Measurement Systems

Application and Design

Measurement Systems

Application and Design

Ernest O. Doebelin

Department of Mechanical Engineering
The Ohio State University

Revised Edition

McGRAW-HILL BOOK COMPANY

New York / St. Louis / San Francisco
Düsseldorf / Johannesburg / Kuala Lumpur
London / Mexico / Montreal / New Delhi
Panama / Paris / São Paulo / Singapore
Sydney / Tokyo / Toronto

MEASUREMENT SYSTEMS
Application and Design

4567890   DODO   79876

This book was set in Modern 8A by The Maple Press Company.
The editors were B. J. Clark and Michael Gardner;
the cover was designed by J. Paul Kirouac, A Good Thing, Inc.;
the production supervisor was Leroy A. Young.
R. R. Donnelley & Sons Company was printer and binder.

Library of Congress Cataloging in Publication Data

Doebelin, Ernest O
    Measurement systems: application and design.

    Includes bibliographies.
    1. Measuring instruments.   I.   Title.
TJ1313.D64   1976   681'.2   74-10881
ISBN 0-07-017336-2

Contents

Chapter 7   Flow Measurement   435

Chapter 8   Temperature and Heat-flux Measurement   502

Preface

The title of this book, "Measurement Systems: Application and Design," is sufficiently broad that one can read into it a wide range of possibilities with regard to content and approach. The author would like to use the preface to define them more sharply, in addition to presenting the way in which the material has been used and the purpose he feels it can serve in engineering education and practice.

Since measurement in one form or another is used regularly by all sorts of people in all sorts of jobs, one must first restrict the scope. This material has been used in connection with courses and laboratories in the mechanical engineering curriculum at The Ohio State University and is thus biased toward this audience. Sufficient material at a suitable level is included for two courses: an introductory treatment useful for a required undergraduate course and more advanced considerations suitable for an elective course at the advanced undergraduate or beginning graduate level. The inclusion of both types of material in a single text is in part a recognition of the variation in curricula from school to school. By including a wide scope of material it is hoped that the individual instructor will be able to select topics and place emphasis so as to best utilize the previous preparation of his particular students.

The area of measurement is, of course, closely allied to that of laboratory teaching, and since this facet of engineering education is the subject of some controversy, there is certainly room for a wide variety of approaches. One can consider the general field to be composed of two parts: the hardware of measurement and the techniques of experimentation. It is difficult completely to separate the two, and some, by choice or because of pressure of time, will design courses to treat both types of material concurrently. At the author's school, three required courses are devoted to this general subject. The first is aimed mainly at developing an understanding of the operating principles of measurement hardware and the problems involved in the analysis, design, and application of such equipment. "Application" here (and in the book title) is not construed to encompass the detailed planning of comprehensive experiments but rather is limited mainly to consideration of the disturbing effect of the measuring instrument on the measured system and the influence of extraneous system variables on the instrument output. The method of presentation of this material is by 1-hr lectures each week and one 4-hr laboratory session each week. Lecturing is divided about equally between generalized theory (static characteristics, dynamic characteristics, etc.) and specific measurement hardware (vibration pikups, recorders, etc.) which may not be covered in the laboratory experiments. Laboratory experiments are preceded by about 1 hr of lecturing specific to the particular measurement area treated in that experiment. Experiments will not here be discussed in detail; however, the philosophy is that of having a relatively small number of experiments so as to allow sufficient time for adequate penetration and understanding. It is felt that it is impossible to cover all the significant hardware. Therefore experiments, while dealing with important specific areas, are designed mainly to serve as vehicles for the illustration of general concepts.

Following the course just described are two courses devoted to experimental analysis of engineering problems. These initially devote lecture time to development of systematic methods of planning, executing, and evaluating experiments. Then small groups of students undertake original projects which involve theoretical analysis, experiment planning, equipment design and construction, measurement-system design, experiment execution, evaluation, and report writing.

For the first course mentioned above, selected material from this book serves as text, while for the second and third courses it becomes a valuable reference. In addition to these three required courses, an additional elective in measurement systems, which essentially extends in breadth and depth from the first required course, is offered. This course has 3 hr of lecture and one 2-hr laboratory period per week. The objective is to develop increased competence in both the design and use of measurement equipment. This is implemented by consideration of both more advanced general concepts and also more sophisticated specific hardware. Experiments are designed mainly to provide familiarity with actual instrumentation equipment and problems involved in its use. Material from this book again serves as the text for this course.

Some explanation of the use of the word "system" in the title of this book may be in order. While the term "system engineering" has come to mean, at least to some, the planning and implementation of complex schemes on a grand scale, no such meaning is intended here. In fact, the author subscribes to the view that one man's component may be another man's system and that this varied use of the word is not objectionable. Thus to the designer of a large-scale data system, who essentially selects hardware from that available to achieve a compatible arrangement that meets the specified requirements, a tape recorder is legitimately considered a component. However, a designer of tape recorders would certainly insist that his machine is a most complex electromechanical system. Since this book is addressed to both these classes of application, the title word "system" is intended to be interpreted in either way, as appropriate.

The word "design" is used in the title to emphasize that many mechanical engineers not only use instrumentation but also are engaged in designing it. While the design of an electronic amplifier has perhaps only limited mechanical aspects (packaging, shock mounting, cooling, manufacturing, etc.), much other equipment, particularly transducers, is as much mechanical as electrical. Although highly specialized electrical aspects of electromechanical system design are handled by electrical engineers, the electrical background of mechanical engineers is adequate to allow them to treat many electrical problems that are closely coupled to the mechanical aspects. Design is intended to consider not only the problems of individual components but also the assembly of available components (transducers, amplifiers, recorders, etc.) into a compatible system capable of meeting required specifications.

Some important features of the text include the following:

1 Consideration of measurement as applied to research and development operations and also to monitoring and control of industrial and military systems and processes.

2 A generalized treatment of error-compensating techniques.

3 Treatment of dynamic response for all types of inputs: periodic, transient, and random, on a uniform basis, utilizing frequency response.

4 Detailed consideration of problems involved in interconnecting components.

5 Discussion, including numerical values, of standards for all important quantities. These give the reader a feeling for the ultimate performance currently achievable.

6 Quotation of detailed numerical performance specifications of actual instruments.

7 Inclusion of significant material on important specific areas such as sound measurement, heat-flux sensors, gyroscopic instruments, hot-wire anemometers, digital methods, random signals, mass flowmeters, amplifiers, and the use of feedback principles.

Since material for both introductory and advanced courses is included, it may be helpful to indicate the division. This is somewhat arbitrary since material considered advanced for a course at a given level in one curriculum might be thought elementary in another.   For example, students who have had a course in dynamic systems analvsis could cover the material on dynamic characteristics very rapidly.   Thus the individual instructor must make the necessary selection.   For those who would appreciate some assistance, the author offers the following suggestions.

Chapters 1 and 2 can be covered quite quickly and easily and should be included in an introductory course.   In Chap. 3 the material on the proofs of the generalized loading equations, which starts at Fig. 3.29, can be omitted in a first course.   The material on dynamic response utilizing Fourier transforms for transient response (Fig. 3.74) and on amplitude-modulated and -demodulated signals (Figs. 3.81 to 3.90) can also be omitted, as can the treatment of random signals.   In the remainder of this chapter the sections Requirements on the Instrument Transfer Function to Ensure Accurate Measurement and Experimental Determination of Measurement-system Parameters should be retained and the others omitted.

In Chap. 4 we begin to deal with specific devices, and the choice of topics to be omitted becomes less clear and somewhat a matter of personal preference.   Sections 4.10 to 4.12 might reasonably be omitted and, in the remaining material, emphasis put on standards and calibration, potentiometers, strain gages, differential transformers, piezoelectric transducers, and accelerometers.   Section 5.7 can be omitted, as can Secs. 6.6, 6.7, and 6.10.   In Chap. 7 the material on hot-wire anemometers, electromagnetic flowmeters, and ultrasonic flowmeters can be left out. Radiation methods in Chap. 8 can be considerably cut, and heat-flux sensors omitted.   Section 9.1 is the only essential part of Chap. 9.   Section 10.1 is essential and would, in fact, usually be assigned much earlier, since it is needed in Chap. 4.   Section 10.2 can be eliminated as can the hydraulic filter and statistical averaging of Sec. 10.3, and also Secs. 10.6 to 10.13.   Chapter 11 may be omitted but most instructors will wish to cover all of Chap. 12, perhaps early in the course if an associated laboratory requires the use of recording equipment.   Since Chap. 13 is short, entirely descriptive, and designed as a unifying conclusion, it should be included in an introductory course.

The author would like to acknowledge his appreciation to other workers in this field whose contributions are evidenced by the voluminous references and bibliography.   Production of the manuscript was made as painless as possible by the faultless typing of Mrs. Maxine Fitzgerald.   The contributions of students over the past 10 years to my understanding of methods of presenting material should not be minimized.   Finally, the forbearance of a long-suffering wife and family is gratefully acknowledged.

Ernest O. Doebelin

Preface to the Revised Edition

Since the first edition of this text emphasized basic principles, concepts, and methods of analysis which remain largely unchanged as time passes, this revision was planned as one of limited scope and low cost devoted mainly to the correction of errors, the addition of certain significant hardware developments, and recognition of the need for a gradual conversion from British to metric (SI) units. Since "new" hardware generally *augments* rather than *replaces* "old", the revision does not delete any material but adds a final chapter devoted to new hardware. Metric conversion is limited to inclusion of a conversion table and use of metric units in the new chapter. This approach recognizes the fact that practicing engineers will need to use *both* British and metric units for some years, thus it is unrealistic to train students only in metric. New references are added both at the end of each chapter (where available "white space" allows) and also after the new chapter.

Ernest O. Doebelin

Measurement Systems

Application and Design

Part One
General Concepts

Chapter 1
Types of Applications of Measurement Instrumentation

1.1

Introduction   As background for our later detailed study of measuring instruments and their characteristics, it will be useful first to discuss in a general way the uses to which such devices are put.   We here choose to classify these applications according to the following scheme:

1. Monitoring of processes and operations
2. Control of processes and operations
3. Experimental engineering analysis

Each of these classes of application will now be described in more detail.

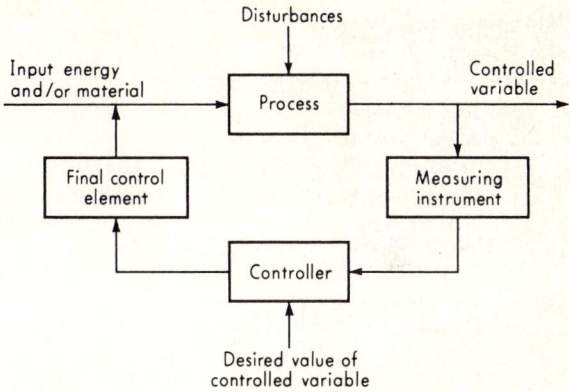

Fig. 1.1.   *Feedback-control system.*

1.2

Monitoring of Processes and Operations   Certain applications of measuring instruments may be characterized as having essentially a monitoring function.   The thermometers, barometers, and anemometers used by the Weather Bureau serve in such a capacity.   They simply indicate the condition of the environment, and their readings do not serve any control functions in the ordinary sense.   Similarly, water, gas, and electric meters in the home keep track of the quantity of the commodity used so that the cost to the user may be computed.   The film badges worn by workers in radioactive environments serve to monitor the cumulative exposure of the wearer to radiations of various types.

1.3

Control of Processes and Operations   Another extremely important type of application for measuring instruments is that in which the instrument serves as a component of an automatic control system. A functional block diagram illustrating the operation of such a system is shown in Fig. 1.1.   It is clear that in order to control any variable by such a "feedback" scheme it is first necessary to measure it; thus all such control systems must incorporate at least one measuring instrument.

Examples of this type of application are endless.   A familiar one is the typical home-heating system employing some type of thermostatic control.   A temperature-measuring instrument (often a bimetallic element) senses the room temperature, thus providing the information necessary for proper functioning of the control system.   Much more sophisticated examples are found among the aircraft and missile control

1. Often give results that are of general use rather than for restricted application.
2. Invariably require the application of simplifying assumptions. Thus not the actual physical system but rather a simplified "mathematical model" of the system is studied. This means the theoretically predicted behavior is *always* different from the real behavior.
3. In some cases, may lead to complicated mathematical problems. This has blocked theoretical treatment of many problems in the past. Today, increasing availability of high-speed computing machines allows theoretical treatment of many problems that could not be so treated in the past.
4. Require only pencil, paper, computing machines, etc. Extensive laboratory facilities are not required. (Some computers are very complex and expensive, but they can be used for solving all kinds of problems. Much laboratory equipment, on the other hand, is special-purpose and suited only to a limited variety of tasks.)
5. No time delay engendered in building models, assembling and checking instrumentation, and gathering data.

Fig. 1.2.   *Features of theoretical methods.*

systems. Here a single control system may require information from many measuring instruments such as pitot-static tubes, angle-of-attack sensors, thermocouples, accelerometers, altimeters, and gyroscopes.

The reader, in attempting to classify applications within his own experience according to the three categories of Sec. 1.1, may find instances where the distinction between monitoring, control, and analysis functions is not clear-cut. Thus the category decided upon may depend somewhat on one's point of view. The data obtained by the Weather Bureau, for instance, serve mainly in a monitoring function for the average person. For fruit growers, however, a report of cold weather may act in a control sense because it signals them to turn on smudge pots and apply other anti-frost measures. Also, present weather data for large areas are correlated and analyzed to form the basis of short- and long-range weather predictions, so that one could say the instruments are supplying data for an engineering analysis. Once one recognizes the possibility of a variety of interpretations, depending on the point of view, the apparent looseness of the classification should not cause any difficulty.

1.4

Experimental Engineering Analysis   In solving engineering problems, two general methods are available: theoretical and experimental.

1. Often give results that apply only to the specific system being tested. However, techniques such as dimensional analysis may allow some generalization.

2. No simplifying assumptions necessary if tests are run on an actual system. The true behavior of the system is revealed.

3. *Accurate* measurements necessary to give a true picture. This may require expensive and complicated equipment. *The characteristics of all the measuring and recording equipment must be thoroughly understood.*

4. Actual system or a scale model required. If a scale model is used, similarity of all significant features must be preserved.

5. Considerable time required for design, construction, and debugging of apparatus.

Fig. 1.3.   *Features of experimental methods.*

Many problems require the application of both methods. The relative amount of each employed depends on the nature of the problem. Problems on the frontiers of knowledge often require very extensive experimental studies since adequate theories are not yet available. Theory and experiment should thus be thought of as complementing each other, and the engineer who takes this attitude will, in general, be a more effective problem solver than one who neglects one or the other of these two approaches.

It may be helpful to summarize quickly the salient features of the theoretical and the experimental methods of attack. This is done in Figs. 1.2 and 1.3.

In considering the application of measuring instruments to problems of experimental engineering analysis, it may be helpful to have at hand a classification of the types of problems encountered. This classification may be accomplished according to several different plans, but one which the author has found meaningful is given in Fig. 1.4.

1.5

Conclusion   Whatever the nature of the application, intelligent selection and use of measurement instrumentation depend on a broad knowledge of what is available and how the performance of the equipment may be best described in terms of the job to be done. New equipment is continuously being developed, but there are certain basic devices that have proved their usefulness in broad areas and will undoubtedly be widely used for many years. A representative cross section of such devices is discussed in this text. These devices are of great interest in

---

1. Testing the validity of theoretical predictions based on simplifying assumptions.

Example: frequency-response testing of mechanical linkage for resonant frequencies.

2. Formulation of generalized empirical relationships in situations where no adequate theory exists.

Example: determination of friction factor for turbulent pipe flow.

3. Determination of material, component, and system parameters; variables; and performance indices.

Examples: determination of yield point of a certain alloy steel, speed-torque curves for an electric motor, thermal efficiency of a steam turbine.

4. Study of phenomena with hopes of developing a theory.

Example: electron microscopy of metal fatigue cracks.

5. Solution of mathematical equations by means of analogies.

Example: solution of shaft torsion problems by measurements on soap bubbles.

Fig. 1.4. *Types of experimental-analysis problems.*

themselves; they also serve as the vehicle for the presentation and development of general techniques and principles needed in handling problems in measurement instrumentation. In addition, these general concepts will be useful in treating any new devices that may be developed in the future.

The treatment is also intended to be on a level that will be of service not only to the user but also to the designer of measurement instrumentation equipment. There are two main reasons for this emphasis. One is that much experimental equipment (including measurement instruments) is often "homemade," especially in smaller companies where the high cost of specialized gear cannot always be justified. The other reason is that the instrument industry is a large and growing one which utilizes many engineers in a design capacity. While the general techniques of mechanical and electrical design as applied to *machines* are also applicable to instruments, in many cases a rather different point of view is necessary in instrument design. This is due, in part, to the fact that the design of machines is mainly concerned with considerations of *power* and *efficiency* whereas instrument design almost completely neglects these areas and concerns itself with the acquisition and manipulation of *information*. Since a considerable number of engineering graduates will work in the instrument industry, their education should include treatment of the most significant aspects of this area.

The planning, execution, and evaluation of experiments are barely touched on in this text. The author feels this extremely important material logically *follows* a treatment of measurement instrumentation. Fortunately, excellent texts treating these matters are available.[1]

Problems

1.1 By consulting various technical journals in the library, find accounts of experimental studies carried out by engineers or scientists. Find three such articles, reference them completely, explain briefly what was accomplished, and attempt to classify them according to one or more categories of Fig. 1.4.

1.2 Give three specific examples of measuring-instrument applications in each of the following areas:

    a. Monitoring of processes and operations

    b. Control of processes and operations

    c. Experimental engineering analysis.

1.3 Compare and contrast the experimental and the theoretical approaches to the following problems:

    a. What is the tolerable vibration level to which astronauts may safely be exposed in launch vehicles?

    b. Find the relationship between applied force F and resulting friction torque T_f in the simple brake of Fig. P1.1.

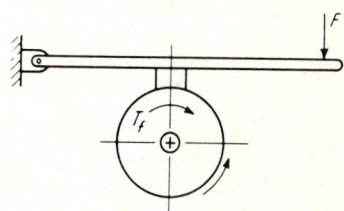

Fig. P1.1

    c. Find the location of the center of mass of the rocket shown in Fig. P1.2 if the shapes, sizes, and materials of all the component parts are known.

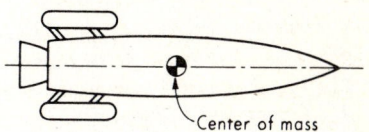

Fig. P1.2　　Center of mass

    d. At what angle with the horizontal should a projectile be launched to achieve the greatest horizontal range?

Bibliography

books

1. K. S. Lion: "Instrumentation in Scientific Research," McGraw-Hill Book Company, New York, 1959.

[1] Hilbert Schenck, Jr., "Theories of Engineering Experimentation," McGraw-Hill Book Company, New York, 1961.

2. C. F. Hix and R. P. Alley: "Physical Laws and Effects," John Wiley & Sons, Inc., New York, 1958.
3. P. K. Stein: "Measurement Engineering," Stein Engineering Services, Inc., Phoenix, Ariz., 1964.
4. C. S. Draper, Walter McKay, and Sidney Lees: "Instrument Engineering," vols. I, II, III, McGraw-Hill Book Company, New York, 1955.
5. R. H. Cerni and L. E. Foster: "Instrumentation for Engineering Measurement," John Wiley & Sons, Inc., New York, 1962.
6. W. M. Cady: "Physical Measurements in Gas Dynamics and Combustion," Princeton University Press, Princeton, N.J., 1954.
7. E. B. Wilson, Jr.: "An Introduction to Scientific Research," McGraw-Hill Book Company, New York, 1952.
8. R. C. Dove and P. H. Adams: "Experimental Stress Analysis and Motion Measurement," Charles E. Merrill Books, Inc., Columbus, Ohio, 1964.
9. D. Bartholomew: "Electrical Measurements and Instrumentation," Allyn and Bacon, Inc., Boston, 1963.
10. E. Frank: "Electrical Measurement Analysis," McGraw-Hill Book Company, New York, 1959.
11. T. G. Beckwith and W. L. Buck: "Mechanical Measurements," Addison-Wesley Publishing Company, Inc., Reading, Mass., 1961.
12. N. H. Cook and E. Rabinowicz: "Physical Measurement and Analysis," Addison-Wesley Publishing Company, Inc., Reading, Mass., 1963.
13. D. P. Eckman: "Industrial Instrumentation," John Wiley & Sons, Inc., New York, 1950.
14. G. L. Tuve: "Mechanical Engineering Experimentation," McGraw-Hill Book Company, New York, 1961.
15. D. M. Considine (ed.): "Process Instruments and Controls Handbook," McGraw-Hill Book Company, New York, 1957.
16. Transducer Compendium, Instrument Society of America, Pittsburgh, Pa.
17. H. K. P. Neubert: "Instrument Transducers," Oxford Univ. Press, London, 1963.
18. C. Lipson and J. Sheth: "Statistical Design and Analysis of Engineering Experiments," McGraw-Hill Book Company, New York, 1973.

periodicals

1. *The Review of Scientific Instruments*
2. *Journal of Scientific Instruments* (Great Britain)
3. *Transactions of Instrument Society of America*
4. *Experimental Mechanics*
5. *Measurement Techniques* (USSR; English translation)
6. *Instruments and Experimental Techniques* (USSR; English translation)
7. *Industrial Laboratory* (USSR; English translation)
8. *Instruments and Control Systems*
9. *Control Engineering*
10. *Journal of Instrument Society of America*
11. *Archiv für Technisches Messen* (Germany)
12. *Journal of Research of the National Bureau of Standards*
13. *Transactions of the Society of Instrument Technology* (Great Britain)
14. *Electromechanical Design*
15. *ASME Jour. of Dyn. Syst., Measurement and Control*
16. *Biomedical Engineering* (London)
17. *Medical Electronics and Data*

Chapter 2

Generalized Configurations and Functional Descriptions of Measuring Instruments

2.1

The Functional Elements of an Instrument   It is possible and desirable to describe both the operation and the performance (degree of approach to perfection) of measuring instruments and associated equipment in a generalized way without recourse to specific physical hardware. The operation can be described in terms of the functional elements of instrument systems, and the performance is defined in terms of the static and dynamic performance characteristics.   This section develops the concept of the functional elements of an instrument or instrument system.

If one examines diverse physical instruments with a view toward

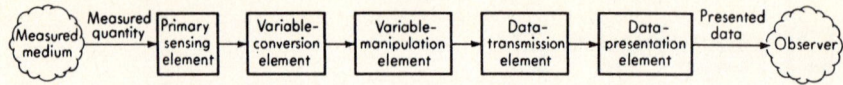

Fig. 2.1.   *Functional elements of an instrument or measurement system.*

generalization, he soon recognizes in the elements of the instruments a recurring pattern of similarity with regard to function.   This leads to the concept of breaking down instruments into a limited number of types of elements according to the generalized function performed by the element. This breakdown can be made in a number of ways, and no standardized universally accepted scheme is at present in use.   We shall present one such scheme which may be of help to the reader in understanding the operation of any new instrument with which he may come in contact and also in planning the design of a new instrument.

Consider the diagram of Fig. 2.1, which represents a possible arrangement of functional elements in an instrument and includes *all* the basic functions considered necessary for a description of any instrument. The *primary sensing element* is that which first receives energy from the measured medium and produces an output depending in some way on the measured quantity.   It is important to note that an instrument *always* extracts some energy from the measured medium.   Thus the measured quantity is *always* disturbed by the act of measurement, making a perfect measurement theoretically impossible.   Good instruments are designed to minimize this effect but it is always present to some degree.

The output signal of the primary sensing element is some physical variable, such as a displacement or a voltage.   For the instrument to perform the desired function, it may be necessary to convert this variable to another more suitable variable while preserving the information content of the original signal.   An element that performs such a function is called a *variable-conversion element*.   It should be noted that every instrument need not include a variable-conversion element while some require several. Also, the "elements" we speak of are *functional* elements, not physical elements.   That is, Fig. 2.1 shows an instrument neatly separated into blocks, which may lead the reader to think of the physical apparatus as being precisely separable into subassemblies performing the specific functions shown.   This is, in general, not the case; a specific piece of hardware may perform *several* of the basic functions, for instance.

In performing its intended task, an instrument may require that a signal represented by some physical variable be manipulated in some way. By manipulation we here mean specifically a change in numerical value according to some definite rule but a preservation of the physical nature of the variable.   Thus an electronic amplifier accepts a small voltage signal as input and produces an output signal that is also a voltage but is some constant times the input. An element that performs such a function will be called a *variable-manipulation element*. Again, the reader should not be misled by Fig. 2.1. A variable-manipulation element does not necessarily *follow* a variable-conversion element; it may precede it, appear elsewhere in the chain, or not appear at all.

When the functional elements of an instrument are actually physically separated, it becomes necessary to transmit the data from one to another. An element performing this function is called a *data-transmission element*. It may be as simple as a shaft and bearing assembly or as complicated as a telemetry system for transmitting signals from missiles to ground equipment by radio.

If the information about the measured quantity is to be communicated to a human being for monitoring, control, or analysis purposes, it must be put into a form recognizable by one of the human senses. An element that performs this "translation" function is called a *data-presentation element*. This function includes the simple *indication* of a pointer moving over a scale and also the *recording* of a pen moving over a chart. Indication and recording may also be performed in discrete increments (rather than smoothly) as exemplified by an optical flat used for measuring flatness of surfaces by light-interference principles and an electric typewriter for recording numerical data. While the majority of instruments communicate with people through the medium of the visual sense, the use of other senses such as hearing and touch is certainly conceivable. Certain methods of recording may present the data in a form not directly detectable by human senses. The magnetic tape recorder is a noteworthy example. For such a case, suitable instruments for extracting the stored information at any desired time and converting it to a form intelligible to man are required.

Before going on to some illustrative examples, let it be emphasized again that Fig. 2.1 is intended as a vehicle for presenting the concept of functional elements and not as a physical schematic of a generalized instrument. A given instrument may involve the basic functions in any number and combination; they need not appear in the order of Fig. 2.1. A given physical component may serve several of the basic functions.

As an example of the above concepts, consider the rudimentary pressure gage of Fig. 2.2. One of several possible valid interpretations is as follows: The primary sensing element is the piston, which also serves the function of variable conversion since it converts the fluid pressure (force per unit area) into a resultant force on the piston face. Force is transmitted by the piston rod to the spring, which converts force into a proportional displacement. This displacement of the piston rod is magnified (manipulated) by the linkage to give a larger pointer displace-

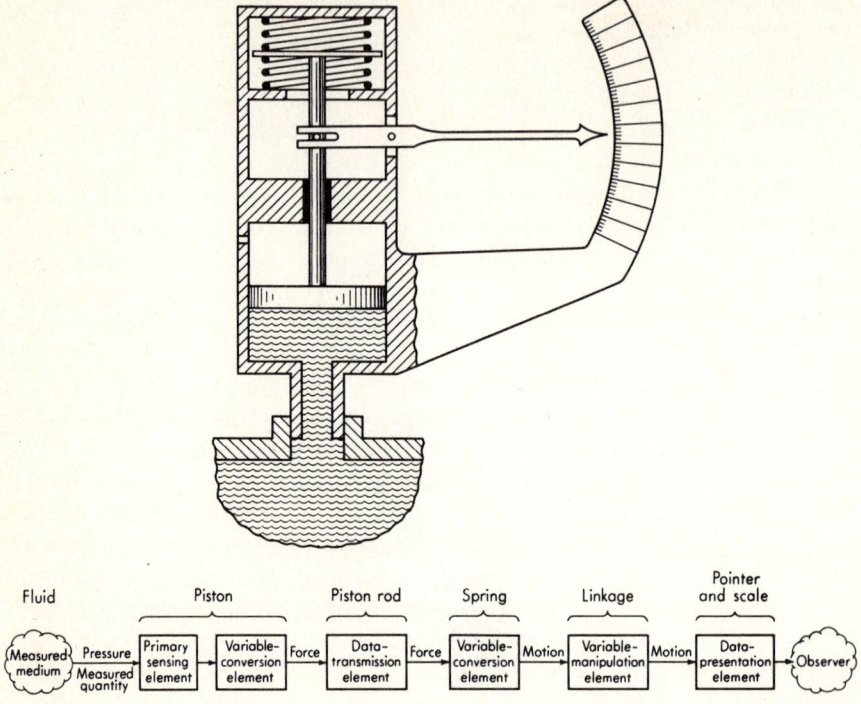

Fig. 2.2. *Pressure gage.*

ment.   The pointer and scale indicate the pressure, thus serving as data-presentation elements.   If it were necessary to locate the gage at some distance from the source of pressure, a small tube could serve as a data-transmission element.

Figure 2.3 depicts a pressure-type thermometer.   The liquid-filled bulb acts as primary sensor and variable-conversion element since a temperature change results in a pressure buildup within the bulb, because of the constrained thermal expansion of the filling fluid.   This pressure is transmitted through the tube to a Bourdon-type pressure gage which converts pressure to displacement.   This displacement is manipulated by the linkage and gearing to give a larger pointer motion.   A scale and pointer again serve for data presentation.

A remote-reading shaft-revolution counter is shown in Fig. 2.4.   The microswitch sensing arm and the camlike projection on the rotating shaft serve both a primary sensing function and a variable-conversion function since rotary displacement is converted to linear displacement.   The microswitch contacts also serve for variable conversion, converting a

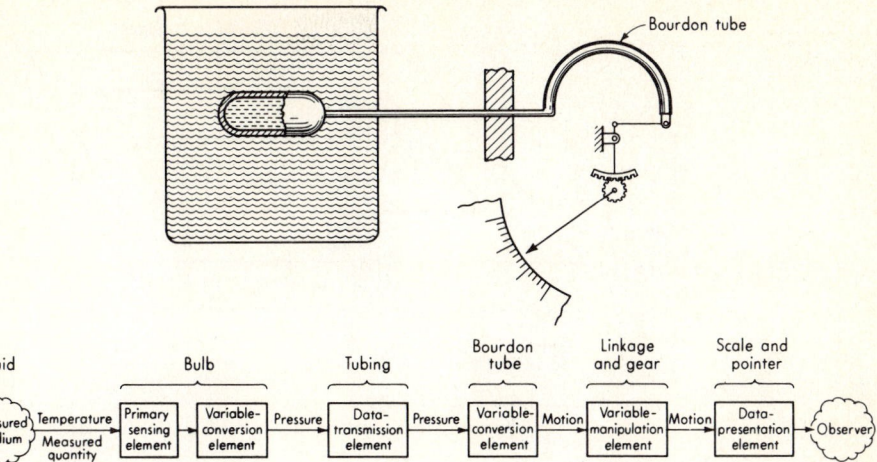

Fluid — Bulb — Tubing — Bourdon tube — Linkage and gear — Scale and pointer

Fig. 2.3. *Pressure thermometer.*

mechanical oscillation into an electrical oscillation (a sequence of voltage pulses). These voltage pulses may be transmitted relatively long distances over wires to a solenoid. The solenoid reconverts the electrical pulses into mechanical reciprocation of the solenoid plunger which serves as input to a mechanical counter. The counter itself involves variable conversion (reciprocating to rotary motion), variable manipulation (rotary motion to decimalized rotary motion), and data presentation.

As a final example, let us examine Fig. 2.5 which illustrates schematically a D'Arsonval galvanometer as used in oscillographs. A time-varying voltage to be recorded is applied to the ends of the two wires

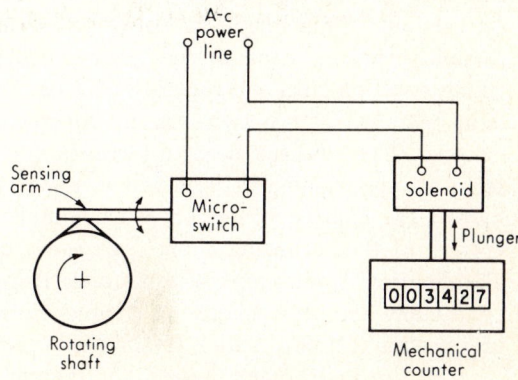

Fig. 2.4. *Digital revolution counter.*

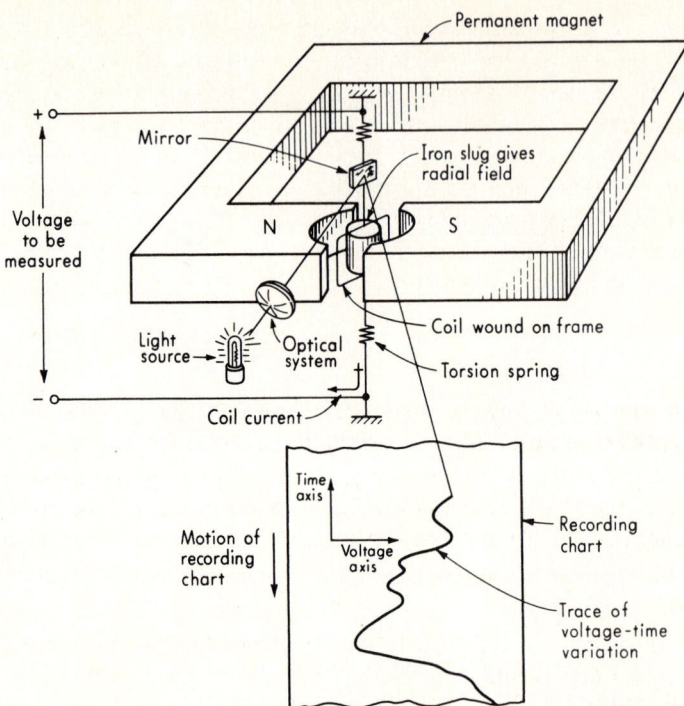

Fig. 2.5. *D'Arsonval galvanometer.*

which transmit the voltage to a coil made up of a number of turns wound on a rigid frame. This coil is suspended in the field of a permanent magnet. The resistance of the coil converts the applied voltage to a proportional current (ideally). The interaction between the current and the magnetic field produces a torque on the coil, giving another variable conversion. This torque is converted to an angular deflection by the torsion springs. A mirror rigidly attached to the coil frame converts the frame rotation into the rotation of a light beam which it reflects. The light-beam rotation is twice the mirror rotation, giving a motion magnification. The reflected beam intercepts a recording chart made of photosensitive material which is moved at a fixed and known rate, giving a time base. The combined horizontal motion of the light spot and the vertical motion of the recording chart generate a graph of voltage versus time. The "optical lever arm" (the distance from the mirror to the recording chart) has a motion-magnifying effect since the spot displacement per unit mirror rotation is directly proportional to it.

In this instrument the coil and magnet assembly would probably be considered as the primary sensing element since the lead wires (which serve a transmission function) are not really part of the instrument and the coil resistance (which acts in a variable-conversion function) is an intrinsic part of the coil.   In any case, the assignment of precise names to specific components is not nearly as important as the recognition of the basic functions necessary to the successful operation of the instrument. By concentrating on these functions and the various physical devices available for accomplishing them, we develop our ability to synthesize new combinations of elements leading to new and useful instruments. This ability is fundamental to all instrument design.

2.2

Active and Passive Transducers   Once certain basic functions common to all instruments have been identified, it is then in order to see if it is possible to make some generalizations on *how* these functions may be performed.   One such generalization is concerned with energy considerations.   In performing any of the general functions indicated in Fig. 2.1, a physical component may act as an *active transducer* or a *passive transducer*.

A component whose output energy is supplied entirely or almost entirely by its input signal is commonly called a *passive transducer*.   The output and input signals may involve energy of the same form (say both mechanical) or there may.be an energy conversion from one form to another (say mechanical to electrical).   (In much technical literature the term transducer is restricted to devices involving energy *conversion*, but, conforming to the dictionary definition of the term, we do not make this restriction.)

An *active transducer*, on the other hand, has an auxiliary source of power which supplies a major part of the output power while the input signal supplies only an insignificant portion.   Again, there may or may not be a conversion of energy from one form to another.

In all the examples of Sec. 2.1 there is only one active transducer, the microswitch of Fig. 2.4; all other components are passive transducers. The power to drive the solenoid comes not from the rotating shaft but from the a-c power line, an auxiliary source of power.   Some further examples of active transducers may be in order.   The electronic amplifier shown in Fig. 2.6 furnishes a good one.   The element supplying the input-signal voltage e_i need supply only a negligible amount of power since almost no current is drawn, owing to negligible grid current and a high R_g. However, the output element (the load resistance R_L) receives significant current and voltage and thus power.   This power must be supplied by the plate battery E_{bb}, the auxiliary power source.   Thus the input *controls* the output but does not actually supply the output power.

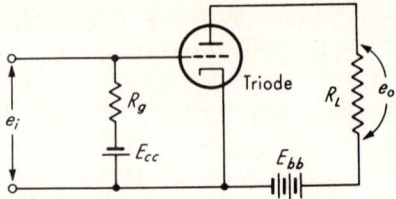

Fig. 2.6. *Triode amplifier.*

Another active transducer of great practical importance, the *instrument servomechanism*, is shown in simplified form in Fig. 2.7. This is actually an instrument *system* made up of components, some of which are passive transducers and some active transducers. When considered as an entity, however, with input voltage e_i and output displacement x_o, it meets the definition of an active transducer and is profitably thought of as such. The purpose of this device is to cause the motion x_o to follow the variations of the voltage e_i in a proportional manner. Since the motor torque is proportional to the error voltage e_e, it is clear that the system can be at rest only if e_e is zero. This occurs only when $e_i = e_{sl}$; since e_{sl} is proportional to x_o, this means that x_o must be proportional to e_i in the

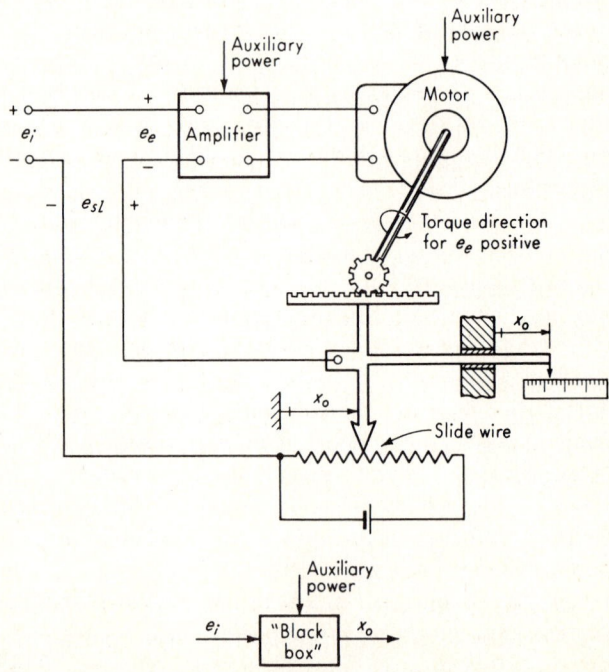

Fig. 2.7. *Instrument servomechanism.*

static case.   If e_i varies, x_o will tend to follow it, and by proper design accurate "tracking" of e_i by x_o should be possible.

2.3

Analog and Digital Modes of Operation   It is possible further to classify how the basic functions may be performed by turning attention to the continuous or discrete nature of the signals that represent the information.   Signals that vary in a continuous fashion and can take on an infinity of values in any given range are called *analog* signals; the devices that produce such signals are called analog devices.   (This is strictly in a macroscopic sense since all physical effects become discrete in atomistic considerations.)   In contrast, signals that vary in discrete steps and can thus take on only a finite number of different values in any given range are described as *digital* signals; the devices that produce such signals are called digital devices.

The majority of present-day measuring instruments are of the analog type.   The only digital device illustrated in this text up to this point is the revolution counter of Fig. 2.4.   This is clearly a digital device since it is impossible for this instrument to indicate, say, 0.79 revolution; it measures only in steps of one revolution.   The importance of digital instruments is increasing, perhaps mainly because of the increasing use of digital computers in both data-reduction and automatic control systems. Since the digital computer works only with digital signals, any information supplied to it must be in digital form.   The computer's output is also in digital form.   Thus any communication with the computer at either the input or output end must be in terms of digital signals.   Since most present-day measurement and control apparatus is of an analog nature it is necessary to have both *analog-to-digital converters* (at the input to the computer) and *digital-to-analog converters* (at the output of the computer).   These devices (which are discussed in more detail in a later chapter) serve as "translators" that enable the computer to communicate with the outside world, which is largely of an analog nature.   Effort is being expended to develop both measuring and control devices that are *inherently* digital in nature so that the somewhat complex converters will not be needed.

2.4

Null and Deflection Methods   A useful classification with regard to the mode of operation of instruments separates devices by their operation on a null or a deflection principle.   In a *deflection-type* device the measured quantity produces some physical effect that engenders a similar

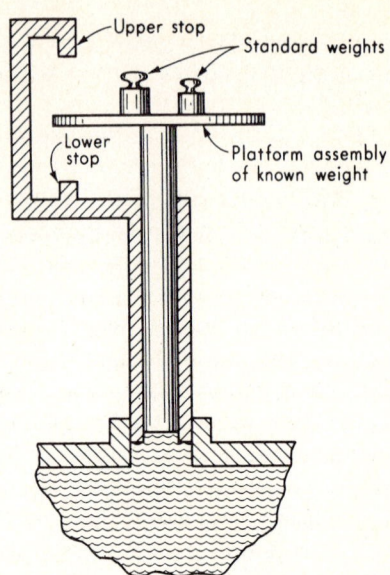

Fig. 2.8.   *Dead-weight pressure gage.*

but opposing effect in some part of the instrument.   The opposing effect is closely related to some variable (usually a mechanical displacement or deflection) that can be directly observed by some human sense.   The opposing effect increases until a balance is achieved, at which point the "deflection" is measured and the value of the measured quantity inferred from this.   The pressure gage of Fig. 2.2 exemplifies this type of device since the pressure force engenders an opposing spring force due to an unbalance of forces on the piston rod (called the force-summing link) which causes a deflection of the spring.   As the spring deflects, its force increases; thus a balance will be achieved at some deflection if the pressure is within the design range of the instrument.

In contrast to the deflection-type device, a *null-type* device attempts to maintain deflection at zero by suitable application of an effect opposing that generated by the measured quantity.   Necessary to such an operation are a detector of unbalance and a means (manual or automatic) of restoring balance.   Since deflection is kept at zero (ideally), determination of numerical values requires accurate knowledge of the magnitude of the opposing effect.   A pressure gage operating on a null principle is depicted in simplified form in Fig. 2.8.   By adding the proper standard weights to the platform of known weight, the pressure force on the face of the piston may be balanced by gravitational force.   The condition of force balance is indicated by the platform remaining at rest between the upper and lower stops.   Since the weights and the piston area are all known, the unknown pressure may be computed.

Upon comparing the null and deflection methods of measurement exemplified by the pressure gages described above, we note that, in the deflection instrument, accuracy depends on the calibration of the spring whereas in the null instrument it depends on the accuracy of the standard weights. In this particular case (and also for most measurements in general) the accuracy attainable by the null method is of a higher level than that of the deflection method. One reason for this is that the spring is not in itself a primary standard of force but must be calibrated by standard weights, whereas in the null instrument a *direct* comparison of the unknown force with the standard is achieved. Another advantage of null methods is the fact that, since the measured quantity is balanced out, the detector of unbalance can be made very sensitive because it need cover only a small range around zero. Also the detector need not be calibrated since it must detect only the presence and direction of unbalance and not the amount. On the other hand, a deflection instrument must be larger, more rugged, and thus less sensitive if it is to measure large magnitudes.

The disadvantages of null methods appear mainly in dynamic measurements. Let us consider the pressure gages again. The difficulty in keeping the platform balanced for a fluctuating pressure should be apparent. The spring-type gage suffers not nearly so much in this respect. By use of automatic balancing devices (such as the instrument servomechanism of Fig. 2.7) the speed of null methods may be improved considerably, and instruments of this type are of great importance.

2.5

Input-Output Configuration of Measuring Instruments and Instrument Systems   Before going on to discuss instrument performance characteristics, it is desirable to develop a generalized configuration which brings out the significant input-output relationships present in all measuring apparatus. A scheme suggested by Draper, McKay, and Lees[1] is presented in somewhat modified form in Fig. 2.9. Input quantities are classified into three categories: desired inputs, interfering inputs, and modifying inputs. *Desired inputs* represent the quantities that the instrument is specifically intended to measure. *Interfering inputs* represent quantities to which the instrument is unintentionally sensitive. A desired input produces a component of output according to an input-output relation symbolized by F_D, where F_D denotes the mathematical operations necessary to obtain the output from the input.

[1] C. S. Draper, Walter McKay, and Sidney Lees, "Instrument Engineering," vol. III, p. 58, McGraw-Hill Book Company, New York, 1955.

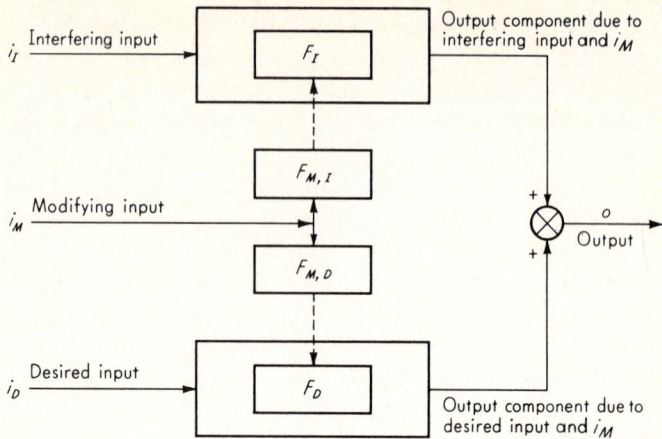

Fig. 2.9. *Generalized input-output configuration.*

The symbol F_D may represent different concepts, depending on the particular input-output characteristic that is being described. Thus F_D might be a constant number K giving the proportionality constant relating a constant static input to the corresponding static output for a linear instrument. For a nonlinear instrument a simple constant is not adequate to relate static inputs and outputs; an algebraic or transcendental *function* is required. To relate dynamic inputs and outputs, differential equations are necessary. If a description of the output "scatter," or dispersion, for repeated equal static inputs is desired, a statistical distribution function of some kind is needed. The symbol F_D encompasses all such concepts. The symbol F_I serves a similar function for an interfering input.

The third class of inputs might perhaps be thought of as being included among the interfering inputs, but a separate classification is actually more significant. This classification is that of modifying inputs. *Modifying inputs* are the quantities that cause a change in the input-output relations for the desired and interfering inputs; that is, they cause a change in F_D and/or F_I. The symbols $F_{M,I}$ and $F_{M,D}$ represent (in the appropriate form) the specific manner in which i_M affects F_I and F_D. These symbols, $F_{M,I}$ and $F_{M,D}$, are again to be interpreted in the same general way as F_I and F_D were.

The block diagram of Fig. 2.9 illustrates the above concepts. The circle with a cross in it is a conventional symbol for a *summing device*. The two plus signs as shown indicate that the output of the summing device is the instantaneous algebraic sum of its two inputs. Since an instrument system may have several inputs of each of the three types and

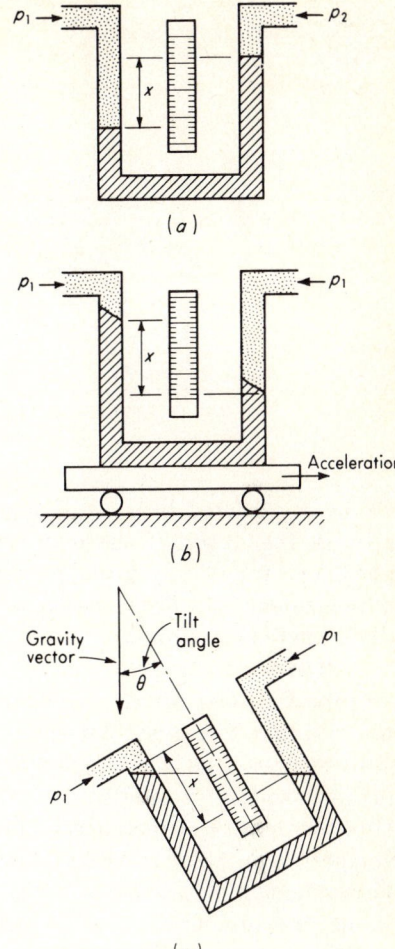

Fig. 2.10.   *Spurious inputs for manometer.*

also several outputs, it may be necessary to draw more complex block diagrams than in Fig. 2.9.   This extension is, however, straightforward.

The above concepts can be clarified by means of specific examples. Consider the mercury manometer used for differential-pressure measurement as shown in Fig. 2.10a.   The desired inputs are the pressures p_1 and p_2 whose difference causes the output displacement x which can be read off the calibrated scale.   Figure 2.10b and c shows the action of two possible interfering inputs.   In Fig. 2.10b the manometer is mounted on some vehicle that is accelerating.   A simple analysis will show that there will be an output x even though the differential pressure might be zero.   Thus if one is trying to measure pressures under such circumstances an error

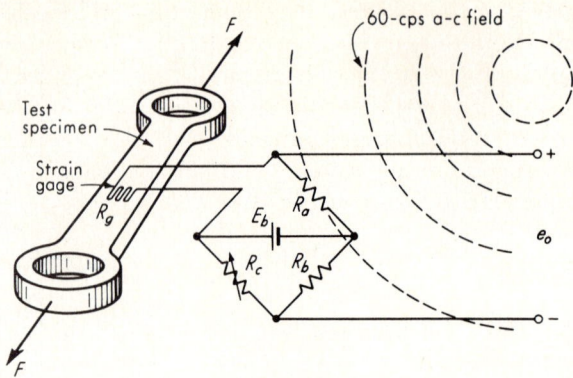

Fig. 2.11.   *Interfering input for strain-gage circuit.*

will be engendered because of the interfering acceleration input. Similarly, in Fig. 2.10c, if the manometer is not properly aligned with the gravity vector it may give an output signal x even though no pressure difference exists. Thus the tilt angle θ is an interfering input. (It is also a modifying input.)

Modifying inputs for the manometer include ambient temperature and gravitational force. Ambient temperature manifests its influence in a number of ways. First, the calibrated scale changes length with temperature; thus the proportionality factor relating $p_1 - p_2$ to x is modified whenever temperature varies from its basic calibration value. Also, the density of mercury varies with temperature, again leading to a change in the proportionality factor. A change in gravitational force due to changes in location of the manometer, such as moving it to another country or putting it aboard a spaceship, leads to a similar modification in the scale factor. It should be noted that the effects of *both* the desired and interfering inputs may be modified by the modifying inputs.

As another example, consider the electrical-resistance strain-gage setup shown in Fig. 2.11. The gage consists of a fine wire grid of resistance R_g firmly cemented to the specimen whose unit strain ϵ at a certain point is to be measured. When strained, the gage's resistance changes according to the relation

$$\Delta R_g = (GF)R_g\epsilon \qquad (2.1)$$

where $\Delta R_g \triangleq$ change in gage resistance, ohms (2.2)

$GF \triangleq$ gage factor, dimensionless (2.3)

$R_g \triangleq$ gage resistance when unstrained, ohms (2.4)

$\epsilon \triangleq$ unit strain, in./in. (2.5)

The resistance change is proportional to the strain; thus if we could measure the resistance, we could compute the strain. The resistance is measured by using the Wheatstone-bridge arrangement shown. When no load F is present, the bridge is balanced (e_o set to zero) by adjusting R_c. Application of load causes a strain, a ΔR_g, and thus unbalances the bridge, causing an output voltage e_o which is proportional to ϵ and can be measured on a meter or oscilloscope. The voltage e_o is given by

$$e_o = -(GF)R_g \epsilon E_b \frac{R_a}{(R_g + R_a)^2} \qquad (2.6)$$

The desired input here is clearly the strain ϵ which causes a proportional output voltage e_o. One interfering input which often results in trouble in such apparatus is the 60-cycle field caused by nearby power lines, electric motors, etc. This field induces voltages in the strain-gage circuit, causing output voltages e_o even when the strain is zero. Another interfering input is the gage temperature. If this varies, it causes a change in gage resistance that will cause a voltage output even if there is no strain. Temperature has another interfering effect since it causes a differential expansion of the gage and the specimen which gives rise to a strain ϵ and a voltage e_o even though no force F has been applied. Temperature also acts as a modifying input since the gage factor is sensitive to temperature. The battery voltage E_b is another modifying input. Both these are modifying inputs since they tend to change the proportionality factor between the desired input ϵ and the output e_o or between an interfering input (gage temperature) and output e_o.

Methods of correction for interfering and modifying inputs. In the design and/or use of measuring instruments a number of methods for nullifying or reducing the effects of spurious inputs are available. We shall briefly describe some of the most widely used.

The *method of inherent insensitivity* proposes the obviously sound design philosophy that the elements of the instrument should *inherently* be sensitive only to the desired inputs. While this is not usually entirely possible, the simplicity of this approach encourages one to consider its application wherever feasible. In terms of the general configuration of Fig. 2.9, this approach requires that somehow F_I and/or $F_{M,D}$ be made as nearly equal to zero as possible. Thus, even though i_I and/or i_M may exist, they cannot affect the output. As an example of the application of this concept to the strain gage of Fig. 2.11, we might try to find some gage material that exhibits an extremely low temperature coefficient of resistance while retaining its sensitivity to strain. If such a material can be found, the problem of interfering temperature inputs is at least partially solved. Similarly, in mechanical apparatus that must maintain

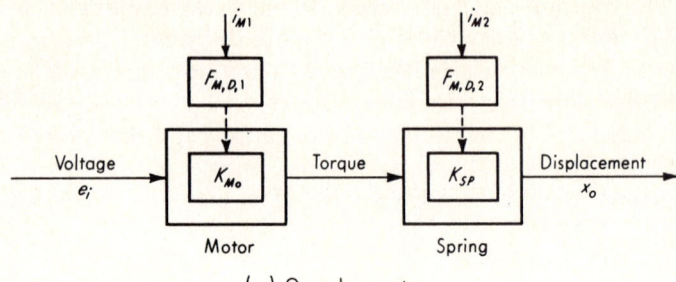

(*a*) Open-loop system

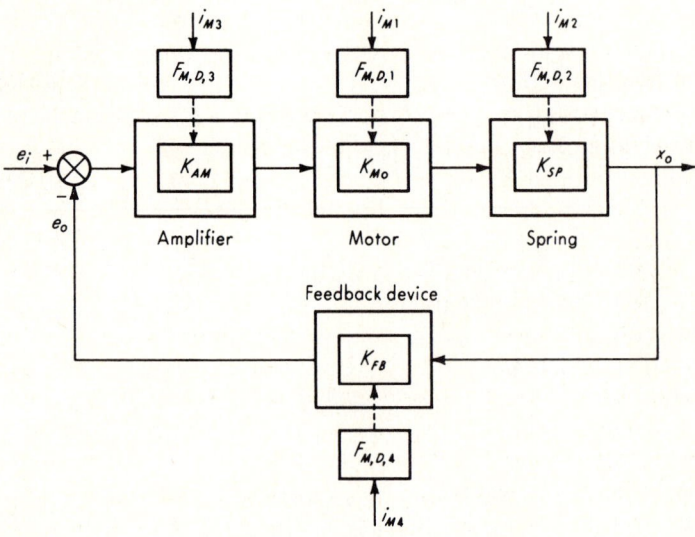

(*b*) Closed-loop or feedback system

Fig. 2.12.   *Use of feedback to reduce effect of spurious inputs.*

accurate dimensions in the face of ambient-temperature changes, the use of a material of very small temperature coefficient of expansion (such as the alloy Invar) may be helpful.

The *method of high-gain feedback* is exemplified by the system shown in Fig. 2.12*b*.   Suppose we wish to measure a voltage e_i by applying it to a motor whose torque is applied to a spring, causing a displacement x_o which may be measured on a calibrated scale.   By proper design, the displacement x_o might be made proportional to the voltage e_i according to

$$x_o = (K_{Mo}K_{SP})e_i \qquad (2.7)$$

where K_{Mo} and K_{SP} are appropriate constants.   This arrangement, shown in Fig. 2.12a, would be called an open-loop system. If modifying inputs i_{M1} and i_{M2} exist, they cause changes in K_{Mo} and K_{SP} which lead to errors in the relation between e_i and x_o. These errors are in *direct proportion* to the changes in K_{Mo} and K_{SP}. Suppose, instead, we construct a system as in Fig. 2.12b. Here the output x_o is measured by the feedback device which produces a voltage e_o proportional to x_o. This voltage is subtracted from the input voltage e_i and the difference applied to an amplifier which drives the motor and thereby the spring to produce x_o. We may write

$$(e_i - e_o)K_{AM}K_{Mo}K_{SP} = (e_i - K_{FB}x_o)K_{AM}K_{Mo}K_{SP} = x_o \qquad (2.8)$$

$$e_i K_{AM}K_{Mo}K_{SP} = (1 + K_{AM}K_{Mo}K_{SP}K_{FB})x_o \qquad (2.9)$$

$$x_o = \frac{K_{AM}K_{Mo}K_{SP}}{1 + K_{AM}K_{Mo}K_{SP}K_{FB}} e_i \qquad (2.10)$$

Suppose now that we design K_{AM} to be very large (a "high-gain" system), so that $K_{AM}K_{Mo}K_{SP}K_{FB} \gg 1$. Then

$$x_o \approx \frac{1}{K_{FB}} e_i \qquad (2.11)$$

The significance of Eq. (2.11) is that the effect of variations in K_{Mo}, K_{SP}, and K_{AM} (due to modifying inputs i_{M1}, i_{M2}, and i_{M3}) on the relation between input e_i and output x_o has been made negligible. *We now require only that K_{FB} stay constant (unaffected by i_{M4}) in order to maintain constant input-output calibration as shown by Eq. (2.11).*

The reader may question whether much really has been gained by this somewhat elaborate scheme since we have merely transferred the requirements for stability from K_{Mo} and K_{SP} to K_{FB}. In actual practice, however, this method often leads to great improvements in accuracy. One reason for this is that, since the amplifier supplies most of the power needed, the feedback device can be designed with low power-handling capacity. This in general leads to greater accuracy and linearity in the feedback-device characteristics. Also, the input signal e_i need carry only negligible power; thus the feedback system extracts less energy from the measured medium than the corresponding open-loop system. This, of course, results in less distortion of the measured quantity due to the presence of the measuring instrument. Finally, if the open-loop chain consists of several (perhaps many) devices, each susceptible to its own spurious inputs, *all* these bad effects can be negated by the use of high amplification and a stable and accurate feedback device.

Before passing on to other methods, it should be mentioned that application of the feedback principle is not without its own peculiar problems. The main one is that of dynamic instability, wherein excessively high amplification leads to destructive oscillations. The study of the design of feedback systems is a whole field in itself, and many texts treating this subject are available.[1]

The *method of calculated output corrections* requires that one be able to measure or estimate the magnitudes of the interfering and/or modifying inputs and that one know quantitatively how they affect the output. With this information available, it is possible to calculate corrections which may be added to or subtracted from the indicated output so as to leave (ideally) only that component associated with the desired input. Thus, in the manometer of Fig. 2.10, the effects of temperature on the calibrated scale's length and on the density of mercury may both be quite accurately computed if the temperature is known. The local gravitational acceleration is also known for a given elevation and latitude so that this effect may also be corrected by calculation. Although theoretically applicable to any form of input, the method of calculated output corrections is probably most used for inputs that are essentially constant.

The *method of signal filtering* is based on the possibility of introducing certain elements ("filters") into the instrument which in some fashion block the spurious signals so that their effects on the output are removed or reduced. The filter may be applied to any suitable signal in the instrument, be it input, output, or intermediate signal. The concept of signal filtering is shown schematically in Fig. 2.13 for the cases of input filtering and output filtering. The application to intermediate signals should be obvious. In Fig. 2.13a the inputs i_I and i_M are caused to pass through filters whose input-output relation is (ideally) zero. Thus i_I' and i_M'' are zero even if i_I and i_M are not zero. The concept of output filtering is illustrated in Fig. 2.13b. Here the output o, though really one signal, is thought of as a superposition of o_I (output due to interfering input), o_D (output due to desired input), and o_M (output due to modifying input). If it is possible to construct filters that selectively block o_I and o_M but allow o_D to pass through, this may be symbolized as in Fig. 2.13b and results in o' consisting entirely of o_D.

The filters necessary in the application of this method may take several forms; they are best illustrated by examples. If a filter is put directly in the path of a spurious input, it can be designed (ideally) to block completely the passage of the signal. If, however, the filter is inserted at a point where the signal contains both desired and spurious components, it must be designed to be selective; that is, it must pass the desired components essentially unaltered while effectively suppressing all others.

It is often necessary to attach delicate instruments to structures

---

[1] E. O. Doebelin, "Dynamic Analysis and Feedback Control," McGraw-Hill Book Company, New York, 1962.

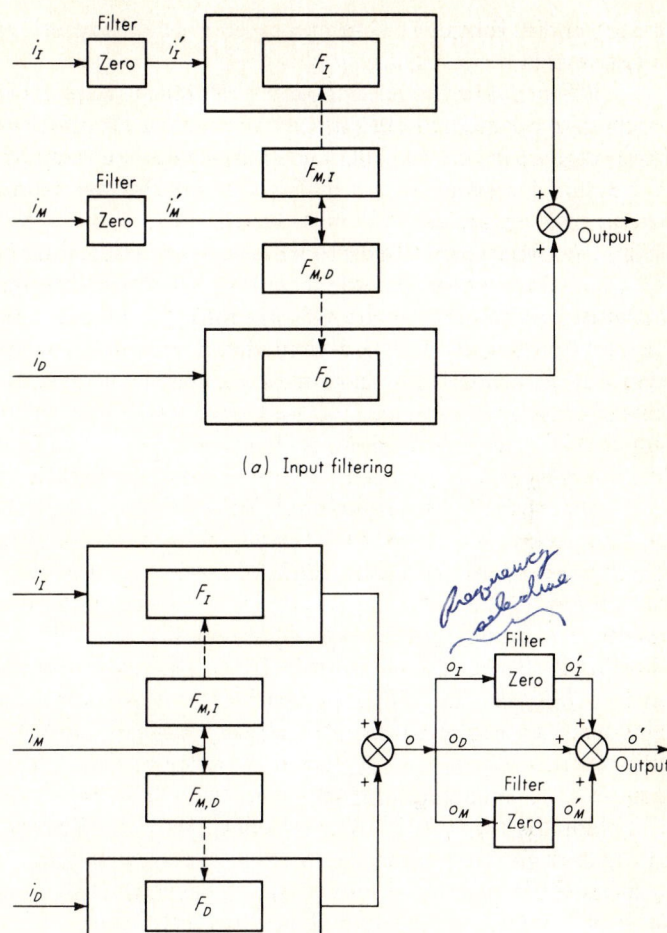

(*a*) Input filtering

(*b*) Output filtering

Fig. 2.13. *General principle of filtering.*

that vibrate. Electromechanical devices for navigation and control of aircraft or missiles are outstanding examples. Figure 2.14a shows how the interfering vibration input may be filtered out by use of suitable spring mounts. The mass-spring system is actually a mechanical filter which passes on to the instrument only a negligible fraction of the motion of the vibrating structure.

The interfering tilt-angle input to the manometer of Fig. 2.10c may be effectively filtered out by means of the gimbal-mounting scheme of Fig. 2.14b. If the gimbal bearings are essentially frictionless, the rotations θ_1 and θ_2 cannot be communicated to the manometer; thus it always hangs vertical.

In Fig. 2.14c the thermocouple reference junction is shielded from ambient-temperature fluctuations by means of thermal insulation. Such an arrangement acts as a filter for temperature or heat-flow inputs.

The strain-gage circuit of Fig. 2.14d is shielded from the interfering 60-cps field by enclosing it in a metal box of some sort. This solution of the problem corresponds to filtering the interfering *input*. Another possible solution, which corresponds to selective filtering of the *output*, is shown in Fig. 2.14e. For this approach to be effective, it is essential that the frequencies in the desired signal occupy a frequency range considerably separated from those in the undesired component of the signal.

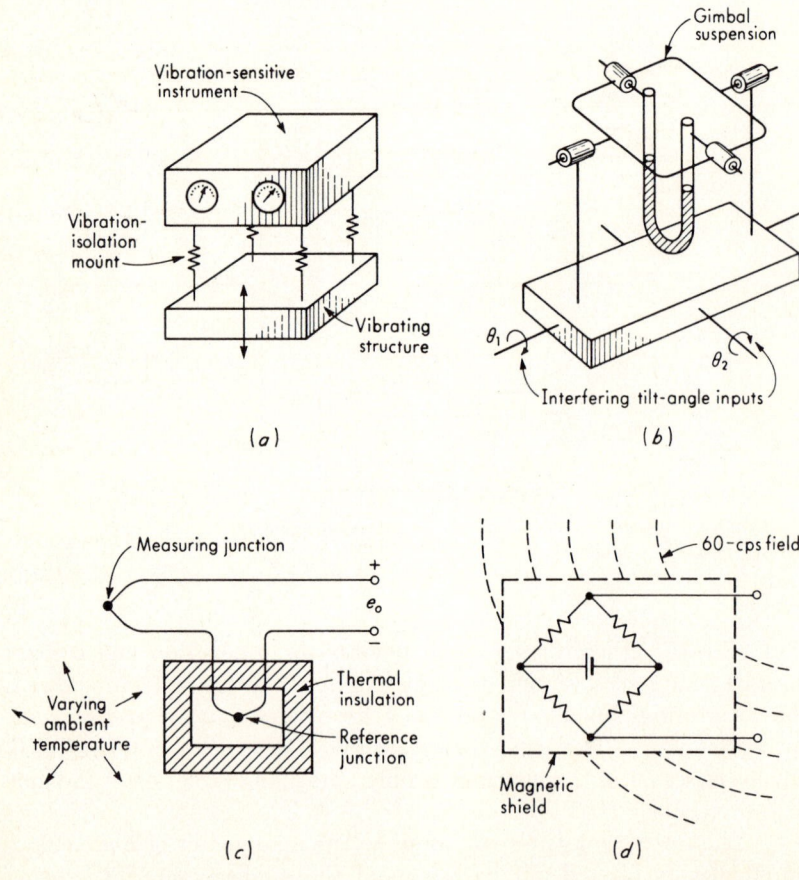

Vibration-sensitive instrument

Vibration-isolation mount

Vibrating structure

(a)

Gimbal suspension

θ_1 θ_2

Interfering tilt-angle inputs

(b)

Measuring junction

e_o

Varying ambient temperature

Thermal insulation

Reference junction

(c)

60-cps field

Magnetic shield

(d)

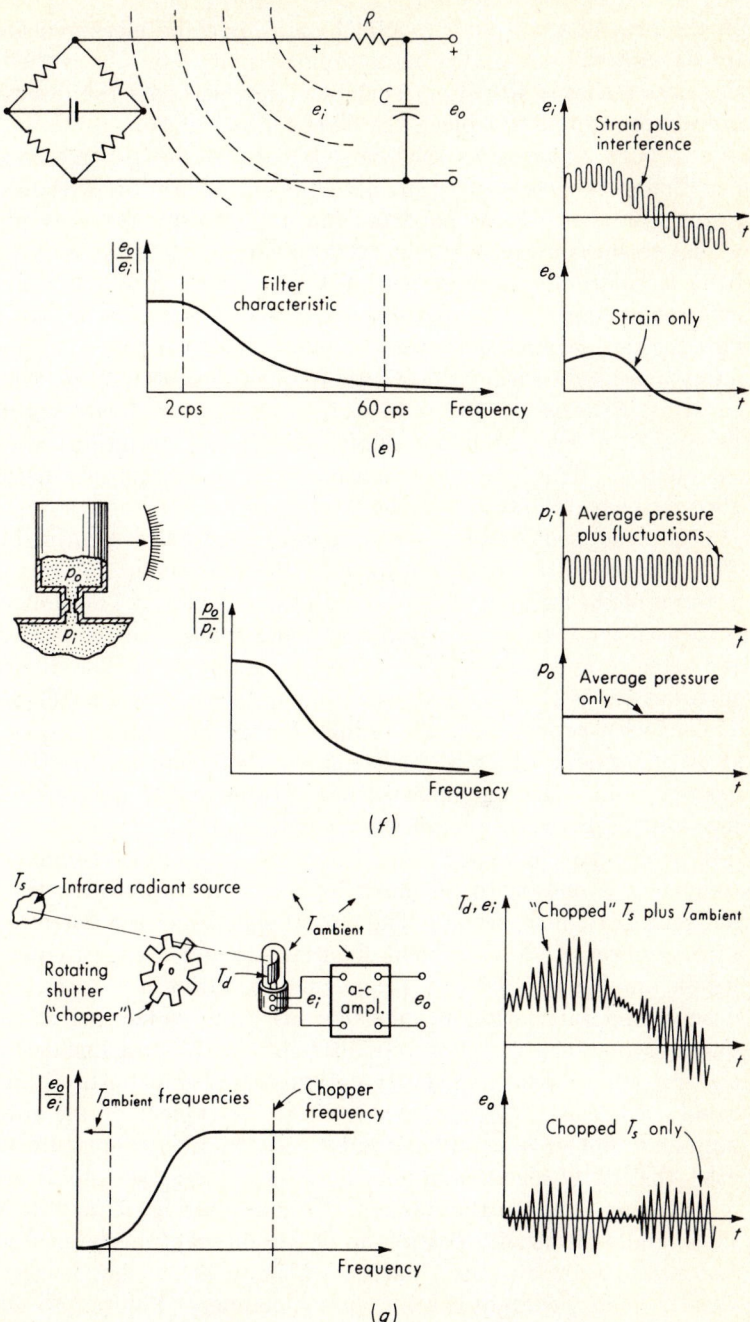

Fig. 2.14. *Examples of filtering.*

In the present example, suppose the strains to be measured are mainly steady and will never vary more rapidly than 2 cps. It is then possible to insert a simple *RC* filter, as shown, that will pass the desired signals but almost completely block the 60-cps interference.

Figure 2.14*f* shows the pressure gage of Fig. 2.2 modified by the insertion of a flow restriction between the source of pressure and the piston chamber. Such an arrangement is useful, for example, if one wishes to measure only the average pressure in a large air tank that is being supplied by a reciprocating compressor. The pulsations in the air pressure may be smoothed by the pneumatic filtering effect of the flow restriction and associated volume. The variation of the output-input amplitude ratio $|p_o/p_i|$ with frequency is similar to that for the electrical *RC* filter of Fig. 2.14*e*. Thus steady or slowly varying input pressures are accurately measured while rapid variations are strongly attenuated. The flow restriction may be in the form of a needle valve, allowing easy adjustment of the filtering effect.

A "chopped" radiometer is shown in simplified form in Fig. 2.14*g*. The purpose of this device is to sense the temperature T_s of some body in terms of the infrared radiant energy that it emits. The emitted energy is focused on a detector of some sort and causes the temperature T_d of the detector and thus its output voltage e_i to vary. The difficulty with such devices is that the ambient temperature, as well as T_s, affects T_d. This effect is serious since the radiant energy to be measured causes very small changes in T_d; thus small ambient drifts can completely mask the desired input. An ingenious solution to this problem interposes a rotating shutter between the radiant source and the detector so that the desired input is "chopped," or modulated, at a known frequency. This frequency is chosen to be much higher than the frequencies at which ambient drifts may occur. The output signal e_i of the detector thus is a superposition of slow ambient fluctuations and a high-frequency wave whose amplitude varies in proportion to variations in T_s. Since the desired and interfering components are thus widely separated in frequency, they may be selectively filtered. In this case one desires a filter that rejects constant and slowly varying signals but faithfully reproduces rapid variations. Such a characteristic is typical of an ordinary a-c amplifier, and since amplification is necessary in such instruments in any case, the use of an a-c amplifier as shown solves two problems at once.

In summing up the method of signal filtering, it may be said that, in general, it is usually possible to design filters of mechanical, electrical, thermal, pneumatic, etc., nature which separate signals according to their frequency content in some specific manner. Figure 2.15 summarizes the most common useful forms of such devices.

The *method of opposing inputs* consists of intentionally introducing

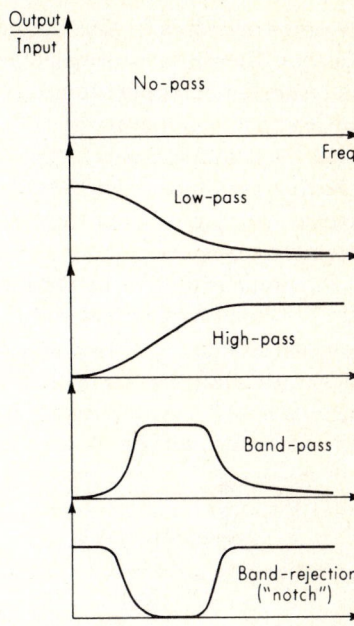

Fig. 2.15. *Basic filter types.*

into the instrument interfering and/or modifying inputs that tend to cancel the bad effects of the unavoidable spurious inputs.   Figure 2.16 shows schematically the concept for interfering inputs.   The extension to modifying inputs should be obvious.   The intentionally introduced input is designed so that the signals o_{I1} and o_{I2} are essentially equal but act in opposite sense; thus the net contribution ($o_{I1} - o_{I2}$) to the output

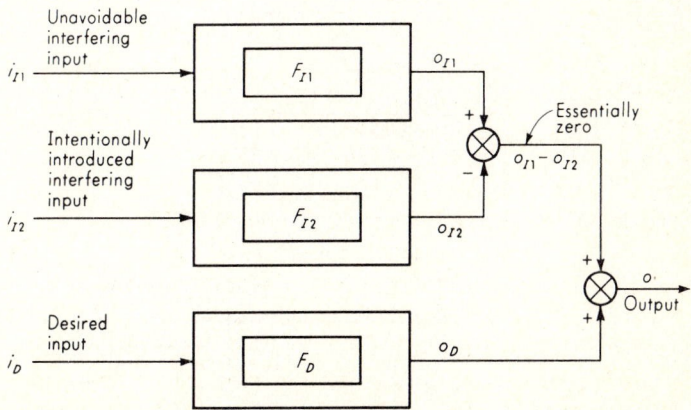

Fig. 2.16. *Method of opposing inputs.*

is essentially zero.   This method might actually be considered as a variation on the method of calculated output corrections; however, the "calculation" and application of the correction are achieved automatically
owing to the structure of the system, rather than by numerical calculation by a human operator.   Thus the two methods are similar; however,
the distinction between them is a worthwhile one since it helps to organize
one's thinking in inventing new applications of these generalized correction concepts.

Some examples of the method of opposing inputs are shown in Fig.
2.17.   A millivoltmeter, as shown in Fig. 2.17a, is basically a *current-
sensitive* device.   However, as long as the total circuit resistance is constant, its scale can be calibrated in voltage since voltage and current are
proportional.   A modifying input here is the ambient temperature, since
it causes the coil resistance R_{coil} to change, thereby changing the propor-

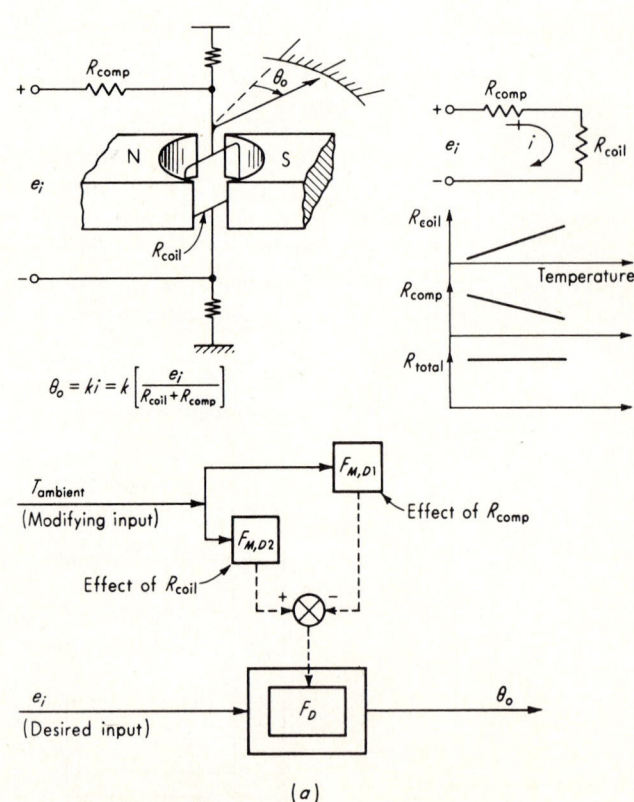

(a)

tionality factor between current and voltage.   To correct for this error, the compensating resistance R_{comp} is introduced into the circuit and its material carefully chosen to have a temperature coefficient of resistance *opposite* to that of R_{coil}.   Thus when the temperature changes, the total resistance of the circuit is unaffected and the calibration of the meter remains accurate.

Figure 2.17*b* shows a static-pressure-probe design due to L. Prandtl. As the fluid flows over the surface of the probe, its velocity must increase since these streamlines are longer than those in the undisturbed flow. This velocity increase causes a drop in static pressure, so that a tap in the surface of the probe gives an incorrect reading.   This underpressure error varies with the distance d_1 of the tap from the probe tip.   Prandtl recognized that the probe support will have a stagnation point (line) along its front edge and that this overpressure will be felt upstream, the effect decreasing as the distance d_2 increases.   By properly choosing the

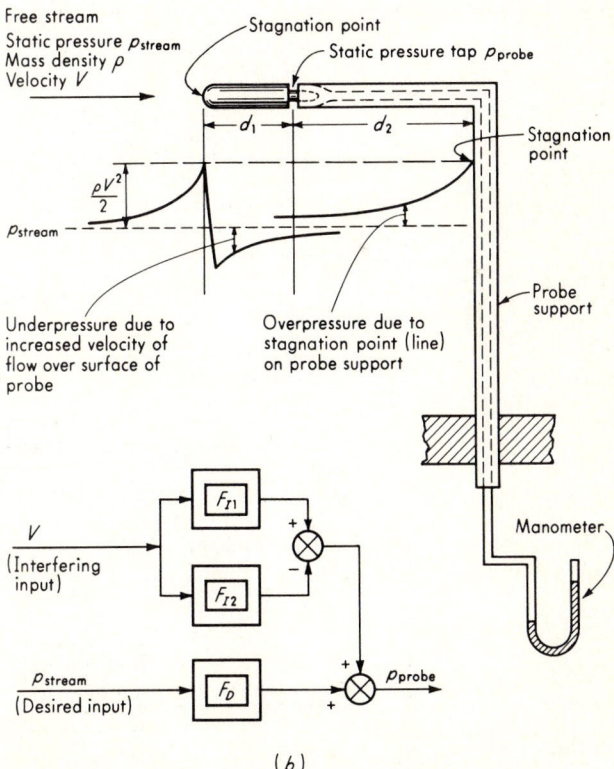

(*b*)

Fig. 2.17.   *Examples of method of opposing inputs.*

distances d_1 and d_2 (by experimental test) these two effects can be made exactly to cancel each other, giving a true static-pressure value at the tap.

A device[1] for the measurement of the mass flow rate of gases is shown in Fig. 2.17c.   The mass flow rate of gas through an orifice may be measured by measuring the pressure drop across the orifice, perhaps by means of a U-tube manometer.   Unfortunately, the mass flow rate also depends on the density of the gas, which varies with pressure and

[1] National Instrument Laboratories, Inc., Washington, D.C.

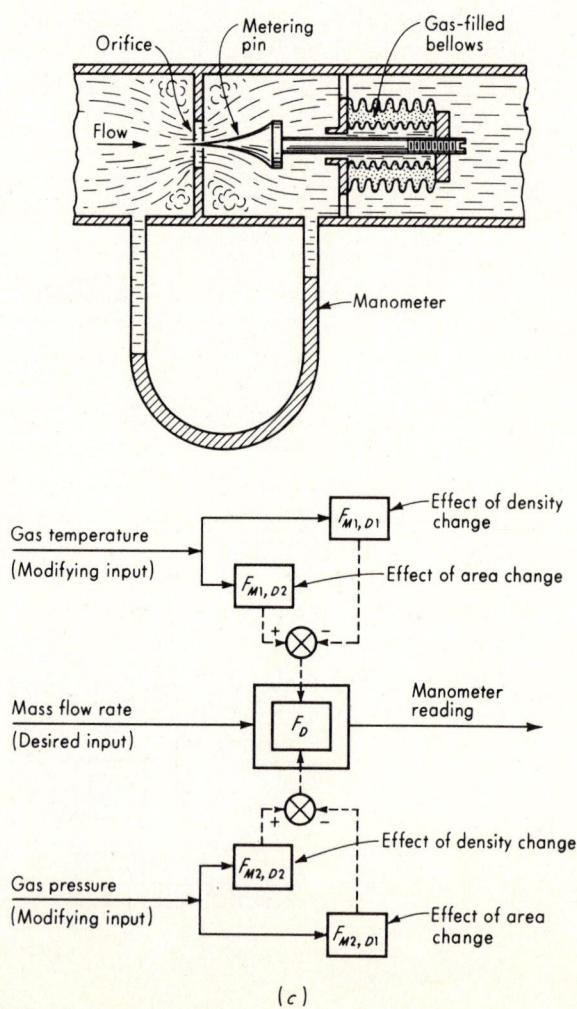

(c)

temperature.   Thus the pressure-drop measuring device cannot usually be calibrated to give mass flow rate, since variations in gas temperature and pressure will give different mass flow rates for the same orifice pressure drop.   The instrument of Fig. 2.17*c* overcomes this problem in an ingenious fashion.   The flow rate through the orifice also depends on its flow area.   Thus if the flow area could be varied in just the right way, this variation could compensate for pressure and temperature changes so that a given orifice pressure drop would *always* correspond to the same mass flow rate.   This is accomplished by attaching the specially shaped metering pin to a gas-filled bellows as shown.   When the temperature drops (causing an increase in density and therefore in mass flow rate) the gas in the bellows contracts, moving the metering pin into the orifice and thereby reducing the flow area.   This returns the mass flow rate to its proper value.   Similarly, should the pressure of the flowing gas

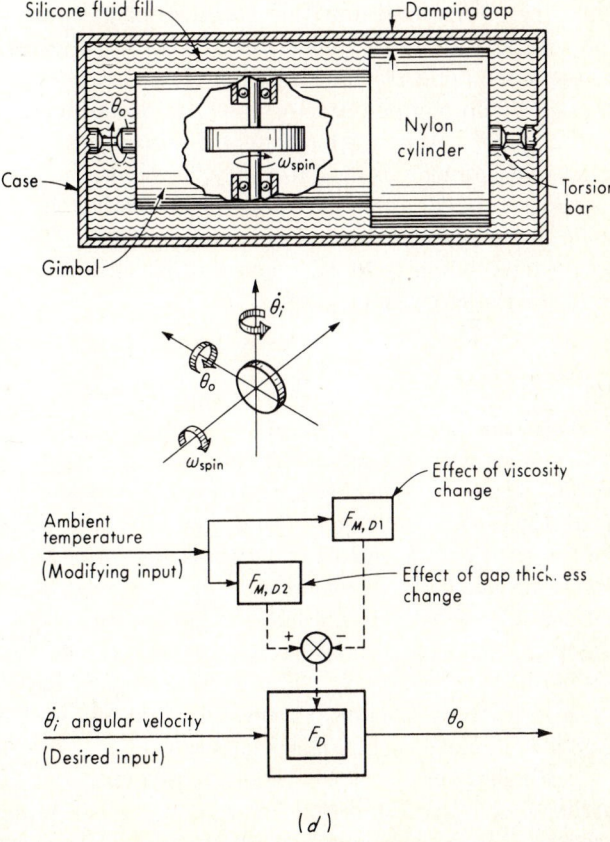

(*d*)

Fig. 2.17. (*Continued.*)

increase, causing an increase in density and mass flow rate, the gas-filled bellows would again be compressed, reducing the flow area and correcting the mass flow rate.   The proper shape for the metering pin is revealed by a detailed analysis of the system.

A final example of the method of opposing inputs is given by the rate gyroscope of Fig. 2.17d.   Such devices are widely used in aerospace vehicles for the generation of stabilization signals in the control system. The action of the device is that a vehicle rotation at angular velocity θ_i causes a proportional rotation θ_o of the gimbal relative to the case.   This rotation θ_o is measured by some motion pickup (not shown in Fig. 2.17d). A signal proportional to vehicle angular velocity is thus available, and this is useful in stabilizing the vehicle.   When the vehicle undergoes rapid motion changes, however, the angle θ_o tends to oscillate, giving an incorrect angular-velocity signal.   To control these oscillations, the gimbal rotation θ_o is damped by the shearing action of a viscous silicone fluid in a narrow damping gap.   The damping effect varies with the viscosity of the fluid and the thickness of the damping gap.   Although the viscosity of the silicone fluid is fairly constant, it does vary with ambient temperature, causing an undesirable change in damping charac- teristics.   To compensate for this, a nylon cylinder is used in the gyro of Fig. 2.17d.   When the temperature increases, viscosity drops, causing a loss of damping.   Simultaneously, however, the nylon cylinder expands, narrowing the damping gap, and thus restoring the damping to its proper value.   By proper choice of materials and geometry, the two effects may be made very nearly to cancel each other over the operating temperature range of the equipment.

2.6

Conclusion   This chapter has attempted to develop useful gen- eralizations with regard to the functional elements and the input-output configurations of measuring instruments and systems.   In the analysis of a given instrument or in the design of a new one, the starting point is the separation of the overall operation into its functional elements. Here one must take a broad view of *what* must be done but not be con- cerned with *how* it is actually to be accomplished.   Once the general functional concepts have been clarified, the details of operation may fruitfully be considered.   The ideas of active and passive transducers, analog and digital modes of operation, and null versus deflection methods give a systematic approach for either analysis or design.

Finally, compensation of spurious inputs and detailed evaluation of performance are facilitated by application of input-output block dia- grams.   These configuration diagrams show clearly which physical analyses must be made to evaluate performance with respect to accurate measurement of the desired inputs and rejection of spurious inputs.   The evaluation of the relative quality of different instruments (or the same instrument with different numerical parameter values) requires the definition of performance criteria against which competitive designs may be compared.   This is the subject of Chap. 3.

Problems

2.1   Make block diagrams such as Fig. 2.1, showing the functional elements of the instruments depicted in the following:

    a. Fig. 2.7.

    b. Fig. 2.8.

    c. Fig. 2.10*a.*

    d. Fig. 2.11.   Take F as input and e_o as output.

    e. Fig. 2.14*g.*   Take T_s as input and e_o as output.

    f. Fig. 2.17*b.*   Take V as input and manometer Δh as output.

    g. Fig. 2.17*d.*   Take θ_i as input and θ_o as output.

2.2   Identify the active transducers, if any, in the instruments of the following:

    a. Fig. 2.8.

    b. Fig. 2.10*a.*

    c. Fig. 2.11.

    d. Fig. 2.17*b.*

    e. Fig. 2.17*c.*

2.3   Consider a man driving a car along a road when he sees the opportunity to pass and decides to accelerate.

    a. If the light waves entering his eyes are considered input and accelerator-pedal travel as output, is the man functioning as an active or a passive transducer?

    b. If accelerator-pedal travel is considered input and car velocity as output, is the automobile engine an active or passive transducer?

2.4   Give an example of a null method of force measurement.

2.5   Give an example of a null method of voltage measurement.

2.6   Sketch and explain two possible modifications of the system of Fig. 2.4 which will allow measurement to $\frac{1}{10}$ revolution.

2.7   Identify desired, interfering, and modifying inputs for the systems of the following:

    a. Fig. 2.2.

    b. Fig. 2.3.

    c. Fig. 2.4.

    d. Fig. 2.5.

2.8   Why is tilt angle in Fig. 2.10*c* a modifying input?

2.9   Suppose in Eq. (2.7) that $K_{MO} = K_{SP} = e_i = 1.0$.   Now let K_{MO} change by 10 percent to 1.1.   What is the change in x_o?   In Eq. (2.10) let $K_{MO} = K_{SP} = K_{FB} = e_i = 1.0$; $K_{AM} = 100$.   Now let K_{MO} change by 10 percent to 1.1. What is the change in x_o?   Investigate the effect of similar changes in K_{AM}, K_{SP}, and K_{FB}.

2.10   The natural frequency of oscillation of the balance wheel in a watch depends on the moment of inertia of the wheel and the spring constant of the (torsional) hairspring.   A temperature rise results in a reduced spring constant, lowering the oscillation frequency.   Propose a compensating means for this effect. Non-temperature-sensitive hairspring material is not an acceptable solution.

Chapter 3

Generalized Performance Characteristics of Instruments

3.1

Introduction   If one is trying to choose, from commercially available instruments, the one most suitable for a proposed measurement, or, alternatively, if one is engaged in the design of instruments for specific measuring tasks, the subject of performance criteria assumes major proportions. That is, to make intelligent decisions, there must be some quantitative bases for comparing one instrument (or proposed design) with the possible alternatives. Chapter 2 has served as a useful preliminary to these considerations since there we developed systematic methods for breaking down the overall problem into its component parts. We now propose to study in considerable detail the performance of measuring instruments and systems with regard to how well they measure the desired inputs and how thoroughly they reject the spurious inputs.

The treatment of instrument performance characteristics has generally been broken down into the subareas of *static characteristics* and *dynamic characteristics*, and this plan will be followed here. The reasons for such a classification are several. First, some applications involve the measurement of quantities that are constant or vary only quite slowly. Under these conditions it is possible to define a set of performance criteria that give a meaningful description of the quality of measurement without becoming concerned with dynamic descriptions involving differential equations. These criteria are called the static characteristics. Many other measurement problems are concerned with rapidly varying quantities; here the dynamic relations between the instrument input and output must be examined, generally by the use of differential equations. Performance criteria based on these dynamic relations constitute the dynamic characteristics.

Actually, static characteristics also influence the quality of measurement under dynamic conditions, but the static characteristics generally show up as nonlinear or statistical effects in the otherwise linear differential equations giving the dynamic characteristics. These effects would make the differential equations unmanageable, and so the conventional approach is to treat the two aspects of the problem separately. Thus the differential equations of dynamic performance generally neglect the effects of dry friction, backlash, hysteresis, statistical scatter, etc., even though they affect the dynamic behavior. These phenomena are more conveniently studied as static characteristics, and the overall performance of an instrument is then judged by a semiquantitative superposition of the static and dynamic characteristics. This approach is, of course, approximate but is a necessary expedient.

3.2

Static Characteristics   We begin our study of static performance characteristics by considering the meaning of the term static calibration.

The meaning of static calibration. All the static performance characteristics are obtained by one form or another of a process called static calibration. It is therefore appropriate to develop at this point a clear concept of what is meant by this term.

In general, static calibration refers to a situation where all inputs (desired, interfering, modifying) except one are kept at some constant values. The one input under study is then varied over some range of constant values, causing the output(s) to vary over some range of constant values.   The input-output relations developed in this way comprise a static calibration *valid under the stated constant conditions of all the other inputs.*   This procedure may be repeated, varying in turn each input considered to be of interest and thus developing a family of static input-output relations.   One might then hope to describe the overall instrument static behavior by means of some suitable form of superposition of these individual effects.   In some cases, if overall rather than individual effects are desired, the calibration procedure would specify the variation of several inputs simultaneously.   It should also be understood that if one examines any practical instrument critically he will find many modifying and/or interfering inputs each of which might have quite small effects and which would be impractical to control.   Thus the statement that *all* other inputs are held constant refers to an ideal situation which can only be approached, but never reached, in practice.   The term *measurement method* has been used to describe the ideal situation while the term *measurement process* describes the (imperfect) physical realization of the measurement method.

The statement that one input is varied and all others held constant implies that all these inputs are determined (measured) independently of the instrument being calibrated.   For interfering or modifying inputs (whose effects on the output should be relatively small in a good instrument), the measurement of these inputs usually need not be at an extremely high accuracy level.   For example, suppose a pressure gage has temperature as an interfering input to the extent that a temperature change of 100°F causes a pressure error of 0.100 percent.   Now, if we had measured the 100°F interfering input with a thermometer which itself had an error of 2.0 percent, the pressure error could actually have been 0.102 percent.   It should be clear that the difference between an error of 0.100 and 0.102 percent is entirely negligible in most engineering situations.   However, when calibrating the response of the instrument to its *desired* inputs one must exercise considerable care in choosing the means of determining the numerical values of these inputs.   That is, if a pressure gage is inherently capable of an accuracy of 0.1 percent one must certainly be able to determine its input pressure during calibration with an accuracy somewhat greater than this.   In other words, it is impossible to calibrate an instrument to an accuracy greater than that of the standard with which it is compared.   A rule often followed is that the calibration standard should be at least about 10 times as accurate as the instrument being calibrated.

While we shall not go into a detailed discussion of standards at this point, it is of utmost importance that the person performing the calibration be able to answer the question: How do I know that this standard is capable of its stated accuracy? The ability to trace the accuracy of a standard back to its ultimate source in the fundamental standards of the National Bureau of Standards is termed *traceability*.

In performing a calibration the following steps are thus necessary:

1. Examine the construction of the instrument and identify and list all the possible inputs.
2. Decide, as best you can, which of the inputs will be significant in the application for which the instrument is to be calibrated.
3. Procure apparatus that will allow you to vary all significant inputs over the ranges considered necessary.
4. By holding some inputs constant, varying others, and recording the output(s), develop the desired static input-output relations.

We are now ready for a more detailed discussion of specific static characteristics. These characteristics may first be classified as either general or special. General static characteristics are of interest in *every* instrument. Special static characteristics are of interest only in a particular instrument. We shall concentrate mainly on general characteristics, leaving the treatment of special characteristics to later sections of the text in which specific instruments are discussed.

Accuracy, precision, and bias. When one makes a measurement of some physical quantity with an instrument and obtains a numerical value, he is usually concerned with how close this value may be to the "true" value. It is first necessary to understand that this so-called true value is, in general, unknown and unknowable, since perfectly exact definitions of the physical quantities that are to be measured are impossible. This can be illustrated by specific example, for instance, the length of a cylindrical rod. When we ask ourselves what we *really* mean by the length of this rod, we must consider such questions as these:

1. Are the two ends of the rod planes?
2. If they are planes, are they parallel?
3. If they are not planes, what sort of surfaces are they?
4. What about surface roughness?

We see that complex problems are introduced when we deal with a real object rather than an abstract geometrical solid. The term *true value*, then, refers to a value that would be obtained if the quantity under consideration were measured by an *exemplar method*,[1] that is, a method agreed

[1] Churchill Eisenhart, Realistic Evaluation of the Precision and Accuracy of Instrument Calibration Systems, *J. Res. Natl. Bur. Std., C*, vol. 67C, no. 2, April–June, 1963.

upon by experts as being sufficiently accurate for the purposes to which the data will ultimately be put.

We must also be concerned over whether we are describing the characteristics of a single reading of an instrument or the characteristics of a measurement process. If we are speaking of a single measurement, the *error* is the difference between the measurement and the corresponding true value, taken positive if the measurement is greater than the true value. When using an instrument, however, we are concerned with the characteristics of the measurement process associated with that instrument. That is, we may take a single reading, but this is a sample from a statistical population generated by the measurement process. If we know the characteristics of the process, we can *put bounds on* the error of the single measurement, although we cannot tell what the error itself is, since this would imply that we knew the true value. We are thus interested in being able to make statements about the accuracy (lack of error) of our readings. This can be done in terms of the concepts of *precision* and *bias* of the measurement process.

The measurement process consists of actually carrying out, as well as possible, the instructions for performing the measurement, which are the measurement method. (Since calibration is essentially a refined form of measurement, these remarks apply equally to the process of calibration.) If this process is repeated over and over again under *assumed* identical conditions, we get a large number of readings from the instrument. Usually these readings will not all be the same, and so we note immediately that we may *try* to assure identical conditions for each trial but it is never exactly possible. The data generated in this fashion may be used to describe the measurement process so that, if it is used in the future, we may be able to attach some numerical estimates of error to its outputs.

If the output data are to give a meaningful description of the measurement process, the data must form what is called a *random sequence*. Another way of saying this is that the process must be in a state of *statistical control*.[1] The concept of the state of statistical control is not a particularly simple one but we shall try to explain its essence briefly. We first note that it is meaningless to speak of the accuracy of an instrument as an isolated device; one must always consider the instrument plus its environment and method of use, that is, the instrument plus its inputs. This aggregate constitutes the measurement process. Every instrument has an infinite number of inputs; that is, the causes that can conceivably affect the output, if only very slightly, are limitless. Such effects as atmospheric pressure, temperature, and humidity are among the more

---

[1] *Ibid.*

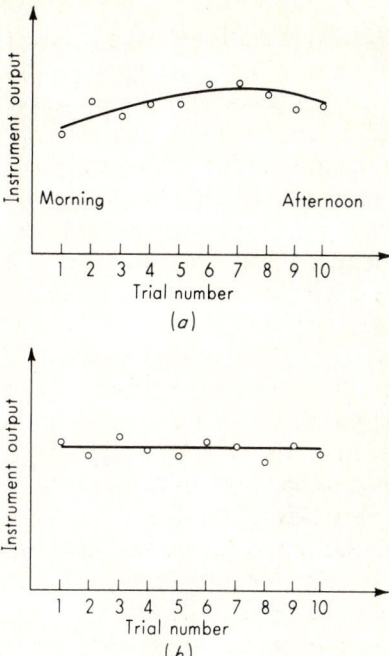

Fig. 3.1. *Effect of uncontrolled input on calibration.*

obvious but if one is willing to "split hairs" he can uncover a multitude of other physical causes which could affect the instrument with varying degrees of severity. In defining a calibration procedure for a specific instrument, one specifies that certain inputs be held "constant" within certain limits. These inputs, it is hoped, are the ones that contribute the largest components to the overall error of the instrument. The remaining infinite number of inputs is left uncontrolled, and it is hoped that each of these individually contributes only a very small effect and that in the aggregate their effect on the instrument output will be of a random nature. If this is indeed the case, the process is said to be in statistical control. Experimental proof that a process is in statistical control is not easy to come by; in fact, *strict* statistical control is unlikely of practical achievement. Thus one can only approximate this situation.

Lack of control is sometimes obvious, however, if one repeats a measurement and plots the result (output) versus the trial number. Figure 3.1a shows such a graph for the calibration of a particular instrument. In this instance it was ascertained after some study that the instrument was actually much more sensitive to temperature than had been thought. The original calibration was carried out in a room without temperature control; thus the room temperature varied from a low in the morning to a peak in the early afternoon and then dropped again in the

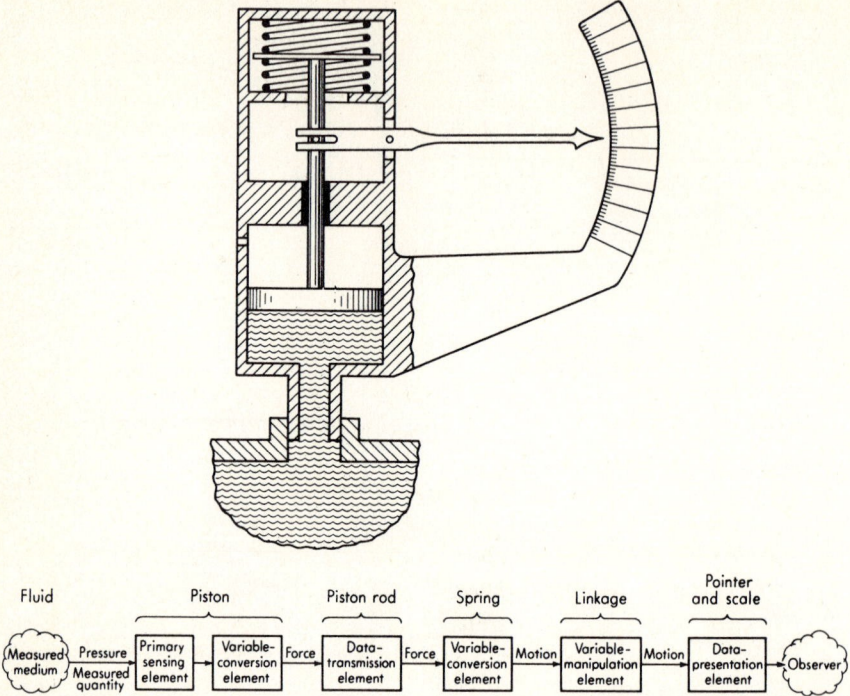

Fig. 3.2. *Pressure gage.*

late afternoon. Since the 10 trials covered a period of about one day, the trend of the curve is understandable. By performing the calibration in a temperature-controlled room, the graph of Fig. 3.1*b* was obtained. For the detection of more subtle deviations from statistical control, the methods of statistical-quality-control charts are useful.[1]

If the measurement process is in reasonably good statistical control and if one repeats a given measurement (or calibration point) over and over, he will generate a set of data exhibiting random scatter. As an example, consider the pressure gage of Fig. 3.2. Suppose we wish to determine the relationship between the desired input (pressure) and the output (scale reading). Other inputs which could be significant and which might have to be controlled during the pressure calibration include temperature, acceleration, and vibration. Temperature can cause expansion and contraction of instrument parts in such a way that the scale reading will change even though the pressure has remained constant.

[1] E. B. Wilson, Jr., "An Introduction to Scientific Research," chap. 9, McGraw-Hill Book Company, New York, 1952.

True pressure $= 10.000 \pm .001$ *psig*
Acceleration $= 0$
Vibration level $= 0$
Ambient temperature $= 70 \pm 1°F$

Trial number	Scale reading, psig
1	10.02
2	10.20
3	10.26
4	10.20
5	10.22
6	10.13
7	9.97
8	10.12
9	10.09
10	9.90
11	10.05
12	10.17
13	10.42
14	10.21
15	10.23
16	10.11
17	9.98
18	10.10
19	10.04
20	9.81

Fig. 3.3.   *Pressure-gage calibration data.*

An acceleration of the instrument, which has a component along the axis of the piston rod, will cause a scale reading even though pressure has again remained unchanged. This input is significant if the pressure gage is to be used aboard a vehicle of some kind. A small amount of vibration may actually be helpful to the operation of an instrument since it may reduce the effects of static friction. Thus if the pressure gage is to be used by attaching it to a reciprocating air compressor (which always has some vibration) it may be more accurate under these conditions than it would be under calibration conditions where no vibration was provided. These examples should illustrate the general importance of carefully considering the relationship between the calibration conditions and the actual application conditions.

Suppose now that we have procured a sufficiently accurate pressure standard and have arranged to maintain the other inputs reasonably close to the actual application conditions. Repeated calibration at a given pressure (say 10 psig) might give the data of Fig. 3.3. Suppose we now

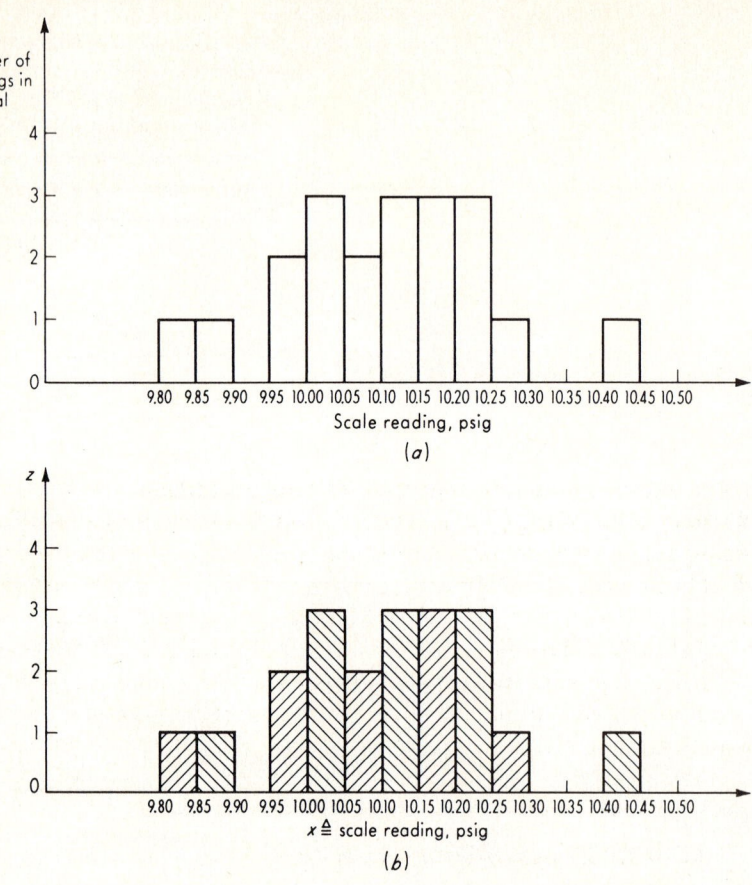

Fig. 3.4.   *Distribution of data.*

order the readings from the lowest (9.81) to the highest (10.42) and see how many readings fall in each interval of, say, 0.05 psig, starting at 9.80. The result can be represented graphically as in Fig. 3.4a. Suppose we now define the quantity Z by

$$Z \triangleq \frac{(\text{number of readings in an interval})/(\text{total number of readings})}{\text{width of interval}}$$

$$(3.1)$$

and plot a "bar graph" with height Z for each interval. Such a "histogram" is shown in Fig. 3.4b. It should be clear from Eq. (3.1) that the area of a particular "bar" is numerically equal to the probability that a particular reading will fall in the associated interval. The area of the entire histogram must then be 1.0 (100 percent = 1.0), since there is

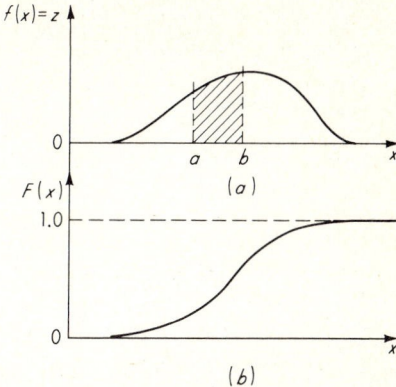

Fig. 3.5. *Probability distribution functions.*

100 percent probability that the reading will fall somewhere between the lowest and highest value, at least based on the data available. If it were now possible to take an infinite number of readings, each with an infinite number of significant digits, we could make the chosen intervals as small as we pleased and still have each interval contain a finite number of readings. Thus the steps in the graph of Fig. 3.4b would become smaller and smaller, the graph approaching in the limit a smooth curve. If we take this limiting abstract case as a mathematical model for the real physical situation, the function $Z = f(x)$ is called the *probability density function* for the mathematical model of the real physical process (see Fig. 3.5a). From the basic definition of Z, it should be clear that

Probability of reading lying between a and $b \triangleq P(a < x < b)$

$$= \int_a^b f(x)\, dx \qquad (3.2)$$

From the infinite number of forms possible for probability density functions, a relatively small number have been found useful mathematical models for practical applications; in fact, *one* particular form is quite dominant. The probability information is sometimes given in terms of the *cumulative distribution function* $F(x)$, which is defined by

$F(x) \triangleq$ probability that reading is less than any chosen

$$\text{value of } x = \int_{-\infty}^x f(x)\, dx \qquad (3.3)$$

and is shown in Fig. 3.5b.

The most useful density function or distribution is the normal or *gaussian* function, which is given by

$$f(x) = \frac{1}{\sqrt{2\pi}\,\sigma}\, e^{-(x-\mu)^2/2\sigma^2} \qquad -\infty < x < +\infty \qquad (3.4)$$

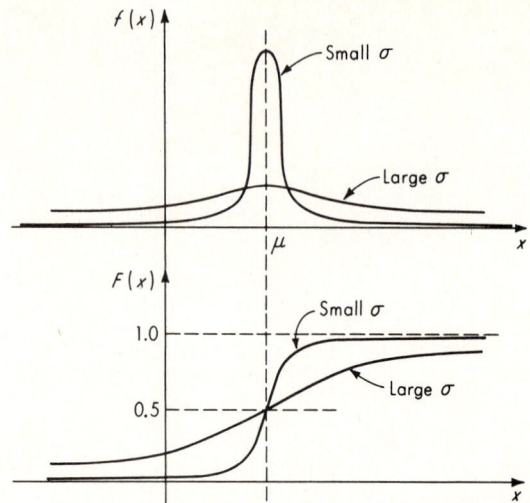

Fig. 3.6. *Gaussian distribution.*

Equation (3.4) defines a whole family of curves depending on the particular numerical values of μ (the mean value) and σ (the standard deviation). The shape of the curve is determined entirely by σ, μ serving only to locate its position along the x axis. The cumulative distribution function $F(x)$ cannot be written explicitly in this case because the integral of Eq. (3.3) cannot be carried out; however, the function has been tabulated by performing the integration by numerical means. Figure 3.6 shows that a small value of σ indicates a high probability that a "reading" will be found close to μ. Equation (3.4) also shows that there is a small probability that very large ($\rightarrow \pm \infty$) readings will occur. This is one of the reasons why a true gaussian distribution can never occur in the real world; physical variables are always limited to finite values. There is *zero* probability, for example, that the pointer on a pressure gage will read 100 psig when the range of the gage is only 20 psig. Real distributions must thus, in general, have their "tails" cut off, as in Fig. 3.7.

Although actual data may not conform *exactly* to the gaussian distribution, they very often are sufficiently close to allow use of the gaussian

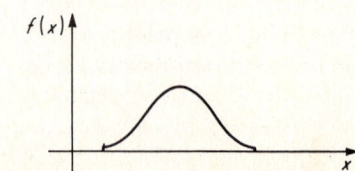

Fig. 3.7. *Nongaussian distribution.*

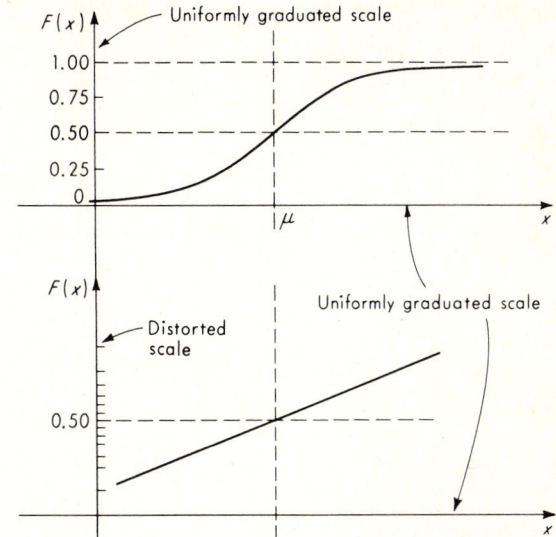

Fig. 3.8.   *Rectification of gaussian curve.*

model in engineering work.   It would be desirable to have available tests
that would indicate whether the data are "reasonably" close to gaussian,
and two such procedures will be explained briefly.   It must be admitted,
however, that in much practical work the time and effort necessary for
such tests cannot be justified and the gaussian model is simply *assumed*
until troubles arise which justify a closer study of the particular situation.

The first method of testing for an approximate gaussian distribution
involves the use of probability graph paper.   If one takes the cumulative
distribution function for a gaussian distribution and suitably distorts the
vertical scale of the graph, the curve can be made to plot as a straight
line, as shown in Fig. 3.8.   (This, of course, can be done with any curvilinear relation, not just probability curves.)   Such graph paper is commercially available and may be used to give a rough qualitative test for
conformity to the gaussian distribution.   For example, consider the data
of Fig. 3.3.   These data may be plotted on gaussian probability graph
paper as follows: First lay out on the uniformly graduated horizontal axis
a numerical scale that includes all the pressure readings.   Now the probability graph paper represents the cumulative distribution, so that the
ordinate of any point represents the probability that a reading will be
less than the abscissa of that particular point.   This probability, in terms
of the sample of data available, is simply the percentage (in decimal form)
of points that fell at or below that particular value.   Figure 3.9 shows

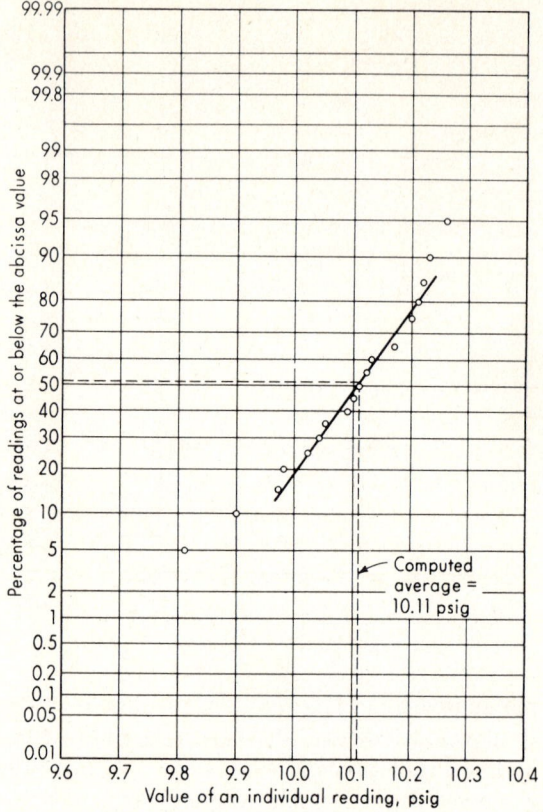

Fig. 3.9.   *Graphical check of gaussian distribution.*

the resulting plot.   Note that it is not possible to plot the highest point (10.42) since the 100 percent point cannot appear on the ordinate scale. For the data to be perfectly gaussian, the following requirements must be met:

1.   All the points must fall exactly on one straight line.
2.   The average of all the data points must intersect this line at an ordinate of exactly 0.50 (50 percent).

The procedure is thus to fit a straight line to the data points (usually by eye) and then see how well the above two requirements are met.   It is advisable not to give much weight to the points at the low and high ends of the plot since these represent the "tails" of the distribution which are not accurately defined unless a very large number of readings have been made.

The results of Fig. 3.9 would be considered by most people to indicate a reasonable approximation to a gaussian distribution. However, this test is obviously only qualitative, and its main usefulness is perhaps in indicating *gross* departures from the theoretical distribution when they exist. Such deviations would lead one to examine the instrument and measurement process more closely before attempting to make any statistical statements about accuracy.

Another method of testing for "normality" (gaussian distribution) involves use of the chi-squared (χ^2) statistical test. This method puts the conclusions on a somewhat more quantitative basis but there is still an element of uncertainty involved, as there must be, since *perfect* gaussian distributions simply do not exist in nature. To use the χ^2 test, one must first develop some additional information. The exact gaussian distribution which is taken as a mathematical model of the real distribution has the parameters μ (mean or average value) and σ (standard deviation). The true values of these numbers cannot be exactly known; they can merely be estimated from the data taken on the real distribution. The estimate of μ is given the symbol $\bar{x}$ and is the ordinary average value

$$\bar{x} \triangleq \frac{\sum\limits_{i=1}^{N} x_i}{N} \qquad (3.5)$$

$$\text{where} \quad x_i \triangleq \text{individual reading} \qquad (3.6)$$
$$N \triangleq \text{total number of readings} \qquad (3.7)$$

The estimate of σ is given the symbol s and the name sample standard deviation and is computed from

$$s \triangleq \sqrt{\frac{\sum\limits_{i=1}^{N} (x_i - \bar{x})^2}{N - 1}} \qquad (3.8)$$

For the data of Fig. 3.3 we have $\bar{x} = 10.11$ psig and $s = 0.14$ psig. For a perfect gaussian distribution, it can be shown that

$$\begin{array}{l} 68 \text{ percent of the readings lie within } \pm\sigma \text{ of } \mu \\ 95 \text{ percent of the readings lie within } \pm 2\sigma \text{ of } \mu \qquad (3.9) \\ 99.7 \text{ percent of the readings lie within } \pm 3\sigma \text{ of } \mu \end{array}$$

Thus, if we assume our real distribution is nearly gaussian, we might predict, for instance, that if more readings were taken 99.7 percent would fall within ± 0.42 psig of 10.11 psig. These estimates $\bar{x}$ and s of μ and σ can be improved by taking more readings. That is, suppose we took 20 more readings and computed $\bar{x}$ from these 20 readings. We would probably get a value different from 10.11. Thus $\bar{x}$ itself exhibits scatter.

We can describe this scatter by determining a standard deviation of the mean, $s_{\bar{x}}$, which would then allow statements such as (3.9) to be made about the closeness of $\bar{x}$ to μ.   It can be shown[1] that

$$s_{\bar{x}} = \frac{s}{\sqrt{N-1}} \qquad (3.10)$$

where s and N are as in Eqs. (3.7) and (3.8).   This shows that we can get a better and better estimate of μ by using larger and larger samples. For our original sample of 20 readings we could make the statement: "There is a probability of 95 percent that $\bar{x}$ does not differ from μ by more than $\pm 0.28/\sqrt{19} = \pm 0.064$ psig."   If we had 80 readings, this could be reduced to ± 0.032 psig (assuming the value of s stays about at 0.14).

Let us return now to the chi-squared test for estimating the approach to normality of a set of data.   The first step is to order the data from the lowest to the highest and then group them so that no group has less than four or five members.   For the test to be carried out at all, there must be at least four groups, and so we see that a sample of at least about 16 readings should be taken.   The larger the sample, the more significant the test will be.   The quantity χ^2 is defined as follows:

$$\chi^2 \triangleq \sum_{i=1}^{n} \frac{(n_0 - n_e)^2}{n_e} \qquad (3.11)$$

where $n_0 \triangleq$ number of readings actually observed in given range (group)
$\quad n_e \triangleq$ number of readings that would be observed in same range
$\qquad$ if distribution were normal, i.e. with $\mu = \bar{x}$ and $\sigma = s$
$\quad n \triangleq$ number of groups

It is necessary to explain how the number n_e is calculated.   We mentioned earlier that tables of the cumulative gaussian distribution are available.[2] These tables allow one to calculate the number n_e as follows: Entries in the table are values of $F(w)$, where

$F(w) \triangleq$ probability that reading falls in range from $-\infty$ to w $\qquad$ (3.12)

$w \triangleq \dfrac{x - \mu}{\sigma}$ $\hfill$ (3.13)

Definition (3.13) puts the tables on a nondimensional basis so that one table serves for all possible values of μ and σ.   As an example, suppose we wish to calculate n_e for the first range, $-\infty$ to 10.00.   This range of x

[1] A. M. Mood, "Introduction to the Theory of Statistics," pp. 133, 159, McGraw-Hill Book Company, New York, 1950.
[2] R. S. Burington, "Handbook of Mathematical Tables and Formulas," 2d ed., p. 257, McGraw-Hill Book Company, New York, 1940.

Group number	Range of x	Range of w	n_0	n_e	$\dfrac{(n_0 - n_e)^2}{n_e}$
1	$-\infty$ to 10.00	$-\infty$ to -0.79	4	4.296	0.020
2	10.00 to 10.095	-0.79 to -0.107	4	4.864	0.153
3	10.095 to 10.15	-0.107 to 0.286	4	3.080	0.274
4	10.15 to 10.215	0.286 to 0.75	4	3.220	0.189
5	10.215 to ∞	0.75 to ∞	4	4.532	0.062

$$\chi^2 = 0.698$$

Fig. 3.10.   *Tabulation for chi-squared test.*

corresponds to a range on w of $-\infty$ to -0.79 since we use Eq. (3.13) with $\mu = \bar{x} = 10.11$ and $\sigma = s = 0.14$ to calculate w. Now the probability that a reading falls in the range $-\infty$ to -0.79 is the same as the probability that it falls in $+0.79$ to $+\infty$ since the gaussian curve is perfectly symmetrical about $w = 0$. Entering the table for $w = 0.79$, we find $P(-\infty < w < 0.79)$ is 0.7852. Thus, $P(0.79 < w < \infty)$ is $(1.0000 - 0.7852) = 0.2148$, and we would expect in a sample of 20 trials that $(20)(0.2148) = 4.296$ readings would fall in the range $-\infty < x < 10.00$. Actually, we found that exactly four readings fell in this range. All other entries in Fig. 3.10 are found in the same manner.

To make the final interpretation of this test, one must know the number of "degrees of freedom." This is numerically equal to the number of groups minus 3, and so in the present example we have 2 degrees of freedom. The significance of the numerical value of χ^2 is given in Fig. 3.11, which is interpreted as follows: If we had a perfectly gaussian distribution with $\mu = 10.11$ and $\sigma = 0.14$ from which we drew a sample of 20 readings, we would *not*, in general, get a χ^2 value of zero. That is, because of the random nature of the readings, any finite sample will not exhibit the same properties as its "parent" population, and the smaller the sample the more likely it is that, just by chance, there is taken a sample that *appears* to be nongaussian. In Fig. 3.11 there is a probability of 5 percent that χ^2 would fall above the upper curve if the distribution were actually gaussian with $\mu = 10.11$ and $\sigma = 0.14$ and similarly a probability of 95 percent that χ^2 would fall above the lower curve. Thus if we compute a value of χ^2 that falls in either crosshatched region it is highly unlikely (though not impossible) that the sample came from the assumed gaussian distribution. (The 5 and 95 percent values were chosen somewhat arbitrarily. Tables[1] for other chosen percentages are available.)

[1] D. V. Huntsberger, "Elements of Statistical Inference," pp. 177, 259, Allyn and Bacon, Inc., Boston, 1961.

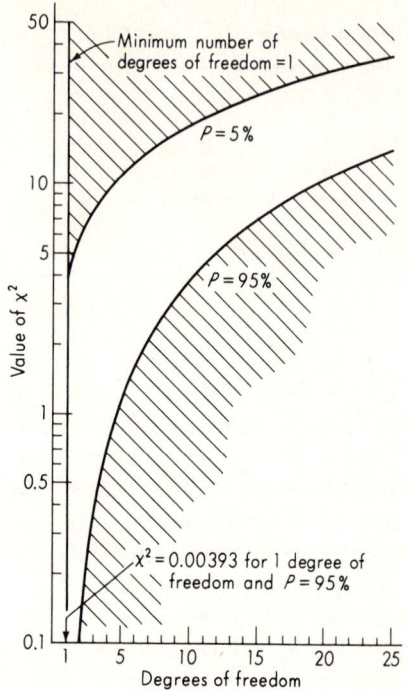

Fig. 3.11. *Chi-squared-test graph.*

If the point does not fall in the crosshatched areas, we can only say that there is no strong evidence of nongaussian behavior. It should be noted that, for small numbers of degrees of freedom, x^2 can vary greatly (0.00393 to 3.84, almost 4 orders of magnitude for 1 degree of freedom) and still fall between the 5 and 95 percent lines. This simply means that for small samples (which give a low number of degrees of freedom) one cannot get a very sensitive test of normality. That is, even if the distribution *is* gaussian, small samples of it can very likely have a wide variation in the n_e's and thus in x^2. However, large samples reduce the allowable range of x^2 considerably (from 14.6 to 37.7 for 25 degrees of freedom). For our present example, $x^2 = 0.698$ for 2 degrees of freedom; thus normality is not disproved.

Most statistics texts recommend that if any of the n_e's are much less than 5 the groups be redefined to prevent this. Since Fig. 3.10 exhibits this problem, it might be in order to make some changes. The new groups and the results are shown in Fig. 3.12. The point ($x^2 = 0.3$, degrees of freedom $= 1$) again falls between the 5 and 95 percent lines and so our conclusions are unchanged.

Having considered the problem of determining the normality of scattered data, we now return to the main business of this section, that is,

Group number	Range of x	Range of w	n_0	n_e	$\dfrac{(n_0 - n_e)^2}{n_e}$
1	$-\infty$ to 10.03	$-\infty$ to -0.572	5	5.66	0.077
2	10.03 to 10.115	-0.572 to 0.0357	5	4.62	0.031
3	10.115 to 10.215	0.0357 to 0.75	6	5.18	0.130
4	10.215 to ∞	0.75 to ∞	4	4.532	0.062

$$\chi^2 = 0.300$$

Fig. 3.12. *Tabulation for chi-squared test.*

definition of the terms accuracy, precision, and bias. Up to now we have been examining the situation wherein a single true value is applied repeatedly and the resulting measured values are recorded and analyzed. In an actual instrument calibration the true value is varied, in increments, over some range, causing the measured value also to vary over a range. Very often there is no multiple repetition of a given true value, the procedure being merely to cover the desired range in both the increasing and the decreasing direction. Thus a given true value is applied, at most, twice if one chooses to use the same set of true values for both increasing and decreasing readings.

As an example, suppose we wish to calibrate the pressure gage of Fig. 3.2 for the relation between the desired input (pressure) and the output (scale reading). Figure 3.13 gives the data for such a calibration over the range 0 to 10 psig. In this instrument (as in most but not all) the input-output relation is ideally a straight line. The *average calibration curve* for such an instrument is generally taken as a straight line which fits the scattered data points best as defined by some chosen criterion. The most common is the least-squares criterion which minimizes the sum of the squares of the vertical deviations of the data points from the fitted line. (The least-squares procedure can also be used to fit curves other than straight lines to scattered data if this should be desired.) The equation for the straight line is taken as

$$q_o = mq_i + b \qquad (3.14)$$

where $q_o \triangleq$ output quantity (dependent variable)   (3.15)
        $q_i \triangleq$ input quantity (independent variable)   (3.16)
        $m \triangleq$ slope of line   (3.17)
        $b \triangleq$ intercept of line on vertical axis   (3.18)

The equations for calculating m and b may be found in several

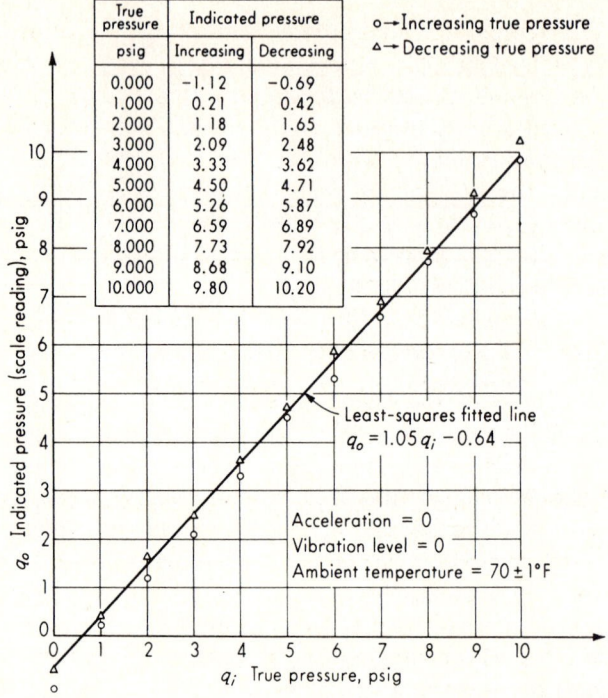

True pressure	Indicated pressure	
psig	Increasing	Decreasing
0.000	−1.12	−0.69
1.000	0.21	0.42
2.000	1.18	1.65
3.000	2.09	2.48
4.000	3.33	3.62
5.000	4.50	4.71
6.000	5.26	5.87
7.000	6.59	6.89
8.000	7.73	7.92
9.000	8.68	9.10
10.000	9.80	10.20

$\circ \rightarrow$ Increasing true pressure
$\triangle \rightarrow$ Decreasing true pressure

Least-squares fitted line
$q_o = 1.05 q_i − 0.64$

Acceleration = 0
Vibration level = 0
Ambient temperature = $70 \pm 1°F$

q_o Indicated pressure (scale reading), psig

q_i True pressure, psig

Fig. 3.13.   *Pressure-gage calibration.*

references[1]:

$$m = \frac{N\Sigma q_i q_o - (\Sigma q_i)(\Sigma q_o)}{N\Sigma q_i^2 - (\Sigma q_i)^2} \qquad (3.19)$$

$$b = \frac{(\Sigma q_o)(\Sigma q_i^2) - (\Sigma q_i q_o)(\Sigma q_i)}{N\Sigma q_i^2 - (\Sigma q_i)^2} \qquad (3.20)$$

where $N \triangleq$ total number of data points   (3.21)

In the present example, calculation gives $m = 1.05$ and $b = -0.64$ psig. Since these values are derived from scattered data, it would be useful to have some idea of their possible variation. The standard deviations of m and b may be found from

$$s_m^2 = \frac{N s_{q_o}^2}{N\Sigma q_i^2 - (\Sigma q_i)^2} \qquad (3.22)$$

$$s_b^2 = \frac{s_{q_o}^2 \Sigma q_i^2}{N\Sigma q_i^2 - (\Sigma q_i)^2} \qquad (3.23)$$

where $s_{q_o}^2 = \frac{1}{N} \sum (m q_i + b - q_o)^2$ (3.24)

[1] H. D. Young, "Statistical Treatment of Experimental Data," p. 121, McGraw-Hill Book Company, New York, 1962.

The symbol s_{q_o} represents the standard deviation of q_o. That is, if q_i were fixed and then repeated over and over, q_o would give scattered values, the amount of scatter being indicated by s_{q_o}. If we assume that this s_{q_o} would be the same for *any* value of q_i, we can calculate s_{q_o} using *all* the data points of Fig. 3.13 and *without* having to repeat any one q_i many times. For the present example, calculation gives $s_{q_o} = 0.23$ psig. Then $s_m = 0.0154$ and $s_b = 0.091$ psig. Assuming a gaussian distribution and the 99.7 percent limits ($\pm 3s$), we could then give m as 1.05 ± 0.05 and b as -0.64 ± 0.27 psig.

In *using* the calibration results, the situation is one where q_o (the indicated pressure) is known and one wishes to make a statement about q_i (the true pressure). The least-squares line gives

$$q_i = \frac{q_o + 0.64}{1.05} \qquad (3.25)$$

However, the q_i value computed in this way must have some $\pm$ error limits put on it. These can be obtained since s_{q_i} can be computed from

$$s_{q_i}{}^2 = \frac{1}{N} \sum \left(\frac{q_o - b}{m} - q_i \right)^2 = \frac{s_{q_o}{}^2}{m^2} \qquad (3.26)$$

which in this example gives $s_{q_i} = 0.22$ psig. Thus if one were using this gage to measure an unknown pressure and got a reading of 4.32 psig, his estimate of the true pressure would be 4.72 ± 0.66 psig if he wished to use the $\pm 3s$ limits.

Another common method of giving bounds on the error uses the *probable error*, e_p. This is defined by

$$e_p \triangleq 0.674s \qquad (3.27)$$

A range of $\pm e_p$ includes the true value 50 percent of the time. In this case the above value would be quoted as 4.72 ± 0.15 psig. It should be clear that when one is using statements of this kind it is extremely important to state whether probable errors or $\pm 3s$ limits are used.

We should note that in computing s_{q_o}, either of two approaches could be used. One might use data such as that in Fig. 3.13 and apply Eq. (3.24) or, alternatively, repeat a given q_i many times and compute s_{q_o} from Eq. (3.8). If s_{q_o} is actually the same for all values of q_i (as assumed above), these two methods should give the same answer for large samples. In computing s_{q_i}, however, the second method is not feasible because one cannot, in general, fix q_o in a calibration and then repeat that point over and over to get scattered values of q_i. This is because q_i is truly an independent variable (subject to one's choice) whereas q_o is dependent (not subject to choice). Thus, in computing s_{q_i}, an approach such as Eq. (3.26) is necessary.

A calibration such as that of Fig. 3.13 allows decomposition of the total error of a measurement process into two parts, the *bias* and the *imprecision*. That is, if one gets a reading of 4.32 psig, the true value is given as 4.72 ± 0.66 psig (3*s* limits), the bias would be −0.40 psig, and the imprecision ±0.66 psig (3*s* limits). Of course, once the instrument has been calibrated, the bias can be removed, and the only remaining error is that due to imprecision. The bias is also called the *systematic error* (since it is the same for each reading and can thus be removed by calibration). The error due to imprecision is called the *random error* or *nonrepeatability* since it is, in general, different for every reading and one can only put bounds on it but cannot remove it. Calibration is thus the process of removing bias and defining imprecision numerically. The *total inaccuracy* of the process is defined by the combination of bias and imprecision. If the bias is known, the total inaccuracy is entirely due to imprecision and can be specified by a single number such as s_{q_i}.

In actual engineering practice the accuracy of an instrument is usually given by a single numerical value; very often it is not made clear just what the precise meaning of this number is meant to be. Often, even though a calibration, as in Fig. 3.13, has been carried out, s_{q_i} is not calculated. The error is taken as the largest horizontal deviation of any data point from the fitted line. In Fig. 3.13 this occurs at $q_i = 0$ and amounts to 0.48 psig. The inaccuracy in this case might thus be quoted as ±4.8 percent of full scale. Note that this corresponds to about ±2s_{q_i} in this case. This practice is no doubt due to the practical viewpoint that when one takes a measurement all he really wants is to say that it cannot be incorrect by more than some specific value; thus the "easy way out" is simply to give a single number. This would be legitimate if the bias were known to be zero (removed by calibration) and if the ± limit given were specified as being ±s, ±2s, ±3s, or ±e_p, since all these terms recognize the random nature of the error. However, if the bias is unknown (and not zero), the quotation of a single number for the total inaccuracy is somewhat unsatisfactory although it may be a necessary expedient.

One reason for this is that if one is trying to estimate the overall inaccuracy of a measurement system made up of a number of components, each of which has a known inaccuracy, the method of combining the individual inaccuracies is different for systematic errors (biases) than for random errors (imprecisions). Thus, if the number given for the total inaccuracy of a given component contains both bias and imprecision in unknown proportions, the calculation of overall system inaccuracy is confused. However, in many cases there is no alternative, and by calculation from theory, past experience, and/or judgment the experimenter must arrive at the best available estimate of the total inaccuracy, or *uncertainty* (as it is sometimes called), to be attached to the reading. In such cases a useful viewpoint is that one is willing to bet with certain odds (say 19 to 1) that the error falls within the given limits. Such limits may then be combined as if they were imprecisions in calculations of overall system error.[1,2]

Irrespective of the precise *meaning* to be attached to accuracy figures provided, say, by instrument manufacturers, the *form* of such specifications is fairly uniform. More often than not, accuracy is quoted as a percentage figure based on the full-scale reading of the instrument. Thus if a pressure gage has a range from 0 to 10 psig and a quoted inaccuracy of ± 1.0 percent of full scale, this is to be interpreted as meaning that no error greater than ± 0.1 psig can be expected for any reading that might be taken on this gage, provided it is "properly" used. The manufacturer may or may not be explicit about the conditions required for "proper use." Note that for an actual reading of 1 psig a 0.1 psig error is 10 *percent of the reading*.

Another method sometimes used gives the error as a percentage of the particular reading with a qualifying statement to apply to the low end of the scale. For example, a spring scale might be described as having an inaccuracy of ± 0.5 percent of reading or ± 0.1 lb$_f$, whichever is greater. Thus for readings less than 20 lb$_f$ the error is constant at ± 0.1 lb$_f$, while for larger readings the error is proportional to the reading.

Combination of component errors in overall system-accuracy calculations. A measurement system is often made up of a chain of components each of which is subject to individual inaccuracy. If the individual inaccuracies are known, how is the overall inaccuracy computed? A similar problem occurs in experiments that use the results (measurements) from several different instruments to compute some quantity. If the inaccuracy of each instrument is known, how is the inaccuracy of the computed result estimated? Or, inversely, if there must be a certain accuracy in a computed result, what errors are allowable in the individual instruments?

To answer the above questions, consider the problem of computing a quantity N, where N is a known function of the n *independent* variables, $u_1, u_2, u_3, \ldots, u_n$. That is,

$$N = f(u_1, u_2, u_3, \ldots, u_n) \qquad (3.28)$$

The u's are the measured quantities (instrument or component outputs) and are in error by $\pm \Delta u_1, \pm \Delta u_2, \pm \Delta u_3, \ldots, \pm \Delta u_n$, respectively. These errors will cause an error ΔN in the computed result N. The Δu's

---

[1] S. J. Kline and F. A. McClintock, Describing Uncertainties in Single Sample Experiments, *Mech. Eng.*, vol. 75, p. 3, January, 1953.

[2] L. W. Thrasher and R. C. Binder, A Practical Application of Uncertainty Calculations to Measured Data, *Trans. ASME*, p. 373, February, 1957.

may be considered as absolute limits on the errors, as statistical bounds such as e_p's or $3s$ limits, or as uncertainties on which we are willing to give certain odds as including the actual error. However, the method of computing ΔN and the interpretation of its meaning are different for the first case as compared with the second and third. If the Δu's are considered as absolute limits on the individual errors and we wish to calculate similar absolute limits on the error in N, we could calculate

$$N \pm \Delta N = f(u_1 \pm \Delta u_1, u_2 \pm \Delta u_2, u_3 \pm \Delta u_3, \ldots, u_n \pm \Delta u_n) \qquad (3.29)$$

and by subtracting N in Eq. (3.28) from $N \pm \Delta N$ in Eq. (3.29) finally obtain $\pm \Delta N$. This procedure is needlessly time-consuming, however, and an approximate solution valid for engineering purposes may be obtained by application of the Taylor series. Expanding the function f in a Taylor series, we get

$$f(u_1 \pm \Delta u_1, u_2 \pm \Delta u_2, \ldots, u_n \pm \Delta u_n) = f(u_1, u_2, \ldots, u_n)$$
$$+ \Delta u_1 \frac{\partial f}{\partial u_1} + \Delta u_2 \frac{\partial f}{\partial u_2} + \cdots + \Delta u_n \frac{\partial f}{\partial u_n}$$
$$+ \frac{1}{2} \left[(\Delta u_1)^2 \frac{\partial^2 f_2}{\partial u_1} + \cdots \right] + \cdots \qquad (3.30)$$

where all the partial derivatives are to be evaluated at the known values of $u_1, u_2, \ldots, u_n$. That is, if the measurements have been made, the u's are all known as numbers and may be plugged into the expressions for the partial derivatives to give other numbers. In actual practice, the Δu's will all be small quantities and thus terms such as $(\Delta u)^2$ will be negligible. Equation (3.30) may then be given approximately as

$$f(u_1 + \Delta u_1, u_2 + \Delta u_2, \ldots, u_n + \Delta u_n) = f(u_1, u_2, \ldots, u_n)$$
$$+ \Delta u_1 \frac{\partial f}{\partial u_1} + \Delta u_2 \frac{\partial f}{\partial u_2} + \cdots + \Delta u_n \frac{\partial f}{\partial u_n} \qquad (3.31)$$

The absolute error E_a is then given by

$$E_a = \Delta N = \left| \Delta u_1 \frac{\partial f}{\partial u_1} \right| + \left| \Delta u_2 \frac{\partial f}{\partial u_2} \right| + \cdots + \left| \Delta u_n \frac{\partial f}{\partial u_n} \right| \qquad (3.32)$$

The absolute-value signs are used because some of the partial derivatives might be negative, and for a positive Δu such a term would *reduce* the total error. Since an error Δu is, in general, just as likely to be positive as negative, to estimate the maximum possible error the absolute-value signs must be used as in Eq. (3.32). The form of Eq. (3.32) is very useful since it shows which variables (u's) exert the strongest influence on the accuracy of the overall result. That is, if, say, $\partial f / \partial u_3$ is a large number compared with the other partial derivatives this means that a small Δu_3 can have a large effect on the total error E_a. If the relative or percentage error E_r is desired, it is clearly given by

$$E_r = \frac{\Delta N}{N} \times 100 = \frac{100 E_a}{N} \qquad (3.33)$$

The computed result may thus be expressed as either $N \pm E_a$ or $N \pm E_r$ percent, and the interpretation is that one is *certain* this error will not be exceeded since this is the way the Δu's were defined.

In carrying out the above computations, questions of significant figures and rounding off will occur. We here briefly review these matters for those not familiar with such procedures. A significant figure is any one of the digits 1, 2, 3, 4, 5, 6, 7, 8, 9; zero is a significant figure except when used to fix the decimal point or to fill the places of unknown or discarded digits. Thus in the number 0.000532 the significant figures are 5, 3, and 2, while in the number 2,076 *all* the digits, including the zero, are significant. For a number such as 2,300 the zeros may or may not be significant. To convey which figures are significant, one should write this as 2.3×10^3 if two significant figures are intended, 2.30×10^3 if three, 2.300×10^3 if four, and so forth.

In computations one often deals with numbers having unequal numbers of significant figures. For example, it may be necessary to multiply 4.62×0.317856. The first number is assumed good to three significant figures while the second is good to six. It can be shown that the product will be good only to three significant figures; therefore to save work, the six-figure number should be rounded off before multiplication. A number of rules have been proposed for this rounding procedure, and we now state one that is widely used:

> To round a number to n significant figures, discard all digits to the right of the nth place. If the discarded number is less than one-half a unit in the nth place, leave the nth digit unchanged. If the discarded number is greater than one-half a unit in the nth place, increase the nth digit by 1. If the discarded number is exactly one-half a unit in the nth place, leave the nth digit unchanged if it is an even number and add 1 to it if it is odd.

To determine to what extent numbers should be rounded, the following rules may be applied.

Addition. For addition, retain one more decimal digit in the more-accurate numbers than is contained in the least-accurate number. (The more-accurate numbers are those with the most significant figures.)

Then round off the result to the same decimal place as the least-accurate number.

Example.

2.635		2.64
0.9		0.9
1.52	$\longrightarrow$	1.52
0.7345		0.73
		5.79 $\longrightarrow$ 5.8

Subtraction.   For subtraction, round off the more-accurate number to the same number of decimal places as the less accurate before subtracting. Give the result to the same number of decimal places as the less-accurate figure.

Example.

$$
\begin{array}{r} 7.6345 \\ -0.031 \end{array} \Big\} \longrightarrow \begin{array}{r} 7.634 \\ -0.031 \\ \hline 7.603 \end{array} \longrightarrow 7.603
$$

Multiplication and division.   For multiplication and division, round off the more-accurate numbers to one more significant figure than the least accurate before computing.   Round the result to the same number of significant figures as the least-accurate number.

Example. $\dfrac{(1.2)(6.335)(0.0072)}{3.14159} \rightarrow \dfrac{(1.2)(6.34)(0.0072)}{3.14} \rightarrow 0.0174 \rightarrow 0.017$

We can now give a step-by-step procedure for computing the overall error:

1. Tabulate all data, each with its $\pm$ error attached.   All errors should be expressed to two significant figures.   (Actually, one significant figure is often adequate.)   The reason for this is that the errors themselves are not generally known very accurately and so it is foolish to carry many significant figures.
2. If the quantity to be computed is N, where $N = f(u_1, u_2, \ldots , u_n)$, compute the partial derivatives $\partial f/\partial u_1$, $\partial f/\partial u_2$, $\ldots$, $\partial f/\partial u_n$ and evaluate each to slide-rule accuracy (three significant figures) by substituting in the basic data u_1, u_2, $\ldots$, u_n.
3. Using Eq. (3.32), compute E_a and round to two significant figures.
4. Compute N from Eq. (3.28) to one more decimal place than the rounded E_a of step 3.   Thus if $E_a = \pm 0.062$, N should be computed as, say, 7.0516.   This value is then rounded to the same number of decimal places as E_a, in this case to 7.052.   In computing N, treat u_1, u_2, etc., as exact numbers; that is, they each have an infinite

number of significant figures.   This viewpoint is necessary because one must be able to compute N to as many significant figures as required by E_a according to the above rule.

5.   The result may then be quoted as

$$7.052 \pm 0.062 \qquad \text{absolute terms}$$

or

$$7.052 \pm 0.88 \text{ percent} \qquad \text{relative terms}$$

When the problem is one in which a certain overall accuracy is known to be required and one wishes to know what component accuracies are needed, the following method may be employed.   It should be apparent that this problem is mathematically indeterminate since there are an infinite number of combinations of individual accuracies that could result in the same overall accuracy.   The means of resolving this difficulty are to be found in the "method of equal effects."   This principle merely assumes that each source of error will contribute an equal amount to the total error.   Mathematically, if

$$\Delta N = \left| \frac{\partial f}{\partial u_1} \Delta u_1 \right| + \left| \frac{\partial f}{\partial u_2} \Delta u_2 \right| + \cdots + \left| \frac{\partial f}{\partial u_n} \Delta u_n \right|$$

then, if each term is assumed to be equal, we may write

$$\left| \frac{\partial f}{\partial u_1} \Delta u_1 \right| = \left| \frac{\partial f}{\partial u_2} \Delta u_2 \right| = \cdots = \left| \frac{\partial f}{\partial u_n} \Delta u_n \right| = \frac{\Delta N}{n} \qquad (3.34)$$

Now the allowable overall error ΔN is known and so are n and $u_1, u_2, \ldots, u_n$.   Thus

$$\frac{\partial f}{\partial u_i} \Delta u_i = \frac{\Delta N}{n}$$
$$\Delta u_i = \frac{\Delta N}{n(\partial f / \partial u_i)} \qquad i = 1, 2, 3, \ldots, n \qquad (3.35)$$

and the allowable error Δu_i in each measurement may be calculated.   (The partial derivatives are evaluated at the known values of $u_i, u_2, \ldots, u_n$.)   If a particular Δu_i turns out to be smaller than what can possibly be achieved by the instruments available it may be possible to relax this requirement if some *other* Δu_i can be made smaller than the value given by Eq. (3.35).   That is, some instruments may give better accuracy than required by Eq. (3.35) while others may be unable to meet the requirements of Eq. (3.35).   In such cases it may still be possible to meet the overall accuracy requirement; this may be checked by the formulas given.

When the Δu's are not considered as absolute limits of error but rather as statistical bounds such as $\pm 3s$ limits, probable errors, or uncertainties, the formulas for computing overall errors must be modified.

It can be shown[1] that the proper method of combining such errors is according to the root-sum square (rss) formula

$$E_{a_{rss}} = \sqrt{\left(\Delta u_1 \frac{\partial f}{\partial u_1}\right)^2 + \left(\Delta u_2 \frac{\partial f}{\partial u_2}\right)^2 + \cdots + \left(\Delta u_n \frac{\partial f}{\partial u_n}\right)^2} \qquad (3.36)$$

The overall error $E_{a_{rss}}$ then has the same meaning as the individual errors. That is, if Δu_i represents a $\pm 3s$ limit on u_i, then $E_{a_{rss}}$ represents a $\pm 3s$ limit on N, and 99.7 percent of the values of N can be expected to fall within these limits. Equation (3.36) always gives a smaller value of error than does Eq. (3.32). Equation (3.35) must also be modified when this viewpoint is taken:

$$\Delta u_i = \frac{\Delta N}{\sqrt{n}\,(\partial f/\partial u_i)} \qquad (3.37)$$

As an example of the above procedures, consider an experiment for measuring by means of a dynamometer the average power transmitted by a rotating shaft. The formula for horsepower can be written as

$$\text{hp} = \frac{2\pi RFL}{550t} \qquad (3.38)$$

where $R \triangleq$ revolutions of shaft during time t
$F \triangleq$ force at end of torque arm, lb_f
$L \triangleq$ length of torque arm, ft   (3.39)
$t \triangleq$ time length of run, sec

A sketch of the experimental setup is shown in Fig. 3.14. The revolution counter is of the type shown in Fig. 2.4 and can be turned on and off with an electric switch. The instants of turning on and off are recorded by a stopwatch. If it is assumed the counter does not miss any counts, the maximum error in R is ± 1 revolution, because of the digital nature of the device (see Fig. 3.15).

There is a related error, however, in determining the time t since perfect synchronization of the starting and stopping of watch and counter is not possible. The stopwatch might be known to be a quite accurate time-measuring instrument but this does not guarantee that it will always measure the time interval intended. In assigning an error to t, then, we are not helped much by the watch manufacturer's guarantee of 0.10 percent inaccuracy if our synchronization error is much larger than this. This synchronization error is certainly not precisely known since it involves human factors. An experiment to determine its statistical characteristics would be a more expensive and involved undertaking than

[1] J. B. Scarborough, "Numerical Mathematical Analysis," 3d ed., p. 429, The Johns Hopkins Press, Baltimore, 1955.

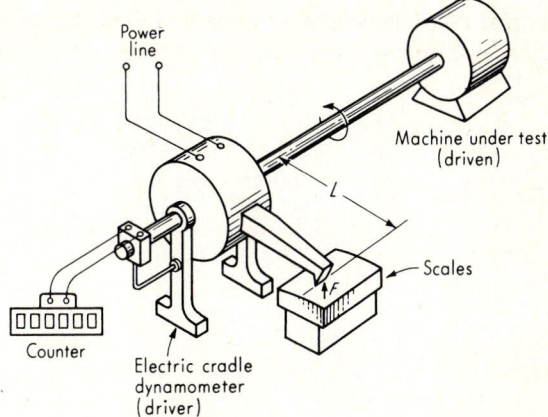

Fig. 3.14. *Dynamometer test setup.*

the power measurement of which it is a part.   We are thus in the rather common position of having to rely on experience and judgment in arriving at an estimate of the proper numerical value, and we begin to appreciate that some of the statistical niceties and fine points of theory considered earlier may appear somewhat academic in such a situation.   They are always useful in terms of the understanding of basic concepts that they develop; however, they cannot be relied upon to give clear-cut answers in situations where the basic data are ill-defined.   In the present case, suppose it is decided that a total starting and stopping error will be taken as ± 0.50 sec.   Whether this is to be considered as an absolute limit or as a $\pm 3s$ limit is somewhat meaningless when the basic number is arrived at in such an arbitrary fashion.

The measurement of the torque arm length L is also subject to

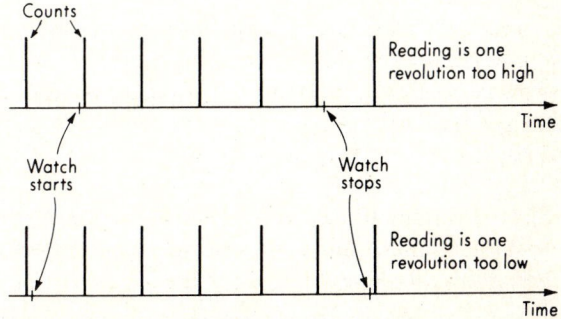

Fig. 3.15. *Revolution-counting error.*

similar vagaries, depending on the care taken in this particular measurement. Suppose we use a fairly rough procedure and decide an on error of ± 0.05 in.

The scales used to measure the force F can be statically calibrated with dead weights, giving a set of data analogous to that of Fig. 3.13. Suppose this is done and an s_{qi} of 0.0133 lb$_f$ is obtained. The $\pm 3s$ limits would then be ± 0.040 lb$_f$. Again, however, the situation is not this simple. When actually used, the scales will be subject to vibration, which may reduce frictional effects and increase precision. At the same time, the pointer on the scale will not stand perfectly still when the dynamometer is running; thus in reading the scale we must perform a mental averaging process which may introduce more error. Such effects are clearly difficult to quantify, and we must again make a decision based partly on experience and judgment. Suppose we assume the two mentioned effects cancel each other and thus take ± 0.040 lb$_f$ as the force measurement error.

If for a specific run the data are

$$
\begin{aligned}
R &= 1{,}202 \pm 1.0 \text{ revolutions} \\
F &= 10.12 \pm 0.040 \text{ lb}_f \\
L &= 15.63 \pm 0.050 \text{ in.} \\
t &= 60.0 \pm 0.50 \text{ sec}
\end{aligned}
\tag{3.40}
$$

the calculation proceeds as follows: In terms of inch units, we have

$$
\text{hp} = \frac{2\pi}{(550)(12)} \frac{FLR}{t} = K \frac{FLR}{t} \tag{3.41}
$$

Then, computing the various partial derivatives to slide-rule accuracy gives

$$
\frac{\partial(\text{hp})}{\partial F} = \frac{KLR}{t} = \frac{(0.000952)(15.63)(1{,}202)}{60} = 0.298 \text{ hp/lb}_f \tag{3.42}
$$

$$
\frac{\partial(\text{hp})}{\partial R} = \frac{KFL}{t} = \frac{(0.000952)(10.12)(15.63)}{60} = 0.00251 \text{ hp/revolution} \tag{3.43}
$$

$$
\frac{\partial(\text{hp})}{\partial L} = \frac{KFR}{t} = \frac{(0.000952)(10.12)(1{,}202)}{60} = 0.193 \text{ hp/in.} \tag{3.44}
$$

$$
\frac{\partial(\text{hp})}{\partial t} = -\frac{KFLR}{t^2} = -\frac{(0.000952)(10.12)(15.63)(1{,}202)}{3{,}600}
$$
$$
= -0.0500 \text{ hp/sec} \tag{3.45}
$$

If we now choose to consider the component errors as absolute limits and wish to compute the absolute limits on the overall error, we use Eq. (3.32) and get

$$
E_a = (0.298)(0.040) + (0.00251)(1.0) + (0.193)(0.050) + (0.05)(0.50) \tag{3.46}
$$

$$
E_a = 0.0119 + 0.00251 + 0.00965 + 0.025 = 0.049 \text{ hp} \tag{3.47}
$$

We now compute a rough value of hp by slide rule as

$$hp = \frac{(0.000952)(10.12)(15.63)(1{,}202)}{60} = 3.02 \qquad (3.48)$$

If we now follow the rule saying that hp should be computed to one more decimal place than E_a, we must calculate hp to four decimal places, which in this case means five significant figures.   Thus

$$hp = \frac{(2.0000)(3.1416)(10.120)(15.630)(1{,}202.0)}{(550.00)(12.000)(60.000)} = 3.0167 \qquad (3.49)$$

which we round off to 3.017.   The result may then be quoted as hp = 3.017 ± 0.049 hp or hp = 3.017 ± 1.6 percent.   If the component errors were considered as having only one significant figure (which might well be the case here where they are in considerable doubt), the above computations can be simplified since fewer significant figures need be carried.   The final result thus might be given more realistically as hp = 3.02 ± 0.05.

If the individual errors are thought of as ±3s limits, then Eq. (3.36) should be used to compute ±3s limits on hp.   Let us carry this out to see the numerical significance.

$$E_{a_{rss}} = \sqrt{(0.0119)^2 + (0.00251)^2 + (0.00965)^2 + (0.025)^2} = 0.029 \text{ hp} \tag{3.50}$$

We see that $E_{a_{rss}}$ is significantly smaller than E_a.   We might say that the error is *possibly* as large as 0.049 hp but *probably* not larger than 0.029 hp. If the individual errors had been accurately known to be ±3s limits, the word "probably" would have precise statistical meaning, otherwise not.

Finally, suppose we wish to measure hp to 0.5 percent accuracy in the previous example.   What accuracies are needed in the individual measurements?   We can use either Eq. (3.35) (if we wish to be conservative) or Eq. (3.37) (if we wish to give ourselves every chance of showing the measurement to be possible).   Using Eq. (3.37), we get

$$\Delta F = \frac{(3.02)(0.005)}{\sqrt{4}\,(0.298)} = 0.025 \text{ lb}_f$$

$$\Delta R = \frac{(3.02)(0.005)}{\sqrt{4}\,(0.0025)} = 3.0 \text{ revolutions}$$

$$\Delta L = \frac{(3.02)(0.005)}{\sqrt{4}\,(0.193)} = 0.039 \text{ in.} \tag{3.51}$$

$$\Delta t = \frac{(3.02)(0.005)}{\sqrt{4}\,(0.05)} = 0.15 \text{ sec}$$

[If we use Eq. (3.35), all these allowable errors are cut in half.]   If it

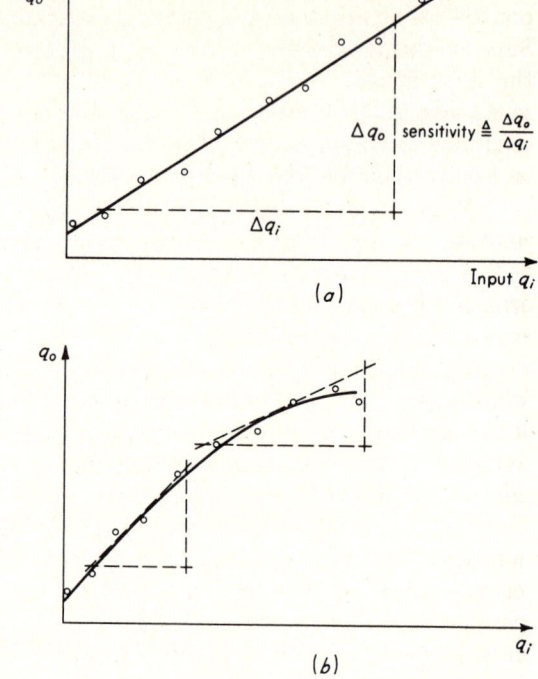

Fig. 3.16. *Definition of sensitivity.*

is found that the best instrument and technique available for measuring, say, F are good only to 0.04 lb$_f$ rather than the 0.025 lb$_f$ called for by Eq. (3.51), this does not necessarily mean that hp cannot be measured to 0.5 percent.   However, it does mean that one or more of the other quantities R, L, and t *must* be measured *more* accurately than required by Eq. (3.51).   Making one or more of these measurements more accurately may offset the excessive error in the F measurement.   The given formulas allow calculation of whether this will be true or not.

Static sensitivity.   When an input-output calibration such as that of Fig. 3.13 has been performed, the *static sensitivity* of the instrument can be defined as the slope of the calibration curve.   If the curve is not nominally a straight line, the sensitivity will vary with the input value, as shown in Fig. 3.16b.   To get a meaningful definition of sensitivity, the output quantity must be taken as the actual physical output, not the meaning attached to the scale numbers.   That is, in Fig. 3.13 the output quantity was plotted as pounds per square inch gage; however, the actual physical output is an angular rotation of the pointer.   Thus to define sensitivity properly, one must know the angular spacing of the pounds-per-square-inch-gage marks on the scale of the pressure gage. Suppose this is 5 angular degrees/psig. Since we have already calculated the slope in psig/psig as 1.05 in Fig. 3.13, we get the instrument static sensitivity as (5)(1.05) = 5.25 angular degrees/psig. In this form the sensitivity allows comparison of this pressure gage with others as regards its ability to detect pressure changes.

While the instrument's sensitivity to its desired input is of primary concern, its sensitivity to interfering and/or modifying inputs may also be of interest. As an example, consider temperature as an input to the pressure gage mentioned above. Temperature can cause a relative expansion and contraction that will result in a change in output reading even though the pressure has not changed. In this sense, it is an interfering input. Also, temperature can change the modulus of elasticity of the pressure-gage spring, thereby giving a change in the pressure sensitivity. In this sense, it is a modifying input. The first effect is often called a *zero drift* while the second is a *sensitivity drift* or *scale-factor drift*. These effects can be numerically evaluated by running suitable calibration tests. To evaluate zero drift the pressure is held at zero while the temperature is varied over a range and the output reading recorded. For reasonably small temperature ranges the effect is often nearly linear, and one can then quote the zero drift as, say, 0.01 angular degree/F°. Sensitivity drift may be found by fixing the temperature and running a pressure calibration to determine pressure sensitivity. Repeating this for various temperatures should show the effect of temperature on pressure sensitivity. Again, if this is nearly linear, one can specify sensitivity drift as, say, 0.0005 (angular degree/psig)/F°.

Figure 3.17 shows how the superposition of these two effects determines the total error due to temperature. If the instrument is used for measurement only and the temperature is known, numerical knowledge of zero drift and sensitivity drift allows correction of the readings. If the instrument is part of a large data-collection system or control system, such corrections may not be feasible; then knowledge of the drifts is used mainly to estimate overall system errors due to temperature.

The generalized input-output configuration of Fig. 2.9 can, for the present example, be made specific, as in Fig. 3.18. New symbology introduced here includes the *transfer function* and the *variable multiplier*. The output of the variable-multiplier symbol is taken as the product of the two inputs. The output of the transfer-function symbol is taken as the input times the function inside the box. In the present case, all the "functions" are merely constants. However, later we shall generalize this concept to include dynamic relations derived from differential equations relating input and output.

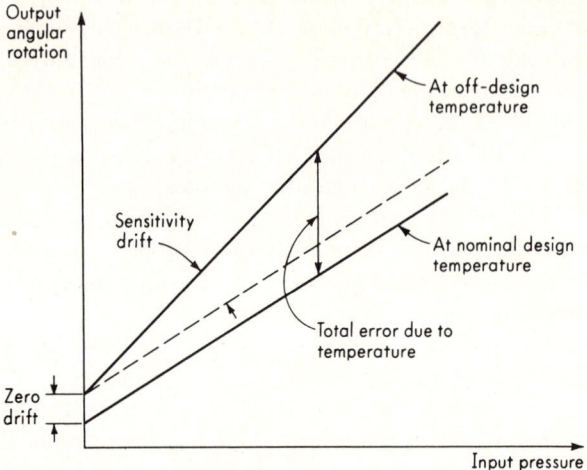

Fig. 3.17.   *Zero and sensitivity drift.*

Linearity. If an instrument's calibration curve for desired input is not a straight line, the instrument may still be highly accurate. There are many applications, however, where linear behavior is most desirable. The conversion from a scale reading to the corresponding measured value of input quantity is most convenient if one merely has to multiply by a fixed constant rather than consult a nonlinear calibration curve or compute from a nonlinear calibration equation. Also, when the instrument is part of a larger data or control system, linear behavior of the parts often simplifies design and analysis of the whole. Thus specifications relating to the degree of conformity to straight-line behavior are common.

Several definitions[1] of linearity are possible. However, the so-called *independent linearity* seems to be preferable in many cases. Here the reference straight line is the least-squares fit, as in Fig. 3.13. The linearity is then simply a measure of the maximum deviation of any calibration points from this straight line. This may be expressed as a percent of the actual reading, a percent of full-scale reading, or a combination of the two. The last method is probably the most realistic and leads to the following type of specification:

$$\text{Independent nonlinearity} = \pm A \text{ percent of reading or}$$
$$\pm B \text{ percent of full scale, whichever is greater} \qquad (3.52)$$

The first part ($\pm A$ percent of reading) of the specification recognizes the

---

[1] L. P. Entin, Instrument Uncertainties: I and II, *Control Eng.*, December, 1959; February, 1960.

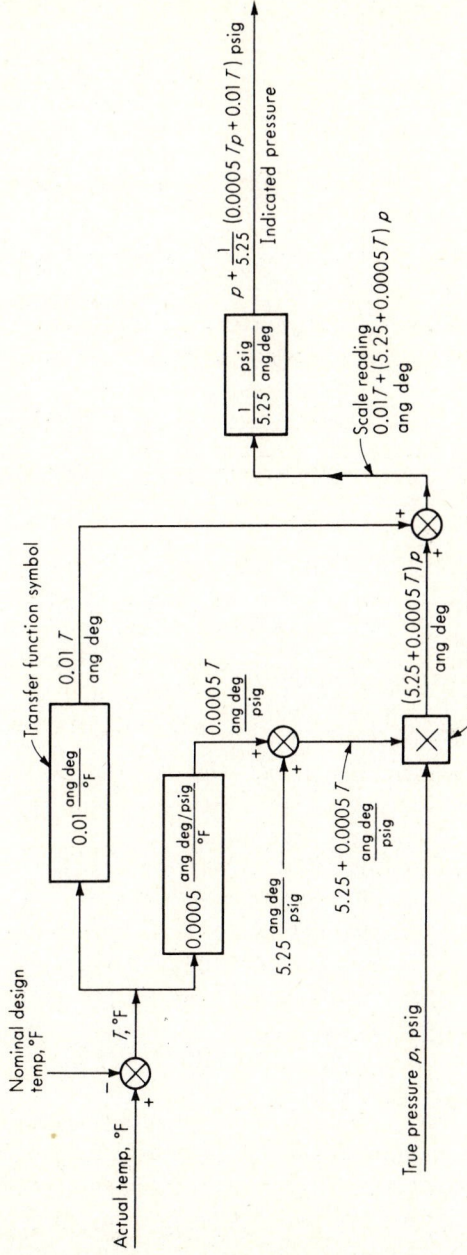

Fig. 3.18. *Block diagram of pressure gage.*

(The −0.64 psig zero bias of Fig. 3.13 is assumed to have been removed by gage adjustment)

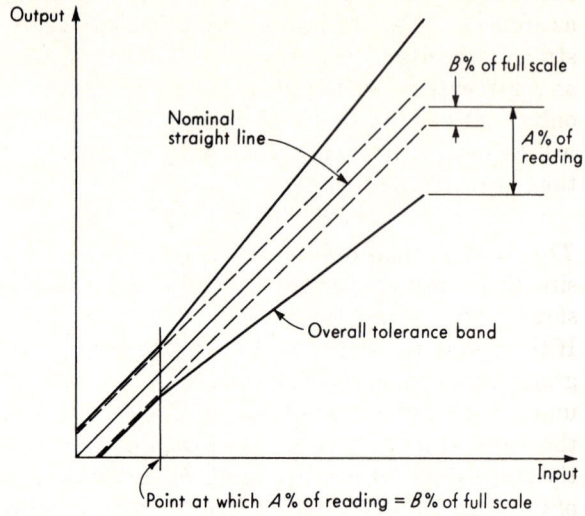

Fig. 3.19.   *Linearity specification.*

desirability of a constant-percentage nonlinearity, while the second ($\pm B$ percent of full scale) recognizes the impossibility of testing for extremely small deviations near zero. That is, if a fixed percentage of reading is specified, the absolute deviations approach zero as the readings approach zero. Since the test equipment should be about 10 times as accurate as the instrument under test, this leads to impossible requirements on the test equipment. Figure 3.19 shows the type of tolerance band allowed by specifications of the form (3.52).

It should be pointed out that in instruments that are considered essentially linear the specification of nonlinearity is equivalent to a specification of overall inaccuracy when the common (nonstatistical) definition of inaccuracy is used. Thus in many commercial linear instruments only a linearity specification (and not an accuracy specification) may be given. The reverse (an accuracy specification but not a linearity specification) may be the case if nominally linear behavior is implied by the quotation of a fixed sensitivity figure.

In addition to overall accuracy requirements, linearity specifications are often useful in dividing the total error into its component parts. Such a division is sometimes advantageous in choosing and/or applying measuring systems for a particular application in which, perhaps, one type of error is more important than another. In such cases, different definitions of linearity may be especially suitable for certain types of systems. The Scientific Apparatus Makers Association standard load-cell (force-measuring device) terminology,[1] for instance, defines linearity as follows: "The maximum deviation of the calibration curve from a straight line drawn between no-load and full-scale load outputs, expressed as a percentage of the full-scale output and measured on increasing load only." The breakdown of total inaccuracy into its component parts will be carried further in the next few sections where hysteresis, resolution, etc., are considered.

Threshold, resolution, hysteresis, and dead space. Consider a situation wherein the pressure gage of Fig. 3.2 has the input pressure slowly and smoothly varied from zero to full scale and then back to zero. If there were no friction due to sliding of moving parts, the input-output graph might appear as in Fig. 3.20a. The noncoincidence of loading and unloading curves is due to the internal friction or hysteretic damping of the stressed parts (mainly the spring). That is, all the energy put into the stressed parts upon loading is not recoverable upon unloading, because of the second law of thermodynamics, which rules out perfectly reversible processes in the real world. Certain materials exhibit a minimum of internal friction, and they should be given consideration in designing highly stressed instrument parts, provided that their other properties are also suitable for the specific application. For instruments with a usable range on both sides of zero, the behavior is as shown in Fig. 3.20b.

If it were possible to reduce internal friction to zero but external sliding friction were still present, the results might be as in Fig. 3.20c and d, where a constant coulomb (dry) friction force is assumed. If there is any free play or looseness in the mechanism of an instrument, a curve of similar shape will result.

Hysteresis effects also show up in electrical phenomena. One example is found in the relation between output voltage and input field current in a d-c generator, which is similar in shape to Fig. 3.20b. This effect is due to the magnetic hysteresis of the iron in the field coils.

In a given instrument a number of causes such as those just mentioned may combine to give an overall hysteresis effect which might result in an input-output relation as in Fig. 3.20e. The numerical value of hysteresis can be specified in terms of either input or output and is usually given as a percentage of full scale. When the total hysteresis has a large component of internal friction, time effects during hysteresis testing may confuse matters since sometimes significant relaxation and recovery effects are present. Thus in going from one point to another in Fig. 3.20e one may get a different output reading immediately after

---

[1] Standard Load Cell Terminology and Definitions: II, Scientific Apparatus Makers Association, Chicago, Jan. 11, 1962.

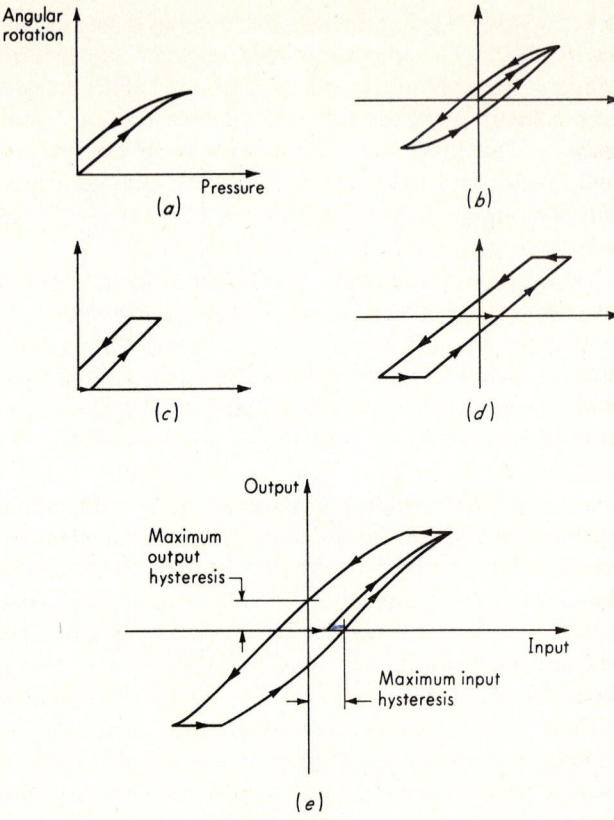

Fig. 3.20. *Hysteresis effects.*

changing the input than if some time elapses before the reading is taken. If this is the case, the time sequence of the test must be clearly specified if reproducible results are to be obtained.

If the instrument input is very gradually increased from zero there will be some minimum value below which no output change can be detected. This minimum value defines the *threshold* of the instrument. In specifying threshold, the first detectable output change is often described as being any "noticeable" or "measurable" change. Since these terms are somewhat vague, to improve reproducibility of threshold data it may be preferred to state a definite numerical value for output change for which the corresponding input is to be called the threshold.

If the input is slowly increased from some arbitrary (nonzero) input value, it will again be found that the output does not change at all until a certain input increment is exceeded. This increment is called the *resolution;* again, to reduce ambiguity, it is defined as the input increment that gives some small but definite numerical change in the output. Thus resolution defines the smallest measurable input *change* while threshold defines the smallest measurable *input*. Both threshold and resolution may be given either in absolute terms or as a percentage of full-scale reading. An instrument with large hysteresis does not necessarily have poor resolution; internal friction in a spring can give a large hysteresis, but even small changes in input (force) cause corresponding changes in deflection, giving high resolution.

The terms dead space, dead band, and dead zone are sometimes used interchangeably with the term hysteresis. However, they may be defined as the total range of input values possible for a given output and may thus be numerically twice the hysteresis as defined in Fig. 3.20*e*. Since none of these terms is completely standardized, one should always be sure which definition is meant.

Scale readability. Since the majority of instruments that have analog (rather than digital) output are read by a human observer noting the position of a "pointer" on a calibrated scale, it is usually desirable for the data taker to state his opinion as to how closely he believes he can read this scale. This characteristic, *which depends on both the instrument and the observer*, is called the scale readability. While this characteristic should logically be *implied* by the number of significant figures recorded in the data, it is probably good practice for the observer to stop and think about this before taking data and then *record* the scale readability he decides upon on the data sheet. It may also be appropriate at this point to suggest that all data, including scale readabilities, be given in decimal rather than fractional form. Since some instrument scales are calibrated in $\frac{1}{4}$'s, $\frac{1}{2}$'s, etc., this requires the data taker to convert to decimal form before recording data. This procedure is considered preferable to recording a piece of data as, say, $21\frac{1}{4}$ and then *later* trying to decide whether 21.250 or 21.3 was meant.

Span. The range of measured variable that an instrument is designed to measure is sometimes called the span. Equivalent terminology also in use states the "low operating limit" and "high operating limit." For essentially linear instruments, the term "linear operating range" is also common. A related term, which, however, implies dynamic fidelity also, is the *dynamic range*. This is the ratio of the largest to the smallest dynamic input that the instrument will faithfully measure. The number representing the dynamic range is often given in decibels, where the decibel (db) value of a number N is defined as db $\triangleq 20 \log_{10} N$. Thus a dynamic range of 60 db indicates the instrument can handle a range of input sizes of 1,000 to 1.

Generalized static stiffness and input impedance. It has been mentioned before that the introduction of any measuring instrument into a measured medium always results in the extraction of some energy from the medium, thereby changing the value of the measured quantity from its undisturbed state and thus making perfect measurements theoretically impossible.   Since the instrument designer wishes to approach perfection as nearly as practicable, some numerical means of characterizing this "loading" effect of the instrument on the measured medium would be helpful in comparing competitive instrument designs.   The concepts of *stiffness* and *input impedance*[1] are intended to serve such a function. While both these terms are useful for both static and dynamic conditions, we here introduce them by considering their static aspects only.

In Fig. 2.1 and subsequent schematic and block diagrams the connection of functional elements by single lines perhaps gives the impression that the transfer of information and energy is described by a single variable only.   Closer examination reveals that energy transfers require the specification of two variable quantities for their description.   The definitions of stiffness and generalized input impedance are in terms of two such variables.   At the input of each component in a measuring system there exists a variable q_{i1} with which one is primarily concerned, insofar as information transmission is concerned.   At the same point, however, there is associated with q_{i1} another variable q_{i2} such that the product $q_{i1}q_{i2}$ has the dimensions of power and represents an instantaneous rate of energy withdrawal from the preceding element.   When these two signals are identified, one can define the generalized input impedance Z_{gi} by

$$Z_{gi} \triangleq \frac{q_{i1}}{q_{i2}} \qquad (3.53)$$

if q_{i1} is an "effort variable."   (The definition of an "effort variable" will be given shortly.)   [At this point we consider only systems where (3.53) is an ordinary algebraic equation.   However, the concept of impedance can easily be extended to dynamic situations and (3.53) then must be given a more general interpretation.]   Using (3.53), we see that the power drain $P = q_{i1}{}^2/Z_{gi}$ and that *a large input impedance is needed to keep the power drain small.*   The concept of generalized impedance (and of course the terminology itself) is a generalization of electrical impedance, and we first give some examples from this perhaps somewhat more familiar field.

Consider a voltmeter of the common type shown in Fig. 2.17.   Suppose this meter is to be applied to a circuit in order to measure an unknown

---

[1] R. G. Boiten, The Mechanics of Instrumentation, *Proc. Inst. Mech. Engrs.* (*London*), vol. 177, no. 10, 1963.

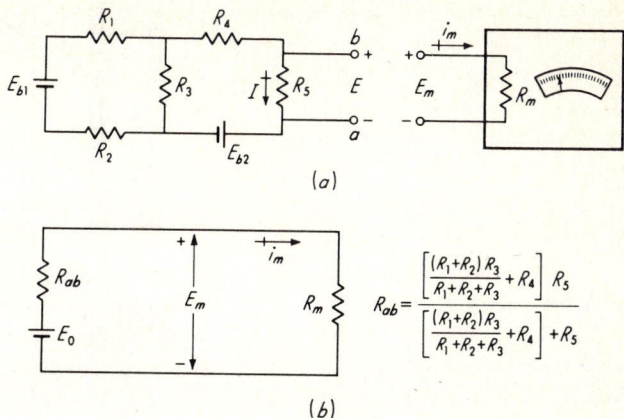

$$R_{ab} = \frac{\left[\dfrac{(R_1+R_2)R_3}{R_1+R_2+R_3} + R_4\right]R_5}{\left[\dfrac{(R_1+R_2)R_3}{R_1+R_2+R_3} + R_4\right] + R_5}$$

(b)

Fig. 3.21. *Voltmeter loading effect.*

voltage E as in Fig. 3.21a. As soon as the meter is attached to the terminals a, b, the circuit is changed and the value of E is no longer the same. For the meter alone, the input variable of direct interest (q_{i1}) is the terminal voltage E_m. If we look for an associated variable (q_{i2}) which when multiplied by q_{i1} gives the power withdrawal we find the meter current i_m meets these requirements. In this example, then, $Z_{gi} = E_m/i_m = R_m$, the meter resistance.

Further to illustrate the significance of input impedance, let us determine just how much error is caused when the meter is connected to the circuit. To facilitate this, we first give without proof a very useful network theorem called Thévenin's theorem.[1] Consider any network made up of linear, bilateral impedances and generators (or batteries). A linear impedance is one whose elements (R,L,C) do not change value with the magnitude of the current or voltage. Most resistances, capacitances, and air-core inductances are linear; iron-core inductances are nonlinear. A bilateral impedance is one that transmits energy equally well in either direction. Resistances, capacitances, and inductances are essentially bilateral. Vacuum tubes are unilateral since they effectively transmit energy only in one direction (from grid circuit to plate circuit, *not* the reverse).

A linear bilateral network is shown in Fig. 3.22a as a "black box" with terminals A, B. A load of impedance Z_l may be connected across the terminals A, B. When the load Z_l is *not* connected, a voltage will, in general, exist at terminals A, B. This is called E_o, the open-circuit

[1] K. Y. Tang, "Alternating Current Circuits," 2d ed., p. 202, International Textbook Company, Scranton, Pa., 1952.

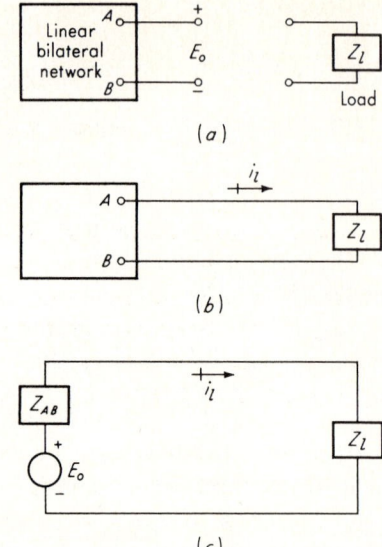

Fig. 3.22. *Thévenin's theorem.*

output voltage of the network.   Also with Z_l *not* connected, it is possible to determine the impedance Z_{AB} between terminals A and B.   When this is done, any batteries or generators in the network are to be replaced by their internal impedance.   If their internal impedance is assumed to be zero, they are replaced by a short circuit (just a wire with no resistance).   Thévenin's theorem then states: If the load Z_l is connected as shown in Fig. 3.22b a current i_l will flow.   This current will be the *same* as the current that flows in the fictitious equivalent circuit of Fig. 3.22c.   Thus the network, no matter how complex, may be replaced by a single impedance (the output impedance) Z_{AB} in series with a single voltage source E_o.

Applying Thévenin's theorem to Fig. 3.21a, we get Fig. 3.21b.   We see here that the value E_m indicated by the meter is *not* the true value E but rather

$$E_m = \frac{R_m}{R_{ab} + R_m} E \qquad (3.54)$$

and that if E_m is to approach E we must have $R_m \gg R_{ab}$.   Thus our earlier statement about the desirability of high input impedance can now be made more specific.   *The input impedance must be high relative to the output impedance of the system to which the load is connected.*   Assuming that it is possible to define generalized input and output impedances Z_{gi} and Z_{go} in nonelectrical as well as electrical systems, we may generalize

Eq. (3.54) to

$$q_{i1m} = \frac{Z_{gi}}{Z_{go} + Z_{gi}} q_{i1u} = \frac{1}{Z_{go}/Z_{gi} + 1} q_{i1u} \qquad (3.55)$$

where $q_{i1m} \triangleq$ measured value of effort variable

          $q_{i1u} \triangleq$ undisturbed value of effort variable

Of course, if we knew both Z_{gi} and Z_{go}, we could correct q_{i1m} by means of Eq. (3.55). However, this would be inconvenient; also Z_{go} is not always known, especially in nonelectrical systems, where definition of *both* Z_{gi} and Z_{go} is not always straightforward. Thus a high value of Z_{gi} is desirable since then corrections are unnecessary and the actual values of either Z_{gi} or Z_{go} need not be known.

To achieve a high value of input impedance for any instrument, not just voltmeters, a number of paths are open to the designer. We shall now describe three of them, using the voltmeter as a specific example. The most obvious approach is to leave the configuration of the instrument unchanged but to change the numerical values of physical parameters so that the input impedance is increased. In the voltmeter of Fig. 3.21 this is simply accomplished by winding the coil in such a way (higher resistance material and/or more turns) that R_m is increased. While this accomplishes the desired result, certain undesirable effects also appear. Since this type of voltmeter is basically a *current*-sensitive rather than a voltage-sensitive device, an increase in R_m will *reduce* the magnetic torque available from a given impressed voltage. Thus if the spring constant of the restraining springs is not changed, the angular deflection for a given voltage (the sensitivity) is reduced. To bring the sensitivity back to its former value, we must reduce the spring constant. Also, because of lower torque levels, pivot bearings with less friction must be employed. These design changes generally result in a less rugged and reliable instrument so that this method of increasing input impedance is limited in the degree of improvement possible before other performance features are compromised. This situation will be found to occur in most instruments, not just in this specific example.

If input impedance is to be increased without compromising other characteristics, different approaches are needed. One of general usefulness employs a change of configuration of the instrument so as to include an auxiliary power source. The concept is that a rugged instrument requires a fair amount of power to actuate its output elements but that this power need *not* necessarily be taken from the measured medium. Rather, the low power signal from the primary sensing element may *control* the output of the auxiliary power source so as to realize a power-amplifying effect.

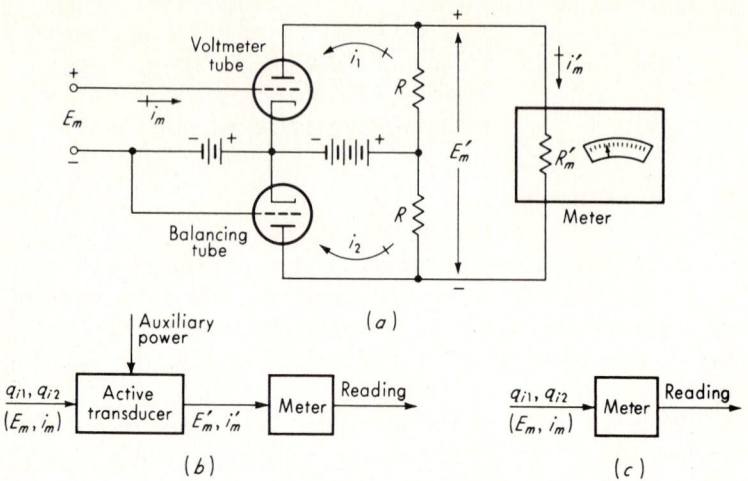

Fig. 3.23.   *Vacuum-tube voltmeter.*

Continuing our voltmeter example, this approach is exemplified by the vacuum-tube voltmeter (VTVM). Such a device is shown in rudimentary form in Fig. 3.23a. When the input voltage is zero (short circuit across the E_m terminals), if the two tubes are perfectly identical and the two resistances R are exactly equal, the meter voltage E'_m must be zero, because of symmetry. If an input E_m is applied, the grid bias on each tube is no longer the same, the currents i_1 and i_2 will no longer be equal, and a meter voltage E'_m will exist. While the meter current i'_m may still be as large as in a conventional meter, the current that determines the input impedance is i_m, the grid current, which can be extremely small. Thus a very high input impedance can be realized while still employing a rugged meter element. A block diagram for the VTVM is given in Fig. 3.23b which may be compared with that for an ordinary meter in Fig. 3.23c.

Still another approach to the problem of increasing input impedance uses the principle of feedback or null balance. For the specific area of voltage measurements, this technique is exemplified by the potentiometer. The most simple form of this instrument is shown in Fig. 3.24. It should be clear that in Fig. 3.24a each position of the sliding contact corresponds to a definite voltage between the terminals a and b. Thus the scale can be calibrated and any voltage between zero and the battery voltage obtained by properly positioning the slider. If one now connects in an unknown voltage E_m and a galvanometer (current detector) as in Fig. 3.24b, if E_m is less than the battery voltage there will be some point on the slider scale at which the voltage picked off the slide-wire just equals

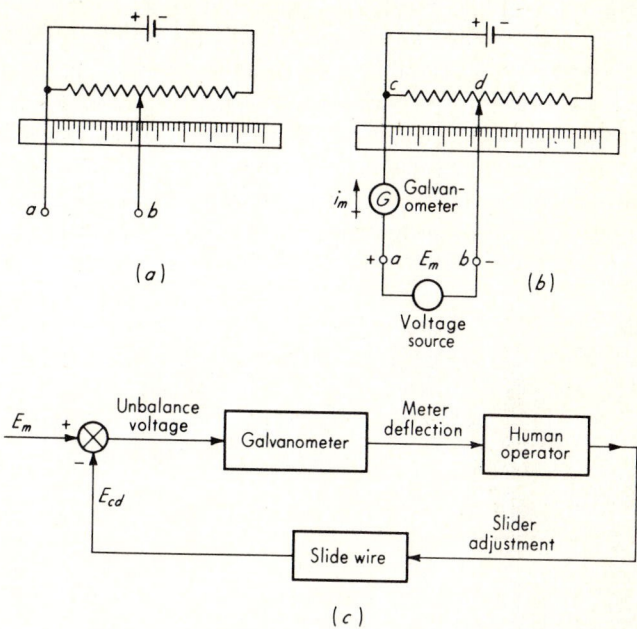

Fig. 3.24. *Potentiometer voltage measurement.*

the unknown E_m. This point of null balance can be detected by a zero deflection of the galvanometer since the net loop voltage a, c, d, b, a will be zero. The unknown voltage can then be read from the calibrated scale.

We should note that under conditions of perfect balance the current drawn from the unknown voltage source is exactly zero, thus giving an infinite input impedance. In actual practice, there must always remain some unknown unbalance current since a galvanometer always has a threshold below which currents cannot be detected. The interpretation of potentiometric voltage measurement as a feedback scheme is given in Fig. 3.24c. While the manual balancing described above is adequate when the unknown voltage is relatively constant, the procedure may be made automatic by use of the instrument servomechanism (in this case called a self-balancing potentiometer) as shown in Fig. 2.7. Then, by providing a pen and recording chart, varying voltages may be accurately measured.

In generalizing the concepts of input impedance, a reasonable starting point might be a listing of q_{i1} and corresponding q_{i2} variables for some common measurement situations. In general, the quantity q_{i1}, which is of primary concern, may be either a *flow variable* or an *effort variable*.

The concepts of flow and effort variables are discussed by Paynter.[1] Briefly, energy transfer across the boundaries of a system may be defined in terms of two variables, the product of which gives the instantaneous power. One of these variables, the flow variable, is an *extensive* variable, in the sense that its magnitude depends on the extent of the system taking part in the energy exchange. The other variable, the effort variable, is an *intensive* variable, whose magnitude is independent of the amount of material being considered. In the literature, flow variables are also called "through" variables, and effort variables are called "across" variables. When q_{i1} is an effort variable, Eq. (3.53) and subsequent developments apply. However, if q_{i1} is a flow variable, the situation is somewhat different. It is then appropriate to define a *generalized input admittance* Y_{gi} as

$$Y_{gi} \triangleq \frac{\text{flow variable}}{\text{effort variable}} = \frac{q_{i1}}{q_{i2}} \qquad (3.56)$$

rather than a generalized input impedance Z_{gi},

$$Z_{gi} = \frac{\text{effort variable}}{\text{flow variable}} \qquad (3.57)$$

We can then write the power drain of the instrument from the measured medium in terms of the measured variable q_{i1} as

$$P = q_{i1}q_{i2} = \frac{q^2_{i1}}{Y_{gi}} \qquad (3.58)$$

and we note that now a large value of input admittance Y_{gi} is required in order to minimize the power drain. A familiar electrical example of this situation is the ammeter. In Fig. 3.25a we are interested in measuring the current by means of an ammeter inserted into the circuit as shown. Applying Thévenin's theorem, we can reduce Fig. 3.25a to Fig. 3.25b. We see that the measured value of the current is given by

$$I_m = \frac{E_{ab}}{R_{ab} + R_m} = \frac{E_{ab}}{1/Y_{ab} + 1/Y_m} \qquad (3.59)$$

whereas the true (undisturbed) value of the current would be

$$I_u = \frac{E_{ab}}{R_{ab}} = \frac{E_{ab}}{1/Y_{ab}} \qquad (3.60)$$

It is now clear that if I_m is to approach I_u we must use an ammeter with $Y_m \gg Y_{ab}$; that is, the meter resistance must be sufficiently *low*, just the *opposite* of that desired in a voltmeter. This result can be generalized

[1] H. M. Paynter, "Analysis and Design of Engineering Systems," p. 18, The M.I.T. Press, Cambridge, Mass., 1960.

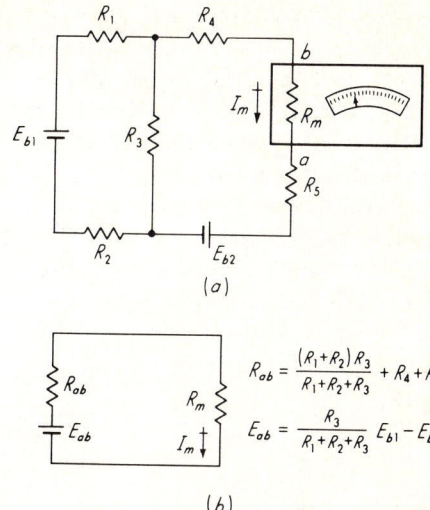

Fig. 3.25.   *Ammeter loading effect.*

$$R_{ab} = \frac{(R_1 + R_2)R_3}{R_1 + R_2 + R_3} + R_4 + R_5$$

$$E_{ab} = \frac{R_3}{R_1 + R_2 + R_3} E_{b1} - E_{b2}$$

to apply to all effort variables (such as voltage) and all flow variables (such as current). Equation (3.55) is applicable to those cases in which the measured variable q_{i1} is an effort variable, and Eq. (3.61) gives the corresponding relationship when q_{i1} is a flow variable.

$$q_{i1m} = \frac{1}{Y_{go}/Y_{gi} + 1} q_{i1u} \qquad (3.61)$$

where $Y_{go} \triangleq$ generalized output admittance of preceding element
$Y_{gi} \triangleq$ generalized input admittance of instrument
$q_{i1m} \triangleq$ measured value of flow variable
$q_{i1u} \triangleq$ undisturbed value of flow variable

For some instruments, in the case of a static input, the *power* drain from the preceding element is zero in the steady state, although some total *energy* is removed in going from one steady state to another. In such an instance the concepts of impedance and admittance are not as directly useful as one would like, and it is appropriate to consider the concepts of *static stiffness* and *static compliance*. These make it possible to characterize the *energy* drain (in the same way that impedance and admittance serve to define the *power* drain) in those situations where impedance or admittance becomes infinite and thus not directly meaningful. The terms stiffness and compliance come from the terminology of mechanical systems which afford some of the best examples of the application of these concepts. However, we shall generalize their definitions to include all types of physical systems, just as we did with impedance and admittance.

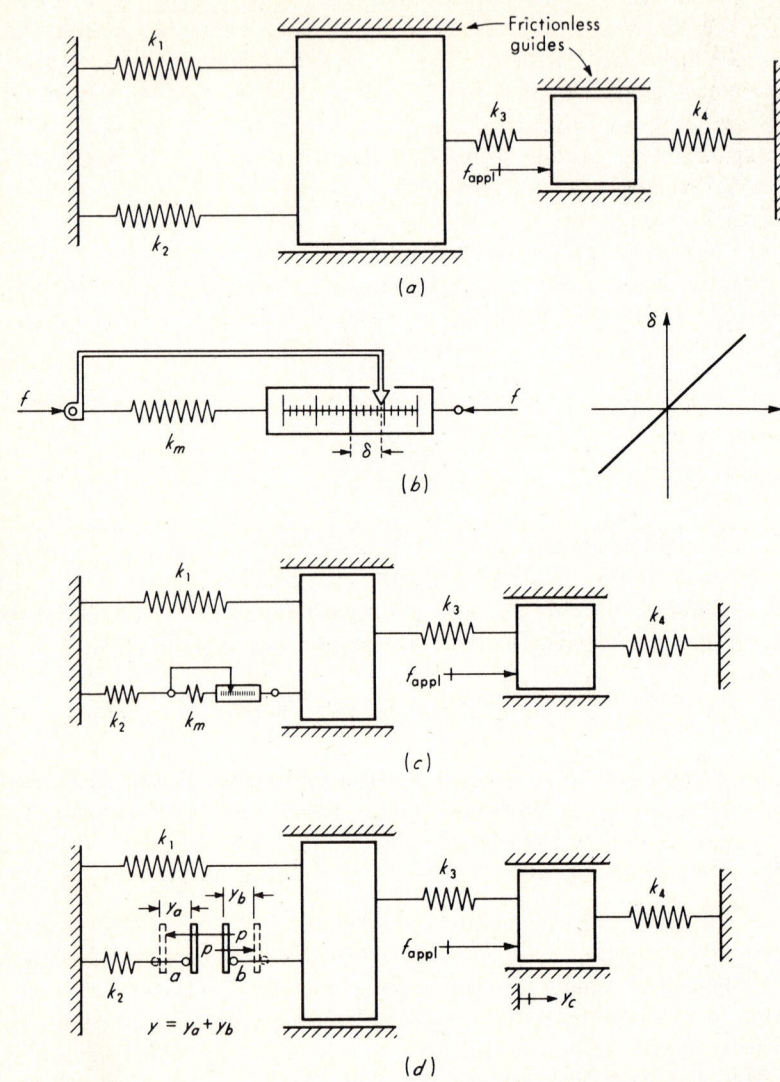

Fig. 3.26. *Force-gage loading effect.*

Consider the system of Fig. 3.26a as an idealized model of some elastic structure under applied load f_{appl}. This load will cause forces in the various structural members. Suppose we wish to measure the force in the member represented by the spring k_2. A common method of force measurement employs a calibrated elastic link whose deflection is proportional to force; thus a deflection measurement allows a force measurement.   Such a device, with spring constant k_m, is shown in Fig. 3.26b.   To measure the force in link k_2 we insert the force-measuring device "in series" with the link k_2 as shown in Fig. 3.26c.   The usual difficulty is encountered here in that the insertion of the measuring instrument alters the condition of the measured system and thus changes the measured variable from its undisturbed value.   We wish to assess the nature and amount of this error and shall use this example to introduce the concept of static stiffness.

The measured variable, force, is an effort variable.   Thus if we try to use the impedance concept, we must find an associated flow variable whose product with force will give power.   Since mechanical power has dimensions ft-lb$_f$/sec, we have

$$\text{Flow variable} = \frac{\text{power}}{\text{effort variable}} = \frac{\text{ft lb}_f/\text{sec}}{\text{lb}_f} = \frac{\text{ft}}{\text{sec}} = \text{velocity} \qquad (3.62)$$

Mechanical impedance is thus given by

$$\text{Mechanical impedance} = \frac{\text{effort variable}}{\text{flow variable}} = \frac{\text{force}}{\text{velocity}} \qquad (3.63)$$

If we now calculate the static mechanical impedance of an elastic system by applying a constant force and noting the resulting velocity, we get

$$\text{Static mechanical impedance} = \frac{\text{force}}{0} = \infty \qquad (3.64)$$

This difficulty may be overcome by using energy rather than power in the definition of the variable associated with the measured variable.   If this is done, a new term for the ratio of the two variables must be introduced, since the use of mechanical impedance as the ratio of force to velocity is well established.   We thus define

$$\text{Mechanical static stiffness} \triangleq \frac{\text{force}}{\text{displacement}} = \frac{\text{force}}{\int(\text{velocity})\,dt} \qquad (3.65)$$

$$\text{since} \quad \text{Energy} = (\text{force})(\text{displacement}) \qquad (3.66)$$

Thus, in general, whenever the measured variable is an effort variable and the static impedance is infinite, instead of using impedance one uses a generalized static stiffness S_g defined by

$$S_g \triangleq \frac{\text{effort variable}}{\int(\text{flow variable})\,dt} \qquad (3.67)$$

If this is done, it can be shown that the same formulas can be used for calculating the error due to inserting the measuring instrument as were used for impedance, except S is used instead of Z.   Thus, Eq. (3.55) becomes

$$q_{i1m} = \frac{S_{gi}}{S_{go} + S_{gi}} q_{i1u} = \frac{1}{S_{go}/S_{gi} + 1} q_{i1u} \qquad (3.68)$$

where $q_{i1m} \triangleq$ measured value of effort variable
$q_{i1u} \triangleq$ undisturbed value of effort variable
$S_{gi} \triangleq$ generalized static input stiffness of measuring instrument
$S_{go} \triangleq$ generalized static output stiffness of measured system

Let us now apply these general concepts to the specific case at hand. The output stiffness of the system of Fig. 3.26a at the point of insertion of the measuring device is simply the ratio of force p to deflection y at the terminals a, b in Fig. 3.26d. This stiffness can be found theoretically by applying a fictitious load p and calculating the resulting y, or if the structure (or a scale model) has been constructed one can obtain the stiffness experimentally by applying known loads and measuring the resulting deflections. A theoretical analysis might proceed as below:

$$\Sigma \text{ forces} = 0$$

$$\begin{cases} p - y_b(k_1) + k_3(y_c - y_b) = 0 & (3.69) \\ f_{\text{appl}} - k_3(y_c - y_b) - k_4 y_c = 0 & (3.70) \end{cases}$$

$$\begin{cases} (-k_1 - k_3)y_b + (k_3)y_c = -p & (3.71) \\ (k_3)y_b + (-k_3 - k_4)y_c = -f_{\text{appl}} & (3.72) \end{cases}$$

using determinants,

$$y_b = \frac{\begin{vmatrix} -p & k_3 \\ -f_{\text{appl}} & -(k_3 + k_4) \end{vmatrix}}{\begin{vmatrix} -(k_1 + k_3) & k_3 \\ k_3 & -(k_3 + k_4) \end{vmatrix}} = \frac{p(k_3 + k_4) + f_{\text{appl}}(k_3)}{(k_3 + k_4)(k_1 + k_3) - k_3^2} \qquad (3.73)$$

The output stiffness is now obtained from Eq. (3.73) by letting f_{appl} be zero:

$$S_{go} = \frac{p}{y} = \frac{p}{y_a + y_b} = \frac{p}{p/k_2 + p(k_3 + k_4)/[(k_3 + k_4)(k_1 + k_3) - k_3^2]} \qquad (3.74)$$

$$S_{go} = \frac{1}{1/k_2 + (k_3 + k_4)/[(k_3 + k_4)(k_1 + k_3) - k_3^2]} \qquad (3.75)$$

The input stiffness of the measuring instrument is given by

$$S_{gi} = \frac{\text{force}}{\text{displacement}} = k_m \qquad (3.76)$$

We may now apply Eq. (3.68) to get

$$\frac{\text{Measured value of force}}{\text{True value of force}} = \frac{k_m}{\dfrac{1}{1/k_2 + (k_3 + k_4)/(k_1 k_3 + k_1 k_4 + k_3 k_4)} + k_m} \qquad (3.77)$$

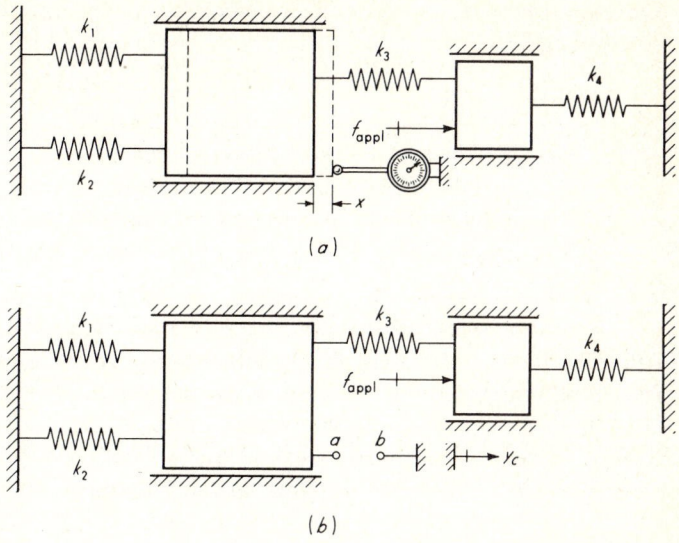

Fig. 3.27. *Displacement-gage loading effect.*

From Eq. (3.68) it is apparent that in general we should like to have $S_{gi} \gg S_{go}$ in order to have the measured value close to the true value. In this example, this requirement corresponds to

$$k_m \gg \frac{1}{1/k_2 + (k_3 + k_4)/(k_1 k_3 + k_1 k_4 + k_3 k_4)} \qquad (3.78)$$

Thus the measuring device must have a sufficiently stiff spring.

    We saw earlier that when the measured variable is not an effort variable, admittance rather than impedance is a more convenient tool. Again, however, under static conditions it may happen that admittance is infinite; thus a concept analogous to stiffness is needed to facilitate the treatment of such situations. For such cases the generalized compliance C_g is defined by

$$C_g \triangleq \frac{\text{flow variable}}{\int (\text{effort variable}) \, dt} \qquad (3.79)$$

    As a mechanical example, suppose we wish to measure the displacement x in Fig. 3.27a by means of a dial indicator. Such indicators generally have a spring load to ensure positive contact with the body whose motion is being measured. This spring load adds a force to the measured system, thereby causing error in the motion measurement. It is clear that the indicator spring load should be as light as possible, but we wish to make more quantitative statements about measurement accuracy.

Again, it is possible to show the applicability of the admittance relations [Eq. (3.61)] if we replace admittance by compliance:

$$q_{i1m} = \frac{1}{C_{go}/C_{gi} + 1} \, q_{i1u} \qquad (3.80)$$

where $q_{i1m} \triangleq$ measured value of flow variable
$q_{i1u} \triangleq$ undisturbed value of flow variable
$C_{gi} \triangleq$ generalized static input compliance of measuring instrument
$C_{go} \triangleq$ generalized static output compliance of measured system

We note that for accurate measurement we require $C_{gi} \gg C_{go}$.

In our example the measured variable is displacement, which is a flow variable. If we try to use admittance concepts, we can find the associated effort variable in the usual way:

$$\text{Power} = (\text{effort variable})(\text{flow variable}) \qquad (3.81)$$

$$\frac{\text{ft-lb}_f}{\text{sec}} = (\text{effort variable})(\text{displacement, ft}) \qquad (3.82)$$

$$\text{Effort variable} = \frac{\text{lb}_f}{\text{sec}} = \text{rate of change of force} \qquad (3.83)$$

The admittance is then

$$Y = \frac{\text{flow variable}}{\text{effort variable}} = \frac{\text{displacement}}{\text{rate of change of force}} \qquad (3.84)$$

and if we apply this definition to the case of a static load on a spring, we get $Y = \infty$. In this case, however, the compliance would be

$$C_g = \frac{\text{flow variable}}{\int (\text{effort variable}) \, dt} = \frac{\text{displacement}}{\text{force}} = \frac{1}{\text{spring constant}} \qquad (3.85)$$

In our example the output compliance of the measured system is the ratio of displacement to force for the terminals a, b of Fig. 3.27b. If we apply a fictitious force p between the terminals a, b a displacement y will occur; it may be computed as follows:

$$\Sigma \text{ forces} = 0$$

$$\begin{cases} p - y(k_1 + k_2) + k_3(y_c - y) = 0 & (3.86) \\ f_{\text{appl}} - k_3(y_c - y) - k_4 y_c = 0 & (3.87) \end{cases}$$

$$\begin{cases} (-k_1 - k_2 - k_3)y + (k_3)y_c = -p & (3.88) \\ (k_3)y + (-k_3 - k_4)y_c = -f_{\text{appl}} & (3.89) \end{cases}$$

$$y = \frac{\begin{vmatrix} -p & k_3 \\ -f_{\text{appl}} & (-k_3 - k_4) \end{vmatrix}}{\begin{vmatrix} (-k_1 - k_2 - k_3) & k_3 \\ k_3 & (-k_3 - k_4) \end{vmatrix}}$$

$$= \frac{p(k_3 + k_4) + f_{\text{appl}}(k_3)}{(k_3 + k_4)(k_1 + k_2 + k_3) - k_3{}^2} = \frac{p(k_3 + k_4) + f_{\text{appl}}(k_3)}{(k_3 + k_4)(k_1 + k_2) + k_3 k_4} \qquad (3.90)$$

We can now get C_{go} from Eq. (3.90) by letting $f_{appl} = 0$:

$$C_{go} = \frac{y}{p} = \frac{k_3 + k_4}{(k_3 + k_4)(k_1 + k_2) + k_3 k_4}$$

$$= \frac{1}{k_1 + k_2 + k_3 k_4/(k_3 + k_4)} \qquad (3.91)$$

If the spring constant of the dial indicator is k_m, the input compliance of the measuring instrument is given by

$$C_{gi} = \frac{1}{k_m} \qquad (3.92)$$

We then have

$$\frac{\text{Measured value of deflection}}{\text{True value of deflection}} = \frac{1}{\dfrac{k_m}{k_1 + k_2 + k_3 k_4/(k_3 + k_4)} + 1}$$

$$(3.93)$$

and k_m must be sufficiently small in order to get accurate displacement measurement.

To illustrate the general applicability of the concepts of impedance, admittance, stiffness, and compliance to measurement problems, Fig. 3.28 has been compiled. In the first column are listed some of the physical quantities commonly measured, each one identified as a flow variable or an effort variable. The appropriate associated variables which give either power or energy when multiplied with the measured variable are then listed. The last four columns indicate the dimensions of the appropriate loading criteria for that particular measurement. For effort variables, both impedance and stiffness are given; which one to use depends on the nature of the specific instrument. The fact that admittance and compliance are *not* given for effort variables does not mean that they could not be defined but merely that there is no need to consider them when the methods explained earlier in this section are used. Similar statements apply to those measured variables that are flow variables.

The basic formulas (3.55), (3.61), (3.68), and (3.80) were given without a detailed proof. At this point we wish to show the justification for these results and also to make clearer the physical meaning of impedance, admittance, stiffness, and compliance for physical systems in general. This discussion will also indicate more clearly how one calculates theoretically or measures experimentally these important system characteristics.

Consider first Fig. 3.29, showing two separate elements which we shall subsequently wish to interconnect in the order shown. Our objective is to determine the characteristics of the *individual* devices that must

Associated variable

Measured variable	Power based	Energy based	Impedance	Admittance	Stiffness	Compliance
Voltage (effort)	Current	Charge	$\dfrac{\text{volts}}{\text{amp}}$		$\dfrac{\text{volts}}{\text{coul}}$	
Current (flow)	Voltage	$\int(\text{voltage})\,dt$		$\dfrac{\text{amp}}{\text{volt}}$		$\dfrac{\text{amp}}{\text{volt-sec}}$
Force (effort)	Translational velocity	Translational displacement	$\dfrac{\text{lb}_f}{\text{fps}}$		$\dfrac{\text{lb}_f}{\text{ft}}$	
Translational displacement (flow)	$d/dt(\text{force})$	Force		$\dfrac{\text{in.}}{\text{lb}_f/\text{sec}}$		$\dfrac{\text{in.}}{\text{lb}_f}$
Torque (effort)	Rotational velocity	Rotational displacement	$\dfrac{\text{ft-lb}_f}{\text{rad}/\text{sec}}$		$\dfrac{\text{ft-lb}_f}{\text{rad}}$	
Rotational displacement (flow)	$d/dt(\text{torque})$	Torque		$\dfrac{\text{rad}}{\text{ft-lb}_f/\text{sec}}$		$\dfrac{\text{rad}}{\text{ft-lb}_f}$
Translational velocity (flow)	Force	$\int(\text{force})\,dt$		$\dfrac{\text{fps}}{\text{lb}_f}$		$\dfrac{\text{fps}}{\text{lb}_f\text{-sec}}$
Rotational velocity (flow)	Torque	$\int(\text{torque})\,dt$		$\dfrac{\text{rad}/\text{sec}}{\text{ft-lb}_f}$		$\dfrac{\text{rad}/\text{sec}}{\text{ft-lb}_f\text{-sec}}$

Associated variable

Measured variable	Power based	Energy based	Impedance	Admittance	Stiffness	Compliance
Translational acceleration (flow)	$\int(\text{force})\,dt$	$\int[\int(\text{force})\,dt]\,dt$		$\dfrac{\text{ft/sec}^2}{\text{lb}_f\text{-sec}}$		$\dfrac{\text{ft/sec}^2}{\text{lb}_f\text{-sec}^2}$
Rotational acceleration (flow)	$\int(\text{torque})\,dt$	$\int[\int(\text{torque})\,dt]\,dt$		$\dfrac{\text{rad/sec}^2}{\text{ft-lb}_f\text{-sec}}$		$\dfrac{\text{rad/sec}^2}{\text{ft-lb}_f\text{-sec}^2}$
Fluid pressure (effort)	Volume flow rate	Volume flow	$\dfrac{\text{lb}_f{}^2/\text{ft}}{\text{ft}^3/\text{sec}}$		$\dfrac{\text{lb}_f{}^2/\text{ft}}{\text{ft}^3}$	
Volume flow rate (flow)	Fluid pressure	$\int(\text{fluid pressure})\,dt$		$\dfrac{\text{ft}^3/\text{sec}}{\text{lb}_f/\text{ft}^2}$		$\dfrac{\text{ft}^3/\text{sec}}{(\text{lb}_f/\text{ft}^2)\text{-sec}}$
Temperature (effort)	Heat-transfer rate per unit temperature difference	Total heat transfer per unit temperature difference	$\dfrac{\text{F}^\circ}{\text{Btu}/(\text{sec-F}^\circ)}$		$\dfrac{\text{F}^\circ}{\text{Btu}/\text{F}^\circ}$	

Fig. 3.28. *Loading parameters for common variables.*

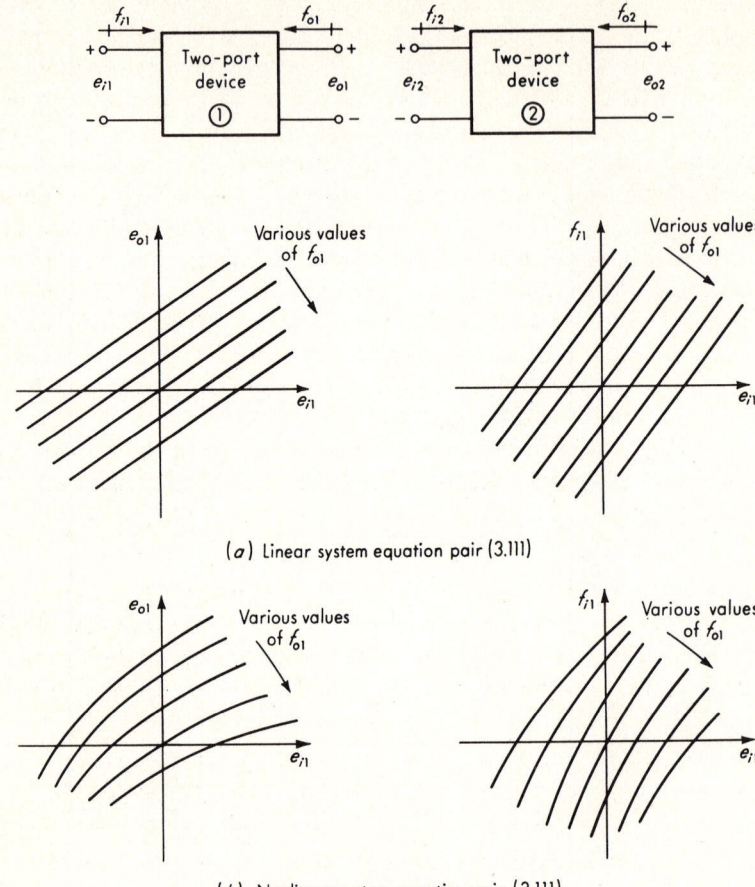

(*a*) Linear system equation pair (3.111)

(*b*) Nonlinear system equation pair (3.111)

Fig. 3.29.   *Generalized loading configuration.*

be known in order to predict their overall operation when connected together. We first assume that each element may be characterized as a "two-port."[1] This means essentially that the only significant energy exchanges that take place between the device and others that might be connected to it occur at two places ("ports"), which we denote as the input and the output. This does not necessarily mean that a third (or fourth) flow of energy through the "boundaries" of the device might not exist but merely that such additional energy flows are assumed constant and unaffected by changes in the two main energy flows. For example, an electronic amplifier has mainly an input voltage (and current) and an

---

[1] Paynter, *op. cit.*, p. 50.

output voltage (and current). However, it actually also has a third "port," the connection to its power supply. However, if the power supply is assumed *constant*, it can be included within the boundaries of the amplifier itself, allowing characterization of such amplifiers as two-port.

In Fig. 3.29, the symbols e represent effort variables whereas f's denote flow variables. Any one of the four variables associated with each of the devices may be considered as being determined by the values of the other variables. That is, we may consider any one variable as a *dependent* variable whose value is determined by the values of several independent variables. It is important to note that the quantities in question (the e's and f's) are all available at the "terminals" of the device and may be measured experimentally *without any knowledge of the internal details of the device. Thus complex devices for which no adequate theory exists may be studied experimentally by such methods.*

Suppose we choose to consider first the quantity e_{o1} as a dependent variable. Then we might at first think that the independent variables would be e_{i1}, f_{i1}, and f_{o1}, that is,

$$e_{o1} = e_{o1}(e_{i1}, f_{i1}, f_{o1}) \qquad (3.94)$$

This, however, is not the case, since e_{i1}, f_{i1}, and f_{o1} are not all independent. If any two of these three are assigned values, the value of the third is determined by the system and is not open to independent choice. The truth of this statement may be demonstrated as follows: Similarly to Eq. (3.94), we could write

$$f_{o1} = f_{o1}(e_{i1}, f_{i1}, e_{o1}) \qquad (3.95)$$
$$e_{i1} = e_{i1}(f_{i1}, f_{o1}, e_{o1}) \qquad (3.96)$$
$$f_{i1} = f_{i1}(e_{i1}, e_{o1}, f_{o1}) \qquad (3.97)$$

We note now, however, from (3.95) that f_{o1} depends on e_{i1} and f_{i1}. Thus it would not be correct to have e_{i1}, f_{i1}, *and f_{o1}* as independent variables in (3.94). Similar inconsistencies can be found in all four formulas. It is thus clear that three independent variables are too many. If we try two instead, we can write

$$e_{o1} = e_{o1}(e_{i1}, f_{i1}) \qquad (3.98)$$
$$\text{or} \quad e_{o1} = e_{o1}(e_{i1}, f_{o1}) \qquad (3.99)$$
$$\text{or} \quad e_{o1} = e_{o1}(f_{i1}, f_{o1}) \qquad (3.100)$$
$$\text{Also} \quad f_{o1} = f_{o1}(e_{i1}, e_{o1}) \qquad (3.101)$$
$$\text{or} \quad f_{o1} = f_{o1}(e_{i1}, f_{i1}) \qquad (3.102)$$
$$\text{or} \quad f_{o1} = f_{o1}(e_{o1}, f_{i1}) \qquad (3.103)$$
$$\text{Also} \quad e_{i1} = e_{i1}(e_{o1}, f_{i1}) \qquad (3.104)$$
$$\text{or} \quad e_{i1} = e_{i1}(e_{o1}, f_{o1}) \qquad (3.105)$$
$$\text{or} \quad e_{i1} = e_{i1}(f_{i1}, f_{o1}) \qquad (3.106)$$
$$\text{Also} \quad f_{i1} = f_{i1}(e_{i1}, e_{o1}) \qquad (3.107)$$
$$\text{or} \quad f_{i1} = f_{i1}(e_{i1}, f_{o1}) \qquad (3.108)$$
$$\text{or} \quad f_{i1} = f_{i1}(e_{o1}, f_{o1}) \qquad (3.109)$$

We can now choose any one of the above equations and immediately find its companion (another equation with the *same* independent variables) and thus define the system in terms of two "input" quantities (the independent variables) and two "output" quantities (the dependent variables).   Enumerating all possibilities, we get

$$e_{o1} = e_{o1}(e_{i1}, f_{i1})$$
$$f_{o1} = f_{o1}(e_{i1}, f_{i1}) \qquad (3.110)$$

$$e_{o1} = e_{o1}(e_{i1}, f_{o1})$$
$$f_{i1} = f_{i1}(e_{i1}, f_{o1}) \qquad (3.111)$$

$$e_{o1} = e_{o1}(f_{i1}, f_{o1})$$
$$e_{i1} = e_{i1}(f_{i1}, f_{o1}) \qquad (3.112)$$

$$f_{o1} = f_{o1}(e_{i1}, e_{o1})$$
$$f_{i1} = f_{i1}(e_{i1}, e_{o1}) \qquad (3.113)$$

$$f_{o1} = f_{o1}(e_{o1}, f_{i1})$$
$$e_{i1} = e_{i1}(e_{o1}, f_{i1}) \qquad (3.114)$$

$$e_{i1} = e_{i1}(e_{o1}, f_{o1})$$
$$f_{i1} = f_{i1}(e_{o1}, f_{o1}) \qquad (3.115)$$

Thus we see that, in general, one can choose values for any two of the four quantities and then the values of the other two are determined. If the physical system is strictly linear, its static input-output characteristics can be displayed graphically for any one of the equation pairs (3.110) to (3.115) in a fashion similar to Fig. 3.29a.   If it is nonlinear in a continuous (smooth) fashion, the curves might be as in Fig. 3.29b.

We are now in a position to derive Eq. (3.55).   Taking the first equation of pair (3.111), we may write

$$de_{o1} = \left. \frac{\partial e_{o1}}{\partial e_{i1}} \right|_{f_{o1} = \text{const}} de_{i1} + \left. \frac{\partial e_{o1}}{\partial f_{o1}} \right|_{e_{i1} = \text{const}} df_{o1} \qquad (3.116)$$

We now define

$$\left. \frac{\partial e_{o1}}{\partial e_{i1}} \right|_{f_{o1} = \text{const}} \triangleq \text{no-load static transfer function} \triangleq K \qquad (3.117)$$

$$\left. \frac{\partial e_{o1}}{\partial f_{o1}} \right|_{e_{i1} = \text{const}} \triangleq \text{generalized output impedance } Z_{go} \qquad (3.118)$$

The physical interpretation here is that we consider the system originally in equilibrium with e_{i1}, f_{i1}, e_{o1}, and f_{o1} all at some constant values.   Now, if e_{i1} changes by de_{i1} and if f_{o1} should stay constant [the simplest case is where f_{o1} is constant *at zero* because device 1 is open circuit (unloaded) at its output], the change in output is given by

$$de_{o1} = K \, de_{i1} \qquad (3.119)$$

If, however, the output of device 1 *is* connected to the input of device 2, then f_{o1} will not be constant and de_{o1} will be different from $K\,de_{i1}$. The amount of "loading error" depends on df_{o1}; to find it, we must know the input impedance of device 2. To define this, we use the second equation of pair (3.112) and apply it to device 2.

$$de_{i2} = \frac{\partial e_{i2}}{\partial f_{i2}}\bigg|_{f_{o2}=\text{const}} df_{i2} + \frac{\partial e_{i2}}{\partial f_{o2}}\bigg|_{f_{i2}=\text{const}} df_{o2} \qquad (3.120)$$

If device 2 has no third device connected to its output, we may take $df_{o2} = 0$ and get

$$de_{i2} = \frac{\partial e_{i2}}{\partial f_{i2}}\bigg|_{f_{o2}=\text{const}} df_{i2} \qquad (3.121)$$

We now define

$$\frac{\partial e_{i2}}{\partial f_{i2}}\bigg|_{f_{o2}=\text{const}} \triangleq \text{ generalized input impedance } Z_{gi} \qquad (3.122)$$

From Fig. 3.29, we note that, when devices 1 and 2 are connected, $e_{o1} = e_{i2}$ and $f_{o1} = -f_{i2}$. We thus get

$$de_{o1} = K\,de_{i1} + Z_{go}\,df_{o1} = K\,de_{i1} - \frac{Z_{go}}{Z_{gi}}\,de_{o1} \qquad (3.123)$$

$$\left(1 + \frac{Z_{go}}{Z_{gi}}\right) de_{o1} = K\,de_{i1} \qquad (3.124)$$

Now, $K\,de_{i1}$ is the value de_{o1} would have if device 2 were not connected to device 1. This corresponds to q_{i1u} of Eq. (3.55). We can thus write

$$de_{o1} = \frac{1}{Z_{go}/Z_{gi} + 1}\,K\,de_{i1} \qquad (3.125)$$

thereby proving Eq. (3.55). Equations (3.61), (3.68), and (3.80) can all be established in similar fashion.

When the physical devices are strictly linear, the small changes de_{o1}, de_{i1}, etc., can be replaced by the actual quantities e_{o1}, e_{i1}, etc., and the partial derivatives are constant for all values of the other independent variable. That is, in Eq. (3.116), for example, the term

$$\frac{\partial e_{o1}}{\partial e_{i1}}\bigg|_{f_{o1}=\text{const}}$$

would be numerically the same for any and all values of f_{o1} that one might choose. One generally chooses the simplest value of f_{o1} with which to work theoretically or experimentally. This is $f_{o1} \equiv 0$. The other partial derivatives are similarly handled. Thus, in linear systems, the various impedances, admittances, transfer functions, etc., are all constant and *independent of the size* of the various signals e_{i1}, e_{o1}, etc., in the devices.

In nonlinear devices, on the other hand, the system terminal

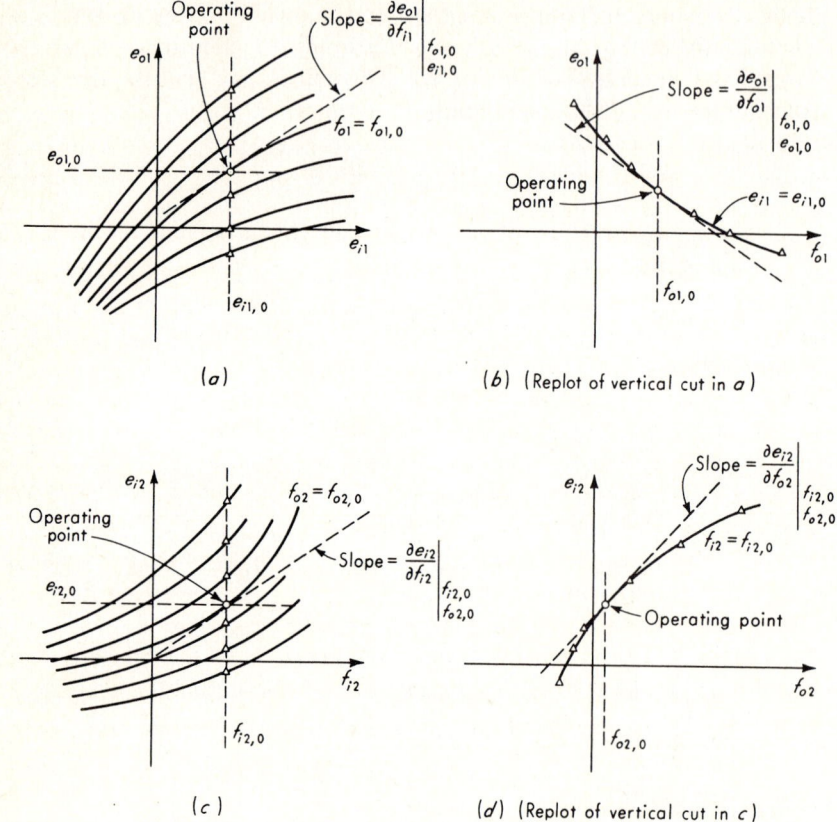

(a)

(b) (Replot of vertical cut in *a*)

(c)

(d) (Replot of vertical cut in *c*)

Fig. 3.30. *Definition of loading parameters.*

characteristics vary with the size of the signal and give accurate results only if the signal variations are limited to *small* excursions from some operating point.   That is, we assume, in Fig. 3.29*b*, for example, that the independent variables f_{o1} and e_{i1} are set at some fixed values $f_{o1,0}$ and $e_{i1,0}$, as shown in Fig. 3.30*a*.   The changes in the output quantities caused by small changes in the input quantities can then be predicted by Eq. (3.116), using the curve slopes of Fig. 3.30*a* and *b* as the numerical values of the partial derivatives.   A similar interpretation of Eq. (3.120) for the second device leads to Fig. 3.30*c* and *d*.   The "operating points" cannot be chosen arbitrarily.   They must correspond to a stable equilibrium condition when the two devices are interconnected, since then $e_{o1,0}$ must equal $e_{i2,0}$ and $f_{o1,0}$ must equal $-f_{i2,0}$.

As an example of the above procedures, consider a situation where one wishes to measure the volume flow rate of a motor-driven pump by means of a flowmeter connected at its discharge line. With no flowmeter attached, the pump is assumed to discharge to atmospheric pressure (0 psig) at a certain flow rate. When the flowmeter is attached, it will cause a pressure drop across itself; thus, if the flowmeter discharges to atmosphere, the pump discharge is now above atmosphere. Depending on the type of pump, this increase in discharge pressure will cause a greater or lesser change in flow rate; thus the flowmeter will not be measuring the no-load flow accurately. A generalization of this problem might be stated as follows: Suppose the pump (with flowmeter attached) is driven at a speed ω_0 with a shaft torque T_0. This will result in pump discharge pressure $p_{p,0}$ and a volume flow rate Q_0. If the pump speed is changed by an amount $d\omega$, the flow rate will change by an amount dQ. How much different will this dQ be from the dQ that would occur for the same $d\omega$ if the flowmeter were not present?

Figure 3.31 illustrates this situation and its analysis by means of admittance techniques. Since the measured variable Q is a flow variable, we search the equation list (3.110) to (3.115) for equations that will allow us to define the output admittance Y_{go} of the pump and the input admittance Y_{gi} of the flowmeter. Also, since we are concerned with a change in speed ω (rather than torque T), we desire an equation containing the flow variable (rather than the effort variable) as one of the independent variables. For the pump, the first equation of pair (3.114) gives

$$dQ = \left.\frac{\partial Q}{\partial \omega}\right|_{p_p = p_{p,0}} d\omega + \left.\frac{\partial Q}{\partial p_p}\right|_{\omega = \omega_0} dp_p \qquad (3.126)$$

We define

$$\left.\frac{\partial Q}{\partial \omega}\right|_{p_p = p_{p,0}} \triangleq \text{no-load static transfer function} \triangleq K_{Q,\omega} \qquad (3.127)$$

$$\left.\frac{\partial Q}{\partial p_p}\right|_{\omega = \omega_0} \triangleq \text{output admittance} \triangleq Y_{go} \qquad (3.128)$$

The numerical values of these parameters may be established by experiments, which give the graphical results of Fig. 3.31c. Turning now to the flowmeter, we use the second equation of pair (3.113) to get

$$-dQ = \left.\frac{\partial Q}{\partial p_p}\right|_{p_d = p_{d,0}} dp_p + \left.\frac{\partial Q}{\partial p_d}\right|_{p_p = p_{p,0}} dp_d \qquad (3.129)$$

We define

$$\left.\frac{\partial Q}{\partial p_p}\right|_{p_d = p_{d,0}} \triangleq \text{input admittance} \triangleq Y_{gi} \qquad (3.130)$$

$$\left.\frac{\partial Q}{\partial p_d}\right|_{p_p = p_{p,0}} \triangleq \text{flowmeter output admittance} \triangleq Y_{go.f} \qquad (3.131)$$

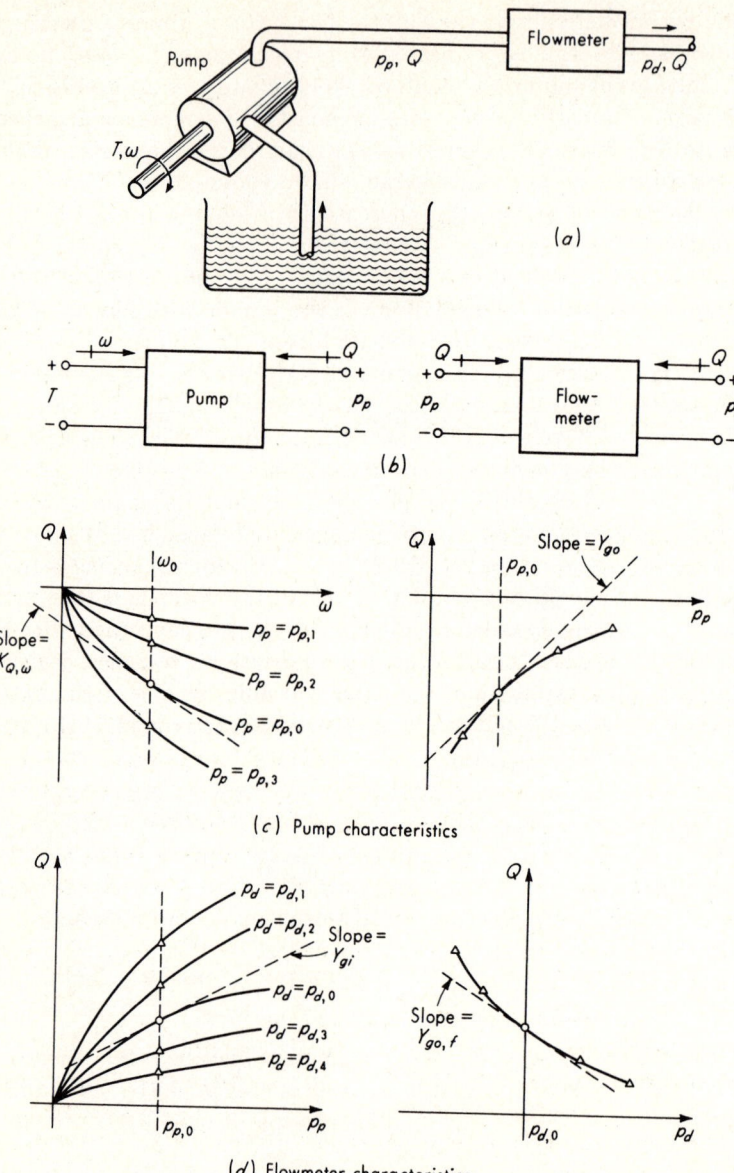

(a)

(c) Pump characteristics

(d) Flowmeter characteristics

Fig. 3.31. *Example of loading analysis.*

Figure 3.31*d* illustrates the method of obtaining these parameters from experimental tests on the flowmeter.

If we assume that the flowmeter output is not attached to the input of some other device but discharges directly to atmosphere, then $p_d =$ const and $dp_d = 0$. Combining Eqs. (3.126) and (3.129), we get

$$dQ = K_{Q,\omega}\, d\omega + Y_{go}\, dp_p = K_{Q,\omega}\, d\omega - \frac{Y_{go}}{Y_{gi}}\, dQ \qquad (3.132)$$

$$\left(1 + \frac{Y_{go}}{Y_{gi}}\right) dQ = K_{Q,\omega}\, d\omega \qquad (3.133)$$

$$dQ = \frac{1}{1 + Y_{go}/Y_{gi}}\, K_{Q,\omega}\, d\omega \qquad (3.134)$$

If the flowmeter were not present, dp_p in Eq. (3.126) would be zero, giving

$$dQ = K_{Q,\omega}\, d\omega \qquad (3.135)$$

Thus the flowmeter causes an "error" (flow change) that depends on the numerical value of Y_{go}/Y_{gi}. If Y_{go}/Y_{gi} is very small compared with 1, the error becomes negligible.

In concluding this section on loading effects, the following comments are appropriate. In every measurement, the instrument input causes a load on the measured-medium output. If the instrument system consists of several interconnected stages (the "elements" of Fig. 2.1), there may, furthermore, be significant loading effects between stages. This is often the case when the "elements" are general-purpose devices such as amplifiers and recorders which are connected together in different ways at different times to create a measurement system suited to a particular problem. In using such a "building block" approach, loading problems must be carefully considered, and methods such as have been outlined above must be used if satisfactory and predictable results are to be obtained. If, on the other hand, the elements are merely functional components (permanently connected together) of a specific instrument, loading between elements may be a necessary consideration for the instrument *designer* but not for the user. The user need concern himself only with the loading situation at the measured-medium–primary-sensing-element interface.

While the impedance-type concepts discussed in this section are extremely useful in studying the disturbing effects of measuring instruments on measured media, some such effects do not lend themselves to this type of approach. For example, if a pitot tube is inserted into a flow field to measure flow velocity, the presence of the tube distorts the velocity field without necessarily extracting any energy or power. Thus, our impedance-type concepts are not directly applicable to such situations.

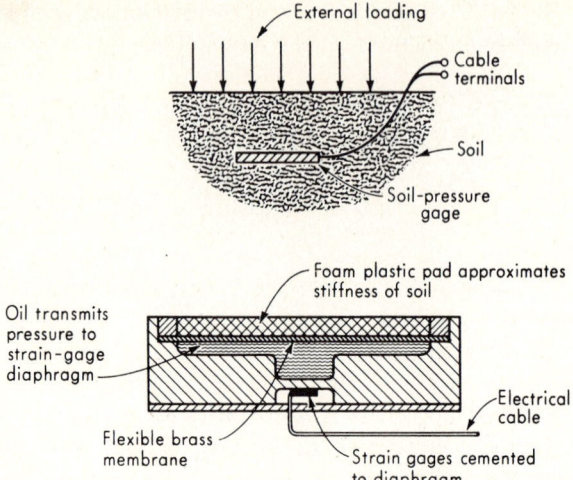

Fig. 3.32. *Soil-pressure gage.*

A related example, in which impedance-type concepts are useful (though not in the way described in this section) is given by Boiten.[1] An instrument for measuring subsoil pressures due to surface loading is useful in soil mechanics studies related to the construction of roads, airfields, etc. Figure 3.32 shows the essential features of such a device. While it is usually desirable that a force-measuring device have the highest possible stiffness, this criterion is *not* correct for this particular application. The reason is that the presence of the pressure gage in the soil causes a distortion of the pressure field if the stiffness of the gage is either higher or lower than the stiffness of the surrounding soil. That is, a very stiff gage will read too high when the external loading is applied, since the soil around it will compress more than the gage; thus the gage carries a disproportionate share of the total load. Similarly, an excessively compliant gage will read too low a pressure. The best gage is one whose stiffness just *matches* that of the soil in which it is to be used and thus does not distort the pressure field in the soil.

The feature distinguishing this example from those treated earlier in this section is that it deals with a distributed-parameter- ("field-") type of model for the physical system whereas our earlier discussions were all in terms of lumped-parameter ("network") models. The pitot-tube example also falls in the distributed-parameter category. While

[1] R. G. Boiten, The Mechanics of Instrumentation, *Proc. Inst. Mech. Engrs.* (*London*), vol. 177, no. 10, 1963.

impedance concepts have been applied to distributed-parameter models, we shall not pursue this subject further in this text.

Concluding remarks on static characteristics. In this section we have attempted to present the most significant static characteristics commonly used in describing instrument performance.   It is not possible to list *all* the specific static characteristics that might be pertinent in a particular instrument; rather, those of general interest were considered. It should also be emphasized that the terminology of the measurement field has not been thoroughly standardized.   Therefore one should be careful to determine the precise definition intended when an apparently familiar term is encountered.

Also, errors not intrinsically associated with the instrument itself, but due to human factors involved in taking readings or incurred by incorrect installation of the instrument, were not considered in detail. However, such errors could certainly be included in our general framework merely by extending the concept of "inputs" to include human factors and installation effects.

3.3

Dynamic Characteristics   We begin our study of the dynamic characteristics of measurement systems by postulating a generalized mathematical model embodying the features pertinent to the characterization of the dynamic relation between any particular input and output.

Generalized mathematical model of measurement system. As in so many other areas of engineering application (vibration theory, circuit theory, automatic-control theory, aircraft stability and control theory, etc.), the most widely useful mathematical model for the study of measurement-system dynamic response is the ordinary linear differential equation with constant coefficients.   We assume that the relation between any particular input (desired, interfering, or modifying) and the output can, by application of suitable simplifying assumptions, be put in the form

$$a_n \frac{d^n q_o}{dt^n} + a_{n-1} \frac{d^{n-1} q_o}{dt^{n-1}} + \cdots + a_1 \frac{dq_o}{dt} + a_0 q_o = b_m \frac{d^m q_i}{dt^m}$$

$$+ b_{m-1} \frac{d^{m-1} q_i}{dt^{m-1}} + \cdots + b_1 \frac{dq_i}{dt} + b_0 q_i \qquad (3.136)$$

where $q_o \triangleq$ output quantity

$\quad q_i \triangleq$ input quantity

$\quad\quad t \triangleq$ time

a's, b's $\triangleq$ combinations of system physical parameters, assumed constant

If we define the differential operator $D \triangleq d/dt$, Eq. (3.136) can be written as

$$(a_n D^n + a_{n-1} D^{n-1} + \cdots + a_1 D + a_0)q_o$$
$$= (b_m D^m + b_{m-1} D^{m-1} + \cdots + b_1 D + b_0)q_i \qquad (3.137)$$

The solution of equations of this type has been put on a systematic basis using either the "classical" method of D operators or the Laplace-transform method. With the D-operator method, the complete solution q_o is obtained in two separate parts as

$$q_o = q_{ocf} + q_{opi} \qquad (3.138)$$

where $\quad q_{ocf} \triangleq$ complementary-function part of solution
$\qquad q_{opi} \triangleq$ particular-integral part of solution

The solution q_{ocf} has n arbitrary constants; q_{opi} has none. These n arbitrary constants may be evaluated numerically by imposing n initial conditions on Eq. (3.138). The solution q_{ocf} is obtained by calculating the n roots of the algebraic *characteristic equation*

$$a_n D^n + a_{n-1} D^{n-1} + \cdots + a_1 D + a_0 = 0 \qquad (3.139)$$

Once these roots $r_1, r_2, \ldots, r_n$ have been found, the complementary-function solution is immediately written by following the rules stated below.

1. Real roots, unrepeated. For each real unrepeated root r one term of the solution is written as Ce^{rt}, where C is an arbitrary constant. Thus, for example, roots -1.7, $+3.2$, and 0 give a solution $C_1 e^{-1.7t} + C_2 e^{3.2t} + C_3$.

2. Real roots, repeated. For each root r which appears p times, the solution is written as $(C_0 + C_1 t + C_2 t^2 + \cdots + C_{p-1} t^{p-1})e^r$. Thus, if there are roots $-1, -1, +2, +2, +2, 0, 0$ the solution is written as $(C_0 + C_1 t)e^{-t} + (C_2 + C_3 t + C_4 t^2)e^{2t} + (C_5 + C_6 t)$.

3. Complex roots, unrepeated. A complex root has the general form $a + ib$. It can be shown that if the a's of Eq. (3.139) are themselves real numbers (which they generally will be since they are physical quantities such as mass, spring, rate, etc.) then if any complex roots occur they will always occur in pairs of the form $a \pm ib$. For each such root pair, the corresponding solution is $Ce^{at} \sin (bt + \phi)$, where C and ϕ are the two arbitrary constants. Thus roots $-3 \pm i4$, $2 \pm i5$, and $0 \pm i7$ give a solution $C_0 e^{-3t} \sin (4t + \phi_0) + C_1 e^{2t} \sin (5t + \phi_1) + C_2 \sin (7t + \phi_2)$.

4. Complex roots, repeated. For each pair of complex roots $a \pm ib$ which appears p times the solution is $C_0 e^{at} \sin (bt + \phi_0) + C_1 t e^{at}$

$\sin (bt + \phi_1) + \cdots + C_{p-1}t^{p-1}e^{at} \sin (bt + \phi_{p-1})$. Roots $-3 \pm i2$, $-3 \pm i2$, and $-3 \pm i2$ thus give a solution $C_0e^{-3t} \sin (2t + \phi_0) + C_1te^{-3t} \sin (2t + \phi_1) + C_2t^2e^{-3t} \sin (2t + \phi_2)$.

The complete complementary-function solution is simply the algebraic sum of the individual parts found from the four rules. Whereas the above method for finding q_{ocf} *always* works, no universal method for finding the particular solution q_{opi} exists. This is because q_{opi} depends on the form of q_i, and one can always define a sufficiently "pathological" form of q_i to prevent solution for q_{opi}. However, if q_i is restricted to functions of prime engineering interest a relatively simple method for finding q_{opi} is available. This is the *method of undetermined coefficients*, which will now be briefly reviewed. Since the method does not work for all q_i, the first question to be answered is whether it will work for the q_i of interest. For a given q_i, the right side of Eq. (3.136) is some known function of time $f(t)$. To test whether this method can be applied, we repeatedly differentiate $f(t)$ and then examine the functions created by these differentiations. There are three possibilities:

1. After a certain-order derivative, all higher derivatives are zero.
2. After a certain-order derivative, all higher derivatives have the same functional form as some lower-order derivative.
3. Upon repeated differentiation, new functional forms continue to arise.

If case 1 or 2 occurs, the method will work. If case 3 occurs, this method will not work, and others must be tried. If the method is applicable, the solution q_{opi} is immediately written as

$$q_{opi} = Af(t) + Bf'(t) + Cf''(t) + \cdots \qquad (3.140)$$

where the right-hand side includes one term for each functionally different form found by examining $f(t)$ and all its derivatives. The constants A, B, C, etc., can be found *immediately* (they do *not* depend on the initial conditions) by substituting q_{opi} [as given in Eq. (3.140)] into Eq. (3.136) and requiring (3.136) to be an *identity*. This procedure always generates as many simultaneous algebraic equations in the unknowns A, B, C, etc., as there are unknowns; thus the equations can be solved for A, B, C, etc.

The operational transfer function. In the analysis, design, and application of measurement systems the concept of the operational transfer function is very useful. The operational transfer function relating

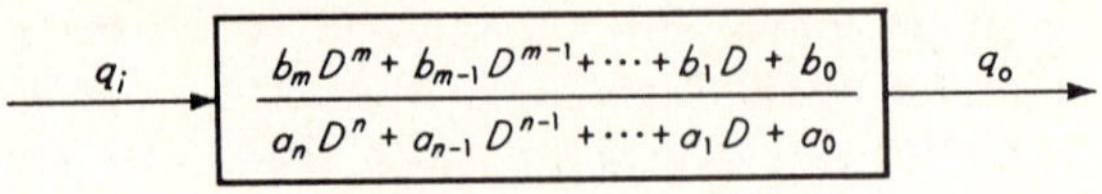

$$q_i \longrightarrow \boxed{\dfrac{b_m D^m + b_{m-1} D^{m-1} + \cdots + b_1 D + b_0}{a_n D^n + a_{n-1} D^{n-1} + \cdots + a_1 D + a_0}} \longrightarrow q_o$$

Fig. 3.33. *General operational transfer function.*

output q_o to input q_i is defined by treating Eq. (3.137) as if it were an algebraic relation and forming the ratio output/input:

$$\text{Operational transfer function} \triangleq \frac{q_o}{q_i}(D)$$

$$\triangleq \frac{b_m D^m + b_{m-1} D^{m-1} + \cdots + b_1 D + b_0}{a_n D^n + a_{n-1} D^{n-1} + \cdots + a_1 D + a_0} \qquad (3.141)$$

In writing transfer functions, one always writes $(q_o/q_i)(D)$, not just q_o/q_i, to emphasize that the transfer function is a *general* relation between q_o and q_i and very definitely *not* the instantaneous ratio of the time-varying quantities q_o and q_i.

One of the several useful features of transfer functions is their utility for graphic symbolic depiction of system dynamic characteristics by means of block diagrams. That is, if we wish to depict graphically a device with transfer function (3.141) we can draw a block diagram as in Fig. 3.33. Furthermore, the transfer function is helpful in determining the overall characteristics of a system made up of components whose individual transfer functions are known. This combination is most simply achieved when there is negligible loading (the input impedance of the second device much higher than the output impedance of the first, etc.) between the connected devices. For this case, the overall transfer function is simply the product of the individual ones since the output of the preceding device becomes the input of the following one. Figure 3.34 illustrates this procedure. When significant loading *is* present, one may apply the impedance concepts of Sec. 3.2 (extended to the dynamic

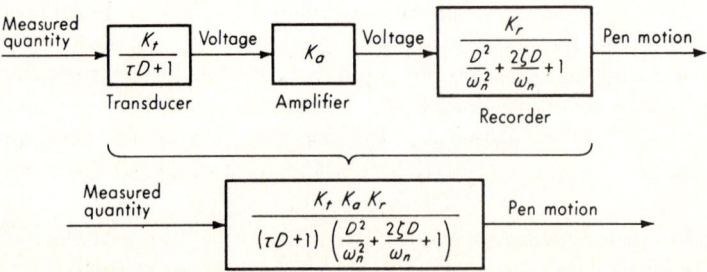

Fig. 3.34. *Combination of individual transfer functions.*

case) or simply analyze the complete system "from scratch" without using the individual transfer functions.

In the technical literature, the Laplace-transform method is in common use for the study of linear systems. When such methods are employed, the *Laplace transfer function* is defined as the ratio of the Laplace transform of the output quantity to the Laplace transform of the input quantity when all initial conditions are zero. Thus, analogous to Eq. (3.141), the Laplace transfer function would be written as

$$\frac{q_o(s)}{q_i(s)} \triangleq \frac{q_o}{q_i}(s) \triangleq \frac{b_m s^m + b_{m-1} s^{m-1} + \cdots + b_1 s + b_0}{a_n s^n + a_{n-1} s^{n-1} + \cdots + a_1 s + a_0} \qquad (3.142)$$

where $s \triangleq \sigma + i\omega$ is the complex variable of the Laplace transform. We note that, so far as the *form* of the transfer function is concerned, one can shift from the Laplace form to the D-operator form (or vice versa) simply by interchanging s and D. Thus, if one encounters a block diagram using the Laplace notation, he can *always* convert to the D notation by a simple substitution. All the methods we shall subsequently develop may then be applied to the operational transfer function.

The sinusoidal transfer function. In studying the quality of measurement under dynamic conditions, we shall be analyzing the response of measurement systems to certain "standard" inputs. One of the most important of such responses is the steady-state response to a sinusoidal input. Here the input q_i is of the form $A_i \sin \omega t$. If one waits for all transient effects to die out (the complementary-function solution of a stable linear system always eventually dies out) it will be seen that the output quantity q_o will be a sine wave of exactly the same frequency (ω) as the input. However, the amplitude of the output may differ from that of the input, and a phase shift may be present. These results are easily shown by obtaining the particular (steady-state) solution by means of the method of undetermined coefficients. Since the frequency is the same, the relation between the input and output sine waves is completely specified by giving their amplitude ratio and phase shift. Both these quantities, in general, change when the driving frequency ω changes. Thus the *frequency response* of a system consists of curves of amplitude ratio and phase shift as a function of frequency. Figure 3.35 illustrates these concepts.

While the frequency response of any linear system may be obtained by getting the particular solution of its differential equation with

$$q_i = A_i \sin \omega t$$

much quicker and easier methods are available. These methods depend on the concept of the sinusoidal transfer function. The sinusoidal

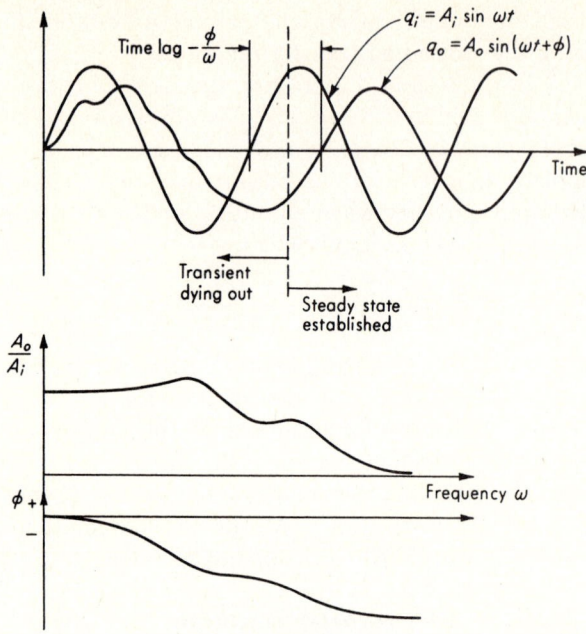

Fig. 3.35. *Frequency-response terminology.*

transfer function of a system is obtained by substituting $i\omega$ for D wherever it appears in the operational transfer function:

$$\text{Sinusoidal transfer function} \triangleq \frac{q_o}{q_i}(i\omega)$$

$$\triangleq \frac{b_m(i\omega)^m + b_{m-1}(i\omega)^{m-1} + \cdots + b_1(i\omega) + b_0}{a_n(i\omega)^n + a_{n-1}(i\omega)^{n-1} + \cdots + a_1(i\omega) + a_0} \quad (3.143)$$

$$\text{where} \quad i \triangleq \sqrt{-1}$$
$$\omega \triangleq \text{frequency, rad/time}$$

For any given frequency ω, Eq. (3.143) shows that $(q_o/q_i)(i\omega)$ is a complex number, which can always be put in the polar form $M\underline{/\phi}$. *We shall prove that the magnitude M of the complex number is the amplitude ratio A_o/A_i while the angle ϕ is the phase angle by which the output q_o leads the input q_i.* (If the output *lags* the input, ϕ is negative.)

The proof of the above statement is most readily demonstrated by means of the rotating-vector or phasor method of representing sinusoidal quantities. By a well-known trigonometric identity, we may write, in general,

$$Ae^{i\theta} = A(\cos\theta + i\sin\theta) = A\cos\theta + iA\sin\theta \quad (3.144)$$

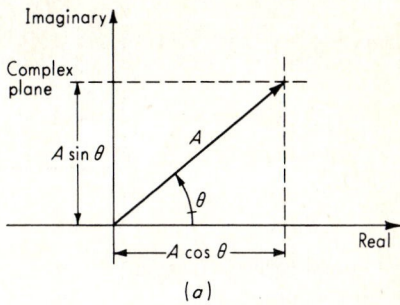

(a)

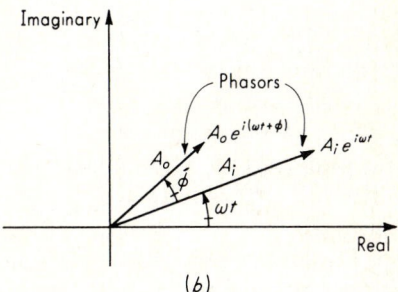

Fig. 3.36. *Phasor representation of sine waves.*

(b)

The complex number represented by the right-hand side can be exhibited graphically as in Fig. 3.36a. If we now apply this general result to the specific problem of representing q_o and q_i, we get

$$\text{For input } q_i \begin{cases} \text{let } A = A_i \text{ and } \theta = \omega t \\ A_i e^{i\omega t} = A_i \cos \omega t + i A_i \sin \omega t \end{cases} \quad (3.145)$$

$$\text{For output } q_o \begin{cases} \text{let } A = A_o \text{ and } \theta = \omega t + \phi \\ A_o e^{i(\omega t + \phi)} = A_o \cos (\omega t + \phi) + i A_o \sin (\omega t + \phi) \end{cases} \quad (3.146)$$

We note that the frequency ω of sinusoidal oscillation is also the angular velocity of rotation of the phasors of Fig. 3.36b. The phasors both rotate at the same angular velocity ω but maintain a fixed angle ϕ between them.

In carrying out our proof, we shall need to be able to differentiate phasor quantities. Since the amplitude A and the quantity i are constants, we have, in general,

$$\frac{d}{dt} (A e^{i\theta}) = \left(i \frac{d\theta}{dt} \right) A e^{i\theta} \quad (3.147)$$

and in particular, if $A = A_i$ and $\theta = \omega t$,

$$\frac{d}{dt} (A_i e^{i\omega t}) = (i\omega) A_i e^{i\omega t} \quad (3.148)$$

or, if $A = A_o$ and $\theta = \omega t + \phi$,

$$\frac{d}{dt} [A_o e^{i(\omega t + \phi)}] = (i\omega) A_o e^{i(\omega t + \phi)} \quad (3.149)$$

Clearly, for any higher derivative, we would get

$$\frac{d^n}{dt^n}\left(A_i e^{i\omega t}\right) = (i\omega)^n (A_i e^{i\omega t}) \qquad (3.150)$$

$$\text{and} \quad \frac{d^n}{dt^n}[A_o e^{i(\omega t + \phi)}] = (i\omega)^n[A_o e^{i(\omega t + \phi)}] \qquad (3.151)$$

Thus, differentiating a phasor quantity n times with respect to time t may be achieved simply by multiplying it by $(i\omega)^n$.

Suppose we now consider Eq. (3.137) for the sinusoidal steady-state case. Then *every* term on each side of the equation will be a sinusoidally varying quantity since repeated differentiation of sine waves gives only more sine waves (or cosines, which can be replaced by sines with a phase angle). We now convert the differential equation (3.137) into a complex algebraic equation by replacing each sinusoidal term by its phasor representation. This is *not* a matter of simple substitution since the sinusoidal terms are not *equal* to the phasor quantities; rather they are *represented* by the phasor quantities. We must thus be careful to show that, when the new phasor (complex-number) equation is satisfied, we are guaranteed that the original system differential equation is also satisfied. We can then perform any desired manipulations on the complex-number equation with assurance that correct results will be obtained. This is done by first replacing the sinusoidal terms by their phasor representations:

$$a_n(i\omega)^n A_o e^{i(\omega t + \phi)} + a_{n-1}(i\omega)^{n-1} A_o e^{i(\omega t + \phi)} + \cdots$$
$$+ a_1(i\omega)A_o e^{i(\omega t + \phi)} + a_0 A_o e^{i(\omega t + \phi)} = b_m(i\omega)^m A_i e^{i\omega t}$$
$$+ b_{m-1}(i\omega)^{m-1} A_i e^{i\omega t} + \cdots + b_1(i\omega)A_i e^{i\omega t} + b_0 A_i e^{i\omega t} \qquad (3.152)$$

This complex-number equation can be satisfied only if the real parts on the left equal the real parts on the right and similarly for the imaginary parts. Thus, if Eq. (3.152) is enforced, we are guaranteed that the equation given by the imaginary parts will also be satisfied. If we obtain the first few terms in this equation, the pattern should be obvious. We have

$$\text{Im}\,[a_0 A_o e^{i(\omega t + \phi)}] = a_0 A_o \sin(\omega t + \phi) = a_0 q_o \qquad \text{lowest-order terms in the original differential equation}$$

$$\text{Im}\,[b_0 A_i e^{i\omega t}] = b_0 A_i \sin \omega t = b_0 q_i$$

$$\text{Im}\,[a_1(i\omega)A_o e^{i(\omega t + \phi)}] = \text{Im}\,\{a_1(i\omega)A_o[\cos(\omega t + \phi)$$
$$+ i\sin(\omega t + \phi)]\} = a_1\omega A_o \cos(\omega t + \phi) = a_1 Dq_o \qquad \text{next terms in original differential equation}$$

$$\text{Im}\,[b_1(i\omega)A_i e^{i\omega t}] = \text{Im}\,[b_1(i\omega)A_i(\cos \omega t + i\sin \omega t)]$$
$$= b_1\omega A_i \cos \omega t = b_1 Dq_i$$

It should be clear now that requiring Eq. (3.152) to hold is *equivalent* to requiring (3.137) to hold, even though they are *not* the same equation.

We now manipulate Eq. (3.152) as follows to prove our final result:

$$[a_n(i\omega)^n + a_{n-1}(i\omega)^{n-1} + \cdots + a_1(i\omega) + a_0]A_o e^{i(\omega t + \phi)}$$
$$= [b_m(i\omega)^m + b_{m-1}(i\omega)^{m-1} + \cdots + b_1(i\omega) + b_0]A_i e^{i\omega t} \qquad (3.153)$$

$$\frac{A_o e^{i(\omega t + \phi)}}{A_i e^{i\omega t}} = \frac{b_m(i\omega)^m + b_{m-1}(i\omega)^{m-1} + \cdots + b_1(i\omega) + b_0}{a_n(i\omega)^n + a_{n-1}(i\omega)^{n-1} + \cdots + a_1(i\omega) + a_0}$$

$$\triangleq \frac{q_o}{q_i}(i\omega) \qquad (3.154)$$

$$\text{Now } \frac{A_o e^{i(\omega t + \phi)}}{A_i e^{i\omega t}} = \frac{A_o}{A_i} e^{i\phi} = \frac{A_o}{A_i}(\cos\phi + i\sin\phi) \qquad (3.155)$$

$$\cos\phi + i\sin\phi = \sqrt{\cos^2\phi + \sin^2\phi}\ \underline{/\phi} = 1\ \underline{/\phi}$$

$$\text{and thus } \frac{q_o}{q_i}(i\omega) = \frac{A_o}{A_i}\ \underline{/\phi} = M\ \underline{/\phi} \qquad (3.156)$$

Equation (3.156) states that at any chosen frequency ω the magnitude of the complex number $(q_o/q_i)(i\omega)$ is numerically the amplitude ratio A_o/A_i while the angle of the complex number is the angle by which the output leads the input. Therefore our desired result is proved.

The zero-order instrument. While the general mathematical model of Eq. (3.136) is adequate for handling any linear measurement system, certain special cases occur so frequently in practice that they warrant separate consideration. Furthermore, more complicated systems can profitably be studied as combinations of these simple special cases.

The simplest possible special case of Eq. (3.136) occurs when all the a's and b's other than a_0 and b_0 are assumed to be zero. The differential equation then degenerates into the simple algebraic equation

$$a_0 q_o = b_0 q_i \qquad (3.157)$$

Any instrument or system that closely obeys Eq. (3.157) over its intended range of operating conditions is defined to be a *zero-order instrument*. Actually, two constants a_0 and b_0 are not necessary, and so we define the static sensitivity (or steady-state "gain") as follows:

$$q_o = \frac{b_0}{a_0} q_i = K q_i \qquad (3.158)$$

$$K \triangleq \frac{b_0}{a_0} \triangleq \text{static sensitivity} \qquad (3.159)$$

Since the equation $q_o = K q_i$ is an algebraic equation, it is clear that, no matter how q_i might vary with time, the instrument output (reading) follows it *perfectly* with no distortion or time lag of any sort. Thus, the zero-order instrument represents ideal or perfect dynamic performance.

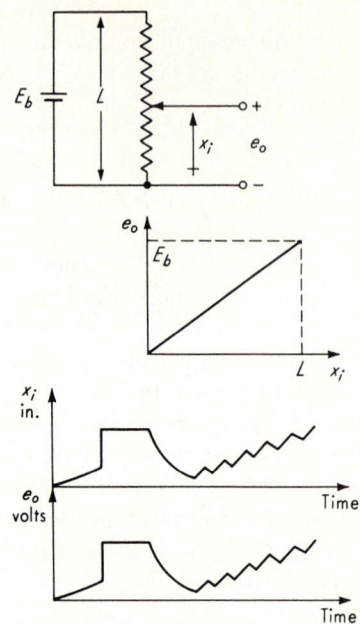

Fig. 3.37. *Zero-order instrument.*

A practical example of a zero-order instrument is the displacement-measuring potentiometer. Here (see Fig. 3.37) a strip of resistance material is excited with a voltage and provided with a sliding contact. If the resistance is linearly distributed along the length L, we may write

$$e_o = \frac{x_i}{L} E_b = K x_i \qquad (3.160)$$

$$\text{where} \quad K \triangleq \frac{E_b}{L} \quad \text{volts/in.}$$

If one examines this measuring device more critically, he will find that it is not *exactly* a zero-order instrument. This is simply a manifestation of the *universal* rule that no mathematical model can *exactly* represent *any* physical system. In our present example, we would find that, if we wish to *use* a potentiometer for motion measurements, we must attach to the output terminals some voltage-measuring device (such as an oscilloscope). Such a device will always draw some current (however small) from the potentiometer. Thus, when x_i changes, the potentiometer winding current will also change. This in itself would cause no dynamic distortion or lag *if the potentiometer were a pure resistance.* However, the idea of a pure resistance is a *mathematical model*, not a real system; thus the potentiometer will have some (however small) inductance and capacitance. If x_i is varied relatively slowly, these parasitic inductance and capacitance effects will not be apparent. However, for sufficiently fast variation of x_i these effects are no longer negligible and cause dynamic errors between x_i and e_o. The reasons a potentiometer is normally called a zero-order instrument are as follows:

1. The parasitic inductance and capacitance can be made very small by design.
2. The speeds ("frequencies") of motion to be measured are not high enough to make the inductive or capacitive effects noticeable.

Another aspect of nonideal behavior in a real potentiometer comes to light when we realize that the sliding contact must be attached to the body whose motion is to be measured. Thus, there is a mechanical loading effect, due to the inertia of the sliding contact and its friction, which will cause the measured motion x_i to be different from that which would occur were the potentiometer not present. This effect is thus different *in kind* from the inductive and capacitive phenomena mentioned earlier since they affected the relation [Eq. (3.160)] between e_o and x_i whereas the mechanical loading has no effect on this relation but, rather, makes x_i different from the undisturbed case.

The first-order instrument. If in Eq. (3.136) all a's and b's other than a_1, a_0, and b_0 are taken as zero, we get

$$a_1 \frac{dq_o}{dt} + a_0 q_o = b_0 q_i \qquad (3.161)$$

Any instrument that follows this equation is by definition a first-order instrument. There may be some conflict here between mathematical terminology and common engineering usage. In mathematics, a first-order *equation* has the general form

$$a_1 \frac{dq_o}{dt} + a_0 q_o = (b_m D^m + b_{m-1} D^{m-1} + \cdots + b_1 D + b_0) q_i \qquad (3.162)$$

where m could have any numerical value. However, through long usage, in engineering one commonly understands a first-order *system* to be defined by Eq. (3.161). Since in technical presentations both words *and equations* are generally employed, confusion on this point is rarely a problem.

While Eq. (3.161) has three parameters a_1, a_0, and b_0, only two are really essential since the whole equation could always be divided through by either a_1, a_0, or b_0, thus making the coefficient of one of the terms numerically equal to 1. The most useful procedure is to divide through

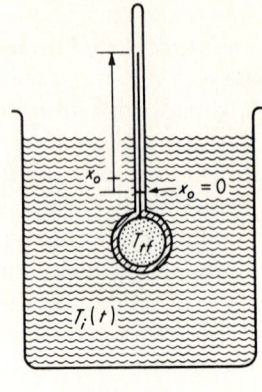

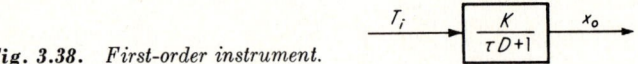

Fig. 3.38.   *First-order instrument.*

by a_0, giving

$$\frac{a_1}{a_0}\frac{dq_o}{dt} + q_o = \frac{b_0}{a_0}q_i \qquad (3.163)$$

which becomes

$$(\tau D + 1)q_o = Kq_i \qquad (3.164)$$

when we define

$$K \triangleq \frac{b_0}{a_0} \triangleq \text{static sensitivity} \qquad (3.165)$$

$$\tau \triangleq \frac{a_1}{a_0} \triangleq \text{time constant} \qquad (3.166)$$

The time constant τ always has the dimensions of time, while the static sensitivity K has the dimensions output/input.   For *any*-order instrument, K is always defined as b_0/a_0 and always has the same physical meaning, that is, the amount of output per unit input when the input is static (constant), because under such conditions all the derivative terms in the differential equation are zero.   The operational transfer function of any first-order instrument is

$$\frac{q_o}{q_i}(D) = \frac{K}{\tau D + 1} \qquad (3.167)$$

As an example of a first-order instrument, let us consider the liquid-in-glass thermometer of Fig. 3.38.   The input (measured) quantity here is the temperature $T_i(t)$ of the fluid surrounding the bulb of the thermometer, and the output is the displacement x_o of the thermometer fluid in the capillary tube.   We assume the temperature $T_i(t)$ is uniform throughout the fluid at any given time but may vary with time in an arbitrary fashion.   The principle of operation of such a thermometer is the thermal expansion of the filling fluid which drives the liquid column up or down in response to temperature changes. Since this liquid column has inertia, mechanical lags will be involved in moving the liquid from one level to another. However, we shall assume this lag is negligible compared with the thermal lag involved in transferring heat from the surrounding fluid through the bulb wall and into the thermometer fluid. This assumption rests (as all such assumptions necessarily must) on experience, judgment, order-of-magnitude calculations, and, ultimately, experimental verification (or refutation) of the results predicted by the analysis. Assumption of negligible mechanical lag allows us to relate the temperature of the fluid in the bulb to the reading x_o by the instantaneous (algebraic) equation

$$x_o = \frac{K_{ex}V_b}{A_c} T_{tf} \qquad (3.168)$$

where $x_o \triangleq$ displacement from reference mark, in.
$T_{tf} \triangleq$ temperature of fluid in bulb (assumed uniform throughout bulb volume), $T_{tf} = 0$ when $x_o = 0$, °F
$K_{ex} \triangleq$ differential expansion coefficient of thermometer fluid and bulb glass, in.^{3}/(in.^{3}-F°)
$V_b \triangleq$ volume of bulb, in.^{3}
$A_c \triangleq$ cross-sectional area of capillary tube, in.^{2}

To get a differential equation relating input and output in this thermometer, we consider conservation of energy over an infinitesimal length of time dt for the thermometer bulb:

$$\text{(Heat in)} - \text{(heat out)} = \text{energy stored}$$
$$UA_b(T_i - T_{tf})\, dt - 0(\text{assume no heat loss}) = V_b\rho C\, dT_{tf} \qquad (3.169)$$

where $U \triangleq$ overall heat-transfer coefficient across bulb wall, Btu/ (in.^{2}-F°-sec)
$A_b \triangleq$ heat-transfer area of bulb wall, in.^{2}
$\rho \triangleq$ mass density of thermometer fluid, lb$_m$/in.^{3}
$C \triangleq$ specific heat of thermometer fluid, Btu/(lb$_m$-F°)

Equation (3.169) involves many assumptions:

1. The bulb wall and fluid films on each side are pure resistance to heat transfer with no heat-storage capacity. This will be a good assumption if the heat-storage capacity (mass) (specific heat) of the bulb wall and fluid films is small *compared* with $C\rho V_b$ for the bulb.
2. The overall coefficient U is constant. Actually, film coefficients and bulb-wall conductivity all change with temperature, but these changes are quite small so long as the temperature does not vary over wide ranges.

3. The heat-transfer area A_b is constant. Actually, expansion and contraction would cause this to vary, but this effect should be quite small.
4. No heat is lost from the thermometer bulb by conduction up the stem. Heat loss will be small if the stem is of small diameter, made of a poor conductor, and is immersed in the fluid over a great length and if the exposed end is subjected to an air temperature not much different from T_i and T_{tf}.
5. The mass of fluid in the bulb is constant. Actually mass must enter or leave the bulb whenever the level in the capillary tube changes. For a fine capillary and a large bulb this effect should be small.
6. The specific heat C is constant. Again, this fluid property varies with temperature but the variation is slight except for large temperature changes.

The above list of assumptions is not complete but should give some appreciation of the discrepancies between a mathematical model and the real system it represents. Many of these assumptions could be relaxed to get a more accurate model but one would pay a heavy price in increased mathematical complexity. The choice of assumptions that are *just good enough* for the needs of the job at hand is one of the most difficult and important tasks of the engineer.

Returning to Eq. (3.169), we may write it as

$$V_b \rho C \frac{dT_{tf}}{dt} + UA_b T_{tf} = UA_b T_i \qquad (3.170)$$

Using Eq. (3.168), we get

$$\frac{\rho C A_c}{K_{ex}} \frac{dx_o}{dt} + \frac{UA_b A_c}{K_{ex} V_b} x_o = UA_b T_i \qquad (3.171)$$

which we recognize to be the form of Eq. (3.163), and so we immediately define

$$K \triangleq \frac{K_{ex} V_b}{A_c} \qquad \text{in./F}° \qquad (3.172)$$

$$\tau \triangleq \frac{\rho C V_b}{UA_b} \qquad \text{sec} \qquad (3.173)$$

Having shown a concrete example of a first-order instrument, let us now return to the problem of examining the dynamic response of first-order instruments in general. Once one has obtained the differential equation relating the input and output of an instrument he can study its dynamic performance by taking the input (quantity to be measured) to be some known function of time and then solving the differential equation for the output as a function of time. If the output is closely proportional to the input at all times, the dynamic accuracy is good. The fundamental

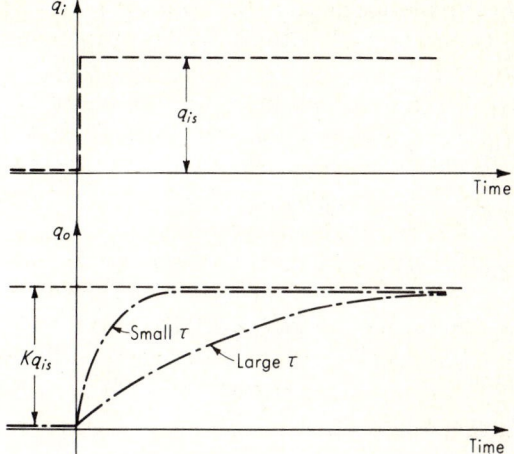

Fig. 3.39. *Step-function response of first-order instrument.*

difficulty in this approach lies in the fact that, in actual practice, the quantities to be measured usually do not follow some simple mathematical function but rather are of a random nature. Fortunately, however, much can be learned about instrument performance by examining the response to certain rather simple "standard" input functions. That is, just as one is not able to analyze the real *system* but rather an idealized model of it, so also one cannot work with the real *inputs* to a system but rather with simplified representations of them. This simplification of inputs (just as that of systems) can be carried out at several different levels, leading to either simple, rather inaccurate input functions that are readily handled mathematically or, alternatively, complex, more accurate representations that lead to mathematical difficulties.

We commence our study by considering several quite simple standard inputs that are in wide use. Although these inputs are, in general, only crude approximations to the actual inputs, they are extremely useful for studying the effects of parameter changes in a given instrument or comparing the *relative* performance of two competitive measurement systems.

Step response of first-order instruments. To apply a step input to a system, we assume that initially it is in equilibrium, with $q_i = q_o = 0$, when at time $t = 0$ the input quantity increases instantly an amount q_{is} (see Fig. 3.39). For $t > 0$, Eq. (3.164) becomes

$$(\tau D + 1)q_o = Kq_{is} \qquad (3.174)$$

It can be shown generally (by mathematical reasoning) or in any specific physical problem, such as the thermometer (by physical reasoning), that the initial condition for this situation is $q_o = 0$ for $t = 0^+$ ($t = 0^+$ means an infinitesimal time after $t = 0$). The complementary-function solution is

$$q_{ocf} = Ce^{(-t)/\tau} \qquad (3.175)$$

while the particular solution is

$$q_{opi} = Kq_{is} \qquad (3.176)$$

giving the complete solution as

$$q_o = Ce^{(-t)/\tau} + Kq_{is} \qquad (3.177)$$

Applying the initial condition,

$$0 = C + Kq_{is}$$
$$C = -Kq_{is}$$

giving finally

$$q_o = Kq_{is}[1 - e^{(-t)/\tau}] \qquad (3.178)$$

Examination of Eq. (3.178) shows that the speed of response depends *only* on the value of τ and is faster if τ is smaller. Thus in first-order instruments one strives to minimize τ for faithful dynamic measurements.

These results may be nondimensionalized by writing

$$\frac{q_o}{Kq_{is}} = 1 - e^{(-t)/\tau} \qquad (3.179)$$

and then plotting q_o/Kq_{is} versus t/τ as in Fig. 3.40a. This curve is then universal for any value of K, q_{is}, or τ that might be encountered. We could also define the measurement error e_m as

$$e_m \triangleq q_i - \frac{q_o}{K} \qquad (3.180)$$

$$e_m = q_{is} - q_{is}[1 - e^{(-t)/\tau}]$$

and nondimensionalize for plotting in Fig. 3.40b as

$$\frac{e_m}{q_{is}} = e^{(-t)/\tau} \qquad (3.181)$$

A dynamic characteristic useful in characterizing the speed of response of any instrument is the *settling time*. This is the time (after application of a step input) for the instrument to reach and stay within a stated plus-and-minus tolerance band around its final value. A small settling time is thus indicative of fast response. It is obvious that the numerical value of a settling time depends on the percentage tolerance band used; one must always state this. Thus one speaks of, say, a 5 percent settling time. For a first-order instrument a 5 percent settling time is equal to three time constants (see Fig. 3.41). Other percentages may be and are used in actual practice.

Knowing now that fast response requires a small value of τ, we can examine any specific first-order instrument to see what physical changes

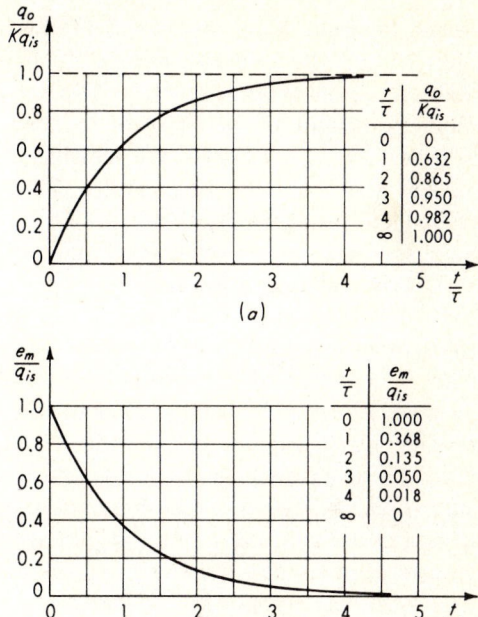

(a)

(b)

Fig. 3.40. *Nondimensional step-function response.*

would be needed to reduce τ. If we use our thermometer example, Eq. (3.173) shows that τ may be reduced by the following:

1. Reducing ρ, C, and V_b
2. Increasing U and A_b

Since ρ and C are properties of the fluid filling the thermometer they cannot be varied independently of one another, and so for small τ we

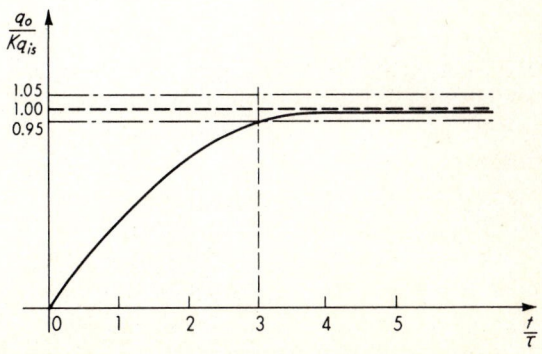

Fig. 3.41. *Settling-time definition.*

search for fluids with a small ρC product. The bulb volume V_b may be reduced but this will also reduce A_b unless some extended-surface heat-transfer augmentation (such as fins on the bulb) is introduced. Even more significant is the effect of reduced V_b on the static sensitivity K, as given by Eq. (3.172). We see that attempts to reduce τ by reducing V_b will result in reductions in K. Thus increased speed of response is traded off for lower sensitivity. This tradeoff is not unusual and will be observed in many other instruments.

The fact that τ depends on U means that one cannot state that a certain *thermometer* has a certain time constant but only that a certain thermometer *used in a certain fluid under certain heat-transfer conditions* (say free or forced convection) has a certain time constant. This is because U depends partly on the value of the film coefficient of heat transfer at the outside of the bulb, which varies greatly with changes in fluid (liquid or gas), flow velocity, etc. For example, a thermometer in stirred oil might have a time constant of 5 sec while the same thermometer in stagnant air would have a τ of perhaps 100 sec. Thus one must always be careful in giving (or using) performance data to be sure that the conditions of use correspond to those in force during calibration or that proper corrections are applied.

Ramp response of first-order instruments. To apply a ramp input to a system, we assume that initially the system is in equilibrium, with $q_i = q_o = 0$, when at $t = 0$ the input q_i suddenly starts to change at a constant rate $\dot{q}_{is}$. We thus have

$$q_i = q_o = 0 \qquad t \leq 0$$
$$q_i = \dot{q}_{is}t \qquad t \geq 0$$

and therefore $\quad (\tau D + 1)q_o = K\dot{q}_{is}t \qquad (3.182)$

The necessary initial condition can again be shown to be $q_o = 0$ for $t = 0^+$. Solution of Eq. (3.182) gives

$$q_{ocf} = Ce^{(-t)/\tau}$$
$$q_{opi} = K\dot{q}_{is}(t - \tau)$$
$$q_o = Ce^{(-t)/\tau} + K\dot{q}_{is}(t - \tau)$$

and applying the initial condition gives

$$q_o = K\dot{q}_{is}[\tau e^{(-t)/\tau} + t - \tau] \qquad (3.183)$$

We again define measurement error e_m by

$$e_m \triangleq q_i - \frac{q_o}{K} = \dot{q}_{is}t - \dot{q}_{is}\tau e^{(-t)/\tau} - \dot{q}_{is}t + \dot{q}_{is}\tau \qquad (3.184)$$

$$e_m = \underbrace{-\dot{q}_{is}\tau e^{(-t)/\tau}}_{\substack{\text{transient error} \\ e_{m,t}}} + \underbrace{\dot{q}_{is}\tau}_{\substack{\text{steady-} \\ \text{state} \\ \text{error} \\ e_{m,ss}}} \qquad (3.185)$$

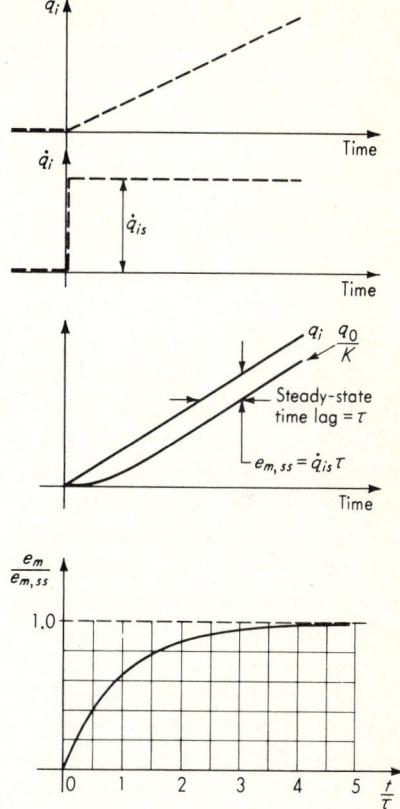

Fig. 3.42.   *Ramp response of first-order instrument.*

We note that the first term of e_m will gradually disappear as time goes by, and so it is called the *transient error*. The second term, however, persists forever and is thus called the *steady-state error*. The transient error disappears more quickly if τ is small. The steady-state error is directly proportional to τ; thus small τ is desirable here also. Steady-state error also increases directly with $\dot{q}_{is}$, the rate of change of the measured quantity. In steady state, the horizontal (time) displacement between input and output curves is seen to be τ, and so one may make the interpretation that the instrument is reading what the input *was* τ sec ago. The above results, together with a nondimensionalized representation, are given graphically in Fig. 3.42.

Frequency response of first-order instruments. Equation (3.143) may be applied directly to the problem of finding the response of first-order systems to sinusoidal inputs. We have

$$\frac{q_o}{q_i}(i\omega) = \frac{K}{i\omega\tau + 1} = \frac{K}{\sqrt{\omega^2\tau^2 + 1}} \, \underline{/\tan^{-1} - \omega\tau} \qquad (3.186)$$

Thus the amplitude ratio is

$$\frac{A_o}{A_i} = \left| \frac{q_o}{q_i} (i\omega) \right| = \frac{K}{\sqrt{\omega^2 \tau^2 + 1}} \qquad (3.187)$$

and the phase angle

$$\phi = \bigg/ \frac{q_o}{q_i} (i\omega) = \tan^{-1} - \omega\tau \qquad (3.188)$$

The ideal frequency response (zero-order instrument) would have

$$\frac{q_o}{q_i} (i\omega) = K\underline{/0^\circ} \qquad (3.189)$$

Thus a first-order instrument approaches perfection if Eq. (3.186) approaches Eq. (3.189). We see this occur if the product $\omega\tau$ is sufficiently small. Thus for *any* τ there will be some frequency of input ω below which measurement is accurate, or, alternatively, if a q_i of high frequency ω must be measured, the instrument used must have a sufficiently small τ. Again we see that accurate dynamic measurement requires a small time constant.

If one were concerned with the measurement of *pure* sine waves only, the above considerations would not be very pertinent since if one knew the frequency and τ he could easily correct for amplitude attenuation and phase shift by simple calculations. In actual practice, however, q_i is often a combination of several sine waves of different frequencies. An example will show the importance of adequate frequency response under such conditions. Suppose we must measure a q_i given by

$$q_i = 1 \sin 2t + 0.3 \sin 20t \qquad (3.190)$$

(where t is in seconds) with a first-order instrument whose τ is 0.2 sec. Since this is a linear system we may use the superposition principle to find q_o. We first evaluate the sinusoidal transfer function at the two frequencies of interest.

$$\frac{q_o}{q_i} (i\omega) \bigg|_{\omega=2} = \frac{K}{\sqrt{0.16 + 1}} \underline{/-21.8^\circ} = 0.93K\underline{/-21.8^\circ} \qquad (3.191)$$

$$\frac{q_o}{q_i} (i\omega) \bigg|_{\omega=20} = \frac{K}{\sqrt{16 + 1}} \underline{/-76^\circ} = 0.24K\underline{/-76^\circ} \qquad (3.192)$$

We can then write q_o as

$$q_o = (1)(0.93K) \sin (2t - 21.8^\circ) \\ + (0.3)(0.24K) \sin (20t - 76^\circ) \qquad (3.193)$$

$$\frac{q_o}{K} = 0.93 \sin (2t - 21.8^\circ) + 0.072 \sin (20t - 76^\circ) \qquad (3.194)$$

Since ideally $q_o/K = q_i$, comparison of Eq. (3.194) with (3.190) shows the presence of considerable measurement error. A graph of these two

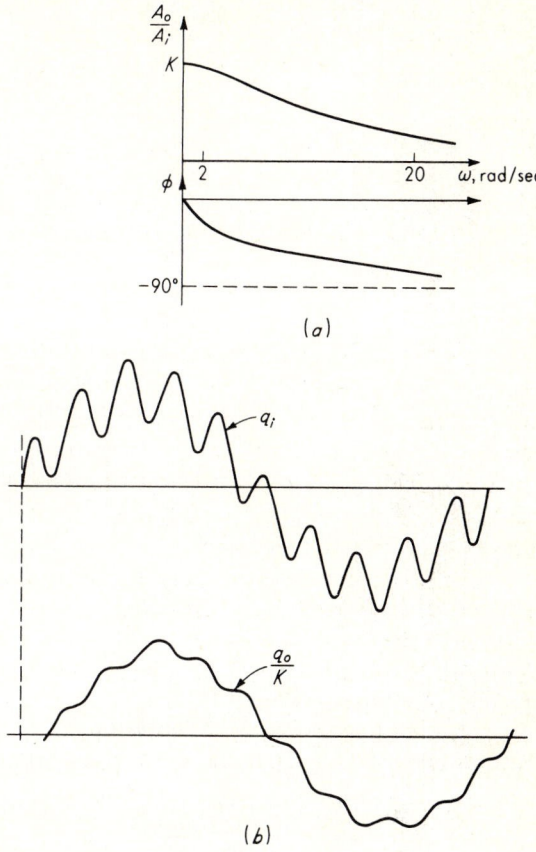

Fig. 3.43.   *Example of inadequate frequency response.*

equations in Fig. 3.43b shows that the instrument gives a severely distorted measurement of the input.   Furthermore, the high-frequency (20 rad/sec) component present in the instrument output is now so small relative to the low-frequency component that any attempts at correction are not only inconvenient but also inaccurate.

   Suppose we now consider use of an instrument with $\tau = 0.002$ sec. We then have

$$\frac{q_o}{q_i}\,(i\omega)\,\Big|_{\omega=2} = \frac{K}{\sqrt{1.6 \times 10^{-5} + 1}}\,\underline{/-0.23°} = 1.00K\underline{/-0.23°} \qquad (3.195)$$

$$\frac{q_o}{q_i}\,(i\omega)\,\Big|_{\omega=20} = \frac{K}{\sqrt{1.6 \times 10^{-3} + 1}}\,\underline{/-2.3°} = 1.00K\underline{/-2.3°} \qquad (3.196)$$

giving $\dfrac{q_o}{K} = 1.00 \sin\,(2t - 0.23°) + 0.3 \sin\,(20t - 2.3°)$ \qquad (3.197)

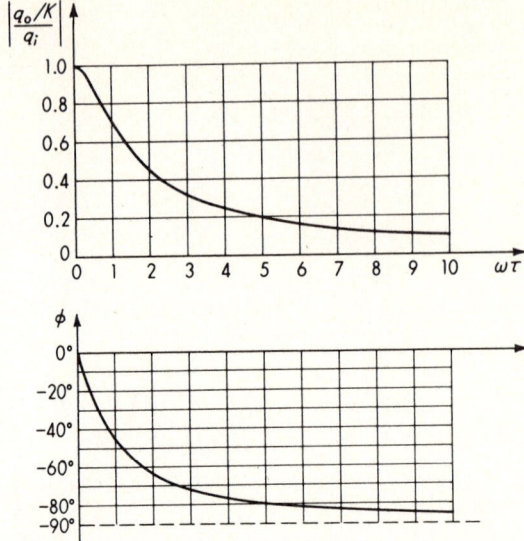

Fig. 3.44. *Frequency response of first-order system.*

Comparison of Eq. (3.190) and (3.197) shows clearly that this instrument faithfully measures the given q_i.

A nondimensional representation of the frequency response of any first-order system may be obtained by writing Eq. (3.186) as

$$\frac{q_o/K}{q_i}\,(i\omega) = \frac{1}{\sqrt{(\omega\tau)^2 + 1}}\;\underline{/\tan^{-1} - \omega\tau} \qquad (3.198)$$

and plotting as in Fig. 3.44.

Impulse response of first-order instruments. The final standard input we shall consider is the *impulse function*. Consider the pulse function $p(t)$ defined graphically in Fig. 3.45a. The impulse function of "strength" (area) A is defined by the limiting process

$$\text{Impulse function of strength } A \triangleq \lim_{T \to 0} p(t) \qquad (3.199)$$

We see that this "function" has rather peculiar properties. Its time duration is infinitesimal, its peak is infinitely high, and its area is A. If A is taken as 1, it is called the *unit* impulse function, $u_1(t)$. Thus an impulse function of any strength A may be written as $Au_1(t)$. This rather peculiar function plays an important role in system dynamic analysis, as we shall see in more detail later.

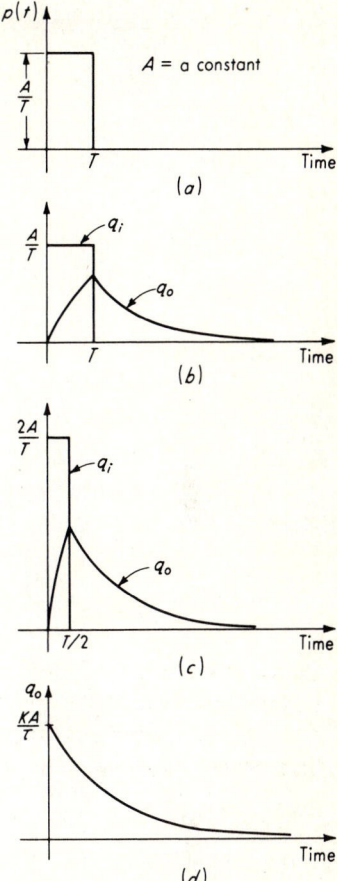

Fig. 3.45. *Impulse response of first-order system.*

We shall now find the response of a first-order instrument to an impulse input. We do this by finding the response to the pulse $p(t)$ and then applying the limiting process to the result. For $0 < t < T$ we have

$$(\tau D + 1)q_o = Kq_i = \frac{KA}{T} \qquad (3.200)$$

Since, up until time T, this is no different from a *step* input of size A/T, our initial condition is $q_o = 0$ at $t = 0^+$, and the complete solution is

$$q_o = \frac{KA}{T}\,[1 - e^{(-t)/\tau}] \qquad (3.201)$$

However, this solution is valid only up to time T. At this time we have

$$q_o\,\Big|_{t=T} = \frac{KA}{T}\,[1 - e^{(-T)/\tau}] \qquad (3.202)$$

Now for $t > T$, our differential equation is

$$(\tau D + 1)q_o = Kq_i = 0 \qquad (3.203)$$
$$\text{giving} \quad q_o = Ce^{(-t)/\tau} \qquad (3.204)$$

The constant C is found by imposing initial condition (3.202),

$$\frac{KA}{T}[1 - e^{(-T)/\tau}] = Ce^{(-T)/\tau} \qquad (3.205)$$

$$C = \frac{KA[1 - e^{(-T)/\tau}]}{Te^{(-T)/\tau}} \qquad (3.206)$$

giving finally
$$q_o = \frac{KA[1 - e^{(-T)/\tau}]e^{(-t)/\tau}}{Te^{(-T)/\tau}} \qquad (3.207)$$

Figure 3.45*b* shows a typical response, and Fig. 3.45*c* shows the effect of cutting T in half. As T is made shorter and shorter, the first part ($t < T$) of the response becomes of negligible consequence so that we can get an expression for q_o by taking the limit of Eq. (3.207) as $T \to 0$.

$$\lim_{T \to 0} \left\{ \frac{KA[1 - e^{(-T)/\tau}]}{Te^{(-T)/\tau}} \right\} e^{(-t)/\tau} = KAe^{(-t)/\tau} \lim_{T \to 0} \frac{1 - e^{(-T)/\tau}}{Te^{(-T)/\tau}} \qquad (3.208)$$

$$\lim_{T \to 0} \frac{1 - e^{(-T)/\tau}}{T} = \frac{0}{0} \qquad \text{an indeterminate form}$$

Applying L'Hospital's rule,

$$\lim_{T \to 0} \frac{1 - e^{(-T)/\tau}}{T} = \lim_{T \to 0} \frac{(1/\tau)e^{(-T)/\tau}}{1} = \frac{1}{\tau} \qquad (3.209)$$

Thus we have finally for the impulse response of a first-order instrument

$$q_0 = \frac{KA}{\tau} e^{(-t)/\tau} \qquad (3.210)$$

which is plotted in Fig. 3.45*d*.

We note that the output q_o is also "peculiar" in that it has an infinite (vertical) slope at $t = 0$ and thus goes from zero to a finite value in infinitesimal time. Such behavior is clearly impossible for a physical system since it requires energy transfer at an infinite rate. In our thermometer example, for instance, to cause the temperature of the fluid in the bulb *suddenly* to rise a finite amount requires an infinite rate of heat transfer. Mathematically, this infinite rate of heat transfer is provided by having the input $T_i(t)$ be infinite, i.e., an impulse function. In actuality, of course, T_i cannot go to infinity; however, if it is large enough and of sufficiently short duration (relative to the response speed of the system) the system may respond very nearly as it would for a perfect impulse.

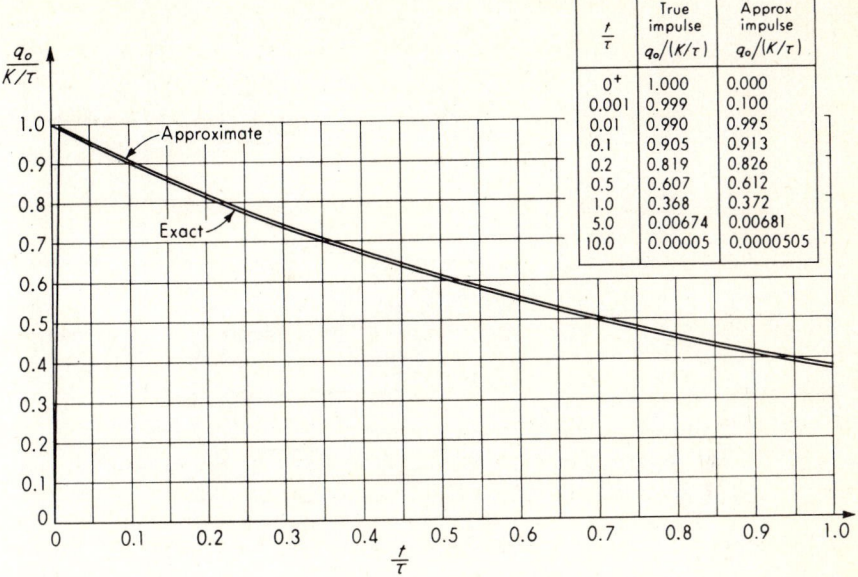

$\dfrac{t}{\tau}$	True impulse $q_o/(K/\tau)$	Approx impulse $q_o/(K/\tau)$
0^+	1.000	0.000
0.001	0.999	0.100
0.01	0.990	0.995
0.1	0.905	0.913
0.2	0.819	0.826
0.5	0.607	0.612
1.0	0.368	0.372
5.0	0.00674	0.00681
10.0	0.00005	0.0000505

Fig. 3.46. *Exact and approximate impulse response.*

To illustrate this, suppose in Fig. 3.45a we take $A = 1$ and $T = 0.01\tau$. The response to this approximate unit impulse is

$$q_o = \frac{100K}{\tau}[1 - e^{(-t)/\tau}] \qquad 0 \leq t \leq T \qquad (3.211)$$

$$q_o = \frac{100K(1 - e^{-0.01})e^{(-t)/\tau}}{\tau e^{-0.01}} \qquad T \leq t \leq \infty \qquad (3.212)$$

Figure 3.46 gives a tabular and graphical comparison of the exact and approximate response, showing excellent agreement. The agreement is quite acceptable in most cases if T/τ is even as large as 0.1. It can also be shown that *the shape of the pulse is immaterial;* as long as its duration is sufficiently short, only its *area* matters. The plausibility of this statement may be shown by integrating the terms in the differential equation as follows:

$$\tau\frac{dq_o}{dt} + q_o = Kq_i \qquad (3.213)$$

$$\int_0^{0^+} \tau\,dq_o + \int_0^{0^+} q_o\,dt = \int_0^{0^+} Kq_i\,dt \qquad (3.214)$$

$$\tau\left(q_o\Big|_{0^+} - q_o\Big|_0\right) + 0$$

$$= K\text{ (area under } q_i \text{ curve from } t = 0 \text{ to } t = 0^+) \qquad (3.215)$$

$$q_o\Big|_{0^+} = \frac{K}{\tau}\text{ (area of impulse)} \qquad (3.216)$$

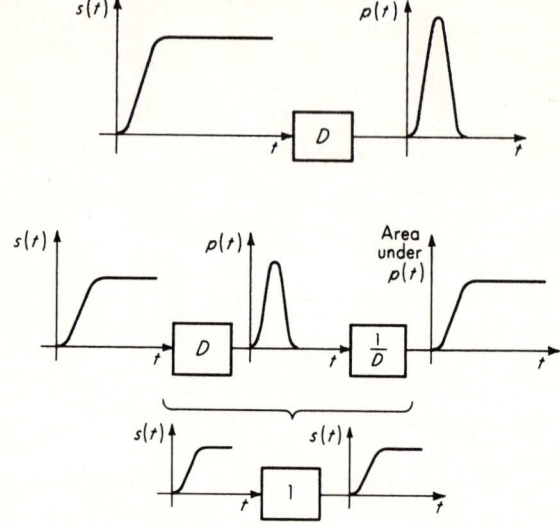

Fig. 3.47. *Approximate step and impulse functions.*

This analysis holds strictly for an exact impulse and is a good approximation for a pulse of arbitrary shape if its duration is sufficiently short. It
should be noted that, since the right side of the differential equation
(3.213) is zero for $t > 0^+$, an impulse (or short pulse) is equivalent to a
zero forcing function and a nonzero initial ($t = 0^+$) condition. That is,
the solution of

$$(\tau D + 1)q_o = 0$$

$$q_o = \frac{K}{\tau} \qquad \text{at } t = 0^+ \qquad (3.217)$$

is exactly the same as the impulse response.

Another interesting aspect of the impulse function is its relation to
the step function. Since a perfect step function is also physically
unrealizable because it changes from one level to another in infinitesimal
time, consider an approximation such as in Fig. 3.47. If this approximate
step function is fed into a differentiating device the output will be a
pulse-type function. As the approximate step function is made to
approach the mathematical ideal more and more closely the output of the
differentiating device will approach a perfect impulse function. *In this
sense, the impulse function may be thought of as the derivative of the step
function,* even though the discontinuities in the step function preclude the
rigorous application of the basic definition of the derivative. In Fig. 3.47
the truth of these assertions is demonstrated by passing the output of the
differentiating device through an integrating device $(1/D)$.

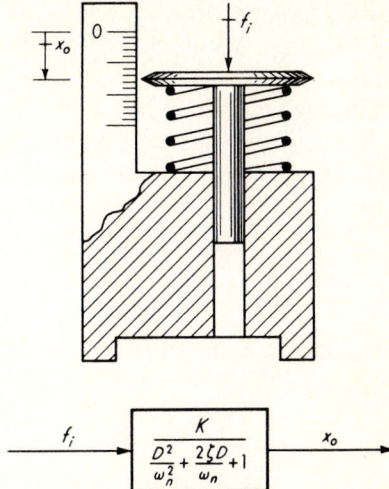

Fig. 3.48. *Second-order instrument.*

The second-order instrument. A second-order instrument is defined as one that follows the equation

$$a_2 \frac{d^2 q_o}{dt^2} + a_1 \frac{dq_o}{dt} + a_0 q_o = b_0 q_i \qquad (3.218)$$

Again, a second-order *equation* could have more terms on the right-hand side, but in common engineering usage, Eq. (3.218) is generally accepted as defining a second-order *system*.

The essential parameters in Eq. (3.218) can be reduced to three:

$$K \triangleq \frac{b_0}{a_0} \triangleq \text{static sensitivity} \qquad (3.219)$$

$$\omega_n \triangleq \sqrt{\frac{a_0}{a_2}} \triangleq \text{undamped natural frequency, rad/time} \qquad (3.220)$$

$$\zeta \triangleq \frac{a_1}{2\sqrt{a_0 a_2}} \triangleq \text{damping ratio, dimensionless} \qquad (3.221)$$

giving
$$\left(\frac{D^2}{\omega_n^2} + \frac{2\zeta D}{\omega_n} + 1 \right) q_o = K q_i \qquad (3.222)$$

The operational transfer function is thus

$$\frac{q_o}{q_i}(D) = \frac{K}{D^2/\omega_n^2 + 2\zeta D/\omega_n + 1} \qquad (3.223)$$

A good example of a second-order instrument is the force measuring spring scale of Fig. 3.48. We assume the applied force f_i has frequency components only well below the natural frequency of the spring itself. Then the main dynamic effect of the spring may be taken into account by adding one-third of the spring's mass to the main moving mass. This total mass we call M. The spring is assumed linear with spring constant K_s lb$_f$/in. Although in a real scale there might be considerable dry friction, we assume perfect film lubrication and therefore a viscous damping effect with constant B lb$_f$/(in./sec).

The scale can be adjusted so that $x_o = 0$ when $f_i = 0$ (gravity force will then drop out of the equation), giving

$$\Sigma \text{ forces } = (\text{mass})(\text{acceleration})$$

$$f_i - B\frac{dx_o}{dt} - K_s x_o = M\frac{d^2 x_o}{dt^2} \qquad (3.224)$$

$$(MD^2 + BD + K_s)x_o = f_i \qquad (3.225)$$

Noting this to fit the second-order model, we immediately define

$$K \triangleq \frac{1}{K_s} \qquad \text{in./lb}_f \qquad (3.226)$$

$$\omega_n \triangleq \sqrt{\frac{K_s}{M}} \qquad \text{rad/sec} \qquad (3.227)$$

$$\zeta \triangleq \frac{B}{2\sqrt{K_s M}} \qquad (3.228)$$

Step response of second-order instruments. For a step input of size q_{is} we get

$$\left(\frac{D^2}{\omega_n{}^2} + \frac{2\zeta D}{\omega_n} + 1\right)q_o = Kq_{is} \qquad (3.229)$$

with initial conditions

$$q_o = 0 \qquad \text{at } t = 0^+$$
$$\frac{dq_o}{dt} = 0 \qquad \text{at } t = 0^+ \qquad (3.230)$$

The particular solution of Eq. (3.229) is clearly $q_{opi} = Kq_{is}$. The complementary-function solution takes on one of three possible forms, depending on whether the roots of the characteristic equation are real and unrepeated (overdamped case), real and repeated (critically damped case), or complex (underdamped case). The complete solutions of Eq. (3.229) with initial conditions (3.230) are, in nondimensional form,

$$\frac{q_o}{Kq_{is}} = -\frac{\zeta + \sqrt{\zeta^2 - 1}}{2\sqrt{\zeta^2 - 1}} e^{(-\zeta + \sqrt{\zeta^2 - 1})\omega_n t}$$

$$+ \frac{\zeta - \sqrt{\zeta^2 - 1}}{2\sqrt{\zeta^2 - 1}} e^{(-\zeta - \sqrt{\zeta^2 - 1})\omega_n t} + 1 \qquad \text{overdamped} \qquad (3.231)$$

$$\frac{q_o}{Kq_{is}} = -(1 + \omega_n t)e^{-\omega_n t} + 1 \qquad \text{critically damped} \qquad (3.232)$$

$$\frac{q_o}{Kq_{is}} = -\frac{e^{-\zeta\omega_n t}}{\sqrt{1 - \zeta^2}} \sin\left(\sqrt{1 - \zeta^2}\,\omega_n t + \phi\right) + 1 \qquad \text{underdamped} \qquad (3.233)$$

$$\phi \triangleq \sin^{-1}\sqrt{1 - \zeta^2}$$

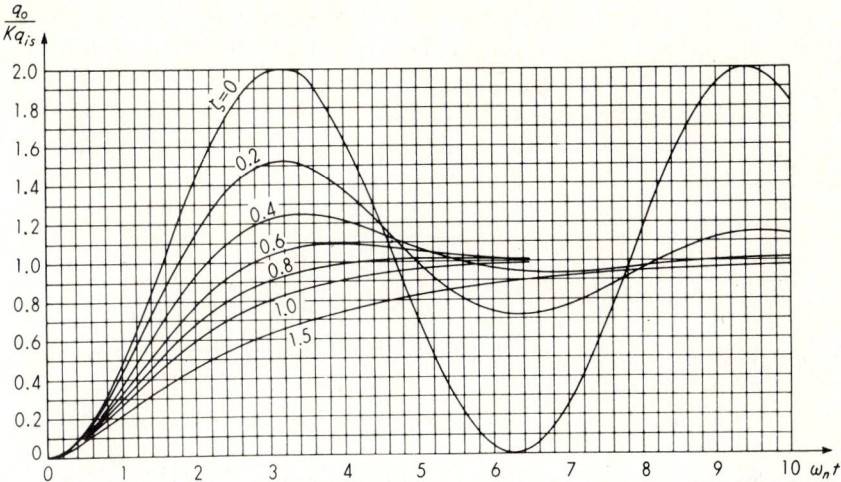

Fig. 3.49.   *Nondimensional step-function response of second-order instruments.*

Since t and ω_n always appear as the product $\omega_n t$, the curves of q_o/Kq_{is} may be plotted against $\omega_n t$, making them universal for any ω_n, as in Fig. 3.49.   This fact also shows that ω_n *is a direct indication of speed of response*.   For a given ζ, doubling ω_n will halve the response time since $\omega_n t$ (and thus q_o/Kq_{is}) achieves the same value at one-half the time. The effect of ζ is not clearly perceived from the equations but is evident from the graphs.   An increase in ζ reduces oscillation but also slows the response in the sense that the first crossing of the final value is retarded. A settling time may actually be a better indication of response speed; however, the optimum value of ζ will then vary with the chosen tolerance band.   For example, if we choose a 10 percent settling time, the curve for $\zeta = 0.6$ gives a settling time of about $2.4/\omega_n$, and this is optimum since ζ either larger or smaller gives a longer settling time.   However, if we had chosen a 5 percent settling time, a ζ between 0.7 and 0.8 gives the shortest value.   In choosing a proper ζ value for a practical application, the situation is further complicated by the fact that the real inputs will not be step functions and their *actual* form influences what will be the best ζ value.   If the actual inputs are quite variable in form, some compromise must be struck.   It will be found that many commercial instruments use $\zeta = 0.6$ to 0.7.   We shall show shortly that this range of ζ gives good frequency response over the widest frequency range.

Terminated-ramp response of second-order instruments.   Under certain circumstances the response of second order instruments to perfect step inputs is misleading.   The best example of this is perhaps found in

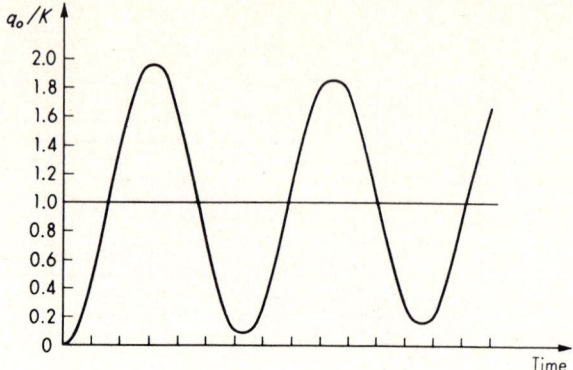

Fig. 3.50.   *Step response of poorly damped system.*

piezoelectric pressure pickups, accelerometers, etc.   While these devices will be discussed in detail later, for the present it is sufficient to state that they usually have an extremely high natural frequency and very little damping ($\zeta < 0.01$, often).   Based on a perfect step input, such an instrument appears highly undesirable because of its large overshoot and strong oscillation (Fig. 3.50).   Actually, these instruments may give excellent response.   The explanation of this apparent inconsistency lies in the fact that *perfect* step inputs do not occur in nature, since a macroscopic quantity cannot change a finite amount in an infinitesimal time.   Thus a more realistic input than the step is the *terminated-ramp input,* defined in Fig. 3.51.   This input has a *finite* slope equal to $1/T$, whereas a step input has an infinite slope.   By letting T get smaller and smaller, one can approach the perfect step input.   For a second-order system we would have mathematically

$$\left(\frac{D^2}{\omega_n{}^2} + \frac{2\zeta D}{\omega_n} + 1\right) q_o = K q_i \qquad (3.234)$$

$$q_i = \begin{cases} \dfrac{t}{T} & 0 \le t \le T \\[2mm] 1.0 & T \le t < \infty \end{cases} \qquad (3.235)$$

$$q_o = \frac{dq_o}{dt} = 0 \qquad \text{at } t = 0^+ \qquad (3.236)$$

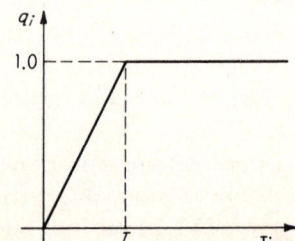

Fig. 3.51.   *Terminated-ramp input.*

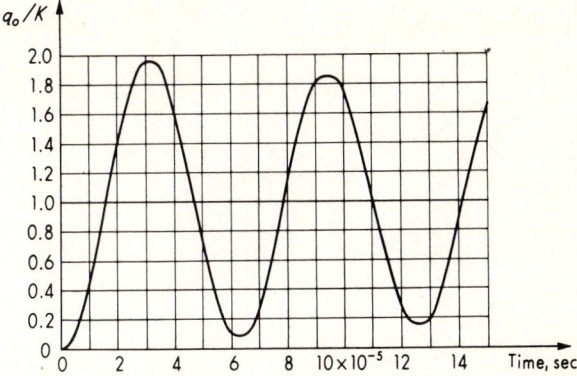

Fig. 3.52. *Step response of poorly damped system.*

Since we are concerned here with lightly damped systems, we obtain the solution only for the underdamped case as

$$\frac{q_o}{K} = \frac{t}{T} - \frac{2\zeta}{\omega_n T} + \frac{1}{\omega_n T \sqrt{1 - \zeta^2}} e^{-\zeta \omega_n t} \sin \left(\sqrt{1 - \zeta^2}\, \omega_n t + \phi \right)$$

$$0 \le t \le T \qquad (3.237)$$

$$\frac{q_o}{K} = \left[\frac{t}{T} - \frac{2\zeta}{\omega_n T} + \frac{1}{\omega_n T \sqrt{1 - \zeta^2}} e^{-\zeta \omega_n t} \sin \left(\sqrt{1 - \zeta^2} \right. \right.$$

$$\left. \omega_n t + \phi \right) \bigg] - \left[\frac{t}{T} - 1 - \frac{2\zeta}{\omega_n T} + \frac{1}{\omega_n T \sqrt{1 - \zeta^2}} e^{-\zeta \omega_n (t - T)} \right.$$

$$\left. \sin \left(\sqrt{1 - \zeta^2}\, \omega_n (t - T) + \phi \right) \right] \qquad T \le t < \infty \qquad (3.238)$$

$$\phi \triangleq 2 \tan^{-1} \frac{\sqrt{1 - \zeta^2}}{\zeta} \qquad (3.239)$$

From Eq. (3.237) we note immediately that, for $0 \le t \le T$, the following is true:

1.   There is a steady-state error of size $2\zeta/\omega_n T$.
2.   The transient error can be no larger than $1/(\omega_n T \sqrt{1 - \zeta^2})$.

Thus if $\zeta = 0$ (no damping) the steady-state error is zero and the "transient" error is a sustained sine wave of amplitude $1/\omega_n T$. *Therefore if ω_n is sufficiently large relative to $1/T$ the transient error can be made very small even if the damping is practically nonexistent.* This result is based on Eq. (3.237) but similar results are obtained from (3.238) for $T \le t \le \infty$,

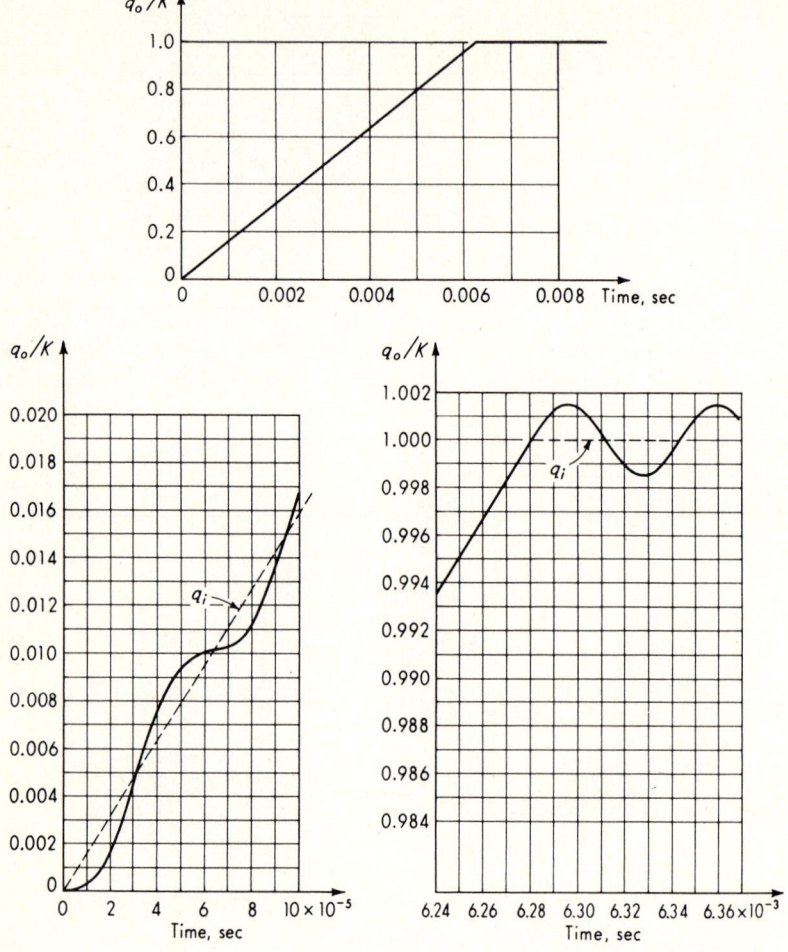

Fig. 3.53. *Terminated-ramp response of poorly damped system.*

since the transient induced at $t = 0$ by the increasing ramp is essentially the same as that induced at $t = T$ by a decreasing ramp. That is, the q_i of Fig. 3.51 is really a superposition of an increasing ramp starting at $t = 0$ and a decreasing ramp starting at $t = T$.

As a numerical example, suppose a pressure pickup with $\zeta = 0.01$ and $\omega_n = 100,000$ rad/sec is subjected to terminated-ramp-type inputs with $T = 0.00628$ sec. The step response of such an instrument is shown in Fig. 3.52 and indicates the severe overshooting and oscillation which could lead one to reject the instrument. Figure 3.53, however,

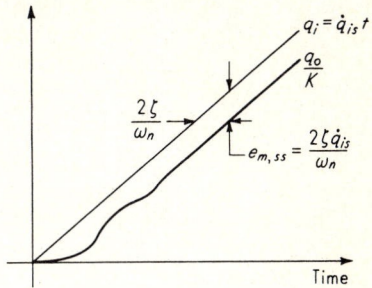

Fig. 3.54. *Ramp response of second-order instrument.*

shows the terminated-ramp response corresponding to the *actual* input. It is clear that the response is almost perfect. The conclusion, then, is that if the instrument is not subjected to any inputs more rapid in variation than that stated above it should prove quite acceptable.

Ramp response of second-order instruments. The differential equation here is

$$\left(\frac{D^2}{\omega_n{}^2} + \frac{2\zeta D}{\omega_n} + 1\right) q_o = K\dot{q}_{is}t \qquad (3.240)$$

$$q_o = \frac{dq_o}{dt} = 0 \qquad \text{at } t = 0^+$$

The solutions are found to be

$$\frac{q_o}{K} = \dot{q}_{is}t - \frac{2\zeta\dot{q}_{is}}{\omega_n}\left[1 + \frac{2\zeta^2 - 1 - 2\zeta\sqrt{\zeta^2 - 1}}{4\zeta\sqrt{\zeta^2 - 1}} e^{(-\zeta + \sqrt{\zeta^2 - 1})\omega_n t}\right.$$

$$\left. + \frac{-2\zeta^2 + 1 - 2\zeta\sqrt{\zeta^2 - 1}}{4\zeta\sqrt{\zeta^2 - 1}} e^{(-\zeta - \sqrt{\zeta^2 - 1})\omega_n t}\right] \qquad \text{overdamped} \qquad (3.241)$$

$$\frac{q_o}{K} = \dot{q}_{is}t - \frac{2\dot{q}_{is}}{\omega_n}\left[1 - e^{-\omega_n t}\left(1 + \frac{\omega_n t}{2}\right)\right] \qquad \text{critically damped} \qquad (3.242)$$

$$\frac{q_o}{K} = \dot{q}_{is}t - \frac{2\zeta\dot{q}_{is}}{\omega_n}\left[1 - \frac{e^{-\zeta\omega_n t}}{2\zeta\sqrt{1 - \zeta^2}} \sin\left(\sqrt{1 - \zeta^2}\,\omega_n t + \phi\right)\right] \qquad (3.243)$$

$$\tan\phi = \frac{2\zeta\sqrt{1 - \zeta^2}}{2\zeta^2 - 1} \qquad \text{underdamped}$$

Figure 3.54 shows the general character of the response. There is a steady-state error $2\zeta\dot{q}_{is}/\omega_n$. Since the value of q_{is} is set by the measured quantity, the steady-state error can be reduced only by reducing ζ and increasing ω_n. For a given ω_n, reduction in ζ results in larger oscillations. There is also a steady-state time lag $2\zeta/\omega_n$. Figure 3.55 gives a set of nondimensionalized curves summarizing system behavior.

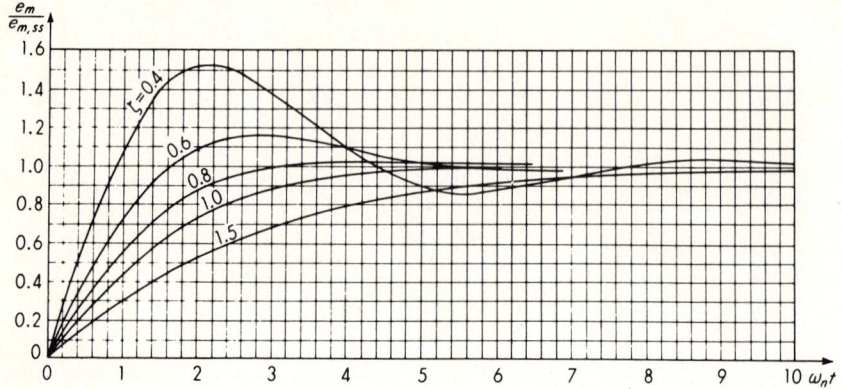

Fig. 3.55.   *Nondimensional ramp response.*

Frequency response of second-order instruments. The sinusoidal transfer function is

$$\frac{q_o}{q_i}(i\omega) = \frac{K}{(i\omega/\omega_n)^2 + (2\zeta i\omega/\omega_n) + 1} \qquad (3.244)$$

which can be put in the form

$$\frac{q_o/K}{q_i}(i\omega) = \frac{1}{\sqrt{[1 - (\omega/\omega_n)^2]^2 + 4\zeta^2\omega^2/\omega_n^2}}\underline{/\phi} \qquad (3.245)$$

$$\phi \triangleq \tan^{-1}\frac{2\zeta}{\omega/\omega_n - \omega_n/\omega} \qquad (3.246)$$

Figure 3.56 gives the nondimensionalized frequency-response curves. Clearly, increasing ω_n will increase the range of frequencies for which the amplitude-ratio curve is relatively flat; thus a high ω_n is needed to measure accurately high-frequency q_i's. An optimum range of values for ζ is indicated by both the amplitude-ratio and phase-angle curves. The widest flat amplitude ratio exists for ζ of about 0.6 to 0.7. While zero phase angle would be ideal it is rarely possible to realize this even approximately. Actually, if the main interest is in q_o reproducing the correct *shape* of q_i and if a time delay is acceptable, we shall show shortly that ϕ need not be zero; rather it should vary *linearly* with frequency ω. Examining the phase curves of Fig. 3.56, we note that the curves for $\zeta = 0.6$ to 0.7 are nearly straight for the widest frequency range. These considerations lead to the widely accepted choice of $\zeta = 0.6$ to 0.7 as the optimum value of damping for second-order instruments. There are exceptions, however, as noted in the section on terminated-ramp response.

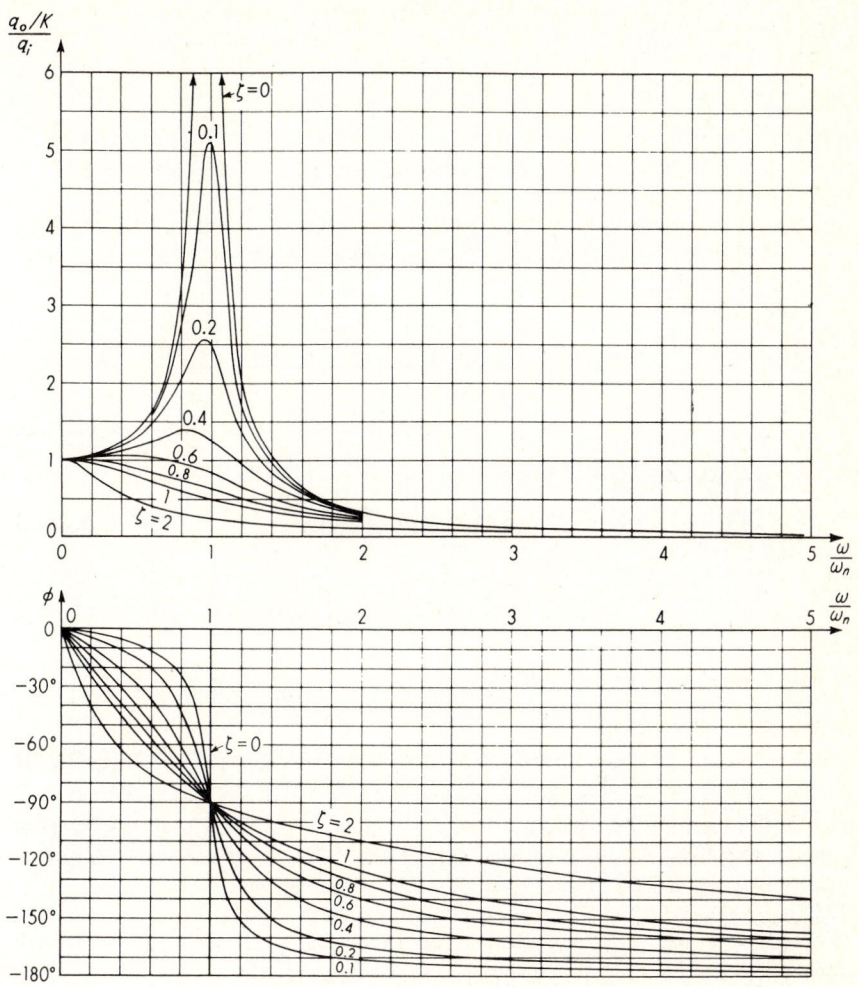

Fig. 3.56. *Frequency response of second-order system.*

Impulse response of second-order instruments. In the section on first-order instruments we showed that the impulse response is equivalent to the free (unforced) response if the initial $(t = 0^+)$ conditions produced by the impulse are taken into account. To find the initial conditions produced by applying an impulse of area A to a second-order instrument redraw the block diagram of Fig. 3.57a as in Fig. 3.57b. (The equivalence of the two diagrams is easily demonstrated by tracing through the signals in Fig. 3.57b to get the differential equation relating q_o to q_i.) In Fig. 3.57c

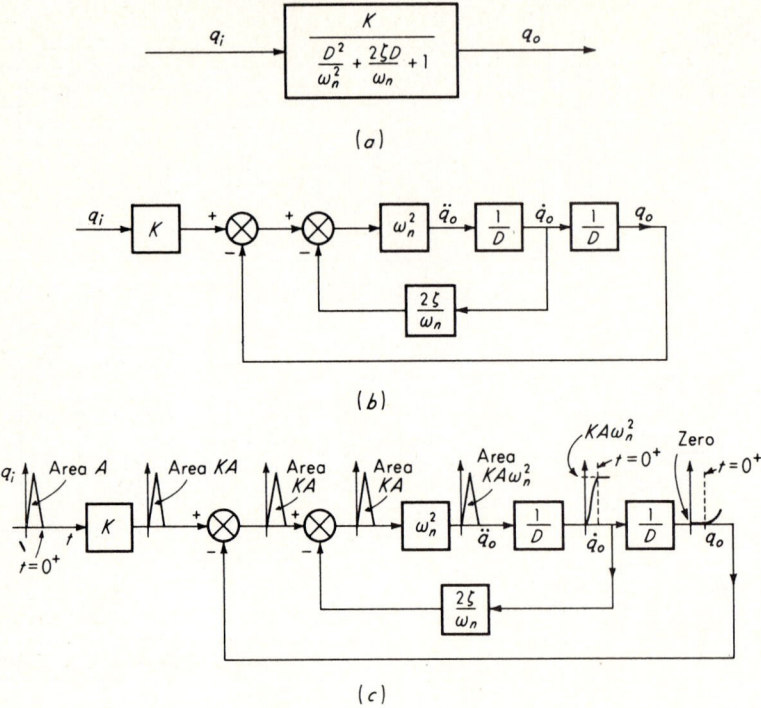

Fig. 3.57.   *Block-diagram analysis of impulse response.*

the impulse is applied at q_i, and the "propagation" of this input signal is traced through the rest of the diagram. This analysis shows that at $t = 0^+$ we have $q_o = 0$ and $\dot{q}_o = KA\omega_n^2$. The differential equation to be solved is then

$$\left(\frac{D^2}{\omega_n^2} + \frac{2\zeta D}{\omega_n} + 1\right) q_o = 0$$

$$q_o = 0 \qquad \frac{dq_o}{dt} = KA\omega_n^2 \qquad \text{at } t = 0^+ \qquad (3.247)$$

The solutions are found to be

$$\frac{q_o}{KA\omega_n} = \frac{1}{2\sqrt{\zeta^2 - 1}} \left[e^{(-\zeta+\sqrt{\zeta^2-1})\omega_n t} - e^{(-\zeta-\sqrt{\zeta^2-1})\omega_n t}\right] \qquad \text{overdamped}$$

$$(3.248)$$

$$\frac{q_o}{KA\omega_n} = \omega_n t e^{-\omega_n t} \qquad \text{critically damped} \qquad (3.249)$$

$$\frac{q_o}{KA\omega_n} = \frac{1}{\sqrt{1-\zeta^2}} e^{-\zeta\omega_n t} \sin\left(\sqrt{1-\zeta^2}\,\omega_n t\right) \qquad \text{underdamped} \qquad (3.250)$$

Figure 3.58 displays these results graphically.

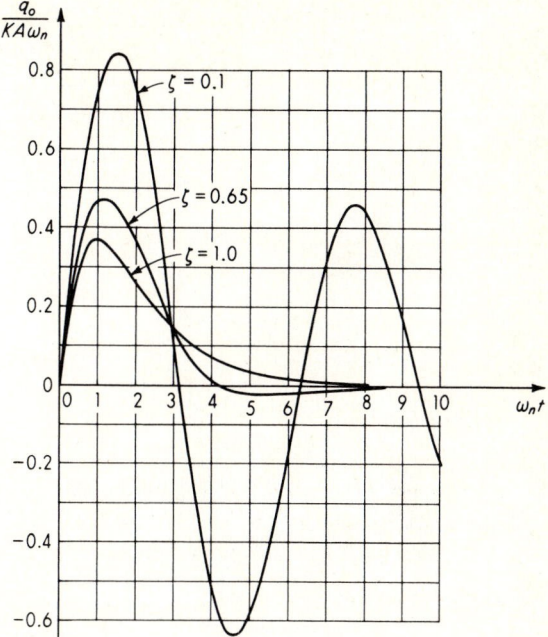

Fig. 3.58. *Nondimensional impulse response of second-order system.*

Dead-time elements. Some components of measuring systems are adequately represented as dead-time elements. A dead-time element is defined as a system in which the output is exactly the same form as the input but occurs τ_{dt} sec (the dead time) later. Mathematically,

$$q_o(t) = Kq_i(t - \tau_{dt}) \qquad t \geq \tau_{dt} \qquad (3.251)$$

This type of element is also called a pure delay or a transport lag. An example of such an effect is found in pneumatic signal-transmission systems. A pressure signal at one end of a length of pneumatic tubing will cause no response at all at the other end until the pressure wave has had time to propagate the distance between them. Because this speed of propagation is the same as the speed of sound, a 1,000-ft length of tubing will have a dead time of about 1 sec, since the speed of sound in standard air is about 1,000 fps.

The response of dead-time elements to the standard inputs is easily found. For steps, ramps, and impulses the results are given in Fig. 3.59. For sinusoidal input we have

$$q_i = A_i \sin \omega t$$

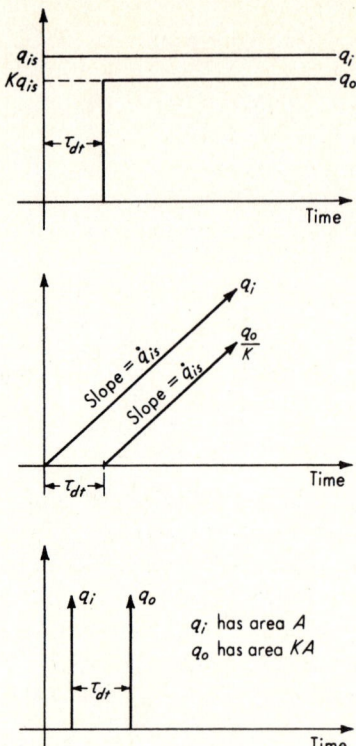

Fig. 3.59. *Dead-time responses.*

and from Eq. (3.251)

$$q_o = KA_i \sin \omega(t - \tau_{dt}) \qquad (3.252)$$

$$q_o = KA_i \sin (\omega t - \omega \tau_{dt}) = KA_i \sin (\omega t + \phi) \qquad (3.253)$$

$$\phi \triangleq -\omega \tau_{dt}$$

$$\text{Thus} \quad \frac{q_o/K}{q_i} (i\omega) = 1\underline{/\phi} = e^{-i\omega \tau_{dt}} \qquad (3.254)$$

The frequency-response curves for a dead-time element are shown in Fig. 3.60.

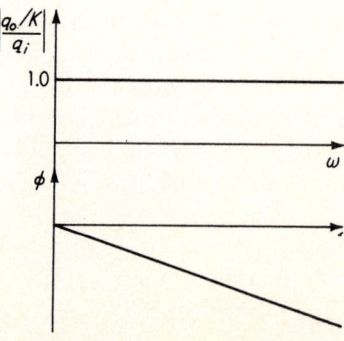

Fig. 3.60. *Dead-time frequency response.*

Logarithmic plotting of frequency-response curves. We shall find the frequency response of measurement systems extremely useful, so that rapid methods for getting the amplitude-ratio and phase-angle curves would be helpful. Certain logarithmic methods are in wide use and will now be explained.

The sinusoidal transfer function of a measurement system can generally be put in the form

$$\frac{q_o}{q_i}(i\omega) = \frac{\left\{\begin{array}{c} K(i\omega)^n(i\omega\tau_1+1)\ \cdots\ (i\omega\tau_m+1)[(i\omega/\omega_{n1})^2+2\zeta_1 i\omega/\omega_{n1}+1] \\ \cdots\ [(i\omega/\omega_{nr})^2+2\zeta_r i\omega/\omega_{nr}+1](e^{-i\omega\tau_{dt1}})\ \cdots\ (e^{-i\omega\tau_{dtp}}) \end{array}\right\}}{\left\{\begin{array}{c} (i\omega\tau_I+1)\ \cdots\ (i\omega\tau_M+1)[(i\omega/\omega_{nI})^2+2\zeta_I i\omega/\omega_{nI}+1] \\ \cdots\ [(i\omega/\omega_{nR})^2+2\zeta_R i\omega/\omega_{nR}+1] \end{array}\right\}}$$

$$(3.255)$$

This follows from the fact that the polynomials in the numerator and denominator of Eq. (3.141) can, in general, be *factored* into terms of the form $(D)^n$, $\tau D+1$, and $D^2/\omega_n^2+2\zeta D/\omega_n+1$. Replacing D by $i\omega$ then gives Eq. (3.255) when dead-time elements are also included.

Since Eq. (3.255) is in the form of a *product* of complex numbers, the use of logarithms suggests itself as a means of replacing multiplication by addition. That is,

$$\frac{q_o}{q_i}(i\omega) = G_1(i\omega)G_2(i\omega)\ \cdots\ G_u(i\omega) \qquad (3.256)$$

where the $G(i\omega)$ functions represent the various terms of Eq. (3.255). The amplitude ratio would be given by

$$\left|\frac{q_o}{q_i}(i\omega)\right| = |G_1(i\omega)|\,|G_2(i\omega)|\ \cdots\ |G_u(i\omega)| \qquad (3.257)$$

A widely used logarithmic method uses the *decibel notation* to express amplitude ratios. An amplitude ratio A is given in decibels (db) by

$$\text{Decibel value} \triangleq \text{db} \triangleq 20\log_{10}A \qquad (3.258)$$

Then

$$20\log_{10}\left|\frac{q_o}{q_i}(i\omega)\right| = 20\log_{10}\left[|G_1(i\omega)|\,|G_2(i\omega)|\ \cdots\ |G_u(i\omega)|\right] \qquad (3.259)$$

$$20\log_{10}\left|\frac{q_o}{q_i}(i\omega)\right| = 20\log_{10}|G_1(i\omega)| + 20\log_{10}|G_2(i\omega)|$$
$$+\ \cdots\ + 20\log_{10}|G_u(i\omega)| \qquad (3.260)$$

Thus, if one gets the amplitude-ratio curves for the individual terms in

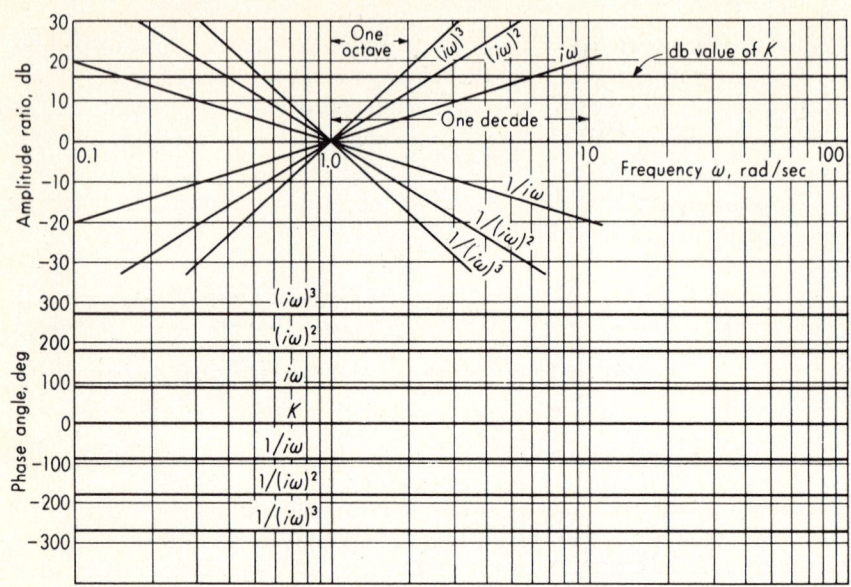

Fig. 3.61. *Integrator and differentiator frequency response.*

Eq. (3.255) in the decibel form, the overall decibel curve is obtained by simple graphical addition. The phase-angle curves are also obtained by simple addition since one adds phase angles when multiplying complex numbers.

We now show how to obtain with a minimum of effort the amplitude-ratio (db) and phase-angle curves for each of the types of terms in Eq. (3.255). To put the curves in their simplest form, we plot against the logarithm of frequency rather than frequency itself. The simplest term is the sensitivity K which is a real number $K/\underline{0°}$; thus its decibel curve is a straight horizontal line through the decibel value of K while its phase-angle curve is a straight horizontal line through zero degrees (see Fig. 3.61). Figure 3.62 gives a convenient table for interconverting decibel and ordinary-number values.

The next type of term is $(i\omega)^n$, where $n = \pm 1, \pm 2, \ldots$. The phase angle of such terms is constant with frequency and is given by $90n°$. The amplitude ratio is ω^n, so that the decibel value is

$$20 \log_{10} \omega^n = 20n \log_{10} \omega$$

Since we plot against $\log_{10} \omega$, the decibel curves become straight lines of slope $20n$ db/decade (see Fig. 3.61). A *decade* is defined as any 10-to-1 frequency range; an *octave* is any 2-to-1 range.

Number A	0	1	2	3	4	5	6	7	8	9
0		−40.00	−33.98	−30.46	−27.96	−26.02	−24.44	−23.10	−21.94	−20.92
0.1	−20.00	−19.17	−18.42	−17.72	−17.08	−16.48	−15.92	−15.39	−14.89	−14.42
0.2	−13.98	−13.56	−13.15	−12.77	−12.40	−12.04	−11.70	−11.37	−11.06	−10.76
0.3	−10.46	−10.16	−9.90	−9.63	−9.37	−9.12	−8.87	−8.64	−8.40	−8.18
0.4	−7.96	−7.74	−7.54	−7.33	−7.13	−6.94	−6.74	−6.56	−6.38	−6.20
0.5	−6.02	−5.85	−5.68	−5.51	−5.35	−5.19	−5.04	−4.88	−4.73	−4.58
0.6	−4.44	−4.29	−4.15	−4 01	−3.88	−3.74	−3.61	−3.48	−3.35	−3.22
0.7	−3.10	−2.97	−2.85	−2.73	−2.62	−2.50	−2.38	−2.27	−2.16	−2.05
0.8	−1.94	−1.83	−1.72	−1.62	−1.51	−1.41	−1.31	−1.21	−1.11	−1.01
0.9	−0.92	−0.82	−0.72	−0.63	−0.54	−0.45	−0.35	−0.26	−0.18	−0.09
1.0	0.00	0.09	0.17	0.26	0.34	0.42	0.51	0.59	0.67	0.75
1.1	0.83	0.91	0.98	1.06	1.14	1.21	1.29	1.36	1.44	1.51
1.2	1.58	1.66	1.73	1.80	1.87	1.94	2.01	2.08	2.14	2.21
1.3	2.28	2.35	2.41	2.48	2.54	2.61	2.67	2.73	2.80	2.86
1.4	2.92	2.98	3.05	3.11	3.17	3.23	3.29	3.35	3.41	3.46
1.5	3.52	3.58	3.64	3.69	3.75	3.81	3.86	3.92	3.97	4.03
1.6	4.08	4.14	4.19	4.24	4.30	4.35	4.40	4.45	4.51	4.56
1.7	4.61	4.66	4.71	4.76	4.81	4.86	4.91	4.96	5.01	5.06
1.8	5.11	5.15	5.20	5.25	5.30	5.34	5.39	5.44	5.48	5.53
1.9	5.58	5.62	5.67	5.71	5.76	5.80	5.85	5.89	5.93	5.98
2.0	6.02	6.44	6.85	7.23	7.60	7.96	8.30	8.63	8.94	9.25
3.0	9.54	9.83	10.10	10.37	10.63	10.88	11.13	11.36	11.60	11.82
4.0	12.04	12.26	12.46	12.67	12.87	13.06	13.26	13.44	13.62	13.80
5.0	13.98	14.15	14.32	14.49	14.65	14.81	14.96	15.12	15.27	15.42
6.0	15.56	15.71	15.85	15.99	16.12	16.26	16.39	16.52	16.65	16.78
7.0	16.90	17.03	17.15	17.27	17.38	17.50	17.62	17.73	17.84	17.95
8.0	18.06	18.17	18.28	18.38	18.49	18.59	18.69	18.79	18.89	18.99
9.0	19.08	19.18	19.28	19.37	19.46	19.55	19.65	19.74	19.82	19.91

db value corresponding to A

Fig. 3.62. *Decibel conversion table.*

Terms of the form $(i\omega\tau + 1)$ or $1/(i\omega\tau + 1)$ give, respectively,

$$\text{db} = 20 \log_{10} \sqrt{(\omega\tau)^2 + 1} \tag{3.261}$$

$$\text{and} \quad \text{db} = -20 \log_{10} \sqrt{(\omega\tau)^2 + 1} \tag{3.262}$$

When $\omega\tau \gg 1$, these become

$$\text{db} \approx 20 \log_{10} \omega\tau = 20 \log_{10} \tau + 20 \log_{10} \omega \tag{3.263}$$

$$\text{and} \quad \text{db} \approx -20 \log_{10} \omega\tau = -20 \log_{10} \tau - 20 \log_{10} \omega \tag{3.264}$$

We see that both of these represent straight lines of slope ± 20 db/decade, and these straight lines will be the high-frequency asymptotes of the actual amplitude-ratio curves. Similarly, for $\omega\tau \ll 1$,

$$\text{db} \approx 20 \log_{10} 1 = 0 \tag{3.265}$$

$$\text{and} \quad \text{db} \approx -20 \log_{10} 1 = 0 \tag{3.266}$$

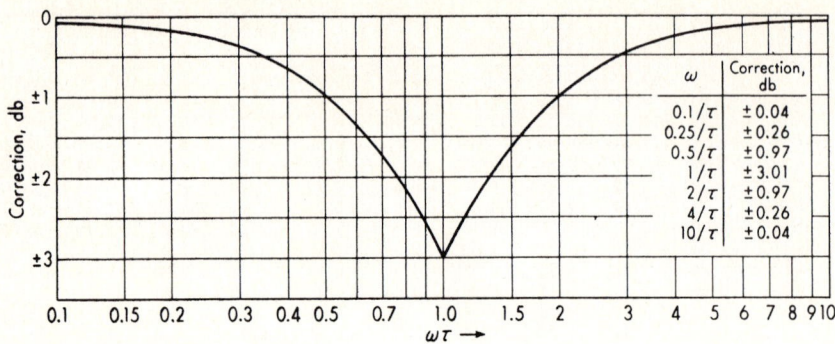

Fig. 3.63.   *First-order-system amplitude-ratio corrections.*

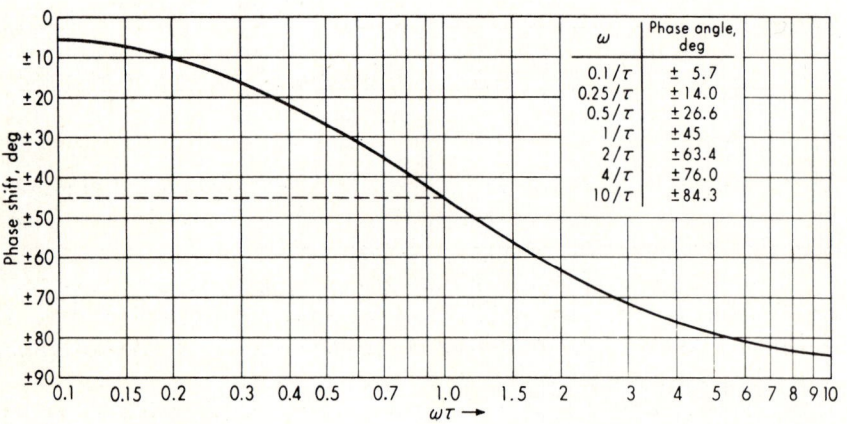

Fig. 3.64.   *First-order-system phase angle.*

so that the low-frequency asymptote is simply the 0-db line.   The two straight-line asymptotes will meet at $\omega\tau = 1$ because this is where (3.263) and (3.264) are zero.   The point $\omega = 1/\tau$ is called the "breakpoint" or "corner frequency."   In plotting curves for such terms, one first locates the breakpoint and then draws the two asymptotes.   The true curve is obtained by correcting the straight-line asymptotes at several points, using the data of Fig. 3.63.   The phase-angle curves may be quickly plotted, using the data of Fig. 3.64.   A numerical example illustrating these methods is given in Fig. 3.65.

Terms of the form $[(i\omega/\omega_n)^2 + 2\zeta i\omega/\omega_n + 1]^{\pm 1}$ have low-frequency asymptotes of 0 db and high-frequency asymptotes of slope ± 40 db/decade.

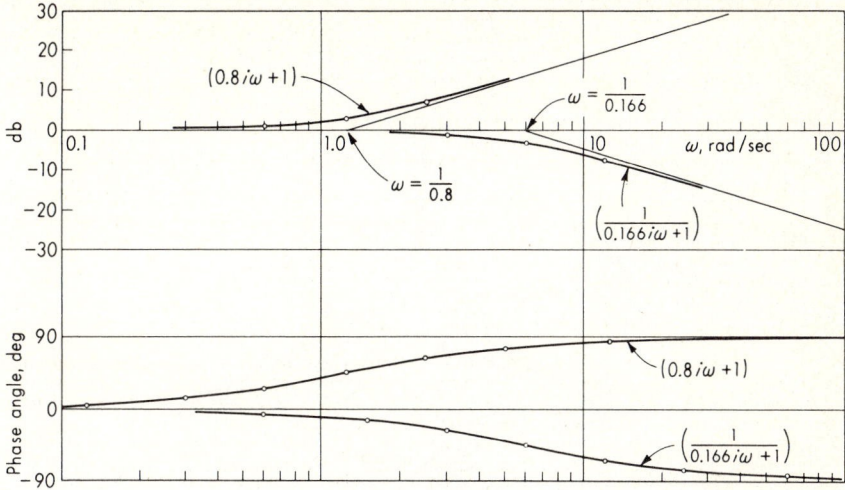

Fig. 3.65.   *Example of first-order terms.*

They intersect at $\omega = \omega_n$. The exact curves for a given ζ are obtained by applying the corrections of Fig. 3.66. The phase-angle curves are obtained from Fig. 3.67. Figure 3.68 gives numerical examples.

The final type of term to be considered is the dead-time term $e^{-i\omega\tau_{dt}}$. Since the amplitude ratio is 1.0 for all frequencies, the decibel curve is simply the 0-db line. The phase-angle curve is easily plotted from $\phi = -\omega\tau_{dt}$ for any given dead time.

To illustrate the procedure for combining the individual terms to obtain the overall frequency-response curves, we consider the following example:

$$\frac{q_o}{q_i}(i\omega) = \frac{4.4(i\omega)}{(i\omega + 1)(0.2i\omega + 1)} \tag{3.267}$$

Figure 3.69 shows the procedure and results.

Response of a general form of instrument to a periodic input.
Our approach to the dynamic response of measurement systems has, to this point, been limited in two ways. First, we considered only rather simple types of instruments (zero-order, first-order, and second-order) and, secondly, we subjected these instruments only to rather simple inputs (steps, ramps, sine waves, and impulses). At this point, by applying more advanced mathematical tools, we begin to remove both these limitations. We shall see that the concept of frequency response plays a central role

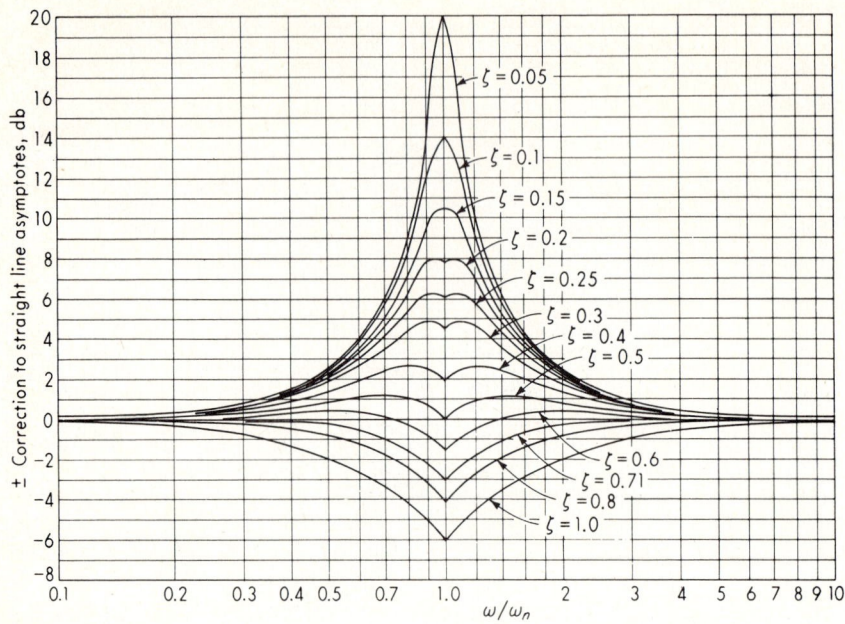

Fig. 3.66.   *Second-order-system amplitude-ratio corrections.*

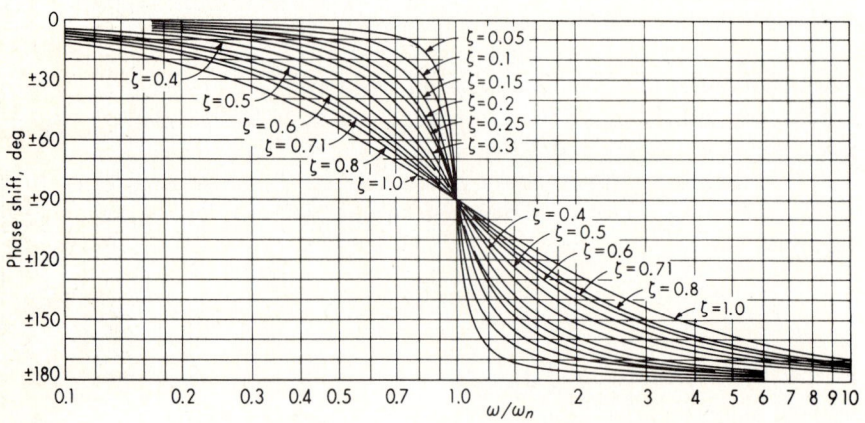

Fig. 3.67.   *Second-order-system phase angle.*

in these developments.   The first step involves the study of the response of a general (linear, time-invariant) instrument to periodic inputs.

By a periodic function we mean one that repeats itself cyclically over and over again, as in Fig. 3.70.   If this function meets the *Dirichlet conditions* (it must be single-valued, finite, and have a finite number of

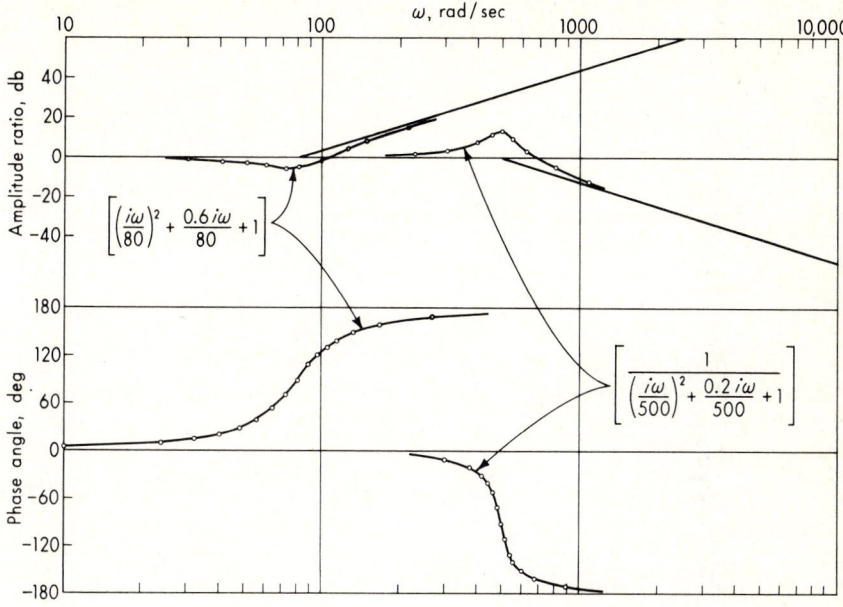

Fig. 3.68.   *Example of second-order terms.*

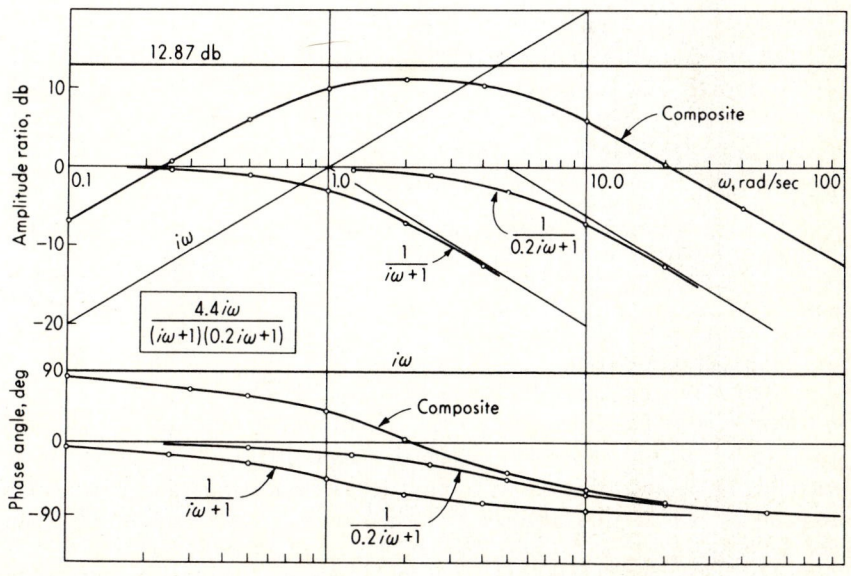

Fig. 3.69.   *Example of frequency-response plot.*

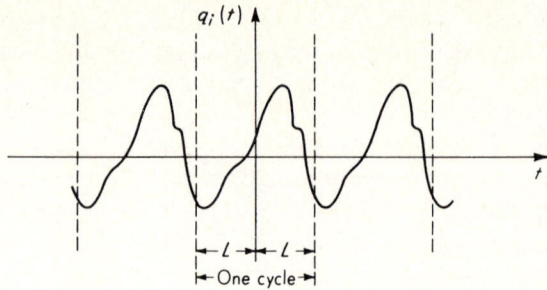

Fig. 3.70. *General periodic function.*

discontinuities and maxima and minima in one cycle), it may be represented by a *Fourier series*.[1] That is,

$$q_i(t) = q_{i,\text{av}} + \frac{1}{L}\left(\sum_{n=1}^{\infty} a_n \cos \frac{n\pi t}{L} + \sum_{n=1}^{\infty} b_n \sin \frac{n\pi t}{L} \right) \qquad (3.268)$$

where $\quad q_{i,\text{av}} \triangleq$ average value of $q_i = \dfrac{1}{2L} \displaystyle\int_{-L}^{L} q_i(t)\, dt \qquad (3.269)$

$$a_n \triangleq \int_{-L}^{L} q_i(t) \cos \frac{n\pi t}{L}\, dt \qquad (3.270)$$

$$b_n \triangleq \int_{-L}^{L} q_i(t) \sin \frac{n\pi t}{L}\, dt \qquad (3.271)$$

(The origin of the t coordinate may be chosen wherever most convenient.) We see then that any periodic function satisfying the Dirichlet conditions can be replaced by a sum of terms consisting of a constant and sine and cosine waves of various frequencies. If one wishes, any *pair* of sine and cosine terms of the *same* frequency can be replaced by a *single* sine wave of the same frequency at some phase angle since

$$A \cos \omega t + B \sin \omega t = C \sin (\omega t + \alpha) \qquad (3.272)$$

$$\text{where} \quad C \triangleq \sqrt{A^2 + B^2} \qquad (3.273)$$

$$\alpha \triangleq \tan^{-1} \frac{A}{B} \qquad (3.274)$$

The Fourier series will usually be an *infinite* series, and to get a *perfect* reconstruction of $q_i(t)$ from the series an infinite number of terms would have to be added. Fortunately, perfect reproduction is not required in engineering applications; thus generally $q_i(t)$ is *approximated* by a truncated (cut off after a certain number of terms) series. Just how many terms

[1] B. J. Ley, S. G. Lutz, and C. F. Rehberg, "Linear Circuit Analysis," chap. 6, McGraw-Hill Book Company, New York, 1959.

to use depends on the form of $q_i(t)$ (if it has very sharp changes more terms are required) and also on the use to which the information is to be put. Often, less than 10 "harmonics" (the first 10 different frequencies) are adequate.

The method of obtaining the desired terms in the Fourier series depends on the nature of $q_i(t)$. If $q_i(t)$ is given as a known mathematical formula, Eqs. (3.268) to (3.271) may be employed. If the required integrations cannot be performed [because of the complexity of $q_i(t)$], one can use one of several approximate numerical integration schemes or he can *plot* the functions to be integrated [such as $q_i(t) \cos(n\pi t/L)$] and use a planimeter to perform the integrations. If this plotting and planimetering (area measuring) are accurately done, the exact values of the desired coefficients will be obtained. This method is also feasible if $q_i(t)$ is not given by a formula but rather by a table of experimental data or a chart from a recording device. An alternative scheme is to choose a definite number of terms (such as 12) in the series and then adjust the coefficients of these terms so that $q_i(t)$ is fitted *exactly* at 12 points but an error exists at all other points. This method of obtaining the approximate Fourier series is in error in two ways:

1. Only a finite number of terms is used.
2. The terms that *are* used are not exact.

Both these types of errors can be reduced by using more terms. Such methods have been reduced to rather simple computation schemes in books on numerical analysis[1] and are thus quite popular. We now give two of these without proof. The *method of 12 ordinates* is given in Fig. 3.71 and the method of 24 ordinates in Fig. 3.72.

There have also been developed various mechanical, electrical, and optical instruments for obtaining the Fourier coefficients from a graphical or electrical record of $q_i(t)$. Some of these will be touched on in a later chapter under the topic Signal Analyzers.

Once the Fourier series for a particular $q_i(t)$ has been found, the *steady-state* response of any instrument to this input may be found by use of frequency-response techniques and the principle of superposition. That is, the response for each individual sinusoidal term is found and then they are added algebraically to get the total response. By use of Eqs. (3.272) to (3.274) all terms in the Fourier series can be put in the form

$$A_{ik} \sin(\omega_k t + \alpha_k)$$

---

[1] J. B. Scarborough, "Numerical Mathematical Analysis," 3d ed., chap. 17, The Johns Hopkins Press, Baltimore, 1958.

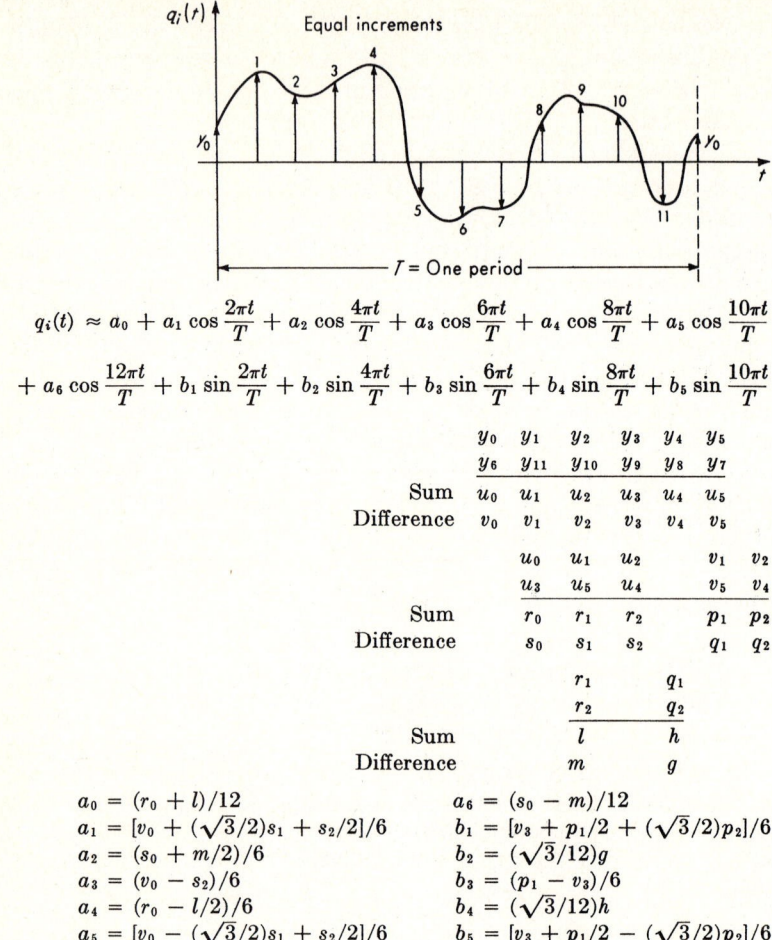

$$q_i(t) \approx a_0 + a_1 \cos \frac{2\pi t}{T} + a_2 \cos \frac{4\pi t}{T} + a_3 \cos \frac{6\pi t}{T} + a_4 \cos \frac{8\pi t}{T} + a_5 \cos \frac{10\pi t}{T}$$

$$+ a_6 \cos \frac{12\pi t}{T} + b_1 \sin \frac{2\pi t}{T} + b_2 \sin \frac{4\pi t}{T} + b_3 \sin \frac{6\pi t}{T} + b_4 \sin \frac{8\pi t}{T} + b_5 \sin \frac{10\pi t}{T}$$

	y_0	y_1	y_2	y_3	y_4	y_5	
	y_6	y_{11}	y_{10}	y_9	y_8	y_7	
Sum	u_0	u_1	u_2	u_3	u_4	u_5	
Difference	v_0	v_1	v_2	v_3	v_4	v_5	

	u_0	u_1	u_2	v_1	v_2
	u_3	u_5	u_4	v_5	v_4
Sum	r_0	r_1	r_2	p_1	p_2
Difference	s_0	s_1	s_2	q_1	q_2

	r_1	q_1
	r_2	q_2
Sum	l	h
Difference	m	g

$$a_0 = (r_0 + l)/12 \qquad\qquad a_6 = (s_0 - m)/12$$
$$a_1 = [v_0 + (\sqrt{3}/2)s_1 + s_2/2]/6 \qquad b_1 = [v_3 + p_1/2 + (\sqrt{3}/2)p_2]/6$$
$$a_2 = (s_0 + m/2)/6 \qquad\qquad b_2 = (\sqrt{3}/12)g$$
$$a_3 = (v_0 - s_2)/6 \qquad\qquad b_3 = (p_1 - v_3)/6$$
$$a_4 = (r_0 - l/2)/6 \qquad\qquad b_4 = (\sqrt{3}/12)h$$
$$a_5 = [v_0 - (\sqrt{3}/2)s_1 + s_2/2]/6 \qquad b_5 = [v_3 + p_1/2 - (\sqrt{3}/2)p_2]/6$$

Fig. 3.71. *Method of 12 ordinates.*

We now define the complex number $Q_i(i\omega_k)$ by

$$Q_i(i\omega_k) \triangleq A_{ik}\underline{/\alpha_k} \qquad (3.275)$$

For example, the constant term -7.2 becomes $7.2\underline{/180°}$, and the term $9.3 \sin(20t + 37°)$ becomes $9.3\underline{/37°}$. When the Fourier series representing $q_i(t)$ is expressed in this form, it is called $Q_i(i\omega)$, the *input-frequency spectrum.* Thus, if

$$q_i(t) = A_{i0} + A_{i1} \sin(\omega_1 t + \alpha_1) + A_{i2} \sin(\omega_2 t + \alpha_2) + \cdots \qquad (3.276)$$

then $\quad Q_i(i\omega) = |A_{i0}|\underline{/0° \text{ or } 180°} + A_{i1}\underline{/\alpha_1} + A_{i2}\underline{/\alpha_2} + \cdots \qquad (3.277)$

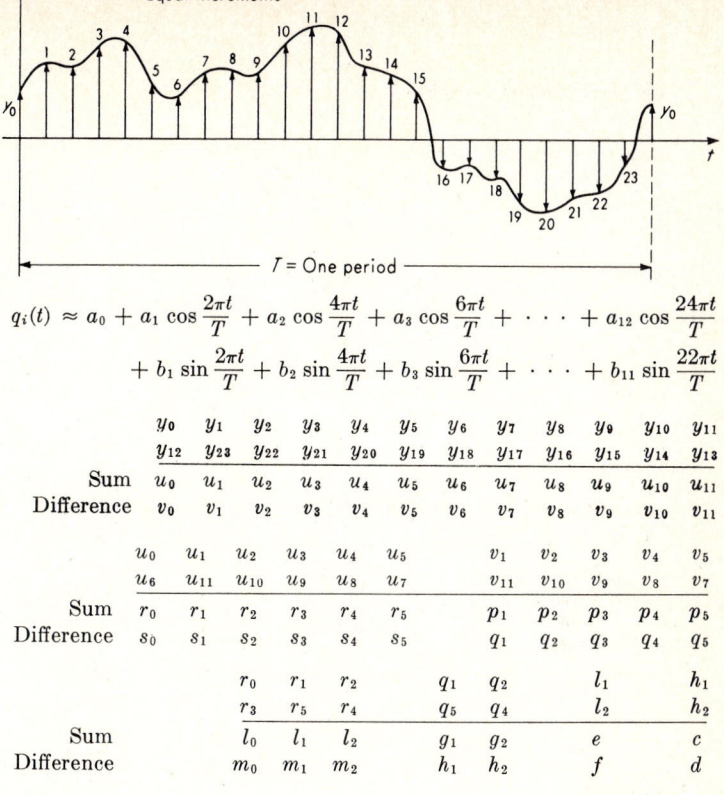

$$q_i(t) \approx a_0 + a_1 \cos \frac{2\pi t}{T} + a_2 \cos \frac{4\pi t}{T} + a_3 \cos \frac{6\pi t}{T} + \cdots + a_{12} \cos \frac{24\pi t}{T}$$

$$+ \, b_1 \sin \frac{2\pi t}{T} + b_2 \sin \frac{4\pi t}{T} + b_3 \sin \frac{6\pi t}{T} + \cdots + b_{11} \sin \frac{22\pi t}{T}$$

	y_0	y_1	y_2	y_3	y_4	y_5	y_6	y_7	y_8	y_9	y_{10}	y_{11}
	y_{12}	y_{23}	y_{22}	y_{21}	y_{20}	y_{19}	y_{18}	y_{17}	y_{16}	y_{15}	y_{14}	y_{13}
Sum	u_0	u_1	u_2	u_3	u_4	u_5	u_6	u_7	u_8	u_9	u_{10}	u_{11}
Difference	v_0	v_1	v_2	v_3	v_4	v_5	v_6	v_7	v_8	v_9	v_{10}	v_{11}

	u_0	u_1	u_2	u_3	u_4	u_5		v_1	v_2	v_3	v_4	v_5
	u_6	u_{11}	u_{10}	u_9	u_8	u_7		v_{11}	v_{10}	v_9	v_8	v_7
Sum	r_0	r_1	r_2	r_3	r_4	r_5		p_1	p_2	p_3	p_4	p_5
Difference	s_0	s_1	s_2	s_3	s_4	s_5		q_1	q_2	q_3	q_4	q_5

	r_0	r_1	r_2		q_1	q_2	l_1	h_1
	r_3	r_5	r_4		q_5	q_4	l_2	h_2
Sum	l_0	l_1	l_2		g_1	g_2	e	c
Difference	m_0	m_1	m_2		h_1	h_2	f	d

In the formulas below, $C = \cos 15° = 0.9659258$, $S = \sin 15° = 0.2588190$.

$$a_0 = (l_0 + e)/24$$
$$a_1 = [v_0 + Cs_1 + (\sqrt{3}/2)s_2 + (1/\sqrt{2})s_3 + (1/2)s_4 + Ss_5]/12$$
$$a_2 = [s_0 + (\sqrt{3}/2)m_1 + (1/2)m_2]/12$$
$$a_3 = [v_0 + (1/\sqrt{2})(s_1 - s_3 - s_5) - s_4]/12$$
$$a_4 = [m_0 + (1/2)f]/12$$
$$a_5 = [v_0 + Ss_1 - (\sqrt{3}/2)s_2 - (1/\sqrt{2})s_3 + (1/2)s_4 + Cs_5]/12$$
$$a_6 = (s_0 - m_2)/12$$
$$a_7 = [v_0 - Ss_1 - (\sqrt{3}/2)s_2 + (1/\sqrt{2})s_3 + (1/2)s_4 - Cs_5]/12$$
$$a_8 = [l_0 - (1/2)e]/12$$
$$a_9 = [v_0 - (1/\sqrt{2})(s_1 - s_3 - s_5) - s_4]/12$$
$$a_{10} = [s_0 - (\sqrt{3}/2)m_1 + (1/2)m_2]/12$$
$$a_{11} = [v_0 - Cs_1 + (\sqrt{3}/2)s_2 - (1/\sqrt{2})s_3 + (1/2)s_4 - Ss_5]/12$$
$$a_{12} = (m_0 - f)/24$$

$$b_1 = [Sp_1 + (1/2)p_2 + (1/\sqrt{2})p_3 + (\sqrt{3}/2)p_4 + Cp_5 + v_6]/12$$
$$b_2 = [(1/2)g_1 + (\sqrt{3}/2)g_2 + q_3]/12$$
$$b_3 = [p_2 - v_6 + (1/\sqrt{2})(p_1 + p_3 - p_5)]/12$$
$$b_4 = (\sqrt{3}/24)c$$
$$b_5 = [Cp_1 + (1/2)p_2 - (1/\sqrt{2})p_3 - (\sqrt{3}/2)p_4 + Sp_5 + v_6]/12$$
$$b_6 = (g_1 - q_3)/12$$
$$b_7 = [Cp_1 - (1/2)p_2 - (1/\sqrt{2})p_3 + (\sqrt{3}/2)p_4 + Sp_5 - v_6]/12$$
$$b_8 = (\sqrt{3}/24)d$$
$$b_9 = [v_6 - p_2 + (1/\sqrt{2})(p_1 + p_3 - p_5)]/12$$
$$b_{10} = [(1/2)g_1 - (\sqrt{3}/2)g_2 + q_3]/12$$
$$b_{11} = [Sp_1 - (1/2)p_2 + (1/\sqrt{2})p_3 - (\sqrt{3}/2)p_4 + Cp_5 - v_6]/12$$

Fig. 3.72. *Method of 24 ordinates.*

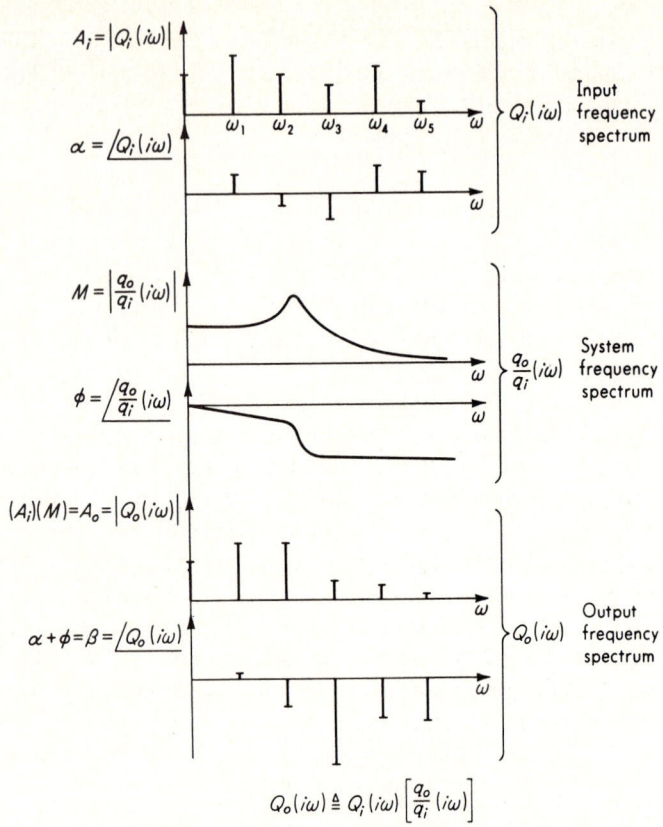

Fig. 3.73. *System response to periodic input.*

Such a spectrum, which exists only at isolated frequencies, is called a *discrete spectrum*.

Now, we recall that the sinusoidal transfer function of the system, $(q_o/q_i)(i\omega)$, is also a complex number for any given frequency. An alternative name for the sinusoidal transfer function is the *system frequency spectrum*. Such a spectrum, which exists at *all* frequencies, is called a *continuous spectrum*. If we pick any frequency ω_k and multiply the complex numbers $Q_i(i\omega_k)$ and $(q_o/q_i)(i\omega_k)$ we get another complex number which we define as $Q_o(i\omega_k)$. If we do this for *all* frequencies and add all the $Q_o(i\omega_k)$, the sum is called $Q_o(i\omega)$, the *output frequency spectrum*, which has the form

$$Q_o(i\omega) = A_{o0}\underline{/0° \text{ or } 180°} + A_{o1}\underline{/\beta_1} + A_{o2}\underline{/\beta_2} + \cdots \qquad (3.278)$$

This is now interpreted as in Eqs. (3.276) and (3.277) to give

$$q_o(t) = \pm A_{o0} + A_{o1}\sin(\omega_1 t + \beta_1) + A_{o2}\sin(\omega_2 t + \beta_2) + \cdots \qquad (3.279)$$

Since $q_i(t)$ is periodic, the frequencies ω_2, ω_3, etc., are all integer multiples of ω_1 and thus $q_o(t)$ will also be a periodic function. For accurate measurement, $q_i(t)$ and $q_o(t)$ must, of course, have nearly identical wave forms. The validity of the statement

$$Q_o(i\omega) = Q_i(i\omega) \left[\frac{q_o}{q_i}(i\omega) \right] \qquad (3.280)$$

used in the above manipulations follows easily from the basic definition of the sinusoidal transfer function, the superposition theorem, and the rules for multiplying complex numbers. Figure 3.73 illustrates the method graphically.

Response of a general form of instrument to a transient input.
By a transient input we mean a $q_i(t)$ that is identically zero for all values of time greater than some finite value t_o, that is, an input that eventually dies out. For transient inputs of specific mathematical form we can usually solve the differential equation and get $q_o(t)$ directly. For q_i's given by experimental data or, more importantly, if we wish to bring out certain important results of a *general* (not restricted to a specific type of q_i) nature, the methods of Fourier transforms[1,2,3] or Laplace transforms[1,2,3] are useful. We now present the methods of applying these techniques, without proof of their validity.

The *direct Fourier transform* $Q_i(i\omega)$ (or the Laplace transform with $s = i\omega$) of the transient input $q_i(t)$ which is zero for $t < 0$ is given by

$$Q_i(i\omega) \triangleq \int_0^\infty q_i(t) \cos \omega t \, dt - i \int_0^\infty q_i(t) \sin \omega t \, dt \qquad (3.281)$$

where ω can take all values from $-\infty$ to $+\infty$. Equation (3.281) is said to transform the input function from the time domain $[q_i(t)]$ to the frequency domain $[Q_i(i\omega)]$. The function $Q_i(i\omega)$ is also called the frequency spectrum of the input and plays the same role for transient inputs as Eq. (3.277) does for periodic inputs. However, whereas $Q_i(i\omega)$ is a *discrete* spectrum for $q_i(t)$ periodic, it is a *continuous* spectrum for $q_i(t)$ transient. That is, if one carries out Eq. (3.281) for a given $q_i(t)$, he will find $Q_i(i\omega)$ to be a complex number which varies with (is a function of) frequency ω and exists for *all* ω, not just at isolated points. As an example, consider the transient input of Fig. 3.74a. Applying Eq.

---

[1] Ley, Lutz, and Rehberg, *op. cit.*

[2] M. F. Gardner and J. L. Barnes, "Transients in Linear Systems," John Wiley & Sons, Inc., New York, 1942.

[3] A. Papoulis, "The Fourier Integral and Its Applications," McGraw-Hill Book Company, New York, 1962.

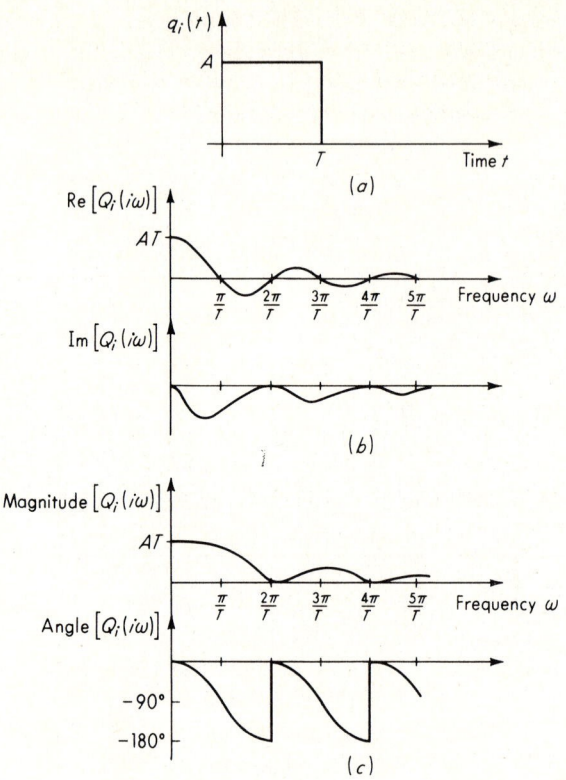

Fig. 3.74. *Frequency spectrum of transient.*

(3.281) we get

$$Q_i(i\omega) = \int_0^T A \cos \omega t \, dt - i \int_0^T A \sin \omega t \, dt \qquad (3.282)$$

$$Q_i(i\omega) = \underbrace{\frac{A \sin \omega T}{\omega}}_{\text{real part}} + i \underbrace{\frac{A}{\omega}(-1 + \cos \omega T)}_{\text{imaginary part}} \qquad (3.283)$$

or, alternatively, $$Q_i(i\omega) = \underbrace{\frac{\sqrt{2}\,A}{\omega}\sqrt{1 - \cos \omega T}}_{\text{magnitude}}\ \underbrace{/\alpha}_{\text{angle}} \qquad (3.284)$$

where $$\tan \alpha = \frac{\cos \omega T - 1}{\sin \omega T} \qquad (3.285)$$

Plots of this frequency spectrum are given in Fig. 3.74*b* and *c*. [While $Q_i(i\omega)$ exists for both positive and negative values of ω, because of symmetry, we can often get the desired results by considering only the range $0 \leq \omega < +\infty$. The stated symmetry consists of the following:

1. $\text{Re } Q_i(-i\omega) = \text{Re } Q_i(+i\omega)$
2. $\text{Im } Q_i(-i\omega) = -\text{Im } Q_i(+i\omega)$
3. $\text{Magnitude } Q_i(-i\omega) = \text{magnitude } Q_i(+i\omega)$
4. $\text{Angle } Q_i(-i\omega) = -\text{angle } Q_i(+i\omega)$

Most of our graphs and calculation methods will employ the range $0 \leq \omega < +\infty$, but $Q_i(-i\omega)$ always exists and can be found from a given $Q_i(+i\omega)$ by application of the above symmetry rules.] These graphs indicate the "frequency content" of the transient input just as the Fourier series indicates the frequency content of a periodic input. Thus we see that if T is small large values of $Q_i(i\omega)$ persist out to higher frequencies than if T is large. Therefore a short-duration pulse is said to have more high-frequency content than a long one. It is important to point out that the concept of frequency content for transients is not as clear-cut as for periodic functions; $q_i(t)$ can *not*, for a transient, be built up by simply adding distinct sine waves because $Q_i(i\omega)$ is now a *continuous function and no distinct frequencies exist.*

A further illustration of this distinction may be found by examining the dimensions of $Q_i(i\omega)$ in both cases. As an example, consider a $q_i(t)$ which is a pressure, $\text{lb}_f/\text{in.}^2$. If this pressure is periodic, Eqs. (3.268), etc., show that $Q_i(i\omega)$ has the same dimensions as $q_i(t)$, that is, $\text{lb}_f/\text{in.}^2$. Now, however, if the pressure is a transient, Eq. (3.281) gives

$$Q_i(i\omega) = \int_0^\infty \underbrace{(\text{lb}_f/\text{in.}^2)}_{} \underbrace{\cos \omega t}_{\text{dimensionless}} (\text{sec}) - i \, (\text{same dimensions}) \qquad (3.286)$$

We see that $Q_i(i\omega)$ now has dimensions of $\text{lb}_f\text{-sec}/\text{in.}^2$ or, reinterpreting this, $(\text{lb}_f/\text{in.}^2)/(\text{rad/sec})$. That is, $Q_i(i\omega)$ is thought of as the amount of signal *per unit frequency increment* rather than the actual amount of signal at a discrete frequency. This is analogous to the concept of distributed (rather than concentrated) loads in strength of materials. When a beam has sand (or water, etc.) piled on it, the applied load at any particular *point* is zero, but over an *area* the load is the force density times the area. Similarly a transient signal has no discrete frequencies but does contain a certain amount of signal within any frequency *band*. Thus, for a transient, $Q_i(i\omega)$ may be thought of as the *density* of signal per frequency bandwidth rather than as the signal itself.

The main purpose of using Eq. (3.281) is to convert functions from the time domain to the frequency domain, perform certain desired operations (which are *easier* or more *revealing* in the frequency domain than in the time domain), and then convert the information back to the time domain since this is the more familiar and directly applicable (in an engineering sense) form. The conversion from frequency domain to time domain is

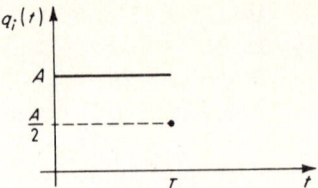

Fig. 3.75. *Single-valued definition of transient.*

accomplished by the *inverse-Fourier-* (or Laplace-) transform formula given by

$$q_i(t) = \frac{2}{\pi} \int_0^\infty \text{Re} \,[Q_i(i\omega)] \cos \omega t \, d\omega \qquad t > 0$$

$$q_i(t) \equiv 0 \qquad t < 0 \tag{3.287}$$

Since these transformations are unique, if a $Q_i(i\omega)$ for a given $q_i(t)$ is found from Eq. (3.281), it should be possible to reconstruct the original $q_i(t)$ from the $Q_i(i\omega)$ by Eq. (3.287). Carrying this out for our example, we get

$$q_i(t) = \frac{2}{\pi} \int_0^\infty \frac{A \sin \omega T}{\omega} \cos \omega t \, d\omega \qquad t > 0$$

$$q_i(t) \equiv 0 \qquad t < 0 \tag{3.288}$$

After some transformations, this can be put in a standard form found in integral tables and gives

$$q_i(t) = \begin{cases} 0 & t < 0 \\ A & 0 < t < T \\ A/2 & t = T \\ 0 & T < t \end{cases} \tag{3.289}$$

This function is shown in Fig. 3.75, and we see that it is practically identical to Fig. 3.74a. Actually, in Fig. 3.74a, we were not mathematically precise in defining $q_i(t)$ at $t = T$ since the graph shows it taking on *all* values between 0 and A. The usual practice for such discontinuities is to define the function as single-valued and equal to the midpoint. The Fourier transform is set up on this basis and thus always gives results similar to (3.289). The Fourier *series* for a periodic function with step discontinuities also behaves in this fashion; that is, it converges to the midpoint. Thus in using numerical schemes such as those of Figs. 3.71 and 3.72, if an ordinate falls right on a discontinuity, the midpoint should be used as the numerical entry in the computation schedule.

To get a better feeling for the above methods and also to show how they are graphically or numerically applied to functions (data) for which mathematical formulas are not available, let us consider the following development. Figure 3.76a shows a typical transient $q_i(t)$ as might be

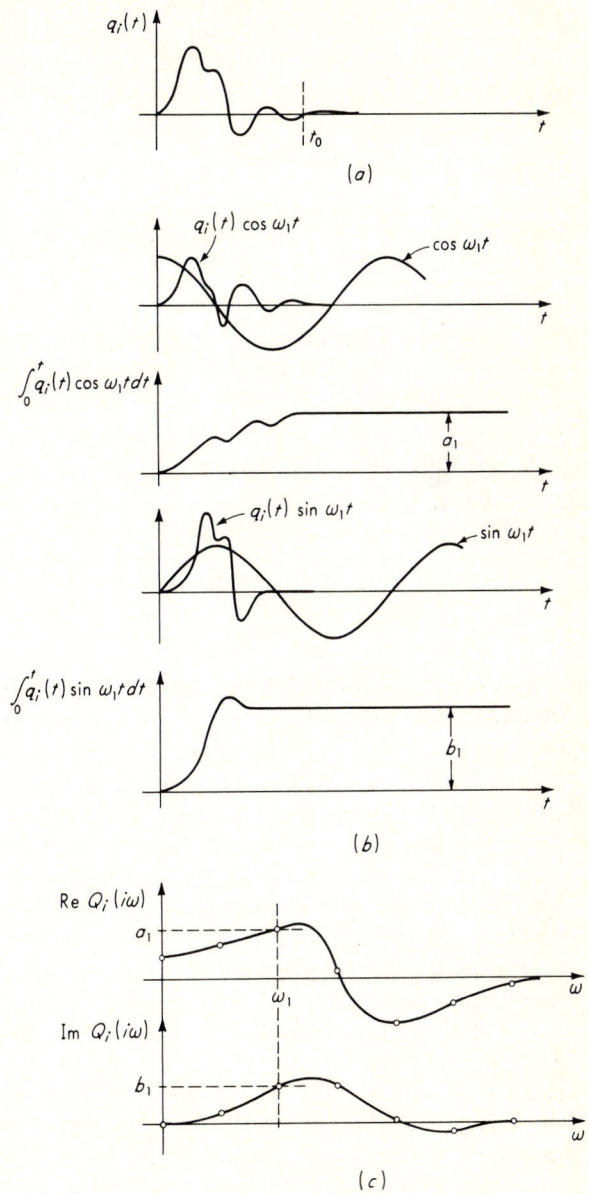

Fig. 3.76. *Graphical interpretation of direct Fourier transform.*

experimentally recorded in, say, a shock test. The first thing we note is that, for all practical purposes, the transient is ended in a finite length of time t_0. Thus, since $q_i(t)$ is a multiplying factor in the integrand and becomes zero for $t > t_0$, we may write Eq. (3.281) as

$$Q_i(i\omega) = \int_0^{t_0} q_i(t) \cos \omega t \, dt - i \int_0^{t_0} q_i(t) \sin \omega t \, dt \qquad (3.290)$$

We obtain $Q_i(i\omega)$ one point (frequency) at a time as follows:

1. Choose a numerical value of ω, say ω_1.
2. Now, $\cos \omega_1 t$ is a perfectly definite curve and may be plotted against t.
3. Multiply $q_i(t)$ and $\cos \omega_1 t$ point by point to get the curve $q_i(t) \cos \omega_1 t$.
4. Integrate, by any suitable numerical, graphical, or machine means, the curve $q_i(t) \cos \omega_1 t$ from $t = 0$ to $t = t_0$. Call the integral (area under curve) a_1.
5. Repeat the above procedure for $q_i(t) \sin \omega_1 t$ and call the integral b_1.
6. $Q_i(i\omega_1)$ is then $a_1 + ib_1$.
7. Repeat for as many ω's as desired to generate the curves for $Q_i(i\omega)$ versus ω.

Figure 3.76 illustrates these procedures.

To appreciate the difference in "frequency content" between a "slow" transient and a "fast" one, we consider Fig. 3.77. Here, $Q_i(i\omega)$ is found (for a high value of ω) for both a slow transient (Fig. 3.77a) and a fast one (Fig. 3.77b). It is clear that $Q_i(i\omega)$ will be nearly zero for ω's at or above the chosen ω_1 for the slow transient; thus its frequency content is limited to lower frequencies. The fast transient, on the other hand, has a nonzero value for $Q_i(i\omega_1)$ and thus "contains" frequencies at and somewhat above this value. For any real-world transient, one can always find *some* ω_1 high enough to make $Q_i(i\omega_1) \approx 0$; that is, all *real* transients are limited in frequency content at the high end. An "unreal" (mathematically possible only) transient is the impulse function which we can easily show to contain *all* frequencies from 0 to ∞ and all in equal "strength." For an impulse of area A

$$Q_i(i\omega) = \int_0^{\infty} A u_1(t) \cos \omega t \, dt - i \int_0^{\infty} A u_1(t) \sin \omega t \, dt \qquad (3.291)$$
$$Q_i(i\omega) = A - i0 = A \qquad \text{for any finite } \omega \qquad (3.292)$$

Thus Fig. 3.78 shows the frequency content of an impulse. This property of an impulse makes it most useful as a "test signal" for investigating unknown systems, since all frequencies will be excited equally and the

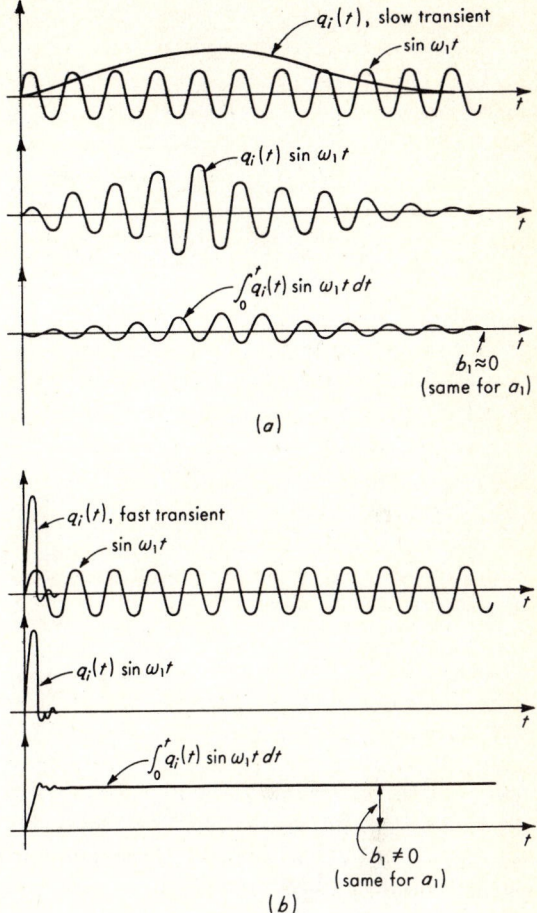

Fig. 3.77. *Frequency content of fast and slow transients.*

true nature of the system will be revealed by its response.   We shall develop this important concept in more detail later.

The process of *inverse* transformation can also be interpreted graphically.   The defining equation (3.287) may, in actual practice, be written as

$$q_i(t) = \frac{2}{\pi} \int_0^{\omega_0} \text{Re}\,[Q_i(i\omega)] \cos \omega t \, d\omega \qquad t > 0$$

$$q_i(t) \equiv 0 \qquad t < 0$$

$$(3.293)$$

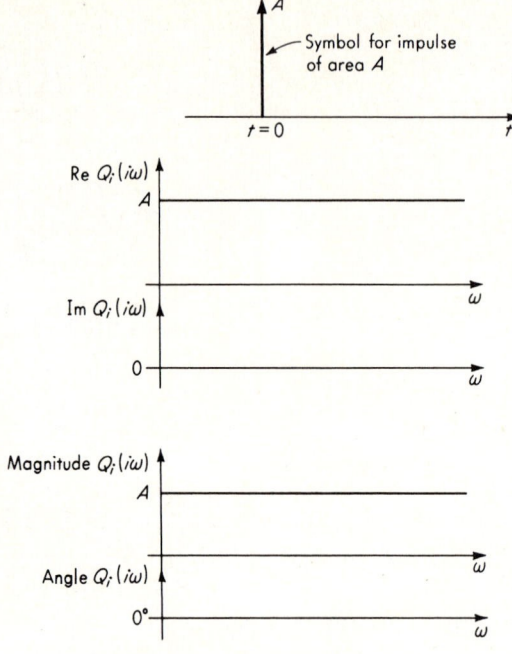

Fig. 3.78.   *Frequency spectrum of impulse.*

since all $Q_i(i\omega)$ representing physical quantities become approximately equal to zero for ω greater than some finite value ω_0. This follows directly from the fact that the $Q_i(i\omega)$ come from the $q_i(t)$ and they cannot contain infinitely high frequencies. A step-by-step procedure for finding $q_i(t)$ one point at a time from a given $Q_i(i\omega)$ is as follows:

1. Choose a numerical value of t, say t_1.
2. Now, $\cos \omega t_1$ is a perfectly definite curve and may be plotted against ω.
3. Multiply $\mathrm{Re}\,[Q_i(i\omega)]$ and $\cos \omega t_1$ point by point to get the curve $\mathrm{Re}\,[Q_i(i\omega)]\cos \omega t_1$.
4. Integrate, by any suitable numerical, graphical, or machine means, the curve $\mathrm{Re}\,[Q_i(i\omega)]\cos \omega t_1$ from $\omega = 0$ to $\omega = \omega_0$. The integral (area under curve) is $(\pi/2)q_i(t_1)$, from Eq. (3.293). Plot $q_i(t_1)$ versus t.
5. Repeat for as many t's as desired to generate the curve $q_i(t)$ versus t.

Figure 3.79 illustrates this procedure.

Since the direct and inverse transformations described above are widely used, methods for speeding up the calculations have been devel-

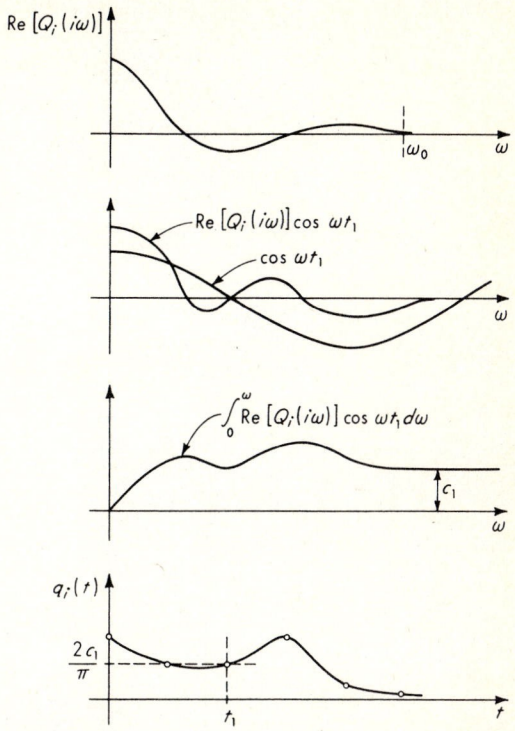

Fig. 3.79. *Graphical interpretation of inverse Fourier transform.*

oped.[1,2,3] Further treatment of these methods can be found in the bibliography entries at the end of this chapter.

The main usefulness of the above transform methods is based on the important result[4] relating the Fourier transform of the input signal $Q_i(i\omega)$, the system frequency response $(q_o/q_i)(i\omega)$, and the Fourier transform of

[1] C. R. Huss and J. J. Donegan, Method and Tables for Determining the Time Response to a Unit Impulse from Frequency-response Data and for Determining the Fourier Transform of a Function of Time, *NACA, Tech. Note* 3598, January, 1956.

[2] C. R. Huss and J. J. Donegan, Tables for the Numerical Determination of the Fourier Transform of a Function of Time and the Inverse Fourier Transform of a Function of Frequency, with Some Applications to Operational Calculus Methods, *NACA, Tech. Note* 4073, October, 1957.

[3] NACA is National Advisory Committee on Aeronautics; name changed in 1958 to National Aeronautics and Space Administration (NASA).

[4] J. A. Aseltine, "Transform Method in Linear System Analysis," McGraw-Hill Book Company, New York, 1958.

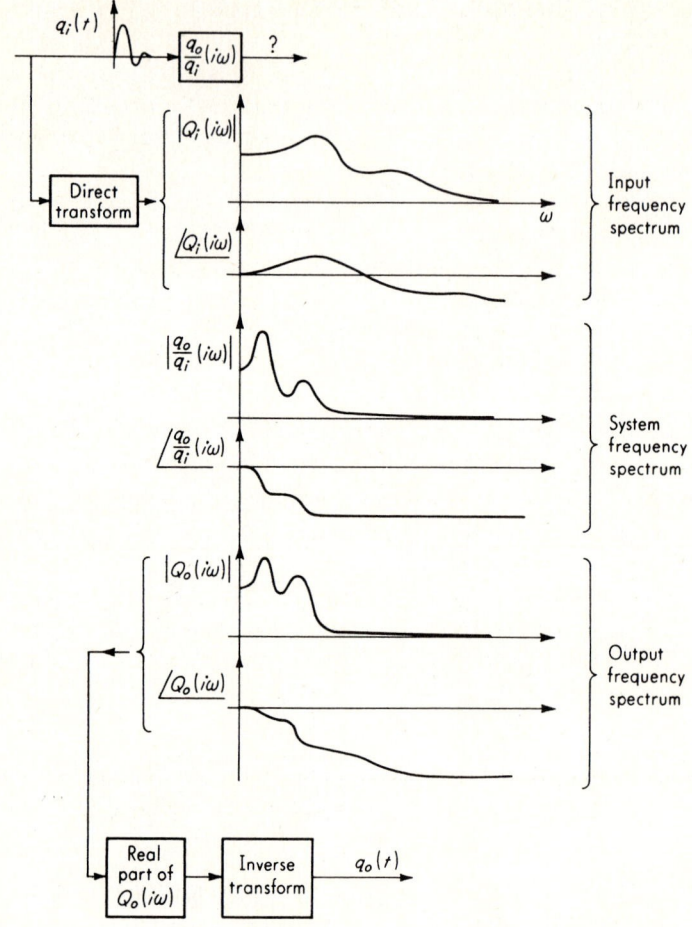

Fig. 3.80. *System response to transient input.*

the output signal $Q_o(i\omega)$:

$$Q_o(i\omega) = Q_i(i\omega)\left[\frac{q_o}{q_i}(i\omega)\right] \qquad (3.294)$$

The restriction on this result is that the system and the input must be such that all derivatives (other than the highest derivative) and integrals of $q_o(t)$ that appear in the differential equation relating $q_o(t)$ and $q_i(t)$ must be zero at $t = 0^+$ (an infinitesimal time after the input is applied). That is, the initial conditions must be zero; the system starts from rest. Many important practical problems meet this requirement. (Also, for nonzero initial conditions, the method may still be applied if the input is suitably modified.[1])

The meaning of Eq. (3.294) is that when the frequency-response curves for a system are known, if a transient input $q_i(t)$ is applied, one can transform $q_i(t)$ to $Q_i(i\omega)$, multiply $Q_i(i\omega)$ and $(q_o/q_i)(i\omega)$ point by point to get $Q_o(i\omega)$, and then inverse-transform $Q_o(i\omega)$ to $q_o(t)$ to get the output or response of the system.   Figure 3.80 illustrates the procedure.

From a measurement-system point of view, Eq. (3.294) has the following important interpretation.   For accurate measurement, $q_o(t) \approx Kq_i(t)$, and since the Fourier transforms are unique [only one possible $F(i\omega)$ for each $f(t)$ and vice versa], this requires $Q_o(i\omega) \approx KQ_i(i\omega)$.   Since $Q_o(i\omega)$ is obtained by multiplying $Q_i(i\omega)$ by $(q_o/q_i)(i\omega)$, this means $(q_o/q_i)(i\omega)$ must be $K\underline{/0°}$ over the entire range of frequencies for which $Q_i(i\omega)$ is not practically zero *but can be anything elsewhere*.   The requirement that $(q_o/q_i)(i\omega)$ be $K\underline{/0°}$ for *all* frequencies for perfect measurement is obvious without the use of transform methods.   The condition that this need be so only for a definite, finite *range* of frequencies (corresponding to the frequency content of the input) is the contribution of the transform methods and is of great practical significance since it puts much more realistic demands on the measurement system.   A further relaxation of these requirements (which allows phase shift) will be developed later in this chapter.

Frequency spectrum of amplitude-modulated signals. Interest in amplitude-modulated signals stems mainly from two considerations:

1.   Physical data that are to be measured and interpreted sometimes are amplitude-modulated.
2.   Certain types of measurement systems intentionally introduce amplitude modulation for one or more benefits this process may supply.

While, in general, the signal that modulates the amplitude of a carrier wave may be of any form (single sine wave, general periodic function, random wave, transient, etc.) and the carrier may also be given different forms (sine wave, square wave, etc.), the process is perhaps most easily understood for a single sine wave modulating a sinusoidal carrier.   The modulation process is basically one of multiplying the signal carrying the information by a carrier wave of constant frequency and amplitude; see

[1] Aseltine, *op. cit.*

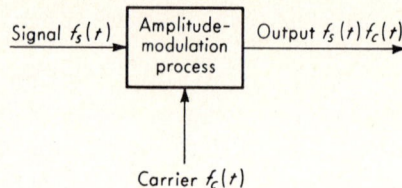

Fig. 3.81. *Amplitude modulation.*

Fig. 3.81.   For our simple example we have

$$\text{Output} = (A_s \sin \omega_s t)(A_c \sin \omega_c t) \qquad (3.295)$$

where $A_s \triangleq$ amplitude of signal
$\omega_s \triangleq$ frequency of signal
$A_c \triangleq$ amplitude of carrier
$\omega_c \triangleq$ frequency of carrier

The frequency ω_c is greater (usually considerably greater) than ω_s.   For such a situation the output has the shape shown in Fig. 3.82a.   The frequency spectrum of such a signal is easily obtained from the following trigonometric identity:

$$\sin \alpha \sin \beta \equiv \tfrac{1}{2} \cos (\alpha - \beta) - \tfrac{1}{2} \cos (\alpha + \beta) \qquad (3.296)$$

Applying this to Eq. (3.295), we get

$$\text{Output} = \frac{A_s A_c}{2} [\cos (\omega_c - \omega_s) t - \cos (\omega_c + \omega_s) t] \qquad (3.297)$$

$$\text{Output} = \frac{A_s A_c}{2} \sin [(\omega_c - \omega_s) t + 90°]$$

$$+ \frac{A_s A_c}{2} \sin [(\omega_c + \omega_s) t - 90°] \qquad (3.298)$$

We see that the frequency spectrum of this signal is a discrete spectrum existing only at the frequencies $\omega_c - \omega_s$ and $\omega_c + \omega_s$, the so-called *side frequencies*.   If such a signal is the input $q_i(t)$ to a measurement system, one can find the steady-state output easily by the methods of Fig. 3.73.

Some applications of these concepts may be appropriate at this point. In a first, rough consideration of the vibration and noise of shafts with gears, one would perhaps expect the important frequencies to be those corresponding to the rotational speeds of the shafts and those corresponding to the tooth-meshing frequencies.   For example, a shaft with a 20-tooth gear running at 200 rps would be expected to generate noise at 200 and 4,000 cps.   However, actual noise measurements in such situations may show the peak noise to occur at frequencies different from those

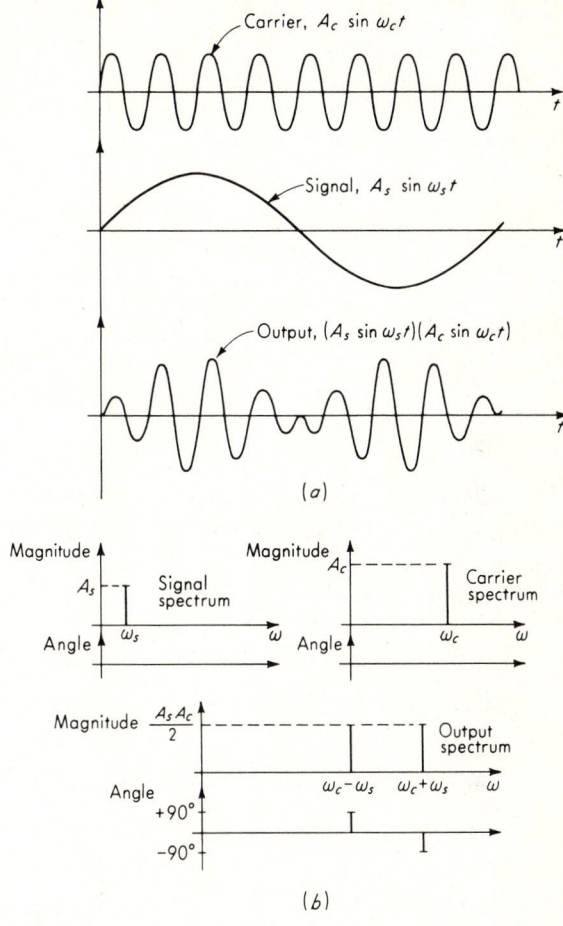

Fig. 3.82.   *Frequency spectrum of amplitude-modulated signals.*

expected from the above crude analysis.[1]   These discrepancies may often be resolved by the application of amplitude-modulation concepts as follows:

For a pair of absolutely true-running gears, the tooth forces (which cause vibration and, thereby, noise) would have a fundamental frequency equal to the tooth-meshing frequency.   These forces would not be pure

---

[1] P. K. Stein, "Measurement Engineering," vol. I, sec. 17, Stein Engineering Services, Inc., Phoenix, Ariz., 1962.

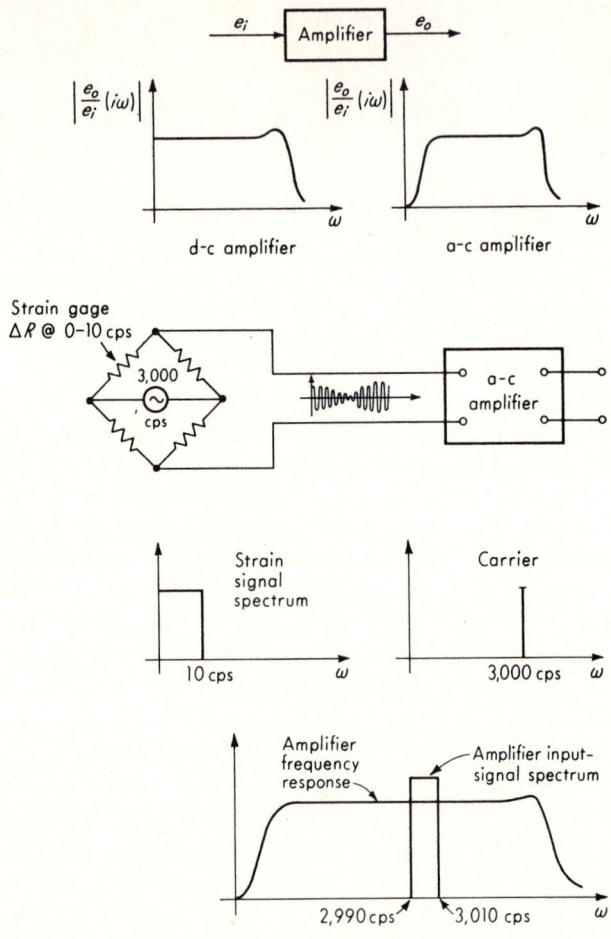

Fig. 3.83. *Application of amplitude modulation.*

sine waves but *would* be periodic. Thus one could get a Fourier series for them. For simplicity, let us assume these forces to be pure sine waves of fixed amplitude. In an actual set of gears there is always some eccentricity or "runout"; that is, the gears are closer together at some points in their rotation than at others. It is postulated that this runout leads to a force amplitude that *varies* as the gear rotates; that is, the tooth-force amplitude is *modulated* as a function of rotational position. If this is so, the frequencies of generated noise (corresponding to tooth-force frequencies) would be expected to be the side frequencies generated by modulating the 4,000-cps tooth-meshing frequency with the 200-cps (once per rotation) runout frequency. These frequencies (3,800 and

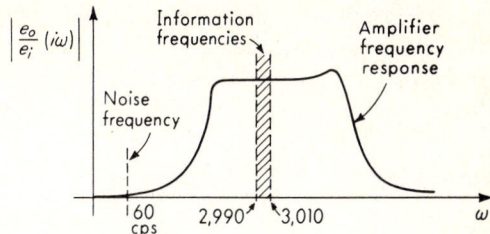

Fig. 3.84. *Noise rejection through amplitude modulation.*

4,200 cps in our example) have actually been measured, confirming the original conjecture. Without the amplitude-modulation concepts, the engineer would have been hard pressed to explain these frequencies in the measured data.

Another interesting example is found in the carrier amplifier. To measure easily and record very small voltages coming from transducers (such as strain gages) requires a very-high-gain amplifier. Because of drift problems, a high-gain amplifier is easier to build as an a-c, rather than a d-c, unit. An a-c amplifier, however, does not amplify constant or slowly varying voltages and so would appear to be unsuitable for measuring static strains. This problem is overcome by exciting the strain-gage bridge with alternating voltage (say 5 volts at 3,000 cps) rather than direct current. Thus when the bridge is unbalanced by strain-induced resistance changes, the output voltage will be a 3,000-cps a-c voltage whose amplitude will be modulated by the strain changes. Thus if we are measuring strains which vary from, say, 0 cps (static) to 10 cps, the amplifier will have an input-frequency spectrum bounded by 2,990 and 3,010 cps. This range of frequencies is easily handled by an a-c amplifier. Figure 3.83 illustrates these concepts. This "shifting" of the information frequencies from one part of the frequency range to another is the basis of many useful applications of amplitude modulation.

As a final example, suppose that the wires leading from the bridge to the amplifier in Fig. 3.83 are subjected to a stray 60-cps field from surrounding a-c machinery and a 60-cps noise, or "hum," is superimposed (additively) on the desired signals. This 60-cps noise could easily be larger than the desired strain signals. With a carrier system, however, this noise may be easily eliminated merely by designing the a-c amplifier so that it does not respond to 60 cps. Since the desired band of frequencies is 2,990 to 3,010 cps, making the low-frequency cutoff of the amplifier greater than 60 cps is not difficult. Figure 3.84 illustrates this situation.

We shall now extend the amplitude-modulation concept to signals other than just a single sine wave. If the modulating signal is a periodic function $f_i(t)$ it may be expanded in a Fourier series to get the output of the modulator as [see Eq. (3.268)]

$$\text{Output} = \left[f_{i,\text{av}} + \frac{1}{L}\left(\sum_{n=1}^{\infty} a_n \cos \frac{n\pi t}{L} + \sum_{n=1}^{\infty} b_n \sin \frac{n\pi t}{L} \right) \right] A_c \sin \omega_c t \quad (3.299)$$

which can be written as

$$\text{Output} = A_0 A_c \sin \omega_c t + (A_1 A_c \cos \omega_1 t \sin \omega_c t$$
$$+ A_2 A_c \cos \omega_2 t \sin \omega_c t + \cdots) + (B_1 A_c \sin \omega_1 t \sin \omega_c t$$
$$+ B_2 A_c \sin \omega_2 t \sin \omega_c t + \cdots) \quad (3.300)$$
$$\text{Now,} \quad \sin \alpha \sin \beta \equiv \tfrac{1}{2}\cos(\alpha - \beta) - \tfrac{1}{2}\cos(\alpha + \beta)$$
$$\text{and} \quad \sin \alpha \cos \beta \equiv \tfrac{1}{2}\sin(\alpha + \beta) + \tfrac{1}{2}\sin(\alpha - \beta)$$

and so

$$\text{Output} = A_0 A_c \sin \omega_c t + C_1 \{ \sin[(\omega_c + \omega_1)t - \alpha_1]$$
$$+ \sin[(\omega_c - \omega_1)t + \alpha_1] \} + \cdots \quad (3.301)$$

where

$$C_1 \triangleq \frac{A_c}{2} \sqrt{A_1{}^2 + B_1{}^2}$$

$$(3.302)$$

$$\alpha_1 \triangleq \tan^{-1} \frac{B_1}{A_1}$$

We see that the spectrum of the output signal is a discrete spectrum containing the frequencies ω_c, $\omega_c \pm \omega_1$, $\omega_c \pm \omega_2$, $\omega_c \pm \omega_3$, etc. That is, each frequency component of the modulating signal produces one pair of side frequencies (see Fig. 3.85). If the output of the modulator is applied to the input of a system with known frequency response the methods of Fig. 3.73 can again be used to find the steady-state output.

If the modulating signal is a transient, the spectrum of the modulator output may be obtained with the help of the modulation theorem[1,2] for Fourier (or Laplace) transforms. If the modulating signal is a transient $f_i(t)$ it will have a Fourier transform $F_i(i\omega)$ which can be obtained in the usual ways. The modulation theorem leads to the following result if $f_i(t)$ is multiplied by the carrier $A_c \sin \omega_c t$ to produce the modulated output:

Fourier transform of modulated output $\triangleq |Q_i(i\omega)| / \underline{Q_i(i\omega)}$ \quad (3.303)

where

$$|Q_i(i\omega)| = \frac{A_c}{2} \text{ magnitude } \{ F_i[i(\omega - \omega_c)] \} \quad (3.304)$$

$$\underline{/Q_i(i\omega)} = \text{angle } \{ F_i[i(\omega - \omega_c)] \} - 90° \quad (3.305)$$

[1] G. A. Korn and T. M. Korn, "Mathematical Handbook for Scientists and Engineers," p. 219, McGraw-Hill Book Company, New York, 1961.
[2] Papoulis, *op. cit.*, p. 15.

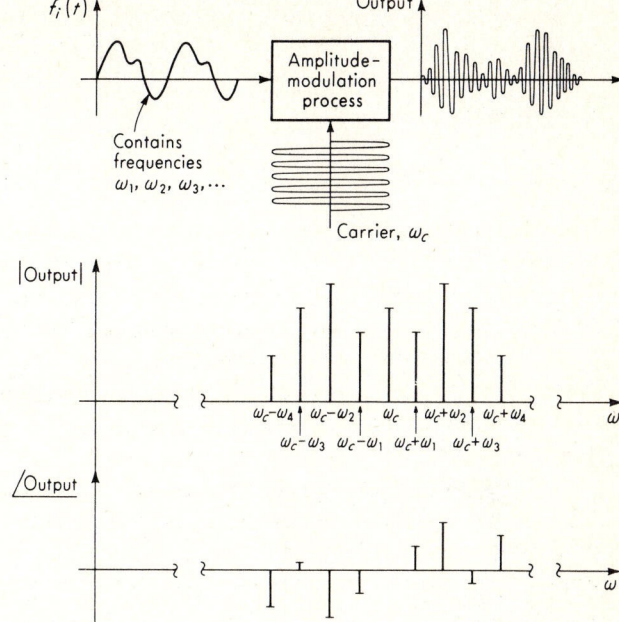

Fig. 3.85. *Frequency spectrum when modulating signal is periodic.*

and $0 \leq \omega < \infty$. Note that the argument (independent variable) of the F_i function is now $i(\omega - \omega_c)$, so that if one wants to find $Q_i(i\,4{,}000)$ and $\omega_c = 3{,}000$, he must evaluate $F_i(i\,1{,}000)$. Also note that, to get $Q_i(i\omega)$ for $0 < \omega < \infty$, one must know $F_i(i\omega)$ for *negative* ω's, since any $\omega < \omega_c$ gives $F_i i(\omega - \omega_c)$ a negative argument. While we have generally worked with positive ω's, the transform for negative ω's always exists and is easily found from the previously given symmetry rules. The spectrum given by Eq. (3.303) will be a continuous one and if the modulated output is applied as an input to some system with known frequency response the corresponding output can be obtained by the methods of Fig. 3.80.

For measurement systems in which amplitude modulation is intentionally introduced to allow the use of carrier-amplifier techniques, the carrier frequency must be considerably greater (usually 5 to 10 times) than any significant frequencies present in the modulating signal. For such a situation the pertinent frequency spectra are as shown in Fig. 3.86. We see again that the amplitude-modulation process shifts the frequency spectrum by the amount ω_c.

When amplitude modulation is intentionally introduced to facilitate data handling in one way or another, it generally plays the role of an

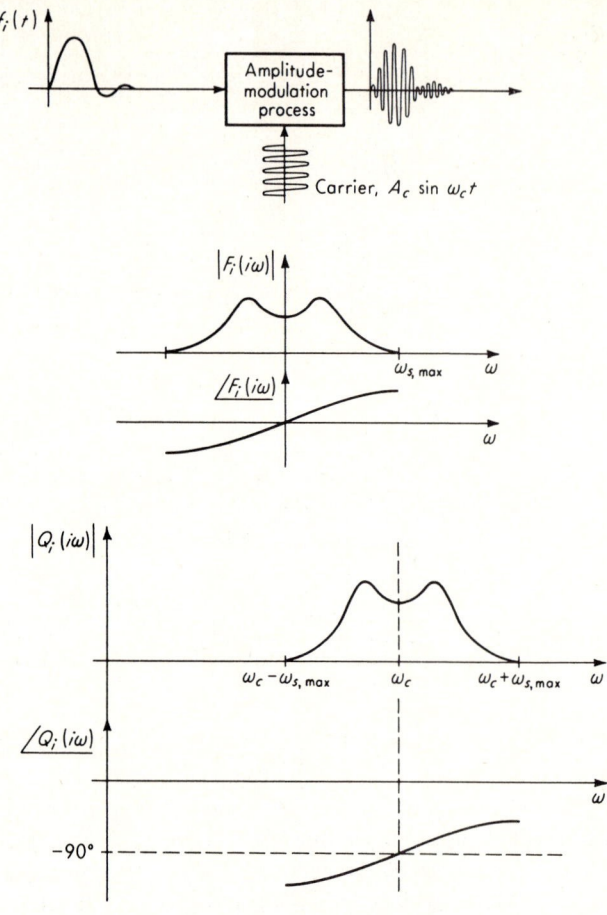

Fig. 3.86. *Frequency spectrum when modulating signal is a transient.*

intermediate step, and the amplitude-modulated signal is not usually considered a suitable final readout. Rather, the original form of the modulating signal (the basic measured data from, say, a transducer) should be recovered. The process for accomplishing this involves *demodulation* (or *detection*, as it is sometimes called) and filtering. Demodulation may be full-wave, half-wave, phase-sensitive, or non-phase-sensitive (Fig. 3.87). We here treat the form giving the best reproduction of the original data, full-wave phase-sensitive demodulation, and consider only the process, not the hardware for accomplishing it. Again it is necessary to consider whether the form of the original signal was single

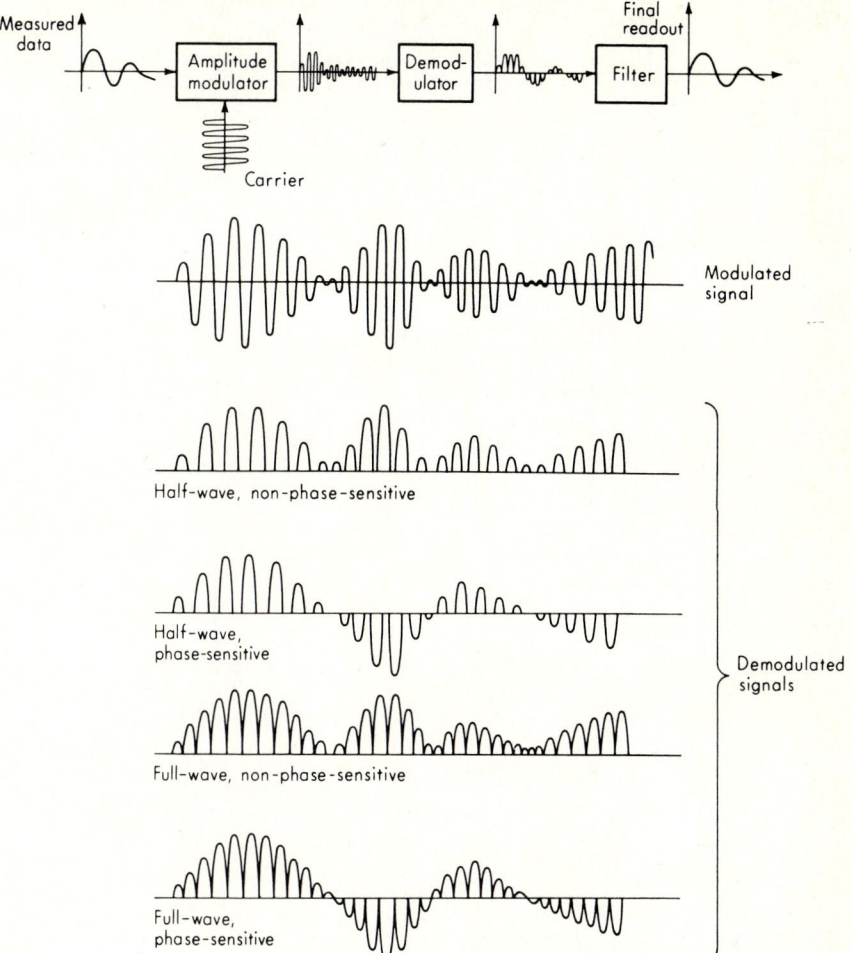

Fig. 3.87.   *Types of demodulation.*

sine wave, periodic function, or transient.   For a single sine wave A_s sin $\omega_s t$ which is modulating a carrier A_c sin $\omega_c t$ the expression for the full-wave phase-sensitive demodulated signal is

$$\text{Demodulator output} = (A_s \sin \omega_s t)|A_c \sin \omega_c t| \qquad (3.306)$$

as seen from Fig. 3.88a.   Now $|A_c \sin \omega_c t|$ is a periodic function and may be expanded in a Fourier series by application of Eq. (3.268).   The results

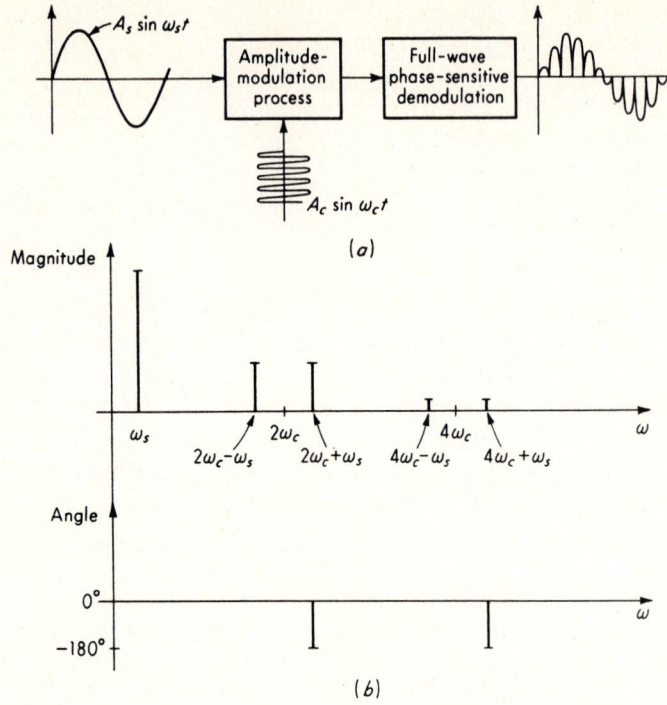

Fig. 3.88. *Frequency spectrum of full-wave phase-sensitive demodulation.*

are

$$|A_c \sin \omega_c t| = \frac{2A_c}{\pi}\left(1 - \tfrac{2}{3}\cos 2\omega_c t - \tfrac{2}{15}\cos 4\omega_c t + \cdots\right.$$

$$\left. + \frac{2}{1 - 4n^2}\cos 2n\omega_c t + \cdots\right) \qquad n = 3, 4, 5, \ldots \qquad (3.307)$$

Equation (3.306) can then be written as

$$\text{Demodulator output} = (A_s \sin \omega_s t)\left[\frac{2A_c}{\pi}\left(1 - \tfrac{2}{3}\cos 2\omega_c t\right.\right.$$

$$\left.\left. - \tfrac{2}{15}\cos 4\omega_c t + \cdots\right)\right] \qquad (3.308)$$

which when multiplied gives

$$\text{Demodulator output} = \frac{2A_c A_s}{\pi}\sin \omega_s t - \frac{4A_c A_s}{3\pi}\sin \omega_s t \cos 2\omega_c t$$

$$- \frac{4A_c A_s}{15\pi}\sin \omega_s t \cos 4\omega_c t + \cdots \qquad (3.309)$$

Now, terms of the form $(\sin \omega_s t)(\cos 2n\omega_c t)$ can, by a trigonometric identity, be written as $[\sin (2n\omega_c + \omega_s)t - \sin (2n\omega_c - \omega_s)t]/2$. We can thus write Eq. (3.309) as

$$\text{Demodulator output} = \frac{2A_cA_s}{\pi} \sin \omega_s t - \frac{2A_cA_s}{3\pi} [\sin (2\omega_c + \omega_s)t$$

$$- \sin (2\omega_c - \omega_s) t] - \frac{2A_cA_s}{15\pi} [\sin (4\omega_c + \omega_s)t - \sin (4\omega_c - \omega_s)t]$$

$$+ \cdots \qquad (3.310)$$

From this we see that the frequency spectrum of the demodulator output signal is a discrete spectrum with frequency content at ω_s, $2\omega_c \pm \omega_s$, $4\omega_c \pm \omega_s$, etc., as shown in Fig. 3.88b. If this signal were an input to a system of known frequency response (such as the filter of Fig. 3.87) the output of this system may be found by the methods of Fig. 3.73. If the output of the filter is to look like the original data, the filter must be designed to *reject* the frequencies $2\omega_c \pm \omega_s$, $4\omega_c \pm \omega_s$, etc., while *passing* with a minimum of distortion the signal frequency ω_s. The design of such a low-pass filter is made simpler if the passband and the rejection band are more widely separated. This is the basis of our earlier statement that carrier frequencies are usually chosen to be 5 to 10 times the highest expected signal frequency.

When the modulating signal is a periodic wave rather than a single sine wave a procedure similar to that just used is employed, except now the modulating signal is *also* expressed as a Fourier series of the form

$$\text{Modulating signal} = A_{s0} + A_{s1} \sin (\omega_s t + \alpha_1)$$
$$+ A_{s2} \sin (2\omega_s t + \alpha_2) + \cdots \qquad (3.311)$$

When this is multiplied by $|A_c \sin \omega_c t|$ as given by Eq. (3.307), we find exactly the same situation as for a single sine wave, but it must be applied for *each* signal frequency (0, ω_s, $2\omega_s$, $3\omega_s$, etc.). The frequency spectrum of the demodulated signal will thus be a discrete spectrum with frequency content at $\omega = 0$, $(\omega_s, 2\omega_c \pm \omega_s, 4\omega_c \pm \omega_s, 6\omega_c \pm \omega_s$, etc.$)$, $(2\omega_s, 2\omega_c \pm 2\omega_s$, $4\omega_c \pm 2\omega_s, 6\omega_c \pm 2\omega_s$, etc.$)$, etc. Figure 3.89 illustrates these concepts. Again, if such a signal is applied to a system of known frequency response the methods of Fig. 3.73 allow calculation of the output.

When the modulating signal is a transient $f_i(t)$ the demodulated signal will be $f_i(t)|A_c \sin \omega_c t|$, which can be written as

$$\text{Demodulated signal} = f_i(t) \left[\frac{2A_c}{\pi} (1 - \tfrac{2}{3} \cos 2\omega_c t - \tfrac{2}{15} \cos 4\omega_c t + \cdots) \right]$$

$$(3.312)$$

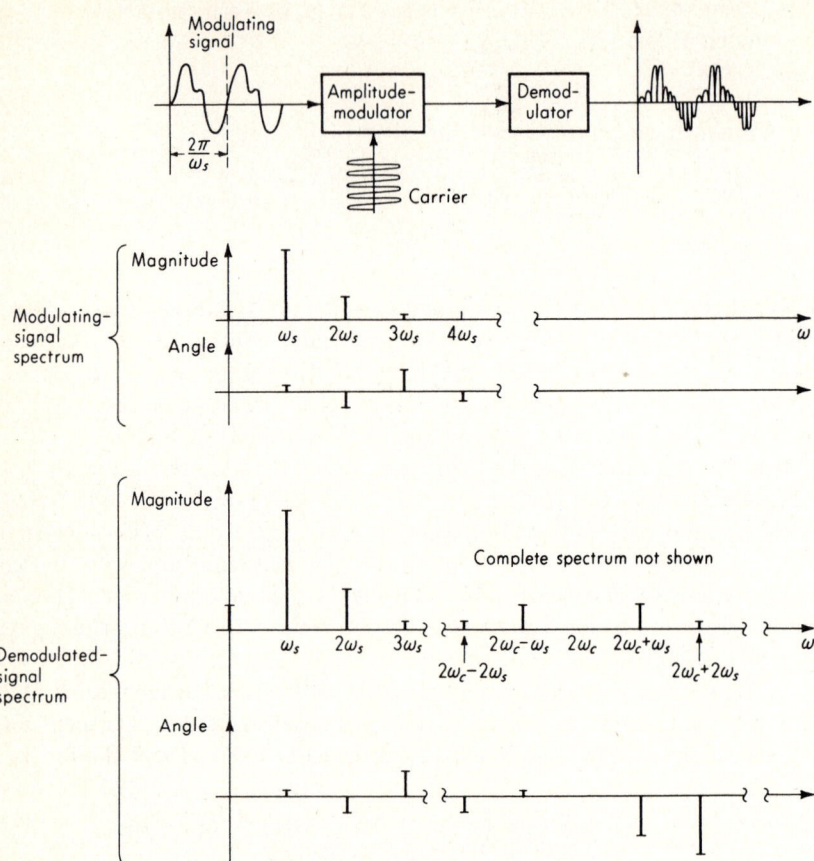

Fig. 3.89.   *Demodulation spectrum for periodic input.*

or, multiplying,

$$\text{Demodulated signal} = \frac{2A_c}{\pi} f_i(t) - \frac{4A_c}{3\pi} f_i(t) \cos 2\omega_c t$$

$$- \frac{4A_c}{15\pi} f_i(t) \cos 4\omega_c t + \cdots \qquad (3.313)$$

Application of the modulation theorem to each of the modulated terms of Eq. (3.313) leads to the result

$$\text{Fourier transform of demodulated signal} \triangleq |Q_i(i\omega)|/\underline{Q_i(i\omega)}$$

where

$$|Q_i(i\omega)| = \frac{2A_c}{\pi} |F_i(i\omega)| + \frac{2A_c}{3\pi} |F_i[i(\omega - 2\omega_c)]|$$

$$+ \frac{2A_c}{15\pi} |F_i[i(\omega - 4\omega_c)]| + \cdots \qquad (3.314)$$

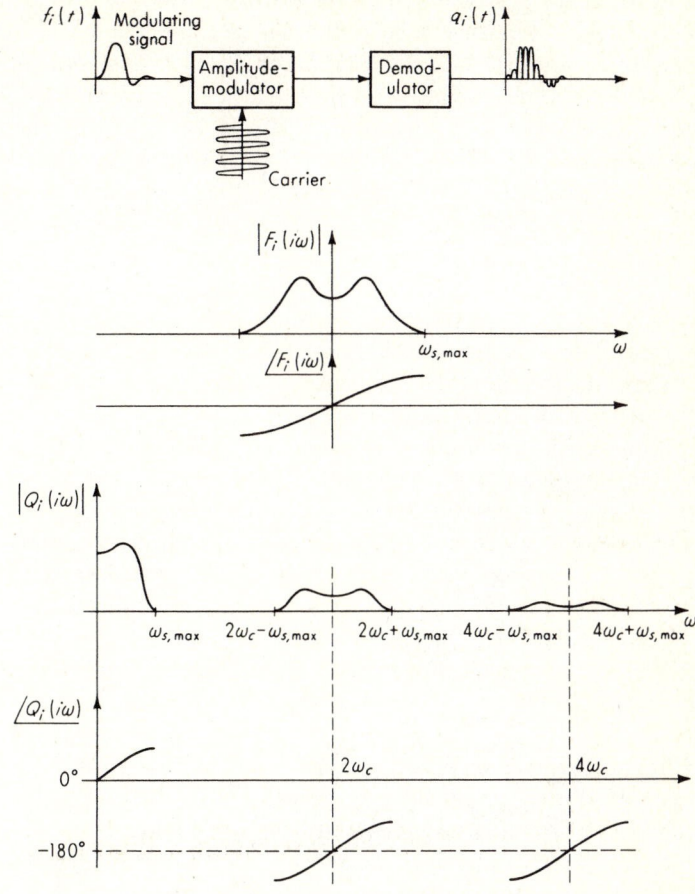

Fig. 3.90. *Demodulation spectrum for transient input.*

and $0 \leq \omega \leq \infty$. The expression for $\underline{/Q_i(i\omega)}$ is not easily given since in Fourier-transforming Eq. (3.313) (each term may be treated separately since superposition is allowed) we find $Q_i(i\omega)$ as a *sum* of complex numbers rather than a product. To find the overall angle of a sum of complex numbers one must express *each* number as $a_i + ib_i$ and then add the a's and b's to get the overall $a + ib$. The angle is then $\tan^{-1}(b/a)$. The angle due to transforming $f_i(t)$ is $\underline{/F_i(i\omega)}$, the angle due to transforming $-f_i(t) \cos 2\omega_c t$ is $-180° + \underline{/F_i[i(\omega - 2\omega_c)]}$, the angle due to transforming $-f_i(t) \cos 4\omega_c t$ is $-180° + \underline{/F_i[i(\omega - 4\omega_c)]}$, etc. These results allow calculation of $\underline{/Q_i(i\omega)}$. The frequency spectrum of such a signal is shown in Fig. 3.90. Since it is a continuous spectrum, the response of a system to such an input may be found by the methods of Fig. 3.80. In Fig. 3.90

it is assumed that the $F_i(i\omega)$ is practically zero for $\omega > \omega_{s,\max}$ and that $\omega_c > \omega_{s,\max}$. For such a situation the phase-angle calculation is greatly simplified, since the individual terms of Eq. (3.314) do not coexist over any frequency range. That is, for $0 < \omega < \omega_{s,\max}$, only the first term is nonzero; for $(2\omega_c - \omega_{s,\max}) < \omega < (2\omega_c + \omega_{s,\max})$, only the second term is nonzero, etc. Thus the overall phase angle is determined by one term only within each of the specified frequency bands.

This concludes our treatment of amplitude-modulated and -demodulated signals. The methods developed can be readily applied to the other common variations such as half-wave demodulation, non-phase-sensitive demodulation, non-sinusoidal-carrier wave forms (square wave, for instance), etc. Non-phase-sensitive demodulation cannot detect a sign change in the modulating signal. Half-wave systems shift the side frequencies of demodulated signals to $\omega_c \pm \omega_s$ rather than $2\omega_c \pm \omega_s$, thus making the filtering problem more difficult. They also have less amplitude than full-wave systems. Nonsinusoidal carriers may be useful in reducing heating effects in resistive transducers such as strain gages. The carrier wave form is such as to give a high ratio of peak value to rms (effective heating) value. The output signal is related to peak value while the power dissipated in the strain gage is related to rms value; thus a high peak/rms ratio increases output for a given allowable heating level. A carrier in the form of a train of high, narrow pulses satisfies this sort of criterion.

Characteristics of random signals. The final class of signals we consider is the so-called random or stochastic type. Such signals are of increasing importance since they serve as more realistic mathematical models of many physical processes than do deterministic signals. By a random signal we shall mean one that can be described only statistically before it actually occurs; that is, it cannot be described by a specific function of time prior to its occurrence. Of course, *all* signals in the real world have some degree of randomness, so that we should be clear that we are *always* dealing with random signals even though we may take a specific time function (periodic, transient, etc.) as a *model* of what is really going on to simplify analysis in some types of problems.

Figure 3.91a shows a time record of a typical random signal such as might be measured by a vibration pickup mounted on a booster-rocket structure subjected to acoustic-pressure forces generated by the rocket exhaust. These pressures are strongly random in that no specific frequencies are apparent in a time record. The stresses caused by such pressures can lead to fatigue failure of the structure. Thus engineers are concerned with means to analyze the effects of such random forcing functions on structures and machinery. Since experimental methods are

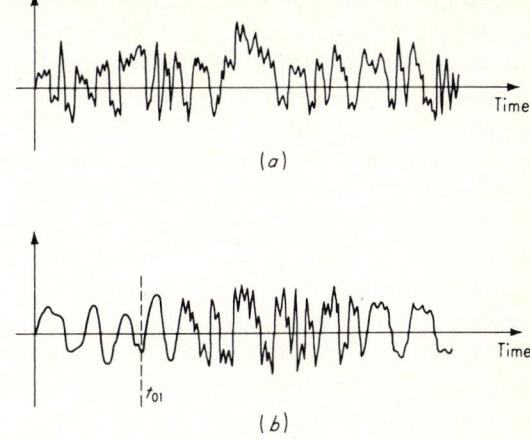

(a)

(b)

Fig. 3.91. *Random signal.*

needed in much of this work, accurate measurement of random signals is important. We shall see that frequency-response techniques are again most useful for this type of problem.

First, we should note that, as far as deciding on the requirements for a measuring system is concerned, *after* a random process has occurred and there is a time record of it, one can define the function as zero for some time $t > t_0$, treat it as a transient, and calculate its Fourier transform to determine its frequency content and thus the required frequency response of the instrumentation. The only problem here is the selection of the cutoff time t_0 so as to have an adequate statistical "sample" of the random function. In Fig. 3.91b, for example, if we based our calculation on the record for $0 < t < t_{01}$ we should miss completely the high-frequency character apparent in the record for $t > t_{01}$. The existence of some minimum valid cutoff time t_0 implies that the random process is *stationary;* that is, its statistical properties (such as average value, mean-square value, etc.) do not change with time. When this is true, there will exist some t_0 corresponding to a chosen level of confidence in the results.

We should be clear that, in dealing with random processes, theoretically an *infinite* record length is needed to give precise results, and results based on finite-length records must always be qualified by statistical statements referring to the *probability* of the result being correct within a certain percentage. This situation leads to an engineering trade-off since long records are desirable from accuracy considerations but are undesirable in terms of the cost involved in obtaining and analyzing them. Also, in many situations the maximum available length of record is limited by the lifetime of the device under study, which in the case of a missile or rocket may be quite short.   If an adequately long record is available, the minimum allowable t_0 can be found by choosing a small t_0 and calculating the Fourier transform.   Then a larger t_0 is chosen and the Fourier transform again evaluated.   If the second transform differs significantly from the first, the first t_0 was not long enough.   By choosing successively longer t_0's one will find some range of t_0 beyond which further increases in t_0 cause no significant change in the transform.   Any t_0 beyond this range would thus be considered acceptably long.   (In actual practice, more refined statistical procedures may be required, and are available,[1] to resolve questions of this sort.)

While the above concepts are adequate for understanding the requirements put on measurement systems for random variables, they do not cover the means available for statistically *describing* the signals.   That is, while the exact form of a random function cannot be predicted ahead of time, certain of its statistical characteristics *can* be predicted; these may be useful in predicting (in a statistical way) the output of some physical system that has the random variable as an input.   We now develop some of the more common methods of statistically describing random signals.   These methods fall mainly into two groups: those concerned with describing the magnitude of the variable and those concerned with describing the rapidity of change (frequency content) of the variable.

If we call a random variable $q_i'(t)$, the *average* or *mean value* $\overline{q_i'(t)}$ is defined by

$$\overline{q_i'(t)} \triangleq \lim_{T \to \infty} \left(\frac{1}{T}\right) \int_0^T q_i(t)\, dt \qquad (3.315)$$

Since this can be thought of as a constant component of the total signal, it is not random and is usually subtracted from the total signal to give a signal with zero mean value.   Also, many real random processes inherently have zero mean value.   For these reasons, from here on we consider only signals $q_i(t)$ with zero mean value.   An indication of the magnitude of the random variable is the *mean-squared value* $\overline{q_i^2(t)}$ given by

$$\overline{q_i^2(t)} \triangleq \lim_{T \to \infty} \left(\frac{1}{T}\right) \int_0^T q_i^2(t)\, dt \qquad (3.316)$$

The mean-squared value has dimensions of $[q_i(t)]^2$.   Thus to get a measure of the size of $q_i(t)$ itself, the *root-mean-square* (rms) value $q_i(t)_{rms}$ is defined by

$$q_i(t)_{rms} \triangleq \sqrt{\overline{q_i^2(t)}} \qquad (3.317)$$

[1] J. S. Bendat, L. D. Enochson, and A. G. Piersol, Analytical Study of Vibration Data Reduction Methods, *NASA*, *N*64-15529, 1963.

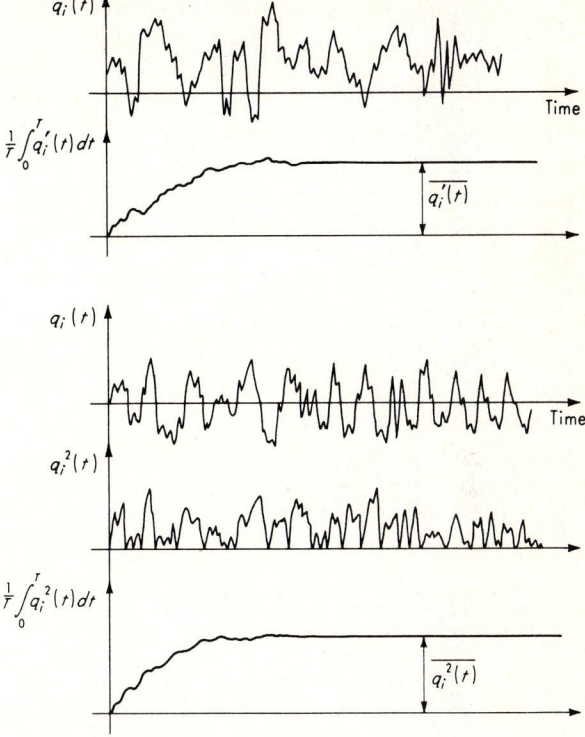

Fig. 3.92. *Average and mean-square values.*

The above definitions are illustrated in Fig. 3.92. The quantities $\overline{q_i^2(t)}$ and $q_i(t)_{\text{rms}}$ give an indication of the overall size of $q_i(t)$ but no clue as to the distribution of "amplitude," that is, the probability of occurrence of large or small values of $q_i(t)$. The specification of this important information is provided by the *amplitude-distribution function* (probability density function) $W_1(q_i)$. To define this function, we consider Fig. 3.93. We define the probability P that $q_i(t)$ will be found between some specific value q_{i1} and $q_{i1} + \Delta q_i$ by

$$\text{Probability } [q_{i1} < q_i < (q_{i1} + \Delta q_i)] \triangleq P[q_{i1}, (q_{i1} + \Delta q_i)] = \lim_{T \to \infty} \frac{\Sigma \, \Delta t_i}{T}$$

$$(3.318)$$

where $\Sigma \, \Delta t_i$ represents the total time spent by $q_i(t)$ within the band Δq_i during the time interval T. We now define $W_1(q_i)$ by

$$\text{Amplitude-distribution function} \triangleq W_1(q_i) \triangleq \lim_{\Delta q_i \to 0} \frac{P[q_i, (q_i + \Delta q_i)]}{\Delta q_i}$$

$$(3.319)$$

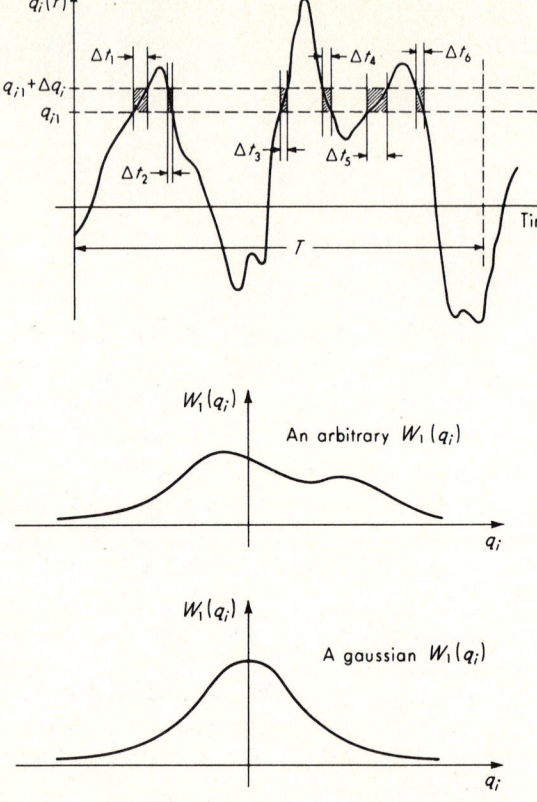

Fig. 3.93. *Probability-density-function definition.*

From this definition it should be clear that

$$W_1(q_i)\,dq_i = \text{probability that } q_i \text{ lies in } dq_i \qquad (3.320)$$

and thus

$$\int_{q_{i1}}^{q_{i2}} W_1(q_i)\,dq_i = \text{probability } q_i \text{ lies between } q_{i1} \text{ and } q_{i2} \qquad (3.321)$$

The function $W_1(q_i)$ can theoretically take on an infinite number of different forms; however, certain forms have been found to be adequate mathematical models for real physical processes. The most common of these is the gaussian or normal distribution given by

$$W_1(q_i) = \frac{1}{\sqrt{2\pi}\,\sigma}\,e^{-q_i{}^2/2\sigma^2} \qquad (3.322)$$

where $\sigma \triangleq$ standard deviation. Whether a given random process closely approximates this form must usually be found experimentally by means of instruments based on Eqs. (3.318) and (3.319). Since the limiting processes in these equations can never be exactly realized in a physical instrument, we again must be satisfied with statements regarding the *probability* that a process is gaussian rather than the *certainty* that it is.

While the quantities $q_i(t)_{rms}$ and $W_1(q_i)$ are usually sufficient to describe the magnitude of a random variable, they give no indication as to the *rapidity* of variation in time. That is, two random processes could both be gaussian with the same numerical value of σ but one could be much more rapidly varying than the other. To describe the time aspect of random variables the concepts of *autocorrelation function* and *mean-square spectral density* (power spectral density) are employed.[1] The autocorrelation function $R(\tau)$ of a random variable $q_i(t)$ is given by

$$R(\tau) \triangleq \lim_{T \to \infty} \left(\frac{1}{T}\right) \int_0^T q_i(t)q_i(t + \tau)\, dt \qquad (3.323)$$

The function $q_i(t + \tau)$ is simply $q_i(t)$ shifted in time by τ sec. Thus, to find $R(\tau)$ one selects a value of τ, say 2 sec, plots against t the functions $q_i(t)$ and $q_i(t + 2)$, multiplies them together point by point, integrates the product curve from 0 to T, and divides by T. This procedure is then repeated for other values of τ to generate the curve $R(\tau)$ versus τ. In actual practice, the shifting of $q_i(t)$ by τ sec is sometimes accomplished by writing $q_i(t)$ on magnetic tape at one point and reading it off at another. The time delay τ is then simply

$$\tau = \frac{\text{distance between read and write heads}}{\text{tape velocity}}$$

The multiplication of $q_i(t)$ and $q_i(t + \tau)$, the integration, and the division by T can all be accomplished by standard electronic-analog-computer components. [Equation (3.323) can also be implemented on a digital computer successfully.] Figure 3.94 illustrates these concepts. For $\tau = 0$, Eq. (3.323) gives $R(0) = \overline{q_i{}^2(t)}$; that is, the autocorrelation function is numerically equal to the mean-square value for $\tau = 0$.

To appreciate the relation between $R(\tau)$ and the rapidity of variation of $q_i(t)$, consider Fig. 3.95 where both a slowly varying and a rapidly varying $q_i(t)$ are shown. For *any* $q_i(t)$, fast or slow, when $\tau = 0$, $q_i(t)q_i(t + \tau)$ is *positive* for all t. Thus integration gives the largest possible value, $\overline{q_i{}^2(t)}$. Any shift ($\tau \neq 0$) will "misalign" the positive and negative parts of $q_i(t)$ and $q_i(t + \tau)$, causing the product curve to be sometimes positive, sometimes negative. Thus integration of this curve gives a smaller value than for $\tau = 0$. If $q_i(t)$ is rapidly varying, it takes

[1] S. H. Crandall, "Random Vibration," John Wiley & Sons, Inc., New York, 1958.

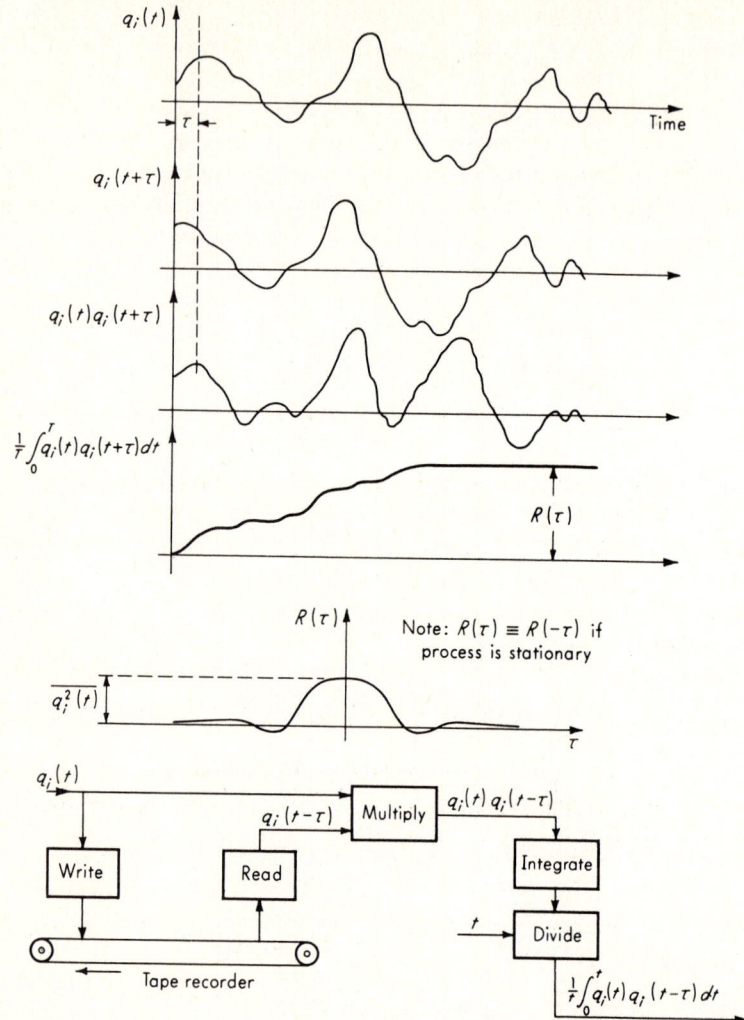

Fig. 3.94. *Autocorrelation-function definition.*

only a small shift (small τ) to cause this misalignment, whereas a slowly varying $q_i(t)$ requires a larger shift before $R(\tau)$ drops off significantly. Thus a sharp peak in $R(\tau)$ at $\tau = 0$ indicates the presence of rapid variation (strong high-frequency content) in $q_i(t)$.

The mean-square spectral density (power spectral density) is another method of determining the frequency content of a random signal. The mean-square spectral density is proportional to the Fourier transform of $R(\tau)$ and conveys in the frequency domain exactly the same information

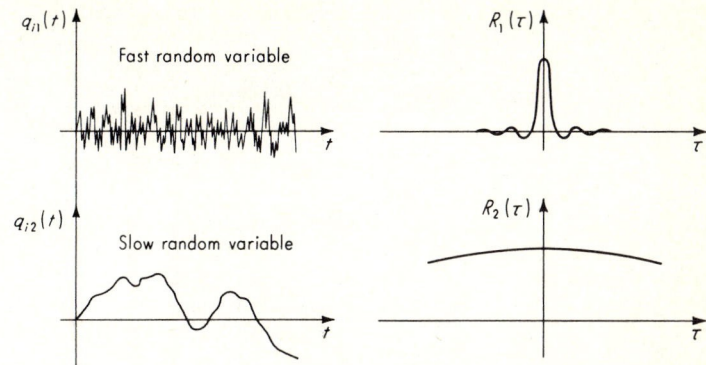

Fig. 3.95. *Frequency significance of autocorrelation function.*

that $R(\tau)$ conveys in the time domain. While they are mathematically related, in actual practice where they must be determined experimentally, one or the other may be preferable. It appears that in random-vibration work the mean-square-spectral-density approach is largely preferred. To develop this concept, let us consider first a periodic function expanded into a Fourier series to give

$$q_i(t) = A_{i1} \sin (\omega_1 t + \alpha_1) + A_{i2} \sin (2\omega_1 t + \alpha_2) + \cdots \qquad (3.324)$$

It is easy to show that the total mean-square value of $q_i(t)$ is equal to the sum of the individual mean-square values for each of the harmonic terms:

$$\overline{q_i^2(t)} = \overline{q_{i1}^2(t)} + \overline{q_{i2}^2(t)} + \cdots \qquad (3.325)$$

Thus the contribution of each frequency to the overall mean-square value is easily found.

We should like now to develop, for a *random* function, a related technique that will show how the total mean-square value is "distributed" over the frequency range. We first note that for a random function no isolated, discrete frequencies exist; thus the frequency spectrum is a continuous one. The concept of mean-square spectral density is perhaps most clearly visualized in terms of the instrumentation used to measure it experimentally. Figure 3.96*a* shows the arrangement necessary to measure the *overall* mean-square value of a signal $q_i(t)$. To find out how much each part of the frequency range contributes to this overall value we simply filter out (with a narrow-band-pass filter of bandwidth $\Delta\omega$) all (ideally) frequencies other than the narrow band of interest and then perform the squaring and averaging operations on what remains, as in

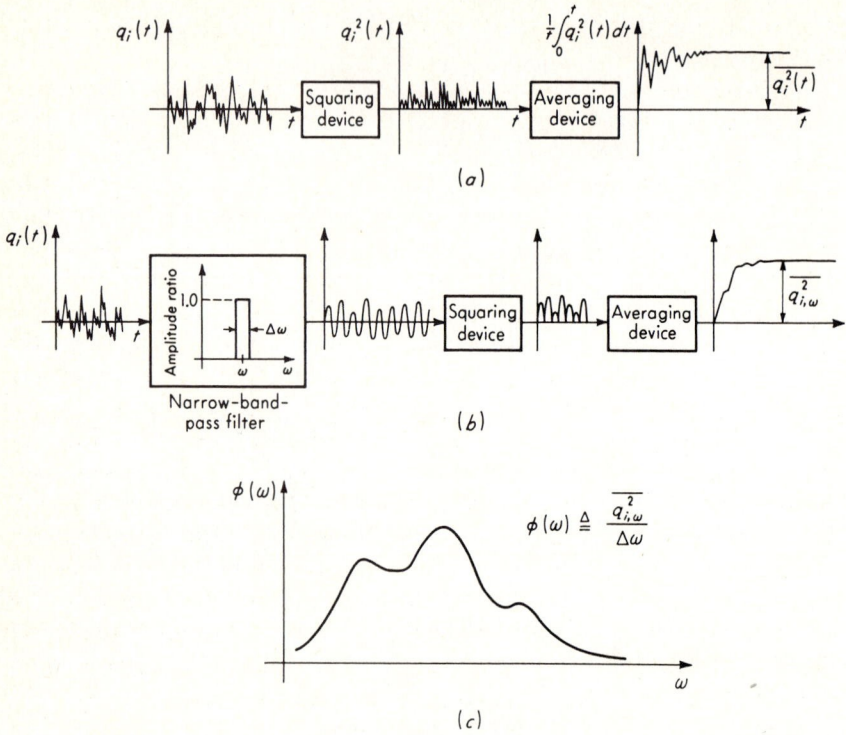

Fig. 3.96. *Mean-square spectral-density definition.*

Fig. 3.96*b*. The results of this operation are thus the mean-square value of that part of the signal $q_i(t)$ lying within the chosen frequency band. We call this value $\overline{q_{i,\omega}^2}$. The filter is adjustable in the sense that we can shift its passband anywhere along the frequency axis, and so we can obtain $\overline{q_{i,\omega}^2}$ for any chosen center frequency ω. A narrow passband is desirable for resolving closely spaced peaks in the spectrum but is undesirable in terms of the increased time required to cover a given frequency range in small steps rather than large. Thus a compromise is needed. The mean-square spectral density $\phi(\omega)$ is defined by

$$\phi(\omega) \triangleq \frac{\overline{q_{i,\omega}^2}}{\Delta\omega} \qquad (3.326)$$

and represents the "density" (amount per unit frequency bandwidth) of the mean-square value since $\phi(\omega)\,\Delta\omega = \overline{q_{i,\omega}^2}$. If we evaluate $\phi(\omega)$ for a whole range of frequencies we can plot it as a curve versus ω as in Fig. 3.96*c*. Note that the dimensions of $\phi(\omega)$ are those of $q_i^2(t)/(\text{rad/sec})$.

The total area under the $\phi(\omega)$-versus-ω curve will be the total mean-square value $\overline{q_i^2(t)}$.

The reader will find that in the literature the term power spectral density is used almost exclusively for the quantity we have called mean-square spectral density, except that power spectral density is π times $\phi(\omega)$. This is a carry-over from communications engineering where the concept was originally developed and where the signal $q_i(t)$ is a voltage applied to a 1-ohm resistor. Under these conditions $\phi(\omega)$ would have dimensions of power/(rad/sec). Actually, however, the concept of $\phi(\omega)$ is a *mathematical* one related to the mean-square value and *not* a physical one related to electrical engineering. In physical applications the dimensions of $\phi(\omega)$ would be $(°F)^2/(rad/sec)$ if $q_i(t)$ is temperature, $g^2/(rad/sec)$ if $q_i(t)$ is acceleration, etc., which are in general *not* power/(rad/sec). However, the term power spectral density seems to be firmly entrenched and probably will continue to prevail. The author here merely suggests that mean-square spectral density might be more appropriate terminology. Thus when dealing with, say, random pressures, one would refer to the "mean-square spectral density of pressure."

Perhaps the main interest in the mean-square spectral density lies in the fact that if a $q_i(t)$ with a known $\phi_i(\omega)$ is applied as input to a linear system of known frequency response, the mean-square spectral density $\phi_o(\omega)$ of the output is easily computed from the relation[1]

$$\phi_o(\omega) \;=\; \phi_i(\omega) \left| \frac{q_o}{q_i}(i\omega) \right|^2 \qquad (3.327)$$

We can thus compute the $\phi_o(\omega)$ curve point by point; see Fig. 3.97. The area under this curve will be the total mean-square value of $q_o(t)$. Furthermore, if the input is gaussian the output will also be gaussian. It is then possible to make statements such as

$$|q_o(t)| > \sqrt{\overline{q_o^2(t)}} \qquad 31.7\% \text{ of the time}$$

$$|q_o(t)| > 2\sqrt{\overline{q_o^2(t)}} \qquad 4.6\% \text{ of the time} \qquad (3.328)$$

$$|q_o(t)| > 3\sqrt{\overline{q_o^2(t)}} \qquad 0.3\% \text{ of the time}$$

Other useful results of this nature can be found in the literature.[2]

A particular form of $\phi(\omega)$ is of great usefulness in practice. This is the so-called *white noise*. For a mathematically perfect white noise $\phi(\omega)$ is equal to a constant for all frequencies; that is, it contains all frequencies in equal amounts (see Fig. 3.98). This makes it useful as a test signal, just as the impulse is useful as a transient test signal because of its

[1] Aseltine, *op. cit.*
[2] Bendat, Enochson, and Piersol, *op. cit.*

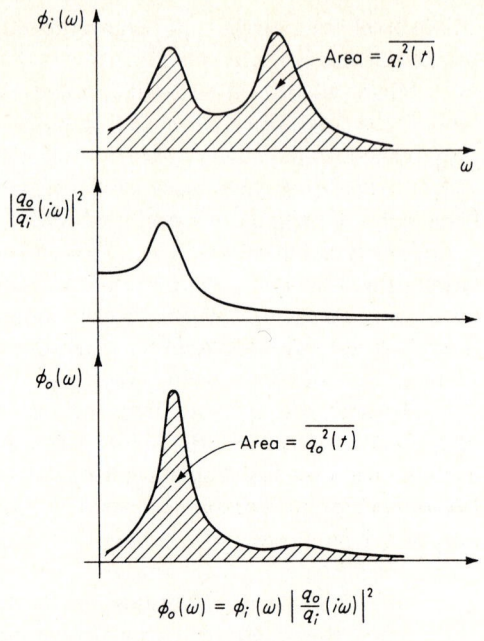

$$\phi_o(\omega) = \phi_i(\omega) \left| \frac{q_o}{q_i}(i\omega) \right|^2$$

Fig. 3.97. *System response to random input.*

uniform frequency content. From Eq. (3.327), if $\phi_i(\omega) = C$

$$\phi_o(\omega) = C \left| \frac{q_o}{q_i}(i\omega) \right|^2 \qquad (3.329)$$

and thus

$$\left| \frac{q_o}{q_i}(i\omega) \right| = \sqrt{\frac{\phi_o(\omega)}{C}} \qquad (3.330)$$

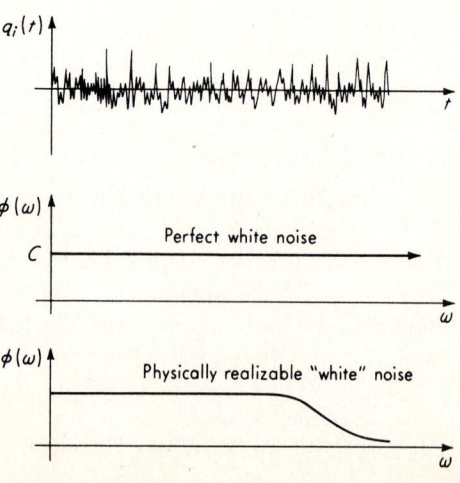

Fig. 3.98. *White noise.*

While it is not possible to build a generator of perfect white noise, the flat range of $\phi(\omega)$ of a practical generator can be quite large, and as long as its flat range extends beyond the frequency response of the system being tested, the "nonwhiteness" will not present any difficulty. If a white-noise generator is available, one can "construct" almost any $\phi(\omega)$ that he wishes simply by passing the white noise through a suitably designed filter according to Eq. (3.329). That is, $|(q_o/q_i)(i\omega)|^2$ for the filter must have the shape desired in $\phi(\omega)$. This technique is widely used in analog-computer simulation studies of aircraft gust response, control-system evaluation, etc., where the frequency characteristics of some random-input quantity are known and it is desired to study their effect on some system.

Most of the previous material has referred to statistical properties of a *single* random variable. Useful practical results may be derived by consideration of two random variables. The *joint amplitude-distribution function* (joint probability density function) of two random variables $q_1(t)$ and $q_2(t)$ is given by

$$W_1(q_1,q_2) = \lim_{\substack{T \to \infty}} \lim_{\substack{\Delta q_1 \to 0 \\ \Delta q_2 \to 0}} \frac{1}{T \, \Delta q_1 \, \Delta q_2} \Sigma \, \Delta t_i \qquad (3.331)$$

where $\Sigma \, \Delta t_i$ represents the total time (during the time T) that $q_1(t)$ and $q_2(t)$ spent *simultaneously* in the bands $q_1 + \Delta q_1$ and $q_2 + \Delta q_2$. Figure 3.99 illustrates these concepts. From the basic definition it should be clear that

$$\text{Probability } (q_{1a} < q_1 < q_{1b}, \, q_{2a} < q_2 < q_{2b}) = \int_{q_{2a}}^{q_{2b}} \int_{q_{1a}}^{q_{1b}} W_1(q_1,q_2) \, dq_1 \, dq_2$$
$$(3.332)$$

Again, $W_1(q_1,q_2)$ can take an infinite variety of forms. The most useful is probably the bivariate gaussian (normal) distribution.[1] The main purpose in experimentally measuring $W_1(q_1,q_2)$ is, just as for $W_1(q_i)$, to determine whether the physical data follow approximately some simple mathematical form such as the gaussian. If this can be proved, many useful theoretical results can be applied. Also, certain calculations can be made directly from $W_1(q_1,q_2)$. For example, if q_1 and q_2 represent the random vibratory motions of two adjacent machine parts, knowledge of $W_1(q_1,q_2)$ allows calculation of the probability that the two parts will strike each other. In actual practice, engineering applications of $W_1(q_1,q_2)$ are somewhat limited because of the difficulty of measuring this function.

[1] A. M. Mood, "Introduction to the Theory of Statistics," p. 165, McGraw-Hill Book Company, New York, 1950.

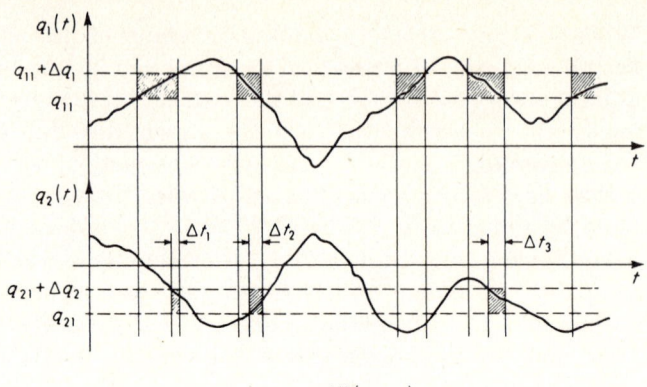

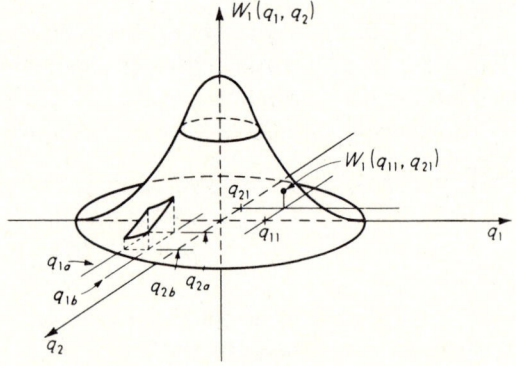

Fig. 3.99. *Bivariate probability density function.*

The *cross-correlation function* $R_{q_1 q_2}(\tau)$ for two random variables $q_1(t)$ and $q_2(t)$ is defined by

$$R_{q_1 q_2}(\tau) = \lim_{T \to \infty} \left(\frac{1}{T}\right) \int_0^T q_1(t) q_2(t + \tau)\, dt \qquad (3.333)$$

An example[1] of its application might be as follows: Suppose a source of vibratory motion $q_1(t)$ exists at one point in a structure and causes a vibratory response motion $q_2(t)$ at another point in the structure. Suppose also that the transmission of vibration from the first point to the second could occur by either (or both) of two mechanisms. The first mechanism is by acoustic (air-pressure) wave propagation through the air separating the two points; the second is by elastic wave propagation through the metallic structure connecting the two points. Since the propagation velocities of waves in air and metal are greatly different, one would expect that an input at q_1 would cause an output at q_2 that would be

---

[1] Bendat, Enochson, and Piersol, *op. cit.*

delayed by different times, depending on whether the transmission was mainly through the air or mainly through the metal. If one knows the transmission path length and the wave velocity, these delays can be calculated. Suppose the air-path delay is 0.01 sec and the structure-path delay is 0.002 sec. If one now experimentally measures $R_{q_1q_2}(\tau)$ and finds a large peak at $\tau = 0.01$ sec and a smaller one at $\tau = 0.002$ sec, he would conclude that the acoustic transmission is responsible for most of the vibration at q_2. Thus, since $R_{q_1q_2}$ is a measure of the correlation ("relatedness") between two signals delayed by various amounts, it can be used as a diagnostic tool for investigating the presence and/or nature of the relation, as in the above example.

Information equivalent to that contained in the cross-correlation function but in a different (and often more practically useful) form is found in the *cross-spectral density* (cross-power spectral density) $\phi_{q_1q_2}(\omega)$ given by

$$\phi_{q_1q_2}(\omega) = C_{q_1q_2}(\omega) - iQ_{q_1q_2}(\omega) \qquad (3.334)$$

where $C_{q_1q_2}(\omega) \triangleq$ cospectrum

$$\triangleq \lim_{T \to \infty} \lim_{\Delta\omega \to 0} \frac{1}{T\,\Delta\omega} \int_0^T (q_{1\Delta\omega})(q_{2\Delta\omega})\,dt \qquad (3.335)$$

$Q_{q_1q_2}(\omega) \triangleq$ quad spectrum

$$\triangleq \lim_{T \to \infty} \lim_{\Delta\omega \to 0} \frac{1}{T\,\Delta\omega} \int_0^T (q_{1\Delta\omega})_{90°}(q_{2\Delta\omega})\,dt \qquad (3.336)$$

$$q_{1\Delta\omega} \triangleq \text{ output of narrow- } (\Delta\omega) \text{ band}$$
$$\text{filter whose input is } q_1(t) \qquad (3.337)$$
$$q_{2\Delta\omega} \triangleq \text{ output of narrow- } (\Delta\omega) \text{ band}$$
$$\text{filter whose input is } q_2(t)$$
$$(q_{1\Delta\omega})_{90°} \triangleq \text{ signal } q_{1\Delta\omega} \text{ with phase shift of } 90° \qquad (3.338)$$

A block diagram of instrumentation necessary to measure $\phi_{q_1q_2}(\omega)$ is given in Fig. 3.100. Note that $\phi_{q_1q_2}(\omega)$ is a *complex* quantity whereas $\phi(\omega)$ is real. Perhaps the main application of the cross-spectral density is in the experimental determination of the sinusoidal transfer function $(q_o/q_i)(i\omega)$ of a linear system. From Eq. (3.327) we see that using the ordinary mean-square spectral density one can find only the magnitude (not the phase angle) of the transfer function if $\phi_i(\omega)$ is known and $\phi_o(\omega)$ is measured. The cross-spectral density determines both magnitude and phase according to the following equation:

$$\frac{q_o}{q_i}(i\omega) = \frac{\phi_{q_iq_o}(\omega)}{\phi_{q_i}(\omega)} \qquad (3.339)$$

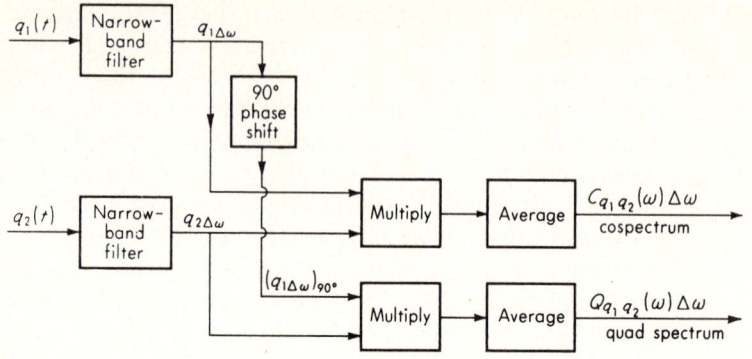

Fig. 3.100.   *Cross-spectral density.*

where $\phi_{q_i q_o}(\omega) \triangleq$ cross-spectral density of q_i and q_o

$\phi_{q_i}(\omega) \triangleq$ mean-square spectral density of q_i

We see that it is necessary to measure one (ordinary) mean-square spectral density and one cross-spectral density.

This concludes our treatment of random signals.   The most important commonly useful results have been presented and the main terminology developed.   Further theoretical and practical details may be found in the literature.   The earlier-referenced work of Bendat, Enochson, and Piersol was found particularly useful from a practical-application viewpoint by the author.

Requirements on the instrument transfer function to ensure accurate measurement.   Having up to this point expressed many different forms of signals in terms of the common denominator of frequency content, we can now give a general statement of the requirements that a measurement system must meet in order to measure accurately a given form of input.

1.   For perfect shape reproduction with no time delay between q_i and q_o and with:

   a.   Periodic inputs. $(q_o/q_i)(i\omega)$ must equal $K/\underline{0°}$ for all frequencies contained in q_i with significant amplitude.

   b.   Transient inputs. $(q_o/q_i)(i\omega)$ must equal $K/\underline{0°}$ for the entire frequency range in which the Fourier transform of $q_i(t)$ has significant magnitude.

   c.   Amplitude-modulated signals.   Same criteria as in parts a and b, depending on whether the modulating signal is periodic or transient.

d. Demodulated signals. $(q_o/q_i)(i\omega)$ for everything following the demodulator should be $K\underline{/0°}$ for all significant frequency bands of the modulating signal and should be zero for all carrier and side frequency bands produced by the modulation process.

e. Random signals. $(q_o/q_i)(i\omega)$ must equal $K\underline{/0°}$ over the entire frequency band where $\phi_{q_i}(\omega)$ is significantly larger than zero.

If the output signal $q_o(t)$ is to be used strictly for measurement rather than as a signal in a feedback-control system, a time delay between q_o and q_i is usually not objectionable as long as the shape of q_i is properly reproduced at q_o. (In control systems, time delay is damaging to stability and therefore is objectionable.) Thus a more easily attained (and usually acceptable) criterion is as follows:

2. For perfect shape reproduction with time delay τ_{dt} between q_i and q_o and with:

a. Periodic inputs.	Same as 1 except that wherever
b. Transient inputs.	$(q_o/q_i)(i\omega) = K\underline{/0°}$ is required now
c. Amplitude-modulated signals.	$(q_o/q_i)(i\omega) = K\underline{/-\omega\tau_{dt}}$ is required.
d. Demodulated signals.	*That is, amplitude ratio is constant*
e. Random signals.	*but phase lag increases linearly with*
	frequency ω.

The validity of the above statements is readily perceived by consideration of Fig. 3.101. To get the output in the frequency domain, the frequency-domain input is always multiplied by the sinusoidal transfer function. If this transfer function is $K\underline{/-\omega\tau_{dt}}$, the output will have a magnitude equal to K times the input magnitude and an angle equal to the angle of the input minus $\omega\tau_{dt}$. This is exactly what we would get if we passed $q_i(t)$ through a pure gain K followed by a dead time τ_{dt}. Thus when we inverse-transform to get from $Q_o(i\omega)$ to $q_o(t)$, we are *bound* to get $Kq_i(t)$ delayed by τ_{dt} sec. Thus, while the *actual* instrument transfer function will be of the form of Eq. (3.255), over the pertinent range of frequencies it must effectively amount to $K\underline{/-\omega\tau_{dt}}$ if accurate wave-form reproduction is to be expected. This is the basis for the selection of $\zeta \approx 0.6$ to 0.7 in second-order instruments since this range of ζ values makes the amplitude ratio most nearly constant and the phase-angle-versus-frequency curve most nearly linear over the widest possible frequency range for a given ω_n.

It should be noted that the above accuracy criteria do not actually state any *numerical* results. That is, no statement of the form "If the amplitude ratio is flat within $\pm x$ percent and the phase angle linear within $\pm y$ percent over the range of frequencies in which $|Q_i(i\omega)|$ (Fourier-series term coefficient or Fourier-transform magnitude) is greater than

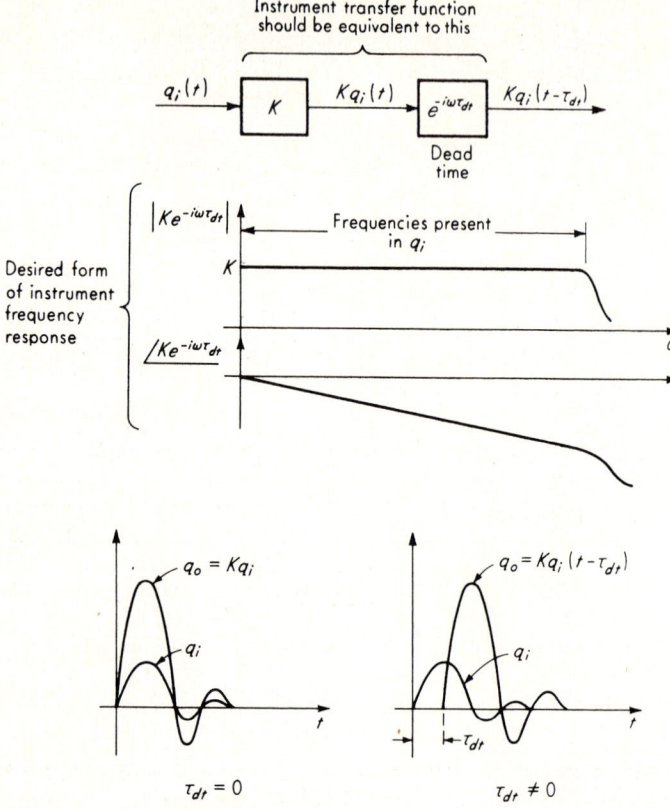

Fig. 3.101.   *Requirements for accurate measurement.*

z percent of its maximum value, then $q_o(t)$ will be within $\pm w$ percent of $Kq_i(t - \tau_{dt})$ at all times" is or *can be* made. While this sort of statement would be exceedingly useful, unfortunately it cannot be made in any *general* sense. For specific forms of $q_i(t)$ and $(q_o/q_i)(i\omega)$ one could investigate mathematically the effect of variations in numerical values of system parameters on dynamic accuracy by means of transform methods or directly from the differential equations. The use of analog and/or digital computers may be most helpful in such studies.

As an example, suppose one has a measuring system whose form is known but some (or all) of its numerical values are open to choice. Suppose it is intended to measure a random variable $q_i(t)$ whose $\phi_{q_i}(\omega)$ is known and may be simulated electrically with a white-noise generator and suitable shaping filters. The arrangement of Fig. 3.102a might be used to study the effect of varying measurement-system parameters on the

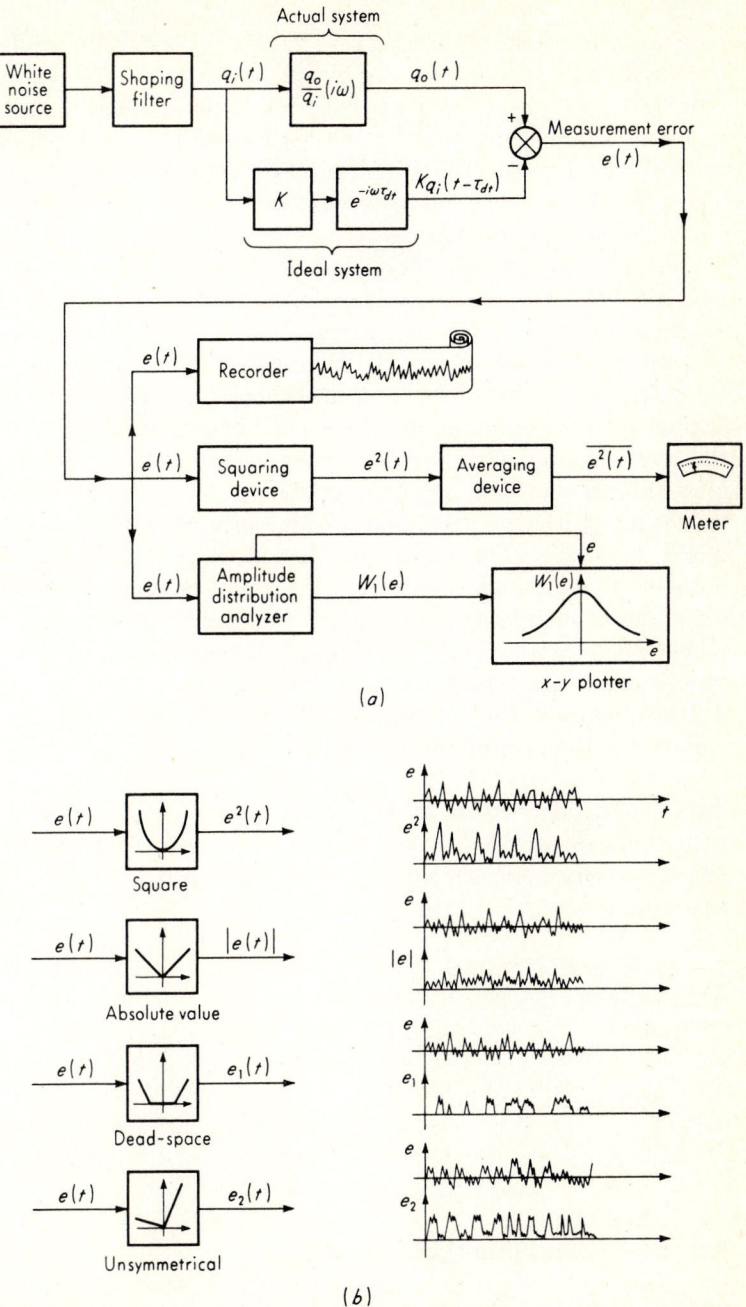

Fig. 3.102. *Experimental system-accuracy analysis.*

measurement error.   Several measures of error are made available to the investigator in this setup.   An overall measure is provided as a single number by the mean-squared error.   A visual display of the error as a function of time is provided by the chart recorder.   A statistical analysis of the error amplitude distribution is provided by a suitable computer. This gives $W_1(e)$ which allows prediction of the probability of errors of a particular magnitude.   Altogether a rather complete picture of the nature of the error is provided.   By changing the parameters of $(q_o/q_i)(i\omega)$ and noting the effect on the various aspects of the error, an optimum measuring system may be found for the particular $q_i(t)$.   Similar arrangements could be used for transient or periodic inputs.

The possibility of defining measures of error that are particularly slanted at some specific application should be kept in mind.   For example, the mean-squared error weights large errors out of proportion to small ones, because of the squaring operation $[2^2 \neq 2(1^2)]$.   If a uniform weighting is desired, the absolute value of error rather than the square might be used.   The absolute value of an electrical signal is easily obtained in an analog computer.   If errors less than a certain threshold value are of *no* consequence, a dead-space element can be used to give an output only when $|e(t)| > e_{\text{threshold}}$.   If negative errors are less important than positive errors, an unsymmetrical element can be used.   Figure 3.102*b* illustrates these concepts.   Because of the versatility of analog computers, the possibilities are endless.

Numerical correction of dynamic data.   Theoretically, if $q_o(t)$ (the actual measured data) is known and if $(q_o/q_i)(i\omega)$ for the measurement system is known one can always reconstruct a *perfect* record of $q_i(t)$ by the following process:

1.   Transform $q_o(t)$ to $Q_o(i\omega)$.
2.   Apply the formula

$$Q_o(i\omega) = Q_i(i\omega) \frac{q_o}{q_i}(i\omega)$$

in the inverse sense as

$$Q_i(i\omega) = \frac{Q_o(i\omega)}{(q_o/q_i)(i\omega)}$$

to find $Q_i(i\omega)$.
3.   Inverse-transform $Q_i(i\omega)$ to $q_i(t)$.

This procedure theoretically will give the exact $q_i(t)$ whether the measurement system meets the $K\underline{/-\omega\tau_{dt}}$ requirements or not.   In actual practice, of course, while the measurement system does not have to meet $K\underline{/-\omega\tau_{dt}}$, it *does* have to respond fairly strongly to all frequencies present in q_i; otherwise some parts of the q_o frequency spectrum will be so small as to be submerged in the unavoidable "noise" present in all systems and thus be unrecoverable by the above mathematical process. This process is tedious and introduces its own errors; therefore one would not use this approach unless absolutely necessary. However, it is a usable alternative in those situations where measurement systems meeting $K\underline{/-\omega\tau_{dt}}$ cannot be constructed with the present state of the art.

An important variation of the above process has been successfully applied in cases where the primary sensor is inadequate but can be cascaded with frequency-sensitive elements whose transfer functions make up the deficiencies in the primary sensor. The above computations are then in a sense automatically and continuously carried out by the compensating equipment to reconstruct $q_i(t)$. This subject is discussed in detail later under the topic Dynamic Compensation.

Experimental determination of measurement-system parameters. While theoretical analysis of instruments is vital to reveal the basic relationships involved in the operation of a device, it is rarely accurate enough to provide usable numerical values for critical parameters such as sensitivity, time constant, natural frequency, etc. Thus calibration of instrument systems is a necessity. We have already discussed static calibration; we here concentrate on dynamic characteristics.

For zero-order instruments the response is instantaneous and so no dynamic characteristics exist. The only parameter to be determined is the static sensitivity K, which is found by static calibration.

For first-order instruments the static sensitivity K is also found by static calibration. There is only one parameter pertinent to dynamic response, the time constant τ, and this may be found by a variety of methods. One common method applies a step input and measures τ as the time to achieve 63.2 percent of the final value. This method is influenced by inaccuracies in the determination of the $t = 0$ point and also gives no check as to whether the instrument is really first-order. A preferred method uses the data from a step-function test replotted semilogarithmically to get a better estimate of τ and also to check conformity to true first-order response. This method goes as follows: From Eq. (3.178) we can write

$$\frac{q_o - Kq_{is}}{Kq_{is}} = -e^{-t/\tau} \qquad (3.340)$$

$$1 - \frac{q_o}{Kq_{is}} = e^{-t/\tau} \qquad (3.341)$$

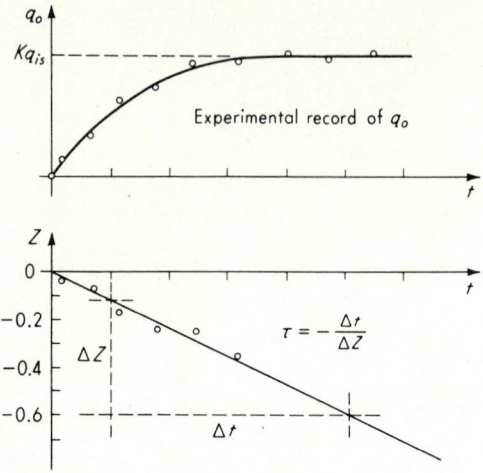

Fig. 3.103. *Step-function test of first-order system.*

Now we define

$$Z \triangleq \log_e \left(1 - \frac{q_o}{Kq_{is}}\right) \qquad (3.342)$$

and then

$$Z = \frac{-t}{\tau} \qquad \frac{dZ}{dt} = \frac{-1}{\tau} \qquad (3.343)$$

Thus if we plot Z versus t we get a straight line whose slope is numerically $-1/\tau$. Figure 3.103 illustrates the procedure. This gives a more accurate value of τ since the best line through *all* the data points is used rather than just two points, as in the 63.2 percent method. Furthermore, if the data points fall nearly on a straight line we are assured that the instrument is behaving as a first-order type. If the data deviate considerably from a straight line we know the instrument is not truly first-order and a τ value obtained by the 63.2 percent method would be quite misleading.

An even stronger verification (or refutation) of first-order dynamic characteristics is available from frequency-response testing, although at considerable cost of time and money if the system is not completely electrical, since nonelectrical sine-wave generators are neither common nor necessarily cheap. If the equipment is available, the system is subjected to sinusoidal inputs over a wide frequency range and the input and output recorded. Amplitude ratio and phase angle are plotted on the logarithmic scales. If the system is truly first-order the amplitude ratio follows the typical low- and high-frequency asymptotes (slope 0 and -20 db/decade) and the phase angle approaches $-90°$ asymptotically. If these characteristics are present, the numerical value of τ is found by finding ω at the

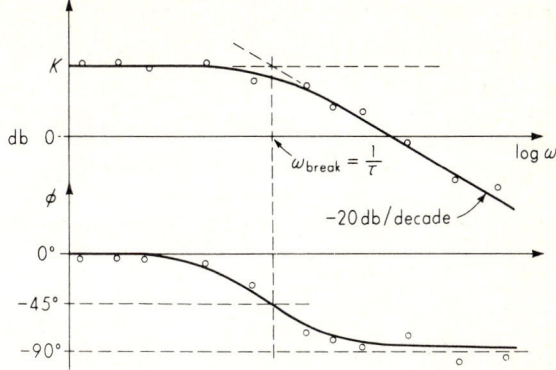

Fig. 3.104.   *Frequency-response test of first-order system.*

breakpoint and using $\tau = 1/\omega_{break}$ (see Fig. 3.104).   Deviations from the above amplitude and/or phase characteristics indicate non-first-order behavior.

For second-order systems, K is found from static calibration, and ζ and ω_n can be obtained in a number of ways from step or frequency-response tests.   Figure 3.105a shows a typical step-function response for an underdamped second-order system.   The values of ζ and ω_n may be found from the relations

$$\zeta = \sqrt{\frac{1}{\left(\dfrac{\pi}{\log_e (a/A)}\right)^2 + 1}} \qquad (3.344)$$

$$\omega_n = \frac{2\pi}{T\sqrt{1 - \zeta^2}} \qquad (3.345)$$

When a system is lightly damped, any fast transient input will produce a response similar to Fig. 3.105b.   Then ζ can be closely approximated by

$$\zeta \approx \frac{\log_e (x_1/x_n)}{2\pi n} \qquad (3.346)$$

This approximation assumes $\sqrt{1 - \zeta^2} \approx 1.0$, which is quite accurate when $\zeta < 0.1$, and ω_n can again be found from Eq. (3.345).   In applying Eq. (3.345), if several cycles of oscillation appear in the record it is more accurate to determine the period T as the average of as many distinct cycles as are available rather than from a single cycle.   If a system is strictly linear and second-order, the value of n in Eq. (3.346) is immaterial; the same value of ζ will be found for any number of cycles.   Thus if ζ is calculated for, say, $n = 1, 2, 4,$ and 6 and *different* numerical values of ζ are obtained, one knows the system is not following the postulated mathe-

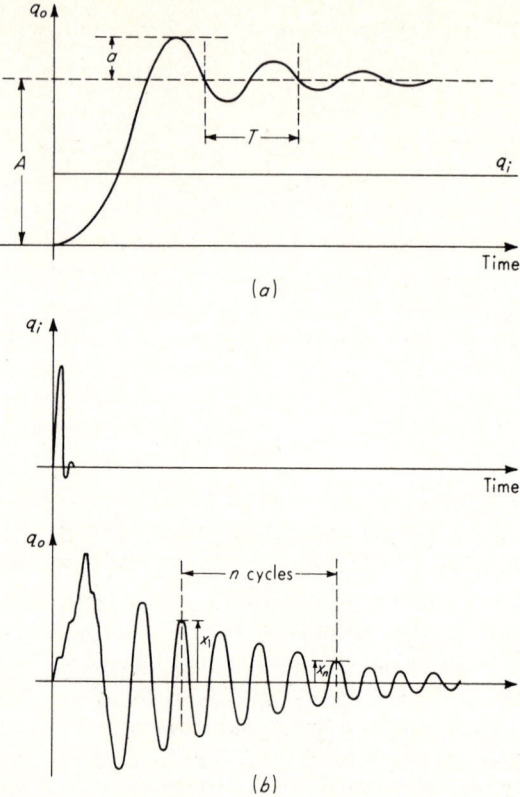

Fig. 3.105.   *Second-order-system step and pulse tests.*

matical model.   For overdamped systems ($\zeta > 1.0$) no oscillations exist, and the determination of ζ and ω_n becomes more difficult.   Usually it is easier to express the system response in terms of two time constants τ_1 and τ_2 rather than ζ and ω_n.   From Eq. (3.231) we can write

$$\frac{q_o}{Kq_{is}} = \frac{\tau_1}{\tau_2 - \tau_1}\, e^{-t/\tau_1} - \frac{\tau_2}{\tau_2 - \tau_1}\, e^{-t/\tau_2} + 1 \qquad (3.347)$$

where
$$\tau_1 \triangleq \frac{1}{(\zeta - \sqrt{\zeta^2 - 1})\omega_n} \qquad (3.348)$$

$$\tau_2 \triangleq \frac{1}{(\zeta + \sqrt{\zeta^2 - 1})\omega_n} \qquad (3.349)$$

To find τ_1 and τ_2 from a step-function-response curve we may proceed as follows[1]:

[1] N. A. Anderson, Step-analysis Method of Finding Time Constant, *Instr. Control Systems*, p. 130, November, 1963.

1. Define the "percent incomplete response" R_{pi} as

$$R_{pi} \triangleq [1 - q_o/Kq_{is}]100$$

2. Plot R_{pi} on a logarithmic scale versus time t on a linear scale. This curve will approach a straight line for large t if the system is second-order. Extend this line back to $t = 0$ and note the value P_1 where this line intersects the R_{pi} scale. Now, τ_1 is the time at which the straight-line asymptote has the value $0.368P_1$.

3. Now plot on the same graph a new curve which is the difference between the straight-line asymptote and R_{pi}. If this new curve is not a straight line, the system is not second-order. If it is a straight line, the time at which this line has the value $0.368(P_1 - 100)$ is numerically equal to τ_2.

Figure 3.106 illustrates this procedure. Once τ_1 and τ_2 are found, ζ and ω_n can be determined from Eqs. (3.348) and (3.349) if desired. Other methods[1] for finding τ_1 and τ_2 are available in the literature. Frequency-response methods may also be used to find ζ and ω_n or τ_1 and τ_2. Figure 3.107 shows the application of these techniques. The methods shown use the amplitude-ratio curve only. If phase-angle curves are available, they constitute a valuable check on conformance to the postulated model.

For measurement systems of arbitrary form (as contrasted to first- and second-order types), description of the dynamic behavior in terms of frequency response is usually desired. Ideally, if one has the amplitude-ratio and phase curves for a wide range of frequency, curve fitting (by cut-and-try or other methods) should give the sinusoidal transfer function in numerical form. The straight-line asymptotes of the decibel plot are particularly helpful in judging the location of time constants and natural frequencies when these are not closely spaced on the frequency axis. Actually, the frequency response in *graphical* form is all that is really needed since we have shown how one can get the response to any input from these graphs.

Since frequency-response testing is expensive and time-consuming, consideration should be given to impulse testing. A perfect impulse input is not needed; the only requirements are the following:

1. The approximate impulse must be of sufficiently short duration so that its Fourier-transform-magnitude curve is flat out to frequencies just beyond the point where the tested system's response practically cuts off.

[1] G. M. Hoerner, Second-order System Characteristics from Initial Step Response, *Control Eng.*, p. 93, December, 1962.

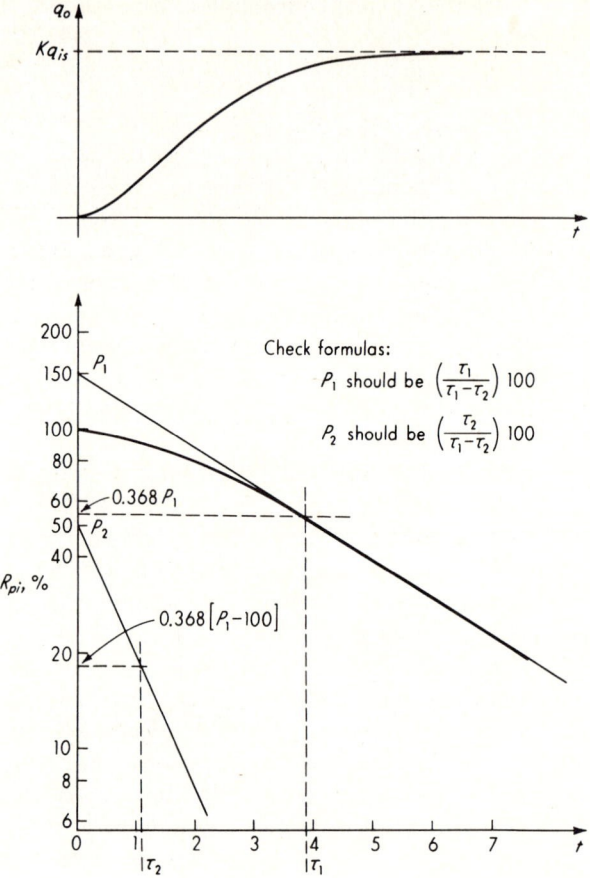

Fig. 3.106. *Step test for overdamped second-order systems.*

2. The approximate impulse must be strong enough to cause an output response that is large enough to be accurately measurable.

The first requirement implies some knowledge of the "unknown" system's characteristics prior to the test, but one can usually estimate the upper bounds of response easily from theory or experience.   The second requirement conflicts with the first since the strength (area) of the pulse is proportional to the duration; thus short durations (dictated by fast-response systems) require very large peak values in order to maintain adequate area. If a suitable impulse test can be set up, one records $q_i(t)$ and $q_o(t)$ and then computes the system frequency response as follows:

$$Q_o(i\omega) = Q_i(i\omega) \frac{q_o}{q_i}(i\omega) \qquad (3.350)$$

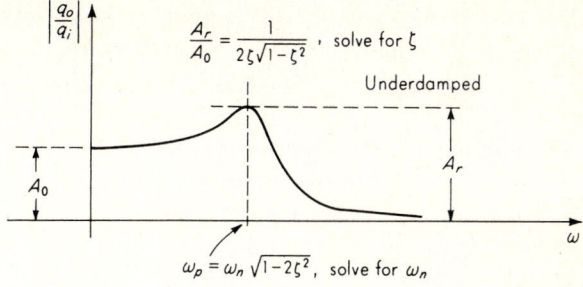

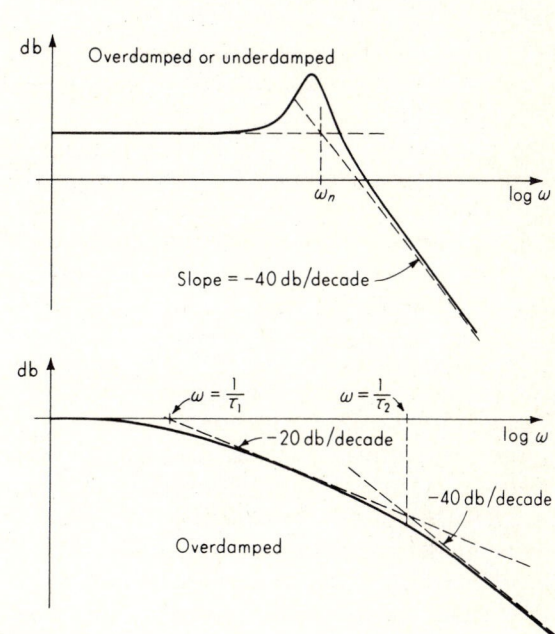

Fig. 3.107. *Frequency-response test of second-order system.*

But, for a "short enough" pulse,

$$Q_i(i\omega) = A_i\underline{/0^\circ} \qquad A_i = \text{area of } q_i(t) \qquad (3.351)$$

for all pertinent frequencies. Thus

$$Q_o(i\omega) = A_i \frac{q_o}{q_i}(i\omega) \qquad (3.352)$$

$$\text{and} \quad \frac{q_o}{q_i}(i\omega) = \frac{Q_o(i\omega)}{A_i} \qquad (3.353)$$

Since $q_o(t)$ is available, one can compute $Q_o(i\omega)$ in the usual way and thus get $(q_o/q_i)(i\omega)$. In fact, if one is satisfied that $q_i(t)$ is fast enough, and if one does not care what the static sensitivity of the system is (since that is easily obtained by static calibration), there is no need to measure $q_i(t)$ at all since A_i in Eq. (3.353) is merely a constant factor in $(q_o/q_i)(i\omega)$.

Loading effects under dynamic conditions. The treatment of loading effects by means of impedance, admittance, etc., was treated in Sec. 3.2 for static conditions. All these results can be immediately transferred to the case of dynamic operation by generalizing the definitions in terms of transfer functions. The basic equations relating the undisturbed value q_{i1u} and the actual measured value q_{i1m} at the input of a device are

$$q_{i1m} = \frac{1}{Z_{go}/Z_{gi} + 1}\, q_{i1u} \qquad (3.55)$$

$$q_{i1m} = \frac{1}{Y_{go}/Y_{gi} + 1}\, q_{i1u} \qquad (3.61)$$

$$q_{i1m} = \frac{1}{S_{go}/S_{gi} + 1}\, q_{i1u} \qquad (3.68)$$

$$q_{i1m} = \frac{1}{C_{go}/C_{gi} + 1}\, q_{i1u} \qquad (3.80)$$

The quantities Z, Y, S, and C were previously considered to be the ratios of small changes in two related system variables under stated conditions. To generalize these concepts, we now define the quantities Z, Y, S, and C as *transfer functions* relating the same two variables under the same conditions except now dynamic operation is to be considered. That is, we must get (theoretically or experimentally) $Z(D)$, $Y(D)$, $S(D)$, and $C(D)$ if we wish to use operational transfer functions and $Z(i\omega)$, $Y(i\omega)$, $S(i\omega)$, and $C(i\omega)$ if we wish to use frequency-response methods.

Usually the frequency-response form is most useful if these quantities must be found experimentally. This means, then, that in finding, say, $Z(i\omega)$, one of the two variables involved in the definition of Z plays the role of an "input" quantity which we vary sinusoidally at different frequencies. This causes a sinusoidal change in the other ("output") variable, and we can thus speak of an amplitude ratio and phase angle between these two quantities, making $Z(i\omega)$ now a complex number that varies with frequency. (If the system is somewhat nonlinear the effective approximate Z now becomes a function also of input amplitude. This situation was adequately described under static conditions in Sec. 3.2.) In Eq. (3.55), for example, both Z_{go} and Z_{gi} would now be complex numbers; if these were known, we could calculate the amplitude and phase of q_{i1m} if the amplitude, phase, and frequency of a sinusoidal q_{i1u} were given. The quantity q_{i1m} would then be the *actual* input (q_i) to the measuring device, and we could calculate q_o if the transfer function $(q_o/q_i)(i\omega)$ were known.   That is,

$$Q_o(i\omega) = \frac{1}{Z_{go}(i\omega)/Z_{gi}(i\omega) + 1} \left[\frac{q_o}{q_i}(i\omega) \right] Q_{i1u}(i\omega) \qquad (3.354)$$

One could thus define a *loaded transfer function* $(q_o/q_{i1u})(i\omega)$ as

$$\frac{q_o}{q_{i1u}}(i\omega) \triangleq \frac{1}{Z_{go}(i\omega)/Z_{gi}(i\omega) + 1} \frac{q_o}{q_i}(i\omega) \qquad (3.355)$$

where $q_o \triangleq$ actual output of measuring device which has no load at *its* output

$q_{i1u} \triangleq$ measured variable value that would exist if measuring device caused *no* loading on measured medium

Equations (3.61), (3.68), and (3.80) may be modified in similar fashion. Also, if differential equations relating $q_o(t)$ and $q_{i1u}(t)$ are desired, we may write

$$\frac{q_o}{q_{i1u}}(D) = \frac{1}{Z_{go}(D)/Z_{gi}(D) + 1} \frac{q_o}{q_i}(D) \qquad (3.356)$$

and then obtain the differential equation in the usual way by "cross-multiplying":

$$\begin{aligned} \{[Z_{go}(D) + Z_{gi}(D)](a_n D^n + a_{n-1}D^{n-1} + \cdots a_1 D \\ + a_0)\}q_o = \{[Z_{gi}(D)](b_m D^m + b_{m-1}D^{m-1} \\ + \cdots + b_1 D + b_0)\}q_{i1u} \qquad (3.357) \end{aligned}$$

An example of the above methods will be helpful.   Consider a device for measuring translational velocity as in Fig. 3.108a.   The unloaded transfer function relating the output displacement x_o and the input (measured) velocity v_i is obtained as follows:

$$B_i(\dot{x}_i - \dot{x}_o) - K_{is}x_o = M_i\ddot{x}_o \qquad (3.358)$$

$$\frac{x_o}{v_i}(D) = \frac{K_i}{D^2/\omega_{ni}^2 + 2\zeta_i D/\omega_{ni} + 1} \qquad (3.359)$$

where $K_i \triangleq$ instrument static sensitivity $\triangleq \dfrac{B_i}{K_{is}}$

$$\text{in./(in./sec)} \qquad (3.360)$$

$\zeta_i \triangleq$ instrument damping ratio $\triangleq \dfrac{B_i}{2\sqrt{K_{is}M_i}} \qquad (3.361)$

$\omega_{ni} \triangleq$ instrument undamped natural frequency

$$\triangleq \sqrt{\frac{K_{is}}{M_i}} \qquad \text{rad/sec} \qquad (3.362)$$

We see that the instrument is second-order and will thus measure v_i accurately for frequencies sufficiently low relative to ω_{ni}.   Suppose we now

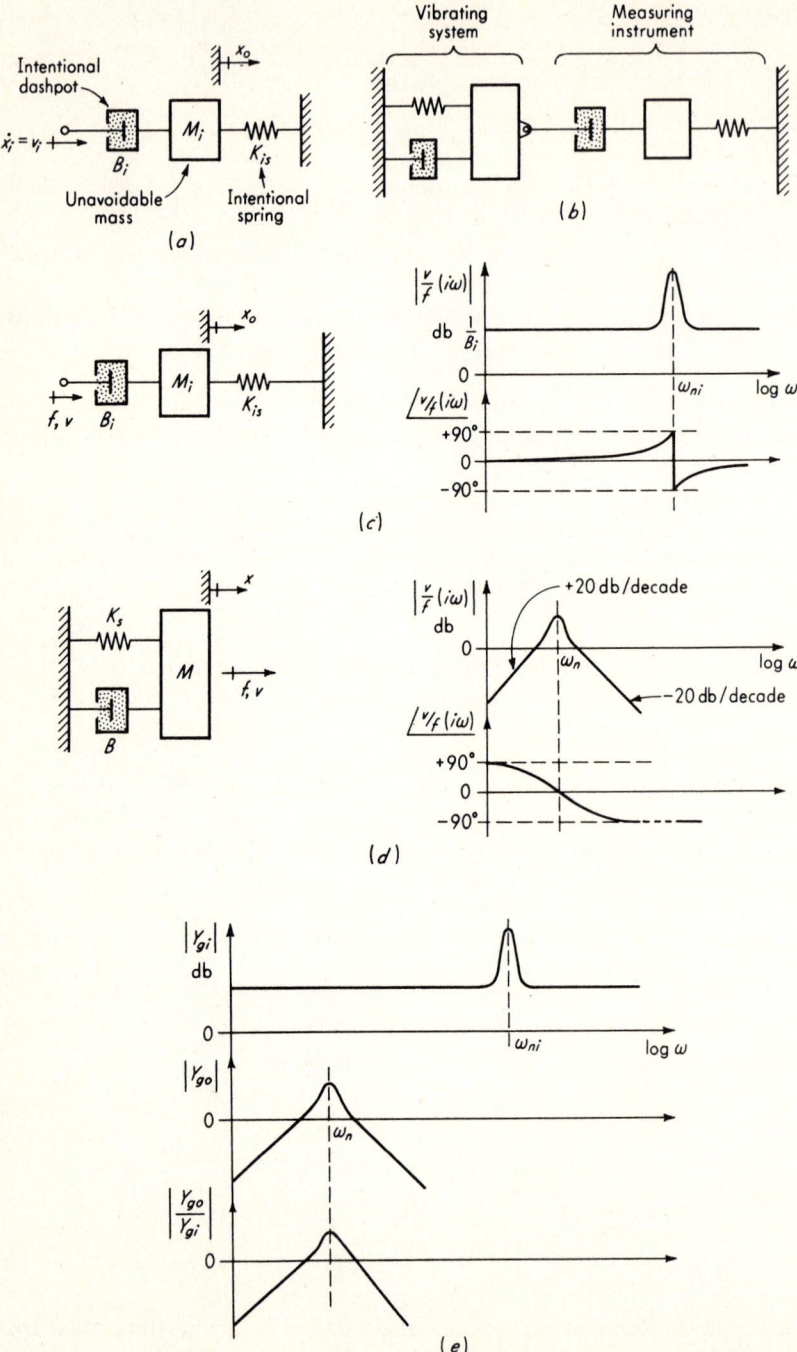

Fig. 3.108. *Example of dynamic-loading analysis.*

attach this instrument to a vibrating system whose velocity we wish to measure, as in Fig. 3.108b.   The presence of the measuring instrument will distort the velocity we are trying to measure.   The character of this distortion may be assessed by application of Eq. (3.61), since the measured quantity is velocity (a flow variable; see Fig. 3.28) and thus admittance is the appropriate quantity to use.   We determine the input admittance $Y_{gi}(D) = (v/f)(D)$ from Fig. 3.108c as follows:

$$f - K_{is}x_o = M_i\ddot{x}_o \qquad (3.363)$$

Also
$$f = B_i(v - \dot{x}_o) \qquad (3.364)$$

and, eliminating x_o, we get

$$Y_{gi}(D) = \frac{v}{f}(D) = \frac{(1/B_i)(D^2/\omega_{ni}^2 + 2\zeta_i D/\omega_{ni} + 1)}{D^2/\omega_{ni}^2 + 1} \qquad (3.365)$$

Figure 3.108c also shows the frequency characteristics of this input admittance.   The output admittance $Y_{go}(D) = (v/f)(D)$ of the measured system is obtained from Fig. 3.108d:

$$f - B\dot{x} - K_s x = M\ddot{x} \qquad (3.366)$$

$$Y_{go}(D) = \frac{v}{f}(D) = \frac{(1/K_s)D}{D^2/\omega_n^2 + 2\zeta D/\omega_n + 1} \qquad (3.367)$$

The frequency characteristic of this output admittance is shown in Fig. 3.108d.   We may now write

$$\frac{x_o}{v_{i1u}}(D) = \frac{1}{Y_{go}(D)/Y_{gi}(D) + 1} \frac{x_o}{v_i}(D) \qquad (3.368)$$

$$\frac{x_o}{v_{i1u}}(D) = \frac{1}{\underbrace{\dfrac{(1/K_s)D}{D^2/\omega_n^2 + 2\zeta D/\omega_n + 1} \dfrac{(D^2/\omega_{ni}^2) + 1}{(1/B_i)(D^2/\omega_{ni}^2 + 2\zeta_i D/\omega_{ni} + 1)} + 1}_{\text{loading effect}}} \frac{K_i}{D^2/\omega_{ni}^2 + 2\zeta_i D/\omega_{ni} + 1} \qquad (3.369)$$

where $x_o \triangleq$ actual output of measuring device
$\quad\quad\quad v_{i1u} \triangleq$ velocity that would exist if measuring device caused no loading

Figure 3.108e shows that in this example the loading effect is most serious for frequencies near the natural frequency of the measured system but approaches zero for both very low and very high frequencies. Since the loading effects can be expressed in frequency terms, they can be handled for all kinds of inputs by using appropriate Fourier series, transform, or mean-square spectral density.

Problems

3.1 For the system of Fig. 2.3:

a. Explain how you would carry out a static calibration to determine the relation between the desired input and the output.

b. The temperature of the air surrounding the capillary tube is an interfering input. Explain how you would calibrate the relation between this input and the output.

c. The elevation difference between the Bourdon tube and the bulb is another interfering input. Discuss means for its calibration.

3.2 Does the system of Fig. 2.4 require calibration? Explain.

3.3 What fundamental difficulties arise in trying to define the true temperature of a physical body?

3.4 Slide a coin along a smooth surface, trying to make it come to rest at a drawn line. Measure the distance of the coin from the line. Repeat 100 times and check the resulting data for conformance to a gaussian distribution, using probability graph paper.

3.5 Using the data generated in Prob. 3.4, apply the chi-squared test for conformance to a gaussian distribution.

3.6 In Eq. (2.6), solve for the strain ϵ in terms of the other parameters; $\epsilon = f(GF, R_g, E_b, R_a, e_o)$. Then take the natural log of both sides; $\ln \epsilon = \ln f$. Now take the differential of both sides so that terms such as $d\epsilon/\epsilon$, de_o/e_o, dR_a/R_a, etc., are formed. This will give the percentage error $d\epsilon/\epsilon$ in ϵ as a function of the percentage errors in the other parameters. If GF, R_g, E_b, R_a, and e_o are all measured to ± 1 percent error, what is the possible error in the computed value of ϵ?

3.7 Is the logarithmic differentiation method of Prob. 3.6 applicable to all forms of functional relations? Explain. Hint: Apply it to the relation $w = \sin x + 5y^3 - 6e^z$.

3.8 The discharge coefficient C_q of an orifice can be found by collecting the water that flows through during a timed interval when it is under a constant head h. The formula is

$$C_q = \frac{W}{t\rho A \sqrt{2gh}}$$

Find C_q and its possible error if:

$W = 865 \pm 0.5\,\text{lb}_m$ $A = \pi d^2/4$ $d = 0.500 \pm 0.001$ in.
$t = 600.0 \pm 2\,\text{sec}$ $g = 32.17 \pm 0.1\%\,\text{ft/sec}^2$
$\rho = 62.36 \pm 0.1\%\,\text{lb}_m/\text{ft}^3$ $h = 12.02 \pm 0.01\,\text{ft}$
considering both the following:

a. The errors are the absolute limits.

b. The errors are $\pm 3s$ limits.

3.9 In Prob. 3.8 if C_q must be measured within ± 0.5 percent for the numerical mean values given, what errors are allowable in the measured data? Use the method of equal effects.

3.10 Static calibration of an instrument gives the data of Fig. P3.1. Calculate the following:

    a. The best-fit straight line

    b. s_m and s_b

    c. s_{q_i}

    d. q_i and its error limits if the instrument is used after calibration and reads $q_o = 5.72$

q_i	q_o Increasing values	q_o Decreasing values
0	−0.07	+0.01
5	1.08	1.16
10	2.05	2.10
15	3.27	3.29
20	4.28	4.36
25	5.41	5.45
30	6.43	6.53
35	7.57	7.61
40	8.66	8.75

Fig. P3.1

3.11 In Fig. 3.21, what percent error may be expected in measuring the voltage across R_5 if $R_1 = R_2 = R_3 = R_4 = R_5 = 100$ ohms and $R_m = 1,000$ ohms? If $R_m = 10,000$ ohms?

3.12 Repeat Prob. 3.11 except now the voltage across R_3 is to be measured.

3.13 In Fig. 3.25, what percent error may be expected in measuring the current through R_5 if $R_1 = R_2 = R_3 = R_4 = R_5 = 100$ ohms and $R_m = 10$ ohms? If $R_m = 1$ ohm?

3.14 Repeat Prob. 3.13 except now the current through R_3 is to be measured.

3.15 In Fig. 3.26, what percent error may be expected in measuring the force in k_2 if $k_1 = k_2 = k_3 = k_4 = 100$ lb_f/in. and $k_m = 1,000$ lb_f/in.? If $k_m = 10,000$ lb_f/in.?

3.16 Repeat Prob. 3.15 except now the force in k_3 is to be measured.

3.17 In Fig. 3.27, what percent error may be expected in measuring the deflection x if $k_1 = k_2 = k_3 = k_4 = 1$ lb_f/in. and $k_m = 0.1$ lb_f/in.? If $k_m = 0.01$ lb_f/in.?

3.18 Repeat Prob. 3.17 except now the motion of the right-hand block is to be measured.

3.19 Using methods similar to those used in proving Eq. (3.55), prove the following:

    a. Eq. (3.61)

    b. Eq. (3.68)

    c. Eq. (3.80)

3.20 A mercury thermometer has a capillary tube of 0.010-in. diameter. If the bulb is made of a zero-expansion material, what volume must it have if a sensitivity of 0.10 in./F° is desired? Assume operation near 70°F. If the bulb is spherical and is immersed in stationary air, estimate the time constant.

3.21 A balloon carrying a first-order thermometer with a 15-sec time constant rises through the atmosphere at 20 fps. Assume temperature varies with altitude at 0.3 F°/100 ft. The balloon radios temperature and altitude readings back to the ground. At 10,000 ft the balloon says the temperature is 30°F. What is the true altitude at which 30°F occurs?

3.22 A first-order instrument must measure signals with frequency content up to 100 cps with an amplitude inaccuracy of 5 percent. What is the maximum allowable time constant? What will be the phase shift at 50 and 100 cps?

3.23 For the spring scale of Fig. 3.48, discuss the tradeoff between sensitivity and speed of response resulting from changes in K_s.

3.24 Derive Eqs. (3.237) to (3.239).

3.25 Find the transfer function of a spring scale (Fig. 3.48) whose mass is negligible. Show that the steady-state time lag for a ramp input is the same whether mass is zero or not.

3.26 Plot decibel and phase-angle curves for the following systems:

    a. $\dfrac{q_o}{q_i}(D) = \dfrac{10D}{10D + 1}$

    b. $\dfrac{q_o}{q_i}(D) = \dfrac{D}{(D + 1)(5D + 1)}$

    c. $\dfrac{q_o}{q_i}(D) = \dfrac{10e^{-2D}}{0.01D^2 + 0.1D + 1}$

    d. $\dfrac{q_o}{q_i}(D) = \dfrac{100D^2}{(D + 1)^2(10D + 1)^2}$

    e. $\dfrac{q_o}{q_i}(D) = \dfrac{5(0.25D^2 + 1)}{0.01D^2 + 0.1D + 1}$

3.27 Find $Q_i(i\omega)$ for the $q_i(t)$ of Fig. P3.2 by the following methods:

    a. The exact analytical method

    b. The method of 12 ordinates

    c. The method of 24 ordinates

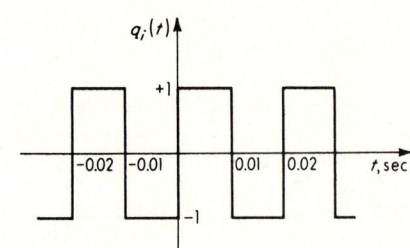

Fig. P3.2

3.28 Repeat Prob. 3.27 for Fig. P3.3.

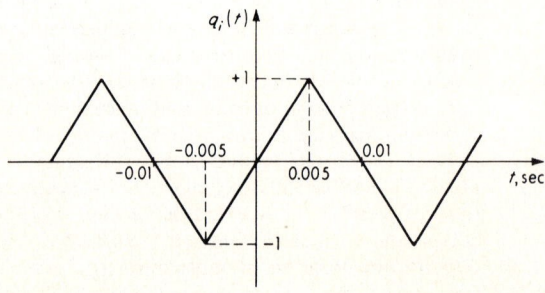

Fig. P3.3

3.29 If the $q_i(t)$ of Prob. 3.27 is the input to a first-order system with a gain of 1 and a time constant of 0.001 sec, find $Q_o(i\omega)$ and $q_o(t)$ for the steady state, using methods a, b, and c of Prob. 3.27.

3.30 Repeat Prob. 3.29 except use $q_i(t)$ from Prob. 3.28.

3.31 In Fig. 3.82 let the carrier be a square wave as in Fig. P3.4. Find the frequency spectrum of the output signal of the modulator.

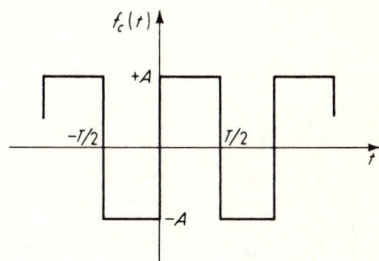

Fig. P3.4

3.32 Repeat Prob. 3.31 if the carrier is a square wave as in Fig. P3.5.

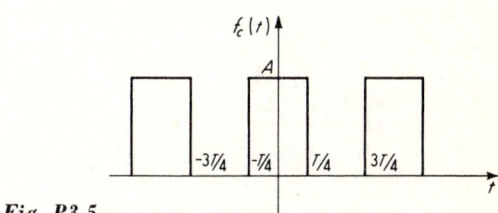

Fig. P3.5

3.33 In an analog-computer study it is desired to simulate a random atmospheric turbulence whose mean-square spectral density $\phi_t(\omega)$ is adequately represented as $10/(1 + 0.0001\omega^2)$, where ω is in radians per second. A white-noise generator having $\phi_{wn}(\omega) = 10$ is available. Select a suitable filter configuration and numerical values to follow the generator and produce the desired $\phi_t(\omega)$. The output of the noise generator should "see" a filter input resistance of 10,000 ohms.

3.34 Tests on a gyroscope show that it can withstand any random vibration along a given axis if the frequency content is between 0 and 1,000 rad/sec and the rms acceleration is less than 80 in./sec^2. This gyro is to be mounted in a rocket where it will be subjected to acoustic-pressure-induced vibration. The transfer function between pressure and acceleration and the mean-square spectral density of pressure are as given in Fig. P3.6. Will this gyro withstand the vibration?

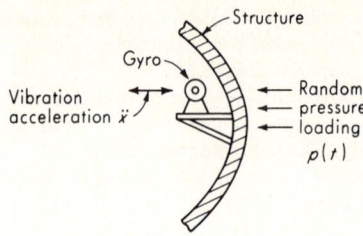

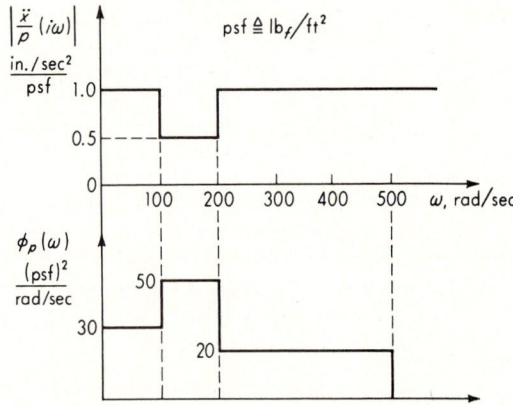

Fig. P3.6

3.35   Derive Eq. (3.344).

3.36   Derive Eq. (3.346).

3.37   Explain how the sinusoidal transfer function of a system may be obtained from measured records of $q_i(t)$ and $q_o(t)$ if q_i is a transient of any shape whatever.

3.38   Reanalyze the force-measuring problem of Fig. 3.26 for *dynamic* operation, assuming the two blocks have masses M_1 and M_2.   That is, get the operational transfer function analogous to Eq. (3.77).

3.39   Reanalyze the displacement-measuring problem of Fig. 3.27 for *dynamic* operation, assuming the two blocks have masses M_1 and M_2.   That is, get the operational transfer function analogous to Eq. (3.93).

3.40   Reanalyze the voltage-measuring problem of Fig. 3.21 for dynamic operation; i.e., replace the batteries with sources of time-varying voltage.   Also, let the voltage-measuring device be an oscilloscope with R_m shunted by a capacitor C_m.

3.41   Reanalyze the current-measuring problem of Fig. 3.25 for dynamic operation; i.e., replace the batteries with sources of time-varying voltage.   Also, let the current-measuring device be a galvanometer which has an inductance L_m in series with R_m.

Bibliography

1. H. E. Koenig and W. A. Blackwell: "Electromechanical System Theory," McGraw-Hill Book Company, New York, 1961.

2. C. L. Cuccia: "Harmonics, Sidebands, and Transients in Communication Engineering," McGraw-Hill Book Company, New York, 1952.
3. T. N. Whitehead: "The Design and Use of Instruments and Accurate Mechanisms," Dover Publications, Inc., New York, 1954.
4. V. L. Lebedev: Random Processes in Electrical and Mechanical Systems, *NASA, Tech. Transl., F*-61, 1961.
5. C. C. Perry: The Least Squares Method, *Machine Design*, p. 210, May 12, 1960.
6. V. R. Boulton: Economics of Instrumentation Precision, *Aerospace Eng.*, p. 30, March, 1961.
7. C. T. Morrow: Averaging Time and Data-reduction Time for Random Vibration Spectra, *J. Acoust. Soc. Am.*, vol. 30, no. 6, p. 572, June, 1958.
8. N. R. Goodman et al.: Frequency Response from Stationary Noise: Two Case Histories, *Technometrics*, p. 245, May, 1961.
9. R. L. Hammon: An Application of Random Process Theory to Gyro Drift Analysis, *IRE Trans. PGANE*, vol. ANE-7, no. 3, September, 1960.
10. D. E. Cartwright et al.: Digital Techniques for the Study of Sea Waves, Ship Motion and Allied Processes, *Trans. Soc. Instr. Tech. (London)*, p. 1, March, 1962.
11. J. T. Broch: Automatic Recording of Amplitude Density Curves, *B & K Tech. Rev.*, B & K Instruments, Cleveland, Ohio, no. 4, 1959.
12. J. T. Broch: Recording of Narrow Band Noise, *B & K Tech. Rev.*, B & K Instruments, Cleveland, Ohio, no. 4, 1960.
13. K. R. Thorson and Q. R. Bohne: Application of Power Spectral Methods in Airplane and Missile Design, *J. Aero/Scope Sci.*, p. 107, February, 1960.
14. J. C. Laurence: Intensity, Scale and Spectra of Turbulence in Mixing Region of Free Subsonic Jet, *NACA, Tech. Notes* 3561, 1955.
15. J. R. Rice et al.: On the Prediction of Some Random Loading Characteristics Relevant to Fatigue, *NASA, CR*-56152, 1964.
16. W. A. Wildhack et al.: Accuracy in Measurements and Calibrations, *NBS, Tech. Notes* 262, 1965.
17. H. W. Maynard: An Evaluation of Ten Fast Fourier Transform (FFT) Programs, US Army Elect. Command, Ft. Monmouth, N.J., 1973.
18. H. McNeill: Digital Data Reduction Methods for Aircraft Engine Noise Analysis, *Sound and Vibration*, April, 1972.
19. S. G. Cline: New Capabilities and Digital Low-Frequency Spectrum Analysis, *Hewlett-Packard Jour.*, June, 1972.
20. Special Issue on the Fast Fourier Transform. *IEEE Trans. on Audio and Electroacoustics*, vol. AV-17, June, 1969.
21. Real Time Signal Processing in the Frequency Domain, Monograph No. 3, Federal Scientific Corp., N.Y., 1972.
22. K. N. Fieldhouse: Techniques for Identifying Sources of Noise and Vibration, *Sound & Vib.*, December, 1970.
23. J. S. Bendat and A. G. Piersol: "Measurement and Analysis of Random Data," J. Wiley & Sons, New York, 1966.
24. G. A. Korn: "Random Process Simulation and Measurement," McGraw-Hill, New York, 1966.

Part Two
Measuring Devices

Chapter 4
Motion
Measurement

4.1

Introduction We commence our study of specific measuring devices with motion measurement since it is based on two of the fundamental quantities in nature (length and time) and also because so many other quantities such as force, pressure, temperature, etc., are often measured by transducing them to motion and then measuring this resulting motion. As indicated in the chapter title, our main interest is in motion (a *changing* displacement). Thus we shall not go extensively into dimensional measurement or gaging of fixed lengths, angles, hole diameters, etc., except as this relates to standards or calibration of motion-measuring devices.

We are also mainly (though not exclusively) concerned with electro-mechanical transducers which convert motion quantities into electrical quantities. The intent is not to present a catalog listing of the myriad physical effects which have been, or might be, used as the basis of a motion transducer but rather to provide sufficient detail for practical application of the relatively small number of transducer types which form the basis of the majority of practical measurements. The above-mentioned catalog-listing type of information is extremely useful to one who has a measurement problem not solvable by one of the standard techniques and who must therefore invent and/or develop a new instrument. Material of this type is available in several references.[1,2]

4.2

Fundamental Standards   The four fundamental quantities of the International Measuring System, for which independent standards have been defined, are length, time, mass, and temperature. Units and standards for all other quantities are *derived* from these. In motion measurement the fundamental quantities are length and time. Prior to 1960 the standard of length was the carefully preserved platinum-iridium International Meter Bar at Sèvres, France. In 1960 the meter was redefined in terms of the wavelength of a krypton-86 lamp as "the length equal to 1,650,763.73 wavelengths in vacuum corresponding to the transition between the energy levels $2p_{10}$ and $5d_5$ of the atom krypton 86."[3] This standard is believed[4] to be reproducible to about 2 parts in 10^8 and can be applied at this precision level to measurements of length in the range of about 10^{-8} to 40 in.[5]

The above National Prototype Standard is not available for routine calibration work. Rather, to protect such top-level standards from deterioration, the National Bureau of Standards has set up National Reference Standards and, below these, Working Standards. Further down the line in accuracy are the so-called Interlaboratory Standards, which are standards sent in to the National Bureau of Standards for

---

[1] K. S. Lion, "Instrumentation in Scientific Research," McGraw-Hill Book Company, New York, 1959.

[2] C. F. Hix, Jr., and R. P. Alley, "Physical Laws and Effects," John Wiley & Sons, Inc., New York, 1958.

[3] A. G. McNish, Fundamentals of Measurement, *Electro-Technol.* (*New York*), p. 113, May, 1963.

[4] W. A. Wildhack, NBS—Source of American Standards, *ISA J.*, p. 45, February, 1961.

[5] L. B. Wilson and H. W. Martin, The Measurements Gap, *Space/Aeron.*, p. 84, March, 1964.

calibration and certification by factories and laboratories all over the country. These last-mentioned standards are the ones usually readily available to the working engineer for calibration of motion transducers.

The fundamental unit of time is the second, which was redefined for scientific use as 1/31,556,925.9747 of the tropical year at 12^h ephemeris time, 0 January 1900,[1] by the International Committee on Weights and Measures in 1956. A serious fault in this definition is that no one can measure an interval of time by direct comparison with the interval of time defining the second. Rather, lengthy astronomical measurements over several years are necessary to relate the current value of the mean solar second to the basic standard. These measurements and calculations result in an estimated probable error of about 1 part in 10^9, which is quite poor compared with the precision implied in the basic definition of the second. To remedy this difficulty, metrologists in 1964 again redefined the second in terms of the frequencies of atomic resonators.[2] Now the second is defined as the interval of time corresponding to 9,192,631,770 cycles of the atomic resonant frequency of cesium 133. Already it is possible for independent laboratories to construct cesium-beam resonators which agree in frequency within a few parts in 10^{11}. The hydrogen maser gives promise of extending this to 10^{13}.

The above short discussion was concerned with the *fundamental* standards of length and time rather than the practical working standards with which most engineers will be concerned. These practical standards and associated calibration procedures will be discussed in each specific section, such as relative displacement, acceleration, etc.

4.3

Relative Displacement, Translational and Rotational   We consider here devices for measuring the translation along a line of one point relative to another and the plane rotation about a single axis of one line relative to another. Such displacement measurements are of great interest as such and also because they form the basis of many transducers for measuring pressure, force, acceleration, temperature, etc., as shown in Fig. 4.1.

Calibration. Static calibration of translational devices can often be satisfactorily accomplished using ordinary dial indicators or micrometers as the standard. When used directly to measure the displacement of the transducer, these devices usually are suitable to read to the nearest

---

[1] McNish, *loc. cit.*

[2] Time Standards, *Instr. Control Systems*, p. 87, October, 1965.

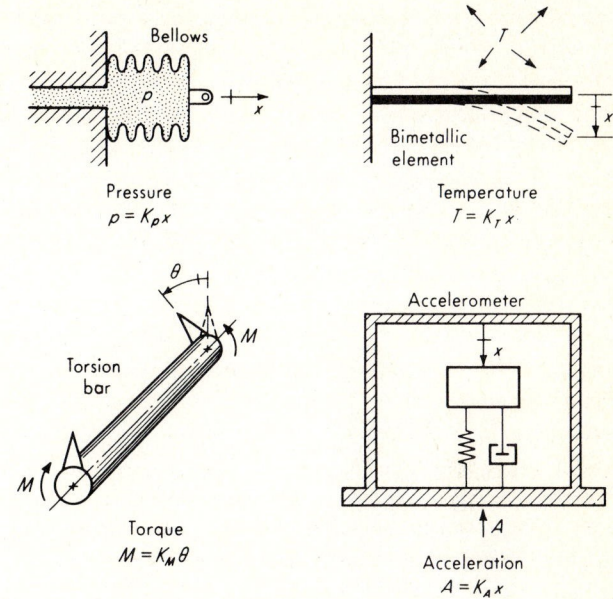

Fig. 4.1. *Applications of displacement measurement.*

0.0001 in. If smaller increments are necessary, lever arrangements (about a 10:1 ratio is fairly easy to achieve) or wedge-type mechanisms (about 100:1) can be employed for motion reduction.[1] The Mikrokator,[2] a unique mechanical gage of high sensitivity, may also be useful in measuring small motions down to a few millionths of an inch.

If accuracy to 0.0001 in. or better is required, such equipment should itself be calibrated against gage blocks, or (for maximum accuracy) gage blocks should be used *directly* to calibrate the transducer. *Gage blocks* are small blocks of hard, dimensionally stable steel or other material, made up in sets which can be stacked up to provide accurate dimensions over a wide range and in small steps. They are the basic working length standards of industry. As purchased from the manufacturer, their dimensions are accurate to ± 8 μin. for working grade blocks, ± 4 μin. for reference grade, and ± 2 μin. for all blocks up to 1 in. (± 2 μin./in. for blocks longer than 1 in.) for master blocks. If these tolerances are too large, the blocks can be sent to the National Bureau of Standards and calibrated against light wavelengths to the nearest 10^{-7} in. Some pre-

[1] H. C. Roberts, "Mechanical Measurements by Electrical Methods," chap. 13, The Instruments Publishing Co., Pittsburgh, Pa., 1951.
[2] C. E. Johansson Gage Co., Dearborn, Mich.

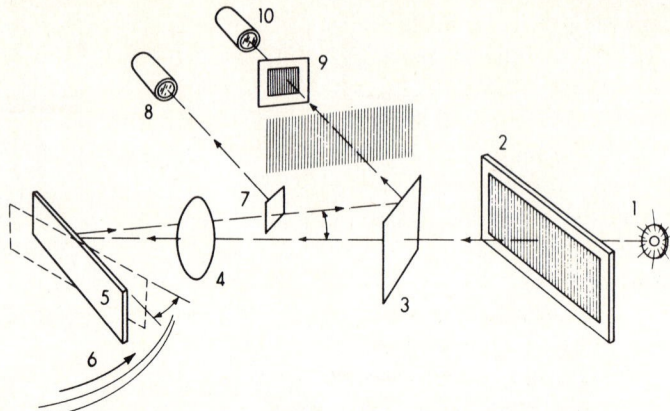

The Midarm system is comprised of an optical unit and an electronic unit. The simplified diagram shows how Midarm operates. Light from a monochromatic light point source (1) passes through a grid (2), a beam splitter (3), a collimating lens (4), and strikes a mirror (5) mounted on the rotating specimen to be tested (6). The image is reflected back into the system where it is directed by a beam splitter (7) to the reference photosensor (8). The image is also reflected by the beam splitter (3) through a second grid (9) to the control photosensor (10). As the test specimen (mirror) rotates, the image of the first grid passes across the second grid which allows minimum and maximum amounts of light to reach the control photosensor. The output voltage of the photosensor has a period of 12.8 arc-sec, the angle subtended by the grid spaces. The Midarm has digital output pulses at 12.8 arc-sec and an analog voltage output of 30 volts/arc-sec.

Fig. 4.2.   *The Midarm system.*

cision-manufacturing operations currently require and use the latter calibration service. When calibrating transducers to very high accuracies it is extremely important to control all interfering and/or modifying inputs such as ambient temperature, electrical excitation to the transducer, etc.

Rotational or angular displacement is not itself a fundamental quantity since it is based on length, and so a fundamental standard is not necessary. However, reference and working standards for angles (and thus angular displacement) are desirable and available. The basic standards (against which other standards or instruments may be calibrated) are called *angle blocks*.[1] These are carefully made steel blocks about ⅝ in. wide and 3 in. long, with a specified angle between the two contact surfaces. Just as for length gage blocks, these angle blocks can

---

[1] C. E. Haven and A. G. Strong, Assembled Polygon for the Calibration of Angle Blocks, *Natl. Bur. Std. (U.S.), Handbook* 77, vol. 3, p. 318, 1961.

be stacked to "build up" any desired angle accurately and in small increments. The blocks can be calibrated to an accuracy of 0.1 second of arc by the National Bureau of Standards.[1]

Recent developments in inertial guidance systems have required angle and angular rate measurements on rotating components to an accuracy approaching or exceeding the capability of National Bureau of Standards calibration. Combinations of optical and electronic principles have led to the development of instruments such as the Midarm[2] system to meet these requirements. This instrument will, with relative convenience, measure angular displacement with an accuracy of 0.05 second of arc and a repeatability of 0.02 second of arc. Figure 4.2 shows a simplified diagram of this instrument.

Rotational transducers rarely require such accuracy for calibration nor can the laborious and expensive techniques necessary to realize these limits be economically justified. Thus most static calibration of angular-displacement transducers can adequately be carried out using more convenient and readily available equipment. Examples[3] of such equipment which should be available in a precision machine shop are the circular division tester (range 360°, microscope reads to 0.1 minute of arc, precision of scale disk ±20 seconds of arc), the optical dividing head (range 360°, scale reads to 1.0 minute of arc, working accuracy ±20 seconds of arc), and the division tester with telescope and collimator (accuracy ±2 seconds of arc). In some applications even cruder devices such as ordinary machine-tool index heads, calibrated dials, etc., may be perfectly adequate.

Resistive potentiometers. Basically, a resistive potentiometer consists of a resistance element provided with a movable contact. The contact motion can be translation, rotation, or a combination of the two (helical motion in a multiturn rotational device), thus allowing measurement of rotary and translatory displacements. Translatory devices have strokes from about 0.1 to 20 in., and rotational ones range from about 10° to as much as 60 full turns. The resistance element is excited with either d-c or a-c voltage, and the output voltage is (ideally) a linear function of the input displacement. Resistance elements in common use may be classified as wire-wound, carbon-film, or conducting-plastic.

If the distribution of resistance with respect to translational or angular travel of the wiper (moving contact) is linear, the output voltage e_o will faithfully duplicate the input motion x_i or θ_i if the terminals at e_o

[1] Independent Standards Laboratory, *Instr. Control Systems*, p. 478, March, 1961.

[2] Razdow Laboratories, Inc., Newark, N.J.

[3] Carl Zeiss, Inc., New York.

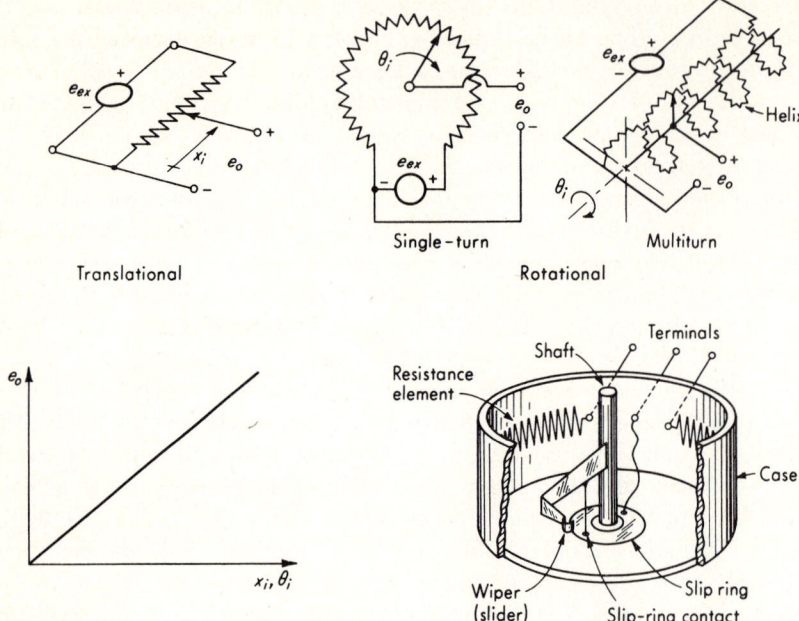

Fig. 4.3. *Potentiometer displacement transducer.*

are open circuit (no current drawn at the output).   (For a-c excitation, x_i or θ_i amplitude-modulate e_{ex}, and e_o does not look like the input motion.) The usual situation, however, is one in which the potentiometer output voltage is the input to a meter or recorder that draws some current from the potentiometer.   Thus a more realistic circuit is as shown in Fig. 4.4. Analysis of this circuit gives

$$\frac{e_o}{e_{ex}} = \frac{1}{1/(x_i/x_t) + (R_p/R_m)[1 - (x_i/x_t)]} \quad (4.1)$$

which becomes for ideal ($R_p/R_m = 0$ for an open circuit) conditions

$$\frac{e_o}{e_{ex}} = \frac{x_i}{x_t} \quad (4.2)$$

Thus for no "loading" the input-output curve is a straight line.   In actual practice, $R_m \neq \infty$ and Eq. (4.1) shows a nonlinear relation between e_o and x_i.   This deviation from linearity is shown in Fig. 4.4.   The maximum error is about 12 percent of full scale if $R_p/R_m = 1.0$ and drops to about 1.5 percent when $R_p/R_m = 0.1$.   For values of $R_p/R_m < 0.1$ the position of maximum error occurs in the neighborhood of $x_i/x_t = 0.67$, and the maximum error is approximately $15(R_p/R_m)$ percent of full scale.

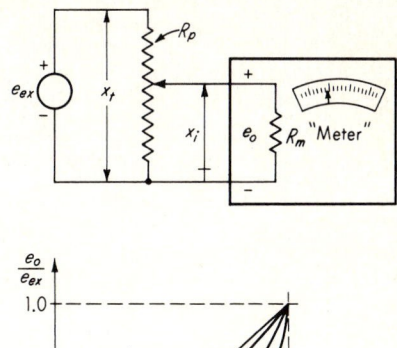

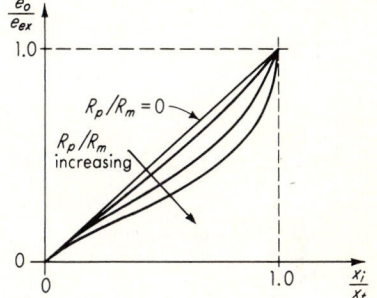

Fig. 4.4. *Potentiometer loading effect.*

We see that to achieve good linearity, for a "meter" of a given resistance R_m, one should choose a potentiometer of sufficiently *low* resistance relative to R_m. This requirement conflicts with the desire for high sensitivity. Since e_o is directly proportional to e_{ex}, it would seem possible to get any sensitivity desired simply by increasing e_{ex}. This is not actually the case, however, since potentiometers have definite power ratings related to their heat-dissipating capacity. Thus a manufacturer may design a series of potentiometers, say single-turn 2-in.-diameter, with a wide range (perhaps 100 to 100,000 ohms) of total resistance R_p but all these will be essentially the same size and mechanical configuration, giving the same heat-transfer capability and thus the same power rating, say about 5 watts at 70°F ambient. If the heat dissipation is limited to P watts, the maximum allowable excitation voltage is given by

$$\text{max } e_{ex} = \sqrt{PR_p} \qquad (4.3)$$

Thus a low value of R_p allows only a small e_{ex} and therefore a small sensitivity. Choice of R_p must thus be influenced by a tradeoff between loading and sensitivity considerations. The maximum available sensitivity of potentiometers varies considerably from type to type and also with size in a given type. It can be calculated from the manufacturer's data on maximum allowable voltage, current, or power and the maximum stroke. The shorter-stroke devices generally have higher sensitivity. *Extreme* values are of the order of 15 volts/deg for short-stroke rotational types ("sector" potentiometers) and 300 volts/in. for short-stroke (about

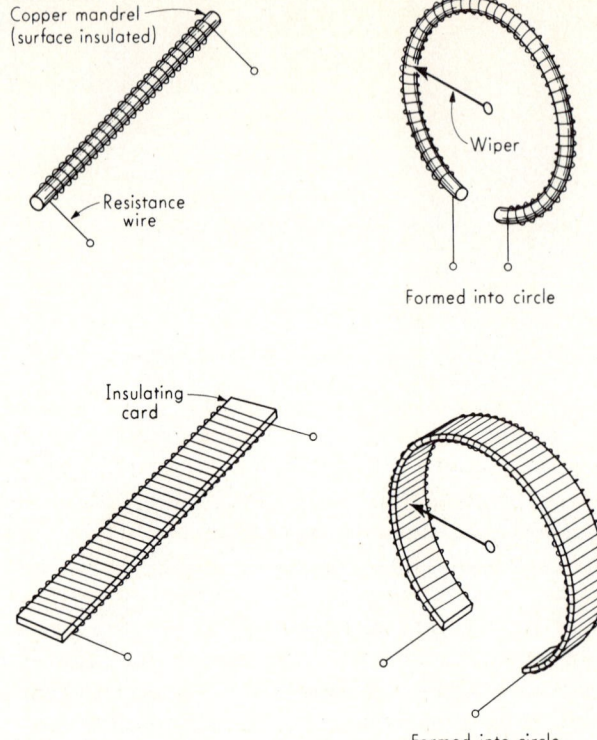

Fig. 4.5. *Construction of resistance elements.*

$\frac{1}{4}$ in.) translational pots. It must be emphasized that these are maximum values and that the usual application involves a much smaller (10 to 100 times smaller) sensitivity.

The resolution of potentiometers is strongly influenced by the construction of the resistance element. An obvious approach is to use a single slide-wire as the resistance, giving an essentially continuous stepless resistance variation as the wiper travels over it. Such potentiometers are available but are limited to rather small resistance values since the length of wire is limited by the desired stroke in a translational device and by space restrictions (diameter) in a rotational one. Resistance of a given length of wire can be increased by decreasing the diameter but this is limited by strength and wear considerations.

To get sufficiently high resistance values in small space the wire-wound resistance element is widely used. The resistance wire is wound on a mandrel or card which is then formed into a circle or helix if a rotational device is desired (see Fig. 4.5). With such a construction the

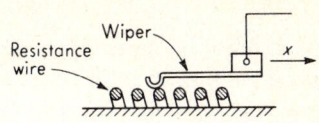

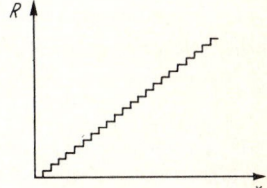

Fig. 4.6. *Resolution of wire-wound potentiometers.*

variation of resistance is not a linear continuous change but actually proceeds in small steps as the wiper moves from one turn of wire to the next (see Fig. 4.6). This phenomenon results in a fundamental limitation on the resolution in terms of resistance-wire size. For instance, if a translational device has 500 turns of resistance wire on a card 1 in. long, motion changes smaller than 0.002 in. cannot be detected. (This is slightly conservative since the wiper, in going from one turn to the next, goes through an intermediate position in which it is touching *both* turns at once.[1,2] The resolution thus actually varies from one position to another, the *worst* value being that given by a simple counting of turns per inch.) The actual practical limit for wire spacing according to current practice is between 500 and 1,000 turns per inch.[3] For translational devices, resolution is thus limited to 0.001 to 0.002 in. while single-turn rotational devices can trade off increased diameter D for increased angular resolution according to the relation

$$\text{Best angular resolution} = \frac{0.12 \text{ to } 0.24}{D} \quad \text{degrees} \quad D \text{ in inches} \quad (4.4)$$

It should be noted that resolution is intimately related to total resistance since the fine wire required to get close wire spacing will naturally have a high resistance. Thus one cannot choose total resistance and resolution independently. If extremely fine resolution and high resistance are required a carbon-film or conductive-plastic resistance ele-

[1] H. Gray, How to Specify Resolution for Potentiometer Servos, *Control Eng.*, p. 129, November, 1959.

[2] C. A. Mounteer, The Effective Resolution of Wire-wound Potentiometers, *Giannini Tech. Notes*, G. M. Giannini Co., Pasadena, Calif., January–February, 1959.

[3] S. A. Davis and B. K. Ledgerwood, "Electromechanical Components for Servomechanisms," p. 53, McGraw-Hill Book Company, New York, 1961.

ment may be indicated.   Carbon-film elements may have a resolution as good as 5×10^{-6} in. (the fact that it is not infinitesimally small is due to *granularity* in the surface) but the overall resolution of the potentiometer is somewhat poorer because of mechanical defects in bearings and wiper springs.   In applying carbon-film devices it is important to keep in mind that the wiper is a relatively high-resistance contact (wire-wound devices have very low contact resistance).   Thus the amount of current drawn from the potentiometer must be kept quite small; otherwise the iR drop across the wiper will cause errors in the output voltage.

Another approach to increased resolution involves the use of multiturn potentiometers.   The resistance element is in the form of a helix, and the wiper travels along a "lead screw."   The number of wires per inch of element is still limited, as mentioned above, but an increase in resolution can be obtained by introducing gearing between the shaft whose motion is to be measured and the potentiometer shaft.   For example, one rotation of the measured shaft could cause 10 rotations of the potentiometer shaft; thus the resolution of measured-shaft motion is increased by a factor of 10.   Multiturn potentiometers are available up to about 60 turns.   For translational devices various motion-amplifying mechanisms could be used in similar fashion.

Most potentiometers used for motion measurement are intended to give a linear input-output relation and are used as purchased, without calibration.   Thus a specification of linearity is essentially equivalent to one of accuracy.   Potentiometers are available in a wide range of linearities and corresponding prices.   Linearity depends greatly on the uniformity of the resistance winding but errors in this can be corrected by adding fixed resistances in series and/or parallel at proper locations on the winding.   This procedure also can correct loading errors so as to give a linear relation for a heavily loaded potentiometer.[1]   The best nonlinearities commercially available range from 1 percent of full scale for $\frac{1}{2}$-in.-diameter single-turn pots through 0.02 percent for a 2-in.-diameter multiturn to 0.002 percent for a 10-in.-diameter multiturn.   The best nonlinearities of translational pots are about 0.05 to 0.10 percent of full scale.   It should be noted that accuracy can be no better (and is generally worse) than one-half the resolution; thus resolution places a limit on accuracy.

Noise in potentiometers refers to spurious-output-voltage fluctuations occurring during motion of the slider and includes the effects of resolution.   In addition, various mechanical and electrical defects produce noise.   In a wire-wound pot, motion of the slider over the resistance wires may cause bouncing of the contact at certain speeds, thus causing

---

[1] Davis and Ledgerwood, *op. cit.*, p. 59.

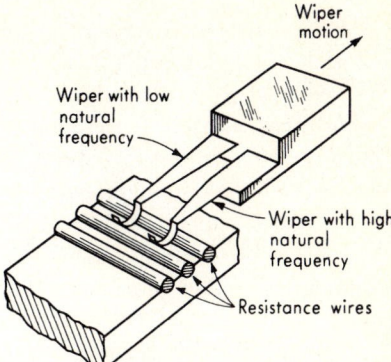

Fig. 4.7. *Antivibration wiper construction.*

intermittent contact.   This phenomenon becomes particularly significant if the speed and wire spacing are such as to produce forces of frequency near the resonant frequency of the spring-loaded contact.   Contacts are sometimes made in the configuration of Fig. 4.7 to overcome this problem. Here the resonant frequency of each section of the wiper is different. Thus if one section is resonating at a certain speed the other will be off resonance and making continuous contact.   Another possibility lies in filling the interior of the potentiometer with a damping fluid to limit resonant amplitudes.   This also generally increases the shock and vibration tolerance of the unit.   Another source of noise is found in dirt and wear products which come between the contact surface and the winding. Even if no dirt or wear products are present, the contact resistance of a moving contact varies during motion, and if any load current is flowing through this contact a spurious iR voltage appears in the output.   This effect also occurs at the slip-ring contact.   Numerical values of noise voltage quoted in specifications generally include all sources of noise and correspond to a definite speed and current.[1]

The dynamic characteristic of potentiometers (considering displacement as input and voltage as output) is essentially that of a zero-order instrument since the impedance of the winding is almost purely resistive at the motion frequencies for which the device is usable.   However, the mechanical loading imposed on the measured motion by the inertia and friction of the potentiometer's moving parts should be carefully considered. The friction is usually mostly dry friction, and the manufacturer generally supplies numerical values of the starting and running friction force or torque.   These values vary over a wide range, depending on the construction of the potentiometer.   Special low-friction rotary pots have starting torques as small as 0.003 oz-in.   More conventional instruments

---

[1] PPMA Conference Report, *Electromech. Design*, p. 8, April, 1964.

may have 0.1 to 0.5 oz-in. or more. Translational pots have friction values from less than 1 oz to over 1 lb. Inertia values for both rotary and translatory pots vary widely with size. A typical $\frac{7}{8}$-in.-diameter single-turn pot has a moment of inertia of 0.12 g-cm^2 while a 2-in.-diameter 10-turn pot has about 18 g-cm^2. Moving masses of translatory pots have weights ranging from fractions of an ounce to several ounces.

Since the measured variable in a pot is displacement (a flow variable) the pertinent loading quantity is the generalized input admittance or compliance. The presence of dry friction (a discontinuous nonlinear effect) prevents one from defining an exact input admittance or compliance. If friction is small, it may be neglected and the system analyzed as if it contained inertia only. We may choose to work with either admittance or compliance. If we choose compliance, the input compliance of a pure mass is given by

$$C_{gi}(D) \triangleq \frac{\text{displacement}}{\text{force}}(D) = \frac{x}{f}(D) \qquad (4.5)$$

and since

$$f = M\ddot{x} = MD^2x \qquad (4.6)$$

we get

$$C_{gi}(D) = \frac{1}{MD^2} \qquad (4.7)$$

or, in frequency-response terms,

$$C_{gi}(i\omega) = \frac{1}{(i\omega)^2 M} \qquad (4.8)$$

For negligible loading effect the term C_{go}/C_{gi}, where C_{go} is the output compliance of the measured system, must be small compared with 1.0. Since $C_{gi}(i\omega)$ approaches infinity at low frequencies, loading is negligible at low frequencies so long as $C_{go}(i\omega)$ does not also approach infinity as $\omega \rightarrow 0$. The range of low frequencies for which $C_{gi}(i\omega)$ is large can be extended by decreasing the moving mass M.

Finally, selection of potentiometers should take into account various environmental factors such as high or low temperatures, shock and vibration, humidity, and altitude. These may act as modifying and/or interfering inputs so as seriously to degrade instrument performance. Under good environmental conditions the life of a potentiometer may be more than 20 million full strokes or rotations.

Resistance strain gages. Consider a conductor of uniform cross-sectional area A and length L, made of a material with resistivity ρ. The resistance R of such a conductor is given by

$$R = \frac{\rho L}{A} \qquad (4.9)$$

If this conductor is now stretched or compressed, its resistance will change because of dimensional changes (length and cross-sectional area) and also because of a fundamental property of materials called *piezoresistance*[1] (pronounced pī-ēzō-resistance) which indicates a dependence of resistivity ρ on the mechanical strain.   To find how a change dR in R depends on the basic parameters, we differentiate Eq. (4.9) to get

$$dR = \frac{A(\rho\,dL + L\,d\rho) - \rho L\,dA}{A^2} \qquad (4.10)$$

Since volume $V = AL$, $dV = A\,dL + L\,dA$.   Also

$$dV = L(1 + \epsilon)A(1 - \epsilon\nu)^2 - AL \qquad (4.11)$$

$$\text{where} \quad \epsilon \triangleq \text{unit strain}$$
$$\nu \triangleq \text{Poisson's ratio}$$

Since ϵ is small, $(1 - \nu\epsilon)^2 \approx 1 - 2\nu\epsilon$ and Eq. (4.11) becomes

$$dV = AL\epsilon(1 - 2\nu) = A\,dL + L\,dA \qquad (4.12)$$

and since $\epsilon \triangleq dL/L$

$$A\,dL(1 - 2\nu) = A\,dL + L\,dA \qquad (4.13)$$
$$-2\nu A\,dL = L\,dA \qquad (4.14)$$

Substituting in Eq. (4.10)

$$dR = \frac{\rho A\,dL + LA\,d\rho + 2\nu\rho A\,dL}{A^2} \qquad (4.15)$$

and thus

$$dR = \frac{\rho\,dL(1 + 2\nu)}{A} + \frac{L\,d\rho}{A} \qquad (4.16)$$

Dividing by Eq. (4.9) gives

$$\frac{dR}{R} = \frac{dL}{L}(1 + 2\nu) + \frac{d\rho}{\rho} \qquad (4.17)$$

and finally   Gage factor $\triangleq \dfrac{dR/R}{dL/L} = \underbrace{1}_{\substack{\text{resistance} \\ \text{change due} \\ \text{to length} \\ \text{change}}} + \underbrace{2\nu}_{\substack{\text{resistance} \\ \text{change due to} \\ \text{area change}}}$

$$+ \underbrace{\frac{d\rho/\rho}{dL/L}}_{\substack{\text{resistance change due} \\ \text{to piezoresistance} \\ \text{effect}}} \qquad (4.18)$$

Thus if the gage factor is known, measurement of dR/R allows measurement of the strain $dL/L = \epsilon$.   This is the principle of the resistance strain gage.   The term $(d\rho/\rho)/(dL/L)$ can also be expressed as $\pi_1 E$, where

$$\pi_1 \triangleq \text{longitudinal piezoresistance coefficient}$$
$$E \triangleq \text{modulus of elasticity}$$

The material property π_1 can be either positive or negative.   Poisson's

---

[1] C. M. Harris and C. E. Crede (eds.), "Shock and Vibration Handbook," vol. 1, p. 16-35, McGraw-Hill Book Company, New York, 1961.

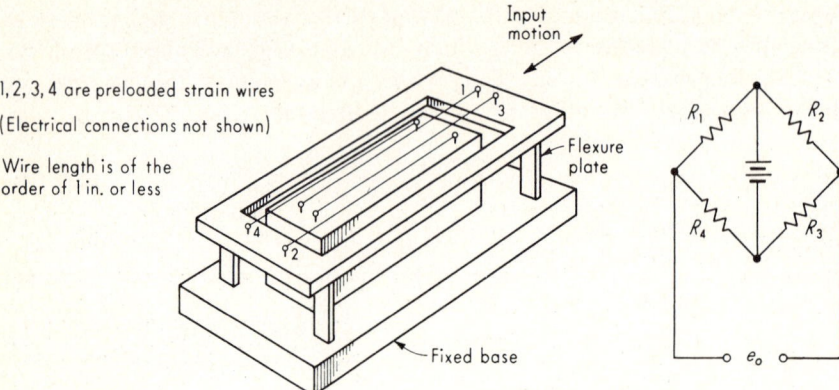

Fig. 4.8.   *Unbonded strain gage.*

ratio is always between 0 and 0.5 for all materials.   The most common type of strain gage uses one of the two alloys Advance (55 percent copper, 45 percent nickel) or Iso Elastic (36 percent nickel, 8 percent chromium, 4 percent manganese, silicon, and molybdenum; remainder iron).   Advance gives a gage factor of about 2 and Iso Elastic about 3.5.

About 1960, strain gages based on semiconductor materials rather than metals began to become commercially available.   While these are somewhat more expensive and more difficult to apply than metallic gages, their outstanding virtue is a very high gage factor of about 130.   From Eq. (4.18) we can see that in the common metallic gages most of the resistance change comes from dimensional changes whereas in semiconductor gages most of it comes from piezoresistance effects.   Ideally the gage factor would be a constant, and for metallic gages it can generally be treated as such.   In semiconductor gages, however, π_1 varies somewhat with strain so that a nonlinear strain/resistance relationship exists. This tends to complicate the interpretation of readings from such gages. Intensive development is rapidly overcoming the disadvantages of semiconductor gages and they are already used in considerable quantities, especially in load cells, accelerometers, and other transducer applications.

For metallic gages two different methods of utilizing the above basic principle are in common use.   These correspond to the *unbonded* and *bonded* strain gage.   In the unbonded gage (shown in simplified form in Fig. 4.8) the resistance wires (about 0.001-in. diameter) are stretched between two frames which can move relative to each other as guided by flexure plates.   Since the wires would buckle if compressive forces were applied, an internal preload greater than any expected external compressive load is employed.   Under these conditions, applied motion to the right increasingly stretches wires 1 and 3 and reduces the tension in wires 2 and 4.   A motion to the left does just the reverse, and so motions in both directions can be measured so long as the preload is not overcome. The resistance wires are generally connected in a bridge circuit (shown in its simplest form in Fig. 4.8).   With the preload present but no external load applied, the bridge is balanced if $R_1/R_4 = R_2/R_3$.   Adjustable resistors are generally provided in the bridge to accomplish this.   An external load will then cause variation in resistance of the wires, unbalancing the bridge and causing an output voltage e_o in proportion to the motion.   The motions directly measurable by gages of this type are very small, of the order of 0.0015 in. full scale.

Unbonded gages are used mainly as elements of force and pressure transducers and accelerometers rather than directly as displacement pickups.   In these applications the strain wires often serve as the necessary spring element in transducing force to deflection, in addition to being the displacement sensor.   The allowable force on the wires is very small, about 0.15 oz for the maximum deflection of 0.0015 in.   Variation in the number, size, and length of wires allows design for a range of force and deflection values.   The resolution of such devices is infinitesimally small since the resistance variation is a smooth change.   Inaccuracy is of the order of 0.15 percent of full scale for a typical[1] unit.   The sensitivity for the recommended 5-volt bridge excitation is 40 mv full-scale output for 0.0024-in. full-scale displacement (30 g full-scale force), that is, 16.7 volts/in. or 0.60 volt/lb$_f$.   Thermal-sensitivity shift is 0.01 percent/F° between -65 and $+250$°F while thermal zero shift is 0.01 percent of full scale/F° between -65 and $+250$°F.   The resistance of the bridge arms is nominally 350 ohms.   With displacement considered as the input and bridge voltage e_o as output, the response of such instruments is essentially instantaneous (zero-order), if wave propagation is neglected.   When measuring the displacement of a system, mechanical loading effects are again determined by the ratio of measured-system-output compliance to instrument-input compliance.   Owing to the use of flexures rather than bearings, friction is negligible, and the instrument input is characterized by a moving mass M and a spring constant K_s determined by the resistance wires and the flexures.   For such a situation, input compliance is given by

$$C_{gi}(D) \triangleq \frac{\text{displacement}}{\text{force}}(D) = \frac{x}{f}(D)$$

where

$$f - K_s x = M\ddot{x} = MD^2 x \qquad (4.19)$$

$$C_{gi}(D) = \frac{1}{MD^2 + K_s} \qquad (4.20)$$

$$C_{gi}(i\omega) = \frac{1}{(i\omega)^2 M + K_s} = \frac{1}{K_s - M\omega^2} \qquad (4.21)$$

[1] Universal transducing cell, Statham Instruments Inc., Los Angeles, Calif.

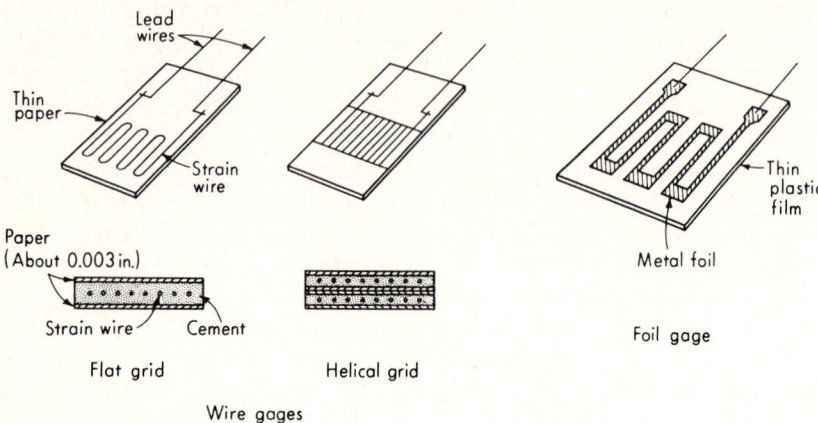

Fig. 4.9. Bonded strain gages.

As $\omega \to 0$, the static case is approached, and $C_{gi} \to 1/K_s$. Thus there will be a loading error for static operation unless the measured system has $C_{go}(i\omega) = 0$ for $\omega = 0$ (an infinitely stiff system). Since large compliance C_{gi} is desired to minimize loading, K_s should be small. Also note that C_{gi} increases with increasing frequency, reaching an infinite peak at $\omega^2 = K_s/M$ and then decreasing for higher frequencies.

Bonded metallic strain gages[1] use elements of wire in a flat grid or (flattened) helical construction or a thin metal foil printed and etched to give a grid-type pattern (see Fig. 4.9). Gages are available in sizes from about 6 in. in length to about $\frac{1}{64}$ in. These gages must be cemented to the surface whose strain is to be measured with a cement suitable to the environmental conditions. Once cemented down, the gages cannot be removed and reused. For wire gages the wire size is about 0.001-in. diameter. Foil gages can be made somewhat thinner; using a foil of about 0.00015-in. thickness and plastic film of 0.001 in. gives an overall thickness of about 0.001 in., somewhat thinner than wire gages.

When the gages are properly cemented down, they effectively become a part of the surface to which they are fastened and undergo essentially the same strain as that surface. They work equally well in both tension and compression since the matrix of cement surrounding the wire or foil completely prevents buckling. With suitable auxiliary electronic equipment, strains down to about 10^{-7} in./in. can be detected. While the useful upper strain limit (about 0.01 in./in.) for most gages is set by the elastic limit of the strain wires, special "post-yield" gages can

[1] C. C. Perry and H. R. Lissner, "The Strain Gage Primer," McGraw-Hill Book Company, New York, 1955.

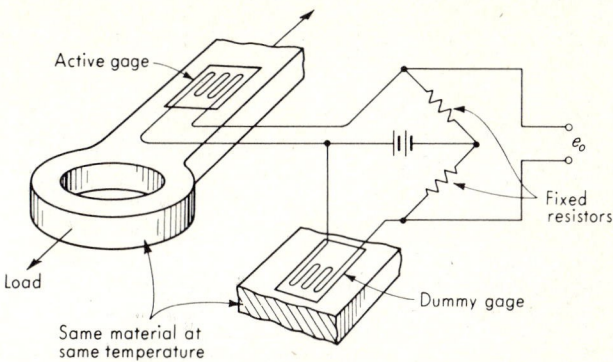

Fig. 4.10.　*Strain-gage temperature compensation.*

be used to measure strains as great as 0.1 in./in.　The gages are mainly sensitive to the component of strain along their longitudinal axis; however, there is some small transverse sensitivity because of the loops at the end of each turn of wire.　This effect is usually less than 1 percent and is reduced even more in foil-type gages by making the end "loops" of greater cross section than the main portion.　The total resistance of individual gages ranges from about 40 to 2,000 ohms, with 120, 350, and 1,000 ohms being common standard values.　Heating of the gages limits the maximum allowable gage current.　This varies with the type of gage and the heat-transfer conditions but is of the order of 0.030 amp.

　　Temperature is an important interfering input for strain gages since resistance changes with *both* strain and temperature.　Since strain-induced resistance changes are quite small, the temperature effect can assume major proportions.　Another aspect of temperature sensitivity is found in the possible differential thermal expansion of the gage and the underlying material.　This can cause a strain and resistance change in the gage even though the material is not subjected to an external load.　These temperature effects can be compensated in various ways.　In Fig. 4.10 a "dummy" gage (identical to the active gage) is cemented to a piece of the same material as is the active gage and placed so as to assume the same temperature.　The dummy and active gages are placed in adjacent legs of a Wheatstone bridge; thus resistance changes due to the temperature coefficient of resistance and differential thermal expansion will have no effect on the bridge output voltage whereas resistance changes due to an applied load will unbalance the bridge in the usual way (see text on bridge circuits).　Another approach to this problem involves special inherently temperature-compensated gages.　These gages are designed to be used on a specific material and have expansion and resistance properties

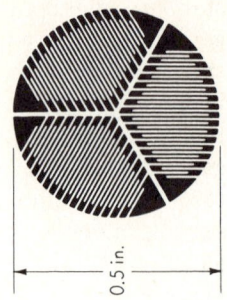

Fig. 4.11. Foil rosette of Baldwin-Lima-Hamilton Corp.

such that the two effects very nearly cancel each other and no dummy gage is required.

Temperature also can act as a modifying input in that it may change the gage factor. With metallic gages this effect is usually quite small except at extremely low or high temperatures. Semiconductor gages are more seriously affected in this way; however compensation is possible. Although the above temperature problems must be carefully considered in each application, strain gages have been successfully employed from liquid-helium temperature (7°R) to the order of 2000°F. However, these extreme (especially the high temperature) applications require special techniques and yield results of lower accuracy than is obtained in routine room-temperature situations.

Whereas the strain in a body is defined at a *point*, the element of a resistance strain gage extends over a finite *area*. Thus what strain is actually measured by such an instrument? Obviously it is some sort of average strain over the sensitive area of the gage. If the strain is uniaxial, if the gage is aligned with this axis, and if the strain gradient (rate of change of strain with distance along the surface) is constant, the average strain indicated by the gage will numerically equal the "point" strain at the midsection of the gage. The midpoint of the gage length is generally marked on the gage by the manufacturer. If the strain gradient is not constant and its form is unknown, the strain value read by the gage cannot be associated with any specific point. For such situations the smallest practical gage should be used in order to reduce this uncertainty. When the direction and magnitude of the maximum strain at a point are completely unknown, it can be shown that strain measurements in three different directions are sufficient to calculate the strain in any direction and thus its maximum value. To facilitate such measurements, strain-gage "rosettes" which combine the necessary three gages into one easily applied assembly have been developed. They are available in both wire and foil types, Fig. 4.11 showing a foil rosette.

To use a strain gage for strain measurement, one must know its resistance and gage factor. The resistance of a gage can be measured by standard techniques.   However, the gage factor of an individual bonded gage cannot be found without cementing it to a simple member (for which strain can be accurately calculated from theory or measured by some independent means), applying known loads or deflections, and measuring the resulting resistance changes.   The difficulty is that this calibrated gage *cannot* now be removed from the calibration member and recemented to a member whose unknown strain is to be measured.   The manufacturer supplies along with a gage both its resistance and its gage factor but the gage factor of that particular gage has *not* itself been determined.   Rather, periodic samples of the gage production are taken and calibrated, and the manufacturer relies on careful statistical quality control to ensure that the gage factors of gages sold do not deviate more than a specified percentage from the quoted figures.   Thus a typical purchased gage might have a resistance of 120 $\pm$ 0.4 ohms and a gage factor of 2.14 $\pm$ 1 percent stated on the package.   The resistance value can be measured more accurately by the user but the typical ± 1 percent uncertainty in the gage factor represents a basic limitation on gage accuracy for strain measurements.

Used directly, the bonded strain gage is useful for measuring only very small displacements (strains).   However, larger displacements may be measured by bonding the gage to a flexible element such as a thin cantilever beam and applying the unknown displacement to the end of the beam, as in Fig. 4.12.   For such an application the gage factor need not be accurately known since the overall system can be calibrated by applying known displacements to the end of the beam and measuring the resulting bridge output voltage.   The configuration shown is temperature-compensated without the need for dummy gages and has four times the sensitivity of a single gage because of judicious application of bridge-circuit properties.   Such transducers may be accurate to 0.1 percent of full scale.

The dynamic response of bonded strain gages with respect to faithfully reproducing as a resistance variation the strain variation of the underlying surface is very good.   The dynamic effects of wave propagation in the cement and strain wires seem to be negligible for frequencies up to at least 50,000 cps, and so a zero-order dynamic model is generally adequate.   The loading effect of the cement and strain wires on the underlying structure is generally negligible except for very thin members, in which the stiffening effect of the strain gage may reduce the measured strain to a value considerably lower than that present without the gage. Compliance techniques can be applied to study this effect.

The voltage output from metallic strain-gage circuits is quite small (a few microvolts to a few millivolts), and so amplification is generally needed.   As an example, consider the measurement of a stress level of

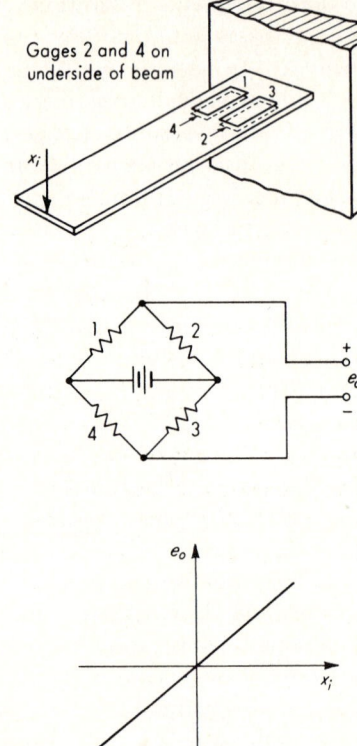

Fig. 4.12. *Beam displacement transducer.*

1,000 psi in steel with a single active gage of 120 ohms resistance and a gage factor of 2.0. If a bridge circuit of all equal arms is used, the maximum allowable bridge voltage for 30-ma gage current is

$$e_{ex} = (240)(0.030) = 7.2 \text{ volts} \qquad (4.22)$$

The strain ϵ is $1,000/30 \times 10^6 = 3.33 \times 10^{-5}$ in./in., so that

$$\Delta R = (\text{gage factor})(\epsilon)(R) = (2)(3.33 \times 10^{-5})(120) = 7.99 \times 10^{-3} \text{ ohm} \qquad (4.23)$$

For the given bridge arrangement,

$$e_o = e_{ex}\left(\frac{1}{4R}\right)\Delta R = \frac{(7.2)(7.99 \times 10^{-3})}{480} = 0.12 \text{ mv} \qquad (4.24)$$

Based on limitations of the gage alone, the smallest detectable strain depends on the thermal or Johnson-noise[1] voltage generated in every

---

[1] E. B. Wilson, Jr., "An Introduction to Scientific Research," p. 116, McGraw-Hill Book Company, New York, 1952.

resistance because of the random motion of its electrons.   This random voltage is essentially a white noise of spectral density $4kTR$ volts^2/cps, where

$$k \triangleq \text{Boltzmann's constant} = 1.38 \times 10^{-23} \text{ joule/K}° \quad (4.25)$$
$$T \triangleq \text{absolute temperature of resistor, }°\text{K} \quad (4.26)$$
$$R \triangleq \text{resistance, ohms} \quad (4.27)$$

Thus if this voltage were measured by a hypothetical noise-free oscilloscope with a bandwidth of Δf cps, the measured rms voltage would be

$$E_{\text{noise,rms}} = \sqrt{4kTR\,\Delta f} \quad \text{volts} \quad (4.28)$$

As an example, a strain gage of $R = 120$ ohms at 300°K over a bandwidth of 100,000 cps would put out an rms noise voltage of 0.45 μv. Comparing this with our earlier calculation of the signal due to 1,000-psi stress, we see that the signal/noise ratio would be $120/0.45 = 267{:}1$. Suppose, however, that we wish to measure 1-psi stress rather than 1,000. The signal is then 0.12 μv, which is less than the noise; therefore the signal would be lost in the noise.   Amplification under these conditions is of no use since the signal and noise are both amplified.   This simple example does not cover other methods that have been developed to reduce this limitation but it should be understood that the limitation is a fundamental one and can be reduced but not overcome.   Similar random fluctuations limit the measurable threshold of all physical variables.[1]   In practical strain-gage measurement systems, Johnson noise of resistances other than the strain gage and other sources of noise in tubes, etc., actually limit the system resolution.

The useful operating life of bonded and unbonded strain gages is heavily influenced by environmental conditions.   Unlike potentiometers, which *wear* out as a result of friction at the sliding contact, failure of strain gages, when it occurs, is often chargeable to fatigue of the metal wires or solder joints because of cyclic stressing.   Since the metals used for strain wires are generally nonferrous, no endurance limit (stress below which failure *never* occurs) exists, and any stress, if repeated often enough, will ultimately cause failure.   Under normal conditions this may take many millions of cycles, and so a long useful life is generally to be expected.   Gradual or sudden failure of the cement is also responsible for some failures.   For long-term applications a protective covering of wax or other material to prevent entrance of humidity may be indicated.

Differential transformers. Figure 4.13 shows schematic and circuit diagrams for translational and rotational linear variable-differential-transformer (LVDT) displacement pickups.   The excitation of such devices is

[1] *Ibid.*

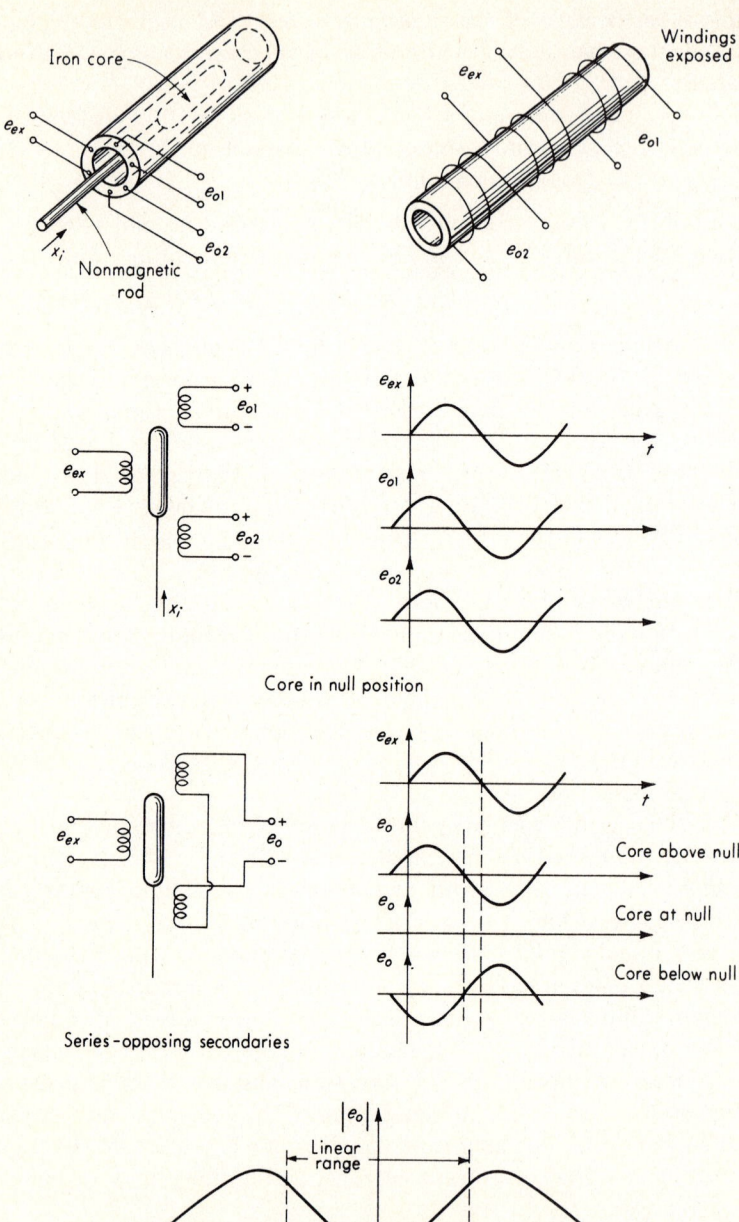

Core in null position

Series-opposing secondaries

normally a sinusoidal voltage of 3 to 15 volts rms amplitude and frequency of 60 to 20,000 cps. The two identical secondary coils have induced in them sinusoidal voltages of the same frequency as the excitation; however, the amplitude varies with the position of the iron core. When the secondaries are connected in series opposition, a null position exists ($x_i \triangleq 0$) at which the net output e_o is essentially zero. Motion of the core from null then causes a larger mutual inductance (coupling) for one coil and a smaller mutual inductance for the other, and the amplitude of e_o becomes a nearly linear function of core position for a considerable range either side of null. The voltage e_o undergoes a 180° phase shift in going through null. The output e_o is generally out of phase with the excitation e_{ex}; however, this varies with the frequency of e_{ex}, and for each differential transformer there exists a particular frequency (numerical value supplied by the manufacturer) at which this phase shift is zero. If the differential transformer is used with some readout system that requires a small phase shift between e_o and e_{ex} (some carrier-amplifier systems require this), excitation at the correct frequency can solve this problem. If the output voltage is applied directly to an a-c meter or an oscilloscope this phase shift is not a problem.

The origin of this phase shift can be seen from analysis of Fig. 4.14. Applying Kirchhoff's voltage-loop law, we get

$$i_p R_p + L_p \left(\frac{di_p}{dt}\right) - e_{ex} = 0 \qquad (4.29)$$

Now the voltage induced in the secondary coils is given by

$$e_{s1} = M_1 \frac{di_p}{dt} \qquad (4.30)$$

$$e_{s2} = M_2 \frac{di_p}{dt} \qquad (4.31)$$

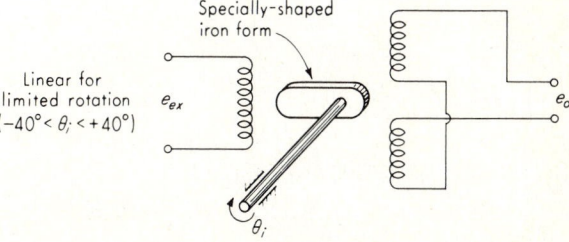

Specially-shaped iron form

Linear for limited rotation ($-40° < \theta_i < +40°$)

e_{ex}

e_o

θ_i

Rotational differential transformer

Fig. 4.13. *Differential transformer.*

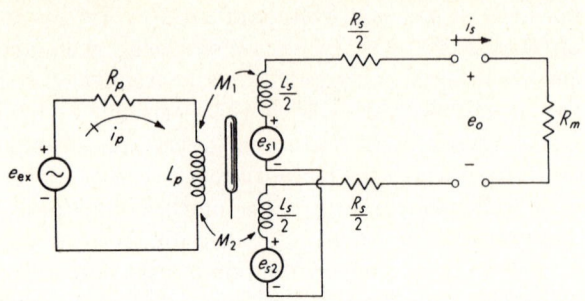

Fig. 4.14. *Circuit analysis.*

where M_1 and M_2 are the respective mutual inductances. The net secondary voltage e_s is then given by

$$e_s = e_{s1} - e_{s2} = (M_1 - M_2) \frac{di_p}{dt} \qquad (4.32)$$

The net mutual inductance $M_1 - M_2$ is the quantity that varies linearly with core motion. If the output is open circuit (no voltage-measuring device attached), we have for a fixed core position

$$e_o = e_s = (M_1 - M_2) \frac{D}{L_p D + R_p} e_{ex} \qquad (4.33)$$

and thus $\qquad \dfrac{e_o}{e_{ex}}(D) = \dfrac{[(M_1 - M_2)/R_p]D}{\tau_p D + 1} \qquad \tau_p \triangleq \dfrac{L_p}{R_p} \qquad (4.34)$

In terms of frequency response,

$$\frac{e_o}{e_{ex}}(i\omega) = \frac{(M_1 - M_2)/R_p}{\sqrt{(\omega\tau_p)^2 + 1}} \underline{/\phi} \qquad \phi = 90° - \tan^{-1} \omega\tau_p \qquad (4.35)$$

thus demonstrating the phase shift between e_o and e_{ex}. If a voltage-measuring device of input resistance R_m is attached to the output terminals, a current i_s will flow, and we can write

$$i_p R_p + L_p D i_p - (M_1 - M_2) D i_s - e_{ex} = 0 \qquad (4.36)$$
$$(M_1 - M_2) D i_p + (R_s + R_m) i_s + L_s D i_s = 0 \qquad (4.37)$$

which lead to

$$\frac{e_o}{e_{ex}}(D) = \frac{R_m(M_2 - M_1)D}{[(M_1 - M_2)^2 + L_p L_s]D^2 \atop + [L_p(R_s + R_m) + L_s R_p]D + (R_s + R_m)R_p} \qquad (4.38)$$

Since the frequency response of $(e_o/e_{ex})(i\omega)$ has a phase angle of $+90°$ at low frequencies and $-90°$ at high, somewhere in between it will be zero,

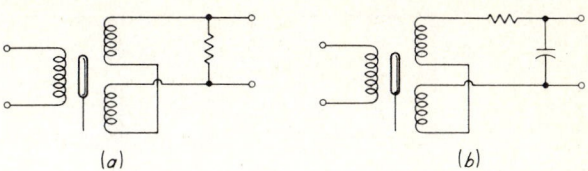

(a)   (b)

Two possible methods for retarding a leading phase angle

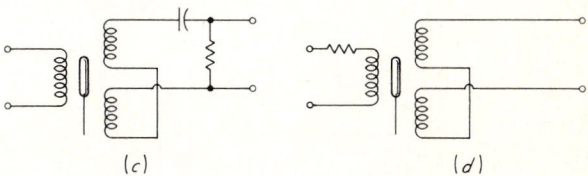

(c)   (d)

Two possible methods for advancing a lagging phase angle

Fig. 4.15. *Phase-angle-adjustment circuits.*

as mentioned earlier. If, for some reason, the excitation frequency cannot be adjusted to this value, the same effect may be achieved for a given frequency by one of the methods[1,2] shown in Fig. 4.15.

While the output voltage at the null position is ideally zero, harmonics in the excitation voltage and stray capacitance coupling between the primary and secondary usually result in a small but nonzero null voltage. Under usual conditions this is less than 1 percent of the full-scale output voltage and may be quite acceptable. Methods of reducing this null when it is objectionable are available. First, the preferred connection shown in Fig. 4.16a should be used if a balanced (center-tapped) excitation-voltage source is available. The grounding shown tends to reduce capacitance-coupling effects. If a center-tapped voltage source is not available, the arrangement of Fig. 4.16b can be used. With the core at the null position and the output-measuring device connected, the potentiometer is adjusted until the minimum null reading is obtained. The values of R and R_p are not critical but should be as low as possible without loading (drawing excessive current from) the excitation source.

The output of a differential transformer is a sine wave whose amplitude is proportional to the core motion. If this output is applied to an a-c voltmeter the meter reading can be directly calibrated in motion units. This arrangement is perfectly satisfactory for measurement of static or very slowly varying displacements except that the meter will give exactly

---

[1] A. Miller, Differential Transformers, *The Right Angle*, The Sanborn Co., Waltham, Mass., August, 1956; November, 1956.

[2] Schaevitz Engineering, Camden, N.J., *Bull.* AA-1A.

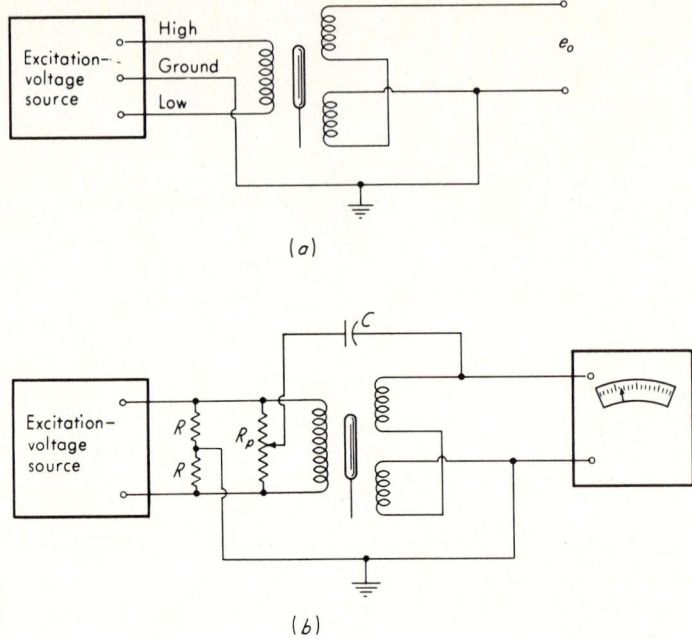

Fig. 4.16. Methods for null reduction.

the same reading for displacements of equal amount on *either* side of the null since the meter is not sensitive to the 180° phase change at null. Thus one cannot tell to which side of null the reading applies without some independent check.   Furthermore, if rapid core motions are to be measured, the meter cannot follow or record them, and an oscillograph or oscilloscope must be used as a readout device.   These instruments record the actual wave form of the output as an amplitude-modulated sine wave, which is usually undesirable; what is desired is an output-voltage record that looks like the mechanical motion being measured.   To achieve the desired results, demodulation and filtering must be performed; if it is necessary to detect unambiguously the motions on both sides of null, the demodulation must be phase-sensitive.   Many different circuits are available for performing these operations.   We show here only one arrangement, which is quite simple.   To use this approach, all four output leads of the LVDT must be accessible (some have the series opposition connection *internal* to the case and would thus not be applicable to the following discussion).

Figure 4.17c shows the circuit arrangement for phase-sensitive demodulation using semiconductor diodes.   Ideally these pass current

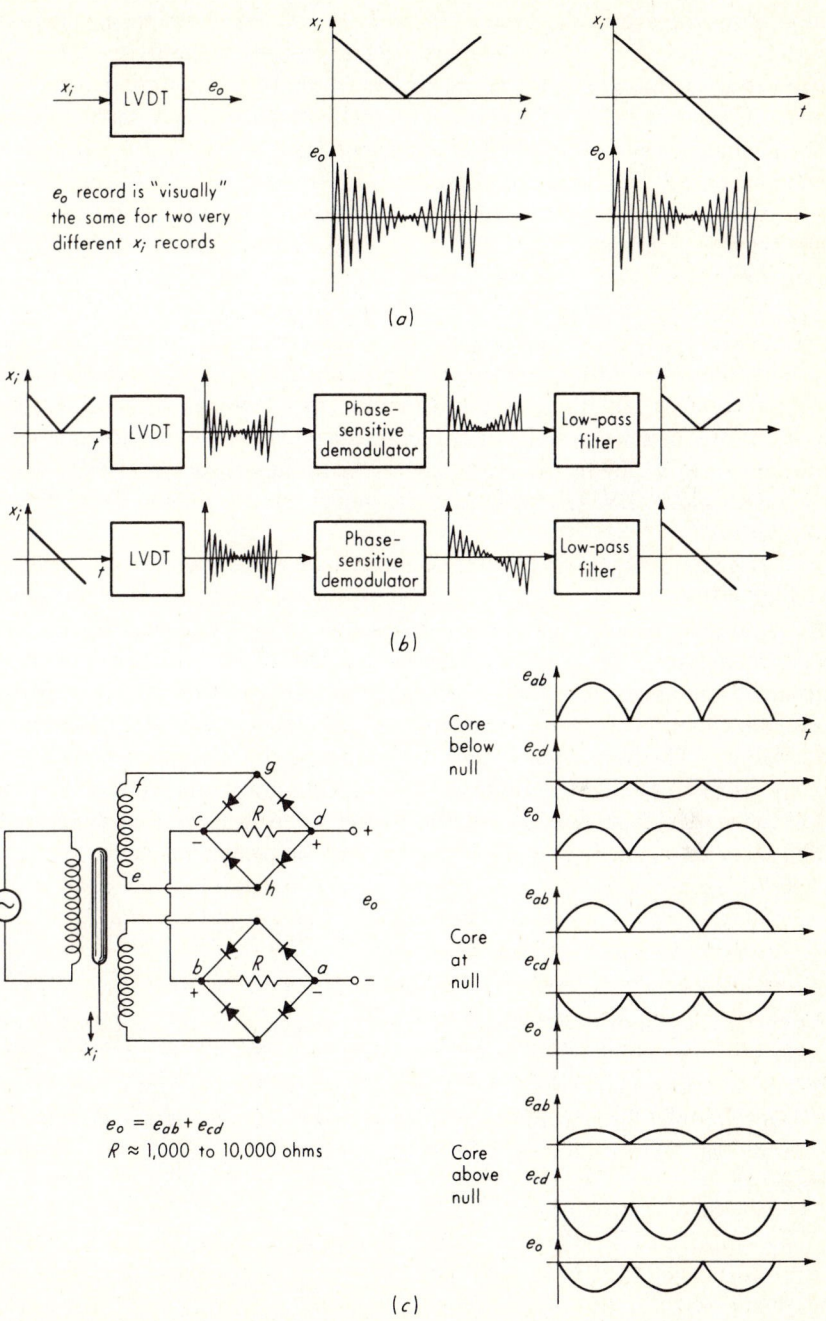

Fig. 4.17. *Demodulation and filtering.*

only in one direction; thus when f is positive and e is negative the current path is *efgcdhe*, while when f is negative and e positive the path is *ehcdgfe*. The current through R is therefore always from c to d. A similar situation exists in the lower diode bridge. For static or very slowly varying core displacements the voltage e_o may be applied directly to a d-c voltmeter. The meter will act as an electromechanical low-pass filter, the needle assuming a position corresponding to the average value of the rectified sine wave e_o. If motions both sides of null are to be measured, a meter with zero in the center of the scale will eliminate the need for switching lead wires when e_o goes negative. When rapid core motions are to be measured, this d-c meter arrangement is useless since the meter movement cannot follow variations more rapid than about 1 cps. It is then necessary to connect e_o of Fig. 4.17c to the input of a low-pass filter which will pass the frequencies present in x_i but reject all those (higher) frequencies produced by the modulation process. The design of such a filter is eased by making the LVDT excitation frequency much higher than the x_i frequencies.

If a frequency ratio of 10:1 or more is feasible, a simple RC filter as in Fig. 4.18a may be adequate. The output of this filter then becomes the input to an oscillograph or oscilloscope. For example, suppose we wish to measure a transient x_i whose Fourier transform has dropped to insignificant magnitude for all frequencies higher than 1,000 cps. Suppose also an LVDT system with an excitation frequency of 10,000 cps is available. The frequencies produced by the modulation process will thus lie in the band 19,000 to 21,000 cps. Suppose that we desire the "ripple" due to frequencies at 19,000 cps and higher to be no more than 5 percent. The filter time constant $\tau_f = R_f C_f$ can then be calculated as

$$0.05 = \frac{1}{\sqrt{[(19,000)(6.28)\tau_f]^2 + 1}} \qquad (4.39)$$

$$\tau_f = 0.00017 \text{ sec} \qquad (4.40)$$

At the highest motion frequency (1,000 cps) this filter has an amplitude ratio of 0.68 and a phase shift of $-47°$; thus it will distort the high-frequency portion of the x_i transient considerably. A more selective (sharper cutoff) filter would help this situation. Consider the double RC filter of Fig. 4.18b. The value of τ_f for a 5 percent ripple is now obtained from

$$0.05 = \frac{1}{[(19,000)(6.28)\tau_f]^2 + 1} \qquad (4.41)$$

$$\tau_f = 0.000037 \text{ sec} \qquad (4.42)$$

Now, at 1,000 cps the amplitude ratio is 0.98 and the phase angle is $-13°$. Since the phase angle of this filter from $\omega = 0$ to $\omega = 6,280$ rad/sec is

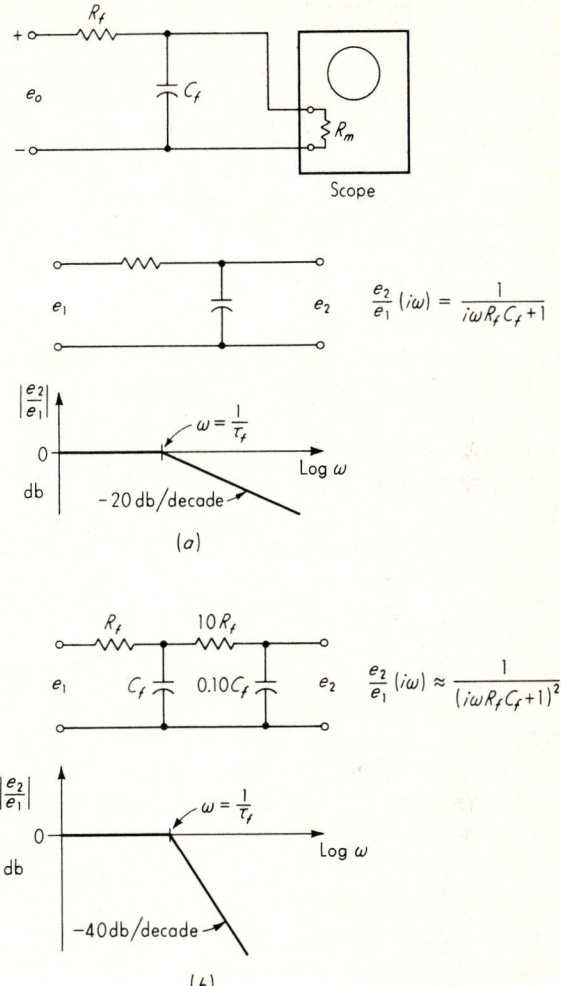

Fig. 4.18. *Filter frequency response.*

nearly linear and the amplitude ratio is nearly flat (1.0 to 0.98), the wave form of the transient will be faithfully reproduced but with a delay (dead time) of about $13/[(57.3)(6{,}280)] = 36\ \mu\text{sec}$. The above calculations give the desired value of τ_f but not R_f and C_f directly. In going from the demodulator circuit to the filter and then to the oscilloscope, transfer functions of the individual elements can be multiplied together only if no significant loading of the successive stages is present. As a rule of thumb the impedance level should go up about 10 to 1 for each successive stage.   A typical oscilloscope input resistance is about 10^6 ohms.   This suggests that in Fig. 4.18b $10R_f$ could be about 10^5 ohms and R_f thus 10^4 ohms.   In Fig. 4.17c the demodulator R can be of the order of 10^3 ohms, and so the overall chain should not show much loading effect and our above calculations should be fairly accurate.   In any case, experimental checks of the system should be performed to verify the final design.   If R_f of 10^4 ohms is satisfactory, C_f will then be $(37 \times 10^{-6})/10^4 = 0.0037$ μf.

The full-range stroke of commercially available translational LVDT's ranges from about ± 0.005 to about ± 3 in., with other sizes available as specials.   The nonlinearity of standard units is of the order of 0.5 percent of full scale, with 0.1 percent possible by selection.   Sensitivity with normal excitation voltage of 3 to 6 volts is of the order of 0.6 to 30 mv per 0.001 in., depending on frequency of excitation (higher frequency gives more sensitivity) and stroke (smaller strokes usually have higher sensitivity).   Some special units have sensitivity as high as 1 to 1.5 volts per 0.001 in.   Since the coupling variation due to core motion is a continuous phenomenon, the resolution of LVDT's is infinitesimal.   Amplification of the output voltage allows detection of motions down to a few microinches.   There is no physical contact between the core and the coil form; thus there is no friction or wear.   There are, however, small radial and longitudinal magnetic forces on the core if it is not centered radially and at the null position.   These are in the nature of magnetic "spring" forces in that they increase with motion from the equilibrium point.   They are rarely more than 0.1 to 0.3 g and are thus often negligible.   Rotary LVDT's have a nonlinearity of about ± 1 percent of full scale for travel of $\pm 40°$ and ± 3 percent for $\pm 60°$.   The sensitivity is of the order of 10 to 20 mv/deg.   The moving mass (core) of LVDT's is quite small, ranging from less than 0.1 g in small units to 5 g or more in larger ones.   There is a small radial clearance (air gap) between the core and the hole in which it moves.   Motion in the radial direction produces a small output signal but this undesirable transverse sensitivity is usually less than 1 percent of the longitudinal sensitivity.

The dynamic response of LVDT's is limited mainly by the excitation frequency, since it must be much higher than the core-motion frequencies so as to be able to distinguish between them in the amplitude-modulated output signal.   For adequate demodulation and filtering, a frequency ratio much less than 10:1 presents problems.   Since few differential transformers are designed to be excited by more than 20,000 cps, the useful range of motion frequencies is limited to about 2,000 cps.   This is adequate for many applications.   The mechanical loading effect on the measured system is mainly mass; thus the compliance analysis of Eq. (4.5), etc., is applicable.   If the small "magnetic spring" force is not negligible, Eq. (4.19), etc., should be used.

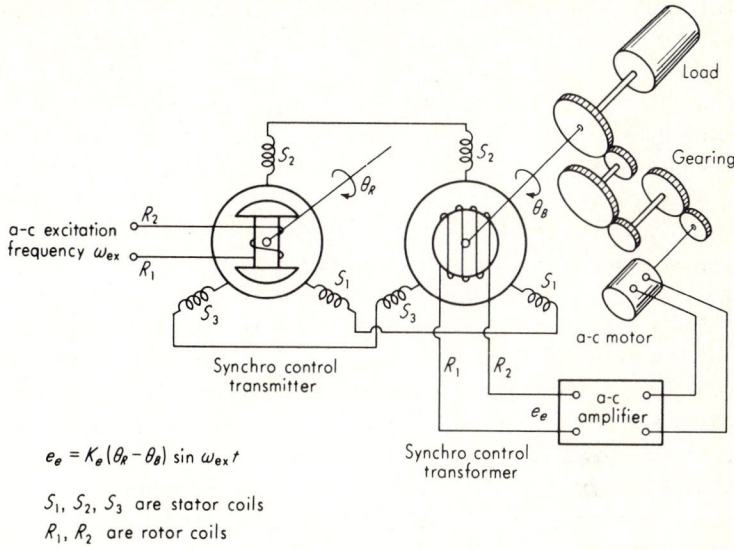

$$e_e = K_e(\theta_R - \theta_B)\sin\omega_{ex}t$$

S_1, S_2, S_3 are stator coils
R_1, R_2 are rotor coils

Fig. 4.19. *A-C servomechanism.*

Synchros and induction potentiometers. The term synchro is applied to a family of a-c electromechanical devices which, in various forms, perform the functions of angle measurement, voltage and/or angle addition and subtraction, remote angle transmission, and computation of rectangular components of vectors. In this section we are concerned only with the angle-measuring function; equipment for performing the other functions is covered in later appropriate chapters.

Synchros for angle measurement are most utilized as components of servomechanisms (automatic motion-control feedback systems) where they are used to measure and compare the actual rotational position of a load with its commanded position, as in Fig. 4.19. To perform this function two different types of synchros, the control transmitter and the control transformer, are used. The error voltage signal e_e is an a-c voltage of the same frequency as the excitation and of amplitude proportional (for small error angles) to the error angle $\theta_R - \theta_B$. Its phase changes by 180° at the null point; thus the direction of the error is detected. When $\theta_R = \theta_B$, the error voltage (and thus the amplifier output and motor input) is zero and the system stays at rest. If a command rotation θ_R is now put in, $e_e \neq 0$ and the motor will rotate so as to return θ_B to correspondence with θ_R.

The physical construction of the control transmitter and control transformer is identical except that the transmitter has a salient-pole

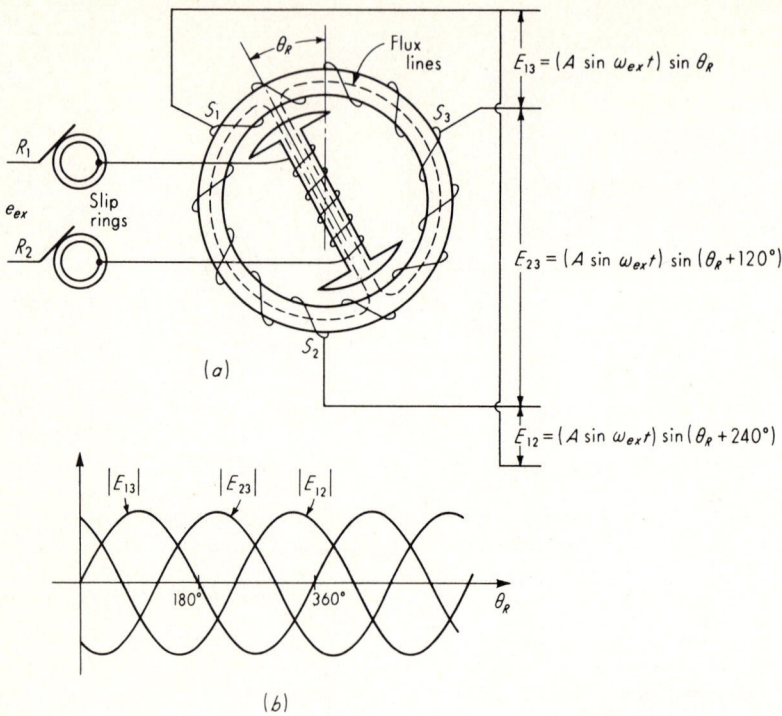

$E_{13} = (A \sin \omega_{ex} t) \sin \theta_R$

$E_{23} = (A \sin \omega_{ex} t) \sin (\theta_R + 120°)$

$E_{12} = (A \sin \omega_{ex} t) \sin (\theta_R + 240°)$

Fig. 4.20. *Synchro.*

("dumbbell") rotor while the transformer has a cylindrical rotor. The construction is similar to that of a wound-rotor induction motor. Figure 4.20*a* shows the coil arrangement of the transmitter alone. Basically, rotation of the rotor changes the mutual inductance (coupling) between the rotor coil and the stator coils. For a given stator coil the open-circuit output voltage is sinusoidal in time and varies in amplitude with rotor position, also sinusoidally, as shown in Fig. 4.20*b*. The three voltage signals from the stator coils uniquely define the angular position of the rotor. When these three voltages are applied to the stator coils of a control transformer, they produce a resultant magnetomotive force aligned in the same direction as that of the transmitter rotor. The rotor of the transformer acts as a "search coil" in detecting the direction of its stator field. If the axis of this coil is aligned with the field, the maximum voltage is induced into the transformer rotor coil. If the axis is perpendicular to the field, zero voltage is induced, giving the null position mentioned above. The output-voltage amplitude actually varies sinusoidally with the misalignment angle, but for small angles the sine and the angle are nearly equal, giving a linear output.

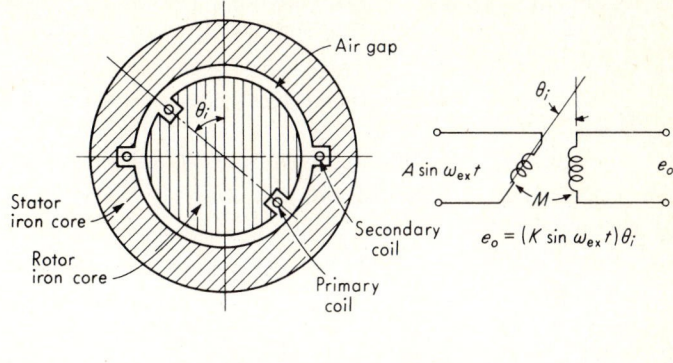

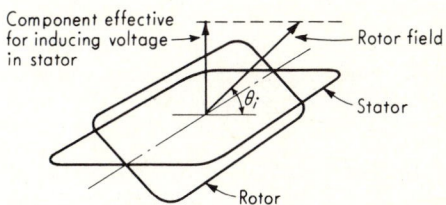

Fig. 4.21.   *Induction potentiometer.*

In an induction potentiometer there is one winding on the rotor and one on the stator. (Additional dummy windings are sometimes used to improve accuracy, however.) Both these windings are concentrated; thus for simplicity we show them as single-turn coils in Fig. 4.21. The primary winding (rotor) is excited with alternating current. This induces a voltage into the secondary (stator). The amplitude of this output voltage varies with the mutual inductance (coupling) between the two coils, and this varies with the angle of rotation. For single-turn (concentrated) coils the variation with angle would be sinusoidal and only a small linear range around null would be obtained. By carefully *distributing* the rotor and stator windings a linear relation for up to $\pm 90°$ rotation may be obtained.

While synchros and induction pots could be designed to work at a variety of excitation frequencies, standard commercial units are generally available only for 60 or 400 cps. The physical size ranges from about $\frac{1}{2}$- to 3-in. diameter. Sensitivities of both synchros and induction pots are of the order of 1 volt/degree rotation while the residual voltage at null is of the order of 10 to 100 mv. For a standard synchro transmitter-transformer pair the misalignment of the two shafts when rotated from an originally established electrical null to any other null position within a complete rotation is of the order of 10 angular minutes. This type of error puts a basic limit on the positioning accuracy of servo systems using synchros. Synchro pairs and induction pots are capable of continuous rotation although the linear range of induction pots is limited to about ± 60 to $\pm 90°$. Within this range the nonlinearity is of the order of 0.25 percent.

Just as in LVDT's, synchros and induction pots require some sort of phase-sensitive demodulation to obtain a signal of the same form as the mechanical-motion input. When used in a-c servomechanisms, the conventional two-phase a-c servomotor itself accomplishes this function without any additional equipment. In a strictly measurement (as opposed to control) application, some sort of phase-sensitive demodulator is needed if an electrical output signal of the same form as the mechanical input is required. The dynamic response is limited by the excitation frequency and demodulator filtering requirements, just as in LVDT's. The mechanical loading of these rotary components on the measured system is mainly the inertia of the rotor.

Variable-inductance and variable-reluctance pickups. Closely related to LVDT's and synchros but in practice distinguished from them by name is a family of motion pickups variously called variable-inductance, variable-reluctance, or variable-permeance (permeance is the reciprocal of reluctance) pickups or transducers. The terminology used for these pickups is not uniform nor necessarily descriptive of their basic principles of operation. We are here concerned mainly with describing some common examples rather than trying to develop a systematic nomenclature.

Figure 4.22*a* shows the arrangement of a typical translational variable-inductance transducer. Outwardly the physical size and shape are very similar to an LVDT. Again there is a movable iron core which provides the mechanical input. However, only two inductance coils are present; they generally form two legs of a bridge which is excited with alternating current of 5 to 30 volts at 60 to 5,000 cps. With the core at the null position, the inductance of the two coils is equal, the bridge is balanced, and e_o is zero. A core motion from null causes a change in the reluctance of the magnetic paths for each of the coils, increasing one and decreasing the other. This reluctance change causes a proportional change in inductance for each coil, a bridge unbalance, and thus an output voltage e_o. By careful construction, e_o can be made a nearly linear function of x_i over the rated displacement range.

Two alternative methods of forming the bridge circuit are shown in Fig. 4.22*b*. The total transducer impedance (Z_1 plus Z_2) at the excitation frequency is of the order of 100 to 1,000 ohms. The resistors R are usually about the same value as Z_1 and Z_2, and the input impedance of

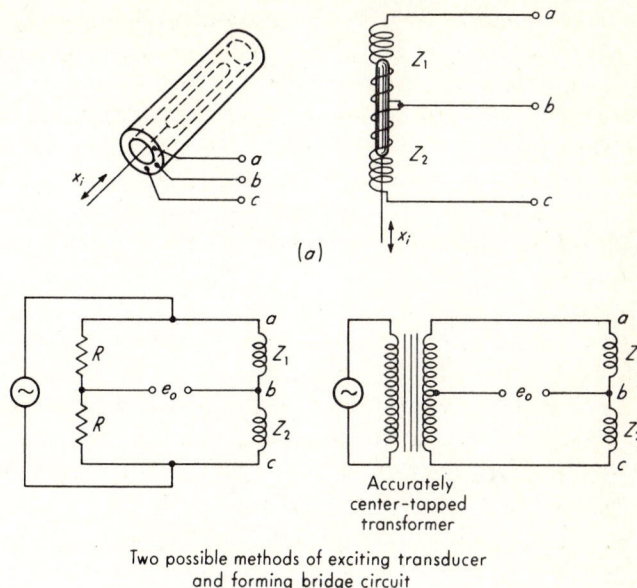

(a)

Accurately
center-tapped
transformer

Two possible methods of exciting transducer
and forming bridge circuit

(b)

Fig. 4.22. *Variable-inductance pickup.*

the voltage-measuring device at e_o should be at least $10R$. If the bridge output must be worked into a low-impedance load, R must be quite small. To get high sensitivity, high excitation voltage is needed; this causes a high power loss (heating) in the resistors R. To solve this problem, a center-tapped transformer circuit may be used. Here the bridge is mainly inductive, and less power is consumed with corresponding less heating.

Such variable-inductance transducers are available in strokes of about 0.1 in. to as much as 200 in. The resolution is infinitesimal, and the nonlinearity ranges from about 1 percent of full scale for standard units to 0.02 percent for special units of rather long stroke. Sensitivity is of the order of 5 to 40 volts/in. Rotary versions using specially shaped rotating cores have a nonlinearity of the order of 0.5 to 1 percent of full scale over a $\pm 45°$ range. The sensitivity is about 0.1 volt/degree.

Figure 4.23 shows another common version of the variable-reluctance principle. This particular application is an accelerometer for measurement of accelerations in the range $\pm 4g$. Since the force required to accelerate a mass is proportional to the acceleration, the springs supporting the mass in Fig. 4.23 deflect in proportion to the acceleration; thus a displacement measurement allows an acceleration measurement. The

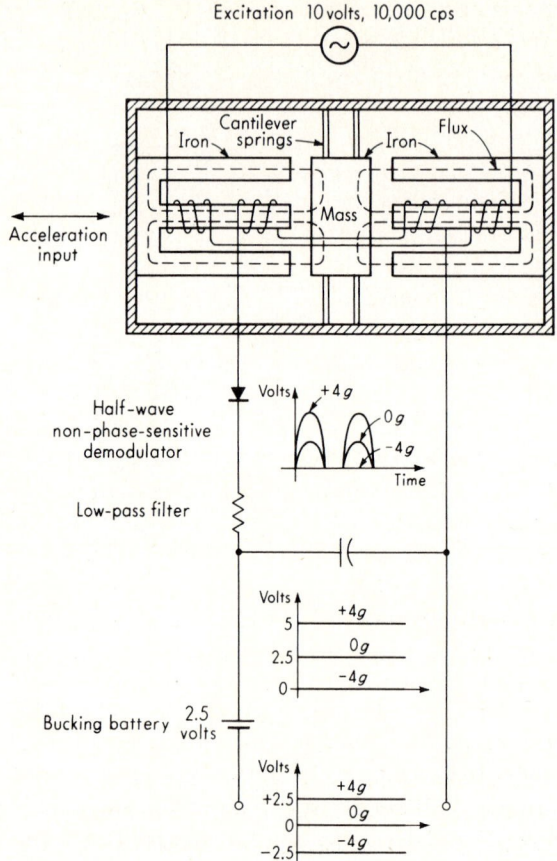

Fig. 4.23. *Variable-reluctance accelerometer.*

mass is of iron and thus serves as both an inertial element for transducing acceleration to force and also as a magnetic circuit element for transducing motion to reluctance.

We shall consider the complete instrument here since it has several features of general interest with regard to displacement measurement. Ordinarily, such an instrument would be constructed so that the iron core would be halfway between the two E frames when the acceleration was zero, thus giving zero output voltage for zero acceleration. However, to detect motion on both sizes of zero (corresponding to plus or minus accelerations) a fairly involved phase-sensitive demodulator would be required. It was desired to save the cost, weight, and space of this demodulator, and so another solution (which can also be used with LVDT's and similar devices) was proposed. With zero-acceleration input the iron core and springs were adjusted so that the core was offset to one side by an amount equal to the spring deflection corresponding to $4g$ acceleration. Thus, with no acceleration applied, the output voltage was not zero but some specific value (2.5 volts in this particular case). Then, when $+4g$ of acceleration was applied the output went to 5.0 volts, and when $-4g$ was applied the output went to zero. In this way a relatively simple demodulator and filter circuit can be used to provide direction-sensitive motion measurement. The main drawback of this scheme (which argues against its use except when necessary) is the loss of linearity. This is because the greatest linearity is found around the null position. Thus for a given total stroke it is better to put one-half of it on each side of null rather than all on one side, as in the above scheme.

Returning to the basic motion-measuring principle of Fig. 4.23, the primary coils set up a flux dependent on the reluctance of the magnetic path. The main reluctance is the air gap. When the core is in the neutral position, the flux is the same for both halves of the secondary coil, and since they are connected in series opposition the net output voltage is zero. A motion of the core increases the reluctance (air gap) on one side and decreases it on the other, causing more voltage to be induced into one half of the secondary coil than the other and thus a net output voltage. Motion in the other direction causes the reverse action, with a 180° phase shift occurring at null. The output voltage is half-wave, non-phase-sensitive rectified (demodulated) and filtered to produce an output of the same form as the acceleration input. If the 2.5-volt output for zero-acceleration input is objectionable, it can be bucked out with a 2.5-volt battery of opposite polarity connected externally to the accelerometer. The actual full-scale motion of the mass in this particular instrument is just a few thousandths of an inch, giving a displacement sensitivity for the variable-reluctance element of almost 1,000 volts/in.

The final variable-reluctance element we shall consider is the Microsyn,[1] a rotary component shown in Fig. 4.24 and widely used in sensitive gyroscopic instruments. The sketch shows the instrument in the null position where the voltages induced in coils 1 and 3 (which aid each other) are just balanced by those of coils 2 and 4 (which also aid each other but oppose 1 and 3). Motion of the input shaft from the null (say clockwise) increases the reluctance (decreases the induced voltage) of coils 1 and 3 and decreases the reluctance (increases the voltage) of coils 2 and 4, thus giving a net output voltage e_o. Motion in the opposite direction causes a similar effect except the output voltage has a 180° phase

---

[1] P. H. Savet (ed.), "Gyroscopes: Theory and Design," p. 332, McGraw-Hill Book Company, New York, 1961.

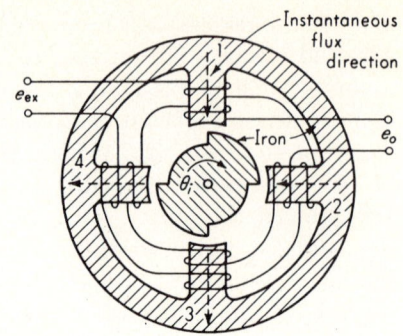

Fig. 4.24. Microsyn.

shift.   If a direction-sensitive d-c output is required, a phase-sensitive demodulator is necessary.

The excitation voltage is 5 to 50 volts at 60 to 5,000 cps.   Sensitivity is of the order of 0.2 to 5 volts/degree rotation.   Nonlinearity is about 0.5 percent of full scale for $\pm 7°$ rotation and 1.0 percent for $\pm 10°$.   The null voltage is extremely small, being less than the output signal generated by 0.01° of rotation; thus very small motions can be detected.   The magnetic-reaction torque is also extremely small.   Since there are no coils on the rotor, no slip rings (with their attendant friction) are needed.

Capacitance pickups.   A rotational or translatory motion may be used in many ways to change the capacitance of a variable capacitor.[1,2,3]   The resulting capacitance change can be converted to a usable electrical signal by means of a variety of circuitry.[1,2,3]   We here consider only a few typical applications.   Capacitance-type pickups tend to be special-purpose devices developed for a particular problem rather than ready-made commercial devices.   The associated electronics is somewhat more complex than for the more common types of transducers.   However, their mechanical simplicity, very small mechanical loading effects, and potential high sensitivity make them attractive in a number of applications, and some commercial general-purpose devices now available are quite convenient to use.

The most common form of variable capacitor used in motion transducers is the parallel-plate capacitor with a variable air gap.   Theory

[1] C. H. Harris and C. E. Crede (eds.), "Shock and Vibration Handbook," vol. 1, p. 14-1, McGraw-Hill Book Company, New York, 1961.

[2] H. C. Roberts, "Mechanical Measurements by Electrical Methods," chaps. 3 and 9, The Instruments Publishing Co., Pittsburgh, Pa., 1951.

[3] R. R. Batcher and W. Moulic, "The Electronic Control Handbook," chaps. 2 and 3, The Instruments Publishing Co., Pittsburgh, Pa., 1946.

gives the capacitance of such an arrangement as

$$C = \frac{0.225A}{x} \qquad (4.43)$$

where $C \triangleq$ capacitance, pf
$A \triangleq$ plate area, in.^{2}
$x \triangleq$ plate separation, in.

For example, the capacitance of an air capacitor with 1-in.^{2} plates separated by 0.01 in. is 22.5 pf. The impedance $1/i\omega C$ of this capacitor at a frequency of, say, 10,000 cps has a magnitude of 708,000 ohms. This very high impedance level of capacitance gages is responsible for some of their main problems with respect to spurious noise voltages, sensitivity to length and position of connecting cables, and requirement for high-input-impedance electronics. From Eq. (4.43) we also note that the variation of capacitance with plate separation x is nonlinear (hyperbolic); thus the percentage change in x from a chosen "neutral" position must be small if good linearity is to be achieved. The sensitivity of capacitance to changes in plate separation may be computed from Eq. (4.43):

$$\frac{dC}{dx} = -\frac{0.225A}{x^2} \qquad (4.44)$$

We note that the sensitivity increases as x decreases. However, the *percentage* change in C is equal to the *percentage* change in x for small changes about *any* neutral position, as shown by

$$\frac{dC}{dx} = -\frac{C}{x} \qquad (4.45)$$

$$\frac{dC}{C} = -\frac{dx}{x} \qquad (4.46)$$

Perhaps the simplest useful circuit is that employed with capacitor microphones (see Sec. 6.10 on microphones for a detailed analysis). Figure 4.25 shows the arrangement. When the capacitor plates are stationary with a separation x_0, no current flows and $e_o = E_b$. If there is then a relative displacement x_1 from the x_0 position, a voltage $e_1 \triangleq e_o - E_b$ is produced and is related to x_1 by

$$\frac{e_1}{x_1}(D) = \frac{K\tau D}{\tau D + 1} \qquad (4.47)$$

where
$$K \triangleq \frac{E_b}{x_0}, \text{ volts/in.} \qquad (4.48)$$

$$\tau \triangleq 0.225 \times 10^{-12} \frac{AR}{x_0} \text{ sec} \qquad (4.49)$$

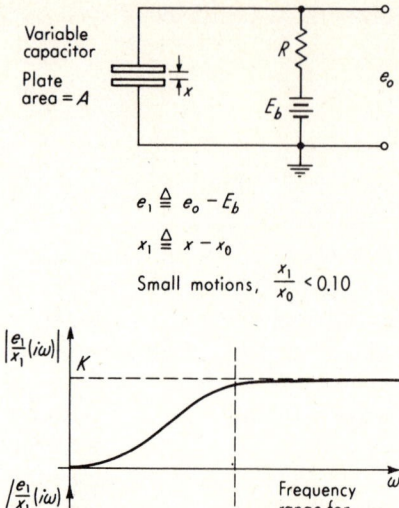

$$e_1 \triangleq e_o - E_b$$

$$x_1 \triangleq x - x_0$$

Small motions, $\dfrac{x_1}{x_0} < 0.10$

Fig. 4.25. *Capacitive transducer.* $\dfrac{e_1}{x_1}(i\omega) = \dfrac{Ki\omega\tau}{i\omega\tau+1}$

Equation (4.47) shows that this arrangement does not allow measurement of static displacements since e_1 is zero in steady state for any value of x_1. For sufficiently rapid variations in x_1, however, the signal e_1 will faithfully measure the motion. This is most easily seen from the frequency response

$$\frac{e_1}{x_1}(i\omega) = \frac{K\tau i\omega}{i\omega\tau + 1} \qquad (4.50)$$

which becomes for $\omega\tau \gg 1$

$$\frac{e_1}{x_1}(i\omega) \approx K \qquad (4.51)$$

Thus e_1 follows x_1 accurately under these conditions. A microphone usually need not measure sound pressures slower than about 20 cps, and so the above arrangement is perfectly satisfactory. To make $\omega\tau \gg 1$ for low frequencies requires a large τ. For a given capacitor and x_0, the value of τ can be increased only by increasing R. Typically, R will be 10^6 ohms or more. Thus to prevent loading of the capacitance transducer circuit the readout device connected to the e_o terminals must have a

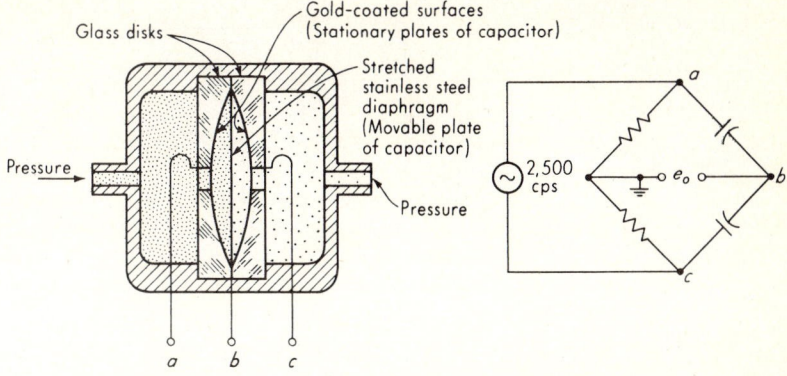

Fig. 4.26. *Differential-capacitor pressure pickup.*

high (10^7 ohms or more) input impedance.   This usually requires use of a cathode-follower type of isolation amplifier.

The use of a variable differential (three-terminal) capacitor with a bridge circuit is shown in the Equibar[1] differential pressure transducer of Fig. 4.26.   Spherical depressions of a depth of about 0.001 in. are ground into the glass disks; then these depressions are gold-coated to form the fixed plates of a differential capacitor.   A thin stainless-steel diaphragm is clamped between the disks and serves as the movable plate.   With equal pressures applied to both ports, the diaphragm is in a neutral position, the bridge is balanced, and e_o is zero.   If one pressure is greater than the other, the diaphragm deflects in proportion, giving an output at e_o in proportion to the differential pressure.   For the opposite pressure difference, e_o exhibits a 180° phase change.   The high impedance level again requires a cathode-follower amplifier at e_o.   A direction-sensitive d-c output can be obtained by conventional phase-sensitive demodulation and filtering.   Balance resistors necessary for initially nulling the bridge are not shown in Fig. 4.26.   This method (as opposed to that of Fig. 4.25) allows measurement of static deflections.   Such differential-capacitor arrangements also exhibit considerably greater linearity than do single capacitor types.[2]

An ingenious method[3] of circumventing the nonlinear relationship [Eq. (4.43)] between x and C is shown in Fig. 4.27.   This technique employs a high-gain feedback amplifier using an approach common in

[1] Trans-Sonics, Inc., Burlington, Mass.
[2] N. H. Cook and E. Rabinowicz, "Physical Measurement and Analysis," p. 142, Addison-Wesley Publishing Company, Inc., Reading, Mass., 1963.
[3] Wayne Kerr Corp., Philadelphia, Pa.

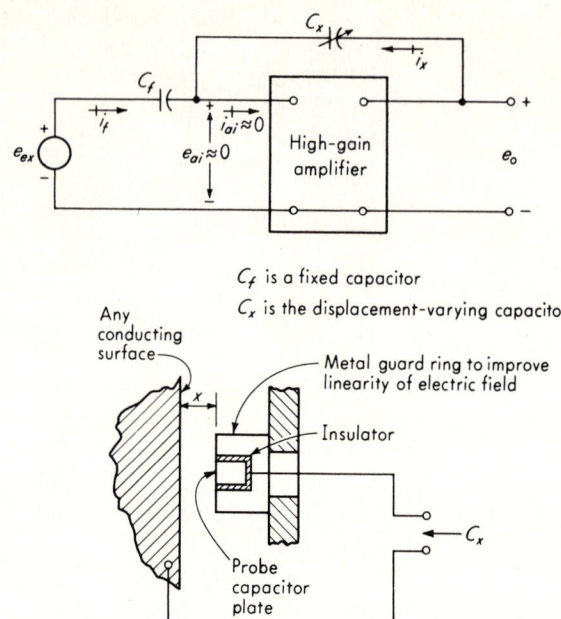

C_f is a fixed capacitor

C_x is the displacement-varying capacitor

Fig. 4.27. *Feedback-type capacitive pickup.*

electronic analog computers.   The assumptions necessary to an analysis of this circuit depend on the following characteristics of so-called "operational" amplifiers:

1. The input impedance is so high that the amplifier input current may be taken as zero relative to other currents.
2. The gain is so high that if the output voltage of the amplifier is not saturated the input voltage is extremely small and may be taken as zero relative to other voltages.   For example, a typical amplifier has linear output for the range ± 100 volts and a gain of 10^8 volts/volt. Thus the maximum input for linear operation is 10^{-6} volt.

Using these assumptions in Fig. 4.27, we can write

$$\frac{1}{C_f} \int i_f \, dt = e_{ex} - e_{ai} = e_{ex} \qquad (4.52)$$

$$\frac{1}{C_x} \int i_x \, dt = e_o - e_{ai} = e_o \qquad (4.53)$$

$$i_f + i_x - i_{ai} = 0 = i_f + i_x \qquad (4.54)$$

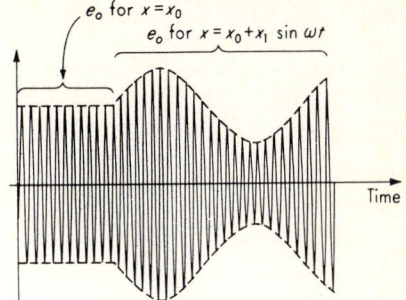

Fig. 4.28. *Wave form for sinusoidal displacement.*

Manipulation then gives

$$e_o = \frac{1}{C_x} \int i_x \, dt = - \frac{1}{C_x} \int i_f \, dt = - \frac{C_f}{C_x} e_{ex} \qquad (4.55)$$

$$e_o = - \frac{C_f x e_{ex}}{0.225A} = Kx \qquad (4.56)$$

Equation (4.56) shows that the output voltage is now *directly* proportional to the plate separation x; linearity is thus achieved for both large and small motions. In the commercial instrument described above, e_{ex} is a 50-kc sine wave of fixed amplitude. The output e_o is also a 50-kc sine wave which is rectified and applied to a d-c voltmeter calibrated directly in distance units.

For vibratory displacements, e_o will be an amplitude-modulated wave as in Fig. 4.28. The average value of this wave after rectification is still the mean separation of the plates and can still be read by the same meter as used for static displacements. The vibration amplitude around this mean position is extracted by applying the e_o signal also to a demodulator and a low-pass filter with cutoff at 10 kc. The output of this filter is applied to a peak-to-peak voltmeter directly calibrated in vibration amplitude and is also available at a jack for connection to an oscilloscope for viewing of the vibration wave form. The instrument is provided with six different probes ranging from 0.0447 to 1.0 in. in diameter of the capacitance plate and covering the full-scale displacement ranges of 0.001, 0.005, 0.01, and 0.5 in., respectively. The overall system accuracy is of the order of 2 percent of full scale. Any flat conductive surface may serve as the second plate of the variable capacitor. Thus in vibrating machine parts the parts themselves may often perform this function. The resolution of 0.5 percent of full scale indicates that (with the 0.001-in. full-scale probe) it is possible to detect motion as small as 5 μin.

The final capacitance-type device we consider here involves the

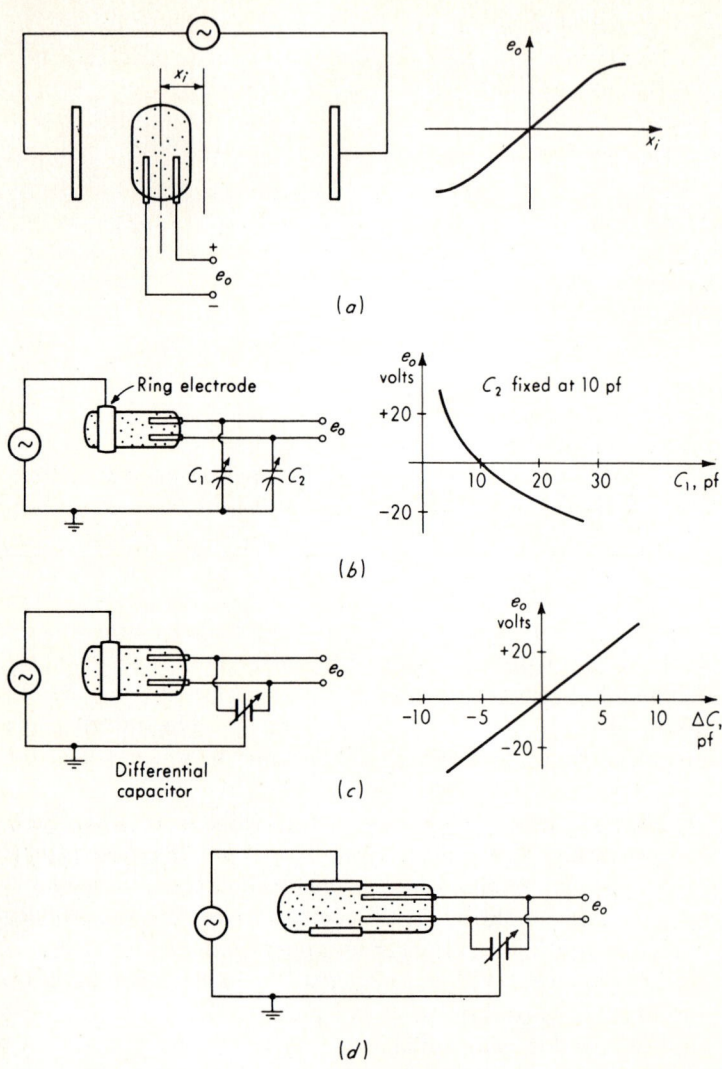

Fig. 4.29. *Ionization transducer.*

use of the T-42 ionization transducer.[1,2]   While this transducer system involves other fundamental physical effects in addition to capacitance, we include it here because the variation of capacitance is used in many

[1] K. S. Lion, Mechanic-Electric Transducer, *Rev. Sci. Instr.*, vol. 27, no. 4, pp. 222–225, April, 1956.

[2] The Decker Corp., Bala-Cynwyd, Pa.

of its practical applications. The basic physical effect employed in the tube shown in Fig. 4.29a is the development of a d-c voltage across the internal electrodes of a glass tube, which contains gas under reduced pressure, when the tube is exposed to an electric field created by the external parallel-plate electrodes connected to a radio-frequency (250 kc in the commercial version) voltage source. This d-c voltage varies with the position x_i, being zero at the neutral position and varying linearly (with a polarity change at null) over a small range on either side. The sensitivity of this arrangement can be as high as several thousand volts per millimeter.

While the basic arrangement of Fig. 4.29a has been employed in several practical transducers, a modified form of the tube which employs capacitance-variation techniques has been found useful for general-purpose motion measurement. Figure 4.29b and c shows these modifications and typical performance. A further refinement leading to increased tube life is pictured in Fig. 4.29d. This arrangement has a sensitivity of about 2 volts/pf capacitance change for an initial capacitance of 10 pf. The output impedance of all these transducer systems is high (of the order of 1 megohm); thus cathode-follower-type amplifiers are needed at the output. Commercial instruments based on Fig. 4.29d plus associated electronics give about 0.2 volt/percent change in capacitance in the preferred range of 5 to 20 pf. Frequency response is from zero to about 1,000 cps. Nonlinearity depends on the capacitor arrangement used and is typically about 1 percent of full scale.

Piezoelectric transducers. When certain solid materials are deformed, they generate within them an electric charge. This effect is reversible in that if a charge is applied the material will mechanically deform in response. These actions are given the name piezoelectric (pī-ēzō-electric) effect.[1] This electromechanical energy-conversion principle is usefully applied in both directions. The mechanical-input/electrical-output direction is the basis of many instruments for measuring acceleration, force, and pressure. It also can be used as a means of generating high-voltage low-current electrical power such as is used in spark-ignition engines and electrostatic dust filters. The electrical-input/mechanical-output direction is applied in small vibration shakers, sonar systems for acoustic detection and location of underwater objects, and industrial ultrasonic nondestructive test equipment.

The materials that exhibit a significant and useful piezoelectric effect fall into two main groups: natural (quartz, rochelle salt) and synthetic (lithium sulfate, ammonium dihydrogen phosphate) crystals

---

[1] Harris and Crede, *op. cit.*, chap. 16, pt. II.

and polarized ferroelectric ceramics (barium titanate). Because of their natural asymmetrical structure, the crystal materials exhibit the effect without further processing. The ferroelectric ceramics must be artificially polarized by applying a strong electric field to the material (while it is heated to a temperature above the Curie point of that material) and then slowly cooling with the field still applied. (The Curie temperature is the temperature above which a material loses its ferroelectric properties; thus it limits the highest temperature at which such materials may be used.) When the external field is removed from the cooled material a remanent polarization is retained and the material will now exhibit the piezoelectric effect.

The piezoelectric effect can be made to respond to (or cause) mechanical deformations of the material in many different modes, such as thickness expansion, transverse expansion, thickness shear, and face shear. The mode of motion effected depends on the shape and orientation of the body relative to the crystal axes and the location of the electrodes. Metal electrodes are plated onto selected faces of the piezoelectric material so that lead wires can be attached for bringing in or leading out the electrical charge. Since the piezoelectric materials are insulators, the electrodes also become the plates of a capacitor. A piezoelectric element used for converting mechanical motion to electrical signals may thus be. thought of as a charge generator and a capacitor. Mechanical deformation generates a charge; this charge then results in a definite voltage appearing between the electrodes according to the usual law for capacitors, $E = Q/C$. The piezoelectric effect is direction-sensitive in that tension produces a definite voltage polarity while compression produces the opposite.

We shall illustrate the main characteristics of piezoelectric motion-to-voltage transducers by considering only one common mode of deformation, thickness expansion. For this mode the physical arrangement is as in Fig. 4.30b. Various double-subscripted physical constants are used to describe numerically the phenomena occurring. The convention is that the first subscript refers to the direction of the electrical effect and the second to that of the mechanical effect, using the axis-numbering system of Fig. 4.30a.

Two main families of constants, the g constants and the d constants, will be considered. For a barium titanate thickness-expansion device the pertinent g constant is g_{33}, which is defined as

$$g_{33} \triangleq \frac{\text{field produced in direction 3}}{\text{stress applied in direction 3}} = \frac{e_o/t}{f_i/(wl)} \qquad (4.57)$$

Thus if one knows g for a given material and the dimension t he can calculate the output voltage per unit applied stress. Typical g values are

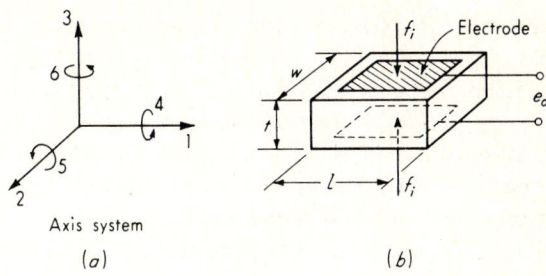

Fig. 4.30.   *Piezoelectric transducer.*

12×10^{-3} (volt/m)/(newtons/m²) for barium titanate and 50×10^{-3} for quartz. Thus, for example, a quartz crystal 0.1 in. thick would have a sensitivity of 0.88 volt /psi, illustrating the large voltage output for small stress typical of piezoelectric devices.

To relate applied force to generated charge the d constants can be defined as

$$d_{33} \triangleq \frac{\text{charge generated in direction 3}}{\text{force applied in direction 3}} = \frac{Q}{f_i} \qquad (4.58)$$

Actually, d_{33} can be calculated from g_{33} if the dielectric constant ϵ of the material is known, since

$$C = \frac{\epsilon w l}{t} \qquad (4.59)$$

$$g_{33} \triangleq \frac{\text{field}}{\text{stress}} = \frac{e_o w l}{t f_i} = \frac{e_o C}{\epsilon f_i} = \frac{Q}{\epsilon f_i} = \frac{d_{33}}{\epsilon} \qquad (4.60)$$

$$d_{33} = \epsilon g_{33} \qquad (4.61)$$

The dielectric constant of quartz is about 4.06×10^{-11} farad/m while for barium titanate it is $1{,}250 \times 10^{-11}$. For quartz, then,

$$d_{11} = \epsilon g_{11} = (4.06 \times 10^{-11})(50 \times 10^{-3}) = 2.03 \text{ pcoul/newton} \qquad (4.62)$$

(The subscripts 11 are used because in quartz the thickness-expansion mode is along the crystallographic axis conventionally called axis 1.) Sometimes it is desired to express the output charge or voltage in terms of deflection (rather than stress or force) of the crystal since it is really the *deformation* that causes the charge generation. To do this, one must know the modulus of elasticity, which is 8.6×10^{10} newtons/m² for quartz and 12×10^{10} for barium titanate.

With the above brief introduction as background, we now proceed to consider piezoelectric elements as displacement transducers. The ultimate purpose is generally force, pressure, or acceleration measurement but we shall consider only the conversion from displacement to voltage. For analysis purposes it is necessary to consider the transducer, connecting cable, and associated amplifier as a unit. The transducer impedance is generally very high; thus the amplifier is usually a high-impedance cathode follower used for isolation purposes rather than voltage gain. The cable capacitance can be significant, especially for long cables. For the transducer alone, if a static deflection x_i is applied and maintained, a transducer terminal voltage will be developed but the charge will slowly leak off through the leakage resistance of the transducer. Since R_{leak} is generally very large (the order of 10^{11} ohms) this decay would be very slow, perhaps allowing at least a quasi-static response. However, when an external voltage-measuring device of low input impedance is connected to the transducer the charge leaks off very rapidly, preventing the measurement of static displacements. Even relatively high-impedance cathode followers do not generally allow static measurements. Some commercially available[1] systems using quartz transducers (very high leakage resistance) and electrometer input amplifiers (very high input impedance) achieve an effective total resistance of 10^{14} ohms which gives a sufficiently slow leakage to allow static measurements.

To put the above discussion on a quantitative basis, we consider Fig. 4.31. The charge generated by the crystal can be expressed as

$$q = K_q x_i \qquad (4.63)$$

where
$$K_q \triangleq \text{coul/in.} \qquad (4.64)$$
$$x_i \triangleq \text{deflection, in.} \qquad (4.65)$$

The resistances and capacitances of Fig. 4.31b can be combined as in 4.31c. We also convert the charge generator to a more familiar current generator according to

$$i_{cr} = \frac{dq}{dt} = K_q \left(\frac{dx_i}{dt} \right) \qquad (4.66)$$

We may then write

$$i_{cr} = i_C + i_R \qquad (4.67)$$

$$e_o = e_C = \frac{\int i_C \, dt}{C} = \frac{\int (i_{cr} - i_R) \, dt}{C} \qquad (4.68)$$

$$C \left(\frac{de_o}{dt} \right) = i_{cr} - i_R = K_q \left(\frac{dx_i}{dt} \right) - \frac{e_o}{R} \qquad (4.69)$$

$$\frac{e_o}{x_i} (D) = \frac{K\tau D}{\tau D + 1} \qquad (4.70)$$

where
$$K \triangleq \text{sensitivity} \triangleq \frac{K_q}{C} \text{ volts/in.} \qquad (4.71)$$
$$\tau \triangleq \text{time constant} \triangleq RC, \text{ sec} \qquad (4.72)$$

[1] Kistler Instrument Corp., North Tonawanda, N.Y.

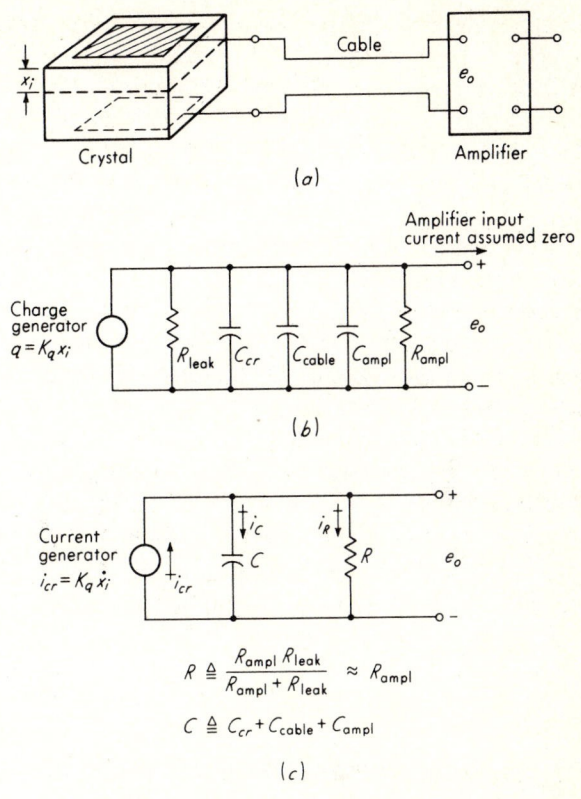

Fig. 4.31. *Equivalent circuit for piezoelectric transducer.*

We see that, just as in the capacitance pickup of Fig. 4.25, the steady-state response to a constant x_i is zero; thus we cannot measure static displacements. For a flat amplitude response within, say, 5 percent the frequency must exceed ω_1, where

$$(0.95)^2 = \frac{(\omega_1 \tau)^2}{(\omega_1 \tau)^2 + 1} \qquad (4.73)$$

$$\omega_1 = \frac{3.04}{\tau} \qquad (4.74)$$

Thus a large τ gives an accurate response at lower frequencies.

The response of these transducers is further illuminated by considering the displacement input of Fig. 4.32. The differential equation is

$$(\tau D + 1)e_o = (K\tau D)x_i \qquad (4.75)$$

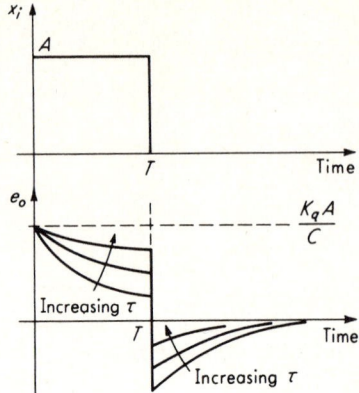

Fig. 4.32. *Pulse response of piezoelectric transducer.*

Since $x_i = A$ for $0 < t < T$, this becomes

$$(\tau D + 1)e_o = 0 \qquad (4.76)$$

Now at $t = 0^+$ the displacement x_i is A and so the charge *suddenly* increases to $K_q A$, and the crystal capacitor voltage and therefore e_o *suddenly* increases to $K_q A / C$. Thus our initial condition is

$$e_o = \frac{K_q A}{C} \qquad \text{at } t = 0^+ \qquad (4.77)$$

Solving Eq. (4.76) with initial condition (4.77) gives

$$e_o = \frac{K_q A}{C} e^{-t/\tau} \qquad 0 < t < T \qquad (4.78)$$

Equation (4.78) holds until $t = T$. At this instant we must stop using it because of the change in x_i. For $T < t < \infty$ the differential equation is

$$(\tau D + 1)e_o = 0 \qquad (4.79)$$

At $t = T^-$ Eq. (4.78) is still valid and

$$e_o = \frac{K_q A}{C} e^{-T/\tau} \qquad (4.80)$$

Now, at $t = T$, x_i suddenly drops an amount A, causing a sudden decrease in charge of $K_q A$ and a sudden decrease in e_o of $K_q A / C$ from its value at $t = T^-$. Thus at $t = T^+$, e_o is given by

$$e_o = \frac{K_q A}{C} (e^{-T/\tau} - 1) \qquad (4.81)$$

which becomes the initial condition for Eq. (4.79). The solution then becomes

$$e_o = \frac{K_q A}{C} (e^{-T/\tau} - 1)e^{-(t-T)/\tau} \qquad T < t < \infty \qquad (4.82)$$

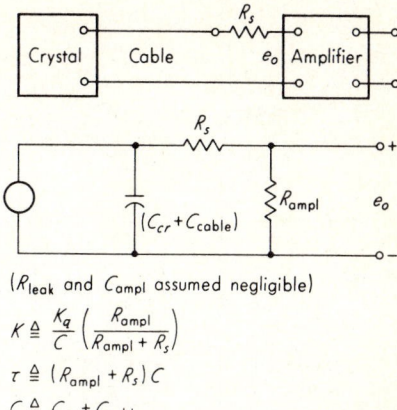

$(R_{leak}$ and C_{ampl} assumed negligible$)$

$$K \triangleq \frac{K_q}{C}\left(\frac{R_{ampl}}{R_{ampl}+R_s}\right)$$

Fig. 4.33. *Use of series resistor to in-crease time constant.*

$$\tau \triangleq (R_{ampl}+R_s)\,C$$

$$C \triangleq C_{cr}+C_{cable}$$

Figure 4.32 shows the complete process for three different values of τ. It is clear that a large τ is desirable for faithful reproduction of x_i. If the decay and "undershoot" at $t = T$ is to be kept within, say, 5 percent of the true value, τ must be at least $20T$. If an increase of τ is required in a specific application, it may be achieved by increasing either or both R and C. An increase in C is easily obtained by connecting an external shunt capacitor across the transducer terminals, since shunt capacitors add directly. The price paid for this increase in τ is a loss of sensitivity according to $K = K_q/C$. This may often be tolerated because of the initial high sensitivity of piezoelectric devices. An increase in R generally requires an amplifier of greater input resistance. If sensitivity can be sacrificed, a series resistor connected external to the amplifier, as shown in Fig. 4.33, will increase τ without the need of obtaining a different amplifier.

Detailed data on static and dynamic performance characteristics of piezoelectric transducers will be deferred to the respective sections on force, pressure, and acceleration measurement where these data will be more meaningful.

Electro-optical devices. The combination of classical optical principles with modern electronic developments has led to a variety of useful electro-optical measuring instruments. The Midarm system of Fig. 4.2 is one example of this class. Here we consider another, the Optron[1] displacement follower, shown in Fig. 4.34.

Light emanating from the phosphorescent spot on the surface of a special cathode-ray tube is sharply focused onto the plane of the edge of

---

[1] Optron Corp., Santa Barbara, Calif.

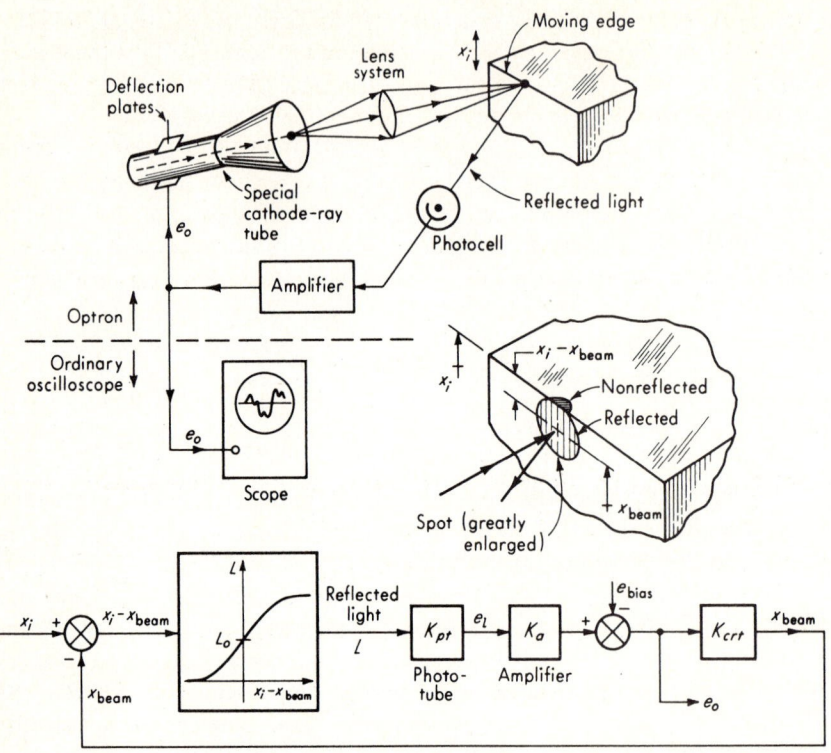

Fig. 4.34. *Optron displacement pickup.*

the body whose motion x_i is to be measured. If the spot is completely above this edge, no light will be reflected back to the photocell, and its output voltage e_l will be zero. The voltage applied to the vertical deflection plates of the special cathode-ray tube is then equal to $-e_{\text{bias}}$, and the spot motion x_{beam} will be $-K_{crt}e_{\text{bias}}$, driving the beam downward. The voltage e_{bias} is sufficient to drive the beam down past the edge. As it comes onto the edge from the top, first a little light and then more and more is reflected back to the photocell. If the spot is completely below the edge, further downward motion produces no more reflected light, saturating this effect. This relation between $x_i - x_{\text{beam}}$ and reflected light is somewhat nonlinear, as shown in the block diagram of Fig. 4.34. As the spot first comes onto the edge from above, the photocell puts out voltage e_l. This is amplified an amount K_a, which is very large, so that the spot moves only a small distance onto the edge before $e_l K_a$ is larger than e_{bias}, and x_{beam} is now positive, driving the beam upward. If it is driven too far up, e_l again becomes smaller than e_{bias} and the spot is again driven down. Thus this feedback system tends at all times to "slave" or "lock on" the point of light to the moving edge. Under these conditions, $x_{\text{beam}} \approx x_i$; since the voltage $e_o = x_{\text{beam}}/K_{crt}$ is also applied to the measuring oscilloscope, its trace is an accurate record of x_i.

The block diagram of Fig. 4.34 shows all effects as occurring instantaneously; thus it is adequate only for an analysis of static behavior. However, this analysis is quite revealing and so we carry it through. The nonlinear relation between $x_i - x_{\text{beam}}$ and reflected light L may be linearized for small excursions about an operating point L_0 as follows:

$$L \approx L_0 + \frac{dL}{d(x_i - x_{\text{beam}})}\bigg|_{(x_i - x_{\text{beam}}) = 0} (x_i - x_{\text{beam}})$$
$$= L_0 + K_L(x_i - x_{\text{beam}}) \qquad (4.83)$$

The phototube, amplifier, and cathode-ray tube are assumed linear although it will be seen that only the cathode-ray tube need be strictly linear. From the block diagram we can write

$$\{[L_0 + K_L(x_i - x_{\text{beam}})]K_{pt}K_a - e_{\text{bias}}\}K_{crt} = x_{\text{beam}} \qquad (4.84)$$

It is possible to choose $L_0 K_{pt} K_a = e_{\text{bias}}$, thus giving

$$(K_L K_{pt} K_a K_{crt} + 1)x_{\text{beam}} = (K_L K_{pt} K_a K_{crt})x_i \qquad (4.85)$$
$$x_{\text{beam}} = \frac{K_L K_{pt} K_a K_{crt}}{1 + K_L K_{pt} K_a K_{crt}} x_i \qquad (4.86)$$

We now by design make $K_L K_{pt} K_a K_{crt} \ggg 1$ (usually by making K_a very large) to get

$$x_{\text{beam}} \approx x_i \qquad (4.87)$$

thus showing that x_{beam} "tracks" x_i. Equation (4.87) holds, whether K_L, K_{pt}, K_a, or K_{crt} is constant or variable (due to time drift and/or nonlinearity). The *only* requirement is that their product at all times should be much greater than 1.0. This is a result of the feedback principle employed. This feedback unfortunately also puts an upper allowable value on $K_L K_{pt} K_a K_{crt}$ (the loop gain) because of stability problems which would become apparent if dynamic effects were taken into account. However, proper design allows a sufficiently high value of loop gain to achieve high accuracy without instability. Since the output of the instrument is the voltage $e_o = x_{\text{beam}}/K_{crt}$, we see that K_{crt} *must* be a constant if e_o is to be in proportion to x_{beam} and thus to x_i. The special cathode-ray tube must therefore have a linear voltage/(spot deflection) characteristic.

The Optron described above must be used in low ambient-light levels, the equipment being covered with a blackout cloth where convenient. An adjustable optical system gives four ranges as follows:

Range	Full scale x_i, in.	Resolution, in.	Working distance, in.
1	0.1	0.0001	0.5
2	0.25	0.0002	2.0
3	1.0	0.001	3.5
4	4.0	0.004	9.0

Nonlinearity is 0.2 percent of full scale over the central 75 percent of range. Nonrepeatability is 0.1 percent of full scale over the same range. The frequency response is from 0 to 5,000 cps with the amplitude ratio down 3 db at 5,000 cps. The output signal is 40 volts full scale.

A more recent version of the Optron uses a lens or telescope to form an optical image of a black-and-white target fastened to the object under study on the photocathode of an image-dissector tube. This tube projects an "electron image" of the target onto an aperture behind which is located a photomultiplier tube. The output of the photomultiplier tube is amplified and drives the deflection coil of the image-dissector tube. The deflection coil positions the electron image on the aperture so that the black-white boundary always splits the aperture. Motion of the target causes motion of the photocathode optical image and the corresponding electron image at the aperture. However, as soon as the image at the aperture starts to deviate from the neutral position the photomultiplier output tends to drive it back by means of the deflection coil.

Although the hardware is different, the basic concept, feedback analysis, and performance are quite analogous to those of the system of Fig. 4.34. An advantage of this system (see Fig. 4.35) over the earlier version is that it is usable in ordinary room light (40 ft-c) and is "passive" in the sense that the Optron system does *not* provide the light source illuminating the target, as did the special cathode-ray tube of Fig. 4.34. The working distance can thus be greatly extended to hundreds of feet or even miles and the instrument applied to such problems as missile tracking. At a working distance of 200 ft the full-scale range of displacement is 10 in. with a resolution of 0.050 in. With special lenses, a working distance of 0.65 in., a full-scale range of 0.050 in., and a resolution of 12 μin. are claimed. The frequency response is flat from 0 to 5,000 cps.

Photographic techniques. The application of still and motion-picture photography often allows qualitative and quantitative analysis of com-

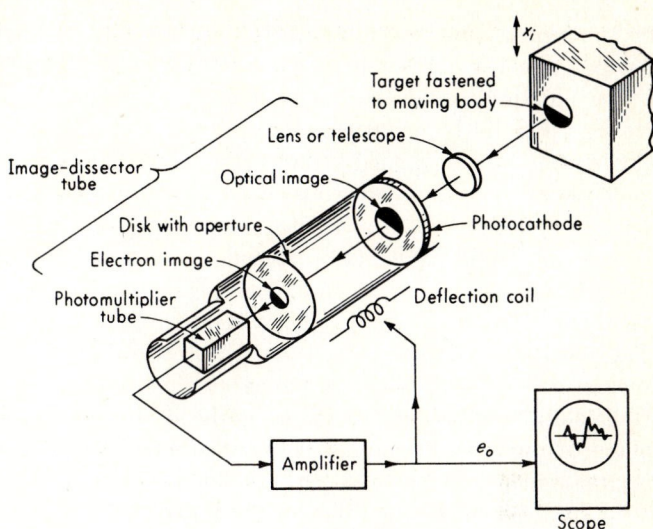

Fig. 4.35. Optron displacement pickup.

plex motions that would be difficult by other methods.[1]   We here touch briefly on some of the more common applications.

Perhaps the simplest application of still photography is the single-flash "stop-action" technique.[2]   The objective is to "freeze" a motion at a particular phase of its occurrence to allow detailed visual study of some physical phenomenon.   The equipment usually employed consists of a still camera, a stroboscopic light source, and some means of triggering a single flash of the strobe light at the desired instant.   If the experiment can be performed in a darkened room, the procedure consists of manually opening the camera shutter, allowing the phenomenon to occur, triggering the light at the desired instant, and then manually closing the shutter. If triggering, focus, and exposure are correct, a photo of the phenomenon, "frozen" at the instant of the light flash, will be obtained.   Such photos can be most helpful in understanding complex physical processes in fluid motion or moving machine parts.   Of course, the effective freezing of the motion depends on the flash duration being sufficiently short compared with the velocity of the motion.   Flash durations of the order of 1 to 3 μsec are readily available.   Thus, for example, a velocity of 1,000 fps will cause a "blurring" of 0.012 to 0.036 in. at the object.   The actual blurring

[1] W. G. Hyzer, "Engineering and Scientific High-speed Photography," The Macmillan Company, New York, 1962.

[2] "Handbook of High-speed Photography," General Radio Co., West Concord, Mass.

on the film will be this value times the image/object ratio of the camera setup. If a 1-in. object shows up on the film as 0.1 in., for example, the above blurring would amount to 0.0012 to 0.0036 in. on the film, which would generally be considered acceptable. If the experiment cannot be carried out in a darkened room, the opening and closing of the shutter must be synchronized with the flash and the open time of the shutter must be short enough so as not to overexpose the film from the room light.

If a displacement-time record is desired, a multiple-flash still-camera technique may be employed. The setup is essentially the same as above except the strobe light flashes repetitively at a known rate. The result is a multiple-exposure photo showing the moving object in successive positions which are separated by known increments of time. By including a calibrated length scale in the photo (preferably in the plane of the motion) numerical values of displacement at specific time intervals may be measured.

High-speed motion-picture photography is used to study motions that occur too rapidly for the eye to analyze properly. This is accomplished by taking the pictures at a high camera picture frequency (frames per second) and then projecting the film at a low projector picture frequency. The lowest usable projector frequency is about 16 frames per second since lower frequencies result in flickering because the human eye's persistence of vision is about 0.06 sec. The highest usable camera picture frequency depends on the construction of the camera. Relatively small, portable cameras are available with picture frequencies up to about 20,000 frames per second, and these are in relatively wide use in industry. Larger, more complex, and expensive cameras are available where the application dictates their higher speed. Their picture frequencies are up to several million per second. The time magnification is defined as the ratio of the camera frequency to the projector frequency; thus if 16 is the projector frequency, magnifications of several hundred thousand to one are achievable while 1,500:1 is not unusual in common industrial practice. Aside from picture frequency, the shutter speed of the camera must be sufficiently high to prevent blurring of the individual frames, just as in still photography. For 16-mm film a blur of 0.002 in. is considered acceptable. The shutter-speed requirement can be greatly relaxed by the use of a synchronized short-duration electronic flash as the light source, since the flash duration then controls blur no matter what the shutter speed. Selection of a proper camera picture rate may be judged roughly by the rule that the projection of the complete motion to be visually analyzed should take about 2 to 10 sec. Thus a motion occurring in 0.001 sec requires a time magnification of 2,000 to 10,000. For vibratory motions the camera picture rate must be several (preferably 5 to 10) times the highest vibration frequency.

The main features of all photographic motion-analysis techniques include their noncontacting nature, the ability to see the overall motion of a body or fluid rather than just that of isolated points, and the capacity to measure quantitatively a general plane motion (any combination of translation and rotation) rather than just a pure rotation or translation, as for potentiometers, differential transformers, etc.

Photoelastic and brittle-coating stress-analysis techniques. Since both these methods are really strain- rather than stress-sensitive and since strain is a small displacement, we include a brief treatment in this section on displacement measurement.

Photoelastic methods[1] depend on the property of birefringence under load exhibited by certain natural or synthetic transparent materials. Birefringence (double refraction) under load refers to the phenomenon wherein light travels at different speeds in a transparent material, depending on the direction of travel relative to the directions of the principle stresses and also depending on the magnitude of the difference between the principal stresses, for a two-dimensional stress field. By constructing models (from suitable transparent materials) of the same shape as the part to be stress-analyzed and shining suitably polarized light through them while they are subjected to loads proportional to those expected in actual service, a pattern of light and dark fringes appears which shows the stress distribution throughout the piece and allows numerical calculation of stresses at any chosen point.

By use of the "frozen stress" technique, the method can be extended to three-dimensional problems. A three-dimensional plastic model is subjected to simultaneous load and high temperature. The load is maintained while the specimen is slowly cooled to room temperature whereupon the load is released. It will be found that a residual stress pattern identical to that produced by the load will be "frozen" into the specimen. Furthermore, the model may now be carefully sliced in various directions to produce flat (two-dimensional) slabs which may be photoelastically analyzed to determine three-dimensional stresses. By combining photoelastic and high-speed photographic techniques, the method may be extended to dynamic studies such as the propagation of shock waves through solid bodies.

A recent extension of photoelastic techniques, reflective photoelasticity, does not require construction of a plastic model. Rather, the metal part itself (with its surface polished or aluminum-painted for reflectivity) is coated with a liquid photoelastic material which hardens

---

[1] M. M. Frocht, "Photoelasticity," vols. I and II, John Wiley & Sons, Inc., New York, 1941, 1948.

and bonds to the surface. Polarized light is directed onto the part and reflects from the shiny undersurface. Dark and light patterns again reveal the stress distribution when a load is applied.

Photoelastic methods, as compared with bonded resistance strain gages, give an overall picture of the stress distribution in a part. This is very helpful in locating and numerically evaluating stress concentrations and in redesigning the part for optimum material use. The method also does not disturb the local stress field as a strain gage might.

In the brittle-coating stress-analysis technique[1] a special lacquerlike material is sprayed on the actual part to be analyzed and the coating allowed to dry. Application of load causes visible cracking of the brittle coating. The direction of the cracking shows the direction of maximum stress while the spacing of the cracks indicates magnitude. Under favorable conditions, numerical values of stress can be calculated from measured crack spacing to about 10 percent accuracy. The main features of the method are its simplicity, low cost, and speed in giving an overall picture of stress distribution. It is often used in conjunction with electric resistance strain gages, the brittle coating locating the points of maximum stress and its direction so that strain gages can be applied at the proper places and in the proper orientation for accurate strain measurement.

Displacement-to-pressure transducer (***nozzle-flapper***). The nozzle-flapper-transducer principle is widely used in precision gaging equipment and also as a basic component of pneumatic and hydraulic measurement and control apparatus. Figure 4.36 shows the general arrangement. Fluid at a regulated pressure is supplied to a fixed flow restriction and a variable flow restriction connected in series. The variable flow restriction is varied by moving the "flapper" to change the distance x_i. This causes a change in output pressure p_o which, for a limited range of motion, is nearly proportional to x_i and extremely sensitive to it. A pressure-measuring device connected to p_o can thus be calibrated to read x_i. Ideally (pressure-containing chambers rigid; fluid incompressible) a sudden change in x_i would cause an instantaneous change in p_o. Actually, the dynamics are approximately those of a linear, first-order system for small changes in x_i. The time constant is determined for gases by the compressibility of the gas; for liquids the elastic deformation of the pressure-sensing device often controls.

We shall analyze the system of Fig. 4.36a for the case of a gaseous medium since a majority of practical applications utilize low-pressure ($p_s \approx 20$ to 30 psig) air as the working fluid. The principle of conserva-

[1] C. C. Perry and H. R. Lissner, "The Strain Gage Primer," chap. 13, McGraw-Hill Book Company, New York, 1955.

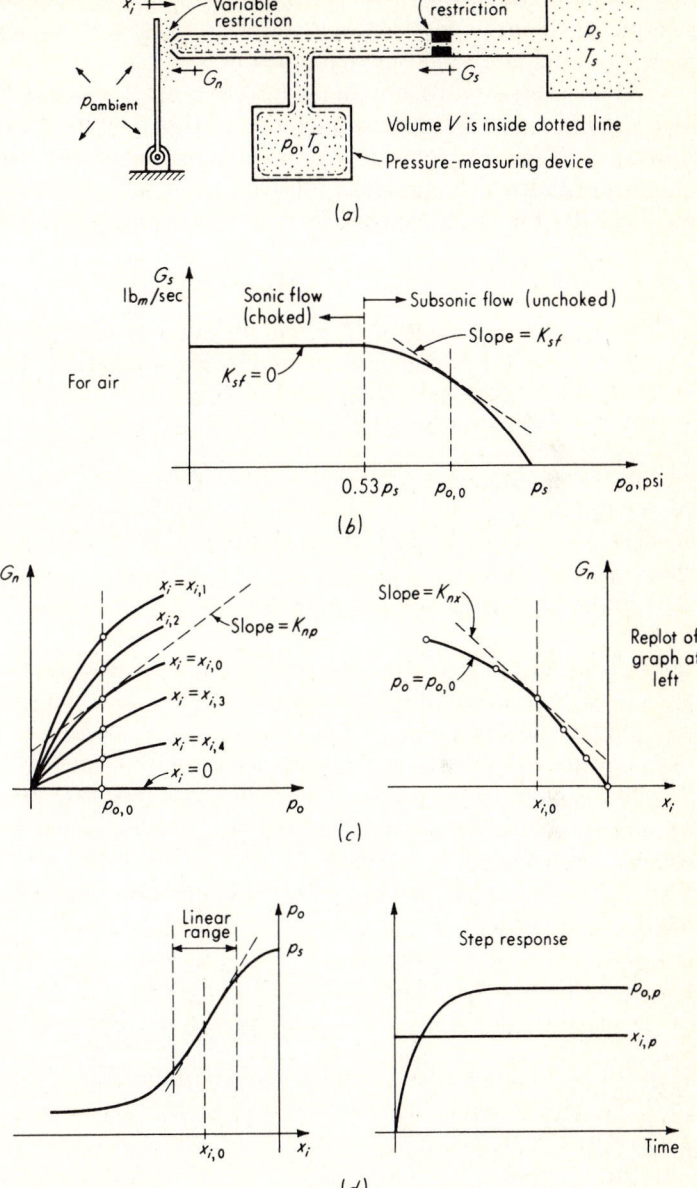

Fig. 4.36. *Nozzle-flapper transducer.*

tion of mass is applied to the volume V by stating that during a time interval dt the difference between entering mass and leaving mass must show up as an additional mass storage in V.

It is necessary to obtain expressions for the mass flow rates G_s and G_n. We assume that supply pressure p_s and temperature T_s are constant. Then G_s depends on p_o only; however, the dependence is nonlinear, and so we employ a linearized (perturbation) analysis which will be valid for small changes from an operating point. We can write

$$G_s = G_s(p_o) \approx G_{s,0} + \frac{dG_s}{dp_o}\bigg|_{p_o = p_{o,0}} (p_o - p_{o,0}) = G_{s,0} + K_{sf}p_{o,p} \qquad (4.88)$$

where
$$G_{s,0} \triangleq \text{value of } G_s \text{ at equilibrium operating point} \qquad (4.89)$$
$$p_{o,0} \triangleq \text{value of } p_o \text{ at equilibrium operating point} \qquad (4.90)$$
$$p_{o,p} \triangleq \text{small change (perturbation) in } p_o \text{ from } p_{o,0} \qquad (4.91)$$
$$K_{sf} \triangleq \text{value of } dG_s/dp_o \text{ at } p_{o,0} \text{ (a constant)} \qquad (4.92)$$

The function $G_s(p_o)$ can be found theoretically from fluid mechanics and thermodynamics (with the help of an experimental orifice-discharge coefficient) or entirely by experiment for a given orifice. Its general shape is given in Fig. 4.36b.

In finding the nozzle mass flow rate G_n, we assume that the process from p_s, T_s to p_o, T_o is a perfect-gas, work-free, adiabatic process. Also, the velocity of the gas at pressure p_o in volume V is assumed zero. Thus the gas at p_o is essentially in a stagnation state, and since the stagnation enthalpy for a perfect-gas, work-free, adiabatic process is constant, the temperature T_o is nearly the same as T_s and remains nearly constant. If T_o may be assumed constant, the nozzle mass flow rate depends only on p_o and x_i. The relationship between G_n, p_o, and x_i can be found from theory (with experimental corrections) or from experiment alone for a specific device. The relationship is again nonlinear, and so a perturbation analysis is in order.

$$G_n(p_o,x_i) \approx G_{n,0} + \frac{\partial G_n}{\partial p_o}\bigg|_{\substack{x_{i,0} \\ p_{o,0}}} (p_o - p_{o,0}) + \frac{\partial G_n}{\partial x_i}\bigg|_{\substack{x_{i,0} \\ p_{o,0}}} (x_i - x_{i,0}) \qquad (4.93)$$

$$G_n \approx G_{n,0} + K_{np}p_{o,p} + K_{nx}x_{i,p} \qquad (4.94)$$

Figure 4.36c shows how K_{np} and K_{nx} could be found from experimental data.

The mass storage in volume V can be treated by using the perfect-gas law $p_o V = MRT_o$. We assume V, R, and T_o to be constant. Then,

$$p_o = \frac{RT_o}{V} M \qquad (4.95)$$

$$p_{o,0} + p_{o,p} = \frac{RT_o}{V} (M_0 + M_p) \qquad (4.96)$$

$$\frac{dp_{o,p}}{dt} = \frac{RT_o}{V} \frac{dM_p}{dt} \qquad (4.97)$$

By conservation of mass during a time interval dt,

$$\text{(Mass in)} - \text{(mass out)} = \text{additional mass stored}$$

$$(G_{s,0} + K_{sf}p_{o,p})\, dt - (G_{n,0} + K_{np}p_{o,p}$$

$$+ K_{nx}x_{i,p})\, dt = dM_p = \frac{V}{RT_o}\, dp_{o,p} \qquad (4.98)$$

If the operating point $p_{o,0}$, $x_{i,0}$ is an equilibrium condition, $G_{s,0} = G_{n,0}$. We then have

$$\frac{V}{RT_o}\frac{dp_{o,p}}{dt} + (K_{np} - K_{sf})p_{o,p} = (-K_{nx})x_{i,p} \qquad (4.99)$$

This is clearly a first-order system, and so we define

$$K \triangleq \frac{-K_{nx}}{K_{np} - K_{sf}} \qquad \text{psi/in.} \qquad (4.100)$$

$$\tau \triangleq \frac{V}{RT_o(K_{np} - K_{sf})} \qquad \text{sec} \qquad (4.101)$$

to give

$$(\tau D + 1)p_{o,p} = Kx_{i,p} \qquad (4.102)$$

and

$$\frac{p_{o,p}}{x_{i,p}}(D) = \frac{K}{\tau D + 1} \qquad (4.103)$$

To improve speed of response (decrease τ) the volume V should be minimized. Since T_o is usually the ambient temperature, R and T_o are not available for adjustment. An increase in $K_{np} - K_{sf}$ will decrease τ but at the expense of sensitivity, as shown by Eq. (4.100).

A relatively crude device of this type made up for student laboratory use had a nozzle diameter of $\frac{1}{32}$ in. and a volume V of the order of 1 in.^{3}. For a supply-orifice diameter of $\frac{1}{32}$ in. and a supply pressure of 25 psig this device had a K (at the most sensitive part of its range) of about 2,000 psi/in. and a τ of about 0.12 sec. It was quite linear over a range of about ± 0.002 in. around $x_{i,0} = 0.004$ in. By changing only the supply-orifice diameter to $\frac{1}{64}$ in. the sensitivity was raised to about 8,000 psi/in. while τ increased to 0.24 sec. The linear range was now about ± 0.0005 in. around $x_{i,0} = 0.0015$ in. Since 1 psi $= 27.7$ in. of water, a water manometer used to read p_o gives a (easily readable) 0.1-in. change for an x_i change of only 0.45×10^{-6} in. when the sensitivity is 8,000 psi/in. This illustrates the great sensitivity of this transducer.

A useful approximate expression for the static sensitivity may be easily obtained by assuming incompressible flow. The results are accurate for liquids and a good estimate for gases if pressure changes are not large. The mass flow through the supply orifice is now

$$G_s = \frac{C_d \pi d_s^2}{4} \sqrt{2\rho(p_s - p_o)} \qquad (4.104)$$

where $C_d \triangleq$ discharge coefficient

$d_s \triangleq$ supply-orifice diameter

$\rho \triangleq$ fluid mass density

and we are neglecting the velocity of approach since the orifice area is very small compared with the upstream passage. The flow area for the nozzle is taken as the peripheral area of a cylinder of height x_i and diameter d_n, the nozzle diameter. This is true only for small values of x_i. The discharge coefficient for this configuration may be different from that for the supply orifice and may vary somewhat with x_i but here we take it to be the same as C_d in Eq. (4.104). We have then

$$G_n = C_d \pi d_n x_i \sqrt{2\rho(p_o - p_{\text{ambient}})} \qquad (4.105)$$

For steady state, $G_n = G_s$ so that (taking $p_{\text{ambient}} = 0$ psig) we get

$$p_o = \frac{p_s}{1 + 16(d_n^2 x_i^2 / d_s^4)} \qquad (4.106)$$

The sensitivity dp_o/dx_i varies with x_i and is found to have its maximum value at $x_i = 0.14 d_s^2 / d_n$. This maximum value is

$$K_{\max} = \frac{2.6 d_n p_s}{(d_s)^2} \qquad \text{psi/in.} \qquad (4.107)$$

Thus we see that large d_n, large p_s, and small d_s lead to high sensitivity.

Digital displacement transducers. More and more, measuring instruments are being required to communicate with digital computers. The amount of raw data generated by large-scale test programs is so great that automated computer reduction of these data to meaningful form is a necessity. Also, feedback-control systems for complex processes are increasingly dependent on digital computers for partial or complete control-action generation. Thus, measuring devices that form a basic part of these overall systems must be compatible with the digital nature of the computer.

There is some question whether any "true" digital transducers exist, since a clear-cut definition is not at present available. Rather, many people prefer to think of all transducers as analog with an analog-to-digital converter "built in" to a greater or lesser extent. We shall not try to resolve this semantic problem here but rather shall simply state that in practice two main methods are used to realize digital signals. The first involves converting the analog variable to a shaft rotation (or translation) and then using one of the many types of *shaft-angle encoder* to generate digital voltage signals. The other approach is to convert the analog variable to an analog voltage which is then converted to a digital voltage by one of the many types of *voltage-to-digital* converters. The shaft-angle encoder and the voltage-to-digital converter are perhaps the closest approach to true digital transducers, the first for motion, the second for voltage. In this section we consider the shaft-angle encoder[1] for transducing analog motion to digital voltage.

While analog voltages can be transmitted with a single pair of wires, digital voltages for input to a computer require a pair of wires for each digit of the signal. That is, a voltage of 564 volts is communicated to a digital computer in three separate pieces. The computer is told that the units digit is 4, the tens digit 6, and the hundreds digit 5. (Actually the computer works on the binary rather than the decimal system, but the concept is the same.) If the computer worked on the decimal system, it would require components that could recognize 10 different states (0 through 9) for each digit. The most simple and reliable arrangement is to use components that recognize only two states. If only two states are needed, one can be "on" and the other "off." Thus the basic components can be essentially "switches" that are either open or closed. This type of hardware leads naturally to representing numbers in the binary (rather than decimal) system since each binary digit (called a *bit*) requires only two states to specify it completely.

Any number can be expressed in binary form by breaking it down into a sum of various powers of 2. For example,

$$10 = 1(2^3) + 0(2^2) + 1(2^1) + 0(2^0) \qquad (4.108)$$

The coefficients 1, 0, 1, 0 are actually sufficient to specify the number completely if it is known that the binary system is being used. Thus the notation used to give the number 10 in binary form is simply 1010. It takes four bits to give this number; thus four "switches" in the computer are needed. To communicate the number 10 to the computer, the first and third switches would be closed (a closed switch corresponds to 1) while the second and fourth would be open (an open switch corresponds to zero). With four switches, the computer could handle only numbers in the range 0 to 15 in steps of 1; thus its "resolution" would be only $100/15 = 6.7$ percent. To get more accuracy (resolution), more switches (bits) must be included in the computer. To handle noninteger numbers, negative powers of 2 are used. Thus

$$9.72 = 1(2^3) + 0(2^2) + 0(2^1) + 1(2^0) + 1(2^{-1})$$
$$\underbrace{+ \ 0(2^{-2}) + 1(2^{-3}) + 1(2^{-4}) + 1(2^{-5}) + \cdots}_{9.71875} \qquad (4.109)$$

We see from this example that a decimal number may not have an exact binary equivalent but that one can come as close as one wishes by adding

[1] P. Barr, Shaft Position Encoders, *Electromech. Design*, p. 165, January, 1964.

more bits.   When noninteger numbers are expressed in binary, a *binary point* is placed just to the right of the 2^0 position; thus 9.72 expressed as closely as possible in nine-bit binary is 1001.10111.

While the computer itself generally works in the pure binary system, other *codes* are sometimes useful where communication between a man (who is most familiar with the decimal system) and the machine is necessary.   The binary-coded decimal (BCD) code is widely used for such purposes.   Here each decimal digit is given by its binary equivalent. Since the numbers 0 through 9 must be representable, this requires four bits (2^3, 2^2, 2^1, 2^0) for each decimal digit.   Thus the decimal number 872.5 would be written as 1000 0111 0010. 0101 in the binary-coded decimal system, requiring a total of 16 bits.   In a pure binary system this number would be 1101101000.1, requiring only 11 bits; thus the binary-coded decimal code trades off ease of conversion for efficiency.   The code given above should not be called *the* binary-coded decimal code since other ways of coding decimal into binary are also in use.

With the above brief introduction as background, we may now consider the operation of a typical shaft-angle encoder.   The principle is most easily visualized for a translational rather than a rotary motion. Figure 4.37 shows such a device in schematic form.   The encoder shown has four tracks (bits) and is divided into conducting and insulating portions, with the smallest increment being 0.01 in.   As the scale moves under the brushes, the respective lamp circuits are made or broken so that the sum of the numbers shown on the readout lamps is at every instant equal to the displacement in hundredths of an inch.   With four bits the maximum travel is 0.15 in.; longer travel requires more tracks (bits).   If resolution finer than 0.01 in. is needed, a more closely spaced black-white pattern can be produced or some motion-amplifying mechanism (gearing, etc.) employed.   If the displacement information is to be used in a digital computer rather than just visually displayed, the brushes are connected to the computer input "switches" rather than to the lamps.

For rotational motions, the pattern of Fig. 4.37 is simply deformed so that the length of the scale becomes the circumference of a circle on a flat disk.   The brushes are then disposed along a radial line on the disk. Figure 4.38 shows a typical disk.   Many detail variations on the basic principle are in use.   To eliminate wear and friction of the brushes and to improve resolution, optical readout using clear and opaque segments, light sources, and photocells is used.   Magnetic readout without contact is also possible.   A typical unit with optical readout has 13 tracks (giving a resolution of $1/2^{13} = 1/8,192$) in a case only 4 in. in diameter and 7 in. long.

One important defect in the arrangement of Fig. 4.37 is that, if the brushes ("reading line") and the grid patterns are not perfectly aligned,

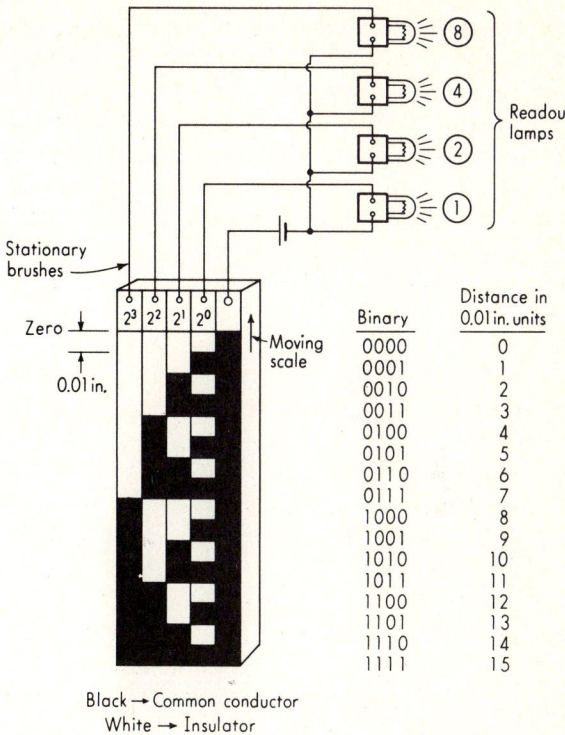

Binary	Distance in 0.01 in. units
0000	0
0001	1
0010	2
0011	3
0100	4
0101	5
0110	6
0111	7
1000	8
1001	9
1010	10
1011	11
1100	12
1101	13
1110	14
1111	15

Black → Common conductor
White → Insulator

Fig. 4.37.   *Digital transducer.*

in moving from one position to the next an error of more than one unit can occur. For example, suppose the reading is 9 (1001 binary) and in going to 10, because of misalignment, the 2^1 digit changes before the 2^0 digit does, giving binary 1011. Now 1011 binary is 11 decimal, and so the reading jumps from 9 to 11. This problem is overcome in practical

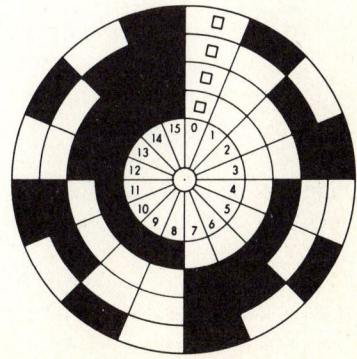

Fig. 4.38.   *Code disk for shaft encoder.*

Ordinary decimal	Cyclic decimal
1	1
2	2
3	3
4	4
5	5
6	6
7	7
8	8
9	9
10	19
11	18
12	17
13	16
14	15
15	14
16	13
17	12
18	11
19	10
20	20
21	21
22	22
23	23
24	24
25	25

Fig. 4.39.   Cyclic decimal system.

encoders either by special brush arrangements or by changing the code to one that changes only one digit in going from one number to the next. The Gray[1] (cyclic binary) code is one such in common use.

The Giannini Datex[2] code combines the cyclic binary code with a binary-coded decimal code as follows: First the ordinary decimal number is converted to "cyclic" decimal according to the rules:

1. An ordinary decimal digit that follows an even digit is not changed.
2. An ordinary decimal digit that follows an odd digit is changed to its 9's complement (9 minus the digit).
3. The first (most significant) digit is not changed.

Figure 4.39 shows the relation between ordinary and cyclic decimal numbers. *Note that any two adjacent numbers differ in only one decimal digit*

[1] J. T. Tou, "Digital and Sampled-data Control Systems," p. 372, McGraw-Hill Book Company, New York, 1959.

[2] Giannini Corp., Monrovia, Calif., *Bull.* 001, 1955.

Binary-coded decimal (Datex)

Ordinary decimal	Cyclic decimal	Brush number	Hundreds digit				Tens digit				Units digit			
			9	10	11	12	5	6	7	8	1	2	3	4
000	000		1	0	0	0	1	0	0	0	1	0	0	0
001	001		1	0	0	0	1	0	0	0	1	1	0	0
009	009		1	0	0	0	1	0	0	0	1	0	0	1
010	019		1	0	0	0	1	1	0	0	1	0	0	1
011	018		1	0	0	0	1	1	0	0	1	1	0	1
099	090		1	0	0	0	1	0	0	1	1	0	0	0
100	190		1	1	0	0	1	0	0	1	1	0	0	0
199	100		1	1	0	0	1	0	0	0	1	0	0	0
200	200		0	1	0	0	1	0	0	0	1	0	0	0

Fig. 4.40. *Binary-coded decimal system.*

for the cyclic decimal system. The next step is to convert the cyclic decimal to binary, using a special binary-coded decimal code. Recall that each decimal digit requires four bits; however, with four bits one can write 16 decimal numbers (0 through 15) while only 10 (0 through 9) are necessary. Thus one can *choose* which of the 16 to use. Based on the desire for error checking, reducing power-supply requirements, and ease of obtaining the 9's complement of any number, it can be shown that the best choice for a binary-coded decimal code here is as follows:

Decimal	Binary-coded decimal (Datex)
0	1000
1	1100
2	0100
3	0110
4	0010
5	0011
6	0111
7	0101
8	1101
9	1001

An encoder based on this system requires 4 brushes for each decimal digit. Thus to handle three-digit decimal numbers, 12 brushes are needed. Figure 4.40 shows some typical conversions. Note that for a unit change in the decimal number only one contact change in the encoder is ever needed, thus preventing the large errors possible in a straight binary system.

The encoders discussed above are all of the so-called "absolute"-position type in that the contact pattern uniquely identifies any given position. A different method, the "incremental" system, merely gives a pulse each time an increment of motion occurs. These pulses are put into a counter which adds pulses for forward motion and subtracts them for reverse motion. The "contents" of the counter are thus at any time an indication of the position. This method uses only one "track" to generate the pulses, and its output is not in the binary form desired for direct computer input; thus a separate pulse-count-to-binary converter is required. The main problem with such systems is that if an error occurs it persists for all later motions.

4.4

Relative Velocity, Translational and Rotational   We consider here devices for measuring the velocity of translation, along a line, of one point relative to another and the plane rotational velocity about a single axis of one line relative to another.

Calibration. The measurement of rotational (angular) velocity is probably more common than that of translational velocity. Since translation can generally be obtained from rotation by suitable gearing or mechanisms, we consider mainly the calibration of rotational devices.

Perhaps the most difficult area of angular rate measurements is the extremely slow rotations associated with inertial guidance equipment. Angular velocities of 1 revolution/day (earth's rate) and less are of interest. Electro-optical devices such as the Midarm system (discussed under displacement calibration) enable measurement (and thus calibration of other less accurate transducers) of these low angular velocities with an accuracy of the order of 0.0002 degree/hr, using a measuring time of 1 min.

For higher angular and linear velocities perhaps the most convenient calibration scheme uses a combination of a toothed wheel, a simple magnetic proximity pickup, and an electronic EPUT (events per unit time) meter (see Fig. 4.41). The angular rotation is provided by some adjustable-speed drive of adequate stability. The toothed iron wheel passing under the proximity pickup produces an electrical pulse each time one tooth passes. These pulses are fed to the EPUT meter which counts them over an accurate time period (say 1.00000 sec), displays the result visually for a few seconds to enable reading, and then repeats the process

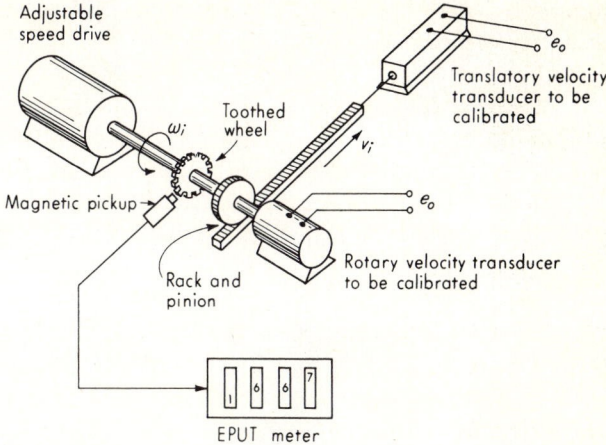

Fig. 4.41.   *Velocity-calibration setup.*

over and over.   The stability of the rotational drive is easily checked by observing the variation of the EPUT meter readings from one sample to another.   The inaccuracy of pulse counting is ± 1 pulse plus the error in the counter time base, which is of the order of 1 ppm.   The overall accuracy achieved depends on the stability of the motion source, the angular velocity being measured, and the number of teeth on the wheel. If the motion source were *absolutely* stable (no change in velocity whatever), very accurate measurement could be achieved simply by counting pulses over a long period of time, since then the average velocity and the instantaneous velocity would be identical.   If the motion source has some drift, however, the time sample must be fairly short.   For example, a shaft rotating at 1,000 rpm with a 100-tooth wheel produces 1,667 pulses in a 1-sec sample period.   The inaccuracy here would be 1 part in 1,667 (the 1-ppm time-base error is totally negligible) or 0.06 percent.   If the shaft rotated at 10 rpm the error would be 6 percent.   Slow rotations can be accurately measured by such means if the toothed wheel is placed on a shaft which is sufficiently geared up from the shaft driving the transducer being calibrated.

The above procedure uses relatively simple equipment and generally provides entirely adequate accuracy.   Other simpler and less accurate procedures can be used if they are adequate for their intended purpose.   These usually consist of simply comparing the reading of a velocity transducer known to be accurate with the reading of the transducer to be calibrated when they are both experiencing the same velocity input.

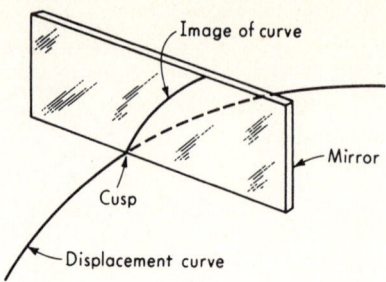

Fig. 4.42.   *Graphical differentiation.*

Adjust mirror until image and curve have
smooth juncture (no cusp). Mirror will then
lie on normal to curve.

Velocity by graphical differentiation of displacement records. If
one has a graphical record of displacement versus time it is possible to
obtain velocity versus time by graphical measurement[1] of instantaneous
slope.   This procedure can be fairly accurate for smooth displacement
curves but becomes increasingly less reliable for rapidly changing displacements.   A mirror used as in Fig. 4.42 to find the normal to the
curve is helpful in establishing the tangent.

*Velocity by electrical differentiation of displacement voltage sig-
nals.* The output voltage of any displacement transducer may be applied
to the input of a suitable differentiating circuit to obtain a voltage proportional to velocity (see Sec. 10.4 on differentiating circuits).   The main
problem is that differentiation accentuates any low-amplitude, high-
frequency noise present in the displacement signal.   Thus a carbon-film
potentiometer would be preferable to the wire-wound type, and demodulated and filtered signals from a-c transducers may cause trouble because
of the remaining ripple at carrier frequency.   Workable systems using
electrical differentiation are possible, however, with adequate attention
to details.

Average velocity from measured Δx and Δt. Often a value of average
velocity over a short distance or time interval is adequate, and a continuous velocity/time record is not required.   A useful basic method
is somehow (optically, magnetically, etc.) to generate a pulse when the
moving object passes two locations whose spacing is accurately known.
If the velocity were constant, any spacing could be used, large spacing
of course leading to greater accuracy.   If the velocity is varying, the

[1] F. A. Willers, "Practical Analysis," p. 158, Dover Publications, Inc., New
York, 1947.

spacing Δx should be small enough so that the average velocity over Δx is not very different from the velocity at either end of Δx. The same technique is applicable to rotational motion.

Figure 4.43a shows the application of a variable-reluctance proximity pickup such as was used in Fig. 4.41. When magnetic material passes close in front of the face of the pickup the reluctance of the magnetic path changes with time, generating a voltage in the coil. These pickups are simple and cheap and give a large output voltage (often several volts) under typical operating conditions. The output voltage increases with velocity and closeness of the external moving iron to the pickup. Display of the two pulses on a single sweep of an oscilloscope with a calibrated time base allows measurement of the average velocity. Greater accuracy may be achieved by applying the voltage pulses to an electronic time-interval meter. Figure 4.43b shows a similar arrangement using electro-optical techniques. Such an arrangement could replace the magnetic pickup in Fig. 4.41 also.

Photography of the motion, using a stroboscopic lamp flashing at a known rate, also provides velocity data of this type.

Mechanical flyball angular-velocity sensor. A classical rotary speed-measuring device still in wide use today, especially as a measuring element of industrial speed-control systems for engines, turbines, etc., is the flyball. Figure 4.44 shows the general arrangement schematically. Since the centrifugal force varies as the square of input velocity ω_i, the output x_o will not vary linearly with speed if an ordinary linear spring is used. For *small* changes in ω_i a linearized model may be used to show that the transfer function between ω_i and x_o is essentially of the form

$$\frac{x_o}{\omega_i}(D) = \frac{K}{D^2/\omega_n^2 + 2\zeta D/\omega_n + 1} \qquad (4.110)$$

The nonlinear static relation between ω_i and x_o for large speed changes may be acceptable in some systems. Where it is not, a nonlinear spring with $F_{\text{spring}} = K_s x_o^2$ can be used to get a linear overall characteristic since, at balance,

$$\text{Centrifugal force} = \text{spring force}$$
$$F_c = K_c \omega_i^2 = F_{\text{spring}} = K_s x_o^2 \qquad (4.111)$$

and thus $x_o = \sqrt{K_c/K_s}\ \omega_i$, a linear relationship.

A variation[1] on this principle uses a pneumatic force-balance system to replace the spring and produces a standard 3- to 15-psig air-pressure signal proportional to ω_i^2. Since this is the standard pressure range of

---

[1] G. C. Carroll, "Industrial Process Measuring Instruments," p. 187, McGraw-Hill Book Company, New York, 1962.

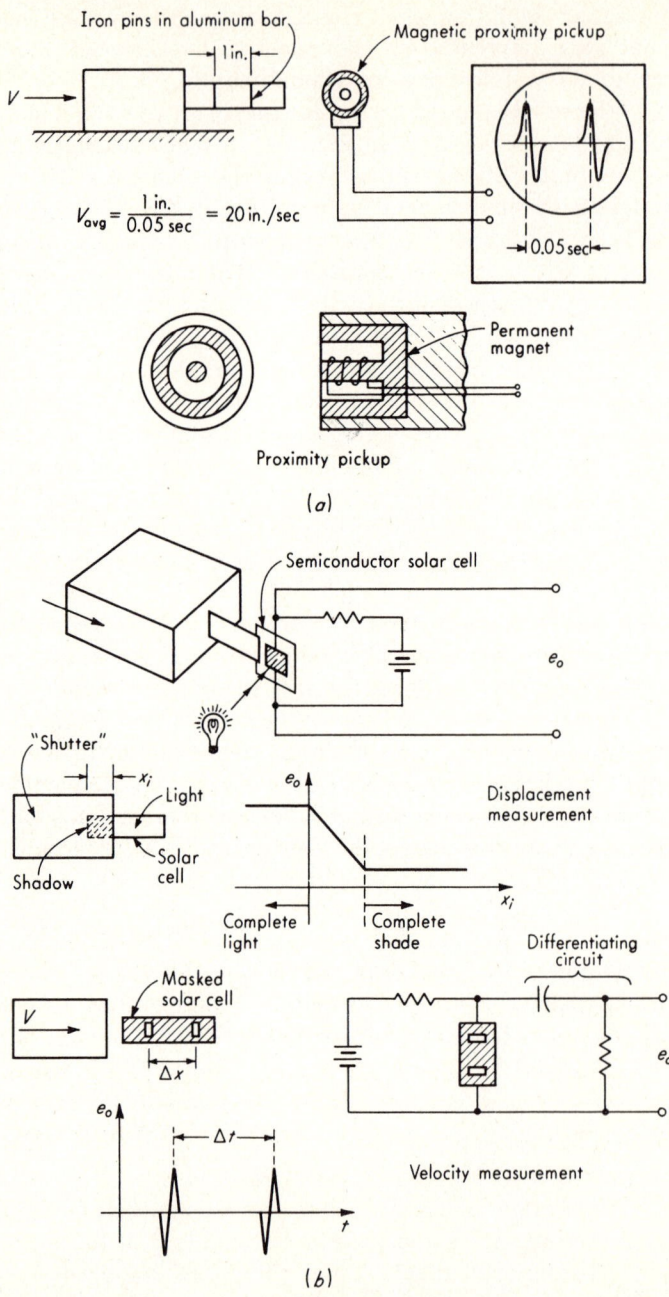

$$V_{avg} = \frac{1\ in.}{0.05\ sec} = 20\ in./sec$$

Iron pins in aluminum bar

Magnetic proximity pickup

Permanent magnet

Proximity pickup

(a)

Semiconductor solar cell

"Shutter"

Light

Solar cell

Shadow

Displacement measurement

Complete light

Complete shade

Masked solar cell

Differentiating circuit

Velocity measurement

(b)

Fig. 4.43. *Velocity measurement as $\Delta x/\Delta t$.*

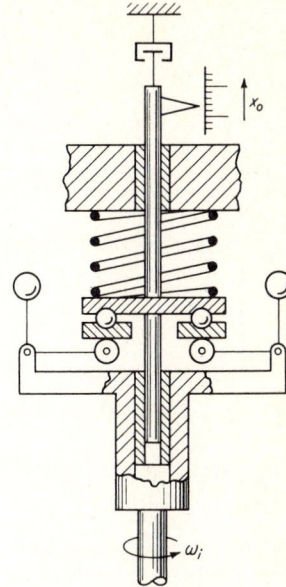

Fig. 4.44. *Flyball velocity pickup.*

industrial-process-control systems, this speed transducer can be directly incorporated into such systems.

Mechanical revolution counters and timers. When continuous reading and an electrical output signal are not required, a variety of mechanical revolution counters (with or without built-in timers) are available. They are generally supplied with a variety of rubber-tipped wheels which transmit by friction the motion to be measured to the counter input shaft.

Magnetic and photoelectric pulse-counting methods. The arrangement of Fig. 4.41, using magnetic pickups (or photocells and light sources with slotted wheels or black-and-white targets) and discussed under Calibration is also often used for measurement since the equipment needed is quite widely available in industry today. If an analog signal (varying d-c voltage) proportional to speed is desired, electronic devices called frequency-to-voltage converters can be connected to the pickup output terminals. However, this arrangement will not generally give as high accuracy as the pulse-counting technique.

Stroboscopic methods. Rotational velocity may be conveniently measured by using electronic stroboscopic lamps which flash at a known and adjustable rate. The light is directed onto the rotating member which itself usually has spokes, gear teeth, or some other feature enabling

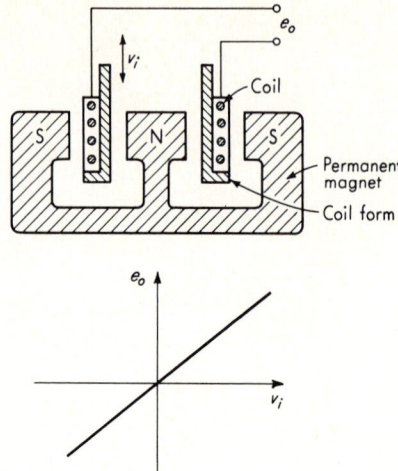

Fig. 4.45. *Moving-coil velocity pickup.*

"lock on."   If not, a simple black-and white paper target can be attached. The frequency of lamp flashing is adjusted until the "target" appears motionless.   At this setting, the lamp frequency and motion frequency are identical, and the numerical value can be read from the lamp's calibrated dial to an inaccuracy of about ± 1 percent of the reading.   The range of lamp frequency of a typical unit[1] is 110 to 25,000 flashes per minute.   Speeds greater than 25,000 rpm can be measured by the following technique.   Synchronism can be achieved at any flashing rate r that is an integral submultiple of the speed to be measured, n.   The flashing rate is adjusted until synchronism is achieved at the largest possible flashing rate, say r_1.   The flashing rate is then slowly decreased until synchronism is again achieved at a rate r_2.   The unknown speed n is then given by

$$n = \frac{r_1 r_2}{r_1 - r_2} \qquad (4.112)$$

For very high speeds, r_1 and r_2 are close together, giving poor accuracy. Accuracy can be improved by reducing the flashing rate below r_2 until synchronism is again achieved.   This procedure can be continued, obtaining synchronism N times ($r_1, r_2, r_3, \ldots, r_N$).   The speed n is then given by

$$n = \frac{r_1 r_N (N - 1)}{r_1 - r_N} \qquad (4.113)$$

This procedure can extend the upper range to about 250,000 rpm.

Translational-velocity transducers (*moving-coil and moving-magnet pickups*). The moving-coil pickup of Fig. 4.45 is based on the

---

[1] Strobotac, General Radio Co., West Concord, Mass.

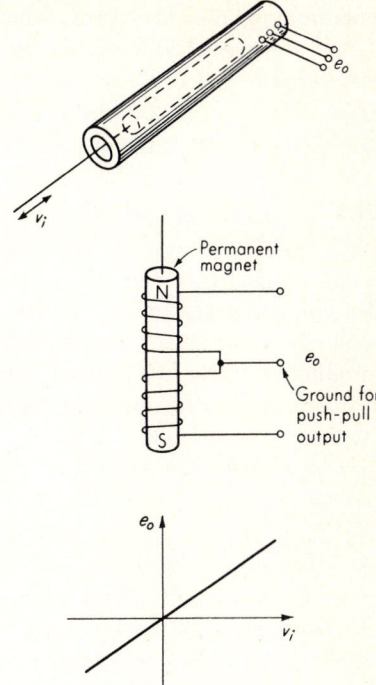

Fig. 4.46. *Velocity pickup.*

law of induced voltage

$$e_o = (Blv_i)10^{-8} \qquad (4.114)$$

where e_o = terminal voltage, volts

        B = flux density, gauss

        l = length of coil, cm

        v_i = relative velocity of coil and magnet, cm/sec

Since B and l are constant, the output voltage follows the input velocity linearly and reverses polarity when the velocity changes sign. Such pickups are widely used for the measurement of vibratory velocities. Since the flux density available from permanent magnets is limited to the order of 10,000 gauss, increase in sensitivity can be achieved only by increase in the length of wire in the coil. To keep the coil small, this requires fine wire and thus high resistance. High-resistance coils require a high-resistance voltage-measuring device at e_o to prevent loading. A typical pickup of about 500 ohms resistance has a sensitivity of 0.15 volt/(in./sec) and a full-scale displacement of 0.15 in. with a nonlinearity of ±1 percent. A more sensitive coil used in a seismometer (instrument to measure earth shocks) has 500,000 ohms resistance and a sensitivity of 115 volts/(in./sec).

The Sanborn[1] LVsyn shown in Fig. 4.46 uses a permanent-magnet

[1] Sanborn Co., Waltham, Mass.

core moving inside a form wound with two coils connected as shown. Units are available in full-range strokes from about 0.5 to 9.0 in. Sensitivity varies from about 0.1 to 0.65 volt/(in./sec), and coil resistance from 2,000 to 32,000 ohms. The nonlinearity is about 1 percent while core weights range from 3.5 to 69 g.

D-C tachometer generators for rotary-velocity measurement. An ordinary d-c generator (using either a permanent magnet or separately excited field) produces an output voltage roughly proportional to speed. By emphasizing certain aspects of design, such a device can be made an accurate instrument for measuring speed rather than a machine for producing power. The basic principle is again Eq. (4.114), which when applied to the rotational configuration of a d-c generator becomes

$$ e_o = \frac{n_p n_c \phi N}{60 n_{pp}} \, 10^{-8} \qquad (4.115) $$

where $e_o \triangleq$ average output voltage, volts
   $n_p \triangleq$ number of poles
   $n_c \triangleq$ number of conductors in armature
   $\phi \triangleq$ flux per pole, lines
   $N \triangleq$ speed, rpm
   $n_{pp} \triangleq$ number of parallel paths between positive and negative brushes

The voltage e_o is a d-c voltage proportional to speed which reverses polarity when the angular velocity reverses. A small superimposed ripple voltage is present because of the finite number of conductors. A typical high-accuracy unit[1] (permanent magnet) has a sensitivity of 7 volts/1,000 rpm, a rated speed of 5,000 rpm, nonlinearity of 0.07 percent over a range 0 to 3,600 rpm, ripple voltage 2 percent of average voltage for speeds above 100 rpm, friction torque of 0.2 in.-oz, rotor inertia of 7 g-cm², output impedance of 2,800 ohms, and a total weight of 3 oz.

A special d-c tachometer[2] of unique design for use where a limited ($\pm 15°$) angular travel is acceptable exhibits a very high sensitivity. A 1-in.-diameter model gives 500 volts/1,000 rpm while a 3-in.-diameter gives 30,000 volts/1,000 rpm. The nonlinearity is ± 9 percent for $\pm 15°$ travel, and the operating torque is 500 g-cm. In this generator the permanent magnet rotates while the coil is stationary, and no commutator is needed because of the limited travel.

[1] General Precision Inc., Kearfott Div., Little Falls, N.J.
[2] Armstrong Whitworth Equipment, Hucclecote, Gloucester, England.

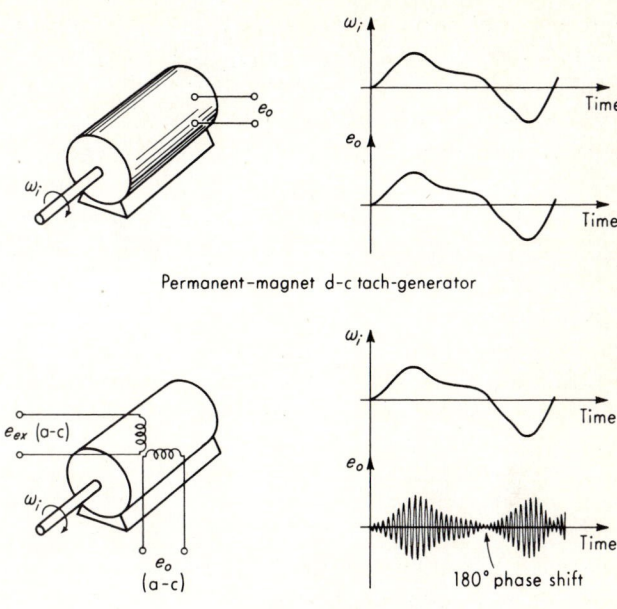

Permanent-magnet d-c tach-generator

A-c tach-generator

Fig. 4.47. *Tachometer generators.*

A-C tachometer generators for rotary-velocity measurement.
An a-c two-phase squirrel-cage induction motor can be used as a tachometer by exciting one phase with its usual a-c voltage and taking the voltage appearing at the second phase as output. With the rotor stationary, the output voltage is essentially zero. Rotation in one direction causes at the output an a-c voltage of the same frequency as the excitation and of an amplitude proportional to the instantaneous speed. This output voltage is in phase with the excitation. Reversal of rotation causes the same action except the phase of the output shifts 180°. While squirrel-cage rotors are sometimes used, the most accurate units employ a drag-cup rotor. This does not change the basic operating characteristics.

A typical high-accuracy unit[1] is excited by 115-volt/400-cps voltage, has a sensitivity of 2.8 volts/1,000 rpm, nonlinearity of 0.05 percent from 0 to 3,600 rpm, negligible rotor friction, rotor inertia of 7 g-cm², and a total weight of 6.7 oz. Most commercial a-c tachometers are designed to be used in a-c servomechanisms which conventionally operate on either 60 or 400 cps, and so they are generally designed for operation at these frequencies. For general-purpose motion measurement, the frequency

[1] General Precision Inc., Kearfott Div., Little Falls, N.J.

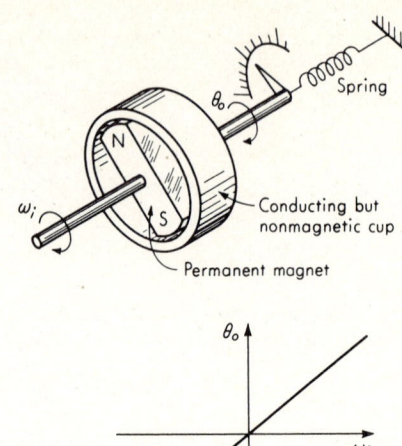

Fig. 4.48.   *Drag-cup velocity pickup.*

response of such units is limited (as are all a-c or "carrier"-type devices) by the carrier frequency to about one-tenth to one-fifth of the carrier frequency. It is usually possible, however, to excite a tachometer designed for 400 cps at considerably higher frequencies if necessary, although some or all of the performance characteristics may change value.

Eddy-current drag-cup tachometer.   Figure 4.48 shows schematically an eddy-current tachometer. Rotation of the magnet induces voltages into the cup which thereby produce circulating eddy currents in the cup material. These eddy currents interact with the magnet field to produce a torque on the cup in proportion to the relative velocity of magnet and cup. This causes the cup to turn through an angle θ_o until the linear spring torque just balances the magnetic torque. Thus in steady state the angle θ_o is directly proportional to ω_i, the input velocity. If an electrical output signal is desired, any low-torque displacement transducer can be used to measure θ_o. Dynamic operation is governed by the rotary inertia of parts moving with θ_o, the spring stiffness, and the viscous damping effect of the eddy-current coupling between magnet and cup, leading to a second-order response of the form

$$\frac{\theta_o}{\omega_i}(D) = \frac{K}{D^2/\omega_n{}^2 + 2\zeta D/\omega_n + 1} \qquad (4.116)$$

Nonlinearity of the order of 0.3 percent can be achieved in such units.

4.5

Relative-acceleration Measurements   Transducers directly sensitive to the relative acceleration of two bodies are not generally commercially available. This is due to the scarcity of physical effects directly producing an electrical signal proportional to relative acceleration and also to the wide availability of ("seismic") transducers for measuring *absolute* acceleration. One possibility (though not widely used) is the a-c tachometer generator described above. If the excitation winding is supplied with direct current rather than alternating current, the output voltage will be a varying direct current proportional to the relative angular acceleration of the rotor and stator. The author does not know of any commercial devices of this kind that are specifically designed and marketed for acceleration measurement. One would thus have to experiment with a given a-c tachometer to see whether it was satisfactory for the particular acceleration measurement.

Other possibilities for relative-acceleration measurement include graphical or electrical differentiation of displacement or velocity signals or records. The double differentiation required of displacement records or voltages can rarely be accurately performed except with very smooth signals. Single differentiation of velocity signals may be practical in some instances.

4.6

Seismic- (Absolute-) Displacement Pickups   Figure 4.49 shows the general construction of a seismic-displacement pickup for translatory or rotary motions. These devices are used almost exclusively for measurement of vibratory displacements in those (many) cases where a fixed reference for relative-displacement measurement is not available. That is, the vibration of a body can be measured with any of the relative-motion transducers discussed earlier in this chapter, but only if one end of the transducer can be attached to a stationary reference. For measurements on moving vehicles, such references are not generally available, and in many other situations measurement of absolute motion is easier and more desirable. The basic principle of seismic- (absolute-) displacement pickups is simply to measure (with any convenient relative-motion transducer) the relative displacement of a mass connected by a soft spring to the vibrating body. For frequencies above the natural frequency, this relative displacement is also very nearly the absolute displacement since the mass tends to stand still.

To obtain a quantitative measure of performance for such systems we analyze the configuration of Fig. 4.49a. The rotational configuration

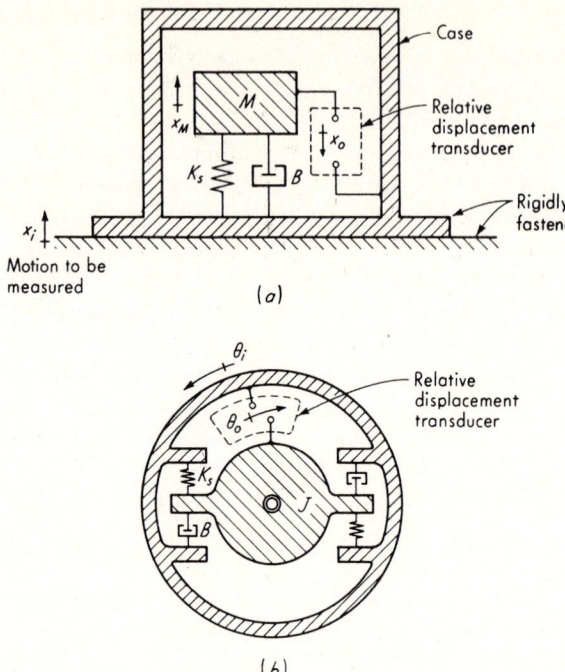

Motion to be
measured

Fig. 4.49.   *Translational and rotational seismic pickups.*

is completely analogous.   Newton's law may be applied to the mass M as
follows:

$$K_s x_o + B\dot{x}_o = M\ddot{x}_M = M(\ddot{x}_i - \ddot{x}_o) \qquad (4.117)$$

where x_i and x_M are the absolute displacements and we have chosen our
reference for x_o such that x_o is zero when the gravity force (weight of M)
is acting along the x axis statically.   Manipulation gives

$$\frac{x_o}{x_i}(D) = \frac{D^2/\omega_n^2}{D^2/\omega_n^2 + 2\zeta D/\omega_n + 1} \qquad (4.118)$$

$$\text{where} \quad \omega_n \triangleq \sqrt{\frac{K_s}{M}}$$

$$\zeta \triangleq \frac{B}{2\sqrt{K_s M}}$$

Since the pickup is intended mainly as a vibration sensor, the frequency
response is of prime interest.

$$\frac{x_o}{x_i}(i\omega) = \frac{(i\omega)^2/\omega_n^2}{(i\omega/\omega_n)^2 + 2\zeta i\omega/\omega_n + 1} \qquad (4.119)$$

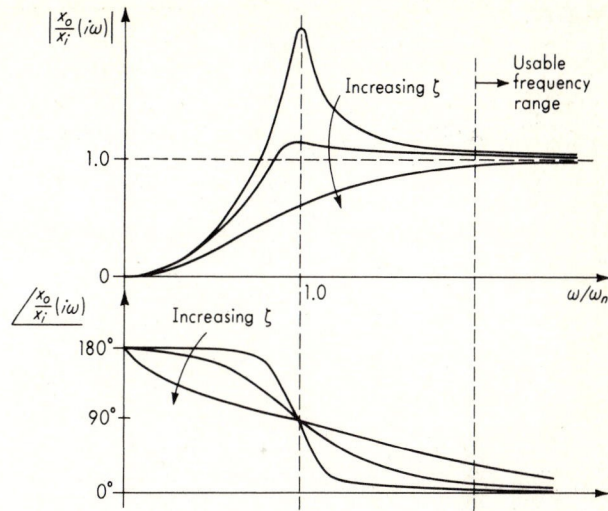

Fig. 4.50. *Seismic-displacement-pickup frequency response.*

This is graphed in Fig. 4.50. We note that there is no response to static displacement inputs and that ω_n should be much less than the lowest vibration frequency ω for accurate displacement measurement. For frequencies much above ω_n, $(x_o/x_i)(i\omega) \rightarrow 1\underline{/0°}$, indicating perfect measurement. The characteristics of the relative-displacement transducer in converting x_o to a voltage e_o must also be considered. Since the force in spring K_s is directly proportional to x_o, if strain gages are used they can be applied directly to this spring, which may be in the form of a cantilever beam. Since a low ω_n is desired, either a large mass or soft spring (or both) is necessary. To keep size (and thereby loading on the measured system) to a minimum, soft springs are preferred to large masses. Intentional damping in the range $\zeta = 0.6$ to 0.7 is often employed to minimize resonant response to slow transients.

4.7

Seismic- (Absolute-) Velocity Pickups The application here is again limited to vibratory velocities, and the basic configuration is exactly the same as in Fig. 4.49. To measure velocity $\dot{x}_i$ rather than displacement x_i, three possibilities will be considered. First, a voltage signal from a displacement pickup may be sent to an electrical differentiation circuit. Second (and this is the most practical), the relative-displacement transducer of Fig. 4.49 is replaced by a relative-velocity transducer

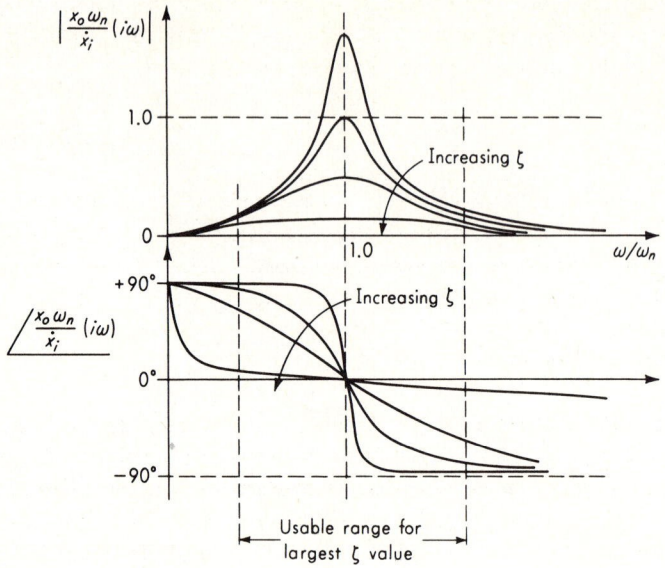

Fig. 4.51.   *Seismic-velocity-pickup frequency response.*

(usually a moving-coil pickup).   Then, since $e_o = K_e \dot{x}_o$, we have

$$\frac{e_o}{\dot{x}_i}(D) = \frac{K_e D^2/\omega_n{}^2}{D^2/\omega_n{}^2 + 2\zeta D/\omega_n + 1} \qquad (4.120)$$

and accurate velocity measurement is possible if $\omega \gg \omega_n$.   Signals from such pickups may be readily integrated electrically to get displacement information.   The third possibility is revealed by rewriting Eq. (4.118) as

$$\frac{x_o}{D x_i}(D) = \frac{D}{D^2 + 2\zeta\omega_n D + \omega_n{}^2} \qquad (4.121)$$

$$\frac{x_o}{\dot{x}_i}(i\omega) = \frac{1}{2\zeta\omega_n - i[(\omega_n{}^2 - \omega^2)/\omega]} \qquad (4.122)$$

Now if we wish x_o to be a measure of $\dot{x}_i$, then $(x_o/\dot{x}_i)(i\omega) \approx$ const.   From Eq. (4.122) we see that this will be the case if $(\omega_n{}^2 - \omega^2)/\omega \approx 0$, since then

$$\frac{x_o}{\dot{x}_i}(i\omega) \approx \frac{1}{2\zeta\omega_n} \qquad (4.123)$$

Now $(\omega_n{}^2 - \omega^2)/\omega \approx 0$ if $\omega \approx \omega_n$.   This would allow measurement only at frequency ω_n.   However, if ζ is made very large, the range of frequencies around ω_n for which $(\omega_n{}^2 - \omega^2)/\omega$ is negligible compared with $2\zeta\omega_n$ is fairly broad.   Figure 4.51 shows this graphically.   While a possibility, this approach is rarely employed in practice. The required large value of ζ reduces the sensitivity [see Eq. (4.123)], and *both* very low and very high frequencies are not accurately measured.

4.8

Seismic- (*Absolute-*) Acceleration Pickups (*Accelerometers*) The most important pickup for vibration, shock, and general-purpose absolute-motion measurement is the accelerometer. This instrument is commercially available in a wide variety of types and ranges to meet correspondingly diverse application requirements. The basis for this popularity lies in the following features:

1. Frequency response is from zero to some high limiting value. Steady accelerations can (except in piezoelectric types) be measured.
2. Displacement and velocity can be easily obtained by electrical integration, which is much preferred to differentiation.
3. Measurement of transient (shock) motions is more readily achieved than with displacement or velocity pickups.
4. Destructive forces in machinery, etc., are often more closely related to acceleration than to velocity or displacement.

The basic accelerometer configuration is again that of Fig. 4.49. The operating principle is as follows: Suppose the acceleration $\ddot{x}_i$ to be measured is constant. Then, in steady state, the mass M will be at rest relative to the case and thus its absolute acceleration will also be $\ddot{x}_i$. If mass M is accelerating at $\ddot{x}_i$ there must be some force to cause this acceleration, and if M is not moving relative to the case, this force can come only from the spring. Since spring deflection x_o is proportional to force, which in turn is proportional to acceleration, x_o is a measure of acceleration $\ddot{x}_i$. Thus absolute-acceleration measurement is reduced to the measurement of the force required to accelerate a known mass (sometimes called the "proof" mass). This dependence on mass leads to problems (mainly in inertial guidance systems, not in vibration measurement) since a mass also experiences forces due to gravitational fields. Thus an accelerometer cannot distinguish between a force due to acceleration and a force due to gravity.

The majority of accelerometers may be classified as either deflection type or null-balance type. Those used for vibration and shock measurement are generally the deflection type whereas those used for measurement of gross motions of vehicles (submarines, aircraft, spacecraft, etc.) may be either type, with the null-balance being used when extreme accuracy is needed.

Deflection-type accelerometers. A large number of practical accelerometers have the configuration of Fig. 4.49 and differ only in details, such as the spring element used, relative-motion transducer used, and type of damping provided. Since the desired input is now $\ddot{x}_i$, we can rewrite Eq. (4.118) as

$$\frac{x_o}{D^2 x_i}(D) = \frac{x_o}{\ddot{x}_i}(D) = \frac{K}{D^2/\omega_n{}^2 + 2\zeta D/\omega_n + 1} \qquad (4.124)$$

where

$$K \triangleq \frac{1}{\omega_n{}^2} \quad \text{in.}/(\text{in.}/\text{sec}^2) \qquad (4.125)$$

Since output voltage $e_o = K_e x_o$ for many motion transducers, Eq. (4.124) has the correct form for the acceleration-to-voltage transfer function also. We see that the accelerometer is an ordinary second-order instrument; thus all our previous work on this type is immediately applicable. The frequency response extends from 0 to some fraction of ω_n, depending on the accuracy required and the damping. Because sensitivity $K = 1/\omega_n{}^2$, high-frequency response must be traded for sensitivity. Since the dynamic characteristics of second-order instruments have already been thoroughly discussed, here we shall discuss mainly the specific characteristics of commercially available instruments.

Accelerometers using resistive potentiometers as their motion pickup are intended mainly for slowly varying accelerations and low-frequency vibration. A typical family[1] of such instruments offers nine models covering the range of $\pm 1g$ full scale to $\pm 50g$ full scale. The natural frequencies range from 12 to 86 cps, and ζ is 0.5 to 0.8 over the temperature range -65 to $+165°F$, using a temperature-compensated liquid damping arrangement. Potentiometer resistance may be selected in the range 1,000 to 10,000 ohms, with corresponding resolution of 0.45 to 0.25 percent of full scale. The potentiometer power rating is 0.5 watt at $+165°F$. The sensitivity to acceleration at right angles to the desired axis (cross-axis sensitivity) is less than ± 1 percent of the sensitivity to the desired direction. The operating life is 2,000,000 cycles. Overall inaccuracy is ± 1 percent of full scale or less at room temperature. This increases to ± 1.8 percent if the temperature is allowed to vary over the design range of -65 to $+165°F$. Size is about a 2-in. cube; weight is about 1 lb.

Unbonded-strain-gage accelerometers use the strain wires as the spring element and also as the motion transducer. They are useful for general-purpose motion measurement and also for vibration up to relatively high frequencies. They are available in a wide range of characteristics, typical[2] values including ± 0.5 to $\pm 200g$ full scale, natural

---

[1] Bourns, Inc., Riverside, Calif.
[2] Statham Instruments Inc., Los Angeles, Calif.

frequency 17 to 800 cps, excitation voltage about 10 volts alternating or direct current, full-scale output ± 20 to ± 50 mv, resolution less than 0.1 percent, inaccuracy 1 percent of full scale or less, cross-axis sensitivity less than 2 percent, and damping ratio (using silicone-oil damping) of 0.6 to 0.8 at room temperature. (Temperature-compensated models are also available.) These instruments can be made quite small and light, a typical size being $\frac{1}{2}$ by $\frac{1}{2}$ by 2 in., with a weight of 26 g.

Bonded-strain-gage accelerometers generally use a mass supported by thin flexure beams, with strain gages cemented to the beam so as to achieve maximum sensitivity, temperature compensation, and insensitivity to cross-axis and angular acceleration. Their characteristics are similar to those of unbonded-gage accelerometers except that size and weight tend to be greater. Silicone-oil damping is again widely used.

A recently developed strain-gage accelerometer[1] using semiconductor materials exhibits many desirable properties. Full-scale range is $\pm 250g$ with full-scale output of ± 250 mv (10 volts d-c excitation). The natural frequency is greater than 10,000 cps with a damping ratio of about 0.06. This low damping is due to material hysteresis and air drag; no intentional damping device is employed. The usable frequency range (amplitude ratio flat to 5 percent) is from 0 to 2,000 cps. In this range, phase shift is very small because of the light damping. The light damping causes no "ringing" problems as long as shock inputs do not contain much frequency content near the 10-kc resonant frequency (see text on terminated-ramp inputs). The cross-axis sensitivity is 1 to 3 percent while nonlinearity and hysteresis are 1 percent of full scale. Temperature compensation is provided to give an operating range of -65 to $+250°F$. The thermal zero shift is 0.02 percent of full scale/F° over this range, and the maximum sensitivity shift is about -7 percent at the extremes of this range. The size is about $\frac{5}{8}$-in. diameter and 1 in. high, with a weight of about 1 oz.

A family of liquid-damped differential-transformer accelerometers[2] exhibits the following characteristics: full-scale ranges from ± 2 to $\pm 700g$, natural frequency from 35 to 620 cps, nonlinearity 1 percent of full scale, full-scale output about 1 volt with excitation of 10 volts at 2,000 cps, damping ratio 0.6 to 0.7 at 70°F, residual voltage at null less than 1 percent of full scale, and hysteresis less than 1 percent of full scale. The size is about a 2-in. cube, with a weight of 4 oz.

A variable-reluctance accelerometer[3] using eddy-current damping has full-scale ranges of ± 1 to $\pm 40g$, natural frequency 16 to 100 cps, damping

---

[1] Endevco Corp., Pasadena, Calif.
[2] Schaevitz Engineering, Camden, N.J.
[3] Honeywell Inc., Boston, Mass.

ratio 0.6 ± 0.2 from -65 to $+250°F$, 25 volts full-scale output with 26-volt/400-cps excitation, hysteresis 0.15 percent of full scale, cross-axis sensitivity 0.5 percent, and nonlinearity of ± 0.25 percent for one-half range and ± 1.6 percent full range. The threshold and resolution are each $0.0001g$. The size is about a 2-in. cube, and the weight is about 1 lb.

Piezoelectric accelerometers are in wide use for shock and vibration measurements. In general they do not give an output for constant acceleration because of the basic characteristics of piezoelectric motion transducers. They do, however, have large output-voltage signals and can have very high natural frequencies (higher than any other type) which are necessary for accurate shock measurements. In general no intentional damping is provided, material hysteresis being the only source of energy loss. This results in a very low (about 0.01) damping ratio but this is acceptable because of the very high natural frequency. The transfer function is a combination of Eqs. (4.70) and (4.124):

$$\frac{e_o}{\ddot{x}_i}(D) = \frac{(K_q/C\omega_n{}^2)\tau D}{(\tau D + 1)(D^2/\omega_n{}^2 + 2\zeta D/\omega_n + 1)} \qquad (4.126)$$

The low-frequency response is limited by the piezoelectric characteristic $\tau D/(\tau D + 1)$ while the high-frequency response is limited by mechanical resonance. The damping ratio ζ of piezoelectric accelerometers is not usually quoted by the manufacturer but can be taken as zero for most practical purposes. The accurate (5 percent high at the high-frequency end and 5 percent low at the low-frequency end) frequency range of such an accelerometer is $3/\tau < \omega < 0.2\omega_n$. Accurate low-frequency response requires large τ, which is usually achieved by use of high-impedance (cathode-follower) amplifiers or charge amplifiers. Some quartz accelerometers and electrometer amplifiers have large enough τ to allow measurement of constant accelerations.

Typical construction of a piezoelectric accelerometer is shown in Fig. 4.52. The "crystal" is preloaded to about 10,000-psi stress by screwing down the cap on the hemispherical spring. This prestressing puts the piezoelectric material at a more linear part of its stress-charge curve. It also allows measurement of acceleration in both directions without the crystal going into tension. When the preload is applied, a voltage of a certain polarity is developed but this soon leaks off to zero. Any further deflection (due to acceleration forces) gives a plus or minus charge, depending on the direction of the motion. Figure 4.53 shows some other constructions designed to minimize cross-axis sensitivity.

Piezoelectric accelerometers are available in a wide range of characteristics; we quote only a few typical examples.[1] A single low-g instru-

---

[1] Endevco Corp., Pasadena, Calif.

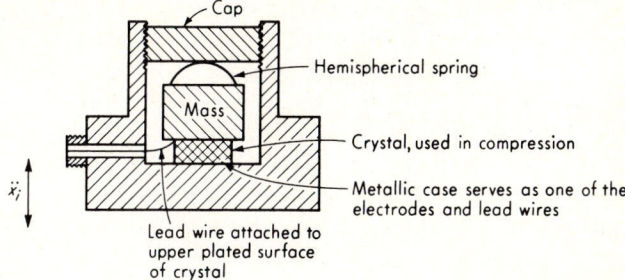

Fig. 4.52. *Piezoelectric-accelerometer construction.*

ment has a sensitivity of 50 mv/g, will measure accelerations from 0.03 to 1,000g with a nonlinearity of 1 percent of full scale, and has a natural frequency of 20 kc, flat frequency response ± 5 percent from 20 to 4,000 cps when used with 100-megohm input impedance, capacitance of 600 pf with 3 ft of cable, cross-axis sensitivity of 5 percent, sensitivity drift of ± 10 percent from -30 to $+230°F$, size about a 1-in. cube, and a weight of 2 oz.

A shock accelerometer has a sensitivity of 5 mv/g, range 0 to 10,000g with 1 percent nonlinearity, natural frequency of 35 kc, flat frequency response ± 5 percent from 0.1 to 7,000 cps with charge amplifier, pulse response ± 5 percent for pulses shorter than 0.66 sec (see Fig. 4.32), no ringing for pulses longer than 0.15 msec, capacitance of 100 pf, cross-axis sensitivity of 5 percent, sensitivity drift ± 1 percent from -100 to $+350°F$, size about a 0.7-in. cube, and a weight of 1 oz.

Some special piezoelectric-accelerometer characteristics available in specific models include water-cooled units usable at 2200°F, triaxial units combining three mutually perpendicular elements in a 1-in.-cube case, high-capacitance (7,000 pf) units insensitive to cable capacitance and giving good low-frequency response with relatively low-input-impedance measuring equipment, and miniature units weighing the order of 1 g for small-object testing.

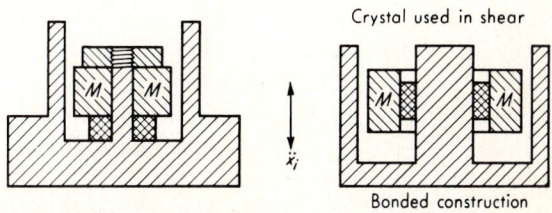

Fig. 4.53. *Piezoelectric-accelerometer construction.*

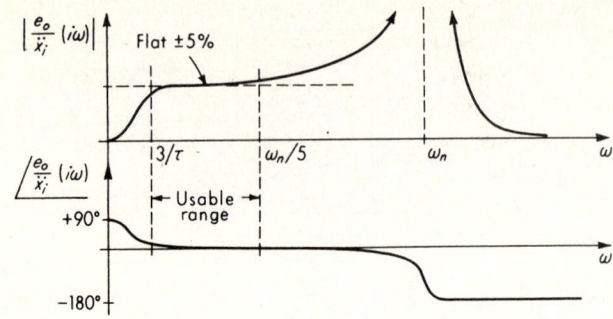

Piezoelectric accelerometer frequency response

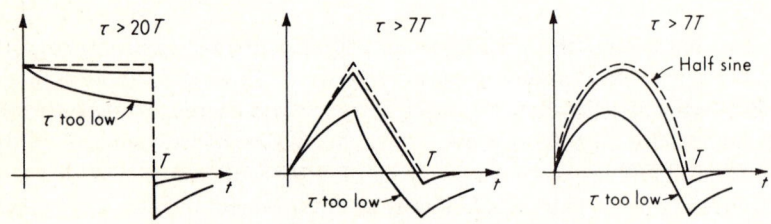

Requirements for accurate peak measurements ±5%

Low-frequency response problems

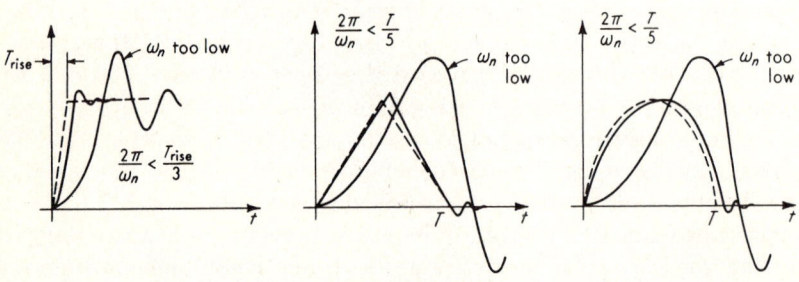

Requirements for accurate peak measurements ±10%

High-frequency response problems

Fig. 4.54. *Piezoelectric-accelerometer response.*

The use of charge amplifiers rather than voltage amplifiers for piezo-electric devices is becoming quite common because of their advantages of insensitivity to cable capacitance and better low-frequency-response characteristics. These instruments are discussed later in the text in the section on amplifiers.

While the general dynamic-response techniques of Chap. 3 are applicable to accelerometer problems, Fig. 4.54 summarizes some results useful in choosing piezoelectric accelerometers for practical applications. The low-frequency problems are peculiar to piezoelectric types whereas the high-frequency problems are common to all accelometers.

Angular accelerometers based on the configuration of Fig. 4.49 b can be constructed by using various pickups, just as in translational accelerometers. An interesting variation on the basic principle is found in the Statham liquid-rotor angular accelerometer.[1] The inertial mass is liquid contained in a circular case having a flexure-mounted paddle. Rotational acceleration of the case causes a fluid pressure on the paddle whose resulting deflection is measured with a suitable motion pickup. A family of such devices consisting of 11 models exhibits the following characteristics: Full-scale range is from ±1.5 to ±3,000 rad/sec², natural frequency 3 to 150 cps, damping ratio 0.7 ± 0.1 at room temperature, cross-axis (angular) sensitivity 2 percent, nonlinearity and hysteresis ±2 percent of full scale, sensitivity to linear acceleration from 0.3 percent of full scale/g for the lowest range to 0.02 percent for the highest, and sensitivity to angular velocity about axes other than the instrument axis of about 0.05 rad/sec² for an angular velocity of 5 rad/sec. The pickup is either an unbonded strain gage (10 volts excitation, ±25-mv full-scale output) or a two-arm inductive bridge (10 volts at 3,000 cps excitation, ±0.75-volt full-scale output). The size and weight range from 9-in. diameter by 3 in. high (8 lb) for the lowest-range instrument to 3-in. diameter by 3 in. high (1½) lb) for the highest range.

Null-balance- (servo-) type accelerometers. So-called servo accelerometers using the principle of feedback have been developed for applications requiring greater accuracy than is generally achieved with instruments using mechanical springs as the force-to-displacement transducer. In these null-balance instruments the acceleration-sensitive mass is kept very close to the zero-displacement position by sensing this displacement and generating a magnetic force which is proportional to this displacement and which always opposes motion of the mass from neutral. This restoring force plays the same role as the mechanical spring force in a conventional accelerometer. Thus one may consider the mechanical spring to have been replaced by an electrical "spring." The advantages derived from this approach are the greater linearity and lack of hysteresis of the electrical spring as compared with the mechanical one. Also, in some cases, electrical damping (which can often be made less temperature-sensitive than mechanical damping) may be employed. There is also the possibility of testing the static and dynamic performance of the device just

---

[1] G. N. Rosa, Some Design Considerations for Liquid Rotor Angular Accelerometers, *Statham Instr. Notes* 26, Statham Instruments Inc., Los Angeles, Calif.

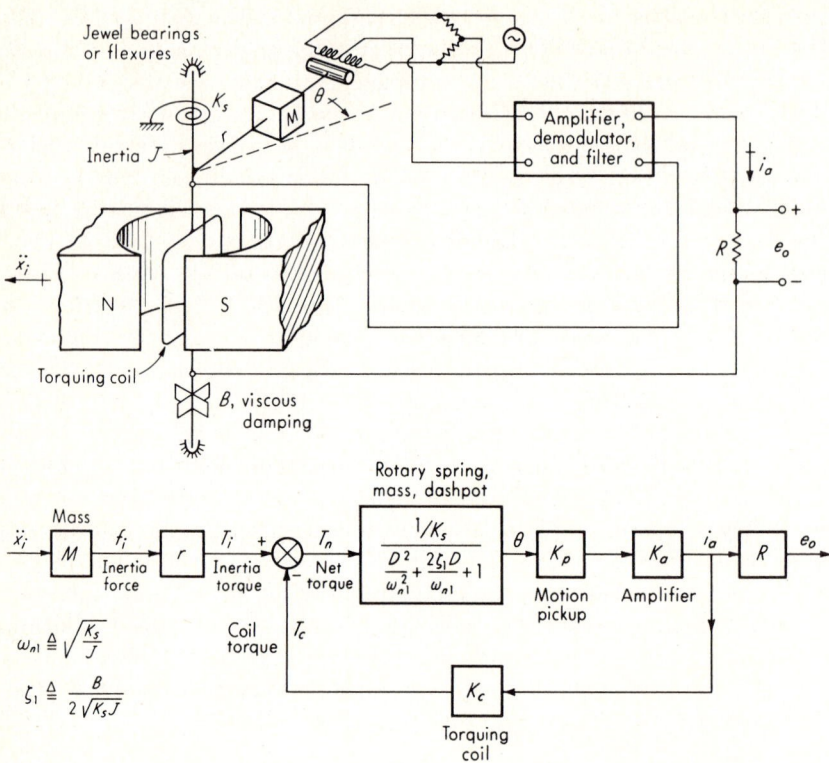

Fig. 4.55.   *Servo-type accelerometer.*

prior to a test run by introducing electrically excited test forces into the system.   This convenient and rapid remote self-checking feature can be quite important in complex and expensive tests where it is extremely important that all systems operate correctly before the test is commenced. These servo accelerometers are usually used for general-purpose motion measurement and low-frequency vibration.   They are also particularly useful in acceleration-control systems since the desired value of acceleration can be put into the system by introducing a proportional current i_a from some external source.

Figure 4.55 illustrates in simplified fashion the operation of a typical instrument designed to measure a translational acceleration $\ddot{x}_i$.   (Angular acceleration can also be measured by these techniques by using an obvious mechanical modification.)   The acceleration $\ddot{x}_i$ of the instrument case causes an inertia force f_i on the sensitive mass M, tending to make it pivot in its bearings or flexure mount.   The rotation θ from neutral is sensed by an inductive pickup and is amplified, demodulated, and filtered to produce a current i_a directly proportional to the motion from null. This current is passed through a precision stable resistor R to produce the output-voltage signal and also is applied to a coil suspended in a magnetic field. The current through the coil produces a magnetic torque on the coil (and the attached mass M) which acts to return the mass to neutral. The current required to produce a coil magnetic torque that just balances the inertia torque due to $\ddot{x}_i$ is directly proportional to $\ddot{x}_i$; thus e_o is a measure of $\ddot{x}_i$. Since a nonzero displacement θ is necessary to produce a current i_a, the mass is not returned exactly to null but comes very close because a high-gain amplifier is used. Analysis of the block diagram reveals the details of performance as follows:

$$\left(Mr\ddot{x}_i - \frac{e_oK_c}{R}\right)\frac{K_pK_a/K_s}{D^2/{\omega_{n1}}^2 + 2\varsigma_1 D/\omega_{n1} + 1} = \frac{e_o}{R} \qquad (4.127)$$

$$\left[\frac{D^2}{{\omega_{n1}}^2} + \frac{2\varsigma_1 D}{\omega_{n1}} + \left(1 + \frac{K_cK_pK_a}{K_s}\right)\right]e_o = \frac{MrRK_pK_a}{K_s}\ddot{x}_i \qquad (4.128)$$

Now, by design, the amplifier gain K_a is made large enough so that $K_cK_pK_a/K_s \gg 1.0$, so that then

$$\frac{e_o}{\ddot{x}_i}(D) = \frac{K}{D^2/{\omega_n}^2 + 2\varsigma D/\omega_n + 1} \qquad (4.129)$$

where

$$K \triangleq \frac{MrR}{K_c} \qquad \text{volts/(in./sec}^2) \qquad (4.130)$$

$$\omega_n \triangleq \omega_{n1}\sqrt{\frac{K_pK_aK_c}{K_s}} \qquad \text{rad/sec} \qquad (4.131)$$

$$\varsigma \triangleq \frac{\varsigma_1}{\sqrt{K_pK_aK_c/K_s}} \qquad (4.132)$$

Equation (4.130) shows that the sensitivity now depends only on M, r, R, and K_c, all of which can be made constant to a high degree. This demonstrates again the usefulness of high-gain feedback in shifting the requirements for accuracy and stability from many components to a few chosen ones where the requirements can be met. As in all feedback systems, the gain cannot be made arbitrarily high because of dynamic instability; however, a sufficiently high gain can be achieved to give excellent performance. Turning to Eq. (4.131), we see that ω_n is increased from the basic spring-mass frequency ω_{n1} by the factor $\sqrt{K_pK_aK_c/K_s}$, another benefit of high-gain feedback. However, ς is decreased by the same factor, and so ς_1 must be made sufficiently high to compensate for this.

A typical accelerometer[1] of this kind, using a flexure pivot, is available in full-scale ranges of ± 10 to $\pm 100g$, natural frequency of 100 to

[1] Systron-Donner Corp., Concord, Calif.

250 cps, damping ratio of from 0.3 to 5, cross-axis sensitivity of 0.1 percent, resolution better than 0.0001 percent of full scale, nonlinearity and hysteresis each better than 0.005 percent of full scale, and has a full-scale output of ± 7.5 volts with considerable current capacity (± 1.2 to ± 12 ma).

Accelerometers for inertial navigation. Inertial navigation is accomplished in principle by measuring the absolute acceleration (usually in terms of three mutually perpendicular components of the total-acceleration vector) of the vehicle and then integrating these acceleration signals twice to obtain displacement from an initial known starting location. Thus instantaneous position is always known without the need for any communication with the world outside the vehicle.   To keep the accelerometers' sensitive axes always oriented parallel to their original starting positions, elaborate stable platforms using gyroscopic references and feedback systems are necessary.   Since the accelerometers are also sensitive to gravitational force, this force must be computed and corrections applied continuously.   Since the inertial navigation system measures absolute motion, systems for navigation over the earth's surface (such as for submarines) must include means for compensating for the earth's own motions.

While accelerometers for such navigation systems must operate on essentially the same basic principles that we have considered above, their extreme performance requirements and the desire to obtain integrals of the acceleration rather than acceleration itself lead to special techniques and configurations.   The desire for compatibility with the required data-processing computers (often digital) also influences the designer's choice of alternatives.   The details of these applications are beyond the scope of this book but may be found in numerous references.[1,2,3,4,5,6]

Mechanical loading of accelerometers on the test object. The attachment of an accelerometer to a vibrating system results in a change in the motion measured as compared with the undisturbed case.   We can apply general impedance principles to this problem to calculate the significance of this effect in any particular instance.   In doing so, a useful

[1] H. B. Sabin, 17 Ways to Measure Acceleration, *Control Eng.*, p. 106, February, 1961.

[2] J. M. Slater and D. E. Wilcox, How Precise Are Inertial Components?, *Control Eng.*, p. 86, July, 1958.

[3] *Sperry Engineering Review*, spring, 1964.

[4] J. M. Slater, Inertial Guidance Notes, North American Aviation Corp., Autonetics Div.

[5] C. F. Savant et al., "Principles of Inertial Navigation," McGraw-Hill Book Company, New York, 1961.

[6] P. H. Savet (ed.), "Gyroscopes: Theory and Design," McGraw-Hill Book Company, New York, 1961.

simplification, which is adequate in most cases, is to regard the entire accelerometer as one rigid mass equal to the total mass of the instrument. This approximation generally holds since accelerometers are used below their natural frequency and there is thus little relative motion of the proof mass and the instrument case.

4.9

Calibration of Vibration Pickups   While the response of vibration pickups to interfering and modifying inputs such as temperature, acoustic noise, and magnetic fields is often of interest, we are here concerned with the response to the desired input of displacement, velocity, or acceleration.   An excellent reference giving a more complete treatment of this subject is the publication "American Standard Methods for the Calibration of Shock and Vibration Pickups 52.2-1959."[1]   We shall here briefly touch on some of the main points only.

The calibration methods in wide use may be classified into three broad types: constant acceleration, sinusoidal motion, and transient motion.   Constant-acceleration methods (which are suitable only for calibrating accelerometers) include the tilting-support method and the centrifuge.   The tilting-support method utilizes the accelerometer's inherent sensitivity to gravity.   Static "accelerations" over the range $\pm 1g$ may be accurately applied by fastening the accelerometer to a tilting support whose tilt angle from vertical is accurately measured. This method requires that the accelerometer respond to static accelerations; therefore most piezoelectric devices cannot be calibrated in this way. The accuracy of the method depends on the accuracy of angle measurement and the knowledge of the local gravity.   The accuracy is of the order of $\pm 0.0003g$.   In the centrifuge method the sensitive axis of the accelerometer is radially disposed on a rotating horizontal disk so that it experiences the normal acceleration of uniform circular motion.   Static accelerations in the range 0 to 60,000g are achievable with an accuracy of ± 1 percent.   The allowable weight of the pickup varies from 100 lb at 100g to 1 lb at 60,000g.

The sinusoidal-motion method is exemplified by the calibration facility of the National Bureau of Standards.[2]   This consists of a modified electrodynamic vibration shaker which has been carefully designed to provide uniaxial pure sinusoidal motion and which is equipped with an

---

[1] Available from the American Standards Association, Inc., 10 E. 40th St., New York, N.Y., 10016.

[2] R. R. Bouche, Improved Standard for the Calibration of Vibration Pickups, *Exptl. Mechanics*, April, 1961.

accurately calibrated moving-coil velocity pickup to measure its table motion. If a motion is known to be purely sinusoidal, knowledge of its velocity and frequency allows accurate calculation of the displacement and acceleration. (The motion frequency is easily obtained with high accuracy by electronic counters.) This technique is thus useful for displacement, velocity, or acceleration pickups. The particular equipment referred to above can calibrate pickups (obtaining both amplitude ratio and phase angle) of a weight up to 2 lb over the frequency range 8 to 2,000 cps. The acceleration range available is 0 to 25g, velocity range is 0 to 50 in./sec, and displacement range is 0 to 0.5 in. Accuracy is ± 1 percent from 8 to 900 cps and ± 2 percent from 900 to 2,000 cps.

At frequencies up to about 200 cps the displacement of a vibrating shake table can be quite accurately measured by viewing a suitable target (a point-light-source-illuminated piece of 320-grit emery cloth cemented to the table is convenient) through a measuring microscope. The illuminated grit crystals generate short straight lines of length equal to the peak-to-peak displacement amplitude in the microscope viewing field. The length of the lines can be measured with an inaccuracy of the order of 0.0001 in. This inaccuracy limits the usable frequency range of the method since the displacement amplitude corresponding to the acceleration levels of practical interest becomes very small as the frequency increases. For example, consider a sinusoidal vibration of 10g peak acceleration at 1,000 cps. The peak-to-peak displacement would be 0.000196 in., and an inaccuracy of 0.0001 in. would represent a 50 percent error. Thus, at high frequencies, more accurate displacement-measuring techniques are needed. The optical interferometer fringe-disappearance method[1] is used at amplitudes of the order of 4×10^{-6} in. and allows calibration at frequencies exceeding 10,000 cps.

Transient-motion calibration methods include the physical pendulum,[2] ballistic pendulum,[2] and drop-test[2] techniques. The latter two methods (used for accelerometers) are of greatest interest and will be briefly discussed. Both employ the concept that the velocity change Δv during a time interval $\Delta t = t_2 - t_1$ is given by

$$\Delta v = \int_{t_1}^{t_2} a \, dt \qquad (4.133)$$

where $a \triangleq$ acceleration. The procedure involves measurement, by some independent means, of the velocity change Δv of a rigid body to which the pickup is attached and simultaneous recording of the output voltage e_o of

---

[1] J. Johansson, Accelerometer Calibration, *Instr. Control Systems*, p. 79, December, 1963.

[2] C. M. Harris and C. E. Crede (eds.), "Shock and Vibration Handbook," vol. I, chap. 18, McGraw-Hill Book Company, New York, 1961.

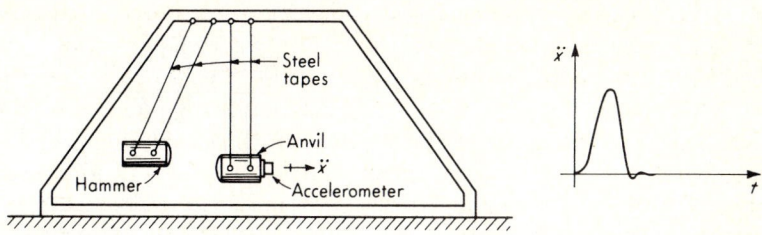

Fig. 4.56. *Ballistic pendulum.*

the pickup. This output voltage is given by

$$e_o = Ka \qquad (4.134)$$

where K is the unknown sensitivity [volts/(in./sec^2)] of the pickup. Thus we may write

$$\Delta v = \frac{1}{K} \int_{t_1}^{t_2} e_o \, dt \qquad (4.135)$$

and thus

$$K = \frac{\int_{t_1}^{t_2} e_o \, dt}{\Delta v} \qquad (4.136)$$

The integral of e_o can be obtained by connecting the output of the accelerometer to an integrating circuit or by recording e_o on paper or film and then measuring the area under the curve with a planimeter.

The physical layout of the ballistic pendulum is shown in Fig. 4.56. The hammer is raised on its suspension tapes and allowed to impact the anvil and attached accelerometer. The velocity change can be measured by means of any suitable relative-velocity transducer, or, alternatively, it may be calculated from the maximum height of rise of the anvil after the impact. The pulse duration can be varied by changing the shape and/or material of the impacting surfaces, typical values being of the order of 0.001 sec.

A drop-test calibrator is shown in Fig. 4.57. The velocity change

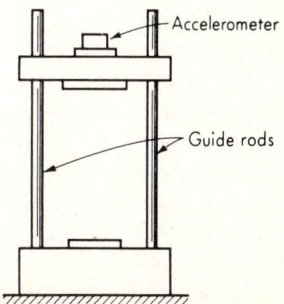

Fig. 4.57. *Drop-test apparatus.*

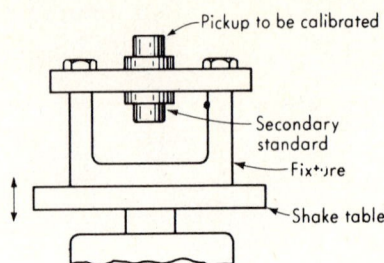

Fig. 4.58.   *Calibration fixture.*

is again measured by means of conventional transducers or may be calculated (assuming low friction) from the height of fall h_1 and the height of rebound h_2 as

$$\Delta v = \sqrt{2gh_1} + \sqrt{2gh_2} \qquad (4.137)$$

Perhaps the main usefulness of the above transient techniques lies in their ability to provide high acceleration values (up to several thousand g's) with large high-frequency content (short pulse duration). Such tests, while in general not as accurate for determining numerical values of instrument characteristics as the frequency-response tests, provide a very severe test of the accelerometer's freedom from internal resonances and should thus be included in the calibration programs for any accelerometers to be used for shock work.

As a final comment on calibration methods, it should be noted that once a pickup has been accurately calibrated (say by the National Bureau of Standards) it becomes a secondary standard against which other pickups may be calibrated by direct comparison. This can readily be done by fastening both pickups to a common rigid fixture mounted on a vibration shake table and then applying the frequencies and amplitudes desired, as in Fig. 4.58.

4.10

Jerk Pickups   In some measurement and control applications the rate of change of acceleration, or jerk, d^3x/dt^3, must be measured. An obvious approach is to apply the electrical output from an accelerometer to a differentiating circuit. A more subtle technique which avoids the noise-accentuating problems of differentiating circuits is applied in the Donner Jerkmeter.[1] By ingenious use of feedback principles, this null-balance instrument provides both acceleration and jerk signals of good quality. The physical configuration is essentially that of Fig. 4.55 with

[1] Systron Donner Corp., Concord, Calif.

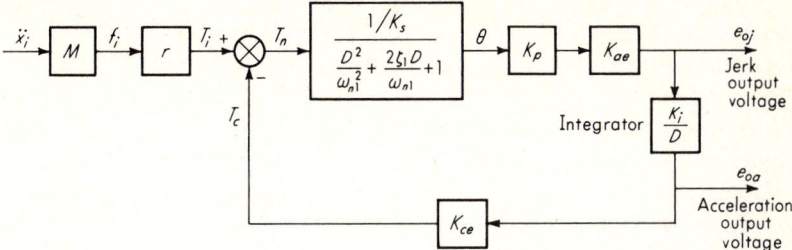

Fig. 4.59. *Jerkmeter block diagram.*

the addition of an electronic integrator. The resulting block diagram is shown in Fig. 4.59. Analysis gives

$$\left(MrD^2 x_i - \frac{K_i K_{ce} e_{oj}}{D} \right) \frac{K_p K_{ae}/K_s}{D^2/\omega_{n1}{}^2 + 2\zeta_1 D/\omega_{n1} + 1} = e_{oj} \qquad (4.138)$$

leading to the differential equation

$$\left(\frac{K_s}{K_i K_p K_{ae} K_{ce} \omega_{n1}{}^2} D^3 + \frac{2\zeta_1 K_s}{K_i K_p K_{ae} K_{ce} \omega_{n1}} D^2 \right.$$
$$\left. + \frac{K_s}{K_i K_p K_{ae} K_{ce}} D + 1 \right) e_{oj} = \frac{Mr}{K_i K_{ce}} D^3 x_i \qquad (4.139)$$

We note that the relationship between jerk input $D^3 x_i$ and voltage output e_{oj} is that of a third-order differential equation. The static sensitivity is easily seen to be $Mr/K_i K_{ce}$ volts/(in./sec³); however, the dynamic behavior is not obvious since we have not previously considered third-order instruments. For any given set of numerical values, the cubic characteristic equation of Eq. (4.139) will have three numerical roots. These will be either three real roots or one real root plus a pair of complex conjugates. In an actual design the latter situation often prevails. This means that the transfer function of Eq. (4.139) will generally be of the form

$$\frac{e_{oj}}{D^3 x_i}(D) = \frac{K}{(\tau D + 1)(D^2/\omega_n{}^2 + 2\zeta D/\omega_n + 1)} \qquad (4.140)$$

where $\qquad\qquad\qquad\qquad\qquad\qquad\qquad\qquad K \triangleq \dfrac{Mr}{K_i K_{ce}}$

and τ, ζ, and ω_n can be found if numerical values for all the constants are given. The frequency response for Eq. (4.140) is easily plotted by using logarithmic methods. Actually, the frequency response for Eq. (4.139) is quite revealing. It is clear that, for sufficiently low frequencies, $e_{oj} \approx (Mr/K_i K_{ce}) \ddot{x}_i$ for *any* values of the system constants since the first three terms involve ω^3, ω^2, and ω as a factor and thus go to zero as $\omega \to 0$. To increase the usable frequency range, the *coefficients* of the ω^3, ω^2, and ω terms must be made small. This can be done by making $K_s/K_iK_pK_{ae}K_{ce}$ small, which corresponds to making the gain of the feedback loop large. A limit is placed on the gain, however, since excessive gain will cause dynamic instability and resultant destruction of the instrument.

The Routh stability criterion[1] shows that for a cubic characteristic equation of the form $a_3D^3 + a_2D^2 + a_1D + a_0 = 0$ stability is assured if all the a's are positive and $a_0a_3 < a_1a_2$. In our case this becomes

$$\frac{K_s}{K_iK_pK_{ae}K_{ce}\omega_{n1}^2} < \frac{2\zeta_1K_s^2}{(K_iK_pK_{ae}K_{ce})^2\omega_{n1}} \quad (4.141)$$

or

$$\zeta_1 > \frac{K_iK_pK_{ae}K_{ce}}{2K_s\omega_{n1}} \quad (4.142)$$

This shows that if $K_s/K_iK_pK_{ae}K_{ce}$ is made small to increase the usable frequency range, the damping ζ_1 must be correspondingly increased to retain adequate stability. This required damping effect can also be obtained by adding proper electrical compensating networks to the circuit. This approach is often used. No matter how the damping is achieved, however, a tradeoff will always be necessary between frequency range and stability.

The resolution, nonlinearity, and hysteresis of a commercially available Jerkmeter are each better than 0.1 percent of full scale. Instruments with full-scale jerk values ranging from ± 0.5 to $\pm 20g/\text{sec}$ and full-scale acceleration values of ± 1 to $\pm 30g$ are available. The full-scale acceleration and jerk output voltages are both ± 7.5 volts, while the instrument weight is about 8 oz and its size 3 in. long by 1.5 in. square.

4.11

Pendulous (Gravity-referenced) Angular-displacement Sensors In a number of applications the measurement of angular displacement relative to the local vertical (gravity vector) is a useful technique. Examples include sensing elements for control systems of road-paving and scraping machines; drainage-tile-laying machines; alignment of construction forms, piles, and bridges; and attitude control of vehicles (such as submarines) and torpedoes or missiles. These relatively simple instruments (basically plumb bobs with electrical output) can sometimes replace more complex and expensive gyroscopic instruments which perform similar functions. Their main disadvantages relative to gyros are their

[1] E. O. Doebelin, "Dynamic Analysis and Feedback Control," p. 175, McGraw-Hill Book Company, New York, 1962.

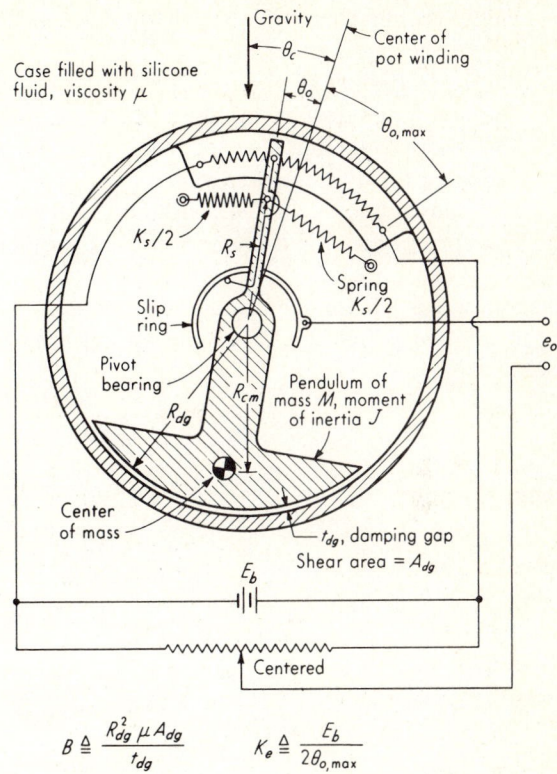

$$B \triangleq \frac{R_{dg}^2 \mu A_{dg}}{t_{dg}} \qquad K_e \triangleq \frac{E_b}{2\theta_{o,max}}$$

Fig. 4.60. *Pendulum displacement sensor.*

sensitivity to interfering translatory acceleration inputs and their dependence on a gravity field. (They will not work in essentially gravity-free space.)

Figure 4.60 shows a typical configuration of a single-axis pendulum-type sensor. The desired input to be measured is the case rotation angle θ_c. Most commercial sensors do not include the springs K_s; we include them here because their presence makes possible interesting and potentially useful dynamic behavior. For the usual case of no springs, K_s is simply set equal to zero. The damping effect is not essential to the theoretical operation of the device but is included in most practical instruments to reduce oscillations at pendulum frequency caused by transient interfering inputs. A variety of electrical displacement transducers may be employed, depending on the required characteristics; a potentiometer is shown for simplicity. The following assumptions are justifiable for most purposes in simplifying the analysis:

1. Angles are small enough so that the sine and the angle are nearly equal and the cosine is nearly 1.
2. The inertia effect of the fluid on the pendulum motion is negligible.
3. The damping effect of the fluid is limited to the damping gap.
4. All dry-friction effects in pot wipers, bearings, and slip rings may be neglected for dynamic analysis.
5. The buoyant force on the pendulum is negligible.
6. The springs provide a linear restoring torque.

The analysis is left for the problems at the end of this chapter. However the results are as follows: When the springs are present ($K_s \neq 0$) we have

$$\frac{e_o}{\theta_c} (D) = \frac{K(D^2/\omega_{n1}^2 + 1)}{D^2/\omega_{n2}^2 + 2\zeta D/\omega_{n2} + 1} \qquad (4.143)$$

where
$$K \triangleq \frac{MgR_{cm}K_e}{R_s^2 K_s + MgR_{cm}} \qquad (4.144)$$

$$\omega_{n1} \triangleq \sqrt{\frac{MgR_{cm}}{J}} \qquad (4.145)$$

$$\omega_{n2} \triangleq \sqrt{\frac{R_s^2 K_s + MgR_{cm}}{J}} \qquad (4.146)$$

$$\zeta \triangleq \frac{B}{2\sqrt{J(R_s^2 K_s + MgR_{cm})}} \qquad (4.147)$$

The frequency response of this system is shown in Fig. 4.61a. Note the "notch-filter" effect at ω_{n1} followed by a resonant peak near ω_{n2}. If the springs are removed (the usual case) we get $\omega_{n1} = \omega_{n2} = \omega_n$ and $K_e = K$, giving

$$\frac{e_o}{\theta_c} (D) = \frac{K(D^2/\omega_n^2 + 1)}{D^2/\omega_n^2 + 2\zeta D/\omega_n + 1} \qquad (4.148)$$

and the frequency response of Fig. 4.61b.

The pendulum sensor is unfortunately sensitive to horizontal accelerations, and so its application is ruled out where such accelerations are large enough to cause a significant output signal. A simple analysis shows that a steady horizontal acceleration A_x will cause an output voltage $K_e A_x/g$ for an instrument with no springs.

A commercial pendulum[1] using a potentiometer motion pickup has a full-scale range of $\pm 8°$, natural frequency of 2 cps, damping ratio of 0.6, and resolution of 0.1° (0.05° if vibration is present). Pendulums are also available for measuring rotation about two mutually perpendicular axes.

---

[1] Honeywell Inc., Minneapolis, Minn.

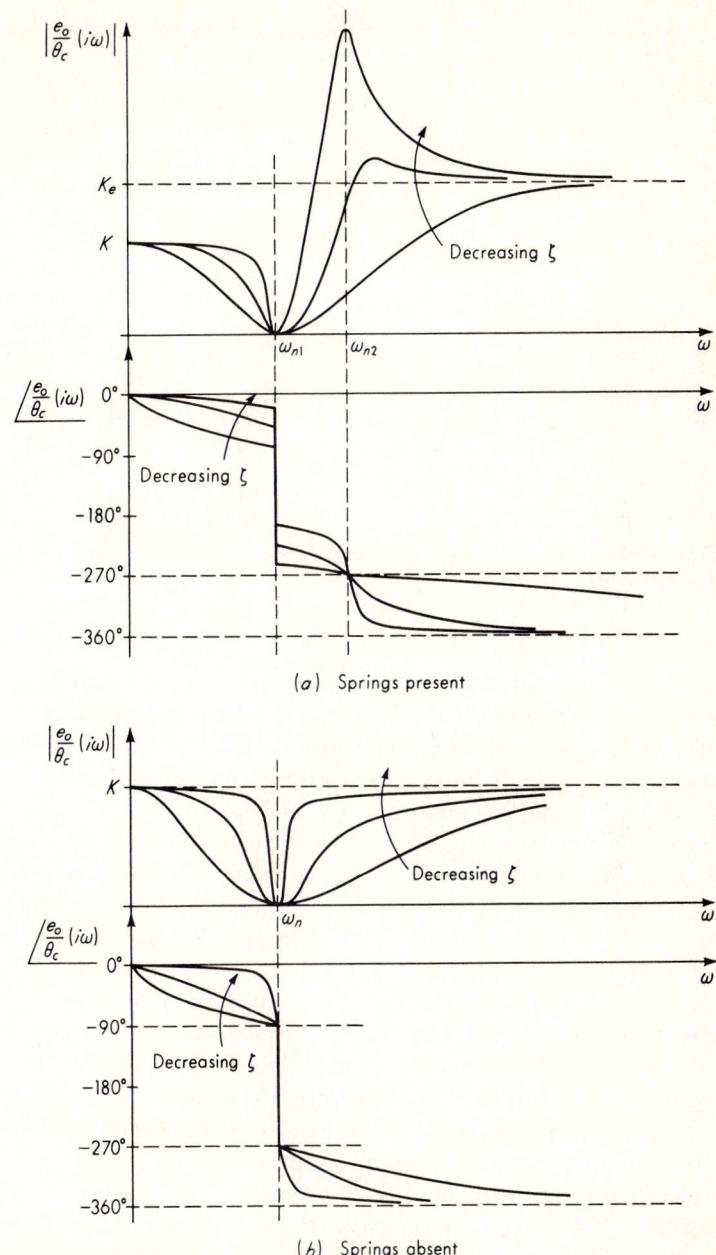

Fig. 4.61. *Frequency response of pendulum sensor.*

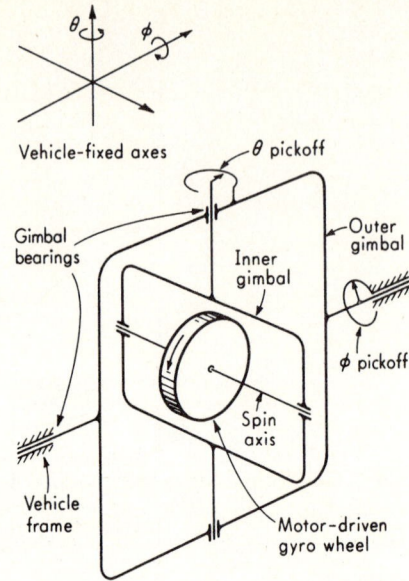

Fig. 4.62. *Free gyroscope (two-axis position gyro).*

4.12
Gyroscopic (Absolute) Angular-displacement and Velocity Sensors   While gyroscopic instruments have been used in limited numbers and rather restricted applications (gyrocompasses in ships and aircraft, turn-and-bank indicators in aircraft, etc.) since around World War I, developments during and after World War II have brought them to an extreme degree of refinement, and their use in large numbers is now common in military systems.[1]   Increased commercial and industrial applications have also resulted from the availability of equipment generated by military development programs.   A recent estimate[2] of the value of total gyro production is placed at 121 million dollars per year.

Perhaps the simplest gyro configuration is the free gyro shown in Fig. 4.62.   These instruments are used to measure the absolute angular displacement of the vehicle to which the instrument frame is attached. A single free gyro can measure rotation about two perpendicular axes, such as the angles θ and ϕ.   This can be accomplished because the axis of the spinning gyro wheel remains fixed in space (if the gimbal bearings are frictionless) and thus provides a reference for the relative-motion transducers.   If the angles to be measured do not exceed about 10°, the readings of the relative-displacement transducers give directly the absolute rota-

---

[1] Sidney Lees (ed.), "Air, Space, and Instruments," p. 32, McGraw-Hill Book Company, New York, 1963.

[2] *Control Engineering*, p. 77, May, 1963.

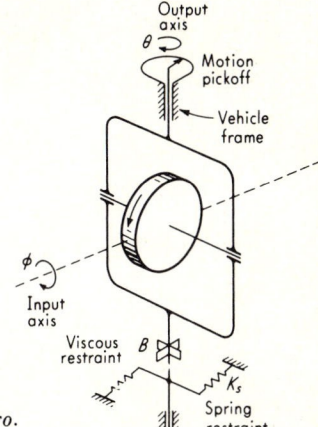

Fig. 4.63. *Single-axis restrained gyro.*

tions with good accuracy. For larger rotations of both axes, however, there is an interaction effect between the two angular motions, and the transducer readings do *not* accurately represent the absolute motions of the vehicle. The free gyro is also limited to relatively short-time applications (less than about 5 min) since gimbal-bearing friction causes gradual drift (loss of initial reference) of the gyro spin axis. A constant friction torque T_f causes a drift (precession) of angular velocity ω_d given by

$$\omega_d = \frac{T_f}{H_s} \qquad (4.149)$$

where H_s is the angular momentum of the spinning wheel. It is clear that a high angular momentum is desirable in reducing drift. A typical drift rate is about 0.5 degree/min for each axis.

Rather than using free gyros to measure two angles in one gyro (thus requiring two gyros to define completely the required three axes of motion), most recent high-performance systems utilize the so-called single-axis or constrained gyros. Here a single gyro measures a single angle (or angular rate); therefore three gyros are required to define the three axes. This approach avoids the coupling or interaction problems of free gyros, and the constrained (rate-integrating) gyros can be constructed with exceedingly small drift. We shall here consider two common types of constrained gyros, the rate gyro and the rate-integrating gyro. The rate gyro measures absolute angular velocity and is widely used to generate stabilizing signals in vehicle-control systems. The rate-integrating gyro measures absolute angular displacement and thus is used as a fixed reference in navigation and attitude-control systems. The configuration of a rate gyro is shown in Fig. 4.63; the rate-integrating gyro is functionally identical except that it has no spring restraint.

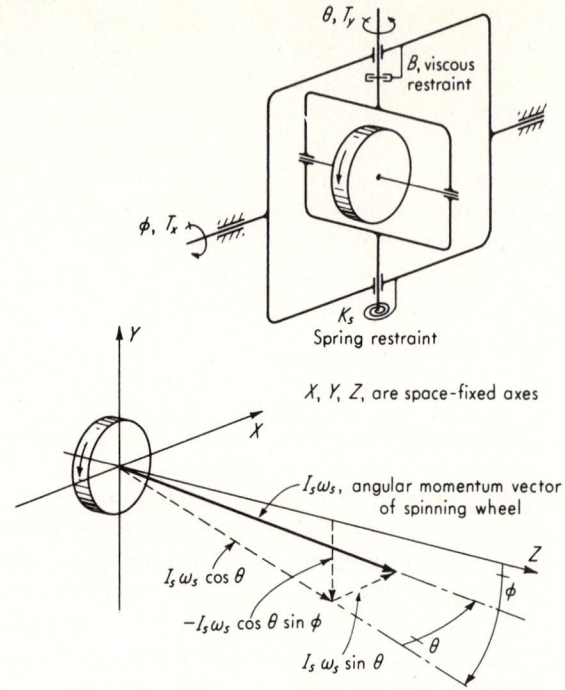

Fig. 4.64. *Gyro analysis.*

While a general analysis of gyroscopes is exceedingly complex, useful results for many purposes may be obtained relatively simply by considering small angles only. This assumption is satisfied in many practical systems. Figure 4.64 shows a gyro whose gimbals (and thus the angular-momentum vector of the spinning wheel) have been displaced through small angles θ and ϕ. We shall apply Newton's law

$$\sum \text{torques} = \frac{d}{dt}(\text{angular momentum}) \qquad (4.150)$$

to the x and y axis components of angular momentum. This angular momentum is made up of two parts, one part due to the spinning wheel and another part (due to the motion of the wheel, case, gimbals, etc.) which would exist even if the wheel were not spinning. The latter part depends on (for the x axis) the angular velocity $d\phi/dt$ and the moment of inertia I_x of everything that rotates when the outer gimbal turns in its bearing. For the y axis it depends on $d\theta/dt$ and the moment of inertia I_y of everything that rotates when the inner gimbal turns in its bearing. The external applied torques T_x and T_y are included to provide for the possibility of bearing friction and also for intentionally applied torques from small electromagnetic "torquers" which are used in some systems to cause desired precessions for control or correction purposes. The inertias I_x and I_y (which are about *space-fixed* axes) actually change when θ and ϕ change, but this effect is negligible for small angles. Also, the exact equations would contain terms in the *products* of inertia as well as the moments of inertia, but these are again negligible due to the small angles and also to the inherent symmetry of gyro structures. With the above qualifications we may write for the x axis

$$T_x = \frac{d}{dt}\left(H_s \sin\theta + I_x \frac{d\phi}{dt}\right) \qquad (4.151)$$

and for the y axis

$$T_y - B\frac{d\theta}{dt} - K_s\theta = \frac{d}{dt}\left(-H_s \cos\theta \sin\phi + I_y \frac{d\theta}{dt}\right) \qquad (4.152)$$

We now assume H_s is a constant (the gyro wheel is driven by a constant-speed motor) and $\cos\theta = 1$, $\sin\theta = \theta$, $\sin\phi = \phi$ to get

$$T_x = H_s\frac{d\theta}{dt} + I_x\frac{d^2\phi}{dt^2} \qquad (4.153)$$

$$T_y - B\frac{d\theta}{dt} - K_s\theta = -H_s\frac{d\phi}{dt} + I_y\frac{d^2\theta}{dt^2} \qquad (4.154)$$

These are two simultaneous linear differential equations with constant coefficients relating the two inputs T_x and T_y to the two outputs θ and ϕ. Writing these equations in operator form, we can treat them as algebraic equations to solve for ϕ and θ as desired. For ϕ we get

$$\phi = \frac{(I_yD^2 + BD + K_s)T_x - (H_sD)T_y}{D^2[I_xI_yD^2 + BI_xD + (H_s^2 + I_xK_s)]} \qquad (4.155)$$

Since ϕ depends on both T_x and T_y, transfer functions may be obtained by considering each input separately and then using superposition. Letting $T_y = 0$, we get

$$\frac{\phi}{T_x}(D) = \frac{I_yD^2 + BD + K_s}{D^2[I_xI_yD^2 + BI_xD + (H_s^2 + I_xK_s)]} \triangleq G_1(D) \qquad (4.156)$$

and letting $T_x = 0$ gives

$$\frac{\phi}{T_y}(D) = -\frac{H_s}{D[I_xI_yD^2 + BI_xD + (H_s^2 + I_xK_s)]} \triangleq G_2(D) \qquad (4.157)$$

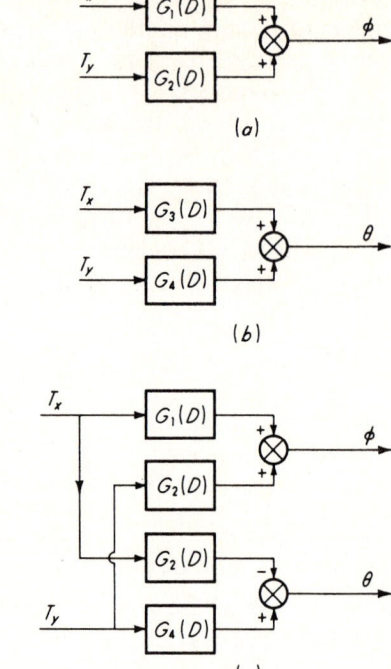

Fig. 4.65.   *Gyro block diagrams.*

(a)

(b)

(c)

This leads to the block diagram of Fig. 4.65a.   Similar analysis for θ gives

$$\frac{\theta}{T_x}(D) = \frac{H_s}{D[I_xI_yD^2 + BI_xD + (H_s{}^2 + I_xK_s)]} \triangleq G_3(D) = -G_2(D)$$
(4.158)

and $$\frac{\theta}{T_y}(D) = \frac{I_x}{I_xI_yD^2 + BI_xD + (H_s{}^2 + I_xK_s)} \triangleq G_4(D)$$
(4.159)

leading to the block diagram of Fig. 4.65b.   An overall block diagram may then be constructed as in Fig. 4.65c.

The above results are of quite general applicability to gyro systems of various configurations as long as the small-angle requirement is met. For single-axis rate and rate-integrating gyros considerable simplification of the above results is possible.   In these applications (see Fig. 4.63) the input is the *motion* ϕ.   A torque T_x also exists, accompanying this motion, but it is usually of no interest since it generally is so small as not to affect the motion ϕ, which is the rotation of a (usually large) vehicle.   The angle θ is an indication of the angle ϕ (rate-integrating gyro) or angular velocity $\dot{\phi}$ (rate gyro); thus we should like to have transfer functions relating θ to ϕ. The torque T_y (neglecting bearing friction) is zero for this application unless a torquer is used for some special purpose. The desired θ-ϕ relation may then easily be obtained by solving Eqs. (4.156) and (4.158) for T_x and setting them equal. The result is

$$\frac{\theta}{\phi}(D) = \frac{H_s D}{I_y D^2 + BD + K_s} \qquad (4.160)$$

For a rate gyro we then have the second-order response

$$\frac{\theta}{D\phi}(D) = \frac{\theta}{\dot{\phi}}(D) = \frac{K}{D^2/\omega_n{}^2 + 2\zeta D/\omega_n + 1} \qquad (4.161)$$

where
$$K \triangleq \frac{H_s}{K_s} \qquad \text{rad/(rad/sec)} \qquad (4.162)$$

$$\omega_n \triangleq \sqrt{\frac{K_s}{I_y}} \qquad \text{rad/sec} \qquad (4.163)$$

$$\zeta \triangleq \frac{B}{2\sqrt{I_y K_s}} \qquad (4.164)$$

A high sensitivity is achieved by large angular momentum H_s and soft spring K_s, although low K_s gives a low ω_n. Natural frequencies of commercially available rate gyros are of the order of 10 to 100 cps. Damping ratio is usually set at 0.3 to 0.7. Large angular momentum is obtained in small size by using high-speed (often 24,000 rpm) motors to spin the gyro wheel. Full-scale ranges of about ± 10 to $\pm 1,000$ degree/sec are readily available. Resolution of a ± 10-degree/sec-range instrument is of the order of 0.005 degree/sec. In some high-performance rate gyros the mechanical spring is replaced by an "electrical-spring" arrangement similar to that used in the servo accelerometer of Fig. 4.55.

To measure all three components (roll, pitch, and yaw) of angular velocity in a vehicle, an arrangement of three rate gyros, such as in Fig. 4.66, may be employed. It should be pointed out that in a rate gyro only the output angle θ must be kept small. This requires use of a very sensitive motion pickoff, but such are available. The input angle ϕ may be indefinitely large since no matter how large it gets the spin angular-momentum vector is always perpendicular (except for a small error due to nonzero θ) to the input angular-velocity vector. A roll-rate gyro in an aircraft thus will function correctly even if the aircraft rolls completely over. Of course the angular-velocity components measured are those about the *vehicle* axes rather than space-fixed axes, but these are usually the velocities desired for stabilization signals in control systems.

To obtain a rate-integrating gyro one merely removes the spring restraint from the configuration of Fig. 4.63. Equation (4.160) then

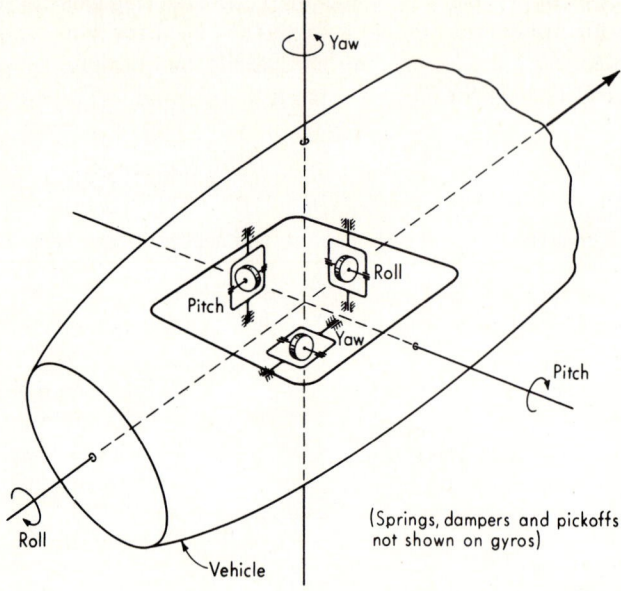

Fig. 4.66.   *Three-axis rate-gyro package.*

becomes

$$\frac{\theta}{\phi}(D) = \frac{K}{\tau D + 1} \qquad (4.165)$$

where

$$K \triangleq \frac{H_s}{B} \qquad \text{rad/rad} \qquad (4.166)$$

$$\tau \triangleq \frac{I_y}{B} \qquad \text{sec} \qquad (4.167)$$

We see that the output angle θ is a direct indication of the input angle ϕ according to a standard first-order response form.   High sensitivity again requires high H_s, a typical value being 5×10^5 g-cm²/sec.   Low damping B also increases sensitivity but at the expense of speed of response, as shown by Eq. (4.167).   For a K of 15 a τ of 0.006 sec is typical of high-performance units.[1]   Increase of K to 440 raises τ to 0.17 sec.   The rate-integrating gyro is the basis of highly accurate inertial navigation systems where it is used as a reference to maintain so-called stable platforms in a fixed attitude within a vehicle while the vehicle moves arbitrarily.   This is done by using the motion signals from the gyros to drive servomechanisms which maintain the platform in a fixed angular orientation.   The

[1] General Precision Inc., Kearfott Div., Little Falls, N.J.

system accelerometers are mounted on this platform, and their double-integrated output is an accurate measure of vehicle motion along the three orthogonal axes since the platform always moves parallel to its initial orientation.   The rate-integrating gyro has been subjected to extremely intensive and extensive engineering development to bring its performance characteristics to remarkable levels while maintaining small size and weight and great resistance to rugged environments.   This has been accomplished by painstaking attention to minute mechanical, thermal, and electrical details which would be completely negligible in a less sophisticated device.   While the performance of the best current instruments is generally classified information, drift rates less than 0.01 degree/hr are published in the open literature.

In this short section we can only indicate a few major concepts while many significant details are neglected.   A few samples[1,2] of the voluminous literature are mentioned here while more will be found in the Bibliography of this chapter.

Problems

4.1 For a steel gage block of 1-in. length, what temperature change is required to cause a length change of 1 μin.?

4.2 Derive Eq. (4.1).

4.3 The output of a potentiometer is to be read by a recorder of 10,000 ohms input resistance.   Nonlinearity must be held to 1 percent.   A family of potentiometers having a thermal rating of 5 watts and resistances ranging from 100 to 10,000 ohms in 100-ohm steps is available.   Choose from this family the potentiometer that has the greatest possible sensitivity and also meets the other requirements.   What is this sensitivity if the potentiometers are single-turn (360°) units?

4.4 If a potentiometer changes resistance because of temperature changes, what effect does this have on motion measurements?

4.5 A 10-in.-stroke wire-wound translational potentiometer is excited with 100 volts.   The output is read on an oscilloscope with a "sensitivity" of 0.5 mv/cm. It would appear that measurements to the nearest 0.0001 in. are easily possible. Explain why this is not so.

4.6 What resolution is possible with a 60-turn wire-wound potentiometer using appropriate gearing?

4.7 Explain methods whereby one can experimentally determine the moment of inertia and the starting and running friction torque of a rotary potentiometer.

4.8 In Fig. P4.1 a potentiometer whose moving part weighs 0.01 lb$_f$ measures the displacement of a spring-mass system subjected to a step input.   The measured natural frequency is 30 cps.   If the spring constant and mass M of the system are unknown, can the true natural frequency be deduced from the above data?   Suppose an additional 0.01-lb$_f$ weight is attached to the poten-

---

[1] B. Lichtenstein, "Technical Information for the Engineer, Gyros," General Precision Inc., Kearfott Div., Little Falls, N.J.

[2] Savet, *op. cit.*

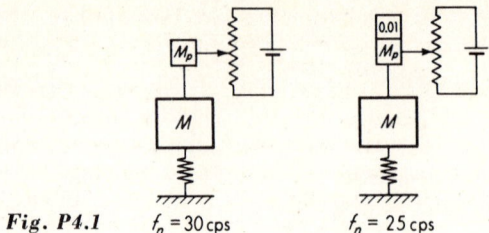

Fig. P4.1 $f_n = 30$ cps $f_n = 25$ cps

tiometer and the test repeated, giving a 25-cps frequency.   Calculate the true natural frequency of the system, that is, the frequency before the potentiometer was attached.

4.9 Find C_{gi} for the system of Fig. 4.8 if a viscous damping effect is included.

4.10 Explain why increasing the cross-sectional area of the end loops in foil-type strain gages reduces transverse sensitivity.

4.11 If, in the discussion following Eq. (4-28), dynamic strains only in the range 0 to 10,000 cps need be measured, explain how and to what extent the noise voltage may be reduced.

4.12 In a Wheatstone bridge, leg 1 is an active strain gage of Advance alloy and 120 ohms resistance, leg 4 is a similar dummy gage for temperature compensation, and legs 2 and 3 are fixed 120-ohm resistors.   The maximum gage current is to be 0.030 amp.

  a. What is the maximum permissible d-c bridge excitation voltage?   (Use this value in the remaining parts of this problem.)

  b. If the active gage is on a steel member, what is the bridge output voltage per 1,000 psi of stress?

  c. If temperature compensation were *not* used, what bridge output would be caused by the active gage increasing temperature by 100F° if the gage is bonded to steel?   What stress value would be represented by this voltage?   Thermal-expansion coefficients of steel and Advance alloy are 6.5×10^{-6} and 14.9×10^{-6} in./(in.-F°), respectively.   The temperature coefficient of resistance of Advance is 6×10^{-6} ohm/(ohm-F°).

  d. Compute the value of a shunt calibrating resistor that would give the same bridge output as 10,000-psi stress in a steel member.

4.13 From Eq. (4.38), find an expression for the frequency at which zero phase shift occurs.

4.14 Perform an analysis similar to that leading to Eq. (4.38), assuming output loaded with R_m, for the following:

  a. The circuit of Fig. 4.15*a*

  b. The circuit of Fig. 4.15*b*

  c. The circuit of Fig. 4.15*c*

  d. The circuit of Fig. 4.15*d*

4.15 In Fig. 4.17*c*, let x_i be a periodic motion with a significant frequency content up to 500 cps, and let the excitation frequency be 10,000 cps.   The output voltage e_o is connected to an oscillograph galvanometer, which is a second-order system with $\zeta = 0.65$ and a natural frequency of 1,000 cps.   Will this combination result in a satisfactory measurement system?   Justify your answer with numerical results.

4.16 Air exhibits a dielectric breakdown at fields of about 50,000 volts/in. What limitation does this impose on the ultimate sensitivity of a capacitance transducer such as in Fig. 4.25?

4.17 In Eq. (4.50), suppose a flat amplitude ratio within 5 percent down to 20 cps is required. What is the minimum allowable τ? If $A = 0.5$ in.^{2} and x_0 is 0.005 in., what value of R is needed?

4.18 A piezoelectric transducer has a capacitance of 1,000 pf and K_q of 10^{-5} coul/in. The connecting cable has a capacitance of 300 pf while the oscilloscope used for readout has an input impedance of 1 megohm paralleled with 50 pf.

    a. What is the sensitivity (volts/in.) of the transducer alone?

    b. What is the high-frequency sensitivity (volts/in.) of the entire measuring system?

    c. What is the lowest frequency that can be measured with 5 percent amplitude error by the entire system?

    d. What value of C must be connected in parallel to extend the range of 5 percent error down to 10 cps?

    e. If the C value of part *d* is used, what will the system high-frequency sensitivity be?

4.19 A piezoelectric transducer has an input

$$x_i = At \qquad 0 \leq t < T$$
$$x_i = 0 \qquad T < t < \infty$$

Solve the differential equation to find e_o. For $t = T^-$, find the error [(ideal value of e_o) − (actual value of e_o)]. Approximate this error by using the truncated series

$$e^{-T/\tau} \approx 1 - \frac{T}{\tau} + \frac{1}{2}\left(\frac{T}{\tau}\right)^2$$

Express this approximate error as a percentage of the ideal value of e_o. What must T/τ be if the error is to be 5 percent? For this value of T/τ, evaluate the error caused by truncating the series. (Use the theorem on the remainder of an alternating series.)

4.20 Analyze the nozzle-flapper displacement pickup of Fig. P4.2, using the simple incompressible relations. Explain the advantages of this configuration.

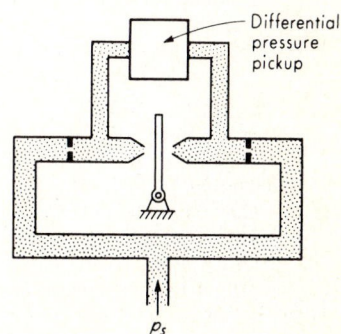

Fig. P4.2 p_s

4.21 Express in binary form the decimal numbers 27, 6,382, 9.125, 9.126.

4.22 Express in decimal form the binary numbers 10111001., 1001001111., 101011.-1100.

4.23 Express in Datex binary-coded decimal the numbers 137, 9,764, and 42.

4.24 Prove Eq. (4.112).

4.25 Derive Eq. (4.113).

4.26 In the variable-capacitance velocity pickup shown in Fig. P4.3, prove that the current i is directly proportional to the angular velocity $d\theta/dt$. Since voltage signals are more readily manipulated, how might the current signal be transduced to a proportional voltage? Does your method of doing this affect the basic operation? What must be required if the basic operation is to be only slightly affected?

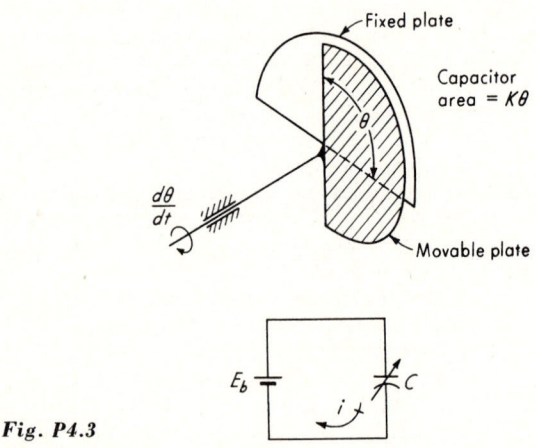

Fig. P4.3

4.27 Construct logarithmic frequency-response curves for a seismic-displacement pickup with $\zeta = 0.3$ and $\omega_n/2\pi = 10$ cps. For what range of frequencies is the amplitude ratio flat within 1 db?

4.28 Make a comprehensive list and explain the action of modifying and/or interfering inputs for the systems of the following:

a.	Fig. 4.3	*g.*	Fig. 4.36
b.	Fig. 4.8	*h.*	Fig. 4.44
c.	Fig. 4.13	*i.*	Fig. 4.45
d.	Fig. 4.25	*j.*	Fig. 4.48
e.	Fig. 4.31	*k.*	Fig. 4.55
f.	Fig. 4.34	*l.*	Fig. 4.63

4.29 Derive $(\theta_o/\theta_i)(D)$ for Fig. 4.49b.

4.30 Construct logarithmic frequency-response curves for a piezoelectric accelerometer with $K_q/C\omega_n{}^2 = 0.001$ volt/(in./sec^2), $\tau = 0.10$ sec, $\omega_n/2\pi = 10,000$ cps, and $\zeta = 0$. Will this accelerometer be satisfactory for shock measurements of half-sine pulses with a duration of 0.05 sec? If not, suggest needed changes.

4.31 Explain, giving a sketch, how the principle of the system of Fig. 4.55 can be adapted to the measurement of angular acceleration. Your device must *not* be sensitive to translational acceleration.

4.32 In the system of Fig. 4.55, if K_s and B are made zero, $(\theta/T_n)(D) = 1/JD^2$. Obtain $(e_o/\ddot{x}_i)(D)$ for this situation. What is the defect in this system? To remedy this, electrical "damping" may be introduced by adding a circuit with a transfer function as shown in Fig. P4.4. Obtain $(e_o/\ddot{x}_i)(D)$ for this arrangement. Why must $\tau_1 > \tau_2$?

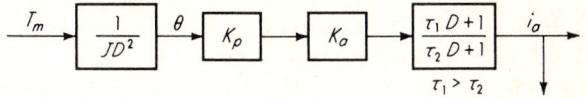

Fig. P4.4

4.33 An accelerometer used in an inertial navigation system has a so-called bias error such that there is a small output signal even when the input acceleration is exactly zero. If this bias signal is equivalent to $10^{-5}g$ of acceleration, what *position* (displacement) error does it cause over a time interval of 2 days? Assume travel along a straight line.

4.34 When a seismic-displacement pickup is used in its proper frequency range, what is an adequate mechanical model (masses, springs, dashpots) for its loading effect on the measured system?

4.35 In the Jerkmeter of Fig. 4.59, replace ω_{n1} and ζ_1 by their values in terms of J, B, and K_s, thus rewriting Eq. (4.139). Suppose a sensitivity $Mr/K_iK_{ce} = 0.05$ volt/(ft/sec³) is required. Assume temporarily that $B = K_s = 0$ and neglect the fact that this system would be unstable. Assume also that J is due mainly to M; thus take $J = Mr^2$.

    a. Find the numerical value of K_pK_{ae}/r needed to give a flat amplitude ratio for $(e_o/D^3x_i)(i\omega)$ within 5 percent over the range 0 to 5 cps.

    b. If $r = 0.1$ ft and $K_p = 57.3$ volts/rad, find K_{ae}.

    c. Suppose now that $M = 0.01$ lb$_m$. Find K_iK_{ce}.

    d. If B and K_s are not zero, find the value of BK_s needed to put the system just on the margin of instability. (Use the Routh criterion.)

    e. Let the design value of BK_s be 10 times the value of part *d*. If $K_s = 0.275$ ft-lb$_f$/rad, what is B?

    f. With the above values of B and K_s, recheck the amplitude ratio at 5 cps. Does it meet the 5 percent requirement?

    g. To correct the situation found in *f*, make $K_{ae} = 2,490$. Recheck the amplitude ratio at 5 cps. Recheck the stability.

    h. To regain the stability lost in part *g*, reduce M to 0.001 lb$_m$. Recheck the amplitude ratio at 5 cps.

    i. Recheck the overall system static sensitivity. How much external amplification is now needed to return to the required 0.05 volt/(ft/sec³)?

4.36 Derive Eqs. (4.143) to (4.147). Give a physical explanation of the "notch-filter" effect. Explain the apparent discontinuity in phase angle.

4.37 Find the steady-state response of the system of Fig. 4.60 to a constant horizontal acceleration.

4.38 A commercial version of the system of Fig. 4.60 has a pendulum made as shown in Fig. P4.5. Where is the center of buoyancy of this pendulum? Where is the center of mass? Will the buoyant force tend to cause an output? Why? Derive a relation showing the requirements for completely unloading the pivot bearing. (This "floating" reduces bearing friction and thus system threshold.)

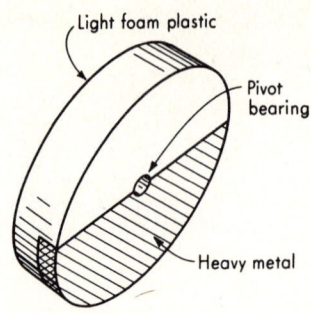

Fig. P4.5   Completely immersed in liquid

4.39 Reduce Eqs. (4.156) to (4.159) to standard form, defining appropriate K's, ζ's, and ω_n's.

4.40 Equation (4.157) can be written as

$$[I_x I_y D^2 + B I_x D + (H_s{}^2 + I_x K_s)]\phi = - \frac{H_s}{D} T_y$$

For a gyro with no spring nor damping, suppose T_y is a unit impulse. Solve Eqs. (4.157) and (4.159) for ϕ and θ. The combined sinusoidal motion of ϕ and θ causes the spin axis to rotate in space. This motion is called nutation. Describe it qualitatively and define its frequency. What is the effect on the above results if damping is present?

4.41 A rate gyro is mounted on a long thin missile, which is quite flexible in bending, as in Fig. P4.6. Bending vibrations cause the slope at the gyro location to go

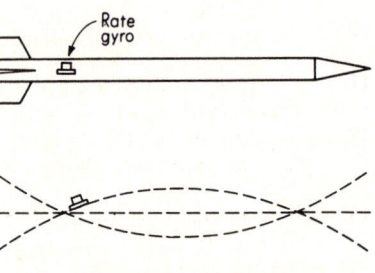

Fig. P4.6   Mode shape of bending vibrations

through sinusoidal oscillations of 0.1° amplitude at 50 cps. What maximum angular velocity will the gyro feel due to vibration? If the gross (rigid-body) rotation of the missile (which is what the gyro is *intended* to measure) is 10 rad/sec, what percent of the total gyro signal is due to the vibration? Where could one relocate the gyro to minimize this problem? If the fastest rigid-body motions expected are 1 cps, what other solution is possible?

4.42 In a rate gyro, the steady-state output/input ratio θ/ϕ is not strictly linear because the angular-momentum vector does not remain perpendicular to the angular-velocity vector ϕ when θ rotates away from zero.

a. What is the maximum allowable θ if this nonlinearity is not to exceed 1 percent?

b. If a rate gyro requires $\omega_n = 100$ rad/sec and if $I_y = 0.0013$ in.-lb$_f$-sec^2, what must the spring constant K_s be?

c. If the maximum input rate $\dot{\phi}$ is 10 rad/sec, what must H_s be if θ_{max} is the value found in part *a*? Use K_s from part *b*.

d. If the spin motor runs at 24,000 rpm and if the wheel is a solid cylinder of a length equal to its diameter and is made of a material with a specific weight of 0.3 lb$_f$/in.^{3}, what are the required dimensions of the wheel? Use all necessary numbers from the previous parts of the problem.

Bibliography

1. J. W. Dally and W. F. Riley: "Experimental Stress Analysis," McGraw-Hill Book Company, New York, 1965.
2. H. G. Buchbinder: Precision Potentiometers, *Electromech. Design*, p. 21, January, 1964.
3. K. H. Hardman: Conductive Plastic Precision Pots, *Electromech. Design*, p. 44, October, 1963.
4. J. F. Blackburn (ed.): Potentiometers, "Components Handbook," chap. 8, McGraw-Hill Book Company, New York, 1948.
5. High Temperature Potentiometers, *Electromech. Design*, p. 41, January, 1960.
6. T. L. Foldvari and K. S. Lion: Capacitive Transducers, *Instr. Control Systems*, p. 77, November, 1964.
7. Electro-optical Moire Fringe Transducer, *Electromech. Design*, p. 26, December, 1964.
8. W. H. Kliever: Measure Position Digitally, *Control Eng.*, p. 107, November, 1956.
9. J. O. Morin: 6 Transducers for Precision Position Measurement, *Control Eng.*, p. 107, May, 1960.
10. J. H. Brown: Measure Motion to 0.0001 In. without Friction or Wear, *Control Eng.*, p. 50, April, 1955.
11. G. E. Bowie: Measurement of Displacements in Contact-Stress Experiments, *ASD Tech. Rept.* 61-450, Wright-Patterson Air Force Base, Ohio, October, 1961.
12. T. G. Baxter: Measurement of Pier Tilt with a Quartz Torsion Fiber Pendulum, *Jet Propulsion Lab. Tech. Rept.* 32-44, Jet Propulsion Laboratory, Pasadena, Calif., 1960.
13. W. H. Faulkner and J. G. Wood: Thickness Measuring Devices for Sheet and Web Materials, *Automation*, p. 67, July, 1962.
14. K. V. Olsen: On the Standardization of Surface Roughness Measurements, *B & K Tech. Rev.*, B & K Instruments, Cleveland, Ohio, no. 3, 1961.
15. Precise Amplitude Measurements, *Mech. Eng.*, p. 65, February, 1963.
16. Length Calibration, National Bureau of Standards, *ISA J.*, p. 75, April, 1963.
17. W. Kinder: Comparator Measures up to 40″ with Accuracy of 0.4 Microinch, *Instr. Control Systems*, p. 123, April, 1963.
18. E. G. Loewen: Positioning System Spaces Lines to within $\frac{1}{10}$ Microinch, *Control Eng.*, p. 95, May, 1963.
19. W. R. Ketterer and R. H. Schuman: Photocell Technique for Linear Measurements, *Electro-Technol.*, p. 120, November, 1963.
20. B. Sternlicht: An Indirect Method of Film-thickness Measurement in Fluid-Film Bearings, *ISA Trans.*, vol. 2, no. 1, p. 28, January, 1963.

21. J. B. Bryan and G. I. Boyadjieff: Measuring Surface Finish, A State-of-the-Art Report, *Mech. Eng.*, p. 42, December, 1963.

22. F. Farago: Measuring the Critical Profile of Barrel Roller Bearings with Microinch Sensitivity, *Gen. Motors Eng. J.*, p. 17, January–February–March, 1964.

23. R. Zito: Nuclear-Resonance Sensing of Mechanical Motion, *Electro-Technol.*, p. 43, June, 1964.

24. K. G. Overbury: Temperature in Length Measurement, Sandia Corp., Albuquerque, N.Mex., February, 1962.

25. D. H. Parkes: The Application of Microwave Techniques to Noncontact Precision Measurement, *ASME Paper* 63-WA-346, 1963.

26. F. H. London: Laser Interferometer, *Instr. Control Systems*, p. 87, November, 1964.

27. J. G. Collier and G. F. Hewitt: Film-thickness Measurements, *ASME Paper* 64-WA/HT-41, 1964.

28. E. V. Sundt: Touchless Tachometry, *Electromech. Design*, p. 36, May, 1964.

29. R. Zito: Velocity Sensing for Spacecraft Docking, *Space/Aeron.*, p. 90, December, 1963.

30. J. Frey: The A-C Tachometer, *Electro-Technol.*, p. 88, August, 1963.

31. R. L. Pike: Measurement of Low Angular Rates, *Instr. Control Systems*, p. 83, December, 1962.

32. A. L. Fisher: Ball and Disk Read Angular Velocity Directly, *Control Eng.*, p. 125, November, 1962.

33. L. E. Bollinger and K. E. Kissell: Measurement of Detonation-wave Velocities, *ISA J.*, p. 170, May, 1957.

34. Shaoue Ezekiel: Towards a Low-level Accelerometer, *NASA*, CR-56941, 1964.

35. P. K. Chapman: A Cryogenic Test-Mass Suspension for a Sensitive Accelerometer, *NASA*, N64-27883, 1964.

36. Pressure Sensitivity of Accelerometers and Cables, Wilcoxon Research, Bethesda, Md.

37. Mercury Drop Measures Space-vehicle Acceleration, *Machine Design*, p. 28, Mar. 29, 1962.

38. H. R. Judge: Performance of Donner Linear Accelerometer Model 4310, *Space Technol. Lab. Tech. Note* 60-0000-09117, Space Technology Laboratories, Los Angeles, Calif., 1960.

39. A. Castle: Accelerometer Scribes Vector-Force Signatures, *Control Eng.*, p. 105, March, 1965.

40. R. R. Bouche: High Frequency Response and Transient Motion Performance Characteristics of Piezoelectric Accelerometers, Endevco Corp., Pasadena, Calif., 1961.

41. C. K. Stedman: Some Characteristics of Gas Damped Accelerometers, Statham Instruments Inc., Los Angeles, Calif., 1958.

42. H. B. Sabin: 17 Ways to Measure Acceleration, *Control Eng.*, p. 106, February, 1961.

43. R. P. Bowen: Calibrating Vibration Pickup Calibrators, *ISA J.*, p. 58, March, 1951.

44. V. B. Corey: Measuring Angular Acceleration with Linear Accelerometers, *Control Eng.*, p. 79, March, 1962.

45. K. E. Pope: A New Double-integrating Accelerometer, *Control Eng.*, p. 97, November, 1958.

46. K. N. Sergeyev: Investigation of Acceleration Pickup Having Filtering Properties, *NASA*, N64-23543, 1964.

47. D. K. Phillips: Balanced Beam Improves Angular Accelerometer, *Control Eng.*, p. 91, July, 1964.

48. S. Rubin: Design of Accelerometers for Transient Measurement, *J. Appl. Mech.*, p. 509, December, 1958.
49. A. Degenholtz: Optical-Wedge Technique for Measuring Angular Vibration, *Machine Design*, p. 167, March 12, 1964.
50. Design and Construction of a Lunar Seismograph, *NASA*, *N*63-18290, 1963.
51. G. B. Foster: Non-contacting Self-calibrating Vibration Transducer, *Instr. Control Systems*, p. 83, December, 1963.
52. B. D. Van Deusen: Analysis of Vehicle Vibration, *ISA Trans.*, vol. 3, no. 2, p. 138, April, 1964.
53. D. F. Wilkes and C. E. Kreitler: The Long Period Horizontal Air Bearing Seismometer, Sandia Corp., Albuquerque, N.Mex., SCTM74-62(13), 1962.
54. J. M. Slater: Exotic Gyros, *Control Eng.*, p. 92, November, 1962.
55. Traverse Meter Uses Gyroscope Sensor, *Machine Design*, p. 163, Apr. 13, 1961.
56. R. H. Cherwin: Ball Bearings for Precision Gyros, *Control Eng.*, p. 79, August, 1959.
57. A. W. Lane et al.: Achieving Extremely Accurate Non-floated Gyros, *Aero/ Space Eng.*, p. 43, January, 1959.
58. H. Stern: Which Rate Gyro to Use, *Control Eng.*, p. 79, February, 1958.
59. C. S. Draper et al.: The Floating Integrating Gyro, *Aeron. Eng. Rev.*, p. 46, June, 1956.
60. Inertial Gyro Test, *Electromech. Design*, p. 18, November, 1960.
61. R. E. Barnaby et al.: Control of Thermal Drift in Floated Gyroscopes, *Sperry Eng. Rev.*, Sperry Gyroscope Co., Great Neck, N.Y., p. 36, September, 1961.
62. W. G. Wing: Fluid Rotor Gyros, *Control Eng.*, p. 105, March, 1963.
63. G. C. Newton: Vibratory Rate Gyros, *Control Eng.*, p. 95, June, 1963.
64. E. H. Ernst: Basic Theory-Particle Gyroscope, *Electro-Technol.*, p. 12, August, 1963.
65. H. L. Kreitzburg: Compensating Gyro Drifts, *Control Eng.*, p. 113, November, 1963.
66. H. W. Knoebel: The Electric Vacuum Gyro, *Control Eng.*, p. 70, February, 1964.
67. Gimbal-less Gyro, *Mech. Eng.*, p. 59, May, 1964.
68. S. Redner and F. Zandman: Experimental Stress Analysis, *Ind. Res.*, p. 67, May, 1965.
69. L. H. Ravitch: Some Applications of Stress Analysis Techniques in Improving Casting Designs, *Gen. Motors Eng. J.*, p. 22, October–November–December, 1958.
70. S. S. Manson: Thermal Stresses, Measurements by Photoelasticity, *Machine Design*, p. 143, Nov. 26, 1959.
71. F. Zandman: Stress Analysis with a Photoelastic Coating, *Metal Progr.*, p. 111, November, 1960.
72. F. B. Stern: Strain Sensitive Ceramic Base Brittle Coatings, *Machine Design*, p. 147, May 29, 1958.
73. G. Gerard and H. Tramposch: Photothermoelastic Investigation of Transient Thermal Stresses in a Multiweb Wing Structure, *J. Aerospace Sci.*, p. 783, December, 1959.
74. W. R. Campbell: Performance Tests of Wire Strain Gages: I. Calibration Factors in Tension, *NACA*, *Tech. Note* 954, 1944.
75. P. K. Stein: Pulsing Strain-gage Circuits, *Instr. Control Systems*, p. 128, February, 1965.
76. P. K. Stein: Strain Gages, "Measurement Engineering," Stein Engineering Services, Inc., Phoenix, Ariz.

77. S. S. Manson: Thermal Stresses in Design-Strain Gage Applications, *Machine Design*, p. 683, Nov. 12, 1959.
78. A. Kaufman: Performance of Electrical-Resistance Strain Gages at Cryogenic Temperatures, *NASA, Tech. Note*, D-1663, 1963.
79. R. H. Kemp et al.: Application of a High-temperature Static Strain Gage to the Measurement of Thermal Stresses in a Turbine Stator Vane, *NACA, Tech. Note* 4215, 1958.
80. S. S. Manson: Thermal Stresses in Design-Strain-Gage Measurements, *Machine Design*, p. 109, Oct. 29, 1959.
81. New Strain Gages for the Space Age, *ISA J.*, p. 50, February, 1959.
82. J. Gunn and E. Billinghurst: Magnetic Fields Affect Strain Gages, *Control Eng.*, p. 109, August, 1957.
83. Semiconductor Strain Gage Handbook, Baldwin-Lima-Hamilton Corp., Waltham, Mass.
84. R. J. Whitehead: Protective Coating for Strain Gages, *ISA J.*, p. 71, March, 1964.
85. R. Shiver and W. Putman: Measuring Dynamic Strain on High-speed Turbine Wheels, *Instr. Control Systems*, p. 118, September, 1962.
86. A. J. Bush: Soldered-cap Strain Gages, *Machine Design*, p. 163, Nov. 8, 1962.
87. C. E. Mathewson: The "Dimensionless" Strain Gage, *Instr. Control Systems*, p. 1870, October, 1961.
88. D. Post: The Moiré Grid-Analyzer Method for Strain Analysis, *Experimental Mechanics*, pp. 368–377, November, 1965.
89. G. R. Sarna: Extending the Range of Seismic Transducers, *Inst. & Cont. Syst.*, February, 1972.
90. W. J. Pastorius: Vibration Analysis by Holography, *Mechanical Engineering*, June, 1972.
91. J. K. S. Walter: Motion Sensor Development, Sandia Labs Report SC-DR-710910, Albuquerque, 1972.
92. Y. T. Li: Air Damped Capacitance Accelerometers and Velocimeters, *Trans. of IECI Group of IEEE Second Transducer Conf.*, May 4, 1970.
93. S. Y. Lee: Signal Transduction with Differential Pulse Width Modulation (Capacitance Transducers), *Trans. of IECI Group of IEEE Sec. Trans. Conf.*, May 4, 1970.
94. D. Olshove: Velocity Measurement with a Piezoresistive Acceleration-Integration System, "Inst. in the Aerospace Ind., Vol. 17," *Inst. Soc. of Amer.*, Pittsburgh, 1971.
95. R. L. Gates: Reviewing the Status of Inertial Sensors, *Control Engineering*, March, 1971.
96. J. K. Emery: Roundness Measurement, *Mechanical Engineering*, October, 1969.
97. R. F. Hill and M. Skunda: Measuring Tool Wear Radiometrically, *Mechanical Engineering*, February, 1972.
98. J. B. Bryan et al: Thermal Effects in Dimensional Metrology, *ASME Paper* 65-*Prod*-13, 1965.
99. R. Iltis: Solid-State Sensors: Strain Gages, *Control Engineering*, January, 1970.
100. R. E. Herzog: Forecasting Failures with Acoustic Emission, *Machine Design*, June 14, 1973.

Chapter 5

Force, Torque, and Shaft Power Measurement

5.1

Standards and Calibration   Force is defined by the equation $F = MA$; thus a standard for force depends on standards for mass and acceleration. Mass is considered a fundamental quantity, and its standard is a cylinder of platinum-iridium, called the International Kilogram, kept in a vault at Sèvres, France. Other masses (such as national standards) may be compared with this standard by means of an equal-arm balance, with a precision of a few parts in 10^9 for masses of about 1 kg. Tolerances on various classes of standard masses available from the National Bureau of Standards vary with the magnitude of the mass and may be found in its publications.[1]

Acceleration is not a fundamental quantity but rather is derived from length and time, two fundamental quantities whose standards were discussed in Chap. 4. The acceleration of gravity, g, is a convenient standard which can be determined with an accuracy of about 1 part in 10^6 by measuring the period and effective length of a pendulum or by determining the change with time of the speed of a freely falling body. The actual value of g varies with location and also slightly with time (in a periodic predictable fashion) at a given location. It may also change (slightly) unpredictably because of local geological activity. The so-called standard value of g refers to the value at sea level and 45° latitude and is numerically 980.665 cm/sec². The value at any latitude ϕ degrees may be computed from

$$g = 978.049(1 + 0.0052884 \sin^2 \phi - 0.0000059 \sin^2 2\phi) \qquad \text{cm/sec}^2$$

$$(5.1)$$

while the correction for altitude h in meters above sea level is

$$\text{Correction} = -(0.00030855 + 0.00000022 \cos 2\phi)h$$
$$+ 0.000072 \left(\frac{h}{1,000}\right)^2 \qquad \text{cm/sec}^2 \qquad (5.2)$$

When the numerical value of g has been determined at a particular locality, the gravitational force (weight) on accurately known standard masses may be computed to establish a standard of force. This is the basis of the so-called "dead-weight" calibration of force-measuring systems. The current National Bureau of Standards capability for such calibrations is an inaccuracy of about 1 part in 5,000 for the range of 10 to 1 million lb_f. Above this range, direct dead-weight calibration is not presently available. Rather, proving rings[2] or load cells of a capacity of 1 million lb_f or less are calibrated against dead weights and then the unknown force is applied to a multiple array of these in parallel. The range 1 to 10 million lb_f is covered by such arrangements with somewhat reduced accuracy. At the low-force end of the scale, the accuracy of standard masses ranges from about 1 percent for a mass of 10^{-5} lb_m to 0.0001 percent for the 0.1 to 10-lb_m range to 0.001 percent for a 100-lb mass. The accuracy of *force* calibrations using these masses must be somewhat

---

[1] T. W. Lashof and L. B. Macurdy, Precision Laboratory Standards of Mass and Laboratory Weights, *Natl. Bur. Std. (U.S.)*, *Circ.* 547, sec. 1, 1954.

[2] Proving Rings for Calibrating Testing Machines, *Natl. Bur. Std. (U.S.)*, *Circ.* C454, 1946.

less than the quoted figures because of error sources in the experimental procedure.

The measurement of torque is intimately related to force measurement; thus torque standards as such are not necessary, force and length being sufficient to define torque. The power transmitted by a rotating shaft is the product of torque and angular velocity. Angular-velocity measurement was treated in Chap. 4.

5.2

Basic Methods of Force Measurement   An unknown force may be measured by the following means:

1.   Balancing it against the known gravitational force on a standard mass, either directly or through a system of levers
2.   Measuring the acceleration of a body of known mass to which the unknown force is applied
3.   Balancing it against a magnetic force developed by interaction of a current-carrying coil and a magnet
4.   Transducing the force to a fluid pressure and then measuring the pressure
5.   Applying the force to some elastic member and measuring the resulting deflection

In Fig. 5.1, method 1 is illustrated by the analytical balance, the pendulum scale, and the platform scale. The analytical balance, while simple in principle, requires careful design and operation to realize its maximum performance. The beam is designed so that the center of mass is only slightly (a few thousandths of an inch) below the knife-edge pivot and thus barely in stable equilibrium. This makes the beam deflection (which in sensitive instruments is read with an optical micrometer) a very sensitive indicator of unbalance. For the low end of a particular instrument's range the beam deflection is often used as the output reading rather than attempting to null by adding masses or adjusting the arm length of a poise weight. This approach is faster than nulling but requires that the deflection-angle unbalance relation be accurately known and stable. This relation tends to vary with the load on the balance, because of deformation of knife edges, etc., but careful design can keep this to a minimum. For highly accurate measurements the buoyant force due to the immersion of the standard mass in air must be taken into account. Also, the most sensitive balances must be installed in temperature-controlled chambers and manipulated by remote control to reduce the effects of the operator's body heat and convection currents. Typically,

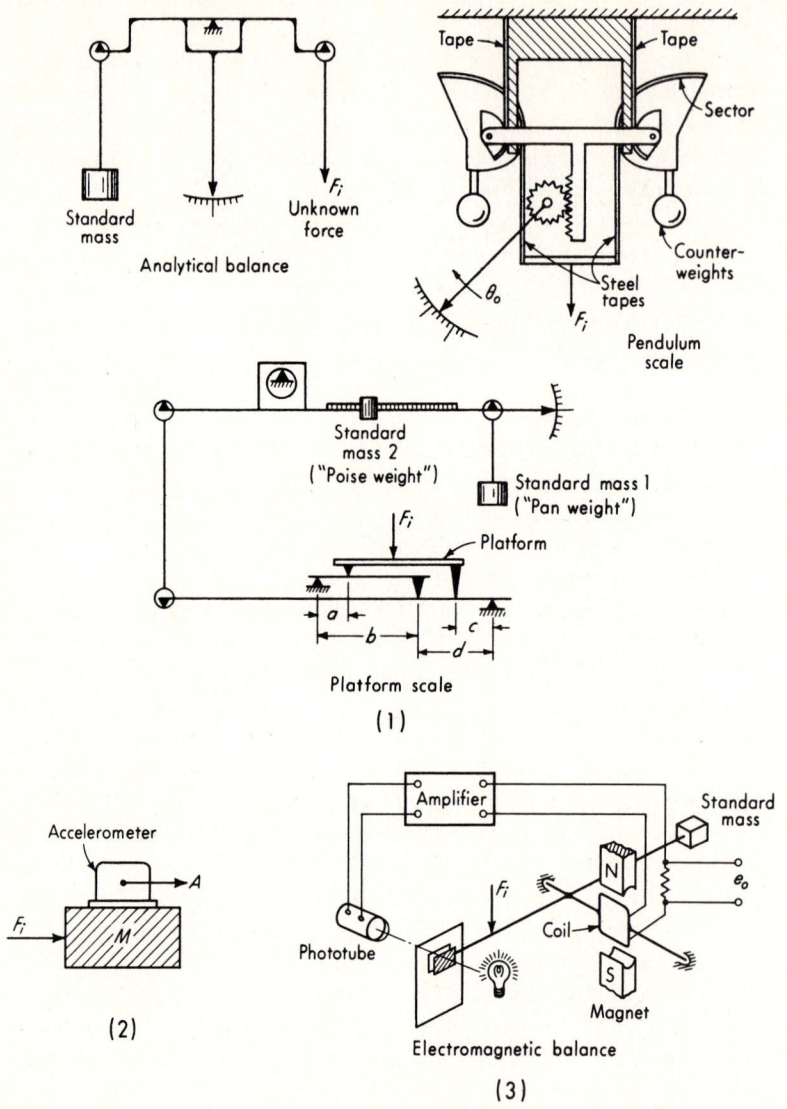

Standard mass

Unknown force F_i

Analytical balance

Tape

Tape

Sector

Counterweights

Steel tapes

θ_0

F_i

Pendulum scale

Standard mass 2 ("Poise weight")

Standard mass 1 ("Pan weight")

F_i

Platform

a b c d

Platform scale

(1)

Accelerometer

A

F_i

M

(2)

Amplifier

Standard mass

e_0

F_i

N

Coil

Phototube

Magnet

S

Electromagnetic balance

(3)

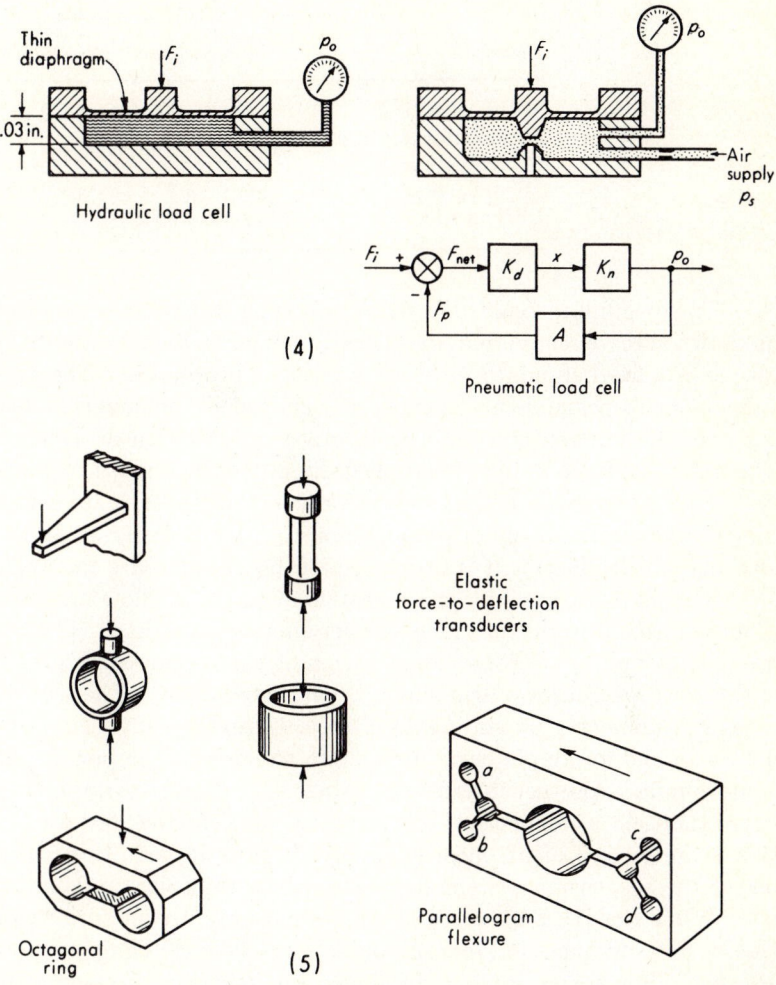

Fig. 5.1. *Basic force-measurement methods.*

a temperature difference of $1/20C°$ between the two arms of a balance can cause an arm-length ratio change of 1 ppm, significant in some applications. Commercially available analytical balances may be classified as follows.[1]

Description	Range, gram	Resolution, gram
Macro analytical	200–1,000	10^{-4}
Semimicro analytical	50–100	10^{-5}
Micro analytical	10–20	10^{-6}
Micro balance	less than 1	10^{-6}
Ultramicro balance	less than 0.01	10^{-7}

The pendulum scale is a deflection-type instrument in which the unknown force is converted to a torque which is then balanced by the torque of a fixed standard mass arranged as a pendulum. The practical version of this principle utilizes specially shaped sectors and steel tapes to linearize the inherently nonlinear torque/angle relation of a pendulum. The unknown force F_i may be applied directly as in Fig. 5.1 or through a system of levers, such as that shown for the platform scale, to extend the range. An electrical signal proportional to force is easily obtained from any angular-displacement transducer attached to measure the angle θ.

The platform scale utilizes a system of levers to allow measurement of large forces in terms of much smaller standard weights. The beam is brought to null by a proper combination of pan weights and adjustment of the poise-weight lever arm along its calibrated scale. The scale can be made self-balancing by adding an electrical displacement pickup for null detection and an amplifier-motor system to position the poise weight to achieve null. Another interesting feature is that, if $a/b = c/d$, the reading of the scale is independent of the location of F_i on the platform. Since this is quite convenient, most commercial scales provide this feature by use of the suspension system shown or others that allow similar results.

While analytical balances are used almost exclusively for "weighing" (really determining the *mass* of) objects or chemical samples, platform and pendulum scales are also employed for force measurements such as those involved in shaft power determinations with dynamometers. All three instruments are intended mainly for static force measurements.

Method 2, the use of an accelerometer for force measurement, is of somewhat limited application since the force determined is the *resultant* force on the mass. Often *several* unknown forces are acting, and they cannot be separately measured by this method.

---

[1] F. Baur, The Analytical Balance, *Ind. Res.*, p. 64, July–August, 1964.

The electromagnetic balance[1,2] (method 3) utilizes a photoelectric null detector, amplifier and torquing coil in a servosystem to balance the difference between the unknown force F_i and the gravity force on a standard mass. This instrument is mainly a competitor of the mechanical analytical balance and is used for the same types of applications. Within its weight range (full scale is presently limited to 1 g or less), its advantages relative to mechanical balances are ease of use, less sensitivity to environment, faster response, smaller size, and ease of remote operation. Also, the electrical output signal is convenient for continuous recording and/or automatic-control applications.

Method 4 is illustrated in Fig. 5.1 by hydraulic and pneumatic load cells. Hydraulic cells[3] are completely filled with oil and usually have a preload pressure of the order of 30 psi. Application of load increases the oil pressure which is read on an accurate gage. Electrical pressure transducers can be used to obtain an electrical signal. The cells are very stiff, deflecting only a few thousandths of an inch under full load. Capacities to 100,000 lb_f are available as standard while special units up to 10 million lb_f are obtainable. Accuracy is of the order of 0.1 percent of full scale; resolution is about 0.02 percent.

The pneumatic load cell shown uses a nozzle-flapper transducer as a high-gain amplifier in a servo loop. Application of force F_i causes a diaphragm deflection x which in turn causes an increase in pressure p_o since the nozzle is more nearly shut off. This increase in pressure acting on the diaphragm area A produces an effective force F_p which tends to return the diaphragm to its former position. For any constant F_i the system will come to equilibrium at a specific nozzle opening and corresponding pressure p_o. The static behavior is given by

$$(F_i - p_o A)K_d K_n = p_o \qquad (5.3)$$

where
$$K_d \triangleq \text{diaphragm compliance, in./}lb_f \qquad (5.4)$$
$$K_n \triangleq \text{nozzle-flapper gain, psi/in.} \qquad (5.5)$$

Solving for p_o, we get

$$p_o = \frac{F_i}{1/K_d K_n + A} \qquad (5.6)$$

Now K_n is not strictly constant but varies somewhat with x, leading to a nonlinearity between F_i and p_o. However, in practice, the product $K_d K_n$

---

[1] L. Cahn, Electromagnetic Weighing, *Instr. Control Systems*, p. 107, September, 1962.

[2] Cahn Instrument Co., Paramount, Calif.

[3] A. H. Emery Co., New Canaan, Conn.

is very large so that $1/K_d K_n$ is made negligible compared with A, giving

$$p_o = \frac{F_i}{A} \quad (5.7)$$

which is linear since A is constant. As in any feedback system, dynamic instability limits the amount of gain that can actually be used. A typical supply pressure p_s is 60 psi, and since the maximum value of p_o cannot exceed p_s, this limits F_i to somewhat less than $60A$. Commercial[1] load cells operating on this general principle (with refinements) have an accuracy of about 0.5 percent of full scale and a deflection under full load of less than 0.001 in. and come in full-scale ranges of 5 to 5,000 lb$_f$. The air consumption is of the order of 0.1 ft^3/min of free air.

While all the previously described force-measuring devices are intended mainly for static or slowly varying loads, the elastic deflection transducers of method 5 are widely used for both static and dynamic loads of frequency content up to many thousand cycles per second. While all are essentially spring-mass systems with (intentional or unintentional) damping, they differ mainly in the geometric form of "spring" employed and in the displacement transducer used to obtain an electrical signal. The displacement sensed may be a gross motion, or strain gages may be judiciously located to sense force in terms of strain. Bonded strain gages have been found particularly useful in force measurements with elastic elements. In addition to serving as force-to-deflection transducers, some elastic elements also perform the function of resolving vector forces or moments into rectangular components. An example, the parallelogram flexure[2] of Fig. 5.1, is extremely rigid (insensitive) to all applied forces and moments except in the direction shown by the arrow. A displacement transducer arranged to measure motion in the sensitive direction will thus measure only that component of an applied vector force which lies along the sensitive axis. The action of this flexure may perhaps be most easily visualized by considering it as a four-bar linkage with pin joints at a, b, c, and d.

Because of the importance of elastic force transducers in modern dynamic measurements, we shall devote a considerable portion of this chapter to their consideration. Although they may differ widely in detail construction, their dynamic-response form is generally the same, and so we shall treat an idealized model representative of all such transducers in the next section.

[1] A. H. Emery Co., New Canaan, Conn.
[2] Flex-Cell, Fluidyne Engineering Corp., Minneapolis, Minn.

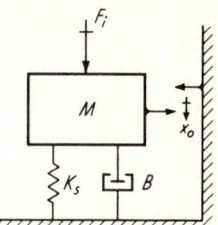

Fig. 5.2.   *Elastic force transducer.*

5.3

Characteristics of Elastic Force Transducers   Figure 5.2 shows an idealized model of an elastic force transducer. The relationship between input force and output displacement is easily established as a simple second-order form:

$$F_i - K_s x_o - B\dot{x}_o = M\ddot{x}_o \qquad (5.8)$$

$$\frac{x_o}{F_i}(D) = \frac{K}{D^2/\omega_n{}^2 + 2\zeta D/\omega_n + 1} \qquad (5.9)$$

where

$$\omega_n \triangleq \sqrt{\frac{K_s}{M}} \qquad (5.10)$$

$$\zeta \triangleq \frac{B}{2\sqrt{K_s M}} \qquad (5.11)$$

$$K \triangleq \frac{1}{K_s} \qquad (5.12)$$

For transducers that do not measure a gross displacement but rather use strain gages bonded to the "spring" K_s, the output strain ϵ may be substituted for x_o if K_s is reinterpreted as force per unit strain rather than force per unit deflection.   In many transducers a distinct and separate "spring" and "mass" cannot be distinguished because the elasticity and inertia are distributed rather than lumped.   In these cases, for design purposes the natural frequency must be calculated from the appropriate formulas[1,2,3] for the geometric shapes involved rather than by employing Eq. (5.10).   Once the transducer is constructed, its lowest natural frequency can generally be found experimentally, as can ζ, which is usually

[1] C. M. Harris and C. E. Crede (eds.), "Shock and Vibration Handbook," vol. 1, chap. 7, McGraw-Hill Book Company, New York, 1961.

[2] J. P. Den Hartog, "Mechanical Vibrations," 4th ed., pp. 431–433, McGraw-Hill Book Company, New York, 1956.

[3] R. K. Mitchell, Some Considerations in the Design of Elastic Force Transducers, M.Sc. Thesis, The Ohio State University, Mechanical Engineering Department, 1965.

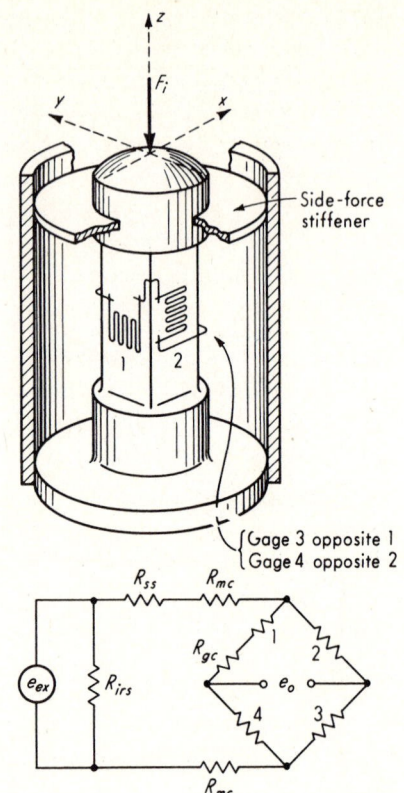

Fig. 5.3   *Strain-gage load cell.*

small and difficult to calculate.   The sensitivity K is generally available theoretically from strength-of-materials or elasticity formulas, whether it relates to a gross deflection or a local unit strain.   Once the transducer is constructed, it should be given an overall calibration relating electrical output to force input since none of the theoretical formulas is sufficiently accurate for this purpose.

Since the dynamic response of second-order instruments has been fully discussed previously, we shall concentrate mainly on details peculiar to specific force transducers.

Bonded-strain-gage transducers.   A typical construction for a strain-gage load cell for measuring compressive forces is shown in Fig. 5.3. (Cells to measure both tension and compression require merely the addition of suitable mechanical fittings at the ends.)   The load-sensing member is short enough to prevent column buckling under the rated load.   Foil-type metal gages are bonded on all four sides; gages 1 and 3 sense the direct stress due to F_i and gages 2 and 4 the transverse stress due to Poisson's ratio μ. This arrangement gives a sensitivity $2(1 + \mu)$ times that achieved with a single active gage in the bridge. It also provides primary temperature compensation since all four gages are (at least for steady temperatures) at the same temperature. Furthermore, the arrangement is insensitive to bending stresses due to F_i being applied off center or at an angle. This can be seen by replacing an off-center force by an equivalent on-center force and a couple. The couple can be resolved into x and y components which cause bending stresses in the gages. If the gages are carefully placed so as to be symmetrical, the bending stresses in gages 1 and 3 will be of opposite sign, and by the rules of bridge circuits the net output e_o due to bending will be zero. Similar arguments hold for gages 2 and 4 and for bending stresses due to F_i being at an angle. The side-force stiffener plate also reduces the effects of angular forces since it is very stiff in the radial (x,y) direction but very soft in the z direction.

The deflection under full load of such load cells is of the order of 0.001 to 0.015 in., indicating their high stiffness. The natural frequency is often not quoted since it is frequently determined almost entirely by the mass of force-carrying elements external to the transducer. This is especially true in the many applications where the load cell is used for weighing purposes. The high stiffness also implies a low sensitivity. To increase sensitivity (in low-force cells where it is needed) without sacrificing column stability and surface area for mounting gages, a hollow (square on the outside, round on the inside) load-carrying member may be employed.

To achieve the high accuracy (0.3 to 0.1 percent of full scale) required in many applications, additional temperature compensation is needed. This is accomplished by means of the temperature-sensitive resistors R_{gc} and R_{mc} shown in Fig. 5.3. These resistors are permanently attached internal to the load cell so as to assume the same temperature as the gages. The purpose of R_{gc} is to compensate for the slightly different temperature coefficients of resistance of the four gages. The purpose of R_{mc} is to compensate for the temperature dependence of the modulus of elasticity of the load-sensing member. That is, although one wishes to measure force, the gages sense strain; thus any change in the modulus of elasticity will give a different strain (and thus a different e_o) even though the force is the same. Since all metals change modulus somewhat with temperature, this effect causes a sensitivity drift. The resistance R_{mc} compensates for this by changing the excitation voltage actually applied to the bridge by just the right amount to counteract the modulus effect.

Two additional (non-temperature-sensitive) resistors are often found in commercial load cells. They are R_{ss}, which is adjusted to standardize the sensitivity for a nominal e_{ex} to a desired value, and R_{irs}, which is used to adjust the input resistance to a desired value.

A family[1] of load cells covering the full-scale range from 10 to 250,000 lb$_f$ has a full-scale output of about 35 mv, nonlinearity as good as 0.05 percent of full scale, hysteresis and nonrepeatability as good as 0.02 percent of full-scale, best temperature-induced zero drift of 0.0015 percent of full scale/F°, and best temperature-induced sensitivity drift of 0.0008 percent/F° over the range 15 to 115°F. These cells all use 350-ohm gages and 12 volts bridge excitation. The 10-lb$_f$ cell is a cylinder about 3.5 in. in diameter and length (weight 2.5 lb$_f$) while the 250,000-lb$_f$ cell is a cylinder 12 in. in diameter by 24 in. long (weight 400 lb$_f$).

When adequate sensitivity cannot be achieved by use of tension/ compression members, configurations employing bending stresses may be helpful. These generally provide more strain per unit applied force but at the expense of reduced stiffness and thus natural frequency. Of the many possibilities, the cantilever beam and proving ring are shown in Fig. 5.4. The cantilever-beam gage arrangement provides four times the sensitivity of a single gage, temperature compensation, and insensitivity to x and y components of force if identical gages and perfect symmetry are assumed. The proving-ring transducer also is inherently temperature-compensated. For F_i as a tension load, gages 1 and 3 are in compression and 2 and 4 are in tension; thus, with the bridge arrangement shown, these effects are all additive, giving a large output.

When maximum output is desired for any strain-gage transducer one should consider the possible use of low-modulus materials (such as aluminum) to increase strain per unit force, several gages (or one high-resistance gage) per bridge leg (if space allows), and the intentional introduction of stress concentrations at the gage locations. However, such techniques also present associated problems. Low modulus reduces stiffness and natural frequency, and some low-modulus materials have excessive hysteresis and low fatigue life. Stress concentrations also lower fatigue life, and their effect may be difficult to calculate for design purposes.

Unbonded-strain-gage transducers. The unbonded strain gage described in Chap. 4 can be used for measurement of small forces by applying them directly to the strain wires and for large forces by using the strain gage to measure displacement of an elastic member. A family[2] of such instruments has full-scale ranges of ± 0.15 oz to $\pm 1,000$ lb$_f$,

[1] Baldwin-Lima-Hamilton Corp., Waltham, Mass.
[2] Statham Instruments Inc., Los Angeles, Calif.

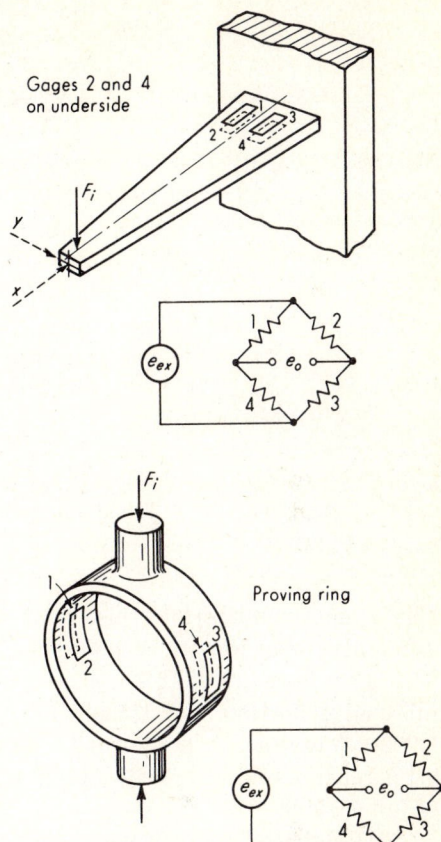

Fig. 5.4. *Beam and ring transducers.*

deflection of 0.015 to 0.0015 in. under full load, full-scale output of ± 20 to ± 53 mv, and nonlinearity and hysteresis less than ± 1 percent of full scale, with temperature-compensated units having a thermal sensitivity shift of 0.01 percent/F° from -65 to $+250$°F and a thermal zero shift of 0.01 percent of full scale/F° from -65 to $+250$°F.

Differential-transformer transducers. A family[1] of load cells which use a differential transformer to measure deflection of a machined-from-solid dual diaphragm spring of alloy steel (see Fig. 5.5) has the following characteristics: full-scale ranges from ± 1 to $\pm 5,000$ lb$_f$, full-load deflection of 0.005 in., full-scale output of 0.2 volt for 10-volt/3,000-cps excitation, nonlinearity 0.2 percent of full scale, repeatability 0.1 percent of full scale, and thermal zero or sensitivity shift of 0.02 percent of full scale/F°.

---

[1] Daytronic Corp., Dayton, Ohio.

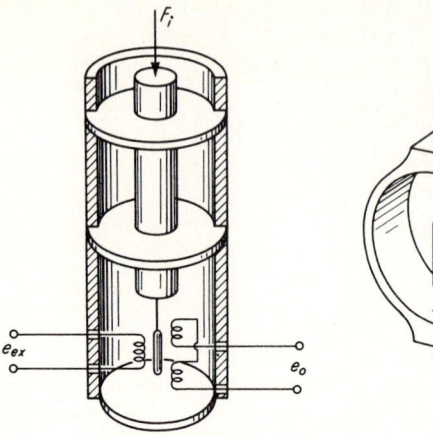

Fig. 5.5. *Differential-trans-former transducer.*

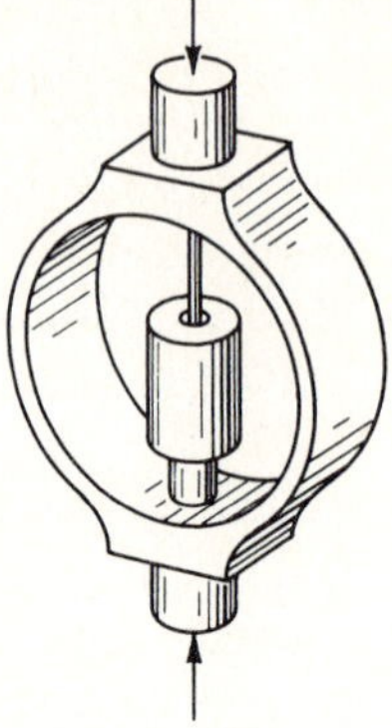

Fig. 5.6. *Proving ring transducer.*

The same manufacturer also makes a proving-ring transducer with differential-transformer measurement of the deflection (see Fig. 5.6). The characteristics of this unit are full-scale range ± 10 to $\pm 10{,}000$ lb$_f$, full-load deflection of 0.025 in., full-scale output of 2 volts for 10-volt/3,000-cps excitation, nonlinearity 0.25 percent of full scale, repeatability 0.1 percent of full scale, and thermal zero or sensitivity shift of 0.03 percent of full scale/F°.

Differential-transformer force transducers are also available[1] in the range ± 1 to ± 100 g with full-load deflection of 0.01 in. and 0.35-volt full-scale output.

Piezoelectric transducers. These force transducers have the same form of transfer function as piezoelectric accelerometers. They are intended for dynamic force measurement only, although some types (quartz pickup with electrometer charge amplifier) have sufficiently large τ to allow short-term measurement of static forces and static calibration. Commercially available pickups are designed primarily for compressive loading and exhibit high stiffness of the order of 5 to 35×10^6 lb$_f$/in.

A quartz-pickup/amplifier combination[2] has five switch-selectable full-scale ranges of 0.125, 1.25, 12.5, 125, and 500 lb$_f$ with corresponding sensitivities of 400, 140, 19, 2, and 0.05 mv/lb$_f$ when a 1-ft cable length is used. The natural frequency of the pickup itself is 60,000 cps, and non-

---

[1] Sanborn Co., Waltham, Mass.
[2] Kistler Instrument Corp., North Tonawanda, N.Y.

linearity is 1 percent. This pickup/amplifier combination will hold a static reading for several minutes, thus allowing static calibration. The good temperature characteristics of quartz give an operating range of -400 to $+500°F$.

A family[1] of force gages using synthetic-ceramic piezoelectric elements covers the full-scale range of 100 to 5,000 lb_f with corresponding sensitivities of 300 to 20 mv/lb_f and stiffness of 5 to 35 $\times$ 10^6 lb_f/in. The natural frequency of the pickups themselves is of the order of 25,000 cps, while an added mass of M lb_m gives frequencies ranging from $7,000/\sqrt{M}$ to $18,300/\sqrt{M}$ cps. Nonlinearity is ± 1 percent and sensitivity changes ± 10 percent over the temperature range -30 to $+230°F$. The low-frequency response is 5 percent down at 2 cps if the pickup output is connected to 1,000-megohms input resistance.

Piezoelectric force pickups tend to be sensitive to side loading, and most manufacturers recommend special precautions to minimize this. Numerical sensitivities to side loads are not often quoted on instrument specification sheets. One manufacturer[2] offering specially designed pickups resistant to side loading quotes transverse sensitivity as less than 7 percent of axial sensitivity.

Variable-reluctance/FM-oscillator digital systems. While the electrical signal from any force transducer can be converted to digital form by suitable equipment, some types of pickups are specifically designed with this in mind. We here discuss briefly one such type[3] used in rocket-engine testing where digital data provide advantages of accuracy and ease of performing computations automatically. The elastic member is a proving ring with a two-arm variable-reluctance bridge displacement transducer. The signal from this transducer is used to change the frequency of a frequency-modulated (FM) oscillator. The frequency change (from some base value) of this oscillator is directly proportional to displacement (and thus to force). A digital reading of force is accomplished by applying the output signal of the oscillator to an electronic counter over a known time interval. The total number of pulses accumulated is thus a digital measure of force. In a typical unit a change of force from minus full scale to plus full scale causes the frequency to change from 10,000 to 12,500 cps. For a 1-sec counting period a full-scale force will thus cause a counter reading 1,250 above the base value of 11,250.

Computing advantages of such a system arise from the desire to know the total impulse of the rocket engine and to be able to add and sub-

---

[1] Endevco Corp., Pasadena, Calif.

[2] Wilcoxon Research, Bethesda, Md.

[3] Daystrom-Wiancko Co., Pasadena, Calif.

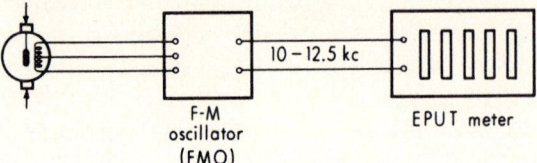

Force measurement

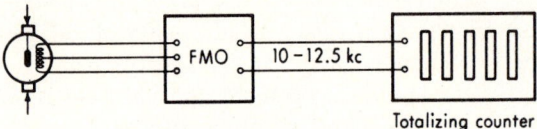

Total impulse measurement

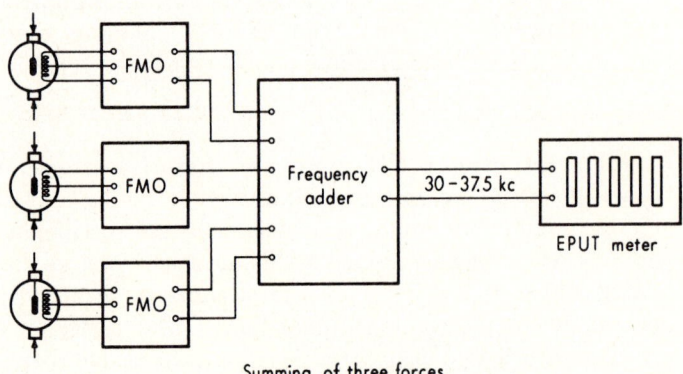

Summing of three forces

Fig. 5.7.   *Digital force measurement, integration, and summing.*

tract various forces in multicomponent test stands.   The total impulse is the integral of force with respect to time; this is simply the total number of counts over the desired integrating time.   When several forces must be added or subtracted, each has its own force pickup and 10,000 to 12,500-cps oscillator.   The oscillator output signals are combined in an electronic adder unit which produces a single output whose frequency is the sum (or difference) of the input frequencies.   The output of the adder unit is a signal of a frequency of 30,000 to 37,500 cps; counting this over a timed interval gives the algebraic sum of the measured forces.   Integration of this sum signal can also be easily accomplished by the same method used for a single signal.   Figure 5.7 shows block diagrams of these systems.

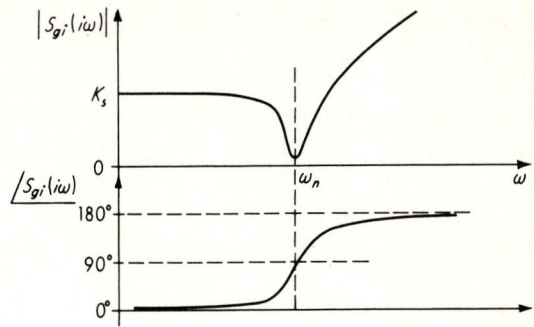

Fig. 5.8.   *Stiffness of force transducers.*

Loading effects.   Since force is an effort variable, the pertinent loading parameter is either stiffness or impedance, and the associated flow variable is either displacement or velocity.   Stiffness is perhaps more convenient for elastic force sensors since impedance would be infinite for the static case.   The generalized input stiffness S_{gi} of the system of Fig. 5.2 is given by

$$S_{gi}(D) = \frac{F_i}{x}(D) = K_s\left(\frac{D^2}{\omega_n{}^2} + \frac{2\zeta D}{\omega_n} + 1\right) \qquad (5.13)$$

Recall that for small error due to loading, S_{gi} must be sufficiently large compared with S_{go}, the generalized output stiffness of the system being measured.   The frequency characteristic $S_{gi}(i\omega)$ is shown in Fig. 5.8 for a particular (small) value of ζ.   Note that, near ω_n, $S_{gi}(i\omega)$ becomes very small.   However, force pickups are generally used only for $\omega \ll \omega_n$; thus $S_{gi}(i\omega)$ is, in most cases, adequately approximated as simply K_s.

5.4

Resolution of Vector Forces and Moments into Rectangular Components   In a number of important practical applications the force or moment to be measured is not only unknown in magnitude but also of unknown and/or variable direction.   Outstanding examples of such situations are "balances" for measuring forces on wind-tunnel models,[1,2] dynamometers (force gages) for measuring cutting forces in

---

[1] P. K. Stein, "Measurement Engineering," vol. 1, p. 431, Stein Engineering Services, Inc., Phoenix, Ariz., 1964.

[2] C. C. Perry and H. R. Lissner, "The Strain Gage Primer," p. 212, McGraw-Hill Book Company, New York, 1955.

machine tools,[1] and thrust stands[2] for determining forces of rocket engines. Elastic force transducers of either the bonded-strain-gage or gross-deflection variety are employed in these applications. Ingenious use of various types of flexures for isolating and measuring the various force components characterizes the design of these devices. Depending on the degree to which the force or moment direction is unknown, force-resolving systems of varying degrees of complexity may be devised. The most general situation (measurement of three mutually perpendicular force components and three mutually perpendicular moment components) is regularly accomplished with high accuracy.

Figure 5.9 shows a six-component thrust stand[2] used in testing rocket engines. The load cells 1, 2, and 3 are mounted at the corners of an equilateral triangle, and load cells 4, 5, and 6 are in the sides of a concentric smaller equilateral triangle. The engine to be tested is rigidly fastened at the common center of both triangles and produces a force of unknown magnitude and direction (which can be expressed in terms of components F_x, F_y, and F_z) and a moment of unknown magnitude and direction (which can be expressed in terms of components M_x, M_y, and M_z). The rocket forces are transmitted from the mounting plate to the rigid foundation through the six load cells and their associated flexures. The action of the suspension system is most clearly seen if one considers each flexure as a pin joint. Pin joints are not actually used because of their lost motion and friction. A static analysis of the force system gives the following results which allow calculation of the rocket forces and moments from the measured load-cell forces and stand dimensions:

$$F_x = F_1 + F_2 + F_3 \qquad (5.14)$$

$$F_y = \frac{(F_5 - F_4) - (F_4 - F_6)}{2} \qquad (5.15)$$

$$F_z = \frac{\sqrt{3}\,(F_5 - F_6)}{2} \qquad (5.16)$$

$$M_x = \frac{-d_1(F_4 + F_5 + F_6)}{2\sqrt{3}} \qquad (5.17)$$

$$M_y = d_2 \frac{(F_1 - F_2) - (F_3 - F_1)}{2\sqrt{3}} \qquad (5.18)$$

$$M_z = d_2 \frac{(F_3 - F_2)}{2} \qquad (5.19)$$

[1] E. G. Loewen, E. R. Marshall, and M. C. Shaw, Electric Strain Gage Tool Dynamometers, *Proc. Soc. Exptl. Stress Anal.*, vol. 8, no. 2, 1951.

[2] The Design of High-accuracy Thrust Stands and Calibrators, WEC-BA-7D, Daystrom-Wiancko Co., Pasadena, Calif., 1961.

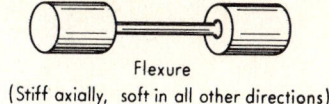

Flexure
(Stiff axially, soft in all other directions)

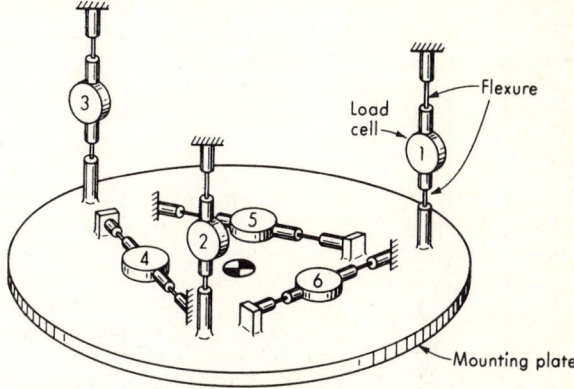

⊔⊔⊔⊔ Denotes common rigid foundation

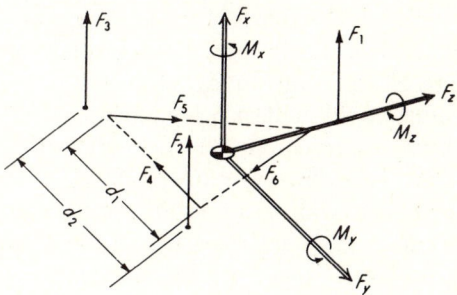

Fig. 5.9. *Six-component load frame.*

The indicated additions and subtractions of forces are (in the thrust stand described above) performed automatically by digital counters and adders since the load cells used are the variable-reluctance/FM-oscillator type described in the previous section.

A combination of bonded strain gages, Wheatstone-bridge circuits, and flexible elements of various geometries has proved a versatile tool in the development of multicomponent-force pickups of small size and high natural frequency. Figure 5.10 shows a beam with three separate bridge circuits of gages arranged to measure the three rectangular components of an applied force. All bridge circuits are temperature-compensated and respond only to the intended component of force; however, the point of

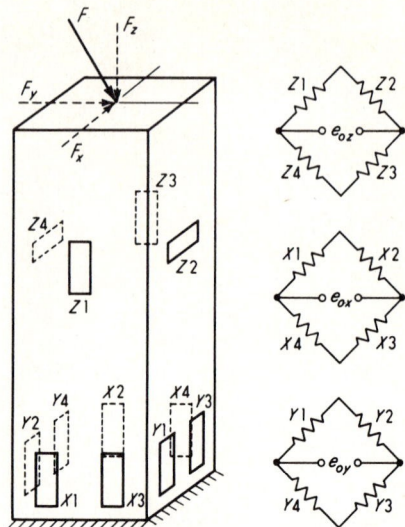

Fig. 5.10. *Resolution of vector forces.*

application of the force must be at the center of the beam cross section. An eccentricity in the y direction, for example, would give F_z a moment arm, causing bending stresses in the y direction that would be indistinguishable from those due to F_y. If side loads F_x and F_y are present, the end of the beam will deflect, causing just such eccentricities; thus beam stiffness must be adequate to keep deflection sufficiently low. This stiffness tends, of course, to reduce sensitivity.

While many strain-gage pickup configurations are possible and have been used, the octagonal ring[1] of Fig. 5.1 has particularly interesting properties. The reference gives information on its design and use.

5.5

Torque Measurement on Rotating Shafts   Measurement of the torque carried by a rotating shaft is of considerable interest for its own sake and also as a necessary part of shaft power measurements. Torque transmission through a rotating shaft generally involves both a source of power and a sink (power absorber or dissipator), as in Fig. 5.11. Torque measurement may be accomplished by mounting either the source or the sink in bearings ("cradling") and measuring the reaction force F and arm length L, or the torque in the shaft itself is measured in terms of the angular twist or strain of the shaft (or a torque sensor coupled into the shaft).

The cradling concept is the basis of most shaft power dynamometers.

[1] N. H. Cook and E. Rabinowicz, "Physical Measurement and Analysis," p. 162, Addison-Wesley, Publishing Company, Inc., Reading, Mass., 1963.

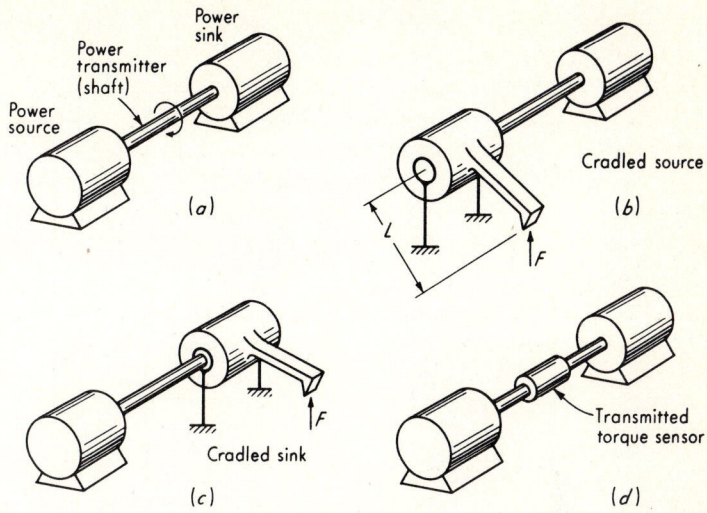

Fig. 5.11. *Torque measurement of rotating machines.*

These are used mainly for measurements of steady power and torque, using pendulum or platform scales to measure F.   A free-body analysis of the cradled member reveals error sources due to friction in the cradle bearings, static unbalance of the cradled member, windage torque (if the shaft is rotating), and forces due to bending and/or stretching of power lines (electric, hydraulic, etc.) attached to the cradled member.   To reduce frictional effects and also to make possible dynamic torque measurements, the cradle-bearing arrangement may be replaced by a flexure pivot with strain gages to sense torque,[1] as in Fig. 5.12.   The crossing point of the

[1] Lebow Associates, Oak Park, Mich.

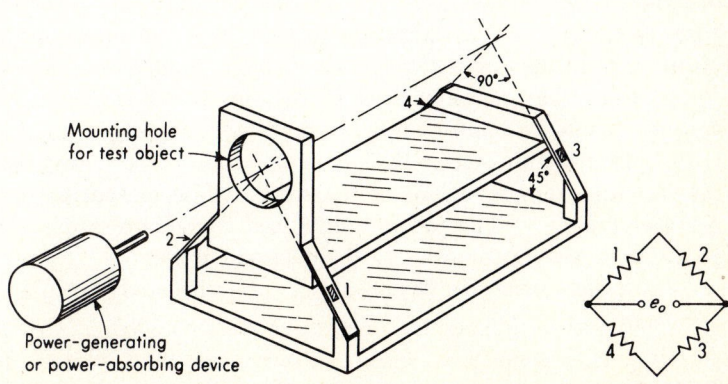

Fig. 5.12. *Strain-gage torque table.*

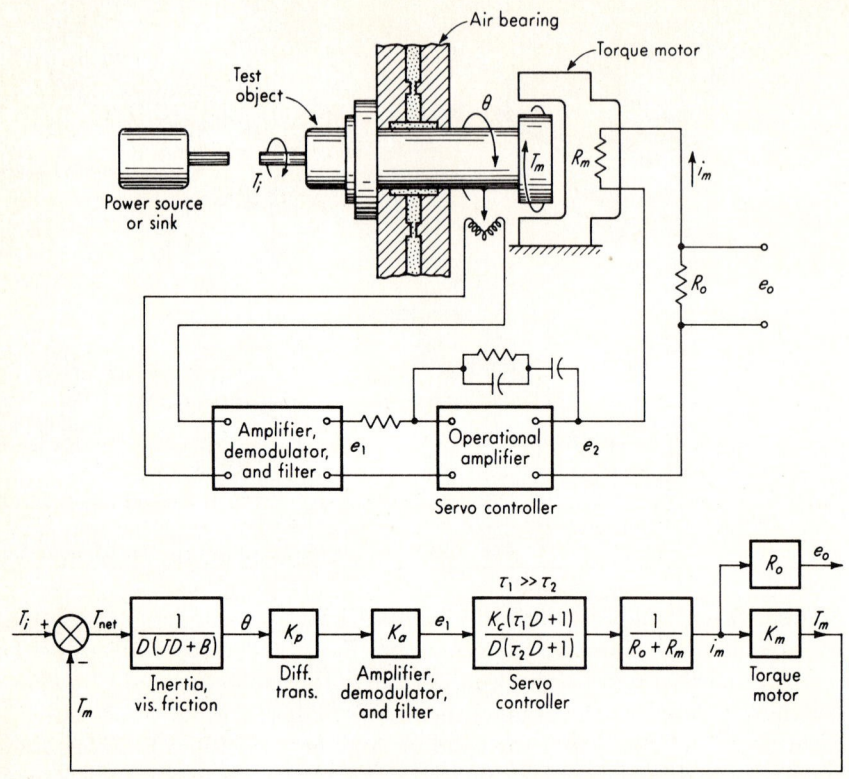

Fig 5.13.   *Feedback torque sensor.*

flexure plates defines the effective axis of rotation of the flexure pivot. Angular deflection under full load is typically less than 0.5°. This type of cross-spring flexure pivot is relatively very stiff in all directions other than the rotational one desired, just as in an ordinary bearing. The strain-gage bridge arrangement also is such as to reduce the effect of all forces other than those related to the torque being measured. Speed-torque curves for motors may be obtained quickly and automatically with such a torque sensor by letting the motor under test accelerate an inertia from zero speed up to maximum while measuring speed with a d-c tachometer.[1] The torque and speed signals are applied to an *X-Y* recorder to give automatically the desired curves.

Another variation (see Fig. 5.13) on the cradle principle is found

---

[1] B. Hall, Motor Tests Using *X-Y* Recorders, *Electro-Technol.*, p. 116, May, 1964.

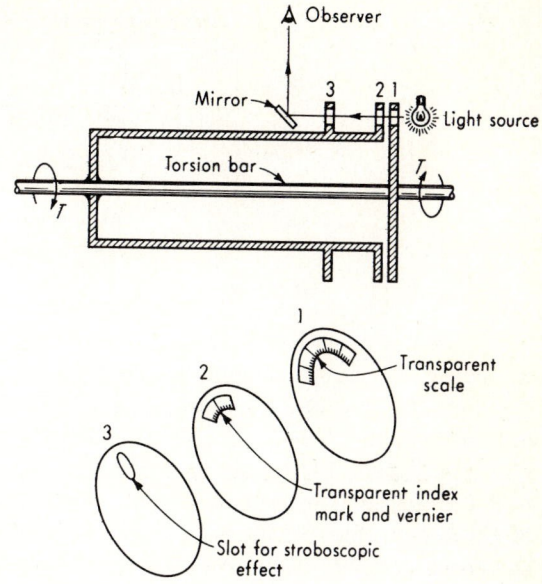

Fig. 5.14. *Torsion-bar dynamometer.*

in a null-balance torquemeter using feedback principles to measure small torques in the range 0 to 10 oz-in. In this device[1] the test object is mounted on a hydrostatic air-bearing table to reduce bearing friction to exceedingly small values. Any torque on the test object tends to cause rotation of the air-bearing table but this rotation is immediately sensed by a differential-transformer displacement pickup. The output from this pickup is converted to direct current and amplified to provide the coil current of a torque motor which applies opposing torque to keep displacement at zero. The amount of current required to maintain zero displacement is a measure of torque and is read on a meter. The servo loop uses integral control[2] to give zero displacement for any constant torque. Approximate derivative control is also used to give stability. The threshold of this air-bearing system is less than 0.0005 oz-in. while the torque/current nonlinearity is 0.001 oz-in. The overall system behaves approximately as a second-order system with a natural frequency of about 10 cps and damping ratio of 0.7 when no test object is present on the table.

---

[1] McFadden Electronics, South Gate, Calif.

[2] E. O. Doebelin, "Dynamic Analysis and Feedback Control," p. 223, McGraw-Hill Book Company, New York, 1962.

Gages 2 and 3 are also at 45° with shaft axis

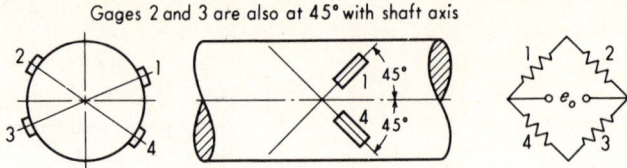

Fig. 5.15.   Strain-gage torque measurement.

The use of elastic deflection of the transmitting member for torque measurement may be accomplished by measuring either a gross motion or a unit strain.   In either case, a main difficulty is the necessity of being able to read the deflection while the shaft is rotating.   Figure 5.14 illustrates a torsion-bar torquemeter using optical methods of deflection measurement.   The relative angular displacement of the two sections of the torsion bar can be read from the calibrated scales because of the stroboscopic effect of intermittent viewing and the persistence of vision. The desire for electrical output signals and for the ability to measure rapidly varying torque has led to the development of strain-gage torquemeters.   The problem of getting bridge power onto the shaft and output signals off the shaft has a number of possible solutions.   If rotation is slow and the total angle turned through is small, one may simply let the connecting wires wrap around the shaft.   For continuous rotation, slip rings and brushes may be used, or a subminiature telemetry system may be attached to the rotating shaft and the signals sent to a stationary receiver by radio.

The strain-gage bridge configuration generally used to measure torque is shown in Fig. 5.15.   This arrangement (assuming accurate gage placement and matched gage characteristics) is temperature-compensated and insensitive to bending or axial stresses.   The gages must be precisely at 45° with the shaft axis, and gages 1 and 3 must be diametrically opposite, as must gages 2 and 4.   Accurate gage placement is facilitated by the availability of special rosettes in which two gages are precisely oriented on one sheet of backing material.   In some cases the shaft already present in the machine to be tested may be fitted with strain gages.   In other cases a different shaft or a commercial torquemeter must be used to get the desired sensitivity or other properties.   Placement of the gages on a square, rather than round, cross section of the shaft (see Fig. 5.16) has some advantages.   The gages are more easily and accurately located and more firmly bonded on a flat surface.   Also, the corners of a square section in torsion are stress-free and thus provide a good location for solder joints between lead wires and gages.   These joints are often a

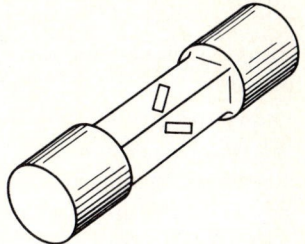

Fig. 5.16. *Square shaft-torque element.*

source of fatigue failure if located in a high-stress region. Also, for equivalent strain/torque sensitivity, a square shaft is much stiffer in bending than a round one, thus reducing effects of bending forces and raising shaft natural frequencies.

The torque of many machines, such as reciprocating engines, is not smooth even when the machine is running under "steady-state" conditions. If one wishes to measure the average torque so as to calculate power, the higher frequency response of strain-gage torque pickups may be somewhat of a liability since the output voltage will follow the cyclic pulsations and some sort of averaging process must be performed to obtain average torque. If exceptional accuracy is not needed, the low-pass filtering effect of a d-c meter used to read e_o may be sufficient for this purpose. In the cradled arrangements of Fig. 5.11 (used in many commercial dynamometers for engine testing, etc.) the inertia of the cradled member and the low frequency response of the platform or pendulum scales used to measure F perform the same averaging function.

Commercial strain-gage torque sensors are available with built-in slip rings and speed sensors. A family[1] of such devices covers the range 50 to 100,000 in.-lb$_f$ with full-scale output of about 40 mv. The smaller units may be used at speeds up to 24,000 rpm, the largest to 4,000 rpm. Torsional stiffness of the 50-in.-lb$_f$ unit is about 4,000 in.-lb$_f$/rad while the 100,000-in.-lb$_f$ unit has 9.5×10^6 in.-lb$_f$/rad. Nonlinearity is 0.1 percent of full scale while temperature effect on zero is 0.002 percent of full scale/F° and temperature effect on sensitivity is 0.002 percent/F° over the range 30 to 150°F.

The dynamic response of elastic deflection torque transducers is essentially slightly damped second-order, with the natural frequency usually determined by the stiffness of the transducer and the inertia of the parts connected at either end. Damping of the transducers themselves is usually not attempted, and any damping present is due to bearing friction, windage, etc., of the complete test setup.

---

[1] Lebow Associates, Oak Park, Mich.

5.6

Shaft Power Measurement (Dynamometers)   The accurate measurement of the shaft power input or output of power-generating, -transmitting, and -absorbing machinery is of considerable interest. While the basic measurements, torque and speed, have already been discussed, their practical application to power measurement will be considered briefly here.   The term dynamometer is generally used to describe such power-measuring systems although it is also used as a name for elastic force sensors.

The type of dynamometer employed depends somewhat on the nature of the machine to be tested.   If the machine is a power generator, the dynamometer must be capable of absorbing its power.   If the machine is a power absorber, the dynamometer must be capable of driving it.   If the machine is a power transmitter or transformer, the dynamometer must provide both the power source and the load.

Perhaps the most versatile and accurate dynamometer is the d-c electric type.   Here a d-c machine is mounted in low-friction trunnion bearings (see Fig. 5.11*b*) and provided with field and armature control circuits.[1,2]   This machine can be coupled to either power-absorbing or power-generating devices since it may be connected as either a motor or a generator.   When it is used as a generator, the generated power is dissipated in resistance grids or recovered for use.   Modern control circuits allow accurate control of dynamometer speed under varying load torque (100 percent change in load torque causes about 0.5 percent steady-state speed change).   Control of dynamometer torque is also possible, although it is less used and less accurate.   A speed change of about 50 percent of top speed may be expected to cause a torque change of 5 percent of maximum torque when torque control is used.   The d-c dynamometer can be adjusted to provide any torque from zero to the maximum design value for speeds from zero to the so-called base speed of the machine.   This is the speed at which the maximum torque develops the maximum design horsepower.   At speeds above base speed, torque must be progressively reduced so as to maintain horsepower less than the design maximum.   The controllability of the d-c dynamometer lends itself particularly to modern automatic load and speed programming applications.

Figure 5.17 illustrates such a situation.   Tape recordings of engine torque and speed measured under actual driving conditions for an auto-

[1] P. S. Potts and P. T. Schuerman, How to Choose Electric Dynamometers, *Machine Design*, p. 102, June 27, 1957.

[2] R. F. Knudsen, A Discussion of Present Day Dynamometers, *Gen. Motors Eng. J.*, p. 18, October–November–December, 1957.

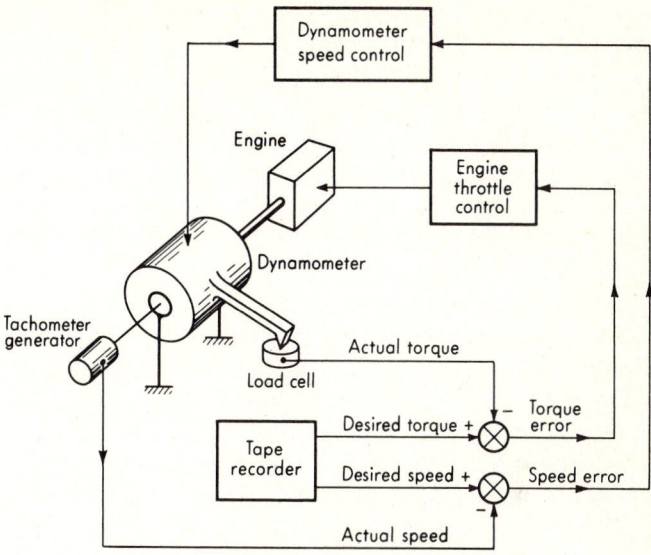

Fig. 5.17. *Servo-controlled dynamometer.*

mobile are used to reproduce these conditions in the laboratory engine test. Two feedback systems are used to control engine speed and torque. A tachometer generator speed signal from the dynamometer is compared with the desired speed signal from the tape recorder; if the two are different, the dynamometer control is automatically adjusted to change speed until agreement is reached. Actual engine torque is obtained from a load cell on the dynamometer and compared with the desired torque from the tape recorder. If these do not agree, the error signal actuates the engine throttle control in the proper direction. Both these systems operate simultaneously and continuously to force engine speed and torque to follow the tape-recorder commands.

   Dynamometers capable only of absorbing power include the eddy-current brake (inductor dynamometer) and various mechanical brakes employing dry friction (Prony brake) and fluid friction (air and water brakes). The eddy-current brake is easily controllable by varying a d-c input, but it cannot produce any torque at zero speed and only small torque at low speeds. However, it is capable of higher power and speed than a d-c dynamometer. The power absorbed is carried away by cooling water circulated through the air gap between rotor and stator. The Prony brake is a simple mechanical brake in which friction torque is manually adjusted by varying the normal force with a handwheel. Torque is available at zero speed but operation may be jerky because

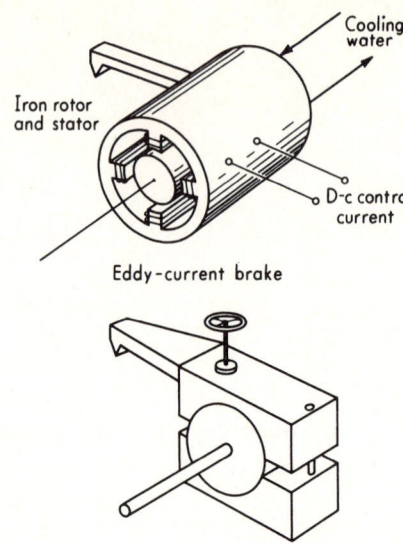

Fig. 5.18. *Absorption dynamometers.*

Prony brake

of the basic nature of dry friction. Water and air brakes utilize the churning action of paddle wheels or vanes rotating inside a fluid-filled casing to absorb power. A flow of air or water through the device is maintained for cooling purposes. No torque is available at zero speed and only small torques at low speeds. High speed and high power can be handled well, however, some water brakes being rated at 10,000 hp at 30,000 rpm.

In all the above power-measurement applications torque and speed are separately measured and then power is manually calculated. This calculation can be performed automatically in a number of ways since the basic operation (multiplication) can be accomplished physically in various ways. An interesting scheme using the properties of bridge circuits is shown in Fig. 5.19. Speed is measured with a d-c tachometer generator, and this voltage is applied as the excitation of a strain-gage load cell used to measure torque. Since bridge output is directly proportional to excitation voltage and also directly proportional to torque, the voltage e_o is actually an instantaneous power signal.

Another ingenious solution[1] of this problem is shown in Fig. 5.20. A torsion bar carrying the torque to be measured has a permanent-magnet a-c generator (alternator) coupled at either end. Each alternator puts out an a-c voltage of amplitude proportional to shaft speed and

---

[1] Waugh Div., Van Nuys, Calif.

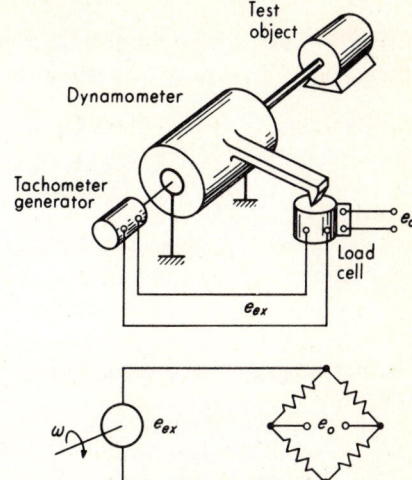

Fig. 5.19. *Instantaneous power measurement.*

frequency equal to shaft speed. When the torsion bar is unloaded, the two alternator rotor-stator positions are mechanically adjusted so that the a-c output voltages are exactly out of phase. If the two alternators are now connected electrically in series, the net output will be zero at any shaft speed so long as no torque is present. When the shaft carries torque, it twists, causing a phase shift between the two alternators and a net

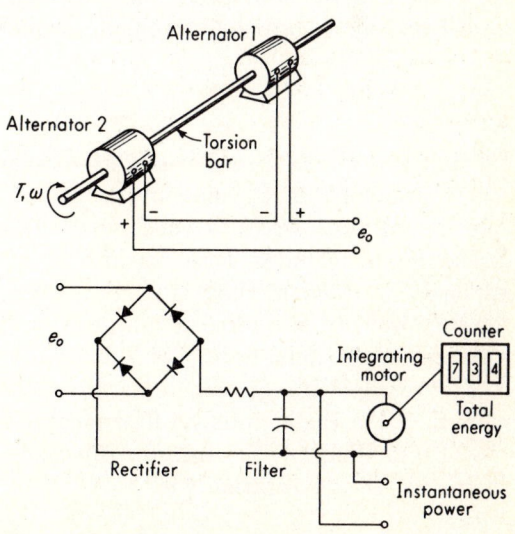

Fig. 5.20. *Alternator power measurement.*

output voltage whose amplitude is proportional to the product of torque and speed. This may be seen from the following analysis:

$$\text{Alternator 1 output} = K_\omega \omega \sin \omega t \qquad (5.20)$$
$$\text{Alternator 2 output} = K_\omega \omega \sin (\omega t + \phi) \qquad (5.21)$$

where
$$K_\omega \triangleq \text{peak volts/(rad/sec)} \qquad (5.22)$$
$$\phi = K_t T \qquad (5.23)$$
$$K_t \triangleq \text{rad/in.-lb}_f \qquad (5.24)$$
$$T \triangleq \text{torque} \qquad (5.25)$$

The net output of the series-connected alternators is

$$\text{Net output} = e_o = K_\omega \omega [\sin \omega t - \sin (\omega t + K_t T)] \qquad (5.26)$$
$$e_o = K_\omega \omega [\sin \omega t - (\sin \omega t \cos K_t T + \cos \omega t \sin K_t T)] \qquad (5.27)$$

The twist angle $\phi = K_t T$ is very small, and so $\cos K_t T \approx 1.0$ and $\sin K_t T \approx K_t T$. Thus

$$e_o = -K_\omega \omega K_t T \cos \omega t \qquad (5.28)$$

and e_o is a sine wave of amplitude proportional to ωT and thus to power. In the instrument described above this a-c voltage is rectified and filtered to produce a proportional direct current of 20 volts full scale. Also, if total energy over a time period is desired, an integrator is available to integrate the d-c voltage. This is in the form of a precise d-c motor whose speed is directly proportional to the voltage applied to it from the horsepower meter. If speed is proportional to horsepower, total revolutions (read by an ordinary mechanical counter) give total energy.

5.7

Gyroscopic Force and Torque Measurement   The use of gyroscopic principles in the measurement of force and torque is not widespread and thus was not considered earlier in this chapter. A short discussion is included here because of general interest and potential application. The method is particularly useful if one wishes to obtain the time integral of a force or torque since the gyro supplies a mechanical rotation proportional to it. This concept has found practical application in a gyroscopic integrating mass flowmeter which will be discussed in more detail in the chapter on flow measurement.

From Chap. 4, the transfer function relating an applied torque T_x to an output axis rotation θ is

$$\frac{\theta}{T_x}(D) = \frac{H_s/I_x}{D[I_y D^2 + BD + (H_s^2/I_x + K_s)]}$$

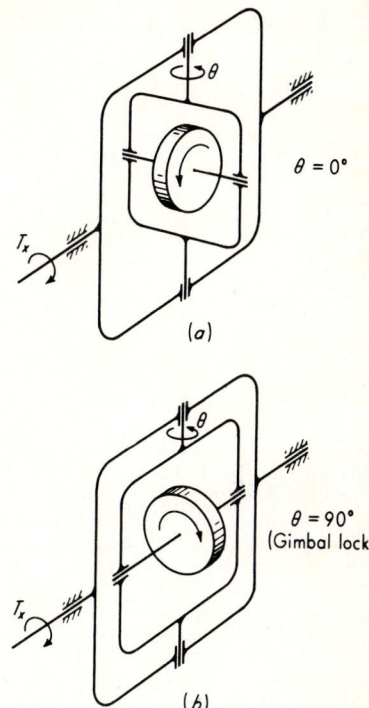

Fig. 5.21. *Gyroscopic torque measurement.*

For a free gyro, as in Fig. 5.21a, B and K_s are effectively zero, giving

$$\frac{\dot{\theta}}{T_x}(D) = \frac{K}{D^2/\omega_n{}^2 + 1} \qquad (5.29)$$

where

$$K \triangleq \frac{1}{H_s} \qquad \text{(rad/sec)/in.-lb}_f \qquad (5.30)$$

$$\omega_n \triangleq \sqrt{\frac{H_s{}^2}{I_x I_y}} \qquad (5.31)$$

While the second-order term of Eq. (5.29) is undamped, small damping (friction) in bearings, etc., is always present to prevent sustained oscillation. Also, if T_x is applied gradually, oscillations will not be started. Thus a constant torque T_x will produce a precessional angular velocity $\dot{\theta}$ in direct proportion according to $\dot{\theta} = KT_x$. This will be true only so long as θ is small, however, and since a constant torque produces a constant *velocity*, θ must eventually become large. This will ultimately (when θ reaches 90°) lead to what is called "gimbal lock." The gyroscopic precession actually depends on the component of the torque vector that is perpendicular to the spin angular-momentum vector. For small angles, this component is directly proportional to torque. For large angles, it is

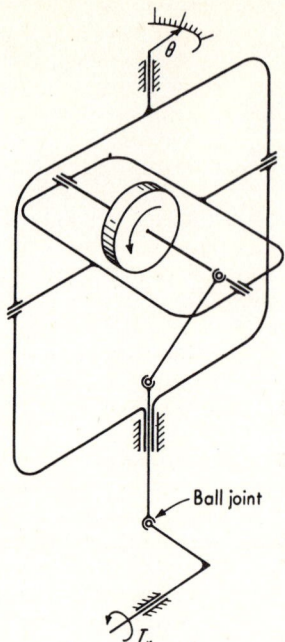

Fig. 5.22. *Solution of gimbal-lock problem.*

proportional to the product of torque and cos θ; this becomes smaller as θ increases and disappears completely at $\theta = 90°$. Thus the precession θ produced by a constant torque T_x gets smaller and smaller as θ approaches 90°. At 90° a torque T_x produces no precession at all; rather, the inner and outer gimbal both rotate together about the x axis.

To prevent gimbal lock and thus achieve a useful torque-sensing instrument, a simple mechanical solution is available. The requirement that torque vector and spin angular-momentum vector always be perpendicular is met by the configuration of Fig. 5.22. The equation $\theta = KT_x$ now holds for *any* angle θ, and one can measure torque by measuring θ or measure the time integral of torque by means of a simple revolution counter attached to the θ shaft.

Problems

5.1 Compute the value of g at your local latitude and altitude. What is the percent deviation from the standard value 980.665?

5.2 What is the percentage change in weight (compared with sea level) of a mass located above the equator at an altitude of 500 miles?

5.3 An object with a volume of 10 in.³ is weighed on an equal-arm balance. The standard mass required for balance is 1 lb$_m$ and has a volume of 3 in.³ What is the value of the correction necessary for air buoyancy?

5.4 A brass balance beam has a length of exactly 1 m at 60°F, and the pivot is perfectly centered to give an equal-arm balance. If one end of the beam comes to 80°F and there is a uniform temperature gradient to the other end at 60°F, what inequality in arm length results?

5.5 What general form of dynamic response would you expect from the systems of Fig. 5.1-1? Why?

5.6 Prove that the reading is independent of location of F_i if $a/b = c/d$ in the platform scale of Fig. 5.1-1.

5.7 If, in Fig. 5.1-2, $M = 1$ lb_m and $A = 20g$, what is the net force on M? If a friction force which is unknown but less than 1 lb_f may be present, what error may be expected in F_i?

5.8 Carry out a simplified, linear dynamic analysis of the system of Fig. 5.1-3.

5.9 Carry out a linear dynamic analysis and an accuracy/stability tradeoff study of the pneumatic load cell of Fig. 5.1-4. Use $(x/F_{net})(D) = K_d/(D^2/\omega_n{}^2 + 2\zeta D/\omega_n + 1)$ and $(p_o/x)(D) = K_n/(\tau D + 1)$.

5.10 In Fig. 5.3 if F_i is eccentric and also angularly misaligned a torque is produced. Does this affect the bridge output? Explain.

5.11 A load cell deflects 0.005 in. under its full-scale load of 1,000 lb_f. It is used to measure force on a machine-tool slide which weighs 500 lb_f. Estimate the highest frequency of force that may be accurately measured.

5.12 Derive Eqs. (5.14) to (5.19).

5.13 From the block diagram of Fig. 5.13, obtain the transfer function $(e_o/T_i)(D)$, assuming τ_2 is small enough to neglect. Investigate the effect of system parameters on dynamic accuracy and stability.

5.14 In Fig. 5.13 prove that $\theta = 0$ for any constant value of T_i.

5.15 Prove that the arrangement of Fig. 5.15 is insensitive to axial or bending stresses.

5.16 Prove that for equivalent strain/torque sensitivity a square shaft is stiffer in bending than a round one.

5.17 A torque sensor with torsional stiffness of 1,000 in.-lb_f/rad is coupled between an electric motor and a hydraulic pump. The moments of inertia of the motor and pump are each 0.01398 in.-lb_f-sec². If the motor has a small oscillatory torque component at 60 cps, will this measuring system be satisfactory? Explain what torsional stiffness is needed if the response at 60 cps is to be no more than 105 percent of the static response. The amount of damping is unknown.

5.18 Suppose the tachometer generator in the system of Fig. 5.19 puts out 6 volts/1,000 rpm and the load cell produces 0.05 mv/(lb_f-volt excitation). What will be the power calibration factor for e_o in horsepower per millivolt if the arm length is 1 ft?

5.19 In the system of Fig. P5.1:

  a. For $F = 0$ and heat off, $R_1 = R_2 = R_3 = R_4$ and $e_o = 0$.

  b. The gage factor of the gages is $+2.0$, and the temperature coefficient of resistance of the gages is positive.

  c. The modulus of elasticity of the beam decreases with increased temperature.

  d. The thermal-expansion coefficient of the gage is greater than that of the beam.

  e. Assume the gage temperature is the same as that of the beam *immediately* beneath it.

At time $t = 0$, an upward force F is applied and maintained constant thereafter. After oscillations have died out, at a later time t_1 the radiant-heat source is

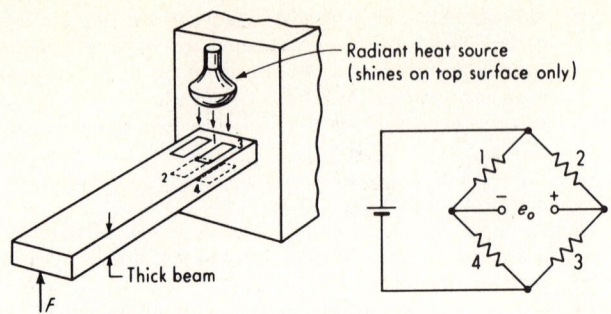

Fig. P5.1

turned on and left on thereafter. Sketch the general form of e_o versus t, justifying clearly by detailed reasoning the shape you give the curve.

5.20 It is necessary to design a strain-gage thrust transducer for small experimental rocket engines which are roughly in the shape of a cylinder 6 in. in diameter by 12 in. long. The following information is given:

 a. Weight of motor and mounting bracket, 20 lb$_f$.

 b. Maximum steady thrust, 50 lb$_f$.

 c. Oscillating component of thrust, ± 10 lb$_f$ maximum.

 d. Oscillating components of thrust up to 100 cps must be measured with a flat amplitude ratio within ± 5 percent.

 e. A recorder with a sensitivity of 0.1 volt/in., frequency response flat to 120 cps, and input resistance of 10,000 ohms is available.

 f. Thrust changes of 0.5 lb$_f$ must be clearly detected.

 g. Gages with a resistance of 120 ohms and a gage factor of 2.1 are available. They are 0.5 × 1.0 in. in size.

 h. An amplifier (to be placed between transducer and recorder) is available with a gain up to 1,000.

Design the transducer so as to require a minimum of amplifier gain. If damping is employed, calculate the required damping coefficient B but do not design the damper. Use the cantilever-beam arrangement of Fig. P5.2.

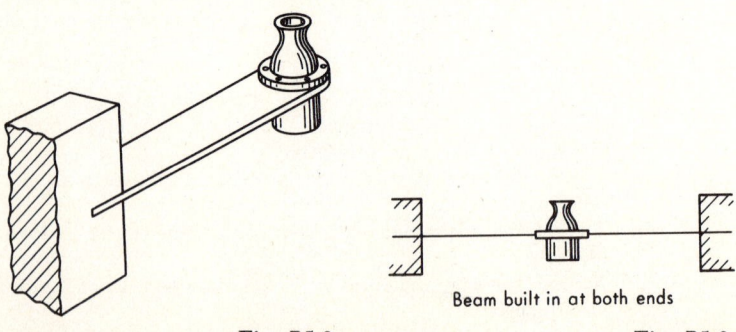

Fig. P5.2   **Fig. P5.3**

5.21 Repeat Prob. 5.20 using the configuration of Fig. P5.3.

5.22   Repeat Prob. 5.20 using the configuration of Fig. P5.4.

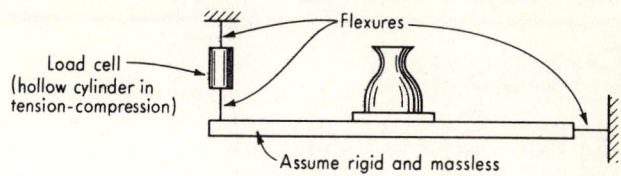

Fig. P5.4

Bibliography

1.   A. Krsek and M. Tiefermann: Optical Torquemeter for High Rotational Speeds, *NASA Tech. Note*, D-1437, October, 1962.
2.   H. E. Lockery: Applying the Strain-gage Torque Transducer, *ISA J.*, p. 65, March, 1962.
3.   Torque-gauge Without Sliprings, *Electromech. Design*, p. 6, November, 1959.
4.   J. Guthrie: Lever-shaft Torque Measurement, *Instr. Control Systems*, p. 116, August, 1964.
5.   D. Ettleman and M. Hoberman: Torquemeters, *Machine Design*, p. 134, Feb. 28, 1963.
6.   O. Dahle: Heavy Industry Gets a New Load Cell, *ISA J.*, p. 32, August, 1959.
7.   F. M. Ryan: Automatic Weighing for Solids, *Control Eng.*, p. 103, September, 1962.
8.   D. W. Kennedy: Weighing Scales Couple to Computer, *Control Eng.*, p. 83, July, 1962.
9.   F. A. Ludewig: Digital Force Transducer, *Control Eng.*, p. 107, June, 1961.
10.   K. Harris: Servo-balanced Supply Tank Measures Nozzle Thrust, *Control Eng.*, p. 115, February, 1960.
11.   Planetary Gearing in Torquemeter Does Away with Sliding Contacts, *Machine Design*, p. 135, Sept. 1, 1960.
12.   E. T. Gay: Precision Weighing with Platform Scales, *Tool Engr.*, June, 1959.
13.   S. Edwards: Dynamic Measurement of Vehicle Front Wheel Loads Using a Special Purpose Transducer, *Gen. Motors Eng. J.*, p. 15, October–November–December, 1964.
14.   S. Hejzlar: Backweighing Error in Scales, *Instr. Control System*, p. 95, February, 1965.
15.   Apparatus Measures Very Small Thrusts, *NASA Brief* 64-10284, 1964.
16.   J. A. Bierlein: Methods of Measuring Thrust, *ARS J.*, p. 128, May–June, 1953.
17.   L. E. Stone: Criteria for Design and Use of an Internal Strain Gage Floating-frame Balance, *ISA Trans.*, p. 152, April, 1965.
18.   R. L. Small: Belt Scales, *ISA J.*, p. 65, May, 1965.
19.   V. C. Plane: Total Impulse Measuring System for Solid-propellant Rocket Engine, Rocketdyne Div., Canoga Park, Calif., *Rept.* R-5638, 1964.
20.   R. W. Postma: Pulse Thrust Measuring Transducer (with Accelerometer Dynamic Compensation), Rocketdyne Div., Canoga Park, Calif., *Rept.* R-6044, 1965.
21.   G. F. Malikov: Computation of Elastic Tensometric Elements (Load Cells), *NASA* TT F-513, 1968.
22.   K. W. Stark: Design and Development of a Micropound Extended Range Thrust Stand, *NASA* TN D-7029, 1971.

Chapter 6
Pressure and Sound Measurement

6.1

Standards and Calibration Pressure is not a fundamental quantity but rather is derived from force and area, which in turn are derived from mass, length, and time, the latter three being fundamental quantities whose standards have been discussed earlier. Pressure "standards" in the form of very accurate instruments are available, however, for calibration of less accurate instruments. However, these "standards" depend ultimately on the fundamental standards for their accuracy. The basic standards[1,2,3] for pressures ranging from medium vacuum (about 10^{-1} mm Hg) up to several hundred thousand pounds per square inch are in the form of precision mercury columns (manometers) and dead-weight piston gages.   For pressures in the range 10^{-1} to 10^{-3} mm Hg the McLeod vacuum gage is considered the standard.   For pressures below 10^{-3} mm Hg a pressure-dividing technique using flow through a succession of accurate orifices to relate the low downstream pressure to a higher upstream pressure (which is accurately measured with a McLeod gage) is presently in use.[4]

This technique can be further improved by substituting a Schulz hot-cathode or radioactive ionization vacuum gage for the McLeod gage. Each of these must be calibrated against a McLeod gage at one point (about 9×10^{-2} mm Hg) but their known linearity is then used to extend their accurate range to much lower pressures.[5]   This procedure is fairly well accepted to about 10^{-7} mm Hg at present, but the great interest in the high vacuums (10^{-12} mm Hg and less) of space environments will undoubtedly lead to continuing improvements.

The inaccuracies of the above-mentioned pressure standards range from about ± 4 percent at 10^{-7} mm Hg to ± 1 percent at 10^{-3} mm Hg to ± 0.1 percent at 10^{-1} mm Hg to a peak of ± 0.001 percent at 1 atm and down again to ± 0.1 percent at 200,000 psi.   Since the above-mentioned pressure standards are also pressure-measuring instruments (of the highest quality and used under carefully controlled conditions), their operating principles and characteristics will not be discussed here since they are adequately covered later.

6.2

Basic Methods of Pressure Measurement   Since pressure can usually be easily transduced to force by allowing it to act on a known area, the basic methods of measuring force and pressure are essentially the same, except for the high-vacuum region where a variety of special methods not directly related to force measurement are necessary.   These special

[1] D. P. Johnson and D. H. Newhall, The Piston Gage as a Precise Pressure-measuring Instrument, *Instr. Control Systems*, p. 120, April, 1962.

[2] Errors in Mercury Barometers and Manometers, *Instr. Control Systems*, p. 121, March, 1962.

[3] 2″ Range Hg Manometer, *Instr. Control Systems*, p. 152, September, 1962.

[4] J. R. Roehrig and J. C. Simons, Calibrating Vacuum Gages to 10^{-9} Torr, *Instr. Control Systems*, p. 107, April, 1963.

[5] J. C. Semons, On Uncertainties in Calibration of Vacuum Gages and the Problem of Traceability, "Transactions of 10th National Vacuum Symposium," p. 246, The Macmillan Company, New York, 1963.

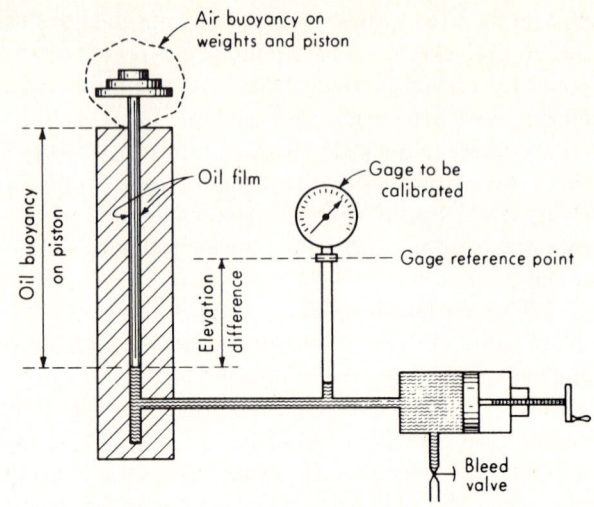

Fig. 6.1.   *Dead-weight-gage calibrator.*

methods will be described in the section on vacuum measurement.   Other than the special vacuum techniques, most pressure measurement is based on comparison with known dead weights acting on known areas or on the deflection of elastic elements subjected to the unknown pressure.   The dead-weight methods are exemplified by manometers and piston gages while the elastic deflection devices take many different forms.

6.3

Dead-weight Gages and Manometers   Figure 6.1 shows the basic elements of a dead-weight or piston gage.   Such devices are used mainly as standards for the calibration of less accurate gages or transducers.   The gage to be calibrated is connected to a chamber filled with fluid whose pressure can be adjusted by means of some type of pump and bleed valve.   The chamber also connects with a vertical piston-cylinder to which various standard weights may be applied.   The pressure is slowly built up until the piston and weights are seen to "float," at which point the fluid "gage" pressure (pressure above atmosphere) must equal the dead weight supported by the piston, divided by the piston area.

For highly accurate results, a number of refinements and corrections are necessary.   The frictional force between the cylinder and piston must be reduced to a minimum and/or corrected for.   This is generally accomplished by rotating either the piston or the cylinder.   If there is no axial relative motion, this rotation should reduce the axial effects of dry friction to zero.   There must, however, be a small clearance between the piston and the cylinder and thus an axial flow of fluid from the high-pressure end to the low.   This flow produces a viscous shear force tending to support part of the dead weight.   This effect can be estimated from theoretical calculations;[1] however, it varies somewhat with pressure since the piston and cylinder deform under pressure, thereby changing the clearance.   The clearance between the piston and cylinder also raises the question of which area is to be used in computing pressure.   The effective area is generally taken as the average of the piston and cylinder areas.   Further corrections are needed for temperature effects on areas of piston and cylinder, air and pressure-medium buoyancy effects, local gravity conditions, and height differences between the lower end of the piston and the reference point for the gage being calibrated.   Special designs and techniques allow use of dead-weight gages for pressures up to several hundred thousand pounds per square inch.

Since the piston assembly itself has weight, conventional dead-weight gages are not capable of measuring pressures lower than the piston weight/area ratio ("tare" pressure).   This difficulty is overcome by the tilting-piston gage[2] in which the cylinder and piston can be tilted from vertical through an accurately measured angle, thus giving a continuously adjustable pressure from 0 psig up to the tare pressure.   The described gage uses nitrogen or other inert gas as the pressure medium and covers the range 0 to 600 psig, having two interchangeable piston-cylinders and 14 weights.   The accuracy is 0.01 percent of reading in the range 0.3 to 15 psig and 0.015 percent of reading in the range 2 to 600 psig.   The tilting feature is used for the ranges 0 to 0.3 and 0 to 2.0 psig; higher pressures are obtained in increments by the addition of discrete weights.

Dead-weight gages may be used for absolute- rather than gage-pressure measurement by placing them inside an evacuated enclosure at (ideally) 0-psia pressure.   Since the degree of vacuum (absolute pressure) inside the enclosure must be known, this really requires an additional independent measurement of absolute pressure.

The manometer, in its various forms, is closely related to the piston gage since both are based on the comparison of the unknown pressure force with the gravity force on a known mass.   The manometer differs, however, in that it is self-balancing, is a deflection rather than a null instrument, and has continuous rather than stepwise output.   The accuracies of dead-weight gages and manometers of similar ranges are quite comparable; however, manometers become unwieldy at high pres-

[1] R. J. Sweeney, "Measurement Techniques in Mechanical Engineering," p. 104, John Wiley & Sons, Inc., New York, 1953.
[2] Ruska Instrument Corp., Houston, Texas.

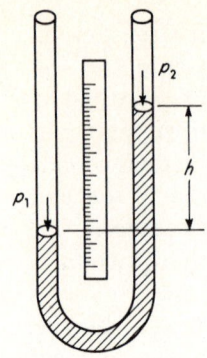

Fig. 6.2. *U-tube manometer.*

sures because of the long liquid columns involved.   The U-tube manometer of Fig. 6.2 is usually considered the basic form and has the following
relation between input and output for static conditions:

$$h = \frac{p_1 - p_2}{\rho g} \qquad (6.1)$$

where $g \triangleq$ local gravity
$\rho \triangleq$ mass density of manometer fluid

If p_2 is atmospheric pressure, then h is a direct measure of p_1 as a gage
pressure.   Note that the cross-sectional area of the tubing (even if not
uniform) has no effect.   At a given location (given value of g) the sensitivity depends only on the density of the manometer fluid.   Water and
mercury are the most commonly used fluids.   To realize the high accuracy possible with manometers, a number of corrections must often be
applied.   When visual reading of the height h is employed, the engraved-
scale's temperature expansion must be considered.   The variation of ρ
with temperature for the manometer fluid used must be corrected and the
local value of g determined.   Additional sources of error are found in the
nonverticality of the tubes and the difficulty in reading h due to the
meniscus formed by capillarity.   Considerable care must be exercised in
order to keep inaccuracies as small as 0.005 in. Hg for the overall
measurement.[1]

A number of practically useful variations on the basic manometer
principle are shown in Fig. 6.3.   The *cistern* or *well-type manometer* is
widely used because of its convenience in requiring reading of only a single
leg.   The well area is made very large compared with the tube; thus the
zero level moves very little when pressure is applied.   Even this small
error is compensated by suitably distorting the length scale.   However,

---

[1] A. J. Eberlein, Laboratory Pressure Measurement Requirements for Evaluating the Air Data Computer, *Aeron. Eng. Rev.*, p. 53, April, 1958.

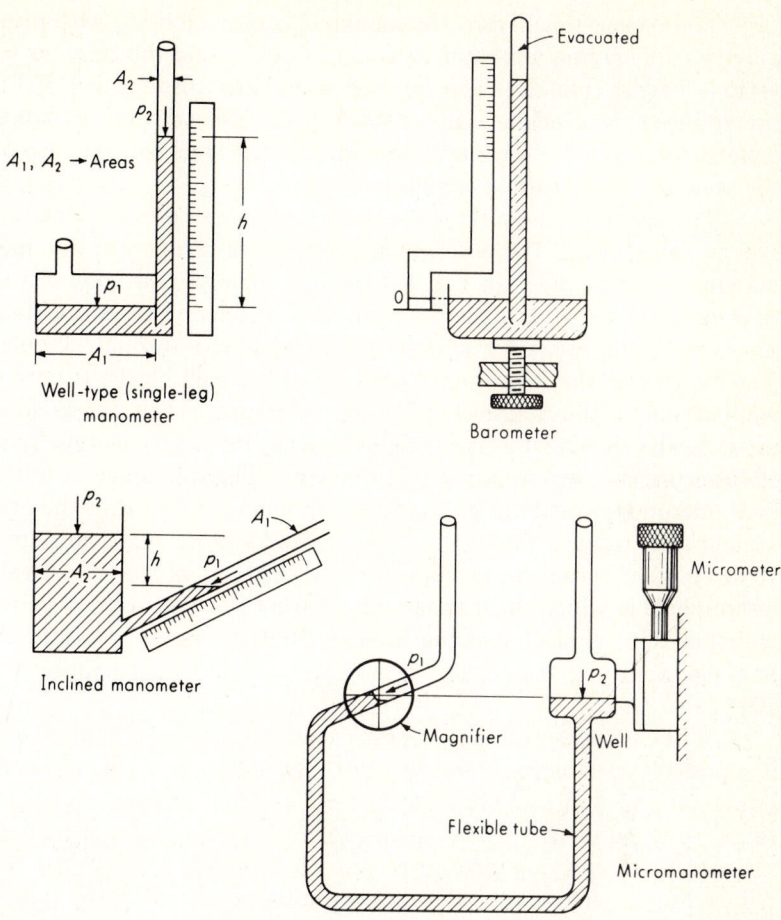

Fig. 6.3.   *Various forms of manometers.*

such an arrangement, unlike a U tube, is sensitive to nonuniformity of the tube cross-sectional area and is thus considered somewhat less accurate.

Since manometers inherently measure the pressure *difference* between the two ends of the liquid column, if one end is at zero absolute pressure then h is an indication of absolute pressure. This is the principle of the *barometer* of Fig. 6.3. Although a "single-leg" instrument, high accuracy is achieved by setting the zero level of the well at the zero level of the scale before each reading is taken. The pressure in the evacuated portion of the barometer is not really absolute zero but rather the vapor pressure of the filling fluid, mercury, at ambient temperature. This is about 10^{-4} psia at 70°F and is usually negligible as a correction.

To increase sensitivity, the manometer may be tilted with respect to gravity, thus giving a greater motion of liquid along the tube for a given vertical height change.   The *inclined manometer* (draft gage) of Fig. 6.3 exemplifies this principle.   Since this is a single-leg device, the calibrated scale is corrected for the slight changes in well level so that rezeroing of the scale for each reading is not required.

The accurate measurement of extremely small pressure differences is accomplished with the *micromanometer*, a variation on the inclined-manometer principle.   In Fig. 6.3 the instrument is initially adjusted so that when $p_1 = p_2$ the meniscus in the inclined tube is located at a reference point given by a fixed hairline viewed through a magnifier. The reading of the micrometer used to adjust well height is now noted. Application of the unknown pressure difference causes the meniscus to move off the hairline, but it can be restored to its initial position by raising or lowering the well with the micrometer.   The difference in initial and final micrometer readings gives the height change h and thus the pressure. Instruments using water as the working fluid and having a range of either 10 or 20 in. of water can be read to about 0.001 in. of water.[1]   In another instrument[2] in which the inclined tube (rather than the well) is moved and which uses butyl alcohol as the working fluid, the range is 2 in. of alcohol, and readability is 0.0002 in.   This corresponds to a resolution of 6 $\times$ 10^{-6} psi.

While manometers are generally read visually by a human operator, it is possible to construct servosystems that will "track" the motion of the liquid column and provide a mechanical and/or electrical signal proportional to the pressure.   Such arrangements allow the use of manometers for measuring varying pressures, are much faster than visual reading, reduce human errors, provide signals usable in control or computing systems, and may provide automatic temperature corrections.   The sensing of the liquid level in the tube may be accomplished with a light source and photocell or by a differential transformer.   The differential transformer has the advantage of allowing use of (nonmagnetic) stainless-steel tubes (in place of glass) for high-pressure work.

Such a "servomanometer"[3] is shown schematically in Fig. 6.4.   The core of the differential transformer is fastened to a small float which rests on the surface of the manometer liquid.   The coils are concentric with the tube and are positioned vertically by a perforated steel tape, $\frac{1}{2}$ in. wide and 0.005 in. thick, running over sprockets.   Whenever the core is not at the null position of the coils, a voltage is applied to the amplifier

[1] Meriam Instrument Co., Cleveland, Ohio.
[2] Flow Corp., Arlington, Mass.
[3] Exactel Instrument Co., Mountain View, Calif.

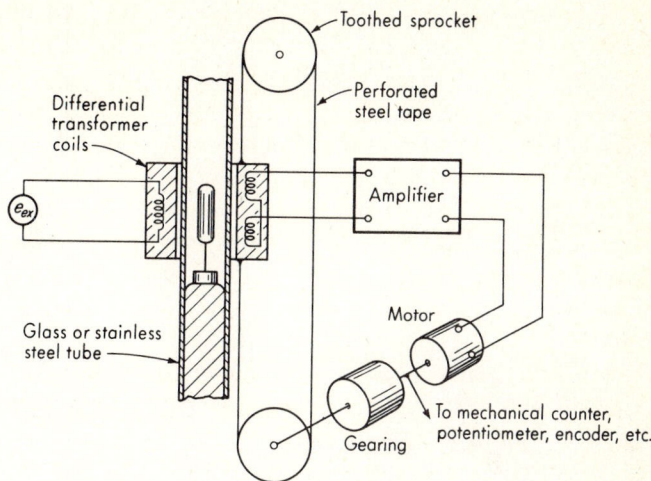

Fig. 6.4. *Servomanometer.*

(and thus to the motor), causing the coils to drive toward null. Thus the coils tend to track the core and thereby the liquid-column position. Since coil motion is kinematically related to motor rotation, this rotation is proportional to pressure and may be read out by a variety of means, depending on the application. Both U-tube and single-leg instruments are available. A U tube requires two servo followers, one for each leg. The subtraction necessary to calculate pressure is performed mechanically by applying the two motor-shaft rotations to a gear differential. Such servomanometers have a resolution of 0.0005 in. and an overall accuracy of a few thousandths of an inch. Following speed is of the order of 100 in./min or more, and automatic temperature correction is provided.

Manometer dynamics. While manometers are used mainly for static measurements, their dynamic response is sometimes of interest. The general problem of the oscillations of liquid columns is an interesting (and rather difficult) question in fluid mechanics and has received considerable attention in the literature.[1,2] Here we take a considerably simplified view of the problem which, however, gives results of practical interest. In the U-tube configuration of Fig. 6.5a the unknown pressures p_1 and p_2 are exerted by a gas whose inertia and viscosity may be considered

[1] J. F. Ury, Viscous Damping in Oscillating Liquid Columns, Its Magnitude and Limits, *Intern. J. Mech. Sci.*, vol. 4, p. 349, 1962.

[2] P. D. Richardson, Comments on Viscous Damping in Oscillating Liquid Columns, *Intern. J. Mech. Sci.*, vol. 5, p. 415, 1963.

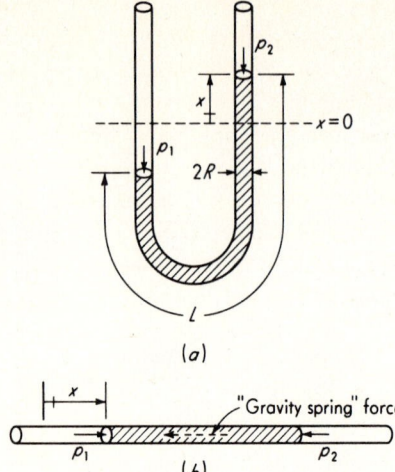

Fig. 6.5. *Manometer model.*

negligible compared with the manometer liquid. If the pressures vary with time, the reading of the manometer varies with time; we are interested in the fidelity with which the manometer reading follows the pressure variation. The motion of the manometer liquid in the tube is caused by the action of various forces. If we consider the manometer liquid in its entirety as a free body and search for forces acting on it, the following forces come to mind:

1. The gravity force (weight) distributed uniformly over the whole body of fluid
2. A drag force due to motion of the fluid within the tube and related to the wall shearing stress
3. The forces on the two ends of the free body due to the pressures p_1 and p_2
4. Distributed normal pressure of the tube on the fluid
5. Surface-tension effects at the two ends of the body of fluid

A detailed analysis of all these effects would lead to rather complex and unwieldy mathematics. Fortunately, useful results may be obtained by a simplified analysis. The initial step is the assumption that the system shown in Fig. 6.5b is dynamically equivalent to that of Fig. 6.5a. The "gravity spring" force of Fig. 6.5b is explained as follows: In Fig. 6.5a, whenever $x \neq 0$, there is an unbalanced gravity force acting on the liquid column, tending to restore the level to $x = 0$. The magnitude of this force is $-2\pi R^2 x \gamma$, where $\gamma \triangleq$ manometer-fluid specific weight, in $\mathrm{lb}_f/\mathrm{in.}^3$. We see that this force is proportional to the displacement x and always opposes it; thus it has all the characteristics of a spring force. When the liquid column is "straightened out" in Fig. 6.5*b*, we must include this "gravity spring" force in our equivalent system if we are to preserve the analogy of the two configurations. In comparing Fig. 6.5*b* with Fig. 6.5*a*, we note that any effects of flow curvature in the 180° bend are lost, but these will probably be small if the diameter of the bend is large compared with the inside diameter of the manometer tube and if the total length L is large compared with the bend length. We shall also neglect any sur-face-tension effects at the ends of the column. This is usually a good assumption if the column is long relative to its diameter.

In addition to the gravity-spring force and the pressure forces due to p_1 and p_2, the liquid column is also subjected to a drag force at the interface between the liquid and the wall of the tube. This drag force is equal to the wall shearing stress times the surface area of the liquid column. The motion of the liquid in the tube may be thought of as an unsteady pipe flow. We shall assume that at any instant of time the wall shearing stress can be computed from the instantaneous velocity of the liquid, *using the formulas commonly used for steady pipe flows.*

The flow of liquid in the tube may occur in the laminar, transition, or turbulent regimes. Let us first assume laminar flow prevails. The pressure drop Δp due to pipe friction for both laminar and turbulent flow is given by

$$\Delta p = f \frac{\gamma L V_{av}^2}{2gd} = f \left(\frac{L}{d}\right) \frac{\rho V_{av}^2}{2g_0} \qquad (6.2)$$

where
$g_0 \triangleq$ mass unit conversion factor, $lb_m/slug$
$\rho \triangleq$ fluid density, lb_m/ft^3
$f \triangleq$ friction factor
$\gamma \triangleq$ fluid specific weight, local, lb_f/ft^3
$V_{av} \triangleq$ average velocity
$g \triangleq$ local gravity acceleration, ft/sec^2
$d \triangleq$ diameter of pipe
$L \triangleq$ pipe length

The wall shearing stress τ_0 is given by

$$\tau_0 = \Delta p \frac{d}{4l} \qquad (6.3)$$

Thus

$$\tau_0 = f \frac{\gamma V_{av}^2}{8g} = f \frac{\rho V_{av}^2}{8g_0} \qquad (6.4)$$

For laminar flow the friction factor is given by

$$f = \frac{64\mu g}{d\gamma V_{av}} = \frac{64}{d V_{av}\rho/\mu g_0} \qquad (6.5)$$

so that
$$\tau_0 = \frac{4\mu V_{av}}{R} \qquad (6.6)$$

where $R \triangleq d/2$.

This result can also be obtained directly from the laminar velocity distribution

$$V = V_c \left[1 - \left(\frac{r}{R} \right)^2 \right] \quad (6.7)$$

where $V \triangleq$ velocity at radius r

$V_c \triangleq$ center-line velocity

The velocity gradient is

$$\frac{dV}{dr} = - \frac{V_c}{R^2} (2r) \quad (6.8)$$

which becomes at the wall

$$\frac{dV}{dr} \bigg|_{r=R} = - \frac{2V_c}{R} \quad (6.9)$$

Shearing stress is given by

$$\tau = \mu \frac{dV}{dr} \quad (6.10)$$

and so the magnitude of the wall shearing stress τ_0 is

$$\tau_0 = \frac{4\mu V_{\text{av}}}{R}$$

since $V_c = 2V_{\text{av}}$ for laminar flow in circular pipes.

We are now in a position to apply Newton's law to the system of Fig. 6.5b. The average flow velocity V_{av} corresponds to $\dot{x}$, the first derivative of x with respect to time. Considering the entire body of liquid as a free body and taking the effective mass of the moving liquid as four-thirds of its actual mass, based on the kinetic energy of steady laminar flow, we can write for motion in the x direction

$$\pi R^2(p_1 - p_2) - 2\pi R^2 \gamma x - 2\pi R L \frac{4\mu \dot{x}}{R} = \frac{4}{3} \frac{\pi R^2 L \gamma}{g} \ddot{x} \quad (6.11)$$

This reduces to

$$\frac{2\ddot{x}}{3g/L} + \frac{4\mu L}{R^2 \gamma} \dot{x} + x = \frac{1}{2\gamma} p \quad (6.12)$$

where we have defined $p \triangleq p_1 - p_2$. In operator form, this becomes

$$\left(\frac{2D^2}{3g/L} + \frac{4\mu L}{R^2 \gamma} D + 1 \right) x = \frac{1}{2\gamma} p \quad (6.13)$$

The operational transfer function relating output x to input p is

$$\frac{x}{p}(D) = \frac{1/2\gamma}{\dfrac{2D^2}{3g/L} + \dfrac{4\mu L}{R^2 \gamma} D + 1} \quad (6.14)$$

which is of the form $\dfrac{x}{p}(D) = \dfrac{K}{D^2/\omega_n^2 + 2\zeta D/\omega_n + 1} \quad (6.15)$

where
$$K \triangleq \frac{1}{2\gamma} \qquad \text{in./psi} \qquad (6.16)$$

$$\omega_n \triangleq \sqrt{\frac{3g}{2L}} \qquad \text{rad/sec} \qquad (6.17)$$

$$\zeta \triangleq 2.45\mu \, \frac{\sqrt{gL}}{R^2\gamma} \qquad (6.18)$$

We note from the above that the manometer is a second-order instrument. The numerical values of the parameters are usually such that $\zeta < 1.0$; that is, the instrument is underdamped.

Since laminar flow was assumed in carrying out the above analysis, we should try to estimate a typical Reynolds number to see under what conditions laminar flow occurs. As a numerical example, we take a mercury manometer with

$$L = 26.5 \text{ in.}$$
$$R = 0.13 \text{ in.}$$
$$\mu = 2.18 \times 10^{-7} \text{ lb}_f\text{-sec/in.}^2$$
$$\gamma = 0.491 \text{ lb}_f/\text{in.}^3$$

Suppose that we wish to check our theoretical results by measuring ζ and ω_n experimentally for a step-function input. Computing ζ and ω_n from Eqs. (6.17) and (6.18) gives $\zeta = 0.007$ and $\omega_n = 4.7$ rad/sec. Since $\zeta = 0.007$ represents a *very lightly* damped system, we can estimate the maximum flow velocity by assuming no damping at all. A second-order system with no damping executes pure sinusoidal oscillations when subjected to a step-function input. Its motion would thus be given by

$$x = X \sin \omega_n t \qquad (6.19)$$

where X is the size of the step function. The velocity $\dot{x}$, which is the same as the average flow velocity, would be

$$\dot{x} = \omega_n X \cos \omega_n t \qquad (6.20)$$

and its maximum value would thus be $\omega_n X$. The Reynolds number for steady pipe flow is given by

$$N_R = \frac{2\gamma R V_{av}}{g\mu} \qquad (6.21)$$

and the critical value for transition from laminar to turbulent flow is 2,100. Since $V_{av} = \omega_n X$, it should be clear that there is a maximum-size step function that can be used without exceeding $N_R = 2,100$. This limiting value X_m is given by

$$2{,}100 = \frac{2\gamma R \omega_n X_m}{g\mu} \qquad (6.22)$$

which in this example gives

$$X_m = \frac{(2,100)(386)(2.18 \times 10^{-7})}{(2)(0.491)(0.13)(4.7)} = 0.30 \text{ in.} \qquad (6.23)$$

Thus, to ensure laminar flow at all times during the oscillation the step input can be no larger than 0.30 in. Suppose we wish to measure ζ and ω_n by simple visual methods—ζ from the size of the first overshoot and ω_n by counting and timing cycles. This requires much larger step inputs for reasonable accuracy. Therefore we must investigate the effect of the presence of turbulent flow on our analysis.

Suppose that we decide that a step input X_m of 5 in. Hg will be sufficiently large to allow accurate measurements of ζ and ω_n. The maximum Reynolds number would then be

$$N_R = \frac{5}{0.30}(2,100) = 35,000 \qquad (6.24)$$

For steady turbulent flow in smooth pipes with $3,000 < N_R < 100,000$ the Blasius equation for friction factor is

$$f = \frac{0.316}{(N_R)^{0.25}} \qquad (6.25)$$

The turbulent wall shearing stress is then given by Eq. (6.4) as

$$\tau_0 = \frac{0.0378 \gamma^{0.75} \mu^{0.25} V_{av}^{1.75}}{g^{0.75} R^{0.25}} \qquad (6.26)$$

Using the numerical values of this particular example, we get

$$\tau_0 = 9.18 \times 10^{-6} V_{av}^{1.75} \qquad \text{lb}_f/\text{in.}^2 \qquad (6.27)$$

For laminar flow the comparable expression is

$$\tau_0 = 6.71 \times 10^{-6} V_{av} \qquad \text{lb}_f/\text{in.}^2 \qquad (6.28)$$

The most significant difference between Eqs. (6.27) and (6.28) is that (6.27) represents a nonlinear relation between shear stress and velocity whereas (6.28) is linear. This means that when the force due to wall shearing stress is substituted into Newton's law the result is a nonlinear differential equation because of the term $(\dot{x})^{1.75}$. This nonlinear equation cannot be solved except by use of analog computers or approximate numerical methods. Thus, the presence of turbulent flow leads to mathematical difficulties.

In working with oscillations of systems with nonlinear damping terms similar to the $(\dot{x})^{1.75}$ of this problem, engineers have developed an approximate method of analysis which is quite useful. This approach is based on the observation that, while the linear damping term $\dot{x}$ and the nonlinear term, such as $(\dot{x})^{1.75}$, are quite different mathematically, the general *form* of the oscillation in the two systems is not radically different in experimental tests. If the linear system is excited by a sinusoidal exciting force it will respond with a sinusoidal motion, whereas the nonlinear systems's motion will not be purely sinusoidal. However, observation shows that the deviation from pure sinusoidal motion is usually quite small. Using these facts as a basis, we might then reason as follows: If a system with nonlinear damping is executing steady oscillations of fixed amplitude, during each cycle the damping force will dissipate a certain amount of energy. If we know from experience that the wave form of the nonlinear oscillation is nearly sinusoidal, we can compute approximately the energy dissipation per cycle. This is done as follows: Suppose there exists a steady oscillation of amplitude Y and frequency ω. If we assume the wave form to be sinusoidal, we can write

$$y = Y \sin \omega t \qquad (6.29)$$

and

$$\dot{y} = Y\omega \cos \omega t \qquad (6.30)$$

Now, in general, the instantaneous power is the product of instantaneous force and instantaneous velocity. The power dissipation due to damping is thus the product of velocity and damping force. If the damping force is a known function of velocity $f(\dot{y})$, we can write

$$\text{Instantaneous power dissipation} = \dot{y}f(\dot{y}) \qquad (6.31)$$

and the energy dissipated per cycle will be given by

$$\int_{\text{one cycle}} \dot{y}f(\dot{y}) \, dt \qquad (6.32)$$

For a linear damping the function $f(\dot{y})$ is just $B\dot{y}$, where B is a constant. The energy dissipation per cycle is thus

$$\int_0^{2\pi/\omega} (Y\omega \cos \omega t)(BY\omega \cos \omega t) \, dt = \pi B\omega Y^2 \qquad (6.33)$$

For the nonlinear damping due to turbulent flow the function $f(\dot{y})$ is $C(\dot{y})^{1.75}$, where C is a constant. The energy dissipation per cycle for this nonlinear damping is

$$\int_0^{2\pi/\omega} (Y\omega \cos \omega t)C(Y\omega \cos \omega t)^{1.75} \, dt \qquad (6.34)$$

This is equal to

$$C(Y\omega)^{2.75} \int_0^{2\pi/\omega} (\cos \omega t)^{2.75} \, dt \qquad (6.35)$$

In evaluating the integral in (6.35), we must use physical reasoning, because when $\cos \omega t$ becomes *negative* the quantity $(\cos \omega t)^{2.75}$ is not defined in terms of real numbers. Physical reasoning, however, tells us that the physical processes occurring during the first quarter cycle $(0 \le t \le \pi/2\omega)$ give exactly the same energy dissipation as those occur-

ring during the other three quarters of the cycle. Thus, we can integrate over only the first quarter and multiply by 4 to get the total energy dissipation. During the first quarter cycle cos ωt is always positive, and so no mathematical difficulties arise. This amounts to saying that, to agree with the known physical facts, integral (6.34) should really be written as

$$\int_0^{2\pi/\omega} |Y\omega \cos \omega t| \, [C(Y\omega)^{1.75}|(\cos \omega t)|^{1.75}] \, dt \qquad (6.36)$$

with the absolute-value signs as shown.

Evaluation of integral (6.36) for one quarter cycle is most easily done by plotting and using a planimeter to get the area under the curve. By defining $\theta \triangleq \omega t$, integral (6.36) can be written as

$$CY^{2.75}\omega^{1.75} \int_0^{2\pi} (\cos \theta)^{2.75} \, d\theta \qquad (6.37)$$

Plotting and planimetering give the energy dissipation per cycle for nonlinear damping as

$$2.50C\omega^{1.75}Y^{2.75} \qquad (6.38)$$

Having obtained the above results, we now define the *equivalent linear damping* as that linear damping which would dissipate exactly the same energy per cycle as the nonlinear damping at a given frequency and amplitude. Thus we set (6.33) equal to (6.38) and get

$$\pi B_e \omega Y^2 = 2.50C\omega^{1.75}Y^{2.75} \qquad (6.39)$$
$$B_e \triangleq \text{equivalent linear damping}$$
$$B_e = 0.796C(\omega Y)^{0.75} \qquad (6.40)$$

Since C is the constant that multiplies $(\dot{y})^{1.75}$, in the manometer problem we have

$$\text{Damping force} = 2\pi RL\tau_0 = \frac{0.237R^{0.75}\gamma^{0.75}\mu^{0.25}L\dot{x}^{1.75}}{g^{0.75}} \qquad (6.41)$$

and thus

$$C = \frac{0.237\gamma^{0.75}R^{0.75}L\mu^{0.25}}{g^{0.75}} \qquad (6.42)$$

Now Eq. (6.18) can be written as

$$\zeta = \frac{2.45\mu L \sqrt{g}}{R^2\gamma \sqrt{L}} = \frac{0.0974B \sqrt{g}}{R^2\gamma \sqrt{L}} \qquad (6.43)$$

since $B = 8\pi\mu L$ for the linear system. We can now define the equivalent linear damping ratio ζ_e by substituting B_e from Eqs. (6.40) and (6.42) in (6.43):

$$\zeta_e \triangleq \frac{0.0184 \sqrt{L}(\mu/\gamma g)^{0.25}}{R^{1.25}} (\omega Y)^{0.75} \qquad (6.44)$$

This shows clearly the dependence of ζ_e on the frequency and amplitude of the oscillation. For turbulent flow the value of ω_n is also somewhat different since the velocity profile tends to be more nearly square rather than parabolic. If the velocity is assumed uniform over the cross section, the effective mass becomes equal to the actual mass since all particles have the same velocity. If the turbulent damping, though larger than the laminar, is assumed to be still quite small, it is reasonable to expect that it will have little effect on the frequency. We shall therefore compute ω_n for turbulent flow by neglecting the nonlinear damping completely where it would appear in Eq. (6.11). We then get

$$\omega_n = \sqrt{\frac{2g}{L}} \qquad (6.45)$$

for turbulent flow.

Many assumptions were made in the above analysis. The formulas for steady pipe flow were used for an unsteady situation. In an oscillating flow, velocity actually goes to zero twice each cycle, no matter how great the amplitude or frequency; thus one wonders whether part of such a cycle is turbulent and part laminar. In the analysis above, turbulent equations were used for the whole cycle. The nonlinear differential equation containing $(\dot{x})^{1.75}$ actually has no closed-form analytical solution. Thus what is the meaning of a ζ_e and ω_n attached to such a process? Such questions and others may be at least partially resolved by more complex analyses or experimental studies. To provide some idea of the degree of validity of our simplified analysis some experimental results will be given. They were obtained at The Ohio State University by undergraduate students who study manometer dynamics in a simple experiment in a measurement course.

The experiment consists in part of suddenly releasing an air pressure applied to a mercury manometer and observing the resulting oscillations. The process is slow enough that ζ_e can be estimated from the size of the first overshoot and ω_n by counting and timing cycles with a stopwatch. The manometers used have the numerical values quoted earlier in this section. A step pressure input of 10 in. Hg ($x = 5$ in.) is used; thus turbulent flow may be expected. If laminar flow were assumed, the theoretical values would be $\zeta = 0.007$ and $\omega_n = 4.7$ rad/sec. For turbulent flow, ω_n becomes 5.4 rad/sec. To calculate ζ_e from Eq. (6.44) the frequency and amplitude of the oscillation must be known. For a step input the frequency is the damped natural frequency rather than ω_n; however, these are practically identical for the small damping present and so we use ω_n. We experimentally measure ζ_e from the first overshoot, and so the proper amplitude to use in the theoretical calculation might be an average of the initial amplitude and that at the first overshoot. Only the initial amplitude (5 in.) is known, however, and so this is used. Equation (6.44) then gives $\zeta_e = 0.082$. By timing and counting cycles (about eight cycles of the decaying oscillation can be easily measured) the experimental value of ω_n is 5.2 rad/sec. This lies between the values calculated for turbulent and laminar flow and thus is not unreasonable since several of the eight cycles used were of quite low amplitude due to decay of the oscillation. The first overshoot is about 4.05 in., giving an experimental value of ζ of 0.067. Therefore the theoretical estimate of 0.082 is fairly good. If we now use the experimental value of ω (5.2) and the average amplitude of the first half cycle (4.53) in Eq. (6.44), the predicted ζ_e is 0.074, which compares even more favorably. Based on even these limited results, a certain amount of confidence in the theoretical predictions is established.

6.4

Elastic Transducers   While a wide variety of flexible metallic elements might conceivably be used for pressure transducers, the vast majority of practical devices utilize one or another form of Bourdon tube,[1,2] diaphragm,[1,3,4] or bellows[1,5] as their sensitive element, as shown in Fig. 6.6. The gross deflection of these elements may directly actuate a pointer/scale readout through suitable linkages or gears, or the motion may be transduced to an electrical signal by one means or another. Strain gages bonded to diaphragms are also widely used to measure local strains that are directly related to pressure.

The Bourdon tube is the basis of many mechanical pressure gages and is also widely used in electrical transducers by measuring the output displacement with potentiometers, differential transformers, etc. The basic element in all the various forms is a tube of noncircular cross section. A pressure difference between the inside and outside of the tube (higher pressure inside) causes the tube to attempt to attain a circular cross section. This results in distortions which lead to a curvilinear translation of the free end in the C type and spiral and helical types and an angular rotation in the twisted type, which motions are the output. The theo-

---

[1] D. M. Considine (ed.), "Process Instruments and Controls Handbook," sec. 3, McGraw-Hill Book Company, New York, 1957.

[2] R. W. Bradspies, Bourdon Tubes, *Giannini Tech. Notes*, Giannini Corp., Duarte, Calif., January–February, 1961.

[3] Pressure Capsule Design, *Giannini Tech. Notes*, November, 1960.

[4] C. K. Stedman, The Characteristics of Flat Annular Diaphragms, *Statham Instr. Notes*, Statham Instruments Inc., Los Angeles, Calif.

[5] R. Carey, Welded Diaphragm Metal Bellows, *Electromech. Design*, p. 22, August, 1963.

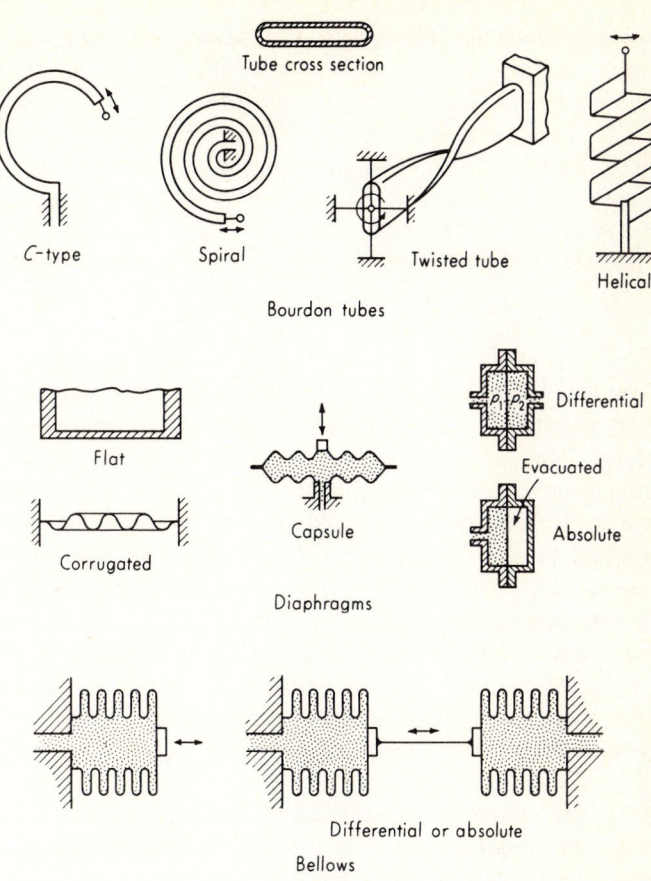

Fig. 6.6.   *Elastic pressure transducers.*

retical analysis of these effects is difficult, and practical design at present still makes use of considerable empirical data.   The C-type Bourdon tube has been used up to about 100,000 psi.   The spiral and helical configurations are attempts to obtain more output motion for a given pressure and have been used mainly below about 1,000 psi.   The twisted tube shown has a crossed-wire stabilizing device which is stiff in all radial directions but soft in rotation.   This reduces spurious output motions due to shock and vibration.

Flat diaphragms are widely used in electrical transducers either by sensing the center deflection with some displacement transducer or by bonding strain gages to the diaphragm surface.   The full-scale deflection at the center must be less than about one-third the diaphragm thickness if nonlinearity of less than 5 percent is desired.   The pressure-deflection formula for a flat diaphragm with edges clamped is

$$p = \frac{16Et^4}{3R^4(1 - \mu^2)} \left[\frac{y_c}{t} + 0.488 \left(\frac{y_c}{t} \right)^3 \right] \quad (6.46)$$

where $p \triangleq$ pressure difference across diaphragm
$E \triangleq$ modulus of elasticity
$t \triangleq$ diaphragm thickness
$\mu \triangleq$ Poisson's ratio
$R \triangleq$ diaphragm radius
$y_c \triangleq$ center deflection

For small deflections, $(y_c/t)^3$ is negligible compared with y_c/t, and linear behavior may be expected since bending stresses predominate.   At larger deflections, a stretching action is added to the bending, stiffening the diaphragm and contributing the $(y_c/t)^3$ term.   If local strain rather than center deflection is measured, a similar nonlinear effect is noted.   For strictly mechanical elements, larger linear deflections than those allowed above are often needed; they may be obtained by using corrugated diaphragms or capsules, as in Fig. 6.6.   Metallic bellows serve a similar function.   Bellows and diaphragms are most often used for relatively low pressures (less than a few hundred pounds per square inch) except for the diaphragm-type electrical transducers, which are available up to several thousand pounds per square inch.

Typical characteristics of electrical pressure pickups. Some numerical characteristics of the most common types of electrical pressure transducers will now be briefly reviewed.   Since they are all basically spring-mass systems with intentional or unintentional damping, their dynamic behavior is of standard second-order form, just as for force transducers.   An important point to note, however, is that under actual operating conditions the values of ω_n and ζ are greatly affected by the configuration of associated piping connections and the characteristics of the fluid medium.   Thus numerical values of ω_n and ζ supplied by instrument manufacturers usually refer to the instrument's behavior in ambient air and may be greatly different under actual operating conditions.   In some cases, instrument volume and tubing flow resistance are large enough so that the pressure felt by the elastic element follows with a first-order lag the pressure to be measured.   This lag may be so large that the second-order spring-mass dynamics is completely overshadowed and the overall instrument response becomes first-order.   These and similar problems will be treated in Sec. 6.6.   Most instruments using various combinations of elastic elements and electrical displacement transducers are available in versions to measure gage, differential, or absolute pressure.

Pressure pickups using resistive potentiometers for motion measurement are not generally intended for measurement of extremely fast pressure changes, and their natural frequencies are not often quoted, although rise time for a step input may be. The rather large motion required by potentiometers leads to a relatively large internal volume and volume change. A family[1] of differential-pressure transducers employing a capsule diaphragm of NI-SPAN-C (an alloy widely used for its constancy of elastic modulus with temperature) has full-scale ranges of 2 to 100 psi, nonlinearity ± 0.6 percent, hysteresis and friction ± 2 percent, and a temperature error ± 1 percent over the range -65 to $+200°F$. The standard potentiometer has 5,000 ohms resistance and a power rating of about 0.8 watt at room temperature. Acceleration sensitivity is of the order of 0.08 percent of full scale/g in the worst direction. The differential-pressure action is obtained by placing the capsule inside a pressure-tight case. One pressure is applied inside the capsule and the other to the space between the capsule and case. The fluid inside the capsule may be any one compatible with NI-SPAN-C; however, the space between the capsule and case contains the potentiometer and electrical leads so that only clean, nonconductive, noncorrosive fluids may be used. The 90 percent rise time for the step input of air pressure in the capsule is of the order of 40 to 70 msec. The internal case volume is 85 cm³, the capsule volume 6 cm³, and the full-scale volume change 4 cm³.

A family[1] of absolute-pressure transducers using a helical Bourdon tube has full-scale ranges of 500 to 10,000 psia, total error due to nonlinearity, friction, hysteresis, resolution, and repeatability of ± 1.2 percent (± 2.2 percent for the temperature range -100 to $+200°F$), and 63 percent response time of 6 msec. Acceleration sensitivity is 0.05 percent/g; the size is a 1-in.-diameter cylinder 2.7 in. long, and weight is 4 oz.

Potentiometer-type pickups generally exhibit a finite operating life. The two pickups mentioned above have a minimum life of about 50,000 and 25,000 full-scale cycles, respectively. Other potentiometer pickups have lives to a million cycles or more.

Pressure pickups using unbonded-strain-gage transducers are available for a wide range of applications. They generally employ the center deflection of a diaphragm as the mechanical input. Both flush-diaphragm and chamber types are made. Full-scale ranges from 0.01 to several thousand pounds per square inch can be obtained. A typical family[2] of flush-diaphragm absolute-pressure transducers has ranges from 5 to 1,000 psia, full-scale output of 56 mv with 7-volts excitation, nonlinearity and hysteresis less than 0.75 percent of full scale, thermal sen-

[1] Bourns Inc., Riverside, Calif.
[2] Statham Instruments Inc., Los Angeles, Calif.

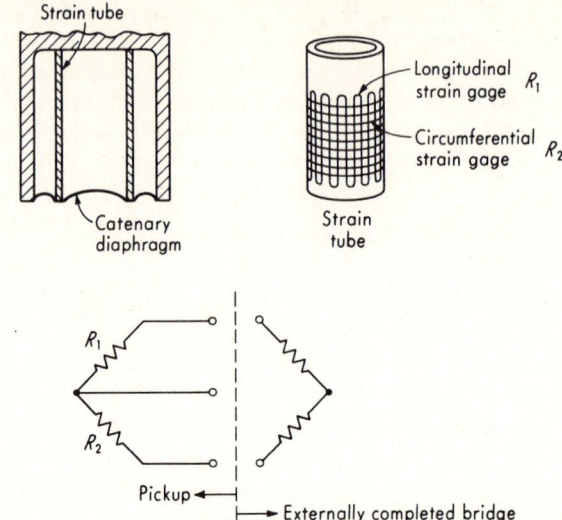

Fig. 6.7.   *Strain-tube pressure pickup.*

sitivity shift of 0.01 percent/F° and thermal zero shift of 0.01 percent of full scale/F° for the range −65 to +250°F, a natural frequency of 3,500 to 25,500 cps, and acceleration sensitivity 0.25 to 0.01 percent of full scale/g.

Bonded-strain-gage pressure pickups employ a number of pressure/strain transducing schemes. The most direct method bonds the gages directly to the diaphragm. Another applies the force from the diaphragm to a proving ring. A third[1] employs a catenary diaphragm to apply compression loads to a hollow strain tube on which the strain gages are mounted (see Fig. 6.7). A typical pickup of the last-mentioned type has full scale of 1,000 psi, nonlinearity ±1 percent of full scale, natural frequency of 45,000 cps, acceleration sensitivity 0.01 percent/g, thermal zero shift 0.02 percent of full scale/F°, and full-scale output of 50 mv. A subminiature evacuated-capsule absolute-pressure pickup[2] having one active flat diaphragm with a four-arm strain-gage bridge has an overall size of 0.25-in. diameter and 0.02-in. thickness. The pressure-sensitive area is 0.028 in.[2], ranges are 2 to 100 psia, natural frequency 20,000 cps, nonlinearity and hysteresis ±1 percent, thermal zero shift 0.1 percent of full scale/F°, thermal sensitivity shift 0.05 percent/F°, and full-scale output of 0.4 to 4 mv. The above-mentioned pickups all use metallic strain gages of the wire or foil type. Pickups using semiconductor gages

[1] Norwood Controls, Norwood, Mass.
[2] Scientific Advances, Inc., Columbus, Ohio.

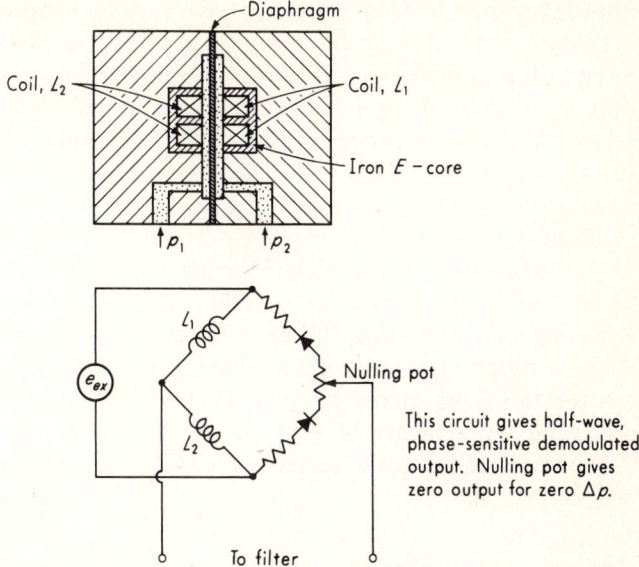

Fig. 6.8. *Variable-inductance pickup.*

are also available; they generally have considerably greater output, of the order of 0.25 volt full scale.

Variable-inductance pressure pickups are available in several forms. Often a magnetic stainless-steel diaphragm serves as the moving "iron" between two E coils arranged in a half-bridge circuit. One such pickup[1] has the interesting feature of interchangeable diaphragms, giving full-scale ranges of ±1, 5, 25, 100, and 500 psi in a single transducer. This pickup can be used for gage or differential pressure since both sides of the diaphragm may be exposed to corrosive liquids or gases. The nonlinearity is 0.5 percent, full-scale output is 1.5 volts at 3,000 cps, thermal zero shift 0.01 percent of full scale/F°, and thermal sensitivity shift 0.02 percent/F°, both from −65 to +250°F. The pressure cavity has a volume of 0.004 in.³ while the full-scale volume change is 0.0003 in.³. Natural frequency for the softest diaphragm is 5,000 cps, with acceleration sensitivity 0.2 percent/g, while the stiffest diaphragm has 40,000 cps, with 0.003 percent/g. Figure 6.8 shows the construction of such a transducer.

Piezoelectric pressure pickups have the same form of dynamic response as piezoelectric accelerometers. Except for certain quartz pickups used with electrometer amplifiers, they do not give an output for

---

[1] Pace Engineering Co., North Hollywood, Calif.

a static pressure.   They generally have very high natural frequencies and little damping.   A flush diaphragm is generally used to apply the pressure force to the piezoelectric element.   A quartz pickup/amplifier combination[1] specifically designed for measurements in shock tubes has switch-selectable full-scale ranges of 10, 100, 1,000, and 5,000 psi, responds to steady pressures, and has full-scale output of 0.5 to 2.4 volts, natural frequency of 150,000 cps, nonlinearity of 1 percent, and acceleration sensitivity of 0.02 psi/g.

A capacitance-type pickup[2] using the capacitor as part of a tuned 25-Mc oscillating circuit has, together with its associated electronics, the following characteristics: full-scale ranges from 5 to 50,000 psi (absolute, gage, or differential units are available), natural frequencies from 33,000 cps for the 5-psi model to over 350,000 cps for the 50,000-psi model, operating temperature of -65 to $+250°F$, nonlinearity and hysteresis 0.75 percent of full scale, repeatability 0.15 percent, and full-scale output of 5 volts.

Analysis of diaphragm-type bonded-strain-gage transducer.   To illustrate some general concepts involved in pressure-transducer design, we here consider the flat-diaphragm type since it lends itself to theoretical calculation.   Such a diaphragm, clamped at the edges and subjected to a uniform pressure difference p, has at any point on the low-pressure surface a radial stress s_r and a tangential stress s_t given by the following formulas (see Fig. 6.9):

$$s_r = \frac{3pR^2\mu}{8t^2}\left[\left(\frac{1}{\mu}+1\right) - \left(\frac{3}{\mu}+1\right)\left(\frac{r}{R}\right)^2\right] \quad (6.47)$$

$$s_t = \frac{3pR^2\mu}{8t^2}\left[\left(\frac{1}{\mu}+1\right) - \left(\frac{1}{\mu}+3\right)\left(\frac{r}{R}\right)^2\right] \quad (6.48)$$

where $\mu \triangleq$ Poisson's ratio

The deflection at any point is given by

$$y = \frac{3p(1-\mu^2)(R^2-r^2)^2}{16Et^3} \quad (6.49)$$

Equations (6.47) to (6.49) all give linear relations between stress or deflection and pressure and are accurate only for sufficiently small pressures.   Equation (6.46) may be used to estimate the degree of nonlinearity.   The stress situation on the diaphragm surface is fortunate because both tension and compression stresses exist simultaneously.

[1] Kistler Instrument Corp., North Tonawanda, N.Y.
[2] Omega Dynamics Corp., Pasadena, Calif.

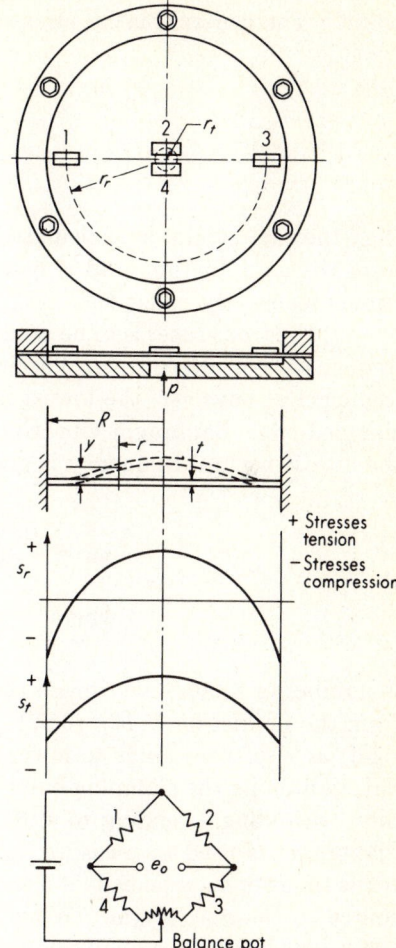

Fig. 6.9. *Diaphragm-type strain-gage pressure pickup.*

This allows use of a four-active-arm bridge in which all effects are additive (giving large output) and also gives temperature compensation. Gages 2 and 4 are placed as close to the center as possible and oriented to read tangential strain since it is maximum (positive) at this point. Gages 1 and 3 are oriented to read radial strain and placed as close to the edge as possible since radial strain has its maximum negative value at that point. The laws of bridge circuits show that the pressure effects on all four gages are additive. In computing the overall sensitivity, Eqs. (6.47) and (6.48) cannot be used directly to determine the strains "seen" by the gages since the diaphragm surface is in a state of *biaxial* stress and *both* the radial and tangential stress contribute to the radial or tangential strain at any point.   The general biaxial stress-strain relation gives

$$\epsilon_r = \frac{s_r - \mu s_t}{E} \qquad (6.50)$$

$$\epsilon_t = \frac{s_t - \mu s_r}{E} \qquad (6.51)$$

Once the gage strains are calculated, the individual gage ΔR's are obtained from the gage factors, and e_o can then be determined from the bridge-circuit sensitivity equations.

If the transducer is to be used for dynamic measurements, its natural frequency is of interest.   A diaphragm has an infinite number of natural frequencies; however, the lowest is the only one of interest here.   For a clamped-edge diaphragm vibrating in a vacuum (no fluid-inertia effects) the lowest natural frequency is given by

$$\omega_n = \frac{10.21}{R^2} \sqrt{\frac{Et^2}{12\rho(1 - \mu^2)}} \qquad \text{rad/sec} \qquad (6.52)$$

where $\rho \triangleq$ mass density of diaphragm material

A number of factors may make the actual operating value of ω_n different from the prediction of Eq. (6.52).   The edge clamping is never perfectly rigid; any softness tends to lower ω_n.   If the diaphragm is not perfectly flat, tightening the clamping bolts may cause a slight (perhaps impercepti-ble) "wrinkling," tending to stiffen the diaphragm and raise ω_n.   If the diaphragm is used to measure liquid pressures, the inertia of the liquid tends to lower ω_n, especially if a small-diameter tube connects the pressure source to the diaphragm.   When it is used with gases, the volume of gas "trapped" behind the diaphragm may act as a stiffening spring, raising ω_n.

A pressure transducer of the above type constructed for use in a transducer research project at The Ohio State University serves as a numerical example for the above discussion.   This transducer used a phosphor-bronze diaphragm with

$$E = 16 \times 10^6 \text{ lb}_f/\text{in.}^2 \qquad R = 1.830 \pm 0.002 \text{ in.}$$
$$\mu = \tfrac{1}{3} \qquad r_t = 0.15 \pm 0.01 \text{ in.}$$
$$\rho = 0.00083 \text{ lb}_f\text{-sec}^2/\text{in.}^4 \qquad r_r = 1.52 \pm 0.01 \text{ in.}$$
$$t = 0.0454 \pm 0.0003 \text{ in.}$$

The strain gages were SR-4 type A-7 wire gages with a gage factor of 1.97 $\pm$ 2 percent and a resistance of 119.5 $\pm$ 0.3 ohms.   The bridge was excited with 7.5 volts direct current, and the transducer was designed for a full-scale range of 10 psi.   The theoretically calculated sensitivity was

0.516 mv/psi. Static calibration gave 0.513 mv/psi and indicated a maximum nonlinearity of 2 percent of full scale. The theoretically calculated natural frequency was 924 cps, and the experimental value (with atmospheric air on both sides of the diaphragm) was 897.

6.5

Force-Balance Transducers   Feedback or null-balance principles may be applied to pressure measurement in a manner similar to that employed for force measurement. Figure 6.10*a* shows a pneumatic/mechanical type, and electromechanical methods are used in Fig. 6.10*b*. High loop gain in these servosystems gives good linearity and accuracy. The block diagrams give the static relations only; a dynamic analysis is necessary to determine the limit set on gain by stability requirements. The operation of these instruments is left for the reader to deduce from the schematic and block diagrams. Previous discussions of feedback-type instruments may be helpful.

6.6

Dynamic Effects of Volumes and Connecting Tubing   We have mentioned the possible strong effect of fluid properties and "plumbing" configurations on the dynamic behavior of pressure-measuring systems. In this section some of these problems will be investigated. It should first be pointed out that, if maximum dynamic performance is to be attained, a flush-diaphragm transducer mounted directly at the point where a pressure measurement is wanted should be used if at all possible (see Fig. 6.11). Any connecting tubing or volume chambers will degrade performance to some extent. The fact that this degradation is studied in this section indicates that in many practical circumstances a flush-diaphragm transducer is not applicable.

Liquid systems, heavily damped, slow-acting. In the system of Fig. 6.12, the spring-loaded piston represents the flexible element of the pressure pickup. For the present analysis, the only pertinent characteristic of the pressure pickup is its volume change per unit pressure change, C_{vp} in.^{3}/psi. This can be calculated or measured experimentally. For systems that are heavily damped (a criterion for judging this will be given shortly) and subjected to relatively slow pressure changes, the inertia effects of both the fluid and the moving parts of the pickup are negligible compared with viscous and spring forces. We shall show that under these conditions the measured pressure p_m follows the desired pres-

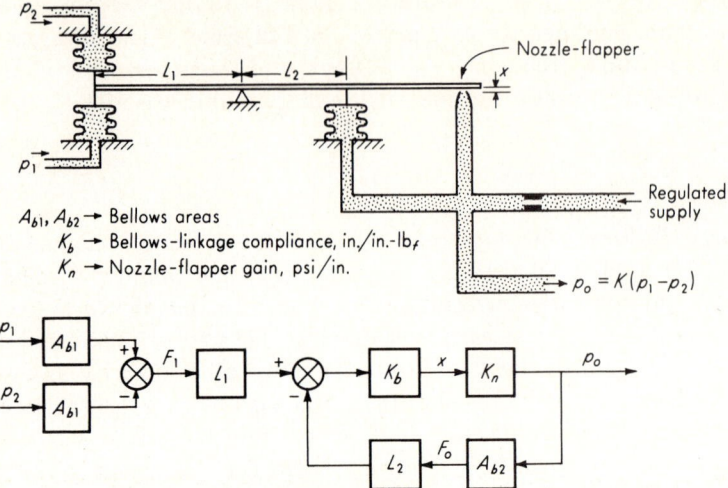

Force-balance differential-pressure transducer

(a)

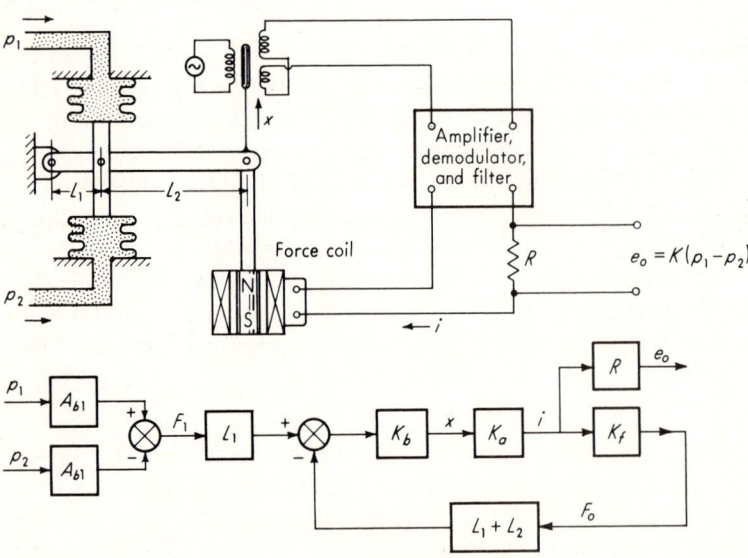

Electromagnetic pressure balance

(b)

Fig. 6.10.   *Null-balance pressure sensors.*

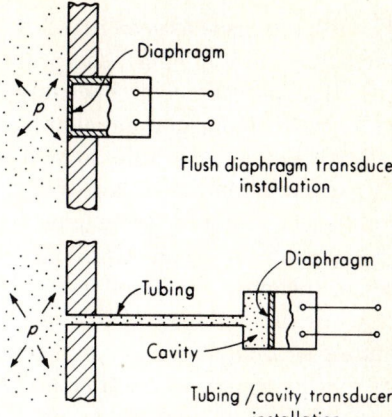

Fig. 6.11. *Transducer installation types.*

sure p_i with a first-order lag. For steady laminar flow in the tube we have

$$p_i - p_m = \frac{32\mu L V_{t,\mathrm{av}}}{d_t^2} \qquad (6.53)$$

where $\qquad \mu \triangleq$ fluid viscosity

$\qquad V_{t,\mathrm{av}} \triangleq$ average flow velocity in tube

While this equation is exact only for steady flow, it holds quite closely for slowly varying velocities. During a time interval dt a quantity of liquid enters the chamber. This is given by

$$dV = \text{volume entering} = \frac{\pi d_t^2}{4} V_{t,\mathrm{av}}\, dt = \frac{\pi d_t^4 (p_i - p_m)\, dt}{128\mu L} \qquad (6.54)$$

Any volume added or taken away results in a pressure change dp_m given by

$$dp_m = \frac{dV}{C_{vp}} = \frac{p_i - p_m}{\tau}\, dt \qquad (6.55)$$

and thus

$$\tau \frac{dp_m}{dt} + p_m = p_i \qquad (6.56)$$

where

$$\tau \triangleq \frac{128\mu L C_{vp}}{\pi d_t^4} \qquad (6.57)$$

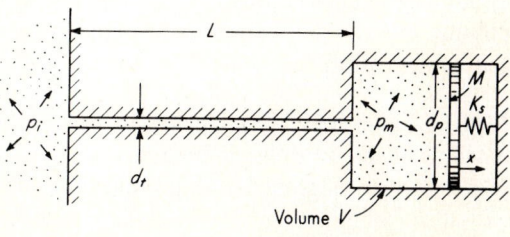

Fig. 6.12. *Transducer/tubing model.*

The response is thus seen to follow a standard first-order form. To keep τ small, the tubing length should be short and the diameter large.

The viewpoint of this analysis is that any sudden changes in p_i will cause much more gradual changes in p_m and thus the pickup spring-mass system will not be able to manifest its natural oscillatory tendencies. Under such conditions an overall first-order response from the tubing/transducer system may be expected. To obtain some numerical estimate of the conditions required for such behavior, we later carry out an analysis which includes inertia effects. This will lead to a second-order type of response, and when ζ of this model is greater than about 1.5, the simpler first-order model may be employed with fair accuracy, at least (in terms of frequency response) for frequencies less than ω_n.

The model used above predicts that, for a change in p_i, p_m starts to change immediately. This cannot be true since a pressure wave propagates through a fluid at finite speed, the velocity of sound, for small disturbances. There is thus a dead time τ_{dt} equal to the distance traversed divided by the speed of sound. For liquids and reasonable tube lengths this delay is usually small enough to ignore completely. The speed of sound, v_s, in a fluid contained in a nonrigid tube is given by

$$v_s = \sqrt{\frac{E_L}{\rho_L}} \sqrt{\frac{1}{1 + 2R_t E_L / t E_t}} \qquad (6.58)$$

where $E_L \triangleq$ bulk modulus of fluid
$\rho_L \triangleq$ mass density of fluid
$R_t \triangleq$ tube inside radius
$t \triangleq$ tube wall thickness
$E_t \triangleq$ tube-material modulus of elasticity

If this dead time is significant in a practical problem, it may be included in the transfer function as

$$\frac{p_m}{p_i}(D) = \frac{e^{-\tau_{dt}D}}{\tau D + 1} \qquad (6.59)$$

with fair accuracy.

Finally, it should also be pointed out that the result [Eq. (6.56)] of this analysis may also be applied to systems using gases rather than liquids if the elastic pressure-sensing element is sufficiently soft so that its volume change per unit pressure change is much larger than that due to gas compressibility.

Liquid systems, moderately damped, fast-acting. When the motions of the liquid and the pickup elastic element are rapid, their inertia is no longer negligible. An analysis[1] of this situation using energy

---

[1] G. White, Liquid Filled Pressure Gage Systems, *Statham Instr. Notes*, Statham Instruments Inc., Los Angeles, Calif., January–February, 1949.

methods is available. In the system of Fig. 6.12, any change in pressure p_m must be accompanied by a volume change; this in turn requires an inflow or outflow of liquid through the tube. If the tube is of small diameter compared with the equivalent piston diameter of the pickup, the tube flow will be at a much higher velocity than the piston velocity, and the kinetic energy of the liquid in the tube may be a large (sometimes major) part of the total system kinetic energy. This increase in kinetic energy is equivalent to adding mass to the piston and ignoring the fluid inertia, and the analysis below calculates just how much mass should be added to give the same effect as the fluid inertia. This added mass lowers the system natural frequency and thereby degrades dynamic response.

To find the equivalent piston/spring configuration for a given transducer, the volume change per unit pressure change must be equal for both systems. This gives

Transducer volume change = equivalent piston volume change

$$pC_{vp} = \frac{\pi^2 d_p{}^4 p}{16 K_s} \qquad (6.60)$$

$$\frac{d_p{}^4}{K_s} = \frac{16 C_{vp}}{\pi^2} \qquad (6.61)$$

Thus d_p and K_s for the equivalent system can have any values that satisfy Eq. (6.61). Also, the natural frequency of each system with no fluid present must be equal. This gives

$$\omega_{n,t} = \sqrt{\frac{K_s}{M}} \qquad (6.62)$$

$$\frac{K_s}{M} = \omega_{n,t}{}^2 \qquad (6.63)$$

where $\omega_{n,t} \triangleq$ transducer natural frequency. Again, any values of K_s and M that satisfy Eq. (6.63) may be used. Thus, to define the equivalent system, only C_{vp} and $\omega_{n,t}$ need be known; they can be found from experiment if theoretical formulas are unavailable.

Since we have just shown how the equivalent system is related to the real system, we can now proceed with an analysis of the equivalent system. The volume change dV is related to the piston motion dx by

$$dV = \frac{\pi d_p{}^2 \, dx}{4} \qquad (6.64)$$

Thus

$$\frac{dV}{dt} = \frac{\pi d_p{}^2}{4} \frac{dx}{dt} \qquad (6.65)$$

and

$$\frac{\pi}{4} d_t{}^2 \, V_{t,\text{av}} = \frac{\pi d_p{}^2}{4} \frac{dx}{dt} \qquad (6.66)$$

where $V_{t,\mathrm{av}}$ = average flow velocity in tube

We then get
$$V_{t,\mathrm{av}} = \left(\frac{d_p}{d_t}\right)^2 \frac{dx}{dt} \qquad (6.67)$$

Next we assume laminar flow in the tube, with the parabolic velocity profile characteristic of steady flow:

$$V_t = V_{t,c}\left[1 - \left(\frac{r}{R}\right)^2\right] \qquad (6.68)$$

where $V_t \triangleq$ velocity at radius r
$V_{t,c} \triangleq$ center-line velocity
$R \triangleq$ tube inside radius

The kinetic energy of an annular element of fluid (density ρ) of thickness dr at radius r is

$$d(KE) = \frac{(2\pi r\, dr)L\rho V_t^2}{2} \qquad (6.69)$$

Substitution of Eq. (6.68) and integration give the fluid kinetic energy as

$$KE = \frac{\pi\rho L V_{t,\mathrm{av}}^2 d_t^2}{6} \qquad (6.70)$$

For a square velocity profile the kinetic energy would be

$$KE_s = \frac{\pi\rho L V_{t,\mathrm{av}}^2 d_t^2}{8} \qquad (6.71)$$

The actual velocity profile will be somewhere between parabolic and square.   Even if laminar flow exists, the velocity profile is nonparabolic except for steady flow.[1]   Turbulent flow gives a rather square profile. We shall assume Eq. (6.70) to hold here; however, Eq. (6.71) can be carried through with equal ease to "bracket" the correct value.   The rigid mass M_e, attached to M, which would have the same kinetic energy as the fluid, is given by

$$\frac{M_e}{2}\left(\frac{dx}{dt}\right)^2 = \frac{\pi\rho L d_p^4}{6d_t^2}\left(\frac{dx}{dt}\right)^2 \qquad (6.72)$$

$$M_e = \frac{\pi\rho L d_p^4}{3d_t^2} \qquad (6.73)$$

The natural frequency of the transducer/tubing system is then

$$\omega_n = \sqrt{\frac{K_s}{M + M_e}} = \sqrt{\frac{1}{M/K_s + M_e/K_s}} = \sqrt{\frac{1}{1/\omega_{n,t}^2 + 16\rho L C_{vp}/3\pi d_t^2}} \qquad (6.74)$$

[1] C. K. Stedman, Alternating Flow of Fluid in Tubes, *Statham Instr. Notes*, Statham Instruments Inc., Los Angeles, Calif., January, 1956.

To keep ω_n as high as possible, L and C_{vp} must be as small as possible and d_t as large as possible. In many practical cases, $M_e \gg M$, allowing simplification of Eq. (6.74) to

$$\omega_n = \sqrt{\frac{3\pi d_t{}^2}{16\rho L C_{vp}}} \qquad (6.75)$$

We next calculate the damping ratio of the transducer/tubing system. The transducer itself is assumed to have negligible damping; thus the only damping is due to the fluid friction in the tube. We again assume the validity of the steady laminar-flow relations to calculate the pressure drop due to fluid viscosity as $32\mu L V_{t,\text{av}}/d_t{}^2$. The force on the piston due to this pressure drop is the damping force $B(dx/dt)$; thus

$$\frac{\pi d_p{}^2}{4} \frac{32\mu L V_{t,\text{av}}}{d_t{}^2} = B\frac{dx}{dt} \qquad (6.76)$$

and, since $V_{t,\text{av}} = (dx/dt)(d_p/d_t)^2$,

$$B = 8\pi\mu L \left(\frac{d_p}{d_t}\right)^4 \qquad (6.77)$$

Then, using the general formula for the damping ratio of a spring-mass-dashpot system, we get

$$\zeta = \frac{B}{2\sqrt{K_s(M + M_e)}} = \frac{64\mu L C_{vp}}{\pi d_t{}^4 \sqrt{1/\omega_{n,t}{}^2 + 16\rho L C_{vp}/3\pi d_t{}^2}} \qquad (6.78)$$

If $M_e \gg M$, this simplifies to

$$\zeta = \frac{16\sqrt{3/\pi}\,\mu\,\sqrt{LC_{vp}/\rho}}{d_t{}^3} \qquad (6.79)$$

The above theory has been partially checked experimentally in an M.Sc. thesis by Fowler.[1] The transducer used was the diaphragm strain-gage instrument whose numerical parameters were given at the end of Sec. 6.4. The liquid used was water, and ω_n and ζ were found from step-function tests using a bursting cellophane diaphragm to obtain a sudden 10-psi release of pressure. Tubing of 0.042- to 1.022-in. inside diameter and 4 to 32 in. in length was studied. Conditions were such that turbulent flow probably existed much of the time; thus use of Eq. (6.71) was indicated. The value of C_{vp} was calculated from theory as 0.00441 in.^{3}/psi while an experimental calibration gave 0.00500. For all cases in which oscillation occurred and thus ω_n could be accurately measured, it was found that ω_n was accurately predicted within about 5 to 10 percent. To show the severity of the performance degradation,

---

[1] R. L. Fowler, An Experimental Study of the Effects of Liquid Inertia and Viscosity on the Dynamic Response of Pressure Transducer-Tubing Systems, M.Sc. Thesis, The Ohio State University, Mechanical Engineering Department, 1963.

the natural frequency (which was 897 cps with no tubing and in air) became about 60 cps with a 1.022-in.-diameter tube 10 in. in length and about 3 cps with a 0.092-in. tube 32 in. long. However, the prediction of ζ was much less satisfactory, being invariably too low. The small-diameter long tubes (which encourage laminar flow) gave the best correlation with theory but even there errors of 100 percent occurred. For example, the 0.042-in. tube 10 in. long had ζ of about 1.0 while theory predicted 0.66. The poor agreement can undoubtedly be charged to turbulent flow and energy losses due to expansion and contraction (end effects) at the tubing ends.

Fowler developed corrections for these effects which resulted in much better agreement. However, the turbulent flow and end effects are nonlinear damping mechanisms; thus the meaning of ζ is confused for both theoretical predictions and experimental measurements. The use of smaller step functions should give lower velocities and thus more nearly laminar flow. This effect was checked for a 0.092-in.-diameter 10-in. long tube. A 10-psi step function gave $\zeta = 0.34$; a 5-psi step function gave 0.22. The theoretical value was 0.065, suggesting that for sufficiently small inputs the theory might be quite accurate. Detailed discussion of these problems may be found in the reference.

When ζ calculated from Eq. (6.78) or (6.79) is 1.5 or greater, the simpler first-order model of Eq. (6.56) may be adequate. To show the relationship of these two analyses, recall that in the second-order form $K/[D^2/\omega_n{}^2 + 2\zeta D/\omega_n + 1]$ the inertia effect resides in the D^2 term. Neglecting this gives $K/[2\zeta D/\omega_n + 1]$. If one calculates $2\zeta/\omega_n$ from Eqs. (6.74) and (6.78), he will find it numerically equal to τ of Eq. (6.57).

Gas systems, with tube volume small fraction of chamber volume. When, in the configuration of Fig. 6.12, a gas is the fluid medium, the compressibility of the gas in the volume V becomes the major spring effect when the pressure pickup is at all stiff. It is reasonable, then, to assume that the volume V is enclosed by rigid walls ($C_{vp} = 0$). The majority of practical problems involve rather low frequencies. This allows treatment as a lumped-parameter system with fluid properties considered constant along the length of the tube for small disturbances. The validity of this viewpoint is shown by noting that pressure-wave propagation in the gas follows the general law of wave motion

$$\lambda = \text{wavelength} = \frac{\text{velocity of propagation}}{\text{frequency}} = \frac{c}{f} \qquad (6.80)$$

Our lumped-parameter assumption says that the gas in the tube moves as a unit, as opposed to wave motion *within* the tube. For any given

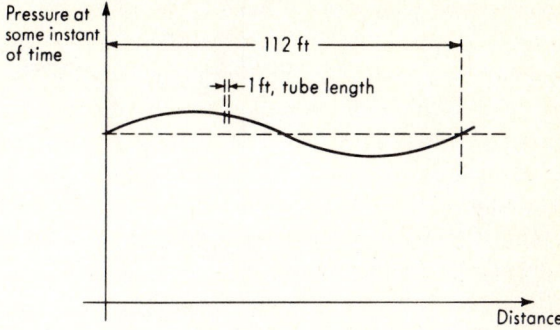

Fig. 6.13. *Justification of lumped-parameter analysis.*

tube length L, for sufficiently low frequencies, the lumped-parameter approach becomes valid. For example, the velocity of pressure waves in standard air is about 1,120 fps. If an oscillation of 10-cps frequency exists, its wavelength must be 112 ft/cycle. This means that the space-wise variation of fluid pressure due to wave motion has a wavelength of 112 ft (see Fig. 6.13). For a tube of, say, 1-ft length, the variation of pressure (and thus of density, etc.) from one end of the tube to the other, due to wave motion, is very small. That is, there is negligible *relative* motion of particles in the tube; they all move together, as a unit. While the above aspect of wave motion is neglected, the dead time due to the finite speed of propagation can be relatively easily taken into account (approximately) by multiplying the second-order transfer function (which we shall obtain shortly) by $e^{-\tau_{dt} D}$, where

$$\tau_{dt} = \frac{L}{v_s} \qquad (6.81)$$

$$\text{Sound velocity} = v_s = \sqrt{\frac{\gamma p}{\rho}} \qquad (6.82)$$

where　$\gamma \triangleq$ ratio of specific heats
$\rho \triangleq$ mass density
$p \triangleq$ average pressure

This dead time is negligible unless very long lines (several hundred feet, as found in some industrial pneumatic control systems) are used. For ambient air, a 100-ft line has a dead time of about 0.1 sec. Thus if ω_n of the second-order response is, for example, 3 rad/sec, the effect of the dead time on the overall response will be of relatively slight importance (see Fig. 6.14).

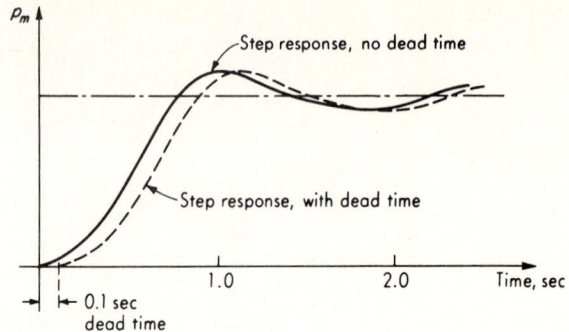

Fig. 6.14. *Effect of small dead time.*

The analysis[1] we carry out below is valid only for small pressure changes, the system becoming nonlinear for large disturbances. Steady-laminar-flow formulas are used for calculating fluid friction. However, the effective mass is taken equal to the actual mass (rather than the four-thirds actual mass given by the parabolic velocity profile). We shall follow the reference in this respect; use of the $\frac{4}{3}$ factor is easily incorporated if one wishes to bracket a more correct value. Numerically the effect is rather small in any case.

The analysis consists merely of applying Newton's law to the "slug" of gas in the tube. We assume that initially $p_i = p_m = p_0$ when p_i changes slightly in some way. From here on, the symbols p_i and p_m are taken to mean the *excess* pressures over and above p_0. The force due to the pressure p_i is $\pi p_i d_t^2/4$. The viscous force due to the wall shearing stress is $8\pi\mu L\dot{x}_t$, where x_t is the displacement of the slug of gas in the tube. If the slug of gas moves into the volume V an amount x_t, the pressure p_m will increase. We assume this compression occurs under adiabatic conditions. The adiabatic bulk modulus E_a of a gas is given by

$$E_a \triangleq - \frac{dp}{dV/V} = \gamma p \qquad (6.83)$$

The displacement x_t causes a volume change $dV = \pi d_t^2 x_t/4$. This in turn causes a pressure excess $p_m = \pi E_a d_t^2 x_t/4V$. The force due to this pressure excess is $\pi^2 E_a d_t^4 x_t/16V$. Newton's law then gives

$$\frac{\pi p_i d_t^2}{4} - 8\pi\mu L\dot{x}_t - \frac{\pi^2 E_a d_t^4 x_t}{16V} = \frac{\pi d_t^2 L\rho}{4} \ddot{x}_t \qquad (6.84)$$

[1] G. J. Delio, G. V. Schwent, and R. S. Cesaro, Transient Behavior of Lumped-constant Systems for Sensing Gas Pressures, *NACA Tech. Note* 1988, 1949.

and, since $p_m = \pi E_a d_t{}^2 x_t / 4V$,

$$\frac{4L\rho V}{\pi E_a d_t{}^2}\, \ddot{p}_m + \frac{128\mu LV}{\pi E_a d_t{}^4}\, \dot{p}_m + p_m = p_i \qquad (6.85)$$

This is clearly the standard second-order form, and so we define

$$\omega_n \triangleq \frac{d_t}{2}\sqrt{\frac{\pi E_a}{L\rho V}} \qquad (6.86)$$

$$\zeta \triangleq \frac{32\mu}{d_t{}^3}\sqrt{\frac{VL}{\pi E_a \rho}} \qquad (6.87)$$

Since E_a and ρ both vary during pressure changes, ζ and ω_n are not really constants; that is, the system is nonlinear. For small-percentage pressure variations around the original equilibrium pressure p_0, however, ζ and ω_n vary only slightly and the behavior is nearly linear. In calculating ζ and ω_n, E_a and ρ are computed by using pressure p_0.

When L becomes very short, Eq. (6.86) predicts a very large ω_n. In practice, this will not occur since even when $L = 0$ there is some air (close to the opening in the volume) which has appreciable velocity and, therefore, kinetic energy. Theory shows that this end effect may be taken into account by using for L in Eq. (6.86) an effective length L_e given by

$$L_e = L\left(1 + \frac{8}{3\pi}\frac{d_t}{L}\right) \qquad (6.88)$$

In most cases the term $(8/3\pi)(d_t/L)$ will be completely negligible compared with 1.0. However, if $L = 0$ (tube degenerates into simply a hole in the side of volume V), one can still compute an ω_n since then $L_e = (8/3\pi)d_t$. The computation of damping for this case is not straightforward and will not be discussed here.

Gas systems, with tube volume comparable to chamber volume. When the volume of the tube becomes a significant part of the total volume of a system, compressibility effects are no longer restricted to the volume chamber alone and the above formulas become inaccurate. More refined analyses[1] give the following formulas for ζ and ω_n:

$$\omega_n = \frac{\sqrt{\gamma p/\rho}}{L\sqrt{1/2 + V/V_t}} \qquad (6.89)$$

$$\zeta = \frac{16\mu L}{d_t^2\sqrt{\gamma p \rho}}\sqrt{\frac{1}{2} + \frac{V}{V_t}} \qquad (6.90)$$

where $V_t \triangleq$ tubing volume

---

[1] J. O. Hougen, O. R. Martin, and R. A. Walsh, Dynamics of Pneumatic Transmission Lines, *Control Eng.*, p. 114, September, 1963.

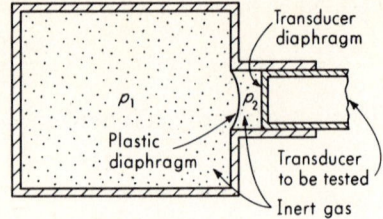

Fig. 6.15. *Step-test apparatus.*

If, in these formulas, $V/V_t \gg \frac{1}{2}$ (tubing volume negligible compared with chamber volume), the term $\frac{1}{2}$ may be neglected and the formulas become identical to (6.86) and (6.87).

Conclusion. The results of this section are to be thought of as practical working relations. The general problem treated here is quite complex and has been the subject of many intricate analyses and experimental studies, some of which will be found in the bibliography of this chapter. Most of the difficulties encountered are in the area of very high frequencies, where the lumped-parameter models used in this section are inadequate and give faulty predictions.

6.7

Dynamic Testing of Pressure-measuring Systems   To determine the regions of accuracy of theoretical predictions or to find accurate numerical values of system dynamic characteristics for critical applications, recourse must be made to experimental testing. This commonly takes the form of impulse, step, or frequency-response tests, with step-function tests being perhaps the most common. A comprehensive review[1] of this subject is available. Here we can mention only a few high points.

For step-function tests of systems in which natural frequencies are not greater than about 1,000 cps, the bursting of a thin diaphragm subjected to gas pressure is often satisfactory. A general rule for step testing is that the rise time of the "step" function must be less than about one fourth of the natural period of the system tested if it is to excite the natural oscillations. Thus a 1,000-cps system requires a step with a rise time of 0.25 msec or less. Figure 6.15 shows schematically the principle of such devices. The pressures p_1 and p_2 are each individually adjustable. The volume containing p_2 is much smaller than that containing p_1; thus when the thin plastic diaphragm is ruptured by a solenoid-actuated

---

[1] J. L. Schweppe et al., Methods for the Dynamic Calibration of Pressure Transducers, *Natl. Bur. Std. (U.S.), Monograph* 67, 1963.

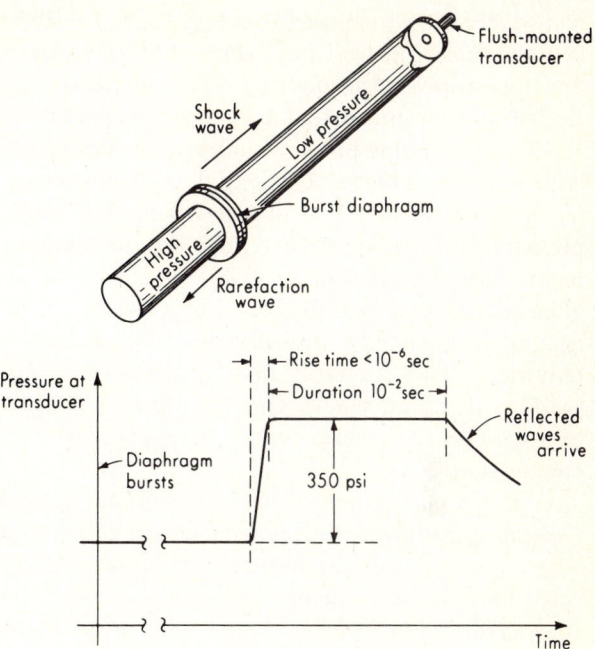

Fig. 6.16.   *Shock tube.*

knife, the pressure p_2 rises to p_1 very quickly.   If a decreasing step function is wanted, p_2 can be made larger than p_1.   Construction and operation of such devices are quite simple, and they have been widely used in their range of applicability.

For pickups of natural frequency greater than 1,000 cps, the simple burst-diaphragm testers are not capable of exciting the natural oscillations, and the pickup output is simply an accurate record of the terminated-ramp pressure input.   To achieve sufficiently short pressure-rise times the shock tube is widely used.   Figure 6.16 shows a sketch of such a device.   A thin diaphragm separates the high-pressure and low-pressure chambers, and the transducer to be tested is mounted flush with the end of the low-pressure chamber.   When the diaphragm is caused to burst, a shock wave travels toward the low-pressure end at a speed that may greatly exceed the speed of sound (5,000 fps is not unusual).   From one side of this shock front to the other there is a pressure change of the order of 2 to 1 over a distance which may be of the order of 10^{-4} in.   (At the same time a rarefaction wave travels from the diaphragm toward the high-pressure end.)   When the shock front reaches the end of the tube where the transducer is mounted, it is reflected as a shock wave with more than twice the pressure difference of the original shock wave.   The transducer is thus exposed to a very sharp ($\sim 10^{-8}$ sec) pressure rise which is maintained constant for a short interval before various reflected waves arrive to confuse the picture. The length of this interval may be controlled to a certain extent by proper proportioning and operation of the shock tube. Some numerical characteristics of a typical shock tube[1] are: high-pressure chamber 7 ft long, low-pressure chamber 15 ft long, tubing inside dimensions 1.4 in. square (wall thickness $\frac{1}{4}$ in.), maximum high pressure 600 psi, operating fluid air, maximum pressure step 350 psi, burst diaphragm 0.001- to 0.005-in.-thick Mylar plastic, and duration of constant pressure 0.01 sec. For a pressure pickup of, say, 100,000-cps natural frequency, a pressure-rise time of less than 0.25×10^{-5} sec is required. This is readily met by the tube described above. The 0.01-sec step duration would give time for about 1,000 cycles of oscillation of a 100,000-cps pickup—more than adequate to determine the dynamic characteristics.

A simple impulse-type test method applicable to flat, flush-diaphragm transducers utilizes a small steel ball dropped onto the diaphragm. The impact excites the natural oscillations which are recorded and analyzed for natural frequency and damping ratio. Although this input is a concentrated force rather than a uniform pressure, the results have been found[2] to correlate quite well with shock-tube tests. Another impulse method[3] uses the shock wave created by discharging 25,000 volts across a spark gap. The spark gap and transducer are located about 3 in. apart in open air. A pressure impulse of 0.2 μsec rise time and 100 psi peak can be obtained.

Figure 6.17 shows one method of constructing a frequency-response tester using liquid as the pressure medium. The vibration shaker applies a sinusoidal force of adjustable frequency and amplitude to the piston/diaphragm to create sinusoidal pressure in the liquid-filled chamber. Such vibration shakers are readily available in industry and cover a wide range of force and frequency. The average pressure about which oscillations take place may be adjusted by regulating the bias pressure on the air side of the cylinder. Since it is not usually possible accurately and reproducibly to predict the pressure actually produced by such an arrangement as "frequency and/or amplitude is varied," it is customary to mount a reference transducer at a location where it will experience the same pressure as the transducer under test. The reference transducer must have a known flat frequency response beyond any frequencies to be tested. This

---

[1] R. Bowersox, Calibration of High-frequency-response Pressure Transducers, *ISA J.*, p. 98, November, 1958.

[2] W. C. Bentley and J. J. Walter, Transient Pressure Measuring Methods Research, *Princeton Univ. Aeron. Eng. Dept. Rept.* 595g, p. 103, 1963.

[3] Omega Dynamics Corp., Pasadena, Calif.

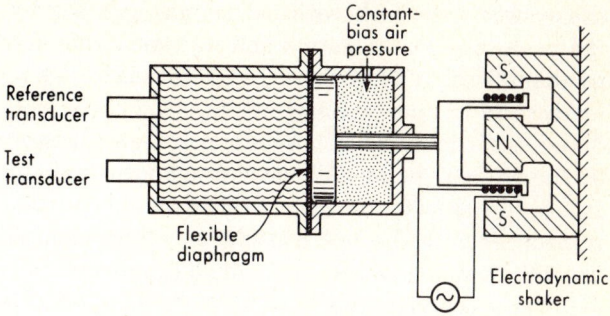

Fig. 6.17.   *Sinusoidal test apparatus for liquid.*

can be determined by some independent method, such as a shock tube. In testing another transducer, one merely calculates the amplitude ratio and phase shift between the reference and test transducer to determine the test-transducer frequency response.

A different approach[1] to frequency testing, using a flow-modulating principle and gas as the fluid medium, is shown in Fig. 6.18.   A chamber is supplied with compressed gas from a constant-pressure source through a small inlet passage.   The gas is exhausted to the atmosphere through an outlet passage whose area is modulated approximately sinusoidally with time.   This is accomplished by rotating a disk containing holes in front of the exhaust port so that outflow periodically is cut on and off.   This produces a periodic (nearly sinusoidal) variation in chamber pressure which is measured by both a reference transducer and the test transducer.   Varying the speed of the rotating disk changes the frequency.   The amplitude of pressure oscillation of such a device drops off with frequency.   For

[1] Bentley and Walter, *op. cit.*, p. 63.

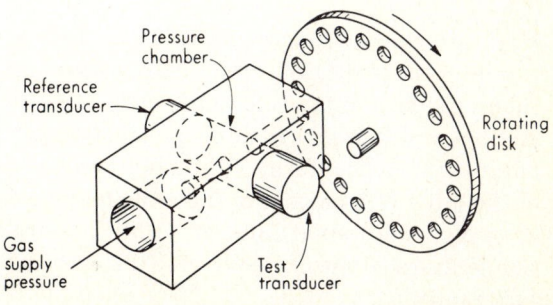

Fig. 6.18.   *Sinusoidal test apparatus for gas.*

the system described, with helium gas at a supply pressure of 121 psia, the peak-to-peak pressure amplitude goes from about 15 psi at 1,000 cps to about 2 psi at 11,000 cps.   In addition to this reduction in amplitude, increase in frequency also leads into the range of resonant acoustical frequencies of the chamber.   When these resonances occur one cannot depend on the pressure being uniform throughout the chamber.   This uniformity is a necessity when using a method based on comparison of a reference transducer with the test transducer.   The acoustic resonant frequencies depend on the chamber size (smaller chambers have higher frequencies) and the speed of sound in the gas (higher sound speed gives higher frequencies).   The use of helium in the above example is based on this last consideration.   A system of the above type was found to be usable for dynamic calibration up to about 10,000 cps.

In summary, it should be pointed out that the dynamic testing of very-high-frequency pressure pickups involves a number of complicating factors.   It has been found extremely difficult to generate high-frequency pressure sine waves that also are of large enough amplitude to give a relatively noise-free transducer output signal.   Small-amplitude pressure waves (such as are applicable to sound-measuring systems) can be relatively easily produced with loudspeaker-type systems but their amplitude is far below the levels needed for pressure pickups whose full-scale range is tens, hundreds, or thousands of pounds per square inch.   Step testing with a shock tube has thus been widely used since the fast rise time and large pressure steps result in a transient input with strong high-frequency content.   The pickups themselves present problems since at the high natural frequencies involved many complex wave-propagation and reflection effects make the response deviate considerably from the simple second-order model.   Also, these pickups generally have little damping ($\zeta \approx 0.01$ to 0.04) which makes them particularly prone to ringing at their natural frequency if any sharp transients occur.

6.8

High-pressure Measurement   Pressures up to about 100,000 psi can be measured fairly easily with strain-gage pressure cells or Bourdon tubes.   Bourdon tubes for such high pressures have nearly circular cross sections and thus give little output motion per turn.   To get a measurable output, the helical form with many turns is generally used.   Inaccuracy of the order of 1 percent of full scale may be expected with a temperature error of an additional 2 percent/100F°.   Strain-gage pressure cells can be temperature-compensated to give 0.25 percent error over a large temperature range.

For fluid pressures above 100,000 psi, electrical gages based on the

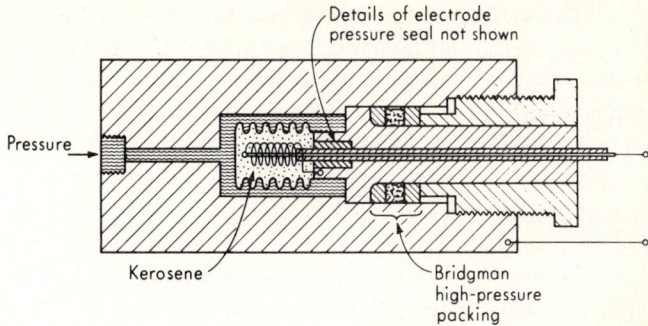

Fig. 6.19.   *Very-high-pressure transducer.*

resistance change of Manganin or gold-chrome wire with hydrostatic pressure are generally utilized.[1]   Figure 6.19 shows a typical gage.   The sensitive wire is wound in a loose coil, one end of which is grounded to the cell body and the other end brought out through a suitable insulator. The coil is enclosed in a flexible, kerosene-filled bellows which transmits the measured pressure to the coil.   The resistance change, which is linear with pressure, is sensed by conventional Wheatstone-bridge methods. Pertinent characteristics of the common wire materials are as follows:

	Pressure sensitivity, (ohm/ohm)/psi	Temperature sensitivity, (ohm/ohm)/F°	Resistivity, ohm-cm
Manganin	1.69×10^{-7}	1.7×10^{-5}	45×10^{-6}
Gold chrome	0.673×10^{-7}	0.8×10^{-6}	2.4×10^{-6}

Although its pressure sensitivity is lower, gold chrome is preferred in many cases because of its much smaller temperature error.   This is particularly significant since the kerosene used in the bellows will experience a transient temperature change when sudden pressure changes occur, because of adiabatic compression or expansion.   The response of the wire resistance to pressure changes is practically instantaneous; however, the accompanying temperature change will cause a transient error if temperature sensitivity is too high.   Gages of the above type are commercially available with full scale up to 200,000 psi and inaccuracy of 0.1 to 0.5 percent.   They have also been used successfully for much higher pressures on a special-application basis.

[1] W. H. Howe, The Present Status of High Pressure Measurement and Control, *ISA J.*, p. 77, March, 1955.

The measurement of local contact pressures between rolling elements in gears, cams, and bearings may possibly be accomplished by depositing a thin strip of Manganin or gold chrome onto the surface as a pressure transducer. Preliminary studies[1] of such a technique, using a Manganin element 0.001 in. wide and 5×10^{-6} in. thick, have been reported.

6.9

***Low-pressure (Vacuum) Measurement*[2]**   Two commonly used units of vacuum measurement are the torr and the micron. One torr is a pressure equivalent to 1 mm Hg at standard conditions; one micron is 10^{-3} torr. Manometers and bellows gages are usable to about 0.1 torr, Bourdon gages to 10 torrs, and diaphragm gages to 10^{-3} torr. Below these ranges, other types of vacuum gages are necessary.

The McLeod gage. The McLeod gage is considered a vacuum standard since the pressure can be computed from the dimensions of the gage. It is not directly usable below about 10^{-4} torr; however, pressure-dividing techniques (see Sec. 6.1) allow its use as a calibration standard for considerably lower ranges. The multiple-compression technique[3] is also being studied to extend its range. The inaccuracy of McLeod gages is rarely less than 1 percent and may be much higher at the lowest pressures.

Of the many variations of McLeod gages, we here consider only the most basic. The principle of all McLeod gages is the compression of a sample of the low-pressure gas to a pressure sufficiently high to read with a simple manometer. Figure 6.20 shows the basic construction. By withdrawing the plunger, the mercury level is lowered to the position of Fig. 6.20a, admitting the gas at unknown pressure p_i. When the plunger is pushed in, the mercury level goes up, sealing off a gas sample of known volume V in the bulb and capillary tube A. Further motion of the plunger causes compression of this sample, and motion is continued until the mercury level in capillary B is at the zero mark. The unknown pressure is then calculated, using Boyle's law, as follows:

$$p_i V = p A_t h \qquad (6.91)$$
$$p = p_i + h\gamma \qquad (6.92)$$
$$p_i = \frac{\gamma A_t h^2}{V - A_t h} \approx \frac{\gamma A_t h^2}{V} \qquad \text{if } V \gg A_t h \qquad (6.93)$$

[1] F. K. Orcutt, Elastohydrodynamic Lubrication-Experimental Investigation, Mechanical Technology Inc., Latham, N.Y., *Rept.* 64TR6, 1964.

[2] S. Dushman, "Scientific Foundations of Vacuum Technique," chap. 6, John Wiley & Sons, Inc., New York, 1949.

[3] W. Kreisman, Extension of the Low Pressure Limit of McLeod Gages, *NASA*, *CR*-52877.

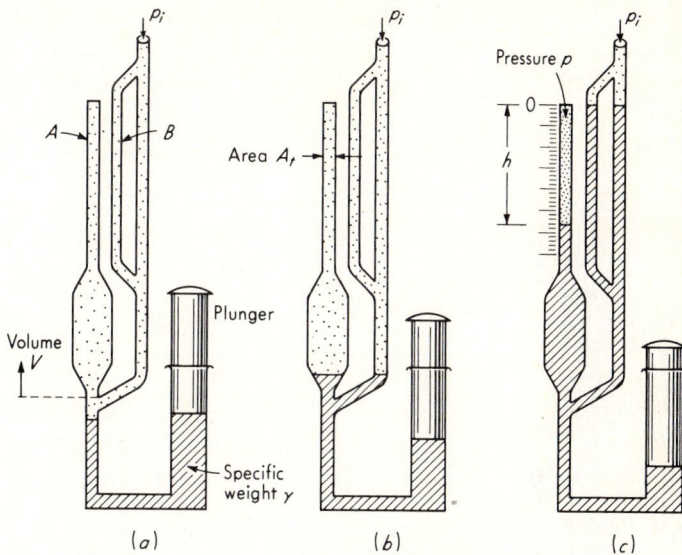

Fig. 6.20. *McLeod gage.*

In using a McLeod gage it is important to realize that if the measured gas contains any vapors that are condensed by the compression process the pressure will be in error. Except for this effect, the reading of the McLeod gage is not influenced by the composition of the gas. Only the Knudsen gage shares this desirable feature of composition insensitivity. The main drawbacks of the McLeod gage are the lack of a continuous output reading and the limitations on the lowest measurable pressures. When it is used to calibrate other gages, a liquid-air cold trap should be used between the McLeod gage and the gage to be calibrated to prevent the passage of mercury vapor.

The Knudsen gage. Although the Knudsen gage is little used at present, it is discussed briefly since it is relatively insensitive to gas composition and thus gives promise of development into a standard for pressures too low for the McLeod gage. In Fig. 6.21 the unknown pressure p_i is admitted to a chamber containing fixed plates heated to absolute temperature T_f, which temperature must be measured, and a spring-restrained movable vane whose temperature T_v must also be known. The spacing between the fixed and movable plates must be less than the mean free path of the gas whose pressure is being measured. The kinetic theory of gases shows that gas molecules rebound from the heated plates with greater momentum than from the cooler movable vane, thus giving a net

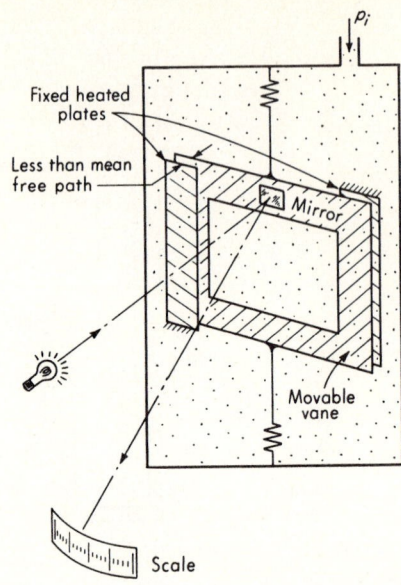

Fig. 6.21. *Knudsen gage.*   Scale

force on the movable vane which is measured by the deflection of the spring suspension.   Analysis shows that the force is directly proportional to pressure for a given T_f and T_v, following a law of the form

$$p_i = KF/\sqrt{T_f/T_v - 1}$$

where F is force and K is a constant.   The Knudsen gage is insensitive to gas composition except for the variation of accommodation coefficient from one gas to another.   The accommodation coefficient is a measure of the extent to which a rebounding molecule has attained the temperature of the surface.   This effect results, for example, in a 15 percent change in sensitivity between helium and air.   Knudsen gages at present cover the range from about 10^{-8} to 10^{-2} torr.

Momentum-transfer (viscosity) gages.   For pressures less than about 10^{-2} torr the kinetic theory of gases predicts that the viscosity of a gas will be directly proportional to the pressure.   The viscosity may be measured, for example, in terms of the torque required to rotate, at constant speed, one concentric cylinder within another.   (For pressures greater than about 1 torr, the viscosity is independent of pressure.)   The variation of viscosity with pressure is different for different gases; thus gages based on this principle must be calibrated for a specific gas.   While gages based on viscosity principles can measure to about 10^{-7} torr, such ranges are characteristic of laboratory-type equipment requiring great care in its use.

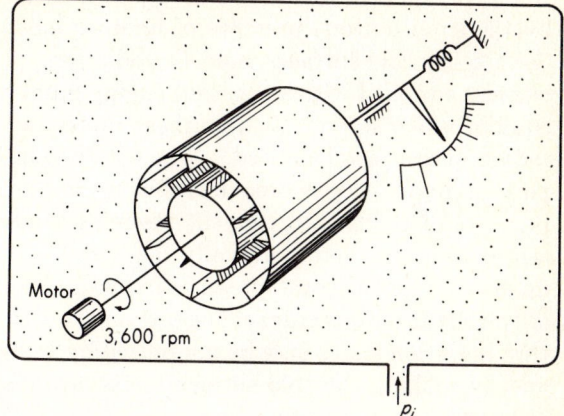

Fig. 6.22. *Momentum gage.*

A typical commercial gage,[1] shown schematically in Fig. 6.22, is calibrated for dry air and covers the range from 0 to 20 torrs. The range from 0 to 0.01 torr occupies about 10 percent of the total scale. The scale is nonlinearly calibrated because most of the range is above 10^{-2} torr and viscosity here is not proportional to pressure. To enable readings above 1 torr, where viscosity tends to become pressure-independent, bladed wheels rather than smooth concentric cylinders are used in this gage. These wheels cause a turbulent momentum exchange which is pressure-dependent above 1 torr, extending the useful range to 20 torrs. To reach the quoted lower limit (10^{-7} torr) of viscosity gages the construction of Fig. 6.22 is not used. Rather, the rate of decay of delicately constructed vibrating systems subjected to the damping effects of the gas is determined and pressure inferred from this.

Thermal-conductivity gages. Just as for viscosity, when the pressure of a gas becomes low enough that the mean free path of molecules is large compared with the pertinent dimensions of the apparatus, a linear relation between pressure and thermal conductivity is predicted by the kinetic theory of gases. For a viscosity gage the pertinent dimension is the spacing between the relatively moving surfaces. For a conductivity gage it is the spacing between the hot and cold surfaces. Again, when the pressure is increased sufficiently, conductivity becomes independent of gas pressure. The transition region between dependence and nondependence of viscosity and thermal conductivity on pressure is approximately the range 10^{-2} to 1 torr for apparatus of a size convenient to construct.

[1] General Electric Co.

The application of the thermal-conductivity principle is complicated by the simultaneous presence of another mode of heat transfer between the hot and cold surfaces, namely, radiation. Most gages utilize a heated element supplied with a constant energy input. This element assumes an equilibrium temperature when heat input and losses by conduction and radiation are just balanced. The conduction losses vary with gas composition and with gas pressure; thus for a given gas the equilibrium temperature of the heated element becomes a measure of pressure, and this temperature is what is actually measured. If the radiation losses are a major part of the total, pressure-induced conductivity changes will cause only a slight temperature change, giving poor sensitivity. Analysis shows that radiation losses may be minimized by using surfaces of low emissivity and by making the cold-surface temperature as low as practical. Since conduction and radiation losses depend on *both* the hot- and cold-surface temperatures, the cold surface may be maintained at a known constant temperature if overall accuracy warrants this measure. A further source of error is in the heat-conduction loss through any solid supports by which the heated element is mounted. The relative importance of the above-mentioned effects varies with the details of construction of the gage. The most common types of conductivity gages are the thermocouple, resistance thermometer (Pirani), and thermistor.

Figure 6.23 shows in schematic form the basic elements of a thermocouple vacuum gage. The hot surface is a thin metal strip whose temperature may be varied by changing the current passing through it. For a given heating current and gas the temperature assumed by the hot surface depends on pressure; this temperature is measured by a thermocouple welded to the hot surface. The cold surface here is the glass tube which usually is near room temperature. The accuracy of such gages is usually not high enough to warrant measurement or correction for changes in room temperature. Thermocouple gages of one type or another are available to measure in the range 10^{-4} to 1 torr.

In the resistance-thermometer (Pirani) gage the functions of heating and temperature measurement are combined in a single element. A typical construction is shown in Fig. 6.24. The resistance element is in the form of four coiled tungsten wires connected in parallel and supported inside a glass tube to which the gas is admitted. Again the cold surface is the glass tube. Two identical tubes are generally connected in a bridge circuit as shown. One of the tubes is evacuated to a very low pressure and then sealed off while the other has the gas admitted to it. The evacuated tube acts as a compensator to reduce the effect of bridge-excitation-voltage changes and temperature changes on the output reading. Current flowing through the measuring element heats it to a temperature depending on the gas pressure. The electrical resistance of the element

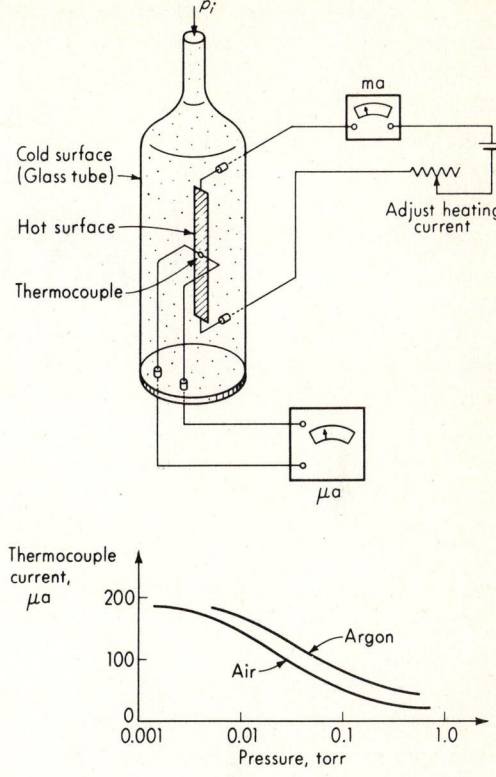

Fig. 6.23. *Thermocouple gage.*

changes with temperature, and this resistance change causes a bridge unbalance. The bridge is generally used as a deflection rather than a null device. To balance the bridge initially, the pressure in the measuring element is made very small and the balance pot set for zero output. Any changes in pressure will cause a bridge unbalance. Of course the gage must be calibrated against some standard. Calibration is nonlinear and varies from one gas to another. Pirani gages cover the range from about 10^{-5} to 1 torr.

Thermistor vacuum gages operate on the same principle as the Pirani gage except that the resistance elements are temperature-sensitive semiconductor materials called thermistors rather than metals such as tungsten or platinum. Thermistor gages are used in the range 10^{-4} to 1 torr.

Ionization gages. An electron passing through a potential difference acquires a kinetic energy proportional to the potential difference. When

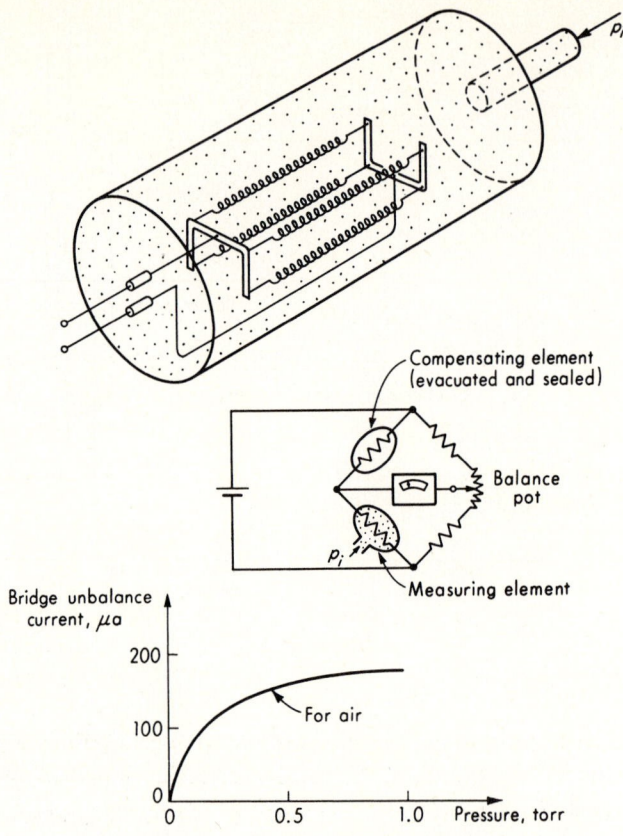

Fig. 6.24.   Pirani gage.

this energy is large enough, if the electron strikes a gas molecule there is a
definite probability that it will drive an electron out of the molecule,
leaving it as a positively charged ion.   In an ionization gage a stream of
electrons is emitted from a cathode.   Some of these strike gas molecules
and knock out secondary electrons, leaving the molecules as positive ions.
For normal operation of the gage the secondary electrons are a negligible
part of the total electron current; thus, for all practical purposes, electron
current i_e is the same whether measured at the emitting point (cathode) or
the collecting point (anode).   The number of positive ions formed is
directly proportional to i_e and directly proportional to the gas pressure.
If i_e is held fixed (as in most gages) the rate of production of positive ions
(ion current) is, for a given gas, a direct measure of the number of gas
molecules per unit volume and thus of the pressure.   The positive ions
are attracted to a negatively charged electrode which collects them and

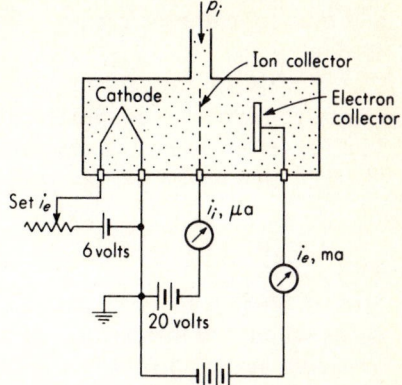

Fig. 6.25.   *Ionization gage.*

carries the ion current.   The "sensitivity" S of an ionization gage is defined by

$$S \triangleq \frac{i_i}{p i_e} \qquad (6.94)$$

where $i_i \triangleq$ ion current, gage output
$i_e \triangleq$ electron current
$p \triangleq$ gas pressure, gage input

According to our usual definition of sensitivity as output/input, the "sensitivity" would be $S i_e$ rather than S.   But the definition of Eq. (6.94) makes "sensitivity" independent of i_e and dependent only on gage construction.   This allows comparison of the "sensitivity" of different gages without reference to the particular i_e being used.   A main advantage of ionization gages in general is their linearity; that is, the sensitivity S is constant for a given gas over a wide range of pressures.

Figure 6.25 shows the basic elements of a hot-cathode ionization gage.   The emission of electrons is due to the heating of the cathode. Some disadvantages of hot-cathode gages are filament burnout if exposed to air while hot, decomposition of some gases by the hot filament, and contamination of the measured gas by gases forced out of the hot filament. Hot-cathode gages cover the range from 10^{-10} to 1 torr.

The Philips cold-cathode gage[1] overcomes the problems associated with a high-temperature filament by the use of a cold cathode and a high accelerating potential (2,000 volts).   A superimposed magnetic field causes the electrons ejected from the cathode to travel in long helical paths to the anode.   The long path results in more collisions with gas molecules

[1] J. M. Lafferty and T. A. Vanderslice, Vacuum Measurement by Ionization, *Instr. Control Systems*, p. 90, March, 1963.

and thus a greater ionization.   Philips gages are used in the range 10^{-5} to 10^{-2} torr.

For the lowest pressures, hot-cathode and cold-cathode gages of the magnetron type[1] are available.   They are useful down to about 10^{-13} torr. Mass spectrometers[1] are employed for even lower pressures; they allow identification of the partial pressures of components in gas mixtures.

6.10

Sound Measurement   The measurement of air-borne and water-borne sound is of increasing interest to engineers.   Air-borne-sound measurements are important in the development of less noisy machinery and equipment, in diagnosis of vibration problems, and in the design and test of sound-recording and-reproducing equipment.   In large booster rockets the sound pressures produced by the exhaust are large enough to cause fatigue failure of metal panels because of vibration ("acoustic fatigue"). Water-borne sound has been applied in underwater direction and range-finding equipment (sonar).   Since most sound transducers (microphones and hydrophones) are basically pressure-measuring devices, it is appropriate to consider them briefly in this chapter.

The basic definitions of sound are in terms of the magnitude of the fluctuating component of pressure in a fluid medium.   The *sound pressure level* is defined by

$$SPL \triangleq \text{sound pressure level} \triangleq 20 \log_{10} \frac{p}{0.0002}$$

$$\text{decibels (db)} \qquad (6.95)$$

where $p \triangleq$ root-mean-square (rms) sound pressure,

$$\text{microbar} (\mu b) \qquad (6.96)$$

$$1 \ \mu b = 1 \ \text{dyne/cm}^2 = 1.45 \times 10^{-5} \ \text{psi} \qquad (6.97)$$

The rms value of the fluctuating component of pressure is employed because most sounds are random signals rather than pure sine waves.   The value $0.0002 \ \mu b$ is an accepted standard reference value of pressure against which other pressures are compared by Eq. (6.95).   Note that when $p = 0.0002 \ \mu b$ the sound pressure level is 0 db.   This value was selected somewhat arbitrarily but represents the average threshold of hearing for human beings, if a 1,000-cps tone is used.   That is, the 0-db level was selected as the lowest pressure fluctuation normally discernible by human beings.   Since 0 db is about 3×10^{-9} psi, the remarkable sensitivity of the human ear should be apparent.   The decibel (logarithmic) scale is used as a convenience because of the great ranges of sound pressure level

---

[1] J. M. Lafferty and T. A. Vanderslice, Vacuum Measurement by Ionization, *Instr. Control Systems*, p. 90, March, 1963.

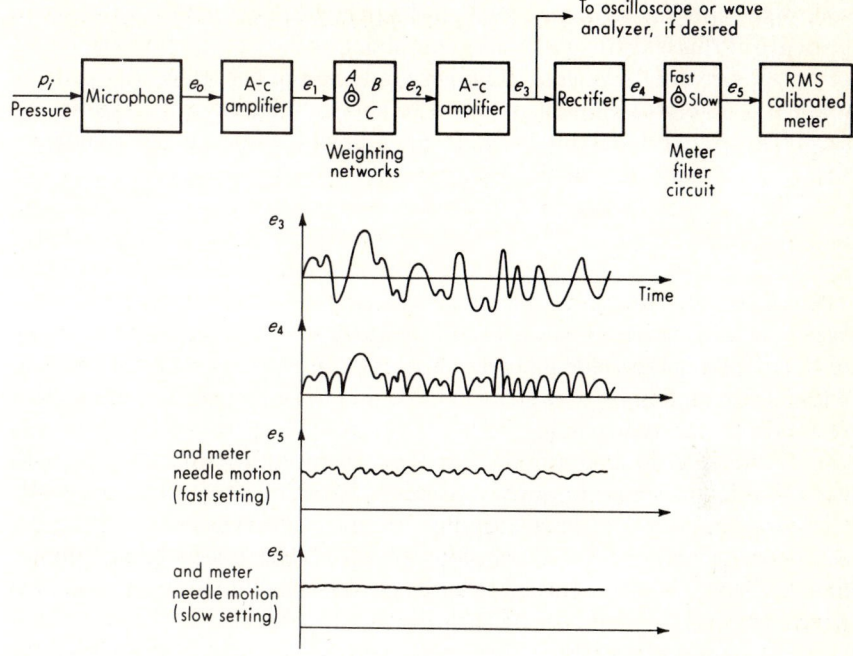

Fig. 6.26. *Sound-level meter.*

of interest in ordinary work. For example, an office with tabulating machines may have an *SPL* of 74 db (1 μb). The average human threshold of pain is 144 db. Sound pressures close to large rocket engines are the order of 170 db (1 psi). One atmosphere (14.7 psi) is 194 db. The range from the lowest to the highest pressures of interest is thus of the order of 10^{-9} to 1, a tremendous range.

The sound-level meter. The most commonly used instrument for routine sound measurements is the sound-level meter. This is actually a measurement *system* made up of a number of interconnected components. Figure 6.26 shows a typical arrangement. The sound pressure p_i is transduced to a voltage by means of the microphone. Microphones generally employ a thin diaphragm to convert pressure to motion. The motion is then converted to voltage by some suitable transducer, usually a capacitance, piezoelectric, or moving-coil type. Microphones usually have a "slow leak" (capillary tube) connecting the two sides of the diaphragm to equalize the average pressure (atmospheric pressure) and prevent bursting of the diaphragm. This is necessary because the (slow) hour-to-hour and day-to-day changes in atmospheric pressure are much greater than the sound-pressure fluctuations to which the microphone must respond. The presence of this leak dictates that microphones will not respond to constant or slowly varying pressures. This is usually no problem since many measurements involve a human response to the sound, and this is known to extend down to only about 10 to 20 cps. Thus the microphone frequency response need go only to this range, not to zero frequency.

The output voltage of the microphone is generally quite small and also at a high impedance level; thus an amplifier of high input impedance and gain is used at the output of the microphone. This can be a relatively simple a-c amplifier since response to static or slowly varying voltages is not required. Capacitor microphones often use for the first stage of the amplifier a cathode follower built right into the microphone housing. This close coupling reduces stray capacitance effects by eliminating cables at the high-impedance end.

Following the first amplifier are the weighting networks. They are electrical filters whose frequency response is tailored to approximate the frequency response of the average human ear. The ear does not interpret equal sound pressure levels at, say, 500 and 1,000 cps as being equally loud. Thus if we are trying to measure the "loudness" of a sound as heard by a person, a flat or uniform frequency response in the measuring instrument is not wanted. Measurements on human hearing have established the shape of the ear's frequency response fairly well; these shapes can be approximated by suitable filter networks. An additional complication is the nonlinearity of the ear's response in that the shape of its frequency-response curve varies with the *amplitude* of the sound. Thus the weighting network for low-level sound should be different from that for high. Most sound-level meters have three different settings (A,B,C) for the weighting networks. The A setting is used for sounds of 55-db level or below, the B for 55 to 85 db, and C for above 85 db. The C setting corresponds to a flat overall system frequency response and thus is used when an actual measurement of sound pressure level (rather than a human response to the sound) is desired. Readings taken on the A or B settings are called *sound level*, not sound pressure level.

The output of the weighting network is further amplified and an output jack provided to lead this signal to an oscilloscope (if observation of the wave form is desired) or to a wave analyzer (if the frequency content of the sound is to be determined). If only the overall sound magnitude is desired, the rms value of e_3 must be found. While true rms voltmeters are available, their expense is usually not justifiable in an ordinary sound-level meter. Rather, the *average* value of e_3 is determined by rectifying and filtering and then the meter scale is *calibrated* to read rms values. This procedure is exact for pure sine waves since there is a precise relation between the average value and the rms value of a sine wave. For nonsinusoidal waves this is not true, but the error is generally small enough to be acceptable for relatively unsophisticated work. The filtering is accomplished both by a simple low-pass *RC* filter and the low-pass meter dynamics. Some meters have a slow and fast response switch which changes the filtering. The "slow" position gives a steady, easy-to-read needle position but masks any short-term variations in the signal. If these short-term variations are of interest, they may be visually observed on the meter by switching it to fast response. While the meter is actually reading the rms value of e_3 (and thus of p_i), it is calibrated in decibels since Eq. (6.95) establishes a definite relation between sound pressure in microbars and decibels.

Microphones. While the design of microphones is a specialized and complex field with a large technical literature, we can here point out some of the main considerations. Frequency response is still of major interest; however, the effects on frequency response of sound wavelength and direction of propagation are aspects of dynamic behavior not regularly encountered in other measurements. The *pressure response* of a microphone refers to the frequency response relating the actual (uniform) sound pressure existing at the microphone diaphragm to the output voltage of the microphone. The pressure response of a given microphone may be estimated theoretically or measured experimentally by one of a number of accepted methods.[1]

What is usually desired is the *free-field response* of the microphone. That is, what is the relation between the microphone output voltage and the sound pressure that existed at the microphone location *before* the microphone was introduced into the sound field? The reason that the microphone distorts the pressure field is that its acoustical impedance is radically different from that of the medium (air) in which it is immersed. In fact, for most purposes, the microphone (including its diaphragm) may be considered as a rigid body. Sound waves impinging on this body give rise to complex reflections that depend on the frequency, the direction of propagation of the sound wave, and the microphone size and shape. When the wavelength of the sound wave is very large compared with the microphone dimensions (low frequencies), the effect of reflections is negligible for any angle of incidence between the diaphragm and the wave-propagation direction, and the free-field response is the same as the pressure response. At very high frequencies, where the wavelength is much smaller than the microphone dimensions, the microphone acts as an

[1] P. V. Bruel and G. Rasmussen, Free Field Response of Condenser Microphones, *B & K Tech. Rev.*, B & K Instruments Inc., Cleveland, Ohio, no. 1, January, 1959; no. 2, April, 1959.

infinite wall and the pressure at the microphone surface [for waves propagating perpendicular to the diaphragm (0° angle of incidence)] is twice what it would be if the microphone were not there. For waves propagating parallel to the diaphragm (90° incidence angle) the average pressure over the diaphragm surface is zero, giving no output voltage. Between the very low and very high frequencies the effect of reflections is quite complicated and depends on sound wavelength (frequency), microphone size and shape, and angle of incidence.

For simple geometrical shapes such as spheres and cylinders, theoretical results are available.[1] Experiments on actual microphones give results such as those shown in Fig. 6.27. Note that for sufficiently low frequencies (below a few thousand cycles per second) there is little change in pressure due to the presence of the microphone; also the angle of incidence has little effect. This flat frequency range can be extended by reducing the size of the microphone; however, smaller size tends to reduce sensitivity. The size effect is directly related to the relative size of the microphone and the wavelength of the sound. The wavelength λ of sound waves in air is roughly $13{,}000/f$ in., where f is frequency in cycles per second. When λ becomes comparable to the microphone-diaphragm diameter, significant reflection effects can be expected. For example, a 1-in.-diameter microphone would not be expected to have good response much above 13,000 cps. (These limitations can be relaxed to some extent by clever use of acoustical-mechanical techniques.[2])

The lower part of Fig. 6.27 shows a curve labeled "random incidence." This refers to the response to a diffuse sound field where the sound is equally likely to come to the microphone from any direction, the waves from all directions are equally strong, and the phase of the waves is random at the microphone position. Such a field may be approximated by constructing a room with highly irregular walls and placing reflecting objects of various sizes and shapes in it. A source of sound placed in such a room gives rise to a diffuse sound field at any point in the room. Microphones calibrated under such conditions are of interest because many sound measurements take place in enclosures which, while not giving perfect random incidence, certainly do not give pure plane waves.

Microphone calibrations may give the pressure response and also the free-field response for selected incidence angles, usually 0 and 90°. Figure 6.28 shows typical curves.

[1] L. Beranek, "Acoustic Measurements," chap. 3, John Wiley & Sons, Inc., New York, 1949.

[2] Gunnar Rasmussen, Miniature Pressure Microphones, *B & K Tech. Rev.*, B & K Instruments Inc., Cleveland, Ohio, no. 1, 1963.

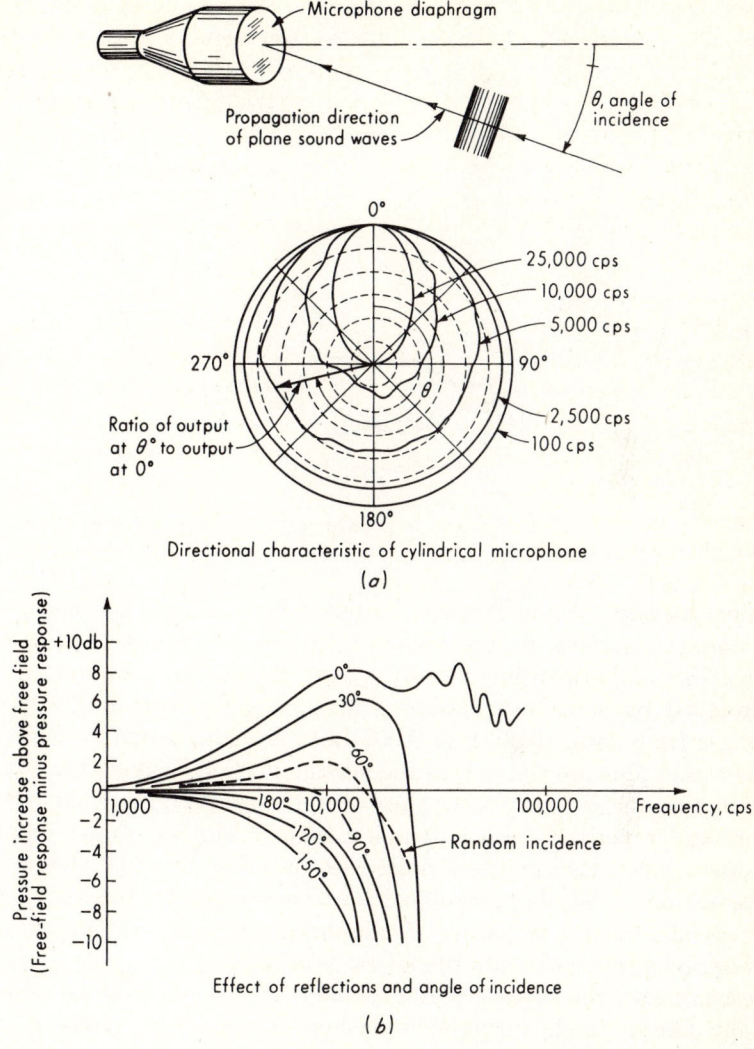

Directional characteristic of cylindrical microphone

(*a*)

Effect of reflections and angle of incidence

(*b*)

Fig. 6.27.   *Microphone response characteristics.*

Pressure response of a capacitor microphone. Of the several types of microphones in common use the capacitor type is generally considered to be capable of the highest performance.   Figure 6.29 shows in simplified fashion the construction of a typical capacitor microphone.   The pressure response is found by assuming a uniform pressure p_i to exist all around the microphone at any instant of time.   This is actually the case for

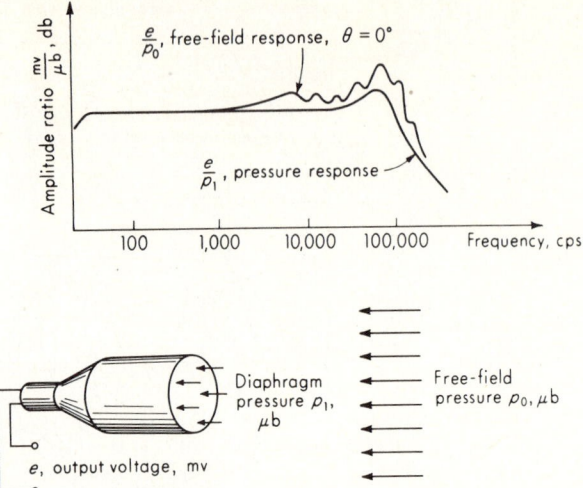

Fig. 6.28. *Free-field and pressure response.*

sufficiently low sound frequencies but reflection and diffraction effects distort this uniform field at higher frequencies, as pointed out earlier.

The diaphragm is generally a very thin metal membrane which is stretched by a suitable clamping arrangement. Diaphragm thickness ranges from about 0.0001 to 0.002 in. The diaphragm is deflected by the sound pressure and acts as the moving plate of a capacitance displacement transducer. The other plate of the capacitor is stationary and may contain properly designed damping holes. Motion of the diaphragm causes air flow through these holes with resulting fluid friction and energy dissipation. This damping effect is utilized to control the resonant peak of the diaphragm response. A diaphragm actually has many natural frequencies; however only the lowest is of interest here. For frequencies near or below the lowest natural frequency the diaphragm behaves essentially like a simple spring-mass-dashpot second-order system and may be analyzed as such.

A capillary air leak is provided to give equalization of steady (atmospheric) pressure on both sides of the diaphragm to prevent diaphragm bursting. For varying (sound) pressures the capillary-volume system results in the varying component of pressure acting *only* on the outside of the diaphragm and thus causing the desired diaphragm deflection.

The variable capacitor is connected into a simple series circuit with a high resistance R and "polarized" with a d-c voltage E_b of about 200 volts. This polarizing voltage acts as circuit excitation and also determines the neutral (zero-pressure) diaphragm position because of the elec-

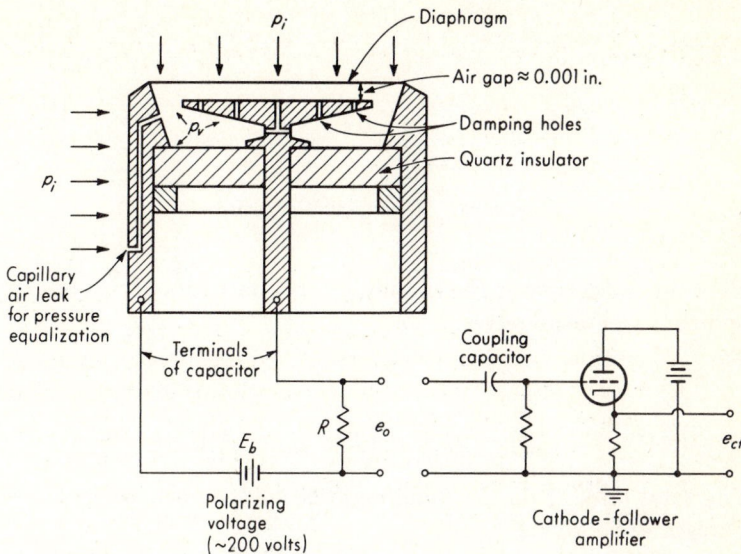

Fig. 6.29. *Capacitor microphone.*

trostatic attraction force between the capacitor plates. For a constant diaphragm deflection no current flows through R and no output voltage e_o exists; thus there is no response to static pressure differences across the diaphragm. For dynamic pressure differences a current *will* flow through R and an output voltage exists. The voltage e_o is usually applied to the input of a capacitance-coupled cathode-follower amplifier. A cathode follower always has a gain less than 1. Thus the purpose of the amplifier is not to increase the voltage level. Rather, the cathode follower has a high input impedance to prevent loading of the microphone, which has a high output impedance. Since the output impedance of the cathode follower is low, its output signal e_{cf} may be coupled into long cables and low-impedance loads without loss of signal magnitude.

The first step in the analysis involves determination of the effective force tending to deflect the diaphragm in terms of the pressure p_i. The relation between p_i and the pressure p_v in the microphone internal volume V may be obtained from Eq. (6.85). We shall neglect the inertia term since in microphones the viscous effect predominates and the filtering effect of the capillary is significant only at low frequencies. Thus we get

$$\tau_l \dot{p}_v + p_v = p_i \qquad (6.98)$$

where $\qquad \tau_l \triangleq$ leak time constant $\triangleq \dfrac{128\,\mu L V}{\pi E_a d_t^{\,4}} \qquad (6.99)$

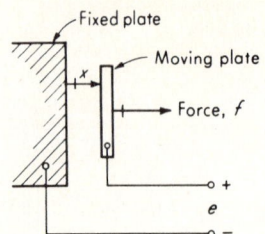

Fig. 6.30.   *Moving-plate capacitor.*

Now the deflection of the diaphragm is due to the *difference* between p_i and p_v. Operationally,

$$p_v = \frac{p_i}{\tau_l D + 1}$$

and thus

$$p_i - p_v = p_i\left(1 - \frac{1}{\tau_l D + 1}\right) = \frac{\tau_l D}{\tau_l D + 1} p_i \qquad (6.100)$$

The total force f_d on the diaphragm is $A_d(p_i - p_v)$, where A_d is the diaphragm area, giving

$$\frac{f_d}{p_i}(D) = \frac{A_d \tau_l D}{\tau_l D + 1} \qquad (6.101)$$

The frequency response of this shows clearly that $f_d \to 0$ as frequency $\to 0$; thus slow pressure changes do not result in forces tending to burst the diaphragm. However, the time constant τ_l must be small enough so that $(f_d/p_i)(i\omega) \approx A_d$ for all frequencies above about 10 cps, the lowest-frequency sound pressures usually of interest.

The next step requires study of the electromechanical-energy-conversion process in a moving-plate capacitor. While the moving "plate" (diaphragm) of the microphone is not flat, we shall analyze the situation for a flat plate for reasons of simplicity. One can always find a flat-plate capacitor that is equivalent to the diaphragm capacitor in the sense that the capacitance variation with plate separation is the same (at least for small motions) for both. Considering Fig. 6.30, we recall that

$$\text{Energy stored by a capacitor} = \frac{q^2}{2C} = \frac{Ce^2}{2} \qquad (6.102)$$

$$\begin{aligned}
\text{where} \quad & q \triangleq \text{charge} \\
& e \triangleq \text{voltage} \\
& C \triangleq \text{capacitance}
\end{aligned}$$

We wish to show that the two plates attract each other with a force f. The capacitance of a parallel-plate capacitor whose area A is large compared with the plate separation x is given very closely by

$$C = \frac{\epsilon A}{x} \qquad (6.103)$$

where $\epsilon \triangleq$ permittivity of material between plates
$= 8.86 \times 10^{-12}$ farad/m for vacuum or dry air

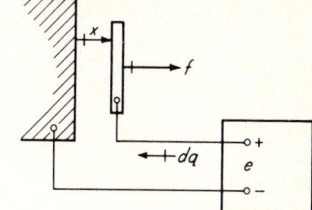

Fig. 6.31. *Capacitor with external circuit.*

We now suppose the capacitor is charged and then open-circuited so that q must remain constant. If the plates are now separated an additional amount dx, we may write

$$\text{Original energy } (x = x_0) = \frac{q^2}{2C_0} = \frac{q^2 x_0}{2\epsilon A} \qquad (6.104)$$

$$\text{Final energy } (x = x_0 + dx) = \frac{q^2}{2C_f} = \frac{q^2(x_0 + dx)}{2\epsilon A} \qquad (6.105)$$

The energy change is thus $q^2 dx/2\epsilon A$. Since energy is conserved in this system, it must have required a force f on the plate to cause the motion dx, since then mechanical work $f\,dx$ would have been done and converted into electrical energy $(q^2\,dx)/2\epsilon A$. The force f may thus be calculated from

$$\frac{q^2\,dx}{2\epsilon A} = f\,dx$$

$$f = \frac{q^2}{2\epsilon A} = \frac{\epsilon A e^2}{2x^2} \qquad (6.106)$$

For air, with e in volts and A and x in any consistent units, this becomes

$$f = 0.99 \times 10^{-12}\,\frac{Ae^2}{x^2} \qquad \text{lb}_f \qquad (6.107)$$

As an example, if $e = 200$ volts, $A = 1$ in.^{2}, and $x = 0.001$ in., the force is 0.04 lb$_f$.

If the capacitor is connected to an external circuit as in Fig. 6.31, we can show Eq. (6.106) still holds as follows: The work done in moving a charge dq through a potential difference e is $e\,dq$. Then, by conservation of energy,

$$f\,dx + e\,dq = d(\text{stored energy}) = d\left(\frac{Ce^2}{2}\right) \qquad (6.108)$$

Then, $\qquad f = -e\,\frac{dq}{dx} + \frac{d}{dx}\left(\frac{Ce^2}{2}\right) = -e\,\frac{d}{dx}\,(Ce) + \frac{d}{dx}\left(\frac{Ce^2}{2}\right) \qquad (6.109)$

$$f = -e\left(C\,\frac{de}{dx} + e\,\frac{dC}{dx}\right) + Ce\,\frac{de}{dx} + \frac{e^2}{2}\,\frac{dC}{dx} \qquad (6.110)$$

$$f = -\frac{e^2}{2}\,\frac{dC}{dx} = -\frac{e^2}{2}\left(-\frac{\epsilon A}{x^2}\right) = \frac{\epsilon A e^2}{2x^2} \qquad (6.111)$$

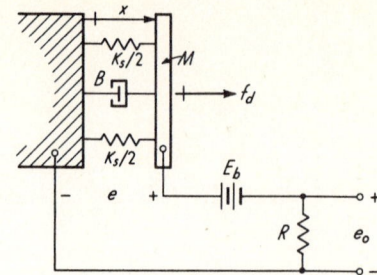

Fig. 6.32. *Microphone model.*

We next model the microphone as in Fig. 6.32. The mass M and spring K_s must be such as to give the same natural frequency as the lowest natural frequency of the diaphragm. The dashpot B must be such as to give the same resonant peak as in the microphone's measured pressure response. The capacitor plate area and air gap (with no external forces acting) must be such as to give the same capacitance as is measured for the microphone under similar conditions. The spring constant K_s and capacitor dimensions must also be such that the force f_d causes a capacitance variation equal (at least for small motions) to that caused in the actual microphone by a pressure difference $p_i - p_v = f_d/A_d$. If all the above conditions are met, the simplified model of Fig. 6.32 will respond essentially the same as the microphone itself. While the equivalent system described is defined in terms of experimental measurements on an existing microphone, microphone designers have available theoretical formulas for estimating these parameters *before* a new microphone is built.

Assuming that the equivalent system is a reasonable model, we can proceed with the analysis. With no force f_d applied and with the capacitor uncharged, the mass will assume an equilibrium position x_{fl}, where x_{fl} is the free length of the springs. If the polarizing voltage E_b is now applied, the moving plate will experience an attractive force and will move to a new position x_0 such that the spring force and electrostatic force just balance (see Fig. 6.33). Now, when pressure force f_d is applied, motion will take place around x_0 as an operating point. To find x_0, we can write

$$K_s(x_{fl} - x_0) = \frac{\epsilon A E_b^2}{2x_0^2} \qquad (6.112)$$

This equation in x_0 has two positive solutions x_0 and x_0' for a practical case. The solution (equilibrium position) x_0' is unstable in the sense that any slight motion away from x_0' results in *further* motion away from this point. The desired (stable) equilibrium position is x_0, where small disturbances

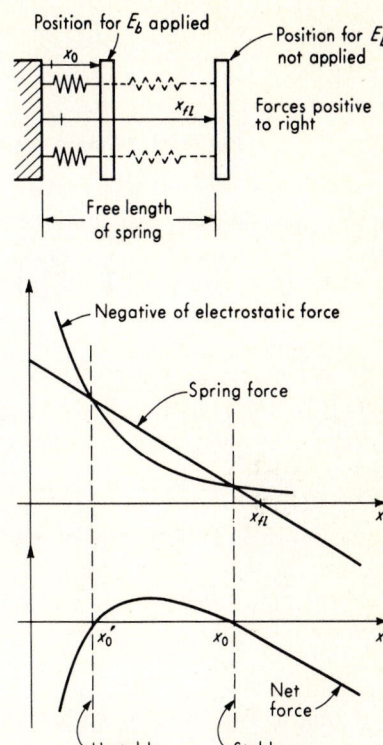

Fig. 6.33. *Determination of equilibrium points.*

from equilibrium give rise to forces tending to restore equilibrium. The microphone must thus be designed to operate at x_0 rather than x_0'.

We now apply Newton's law to the mass M to get

$$-B\frac{dx}{dt} + K_s(x_{fl} - x) - \frac{\epsilon A e^2}{2x^2} + f_d = M\frac{d^2x}{dt^2} \qquad (6.113)$$

The electrostatic-force term makes this differential equation nonlinear. For small changes in e and x from the equilibrium operating point this nonlinear term may be linearized approximately with good accuracy. This may be done by employing only the linear terms of a Taylor-series expansion of the nonlinear function. That is, if in general

$$z = z(x,y)$$

then $\qquad z \approx z(x_0,y_0) + \frac{\partial z}{\partial x}\bigg|_{\substack{x=x_0\\y=y_0}} (x - x_0) + \frac{\partial z}{\partial y}\bigg|_{\substack{x=x_0\\y=y_0}} (y - y_0) \qquad (6.114)$

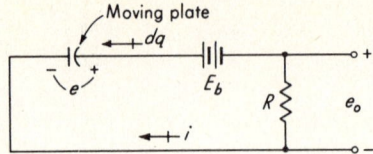

Fig. 6.34. *Circuit analysis.*

In this specific case, the nonlinear function is e^2/x^2; thus

$$\frac{e^2}{x^2} \approx \frac{E_b{}^2}{x_0{}^2} + E_b{}^2 \left(-\frac{2}{x_0{}^3}\right)(x - x_0) + \frac{1}{x_0{}^2} 2E_b(e - E_b) \qquad (6.115)$$

$$\text{We now define} \quad x_1 \triangleq x - x_0$$
$$e_o \triangleq e - E_b$$

to get

$$\frac{e^2}{x_2} \approx \frac{E_b{}^2}{x_0{}^2} - \frac{2E_b{}^2}{x_0{}^3} x_1 + \frac{2E_b}{x_0{}^2} e_o \qquad (6.116)$$

Also

$$K_s(x_{fl} - x) = K_s(x_{fl} - x_1 - x_0) = -K_s x_1 + \frac{\epsilon A E_b{}^2}{2x_0{}^2} \qquad (6.117)$$

Equation (6.113) may now be written as

$$-B \frac{dx_1}{dt} - K_s x_1 + \frac{\epsilon A E_b{}^2}{2x_0{}^2} - \frac{\epsilon A}{2}\left(\frac{E_b{}^2}{x_0{}^2} + \frac{2E_b}{x_0{}^2} e_o - \frac{2E_b{}^2}{x_0{}^3} x_1\right)$$
$$+ f_d = M \frac{d^2 x_1}{dt^2} \qquad (6.118)$$

Bringing in Eq. (6.101), we may write

$$\left[MD^2 + BD + \left(K_s - \frac{\epsilon A E_b{}^2}{x_0{}^3}\right)\right] x_1 + \frac{\epsilon A E_b}{x_0{}^2} e_o = \frac{A_d \tau_l D}{\tau_l D + 1} p_i \qquad (6.119)$$

This equation contains two unknowns, x_1 and e_o; thus an additional equation must be found before a solution can be effected. This can be found from an analysis of the circuit of Fig. 6.34 as follows:

$$e_o = e - E_b = iR = -\frac{dq}{dt} R \qquad (6.120)$$

$$q = Ce = \frac{\epsilon A e}{x} \qquad (6.121)$$

Equation (6.121) may be linearized as

$$q \approx \frac{\epsilon A E_b}{x_0} - \frac{\epsilon A E_b}{x_0{}^2} x_1 + \frac{\epsilon A}{x_0} e_o \qquad (6.122)$$

Then, approximately,

$$\frac{dq}{dt} = -\frac{\epsilon A E_b}{x_0{}^2} \frac{dx_1}{dt} + \frac{\epsilon A}{x_0} \frac{de_o}{dt} \qquad (6.123)$$

and

$$\frac{e_o}{R} = -\frac{dq}{dt} = \frac{\epsilon A E_b}{x_0{}^2} \frac{dx_1}{dt} - \frac{\epsilon A}{x_0} \frac{de_o}{dt} \qquad (6.124)$$

thus finally giving

$$-\frac{\epsilon A E_b R}{x_0{}^2} D x_1 + \left(1 + \frac{\epsilon A R}{x_0} D\right) e_o = 0 \qquad (6.125)$$

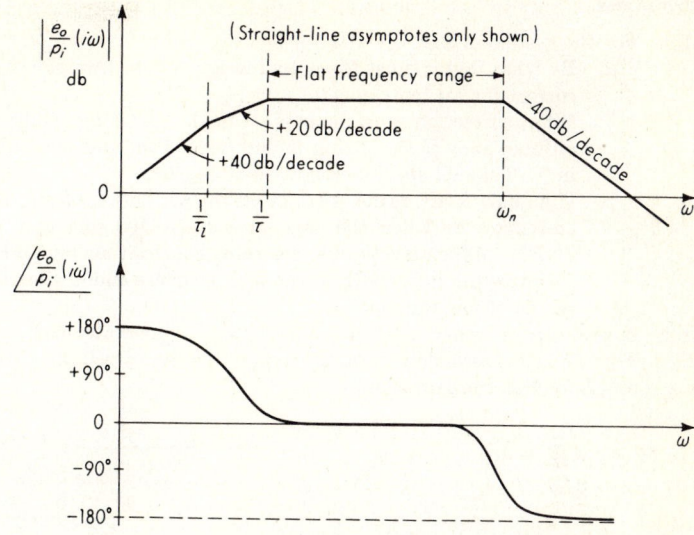

Fig. 6.35. *Microphone frequency response.*

Since we are primarily interested in e_o rather than x_1, Eq. (6.125) may be combined with (6.119) to eliminate x_1 and get

$$\left[\frac{M\tau_e}{K_e} D^3 + \left(\frac{M}{K_e} + \frac{B\tau_e}{K_e}\right) D^2 + \left(\frac{B}{K_e} + \tau_e + \frac{\tau_e^2 E_b^2}{x_0^2 R K_e}\right) D + 1\right] e_o$$

$$= \frac{A_d E_b \tau_e}{K_e x_0} \frac{\tau_l D^2}{\tau_l D + 1} p_i \qquad (6.126)$$

where

$$K_e \triangleq K_s - \frac{\epsilon A E_b^2}{x_0^3} \qquad (6.127)$$

$$\tau_e \triangleq \frac{\epsilon A R}{x_0} \qquad (6.128)$$

The cubic left-hand side is not readily factored until numerical values are known. In general, one gets two complex roots and one real root. This leads to a transfer function of the form

$$\frac{e_o}{p_i} (D) = \frac{K D^2}{(\tau_l D + 1)(\tau D + 1)(D^2/\omega_n^2 + 2\zeta D/\omega_n + 1)} \qquad (6.129)$$

The frequency response of the microphone is then as shown in Fig. 6.35. The sensitivity in the flat range is typically of the order of 1 to 5 mv/μb while the low-frequency cutoff is about 1 to 10 cps, though lower values are possible. The upper limit of frequency can be extended well beyond the range of human hearing; 100,000 cps is not unattainable.

Problems

6.1 For the system of Fig. 6.1:

    a. By what factor must the actual weight of steel weights be multiplied to correct for air buoyancy?

    b. What correction must be applied to the platform weight to account for oil buoyancy if the piston is immersed 5 in. and has a diameter of 0.2 in.? Take oil specific weight as 50 lb_f/ft^3.

    c. If, in part *b*, air, rather than oil, is the pressure medium, what would the correction be when the gage pressure is 100 psig and temperature is 70°F? Make an estimate, assuming constant air temperature and pressure varying linearly from the high-pressure end of the piston to atmospheric at the top end.

6.2 A well-type mercury manometer used to measure water flow rate is shown in Fig. P6.1 for zero flow rate ($p_1 = p_2$). Derive a relation between $p_1 - p_2$ and h for this configuration.

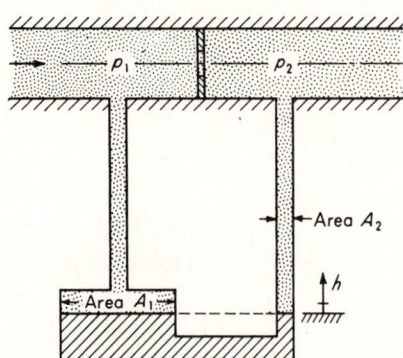

Fig. P6.1

6.3 For the inclined manometer of Fig. 6.3, derive a relation between $p_1 - p_2$ and displacement reading along the calibrated scale.

6.4 Estimate the largest step change that will give linear behavior in a water manometer with $L = 26.5$ in. and $R = 0.13$ in. What are ζ and ω_n for this manometer? If a step change five times the value found above is applied, estimate ζ_e and ω_n for this situation.

6.5 In Eq. (6.44), get an expression for $d\zeta_e/\zeta_e$ by taking the log of both sides and then differentiating. If L, μ, γ, g, R, ω, and Y are each in error by 1 percent, what is the percent error in ζ_e?

6.6 From Eq. (6.46), plot p versus y_c/t for $0 \le y_c/t \le 1$ and $E = 25 \times 10^6$ psi, $t = 0.04$ in., $R = 4.0$ in., and Poisson's ratio $= 0.26$.

6.7 Design a pressure pickup and bridge circuit such as that in Fig. 6.9 to meet the following requirements:

    Maximum pressure $= 100$ psig

    Natural frequency in vacuum $= 10$ cps minimum

    Maximum nonlinearity by Eq. (6.46) $= 3$ percent

    Full-scale output $= 10$ mv minimum

    Diaphragm material is stainless steel.

    Strain gages with 350 ohms resistance, gage factor of 2, and size 0.3 by 0.3 in. are to be used.

6.8   A pressure pickup as in Fig. 6.9 has the following characteristics:

$R = 3.0$ in. $E = 28 \times 10^6$ psi   Gage resistance = 120 ohms

$r_r = 2.5$ in. $\mu = 0.26$ Gage factor = 2.0

$r_t = 0.5$ in. $\gamma = 0.3$ lb$_f$/in.^{3} Battery voltage = 5.0 volts

$t = 0.05$ in.

    a. Calculate the sensitivity in mv/psi.

    b. What is the natural frequency in vacuum?

    c. Based on Eq. (6.46), what is the maximum allowable pressure for 2 percent nonlinearity?   What is the voltage output at this point?

6.9   Explain in words the operation of the system of Fig. 6.10*a*.   Derive an equation relating p_o to $p_1 - p_2$ and show how linearity is achieved even if K_n varies.

6.10   Explain in words the operation of the system of Fig. 6.10*b*.   Derive an equation relating e_o to $p_1 - p_2$.

6.11   Perform a linearized dynamic analysis of the system of Fig. 6.10*a* and discuss stability/accuracy tradeoff.

6.12   Perform a linearized dynamic analysis of the system of Fig. 6.10*b* and discuss stability/accuracy tradeoff.

6.13   From Eq. (6.57), compute τ for a system with $C_{vp} = 0.4$ cm^3/psi, $d_t = 0.10$ in., $L = 10$ ft, and $\mu = 0.001$ lb$_f$-sec/ft^2.   Using Eq. (6.58), find the dead time associated with this system if the tube is of steel with a wall thickness of 0.02 in., $E_L = 100,000$ psi, and the fluid specific weight is 0.03 lb$_f$/in.^{3}.   Is this dead time significant relative to τ?

6.14   A pressure transducer has a natural frequency in vacuum of 5,000 cps and $C_{vp} = 0.0003$ in.^{3}/psi.   It is used with a liquid of specific weight 0.04 lb$_f$/in.^{3} and viscosity 0.0005 lb$_f$-sec/ft^2.   The tubing inside diameter is 0.2 in., and its length is 5 ft.   Find ω_n and ζ of the combined transducer/tubing system.

6.15   The pressure pickup of Prob. 6.14 has an internal volume of 0.004 in.^{3}.   If it is used with the same tubing as in Prob. 6.14 but if the fluid medium is changed to air at 100 psia and 100°F, what are the values of ζ and ω_n?

6.16   If the transducer of Prob. 6.15 is used with tubing of 1-in. length and 0.1-in. inside diameter, what will ω_n and ζ be?

6.17   Compute the resistance change of 100-ohm coils of Manganin and gold chrome for 50,000-psi pressure and 100°F temperature changes.

6.18   Design a capillary leak for a microphone such that frequencies of 10 cps and above will be measured with an amplitude-ratio error of no more than 10 percent.   Assume standard atmospheric air and a leak length L of 1 in.   Find the required leak diameter d_t.   The microphone has an internal volume of 0.5 in.^{3}.   What will be the amplitude ratio for atmospheric pressure drifts of 2-cph frequency?

Bibliography

1.   J. B. Damrel: Quartz Bourdon Gage, *Instr. Control Systems*, p. 87, February, 1963.

2.   W. R. Myers: The Electromanometer, *Instr. Control Systems*, p. 116, April, 1962.

3.   R. R. Koolman: Reference Pressure Cells, *Instr. Control Systems*, p. 123, February, 1962.

4.   Dead-weight Testers, *Instr. Control Systems*, p. 126, April, 1962.

5.   R. C. Schumacher: Automatic Pressure Calibration Systems, *Instr. Control Systems*, p. 83, February, 1964.

6.   H. Norville: A Dead Weight Pressure Balance with Extended Range to 5000 psig, *NASA*, N-64-33684, 1964.

7. S. Siegel: Pressure Calibration Circuits, *Instr. Control Systems*, p. 116, February, 1965.

8. R. C. Cerni: Measuring 100 Pressures in 15 Seconds, *Control Eng.*, p. 89, July, 1964.

9. R. I. Kreisler: Rocket Propellant Manometer System, *ISA J.*, p. 55, October, 1962.

10. W. W. Willmarth: Wall Pressure Fluctuations in a Turbulent Boundary Layer, *NACA, Tech. Note* 4139, 1958.

11. P. K. Stein: Measuring Fluctuating Pressure, *Instr. Control Systems*, p. 156, September, 1964.

12. Y. T. Li: Pressure Transducers for Missile Testing and Control, *ISA J.*, p. 81, November, 1958.

13. L. R. Voss et al.: Recent Developments in Balanced-diaphragm Pressure Transducers, *ISA J.*, p. 348, September, 1955.

14. A. F. Welch et al.: Auxiliary Equipment for the Capacitor-type Transducer, *ISA J.*, p. 548, December, 1955.

15. E. J. Rogers: Semiconductor Pressure Transducer Features Mechanical Compensation, *Instr. Control Systems*, p. 128, April, 1963.

16. D. B. Clark: Rare-earth Pressure Transducers, *Instr. Control Systems*, p. 93, February, 1963.

17. J. C. Sanchez: Semiconductor Strain-gage Pressure Sensors, *Instr. Control Systems*, p. 117, November, 1963.

18. Y. T. Li: Two-cylinder Transducer Has Straight Line Response, *Control Eng.*, p. 151, April, 1962.

19. Y. Kobashi et al: Improvements of a Pressure Pickup for the Measurement of Turbulence Characteristics, *J. Aerospace Sci.*, p. 149, February, 1960.

20. T. Wrathall: Measuring Impact Pressures on Re-entering Missile Nose Cones. *ISA J.*, p. 54, October, 1959.

21. G. E. Reis: Theoretical Examination of Variable Reluctance Diaphragm Gage, Sandia Corp., Albuquerque, N.Mex., *SCR-162*, 1960.

22. Capacitive Pressure Sensors, *Instr. Control Systems*, p. 119, May, 1962.

23. P. Smelser: Pressure Measurements in Cryogenic Systems, National Bureau of Standards, Boulder, Colo.

24. H. Chelner: High Frequency Semiconductor Probe Pressure Transducer, *AIAA Paper* 64-508, 1964.

25. J. H. Thomson: Torsion Bar Pressure Transducer, *Electromech. Design*, p. 46, June, 1964.

26. D. S. Johnson: Design and Application of Piezoceramic Transducers to Transient Pressure Measurements, *NASA*, N-63-18139, 1962.

27. R. E. Engdahl: Pressure Measuring Systems for Closed Cycle Liquid Metal Facilities, *NASA, CR*-54140.

28. J. A. Haner: Pressure/Displacement Transducer, *Instr. Control Systems*, p. 107, November, 1964.

29. D. D. Keough et al: Piezoresistive Pressure Transducer, *ASME Paper* 64-WA/PT-5, 1964.

30. Pressure Transducer $\frac{3}{8}$-inch in Size Can be Faired into Surface, *NASA, Brief* 64-10021, 1964.

31. Welded Pressure Transducer Made as Small as $\frac{1}{8}$-inch Diameter, *NASA, Brief* 63-10429, 1963.

32. Improved Variable-reluctance Transducer Measures Transient Pressure, *NASA, Brief* 63-10321, 1963.

33. H. B. Jones et al.: Transient Pressure Measurements in Liquid Propellant Rocket Thrust Chambers, *ISA Trans.*, p. 117, April, 1965.

34. R. L. Ledford and W. E. Smotherman: Miniature Transducers for Pressure and Heat Transfer Rate Measurements in Hypervelocity Wind Tunnels, *ISA Trans.*, p. 133, April, 1965.

35. P. S. Lederer and R. O. Smith: An Experimental Technique for the Determination of the Fidelity of the Dynamic Response of Pressure Transducers, *Natl. Bur. Std. (U.S.), Rept.* 7862, 1963.

36. R. O. Smith: A Liquid-medium Step-function Pressure Calibrator, *ASME Paper* 63-WA-263, 1963.

37. J. L. Schweppe: Calibration of Pressure Transducers with Aperiodic Input-function Generators, *ISA Trans.*, p. 72, January, 1964.

38. W. C. Bentley and J. J. Walter: Dynamic Response Testing of Transient Pressure Transducers for Liquid Propellant Rocket Combustion Chambers, *NASA, CR*-51995, 1963.

39. W. E. Amend: Dynamic Performance of Pressure Transducers in Shock and Detonation Tubes, *NASA, N*-65-13313.

40. D. Baganoff: Pressure Gauge with One-tenth Microsecond Risetime for Shock Reflection Studies, *Rev. Sci. Instr.*, p. 288, March, 1964.

41. E. L. Davis: The Measurement of Unsteady Pressures in Wind Tunnels, *AGARD Rept.* 169, March, 1958.

42. R. Oldenburger and R. E. Goodson: Hydraulic Line Dynamics, *NASA, CR*-52148, 1963.

43. T. R. Stalzer and G. J. Fiedler: Criteria for Validity of Lumped-parameter Representation of Ducting Air-flow Characteristics, *ASME Trans.*, p. 833, May, 1957.

44. F. Nagao and M. Ikegami: Errors of an Indicator Due to a Connecting Passage, *Bull. JSME*, vol. 8, no. 29, 1965.

45. A. L. Ducoffe and F. M. White: The Problem of Pneumatic Pressure Lag, *ASME Trans.*, p. 234, June, 1964.

46. A. S. Iberall: Attenuation of Oscillatory Pressures in Instrument Lines, *Natl. Bur. Std. (U.S.), Res. Paper* RP2115, July, *ASME Trans.*, 1950.

47. F. Nagao et al.: Influence of the Connecting Passage of a Low Pressure Indicator on Recording, *Bull. JSME*, vol. 6, no. 21, 1963.

48. C. B. Schuder and G. C. Blunck: The Driving Point Impedance of Fluid Process Lines, *ISA Trans.*, p. 39, January, 1963.

49. R. P. Benedict: The Response of a Pressure-sensing System, *ASME Paper* 59-A-289, 1959.

50. R. J. Martin and D. S. Moseley: Analysis of the Effect of Pulsations on the Response of Mercurial-type Differential-pressure Recorders, *ASME Trans.*, p. 1343, October, 1958.

51. J. E. Broadwell and A. G. Hammitt: Transient Response of Fluid Systems, *J. Aerospace Sci.*, July, 1962.

52. A. F. D'Souza and R. Oldenburger: Dynamic Response of Fluid Lines, *ASME Paper* 63-WA-73, 1963.

53. R. Oldenburger and R. E. Goodson: Simplification of Hydraulic Line Dynamics by Use of Infinite Products, *ASME Paper* 62-WA-55, 1962.

54. J. D. Regetz: Experimental Determination of the Dynamic Response of a Long Hydraulic Line, *NASA, Tech. Note* D-576, 1960.

55. W. Lewis et al.: Study of the Effect of a Closed-end Side Branch on Sinusoidally Perturbed Flow of Liquid in a Line, *NASA, Tech. Note* D-1876, 1963.

56. I. Taback: The Response of Pressure Measuring Systems to Oscillating Pressures, *NACA, Tech. Note* 1819, 1949.

57. F. T. Brown: The Transient Response of Fluid Lines, *ASME Paper* 61-WA-143, 1961.

58. C. B. Schuder and R. C. Binder: The Response of Pneumatic Transmission Lines to Step Inputs, *ASME Trans.*, p. 578, December, 1959.
59. High Pressure Measurement, *Mech. Eng.*, p. 76, February, 1963.
60. D. H. Newhall and L. H. Abbott: High Pressure Measurement, *Instr. Control Systems*, p. 232, February, 1961.
61. W. H. Howe: High-pressure Measurement and Control, *Control Eng.*, p. 53, April, 1955.
62. A. J. Yerman: The Tunnel Diode as an FM Hydrostatic Pressure Sensor, *ASME Paper* 63-WA-264, 1963.
63. R. J. Melling: Ionization Vacuum Gage Measures Absolute Pressures up to 1 mm Hg, *Instr. Control Systems*, p. 119, September, 1964.
64. Vacuum Instrumentation, *Instr. Control Systems*, p. 110, September, 1964.
65. J. M. Lafferty and T. A. Vanderslice, Vacuum Measurement by Ionization, *Instr. Control Systems*, p. 90, March, 1963.
66. A. P. Flanick and J. Ainsworth: A Thermistor Pressure Gage, *NASA, Tech. Note* D-504, 1960.
67. J. M. Benson: Calibrating Thermal Conductivity Gauges, *Instr. Control Systems*, p. 115, September, 1964.
68. J. P. Walsh: Molecular Vacuum Gages, *Instr. Control Systems*, p. 106, August, 1963.
69. W. Kreisman: Extension of the Low Pressure Limit of McLeod Gages, *NASA, CR*-52877, 1963.
70. M. P. Hnilicka: Extreme High Vacuum, *Ind. Res.*, p. 36, September, 1964.
71. J. M. Benson: Thermal Conductivity Vacuum Gages, *Instr. Control Systems*, p. 98, March, 1963.
72. D. Alpert: Theoretical and Experimental Studies of the Underlying Processes and Techniques of Low Pressure Measurement, *NASA, N*-64-17582, 1964.
73. P. J. Bryant et al.: Extreme Vacuum Technology, *NASA, CR*-84, 1964.
74. Vacuum Instrumentation, *Instr. Control Systems*, p. 113, October, 1964.
75. Precision Gage Measures Ultrahigh Vacuum Levels, *NASA, Brief* 63-10597, 1963.
76. Absolute Pressure Gage Feasibility Study, *NASA, CR*-58075, 1963.
77. J. Gavis: Vacuum Gage Systems, *NASA, N*-64-28208, 1964.
78. R. W. Roberts: Ultrahigh Vacuum Technology, General Electric Co., Schenectady, N.Y., *Rept.* 64-RL-3644C, 1964.
79. R. W. Roberts: An Outline of Vacuum Technology, General Electric Co., Schenectady, N.Y., *Rept.* 64-RL-3394C, 1964.
80. L. T. Melfi and P. R. Yeager: A Method for Calibration of Gas-composition Sensitive Pressure Gages in Condensible Vapors, *NASA, Tech. Note* D-2567, 1965.
81. F. Feakes et al.: Gauge Calibration Study in Extreme High Vacuum, *NASA, CR*-167, 1965.
82. S. W. Athey: Acoustics Technology, A Survey, *NASA* SP-5093, 1970.
83. D. O. Conn, III: The Audio Dosimeter—A System for Measuring Personal Noise Exposure, *Sound & Vib.*, September, 1972.
84. The Electret-Condenser Microphone: A New Quality Microphone for Acoustical Measurements, *GR Today*, General Radio Co., Concord, Mass., Autumn, 1972.
85. H. C. Sommer: Description and Use of a Measurement System for Air Bag Acoustic Transient Data Acquisition and Analysis, *Aero. Med. Res. Lab. Dept.* AMRL-TR-73-8, Wright Patterson AF Base, Ohio., 1973.
86. T. W. Nyland and R. E. Chase: High-Temperature Transient Pressure Transducer for Use in Liquid-Metal Systems, *NASA* TN D-5589, 1969.

Chapter 7
Flow
Measurement

7.1

Local Flow Velocity, Magnitude, and Direction   In many experimental studies of fluid flow phenomena it is necessary to determine the magnitude and/or direction of the flow-velocity vector at a "point" in the fluid and how this varies from point to point.   That is, a description of the flow field is desired.   While the conditions at a mathematical *point* are not susceptible to direct measurement, the *average* conditions over a small area or volume can be determined with suitable instruments.

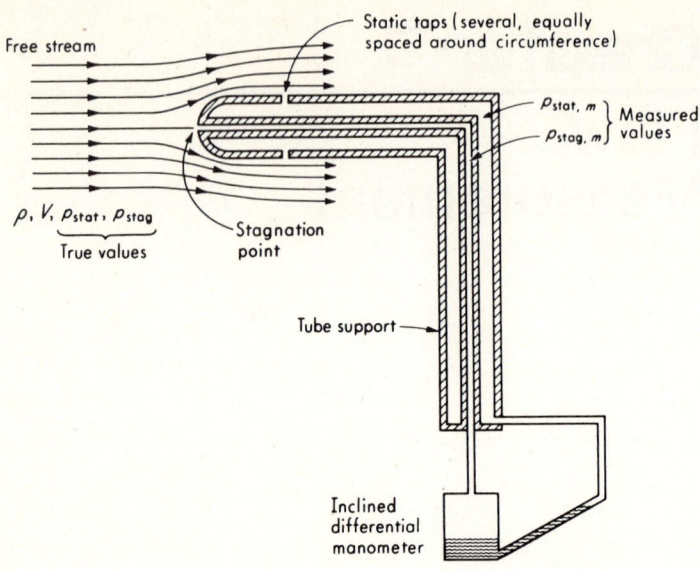

Free stream

Static taps (several, equally spaced around circumference)

$p_{stat, m}$ ⎫ Measured
$p_{stag, m}$ ⎬ values

$\rho, V, p_{stat}, p_{stag}$
True values

Stagnation point

Tube support

Inclined differential manometer

Fig. 7.1. *Pitot-static tube.*

Velocity magnitude from pitot-static tube. In some situations the direction of the velocity vector is known with sufficient accuracy without taking any measurements. If the direction is not known, it may be found in several ways discussed later. Let us assume the direction is known, so that a pitot-static tube may be properly aligned with this direction, as in Fig. 7.1. Assuming steady one-dimensional flow of an incompressible frictionless fluid, we can derive the well-known result

$$V = \sqrt{\frac{2(p_{stag} - p_{stat})}{\rho}} \qquad (7.1)$$

where $V \triangleq$ flow velocity
$\rho \triangleq$ fluid mass density
$p_{stag} \triangleq$ stagnation or total pressure, free stream
$p_{stat} \triangleq$ static pressure, free stream

In an actual pitot-static tube, deviations from the ideal theoretical result of Eq. (7.1) arise from a number of sources. If ρ is accurately known, the errors can be traced to inaccurate measurement of p_{stag} and p_{stat}.

The static pressure is usually the more difficult to measure accurately. The difference between true (p_{stat}) and measured ($p_{stat,m}$) values of static pressure may be due to the following:

1. Misalignment of the tube axis and velocity vector.   This exposes the static taps to some component of velocity.
2. Nonzero tube diameter.   Streamlines next to the tube must be longer than those in undisturbed flow, indicating an increase in velocity. This is accompanied by a decrease in static pressure, making the static taps read low.   A similar (and possibly more severe) effect occurs if a tube is inserted in a duct whose cross-sectional area is not much larger than that of the tube.
3. Influence of stagnation point on the tube-support leading edge. This higher pressure causes the static pressure upstream of the leading edge also to be high.   If the static taps are too close to the support, they will read high because of this effect.   Note that this error and that of item 2 above tend to cancel.   By proper design, effective cancellation may be achieved[1] (see also Prandtl pitot tube. Fig. 2.17*b*).   Figure 7.2 shows the nature of both these errors as revealed by experimental tests.[2]

An important application of the pitot-static tube is found in aircraft and missiles.   Here the stagnation- and static-pressure readings of a tube fastened to a vehicle are used to determine the airspeed and Mach number while the static reading alone is used to measure altitude.   If altitude is to be measured with an error of 100 ft, the static pressure must be accurate to 0.5 percent.[1]   To achieve this accuracy, methods for compensating errors of the type mentioned in items 1, 2, and 3 above have been developed and reported.[3]   An interesting and useful result of these studies is a simple method for reducing error due to angular misalignment. It was found that by locating the static-pressure taps as in Fig. 7.3 the measurement was essentially insensitive to angle of attack for the range $-2° < \alpha < 12°$ and Mach numbers in the range 0.4 to 1.2.   While this method, as shown, is effective only for misalignment in the particular plane shown, it can be extended to arbitrary directions of misalignment by designing the probe with a single vane to rotate about its longitudinal axis and automatically locate the taps 37.5° from the cross-flow stagnation point.[4]   It is also possible to design multiple-vaned probes[4] mounted on a gimbal system with complete rotational freedom.   These probes

---

[1] V. S. Ritchie, Several Methods for Aerodynamic Reduction of Static-pressure Sensing Errors for Aircraft at Subsonic, Near-sonic, and Low Supersonic Speeds, *NASA, Tech. Rept.* R-18, 1959.

[2] R. G. Folsom, Review of the Pitot Tube, *Trans. ASME*, p. 1450, October, 1956.

[3] Ritchie, *op. cit.*

[4] F. J. Capone, Wind-tunnel Tests of Seven Static-pressure Probes at Transonic Speeds, *NASA, Tech. Note* D-947, 1961.

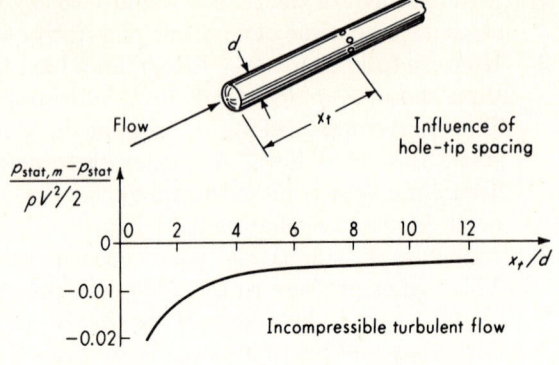

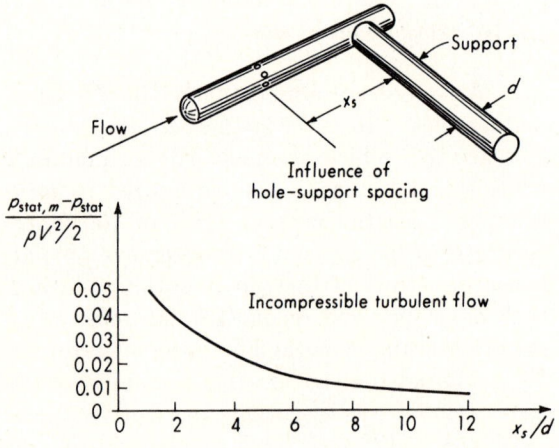

Fig. 7.2. *Static-pressure errors.*

act in a fashion similar to a weather vane (except that they have angular freedom about two axes) and thus align themselves with the velocity vector. A conventional ring of evenly spaced static taps then gives accurate readings. By measuring the rotations of the gimbals, such probes also provide information on the *direction* of the velocity.

While errors in the stagnation pressure are likely to be smaller than those in the static pressure, several possible sources of error are present, namely:

1.   Misalignment. This situation prevents formation of a true stagnation point at the measuring hole since the velocity is not zero. Tubes

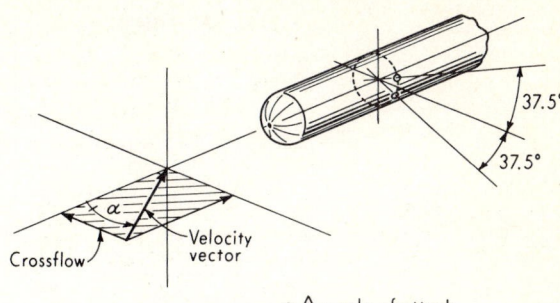

$\alpha \triangleq$ angle of attack

Fig. 7.3. *Probe insensitive to misalignment*

of special design have been developed which exhibit considerable tolerance to misalignment.[1,2] Figure 7.4 shows an example of such a tube which has an error less than 1 percent of the velocity pressure $\rho V^2/2$ for misalignments up to $\pm 38°$ for velocities from low subsonic to Mach 2. Conventional tubes not specifically designed for misalignment insensitivity may show 1 percent errors at only 5 or 10°.

[1] W. Gracey, D. E. Coletti, and W. R. Russell, Wind-tunnel Investigation of a Number of Total-pressure Tubes at High Angles of Attack, Supersonic Speeds, *NACA, Tech. Note* 2261, January, 1951.

[2] W. Gracey, W. Letko, and W. R. Russell, Wind-tunnel Investigation of a Number of Total-pressure Tubes at High Angles of Attack, Subsonic Speeds, *NACA, Tech. Note* 2331, April, 1951.

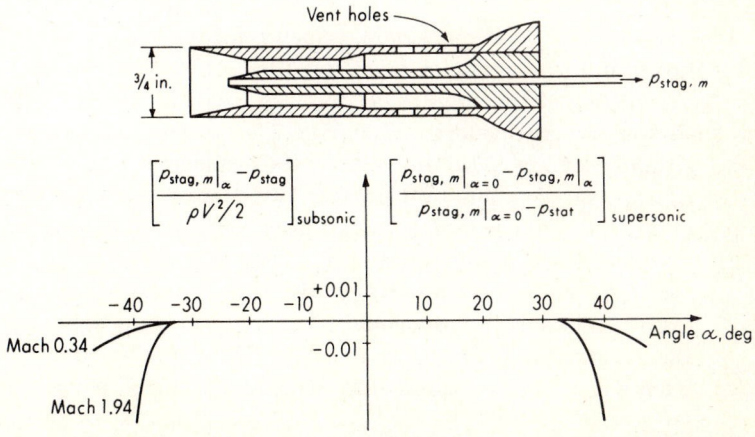

Fig. 7.4. *Special stagnation probe.*

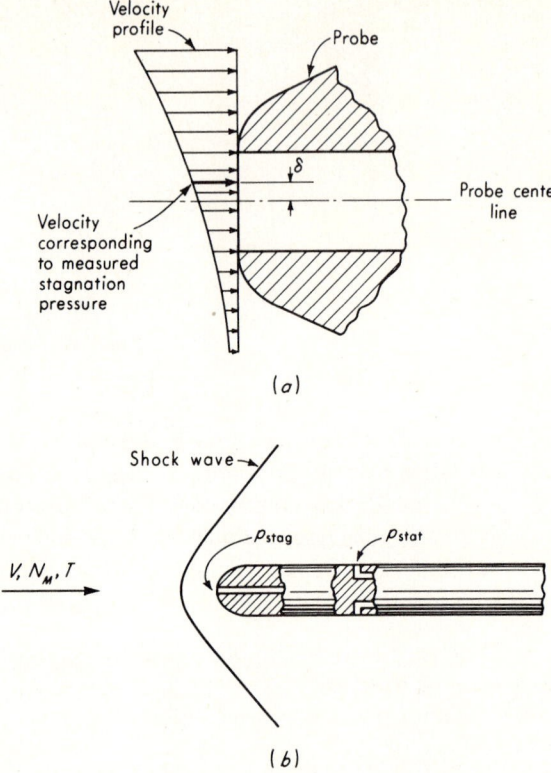

Fig. 7.5. *Nonuniform-velocity profile and supersonic probe.*

2. Two- and three-dimensional velocity fields. When the velocity is not uniform, a probe of finite size intercepts streamlines of different velocities and the stagnation pressure measured corresponds to some sort of average velocity (see Fig. 7.5a). For the two-dimensional situation of Fig. 7.5a, if one knew the displacement δ, he could assign the measured stagnation pressure (and thus velocity) to a specific point in the flow. Some limited data[1] on this problem are available.

3. Effect of viscosity. Equation (7.1) assumes the fluid to be frictionless. At sufficiently low Reynolds number the viscosity of the fluid exerts a noticeable additional force at the stagnation hole, causing the stagnation pressure to be higher than predicted by Eq. (7.1). This effect can be taken into account by introducing a correction

---

[1] Folsom, *op. cit.*, p. 1451.

factor C as follows:

$$p_{stag,m} = p_{stat} + \frac{C\rho V^2}{2} \qquad (7.2)$$

For negligible viscosity effects, $C = 1.0$ and Eq. (7.2) is the same as (7.1). For a given probe, the factor C is a function of Reynolds number only and may be found theoretically for simple probe shapes such as spheres and cylinders. A typical result[1] for a cylindrical probe is

$$C = 1 + \frac{4}{N_R} \qquad 10 < N_R < 100 \qquad (7.3)$$

where   Reynolds number $\triangleq N_R \triangleq \dfrac{V\rho r}{\mu}$

$r \triangleq$ probe radius

$\mu \triangleq$ fluid viscosity

Equation (7.3) shows that the effect is about 4 percent of the velocity pressure $\rho V^2/2$ at $N_R = 100$. At $N_R = 10$, however, the effect is 40 percent. Theory and experimental tests[1] show that viscosity corrections are rarely needed for $N_R > 500$, no matter what the shape of the probe.

When a pitot-static tube is used in a compressible fluid, Eq. (7.1) no longer applies, although it may be sufficiently accurate if the Mach number is low enough. For subsonic flow (Mach number $N_M < 1$) the velocity is given by[2]

$$V = \sqrt{\frac{2k}{k-1} \frac{p_{stat}}{\rho_{stat}} \left[\left(\frac{p_{stag}}{p_{stat}} \right)^{(k-1)/k} - 1 \right]} \qquad (7.4)$$

where $k \triangleq \dfrac{\text{specific heat at constant pressure}}{\text{specific heat at constant volume}} = \dfrac{C_p}{C_v} \qquad (7.5)$

Measurement of free-stream density ρ_{stat} requires knowledge of static temperature, which may itself be a difficult measurement. Equation (7.4) may be rewritten as

$$p_{stag} = p_{stat} \left[1 + \frac{k-1}{2} \left(\frac{V}{c} \right)^2 \right]^{k/(k-1)} \qquad (7.6)$$

where $c \triangleq$ acoustic velocity $= \sqrt{\dfrac{kp_{stat}}{\rho_{stat}}} = \sqrt{kgRT} \qquad (7.7)$

and $g \triangleq$ gravitational acceleration

$T \triangleq$ free-stream static temperature

$R \triangleq$ gas constant

[1] Folsom, *op. cit.*, p. 1453.

[2] R. C. Binder, "Advanced Fluid Dynamics and Fluid Machinery," p. 51, Prentice-Hall, Inc., Englewood Cliffs, N.J., 1951.

The right side of Eq. (7.6) may be expanded in a power series to give

$$p_{stag} = p_{stat} + \left(\rho_{stat} \frac{V^2}{2} \right) \left(1 + \frac{N_M^2}{4} + \frac{2 - k}{24} N_M^4 + \cdots \right) \quad (7.8)$$

where
$$N_M \triangleq \frac{V}{c} \quad (7.9)$$

Since the Mach number of an incompressible fluid is zero, Eq. (7.8) shows that p_{stag} is higher for compressible than for incompressible flow. Also, if N_M is sufficiently small, Eq. (7.8) is closely approximated by Eq. (7.1).

For supersonic flow ($N_M > 1$) a compression shock wave forms ahead of the pitot tube. Between this shock wave and the tube end the velocity is subsonic. This subsonic velocity is then reduced to zero at the tube stagnation point (see Fig. 7.5b). Analysis[1] gives the following formula for computing the free-stream Mach number and thereby the velocity:

$$\frac{p_{stag}}{p_{stat}} = N_M^2 \left(\frac{k + 1}{2} \right)^{k/(k-1)} \left[\frac{2kN_M - k + 1}{N_M^2(k + 1)} \right]^{1 - 1/(k-1)} \quad (7.10)$$

The measurement of stagnation and static pressures may be combined in a single probe, as in Fig. 7.1, or two separate probes, one for stagnation and one for static, may be employed. Figure 7.6 shows several examples[2] of commonly used forms. The wedge static-pressure probe of Fig. 7.6a also can be used to measure velocity direction in a single plane. When the two static taps read equal pressures the wedge is aligned with the flow. This probe is usable for both subsonic and supersonic flow. At Mach 0.9 the sensitivity to misalignment is about 1.5 in. of water per angular degree. The total (stagnation) probes of Fig. 7.6b and c are also intended for both sub- and supersonic flow. The simple tube is insensitive to misalignment up to about $\pm 20°$ while the venturi shielded tube is good to $\pm 50°$. The boundary-layer probe is usable up to Mach 1.0 and is insensitive to misalignment up to $\pm 5°$. Boundary-layer thickness can be measured with such a probe with an error of the order of 0.002 in. The probe and associated pressure-measuring equipment have a long time lag because of the small flow passage (0.001 in.) at the probe tips.

Velocity direction from yaw tube, pivoted vane, and servoed sphere. In addition to laboratory studies of flow processes in fluid machinery, ducting, etc., flow-velocity direction information is also of

---

[1] Binder, *op. cit.*, p. 52.

[2] Aero Research Instrument Co., Chicago, Ill.

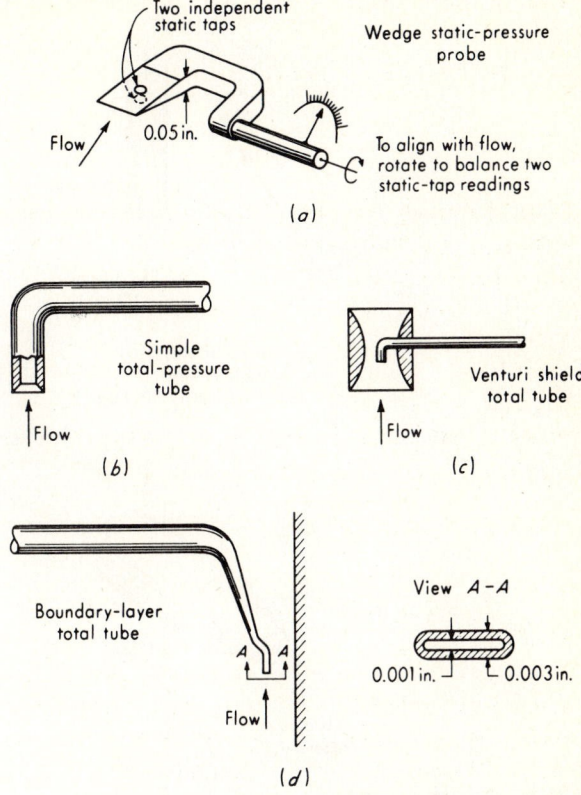

Fig. 7.6.   *Pressure probes.*

interest in flight vehicles[1] where angle-of-attack measurements are utilized in attitude measurement and control, stability augmentation, and gust alleviation systems.

So-called yaw tubes[2] of one form or another are conventionally employed to determine the direction of local flow velocity. Perhaps the simplest form, useful for finding the angular inclination in one plane only, is shown in Fig. 7.7a. Taps 1 and 3 are connected to a differential-pressure instrument that reads zero when the tube is aligned with the flow. A central tap 2 is often included to read the stagnation pressure after alignment is attained (valid only if the angle of attack is zero). The

[1] H. H. Koelle, "Handbook of Astronautical Engineering," p. 13-33, McGraw-Hill Book Company, New York, 1961.

[2] Aero Research Instrument Co., Chicago, Ill.

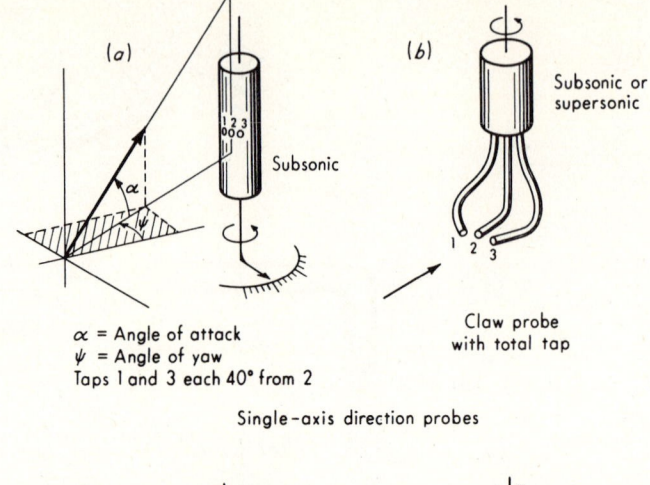

α = Angle of attack
ψ = Angle of yaw
Taps 1 and 3 each 40° from 2

Single-axis direction probes

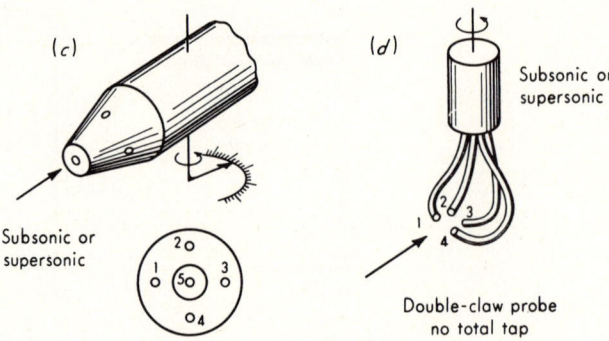

Two-axis direction probes

Fig. 7.7.   *Flow-direction probes.*

claw tube of Fig. 7.7b operates on similar principles but may be used in regions where the flow direction changes greatly since its sensing holes may be located very close together.   The two-axis probes of Fig. 7.7c and d could conceivably be designed to allow rotation about each axis; however, the complexity and size of such a design are generally prohibitive.   Thus probe operation consists of rotation about the probe axis to balance taps 1 and 3.   Then pressures 2 and 4 are each measured, and calibration charts give the angle of attack.   Tap 5 does not read stagnation pressure directly; this can be obtained from calibration charts.   Any of these probes may be made automatically self-aligning by using the pressure difference $p_1 - p_3$ as the error signal in a servosystem which

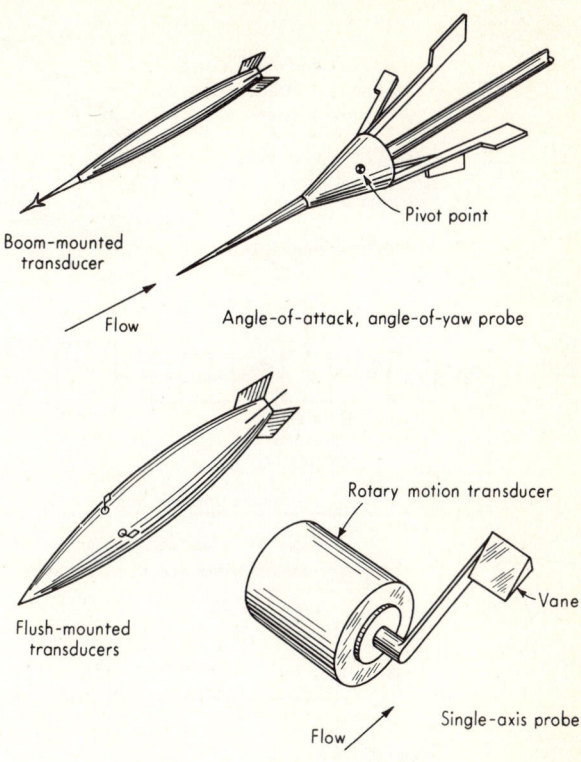

Boom-mounted transducer

Pivot point

Flow

Angle-of-attack, angle-of-yaw probe

Rotary motion transducer

Vane

Flush-mounted transducers

Flow

Single-axis probe

Fig. 7.8.   *Vane-type probes.*

rotates the probe until a null is achieved.   The details of a system of this type are shown in Fig. 7.9.

Determination of angles of attack and yaw aboard flight vehicles is often accomplished with vane-type probes as in Fig. 7.8.   These devices are essentially one- or two-axis weather vanes with suitable damping to reduce oscillation and with motion pickups to provide electrical angle signals.   A dynamic analysis of these devices is available.[1]   Limitations of this type of device for certain high-speed, high-altitude applications have led to the development of the servoed-sphere type of sensor shown in Fig. 7.9.   The one shown was developed[2] for the X-15 rocket research aircraft.   A servo-driven sphere is continuously and automatically aligned with the velocity vector by means of two independent servosystems using the differential-pressure signals $p_1 - p_2$ and $p_3 - p_4$ as

[1] G. J. Friedman, Frequency Response Analysis of the Vane-type Angle of Attack Transducer, *Aero/Space Eng.*, p. 69, March, 1959.

[2] Northrop Corp., Nortronics Div., *Rept.* NORT60-46.

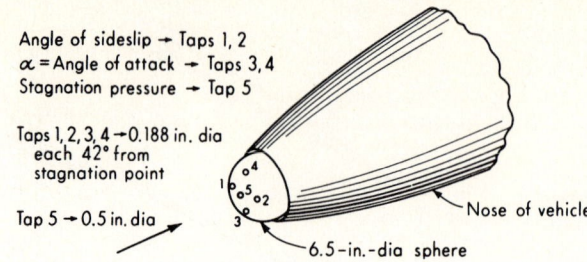

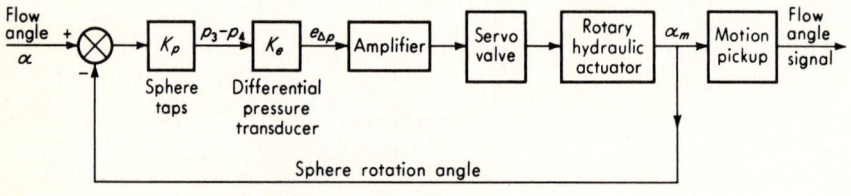

Angle-of-attack servo system
(Angle-of-sideslip system functionally identical)

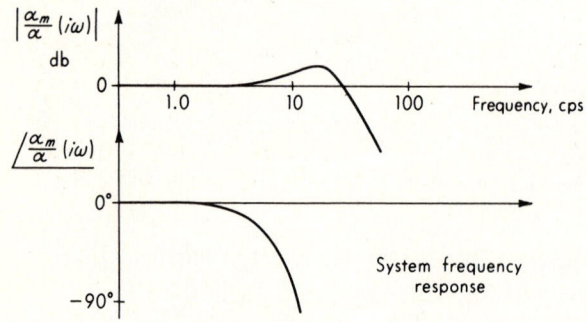

Fig. 7.9. *Servoed-sphere probe.*

error signals. A fifth tap measures the stagnation pressure. Block diagrams and frequency response of a single axis are given in Fig. 7.9. The angle-of-attack axis is designed for the range -10 to $+40°$ while the sideslip axis covers $\pm 20°$. Static inaccuracy of angle measurement is $0.25°$.

Dynamic wind-vector indicator.[1] Figure 7.10 shows a transducer that measures the magnitude and direction of flow velocity in terms of the drag force exerted on a hollow sphere. The drag force F_d on a body is

[1] Flow Corp., Cambridge, Mass.

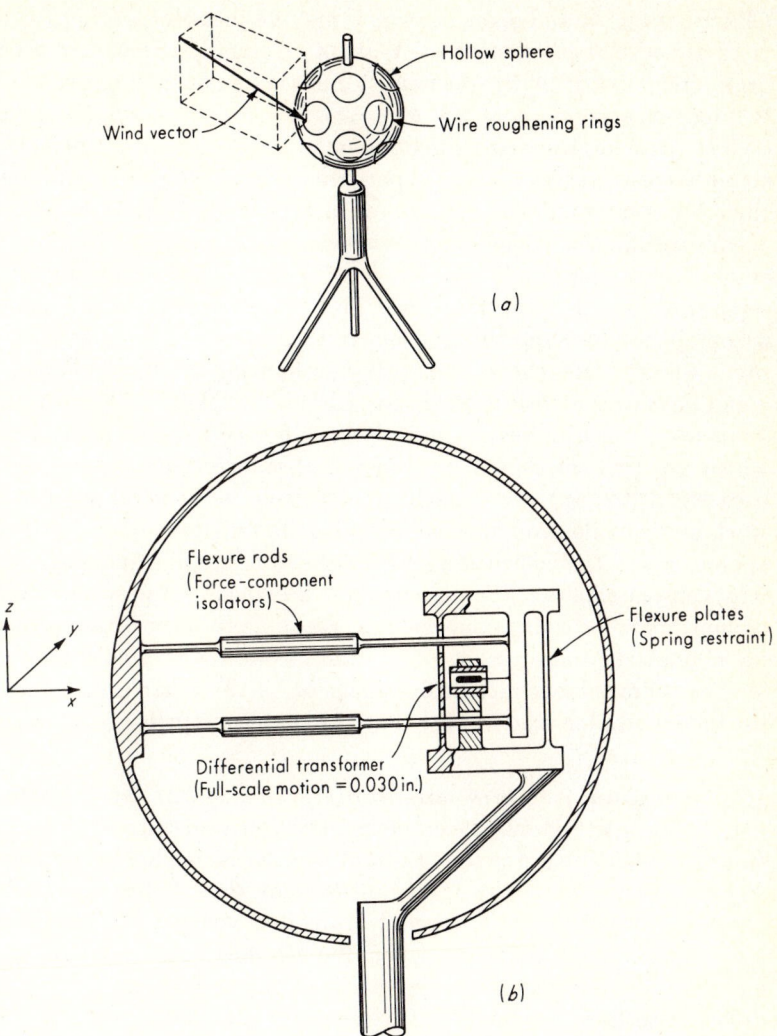

Fig. 7.10.   *Wind-vector indicator.*

given by

$$F_d = C_d \frac{A \rho V^2}{2} \qquad (7.11)$$

where $C_d \triangleq$ drag coefficient of body
= 0.567 for these transducers
$A \triangleq$ projected area of body

Clearly, if C_d, A, and ρ are known, V may be measured by measuring F_d. If all directional components of V are to be equally effective in producing drag force, a body with spherical symmetry must be employed. If this is done, measurement of the x, y, and z components of F_d completely defines the magnitude and direction of V. Since the drag coefficient of a smooth sphere is somewhat dependent on the Reynolds number (and thus V), wire roughening rings are attached to the sphere surface to ensure turbulence. The drag coefficient of the roughened sphere is constant over the entire design range of V for a given transducer. The separation of the total drag force into three rectangular components is accomplished by mounting the sphere on a force-resolving flexure assembly. Figure 7.10*b* shows the x-axis mechanism of this structure. (The y and z axes are identical except for their orientation.) The force components are applied to flexure-plate springs to produce proportional motions which are then measured by differential transformers. (A later model uses strain gages.) A transducer designed for general meteorological work and missile ground support at launching sites uses a 1-ft-diameter sphere, covers the velocity range 0 to 100 mph, has flat frequency response 0 to 6 cps, and has a static inaccuracy of 2 percent of full scale. Auxiliary computing and display equipment is available to show total vector velocity magnitude and angular orientation in two planes visually on oscilloscopes. Transducer models designed for use in water are also available for oceanographic studies such as those of ocean currents.

Hot-wire and hot-film anemometers. Hot-wire anemometers are commonly made in two basic forms, the constant-current type and the constant-temperature type. Both utilize the same physical principle but in different ways. In the constant-current type, a fine resistance wire carrying a fixed current is exposed to the flow velocity. The wire attains an equilibrium temperature when the i^2R heat generated in it is just balanced by the convective heat loss from its surface. The circuit is designed so that the i^2R heat is essentially constant; thus the wire temperature must adjust itself to change the convective loss until equilibrium is reached. Since the convection film coefficient is a function of flow velocity, the equilibrium wire temperature is a measure of velocity. The wire temperature can be measured in terms of its electrical resistance. In the constant-temperature form, the current through the wire is adjusted to keep the wire temperature (as measured by its resistance) constant. The current required to do this then becomes a measure of flow velocity.

For equilibrium conditions one can write an energy balance for a hot wire as

$$I^2R_w = K_c hA(T_w - T_f) \qquad (7.12)$$

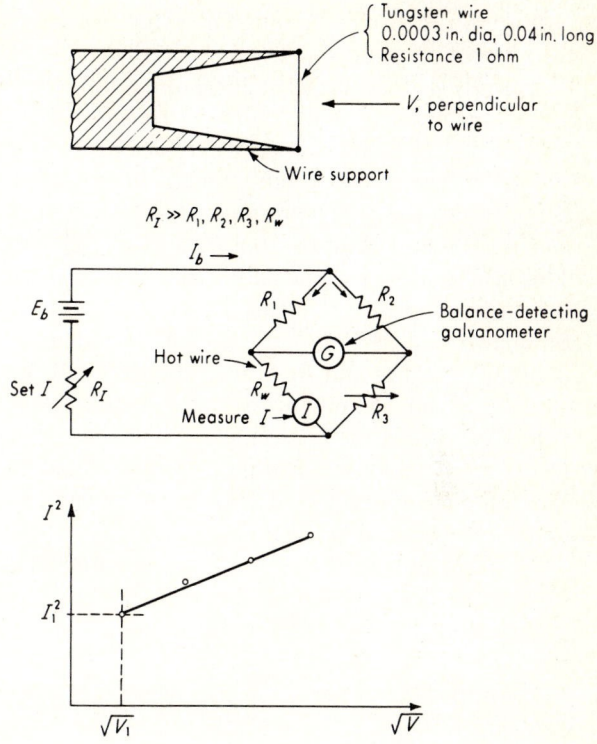

Fig. 7.11. *Hot-wire anemometer.*

where

$I \triangleq$ wire current
$R_w \triangleq$ wire resistance
$K_c \triangleq$ conversion factor, thermal to electrical power
$T_w \triangleq$ wire temperature
$T_f \triangleq$ temperature of flowing fluid
$h \triangleq$ film coefficient of heat transfer
$A \triangleq$ heat-transfer area

Now h is mainly a function of flow velocity for a given fluid density. For a range of velocities, this function has the general form

$$h = C_0 + C_1 \sqrt{V} \qquad (7.13)$$

For the measurement of average (steady) velocities, the constant-temperature mode of operation is often used. Figure 7.11 shows a possible circuit arrangement. For accurate work a given hot-wire probe must be calibrated in the fluid in which it is to be used. That is, it is exposed to *known* velocities (measured accurately by some other means) and its output recorded over a range of velocities. In the circuit of Fig. 7.11, the current through R_w stays essentially constant even when R_w changes because R_I is of the order of 2,000 ohms while R_1, R_2, R_3, and R_w are much less, of the order of 1 to 20 ohms. In calibration, V is set at some known value V_1. Then R_I is adjusted to set hot-wire current I at a value low enough to prevent wire burnout but high enough to give adequate sensitivity to velocity. The resistance R_w will come to a definite temperature and resistance. The resistor R_3 is then adjusted to balance the bridge. This adjustment is essentially a measurement of wire temperature, which is held fixed at all velocities. The first point on the calibration curve is thus plotted as $I_1{}^2$, $\sqrt{V_1}$. Now V is changed to a new value, causing wire temperature and R_w also to change and thus unbalancing the bridge. Then R_w, and thereby wire temperature, are restored to their original values by adjusting I (by means of R_I) until bridge balance is restored (R_3 is *not* changed). The new current I and the corresponding V may then be plotted on the calibration curve and this procedure repeated for as many velocities as desired.

Once calibrated, the probe can be used to measure unknown velocities by adjusting R_I until bridge balance is achieved, reading I, and getting the corresponding V from the calibration curve. This assumes that the measured fluid is at the same temperature and pressure as for the calibration. Correction methods for varying temperature and pressure are fairly simple but will not be gone into here. For the above constant-temperature mode of operation, Eqs. (7.12) and (7.13) can be combined to give

$$I^2 = \frac{K_c A (T_w - T_f)(C_0 + C_1 \sqrt{V})}{R_w} \triangleq C_2 + C_3 \sqrt{V} \qquad (7.14)$$

indicating that the calibration curve of Fig. 7.11 should be essentially a straight line. This is borne out by experimental tests.

While the above described measurement of steady velocities is of some practical interest, perhaps the main application of hot-wire instruments is the measurement of rapidly fluctuating velocities, such as the turbulent components superimposed on the average velocity. Both constant-current and constant-temperature techniques are used; we first consider the constant-current operation. Figure 7.12 shows the basic arrangement. The current can again be assumed constant at a value I even if R_w changes, since $R_I \gg R_w$. Let us suppose the velocity is constant at a value V_0. This will cause R_w to assume a constant value, say R_{w0}, and a voltage $I R_{w0}$ will appear across R_w. Now, we let the velocity V fluctuate about the value V_0 so that $V = V_0 + v$, where v is the fluctuating component. This will result in R_w varying so that $R_w = R_{w0} + r_w$, where r_w is the varying component. Now, during a time interval dt,

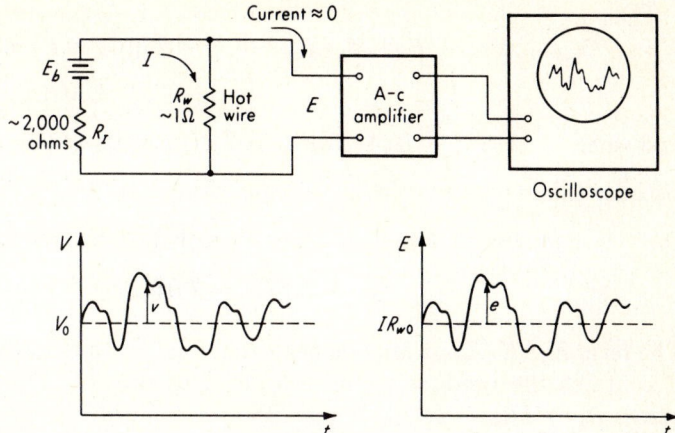

Fig. 7.12.   *Velocity-fluctuation measurement.*

we may write for the wire

(Electrical energy generated) − (energy lost by convection)

$$= \text{energy stored in wire} \qquad (7.15)$$

The energy lost by convection is given by

$$K_c A (T_w - T_f)(C_0 + C_1 \sqrt{V}) \, dt \qquad (7.16)$$

while the wire temperature T_w may be related to its resistance by

$$T_w = K_{tr}(R_{w0} + r_w) \qquad (7.17)$$

where K_{tr} is the reciprocal of a temperature coefficient of resistance. The term $C_0 + C_1 \sqrt{V}$ may be approximately linearized for small changes in V with good accuracy as follows:

$$f(V) = C_0 + C_1 \sqrt{V} \approx (C_0 + C_1 \sqrt{V_0}) + \left. \frac{\partial f}{\partial V} \right|_{V = V_0} (V - V_0)$$

$$(7.18)$$

$$C_0 + C_1 \sqrt{V} \approx (C_0 + C_1 \sqrt{V_0}) + K_v v \qquad (7.19)$$

Equation (7.15) then becomes

$$I^2(R_{w0} + r_w) \, dt - K_c A (T_w - T_f)(C_0 + C_1 \sqrt{V_0} + K_v v) \, dt = M C d T_w$$

$$(7.20)$$

where $M \triangleq$ mass of wire

$C \triangleq$ specific heat of wire

Now,

$$I^2R_{w0} + I^2r_w - K_cA[K_{tr}(R_{w0} + r_w) - T_f](C_0 + C_1\sqrt{V_0}$$
$$+ K_vv) = MCK_{tr}\frac{dr_w}{dt} \tag{7.21}$$

and since $I^2R_{w0} - K_cA(K_{tr}R_{w0} - T_f)(C_0 + C_1\sqrt{V_0}) = 0$ (7.22)

because this represents the initial equilibrium state, we get

$$I^2r_w - K_cAK_{tr}r_w(C_0 + C_1\sqrt{V_0}) - K_cAK_{tr}r_wK_vv$$
$$- K_cA(K_{tr}R_{w0} - T_f)K_vv = MCK_{tr}\frac{dr_w}{dt} \tag{7.23}$$

The term $K_cAK_{tr}K_vr_wv$ may be neglected relative to the other terms since it contains the product r_wv of two small quantities.   Now the voltage across R_w is $IR_w = I(R_{w0} + r_w)$.   The fluctuating component of this is Ir_w, which we shall call e.   Equation (7.23) then leads to

$$\frac{e}{v}(D) = \frac{K}{\tau D + 1} \tag{7.24}$$

where $K \triangleq \dfrac{-K_vK_cAI(K_{tr}R_{w0} - T_f)}{K_cK_{tr}A(C_0 + C_1\sqrt{V_0}) - I^2}$ volts/fps   (7.25)

$$\tau \triangleq \frac{MCK_{tr}}{K_cK_{tr}A(C_0 + C_1\sqrt{V_0}) - I^2} \qquad \text{sec} \tag{7.26}$$

We see that the voltage follows the flow velocity with a first-order lag. The time constant τ cannot be reduced much below 0.001 sec in actual practice, which would limit the flat frequency response to less than 160 cps.   This is quite inadequate for turbulence studies since frequencies of 50,000 cps and more are of interest.   This limitation is overcome by use of electrical dynamic compensation.   Circuits whose frequency response just makes up the deficiency in the hot wire itself are employed as in Fig. 7.13.[1]   The overall system then has a flat response to almost 100,000 cps.   The main difficulty in applying this technique is that the correct compensation depends on τ whose value is not known and varies with flow conditions.   The next paragraph explains a method[1] of solving this difficulty.

The basic idea of the scheme is to force a square-wave current through the hot wire while it is exposed to the flow to be studied (see Fig. 7.14).   We shall show that the output-voltage response to this *current* signal has exactly the same time constant as the response to the flow-*velocity* signal.   Thus if the compensation can be adjusted to be correct for the current signal it will also be correct for the velocity input. The "correctness" of the adjustment may be judged by the degree to

[1] Flow Corp., Cambridge, Mass.

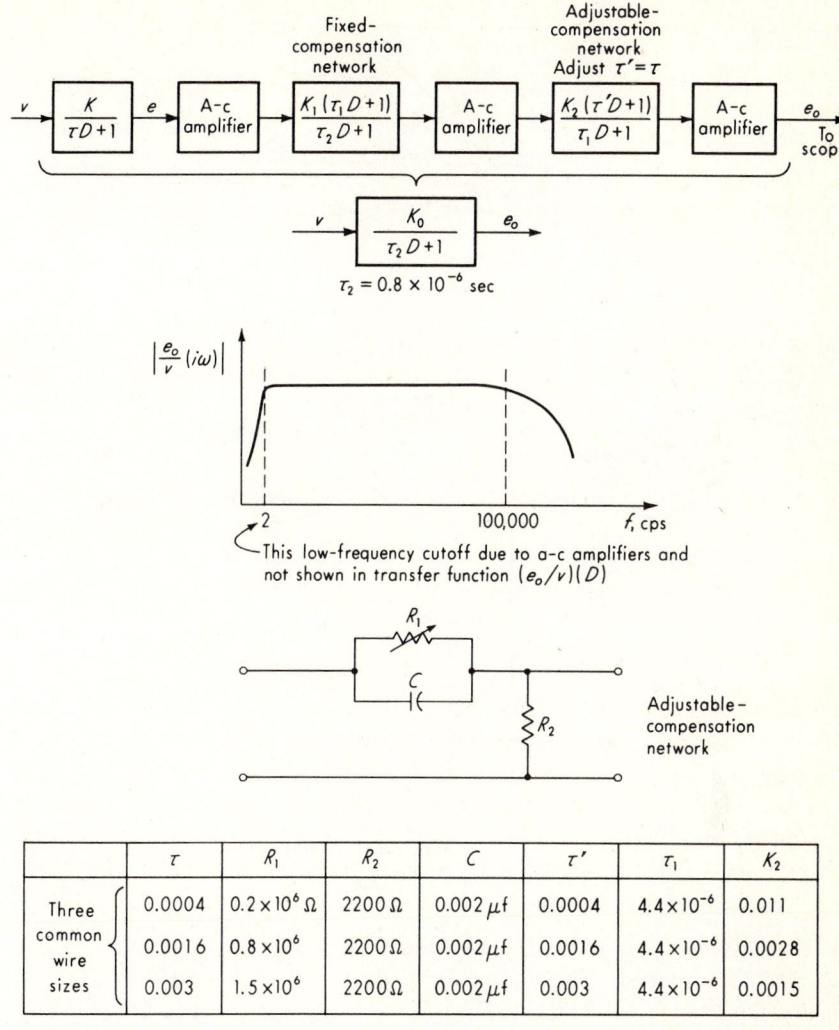

Fig. 7.13. *Dynamic compensation.*

which the output voltage corresponds to a square wave. In the circuit of Fig. 7.14 a good approximation to linear behavior may be expected for small input signals (current or velocity); thus the superposition principle will apply, and the effects of current and velocity inputs may be considered separately. If the square-wave current is turned off, R_I adjusted to give the desired hot-wire current I_0, and R_b adjusted to balance the bridge, then $R_a/R_{w0} = R_r/R_b$, and the voltage $E_{B1,B2} = 0$. Now we

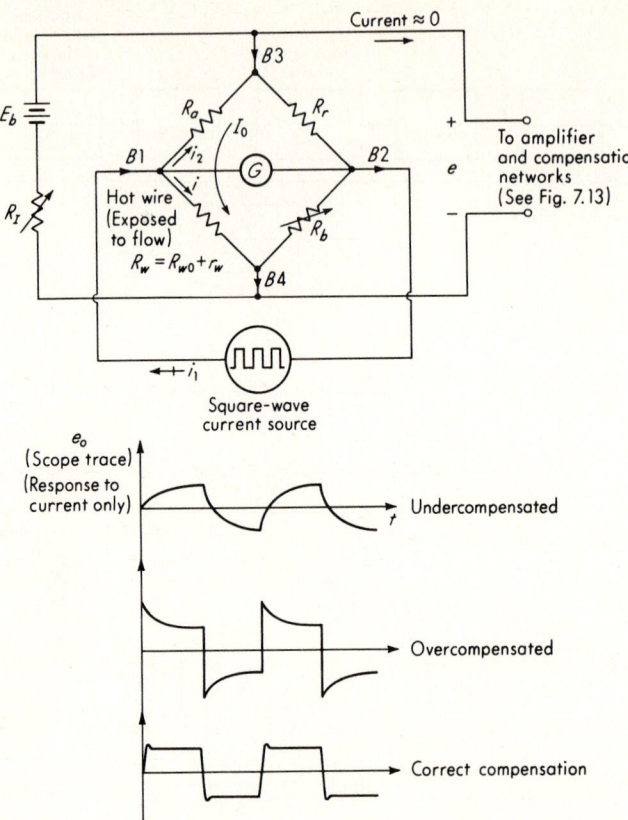

Fig. 7.14. *Compensation adjustment scheme.*

let the square-wave current i_1 be turned on, causing a current i, which we calculate to be $i_1(R_a + R_r)/(R_a + R_r + R_w + R_b)$, to flow through R_w and a current $i_2 = i_1(R_w + R_b)/(R_a + R_r + R_w + R_b)$ to flow through R_a and R_r. Equation (7.15) may be applied to this situation to give

$$(I_0 + i)^2(R_{w0} + r_w) - K_c A[K_{tr}(R_{w0} + r_w) - T_f](C_0$$
$$+ C_1 \sqrt{V_0}) = MCK_{tr}\frac{dr_w}{dt} \qquad (7.27)$$

Now, since $I_0^2 R_{w0} = K_c A(K_{tr}R_{w0} - T_f)(C_0 + C_1 \sqrt{V_0})$, and neglecting $2I_0 i r_w$ and $i^2(R_{w0} + r_w)$ since they are products of small quantities, Eq. (7.27) reduces to

$$\frac{r_w}{i}(D) = \frac{K_i}{\tau D + 1} \qquad (7.28)$$

where

$$\tau \triangleq \frac{MCK_{tr}}{K_cK_{tr}A(C_0 + C_1\sqrt{V_0}) - I_0{}^2} \qquad \text{sec} \qquad (7.29)$$

$$K_i \triangleq \frac{2I_0R_{w0}}{K_cK_{tr}A(C_0 + C_1\sqrt{V_0}) - I_0{}^2} \qquad \text{ohms/amp} \qquad (7.30)$$

Let us now calculate e, the varying component of the voltage appearing across $B3$ and $B4$, which will be the input to the amplifiers and compensating networks.

$$e = -R_ai_2 + (R_{w0} + r_w)i + I_0r_w \approx -R_ai_2 + R_{w0}i + I_0r_w \qquad ir_w \approx 0 \qquad (7.31)$$

$$e = -R_a\frac{R_w + R_b}{R_a + R_r}i + R_{w0}i + I_0r_w$$

$$= \frac{-R_aR_w - R_aR_b + R_{w0}R_a + R_{w0}R_r}{R_a + R_r}i + I_0r_w \qquad (7.32)$$

Now $R_{w0}R_a \approx R_wR_a$ and $R_{w0}R_r = R_aR_b$ (balanced-bridge relation); thus

$$e \approx I_0r_w \qquad (7.33)$$

Thus, finally,

$$\frac{e}{i}(D) = \frac{K_e}{\tau D + 1} \qquad (7.34)$$

where

$$K_e \triangleq I_0K_i \qquad (7.35)$$

We see now that the response of the voltage e to impressed current signals has the identical time constant τ as the response to flow-velocity signals. Thus the compensating networks may be adjusted to optimize the response to current inputs and ensure optimum response for flow-velocity inputs. Since this adjustment is made while the probe is exposed to flow, the output will contain a superposition of current response and velocity response, resulting in a sometimes confusing picture, rather than the simple wave forms of Fig. 7.14. Usually, however, the compensation adjustment can be made satisfactorily.

The operation of the constant-temperature type of instrument for steady velocities was explained earlier in relation to Fig. 7.11. This mode of operation can be extended to measure both average and fluctuation components of velocity by making the bridge-balancing operation automatic, rather than manual, through the agency of a feedback arrangement. A simplified functional schematic of such a system is shown in Fig. 7.15. With zero flow velocity and the bridge excitation shut off ($i_w = 0$), the hot wire assumes the fluid temperature. The variable resistor R_3 is then manually adjusted so that $R_3 > R_w$, thereby unbalancing the bridge. When the excitation current is turned on, the unbalanced bridge produces an unbalance voltage e_e which is applied to the input of a high-gain current amplifier supplying the bridge excitation current. The current now flowing through R_w increases its temperature and thus its

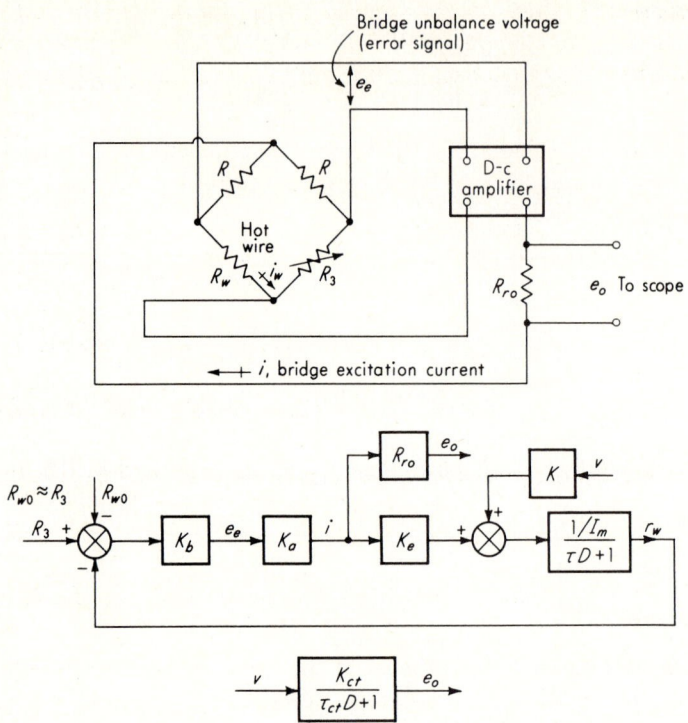

Fig. 7.15. *Constant-temperature anemometer.*

resistance. As R_w increases, it approaches R_3 and the bridge-unbalance voltage e_e decreases. If the current amplifier had an infinite gain, the bridge current required to heat R_w to match R_3 precisely could be produced with an infinitesimally small error voltage e_e and thus perfect bridge balance would be attained automatically. The current i (which is a measure of flow velocity V) produces an output voltage in passing through the readout resistor R_{ro}. Since an actual amplifier has finite gain, the bridge-unbalance voltage cannot go to zero, but it can be extremely small. This also means that R_w will be very close, but not exactly equal, to R_3.

The above described self-adjusting equilibrium will come about at *any* steady-flow velocity, not just zero, the equilibrium current in each case being a measure of velocity. If the velocity is rapidly fluctuating, perfect continuous bridge balance requires an infinite-gain amplifier. Lacking this, practical systems exhibit some lag between velocity and current. System dynamic response for small velocity fluctuations can be obtained by superimposing previously obtained results for response to velocity (wire-current constant) and response to current (flow-velocity constant), since in the constant-temperature mode both current and velocity are changing simultaneously. The effect of velocity on r_w (the varying component of R_w) is obtained from Eq. (7.24) as

$$\frac{r_w}{v}(D) = \frac{K/I_m}{\tau D + 1} \qquad (7.36)$$

where I_m is the constant current about which fluctuations take place. From Eq. (7.34), the effect of current on r_w is given by

$$\frac{r_w}{i}(D) = \frac{K_e/I_m}{\tau D + 1} \qquad (7.37)$$

The total effect of i and v on r_w is then, by superposition,

$$(\tau D + 1)r_w = \frac{Kv + K_e i}{I_m} \qquad (7.38)$$

Now a change in r_w causes a bridge-unbalance voltage change according to

$$e_e = \frac{I_m R}{R + R_{w0}} r_w \triangleq K_b r_w \qquad (7.39)$$

We assume that the amplifier has no lag and produces output current proportional to input voltage as given by

$$i = K_a e_e \qquad (7.40)$$

The block diagram of Fig. 7.15 embodies the above relations which may now be manipulated to give the relation between input v and output e_o:

$$\left(\frac{K_e e_o}{R_{ro}} + Kv\right)\frac{(1/I_m)(-K_b K_a R_{ro})}{\tau D + 1} = e_o \qquad (7.41)$$

This gives finally

$$\frac{e_o}{v}(D) = \frac{K_{ct}}{\tau_{ct}D + 1} \qquad (7.42)$$

where

$$\tau_{ct} \triangleq \frac{\tau}{1 + K_e K_b K_a/I_m} \qquad (7.43)$$

$$K_{ct} \triangleq \frac{-KK_b K_a R_{ro}/I_m}{1 + K_e K_b K_a/I_m} \qquad (7.44)$$

Note that τ_{ct}, the time constant of the constant-temperature anemometer system, is always less than τ (the time constant of the wire itself) and in actual practice is *much* less since a very high value of amplifier gain K_a is used. As in all feedback systems, too high a loop gain will cause instability; however, careful design allows sufficiently high gain to make τ_{ct} of the order of $\frac{1}{100}$ of τ or less. A typical instrument has flat (within 3 db) frequency response to 17,000 cps when the average flow velocity is 30 fps, 30,000 cps for 100 fps, and 50,000 cps at 300 fps.

Both constant-current and constant-temperature anemometers are in use; it may be helpful to list their comparative features. In the constant-current type, the current must be set high enough to heat the wire considerably above the fluid temperature for a given average velocity. If the flow should suddenly drop to a much lower velocity or come to rest, the hot wire will burn out since the convection loss cannot match the heat generation before the wire temperature reaches the melting point. The constant-temperature type does not have this drawback because the feedback system *automatically* sets wire current to maintain the desired (safe) wire temperature for every velocity. A further advantage of the constant-temperature method lies in the nature of the dynamic compensation. In the constant-current method the compensating network must be reset (using the square-wave current) whenever the average velocity changes appreciably. Furthermore, if velocity fluctuations about the average are large (more than, say, 5 percent of the average velocity) the dynamic compensation will not be complete since the value of τ varies with V and thus the compensating network time constant τ' should be continuously and instantaneously varied, which it is not. The feedback arrangement of the constant-temperature system provides the proper compensation for velocity fluctuations of *any* size, so long as the amplifier gain is large enough to maintain nearly perfect instantaneous bridge balance. The main disadvantages of the constant-temperature scheme are the higher noise level in the electronics (which prevents measurement of very small velocity fluctuations) and the difficulty in designing sufficiently high-gain d-c amplifiers without causing instability and drift problems. Note that since the constant-temperature system uses a d-c amplifier it is usable down to zero frequency (steady velocity), whereas the constant-current (which uses a-c amplifiers) is good only to about 1 cps.

In the above analyses the fluid density ρ was assumed constant. If ρ varies, the anemometer actually measures the product ρV, that is, the local mass flow rate. Thus in the calibration curve of Fig. 7.11, a more general relationship could have been shown; that is, I^2 is really proportional to $\sqrt{\rho V}$. Thus in compressible flows where ρ varies significantly, one would have to know ρ in order to reduce anemometer readings to velocity values.

For both constant-current and constant-temperature instruments, the output voltage (assuming constant ρ) may be taken directly proportional to velocity fluctuations as long as they are a small percentage of the average velocity. Large velocity changes are inaccurately measured by the constant-current method but are properly handled by the constant-temperature scheme. However, the instrument output voltage is *not* now proportional to velocity but rather follows the basic relationship of Fig. 7.11; that is, I^2 varies linearly with $\sqrt{V}$. Since the output voltage e_o of Fig. 7.15 is proportional to I, we see that e_o varies linearly with $(V)^{\frac{1}{4}}$. It is possible to construct an electrical computing circuit (function generator)

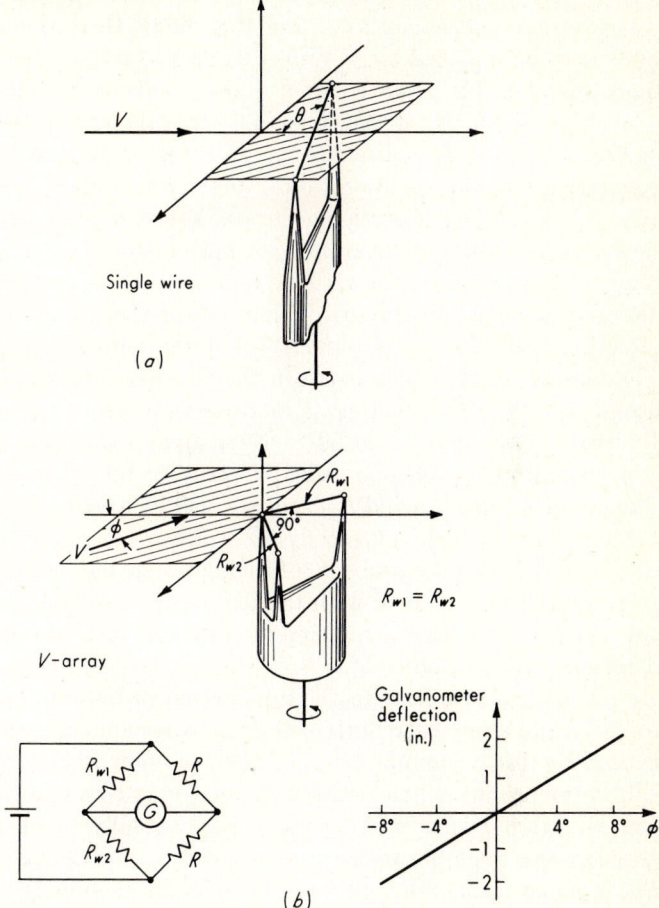

Fig. 7.16. *Flow-direction measurement.*

that will produce an output voltage proportional to the fourth power of its input voltage. If e_o is applied to the input of such a circuit, the output voltage will be proportional to velocity for both large and small changes, thus giving an easily interpretable record.

The hot-wire anemometer may be used to measure the direction of the average flow velocity in several ways.[1] It has been found that a single wire, as in Fig. 7.16a, responds essentially to the component of

---

[1] H. H. Lowell, Design and Applications of Hot-wire Anemometers for Steady-state Measurements at Transonic and Supersonic Speeds, *NACA, Tech. Note* 2117, 1950.

velocity perpendicular to it, if the angle between wire and velocity vector is between 90 and about 25°. For this range, then, the V in our derivations may be replaced by $V \sin \theta$. (For $\theta < 25°$ the heat loss is greater than predicted by $V \sin \theta$; for $\theta = 0$ it is about 55 percent of that for $\theta = 90°$.) With the arrangement of Fig. 7.11 and a rotatable probe as in Fig. 7.16a, the flow-direction angle (in a single plane) could be found by determining the probe-rotation angle which gives a maximum value of I. This method is quite inaccurate, however, since $\sin \theta$ changes very slowly with θ when θ is near 90°. A better procedure, if the flow angle is roughly known, is as follows: The wire is set at about 50° from the flow direction and I is measured. The probe is then rotated in the opposite direction until an angle is found at which the same I as before is measured. The bisector of the angle between the two locations then determines the flow direction. This method is more accurate since the rate of change of I with θ is a maximum near 50°. Even greater convenience and accuracy is achieved by use of a so-called V array of hot wires, as in Fig. 7.16b. The two hot wires R_{w1} and R_{w2} are assumed identical and form a V with an included angle which is typically 90°. They are connected into a bridge as shown. When the probe is rotated, a bridge null occurs when the flow-velocity vector is aligned with the bisector of the V. In a specific case,[1] the sensitivity of this arrangement was sufficient to determine velocity direction within about 0.5°.

Practical problems in the application of hot-wire anemometers are found in the limited strength of the fine wires and the calibration changes caused by dirt accumulations. Unless the flow is very clean, significant calibration changes can occur in a relatively few minutes of operation. Larger dirt particles striking wires may actually break them. At high speeds, wires may vibrate because of aerodynamic loads and flutter effects. While some measurements have been made in liquids, the majority of applications have been in gases.

A variation of the hot-wire anemometer intended to overcome some of its problems is the hot-film transducer. Here the resistance element is a thin film of platinum deposited on a glass base. The film takes the place of the hot wire; the required circuitry is basically similar to the constant-temperature hot-wire approach. The film transducers have great mechanical strength and may also be used at very high temperatures by constructing them with internal cooling-water passages. Various configurations of sensors are possible; Fig. 7.17 shows two possibilities.[2,3]

In addition to measurement of velocity magnitude and direction, hot-

---

[1] *Ibid.*
[2] DISA Electronik A/S, Herlev, Denmark.
[3] Thermo-Systems Inc., Minneapolis, Minn.

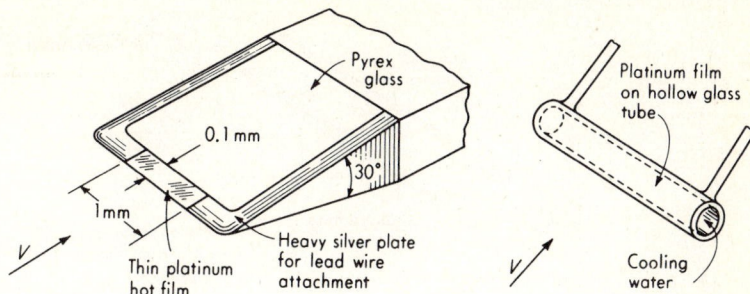

Fig. 7.17. *Hot-film anemometer.*

wire and hot-film instruments may also be adapted to measurements of fluid temperatures, turbulent shear stresses, and concentrations of individual gases in mixtures.[1] Furthermore, if several quantities (such as, say, velocity and temperature) vary simultaneously, it may be possible to extract information about each separate quantity from a hot wire with suitable auxiliary equipment.[1]

Hot-film shock-tube velocity sensors. In shock-tube experiments the propagation velocity of the shock wave down the tube must often be measured.   Of the various means available for making this measurement, hot-film temperature sensors are in wide use.   The passage of the shock wave past a particular section of the tube is accompanied by a step change in gas temperature.   By locating thin resistance films flush with the inside of the tube, the instant of wave-front passage may be detected as a temperature (and therefore resistance) change.   If two such film sensors are mounted a known distance apart, the average wave velocity may be computed from the time interval between the two sensor responses.   The films are operated at constant current by the simple circuit of Fig. 7.18, and a differentiating circuit is used to sharpen the pulses for greater timing accuracy.   With systems of this type,[2] shock-wave velocities have been measured with an accuracy of 1 percent for shock Mach numbers as high as 7.5 to 10.

7.2

Gross Volume Flow Rate   The total flow rate through a duct or pipe must often be measured and/or controlled.   Many instruments

---

[1] S. Corrsin, Extended Applications of the Hot-wire Anemometer, *NACA, Tech Note* 1864, 1949.

[2] Shock Tubes, Handbook of Supersonic Aerodynamics, sec. 18, *NAVORD Rept.* 1488, vol. 6, pp. 543, 558.

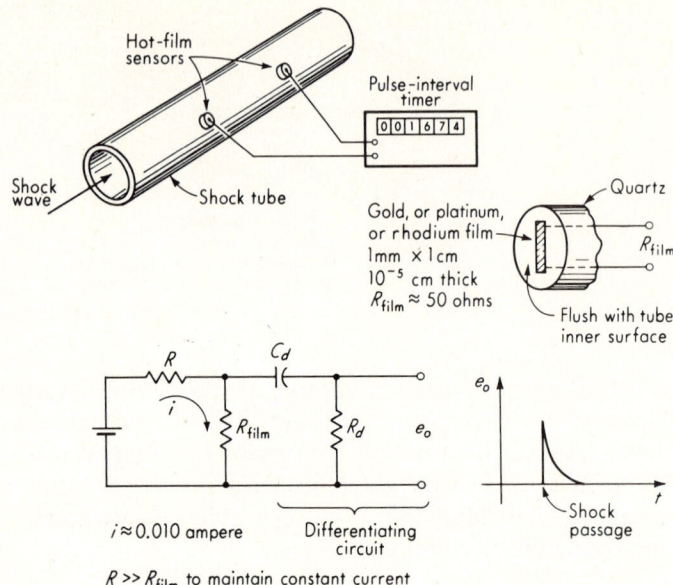

Fig. 7.18.   *Thin-film velocity sensor.*

(flowmeters) have been developed for this purpose. They may be classified in various ways; a useful overall classification divides devices into those which measure volume flow rate (ft³/time) and those which measure mass flow rate (lb$_m$/time). The total flow (ft³ or lb$_m$) occurring during a given time interval may also be of interest. This measurement requires integration with respect to the time of the instantaneous flow rate. The integrating function may also be performed in various ways, sometimes being an integral part of the flowmetering concept and other times being performed by a general-purpose integrator more or less remote from the flowmeter.

Calibration and standards. Flow-rate calibration depends on standards of volume (length) and time or mass and time. Primary calibration is, in general, based on the establishment of steady flow through the flowmeter to be calibrated and subsequent measurement of the volume or mass of flowing fluid that passes through in an accurately timed interval. If steady flow exists, the volume or mass flow rate may be inferred from such a procedure. Any stable and precise flowmeter calibrated by such primary methods then itself becomes a secondary flow-rate standard against which other (less accurate) flowmeters may be conveniently calibrated. As in any other calibration, significant deviations of the conditions of use from

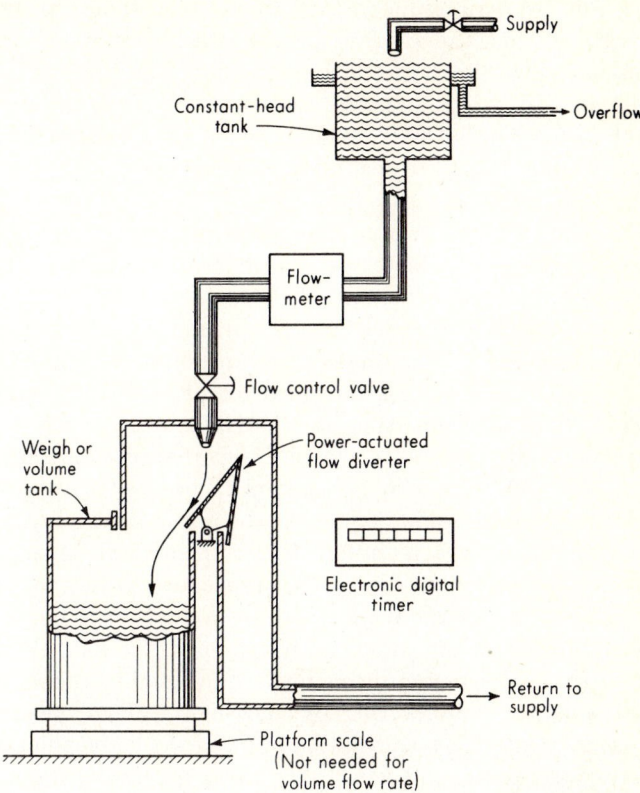

Fig. 7.19. *Flowmeter calibration setup.*

those at calibration will invalidate the calibration. Possible sources of error in flowmeters include variations in fluid properties (density, viscosity, and temperature), orientation of the meter, pressure level, and particularly flow disturbances (such as elbows, tees, valves, etc.) upstream (and to a lesser extent downstream) of the meter.

A typical calibration setup for precise primary calibration of flow-meters using liquids is shown in Fig. 7.19. A constant-head tank maintains a fixed inlet pressure to the flowmeter under test, irrespective of flow rate. (A constant-head tank may be impractical, because of the excessive elevation required, for very high inlet pressures, or because of toxicity, volatility, or flammability of certain liquids. Then a closed system using a pump to supply pressure is used. Manual or automatic control of the flow rate to maintain constancy during a given run is then necessary.) The flow rate through the meter is adjusted to the various desired values with a flow-control valve. If necessary, this valve is manipulated during the run to maintain constant flow. Constancy of flow is generally observed from the reading of the meter under calibration. Until a constant flow rate is established, the liquid is diverted from the weigh or volume tank which must be emptied (volume tank) or weighed (weigh tank) before flow into it is started. There are available volume tanks, called prover tanks, that are accurate to 0.05 to 0.1 percent of nominal volume. Platform scales accurate to 0.05 to 0.1 percent are used for mass flow measurements. When the flow diverter is suddenly moved to the tank position, a switch starts the electronic timer as the diverter passes the mid-position. Flow is continued until the tank is filled, at which time the motion of the diverter through the mid-position to the return position stops the timer. The weight or volume of liquid accumulated during the timed interval is then determined to calculate the volume or mass flow rate. With extreme attention to details, calibrations of the above type result in overall flow-rate errors of the order of a few tenths of 1 percent.[1]

The calibration of flowmeters to be used with gases can often be carried out with liquids as long as the pertinent similarity relations (Reynolds number) are maintained and theoretical density and expansion corrections are applied. If this procedure is felt to be of insufficient accuracy, a direct calibration with the actual gas to be used can be carried out by means of the *gasometer* system of Fig. 7.20. Here the gas flowing through the flowmeter during a timed interval is trapped in the gasometer bell and its volume thereby measured. Temperature and pressure measurements allow calculation of mass and conversion of volume to any desired standard conditions. By filling the bell with gas, raising it to the top, and adding appropriate weights, such a system may also be used as a gas *supply* to drive gas through a flowmeter as the bell gradually drops at a measured rate.

When the above primary calibration methods cannot be justified, comparison with a secondary standard flowmeter connected in series with the meter to be calibrated may be sufficiently accurate. Turbine flowmeters and their associated digital counting equipment have been found particularly suitable for such secondary standards. With attention to detail, such standards can closely approach the accuracy of the primary methods themselves. The Navy Primary Standards Laboratory at Pensacola, Florida, has such a system[2] with an inaccuracy of the order of 0.2 percent. For gas flow, a flow nozzle discharging air to the atmosphere can be very accurately calibrated[3] for mass flow rate by means of pitot-

[1] M. R. Shafer and F. W. Ruegg, Liquid-flowmeter Calibration Techniques, *Trans. ASME*, p. 1369, October, 1958.

[2] R. P. Bowen, Designing Portability into a Flow Standard, *ISA J.*, p. 40, May, 1961.

[3] R. J. Sweeney, "Measurement Techniques in Mechanical Engineering," p. 220, John Wiley & Sons, Inc., New York, 1953.

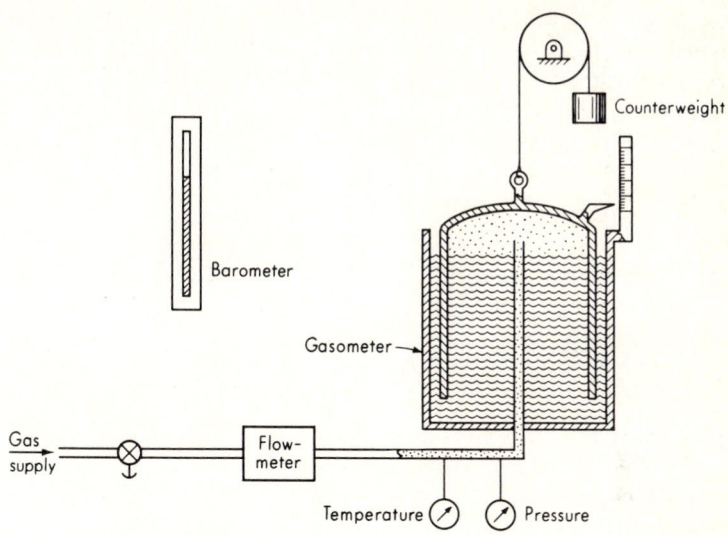

Fig. 7.20. *Gas-flow calibration.*

tube traverses at the discharge. This nozzle can then be connected in series with any flowmeter to be calibrated and used as an accurate standard. This can be done with very small error up to Mach number about 0.9.

Constant-area, variable-pressure-drop meters ("obstruction" meters). Perhaps the most widely used flowmetering principle involves placing a fixed-area flow restriction of some type in the pipe or duct carrying the fluid. This flow restriction causes a pressure drop which varies with flow rate; thus measurement of the pressure drop by means of a suitable differential pressure pickup allows flow-rate measurement. In this section we shall briefly discuss the most common practical devices that utilize this principle: the orifice, the flow nozzle, the venturi tube, the Dall flow tube, and the laminar-flow element.

The *sharp-edge orifice* is undoubtedly the most widely used flowmetering element, mainly because of its simplicity, low cost, and the great volume of research data available for predicting its behavior. A typical flowmetering setup is shown in Fig. 7.21. If one-dimensional flow of an incompressible frictionless fluid without work, heat transfer, or elevation change is assumed, theory gives the volume flow rate Q_t (ft³/sec) as

$$Q_t = \frac{A_{2f}}{\sqrt{1 - (A_{2f}/A_{1f})^2}} \sqrt{\frac{2(p_1 - p_2)}{\rho}} \qquad (7.45)$$

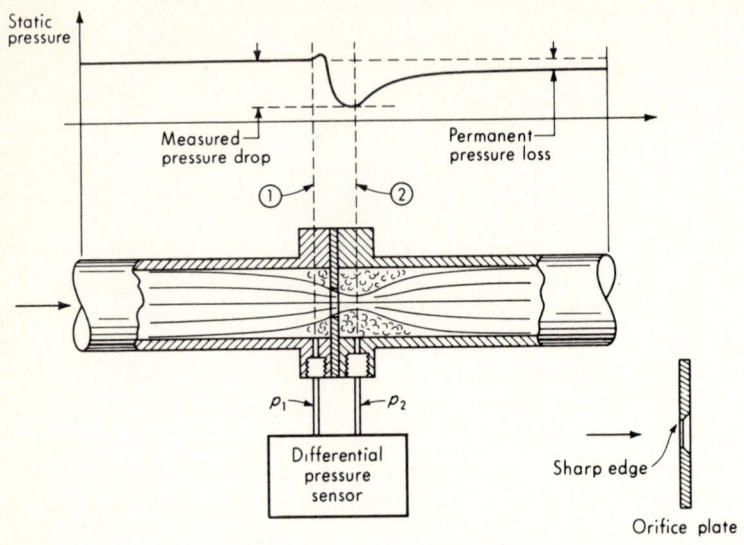

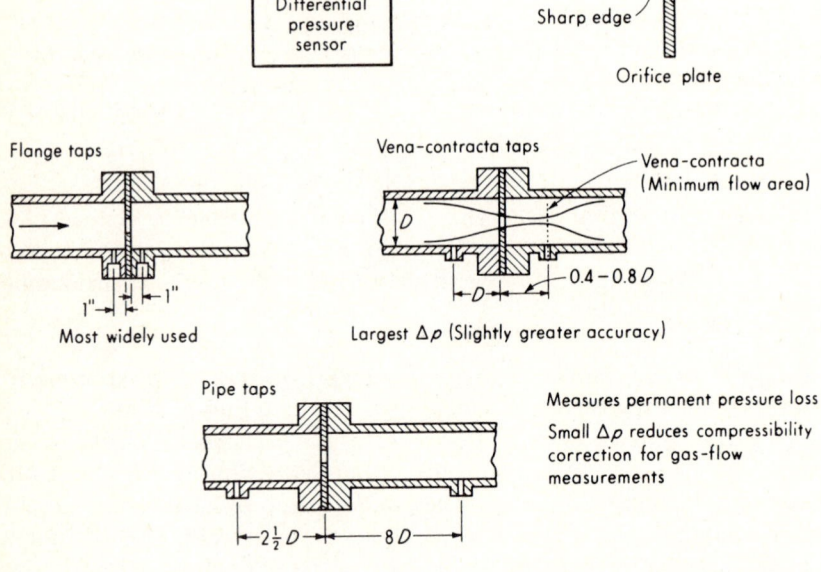

Fig. 7.21.   *Orifice flowmetering.*

where $A_{1f}, A_{2f} \triangleq$ cross-section flow areas where p_1 and p_2 are measured, ft²

$\rho \triangleq$ fluid mass density, slug/ft³

$p_1, p_2 \triangleq$ static pressures, lb$_f$/ft²

We see that measurement of Q requires knowledge of A_{1f}, A_{2f}, and ρ and measurement of the pressure differential $(p_1 - p_2)$. Actually, the real situation deviates from the assumptions of the theoretical model suffi-

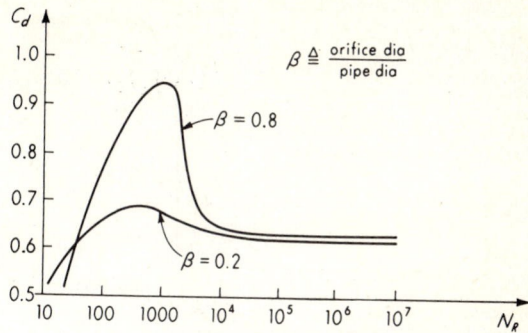

Fig. 7.22. *Variation of discharge coefficient.*

ciently to require experimental correction factors if acceptable flowmetering accuracy is to be attained.   For example, A_{1f} and A_{2f} are areas of the actual flow cross section, which are *not*, in general, the same as those corresponding to the pipe and orifice diameters, which are the ones susceptible to practical measurement.   Furthermore, A_{1f} and A_{2f} may change with flow rate because of flow geometry changes.   Also, there are present frictional losses that affect the measured pressure drop and also lead to a permanent pressure loss.   To take these factors into account, an experimental calibration to determine the actual flow rate Q_a by methods such as that of Fig. 7.19 is necessary.   A discharge coefficient C_d may then be defined by

$$C_d \triangleq \frac{Q_a}{Q_t} \qquad (7.46)$$

and thus

$$Q_a = \frac{C_d A_2}{\sqrt{1 - (A_2/A_1)^2}} \sqrt{\frac{2(p_1 - p_2)}{\rho}} \qquad (7.47)$$

where $A_1 \triangleq$ pipe cross-section area
$A_2 \triangleq$ orifice cross-section area

The discharge coefficient of a given installation varies mainly with the Reynolds number N_R at the orifice.   Thus the calibration can be performed with a single fluid, such as water, and the results used for any other fluid as long as the Reynolds numbers are the same.   Variation of C_d with N_R follows typically the trend[1] of Fig. 7.22.   While the above discussion would seem to indicate that each installation must be individually calibrated, this is fortunately not the case.   If one is willing to

---

[1] "Handbook of Measurement and Control," p. 96, The Instruments Publishing Co., Pittsburgh, Pa., 1954.

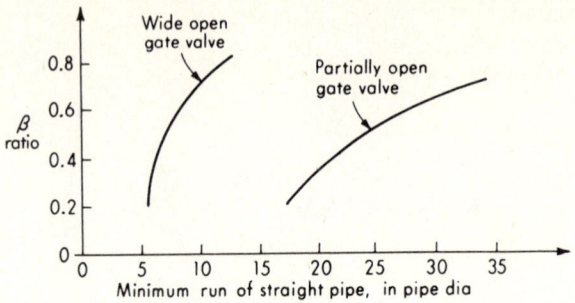

Fig. 7.23. *Effect of upstream disturbances.*

construct the orifice according to certain standard dimensions and also locate the pressure taps at specific points, then quite accurate (about 0.4 to 0.8 percent error) values of C_d may be obtained from tables or charts[1] which have been compiled based on many past experiments. Such data are available for pipe diameters 2 in. and greater, β (see Fig. 7.22) ratios of 0.2 to 0.7, and Reynolds number above 10,000. Installations exceeding these limits should be individually calibrated if high accuracy is required. It should also be noted that the standard calibration data assume no significant flow disturbances such as elbows, bends, tees, valves, etc., for a certain minimum distance upstream of the orifice. The presence of such disturbances close to the orifice can invalidate the standard data, causing errors of as much as 15 percent. Information on the minimum distances is available[3,4]; Fig. 7.23 shows a typical example. If the minimum distances are not feasible, straightening vanes[2,3,4] may be introduced ahead of the flowmeter to smooth out the flow.

Since flow rate is proportional to $\sqrt{\Delta p}$, a 10:1 change in Δp corresponds to only about a 3:1 change in flow rate. Since a given Δp-measuring instrument becomes quite inaccurate below about 10 percent of its full-scale reading, this nonlinearity typical of all obstruction meters (other than the laminar-flow element) restricts the accurate range of flow measurement to about 3 to 1. That is, a meter of this type cannot be used accurately below about 30 percent of its maximum flow rating. The square-root nonlinearity also causes difficulties in pulsating flow measure-

---

[1] Fluid Meters, Their Theory and Application, American Society of Mechanical Engineers, New York, 1937.

[2] "Handbook of Measurement and Control," p. 96.

[3] Fluid Meters, Their Theory and Application.

[4] P. S. Starrett and P. F. Halfpenny, The Effect of Non-standard Approach Sections on Orifices and Venturi Meters, Lockheed California Co., LR17905, 1964.

ment,[1,2] where the average flow rate (the rate to be measured) has a fluctuating component superimposed on it. Let us consider, as a simple example, a flow Q where

$$Q = Q_{av} + Q_p \sin \omega t \qquad Q_p < Q_{av} \qquad (7.48)$$

and a flowmeter such that $\Delta p = KQ^2$. The Δp presented as input to the pressure-measuring system is then

$$\Delta p = K(Q_{av}^2 + 2Q_{av}Q_p \sin \omega t + Q_p^2 \sin^2 \omega t) \qquad (7.49)$$

If the Δp instrument has a low-pass filtering characteristic, it will tend to read the average value of Δp. This is seen to be

$$\Delta p_{av} = K\left(Q_{av}^2 + \frac{Q_p^2}{2}\right) \qquad (7.50)$$

Thus if one takes a measured Δp_{av} and computes the corresponding Q_{av} from it, using $Q_{av} = \sqrt{\Delta p_{av}/K}$, he will get a flow rate *higher* than actually existed. A further difficulty caused by the nonlinearity occurs when flow rate must be integrated to get total flow during a given time interval. The Δp signal must then be square-rooted before integration or this compensation included in the integrating device.

The orifice has the largest permanent pressure loss of any of the obstruction meters (other than the laminar-flow element); this is one of its disadvantages since it represents a power loss that must be replaced by whatever pumping machinery is causing the flow. The permanent pressure loss is given approximately by $\Delta p(1 - \beta^2)$, where Δp is the differential pressure used for flow measurement. Thus for the usual range of $\beta(0.2$ to $0.7)$ the permanent pressure loss ranges from $0.96\Delta p$ to $0.51\Delta p$. The actual power loss may, in fact, be quite small since the Δp recommended[3] for conventional flowmetering of liquids is only 20 to 400 in. of water (0.72 to 14.4 psi).

Orifice discharge coefficients are quite sensitive to the condition of the upstream edge of the hole. The standard orifice design requires that this edge be very sharp and also that the orifice plate be sufficiently thin relative to its diameter. Wear (rounding) of this sharp edge by long use, particularly if the fluid contains abrasive particles, can cause significant changes in the discharge coefficient. Flows that contain suspended solids

---

[1] A. K. Oppenheim and E. G. Chilton, Pulsating Flow Measurement—A Literature Survey, *Trans. ASME*, p. 231, February, 1955.

[2] T. Isobe and H. Hattori, A New Flowmeter for Pulsating Gas Flow, *ISA J.*, p. 38, December, 1959.

[3] L. K. Spink, "Principles and Practice of Flow Meter Engineering," The Foxboro Co., Foxboro, Mass., 1959.

also cause difficulty since the solids tend to collect behind the "dam" formed by the orifice plate and cause irregular flow. This problem can often be solved by use of an "eccentric" orifice in which the hole is at the bottom of the pipe rather than on the center line. This allows the solids to be continuously swept through. Liquids containing traces of vapor or gas may be metered if the orifice is installed in a vertical run of pipe with the flow upward. Gases containing traces of liquid may be similarly handled except that the flow should be downward.

When compressible fluids are metered, Eq. (7.47) is no longer correct. By assuming an isentropic process between states 1 and 2, the following relation may be derived[1] for compressible fluids:

$$W = C_d A_2 \sqrt{\frac{2gk p_1}{(k-1)v_1}} \sqrt{\frac{(p_2/p_1)^{2/k} - (p_2/p_1)^{(k+1)/k}}{1 - \beta^4 (p_2/p_1)^{2/k}}} \qquad (7.51)$$

where $W \triangleq$ weight flow rate, lb_f/sec
$k \triangleq$ ratio of specific heats (1.4 for air)
$g \triangleq$ local gravity
$v_1 \triangleq$ specific volume at state 1, ft^3/lb_f

The discharge coefficient C_d is the same for liquids or gases as long as the Reynolds number is the same. In many practical gas-flow installations the meter pressure drop is so small that the pressure ratio p_2/p_1 is 0.99 or greater. Under these conditions the simpler incompressible relation of Eq. (7.47) may be used with an error less than 0.6 percent if, for example, $\beta = 0.5$ and $k = 1.4$. Modified to give weight rather than volume flow rate, this is

$$W = C_d A_2 \left(\sqrt{1 - \left(\frac{A_2}{A_1}\right)^2} \right)^{-1} \sqrt{\frac{2g(p_1 - p_2)}{v_1}} \qquad (7.52)$$

For $p_2/p_1 < 0.99$ the error becomes greater, being 6 percent at $p_2/p_1 = 0.9$, 12 percent at 0.8, 19 percent at 0.7, and 26 percent at 0.6 (if $\beta = 0.5$ and $k = 1.4$). For such situations the more complex compressible formula must obviously be used. In flow nozzles and venturis the flow process is close enough to isentropic to allow theoretical calculation of compressibility effects from Eq. (7.51). In orifices, however, deviation from isentropic conditions is significant (greater turbulence), and an experimental compressibility factor Y is used in the equation:

$$W = Y \left[C_d A_2 \left(\sqrt{1 - \left(\frac{A_2}{A_1}\right)^2} \right)^{-1} \right] \sqrt{\frac{2g(p_1 - p_2)}{v_1}} \qquad (7.53)$$

[1] D. P. Eckman, "Industrial Instrumentation," p. 270, John Wiley & Sons, Inc., New York, 1950.

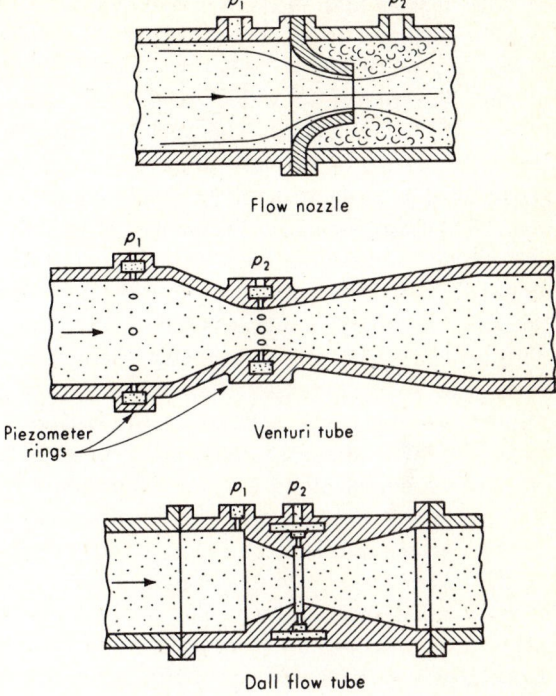

Fig. 7.24. *Variable-pressure-drop meters.*

For flange taps or vena-contracta taps[1]

$$Y = 1 - (0.41 + 0.35\beta^4) \frac{p_1 - p_2}{p_1} \frac{1}{k} \qquad (7.54)$$

while for pipe taps[1]

$$Y = 1 - [0.333 + 1.145(\beta^2 + 0.7\beta^5 + 12\beta^{13})] \frac{p_1 - p_2}{p_1} \frac{1}{k} \qquad (7.55)$$

These empirical formulas are accurate to ± 0.5 percent if $0.8 < p_2/p_1 < 1.0$ and the flowing fluid is a gas or vapor other than steam. For steam the accuracy is ± 1.0 percent. In sizing orifices for gas measurement, a useful rule of thumb is that the maximum Δp (in inches of water) should not exceed the upstream gage pressure in pounds per square inch.

The *flow nozzle, venturi tube,* and *Dall flow tube* (Fig. 7.24) all operate on exactly the same principle as the orifice, the significant differences lying in the numerical values of certain characteristics. Discharge coefficients

---

[1] Fluid Meters, Their Theory and Application.

of flow nozzles and venturis are larger than those for orifices and also exhibit an opposite trend with Reynolds numbers, varying from about 0.94 at $N_R = 10,000$ to 0.99 at $N_R = 10^6$. The Dall-flow-tube coefficient is more like that of an orifice, for $\beta = 0.7$, for example, going from about 0.68 at $N_R = 100,000$ to 0.66 at $N_R = 10^6$. Individual calibrations are generally needed on all these devices since their complicated shapes (compared with an orifice) make accurate reproduction difficult.

When comparing the permanent pressure losses of the various devices one should require that each device be producing the same measured Δp, since this would keep the accuracy constant. On this basis, the permanent pressure loss of a flow nozzle is practically identical with that of an orifice. This is because, to get the same Δp, the flow nozzle must have a smaller β ratio, and losses increase with decreasing β ratio. The venturi tube also requires a smaller β for a given Δp, but because of its streamlined form, its losses are low and nearly independent of β. The permanent pressure loss is of the order of 10 to 15 percent of the measured Δp over the range $0.2 < \beta < 0.8$; thus a venturi gives a definite improvement in power losses over an orifice and is often indicated for measuring very large flow rates, where power losses, though a small *percentage*, become economically significant in absolute value. The initial higher cost of a venturi over an orifice may thus be offset by reduced operating costs. The Dall flow tube has the unexpected (though desirable) features of a high measured Δp (similar to an orifice) and a low permanent pressure loss (similar to, and sometimes better than, a venturi). These apparently inconsistent virtues have been checked experimentally but are not fully explained theoretically.[1] The permanent pressure loss of a Dall tube is of the order of 50 percent or less of that of a venturi tube with the same Δp. Other factors to consider in choosing among the orifice, flow nozzle, venturi, or Dall tube include freedom from pressure-tap clogging due to suspended solids (venturi is best), loss of accuracy due to wear (venturi, flow nozzle, and Dall tube are better than an orifice), accuracy (venturi, when calibrated, is best), cost, and ease of changing the flow element to a different size.

Laminar-flow elements differ from the metering devices discussed above in that they are specifically designed to operate in the laminar-flow regime. Pipe flows generally are considered laminar if Reynolds number N_R is less than 2,000; however, in laminar-flow elements, considerably lower values are often designed for to ensure laminar conditions. The simplest form of laminar-flow element is merely a length of small-diameter (capillary) tubing.[2] For $N_R < 2,000$, the Hagen-Poiseuille viscous-

[1] I. O. Miner, The Dall Flow Tube, *Instr. Engr.*, p. 45, April, 1957.
[2] L. M. Polentz, Capillary Flowmetering, *Instr. Control Systems*, p. 648, April, 1961.

flow relation gives for incompressible fluids

$$Q = \frac{\pi D^4}{128\,\mu L}\,\Delta p \qquad (7.56)$$

where $Q \triangleq$ volume flow rate, ft³/sec
$D \triangleq$ tube inside diameter, ft
$\mu \triangleq$ fluid viscosity, lb$_f$-sec/ft²
$L \triangleq$ tube length between pressure taps, ft
$\Delta p \triangleq$ pressure drop, lb$_f$/ft²

One usually designs for $N_R \lesssim 1,000$ in such a device. Extremely small flows can be measured in this way; a 3-ft length of 0.004-in.-diameter tubing measuring Δp with a 2-in. water inclined manometer gives a threshold sensitivity of about 0.000175 in.³/hr when hydrogen is flowing.[1]

A single capillary tube is capable of handling only small flow rates at laminar Reynolds numbers. To increase the capacity of laminar-flow elements, many capillaries in parallel (or their equivalent) may be employed. One commercially available variation[2] uses a large tube (about 1 in. in diameter) packed with small spheres. The passages between the spheres give the same effect as many capillary tubes. This particular instrument is designed to give a Reynolds number of 20 or less. A 5-in. length of tube gives a Δp of 20 in. of water for a 2 cm³/min flow rate of air at 14.7 psia and 70°F. Another approach[3] uses a "honeycomb" element (see Fig. 7.25) with triangular members a few thousandths of an inch on a side and a few inches long. These devices have been used mainly to measure flow of low-pressure air, and standard models are available in ranges from 50 to 2,000 ft³/min at pressure drops of 4 to 8 in. of water. All the above laminar elements have the advantages accruing from a linear (rather than square-root) relation between flow rate and pressure drop. These are principally a large accurate range of as much as 100 to 1 (compared with 3 or 4 to 1 for square-root devices), accurate measurement of average flow rates in pulsating flow, and ease of integrating Δp signals to compute total flow. The laminar elements also can measure reversed flows with no difficulty. They are also usually less sensitive to upstream and downstream flow disturbances than other devices that have been discussed. Their disadvantages include clogging due to dirty fluids, high cost, large size, and high pressure loss (*all* the measured Δp is lost).

[1] Polentz, *loc. cit.*

[2] A. R. Hughes, New Laminar Flowmeter, *Instr. Control Systems*, p. 98, April, 1962.

[3] Meriam Instrument Co., Cleveland, Ohio, *Tech. Note* 2A.

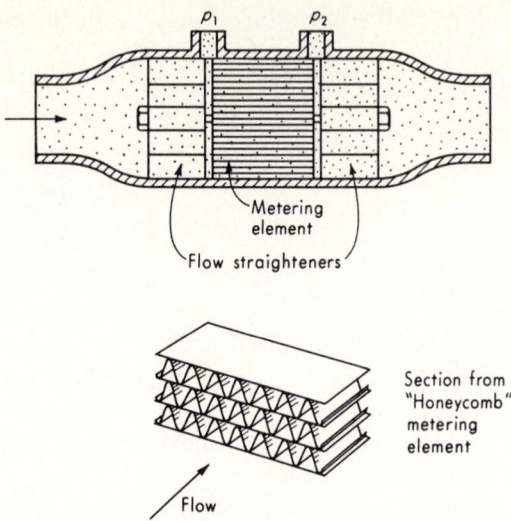

Fig. 7.25. *Laminar-flow element.*

Constant-pressure-drop, variable-area meters (rotameters). A rotameter consists of a vertical tube with tapered bore in which a "float" assumes a vertical position corresponding to each flow rate through the tube (see Fig. 7.26). For a given flow rate the float remains stationary since the vertical forces of differential pressure, gravity, viscosity, and buoyancy are balanced. This balance is self-maintaining since the meter flow area (annular area between the float and tube) varies continuously with vertical displacement; thus the device may be thought of as an

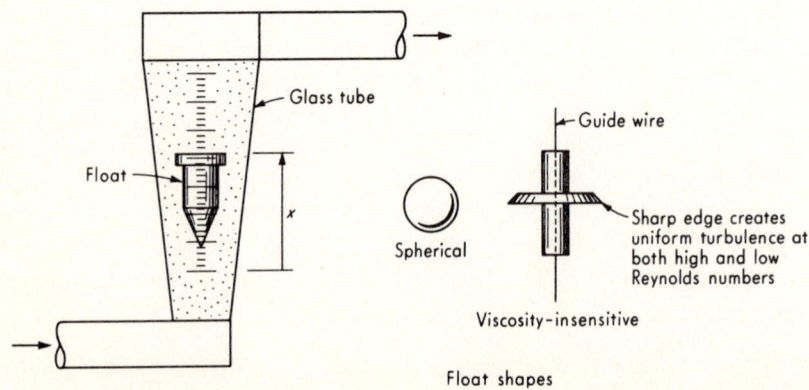

Fig. 7.26. *Rotameter.*

orifice of adjustable area.   The downward force (gravity minus buoyancy) is constant and so the upward force (mainly the pressure drop times the float cross-section area) must also be constant.   Since the float area is constant, the pressure drop must be constant.   For a *fixed* flow area, Δp varies with the square of flow rate, and so to keep Δp *constant* for differing flow rates the area must vary.   The tapered tube provides this variable area.   The float position is the output of the meter and can be made essentially linear with flow rate by making the tube area vary linearly with the vertical distance.   Rotameters thus have an accurate range of about 10 to 1, considerably better than square-root-type elements.   Assuming incompressible flow and the above described simplified model, one can derive the result

$$Q = \frac{C_d(A_t - A_f)}{\sqrt{1 - [(A_t - A_f)/A_t]^2}} \sqrt{2gV_f \frac{w_f - w_{ff}}{A_f w_{ff}}} \qquad (7.57)$$

where $Q \triangleq$ volume flow rate, ft³/sec
$C_d \triangleq$ discharge coefficient
$A_t \triangleq$ area of tube, ft²
$A_f \triangleq$ area of float, ft²
$g \triangleq$ local gravity, ft/sec²
$V_f \triangleq$ volume of float, ft³
$w_f \triangleq$ specific weight of float, lb$_f$/ft³
$w_{ff} \triangleq$ specific weight of flowing fluid, lb$_f$/ft³

If the variation of C_d with float position is slight and if $[(A_t - A_f)/A_t]^2$ is always much less than 1, Eq. (7.57) has the form

$$Q = K(A_t - A_f) \qquad (7.58)$$

and if the tube is shaped so that A_t varies linearly with float position x, then $Q = K_1 + K_2 x$, a linear relation.   The floats of rotameters may be made of various materials to obtain the desired density difference [$w_f - w_{ff}$ in Eq. (7.57)] for metering a particular liquid or gas.   Some float shapes, such as spheres, require no guiding in the tube; others are kept central by guide wires or by internal ribs in the tube.   Certain shapes of floats have been found to reduce viscosity effects.   The tubes are often made of high-strength glass to allow direct observation of the float position.   Where greater strength is required, metal tubes can be used and the float position detected magnetically through the metal wall.   If a pneumatic or electrical signal related to the flow rate is desired, the float motion can be measured with a suitable displacement transducer.

Turbine meters.   If a turbine wheel is placed in a pipe containing a flowing fluid, its rotary speed depends on the flow rate of the fluid.   By

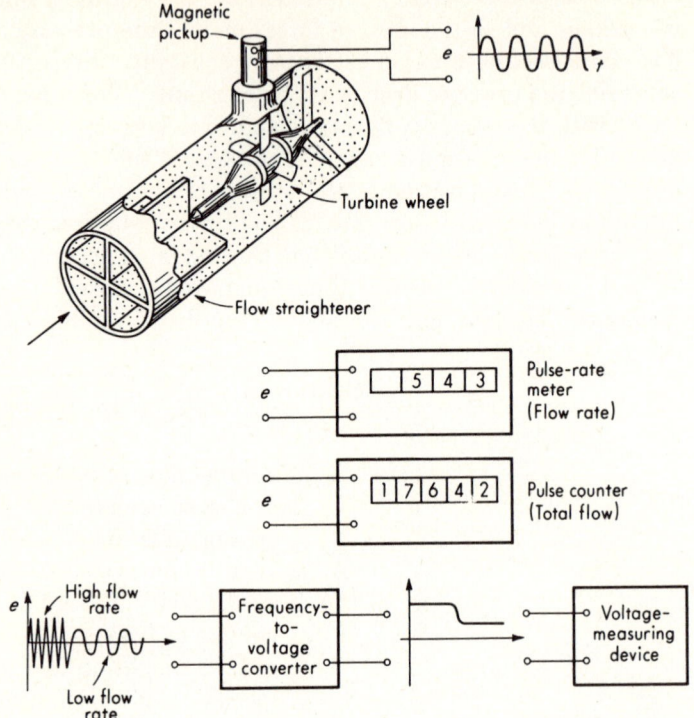

Fig. 7.27.   *Turbine flowmeter*

reducing bearing friction and other losses to a minimum, one can design a turbine whose speed varies linearly with flow rate; thus a speed measurement allows a flow-rate measurement.   The speed can be measured simply and with great accuracy by counting the rate at which turbine blades pass a given point, using a magnetic proximity pickup to produce voltage pulses.   By feeding these pulses to an electronic pulse-rate meter one can measure flow rate, and by accumulating the total number of pulses during a timed interval the total flow is obtained.   These measurements can be made very accurately because of their digital nature.   If an analog voltage signal is desired, the pulses can be fed to a frequency-to-voltage converter with, however, some loss in accuracy.   Figure 7.27 shows a flowmetering system of this type.

Dimensional analysis[1] of the turbine flowmeter shows that (if bearing

[1] H. M. Hochreiter, Dimensionless Correlation of Coefficients of Turbine-type Flowmeters, *Trans. ASME*, p. 1363, October, 1958.

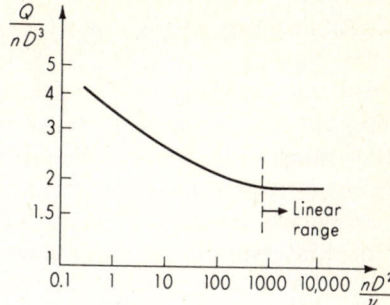

Fig. 7.28. *Turbine-flowmeter character-istic.*

friction and shaft power output are neglected) the following relation should hold:

$$\frac{Q}{nD^3} = \text{some function of } \frac{nD^2}{\nu} \qquad (7.59)$$

where $Q \triangleq$ volume flow rate, in.³/sec
$n \triangleq$ rotor angular velocity, rps
$D \triangleq$ meter bore diameter, in.
$\nu \triangleq$ kinematic viscosity, in.²/sec

Actually, the effect of viscosity is limited mainly to low flow rates, high flow rates being in the turbulent regime where viscosity effects are secondary. For negligible viscosity effects, a simplified analysis[1] based on strictly kinematic relationships gives the following result:

$$\frac{Q}{nD^3} = \frac{\pi L}{4D} \left[1 - \alpha^2 - \frac{2m(D_b - D_h)t}{\pi D^2} \sqrt{1 + \left(\frac{\pi D_b}{L}\right)^2} \right] \qquad (7.60)$$

where $L \triangleq$ rotor lead, in.
$\alpha \triangleq D_h/D$
$m \triangleq$ number of blades
$D_b \triangleq$ rotor-blade-tip diameter, in.
$D_h \triangleq$ rotor-hub diameter, in.
$t \triangleq$ rotor-blade thickness, in.

Equation (7.60) gives $Q = Kn$, where K is a constant for any given meter and is independent of fluid properties. This thus represents the ideal situation. Deviations from this ideal may be found from experimental calibrations,[2] such as are shown for a meter with $D = 1$ in. in Fig. 7.28. We see for sufficiently high values of nD^2/ν that Q/nD^3 becomes essentially constant as predicted by Eq. (7.60). In the particular case shown, $Q/nD^3 = 1.92$ for at least at 10-to-1 range of nD^2/ν; thus this would be a

[1] H. M. Hochreiter, Dimensionless Correlation of Coefficients of Turbine-type Flowmeters, *Trans. ASME*, p. 1363, October, 1958.
[2] *Ibid.*

useful linear operating range for this meter. The meter could be used at lower flow rates by applying corrections obtained from Fig. 7.28. However, this is usually not done since turbine meters are available in a wide range of sizes, each being linear over a different flow range. If the total flow range to be accommodated is about 10 to 1 or less, one can usually select a turbine meter that is linear in the desired range. Linearity is particularly desirable if one is totalizing pulses to get a total flow over a timed interval during which flow rate fluctuates.

Commercial turbine meters are available with full-scale flow rates ranging from about 0.1 to 30,000 gpm for liquids and 0.1 to 15,000 ft^3/min for air. Nonlinearity within the design range (usually about 10 to 1) can be as good as 0.05 percent in the larger sizes. The output voltage of the magnetic pickups is of the order of 10 mv rms at the low end of the flow range and 100 mv at the high. Pressure drop across the meter varies with the square of flow rate and is about 3 to 10 psi at full flow. Turbine meters can follow flow transients quite accurately since their fluid/mechanical time constant is of the order of 2 to 10 msec.[1] If a frequency-to-voltage converter is used to get an analog voltage output, however, its response may be somewhat slower than this since the operating frequencies of turbine meters are of the order of 100 to 2,000 cps. That is, the frequency-to-voltage converter requires low-pass filtering which rejects frequencies somewhat below the turbine operating frequency and is thus limited in transient response also. For example, if the turbine is putting out 500 cps, the low-pass filter will have to cut off at *least* at 500 cps and probably considerably lower. If a first-order filter is designed to attenuate 20 db at 500 cps, it will have a time constant of 0.0032 sec, thus adding this much to the lag of the overall system.

Positive-displacement meters. These meters are actually positive-displacement fluid motors in which friction and inertia have been reduced to a minimum. The flow of a fluid through volume chambers of definite size causes rotation of an output shaft. This type of meter usually is used where the total flow and not the instantaneous flow rate is of interest. A simple mechanical counter records the total number of rotations, which is proportional to the total flow. The accuracy of such devices is of the order of 1.5 percent, with the pressure drop being about 5 psi at maximum flow.

Metering pumps. A variable-displacement positive-displacement pump, if properly designed, can serve both to *cause* a flow rate and also simultaneously to *measure* it. The principle again is merely that a positive-displacement machine, except for leakage and compressibility, delivers a

[1] J. Grey, Transient Response of the Turbine Flowmeter, *Jet Propulsion*, p. 98, Feb. 1956.

definite flow rate of fluid at a given speed. In most pumps of this kind the operating speed is fixed and the flow rate is varied by changing pump displacement, usually with some form of mechanical linkage. Since these pumps are often used in automatic control systems, many are designed to accept pneumatic or electrical input signals which adjust the pump displacement in a linear fashion. The flow rate of such a system can be set with an accuracy of the order of 1 percent.

Electromagnetic flowmeters. The electromagnetic flowmeter is an application of the principle of induction, shown in Fig. 7.29*a*. If a conductor of length *l* moves with a transverse velocity *v* across a magnetic field of intensity *B* there will be forces on the charged particles of the conductor that will move the positive charges toward one end of the conductor and the negative to the other. Thus a potential gradient is set up along the conductor, and there is a voltage difference *e* between its two ends. The quantitative relation among the variables is given by the well-known equation

$$e = Blv \qquad (7.61)$$

where $B \triangleq$ field flux density, volt-sec/ft^2
$l \triangleq$ conductor length, ft
$v \triangleq$ conductor velocity, fps

If the ends of the conductor are connected to some external circuit that is stationary with respect to the magnetic field, the induced voltage will, in general, cause a current *i* to flow. This current flows through the moving conductor, which has a resistance *R*, causing an *iR* drop, so that the terminal voltage of the moving conductor becomes *e-iR*.

We consider now a cylindrical jet of conductive fluid with a uniform velocity profile, traversing a magnetic field as in Fig. 7.29*b*. In a liquid conductor the positive and negative ions are forced to opposite sides of the jet, giving a potential distribution as shown. The maximum voltage difference is found across the ends of a horizontal diameter and is BD_pv in magnitude. In a practical situation, the magnetic field is of limited extent, as in Fig. 7.29*c*; thus no voltage is induced in that part of the jet outside the field. Since these parts of the fluid are, however, still conductive paths, they tend partially to "short circuit" the voltages induced in the section exposed to the field; thus the voltage is reduced from the value BD_pv. If the field is sufficiently long, this effect will be slight at the center of the field length. A length of 3 diameters is usually sufficient.[1]

In a practical flowmeter, the "jet" is contained within a stationary

---

[1] I. C. Hutcheon, Electrical Characteristics of the Magnetic Flow Detector Head, *Instr. Eng.*, p. 1, April, 1964.

pipe. The pipe must be nonmagnetic to allow the field to penetrate the fluid and usually is nonconductive (plastic, for instance) so that it does not provide a short-circuit path between the positive and negative induced potentials at the fluid surface. This nonconductive pipe has two electrodes placed at the points of maximum potential difference. These electrodes then supply a signal voltage to external indicating or recording apparatus. Because it is impractical to make the entire piping installation nonconductive, a short length (the flowmeter itself) of nonconductive pipe must be coupled into an ordinary metal-pipe installation. Since the fluid itself is conductive, this means that there is a conductive path between the

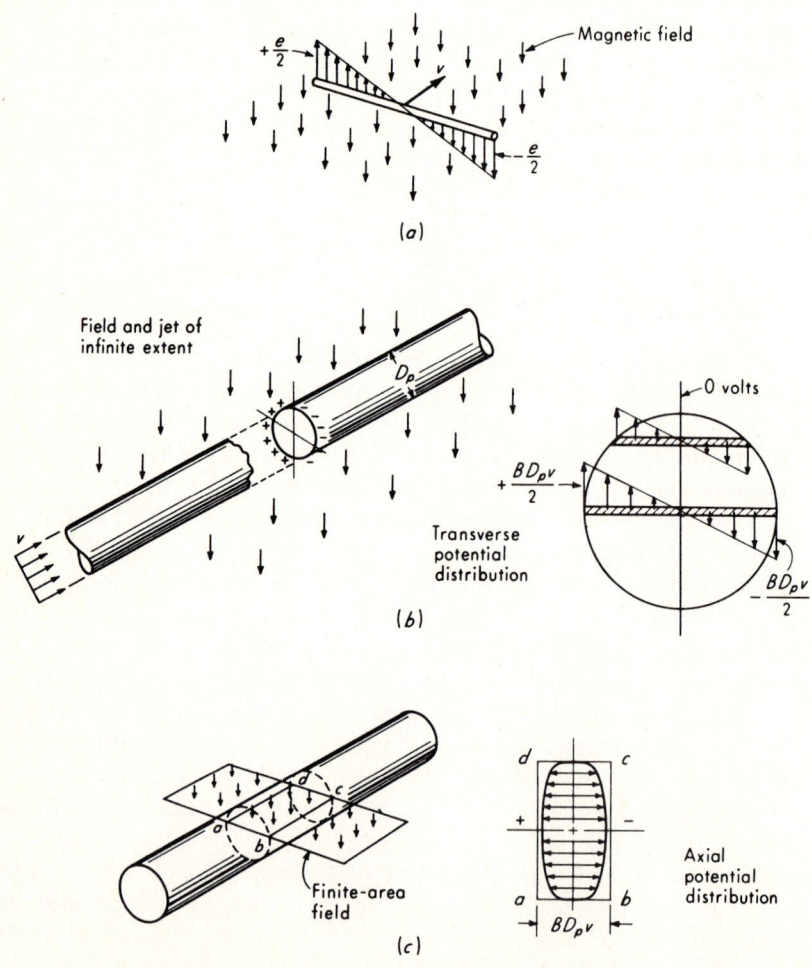

electrodes. In Fig. 7.29d this path is shown split into two equal parts $R_1/2$ and containing the signal source $e = BD_p v$. This resistance is not simple to calculate since it involves a continuous distribution of resistance over complex bodies. It can, however, be estimated from theory, and once a device is built it can be directly measured. The magnitude of this source resistance determines the loading effect of any external circuit connected to the electrodes.

The magnetic field used in such a flowmeter could conceivably be either constant or alternating, giving rise to a d-c or an a-c output signal, respectively. The d-c form of meter is little used for a number of reasons. First, many hydrogen-bearing or aqueous solutions exhibit a polarization effect wherein positive ions migrate to the negative electrode and disassociate, forming an insulating pocket of gaseous hydrogen. An a-c field inhibits this action. Secondly, a d-c field may distort the fluid-velocity profile by magnetohydrodynamic action. Above we assumed a uniform (square) velocity profile, whereas actual flows are never of exactly this form. It has been shown,[1] however, that, for *any* velocity profile that is symmetrical about the center line of the pipe, the voltage generated will correspond to the *average* velocity of the flow. However, *unsymmetrical*

[1] A. Kolin, An Alternating Field Induction Flowmeter of High Sensitivity, *Rev. Sci. Instr.*, vol. 16, p. 109, May, 1945.

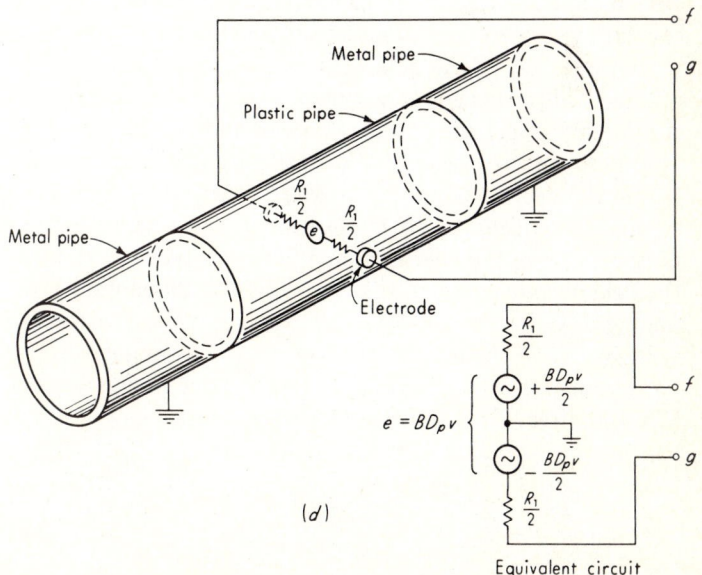

Fig. 7.29. *Electromagnetic flowmeter.*

profiles do not give correct average velocity readings and magnetohydro-dynamic action can give such profiles. An a-c field (60 cps is generally adequate) has little effect on velocity profiles because fluid inertia and friction forces at 60 cps are sufficient to prevent any large fluid motions. Finally, since the flow-related voltage signals of such meters are quite small (a few millivolts), interfering voltage inputs due to thermocouple-type effects and galvanic action of dissimilar metals used in meter con-struction may be of a magnitude similar to the desired signal. Since these interfering inputs are generally drifts of very-low-frequency content, a 60-cps a-c system can use high-pass filtering to wash these out completely. Furthermore, the small flow signals require high amplification, which is more easily, cheaply, and reliably accomplished with a-c than with d-c amplifiers.

While a-c systems predominate, d-c types have been used in metering liquid metals, such as mercury. Here, no polarization problem exists. Also, an insulating pipe liner is not needed, since the conductivity of the liquid metal is very good relative to an ordinary metal (stainless-steel) pipe. This means that a metal pipe is not very effective as a "short circuit" for the voltage induced in a liquid-metal flow. Also, no special electrodes are necessary, the output voltage being tapped off the metal pipe itself at the points of maximum potential difference.

In a typical a-c system[1] shown in simplified form in Fig. 7.30, the field B is provided by coils operating from a 60-cps source. Since the output signal is directly proportional to B, it would seem that B must be regulated very closely to maintain constant flowmeter sensitivity. The power supplied to the field coils is not small (600 va for a typical 3-in.-diameter flowmeter), and currents of this magnitude are not easily regulated. Most systems therefore employ an instrument servo feedback system arranged to cancel the effect of any changes in field, rather than trying to keep B constant. In Fig. 7.30 a signal proportional to the field current (and thus to B) is obtained from a current transformer whose primary carries the field current. The secondary current (proportional to the field current) is applied as excitation to the motion-measuring potenti-ometer of an instrument servo whose input voltage is the flowmeter output voltage. The output of the servo is the angular rotation θ_p of the potentiometer, which we shall show is a direct indication of flow velocity v. When the servo is balanced, the error voltage $e_e \approx 0$ and thus

$$vD_pB = BK_BR_p\theta_p \qquad (7.62)$$

$$\theta_p = \frac{D_p}{K_BR_p}v \qquad (7.63)$$

[1] I. C. Hutcheon, Some Problems of Magnetic Flow Measurement, *Instr. Eng.*, p. 1, April, 1960.

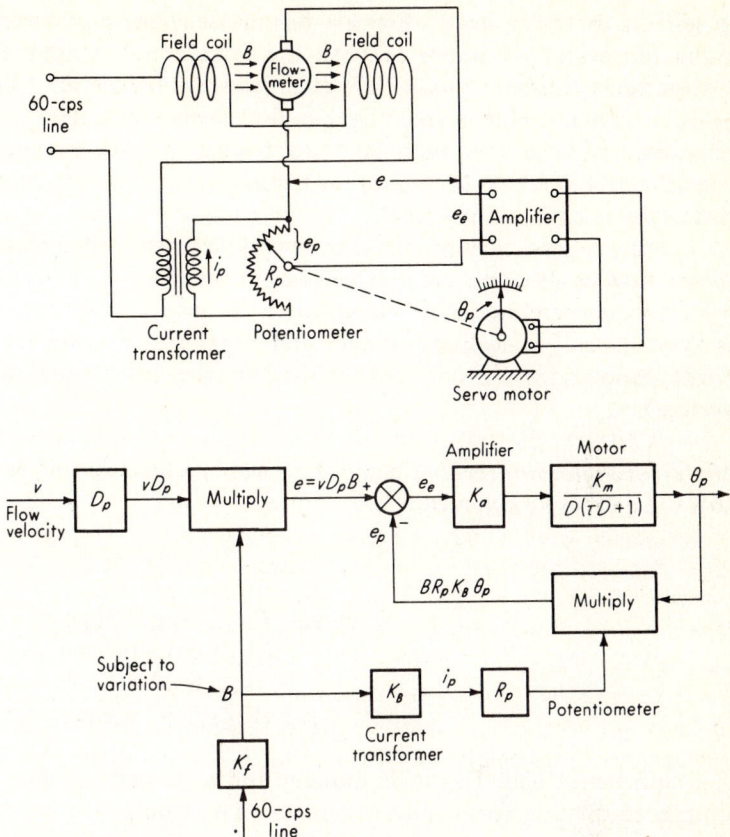

Fig. 7.30. *Magnetic flowmeter servosystem.*

We see that changes in B have no effect on the indication of such a system. A typical system produces a voltage e of about 3 mv rms if $D = 3$ in. and tap water is flowing at 100 gpm. The resistance between the electrodes is given by theory[1] as approximately $1/\sigma d$, where $\sigma \triangleq$ fluid conductivity and $d =$ electrode diameter. For tap water $\sigma \approx 200 \times 10^{-6}$ mho/cm, and so if d is, say, $\frac{1}{4}$ in. (0.64 cm), there is a resistance of about 7,800 ohms as the internal resistance of the voltage source producing e. The signal cable and amplifier input act as a load on this source and must be of sufficiently high impedance.

A limitation of electromagnetic flowmeters is the conductivity of the metered fluid. This must be sufficiently high so that the external circuitry

[1] V. P. Head, Electromagnetic Flowmeter Primary Elements, *Trans. ASME,* p. 662, December, 1959.

is not an excessive load. For a-c flowmeters one must consider both capacitance and resistance of both the fluid and the external circuit. Capacitance causes a phase shift that leads to balancing errors in the servo. These problems can be handled fairly routinely if the fluid conductivity is greater than 20×10^{-6} mho/cm; lower values require special care but 0.1×10^{-6} mho/cm is feasible,[1] with future developments promising even lower values.

Some general features of electromagnetic flowmeters include the lack of any flow obstruction; ability to measure reverse flows; insensitivity to viscosity, density, and flow disturbances as long as the velocity profile is symmetrical; wide linear range; and rapid response to flow changes (instantaneous for a d-c system; limited by the field frequency in an a-c system).

Drag-force flowmeters. A body immersed in a flowing fluid is subjected to a drag force F_d given by

$$F_d = \frac{C_d A \rho V^2}{2} \qquad (7.64)$$

where $C_d \triangleq$ drag coefficient
$A \triangleq$ cross-section area, ft^2
$\rho \triangleq$ fluid mass density, slug/ft^3
$V \triangleq$ fluid velocity, fps

For sufficiently high Reynolds number and a properly shaped body the drag coefficient is reasonably constant. Therefore, for a given density, F_d is proportional to V^2 and thus to the square of volume flow rate. The drag force can be measured by attaching the drag-producing body to a strain-gage force-measuring transducer. One type[2] uses a cantilever beam with bonded strain gages (see Fig. 7.31a). A hollow-tube arrangement with the gages on the outside serves to isolate the gages from the flowing fluid. If the drag body is made symmetrical, reversed flows can be measured. A main advantage of this class of flowmeters is the high dynamic response. The type just described is basically second-order with a natural frequency of 70 to 200 cps. When such a meter is used for dynamic-flow studies it is desirable to linearize the square-root characteristic. This can be done by feeding the strain-gage voltage signal to a square-root computing element whose output will then be linear with flow rate. Another[3] flow transducer of the drag type uses internal flow

[1] D. R. Lynch, A Low-conductivity Magnetic Flowmeter, *Control Eng.*, p. 122, December, 1959.
[2] Ramapo Instrument Co., Bloomingdale, N.J.
[3] Dynamic Instrument Co., Cambridge, Mass.

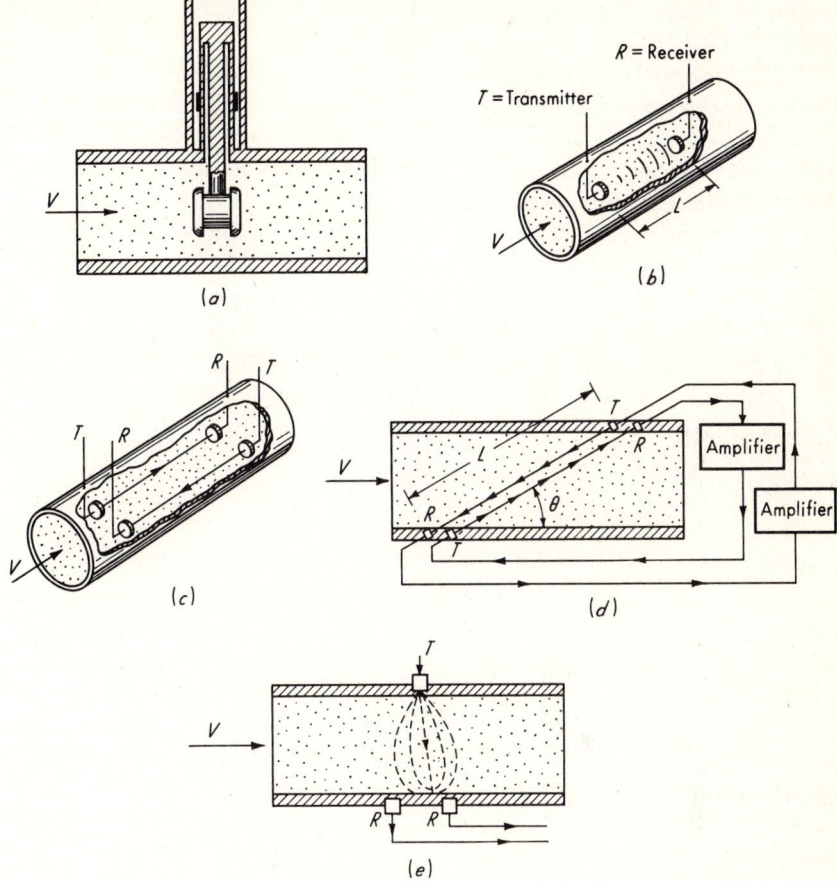

Fig. 7.31. *Drag-force and ultrasonic flowmeters.*

through an orifice plate with many holes. The force on the orifice plate is measured by an unbonded-strain-gage transducer. The natural frequency of this flowmeter is 1,500 cps.

Ultrasonic flowmeters. Small-magnitude pressure disturbances are propagated through a fluid at a definite velocity (the speed of sound) *relative to the fluid.* If the fluid also has a velocity, the *absolute* velocity of pressure-disturbance propagation is the algebraic sum of the two. Since flow rate is related to fluid velocity, this effect may be used in several ways as the operating principle of an "ultrasonic" flowmeter. The term ultrasonic refers to the fact that, in practice, the pressure disturbances usually are short bursts of sine waves whose frequency is above the range audible to human hearing, about 20,000 cps.   A typical frequency might be 10 Mc (10^7 cps).

The various methods of implementing the above phenomenon all depend on the existence of transmitters and receivers of acoustic energy. A common approach is to utilize piezoelectric crystal transducers for both these functions.   In a transmitter, electrical energy in the form of a short burst of high-frequency voltage is applied to a crystal, causing it to vibrate.   If the crystal is in contact with the fluid, the vibration will be communicated to the fluid and propagated through it.   The receiver crystal is exposed to these pressure fluctuations and responds by vibrating. The vibratory motion produces an electrical signal in proportion, according to the usual action of piezoelectric displacement transducers.

Figure 7.31*b* shows the most direct application of these principles. With zero flow velocity the transit time t_0 of pulses from the transmitter to the receiver is given by

$$t_0 = \frac{L}{c} \qquad (7.65)$$

where $L \triangleq$ distance between transmitter and receiver
$c \triangleq$ acoustic velocity in fluid

For example, in water, $c \approx 5,000$ fps, and so, if $L = 1$ ft, $t_0 = 0.0002$ sec. If the fluid is moving at a velocity V, the transit time t becomes

$$t = \frac{L}{c + V} = L\left(\frac{1}{c} - \frac{V}{c^2} + \frac{V^2}{c^3} - \cdots\right) \approx \frac{L}{c}\left(1 - \frac{V}{c}\right) \qquad (7.66)$$

and if we define $\Delta t \triangleq t_0 - t$

$$\Delta t \approx \frac{LV}{c^2} \qquad (7.67)$$

Thus, if c and L are known, measurement of Δt allows calculation of V. While L may be taken as constant, c varies, for example, with temperature and since c appears as c^2 the error caused may be significant.   Also, Δt is quite small since V is a small fraction of c.   For example, if $V = 10$ fps, $L = 1$ ft, and $c = 5,000$ fps, $\Delta t = 0.4$ μsec, a very short increment of time to measure accurately.   Since the measurement of t_0 is not directly provided for in this arrangement, the modification of Fig. 7.31*c* may be preferable.   If t_1 is the transit time with the flow and t_2 is the transit time against the flow, we get

$$\Delta t \triangleq t_2 - t_1 = \frac{2VL}{c^2 - V^2} \approx \frac{2VL}{c^2} \qquad (7.68)$$

This Δt is twice as large as before and also is a time increment that physically exists and may be directly measured.   However, the dependence on c^2 is still a drawback.

In Fig. 7.31*d* two self-excited oscillating systems are created by using the received pulses to trigger the transmitted pulses in a feedback arrangement. The pulse repetition frequency in the forward propagating loop is $1/t_1$ while that in the backward loop is $1/t_2$. The frequency difference $\Delta f \triangleq 1/t_1 - 1/t_2$ can be measured by multiplying the two signals together to get a beat frequency. Since $t_1 = L/(c + V \cos \theta)$ and $t_2 = L/(c - V \cos \theta)$ we get

$$\Delta f = \frac{2V \cos \theta}{L} \qquad (7.69)$$

which is independent of c and thus not subject to errors due to changes in c. The above analysis assumes a square velocity profile which, of course, does not occur in practice. For actual profiles, V can be replaced by V_{av} as long as the profiles are symmetrical about the pipe center line. A commercial system[1] of this type has an accuracy ± 2 percent of full scale, a linear range 20 to 1, and a capacity to 25 fps.

Another approach,[2] shown in Fig. 7.31e, senses the deflection of an ultrasonic beam propagated transversely to the flow. With no flow, the two receivers get equal signals; with a flow present, the beam deflects, giving one receiver a stronger signal and the other a weaker signal. Electronic circuitry forms the ratio of these two signals and this ratio is a linear function of flow rate. This arrangement is sensitive to changes in sound velocity but these are mainly dependent on temperature and are compensated for by a thermistor placed on the outside wall of the pipe.

Ultrasonic flowmeters are presently used mainly for liquids. The practical versions (Fig. 7.31d and e) have no flow obstruction and are relatively insensitive to viscosity, temperature, and density variations. Their complexity and relatively high cost somewhat limit industrial application at present.

7.3

Gross Mass Flow Rate In many applications of flow measurement mass flow rate is actually more significant than volume flow rate. As an example, the range capability of an aircraft or liquid-fuel rocket is determined by the *mass* of fuel, not the volume. Flowmeters used in fueling such vehicles should thus indicate mass, not volume. In chemical process industries also, mass flow rate is often the significant quantity.

Two general approaches are used to measure mass flow rate. One involves the use of some type of volume flowmeter, some means of density measurement, and some type of simple computer to compute mass flow

[1] Fischer and Porter Co., Hatboro, Pa.
[2] H. E. Dalke and W. Welkowitz, A New Ultrasonic Flowmeter for Industry, *ISA J.*, p. 60, October, 1960.

rate. The other, more basic approach is to find flowmetering concepts that are inherently sensitive to mass flow rate. Both methods are currently finding successful application in various detail forms.

Volume flowmeter plus density measurement. Some basic methods of fluid-density measurement are shown in Fig. 7.32. In Fig. 7.32a a portion of the flowing liquid is bypassed through a still well. The buoyant force on the float is directly related to density and may be measured in a number of ways, such as the strain-gage beam shown. Buoyant force is used in a different way in Fig. 7.32b. For each liquid density the system of three floats assumes a unique angular orientation θ. By proper choice of float densities, volumes, and angular locations, a very closely linear relation between density and θ is achieved.[1] The angle θ is measured by an angular position transducer to obtain an electrical signal. In Fig. 7.32c a definite volume of flowing liquid contained within the U tube is continuously weighed by a spring and pneumatic displacement transducer. Flexible couplings isolate external forces from the U tube. A pneumatic force-balance feedback system can also be used to measure the weight.[2] This minimizes deflection and thus reduces errors due to variable spring effects of flexible couplings and flexure pivots.

Figure 7.32d shows a method of measuring gas density using a small centrifugal blower (run at constant speed) to pump continuously a sample of the flow. The pressure drop across such a blower is proportional to density and may be measured with a suitable differential pressure pickup. Ultrasonic volume flowmeters often use an ultrasonic density-measuring technique when mass flow rate is wanted. In Fig. 7.32e the crystal transducer serves as an acoustic-impedance detector. Acoustic impedance depends on the product of density and speed of sound. Since a signal proportional to the speed of sound is available from the volume flowmeter, division of this signal into the acoustic-impedance signal gives a density signal. The attenuation of radiation from a radioisotope source depends on the density of the material through which the radiation passes (see Fig. 7.32f). Over a limited (but generally adequate) density range the output current of the radiation detector is nearly linear with density, for a given flowing fluid.[3] For gas flow, indirect measurement of density by means of computation from pressure and temperature signals (Fig. 7.32g) is also common. Figure 2.17c shows an ingenious method of accomplishing this.

In computing mass flow rate from volume flowmeter and densitometer (density-measuring device) signals, the necessary form of com-

---

[1] Potter Aeronautical Corp., Union, N.J.
[2] Halliburton Co., Duncan, Okla.
[3] Industrial Nucleonics Corp., Columbus, Ohio.

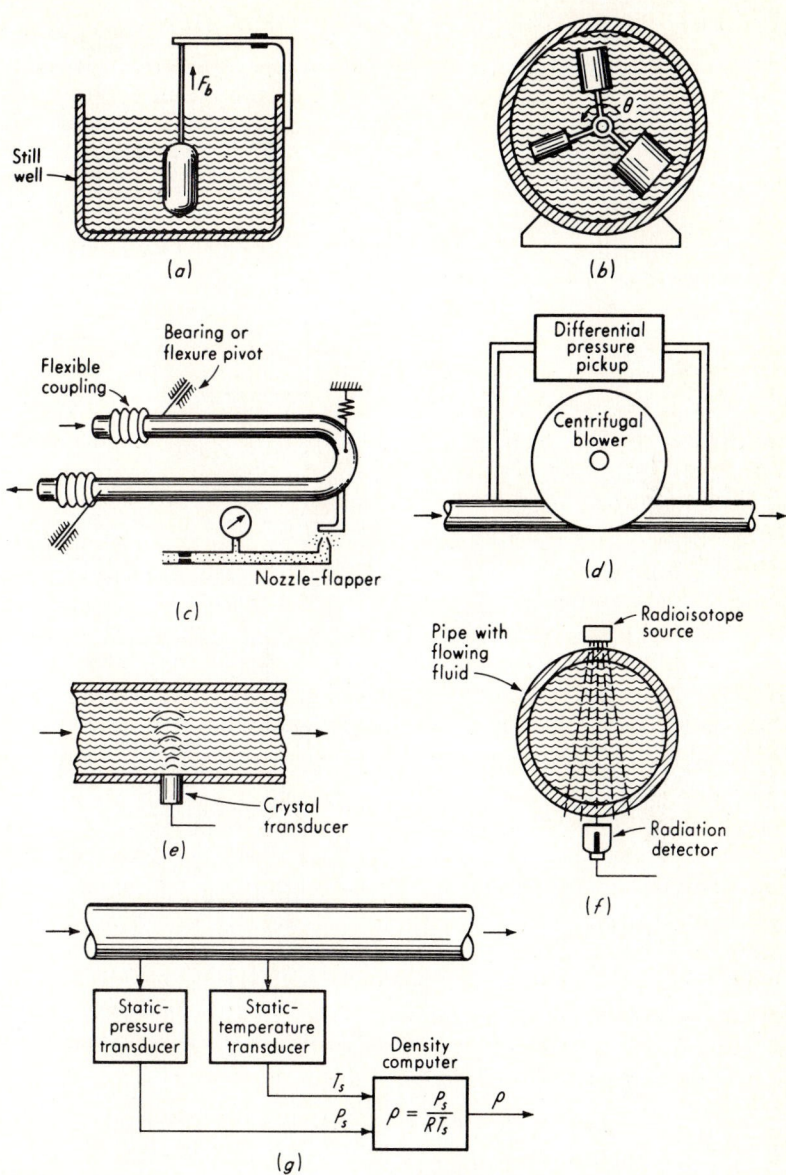

Fig. 7.32. *Fluid-density measurement.*

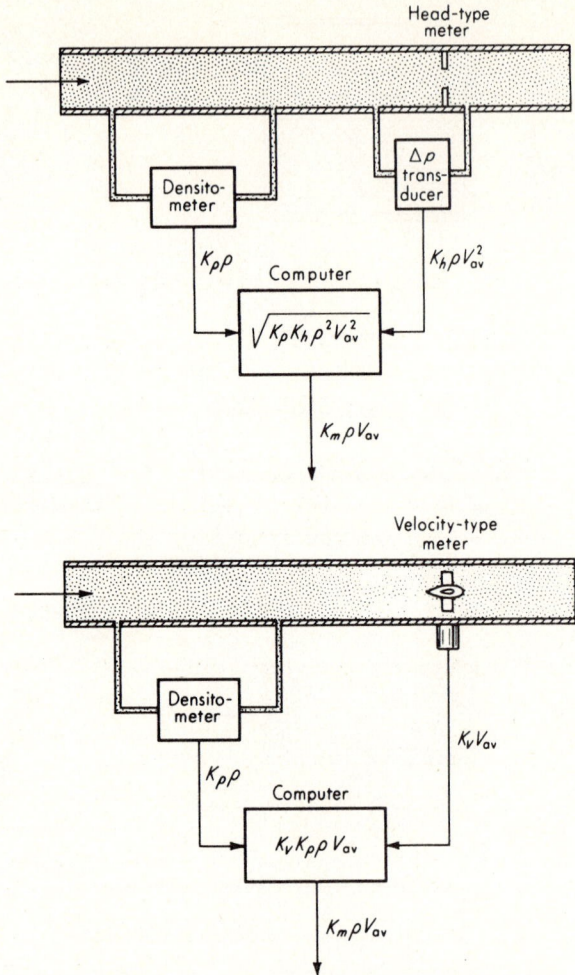

Fig. 7.33.   *Computed mass flow measurement.*

puter varies somewhat, depending on the type of flowmeter.   For so-called "head" meters (those producing a differential pressure or electrical signal proportional to ρV_{av}^2, such as an orifice) the computer multiplies the ρ signal by the ρV_{av}^2 to form $\rho^2 V^2$ and then takes the square root of this to get ρV, which is proportional to the mass flow rate.   For "velocity" flowmeters, such as the turbine and electromagnetic types, the available signal is proportional to V_{av}; thus the computer must simply multiply this by the ρ signal, a square-root operation being unnecessary.   Figure 7.33 shows these concepts.

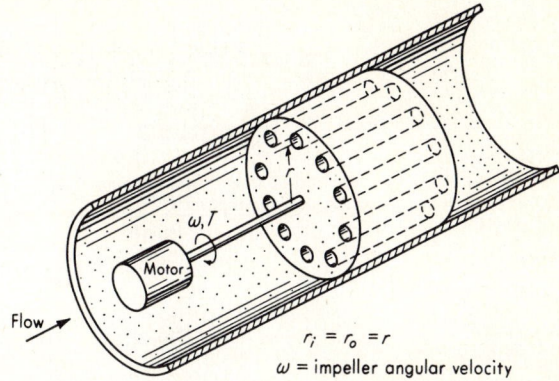

Fig. 7.34. *Angular-momentum element.*

Direct mass flowmeters. While the above indirect methods of mass-flow-rate measurement are often satisfactory and are in wide use, it is possible to find flowmetering concepts that are more directly sensitive to mass flow rate. These may have advantages with respect to accuracy, simplicity, cost, weight, space, etc., in certain applications. We shall discuss briefly some of the more common principles in terms of the practical hardware through which they have been realized.

Perhaps the most widely used principle depends on the moment-of-momentum law of turbomachines. Fluid mechanics shows that, for one-dimensional, incompressible, lossless flow through a turbine or impeller wheel, the torque T exerted by an impeller wheel on the fluid (minus sign) or on a turbine wheel by the fluid (plus sign) is given by

$$T = G(V_{ti}r_i - V_{to}r_o) \qquad (7.70)$$

where $\quad G \triangleq$ mass flow rate through wheel, slug/sec
$\qquad V_{ti} \triangleq$ tangential velocity at inlet, fps
$\qquad V_{to} \triangleq$ tangential velocity at outlet, fps
$\qquad r_i \triangleq$ radius at inlet, ft
$\qquad r_o \triangleq$ radius at outlet, ft
$\qquad T \triangleq$ torque, ft-lb$_f$

Consider now the system of Fig. 7.34. The flow to be measured is directed through an impeller wheel which is motor-driven at constant speed. If the incoming flow has no rotational component ($V_{ti} = 0$) and if the axial length of the impeller is enough to make $V_{to} = r\omega$, the driving torque necessary on the impeller is

$$T = r^2\omega G \qquad (7.71)$$

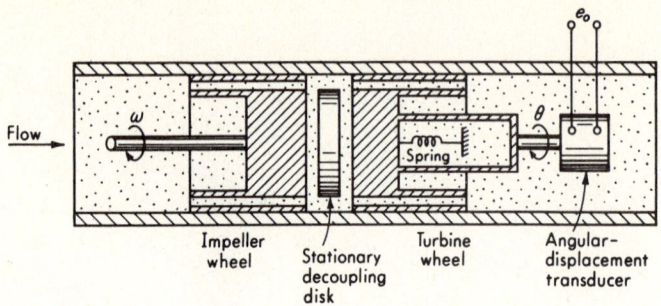

Fig. 7.35. *Angular-momentum mass flowmeter.*

Since r and ω are constant, the torque T (which could be measured in several ways) is a direct and linear measure of mass flow rate G. However, for $G = 0$, torque will *not* be zero, because of frictional effects; furthermore, viscosity changes would also cause this zero-flow torque to vary. A variation on this approach is to drive the impeller at constant *torque* (with some sort of slip clutch). Then, impeller *speed* is a measure of mass flow rate according to

$$\omega = \frac{T/r^2}{G} \qquad (7.72)$$

The speed ω is now nonlinear with G but may be easier to measure than torque. If a magnetic proximity pickup is used for speed measurement, the time duration t between pulses is inversely related to ω; thus measurements of t would be linear with G.

A further variation, used in several commercial instruments, is shown in Fig. 7.35. A constant-speed motor-driven impeller again imparts angular momentum to the fluid; however no torque or speed measurements are made on this wheel. Close by, downstream, a second ("turbine") wheel is held from turning by a spring restraint. For the impeller, $V_{to} = r\omega$; furthermore, this becomes V_{ti} for the turbine. Since the turbine cannot rotate, if it is long enough axially, the angular momentum is removed and V_{to} for the turbine is zero. Then the torque on the turbine is

$$T = r^2\omega G \qquad (7.73)$$

If the spring restraint is linear, the deflection θ is a direct and linear measure of G and can be transduced to an electrical signal in a number of ways. The decoupling disk reduces the viscous coupling between the impeller and turbine so that, at zero flow rate, a minimum of viscous torque is exerted on the turbine wheel.

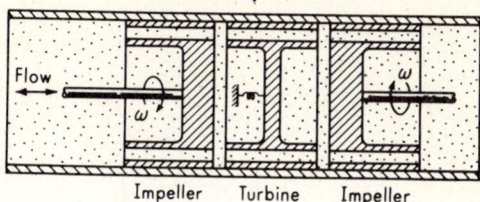

Fig. 7.36. *Bidirectional flow-meter.*

The arrangement of Fig. 7.35 handles flow in one direction only. To make the meter bidirectional, two counterrotating impellers are used, as in Fig. 7.36. The reverse flow causes a reversed turbine-deflection signal. Also, at zero flow, the viscous coupling effects tend to cancel. A drawback is the increased pressure drop due to more obstructions in the flow.

When total mass flow over a time interval, rather than instantaneous rate, is desired, the output signal must be integrated. An ingenious method[1] for doing this, using the basic arrangement of Fig. 7.35, is shown schematically in Fig. 7.37. The torque signal is brought through the

[1] General Electric Co., West Lynn, Mass.

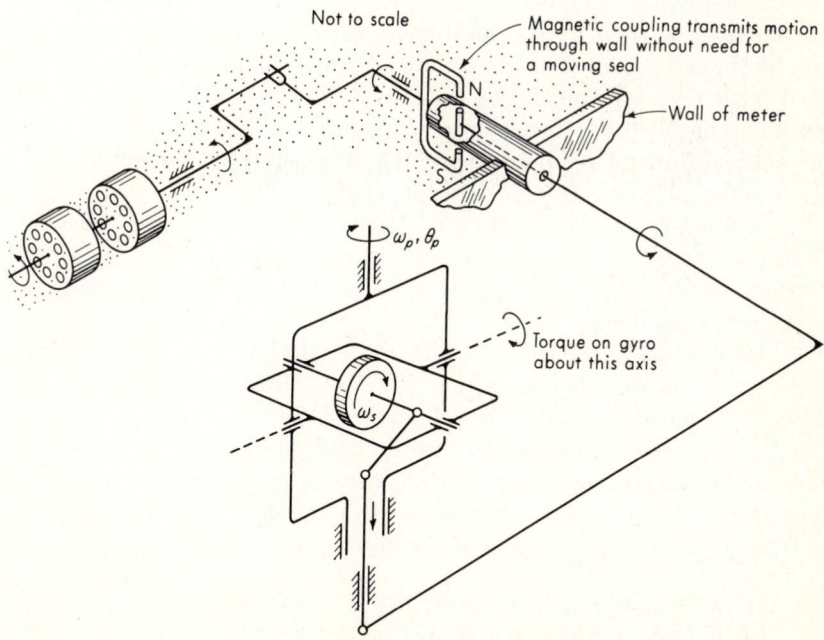

Fig. 7.37. *Gyro integrating flowmeter.*

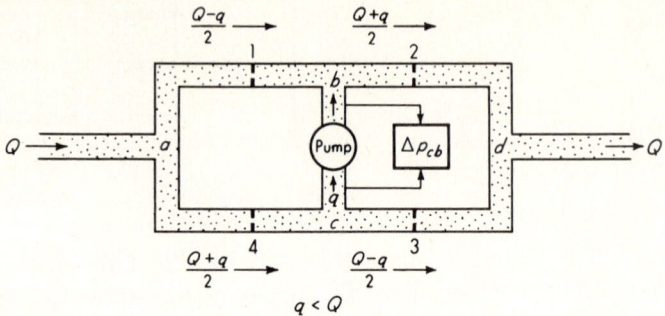

Fig. 7.38.   *Bridge-circuit flowmeter.*

wall of the flowmeter by a magnetic coupling and then applied through suitable linkages to a gyro as shown. The precession velocity ω_p is proportional to the torque and inversely proportional to the gyro spin angular velocity ω_s. If the flowmeter impeller motor and the gyro spin motor are driven from the same a-c source, any frequency changes will affect both speeds equally. Since torque is directly proportional to impeller speed whereas ω_p is inversely proportional to ω_s, the effect on ω_p of an a-c-source frequency change is made zero. If ω_p is directly proportional to G, then θ_p, the total angle turned through in a given time, will be the integral of G or the total mass flow. This total angle is measured with a simple revolution counter. Meters of this type for both gases and liquids have an accurate operating range from 10 to 133 percent of nominal flow rating. Accuracy in this range is ± 1 percent of reading while repeatability is 0.25 percent of reading.

A number of other meters use an angular-momentum principle in different ways. They include the vibrating gyroscope meter,[1] the Coriolis meter,[1] the rotating gyroscope meter,[1] the twin-turbine meter,[1] and the S-tube meter.[2]

An interesting mass flowmeter[3] for liquids, based on quite a different principle, is shown schematically in Fig. 7.38. For a given fluid the pressure drop across an orifice is proportional to ρQ^2. In Fig. 7.38, four identical orifices are connected into a "bridge circuit." A positive-displacement pump of fixed displacement runs at constant speed and volume flow rate q. The pressure rise across this pump is Δp_{cb}, which pressure difference is the output signal of the flowmeter. The flow rates through

[1] C. M. Halsell, Mass Flowmeters, *ISA J.*, p. 49, June, 1960.
[2] J. Haffner, A. Stone, and W. K. Genthe, Novel Mass Flowmeter, *Control Eng.*, p. 69, October, 1962.
[3] Flo-Tron Inc., Paterson, N.J.

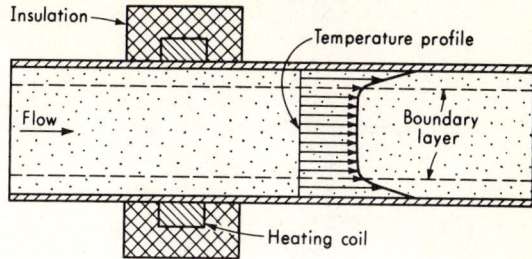

Fig. 7.39. *Boundary-layer principle.*

the individual orifices must be as shown, because of symmetry. The output signal Δp_{cb} is

$$\Delta p_{cb} = K\rho \left(\frac{Q+q}{2}\right)^2 - K\rho \left(\frac{Q-q}{2}\right)^2 = Kq\rho Q = K_1 G \qquad (7.74)$$

Thus Δp_{cb} is linear with mass flow rate G. The "constant" K_1 includes the orifice discharge coefficient; thus all orifice flow rates must be maintained at high enough Reynolds numbers so that the discharge coefficients of all orifices are equal and do not vary when Q varies. Various other arrangements of orifices, pumps, and pressure pickups have been devised; they give the same overall result but have relative advantages and disadvantages in other respects.

Our final example of mass flowmeters uses heat-transfer principles. In Fig. 7.39 an electric heating coil is transferring heat to a fluid flowing inside a pipe. If the pipe wall is a good thermal conductor and quite thin and if heat losses are minimized by insulation, the temperature drop across the boundary layer for turbulent flow is given by

$$\Delta T = \frac{K_1 P_h}{h} \qquad (7.75)$$

where $K_1 \triangleq$ constant (conversion factor and heat-transfer area)
$P_h \triangleq$ heater power
$h \triangleq$ film conductance of boundary layer

For turbulent flow the film conductance is given by

$$h = 0.023 \left[\frac{k^{0.6}c^{0.4}}{D^{0.2}\mu^{0.4}}\right] G^{0.8} \qquad (7.76)$$

where $k \triangleq$ fluid thermal conductivity
$c \triangleq$ fluid specific heat
$D \triangleq$ pipe diameter
$\mu \triangleq$ fluid absolute viscosity
$G \triangleq$ mass flow rate

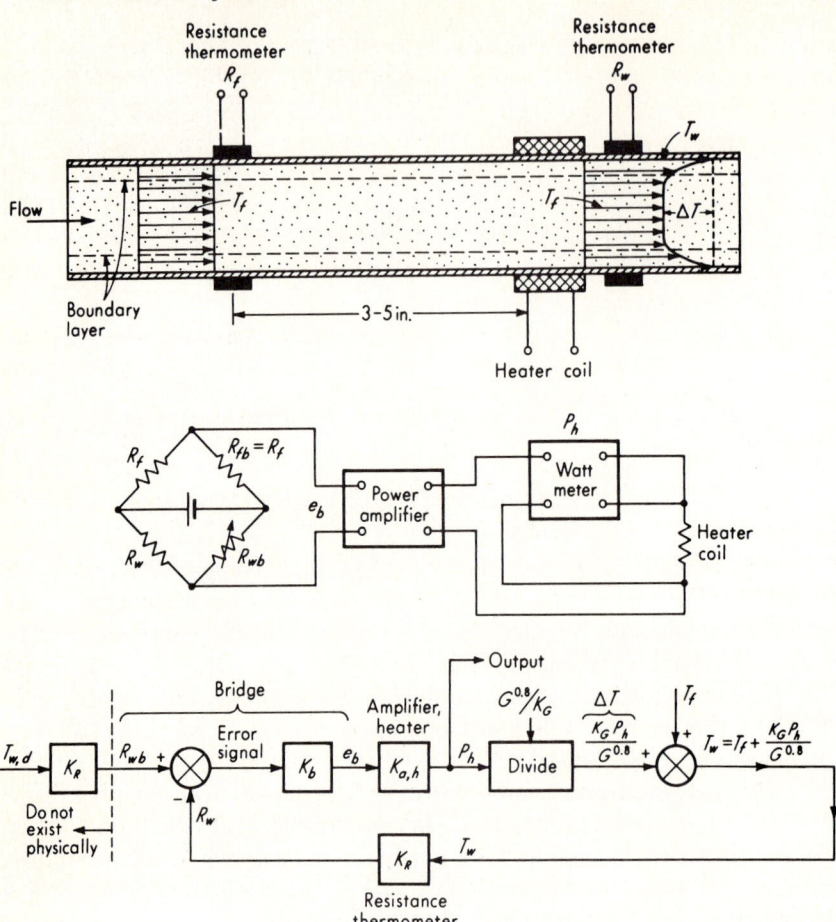

Fig. 7.40. *Boundary-layer flowmeter.*

In Eq. (7.76), if the bracketed quantity is constant or can be compensated for, we see that h is given by $h = K_2 G^{0.8}$. Then from Eq. (7.75) we get

$$P_h = \frac{\Delta T K_2}{K_1} G^{0.8} \qquad (7.77)$$

If the heater power P_h is adjusted to keep ΔT always constant, P_h becomes a direct (and almost linear) measure of mass flow rate G. This is the principle of the boundary-layer (electrocaloric) flowmeter.[1]

It would seem that to measure ΔT one would require a temperature

[1] J. H. Laub, Measuring Mass Flow with the Boundary-layer Flowmeter, *Control Eng.*, p. 112, March, 1957.

probe in the core of the flowing fluid. Fortunately this complication is unnecessary; the pipe-wall temperature 3 to 5 in. upstream of the heater is very close to the fluid core temperature downstream of the heater. The adjustment of heater power to maintain a constant ΔT (about 2°F is used) is accomplished continuously and automatically by a feedback system shown simplified in Fig. 7.40. A wattmeter reading P_h can be calibrated in mass-flow-rate units for a given fluid. Its reading is the output signal of the flowmeter. A typical value of P_h is about 40 watts if 12,000 lb/hr of water is flowing in a 1.5-in.-diameter pipe.

The feedback-system operation may be explained as follows: If the incoming fluid has a fixed temperature T_f, the resistance R_f will be constant. The bridge resistor R_{fb} will also be constant and made equal to R_f. Suppose that R_{wb} is set equal to R_w; then the bridge is balanced and no heater power is being supplied. If the temperature coefficient of R_w is known, the change in R_w corresponding to the desired ΔT can be calculated. Now R_{wb} is changed by this amount, unbalancing the bridge and thus turning on the heater power. This raises R_w toward the new value of R_{wb}, tending to rebalance the bridge. Perfect balance ($R_w = R_{wb}$) cannot be achieved, since then the heater power would have to be zero. If the gain of the feedback system is high, however, nearly perfect bridge balance is possible without making the heater power zero. With the bridge in this condition, an increase, say, in mass flow rate G results in a momentary decrease in T_w (and thus R_w), unbalancing the bridge in the direction to increase heater power, increase T_w, and thus maintain the desired ΔT. A decrease in G results in the opposite action, again tending to maintain ΔT fixed. If ΔT is maintained fixed for all flow rates, Eq. (7.77) shows that the wattmeter reading (system output signal) is a direct indication of mass flow rate.

Problems

7.1 Water flows in a 1-in.-diameter pipe at 10 fps. If a pitot-static tube of 0.5-in.-diameter is inserted, what velocity will be indicated? Assume one-dimensional frictionless flow. Find the pitot-static-tube diameter needed to reduce the above error to 1 percent.

7.2 For the system of Fig. 7.10, what diameter sphere is needed to obtain a 1-lb force from a 50-mph wind when atmospheric pressure is 14.7 psi and temperature is 70°F? If the 50-mph wind is assumed to be the full-scale value and if 0.05-in. full scale deflection of the differential transformer is desired, what must be the flexure-plate spring constant? If the total moving mass is assumed to be due to the spherical shell alone and if the shell is made of $\frac{1}{16}$-in.-thick aluminum, estimate the usable frequency range of this instrument.

7.3 If, in the system of Fig. 7.15, the amplifier lag is not neglected, so that $(i/e_e)(D) = K_a/[(\tau_1 D + 1)(\tau_2 D + 1)]$, find the value of K_a that will put the feedback system just on the margin of instability (use the Routh criterion). If $\tau = 0.001$ sec and $\tau_1 = \tau_2 = 0.000001$, what is the maximum allowable value of $K_b K_a K_e/I_m$

for marginal stability? If a value of $K_b K_a K_e / I_m$ about one-fifth of that giving marginal stability can be used and if Eq. (7.43) is assumed applicable under these conditions, what percentage improvement of τ_{ct} as compared with τ may be achieved?

7.4 The frequency response of a hot-wire anemometer system also determines its ability to resolve *spacewise* variations in velocity. That is, at a given instant of time, if the velocity pattern in a flow were "frozen," the velocity component in a certain direction would be different at different stations along the line of travel of the gross flow. When this velocity structure is swept past a hot wire by the average velocity, it requires adequate frequency response to resolve the spacewise velocity variations. Consider a simple spacewise variation wherein velocity deviation from average is given by $v = v_0 \sin (2\pi x / \lambda)$, where λ is the wavelength in inches and x is displacement in inches along the flow direction. If the average flow velocity is V_0 in./sec, find an expression for the smallest wavelength λ_{min} that can be resolved by a system whose flat frequency response extends to f_0 cps. Plot λ_{min} versus V_0 for a system with f_0 of 100,000 cps.

7.5 Analyze the error in flow-rate measurement caused by thermal expansion of an orifice plate.

7.6 A capillary-tube laminar-flow element is needed to measure water flow of 0.01 in.³/min at 70°F. A flowmeter pressure drop of 3 in. of water is desired. If the element is designed for a Reynolds number of 500, what length and diameter of tubing are needed?

7.7 Using the simplified model discussed in the text, derive Eq. (7.57).

7.8 Using the assumptions discussed following Eq. (7.57), one can write for the weight flow rate

$$W = K_1 (A_t - A_f) \sqrt{w_{ff}(w_f - w_{ff})} \qquad \text{lb}_f/\text{sec}$$

where K_1 includes all the other constants in Eq. (7.57). To make the weight-flow indication relatively insensitive to changes in fluid density w_{ff}, the float density w_f should be twice the density of the flowing fluid. Show the truth of this statement. Hint: Set $\partial W / \partial w_{ff} = 0$.

7.9 Classify, according to the categories of Sec. 2.5, the compensation method used in the system of Fig. 7.30.

7.10 Outline the procedure you would use to design a drag-force flowmeter of the type shown in Fig. 7.31*a*. The given specifications include static sensitivity, dynamic response, flow-velocity range to be covered, allowable size, and fluid-density range to be covered.

7.11 Perform a dynamic analysis on the system of Fig. 7.32*a* to obtain the transfer function relating fluid density as an input to strain-gage bridge voltage as an output. What is the effect of changes in liquid level in the still well on the output signal? What is the effect of thermal (volume) expansion of the float? If the entire assembly is aboard a vehicle that is accelerating vertically, explain the effect on the output.

7.12 Perform a static analysis on the system of Fig. 7.32*b*, obtaining a relation between fluid density as an input and rotation angle θ as an output. List modifying and/or interfering inputs for this instrument.

7.13 Perform a static analysis on the system of Fig. 7.32*c* to get a relation between fluid density as an input and nozzle-flapper pressure as output. List modifying and/or interfering inputs for this instrument.

7.14 Modify the system of Fig. 7.32*c* using a feedback principle (null-balance system) to keep vertical deflection nearly zero for all densities. A bellows may be used to provide the rebalancing force.

7.15 For the system of Fig. 7.35, suppose that the full-scale flow rate is 10 lb_m/sec, an impeller speed of 100 rpm is used, and $r = 1.0$ in. If full-scale transducer rotation is to be 20°, what torsional spring constant is required? If this spring constant must be increased to improve dynamic response, what design changes are possible to achieve this?

7.16 Intuitively, what would you expect the dynamic response of the system of Fig. 7.35 to be? What design parameters would have a major influence on this response and in what way? Devise an experimental technique for subjecting this instrument to an approximate step input.

7.17 List modifying and/or interfering inputs for the system of Fig. 7.40.

Bibliography

1. Standards for Discharge Measurement with Standardized Nozzle and Orifices, *NACA*, *TM*-952, 1940.
2. V. P. Head: Electromagnetic Flowmeter Primary Elements, *ASME Trans.*, p. 660, December, 1959.
3. D. R. Lynch: A Low-conductivity Magnetic-Flowmeter, *Control Eng.*, p. 122, December, 1959.
4. V. Cushing et al.: Development of an Electromagnetic Induction Flowmeter for Cryogenic Fluids, *NASA*, N-64-20708, 1964.
5. J. A. Shercliff: "The Theory of Electromagnetic Flow Measurement," Cambridge University Press, New York, 1962.
6. Seawater Voltage Tells Submarine Speed, *Machine Design*, p. 28, Feb. 18, 1965.
7. S. Blechman: Techniques for Measuring Low Flows, *Instr. Control Systems*, p. 82, October, 1963.
8. E. W. Miller: Turbine Gas-flow Sensor, *Instr. Control Systems*, p. 105, January, 1962.
9. H. J. Evans: Turbine Flowmeter for Gases, *Instr. Control Systems*, p. 103, March, 1964.
10. R. D. Wood: Steam Measurement by Orifice Meter, *Instr. Control Systems*, p. 135, April, 1963.
11. R. B. Crawford: A Broad Look at Cryogenic Flow Measurement, *ISA J.*, p. 65, June, 1963.
12. D. Shichman and B. S. Johnson: Tap Location for Segmental Orifices, *Instr. Control Systems*, p. 102, April, 1962.
13. P. J. Klass: Laser Flowmeter, *Aviation Week*, Jan. 11, 1965.
14. R. W. Henke: Positive Displacement Meters, *Control Eng.*, p. 56, May, 1955.
15. T. Isobe and H. Hattori: A New Flowmeter for Pulsating Gas Flow, *ISA J.*, p. 38, December, 1959.
16. A. K. Oppenheim and E. G. Chilton: Pulsating-flow Measurement—A Literature Survey, *ASME Trans.*, p. 231, February, 1955.
17. J. R. Musham and B. G. Lewis: Direct Reading Flow Rate Meter for Low Flow Rates, *Control Eng.*, p. 115, December, 1961.
18. L. Gess: Common Troubles with Head Flowmeters, *ISA J.*, p. 58, February, 1958.
19. R. Shapcott: How to Select Flowmeters, *ISA J.*, p. 272, July, 1957.
20. H. E. Wingo: Thermistors Measure Low Liquid Velocities, *Control Eng.*, p. 131, October, 1959.
21. W. D. Hamilton: Flow Elements from Tubing Elbows, *ISA J.*, p. 61, July, 1963.
22. E. G. Keshock: Comparison of Absolute- and Reference-system Methods of Measuring Containment Vessel Leakage Rates, *NASA*, *Tech. Note* D-1588, 1964.

23. Ball Bearing Used in Design of Rugged Flowmeter, *NASA, Brief* 64-10170, 1964.
24. Meter Accurately Measures Flow of Low-conductivity Fluids, *NASA, Brief* 63-10280, 1963.
25. E. L. Upp: Flowmeters for High-pressure Gas, *Instr. Control Systems*, p. 151, March, 1965.
26. Turbine Flow Sensors, *Instr. Control Systems*, p. 123, March, 1965.
27. A. Haalman: Pulsation Errors in Turbine Flowmeters, *Control Eng.*, p. 89, May, 1965.
28. R. Siev: Mass Flow Measurement, *Instr. Control Systems*, p. 966, June, 1960.
29. Heat Transfer Flowmeter Has No Pressure Drop, *Space/Aeron.*, p. 259, January, 1964.
30. Mass Flow by Temperature Measurement, *Instr. Control Systems*, p. 95, March, 1964.
31. C. M. Holsell: Mass Flowmeters, *ISA J.*, p. 49, June, 1960.
32. L. N. Mortenson: Mass Flowmeter Calibration, *Instr. Control Systems*, p. 133, March, 1964.
33. J. Haffner et al.: Novel Mass Flowmeter, *Control Eng.*, p. 69, October, 1962.
34. G. T. Gebhardt: What's Available for Measuring Mass Flow, *Control Eng.*, p. 90, February, 1957.
35. G. F. Battista: The Use of Momentum Effects in Liquid Flow Measurement, U.S. Naval Air Turbine Test Station, Trenton, N.J., *NATTS-ATL-TN-26*, 1963.
36. G. Bloom: Low Flow Mass Flow Meter, *Instr. Control Systems*, p. 117, March, 1965.
37. E. C. Evans and G. W. Ray: Gas Mass Flow Rate Measurement to 0.1%, *Instr. Control Systems*, p. 105, March, 1965.
38. J. W. Freshour: Mass Flow Measurement of Cryogens, *Instr. Control Systems*, p. 97, March, 1965.
39. J. C. Pemberton: Flow Measurement in Rotating Machinery, *Instr. Control Systems*, p. 105, March, 1964.
40. T. J. Larson and L. D. Webb: Calibrations and Comparisons of Pressure-type Airspeed-Altitude Systems of the X-15 Airplane from Subsonic to High Supersonic Speeds, *NASA, Tech. Note* D-1724, 1963.
41. W. Gracey et al.: Wind-tunnel Investigation of a Number of Total-pressure Tubes at High Angles of Attack, *NACA, Tech. Note* 2261, 1951.
42. F. J. Capone: Wind-tunnel Tests of Seven Static-pressure Probes at Transonic Speeds, *NASA, Tech. Note* D-947, 1961.
43. R. S. Ritchie: Several Methods for Aerodynamic Reduction of Static-pressure Sensing Errors for Aircraft at Subsonic, Near-sonic, and Low Supersonic Speeds, *NASA, Tech. Rept.* R-18, 1959.
44. A. O. Pearson and H. A. Brown: Calibration of a Combined Pitot-static Tube and Vane-type Flow Angularity Indicator at Transonic Speeds and at Large Angles of Attack or Yaw, *NACA, RM*-L52F24, 1952.
45. J. M. Savino and A. J. Hilovsky: On the Use of Single Total- and Static-pressure Probes to Measure the Average Mass Velocity in Thin Rectangular Channels, *NASA, Tech. Note* D-2212, 1964.
46. W. Gracey: Measurement of Static Pressure on Aircraft, *NASA, Rept.* 1364, 1958.
47. W. H. Reed: Dynamic Response of Rising and Falling Balloon Wind Sensors with Application to Estimates of Wind Loads on Launch Vehicles, *NASA, Tech. Note* D-1821, 1963.
48. W. F. Van Tassell and C. E. Covert: Relaxation Effects on the Interpretation of Impact-probe Measurements, *J. Aerospace Sci.*, p. 147, February, 1960.

49. F. S. Sherman: New Experiments on Impact-pressure Interpretation in Supersonic and Subsonic Rarefied Air Streams, *NACA, Tech. Note* 2995, 1953.

50. New Anemometer Has Fast Response, Measures Dynamic Pressure Directly, *NASA, Brief* 63-10530, 1963.

51. L. V. Baldwin and V. A. Sandborn: Hot-wire Calorimetry: Theory and Application to Ion Rocket Research, *NASA, Tech. Rept.* R-98, 1961.

52. H. P. Grant: Hot-Wire in Liquid Flow Measurement, *Flow Corp.*, Cambridge, Mass., *Bull.* 89.

53. L. Kovasznay: Calibration and Measurement in Turbulence Research by Hot-wire Method, *NACA, TM*-1130, 1947.

54. H. L. Dryden and A. M. Kuethe: The Measurement of Fluctuations of Air Speed by the Hot-wire Anemometer, *NACA, Rept.* 320, 1929.

55. W. G. Spangenberg: Heat-loss Characteristics of Hot-wire Anemometers at Various Densities in Transonic and Supersonic Flow, *NACA, Tech. Note* 3381, 1955.

56. L. Kovasznay: Development of Turbulence-measuring Equipment, *NACA, Rept.* 1209, 1954.

57. V. A. Sandborn and J. C. Laurence: Heat Loss from Yawed Hot Wires at Subsonic Mach Numbers, *NACA, Tech. Note* 3563, 1955.

58. C. E. Shepard: A Self-excited, Alternating Current, Constant-temperature Hot-wire Anemometer, *NACA, Tech. Note* 3406, 1955.

59. W. G. Rose: Some Corrections to the Linearized Response of a Constant-temperature Hot-wire Anemometer Operated in a Low-speed Flow, *ASME Paper* 62-WA-11, 1962.

60. G. P. Katys: "Continuous Measurement of Unsteady Flow," The Macmillan Company, New York, 1964.

61. R. R. Dowden: Fluid Flow Measurement, A Bibliography, *British Hydromechanics Res. Assoc.*, Cranfield, Bedford, England, 1972.

62. Benedict, R. P.: "Fund. of Temp., Pressure and Flow Meas.," Wiley, New York, 1969.

63. The Use of Laser Doppler Velocimeter for Flow Measurements, *Proc. of Project Squid Workshop*, Purdue Univ., 1972.

64. L. C. Lynworth et al: Ultrasonic Mass Flowmeter for Army Aircraft Engine Diagnostics, *USAA MRDL Tech. Rept.* 72-66, Fort Eustis, Va., 1973.

65. J. W. Tanney: Fluidic Velocity Sensor, *Insts. and Cont. Syst.*, June, 1969.

66. Thin-Film Probes for Measurement of Instantaneous Blood Flow Velocity, *Disa Information*, Disa Electronics, Herlev, Denmark, October, 1970.

67. M. R. Davis and P. O. A. L. Davies: The Physical Characteristics of Hot-Wire Anemometers, *Inst. of Sound and Vib.*, *Rept.* No. 2, Univ. of Southhampton, England, 1968.

68. F. J. Resch: Use of the Dual-sensor Hot-Film Probe in Water Flow, Disa Information, *Disa Electronics*, Herlev, Denmark, March, 1973.

69. R. T. Lakey, Jr. and B. S. Shiralkar: Transient Flow Measurements with Sharp-Edge Orifices, *ASME Paper* 71-FE-30, 1971.

70. J. A. Breman et al: An Evaluation of Positive Displacement Cryogenic Volumetric Flowmeters, *NBS Tech Note* 605, 1971.

Chapter 8

Temperature and Heat-flux Measurement

8.1

Standards and Calibration   The International Measuring System sets up independent standards for only four fundamental quantities: length, time, mass, and temperature.   Standards for all other quantities are basically derived from these.   We have previously discussed the standards of length, time, and mass; let us now consider the temperature standard.[1]   It should first be noted that temperature is fundamentally

[1] A. G. McNish, Fundamentals of Measurement, *Electro-Technol.* (*New York*), p. 114, May, 1963.

different in nature from length, time, and mass in that it is an intensive quantity whereas the others are extensive. That is, if two bodies of like length are "combined" the total length is twice the original; the same is true for two time intervals or two masses. However, the combination of two bodies of the same temperature results in exactly the same temperature. Thus the idea of a standard unit of mass, length, or time that can be indefinitely divided or multiplied to generate any arbitrary magnitude of these quantities cannot be carried over to the concept of temperature. Also, even though statistical mechanics relates temperature to the mean kinetic energies of molecules, these kinetic energies (which are dependent only on mass, length, and time standards for their description) are not at present measurable. Thus an *independent* temperature standard is required.

The fundamental meaning of temperature, just as for all basic concepts of physics, is not easily given. For most purposes the zeroth law of thermodynamics gives a useful concept. This is that, for two bodies to be said to have the same temperature, they must be in thermal equilibrium; that is, when thermal communication is possible between them, no change in the thermodynamic coordinates of either occurs. The zeroth law says that two bodies each in thermal equilibrium with a third body are in thermal equilibrium with each other. Then, by definition, the bodies are all at the same temperature. Thus if one can set up a reproducible means of establishing a range of temperatures, unknown temperatures of other bodies may be compared with the standard by subjecting any type of "thermometer" successively to the standard and the unknown temperatures and allowing equilibrium to occur in each case. That is, the thermometer is calibrated against the standard and thereafter may be used to read unknown temperatures.

In choosing the means of defining the standard temperature scale, one could conceivably employ any of the many physical properties of materials that vary reproducibly with temperature. For instance, the length of a metal rod varies with temperature. To define a temperature scale numerically one must choose a reference temperature and also state a rule for defining the difference between the reference and other temperatures. (Mass, length, or time measurements do *not* require universal agreement on a reference point at which each quantity is assumed to have a particular numerical value. Every centimeter, for example, in a meter is the same as every other centimeter.)

Suppose we take a copper rod 1 m long, place it in an ice-water bath which we have taken as our reference temperature source, and measure its length. Let us choose to call the ice-bath temperature 0°. We are now free to define any rule we wish to fix the numerical value to be assigned to all lower and higher temperatures. Suppose we decide that each additional 0.01 mm of expansion will correspond to $+1.0°$ on our temperature scale and each 0.01 mm of contraction to $-1.0°$. If the expansion phenomenon were reproducible, such a temperature scale would, in principle, be perfectly acceptable as long as everyone adhered to it. Would it be correct to say that each degree of temperature on this scale was "equal" to every other degree? That depends on what one means by "equal." If "equal" means that each degree causes the same amount of expansion of the copper rod, then all degrees are equal. If, instead, one considers the expansion of, say, iron rods, then equal amounts of expansion would not, in general, be caused by a 1° (copper scale) change from -6 to $-5°$ as by a 1° change from 100 to 101°. Or, consider conduction heat transfer in, for example, silver. If a temperature difference from 100 to 200° causes a given heat-transfer rate, will the same rate be caused by a temperature difference from -50 to $+150$? The answer is, in general, no.

The point of the above discussion is that, while our arbitrarily defined temperature scale is, in principle, as good as any other such scale based on some material property, its graduations have no particular significance with regard to physical laws *other* than the one used in the definition. One measures temperature for some reason, such as computing thermal expansion, heat-transfer rate, electrical conductivity, gas pressure, etc. The forms of the equations used to make such calculations depend on the nature of the standard used to define temperature. A temperature scale that gives a simple form to thermal-expansion equations may give complex forms to all other physical relations involving temperature. Since this difficulty is common to *all* standards based on the properties of a particular substance, a way of defining a temperature scale independent of *any* substance is desirable.

The thermodynamic temperature scale[1] proposed by Lord Kelvin in 1848 provides the theoretical base for a temperature scale independent of any material property and is based on the Carnot cycle. Here a perfectly reversible heat engine transfers heat from a reservoir of infinite capacity at temperature T_2 to another such reservoir at T_1. If the heat taken from reservoir 2 is Q_2 and that supplied to reservoir 1 is Q_1, for a Carnot cycle, $Q_2/Q_1 = T_2/T_1$; this may be taken as a *definition* of temperature ratio. If, also, a number is selected to describe the temperature of a chosen fixed point, the temperature scale is completely defined. At present, the fixed point is taken as the triple point (the state at which solid, liquid, and vapor phases are in equilibrium) of water because this is the most reproducible state known. The number assigned to this point is

---

[1] F. W. Sears, "Thermodynamics, Kinetic Theory and Statistical Mechanics," p. 116, Addison-Wesley Publishing Company, Inc., Reading, Mass., 1950.

273.16°K (°K = degrees Kelvin) since this makes the temperature
interval from the ice point (273.15°K) to the steam point equal to 100K°.
This would thus coincide with the previously established centigrade (now
called Celsius) scale as a matter of convenience.[1]

While the Kelvin absolute thermodynamic scale is ideal in the sense
that it is independent of any material properties, it is not physically
realizable since it depends on an ideal Carnot cycle.   Fortunately it can
be shown[2] that a temperature scale defined by a constant-volume or
constant-pressure gas thermometer using an ideal gas is *identical* with the
thermodynamic scale.   A constant-volume gas thermometer keeps a
fixed mass of gas at constant volume and measures the pressure changes
caused by temperature changes.   The perfect-gas law then gives the fact
that temperature ratios are identical to pressure ratios.   The constant-
pressure thermometer keeps mass and pressure constant and measures
volume changes caused by temperature changes.   Again the perfect-gas
law says that temperature ratios are identical to volume ratios.   These
ratios are identical with those of the thermodynamic scale; thus if the
same fixed point (the triple point of water) is selected for the reference
point, the two scales are numerically identical.   However, there is now
the problem that the ideal gas is a mathematical model, not a real substance, and therefore the gas thermometers described above cannot
actually be built and operated.

To obtain a physically realizable temperature scale, *real* gases must
be used in the gas thermometers; the readings must be corrected, as well
as possible, for deviation from ideal-gas behavior; and then the resulting
values accepted as a definition of the temperature scale.   The corrections
for non-ideal-gas behavior are obtained for a constant-volume gas thermometer as follows: The thermometer is filled with a certain mass of gas
and mercury is added until the desired volume is achieved (see Fig. 8.1).
Suppose that this is done with the system at the ice-point temperature.
The gas pressure is measured; let us call it p_{i1}.   The system is then raised
to the steam-point temperature, causing volume expansion.   By adding
more mercury, however, the volume can be returned to the original value.
The pressure will now be higher; we shall call it p_{s1}.   For an ideal gas, the
ratio of the steam-point and ice-point temperatures would also be given
by the pressure ratio p_{s1}/p_{i1}.   If one repeats this experiment but uses a
different mass of gas, thus giving different ice-point and steam-point
pressures, p_{i2} and p_{s2}, he finds that $p_{s1}/p_{i1} \neq p_{s2}/p_{i2}$.   This is a manifestation of the nonideal behavior of the gas; an ideal gas would have
$p_{s1}/p_{i1} = p_{s2}/p_{i2}$.

[1] *Ibid.*, p. 8.
[2] *Ibid.*, p. 116.

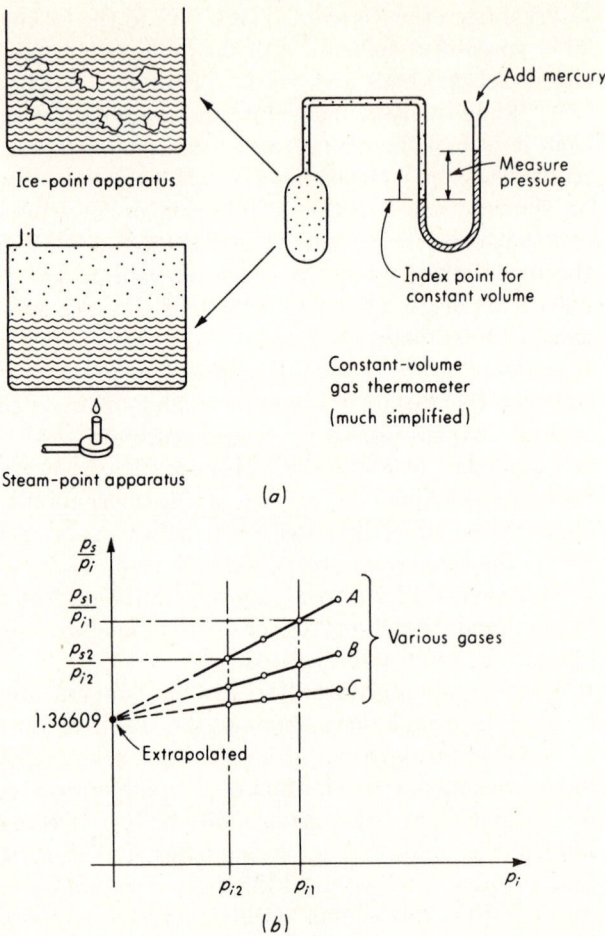

Fig. 8.1. *Gas-thermometer temperature scale.*

Real gases approach ideal-gas behavior if their pressure is reduced toward zero; thus one repeats the above experiment with successively smaller masses of gas, generating the curve A of Fig. 8.1b. Since one cannot use zero mass of gas, the zero-pressure point on this curve must be obtained by extrapolation. This zero-pressure point is taken as the true value of the pressure ratio corresponding to the steam-point/ice-point temperature ratio. If this experiment is repeated with *different* gases (B, C in Fig. 8.1b), all the curves intersect at the same point, showing that the procedure is independent of the type of gas used. Actual results give the numerical value $p_s/p_i = 1.36609 \pm 0.00004$. If one takes $T_s/T_i = p_s/p_i$, choice of a numerical value for any chosen reference point (such as calling $T_i = 273.15°K$) completely fixes the entire temperature scale. Such a scale, unfortunately, is not practical for day-to-day temperature measurements since the procedures involved are extremely tedious and time-consuming. Also, gas thermometers actually have a lower precision (repeatability) than some other temperature-measuring devices, such as resistance thermometers. This situation led to the acceptance in 1927 of the International Practical Temperature Scale (IPTS) which, with some minor revisions, is the temperature standard today.

The International Practical Temperature Scale is set up to conform as closely as practical with the thermodynamic scale. At the triple point of water the two scales are in exact agreement, by definition. Five other primary fixed points are used. They are the boiling points of liquid oxygen ($-182.970°C$), water ($100°C$), and sulfur ($444.600°C$) and the melting points of silver ($960.8°C$) and gold ($1063.0°C$), all at standard atmospheric pressure. The thermodynamic temperatures of these states were all determined as accurately as possible by gas thermometry. Interpolation between these fixed points is accomplished by use of a platinum resistance thermometer for the range -182.970 to $630.5°C$ and by a platinum/platinum–10 percent rhodium thermocouple for 630.5 to $1063.0°C$. (The melting point of antimony, $630.5°C$, is a secondary fixed point.) The equation[1] used to calculate Celsius temperature from the measured resistance of a platinum thermometer is, for the range $-182.970°$ to $0°C$,

$$t = \frac{R_t - R_0}{R_{100} - R_0} 100 + \delta \left(\frac{t}{100} - 1\right) \frac{t}{100} + \beta \left(\frac{t}{100} - 1\right)\left(\frac{t}{100}\right)^3 \qquad (8.1)$$

where $t \triangleq$ Celsius (centigrade) temperature
$R_t \triangleq$ resistance at t
$R_0 \triangleq$ resistance at ice point
$R_{100} \triangleq$ resistance at steam point

For the range 0 to $630.5°C$, the equation is

$$t = \frac{R_t - R_0}{R_{100} - R_0} 100 + \delta \left(\frac{t}{100} - 1\right) \frac{t}{100} \qquad (8.2)$$

The constant δ is a characteristic of a particular thermometer and is found by solving for δ in Eq. (8.2) when t is the sulfur point, $444.600°C$.

---

[1] R. P. Benedict, Temperature and Its Measurement, *Electro-Technol.* (*New York*), p. 71, July, 1963.

The constant β in Eq. (8.1) is similarly obtained when t in Eq. (8.1) is the oxygen point, $-182.970°C$. From 630.5 to 1063.0°C the thermocouple interpolation formula is

$$E = a + bt + ct^2 \qquad (8.3)$$

where E is the net emf of the thermocouple with one junction at the ice point and the other at t. The constants a, b, and c are found by solving the set of three simultaneous equations formed when Eq. (8.3) is applied at the antimony, silver, and gold points. When these constants have been found for the particular thermocouple, temperature t can be calculated from measured voltage E by Eq. (8.3).

For temperatures below the oxygen point the International Practical Temperature Scale is not defined. However, the National Bureau of Standards has standards extending down to about 2°K, although they are not all internationally accepted. They include the acoustical interferometer, the helium-vapor-pressure thermometer, and the platinum resistance thermometer.

Above the gold point the International Practical Temperature Scale *is* defined and uses a narrow-band radiation pyrometer ("optical" pyrometer) and the Planck equation to establish temperatures. The formula used is

$$\frac{J_t}{J_{Au}} = \frac{e^{1.438/\lambda(t_{Au}+273.15)} - 1}{e^{1.438/\lambda(t+273.15)} - 1} \qquad (8.4)$$

where $t_{Au} \triangleq$ gold-point temperature, 1063.0°C
$\lambda \triangleq$ effective wavelength of pyrometer

The quantity J_t/J_{Au} is measurable with the pyrometer and is the ratio of the spectral radiance of a blackbody at temperature t to one at temperature t_{Au}. Since λ can be determined for a given pyrometer, Eq. (8.4) allows calculation of t when J_t/J_{Au} has been measured. In principle, this method can be applied to arbitrarily high temperatures but in practice few reliable results above 4000°C are known.

The highest meaningful temperatures, existing in the interior of stars and for short times in atomic explosions, are inferred from kinetic theory to be in the range 10^7 to $10^{9°}K$. Definition of temperature, much less measurement, is difficult at these extremes, although spectroscopic methods have given useful results. At the other extreme, temperatures of $10^{-6°}K$ have been produced by using the concept of nuclear cooling. Magnetic susceptibility of certain materials has been used to measure temperatures in the extremely low ranges.

The question of the accuracy of temperature standards may be considered from two viewpoints. First, how closely can the International

Practical Temperature Scale be reproduced and second, how closely does it agree with the thermodynamic absolute scale?   The highest reproducibility of the International Practical Temperature Scale occurs at the triple point of water, which can be realized with a precision of a few ten-thousandths of a degree, giving an accuracy of about 1 ppm.   For either lower or higher temperatures the accuracy falls off.   At 100°K it is about 1 part in 2×10^4, at 10°K about 1 part in 10^3, and at 1°K about 1 part in 300.   Above the triple point, accuracy is about 1 part in 10^5 at 800°K, drops sharply to about 1 part in 5,000 at 1500°K, and then gradually decreases to 1 part in 500 at 4000°K and 1 part in 15 at 10,000°K.   The above statements refer to the International Practical Temperature Scale within its defined range and to the best available standards outside this range.[1,2]   The question of agreement between the various empirical scales (such as the International Practical Temperature Scale) and the absolute thermodynamic scale involves the fact that, in general, the thermodynamic scale is considerably less reproducible than the empirical scales.   For example, the steam-point temperature is reproducible to 0.0005° with a platinum resistance thermometer but to only 0.02° with a gas thermometer.   The disagreement between the International Practical Temperature Scales and the absolute thermodynamic scale has been estimated in centigrade degrees as[3]

$$\frac{t}{100}\left(\frac{t}{100} - 1\right)[0.04106 - 7.363(10^{-5})t] \qquad 0° < t < 444.6°C \qquad (8.5)$$

where t is the Celsius temperature.   The error is seen to be zero at $t = 0$ and 100°C and has a maximum value of about 0.14C° near $t = 400$°C.

   Calibration of a given temperature-measuring device is generally accomplished by subjecting it to some established fixed-point environment, such as the melting and boiling points of standard substances, or by comparing its readings with those of some more accurate (secondary standard) temperature sensor which has itself been calibrated.   The latter is generally accomplished by placing the two devices in intimate thermal contact in a constant-temperature-controlled bath.   By varying the temperature of the bath over the desired range (allowing equilibrium at each point), the necessary corrections are determined.   Accurate resistance thermometers, thermocouples, or mercury-in-glass expansion thermometers are generally useful as secondary standards.   Fixed-point standards using the melting points of various metals and the triple point of water are commercially available.

[1] McNish, *op. cit.*, p. 113.

[2] "High Temperature Technology," p. 34, McGraw-Hill Book Company, 1960.

[3] "Temperature, Its Measurement and Control in Science and Industry," vol. 2, p. 93, Reinhold Publishing Corporation, New York, 1955.

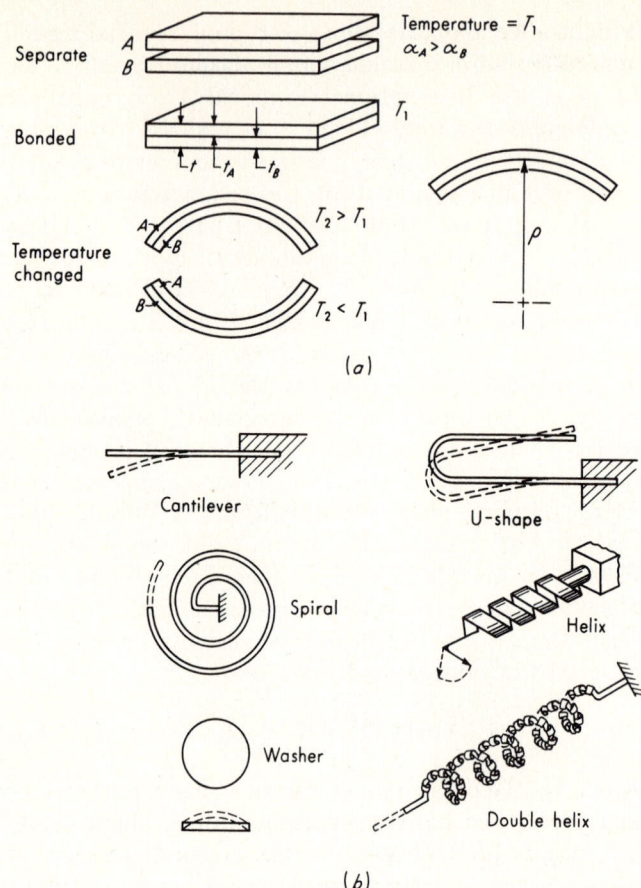

Fig. 8.2. *Bimetallic sensors.*

8.2

Thermal-expansion Methods   A number of practically important temperature-sensing devices utilize the phenomenon of thermal
expansion in one way or another. The expansion of solids is employed
mainly in bimetallic elements by utilizing the differential expansion of
bonded strips of two metals. Liquid expansion at essentially constant
pressure is used in the common liquid-in-glass thermometers. Restrained
expansion of liquids, gases, or vapors results in a pressure rise which is
the basis of pressure thermometers.

Bimetallic thermometers. If two strips of metals A and B with different thermal-expansion coefficients α_A and α_B but at the same temperature (Fig. 8.2a) are firmly bonded together, a temperature change causes

a differential expansion and the strip, if unrestrained, will deflect into a uniform circular arc.   Analysis[1] gives the relation

$$\rho = \frac{t[3(1 + m)^2 + (1 + mn)(m^2 + 1/mn)]}{6(\alpha_A - \alpha_B)(T_2 - T_1)(1 + m)^2} \tag{8.6}$$

where $\rho \triangleq$ radius of curvature

$t \triangleq$ total strip thickness, 0.0005 in. $< t <$ 0.125 in. in practice

$n \triangleq$ elastic modulus ratio, E_B/E_A

$m \triangleq$ thickness ratio, t_B/t_A

$T_2 - T_1 \triangleq$ temperature rise

In most practical cases, $t_B/t_A \approx 1$ and $n + 1/n \approx 2$, giving

$$\rho \approx \frac{2t}{3(\alpha_A - \alpha_B)(T_2 - T_1)} \tag{8.7}$$

Combination of this equation with appropriate strength-of-materials relations allows calculation of the deflections of various types of elements in practical use.   The force developed by completely or partially restrained elements can also be calculated in this way.   Accurate results require the use of experimentally determined factors[2] which are available from bimetal manufacturers.

Since there are no practically usable metals with negative thermal expansion, the B element is generally made of Invar, a nickel steel with a nearly zero [1.7×10^{-6} in./(in.-C°)] expansion coefficient.   While brass was originally employed, a variety of alloys are now used for the high-expansion strip, depending on the mechanical and electrical characteristics required.   Details of materials and bonding processes are in some cases considered trade secrets.   A wide range of configurations has been developed to meet application requirements (Fig. 8.2b).

Bimetallic devices are used for temperature measurement and also very widely as combined sensing and control elements in temperature-control systems, mainly of the on-off type.   They are also used as overload cutout switches in electrical apparatus by allowing the current to flow through the bimetal, heating and expanding it and causing a switch to open when excessive current flows.   Further applications are found as temperature-compensating devices[3] for various instruments that have temperature as a modifying or interfering input.   The mechanical motion proportional to temperature is used to generate an opposing

[1] S. G. Eskin and J. R. Fritze, Thermostatic Bimetals, *Trans. ASME*, p. 433, July, 1940.

[2] General Plate Division, Attleboro, Mass., *Bull*. PR750.

[3] R. Gitlin, How Temperature Effects Instrument Accuracy, *Control Eng.*, April, May, June, 1955.

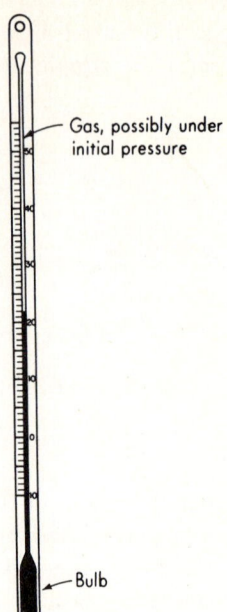

Gas, possibly under initial pressure

Bulb

Fig. 8.3.   *Liquid-in-glass thermometer.*

compensating effect.   The accuracy of bimetallic elements varies greatly, depending on the requirements of the application.   Since the majority of control applications are not extremely critical, requirements can be satisfied with a rather low-cost device.   For more critical applications, performance can be much improved.   The working temperature range is about from -100 to $1000°F$.   Inaccuracy of the order of 0.5 to 1 percent of scale range may be expected in bimetal thermometers of high quality.

Liquid-in-glass thermometers.[1,2]   The well-known liquid-in-glass thermometer is adaptable to a wide range of applications by varying the materials of construction and/or configuration.   Mercury is the most common liquid used at intermediate and high temperatures; its freezing point of $-38°F$ limits its lower range.   The upper limit is in the region of $1000°F$ and requires use of special glasses and an inert-gas fill in the capillary space above the mercury (see Fig. 8.3).   Compression of the gas helps to prevent separation of the mercury thread and raises the liquid boiling point.   For low temperatures, alcohol is usable to $-80°F$,

[1] J. F. Swindells, Calibration of Liquid-in-glass Thermometers, *Natl. Bur. Std.* (*U.S.*), *Circ.* 600, 1959.
[2] M. F. Behar, "Handbook of Measurement and Control," p. 25, The Instruments Publishing Co., Pittsburgh, Pa., 1954.

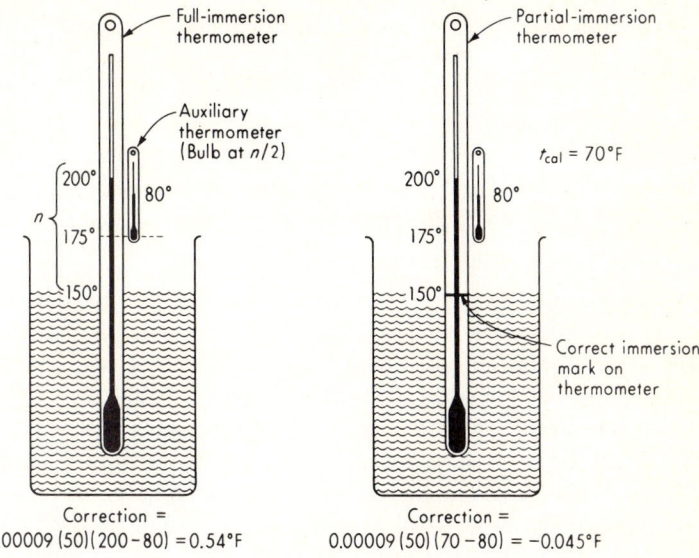

Fig. 8.4. *Full- and partial-immersion thermometers.*

toluol to $-130°$, pentane to $-330°$, and a mixture of propane and propylene giving the lower limit of $-360°$.

Thermometers are commonly made in two types: total immersion and partial immersion. Total-immersion thermometers are calibrated to read correctly when the liquid column is completely immersed in the measured fluid. Since this may obscure the reading, a small portion of the column may be allowed to protrude with little error. Partial-immersion thermometers are calibrated to read correctly when immersed a definite amount and with the exposed portion at a definite temperature. They are inherently less accurate than full-immersion types. If the exposed portion is at a temperature different from that at calibration a correction must be applied. Corrections for full- and partial-immersion thermometers used at conditions other than those intended are most accurately determined by the use of a special "faden" thermometer[1] designed to measure the average temperature of the emergent stem. If such a thermometer is not available, the correction may be estimated by suspending a small auxiliary thermometer close to the stem of the thermometer to be corrected, as in Fig. 8.4. This auxiliary thermometer estimates the mean temperature of the emergent stem. When a partial-immersion thermometer is used at correct immersion but with a surround-

---

[1] Swindells, *op. cit.*

ing air temperature different from its original calibration condition, the correction may be calculated from (for mercury-in-glass)

$$\text{Correction} = 0.00009n(t_{cal} - t_{act}) \qquad \text{F}° \qquad (8.8)$$

where $\quad n \triangleq$ number of scale degrees equivalent to emergent stem length, F°

$t_{cal} \triangleq$ air temperature at calibration, °F

$t_{act} \triangleq$ actual air temperature at use, °F (from auxiliary thermometer)

When a total-immersion thermometer is used at partial immersion the same formula may be used except that $t_{cal} - t_{act}$ is replaced by (main-thermometer reading) − (auxiliary-thermometer reading). For Celsius thermometers the constant 0.00009 becomes 0.00016.

The accuracy obtainable with liquid-in-glass thermometers depends on instrument quality, temperature range, and type of immersion. For full-immersion thermometers the best instruments, when calibrated, are capable of errors as small as 0.4F° (range −328 to 32°F), 0.05F° (range −69 to 32°F), 0.04F° (range 32 to 212°F), 0.4F° (range 212 to 600°F), and 0.8F° (range 600 to 950°F). Errors in partial-immersion types may be several times larger even after corrections have been applied for air-temperature variations. All the above figures refer to the ultimate performance attainable with the best instruments and great care in application. Errors in routine day-to-day measurements may be much larger.

For measuring small temperature changes from some chosen value with high accuracy and sensitivity the Beckman thermometer (Fig. 8.5) is available. Its high sensitivity requires a very large bulb. An impractically long capillary would also be required in order to cover a sizable temperature range with a single instrument. This difficulty is overcome by providing means for transferring mercury from the main bulb to an auxiliary bulb, so that at any chosen temperature the mercury thread is within the range of the capillary scale. The usual range is only 5 or 6C°, and the scale may be graduated in 0.01C° intervals. Measurement of the difference between the original set temperature and the measured temperature may be made with an error as little as 0.002 to 0.005C° if sufficient care is taken.

Pressure thermometers.[1,2] Pressure thermometers consist of a sensitive bulb, an interconnecting capillary tube, and a pressure-measuring

---

[1] D. M. Considine (ed.), "Process Instruments and Controls Handbook," p. 68, McGraw-Hill Book Company, New York, 1957.

[2] Behar, *op. cit.*, p. 29.

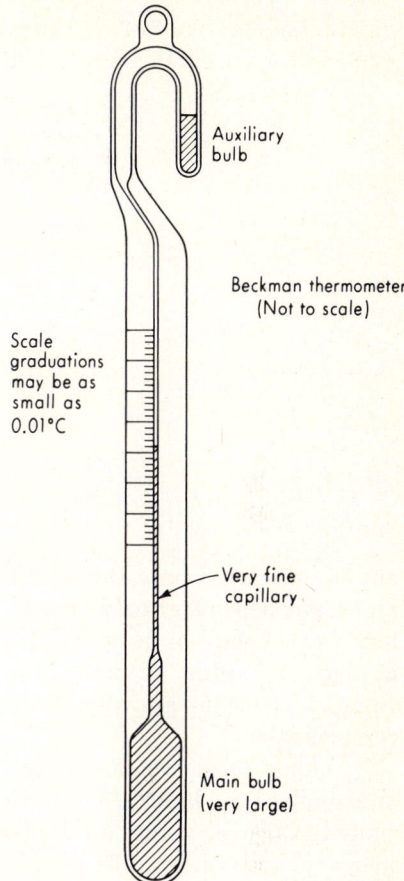

Auxiliary
bulb

Beckman thermometer
(Not to scale)

Scale
graduations
may be as
small as
0.01°C

Very fine
capillary

Main bulb
(very large)

Fig. 8.5. *Beckman thermometer.*

device such as a Bourdon tube, bellows, or diaphragm (Fig. 8.6). When the system is completely filled with a liquid (mercury and xylene are common) under an initial pressure, the compressibility of the liquid is often small enough relative to the pressure gage $\Delta V/\Delta p$ that the measurement is essentially one of volume change. For gas or vapor systems the reverse is true, and the basic effect is one of pressure change at constant volume.

Capillary tubes as long as 200 ft may be used for remote measurement. Temperature variations along the capillary and at the pressure-sensing device generally require compensation, except in the vapor-pressure type, where pressure depends only on the temperature at the liquid's free surface, located at the bulb. A common compensation scheme using an auxiliary pressure sensor and capillary is shown in Fig. 8.7. The motion of the compensating system is due to the interfering effects only

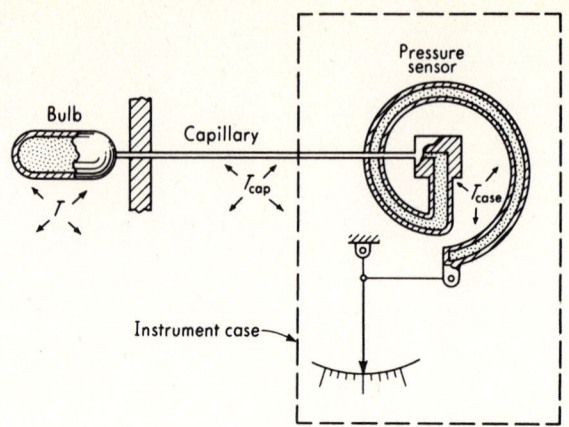

Fig. 8.6. *Pressure thermometer.*

and is subtracted from the total motion of the main system, resulting in an output dependent only on bulb temperature. The "trimming" capillary (which may be lengthened or shortened) allows the volume to be changed to attain accurate case compensation by experimental test. Bimetal elements are also used to obtain case and partial capillary compensation.

Liquid-filled systems cover the range −150 to 750°F with xylene or a similar liquid and −38 to 1100°F with mercury. Response is essentially linear over ranges up to about 300°F with xylene and 1000°F with mercury. Elevation differences between the bulb and pressure sensor different from those at calibration may cause slight errors. Gas-filled

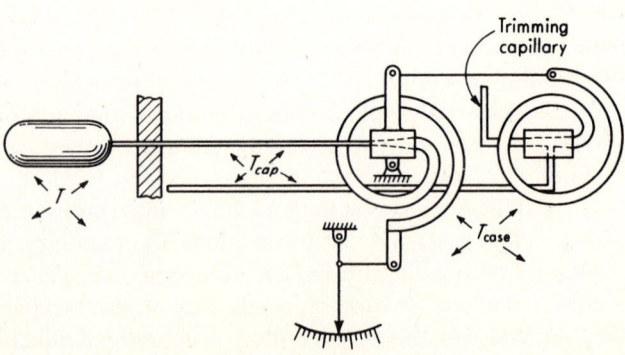

Fig. 8.7. *Compensation methods.*

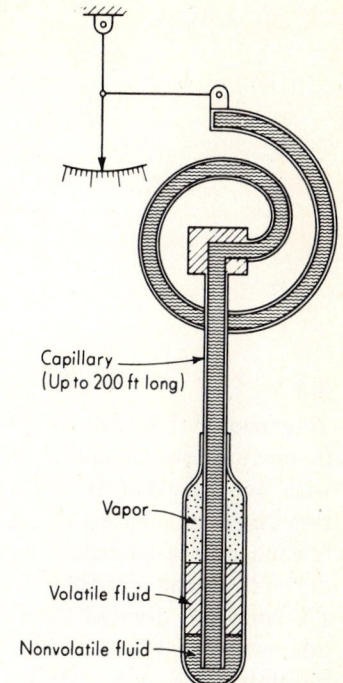

Capillary
(Up to 200 ft long)

Vapor

Volatile fluid

Nonvolatile fluid

Fig. 8.8. *Vapor-pressure thermometer.*

systems operate over the range −400 to 1200°F with linear ranges as great as 1000°F; errors due to capillary temperature variations are usually small enough not to justify compensation. Case compensation is accomplished with bimetal elements. Vapor-pressure systems are usable in the range −40 to 600°F. The calibration is strongly nonlinear; special linearizing linkages are needed if linear output is required. Characteristics of the system vary, depending on whether the bulb is hotter than, colder than, or equal in temperature to the rest of the system, since this determines where liquid and vapor will exist. The most versatile arrangement is shown in Fig. 8.8, where the volatile-liquid surface is *always* in the bulb. Capillary and case corrections are not needed in such a device since the vapor pressure of a liquid depends only on the temperature of its free surface. Commonly used volatile liquids include ethane (vapor pressure changes from 20 to 600 psig for a temperature change from −100 to +80°F), ethyl chloride (0 to 600 psig for 40 to 350°F), and chlorobenzene (0 to 60 psig for 275 to 400°F). The accuracy of pressure thermometers under the best conditions is of the order ±0.5 percent of the scale range. Adverse environmental conditions may increase this error considerably.

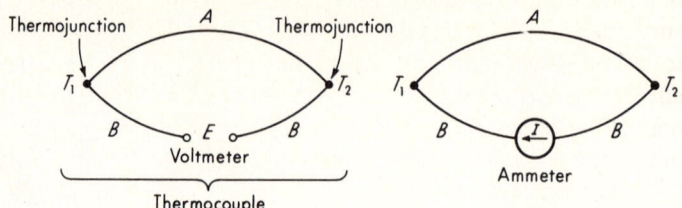

Fig. 8.9.   Basic thermocouple.

8.3

Thermoelectric Sensors (Thermocouples)   If two wires of different materials A and B are connected in a circuit as shown in Fig. 8.9, with one junction at temperature T_1 and the other at T_2, an infinite-resistance voltmeter detects an electromotive force E, or if an ammeter is connected a current I is measured.   The magnitude of the voltage E depends on the materials and the temperatures T_1 and T_2.   The current I is simply E divided by the total resistance of the circuit, including the ammeter resistance.   If current is allowed to flow, electrical power is developed; this comes from a heat flow from the surroundings to the wires. A direct conversion of heat energy to electrical energy is thus obtained. The effect is reversible, so that forcing a current from an external source through a thermoelectric circuit will cause heat flows to and from the circuit.   While here we are concerned with this thermoelectric effect only as a means of sensing temperatures, modern developments in materials have made the principle of practical application in electric power generation, heating, and cooling, though at present only on a small scale.

The overall relation between voltage E and temperatures T_1 and T_2, which is the basis of thermoelectric temperature measurement, is called the Seebeck effect.   The temperatures T_1 and T_2 refer to the junctions themselves, whereas when using a thermocouple one is trying to measure the temperature of some body in contact with the thermojunction.   These two temperatures are not exactly the same if current is allowed to flow through the thermojunction, since then heat is generated or absorbed at the junction, which must thus be hotter or colder than the surrounding medium whose temperature is being measured.   This heating and cooling are related to the Peltier effect.[1]   If the thermocouple voltage is measured with a potentiometer, no current flows and Peltier heating and cooling are not present.   When a millivoltmeter is used, current flows, and heat

---

[1] P. H. Dike, "Thermoelectric Thermometry," Leeds & Northrup Co., Philadelphia, Pa., 1954.

is absorbed at the hot junction (requiring it to become cooler than the surrounding medium) while heat is liberated at the cold junction, making it hotter than its surrounding medium. These heating and cooling effects are proportional to the current and fortunately are completely negligible[1] when the current is that produced by the thermocouple itself in a practical millivoltmeter circuit.

Another reversible heat-flow effect, the Thomson effect,[1] influences the temperature of the conductors between the junctions rather than the junctions themselves. When current flows through a conductor having a temperature gradient (and thus a heat flow) along its length, heat is liberated at any point where the current flow is in the same direction as the heat flow, while heat is absorbed at any point where these are opposite. Since this effect also depends on current flow, it is not present if a potentiometer is used. Even if a millivoltmeter is used, the effect of the heat flows on conductor temperature is completely negligible. Finally it should be noted that in any current-carrying conductor I^2R heat is generated, raising the circuit temperature above its local surroundings. Again, potentiometric voltage measurements are not susceptible to this error. Errors in millivoltmeter circuits are usually negligible also but can be estimated if heat-transfer conditions are known.

The above physical effects can be analyzed[2] on a macroscopic scale by classical thermodynamics, with fewer assumptions by irreversible thermodynamics, and qualitatively on a microscopic basis by solid-state physics. Thermodynamic approaches are based on the two experimentally observed reversible energy-conversion processes, the Peltier and Thomson effects, and neither require nor give any explanation of the basic atomic mechanisms. The total emf produced is made up of a part due to the Peltier effect, which is localized at each junction, and a (usually much smaller) part caused by the Thomson effect, which is distributed along each conductor between the junctions. The Peltier emf's are assumed proportional to the junction temperature while the Thomson emf's are proportional to the difference between the squares of the junction temperatures. For the total voltage, the equation takes the form

$$E = C_1(T_1 - T_2) + C_2(T_1^2 - T_2^2) \qquad (8.9)$$

Copper/constantan thermocouples, for example, give

$$E = 37.5(T_1 - T_2) - 0.045(T_1^2 - T_2^2) \qquad (8.10)$$

where $\quad E \triangleq$ total voltage, μv

$\qquad T_1, T_2 \triangleq$ absolute junction temperatures, °K

---

[1] *Ibid.*

[2] R. R. Heikes and R. W. Ure, "Thermoelectricity," Interscience Publishers, Inc., New York, 1961.

Unfortunately the assumptions made in the analyses leading to Eq. (8.9) are not exactly satisfied in practice; thus equations such as (8.10) can *not* usually be used to predict accurately temperatures from measured voltages. Rather, a given thermocouple material must be calibrated over the complete range of temperatures in which it is to be used. In this calibration only the overall voltage is of interest, and the separate contributions of Peltier and Thomson effects are not determined. Temperature measurement by thermoelectric means is thus based entirely on empirical calibrations and the application of so-called thermoelectric "laws" which experience has shown to hold. These laws, quoted below, are adequate for analysis of most practical thermocouple circuits. In those cases where the circuit configuration does not lend itself to direct application of these laws, alternative approaches[1] are available.

The laws of thermocouple behavior may be stated as follows:

1. The thermal emf of a thermocouple with junctions at T_1 and T_2 is totally unaffected by temperature elsewhere in the circuit if the two metals used are each homogeneous (Fig. 8.10a).

2. If a third homogeneous metal C is inserted into either A or B (see Fig. 8.10b), as long as the two new thermojunctions are at like temperatures, the net emf of the circuit is unchanged irrespective of the temperature of C away from the junctions.

3. If metal C is inserted between A and B at one of the junctions, the temperature of C at any point away from the AC and BC junctions is immaterial. So long as the junctions AC and BC are both at the temperature T_1, the net emf is the same as if C were not there (Fig. 8.10c).

4. If the thermal emf of metals A and C is E_{AC} and that of metals B and C is E_{CB}, the thermal emf of metals A and B is $E_{AC} + E_{CB}$ (Fig. 8.10d).

5. If a thermocouple produces emf E_1 when its junctions are at T_1 and T_2, and E_2 when at T_2 and T_3, it will produce $E_1 + E_2$ when the junctions are at T_1 and T_3 (Fig. 8.10e).

These laws are of great importance in the practical application of thermocouples. The first one states that the lead wires connecting the two junctions may be safely exposed to an unknown and/or varying temperature environment without affecting the voltage produced. Laws 2 and 3 make it possible to insert a voltage-measuring device into the circuit actually to measure the emf rather than just talking about its

---

[1] P. Stein, "Measurement Engineering," vol. I, chap. 18, Stein Engineering Services, Inc., Phoenix, Ariz., 1964.

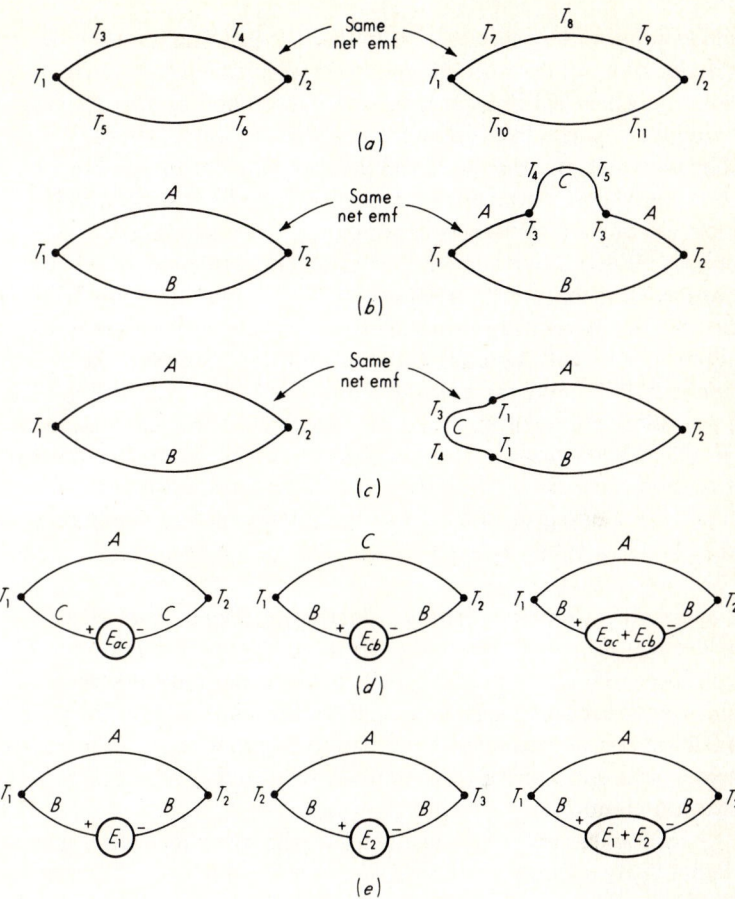

Fig. 8.10. *Thermocouple laws.*

existence. That is, the metal C represents the internal circuit (usually all copper in precise instruments) between the binding posts of a milli-voltmeter or potentiometer. The instrument can be connected in two ways, as shown in Fig. 8.10b and c. Law 3 also shows that thermocouple junctions may be soldered or brazed (thereby introducing a third metal) without affecting the readings. The fourth law shows that all possible pairs of metals need not be calibrated since the individual metals can each be paired with *one* standard (platinum is used) and calibrated. Any other combinations can then be *calculated;* calibration is not necessary.

In considering the fifth law we should note that, in using a thermo-couple to measure an unknown temperature, the temperature of one of the thermojunctions (called the reference junction) must be known by

some independent means. A voltage measurement then allows one to
get the temperature of the other (measuring) junction from calibration
tables. These calibration tables were obtained by maintaining the reference junction at a fixed known value (usually 32°F, the ice point), varying
the measured junction over the desired range of temperatures (known by
some independent means), and recording the resulting voltages. Thus
most calibration tables are based on the reference junction being at the ice
point. When a thermocouple is used, the reference junction may or may
not be at the ice point. If it is, the calibration table may be used directly
to find the measuring-junction temperature. If it is not, the fifth law
allows use of the standard table as follows: Suppose the reference junction is at 70°F and the voltage reading is 1.23 mv. In Fig. 8.10e we take
$T_1 = 32°F$, $T_2 = 70°F$, and T_3 is unknown. We can look up E_1 directly
in the table; suppose it is 0.71 mv. Now E_2 is the measured value
1.23 mv; thus $E_1 + E_2 = 1.94$ mv. The unknown temperature can be
found by looking up the temperature value corresponding to 1.94 mv in
the standard table; it is 100°F.

Common thermocouples. Thermojunctions formed by welding,
soldering, or merely pressing the two materials together give identical
voltages. If current is allowed to flow, the currents may be different
since the contact resistance differs for the various joining methods.
Welding (either gas or electric) is most widely used although both silver
solder and soft solder (low temperatures only) are used in copper/con-
stantan couples.

While many materials exhibit the thermoelectric effect to some
degree, only a small number of pairs are in wide use. They are platinum/
rhodium, Chromel/Alumel, copper/constantan, and iron/constantan.
Each of these pairs exhibits a combination of properties that suit it to a
particular class of applications. Since the thermoelectric effect is somewhat nonlinear, the sensitivity varies with temperature. The maximum
sensitivity of any of the above pairs is about $60\mu V/C°$ for copper/con-
stantan at 350°C. Platinum/platinum-rhodium is the least sensitive:
about $6\mu V/C°$ between 0 and 100°C.

The accuracy of the common thermocouples may be stated in two
different ways. If one uses standard thermocouple wire (which is *not*
individually calibrated by the manufacturer) and makes up a thermocouple to be used without calibration, he is relying on the wire manufacturer's quality control to limit deviations from the published calibration tables. These tables give the *average* characteristics, not those of a
particular batch of wire. Platinum/platinum-rhodium is the most accurate; error is of the order of ± 0.25 percent of reading. Copper/con-
stantan gives ± 0.5 percent or $\pm 1.5F°$ (whichever is larger) between -75

and 200°F and ±0.75 percent between 200 and 700°F. Chromel/Alumel gives ±5F° (32 to 660°F) and ±0.75 percent (660 to 2300°F). Iron/constantan has ±66μV below 500°F and ±1.0 percent from 500 to 1500°F. If higher accuracies are needed, the individual thermocouple may be calibrated. An indication of the achievable accuracy is available from National Bureau of Standards listings[1] of the results the Bureau will guarantee. At the actual calibration points the error ranges from 0.05 to 0.5C°. Interpolated points are less accurate: 0.1 to 1.0C°, except platinum/platinum-rhodium at 1450°F, 2.0 to 3.0C°. The realization of this potential accuracy in applying such calibrated thermocouples to practical temperature measurement is, of course, dependent on the application conditions and is rarely possible.

Platinum/platinum-rhodium thermocouples are used mainly in the range 0 to 1500°C. The main features of this combination are its chemical inertness and stability at high temperatures in oxidizing atmospheres. Reducing atmospheres cause rapid deterioration at high temperatures as the thermocouple metals are contaminated by absorbing small quantities of other metals from nearby objects (such as protecting tubes). This difficulty, causing loss of calibration, is unfortunately common to most thermocouple materials above 1000°C.

Chromel ($Ni_{90}Cr_{10}$)/Alumel ($Ni_{94}Mn_3Al_2Si_1$) couples are useful over the range −200 to +1300°C. Their main application, however, is from about 700 to 1200°C in nonreducing atmospheres. The temperature/voltage characteristic is quite linear for this combination (see Fig. 8.11).

Copper/constantan ($Cu_{57}Ni_{43}$) is used at temperatures as low as −200°C; its upper limit is about 350°C because of the oxidation of copper above this range. Iron/constantan is the most widely used thermocouple for industrial applications and covers the range −150 to +1000°C. It is usable in oxidizing atmospheres to about 760°C and reducing atmospheres to 1000°C.

Thermocouple manufacturers have a wealth of experience concerning the application of thermocouples to diverse temperature-measuring problems and should be consulted if special types of problems are foreseen in a particular case.

Reference-junction considerations. For the most precise work, reference junctions should be kept in a triple-point-of-water apparatus[2] whose temperature is 0.01 ± 0.0005C°. Such accuracy is rarely needed, and an ice bath is much more commonly used. A carefully made ice bath is reproducible to about 0.001C° but a poorly made one may have an error of

---

[1] Thermocouple Calibration, *Instr. Control Systems*, p. 1663, September, 1961.
[2] Transonics Inc., Lexington, Mass.

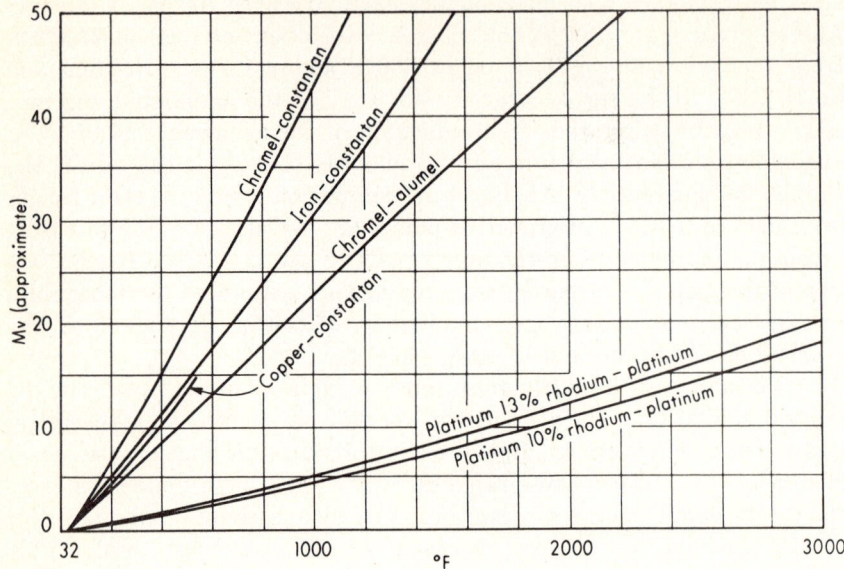

Fig. 8.11. *Thermocouple temperature/voltage curves.*

$1C°.$[1]   Figure 8.12 shows one method of constructing an ice-bath reference junction.   The main sources of error are insufficient immersion length and an excessive amount of water in the bottom of the flask.   Automatic ice baths that use the Peltier cooling effect as the refrigerator, rather than relying on externally supplied ice (which must be continually replenished), are available with an accuracy of $0.05C°.$[1]   These systems use the expansion of freezing water in a sealed bellows as the temperature-sensing element that signals the Peltier refrigerator when to turn on or off by displacing a microswitch.

Since low-power heating is more easily obtained than low-power cooling, some reference junctions are designed to operate at a fixed temperature higher than any expected ambient.   A feedback system operates an electrical heating element to maintain a constant temperature in an enclosure containing the reference junctions.   Since the reference junction is not at 32°F the thermocouple-circuit net voltage must be corrected by adding the reference-junction voltage before the measuring-junction temperature can be found.   This correction is, however, a constant.

In some situations the reference junction is allowed to assume

[1] C. L. Feldman, Automatic Ice-point Thermocouple Reference Junction, *Instr. Control Systems*, p. 101, January, 1965.

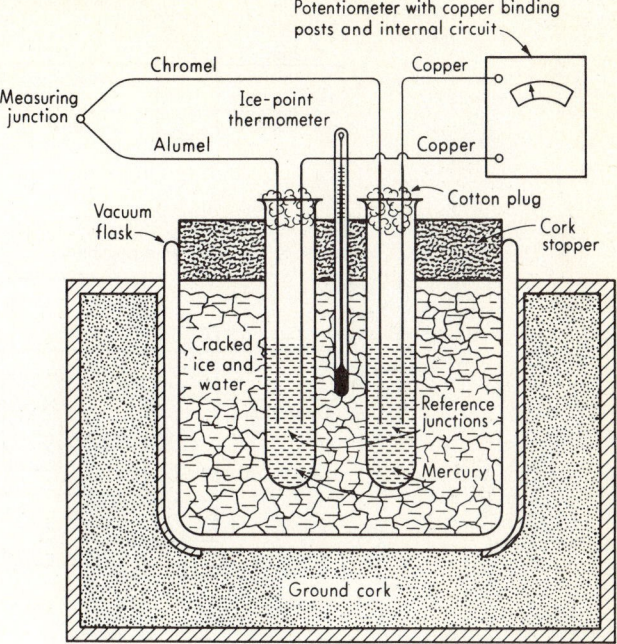

Fig. 8.12.   *Ice-bath reference junction.*

ambient temperature.   Knowledge of the ambient temperature then allows correction of the net measured voltage.   This correction is made automatic in some potentiometers that are intended to measure voltages from a specific type of thermocouple.   Ambient temperature is sensed by a bimetal element or a temperature-sensitive resistor.   These adjust a compensating voltage in the circuit so that the voltage indicated by the instrument is the same as it would be if the reference junction were at 32°F, thus allowing a direct reading of the measuring-junction temperature.

Another approach uses two temperature-controlled ovens and an auxiliary thermocouple to provide an equivalent reference junction at 32°F.[1]   Figure 8.13*b* shows the arrangement for an iron/constantan thermocouple.   In Fig. 8.13*a* if the reference junctions are kept at 130°F, when the measuring junction is at 32°F the output voltage is 2.82 mv.   To make the system act as if the reference junctions were at 32°F an opposing voltage of 2.82 mv must be put in series.   This is done by connecting a Chromel/constantan thermocouple as shown in Fig. 8.13*b*.   This produces an opposing voltage which will be 2.82 mv if the second oven is set at

[1] Pace Engineering Co., North Hollywood, Calif.

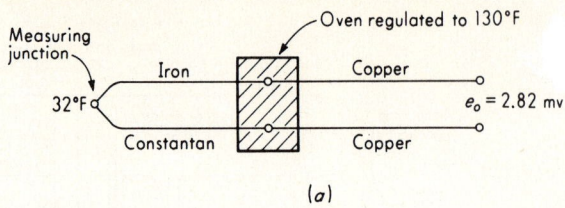

(a)

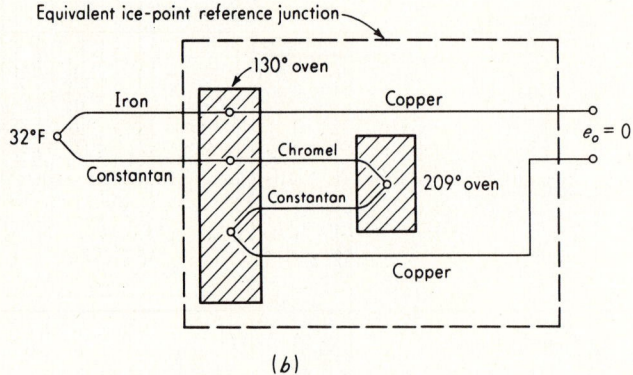

(b)

Fig. 8.13.   *Oven reference junctions.*

209°F. Systems of this type maintain a reference temperature within ± 0.2F° for long-term unattended operation.

Special materials, configurations, and techniques. Increasing interest in high-temperature processes in jet and rocket engines and nuclear reactors has led to requirements for reliable temperature sensors in the range 2000 to 4500°F. New thermocouples developed for these applications include rhodium-iridium/rhodium,[1] tungsten/rhenium,[1] and boron/graphite.[2] Rhodium-iridium is usable to about 4000°F under proper conditions and has a sensitivity of the order of $6\mu V/C°$. Various alloys of tungsten and rhenium may be used up to 5000°F under favorable conditions and have about the same sensitivity at the highest temperatures as rhodium-iridium. Boron/graphite has a high sensitivity (about $40\mu V/C°$) and is usable for short times up to 4500°F.

An alternative solution to high-temperature problems may be found in various cooling schemes. Two such[3] in actual use are shown in Fig. 8.14. In Fig. 8.14a the hot-gas flow whose temperature is to be measured

---

[1] P. D. Freeze, Review of Recent Developments of High-temperature Thermocouples, *ASME Paper* 63-WA-212, 1963.

[2] Astro Industries Inc., Santa Barbara, Calif., *Bull.* BGT-1, 1963.

[3] *NASA, SP*-5015, p. 128, 1964.

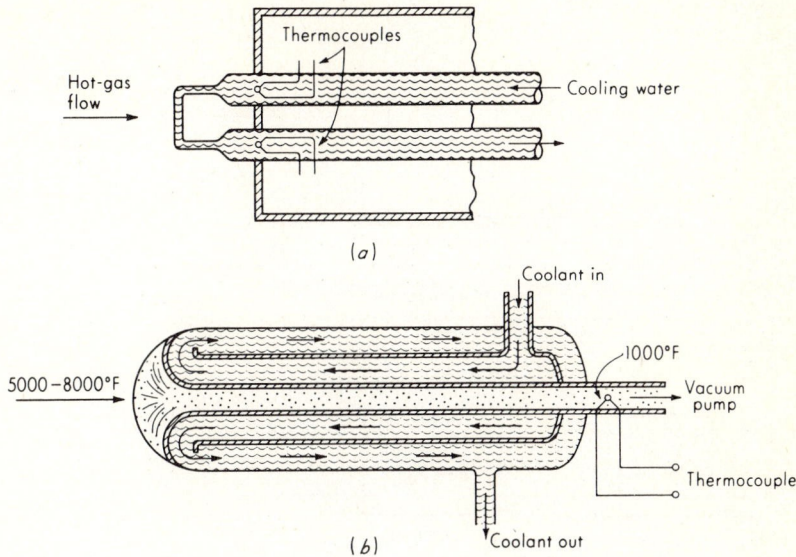

Fig. 8.14. *Cooled thermocouples.*

impinges on a small tube carrying cooling water, causing a temperature rise of about 100F°. If heat-transfer coefficients are known, measurements of water flow rate, temperature, and temperature rise allow calculation of the hot-gas temperature. Figure 8.14*b* shows another approach wherein the hot gas is aspirated through a heat exchanger, cooling it to about 1000°F. Knowledge of heat-transfer characteristics and flow rates again allows calculation of the hot-gas temperature. Such methods have been used in the range 5000 to 8000°F.

Figure 8.15 shows in simplified fashion the principle of a pulse-cooling technique[1] which allows use of Chromel/Alumel thermocouples (melting point 2550°F) to measure temperatures up to 7000°F. The measuring junction is kept at a low temperature by a cooling air flow. When this flow is shut off by a solenoid valve the thermocouple starts to heat up, following the first-order equation

$$\tau \frac{dT_{tc}}{dt} + T_{tc} = T_{\text{gas}} \qquad (8.11)$$

where $\tau \triangleq$ thermocouple time constant (assumed known)
$T_{tc} \triangleq$ thermocouple temperature
$T_{\text{gas}} \triangleq$ hot-gas temperature

---

[1] A. F. Wormser and R. A. Pfuntner, Pulse Technique Extends Range of Chromel-Alumel to 7000°F, *Instr. Control Systems*, p. 101, May, 1964.

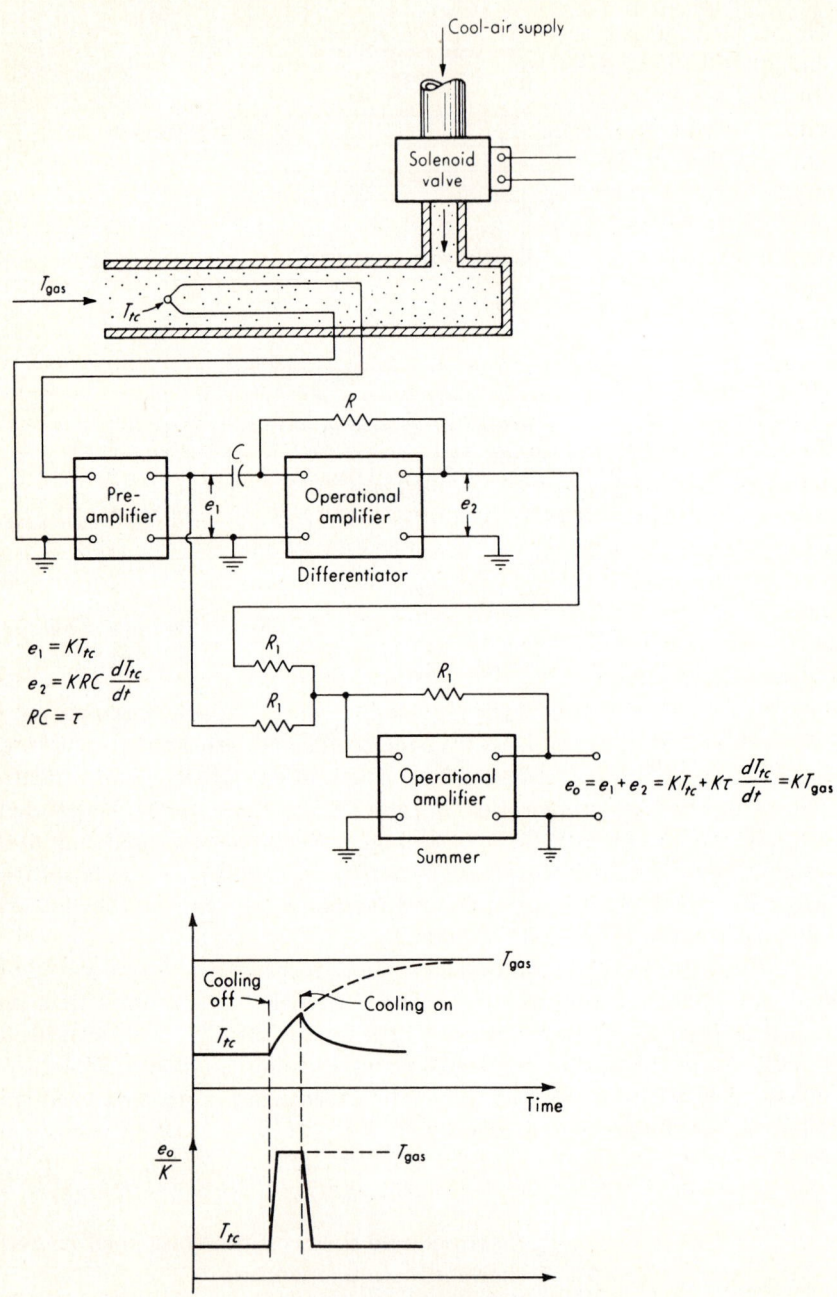

Fig. 8.15.   *Pulsed-thermocouple technique.*

This equation shows that T_{gas} can be *computed* any time after the cooling is shut off if dT_{tc}/dt is known. A voltage proportional to dT_{tc}/dt can be obtained by use of the differentiating circuit shown and, when summed with a voltage proportional to T_{tc}, provides a signal proportional to T_{gas}. This signal is theoretically available immediately after the cooling is shut off; however, in practice, the cooling is left off a finite interval during which the value of T_{gas} is recorded. The cooling is turned on again before the thermocouple is overheated. In the actual system of the quoted reference, additional computing elements also compute the numerical value of τ; thus this need not be known beforehand.

The measurement of rapidly changing internal and surface temperatures of solid bodies may be accomplished with arrangements such as those in Fig. 8.16. The main requirements in such applications are that the thermojunction be of minimum size and be precisely located and that any materials placed into the wall have thermal properties identical with those of the wall so that temperature distributions are not distorted. In Fig. 8.16a[1] the thermojunction is formed by drawing an abrasive tool, such as a file or emery cloth, across the end of the sensing tip. This action flows metal from one thermocouple element to the other since the 0.0002-in. mica insulation is easily bridged over, thus forming numerous microscopic hot-weld thermojunctions. Subsequent erosion or abrasive action forms new thermal junctions continuously as the tip wears away. Such thermocouples have time constants as small as 10^{-5} sec and are available in materials usable to 5000°F and 10,000-psi pressure. In Fig. 8.16b[2,3] two thermojunctions are formed by plating a thin rhodium film over the end of a coaxial pair of thermocouple metals. Since the rhodium/metal A and rhodium/metal B junctions are at the same temperature, the third metal (rhodium) has no effect. The plating is performed by vacuum evaporation and results in a rhodium layer 10^{-4} to 10^{-5} in. in thickness. Theoretical calculations indicate the time constant of such a probe is of the order of 0.3 μsec.

Thermocouples in common use are made from wires ranging from about 0.020 to 0.1 in. in diameter, the larger diameters being required for long life in severe environments. Since speed of response, conduction and radiation errors, and precision of junction location all are improved by the use of smaller wire, very-fine-wire thermocouples are used in special applications requiring these attributes and where lack of ruggedness is not a

---

[1] Nanmac Corp., Indian Head, Md.

[2] D. Bendersky, A Special Thermocouple for Measuring Transient Temperatures, *Mech. Eng.*, p. 117, February, 1953.

[3] MO-RE', Inc., Bonners Springs, Kans.

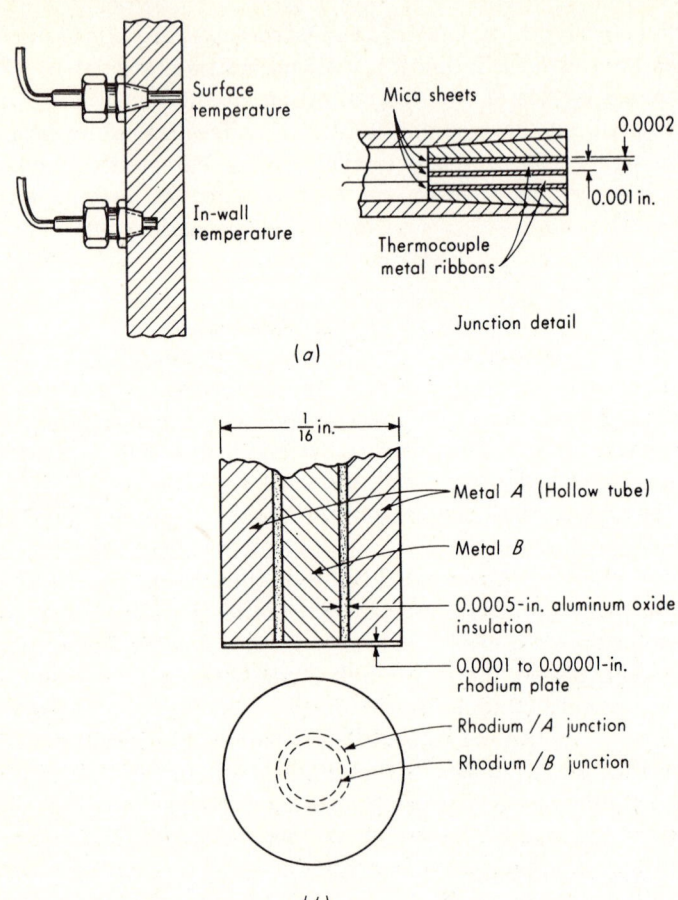

Fig. 8.16.   *High-speed thermocouples.*

serious drawback.   Such couples are available ready-made[1,2] in most common materials and wire sizes from 0.0005 to 0.015 in. in diameter.   The
time constant of an iron/constantan couple of 0.0005-in.-diameter wire
for a step change from 200 to 100°F in still water is about 0.001 sec.

Several thermocouples may be connected together in series or parallel
to achieve useful functions (Fig. 8.17).   The series connection with all
measuring junctions at one temperature and all reference junctions at
another is used mainly as a means of increasing sensitivity.[3]   Such an

[1] Omega Engineering Inc., Springdale, Conn.
[2] Baldwin-Lima-Hamilton, Waltham, Mass., *Tech. Data* 4336-1.
[3] R. A. Schnurr, Thermopiles Aid in Measuring Heat Rejection, *Gen. Motors Eng. J.*, p. 8, April–May–June, 1963.

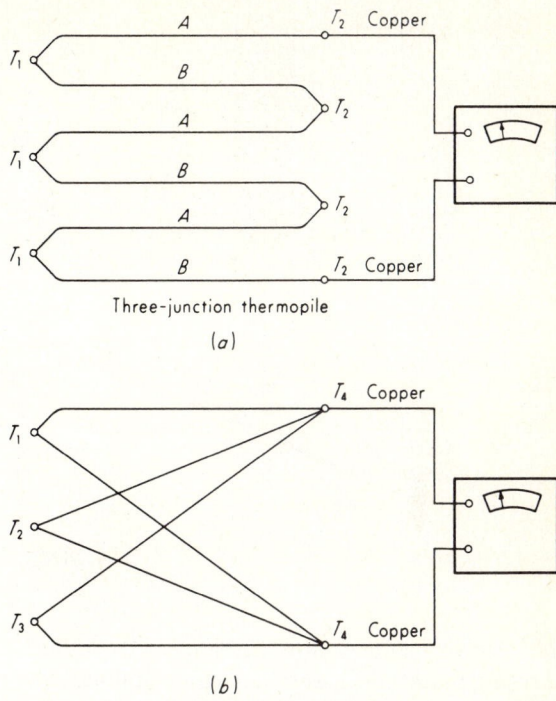

Three-junction thermopile

(a)

(b)

Fig. 8.17.   *Multiple-junction thermocouples.*

arrangement is called a thermopile and for n thermocouples gives an output n times as great as a single couple.   A commercially available[1] Chromel/constantan thermopile has 25 couples and produces about 1 mv/F°.   Common potentiometers can resolve 1 μv, thus making such an arrangement sensitive to 0.001F°.   The parallel combination generates the same voltage as a single couple if all measuring and reference junctions are at the same temperatures.   If the measuring junctions are at different temperatures and the thermocouples are all the same resistance, the voltage measured is the average of the individual voltages.   The temperature corresponding to this voltage is the average temperature only if the thermocouples are linear over the temperature range being measured.

8.4

Electrical-resistance Sensors   The electrical resistance of various materials changes in a reproducible manner with temperature, thus forming the basis of a temperature-sensing method.   Materials in actual use

[1] Science Products Corp., Dover, N.J.

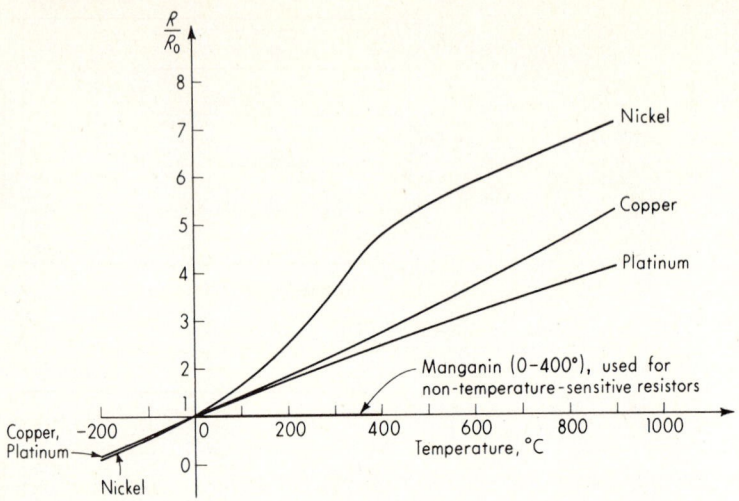

Fig. 8.18.   *Resistance/temperature curves.*

fall into two main classes: conductors (metals) and semiconductors.   Conducting materials historically came first and have traditionally been called resistance thermometers.   Semiconductor types appeared more recently and have been given the generic name thermistor.   Any of the various established techniques of resistance measurement may be employed to measure the resistance of these devices, with the Wheatstone bridge being the most common.

Conductive sensors (resistance thermometers).   The variation of resistance R with temperature T for most metallic materials can be represented by an equation of the form

$$R = R_0(1 + a_1T + a_2T^2 + \cdots + a_nT^n) \qquad (8.12)$$

where R_0 is the resistance at temperature $T = 0$ (see Fig. 8.18).   The number of terms necessary depends on the material, the accuracy required, and the temperature range to be covered.   Platinum, nickel, and copper are the most commonly used and generally require, respectively, two, three, and three of the a constants for a highly accurate representation.   Tungsten and nickel/iron alloys are also in use.   Only constant a_1 may often be used since quite respectable linearity may be achieved over limited ranges.   Platinum,[1] for instance, is linear within ±0.4 percent over the ranges -300 to $-100°F$ and -100 to $+300°F$, ±0.3 percent from

[1] Platinum Resistance Thermometers, Transducer Handbook 1, Trans-Sonics Inc., Burlington, Mass.

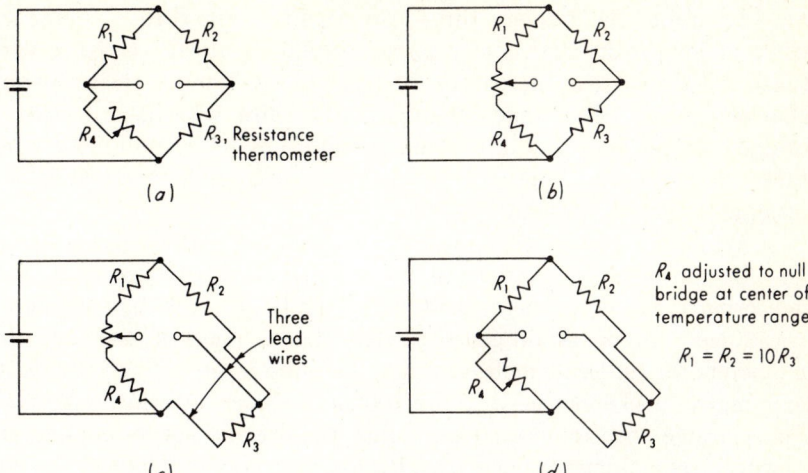

Fig. 8.19.   *Resistance-thermometer bridge circuits.*

0 to 300°F, ±0.25 percent from −300 to −200°F, ±0.2 percent from 0 to 200°F, and ±1.2 percent from 500 to 1500°F.

Sensing elements are made in a number of different forms.[1]   For measurement of fluid temperatures the winding of resistance wire may be encased in a stainless-steel bulb to protect it from corrosive liquids or gases.   Open-type pickups expose the resistance winding directly to the fluid (which must be noncorrosive) and give faster response.   Various flat grid windings are available for measuring surface temperatures of solids. These may be taped, welded, or cemented onto the surface.   Thin deposited films of platinum[2] are also used in place of wire windings.   Surface-temperature transducers affixed to bodies may exhibit spurious output due to interfering strain inputs.[3]   These strains may be due to loading of the structure or differential expansion.

Bridge circuits used with resistance temperature sensors may employ either the deflection mode of operation or the null (manually or automatically balanced) mode.   If the null method is used, resistor R_4 in Fig. 8.19a is varied until balance is achieved.   When the highest accuracy is required, the arrangement of Fig. 8.19b is preferred since the (variable and unknown) contact resistance in the adjustable resistor has no influence on the resistance of the bridge legs.   If long lead wires subjected to temperature varia-

[1] Platinum Resistance Thermometers, Tranducer Handbook   1, Trans-Sonics Inc., Burlington, Mass.

[2] Microdot Inc., South Pasadena, Calif.

[3] A. B. Kaufman, Bonded-wire Temperature Sensors, *Instr. Control Systems*, p. 103, May, 1963.

tions are unavoidable, errors due to their resistance changes may be canceled by use of the Siemens three-lead circuit of Fig. 8.19c. Three lead wires of identical length and material exhibit identical resistance variations, and since one of these leads is in each of legs 2 and 3, their resistance changes cancel. Resistance change in the third wire has no effect on bridge balance since it is in the null detector circuit for null-mode operation. For deflection operation its effect is negligible if the indicating instrument draws little current.

While the resistance/temperature variation of the sensing element may be quite linear, the output voltage signal of a bridge used in the deflection mode is not necessarily linear for large percentage changes of resistance. Unlike strain gages, the resistance change of resistance thermometers for full-scale deflection may be quite large. Typically, a 500-ohm platinum element may exhibit 100 ohms change over its design range. For a bridge with four equal arms this would cause severe nonlinearity; however, by making the fixed arms R_1, R_2 of considerably higher resistance (say 10:1) than R_3 and R_4 and by balancing the bridge at the middle of the temperature range rather than at one end, good linearity may be achieved (Fig. 8.19d). Typically,[1] a platinum element covering a range from 0 to 100°C using the 10:1 resistance ratio mentioned above gives a nonlinearity of only 0.5C°. For nickel elements, whose resistance/temperature variation is quite nonlinear, this nonlinearity and the bridge nonlinearity can be made nearly to cancel by proper design[2] since the two effects are of opposite directions.

Resistance-thermometer bridges may be excited with either a-c or d-c voltages. The direct or rms alternating current through the thermometer is usually in the range 2 to 20 ma. This current causes an I^2R heating which raises the temperature of the thermometer above its surroundings, causing the so-called self-heating error. The magnitude of this error depends also on heat-transfer conditions and is usually quite small. A 450-ohm platinum element of open construction carrying 25-ma current has a self-heating error of 0.2F° when immersed in liquid oxygen. Actually, by using an unsymmetrical pulse type of excitation voltage whose rms (heating) value is small compared with its peak value, quite large instantaneous currents (and thus large peak output voltages) may be obtained without significant self-heating. Such pulse excitation voltages (Fig. 8.20) can be obtained by commutating a d-c source; this also allows time sharing of the bridge among several resistance sensors. As much as

---

[1] Temperature Recording from Platinum Resistance Sensors, Brush Instruments, Cleveland, Ohio.

[2] D. R. Mack, Linearizing the Output of Resistance Temperature Gages, *Exptl. Mechanics*, p. 122, April, 1961.

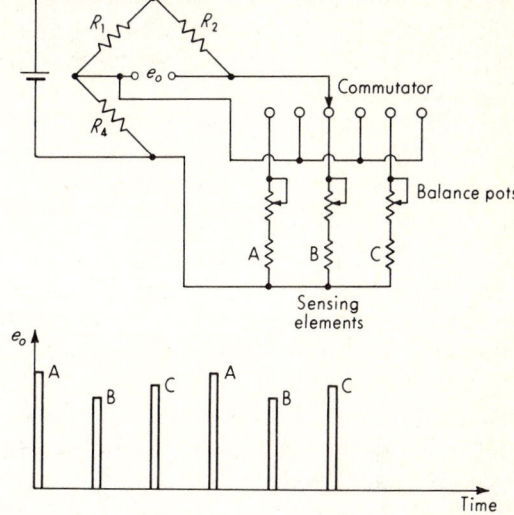

Fig. 8.20. *Pulse-excitation technique.*

5 volts full-scale bridge output signal can be obtained from resistance sensors used in this way.[1]

Resistance-thermometer elements range in resistance from about 10 ohms to as high as 25,000 ohms.   Higher-resistance elements are less affected by lead-wire and contact resistance variations, and since they generally produce large voltage signals, spurious thermoelectric emf's due to joining of dissimilar metals are also usually negligible.   Platinum is used from −450 to 1850°F, copper from −320 to 500°F, nickel from −320 to 800°F, and tungsten from −450 to 2000°F.   Average temperatures may be measured using resistance thermometers as in Fig. 8.21*a*, while differential temperature is sensed by the arrangement of Fig. 8.21*b*.[2] Differential-temperature measurements to an accuracy of 0.05C° have been accomplished in a nuclear-reactor-coolant heat-rise application.[3]

Semiconductor sensors (*thermistors*). The earlier types of semi-conductor resistance temperature sensors were made of manganese, nickel, and cobalt oxides which were milled, mixed in proper proportions with binders, pressed into the desired shape, and sintered.   These were given

---

[1] Platinum Resistance Thermometers, Transducer Handbook 1.

[2] Temperature Recording from Platinum Resistance Sensors.

[3] B. G. Kitchen, Precise Measurement of Process Temperature Differences, *ISA J.*, p. 39, February, 1959.

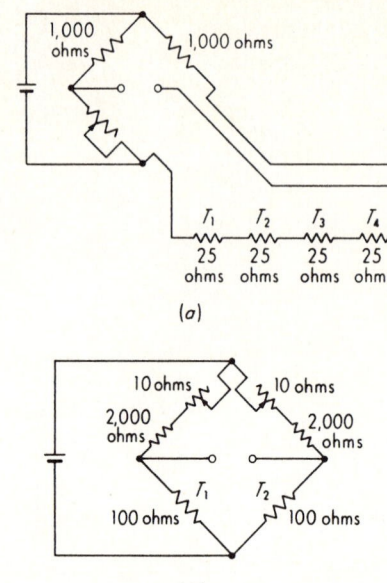

Fig. 8.21. *Average- and differential-temperature sensing.*

(b)

the name thermistor and are in wide use today. Compared with conductor-type sensors (which have a small positive temperature coefficient), thermistors have a very large negative coefficient. While some conductors (copper, platinum, tungsten) are quite linear, thermistors are very non-linear. Their resistance/temperature relation is generally of the form

$$R = R_0 e^{\beta(1/T - 1/T_0)} \qquad (8.13)$$

where R ≜ resistance at temperature T, ohms
R_0 ≜ resistance at temperature T_0, ohms
β ≜ constant, characteristic of material, °K
e ≜ base of natural log
T, T_0 ≜ absolute temperatures, °K

The reference temperature T_0 is generally taken as 298°K (25°C) while the constant β is of the order of 4000. By computing $(dR/dT)/R$ we find the temperature coefficient of resistance to be given by $-\beta/T^2$ ohms/(ohm-C°). If β is taken as 4000, the temperature coefficient at room temperature (25°C) is -0.045, compared with $+0.0036$ for platinum. While the exact resistance/temperature relation varies somewhat with the particular material used and the configuration of the resistance element, Fig. 8.22 shows the general type of curve to be expected.

Thermistors are commercially available in the form of beads, probes,

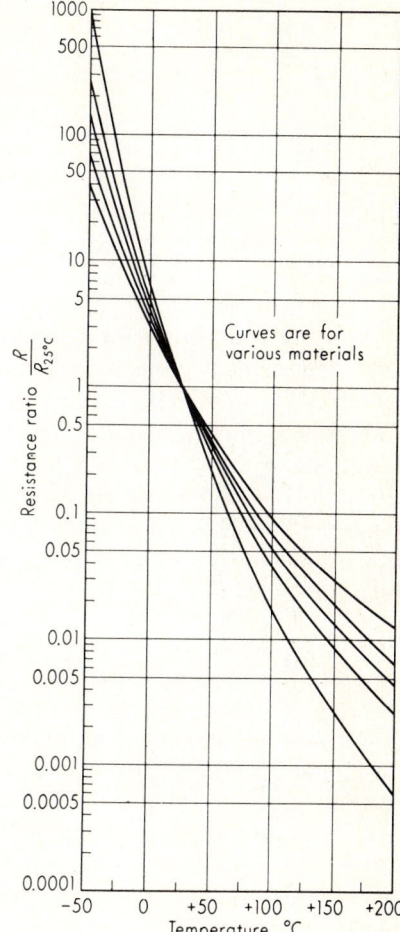

Fig. 8.22. *Thermistor resistance/temperature curves.*

disks, and rods as shown in Fig. 8.23. Beads, much used for temperature measurement, may be bare but are more often glass-coated. They may be as small as a few thousandths of an inch in diameter, giving fast response. Resistance at 25°C can vary over a wide range, from 500 ohms to several megohms. The usable temperature range is from about −420 to 1200°F; however, a single thermistor is not ordinarily used over such a large range. Glass probes have a diameter of about 0.1 in. and a length varying from ¼ to 2 in., are also widely used in temperature measurement, and have resistance properties similar to beads. Disks and rods are used more as temperature-compensating devices, time delay elements, and voltage and power controls in electronic circuits.

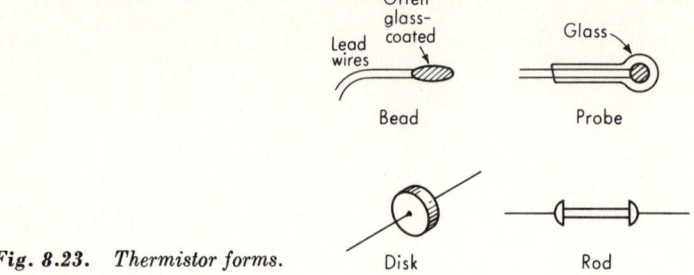

Fig. 8.23.   *Thermistor forms.*

Other semiconductor temperature sensors include carbon resistors and silicon[1] and germanium[2] crystal elements. Carbon resistors are merely the commercial carbon-composition elements commonly used as resistance elements in radios and other electronic circuitry. The 0.1- to 1-watt rated resistors with room-temperature resistance of 2 to 150 ohms are widely used for the measurement of cryogenic temperatures in the range 1 to 20°K. From about 20°K downward these elements exhibit a large increase in resistance with decrease in temperature given by the relation[3]

$$\log_{10} R + \frac{K}{\log_{10} R} = A + \frac{B}{T} \qquad (8.14)$$

where R is the resistance at Kelvin temperature T, and A, B, and K are constants determined by calibration of the individual resistor. Reproducibility of the order of 0.2 percent is obtained in the range 1 to 20°K.

Silicon, with varying amounts of boron impurities, can be designed to have either a positive or negative temperature coefficient over a particular temperature range. The resistance/temperature relation is quite nonlinear. A typical element shows a resistance change (from the nominal value at 25°C) of −80 percent at −150°C to +180 percent at +200°C. The temperature coefficient near room temperature is of the order of +0.7 percent/C°. Germanium, doped with arsenic, gallium, or antimony, is used for cryogenic temperatures, where it exhibits a large decrease in resistance with increasing temperature. The relation is quite nonlinear but very reproducible, giving precise measurements within 0.001 to

[1] J. R. Pies, A New Semiconductor for Temperature Measuring, *ISA J.*, p. 50, August, 1959.

[2] J. S. Blakemore, Germanium for Low-temp Resistance Thermometry, *Instr. Control Systems*, p. 94, May, 1962.

[3] L. G. Rubin, Temperature, *Electron. Progr.*, Raytheon Corp., p. 1, autumn, 1963.

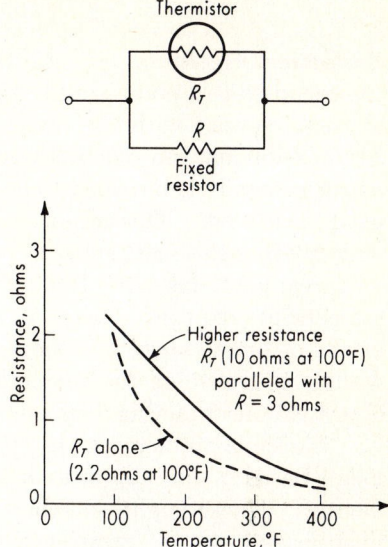

Fig. 8.24. *Thermistor linearization.*

0.0001K° near 4°K when adequate care is taken in technique. Commercially available elements cover a range from about 0.5 to 100°K, a typical unit changing resistance from 7,000 ohms at 2°K to 6 ohms at 60°K.

Circuitry[1,2] for applying the various types of semiconductor resistance sensors to temperature measurement, control, and compensation problems is essentially the same as for conductive sensors, although the greater nonlinearity makes wide temperature ranges less convenient. One technique[3,4] for reducing this nonlinearity is to shunt the thermistor with an ordinary resistor as shown in Fig. 8.24. The stability (variation of resistance/temperature characteristic with time) of early semiconductor elements was inferior to that of conductive elements. While it is unlikely that they will ever approach the excellent stability of platinum, modern semiconductor elements have quite acceptable stability for many applications. This is achieved by proper aging of the elements prior to sale by the manufacturer and continuing improvements in design and production techniques.

[1] D. S. Saulson, The Thermistor Bridge, *Electro-Technol.* (*New York*), p. 73, September, 1961.

[2] O. Schwelb and G. C. Temes, Thermistor-Resistor Temperature-sensing Networks, *Electro-Technol.* (*New York*), p. 71, November, 1961.

[3] F. Bennett, Designing Thermistor Temperature-correcting Networks Graphically, *Control Eng.*, November, 1955.

[4] R. W. Smith, An Evaluation of Thermistors, *Gen. Motors Eng. J.*, p. 14, October–November–December, 1960.

8.5

Radiation Methods   All the temperature-measuring methods discussed up to this point require that the "thermometer" be brought into physical contact with the body whose temperature is to be measured. Also, except for the pulsed thermocouple of Fig. 8.15, the temperature sensor generally is intended to assume the same temperature as the body being measured. This means that the thermometer must be capable of withstanding this temperature, which in the case of very hot bodies presents real problems, since the thermometer may actually melt at the high temperature required. Also, for bodies that are moving, a noncontacting means of temperature sensing is most convenient. Furthermore, if one wishes to determine the temperature variations over the surface of an object, a noncontacting device can readily be "scanned" over the surface.

To solve problems of the type mentioned above, a variety of instruments based in one way or another on the sensing of radiation have been devised. These might, in general, be called radiometers; however, common usage employs terms such as radiation pyrometer, radiation thermometer, optical pyrometer, etc., to describe a particular type of instrument. Since this terminology is not standardized, one must inquire into the basic operating principle of a given instrument to be sure what its characteristics are, rather than relying on the name given the instrument.

Other important applications of infrared radiation include missile guidance, satellite attitude sensing, and infrared spectroscopy. In missile guidance (the Sidewinder missile is an outstanding example) the missile is designed to "home" on the infrared radiation emitted by the target, often the hot jet exhaust of the target aircraft's engine. A scanning system in the missile locates the target and produces error signals that steer the missile into the target. For satellite attitude sensing[1] the infrared sensors are able to distinguish the radiation from the earth, the moon, or a planet from the background of space and thus generate accurate orientation signals for control purposes. Infrared spectroscopy[2] involves the use of infrared principles for the analysis of gases, liquids, and solids to identify and determine the concentration of molecules or molecular groups.

Radiation fundamentals. Radiation-temperature sensors operate with electromagnetic radiation whose wavelengths lie in the visible and infrared portions of the spectrum. The visible spectrum is quite narrow: 0.3 to 0.72 μ (1 μ = 10^{-6} m). The infrared spectrum is generally defined as the range from 0.72 to about 1,000 μ. Bordering the visible spectrum

---

[1] Barnes Engineering Co., Stamford, Conn., *Bull.* 14-003 and 0-014, 1962.

[2] D. M. Considine (ed.), "Process Instruments and Controls Handbook," pp. 6–67, McGraw-Hill Book Company, New York, 1957.

on the low-wavelength side are the ultraviolet rays, while microwaves border the infrared spectrum on the high side. Radiation-temperature-sensing devices utilize mainly some part of the range 0.3 to 40 μ.

Physical bodies (solids, liquids, gases) may emit electromagnetic radiation or subatomic particles for a number of reasons. As far as temperature sensing is concerned, we need be concerned only with that part of the radiation caused solely by temperature. Every body above absolute zero in temperature emits radiation dependent on its temperature. The ideal thermal radiator is called a blackbody. Such a body would absorb completely any radiation falling on it and also, for a given temperature, emits the maximum amount of thermal radiation possible. The law governing this ideal type of radiation is Planck's law, which states that

$$W_\lambda = \frac{C_1}{\lambda^5(e^{C_2/\lambda T} - 1)} \qquad (8.15)$$

where $W_\lambda \triangleq$ hemispherical spectral radiant intensity, watts/(cm²-μ)
$C_1 \triangleq$ 37,413, (watts-μ^4)/cm²
$C_2 \triangleq$ 14,388, μ-°K
$\lambda \triangleq$ wavelength of radiation, μ
$T \triangleq$ absolute temperature of blackbody, °K

The quantity W_λ is the amount of radiation emitted from a flat surface into a hemisphere, per unit wavelength, at the wavelength λ. Equation (8.15) thus gives the distribution of radiant intensity with wavelength; that is, a blackbody at a certain temperature emits *some* radiation per unit wavelength at every wavelength from zero to infinity, but not the same amount at each wavelength. Figure 8.25 shows the curves obtained from Eq. (8.15) by fixing T at various values and plotting W_λ versus λ. The curves exhibit peaks at particular wavelengths, and the peaks occur at longer wavelengths as the temperature decreases. The area under each curve is the total emitted power and increases rapidly with temperature. Equations giving the peak wavelength λ_p and the total power W_t are

$$\lambda_p = \frac{2,891}{T} \qquad \mu \qquad (8.16)$$

and
$$W_t = 5.67 \times 10^{-12}\, T^4 \qquad \text{watts/cm}^2 \qquad (8.17)$$

Figure 8.26 shows the wavelength range over which 90 percent of the total power is found for various temperatures. Note that lower temperatures require measurement out to longer wavelengths.

While the concept of a blackbody is a mathematical abstraction, real physical bodies can be constructed to approximate closely blackbody behavior. Such radiation sources are needed for calibration of radiation thermometers and generally take the form of a blackened conical cavity of about 15° cone angle. The temperature is adjustable, automatically con-

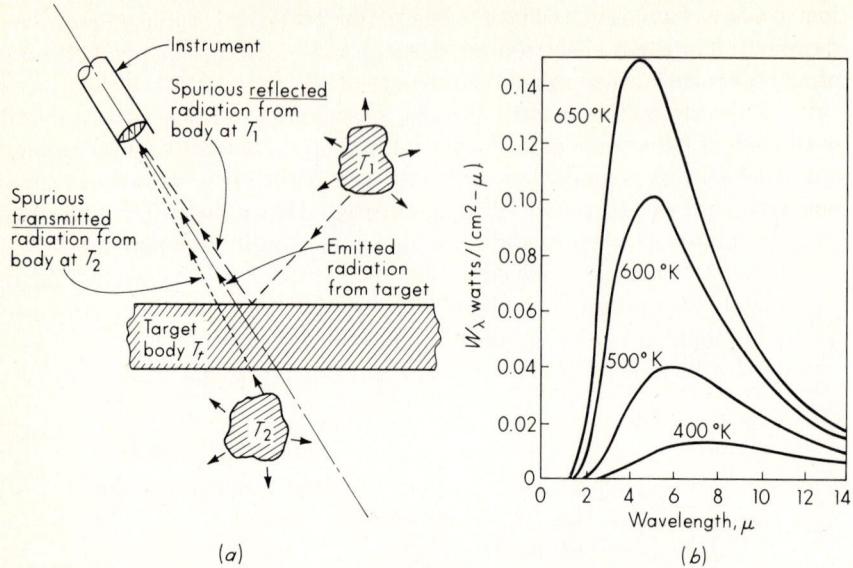

(a)

Fig. 8.25. (a) *Measurement problems with targets that are reflective and/or translucent* (b) *Blackbody radiation.*

trolled for constancy, and measured by some accurate sensor such as a platinum resistance thermometer. A typical unit[1] covers the range 500 to 1000°K with 1K° accuracy and emittance 0.99 ± 0.01 (blackbody has 1.00). While it is possible to construct a nearly perfect blackbody, the bodies whose temperatures are to be measured with some radiation-type

[1] Infrared Industries Inc., Riverside, Calif.

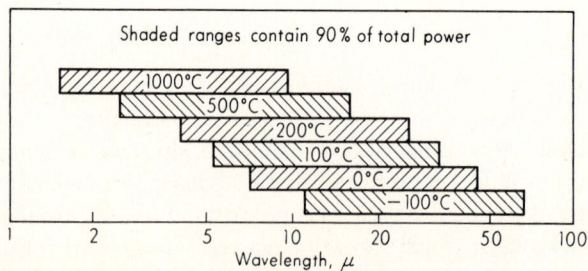

Fig. 8.26. *Power/wavelength distribution.*

instrument often deviate considerably from such ideal conditions. The deviation from blackbody radiation is expressed in terms of the emittance of the measured body. Several types of emittance have been defined to suit particular applications. The most fundamental form of emittance is the hemispherical spectral emittance $\epsilon_{\lambda,T}$. Let us call the *actual* hemispherical spectral radiant intensity of a real body at temperature T $W_{\lambda a}$ and assume it can be measured. Then $\epsilon_{\lambda,T}$ is defined as

$$\epsilon_{\lambda,T} \triangleq \frac{W_{\lambda a}}{W_\lambda} \qquad (8.18)$$

where W_λ is the blackbody intensity at temperature T. Emittance is thus dimensionless and always less than 1.0 for real bodies. In the most general case it varies with both λ and T. With the definition of Eq. (8.18), the radiation from a real body may be written as

$$W_{\lambda a} = \frac{C_1 \epsilon_{\lambda,T}}{\lambda^5 (e^{C_2/\lambda T} - 1)} \qquad (8.19)$$

Similarly, the total power W_{ta} of an actual body is given by

$$W_{ta} = C_1 \int_0^\infty \frac{\epsilon_{\lambda,T}}{\lambda^5 (e^{C_2/\lambda T} - 1)} \, d\lambda \qquad (8.20)$$

and if we assume that W_{ta} can be measured experimentally we may define the hemispherical total emittance $\epsilon_{t,T}$ by

$$\epsilon_{t,T} \triangleq \frac{W_{ta}}{W_t} \qquad (8.21)$$

where W_t is the blackbody total power at temperature T. Thus if $\epsilon_{t,T}$ is known, the total power of a real body is given by

$$W_{ta} = 5.67 \times 10^{-12} \, \epsilon_{t,T} T^4 \qquad \text{watts/cm}^2 \qquad (8.22)$$

If a body has $\epsilon_{\lambda,T}$ equal to a constant for all λ and at a given T, it is called a *graybody*. In this case we see that $\epsilon_{\lambda,T} \equiv \epsilon_{t,T}$. Also the curves of $W_{\lambda a}$ versus λ have exactly the same shape as for W_λ. Since many radiation thermometers operate in a restricted band of wavelengths, the hemispherical band emittance $\epsilon_{b,T}$ has been defined by

$$\epsilon_{b,T} \triangleq \frac{\int_{\lambda_a}^{\lambda_b} [\epsilon_{\lambda,T}/\lambda^5 (e^{C_2/\lambda T} - 1)] \, d\lambda}{\int_{\lambda_a}^{\lambda_b} [1/\lambda^5 (e^{C_2/\lambda T} - 1)] \, d\lambda} \qquad (8.23)$$

This is seen to be just the ratio of the total powers, actual and blackbody, within the wavelength interval λ_a to λ_b for bodies at temperature T. If the actual power can be measured directly, $\epsilon_{b,T}$ can be found without knowing $\epsilon_{\lambda,T}$. For a graybody, $\epsilon_{b,T} \equiv \epsilon_{\lambda,T}$.

If a radiation thermometer has been calibrated against a blackbody source, knowledge of the appropriate emittance value allows correction

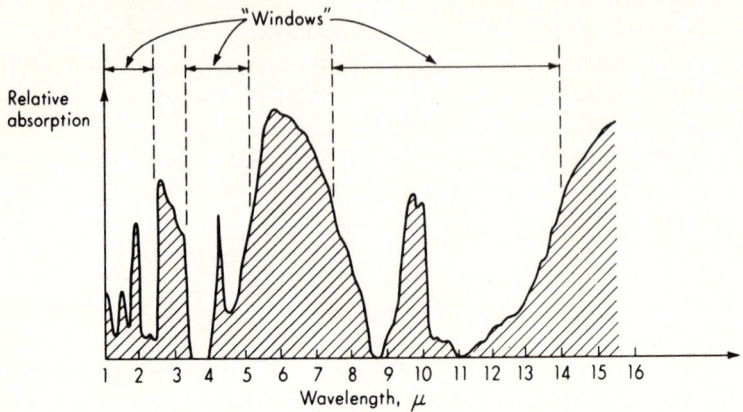

Fig. 8.27. *Atmospheric absorption.*

of its readings for nonblackbody measurements.   Unfortunately, emittances are not simple material properties such as densities but rather depend on size, shape, surface roughness, angle of viewing, etc.   This leads to uncertainties in the numerical values of emittances, which are one of the main problems in radiation-temperature measurement.

Another source of error is the losses of energy in transmitting the radiation from the measured object to the radiation detector.   Generally the optical path consists of some gas (often atmospheric air) and various windows, lenses, or mirrors used to focus the radiation or protect sensitive elements from the environment.   In atmospheric air the attenuation of radiation is due mainly to the resonance-absorption bands of water vapor, carbon dioxide, and ozone and the scattering effect of dust particles and water droplets.   The combined absorption effect of H_2O, CO_2, and O_3 is roughly as shown in Fig. 8.27.   Since the absorption varies with wavelength, a radiation thermometer can be designed to respond only within one of the "windows" shown, thus making it insensitive to these effects. Since the absorption varies with the thickness of the gas traversed by the radiation, the effect is not an instrument constant and cannot thus be calibrated out.   The lenses used in infrared instruments must often be made of special materials, since glasses normally used for the visible spectrum are almost opaque to radiation of wavelength longer than about 2 μ.   Figure 8.28 shows the variation of transmission factor of various materials with wavelength.   While infrared radiation follows the same optical laws used for lens and mirror design as visible light, some materials useful for infrared wavelengths (arsenic trisulfide, for example) are opaque to visible-light wavelengths.

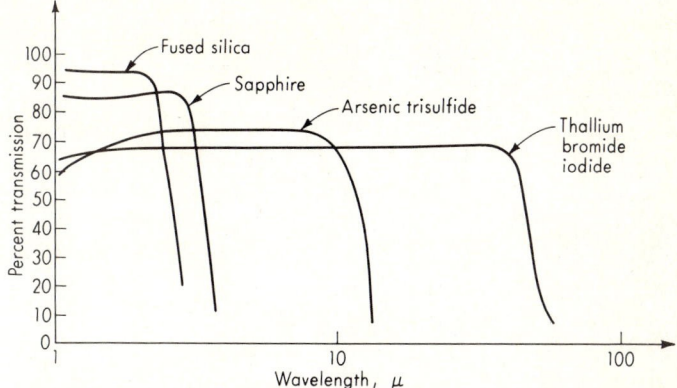

Fig. 8.28. *Optical-material spectral transmission.*

Radiation detectors. In all radiation thermometers (other than the disappearing-filament optical pyrometer) the radiation from the measured body is focused on some sort of radiation detector which produces an electrical signal.   Detectors may be classified as thermal detectors or photon detectors.   Thermal detectors are blackened elements designed to absorb a maximum of the incoming radiation at all wavelengths.   The absorbed radiation causes the temperature of the detector to rise until an equilibrium is reached with heat losses to the surroundings.   Thermal detectors actually measure this temperature, using a resistance thermometer, thermistor, or thermocouple (thermopile) principle.

Resistance-thermometer and thermistor elements are made in the form of thin films or flakes and are called bolometers.   Performance criteria for both thermal and photon detectors include the time constant (most detectors behave roughly as first-order systems), the responsivity (volts of signal per watt of incident radiation), and the noise-equivalent power (N.E.P.).   The noise-equivalent power gives an indication of the smallest amount of radiation that can be detected, which is limited by the inherent electrical noise level of the detector.   That is, with no radiation whatever coming in, the detector still puts out a small random voltage due to various electrical noise sources within the detector itself.   The amount (watts) of incoming radiation required to produce a signal just equal in strength to the noise (signal/noise ratio of 1) is called the noise-equivalent power.   A low value of noise-equivalent power is thus desirable.   An evaporated-nickel-film bolometer of about 35 mm^2 area has a resistance of about 100 ohms, a time constant of 0.004 sec, responsivity of 0.4 volt/watt, and noise-equivalent power of 3×10^{-9} watt.   A thermistor bolometer of 0.5 mm^2 area might have 3 megohms resistance,

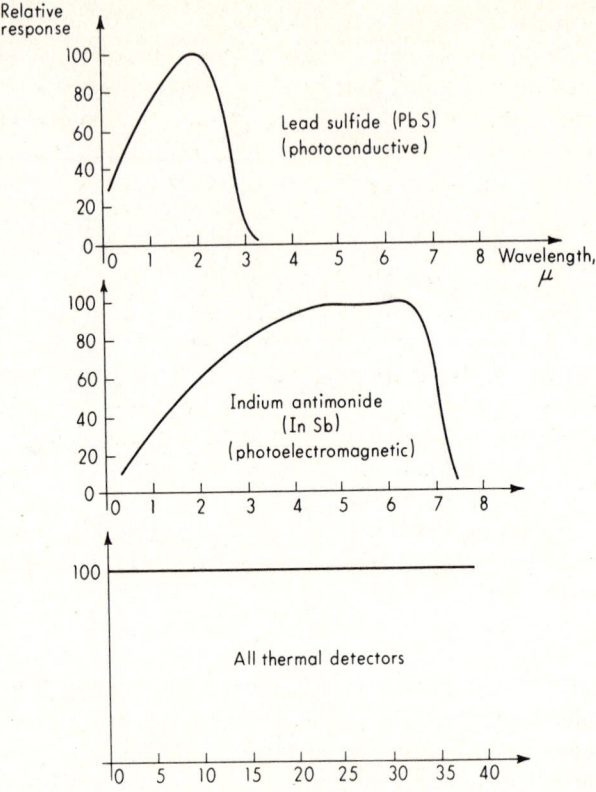

Fig. 8.29.   *Radiation-detector spectral sensitivity.*

a time constant of 0.004 to 0.030 sec, responsivity of 700 to 1,200 volts/watt, and noise-equivalent power of 2×10^{-10} watt. Thermopiles of an area from 0.4 to 10 mm² have resistance of the order of 10 to 100 ohms, time constants in the range 0.005 to 0.3 sec, responsivity of 3 to 90 volts/watt, and noise-equivalent power of 2×10^{-11} to 7×10^{-10}.

In the various types of photon detectors the incoming radiation (photons) frees electrons in the detector structure and produces a measurable electrical effect. These events occur on an atomic or molecular time scale rather than on the gross time scale involved in the heating and cooling of thermal detectors. A much higher response speed is thus possible. However, photon detectors have a sensitivity that varies with wavelength; thus incoming radiation of different wavelengths is not equally treated. Typical spectral response of some common types is shown in Fig. 8.29. Photon detectors commonly in use operate in the photoconductive, photovoltaic, or photoelectromagnetic (PEM) modes.

Photoconductive types exhibit an electrical resistance that changes

Two matched bolometer
elements

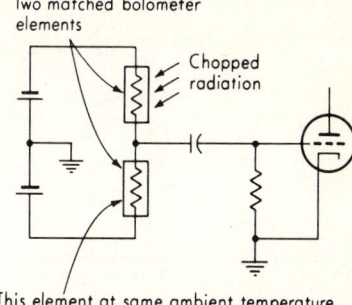

Chopped
radiation

This element at same ambient temperature
but shielded from radiation to provide
ambient-temperature compensation

Simple circuit for a high-resistance bolometer

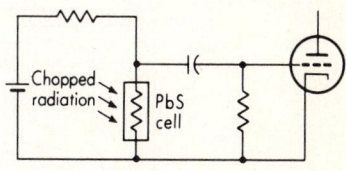

Chopped
radiation

PbS
cell

Fig. 8.30. *Basic detector circuits.*   Simple circuit for a lead sulfide cell

with the incoming radiation level.   Photovoltaic cells, also called barrier photocells, employ a photosensitive barrier of high resistance, deposited between two layers of conducting material.   A potential difference between these two layers is built up when the cell is exposed to radiation. In photoelectromagnetic detectors the Hall effect is utilized.   A semi-conductor crystal is subjected to a strong magnetic field and radiation applied to one side.   A potential difference is developed across the ends of the crystal.   Lead sulfide photoconductive cells are by far the most used type, typical units of 1 to 35 mm² area having resistances of 10^5 to 2×10^6 ohms, time constants of 2 to 0.04 msec, responsivity of 5,000 to 150,000 volts/watt, and noise-equivalent power of 4×10^{-11} to 4×10^{-12} watt.   An indium antimonide photoelectromagnetic cell of 100 ohms resistance may have a time constant less than 1 μsec, responsivity of 1 volt/watt, and noise-equivalent power of 10^{-9} watt.

Some type of circuit must be employed to realize a usable electrical signal (generally a voltage) from a radiation detector.   Thermopile devices generally work with an uninterrupted stream of radiation and require no circuitry other than the usual reference junction, which is commonly left at ambient temperature.   Bolometers and photoconductive cells often employ a chopper to interrupt the radiation at a fixed rate of the order of several hundred cycles per second.   This leads to an a-c-type electrical signal and allows use of high-gain a-c amplifiers. Typical circuits for such arrangements are shown in Fig. 8.30.

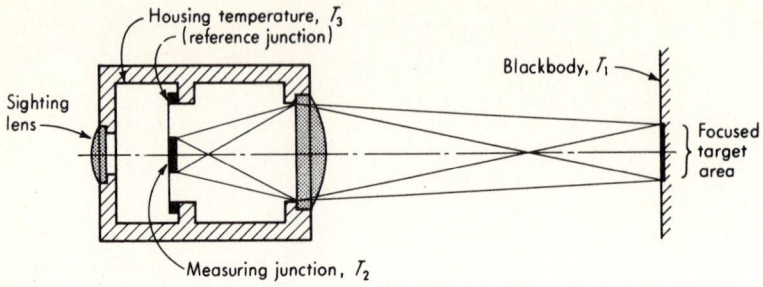

Lens-type radiation thermometer

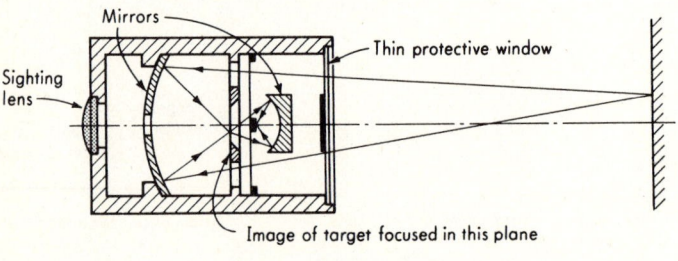

Mirror-type radiation thermometer

Fig. 8.31. *Lens- and mirror-type radiation thermometers.*

Unchopped (d-c) broadband radiation thermometers. We begin our study of complete radiation-sensing instruments by consideration of the most common type used in day-to-day industrial applications.[1] These instruments use a blackened thermopile as detector and focus the radiation by means of either lenses or mirrors. Figure 8.31 shows in simplified fashion the construction of this class of instruments. The reference of footnote 1 gives a very complete analysis of such devices.

Basically, for a given source temperature T_1, the incoming radiation heats the measuring junction until conduction, convection, and radiation losses just balance the heat input. The measuring-junction temperature is usually less than 40°C above its surroundings even if the source is incandescent. An oversimplified analysis gives

$$\text{Heat loss} = \text{radiant heat input}$$
$$K_1(T_2 - T_3) = K_2 T_1^4 \qquad (8.24)$$

[1] T. R. Harrison, "Radiation Pyrometry and Its Underlying Principles of Radiant Heat Transfer," John Wiley & Sons, Inc., New York, 1960.

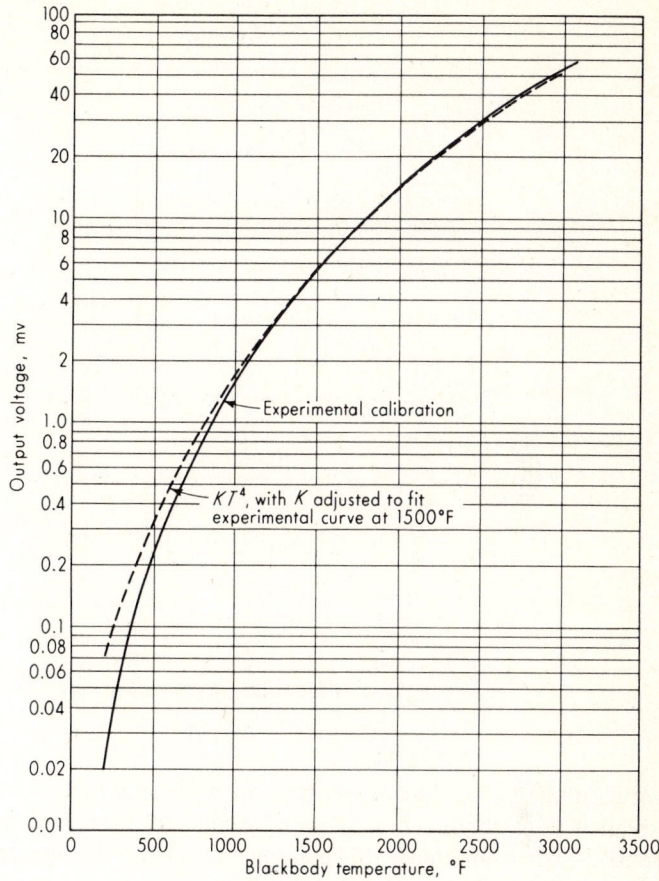

Fig. 8.32.   *Theoretical and experimental calibration curves.*

If the thermocouple voltage is proportional to $T_2 - T_3$, the voltage output should be proportional to T_1^4.   Figure 8.32 shows an actual calibration curve of such an instrument together with the ideal relationship.   For high temperatures the agreement is quite close.   The temperatures T_2 and T_3 are both influenced by the environmental temperature; thus compensation must generally be provided for this, particularly if the instrument is intended to measure low temperatures.   This compensation may include a thermostatically controlled housing temperature.

The thermopiles used may have from 1 or 2 to 20 or 30 junctions. A small number of junctions has less mass and thus faster response, but lower sensitivity limits application to high temperatures.   Response is

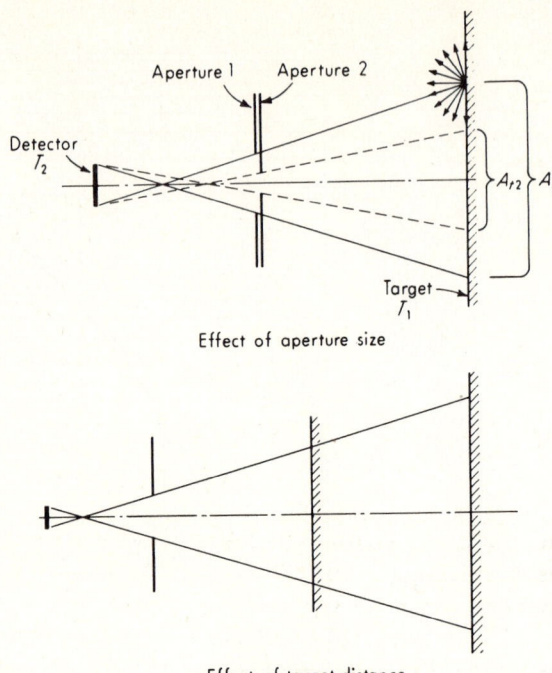

Effect of aperture size

Effect of target distance

Fig. 8.33. *Effect of aperture size and target distance.*

roughly that of a first-order system, with time constants ranging from about 0.1 sec (high-temperature systems) to 2 sec (low-temperature systems). Instruments of this class are available to measure temperatures as low as 0°F; that is, the thermopile is actually cooler than the ambient temperature and reads a negative voltage. Theoretically there is no upper limit to the temperatures that can be measured in this way. Commercial instruments usable to 3200°F are readily available.

Conceivably, an instrument of the above type could be constructed with no focusing means, i.e., no lens or mirror. A simple diaphragm with a circular aperture (Fig. 8.33) would define the target from which radiation is received. To define smaller target areas for a given target distance, a smaller aperture could be used; however, there would be a proportionate loss in incoming radiation and thus sensitivity. The reading of such an instrument is independent of the distance between the target and the instrument, since the amount of radiation received is limited by the solid angle of the cone defined by the aperture and detector and this is always the same. However, as the target distance increases, the target area necessary to fill the cone increases. If, because of non-

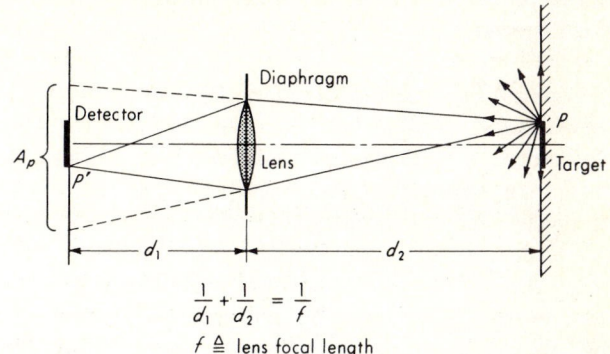

$$\frac{1}{d_1} + \frac{1}{d_2} = \frac{1}{f}$$

$f \triangleq$ lens focal length

Fig. 8.34. *Advantages of focusing.*

uniformity of target temperature or small target size, one wishes to restrict the target area to small values, a very small aperture is required, giving very low sensitivity.

The basic purpose of lens or mirror systems is to overcome this restriction, thus allowing the resolution of small targets without loss of sensitivity. Thus, in Fig. 8.34, radiation emanating from the point P on the target is focused on the corresponding point P' of the detector. If a simple diaphragm, rather than a focused lens, had been used, the same amount of energy would be spread out over the area A_p, with the detector receiving only a fraction of the radiation. Commercial lens- or mirror-type instruments generally have parameters such that targets 2 ft or more distant are adequately focused with a fixed-focus lens, and the instrument calibration is independent of distance as long as the target fills the field of view. Minimum target size to fill the field of view is of the order of one-twentieth of the target distance for common instruments. For targets closer than 2 ft, focusing is necessary and may affect the calibration, depending on instrument construction and closeness of target. Target diameters of 0.1 to 0.3 in. at target distances of 4 to 12 in. are available. Since the focal length of a lens depends on the index of refraction, which in turn varies with wavelength, all wavelengths are not focused at the same point. In particular, if one focuses a lens visually (using visible light), the longer infrared wavelengths, which contribute a large portion of the total energy at lower temperatures, will be out of focus. Such an instrument may thus have to be focused by adjusting for maximum thermopile output rather than for sharpest visual definition. Another effect of lenses is selective transmission, as shown in Fig. 8.28. The use of mirrors rather than lenses is an attempt to alleviate some of these problems. However, instruments of both types are widely used with success.

Chopped (a-c) broadband radiation thermometers. A number of advantages accrue when the radiation coming from the target to the detector is periodically interrupted (chopped) at a fixed frequency; therefore many infrared systems employ this technique. When high sensitivity is needed, amplification is required, and high-gain a-c amplifiers are easier to construct than their d-c counterparts. This is usually the main reason for using choppers. Additional benefits related to ambient-temperature compensation and reference-source comparison may also be obtained. Systems employing thermal (broadband) and photon (restricted-band) detectors and choppers are in common use. We here consider those using thermal detectors.

The time constants of adequately sensitive thermopile detectors are generally too long to allow efficient use of chopping; thus the faster bolometers, usually the thermistor type, are employed. We shall consider two specific forms of this class of instruments: the blackened-chopper type and the mirror-chopper type. Figure 8.35 shows the basic elements of a blackened-chopper radiometer. A mirror focuses the target radiation on the detector; however, this beam is interrupted periodically by the chopper rotating at constant speed. Thus the detector alternately "sees" radiation from the target and the radiation from the chopper's blackened surface. For high target temperatures, sufficient accuracy may be achieved by leaving the chopper temperature at ambient. Higher accuracy, particularly at low target temperatures, is obtained by thermostatically controlling the chopper temperature. An arrangement similar to that of Fig. 8.30 is used for the detector circuit. The output voltage of the detector circuit is essentially as shown in Fig. 8.35. By amplifying this in an a-c amplifier the mean value (which is subject to drift) is discarded, and only the difference between the target and chopper radiation levels is amplified. If the chopper radiation level is considered as a known reference value, the target radiation and thus its temperature may be inferred. To provide a d-c output signal related to target temperature and suitable for recording or control purposes the a-c amplifier is followed by a phase-sensitive demodulator and filter circuit. The necessary synchronizing signal for the demodulator may be generated by placing a magnetic proximity pickup near the chopper blades. While the response time of the detector itself may be of the order of a millisecond, the chopper frequency and necessary demodulator filter time constant greatly reduce the overall system speed. High chopper speeds allow faster overall response but reduce sensitivity if the detector time constant is too large, since the detector does not have time to reach equilibrium during the time either the target or chopper is in view.

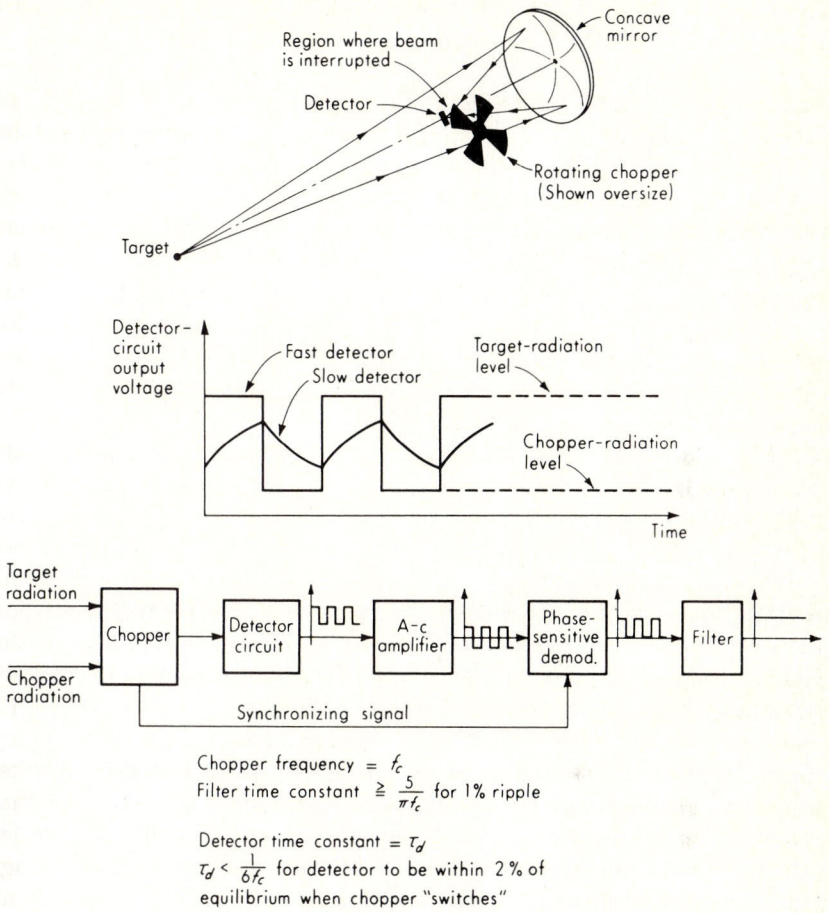

Fig. 8.35. *Blackened-chopper system.*

A typical instrument[1] uses a square thermistor detector, thus giving rise to a square field of view of size 1° by 1°. Sighting and focusing in the range 2 ft to infinity are accomplished with an attached optical telescope. The standard chopping frequency is 180 cps, leading to an overall system time constant of about 0.008 sec. Standard temperature range is from ambient to 1300°C.

In the mirror-chopper instrument[2] of Fig. 8.36 two thermistor detectors are used. Also, an accurate blackbody source whose temperature is automatically controlled and accurately measured (say by a

---

[1] Servo Corp. of America, New Hyde Park, N.Y.
[2] Barnes Engineering Co., Stamford, Conn.

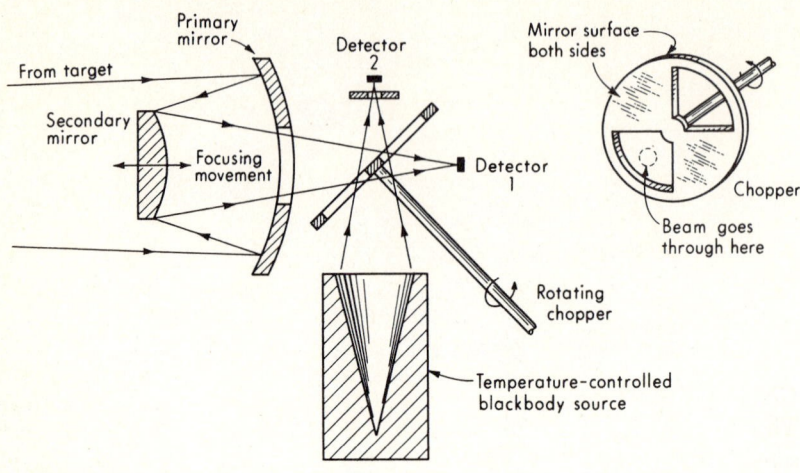

Fig. 8.36.   *Mirror-chopper system.*

thermocouple) is provided within the radiometer.   A chopper operating at 77 cps alternately exposes each of the two detectors to the target radiation and the blackbody source of known temperature.   In Fig. 8.36 detector 1 is receiving target radiation while 2 is receiving blackbody. When the chopper disk rotates 90° so that a solid sector is in the line of sight the number 1 detector receives blackbody radiation reflected from the rear mirror surface of the chopper while number 2 receives target radiation reflected from the front surface.   Circuitry similar to that of Fig. 8.35 can be employed to develop a d-c output signal related to target temperature.   Such a system can be used for very accurate static measurements by using it in a null method of operation.   The blackbody source temperature is adjusted until no output is obtained.   Then the target and blackbody are at identical temperatures, provided the target emittance is 1.0.   If the emittance is not 1.0 but is known, a correction may be applied.   Such a null method makes the reading independent of detector sensitivity (which may drift) and amplifier gain.   A system of this type employing 1.5- by 1.5-mm square thermistor detectors has a 0.5 by 0.5° field of view (1-in.-square target at 10 ft), an overall system time constant of 0.016 sec (bandwidth 10 cps), and will detect (signal/noise ratio = 1.0) a target temperature change of 0.4C°.   By sacrificing response speed for sensitivity, heavier filtering can be switched in to give a time constant of 0.16 or 1.6 sec, with corresponding increase of resolution to 0.14 and 0.04C°.   The instrument can be focused on targets from 4 ft to infinity.

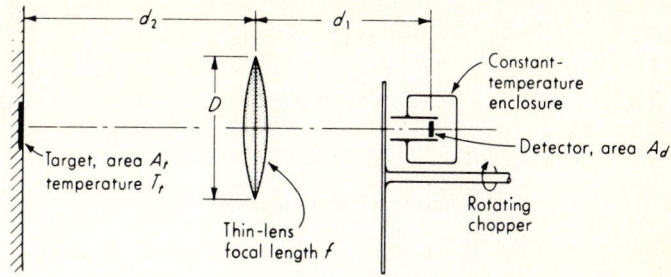

Fig. 8.37.   *Photon-detector system.*

Chopped (a-c) selective band (photon) radiation thermometers.
The use of photon detectors allows faster response speeds and may reduce
sensitivity to ambient temperature. Since such detectors respond
directly to the incident photon flux, rather than detecting a temperature
change, the disturbing influence of ambient-temperature changes is
restricted mainly to changes in the responsivity of the detector. When
radiation is expressed in terms of photon flux rather than watts, the
formulas are somewhat modified. Equation (8.15) becomes

$$N_\lambda = \frac{2\pi c}{\lambda^4 (e^{C_2/\lambda T} - 1)} \qquad (8.25)$$

where $N_\lambda \triangleq$ hemispherical spectral photon flux, photons/(cm²-sec-μ)
$c \triangleq$ speed of light, 3×10^{10} cm/sec

The peak of the photon-flux curves occurs at a different wavelength from
that of the radiant intensity. It is given by

$$\lambda_{p,p} = \frac{3,669}{T} \qquad \text{microns} \qquad (8.26)$$

The total photon flux for all wavelengths is

$$N_t = 1.52 \times 10^{11} T^3 \qquad \text{photons/(cm²-sec)} \qquad (8.27)$$

Figure 8.37 shows the basic arrangement of an instrument[1] using a
photon detector and a chopper. Basic optics gives

$$\frac{1}{d_1} + \frac{1}{d_2} = \frac{1}{f} \qquad (8.28)$$

and

$$A_t = \frac{A_d d_2^2}{d_1^2} \qquad (8.29)$$

[1] Infrared Thermometry, Infrared Industries, Santa Barbara, Calif.

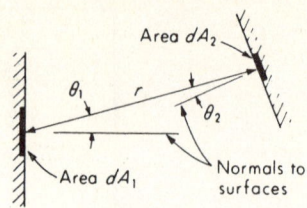

Fig. 8.38.   *Basic radiation configuration.*

Combining these gives

$$A_t = \frac{A_d(d_2 - f)^2}{f^2} \qquad (8.30)$$

Now if $d_2 \gg f$, we get approximately

$$A_t = \frac{A_d d_2^2}{f^2} \qquad (8.31)$$

If the detector is a square with side L_d, the resolved target will be a square of side $L_d d_2/f$.   For example, a detector 1 mm square used with a lens with a focal length of 75 mm requires a target of size $d_2/75$ to fill exactly the field of view.

For the general configuration of Fig. 8.38, basic radiation laws[1] give

$$\frac{\text{Radiation incident on } dA_2}{dA_2} = \frac{\cos \theta_1 \cos \theta_2}{\pi r^2} \; (\text{radiation emitted from } dA_1)$$

$$(8.32)$$

We can apply this to the configuration of Fig. 8.37 to find the radiation received over the area of the lens from the target.   If d_2 is large compared with the size of target and lens (usually true), $\cos \theta_1 \approx \cos \theta_2 \approx 1$, and dA_1 and dA_2 may be replaced by A_1 and A_2 in Eq. (8.32).   To simplify the analysis, let us assume that the detector has uniform spectral response from $\lambda = 0$ to $\lambda = 6.9$ and zero response beyond and that the target is a graybody with emittance ϵ.   The radiation emitted by the target is then

$$A_t \epsilon \int_0^{6.9} N_\lambda \, d\lambda = A_t \epsilon E(T_t) \qquad \text{photons/sec} \qquad (8.33)$$

where
$$E(T_t) \triangleq \int_0^{6.9} N_\lambda \, d\lambda \qquad \text{a function of } T_t$$

The radiation received at the lens is

$$\frac{A_t \epsilon E(T_t)(\pi D^2/4)}{\pi d_2^2} = \frac{A_t \epsilon E(T_t) D^2}{4 d_2^2} \qquad (8.34)$$

If the lens transmits $100 K_{tr}$ percent of the flux incident on it (perfect transmission has $K_{tr} = 1.0$) and if the system is focused so that the target

[1] A. I. Brown and S. M. Marco, "Introduction to Heat Transfer," p. 237, McGraw-Hill Book Company, New York, 1951.

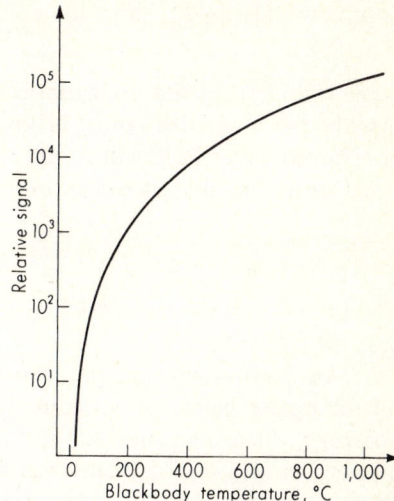

Fig. 8.39. *Indium antimonide system-response curve.*

image just fills the detector area, the photon flux density at the detector is given by

$$N_d \triangleq \frac{K_{tr}A_t\epsilon E(T_t)D^2}{4d_2{}^2A_d} \qquad \text{photons/(cm}^2\text{-sec)} \qquad (8.35)$$

For the indium antimonide photoelectromagnetic detector the output voltage is proportional to N_d. Since for a focused target $A_t/A_d d_2{}^2 = 1/f^2$, Eq. (8.35) can be written as

$$N_d = \frac{K_{tr}D^2}{4f^2}\,\epsilon E(T_t) \qquad (8.36)$$

The output voltage of the overall instrument is $K_{dr}K_aN_d$, where $K_{dr} \triangleq$ detector responsivity, volts/[photons/(cm^2-sec)], and K_a is the amplifier gain, volts/volt. Thus

$$\text{Instrument output voltage} \triangleq e_o = \frac{K_{dr}K_aK_{tr}D^2}{4f^2}\,\epsilon E(T_t) \qquad (8.37)$$

Note that the first factor in Eq. (8.37) is a constant of the instrument while the second [$\epsilon E(T_t)$] is a function of target temperature and emittance only. Also, as long as the target is focused, the reading is independent of the distance from the target. The variation of $E(T_t)$ with T_t for a black-body target can be found by experimental calibration of the overall instrument. Its general shape is shown in Fig. 8.39. Since Eq. (8.27) shows that the total photon flux varies as T^3 and since the detector output is roughly proportional to total flux, the instrument output signal varies approximately as T^3; thus, for a graybody

$$e_o \approx K\epsilon T^3 \qquad (8.38)$$

Since the instruments are calibrated against blackbody sources, for non-blackbodies a value of ϵ must be known to find T.   If the value of ϵ used is in error, an error in T will result.   Because of the third power law, however, errors in ϵ do not cause proportionate errors in T.   Rather

$$\frac{T_{\text{actual}}}{T_{\text{assumed}}} = \left(\frac{\epsilon_{\text{assumed}}}{\epsilon_{\text{actual}}}\right)^{\frac{1}{3}} \qquad (8.39)$$

Thus if we assume $\epsilon = 0.8$ when it is really 0.6, the temperature error is only 10 percent.

An instrument[1] of the above general class which, however, uses mirror optics has a range from 100 to 2000°F (8000°F with calibrated aperture), focusing range 4 ft to infinity (18 to 60 in. optional), 0.5° field of view, output signal 10 mv full scale, and an overall time constant (in chopped mode) of 2, 20, or 200 msec.   For the study of very rapid transients, the chopper may be turned off and the system operated with just the detector and a-c amplifier.   Transients as brief as 10 μsec may thus be measured.   Other models of this manufacturer, using lens optics and lead sulfide detectors, have target sizes as small as 0.009 in. at a target distance of 2.8 in.

Another instrument[2] of this class accepts radiation only in the wavelength band 4.8 to 5.6 μ.   This band avoids the absorption bands of atmospheric water vapor and carbon dioxide, thus removing the effect of these variables on instrument response.   The measurement of the surface temperature of glass is also facilitated since in this spectral range the emittance of glass is high and independent of thickness.   This instrument has a range of 100 to 1000°F, target size equal to its distance divided by 57, time constant 0.2 to 0.5 sec, focusing range 17 in. to infinity, calibration accuracy 2 percent of range or 20°F (whichever is larger), resolution 0.25 percent of range or 3°F (whichever is larger), and repeatability of 0.5 percent of range or 10°F (whichever is larger).

Monochromatic-brightness radiation thermometers (optical pyrometers).   The classical form of this type of instrument is the disappearing-filament optical pyrometer.   It is the most accurate of all the radiation thermometers; however, it is limited to temperatures greater than about 700°C since it requires a visual brightness match by a human operator.   This instrument is used to realize the International Practical Temperature Scale above 1063°C.

---

[1] Infrared Industries, Santa Barbara, Calif.
[2] Ircon, Inc., Chicago, Ill.

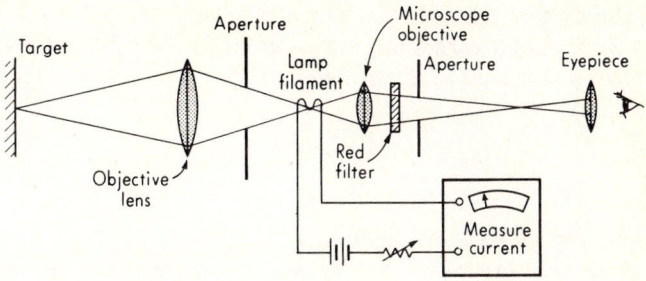

Fig. 8.40. *Disappearing-filament optical pyrometer.*

Monochromatic-brightness thermometers utilize the principle that, at a given wavelength λ, the radiant intensity ("brightness") varies with temperature as given by Eq. (8.15). In a disappearing-filament instrument (Fig. 8.40) an image of the target is superimposed on a heated tungsten filament. This tungsten lamp, which is very stable, has been previously calibrated so that when the current through the filament is known the brightness temperature of the filament is known. (Such a calibration is basically obtained by visually comparing the brightness of a blackbody source of known temperature with that of the tungsten lamp.) A red filter which passes only a narrow band of wavelengths around 0.65 μ is placed between the observer's eye and the tungsten lamp and target image. The observer controls the lamp current until the filament disappears in the superimposed target image. Then the brightness of the target and lamp are equal, and one can write

$$\frac{\epsilon_{\lambda_e} C_1}{\lambda_e{}^5 (e^{C_2/\lambda_e T_t} - 1)} = \frac{C_1}{\lambda_e{}^5 (e^{C_2/\lambda_e T_L} - 1)} \qquad (8.40)$$

where $\epsilon_{\lambda_e} \triangleq$ emittance of target at wavelength λ_e
    $\lambda_e \triangleq$ effective wavelength of filter, usually 0.65 μ
    $T_t \triangleq$ target temperature
    $T_L \triangleq$ lamp brightness temperature

For T less than about 4000°C, the terms $e^{C_2/\lambda_e T}$ are much greater than 1, allowing Eq. (8.40) to be simplified to

$$\frac{\epsilon_{\lambda_e}}{e^{C_2/\lambda_e T_t}} = \frac{1}{e^{C_2/\lambda_e T_L}} \qquad (8.41)$$

Then
$$\epsilon_{\lambda_e} = e^{-(C_2/\lambda_e)(1/T_L - 1/T_t)} \qquad (8.42)$$

and finally
$$\frac{1}{T_t} - \frac{1}{T_L} = \frac{\lambda_e \ln \epsilon_{\lambda_e}}{C_2} \qquad (8.43)$$

If the target is a blackbody ($\epsilon_{\lambda_e} = 1.0$), there is no error since $\ln \epsilon_{\lambda_e} = 0$ and $T_t = T_L$. If ϵ_{λ_e} is not 1.0 but is known, Eq. (8.43) allows calculation of the needed correction.   The errors caused by inexact knowledge of ϵ_{λ_e} for a particular target are not as great for an optical pyrometer as for an instrument sensitive to a wide band of wavelengths.   The percent error is given by

$$\frac{dT_t}{T_t} = -\frac{\lambda_e T_t}{C_2}\frac{d\epsilon_{\lambda_e}}{\epsilon_{\lambda_e}} \qquad (8.44)$$

Thus, for a target at 1000°K, a 10 percent error in ϵ_{λ_e} results in only a 0.45 percent error in T_t.   The use of a monochromatic red filter aids the operator in matching the brightness of target and lamp since color effects are eliminated.   Also, the target emittance need be known only at one wavelength.   If ϵ_{λ_e} is exactly known, temperatures can be measured with optical pyrometers with errors of the order of 3C° at 1000°C, 6C° at 2000°C, and 40C° at 4000°C.   With special optical systems, targets as small as 0.001 in. can be measured at distances of 5 or 6 in.

Because of its manual null-balance principle, the optical pyrometer is not usable for continuous-recording or automatic-control applications. To overcome this drawback, automatic brightness pryometers[1,2,3] have been developed.   One model[3] of such a device uses a mirror-chopper arrangement to produce a square wave of radiation flux in which the target radiation and standard lamp radiation are alternately applied to a photomultiplier tube.   The standard lamp is left at a fixed brightness; thus a null method is not used.   A red filter with $\lambda_e = 0.653\ \mu$ is employed. This instrument has a range of 700 to 3000°C (though not in a single instrument), inaccuracy of 1 percent of span plus 0.3 percent of measured temperature, repeatability 0.3 percent of measured temperature, resolution 1°F, time constant 0.3 sec, target size 0.4 in. at 18 in., and a recorder output of 50 or 100 mv full scale.   The above specifications refer to an instrument for commercial use; refined models for standards laboratories are expected to exceed the performance of manual pyrometers shortly and will probably replace them for the most accurate work.

Another class of brightness pyrometer which does not use a reference lamp source is also available.   They are merely chopper-type radiation thermometers using photon detectors and narrow-band optical filters. One such instrument[4] has a range of 1400 to 8300°F, 0.5° field of view, focusing range 24 in. to infinity, time constant 0.1 sec (0.01 sec optional), repeatability 0.25 percent of span, inaccuracy 1 percent of span, and

---

[1] S. Ackerman and J. S. Lord, Automatic Brightness Pyrometer Uses a Photomultiplier "Eye," *ISA J.*, p. 48, December, 1960.

[2] J. S. Lord, Brightness Pyrometry, *Instr. Control Systems*, p. 109, February, 1965.

[3] Instrument Development Laboratories, Attleboro, Mass., *Bull.* 614.

[4] Infrared Industries, Santa Barbara, Calif.

recorder output 10 to 100 mv full scale.   This instrument uses a filter centered at 0.80 μ with a bandwidth of 0.06 μ.

Two-color radiation thermometers. Since errors due to inaccurate values of emittance are a problem in all radiation-type temperature measurements, considerable attention has been given to possible schemes for alleviating this difficulty.   Although no universal solution has been found, the two-color concept has met with some practical success.   The basic concept requires that W_λ be determined at two different wavelengths and then the ratio of these two W_λ's be taken as a measure of temperature.   For the usual conditions of practical application, the terms $e^{C_2/\lambda T}$ are much greater than 1.0, and we may write with close approximation

$$W_{\lambda 1} = \frac{\epsilon_{\lambda_1} C_1}{\lambda_1{}^5 e^{C_2/\lambda_1 T}} \qquad (8.45)$$

$$W_{\lambda 2} = \frac{\epsilon_{\lambda_2} C_1}{\lambda_2{}^5 e^{C_2/\lambda_2 T}}$$

Then

$$\frac{W_{\lambda 1}}{W_{\lambda 2}} = \frac{\epsilon_{\lambda_1}}{\epsilon_{\lambda_2}} \left(\frac{\lambda_2}{\lambda_1}\right)^5 e^{(C_2/T)(1/\lambda_2 - 1/\lambda_1)} \qquad (8.46)$$

For a graybody, $\epsilon_{\lambda_1} = \epsilon_{\lambda_2}$; thus

$$\frac{W_{\lambda 1}}{W_{\lambda 2}} = \left(\frac{\lambda_2}{\lambda_1}\right)^5 e^{(C_2/T)(1/\lambda_2 - 1/\lambda_1)} \qquad (8.47)$$

and we see that the ratio $W_{\lambda 1}/W_{\lambda 2}$ is independent of emittance as long as it is numerically the same at λ_1 and λ_2.   The wavelengths λ_1 and λ_2 are usually both in the visible range and are generally varied depending on the temperature range of the particular instrument.   In one commercial instrument[1,2] the two filters are mounted on a rotating wheel so that the incoming radiation passes alternately through each on its way to a photon detector.   Special electronic circuitry performs operations equivalent to taking the ratio of $W_{\lambda 1}$ and $W_{\lambda 2}$.   Instruments covering the range 1400 to 4000°F are available as standard models.   A recorder output of 100 mv full scale is provided.

8.6

A Digital Temperature-sensing System   While the analog outputs of the various temperature sensors described up to this point can be transformed to digital form by using any of several available analog-to-

---

[1] T. P. Murray and V. G. Shaw, Two-color Pyrometry in the Steel Industry, *ISA J.*, p. 36, December, 1958.

[2] Shaw Instrument Corp., Pittsburgh, Pa.

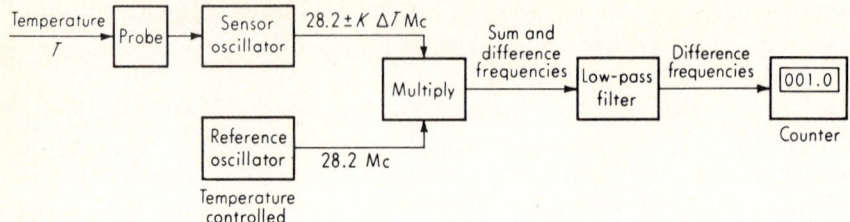

Fig. 8.41. *Digital thermometer.*

digital conversion techniques, we here describe briefly a system available as a package which accepts temperature inputs and provides digital signals and/or displays as outputs.

Electronic oscillators using piezoelectric quartz crystals as the resonant element that establishes the frequency of oscillation have been widely used for many years. For the most critical applications it has been necessary to place the crystal in a temperature-controlled oven, since the natural frequency of the crystal varies with temperature, causing drifts in oscillator frequency. This difficulty is turned to good advantage in the quartz thermometer where the crystal is placed in a probe which serves as a temperature-sensing element. Changes in probe temperature cause a frequency change in proportion. By applying the oscillator voltage to an electronic counter for a definite time interval, a direct digital reading of temperature is obtained.

A commercial version[1] of the above principle has a block diagram as in Fig. 8.41. The crystals used have a temperature coefficient of 35.4 ppm/C° which is constant within ±0.05 percent of range over the span −40 to +230°C, giving a very linear response. If the sensor oscillator is designed for a frequency of 28.2 Mc when the probe is at 0°C, a 1C° change in temperature will cause a frequency change of 1,000 cps. Were this frequency change applied to a four-digit electronic counter for a sample period of 0.01 sec, the counter would read 001.0°C, and such an arrangement would have a resolution of 0.1C°. To obtain the frequency *change* as a usable signal, the sensor oscillator signal is multiplied with the signal of a 28.2-Mc reference oscillator which is temperature-controlled and thus of fixed frequency. The output of the multiplier (amplitude modulator) contains the sum and difference frequencies of the two input signals. By filtering out the high-frequency (sum) component, only the difference frequency signal remains and may be sent to the counter.

Probes used in the above system are ⅜ in. in diameter and either ¾ or

[1] Hewlett-Packard Co., Dymec Div., Palo Alto, Calif.

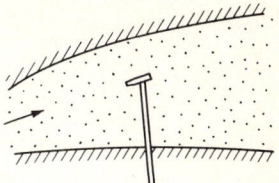

Fig. 8.42.   *Probe configuration.*

9 in. long, with a time constant of 1 sec in water with a velocity of 2 fps. System resolution can be increased to 0.0001C° by increasing the sampling period to 10 sec. Differential temperature can be measured by multiplying the signals from two probes rather than one probe and the reference oscillator. Self-heating of the probe is 10 μw and gives less than 0.01C° error in water at 2 fps. A binary-coded-decimal output signal is also available.

8.7

Temperature-measuring Problems in Flowing Fluids   In attempting to measure the static temperature of flowing fluids (particularly high-speed gas flows), one encounters certain types of problems irrespective of the particular sensor being used. These have to do mainly with errors caused by heat transfer between the probe and its environment and the problem of measuring static temperature of a high-velocity flow with a stationary probe.

Conduction error. Let us consider the so-called conduction error first. Figure 8.42 shows a common situation. A probe has been inserted into a duct or other flow passage and is supported at a wall. In general, the wall will be hotter or colder than the flowing fluid; thus there will be heat transfer, and this leads to a probe temperature different from that of the fluid. We shall analyze a simplified model of this arrangement to find what measures can be taken to reduce and/or correct for the error to be expected in such a case. Figure 8.43 shows the simplified model to be used in analyzing this situation. A slender rod extends a distance L from the wall. We assume the rod temperature T_r is a function of x only; it does not vary with time or over the rod cross section at a given x. A fluid of constant and uniform temperature T_f completely surrounds the rod and exchanges heat with it by convection. For a steady-state situation

Heat in at x = (heat out at $x + dx$) + (heat loss at surface)

$$q_x = q_{(x+dx)} + q_l$$

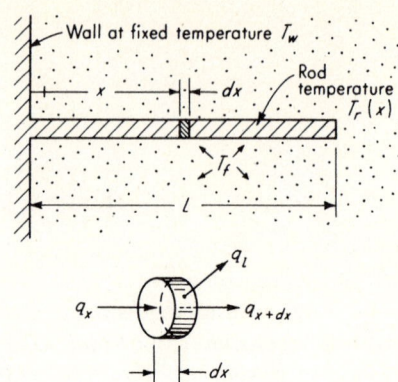

Fig. 8.43. *Conduction-error analysis.*

One-dimensional conduction heat transfer gives

$$q_x = -kA \frac{dT_r}{dx} \qquad (8.48)$$

where $k \triangleq$ thermal conductivity of rod
$A \triangleq$ cross-section area

Then $$q_{(x+dx)} = q_x + \frac{d}{dx}(q_x)\, dx = -kA \frac{dT_r}{dx} + \frac{d}{dx}\left(-kA \frac{dT_r}{dx}\right) dx$$

$$(8.49)$$

Now if k and A are assumed constant

$$q_{(x+dx)} = -kA \frac{dT_r}{dx} - kA \frac{d^2T_r}{dx^2}\, dx \qquad (8.50)$$

We assume the heat loss by convection at the surface to be given by

$$q_l = h(C\, dx)(T_r - T_f) \qquad (8.51)$$

where $h \triangleq$ film coefficient of heat transfer
$C \triangleq$ "circumference" of rod (it need not be circular)
$C\, dx =$ surface area

We then have $$\frac{d^2T_r}{dx^2} - \frac{hC}{kA} T_r = -\frac{hC}{kA} T_f \qquad (8.52)$$

If we take h and C as being constants, Eq. (8.52) is a linear differential equation with constant coefficients and is readily solved for T_r as a function of x. We need two boundary conditions to accomplish this. Clearly, $T_r = T_w$ at $x = 0$ is one such condition. The simplest assumption at $x = L$ is an insulated end; this gives $dT_r/dx = 0$ at $x = L$. Even if the end is not insulated, if L is quite large we intuitively see that the variation of T_r with x must be as in Fig. 8.44; thus $dT_r/dx \approx 0$ for $x = L$. Using

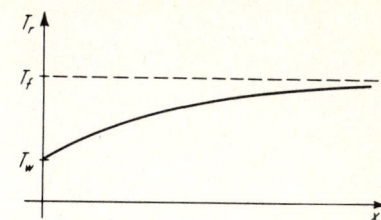

Fig. 8.44. *Temperature profile.*

these two boundary conditions with Eq. (8.52) gives

$$T_r = T_f - (T_f - T_w) \left[\left(1 - \frac{e^{mL}}{2 \cosh mL} \right) e^{mx} + \frac{e^{mL}}{2 \cosh mL} e^{-mx} \right]$$
$$(8.53)$$

where
$$m \triangleq \sqrt{\frac{hC}{kA}} \qquad (8.54)$$

Since the temperature-sensing element (thermocouple bead, thermistor, etc.) is generally located at $x = L$, we evaluate Eq. (8.53) there to get

$$\text{Temperature error} = T_r - T_f = \frac{T_w - T_f}{\cosh mL} \qquad (8.55)$$

Equation (8.55) may be used in two ways: to indicate how to design a probe support to minimize the error and also to allow one to calculate and correct for whatever error there might be. It is clear that the error is reduced if T_w is close to T_f. Insulating or controlling the temperature of the wall encourages this. The term $\cosh mL$ will be large if m and L are large. Thus the probe should be immersed (L is called the immersion length) as far as practical. We see that, to make m large, h should be large (high rate of convection heat transfer) and k should be small (the probe support made of insulating material). The term C/A depends on the shape of the rod. For the usual circular cross section, $C/A = 2/r$, where r is the rod radius. Thus we see that the rod should be of small cross section to reduce error.

If the boundary condition at $x = L$ is changed to a more realistic (and complicated) one in which there is convection heat transfer with a film coefficient h_e at the end, one gets at $x = L$

$$T_r - T_f = \frac{T_w - T_f}{\cosh (mL) + (h_e/mk) \sinh mL} \qquad (8.56)$$

Since the error predicted by (8.56) is less than that of (8.55), the use of the simpler relation is conservative.

Radiation error. Additional error is caused by radiant-heat exchange between the temperature probe and its surroundings. This occurs

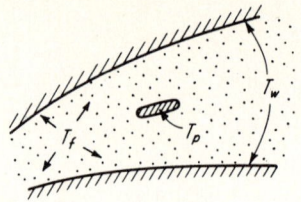

Fig. 8.45. *Radiation-error analysis.*

simultaneously with the previously studied conduction losses but we here consider it separately for simplicity. We also assume radiation exchange only between the probe and the surrounding walls, neglecting radiation of the gas itself or the absorption by the gas of radiation passing through it. Neglecting conduction losses, we may consider the probe as in Fig. 8.45. For steady-state conditions

$$\text{Heat convected to probe} = \text{net heat radiated to wall} \qquad (8.57)$$

$$hA_s(T_f - T_p) = 0.174\epsilon_p A_s \left[\left(\frac{T_p}{100} \right)^4 - \left(\frac{T_w}{100} \right)^4 \right] \qquad (8.58)$$

where $h \triangleq$ film coefficient at probe surface, Btu/(hr-ft^2-°F)
         $A_s \triangleq$ probe surface area
         $\epsilon_p \triangleq$ emittance of probe surface
         $T_p \triangleq$ probe absolute temperature, °R
         $T_w \triangleq$ wall absolute temperature, °R

Equation (8.58) assumes the radiation configuration described as "a small body completely enclosed by a larger one." The error due to radiation is given by

$$\text{Temperature error} = T_p - T_f = \frac{0.174\epsilon_p}{h} \frac{T_w{}^4 - T_p{}^4}{10^8} \qquad (8.59)$$

By insulating the wall or controlling its temperature, error can be reduced by making the difference between T_w and T_p as small as possible. A probe surface of low emittance ϵ_p (a shiny surface) will further reduce such errors, as will a high value of heat-transfer coefficient h. To obtain a high value of h when the fluid velocity is low, the aspirated type of probe may be used. Here a high local velocity is induced at the probe by connecting a vacuum pump into the probe tubing. Equation (8.59) may also be used to calculate corrections if numerical values of the needed quantities are available.

Probes with some form or another of radiation shield are widely used to reduce radiation errors. Figure 8.46 shows a probe with a single shield. The principle of all radiation shields is to interpose between the probe and the wall a body (the shield) whose temperature is closer to the fluid temperature than is the wall. Thus the probe "sees" the shield rather than

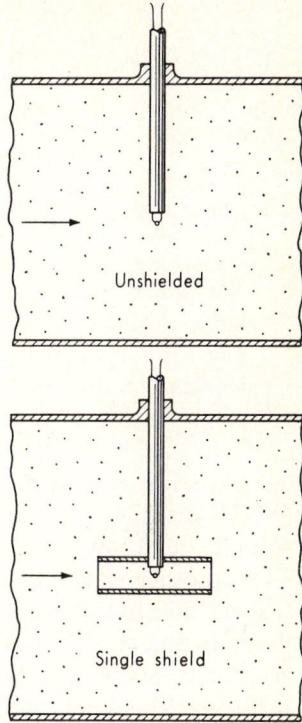

Fig. 8.46. *Radiation shielding.*

the wall, and if the shield is close to fluid temperature the probe radiation-heat loss will be small. It is easy to show that, for pure-radiation-heat transfer, interposing a single screen between a body and its surroundings will reduce the heat loss to one-half the former value, since the screen comes to a temperature $T_{\text{screen}}^4 = (T_{\text{body}}^4 + T_{\text{surroundings}}^4)/2$. For n screens the heat loss is reduced to $1/(n + 1)$ of the unscreened value. All these results are for the simplest case where the screens and surroundings completely enclose the body. For actual probe shields, various geometrical and emittance factors complicate the situation. Also, additional heating of the shield by convection raises its temperature and reduces probe error. Experimental tests[1] with concentric circular cylinder shields (Fig. 8.47) have shown the following:

1. A significant decrease in error may be achieved by adding more shields, at least up to about four.
2. Little is gained by increasing the length/diameter ratio beyond 4:1 in attempting to reduce the unshielded angles at the open ends.

[1] W. J. King, Measurement of High Temperatures in High-velocity Gas Streams, *Trans. ASME*, p. 421, July, 1943.

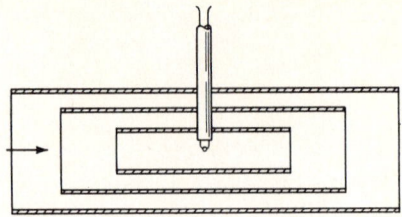

Fig. 8.47. *Multiple radiation shield.*

3. For multiple shields the spacing between shields must be sufficiently
    large to prevent excessive conduction heat transfer between shields
    and to allow high enough flow velocity for good convection from gas to
    shields. A double shield with only $\frac{1}{32}$-in. spacing acted almost like a
    single shield.

To illustrate the need for shielding, an unshielded probe exposed to
1800°F gas flow at 270 lb$_m$/(ft²-min) may exhibit an error of 160F°. A
suitable quadruple shield can reduce this to about 20F°.

Another shielding technique employs an electrically heated shield
with an additional temperature sensor fastened to the shield. The heat
input to the shield is adjusted until the probe sensor and shield sensor
register identical temperatures. At this point, probe and shield should
both be at the fluid temperature, with the heat loss from the shield to the
cooler wall being replaced by the shield heater.

Velocity effects. It is often necessary to determine the static temperature of a flowing gas, since its physical properties depend on this temperature. To measure this temperature directly with a probe, however,
requires that the probe be stationary with respect to the fluid; thus it
must be moving at the same velocity as the fluid. Since this is usually
impractical, various indirect methods of measuring static temperatures
are in use. If one can measure static pressure and either density, sound
velocity, or index of refraction, formulas allow calculation of static
temperature. Experimental techniques based on each of these principles
have been developed. However for routine measurements a different
approach is generally employed. This involves placing a stationary probe
in the stream and calculating the static temperature from the readings of
this probe, using suitable corrections. Ideally, if a perfect gas is deceler-
ated from free-stream velocity to zero velocity adiabatically (not necessarily isentropically) the temperature rises from the free-stream static
temperature T_{stat} to the so-called stagnation or total temperature T_{stag},
where

$$\frac{T_{stag}}{T_{stat}} = 1 + \frac{\gamma - 1}{2} N_m^2 \qquad (8.60)$$

where $\qquad \gamma \triangleq c_p/c_v$, ratio of specific heats

$\qquad\qquad N_m \triangleq$ Mach number

$\qquad T_{stag},\ T_{stat} \triangleq$ absolute temperatures

This result holds for both subsonic and supersonic flow because the shock wave that forms ahead of a probe in supersonic flow affects only the entropy and not the total enthalpy of the gas. For air, Eq. (8.60) becomes

$$\frac{T_{stag}}{T_{stat}} = 1 + 0.2N_m{}^2 \qquad (8.61)$$

and if $N_m < 0.22$, T_{stag} is within 1 percent of T_{stat}. Thus for sufficiently low velocities a stationary probe can be used to read static temperature directly. For higher Mach numbers, T_{stat} can be calculated from Eq. (8.60) if γ and N_m are known. Measurement of Mach numbers with a pitot-static tube is discussed in Chap. 7.

Unfortunately, real temperature probes do not attain the theoretical stagnation temperature predicted by Eq. (8.60). Even if the conduction and radiation errors discussed earlier in this section are corrected for, there remain further deviations of the actual situation from the assumed ideal. Correction for these effects is generally accomplished by experimental calibration to determine the recovery factor r of the particular probe. This is defined by

$$r \triangleq \frac{T_{stag,ind} - T_{stat}}{T_{stag} - T_{stat}} \qquad (8.62)$$

where $\qquad T_{stag,ind} \triangleq$ temperature actually indicated by probe

If r is assumed to be a known number for a given probe, combination of Eqs. (8.60) and (8.62) gives

$$T_{stat} = \frac{T_{stag,ind}}{1 + r[(\gamma - 1)/2]N_m{}^2} \qquad (8.63)$$

A probe that measures T_{stag} exactly would have a recovery factor of 1.0 while one that measures T_{stat} exactly would have $r = 0$.

A possible apparatus[1] for determination of r is shown in Fig. 8.48. The flow velocity in the stagnation chamber is $\frac{1}{100}$ of the nozzle flow velocity; thus measurement of tank temperature and pressure is accurately carried out under essentially zero velocity conditions. By careful design to minimize friction and heat transfer, the nozzle can be made to provide an almost perfect isentropic expansion. The validity of this assumption has been checked experimentally. For an isentropic process

[1] H. C. Hottel and A. Kalitinsky, Temperature Measurements in High-velocity Air Streams, *J. Appl. Mech.*, p. A-25, March, 1945.

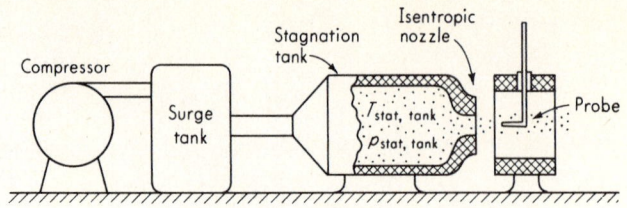

$T_{stat,tank}$ kept near room temperature to minimize
conduction and radiation errors

Fig. 8.48. *Recovery-factor calibration setup.*

from tank to nozzle,

$$T_{stag,nozzle} = T_{stat,tank} \qquad (8.64)$$

and

$$p_{stag,nozzle} = p_{stat,tank} \qquad (8.65)$$

In a free jet

$$p_{stat,nozzle} = p_{atmosphere} \qquad (8.66)$$

The nozzle Mach number N_m can now be computed from the standard pitot-tube formulas since $p_{stat,nozzle}$ and $p_{stag,nozzle}$ are both known. However, the actual use of a pitot tube (with its attendant errors) is avoided since $p_{stag,nozzle}$ is obtained by measurement of $p_{stat,tank}$, and $p_{stat,nozzle}$ is obtained from a barometer reading of $p_{atmosphere}$. Once N_m is known, $T_{stat,nozzle}$ can be computed from

$$T_{stat,nozzle} = \frac{T_{stag,nozzle}}{1 + [(\gamma - 1)/2]N_m{}^2} = \frac{T_{stat,tank}}{1 + [(\gamma - 1)/2]N_m{}^2} \qquad (8.67)$$

The reading of the probe itself supplies $T_{stag,ind}$. Thus one can now compute r from its definition

$$r = \frac{T_{stag,ind} - T_{stat,nozzle}}{T_{stag,nozzle} - T_{stat,nozzle}} \qquad (8.68)$$

For bare thermocouple sensors the numerical value of r usually lies in the range 0.6 to 0.9, depending on the form of the junction (butt-welded, twisted, or spherical bead) and the orientation (wire parallel to flow or transverse to flow). To get a high value of r and one relatively independent of flow conditions such as velocity magnitude and direction, sensors (usually thermocouples or resistance thermometers) are built into probes that have been specifically designed to approach ideal stagnation conditions. Figure 8.49 shows two examples[1] of such probes.

[1] R. W. Ladenburg et al., "Physical Measurements in Gas Dynamics and Combustion," p. 186, Princeton University Press, Princeton, N.J., 1954.

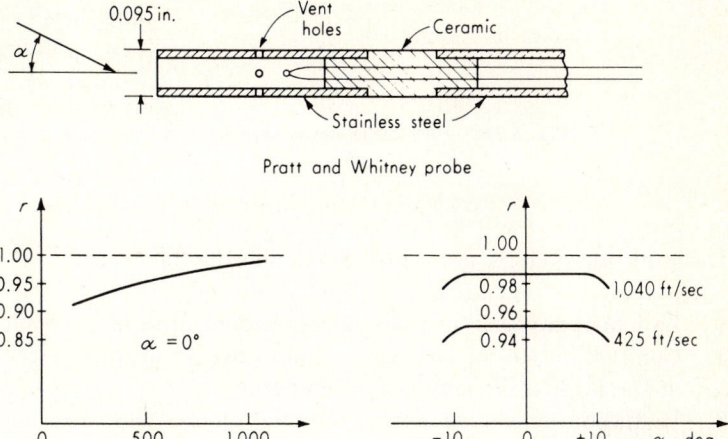

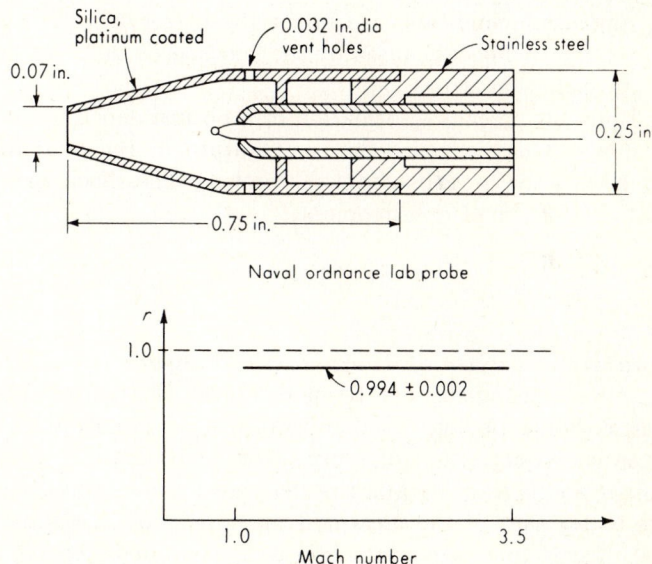

Fig. 8.49. *Stagnation-temperature probes.*

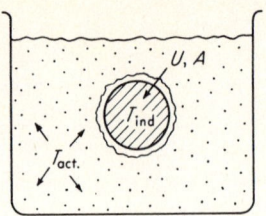

Fig. 8.50. *First-order sensor model.*

Desirable characteristics of probes include the following:

1. Low heat capacity in the sensing element for a fast response.
2. Conduction loss of lead wires minimized by exposing enough length of the lead to the stagnation temperature.
3. Radiation shield of low thermal conductivity and low surface emittance.
4. Vent holes provided to replenish continuously the fluid in the stagnation chamber; otherwise it would be cooled by conduction and radiation. This flow must be kept small enough, however, so that stagnation conditions are essentially preserved. The increased convection coefficient caused by the flow speeds the response and reduces the radiation error.
5. Blunt shape causes formation of a normal shock wave in supersonic flow. This increases the temperature in the boundary layer and reduces the heat loss from the probe. The shock wave also reduces the influence of misalignment.

8.8

Dynamic Response of Temperature Sensors   Since the conversion from sensing-element temperature to thermal expansion, thermoelectric voltage, or electrical resistance is essentially instantaneous, the dynamic characteristics of temperature sensors are related to the heat-transfer and storage parameters that cause the sensing-element temperature to lag that of the measured medium. When a sensing element is used "bare" (not in a protective well), the model of Fig. 8.50 is often adequate. Here heat losses are neglected, resistance to heat transfer is lumped in a single element, and energy storage is lumped in a single element. Conservation of energy gives

$$UA(T_{act} - T_{ind})\,dt = MC\,dT_{ind} \qquad (8.69)$$

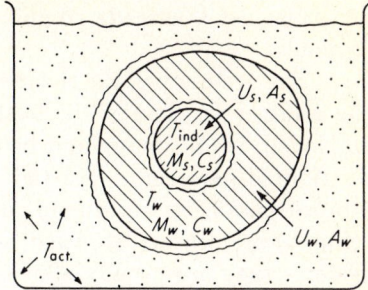

Fig. 8.51.   *Second-order sensor model.*

where $U \triangleq$ overall heat-transfer coefficient, Btu/(sec-F°-in.²)

$A \triangleq$ heat-transfer area

$T_{act} \triangleq$ actual temperature of surrounding fluid

$T_{ind} \triangleq$ temperature indicated by sensor

$M \triangleq$ mass of sensing element

$C \triangleq$ specific heat of sensing element

This leads to
$$\frac{T_{ind}}{T_{act}} (D) = \frac{1}{\tau D + 1} \qquad (8.70)$$

where
$$\tau \triangleq \frac{MC}{UA} \qquad (8.71)$$

Clearly, speed of response may be increased by decreasing M and C and/or increasing U and A. Since U, in general, depends on the surrounding fluid and its velocity, τ is not a constant for a given sensor but rather varies with how it is used.

Since temperature sensors are often enclosed in protective wells or sheaths, a thermal model taking into account heat-transfer resistance and energy storage in the well is of practical interest. Figure 8.51 shows such a configuration. Analysis gives

$$\frac{T_{ind}}{T_{act}} (D) = \frac{1}{\tau_w\tau_s D^2 + (\tau_w + \tau_s + M_sC_s/U_wA_w)D + 1} \qquad (8.72)$$

where $\tau_w \triangleq M_wC_w/U_wA_w$, time constant of well alone

$\tau_s \triangleq M_sC_s/U_sA_s$, time constant of sensor alone

We see that the addition of a well changes the form of response to second-order and increases the lag. The term M_sC_s/U_wA_w is called the coupling term between the well and the sensor. If it is small compared with $\tau_w + \tau_s$, we have approximately

$$\frac{T_{ind}}{T_{act}} (D) = \frac{1}{\tau_w\tau_s D^2 + (\tau_w + \tau_s)D + 1} = \frac{1}{\tau_w D + 1}\frac{1}{\tau_s D + 1} \qquad (8.73)$$

which is just a cascade combination of the sensor and well individual dynamics.

The accuracy of the theoretical model can be increased by increasing the number of "lumps" of heat-transfer resistance and energy storage employed, the ultimate limit being an infinite number corresponding to a distributed-parameter (partial differential equation) rather than a lumped-parameter (ordinary differential equation) approach. When temperature sensors are used as measuring devices in feedback-control systems they are usually allowed to contribute no more than 30° phase lag at the frequency where the entire open-loop lag is 180°. Under such conditions they are usually adequately modeled as simple first-order systems. The best[1] time constant to use for such a model is determined by an experimental ramp-input test and is numerically the steady-state time lag observed in such a test.

When greater accuracy is needed in utilizing the results of experimental tests to determine sensor dynamics, a model[2] using three time constants and a dead time may be employed. The transfer function is then

$$\frac{T_{ind}}{T_{act}}(D) = \frac{e^{-\tau_{dt}D}}{(\tau_1 D + 1)(\tau_2 D + 1)(\tau_3 D + 1)} \qquad (8.74)$$

Numerical values of τ_1, τ_2, τ_3, and τ_{dt} may be obtained from step-function response tests. For example, a thermocouple used in a heavy-duty stainless-steel well had

$$\frac{T_{ind}}{T_{act}}(D) = \frac{e^{-2.6D}}{(21.6D + 1)(2.9D + 1)(2.1D + 1)} \qquad (8.75)$$

where the time constants are in seconds.

When accurate numerical values are needed, experimental tests are generally required to determine temperature probe dynamics. For simple bare thermocouples, however, extensive research and testing have provided semiempirical formulas which allow calculation of the time constant with fair accuracy. One such relation[3] useful for temperatures from 160 to 1600°F, wire diameter 0.016 to 0.051 in., mass velocity 3 to 50 $lb_m/(ft^2\text{-sec})$, and static pressure of 1 atm is

$$\tau = \frac{3{,}500\rho c d^{1.25} G^{-15.8/\sqrt{T}}}{T} \qquad \text{sec} \qquad (8.76)$$

[1] G. A. Coon, Responses of Temperature-sensing-element Analogs, *Trans. ASME*, p. 1857, November, 1957.

[2] J. R. Louis and W. E. Hartman, The Determination and Compensation of Temperature-sensor Transfer Functions, *ASME Paper* 64-WA/AUT-13.

[3] R. J. Moffat, How to Specify Thermocouple Response, *ISA J.*, p. 219, June, 1957.

where $\rho \triangleq$ average density of two thermocouple materials, lb_m/ft^3

$\qquad c \triangleq$ average specific heat of two thermocouple materials, Btu/ $(\text{lb}_m\text{-F}°)$

$\qquad d \triangleq$ wire diameter, in.

$\qquad G \triangleq$ flow mass velocity, $\text{lb}_m/(\text{ft}^2\text{-sec})$

$\qquad T \triangleq$ stagnation temperature, °R

Within the above restrictions, this formula will predict τ for butt-welded junctions within about 10 percent.   Another such result,[1] based on tests for a Mach-number range of 0.1 to 0.9 and a Reynolds-number range of 250 to 30,000, gives

$$\tau = \frac{4.05\rho c d^{1.50}\{1 + [(\gamma - 1)/2]N_m{}^2\}^{0.25}}{p^{0.5}N_m{}^{0.5}T^{0.18}} \qquad (8.77)$$

$\qquad\qquad$ where $\qquad \gamma \triangleq$ ratio of specific heats

$\qquad\qquad\qquad\qquad N_m \triangleq$ Mach number

$\qquad\qquad\qquad\qquad p \triangleq$ static pressure, atm

This reference also presents a comprehensive analysis of conduction and radiation errors and the effects of differences in the thermal properties of the two metals used in a thermocouple.

Dynamic compensation of temperature sensors. When environmental conditions require a rugged temperature sensor, the mass may be so high as to cause a sluggish response.   It may be possible to obtain an improved overall measuring-system response by cascading an appropriate dynamic compensation device with the sensor.   Such schemes have been applied in practice with considerable success and may be implemented in a number of ways.[2,3]   An RC network such as that shown for hot-wire anemometer compensation in Chap. 7 may be used if the sensor is essentially first-order.   Second-order sensors may also be compensated;[3] operational-amplifier networks provide a convenient means of implementation.   Since the compensation is correct only for specific sensor dynamics, changes in numerical values or form of transfer function caused by changes in operating conditions can lead to loss of compensation.

[1] M. D. Scadron and I. Warshawsky, Experimental Determination of Time Constants and Nusselt Numbers for Bare-wire Thermocouples in High-velocity Air Streams and Analytic Approximation of Conduction and Radiation Errors, *NACA, Tech. Note* 2599, 1952.

[2] C. E. Shepard and I. Warshawsky, Electrical Techniques for Compensation of Thermal Time Lag of Thermocouples and Resistance Thermometer Elements, *NACA, Tech. Note* 2703, 1952.

[3] Louis and Hartman, *op. cit.*

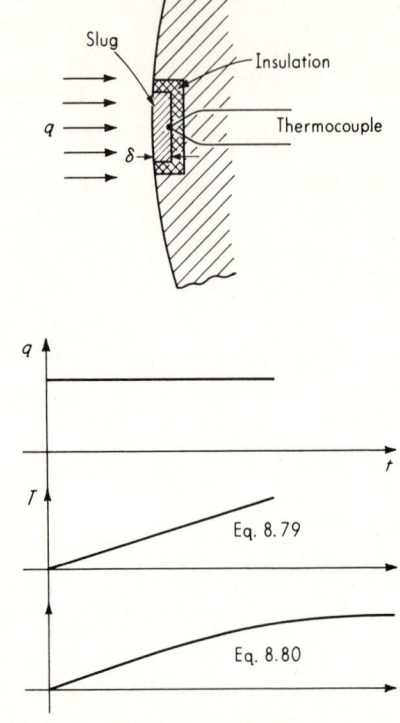

Fig. 8.52. *Slug-type heat-flux sensor,*

Increased speed of response is, in general, traded off for overall sensitivity in such compensation schemes. This can be made up by additional amplification but only up to a point; then the inherent noise level prevents further improvement. However, improvement of 100:1 or more is often possible.

8.9

Heat-flux Sensors   In recent years, requirements for measurement of local convective, radiative, or total heat-transfer rates in missile structures have led to the development of several types of heat-flux sensors. We here briefly review the operating principles and characteristics of the most common types.

Slug-type sensors. In Fig. 8.52 a slug of metal is buried in (but insulated from) the surface across which the heat-transfer rate is to be measured. Neglecting losses through the insulation and the thermocouple wires, one may write

$$\text{Heat transferred in } = \text{ energy stored}$$
$$Aq\, dt = Mc\, dT \qquad (8.78)$$

where $A \triangleq$ surface area of slug, in.^{2}
$q \triangleq$ local heat-transfer rate, Btu/(sec-in.^{2})
$M \triangleq$ mass of slug, lb$_m$
$c \triangleq$ specific heat of slug, Btu/(lb$_m$-F°)
$T \triangleq$ slug temperature, °F

Then
$$q = \frac{Mc}{A}\frac{dT}{dt} \qquad (8.79)$$

and thus q may be determined by measuring dT/dt if Mc/A is known. Since the thermocouple reads T rather than dT/dt, a graphical, numerical, or electrical differentiation must be carried out to get q. For greater accuracy, the heat losses may be taken into account by modifying Eq. (8.79) to give

$$q = \frac{Mc}{A}\frac{dT}{dt} + K_l\,\Delta T \qquad (8.80)$$

where $K_l \triangleq$ loss coefficient, Btu/(in.^{2}-sec-F°)
$\Delta T \triangleq$ temperature difference between slug and casing (usually taken as temperature rise of slug by assuming constant casing temperature)

The numerical values of Mc/A and K_l for a given sensor are determined by calibration and supplied by the manufacturer. Equation (8.79) predicts that, for a constant q, T increases linearly with time and without limit. Actually, the unavoidable heat losses eventually make dT/dt approach zero, as shown by the more correct Eq. (8.80).

The analysis of Eq. (8.78) assumes the slug is at all times at uniform temperature T throughout. This is not actually the case; thus there is a time-lag effect which has been evaluated[1] on the basis of a step input of q. A partial-differential-equation analysis leads to

$$q_m \approx q(1 - 2e^{-\pi^2\alpha t/\delta^2}) \qquad (8.81)$$

where $q_m \triangleq$ measured flux, using temperature at back surface of slug
$q \triangleq$ actual flux
$\alpha \triangleq$ thermal diffusivity $= k/\rho c$
$\delta \triangleq$ slug thickness
$k \triangleq$ thermal conductivity
$\rho \triangleq$ mass density

We see that a fast response requires a small value of $\delta^2\rho c/k$.

[1] Heat Technology Laboratory, Inc., Huntsville, Ala., *Rept. HTL*-ER-4, p. 4, 1962.

Since the materials of which the sensor is made can withstand only a certain maximum temperature rise ΔT_{max}, a slug can be exposed to a given heat-transfer rate q for only a limited time t_{max}. Neglecting losses, Eq. (8.79) can be integrated to give the slug thickness δ required for a given q, ΔT_{max}, and t_{max} as

$$\delta = \frac{qt_{max}}{\rho c \, \Delta T_{max}} \qquad (8.82)$$

For convective heat transfer from a gas of fixed temperature T_g, the heat flux into the slug is $h(T_g - T)$, where h is the film coefficient. Neglecting losses, we may write

$$h(T_g - T) \, dt = \rho c \delta \, dT \qquad (8.83)$$

which leads to

$$\delta = \frac{ht_{max}}{\rho c \, \log_e \left| \dfrac{1}{1 - \Delta T_{max}/(T_g - T_i)} \right|} \qquad (8.84)$$

where T_i is the initial slug temperature.

A more refined analysis[1] for the case of constant q (which takes into account that the *front* surface will overheat before the back) shows that there is an optimum value of δ in the sense that the linear (steady-state) part of the rear-surface response is the longest possible before the front surface overheats. This optimum value of δ is given by

$$\delta_{opt} = \frac{kT_{f,max}}{1.366q} \qquad (8.85)$$

where $T_{f,max}$ is the maximum allowable front-surface temperature. The time interval of linear response is found to be

$$\Delta t_{linear} = \frac{0.366k^2 T_{f,max}^2}{\alpha q^2} \qquad (8.86)$$

Figure 8.53 illustrates these concepts.

Steady-state or asymptotic sensors (Gardon gage). Figure 8.54 shows the essential features of this type of sensor which was first proposed by R. Gardon.[2] A thin constantan disk is connected at its edges to a large copper heat sink while a very thin (< 0.005-in.-diameter) copper

---

[1] R. H. Kirchhoff, Calorimetric Heating-rate Probe for Maximum-response-time interval, *AIAA J.*, p. 967, May, 1964.

[2] R. Gardon, An Instrument for the Direct Measurement of Intense Thermal Radiation, *Rev. Sci. Instr.*, p. 366, May, 1953.

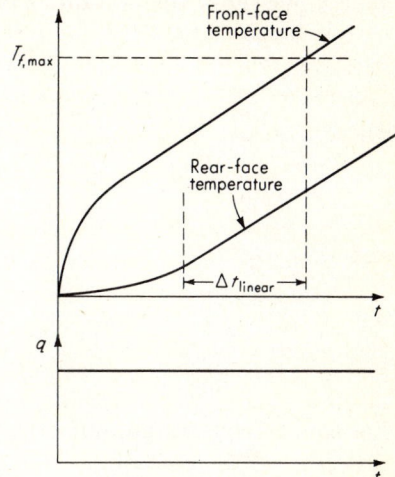

Fig. 8.53.   *Slug-type-sensor response.*

wire is fastened at the center of the disk.   This forms a differential thermocouple between the disk center and its edges.   When the disk is exposed to a constant heat flux, an equilibrium temperature difference is rapidly established which is proportional to the heat flux.   Since the thermocouple signal is now directly proportional to the heat flux, no differentiating process (such as is required in a slug-type sensor) is necessary.   Furthermore, loss corrections are generally not needed nor is a thermocouple reference junction required.   Instrument response[1] is

[1] Heat Technology Laboratory, *Rept. HTL*-ER-4, p. 4.

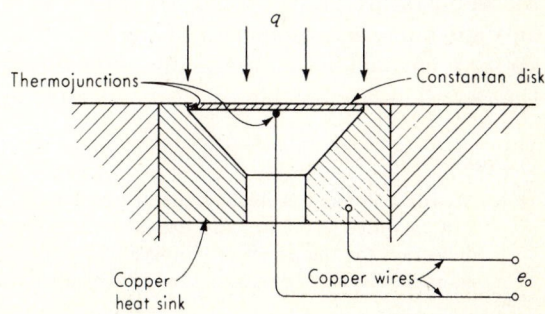

Fig. 8.54.   *Gardon gage.*

approximately of the first-order type; thus

$$\frac{e_o}{q}(D) = \frac{K}{\tau D + 1} \qquad (8.87)$$

where

$$K \triangleq \frac{d^2 K_e}{16 \delta k} \qquad (8.88)$$

$$\tau \triangleq \frac{\rho c d^2}{16 k} \qquad (8.89)$$

and

$d \triangleq$ diameter of disk

$\delta \triangleq$ thickness of disk

$K_e \triangleq$ thermocouple sensitivity, mv/F°

$k \triangleq$ thermal conductivity of disk

$c \triangleq$ specific heat of disk

For copper/constantan, numerical values are

$$\frac{e_o}{q}(D) = \frac{0.0308(d^2/\delta)}{(5.96d^2)D + 1} \qquad (8.90)$$

where d and δ are in inches, e_o is in millivolts, q is in Btu/(sec-ft²), and $5.96d^2$ is in seconds. Typical commercial units are available for full-scale heat fluxes of 15 to 300 Btu/(sec-ft²), produce 10 mv full-scale output, and have time constants of 0.07 to 0.2 sec.

Application considerations. The introduction of the sensor into the wall locally alters the thermal properties of the wall and causes the measured heat flux to differ from that which would occur if the sensor were not present.[1] It is thus desirable to match, insofar as feasible, the thermal properties of sensor and wall. For a Gardon gage, there is a radial temperature gradient over the disk which, if excessive, causes a variation in local convection coefficient and thus an error. By sacrificing sensitivity (and then recovering it by external amplification if necessary) the temperature gradient and associated error may be reduced. When only the radiation component of the total flux is desired, the front of the sensor is covered with a thermally isolated sapphire window which passes the radiation flux but blocks the convective flux.

Problems

8.1 An Invar/brass cantilever bimetal (see Fig. 8.2b) has $t = 0.05$ in., a length of 2 in., a width of 0.5 in., and has $t_A = t_B$ and $n + 1/n \approx 2$. Estimate the end deflection for temperature changes of 30 and 60°C. If the end is held fixed, estimate the force developed for temperature changes of 30 and 60°C.

[1] D. R. Hornbaker and D. L. Rall, Thermal Perturbations Caused by Heat-flux Transducers and Their Effect on the Accuracy of Heating-rate Measurements, *ISA Trans.*, p. 123, April, 1964.

8.2 A Beckman thermometer is to have a sensitivity of 10 in./°C when using mercury in the neighborhood of room temperature. Obtain an expression relating capillary cross-section area and bulb volume to meet this requirement. Obtain expressions for the time constant if the bulb is spherical and also if it is cylindrical with a length equal to 5 diameters. Also find the ratio of these two time constants.

8.3 Sketch and explain the operation of a bimetallic compensator to replace the auxiliary pressure sensor in Fig. 8.7. Can both case and capillary compensation be obtained? Explain.

8.4 Analyze the system of Fig. 8.14*a* to obtain a steady-state relation between hot-gas temperature as an input and thermocouple voltage as an output.

8.5 Develop equations to estimate the dynamic response of the system of Fig. 8.14*a*.

8.6 Repeat Prob. 8.4 for the system of Fig. 8.14*b*.

8.7 In Fig. 8.15 if $\tau = 2.8$ sec, $T_{gas} = 5000°F$, and the thermocouple damage limit is 2000°F, how long can the cooling be left off if the steady-state cooled thermocouple temperature is 500°F?

8.8 A resistance-thermometer circuit as in Fig. 8.19*d* has $R_1 = R_2 = 10,000$ ohms and is to cover the temperature range 0 to 400°C. The thermometer element is platinum with a resistance of 1,000 ohms at 200°C. Plot the curve of bridge output voltage (open circuit) versus input temperature, using the data of Fig. 8.18. See Chap. 10 for bridge-circuit equations. Bridge excitation is 20 volts.

8.9 A 500-ohm resistance thermometer carries 5-ma current. Its surface area is 0.5 in.^{2}, and it is immersed in stagnant air, so that the heat-transfer coefficient is $U = 1.5$ Btu/(hr-ft^2-F°). Find its self-heating error. What would be the error in water with $U = 100$ Btu/(hr-ft^2-F°)?

8.10 A pulse-excited resistance thermometer (see Fig. 8.20) has an excitation voltage in the form of a rectangular pulse of 100 volts height and 0.1-sec duration. The pulse is on for 0.1 sec and off for 0.9 sec in a repetitive cycle. Compute the ratio of peak/rms voltage for this pulse. What average heating power would this voltage pulse produce in a 500-ohm resistor?

8.11 For blackbody radiation, what surface temperature is needed to radiate 1 hp/in.^{2}?

8.12 Estimate the percentage of total power found above 10μ for blackbody radiation at $T = 400°K$.

8.13 Explain the disadvantage of a large time constant in a thermal-radiation detector using a chopper.

8.14 Derive Eq. (8.53).

8.15 Consider a subsonic air flow in a duct. Pitot-static-tube measurements give a static pressure of 100 psia and a stagnation pressure of 129.1 psia. A temperature probe with a recovery factor $r = 0.80$ extends from a 100°F wall a distance of 1 ft into the flow. The probe thermocouple reads 400°F. The probe support has a radius of 0.02 ft and a thermal conductivity of 100 Btu/(hr-ft-F°). The end of the probe may be assumed insulated, and the surface convection coefficient is 10 Btu/(hr-ft^2-F°). Radiation effects are negligible. Calculate the static temperature of the flow.

8.16 Derive Eq. (8.72).

8.17 Make and analyze a third-order model of a temperature sensor analogous to the second-order model of Fig. 8.51.

8.18 A butt-welded 0.03-in. bare-wire copper/constantan thermocouple is used to measure the temperature (near 100°F) of atmospheric air flowing at 100 fps. Estimate the time constant using both formulas available.

8.19   A copper slug-type heat-flux sensor is 0.2 in. thick.   Plot its time response $(q_m$ versus $t)$ for a step change in q.   Is this a first-order instrument?   Explain. How long must one wait before q_m is 95 percent of q?

8.20   Derive Eq. (8.82).

8.21   Derive Eq. (8.84).

8.22   Compare the sensitivity and response speed of Gardon gages of like diameter and thickness but made of the following:

  a.   Constantan disk, copper heat sink

  b.   Copper disk, constantan heat sink

  c.   Iron disk, constantan heat sink

  d.   Constantan disk, iron heat sink

8.23   Sketch and explain a test setup for evaluating the step-function response of temperature sensors exposed to air flows of different velocities.

8.24   Sketch and explain a test setup for static calibration of heat-flux sensors.

8.25   Sketch and explain a test setup for step-function testing of heat-flux sensors.

Bibliography

1.   H. F. Stimson: International Temperature Scale of 1948, Text Revision of 1960, *Natl. Bur. Std. (U.S.)*, *Monograph* 37, 1961.

2.   J. Nicol and C. J. Rauch: Below One Degree, *Ind. Res.*, p. 60, September, 1964.

3.   J. R. Van Orsdel et al.: Development of a Vapor-pressure-operated High Temperature Sensor Device, *NASA*, *CR*-50001, 1964.

4.   C. F. Alban and C. C. Perry: Maximum-work Bimetals, *Machine Design*, p. 143, Apr. 16, 1959.

5.   C. F. Alban and C. C. Perry: Adjusting Performance of Thermostatic Bimetals, *Machine Design*, p. 195, May 14, 1959.

6.   C. F. Alban and C. C. Perry: Optimum Design of Thermostatic Bimetal Elements, *Machine Design*, p. 119, Feb. 21, 1957.

7.   J. M. Benson and R. Horne: Surface Temperature of Thin Sheets and Filaments, *Instr. Control Systems*, p. 115, October, 1962.

8.   C. E. Moeller: Do Shields Improve Thermocouple Response?, *ISA J.*, p. 56, August, 1960.

9.   J. L. LeMay: More Accurate Thermocouples with Percussion Welding, *ISA J.*, p. 42, March, 1959.

10.   L. E. Bollinger: Thermocouple Measurements in an RF Field, *ISA J.*, p. 338, September, 1955.

11.   J. C. Lachman and F. W. Kuether: Stability of Rhenium/Tungsten Thermocouples in Hydrogen Atmospheres, *ISA J.*, p. 67, March, 1960.

12.   A. R. Driesner et al.: High Temperature W/W-25 Re Thermocouples, *Instr. Control Systems*, p. 105, May, 1962.

13.   F. W. Kuether and J. C. Lachman: How Reliable Are the Two New High-temperature Thermocouples in Vacuum?, *ISA J.*, p. 67, April, 1960.

14.   J. J. Van Drasek and B. A. Short: Conversion Formulas for Copper-Constantan Thermocouples, *Instr. Control Systems*, p. 106, February, 1965.

15.   G. E. Reis et al.: A Thermocouple Unit for Measuring Transient Temperatures at Specified Locations in Metal Bodies, Sandia Corp., Albuquerque, N.Mex., *SCR*-3, 1958.

16.   H. C. Jordan: Welded Thermocouple Junctions, *Instr. Control Systems*, p. 988, June, 1960.

17.   Simple Circuit Continuously Monitors Thermocouple Sensor Continuity, *NASA*, *Brief* 63-10567, 1963.

18. Thermocouple Calibration, *Instr. Control Systems*, p. 1663, September, 1961.
19. M. B. Dow: Comparison of Measurements of Internal Temperatures in Ablation Material by Various Thermocouple Configurations, *NASA, Tech. Note* D-2165, 1964.
20. R. Dutton and E. C. Lee: Surface-temperature Measurement of Current-carrying Objects, *ISA J.*, p. 49, December, 1959.
21. J. Nanigian: Thermal Properties of Thermocouples, *Instr. Control Systems*, p. 87, October, 1963.
22. Unusual Thermocouples and Accessories, *Instr. Control Systems*, p. 110, June, 1963.
23. E. G. Weissenberger: Metal Sheathed Thermocouples, *Instr. Control Systems*, p. 109, May, 1963.
24. C. E. Moeller: Special Thermocouple Solves Surface Temperature Problem, *ISA J.*, p. 47, June, 1959.
25. C. M. Stover: Method of Butt Welding Small Thermocouples 0.001 to 0.01 Inch in Diameter, *Rev. Sci. Instr.*, vol. 31, no. 6, p. 605, June, 1950.
26. C. M. Stover: Method of Making Small Pointed Thermocouples, *Rev. Sci. Instr.*, vol. 32, no. 3, p. 366, March, 1961.
27. G. E. Reis: Temperature Measurements on High Speed Missiles, Sandia Corp., Albuquerque, N.Mex., *SCR*-73, 1956.
28. O. Schwelb and G. C. Temes: Thermistor-Resistor Temperature Sensing Networks, *Electro-Technol. (New York)*, p. 71, November, 1961.
29. D. S. Saulson: The Thermistor Bridge, *Electro-Technol. (New York)*, p. 73, September, 1961.
30. J. S. Blakemore: Germanium for Low-temp Resistance Thermometry, *Instr. Control Systems*, p. 94, May, 1962.
31. J. R. Pies: A New Semiconductor for Temperature Measuring, *ISA J.*, p. 50, August, 1959.
32. J. M. Janicke: Direct-reading Platinum Thermometer, *Instr. Control Systems*, p. 129, May, 1965.
33. H. N. Norton: Resistance Elements for Missile Temperatures, *Instr. Control Systems*, p. 993, June, 1960.
34. A. R. Anderson and T. M. Stickney: Ceramic Resistance Thermometers as Temperature Sensors Above 2200°R, *Instr. Control Systems*, p. 1864, October, 1961.
35. A. B. Kaufman: Cryogenic Characteristics of Alloy Wires, *Instr. Control Systems*, p. 119, March, 1964.
36. H. C. Tsien: Piston Zone Temperature Measurement, *Instr. Control Systems*, p. 105, May, 1964.
37. R. S. Benson: Measurement of Transient Exhaust Temperatures in I.C. Engines, *The Engineer*, Feb. 28, 1964.
38. T. Coor and L. Szmanz: Digital Thermometer, *Instr. Control Systems*, p. 125, May, 1965.
39. E. W. Jones: Calibration Techniques for Thermistors, *Instr. Control Systems*, p. 123, May, 1965.
40. D. B. Schneider: The Thermistor Thermometer, *Instr. Control Systems*, p. 119, May, 1965.
41. P. W. Montgomery and R. L. Lowery: Jet Temperature by IR Pyrometry, *ISA J.*, p. 61, April, 1965.
42. R. J. Thorn and G. H. Winslow: Recent Developments in Optical Pyrometry, *ASME Paper* 63-WA-224, 1963.

43. R. H. Tourin: Recent Developments in Gas Pyrometry by Spectroscopic Methods, *ASME Paper* 63-WA-252, 1963.

44. D. A. McGraw and R. G. Mathias; Application of Radiation Pyrometry to Glass-forming Processes, *Ceram. Age*, August, 1962.

45. R. K. McDonald: Infrared Radiometry, *Instr. Control Systems*, p. 1527, September, 1960.

46. R. A. Hanel: The Dielectric Bolometer, A New Type of Thermal Radiation Detector, *NASA, Tech. Note* D-500, 1960.

47. R. A. Hanel: A Low-resolution Unchopped Radiometer for Satellites, *NASA Tech. Note* D-485, 1961.

48. R. W. Reynolds: Infrared-radiation Reference Sources, *Electro-Technol. (New York)*, p. 46, January, 1963.

49. G. Conn and D. Avery: "Infrared Methods," Academic Press Inc., New York 1960.

50. F. Schwarz: Infrared Detectors, *Electro-Technol. (New York)*, p. 116, November, 1963.

51. Calibration of Thermopiles, *NASA*, N-64-28205, 1964.

52. Pyroelectric Detection Techniques and Materials, *NASA*, CR-44, 1964.

53. D. Greenshields: Spectrometric Measurements of Gas Temperatures in Arc-heated Jets and Tunnels, *NASA, Tech. Note* D-1960, 1963.

54. D. R. Buchele: Nonlinear-averaging Errors in Radiation Pyrometry, *NASA, Tech. Note* D-2406, 1964.

55. M. Weiss: High Temperature Ultraviolet Radiometer, *Instr. Control Systems*, p. 95, May, 1964.

56. E. W. Bivans: Measuring Infrared Detector Noise, *Electron. Design*, Aug. 2, 1962.

57. Reference Blackbody Is Compact, Convenient to Use, *NASA Brief* 63-10004, 1963.

58. Lunar Surface Temperature Instrument, *NASA*, N-64-10097, 1964.

59. A. J. Metzler and J. R. Branstetter: Fast Response, Blackbody Furnace for Temperatures to 3000°K, *Rev. Sci. Instr.*, vol. 34, no. 11, p. 1216, November, 1963.

60. H. C. Ingrao et al.: Ferroelectric Bolometer for Space Research, *NASA*, CR-55542, 1964.

61. E. M. Wormser: Radiation Thermometer with In-line Blackbody Reference, *Instr. Control Systems*, p. 101, December, 1964.

62. T. P. Murray and V. G. Shaw: Two-color Pyrometry in the Steel Industry, *ISA J.*, p. 36, December, 1958.

63. Calibration of Optical Pyrometers, *Instr. Control Systems*, p. 84, May, 1962.

64. B. Bernard: Flame Temperature Measurements, *Instr. Control Systems*, p. 113, May, 1965.

65. A. G. Gaydon: "The Spectroscopy of Flame," John Wiley & Sons, Inc., New York, 1957.

66. R. Looney: Method for Presenting the Response of Temperature Measuring Systems, *ASME Trans.*, p. 1851, November, 1957.

67. W. J. King: Measurement of High Temperatures in High-velocity Gas Streams, *ASME Trans.*, p. 421, July, 1943.

68. W. M. Rohsenow and J. P. Hunsaker: Determination of the Thermal Correction for a Single-shielded Thermocouple, *ASME Trans.*, p. 699, August, 1947.

69. T. M. Stickney: Recovery and Time-response Characteristics of Six Thermocouple Probes in Subsonic and Supersonic Flow, *NACA, Tech. Note* 3455, 1955.

70. R. C. Turner and G. D. Gordon: Thermocouple for Vacuum Tests Minimizes Error, *Space/Aeron.*, p. 256, January, 1964.
71. D. Wald: Measuring Temperature in Strong Fields, *Instr. Control Systems*, p. 100, May, 1963.
72. L. M. K. Boelter et al.: Thermocouple Conduction Error Observed in Measuring Surface Temperatures, *NACA, Tech. Note* 2427, 1951, and *Tech. Note* 1452, 1948.
73. M. Sibulkin: A Total-temperature Probe for High-temperature Boundary-layer Measurements, *J. Aerospace Sci.*, p. 458, July, 1959.
74. J. C. Faul: Thermocouple Performance in Gas Streams, *Instr. Control Systems*, p. 104, December, 1962.
75. I. Fruchtman: Temperature Measurement of Hot Gas Streams, *AIAA J.*, vol. 1, no. 8, p. 1909, 1963.
76. D. L. Goldstein and R. Scherrer: Design and Calibration of a Total-temperature Probe for Use at Supersonic Speeds, *NACA, Tech. Note* 1885, 1949.
77. R. Sandri et al.: On the Measurement of the Average Temperature of a Fluid Stream in a Tube by Means of a Special Type of Resistance Thermometer, National Research Laboratory, Ottawa, Canada, *Mech. Eng. Div. Rept. MI-826*, April, 1962.
78. R. D. Wood: A Heated Hypersonic Stagnation-temperature Probe, *J. Aerospace Sci.*, p. 556, July, 1960.
79. R. P. Benedict: Temperature Measurement in Moving Fluids, *Electro-Technol. (New York)*, p. 56, October, 1963.
80. R. V. DeLeo et al.: Measurement of Mean Temperature in a Duct, *Instr. Control Systems*, p. 1659, September, 1961.
81. C. F. Hansen et al.: Investigation of Heat Conduction in Air, *NASA, Tech. Rept.* R-27 (Nickel Film Surface Temperature Detectors), 1959.
82. R. P. Benedict: High Response Aerosol Probe for Sensing Gaseous Temperature in a Two-phase, Two-component Flow, *ASME Paper* 62-WA-317, 1962.
83. I. Warshawsky: Measurements of Rocket Exhaust-gas Temperatures, *ISA J.*, p. 91, November, 1958.
84. M. G. Holland et al.: Temperature Measurement from 2°K–400°K, *Instr. Control Systems*, p. 89, May, 1962.
85. J. Grey: Thermodynamic Methods of High-temperature Measurement, *ISA Trans.*, p. 102, April, 1965.
86. T. A. Perls and J. J. Hartog: Pyroelectric Transducers for Heat-transfer Measurements, *ISA Trans.*, p. 21, January, 1963.
87. D. L. Johnson: The Design and Application of a Steady-state Heat Flux Transducer for Aerodynamic Heat-transfer Measurements, *ISA Trans.*, p. 46, January, 1965.
88. Simple Transducer Measures Low Heat-transfer Rates, *NASA, Brief* 64-10122, 1964.
89. F. C. Stempel and D. L. Rall: Direct Heat Transfer Measurements, *ISA J.*, p. 68, April, 1964.
90. E. A. Laumann: A Steady-state Heat Meter for Determining the Heat-transfer Rate to a Cooled Surface, *NASA, N-63-18868*, 1963.
91. L. R. Hunt and R. R. Howell: Experimental Technique for Measuring Total Aerodynamic Heating Rates to Bodies of Arbitrary Shape with Results to Mach 7, *NASA, Tech. Note* D-2446, 1964.
92. R. J. Conti: Heat-transfer Measurements at a Mach Number of 2 in the Turbulent Boundary Layer on a Flat Plate Having a Stepwise Temperature Distribution, *NASA, Tech. Note* D-159, 1959.

93.   R. A. Jones and J. L. Hunt: Use of Temperature-sensitive Coatings for Obtaining Quantitative Aerodynamic Heat-transfer Data, *AIAA J.*, p. 1354, July, 1964.

94.   P. H. Rose and J. O. Stankevics: Heat Transfer Measurements in Partially Ionized Gases, *NASA, CR*-59768, 1964.

95.   C. H. Liebert et al.: Application of Various Techniques for Determining Local Heat-transfer Coefficients in a Rocket Engine from Transient Experimental Data, *NASA, Tech. Note* D-277, 1960.

96.   D. R. Beck and F. Kreith: A New Steady State Calorimeter for Measuring Heat Transfer through Cryogenic Insulation, *NASA, N*-64-14283, 1964.

97.   L. Bogdon: High-temperature, Thin-film Resistance Thermometers for Heat Transfer Measurement, *NASA, CR*-26, 1964.

98.   J. C. Cook and H. S. Levine: Calorimeter and Accessories for Very High Thermal Radiation Flux Measurements, *Rev. Sci. Instr.*, October, 1960.

99.   L. Bogdan: Measurement of Radiative Heat Transfer with Thin-film Resistance Thermometers, *NASA, CR*-27, 1964.

100. R. C. Bachmann et al.: Investigation of Surface Heat-flux Measurements with Calorimeters, *ISA Trans.*, p. 143, April, 1965.

101. Bibliography of Temp. Meas. 1953–1969, *NBS* SP-373, 1972.

102. R. P. Benedict: "Fund. of Temp., Pressure and Flow Meas.," Wiley, New York, 1969.

103. G. D. Nutter: Recent Advances and Trends in Radiation Thermometry, *ASME Paper* 71-WA/Temp-3, 1971.

104. J. Geist: Fundamental Principles of Absolute Radiometry and the Philosophy of This NBS Program (1968–1971), *NBS Tech. Note* 594-1, 1972.

105. D. R. Buchele and D. J. Lesco: Pyrometer for Measurement of Surface Temperature Distribution on a Rotating Turbine Blade, *NASA* TMX-68113, 1972.

106. J. R. Branstetter: Some Practical Aspects of Surface Temperature Measurement by Optical and Ratio Pyrometers, *NASA* TN D-3604, 1966.

107. R. A. Pease: Using Semiconductor Sensors for Linear Thermometers, *Inst. and Cont. Sys.*, June, 1972.

108. R. P. Benedict and R. J. Russo: A Note on Grounded Thermocouple Circuits, *ASME Paper* 71-WA/Temp-1, 1971.

109. C. E. Moeller et al: NASA Contributions to Development of Special-Purpose Thermocouples, *NASA* SP-5050, 1968.

110. G. Cataland and H. H. Plumb: Low Temperature Thermometry: Interim Report, *NBS Tech. Note* 765, 1973.

111. N. R. Keltner: Heat Transfer in Intrinsic Thermocouples—Application to Transient Temperature Measurement Errors, *Rept.* SC-RR-720719, Sandia Labs, Albuquerque, 1973.

112. R. P. Shreeve and D. W. Peecher: Stagnation Temperature Measurement at High Mach Number Using Very Small Probes, *Boeing Res. Lab. Rept.* DI-82-0945, 1970.

113. D. J. Baines: Selecting Unsteady Heat Flux Sensors, *Inst. and Cont. Syst.*, May, 1972.

Chapter 9
Miscellaneous Measurements

9.1

Time, Frequency, and Phase-angle Measurement The fundamental standard of time was discussed in Sec. 4.2. The United States Frequency Standard is a cesium-beam resonator whose precision is of the order of 1 part in 10^{11}. By radio-broadcasting signals related to the frequency of the standard, the National Bureau of Standards makes these frequency and time standards available to any other laboratory equipped to receive the signals. Radio station WWV broadcasts such signals with a precision of 1 part in 10^8. Owing to errors in transmission, the signals as remotely received have a precision of about 1 part in 10^7. Recently

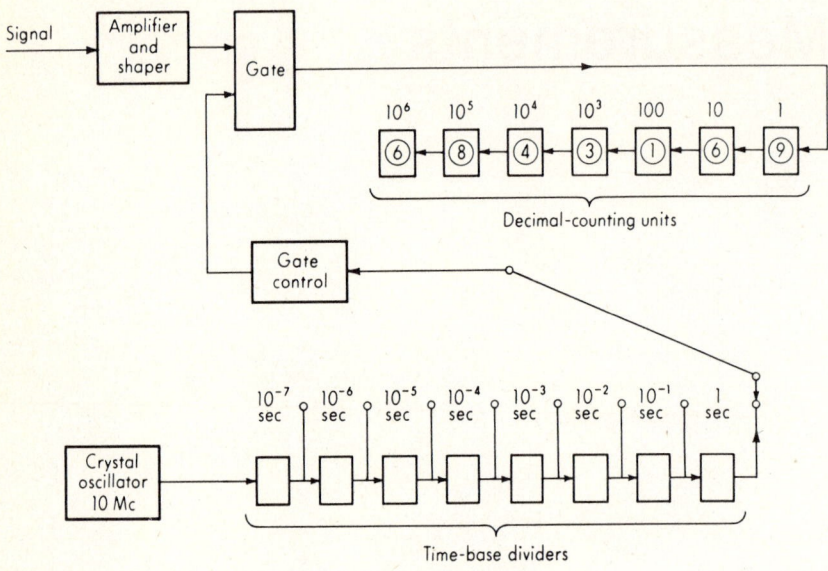

(*a*)   Frequency measurement

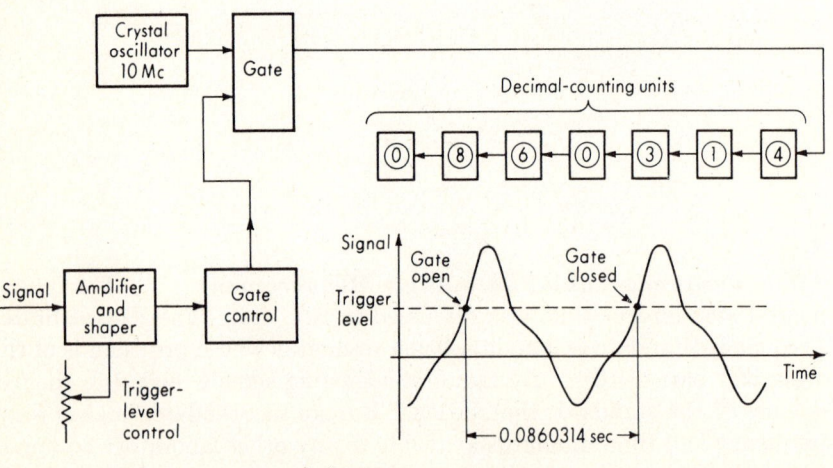

(*b*)   Period measurement

Fig. 9.1.   *Basic counter applications.*

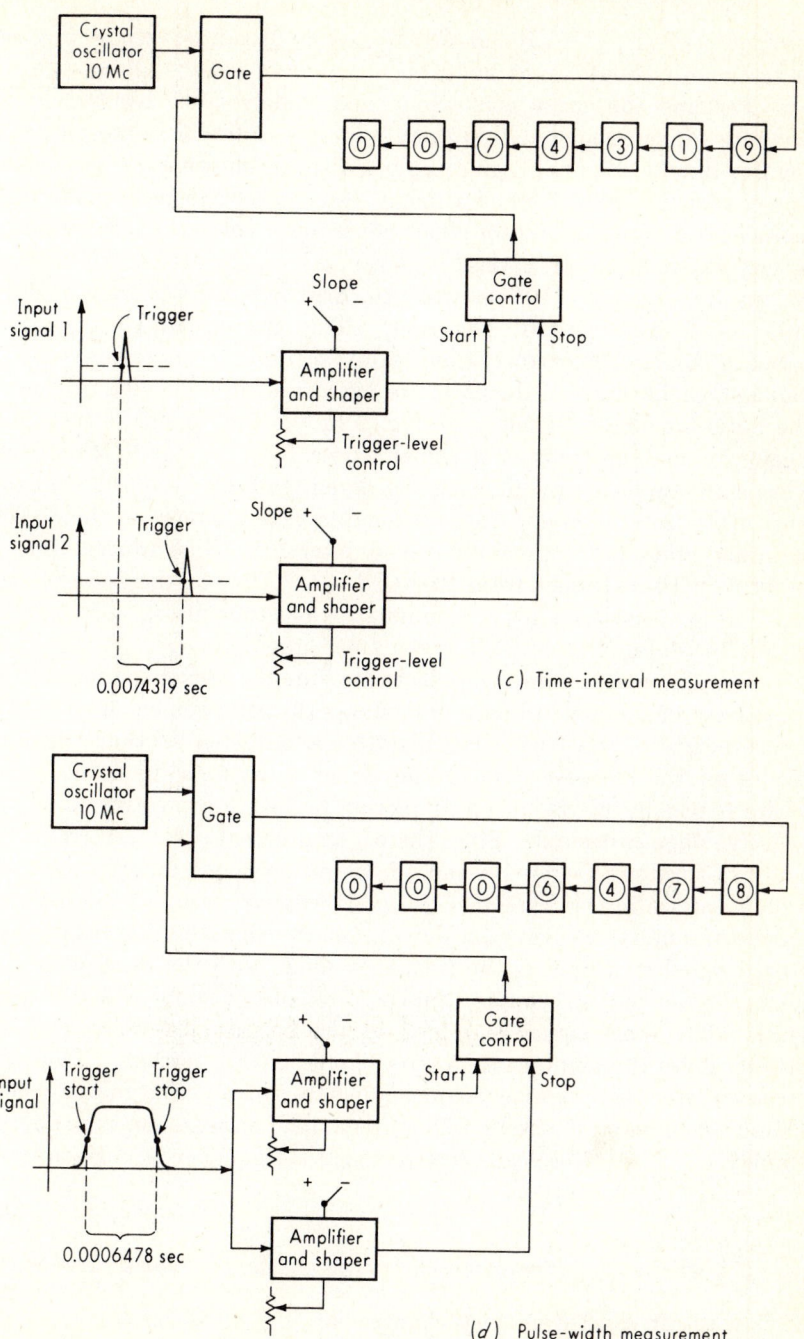

(c) Time-interval measurement

(d) Pulse-width measurement

Fig. 9.1. (*Continued*).

two new stations, WWVB and WWVL, began broadcasts whose precision as received is about 1 part in 10^{10}.

Perhaps the most convenient and widely used instrument for accurate measurement of frequency and time interval is the electronic counter-timer. Figure 9.1 gives a block diagram showing the basic operation of such devices. The instrument's time and frequency standard is a piezoelectric crystal oscillator which generates a voltage whose frequency is very stable since the crystal is kept in a temperature-controlled oven. A typical frequency is 10^7 cps while the drift in frequency may be of the order of 3 parts in 10^7 per week. This gradual drift can, over a period of time, cause errors; thus highly accurate measurements require periodic recalibration of the oscillator against a suitable standard such as the radio-broadcast signals. In Fig. 9.1*a* the instrument is set up for frequency measurement of a signal whose frequency is 6,843,169 cps. This is accomplished by allowing the signal (suitably "shaped" to define each cycle more precisely) to go through a gating circuit to the decimal counting units for a precisely timed interval. This interval may be selected in 10-to-1 steps from 10^{-7} to 1 sec. Thus in the 1-sec interval used in Fig. 9.1*a* the counters accumulate 6,843,169 pulses. This mode of operation is also called EPUT (events-per-unit-time).

Sometimes it is more desirable to measure the period (rather than the frequency) of a signal. Figure 9.1*b* shows the arrangement used for this measurement. The trigger-level control is adjusted so that triggering occurs on the steepest part of the signal wave form to reduce error. There is usually provision for triggering on either a positive slope or a negative slope as desired. Since there is an inherent potential error of ± 1 count in turning the gate on and off, frequency-mode measurements are more accurate for high-frequency signals whereas period-mode measurements are more accurate for low-frequency signals. For example, a 10-cps signal measured in the frequency mode with the usual 1-sec time interval gives only 10 counts; thus an error of ± 1 count is a 10 percent error. The same signal measured in the period mode with a 10-Mc counter gives 10^6 counts and an error of only 0.0001 percent. Thus for a given counter there is some frequency below which period measurements should be used and above which frequency measurements should be employed. For a 1-sec sampling period this frequency f_0 is given by

$$f_0 = \sqrt{f_c} \qquad (9.1)$$

where $f_c \triangleq$ frequency of crystal oscillator ("clock" frequency)

Thus a 10-Mc clock-frequency counter has $f_0 = 3,160$ cps.

Measurements of the time interval between two events are very important in many experimental studies. The basic building blocks

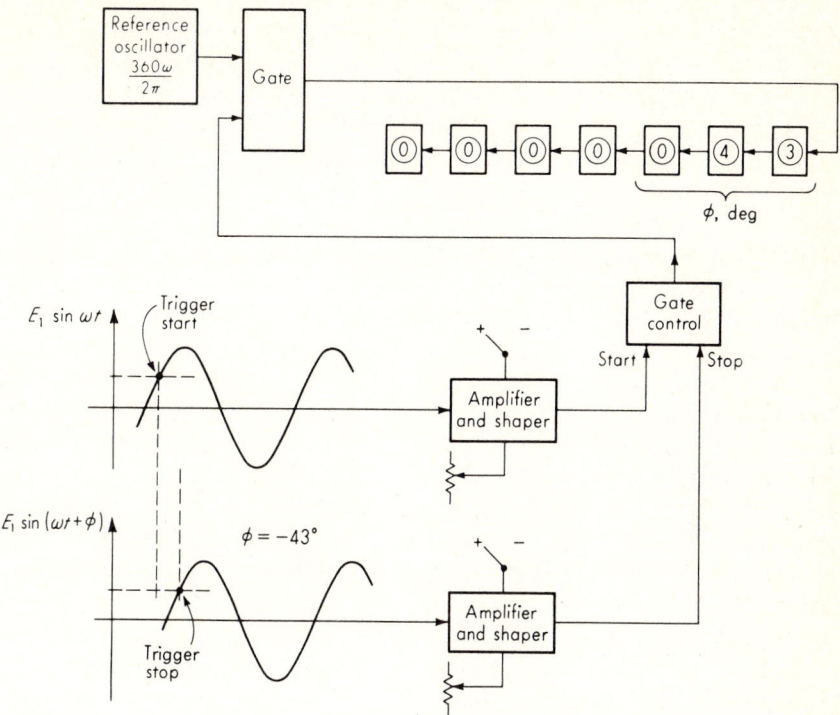

Fig. 9.2.   *Phase-angle measurement.*

described above can be interconnected in a slightly different fashion, as in Fig. 9.1c and d, to accomplish this.   In Fig. 9.1c two separate events have been transduced to electrical pulses; one event pulse is used to open the gate, and the other to close it, thereby timing the interval between them. Considerable versatility in triggering is obtained by providing trigger-level and slope controls on each input.   By using the above arrangement, but only one input signal (Fig. 9.1d), the widths of pulses may be determined.   Additional circuits are sometimes provided to send to an oscilloscope pulses that show the exact point on the incoming signals at which triggering is initiated.   These are helpful in adjusting the trigger-level and slope controls and in interpreting the resulting information.

Measurement of the phase angle between two sinusoidal signals of the same frequency is often required.   The experimental determination of the frequency response of some physical system is a good example of this type of measurement.   A general-purpose digital counter-timer can be used for such measurements, as shown in Fig. 9.2.   To use this method the amplitude of the two signals must be made equal and the triggering

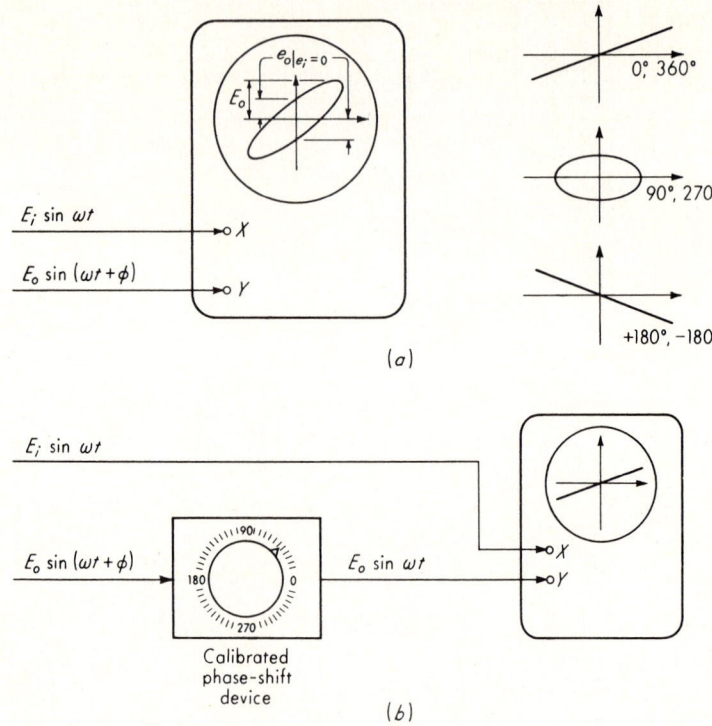

Fig. 9.3. *Phase angle from Lissajous figure.*

point of the two channels adjusted to be the same. Then the phase angle can be read directly with a resolution of 1° for the setup shown, or 0.1° if the reference frequency is set at $3,600f_s$.

Another common method of phase-angle measurement involves cross-plotting the two sinusoidal signals against each other, using an X-Y plotter for very low frequencies and an oscilloscope for high frequencies. The cross plot can be shown to be an ellipse, and suitable measurements on this ellipse give the phase angle (see Fig. 9.3a). We have

$$e_i = E_i \sin \omega t \qquad (9.2)$$

and

$$e_o = E_o \sin (\omega t + \phi) \qquad (9.3)$$

If we set $t = 0$, $e_i = 0$, and $e_o = E_o \sin \phi$. Then

$$\sin \phi = \frac{e_o \big|_{e_i = 0}}{E_o} \qquad (9.4)$$

Since $e_o \big|_{e_i = 0}$ has two values $(+,-)$, the quadrant of ϕ is ambiguous;

however this can usually be resolved by visual observation of the two sine waves plotted against time (say on a dual-beam oscilloscope) or from knowledge of the system characteristics. The direction of travel of the "spot" as it plots the ellipse also resolves this difficulty but may be hard to detect at high frequencies. An alternative method employing the same basic principle but a null technique is shown in Fig. 9.3*b*. Here the calibrated phase-shift circuit is adjusted until the ellipse degenerates into a straight line (0° phase shift). The phase angle ϕ is then read directly from the phase-shifter dial.

9.2

Liquid Level　　Measurement and/or control of liquid level in tanks is an important function in many industrial processes and also in more exotic applications such as the operation and fueling of large liquid-fuel rocket motors. Figure 9.4 illustrates the more common methods of accomplishing this measurement.

The simple float of Fig. 9.4*a* can be coupled to some suitable motion transducer to produce an electrical signal proportional to the liquid level. Figure 9.4*b* shows a "displacer" which has negligible motion and measures the liquid level in terms of buoyant force by means of a force transducer. Since hydrostatic pressure is directly related to liquid level, the pressure-sensing schemes of Fig. 9.4*c* and *d* allow measurement of the liquid level in open and pressure vessels, respectively. In the "bubbler" or purge system of Fig. 9.4*e* the gas pressure downstream of the flow restriction is the same as the hydrostatic head above the bubble-tube end. The flow of gas is quite small; a bottle of nitrogen used as a source of pressurized gas may last six months or more.

Capacitance variation has been employed in various ways for level sensing. For essentially nonconducting liquids (conductivity less than 0.1 μmho/cm³) the bare-probe arrangement of Fig. 9.4*f* may be satisfactory since the liquid resistance R is sufficiently high. For conductive liquids the probe must be insulated as in Fig. 9.4*g* to prevent short-circuiting of the capacitance by the liquid resistance. The measurement of the capacitance between the terminals *ab* may be accomplished in several ways. However high-frequency a-c (radio-frequency) methods offer significant advantages. Capacitance level-sensing techniques have been used with many common liquids, powdered or granular solids, liquid metals (high temperatures), liquefied gases (low temperatures), corrosive materials such as hydrofluoric acid, and in very high-pressure processes.

Figure 9.4*h* illustrates the use of radioisotopes for level measurement. Since the absorption of beta-ray or gamma-ray radiation varies with the thickness of absorbing material between the source and the detector, a

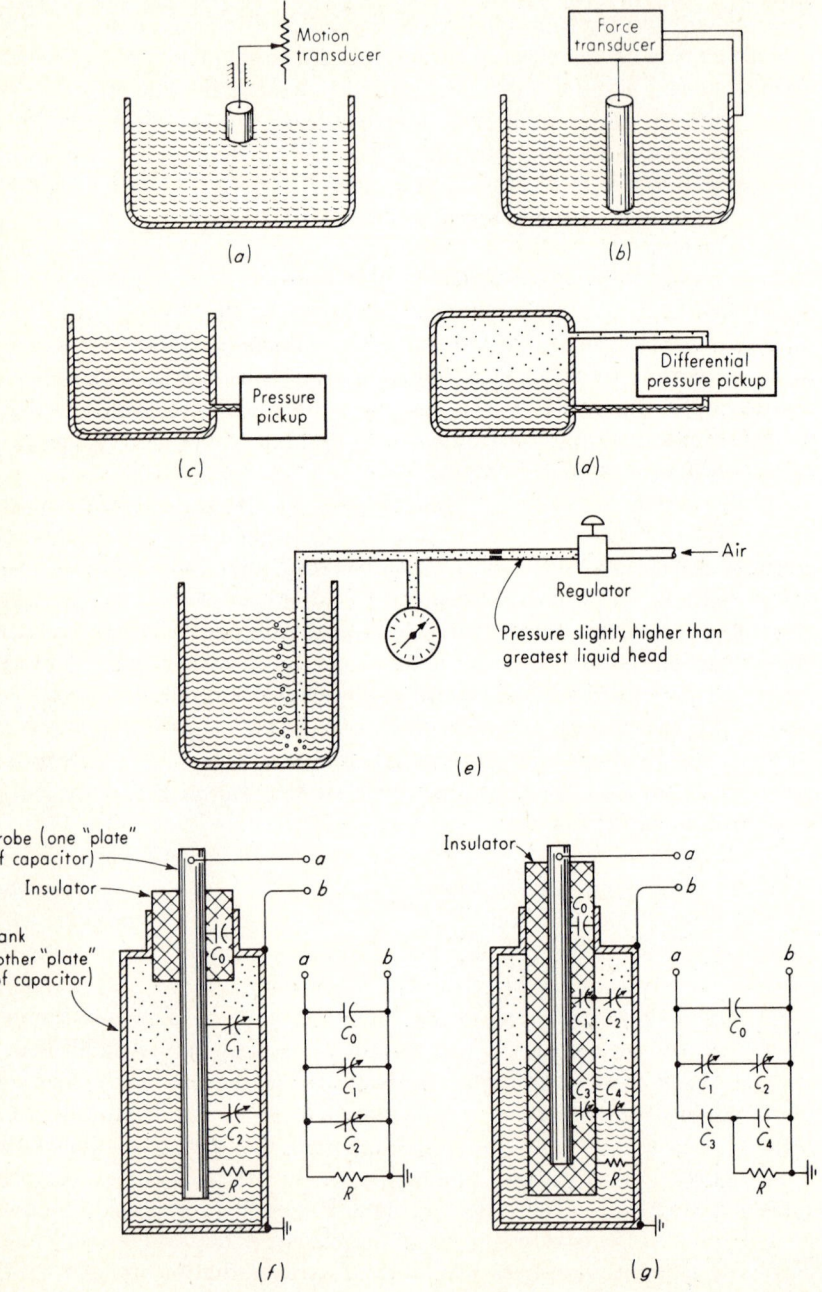

Fig. 9.4.   *Liquid-level measurement.*

signal related to tank level may be developed. For analyzing such arrangements one may use the law

$$I = I_o e^{-\mu \rho x} \qquad (9.5)$$

where $I \triangleq$ intensity of radiation falling on detector
   $I_0 \triangleq$ intensity at detector with absorbing material not present
   $e \triangleq$ base of natural logarithms
   $\mu \triangleq$ mass absorption coefficient (constant for given source and absorbing material), cm²/g
   $\rho \triangleq$ mass density of absorbing material, g/cm³
   $x \triangleq$ thickness of absorbing material, cm

The gamma-ray source cesium 137 has been widely used for liquid-level measurements and has $\mu = 0.077$ cm²/g for water or oil, 0.074 for aluminum, 0.072 for steel, and 0.103 for lead. For gaging a tank of water, then, if a vertical radiation path (rather than the angled one of Fig. 9.4h) is assumed, the variation of I with liquid height h is given by (neglecting absorption of air path)

$$I = I_0 e^{-0.077h} \qquad (9.6)$$

This exponential relation of I and h is nearly linear only for sufficiently small values of h. For h as large as, say, 100 cm, the nonlinearity is quite apparent. This can be overcome by using either a radiation source or detector in the form of a strip oriented vertically rather than a "point" source or detector. Such arrangements are nonlinear for small values of $\mu \rho x_{max}$. Therefore point-to-point configurations are indicated for small ranges, whereas larger ranges require the more complex strip-to-point

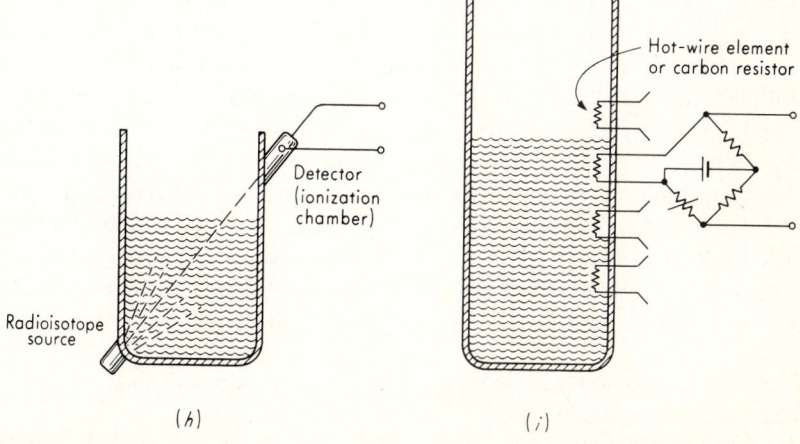

Hot-wire element or carbon resistor

Detector
(ionization
chamber)

Radioisotope
source

(h) (i)

Fig. 9.4. (*Continued*)

type. For a strip source (or detector), the strength (sensitivity for a detector), can be "tailored" to vary in just the right way with position along the strip to give a linear tank-level/detector-signal relation.

Figure 9.4*i* shows the method of using hot-wire or carbon resistor elements for the measurement of liquid level in discrete increments. The basic concept is that the heat-transfer coefficient at the surface of the resistance element changes radically when the liquid surface passes it. This changes its equilibrium temperature and thus its resistance, causing a change in bridge output voltage. By locating resistance elements at known height intervals, the tank level may be measured in discrete increments. Such arrangements have been used in filling fuel tanks of large rocket engines with cryogenic liquid fuels.

9.3

Humidity   Knowledge of the amount of water vapor in the air is important to the operation and/or automatic control of many industrial processes. This information may be gathered and presented in a number of ways, depending on the needs of the particular process and the measuring instrumentation used. In common use are relative humidity (ratio of water partial pressure to saturation pressure), dew-point temperature, mixing ratio or specific humidity (mass of water per unit mass of dry gas), and volume ratio (parts of water vapor per million parts of air).

The ultimate standard for calibration of humidity-measuring devices is the National Bureau of Standards standard gravimetric hygrometer. This is a strictly laboratory apparatus in which the water vapor in an air sample is absorbed by suitable chemicals and then very carefully weighed. It directly determines the mixing ratio in grams per kilogram, covers the range 0.30 to 20.0, and has an uncertainty (systematic error plus three standard deviations) of about 0.1 percent of the reading. For less critical calibrations the National Bureau of Standards uses its two-pressure humidity generator.[1] This equipment generates air/water mixtures in the dew-point range $-70°C$ (uncertainty $\pm 1.2C°$) to $+25°C$ ($\pm 0.1C°$), relative-humidity range 10 to 98 percent at temperatures ranging from $-55°C$ (relative-humidity uncertainty ± 2.5 percent) to $+40°C$ (relative-humidity uncertainty 0.5 percent), mixing ratio in the range 0.0013 g/kg (uncertainty ± 0.0003 g/kg) to 20 g/kg (uncertainty ± 0.5 percent), and volume ratio 2 ppm (± 0.5 ppm) to 30,000 ppm (± 0.5 percent).[2]

---

[1] A. Wexler and R. D. Daniels, Pressure-Humidity Apparatus, *J. Res. Natl. Bur. Std.*, vol. 48, p. 269, 1952.

[2] Humidity Calibration Service, *Instr. Control Systems*, p. 123, November, 1964.

Classically, relative humidity has been found from psychrometric charts and the temperature readings of two thermometers. One, the dry-bulb thermometer, reads the ordinary air temperature while the other, the wet-bulb, is intended to read the temperature of adiabatic saturation. The latter measurement requires that the bulb be kept wet and a suitably high (about 1,000 fpm) air velocity be maintained over the wet bulb. While these operations may be automated to a certain extent, the complexity of the calculations equivalent to the psychrometric chart hinders development of this technique into a continuous-reading instrument. Furthermore, the evaporation process at the wet bulb adds moisture to the air, thus disturbing the measured medium.

For continuous recording and/or control of relative humidity, electrical transducers of the Dunmore type are widely used. These were first developed about 1944 by F. W. Dunmore of the National Bureau of Standards and are basically a resistance element which changes resistance with relative humidity. The resistance element is constructed of a dual winding of noble-metal wires on a plastic form with a definite spacing between them. When the windings are coated with a lithium chloride solution a conducting path is formed between the windings. The electrical resistance of this path is found to vary reproducibly with the relative humidity of the surrounding air and may thus be used as a sensing element. Bridge-type resistance-measuring circuitry with a-c excitation is normally employed. The resistance/relative-humidity relation is quite nonlinear, and a single transducer generally can cover only a small range of the order of 10 percent relative humidity. Where large ranges (as great as 5 to 99 percent relative humidity) are needed, seven or eight of the transducers, each designed for a specific part of the total range, are combined in a single package. A single narrow-range sensing element may have an inaccuracy of the order of 1.5 percent relative humidity, resolution about 0.15 percent relative humidity, time constant as small as 3 sec, and size 1 in. in diameter by 2 in. long. Since these units are also sensitive to temperature, some form of temperature compensation may be required. The sensors do not add or subtract moisture or heat from their environment in significant amounts and may thus be used in sealed areas. Working temperatures in the range of -40 to $+150°F$ are possible.

Dew-point temperature can be determined by noting the temperature of a polished metal surface (mirror) when the first traces of condensation ("fogging") appear. Commercial devices in which this operation has been completely automated by means of a feedback system are available. Figure 9.5 shows the operation of a typical system.[1] The mirror (0.13-in.-diameter by 0.003-in.-thick rhodium-plated copper) is cooled by a CO_2

[1] Burton Manufacturing Co., Los Angeles, Calif.

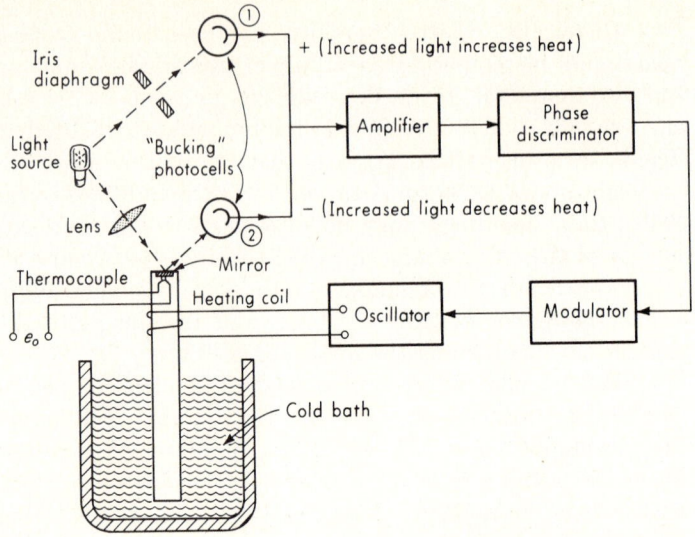

Fig. 9.5. *Automatic dew-point sensor.*

and acetone bath and heated by an induction coil. Initial adjustment of the system by a human operator consists of setting the opening of the iris diaphragm so that the difference in light received by photocells 1 and 2 is just sufficient to produce enough heating to maintain the lightest discernible fog on the mirror, which is observed with a 5-power magnifier. Since the light received by photocell 2 is reduced as the fogging of the mirror increases, the heating power is increased if the fog builds up and is decreased if the fog disappears. The system thus tends at all times to maintain automatically the degree of fog initially set into it. Dew-point temperature (mirror temperature) is continuously measured by the mirror thermocouple. An accuracy of $\pm 1C°$ over the dew-point range -58 to $+65°F$ is attained by the instrument described above.

The lithium chloride sensor mentioned above under relative humidity can be modified to give a signal related to dew-point temperature.[1] The dual wire windings are supplied with a-c power, causing a heating of the lithium chloride film. Lithium chloride shows a very sharp decrease in electrical resistance when *its* relative humidity increases above about 11 percent. Thus, when surrounded by moist air the lithium chloride momentarily absorbs moisture, its resistance drops, allowing more current to flow and more heat to be generated. This raises the temperature, driving off excess moisture, increasing the resistance, and reducing the heating.

[1] E. J. Amdur, Humidity Sensors, *Instr. Control Systems*, p. 93, June, 1963.

The sensor itself thus regulates its temperature so that the relative humidity of the lithium chloride element stays near 11 percent no matter what the moisture content of the surrounding air.   It has been established that the temperature attained by the lithium chloride element, while not equal to the dew-point temperature, is directly related to it.   Thus by measuring this temperature with some appropriate sensor, the dew-point temperature may be established.   Probes of this type cover a dew-point range of -50 to $+160°F$ with an error of the order of 1 or 2F°.

The final humidity sensor considered here is the electrolytic type.[1] Here a continuous flow of sample gas (100 cm³/min, regulated ± 2 percent) is passed through an analyzer tube.   This tube has two platinum wires wound in a double helix on the inside of the tube.   The space between the wires is coated with a strong desiccant (phosphorous pentoxide) and a d-c potential applied to the wire ends.   When moisture in the sample gas is taken up by the P_2O_5, the water is electrolyzed into hydrogen and oxygen gas and a measurable electrolysis current flows.   Such instruments have an inaccuracy of the order of 5 percent of full scale, ranges of 0 to 100 to 0 to 10,000 ppm, resolution better than 1 ppm, and a time constant of the order of 30 sec.   They have also been adapted to the measurement of the water content of liquids and solids.

9.4

Chemical Composition   In years past the chemical composition of materials was necessarily determined by taking a sample to a laboratory and performing the required chemical tests, usually with somewhat tedious procedures.   Today, many important measurements of this type are made on a relatively continuous and automatic basis without the need for a human operator.   The need for such measuring systems is due largely to the desire automatically to control product quality directly in terms of its chemical composition rather than inferring it from measurements of temperature, pressure, flow rate, etc.   Even in manually controlled situations the desire to increase production rates while also maintaining or improving quality leads to a need for rapid analysis methods.   Rapid and accurate analysis techniques are also very useful in research and development problems.   These needs of industry have led to the development of a wide variety of instruments for measuring various aspects of chemical composition and related quantities.

Some examples of measurements of the above type include analysis of products of combustion, monitoring of the composition of dissolved gases in oil-well drilling mud, detection of alcohol contaminant in a heavy

---

[1] Consolidated Electrodynamics, Pasadena, Calif., *Bull.* 26303, June, 1963.

hydrocarbon liquid stream, detection of explosive solvents in the atmosphere within a uranium-extraction kettle, measurement of pH of industrial-waste effluent to control river pollution, determination of alloying constituents in metals, outgassing of materials under high vacuum, rocket-borne instruments for analyzing atmospheric gases at high altitudes, air-pollution studies, and analysis of anesthetic gases in blood. A wide variety of techniques[1] have been developed for handling such problems. A discussion of these methods adequate for their selection and use is beyond the scope of this text. However the references given will serve this purpose for the interested reader.

Problems

9.1 Derive Eq. (9.1).

9.2 Prove that a plot of $A_o \sin (\omega t + \phi)$ against $A_i \sin \omega t$ is an ellipse.

9.3 Analyze the system of Fig. 9.4a to obtain the transfer function relating liquid level h_i to float motion x_o. Neglect dry-friction effects. If the liquid level increases very slowly and the float motion is subject to a dry-friction force F_f, develop a formula to estimate the maximum steady-state error.

9.4 Assume the force transducer in Fig. 9.4b is of the elastic deflection type and obtain the transfer function relating liquid level h_i to force-transducer deflection x_o.

9.5 Discuss the effect of liquid density changes on the accuracy of liquid-level measurement in the systems of Figs. 9.4a and 9.4b.

9.6 For the system of Fig. 9.4a, discuss the effect on static and dynamic behavior of using a float that is a body of revolution but *not* a cylinder.

9.7 Repeat Prob. 9.6 for the system of Fig. 9.4b.

9.8 Discuss interfering and/or modifying inputs for the system of Fig. 9.4d. Assume the pressure pickup itself to be insensitive to such inputs.

9.9 Repeat Prob. 9.8 for the system of Fig. 9.4e.

Bibliography

1. S. J. Goldwater: Phase-angle Measurement in Control Systems, *Trans. Soc. Instr. Tech. (London)*, p. 100, June, 1960.

2. R. J. A. Paul and M. H. McFadden: Measurement of Phase and Amplitude at Low Frequencies, *Electron. Eng.*, vol. 31, no. 373, March, 1959.

3. F. J. Huddleston: Frequency Response by Sum or Difference, *Control Eng.*, p. 113, October, 1957.

4. Timers, *Electromech. Design*, p. 51, March, 1961.

5. Electric Timing Motors, *Electromech. Design*, p. 59, May, 1961.

6. Electronic Tuning Fork Beats Time for Accuracy, *Machine Design*, p. 30, October 27, 1960.

7. Time Interval Measurement, *Instr. Control Systems*, p. 125, September, 1962.

8. P. Young: 1 Nanosecond Time Interval Counter, *Instr. Control Systems*, p. 105, January, 1965.

[1] D. M. Considine (ed.), "Process Instruments and Controls Handbook," sec. 6, McGraw-Hill Book Company, New York, 1957.

9. A. MacMullen: Sources of Error in Phase Measurement, *Instr. Control Systems*, p. 91, January, 1965.

10. A New Approach to Precision Time Measurements, *Gen. Radio Experimenter*, General Radio Co., West Concord, Mass., February–March, 1965.

11. Correlating Time from Europe to Asia with Flying Clocks, *Hewlett-Packard J.*, Hewlett-Packard Co., Palo Alto, Calif., April, 1965.

12. Level Measurement and Control, *Instr. Control Systems*, p. 148, March, 1965.

13. N. Z. Alcock, and S. K. Ghosh: Minimizing Measurement Errors in Nuclear Gages, *Control Eng.*, p. 87, May, 1961.

14. F. W. Hannula: Use Capacitance for Accurate Level Measurement, *Control Eng.*, p. 104, November, 1957.

15. R. C. Muhlenhaupt and P. Smelser: Carbon Resistors for Cryogenic Liquid Level Measurement, *Natl. Bur. Std. (U.S.)*, *Tech. Note* 200, 1963.

16. W. A. Olsen: An Integrated Hot Wire–Stillwell Liquid Level Sensor System for Liquid Hydrogen and Other Cryogenic Fluids, *NASA, Tech. Note* D-2074, 1963.

17. Liquid Hydrogen Level Sensors, *Instr. Control Systems*, p. 129, May, 1964.

18. R. L. Rod: Propellant Gaging and Control, *Instr. Control Systems*, p. 119, October, 1962.

19. G. H. Burger: Reliable Level Measurements for Liquid Metals, *Control Eng.*, p. 131, July, 1959.

20. F. Marton: Level Measurement and Control, *Instr. Control Systems*, p. 107, January, 1965.

21. E. Ulicki: Propellant Gaging System for Apollo Spacecraft, *Space/Aeron.*, p. 68, October, 1964.

22. D. D. Kana: A Resistive Wheatstone Bridge Liquid Wave Height Transducer, *NASA, CR*-56551, 1964.

23. L. Siegel: Nuclear and Capacitance Techniques for Level Measurement, *Instr. Control Systems*, p. 129, July, 1964.

24. N. H. Roos: Level Measurement in Pressurized Vessels, *ISA J.*, p. 55, May, 1963.

25. Wire Matrix Gages Zero-*g* Liquids, *Machine Design*, p. 10, Feb. 16, 1961.

26. C. L. Pleasance: Accurate Volume Measurement of Large Tanks, *ISA J.*, p. 56, May, 1961.

27. Moisture and Humidity, *Instr. Control Systems*, p. 121, October, 1964.

28. D. J. Fraade: Measuring Moisture in Gases, *Instr. Control Systems*, p. 100, April, 1963.

29. R. E. Fishburn: Measurement and Control of Humidity, *Automation*, p. 61, January, 1963.

30. R. M. Atkins: Wet/Dry Bulb Thermistor Hygrometer with Digital Indication, *Instr. Control Systems*, p. 111, April, 1964.

31. H. Hellivig: Frequency Standards and Clocks: A Tutorial Introduction, *NBS Tech. Note* 616, 1972.

32. E. H. Schulte: Carbon Resistors for Multipoint Level Sensing, *Cryogenic Tech.*, September/October, 1970.

33. B. E. Dozer: Liquid Level Measurement for Hostile Environment, *Inst. Tech.*, February, 1967.

34. O. W. Schoen: A Continuously-Variable Humidity Reference, *Inst. and Cont. Sys.*, October, 1972.

35. Survey of Humidity and Moisture Instrumentation, *Inst. and Cont. Sys.*, January, 1972.

Part Three
Manipulation, Transmission, and Recording of Data

Chapter 10
Manipulating, Computing, and Compensating Devices

The information or data generated by a basic measuring device generally require "processing" or "conditioning" of one sort or another before they are finally presented to the observer as an indication or record. Devices for accomplishing these operations may be specific to a certain class of measuring sensors or they may be quite general-purpose. In this chapter we briefly consider those devices most often needed in building up measurement systems.

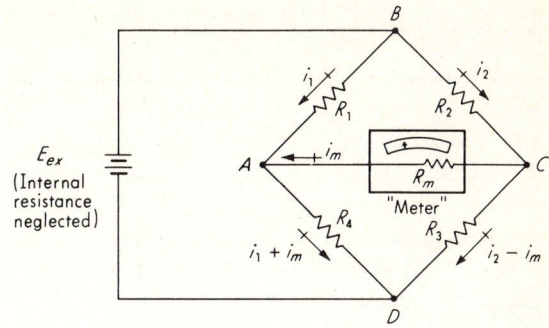

Fig. 10.1. *Basic Wheatstone bridge.*

10.1

Bridge Circuits   Bridge circuits of various types are widely used for the measurement of resistance, capacitance, and inductance. Since we have seen that many transducers convert some physical variable into a resistance, capacitance, or inductance change, bridge circuits are of considerable interest. While capacitance and inductance bridges are important, the simpler resistance bridge is in the widest use, and we shall thus concentrate on it here. Adequate technical literature on all types of bridge circuits is readily available.[1]

Figure 10.1 shows a purely resistive (Wheatstone) bridge in its simplest form. The excitation voltage E_{ex} may be either d-c or a-c voltage; we here consider only direct voltage. In measurement applications, one or more of the legs of the bridge is a resistive transducer such as a strain gage, resistance thermometer, or thermistor. The basic principle of the bridge may be applied in two different ways: the null method and the deflection method. Let us assume that the resistances have been adjusted so that the bridge is balanced; that is, $e_{AC} = 0$. (It is easily shown that this requires $R_1/R_4 = R_2/R_3$.) Now we let one of the resistors, say R_1, change its resistance. This will unbalance the bridge and a voltage will appear across AC, causing a meter reading. The meter reading is an indication of the change in R_1 and can actually be used to compute this change. This method of measuring the resistance change is called the *deflection method*, since the meter deflection indicates the resistance change. In the *null method*, one of the resistors is manually adjustable. Thus if R_1 changes, causing a meter deflection, R_2 can be manually adjusted until its effect just cancels that of R_1 and the bridge

[1] E. Frank, "Electrical Measurement Analysis," chaps. 10 and 13, McGraw-Hill Book Company, New York, 1959.

is returned to its balanced condition. The adjustment of R_2 is guided by the meter reading; R_2 is adjusted so that the meter returns to its null or zero position. In this case the numerical value of the change in R_1 is directly related to the change in R_2 required to effect balance.

Both the null and deflection methods are used in practice. In the deflection method a calibrated meter is needed, and if the excitation E_{ex} changes, an error is introduced, since the meter reading is changed by changes in E_{ex}. With the null method, a calibrated variable resistor is needed, and since there is no meter deflection when the final reading is made, no error is caused by changes in E_{ex}. The deflection method gives an output voltage across terminals AC that almost instantaneously follows the variations of R_1. This output voltage can be applied to an oscilloscope (rather than the meter shown in Fig. 10.1) and thus measurements of rapid dynamic phenomena are possible. The null method, on the other hand, requires that the balancing resistor be adjusted to null the meter before a reading can be taken. This adjustment takes considerable time if done manually; even when an instrument servomechanism makes the adjustment automatically, the time required is much longer than is allowable for measuring many rapidly changing variables. Thus the choice of the null or the deflection method in a given practical case depends on the speed of response, drift, etc., required by the particular application.

In order to obtain quantitative relations governing the operation of the bridge circuit, a circuit analysis is necessary. The following information is desired:

1. What relation exists among the resistances when the bridge is balanced ($e_{AC} = 0$)? The answer to this has already been given as $R_1/R_4 = R_2/R_3$.
2. What is the sensitivity of the bridge? That is, how much does the output voltage e_{AC} change per unit change of resistance in one of the legs?
3. What is the effect of the meter internal resistance on the measurement?

We shall consider the question of bridge sensitivity first for the case where the "meter" has a very high internal resistance R_m. If this is the case, the meter current i_m will be negligible compared with the currents in the legs. This situation is closely approximated in practice in the following cases:

1. The "meter" is an oscilloscope. The internal or input resistance of a typical oscilloscope is of the order of 1 million ohms. When

the legs of the bridge are each of the order of 10,000 ohms or less, the current in the oscilloscope will be of the order of $\frac{1}{100}$ or less of the currents in the legs and may thus be neglected. In a typical strain-gage bridge, $R_1 = R_2 = R_3 = R_4 = 120$ ohms, and so the oscilloscope current would be extremely negligible.

2. The voltage e_{AC} is applied to the input of an electronic amplifier. Here again, the input resistance of amplifiers is often much higher than the resistance of the legs in the bridge, and thus the current may be treated as effectively zero. (Actually, the input to an oscilloscope is through an amplifier; thus cases 1 and 2 are not really different.)

3. The voltage e_{AC} is measured with a potentiometer. In the potentiometer method of voltage measurement the unknown voltage (such as e_{AC}) is bucked against a known and adjustable voltage of opposite polarity supplied by the potentiometer. The known voltage is adjusted until a galvanometer indicates the two bucking voltages are equal. The value is then read from the calibrated dial of the potentiometer. When the potentiometer is balanced, no current is being drawn from terminals AC. (Actually the potentiometer cannot be *perfectly* balanced since any galvanometer has a threshold sensitivity below which it cannot detect the presence of current.)

Since it appears the condition of $i_m = 0$ is quite closely approximated in many practical cases, it will be worthwhile to study this situation. We have

$$i_1 = \frac{E_{ex}}{R_1 + R_4} \qquad (10.1)$$

$$i_2 = \frac{E_{ex}}{R_2 + R_3} \qquad (10.2)$$

$$e_{AB} = \text{voltage rise from } A \text{ to } B = i_1 R_1 = \frac{R_1}{R_1 + R_4} E_{ex} \qquad (10.3)$$

$$e_{CB} = \frac{R_2}{R_2 + R_3} E_{ex} \qquad (10.4)$$

and finally

$$e_{AC} = e_{AB} + e_{BC} = e_{AB} - e_{CB} = \left(\frac{R_1}{R_1 + R_4} - \frac{R_2}{R_2 + R_3} \right) E_{ex} \qquad (10.5)$$

Thus we see that the output voltage is a linear function of the bridge excitation E_{ex} but, in general, a *nonlinear* function of the resistances R_1, R_2, R_3, and R_4. If the bridge is initially balanced and then R_1, say, begins to change, the output voltage signal will *not* be directly proportional to the change in R_1. For certain practically important special cases, however, perfect linearity is possible. The best example of this is found in many strain-gage transducers in which, at the balanced condition, $R_1 = R_2 = R_3 = R_4 = R$. Also, the resistance changes are such that $+\Delta R_1 = -\Delta R_2 = +\Delta R_3 = -\Delta R_4$. We may then write

$$e_{AC} = \left[\frac{R_1 + \Delta R_1}{(R_1 + \Delta R_1) + (R_4 + \Delta R_4)} - \frac{R_2 + \Delta R_2}{(R_2 + \Delta R_2) + (R_3 + \Delta R_3)} \right] E_{ex}$$

(10.6)

$$e_{AC} = \frac{\Delta R_1}{R} E_{ex} \qquad (10.7)$$

Clearly, Eq. (10.7) shows a strictly linear relationship of e_{AC} with ΔR_1.

Even when the above symmetry does not exist, the bridge response is very nearly linear as long as the ΔR's are small percentages of the R's. In strain gages, for example, the ΔR's rarely exceed 1 percent of the R's. Since the case of small ΔR's is of practical interest, we shall work out an expression for bridge sensitivity that is a good approximation for such a situation. From Eq. (10.5), $e_{AC} = f(R_1, R_2, R_3, R_4)$, and thus for small changes from the null condition we may write approximately

$$\Delta e_{AC} = e_{AC} = \frac{\partial e_{AC}}{\partial R_1} \Delta R_1 + \frac{\partial e_{AC}}{\partial R_2} \Delta R_2 + \frac{\partial e_{AC}}{\partial R_3} \Delta R_3 + \frac{\partial e_{AC}}{\partial R_4} \Delta R_4 \qquad (10.8)$$

Now,

$$\frac{\partial e_{AC}}{\partial R_1} = E_{ex} \frac{R_4}{(R_1 + R_4)^2} \qquad \text{volts/ohm} \qquad (10.9)$$

$$\frac{\partial e_{AC}}{\partial R_2} = -E_{ex} \frac{R_3}{(R_2 + R_3)^2} \qquad (10.10)$$

$$\frac{\partial e_{AC}}{\partial R_3} = E_{ex} \frac{R_2}{(R_2 + R_3)^2} \qquad (10.11)$$

$$\frac{\partial e_{AC}}{\partial R_4} = -E_{ex} \frac{R_1}{(R_1 + R_4)^2} \qquad (10.12)$$

The partial derivatives are taken as constants; thus Eq. (10.8) shows a linear relation between e_{AC} and the ΔR's.

We have explained above, in a qualitative fashion, that if the meter resistance is "high enough" the terminals AC may be thought of as an open circuit (no current i_m). It would be useful to have a more quantitative method of deciding whether the meter resistance was "high enough" and also, if it were not, how to correct for it. This will now be done.

By using Thévenin's theorem, the bridge circuit and the "meter" that loads it may be represented as in Fig. 10.2. Since we have been calling the bridge output voltage under assumed open-circuit conditions e_{AC}, this becomes the E_o of Fig. 3.22. Let us call the bridge output

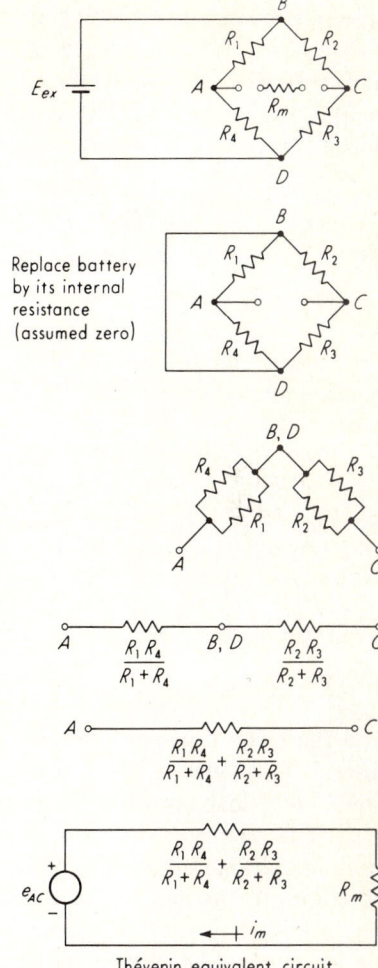

Fig. 10.2.   *Thévenin analysis of bridge.*

Thévenin equivalent circuit

under the actual loaded condition e_{ACL}.   We can then immediately write

$$i_m = \frac{e_{AC}}{R_{\text{total}}} = E_{ex}\frac{R_1/(R_1 + R_4) - R_2/(R_2 + R_3)}{R_m + R_1R_4/(R_1 + R_4) + R_2R_3/(R_2 + R_3)} \qquad (10.13)$$

Knowing i_m, we can now compute the actual voltage e_{ACL} across the meter under the condition where the meter draws current, since the voltage across the meter will be the product of the current i_m and the meter resistance R_m.   Carrying this out and simplifying, we get

$$e_{ACL} = \frac{E_{ex}(R_1R_3 - R_2R_4)}{(R_1 + R_4)(R_2 + R_3) + [(R_1 + R_4)R_2R_3 + R_1R_4(R_2 + R_3)]/R_m} \qquad (10.14)$$

Now
$$e_{AC} = \frac{E_{ex}(R_1 R_3 - R_2 R_4)}{(R_1 + R_4)(R_2 + R_3)} \qquad (10.15)$$

and if we wish to display the effect of the meter resistance on the bridge output voltage we can form the ratio of e_{ACL} to e_{AC}. After some manipulation this can be shown to be

$$\frac{e_{ACL}}{e_{AC}} = \frac{1}{1 + (1/R_m)[R_2 R_3/(R_2 + R_3) + R_1 R_4/(R_1 + R_4)]} \qquad (10.16)$$

We now have a quantitative way of assessing the effect of the meter resistance R_m on the bridge output. We see that, if $R_m = \infty$, $e_{ACL} = e_{AC}$, as one would expect. If R_m is not infinite, there will be a *reduction* in the output signal, and the magnitude of this reduction depends on the relative values of R_m and the bridge "equivalent resistance" R_e which is defined as

$$R_e \triangleq \frac{R_2 R_3}{R_2 + R_3} + \frac{R_1 R_4}{R_1 + R_4} \qquad (10.17)$$

In terms of R_e, Eq. (10.16) becomes

$$\frac{e_{ACL}}{e_{AC}} = \frac{1}{1 + R_e/R_m} \qquad (10.18)$$

Thus, if $R_m = 10R_e$,

$$\frac{e_{ACL}}{e_{AC}} = \frac{1}{1.1} = 0.91 \qquad (10.19)$$

and there is a 9 percent loss in signal due to the noninfinite meter resistance. This type of loss is usually referred to as a *loading effect;* that is, the meter "loads down" the bridge and reduces its sensitivity.

The theory developed above is useful in assessing the effects of various parameters on the bridge sensitivity and could actually be used to compute the sensitivity if all quantities were known exactly. It is preferable, however, to calibrate the bridge directly by introducing a known resistance change and noting the effect on the bridge output. This known resistance change is usually introduced by means of the arrangement shown in Fig. 10.3. The resistance R_c of the calibrating resistor is accurately known. If the bridge is originally balanced with the switch open, when the switch is closed the resistance in leg 1 will change and the bridge will be unbalanced. The output voltage e_{AC} is read on the meter, and the resistance change ΔR that caused this voltage is computed from

$$\Delta R = R_1 - \frac{R_1 R_c}{R_1 + R_c} \qquad (10.20)$$

The bridge sensitivity is then

$$S \triangleq \frac{e_{AC}}{\Delta R} \qquad \text{volts/ohm} \qquad (10.21)$$

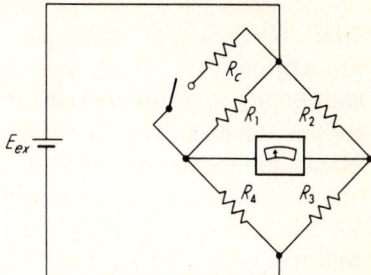

Fig. 10.3. *Shunt calibration method.*

This procedure gives an overall calibration, since the values of all the resistors and the battery voltage are taken into account.

Figure 10.1 shows a bridge circuit with the bare essentials. Often additional features are necessary or desirable for the convenience of the user. Figure 10.4 shows a versatile arrangement providing the following capabilities:

1. Variation of overall sensitivity without the need to change E_{ex}
2. Provision for adjusting the output voltage to be precisely zero when the measured physical quantity is zero even if the legs are not exactly matched
3. Shunt resistor calibration

Commercial transducers may also include additional temperature-sensitive resistors to achieve temperature compensation.

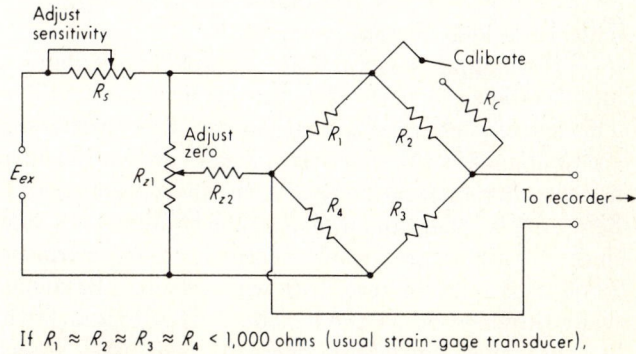

If $R_1 \approx R_2 \approx R_3 \approx R_4 < 1,000$ ohms (usual strain-gage transducer), then
$$R_{z2} \approx 100 \, R_1$$
$$R_{z1} \approx 25,000 \text{ ohms}$$

Fig. 10.4. *Bridge with sensitivity, balance, and calibration features.*

10.2

Amplifiers   Since the electrical signals produced by most transducers are at a low voltage and/or power level, it is often necessary to amplify them before they are suitable for transmission, further manipulation, indication, or recording. While the design of amplifiers is beyond the scope of this text, the criteria to be applied in choosing and using an amplifier for a specific measuring-system application are of interest and will be briefly reviewed here.

General performance characteristics of amplifiers. The dynamic response of an amplifier must equal or exceed that of the transducer feeding it and is usually specified as the essentially flat range of frequency response. In order not to draw much current from the transducer (thereby "loading" it and causing loss of sensitivity and/or linearity), the input impedance of the amplifier must be sufficiently high relative to the transducer output impedance. Such questions should be studied for both static and dynamic conditions in critical applications, using the general impedance methods developed in Chap. 3. Similarly, the output impedance of the amplifier must be sufficiently low relative to the input impedance of the following device (often a recorder) that loading effects at this point are not excessive. Amplifier output impedance can also be interpreted as current-supplying ability and is sometimes quoted as such. For example, a particular amplifier may be quoted as supplying a full-scale voltage of ± 10 volts to any load with resistance R_L greater than 1,000 ohms and a full-scale current of 0.010 amp to any load with resistance less than 1,000 ohms. For such an amplifier we note that the voltage gain is independent of load resistance for $R_L > 1,000$ but decreases progressively for $R_L < 1,000$, full-scale voltage output being only 1 volt, for example, if $R_L = 100$ ohms. In general, a low output impedance indicates a high current capacity.

If the input terminals of an amplifier are short-circuited so that the input voltage is exactly zero, the output voltage will *not* be precisely zero. This defect is charged to two sources, zero drift and noise. Zero drift is a slow change in the short-term mean value of the output voltage whereas noise is a fast fluctuation around the short-term mean value (see Fig. 10.5). Zero drift is generally large until the equipment is thoroughly warmed up since its origins are largely thermal. Once warm-up is over, a certain amount of random zero drift will remain. Balancing controls are generally provided to set the output voltage to zero with the input shorted; however, they must be periodically readjusted to compensate for drift. Since the drift after warm-up is quite slow, it interferes mainly with the measurement of slowly varying quantities over long time periods. Noise has its origins in a number of random processes, such as Johnson noise in

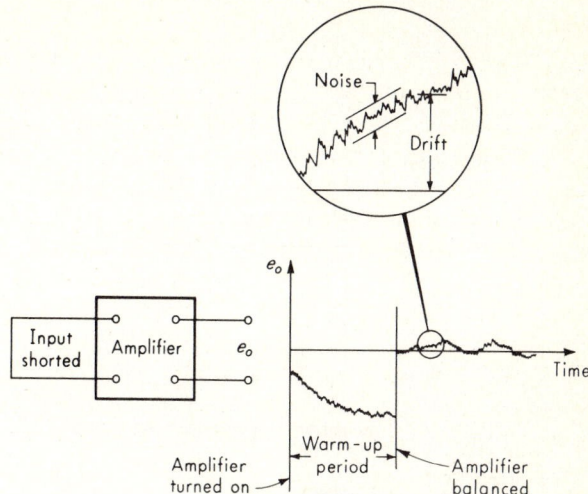

Fig. 10.5. *Amplifier noise and drift test.*

resistors; shot, partition, and gas noise in electron tubes; and transistor noise. It also arises from nonrandom sources such as power-line (60 cps) hum and chopper action in amplifiers that use choppers. Noise limits amplifier behavior mainly with respect to the threshold or smallest signal voltages that can be detected.

The overload recovery time of an amplifier is mainly of interest in multiplexed systems in which a given amplifier is time-shared by several transducers. Here the amplifier input is switched periodically from one transducer to another. This switching action may introduce large transient overloads into the amplifier, and a certain settling time is required before the amplifier output again reads correctly. A long recovery time prevents the amplifier from being switched in a rapid cycle and thus reduces the number of transducers that can be sampled by an amplifier in a given time.

In many measurement systems the amplifiers are located remotely from the transducers for one reason or another. This means that long connecting cables are necessary (several hundred feet is not unusual). A number of problems are introduced by such remote installations, the main ones being inductive pickup of noise voltages in the long cables (usually from nearby 60-cycle power lines, motors, etc.) and the inability to provide an identical ground reference point for two locations several hundred feet apart. Both these phenomena result in spurious voltages appearing at the amplifier input, and these can be very large relative to the transducer signal if adequate precautions are not taken. Shielding of the

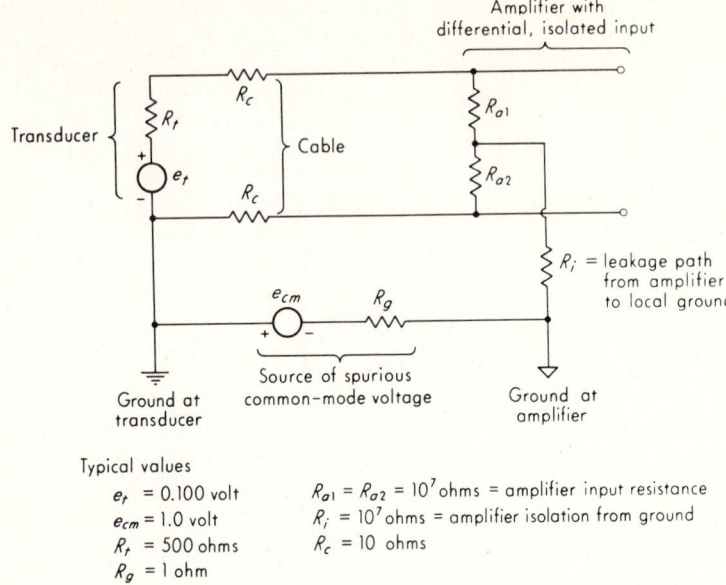

Typical values

e_t = 0.100 volt	$R_{a1} = R_{a2} = 10^7$ ohms = amplifier input resistance
e_{cm} = 1.0 volt	R_i = 10^7 ohms = amplifier isolation from ground
R_t = 500 ohms	R_c = 10 ohms
R_g = 1 ohm	

Fig. 10.6. *Differential input configuration.*

signal cables is helpful but the most critical applications also require the amplifier to have certain characteristics.

The performance specification generally used to indicate the amplifier's ability to reject the above-mentioned spurious voltages is the common-mode rejection ratio (CMRR). This is the ratio by which the undesired signals are attenuated relative to the desired signals. It depends on both the amplifier and the transducer and cable circuitry; thus any numerical value quoted for an amplifier assumes a certain input configuration. The common-mode rejection ratio also varies with the frequency of the noise voltages. Thus an amplifier may have a common-mode rejection ratio of 10^6 at direct current but only 10^4 at 60 cps. Some idea of the frequency content of the noise voltages is therefore necessary in order to evaluate an amplifier adequately. A full discussion of these questions is not possible here; however, adequate literature[1,2] is available. The amplifier features that are most significant in obtaining a desirable high value of the common-mode rejection ratio are differential

---

[1] A. S. Buchman, Noise Control in Low Level Data Systems, *Electromech. Design*, p. 64, September, 1962.

[2] Instrumentation Grounding and Noise Minimization Handbook, Consolidated Systems Corp., Pomona, Calif., AD612-027, *Tech. Rept.* AFRPL-TR-65-1, January, 1965.

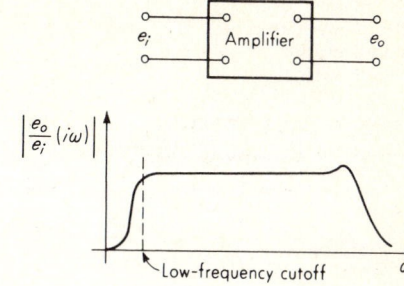

Fig. 10.7. *A-C-amplifier frequency response.*

input and "floating" (highly isolated from ground) input. These can be employed separately or, in the most critical applications, in conjunction in order to obtain a high common-mode rejection ratio. Figure 10.6 shows the circuit configuration when both methods are used. The spurious-voltage source e_{cm} can be shown[1] to act identically whether it is due to "ground loop" (lack of identical ground at two remote points) or inductive pickup in the cable; thus for simplicity Fig. 10.6 shows it only in the ground loop. Circuit calculations for the typical numerical values shown give[1]

$$\text{Desired signal at amplifier input} \approx 0.100 \text{ volt} \qquad (10.22)$$
$$\text{Spurious common-mode signal at amplifier input} \approx 16.5 \times 10^{-6} \text{ volt}$$
$$(10.23)$$

Thus the percentage error in this case is 0.0165 percent. The above calculations assumed purely resistive circuit elements. However similar calculations can be made for more general impedances.

A-C amplifiers. We begin our discussion of specific amplifier types by considering a class we here call "a-c." These amplifiers are closely related to the amplifiers ordinarily used in radios, television sets, and sound systems; however, the specifications for instrumentation types are generally more stringent. Vacuum-tube, transistor, and hybrid (partly tube, partly transistor) designs are in common use. We choose the name a-c based on the frequency response of such instruments, shown in Fig. 10.7. Note that the amplifier is incapable of handling steady or low-frequency input signals. This is a disadvantage if slowly varying quantities must be measured but also makes such amplifiers relatively drift-free since drift is due to slowly changing effects that cannot pass through the amplifier to its output, owing to the capacitive coupling that blocks direct current.

[1] A. S. Buchman, Noise Control in Low Level Data Systems, *Electromech. Design*, p. 64, September, 1962.

Physical quantity
to be measured,
frequency f_s

Transducer
excitation
frequency f_c

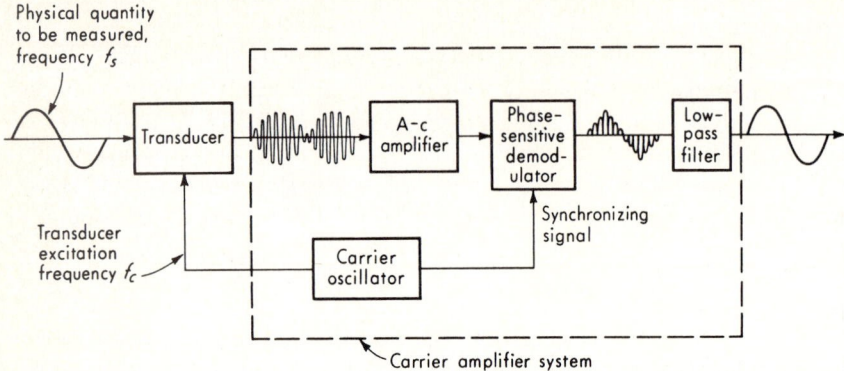

Fig. 10.8.   *Carrier amplifier system.*

A typical tube-type portable battery-operated instrument[1] especially useful with strain-gage transducers has three stages of amplification. By using one, two, or all three stages a gain of 15, 150, or 2,000 may be obtained. Frequency response is flat ±5 percent from 5 to 25,000 cps while the noise level is equivalent to 10 μv at the input. Input impedance is 2 megohms; output impedance is 27,000 ohms. The high output impedance indicates this amplifier is designed to drive a load requiring little current, such as an oscilloscope.

An a-c amplifier[2] using silicon field-effect transistors has adjustable gain (10 to 1,000), frequency response flat ±1 db from 5 to 60,000 cps, input impedance 10^8 ohms shunted by 10^{-10} farad, output impedance 1,000 ohms (1-ma current available), noise level under 10 μv referred to the input, full-scale output of 1.4 volts, rise time (10 to 90 percent) of 3 μsec, and recovery time of 3 sec for a 500-volt input pulse.

Carrier amplifier systems. To extend the advantages of simplicity and lack of drift characteristic of a-c amplifiers to the measurement of steady (d-c) signals, a number of approaches have been developed. The so-called carrier amplifier systems, much used with strain-gage trans-ducers, are one such scheme. Such systems work only with transducers (such as strain-gage bridges and differential transformers) that are excited by an a-c voltage. Figure 10.8 shows the amplitude-modulation principle used; it requires that the a-c amplifiers have a flat frequency response only over the band $f_c \pm f_s$. Because of demodulation and filtering require-

[1] Model BA-1, Ellis Associates, Pelham, N.Y.
[2] Sensonics, Inc., Kensington, Md.

ments, the carrier frequency f_c must be 6 to 10 times the highest signal frequency f_s.

A typical system[1] uses a carrier oscillator of a frequency of 5,000 cps and an amplitude of 0.1 to 5 volts rms, has a flat frequency response ± 5 percent from 0 to 1,000 cps, gain of about 3,000, input resistance of 350 ohms, output resistance of 12 ohms, maximum output voltage and current of 1.5 volts and 0.100 amp, nonlinearity ± 2 percent of full scale, noise less than 1 percent of full-scale output, and zero drift less than 0.5 percent of full scale over a 10-hr period at constant temperature. After $\frac{1}{2}$-hr warm-up, 10F° temperature changes cause less than 2 percent zero drift per hour.

D-C amplifiers. While the carrier system above will amplify "d-c" (constant) physical variables acting on a suitable transducer, it is not generally considered a d-c amplifier because it will not amplify a d-c voltage coming from an arbitrary transducer. Since this latter capability is much desired, various types of so-called d-c amplifiers have been developed.

Although the fundamental principles of both tube and transistor amplification inherently permit d-c operation, practical problems of drift and interstage coupling limited the use of such "true" d-c amplifiers in the past. Increasing demands of instrumentation systems and new developments in components and circuitry have led to the availability today of practical "true" d-c amplifiers as well as various forms of chopper and chopper-stabilized amplifiers. Chopper and chopper-stabilized types, while perhaps not accurately classified as pure d-c instruments, may practically be so considered since (on a black-box, input/output basis) they accept d-c inputs and produce amplified d-c outputs.

In a *chopper amplifier* the input signal (which may have a frequency content from zero to about one-tenth of the chopper frequency) is chopped to produce a square-wave voltage whose amplitude is proportional to the incoming signal. This modulation shifts all frequencies to a band around the chopper frequency; thus the chopped signal can be amplified with an ordinary a-c amplifier. Phase-sensitive demodulation and low-pass filtering reconstitute the input signal in amplified form at the output. Figure 10.9 shows a half-wave circuit which uses a chopper for both modulation and demodulation. More sophisticated circuits using two choppers and/ or feedback techniques result in improved performance.[2]

A very-high-gain chopper amplifier[3] using a 400-cps chopper fre-

---

[1] Honeywell Heiland Div., Denver, Colo.

[2] P. F. Howden, A Review of Chopper Amplifiers, *Electro-Technol.* (*New York*), p. 64, June, 1963.

[3] Honeywell Philadelphia Div., Philadelphia, Pa.

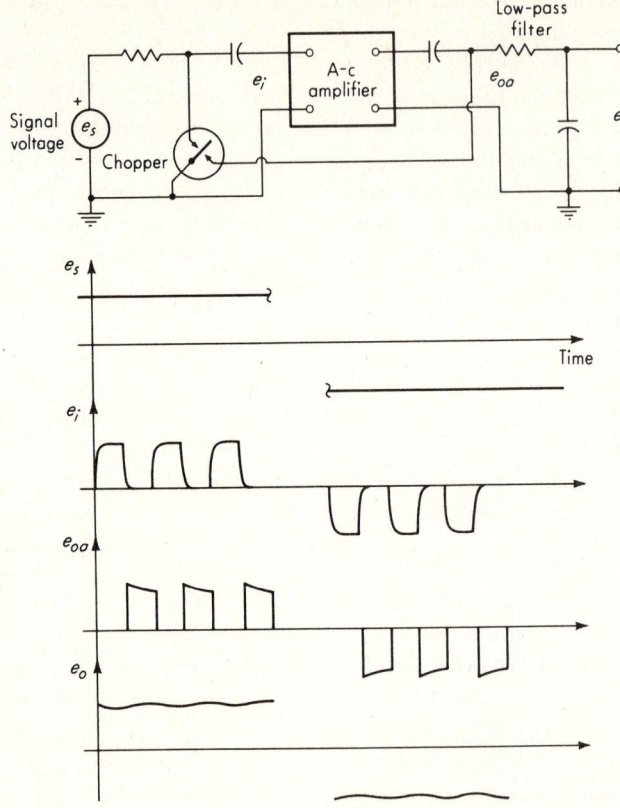

Fig. 10.9.   *Chopper amplifier.*

quency has a gain adjustable from 50 to 100,000, output voltage of 5 volts into a 10,000-ohm load, maximum source resistance of 300,000 ohms, frequency response flat ± 3 db from 0 to 30 cps (for a gain of 50, reduced at higher gain), gain stability ± 0.1 percent, zero drift ± 0.5 μv (referred to the input) per month, nonlinearity 0.1 percent, noise 0.5 μv peak to peak with the input shorted, ripple ± 0.1 percent, and a common-mode rejection ratio of 10^8 at d-c and 60 cps.

By using electronic modulation techniques rather than electromechanical choppers, a sinusoidal rather than square-wave signal is obtained and much higher "chopping" frequency is possible, thus extending the frequency response of the overall system.   A system of this type[1] designed to drive oscillograph galvanometers uses a 50,000-cps modulator, has a flat response of 0 to 5,000 cps and an input resistance of 35,000 ohms, pro-

[1] Alleghany Instrument Co., Cumberland, Md.

vides ± 60-ma output current into 5- to 60-ohm loads, and has 10 μv rms noise referred to the input, nonlinearity 0.5 percent of full scale, and drift of 50 μv/hr referred to the input.

By careful selection of components and attention to compensation of drift-producing inputs it is today possible to produce *pure d-c amplifiers* (completely devoid of choppers or modulators) with quite respectable drift characteristics, particularly if the operating environment is relatively constant-temperature. Amplifiers of this type range from rather simple low-cost units of limited performance to those incorporating ingenious compensation schemes and meeting critical specifications. A typical unit[1] of the latter sort using silicon semiconductors has a frequency response flat within ± 1 db from 0 to 20,000 cps, input impedance of 10^7 ohms, output impedance of 0.1 ohm, output voltage and current of ± 10 volts and ± 100 ma, nonlinearity ± 0.01 percent for direct current to 2,000 cps, gain of 1,000, common-mode rejection 10^6 for direct current to 60 cps, noise referred to the input of 4 μv rms for full bandwidth (1 μv peak to peak for 10 cps bandwidth), recovery time for 10-volt overload of 0.01 sec, constant-temperature zero drift after 30-min warm-up ± 0.02 percent for 200 hr, temperature-induced zero drift ± 0.001 percent/C° referred to the output ± 1 μv referred to the input, and an operating temperature of 0 to 50°C.

Another widely used d-c amplifier in instrumentation systems is the *chopper-stabilized d-c amplifier*. This instrument combines an essentially drift-free chopper amplifier (similar to Fig. 10.9) with a pure d-c amplifier in such a way that the drift of the overall system is reduced by a factor equal to the gain of the chopper amplifier. The arrangement also allows a frequency response that is not limited by the chopping frequency, since high-frequency signals are made to bypass the chopper amplifier and go directly through the pure d-c amplifier. Thus the wide bandwidth of the pure d-c amplifier is combined with the lack of drift of the chopper amplifier to produce a "hybrid" unit of highly desirable characteristics.

To illustrate these points briefly, we consider Fig. 10.10*a*, where a three-stage amplifier is shown with the drift voltage of each stage referred to its input. We can easily show that

$$\frac{e_o}{K_1 K_2 K_3} = e_i + e_{\text{drift},1} + \frac{e_{\text{drift},2}}{K_1} + \frac{e_{\text{drift},3}}{K_1 K_2} \qquad (10.24)$$

and thus compare the output due to desired input e_i with that due to drift. Note that the drift of the first stage is most important, since its effect at the output is unattenuated relative to e_i. Drift of stages 2 and 3, however, is reduced by factors K_1 and $K_1 K_2$, respectively. Thus if a low-drift first stage of relatively high gain can be constructed, the overall

---

[1] Dana Laboratories, Irvine, Calif.

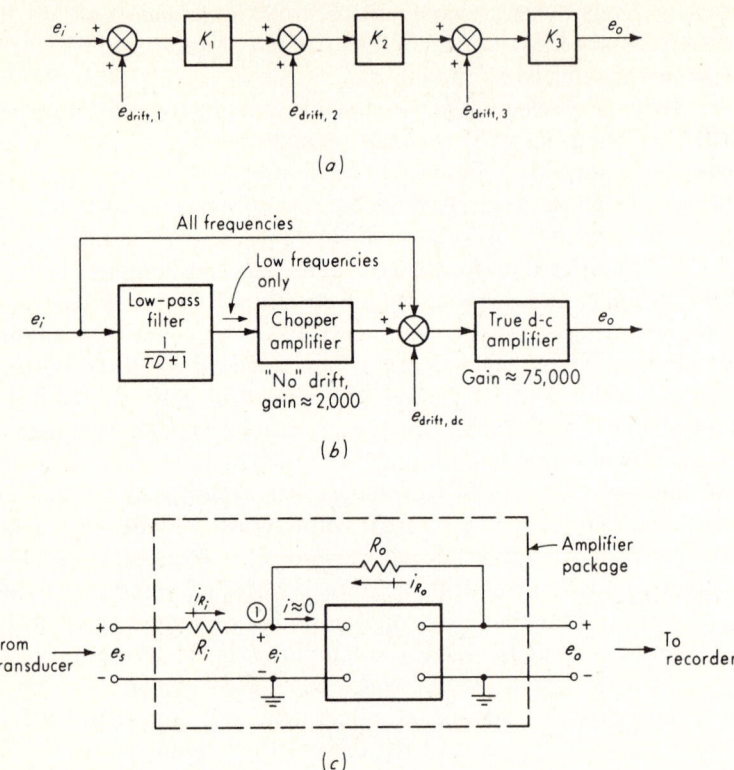

(*a*)

(*b*)

(*c*)

Fig. 10.10.   *Chopper-stabilized amplifier.*

system drift will be small.   Figure 10.10*b* shows in block-diagram form
how this is realized in a typical chopper-stabilized d-c amplifier.   All frequencies present in the signal e_i pass through the true d-c amplifier while
only low-frequency components pass through the chopper amplifier and
then through the d-c amplifier.   Using the typical numerical values shown
and assuming the first (chopper) stage has negligible drift, we see that the
overall drift is reduced to 1/2,000 of the drift of the d-c amplifier itself.
The gain for low-frequency signals is $(2,000) (75,000) = 150 \times 10^6$ while
the high-frequency gain is 75,000.   Since drift is a low-frequency phenomenon, the first-stage gain of 2,000 is effective in reducing drift to
1/2,000 of its former value, as noted above.

The amplifiers are not actually used in the "open-loop" configuration shown in Fig. 10.10*b* because such high gain is not needed, and by
closing a feedback loop around the amplifier one can trade off some of
this gain for other desirable properties such as linearity and control of
input and output impedances.   Thus the package sold as a chopper-

stabilized instrumentation amplifier has the configuration of Fig. 10.10c, where the resistances R_i and R_o are chosen to give a closed-loop (e_s to e_o) gain of the order of a few thousand or less. We may analyze this circuit as follows: Assuming a high input resistance of the open-loop amplifier makes the current $i \approx 0$ and summing currents at the node 1 give

$$\frac{e_s - e_i}{R_i} = - \frac{e_o - e_i}{R_o} \qquad (10.25)$$

Because of the very high open-loop gain (150×10^6 for low frequencies), the signal e_i can produce full-scale e_o (say 10 volts) without ever being greater than a few microvolts at most. Thus e_i is always negligible compared with e_s (which is of the order of millivolts) and e_o (which is of the order of volts). Equation (10.25) then becomes, to a very good degree of approximation,

$$\frac{e_o}{e_s} = - \frac{R_o}{R_i} \qquad (10.26)$$

and thus the gain becomes a function of R_o and R_i only. Their numerical values can be very precisely set and maintained, giving very stable and accurate gain.

The above discussions and diagrams are simplifications intended to make clear the principles involved but, of course, neglect many details that must be considered in arriving at a practical amplifier. A typical differential-input chopper-stabilized instrument[1] has a gain of 1,000, nonlinearity ± 0.1 percent of full scale, input impedance of 10^8 ohms shunted by 700 pf, full-scale output of ± 10 volts and ± 10 ma, output resistance less than 25 ohms, drift ± 2 μv referred to the input plus ± 0.01 percent of full scale at 25°C, noise 3.5 μv rms referred to the input plus 150 μv referred to the output from 0.05 to 5,000 cps, frequency response flat within 1 db from 0 to 5,000 cps, recovery time from 500 percent overload of 300 μsec, and common-mode rejection of 10^7 at direct current and 10^6 at 60 cps.

Operational amplifiers. Operational amplifiers are actually open-loop pure d-c or chopper-stabilized amplifiers, as just described above, with very high gain. By closing appropriate feedback loops around them one can build up many kinds of useful active circuits. Such amplifiers are the basic building blocks of electronic analog computers,[2] for instance. They are also extremely useful in building special-purpose instrumentation equipment such as amplifiers, filters, integrators, etc., since a person with little electronics background can purchase these amplifiers ready-

[1] Astrodata, Anaheim, Calif.
[2] A. S. Jackson, "Analog Computation," McGraw-Hill Book Company, New York, 1960.

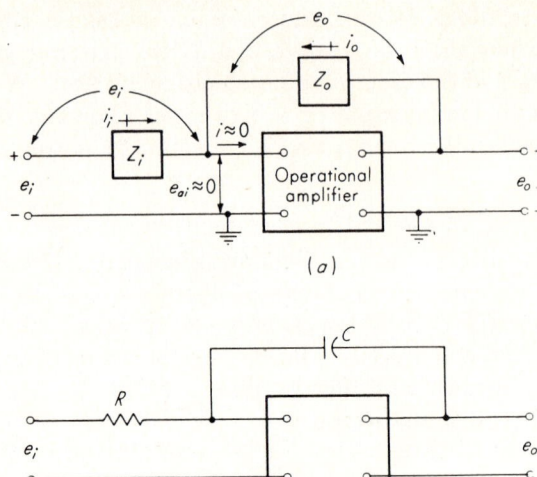

Fig. 10.11.   *Operational amplifier.*

made and then add various relatively simple passive elements (resistors, capacitors, diodes, etc.) to obtain the desired functional characteristics.[1] Both linear and nonlinear operations are conveniently performed.   To show the general linear operations, we consider Fig. 10.11a.   Using assumptions as in Fig. 10.10c, we get

$$\frac{e_o}{e_i}(D) = -\frac{Z_o}{Z_i} \qquad (10.27)$$

where

$$Z_o \triangleq \frac{e_o}{i_o}(D) = \text{operational impedance} \qquad (10.28)$$

$$Z_i \triangleq \frac{e_i}{i_i}(D) = \text{operational impedance} \qquad (10.29)$$

As an example, in Fig. 10.11b, $Z_o = 1/CD$ and $Z_i = R$, giving

$$\frac{e_o}{e_i}(D) = -\frac{1}{RCD} \qquad (10.30)$$

$$e_o = -\frac{1}{RC}\int e_i\,dt \qquad (10.31)$$

Thus this circuit performs the useful operation of integrating the signal e_i with respect to time.   A wide variety of other operations can be per-

[1] A. S. Jackson, "Analog Computation," McGraw-Hill Book Company, New York, 1960.

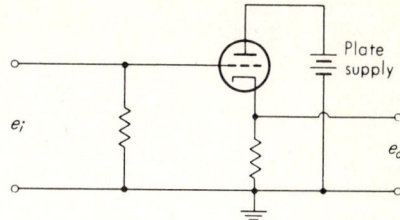

Fig. 10.12.   *Cathode follower.*

formed, some of which will be discussed in the sections on filters and dynamic compensation.

Cathode followers, emitter followers. When a high-impedance transducer such as a piezoelectric crystal or capacitance displacement pickup must be coupled into a recording system of some type, the amplifier used must have a very high-input impedance if loading is to be minimized. If the amplifier is to supply appreciable current to the recorder, its output impedance must be low. The cathode-follower circuit of Fig. 10.12 provides such impedance-transformation properties. However, its gain is always less than 1 (often about 0.9); thus, if voltage amplification is needed, additional amplification must be provided. While the basic circuit of Fig. 10.12 has d-c response, many applications do not require this; thus a coupling capacitor is used at the input to reduce drift and biasing problems.

A typical unit[1] has a gain of 0.95, flat frequency response of 0.02 to 10^6 cps, input impedance of 2×10^9 ohms shunted by 12×10^{-12} farad, output impedance of 290 ohms, output voltage of 30 volts maximum, output current of 3 ma into 10,000 ohms load, and noise 75 μv rms. The transistor version of the cathode follower is called an emitter follower or source follower. Typical specifications[1] are a gain of 0.994, frequency response flat -3 db from 1 to 100,000 cps, input impedance of 10^8 ohms shunted by 20×10^{-12} farad, output impedance under 500 ohms, output voltage of 6 volts maximum, and noise of 300 μv at the output with the input shorted.

Charge amplifiers. Increasing use of piezoelectric accelerometers, pressure pickups, and load cells has led to the development of an amplifier type that offers some advantages over the usual voltage amplifier in certain applications. Such a *charge amplifier* is shown connected to a piezoelectric transducer in Fig. 10.13. The idealized form is shown in Fig. 10.13a where we note that an operational amplifier is used with a

[1] Columbia Research Laboratories, Woodlyn, Pa.

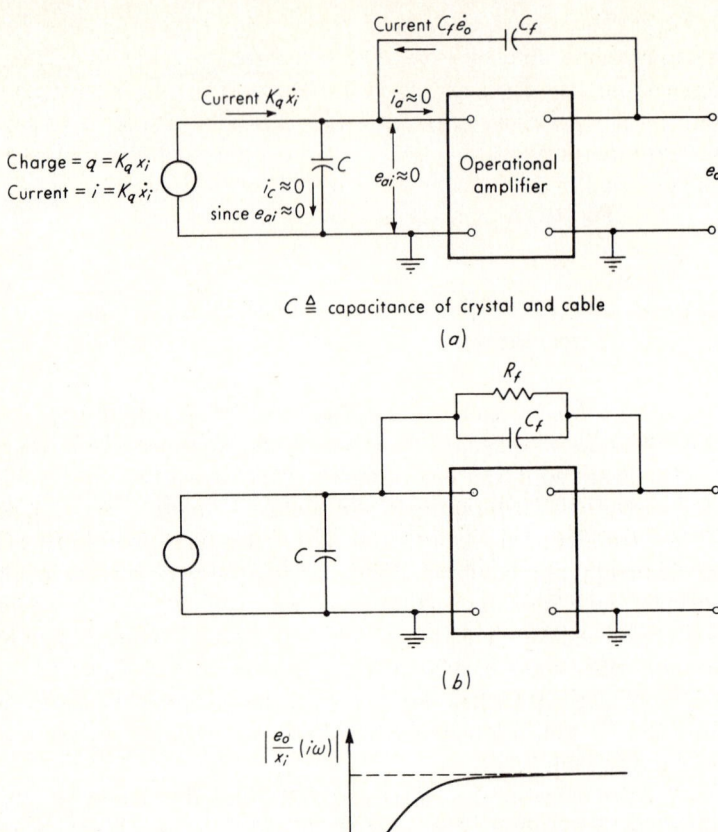

Fig. 10.13. *Charge amplifier.*

capacitor C_f in the feedback path. Assuming, as usual, that the input voltage e_{ai} and current i_a of the operational amplifier are small enough to take as zero, we get

$$K_q Dx_i = -C_f De_o \qquad (10.32)$$

$$e_o = -\frac{K_q x_i}{C_f} \qquad (10.33)$$

Equation (10.33) indicates that e_o would be instantaneously and linearly related to displacement x_i without the usual loss of steady-state response associated with piezoelectric transducers and voltage amplifiers. Unfortunately this advantage is not realizable since a system constructed as in Fig. 10.13*a* would, because of noninfinite input resistance of the operational amplifier and leakage of C_f, exhibit a steady charging of C_f by the leakage current until the amplifier saturated. To overcome this problem, in the practical circuit of Fig. 10.13*b* a feedback resistance R_f is included to prevent this small leakage current from developing a significant charge on C_f. Analysis of this new circuit gives

$$\frac{e_o}{x_i}(D) = \frac{K\tau D}{\tau D + 1} \qquad (10.34)$$

where

$$K \triangleq \frac{K_q}{C_f} \qquad \text{volts/in.} \qquad (10.35)$$

$$\tau \triangleq R_f C_f \qquad \text{sec} \qquad (10.36)$$

Equation (10.34) is of identical form with the transfer function of a piezo-electric transducer and a *voltage* amplifier and exhibits the same loss of static and low-frequency response. The advantages of the charge amplifier are found in Eqs. (10.35) and (10.36). We note that both the sensitivity K and time constant τ are now independent of the capacitance of the crystal itself and the connecting cable whereas with a voltage amplifier neither of these advantages is obtained. Thus long cables (often several hundred feet in practical setups) do not result in a reduced sensitivity or a variation in frequency response. These advantages and others[1] are sufficient to make the charge amplifier of practical interest in many systems. Disadvantages[2] that may arise in certain applications include a possibly poorer signal/noise ratio and a reduction in natural frequency of the transducer due to loss of stiffness caused by what amounts to a short circuit across the crystal.

When used with quartz-crystal transducers,[3] the value of C_f is from 10 to 100,000 pf and R_f is 10^{10} to 10^{14} ohms. For $C_f = 100,000$ pf and $R_f = 10^{14}$ ohms, $\tau = 10^6$ sec, showing that practically d-c response, allowing static calibration and measurement, is possible under these conditions. For ceramic-type transducers, C_f is from 10 to 1,000 pf and R_f from 10^8 to 10^{10} ohms, making the maximum τ about 10 sec and static measurements thus usually impractical.[4] A typical charge amplifier[4] for use with quartz-crystal transducers has adjustable gain from 0.01 to 100 mv/pcoul, output voltage to high impedance load $+10$, -5 volts, output current to low impedance load ± 10 ma, output impedance of 100 ohms, R_f switch selectable as 10^{11} or 10^{14} ohms, frequency response from practically direct cur-

[1] D. Pennington, Charge Amplifier Applications, Endevco Corp., Pasadena, Calif., April, 1964.

[2] Wilcoxon Research, Bethesda, Md., *Wilcoxon Res. Bull.* 5.

[3] Kistler Instrument Corp., Clarence, N.Y., *Tech. Notes* 133762 and 130662.

[4] Kistler Instrument Corp., *Tech. Notes* 133762 and 130662.

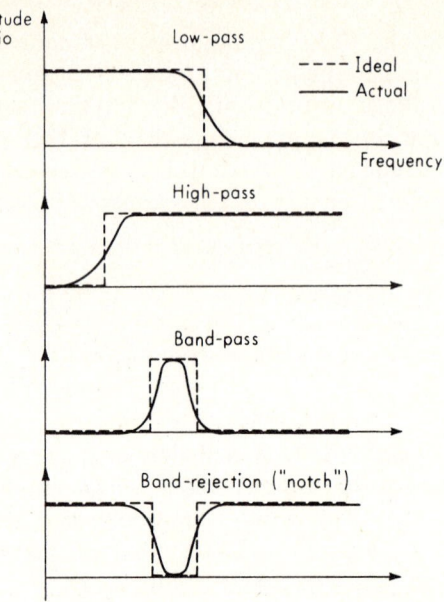

Fig. 10.14.   Basic filter characteristics.

rent to 150,000 cps, nonlinearity 0.1 percent, and output noise of 2 mv with the input shorted.

Magnetic amplifiers. Magnetic amplifiers are not at present widely used as general-purpose instruments in research and development laboratories or data-gathering systems. For special-purpose applications in measurement and control systems their reliability, high isolation, and other advantages make them attractive, and numerous successful applications exist.[1,2]

10.3

Filters   The use of frequency-selective filters to pass the desired signals and reject spurious ones has been discussed before.   Figure 10.14 summarizes the most common frequency characteristics used.   Filters may take many physical forms; however, the electrical form is most common and highly developed with regard to both theory and practical realization.   By use of analogies, the material on electrical filters may suggest

[1] B. A. Mazzeo, A Low-level, High-accuracy D-C Magnetic Amplifier, *Elec. Mfg.*, November, 1958.

[2] Acromag Design Manual, Acromag Inc., Detroit, Mich.

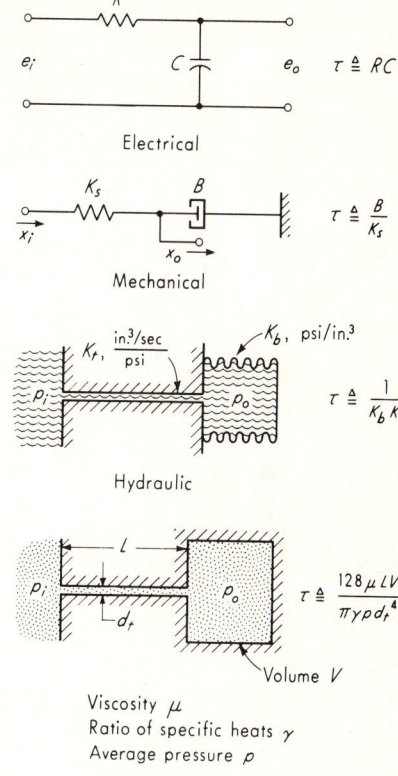

$$\tau \triangleq RC$$

Electrical

$$\tau \triangleq \frac{B}{K_s}$$

Mechanical

$$\tau \triangleq \frac{1}{K_b K_f}$$

Hydraulic

$$\tau \triangleq \frac{128 \mu LV}{\pi \gamma p d_f{}^4}$$

Viscosity μ
Ratio of specific heats γ
Average pressure p

Fig. 10.15. *Low-pass filters.*

Pneumatic

the configurations of mechanical, hydraulic, acoustical, etc., systems that will provide the desired filtering action in specific problems.

Low-pass filters. The simplest low-pass filters commonly in use are shown in several different physical forms in Fig. 10.15. They all have identical transfer functions given by

$$\frac{e_o}{e_i}(D) = \frac{x_o}{x_i}(D) = \frac{p_o}{p_i}(D) = \frac{1}{\tau D + 1} \qquad (10.37)$$

Since these are all simple first-order systems the attenuation is quite gradual with frequency: 6 db/octave. This does not give a very sharp distinction between the frequencies that are passed and those that are rejected. By adding more "stages" (see Fig. 10.16*a*) the sharpness of cutoff may be increased. The use of inductance elements (Fig. 10.16*b*) may also lead to better filtering action. When inserting a filter into a

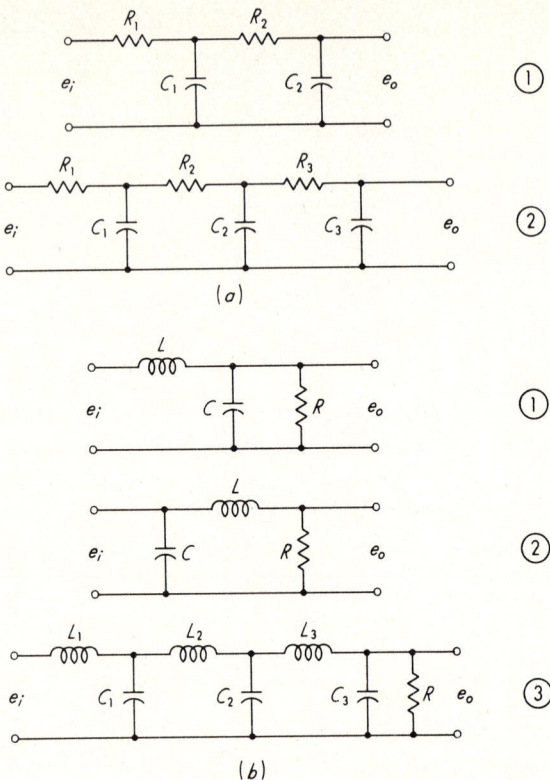

Fig. 10.16.   *Sharper-cutoff low-pass filters.*

system, it is necessary to take into account possible loading effects by use of appropriate impedance analysis.

High-pass filters. Figure 10.17 shows the simplest high-pass filters, which all have the transfer function

$$\frac{e_o}{e_i}(D) = \frac{x_o}{x_i}(D) = \frac{x_o/K_{px}}{p_i}(D) = \frac{\tau D}{\tau D + 1} \qquad (10.38)$$

Again, the attenuation is quite gradual, and more complex configurations are needed to obtain a more sharply defined cutoff (Fig. 10.18).

Band-pass filters. By cascading a low-pass and a high-pass filter one can obtain the band-pass characteristic (Fig. 10.19). To sharpen the rejection on either side of the passband one can simply use the sharper low- and high-pass sections mentioned above.

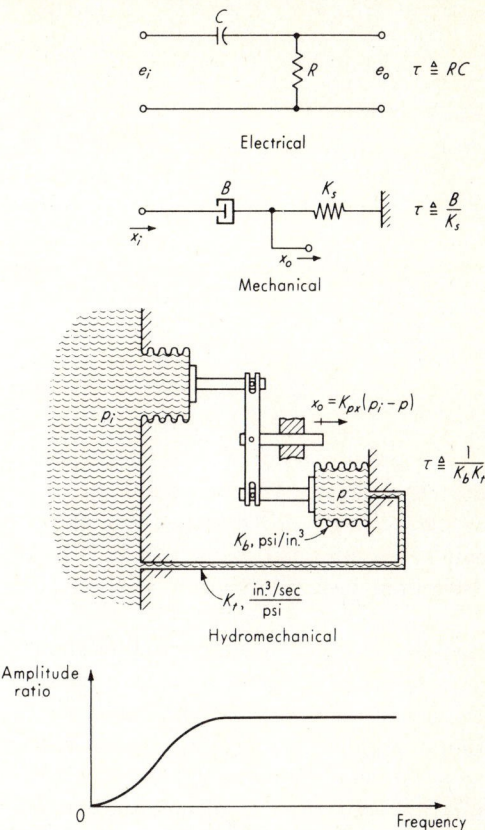

Fig. 10.17. *High-pass filters.*

Band-rejection filters. A common application of a band-rejection filter is found in the input circuits of self-balancing potentiometer and *X-Y* recorders. These instruments are subject to interfering 60-cps noise voltages. Since the frequency response of the overall recorder is good only to a few cycles per second, a band-rejection filter tuned to 60 cps may be employed without distorting any desired signals. Such a filter prevents noise signals from saturating the recorder's amplifiers and preventing the proper amplification of the desired signals.

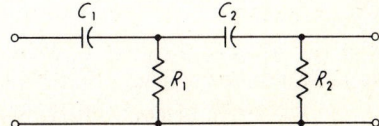

Fig. 10.18. *Sharper-cutoff high-pass filter.*

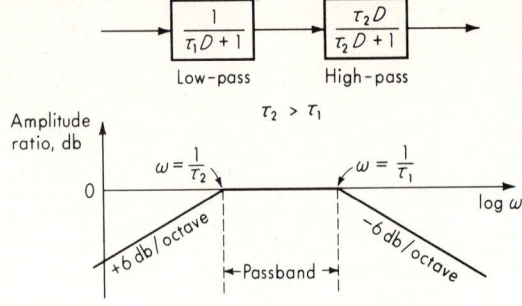

Fig. 10.19. *Band-pass filter*

Passive networks commonly used for rejection of a band of frequencies include the bridged-T and the twin-T network (Fig. 10.20). While the bridged-T does not completely reject any frequency, the twin-T can be so designed, as shown in Fig. 10.20. Equations and charts for designing these filters are available.[1]

Active filters. The filter circuits shown up to this point have all been passive networks; that is, they have no power source within them. By the use of amplifiers as buffer devices, in feedback schemes, or in operational amplifier configurations, many desirable features difficult to obtain in other ways may be achieved. While we shall not discuss here the internal construction of such active filters, we shall describe briefly the operating characteristics of some particular general-purpose laboratory-type units.[2] The frequency characteristics of these filters are adjustable over a wide range, including very low frequencies which are difficult to achieve with passive filters. Their high input impedance and low output impedance allow them to be inserted into a system without causing loading problems. One unit provides a band-pass characteristic with the low-frequency and the high-frequency cutoffs separately and continuously adjustable over the range 0.02 to 2,000 cps. The transfer function is approximately

$$\frac{e_o}{e_i}(D) = \frac{\tau_1^4 D^4}{(\tau_1^2 D^2 + 1.2\tau_1 D + 1)^2(\tau_2^2 D^2 + 1.2\tau_2 D + 1)^2} \tag{10.39}$$

where $1/2\pi\tau_1$ is the low cutoff frequency and $1/2\pi\tau_2$ is the high. We see that the attenuation rate of this filter at the edges of the passband is

[1] J. E. Gibson and F. B. Tuteur, "Control System Components," p. 43, McGraw-Hill Book Company, New York, 1958.
[2] Krohn-Hite Corp., Cambridge, Mass.

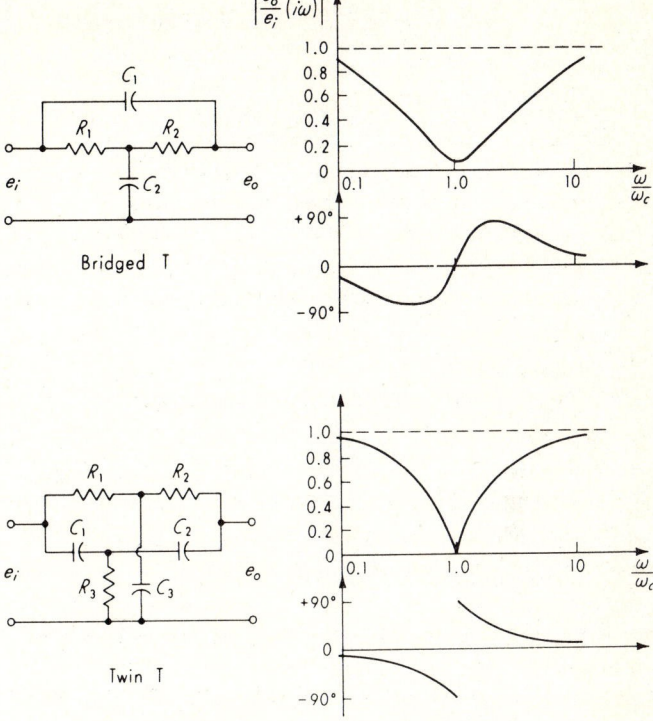

Fig. 10.20. *Band-rejection filters.*

24 db/octave. With the input shorted the noise level at the output is 0.1 to 1.0 mv. This noise may be attributed mainly to the active elements and limits the use of active filters for very low-level signals.

A band-rejection filter of the same manufacturer has a transfer function of approximately

$$\frac{e_o}{e_i}(D) = \frac{1}{(\tau_1^2 D^2 + 1.2\tau_1 D + 1)^2} + \frac{\tau_2^4 D^4}{(\tau_2^2 D^2 + 1.2\tau_2 D + 1)^2} \qquad (10.40)$$

where the low cutoff frequency is $1/2\pi\tau_1$ and the high $1/2\pi\tau_2$. Both are independently adjustable over the range 0.02 to 2,000 cps. The unit may also be set to give a sharp single-frequency null over the range 0.1 to 500 cps. The attenuation rate is again 24 db/octave.

A hydraulic band-pass filter for an oceanographical transducer. While filtering with electrical networks is most common, in some instances other physical forms present advantages. This section illustrates this

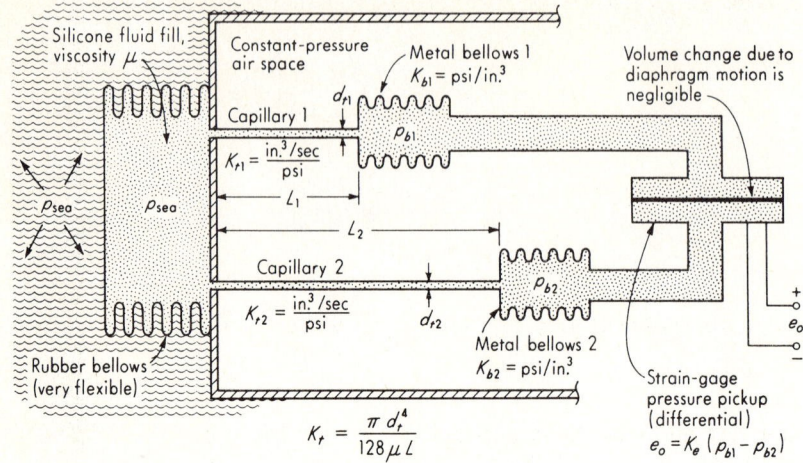

Fig. 10.21. *Hydraulic band-pass filter.*

with an example of a filter that has been successfully constructed and used.[1]

The Scripps Institution of Oceanography at La Jolla, California, uses pressure transducers in its studies of ocean-wave phenomena. A particular study required measurements of waves whose frequencies are lower than those of ordinary gravity waves (which one observes visually) and higher than those due to tides. These waves of intermediate frequency are of rather low amplitude relative to those due to tides and gravity and thus are rather difficult to measure with a pressure pickup which treats all frequencies about equally. The band-pass filter and pressure pickup of Fig. 10.21 solves this problem since it is "tuned" to the frequency range of interest, which is about 0.001 cps. Such low frequencies are very difficult to handle with electrical circuits but the hydraulic filter shown gives very good results with quite simple and reliable components.

In use, the pressure transducer is located underwater, often buried in a foot of sand for temperature insulation, with a "snorkel" tube extending up through the sand to sense water pressure. This pressure is directly related to the height of the waves passing overhead; thus a record of pressure-transducer output voltage is a record of wave activity. Analysis of the system gives

$$\frac{e_o}{p_{\text{sea}}}(D) = \frac{K_e(\tau_2 - \tau_1)D}{(\tau_1 D + 1)(\tau_2 D + 1)} \qquad (10.41)$$

[1] F. E. Snodgrass, Shore-based Recorder of Low-frequency Ocean Waves, *Trans. Am. Geophys. Union*, p. 109, February, 1958.

where $K_e \triangleq$ sensitivity of differential pressure pickup, mv/psi

and
$$\tau_1 \triangleq \frac{1}{K_{t1}K_{b1}} \qquad \text{sec} \qquad (10.42)$$

$$\tau_2 \triangleq \frac{1}{K_{t2}K_{b2}} \qquad \text{sec} \qquad (10.43)$$

One can easily show that the frequency ω_p of peak response is given by

$$\omega_p = \sqrt{\frac{1}{\tau_1\tau_2}} \qquad \text{rad/sec} \qquad (10.44)$$

the amplitude ratio M_p at this frequency is

$$M_p = \frac{K_e(\tau_2 - \tau_1)}{\tau_2 + \tau_1} \qquad (10.45)$$

and the phase angle at ω_p is zero.

Filtering by statistical averaging. All the filters mentioned above are of the frequency-selective type and of course require that the desired and spurious signals occupy different portions of the frequency spectrum. When signal and noise contain the same frequencies, such filters are useless. A basically different scheme may be usefully employed under such circumstances if the following is true:

1. The noise is random.
2. The desired signal can be caused to repeat itself over and over.

If these two conditions are fulfilled, it should be clear that, if one adds up the ordinates of several samples of the total signal at like values of abscissa (time), the desired signal will reinforce itself while the random noise will gradually cancel itself. This will occur even if the frequency content of signal and noise occur in the same part of the frequency spectrum. It can be shown that the signal/noise ratio improves in proportion to the square root of the number of samples used. Thus theoretically the noise can be eliminated to any desired degree by adding up a sufficiently large number of signals. In practice, various factors prevent realization of theoretically optimum performance.

While the above procedure could be carried out manually, it may also be automated to increase speed, convenience, and accuracy. One such system[1] samples the incoming signal at 512 time points, converts the ordinates to digital values, and stores them in a digital memory. Further

---

[1] Noise Reduction by Digital Signal Averaging, Signal Averaging by Waveform Totalling, Signal Averaging by Waveform Comparison, Northern Scientific Inc., Madison, Wis.

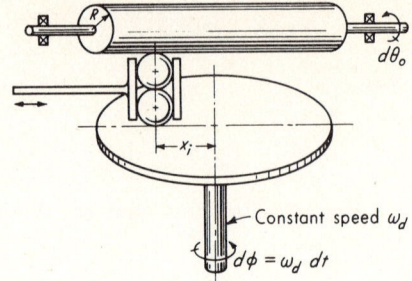

Fig. 10.22.   *Ball-disk integrator.*

repetitions of the signal are similarly sampled with the respective ordinate values being totalized.   The contents of the digital memory are presented on an oscilloscope screen, and one can actually watch the true signal "emerge" from the noise as one sample after another is put into the system.

10.4

Integration and Differentiation   Often in measurement systems it is necessary to obtain integrals and/or derivatives of signals with respect to time.   Depending on the physical nature of the signal, various devices may be most appropriate.   Accurate differentiation is generally harder to accomplish than integration since differentiation tends to accentuate noise (which is usually high frequency) whereas integration tends to smooth noise.   Thus second and higher integrals may easily be achieved while derivatives present real difficulties.

Integration. If the signal to be integrated is already a mechanical displacement or is easily transduced to one, the *ball-and-disk integrator* of Fig. 10.22 may be used.   Assuming rigid bodies and no slippage, we can show that

$$\frac{\theta_o}{x_i}(D) = \frac{\omega_d}{R}\left(\frac{1}{D}\right) \qquad (10.46)$$

and thus the rotation angle θ_o is proportional to the first time integral of the displacement x_i.   A typical unit[1] has a maximum ω_d of 500 rpm, $x_i = \pm 0.75$ in., maximum input force of 2 oz, output torque of 3 in.-oz, reproducibility 0.01 percent of full scale, accuracy 0.05 percent of full scale, and expected life of 10,000 hr.   This unit uses a precision-lapped tungsten carbide disk with 1-μin. surface finish and hardened-tool-steel roller and balls.

Two electromechanical means of obtaining integrals are shown in

---

[1] Librascope Div., Glendale, Calif.

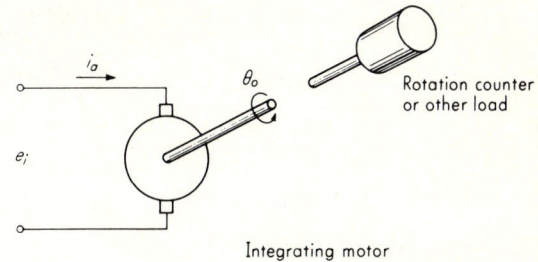

Integrating motor

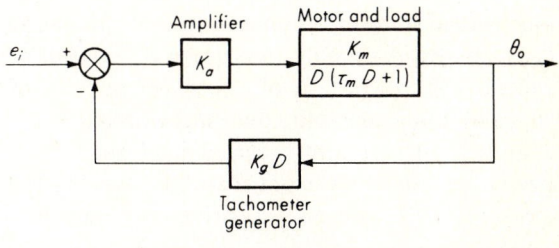

Velocity servo

Fig. 10.23. *Electromechanical integrators.*

Fig. 10.23: the *integrating motor* and the *velocity-servo integrator*. Both these accept electrical signals as input and produce mechanical rotations in proportion to the time integral of the input voltage. The integrating motor is essentially a d-c motor with permanent magnet field in which friction, iron losses, and brush-contact voltage drop have been reduced to extremely low levels, resulting in an input-voltage/output-speed characteristic that is very linear over a wide range of input voltage. For a d-c motor with constant field,

$$\text{Armature current} = i_a = \frac{e_i - e_m}{R} \qquad (10.47)$$

where

$$e_m \triangleq \text{motor back emf} = K_e \dot{\theta}_o \qquad (10.48)$$
$$R \triangleq \text{armature resistance} \qquad (10.49)$$

Motor torque $T_m = K_{mt} i_a$, where K_{mt} is the motor-torque constant. Thus if rotor inertia is J, we have

$$T = J\ddot{\theta}_o \qquad (10.50)$$

$$K_{mt} \frac{e_i - K_e \dot{\theta}_o}{R} = J\ddot{\theta}_o \qquad (10.51)$$

$$\frac{\theta_o}{\int e_i \, dt} (D) = \frac{1/K_e}{\tau D + 1} \qquad (10.52)$$

$$\tau \triangleq \frac{RJ}{K_{mt} K_e} \qquad (10.53)$$

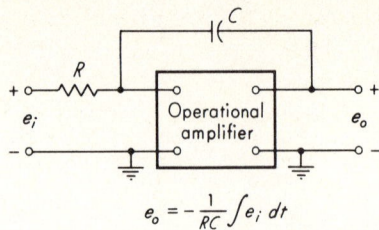

Fig. 10.24. *Electronic integrator.*

$$e_o = -\frac{1}{RC}\int e_i\, dt$$

We see that the rotation angle θ_o (which can be counted by a simple mechanical counter) is a measure of the time integral of e_i with a first-order lag. A family of such instruments[1] has full-scale input voltage ranging from 1.5 to 24 volts, R of 2.8 to 700 ohms, τ of about 0.01 sec, full-scale speed of 1,885 to 1,260 rpm, and starting voltages of 4.2 to 79 mv. For a motor without any external load the nonlinearity is better than 0.3 percent of full scale from 5 to 200 percent of full-scale voltage. These motors can be used only to drive very light loads, 1.8 to 12.4 g-cm at full voltage.

For greater accuracy and the ability to drive loads requiring greater power output, the velocity-servo integrator may be employed. Analysis of the block diagram of Fig. 10.23 gives

$$\frac{\theta_o}{\int e_i\, dt}\,(D) = \frac{1/K_g}{\tau D + 1} \qquad (10.54)$$

where

$$\tau \triangleq \frac{\tau_m}{1 + K_a K_g K_m} \qquad (10.55)$$

and $1/K_g \approx K_a K_m/(1 + K_a K_m K_g)$ since K_a is very large. Such integrators achieve accuracies of 0.1 percent and better.[2]

Figure 10.24 shows the operational-amplifier type of integrator extensively used in general-purpose electronic analog computers. By use of high-quality chopper-stabilized operational amplifiers an integrator of quite low drift can be constructed in this way. Accuracies of the order of 0.1 percent for short-term operation and 1 percent over 14 hr are typical of high-quality electronic integrators of this type.[2] If higher integrals are desired, such units may be cascaded; however, drift becomes more troublesome. In addition to providing a closer approximation to true integration than the passive networks discussed in the following paragraph, the presence of the amplifier (with its own power supply) means that power can be supplied to the device following the integrator without taking any signifi-

[1] Electro Methods Ltd., Stevenage, England.
[2] W. H. Barr, Integrators, *Electromech. Design,* p. 57, October, 1961.

cant power from the device supplying the integrator. That is, operational-amplifier circuits can generally have a high input impedance and low output impedance.

All the low-pass filters of Fig. 10.15 may be used as *approximate integrators* for input signals within a restricted frequency range. This can be shown as follows:

$$\frac{e_o}{e_i}(i\omega) = \frac{1}{i\omega\tau + 1} \qquad (10.56)$$

Now if $\omega\tau \gg 1$

$$\frac{e_o}{e_i}(i\omega) \approx \frac{1}{i\omega\tau} \qquad (10.57)$$

and thus

$$\frac{e_o}{e_i}(D) \approx \frac{1}{\tau D} \qquad (10.58)$$

$$e_o \approx \frac{1}{\tau} \int e_i\, dt \qquad (10.59)$$

Thus, if the frequency spectrum of the input signal is such that $\omega\tau \gg 1$ for all significant frequencies, a good approximation to the desired integrating action is obtained. For a given τ the approximation improves as ω increases. It appears as if any ω can be accommodated by choosing τ sufficiently large. However, large τ decreases the magnitude of the output; thus this can be carried only as far as the noise level of the system permits.

If a signal is available in digital form, it may be integrated by a general-purpose digital computer by programming it for one of the approximate numerical-integration schemes such as Simpson's rule. Another type of *digital integration*, which can be carried out without use of a general-purpose computer, involves use of pulse-totalizing methods. Here the analog voltage signal is converted to a periodic voltage signal whose frequency is proportional to the input-signal amplitude (voltage-to-frequency converter). This periodic signal is then applied to an electronic counter. Thus the reading of the counter at any time is proportional to the time integral of the input signal up to that time.

Differentiation. For mechanical displacement signals the various velocity pickups, accelerometers, jerk pickup, tachometer generator, and rate gyro of Chap. 4 may be considered as differentiating devices.

All the high-pass filters of Fig. 10.17 may be used as *approximate differentiators* for input signals within a restricted frequency range, as shown by the following analysis:

$$\frac{e_o}{e_i}(i\omega) = \frac{i\omega\tau}{i\omega\tau + 1} \qquad (10.60)$$

Now if $\omega\tau \ll 1$

$$\frac{e_o}{e_i}(i\omega) \approx i\omega\tau \qquad (10.61)$$

$$\frac{e_o}{e_i}(D) \approx \tau D \qquad (10.62)$$

$$e_o \approx \tau \frac{de_i}{dt} \qquad (10.63)$$

We note here that for a given τ the approximation improves for lower values of ω. Again τ may be reduced to extend accurate differentiation to higher frequencies. However, small τ reduces sensitivity; thus noise level is limiting just as in the approximate integrators.

Use of operational amplifiers results in both approximate and "exact" differentiators of improved performance relative to the passive high-pass filters discussed above. Figure 10.25 shows some of these circuits. In Fig. 10.25*a*, analysis of this "exact" differentiator gives

$$\frac{e_o}{e_i}(D) = -RCD \qquad (10.64)$$

This circuit is rarely useful because the ever-present noise (generally of high frequency relative to the desired signal) will completely swamp the desired signal at the output. All exact differentiators must suffer from this problem. It can be alleviated only by shifting to approximate differentiators which include low-pass filters to take out the effects of high-frequency noise. Figure 10.25*b* shows a common scheme which, when analyzed, gives

$$\frac{e_o}{e_i}(D) = -\frac{R_2 CD}{R_1 CD + 1} \qquad (10.65)$$

This gives an accurate derivative for frequencies such that $R_1 C\omega \ll 1$ and amplifies high-frequency noise only by an amount R_2/R_1. To attenuate noise, one must use a second-order-type low-pass filter, such as given by the circuit of Fig. 10.25*c*. Analysis of this gives

$$\frac{e_o}{e_i}(D) = -\frac{R_2 C_1 D}{(R_2 C_2 D + 1)(R_1 C_1 D + 1)} \qquad (10.66)$$

Figure 10.25*d* shows an actual circuit[1] designed for measuring the rate of charging or discharging of batteries and using a solid-state operational amplifier. Analysis gives

$$\frac{e_o}{e_i}(D) = -\frac{10 R_2 C_1 D}{(R_1 C_1 D + 1)[R_2(10 C_3 + C_2)D + 1]} \qquad (10.67)$$

---

[1] The Lightning Empiricist, Philbrick Researches Inc., Boston, Mass., October, 1963.

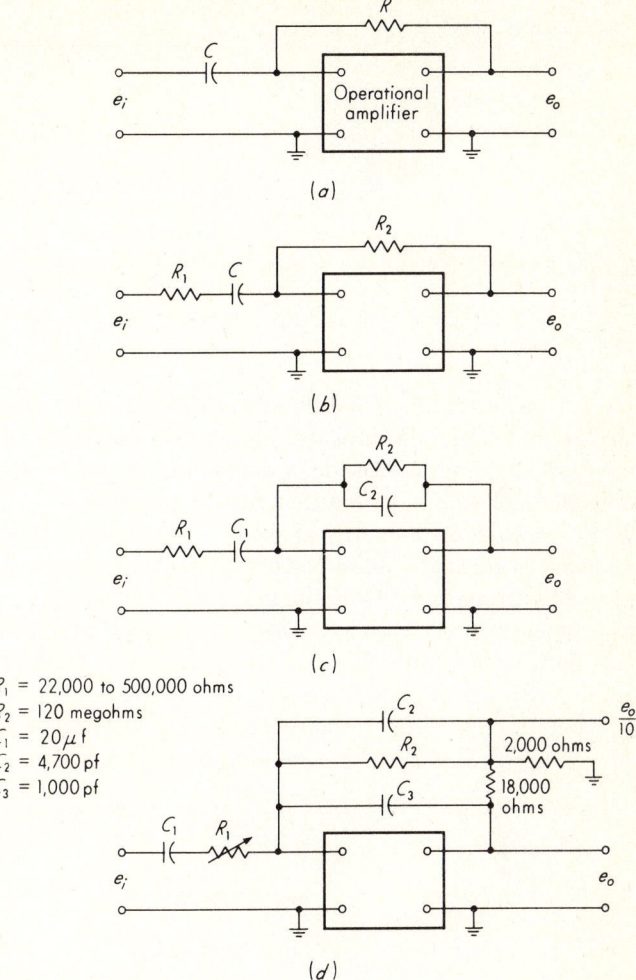

Fig. 10.25. *Electronic differentiators.*

The output is read on a meter which may be connected to e_o or $e_o/10$, depending on the size of the output. For the numerical values given and $R_1 = 22,000$

$$\frac{e_o}{e_i}(D) = -\frac{24,000D}{(0.44D + 1)(1.764D + 1)} \qquad (10.68)$$

We note that for $De_i = 10$ mv/min the output is 4 volts. If, say, $\frac{1}{2}$ mv of 60-cps noise is present at the input, the output noise is only 41 mv, which is about 1 percent of the desired output.

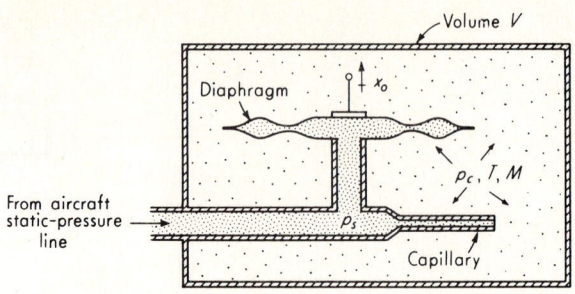

Fig. 10.26.   *Rate-of-climb sensor.*

A final example of a differentiator using nonelectrical methods is the aircraft rate-of-climb indicator shown in Fig. 10.26.   Since atmospheric pressure varies with altitude, a device that measures rate of change of atmospheric pressure can indicate rate of climbing or diving.   While actual design requires a more critical study,[1] we here consider a simplified linear analysis to show the main features.   Static pressure p_s corresponding to aircraft altitude is fed from the vehicle's static pressure probe to the input tube of the rate-of-climb indicator.   Leakage through a capillary tube into the chamber of volume V occurs at a mass flow rate assumed to be $K_c(p_s - p_c)$ lb$_m$/sec.   Air in the chamber follows the perfect-gas law $p_c V = MRT$.   Motion of the output diaphragm is according to $x_o = K_d(p_s - p_c)$.   Assuming K_c and T to be constant, analysis gives

$$\frac{x_o}{Dp_s}(D) = \frac{K}{\tau D + 1} \qquad (10.69)$$

where

$$K \triangleq \frac{K_d V}{RTK_c} \qquad \text{in./(psi/sec)} \qquad (10.70)$$

$$\tau \triangleq \frac{V}{RTK_c} \qquad \text{sec} \qquad (10.71)$$

Thus x_o, which may be measured with any displacement transducer, is an indication of rate of change of p_s, and thereby a measure of rate of climb, if pressure is assumed to be a linear function of altitude.   Since this is not exactly true, various compensating devices are needed in a practical instrument for this and other spurious effects.

10.5

Dynamic Compensation   Sometimes it is not possible to obtain the desired behavior from a measuring device solely by adjusting its own

[1] D. P. Johnson, Aircraft Rate-of-climb Indicators, *NACA, Rept.* 666, 1939.

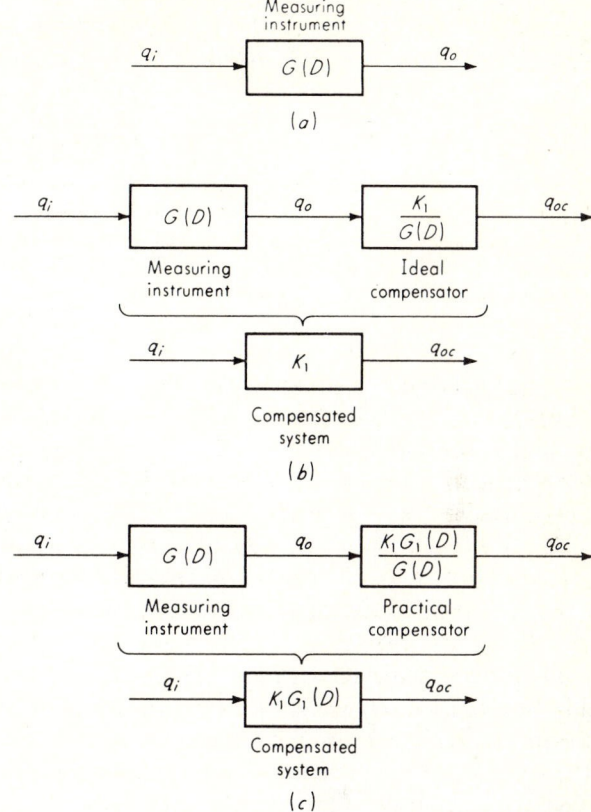

Fig. 10.27. *Generalized dynamic compensation.*

parameters. To get fast response from a thermocouple, for example, very fine wire must be used. Perhaps the vibration and temperature environment might be so severe that such a fine-wire thermocouple would be destroyed before any readings could be obtained. For this and similar situations the concept of dynamic compensation may provide a solution.

Figure 10.27 shows the general arrangement by which dynamic compensation may be employed. Ideally an instrument with transfer function $G(D)$ is cascaded with a compensator $K_1/G(D)$ and thus (if negligible loading is assumed) the overall system now has *instantaneous response* since its transfer function is just the constant K_1. This result is, of course, too good to be true, the practical difficulty lying in the construction of the compensator $K_1/G(D)$, which is generally not physically realizable, because of the need for perfect differentiating effects. While perfect compensation for instantaneous response is *not* possible, very great improve-

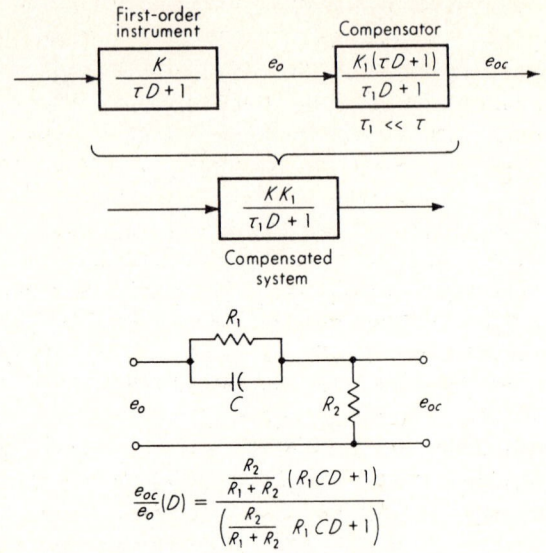

Fig. 10.28. *Dynamic compensation for first-order system.*

ments *may* be achieved with the scheme of Fig. 10.27c.   Here the undesirable dynamics $G(D)$ are *replaced* with more desirable ones, $G_1(D)$.   This
technique has, for example, been used with good success in speeding up
the response of temperature-sensing elements and hot-wire anemometers.
These are basically first-order instruments; thus $G(D) = K/(\tau D + 1)$.
While the compensator can, in general, take any suitable physical form,
because most sensors produce an electrical output, most compensators in
use are electrical circuits.   The compensator generally used for first-order
systems takes the form shown in Fig. 10.28.   Note that any increase in
speed of response $(\tau_1 \ll \tau)$ is paid for by a loss of sensitivity in direct
proportion, since $K_1 = R_2/(R_1 + R_2) = \tau_1/\tau$.   If this loss of sensitivity
is not tolerable, additional amplification is needed.   It is usually placed
between the sensor and the compensator because it will then also serve to
unload the two circuits from each other.   While several stages of such
compensation may sometimes be used, such staging cannot be carried
beyond a certain point because of additional noise introduced by the
amplifiers and accentuated by the compensators.   However, a response
speedup of the order of 100:1 is feasible and has been achieved for thermocouples and hot-wire anemometers.

The concept of dynamic compensation is theoretically applicable to
any-order system.   A good example of a more complex application is
found in the "equalization" of vibration shaker systems.   Figure 10.29a

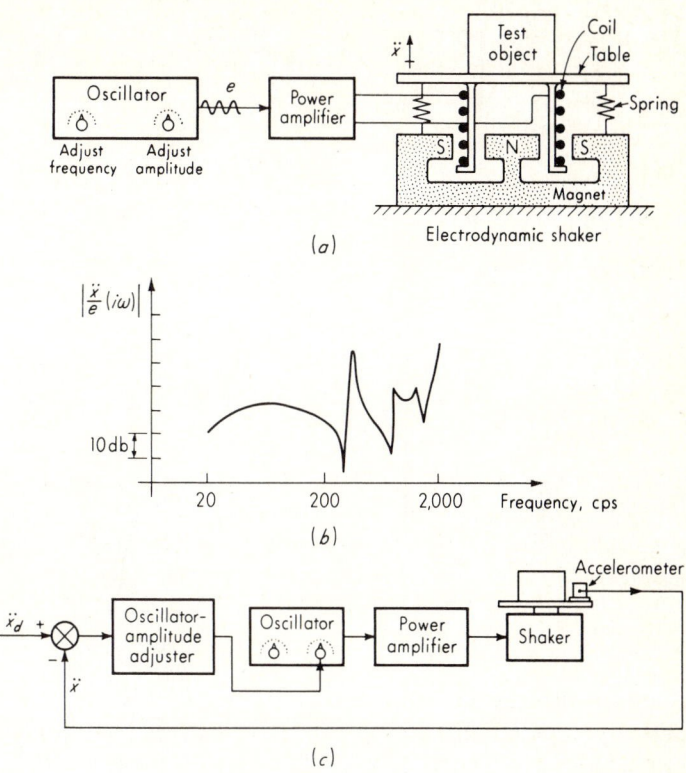

Fig. 10.29. *Vibration shaker systems.*

shows the basic arrangement of a shaker system for sinusoidal vibration testing. Many tests involve "sweeping" the frequency of the oscillator through a certain range while maintaining the acceleration amplitude constant at the test object. While it is not difficult to maintain constant the amplitude of oscillator output voltage e while sweeping through the frequency range, the acceleration $\ddot{x}$ will not be constant since the transfer function $(\ddot{x}/e)(i\omega)$ is not constant over this range. In fact, because of various resonances in the electromechanical shaker, test fixtures, and the test object itself, severely distorted frequency response is not uncommon (Fig. 10.29b). This difficulty can be overcome by use of a feedback scheme as in Fig. 10.29c. Here the actual acceleration $\ddot{x}$ is compared with the desired value $\ddot{x}_d$; if they differ, the amplitude of the oscillator is adjusted to obtain correspondence. This adjustment is performed automatically and continuously as the frequency range is swept.

When random vibration testing (see Fig. 10.30) rather than pure sinusoidal is desired, the above approach is not directly applicable since

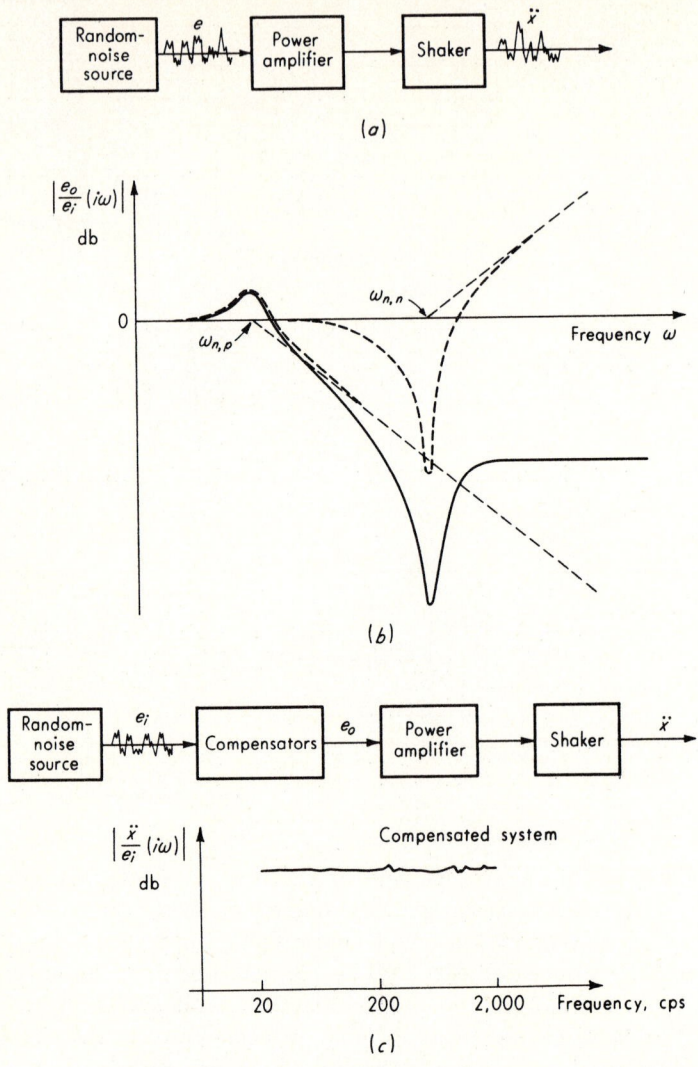

Fig. 10.30. *Dynamic compensation for vibration shaker.*

the signal $\ddot{x}$ is now a random signal and one cannot adjust the noise source in any simple fashion to force $\ddot{x}$ to have the desired frequency spectrum (the spectrum of e). One approach is to provide dynamic compensation such that the transfer function $(\ddot{x}/e)(i\omega)$ *is* flat over the desired frequency range. Then the spectrum at $\ddot{x}$ will be the same as that put in at e. The necessary dynamic compensation here is one which can put "peaks" where there are "notches" and notches where there are peaks in the curve of Fig.

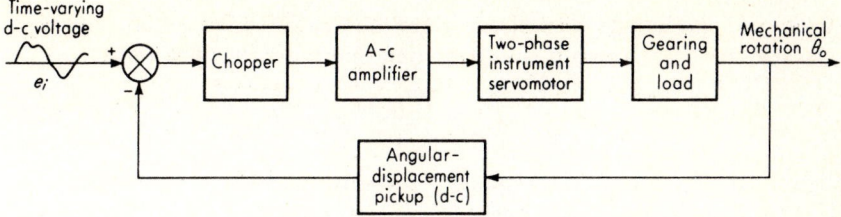

(*a*) D-c voltage-input position servo

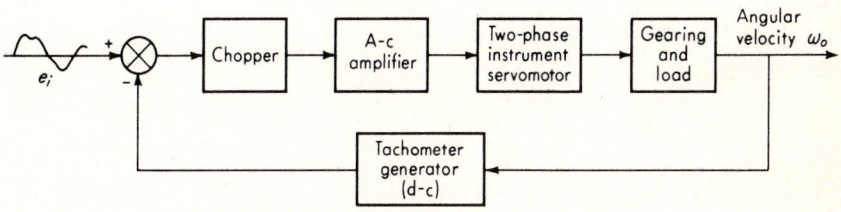

(*b*) D-c voltage-input velocity servo

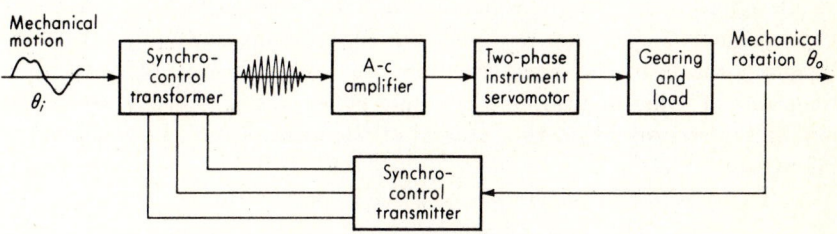

(*c*) Motion-input position servo (all a-c)

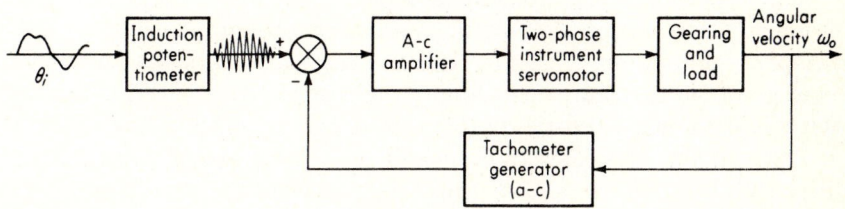

(*d*) Motion-input velocity servo (all a-c)

Fig. 10.31. *Instrument servomechanisms.*

10.29*b*.   Thereby the overall curve can be made relatively flat.   A compensator that will provide one peak and one notch has the form

$$\frac{e_o}{e_i}(D) = \frac{D^2/\omega_{n,n}^2 + 2\zeta_n D/\omega_{n,n} + 1}{D^2/\omega_{n,p}^2 + 2\zeta_p D/\omega_{n,p} + 1} \qquad (10.72)$$

Figure 10.30*b* shows the frequency response of such a compensator in which the peak occurs at a lower frequency than the notch.   The reverse is also possible if needed.   Since ζ_n and ζ_p are also adjustable, the compensator can be "tailored" to cancel out exactly the undesired shaker-system dynamics.   When several peaks and notches are present (as in Fig. 10.29*b*), several compensators are used; as many as 10 are not uncommon.

10.6

Instrument Servomechanisms   Measurement systems often require the conversion of a low-power electrical signal or a low-power mechanical motion into an accurately proportional and (relative to the input signal) high-power mechanical motion.   When frequency response beyond about 5 cps is not required, this function may often be performed by an appropriate instrument servomechanism.   Figure 10.31 shows block diagrams of some of the most common types.   The velocity servos are generally used to obtain the integral of the input signal, as explained in Sec. 10.4.

The d-c-input position servo, while important in its own right, also is the basis of practically all self-balancing potentiometer recorders and *X-Y* plotters.   Thus we shall explain its operation in more detail.   Most instrument servos utilize a-c amplifiers and two-phase a-c instrument servomotors even if the input signal is d-c.   The use of a-c amplifiers is based on their freedom from drift and reasonable cost, while the use of two-phase a-c motors relates to their low friction (no brushes are needed as in a d-c motor) and controllability.   Figure 10.32 briefly summarizes the operating characteristics of this type of motor.   One of the phases is of fixed amplitude.   The amplitude of the other phase (which must be displaced in phase by ± 90 electrical degrees from the fixed phase) controls the direction and amount of torque developed.   When the controlled phase reverses polarity (goes from $+90$ to $-90°$ or vice versa) the torque reverses.

The schematic and graphs of Fig. 10.33 show how a d-c signal is converted to alternating current by a chopper and how a reversal in polarity of the d-c error signal e_{aa} results in a $180°$ phase shift (from $+90$ to $-90°$ or vice versa) in the motor-control phase voltage e_{dd}, thereby causing the required reversal in the direction of torque.   We see that whenever

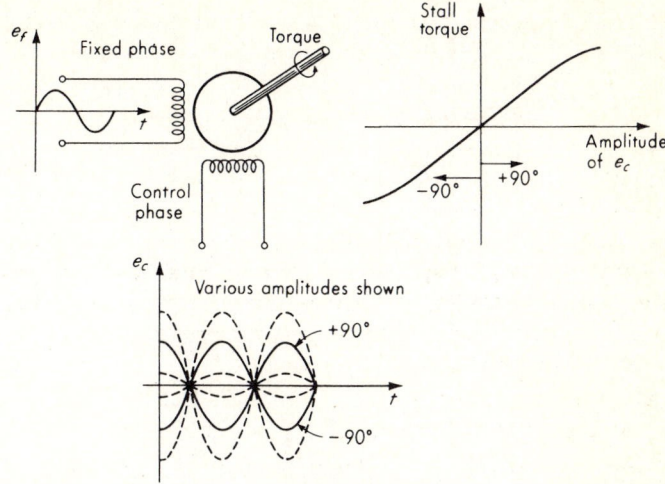

Fig. 10.32. *Two-phase servomotor.*

$e_i \neq e_p$ there will be an error voltage e_{aa} which is converted to alternating current and amplified so that it tends to drive the motor in a direction to change e_p until it equals e_i. If e_i is changing, if the amplifier gain is high enough, e_p (and therefore output motion θ_o) will "track" e_i with very little error. Instrument servos regularly achieve high static accuracy, having errors as small as 0.1 to 0.2 percent of full scale in positioning θ_o as a linear function of e_i. However, their frequency response is limited by inertia of moving parts to about 5 cps or less.

10.7

Addition and Subtraction   The addition or subtraction of mechanical-motion signals is generally accomplished by use of gear differentials or summing links; see Fig. 10.34a. Forces or pressures are summed and transduced to displacement by the schemes shown in Fig. 10.34b. The spring restraints that transduce force to displacement may be removed if a feedback system using a null-balance force to return deflection to zero is employed. Summing of voltage signals is accomplished by the simple series circuit or the operational-amplifier circuit shown in Fig. 10.34c. Subtraction rather than addition in all the above devices is obtained by simply reversing the sense of the input to be subtracted. Addition is the basic operation of digital computers; thus addition or subtraction of digital signals in binary form is easily accomplished with such equipment. Equipment for interconverting numbers in binary and decimal form is also available. Digital signals in the form of pulse

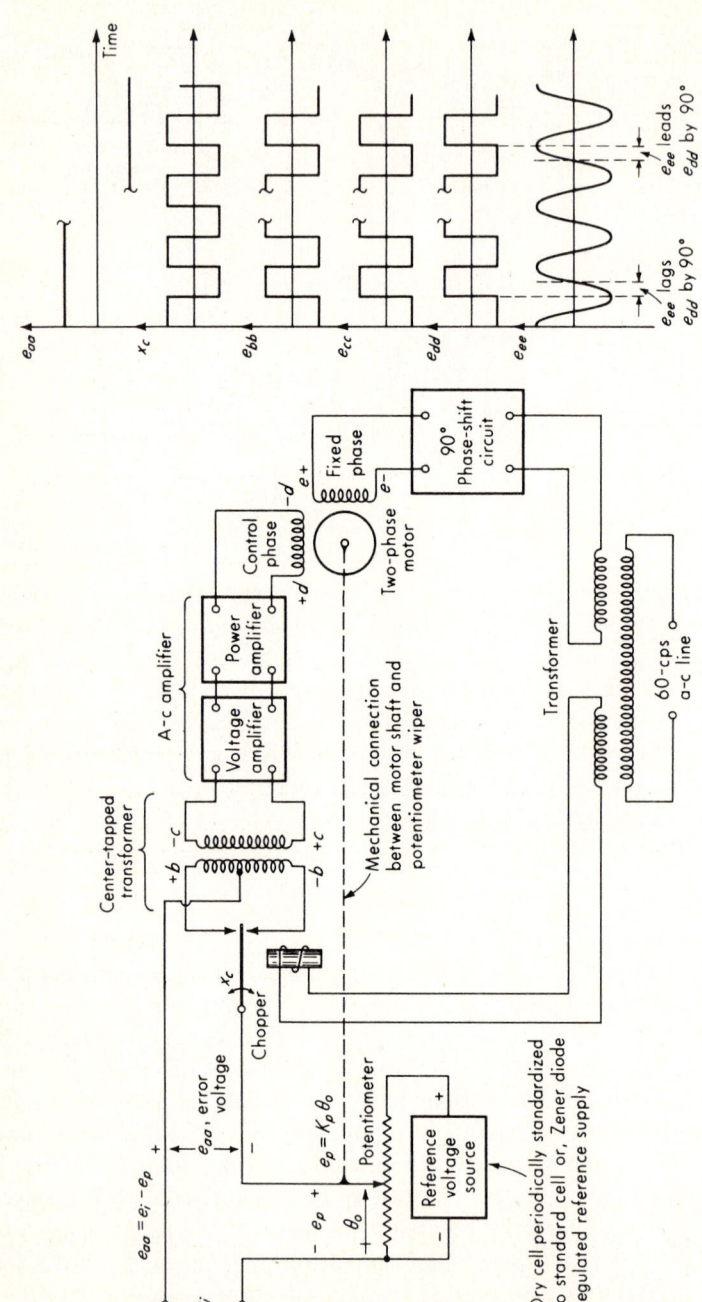

Fig. 10.33. Position servo.

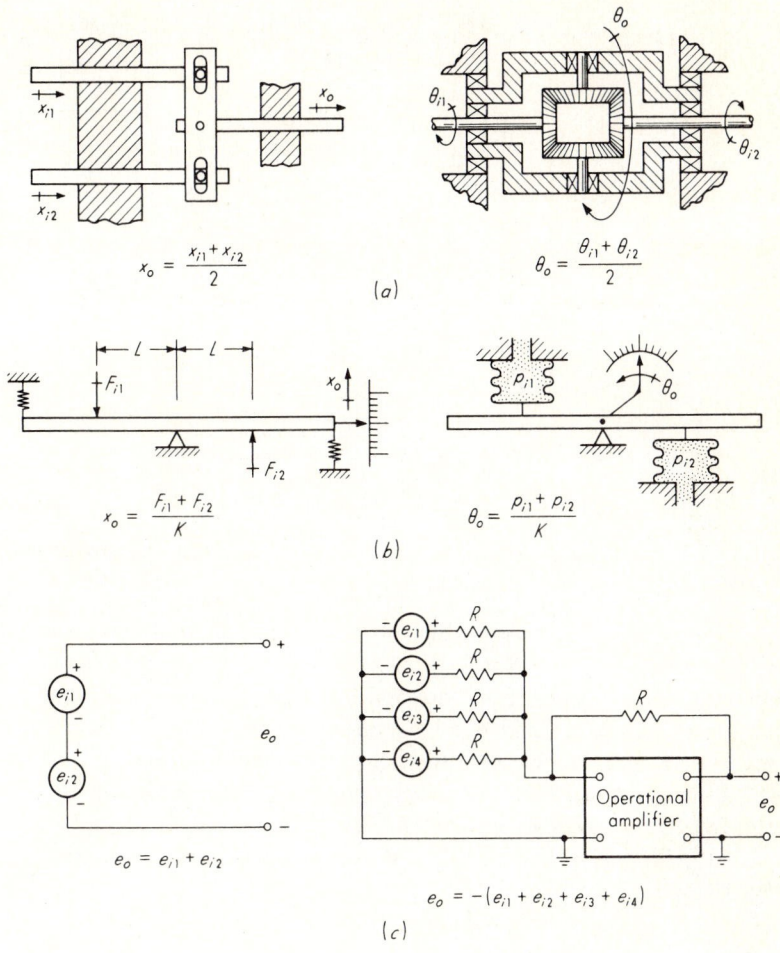

$$x_o = \frac{x_{i1} + x_{i2}}{2}$$

(a)

$$\theta_o = \frac{\theta_{i1} + \theta_{i2}}{2}$$

$$x_o = \frac{F_{i1} + F_{i2}}{K}$$

(b)

$$\theta_o = \frac{p_{i1} + p_{i2}}{K}$$

$$e_o = e_{i1} + e_{i2}$$

(c)

$$e_o = -(e_{i1} + e_{i2} + e_{i3} + e_{i4})$$

Fig. 10.34. *Addition and subtraction.*

rates may also be added and subtracted to obtain pulse rates that are the sum or difference of the input pulse rates.

10.8

Multiplication and Division When data manipulation requires multiplication or division of two variable signals a number of techniques are available, depending on the physical nature of the signals.[1] We here consider a few of the most common.

[1] S. A. Davis, 31 Ways to Multiply, *Control Eng.*, p. 36, November, 1954.

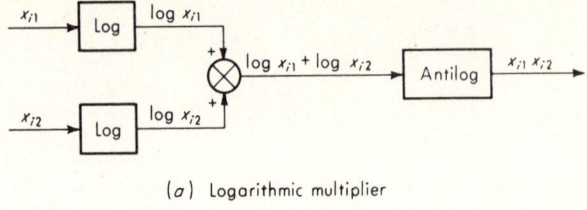

(*a*) Logarithmic multiplier

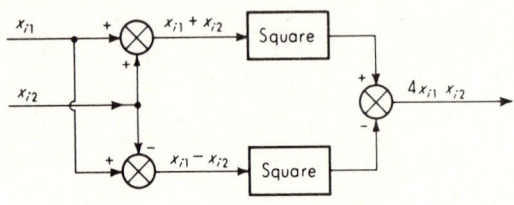

(*b*)   Quarter-squares multiplier

Fig. 10.35.   *Multiplier methods.*

Two general methods which may be implemented either mechanically or electrically are the logarithmic method and the quarter-squares method.   In the logarithmic method one first obtains logarithms of the two input signals x_{i1} and x_{i2}.   For mechanical signals this can be done with special noncircular gearing[1]; electromechanical types use specially wound potentiometers, while electronic units use diode function generators.   Once the two logarithms are obtained, they are added (subtracted if division is wanted) and the antilog of the sum taken (see Fig. 10.35*a*).

$$\text{Antilog} (\log x_{i1} + \log x_{i2}) = x_{i1}x_{i2} \qquad (10.73)$$

The devices for taking the antilog are essentially inversions of the devices for taking the log.

In the quarter-squares method one must have available devices for adding, subtracting, and squaring.   Squaring can be done mechanically with linkages, cams, or noncircular gears; electromechanically with nonlinear potentiometers; and electronically with special nonlinear resistors[2] or diode function generators.   If the needed operations can be performed, the action of the multiplier is given by Fig. 10.35*b*.

An all-mechanical multiplier[3] based on a similar triangles principle

[1] A. E. Lockenvitz et al., Geared to Compute, *Automation*, p. 37, August, 1955.

[2] Quadratron, Bourns Inc., Riverside, Calif.

[3] G. W. Michalec, Analog Computing Mechanisms, *Machine Design*, p. 157, Mar. 19, 1959.

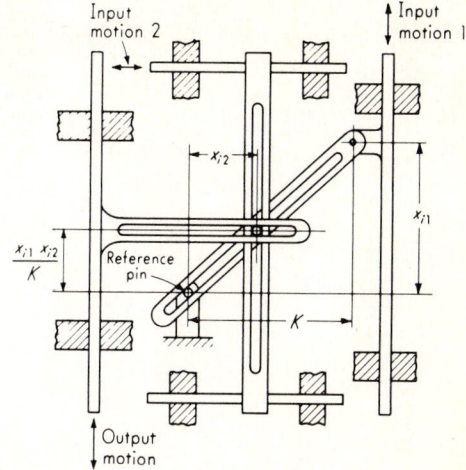

Fig. 10.36. *Mechanical multiplier.*

is shown in Fig. 10.36. If the motions to be multiplied are at a low power level, instrument servos may be used to drive the x_{i1} and x_{i2} input members. Similarly, if nonmotion quantities such as temperatures, pressures, etc., are to be multiplied, they may be transduced to voltages and instrument servos again used to provide the required multiplier input motions.

A Wheatstone bridge may be used for multiplication if one of the signals can be transduced to a voltage (which is used as the bridge excitation voltage) and the other can be transduced to a resistance change (see Fig. 10.37). The bridge output is then proportional to the product $q_{i1}q_{i2}$ for small percentage resistance changes.

Servo multipliers have been widely used in general-purpose analog computing installations and also find application in control and data systems in which two voltage signals must be multiplied. As shown in Fig. 10.38a, one voltage is applied to an instrument servo which positions a potentiometer wiper in direct proportion. The other voltage serves as the potentiometer excitation; thus the output voltage is proportional

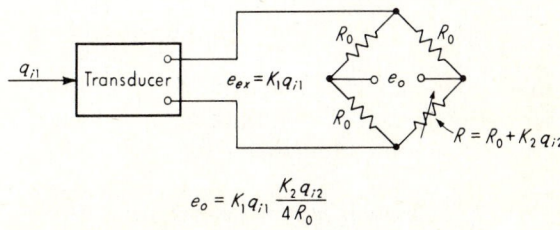

$$e_o = K_1 q_{i1} \frac{K_2 q_{i2}}{4 R_0}$$

Fig. 10.37. *Wheatstone-bridge multiplier.*

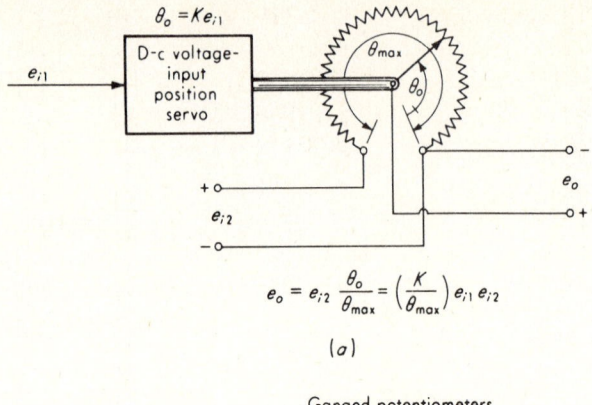

$$e_o = e_{i2} \frac{\theta_o}{\theta_{max}} = \left(\frac{K}{\theta_{max}}\right) e_{i1} e_{i2}$$

(a)

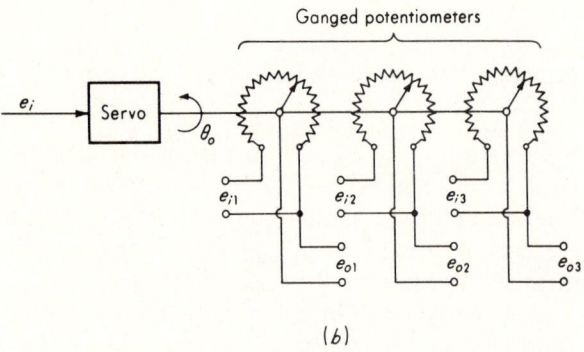

(b)

Fig. 10.38.   *Servo multiplier.*

to the desired product. Such multipliers are particularly useful if several signals e_{i1}, e_{i2}, e_{i3}, etc., are each to be multiplied by one other signal e_i. Then "ganged" potentiometers all driven from the same servo shaft give the desired products (see Fig. 10.38b). Servo multipliers achieve high accuracy, errors being of the order of 0.1 percent for static operation. The frequency response of the servos limits their accurate dynamic operation to less than about 5 cps.

When multiplication or division of rapidly varying voltage signals is required, all-electronic techniques employing the quarter-square scheme (with diode function generators)[1] or the time-division method[1] are available. We shall not go into the details of these methods but merely quote typical performance specifications. A typical quarter-square multiplier[2] accepts full-scale inputs of ± 100 volts and produces ± 100 volts output,

[1] A. S. Jackson, "Analog Computation," McGraw-Hill Book Company, p. 474, New York, 1960.

[2] Comcor Inc., Denver, Colo.

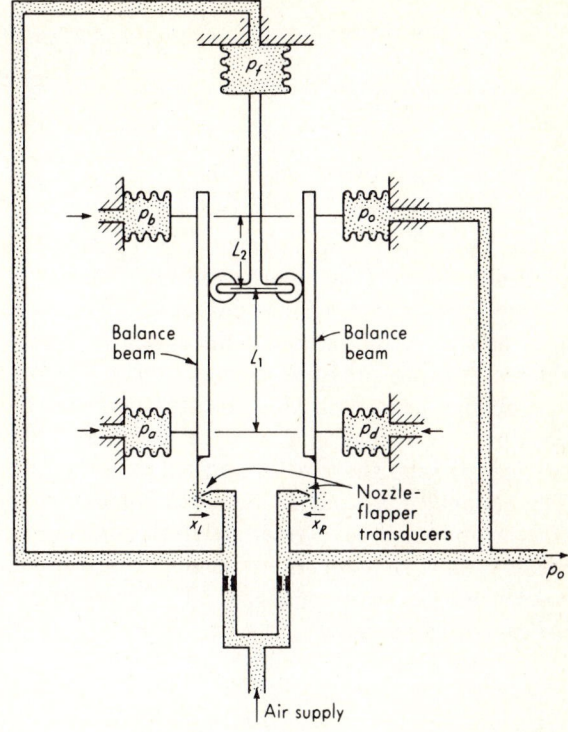

Fig. 10.39. *Sorteberg force bridge.*

has ± 100 mv static error, noise level less than 20 mv peak to peak, frequency response flat within 3 db from 0 to 20,000 cps, phase shift of 2° at 1,000 cps, zero error less than ± 20 mv with either input zero (± 2 mv with both inputs zero), and drift of 1.6 mv/F° from 70 to 95°F.   A time-division multiplier[1] accepts full-scale inputs of ± 100 volts and produces ± 100 volts output, has a static inaccuracy of 0.01 percent of full scale for single-quadrant multiplication (0.05 percent in all four quadrants), noise level less than 100 mv peak, frequency response flat within 1 db from 0 to 700 cps, phase shift less than 1° at 100 cps, zero error of 40 mv with one input zero, and drift less than 100 mv for 8 hr.

Multiplication and division of pneumatic pressure signals may be accomplished by devices such as the Sorteberg force bridge[2] of Fig. 10.39. Let us assume that for some chosen initial equilibrium condition (say $p_a = p_b = p_d = p_o$, $L_1 = L_2$, and $p_f = p_{f0}$) both balance beams are

---

[1] Donner Scientific Co., Concord, Calif.
[2] Sorteberg Controls Co., South Norwalk, Conn.

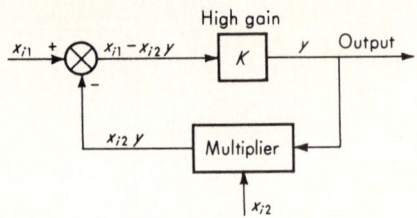

Fig. 10.40. *Feedback inversion of multiplier.*

vertical. Now, if we assume the gain of the nozzle-flapper transducers is very high, only a minute deflection x_L or x_R can cause a large change in p_f or p_o. Examination of the action of the system will show that any change in input pressure p_d that causes a beam deflection x_R will result in a change in p_o such as to return the beam to very nearly its original position. A change in p_a and/or p_b causes a change in x_L which in turn changes p_f (and thereby L_1 and L_2) so as to return the beam to balance. The change in L_1 and L_2 tends to unbalance the right-hand beam but this again tends to correct itself through a change in p_o. The overall result is thus that for any changes in the three inputs p_a, p_b, and p_d the beams always stay balanced. If this is true, we have (assuming all bellows of equal area)

$$p_b L_2 = p_a L_1 \qquad (10.74)$$
$$p_o L_2 = p_d L_1 \qquad (10.75)$$

and thus

$$p_o = \frac{p_b p_d}{p_a} \qquad (10.76)$$

We see that the system thus multiplies p_b and p_d and divides the product by p_a. Inaccuracy of this system is of the order of 1 percent of full scale, resolution 0.1 percent, and hysteresis 0.25 percent. Dynamic response is approximately first-order with a time constant of 4 to 30 sec, depending on the length of the transmission tubing (1 to 200 ft).

To perform the operation of division, any multiplier may be connected into a high-gain feedback loop as shown in Fig. 10.40, and this procedure is regularly used with mechanical, electromechanical, and electronic multipliers. From the block diagram we have

$$K(x_{i1} - x_{i2}y) = y \qquad (10.77)$$
$$x_{i1} - x_{i2}y = \frac{y}{K} \qquad (10.78)$$

and if gain K is very high

$$x_{i1} - x_{i2}y \approx 0$$

giving the output y as

$$y \approx \frac{x_{i1}}{x_{i2}} \qquad (10.79)$$

Dividers have essentially the same performance characteristics as

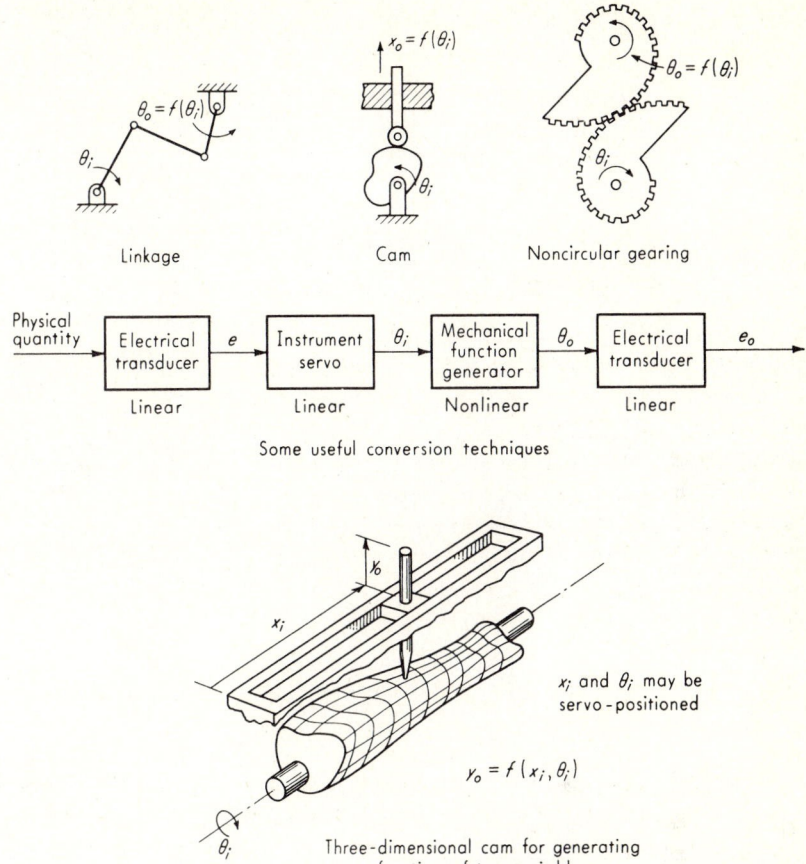

Fig. 10.41. *Mechanical function generation.*

multipliers except that, since division by zero is allowed neither mathematically nor physically, the signal representing the divisor cannot change sign (go through zero). Also, very small values of the divisor overload the device.

10.9

Function Generation   When one needs to generate a specific nonlinear function of a mechanical-motion signal, the use of cams, linkages, and noncircular gears allows great freedom since almost any reasonable function can be adequately approximated by one or a combination of these methods. The use of instrument servos also allows these methods

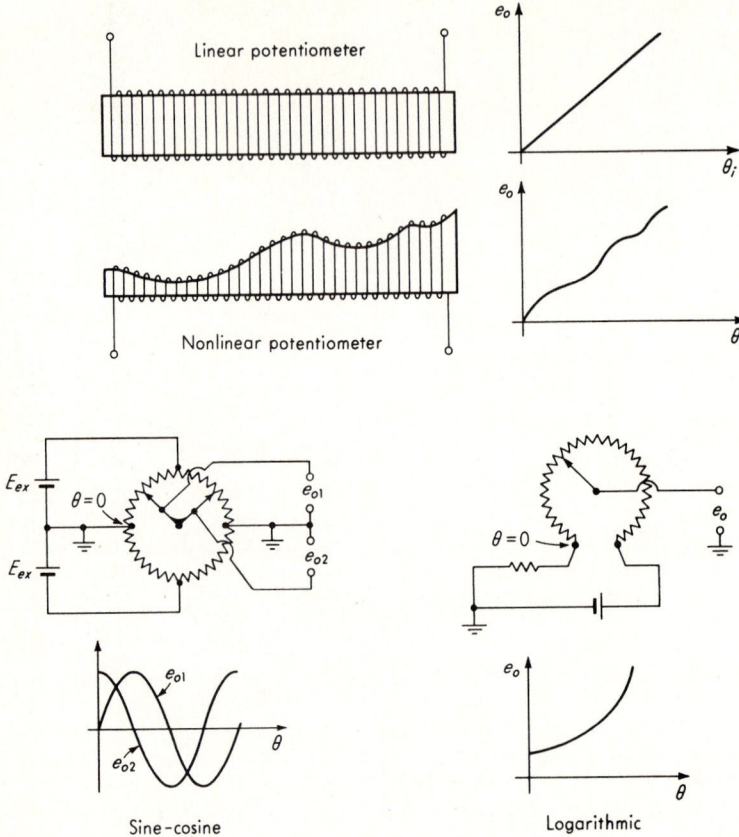

Fig. 10.42. *Nonlinear potentiometers.*

to be employed with electrical signals. If an electrical output is wanted, a motion transducer can be attached to the mechanical output member. Figure 10.41 illustrates these concepts.

Nonlinear potentiometers are also widely used in function generation. They are constructed in basically the same manner as potentiometer displacement transducers except that a specific *nonlinear* relation between θ_i and e_o is wanted rather than the linear relation desired for a motion transducer; see Fig. 10.42. A wide variety of functions is possible by distributing the resistance winding in a proper nonlinear fashion on the mandrel. Techniques have also been developed for constructing nonlinear potentiometers, using conducting plastic or deposited-film (rather than wire-wound) resistance elements. While functions of rather arbitrary form are available as special items, a small number of basic

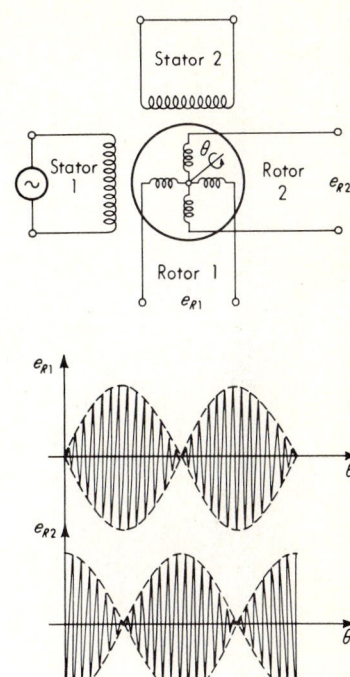

Fig. 10.43. *Resolver.*

functions are so commonly used that they are obtainable ready-made as stock items. These include sine and cosine over 360°, sine or cosine over 360°, 180° sine, 90° sine, ±75° tangent, square function, and logarithmic function. The conformity of the voltage-output/rotation-input relation to the theoretical function is of the order of 0.3 to 2 percent of full scale, depending on the type of function and the instrument quality.

When very accurate sine and/or cosine functions are needed (as in navigation and fire-control computers where resolution and composition of vectors must be performed), the use of resolvers rather than nonlinear potentiometers may be indicated. Resolvers are small a-c rotating machines similar to synchros. In general they have two stator windings and two rotor windings (see Fig. 10.43). If one of the stator windings is excited with an a-c signal of constant amplitude (60 or 400 cps is commonly used) and the other is short-circuited, rotation of the rotor through an angle θ_i from a null position gives at the two rotor windings a-c signals whose amplitudes are respectively proportional to sine θ_i and cosine θ_i. Other important computing functions[1] such as converting vehicle rotation angles to earth coordinates in navigation systems can also be performed

---

[1] "Resolvers," Ford Instrument Co., Long Island City, N.Y.

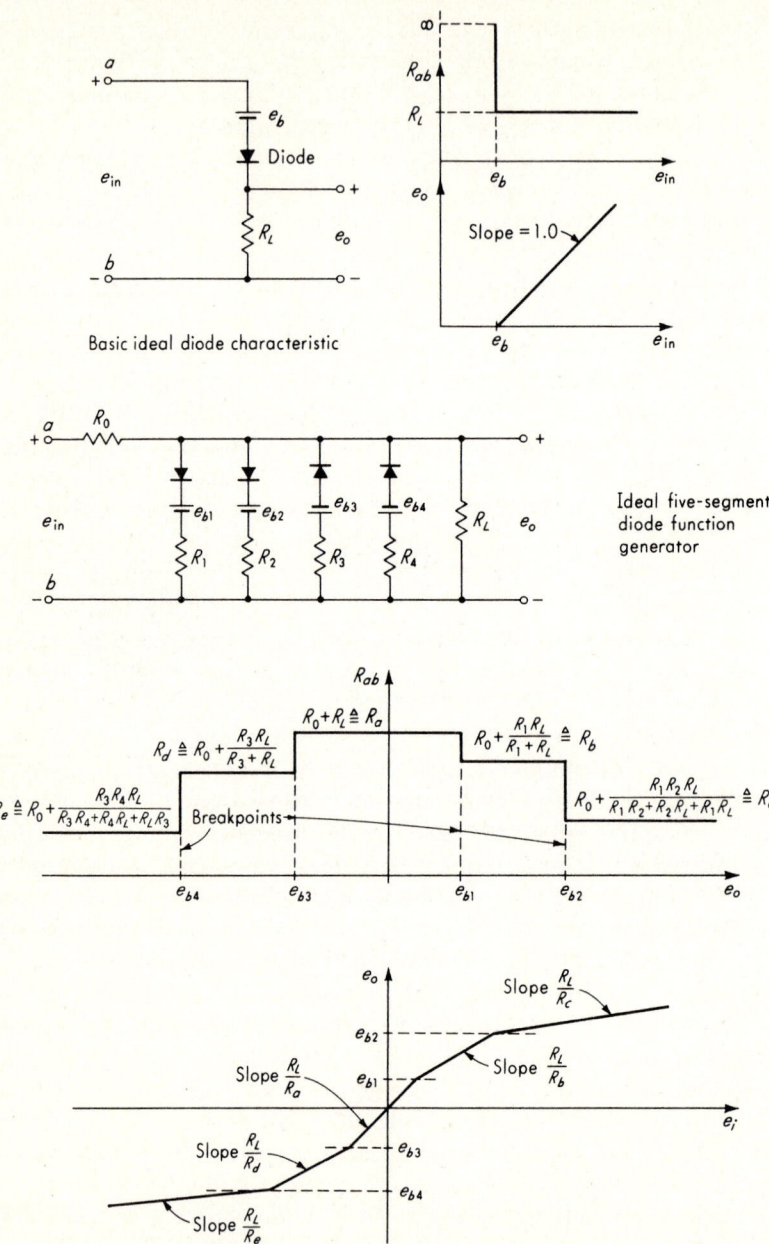

Basic ideal diode characteristic

Ideal five-segment diode function generator

Fig. 10.44.   *Diode function generator.*

by resolvers. A typical high-accuracy resolver[1] has an excitation voltage of 26 volts maximum at 400 cps, open-circuit output voltage of 0 to 26 volts, residual null voltage of 1 mv maximum, and a maximum deviation from the desired functional relation of 0.01 percent.

All the function generators discussed up to this point involve moving parts and are thus limited in speed of response. If higher speed is needed, all-electronic methods are available. The most widely used and versatile device is the diode function generator.[2] It is available in general-purpose forms which can be adjusted to fit almost any single-valued function one can draw on a piece of paper. Figure 10.44 shows the operating principle of such devices. Ideally the diode (either vacuum tube or semiconductor) is assumed to have zero resistance in the forward direction and infinite resistance in the reverse; thus it may be thought of as a switch that is open or closed, depending on the polarity of the voltage across it. While the circuit of Fig. 10.44 allows generation of monotonic functions only, com-mercially available units using operational amplifiers can sum two monotonic functions (one increasing, one decreasing) to obtain functions with both positive and negative slopes. Such units also allow continuous adjustment of breakpoint locations and segment slopes over wide ranges. Some units[3] also blend the straight-line-segment breakpoints with a tangent parabola of adjustable curvature. This allows generation of smoother curves with more continuous derivatives. Commercial diode function generators generally have 10 to 20 straight-line segments. A typical unit[4] having 10 segments has a full-scale output of ± 100 volts, maximum slope of 10 volts/volt, phase shift of 0.8 to 1.4° at 100 cps, fre-quency response down 3 db at 7,000 cps, and noise level of 20 to 100 mv.

The use of specially prepared nonlinear resistors[5] together with conventional operational amplifiers allows generation of many common functions with a relatively small amount of equipment. Being all-electronic, such methods allow high-speed operation.

10.10

Amplitude Modulation and Demodulation   We have seen earlier in the text a number of examples of measurement systems in which interconversion between a-c and d-c signals was necessary and/or

---

[1] "Resolvers," Ford Instrument Co., Long Island City, N.Y.

[2] E. J. Galli, How Diodes Generate Functions, *Control Eng.*, p. 109, March, 1959; and p. 107, February, 1960.

[3] G. A. Philbrick Corp., Boston, Mass.

[4] Comcor Inc., Denver, Colo.

[5] Specifications and Applications of the Douglas Quadratron, Douglas Aircraft Co., Santa Monica, Calif., Feb. 27, 1963.

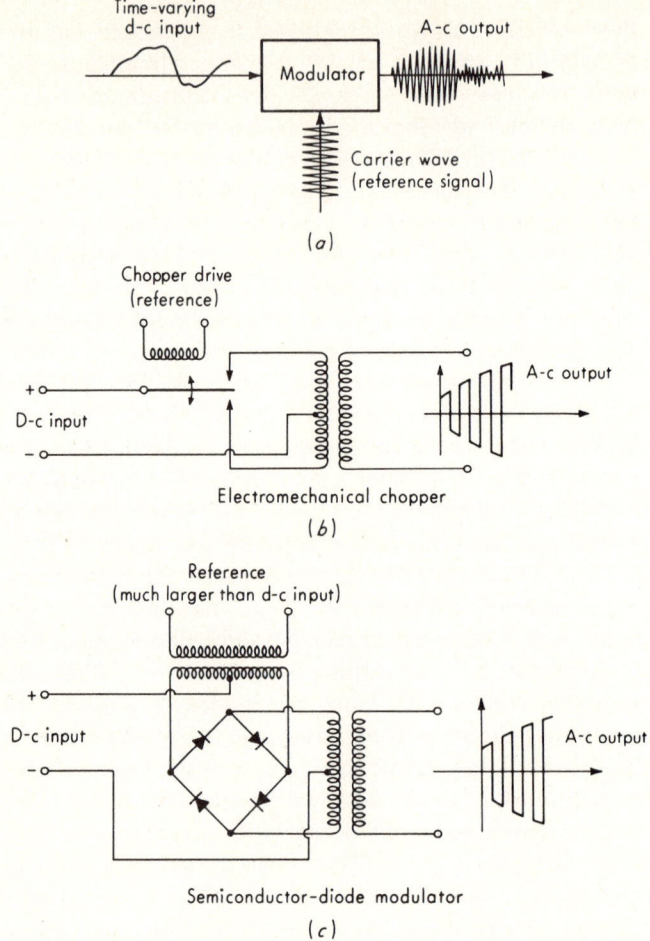

Fig. 10.45. *Amplitude modulation.*

desirable. The conversion from direct current to alternating current is a form of amplitude modulation whereas conversion from alternating to direct current is called demodulation or detection. We here give briefly a few examples of the hardware needed to accomplish these functions.

The process of modulation may be performed by a wide variety of devices[1]; however, all may be represented in block-diagram form as in Fig. 10.45a. In general, the frequency spectrum of the d-c input signal

---

[1] B. T. Barber, Servo Modulators, *Control Eng.*, August, October, November, December, 1957.

must not go beyond about 10 to 20 percent of the carrier frequency for proper operation.   Electromechanical choppers (vibrators) are widely used as modulators.   The carrier is a square wave, often at 60 or 400 cps and limited to less than about 1,000 cps.   The output is often transformer-coupled to the following circuitry to provide electrical isolation (see Fig. 10.45b).   When higher carrier frequencies are required, all-electronic modulators are available.   Tube, transistor, and diode types are usable up to about 10,000 cps.   Figure 10.45c shows a ring-type diode modulator. Here the a-c reference signal serves to "switch" the diodes from their conducting to nonconducting states, thus giving an action quite similar to the chopper but without moving parts.

In most measurement systems, phase-sensitive demodulation is required, if modulation was performed earlier, so as to recover the algebraic sign of the original d-c information.   This requires that the reference signal used to drive the modulator must also be used in the demodulator to ensure proper "synchronization."   For "square-wave"-type modulators and demodulators such as the examples of Figs. 10.45 and 10.46, if there were no time delays, attenuation, or phase shifts, the d-c signal recovered at the demodulator would be identical with that originally put into the modulator and no filtering would be required.   This, of course, is not possible, and so a low-pass filter is generally required at the demodulator output to remove high-frequency components introduced by imperfect modulation and demodulation.   Demodulation may be accomplished with a number of devices[1]; Fig. 10.46 shows two analogous to the modulators of Fig. 10.45.

10.11

Voltage-to-Frequency and Frequency-to-Voltage Converters
The conversion of a d-c voltage input to a periodic-wave output whose frequency is proportional to the d-c input may serve several useful functions in measurement systems.   Such devices, called voltage-controlled oscillators, are widely used in FM/FM telemetry systems, since the voltage-to-frequency conversion process is a form of frequency modulation.   They are also used in the integrating digital voltmeter where a d-c signal is converted into a periodic wave of proportional frequency. This wave is then applied to an electronic counter for a fixed time interval, giving a reading proportional to the average d-c voltage over the time interval.   The recording of d-c voltages on magnetic tape recorders is also accomplished through the use of frequency modulation.

[1] J. E. Gibson and F. B. Tuteur, "Control System Components," p. 249, McGraw-Hill Book Company, New York, 1958.

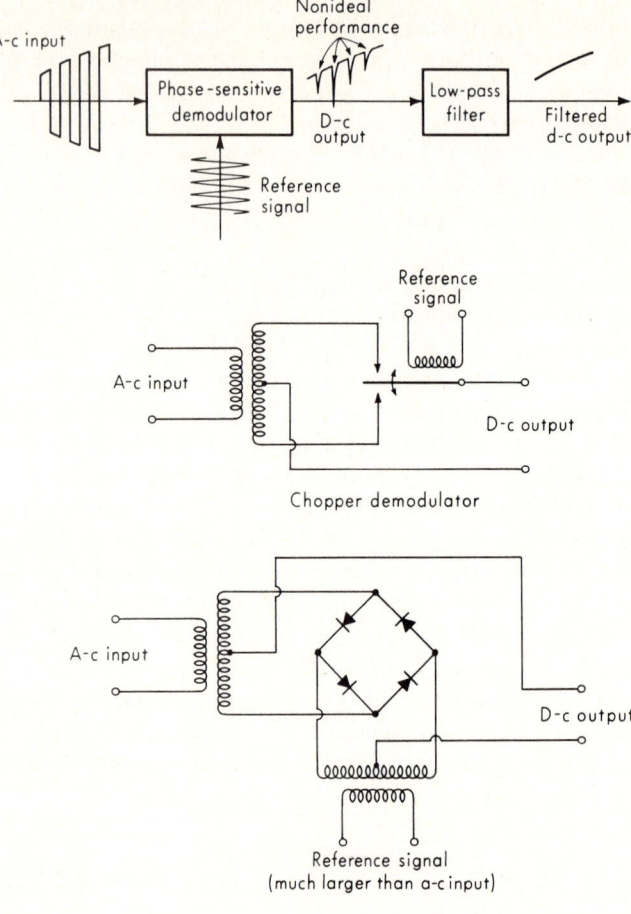

Fig. 10.46.   *Phase-sensitive demodulation.*

Voltage-controlled oscillators[1] are generally of the resistance-capacitance phase-shift type or the multivibrator type. For zero input these oscillators produce a given output frequency, called the center frequency, which commonly is from 400 to 70,000 cps. Variation of the d-c input then results in frequency variation of ± 7.5 or ± 15 percent around the center frequency. Input variation is ordinarily ± 2.5 or 0 to $+5$ volts direct current; however, some units incorporate their own amplification so that ± 10-mv input causes full output-frequency deviation. Non-

---

[1] M. H. Nichols and L. L. Rauch, "Radio Telemetry," p. 253, John Wiley & Sons, Inc., New York, 1956.

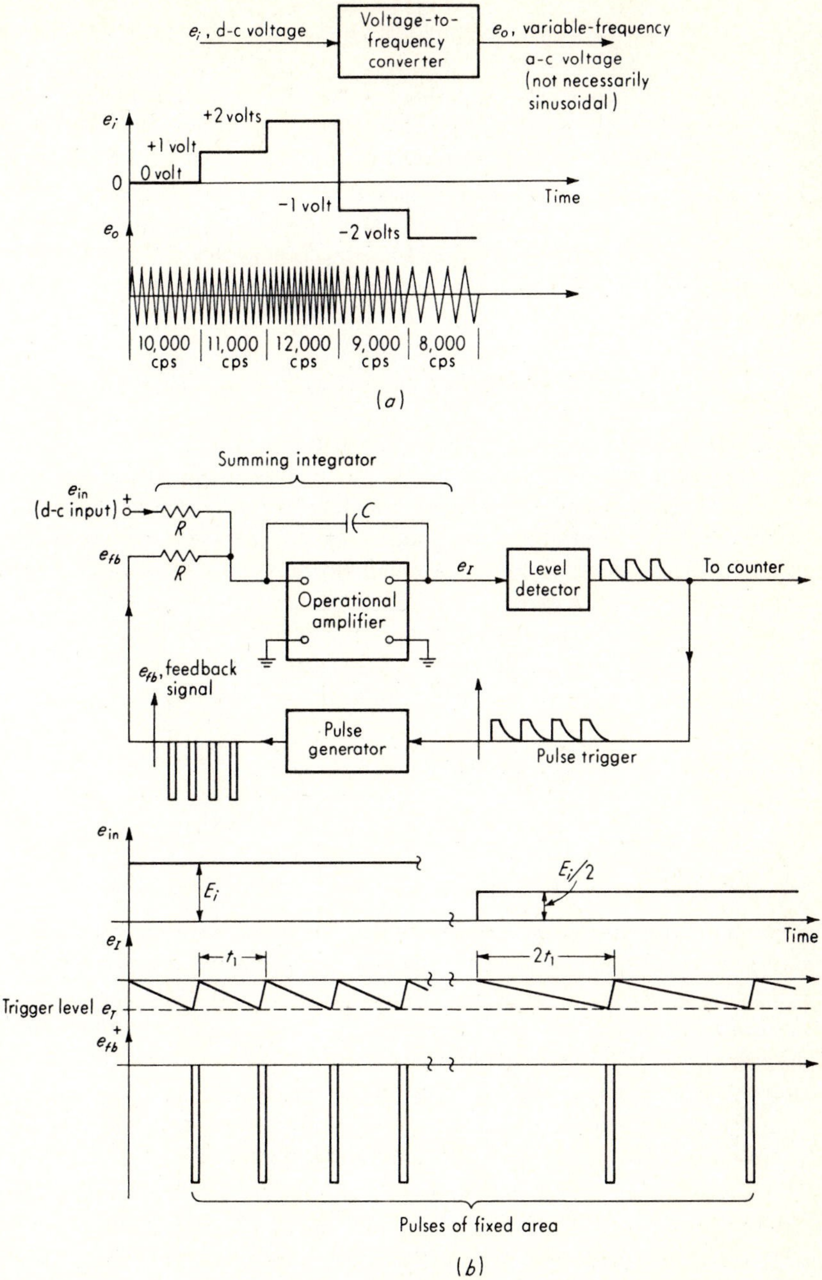

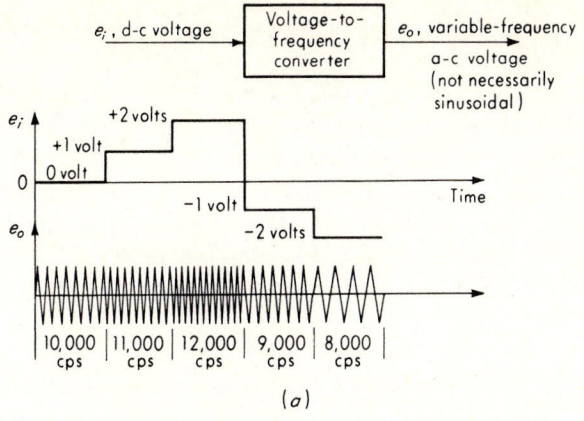

e_i, d-c voltage $\rightarrow$ Voltage-to-frequency converter $\rightarrow$ e_o, variable-frequency a-c voltage (not necessarily sinusoidal)

e_i

+2 volts

+1 volt

0 volt

0

−1 volt

−2 volts

Time

e_o

| 10,000 cps | 11,000 cps | 12,000 cps | 9,000 cps | 8,000 cps |

(a)

Summing integrator

e_{in} (d-c input) +

R

e_{fb}

R

C

Operational amplifier

e_I

Level detector

To counter

e_{fb}, feedback signal

Pulse generator

Pulse trigger

e_{in}

E_i

$E_i/2$

Time

e_I

Trigger level e_r

t_1

$2t_1$

e_{fb}

Pulses of fixed area

(b)

Fig. 10.47. *Voltage-to-frequency converter.*

linearity of the output-frequency/input-voltage characteristic is of the order of 0.1 to 1.0 percent.

Figure 10.47*b* shows a voltage-to-frequency converter used in an integrating digital voltmeter.[1] This device produces an output frequency proportional to the input d-c voltage, down to and including 0-volt input. That is, zero input produces zero-frequency output; thus there is no "center frequency" as described above. The operation of this instrument may be explained as follows: Using the usual operational-amplifier network analysis assumptions, we get the equation of the summing integrator as

$$e_I = - \frac{1}{RC} \int (e_{in} + e_{fb}) \, dt = - \frac{1}{RC} \int e_{in} \, dt - \frac{1}{RC} \int e_{fb} \, dt \qquad (10.80)$$

Now, if a constant input voltage of magnitude E_i is applied, the output e_I of the integrator will be $(-E_i/RC)t$, a negative-going ramp. This voltage is applied to a level detector that will produce a trigger pulse whenever e_I goes through a set value, say e_T. It is clear that the time required for e_I to reach e_T is inversely proportional to E_i. The trigger pulse triggers a pulse generator which produces a pulse of short duration and high amplitude whose area is accurately maintained constant. The area of this pulse, which is applied at the e_{fb} input terminal of the summing integrator, is calculated to just return e_I to zero. That is,

$$e_I \text{ due to } e_{fb} = - \frac{1}{RC} \int e_{fb} \, dt = |e_T| \qquad (10.81)$$

$$e_I \text{ due to } e_{fb} = - \frac{1}{RC} \text{ (pulse area)} = |e_T| \qquad (10.82)$$

and thus $\qquad\qquad\qquad |\text{Pulse area}| = RC \, |e_T| \qquad (10.83)$

Since e_T is fixed and known, the pulse generator can be designed to produce pulses of the required area. When the pulse is over, the negative-going ramp due to E_i starts again and the process repeats itself over and over, producing pulses at a rate proportional to E_i. The commercial unit described above has an output frequency range of 0 to 10,000 cps and a short-term inaccuracy of ± 0.1 percent of full scale.

Frequency-to-voltage converters accept a periodic (not necessarily sinusoidal) input signal and produce a d-c output in direct proportion to input frequency. They are used, for example, to get a d-c analog signal from a-c tachometers or turbine flowmeters. When the input is non-sinusoidal the "frequency" is interpreted as the fundamental frequency or repetition rate of the wave form. Figure 10.48 shows in block-diagram form the mode of operation of a typical frequency-to-voltage converter.

---

[1] Hewlett-Packard Co., Palo Alto, Calif.

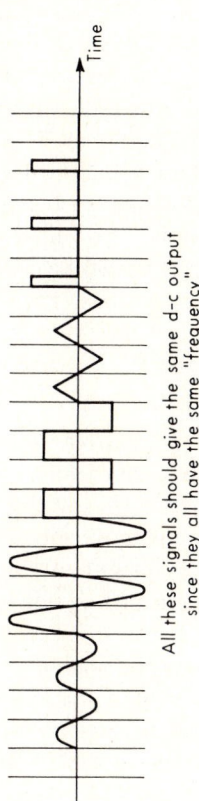

All these signals should give the same d-c output
since they all have the same "frequency"

Time

"Frequency" input

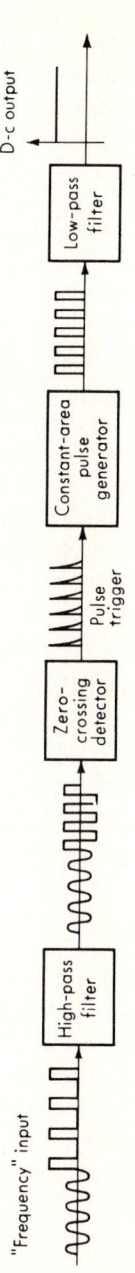

High-pass
filter

Zero-
crossing
detector

Pulse
trigger

Constant-area
pulse
generator

Low-pass
filter

D-c output

Fig. 10.48. *Frequency-to-voltage converter.*

Input wave forms are passed through a high-pass filter so that pulses that do not themselves go negative can be handled.   The characteristic of a signal of any wave form that determines its repetition rate is the number of zero crossings per unit time, and so these are detected and a trigger pulse generated for each one.   These pulses then trigger constant-area pulses at a rate proportional to input frequency.   The average value of this pulse train is thus proportional to frequency and is obtained by low-pass filtering.   This filter gives the system approximately a first-order type of response since its dynamics are the slowest of the entire system.

A commercial frequency-to-voltage converter[1] handles the frequency range 80 to 100,000 cps in nine ranges, accepts sinusoidal input signals from 5 mv rms (5 cps to 10 kc), 30 mv rms (10 to 100 kc) to 10 volts rms, pulse inputs of 25 mv 0 to peak (5 to 20,000 pps), 75 mv 0 to peak (20,000 to 100,000 pps), 5 μsec minimum pulse width, and has nonlinearity 0.025 percent of full scale, long-term drift less than 0.1 percent per week, and an effective time constant of 50 divided by the full-scale frequency range (200, 500, 1,000, 2,000, 5,000, 10 kc, 20 kc, 50 kc, 100 kc).   The longest time constant is thus $50/200 = 0.25$ sec.

10.12

Analog-to-Digital and Digital-to-Analog Converters   The increasing use of digital computers in measurement and control systems and the scarcity of true digital measuring devices lead to a need for analog-to-digital converters to allow analog sensors to communicate with the computer.   Sometimes the digital output of the computer must be used in an analog system; thus digital-to-analog converters are also necessary.

Analog-to-digital converters may be classified broadly as shaft-angle-to-digital or voltage-to-digital types.   The shaft-angle-to-digital type simply uses one of the encoders discussed at the end of Sec. 4.3 to convert the analog shaft angle into coded digital signals.   If the analog signal is not already a shaft rotation, it can be converted to one by using a suitable transducer and position servo (see Fig. 10.49).   Such systems can achieve very high accuracy but are limited in speed of response by the capability of the servos (less than 5 cps).

Voltage-to-digital converters take a number of different forms.[2]
Figure 10.50 illustrates the operation of a time-base encoder in simplified form.   Basically, the analog input signal is compared with a recurrent

---

[1] Vidar Corp., Mountain View, Calif.

[2] J. T. Tou, "Digital and Sampled-data Control Systems," p. 368, McGraw-Hill Book Company, New York, 1959.

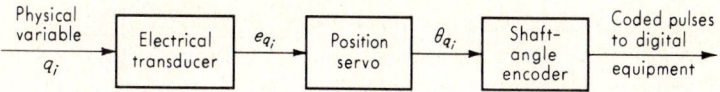

Fig. 10.49. *Shaft-angle-to-digital converter.*

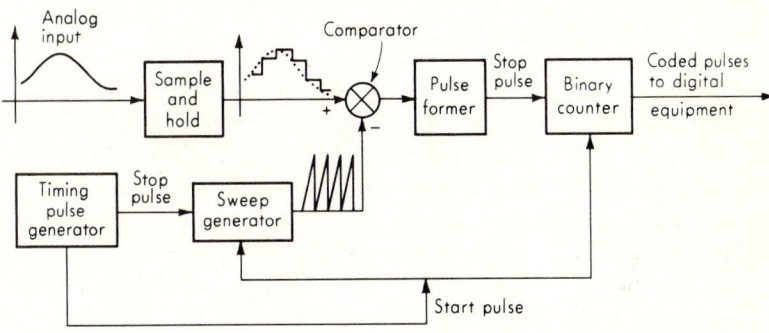

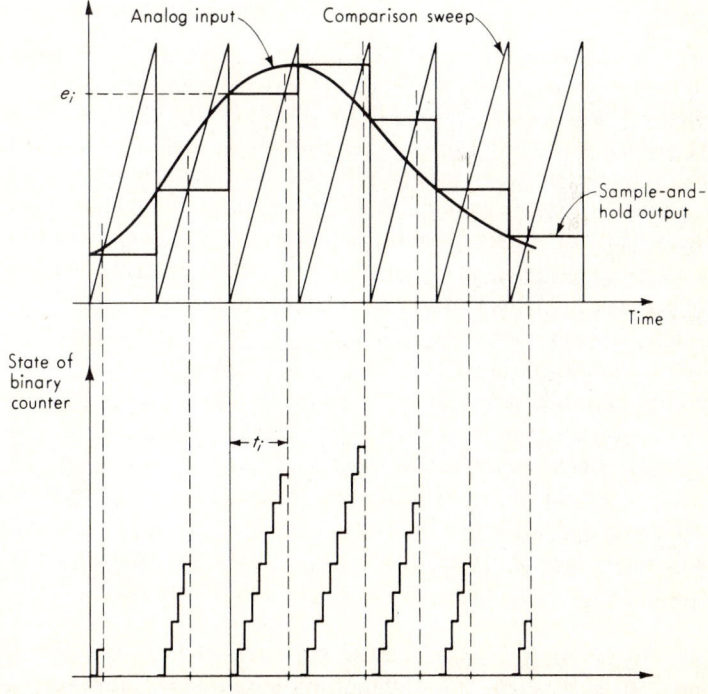

Fig. 10.50. *Time-base voltage-to-digital converter.*

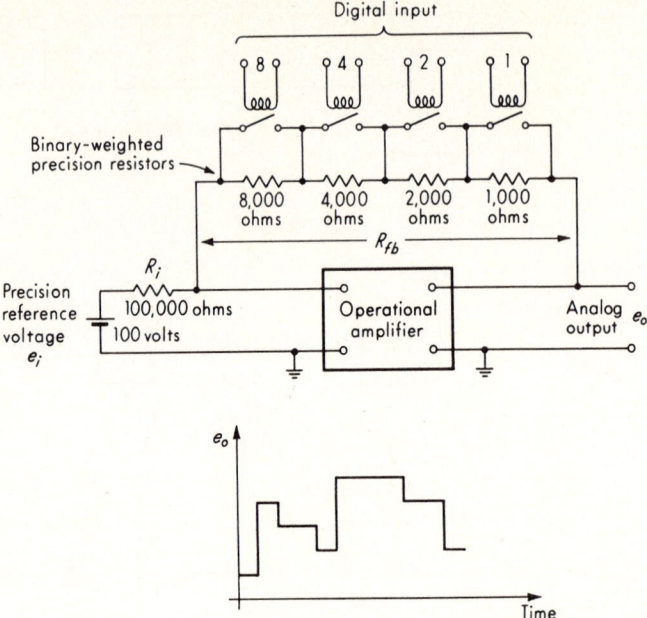

Fig. 10.51.   *Digital-to-analog converter.*

sweep voltage whose peak value is larger than any analog-signal voltage expected. To make the readings of analog input occur at constant time intervals, the input is generally passed through a sample-and-hold device as shown. The output of the sample-and-hold is then compared with the sweep. A binary counter is started at the beginning of each sweep cycle and is stopped at the instant the sweep coincides with the sampled analog input. This counter generates coded pulses representing a binary number which increases in magnitude at a uniform rate. If such a counter is left on for a time interval t_i, as shown in Fig. 10.50, its "state" (binary number represented by its contents) will be proportional to t_i; since t_i is proportional to the sampled analog voltage e_i, the analog voltage will have been converted to its binary digital equivalent. The counter is reset to zero at the end of each cycle; thus the process repeats itself and generates a sequence of digital numbers representing the value of the analog signal at equally spaced intervals of time. All-electronic equipment of this type can perform many thousand such conversions per second.

When digital signals must be converted to analog, the "switch" openings or closures that define the digital signal must somehow be converted into an equivalent voltage. Figure 10.51 shows one method of accomplishing this by using an operational amplifier. The switches shown are closed when a particular digit is not present and open when it is. The feedback resistance at any time is thus the sum of the precision resistors that are not shorted out. For example, if the digital number 13 is present the switches for digits 1, 4, and 8 would be open, giving

$$e_o = -e_i \frac{R_{fb}}{R_i} = -100 \frac{13,000}{100,000} = -13 \text{ volts} \qquad (10.84)$$

The system shown can handle numbers up to 15 in steps of 1 but obviously can be expanded to get greater resolution. The analog output voltage is seen to vary in stepwise fashion but these steps can be smoothed by filtering if desired.

10.13

Signal and System Analyzers   In the analysis and design of many devices and systems it is necessary to have accurate knowledge about the characteristics of the inputs to the system. Once a system has been built, its performance is often checked by studying its output. Equipment for carrying out such studies may be characterized as *signal-analysis equipment*. Closely related to this is the problem of experimentally defining the characteristics (transfer function, frequency response, etc.) of a physical system which may be too complex to analyze accurately by theory alone. Equipment for such investigations might be called *system-analysis equipment* and generally utilizes coordinated simultaneous measurements of both the system input signal and the output signal, together with suitable data processing to obtain conveniently the desired system characterization.

Perhaps the most widely used signal-analysis equipment is that which measures the frequency spectrum of a fluctuating physical quantity. The most common applications are in the field of sound and vibration where the frequency spectrum of a sound pressure, stress, acceleration, etc., may be very useful in diagnosing faults in an operating machine or system. These can be traced back to their origin by noting peaks in the frequency content at certain frequencies and then finding the machine parts that run at speeds that would produce such frequencies. While machines containing rotating and/or reciprocating parts give rise to sound and vibration signals having strong peaks at certain frequencies, a glance at an oscilloscope screen showing such a signal will generally not indicate a simple periodic wave form. Rather, the appearance will be that of a more-or-less random variation. Since this is the case, we would expect that the proper description of such a signal in the frequency domain would be in terms of the mean-square spectral density (power spectral density). While this is theoretically true, many problems of sound and vibration analysis do not require that a true mean-square spectral-density analysis be performed. Rather, somewhat simpler equipment is employed which gives sufficient information for most purposes.

Figure 10.52 shows the functional operation of a typical system. The incoming signal is passed through a narrow band-pass filter with center frequency ω_c and effective bandwidth $\Delta\omega$. This filter is tunable, so that ω_c may be varied smoothly and continuously over a given frequency range, say 20 to 20,000 cps, by manual turning of a knob or automatic drive from a constant-speed motor. This filtering operation appears to be exactly the same as that which would be performed in a true mean-square spectral-density (MSSD) analysis. However a difference exists with regard to the filter bandwidth $\Delta\omega$. In a true mean-square spectral-density analysis the output of the mean-squaring operation is divided by $\Delta\omega$. In practical mean-square spectral-density systems this division need not actually be carried out since $\Delta\omega$ is held constant for all ω_c's and thus the output recorder simply has its scale adjusted to take account of the division by a constant and known $\Delta\omega$. Circuits for obtaining a constant, small $\Delta\omega$ over a wide range of frequencies are quite complex and costly; thus many practical analyzers do not attempt to measure a true mean-square spectral density.

A filter which *can* be constructed fairly easily is the so-called constant-percentage bandwidth type. Here $\Delta\omega$ is not constant but rather varies in direct proportion to ω_c. Thus if $\Delta\omega$ is, say, 1 cps when ω_c is 100 cps it will be 10 cps when ω_c is 1,000 cps; that is, the bandwidth is 1 percent of the center frequency. Now with such a filter one can still compute a true mean-square spectral density by dividing a reading at a given ω_c by the proper $\Delta\omega$. However *this* division is not now possible by a simple fixed scaling of the output recorder, and most systems of this type do not perform it at all. The output record of such an analyzer is *not* a plot of mean-square spectral density, and even though it could be replotted manually to give it (since $\Delta\omega$ is usually known as a function of ω_c) this is not generally done since the uses to which the record is to be put do not require this level of sophistication. Thus the analyzer of Fig. 10.52 performs operations *similar* to those of a true mean-square spectral-density analysis but there is a definite difference which must be remembered. In particular, equal recorded output levels at a low and at a high frequency do *not* mean that the input signal has equal frequency content at these two points, because the true measure of frequency content is mean-square spectral density which requires division by a larger $\Delta\omega$ at the high frequency than at the low.

For applications that *do* require an accurate measurement of mean-

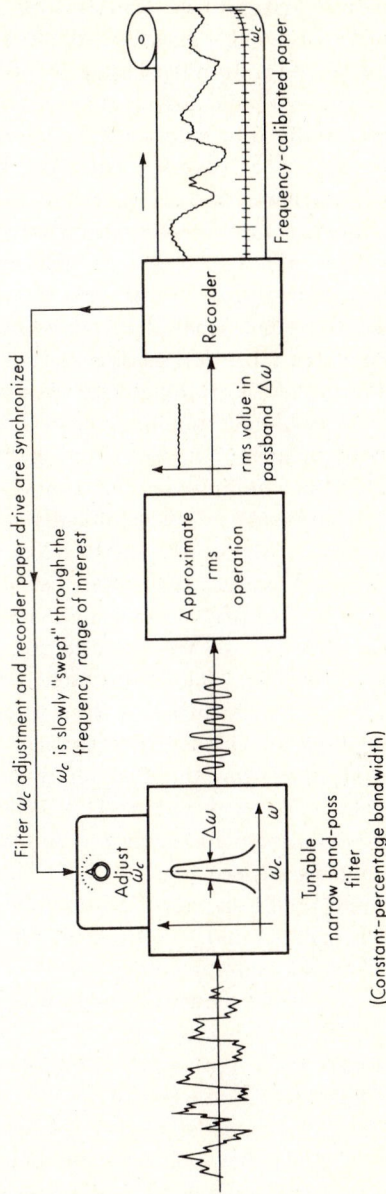

Filter ω_c adjustment and recorder paper drive are synchronized

ω_c is slowly "swept" through the frequency range of interest

Adjust ω_c

Tunable narrow band-pass filter

(Constant-percentage bandwidth)

$\Delta \omega$

ω_c

ω

rms value in passband $\Delta \omega$

Approximate rms operation

Recorder

Frequency-calibrated paper

ω_c

Fig. 10.52. *Frequency-spectrum analyzer.*

square spectral density, suitable analysis equipment is available, though rather costly. The block diagram of such analyzers is quite similar to that of Fig. 10.52 except that the passband $\Delta\omega$ is kept constant for all ω_c's. The value of $\Delta\omega$ can usually be selected to suit the needs of a particular problem; 2 to 100 cps bandwidths are generally available. Also, a mean-square rather than root-mean-square operation is performed, and greater pains are taken to make it accurate. The range of ω_c that can be covered is of the order of 2 to 20,000 cps. If lower or higher frequencies must be studied, a tape-recorder approach using tape speedup or slowdown is possible. Systems are also commercially available for obtaining the cross spectral density of two random signals. Their operation follows essentially the block diagram of Fig. 3.100.

While most sound and vibration analyzers describe the signal in the frequency domain, equipment working in the time domain is also available. A commercially available time-delay correlator[1] computes autocorrelation and cross-correlation functions and automatically plots the required curves. The time delay τ is continuously adjustable from 0 to 17 msec and is accomplished all-electronically without the use of a magnetic-tape loop device. In operation, the value of τ is slowly swept from zero to the maximum value, the sweep period ranging from 1 to 200 min. The averaging time [time over which the integral that defines $R(\tau)$ is computed] can be adjusted from 0.2 to 20 sec, longer averaging times requiring slower τ sweeps. The instrument puts out two d-c voltages, one proportional to τ and the other to $R(\tau)$; thus they can be applied to an X-Y plotter to plot $R(\tau)$ versus τ directly and automatically.

Commercial equipment for the frequency analysis of transient signals is also available. One such analyzer[2] is a direct analog mechanization of the Fourier-transform integrals for computing the frequency spectrum of a transient signal (see Fig. 10.53). The transient input $f(t)$ is multiplied by sine and cosine signals of adjustable frequency; the product curves are then integrated and the integrator outputs applied to d-c meters which read the real and imaginary parts of $F(i\omega)$ directly after the signals settle to their final values. Points on a plot of $F(i\omega)$ versus ω are obtained one frequency at a time by setting the desired ω into the sine and cosine generator and initiating the computing cycle. The input transient $f(t)$ must thus be reproducible since it is needed for each frequency computation. This is usually accomplished by tape recording $f(t)$ and then playing it into the analyzer over and over again, once for each frequency. A standard unit of this type can be used over the frequency range 0.01 to 1,000 cps. If $F(i\omega)$ is wanted in the $M/\underline{\phi}$ form

[1] Honeywell Denver Div., Denver, Colo.
[2] Weston-Boonshaft and Fuchs, Hatboro, Pa.

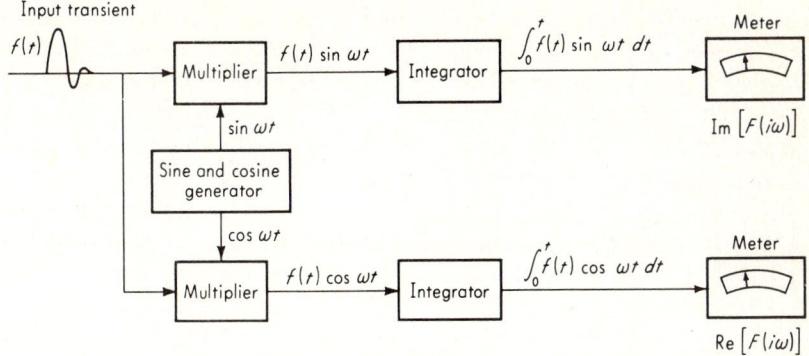

Fig. 10.53. *Fourier-transform transient analyzer.*

rather than the $a \pm ib$ form, a phase and magnitude computer is available to perform this operation.

In some studies of random signals it is desired to know the form of the amplitude-distribution function $W_1(q_i)$. Equipment based essentially on the definition of $W_1(q_i)$ associated with Fig. 3.93 is commercially available.[1,2] This instrument produces d-c voltages proportional to q_i and $W_1(q_i)$ when the signal to be analyzed is applied to its input. The value of q_i can be set manually or automatically swept slowly through a range of q_i while an X-Y plotter plots the desired curve of $W_1(q_i)$ versus q_i. Input signals in the frequency range 0 to 10,000 cps can be handled, while the range of q_i covered is ± 5 rms values.

To determine the transfer function of an unknown system experimentally (*system* analysis) one must measure and analyze both the input signal and the output signal of the system. Equipment is available to carry out these analyses for sinusoidal, transient, or random input signals. Sinusoidal (frequency-response) testing is the most common approach, and many analyzers of this type are available. Figure 10.54a shows the operation of a particular commercial unit.[3] The oscillator provides both the input signal for the system under test and the sine and cosine waves needed at the multipliers. Its frequency can be set over the range 0.01 to 1,000 cps on a standard unit. The frequency response $G(i\omega)$ of the unknown system is obtained one point at a time by setting the oscillator at the desired frequency ω and waiting for transients to die out. Then the two meters can be read to get $(A_o \sin \phi)/2$ and $(A_o \cos \phi)/2$, which

[1] B & K Instruments Inc., Cleveland, Ohio.

[2] H. L. Fox, Probability Density Analyzer, Bolt, Beranek and Newman Inc., Cambridge, Mass., *Rept.* 895, 1962.

[3] Weston-Boonshaft and Fuchs, Hatboro, Pa.

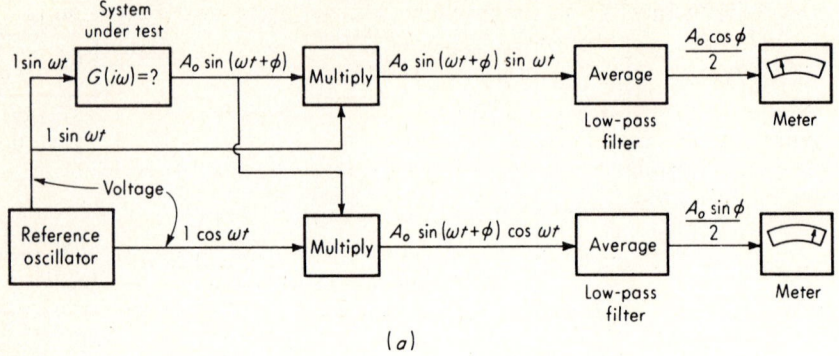

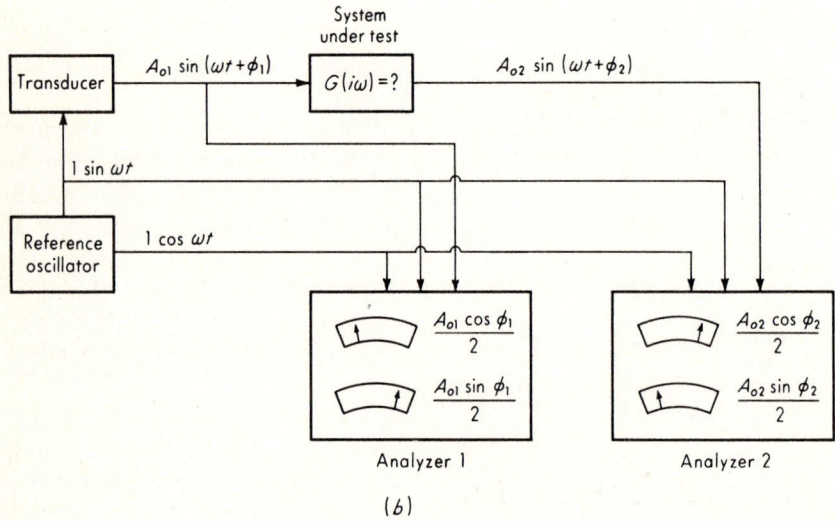

Fig. 10.54.   *Frequency-response analyzer.*

allows calculation of A_o and ϕ either by hand or by an additional computer which is available.   The operation of this analyzer may be understood by writing

$$[A_o \sin (\omega t + \phi)] \sin \omega t = A_o (\sin \omega t \cos \phi + \cos \omega t \sin \phi) \sin \omega t$$

$$(10.85)$$

Then

Average value of $(A_o \cos \phi \sin^2 \omega t + A_o \sin \phi \sin \omega t \cos \omega t)$

$$= \frac{A_o \cos \phi}{2} + 0 \qquad (10.86)$$

since the integral of sin ωt cos ωt is zero over any number of complete cycles and thus its average value is zero. A similar analysis leads to $(A_o \sin \phi)/2$. If the system being tested is nonlinear, its output will not be a pure sine wave $A_o \sin (\omega t + \phi)$ but, rather, will also contain harmonics of the fundamental frequency ω. An analyzer of the type just described completely rejects such harmonics and measures only the fundamental component. This is a result of the general relation

$$\int_{-\pi}^{\pi} \cos mx \cos nx \, dx = \int_{-\pi}^{\pi} \sin mx \sin nx \, dx$$

$$= \int_{-\pi}^{\pi} \cos mx \sin nx \, dx = 0 \qquad \text{if } n \neq m \qquad (10.87)$$

Such behavior is particularly useful because it corresponds to determining the *describing function*[1] when the system is nonlinear. Describing function analysis is a powerful tool in nonlinear-system studies. When the input to the system under study is not an electrical voltage, the oscillator output must be transduced to the appropriate form. Since the transducer may not have perfect dynamics, *two* signals must now be analyzed by using the above procedures on each and then dividing the two complex numbers obtained to get $G(i\omega)$. This can be done sequentially, using one analyzer twice, or simultaneously if two analyzers are available (see Fig. 10.54b).

System analysis by transient techniques utilizes the procedure of Fig. 10.53 except that the frequency analysis must be performed on *two* transients one of which is the input $f_i(t)$ to the system and the other $f_o(t)$ which is the output. When $F_i(i\omega)$ and $F_o(i\omega)$ have been obtained over the frequency range of interest, one simply divides, point by point, the complex number $F_o(i\omega)$ by $F_i(i\omega)$ to get $G(i\omega)$, the frequency response of the unknown system. System analysis by random techniques utilizes the hardware arrangement of Fig. 3.100 and Eq. (3.39) to obtain the system frequency response from a measured spectral density and cross spectral density. Commercial systems that do this automatically and plot $G(i\omega)$ directly on an X-Y plotter are available.

A piece of equipment useful in performing a number of the analyses described in this section is the so-called tracking filter or synchronous filter. This is a narrow band-pass filter whose center frequency is adjusted electronically rather than by mechanical rotation of a knob. Such filters have two inputs: the signal to be filtered and the tuning signal. The filter is constructed so that its center frequency is equal to the frequency of the tuning signal, even if the tuning signal is changing frequency very rapidly. Commercial units[2] can "track" the tuning signal at rates

---

[1] E. O. Doebelin, "Dynamic Analysis and Feedback Control," p. 207, McGraw-Hill Book Company, New York, 1962.

[2] Spectral Dynamics Corp., San Diego, Calif.

in excess of 20,000 cps/sec.   Such filters are useful in studying the sound and vibration of machines whose operating speed changes while the analysis is in progress.   By picking off the operating speed with a magnetic pickup or a-c tachometer and using it as the tuning signal of the tracking filter, one can "slave" the center frequency to the machine operating speed.   Equipment[1] is also available to tune the filter to any desired multiple of the tracking signal, continuously over a wide range. Thus if one wishes to observe, say, the third harmonic of operating speed, he can easily do so even if the speed varies during the analysis.

Problems

10.1   Derive the balanced-bridge relationship $R_1/R_4 = R_2/R_3$.

10.2   For a Wheatstone bridge, show that if $R_1 = R_2 = R_3 = R_4$ at balance, and if $\Delta R_2 = \Delta R_3 = 0$ and $\Delta R_1 = -\Delta R_4$, the output voltage is a perfectly linear function of ΔR_1, no matter how large ΔR_1 gets.

10.3   Discuss qualitatively the effect on bridge operation, for both the null method and the deflection method, of the excitation voltage source having an internal resistance.

10.4   In the system of Fig. 10.4, what considerations determine the numerical value of R_s?

10.5   In a Wheatstone bridge, $R_1 = 3,000$, $R_4 = 4,000$, $R_2 = 6,000$, and $R_3 = 8,000$ ohms at balance.   Find the open-circuit output voltage if $\Delta R_1 = 30$, $\Delta R_2 = -20$, $\Delta R_3 = 40$, $\Delta R_4 = -50$, and $E_{ex} = 50$ volts.   If the bridge output is connected to a meter of 20,000 ohms resistance, what will the output voltage now be?

10.6   Derive the results of Eqs. (10.22) and (10.23).

10.7   Derive Eq. (10.34).

10.8   Why does a charge amplifier essentially amount to a short circuit across the crystal?

10.9   Derive the transfer functions of the circuits of the following:
      a.   Figure 10.16*a*1
      b.   Figure 10.16*a*2
      c.   Figure 10.16*b*1
      d.   Figure 10.16*b*2
      e.   Figure 10.16*b*3

10.10   Derive the transfer function of the hydromechanical filter of Fig. 10.17.

10.11   In the circuit of Fig. 10.17, let e_i be supplied by a sinusoidal generator with an internal resistance of 1,000 ohms and an open-circuit voltage of 10 volts peak to peak.   Also let $C = 10$ μf and $R = 10$ ohms.   If e_o is open-circuit, what voltage will actually appear at the e_i terminals for frequencies of 0, 100, 1,000, 10,000, and 100,000 cps?

10.12   Derive the transfer function of the filter of Fig. 10.18.

10.13   Explain how a notch filter can be used in the feedback path of a high-gain feedback system to construct a band-pass filter.

10.14   Plot logarithmic frequency-response curves of Eq. (10.39) if $1/2\pi\tau_1 = 100$ cps and $1/2\pi\tau_2 = 200$ cps.

10.15   Explain how you would plot frequency-response curves of Eq. (10.40) if numerical values were given.

---

[1] Spectral Dynamics Corp. San Diego, Calif.

10.16 Derive Eqs. (10.41) to (10.45). Show also that the phase angle at ω_p is zero.

10.17 Sketch the configuration of a hydraulic band-pass filter which has an amplitude-ratio attenuation of 40 db/decade on either side of the passband. This is twice the attenuation rate of the system of Fig. 10.21. Only components of the type used in Fig. 10.21 are allowed. Short "transition" regions of slope ±20 db/decade are allowed between the flat response portion and ±40 db/decade portions. You must derive the transfer functions to prove your "invention" works as claimed.

10.18 Derive Eq. (10.46).

10.19 Derive Eq. (10.54).

10.20 Derive Eq. (10.64).

10.21 For the system of Fig. 10.25a, let $e_i = e_{\text{signal}} + e_{\text{noise}}$, where $e_{\text{signal}} = 10 \sin 20t$ and $e_{\text{noise}} = 0.1 \sin 377t$. What is the signal/noise ratio before and after the differentiation?

10.22 The input to a differentiator with transfer function $(e_o/e_i)(D) = D$ is a random signal with a constant mean-square spectral density of 0.001 volt^2/cps from 0 to 10,000 cps and zero elsewhere. Calculate the rms voltage at the input and at the output of the differentiator.

10.23 Derive Eq. (10.66).

10.24 Derive Eq. (10.67).

10.25 Using the system of Eq. (10.68) and the e_i of Prob. 10.21, compute the signal/noise ratio at both the input and the output.

10.26 Derive Eq. (10.69).

10.27 Discuss sensitivity/response-speed tradeoffs in the system of Fig. 10.26.

10.28 Derive the transfer function of the compensating circuit of Fig. 10.28.

10.29 Design a compensating network to speed up by a factor of 10 the response of a thermocouple with a time constant of 1 sec. Thermocouple resistance is 10 ohms and full-scale output is 5 mv. The amplifier/recorder available has maximum full-scale sensitivity of 0.1 mv and an input resistance of 100,000 ohms.

10.30 List and explain the action of all effects that tend to degrade the static accuracy of the system in Fig. 10.33.

10.31 Derive the equation of the operational-amplifier summing circuit of Fig. 10.34c.

10.32 Derive the operating equation of the mechanical multiplier of Fig. 10.36.

Bibliography

1. A. Miller: Bridge Circuits, *The Right Angle*, Sanborn Co., Cambridge, Mass., May and August, 1954.

2. P. R. Perino: The Effect of Transmission Line Resistance in the Shunt Calibration of Bridge Transducers, *Statham Instr. Notes* 36, Statham Instruments Inc., Los Angeles, Calif., November, 1959.

3. P. Pohl: Signal Conditioning for Semiconductor Strain Gages, *ISA J.*, p. 33, June, 1962.

4. Another Look at the Wheatstone Bridge, *Electromech. Design*, p. 36, February, 1965.

5. A. Baracz: Graph Finds Temperature Sensing Bridge Response, *Control Eng.*, p. 85, October, 1961.

6. R. B. F. Schumacher: Differential High-resistance Bridge, *ISA J.*, p. 65, April, 1965.

7. P. Perino: System Considerations for Bridge Circuit Transducers, *Statham Instr. Notes* 37, Statham Instruments Inc., Los Angeles, Calif., September, 1964.

8. G. White: Temperature Compensation of Bridge Type Transducers, *Statham Instr. Notes* 5, Statham Instruments Inc., Los Angeles, Calif., October, 1948.

9. B. B. Helfand: Summation and Averaging of Multiple Measurements by Parallel Transducer Operation, *Statham Instr. Notes* 16, Statham Instruments Inc., Los Angeles, Calif., July, 1950.

10. B. B. Helfand and J. Burns: Calibration of Resistance Bridge Transducer Circuits Under Temperature Extremes, Statham Instruments Inc., Los Angeles, Calif., *Statham Instr. Notes* 14.

11. H. E. Darling: Magnetic Amplifiers for Instrumentation, *ISA J.*, p. 58, January, 1960.

12. J. J. Rado: Input Impedance of a Chopper-modulated Amplifier, *Electro-Technol.* (*New York*), p. 140, June, 1962.

13. J. DiRocco: Potentiometric Amplifiers Improve Impedance Buffering, *Control Eng.*, p. 87, July, 1962.

14. J. Minck and E. Smith: Noise Figure Measurement, *Instr. Control Systems*, p. 115, August, 1963.

15. W. R. Williams and R. C. Hawes: Vibrating Reed Electrometer, *Instr. Control Systems*, p. 112, November, 1963.

16. A. Pearlman: Selecting and Testing Solid-state Operational Amplifiers, *Instr. Control Systems*, p. 121, February, 1965.

17. R. D. Moore: Lock-in Amplifiers for Signals Buried in Noise, *Electronics*, June 8, 1962.

18. C. T. Stelzried: Loaded Parallel-T *RC* Filters, *Control Eng.*, p. 113, May, 1961.

19. G Cocquyt: Evaluating Bridged-T Networks for AC Systems, *Control Eng.*, p. 77, December, 1963.

20. A. I. Zverev: Introduction to Filters, *Electro-Technol.* (*New York*), p. 61, June, 1964.

21. W. Gile: Solid-state Low-frequency Filter, *Electro-Technol.* (*New York*), p. 34, September, 1964.

22. A. W. Langill: Designing Passive Compensators, *Electro-Technol.* (*New York*), p. 26, January, 1965.

23. Miniature Servo Packages, *Electromech. Design*, p. 70, June, 1960.

24. J. B. Heaviside: Sources of Error in AC Servos, *Control Eng.*, p. 85, February, 1964.

25. H. J. Huttenlocker et al.: Instrument Servomechanism Systems, *Electromech. Design*, p. 37, July, 1964.

26. Miniature Servo Packages, *Electromech. Design*, p. 202, May, 1962.

27. Specifying an Instrument Servomechanism, *Electromech. Design*, p. 32, November, 1959.

28. M. Richter: A Simplified Technique in Instrument Servo Analysis, *Electromech. Design*, p. 36, February, 1962.

29. A. Svoboda: "Computing Mechanisms and Linkages," McGraw-Hill Book Company, New York, 1948.

30. T. R. Fredriksen: A Way to Design Low-loss Nonlinear Networks, *Control Eng.*, p. 117, June, 1962.

31. A. J. Baracz: How to Design a Compensating Bridge, *Control Eng.*, p. 81, March, 1965.

32. F. M. Ryan: Special Purpose Analog Computers, *Control Eng.*, p. 103, May, 1963.

33. J. T. Nichols: Zener-regulated Power Supplies, *Instr. Control Systems*, p. 2242, December, 1961.

34. J. Nagy: Zener Diode Power Supplies, *ISA J.*, p. 65, July, 1964.

35. G. A. Korn: "Minicomputers for Engineers and Scientists," McGraw-Hill, New York, 1973.

Chapter 11

Data
Transmission

When components of a measurement system are located more or less remotely from one another, it becomes necessary to transmit information between them by some sort of communication channel. There are also cases where, even though components are close together, transmission problems arise because of relative motion of one part of the system with respect to another. We shall briefly examine questions of this sort and some of the equipment commonly used to solve such problems.

The transmission of information is amenable to mathematical analysis totally disassociated from any hardware considerations, and there is a large body of technical literature on this subject. This science of communication has been extremely useful in putting the design of hardware on a rational basis, showing the tradeoffs in competitive systems, and putting theoretical limits on what can possibly be done. Its consideration, however, is beyond the scope of this text, and we shall restrict ourselves to rather qualitative, hardware-oriented discussions.

11.1

Cable Transmission of Analog Voltage Signals   Perhaps the most common situation is that in which a simple cable is used to transmit an analog voltage signal from one location to another. The accurate analysis of a cable or transmission line involves the use of a distributed-parameter (partial differential equation) approach since the properties of resistance, inductance, and capacitance are not lumped or localized.[1] Figure 11.1a shows the model generally used for such an analysis. An approximation suitable for low frequencies is the lumped network of Fig. 11.1b. The shunt conductance has been neglected here since, in practice, it is generally negligible. If the line is not too long and frequencies not too high, the crude model of Fig. 11.1c may be adequate. The inductance has been totally neglected, and R represents the total resistance of both conductors in the cable while C represents the total capacitance between them. These can be numerically calculated from the values per foot of length given by cable manufacturers. Typically, resistance per foot might be of the order of 0.01 ohm while capacitance would be about 30 pf/ft. Thus a 1,000-ft length of two-conductor cable has $R = 20$ ohms and $C = 30,000$ pf. The actual frequency response of a length of cable can be measured experimentally to get its exact characteristics. Figure 11.1d and e show some typical results.[2] The use of cables up to 7,000 ft long to transmit low-level (± 10 mv full scale) data has been accomplished.[2] However, even short cables (less than 10 ft) can cause difficulties in high-impedance transducers such as the piezoelectric type. The use of charge amplifiers rather than voltage amplifiers may be helpful in such cases.

11.2

Cable Transmission of Digital Data   When data must be transmitted very long distances (100 miles is not unusual), analog signals tend

[1] H. H. Skilling, "Electric Transmission Lines," McGraw-Hill Book Company, New York, 1951.

[2] R. L. Smith, Transmission of Low-level Voltage Over Long Telephone Cables, *NASA, Tech. Note* D-1320, p. 14, January, 1963.

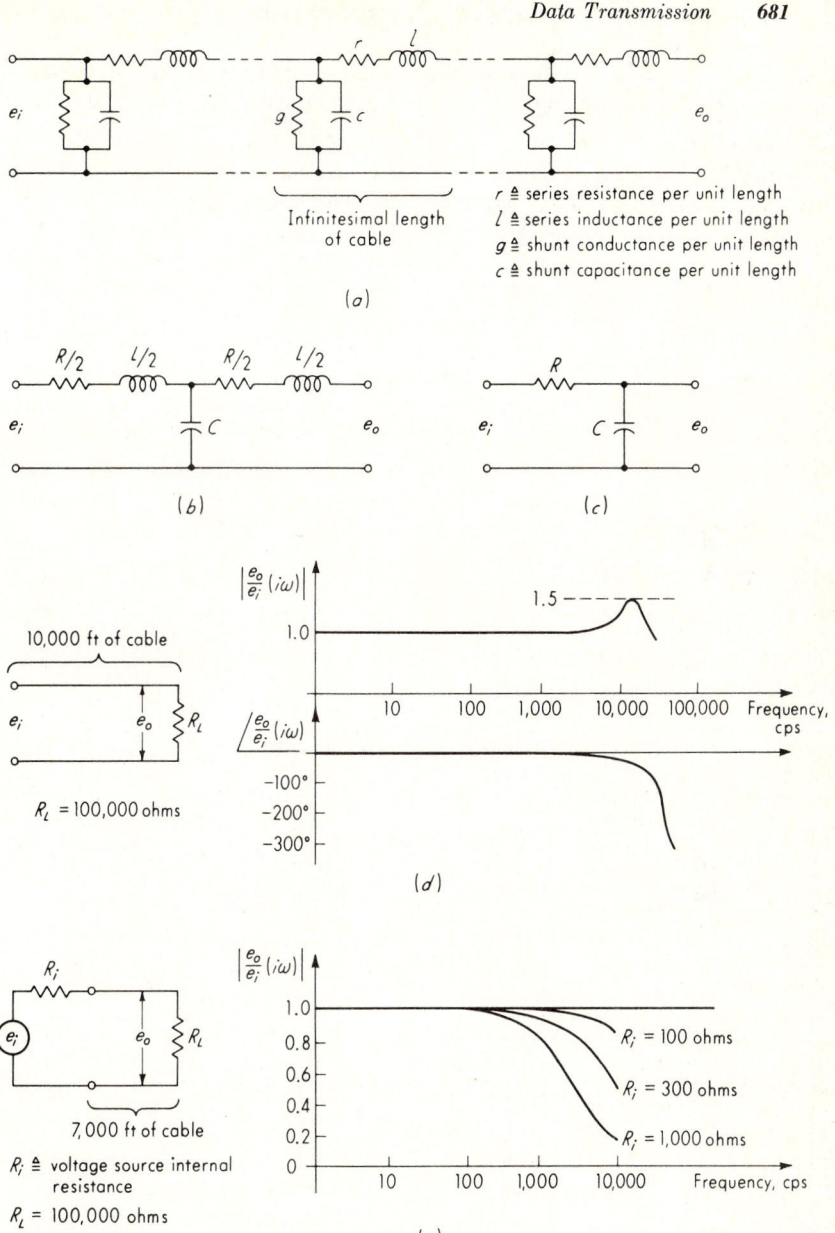

Fig. 11.1.   *Cable models and response.*

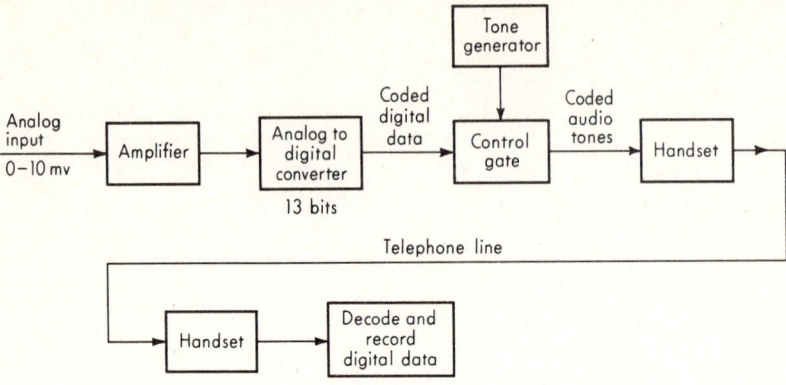

Fig. 11.2.   *Telephonic digital data transmission.*

to be corrupted by the response characteristics of the transmission line and the pickup of spurious noise voltages from a number of sources. Under such conditions it may be desirable to convert the analog data to some digital form, transmit it in digital form, and then reconvert to analog form if desired.   A time-honored example of digital transmission is the telegraph system, in which letters and numbers are represented by a system of coded pulses (dots and dashes).

In many cases the information can be transmitted over lines that have already been installed for other purposes.   Electric power systems, for example, transmit information signals over their power transmission lines simultaneously with the transmission of 60-cycle power.   It is simply necessary to keep the frequencies used sufficiently separated to allow easy filtering for elimination of unwanted signals.   Telephone lines are also in wide use for data transmission.   An interesting system[1] shown in block-diagram form in Fig. 11.2 uses ordinary telephone handsets together with auxiliary equipment to transmit and receive data.   The telephone line is not leased; one can use *any* telephone to call any other in the usual way.   Once voice contact is established, the handsets are placed into an "acoustic coupler."   Here tones at 1,400 cps (representing binary 1) and 2,100 cps (representing binary 0) produced by a tone generator actuated from the analog-to-digital converter are coupled into a loudspeaker which the telephone "hears."   These sounds are transmitted over the telephone line in the usual way to be received by a similar system (but working in reverse fashion) at the other end.   Frequency response of

[1] M. L. Klein, Telephonic Transmission of Data, *Instr. Control Systems*, p. 99, June, 1962.

such a system is quite low but accuracy of transmission is the order of 0.1 percent.

The high accuracy of digital data transmission as compared with analog is due to the fact that the size or precise shape of a pulse in a digital system is not particularly important. Rather, the system operates on the presence or absence of some sort of pulse. Thus even rather severe degradation of pulse shape by the transmission medium will not affect the accuracy of a digital system *at all* as long as the presence or absence of a pulse can be detected.

11.3

FM/FM Radio Telemetry   When interconnecting wires are not possible or desirable, data may be transmitted by radio. A number of different schemes[1,2] are in use; we here consider only one which is widely employed. The word telemetry means simply measurement at a distance and includes all forms of such systems, irrespective of the means of transmission or physical nature of the hardware.

Radio telemetry probably received its greatest impetus from the requirements of aircraft and missile flight testing during and after World War II. Considerable standardization based on the requirements of such systems has been accomplished, and our discussion here reflects this emphasis. Figure 11.3 shows the widely used FM/FM system of radio telemetry. The symbol FM/FM refers to the fact that two frequency-modulation processes are employed. In the first process, time-varying d-c voltages are converted to proportional frequencies, using voltage-to-frequency converters as described in Sec. 10.11. When used in FM/FM telemetry systems these converters are generally called subcarrier oscillators (SCO's). Instead of voltage-to-frequency converters, subcarrier oscillators may also be inductance-controlled. The inductance of a variable-inductance transducer forms part of the oscillator circuit, and changes in inductance cause proportional changes in frequency from the center frequency (see Fig. 11.4).

The standard FM/FM system of Fig. 11.3 has 18 available channels; thus 18 different physical variables may be measured and transmitted simultaneously. Figure 11.5 shows the characteristics of these 18 channels. Note that the center frequencies range from 400 to 70,000 cps. Such low frequencies cannot be practically transmitted by radio propaga-

[1] C. M. Harris and C. E. Crede (eds.), "Shock and Vibration Handbook," vol. 1, pp. 19–76, McGraw-Hill Book Company, New York, 1961.

[2] M. H. Nichols and L. L. Rauch, "Radio Telemetry," John Wiley & Sons, Inc., New York, 1956.

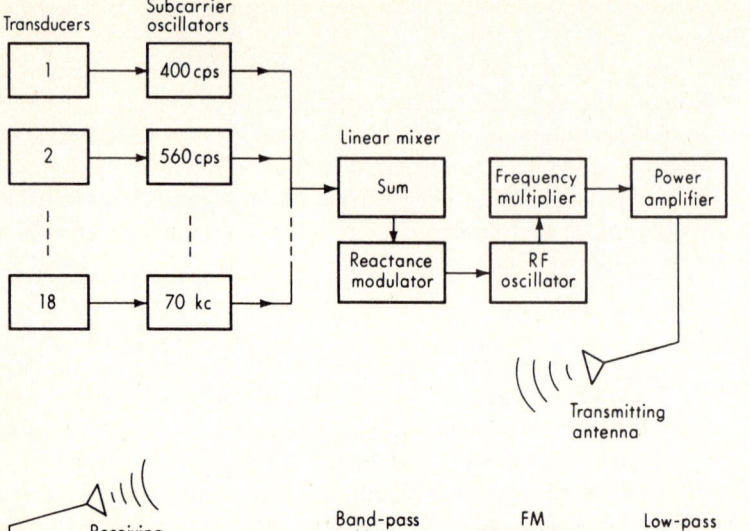

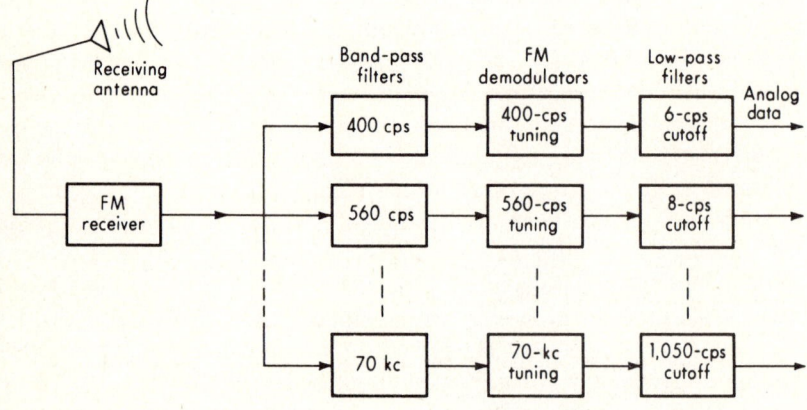

Fig. 11.3.   *FM/FM radio telemetry.*

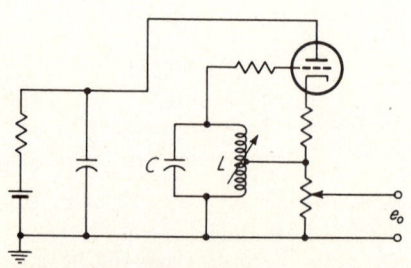

Fig 11.4.   *Variable-inductance subcarrier oscillator.*

L is variable-inductance transducer
C is capacitor which sets center frequency

Band	Center frequency, cps	±Full-scale frequency deviation, %	Overall frequency response, d-c to ____cps
1	400	7.5	6.0
2	560	7.5	8.4
3	730	7.5	11.0
4	960	7.5	14
5	1,300	7.5	20
6	1,700	7.5	25
7	2,300	7.5	35
8	3,000	7.5	45
9	3,900	7.5	59
10	5,400	7.5	81
11	7,350	7.5	110
12	10,500	7.5	160
13	14,500	7.5	220
14	22,000	7.5	330
15	30,000	7.5	450
16	40,000	7.5	600
17	52,500	7.5	790
18	70,000	7.5	1,050

Optional bands

	Band	Center frequency, cps	±Full-scale frequency deviation, %	Overall frequency response, d-c to ____cps
Omit 15 and B	A	22,000	15.0	660
Omit 14, 16, A, C	B	30,000	15.0	900
Omit 15, 17, B, D	C	40,000	15.0	1,200
Omit 16, 18, C, E	D	52,500	15.0	1,600
Omit 17 and D	E	70,000	15.0	2,100

Fig. 11.5. *Telemetry channel characteristics.*

tion since they would require antennas of immense size because the size of an antenna must be of the order of the wavelength to be transmitted [wavelength in meters = 3×10^8/(frequency in cps)]. Thus an additional frequency modulation to boost all frequencies into the radio-frequency range is employed. Rather than use a separate radio-frequency transmitter for each of the 18 channels (which is wasteful of the crowded RF spectrum and also requires much more equipment), the 18 channels are "mixed" (added) and sent out together over one radio-frequency channel. Two such channels, 217.550 and 219.450 Mc, are available for radio telemetry. The frequency deviation caused by any of the sub-carrier-oscillator frequency variations cannot exceed ±125 kc around either of these two frequencies. (Other radio frequencies in the range 216 to 235 Mc may also be used if available.) When the radio signals are received, the 18 channels must be reseparated by suitable band-pass filters, FM-demodulated and low-pass-filtered to reconstruct the original analog data. We see from Fig. 11.5 that by using optional band E a channel with frequency response as great as 0 to 2,100 cps is available. This is adequate for many purposes. Systems of the above type have ranges up to about 400 miles and accuracies of the order of ± 2 percent of full scale. For data of relatively low frequency content, any one of the channels may be time-shared by several transducers if a commutator is used to sample each transducer periodically. If a very large number of low-frequency signals must be telemetered, other methods such as pulse-duration modulation (PDM/FM) may be indicated. More sophisticated telemetry systems can be used over ranges of millions of miles, as in the space probes to Venus.

Radio telemetry is also useful over very short distances when the relative motion of the measuring device and the readout equipment prevent a suitable direct connection. Good examples of such situations are found in measurements on rotating machinery, where slip-ring techniques are not feasible because of high speeds or inaccessibility, and in physiological measurements on test animals or human beings, in which restriction of motion due to connecting wires is not tolerable. Miniaturization of the telemetry components and improvement of shock resistance by use of semiconductor devices now make possible many such applications formerly not feasible.

11.4

Pneumatic Transmission   Transmission of pressure signals in industrial pneumatic control systems is regularly accomplished over distances of several hundred feet. Pneumatic-transmission-line dynamics is analogous to that of electrical cables but, of course, at a much lower frequency. Adequate simplified models employing a dead time equal to the acoustic transmission time and either a first-order or second-order system are available and were discussed in Chap. 6 [see Eqs. (6.89) and (6.90)].

11.5

Synchro Position Repeater Systems   Figure 11.6 illustrates synchro position repeater systems used for transmitting low-power mechanical motion over considerable distances with only a three-wire interconnecting cable. Whenever the two angles θ_i and θ_o are not identical, an electromagnetic torque is exerted on the rotor of *each* machine, tending to bring the shafts into alignment. Thus, if the transmitter shaft θ_i is turned, the receiver shaft θ_o will follow accurately so long as there is no appreciable torque load on the θ_o shaft. Accuracy of such systems

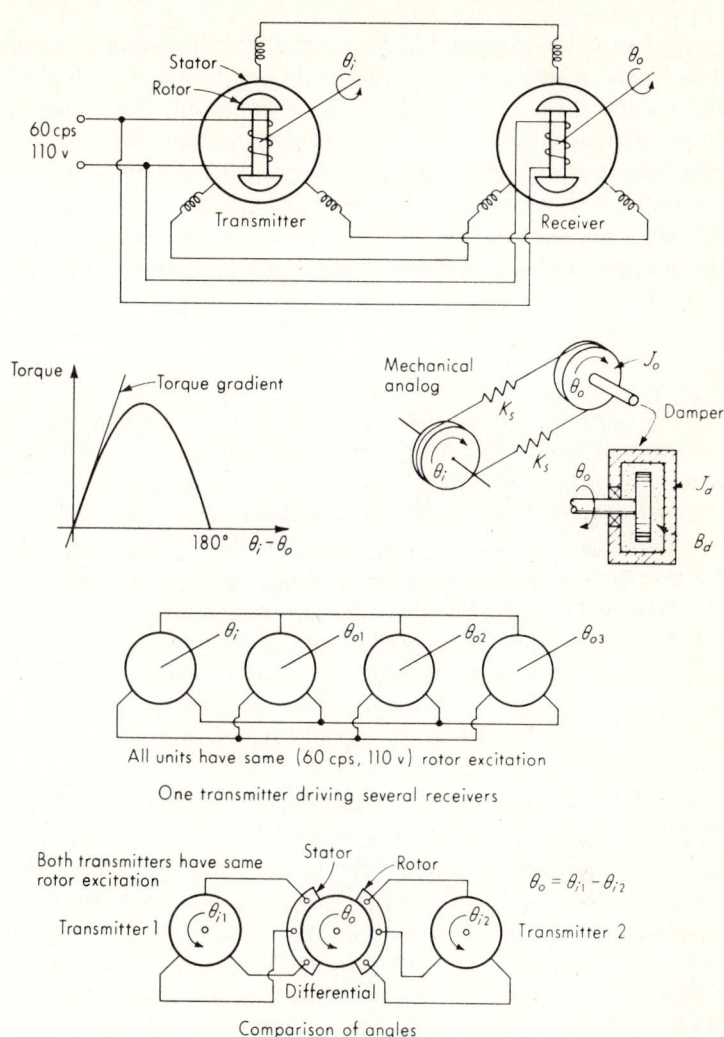

Fig. 11.6. *Torque-synchro angle transmission.*

depends on the torque gradient (torque per unit error angle) of the transmitter/receiver system. A typical value might be 0.35 in.-oz/ degree. The electrical system serves only to transmit power from the θ_i to the θ_o shaft; all the mechanical work taken out at θ_o must be provided as mechanical work at θ_i. The torque gradient is reduced as the resistance of the connecting cable is increased; thus long cables result in reduced accuracy. In a typical unit, 10 ohms resistance in each of the three wires results in a 50 percent loss of torque. The dynamic response of these systems is essentially second-order, a mechanical analog being as shown in Fig. 11.6. A damper is sometimes put on the θ_o shaft to reduce oscillations since little inherent damping is present. When one transmitter drives several receivers (all units of identical size) the torque available at each receiver is $2/(N + 1)$ times the torque for a single pair, where N is the number of receivers. The synchro differential shown in Fig. 11.6 is useful for comparing two rotations at a location remote from either. Its static and dynamic behavior is essentially the same as for a transmitter/receiver pair.

11.6

Slip Rings   When transducers must be mounted on the rotating members of machines, some means must be provided to bring excitation power into the transducer and to take away the output signal. Some transducers (such as synchros) are themselves rotating "machines" in which such data and/or power transmission between a rotating and a stationary member is necessary. When only a small relative motion is involved, continuous flexible conductors (often in the form of light coil springs) can be used. In some cases of limited rotation through a few revolutions, the connecting wires can simply be allowed to wind or unwind on the rotating shaft. However, continuous high-speed rotation requires slip rings, radio telemetry, or some form of magnetic coupling between rotating and stationary parts.

Figure 11.7 shows the common forms of slip rings.[1] Rings are made of coin gold, silver, or other noble metals and alloys. Block-type brushes are often sintered silver graphite while wire-type brushes are alloys of platinum, gold, etc. An important consideration in slip rings used to transmit low-level instrumentation signals is the electrical noise produced at the sliding contact. One component of this noise is due to thermocouple action if the brush and ring are of different materials. The other main effect is a random variation of contact resistance due to surface roughness, vibration, etc. If the contact carries current, a variation in contact resistance causes a noise voltage to appear at the contact. A high-quality miniature sliding slip ring may exhibit a contact-resistance variation of the order of 0.05 ohm peak to peak and 0.005 ohm rms.[2]

While slip rings have been successfully operated at about 100,000 rpm, applications above 10,000 rpm generally require extreme care because of heating and vibration problems. A particular slip-ring assembly[3] usable

---

[1] A. J. Ferretti, Slip Rings, *Electromech. Design*, p. 145, July, 1964.

[2] E. J. Devine, Rolling Element Slip Rings for Vacuum Application, *NASA, Tech. Note* D-2261, p. 11, 1964.

[3] Ferretti, *op. cit.*, p. 159.

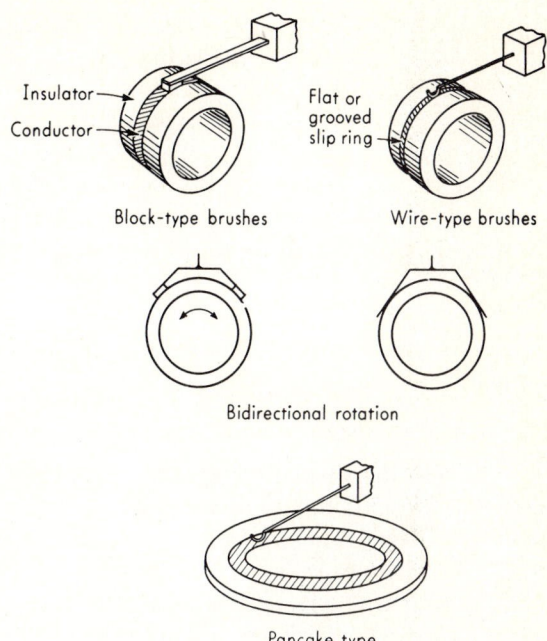

Block-type brushes   Wire-type brushes

Bidirectional rotation

Pancake type

Fig. 11.7. *Slip-ring configurations.*

to 100,000 rpm and intended for strain-gage work had peak-to-peak noise voltage of 0.02 mv at 52,000 rpm and 0.40 mv at 100,000 rpm. This assembly used liquid cooling and lubrication of slip rings and bearings and gave a brush life of 30 hr at 35,000 rpm. At 52,000 rpm the noise level in a typical strain-gage circuit gave a signal/noise ratio of about 150:1. Hard gold rings of $\frac{1}{4}$-in.-diameter were used with two cantilevered wire-tuft brushes per ring.

When slip rings are used with strain-gage circuits, particular care must be taken since the resistance variation of the sliding contact may be comparable with the small strain-gage resistance change to be measured. If possible, a full bridge on the rotating member should be employed so that the sliding contacts can be taken out of the bridge circuit. This arrangement (Fig. 11.8) greatly reduces the effects of slip-ring resistance variations. For the most demanding applications, more complex schemes are available[1,2] to reduce noise to even lower levels.

[1] C. C. Perry and H. R. Lissner, "The Strain Gage Primer," p. 186, McGraw-Hill Book Company, New York, 1955.

[2] P. K. Stein, "Measurement Engineering," vol. II, chap. 29, Stein Engineering Services, Inc., Phoenix, Ariz.

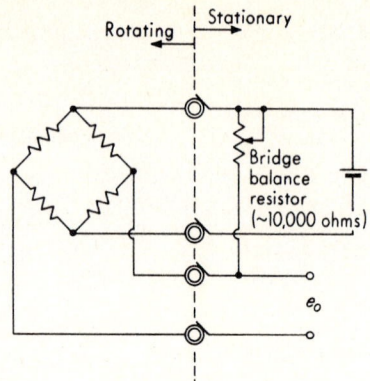

Fig. 11.8. *Bridge-circuit slip-ring configuration.*

A rotating disk dipping into a mercury pool (see Fig. 11.9) can perform the same function as a conventional slip ring. A commercially available device[1] is usable from 0 to 10,000 rpm, has a contact resistance of 0.005 ohm, contact-resistance variation of ± 0.00025 ohm for 0 to 600 rpm, and no measurable resistance variation from 600 to 10,000 rpm, is compensated for self-generated thermoelectric voltages, and can be made with 2 to 160 terminals.

Another alternative to slip rings is the rotary transformer.[2] Here signal or power voltages (they must be a-c) are transferred through an annular air gap between a concentrically rotatable primary and secondary coil.

Ordinary sliding slip rings may not operate properly in the high-vacuum environment of space. Preliminary research[3] using a thrust-type ball bearing as the signal-transfer mechanism indicates that rolling contact slip rings may provide a solution for such problems. A particular

[1] Meridian Laboratory, Lake Geneva, Wis.
[2] Data Tech, Cambridge, Mass.
[3] Devine, *op. cit.*

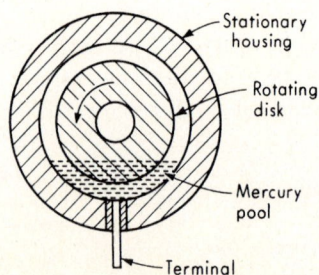

Fig. 11.9. *Mercury-pool slip ring.*

test at 2,000 rpm and vacuum of 2×10^{-9} torr gave operation for over 100 million revolutions at a resistance variation of 0.002 ohm rms.

Problems

11.1 Find a general expression for $(e_o/e_i)(D)$ for the system of Fig. 11.1*b*. If $R = 20$ ohms, $C = 0.3$ μf, and $L = 0.2$ millihenry, plot the logarithmic frequency-response curves.

11.2 A synchro repeater system has one transmitter and five receivers. The torque gradient of a single pair of devices with very short cable connections is 0.5 in.-oz/degree, and 10 percent of this is lost for each ohm of cable resistance. Each receiver drives a dial with 0.05 in.-oz of friction. If the allowable error is 0.5° and cable resistance is 0.05 ohm/ft, find the maximum allowable cable length.

Bibliography

1. R. H. Cerni and L. E. Foster: "Instrumentation for Engineering Measurement," chap. 5, John Wiley & Sons, Inc., New York, 1961.
2. J. D. Tate: Synchro Systems, *Machine Design*, p. 150, June 8, 1961.
3. R. J. Barber: 21 Ways to Pick Data Off Moving Objects, *Control Eng.*, p. 82, October, 1963; p. 61, January, 1964.
4. F. W. Hannula: Transmitting Test Information, *Control Eng.*, p. 173, September, 1959.
5. E. D. Lucas: Techniques for Radio Telemetry, *Control Eng.*, p. 71, December, 1962.
6. E. A. Ragland and D. E. Wassall: The Digital Answer to Data Telemetering, *Control Eng.*, p. 95, August, 1957.
7. E. H. Krause: Telemetering for Interplanetary Flight, *ISA J.*, p. 478, October, 1957.
8. C. I. Cummings and A. W. Newberry: Radio Telemetry, *ARS J.*, p. 141, May–June, 1953.
9. E. H. de Grey and J. G. Bayly: Measuring through Vessel Walls, *ISA J.*, p. 82, May, 1963.
10. L. W. Gardenhire: Evolution of PCM Telemetry, *Instr. Control Systems*, p. 87, April, 1965.
11. M. K. Stark: Short Range Telemetry System Provides Test Data on Rotating Parts, *Gen. Motors Eng. J.*, p. 23, January-February-March, 1965.
12. J. Valentich: Simple Slip Rings for Strain Gage Measurement, *Machine Design*, p. 154, Jan. 7, 1960.
13. J. Valentich: Broadcasting Power to Shaft-Mounted Sensors, *Machine Design*, August 9, 1973.
14. Liquid Metal Contacts Outrun Slip Rings, *Electromechanical Design*, November, 1973.
15. D. J. Lesco et al: On-the-Shaft Data Systems for Rotating Engine Components, *NASA* TMX-68112, 1972.
16. R. T. Troutner: Acoustic Telemetry, *Oceanology Intl.*, January, 1970.
17. Medical and Biological Applications of Space Telemetry, *NASA* SP-5023, 1965.

Chapter 12

Voltage Indicating and Recording Devices

The majority of signals in measurement systems ultimately appear as voltages. Since voltage cannot be seen, it must be transduced into a form intelligible to a human observer. The form in which the data are presented is generally that of a pointer moving over a scale, a pen writing on a chart (including light beams writing on photosensitive paper and electron beams writing on cathode-ray tubes), visual presentation of a set of ordered digits, or printout of digital data by a typewriter or similar device. We shall consider the most common types of such indicating and/or recording devices.

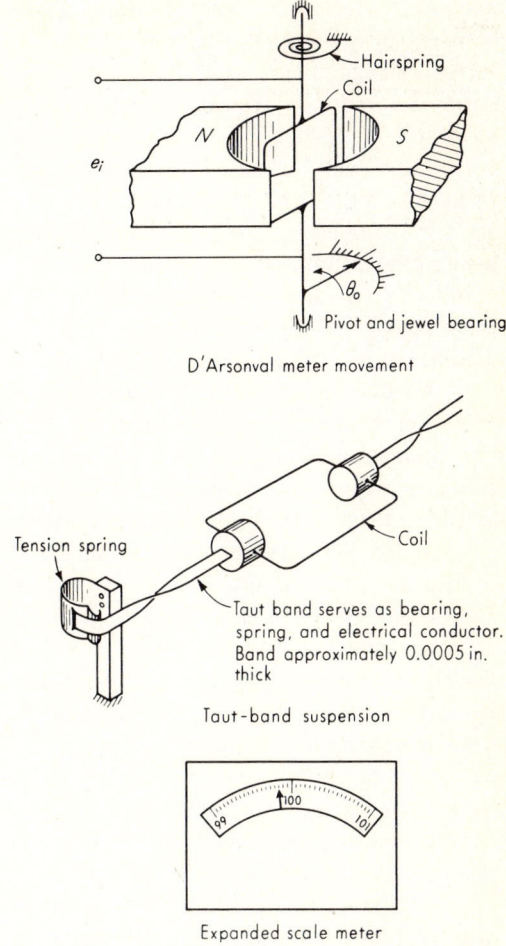

D'Arsonval meter movement

Taut-band suspension

Expanded scale meter

Fig. 12.1.   *D-C analog meters.*

12.1

Analog Voltmeters and Potentiometers   The most widely used meter movement for d-c and (with rectifiers) a-c measurement in electronics and instrumentation work is the classical D'Arsonval movement (see Fig. 12.1). This is basically a current-sensitive device but is used to measure voltage by maintaining circuit resistance constant by means of compensating techniques (see Fig. 2.17a). Relatively recent improvements on this basic configuration include taut-band suspension (rather than pivot-and-jewel bearings), individually calibrated scale divisions, and expanded-scale instruments. Taut-band suspension completely eliminates bearing friction, reduces inertia and temperature effects, increases ruggedness, and results in less loading on the measured circuit since the reduced friction requires less power drain. The increased accuracy made possible by taut-band construction can be provided at reasonable cost through the use of automatic calibration systems which print an individual scale for each and every instrument. Expanded-scale instruments use a precision voltage-suppression circuit to measure a small variation around a larger voltage. Thus if one needs to measure a 100-volt signal accurately it is possible to do this with a meter whose scale goes from 99 to 101 volts. Static inaccuracies of 0.1 percent are attainable in a rugged and portable instrument with these methods.

Vacuum-tube voltmeters (see Fig. 3.23) still use the D'Arsonval meter movement but precede it by amplifier circuits. These increase the input impedance and overall sensitivity. Such instruments generally accept a wide range of d-c and a-c input voltages and have static error of the order of 1 to 3 percent of full scale.

When a-c (not necessarily sinusoidal) voltages are to be measured with a D'Arsonval movement, it is necessary to perform rectification. Depending on the circuitry used, a meter may be sensitive to the average, peak, or rms value of the input wave form. It is common practice to calibrate the scale of the meter to read rms value no matter what quantity is fundamentally sensed. This procedure is accurate only if pure sinusoidal wave forms are being measured since, in this case only, the peak, average, and rms values are all related by fixed constants and can thus be included in the scale calibration. For nonsinusoidal wave forms, peak- or average-sensing meters will not read the correct rms value. In some cases, peak or average value is actually what is wanted; however, rms is most often desired. A true rms voltmeter is complex and expensive; thus peak- and average-sensing meters calibrated to read rms are in wide use and are generally satisfactory except in the most critical applications.

Figure 12.2 shows circuits for peak, average, and rms meters. In the peak circuit the capacitor is charged to the peak value of a periodic input voltage. This charge cannot leak off rapidly because of the one-way conduction of the diodes and the high input impedance of the voltmeter (vacuum-tube voltmeters often use peak sensing). The voltage across the meter thus stays near the peak value of the input with only slight fluctuations due to diode reverse leakage and meter noninfinite impedance. The meter reads the *largest* peak, whether positive or negative. In the average-reading circuit the input is full-wave-rectified, and the low-pass filtering characteristic of the meter movement is used to

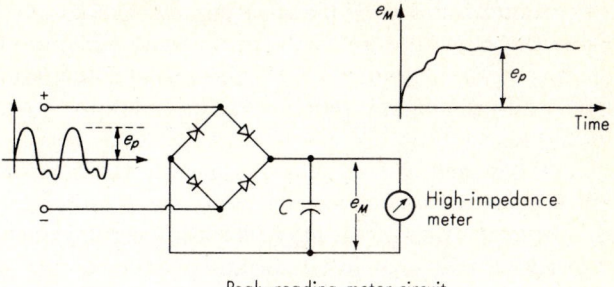

Peak-reading meter circuit

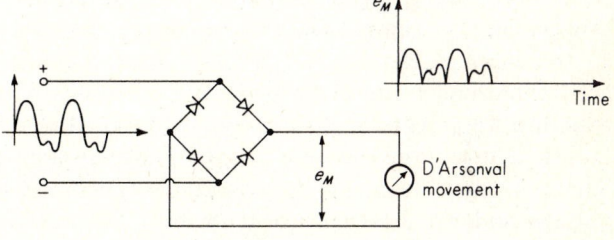

Average-reading meter circuit

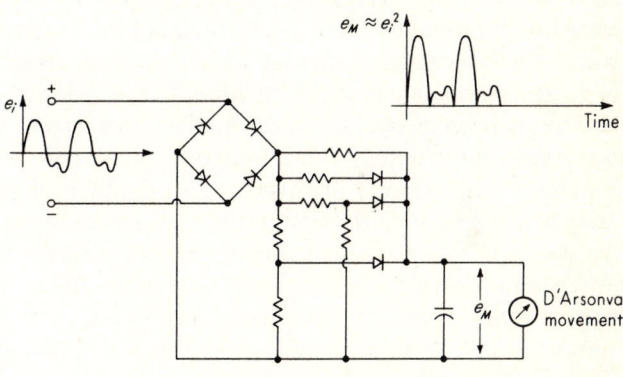

RMS-reading meter circuit

Fig. 12.2. *Peak, average, and rms circuits.*

extract the average value.   The rms-reading circuit[1] approximates the required square-law parabola with a few straight-line segments in the fashion of a diode function generator.   The average voltage on the capacitor is used to provide a variable bias on the diodes in the function gener-

---

[1] C. G. Wahrman, A True RMS Instrument, *B & K Tech. Rev.*, B & K Instruments, Cleveland, Ohio.

ator, thereby obtaining higher accuracy than possible in a fixed-bias unit using the same number of diodes. The averaging required in obtaining an rms value is performed by the meter's low-pass filtering characteristic while the square-root operation is simply obtained by meter-scale distortion.

When highly accurate rms measurements of nonsinusoidal signals are required (random signals are a good example), methods based on the heating power of the wave form are employed since heating power is directly proportional to the mean-squared voltage. One such true rms voltmeter[1] amplifies the incoming signal and applies it to a highly stable resistance heating element bonded to silica and surrounded by an inert atmosphere. A thermopile of 45 copper/constantan junctions is attached to the heating element to measure its temperature. Conditions are such that highly linear heat-transfer processes occur and thus heater equilibrium temperature is closely proportional to the square of current, and therefore to the square of voltage, since heater resistance is constant. The thermopile output voltage is thus an accurate measure of rms input voltage and can easily be read on a D'Arsonval millivoltmeter. The described instrument accepts input signals in the frequency range 2 to 250,000 cps with response flat ± 0.2 db and has full-scale voltage ranges from 0.5 mv to 250 volts. The averaging time (related to the time over which the integration is carried out in the exact mathematical definition of mean-square value) of this meter is 16 sec. Since the *exact* determination of the mean-square value of a random signal requires an infinite integrating (averaging) time, long averaging times are needed for high accuracy. However, they also make the meter sluggish in reaching its final value; thus a switch-selectable 0.5-sec averaging time is provided for use in situations where the 16-sec value is not needed. Voltmeters for random signals must be able to handle peaks that are large compared with the rms value. This is specified by the peak factor of the meter. Large peak factors (ratio of peak to rms value) are desirable; the meter described above has a value of 10.

When the most accurate measurements of d-c voltage are required, potentiometers rather than deflection meters are employed. The potentiometer is a null-balance instrument in which the unknown voltage is compared with an accurate reference voltage which can be adjusted until the two are equal. Since, at the null point, no current flows, errors due to IR drops in lead wires are eliminated. Such IR drops are always present when a D'Arsonval-type meter is used to measure voltage directly. Figure 12.3a shows the basic potentiometer circuit. We see that a galvanometer (just a very sensitive D'Arsonval movement) is used as a

---

[1] Flow Corp., Cambridge, Mass., *Bull.* 59, 1960.

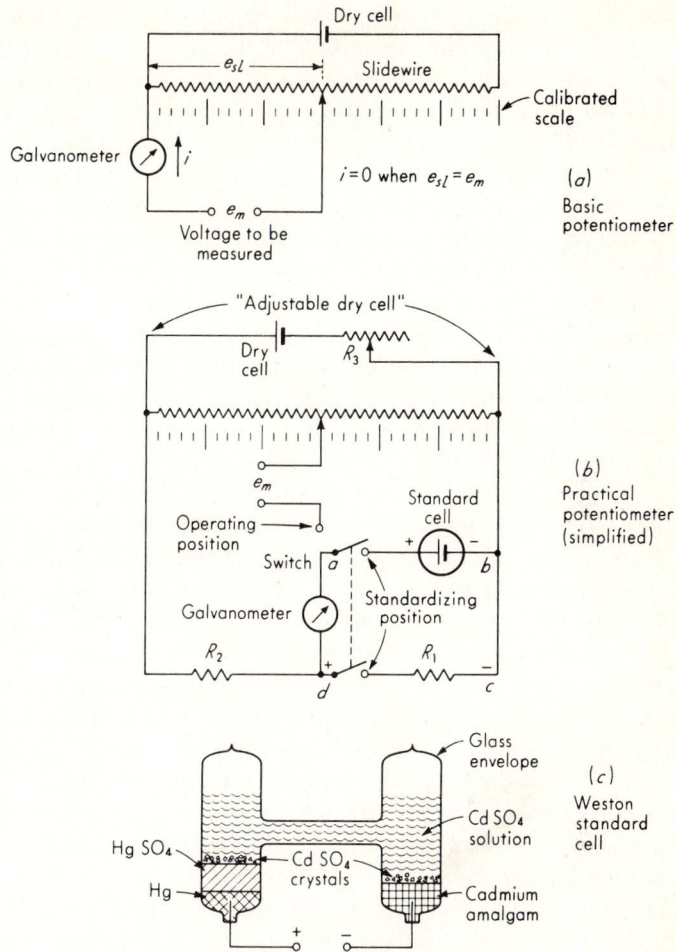

Fig. 12.3. *Manually-balanced potentiometer.*

null detector. It detects the presence or absence of current by deflecting whenever the unknown and reference voltages are unequal. However, it need not be calibrated since it must indicate only the presence of current, not its numerical value. The basic circuit of Fig. 12.3a is not practical since the accuracy of the reference voltage picked off the slide-wire is directly influenced by changes in the dry-cell voltage. Since the dry cell supplies power to the slide-wire, its voltage is bound gradually to drop off. This problem is solved in the practical circuit of Fig. 12.3b by inclusion of an additional component, the standard cell.

Figure 12.3c shows the Weston cadmium saturated standard cell which is the basic working standard of voltage.   Its terminal voltage is 1.018636 volts and is reproducible to the order of 0.1 to 0.6 ppm.   Its accuracy in terms of the fundamental mass, length, and time standards can be established to only about 10 ppm, however.   Its temperature coefficient is -40 $\mu v/C°$; thus close temperature control must obviously be employed in the most exacting situations.   A standard cell cannot be substituted for the dry cell of Fig. 12.3a since its accuracy is destroyed if any appreciable current is drawn from it over a time interval.   It must thus be used as an intermittent reference against which the slide-wire excitation voltage can be checked whenever desired.   The *unsaturated* Weston cell is used in practical instruments since it is more portable. Its terminal voltage varies from one unit to another.   However its drift at constant temperature is only about -0.003 percent per year; thus it is perfectly adequate for most purposes.   Its temperature coefficient is about -10 $\mu v/C°$.

The operation of the circuit of Fig. 12.3b is as follows: When the slide-wire scale on the potentiometer was originally calibrated at the factory, slide-wire excitation-adjusting resistor R_3 was set at a fixed value and resistor R_1 was adjusted until, when loop $abcd$ was completed by the switch, no current flowed in the galvanometer.   This means that the voltage drop across R_1 was just equal to the standard-cell voltage.   Now the slide-wire, R_1, and R_2 are all fixed and stable resistors; thus if the voltage across R_1 is at its calibration value the slide-wire excitation voltage must also be at its calibration value.   Thus, whenever we wish to check the calibration (this is called standardization) we merely complete the loop $abcd$ momentarily (so as not to draw much current from the standard cell) and note whether the galvanometer deflects.   If it does, we adjust the slide-wire excitation with R_3 until deflection ceases.   We are then assured that the slide-wire excitation is at its original calibration value.   The resistor R_2 is merely a current-limiting resistor to prevent drawing large current from the standard cell through the slide-wire path, which is fairly low-resistance.

Fairly common and inexpensive potentiometers which can be read to the nearest microvolt are in wide use.   More sophisticated instruments intended for the most accurate calibration work provide greater accuracy and sensitivity.   One such commercially available instrument[1] measures in three ranges: 0 to 1.611110 volts in steps of 1.0 μv, 0 to 0.1611110 in steps of 0.1 μv, and 0 to 0.01611110 in steps of 0.01 μv.   The total parasitic thermoelectric voltage is less than 0.1 μv.   The limit of error on the high range is ± 0.003 percent of reading ± 0.1 μv, while on the medium and low ranges it is ± 0.005 percent of reading ± 0.1 μv.   These values

---

[1] Honeywell Inc., Philadelphia, Pa.

approach the level of the National Standards achieved by the National Bureau of Standards, which are about 0.001 percent from 0.01 to 1,000 volts.

12.2

Digital Voltmeters and Printers   A digital voltmeter[1,2] accepts analog voltage inputs and produces a direct visual display of the voltage reading in decimal digits.   All such instruments are essentially made up of some sort of analog-to-digital converter (see Sec. 10.12) plus some means of visual display.   They are available in a wide range of capabilities to measure d-c and a-c voltages from less than a millivolt to more than 1,000 volts.   Integrating digital voltmeters (see Sec. 10.11) are also included in this class.   The time required to convert an analog voltage to digital form and display it varies widely with the type of conversion, ranging from about 1 μsec to several seconds.   Inaccuracy of digital voltmeters is generally determined by the stability of the reference voltage (standard-cell or Zener-diode reference) plus the inherent error of ± 1 unit in the least significant digit.   Five-digit units are available, thus giving a resolution of 0.001 percent of full scale.   Accuracy of the order of 0.01 percent of full scale on a given range is attainable.   Since coded digital signals must be produced in a digital voltmeter, many instruments provide these as outputs that can be sent to a digital printer or recorder.

Digital recorders or printers are electromechanical or electro-optical devices that print digital characters (usually the decimal digits 0 through 9 and a plus or minus sign) on a moving strip of paper.   A line of printing may contain any number of digits, depending on the resolution of the equipment feeding the printer.   Electromechanical parallel input printers[3] have a top speed at present of about 40 lines per second.   Electro-optical printers[4] with serial input have speeds of 135 lines per second.   When digitally recorded data must later be reproduced and/or processed, punched cards, punched paper tape, or digital magnetic tape may be employed.

12.3

Self-balancing Potentiometers and *X-Y* Recorders   In considering analog indicating and recording instruments capable of producing

[1] Digital Voltmeters, *Electromech. Design*, p. 58, December, 1960.
[2] DVM, *Instr. Control Systems*, p. 101, June, 1964.
[3] Franklin Electronics, Bridgeport, Pa.
[4] Century Electronics, Tulsa, Okla.

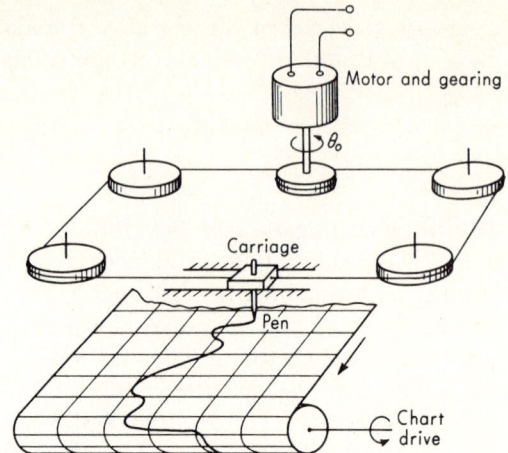

Fig. 12.4. *Self-balancing potentiometer.*

a permanent visual record of the time variation of a voltage, a classification with regard to static accuracy and speed of response is useful.   The majority of recording tasks of this type are performed by three classes of instruments.   These are:

Recorder	Static error, percent of full scale	Frequency response, cps
Self-balancing potentiometer	0.2	Less than 5
Galvanometer oscillograph	2	Up to 10,000
Cathode-ray oscilloscope	3	Up to 10^9

In this section we discuss self-balancing potentiometers and the closely related X-Y plotters.

Almost all self-balancing potentiometers use the instrument-servo-mechanism principle of Fig. 10.33.   To make this into a recording instrument it is only necessary to convert the output angle θ_o to translation of a carriage to which a pen is attached (usually by means of a piano-wire and pulley drive) and then pass calibrated chart paper, from a roll under the pen, at a fixed and known speed to establish the time base (see Fig. 12.4).   The pen will then trace the variation of the voltage e_i with time.   Most such recorders now use a Zener-diode reference supply voltage rather than a standard cell.   This semiconductor circuit provides an accurate and stable reference voltage and, unlike a standard cell, also can supply current to excite the slide-wire.   With such units the inconvenience of periodic standardization is eliminated.

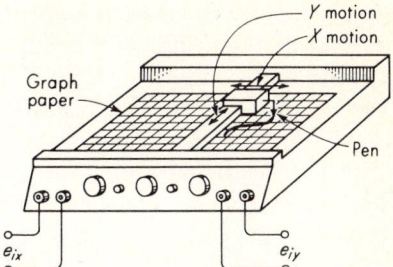

Fig. 12.5. *X-Y plotter.*

Self-balancing potentiometers come in a wide range of capabilities. A general-purpose laboratory type[1] exemplifies the limits of performance presently attainable with this class of instrument. Full-scale input voltage may be switch-selected from 0.1 mv to 100 volts in 19 ranges. Chart width is 6 in. with a static accuracy ± 0.25 percent of full scale or 1 μv, whichever is greater. Frequency response is flat within 1 percent to 5 cps for inputs that are 10 percent of full scale; full-scale inputs result in reduced frequency response. Time for full-scale pen travel is less than 0.5 sec. Chart speeds are thumbwheel-selected from 10 in the range 1 in./sec to 10 min/in.

When multichannel recording is required, two general approaches are used. For slowly varying data the pen is replaced with a print wheel having numbers, say, 1 to 24. A sampling switch connects each of 24 input signals to the potentiometer input in turn. When, for example, channel 9 is connected, the potentiometer drives to the correct chart position, prints a 9, and then goes on to channel 10, etc. For continuous recording, a separate servosystem and pen are provided for each channel but they all write on the same chart. To prevent mechanical interference and still allow each channel to use the full chart width, each pen is displaced by about $\frac{1}{16}$ in. from its neighbors. This causes a slight time displacement from channel to channel but this can be corrected for, if necessary. Up to four channels of such overlapping recording are available with 10-in. chart width.[2]

Often it is desired to cross-plot one variable against another rather than against time. This can be done by employing two independent servosystems to drive a pen over a stationary chart paper. Such arrangements are called *X-Y* plotters. Figure 12.5 shows their typical configuration. They are available for paper sizes up to several feet on a side; however the most common size accepts standard $8\frac{1}{2}$- by 11- or 11- by 17-in.

[1] Honeywell Inc., Philadelphia, Pa.
[2] Texas Instruments, Houston, Texas.

graph paper. Since each axis operates on essentially the same principle as the self-balancing potentiometer, their static and dynamic characteristics are quite similar. Many X-Y plotters also provide for making one of the axes a time base if it is desired to plot a single variable against time. This is done by generating a ramp-function voltage by charging a capacitor with a constant current, thereby making the capacitor voltage increase linearly with time. This voltage is then applied to the input of one of the servosystems and causes that axis to translate at a constant speed. One commercially available 11- by 17-in. instrument[1] has 17-d-c voltage ranges on each axis ranging from 0.1 mv/in. to 20 volts/in. and 12 a-c ranges from 5 mv/in. to 20 volts/in. Static error of d-c ranges is 0.2 percent of full scale and for a-c ranges (20 cps to 100 kc) it is 0.5 percent. Nonlinearity is 0.1 percent for d-c ranges and 0.2 for a-c. A time sweep giving eight speeds from 0.5 to 100 sec/in. may be applied to either axis.

12.4

Galvanometer Oscillographs   While the D'Arsonval movement of Fig. 12.1 as applied to meters has a very limited frequency response (less than 1 cps), it is possible to miniaturize it so as to obtain rotational natural frequencies of the order of 10,000 cps. D'Arsonval movements of this type are called galvanometers and are the basic sensing elements of galvanometer oscillographs. We shall analyze the galvanometer to determine its performance characteristics.

Figure 12.6 shows schematically the construction of a typical galvanometer. Viscous damping B may or may not be present in a given design; we carry it along for generality. The galvanometer is an electromechanical transducer, and we shall write two equations, one electrical and one mechanical, to analyze its behavior. Basically, input voltage e_s from the signal source causes a current to flow in the coil. There is then a current-carrying conductor in a magnetic field; thus the coil experiences an electromagnetic force which, since it has a lever arm, causes a torque. This torque tends to rotate the coil until it is just balanced by the restoring torque of the torsion springs. For a constant e_s the output pointer comes to rest at a definite value of θ_o. By proper design the static relation between θ_o and e_s can be made quite linear.

Writing a Kirchhoff voltage-loop law for the electrical circuit, we have

$$i_g(R_s + R_g) + L_g \frac{di_g}{dt} + HNlb \frac{d\theta_o}{dt} - e_s = 0 \qquad (12.1)$$

[1] Hewlett-Packard Co., Moseley Div., Pasadena, Calif.

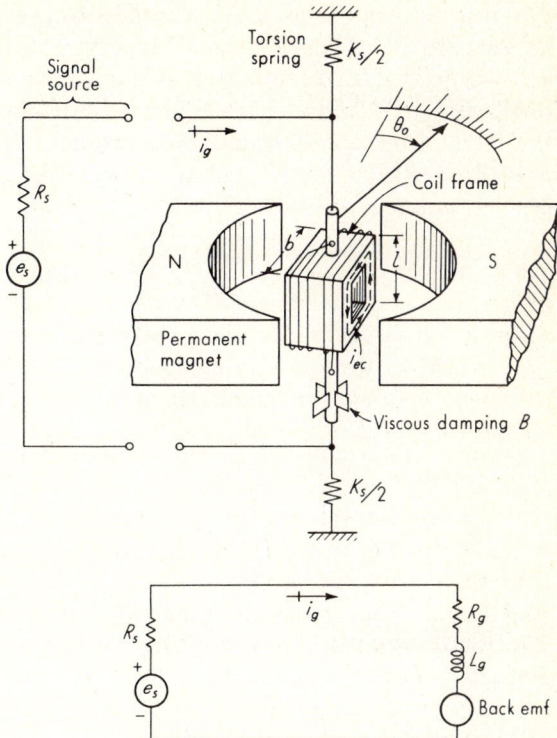

Fig. 12.6.   *Galvanometer.*

where $HNlb(d\theta_o/dt)$ = back emf of coil acting as generator

$H \triangleq$ flux density

$N \triangleq$ number of turns on coil

$l \triangleq$ length of coil

$b \triangleq$ breadth of coil

$L_g \triangleq$ inductance of coil

$R_g \triangleq$ resistance of coil

$R_s \triangleq$ resistance of signal source

We may note that this analysis is comparable to that of a d-c motor since the devices are basically the same.   Now we can apply Newton's law to the rotational motion of the coil:

$$\Sigma \text{ torques} = J \frac{d^2\theta_o}{dt^2}$$

The main electromagnetic torque is given by $HNlbi_g$; however, a more subtle effect also produces an additional torque.   If the *frame* on which the coil is wound is itself a conductor, a voltage $Hlb(d\theta_o/dt)$ will be induced in it (since it is a "one-turn coil") and an eddy current $i_{ec} = (Hlb/R_f)(d\theta_o/dt)$ will flow in the frame, where R_f is the resistance of the frame. This eddy current is also in the magnetic field and thus the frame feels an electromagnetic torque $-Hlbi_{ec}$. The minus sign is needed since such induced voltages always set up effects that oppose the original motion. Newton's law then reads

$$HNlbi_g - \frac{(Hlb)^2}{R_f}\frac{d\theta_o}{dt} - K_s\theta_o - B\frac{d\theta_o}{dt} = J\frac{d^2\theta_o}{dt^2} \qquad (12.2)$$

where $B \triangleq$ viscous damping coefficient
$\quad\quad K_s \triangleq$ torsional spring constant
$\quad\quad J \triangleq$ moment of inertia of moving parts about axis of rotation

Equations (12.1) and (12.2) each contain two unknowns, i_g and θ_o; thus they must be solved simultaneously. Since our main interest is the transfer function relating input e_s to output θ_o, i_g is of little interest, and we do not bother to solve for it. A simplification is possible in Eq. (12.1) since experience shows that, in terms of frequency response, within the useful operating frequency range of the galvanometer the effect of inductance is invariably negligible. We thus drop this term and solve Eq. (12.1) for i_g.

$$i_g = \frac{e_s - HNlb(d\theta_o/dt)}{R_s + R_g} \qquad (12.3)$$

This may now be substituted in Eq. (12.2), which gives, after some manipulation,

$$\frac{\theta_o}{e_s}(D) = \frac{K}{D^2/\omega_n^2 + 2\zeta D/\omega_n + 1} \qquad (12.4)$$

where $\quad\quad K \triangleq \dfrac{HNlb}{K_s(R_s + R_g)} \qquad$ rad/volt $\qquad\qquad (12.5)$

$$\omega_n \triangleq \sqrt{\frac{K_s}{J}} \qquad \text{rad/sec} \qquad\qquad (12.6)$$

$$\zeta \triangleq \frac{B + (Hlb)^2/R_f + (HNlb)^2/(R_s + R_g)}{2\sqrt{K_sJ}} \qquad (12.7)$$

We see that the galvanometer is a second-order instrument. One can increase sensitivity by increasing H, N, l, and b or by decreasing K_s, R_g, and R_s. The flux density H is usually at the maximum value possible with ordinary permanent magnets. Increases in N, l, and b result in increases in R_g since a longer total length of wire in the coil results; thus the net effect on sensitivity is not obvious. Decreases in K_s increase K directly; however, speed of response is lost since ω_n decreases. Note also that sensitivity varies with R_s; thus changing from one trans-

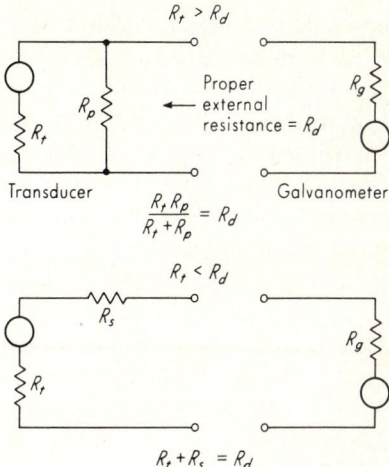

Fig. 12.7. *Damping networks.* $R_f + R_s = R_d$

ducer to another with a different resistance results in a loss of calibration. This is due to the fact that the galvanometer is a current-sensitive device. Equation (12.7) shows that the mechanical viscous damping B can be completely eliminated and the system is still damped.   The damping that remains is of electromagnetic origin and is caused by the back emf proportional to $d\theta_o/dt$ (mechanical-viscous-damping torque is also proportional to $d\theta_o/dt$).   In low-frequency, high-sensitivity galvanometers the electrical damping is adequate to obtain the optimum ζ value of 0.65, and no intentional mechanical damping B is provided.   Also such galvanometers may have the coil frame slotted so that $R_f = \infty$.   Then

$$\zeta = \frac{(HNlb)^2}{2 \sqrt{K_s J} (R_s + R_g)} \qquad (12.8)$$

The most important feature of this result is that ζ depends on R_s, the source resistance.   *Thus such a galvanometer can be properly damped for only one value of external resistance.*   This does not mean it can be used only with transducers that have this value of R_s, but it does mean that a suitable resistance network may have to be interposed between the transducer and galvanometer.   If the transducer resistance is too high, a shunt resistor is needed; if too low, a series resistor (see Fig. 12.7) is required.   For high-frequency, low-sensitivity galvanometers the electromagnetic damping is inadequate, and intentional viscous damping is provided.   The electromagnetic damping is a small percentage of the total, and so such galvanometers can generally be used with any external resistance in a wide range.

   To construct a recording oscillograph it is necessary to provide chart paper moving at a known speed to give a time base, and a means of writing the galvanometer motion on this paper. A number of writing methods have been developed and are in use. The most obvious is simply to mount an ink pen at the end of an arm and let it move over the chart paper. To get straight-line motion from the rotation θ_o, special linkages have been developed. Another popular method also uses a mechanical arm but replaces the pen with a heated stylus. This requires use of a special heat-sensitive paper but eliminates the possible clogging and skipping of ink pens. Any oscillograph using a mechanical arm and a pen or stylus has so much inertia (due to the long arm) and friction (due to contact with the paper) that its frequency response is limited to 100 or 200 cps or less. This is adequate for many applications, and such instruments are in wide use. They generally come provided with amplifiers since their galvanometers are designed quite stiff to be accurate in the face of considerable friction.

In general, such instruments do not have interchangeable galvanometers of various sensitivities and response speeds. Since the input signal goes to an amplifier, the proper damping is provided there, and the signal source resistance usually has no effect on damping. Oscillographs of this type generally have a full-scale motion of 2 or 3 in. and a nonlinearity of 1 or 2 percent. Multichannel oscillographs of this class are generally limited to six or eight channels side by side since the mechanical arms cannot overlap. Recent developments utilizing pen-position feedback systems and pressurized inking systems have improved accuracy to about $\frac{1}{2}$ percent.

To realize the high frequency response (up to 10,000 cps) mentioned earlier, inertia and friction must be drastically reduced. This is accomplished by replacing the mechanical writing arm with a light beam. A tiny mirror is rigidly fastened to the moving coil and a light beam reflected from it. When the coil turns, the light beam, which is focused as a spot on the moving chart paper, deflects over the paper, leaving a trace (see Fig. 2.5). Until recently, ordinary photographic-type recording paper was used and records were not available until time-consuming darkroom work had been carried out to process the records. Today, there are available automatic processors attached directly to the oscillograph to make records quickly accessible. Even more convenient are the new recording processes using special papers and/or light sources that require no processing at all. For slowly varying signals at low chart speeds these processes give an immediately visible trace, just as in ink recording. For high-speed recording the trace becomes visible in 15 or 20 sec if the paper is exposed to an ordinary fluorescent lamp. These records will last for years if kept away from sunlight and can be subjected to a simple liquid fixing process if absolute permanence is required.

Oscillographs using light-beam galvanometers generally provide an

Undamped natural frequency, cps	Flat (±5%) frequency response, cps	External resistance for optimum damping, ohms	Coil resistance, ohms	Current sensitivity, in./ma	Maximum deflection for ±2% nonlinearity, in.
		Electromagnetic damped types			
40	0–24	120	20	136	8
40	0–24	350	35	225	8
100	0–60	120	32	91	8
100	0–60	350	67	160	8
200	0–120	120	53	44	8
400	0–240	120	116	12	8
		Fluid damped types			
1,000	0–600	20–1,000	37	0.356	8
1,650	0–1,000	20–1,000	25	0.107	8
3,300	0–2,000	20–1,000	31	0.039	6
5,000	0–3,000	20–1,000	37	0.023	3.5
8,000	0–4,800	20–1,000	33	0.027	2.0

Fig. 12.8.   *Galvanometer family.*

entire family of interchangeable units covering the range from low frequency, high sensitivity to high frequency, low sensitivity.   Figure 12.8 shows a typical selection.[1]   The galvanometers can be easily and quickly removed from the magnet block and replaced by others suited to the particular job requirements.   Since the light beams do not interfere with each other, as do mechanical arms, each channel of a multichannel instrument can use the entire chart width.   Optical galvanometers also allow greater deflections, full scale being 4 to 8 in. in commercial instruments.   Up to 60 channels may be recorded on one 12-in.-wide chart paper.   Most instruments provide pushbutton selection of chart drive speeds; speeds up to about 160 in./sec are available.   The time base provided by the paper drive is usually accurate only to about 3 to 5 percent.   Thus if accurate time measurement is required, a timing trace (say the accurate 60-cps power-line frequency) can be put on one of the galvanometers.   Accurate time-line generators internal to the oscillograph which print (optically) an accurate time grid at, say, 0.01-sec intervals on the paper are also available.

The optical galvanometer recording principle has been adapted[2]

[1] Honeywell Heiland Div., Denver, Colo.
[2] Sanborn Co., Waltham, Mass.

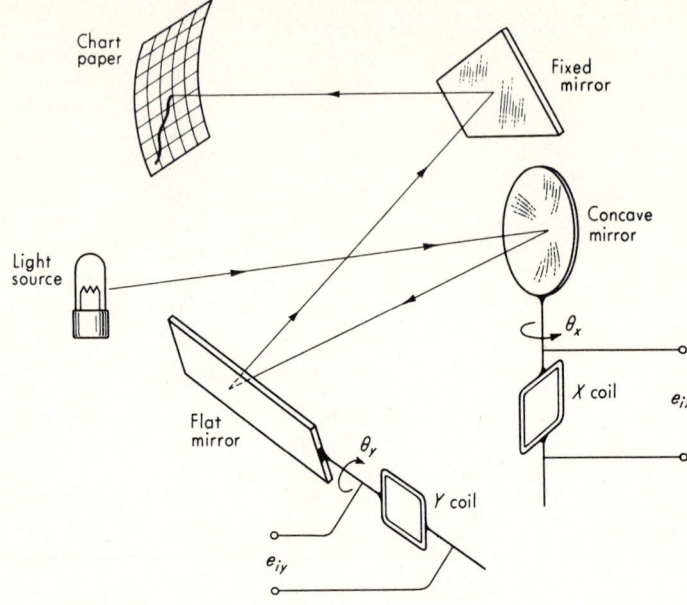

Fig. 12.9. *Galvanometer X-Y plotter.*

to *X-Y* recording as shown in Fig. 12.9. This instrument provides the *X-Y* recording function at speeds intermediate between self-balancing potentiometer types and cathode-ray oscilloscopes. The instrument shown employs ultraviolet-sensitive paper which requires no processing. Chart size is 8 by 8 in. with nonlinearity of 1 percent of full scale. Frequency response is flat from d-c to 100 cps at full-scale deflections.

12.5

Cathode-ray Oscilloscopes Figure 12.10 shows in simplified fashion the functional operation of a typical cathode-ray oscilloscope. A focused narrow beam of electrons is projected from an electron gun through a set of horizontal and vertical deflection plates. Voltages applied to these plates create an electric field which deflects the electron beam and causes horizontal and vertical displacement of its point of impingement on the phosphorescent screen. By proper design this displacement can be made closely linear with deflection-plate voltage. The phosphorescent screen emits light which is visible to the eye and may also be photographed for a permanent record.

The most common mode of operation is that in which one desires a plot of the input signal against time. This may be accomplished by

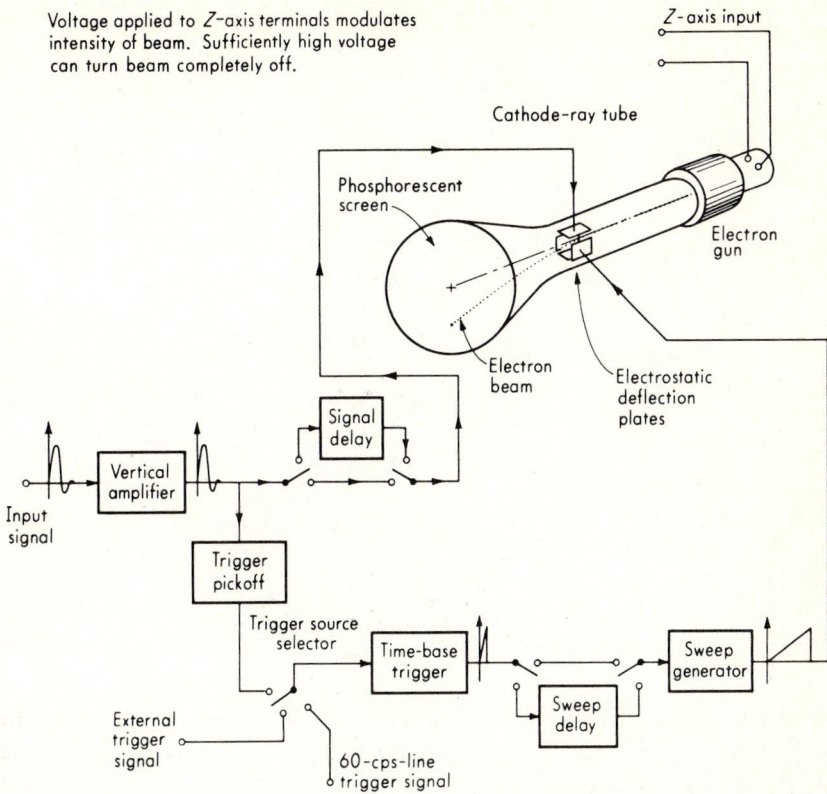

Voltage applied to *Z*-axis terminals modulates intensity of beam. Sufficiently high voltage can turn beam completely off.

Z-axis input

Cathode-ray tube

Phosphorescent screen

Electron gun

Electron beam

Electrostatic deflection plates

Signal delay

Vertical amplifier

Input signal

Trigger pickoff

Trigger source selector

Time-base trigger

Sweep delay

Sweep generator

External trigger signal

60-cps-line trigger signal

Fig. 12.10. *Cathode-ray oscilloscope.*

driving the horizontal deflection plates with a ramp voltage, thus causing the spot to sweep from left to right at a constant speed. To ensure that the sweep and the input signal applied to the vertical deflection plates are properly synchronized, the triggering of the sweep can be initiated by energizing the trigger circuit from the leading edge of the input signal itself (see Fig. 12.11). This results in a loss of the first instants of the input signal on the screen, but this is generally not serious since only about 1 mm of deflection is needed to cause triggering. In those cases where this loss is objectionable, oscilloscopes with signal delay are available. These delay the application of the input signal to the vertical deflection plates so that the sweep starts *before* the rise of the input signal on the screen. Thus the complete input signal is recorded. Most oscilloscopes also provide for triggering from either positive-going or negative-going voltages, and the instant of triggering can be adjusted from the minimum 1 mm level upward to any point on the input wave

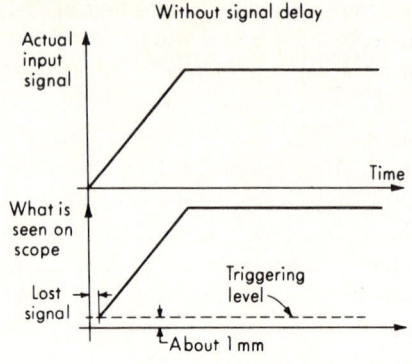

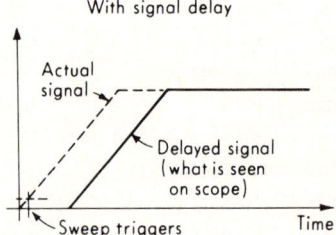

Fig. 12.11. *Signal delay.*

form.   Triggering can also be controlled from external signals or the 60-cps power-line signal.   When external trigger signals which are conveniently available occur somewhat before the input signal of interest, a sweep-delay feature may be useful; some instruments provide this capability.   Since the deflection sensitivity of the cathode-ray tube itself is only of the order of 0.1 cm/volt, oscilloscopes generally include amplifiers so that the instrument can directly handle millivolt-level input signals.   Oscilloscopes are also useful for X-Y plotting.   For such operation the horizontal deflection plates are merely disconnected from the sweep generator and connected to an amplifier identical to the vertical amplifier.

Cathode-ray tubes are available with a number of different phosphors on the screen.   The choice of phosphor controls the intensity of light available for visual observation or photographic recording and also the persistence of the trace after the electron beam has moved on.   Both long- and short-persistence phosphors are available.   Long-persistence phosphors are useful in visual observation of transients since the entire trace is visible long enough for an observer to note its characteristics. Persistence for several seconds is possible.   When the moving-film method of trace photography is used, a very short-persistence phosphor is necessary to prevent blurring. (In this method the electron beam is deflected vertically only, while the film is moved horizontally in front of the screen at a fixed and known velocity.) Persistence of less than 1 μsec is available. The widely used P2 phosphor has a persistence of between 10^{-4} and 10^{-5} sec. For photography of the highest-speed traces the P11 phosphor produces more usable light and is often recommended. Dual-persistence phosphors (P7 is common) provide either a long or short persistence, depending on the color of the filter used over the scope screen. The most common method of photographing oscilloscope traces uses a still camera and the 10-sec Polaroid[1] film process. A common method of photographing transients uses a double exposure to record both the trace and the grid lines. With the grid-line illumination turned off and the camera shutter held open, the transient is triggered, thus recording its image on the film. The shutter is then closed. Now the grid lines are turned on and the shutter snapped in the normal manner (say $\frac{1}{25}$ sec at F:16) to superimpose the grid lines on the picture. This procedure is necessary since the illumination from the grid lines is so great that it would completely fog the film if left on during the long time that the shutter is left open to catch the transient. Because of the rapidity and ease of making trial runs with Polaroid film, it is generally best to determine camera settings by trial and error rather than attempting to calculate them.

To obtain multichannel capability in oscilloscopes, two approaches are used. The dual-beam oscilloscope has two separate electron beams in one cathode-ray tube, with separate deflection plates and amplifiers for each beam. In some units both beams use the same sweep system; thus the two traces are plotted against the same time base. Completely independent beams allowing different time bases on each trace are also available. The other approach uses a single-beam cathode-ray tube and a high-speed electronic switch to time-share the beam among several input signals. Such multitrace systems are available to give up to four traces on a single screen.

Versatility of operation is achieved in plug-in-type oscilloscopes by providing a wide variety of functional plug-in units for a single main frame. Typical plug-ins available include dual-trace and four-trace units, operational amplifiers, carrier amplifiers, spectrum analyzers, high-gain amplifiers, and time bases with special features.

While most laboratory oscilloscopes have a 5-in. diameter screen, larger screens up to about 21 in. are available. These are useful for viewing by large groups, presentation of many channels of data by bar-graph-type displays, etc. Large-screen scopes usually cannot attain the

---

[1] Polaroid Corp., Cambridge, Mass.

high-frequency capability of the 5-in. types. A special 17-in. display system achieving high accuracy by electronic generation of amplitude and time grid lines ("electronic graph paper") has been developed.[1] As many as eight variables can be plotted simultaneously against time by this system with an accuracy of 0.5 percent or better. The X-Y mode of operation is also possible with this equipment. By using suitable function generators in the grid-line system, useful distorted plotting scales (such as semilog, log-log, etc.) may be obtained.

While the Polaroid photography process makes recording very convenient, a *permanent* retention of the trace on the scope screen has certain advantages. This feature has recently become available in the so-called storage oscilloscope. Special cathode-ray tubes are now available that will retain a trace for long periods until it is erased electrically by pushing a button. Oscilloscopes using such storage tubes cannot achieve as high speed performance as conventional types, however they allow one to examine traces visually with ease; when a desired trace is noted, it can then be photographed.

The performance specifications of oscilloscopes cover such a wide range that one must really consult manufacturers' catalogs to appreciate the versatility of this instrument. We can, however, quote some limits of performance as presently available. Voltage sensitivities as high as 0.1 to 0.001 cm/μv with frequency response to 5,000 and 50,000 cps, respectively, can be obtained. Lower sensitivity results in higher frequency response, the upper limit around 1,000 Mc having a sensitivity of the order of 0.1 cm/volt. The accuracy of oscilloscope voltage and time scales is of the order of 2 or 3 percent at best, with operation at the limits of sensitivity and/or speed ranges resulting in reductions in these values. Increased accuracy of time measurements can be obtained in dual-beam or dual-trace scopes by applying an accurately known timing voltage, such as that from a crystal oscillator, to one channel. In single-trace instruments one can use the Z axis (intensity modulation) in a similar way to turn the beam on and off at known time intervals. This produces a dashed-line trace, with the dashes occurring at known time intervals. Increase in voltage (vertical deflection) accuracy is attainable by switching the beam rapidly between an accurately known reference voltage and the unknown voltage. An elaboration of this approach is actually what is done in the "electronic graph paper" system mentioned earlier.

To indicate the type of performance to be expected from a general-purpose oscilloscope suitable for a wide range of mechanical engineering studies, we quote the following specifications of a typical dual-beam

---

[1] G. A. Philbrick Researches, Boston, Mass.

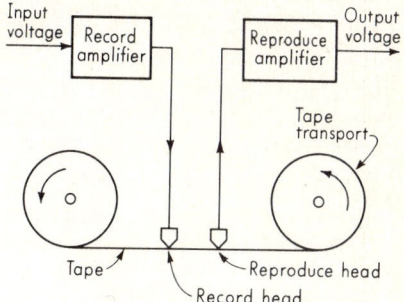

Fig. 12.12. *Tape recorder/reproducer.*

instrument.[1] The vertical-deflection factors (sensitivity) are selectable from 17 in the range 0.1 mv/cm to 20 volts/cm. Frequency response is from d-c to 50,000 cps at 0.1 mv/cm to d-c to 1 Mc at 0.2 volt/cm and higher. Both beams share the same time base, which is selectable from 21 in the range 5 sec/cm to 1 μsec/cm. The input amplifiers offer both single-ended and differential input and may be either a-c- or d-c-coupled. On a-c coupling the frequency response is flat beyond about 2 cps. The input impedance is 1 megohm paralleled by 47 pf. For high-sensitivity X-Y plotting only one beam is needed, and the vertical amplifier for the unused beam is "borrowed" for the horizontal-deflection system.

12.6

Magnetic Tape Recorder/Reproducers The magnetic tape recorder/reproducer has a number of unique features not shared by other recording devices discussed previously. These are derived mainly from its ability to record a voltage, store it for any length of time, and then reproduce it in electrical form essentially identical to its original occurrence. Recording methods used with tape recorders include the direct, FM, PDM (pulse-duration modulation), and digital techniques.[2] We shall consider briefly the direct and FM modes of operation.

Figure 12.12 shows a functional diagram of a tape recorder/reproducer, and Fig. 12.13 shows a closeup of the record and reproduce heads. A current i proportional to the input voltage is passed through the winding on the record head, producing a magnetic flux $\phi = K_\phi i$ at the recording gap. The tape (thin plastic coated with iron oxide particles) passes under the gap, and the oxide particles retain a state of permanent mag-

---

[1] Tektronix Inc., Beaverton, Ore.
[2] P. J. Weber, The Tape Recorder as an Instrumentation Device, Ampex Corp., Redwood City, Calif., 1963.

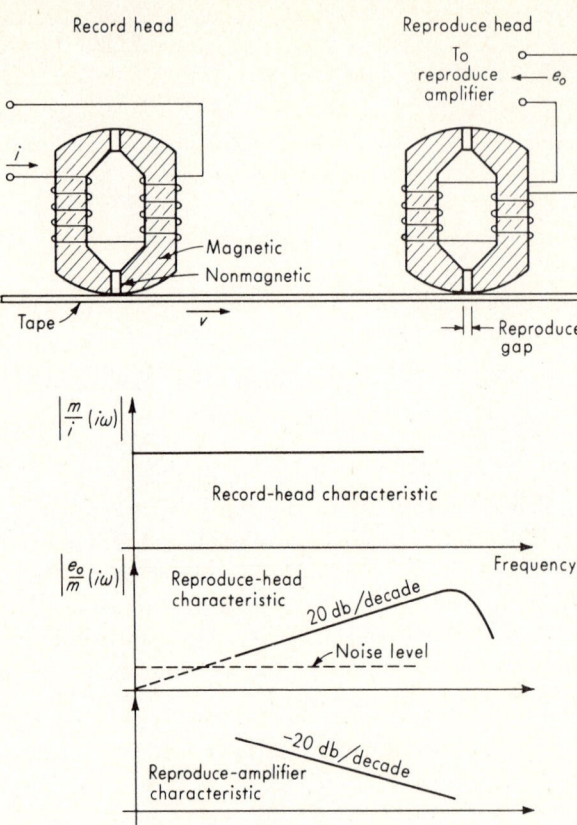

Fig. 12.13.   *Record and reproduce heads.*

netization proportional to the flux existing at the instant the particle leaves the gap. (Actually the applied flux and induced magnetization are not proportional because of the nonlinearity of the magnetic-hysteresis curve. Effectively, however, a close linearity is obtained by a high-frequency bias technique.[1]) Thus, with a sinusoidal input signal $i = i_0 \sin 2\pi f t$ and a tape speed of v in./sec, the intensity of magnetization along the tape will vary sinusoidally with distance x according to

$$\text{Magnetization} \triangleq m = K_m K_\phi i_0 \sin\left(\frac{2\pi f}{v} x\right) \qquad (12.9)$$

$$\text{where} \quad m = K_m \phi$$

The wavelength of the magnetization variation is then v/f in. For example, a 60-cps signal at a 60-in./sec tape speed gives a wavelength

[1] *Ibid.*

of 1 in.   If the tape with this signal on it is then passed under the reproduce head, a voltage proportional to the rate of change of flux bridging its gap will be generated in its coil.   Note that since the output voltage depends on the rate of change of flux, if a d-c current at the input had produced a constant tape magnetization, the reproduce head would have given *zero* output.   Thus the technique described above, the so-called direct recording process, can be used with varying input signals only, about 50 cps being the usual lower limit of frequency.   Furthermore, since the reproducing head has a differentiating characteristic, the reproduce amplifier must have an integrating characteristic in order that the system output be proportional to the input.   An upper frequency limit also exists because at sufficiently high frequencies, for a given reproduce gap and tape speed, one wavelength of magnetization will become equal to or less than the gap width.   Then the average magnetization in the gap will be zero and no output voltage will be generated. For example, at the fastest currently available tape speed, 120 in./sec, and a gap width of 0.00008 in., this occurs at 1.5 Mc.   Actually, the system is usable to only about half this frequency with reasonable accuracy.   The frequency range of the direct recording process is approximately within the band 50 to 600,000 cps, with some machines going to several megacycles.   The direct recording process does not give particularly high accuracy.   This is essentially limited by the signal/noise ratio which is of the order of 25 db (about 18:1).   The rather high noise level is due to minute defects in the tape surface coating to which the direct recording process is sensitive.

When more accurate recording and response to d-c voltages are required, the FM system is generally employed.   Here the input signal is used to frequency-modulate a carrier which is then recorded on the tape in the usual way.   Now, however, only the *frequency* of the recorded trace is significant, and tape defects causing momentary amplitude errors are of little consequence.   The frequency modulators used here are similar in principle to those discussed under voltage-to-frequency converters and subcarrier oscillators in Chaps. 10 and 11.   However the frequency deviation for tape recorders is ±40 percent about the carrier frequency.   The reproduce head reads the tape in the usual way and sends a signal to the FM demodulator and low-pass filter where the original input signal is reconstructed.   The signal/noise ratio of an FM recorder may be of the order of 40 to 50 db (100:1 to 330:1), indicating the possibility of inaccuracies smaller than 1 percent.   By using sufficiently high carrier frequencies (432 kc), the flat (±1 db) frequency response of FM recorders may go as high as 80,000 cps at 120-in./sec tape speed.   To conserve tape when high-frequency response is not needed, a range of tape speeds is generally provided.   When the tape speed is changed, the carrier frequency is changed in direct proportion. This makes the recorded wavelength of a given d-c input signal the same, no matter what tape speed is being used, since ± 40 percent full-scale frequency deviation is used in all cases. Signals may be recorded at one tape speed and played back at any of the others without change in magnitude but with a compression or expansion of the time scale. A common set of specifications might be as follows:

Tape speed, in./sec	Carrier frequency, kc	Flat frequency response ± 0.5 db, cps	RMS signal/noise ratio
120	108	0–20,000	50
60	54	0–10,000	50
30	27	0–5,000	49
15	13.5	0–2,500	48
$7\frac{1}{2}$	6.75	0–1,250	47
$3\frac{3}{4}$	3.38	0–625	46
$1\frac{7}{8}$	1.68	0–312	45

Multichannel tape recorders are available with up to 14 channels on one 1-in.-wide tape. Input to tape recorders is generally at about a 1-volt level, and so most transducers require amplification before recording. The maximum time-base change of about 60:1 ($120/1\frac{7}{8}$) shown in the above table can be even further increased by rerecording the signal. For example, the original signal can be recorded at 120 in./sec. Then it is played back at $1\frac{7}{8}$ with the output of the reproduce amplifier feeding the input of the record amplifier of another machine running at 120 in./ sec. If this tape is now played at $1\frac{7}{8}$, an overall slowdown of 4,096:1 is achieved. An example application of tape slowdown is the recording of a 20,000-cps signal on tape at 120 in./sec and playback at $7\frac{1}{2}$ in./sec into an optical oscillograph (frequency response to 2,000 cps) for a permanent record. Tape slowdown is also used when digital computations are to be performed on high-speed analog data. The digital equipment is very accurate but cannot handle rapidly varying inputs; thus one can analog-tape-record the data and then play it into the digital processing equipment at reduced speed. Tape speedup is useful in the processing (spectrum analysis, autocorrelation, etc.) of low-frequency signals in electronic equipment designed for high frequencies. The storage and playback feature of tape recording is widely used in simulation. A particular environmental condition, such as vibration in an aircraft, is measured and tape-recorded in the actual environment. The tape is then brought into the simulation laboratory where the environmental parameter (say vibration) is recreated by playing the tape into a vibration shaker.

Problems

12.1 Calculate the ratio peak value/rms value and average value/rms value for the following:
  - *a.* Direct current
  - *b.* A sine wave
  - *c.* A square wave
  - *d.* A half-wave-rectified sine wave
  - *e.* A full-wave-rectified sine wave
  - *f.* A train of rectangular pulses that are on 10 percent of the time and off 90 percent of the time.

12.2 Derive Eqs. (12.4) to (12.7).

12.3 Solve for $(i_g/e_s)(D)$ in the system of Fig. 12.6, neglecting inductance.

12.4 Taking inductance into account in the system of Fig. 12.6, find the following:
  - *a.* $(\theta_o/e_s)(D)$
  - *b.* $(i_g/e_s)(D)$

Show the possible shapes of frequency-response curves for these transfer functions.

Bibliography

1. Notes on the Julie Ratiometric Method of Measurement, Julie Research Laboratories, New York, 1964.

2. M. H. Aronson: "Handbook of Electrical Measurements," The Instruments Publishing Co., Pittsburgh, Pa., 1961.

3. L. W. Dean: Potentiometer Specifications, *Instr. Control Systems*, p. 73, January, 1965.

4. S. A. Davis: Analog Voltmeters, *Electromech. Design*, p. 48, November; p. 44, December, 1963.

5. R. Bergeson: Feedback Stiffens D'Arsonval Movement, *Control Eng.*, p. 121, September, 1964.

6. J. W. Martin: Error Analysis in Measuring RMS Voltages, *Electro-Technol.* (*New York*), p. 38, April, 1965.

7. R. J. Erdman: DC Microvolt Measurements, *Instr. Control Systems*, p. 91, January, 1964.

8. R. T. Hood: Measuring Current in High-energy Arc Jets, *Instr. Control Systems*, p. 99, January, 1964.

9. J. F. Keithley: Electrometer Measurements, *Instr. Control Systems*, p. 74, January, 1962.

10. F. C. Martin: RMS Measurement of AC Voltages, *Instr. Control Systems*, p. 65, January, 1962.

11. W. H. Schaeffer: The Six-dial Thermofree Potentiometer, *Instr. Control Systems*, p. 283, February, 1961.

12. A. Miller: Design Considerations of D'Arsonval Galvanometer-Power Amplifier Systems, *The Right Angle*, The Sanborn Co., Waltham, Mass., August, November, 1958.

13. "Typical Oscilloscope Circuitry," Tektronix Inc., Beaverton, Ore., 1961.

14. A. L. Ispas: Interpretation of Magnetic Tape Recorder Specifications, *Instr. Control Systems*, p. 97, July, 1964.

15. P. J. Weber: The Tape Recorder as an Instrumentation Device, Ampex Corp., Redwood City, Calif., 1963.

16. P. A. Mohr: Magnetic Tape Systems for Data Recording, *Automation*, p. 72, February, 1958.

17. History of Magnetism, *Readout*, Ampex Corp., Redwood City, Calif., August–September, 1961.

18. R. E. Morley: Time Compression Disk, *Instr. Control Systems*, p. 108, July, 1964.

19. J. McElwain: Long-term Magnetic Tape Recording, *Instr. Control Systems*, p. 111, July, 1964.

20. E. D. Lucas: Miniature Tape Recorders, *Control Eng.*, p. 53, December, 1964.

21. G. H. Schulze: Tape Recording Errors, *ISA J.*, p. 61, May, 1964.

22. D. R. Davis and C. K. Michener: Graphic Recorder Writing Systems, *Hewlett-Packard Jour.*, October, 1968.

23. L. Brunetti: A More Rugged, Cleaner Writing Oscillographic Ink Recorder, *Hewlett-Packard Jour.*, May, 1973.

24. W. R. McGrath and A. Miller: Fine-line Thermal Recording on Z-fold Paper, *Hewlett-Packard Jour.*, February, 1972.

25. C. A. Donaldson: The New-Generation Oscilloscopes, *Machine Design*, January 27, 1972.

26. C. A. Conaldson and C. A. Gustafason: Easier and Brighter Display of High-Frequency Signals, *Hewlett-Packard Jour.*, May, 1968.

27. J. Johnson: Sampling Oscilloscope Techniques, *Electro-Technology*, September, 1968.

28. W. Farnbach: A Scrutable Sampling Oscilloscope, *Hewlett-Packard Jour.*, November, 1971.

29. Introducing the New Generation of Oscilloscope, *Tekscope*, October, 1969, Tektronix Inc., Beaverton, Ore.

30. A New World of Measurements for the Oscilloscope, *Tekscope*, January, 1971, Tektronix Inc., Beaverton, Ore.

31. The Oscilloscope with Computing Power, *Tekscope*, March/April, 1973, Tektronix Inc., Beaverton, Ore.

32. Fiber-Optic Cathode-Ray-Tube Visicorder Records DC to 1 MHz Responses, *Elect. Inst. Digest*, January, 1967.

33. P. Lowe: Graphic Data Recording: Fiber-Optic CRT, *Digital Design*, October, 1973.

34. R. L. Dudley and V. L. Laing: A Self-Contained, Hand-Held Digital Multimeter, *Hewlett-Packard Jour.*, November, 1973.

35. R. Zaphiropoulos: Nonimpact Printers, *Datamation*, May, 1973.

36. Statos Electrostatic Recorder, REC 1040, Varian Assoc., Palo Alto, Calif., 1967.

Chapter 13
Large-scale
Data Systems

The measurement demands generated by experimental test programs of complex systems such as rockets have led to the development of sophisticated systems for the gathering and processing of measured data. In order to make test results quickly available, analog and/or digital computers may process measurements as they are made and present significant results to the test operator, who may then modify test conditions more nearly to meet requirements. Even if such on-line processing is not employed, the vast quantities of information generated by the transducers must be made available, in intelligible form, as quickly as possible. Such large-scale information-processing systems are also becoming more common in industrial-production operations.   To illustrate the nature of these systems, we shall briefly examine a typical example.   The example chosen is the central facility for recording and processing data from vibration, heat-transfer, and rocket-testing studies at the NASA Lewis Research Center in Cleveland, Ohio.   The information presented below is based on the publication, A Central Facility for Recording and Processing Transient-type Data, *NASA, Tech. Note* D-1320, 1963.   Since we shall discuss only the main features, those interested in more detail should consult this reference which gives a complete account of the problems involved in setting up such an operation.

Figure 13.1 shows a simplified overall block diagram of the data system.   Transducer input signals at the 0 to 10-mv level may be accepted from six different test facilities at various scattered locations at the Lewis Research Center.   When a test facility wishes to utilize the central data-processing system, a phone call is made; if the system is not in use, the necessary connections are established.   Transducer signals are transmitted over a 200-pair shielded telephone cable for distances up to 7,000 ft at the 0 to 10-mv level.   (Figure 11.1*d* and *e* show frequency-response tests of this cable.)   Of the 200 available signal lines, 128 are used for a digital recording system and 24 for an analog system.   The digital system features high accuracy and compatibility with digital-computer data processing but is of limited frequency response.   It also provides for recording a large number of channels by use of time sharing.   High frequency response is the main feature of the analog system.   Various interconnections are possible between the analog and digital systems to provide a maximum of versatility and reliability.

Tracing through the digital system, we first encounter the premultiplexer.   This is a rotating electromechanical sampling switch which samples the 128 input signals, 8 at a time, at a rate of 500 samples per second.   If desired, it is also possible to sample 1, 2, 3, 4, 5, 6, 7, 8, 32, 64, or 128 inputs.   The sample rate per signal is in each case 4,000/(number of signals) per second.   Thus the maximum rate, if only one signal is sampled, is 4,000 per second.   Sampling of signals can lead to difficulties if the number of samples per cycle of signal is not sufficiently great.   For example, if a 60-cps sine-wave signal is sampled at exactly 60 samples per second, when the signal is reconstructed from the sampled points it will appear to be a d-c voltage.   The Shannon sampling theorem[1] shows that the absolute minimum number of samples must be two per cycle of the highest frequency present in the signal.   Actually, somewhat more than two samples per cycle must be used to achieve good accuracy when the signal is reconstructed.   Rational methods of establishing the required

[1] L. W. Gardenhire, Selecting Sample Rates, *ISA J.*, p. 59, April, 1964.

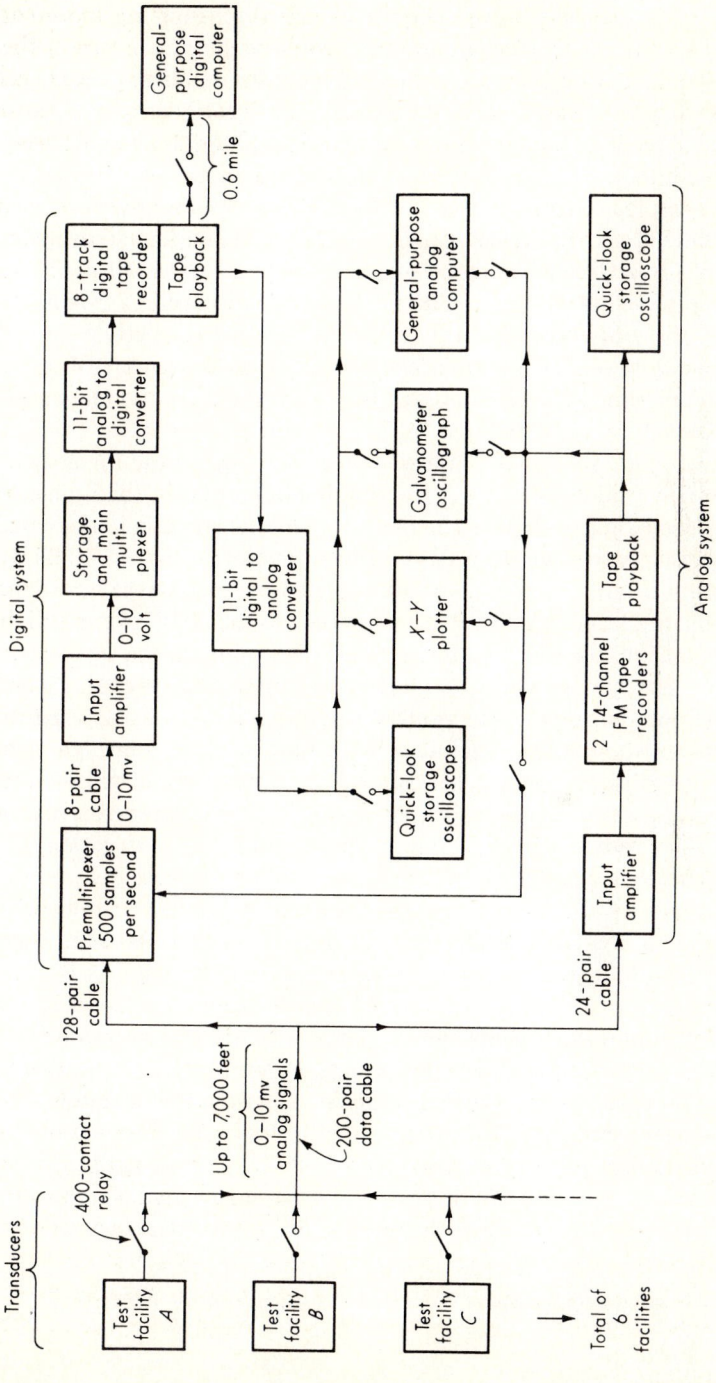

Fig. 13.1. *Large-scale data system.*

sampling rate have been developed[1] but their discussion is beyond the scope of this text. Depending on the accuracy required and the method of reconstructing the signal from the samples, the required sample rate varies over a wide range.[2] If one arbitrarily selects 10 samples per cycle as a reasonable rate, the above system can handle one signal of 400-cps frequency content, two signals of 200-cps content, etc.

The sampled 0 to 10-mv signals are boosted to 0 to 10-volt range by d-c amplifiers of the differential-input type. Their frequency response is flat (-3 db) to 3,000 cps, and the step response time to be within 0.1 percent of final value is 0.5 msec. Since signals are switched in a 2-msec (500 cps) cycle, the amplifiers are sufficiently fast to read correctly before the next sample is taken. Common-mode rejection at 60 cps is $10^6:1$ for a 1,000-ohm unbalanced line, which is quite important in this system because of the extremely long cable.

The storage and main multiplexer unit stores simultaneously the eight sampled analog outputs of the amplifiers and feeds them one at a time to the analog-to-digital converter. This is an 11-bit successive-approximation type which can convert from analog to digital in 22 μsec. Conversion is accurate to 0.05 percent $\pm \frac{1}{2}$ the least-significant bit. Since 2^{11} is 2,048, the least-significant bit represents about 0.05 percent of full scale also. Digitally coded pulses are recorded on eight tracks of a tape recorder at either 15 or 60 in./sec. Playback from this recorder can be sent to a number of plotting or processing equipments. A standard telephone-type cable sends the digital pulses about 0.6 mile to a high-speed, stored-program, general-purpose digital computer. Pulse width at the sending end is about 25 μsec and degrades to 35 μsec at the receiving end. The digital computer can, of course, be programmed to perform any processing operations desired.

To provide quick-look checks on the performance of the digital system and readily available analog records of selected variables, the digital signal is converted back to analog in a digital-to-analog converter. This signal can then be sent to a storage-type oscilloscope, X-Y plotter, five-channel galvanometer oscillograph, or a general-purpose electronic analog computer. This computer has 12 integrator-summers, 24 summers, 8 electronic multipliers, 60 servoset coefficient potentiometers, and 6 servoset function generators. The storage oscilloscope may be used in four modes: One to four selected data channels may be plotted against time in a single sweep, any data channel may be displayed against any other channel in an X-Y plot, all data may be displayed at a fast-recurring

---

[1] L. W. Gardenhire, Selecting Sample Rates, *ISA J.*, p. 59, April, 1964.
[2] *Ibid.*

sweep to produce a vertical bar-graph display, and all data may be displayed during a single slow sweep.

In the analog system, the 0 to 10-mv transducer signals are boosted in d-c amplifiers with a frequency response of 0 to 10,000 cps. These signals are then applied to two 14-channel FM tape recorders. Twelve channels of each are used for data and two for identification and calibration information. The accuracy with which a d-c voltage can be recorded and played back is 1 percent of full scale. Frequency response of the overall system is good to 10,000 cps. Analog information recorded at maximum tape speed can later be played back at $\frac{1}{2}$, $\frac{1}{4}$, $\frac{1}{8}$, $\frac{1}{16}$, or $\frac{1}{32}$ of this speed into the digital system for desired processing. This is necessary since the digital system cannot handle high-speed data. Information from the analog system also may be sent to storage oscilloscopes, X-Y plotters, oscillographs, and the analog computer.

Problems

13.1 A 60-cps sinusoidal signal is sampled once every $\frac{1}{60}$ sec. If the sampled function is reconstructed by linear interpolation, describe the reconstructed function.

13.2 A 10-sec/cycle sine wave is sampled once every 11 sec. If the sampled function is reconstructed by linear interpolation, describe the reconstructed function. What fictitious frequency has been introduced by the sampling process?

Bibliography

1. A. T. Snyder: Airborne Recorder and Computer Speed Flight-test Data Processing, *ISA J.*, p. 44, July, 1958.
2. E. J. Kompass: Information Systems in Control Engineering, *Control Eng.*, p. 103, January, 1961.
3. J. P. Knight et al.: Low-level Data Multiplexing, *Instr. Control Systems*, p. 86, August, 1963.
4. L. W. Gardenhire: Selecting Sample Rates, *ISA J.*, p. 59, April, 1964.
5. W. T. Botner: Digital Data Gathering System for Blowdown Wind Tunnel, Sandia Corp., Albuquerque, N.Mex., *Rept.* SCR-23, 1958.
6. J. K. Slap: Recording and Processing Test Data, *Control Eng.*, p. 177, September, 1959.
7. E. Pacini: How Raytheon Cut Test Analysis Time from Days to Hours, *Instrumentation*, Honeywell Inc., Philadelphia, Pa., vol. 17, no. 1, 1964.
8. P. Westercamp: Computing Power Station Performance, *Control Eng.*, p. 72, December, 1963.
9. Digital Data System Takes 15,625 Engine Samples a Second in Saturn Rocket Static Tests, *Control Eng.*, p. 19, October, 1963.
10. W. C. Hixson: Instrumentation for the Pensacola Centrifuge Slow Rotation Room I Facility, *NASA*, CR-53341, 1964.
11. K. C. Sanderson: The X-15 Flight Test Instrumentation, *NASA*, *TM X*-56000, 1964.

12. J. D. Jones: High-speed Low-level Data Acquisition, *Instr. Control Systems*, p. 96, April, 1965.
13. H. Gruen and B. Olevsky: Increasing Information Transfer, *Space/Aeron.*, p. 40, February, 1965.
14. S. H. Boyd: Digital-to-Visible Character Generators, *Electro-Technol.* (*New York*), p. 77, January, 1965.
15. W. Clifford: Digital Voltmeter Data Systems, *Instr. Control Systems*, p. 105, December, 1964.
16. W. E. Schilke: The Analysis of Transmission and Vehicle Field Test Data Using a Digital Computer, *Gen. Motors Eng. J.*, p. 19, October-November-December, 1964.
17. G. A. Korn: "Minicomputers for Engineers and Scientists," McGraw-Hill, New York, 1973.
18. A Computer for Instrumentation Systems, *Hewlett-Packard Jour.*, March, 1967.
19. K. A. Fox et al: A Human Interface for Automatic Measurement Systems, *Hewlett-Packard Jour.*, April, 1972.
20. P. J. Torpey: Minicomputerizing Analog Data Collection, *Control Eng.*, June, 1970.
21. J. V. Wait: Sampled Data Reconstruction Errors, *Inst. and Cont. Syst.*, June, 1970.
22. J. V. Dirocco: Signal Conditioning for Analog-to-Digital Conversion in Instrumentation Systems, *Elect. Inst. Digest*, May, 1970.
23. D. P. Allen: How to Choose Data Acquisition Systems, *Control Eng.*, November, 1969.
24. R. K. Kaminski: Computer Diagnosis of the VW, *Inst. Technology*, September, 1972.

Chapter 14
Significant Developments: 1965–1975

Since the first edition of this text emphasized basic principles, concepts, and methods of analysis that remain largely unchanged as time passes, the revision is devoted mainly to the correction of errors and addition of significant hardware developments. These functions are served at minimum increase in cost by simply gathering all new hardware discussions into a new final chapter. This approach is feasible since in almost every case, *deletion* of "old" hardware material is neither necessary nor desirable because new hardware rarely *displaces* old—it merely augments it. For example, the invention of the laser-doppler flow

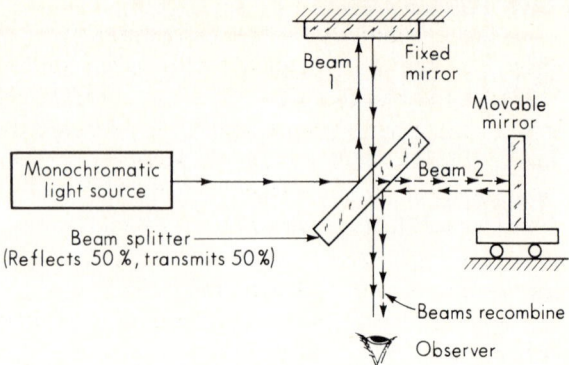

Fig. 14.1.   *Original Michelson interferometer.*

measuring scheme does not make pitot tubes and hot-wires passé, it simply provides yet another method of flow measurement with certain new features.

14.1

Motion Measurement   While lasers have been in existence for over 10 years, their development from laboratory curiosities into devices with "everyday" practical uses is relatively recent.   Today (1975) they are used in eye surgery, computer memories, cutting of microcircuit chips, trimming microresistors and sealing microcircuit packages, balancing gyroscope rotors by precisely blasting out tiny amounts of mass while the rotor is spinning at high speed, cutting fabric for clothing, "machining" small parts made of difficult materials, and various significant measurement applications such as interferometry, holography, and flow sensing.   Here we briefly treat the *laser interferometer*, a very precise motion measuring system.

The use of light-interference principles as a measurement tool can certainly be considered a part of classical physics.   Michelson, in the 1890s, used the scheme of Fig. 14.1, which is the basis of all subsequent developments.   Using the wave model of light, we would expect the observer to see cycles of light and darkness as the motion of the movable mirror shifted the phase of beam two with respect to fixed beam one, causing alternate reinforcement and interference of the two beams.   If we know the light wavelength to be, say, 0.5×10^{-6} m, then each 0.25×10^{-6} m of mirror movement corresponds to one complete cycle (light to dark to light) of illumination.   By counting the number of illumination cycles, we can calculate the distance between any two positions of the movable mirror.

This seemingly simple measurement principle is fraught with difficulties that have kept it a tedious standards-laboratory procedure of limited application (rather than a practical "machine-shop" tool) until the recent appearance of several laser-based developments. The laser itself provides a much higher quality monochromatic (single frequency) light source than did earlier "lamps," since its light is "coherent" (stays in phase with itself) over much greater distances and its frequency is very stable and precisely known (to about one part in 10^7). When light travels through air its frequency is unaffected, but its wavelength changes whenever air pressure, temperature, or humidity cause alterations in the air's refractive index and thus in the speed of light. The pressure effect is about 4 parts in 10^7 per mm of mercury, while the temperature effect is nearly 11 parts in 10^7 per degree centigrade. Many applications do not warrant corrections for these effects, but if they are deemed necessary, some commercial systems[1] provide temperature, pressure, and humidity sensors that automatically scale the instrument readout to provide continuous correction. Since the lengths of objects being measured are *themselves* temperature dependent, automatic sensing of this temperature and correction of readings to some standard (often 20°C) is of practical interest and is provided on some systems.[1]

Clever optical, mechanical, and electronic design have built upon the above simple principles to provide measuring systems of much improved portability, precision, range of applicability, and ease of use. Figure 14.2 shows one such system which we will briefly discuss (Refs. 2 and 3 give more details). The Helium-Neon laser used is unusual in that it produces light at two distinct optical frequencies f_1 and f_2, both in the neighborhood of 5. $\times 10^{14}$ Hz, but separated in frequency by about 2 MHz and of opposite polarizations. As the beam leaves the laser it is split in half, one half going directly to a polarizer and photodetector to create an electrical reference signal, while the other half proceeds out to the external optics. The reference-beam polarizer gives the two frequencies *equal* polarizations so that they may exhibit ordinary constructive and destructive interference. This interference can be thought of in terms of the intensity-versus-time graph of Fig. 14.2. There, two waves initially "in phase" get slightly "out of phase" in one cycle, since the frequencies are slightly different. (Note that the time shift per cycle is a *very* small fraction of one cycle; thus, the waves shown are nearly identical and interfere *constructively*, giving a high illumination to the reference beam photodetector and a large electrical output.)

---

[1] Hewlett-Packard Model 5510A.

[2] A Two-Hundred-Foot Yardstick with Graduations Every Microinch, *Hewlett-Packard Jour.*, August, 1970.

[3] Remote Laser Interferometry, *Hewlett-Packard Jour.*, December, 1971.

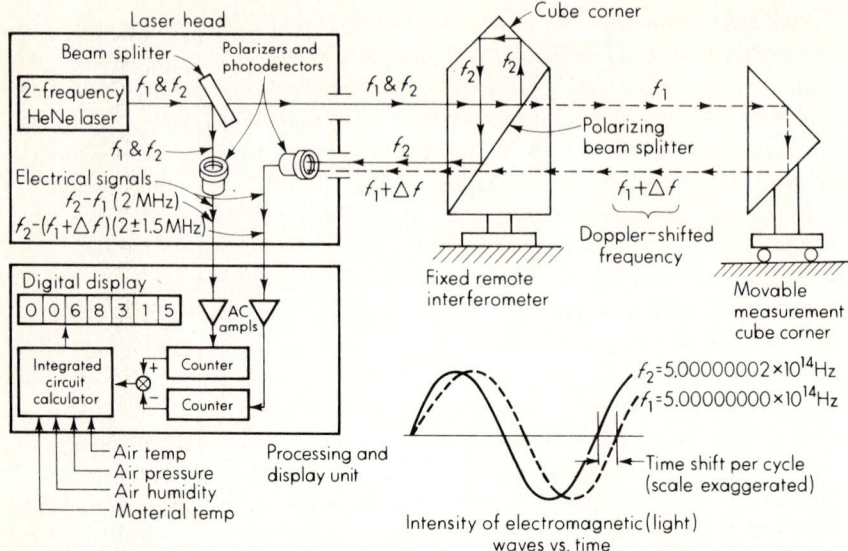

Fig. 14.2.   *Modern laser interferometer.*

Destructive interference (giving low illumination and small electrical output) will occur 1.25×10^8 cycles $(0.25 \times 10^{-6}$ sec) later when the time shift per cycle has accumulated a total phase shift of one-half cycle. The "light" on the reference photodetector thus "flickers" at a 2-MHz rate, producing a 2-MHz electrical signal.

Turning our attention to the measuring beam, it proceeds out from the laser and first encounters the fixed remote interferometer's polarizing beam splitter, which efficiently reflects the f_2 component around a cube corner (prism) and transmits the f_1 component to the measurement cube corner. If this cube corner is stationary, no frequency change occurs in f_1 between the incident and reflected beam. However, if it moves, the reflected beam exhibits a doppler shift of frequency proportional to the velocity of motion [about 3.3 MHz/(m)(sec.)]. When this $f_1 + \Delta f$ beam is recombined with the f_2 beam in the remote interferometer and impinged on the photodetector, we again get interference effects that produce an electrical signal of frequency 0.5 to 3.5 MHz, depending on the velocity. The reference and measurement signals are a-c amplified and applied to a subtracting counter that reads zero if there is no motion, but which accumulates counts in proportion to the distance traveled from a chosen reference position. For example, a motion at 1 cm/sec for 1 sec (total motion 1 cm) gives 33,000 counts, and thus a resolution of about 3×10^{-7} m.

The "a-c" interferometer described above exhibits a number of advantages relative to older "fringe-counting" instruments. The Doppler technique allows use of simpler and more reliable a-c (rather than d-c) amplifiers, gives unambiguous indication of the *direction* of motion, and also allows *velocity* measurement in addition to displacement. Locating the interferometer remote from the laser head reduces thermal deformation problems due to the laser's heat. By varying the external optical hardware, the system[1] can be adapted to measure length, flatness, straightness, and squareness to high precision (a few ppm) over long distances (up to 200 m). In multiaxis machine tools, a single laser beam may be suitably reflected to serve the measurement needs of up to eight axes of motion.

14.2

Flow Measurement   In the area of flow measurement, the most important recent advance is probably the *laser doppler velocimeter*. This instrument measures local flow velocity and is thus in competition with pitot tubes and hot-wire anemometers. Its main advantages relative to these more well-established techniques include:

1. Direct measurement of velocity rather than its inference from pressure (pitot tube) or heat-transfer coefficient (hot-wire).
2. No "physical object" need be inserted into the flow, thus flow is undisturbed by measurement.
3. Sensing volume can be very small (a cube 0.2 mm on a side is not unusual).
4. Very high-frequency response (to megahertz range) is possible.

Disadvantages involve the need for transparent flow channels, the necessity for tracer particles in the fluid, and the cost and complexity of the apparatus. In brief, the operating principle involves focusing laser beams at the point where velocity is to be measured and then sensing with a photodetector the light scattered by tiny particles carried along with the fluid as it passes through the laser focal point. The velocity of the particles (assumed to be identical to the fluid velocity) causes a doppler shift of the scattered light's frequency and produces a photo-detector signal directly related to velocity.

Actually, *artificial* tracer particles are often not necessary; the microscopic particles normally present in tap water or atmospheric air are sufficient. Under extreme conditions, particles may not perfectly follow the flow, but studies have shown highly accurate following in

[1] Hewlett-Packard 5526A.

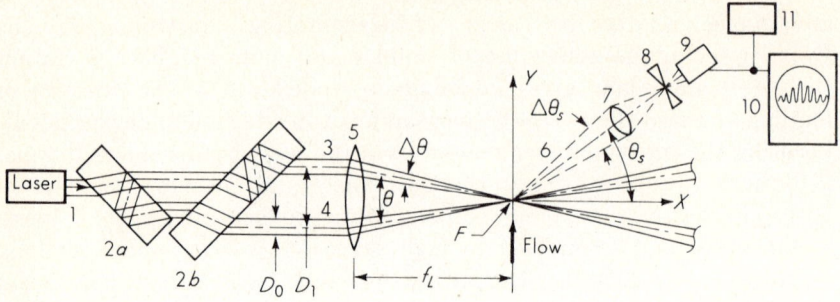

Fig. 14.3.   *Layout of laser doppler velocimeter.*

$$\Delta X = \frac{4\lambda}{\pi \Delta\theta \sin(\theta/2)}$$

Sensing volume (typically contains about 10 fringes)

$$\Delta Y = \frac{4\lambda}{\pi \Delta\theta \cos(\theta/2)}$$

$$\Delta Z = \frac{4\lambda}{\pi \Delta\theta}$$

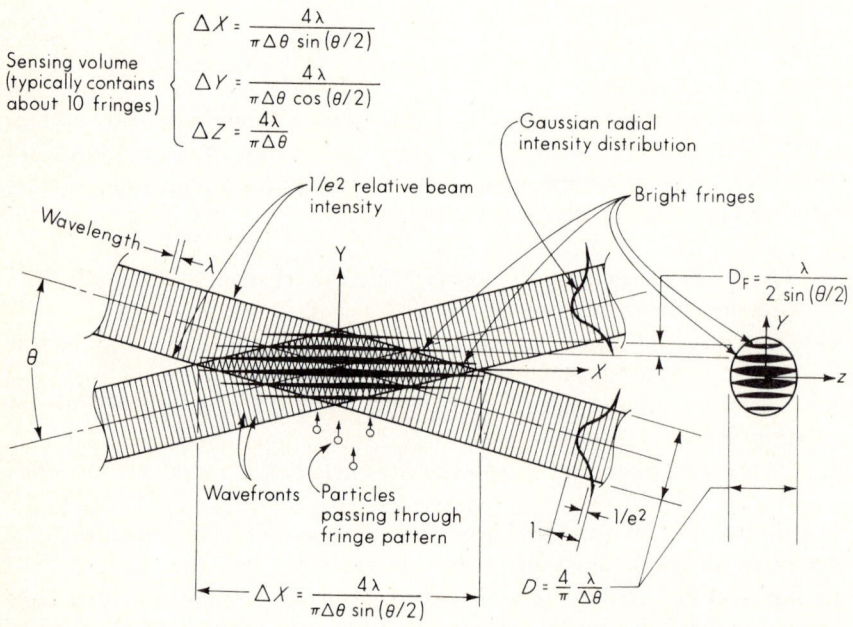

Fig. 14.4.   *Interference fringes in sensing volume.*

many practical cases.   Laser doppler velocimeters (LDV) have been operated in several different configurations; Figs. 14.3 to 14.5[1] give some details on the popular dual-scatter or "fringe" mode.   While a careful physical explanation of the device rests on a detailed study of the doppler-shift effect, many workers in the field find the interference-fringe explanation of Fig. 14.4 useful.   The frequency f of the electrical signals (shown

[1] D. B. Brayton et al, Project Squid Conf. on Laser Doppler Velocimeter, Purdue Univ., March 9–10, 1972, p. 52–100.

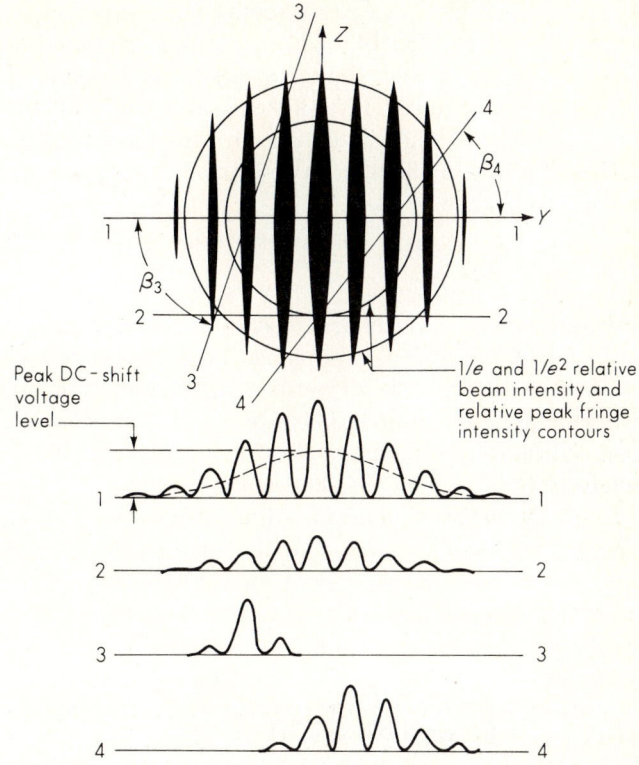

Signal amplitude vs. particle position near $X = 0$ for a number of particle trajectories. The indicated fringe width is proportional to the local peak fringe intensity.

Fig. 14.5. *Electronic signals from velocimeter.*

in Fig. 14.5) produced by a particle moving across the dark and light fringe pattern with a velocity component V normal to the fringes is given by

$$f = \frac{2V \sin \dfrac{\theta}{2}}{\lambda} \qquad (14.1)$$

For typical laser-light wavelength $\lambda = 5 \times 10^{-5}$ cm and $\theta = 30°$, we get $f = 10340\ V$ Hz, where V is in cm/sec. Note that the method measures the velocity *component* perpendicular to the fringe pattern. One can rotate the fringe pattern 90° to measure the other component of a two-dimensional velocity vector. *Simultaneous* measurement of two perpendicular components has been accomplished using polarization tech-

niques.[1]   The simplest signal processing involves visual observation of
the waveforms of Fig. 14.5 on an oscilloscope, where the frequency is
found by counting and timing cycles.   Since these "frequency-burst"
signals are often noisy, intermittent, and confused by the presence of
more than one particle at a time in the sensing volume, sophisticated
electronic processors have been developed that greatly increase the speed,
accuracy, and ease of use.

14.3

Temperature   The first edition of this text only briefly mentioned
infrared imaging systems, and since these have now been developed into
convenient (though still expensive) commercially available instruments
with wide practical applications, we wish to describe them in some detail.
Most of these systems have the function of providing a television-like
visual display in which the shades of gray represent various temperature
levels of the surface of some two-dimensional object (target) on which the
infrared camera is focused.   While these shades of gray are accurately
related to infrared energy levels emitted from the target, recall that all
infrared temperature sensing requires knowledge of the emittance of the
target surface to convert the detector signal to degrees of temperature
(see pages 541–544).

Figure 14.6 shows the arrangement of the camera for a unit of
Swedish manufacture.[2,3]   Scanning of the target surface is accomplished
by focusing the radiation first on a plane mirror oscillating about a
horizontal axis at 16 Hz.   This scans the line of sight vertically over the
target surface.   The image from the plane mirror is focused on an eight-
sided prism (rotating at 200 rps) which provides the horizontal scanning
and results in a "picture" with a frame rate of 16 frames/sec and 100 lines
per picture.   An indium antimonide (InSb) detector (cooled by liquid
nitrogen to reduce its noise level) produces an electrical signal proportional to the incident radiant flux.   To produce a thermal image of the
target on a TV picture tube, horizontal and vertical deflection signals are
picked off the scanning-system motor shafts to position the electron beam,
while the infrared detector signal (video signal) modulates the beam
intensity.   Thus the TV tube displays 100-line picture whose local
intensity (shades of gray) represents target temperature.

Using suitable optics, infrared scanners can produce full-scale
thermal images of objects as small as 0.6 mm square at an 8-mm working

[1] D. B. Brayton et al, *op cit.*

[2] AGA Corp., Secancus, New Jersey.

[3] Borg, Sven-Bertal, Thermal Imaging with Real-Time Picture Presentation,
Applied Optics, vol. 7, page 1697, September, 1968.

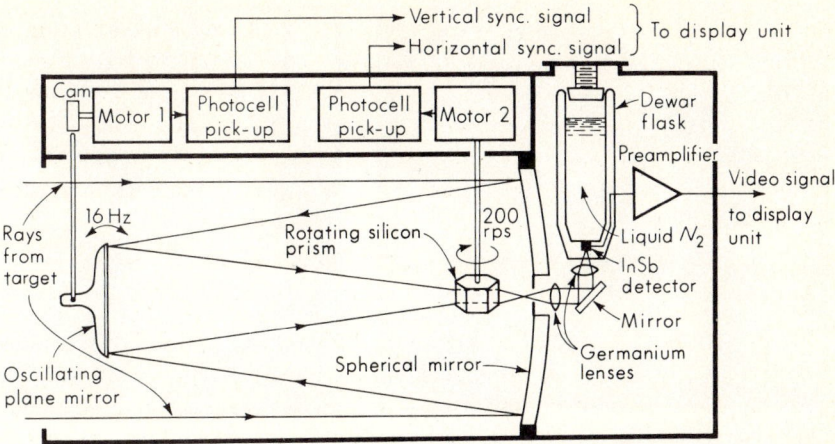

Vertical sync. signal
Horizontal sync. signal
} To display unit

Fig. 14.6. *Infrared camera.*

distance with a "spot size" of 0.01 mm. Temperature differences as small as 0.1°C can be resolved for a target at 30°C. Of course larger targets may be scanned by increasing the working distance; a typical 5° × 5° field of view gives a target size of 8.8 × 8.8 m at a range of 100 m. Figure 14.7 shows the displays produced by a commercial infrared micro-scanner[1] when viewing a 1.5-mm square target area on a metallurgical sample. The center photo shows the "thermogram" with the calibrated gray scale at its left. Digits below give the "temperature" corresponding to black and the range of "temperature" from black to white on the gray scale. An oscilloscope-type trace of the temperature profile across any desired horizontal scan line is also shown. The black line on the thermogram shows its location on the target, while the white trace below displays the temperature profile. A manual knob allows positioning of this trace at any point on the target. The photo at the right shows an "isometric" display obtained by successively displacing horizontally each of the 64 temperature profiles produced by the scanning process. This gives a vivid three-dimensional picture of temperature variations over the surface. Another useful display made (not shown) shows isotherms[2] (lines of constant temperature) as bright white lines on the thermogram. A further refinement[3] gives a 10-color display of isotherms, each color being associated with a different known temperature.

The noncontacting nature of the sensing method, together with the display of the entire surface-temperature distribution over an object,

---

[1] Model RM-50, Barnes Engineering, Stamford, Conn.
[2,3] AGA Thermovision, AGA Corp., Seacaucus, N.J.

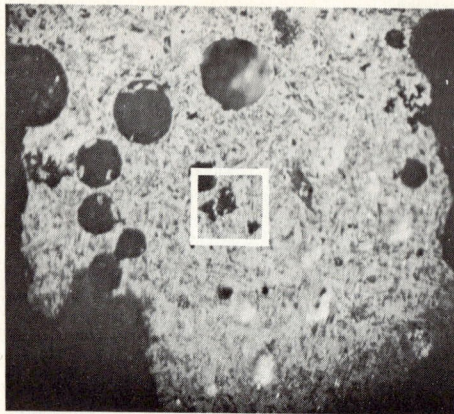

Photograph of metallurgical sample. Outline indicates area shown in thermogram.

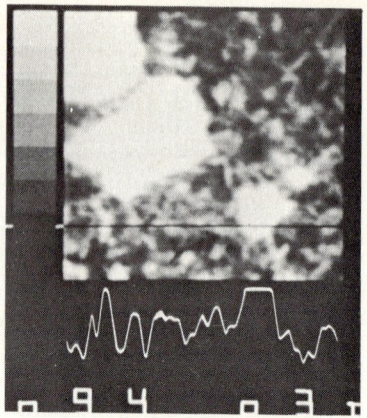

Thermogram (40 X objective, white: hot) of area outlined in photo shows infrared emissivity variation across surface.

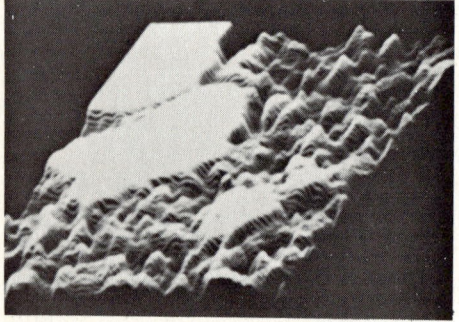

Isometric display of metallurgical sample.

Fig. 14.7.   *Displays from infrared microscanner.*

gives infrared imaging systems unique application possibilities. These include medical diagnosis (cancer detection, peripheral circulation problems, etc.), hot-spot detection in electrical power transmission equipment, surveys of earth and sea temperature from aircraft and satellites, nondestructive testing of products for poor bonding, cracking, wear, heat generation, etc., and biological studies of plant development and insect physiology.

Figure 14.8 shows the physical layout of the Barnes RM-50 Infrared Microscanner mentioned earlier. Basically the system is an infrared microscope with an optical chopper and two-axis scanning system inserted in the optical path. The chopper periodically interrupts the radiant beam so that the detector alternately sees the target and the blackened chopper vane. Since the chopper vane's temperature is accurately known

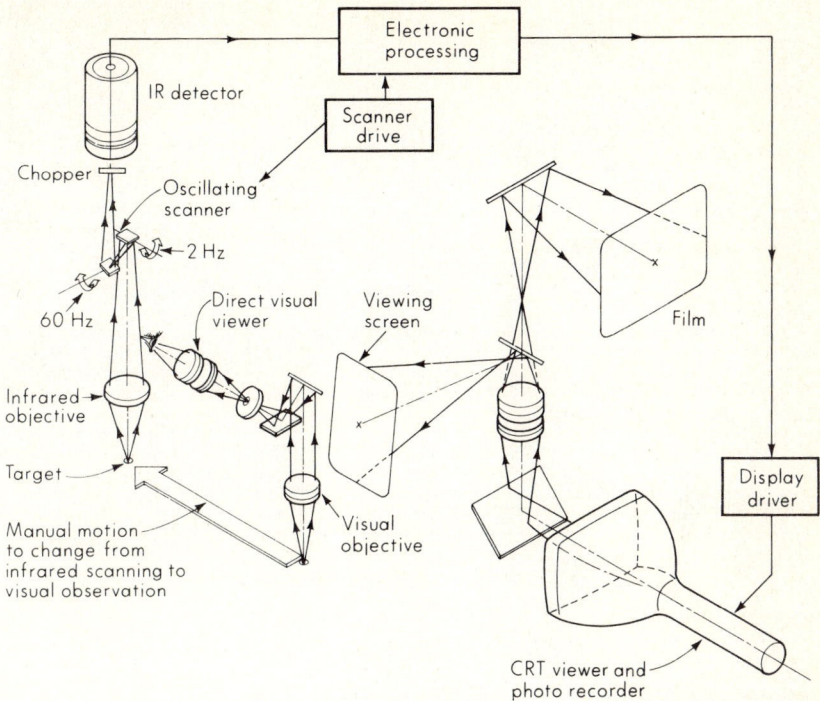

Fig. 14.8.   *Layout of infrared microscanner.*

(and controlled), the detector puts out an alternating electrical signal with an amplitude accurately related to target radiation.   This signal is processed to form a thermal image on the cathode-ray-tube screen. Target scanning uses two oscillating mirrors (x axis at 60 Hz, y axis at either 1 or 2 Hz) driven by small-torque motors to scan the beam over the target area.   This produces an image composed of a raster of 64 horizontal lines at a rate of 1 frame/sec when 2-Hz y scanning is used. To improve resolution (at the expense of frame rate), the 1-Hz scanning gives 128 lines at 2 sec/frame.

14.4

Data Manipulation, Transmission, and Recording   Since electronics plays a large role in this general area, we should begin by noting that while the "transistor revolution" was just well under way in 1965, the use of vacuum tubes in general-purpose electronics is essentially unheard of in 1975.   In fact, discrete transistor technology has in many areas given way to integrated circuits (microcircuits),

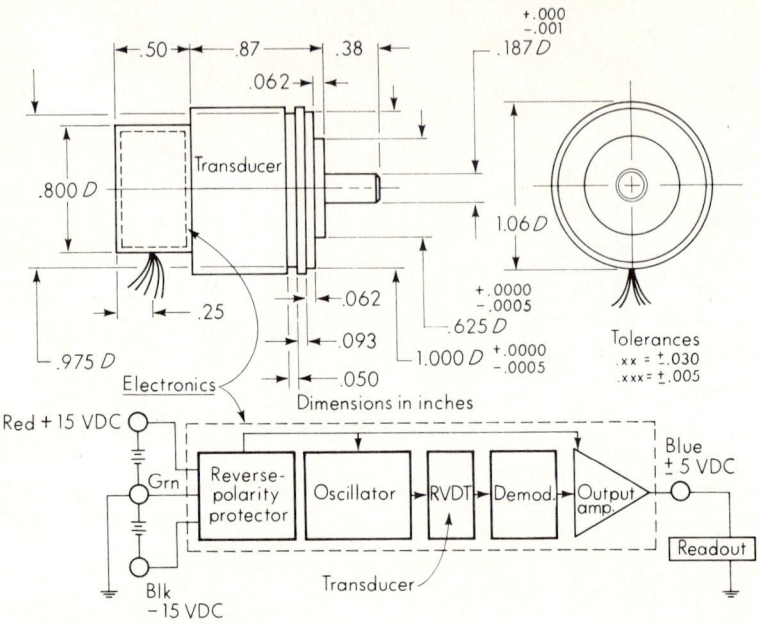

Fig. 14.9.   *Integrated electronics in differential transformer.*

resulting in further miniaturization, reduced cost, and often, improved performance. Thus while the general *functions* described in Chap. 10 (amplification, filtering, analog computation, modulation and demodulation, etc.) have not changed, the details of the hardware used to accomplish them are now quite different. Figure 14.9 shows an example of the possibilities inherent in these miniaturization techniques. The complete carrier-amplifier system (see page 616) needed to operate a rotary variable-differential transformer[1] (RVDT) displacement transducer is enclosed in a space smaller than the transducer itself. Many other types of transducers are now available with miniature amplifiers or other needed signal processing "built in," thus providing large (± 5 V) output signals for transmission. This is often preferable to remote amplification, since transmission of millivolt-level signals from transducer to amplifier encounters many noise problems. Signal processing also makes increasing use of digital rather than analog techniques. This trend is encouraged by the desirable features (accuracy, lack of drift, computer compatibility) of digital methods, together with the decreased cost and improved performance of microcircuit digital hardware.

   In the area of recording, the *digital-memory waveform recorder*

---

[1] Schaevitz Engineering, Camden, N.J.

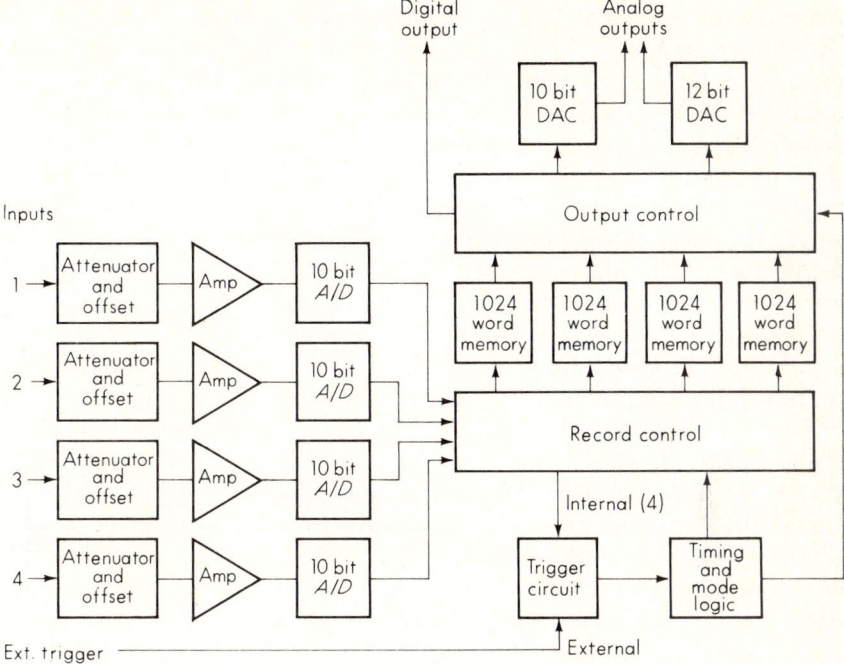

Fig. 14.10. *Digital-memory waveform recorder.*

provides capabilities difficult or impossible to achieve by earlier methods. Figure 14.10 shows a block diagram for a unit[1] capable of recording four independent signals simultaneously. The practical and economic implementation of this concept rests on the ready availability, small size, and reasonable cost of digital memories and associated digital-processing hardware brought about by explosive growth and intense competition in the minicomputer field. Considering a single channel, the analog voltage input signal is digitized in a 10-bit A/D converter with a resolution of 0.1 percent (1 part in 1,024) and frequency response to 25 kHz. The total digital memory of 4,096 words can be used for a single channel, 2,048 can be used for each of two channels, or 1,024 for each of four channels. The analog input voltage is sampled at adjustable rates (up to 100,000 samples/sec), and the data points are read into the memory; a maximum of 4,096 points are storable in this particular instrument. (Sampling rate and memory size must be selected to suit the duration and waveform of the physical event being recorded.)

Once the sampled record of the event is captured in the memory,

---

[1] Model 1015, Biomation Co., Cupertino, Calif.

many useful manipulations are possible, since the memory can be read out without erasing it (nondestructive readout).   By reading the memory out *slowly* through a digital-to-analog converter (DAC), the original event (which could have been extremely rapid) is reproduced as a slowly changing voltage, easy to record permanently in large size on a (slow) *X-Y* plotter.   If the memory is read out *rapidly and repetitively*, an input event which was a "one-shot" transient becomes a repetitive waveform that now may be easily observed on an ordinary (not storage) oscilloscope.   The digital memory may also be read out directly (without going through the DAC) to, say, a computer where a stored program can manipulate the data in almost any desired way.

"Pretrigger recording" allows the device to record the input signal *preceding* the trigger point, a unique and often useful capability.   In "ordinary" triggering, the recording process is *started* by the rise of the input signal (or some external triggering signal) above some preset threshold value.   If this threshold is set too low, random noise will trigger the system inadvertantly; too high a threshold prevents recording of the initial rise of the desired signal.   The digital recorder can be set to record "continuously" (new data coming into memory pushes old data out, once memory is full) until the trigger signal is received, whereupon the recording process is stopped, thus freezing in the memory the data received *prior* to the trigger signal.   An adjustable trigger delay allows operator control of the stop point so that the trigger may occur near the beginning, middle, or end of the stored information.

Signal and system analysis by experimental means (see p. 669) is in much wider use in many areas of engineering now than in 1966.   The *real-time spectrum analyzer* has contributed much to the growth of this important engineering tool.   While the analysis of signals by the methods outlined in Sec. 10.13 can be speeded up by tape recording and replaying into the analyzer at increased tape speed, this method is inconvenient and severly limited in speedup capability.   Use of the same digital-memory technique employed in the digital-memory waveform recorder just described above allows analysis speedup of as much as 500,000 to 1 for slow signals, and can give a complete frequency-spectrum analysis in as little as 0.05 sec.   Such speed (20 complete spectra in 1 sec.) allows one to "track" the changes in frequency content occurring in time-varying signals and gives rise to the "real-time" terminology.   That is, one observes spectral changes "as they occur," i.e., in "real time."

Figure 14.11 shows a simplified block diagram of such an analyzer.[1] The input signal (which can be periodic, transient, or random) is first low-pass filtered (anti-aliasing filtered) to prevent the subsequent

---

[1] Federal Scientific Corp., N.Y.

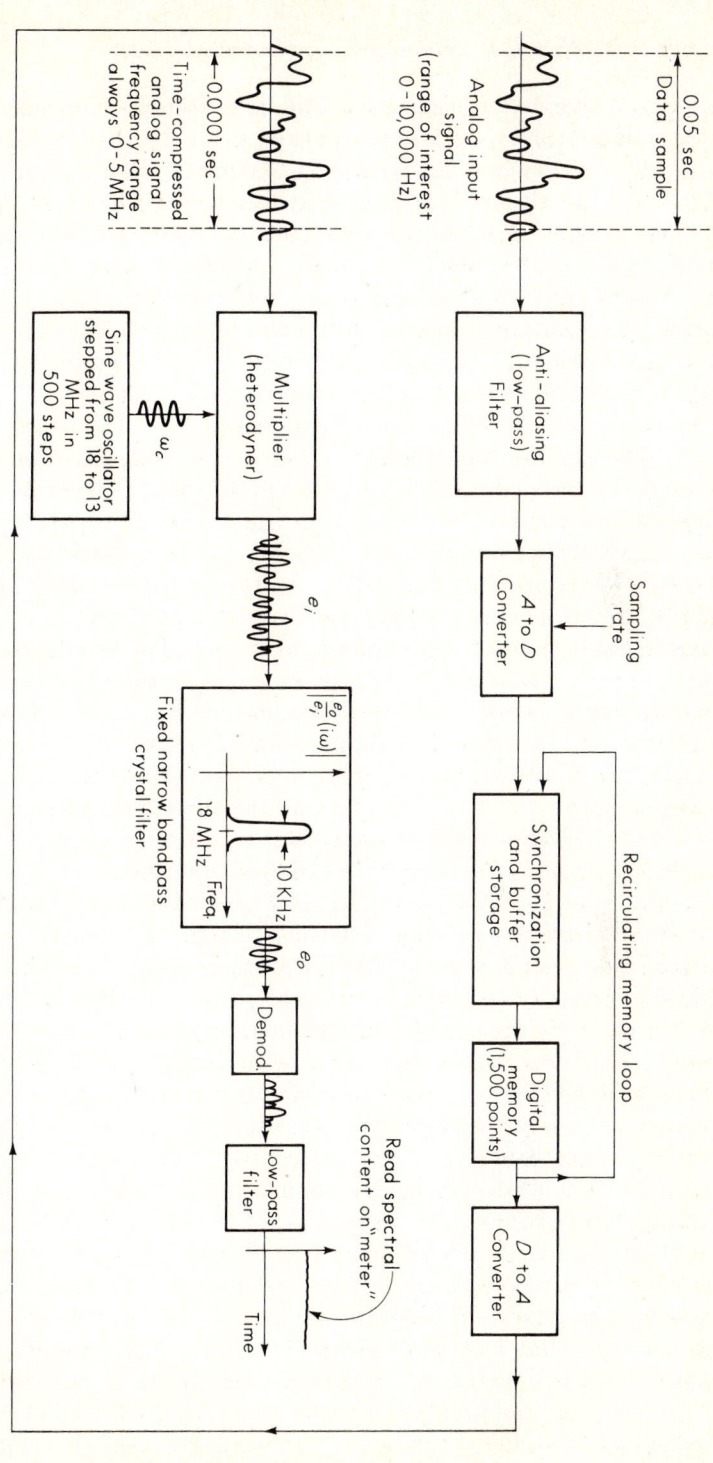

Fig. 14.11. Real-time spectrum analyzer.

sampling process from producing spurious frequencies not really present in the unsampled input. For example, if we set the instrument to analyze the frequency range from 0 to 10,000 Hz, the anti-aliasing filter will increasingly attenuate all frequencies above 10,000 Hz (at 20,000 Hz the attenuation is 50 dB, about a 300-to-1 reduction in amplitude). The filtered signal is periodically sampled at a rate three times the selected upper frequency (30,000 samples per second for a range of 0 to 10,000 Hz), digitized, and entered into the 1,500-word memory. Note that for this example it takes 0.05 sec. to fill the memory. The earlier-quoted calculation rate of 20 independent spectra per second cannot be realized for signals whose spectrum lies below the range of 0 to 10,000 Hz. That is, they are just too slow to accumulate a full memory (needed to get accurate values) in 0.05 sec. For analysis ranges of 0 to 10,000 Hz or greater, the 20 spectra per second rate can be attained or bettered. Very-low-frequency data *can* be analyzed; however, the time needed to accumulate a 1,500-point sample fundamentally limits the speed with which independent spectra can be determined by *any* technique. A typical instrument has 0 to 10 Hz as its lowest range and requires 50 sec. to fill the memory at this setting. Once the memory is full, the spectrum *is* computed every 0.05 sec., however only one or two points in the memory are "refreshed" with new data for each new spectrum.

When the memory is full, it is "recirculated" by nondestructively reading it out at a very high rate. Full memory is read out in 100 μsec (microseconds), irrespective of the rate at which it was loaded. Our 0 to 10,000 Hz data now becomes 0 to 5 MHz data (a 500-to-1 speedup), since the original data sample occupied 0.05 sec. of time while the time-compressed version leaving the memory lasts only 0.0001 sec. The "recirculation" can be done in either of two ways. One can permanently capture a certain time interval and "replay" this fixed data sample over and over again by shifting the data through the memory, thus filling the "empty" locations at the beginning with data "pushed out" of the memory at its output. This technique *must be used* to analyze transients, which occur only once. For continuous signals, a continuously refreshed or updated spectrum can be calculated by discarding the oldest data point in storage when the memory is full and then entering the newest incoming sample in its place.

Since the actual narrow-band-pass filtering is done by an analog filter, the speeded up digital data is reconverted to analog by a digital-to-analog converter. The output of this device (if displayed on an oscilloscope) would look just like the original input record, except it would occur much faster. A heterodyne technique (also used in many "slow" spectrum analyzers) allows use of a single fixed-band-pass filter for examination of the entire range of frequencies. Quartz crystal filters (with center frequency of 18 MHz and bandwidth of 0.01 MHz) are typically used. The "trick" (heterodyning) consists of multiplying the data signal with a sine wave whose frequency ω_c can be accurately and quickly stepped through a range of values. This amplitude modulation (see pp. 161–173) shifts any frequency content of the data signal at, say, frequency ω_s, to $\omega_c + \omega_s$. For example, to find the frequency content at 1 MHz in the speeded up signal (2,000 Hz in the real signal), the oscillator is set at 17 MHz, thus shifting the 1 MHz data to 18 MHz, the filter's pass-band frequency. The filter output is now proportional to the frequency content around 1 MHz and is demodulated and low-pass filtered to get a "d-c" voltage which can be read on a meter or plotted. Any desired frequency within the 0 to 5 MHz range can be examined in this way by setting the oscillator frequency. Due to fast filter-settling time, a single frequency can be measured in one recirculation of the memory (0.0001 sec.); thus a frequency spectrum with 500 points is calculated and displayed on an oscilloscope in 0.05 sec. by stepping the oscillator through 500 frequencies.

This type of spectrum analyzer uses a constant-bandwidth (*not* constant-percentage bandwidth) filter. The filter bandwidth relative to the real (*not* speeded up) data is the crystal filter bandwidth (10,000 Hz) divided by the speed up ratio. For the 0 to 10,000 Hz range the bandwidth is thus 20 Hz, while the 0 to 10 Hz range has a 0.02 Hz bandwidth. These narrow bandwidths allow resolution of any closely spaced peaks that might be present in the frequency spectrum. When *random* signals are being frequency analyzed, a spectrum computed from a single 1,500-point data sample will not be very reliable statistically, and successive spectra exhibit wide variation. If the random data is essentially stationary, a true spectrum exists and can be obtained by *averaging* many individual spectra. Real-time analyzers are available with accessory or built-in averagers to perform this function. Reasonable statistical quality requires averaging of about 64 spectra; this takes 3.2 sec. for data in the 0 to 10000 Hz range and 3,200 sec. for the 0 to 10 Hz range.

Another approach to real-time spectrum analysis utilizes a completely digital computer method based on the fast Fourier transform (FFT) computing algorithm. Here the analog data is again digitized and put in memory, but now the spectrum is *calculated* from the sampled time function rather than *measured* by physical filtering. The keys to the practicality of this approach are the availability of powerful and inexpensive minicomputers and the invention of the FFT algorithm.[1] Page 151 shows the definition of the Fourier transform. When the time

---

[1] Special Issue on FFT: *IEEE Trans. on Audio and Electroacoustics*, vol. AV-15, No. 2, June, 1967.

function $q_i(t)$ is given by sampled data points, various numerical methods for computing the frequency spectrum $Q_i(i\omega)$ are possible. The FFT algorithm is a clever invention which performs the needed calculations much faster than the obvious "brute-force" approach, allowing "real-time" analysis. Due to the flexible programming possible in a digital computer, once the time function is sampled and stored, a wide variety of useful calculations such as averaging, amplitude distribution function, correlation functions, and system transfer function (which requires simultaneous measurement and processing of the input $q_i(t)$ and output $q_o(t)$ of system) are possible.[1]

Annotated Bibliography

In this revised edition, the bibliographies at the end of each chapter were there augmented and updated using the space available. The bibliography section of Chap. 14 will be used to highlight a few selected topics that deserve special attention because of their unique character and/or practical importance. Each entry includes a brief description to help the reader judge the item's significance.

General

1. *Biomedical Instrumentation.* Almost all physical variables (flow, pressure, temperature, voltage, etc.) occur and are of medical interest in living organisms. Basic principles of transducer and measurement-system design as presented in this text are directly applicable to this area of application, just as they are to any other. To provide background on some of the problems and techniques peculiar to this area, the following brief bibliography is offered.

L. Cromwell et al.: "Biomedical Instrumentation and Measurements," Prentice-Hall, Englewood Cliffs, N.J., 1973.

L. A. Geddes and L. E. Baker: "Principles of Applied Biomedical Instrumentation," John Wiley & Sons, New York, 1968.

F. Alt (ed.): "Advances in Bioengineering and Instrumentation," Plenum Press, New York, 1966.

Biomedical Measurements, *Measurements & Data*, January/February, 1969.

J. A. Webb Jr. and V. D. Gebbern: Design of a Specialized Computer for On-Line Monitoring of Cardiac Stroke Volume, *NASA Tech. Note* D-6658, 1972.

---

[1] Time/Data Analyzer, General Radio Co., Concord, Mass.

L. J. Little: The Design and Analysis of a Human Body Motion Measurement System, *US Army Missile Command Dept.* RG-72-19, Redstone Arsenal, Ala.

R. T. McDonald: Development of Respiration-Rate Transducers for Aircraft Environments, *NASA Tech. Note* D-4217, 1967.

J. L. Fanton: A Computer-Aided Hospital System for Cardial Catherization Procedures, *Hewlett-Packard J.*, January, 1972.

J. Gershon-Cohen et al.: Medical Thermography, *The Radiologic Clinics of N. America*, vol. 3, No. 3, December, 1965.

2. *Fluidic Sensors.* Fluidics refers to a technology, development of which began around 1960, that utilizes fluid effects (without moving mechanical parts) to accomplish a wide variety of measurement and control functions. While the technology has not on the whole lived up to the original glowing claims, certain useful devices and systems have been produced, and development continues. We here list a few references aimed mainly at the measurement applications.

K. Foster and G. Parker: "Fluidics: Components and Circuits," John Wiley & Sons, New York, 1970.

H. L. Fox: Direct Fluidic Sensors, *Inst. Technology*, September, 1967.

G. P. Wachtell: Fluidic Vortex Angular Rate Sensor, *USAAVLABS Rept.* 70-25, Fort Eustis, Va., 1970.

B. B. Beeken: Acoustic Fluidic Sensor, *Inst. and Cont. Sys.*, February, 1970.

W. W. Kaniuka: Fluidic Angular Acceleration Sensor, *NADC-AM-6919*, Naval Air Dev. Center, Johnsville, Pa., 1969.

T. W. Bermel: Fluidic Noncontact Sensors, *Automation*, August, 1969.

R. B. Bailey: A Study of All-Fluid High-Temperature Sensing Probes, *NASA CR*-90092, 1967.

3. *Mechanical Signature Analysis.* The use of various signal-analysis techniques (frequency-spectrum analysis, etc.) on machinery vibration signals to detect and predict incipient failures.

R. L. Bannister: Signature Analysis of Turbomachinery, *Sound & Vib.*, September, 1971.

4. *Nondestructive Testing.* An extremely wide range of measurement and evaluation techniques useful for determining properties of, or detecting defects in, materials and products.

A. Vary: Nondestructive Evaluation Technique Guide, *NASA SP*-3079, 1973.

Motion Measurement

1. C. Menadier: The Fotonic Sensor, *Instruments and Control Systems*, June, 1967. (A noncontacting fiber-optic displacement sensor with micrometer sensitivity and 100-kHz frequency response.)

2. Videostrobe System. Unilux Inc., Hackensack, N. J. Real-Time Troubleshooter, *Mechanical Engineering*, February, 1973. (Combination of strobe light and videotape to provide functions of high-speed cinematography with convenience of "instant replay.")

3. Holography. "Holography and Optical Filtering," *NASA SP*-299, May, 1971. (Laser-based optical method for motion measurements over entire surface of an object.)

4. Moire Fringes. A. L. Browne: Fluid-Film Thickness Measurements with Moire Fringes, *General Motors Res. Lab.*, *GMR*-1186, March, 1972. (Optical method for measuring surface contours of objects, or fluid-film thickness.)

5. Fatigue Life Strain Gage. J. Dorsey: Engineering Concepts in Fatigue Life Gage Use, *AN*-127 *Micro-Measurements*, Romulus, Mich., 1971. (Electrical resistance of gage is related to accumulated fatigue loading.)

6. Electro-Fluidic Rate Sensor. Humphrey Inc., San Diego, Calif., 1972. (Replacement for rate gyro, using piezoelectrically pumped gas stream and hot-wire flow sensor.)

Pressure

1. R. L. Gupta: Dynamic Compensation of Cardiac Catheters, *ASME Paper* 73-Wa/Bio-15, 1973.

2. W. M. Shay: The Dynamic Effects of a Pressure Transducer Due to Recessing, *Univ. of Calif.*, *Livermore Lab Rept.* UCRL-51156, 1971.

3. J. S. Hilten: A Simple Hydraulic Sinusoidal Pressure Calibrator, *NBS Tech. Note* 720, 1972.

4. J. T. Brock: "Acoustic Sound Measurements," Bruel and Kjaer Instruments, Cleveland, 1971.

Flow

1. R. R. Dowden: Fluid-Flow Measurement, A Bibliography, *British Hydromechanics Res. Assoc.*, Cranfield, Bedford, England, 1972. (Extensive coverage of the years 1950–70 with significant earlier papers also included.)

2. Vortex-Shedding Flowmeter. Eastech Inc., S. Plainfield, N.J. (A bluff-body inserted transversely across the pipe creates flow oscillations whose frequency, detected by buried hot-film sensors, is linearly related to volume flow rate.)

Temperature

1. S. S. Fam: Ultrasonic Thermometry, *Inst. and Cont. Sys.*, October, 1969. (Use of the velocity of sound as a means of temperature measurement.)

2. Fluidic Sensors. R. G. Bailey: A Study of All-Fluid High-Temperature Sensing Probes, *NASA CR*-90092, 1967. (Frequency of fluid oscillator is related to temperature of fluid.)

3. Bibliography of Temperature Measurements, 1953–1969, *NBS SP*-373, 1972.

4. J. Salzman: Determination of Gas Temperatures from Laser-Raman Scattering, *NASA Tech. Note* D-6336, 1971.

5. G. E. Glawe: A Small Combination Sensing Probe for Temperature, Pressure, and Flow Direction, *NASA Tech. Note* D-4816, 1968.

6. G. E. Glawe: A Cooled-Gas Pyrometer for Use in Hypersonic Engine Testing, *NASA TM* X-2715, 1973.

Computing Devices, Data Systems

The most significant development in this general area is probably the availability of powerful, small, and low-cost digital minicomputers and associated interfacing and peripheral equipment. Since these machines are programmable, they can be adapted to almost any computing or control task. Once the system is put into service, errors, oversights, and/or improvements may be easily accommodated by changes in software (programming), rather than by requiring expensive and time-consuming hardware changes. The literature of this area is voluminous; consult items 17–24 of Chap. 13's bibliography for a start.

APPENDIX A

National Bureau of Standards Accuracy Charts

Figures A-1 through A-7 (taken from *NBS Technical Note* 262, "Accuracy in Measurements and Calibrations," 1965) show the limits of accuracy attainable under ideal laboratory conditions for several basic physical quantities. Technical Note 262 is the most recent such compilation available from NBS in 1975. A recent redefinition of mission for NBS has caused them to concentrate more on the problems of improving practical measurement capabilities, and less on refinements in the ultimate standards. They thus have chosen, at this time, not to provide an updated version of TN 262.

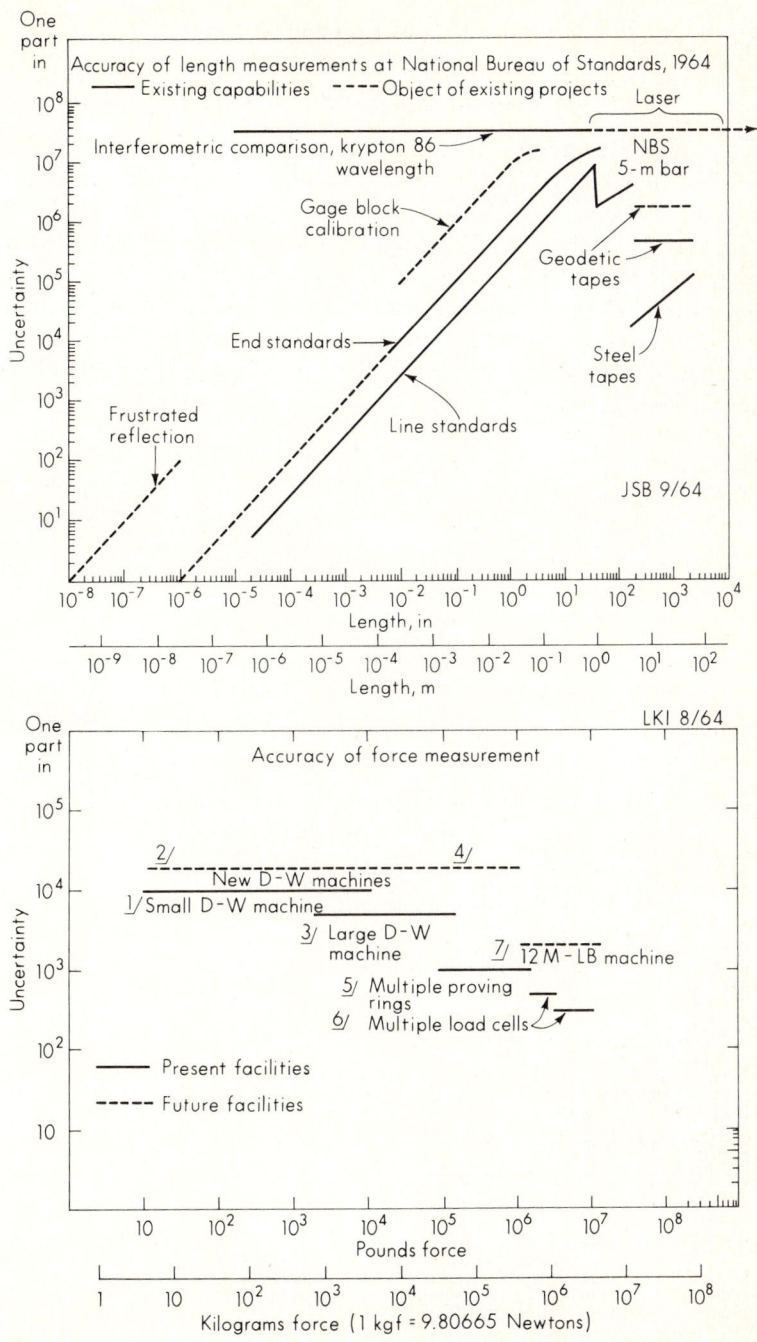

Fig. A-1.

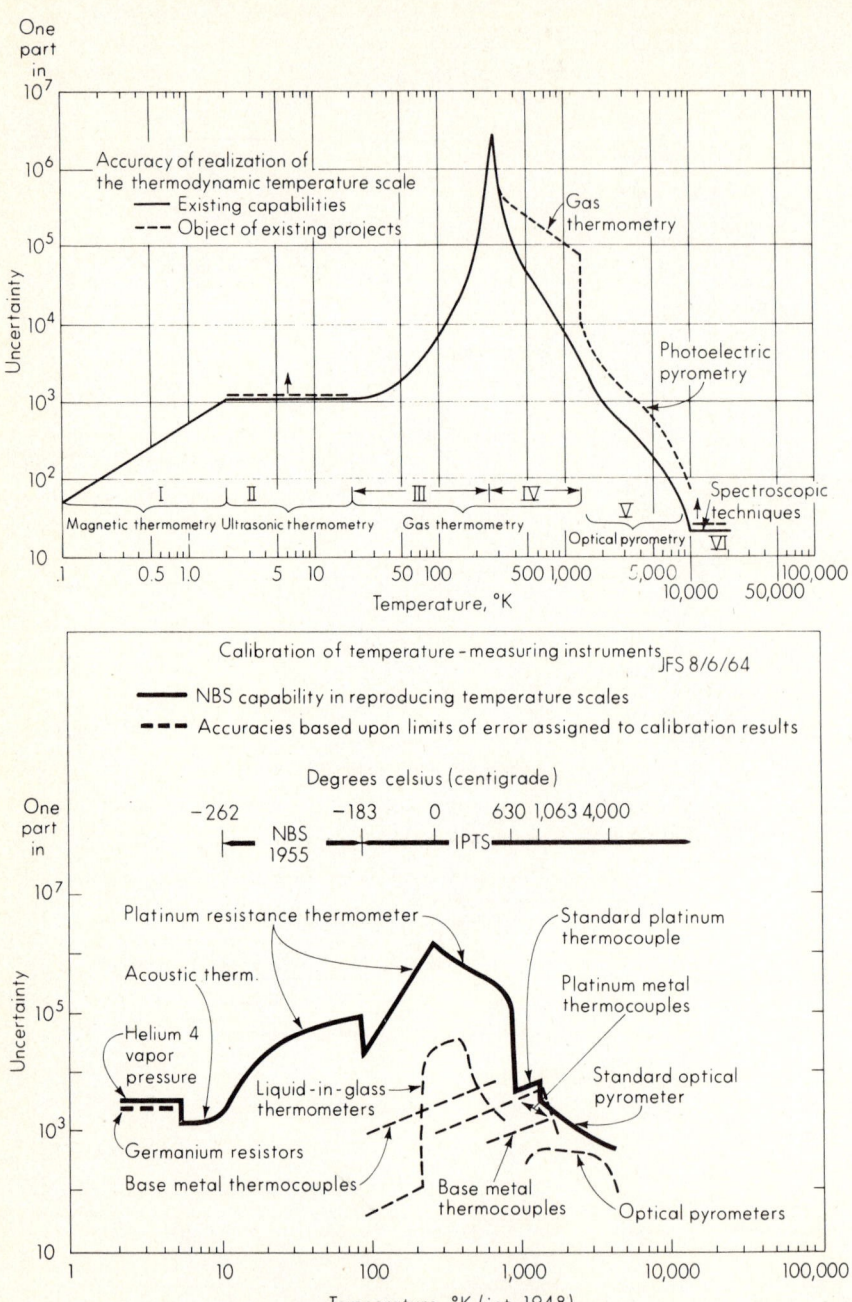

Fig. A-2.

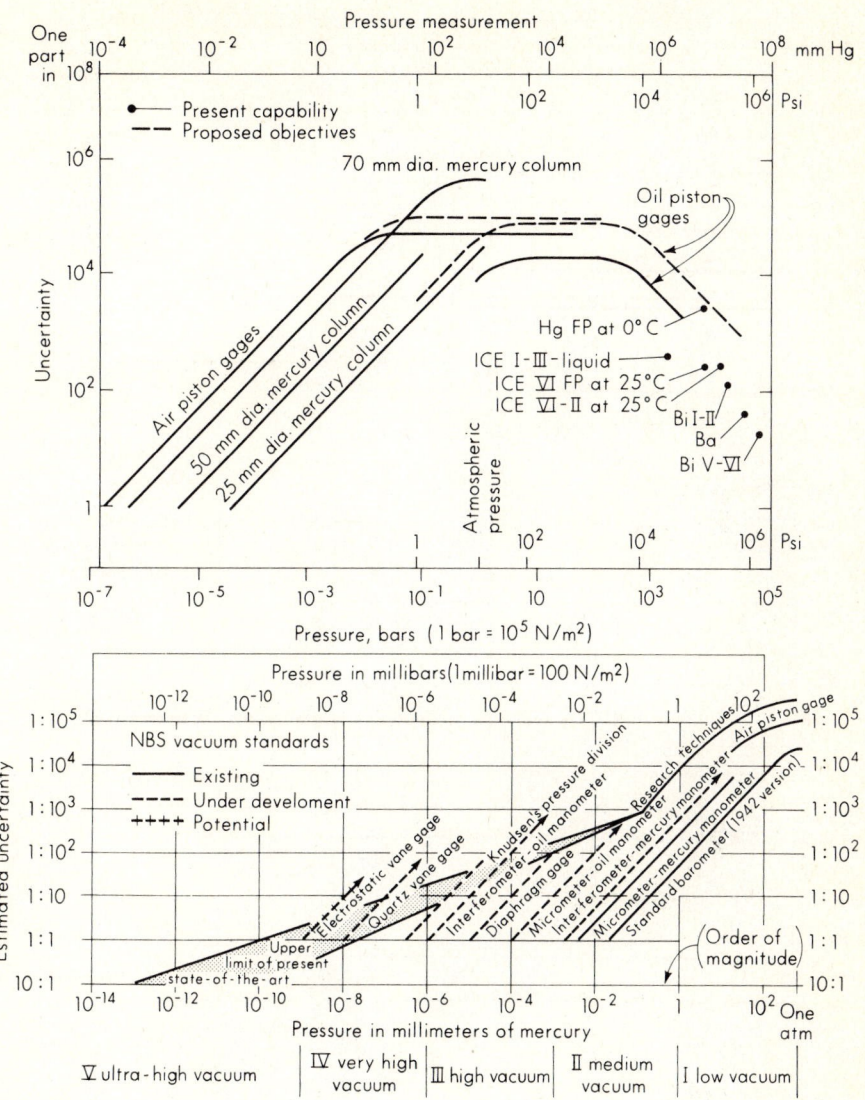

Fig. A-3.

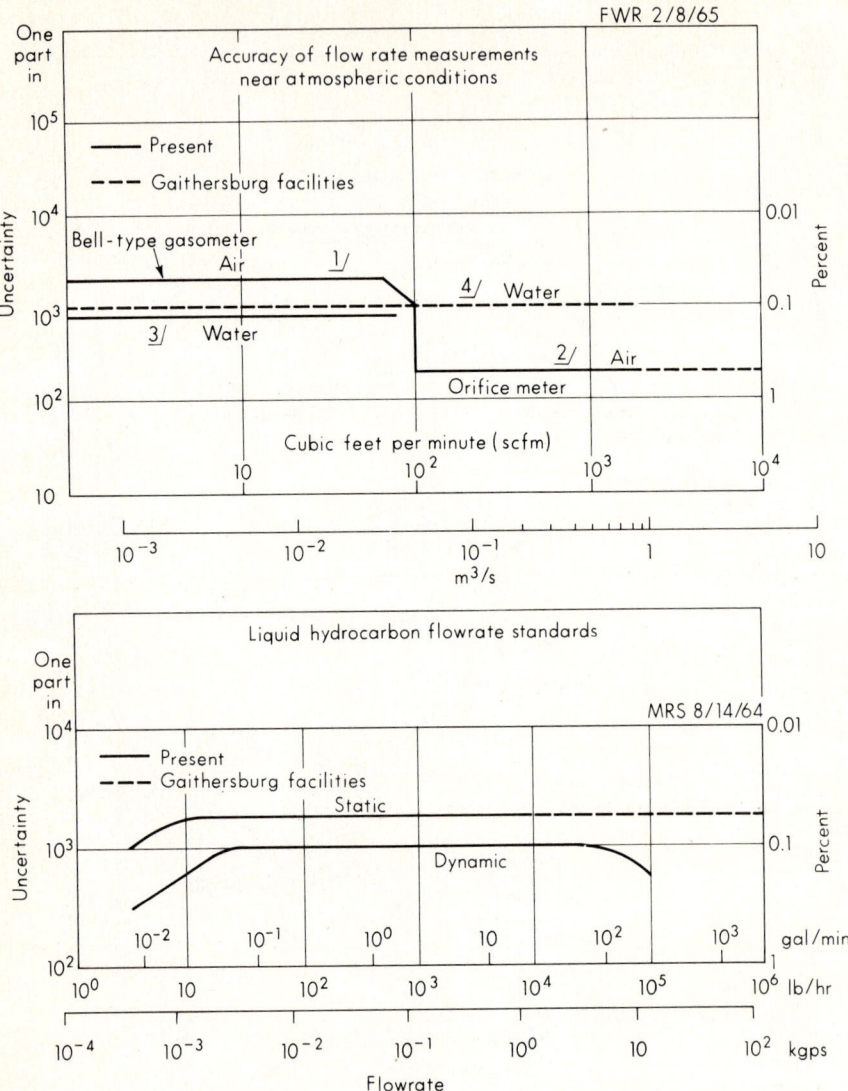

Fig. A-4.

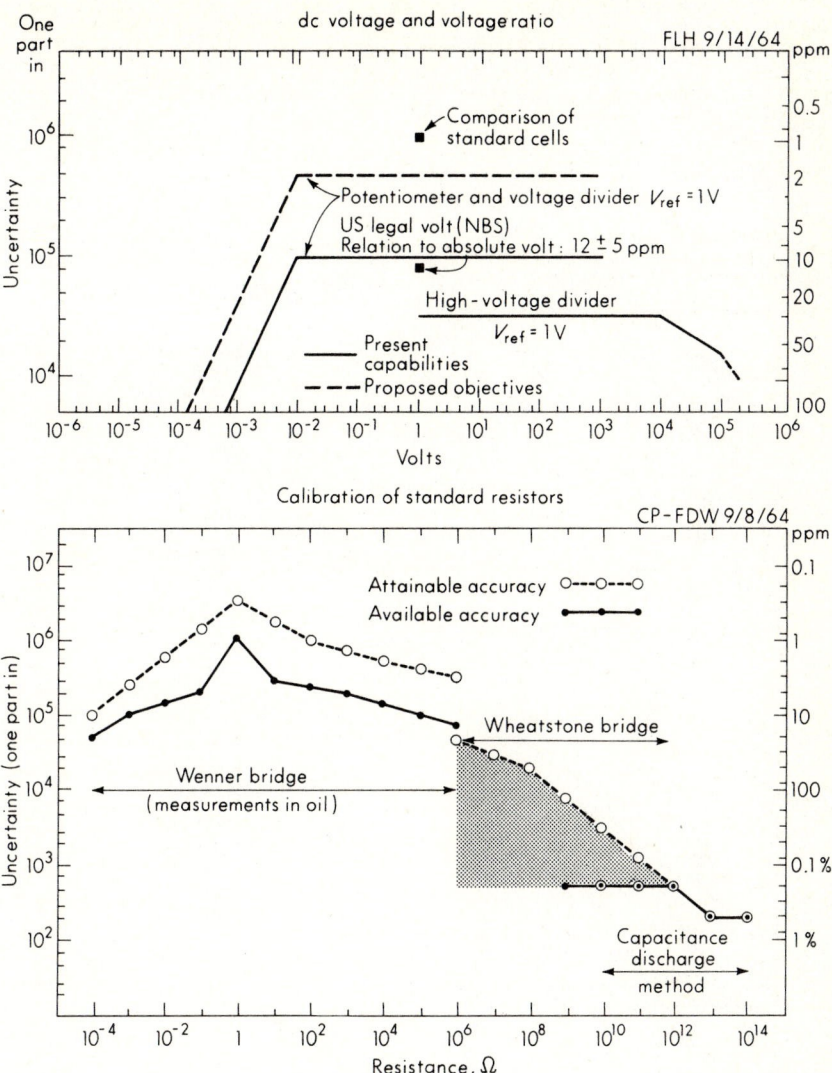

Fig. A-5.

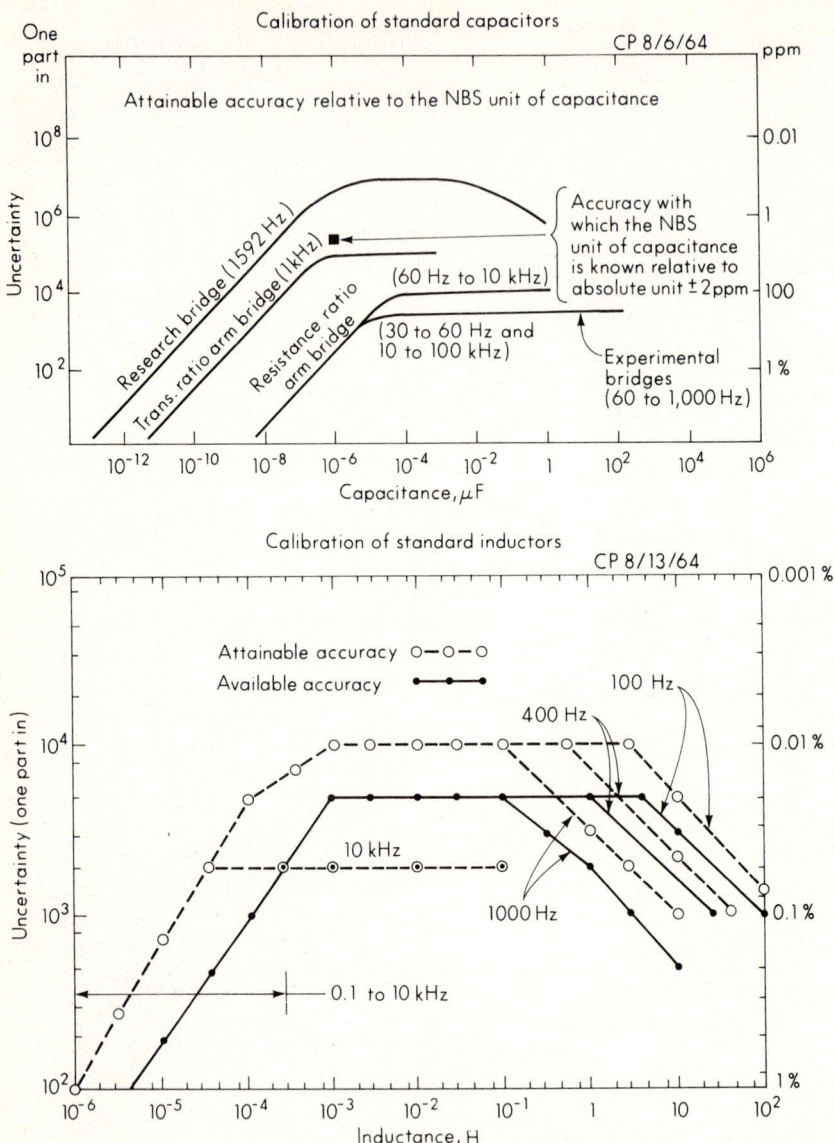

Fig. A-6.

Improvements in the accuracy of the US frequency standard
(USFS)

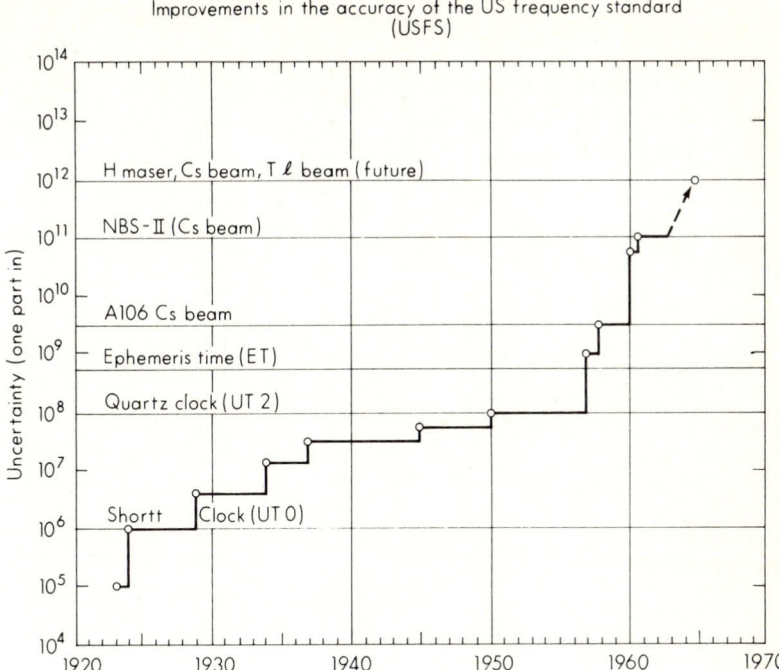

Time interval

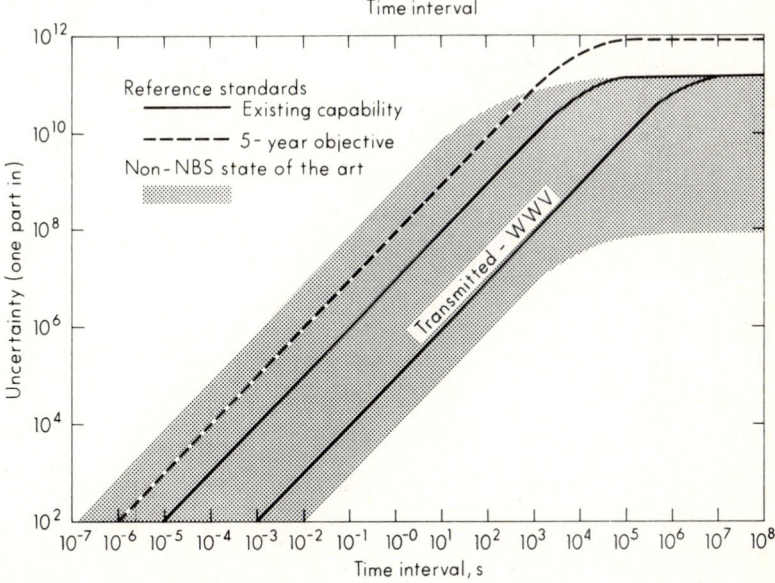

Fig. A-7.

APPENDIX B

British to Metric (SI) Conversion Table

Multiply	By	To Get
Acceleration		
foot/second^2	0.3048	meter/second^2
Area		
foot^2	0.09290304	meter^2
Density		
slug/foot^3	515.379	kilogram/meter^3
lb$_m$/foot^3	16.018463	kilogram/meter^3
Energy		
Btu (mean)	1055.87	joule
calorie (mean)	4.19002	joule
foot-lb$_f$	1.3558179	joule
Force		
lb$_f$	4.4482216	newton
Length		
foot	0.3048	meter
Mass		
lb$_m$	0.453592	kilogram
slug	14.593903	kilogram
Power		
horsepower	745.6999	watt (1 watt $\triangleq$ 1 joule/sec)
Pressure		
inch of mercury (32°F)	3386.389	pascal (1 pascal $\triangleq$ 1 newton/meter^2)
inch of water (39.2°F)	249.082	pascal
lb$_f$/ft^2	47.88026	pascal
lb$_f$/inch^2	6894.7572	pascal

Temperature

°Kelvin = (5/9)(°Fahrenheit + 459.67)
°Celsius = (5/9)(°Fahrenheit − 32)

Index

Index

WORLD POLITICS SINCE 1945

WORLD POLITICS SINCE 1945

SEVENTH EDITION

PETER CALVOCORESSI

LONGMAN
LONDON AND NEW YORK

Addison Wesley Longman Limited
Edinburgh Gate
Harlow, Essex CM20 2JE, England
and Associated Companies throughout the world.

*Published in the United States of America
by Longman Publishing, New York*

First published 1968
Second edition 1971
Third edition 1977
Fourth edition 1982
Fifth edition 1987
Sixth edition 1991
Seventh edition 1996
ISBN 0 582 27796-5 PPR

British Library Cataloguing in Publication Data

A catalogue record for this book is
available from the British Library

Library of Congress Cataloging-in-Publication Data

Calvocoressi, Peter.
    World Politics since 1945 / Peter Calvocoressi. -- 7th ed.
        p.   cm.
    Includes bibliographical references and index.
    ISBN 0-582-27796-5 (PPR)
    1. World Politics--1945- I. Title.
D843.C25 1996
909.82--dc20

                                                                      96-21988
Printed in China   CIP

GCC / 01

CONTENTS

LIST OF MAPS

FOREWORD TO THE SEVENTH EDITION

With this seventh edition this book covers fifty years. The main additions in this edition encompass the immediate aftermath of the disintegration of the Soviet empire in Europe and the Soviet Union itself; the convulsive collapse of the Yugoslav federation; the Gulf War against Iraq and its various sequels; the Maastricht Treaty for the furthering of European union; the tensions in China between economic liberalization and the stringencies of communism in the penumbra of Deng Xiaoping's prolonged rule from the edge of the grave; mixed fortunes in Africa from optimism in South Africa, Ghana and even Angola to grimness or worse in Nigeria, Somalia, Rwanda and Liberia; thin times for the UN, the expiring GATT and internationalism generally.

This is the half century of the Cold War. It begins with the total subjugation of Germany and Japan. For some, the Second World War was an ideological clash which ended with the defeat of fascism and in this perspective it cleared the way for a complementary war between democracy and communism. But more correctly the Second World War was a war against the extra-national ambitions of Germany and Japan whose defeat exalted the power of the United States and the Soviet Union to such a degree that they were dubbed superpowers – a new political category. The Cold War was the clash, ideological in expression, material in substance, between these superpowers.

The ambitions of Germany and Japan were nationalist and predatory but they were also part of an inescapable current compelling the nation state into aggression since the state was no longer a self-sufficient entity and was driven by necessity as well as emotions to extend its range beyond its borders. The United States and the Soviet Union in their turn were not exempt from this necessity although they differed in the means by which they chose to comply with it. The Cold War was about power in the world beyond the borders of the protagonists but it was not a war by either one to take territory from the other.

The Cold War was an immensely confusing phenomenon. It was waged with a rhetorical ideological intensity which was largely rubbish and with weapons which were both terrifyingly and unprecedentedly destructive and in strictly military terms useless. Further, it was assumed that the two superpowers were not only in a class of their own but were also roughly equals: which was never true. Although the armoury of the Soviet Union was in one or two sectors superior and in others sufficient to counter the offensive capacity of the United States, the overall balance between the superpowers was decisively advantageous to the United States which was immeasurably stronger in the crucial economic strength needed to create, sustain and develop the weapons and in the political and economic skills needed to manage the complexities of a modern state. Appearance and reality were uncommonly at odds.

The Cold War was rooted in a distrust which was magnified by misunderstandings and miscalculation into a great fear. This exaggerated fear was the true begetter of the Cold War. The United States and the USSR were profoundly divided in political and economic philosophy but neither intended to make war on the other although each feared that the other did. These fears were irrational. The United States feared further westward advances by the Soviet forces after the German collapse and successful communist subversion of western European states. In fact the Soviet forces were exhausted and the Soviet Union itself in ruins; and no western state was anywhere near the verge of a communist takeover whether by elections or subversion – and had it been so its national army would quickly have extinguished any attempt at a coup. On its side the USSR, in 1945 and for many years to come, feared a concerted western attempt to destroy it. Stalin rightly perceived strong western animosity but was entirely wrong in supposing that the United States and its allies would resort to war against him or could devise some other means to loosen his hold on his newly acquired satellites in central and eastern Europe. The Cold War was an expression of deep antagonism in the realms of ideas and behaviour, unaccompanied however by territorial disputes and conducted therefore as a slanging match. And since it was not territorial it was potentially global.

Nuclear weapons added to the confusion. It has frequently been said that the possession by both sides of nuclear weapons in what was believed to be roughly equivalent force prevented the Cold War from becoming something worse: that each superpower was as much afraid of the weaponry as of its enemy. Clearly, nuclear weapons were peculiarly frightening and to that extent a deterrent – a deterrent greatly enhanced by the much advertised risks of victory only at the cost of self-destruction. But no weapon is dangerous until somebody uses it and it can be argued that nuclear weapons were never likely to be used by the superpowers against one another. The earliest nuclear weapons were weapons of mass destruction and intimidation and for a short time the United States alone had them or could make them. The United States used them against Japan and later pondered their use against China but their use against the USSR in the first postwar years was inconceivable. When the USSR too acquired them the risks of mutual mass maiming forced both sides into developing more and more accurate weapons capable of hitting more narrowly selected targets – medium-range weapons and then tactical or battlefield weapons – in order to make the use of nuclear power in war rational. But the attempt failed. No field commander likes weapons which contaminate the zone into which he proposes to advance and which are more likely to produce anarchy than victory (anarchy being the one thing which scares leaders as much as defeat). Since the deterrent effect of a lesser nuclear weapon depends on the implicit threat to use the next larger one in the chain, the threat to use the lesser depends on the threat ultimately to use the larger which the lesser was brought into being to prevent. The development of a chain of nuclear weapons did not eliminate the bluff which, although present in most international manoeuvring, was in the case of nuclear weapons too large for their credibility. In the end the Cold War destroyed not the living space or the peoples of the superpowers but their economies.

The predominance of a single issue such as the Cold War obscures or distorts other issues. Three of these had specially wide implications. The Manichaean clash of the superpowers stifled the fragile mechanisms of international order. The Cold War began at the same time as the UN came into existence as the new and hopefully improved version of the League of Nations and it immediately incapacitated the UN which was designed for a different world – for a world in which disputes between states were deemed to be, if inevitable in some measure, nevertheless capable of being minimized, managed, sanitized. The impact of the Cold War was not simply to castrate the Security Council by the use of the veto by its principal members. The Cold War made the UN largely irrelevant and inoperative in any issue which might be interpreted as an aspect of the Cold War, so that the wonder was not that the UN was diminished but that it survived. It was, however, much damaged during the Cold War and resurfaced in the last decade of the century in a world which, unlike the world of 1945, placed no high hopes in it. The conduct of international affairs by peaceful means and rational discourse was heavily discounted and the United States, the surviving superpower, emerged from the Cold War uncertain how far to intervene in world affairs as *primus inter pares* and how far as despot.

The crises in Somalia and the Gulf exposed these uncertainties and the war in Bosnia tipped the balance, at first hesitantly and then smartly, away from international co-operation and towards a more effective national assertiveness tempered only by a *zeitgeistlich* figleaf. The United

States moreover was hampered by a legacy of the Cold War which had atrophied thinking about anything except the Cold War: it did not have policies for this or that part of the world but only policies for combating the USSR there. With the end of this determinant the United States was short of policies relevant to a world without a Cold War.

A second feature of the late twentieth century which was obscured by the Cold War was the problem of access by one state to raw materials lying within the borders of another state but essential to the wellbeing of the first state: pre-eminently oil. The surest way of securing these commodities – occupation of the valuable lands – was no longer respectable or reliable. Colonies, even mandates, were tabu but the determination of powerful states to sustain their economies and citizens remained. In this half century Middle East oil passed out of the legal possession of foreign states or corporations. This was a shift in the balance of economic power but there was no comparable shift in the balance of military power: the latter shifted only in the sense that it moved off-shore, not in the sense that the strong became relatively less strong. When in 1990 Iraq invaded Kuwait with the aim of annexing the latter's oilfields and bank balances, Saddam Hussein did two distinct things. He committed an act of aggression in breach of the UN Charter and he threatened to upset the balance of power and production in Middle East oil. His aim alarmed purchasers of this oil whose flow and price had until recently been under their control. By his aggression he gave the UN just cause for war and by upsetting the oil régime he spurred the United States into enlarging the war into a war to overthrow him and his régime. The war demonstrated that the borders of a state containing internationally valuable commodities were peculiarly liable to be violated, be that state Kuwait or Iraq; but it did nothing to illumine the problem of managing without war the relations between a state with such commodities but inferior force and a state in need of the commodities but with no right to them: a conundrum not in the unequal distribution of power but in the distribution of incommensurable powers.

The third question of special perplexity affected the state itself. In much of the world – but far from all of it – the state had become the undeniably prime ingredient in the international fabric. It was a European artefact which European empires had prevented from taking shape in other parts of the world until the retreat of these empires from Asia and Africa allowed other peoples to copy the European pattern. Shortly after mid-century states suddenly became much more numerous. Yet the new states acquired the lesser attributes of statehood – national anthems, central banks – without sufficient awareness of the weaknesses which had always plagued older states: ethnic diversity in what were misleadingly called nation states, economic insufficiency, military weakness, institutional weakness. At the same time older states which enjoyed the manifest benefits of statehood – geographical and legal definition, loyalty, settled government, welfare provision – were casting doubt on the adequacy of their condition and experimenting with superstatal arrangements (the European Union, ASEAN, a plethora of economic associations from free-trade areas to economic unions). In the last quarter of the century almost the entire population of the globe was living in states but almost all of these states belonged to one or more international organizations, beginning with the UN where the newer states in particular stationed their most senior diplomatic representatives. These international organizations were controlled by states through their executive bodies – the UN Security Council, the EU's Council of Ministers, etc. – whose thinly veiled dominance and financial contributions reflected the continuing primacy of the state. No less persistent, however, were the failings of the nation state as exemplified in eastern Europe, in Somalia and Liberia and Rwanda, and in Sri Lanka and Burma and elsewhere. These atrocious encounters were not new but their tools were new and so was their universal visibility on television and in the press. Even in Europe, the country of their birth and maturing, many states failed their citizens in wars and other upheavals throughout the century.

The general outcome has been increasing dissatisfaction with the state without its dethronement. International affairs remain very largely the affairs of states in an international context: states remain the essence, international is the adjectival additive. The 'international community' which comes too readily to the lips of politicians does not exist. The UN is only the first worldwide organization in history, for earlier organizations from the post-Napoleonic Concert of Europe to the League of Nations were regional bodies with limited membership, limited authority, limited aims and no power. Their scope before 1945 grew from Europe itself to Europe plus Latin America and not much more: a journey from Vienna to Geneva. They were collective experiments in the management of states' conflicts by diplomacy, arbitration and the elaboration of international law. The UN began as a similar organization. It was transformed soon after its inauguration by the emergence of dozens of new sovereign states all over the world; but it had little more power (if any) than its predecessor and considerably more cultural diversity. This cultural diversity is a principal obstacle to its being accorded more power or any automotive force. The culture which it seeks to spread is pacific, co-operative, rational, legalistic and (to a degree rarely admitted) based on the European tradition and European understanding of these terms.

By contrast the history of the world, including Europe, since 1945 demonstrates that the world is a patchwork of rampant warrior cultures.

In this imbroglio one European experiment was suggestively different. Regardless of where its bounds may be set, the European Union is regional and can never be global. It embraces an area which by world standards is comparatively small and possesses an appreciable cultural homogeneity (although that homogeneity is bound to be attenuated by the Union's extension from west to east). It bids fair to become more active and effective in European affairs than the UN and, significantly, its members have by the Treaty of Maastricht cautiously broached the question of giving the Union some military power. J. B. S. Haldane once wrote an essay on the 'Importance of Being the Right Size'. What applies in biology may also apply in world affairs. The Roman empire was by modern standards quite small, about the size that the European Union is likely to be.

The post-Cold War world was more than a world without a Cold War. I cite three examples. The disintegration of the Soviet Union posed puzzles about the capabilities of a Russian state, much the most powerful of the Soviet Union's successor states, but Russia was for the time being a less immediate puzzle than China, the oldest and largest imperial state in the history of the world. Brought to the verge of dismemberment more than once in the nineteenth and twentieth centuries, China's huge expanse and population, its renewed coherence and apparent self-confidence, forecast an assertive and menacing role in international affairs. While the United States and Japan seemed to have inherited the earth, China covered much more of it.

Second, the post-Cold War world was not a world purged of nuclear weapons. The Cold War superpowers had negotiated limits to their nuclear armouries and the post-Cold War superpowers seemed less inclined to conduct their relations in the nuclear mode: Japan indeed still had no nuclear weapons. But these weapons, or the power to make or acquire them, were no longer limited to superpowers. There were signs of restraint in Latin America where Brazil and Argentina were mutually distrustful and capable of constructing at least a nuclear capacity, but in the Middle East and southern Asia capacity was alive and spreading and restraint was inconspicuous. Nuclear proliferation was no longer a threat but a fact, and the threat of what was termed a regional nuclear conflict was the more alarming since such a conflict was a contradiction in terms: a prospective nuclear war anywhere must involve more than the prospective belligerents. Yet no regional international organization, nor the UN debilitated by the Cold War, was adequately equipped to moderate conflicts with an inherent nuclear threat or to manage such conflicts on the point of explosion. The world was not necessarily safer because the Cold War petered out.

Nor, third, more prosperous. The world was most truly international in economic terms – by the reciprocal benefits of commerce, by the value of transnational investment to both recipient and provider, by the political dividends of rising standards of living and orderliness, by universal revulsion against gross poverty – but it was at an inchoate stage in the conduct of economic relations between states and between corporations in different states or cultures. Economic power in natural resources and in acquired skills was most unequally distributed; the economic ground rules and the ethos of competition varied from place to place; instruments, such as the World Bank, the IMF and the WTO (successor to the GATT), suffered not only from the limitations of all international bodies controlled by a diversity of national governments but also from disagreement over macro-economic theory, strategies and priorities.

To conclude: I repeat one warning and add another. I again ask readers to be on the lookout for matters which, because they are too speculative for a work of this kind, do not figure in it: for example, the impact of nuclear weapons which, by undermining the rationale of the resort to war – by turning war into suicide unless it succeeds in being genocide – is forcing states to conduct their intercourse with one another up to and beyond the point of aggression by means other than war. How to do this is a notion not easily assimilated. My second warning is to resist the temptation to read current history as an instalment of future history, as an invitation to discern the future. It is far more important to appreciate the interdependence of the present with the past. There is necessarily not much pre-1945 history in this book (if only for reasons of space) but all of it is imbued with the past and in some sections (Bosnia, for example) a past which is both remote and present. The history of the past fifty years is the product of the past and of what people know or fail to know about it. This book is meant to be an aid to understanding things which have been going on for a long time. With all its necessary limitations and unnecessary imperfections it is an attempt to write history.

Peter Calvocoressi
January 1996

PART ONE

WORLD POWER
AND
WORLD ORDER

1 THE SUPERPOWERS

THE COLD WAR: THE FIRST STAGES

The Cold War of the two postwar superpowers was not an episode like other wars which have beginnings and ends, winners and losers. The term 'cold war' was invented to describe a state of affairs. The principal ingredient in this state of affairs was the mutual hostility and fears of the protagonists. These emotions were rooted in their several historical and political differences and were powerfully stimulated by myths which at times turned hostility into hatred. The Cold War dominated world affairs for a generation and more.

President Franklin Roosevelt had believed, or perhaps only hoped, that he could persuade Joseph Stalin not to create a separate Soviet sphere of influence over eastern Europe but to co-operate instead with the United States in creating a global economic order based on free trade and beneficial to all involved, not least to the USSR. Wartime Lend-Lease to the USSR had been a first step; the postwar Marshall Plan was to be a somewhat forlorn last hope, for even after Roosevelt's death in April 1945 there were some in Washington who preferred a policy of resuscitating western Europe without military confrontation with the USSR. But to most Americans the USSR seemed dedicated to the conquest of Europe and the world for itself and for communism and was capable of achieving, or at least initiating, this destructive and evil course by armed force abetted by subversion. On this view the necessary riposte by the United States was military confrontation in alliance with Europeans and others and on the assumption that Soviet hostility was ineradicable or at least much more than temporary. Seen from Moscow the western world (which included half of Europe as well as the United States) was inspired by capitalist values which demanded the destruction of the USSR and the extirpation of communism by any means available but above all by force or the threat of irresistible force. Both these appreciations were absurd. When the Second World War ended the USSR was incapable of further military exertion, while the communist parties beyond its immediate sphere were unable to achieve anything of significance. The western powers, while profoundly mistrustful of the USSR and hostile to its system and beliefs, had no intention of attacking it and were not even prepared to disturb the dominance of central and eastern Europe secured by its armies in the last year of the war. Each side armed itself to win a war which it expected the other to begin but for which it had no stomach and no plans.

The focus of the Cold War was Germany where confrontation over Berlin in 1948–49 came close to armed conflict but ended in victory for the western side without a military engagement. This controlled trial of strength stabilized Europe which became the world's most stable area for several decades, but hostilities were almost simultaneously carried into Asia, beginning with the triumph of communism in China and war in Korea. These events led in turn to an acceleration of the independence and rearmament of West Germany within a new Euro–American alliance and to a succession of conflicts in Asia, of which the Vietnam War was the most devastating. At no point did the protagonists directly engage each other but both sought to extend their influence and win territorial advantage in adjacent parts of the world, notably the Middle East and – after its decolonization – Africa. None of these excursions was decisive and for nearly half a century the chief outward expression of the Cold War was not advances or retreats but the accumulation and refinement of the means by which the two sides tried to intimidate each other: that is to say, their arms race. The slackening of the Cold War resulted from the combined effect of the huge cost of these armaments and the gradual waning of the myths which underlay it.

In the summer of 1945 it was known both in Washington and Moscow that Japan was ready to acknowledge defeat and abandon the war which it had begun by the attack on Pearl Harbor in 1941. In July the Americans experimentally exploded the first nuclear weapon in the history of mankind and in August they dropped two bombs on Hiroshima and Nagasaki. Japan surrendered forthwith and this clinching of the imminent American victory deprived the Russians of all but a token share in the postwar settlement in the Far East. The Second World War ended with an act which contained the two central elements in the Cold War: the advent of nuclear weapons and Russo–American rivalry.

In the European theatre this conflict remained for a short time veiled. The organs and habits of wartime collaboration were to be adapted to the problems of peace, not discarded. The Russian spring offensive of 1944 had set the USSR on the way to military dominance and political authority in Europe unequalled since Alexander I had ridden into Paris in 1814 with plans for a concert of victors which would order the affairs of Europe and keep them ordered. The nature of mid-twentieth-century great power control was a matter for debate – how far the powers were collectively to order the whole world, how far each was to dominate a sector. The Russians and the British, with the reluctant assent of President Franklin D. Roosevelt (and the dissent of his Secretary of State, Cordell Hull), discussed the practical aspects of an immediate division of responsibilities, and in October 1944, at a conference in Moscow which the president was unable to attend owing to the American election campaign, these dispositions were expressed in numerical terms. About western Europe no questions arose because western control was uncontested. Poland was not included because, in parts effectively, in parts imminently, it was under Stalin's military control. Elsewhere the realities were expressed as percentages. The Russian degree of influence in Romania was described as 90, in Bulgaria and Hungary as 80, in Yugoslavia 50, in Greece 10. In practice these figures, although expressed as a bargain, described a situation: 90 and 10 were polite ways of saying 100 and 0, and the diagnosis in the two extreme cases of Romania and Greece was confirmed when the British took control in Greece without Russian protest and the USSR installed a pro-communist régime in Romania with only perfunctory British protest. Bulgaria and Hungary went the same way as Romania for military reasons. Yugoslavia appeared to fall within the Russian sphere but soon fell out of it. Europe became divided into two segments appertaining to the two principal victors, the United States and the USSR. These two powers continued for a while to talk in terms of alliance, and they were specifically pledged to collaborate in the governance of the German and Austrian territories which they and their allies had conquered.

The position of the USSR in these years was one of great weakness. For the USSR the war had been a huge economic disaster accompanied by loss of life so grievous that its full extent was probably not disclosed. The Russian state was a land power which had expanded generation by generation within a zone which presented a persistent German threat. In the Soviet phase of history Russian external politics were further characterized by a diplomacy which led to isolation and so, in 1941, to the threshold of total military defeat. The USSR had been saved by its extraordinary geographical and spiritual resources and by the concurrent war in the west in which the Germans were already engaged before they attacked the USSR and which became graver for them when, shortly afterwards, Hitler gratuitously declared war on the United States of America. For Stalin, however, the anti-fascist alliance of 1941–45 can hardly have appeared to be more than a marriage of convenience and limited span; nor did it look any different when seen from the western end, at any rate by governments, if not in the popular view. With the war over, the purpose of the alliance had been achieved, and there was little in the mentalities and traditions of the allies to encourage the idea that it might be converted into an entente: on the contrary, the diplomatic history of all parties up to 1941 and their respective attitudes to international political, social and economic problems suggest exactly the reverse.

The elimination, permanent or temporary, of the German threat coincided with the explosion of the first American nuclear bombs. For the first time in the history of the world one state had become more powerful than all other states put together. The USSR, no less than the most trivial state, was at the mercy of the Americans if they should be willing to do to Moscow and Leningrad what they had done to Hiroshima and Nagasaki. There were reasons for supposing that they were not so willing, but no government in the Kremlin could responsibly proceed upon this assumption. Stalin's only prudent course in this bitterly disappointing situation was to combine the maximum strengthening of the USSR with a nice assessment of the safe level of provocation of the United States, and to subordinate everything, including postwar reconstruction, to catching up with the Americans in military technology. He possessed a large army, he had occupied large areas of eastern and central Europe, and he had natural allies and servants in communists in various parts of the world.

At home his tasks were immense: they included the safeguarding of the USSR against a repetition of the catastrophe of 1941–45, and the resurrection of the USSR from the catastrophe which had cost something like 25 million lives, the destruction or displacement of a large part of its industry and the distortion of its industrial pattern to the detriment of all except war production, and the devastation and depopulation of its cultivated land so that food production was almost halved. To a man with Stalin's past and temperament the tasks of restitution included the reassertion of party rule and communist orthodoxy and the reduction therefore of the prominence of the army and other national institutions and the reshackling of all modes of thinking outside the prescribed doctrinal run. In matters of national security, in economic affairs and in the life of the spirit the outlook was grim. At home artists and intellectuals were regimented, victorious marshals were slighted and the officer class persistently if quietly purged, while the first postwar five-year plan prescribed strenuous tasks for heavy industry and offered little comfort to a war-weary populace.

Externally Stalin made it clear that the protective acquisitions of 1939–40 were not for disgorging (the three Baltic states, the eastern half of Poland which the Russians called Western Ukraine and Western Byelorussia, Bessarabia and northern Bukovina, and the territory exacted from Finland after the winter war); elsewhere in eastern and central Europe all states must have governments well disposed to the USSR, a vague formula which seemed to mean governments which could be relied upon never again to give facilities to a German aggressor and which came after 1947 or thereabouts to mean governments reliably hostile to the United States in the Cold War. Such governments must be installed and maintained by whatever means might be necessary. During the wartime conferences between Stalin, Roosevelt and Churchill parts of the USSR were still occupied by German forces, and during the war and its psychological aftermath (a period of indefinable duration) Stalin was no doubt obsessed with his German problem. The Cold War first substituted the Americans for the Germans as the main enemy but then, after the rearmament of Western Germany, combined the two threats as a new American–German one. These developments, to which Stalin himself contributed by his actions in eastern and central Europe, may nevertheless have been a disappointment to him if, as seems possible, he had entertained at one time a very different prospect of Russo–American relations.

To Stalin during the war the Americans were personified by President Franklin D. Roosevelt who made no secret of his desire to get on well with the USSR or of his distrust of British and other western imperialisms. Moreover Roosevelt wanted a Russian alliance against Japan and did not seem at all likely to do the one thing which Stalin would have feared; namely, to keep troops permanently in Europe and make the United States a European power. On the contrary, Roosevelt was not much interested in postwar Europe and showed, for example, little of Churchill's concern about what was going to happen in Poland and Greece. Whereas Russo–British relations came near to breaking-point over Poland, Russo–American relations did not; and Stalin, whether out of genuine lack of interest or calculated diplomacy, avoided serious disagreement with Roosevelt over problems of world organization (such as the representation of Soviet republics in the UN) in which Roosevelt was seriously interested. On the basis that the United States would be remote from Europe and in some degree friendly towards the USSR Stalin was prepared to moderate his support for European communists in order not to alarm the United States. He did not foresee that by abandoning Greece to Britain he was preparing the way for the transfer of Greece to the United States within three years. His failure to help the Greek communists may have been principally a result of a calculation that they were not worth helping, but he may also have reflected that helping them would alarm and irritate Americans in Roosevelt's entourage. He was continuing a line of policy applied in Yugoslavia during the war when he restrained the Yugoslav communists' desire to plan and prosecute their social revolution while the war was still in progress, and urged them to

1.1   The Cold War division of Europe

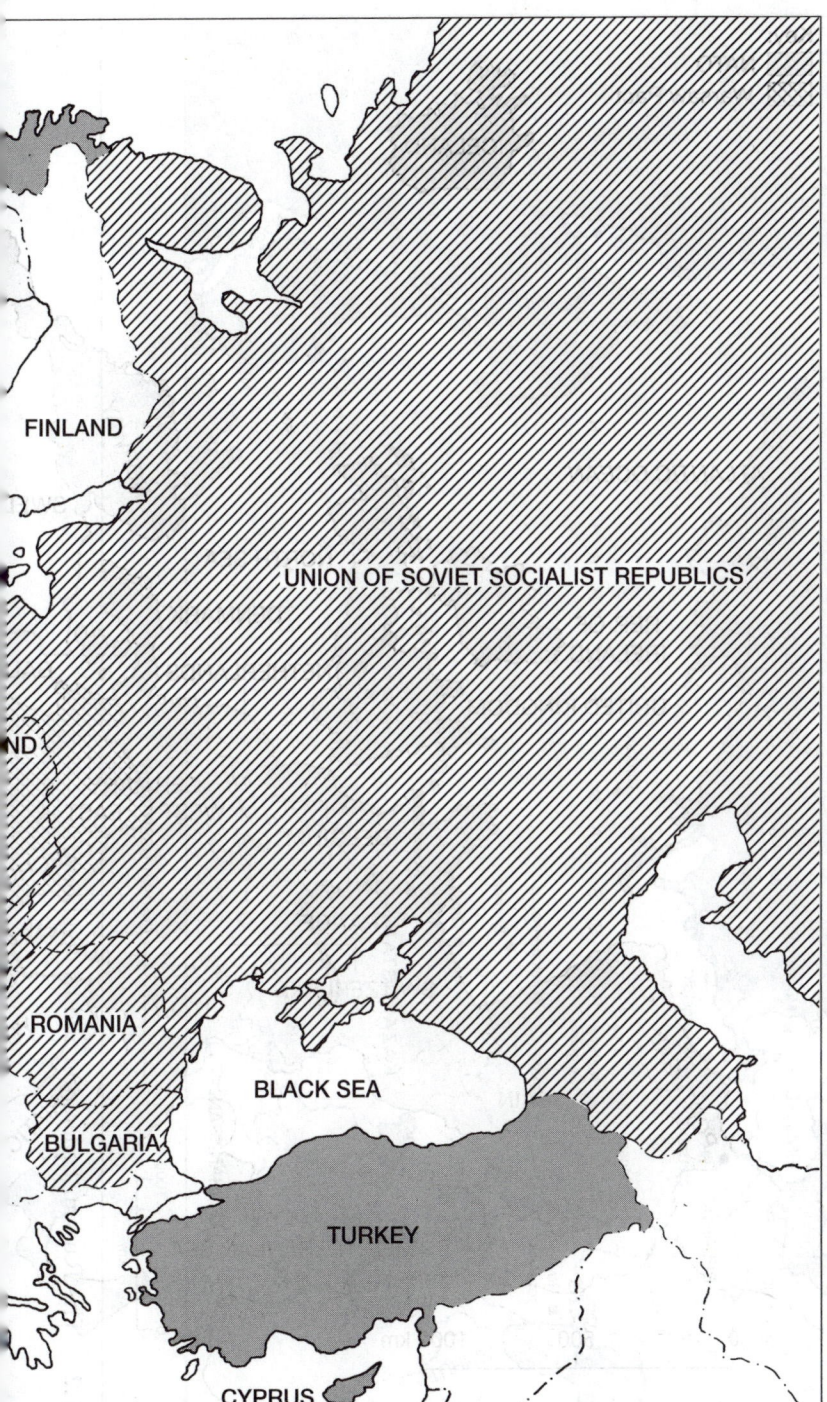

FINLAND

UNION OF SOVIET SOCIALIST REPUBLICS

ND

ROMANIA

BLACK SEA

BULGARIA

TURKEY

CYPRUS

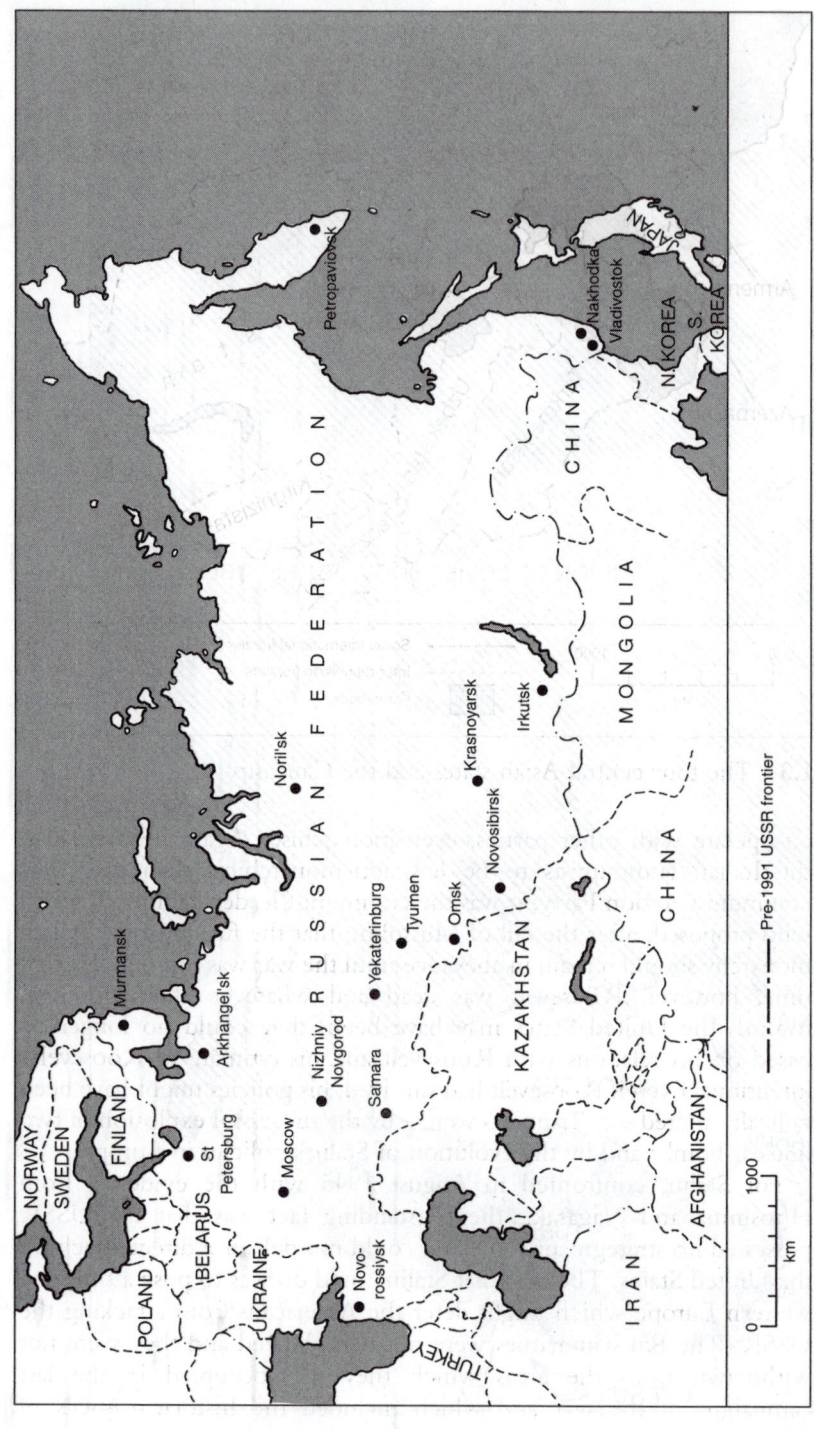

1.2   The Russian Federation and its neighbours

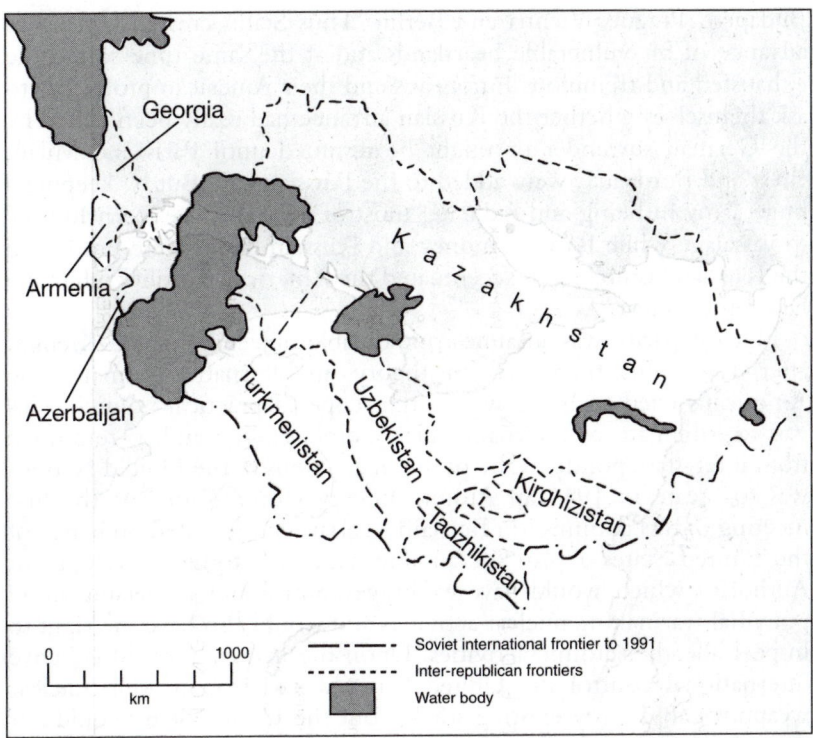

Georgia

Armenia

Azerbaijan

Turkmenistan

Uzbekistan

Kazakhstan

Kirghizistan

Tadzhikistan

| 0 | | | 1000 | |
| | | km | | |

- – - – - Soviet international frontier to 1991
- - - - - - - Inter-republican frontiers
Water body

1.3   The four central Asian states and the Caucasus

co-operate with other parties, even monarchists. Again he persuaded the Italian communists to be less anti-monarchical than the non-communist Action Party; it was the communist leader Palmiro Togliatti who proposed, after the fall of Mussolini, that the future of the Italian monarchy should remain in abeyance until the war was finished. By that time, however, Roosevelt was dead and whatever Stalin's policies towards the United States may have been, they could no longer be based on his relations with Roosevelt and his estimate of Roosevelt's intentions. Even if Roosevelt had survived, his policies might have been radically altered – as Truman's were – by the successful explosion of two nuclear bombs and by the evolution of Stalin's policies in Europe.

For Stalin, confronted in August 1945 with the evidence from Hiroshima and Nagasaki, the outstanding fact was that the USSR possessed no strategic air force and could not deliver a direct attack on the United States. The best that Stalin could do was to pose a threat to western Europe which might deter the Americans from attacking the USSR. The Russian armies were not demobilized and they were not withdrawn from the areas which they had occupied in the last campaigns of the war and which included the historic capitals of

Budapest, Prague, Vienna and Berlin. Thus Stalin created a glacis in advance of his vulnerable heartlands and at the same time scared the exhausted and tremulous Europeans and their American protectors to ask themselves whether the Russian advance had really been halted by the German surrender or might be resumed until Paris and Milan, Brest and Bordeaux, were added to the Russian bag. But by keeping a huge army in being and reducing most of central and eastern Europe to vassalage while Russian money and Russian brains were producing the Russian bomb, Stalin accentuated the American hostility which he had cause to fear.

Nuclear parity was an almost inescapable objective for the Kremlin after 1945, but there was in theory an alternative which some Americans tried to bring within the scope of practical politics. This was to sublimate or internationalize atomic energy and so remove it from inter-state politics. One of the first actions of the United Nations was to create in 1946 an Atomic Energy Commission. At the first meeting of this commission Bernard M. Baruch presented on behalf of the United States a plan for an international Atomic Development Authority which would have exclusive control and ownership in all potential warmaking nuclear activities and would also have the right to inspect all other atomic activities. Upon the inauguration of effective international control the United States would stop making nuclear weapons and destroy existing stocks. But the United States could not destroy its advanced technological knowledge and it would therefore retain a huge advantage over the USSR which, by accepting the Baruch Plan, would inhibit its own advances in nuclear physics. In addition the USSR disliked the plan because it involved the abrogation in this field of the veto, the principal symbol and guarantee of national sovereignty as opposed to international government, which the USSR was even less disposed to countenance than any other state. A. A. Gromyko proposed instead a treaty banning the use of nuclear weapons, the destruction of existing stocks, and an international control commission subordinate to the Security Council (and therefore subject to the veto); he opposed the creation of a new international authority and was only prepared to allow international inspection of those plants which had been declared to have ceased nuclear production and were offered for inspection by the government of the country in which they were situated. These positions were irreconcilable and although the debate continued for a while the UN Atomic Energy Commission eventually decided in 1948 to adjourn indefinitely. The Russian rejection of the Baruch Plan was an additional factor in persuading the administration of Harry S Truman, who had become president on the death of Roosevelt in April 1945, that the USSR was no longer an ally but an adversary.

American policy-makers had more freedom of choice than their Russian counterparts. The nuclear bomb was a military weapon which had been used to bring the war with Japan to an end in a particular way and it was also a political weapon which was used to shackle Russian power. The war left the United States a major power in Europe with considerable military forces in Europe and nuclear weapons in the background. Endowed with mighty technical superiority, which they may have regarded as permanent, the Americans were in a position in which they could strike or threaten or wait and see. To strike – to start a preventive war – was in practice impossible because they were unable to summon up the will to do so. A preventive war is a war undertaken to remove a threat by a people which feels threatened, and the Americans were not threatened by the Russians and did not feel threatened by them. In these circumstances a preventive war by the Americans was an abstract intellectual concept. (For the Russians it was substantive but also suicidal.) The Americans therefore pursued policies which combined threats with waiting to see.

All weapons have political implications and the biggest weapons have the biggest implications. A weapon which is too frightful to use – or turns out to be, in military terms, useless except in the most exceptional circumstances – has the most implications, since its possessors will want to put it to political use in order to compensate for the anomalous limitations on its military usefulness. Truman's position was automatically different from Roosevelt's as soon as Hiroshima had been destroyed. The question was not whether he was to make political use of the new weapon, but to what political end he should use it. The relationship between the United States and the USSR had been altered: how should this alteration be exploited? The context in which this question first arose was not Asian but European, for it was in Europe that the principal political issues were arising. The United States, unlike the USSR and Britain, disliked the idea of spheres of influence. It also disliked the prospect of exclusive Russian control over half Europe and believed that the Russians were moving towards such control in contravention of pledges to instal democratic governments in the countries liberated from German rule. Since Russian designs were uncertain – Stalin seemed content at first to accept coalition government in these countries and to interpret democratic as meaning anything which excluded fascists (only a moderately ambiguous term in the prevailing circumstances) – the United States wished to exert pressure on Moscow in two directions: to make the coalitions more rather than less democratic in the sense of being proportionately representative of the popular will, and to allow American and British representatives on the Control Commissions in liberated countries to have more rather than less authority in relation to their Russian colleagues. And the pressure to be used in the pursuit of these aims included the fact that the nuclear bomb should make Russian leaders think twice before adopting opposite policies.

On the American side that intangible but none the less real factor called the climate of opinion was changing at this juncture for more reasons than the accretion of nuclear power. New men impart new ideas and new ways of handling existing problems. Truman was a very different man from Roosevelt and conscious of the differences: an American of some eminence but in no sense a world figure, a man respected for the qualities which go with simpleness and directness rather than with subtlety, a man in whom political courage would have to take the place of political sophistication, a man typically American in his attachment to a few basic principles and ideologies where the less typical Roosevelt had generally preferred the modes of thought of the pragmatist. Truman in the last resort played politics by precept and not by ear, and where Roosevelt had been concerned with the problem of relations between two great powers Truman was more influenced by the conflict between communism and an even vaguer entity called anti-communism. Again, so far as these generalizations are pardonable, Truman was the more representative American of the late 1940s and more inclined to regard the meeting of Americans and Russians in the middle of Europe as a confrontation of systems and civilizations rather than of states. Reports of indiscipline and barbarities by Russian troops, often described in this context as Asian or mongoloid, increased this propensity. (Europeans, too, if they had to choose between a host of rapists and a host of seducers, not only preferred the latter but also stood in fear of the former as something alien as well as terrifying.)

The United States had some cause to hope that the moderating implications of Hiroshima for Europe were not lost on the Russians. Elections in Hungary in October and November 1945 were held without coercion and gave the communists only a minor share in the government of the capital and the state. Elections in Bulgaria were postponed on American insistence and against Russian wishes. In Romania the United States aligned itself with the anti-communists and the king against the prime minister, Petru Groza, whom the Russians had installed when they entered the country in 1944. But the crucial testing ground was Poland where coalition government, under a socialist prime minister, was maintained until 1947 but gradually transformed in that and the following year – the years which saw the prevailing of the east wind over the west in central and eastern Europe, the failure of the American attempt to resist the partition of Europe into spheres of influence and the formalization of the Cold War.

In these years American policy veered away from the attempt to maximize American influence throughout Europe in favour of the lesser aim of giving the USSR to understand that further territorial advances in Europe were forbidden. This prohibition was to be enforced and institutionalized by a series of open political arrangements, backed by military dispositions. The overwhelming power of the United States would be used obstructively but not destructively. In so far as the USSR presented a material menace, it would be contained by physical barriers; in so far as it presented an ideological menace, it would be countered by democratic example, money and the seeds of decay which westerners discerned in the communist system (as Marxists did in capitalism). American policy was also constructive and reconstructive. The UN Relief and Reconstruction Agency (UNRRA) was mainly financed by American money; its help was enormous, especially to the USSR and Yugoslavia. In March 1947 the United States took over Britain's traditional and now too costly role of keeping the Russians out of the eastern Mediterranean: Truman took Greece and Turkey under the American wing and promised material aid to states threatened by communism. Three months later the United States inaugurated the Marshall Plan to avert an economic collapse of Europe which could, it was feared, leave the whole continent helplessly exposed to Russian power and communist lures. The offer of American economic aid was made to the whole of Europe including the USSR but Moscow refused for itself and for its satellites. For a second time – the Baruch Plan being the first – the Russians rejected a generous overture from Washington rather than accept collaboration which would have enabled Americans and others to go up and down in the USSR and observe its true plight. Separate development of western and eastern Europe was affirmed and the American initiative in restoring western Europe killed Russian hopes (if any) of victories in the west.

Thus a line was drawn. The coup which substituted communist for coalition rule in Prague the next year made it starker. Gloom and fear descended upon European politics. Although there was no fighting except of a local, sporadic and unacknowledged kind, everybody concurred in describing the situation as a war. Then, in Germany, where the major powers had joined hands in 1945 and had established a joint administration, the struggle for the only important piece of Europe outside the two camps produced a challenge and a counter-challenge which seemed bound to lead to shooting.

At the end of the war the United States, the USSR and Britain were in apparent agreement on two fundamental propositions about Germany, both of which they failed to maintain. The first of these was that Germany should be kept under constraint, and the second was that it should remain a single unit. Within a decade Germany became divided into two separate states, each – particularly the western one – playing an increasingly effective part in international politics. The reasons for this outcome included the inability of the victors to agree on other aspects of the German problem, the intrinsic importance of

Germany itself, and external circumstances, of which the Korean War was the most important. So Germany too was partitioned, like Europe as a whole, as a consequence of Russo–American rivalry in the conference chamber and on the ground in Europe and beyond. The bipolarity of postwar power politics led inevitably to delineation and demarcation – even, in a celebrated and late episode, to the building of a wall, a tactic reminiscent of such antique devices as the attempt to bar the Golden Horn to ships or the isthmus of Corinth to armies.

The principal victors were initially agreed that Germany should be disarmed and denazified, divided administratively into zones of occupation but treated economically as a single unit which would pay for imported necessities out of current production. Dismemberment, which had been discussed and may have lingered on in some French minds, was tacitly abandoned without being officially repudiated, and the territorial amputations suffered by Germany were the loss of east Prussia to the USSR and of all other territories beyond the Oder and Western Neisse rivers which were left under Polish administration pending the final delimitation of the German state by the peace conference which never took place. Churchill opposed the designation of the Western as opposed to the Eastern Neisse as the western limit of Poland's sphere (the Western Neisse flows northward into the Oder at a point where the upstream line of the Oder turns sharply eastward) but he felt unable to persist in a seemingly anti-Polish attitude. The Potsdam decision on Germany's frontiers, although expressed in provisional terms, was in reality a victory for the Russians. Germany lost nearly one-quarter of its pre-1938 territory.

The victors were divided in their views on the economic policy of the occupation and on the future structure of the German state. The general principles of economic unity and the balancing of imports and production, adopted at Potsdam, were traversed by the problem of reparations which had been inconclusively discussed at Yalta and then shelved at Potsdam. It was agreed at Yalta that the sum of $20,000 million should be taken as a basis for further discussions, half of this sum being claimed by the USSR for itself and Poland. At Potsdam the Russians, whose need for reparations in kind or cash was intense, secured agreement for removals from their zone of occupation to meet Russian and Polish reparation claims but nothing was settled about the extent of these claims. The western allies were likewise to be entitled to dismantle and remove property in their zones in order to meet their claims and those of the remaining allies. This arrangement was bound to make nonsense of the principle of economic unity since the various zones were economically dissimilar both in regard to manufactures and to agricultural production. It also made nonsense of the principle of paying for imports out of current production, since it permitted the occupiers to destroy the sources of production. Germany could be looted or milked, but it could not for long be both looted and milked. The western allies soon found that dismantling left them with an obligation to provide their zones with imported goods which had to be paid for by their own taxpayers since German production was unable to foot the bill. Moreover, Russian dismantling and removals on a vast scale, coupled with severe shortages in the USSR itself, placed upon the western occupiers and their taxpayers the additional burden of supplying the Russian zone with essential foods and goods. Although the Americans might have been willing to help the Russians directly with equipment for reconstruction, they resented the roundabout way in which the Russians helped themselves to German equipment at eventual American expense, while for their part the Russians were happy to take from Germany what their *amour propre* prevented them from accepting from the Americans.

This conflict was accompanied by related disagreements about the political structure of Germany and, subsequently, about its political attachments. The British were pragmatically predisposed in favour of a unitary rather than a federal structure, for economic rather than political reasons. The main Russian concern was strategic: to maintain their position in eastern Germany. This was the essential minimum to which they subsequently adhered unfalteringly. It was strengthened, in November 1945, by the first significant political event in the reviving postwar Europe – the elections in Austria in which the communists were decisively defeated. If, compatibly with the aim of holding fast to eastern Germany, Russian power could be extended to the whole of Germany, so much the better; but this wider aim can never have seemed more than problematical after the first few disordered months of peace. So long, however, as it remained a possibility, the Russians favoured a strong central government for Germany in the hope that it would be captured by the Socialist Unity Party (SED – an attempt to create a left-wing party under communist control and prevent the operation of a distinct socialist party). They only abandoned this sub-policy after it became evident that a united Germany would no more be a communist Germany in the 1940s than it had been after 1918. They thereupon switched not to a federal solution for Germany as a whole but to a two-Germanys policy.

The supporters of federation, meaning a federation with a weak central government, were the French. Incapable themselves of imposing any control over Germany and dubious of their allies' capacity to do so for long, the French wanted a weak German state, disarmed and disabled by internal political fragmentation. They also wanted coal for their own reconstruction plans and the Saar. Their changes in policy came in stages: first, Georges Bidault, when foreign minister, reluctantly acknowledged the hostility between the USSR and the western allies, threw France's lot in with the latter and accepted the establishment of a new western German state, including the French zone; next, Robert Schuman sought, in a wider European context, to win German friendship rather than insure against German hostility, and René Pleven, in the same context, accepted a modified rearmament of Germany; finally de Gaulle, developing Schuman's policy of reconciliation without the European framework, concluded with Schuman's old colleague, Konrad Adenauer, a bilateral Franco–German alliance.

In the three years after the Potsdam conference the occupiers, having failed to produce a coherent German policy, drifted away from the notion of Germany as something to be constrained to the notion of Germany as something to be acquired, from a basically collaborative to a basically competitive position. Two conferences of foreign ministers in 1947, at Moscow in March and at London in November, failed to elaborate the peace treaty which was supposed to emerge. In the same year, which was also the year of the Truman Doctrine and the Marshall Plan, the American and British zones were fused (January) and endowed with an economic council of fifty-four members (May). In the next year the Americans and the British moved determinedly towards turning their joint zone into a solvent and autonomous parliamentary democracy. The economic council was doubled in size and given a second chamber; a plan was produced for the internationalization of the Ruhr in order to counter the fears of those who found it difficult to stomach the reappearance of a sovereign German state; in June the Americans and the British devalued the mark in their zones, a long discussed and necessary reform which the Russians had obstructed by appealing to the principle of economic unity; in September a constituent assembly met in Bonn. These steps were supplemented in April 1949 when the French zone was joined with the Anglo–American. The western occupiers invigorated this joint zone (and embryonic state) with American financial aid and a reform of its currency; and they prepared to extend the new mark to their sectors of Berlin. At this point the Russians decided to challenge the whole western policy of separate development of a western German state. They chose Berlin where their position was special and strong.

Berlin had been excluded from the zonal system and placed under a separate, joint allied authority, the Kommandatura. For practical purposes the city was divided into four sectors but these sectors did not have the administrative autonomy of the zones. But the Russian position in Berlin was distinct for two reasons. It was the Russians who first entered the city, occupying it a few days before the German surrender and setting about the business of clearing rubble, organizing rations, installing new local authorities and establishing a police force before the arrival of American or British units; and secondly, the drawing of the zonal boundaries left Berlin an enclave within the Russian zone and separated by 260 km from the nearest point under British control. Subsequently there was much debate about the lack of foresight and political sense of the western allies in allowing the Russians to reach Berlin first and in accepting the virtual isolation of the city without even securing clearly and formally defined rights of access to it. Although something must be allowed for the temper and exigencies of wartime collaboration (which had to be preserved to the last at almost any cost), there can be little doubt that the Americans and the British would have driven a different bargain if they had realized that they were in effect handing Berlin over to the Russians subject only to the right to be imprisoned within it.

Berlin was the central point of a Russian attempt to gain control of Germany which, having started with auspicious expedition, quickly ran into trouble. The socialists refused to submerge themselves in a single party with the communists and promoted instead an anti-Russian coalition which, in elections in October 1946, thwarted the Russian design to place the city's administration in communist hands. In 1947 Ernst Reuter, an ex-communist socialist, was elected mayor in a symbolic contest in which the non-Russian occupiers were clearly, if still discreetly, aligned with Reuter against the Russians and the communists. The independent political life of the city had revived before the Russians had been able to impose a throttling substitute so that, while the Russian strategic position remained strong, the Russian political position had not thriven commensurately and the western occupiers found themselves indebted to the anti-Russian activities of Berliners. In return for this uncovenanted aid the western occupiers later felt obliged to commit themselves to the maintenance of the independence of Berlin from the Russian zone and its successor, the East German state or German Democratic Republic.

The steps taken by the western occupiers in 1947–48 to establish a West German state threatened the Russian ambition to keep Germany whole and turn it communist. They also foreshadowed the revival of an independent German power in world politics, armed and hostile to the USSR. The Russians decided to make a major issue of these developments and to resort to force to stop them. They cut the road, rail and water routes by which the western occupiers communicated with Berlin and stopped food, electricity, gas and other necessities from being supplied regularly to the western sectors from the east. The legal right to use the routes uninterruptedly was vague – and also irrelevant in what was clearly a trial of strength. The western occupiers, having considered and rejected the possibility of asserting their rights by sending an armed convoy to force its way along the road from the British zone to the boundary of the city decided instead to pierce the Russian siege by air, thus placing the Russians in the position of having to fire a first shot. They also imposed a counterblockade on the Russian zone, and the Americans moved part of their long-range bomber force to airfields in England. Between July 1948 and May 1949 the American and British air forces carried over 1.5 million tons of food, fuel and other goods into Berlin (the highest load in one day exceeded 12,000 tons), thus ensuring the needs of the entire civilian population of the blockaded sectors as well as their western guardians. This doubly extraordinary feat – extraordinary for what it did and extraordinary for doing it without leading to open hostilities – defeated the Russians, who abandoned the blockade in May after 318 days in return for a promise of one more conference on Germany, which was held in Paris but achieved nothing.

The western victory over Berlin was followed by the transformation of the western zone of Germany into a sovereign state and an armed member of the Euro–American alliance against the USSR. After elections in August the German Federal Republic came into being on 20 September 1949 with its capital in Bonn and Dr Konrad Adenauer as its chancellor. Adenauer thereby tacitly agreed to postpone German reunification by joining the western camp. An occupation statute and a series of agreements (the Petersburg agreements) defined the relations between the new state and the western powers and imposed certain restrictions on its sovereignty, but these detailed provisions were of no importance compared with the transcendent fact that the greater part of Germany had been removed from the joint control of its conquerors and attached to a new anti-communist western alliance. Exactly a year after the inauguration of the Federal Republic its rearmament became a live issue: as a result of the outbreak of war in Korea in June 1950 the Americans persuaded themselves and, with more difficulty, their British and French allies and Adenauer (who was initially opposed) that the Federal Republic should contribute to the armament of the west. The western alliance which was created to wage the Cold War came into existence on 4 April 1949, during the blockade of Berlin. Two years earlier such an alliance would have seemed to most Europeans and Americans impossible because of the strength of the communist parties of France and Italy, but during 1947 communists were excluded from the government of these two countries and the belief that they were ungovernable without communist participation was proved false. The North Atlantic Treaty was an association of twelve states which declared that an armed attack on any one of them in Europe or North America would be regarded as an attack on them all, and that each would in such an event go to the help of the ally attacked by taking such action, including the use of force, as it deemed necessary. The area covered by the treaty was defined as the territories of any signatory in Europe or North America, Algeria and islands, vessels or aircraft of any signatory in the Atlantic north of the tropic of Cancer; the treaty would also be brought into operation by an attack on the occupation forces of any signatory in Europe. Greece and Turkey joined the alliance in 1952 and the German Federal Republic in 1955. The creation of Nato was a affirmation of the dissolution of the wartime alliance. It was a defensive gesture by the principal western powers based on fear of Russian aggression, revulsion against the fact and the nature of Russian domination in eastern Europe, frustration turning to hostility in German affairs, the exposure of western Europe as a result of war damage and demobilization, and the failure to internationalize the control of atomic energy.

In 1945 the American warmaking capacity had been supreme even without nuclear weapons, but in the next years a new pattern was created by American demobilization. While American supremacy was guaranteed by the nuclear bomb, the Russians, by not demobilizing, established a superiority in mobilized land power in Europe. Thus, all future attempts to disarm were bedevilled by the impossibility of comparing like with like; the defence of western Europe became dependent on nuclear power and nuclear strategy and ultimately the collective defence of western Europe provoked dissension about inter-allied control over nuclear weapons. Whereas in 1945 there had been some qualms and, on the Russian side, some hopes of an American retreat from Europe, four years later the United States was formally committed to a dominant role in European affairs for the next twenty years. Realizing too late what had been brought about, Stalin proposed in 1948 the withdrawal of all foreign troops from Germany, but his offer was regarded as a mere device to make the Americans take a long journey from which they would not return while the Russians remained within striking distance of Germany. So long as Germany was debatable ground the United States would remain on it. Hence western Germany's eventual place in Nato alongside its recent enemies.

The Cold War was a short episode in the history of Europe but it assumed at the time an air of permanence owing to the metaphors of frigidity and rigour in which it was discussed. Its two principal features were apparent in 1946 in the speeches by Churchill at Fulton, Missouri, in February in Truman's presence and by Truman's Secretary of State, James F. Byrnes, at Stuttgart in September. These speeches showed that the tripartite wartime alliance was being replaced by a new pattern of two against one and that the United States, so far from turning its back on Europe (and in spite of the reduction of American forces in Europe from 2.5 million men to fewer than half a million at the date of the

Fulton speech), regarded Europe as an essential American sphere of influence. Although Truman had to accept virtual exclusion from central and eastern Europe, he secured by the Truman Doctrine of March 1947 a foothold in the Balkans and the Middle East at the time when he was preparing to consolidate anti-communist and anti-Russian positions in western Europe by a combination of economic aid and military alliance – embodied in the Marshall Plan and the North Atlantic Treaty. These were the beginnings of the policy of containment, designed to curb Russian power and change the Russian mood, but little more than a year after the signing of the North Atlantic Treaty this essentially European policy was complicated by a distant event, the outbreak of war in Korea, which became a drain on the forces available for containment in Europe and converted containment from a European to a more nearly global policy.

The Korean War also embittered the atmosphere. In the United States it was treated as evidence to fortify the myth of a masterly communist conspiracy to conquer the world. Senator Joseph McCarthy, alleging that this conspiracy extended into the United States government itself and other centres of influence, conducted a buoyant and repellent smear campaign in which he and his associates intimidated important segments of the public service by denouncing as communists (or homosexuals) anybody who did not subscribe to their extreme views of how loyal Americans ought to think: a number of Americans were driven into exile and some to suicide, and the formulation and conduct of American external policies were corrupted and demoralized before McCarthyism was anaesthetized by a few bold individuals, by its own excesses and by the residual good sense of the American people – without much help from their more supine elected leaders. This atmospheric pressure affected the American election campaign of 1952 in which the Republicans, in their bid to recapture the presidency for the first time since 1932, adopted General Eisenhower as their candidate. The principal Republican spokesman on foreign affairs was John Foster Dulles, soon to be Secretary of State.

McCarthyism apart there were grounds for questioning the Democrats' foreign policies. The United States was engaged in a grievous war; the USSR was not; containment seemed to mean peace for the Russians who, although prevented from expanding, remained unconstrained in their treatment of their satellites, whose fate bore uneasily on the American conscience. In his election speeches Dulles gave the impression that the Republicans would come to the aid of the enslaved people of eastern Europe and somehow liberate them from Russian domination. Containment was decried as negative and immoral. The Republicans won the election and the policy of liberation was rapidly forgotten. Instead Dulles proceeded with the policy of containment, filling the gap between Nato and the American position in Japan by fostering Seato (the South East Asia Treaty Organization) and the Baghdad Pact. He also tried to escape from the frustrations of containment, which he had criticized for being a series of responses to Russian initiatives, by evolving a strategy of massive retaliation to be applied at times and places of American choosing. But when in Indo-China in 1954 the United States had the choice between massive retaliation and acquiescence in an ally's defeat it chose the latter and so acknowledged that massive retaliation contained a large element of bluff.

In Europe the United States, having successfully annexed western Germany to Nato, accepted as a corollary the impossibility of dislodging the Russians from eastern Germany, which was turned into a satellite communist adjunct of the Russian empire in Europe. After passing through similar stages – an economic council, a parliament, a constitution, the election of a president (Wilhelm Pieck) and prime minister (Otto Grotewohl) – the eastern zone became in March 1954 a separate state under the name of the German Democratic Republic. The integration of western Germany into the western camp involved the end of the occupation and the negotiation of agreements which would both ally the Federal Republic with other western states and allow the latter some control over German rearmament. The three principal western powers offered to terminate their occupation of the Federal Republic if it would join a European Defence Community in which national forces would be subject to international control, and in May 1952 a convention was signed at Bonn ending the occupation and a European Defence Treaty was signed the next day. Elections in 1953 gave Adenauer's Christian Democratic Union and its Bavarian counterpart, the Christian Social Union, a two-to-one victory and in 1954 the Federal Republic ratified the European Defence Treaty. The French parliament, however, refused to ratify a treaty which restored German military might without a countervailing British commitment, whereupon Britain jettisoned part of its traditional aversion to meaningful peacetime associations and promoted the Western European Union – WEU – concluded by the Treaty of Paris in October 1954 and comprising Britain, France and the Benelux countries (which had been associated by treaty since 1948 together with Italy and the Federal Republic). The end of the occupation of West Germany was confirmed and the Federal Republic joined Nato. With the ratification of these agreements in May 1955 the Federal Republic became an almost fully fledged member of the western alliance. It forswore the manufacture of nuclear, bacteriological and chemical weapons and accepted a form of inspection over industrial concerns. In return it got a reiterated pledge on reunification, the recognition of the government in Bonn as the government of the whole of Germany, and the privilege of contributing twelve divisions to Nato's forces.

These developments were strenuously opposed by the Russians who tried a variety of expedients to prevent the adherence of the Federal Republic to Nato. In 1952 they were willing to accept a degree of German rearmament if it were accompanied by neutralization; they proposed, to a sceptical west, a mutual withdrawal from Germany. In June 1953, however, shortly after Stalin's death, risings in the eastern sector of Berlin and in cities in the eastern zone, directed against too much work for too little pay and against political imprisonments, found the east German régime so helpless that it had to be helped and preserved by Russian troops. The USSR became therefore committed to the maintenance of the men whom it had salvaged, and the rising and its suppression strengthened the argument for holding on to what it had.

FROM STALIN'S DEATH TO CUBA

Upon Stalin's death in March 1953 Churchill thought he saw an opportunity to arrest the collision course of the two alliances. In keeping with his own predilections in international diplomacy he proposed a personal meeting of heads of government, but the temper of the times was inimical, the Americans (and many in Britain) were cool, the West Germans suspicious, and Churchill himself soon afterwards suffered a stroke. The riots in eastern Germany in June encouraged those in the west who preferred to wait for the USSR to get into deeper trouble, while in the USSR Stalin's death was followed by an interlude of three years.

The Communist Party had held its nineteenth congress in the previous year after an unconstitutional delay of thirteen years, probably occasioned by the simple fact that the party leaders needed time after the war to set many houses in order. Although nobody knew how nearly Stalin's death was approaching, the succession was inevitably uppermost in all minds. By his handling of the congress's business Stalin indicated a definite preference for G. M. Malenkov who, having outlived A. A. Zhdanov, looked like outpointing his most serious rival N. S. Khrushchev. Zhdanov's death in 1948 had been followed in 1949 by a purge of his associates; the older men had been regressing for some years and the two most eminent among them, Molotov and Mikoyan, had lost their ministerial posts (though not their other positions) in 1949; Lavrenti Beria too, the police chief, seemed somewhat less favoured and less powerful in the early 1950s, despite his control over a police force of 1.5 million men and a militia of 300,000. Then in

January 1953 nine doctors, seven of them Jews, were accused of complicity in the death of Zhdanov. This so-called doctors' plot, which was itself declared to be a baseless plot after Stalin's death, combined anti-Semitism with an attack on Zhdanov's enemies, and it was no secret that Zhdanov's principal enemy was the man who had most markedly profited by his death, Malenkov. When therefore Stalin died in March Malenkov's position was less promising than it had seemed a year earlier, but it was still strong enough to ensure his succession to the top posts in both government and party. The evidence suggests that Malenkov's initial victory was secured in alliance with Beria but it did not last long. Malenkov and Beria may have had some similar ideas, notably in helping the consumer industries at the expense of heavy and armaments industries, but Beria was an exceptionally unpopular and dangerous man, personally and *ex officio*, and for Malenkov the support of the country's chief policeman was offset by the hostility of the armed services which disliked both Beria's private army and Malenkov's economic policy: Beria was killed either in June or December.

Soon after the end of the war Stalin, who did not see himself as a Bonaparte and did not want any Bonapartes around, had set about putting the army and its leaders safely back in a position of subordination to the civil power but in the struggle for power after his death the army was inevitably a major counter and Khrushchev, who had military friends from his days as a commissar on the Stalingrad front, decided to use it. At first he did not have to. The devolution of Stalin's entire position on any one man was more than any of the principal civilian leaders, except Malenkov himself and possibly Beria, was prepared to tolerate. Power was almost immediately divided. Malenkov was forced to choose between being head of the government and head of the party. He chose the former and ceded the latter post to Khrushchev. The antagonism of the two men was thus institutionalized. Two teams of five faced each other. Malenkov and four others formed the top layer of government while Khrushchev and four others constituted the party secretariat. This position lasted until 1955 when Khrushchev defeated Malenkov, partly by reviving rumours of Malenkov's complicity in Zhdanov's death and the subsequent purge and by accusing Malenkov of conspiring with Beria to establish personal instead of collective rule on Stalin's death – charges which alienated party feeling from Malenkov – and partly by manufacturing a war scare which created the alliance between himself and the army. In February 1955 Khrushchev secured the removal of Malenkov and the substitution at the head of the government of Bulganin who was destined to stay there as long as Khrushchev felt it inopportune to claim the post for himself. Bulganin was succeeded at the ministry of defence by Marshal G. K. Zhukhov. Other changes were made at top ministerial level where there seemed to be a shift from political veterans to technical experts, although the chopping and changing of these years more probably reflected uncertainties and inconsistencies in economic planning.

Khrushchev's personal pre-eminence lasted from 1957 to 1964 but was never as secure as it seemed to outsiders. It was won in spite of mistakes which were not forgotten, notably his failure when put in charge of agriculture by Stalin. His policy of exploiting virgin lands in Kazakhstan was radical and sound but disastrously applied in the short run. His political acumen and agility enabled him to survive this setback and, for some years thereafter, the disfavour and machinations of his colleagues who, after forcing Malenkov into the shadows, discovered that Khrushchev was at least as keen on personal authority and impatient with the committee system. But when the elder statesmen in the party tried in 1957 to remove him, he outwitted them and strengthened his own position until he forfeited it through inexperience and waywardness.

In external affairs Khrushchev's term consisted of a short and emollient prelude when he was manoeuvring against domestic rivals and a longer period which displayed his erratic, if agreeably extrovert, temperament. This latter period included major events: risings in Poland and Hungary, the launching of the first sputnik, the building of the Berlin wall, irremediable quarrels with China and his attempt to instal nuclear missiles in Cuba.

In the uncertain years immediately after Stalin's death Russian foreign policy was cautious and comparatively amiable. The German and Austrian problems were brought to the conference table, as were also Korea, where an armistice had been signed in July 1953, and Indo-China. Bulganin and Khrushchev made their peace with Tito, surrendered Porkkala in Finland and Port Arthur, put forward new disarmament proposals, visited India, Burma, Afghanistan (the first non-communist recipient of Russian aid) and Britain, and repaired in July 1955 to a meeting at Geneva with the American president and the British and French prime ministers. This meeting was a demonstration in favour of relaxing the Cold War. It produced some euphoric notions – a non-aggression treaty between Nato and the Warsaw Pact, proposed by the USSR; a free-inspection zone, proposed by Eden; and an open-skies survey, proposed by Eisenhower. An ancillary conference of foreign ministers, designed to give point and precision to the Geneva atmosphere, was a failure and this first attempt to thaw the Cold War was brought to nought by the Polish and Hungarian revolts of 1956. But the leaders had met and had set an example of decent manners and the pursuit of tolerance. In the late 1950s the Russian armed forces were cut from 5.8 to 3.6 million. A further cut of 1.2 million, announced in 1960, was postponed, presumably as a result of military pressure which became more potent after the failure of the summit conference in Paris in 1960.

In relation to Germany Stalin's successors toyed with schemes for reunification, evacuation and neutralization, but in the knowledge that the Americans were committed to two propositions unacceptable to the USSR: reunification by means of free elections and not by sticking the two Germanys together (which the Russians wanted and which entailed treating the Federal Republic and the much smaller and undemocratically constituted Democratic Republic as equals), and freedom for the reunified state to make alliances (namely, to join Nato). At a conference in Berlin at the beginning of 1954 Eden and Molotov produced plans which demonstrated the impossibility of reaching agreement. Eden proposed reunification in five stages: free elections, a constituent assembly, a constitution, an all-German government and a peace treaty. Molotov seemed ready to agree to elections on certain conditions but wanted also a fifty-year European security treaty (with, it was later explained, the United States as a party) with a ban on joining other alliances; that is to say, he dropped the earlier Russian method of reunification in a bid to secure the dissolution of Nato. When this plan failed the USSR even suggested that it should join Nato. In 1955 Bulganin and Khrushchev agreed to the evacuation and neutralization of Austria, and by the State Treaty of that year Austria recovered its full sovereignty within its January 1938 borders, subject only to two prohibitions: no *Anschluss* with Germany and no alliance with either side in the Cold War. (Since by the Warsaw treaty of the same year the USSR acquired the right to station troops in Hungary and Romania, it lost nothing strategically by renouncing its postwar rights in occupied Austria and the concomitant rights of access through adjacent territories.) But Bulganin and Khrushchev secured no comparable arrangement for Germany, even though they recognized the Federal Republic and exchanged ambassadors with it. The attempt to stop the rearmament of the Federal Republic as a part of the anti-Russian alliance had failed. In the same year the Russians created a counterpart to Nato by the Warsaw treaty and in 1956 the German Democratic Republic became a member of it.

In the same year the congress of the Communist Party of the Soviet Union, assembled for its twentieth session, was astonished to hear, first from Mikoyan and then from Khrushchev, wide-ranging and vehement denunciations of Stalin and Stalinism reaching back to what Lenin's wife had said more than thirty years earlier and to the murder of Kirov in 1934. This repudiation of the past, which did not long remain secret and included a specific undertaking to revise the USSR's relations with its satellite neighbours, encouraged anti-Russian feeling and contributed to risings in Poland and Hungary. In Poznan there were strikes for better wages in June and admissions of social unrest. At the same time a conflict arose within the Polish Communist Party between the faction of Boleslaw Bierut, who had died earlier in the year, and the more nationalist or Titoist faction led by Wladyslaw Gomulka, who had recently been released from a prison to which he had been consigned after being disgraced in 1949. In July Khrushchev, Bulganin and other Russian leaders went suddenly to Warsaw and vehemently invaded discussions in the central committee of the Polish party. They were, however, unable to prevent the victory of the Gomulka faction. Gomulka was appointed first secretary and the Russians, discovering that they must choose between allowing Gomulka to take over the government and using force to prevent it, chose the former course and accepted changes which included the dismissal of the defence minister, the Russian Marshal Rokossovsky.

In Hungary the nature of the disturbances and their outcome were different. In July the established rulers, Matyas Rakosi and Erno Gerö, went to Moscow to urge reforms in order to forestall trouble. In October demonstrators demanded better wages and liberty. Hungarian police and Russian troops failed to prevent these demonstrations from turning into an anti-communist revolution. Imre Nagy, who had been prime minister from Stalin's death to 1955, was reinstated. Mikoyan and Suslov arrived from Moscow to direct operations and decided to back Janos Kadar, the reasonably well-regarded first secretary of the Communist Party who represented a compromise between the Rakosi/Gerö team and Nagy, but the revolution gained strength and at the same time the Russians found themselves confronted with the risks and opportunities of a war in the Middle East provoked by the Anglo–French attack on Nasser. After withdrawing their troops from Budapest for tactical reasons they resorted to full-scale military measures to suppress the revolution. Faced with this turn of events Kadar sided with the Russians while Nagy named a new coalition government, promised free elections, proposed to take Hungary out of the Warsaw Pact and appealed to the outside world for help. With the western powers enmeshed at Suez and the USSR vetoing UN action, the revolution was extinguished in the first week of November. The reality of Russian power was underlined by the fact that the American administration not only took no action but never looked as though it might.

These events were a setback both for Khrushchev and for the policy of east–west *rapprochement*. Both, however, recovered. In June 1957 Khrushchev was attacked by Malenkov in the presidium of the Communist Party and defeated on a vote, but he resiliently summoned a meeting of the central committee which expelled Malenkov from the presidium and, for good measure, disgraced the veteran Voroshilov and

Kaganovich and virtually exiled even Molotov, the oldest of the old bolsheviks and the man who had been second only to Stalin among the civilians who had directed the war against the Germans. In October Zhukov too was removed and replaced by Marshal Malinovsky. This year saw the triumph of Khrushchev over his adversaries and over the doctrine of collective leadership. In March 1958 he became prime minister as well as first secretary and he remained predominant until his unexpected fall in October 1964. He reaped the benefit, in external affairs, of the dramatic appearance – in August and October 1957 – of the first intercontinental ballistic missile and the first earth satellite (sputnik). From a platform thus fortified, and observing the alarm in the United States at the thought that the American technological lead had been eliminated, Khrushchev adopted peaceful coexistence as a general description of his intentions. Peaceful coexistence was a benevolent and reassuring (but not new) political slogan of vague import and useful variability. By it Khrushchev reverted to the belief that communism, while remaining unshakeably hostile to capitalism, would prevail over it without war. (The reassertion of this doctrine was intended, among other things, to enlist the sympathies of the emerging Third World.)

Khrushchev's problems in central Europe in 1956 and at home in 1957 were peripheral to the Cold War but they were followed by critical developments in Germany and in Sino–Soviet relations which were directly pertinent to it. Throughout 1958 exchanges about and between the two Germanys had been clouding the atmosphere and the Poles were urging the USSR to find a way of preventing the Federal Republic from becoming a nuclear power and from making mischief in central Europe in association with Nato. Khrushchev was anxious to secure wider international recognition for the German Democratic Republic in order to stabilize the map of Europe and his own frontiers and so facilitate the reduction of the USSR's costly military commitments abroad and pursue his policy of *rapprochement*. He chose to begin with threats and when these did not work reverted to blandishments. He threatened in November 1958 to transfer the USSR's authority in Berlin to the German Democratic Republic unless a solution of the German problems were found within six months, but when the western occupiers contented themselves with controverting the USSR's right to act as it proposed, Khrushchev toned down his ultimatum and then let it die at the end of May. The failure of this gambit, coupled with Khrushchev's growing conviction that the United States did not intend to attack the USSR, and with the presentation to the twenty-first congress of the Communist Party in January 1959 of a seven-year economic plan which depended on switching funds from guns to butter, led to the second serious attempt to thaw the Cold War. After a visit to Moscow by Vice-President

Richard Nixon, Khrushchev visited the United States, conferred privately with Eisenhower at Camp David, presented a plan for general and complete disarmament in four years to the General Assembly of the UN, and announced the second major cut in Russian military manpower. The centrepiece of the détente of 1959–60 was to have been a second summit conference in May 1960 but it was ruined by the shooting down of an American reconnaissance aircraft over Russian territory on 1 May. Reconnaissance flights by U-2 aircraft flying at great heights between bases in Norway and Pakistan provided the United States with militarily valuable information at no political risk so long as the aircraft were not intercepted and their missions not made public by either side. The American president either did not know about the flights or had not thought of cancelling them in the pre-conference weeks, and the Russian government either had not thought of telling its defences to stop trying to shoot them down in this delicate period or – a plausible alternative view – had ordered them to do so. False statements in Washington about the aircraft's mission only compounded the American discomfiture because they were quickly exposed by the Russians who had captured the pilot alive and with his spy kit.

After the consequent failure of the Paris conference Khrushchev repeated, in Warsaw and Moscow, his belief in the policy of *rapprochement*, but for the time being practical progress had been halted by the U-2 as surely as it had been halted in 1956 by the Hungarian revolution and it was arguable that Khrushchev had deliberately engineered this stop on his own policies, possibly in response to pressures from military and pro-Chinese lobbies. His *rapprochement* with the United States affronted the Chinese, who did not share his view about American aggressive intentions, resented and feared Russo–American confabulations, and refused to be mollified when he went to Beijing on his return from the United States. These views found some echo in the Kremlin. Further, Khrushchev's defence policy of relying on nuclear missiles and cutting non-nuclear forces was too bold for some of his colleagues and advisers. Although a Rocket Force Command was established under Marshal Nedelin (later succeeded by Marshal Moskalenko), the second cut was cancelled and at the twenty-first congress in October 1961 Malinovsky stated that he did not see eye to eye with Khrushchev. (This tussle was resumed in 1963–64 when cuts were again proposed and again opposed: Khrushchev was forced to promise that the cuts would be reasonable but this persistence was probably one of the causes of his downfall.) Finally, Khrushchev had discovered that his attempt to secure recognition of the German Democratic Republic by raising the Berlin question could be turned against him by the Americans, who proceeded to couple a settlement in Berlin with a general Russian withdrawal in Europe for which neither Khrushchev himself nor the ruling group in Moscow as a whole was yet ready.

After the dismal experiences of the 1950s the next decade opened with mixed signals. The Russo–Chinese quarrel had become public property (see Chapter 3) and was held to be an incentive to Russo–American accords. In Washington the Eisenhower era closed and was succeeded by the short presidency of John F. Kennedy, whose youth and intellect seemed to promise that the 1960s would be different from the 1950s and less grim. A cease-fire was arranged in Laos, but in an encounter with Khrushchev in Vienna Kennedy made no great mark and may even have left Khrushchev with the impression that the omens were propitious for an anti-American stroke. At any rate Khrushchev gave permission for a new essay in Berlin.

The government of the German Democratic Republic was threatened with collapse. Its citizens were escaping from it at the rate of 1,000 a day, which was economically and psychologically ruinous. Its boss, Walter Ulbricht, had to act urgently in order to maintain his régime, while Khrushchev was probably persuaded that if he did not support Ulbricht the crisis in the German Democratic Republic would lead to a war in Germany. He accordingly gave his consent to the erection of a wall between the eastern sector and western sector in Berlin so that the eastern sector should become a part of the German Democratic Republic and the western sector might be made too uncomfortable for continued western occupation. In the night of 12–13 August the wall was built and the flow of refugees virtually stopped. Kennedy had told Khrushchev a few weeks earlier in Vienna that the United States remained committed to the use of all necessary forces to defend the status and freedom of Berlin. The building of the wall was a provocative act which the United States accepted and which may have influenced Khrushchev's estimate of how far it was safe to provoke the United States: he himself was to venture much further than the East Germans in the next year in Cuba.

Kennedy inherited from his predecessor a Cuban problem which was initially a chapter in the relations between the United States and Latin America but not a chapter in the Cold War. Its origins are described in Part Six of this book. In April 1961 Kennedy, pursuing a venture designed by the Eisenhower régime, aided an attempt by refugees to invade Cuba and overthrow Fidel Castro. It was immediately and totally unsuccessful. At some point thereafter Khrushchev, who was already giving Castro financial and diplomatic aid, decided on an audacious throw. Instead of merely helping Castro to remain in power, he decided to use Cuba to help the USSR, to convert it into a Russian base directly threatening the United States

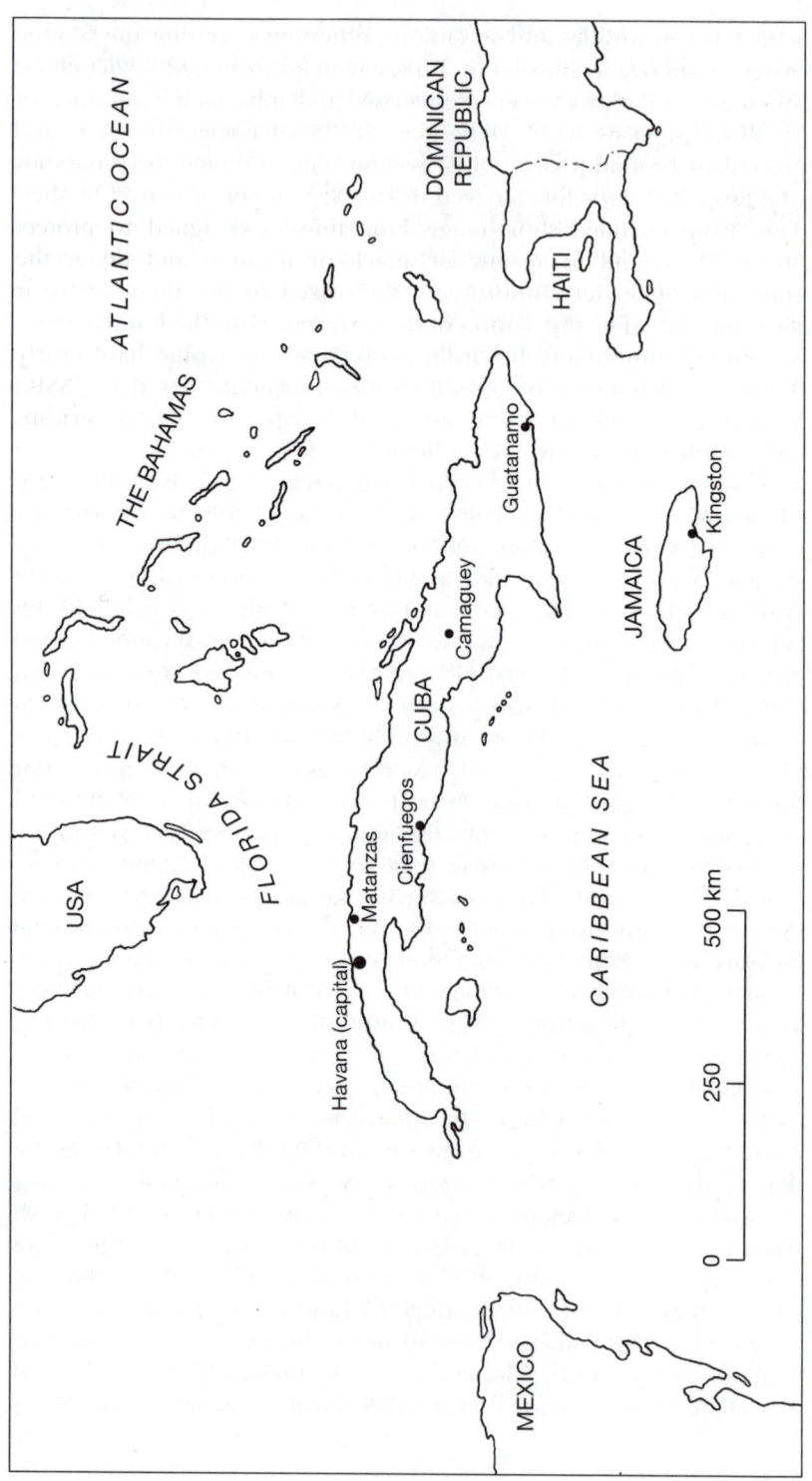

1.4   Cuba and its relationship to the United States and Jamaica

with Russian missiles and so (among other aims) getting the United States to remove its missiles in Turkey which threatened Soviet cities. Surface-to-air missiles were despatched to Cuba in the summer of 1962, followed by MiG 21 fighters, Il. 28 jet nuclear bombers and ground-to-ground (that is, offensive) missiles, of which forty-two out of a projected sixty-four arrived in late September or early October. This array included short-range Frog missiles designed to protect Soviet SS 4s and 5s against air attack or invasion and under the command of Soviet commanders authorized to fire them on their own initiative. The installation of these weapons put the United States for the first time under fire from close range and would have nearly doubled the number of bases or cities threatened by the USSR. Within three weeks of the beginning of the operation the Americans became aware of it, although it was not at first clear to them that the Russians were doing more than strengthening the defence of Cuba. The Russians assured Washington that this was indeed the case and that they had no offensive intentions, and in spite of disturbing tales from refugees close reconnaissance produced no contraverting evidence until the middle of October. On 14 October photographs showed a launching pad and one missile. Kennedy decided at once on his objective: the complete removal of Russian nuclear weapons from Cuba. The vital problem was how to achieve this objective without starting a nuclear war. It was assumed in Washington that action must be taken within about ten days. The obvious retort was an air strike but the objections were serious. Apart from the strong disinclination to open hostilities by using nuclear weapons in whatever circumstances, the president and his advisers were heavily conscious of the dangers of escalation and also of a Russia counterstroke against Berlin. Many of the Russian weapons were still on their way to Cuba by sea, and the secretary of defence, Robert McNamara, supported by the attorney-general, Robert Kennedy, proposed a naval blockade to prevent them reaching their destination and to force the Russians to remove what had already arrived. After long debate this plan, which had immediately appealed to the president, was adopted. It was explained by him to the public in a televised address and to his allies by special emissaries, and American ships of war moved into the path of the vessels bearing the Russian missiles westward. The first Russian response was to repeat that the weapons were defensive and to denounce the blockade. A clash seemed imminent. Then the president, on the suggestion of his close friend the British ambassador, Lord Harlech, moved his line of interceptors southward in order to give Khrushchev a little extra time to think and act. Khrushchev decided not to accept the challenge. The leading ships were reported slackening speed. One innocuous tanker was allowed to pass through the line unsearched and proceed upon its way.

The other ships turned back. In the Security Council Adlai Stevenson confronted the Russian and other delegates with photographic proof of the threat against which the United States had acted.

In his choice of method and in the ensuing diplomatic exchanges Kennedy was careful to leave Khrushchev the openings for the retreat which the Americans aimed to enforce. The dramatic element was intense to the end. Khrushchev signified his surrender in a letter to Kennedy in which he averred once more that the deliveries of arms were a defensive measure; he said that they had been completed and that, if the United States would promise not to invade Cuba and would raise the blockade, the USSR would see no need for a Russian presence in Cuba. Here was the recognition of the Monroe Doctrine which the Americans wanted. But almost at once a second message arrived from Moscow. In it Khrushchev demanded not only an American undertaking not to attack Cuba but also the removal of American missiles from Turkey in return for the removal of Russian missiles from Cuba. Kennedy wanted no bargaining, only a clear-cut decision on the Cuban issue in isolation. After a brief period of consternation the attorney-general suggested that the second letter was in reality an earlier one which had delayed and should be ignored. The president accordingly replied to the first letter, accepting its general tenor and agreeing to open negotiations on the basis that all construction work in Cuba be stopped. This message, sent on 27 October, was accepted on the next day by Khrushchev who agreed to ship the Russian missiles back to the USSR. Throughout these negotiations the UN Secretary-General, U Thant, played a role as crucial as it was unobtrusive. The critical issue was whether Soviet vessels approaching Cuba would stop before reaching the point at which the American president had publicly committed himself to forcing them to do so. A fateful clash was averted by, above all, U Thant who – besides urging restraint and negotiation in a general way on both sides – first suggested to Khrushchev that he order his ships not to cross the American interception line. Khrushchev promptly agreed and U Thant so informed Kennedy. Khrushchev also agreed, in response to U Thant's messages, to remove Soviet missiles and bombers from Cuba under UN supervision. U Thant then flew to Havana where he found Castro much less accommodating, partly because he feared an American invasion but no less because Khrushchev had omitted to inform him of what he had agreed with U Thant. Castro had already announced that he would broadcast that day and could not be dissuaded from doing so for more than one day, but on U Thant's pleas he toned down his speech, particularly the inflammatory remarks which he had intended to cast at Moscow for accepting without recourse to him the presence in Cuba of a UN supervisory mission. U Thant later praised the statesmanship and diplomacy of Kennedy and Khrushchev but theirs were not greater than his.

The clash off Cuba, like the conflict over Berlin sixteen years earlier, brought the protagonists in the Cold War face to face and once again no shot was fired. Khrushchev's attempt to leap into equality failed because it was insane and he himself was ousted, peaceably, by his own colleagues. The United States was becoming drawn deeply into Vietnam following the Bay of Tonkin incident in August. From the mid-1960s, the Cold War relapsed into a matter of static postures, marginally shaken by peripheral adventures (in Ethiopia, in Afghanistan) and maintained by the arms race.

THE ARMS RACE

In 1946 the United States propounded the Baruch Plan for the complete international ownership and control of the sources of nuclear energy and the transfer of American stocks to an international body. The UN Atomic Energy Committee reported that such a plan was technically practicable. It was, however, politically unattainable. The USSR presented counter-proposals for a ban on the manufacture and use of nuclear weapons and the immediate destruction of existing ones (which were exclusively American). The Russian demand for immediate destruction was unacceptable to the Americans who insisted that the creation of international machinery must come first. The American and Russian plans were also irreconcilable in other respects. The USSR accepted the principle of international control but rejected international ownership. The USSR conceded that an international authority created to supervise international control should proceed in some matters by majority vote and without a veto, but it insisted that any proposed enforcement action must be subject to a veto. In the Russian view an international convention should be reinforced by domestic legislation in each of the states signing the treaty but not by any transfer of sovereign powers to an international organ empowered to carry out inspections and to observe whether the convention was being honoured. Inspection was not entirely ruled out by the USSR but was to be limited to the inspection of proclaimed nuclear establishments, excluding any search for clandestine activities. These attitudes reflected the strategic realities of the moment. So did contemporary controversy over the reduction of non-nuclear weapons, the United States wishing to link disarmament of this kind with an agreement on nuclear weapons, while the USSR urged a proportionate reduction of forces (by one-third) which would alter the level of armaments without disturbing the relative strengths of states in these types of armament.

In 1949, the year of the conclusion of the North Atlantic Treaty, the USSR exploded its first nuclear weapon and in the next year it abandoned the UN Disarmament Commission (created in 1948 by fusing the Atomic Energy and Conventional Armaments Committees). In 1952–53 the United States and the USSR exploded their first thermonuclear or hydrogen bombs within nine months of each other. Both rapidly developed their means of delivery so that mutual deterrence ruled by the end of the 1950s, and in 1961 both powers put a man into space – the Russian Yuri Gagarin leading the American John Glenn by six months. American stocks were at all times much greater than Russian stocks and American superiority was enhanced in the early 1960s as Polaris and Minuteman missiles came into service. Unfounded American fears of a missile gap in the USSR's favour had the effect of spurring the United States into extending what was a gap in its own favour.

The Baruch Plan was soon, if tacitly, dropped. The USSR continued to maintain its opposition to anything which could be construed as international intervention in its affairs and to cling to its advantage in non-nuclear weaponry. The United States, Britain and France proposed in 1952 quantitative limits for the armed manpower of all states, and two years later Britain and France produced a new graduated plan which was designed to reconcile the differing American and Russian priorities in a step-by-step disarmament process. The USSR countered with a programme which began with a reduction in conventional forces and then in nuclear stocks and led to the elimination of bases on foreign soil, a cut-off in nuclear weapons production and a conference on a test ban treaty. The USSR also accepted quantitative manpower ceilings, thereby embarrassing the United States whose worldwide commitments demanded larger forces than those envisaged. The United States proposed in return higher ceilings (which would nevertheless have required some reduction in American forces) and an 'open skies' inspection licence whereby each side would keep the other under permanent observation from aircraft or satellites in orbit round the globe, but continued to press for an international control organ – even if subject to a veto – and rejected the idea of a ban on the use of nuclear weapons and destruction of existing stocks. In the American view the time for prohibiting the use of nuclear weapons had gone by. The attempt, initiated by the Baruch Plan, to insulate the science of war from the latest advances in physics had to be abandoned.

The pretence that these disarmament plans were serving any useful or practical purpose wore thin and palliatives were sought in schemes for disengagement, demilitarization and other forms of arms control.

Disengagement – that is, putting a distance between opposing war machines by a mutual retreat from advanced positions – was attractive for a variety of reasons: it might minimize the risk of unpremeditated clashes, it might prove to be a successful experiment in local disarmament which could be repeated on a larger scale, and it might lessen political tensions in the centre of Europe and so lead towards a solution of the German problem. In 1955 Eden proposed limitations on forces in Germany and (unnamed) neighbouring states together with a system of inspection and verification controlled by a reunited German state and its four previous occupiers. Eden also proposed a European experiment in demilitarization, beginning with a zone along Germany's eastern borders, and added a few days later as an experiment in arms control a plan for mixed inspection teams on both sides of the division between eastern and western Europe. These ideas were not well received by the Americans or the West Germans whom Eden apparently failed to consult in advance and who objected to the implied recognition of the German Democratic Republic. Russian proposals presented by Gromyko in 1956 and 1957 for a zone of limitation and inspection were rejected on the same grounds. Similar plans were advanced by the leader of the British parliamentary opposition, Hugh Gaitskell. He proposed a gradual thinning out of foreign forces in the two Germanys, Poland, Czechoslovakia and Hungary and a ban on nuclear weapons throughout the same zone; the two Germanys would be reunited, the Federal Republic would withdraw from Nato and the three eastern states from the Warsaw military alliance. Finally, in 1957 the Polish foreign minister Adam Rapacki, with Russian backing, proposed at the General Assembly (and later elaborated in written form) a plan to prohibit the manufacture and presence of nuclear weapons in both Germanys. He promised that Poland would follow suit, and the Czechoslovak government promised to do so too. The Rapacki plan dealt only with nuclear weapons in a specified area and did not overtly attack the surrounding political issues – the reunification of Germany, the freedom of Germany to join alliances, the presence of considerable American and Russian forces in central European states, the balance of American and Russian power in Europe which would be unsettled by the removal of American nuclear forces without a corresponding withdrawal of Russian non-nuclear forces. It was rejected by the United States in May 1958. Disengagement was thereupon dropped from the political agenda. It had been defeated by the entirely opposite trend of arming the two Germanys as separate contributions to the strength of the forward positions of the two rival blocs, by western fears of the USSR's large non-nuclear forces, by the

Nato policy of putting nuclear weapons into forward positions, and by inexperience and distrust of methods of inspection and verification. The fortification of the alliances had priority over the dismantling of them.

There was, however, some mutual inclination to restrict nuclear tests. The development of varieties of nuclear weapons, large and small, substantiated the need for tests, and these tests produced worldwide alarm about the effects of radioactive fall-out, especially after American tests in the Pacific in 1954 which were believed to have killed some Japanese fishermen, poisoned vast numbers of fish and infected an area of 18,000 sq. km. In 1957 there was a cessation of nuclear tests by tacit agreement – partly a consequence of the conclusion of series of American and Russian tests at this time – and some preliminary examination of the possibility of a more formal and permanent ban. But there was a serious gap between what was acceptable to the United States and what to the USSR: on the constitution of a control organ and the voting within it, on the manning of control posts by observers of foreign nationality, and on the number of on-site inspections to be carried out in any year in a specified area in order to ensure compliance with the ban. On all these matters the gaps between the protagonists were narrowed during discussions in 1958–60 but insufficiently to produce agreement before the destruction of the American U-2 reconnaissance aircraft over the USSR in May 1960 and the abandonment of the summit conference of that month temporarily ruffled the Russo–American *rapprochement*.

Less fruitfully, although with every appearance of seriousness, both the United States and the USSR sponsored unreal plans for general and complete disarmament and spent much time talking about them. The Russian proposals focused on such favourite objectives as the withdrawal of foreign troops and bases rather than on the less spectacular topics of control and verification with which American and British negotiators were more particularly concerned. The Russian plan, presented by Khrushchev to the General Assembly of the UN in September 1959, was followed in March and June 1960 by two American statements. A ten-nation disarmament conference at Geneva in 1960 was short-lived but also the occasion for a joint Russo–American initiative in the shape of a series of recommendations elaborated by John J. McCloy and V. Zorin. No conference was held in 1961. The Americans and Russians agreed a set of principles but failed to agree on a statement about controls submitted by the Americans; both sides produced draft treaties on general and complete disarmament (the Russian was a revision of an earlier draft) in time for the autumn session of the General Assembly.

The high-water mark of these negotiations was the McCloy–Zorin recommendations. These consisted of a set of principles to govern continuing disarmament negotiations, beginning with the acceptance of general and complete disarmament as an ultimate goal. This document was in effect an agreed statement of what had to be agreed. It clarified the issues but did not resolve them. It predicted the need to establish reliable procedures for the peaceful settlement of disputes and the maintenance of peace; the retention by each state of adequate non-nuclear forces for the preservation of law and order; the disbanding of superfluous forces and the elimination of all nuclear, chemical and bacteriological weapons, all means of delivering weapons of mass destruction, all military training institutions and all military budgets; an agreed sequence of stages in the disarmament process, subject to verification at the conclusion of each stage; balanced measures to ensure that neither side secured a temporary advantage as disarmament proceeded; and strict and effective international control of the process and thereafter through an International Disarmament Organization within the framework of the UN. This visionary package required agreement on the order in which different operations would be carried out, agreement on the equivalence of different types of armament, and agreement on the nature and modes of operation of an inspectorate. Agreement on these and other points required a degree of trust which no political leader felt or could perhaps have evinced without incurring plausible charges of gambling with national security. The Russians, for example, accepted at one stage the principle of inspection, but it transpired that what was to be inspected was weapons destroyed and not weapons surviving; they were reluctant to disclose what was left through fear of having it attacked and destroyed. Ingenious schemes for circumventing this difficulty, such as Professor Louis Sohn's proposal to divide national territory into zones and give inspectors the right to search only a limited number of zones at defined intervals, did not suffice to overcome nationalist conservatism.

The crucial feature of the revived disarmament discussion was the problem of inspection and verification (an element which had not obtruded in earlier negotiations such as those which preceded the Washington Treaty of 1922, though it had engaged the attention of the disarmament conference of 1932, which proposed a Permanent Disarmament Commission with powers of inspection but no powers of enforcement). The evolution of nuclear weapons had enormously increased the dangerous consequences of allowing a party to a disarmament agreement to cheat and to conceal; he might by doing so gain the mastery of the world. But concealment was also regarded as one of the conditions of survival. To any power inspectors were potential spies who were licensed to discover and might then reveal how the inspected territory could be denuded of its defences at a blow. The impulses of politicians who took up disarmament for one reason or another were therefore continually countered by more cautious and short-term questionings which prevented the conclusion of any except partial agreements.

Partial agreements were, however, made. The cessation of nuclear tests in the atmosphere was an early example, which was matched in the same year by an agreement to neutralize the Antarctic continent. Tests were resumed by the Russians unilaterally in September 1961, chiefly because they had proceeded to a point at which they had something new which needed testing, but 1963 produced a new and more extensive test ban agreement (as well as the installation of a direct and permanently open, or 'hot', line of communication between the White House and the Kremlin). When the Russians announced the resumption of tests, the Americans and British offered to conclude a treaty banning atmospheric tests without any provision for inspection, but the Russians were not to be deterred from their new series of tests and shortly afterwards the Americans resumed testing too. At the end of the year the Russians proposed a treaty banning all tests, subterranean as well as atmospheric, without inspection; this proposal was immediately rejected, and the three powers reported to the UN their failure to agree and the abandonment of their attempts to do so.

Alarmed by this breakdown the UN resolved to convene yet another disarmament conference, this time with eighteen members, including eight neutrals, an innovation inasmuch as neutrals had not previously participated in test ban discussions but only in discussions on general and complete disarmament. The French, who were in the earlier stages of developing a nuclear armoury of their own and had therefore no wish to ban tests, took no part in the proceedings, but the neutrals (Brazil, Burma, Egypt, Ethiopia, India, Mexico, Nigeria, Sweden) proved themselves valuably persistent in devising compromises and keeping the discussions going. The debate on inspection was resumed between the nuclear powers and was narrowed down to a question of numbers, the western powers ultimately refusing to go below seven a year and the Russians refusing to concede more than two or three. In the early part of 1963 the failure of the talks was generally expected but was avoided and then converted into success largely by the unobtrusive pertinacity of the British prime minister Harold Macmillan. Early in July Khrushchev hinted in a public speech that a partial ban might be agreed, and later in that month Averell Harriman and Lord Hailsham went to Moscow, where the three powers agreed on the terms of a treaty banning nuclear tests in the atmosphere, in outer space and under water for an unlimited period but subject to the right of any party to withdraw from its undertakings if its supreme interests were jeopardized by extraordinary events connected with the subject matter of the treaty. Many other states adhered to the treaty. They did not include China or France.

The conclusion of this treaty raised the question of what to try next. It gave a fillip to the partial approach and therefore to the search for parts ripe for tackling. Lyndon B. Johnson, who had become president in the United States upon the assassination of Kennedy in November 1963, listed some topics early in the next year. They included an anti-proliferation programme to include a total ban on nuclear tests, a ban on the transfer of nuclear materials to non-nuclear states and inspection of peaceful nuclear activities; a chain of observer posts to guard against surprise attack; a verified freeze on missiles; a verified cessation of the production of fissile material; and a ban on the use of force to alter boundaries or otherwise transfer territory from one state's control to another. Proposals from other quarters included a bomber bonfire (applying to B47s and Tu-16s) and percentage cuts in military budgets. In Poland Gomulka recast and revived the Rapacki plan by proposing, in December 1963, a nuclear freeze in Europe, which, however, the Americans disliked partly because they attached little value to controls over the location of weapons unaccompanied by controls over their production and partly out of deference to the suspicions of their German allies.

These plans bore no immediate fruit for three reasons. First, the achievement of the test ban treaty was as much as the leaders on either side could for the time being digest and make palatable in their own countries. More might become possible, but not too soon. Secondly, discussions within Nato about joint control over nuclear dispositions and nuclear weapons roused eastern fears of a nuclear Germany. The United States was caught between the desire for a continuing dialogue and *rapprochement* with the USSR and the desire to accord to its most effective continental ally the status and the authority in allied councils and operations which its material contributions to the alliance warranted. To the Russians the proposed Nato multilateral force (MLF – see Chapter 6) was a form of proliferation of nuclear weapons, although the Americans had devised it as an anti-proliferation policy to satisfy the Federal Republic and also France with something less than independent nuclear forces. And thirdly, the increasing American involvement in Vietnam put an additional strain on Russo–American relations. The American attempt to buttress an independent and non-communist South Vietnamese state entailed war against the Vietcong which ranked in communist terminology as a national liberation movement, and war against the communist state of North Vietnam, and something like war against the vastly more important communist state of China. For the Russians to fail to support President Ho Chi Minh of North Vietnam would be a betrayal of communist solidarity, dangerous at any time to Russian standing in the communist world and doubly dangerous at a time when Russian pre-eminence and doctrinal purity were being assailed by the Chinese. Further, for the Russians to fail to support the Vietcong was again doubly dangerous, for if the Vietcong lost communists would blame the Russians, whereas if the Vietcong won without Russian help Chinese influence might rout Russian influence in Asia and the Chinese would be proved right in their contention, against the Russians, that wars of national liberation could be fought without escalation into nuclear war. When by 1965 the Americans had become unconcealed principals in the war and not merely adjutants of the South Vietnamese, the scale and the nature of the fighting were changed and the Americans had to face worldwide protests against the ensuing horrors. The Russians joined in the outcry. Although, therefore, the evolution of nuclear strategies and the resurgence of China and the passage of time had combined to put an end to the bipolar Cold War, the principal adversaries were prevented by particular crisis and by habit from acknowledging the fact and acting on it more than very tentatively; only de Gaulle did both.

Nevertheless, whether by understanding or as a result of blinder forces, a significant degree of stability and tolerance had evolved on the biggest question of the time: whether the Cold War would engender a nuclear war. In the nuclear age peace could be preserved so long as each side relied on the threat of retaliation to keep the other side from attacking first. But one side might conclude that the only way to avert an attack was to attack first. It would then, using the modern version of the pre-nuclear blitzkrieg, adopt a counter-force strategy and build up, advertise and possibly use its power to destroy the enemy's warmaking capacity at a single blow. In order to neutralize this strategy it was necessary to protect aircraft and missiles sites so effectively as to make it incredible that they could all be destroyed by a first-strike surprise attack. Their protection by anti-aircraft and fighter defences had been rendered obsolete by the enormous increase in the destructive power of each single bomb and by the vastly increased speed of missiles. In its place two new and extremely expensive forms of defence were developed: missiles were placed in dispersed, hardened sites and, in the case of Polaris, under the sea in submarines; and an early warning system gave the defence enough time to get aircraft into the air and to save them from being destroyed on the ground. Before the appearance of the intercontinental missile an early warning of two to three hours was achieved by the Americans, and when this was rendered inadequate it was improved to half an hour, at which point a part of the bomber force was kept always in the air and a part at fifteen minutes' readiness on the ground. These measures to ensure the survival of an effective part of the retaliatory force, bombers and missiles, caused both antagonists to concentrate on second-strike strategies as a means of security. It was essential to the success of this attitude that each side should be seen by the other to have adopted it, and it was a consoling feature of the later years of the Cold War that each side succeeded (with the help of the fact that a first-strike force differed in size and composition from a retaliatory force) in transmitting to the other tacit messages to this effect. The probability of a surprise attack diminished.

Nuclear forces could, however, also be used in reply to a non-nuclear attack. The use of nuclear weapons was not ruled out by the reticence and deterrence of the two principal nuclear powers in relation to each other. Both could and did threaten to use them offensively in other contexts. In January 1954 Eisenhower and Dulles spoke on separate occasions of massive and instant retaliation. The last phase of the French war in Vietnam had begun and the Americans were faced with a choice between helping the French by a nuclear strike or letting them be finally beaten. The Americans tried to make political use of their nuclear armoury while knowing that they would not use it militarily; in a display of brinkmanship Dulles used tough language in order to scare the Russians and the Chinese and prevent them moving into the battle area – and perhaps in order to scare the British too and so get them to restrain him from a course from which he wanted to be diverted in any case. In 1956 Khrushchev, faced simultaneously with a war in the Middle East and a revolution in Hungary, made vague threats of using nuclear weapons in unspecified places in order to terrify the British and French governments and peoples and gain credit in the Arab world, and after the U-2 episode in 1960 he threatened smaller nations with nuclear attack if they facilitated American reconnaissance activities. But in the event both sides reserved their nuclear armaments for each other and became increasingly concerned not merely to keep nuclear weapons out of use but also to keep them out of other nations' hands. Yet one of the more sinister consequences of the Cold War was the postponement, throughout the 1950s and into the 1960s, of the realization of this common interest until after two other powers, France and China, had become nuclear powers and a number of others had acquired the capability and, in the absence of an international control system, were also developing the will to follow suit. During the Cold War the two protagonists had developed an increasing sobriety in relation to one another, and even a sort of fearful intimacy. There was, however, no reason to suppose that other possessors of nuclear weapons would acquire sobriety with might, or – if, for example, they were Israel, Egypt or Iraq – would develop this intimate understanding of the permitted limits of nuclear politics; nor was there reason to suppose that the two giant powers would find it as easy to control the conflicts of others as to control their own.

These two problems – how to prevent a nuclear war between nuclear powers and how to prevent more states graduating to the nuclear élite –

were aspects of the wider problem of arms control which superseded more traditional approaches to disarmament during the 1960s. Arms control, the regulation as opposed to the elimination of the use of weapons of war, was not new. It had been applied particularly to naval weapons by, for example, the Rush–Bagot treaty of 1817 which barred navies from the Great Lakes, and by the Washington treaties of 1922 and 1930 which sought to limit the naval power of one state in proportion to that of another. Interest in similar schemes was revived in the 1960s by scepticism about general disarmament, and academic discussion of arms control was taken up by politicians in Washington and Moscow who were feeling the need to communicate with one another (as in the Cuba missiles crisis) or to co-operate with one another (as in the Middle East crises of the decade). The idea that major powers had more in common than the need to avoid mutual annihilation was further fortified by their common interest in their own superiority. Neither of them wanted to see nuclear weapons in any other states' armouries and in 1968 they, together with Britain, concluded a Nuclear Non-Proliferation Treaty to which they invited all other states to adhere. The object of this treaty was to freeze the existing nuclear hierarchy, keeping France and China in the grade of second-rate nuclear powers and every other state permanently excluded from acquiring nuclear weapons. France and China predictably refused to sign. From the non-nuclear but potentially nuclear states there were sounds of discontent. Before subscribing to such an act of abnegation they objected that they should not be asked to forgo modern weapons without a better prospect of escaping embroilment in a nuclear war and they urged the nuclear powers to balance their enthusiasm for non-proliferation by a more serious attempt to control their own arms race.

The treaty came into force in 1970 and review conferences were held quinquennially, ineffective and acrimonious. Twenty years later over one hundred states had signed the treaty, but more significant were the number and identity of those which had not. They included France and China, which were certainly nuclear powers; South Africa and Israel, which were generally thought to be; and India and Pakistan, Argentina and Brazil, all of them some way along the road. At this date there were in the world several hundred nuclear reactors, including some which produced plutonium as a by-product of their relatively inoffensive activities. A small proportion of this material sufficed to make a bomb a year and newer techniques – in, for example, the enrichment of uranium to weapons-grade standards and the production of plutonium in fast-breeder reactors – were facilitating the manufacture of nuclear weapons. In these circumstances advanced powers which were helping developing states to build reactors were inevitably engaging in proliferation and so were in breach of the treaty, as were all nuclear powers which had failed to make serious efforts to reduce their nuclear armouries. Hopes for a more comprehensive ban were vain so long as the principal nuclear powers continued to develop new weapons which they needed to test. Progress was limited to an agreement between the superpowers, concluded in 1974 but not ratified until 1990, limiting the scale of underground tests, and a further agreement of 1976, also ratified in 1990. South Africa adhered to the NPT in 1991. So too did France and China. Brazil and Argentina renounced the production of long-range nuclear weapons. North and South Korea agreed in 1992 to make the Korean peninsula a nuclear-free zone but North Korea's induction into the NPT régime was jerky (see p. 558). A resumption of testing by China in 1993 imperilled these meagre advances. The 1968 treaty was renewed indefinitely by 178 states in 1995 but not without criticism of nuclear states, notably France and Britain, for their minimal compliance with their treaty obligations. In the same year Jacques Chirac resumed French tests in the Pacific shortly after his election as president. The end of the Cold War laid bare the fact that the Cold War and nuclear weapons were distinct phenomena. The Cold War might be over but nuclear proliferation was not. During the Cold War nuclear states had tried to prevent proliferation in return for promises to reduce their own nuclear armouries, but having done little in the latter respect had achieved nothing in the former.

Alarm at the spread of nuclear weapons was accompanied by alarm at the development of weapons technology. The MIRV (multiple independent re-entry vehicle), whose separately controlled multiple warheads greatly increased the threat from each missile; and ABM (anti-ballistic missile) systems, which could counter a first strike and so destroy the deterrent stability which rested upon the presumptive efficacy of first strike, were enormously increasing the cost of the arms race. At the same time the deadliness of new missiles which were alleged to be capable of landing within a few dozen yards of a target, combined with new defences which could destroy incoming missiles only at the cost of removing the deterrent factor which was designed to prevent their discharge in the first place, inclined the superpowers to talk about the control of the use and development, as well as the proliferation, of nuclear weapons. Strategic arms limitation talks (SALT) were begun in 1969.

Strategic arms may be defined in this context as weapons which can reach targets in an adversary's territory from bases or launching sites in one's own territory or on the high seas. They include missiles or bombs carried by long-range aircraft as well as missiles launched from static or mobile land sites or submarines. The category of strategic arms is therefore a complex one and it is further complicated by the fact that a single missile with many warheads, each of which can be independently directed to a different target, is far from being the equivalent of a missile with a single warhead that can hit one target only. There are furthermore two distinct and incompatible ways of calculating the effectiveness of a nation's strategic armoury. On the one hand it may be assessed by the number of enemy targets theoretically vulnerable, an assessment which involves counting the number of independently targeted warheads deployed; on the other hand it may be assessed by the weight of explosive which can be delivered by all available launchers and aircraft on a single occasion. Finally, the category of strategic arms is not a closed one since there is argument about whether to include nuclear weapons of shorter range which are nevertheless brought within range of the enemy by being based on intermediate territory or capable of being sent there at short notice.

Given these complexities it is remarkable that, by an exercise of political will over technical intractabilities, two agreements were signed in May 1972 (and two minor agreements in the previous year). In negotiations in 1970 at Vienna and Helsinki, which occupied five months of the year, the Americans took the initiative by proposing a total ban on mobile land sites, a special limitation on particularly potent weapons and an overall numerical limit on the sum of the land and sea sites and bombers which either country might possess. Next year the two countries reached agreement on a Seabed Treaty banning the placing of nuclear weapons on the ocean floor (the treaty to be open for signature universally) and updated their agreement of 1963 for a hot line between Washington and Moscow by modifying it to take account of the new means of communication by satellites. The United States then proposed a standstill on the deployment of intercontinental land-based missiles and all submarine-based missiles. Concurrently the negotiators tackled the defensive as distinct from the offensive aspects of strategic nuclear war by trying to set limits to the deployment of anti-ballistic missile (ABM) systems, but the familiar problem of distinguishing a defensive from an offensive weapon bedevilled progress for a time as it was impossible to say that every missile or launcher in an ABM system could never be used except for defensive purposes.

Notwithstanding these conundrums an ABM treaty and an interim agreement on offensive missiles were concluded in 1972, thus marking some progress in the regulation of both defensive and offensive weapons systems. The USSR had at this point an incomplete ABM system round Moscow and the United States was planning two systems for the protection of its intercontinental launching sites. The ABM treaty, of indefinite duration although subject to quinquennial review, permitted each party to deploy two systems, the one for the defence of its capital and the other for the defence of some part of its intercontinental missiles, the centres of the two systems to be not less than 1,300 kilometres apart and the radius of each no more than 150 kilometres; each system might contain 100 launchers, all of them static and capable of firing one warhead only. The agreement on offensive missiles was much barer. It was of limited duration, would expire in 1977 and did no more than impose a freeze on new construction subject to a proviso permitting the substitution of more modern for obsolescent equipment on land and in submarines. The United States and the USSR also agreed to begin in November a second round of SALT talks.

SALT 2 was to be concerned with what had been omitted from the 1972 agreements. This was much. The United States maintained its plea, rejected by the USSR, to ban mobile land-based launchers totally. The Russians had tried but failed to include in the interim agreement specific provisions about long-range bomber aircraft which were still a significant part of the American armoury, although not of the Russian: the Americans had over 500 such aircraft and the Russians, who wanted each aircraft to count as one launcher in an overall total, had 140. There had been no agreement about shorter-range aircraft based outside the United States: the Americans had about 1,300 such aircraft, capable of carrying nuclear weapons, 500 of them in Europe. Above all there was the problem of MIRVs. The USSR had, so far as was known, no operational MIRVs, although it had a larger total than the United States of intercontinental missiles. The United States, however, had begun deploying MIRVs on land (Minuteman 3) in 1970 and at sea (Poseidon) in 1971. By 1972 the USSR was believed to have 2,090 strategic missiles capable of hitting that number of targets, whereas the United States had 1,710 such missiles capable of reaching 3,550 targets and, within a few years as re-equipment with MIRVs proceeded, over 7,000 targets. The Russians were expected to begin deploying MIRVs on their operational sites in 1975, at which point the effectiveness of the Russian force in terms of targets reachable would begin to multiply while the American, already largely re-equipped, would be becoming static. So in order to retain their superiority and match the expanding number of warheads sprouting from Russian launchers the Americans would have to increase the number of their launchers above the total frozen by the interim agreement. If they did not do this after 1977 the Russians would, it was assumed, have almost twice as many American targets in their sights by 1980 as the Americans had Russian. The Americans therefore were primarily concerned to set limits to the land-based missile capacity of each side. The Russians countered by proposing the elimination of American forward bases (aircraft bases in Europe and submarine bases in Spain and Scotland), a limitation on the number of aircraft carriers to be permitted in European waters (the Russians had no carriers of the conventional type but only carriers with helicopters or vertical take-off aircraft), and the relegation of nuclear-armed submarines to parts of the ocean from which they could not hit enemy territory. Progress in SALT 2 was in consequence laborious and negligible until Nixon visited

Moscow in mid–1974 when the impasse was slightly shifted by three minor accords: a modification of the ABM treaty, an agreement to ban underground weapons tests of 150 kilotons and over from March 1976, and agreement that a new SALT treaty would extend to 1985. This last agreement was reflected in November when Gerald Ford, who had stepped into the presidency when Nixon resigned in August, met Brezhnev at Vladivostok in an attempt to prevent the steam from going out of Russo–American détente in general and SALT in particular. Ford and Brezhnev agreed that the talks should continue on the basis that each side might have up to 2,400 strategic launchers of all kinds (a somewhat high ceiling) of which no more than 1,320 might be fitted with MIRVs. The next year was devoted to the attempt to transpose this core into a formal agreement but with no success. The political will was there, fortified by the wish of both leaders to reach agreement before, in Brezhnev's case, the twenty-fifth congress of the Communist Party of the Soviet Union in February 1976 and, in Ford's, the presidential campaign which would take up most of that year. But the complexities and technicalities, themselves in constant flux, overpowered the negotiators. Discussions broke down early in 1977 but a new president, Jimmy Carter, tabled fresh proposals later in the year and then put a stop to the development of the neutron warhead and the long-range B-1 bomber. In 1979 a SALT 2 treaty was signed.

It was to run to 1985 when a third treaty was intended to continue the process. It contained significant limitations. At its core was the restriction of nuclear delivery systems to 2,400 on each side, declining to 2,250 in 1985. Within this overall limitation there were interlocking sub-limitations in more precisely defined categories: MIRV'd systems together with aircraft carrying cruise missiles; land-based and sea-based MIRV'd missiles; and land-based MIRV'd missiles alone. Missiles over a certain size, of which the United States had none and the USSR had 308, were banned. There was a further ban on air-launched missiles delivered otherwise than from aircraft. Limits were set on the number of warheads per missile and the number of cruise missiles per aircraft. It was agreed that any launcher which fired a MIRV'd missile remained, for the purpose of the treaty, a MIRV'd launcher whatever weapons might be fixed to it. Restrictions, albeit modest, were placed on the modernization of weapons; they affected neither submarine delivery systems nor land-based intercontinental ballistic missiles and so permitted the deployment, on the American side, of Trident 1 and 2 and the MX mobile missile system, and their rough Russian equivalents. The parties to the treaty undertook to give notification of certain tests and stocks. A non-circumvention clause aimed to prevent either side from using a third party to get round its treaty obligations; this clause alarmed European members of Nato who feared that the

United States might interpret it as a prohibition on the transfer of new technology. A protocol to the treaty pointed the way towards further restrictions on the deployment, although not on the testing, of new and improved weapons: for example, cruise missiles (other than cruise missiles delivered from aircraft) with an extended range. That these provisions were more effective than cosmetic was proved by the opposition which they provoked among those on either side who put the preservation or attainment of military preponderance ahead of the control of the arms race. In the United States the Senate Foreign Relations Committee recommended ratification of the treaty but public protest and expert disquiet (which concentrated on what the treaty did not secure) and then the Russian invasion of Afghanistan killed it. Non-ratification was a prominent feature in Ronald Reagan's successful bid for the presidency in 1980.

The political atmosphere had become uncongenial to such agreements. The late 1970s was a time of increasing mistrust. On the American side the Russo–Cuban intervention in Angola in 1975 began a series of Russian moves – Ethiopia, Vietnam and Afghanistan were other and widely separated theatres of menacing Russian activity – which hardened American distrust and sharpened the American concern for rearmament rather than arms control. At this time too official American estimates of the size of the Russian defence budget were drastically increased. The passage of the warship *Kiev* from the Black Sea to the Mediterranean, an incident which would hardly have ruffled the waters a few years earlier, led to charges that the USSR had broken the Montreux Convention of 1936 (on the assumption, which was contested, that the *Kiev* was an aircraft carrier within the meaning of that instrument). In 1979 a more serious scare was caused, and deliberately publicized, by the discovery of a Russian brigade in Cuba. President Carter had stopped American intelligence flights over Cuba as a conciliatory gesture, but these were resumed when Cuba was suspected of taking a hand in revolution in Nicaragua. What they revealed was a Russian training unit which had been in Cuba for some years, but the temper of the times converted it into a new threat to the United States.

On the Russian side the flourishing entente between the United States and China, and Deng Xiaoping's attempt to turn it into a triple alliance with Japan, aggravated Moscow's enduring fears of encirclement, which became more acute as the USSR became increasingly dependent on American grains to feed its beasts and so its peoples. The USSR alleged, and perhaps believed, that its troubles in Afghanistan and Poland were accentuated by American undercover interference. Its ageing leadership (procedures for an orderly succession had never been evolved) was plagued at home by an economy which failed to secure basic amenities and was consigning perhaps 20 per cent of GNP to armaments and defence.

Concurrently with this worsening of the political climate were technical advances which were altering the nature of the arms race. The nuclear deterrent had been a crude and inflexible weapon, whose effectiveness depended on its crudeness and inflexibility. It was aimed at cities and their inhabitants. On the American side there were until the late 1970s only two courses that an American president could take in an emergency. He could order an attack on all the targets on which his weapons were trained; or he could order the destruction of all these targets except Moscow. He had no third choice, and because he had none his armoury posed a most fearful threat to the USSR. It was an effective deterrent – so long as it was possible to believe that the president would use it.

But during the 1970s this credibility began to wear thin. Would a president nerve himself unhesitatingly to order so vast a massacre within the minutes available to him for a decision? The very question undermined the strategy of deterrence. It also revived American fears that the USSR would be tempted to revert to a first-strike strategy. At the same time the refinement of, in particular, targeting techniques enabled strategists to think once more in terms of waging war instead of deterring it. The aggressor might attack a single pinpointed military target, such as a missile site or a command headquarters or a bunker containing a head of state; each side might then exchange attacks on such single targets and carry on a war for weeks or months. Surviving a war, even winning a war – concepts which had seemed nonsense in the nuclear age – became possibilities.

A special element in the contest between the superpowers, making that contest global, was the growth of Russian naval power in the 1970s. If a great power is a power that can act in any quarter of the globe – and that is as good a definition as any – then sea power is crucial. The United States was without question a great naval power capable of sailing all the world's oceans and of commanding passage through all but the most private waterways. The USSR was not such a power but was determined to become one. Because it was aiming to be what it was not, its efforts roused considerable alarm. The world was not used to seeing Russian fleets in many oceans, but the universalization of Russian power required this to happen. On land the USSR had advanced little since 1945. Its hold on eastern Europe, although sometimes troubled, remained undisputed and with it the power to pose a threat to western Europe whose credibility was a constant conundrum. It had established in the 1950s its rights to be considered a power in the Middle East, although in the 1970s the limits of this power were exposed by its eviction from Egypt and by its negative role in the diplomacy that followed the war of 1973. Its sallies into the Congo in 1960 and the Caribbean in 1962 were failures, its role in the wars in South-east Asia marginal, the poor performances of communists in Portugal and Greece in 1974 disappointing, the illness and absences of Brezhnev and the uncertainty of the succession a source of hesitations. Its economy and the quality of life of its peoples remained vulnerable to the vagaries of the weather and of an unwieldy bureaucratic system, both of which could perpetrate massive shocks. For all its vast armed strength the USSR was in one vital respect a power of a different kind from the United States, a power still confined whereas American power roamed free. The seaways, and the under-seaways, offered one escape from this inferiority.

By the mid-1970s the Russian navy far surpassed the American in the number of its submarines but was distinctly inferior in every other respect. American naval manpower, including marines, was 733,000 against 500,000 Russian equivalents. The United States had 15 aircraft carriers, the USSR none other than helicopter carriers; of cruisers and destroyers with nuclear weapons the United States had 110 and the USSR 79. But the USSR had 265 submarines against 75, albeit that in nuclear-powered submarines the balance was more even at 75 Russian to 64 American. Such figures ignore a great deal, and a closer comparison of the two navies would have to take account of the age of vessels, their armaments, each nation's reserves, research expenditure and other indicators of comparative effectiveness. Political effectiveness moreover is not the same as military effectiveness. What would happen if the two navies engaged each other in combat was a nearly academic question, but the effect of the appearance of a Russian flotilla of any size in, for example, the Mediterranean was not. This fleet varied between twenty-five and sixty surface and submarine vessels, sometimes larger in numbers and sometimes smaller than the US Sixth Fleet but without air cover, smaller than the French, much smaller than the Italian and trivial beside the combined forces of Nato in this theatre. It was none the less a portent and it made a political point. It had its effect on the conduct of relations between the USSR and Algeria and Libya, two strategically placed anti-western but not pro-Russian states. It reminded western governments that their fears, acute in the late 1940s, of Russian bases in Yugoslavia and Albania might yet one day be realized, with incalculable consequences for Mediterranean politics. A foretaste of these consequences was provided by the prime minister of Malta, Dom Mintoff, whose search for money for his impoverished island led him to demand from Britain greatly increased fees for the use of Malta's harbour, with more than a hint that if Britain did not care to use it at the going price the Russians would. As a result Mintoff secured in 1972 a rental of £14 million a year for seven years, of which Britain's Nato partners were to provide £8.74 million, and additional lump sums of £2.5 million from Italy and £7 million from other members of Nato.

The election of Ronald Reagan to the presidency of the United States in 1980 owed a great deal to the feeling that the United States ought to be making better use of its power instead of striking deals with an adversary who was getting stronger and was not to be trusted. Reagan was not the choice of the nation's political leaders. He was the choice, in the first place, of a group of reactionary conservatives with enough money to buy for him the candidature of the Republican Party and, secondly, of the people at large who liked his personality and simplicity and responded to his forthright reduction of public issues to contests between good and evil in which it behoved the good to exert themselves with more faith than calculation. Reagan retained his popularity and won re-election in 1984 by a landslide. He had powerful support in western Europe which likewise moved to the right in the early 1980s. In Britain Thatcher's electoral victory in 1979 brought to power a prime minister who was temperamentally in favour of tough talking and acting (and was also committed to monetarist economic policies which she believed to be the same as his); her success at the polls was repeated in 1983, when the Labour Party's espousal of unilateral disarmament was a major element in a disastrous campaign, and again in 1987. In West Germany conservatives triumphed when Helmut Kohl ended thirteen years of socialist rule in 1982 and won further convincing victories in 1983 and 1987. Left-wing parties lost power in Norway and Belgium (1981) and the Netherlands (1982) and although France elected a socialist president in 1981 François Mitterrand proved to be more explicit than his predecessor in his support for the deployment of cruise and Pershing II missiles in Europe, even though he was also critical of American policies in Africa and Central America and went so far as to give encouragement to Reagan's enemies in Nicaragua. Until the unforeseen advent of Gorbachev in 1985 therefore the western alliance was at least superficially united behind the tough attitudes expressed by Reagan, and disarmament at all levels from the SALT process to non-nuclear forces appeared to be stalled.

But there were currents of a different kind. The shift to the right in Europe cleared the ground on the left for criticism of American policies more trenchant than left-wing parties were wont to express when in government. Reagan's intemperate bellicosity towards the USSR, however congenial in the United States at this stage, found fewer echoes among Europeans who, much as they disliked and distrusted the USSR, were convinced of the need for a *modus vivendi* with it and did not believe that the threats and abuse to which the president seemed naturally drawn were a sensible way forward. They consorted ill with his affirmation of a two-track policy – strength plus negotiation – with the result that his readiness to negotiate was regarded as shallow and insincere, particularly after his unbridled speech at Orlando, Florida, in March 1985 when he stigmatized the USSR as an evil empire destined for the dustbin of history. His apparent belief that the USSR could be forced to disarm by an escalation of American armaments – a bigger Trident programme, the resurrection of the discarded B-1 bomber, MX missiles, all of which were approved by Congress and, in the case of MX, later expanded – dismayed many among his allies as both dangerous and silly, his espousal of his 'Star Wars' programme doubly so. The United States was also developing to the point of production a greatly more potent anti-tank weapon, the Enhanced Radiation Warhead (or neutron bomb), which would be able to devastate a Russian armoured advance, while in the background were laser beams, chemical weapons and other alternatives to nuclear weapons.

To these complexities were added political and economic squabbles within the western alliance and difficulties of a more technical kind between the two alliances, Nato and the Warsaw Pact. Both trends hampered the pursuit of agreements on disarmaments and arms control.

The politico-economic difficulties were transient. When General Jaruzelski imposed martial law in Poland and imprisoned many of his political opponents Reagan, treating Jaruzelski as a mere Kremlin puppet, imposed sanctions on the USSR as well as Poland and did so without consulting his European allies who regarded these moves as ineffective or worse. He in his turn regarded them as soft on the USSR and their proposals for a mere freeze on Polish borrowing as ludicrously feeble. A more serious conflict arose when Reagan moved to obstruct the building of a pipeline for conveying gas from the USSR to Europe – a project of significance for western Europe as well as the USSR but regarded in the United States as malign strengthening of the USSR, which would double its exports of gas by the end of the century and so earn much needed foreign currency. In using events in Poland to impose sanctions against the USSR Reagan threatened those European businesses which were participating in the construction of the pipeline and had subsidiaries or operations in the United States. This threat to extend the effects of sanctions to European firms was resented as impertinent and illegal. A compromise was reached at Versailles in June 1982 but it immediately broke down, whereupon Reagan – against the advice of his secretary of state, Alexander Haig, who resigned – imposed sanctions which the Europeans ignored. By the end of that year the affair had fizzled out, governments on both sides of the Atlantic being concerned to secure the planned deployment of cruise and Pershing without too much fuss. For the same reason – the solidarity of the alliance – disagreements over other matters were kept down to the muttering level (over American partiality for South Africa and the consequent stalling of negotiations on Namibia, over American policies in the Middle East and Central America), and even the American invasion without notice of a member of the Commonwealth, Grenada, was handled by Thatcher with decorous restraint.

The second complicating factor was the more technical. It had been convenient to keep discussions on nuclear and non-nuclear weapons separate, and to divide nuclear weapons into distinct categories – long-range, intermediate and battlefield. But this way of handling the broad field of disarmament was proving unreal. At the core of SALT was the attempt to set limits to the delivery systems which either side might possess. The fixing of numbers was a matter for haggling. More difficult was the definition of the systems which the treaty should cover. The weapons on either side were not precisely comparable, nor was it any longer easy to agree which were strategic and which not. The Americans wanted the treaty to include the Russian Backfire bomber which, with its in-air refuelling service, was arguably a strategic weapon capable of hitting American cities; and also those categories of mobile missile which, although classified as intermediate, were easily convertible into an SS.16 strategic missile. The Russians were, in the nature of the case, as much concerned about Nato's intermediate-range weapons in Europe as with the ultimate deterrent based in the United States. The USSR was deploying SS.20s targeted on European cities at the rate of about one a week and striving to delay the renovation of Nato's counter-weapons in the European theatre. Thus the limitation of strategic arms – the province of the bilateral SALT talks – could no longer be divorced from the development and deployment of theatre weapons, which involved all the Nato allies and not merely the United States. The regulation of intermediate weapons (INF) became therefore interlocked with the SALT process, now renamed START. It also assumed from the mid-1980s a greater momentum because it offered better prospects of agreement.

Throughout the 1970s Nato's nuclear capacity in Europe lay largely in its aircraft delivering air-to-ground missiles – the British Vulcan and the American F.111 – but in the face of the deployment of the SS.20 Nato resolved in 1979 to deploy new intermediate weapons (464 cruise missiles, non-ballistic, low-flying and very accurate, and 108 Pershing II ballistic missiles) in five European countries, thereby reducing by a factor of five the time needed for a retaliatory strike against targets in the USSR. This decision, urged on the Americans by the Europeans and particularly by West Germany, was to some extent a miscalculation. The SS.20, which began to replace the SS.4 and SS.5 from 1976, was a mobile launcher with three warheads and a range of 5,000 km: the SS.4

and SS.5, in service from 1959 to 1961 respectively, had single warheads with ranges of 2,000 and 4,000 km. Together with the TU-26 (Backfire) aircraft the SS.20 greatly strengthened the Russian position in Europe. The European members of Nato wanted a similar weapon within the theatre in order to square up to the SS.20, reassert the credibility of Nato's strategy in Europe and reknit the Euro–American linkage which, Europeans feared, was becoming frayed. But European leaders underestimated the opposition of their electorates to this proliferation of nuclear weapons and the USSR was able to play on this dissension in the hope of getting Nato's 1979 resolve reversed. In 1981 Nato, in order to placate the critics in its midst, adopted the 'zero option' for INF – that is to say, their total elimination – in the conviction that the USSR would reject it, which it immediately did. The manoeuvring continued with an offer by Moscow to withdraw SS.20s in return for the abandonment of the cruise/Pershing II deployment and a later offer to destroy considerable numbers of them for the same consideration. The key to the manoeuvring was in Bonn and it ended when the Bundestag confirmed the acceptance of West Germany's quota of Pershing IIs. Their arrival at the end of 1983 marked the failure of the Russian attempt to stop the programme. Moscow strengthened its nuclear forces in East Germany and Czechoslovakia and broke off all negotiations, including the START talks on strategic weapons as well as the INF negotiations.

At this point the USSR was confronted with a fresh development which it was even more anxious to obstruct than the deployment of cruise and Pershing II. This was Reagan's Strategic Defence Initiative (SDI), unveiled in 1982. SDI emphasized the distinction between offensive and defensive nuclear weapons. It consisted of startling new proposals for defence against nuclear attack which, if practicable, would destroy nuclear missiles after they had been fired and so provide invulnerability against a first strike. The president proposed to spend $26,000 million on research over an initial five years. This stupendous cost and the implications for the American economy and budget (which the president had promised to reduce) added to the air of fantasy implicit in its nickname of Star Wars. There were also more sober doubts. Many experts thought the proposals intrinsically unworkable or even absurd. They were certainly untestable to any precise degree except in war. They were open to the objection that by the time they were developed counter-measures would also have been developed. They seemed to entail contravention or denunciation of the ABM treaty of 1972 or a request to amend it which would give the USSR a chance, by refusing, to provoke fresh friction in the relations between the United States and its European allies. (Washington began at this time to claim, correctly, that the ABM treaty had already been broken by the USSR.) Most important: the SDI aimed to replace deterrence by defence and so was bound to undermine the deterrence which, however crude and unappealing, had greatly helped to keep the peace between the superpowers for a long generation. In western Europe governments were unwilling to be too openly critical, partly because they did not want to add to discord within Nato but also because Reagan invited them to share in the research and they sensed, mistakenly, that there was a great deal of money to be made by accepting.

The disarmament scene was transformed by a change in leadership in Moscow and a change of heart in Washington. Gorbachev was a new broom who needed above all to cut costs in order to save the USSR from catastrophe and Reagan, whether coincidentally or not, decided to exchange the role of Great Scourge for that of Great Peacemaker. In 1985 the two leaders met in Geneva. Nothing of great consequence was agreed but they met and – as when Napoleon and Alexander met on their raft at Tilsit – there was a kind of humanity about the mere event which cheered onlookers and was deemed to be a part of statecraft. More potent was the stark fact that the USSR was incapable of financing the Cold War and the United States too was heading towards insolvency. As both superpowers took account of the depreciation of their super-status in the world, they were forced to acknowledge the need to disarm. Nor were they any longer prevented by *raison d'état* from doing so. The myths which had played a major part in inflaming the Cold War and sustaining it were grown stale and threadbare. Hostility remained but fear had receded.

The position on disarmament when Gorbachev came to power in 1985 was as follows. The reduction of non-nuclear forces in Europe had been under discussion in Vienna since 1974 between all the members (twenty-two) of Nato and the Warsaw Pact. These discussions were designated MBFR (Mutual and Balanced Force Reductions). Proposals for thinning out conventional forces in central Europe were first made by Nato in 1969 and 1970, and in 1973 Nato proposed a 15 per cent cut by both sides in the central sector to be followed by a reduction to 700,000, which proportionately would be a bigger cut by the Warsaw alliance than by Nato. The Warsaw Pact responded by proposing initial cuts of 20,000 by both sides, followed by a 5 per cent reduction and then a 10 per cent reduction. This scheme, more proportional than numerical, reflected the Warsaw Pact's advantage in numbers, which it designed to keep. Nato on the other hand favoured a mainly numerical approach leading to a parity in numbers which it did not possess. After these opening gambits nothing much happened and nobody seemed to be in a hurry that it should, until 1979 when Brezhnev – possibly in order to sow doubts among the Nato allies about the wisdom of Nato's programme for modernizing its theatre weapons – announced the impending withdrawal of 1,000 tanks and 20,000 men from East Germany. The talks continued, however, to be bedevilled by inadequate and suspect statistics and by the problem of verification. They were wound up in 1989 after nearly 500 sessions and were submerged in a new Vienna colloquy in which all European states were involved and the subject matter was extended to the entire area from the Atlantic to the Urals. This successor conference was dubbed CFE (Conventional Forces in Europe). An even larger group of thirty-five states – thirty-three in Europe plus the United States and Canada – deriving from the Helsinki Final Act in 1975 and its review conference had since 1983 been discussing security and co-operation in Europe (CSCE) and it convoked in 1986 in Stockholm a Conference on Confidence- and Security-Building Measures and Disarmament in Europe (CDE). Whereas the function of the MBFR/CFE negotiations was to reduce the size of the armed forces of Nato and the Warsaw Pact in central Europe and beyond, the CDE was concerned with the movements of these forces and the framing of rules to obviate surprise attack (and the fear of surprise attack) through such measures as advance notification of movements involving more than a given number of men or tanks. Both the CFE and CDE negotiations made satisfactory progress in the closing years of the decade.

In the nuclear field both strategic and intermediate weapons were under discussion, the former wholly, the latter primarily, in the province of the superpowers. At their first meeting, at Geneva, Reagan and Gorbachev had declared themselves in favour of halving their long-range or strategic nuclear weaponry and finding an interim agreement on INF. Gorbachev had inherited his predecessors' failure to prevent the deployment of new Nato INF in Europe. He was anxious to use his bargaining position over strategic weapons to halt Reagan's SDI. After Geneva he tried to accelerate the disarmament process with a surprising proposal for the elimination of all nuclear weapons of all categories by the year 2000. The United States responded with a mixed package which repeated the Geneva target on strategic weapons and coupled it with the removal of all INF from Europe in three years and the continuing attempt to remove the imbalances in the two sides' conventional forces. Gorbachev declared himself flexible on INF and willing to accept a reduced cut in strategic weapons – 30 per cent instead of 50 per cent – and he also modified his tactics on SDI by implying that some research might be permissible provided the United States undertook to observe strictly the ABM treaty for the next fifteen to twenty years. He thereby set the development of SDI firmly within the existing ABM régime and argued for a precise term during which it might not be amended. The United States, while maintaining the proposal to reduce strategic weapons by half, proposed that the ABM treaty be guaranteed for seven and a half years on condition that SDI weapons might be deployed from 1992 (they could not in fact be ready for deployment before that date) and that after 1992 SDI technology would be shared with the USSR. The essence of the American position was to secure freedom of action after a given period during which the constraints in the ABM treaty would be observed. This was the position when Reagan and Gorbachev met again at Reykjavik in Iceland.

Reagan went to Reykjavik in order to affirm and broaden the goodwill established at Geneva. Gorbachev had precise plans: to stop SDI; affirm the strict letter of the ABM treaty for a fixed term and, if possible, strengthen its provisions on testing by getting the United States to agree that while laboratory tests might be permissible, tests in space were not; remove all INF from Europe; and negotiate the reduction of strategic weapons by 50 per cent or at least 30 per cent. He was prepared to limit INF in Asia to 100, and to discuss – and meanwhile freeze – short-range weapons (less than 500 kilometres). He rejected Reagan's offer to share SDI technology. On his side Reagan wanted all INF to be eliminated in Asia as well as Europe. He was not prepared to yield anything on SDI. Surprise at the progress marked by these exchanges was turned to astonishment by a new proposal which appeared to be approved by both leaders, although some doubt was later cast on Reagan's endorsement of it. This proposal was for the complete elimination of nuclear weapons of all kinds in ten years, this process to be geared to the elimination of 50 per cent of strategic weapons in the first five years and the remaining 50 per cent in the next five. The confused euphoria emanating from Reykjavik was dampened when it became apparent that without concessions on SDI Gorbachev would sign nothing. He had failed in his main aim.

He returned to the attack in 1987 by repeating his Reykjavik offer on INF (zero in Europe, 100 in Asia) but without linking it to agreement on SDI, to which the United States replied by agreeing on condition that the USSR's surviving INF should be out of range of western Europe and Japan. This *rapprochement* brought up the question of short-range weapons which had been skirted at Reykjavik, and Gorbachev sprang another surprise by proposing their total elimination (the 'third zero'). He also agreed to forgo the 100 INF in Asia. The outcome was an INF treaty concluded in Washington at the end of the year, and in force from June 1988, for the destruction of all INF weapons by 1991, the USSR being thereby required to destroy more than twice as many missiles and launchers as Nato.

A START I compact was completed after a ten-year gestation when it was signed by Presidents Bush and Gorbachev in 1991. It was followed two years later by START II, signed by Bush and Yeltsin to come into force provided START I did so. By START II the United States and Russia were to reduce their stocks of land-based strategic nuclear weapons from 12,600/11,000 to 3,500/3,000, sea-based weapons to 1,750 each and eliminate all MIRV'd weapons – all over ten years. Some variations, entailing concessions by the United States, were agreed to take account of particular Russian objections. The implementation of these agreements was complicated when the one party to them split into numerous states, four of which – Russia, Ukraine, Belarus and Kazakhstan – possessed on their territory weapons within the scope of the agreements. The last three of these states separately undertook to implement the agreements, to clear their territory of all relevant weapons within seven years and also to adhere to the Non-Proliferation Treaty. (Ukraine prevaricated but fell into line: see p. 77.) Between the signature of the two treaties, both superpowers took unilateral decisions which further reduced their strategic and non-strategic armouries.

In 1993 a Chemical Weapons Convention, fruit of nearly twenty years' discussions in the UN Disarmament Conference, was signed by 130 states to come into force in 1995 if then ratified by half of them. It provided for the destruction of all stocks by the end of ten years, forbade the possession, acquisition or use of chemical weapons, and instituted a strict system of inspection and verification. Like START, this convention created big and costly practical problems of destruction. Weapons invented to destroy were not themselves easily destroyed.

An agreement on Conventional Forces in Europe (CFE) came into force in 1992 covering the area from the Atlantic to the Urals and signed by twenty-two states. It was bolstered by an Open Skies Treaty, likewise elaborated by the CSCE and designed to facilitate the monitoring of the new and improved confidence- and security-building measures.

The arms race had been an economic catastrophe sustained by the psychology of the Cold War. The abatement of the Cold War created new problems for both sides: for the Russian side, how to preserve the semblance of superpower status; on the western side, what to do about Nato. Nato had owed its effectiveness to its huge – mainly American – power and this power had masked doubts about the credibility of its successive strategies. In its first, formative years Nato had adopted, at Lisbon in 1952, a programme to create forces, including German units, capable of fighting a prolonged non-nuclear war, but these forces never came into existence and Nato was refashioned as a link in a chain or ladder whose credibility depended on the existence in the chain of nuclear weapons of rising malignity. This doctrine of flexible response went through a number of phases. The invention of short-range nuclear weapons enabled Nato to plan for smaller forces with more ferocious weapons but once again the targets (set in 1957) were not reached. Nato's role was, however, set as the role of a delaying agency. Whether with non-nuclear weapons or short-range nuclear weapons or intermediate nuclear weapons Nato would resist an attack and cause the attacker to think again before carrying on with his operations. Although this vision of a future conflict was not demonstrably ridiculous it was difficult to see a war developing in this way and Nato's military function was therefore always ambiguous. It relied on the assumption that the aggressor's first step or steps would be taken without much thought but that the same thoughtless aggressor would quickly be made to think better. But whatever the practicalities were likely to be, the theory demanded a graduated response which, given Nato's make-up, was a graduated nuclear response. When Reagan was persuaded at Reykjavik to approve zero options for both INF and strategic weapons he was removing the steps in the ladder – and at the same time dismantling the American strategic nuclear umbrella which Europeans had learnt to regard as the ultimate guarantee of their security. The European dilemma was intensified when Gorbachev proposed to apply the zero option to short-range or battlefield nuclear weapons also. The response could no longer be flexible and a Russian attack would start with considerable advantages, offset only by the counter-fear of an all-out American riposte with what might be left of the strategic missile force. The reversal of the arms race, coupled with the simultaneous disappearance of the Warsaw Pact, left Nato without a strategy, without an obvious foe and perhaps without a purpose.

The weakening of the United States

The 1980s were the decade in which the superpowers ceased to be regarded as so far above and beyond all other states as to constitute a distinct species. This was pre-eminently true of the USSR which no longer looked like a match for the United States, was no longer able to dominate central or eastern Europe, was in acute economic decline and was threatened with disintegration. The depreciation of the United States, although of a very different order, was hardly less startling, given its unchallenged superiority over the past generation and its continuing massive resources and massive industrial and agricultural output.

The decline in American power and prestige was self-inflicted, a consequence of mistakes and misjudgements in economic management and external policies. Fumbling in foreign affairs was most unhappily displayed in Central America, where Reagan failed in his declared intention to pacify the small republic of El Salvador and make it safe for a decent right-wing democracy. In Panama the United States subsidized a known drug dealer and, when his crimes became too blatant, failed to unseat him by bribery and resorted to a military invasion on a flimsy pretext. In Nicaragua Reagan failed to overthrow the Sandinista régime in spite of waging an expensive vicarious war and resorting to clear breaches of international law, for which the United States was censured by the International Court of Justice; the government of President Daniel Ortega was later defeated not by arms but by the economic sanctions which led Nicaraguans to vote against Ortega in elections of surprising democratic rectitude. (For these matters see Chapter 27.) These failures against weak neighbours in the American continent itself betokened an inadequate grasp on the uses of power. The United States was the world's greatest military and industrial power and yet unable to operate efficiently as a regional power.

In the Middle East the United States, confronted with the loss of its ally, the Shah of Iran, and the seizure of hostages from the American embassy by the new régime, became mired in contradictions and subterfuge: the moral imperative to rescue the hostages clashed with the publicly reiterated determination not to let terrorism pay. During the elections of 1980 the release of the hostages was obstructed to Carter's detriment and Reagan's benefit. It was then effected by secret negotiations involving the supply of arms to Iran via Israel from 1980 to 1986 when public scandal put an end to it. High prices were exacted for the arms and the money was used to arm the Nicaraguan Contras by devious means by which the conspirators – who included President Reagan and Vice-President Bush – deceived the Congress and people: a combination of the illicit and hypocritical which the president and his advisers sought to justify on vaguely ideological grounds and by assuming that supposed moderates in the new Iranian régime could be divided from immoderate intransigents.

The second and associated venture arose from the determination to go on arming and funding the Contra opposition to the Nicaraguan government covertly and in defiance of the Boland amendment whereby in 1984 Congress had resolved against providing further military aid. The approach to Iran, pursued in collaboration with Israeli and other arms dealers, involved supplying Iran with weapons with which to wage its war with Iraq; no arms would go directly to Iran from the United States but from stocks in Israel which the United States would then replenish. Senior American officials travelled in secret to Teheran with a consignment of parts for missile launchers. But both sides were expecting more than they could reasonably hope to get. Iran raised its demands not only for weaponry but also for political concessions such as an Israeli withdrawal from southern Lebanon and the Golan Heights and the release of prisoners held in Kuwait on charges of terrorism in Lebanon. One American hostage was released in exchange for more spare parts, a second for yet more weapons. There were many clandestine meetings and much mutual deception by overstatement. The proceedings became public knowledge. Reagan dismissed his national security adviser, Admiral John Poindexter, and his principal subordinate in these matters, Colonel Oliver North, who managed to destroy much documentation before quitting his office. Reagan's first public statements were untrue, and although he knew that the Boland amendment was being bypassed in numerous ways his positive knowledge of the Iran–Contra operations remained unproved throughout the remainder of his presidency. In later testimony he more than a hundred times pleaded an inability to remember crucial matters, so that it was not clear whether he had not been told or had not understood or whether neither of these evasions was valid. His reputation for straightforwardness, competence and application to business did not recover. The importance of the Iran–Contra affair was not that it weakened the United States militarily but that it caught the attention of the world and raised questions about the dependability of the United States in world affairs. The smell of scandal was the more harmful since it came not much more than a decade after the astonishing behaviour of Richard Nixon who had been forced in 1974 by the exposure of his chicanery and mendacity to resign the presidency and transfer it to a congressman of mediocre capacity, Gerald Ford, in exchange for exemption from impeachment.

For the Euro–American alliance these American shortcomings were troubling since the position of the United States in the alliance was uniquely important: no anti-Russian alliance without the United States was conceivable so long as the Cold War prevailed. The alliance had always suffered strains, although these were usually repaired by the crassness of Russian policies (before the advent of Gorbachev) in Europe and beyond. The most serious rift in the 1960s was the Franco–American acerbity which half-removed France from the alliance, but soon after the departure of de Gaulle French units again took part in Nato's naval exercises in the Mediterranean (1970). The 1970s were nevertheless often uneasy. Nato was no longer a preponderantly American host with European facilities: in European theatres and waters the European allies were providing 75 per cent of the air forces, 80 per cent of naval forces and 90 per cent of land forces. The American contribution was symbolic and financially crucial, so that Americans were moved to argue that Europe could not have this costly American military umbrella and at the same time obstruct American policies in other directions. The war in Vietnam had been fiercely criticized in Europe, but the war in the Middle East in 1973 raised ill will to governmental level. The United States was angered by European refusals to permit the use of airfields for the air lift to Israel and by their hurried truckling to Arab threats of an oil boycott.

Europeans retorted by pointing out that they depended on the Middle East for 80 per cent of their oil, the United States for 5 per cent, and by chiding Washington for making policy on the Middle East without consulting its allies and then expecting these to assist it. Europeans were further estranged when Washington appeared to be toying with schemes for assuring the flow of oil by force of arms and they were reluctant to attend a consumers' conference proposed by the United States as a way of putting pressure on Arab producers. For similar reasons France refused to join an Energy Authority created within the OECD or to take part in a consumers' oil-sharing agreement. Europeans preferred a conference between consumers and producers, negotiation rather than confrontation.

At this point the alliance was virtually in abeyance and matters were not improved by the Turkish invasion of Cyprus in 1974. Greece, blaming the United States for not taking a firmer stand against Turkey's excessive exploitation of the Greek dictatorship's inept interference in Cyprus, withdrew from active participation in Nato operations – a protest caused by events in Cyprus but also grounded in a more pervasive anti-Americanism which had grown with American benevolence towards the dictators throughout 1967–74. This Greek hostility was not offset by any countervailing Turkish sentiment, since the US Congress, taking the Greek side, voted in December 1974 to cut off aid to Turkey whereupon the Turkish government took control of twenty-four American military installations in Turkey, concluded a treaty of friendship with the USSR and accepted a large Russian loan.

At the diametrically opposite corner of Nato's territory two other members were engaged in a different conflict which too had implications for Nato installations. In 1972 the Icelandic parliament resolved to extend fishing limits to 50 miles. This act, which particularly affected Britain and West Germany, was a unilateral alteration of treaty dispositions of 1961. At the same time the Althing rejected in advance recourse to the International Court of Justice (which however ruled in August 1972 that British and West German vessels had the right to fish to a 12 miles limit). West Germany and Iceland compromised the resulting dispute in 1975 but with Britain, whose interests were more severely affected, Iceland's action led to armed clashes as Britain provided its fishing vessels with naval protection against the armed Icelandic coastguards trying to drive them away or destroy their gear. A two-year agreement was reached at the end of 1973, limiting the areas in which British vessels might fish and the type of vessel that might be used. This was a way of limiting the catch. Iceland, however, also declared that in 1975 it would extend its exclusive fishing rights to 200 miles. The dispute remained legally unresolved but the tensions were reduced in 1976 by a considerable British abandonment of reasonably well-founded rights. For Iceland the episode was an unyielding assertion of vital economic claims assisted by the advantages of operating in home waters; the real embarrassment of the British as a whole (as opposed to the fishing community); the likelihood that the current conference on the law of the sea would in any case recommend substantial extensions of normally accepted fishing limits; and the Nato connection which could be used to bring pressure on Britain by the United States which did not wish to see Nato's strategic installations in the northern sector imperilled by Icelandic action like that of Greece or Turkey.

In the midst of these aggravating conflicts and policy disputes Nixon upbraided his allies for ganging up on the United States. Kissinger, no less irritated but more constructive, proposed in 1973 a new Atlantic Charter to define the common aims of the United States, western Europe and Japan (added because of the economic conflicts which the United States had with both Europeans and Japanese). The United States, Kissinger said, was prepared to defend western Europe and continued to approve European unity, but it objected to Europeans concerting, among themselves and without consulting Washington, policies objectionable to the United States – which was pretty much what Europeans were objecting to in reverse. The seventies closed with the Russian invasion of Afghanistan which evoked in Washington proposals for action which many Europeans regarded as futile, while in relation to the Middle East European leaders, sceptical about the Camp David accords, set themselves to devise a European policy which, although expressed as a sequel to Camp David, was more precisely an alternative to an American policy which had in their eyes failed.

Then there was the question of paying for the alliance. For years the United States had borne a heavily disproportionate share of the costs and had regarded this burden as both inevitable and just so long as Europe was prostrate after the Second World War, but Europe recovered (with massive American aid) while the United States began to feel the unfamiliar pains of economic overstrain. The Europeans had to acknowledge the unfairness of the economic burden-sharing but when in 1977 Nato resolved that all members should increase their defence spending by 3 per cent a year in real terms many of the Europeans failed to live up to their promises. In this context the question of leadership was an added irritant. There could be only one leader – the president of the United States. But respect for the presidency was calamitously impaired when Lyndon Johnson decided not to run for a further term because he felt that he had failed. Every president after Johnson contributed, sometimes powerfully, to the decline in the standing of the presidency among Europeans and so boosted the latent anti-Americanism which lay not far beneath the surface of European opinion. Yet this anti-Americanism was not so much a foundation for an alternative policy as an emotional luxury.

In retrospect the Helsinki conference of 1972–75 on Security and Co-operation in Europe (CSCE) proved a significant twist in the pattern of European affairs. It was attended by thirty-five states, including the United States and Canada which were accepted as rightful and essential members of the European polity. When the conference ended it was projected forward by providing for periodic re-assembly in review conferences. Its participants wanted different things but got enough of what they wanted to give this wide forum – much the most comprehensive assembled in Europe since the war – a semi-permanency which presented a challenge to the prevailing bipolarity fashioned by the superpowers and to the exceptional position of each superpower *in suo orbe*.

The instigator of the Helsinki conference was the USSR. The west concurred after insisting on the inclusion of the United States and Canada. The Russian purpose was to secure general endorsement of the post-Hitlerian frontiers in Europe which no peace conference had ever ratified; and, secondly, to discuss security in terms of military bases and troop levels. The western approach to the conference was at first a mixture of boredom and cynicism, but towards the end it decided to use the occasion to exact conditions from the USSR. The west and the neutrals combined to reject a Russian proposal to declare Europe's frontiers immutable; the conference declared only that they should not be altered by force. Further, the west and the neutrals insisted on a wider interpretation of security, embracing not only military dispositions but also mutual understanding. Consequently the Helsinki Final Act contained declarations – not legally binding but nevertheless formal and normative – concerning governmental and non-governmental contacts for economic, social and technical co-operation and what was called co-operation in humanitarian fields. This phrase opened the door to discussion about human rights and breaches thereof. A Russian attempt to limit the ambit of these declarations to discussions between systems, as opposed to states, was defeated. Had it been accepted, the Final Act would have permitted abstract debate about the relative merits and vices of the capitalist and communist systems but not criticism of the policies and practices of particular states. The USSR, which had initiated the Helsinki conference, was out-manoeuvred at it and came to regret that it had ever set this particular ball rolling.

The Helsinki conference also provided for periodic reviews of the implementation of its undertakings. The first review took place in Belgrade in 1977 in a climate of considerable animosity. It achieved nothing except agreement to meet again. This next meeting began in Madrid in 1980 after protracted Russian attempts to abort it. The west predictably made great play with the Russian invasion of Afghanistan and the plight of dissidents in the USSR. France proposed yet another conference on disarmament in Europe and Brezhnev spoke warmly of confidence-building measures which he was willing to extend to all Europe up to the Urals. There was nothing new in any of these proposals which evinced only a determination to keep talking and to find something to say. But Helsinki's greater effect was outside these review conferences, for the Helsinki process coincided with a ferment in central Europe, amounting in Poland to a political revolution. The Helsinki declarations and the Helsinki watch committees set up in many countries (including the USSR) to monitor the behaviour of the signatories, contributed to this turbulence of hopes, unquantifiably but far from insignificantly, and after 1989 the CSCE became the pan-European forum which was felt to be needed after the demolition of Russian and communist power in central and eastern Europe and the consequent confusion over the continuing purposes and usefulness of Nato. In 1990 the CSCE participants established a standing headquarters as a step away from armed bipolarity and towards a new, if dimly perceived, order in which confrontation between Europeans and also between the superpowers in Europe would be relegated, if not to the dustbin, at any rate to the pages of history. The CSCE was bidding to become the regional organization for Europe in terms of article 51 of the UN Charter and so a way to moderate and – since the United States was a member – retain American responsibilities and interest in Europe. In 1994, by which date it had fifty-three members, the CSCE was renamed OSCE (O for Organization). But it possessed no armed forces. Nato did but was bereft of enemies, a strange situation for a military alliance. Nato's members wanted to preserve the alliance, its former enemies wanted to join it. For the United States Nato was a symbol of American concerns about Europe and continuing involvement in European affairs; a warning that the revival of hostile Russian power could not be permanently discounted; a form of internationalism more acceptable to the American Congress and people than membership of the UN. Nato was a firmly established organization which the OSCE was not. Therefore the United States wanted to adapt Nato to circumstances by enlarging it. Its allies did not disagree. But embracing new members from eastern Europe was hard to effect without antagonizing Russia (which claimed membership for itself if any other state were admitted).

In 1993 the German minister for defence Volker Ruhe publicly supported an association of Nato (and WEU) with Poland, Hungary and the Czech and Slovak Republics. Yeltsin announced in Warsaw that Russia had no objection to Polish association but a few weeks later he formally informed the United States, Germany, France and Britain that there must be no eastward expansion of Nato; his perennial suspicions of an American intention to use an enlarged Nato as a threat to Russia were sharpened by American keenness to use Nato forces against the Serbs in Bosnia (see Chapter 8).

The United States devised a plan under the title Partnership for Peace for associating former communist enemies with Nato, largely in order to delay the question of membership: more than twenty countries in Europe and ex-Soviet Central Asia accepted the invitation. A Nato Participation Act empowered the president of the United States to transfer surplus arms to these countries.

Yeltsin was pulled two ways. In 1995 he declared his willingness to join the partnership but in the same year vociferous nationalism in Russia and the approach of elections induced him to protest against Nato exercises in conjunction with some of its new partners near Russia's borders. He also protested against American insistence that military operations in Bosnia be conducted under Nato's command (thus excluding Russia). These cross-currents coincided with internal discomfort in Nato when the newly appointed Secretary-General, Willy Claes, was forced to resign upon being charged in Belgium with bribery and corruption. Some Nato members formed regional forces which, ambiguously, were presented as a reinforcement of Nato's European pillar, reinvigoration of the WEU and a beginning of a military arm for the EU. Such was the Franco–German force which was joined by Belgium, Luxembourg and Spain and two (land and sea) Mediterranean forces established by France, Italy, Spain and Portugal.

In economic affairs Reagan inherited a perplexity of problems and made them worse. American supremacy in the world was before all else economic and its fame throughout the century had derived from the incomparable performance of American capitalism in times of adversity as well as times of ease. One of the main elements in the post-1945 international order was the economic system devised in 1944 at Bretton Woods and comprising the World Bank and the International Monetary Fund (IMF). This system presupposed the ascendancy of the US dollar and enshrined the importance of stable exchange rates with the dollar itself pegged to – or determining – the price of gold (at $35 the ounce). In 1945 the strength of the American economy and therefore of the dollar were axiomatic and within the Bretton Woods span of years the United States exported capital on a huge scale, partly in pursuit of foreign policies which demanded massive expenditure on forces overseas and ultimately in waging war in Vietnam, and partly in capital disbursements by corporations investing in and buying up foreign enterprises. At the same time the United States began, intermittently from 1959, to run deficits on its foreign trade and to seek to finance all these operations with no – or, after 1968, only minor – contributions from American citizens in the shape of higher taxes. The consequent growth in the 1960s of piles of Eurodollars (dollars outside the United

States) added both to the uncertainties about the continuing strength of the dollar and to the difficulties of managing the American currency and to the very existence of the ruling world economic order. In the 1970s American industrial growth slid behind the Japanese, German and French and even the Italian, while sharp rises in the price of oil in 1973 and 1979 – coinciding with the conversion of the United States into a net importer of oil and clashing with a habit of cheap energy – compounded the discomfiture of the dollar and the collapse of the system.

The biggest shock to the Bretton Woods order was the soaring price of oil which was itself a function of, first, the transfer of Middle East oil to Middle Eastern ownership and political manipulation and, secondly, war and revolution in the Middle East. OPEC, the cartel of mainly Middle Eastern producers and exporters created in 1961, did little to alter prices in its first ten years but prices were edging upwards in the early 1970s and were increased tenfold during that decade which included the war of 1973 and the overthrow of the shah's régime in Iran in 1979. This economic revolution reversed itself as it drove major oil companies to prospect and produce outside the Middle East with the result that the oil price collapsed around 1988 (as did OPEC's power and cohesion). But during the 1970s and early 1980s governments did not foresee this outcome. They were more harassed by the collapse of the Bretton Woods monetary system, particularly the devaluation of the dollar by one-third in terms of gold. This was perhaps the most enduringly damaging consequence of the Vietnam War and Washington's financing of it by borrowing – damaging worldwide as well as to the United States, since it spelt an end to exchange stability and to growth. In the ensuing Reagan–Bush decade the world's economically leading states – the Group of Five, from 1986 the Seven – tried to establish an interim system. The leadership was initially European (Franco–German with Valéry Giscard d'Estaing and Helmut Schmidt), then American. The Plaza (1985) and Louvre (1987) agreements were an attempt by central banks to manage the devaluation of the dollar by financial measures which were unsuccessful because they were unaccompanied by similarly co-ordinated fiscal measures by governments. The very large sums expended by the banks were largely spent in vain. At the end of 1987 crashing prices on Wall Street advertised the extent of the malady and of public and private fears that the administration had lost control of budget and trade balances. Growth which had averaged about 3 per cent a year from 1945 to the beginning of Reagan's presidency was almost zero. Domestic saving and investment were at their lowest recorded levels and unemployment wavered between between 10 and 7.5 per cent over the twelve years of Republican rule. The Group of Seven sidestepped the predominantly American anxieties over the overvalued yen, Japan's almost unimaginable trade surpluses, hyperinflation in Russia and other menacing consequences of the collapse of communist rule in half Europe and the disintegration of the Soviet Union.

In this context Reagan's policies of deficit financing (not confined to the United States) and tackling inflation by cutting output and jobs contributed to the decline of the United States' unique position in the world. His determination to boost American pride and self-confidence to his own cheery level was accompanied by an equal determination to reduce taxes. Since his way to pride was papered with unprecedentedly costly expenditure on weapons of war his term of office was marked by massive borrowing and the neglect of social services. His promise on entering the White House to balance the budget was even rasher than such promises usually are. He seemed to believe that the gap between spending and revenue would evaporate because lower taxes, in association with monetarist controls, would produce higher profits and so higher tax yields to bridge the gap. But low taxes and tight money did not lay these golden eggs. Deficits grew both absolutely and as a percentage of GNP, after 1982 monetary constraints were relaxed and interest rates lowered, growth continued but so too did the gap, and the only salvation lay in pulling in foreign money to finance current government spending and domestic investment: budget deficits of tens of millions of dollars were half financed by Japanese and other foreign investors who might change their minds unless tempted by higher and higher interest rates. The dollar was allowed to rise but rose so strongly that its exchange value ceased to be credible and it fell even more spectacularly than it had risen.

When Reagan left office in 1989 the United States had swung in less than a decade from being the world's biggest creditor to being its biggest debtor. Its external debt, which exceed $660,000 million, had risen by 25 per cent in a single year. External deficits reached $12,000 million a month and interest on external debts was costing $50,000 million a year. Export business was crippled and foreign assets were being sold off. Within the United States capital resources shrank by $500,000 million as shares in corporations were extinguished or replaced by debt (particularly through the invention of junk bonds). Half the population was worse off in real terms than it had been in 1980. Personal savings had fallen below 15 per cent (half the Japanese rate); higher education in technology and science was in decline; the economic infrastructure was in decay and so were inner cities where housing and infant mortality approximated to the black spots of the Third World and crime and drugs were alarmingly prevalent. Corruption in the public sector was widespread up to the level of the cabinet. These ills were remediable but they required a stark reversal of attitudes and a robust exercise of political will. Given the strength of manufacturing industry, budget deficits could be handled by modest increases in taxation; social decay could be tackled by abandoning Reagan's view that government must be minimized – itself an abdication of responsibility. For the time being, however, the United States seemed to be no longer the sole front-runner in a world in which Japan and the European Community were the thrusting societies. The confidence which Reagan had given Americans through military might and talk was being jeopardized by economic and social malfunction and mediocre leadership. Reagan armed the United States against the USSR but disarmed it against Japan, an equally aggressive adversary although one whose weapons were not military.

Budgetary and external deficits, whatever their narrowly economic significance, sapped domestic and foreign confidence and destroyed the standing of the dollar as a world currency on its own or a currency of last resort. In this political and psychological climate the dollar had a predisposition to sink against the major currencies and particularly against the yen which had an even greater predisposition to rise for a special reason: that the Japanese were saving twice as much per head as were the Americans and were not using their savings to buy American or other foreign goods. Although Japanese spending habits were to a large extent cultural, Americans believed them to be controlled – more comprehensively than was probably the case – by the covert protectionism of Japanese governments using bureaucratic obstructionism and deviousness to make life for foreign exporters impossible.

The Bush administration began with a general weakness and a rare rebuff. For good or ill it was not Reaganite but sub-Reaganism presided over by a vice-president who had not been able to emerge from the vice-presidential shadows either at home or abroad. The rebuff was the refusal of the Senate to approve the new president's nominee as Secretary of Defense. Further, in the months preceding mid-term elections in 1990 the president became seriously and damagingly embroiled with both parties in Congress over ways of reducing the budget deficit. Added to the kinds of conflict which stirred sharp emotions – drugs and crime, abortion, ethics in government – the crisis over the budget demonstrated not only Bush's vacillating character but also two other phenomena, little appreciated outside the United States. The first of these was the fact that, although the American president was commonly described as the most powerful man in the world, the American constitution provided that power and decision were to be shared between the president and the representatives of the people and that the American president had, within his own democracy, less power than many heads of government in other democracies, let alone open or covert autocracies. Secondly and contrastingly, the long-standing growth of the central bureaucracy was turning the Congress from partnership with the executive into a more entrenched opposition to the president, from whichever party he might come.

External affairs were more congenial to Bush than domestic. He filled the office of president at the climax of his country's greatest postwar triumph – the liberation of central and eastern Europe from communist rule and the elimination of the world's second superpower. He scored resounding successes in the war against Iraq in 1991 and proclaimed the advent of a new world order. Yet these exertions and proclamations were not enough to win him a second term in the face of domestic discontents. In 1992 the Democrats regained the White House with Bill Clinton as president. Clinton's victory lasted only two years, effaced by a Republican counter-attack in his first mid-term elections. He took office after winning merely 43 per cent of the popular vote. He was intelligent and articulate, forthright in speech if not always consistent in what he said. He eschewed rodomontade and appeared ready to face responsibilities but he made a shaky start from which he never wholly recovered.

Clinton's first concern was to win congressional and popular approval for a budget which would combine hard measures for reducing a frightening federal deficit with applause or at least resignation: but he won in the Senate by no more than the casting vote of his vice-president. This precarious beginning was accompanied by a number of ill-judged appointments to high office and the revival of scandalous allegations against his private life and his earlier business affairs. It was followed by the defeat of an over-ambitious attempt to introduce nationwide health care, irrespective of age. In foreign affairs he appeared strangely unprepared, uncertain and therefore inept. Although the economy was recovering the politically important middle classes felt little benefit since, paradoxically, economic recovery proceeded simultaneously with a continuing decline in middle-class incomes. Mid-term elections in 1994 dealt the Democrats one of their severest reverses of the century, including the loss for the first time in forty years of a majority in the House of Representatives as well as the Senate. The cause were largely domestic but external affairs contributed to a mood of perplexity and apprehension rising in places to a resentful and censorious exasperation. At first sight this was strange since the United States had only recently marked with triumph the dissolution of the Soviet Union – an unambiguous victory in an all-consuming global conflict. But world communism had provided a compass in world affairs and its disappearance left much of American foreign policy directionless. This aphasia roughly coincided with the removal of the second main assumption about the position and purpose of the United States in the world: that in the international economy the United States outstripped every other state. From one angle the end of the Cold War was victory for capitalism over communism, leaving capitalism without a rival. But capitalism was triumphant without being in good health and the role of the United States in a world capitalist system was increasingly ambiguous from the mid-1970s. Domestically, promising economic growth in the early 1990s was countered by large budgetary and external deficits, by the failure of political will to tackle them except by promulgating long-term visions of financial rectitude, and by the awareness that growth was funded not by savings but by foreign (particularly Japanese) capital which could not be relied on indefinitely. Although capitalism had, in popular parlance, defeated communism under American leadership, the United States seemed hazardously uncertain about how a modern capitalist system, domestic or international, should be managed.

THE DISINTEGRATION OF THE USSR

The problems of the second superpower, the USSR, were very different from those of the United States. Its economic problems were of a different order of hopelessness, its society more corrupt, cruel and inefficient, its very existence as a union in question. After Khrushchev's dismissal in 1964 Leonid Brezhnev had emerged as his successor and remained at the head of affairs until he died in 1982 after a long but – in domestic matters – stagnant term and a slow personal decline. He had three successors in three years: Yuri Andropov who died in 1984, Konstantin Chernenko who died in 1985 and Mikhail Gorbachev with whom the long-awaited next generation at last reached the top. By this time the USSR no longer even looked like a match for the United States. Its empire in central and eastern Europe was unsustainable and the Union itself was threatened with disintegration on all sides of its Slav core. It was not without material resources but, mismanaged and under-equipped (except in certain heavy industries), they sufficed neither to feed its people or provide an acceptable standard of living, nor to make the USSR a world power. Gorbachev – intelligent, courageous and politically exceptionally agile – embarked upon a revolutionary course of economic and political reforms under the twin banners of *glasnost* and *perestroika*: *glasnost*, meaning openness and in particular an end to the pervasive falsification of economic performance; *perestroika*, meaning the restructure of the economy in the broadest sense of that word. He insisted that *perestroika* was not to be achieved without *glasnost* and that *glasnost* entailed not only inroads on censorship and habits of subservience but also reform of the entire political system, including the abolition of the Communist Party's monopoly of power and its control over the institutions of the state and the machinery of the economy. *Glasnost*, however unpalatable to some, was easy to understand. *Perestroika*, however, was a more ambiguous concept since it betokened change without specifying the pace of change or defining the new system into which the old system was to be changed. First steps included greater independence for co-operatives and for the managers of state enterprises, and the introduction of regulation by market forces to some extent. Even if it were agreed (which it was not) that there was a wrong system and a right one, there would still be difficulties in getting from the one to the other. How far, for example, should prices be allowed to find their own levels, if that meant that they would soar, putting goods out of reach of purchasers and leaving producers without a market? Should the prices of some things – food, for example – be controlled and, if so, by whom and on what principles and for how long? The government was caught between conflicting needs to let prices rip and to moderate their inevitable rise. While there was all but universal agreement on the need for a new economic order, there was no agreement on what it was to be: Lenin's New Economic Policy (which promoted small retailers and businesses in the hope of attracting foreign capital and skills), or the state of affairs before the NEP, or a new mixture of private capitalism and free trading and state socialism, or a plunge into something hardly distinguishable from western capitalism. The development of *perestroika* was therefore tentative and shapeless. It was contested in varying degrees; was obstructed by the thousands of people whose posts were likely to be put at risk by it; and was further complicated by the state of the economy which continued to go generally backward and was as fit for surgery as a patient with a weak heart. The economy suffered adventitious blows from the slump in the price of oil, the disaster to the nuclear reactor at Chernobyl in 1986 and an exceptionally grim earthquake in Armenia in 1988.

A first basic law for economic reform in 1987 began the process of decentralization, deregulation of prices, and financial rewards for enterprises but gave industry little freedom to shop around for its supplies and made little alteration to the discredited system of central targeting. These partial measures were extended a year later, although still in a tentative and experimental mode and limited to special zones and undertakings below a prescribed size. They were hampered by the lack of trained managers and by the inertia or opposition of the *nomenklatura*, the privileged and ossified administrators who, having survived the Brezhnev years, were in no mood to lose their jobs and perks in the aftermath.

Gorbachev's political aims did not include diminished executive authority at the centre. The road to power in the USSR might be made freer but the power at the end of the road was to be no less comprehensive and perhaps more so. As with the development of *perestroika* in industry and commerce, so with the accompanying political reforms, Gorbachev was not so much a strategic thinker as a nimble tactician who remained in control of the processes which he inaugurated by quick perception and quick moves which kept him ahead – if only just ahead – of events. At an extraordinary party conference in 1988 he promised to convoke a Congress of People's Deputies of 2,250 members, some chosen by special groups of which the Communist Party was one but the greater number (1,500) in territorial constituencies where the voters were promised a choice of candidates. All candidates were required to issue election manifestos. In elections which followed in the next year there was no contest in 384 constituencies, while in 271 no candidate got half the votes cast. The Congress elected a smaller Supreme Soviet of 750 members which met when the Congress dispersed. At the same time and by something like a coup Gorbachev secured the disappearance of hundreds of party regulars likely to obstruct his plans, purged the Central Committee of the Communist Party and had himself elected president of the USSR in the place of Andrei Gromyko who obligingly resigned. His powers as president were to be considerable but not absolute. The Supreme Court might declare his acts to be unconstitutional; two-thirds of the Supreme Soviet might override a presidential veto of new laws. Although he side-lined the Communist Party, he was obliged to accord prominence at the centre to the republics by instituting a Council of the Federation consisting of himself and all the presidents of the fifteen republics, and by abolishing the Council of Ministers – stronghold of centripetalism – and replacing it by a less prestigious and less powerful cabinet of experts. The resignation at the end of 1990 of his foreign minister Eduard Shevardnadze – a close colleague, successful minister and known advocate of liberal reforms – weakened Gorbachev and reinforced suspicions that he was being forced to veer to the conservative side and might even be becoming a captive of military leaders fearful of losing essential defence installations in dissident republics: the armed forces were, with the KGB, the principal outward and visible sign of centralized power as opposed to the centrifugal ambitions of Baltic and other dissidents or the champions of greater autonomy in the Russian Republic (RSFSR) and the Ukraine. Gorbachev's weakened position was exposed when, seeking approval for his nominee for the new post of vice-president of the Union – for which Shevardnadze had seemed destined – his candidate was rejected by the Congress of People's Deputies and only subsequently approved after a second, strictly unconstitutional, vote.

None of these shifts and changes produced an economic policy. As the economy crumbled two competing strategies imposed themselves: to move fast or very fast. To the bolder or more desperate spirits the situation required a dash for change, damning the consequences and hoping for the best. The protagonists of this strategy produced a

500-day plan for installing a mixed economy in four stages. The first comprised selling government and Communist Party properties and enterprises, dissolving all state and collective farms, giving occupiers their plots or apartments, and cutting the budget by 5 million roubles in three months (including sharp cuts in the costs of defence and the KGB). The prime aim of this stage was to get hoarded money – money hoarded because there was nothing on which to spend it – into circulation and to do so before prices were freed and massive inflation let loose. In the second stage prices would be gradually freed and interest rates raised, but the prices of basic foods would remain controlled. In the third and longest stage (days 250–400) half the manufacturing and service enterprises would be sold and a stock exchange and other exchanges would be established. Finally, all these measures would be accelerated and 90 per cent of retail business would be put up for sale. This programme was opposed by, among others, the prime minister Nicolai Ryzhkov on the grounds that the time limits were too rigid and unreal; that more palliatives were needed for the poorer classes, for pensioners and students; and that the bureaucratic machinery for doing so much so quickly did not exist with the result that there would be more chaos than reform. The objectors argued also that the economic disruption which they foresaw would add to the anarchic forces which were threatening to dissolve the Union. Gorbachev, who seemed to favour first one side and then the other and who did not want to lose his prime minister, tried to force the discordant groups to find a compromise. He failed. The Supreme Soviet preferred to give him emergency powers, shuffling off to the presidency the task of finding an answer and imposing it by decree. The outcome was a new plan, duly endorsed by the Supreme Soviet but so vague as to leave the future not only hazardously obscure but apparently beyond the control of the government. There were rival schemes but no coherent policy.

The political and economic transformation of the USSR was bedevilled by the simultaneous upsurge of dissidence, amounting in places to separatism. Of the USSR's fifteen republics (see Maps 1.2 and 1.3, pp. 10–11) only three were preponderantly Slav: the Russian (which included Siberia), the Ukraine and White Russia (Byelorussia). Ethnic Slavs accounted for not much more than half the population, and the solidarity of the Ukraine with other Slavs could not be taken for granted since the Ukraine had a history of oscillating between subjection to Moscow (or Warsaw or Vilna) and bouts of asserting its independence up to and into the twentieth century. The remaining twelve republics – three Baltic republics, Moldavia, three in the Caucasus and five in central Asia – all had grievances and disruptive separatist aspirations.

The most urgent problem arose in the *Baltic republics* which began to glimpse the recovery of their prewar independence and to claim it. In Lithuania the Sajudis movement came into being in 1988 and, like similar movements in Czechoslovakia, Hungary and elsewhere, evolved from a popular movement into a political party. Its aims were simple: independence. Together with Latvia and Estonia, Lithuania had been overrun by Stalin's armies in 1944–45 and incorporated into the USSR. Many Lithuanians had fled westward, others – perhaps a quarter of a million – were deported or killed. In the Lithuanian republic thus formed, Poles and Russians came to number a fifth of the population of 3.7 million. (In Latvia and Estonia, with 2.7 and 1.5 million, Russians numbered about one–third.) Economically Lithuania's chief features were, on the one hand, an exportable agricultural surplus and, on the other, all but total dependence on the USSR for oil and gas. The economy as a whole suffered from the malformation, stultification and corruption which characterized the USSR. In 1990 elections were held for the republic's Supreme Soviet. Candidates approved by Sajudis won a majority which immediately declared Lithuania independent, repeating the similar declaration of 1918. To Gorbachev this was a challenge which had to be resisted because of its repercussions not only in the other Baltic republics but in all parts of the USSR. He was prepared to give the republics greater freedom in a relaxed USSR, to give them the constitutional right to secede, but he was not prepared to accept a unilateral act which could destroy the Union before he had had time to reform it. In addition, the Baltic republics formed part of an overall defensive system which neither Gorbachev nor his military chiefs could afford to dismantle in a hurry. Gorbachev sent troops into Lithuania and imposed an economic blockade which forced Lithuania's new leaders to moderate, if not their aims, at least their timetable – but not before they had found allies in the heart of the Union, in Moscow itself, where Gorbachev's principal Russian antagonist Boris Yeltsin had won control of the vast Russian republic (RSFSR). All three Baltic republics declared independence in 1991 but the joys of independence were tempered by material hardship as economic links with the USSR and trading preferences were broken, inflation soared and industrial and agricultural output were crippled: in Lithuania, for example, they were halved in three years. None of the three joined the new Commonwealth of Independent States (CIS). All left the rouble zone in 1992. In the years that followed they fared rather better than other parts of the USSR in fending off hyperinflation but Lithuania reverted to left-wing government in 1992 and Estonia in 1995. Latvia, by contrast, surprised itself and others by failing only narrowly in 1995 to give first place to the party of the extreme right led by an immigrant from Germany who spoke no Latvian. Although commonly bracketed together, two of the three – Estonia and Latvia – had historical and other links primarily with Finland, Sweden and Germany while

Lithuania, the largest of the three, was more closely linked with Poland and Russia.

Ukraine, after Russia the largest, most populous and most productive of the Soviet republics, occupying a strategic position between Russia and the Black Sea, peopled by 100 ethnic groups of which besides Ukrainians the most prominent were Great Russians (100 million) and Poles, declared itself an independent state and nuclear-free zone in 1990, a declaration subsequently affirmed by referendum. A focus of revolt against the centralizing thrust of Tsarist Russia in the eighteenth and nineteenth centuries, it had sustained a precarious independence between 1917 and 1920 but was then invaded by a resuscitated Poland (which captured Kiev) before being absorbed into the Soviet Union. The western part of the post-Second World War republic was added to it only in 1939. The nationalist Rukh (or Movement), revived in 1989, won 100 seats in elections in that year for the Supreme Soviet of the Ukraine, whereupon communist leaders including Leonid Kravchuk transformed themselves from communists into nationalists and won enough nationalist support for Kravchuk to win the presidency of the new state in 1991.

Kravchuk's general policy was to assert Ukraine's independence against Russia; his particular concerns, which interlocked, were the Soviet Black Sea fleet and the Crimea. He was a leading proponent of the CIS. For Gorbachev this was a means to salvage as much unity as possible from the Soviet disintegration. For Kravchuk it was a means to minimize and progressively erode the interdependence of the former Soviet republics. While Kravchuk agreed in 1992 to the removal of tactical nuclear weapons for destruction in Russia, he argued that the Black Sea fleet, a force of some 300 ageing vessels and a relatively small part of the Soviet navy, was not a strategic force and so not assignable to Russia, although he was prepared to partition it with Russia. The problem was accentuated by the fact that the fleet's bases were in the Crimea which in 1992 asserted its independence from both Russia and Ukraine – but abandoned this claim in return for substantial devolution of powers from Kiev as an autonomous republic within the Ukraine. Russia and Ukraine made and unmade a series of agreements on the fleet and the use of the naval base at Sebastopol: to control the fleet jointly, to divide it equally, to transfer all of it to Russia for cash. After holding out for much needed cash, Ukraine adhered in 1993 to the Russo–American START I treaty of 1991 and undertook to despatch half of its nuclear weapons to Russia for destruction, a promise later enlarged to the surrender of all these weapons by 1999. Ukraine also obtained promises of cheap fuel from Russia and guarantees of its territorial integrity from Russia, the United States and France.

The *Crimea*, annexed by Tsarist Russia in 1783, had a substantial Tatar population which was deported *en masse* during the Second World War on charges of collaboration with the Germans. Absolved of these accusations after the war they were allowed back but many remained in exile from their ancestral lands. In 1954 the Crimea was transferred from the RSFSR to the Ukrainian Soviet Republic but neither then nor after that change did the Tatars recover their prewar autonomy. By the early 1990s they constituted 10 per cent of the population. Ukrainians numbered 25 per cent, Russians 62 per cent. The Russian language was all but universal. With the break-up of the Soviet Union the Crimea became an autonomous republic within the Ukraine. A presidential election in 1994, sanctioned by the government in Kiev in the false expectation of a victory by its favoured candidate, was won by Yuri Meshkov who advocated reunion with Russia. His success was an embarrassment for Moscow which, by a tripartite treaty with the United States and Ukraine, had simultaneously guaranteed the territorial integrity of the latter in return for the surrender of all its nuclear weapons. But Meshkov proved incompetent as well as awkward and Ukraine stripped him of his powers and then abolished his post without provoking any protest from Moscow.

Elections in the Crimea were followed by elections in the Ukraine itself where a plethora of parties, and candidates belonging to no party, emphasized the divisions between Ukrainians and Russians and the Ukrainians' difficulties in asserting their separate national identity without abandoning to Russia the eastern half of the state and the whole Crimea. In yet further elections for the Ukrainian presidency Kravchuk lost to Leonid Kuchma who was the clear choice of the eastern part of the state, an advocate of renewed close links with Russia and the beneficiary of general discontent with the economy. Kuchma's economic strategies had mixed yields. Inflation was reduced, the level of real wages maintained, reserves built up and interest on debts to Russia paid. But land reform stagnated, subsidies to (unsellable) state industries and farms continued and after a bout of retrenchment budget deficits soared. The minority of reformers put their trust in lower taxes and interest rates to entice capital out of private hoards and foreign bank accounts.

North west of Ukraine the Belorussian Republic of the USSR became the state of *Belarus* – overwhelmingly Russian but with small scattered Polish minorities, its independence threatened less by ethnic divisions than by its nature as an artificial creation of the Soviet régime. In a referendum in 1995 its people voted emphatically for something approaching reunion with Russia: President Alexander Lukashenko stressed Slav unity. West of Ukraine the Moldavian SSR become independent *Moldova* with a mixed population of Romanians, Ukrainians, Russians and Gagauz Turks. Moldavia had enjoyed brief independence towards the end of the Middle Ages before becoming a part of the Ottoman empire. In the nineteenth century parts of it were handed to Russia (1812) and Romania (1859). Its Romanians were deterred from union with Romania after 1989 by the extreme unpleasantness of Ceauşescu's legacy and by the lower standards of living in that country. They affirmed their independence at elections in 1994. Russians proclaimed a separate Dnester Republic but won no recognition for it.

All three Caucasian republics became independent in 1991–92. *Georgia*, an ethnic and religious maze which had lost its independence when its last king gave it to the Russian Tsar in 1800 (in much the same way as the last Medici Duke of Tuscany had bequeathed his duchy to the Habsburg emperor), was governed by Russian proconsuls until 1917 when it received its independence with some help from British troops. It was reconquered a few years later by the Soviet Union. It declared independence once more in 1989. An earlier upheaval in 1972 (which brought Eduard Shevardnadze to power there) was a foretaste of anti-Russian demonstrations in the late 1980s which were suppressed with considerable, probably unnecessary, force. It prompted speculation that army officers hostile to Gorbachev had deliberately inflamed the situation at a moment when Gorbachev was out of the USSR. In 1992 President Zviad Gamsakhouria, a fervent nationalist, annulled the autonomous powers and privileges which had been enjoyed by various minorities within Georgia. Chief among these were the Abkhaz, Turks but Christians in the north west of the country. They constituted 17 per cent of the population and had been incorporated into the republic only in 1930. They took to arms to win independence. Gamsakhouria was forced to flee, whereupon Shevardnadze was invited to return from Moscow and became president (he was re-elected in 1995). He grudgingly joined the CIS to secure Russian support and conceded to Russia the right to keep forces in Georgia and to help train its army: for Russia Georgia's Black Sea ports and Caucasian lines of communication were not unimportant. In 1994 Gamsakhouria committed suicide. The Abkhaz declared independence from Georgia.

In neighbouring *Armenia*, mostly Christian, and *Azerbaijan*, mostly Muslim and Turkish, fighting between the two antedated independence by a number of years. At the core of their conflict was the region of Nagorno-Karabakh which had been created in 1921 as a region of Azerbaijan but in which Armenians outnumbered Azeris by two to one. First Armenians, then Azeris and then again Armenians had the better of it. Both blamed Moscow for their troubles. When in 1991 Armenians had seized Azeri areas and put Azeris to flight these appealed for help to Turkey, their co-religionists and kinsmen with a pronounced history of oppressing Armenians in Turkey. But the Azerbaijani army overthrew the post-communist and pro-Turkish president Abulfaz Elchibey and restored the former KGB chief Gaidar Aliyev. He turned his attention to exploiting Azerbaijan's oil in partnership with foreign companies and planning pipelines in negotiations with foreign states anxious for access and transit dues.

In Asia serious discontent surfaced as early as 1986 in riots in Uzbekistan, notorious for corruption in the days of Brezhnev's son-in-law Yuri Chubanov. In 1990 the five Asian republics of the USSR achieved independence, although not all their 50 million inhabitants – notably in Kazakhstan – were happy with this outcome (see further the Note on Central Asia at the end of Part Four).

The progressive disintegration of the USSR was one of Gorbachev's failures. He had hoped to retain a form of unity but the spirit of the times was against him. The CIS was little more than a transitional device which might lead either to progressive dissolution or alternatively to the reanimation in some measure of the formal and informal links which had existed in the USSR. For the time being there was more distrust of Russian power than awareness of the need for Russian products, services or favours.

A second failure of Gorbachev was the devaluation and demotion (in many places the virtual disappearance) of the Communist Party of the Soviet Union through which he had hoped to exercise power and introduce reforms in default of any other political base. Gorbachev's great achievements had been his boldness in abandoning the Soviet empire in Europe and his initiatives in the mutual disarmament of the Cold War. But he had also wished to preserve the Soviet Union in some modified form and to preserve a reformed Communist Party. In both of these aims he prevaricated and so opened the way for a more single-minded leader – Boris Yeltsin, a Russian nationalist with little concern for the Union and no use for the Communist Party which in Russia he dissolved. Gorbachev had divested the Communist Party of its monopoly of power, carried first instalments of privatization of land and industry, and made the first moves towards a market economy. But although a convinced reformer and modernizer he was no democrat. He did not believe that reform in the USSR could be effective otherwise than through a benevolent autocracy. He secured extended powers for the Soviet president; reassumed the post of General-Secretary of the Communist Party which he had earlier relinquished; created new advisory bodies which, whether by confusion or design, were too numerous to be coherent; and seemed by the first half of 1990 to be veering towards the conservatives. But his economic policy was confused and the economic situation catastrophic. No believer in market economics, he allowed a dozen economic reform plans to follow one another without being put into operation and he failed to press any sustained programme to reform food production.

The difficulties were huge: no foreign exchange, a budget deficit equal to one-third of GDP, nearly all industries losing money, the dead hand of state monopolies still in place, little foreign aid, much but poor foreign advice. In March 1991 proposals for a new Soviet Union of sovereign republics were defeated by popular vote in six of them: the three Baltic republics, Moldova, Georgia and Armenia. In July a new Soviet constitution with restricted powers for the central government was published, in response to which a group of military, KGB and civilian reactionaries staged a coup against Gorbachev who was away in the Crimea. The coup collapsed after three weeks through bad planning but the victor was not Gorbachev who, besides the humiliation involved, was being squeezed between advocates of more drastic constitutional and economic reform on the one hand and on the other by conservatives who feared for their careers and lifestyles and shuddered at the evaporation of Soviet power. Before the year ended Gorbachev was eliminated, resigning the presidency of the Soviet Union which (created in 1923 six years after the bolshevik revolution) had ceased to exist.

The history of Europe in the twentieth century offers no more startling contrast than that between the brevity and the achievements of Gorbachev's rule. He confronted the most daunting problems from the impossible continuance of the Cold War to the internal impossibilities of the Soviet Union and empire. Bold and intelligent though he was, Gorbachev was limited not only by the constraints of his situation but also by the isolation inseparable from an autocratic temperament and his failure to seek the support of close or even competent advisers. These limitations led to vacillation and ultimately to his fall but he did enough to ensure that his departure did not restore the old order.

Yeltsin's power base was not the USSR but Russia. He had acted with courage during the anti-Gobachev coup, risking his life in confronting the plotters and increasing his personal popularity in Russia which had been growing since 1990. In that year the Congress of People's Deputies elected under Gorbachev's reforms of 1988 (and containing an overwhelming majority of professed recent communists) was replaced by a Russian parliment which chose Yeltsin as its president. A few weeks before the coup of 1991 he was elected president of the Russian Republic by direct popular vote: he got 57 per cent of the vote, a narrow majority but much more than any other candidate. This victory turned out to be the high point of his career. By it he stamped his personality on the Russian scene but he failed thereafter to display any mastery over either of the two major problems on that scene: economic policy and nationalities. Economic reform he half supported, barely commended to the public which suffered from it without understanding the case for it, and allowed to be subverted by the institutions of the state itself (including the central bank). His similar ambiguity over the rights of nationalities led to the bungling of the Chechnyan revolt which did him as much harm as any other episode in his years as president.

Yeltsin was given special powers to formulate and implement economic reforms. These, propounded by Yegor Gaidar, Gennady

Bubulis and other adventurous advisers, comprised severe cuts in government spending, the privatization of state enterprises of all kinds for which purchasers could be found, the dismantling of much of the central bureaucracy and a pitiless face to lame ducks. But this programme proved not only painful but also far more protracted than anticipated and was hamstrung by the independent central bank which was outside the executive's control and continued to print money to rescue or simply to please flagging undertakings and their managers. Inflation soared into four figures per cent, production collapsed and the reformers appeared to be benefiting nobody except a handful of enterprising delinquents expert at finding rich pickings in the prevailing confusion. Former allies of Yeltsin, among them Vice-President Alexander Rutskoi, former air force general and hero of the war in Afghanistan, recoiled, judging that Yeltsin and Gaidar – an uncompromising advocate of the sharp shock – were trying to do in five years what should be spread over twenty. Critics formed the Civic Union which became a principal group in the Congress and joined forces with the ex-communists who were even more hostile to Yeltsin than they had been to Gorbachev. Yeltsin was forced to withdraw his nomination of Gaidar as prime minister and appointed instead Victor Chernomyrdin who was expected to be an amenable mediator between Yeltsin and the Congress chairman Ruslan Khasbulatov, another former ally of Yeltsin turned adversary.

Opposition to Yeltsin was not confined to his economic policies and advisers. His strength of character offended as many people as it attracted. His assertiveness verged on the authoritarian. Although he could be stout in defence of a broad principle or belief he gave an uncertain lead over policies either because he was naturally devious or because he was out of his depth. There was in the Congress (whose term ran until 1995) a hard core of Yeltsin loyalists but a comparable number of opponents with a middle group which – probably the largest but also the least coherent – accepted the need for substantial reform but saw with alarm and dismay the suffering involved and wanted less decontrol of prices (especially of necessities), less privatization and above all less speed. This group, however, offered no convincing alternative programme and was influential mainly in preventing the anti-Yeltsin groups from mustering the two-thirds required to impeach or displace him. Given that the constitution forbade the president to dissolve the Congress before the end of its term, there was stalemate and indecision.

Yeltsin continued to command popular approval, which was reaffirmed in a referendum in 1993. He had no adversary of equal stature or eminence, neither Rutskoi nor Khasbulatov having much popular appeal; and he had at least adequate support in the armed forces, perhaps two-thirds among senior officers and more lower down the line. On the other hand he had created no political party of which he could be described as the leader; he was powerful but isolated; he was less determined than, for example, Walesa in Poland; and his conflict with the Congress distracted him from tackling a frightening economic slide. Prices continued to soar, the rouble lost in one year nine-tenths of its value against the dollar, foreign aid was hard to come by and over the years 1989–93 output nearly halved. In 1993 Yeltsin recalled Gaidar, suspended Rutskoi and, in a move almost certainly illegal, dissolved the Congress. The battle between Yeltsin and his enemies became violent, Yeltsin ordered the bombardment of the parliament buildings and put Rutskoi and Khasbulatov in prison.

The election of a new Congress was contested by three main forces: reformers, communists/conservatives and nationalists. Yeltsin himself adopted an aloofness which was not easily distinguishable from indecisiveness. He neither formed a party of his own nor clearly supported any other and so stationed himself autocratically outside and above the party battle. Half the new Congress was elected by proportional representation of lists, half by personal contests in constituencies. The reformers, who took the name Russia's Choice and counted Gaidar as their most prominent leader, debated among themselves instead of presenting a united front to the electorate, came only second (15 per cent of the vote) in the first section but won more constituencies than any other party. The nationalists were led by Vladimir Zhirinovski who had come third in the presidential election of 1991 and campaigned on a demagogic mixture of xenophobia and populism – from the restoration of Russian power and territory (including Alaska) to cheap vodka. His main appeal, which seemed to be irrespective of age, was to people who were fearful of impending chaos and national humiliation. The communists took 12 per cent of the vote in the first section and came third in the second. Together these three groups won half – but only half – the seats in each section.

Yeltsin wanted a strong, centralized Russian state with an effective economic base and coherent within the borders inherited from the USSR. But Russia, like the RSFSR before it, contained twenty-one autonomous regions and about fifty internal border disputes. The calamitous economic and confused political situation fed disintegration, forcing ethnic, local or merely makeshift groups to take economic management and political authority into their own hands willy nilly. In 1992 Yeltsin secured the agreement of all but two of the autonomous regions (Chechens and Tatars) for a Russian federation in which these enclaves and the cities of Moscow and (renamed) St Petersburg would be distinct units with extensive powers over their own affairs.

The Chechens were already in a state of defiance. They had been deported from their Caucasian homelands during the Second World War *en masse* – about 250,000 people, of whom four in five died – and the survivors remained in exile until after the death of Stalin. In the succeeding decades they acquired a reputation as Russia's foremost arms and drugs dealers with an established position in Moscow's underworld.

In 1991 Dzokhar Dudayev, a Soviet general who had seen service in Afghanistan but was adventitiously on leave in his native Chechnya, was propelled into the leadership of a group which, not content with autonomy, proclaimed independence with Dudayev as president. After an unsuccessful attempt to reinstate the regional government by force, Yeltsin withdrew Russian troops and imposed an economic embargo which was however ineffective. He was alarmed about the possibility of similar revolts in other ethnic regions and, more specifically, about Russia's communications – rail and pipelines – through the Caucasus to the Caspian Sea and the oil refinery in the Chechen capital Grozny which was the second largest in Russia. For their part Dudayev and his supporters feared that Russian troops sent to Georgia to help Shevardnadze against Gamsakhouria (who alone recognized Chechen independence) would be turned against Chechnya. Sporadic fighting continued on the borders of Chechnya until 1994 when Yeltsin resolved to extinguish the Dudayev régime, first by helping a rival Chechen group and then by full military action. The first was a failure, the latter a catastrophe, a tragedy and far from a final solution. Bungled but ferocious the Russian action was reported to have cost more lives in a few weeks than the Russians had suffered in ten years in Afghanistan. It united Chechens and Russians living in Chechnya against Yeltsin's Russia; was greatly unpopular in Russia where it weakened Yeltsin's position; and it was widely reported and reprobated around the world. The Chechens retaliated by raids into Russia and seized 1,500 hostages. Russian attempts to rescue them by force cost many lives, the destruction of an entire hospital and an agreement whereby the Chechens were allowed to retreat to their own lands by bus, taking some of their hostages with them. These reverses and excesses posed in Moscow the larger question of Russia's control of the whole area loosely designated the Caucasus, an area the size of France which included fifty languages as well as varying brands of Christianity and Islam.

This small but grisly and continuing war had political and economic repercussions. It raised questions about the mishandling of the situation over a number of years, the misjudgements of the military action in 1994–95, disagreements between Yeltsin and his senior military advisers and the balance of power between them. The cost of the war further disabled the precarious economy and set back the international will to help it. During 1994 inflation had been reduced at one point to 50 per cent a year but it was rising again in 1995 towards 200 per cent instead of falling to a forecast 10–25 per cent. The budget, burdened with continuing credits to bankrupt businesses and farms, showed little sign of being contained within the 7.7 per cent of GNP required by the IMF as a precondition for the support which alone could rescue it. In the industrial sector, half the workforce and two-thirds of businesses had been privatized, albeit by transferring them to their managers and workers at knockdown prices (half of the shares were allotted to managers and workers and a further 30 per cent auctioned, except in defence industries, oil and gas). In agriculture almost all collective farms were converted into limited companies but production and distribution remained inefficient or became chaotic. Only service sectors of the economy registered promising advances.

Russia's central economic problem was two problems: the demolition of an (incompetent) command economy and the substitution of a peace for a war economy. The first entailed changes in attitudes and institutions; the second, switches in production and heavy costs in re-equipment. Both caused confusion, hardship and a steep decline in output. Even with foreign aid these conversions were political, administrative and financial nightmares. As the scale of the problems – in money and over time – became apparent, foreign lenders became warier. Western aid was conditioned by three aims or expectations: that Russia should become a national instead of an imperial state, a democracy and a market economy with a predominance of private property. Russian interference in Georgia and Tajikistan was overlooked but the assault on the Chechens caused greater uneasiness. Yeltsin's bombardment of the parliament was overlooked because Yeltsin seemed a more reliable ruler than any likely successor, but not without uneasiness over the present and future roles of military leaders to whom Yeltsin appeared increasingly indebted. The third, economic condition was the most perplexing, not least because it directly involved the supply of cash credits in very large amounts. To the political need to promote stability and democracy in Russia through economic aid was added the prospect, ultimately, of profits from a country with considerable natural resources and an evident need for western technology and advice.

Germany gave the lead in promising financial aid. It was followed by the United States, the European Union with its newly created European Bank for Reconstruction and Development (created in 1989 on French initiative), the Group of Seven, and the World Bank and IMF. Promises made to Gorbachev were renewed and supplemented when he was displaced by Yeltsin. In round figures commitments had amounted to $100,000 billion plus the rescheduling of a similar amount of existing debt. (Russia had assumed responsibility for the whole of the Soviet Union's external debt.) But the scope of the programme was dubiously commensurate with the needs and much of the aid was not disbursed because of the waywardness of Russia's politics and its economic policies. Yeltsin himself became ambivalent as his opponents accused him of adopting policies which were unduly subservient to foreigners and harsh on Russians. Donors were divided among themselves over the pace at which the old economic system should be replaced.

The principal ingredients of a new system were to be the privatization of state-owned undertakings, the removal of governmental price-fixing and the rigorous restriction of the supply of credit by the central bank. The fast track with its sharp shocks might shorten the process but only by inflicting severe hardships, even starvation, on large numbers of people. A more gradual programme might, however, turn out to be a self-elongating process which never reached its end. Either needed foreign helping hands. Donors tended to prefer the first course, although some – notably the UN's Economics Commission for Europe – opposed the more desperate remedies advocated by the IMF, the World Bank and most governments and experts in western countries not only because of their severity but also on grounds that the short sharp transformation of the economy required the creation of new institutions, new skills and new habits which could not be swiftly conjured up out of very thin air. The transfer of assets from the state to a new or vastly expanded private sector, initially blocked by practical and legal problems (shortages of private capital, non-existence of appropriate assignees), gained momentum but was accompanied by slick profiteering, corruption, conspicuous consumption by the few and widespread poverty and disillusion. Yeltsin, whose health suffered in 1995 as severely as his popularity, seemed no longer to be the man to handle Russia's enormous problems. His inconsistencies, hesitations and growing inconsequence recalled the false start of Russia's last reformer-autocrat, Tsar Alexander II, a century and a half earlier.

In elections at the end of 1995 the most successful party was the Communist Party led by Gennadi Zyuganov. Zhirinovski's ultra-nationalists came second. The principal reformers were divided between Grigori Yavlinski and Yegor Gaidar, leaders respectively of reform in the Gorbachev and Yeltsin periods. Most of the forty or so parties failed to reach the 5 per cent threshold required for entering the parliament.

Post-Soviet Russia was not a world power but it was a regional power of sorts and likely in the medium or longer term to become a more considerable regional power. The spread of that region in Europe and Asia remained to be determined, as it had been after the death of Catherine II and after the bolshevik revolution. It joined the World Bank, the IMF and other international economic bodies and applied at the end of 1994 to join the WTO (the successor to GATT). It was already spending over $100 million a year on foreign goods and services and was becoming one of the world's most potent magnets for foreign exporters. But politically it was unstable so long as Yeltsin embodied authority but not ideas; and his authority was waning as the ideologies of communism and nationalism gathered strength and might even coalesce against him.

2 JAPAN

The postwar history of Japan is an object lesson in what international politics are really about. In 1945 Japan was prostrate, with its military power annihilated and its national symbol, the emperor, nullified. Within a generation Japan regained the status of a great power and it did so not by replacing its armouries but by rebuilding its industries, regaining its foreign trade and reconstituting its reserves of cash and currencies. Only after these achievements did it begin to contemplate and then to refashion its lost military might. It was the one power in the world that could be referred to as a great power but had no nuclear capacity and it was evidently more powerful than some powers – Britain, France, India – which had made nuclear explosions.

Furthermore Japan in this period lacked not only the military trappings with which states are wont to make their mark in the world; it was also conspicuously short of primary resources. It had no – or virtually no – oil, uranium, aluminium or nickel; very little coal, iron ore, copper or natural gas; only half its requirements of lead and zinc. These shortages were made all the more acute by the great expansion of industry, an expansion which paradoxically was both essential to Japan's recovery and yet exacerbated its dependence on foreign materials. This need to secure primary products, whether by participating in exploration or by establishing commercial-political control in the places where they lay, became a major guideline of Japanese foreign policy.

In the immediate postwar years the most striking things about Japan were the physical devastation of a number of its cities, unemployment somewhere between 10 and 15 million, the almost total inadequacy of ordinary means of transport, and the American occupation.

The American share in the defeat of Japan had been so overwhelming that the United States could reduce postwar allied co-operation in the occupation to a not very polite figment. Two bodies were created: the Allied Council for Japan, located in Tokyo and consisting of the United States, the USSR, China and a representative of Britain and the Pacific members of the Commonwealth; and the Far Eastern Commission, located in Washington, with eleven members. But in fact Japan was ruled by the supreme commander, who was General MacArthur and

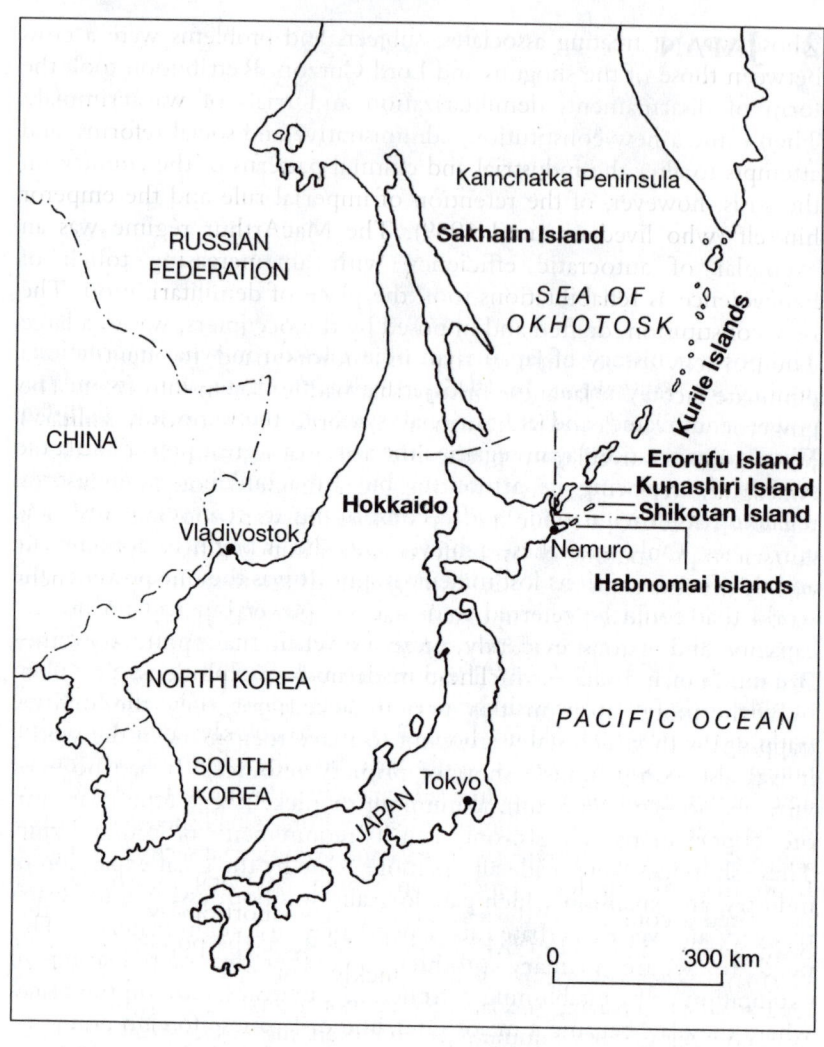

2.1 Japan and its neighbours

whose ways of treating associates, subjects and problems were a cross between those of the shoguns and Lord Curzon. Retribution took the form of disarmament, demilitarization and trials of war criminals. Then came a new constitution, administrative and social reforms, and attempts to alter the industrial and cultural patterns of the country on the basis, however, of the retention of imperial rule and the emperor himself (who lived on until 1989). The MacArthur régime was an exemplar of autocratic efficiency with an increasing touch of benevolence as rehabilitations took the place of demilitarization. The new constitution, drafted and imposed by the occupiers, was to a large extent backward-looking in that its authors tried to identify and eliminate factors in Japanese history that had led Japan into error. The emperor was cut down to human size, over 200,000 (mostly military) persons were barred from public life, the prime minister and all his colleagues were to be civilians, the great financial conglomerates or *zaibatsu* were to be broken up. More forward-looking was the American programme of land reform, again imposed from outside and one of the few examples of its kind in the twentieth century anywhere in the world to eventuate in practice as well as on paper.

None of this stood in the way of revival when the opportunity came, and much of it was helpful. The purge may have eliminated a number of able men but it cleared the way to the top for many more who, without the purge, would not have got there so quickly: some European bureaucracies and businesses would have benefited from such a purge. The elimination of big, often absentee, landowners facilitated the modernization and re-equipment of agriculture, established a rich rural sector alongside the reviving industrial and commercial sectors, and gave Japan the efficient food production which was vital for so densely populated a country. At the same time a new abortion law halved the birthrate over a period of five years and stabilized the population. As in Germany, the Americans were quickly converted from exacting reparations to repairing the Japanese economy, first in the general Cold War context of anti-communism and then more specifically and far more vigorously by the Korean War: in the course of 1948–51 the United States dispensed to Japan postwar relief equal to twice the war reparations exacted from it. War in Asia – the Korean War and later the wars in Vietnam – gave Japan the opportunity and the boost which transformed its fortunes in an astonishingly short time. Like the United States in the Second World War, Japan became an arsenal of war and a war-fuelled economy with the added advantage that it was not itself a belligerent. It constructed a powerful new economy on the basis of low-interest loans for industry, subsidies for public services, high levels of saving, the revival of the prewar *zaibatsu* and a form of guided capitalism in which the state regulated priorities and the allocation of resources without seeking managerial control of operations. At the outset the autocratic rule of General Douglas MacArthur permitted no interference by strikes or unions to the process of making goods and money and together with the discipline and determination of the Japanese people shaped a harsh capitalist culture which sent the weak to the wall but encouraged that adventurousness and vision which had characterized the nineteenth-century English merchant and industrialist before he was turned into a conservative twentieth-century financier. Finally, it was a condition of Japan's success that its new leaders should co-operate closely with the Americans who ruled in Tokyo and Washington.

'New' is hardly the word to describe the first of Japan's postwar prime ministers, Shigeru Yoshida, who was already 70 when he was installed in 1948 in the office which he held for six years, but Yoshida understood the constraints and the possibilities and had no inhibitions about tackling Japan's alarming postwar inflation by the most familiar deflationary devices, killing off weak businesses and adding to the unemployed. His overall purpose was to restore Japan's power in the world but not the power of the military in Japan. He struck lucky. The Korean War came (1950–53) and prosperity bloomed. MacArthur was removed by Truman but Dulles took charge of the business of turning the occupation of a defeated foe into a Japanese–American alliance. In 1951 the United States organized a peace treaty which was signed by the two countries and forty-seven others – the USSR, China, India and Burma being among the absentees. By the treaty of Portsmouth Japan renounced Korea, which it had ruled since 1910; Taiwan and the Pescadores, which had been Japanese for over half a century; the islands in the Pacific which had been administered under mandate since the end of the First World War, after being seized from Germany; all its rights and claims in China and Antarctica; and southern Sakhalin and the Kuriles. The status of these last territories remained ambiguous because the USSR, which had occupied them in 1945, was not a party to the treaty. They were formed in 1947 into a distract or *oblast* of the Russian Soviet Federal Socialist Republic: the island of Sakhalin, stretching northward from Japan's northern tip, runs offshore of the RSFSR for nearly 970 km; the Kurile archipelago, comprising thirty comparatively large volcanic islands and as many smaller ones, runs north east from Japan to the Kamchatka peninsula. The Ryuku islands (including Okinawa) and the Bonins, which had been annexed by Japan at the end of the nineteenth century and were occupied by the United States during and after the Second World War, were gradually recovered by Japan in 1968–72 subject to the continuing presence of American troops and military installations.

Japan's territorial losses were considerable but it was not required to pay any war reparations and it suffered nothing like the bisection of Germany's heartlands. On the same day as the signing of the peace treaty Japan and the United States signed a security treaty which gave the latter the right to station forces throughout Japan for purposes defined as the maintenance of international peace and security in the Far East, the defence of Japan against external aggression and the suppression of rebellion or disorders instigated by an outside power. No such rights were to be accorded to any other state except with American consent. The American occupation came formally to an end in April 1952. What had begun as a crusade to make Japan safe for democracy in the western style was superseded by the recruitment of Japan to the anti-communist side in the Cold War.

Japan's constitution forbade the creation of armed forces but this ban had already been partially circumvented in 1950 by the creation of a National Police Reserve and Self Defence Forces which looked and lived very like an army. These forces gradually expanded to the comparatively modest figure of 250,000 where they stayed. Defence expenditure was kept below 1 per cent of GNP (which was however rising steeply) and around 6 to 7 per cent of government expenditure, again a comparatively modest figure. By the 1970s there were the beginnings of a debate about whether Japan should go nuclear: its weight in the world pointed affirmatively in that direction but there were special political as well as the constitutional obstacles. Japan signed the Partial Test Ban Treaty – although this was an act of supererogation if the constitution were to be taken at its face value – and public opinion in the land of Hiroshima and Nagasaki was ultra-sensitive to the exercise of the nuclear option. An incident in 1954 had dramatized these feelings. In March of that year a Japanese fishing vessel, the *Fukuryu Maru* or *Fortunate Dragon*, a few miles outside an area closed to fishing on account of American nuclear tests on the island of Bikini, was caught in the fall-out of an H-bomb. Before this terrifying fact was realized the *Fukuryu Maru* had returned to port and sold part of its catch. Panic followed. One member of the crew died. Later the United States paid $2 million by way of damages or conscience money for this terrible accident. It also had to pay a political price as Japanese opinion gathered hostility to the United States and to Yoshida as the symbol of the Japanese–American alliance.

This episode coincided with the conclusion in March 1954 of new defence, financial and commercial agreements between the United States and Japan, which included a Mutual Defence Agreement providing for mutual assistance against communism and an expansion and reorganization of Japan's pseudo-military forces. In April Yoshida, whose régime was disfigured by scandals as well as charges of undue subservience to Washington, suffered a parliamentary defeat which he refused to recognize as such. He set out on a tour of western Europe, Canada and the United States, but on his return rifts in his own party had grown too large for repair and in December he was replaced by an old rival, Ichiro Hatoyama, whose foreign minister Mamoru Shigemitsu expressed the intention to restore normal relations with the USSR and China. But Hatoyama did not last long and was succeeded by Nobusuke Kishi, another of the Liberal Democratic Party's numerous but not harmonious chiefs. Kishi preferred to pursue the policy, initiated by his predecessors, of restoring relations with Japan's former enemies in South-east Asia rather than the Hatoyama–Shigemitsu approach to the USSR and China which, in view of Japan's continuing attachment to the United States, was still too hot a potato. Kishi also negotiated in 1960 a revised version of the Security Treaty of 1951, but the new treaty – and especially a clause making it last for ten years – was unpopular. The government was accused of involving Japan in the Cold War by allowing American nuclear weapons to be held on Japanese territory. This criticism was given point by the notorious flight of the U-2 reconnaissance aircraft which was shot down by the Russians in May. There were disorderly scenes in the Japanese parliament and outside it, and although the new treaty was ratified in June a projected visit by Eisenhower had to be cancelled and Kishi resigned before the end of the year.

The 1960s were the years when Japan impressed itself on the rest of the world by annual growth rates of 10 per cent or more; when the alternation of boom and slump which had characterized the 1950s seemed to have gone for good; when the new shape of Japanese industry with its emphasis on heavy goods and chemicals in place of textiles became apparent; when Japan's investment and performance in the most advanced technology captivated the world; and when its admission to the ranks of the OECD publicly designated it as one of the world's economic heavyweights. In 1962 Japan concluded with China a five-year commercial agreement on a barter basis. The vastness of China and its population mesmerized some Japanese industrialists but the government remained inhibited by Washington's hostility to China and in any case the present gains to be made in trade with China were small. Japan's trade with Taiwan was substantially larger and the 1962 agreement was no more than a gesture towards a vague future. More concretely Japan embarked on a policy of creating a region of economic co-operation in South-east Asia and the Pacific rimlands. Already in the 1950s Japan had paved the way with agreements for the payment of reparations to Burma (1956), the Philippines (1956) and Indonesia (1958), and in 1967 the prime minister, Eisaku Sato, undertook a major tour of South-east Asian capitals, preceded and followed by visits to South Korea, Australia and New Zealand. He was the first Japanese prime minister to visit these last two countries. Japan's course was not easy. Besides being a former imperialist aggressor with an unforgotten reputation for peculiar cruelty, Japan was a rapidly developing country in a largely underdeveloped region. Sato lavished loans for development and, in the case of Vietnam, for postwar reconstruction, while by supporting ventures such as the Asian Development Bank and the Agricultural Fund for South-east Asia he hoped to stress Japan's pacific amiability in contrast not only to its past but also to the militaristic policies of the United States evinced by the Seato alliance and the war in Vietnam. Japan was at pains to supply its poorer neighbours with high-grade capital and consumer goods rather than drain them of their natural resources in the classical colonial mode. Further afield, in Australia, Japan became the leading purveyor of investment funds exceeding the sum of British, American and German funds; Australia was by the mid-1960s importing more goods from Japan than from Britain and selling more of its mineral and agricultural products (including wool) to Japan than to any other country. In its smaller way New Zealand was turning in the same direction. Even Canada, another Pacific state, although more firmly fixed in the North American economy, considerably increased its trade with Japan and its loans to and investment in the South-east Asian sector of what was becoming a vast Asian–Pacific economic zone dominated by Japan. The achievements of the 1960s were fittingly crowned by a spectacular demonstration in Tokyo in 1970 called Expo 70.

But shortly after Expo 70 Japan suffered two serious setbacks. Its industrial recovery and commercial expansion had depended on a strong dollar (which made Japanese exports to its largest market exceptionally profitable) and cheap oil. In 1971 Nixon devalued the dollar and in 1973 war in the Middle East created an economic crisis in Japan. Japan relied entirely on imported oil, 85 per cent of which was imported from the Middle East. The war so reduced supply that Japanese stocks were reduced to a few days' consumption and when the flow was resumed by urgent, even grovelling, diplomacy the price had quadrupled. Ruthless economies cut consumption by half but many businesses went under, unemployment rose sharply and the employed had to accept strict wage restraint. A second oil shock occurred a few years later with the fall of the shah in Iran in 1979 but in the interval Japan, uniquely among industrialized countries, had regained its economic health. It did so by abandoning old and moribund industries without compunction and by experimenting and investing in automation and robots, learning from (mainly) American researchers, applying the lessons with a massive investment of funds provided to industry at cheap rates by the Japanese government, and extensive retraining of the workforce again with government funds. In the next decade the same combination of industrialists, experts and government seized for Japan the world's leading place in electronics. In the same period the United States was applying its knowledge and its funds in going to the moon and creating huge, but largely useless, armaments.

Japan's prosperity made it a voracious consumer of the world's products: one estimate in the 1970s had it that by 1981 Japan would need one-tenth of the world's total exports and in oil more than one-tenth of the world's total production. But Japan had no sure way of securing its needs. Britain in the nineteenth century and then the United States had had a similar problem and had solved it by a variety of means which go under the name of imperialism. The essence of imperialism from this viewpoint was not the domination of an area for glory's sake but domination in order to secure materials, whether by taking them, or by ensuring that the producers go on producing them and not something else, or by encouraging a bigger output. The means included investment and so partial or total ownership of minerals, crops or manufacturing enterprises. Japan had plenty of money to invest but there were difficulties about investing it. Apart from its specially delicate standing in the region the whole idea of foreign investment had become suspect. However welcome investment funds might be in purely financial terms, there was a wide and nervous awareness of possible conflicts of interest between investor and recipient and a legacy of hostility to the foreign investor who was presumed to be distorting and retarding a developing economy and indeed to be intent on doing so. In the special case of oil Japan's problem was aggravated by the fact that most known investment opportunities had been pre-empted by the United States or western Europeans. Nevertheless Japan, accelerating in the 1970s a trend begun in the 1960s, placed considerable sums abroad, particularly in Malaysia, Indonesia, Thailand and the Philippines. Anxious to reduce its dependence on Middle Eastern oil it engaged in exploration or investment in Indonesia, New Guinea, Australia and Nigeria. Japan's vulnerability in terms of oil was compounded by the fact that Middle Eastern oil bound for Japan passed through the narrow Malacca Strait and so was at the mercy of any unfriendly power in Malaya or Sumatra. The oil hunger of those years focused attention on certain small islands in the South China Sea: the Paracels, seized by China in 1974 by evicting a small South Vietnamese force, and the Spratlys further south which were claimed by China, Vietnam, the Philippines, the Netherlands and France and some of which were occupied by China in 1988.

Besides his devaluations of the dollar in 1971 Nixon shocked Japan when, without warning to Tokyo, he announced that he had accepted an invitation to visit China. The Japanese government, whose policies had been moulded and constricted by the American alliance and by American policies which, in Asia, were based on hostility to China, was seized with equal astonishment and resentment. What seemed to the rest of the world a sensible move to take some heat out of an overheated quarrel betokened in Japan a reversal of alliances not unconnected with commercial and economic rivalry. Japan feared not only a political *volte face* by Washington but also the closing of American markets to Japanese goods, partly at the instance of the American textile lobby but more generally too in order to check the big Japanese surplus in the balance of trade between the two countries. Proposals by Japan to restrict Japanese exports to the United States had produced no result and Nixon then took unilateral action, including a 10 per cent surcharge designed primarily to hit Japanese trade. Japanese anger was increased when it became apparent that the rate of growth in 1971 had been reduced to 6 per cent, and this interruption in Japan's expansion was blamed on American policy and ill will. The undervalued yen, a source of complaint from all Japan's trading partners, was allowed to float in August 1971 and was in consequence revalued by 16 per cent by the end of the year. Relations between Japan and the United States became, temporarily, bad. They were not sensibly improved by the return of Okinawa to Japanese sovereignty as the agreement to do this set no limit on the use of the island by American troops and was imprecise about its use as a nuclear base. The emperor visited the United States, his first journey outside Japan since 1921, but the visit was treated as a curiosity rather than a political event. Japan joined the United States' attempt to get the United Nations to retain Taiwan as a separate member when China was admitted, but the attempt failed and Japan's part in it seemed to many Japanese to be misplaced loyalty to a shifty ally. Nevertheless the drama of Nixon's visit to Beijing exceeded its immediate consequences and neither Tokyo nor Washington wished their economic disagreements to degenerate into serious political conflict. New trade agreements were signed in 1972 and in September President Nixon and Prime Minister Eisaku Sato met at Honolulu and Japan promised to make massive purchases of American (and other foreign) goods to redress the imbalance in its foreign trade. Japan's balance of payments and reserves rose again in 1972.

Sato did not long survive the Honolulu meeting. Brother to Kishi and no less pro-American he had suffered a serious decline in favour, in the country and in his party, as a result of the buffetings of the previous year and he gave way to Kakuei Tanaka who immediately went to Beijing and restored diplomatic relations with China. This was no reversal of policy, for Tanaka was doing only what the United States had done the previous year and many other countries were hastening to copy. China was neither a possible ally for Japan nor a substitute for the United States as a trading partner; nor was Tanaka markedly pro-Chinese. He was in fact too unpopular to carry through new policies. He retired temporarily into the background during 1972 and was forced to resign in 1974 on being implicated in scandals surrounding the bribery of high government personages by the Lockheed Aviation

Corporation. His successor. Takeo Miki survived only until 1976 when the Liberal Democratic Party (LDP) lost for the first time its overall majority in parliament. Miki's vociferous condemnation of corruption – by implication, of his rivals – made him few friends in the party and the 71-year-old Takeo Fukuda took his place. But Miki got his revenge two years later when the party deposed Fukuda and installed Masayoshi Ohira. In an election in 1979 Ohira disappointed the party but survived the vengeful intrigues of its other chieftains for a short space. In 1980 the party found yet another leader in Zenko Suzuki. Throughout these baronial feudings the party retained its dominant position. What made Japan uniquely strange in this generation was the combination of its worldwide economic thrust with static, geriatric political leadership embodied in the LDP.

That direction continued through the next decade in spite of rebuffs: a decline in electoral favour, severe financial scandals, conflict over fiscal reforms, some sparring with the United States, and a diminution of Japan's hitherto unquenchable economic expansion. Elections in 1983, which registered some reverses for the LDP, were followed three years later by a triumphant campaign under the leadership of Yasuhiro Nakasone who was in consequence accorded an extended hold on the premiership. He was succeeded at the end of 1987 by Noboru Takeshita who, unlike Nakasone, succeeded in imposing on an unwilling parliament a series of reforms to an obsolete tax system – in particular an unpopular consumption tax (3 per cent) – with the aim of redressing the fiscal balance to the advantage of the urban middle class. Simultaneously scandals exploded over the exceptionally lavish bribery of political parties and personalities by the real-estate conglomerate Recourse Cosmos, and in 1989 Takeshita was forced to resign. His immediate successor quickly fell victim to scandals of a more personal nature but the party succeeded in finding in Toshiki Kaifu a prime minister – the thirteenth since the end of the American occupation – sufficiently reputable to hold at bay the challenge of the Japanese Socialist Party (JSP) led by Takako Doi. In elections in 1989 the LDP lost for the first time its majority in the upper house but since only half of the seats were in issue it remained the largest party, with 109 seats out of 252. The JSP won 67 seats. In the next year the LDP retained its majority in the lower house. Kaifu, more durable than had been expected, rebuilt the party's traditional base among farmers, small businesses and women, with the help of a new boom and continuing surpluses in overseas trade.

The comparatively extrovert Nakasone had been second only to Thatcher in praise of Ronald Reagan, but he also visited Moscow, took the initiative in soothing anti-Japanese susceptibilities in South Korea, visited China and South-east Asia, increased Japan's already large aid to the Third World and cancelled some of its more intractable debts. Substantial changes in external affairs were, however, small. A soothing speech by Gorbachev in Vladivostok in 1986, two visits by Shevardnadze to Tokyo in 1986 and 1988 and a visit by Gorbachev in 1991 produced no more than talk about a willingness to discuss a formal resolution of the disputed Northern Territories. A Russian proposal, first made in 1956, to cede two of the four larger Kurile Islands was repeated by Gorbachev (and later by Yeltsin) but rejected by Japan: these large islands, allotted to the USSR at the Yalta conference in 1945, had never been part of Russia and in Japan's view were not included in Japan's renunciation of the Kurile chain by the Treaty of San Francisco of 1952. Japan refused to consider economic aid for Russia unless all four islands were returned.

In response to American criticism of its trading practices Japan modified some import barriers and eased some of the obstacles in the way of foreign bids for construction contracts in Japan, but the balance of trade between the two countries remained above $50,000 million. Under the Omnibus Trade and Competitiveness Act of 1988 the US Congress gave the president authority to designate foreign countries as 'unfair traders' and required him to negotiate the removal of unfair practices and, if he failed, to retaliate. These measures were an indication of persistent American irritation with Japan and also of the fear that Japan's surging pre-eminence in industries from automobiles to computers was being extended into super-conductors and the defence industries. But so long as the Reagan and Bush administrations failed to balance government spending by cutting programmes or raising taxes they remained dependent on heavy borrowing. Since Japan was much the biggest lender, American threats of commercial retaliation had limited substance.

Japan's relatively disarmed condition held great advantages and great risks. On the credit side was an enormous saving: even as late as 1980 Japan's defence spending was less than 1 per cent of GNP (but Japan's GNP was not only one of the largest in the world but also growing by 10 per cent a year). It was also a significant factor in Japan's phenomenal industrial success, since the absence of pressures from military quarters enabled Japan to plan its industrial production and training without the distortions introduced – in the United States, for example, and the USSR – by military exigencies. Within Japan's military-industrial complex the emphasis was more industrial than military. Japan was naturally reluctant to change a pattern which was making it the world's leading economic power, overtaking the United States in much the same way as the United States had overtaken Britain a hundred years earlier. By 1980 Japan had not only surpassed West Germany and Switzerland in many of their leading products but was making more steel and cars than the United States and, by installing more modern equipment, had achieved twice the American output per man. Japanese capitalism proved itself the supreme success of one brand of capitalism and so to some extent a critique of another. It was centrally planned, guided and financed and was taken to include education and training and to be an important element in the broad picture of neo-Japanese culture. Even more successful than the France of the Monnet Plan or the German *Wirtschaftwunder* it allied private business with government and labour. It distrusted the blind forces of the market and the simplistic nostrums in which Reagan's America and Thatcher's Britain were asked to put their faith. Its outward and visible signs included the highest GDP per head in the world, a number of industries which exported 90 per cent of their output and much the largest provision of aid to the rest of the world. In some aspects it was as distinct from American capitalism as was either from communism.

Initially these triumphs owed much to the American alliance which provided the security behind which Japan's energies could be concentrated on the creation of wealth. The efficacy of that alliance was called in question by the American defeat in Vietnam and by Washington's evident incapacity (sometimes interpreted as a failure of will) to redeem its hostages in Iran. Self-defence, militarily adequate in the context of the alliance, had to be re-examined. It was in effect reinterpreted. Already in the 1950s the concept of self-defence had been held to include the pursuit of an enemy to his bases and attacks on those bases. It was also extended to the protection of vital sea routes and, in 1980, to joint naval exercises with other powers. Self-defence was adjudged to be a relative term whose meaning changed with the development of an enemy's military technology and one's own strategic needs. Although Japan's revulsion against nuclear arms was confirmed by its adherence in 1970 to the Nuclear Non-Proliferation Treaty of 1968, the passing of the generation which had known war and defeat was marked by growing support for larger defence budgets. The crisis caused by Iraq's annexation of Kuwait in 1990 gave a precise focus to the debate about Japan's responsibilities in world affairs. A purely financial contribution to the international muster against Iraq seemed unheroic, but a military contribution was unconstitutional and unpopular and Kaifu was obliged to abandon an attempt to change the constitution to permit it. Japan's non-military contribution was nevertheless massive: $9 billion for the general war coffers; a further $2 billion specifically for Egypt, Jordan and Turkey; and mine-sweepers, medical supplies and civilian aircraft to carry refugees. By a narrow majority the parliament later endorsed the use of armed forces in UN peace-keeping operations and Japanese units were sent to UNTAC in Cambodia and in 1993 beyond Asia to Mozambique.

By 1991 Kaifu's time was up. Under his premiership the LDP had done better than expected in elections at the end of 1991 but had nevertheless lost ground. The economy was still growing but at the rate of 5 per cent, low by Japanese standards. Stock prices and property values were slumping. He was replaced by Kiichi Miyazawa, another LDP nominee, who weakly confronted the gravest Japanese recession for twenty years and more vigorous complaints from the United States over Japanese ingenuity in keeping foreigners out of Japanese markets. But more was at stake than the distribution of political offices in what was in effect, if not in appearance, a one-party system. The end of the Cold War distorted postwar Japan's political system and posed testing questions about Japanese security and economic policies. The LDP's monopoly of office was a function of the dominance of the single issue of anti-communism in the decades following the Second World War. In that period the United States had impressed democratic forms on Japan and had impressed Japan into the Cold War against communism. The United States undertook by treaty the defence of Japan and, so long as the Cold War lasted, tolerated, if with increasing irritation, Japan's refusal to conduct its external economic affairs in accordance with the rules of openness enshrined in the Bretton Woods agreements. The LDP enjoyed power in a multi-party system on the terms that it would cede power to no other party and particularly not to the JSP. Putting profit before pride Japan accepted a ban on raising (and paying for) its own armed forces and made prosperity the main business of government in the knowledge that national security was guaranteed by the United States. The LDP developed into a cabal of conservative leaders appropriating the highest political posts among themselves (much as leading Roman families had rotated the Papacy and its prizes in the late Middle Ages before its removal to Avignon). It left administration and even policy-making to senior civil servants and filched scandalously large rewards for not interfering too far in government or with business.

This state of affairs was challenged in the early 1990s not from outside the LDP but from within it. In 1992 Mori Hasokawa left the LDP and formed the New Japan Party: he was an aristocrat and grandson of Prince Fumimaro Konoye, one of the outstanding figures in prewar Japan. In the next year two other defectors, Ichiro Ozawa and Tsutomu Hata, formed the Renewal Party: they were members of the largest faction in the LDP, led by Shin Kanemaru whose ascent to the top office seemed barred by the spectacular scale of his involvement in bribery. After elections in 1993 the LDP remained the largest party in parliament but lost a striking number of seats. The JSP, the main opposition party, fared even worse and LDP dissidents formed the first non-LDP government in Japan's postwar history with Hasokawa as prime minister. But Hasokawa was himself not untainted by financial scandal. The coalition which he led was divided over taxation, electoral reforms and defence policy. He was unable to make more than limited progress with plans for less bureaucracy, decentralization of government and the opening of Japanese markets to foreign competition. He was replaced in less than a year by Hata who had served as foreign minister but was more pertinently regarded as the mouthpiece of Ozawa who was himself regarded as, by inclination, even more an autocrat than a wheeler-dealer, a presage perhaps of a new, or new-old, style of government in Tokyo. Hata's government proved the most short-lived in postwar Japan. The JSP immediately withdrew promised support and did a deal with its inveterate enemies in the LDP to oust the interloping reformers. The LDP recovered power at the price of allowing a socialist, Tomiichi Murayama, to become prime minister for the first time. The economic programme of an uneasy coalition was then wrecked by the devastating earthquake at Kobe in 1995 whose cost fell mainly on the government and Murayama resigned in the first week of 1996, having served the latter half of his eighteen months as a prime minister in evident decline.

The advent of President Clinton to the White House seemed to foreshadow sharper and more public controversy with Japan (as with the EC) over economic affairs but whatever Clinton's intentions his position was no stronger than that of Reagan or Bush. The imbalance in the trade between the two countries was still growing, American attempts to gain access to Japanese markets by setting guaranteed targets for selected products was a failure, and the Clinton administration was forced back to the crude expedients of retaliation through tariffs or depressing the value of the dollar against the yen – the one a device generally reprobated by the United States and the other unwelcome to American importers and by no means certain to achieve its aims. In thirty years the value of American exports to Japan had doubled but the value of trade in the opposite direction had quadrupled – in the Japanese view because American exporters (of, for example, cars) neglected to find out what Japanese buyers wanted, in the American view because Japan used complicated bureaucratic procedures to thwart the operations of market forces. At a loss to find a remedy the United States alternated trade talks with bluster and threats to impose punitive retaliatory super-tariffs. Talks in 1994 produced undisguised stalemate and illustrated the worldwide dangers of bad relations between the United States and Japan as the value of the dollar fell steadily; and in 1995 the United States reverted to threats. Nevertheless, neither side seemed to have the stomach for more than rhetorical brinkmanship.

The half century after the Second World War witnessed no transformation more remarkable than Japan's but by the end of that period doubts were beginning to nag. The twin supports of Japan's rise to global economic power were its financial system and educational ethos and the outward sign of their success was unprecedented economic growth and confidence. Yet in the early 1990s growth slumped to zero, the financial system was tremulous with debt, and critics in Japan and beyond were wondering whether the stresses of achievement in schools and universities were not killing personal initiative, industrial inventiveness and social resilience. Success had been won under the aegis of the Liberal Democrat Party which had become thoroughly and obviously corrupt and in alliance with the United States which, once delighting in Japan as a major anti-communist partner of its own making, was coming to fear that it had a tiger by the tail. Most disturbing was a banking system with bad debts of $500 billion or thereabouts, an economic miracle of the wrong kind.

3 CHINA

THE TRIUMPH OF MAO

In 1949, a little over a quarter of a century after its birth, the Communist Party of China won the ancient capital city of Beijing. A movement had become a government. The defeated Kuomintang was reduced to the island of Taiwan where its leader Jiang Kaishek, like the last of the Ming emperors 300 years earlier, clothed himself with imperial pretences reminiscent of the French nobles and Italian bankers who wandered round western Europe in the later Middle Ages calling themselves emperors of Byzantium. The new lord of China, Mao Zedong, was able to declare in January 1950 with only slight exaggeration that all China was his except Tibet.

The vast lands and imperial traditions which fell to Mao after the Second World War had been going begging since just before the First. The decline of the last (Manchu) dynasty, which had exposed China to foreign intervention and had almost occasioned its partition in the latter part of the nineteenth century, had culminated in the revolution of 1911. Revolutionary groups and secret societies, partly indigenous and partly fomented among overseas Chinese, effected the disintegration of the *ancien régime*. Faced thereafter with the task of re-establishing the cohesion, dignity and power of China they failed, so that fifty years later China was still potentially a great power but not actually one.

The most notable of these successor groups was that of Sun Yat-sen (died 1925), a nationalist, a democrat and a socialist who wanted to reform China through his instrument the Kuomintang. The Kuomintang established a government in the southern capital of Nanking but it never succeeded in bringing the whole country under its obedience. The collapse of the old régime was followed, especially in the north, by the appearance of war lords who created autonomous fiefs for themselves and engaged in civil wars. Even in the south, where the Kuomintang imposed a degree of order and stability, the huge tasks of administrative reform and modernization were tackled but not mastered.

The Kuomintang looked around for outside help and accepted from the new and no less revolutionary government of the USSR the advisory services of Michael Borodin. Under the pressure of circumstances the Kuomintang began to adopt in its fight against the war lords some of the methods (the organization of the party and of party–state relations), though not the doctrines, of European communism, and it also entered into an alliance with the Chinese Communist Party which had been founded in Shanghai in 1922. The communists consisted of small groups in certain southern provinces. They became a part of the Kuomintang under the overall leadership of Jiang Kaishek, but disputes between communist and other Kuomintang leaders (and between left and right non-communist factions within the Kuomintang) were endemic. There was also confusion among the communists themselves because communist committees in Shanghai gave ill-judged instructions to leaders active in the countryside, and still more because Moscow attempted to direct matters without adequate knowledge of Chinese history and conditions. At one moment Moscow had different emissaries in China advocating incompatible policies. In this atmosphere Mao Zedong gradually became one of the principal leaders in the countryside, often at odds with his superiors in Shanghai on political and military tactics and himself still evolving and changing his own ideas on these subjects. Disputes between communists and non-communists led to fighting. In 1931 the Japanese attack in Manchuria forced Jiang Kaishek to send troops to the north, but the confusion in the communists' ranks prevented them from taking advantage of this distraction of the non-communists and in 1934 they were decisively defeated as a result of adopting ill-chosen tactics which Mao had argued against. The communists set out on their strenuous and myth-making Long March from south-eastern China, first westward and then northward to the province of Yenan in the far north-west, where they preserved their movement and bided their time. It was during the Long March that Mao rose to ultimate authority. He had been moving away from the classic Marxist strategy of achieving power by fighting the battles of the industrial proletariat and towards annexing communism to peasant indignation (periodically a mighty subversive force in Chinese history); he proposed to set up a small peasant republic and to create peasant armies which would one day be strong enough to seize towns.

Meanwhile the Kuomintang, having disposed of the communists and made headway against the war lords (it took Peking in 1928 and secured Manchuria the next year by agreement with General Chang Hsueh-liang, the Young Marshal), was attacked by Japan, first in Manchuria in 1931 and then in China proper in 1937. From this date until 1945 the Kuomintang and the communists were rivals for the honours of effecting the revival of China as a great power and for the prize of ruling

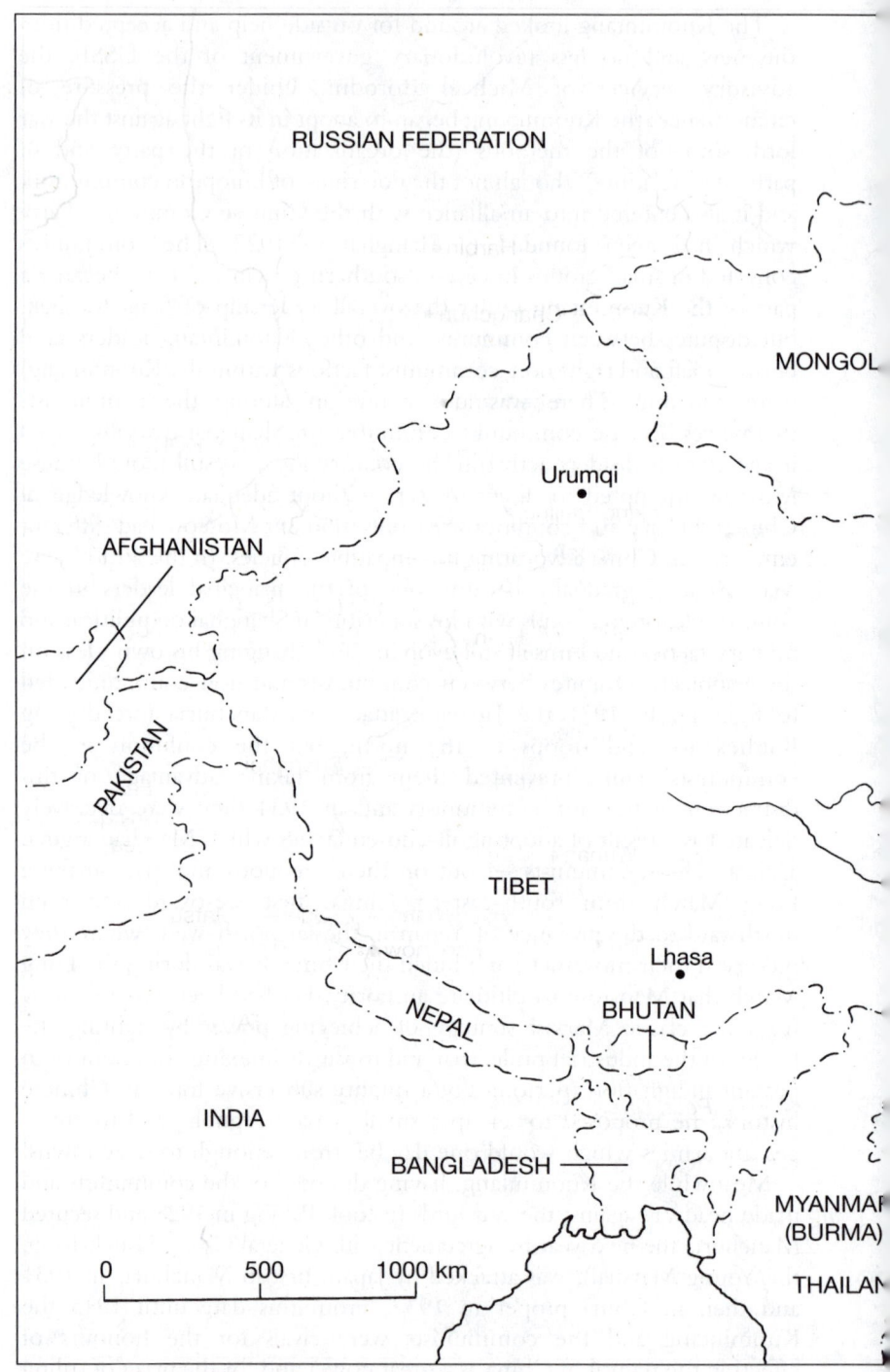

3.1   China and neighbouring countries

Harbin •

Changchun •

Shenyang •

N & S KOREA

JAPAN

Hohhot •

Peking/Beijing •
(capital)

Huanghe
(Yellow River) • Jinan

YELLOW
SEA

ua •

hou

Xian •

•Zhenchou

Hefei •

Shanghai

East
China
Sea

Wuhan •

gdu •

Changjiang (Yangtze)

Changsa •

•Nanchang
·Foochow

Quemoy Matsus

Tachens

•Guiyang

Amoy

TAIWAN

Zhujiang
(Pearl River) Canton

Pescadore
Is.

ng

•Nanning

Hong
Kong

PHILIPPINES

VIETNAM

South
China
Sea

AOS

HAINAN

it after the Japanese had been defeated. The communist remnants which had reached Yenan grew rapidly into an army of about 100,000, attracting recruits from the peasantry and from youth generally on a nationalist anti-Japanese ticket. In 1941 the Japanese attack on the United States at Pearl Harbor merged this Far Eastern war into a general war which temporarily overlaid China's civil discords. The Kuomintang acquired powerful allies, but its fortunes were not destined to revive. It had failed to arrest economic collapse, its leadership remained narrow and cliquish, its administration became corrupt, high-handed and over-reliant on secret police, and finally its armies disintegrated. By contrast the communists rose in popular esteem. Their sojourn in the wilderness had given them glamour and concealed the seamy side of their own methods of rule. They had the reputation of fighting the Japanese more seriously than the Kuomintang and when the war ended they were ready to return from the outskirts to the centre. Only a few years later they were the government of China.

When Jiang Kaishek was chased out of China a generation of civil conflict came to an end and the communists were presented with the opportunity to do what no faction had been able to do since the fall of the empire, and what the empire itself had failed in and been reproached for: to make China strong and healthy enough to maintain its integrity and independence. In the past, threats to the integrity of China had come from the imperial powers of western Europe sailing across the seas, from the Russians approaching over land, and from the Japanese. By 1945 the western European powers were no longer in a position to menace anybody in the Far East, and Japan had been all but annihilated. The Russians, however, had just reappeared on the scene, while the Americans, traditional upholders of the integrity of China against interlopers, were despairing of the Kuomintang and were about to revert to an Asian policy centred on Japan such as they had pursued at the opening of the century.

The history of continuous Russian expansion into Asia begins after the Crimean War. That defeat in Europe inaugurated essays in liberalism at home and in Far Eastern adventure, both of which were sporadic. The tsars were more or less permanently preoccupied with the problems of Poland, the Balkans and the Straits, but there was also a party at St Petersburg which made a speciality of the Far East. The 1860s and 1870s were a period of expansion; the great Russian proconsul Muraviev established the Amur and Maritime provinces in 1858 and 1860 respectively; Vladivostok was founded in 1861; Sakhalin was acquired (but Alaska sold); the Kazakhs and the central Asian Khanates were subjected. By the 1880s Russia, having reached the borders of Persia and Afghanistan, had been brought face to face with the British in India, while further north a warm water Pacific port and a share in the China trade stimulated imperial and financial imaginations.

War between the Japanese and the Chinese in Korea in 1894–95 brought Japan considerable gains (including Taiwan, the Pescadores and, temporarily, the Shantung peninsula and Port Arthur) and served noticed of the decline of Chinese power. Russia, France and Germany intervened in defence of China, although Britain, suspicious of the Russian advance in Asia, now began its pro-Japanese policy as a counter. The Russians, having with the French helped China to pay its war indemnity to Japan, entered into an alliance with China, took Port Arthur and planned the Chinese Eastern and South Manchurian railways. The Russians and the Japanese were now vying with each other for Manchuria. In 1904 Japan attacked Port Arthur and in the ensuing Russo–Japanese War the Russians were defeated and forced back into Manchuria and Mongolia. Japan's standing in Asia and the world was immensely increased and the Russian advance across the top of Asia was checked by a Japanese counter-force in Manchuria. The First World War temporarily eliminated Russian power, effected no Chinese revival, distracted other Europeans from Asian affairs, and left Japan in a situation where, far from curbing the Russians, it had itself to be curbed – and could be curbed by no power except the United States. The Russians recovered positions in Outer Mongolia and Central Asia in the 1920s and came once more into conflict with the Japanese in Manchuria in the 1930s, but they did not seriously return to the struggle for Manchuria until the final week of the Second World War.

For most of the war years Stalin was too much occupied with the Germans to pay much attention to Asia, let alone intervene there. The USSR had signed a neutrality pact with Japan in 1941 and before Pearl Harbor Japan sought and received assurances that the Russians would abide by this pact. Up to the Teheran conference of 1943 Stalin asked for nothing in the east, being in no position to ask for prizes until he was ready to give a helping hand against Japan. At the Moscow conference in October 1944 there were hints of a change in the Russian attitude, and at Yalta in 1945 Stalin's terms were set out and accepted by allies who were anxious to avoid quarrels, still overrated Japan's will and capacity to fight on, and were – except for a few initiates – ignorant of the nuclear weapons about to be used. Stalin asked his allies to reverse the verdict of the Russo–Japanese War of 1904–5 and also to guarantee him certain positions at the expense not of Japan but of China. The positions lost by the Russians in Manchuria and Sakhalin in 1905 were to be restored; the Kuriles were to go to the USSR too; and the USSR's virtual annexation of Outer Mongolia was to be recognized.

Stalin's policy in Asia was to nibble at the Chinese circumference and to prevent the establishment of any powerful central Chinese government. By declaring war on Japan he secured rights, and planted troops, in Manchuria. He was also able to carry off Japanese industrial plant, although not without incurring Chinese resentment. He had already engineered an anti-Chinese revolt in the Ili valley in Xinjiang (Sinkiang), where a secessionist republic of Eastern Turkestan sprouted under Russian protection in 1944. Outer Mongolia, nominally under Chinese suzerainty but effectively in the Russian sphere since 1921 and conceded to the USSR by the western allies at Yalta, was abandoned by the Chinese (who had no choice in the matter) by the Russo–Chinese treaty of August 1945. It was not recovered for China when Mao succeeded Jiang in 1949.

Stalin's declaration of war on Japan had one further notable consequence in the bisection of Korea, for when the Japanese surrendered it was arranged, as a matter of convenience, that Japanese forces north of the 38th parallel should surrender to the Russians and those south of that line to the Americans.

Russian intervention in the Far East was nicely timed. As the Russians knew, the Japanese were anxious for peace; they were taking soundings in Moscow in the hope of securing Russian mediation to end the war. The first nuclear bomb was dropped on 6 August 1945 on Hiroshima by the Americans who also knew that there was an active peace party in the Japanese cabinet. The Russian declaration of war followed two days later. On 9 August the second bomb was dropped on Nagasaki. Hostilities ended on 15 August (on which day the Russians signed their treaty with Jiang Kaishek).

Whereas Stalin wanted a weak China, worked to that end and may well have considered himself successful until the eve of his death, the Americans wanted a strong China, failed through no fault of their own during the years of transition from war to peace, and later began to see the prospect of China strong indeed but not at all to their liking.

American policies in the Far East, which began to take shape at the beginning of the century, were conditioned at the start by the acquisition of the Philippine Islands from Spain in 1898 and by suspicion of the European powers which, having turned most of South-east Asia into colonial terrain, seemed about to carve up China too. With its interests and its moral sense in happy accord Washington claimed for itself any rights or privileges which any European power succeeded in extracting from the Chinese (the policy of the open door, that is, no commercial preferences between foreigners from different states) and stood up at the same time for the integrity and independence of China.

The American entry on this scene was unexpected and unpremeditated. The war of 1898 against Spain, which began in a muddled way in Cuba, had the incidental result of substituting American for Spanish rule in the Philippines. There ensued a considerable public debate on the politics and morality of expansionism, with the lurking background knowledge that if the United States did not assert its power in the Pacific, either the British or the Germans or the Japanese would. This debate coincided with the manifest decline of China.

The British, alarmed by the trend towards annexations, leases and special commercial privileges for particular groups of foreigners, had sought American co-operation to put a stop to this colonialist rat race in China, but Washington was not at this point interested and the British thereupon entered a game which they despaired of preventing and began themselves to mark out enclaves and take leases.

The change in the American attitude came at the end of the nineteenth century with the appointment of John Hay as secretary of state. Hay's first set of notes, condemning spheres of influence and advocating the open door, had little practical effect, although they were popular in the United States; a second set, at the time of the Boxer rebellion and the allied intervention under German command to which the rebellion gave rise in 1900, testified to increasing American concern; so did Washington's mediation in the Russo–Japanese War and the signing of the treaty of Portsmouth on American soil. But it was some time before American and British policies were brought into harmony. Britain, seeking an ally in eastern waters and apparently rebuffed by the United States, had turned to Japan, while the United States was becoming increasingly the protector of China and the enemy at sea of Japan. This discrepancy continued, not without damage to Anglo–American relations, until after the First World War when the Washington treaties of 1921–22 in effect eliminated the Anglo–Japanese alliance, established a three-power naval ratio of 5:5:3, and secured a nine-power guarantee of the integrity of China and the open door. (At the London conference of 1930 the Japanese ratio was raised from 3 to 3.5.) In addition Secretary of State Charles Evans Hughes forced the Japanese to withdraw their troops from the Shantung peninsula and Siberia, but in the 1930s the Americans, undecided whether strong action would check or strengthen the expansionist factions in Japanese politics, became less effective and there was little reaction from Washington to the Japanese conquest of Manchuria, the proclamation of the state of Manchukuo in 1932 or even to the Japanese occupation of Shanghai later in the same year. Japan withdrew from the Washington and London naval treaties at the end of 1936, claiming parity with the United States and Britain, and embarked on full-scale war with China in 1937. At the same time Japan challenged the European powers by propounding a 'new order' for all eastern Asia, and when the Europeans (western and Russian) had made it plain that they wanted no troubles in the east, Washington began to negotiate with Tokyo.

The outbreak of war in Europe in 1939 gave Japan a freer hand against China and new opportunities in South-east Asia. A new and more bellicose government, installed in 1940, decided to attack Indo-China as a first step but hoped to avoid American intervention. This government fell when Hitler attacked the USSR. The foreign minister, Yosuke Matsuoka, who wanted to join in the attack, was dropped, and General Hideki Tojo, who wanted to attack the United States, became the most influential member of the cabinet. He also became prime minister in October 1941 and gave the order for the attack on Pearl Harbor in December.

Until the turn of the tide in the middle of 1943 the Americans were in danger of being evicted from the Far East altogether, but the turning point of the Battle of Midway and the final victory over Japan in 1945 brought them back with a unique preponderance such as no power had ever previously enjoyed.

The war forced Washington to reconsider its Far Eastern policy. Everything combined to make Americans pro-Chinese – traditional policies, missionary connections, Japanese behaviour, and the appealing (in both senses) Madame Jiang – but the Kuomintang government was not successful and was ceasing to be worthy. Nevertheless Washington, which renounced extraterritorial rights in China in 1943 as a gesture, determined that China should emerge from the war fully sovereign in the full extent of its ancient territories (including Manchuria, Taiwan and the Pescadores) and a major power with a permanent place in the Security Council.

China was to be the United States of the Asian continent, a vast, united and liberal-democratic power: a view which Churchill, among others, regarded as romantic moonshine. When the shortcomings of the Kuomintang became more and more obvious, General Joseph Stilwell was sent to keep an eye on Jiang Kaishek and to bolster and, if possible, reform the Kuomintang, but the general's comments on the Kuomintang were so critical that he was recalled in October 1944 at the instance of Jiang's friends. General Patrick Hurley, who had been sent to China in August as President Roosevelt's personal representative with the Generalissimo, became the principal vehicle of American policy with the new task of bringing the Kuomintang and the communists together again in a coalition. General Hurley and General George C. Marshall, who succeeded him in November 1945, made only slight headway. The Kuomintang and the communists were ineradicably suspicious of each other – the communists had had earlier experience of the Kuomintang attacking them under cover of a truce – and discussions for an accommodation broke down on the issue of the amalgamation of their two armies. When the Japanese surrendered the Americans lifted Kuomintang forces by air to take control in north-eastern China (in the event a strategic mistake), while the Russians who had just entered Manchuria gave the communists arms and opportunities in that area. Until the latter part of 1947 there was an uneasy truce, but civil war was only suspended. The Americans, by now thoroughly distrustful of the Kuomintang, withdrew their units from China but were uncertain whether to withdraw all forms of support. General Albert C. Wedemeyer, sent to China in July 1947, recommended in September extensive American aid on the basis that the Kuomintang should introduce extensive reforms under American supervision, but by this time China was approaching economic and financial collapse and rioting was becoming common. Nevertheless interim aid to China was approved before the end of the year and the China Aid Act was passed in April 1948. This was supplemented by a Sino–American agreement and the Kuomintang made a belated effort to put its house in order.

But it was too late. The communists were winning battles in steady succession. All Manchuria was theirs by the end of 1948 and the next year became a roll-call of famous cities conquered for communism. Washington at last wrote off the Kuomintang. Jiang, who had made the characteristic blunder of over-committing himself in the last stages of the civil war, resigned the presidency at the beginning of 1949 and the Kuomintang asked – in vain – for mediation by the United States, the USSR, Britain and France. The communists had no further interest in coalitions or accommodations. A People's Republic was proclaimed in September. On 18 December 1949 Mao arrived in Moscow as a head of state.

The victory of the Chinese communists created problems for both the Americans and the Russians. During the remainder of 1949 and the first half of 1950 the Americans were moving towards diplomatic recognition of the new régime, but possible negotiations to this end were first held up by the arrest of the American consul-general in Mukden and his imprisonment with four of his colleagues for a month at the end of 1949, and were then rendered impossible by war in Korea. At some point in this period the United States decided that Taiwan was a necessary part of a line of strategic bases and must therefore not be allowed to fall into communist hands. This decision was made public and it combined with the Korean War to make the United States keep Mao's régime out of the Chinese seat at the UN and to make the Chinese believe, first, that Taiwan was being held as a temporary refuge for Jiang Kaishek pending an attempted reconquest of the mainland and, secondly, that the UN was an American tool. The Korean War also led the Chinese into an over-confident assessment of their military strength, expressed in the picturesque but inaccurate description of the United States as a paper tiger.

Mao's victory soon proved as awkward for the Russians as for the Americans. The world at large supposed that the interests and policies of the two principal communist powers must be closely allied. This assumption owed something to the Cold War, in which context the new communist state came to birth; it was strengthened by the incidence of the war in Korea in 1950–54 between communists and anti-communists; and it led to bewilderment and amazement when it transpired a few years later that Moscow and Beijing were quarrelling bitterly. In fact the interests of the USSR and China were partly congruent and partly conflicting, and the communist parties of the two countries had no close emotional bonds; in Moscow moreover the Chinese party may well have been regarded as only doubtfully communist in its doctrine and in its loyalty to the world movement dominated by the Russians. After the death of Stalin the divergence in policies became more obvious, partly because of the removal of the undoubted leader of the communist world and partly because the passage of time allowed the differences in the circumstances of the two countries to make themselves increasingly felt.

Stalin, as already stated, wanted no powerful China. So long as the government of China was non-communist it was comparatively easy for him to pursue a policy of alliance coupled with pinpricks. The treaty of August 1945 with Jiang had pins stuck into it, since the Kuomintang acknowledged in effect the absorption of Outer Mongolia into the Soviet sphere and failed to regain in Manchuria rights which now passed from Japan to the USSR. The Chinese communists were, for Stalin, another pin with which to lacerate the Kuomintang, until in 1949 they took the place of the Kuomintang and so ceased to be a mere pin. When Mao arrived in Moscow, Stalin had to devise a completely new China policy which would take into account both the USSR's desire to dominate the Asian heartlands and the inescapable expectation of fraternal love and assistance between communist parties. Given the size and potentialities of China, this was a problem which had never previously come up in Moscow.

Mao had to learn too. He knew little about the USSR (or any other major power) before a visit to Moscow where he was made to hang around for two months before a new Russo–Chinese treaty was concluded on 15 February 1950. This was a business measure dealing with railways, credits and similar matters. The Russians gave the Chinese a handsome present consisting of the Changchun railway and its installations. The Chinese also acquired the right to administer Dalny, though the ultimate fate of this port was to be reconsidered upon the conclusion of a treaty of peace with Japan. The USSR provided China with a credit of $300 million over five years at 1 per cent interest and repayable in 1954–63. Further agreements were concluded a month later for joint exploitation of oil in Sinkiang and joint operation of air routes in central Asia, and a general commercial agreement was signed in April. These agreements were the start of co-operation between the two countries which lasted until the late 1950s and were of vital importance to China during the years of experiment and incipient modernization. They were confirmed during a visit to Moscow by Zhou Enlai in August–September 1952 shortly before the announcement of China's first five-year plan, the formation of a powerful National Planning Commission under Gao Gang and the creation of a Ministry of Higher Education (all in November–December of that year). With the plan, the commission and the drive for higher education the Russo–Chinese agreements were an integral part of Mao's design for a new China.

CHINA AND THE SUPERPOWERS

Nine months after the proclamation of the new Chinese republic war broke out in Korea. This unforeseen and, for the Chinese, extremely inopportune event had a deep effect on Asian politics and on the relations between Asians and non-Asians involved in Far Eastern affairs. It forced the Chinese to look to their defences when they should have been concentrating on their internal affairs; it dominated American policy at a moment when that policy was in the making; it raised questions concerning consultation and co-operation between the principal communist powers; it threatened to import a cold war into Asia, boosted Indian neutralism and caused Americans for a while to equate neutralism and pacifism with indifferentism and moral perversion.

The history of Korea is like the history of any small country which finds itself wedged between more powerful neighbours. Korea is a peninsula which stretches into the sea from the south-eastern corner of Manchuria until it nearly reaches the southernmost of Japan's principal islands. For a thousand years it was ruled by two dynasties separated by a brief Mongol conquest. It suffered Japanese and Manchu incursions in the sixteenth and seventeenth centuries but survived until the end of the nineteenth, by which time it had become a pawn in Sino–Japanese–Russian conflicts. Japan's victories in the war with China in 1894–95 and with Russia in 1904–5 gave Japan a free hand in Korea which was annexed in 1910. It failed in 1919 to recover its independence although a provisional government was established under the presidency of Syngman Rhee, a Korean who had acquired a doctorate in philosophy at Princeton. Independence was, however, promised by the Cairo declaration of 1943 by Roosevelt, Churchill and Jiang Kaishek.

When therefore the war came to an end in 1945 there was no dispute over the status of Korea, but at the moment of independence an accident deprived it of unity. The Japanese having surrendered partly to the Americans and partly to the Russians, Korea was divided into two pieces along the 38th parallel. This famous line came into existence as a result of negotiations between army officers of relatively junior rank; it was not a creation of ministerial decisions. But administrative convenience hardened into political fact and thereafter all attempts to equip Korea with a single government failed. The cause of this failure was the presence of Russian as well as American troops. Contemporaneously in Poland, for which rival governments also contended, there was only one occupying army and so only one possible answer. In Korea there were two armies and so no answer.

In 1947 the United States took the problem to the United Nations which appointed a commission (UN Temporary Commission on Korea – UNTCOK) to effect unity through elections. Elections were held in the south in May 1948 but the commission was prevented from operating in the north and the result of its activities was the creation of a government which claimed to be the government of all Korea but had in fact no authority or existence north of the 38th parallel. The head of this government was Syngman Rhee, now an elderly, rough, reactionary but legendarily essential and, as it turned out, almost irremovable father figure. In 1949 the Russians (who had nurtured a rival government in the north) and the Americans both withdrew their armed forces. Korea was now a country with two governments, overhung by the Russians in Manchuria and Siberia and by the Americans in Japan and vaguely conscious of the emergence of a new China on its borders.

The first half of 1950 was occupied by new elections in South Korea and propaganda in North Korea for reunification either by elections throughout the country (from which Syngman Rhee and others were to be debarred) or by a merger of the two parliaments. When it became apparent that these gambits were unavailing northern troops crossed the border on 25 June, capturing the southern capital of Seoul the next day. It is doubtful in the extreme whether North Korean troops would have taken this action if Dean Acheson had not, in January, excluded Korea (and Taiwan) from the defence perimeter which, by implication, the United States was ready to fight to retain, and it is also extremely doubtful whether the North Koreans would have crossed the parallel without feeling assured of Russian approbation. Kim Il Sung, North Korea's prime minister (1948–72) and president (1972–94) had spent the war years in Moscow and was Moscow's nominee for his jobs. What was perhaps unforeseeable in June was that the American attitude of January would be immediately reversed by the invasion and that South Korea and Taiwan would prove to be parts of the American defence perimeter after all.

The Security Council met at once at the request of the United States and passed, in the absence of the Russian member, a resolution requiring a cease-fire and the return of northern troops to their own side of the border. All members of the UN were asked to support these measures and President Truman thereupon instructed General Douglas MacArthur in Tokyo to give support to the South Koreans. The president also ordered the US Seventh Fleet to insulate Taiwan. Two days later (27 June) a second resolution of the Security Council called on all members to help South Korea to repel the attack made upon it and to restore international peace. This resolution proceeded upon the basis, subsequently challenged by the Russians, that the fighting between North and South was an international threat, although it was not made clear whether it was international because North and South were two different states or because an admittedly internal and civil war was deemed to have international repercussions. On 6 July China protested against illegal intervention in Korean affairs.

At first the fighting went rapidly in favour of the North. The South Koreans and the UN forces which came to their help were driven to the very tip of the Korean peninsula, but in September General MacArthur, who was in command of the UN forces, landed troops at Inchon 240 km to the north and as a result of this bold stroke South Korean and other units crossed the 38th parallel in October and pressed on to the Manchurian border. The war seemed to be over. President Truman flew to Wake Island to congratulate and confer with General MacArthur who gave his opinion that neither the Russians nor the Chinese would intervene. He was right about the Russians, who were now anxious to discountenance the war, but wrong about the Chinese, who attacked on 26 November. Exactly a month later they were across the 38th parallel and the South Koreans and their allies were once more in full retreat.

The Chinese attack was a forestalling action. The Chinese, who remembered the Japanese attack on their country via Korea in 1931, had reason to suspect that the new American power in Japan was about to repeat that performance. American aid to Jiang under the China Aid Act, General MacArthur's visit to Jiang on Taiwan soon after the outbreak of war in Korea, the crossing of the 38th parallel by the Americans in October and their approach to the Yalu River, the open debate on American strategic interests in the Pacific islands and on the possibility of a return by Jiang to the mainland – all these things

3.2 Korea showing division between North and South

combined to alert the new regime in Beijing and persuade it that the Americans intended an anti-communist campaign like the similar, albeit unsuccessful, anti-communist interventions in Russia after the 1917 revolution. So the Chinese struck first.

Chinese intervention altered the nature of the war and gave rise to a fierce debate about how it ought to be prosecuted. From the end of June to the end of November the war, although waged on one side primarily by the Americans, could be represented as an international punitive expedition. After November it became more and more a Sino–American conflict. General MacArthur wished to recognize this fact and to wage open war on China, using the most relevant and effective military means. This meant, in particular, following Chinese aircraft across the Manchurian frontier instead of breaking off pursuit when this line was reached in combat, and it meant also bombing Chinese fixed installations on the Yalu River and elsewhere. Logically this attitude to the war could lead to the use of nuclear bombs in the heart of China itself.

But General MacArthur's views did not carry conviction in Washington. The chiefs of staff shrank from the prospect of embarking on a war with China which might drag on for years, while the president and his civilian advisers were extremely loath to re-enter Chinese politics (from which the United States had very recently extricated itself) and well aware of an almost worldwide disapproval of any such adventure. Washington's allies became alarmed (Attlee flew to Washington to express this alarm, especially about the possible use of nuclear bombs in Asia for a second time) and the neutrals began to give neutralism a markedly anti-American inclination; distrust of the United States in world affairs received in these months a fillip which was very slow to fade away.

As early as July 1950 Nehru, whose government had supported both the Security Council's resolutions, made approaches to Stalin and Acheson to put a stop to the fighting. His intervention was received with bare courtesy but in December India and other neutrals appealed to both sides not to cross the 38th parallel and during that month the Indian representative at the UN, Sir Benegal Rau, had a series of conversations in New York with an emissary from Beijing, General Wu Xiuquan. The latter, however, left New York before the end of the year with nothing achieved; a UN cease-fire committee was rebuffed by Beijing; the time for mediation was not yet.

On the first day of 1951 the Chinese launched their second offensive. UN forces were quickly forced out of Seoul and a fresh UN appeal for a cease-fire was rejected. But the Chinese advance petered out. At the UN China was declared an aggressor and in Washington in March a telegram from General MacArthur was read to the Senate in which the general in effect urged the United States to strike at China and not simply to accept the re-establishment of the 38th parallel as a boundary line. These were the alternatives, and both the declared aims of the UN and the arbitrament of war favoured the second. But it was now the Chinese turn to retreat, the UN recovered Seoul on 14 March and the South Koreans once more crossed the 38th parallel on the twenty-fifth. General MacArthur took the opportunity to issue on his own initiative a peremptory challenge to his adversaries to accept an armistice, coupled with an implied threat of massive retaliation if they did not. The challenge was ignored and President Truman warned General MacArthur that he had exceeded his powers and pursued a policy not approved by his government and commander-in-chief; he was told to stick to his own business. By way of reply the general tried once more to appeal to the Congress and people of the United States over the heads of his military and civilian superiors. He sent to the Republican leader in the House of Representatives, Joseph W. Martin, a letter which was read in the House and which recommended the strongest measures against China, including the use of Jiang's troops. On 11 April he was dismissed by President Truman.

The dramatic dismissal of General MacArthur caused such a stir that its significance was not immediately assimilated. What it meant was that the Korean War had to be brought to an end by compromise and mediation. The American government rejected the alternative of complete victory obtainable only by the military defeat of China. Yet it took more than two years more of fighting and negotiation before an armistice was eventually signed in July 1953, more than three years after the initial act of aggression in June 1950.

The dragging out and winding up of the war can be briefly recorded. Fresh Chinese attacks in April 1951 were soon held and both sides began to feel their way towards a truce. A broadcast in the United States at the end of June by the Russian member of the Security Council, J. A. Malik, led to truce negotiations which were begun at Kaesong in July and later transferred to Panmunjon. These negotiations were tedious, long, fruitless and punctuated by fears of a renewal of full-scale operations and by accusations against the Americans of recourse to bacteriological warfare. The most intractable issue was the fate of prisoners of war, many of whom in Southern hands were alleged to be unwilling to return to the North, but an exchange agreement was eventually signed in June 1953 (shortly after the death of Stalin, although no connection between the events can be definitely proved). The agreement was then wrecked by Syngman Rhee who released prisoners rather than turn them over to the North Koreans, whereupon the Chinese launched a major offensive. Notwithstanding these turbulent episodes an armistice agreement was signed in July.

The Geneva conference which opened in April 1954 failed to produce a final settlement and Syngman Rhee then went to Washington to try to persuade the United States to sanction a joint invasion of China by South Koreans and Jiang Kaishek's forces. He argued that the régime in China was on the verge of collapse but he failed to win over Congress or Eisenhower or Dulles to his view. In the following year American and Chinese troops were gradually withdrawn. Korea remained bisected but it was clear that the war was over.

The year 1954 was something of a landmark in Asia's postwar history. The Geneva conference, although it failed to produce agreement on Korea, demonstrated that the country would revert to the position that ruled in 1949, bisected and freed from foreign occupation. In the same year the United States decided not to intervene at Dien Bien Phu. But also in the same year the United States concluded treaties with Pakistan and Japan and created Seato. The acceptance of the *status quo ante* in Korea and the refusal to engage in war in Indo-China did not betoken an American withdrawal from Asia. For the next eleven years – until the beginning in 1965 of the American attack on North Vietnam – the United States tried to play a major role in eastern and south-eastern Asia by displaying but not using military power. The principal object of American policy was to halt the territorial progress of communism by conquest or subversion. In Korea aggression had failed to annex southern Korea to a communist northern half, but in the American view communism was making dangerous strides elsewhere in Asia by means of subversion. In fact the opposite was the case in Malaya where the tide was running against the communist insurgents, but the growth of Chinese power, still supported by Russian aid at this date, made Washington tremble for the successor states in Indo-China and for Indonesia. American policy was thus anti-Chinese because China had become the fount of a new wave of communism, and by the same token it was ideological, claiming virtue for the stand against the evils of communism and imputing wickedness to those who refused to fight the good fight or at least applaud it.

The events of the early 1950s strengthened and prolonged American links with the remnants of the Kuomintang. On the outbreak of the Korean War Truman gave Jiang Kaishek an undertaking to keep the new Chinese régime away from Taiwan and the Pescadores. In view of China's total naval incapacity, this commitment was easy to honour but it involved Washington in courses not approved by its principal European allies and it contained an awkward ambiguity. When Jiang was evicted from the mainland he had retained control of some small islands off the coast. Did the American umbrella cover these islands as well as Taiwan itself? The question was partly a question of the extent of Washington's unwritten commitments to Jiang, became to some extent a question of asserting American steadfastness after the vacillations over Dien Bien Phu, and was often discussed in terms of whether these islands were necessary for the defence of Taiwan or not.

The offshore islands were the Tachens, Quemoy and the Matsus, situated off Amoy and Foochow and forming an offshore screen similar to the advance guard which the Pescadores provided for Taiwan on the other side of the Taiwan Strait. In 1954 resolutions in favour of the liberation of Taiwan started to be manufactured and emitted by Beijing and provoked retorts from Washington showing that any attempt to attack Taiwan would have to reckon with the American fleet. In September the Chinese began to bombard Quemoy as a retort, it seemed, to the creation of Seato by the Manila Pact. The Kuomintang fired back and for several weeks this fire and counterfire looked like the beginning of a more serious encounter. In November thirteen American airmen, captured during the winter of 1952–53 when their aircraft came down in Manchuria, were sentenced in Beijing to terms varying between life and four years for espionage, and in December the United States and the Kuomintang concluded a new treaty which declared the defence of Taiwan and the Pescadores to be a common interest. Whether the retention of Quemoy and the Matsus was essential to the defence of Taiwan and the Pescadores was left unmentioned. In 1955 Jiang abandoned the Tachens under fire from the mainland.

The Chinese agreed to receive a visit from Dag Hammarskjöld who invited himself to Beijing to talk about the American airmen, but the Geneva summit conference in the summer of 1955 caused the Chinese to reflate the crisis. In July they started another major bombardment of Quemoy. The Americans had meanwhile apparently decided – whether on strategic or general political grounds – that no more territory should be ceded. The new bombardment served to make this resolve clear and after it had died away like its predecessor there followed a period of easier Sino–American relations marked by exchanges at Geneva, ambassadorial conferences in Warsaw and the release of the American prisoners. By the middle of 1958 there was even talk of American recognition of Beijing, but a statement from Washington in August showed that the Eisenhower–Dulles administration contemplated no such step.

This statement was immediately followed by a fresh bombardment of Quemoy. During the 1955–58 lull Jiang had moved troops into Quemoy until about one-third of all his forces was stationed just offshore from the mainland. In September 1958 Beijing demanded the surrender of the islands and was met by a declaration by Dulles to the effect that the Americans would fight to protect Taiwan and another by Eisenhower defining Quemoy and the Matsus as necessary for its defence. American escorts for Jiang's troops sailed within a few miles of the coast. War was feared. Washington's allies and sections of American opinion too became alarmed, and after a few days the crisis was deflated by the resumption of the Sino–American ambassadorial talks in Warsaw. Unpopular and hazardous though Dulles's policy of brinkmanship (that is, using military power to make political threats) had been, it had achieved during these years the object of showing that the American decision not to intervene in Indo-China in 1954 did not betoken a general American failure of will. This determination was to be further sharpened and displayed in South Vietnam where the Americans first gave the anti-communist régime economic and military aid and then engaged themselves directly in war on the Asian mainland. But this engagement changed the nature of American power and policies in the Far East. The defeat of Japan in 1945 had given the United States total dominance in the Pacific. The conflict between Truman and MacArthur and its outcome showed that Pacific dominance was still the basis of policy, buttressed as it was by garrisons in the Japanese and Philippine islands and by alliance with Australia and New Zealand (concluded in 1951 as the price of these countries' consent to the peace treaty with Japan). But the collapse of the French attempt to resume empire in Indo-China, coming as it did after the triumph of communists in China, led the United States towards a new policy of Asian rather than Pacific dominance.

Communist China's conflicts with the United States coincided nearly with a sharp decline in its relations with the USSR in the course of quarrels which developed in the late 1950s and were public news by 1960. One of the most startling consequences of the death of Stalin was the clash between the incompatible temperaments of Khrushchev and Mao which was superimposed upon the sometimes divergent interests of the Russian and Chinese empires and exacerbated by doctrinal dispute.

When Mao visited Stalin in 1949 there was an element of obeisance about this encounter between the Chinese leader and the man who, hyphenated with Marx-Engels-Lenin, was much more than just a Russian leader. Stalin's power had endured for a quarter of a century, his prestige had been enormously increased by the Second World War, he was a legendary figure, a little more than mortal man, a little like a long-lived and successful Chinese emperor. Mao, however, had disliked the Soviet Union since early encounters with it before the Second World War and, if not yet in 1949, became disposed to discern something superhuman in himself.

When Stalin died in March 1953, there was at first no agreement in the USSR about his successor and even a short-lived view that he had no single successor. A composite Stalin – Bulganin/Khrushchev – quickly replaced the committee rule of the Malenkov interlude and visited Beijing in 1954. Then Khrushchev emerged as the new Russian autocrat and was assumed by many to be *ipso facto* the chief of world communism as well. To Mao, however, this proposition was not self-evident; there was no rule which said that the world's leading communist had to be a Russian nor any disposition in Mao's scholarly mind to accept a rumbustious political boss in that role. During his visit to Beijing in 1954 Khrushchev had failed to establish with Mao either a hierarchical or a personal relationship.

For a while relations remained equable. China still needed the economic and technical help of the USSR and remained alive to the need for military support which the outbreak of the Korean War had dramatized. The Russians continued to help the Chinese to modernize their army; the Chinese Fourth Field Army which had entered Korea in 1950 was part of the force which had defeated the Kuomintang and which the Chinese were anxious to transform into a more modern instrument. Resentments, such as the feeling that the Russians had done very little to help the Chinese communists before they won power, were kept in the background. In 1954, with the Korean War clearly over, the Russians left Port Arthur and handed over its installations to China – two years later than the date set in the Russo–China treaty of 1950. They also transferred to China their half share in the joint companies formed in 1945 to exploit oil, non-ferrous metals and civil aviation in Jinjiang and to operate ship building and repair in Dairen. Finally, they entered into new agreements to give economic and technical aid in the shape of financial credits and the provision of skilled technicians and know-how.

The deterioration in Sino–Russian relations set in about 1956 and acquired a concrete edge in the two following years. At the twentieth congress of the Soviet Communist Party in February 1956 Mikoyan, followed by Khrushchev, set about the demolition of Stalin's memory. De-Stalinization involved points of doctrine on which Mao could fairly claim to be heard, but he was not consulted and was perhaps offended by the Russians' evident assumption that such matters could be settled by Russians alone. Khrushchev was still a new man, or at any rate unconfirmed in his new apogee, and it would have been at least becoming for him to consult an elder like Mao. In external affairs too Khrushchev was showing a dangerous unsureness. One of his first initiatives had been to try to repair the breach with Yugoslavia. Chinese attitudes to Yugoslavia were unsettled in the late 1950s. On the one hand Yugoslavia's independence of Moscow appealed to the Chinese; on the other hand they harboured heretical notions. It would seem that the independence prevailed at first over the heresies, but that from about 1957 the heresies seemed to the Chinese more serious, especially when the independence became less pronounced and Moscow gravitated towards a Yugoslav policy of coexistence, of supping with the devil when needs must, a policy which the committedly anti-American Chinese considered reprehensible, inexpedient and silly. Furthermore, in 1955 Bulganin and Khrushchev demonstrated support for non-communists, notably Asian non-communists, by their visits to India, Burma and Afghanistan. Chinese doubts were next reinforced towards the end of 1956 when revolt occurred first in Poland and then in Hungary. In the Chinese view the Russians mishandled, or nearly mishandled, both emergencies. In the Polish case the Chinese intervened in support of the more independent communist Gomulka and cautioned the Russians against the use of military force; in the case of the anti-communist Hungarian rising they urged the Russians not to withdraw their troops prematurely. But these disagreements were not lethal, and in the next year the Chinese were advising the Poles to be more amenable to the Russians.

In 1957 in a new agreement on technical aid, Khrushchev (according to the Chinese at a later date) promised China samples of nuclear material and information about the construction of nuclear weapons. This, moreover, was the year in which the Russians perfected the first intercontinental missile and launched the first sputnik. The whole world drew exaggerated conclusions from this achievement. The Russians were thought to be overhauling the Americans, perhaps even to have done so, and the Chinese expected Khrushchev to exploit this marvellous advantage. The communists had long possessed superiority in sheer numbers; now they were ahead in technology as well. The east wind, in Mao's phrase, was prevailing over the west. The Russian nuclear armoury could be used to pin the Americans to the wall, while the communist states helped their friends to power throughout the underprivileged world; in Asia, Africa and Latin America there were revolutionary movements eager to discard the yoke of capitalist imperialism with the help especially of the Chinese, whose own experiences between 1922 and 1949 had taught them more than anybody else knew about the strategy of revolution in poor, backward, agricultural countries. The Chinese, preparing for the Great Leap Forward and their second five-year plan, felt that they could also make a fresh bid to take Quemoy and make themselves felt in communist and Asian affairs.

There were several points in the Chinese analysis with which Khrushchev and at least some of his senior colleagues disagreed. They may well have been the only persons in the world who knew at this time that the Russian sputnik had not put the USSR ahead of the United States, and therefore they did not believe that they could immobilize the United States by the threat of nuclear annihilation. To some extent they were caught in the toils of their own jubilation, for the more they magnified their technical successes, the more it was supposed that they had won for themselves a much freer hand in international affairs than was really the case. While the Chinese thought them capable of preventing nuclear war, they feared it. The Chinese concluded from the Russian achievements that nuclear war had become much less likely and that therefore the communist powers could afford to try more adventurous policies. They also maintained the orthodox Marxist position that war in some form at some day was inevitable because imperialism made it so.

The Russian view of war had changed, more because the Russians approached the topic from a different angle than because of any radical revision of doctrine. They had diluted the basic doctrine of the inevitability of war not so much for reasons of pure logic but rather because their fear and awareness of the frightful consequences of the use of nuclear weapons had led them to discard so baleful a belief; they were thinking specifically of nuclear war and not of war in general. The nuclear danger also caused them to differ from the Chinese in their approach to non–nuclear wars: both Moscow and Beijing endorsed wars of national liberation, but Moscow was more worried than Beijing about the risks of escalation to nuclear war and therefore more cautious in any particular case. These divergences were a further consequence of the death of Stalin which, in this as in other subjects, released a stifled debate inside the USSR. In March 1954 Malenkov had pronounced that a world war would be likely to destroy all civilization; during his term of office as prime minister he stressed the imbecility of war and coupled it with his policy of increasing the supply of consumer goods. Khrushchev, while still in competition with Malenkov, attacked the latter for defeatism and for advocating coexistence with capitalists and put forward a hard policy of building up the strength of the USSR. When, however, he himself became prime minister he set out to assuage the asperities of the Cold War because the USSR was technically behind the United States and because he found the Russian forces ill organized. At the twentieth congress of the Communist Party in February 1956 he said that war was not inevitable and might not be essential to the worldwide triumph of socialism. This view was repeated in the declaration issued in November 1957 at the end of the Moscow conference of communist parties (the Chinese included), and discussion seemed from this point to be shifting to the question whether all war, and not only nuclear war, had become inexpedient. In 1958, however, the Chinese view that the east wind was prevailing over the west turned a theoretical debate into a live tactical issue with the Chinese expecting the Russians to take positive action incompatible with the general trend of the debate since 1953.

At the same time a further cause of dispute emerged. In general the Russians and the Chinese were in accord on the need to turn Asia, Africa and Latin America away from the capitalist camp, but they differed over the means. The Chinese, intent upon multiplying the number of states under communist rule, wanted to help only communists, whereas the Russians, taking the more pragmatic attitude that any anti-western régime was an advantage, were willing to help bourgeois revolutionary movements where communists were non-existent or unlikely to succeed. At the conference of communist parties in Moscow in December 1960 this disagreement was temporarily smothered by the adoption of a compromise formula: national democracies were to be helped if they were evolving in a socialist direction.

The Russo–Chinese alliance, already ruffled by suspicion and friction in 1956–57, was completely unhinged in the next two years, during which the USSR showed itself indifferent to vital Chinese interests, or even hostile to them. Jiang's immunity on Taiwan under the protection of the American Seventh Fleet was an affront about which Beijing could in fact do nothing since it possessed no fleet of its own, but his possession of Quemoy and the Matsu islands just off the Chinese coast was an even less supportable and also a less irremediable taunt. Beijing's resolve to make a bid for Quemoy and the Matsus by a policy of bang and bluff having repeatedly failed because Dulles refused to be rattled, part of this failure was ascribed by Beijing to the luke-warmness of Russian support. The Russians, while sympathizing with Beijing's feelings about irredeemed Chinese territory, were wary of Pacific entanglements and determined not to be drawn into a war for the reconquest of Taiwan. They refused to set up a joint command for the Far East and made demands which amounted in Chinese eyes to impermissible infringements of Chinese sovereignty. Khrushchev seems to have been willing to establish nuclear bases in China but only on the basis that they would be Russian and that there would be no Chinese finger on the trigger. Some Chinese leaders, including the minister of defence, Marshal Peng Dehuai, apparently considered the price worth paying, but an opposite view prevailed. Marshal Peng was dismissed and the Russian attempt (if such it was) to create in eastern Asia the same sort of strategic position as the American position in western Europe failed. To make matters worse the Russians poured cold water on the Great Leap Forward at a time when their co-operation was essential to its prospects, adopted a neutral attitude in China's disputes with India in 1959, continued to give large amounts of aid to Indonesia even though Beijing was being driven to protest against the Indonesian government's behaviour to its Chinese population, and set out to improve relations with the United States. The Chinese were being forced to reassess the attitudes of the USSR and the balance of the major forces in the world.

Events in the Middle East may have contributed to this rethinking. In July 1958 the Iraqi monarchy was overthrown and the king, his uncle, his prime minister Nuri es-Said and other notabilities murdered.

For a time it seemed to some observers that the Americans and the British would take up arms against this revolution and that a war of major or minor dimensions would begin. The Chinese were certainly interested in these events. At a later stage they demanded to be included in any conference assembled to deal with the situation, and their renewed bombardment of Quemoy began shortly after the Iraqi coup. They may have calculated that the Russians could be induced to use the nuclear threat against the Americans or even to become involved in fighting. If they were thinking in these euphoric terms, their disillusionment was sharp and they were forced to the profoundly melancholy view that, on the contrary, the Russians were engaging in a conspiracy with the Americans to dominate the world and prevent China from becoming a nuclear power.

When the world learned on 3 August 1959 that Khrushchev was going to the United States and would confer with Eisenhower at Camp David, the Chinese concluded that Khrushchev had rejected Beijing's thesis that the USSR should put forth its strength rather than negotiate from it. Although there had been some signs of Chinese endorsement for the Khrushchev policy, a change of mind in Beijing, coincident with changes in the top ranks of the Chinese Communist Party, had produced an unequivocally tough tone towards the United States and warnings against the naïve amateurishness of those who imagined that it was possible to lie down with imperialist lions. And not only was the Camp David spirit not shared by the Chinese leaders; the Camp David negotiations appeared to leave out Chinese interests as though China were nothing more than an impediment to a Russo–American *rapprochement*. In October, shortly after his return from the United States, Khrushchev went to Beijing, a visit which had the almost unparalleled consequence of producing no communiqué. If his hosts asked him what he had said to Eisenhower about Taiwan, it is unlikely that he had a palatable reply, and in the new year both the foreign minister and the prime minister of China made statements which amounted to declarations of no confidence in Khrushchev's foreign policy and proceedings. At a meeting of the presidium of the World Peace Council (a communist front organization) in Rome in January 1960 the Russian delegation attacked China, and at a communist conference in Warsaw in February the Chinese were observers only, although Outer Mongolia and North Korea were full participants. In April the ninetieth anniversary of Lenin's birth provided both sides with an occasion to expound their views at intemperate length, and a powerful Chinese propaganda campaign brought the quarrelling into the open. Mao, emerging from semi-retirement, made five separate statements explaining the Chinese attitude and pouring scorn on the folly of trusting the Americans.

It was no doubt the Chinese hope to convert or unseat Khrushchev, and the Russian leader, preparing for the Paris summit conference with the United States in mid-May, may have been in some danger. The Supreme Soviet met on 5 May, there was a reshuffle in the party secretariat, and there were rumours of splits and cabals. But Khrushchev went to Paris and, despite the U-2 incident and the failure of the conference, repeated his belief in peaceful coexistence on and after his return home. At a communist conclave in Bucharest in June, Khrushchev personally attacked the Chinese in the course of a meeting which was supposed to restore harmony and in August Russian technicians in China, to the number of about 12,000, were ordered to pack up and return home, bringing with them the plans on which they had been working. This bitterly unfriendly act, coinciding with the domestic setbacks of the great famine years, seemed to the Chinese tantamount to an invitation to the Americans to invade China and to the Kuomintang to incite a rising against the communist régime.

Correspondence and propaganda continued with increasing acerbity and in November 1960 eighty-one communist parties attended another general conference held in secret in Moscow to assuage the discord. After vituperative debate this meeting produced a communiqué which too thinly papered a too large crack. The Russians had the better of the argument when it came to counting heads. By this date the Russo–Chinese alliance was non-existent and doctrinal solidarity a farce. The withdrawal of the Russian technicians was a cancellation of the economic co-operation initiated immediately after the establishment of the Chinese People's Republic, and Khrushchev's approaches to Eisenhower had revealed the strict limits of Russian support for China in external affairs. The attempt to operate a Russo–Chinese alliance in world affairs had run on to the rocks of Khrushchev's American policy.

But at the end of 1964 Khrushchev was overthrown and the second triumvirate which took over power *ad interim* (like the earlier collective leadership after Stalin's death) attempted a reconciliation. Kosygin went twice to Beijing. The Chinese, however, refused to attend a conference of communist parties in Moscow in May 1965 and subsequently frustrated Moscow's plans for a world communist conference. The Russian embassy in Beijing was attacked in 1967 and the next year Beijing condemned the Russian invasion of Czechoslovakia and Brezhnev's doctrine on the right and duty of the USSR to act outside its borders for the defence of socialism and the socialist bloc as a whole. In 1969 there were incidents on the Ussuri River, where the possession of a few islands had long been a matter for (not very heated) dispute. Men were killed and there was talk about the possibility of a Russian pre-emptive strike against China before its nuclear capacity assumed deterrent proportions. Russian forces in the east were increased in the 1970s and 1980s from thirty-three to fifty-six divisions, and from 1979 the Russian threat to China was magnified by the establishment of

Russian naval and air bases in Vietnam. Negotiations in Beijing in 1969–70, to which the USSR sent one of its most senior and able diplomats, were lengthy but unproductive, and renewed talks in 1979 were wrecked by the Russian invasion of Afghanistan (with which China has a border).

Some compensation for the collapse of a grand communist alliance was provided for China by its readmission to the comity of nations and an unexpected relaxation of its relations with the United States. Canada and Italy established diplomatic relations in 1970. In the same year a majority in the General Assembly of the UN – but not the requisite two-thirds – voted in favour of China's membership and at the end of the year Mao's régime was admitted to be the rightful incumbent of China's permanent seat in the Security Council. China came into the UN on its own terms which included the rejection of an American–Japanese proposal to retain Taiwan as a separate member alongside China.

In Washington Nixon had decided to jettison the two Chinas policy if that were necessary in order to restore relations with Beijing, thus reversing a basic tenet of American policy. At the Geneva conference of 1954, Zhou had proposed discussions for a reduction of tensions. A first meeting took place in Geneva in August 1955 after China had released the thirteen captured American airmen, but American recognition of Taiwan as a separate state proved an insuperable obstacle and the talks were broken off by the United States at the end of 1957. They were resumed in Warsaw the next year and the United States began to give visas to American newsmen who wanted to go to China, but the three months' crisis over Quemoy and the Matsu islands intervened to freeze attitudes on both sides. In 1960 Zhou proposed a Pacific non-aggression pact. For several years there was a stalemate, although contacts were maintained. In 1966 the American secretary of state Dean Rusk said that the United States would not try to overthrow the government in Beijing by force but however Beijing might interpret this disavowal of Jiang's persistent hope, the war in Vietnam put an end to even the mildest sign of normal relations between Beijing and Washington until the American withdrawal from South-east Asia revived the situation created in the 1950s by the truce in Korea. The Americans were acknowledging the defeat of their efforts to be a continental Asian power. In 1971 an American ping-pong team which had been playing in various Asian countries was invited to China, the first recorded appearance of this sport in high politics. This move was followed by a further relaxation of the American trade embargo (there had been some relaxation in 1969). In July Nixon's adviser on national security, Henry Kissinger, went secretly to Beijing and it was then revealed that Nixon himself would follow the next year. This was a sensational piece of news and the Nixon visit, duly accomplished with the secretary of state William Rogers in attendance, heralded a stream of other highly placed visitors anxious to make their peace with China and recognize and trade with it.

The Mao–Nixon détente was a political demonstration made possible by the ending of the Vietnam War and valuable to both countries as a means of addressing a warning to the USSR – by the Americans, not to take Russo–American détente for granted; by the Chinese, not to make trouble on the Sino–Soviet frontiers. The demonstration had no precise content (and a visit by President Ford to Beijing at the end of 1975 gave it none) but it jolted thinking about international affairs at the great power level. There was now another piece on the board and it served to confuse a game which, by all the rules of a bipolar world, allowed for two players only. In military terms, however, China was still well below superpower standards. It could not touch the United States or greatly harm the USSR. (The Sino–Russian front had been quiet since incidents on the Ussuri in 1969 but the Russian armies in Mongolia had been built up from fifteen divisions to forty-four.) By 1975 China had set off some twenty nuclear explosions and possessed a stock of nuclear weapons estimated at 300. Some of these were intermediate-range missiles but most could be delivered only by aircraft whose range was 2,400 km. By the end of the decade, however, there were good grounds for the belief that China possessed not only the missiles but also the delivery systems needed to attack targets in the European areas of the USSR.

China set out to become both an independent nuclear power and also the world leader of a new international of the underprivileged. Zhou Enlai, who had steadily built up China's diplomatic position in Asia from the Bandung conference in 1955 onwards, made a tour of African countries in 1963–64. China's exclusion from the United Nations, where the seat specifically allotted to China continued to be occupied by the Kuomintang's rump, may have contributed to its wish to assert its claim to a leading world role outside the UN and in defiance of the UN's principal members, even though its comparatively modest defence expenditure and its industrial growth pains made China a power of the future rather than a power to be reckoned with in the present. What China achieved in these years of travail was to project so menacing an image of its future power as to make the world take it very seriously in the present and even to be more afraid of China than of any other country in the world. This fear was not merely a consequence of adept Chinese diplomacy; it was also a product of the mysteriousness with which China was cloaked by the outside world's own determination to treat it as not just different but out of this world. It was a product in particular of confusion about Chinese attitudes towards nuclear war and a belief that China's leaders regarded such a war with equanimity on the grounds that China's vast size and few large cities would enable it to survive nuclear attack. The Chinese were in fact well aware that a nuclear war would be a universal disaster and that China and the Chinese Communist Party would be among the victims, and although they clung to the not implausible thesis that wars were inevitable they did not apparently regard nuclear war as inevitable. Like others they hoped to prevent a nuclear war by a policy of deterrence but, unlike others, they could not in the 1950s and 1960s do the deterring themselves. China in this period was exposed to nuclear threats or preventive war in the same way as the USSR had been exposed between 1945 and 1949.

In this situation the Chinese, like the Russians before them and regardless of whether their ultimate intentions were malevolent or peaceable, had to resort to minor non-nuclear deterrents while being careful to avoid any disastrous provocation of a nuclear power. They challenged and exposed Asia's largest non-nuclear state, India, and came to terms with lesser countries which might have been tempted into an enemy camp. Having taken Tibet because they could and postponed the conquest of Taiwan because they must, they pursued a policy of limited activity which fitted their limited capacities. The decision of North Vietnam to take an active part in war in the south was, it may be presumed, approved by the Moscow conference of eighty-one communist parties and by the government of the USSR, which could not resist appeals to communist solidarity and could have no objection to endorsing and even assisting a modest guerrilla war likely to prove embarrassing to the United States. When in 1963 the United States decided to take a leading part in the war and to raise the level of warfare in order to preserve an independent South Vietnamese state, they in effect served a no-aggression notice on China. The Chinese, faced with a conflict between two principles – the principle that a nuclear power must not be provoked, and the principle that a national liberation movement must be helped – opted for the former.

This dominant principle was further exemplified in Europe. Albania, one of China's few friends, received little more than the rhetorical support which the Russians could be expected to put up with. In any case the Chinese were only marginally interested in Europe. They were more interested in Africa, which presented in their view excellent prospects for revolution. More broadcasting time was devoted to African listeners than even to southern Asians, but the results were disappointing for by the time that China was ready to play a full part in world affairs, most of the nationalist movements in Africa had won power and independence and, being intent on retaining their power, were anything but insurrectionary and were suspicious of Chinese intervention in their affairs. The fall of Ben Bella in Algeria in 1965 was a special setback, similar to the rout and slaughter of the Indonesian communists in the same year.

China was on the way to becoming a regional nuclear power but also was searching in vain for a more than regional role. Its emergence as a future giant power gave notice that the bipolar world of the Cold War was destined to be short-lived. China's first nuclear explosion in October 1964 was followed by a second in May 1965. A year later a first thermonuclear weapon, probably one capable of use in submarines (of which China had thirty, received from the USSR), was exploded. China's first guided missile test occurred in October 1966 and its first hydrogen bomb was exploded in June 1967. It would soon be formidable for a thousand miles around; it could be expected to have a wide range, if limited stocks, of nuclear weapons some time in the 1970s; and it might spring a surprise by developing ahead of expectations a submarine nuclear armoury to threaten the United States and Latin America from the Pacific Ocean. But unlike the United States and the USSR it was not becoming the centre of a group. It had failed to detach more than a few, comparatively insignificant communist parties from the main body of international communism which, if faced with a choice, persisted in choosing Moscow rather than Beijing, even while reducing the intensity of Muscovite control; only the distant and ineffective Albanians and New Zealanders stuck staunchly at Beijing's side. Elsewhere China caught the fancy of sundry malcontents in France or Egypt or Zanzibar, but these were countries with nationalist rather than internationalist preoccupations and with discontents which were peculiar to each of them and provided no basis for common politics.

There is always a certain grandeur about isolation. Britain, Japan and the United States had all at various times dallied with its lures. Communist China made a virtue of its isolation and discounted the dangers by dwelling on a more distant future in which it would ultimately circumvent and discomfit the major powers whose hostility it had to bear in the present. China was used to having powerful enemies. Britain and Germany had been succeeded in this role by Japan, and Japan in 1945 by the United States – especially after the outbreak of the Korean War. The USSR, superficially a natural ally, had turned out within a decade to be an enemy, a foreign power upon whose goodwill China had mistakenly, if for a short space, allowed itself to become over-dependent. In this situation China's leaders seemed to veer towards a nationalism even more intense than might have been expected from a half century of impotence and revolution, and to seek reassurance in their country's vast size and splendid history, their faith in the revolution which they had made and an optimistic view of world politics. In their eyes Asia, Africa and Latin America were revolutionary, anti-colonial domains where their major enemies – the United States, the USSR and the principal western European states – would get into trouble because of their archaic political attitudes and economic contradictions. Western Europe and northern America would also develop similar revolutionary movements in which the bourgeoisie would be threatened and would finally be supplanted by the proletariat. Meanwhile China must assemble and develop its resources and – in the view of Mao himself and some of his associates – preserve the ardour of its revolution.

RESURRECTION

Mao's government was pledged to end corruption in the public services, reverse the economic slide which had overwhelmed the Kuomintang, make China a modern industrial power and introduce sweeping reforms in land-holding and agriculture. These were huge tasks requiring money, authority and peace, and it was difficult to say which was the most pressing. Corruption, waste and bureaucracy were attacked in the Three-Anti campaign begun in Manchuria in August 1951 and extended to the whole of China two months later. A Five-Anti campaign followed, directed against bribery, tax evasion, fraud, the theft of state property and the betrayal of economic secrets. These campaigns seemed to betoken, on the part of the government, a real concern to secure the approbation of the Chinese people and, conversely, an equal concern to ensure that the people should not only behave correctly but think correctly. The campaigns were pursued by means of public meetings, confessions, purges, delation and executions. They were indiscriminate and were used for attacks on unpopular or richer classes such as missionaries, merchants and private entrepreneurs. Landlords and kulaks were also singled out, partly as an acknowledgement of the debt owed to the poorer peasants who had ensured the survival and eventual success of the communists. The dispossession of landlords and kulaks, which began in 1950, was accelerated and made horrible by the panic which spread in China during the first months of the Korean War and claimed 2 million victims or more: the newly installed rulers, like the leaders of the French Revolution facing the armies of *émigrés* and hostile powers, lived in expectation of a counter-revolutionary uprising which would be supported and exploited by the United States.

Collectivization began slowly in 1951. The example of the USSR in the 1930s, and the inadvisability of disillusioning and antagonizing the peasants who had just become proprietors in place of the vanished landlords, dictated caution. In 1953–54 Mao was ill and Gao Gang – the semi-autonomous lord of Manchuria, pro-Moscow, pro-heavy industry – was temporarily in charge. (He committed suicide soon afterwards.)

On his return Mao decided to step up the pace of change, particularly in agriculture. After two poor years the 1955 harvest promised well; he did not want to leave the new peasant proprietors undisturbed for long for fear that a new class of kulaks might emerge from their ranks; the first five-year plan, covering the years 1952–57 (but not published until about half its course was run) was inadequately primed by the Russian pump and needed a further boost that could come only from brigading and driving the peasants into more efficient methods and greater productivity. The basic aim was to sap the peasant's resistance to the state; so long as he owned his land he would keep the will to fight, but as soon as he lost it the spirit would go out of him.

In the earlier years the approach to communal ownership had been cautiously prepared through a series of staging points beginning with co-operation on specified tasks at certain seasons; proceeding thence to a distribution of rewards partly on the basis of the size of a man's holding and partly on the amount of work done by him; and so leading to a final phase in which the ownership of the land would be transferred to the communal group (not the state), rewards would be calculated entirely on the basis of work done, and the group's affairs would be regulated by general meetings and elected committees. From the latter part of 1955 this progression was enormously accelerated and in something like two or two and a half years it was all but completed. The speed of this vast economic and social revolution was characteristic of the methods and outlook of China's new leaders but it also sharpened the discomforts and resentments (especially among the better-off peasants) which such a programme would have evoked in any case.

During these years Mao, whose rapid collectivization was bound to offend the more conservative, also ran the risk of disgruntling the more radical faction by wooing the intellectuals who, though suspect because of their western training and thinking, were very useful to the régime. Mao, perhaps influenced by events in Hungary, wanted to initiate a genuine debate which would engender a genuine conviction of the correctness of his policies. The intellectuals, however, were extremely chary of beginning, even after Mao, by coining the slogan of the Hundred Flowers, virtually pressed them to believe that the régime wanted them to think for themselves and express their views with more freedom and less regard for conformity than they had hitherto dared to use. When criticism was evoked, it proved to be too exuberant. The inevitable disillusion which was eroding the high hopes and self-reliant optimism of the morrow of victory had been sharpened by inflation, labour troubles, and shortages of food and consumer goods, and in this atmosphere the hoped-for discussion about how to progress along the communist path extended to more radical questioning of basic communist tenets. A few months after the first mention of the Hundred Flowers the debate released by that slogan was abruptly closed.

It was succeeded early in 1958, at the commencement of the second five-year plan, by the Great Leap Forward and the vigorous introduction of the commune system in country and town. The Great Leap Forward (preceded by an earlier and unsuccessful Little Leap during 1956–57) was an incitement to greater exertions associated particularly with Liu Shaoqi, a leader of the more impetuous school who succeeded at the end of 1958 to the highest position in the state upon the unexpected retirement of Mao from the presidency. The Great Leap Forward was a short cut to greater production. After some experimentation it was ordained for the whole country in the autumn of 1958. The principal objects were to mobilize labour, set women free for industrial jobs, establish local industries as an adjunct to the larger units of production, and provide country people with an elementary introduction to industrial processes. One of the most publicized items in the programme was the making of steel in back yards (a system which produced much but poor steel), but the most important feature was the communes. These began to be formed at the beginning of 1958, were announced and explained in April, and were rapidly established throughout rural areas to the accompaniment of a propaganda barrage designed to drown criticism in a wave of enthusiasm. Property was brought into communal ownership and individuals were told to look to the community for free food, services and entertainment. In many places results went ludicrously and tragically far; even cottages, trees, fowls and small tools were turned into communal property. The vast size of China made radical reform almost impossibly difficult since the central government had only faulty machinery for ensuring that its wishes were carried out sensibly. It operated in fact through cadres, a non-communist device added by the communists to the system of government in China. These cadres were the links between the central government and the people. They were responsible for much of the inefficiency, crudity and brutality with which new policies were implemented, but without them government would have broken down and, in reverse, government would have been without any means of discovering the mood of the people.

By bad luck the inefficiencies of the Great Leap Forward were magnified by natural disasters which produced great hardships and famines, and the whole experiment, in so far as it was meant to rationalize and boost agriculture production, was abandoned. It was a piece of wilfulness on Mao's part which lacked both the necessary labour and the necessary enthusiasm on the ground, a vast flop which was for a time concealed by false claims. Although the communes survived as new elements in society and in government, by 1960 the small team was once more the basis of the rural economy. The revolution moreover lost face seriously during the great famine years.

It had exacerbated an inevitable crisis by grave overestimates of crop production and it had experimented unsuccessfully with a new communal pattern which the Russians had explicitly and correctly derided.

In the 1960s the revolution began to devour its children. China's ruling group was disrupted with a violence more familiar in the USSR than in China, where the only major communist figures to have been purged had been Gao Gang in 1954 and Peng Dehuai in 1959, both suspected of being too Russophile. Mao, approaching the end of his life, became obsessed by the fear that his life's work was being eroded by bourgeois backsliding and compromise. He determined to displace all those leaders whose steadfastness and fervour were in doubt and to revitalize party and populace by turning to youth. After some years of reflection and preparation he revealed in 1965 to his principal colleagues plans for a cultural revolution and appointed the mayor of Beijing, Peng Zhen, to direct it. Peng Zhen, however, was soon at cross purposes with Jiang Qing, Mao's third wife, and sometime in the first half of 1966 he was dismissed. In May 1966 Mao reappeared in public after an effacement of six months and a few weeks later the Cultural Revolution got publicly under way with a series of rallies, demonstrations and denunciations and a much publicized swim by Mao in the Yangtse near the provincial capital of Wuhan. It so happened that at this time there was a major breakdown in the educational system which had forced the authorities to postpone for a year all entries into universities and similar institutions and so discharged millions of young people into temporary purposelessness. They were turned into Red Guards who were to replace the communist youth organization (which had supposedly gone flabby) and go to work to boost production in field and factory. From these useful, if unacademic, pursuits they were diverted to an ideological crusade and, in an exuberance of anti-revisionist spirits, demonstrated against revolutionary insufficiencies and pre-revolutionary attitudes and symbols, assaulting individuals and destroying property in a movement which spread so extensively that it disrupted communications and brought factories to a halt. Some 20 million young people were said to be involved, most of them in their teens.

'Cultural Revolution' was the misleading name for a reign of terror accompanied by the most revolting cruelty. It had its sources in internal and external problems – economic planning, devolution, consolidation versus forcing the pace of progress, attitudes to the Russians – which had troubled the party in the 1950s and disrupted it in the 1960s. The revolution split the party at all levels. Hundreds or thousands of leaders, from President Liu Shaoqi to much humbler officials, lost their posts, were tortured and killed. Inevitably the army advanced in power. Lin

Biao, who had succeeded Peng Dehuai in 1959 and had invented and distributed the famous little red book, sided with Mao, thus ensuring Mao's victory and confirming the defeat of the pro-Russian faction once represented by Peng. Lin Biao was proclaimed Mao's eventual heir. But in 1971 he disappeared. Rumour had it that he fled to the USSR in an aircraft that crashed in Mongolia, killing him. Two years later at the tenth congress of the Chinese Communist Party he was openly attacked. He was accused of plotting to assassinate Mao and cement an alliance with the USSR. These charges were repeated at the joint trial in 1980 of the leaders of the Cultural Revolution and Lin's one-time closest military associates.

Mao Zedong died in 1976. He was an extreme example of the individual in whom ideology occludes the intellect, common sense and human emotions and helps a naturally authoritarian man to monstrous abuses of power. After a singularly long and arduous struggle which culminated in the victory of 1949 he saw his visions coming to grief. In order to preserve them and his own power he turned to policies which created chaos and brutality on a huge scale, ruined rural China by devastation and depopulation, and destroyed much of the material and spiritual heritage of the oldest civilization in the world. Zhou Enlai died in the same year as Mao and these two deaths released a struggle for power between three main groups. The first beneficiary was Hua Guofeng who was Mao's choice for the leadership and succeeded Zhou as (acting) prime minister. This was a setback for Zhou's 72-year-old deputy and presumed heir Deng Xiaoping. Deng, general secretary of the Communist Party in 1956, had been purged during the Cultural Revolution but reappeared in 1974 and in the next year was appointed deputy chairman of the party, first deputy prime minister and chief of staff of the army. Demonstrations in Beijing in Deng's favour led to his disgrace and the further promotion of Hua to be first deputy chairman of the party (next to Mao) and prime minister. When Mao himself died Hua, temporarily free from attack from the right, moved swiftly against the radical left. The so-called Gang of Four, led by Jiang Qing, were arrested a few weeks later and Hua became chairman of the party. But in the following year Deng made a comeback and for the next three years Hua and Deng shared the principal posts and the limelight, until in 1980 Deng ousted Hua after a campaign in which Hua's position as Mao's successor was undermined in the name of democratic freedom and Mao himself was criticized for having acted autocratically by nominating his successor. At the same time the Gang of Four were brought to trial and linked with Lin Biao's abortive coup of 1971. The effect of the trial, and presumably its purpose, was to represent the Gang of Four as a conspiratorial group in the tradition of China's secret societies; to discredit the armed forces as treasonable and incompetent;

and indirectly to implicate Mao himself in the economic and social disasters of the Cultural Revolution.

In Chinese communist terms Deng was a conservative pragmatist, even a reactionary. His victory meant a return to traditional types of competitive examination, the rehabilitation of the intelligentsia and other victims of the radical years, the restoration of the profit motive, higher prices for peasants and higher wages in industry and commerce, and wider openings to the west and Japan and even to the USSR. But although a revisionist and modernizer, Deng was no liberal or any sort of democrat. He wanted to modernize the Chinese economy without relaxing the political grip of the Communist Party. The Cultural Revolution had set back modernization and in his declining years Mao had failed in his primary task of bringing China back to its rightful place as the greatest power in the world. Deng and his colleagues elaborated programmes for the modernization of industry, the armed forces, agriculture and science and technology. Large sums were borrowed and spent but programmes were cut back when the financial establishment (bankers returning to posts which they had held before the Cultural Revolution or even before Mao's triumph over Jiang Kaishek) discovered that foreign funding was outrunning foreign revenues (exports, remittances from overseas Chinese, tourism). Growth was strong – with some poor spots in energy production from all sources and grains – but so too was demand. Inflation rose therefore to 20–30 per cent a year and corruption was magnified as the gap grew between official prices, unchanged for years, and prices on black markets. Price reform was urgent but anticipation of it further accentuated the gap, caused runs on banks as depositors rushed to change their money into goods, and so forced the government to postpone price reform. As in the USSR the problem of grafting market mechanisms into a command economy generated economic breakdown and bitter disputes between political leaders.

Deng was himself caught in these dilemmas. He both saw the need for liberal measures and feared them. He was alive to his dependence on the army and its ageing and conservative chiefs who had thrown him out in 1966 and restored him in 1975; he much enlarged the armed police force. He oscillated between an innate authoritarianism and a readiness to loosen the reins in order to foster economic progress – and an equal readiness to tighten the reins when relaxation encouraged political protest. His policies engendered therefore their own reversal. He tolerated and seemed to encourage the efflorescence of protest which in 1978–79 took the form of affixing popular complaints and demands on Democracy Wall in Beijing and he was able to combine for several years substantial economic expansion with tolerable political freedoms. But about 1985–86 economic advance faltered while popular demands became on the contrary more pressing. The price of economic progress was rising prices, rising inflation (to around 10 per cent), breakdowns and bottlenecks in communications, shortages of capital and energy, and corruption. Economic growth, instead of relieving want, was making it worse and fuelling indignation. Deng imposed import controls and devalued the currency in an attempt to restrain the economy and, in the face of rising student clamour for more and quicker changes, he relaxed censorship and permitted more open political debate. But in 1987, scared by the outcome of this liberalizing trend, he briefly changed course, muzzled liberal academics and compelled the Communist Party's comparatively liberal and comparatively young general secretary, Hu Yaobang, to resign (as he himself had been compelled by Mao to resign in 1966). After a few months Deng again changed course and reverted to a relatively liberal stance. At the end of the year he fulfilled his frequently announced intention to retire, taking with him into retirement a clutch of old conservatives and leaving at the head of affairs two younger men – Hu's successor Zhao Ziyang and a new prime minister, Li Peng. Deng's retirement was more apparent than real and not very apparent.

In 1989 Hu died. His funeral was made the occasion for massive demonstrations in which students from Beijing's universities were specially prominent, voicing protests against the slow pace of change, economic failure and persistent corruption. They were joined by discontented intellectuals and by workers, and suddenly they were making a big impact not only in the capital but also in some eighty other cities all over China. They posed a threat, not necessarily to the Communist Party, but to its elderly ruling clique. To the octogenarian Deng, however, and to others of his generation these two threats were indistinguishable and amounted also to a threat to China itself. Deng equated the party with the revolution which it had made, and the revolution with China. Anybody who opposed the party was therefore a traitor to his country. At this critical moment Zhao, who shared Hu's belief that economic reforms required some political reforms too, happened to be away in Korea. On his return to Beijing he took a conciliatory, even sympathetic and apologetic, line but behind the scenes his more conservative adversaries persuaded the mutable Deng to their side. The students were attacked and maligned in official publications. Over-optimistically they for their part refused to be intimidated. The resulting confrontation was briefly frozen because of the impending arrival of Gorbachev in Beijing. In Tiananmen Square, which was permanently occupied by huge but well-ordered crowds, the mood turned to hostility against Deng personally. Deng resolved to use force to dispel the crowds. The first moves by troops were blocked by unarmed civilians, the temper of the soldiers appeared uncertain or even friendly to the demonstrators, and two operations were called off; but eventually the drama which had riveted the attention of much of the world from the end of April to the first days of June was brought to its close by massacre. Governments all over the world expressed outrage, loudly but briefly.

These events checked Deng's external policies. A decade earlier Deng had visited Tokyo and Washington and some years later he had received Thatcher in Beijing and sent his foreign minister to Moscow. An eight-year commercial agreement with Japan, concluded in 1978, was followed by a full peace treaty. With the United States full diplomatic relations were established in 1978: Washington agreed to shift formal recognition from Taiwan to Beijing and remove its troops from Taiwan, while China watered down its demand for the complete abrogation of Washington's treaty relations with Taiwan. Reagan angered China shortly after his election by trying to square the new Sino–American concord with a revival of sales of arms to Taiwan but he was quickly forced to abandon this impossible stance. Bush pursued the Nixon–Reagan policy of making friends and contracts in China, overlooking brutal Chinese repression of risings in Tibet in 1987 and 1989 and a personal rebuff when the Chinese government prevented the outspoken critic Fang Lizhi from accepting an invitation to meet Bush at the American embassy in Beijing. At the same time Deng cautiously improved relations with the USSR. Whereas in the 1970s he had advertised his discontent with Moscow by despatching Hua on visits to Romania and Yugoslavia and by publicly advocating a Sino–Japanese–American alliance against the USSR, he responded amicably to Gorbachev's first indications in 1986 that the USSR was prepared to make concessions on border disputes, withdraw troops from Mongolia and Afghanistan and end support for the Vietnamese occupation of Kampuchea. The Chinese foreign minister visited Moscow – the first such visit for over thirty years – and Gorbachev was invited to Beijing. Before going there in 1989 he announced the withdrawal of 500,000 troops from China's borders and a reduction by two-thirds of the USSR's 50,000 troops in Mongolia. Prime Minister Li Peng visited six South-east Asian countries in 1990. Visits to China in 1992 by the Emperor Akihito and Boris Yeltsin put China undisputedly back on the world map in the year in which it had to be carefully wooed over Yugoslavia and other international issues because of its veto in the Security Council.

China, like India almost half a century earlier, had an irridentist programme. Its most substantial item was Taiwan where President Jiang Jingkuo, the son of Jiang Kaishek, brought martial law to an end in 1987 and was succeeded by a native Taiwanese, Lee Tenghui who, although still insistent on Taiwan's independence, promoted economic links with China and Taiwanese investment in China and permitted elections for the first time since 1948. Beijing maintained its policy of integration but with a reticence dictated by the facts. Over Hong Kong, however, and Macao the facts allowed a more insistent tone. Hong Kong had been acquired by Britain in three pieces: the island in 1841, supplemented by Kowloon across the water in 1860 and by the New Territories (which comprised 89 per cent of the colony) leased in 1898 for ninety-nine years. When the Second World War ended the colony's population was about half a million. A move by the Kuomintang to occupy it was thwarted, the victorious communists mysteriously stopped short of a similar attempt a few years later and the British returned to an increasingly prosperous exemplar of capitalist commerce and finance with, however, a deadline set by the term of the lease of the New Territories. No Chinese régime accepted the validity of any of the three nineteenth-century transactions. For China the whole of the colony was and always had been sovereign Chinese territory. For the practical realization of this situation Mao and his successors were prepared to take a long view which would be automatically shortened by the passing of the years. Hong Kong was no use as a base for an attack on China either by the Kuomintang or the United States. Meanwhile it was a window on the outside world, increasingly valuable as China recovered from war and civil war and the Cultural Revolution.

The third and smallest item of irredenta was Macao, a peninsula and two islands on the other, western side of the Pearl River estuary, six and a half square miles in extent. Portuguese traders first occupied Macao in the fifteenth century and it ranked as a province of Portugal from the conclusion in 1887 of a treaty with China. More than 90 per cent of its population of half a million were Chinese. Riots in 1966 demonstrated the precariousness of Portuguese rule and resulted in the humiliation of the Portuguese governor and *de facto* Portuguese–Chinese power-sharing. China promulgated in 1993 a Basic Law for Macao to come into operation on the inauguration in 1999 of the Macao Special Administrative Region of China. A similar course was plotted for Hong Kong.

The Macao riots of 1966 were replicated a year later in Hong Kong, synchronized with verbal and physical attacks on the British in Beijing. This aggressiveness alarmed the business community in Hong Kong and the colony's democratic élite but the governor and police were more resolute than those of Macao and the *status quo* was substantially undisturbed. But violence in the colony and the Chinese capital served notice on the British of the approach of 1997 with the extra complication that the need to resolve the future of the colony had somehow to be reconciled with the need to improve relations with a resurgent China. By the mid-1970s the defeat of the Gang of Four and the restoration of Deng made *rapprochement* with China practicable as well as desirable and at the end of that decade the governor of Hong Kong went to Beijing to initiate talks. Deng, although intransigent on Chinese sovereignty, introduced the notion of Hong Kong as a special region in which communism and capitalism would coexist: two systems in one state.

But the advent of Thatcher blocked agreement and embittered relations. Deng and Thatcher were equally obdurate but Deng was the more realistic, better informed (at least initially) and held practically all the cards. Thatcher, deeply averse to any surrender of British sovereignty and fresh from her triumph in the Falklands, toyed with the idea of a condominium and, even more unrealistic, a surrender of sovereignty so circumscribed as to be no surrender. Proposals for a continuation of British administration after 1997 on the grounds that only thereby could the stability and prosperity of Hong Kong be guaranteed irritated Deng who, after an abrasive meeting of the two leaders in Beijing in 1982, declared that failing agreement within two years China would produce its own unilateral plans for the future of the area. In the ensuing impasse confidence and share prices in Hong Kong slumped, the Hong Kong dollar lost a third of its exchange value in a year and panic loomed. The one concrete outcome of Thatcher's visit to Beijing was Deng's two-year deadline. Before it was out the British foreign secretary announced in Hong Kong and Beijing the abandonment of Britain's proposal for continuing British administration. This retreat cleared the way for the resumption of talks which produced the Joint Declaration of 1984.

This declaration was a compendium which included a statement of principles by China and agreement that China would produce and promulgate a Basic Law for Hong Kong to come into force in 1997 when sole British administration would be replaced by sole Chinese administration. The declaration created a Joint Liaison Group with a vaguely supervisory role in a transitional period before and after 1997 – originally a Chinese proposal but modified in negotiation to restrict the group's authority, delay its inauguration and extend its life to the year 2000. In 1985 China duly established a committee of eighty-two persons, including twenty-three from Hong Kong, to draft a Basic Law. Published in 1988 and approved by the People's Congress in Beijing in 1990, it promised Hong Kong a special status for fifty years, an elective element in the legislative council of one-third from 1997 and up to half by 2003, and required Britain not to exceed these limits before handing over. The publication of this grudging enactment, which disappointed the more optimistic illusions fostered by the Joint Declaration, caused dismay in Hong Kong where fear of the future had been exacerbated by the Tiananmen massacre.

Shortly afterwards a new British governor, Christopher Patten, a Conservative minister who had lost his parliamentary seat at the general election of 1991, arrived to preside over the colony's last years. Patten's hands were tied by the events of 1982–84 which had shortened Deng's temper, introduced the two-year deadline and secured for China the right to legislate for Hong Kong's future with only the vaguest constraints from a set of principles of its own devising. Within these constraints Patten's tasks were to ensure as amicable a transfer of power as possible (London having at least one eye on Anglo–Chinese relations in the indefinite future) and to allay Hong Kong's fears. These fears were already manifest in the rate of emigration which doubled or trebled around 1990, fuelled by the prospects of communist rule and by British niggardliness in admitting fugitives from the colony to Britain. Until 1962 all in the colony had a right to British citizenship and entry into Britain but these rights had been abrogated and Britain was proposing to restrict entry to 50,000 individuals and their dependants. These select few were to be picked on a complicated points system which could not conceal the fact that Britain was fulfilling its obligations not to the people of Hong Kong but to an élite of successful businessmen and colonial servants.

Patten confronted China's constitutional plan with proposals of his own to come into operation at imminent local elections and at the election of a new legislative council in 1995. He proposed to reduce the voting age from twenty-one to eighteen (the voting age in China), so creating an electorate of 3–4 million; to introduce single-seat, single-vote constituencies; to make the two councils for Hong Kong/Kowloon and the New Territories and the nineteen district boards wholly instead of two-thirds elective; and to reform the sixty-seat legislative council by making twenty instead of eighteen members directly elected, ten by local councillors who had themselves been elected to their councils and the remaining ten by functional constituencies with a broadened franchise.

For Hong Kong these changes had the advantage of extending, however belatedly, its fledgling democracy but the disadvantage of giving offence to its impending Chinese masters, whose response was vociferous hostility and charges that the governor was in breach of the Joint Declaration. His policy was narrowly approved by his legislative council. It was a gamble on the chance that by 1997 China would swallow a little democracy in a small corner of its huge territories whose rich material gains would dwarf ideological and constitutional niceties. But elections in 1995 were ignored by two-thirds of the electorate and China reiterated its intention to dissolve the new legislature in 1997. Patten's gamble seemed a valiant irrelevance.

Farther afield – more than 500 miles from China's mainland – China laid claim to islands in the South China Sea (see p. 530) as it developed its naval and, with Russian and Israeli help, air power.

These and all other matters of consequence were overshadowed by the uncertainties caused by the longevity of Deng Xiaoping. Deng resigned his last formal office – Chairman of the Central Military Commission – in 1991 when he was probably 87 years old but he remained a dominating presence. His policy of mending foreign fences was powerfully assisted by evidence of China's economic progress. Growth rates which had been 4–6 per cent in the late 1980s reached 13 per cent in the early 1990s, mainly due to developments by private enterprise in coastal areas. Devaluation of the currency in 1990 was rewarded by handsome surpluses on foreign trade. Yet Deng's dash for growth had question marks against it. What might happen when he died was guesswork; his policy of liberalizing the economy while maintaining authoritarian party rule might become an unsustainable contradiction; his faith in market forces ignored the changes which modern communications had brought into markets by making them as attractive to disruptive speculators as to legitimate and honest traders. Post-Maoist, post-communist China was an authoritarian state which contained a thrusting and thriving neo-capitalist economy and also a huge rural population of family farmers: the last a survival from before Mao's revolutions and the other grafted on to them. It was unclear whether this strange, perhaps unique, complexity was good ground for success or not.

The immediate outlook was unpromising. In 1993 tensions turned into serious riots following tax increases, free market pricing, persistent inflation between 25 and 30 per cent, and conflict between the centre and the provinces, cities and the countryside. A massive migration to the cities was fostering crime, corruption and tax dodging. High spirits were mixed with fears and discontents. Whether communism was to be succeeded by a Chinese capitalism or the more familiar Chinese anarchy was an open question. Yet China was approaching the turn of the century as one of the world's largest economies measured by purchasing power, as the country with the world's longest imperial pedigree and as a nuclear power. China's economy was not only growing but being transformed. By the mid-1990s three-quarters of its exports were manufactured goods, a development which caused some unease to its Japanese and Korean neighbours. It had a surplus on trade with the United States second only to Japan's and the value of its current construction contracts with American corporations exceeded $100 billion a year.

With an eye to these present realities and future possibilities President Bush regularly vetoed congressional wishes to deny China

most-favoured-nation treatment in consequence of the Tiananmen
killings and reports of persistent torture and inhuman labour camps.
President Clinton was less circumspect but, in terms of upholding
human rights, no more effective. By invoking human rights he not
only irritated Chinese leaders but gave them opportunities to display
patriotic anti-Americanism as they manoeuvred for power or survival
in the post-Deng era. His threats to impose super-tariffs on Chinese
exports in retaliation for Chinese disregard of American copyrights
and other intellectual property were no more effective against China
than they were against Japan: that is to say, they had marginal and
probably temporary effect.

4 WORLD ORDER

So long as the state remained the basic element in international society the preservation of world order and the prevention of war could only be secured by the more powerful of these states. They had a choice of methods. Each major power might assume primary or exclusive responsibility in a given region, or all the major powers might together supervise and police the whole globe, or these same powers might equip and finance an association of other states to do the job vicariously on their behalf. After the Second World War international organization and international co-operation were theoretically based on the second of these methods after the first had been unsuccessfully advocated in some quarters; but the circumstances necessary for the success of the second method did not materialize, so that practice approximated rather to an adaptation of the third, imperfectly acknowledged and precariously pursued.

The forms of international organization were discussed during the war by the principal victors-to-be. Churchill and Roosevelt both inclined to a regional pattern, and Churchill elaborated a scheme for a number of local federations to be grouped in three regions under a supreme global council. Power would be concentrated in the three regions − European, American, Pacific − rather than above or below. This pattern did not appeal to Stalin whose suspicions of Churchill, based on his mistrust of the British governing class and on disputes over the timing of the opening of a second front in western Europe, were sharpened by proposals which included the creation of Balkan and Danubian federations in an area of special concern to the USSR: Stalin wanted untrammelled sovereignty and, so far as the two were not incompatible, a continuing association with his allies in order to avoid a return of the USSR's prewar isolation. On the western side too there was opposition to regionalism, especially among professional politicians like Cordell Hull and Eden who feared that it would produce autarkic blocs each dominated by a particular major power and would in particular revive American isolationism. At Moscow, in October 1943, the foreign ministers of the three allies adopted the principle of a global organization based on the sovereign equality of all states and, with a minimum of incompatibility, laid the foundations for a new world organization which was to perpetuate the alliance of democracy and communism against fascism and give to the principal representatives of the former the charge for keeping the peace by the joint exercise of their allied power, this joint charge being coupled with a guarantee for each of them of immunity from any substantive interference by lesser states which were to recognize their superior power and undiluted sovereignty.

The United Nations was, in form, a revised version of the League of Nations. The principal organs of the two bodies were very similar. The authors of the UN Charter aimed, not to devise a new kind of organization, but to retain a familiar framework and insert into it more effective machinery for the prevention of war. The Covenant of the League had not proscribed war. It had bound its signatories to pause before resorting to war and to attempt to resolve their differences by one of three recommended processes. If this interposition of reason failed, there was no covenanted ban on the resort to war, and international sanctions were only applicable in the event of a resort to war in defiance of the preconditions laid down by the Covenant. In 1928 a much more radical attempt to prevent war was made by the signatories of the Kellogg–Briand Pact, who engaged themselves to dispense with war altogether except for certain limited purposes, namely the defence of the Pact itself and of the Covenant and of existing treaties, and in the exercise of the right of self-defence (the justification for the exercise of this right being left to the state claiming it): the United States and Britain also attached conditions relating respectively to the Monroe Doctrine and the defence of the British empire. The UN Charter fell somewhere between the conceptions of war as permissible subject to a pause for taking counsel and war as impermissible subject to defined exceptions. The Charter went far towards banning war except in defence of the Charter, or in pursuance of the obligations contained in it, or in self-defence but it did not totally proscribe war. It explicitly sanctioned not only the use of international force but also the use of national force, by one state or an alliance, in self-defence. The Charter vested considerable authority in the Security Council which was empowered to determine whether a given situation contained a threat to international peace or a breach of the peace or an act of aggression and, if it so determined, to require all UN members to take action against the delinquent (except to use force, a sanction which remained voluntary to each member). On the other hand this collective authority was offset by the procedural obstacles to reaching in the first place a collective decision in the Council, namely, a majority of the Council and the assent of all its five permanent members. In the absence of such a decision it was improper for any UN member to reach an opposite decision or take measures of the kind envisaged in the Charter; whereas an affirmative decision of the Council automatically placed all members under obligation, a failure to reach a decision precluded all action under the Charter. Consequently, although the Council was in this field sovereign over the members, each of the permanent members was sovereign over the Council.

Like the League, the UN was designed as an association of sovereign states (notwithstanding that its ban on interference in the internal affairs of a member – article 2(7) – did not extend to measures for peace enforcement under chapter VII), and like the League it attempted to assert a degree of collective judgement and a field for collective action against its constituent sovereign members in a period when these members had been massively strengthened by the growth of modern technology and of modern ways of influencing people. The state had turned the industrial and the democratic revolutions of the nineteenth century to its own advantage by annexing modern armaments and popular chauvinism to its purposes. Neither the League nor the UN was able to steal the control of armaments from sovereign states nor to create in the peoples of the world an attachment to international organizations which exceeded their purely national patriotism. Besides these general handicaps the UN saw its peace-keeping machinery rendered inoperative at an early stage of its existence. The efficacy of this machinery depended upon the unanimity of the major powers in the Security Council and the provision by all members of forces adequate for the execution of the Council's decisions. The unanimity of the major powers faded in the first breath of peace, so that the veto became a common tactical instrument instead of a weapon of last resort and the Charter's prescription for raising international forces – a series of bilateral agreements between the Security Council and members – was never implemented because the body appointed to negotiate these agreements, the Military Staff Committee, never agreed even the general nature and size of the forces required.

The veto given to the permanent members of the Security Council was a special feature of the UN. In the Council of the League every member had a veto. The authors of the UN Charter had to decide how far to depart from this unanimity rule. They decided to introduce majority voting as a general practice but subject to limited exceptions: in the Security Council, but not in other organs, special power was accorded special privileges with the result that major powers were able to prevent action against themselves or their friends, although they were not entitled to prevent discussion and criticism. No permanent member of the Council has ever objected to this principle, although particular permanent members have objected to the use of the privilege by other permanent members. During the late 1940s and 1950s the USSR, being in a semi-permanent minority in the Council, used the veto to such an extent that other members complained of a breach of the spirit, if not the provisions, of the Charter. Russians argued in return that the UN had become a tool of American policies and that these policies were basically anti-Russian – as evidenced by Truman's use of his nuclear monopoly for political purposes, by speeches of prominent western leaders beginning with Churchill at Fulton and Byrnes at Stuttgart in February and September 1946. The USSR, after an initial attempt to use the UN for its own political purposes by raising or supporting complaints about the Dutch in Indonesia, the British and French in Syria and Lebanon, the British in Egypt and Greece and western tolerance of Franco's fascism in Spain, fell back upon the veto to meet western counter-charges and then in January 1950 ceased to attend meetings of the Security Council. This retreat from the UN was, however, reversed partly because it enabled the Security Council to initiate action in Korea in June 1950 and partly because the expansion of the UN – notably in 1955 and 1960, the years respectively of a package deal admitting an assortment of sixteen blocked candidates and of the major African afflux – altered the character of the organization and offered the USSR political opportunities outweighing its fundamentally minority position in the Security Council. (It was in this period that the USSR ceased to attack the leaders of new states as bourgeois stooges and began instead to make friends with them.)

American hostility to the frequent Russian use of the veto led to two attempts to circumvent it by transferring to the General Assembly some part of the Security Council's authority. In 1948 an *ad hoc* committee of the Assembly, popularly called the Little Assembly, was established as a means of keeping the Assembly in permanent session, but the Little Assembly's powers were circumscribed and it never became an organ of any importance. The Russians were not alone in regarding it as a contravention of the Charter and it faded away. More important was the adoption in 1950 of the Uniting for Peace resolution. This resolution was sponsored by the United States which saw that the UN operations in Korea had been made possible only by the absence of the Russian member and his veto from the Security Council and which sought to ensure that their presence on a future occasion would not fatally obstruct similar action. The resolution, which was adopted by the Assembly by fifty votes to five with two abstentions, introduced machinery for calling an emergency session of the Assembly at short notice; asserted the right of the Assembly to pass judgment on threats to peace, breaches of the peace and acts of aggression when the Security Council was prevented from doing so; created a Peace Observation Committee of fourteen, available for missions of exploration and elucidation in trouble spots; also created a (stillborn) Collective Measures Committee of fourteen to study international peace-keeping machinery; and asked members to earmark forces for service at short notice in UN peace-keeping operations. The Assembly twice stigmatized China as an aggressor in Korea under this procedure and again used it in 1956 to denounce the Anglo–Franco–Israeli attack on Egypt and raise an international force to supplant these aggressors. But the legality of the Uniting for Peace resolution was always in dispute. The USSR and other members attacked it and refused to pay for operations set in motion by the Assembly within its terms of reference. In 1962 the International Court was asked to advise on this issue and declared by nine votes to five that the Security Council's responsibility for peace-keeping was primary but not exclusive.

Thus disputes of a legal nature, reflecting fundamental political disagreements, threatened to thwart the hopes of those who had sought in 1945 to produce a document and an organization which would secure peace and order. Disharmony among the permanent members made a mockery of the Security Council's name and turned the Council itself into an arena for the public display of wrangling and propaganda charges. Open diplomacy became as discredited as secret diplomacy had once been suspect. Likewise the General Assembly became noted for the marshalling of block votes and the trading of unattached ones, especially after the afflux of new members had increased the number of those who were likely to have no direct interest in a particular issue and who would therefore vote either for ulterior reasons or not at all. Recourse to the UN became in consequence something of a gamble owing to the unpredictability of the attitudes of many members.

Disappointment with the functioning of the central organs of the UN led to a recrudescence of interest in regionalism. Although the authors of the Charter had come down on the side of centralism as opposed to regionalism, they had not totally excluded the latter from their design. The Charter recognized regional organizations in two ways. It affirmed by article 51 the right of regional self-defence and in effect sanctioned a regional collective system for defence which was an alternative to the machinery of the Security Council. This article covered regional alliances such as Nato, whose primary purpose was not the maintenance of peace and order within their area but the defence of it from outside threats. Secondly, the Charter made provision in article 52 for regional organizations intended to police the region and resolve disputes arising within it. The Organization of American States was, however, the only organization of this kind to achieve any significant claim to effectiveness in the first thirty years of the UN's existence, with the result that regionalism did not in this period offer any substantial alternative to the UN's central organs as a means of keeping the peace within a region.

But although the Security Council failed to function as anticipated by its authors and although regional organizations failed to step into the gap, the UN became continuously active in security operations, developed a variety of experimental techniques and even engaged – in the Congo – in a major operation in which it deployed a total of nearly 100,000 men over four years. UN intervention in dangerous situations ranged from comparatively modest missions to establish facts, lower tension and gain time, through more complex mediatory operations involving military units but not the use of military force, to military expeditions prepared not only to defend themselves but also to attack others. The line between one type of operation and another cannot be precisely drawn. Thus UN intervention in Kashmir, Palestine and the Suez War could be classified as mediatory and in Korea and the Congo as military, whereas intervention in Cyprus could arguably be placed in either category, or in one category to begin with and in another at a later stage. But classification is no index of activity and the UN's activity in peace-keeping is not to be denied. Where the League had been criticized for passivity, the UN came to be criticized for doing too much.

In the first or pre-Korean phase of this activity the Security Council was faced with situations arising out of the war: the lingering of the Russians in Iran beyond the term set by wartime agreements, and the slowness of British and French troops to leave Syria and Lebanon. These matters were debated in the Council and resolved outside it without further ado. Precedents of recourse to the Council were quickly set. The Council declined to act on a Russian complaint of British interference in Greek affairs through British troops in Greece, but it later investigated Greek complaints of foreign aid to Greek rebels by Yugoslavia, Bulgaria and Albania. It despatched teams of observers who, after being denied access to the non-Greek sides of the frontiers in question, issued a report condemning Greece's enemies. The Council did not put an end to the fighting in progress (which was only stopped when American aid had re-equipped the Greek army and restored its morale and when Yugoslavia, after the breach with the USSR and its communist neighbours in 1948, stopped helping the rebels), but the UN had set a precedent for on-the-spot investigations and could claim that it had contributed to elucidating and holding a potentially dangerous situation until it was eliminated by other means. This function was further exemplified in Indonesia and Kashmir. In Indonesia the Council succeeded in establishing a cease-fire which temporarily halted the first Dutch police action against the Indonesian nationalists and secured a temporary agreement between the two sides through a conciliation committee. Although these achievements were at first transitory, the eventual transfer of sovereignty from the Netherlands to the Indonesian republic at the end of 1949 was mediated by international intervention. In Kashmir too the UN negotiated a cease-fire and succeeded in putting a stop to fighting, even though it failed to secure a withdrawal of Pakistani or Indian troops which had entered Kashmir or to resolve the underlying political dispute between Pakistan and India. Kashmir was the first clear example of a paradox which was later to become explicit and teasing: the fact that a cease-fire and the immobilization of hostilities could obstruct the solution of basic disputes by relieving the contestants of the urgency to come to terms in order to save lives and money. UN observers, sent to Kashmir in 1949, were still there nearly fifty years later.

In Palestine, the final pre-Korean illustration of UN security operations, UN emissaries negotiated a cease-fire, repaired it when it was broken and helped to secure armistice agreements. The Arabs and Israel did not, however, make peace and a UN Truce Supervisory Organization found itself established in the Middle East for more years than its originators had contemplated, subjecting both sides to the hazards of having their infractions of the truce exposed by an impartial body of observers.

The Korean War which, like other events referred to in this summary of UN activities, is described in more detail elsewhere in this book, was a test of a different kind since it arose out of an act of aggression to which there could be no effective response except the use of force. Chapter VII of the Charter was invoked and the Security Council, in the absence of the USSR, in effect endorsed American action and commissioned the United States, which had forces available in Japan and the surrounding waters, to meet force with force. The Korean War became therefore a war conducted by an American general responsible to the American president acting as the agent of the UN. Fifteen other states, mostly allied with the United States for other reasons, sent units to the battlefield but half the ground forces engaged, 93 per cent of air forces and 86 per cent of naval forces were American, and when the Chinese invaded Korea and the war gradually evolved into a trial of strength between the United States and China (with the possibility of an even larger clash between the United States and the USSR whose aid for the communist side was not far in the background) many UN members began to feel that the UN's motives in going to war had been lost sight of and that no future UN operation ought to be conducted in the same way. After the Korean War it was an axiom that major powers should not be invited to make a fighting contribution to UN operations – an axiom only attenuated in Cyprus by the obvious advantage of using the British forces already present in the island. Korea became therefore an exception. But its part in typing the UN as an organization prepared to take action, as opposed to the League's failure to take action in Manchuria twenty years earlier, is not to be underestimated.

The next important operation – the UN intervention at Suez in 1956 – was a combined operation by medium powers set in motion by the General Assembly and placed under the executive control of the secretary-general. The force, which was recruited and despatched with astonishing speed, was equipped for self-defence but not for attack: ten nations contributed troops. Its arrival and deployment was secured in advance by agreement with Egypt (necessary, since the operation fell within chapter VI of the Charter on peaceful settlement of disputes and was not an enforcement action under chapter VII) and by the knowledge that the British and French governments would not use the power which they had to oppose it. The Anglo–French attack on Egypt having been halted by the United States, the UN was used to lever the Anglo–French forces out, to patrol the troubled areas along the Suez Canal and incidentally to clear the canal which the Egyptians had blocked when they were attacked. Peace was restored through the agency of the UN after the United States had displayed a determination, which neither Britain, France nor Israel could gainsay, to stop the fighting. The role of the UN in this crisis was performed within the framework of collective security but was in fact something different. The proponents of collective security, whether in the League or the UN, envisaged the mustering, under pre-existing commitments, of overwhelming force to deter or stop a transgressor. At Suez in 1956 the transgressors were not stopped by such a collective show of force but by the United States. The collective force sent to the area could not have fought the Anglo–French or Israeli aggressors, nor was it meant to. It was intended not to push the intruders out but to keep the major powers out, and the name found for this type of activity has been 'preventive diplomacy'. Hammarskjöld's aim was to forestall by a UN presence the incursion into the conflict zone of the Americans and Russians, and his experiences at Suez fashioned his policy in the Congo four years later. Some of the limitations on action of this kind were demonstrated when, in 1967, the UN force was removed on Egypt's abrupt demand. Others were demonstrated simultaneously with the war of 1956 when the USSR refused to allow even a personal visit to Budapest by Hammarskjöld.

Between Suez and the Congo, a period of less than four years, the UN was again invoked in the Middle East and was required to use its techniques as observer and mediator. In 1958 the Lebanese government complained of interference in its internal affairs by the United Arab Republic and called in American troops. The UN despatched a few groups of mobile observers to check and report on what was happening along Lebanon's borders and subsequently enlarged this mission to enable it to replace the American units, whose presence had become an embarrassment to all concerned, including the Americans. The UN thereby shed light on a confused situation, deflated it and finally smoothed the way for a return to normality. This was an adaptation of previous experiences. It was followed, more or less, in Laos in 1959; in West Irian in 1962–63 when a UN presence eased the transfer from the Netherlands to Indonesia; in North Borneo and Sarawak in 1963 when UN investigators reported that the inhabitants of these territories were not, as Indonesia alleged, opposed to the creation of Malaysia; in Yemen in 1963–64 when a situation as obscure as it was contentious was to some extent clarified by UN observers; and in discovering the wishes of the inhabitants of the trust territories of Togoland and the Cameroons when the administering powers were about to withdraw between 1956 and 1960.

The involvement of the UN in the Congo was a consequence of the precipitate Belgian withdrawal and the mutiny a few days later of the Congolese army, upon which the Belgians returned to protect their nationals, Katanga purported to secede and the Congolese government turned to the UN for help in keeping order, securing essential services, getting foreign forces out and holding the new state together. This involvement had therefore from the start a mixture of international and internal aspects. It was international in so far as it aimed at removing the Belgians and pre-empting the Americans and the Russians – a further instance of preventive diplomacy. But it was also internal in so far as the UN forces were filling the gap created by the mutiny or inadequacy of the Congolese government's own forces and, besides securing law and order, became involved in maintaining the integrity of the new state by preventing the secession of part of it. Hammarskjöld used his right under article 99 of the Charter to bring the matter to the attention of the Security Council. In the four years in which the UN then operated in the Congo a force which reached a strength of 20,000 and averaged 15,000 was used (at a total cost of over $400 million, of which 42 per cent was paid by the United States and none by the USSR) to keep order, prevent civil war and, more contentiously, to force Katanga to acknowledge the authority of the Congolese state. Its operations were based on chapter VI of the Charter. Chapter VII was never explicitly invoked, although shifting circumstances produced a situation very like enforcement action under chapter VII. These operations were initiated by the Security Council, but they were conducted by the secretary-general and medium powers.

Once again only medium states were asked to supply combat units and the secretary-general was given executive control. He created an *ad hoc* advisory committee of representatives of the states with troops engaged, in order to help him to interpret the general directives given by the Security Council and apply them to the circumstances of the moment. Acting in this way the UN became the prime factor in

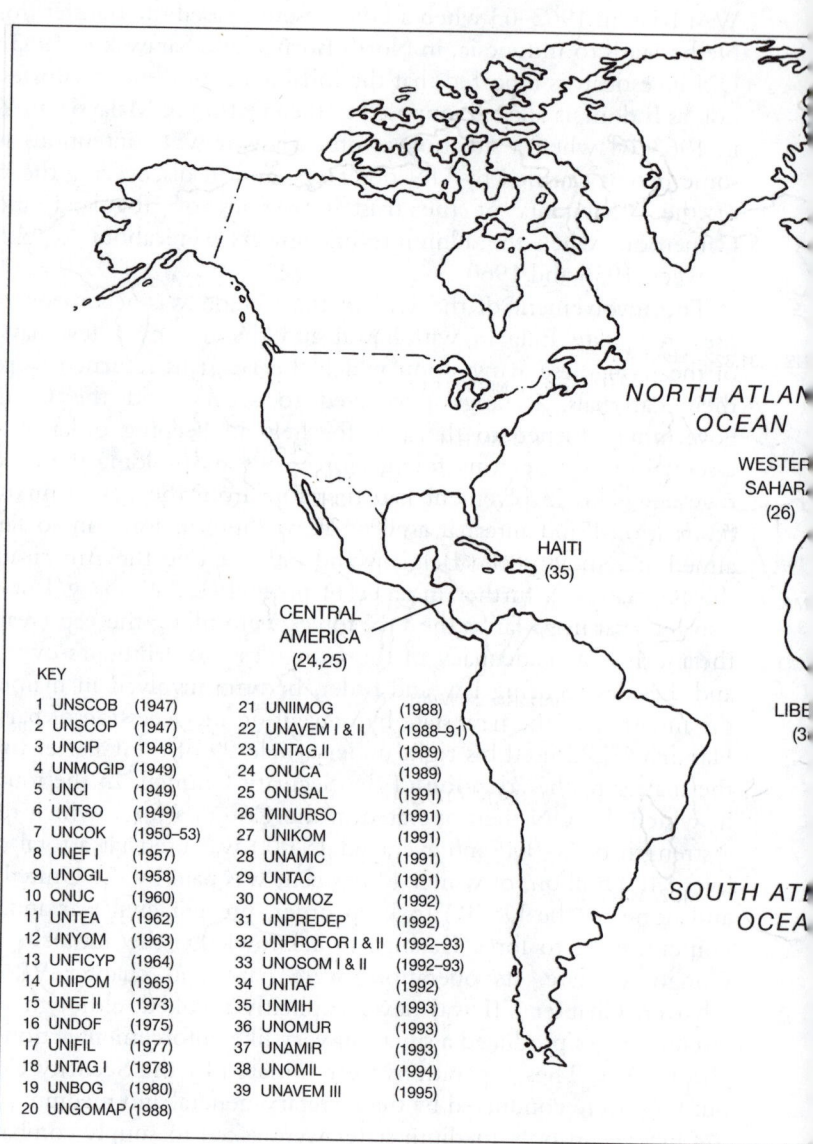

NORTH ATLANTIC
OCEAN

WESTERN
SAHARA
(26)

HAITI
(35)

CENTRAL
AMERICA
(24,25)

LIBE
(3

SOUTH ATL
OCEA

KEY

1	UNSCOB	(1947)	21	UNIIMOG	(1988)
2	UNSCOP	(1947)	22	UNAVEM I & II	(1988–91)
3	UNCIP	(1948)	23	UNTAG II	(1989)
4	UNMOGIP	(1949)	24	ONUCA	(1989)
5	UNCI	(1949)	25	ONUSAL	(1991)
6	UNTSO	(1949)	26	MINURSO	(1991)
7	UNCOK	(1950–53)	27	UNIKOM	(1991)
8	UNEF I	(1957)	28	UNAMIC	(1991)
9	UNOGIL	(1958)	29	UNTAC	(1991)
10	ONUC	(1960)	30	ONOMOZ	(1992)
11	UNTEA	(1962)	31	UNPREDEP	(1992)
12	UNYOM	(1963)	32	UNPROFOR I & II	(1992–93)
13	UNFICYP	(1964)	33	UNOSOM I & II	(1992–93)
14	UNIPOM	(1965)	34	UNITAF	(1992)
15	UNEF II	(1973)	35	UNMIH	(1993)
16	UNDOF	(1975)	36	UNOMUR	(1993)
17	UNIFIL	(1977)	37	UNAMIR	(1993)
18	UNTAG I	(1978)	38	UNOMIL	(1994)
19	UNBOG	(1980)	39	UNAVEM III	(1995)
20	UNGOMAP	(1988)			

4.1 Map of United Nations missions (see Appendix, p. 838, for key
to missions)

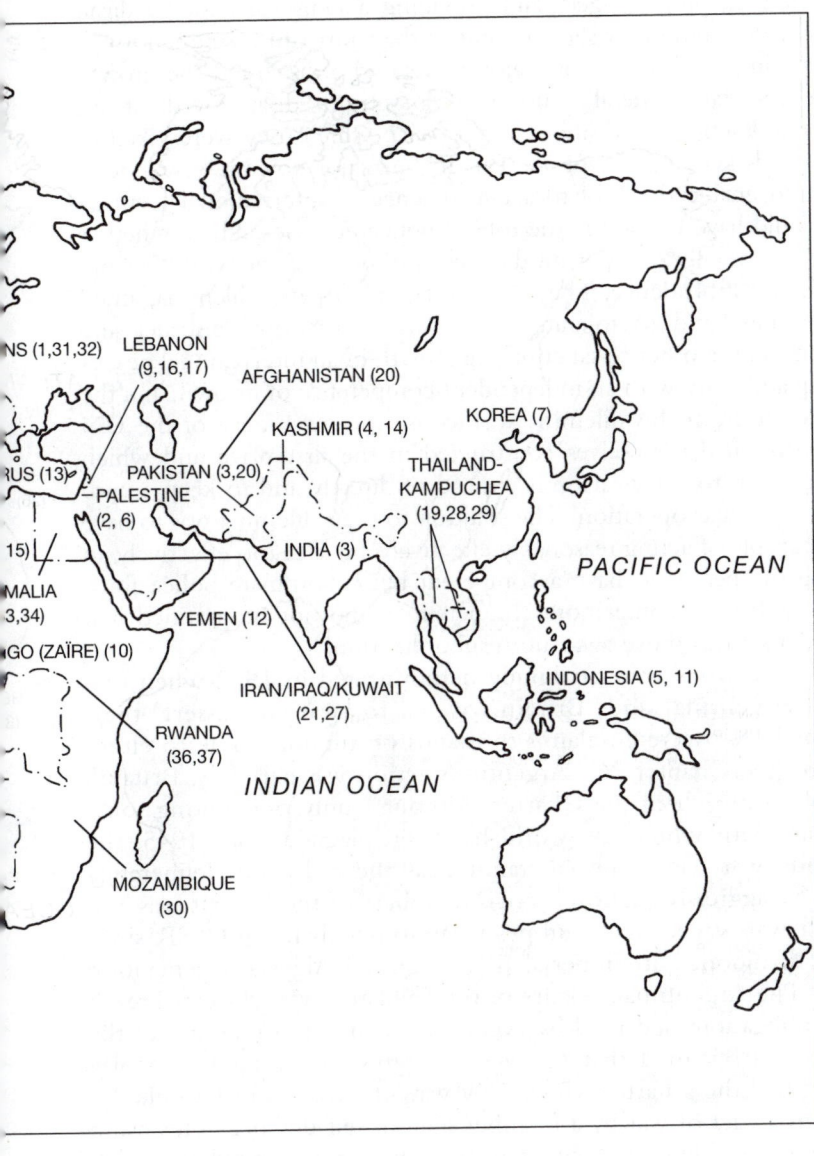

NS (1,31,32)   LEBANON
(9,16,17)   AFGHANISTAN (20)

KASHMIR (4, 14)   KOREA (7)

RUS (13)   PAKISTAN (3,20)   THAILAND-
KAMPUCHEA
(19,28,29)

PALESTINE
(2, 6)

INDIA (3)

15)

PACIFIC OCEAN

MALIA
3,34)   YEMEN (12)

GO (ZAÏRE) (10)

INDONESIA (5, 11)

IRAN/IRAQ/KUWAIT
(21,27)

RWANDA
(36,37)

INDIAN OCEAN

MOZAMBIQUE
(30)

preserving the unity and integrity of the new state and in preventing foreign intervention and perhaps a clash between major powers; it also alleviated human distress by containing civil war, minimizing bloodshed, helping refugees and providing a range of basic medical, administrative and other services; but in the short run it strained itself by incurring the hostility of major powers who distrusted the growth of the secretary-general's authority, sometimes disapproved of his objects and actions, and jibbed at the price which they were asked to pay towards keeping the peace. But if the major powers were not to allow and enable the UN to act in defence of international security, they would have to assume the role of policemen themselves, whether jointly or severally – or permit degrees of disorder in the world beyond the bounds of prudence. They were in fact placed in a dilemma, afraid on the one hand to tolerate too much international disorder and reluctant on the other to sanction the growth of an international peace-keeping authority with an independent competence of its own; and the only escape from this dilemma was to use the machinery of the UN which they had themselves constructed in the first place and which enabled them to act vicariously instead of directly and to keep a curb on any particular operation. The solution to their dilemma was to have a UN capable of acting reasonably effectively, but not too effectively, in keeping the peace; to have a competent but subordinate police force whose wish to become more competent would not be gratified if it entailed a serious move away from subordination.

The limitations were strikingly demonstrated in 1982 when two states, Argentina and Britain, went to war to assert their irreconcilable sovereign claims to islands of minimal value to either of them (see Chapter 26). Argentina's action was certainly, Britain's arguably, a breach of the Charter. Although only one among some 150 wars with which the world has been plagued since 1945, the Falklands War had a special standing as the only war deliberately waged by aggressive action by two members of the UN. It was also the first war waged by a European state other than the USSR since France abandoned its imperial pretensions in Algeria a generation earlier. The Argentinian seizure of the Falklands was a blatant breach of the obligations accepted by every state which is a member of the UN. Britain claimed that this act of aggression brought into play article 51 of the Charter which – by way of exception to article 2 – sanctions an act of war by a member state in self-defence. This claim is not without blemish. Neither in the Charter nor anywhere else is self-defence defined, so that there exists a grey area between self-defence and retaliation. The British government declared promptly its intention to recover the islands, if necessary by force, and wasted no time in assembling the means to do so. On the other hand, if self-defence is to be strictly interpreted, it could be held to mean no more than hitting back at the time of attack with the forces available at the point of attack. The question is a moot one, less clear than statements from the British side chose to portray it. But whatever the true legal judgment on this matter, it can more confidently be asserted that political considerations were uppermost and that, whatever the advice of any number of competent lawyers might have been (if sought), the British government would not have acted otherwise than it did.

The war for the Falklands was a setback for the UN as an organization and for those aspirations to world order which it embodied. For this setback the initial aggressors were overwhelmingly to blame, but the British government did not wholly escape the embarrassment of demonstrating that in a crisis a powerful state will not welcome UN diplomacy and will subordinate the rule of law and its treaty obligations under the Charter to its own assessment of national advantage and prestige. This was in 1982 no great surprise but it was not what the generation of 1945 had hoped for.

A more serious blow to the rule of law in international affairs followed when, in 1986, the United States delivered a heavy naval and air attack on Libya. The occasion for the attack was the explosion ten days earlier of a bomb in a night club in West Berlin. Two persons were killed. The perpetrators were Palestinians. Messages sent by the Libyan People's Bureau (embassy) in East Germany were intercepted and deciphered and disclosed official Libyan involvement. The precise nature of this involvement was uncertain since published English versions of these messages appeared to be not so much translations as paraphrases. Nevertheless Libyan support for Palestinian groups prepared to use violence was uncontestable. The American administration sought agreement for the use of aircraft based in Britain and the right for them to fly over France: the first was granted, the second not. The attack was at first described as reprisal or retaliation but later as self-defence within article 51 of the UN Charter. It was an act of war in clear violation of the treaty obligations accepted by all signatories of the Charter (the appeal to article 51 was so vapid as merely to add hypocrisy to illegality) and also in contravention of the laws of war which, for a thousand years before the Charter, had prescribed that the force used in a punitive venture must be not disproportionate to the offence which gave rise to it. The action was very widely applauded by the American people who, like their president, felt outraged and frustrated by such crimes as the Berlin bombing, the hijacking of the *Achille Lauro* (see p. 420) and more. Unable to get at the small gangs responsible for these crimes the US administration decided to label certain states (Libya, Syria) as 'terrorist' organizations and so claim the right to attack them. The attack was a political act undertaken in defiance of international rules and obligations but in the belief, or hope, that its political expediency would muffle and condone its illegality. This it was not well calculated to do. It had no logical or obvious effect on those small but fervent groups which, needing few arms and little money, could remain in business whether Libyan cities were bombed or not. Qadafi himself was neither killed nor overthrown and did not appear to be intimidated; domestic rivals who had been encouraged to intrigue against him by Libya's economic problems dropped their opposition at least for a time. Other Arab leaders, including those who disliked Qadafi as an upstart and a nuisance, rallied to him; an American attempt to enlist President Mubarak in the raid seriously, perhaps permanently, weakened his position (as Carter had unintentionally but fatally undermined Sadat at Camp David); Arab public opinion gave Qadafi undeserved sympathy as a victim of brute force. Washington's European allies, with the notable exception of the British prime minister and some of her colleagues, were aghast at what they regarded as political folly but muted in their public criticism of Reagan because of their own feebleness in response to Qadafi's lethal meddlings; they were reluctant to apply (alternative and legitimate) economic measures since economic sanctions against Libya would undermine their refusal to use such sanctions against South Africa. In Britain the prime minister's willingness to allow the use of British bases provoked questions about the secret Anglo–American agreement governing their use, particularly for operations which were in nobody's mind when the agreement was made and which had nothing to do with Nato. Although like many acts of violence, the American attack changed little it accentuated some trends of which the more important were, first, the declining ability of the United States to solve rather than aggravate the turmoils of the Middle East and, secondly, the declining ability of Britain to stave off a choice between a junior – even humiliating and now unpopular – role in an Anglo–American partnership and a more wholehearted partnership with the European Community. In the last resort the American action was a massive political miscalculation. On the one hand was a widespread hankering, to which the American administration responded, to find a way to police international and cross-national affairs – even some readiness to put order before law. But in assuming the police role Reagan flew in the face of the fact that, not only the communist bloc, but neither the Arab world nor Asia nor Africa nor most strikingly – western Europe was prepared to entrust this role to the United States. After the attack they were the less willing. So the action lacked not only legitimacy but also *post hoc* endorsement and defenders. American action in Nicaragua and Panama (see Chapter 27) underlined the difficulty of finding a policeman willing to undertake rough work and at the same time stay within the law.

The revolution in the USSR effected by Gorbachev changed this situation by smothering the Cold War, inducing the superpowers to co-operate in international affairs instead of opposing one another as a matter of principle, and by introducing therefore for the first time since 1945 the prospect of making the Charter work as it had been intended to work in the business of keeping the peace and upholding the law. The first test came with Iraq's invasion and annexation of Kuwait in 1990. This was a blatant act of aggression and it raised the question which had been raised on a number of earlier occasions, including Argentina's occupation of the Falkland Islands and Iraq's own attack on Iran ten years earlier: whether counter-action would be taken through or outside the UN. President Bush decided to do both. He despatched large armed forces to Saudi Arabia and he resorted to the UN to impose sanctions against Iraq. Both these undertakings were international in the sense that numerous states participated in both, but only the latter was action by the United Nations. The former was action initiated and led by the United States independently of the Charter. It did not conflict with the terms of the Charter but would do so if these forces were used to make war otherwise than in support of UN resolutions, at UN request and under a system of command established by the UN.

The Security Council took action under chapter VII of the Charter. This chapter comes into operation when the Council itself determines the existence of a threat to peace, a breach of the peace or an act of aggression. If the Council does so, it may prescribe action of two kinds: measures not involving the use of armed force (article 41) and action by air, sea or land forces (article 42 – blockade being specified under this article). In the case of measures under article 41 the Council may call on members of the UN to apply these measures and (by article 25) they are obliged to comply. There is no such compulsion in the case of measures under article 42.

In a series of resolutions adopted during August the Security Council, unanimously, demanded Iraq's immediate withdrawal from Kuwait and a negotiated settlement; imposed, by thirteen votes to two, a commercial, financial and military embargo (medical supplies specifically excepted and rules permitting the supply of food on humanitarian grounds to be worked out by a committee of the Council); declared, unanimously, that the annexation of Kuwait was null and void; demanded, unanimously, that Iraq allow and facilitate the immediate departure from Kuwait and Iraq of all nationals of third states; and authorized, by thirteen votes to two, the use of force to make the embargo effective by, if necessary, stopping and inspecting merchant ships. These first resolutions established sanctions against Iraq and the use of force to monitor them, but not the use of force for any other purpose. Further resolutions defined and regulated UN actions, culminating in a twelfth resolution on 29 November which, by authorizing the use after 15 January 1991 of all measures necessary to achieve the aims set out in the earlier resolutions, sanctioned recourse to war.

The governments of the United States and Britain maintained that, these resolutions apart, they were entitled under article 51 of the Charter to use military force against Iraq but this contention was almost certainly false. Article 51 is one of the least satisfactory articles of the Charter. It expressly safeguards the state's inherent right to defend itself but it makes no attempt to define the line between self-defence and retaliation and, by using the phrase 'individual and collective self-defence', it perpetrates an ambiguity. There are two distinct issues to be met before article 51 may legitimately be invoked: first, the question whether the action proposed or undertaken is defence or retaliation (which was the question posed in the Falklands case) and, secondly, the meaning and scope of the term 'collective' in article 51. Collective self-defence, besides being a contradiction in terms, implies that an attack on one UN member may be resisted by forces other than those of the state attacked. They are presumably the forces of allies, but it is open to question whether the alliance needs to be in existence at the time of the attack or may be concluded after it. In considering this conundrum it is necessary to revert to the circumstances in which article 51 was drafted and adopted at San Francisco in 1945. It was a late addition to the draft Charter and was designed to meet the worries of members of existing regional alliances (the Organization of American States and the Arab League) who feared that by signing the Charter they would invalidate the arrangements in their regions for mutual help against aggression. It is evident that an attack on a UN member may be opposed by the victim's regional partners in accordance with the provisions of the regional alliance, but it is not evident – nor is it easy to argue – that states outside the relevant regional partnership at that time may intervene by force. The broader interpretation would permit any state to throw itself into a dispute simply by declaring itself an ally of the victim of aggression, with the result that every threat to or breach of the peace would become an invitation to military action by self-appointed white knights – or red knights or black knights – on the single condition that the victim of aggression was prepared to welcome that action.

In terms of world order and the usefulness of the UN Iraq's attack on Kuwait served to reanimate the mechanisms of the UN, testing its strengths and exposing its weaknesses. The use of the Charter by the United States was made possible by broad accord between the permanent members of the Security Council and particularly between the United States and the USSR, but few in Washington believed that action under chapter VII would by itself restore the independence of Kuwait or forestall further possible aggression by Iraq – against, for example, Saudi Arabia. The United States therefore established in Saudi Arabia a powerful force which, while it might by its presence in the area reinforce the coercion applied by the UN, was evidently capable of attacking Iraq and not unlikely to be used in order to overthrow the Iraqi régime as well as to restore the Kuwaiti. This display of American purpose and power was a unilateral act which was given international support by securing the participation of the host country and of numerous other states in and beyond the Middle East. It was also an expression of no confidence in the efficacy of the mechanisms and procedures of the UN.

The attempt made in 1945 by the United Nations to shift the making of war from the unsafe hands of the nation state to the safer hands of the community of nations had been obstructed by two things: the Cold War and the developing nature of war. The Cold War undermined the existence of a global community of nations and negated the Charter's provisions for assembling and using international force. The development of nuclear, chemical and other new weapons superposed on the question who should be authorized to use armed force the question whether the use of such force was acceptable or justifiable or even efficient in relation to specific political or military purposes. The crisis over Kuwait was prolonged in a way barely imaginable in the previous century by perplexity over the uses of armed force and a rising undercurrent of opinion against the resort to what war had become. There were, as there ever had been, three main ways of overcoming an enemy: starve him, scare him or beat him. Chapter VII of the Charter is the modern version of the siege, using deprivation to secure surrender but keeping in reserve, as did besiegers, direct force to back it and clinch it. But an embargo in accordance with chapter VII is, like siege warfare, likely to be slow and even ultimately ineffective: besiegers sometimes marched away. The American host assembled in Saudi Arabia was an alternative, a second string. It was intended to scare Saddam Hussein or, if he were not scared, to beat him. It represented the view that if measures under chapter VII failed or did not work quickly enough, the United States and those of like mind would resort to force rather than accept rebuff.

As the crisis developed month by month so did the dilemma between the two aims of chastising the law-breaker and resolving the crisis: between recourse to war and recourse to diplomacy. Saddam Hussein's violation of the Charter created the crisis. The first of the Security Council's resolutions required him to purge his offence by retreating from Kuwait. It required also discussion of matters in dispute between Iraq and Kuwait. It said nothing about the relative timing of these two requirements. Saddam Hussein proclaimed repeatedly that he would not budge from Kuwait, although he expressed willingness to participate in a general conference on Middle Eastern affairs with an agenda from which Kuwait would not be excluded. The United States and some of its associates refused to consider any matters before a total and unconditional Iraqi withdrawal. This was tantamount to the rejection of diplomacy in parallel with retreat. It was a high-risk policy which gave to surrender an absolute priority over negotiation, risking a war which – although caused in a primary sense by Iraq's aggression – would be caused in a second sense by the equal obduracy of the protagonists. In such a contest Bush suffered a double disadvantage. As chief in a democracy and chief of an *ad hoc* alliance he was constrained to a far greater extent than his autocratic adversary who could take much less heed, or none at all, of popular opinion or the misgivings of associates. The overwhelming power of the United States rendered the use of that power obnoxious to much public opinion and barely reconcilable with the well-established rule of law which, for well over a thousand years, had decreed that the use of force must be proportional to the object to be achieved. The tilt towards war was accentuated by the American tactics of reinforcing a policy of sanctions with a deployment of vast armed forces whose cost was unsustainable over a long run and which could not be brought back home without a commensurately unequivocal victory: Bush adopted measures which were economically and politically unsustainable.

That the United States should play the leading part in the UN's undertakings was proper and desirable. But Bush did both more and less. He conducted simultaneously an American operation which overshadowed the UN undertaking almost to the point of obliterating it; and in the American operation he relied on force and the threat of force to the exclusion of diplomacy, even insisting that the direct bilateral talks in Washington and Baghdad which he proposed in December must stop short of anything that might be labelled negotiation. He arrived therefore at a position in which he was demanding unconditional observance of UN resolutions which themselves demanded no such thing. During the Cold War the UN's role in keeping world order and upholding international law had been effaced, but the Cold War had provided the UN with an alibi for its ineffectiveness. The Kuwait crisis, the first serious crisis after the end of the Cold War, reanimated the UN – the United States was among the first to hail this change – but it showed also that the UN's new freedom to perform its central role in world affairs might still be cramped by its more powerful members and would have to be performed without benefit of the excuses for failure provided by the Cold War. Freedom of manoeuvre for the UN could turn out to be useless or – worse –

catastrophic if it were to mean freedom only to demand unconditional surrender. By pressing UN resolutions beyond the objectives explicitly envisaged in them, and by conducting supposedly UN operations without reference to the UN, the United States abused the UN and weakened it while nevertheless fulfilling some of its aims. American aims – the overthrow of a barbarous régime which had among other things violated the Charter and the laws of nations – were widely approved but the methods used demonstrated something different: the power and will of a single national government in the context of a threat to its own national interests. International action and order were subordinated to a national purpose with consequences for the UN which were at best ambivalent.

The Gulf War had a second and potentially more far-reaching consequence. By resolution 688 of 1991 the Security Council asserted that the situation in Iraq was a threat to international peace and demanded access for humanitarian organizations to parts of Iraq where minorities were being abused. It also demanded the right to patrol and monitor these areas. Since the war was at this point over in the sense that Iraq had been defeated, the Council was implicitly raising a semi-concealed issue – the conflict between the UN's obligations to standards of behaviour – in, for example, articles 1(3) and 55 of the Charter – and the prohibition article 2(7) against intervention in the essentially domestic matters of a state except in the circumstances defined in chapter VII of the Charter. In pursuance of Resolution 688 an international force provided by the United States, Britain and France maintained the pressures on Iraq, defining Kurdish and Shi'ite areas in northern and southern Iraq respectively as no-fly areas for Iraqi aircraft and using Turkish air bases for their operations in the north. (The Turks were anxious to prevent Kurdish refugees from entering Turkey and Kurdish armed groups from reinforcing Kurds in Turkey. Turkey even invaded northern Iraq on the plea that it was in chaos and a threat to stability in south-eastern Turkey.) Resolution 688 had a more contentious passage in the Security Council than earlier resolutions for waging war against Iraq's invasion of Kuwait. Three members of the Council opposed it and two abstained. Nevertheless it raised an issue which would not have been formally ventilated during the Cold War and was made precise only by the Gulf War. Whether or not intervention of this kind and on these premises would become a legitimate precedent would depend on future actions of the Security Council in similar circumstances.

The relevant circumstances included three factors beyond the facts of any particular case: the development of international law and its interpretation; the volume of simultaneous demands on the UN; and the practices of the UN in relation to intervention to keep the peace, supply humanitarian aid and enforce humanitarian standards. The law relating to armed conflict was reviewed and recodified after the Second World War by the four Geneva Conventions of 1949 and the two Geneva Protocols of 1972. The term 'armed conflict' itself implied an extension of the law beyond a state of war. These conventions and protocols updated the law in detail and prescribed penal sanctions for 'gross breaches' but failed to establish effective mechanisms for enforcement. Of the two protocols, on international and internal armed conflict respectively, the second was fiercely mutilated before being adopted and ratification of both by the world's more powerful states was patchy. Most states approved their provisions but feared their application to themselves or their friends in an uncertain future.

Demands on the UN had increased dramatically from the late 1980s. In the five years from 1988 the UN's peace-keeping and supervisory operations quadrupled in volume and their cost soared. At the peak of UN involvement in the conflicts in Yugoslavia, where thirty-two states were contributing 25,000 troops, the forces deployed by the UN throughout the world reached more than 50,000 in place of 10,000 on average in previous decades, and their cost was about $3 billion. The UN was hard pressed to recruit the numbers required and its members unwilling to shoulder their prescribed share of the cost. This increase in the demand for world order was occasioned not only by the opportune cessation of the Cold War but also by the redefinition of what sort of order ought to be secured and how the UN should set about the task. World order no longer meant no more than stopping wars between states. It also meant responsibility for domestic order at some undefined level where disorder either threatened international peace or grossly offended against humanitarian norms established by international law and conventions.

In the aftermath of the Gulf War the plight of the Kurds in northern Iraq and of Shi'ites in the south cried out for international intervention partly because of Saddam Hussein's atrocious record against them and partly also because they had been encouraged by the United States to rebel unsuccessfully against their government: two million Kurds were in flight in harsh terrain and dreadful weather and were being refused asylum over the Iraqi–Turkish frontiers. This situation raised the question of the right of the UN to intervene by force in a member state for humanitarian purposes and did so at a time when other disasters, notably in Somalia and Yugoslavia, were demanding attention from the UN and its new secretary-general, Boutros Boutros-Ghali.

The Security Council's Resolution 688, taken without invoking chapter VII of the Charter and so without ostensibly side-stepping article 2(7), to use air power and (in the north) ground troops to protect threatened minorities within Iraq was a breach of article 2(7) unless, either the situation fell within chapter VII – in which case article

2(7) did not by its own formulation apply – or the circumstances giving rise to the intervention could be classed as falling outside Iraq's essentially domestic jurisdiction. For the latter proposition a series of arguments, all forceful but inconclusive, could be adduced. The first was that article 2(7) was at variance with other provisions of the Charter – for example articles 1(3) and 55 – which imposed on the UN a commitment to uphold humanitarian principles. The second was that the word 'domestic' was to be interpreted by the import rather than the location of the enormities to be remedied: that, for example, acts verging on genocide might not properly be termed domestic. A third was that acts in direct contravention of international conventions signed by the peccant state created a right of international intervention.

International intervention in Iraq was a reaction to circumstances, opportunity and emotions (including compassion and shame). It was not clearly legitimate. The protection of the Kurds and Shi'ites was adventurous. It was also relatively easy since the UN was already at war with and had defeated Iraq, and the costs were not impossibly incalculable. Whether the Kurdish venture might come to be accepted as part of the law of the Charter could be determined only with time. On the one hand the International Red Cross reported at this juncture that nine in ten of war victims throughout the world were civilians and refugees, that refugees had reached 17 million of whom 7 million were children, and that half the severe casualties of wars were under 18 years old. On the other hand any readiness for more intervention was countered by failures in Somalia and Yugoslavia.

In Somalia there was anarchy following the overthrow of Siad Barre, no government with which to negotiate the usual preliminaries to the despatch of a peace-keeping mission and a conflict between humanitarian and political aims – the relief of famine as distinct from stopping the fighting and restoring the integrity of the country. The former required not only the provision of food and medicines but also a force to protect those supplying them; the latter required negotiation with warring factions and their disarming. The United States endorsed and took part in UN operations but also despatched forces of its own and pursued ends of its own: it was not easy to tell whether these were UN or US operations or a bit of both (see Chapter 22). In Yugoslavia (see Chapter 8) the confusion was worse. Although the disintegration of that state began with conflicts between Serbs and Croats, the crux internationally was the attempt of Serbs to conquer and annex parts of Bosnia. International concern was twofold: to prevent the fighting from spreading throughout Yugoslavia and beyond, and to succour civilian victims and refugees from the usual consequences of wars and the unusually horrible atrocities of this one. These aims were in conflict in so far as humanitarian intervention required some measure of compliance from the combatants who simultaneously resented external attempts to stop them from carrying on their warfare.

UN missions could be divided into two broad categories which shaded into one another: first, observer missions whose main purpose was to prepare, invigilate and report on elections; and peace-keeping missions sent to a scene of violence which had been halted by a truce with the purposes of maintaining the truce and aiding victims of the violence. The latter were normally empowered to defend themselves but not otherwise to use force and they presupposed a cessation of violence. In Yugoslavia, however, the despatch of UN missions preceded any effective or believable truce. Secondly, the UN became directly involved (from September 1991) in the wake of EC intervention, partly because President Bush encouraged EC leadership in place of UN action which would inevitably and perhaps deeply involve the United States, and partly because some European leaders – notably German and Italian – saw an opportunity for the EC to play a significant international role. The UN's first measure was an embargo on the delivery of arms to all the warring parties, a measure which favoured the Serbs who were better armed than others. The UN also nominated Cyrus Vance, formerly US secretary of state, to co-operate with the EC in its diplomatic attempts to stop the fighting by negotiation. In 1992 the UN raised and despatched a Protection Force (UNPROFOR) to succour victims of the fighting and protect aid workers, first in Croatia, later in Bosnia-Herzegovina, later still to establish a UN presence in Macedonia. The UN imposed economic sanctions on Serbia which were for a time evaded (with the help of Greece and Cyprus) but did considerable damage to the Serbian economy and inflamed popular emotions and irridentist nationalist rhetoric. The UN also assembled a committee of experts to collect evidence on breaches of the Geneva Convention and Protocols. The Security Council resolved in 1993, under chapter VII of the Charter, to create an *ad hoc* tribunal to hear charges of serious violations of international humanitarian law in Yugoslavia from 1991.

In addition to the pre-emptive EC intervention, which was ill-conceived (see Chapter 8), the UN was hampered by serious and genuine differences of opinion among its leading members. President Clinton, who succeeded President Bush in mid-crisis, was anxious to keep his distance or – if that were to prove impolitic – to intervene only through air power. He was revolted by the enormities perpetrated by (although not exclusively) Serbs against Muslims, uneasy about repercussions in the Muslim world and attracted by the calculation that air strikes could be used to compel a cease-fire and a negotiated peace without costing American lives and without irreversible entanglement (as in Vietnam). Europeans on the other hand were opposed to the use of air power on the grounds that it must abort their diplomacy, could not be an effective substitute for ground operations in a bitter war in wooded mountainous country and would abruptly end all relief work. All parties were reluctant to face the fact that a negotiated peace could not be a just peace, that justice for the Muslims and perhaps the Croats too could not be achieved without escalating the war against the Serbs, who had won it. Hence the proffering of a sequence of partition plans which not everybody wanted to accept, and threats of military action which not everybody wanted to carry out.

Boutros-Ghali, presenting in 1992 an *Agenda for Peace*, laid bare the conclusion that, whatever its rules or restraints or practices, no international organization could live up to the expectations embodied in the UN Charter if it were half-heartedly supported by its members, underfunded, sparsely equipped, belatedly informed of likely troubles ahead and uncertain of its capacities and purposes: Angola and Cambodia provided cautionary tales. At the time when he was writing this report the UN was running seventeen separate missions of widely different cost but the total annual cost to the UN's nearly two hundred members was no more than 1 per cent of the defence expenditure of the United States, 10 per cent of the British. This was the negative side of the picture. The positive side was the simple fact that wars are impossible without arms but that arms were being manufactured and traded in plenty. Most of the arms trade was conducted by governments anxious to reduce their heavy military expenditure by selling to other states, and in competition with one another, a surplus on current production or an obsolescent overstock. There was also a black market in arms estimated to be worth billions of dollars a year. In 1991 the UN resolved to establish a register of arms transfers but progress was obstructed by disputes over what arms to include and then brought to a halt when China, in protest against American and French sales of arms to Taiwan, boycotted the proceedings.

There was a further conundrum. That the UN was an organization for keeping the peace between states was not in doubt. That it was also an organization for keeping the peace within states, or intervening in civil wars, or ensuring certain standards of behaviour within states, was much doubted – except in so far as any given situation could be clearly classified as a threat to international peace. Yet civil wars and domestic anarchy or tyranny were by the 1990s no longer off the international agenda as matters for concern and possibly action in their own right. Parts of the UN Charter at least implied that the UN was committed to the protection of human rights and it had created or sponsored a formal framework through a series of (mainly declaratory) instruments of three kinds: general, regional and specific. The first category included the Universal Declaration of Human Rights (1948) and the

Covenants on Civil and Political Rights and on Economic, Social and Cultural Rights (1966). Regional bodies had adopted similar instruments – for example, the European Convention on Human Rights (1953), the OAU's Banjul Charter (1981). In the third category were conventions on specific derelictions such as the Genocide Convention of 1948. To this framework non-official bodies – Amnesty International, the International Commission of Jurists, the Quakers – had contributed pressures, ideas and drafts, but implementation lagged far behind. In 1992 the UN created the office of High Commissioner for Human Rights, an acknowledgement of the *Zeitgeist* and of the growing impact of inhumane activities on international affairs and the international agenda. Yet international action in the cause of justice did not necessarily add to world order, for as in national affairs the claims of justice were not coterminous with the pursuit of stability.

There was yet another source of confusion. In so far as American foreign policies rested on a general principle, that principle was not so much the assertion of international law and order but the promotion, by force if practicable, of democracy, a Wilsonian crusade which could run foul of the law of the UN Charter (see for example the case of Haiti in Chapter 28).

World order is commonly measured by the sum of international and civil wars but financial turmoil, if seemingly less calamitous, may jeopardize world order no less than armed conflict. And just as the international political system was proving too weak to cope with the swelling range of armed conflict, the international economic order – both as provider and as regulator – was showing signs of lagging behind the pace of change. At the date of the Bretton Woods conference in 1944 international economic order meant the co-ordination of distinct national economies or some of them for certain limited purposes, but fifty years later a genuinely worldwide economy had come into being, particularly in the sphere of finance, with a life of its own. The revolution in communications technology had created financial markets which were capable of handling vast numbers of transactions and which never closed: at no hour of the day or night was it impossible to find somewhere to buy or sell currencies or commodities or speculate in futures. A significant part of this business was done by operators using borrowed money or, in extreme cases, phantom money. Attempts by powerful central banks to impose checks and regulations were too easily countered by the intervention in the markets not of governments but of speculators whose interests were diametrically opposed since they thrived on instability rather than the stability which governments, industry and the world of commerce wished for.

In the same years the volume of capital which nourished speculation and made its movements dangerous had also become a necessity for rich and poor countries alike. All faced irresistible needs which could, if not met, produce disasters − whether a collapse of investment in research and production and so in employment, or a collapse in support for the (multiplying) poor. Rich countries, whose standards of living depended on exports, found that trade among themselves was growing more slowly than their trade with poor countries. But the ability of the latter to go on buying the products or services of the former was manifestly limited. They were not themselves creating the domestic capital, the savings or the financial institutions necessary to attract foreign capital to underpin their own development and so in turn sustain the growth of the richer parts of the world.

The magnitude of the problem was illustrated by the cost to Germany alone of rescuing and rehabilitating its eastern Länder after reunification: $100 billion a year for a comparatively small territory with a population of less than 20 million. Rescue plans for the remaining Soviet satellites, let alone for Russia itself and Ukraine, were grievously understated and then unmet; and Europe was only a small part of a world where India, China, South Africa and many others were embarking on ambitiously costly expansion without the necessary capital or likely ways of attracting it. If at the end of the century there was a need to reconsider and redefine the UN's role in keeping the peace, there was a no less urgent need to review and reinforce the operations of the World Bank and the IMF. World order was an empty phrase without the capital to sustain it, the conditions for that capital to fructify, and the regulation to keep a global economic system under the control of responsible national or international government rather than of predators flogging the troubled waters. (In much of West Africa, for example, organized internationalism was most evidently exemplified by illegal, even criminal, trading.)

By the closing stages of the twentieth century, world order was not an alternative to several and separable regional orders: the alternative was world disorder.

5 A THIRD WORLD – AND A FOURTH

NEUTRALISM AND REALIGNMENTS

One of the biggest changes that can occur in a world divided politically into sovereign states is the multiplication of these states. This happened in Europe with the dissolution of the Ottoman and Habsburg empires. A generation later it happened worldwide upon the dissolution of European empires outside Europe. This process was protracted but much the greater part of it was consummated during the twenty to twenty-five years after the end of the Second World War. This period was dominated by the Cold War in Europe and that conflict gave the nations emerging into statehood one of their initial basic characteristics.

Both the protagonists in the Cold War – the United States hardly less than the USSR – were uninhibited in their hostility to European colonialism, but as the Cold War created the Euro–American alliance embodied in Nato American hostility to the British, French and other European presences in Asia and Africa was transformed. While commercially it was frequently intensified by competition in areas hitherto dominated by the colonialists, governmentally it became muted as a result of the priority given to the need for European allies and bases. To nationalists in Asia and Africa this change of mood amounted to something between evasiveness and betrayal. It placed the United States alongside, if not actually in, the ranks of the imperialists, and it was the beginning of a decline in the high standing of the United States in the minds of what was coming to be called the Third World. This phrase – a conscious echo of the antonyms Old and New World – was first proposed by Dag Hammarskjöld to designate the poor countries of Asia and Latin America.

It was a Third World because it rejected the notion of a world divided into two, a world in which only the United States and the USSR counted and everybody else had to declare for the one or the other. It feared the power of the superpowers, exemplified and magnified by nuclear weapons. It distrusted their intentions, envied (particularly in the American case) their superior wealth and rejected their insistence that, in the one case in democratic capitalism and in the other in communism, they had discovered a way of life which others need do no more than copy. Nationalist leaders, although anti-European in the nature of things, had at least one characteristic in common with their retreating masters: their temper was pragmatic. Moscow's rigid communist dogmatism, and Washington's increasingly rigid anti-communism, offended them. Above all they felt beholden neither to the United States nor to the USSR for their independence from European rule, which they attained with unexpected speed and ease.

The decision of the newly emerging states of Asia and Africa, with few exceptions, to throw in their lot with neither superpower was much influenced by one man, Jawaharlal Nehru. Nehru was a world figure before becoming in 1947 prime minister of the most populous of the new states, and he held that office uninterruptedly for seventeen years. He was a pragmatic and eclectic patrician who had imbibed western liberal and democratic values and was also attracted by the USSR's record in auto-industrialization. He was repelled by Stalin's tyranny and police rule, but also by the crudities and stupidities of McCarthyism in the United States and by the arrogant and moralistic division of the world into communists and anti-communists. (Parenthetically, and with hindsight, it is worth emphasizing the worldwide impact of McCarthyism, a domestic upheaval in the United States which seemed to betoken a sharp swing to the right in American politics, coupled with a myopically over-simplified view of world politics. McCarthy's indiscriminate charges of treason and conspiracy flourished on the shocks of the Korean War. In the United States the mood and the methods induced by these shocks were mastered when the peak of the war passed but the damage to the American image abroad persisted for much longer.)

Nehru was the principal creator of the post-imperial Commonwealth as an association of monarchies and republics of all races whose links were not ideological but historical and accidental. When he decided that India should remain in the British Commonwealth (as it was still called at this date) he did so upon the conditions that India should become a republic and that it should have the right to conduct without cavil a foreign policy distinct from, conceivably even at odds with, the foreign policies of Britain and its other Commonwealth associates. Thus he stressed the political independence which all new states needed to assert, while retaining links which had economic, cultural and sentimental value. His example was widely followed. Although Burma severed these links with Britain in 1948, no other British possession did so and by 1990 the Commonwealth had fifty sovereign members (including Pakistan which resigned in 1972 but rejoined in 1989).

Nehru's insistence that each member of the Commonwealth should be free to devise and pursue its own foreign policy was crucial, for it meant that neither the Commonwealth as a whole nor its several members need follow Britain's example in taking the American side in the Cold War. This was the beginning of the Third World's neutralism or non-alignment, to which France's former colonies also adhered in the 1960s. In western Europe there had been at the close of the Second World War a hankering after a similarly independent – and mediatory – stance between the United States and the USSR (a so-called Third Force, championed for example by Georges Bidault), but it quickly shrivelled under the impact of heavy-handed Russian measures such as the Prague coup of 1948. Europe became the seat of the Cold War. But the rest of the world believed that it lay outside the arena.

Its attitudes passed through a number of phases. They were rooted in the concept of neutrality. Neutrality was a general declaration of intent to remain out of any war which might occur, but it had not proved very useful to its various adherents during the Second World War and in any case the new states were not thinking of a shooting war and how to keep out of it, but of the Cold War and how to behave in regard to it. Neutralism and non-alignment, therefore, as distinct from neutrality, were the expression of an attitude towards a particular and present conflict: they entailed, first, equivalent relations with both sides and, secondly – in the phase called positive neutralism – attempts to mediate and abate the dangerous quarrels of the great. In its more negative phase non-alignment involved a reprobation of the Cold War, an assertion that there were more important matters in the world, an acknowledgement of the powerlessness of new states, and a refusal to judge between the two giant powers.

The positive phase of neutralism represented the desires of new states to evade the Cold War but not to be left out of world politics. If at first sight the postwar bipolar world seemed to leave as little scope for small powers as in the days of the great struggles between the Roman and Persian empires, on second thoughts it seemed that the neutralists might nevertheless play a gratifyingly honourable and sensible part. When Africa as well as Asia became independent the number of neutralists and the space they occupied round the globe became considerable. They might at the very least prevent the Cold War from spreading to these areas; by merely setting limits to bipolar commitment they could reduce the occasions and areas of conflict. They could too, by virtue of their combined importance, cause the great powers to woo them, thus becoming a kind of lightning conductor in world politics. More positively still, they might exert influence by the time-honoured method of holding conferences to publicize their views or by the newer method of arguing and voting in the General Assembly of the

United Nations. In this last respect the Indian voice was again decisive. The new states had hesitated at first in their attitudes towards the UN, not knowing whether it might turn out to be dominated by its European members, as the League of Nations had been, or by the west or by the great powers. They feared that the new organization might be used to buttress colonialism or to subserve the purposes of the Cold War, in either of which events they would have had little use for it, but after a little experience they decided otherwise and India in particular became prominent in its discussions and its field commissions and supplied for emergency operations units without which those operations could hardly have been contemplated (especially after Hammarskjöld developed the principle that major powers must not contribute fighting units to UN forces).

It is impossible to assess the precise effects of anti-colonialism, the Cold War and neutralism upon one another, but it is possible to show that the policies and actions of the neutralists had some effect on states outside their ranks. In the first few years after the end of the Second World War American attitudes to Asians were deeply, and adversely, affected by two things: the need to rescue western Europe and the call to fight against an aggressor in Korea. The European Recovery Programme absorbed a great deal of American talent and American attention as well as American money and it may have made Americans less critical of Europeans and particularly of the European colonialism which they had in the past so unequivocally decried. To Asians emerged or emerging from colonialism the American voice seemed to have become muted by concern for war-wrecked Europe and also by the need to find sure and strong allies against a Russian communist threat: in other words the Cold War was perverting the American attitude on colonialism and even carrying the United States, spiritually and physically, into the imperialist camp. To the Americans the war in Korea was a major event in the conflict between communism and anti-communism, in which too few people gave too little help (no more than 10 per cent of the combat effort) and some, notably Asians, indulged in ill-timed and ill-considered criticism. The American attitude to Asian neutralism was one of righteous indignation. Thus the events of these years made Asians dub Americans imperialists and Americans dub Asians traitors. Yet the American mood changed after a few years and it did so partly because the neutralists' behaviour and activities at the UN and elsewhere showed the inadequacy of judging them by a simple criterion of communism and anti-communism.

The Russians too were made to revise their opinions. As anti-imperialists they had been equivocal, supporting communist movements but doubtful about others. What struck them most about the leaders of new states was their bourgeois character, and they attacked them accordingly as western stooges. Men like Nehru and Nasser seemed to the Russians at first no better than any western European politician who joined Nato. But the arrival of such leaders at the UN in increasing numbers converted the Russians to the idea that they constituted, and must therefore be treated as, a separate group midway between the communist bloc and the USSR's enemies. This transition was easier for the Russians than the Americans because the adherence of strange and distant countries to the communist bloc was in any case an unreal and unsatisfactory way of extending Russian influence (perhaps, as a short experience with Iraq showed, even a way of making things more difficult for Moscow), whereas the Americans were used to picking up allies all over the world and holding them together by their easy familiarity with air and sea power. In any event the neutrals succeeded in getting both the Russians and the Americans to accept them in world politics in the role which they had chosen for themselves.

To be effective non-alignment, negative or positive, presupposed solidarity among the non-aligned. The new states were weak and aware of their weakness which – no less than their repudiation of the Cold War – was their hallmark. The two characteristics reinforced each other. Their weakness made them wary of too close an association with a single major power and so obliged them to seek strength by unity among themselves. Many of them were far from being nations, and such political unity as they possessed had been a function of xenophobia. Their governments were metamorphosed liberation movements which had to create the broadest possible consensus in order to prevent the new state from disintegrating or becoming ungovernable. In so far as this problem impinged on external policies it suggested the advisability of the broadest range of contacts and friendships among foreign states and the need to eschew any one precise and discriminatory alignment. Economic needs pointed the same way: no new state was so important to the world's rich powers as to be able to command from one rich power all the aid it needed; better therefore not to contract an alliance with one power which would rule out the possibility of getting aid from others. (This argument was not conclusive. Many of the small states which emerged from French rule in Africa were so weak that they had no choice but to take what they could get from France.) Similarly in the field of defence, while there was a superficial argument in favour of attachment to a particular strong protector, it was also observable that major powers wanted to keep out of the sort of local disputes in which new states wanted help – as opposed to local aspects of global conflict, in which new states did not want to get involved.

The search for solidarity preceded independence among both Asians and Africans. The first notable postwar Asian conference – the Asian Relations Conference held in New Delhi in March 1947 – assembled twenty-eight delegations of which only eight came from sovereign states. Its motive force was a desire to ensure that the United Nations should not become an organization dominated by European or white states and viewpoints, such as the League of Nations had been, but the tone of the discussions was not markedly anti-colonial. The conference was a gathering of Asians to discuss Asian problems including land reform, industrialization, Asian socialism and the application of non-violence in international affairs. The conference established a permanent organization which existed for eight years but did not do much else. Very soon afterwards India, Pakistan, Burma and Ceylon became independent. They did so in a world which had expected peace but not got it. There were guerrilla wars and insurrections in Burma, Malaya and the Philippines, and open fighting in Indonesia, Indo-China and Palestine. The Cold War too was beginning. In domestic politics violence claimed notable victims in Burma in June 1947 with the assassination of Aung San and six colleagues, and an even more notable victim in January 1948 when Gandhi was killed.

In January 1949 another Asian conference assembled in New Delhi. The Soviet Asian republics, which had attended the 1947 meeting, were not this time asked, and Turkey refused an invitation. Otherwise Asia, including the Middle East, was very fully represented and Australia and New Zealand sent observers. The immediate occasion for the conference was Indonesia where an Asian liberation movement was being threatened with extinction by the Dutch and where to Asian eyes the UN seemed bent on facilitating the reimposition of white colonial rule. In the previous December the Dutch had resorted to their second police action and had captured and imprisoned a number of Indonesian leaders. The conference demanded their release and also the immediate establishment of an interim government and independence for Indonesia by 1950 (the Security Council shortly afterwards chose 1960 as a suitable date). Like its predecessor this conference created a permanent organization which proved ineffective, partly because a number of Asian states were becoming jealous of India's predominance and did not wish to see it institutionalized. Owing to the Indonesian issue the conference had a clear anti-colonial note, but it was divided between friends of the west and neutralists. This division was accentuated in the following months when different Asian leaders took up different attitudes towards the two outstanding Asian events of the year, the victory of Mao Zedong and communism in China and the war in Korea. Asian solidarity was proving difficult to achieve, even on an anti-colonialist programme; the British and French campaigns in Malaya and Indo-China did not evoke the same united protest as the Dutch proceedings in Indonesia, partly because of the strong communist flavour in the Malayan and Vietnamese anti-colonialist movements.

In the 1950s Asian solidarity and neutralism waxed and then wore thin. Some Asian states, putting their economic and strategic needs before their neutralism, signed not only commercial but even defence treaties with the United States or the USSR. India, by its treaty of 1954 with China embodying the Panch Shila, maintained its principles but in the same year Pakistan, Thailand and the Philippines concluded military agreements with the United States, while Afghanistan became the first non-communist country to receive Russian aid and the USSR, which already had a trade agreement with India and was about to conclude another with Burma, intensified the diplomatic and economic wooing of Indonesia which was to lead Sukarno to visit Moscow in 1956. The great powers were taking a gratifying interest in Asian affairs but one consequence of this interest was to make it more difficult for Asians to maintain a common attitude towards the great powers or to keep their distance as pure neutralism required.

During 1954 preparations were made for a conference, originally suggested by Ceylon and taken up by Sukarno and Nehru. This conference assembled at Bandung in April 1955. It was a grand assembly to stimulate co-operation among Asians and put Asia on the map. The background to it comprised the treaty between the United States and Taiwan (a consequence of recent crises in the Formosa Strait and offshore islands), and the creation of Seato and the Baghdad Pact. The USSR and China welcomed what looked at first like an anti-western conference, while Washington's friends – Thailand and the Philippines – were half inclined not to go. Israel was excluded on account of Arab opinion. The twenty-nine participants included six from Africa (Egypt, Libya, Sudan, Ethiopia, Liberia, Ghana), so that Bandung became the prototype of Afro–Asian as opposed to purely Asian solidarity. It was an assembly of the needy and the indignant, not a concentration of power. Its members were divided among themselves even on the issue of non-alignment, but the timing was propitious. The Cold War in Europe had, since the Berlin blockade of 1948–49 and the growth of Russian nuclear power to match the American, lapsed into a quiescent stalemate but not into a thaw. Both sides were looking elsewhere and were competing for the allegiance of states in other continents with the vague intention of building up a new preponderance by additional alliances, or of turning the enemy's flank by carrying influence and bases into new terrain. The Russians and the Chinese hoped to advance communism by exploiting anti-western nationalisms, while the Americans hoped to exploit fears of communism and of China and so create new, and if necessary heavily subsidized, military groups. American policy, freshly illustrated by the signing of the Manila Pact, ran counter to the spirit of Bandung. Zhou Enlai, on the other hand, who put in a personal appearance at Bandung, went some way towards showing that Chinese communism was reconcilable with other Asian nationalisms and that at least one Chinese leader was more sensible and amenable than some current pictures of the new China suggested. The Russians had already, by accident or astuteness, taken a number of steps which brought them into closer accord with the Asian mood. The proposal to neutralize Austria was welcome to Asian neutralists, and gestures like the return of Port Arthur to China and of Porkkala to Finland heartened those who hoped that Stalin's death had changed the face of world politics. In 1955 Bulganin and Khrushchev visited Asia with tremendous acclaim (Khrushchev paid a second visit early in 1960) and the Russian campaign to win over the neutralists was so well launched that even the suppression of the Hungarian revolt of 1956 only dented it (the Anglo–French attack on Suez being invaluable to the Russians in saving their new reputation at this juncture).

For the neutralists themselves the principal achievements of the Bandung conference were that they had met and got to know one another (most of them were new to international politics); that they had laid the foundations for joint action at the UN and, through solidarity, increased their security, their status and their diplomatic weight in the world; that they had attracted new men like Nasser to the group and made it bigger; that they were making the giant powers take them seriously and treat their policies as respectable (a trend which was fortified by the admission of sixteen new members to the UN by the package deal of 1955 and still further by the big increase in African membership in 1960); and finally that they had seen one of the leaders of the new, formidable China, had found him not at all frightening and had perhaps inducted China into their pacific circle. In the summer of 1956 Nehru and Nasser visited Tito at Brioni in Yugoslavia. This event was a sign of the development of Asia neutralism into a worldwide association. With an Asian, an African and a European leading them the neutralists became more ambitious in international affairs and hoped to be able to bring pressure to bear on the giant powers in Cold War matters, but this association was already passing the peak of its influence, partly because of the activities of those who wanted to turn it into an alliance of communists and black men against non-communist whites by emphasizing anti-colonialism in place of neutralism. Non-alignment became in practice anti-western non-alignment, particularly with the Afro–Asian People's Solidarity Movement which sponsored a variety of conferences in the late 1950s.

In September 1961 a conference of the unaligned was held in Belgrade. Whereas Bandung had been an exploratory conference, Belgrade had about it an atmosphere of crisis. The background to it included French nuclear tests in the Sahara and the resumption of

Russian tests, the Bay of Pigs and the Berlin wall, the Franco–Tunisian clash over Bizerta and the grinding crisis in the Congo. A new conflict between India and China seemed to be emerging, a conflict between the USSR and China certainly was. Bandung's twenty-nine participants had been overwhelmingly Asian and not overwhelmingly anti–western. The only serious conflict over admissibility had been the Arab veto on Israel. At Belgrade the African representation reflected the division of African states between radicals and moderates, Latin American participants were selected with a similar bias and the Europeans included Yugoslavia and Cyprus but not the traditional neutrals, Sweden and Switzerland. An attempt by Nehru to concentrate on peace rather than anti-colonialism and to get Russo–American talks going met with little success, and a number of delegations displayed a partisan indifference to the nuclear explosion which the Russians set off on the eve of the conference. A proposal to fix a terminal date for colonial rule throughout the world at two to six years was enlarged during the debates to a demand for its immediate and total abolition.

After a pause plans for another conference led to a meeting in Cairo, more African than Asian, in October 1964 and to a project for a conference in Algiers in 1965. This plan was, however, vitiated by the fall of Ben Bella a few days before a preliminary meeting in June and by increasing embarrassment among the likely participants at the prospect of a local Sino–Soviet conflict. The Chinese wished to exclude the USSR and assume the leadership of the underprivileged but at a second preliminary meeting in October the invitation to the USSR was approved, whereupon the Chinese threatened to stay away. In these circumstances a majority thought it better to have no conference at all and the plans were allowed to expire. The movement seemed to be wilting but it was reanimated during 1967 by visits by Tito to Asian and African countries and in 1969 a conference in Belgrade gave it fresh impetus. At Havana ten years later ninety-two full members attended (there were two absentees). Whereas the original members had been non-aligned in their policies and their sympathies, the much wider flock of the 1970s contained a number of states which, although non-aligned by policy, had definite pro-western or pro-communist sympathies.

African solidarity and non-alignment, which began to join forces with the Asian current at Bandung in 1955, had its own remoter origins. Pan-Africanism began as an assertion of the distinctiveness and value of an explicitly African or (by extension to those lands to which the slaves had been consigned) Negro culture. As such it was primarily Caribbean and West African, but it became also part of the wider movement for colonial emancipation in which nationalists from all parts of Africa could consort with and seek strength from one another. Thirdly, there were those (Nkrumah, for example) who saw that political freedom was not the whole of freedom, that economic dependence would persist after the winning of sovereign statehood, and that Africa might stand on its own feet economically only by developing its continental resources in common. This third aspect of pan-Africanism pointed logically to a political union or at least a federation and it was therefore in conflict with the creation of new sovereign states committed to the preservation of their integrity as well as their independence.

Six pan-African conferences had been held between 1900 and 1945. The first of them and the four which followed in the 1920s were predominantly Caribbean and North American but the last was dominated by African leaders from Africa itself. All of them were meetings of personalities. With the beginnings of independence came meetings of African parties and African governments. The former created an All African People's Organization which, at conferences in Accra in 1958, Tunis in 1960 and Cairo in 1961, discussed schemes for African unity or an African commonwealth on the basis that co-operation between governments was not enough. But the third meeting was the last. As the tally of independent states grew, the states' system took hold. Nkrumah continued to beat the drum for a union government until his fall in 1966 but this theme, although a standard item at conferences of the OAU (Organization for African Unity) for some years, attracted declining support both because it was regarded as unpractical and because it became increasingly identified with a left-wing radical minority.

At the date of the first meeting of independent African states, held in Accra in 1958, there were nine independent states in the continent. One of them, South Africa, declined the invitation to Accra. The others were Ethiopia, Liberia, Egypt, Morocco, Tunisia, Libya, Ghana and Guinea. They were chiefly concerned with anti-colonialism, the racial and nationalist struggles in South Africa and Algeria, and the problem of achieving some sort of African unity while at the same time respecting the independence and integrity of African states. This conference was followed in May 1959 by the Declaration of Conakry whereby Ghana and Guinea formed a union which was declared to be the starting point of a wider African union. This step was an unpremeditated retort to the ostracizing of Guinea by France, a practical demonstration of Nkrumah's pan-African principles and a lifeline for Guinea. It was followed in July of the same year by the Declaration of Saniquellie which, primarily on Liberian insistence, emphasized the independence and integrity of existing states.

By 1960, when the second conference of independent African states assembled in Addis Ababa, their number had almost doubled and their unity was about to be tested by the special strains of the Congo as well as by inherited border disputes. Fifteen states were represented. Active border disputes involved Ethiopia and Somalia, Ghana and Togo, Guinea and the Cameroons. The first of these disputes led to fighting on a not inconsiderable scale, but the others did not. More serious for the prospects of African unity was a contest between Ghana and Nigeria in which Ghana urged the case for immediate steps to unity and Nigeria argued in favour of a slow approach to something like a federation. This dispute was spiced with some bitterness between the protagonists since the Nigerians resented Nkrumah's assumption of leadership and distrusted his aims, while Nkrumah feared that Nigeria intended to throw the influence of its vast size on to the side of conservatism versus socialism and of Nigerian nationalism versus pan-Africanism. In the Congo the independent African states tried, both at the UN and in a conference at Leopoldville in August 1960, to present a united front and play a constructive and pacificatory role, but they were not successful.

From this point the independent African states began to form separate groups which were later reassembled in one organization by the founding of the Organization for African Unity. The largest of these was the Brazzaville group consisting of all the former French colonies except Guinea, with the addition of Mauritania (whose claim to be independent and not a part of Morocco was accepted by the group). The Brazzaville group began as an *ad hoc* meeting at Abidjan in October 1960 when the principal topic for discussion was Algeria, but at Brazzaville in December and at further meetings during 1961 at Dakar, Yaoundé and Tananarive it developed into a permanent association, discussed ways of perpetuating the co-operation and common services which had existed in the colonial period, set up an organization for economic co-operation, and considered joint institutions and defence arrangements. This group was neither pan-African nor regional, but an expression of common needs and a common outlook.

A second group took shape at a conference at Casablanca in January 1961. This group consisted of six independent African states plus the Algerian revolutionaries and Ceylon. The six African states were Morocco, Egypt and Libya (which soon afterwards transferred to the Brazzaville group) and Ghana, Guinea and Mali (which had joined the Ghana–Guinea union in the previous year). The Casablanca group opposed the independence of Mauritania and was pro-Lumumbist in the Congo, although at its second conference in May in Cairo Nkrumah strongly opposed proposals to withdraw troops from the UN force and switch them to Gizenga and prevailed upon his associates. This group too established permanent organs, including political, economic and cultural committees, a supreme command, and a headquarters at Bamako in Mali.

In August 1961 no fewer than twenty states assembled in conference at Monrovia. They included the whole of the Brazzaville group, Libya and a majority of former British territories. The Monrovia group thus subsumed the Brazzaville group and, owing to the prominence of Nigeria, acquired a specifically anti-Ghanaian and anti-Nkrumah flavour. The movement for African unity seemed to have been blocked by current problems (the Congo and, to a lesser extent, Mauritania) and by personalities. Nevertheless the idea remained alive. Even if Nkrumah's vision of a union extending into every part of the continent was unacceptable or impracticable, lesser unions might be attempted. The Ghana–Guinea union, with or without Mali, had proved of little practical consequence, but it had been a political demonstration rather than a true regional association. In the north-west Morocco, Tunisia and Algeria espoused federation prematurely, at a meeting in Tangier in April 1958. In east and central Africa there was talk of a federation between Kenya, Uganda, Tanganyika, Zanzibar, Malawi, Zambia and Rhodesia – with possible extensions in some barely visible future to Rwanda, Burundi, Mozambique and even South Africa. A Pan-African Freedom Movement of East and Central Africa (Pafmeca) came into existence in 1958, was enlarged four years later by adding 's' for South as its penultimate letter and was dissolved in 1963; these were associations for self-help in the struggle for liberation.

The French connection was at the base of a number of interstate organizations: the Entente Council (Ivory Coast, Niger, Upper Volta, Togo and Dahomey; see Chapter 20); the Senegal River Association (Senegal, Mali, Guinea, Mauritania; *ibid.*); a West African and a Central African Customs Union. More important than any of these was the African and Malagasy Economic Union (UAMCE) founded in 1965 by thirteen formerly French and Belgian territories and converted into the African and Malagasy Common Organization (OCAM) whose charter, signed at Tananarive in 1966, declared it to be open to all African states – provided existing members all accepted each newcomer. OCAM created a number of useful agencies which were sometimes more effective than those of the OAU, but politically its members were often divided. In the 1960s it was seen as a weapon for Houphouet-Boigny of the Ivory Coast against Nkrumah and in support of Tshombe; in the Nigerian civil war, in which it tried in vain to mediate, Ivory Coast and Gabon recognized Biafra while the remainder were strongly anti-separatist; some members had diplomatic relations with China, others with Taiwan. As the colonial period receded the common French inheritance became a weaker link. There were a number of absentees from the eighth congress held at Lomé in 1972 and Zaïre, feeling that it was not getting enough out of membership, resigned from the organization.

Although the Congo had demonstrated the difficulty of preserving unity among independent African states, it had no less demonstrated the advantages of doing so if at all possible, and a conference at Lagos in 1962 produced a draft charter for an organization of African states. At a further conference in Addis Ababa in 1963 the Organization for African Unity was born with an initial membership of thirty-two. The OAU was not a collective security organization as envisaged and endorsed by article 51 of the charter of the UN but an organization for the promotion of African unity and collaboration and for the eradication of colonialism. It consisted of an annual assembly of heads of state, a council of ministers and a secretariat. A projected commission of mediation, conciliation and arbitration did not materialize, although these functions were in fact performed: in border disputes between Morocco and Algeria, between Somalia and Ethiopia and Somalia and Kenya, and between Ghana and Upper Volta. In the last case, which rose out of the construction by Ghana of a school on territory claimed by Upper Volta, Ghana conceded the claim at a meeting of the OAU's council of ministers. In the other cases the OAU provided mediators and commissions of inquiry which helped to appease the disputes.

The establishment of this organization epitomized two processes which had been going on for a generation or more and had gathered force in the twenty years after the end of the Second World War. Africans ceased to be cut off from each other and they ceased to be cut off from world affairs. Their emancipation had a great variety of causes: the essential liberalism (reinforced by weariness) of the principal colonial powers, the growth of the movement for human rights, American and Russian attacks on colonialism, the Gandhian example, the development of roads and of international airways. While this process was taking place a new class of African, the politician, the lawyer, the intellectual, the *évolué*, was taking the place of the chief and was at the same time rejecting the models prepared for his country by the French and the British. The French had assumed that their colonies would grow into worthy pieces of France, but they had hardly noticed that their doctrine worked neither in terms of government, which was paternalistic and white, nor in terms of society, where low wages and even forced labour were too long tolerated and the favoured few tended not to become leaders in their society but to be extracted from it. The British, who had based themselves originally on paternalism and chiefs, realized the limitations of this model but planned to substitute for it an inapplicable British parliamentary system to be worked by an élite. Consequently the new states – albeit that many of their first leaders, themselves a western-educated and relatively affluent élite, had originally insisted on western democratic institutions as the best available and had expected them to work without essential modification

– found that they had to innovate in theory as well as in practice. They had to find administrators, public servants, economists, teachers, doctors, accountants and trade-union leaders and at the same time construct institutions and develop conventions which would reconcile the Africans' traditions with their thirst for modernity and enable them to enjoy the fruits of efficiency, liberty and justice. They looked at the outside world with a mixture of admiration and suspicion, ready to take the best of what could be learned but convinced that however much they adopted they would evolve a distinct African way of doing things and a separate African culture. This community of aim gave the new states of Africa points of contact with one another which were implicit rather than advertised and which gave their organization a kind of initial cohesion which was not to be found in the Organization of American States or the Arab League or even the European Economic Community. But for all this the OAU was an association of sovereign states in a continent which, in the wake of decolonization, was becoming more fragmented than it had been under foreign rule. The OAU established particularism in Africa in the same way as the UN accepted it in the world at large.

The creation of the OAU represented not only a negation of federal ideas but also an emphasis on specifically African issues. The Charter of the OAU and its founding conference stressed the sacrosanctity of existing frontiers and the role of the new organization in the peaceful settlement of disputes between African states. To some extent therefore it derogated from the concept of Afro–Asian solidarity. The two continents were becoming increasingly concerned each with its own affairs; some of their common concerns faded as the anti-colonial struggle passed into history. Even within each continent solidarity came under strain. In Asia China's attack on India destroyed what was left of the Panch Shila, while India's unpreparedness compelled it to approach the giant powers from which, as leaders of the non-aligned, it had tried to keep at arm's length. India lost a degree of its detachment. After the Chinese invasion in 1962 Ceylon, Burma, Cambodia and Indonesia, in company with Egypt and Ghana, tried to use their good offices to effect a Sino–Indian reconciliation but their efforts were of little effect and were welcomed neither in New Delhi nor Beijing.

POVERTY

Yet there remained a powerful bond – poverty, and the realization that political independence and sovereignty did not remove economic dependence. In the same year as the Chinese invasion of India, and midway through the Congo crisis which threatened to split African opinion, the first UN Conference on Trade and Development (UNCTAD) was held in Geneva. This was not an Afro–Asian affair but something larger. It was a point of junction between the Afro–Asians on the one hand and a posse of other states, mostly Latin American, which were not only poor in comparison with the developed industrial world but also found themselves obliged to live in an economic system devised by and in the interests of the rich.

Before the Second World War ended the world's two major trading countries, the United States and Britain, agreed, at Bretton Woods, to apply to the international economy the principles of free trade, non-discrimination and stable rates of exchange. They created for these purposes two new organizations: an International Trading Organization and the International Monetary Fund, the one to clear the channels of trade of physical obstacles (tariffs and quotas), the other to provide the finance for international trade and its expansion. The latter came out of the Bretton Woods conference in an American rather than a British form: in particular Britain wanted but failed to get an international currency, a variable volume of credit geared to the expansion of trade, and much larger initial reserves than the $25,000 million with which the Fund actually started. The former never came into existence but was replaced by the General Agreement on Tariffs and Trade (GATT) which, if it lacked the institutional solidity of the projected ITO, was nevertheless pledged to the same objectives, which it pursued by a series of meshed bilateral bargainings to reduce tariffs, abolish quotas, rule out new or extended preferences and assure to all every preference available to any.

The prime object of the GATT was the reduction of obstacles to trade in manufactured goods, particularly the reduction of tariffs. The average industrial tariff when the GATT came into force was over 40 per cent, but by 1980 this average had been reduced to not much more than 4 per cent and the volume of world trade had quintupled in twenty-five years. It was in the nature of the GATT process that each round of negotiations be more difficult to conclude than the last and the Uruguay Round inaugurated in 1986 – the eighth in a sequence begun in Geneva in 1947 – did not reach agreement for seven years. Besides the ever increasing number of participants, negotiations were complicated by the shift from manufactured to agricultural products, financial services, so-called intellectual properties (patents, royalties) and the larger export earners of richer countries (civil aircraft, film, cassette and television products). Most intractable for political as well as economic reasons were agricultural subsidies on which agreement was reached only at the cost of abandoning parts of the Round's ambitious programme.

The principal contestants were the European Community acting as a single unit (subject however to the approval of the Council of Ministers and its individual members) and the United States with the support of fifteen other major food exporters collectively denominated the Cairns Group. The United States began by demanding, as a condition for subscribing any of the Round's voluminous packet of provisions, that the Community reduce its export subsidies to cereals by 90 per cent of current rates over ten years. The Community offered cuts of 30 per cent from levels prevailing in 1986. Attempts to close the considerable gap created by the unrealistic opening stance of each side were complicated by the impact of the GATT's debate on the Community's Common Agricultural Policy (CAP) which the members of the Community were painfully and slowly unravelling in the face of strong opposition from their farmers. The Community had reached agreement on reductions in subsidies which, they claimed, would have to be renegotiated – a horrifying prospect – if American demands in the GATT process were accepted. The United States had little sympathy with this argument, particularly in the approach to the presidential election of 1992. They responded with threats to impose tariffs up to 200 per cent (in effect total proscription) on selected European commodities, beginning with French wines.

A further source of mutual exasperation was a quarrel, outside the GATT Round, over the volume of exports and the production of oilseeds, although this issue was narrowed to a sticking point of half a million tons of Community oilseeds valued at a mere $100 million. A bargain was struck between the EC and the United States by the Blair House agreement of 1992 under which EC subsidies for cereal exports were to be cut by 21 per cent over six years. But European farmers remained unreconciled and French farmers in particular put pressure on their government to reject the deal and therewith the entire Round (notwithstanding that these farmers were not on the whole cereal growers, a sector which had been largely taken over by agribusinesses). A contest in obduracy was kept up until the eve of the date fixed by the US Congress for allowing the president to approve the Round's Final Act without Congressional endorsement of its details: Congress might approve or reject but not move amendments to anything accepted by the president before that date. The final stages of the Round were dramatized as a conflict between the United States and the EC – predominantly France with varying degrees of sympathy from other EC members. France's principal aims were to protect a corner of French agriculture, to ward off American attacks on export subsidies and to try to win protection for European against American film-makers (a booming industry) and aircraft manufacturers. These aims were largely secured either by concessions or postponement to a later Round.

When in 1994 the Final Act of the Round was adopted (subject to necessary ratifications) much had been achieved, even if at some cost in postponing the more intractable issues. The agreement gave promise of massive gains in commerce and employment. Measured in money the benefits worldwide were, very speculatively, put at $5,000 billion after ten years but with relatively little immediate impact and more for rich than poor countries. The GATT was also to be transformed into a World Trade Organization (WTO) in 1995. The Uruguay Round marked a shift away from tariffs and quotas to financial and other services and rights in intellectual properties. Its conclusion coincided with attempts to integrate ex-Soviet and satellite economies with western capitalist economies and for the latter to find a *modus vivendi* with surviving command economies, in particular China's. The WTO was therefore economically and politically broader than the GATT. It was, secondly, more likely to highlight conflicts of interest between rich and poor. The United States and France were united in wishing to import social and environmental criteria into the world's commercial and economic order. This thrust was unobjectionable so long as it was restricted to such issues as prison or child labour but developing countries feared that the rich, by enforcing standards unacceptable to low-wage countries, might introduce a new kind of protectionism under the cloak of social justice.

The Uruguay Round marked a victory, but not necessarily an entrenched victory, for universal free trade in an increasingly protectionist climate. If protectionism was no longer a sensible policy for the state, it retained nevertheless some allure for associations of states. The European Community exemplified the rejection of state protectionism but was less convincing in its rejection of the protectionist association. The American position on the Round was softened only after President Clinton had won Congressional approval for the addition of Mexico to a regional North American Free Trade Area (NAFTA – see Chapter 27) and had made a personal appearance at a conference in Seattle to applaud APEC – Asian Pacific Economic Co-operation forum – an Australian initiative of 1989 to foster a large regional block comprising the South Pacific, South-east Asia, North Asia and North America. With NAFTA the United States, conscious of losing its imposing economic dominance throughout the world, was building on its free trade agreement with Canada (1989) to create a powerful regional position including not only Mexico but eventually parts of South America.

Japan had similar visions for a Japanese sphere of economic influence in the three continents which formed the Pacific rim. At a second APEC meeting in Jakarta in 1994 seventeen of the eighteen states which attended agreed to abolish all barriers to imports from their associates by 2010 in the case of the five major states and by 2020 by the rest: only Malaysia refused to accept either date. Although Clinton once more appeared in person there were unspoken doubts about whether the United States was a natural member of the Asian–Australasian group or an eager intruder into an organization whose members might account for half of all international trade by 2010.

Besides the IMF and GATT the Bretton Woods conference created the International Bank for Reconstruction and Development, later known as the World Bank. The World Bank was founded to assist in the postwar reconstruction of Europe. When this task was largely taken over by the Marshall Plan the Bank gradually turned its attention to development and to the rest of the world. Like the IMF, with which the Bank was allied by their common location in Washington and by the requirement that the Bank must be a member of the Fund, the Bank was governed by a professional board where the larger contributors to its capital funds carried commensurate weight. Besides these funds the Bank raised money by floating its own bonds on international stock exchanges. It made loans to creditworthy states at commercial rates of interest, for limited periods (at the end of which the funds could be lent again elsewhere) and principally for economic infrastructure. Its strictly conservative policies were necessary in order to enable it to borrow in money markets at the best available rates but these policies restricted its initial activities to enhancing the rich rather than aiding the poor. For the latter purpose it established the International Finance Corporation which was designed to help the poorer states to get finance from the private banking sector and, in 1960, the International Development Association which made fifty-year interest-free loans to poorer states. These IDA loans were financed mainly by western governments, were used mainly for economic infrastructure and were expected to produce a reasonable, if belated, return. In development therefore the Bank's principal function was that of mediator or procurer. Until the 1970s its activities were restrained but in that decade the scale of its operations was magnified by a factor of ten or more and this expansion was accompanied by a new doctrine which, in order to justify lending to less conventionally creditworthy recipients, emphasized the potential economic value of poor societies, once they were developed or developing.

The parallel function of the International Monetary Fund (IMF) was to facilitate commerce by ensuring stable exchange rates, thus avoiding the financial anarchy of the 1930s when competitive devaluations engendered worldwide recession by halting trade. The Fund was created as the guardian of the Bretton Woods system of fixed exchange rates – fixed in relation to a dollar which was itself convertible into gold. The Fund, like the Bank, was originally funded by members' contributions which both measured those members' voting weight and their drawing rights. It made short-term loans, principally to cushion balance of payments difficulties and principally to developed states, and it provided a monitoring and forecasting service on national and international economic affairs. Its basis was called in question by the almost simultaneous abrogation in 1971 of the dollar's convertibility into gold and the pegging of other currencies to the dollar. The context in which it operated was then transformed by the drastic oil price rises of 1973 and 1979 which created massive imbalances (particularly for states which were oil importers and relatively underdeveloped), while the consequent flood of petrodollars made floating currencies float more wildly. The Fund was all but swamped, lost prestige and influence, and was forced to reconsider its ambit and its methods. Ostensibly worldwide in its purposes, the Fund – like the Bank – had operated as an adjunct of an economic system created by and largely for the developed capitalist world, but from the 1970s it was impelled to take a wider view as the developing (and now independent) Third World clamoured to be treated as part of the world's economic problems and the richer countries began to realize the extent of their economic involvement with the poorer.

The trading system designed at Bretton Woods presupposed a certain community of interests between all trading nations and it also supposed that tariffs and quotas were the principal barriers to commerce between states. But neither assumption was true of the economically weaker states. Although they needed to enter the international economy, they needed also to be protected in it; freedom worked against them. Further, their main problems were not tariffs or quotas but the instability of world prices for their products and the difficulty of getting into foreign markets to sell them. They were for the most part not only very poor – with an average annual income per head of $100–150, compared with over $1,000 in the Nato countries, even with Greece and Turkey included – but also ill equipped for international economic competition. Many of them had inelastic one-crop economies. Their products were primary products, for which the demand (except in the case of oil) was rising less quickly than world income. Their customers were making synthetic or substitute products and, especially in the case of agriculture, were susceptible to domestic protectionist lobbies anxious to close the market to imports from overseas. Although the Korean and Vietnam wars produced booms in commodity prices from which a number of these countries benefited, they did so only temporarily. Another palliative was aid, that is to say, cash, credits, goods or skills given free or transmitted at less than the ruling market price. Considerable benefits were transferred in this way and the donors, totting up the yearly sacrifices, congratulated themselves on their generosity.

The recipients, however, thought otherwise. They came to the conclusion that aid was the wrong answer to their problems. Apart from the fact that there was too little of it, and apart too from the realization that most of this aid was given for political purposes in the Cold War, aid was condemned for a variety of reasons: because the burden of interest payments and capital repayments became a sizeable charge on export earnings; because aid was frequently tied so that the recipient, instead of using it to buy what he wanted where it could be had most cheaply, was obliged to accept schemes not at the top of his list of priorities or buy goods from the donor instead of more cheaply elsewhere; because aid perpetuated an economic pattern created in colonial times, when colonies were kept to the business of producing raw materials for their owners' needs; because, therefore, aid impeded the essential business of diversifying the post-colonial economy and making a start with industrialization in order to enable the new state to create capital. For all these reasons the weaker countries quickly found that aid was no substitute for what they really wanted – a change in the rules governing the international economy and particularly guaranteed prices for their products and ease of access to the world's more affluent markets where, if at all, these products must be sold. The UN, which first set aside special funds for developing Third World countries (SUNFED) in 1960, asked the richer countries to contribute 0.7 per cent of their GNP to help the poor to grow at the rate of 5 per cent a year and convened a special conference, planned in 1962 and held in Geneva in 1964. This conference was intended to be a single event but turned into a permanent institution whose main thrust was for higher and more stable commodity prices. In 1967 the poorer states of Asia, Africa and Latin America established the Group of 77 which held its first meeting in Algiers in 1967 and chose Raul Prebisch of Argentina as its director. These proceedings transformed the international scene by demanding for economic relations between states as much attention as their political relations. The Group of 77 stressed the unfairness of an economic order dominated by purchasers and consumers of raw materials and by the volatility of the world prices of these materials and argued in favour of commercial preferences as opposed to the GATT's rules of equal treatment for the economically weak and strong, for transfers of technology at bargain prices and for cheap borrowing. But although the Group of 77 changed the way in which people thought about international affairs it had much less success in altering the way they behaved. It became divided between so-called moderates and radicals and was further divided when the oil price rises of the 1970s served the oil producers well but impoverished the Group's oil importers. The emergence of OPEC as a distinct group illustrated this cleavage. OPEC maintained a general Third World stance while at the same time gravitating into the western economic order as an operator rather than a rebel. When in 1974 the UN proclaimed the need for a New International Economic Order the declaration testified to the number of votes commanded by the Third World in the General Assembly but the sequel testified to the limited effectiveness of national votes in economic affairs. With its overwhelming economic power the west was able to fend off the complaints and campaigns of the Third World and keep western economic relief at levels of charity to be set by the west itself.

The European Community took note of those problems with benevolent but measured intent. When the treaty of Rome was being negotiated France insisted on the insertion of special provisions for an associate status. It had in mind its own African territories, then all still colonies. Part IV of the treaty therefore provided that the Community might enter into arrangements with non-European states, by which these would be accorded the benefit of tariff reductions and quota extensions while being entitled at the same time to impose tariffs of their own to protect their infant industries (provided that these tariffs did not discriminate between members of the Community). The Community also established a Development Fund which disbursed $581 million in a first quinquennium and $730 million in a second. In 1964, by the first Convention of Yaoundé, eighteen former French colonies took advantage of these provisions and in 1969 three former British territories – Kenya, Uganda, Tanzania – were granted the same privileges for one year (until, that is, a revised Yaoundé Convention should come into operation in 1971). Nigeria too negotiated an agreement with the Community. Although at first suspicious of these arrangements as a form of neo-colonialism Nigeria could not afford to see its neighbours trading on better terms than itself with the Community, to which two-thirds of its exports were despatched. Again, the relations between the Community and its African associates could be attacked on the grounds that, despite their tangible advantages, they tended to perpetuate the colonial pattern instead of modernizing and diversifying the post-colonial economy. Further, some embarrassment was caused by the fact that a minority of the members of the UNCTAD lobby had, by securing a special position for themselves, broken the solidarity of the weaker nations. In 1974 a new convention for five years was signed at Lomé, this time between the EC and forty-six African, Caribbean and Pacific (ACP) states. It made provision for non-reciprocal tariff reductions, created an aid fund of $1,600 million and a scheme for stabilizing export prices, and promised the Commonwealth sugar producers access to the EEC for all their sugar at the prices assured by the Commonwealth Sugar Agreement. By Lomé II, which followed in 1979, the European Community undertook to take exports from the

ACP countries to the value of $15,000 million a year and to provide annually aid rising from an initial base of $850 million, most of it designed for roads, education, hospitals, water and electricity. Lomé III and V followed at quinquennial intervals but by 1990 aid flowing from the conventions, while certainly helpful, was equivalent to no more than half of the aggregate given by members of the Community separately and bilaterally. Recipients were irked by the concomitant dependence and also by the costs and bureaucratic formalities of operating the conventions. In 1995, when Lomé V was concluded after protracted negotiations, beneficiaries numbered 70. The Convention's preferences were permitted deviations from the GATT's regulations but were not likely to survive the transformation of the GATT into the new World Trade Organization.

While money and other benefits flowed to the poorer countries from a variety of sources, their debts grew and, most depressingly, so too did the cost of borrowing. From the 1970s debtor states became more pressingly concerned to reschedule their debts than to attract new money on crippling terms. At UNCTAD IV in Nairobi in 1976 the Third World presented a comprehensive plan for the rescheduling of its debts, technical aid, the promotion of manufacturing industries and the diversification of one-crop economies. The plan was severely mangled by the richer countries who – particularly the former colonial powers among them – were irked by the view that they were not so much commercial creditors as moral debtors. Largely in vain they pointed to the fact that the expansion in the 1950s and 1960s of the world market had been halted and reversed. The effects of world recession on the attitudes and capabilities of the rich were even more in evidence at UNCTAD's fifth conference in Manila in 1979. The desperate poverty, in some areas famine, in the Third World; the prognosis of a world population doubled by the end of the century, with cities under siege and the outbreak of wars for raw materials; a Third World external debt of $300 billion or more; and western aid in decline and western protectionism rising – all these pointers produced a sense of gloomy despair which the conference could find no way of alleviating. By this date the concept of a Third World was becoming obsolete. A new group of countries had come into being whose wealth distinguished them dramatically from the world's poor and whose solidarity enabled them to play a forceful role in international affairs. This was the Fourth World of OPEC (Organization of Petroleum Exporting Countries) whose members owed their wealth to oil and their political and economic clout to the fact that, although very different in geographical extent and population, they were few enough and united enough to subordinate their differences to common action. In the 1970s they raised the price of oil so steeply as to make all their customers, rich and poor, tremble

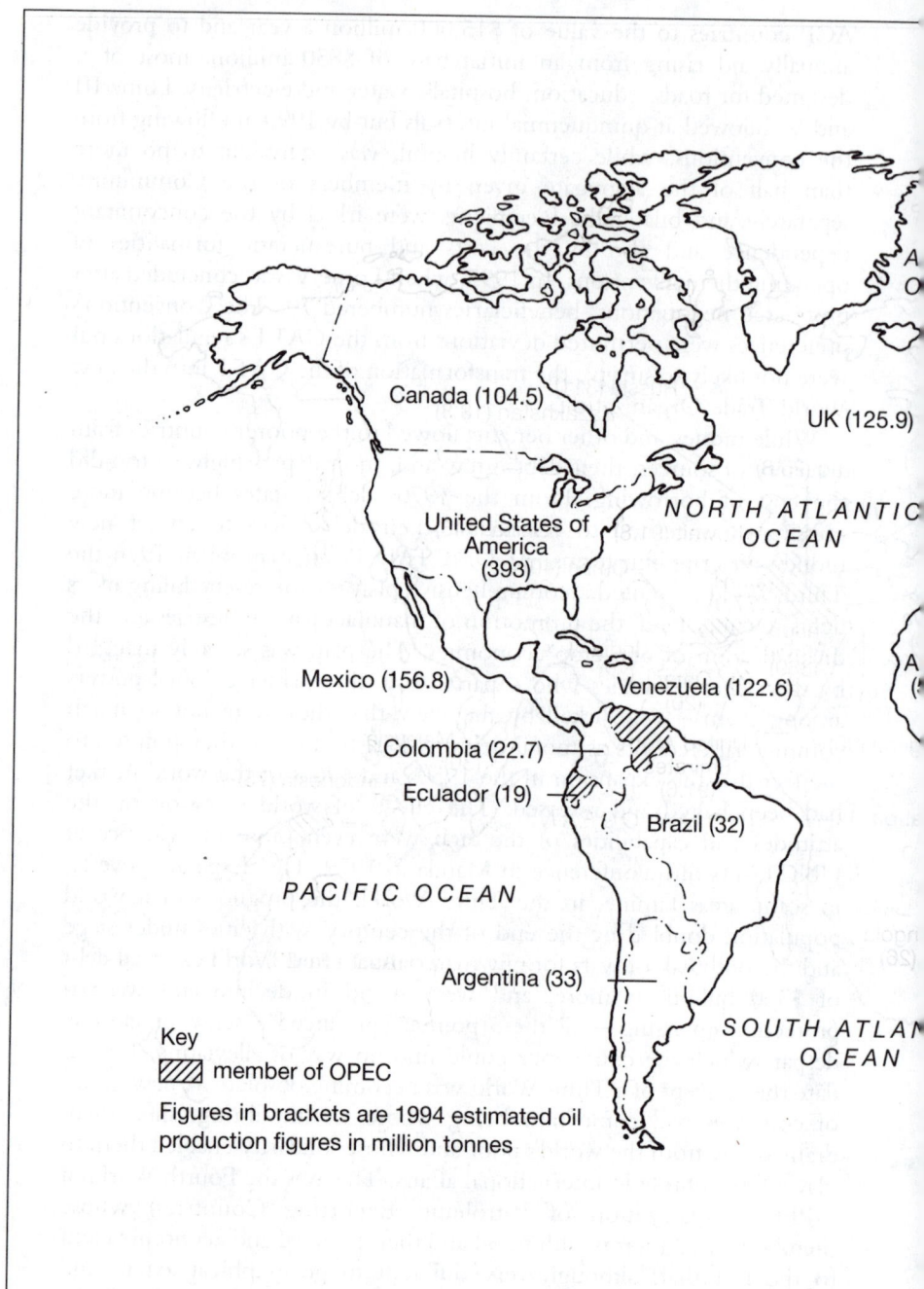

Canada (104.5)

UK (125.9)

NORTH ATLANTIC
OCEAN

United States of
America
(393)

Mexico (156.8)

Venezuela (122.6)

Colombia (22.7)

Ecuador (19)

Brazil (32)

PACIFIC OCEAN

Argentina (33)

SOUTH ATLAN
OCEAN

Key

member of OPEC

Figures in brackets are 1994 estimated oil
production figures in million tonnes

5.1   The world's major oil producers

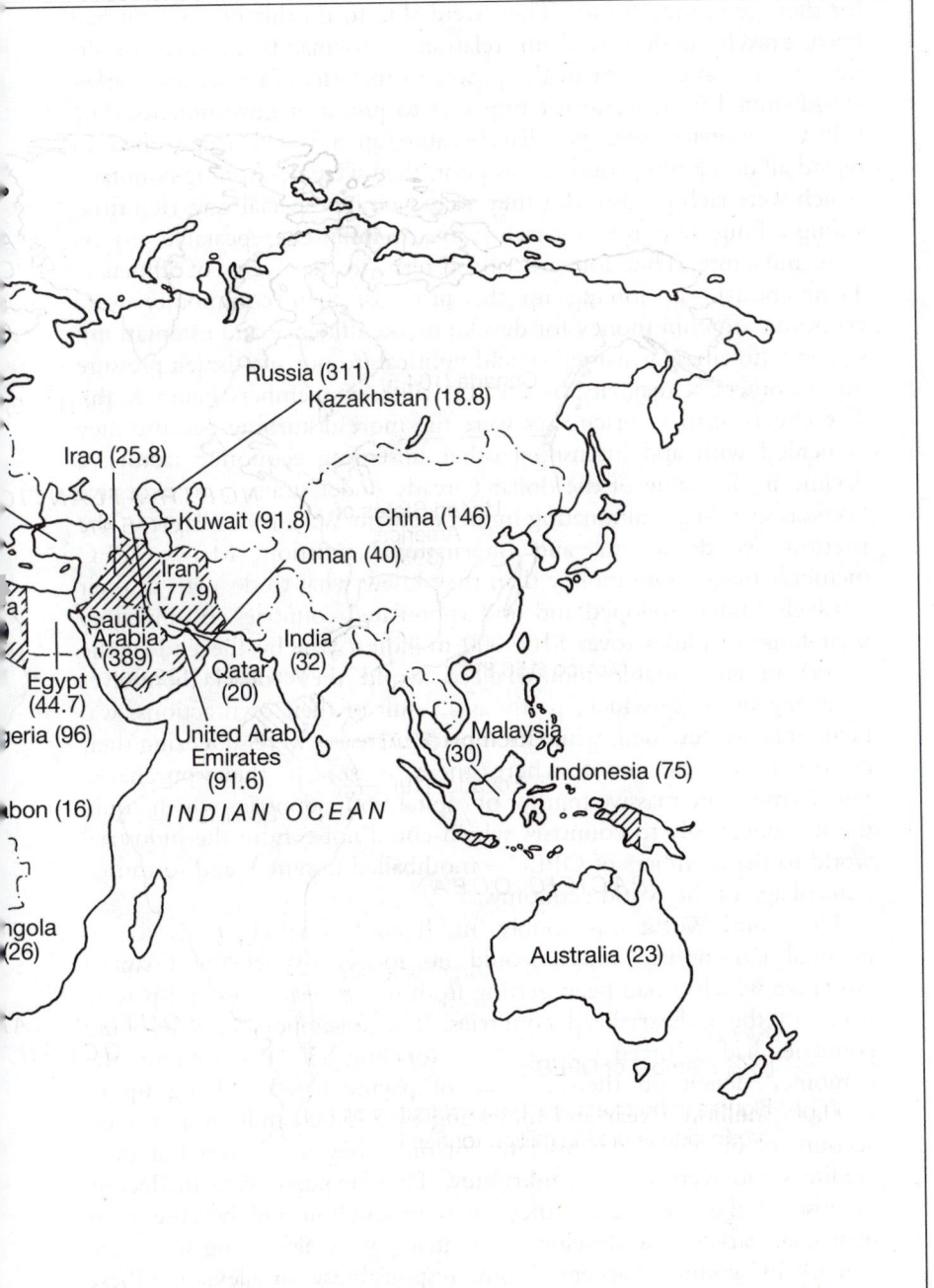

Russia (311)
Kazakhstan (18.8)
Iraq (25.8)
Kuwait (91.8)   China (146)
Iran (177.9)   Oman (40)
Saudi Arabia (389)   India (32)
Egypt (44.7)   Qatar (20)
geria (96)   United Arab Emirates (91.6)   Malaysia (30)
bon (16)   INDIAN OCEAN   Indonesia (75)
ngola (26)   Australia (23)

for their economic future. They were able to do this because oil had been gravely underpriced in relation to demand and because its ownership – and therewith the power to fix prices in a sellers' market – had shifted from western companies to producer governments. The OPEC countries were peculiar because, in a world accustomed to regard all developing countries as poor, they were developing countries which were rich in cash. But they were rich in a special way, rich from selling a finite resource, not rich from the infinitely repeatable process of manufacture. They foresaw a fixed term to their years of affluence. Their motives in forcing up the price of their commodity were economic (to coin money for development, affluence and ostentation); strategic (to amass armaments); and political (to put anti-Israeli pressure on customers, a majority of OPEC's thirteen members being Arab). The effects of these price rises were the more disturbing because they coincided with and intensified other unsettling economic factors: a decline in the value of the dollar (already under strain from excessive overseas spending, culminating in the Vietnam War), the collapse of the Bretton Woods system, and international inflation. Many OPEC members made more money than they knew what to do with in their relatively underdeveloped and underpopulated countries. By investing their huge surpluses (over $100,000 million a year by the end of the 1970s) in an unstable industrialized world they found themselves acquiring property which, partly as a result of their own actions, was being steadily devalued, while their price increases were squeezing their customers to the point where these must cut their purchases. Furthermore the massive transfer of capital from countries which could use it productively to countries which could not – from the industrial world to the members of OPEC – mothballed resources and so caused a shrinkage of the world economy.

The Third World was doubly hit. It could no longer pay for an essential commodity and it could no longer expect the financial assistance which it had been getting from the governments and private banks of the industrialized countries. The oil-importing developing countries had incurred foreign debts totalling $300,000 million; the combined deficit on their balance of payments was edging up to $100,000 million a year and increasing by $25,000 million a year on account of oil alone. At this rate not only they themselves but their creditors too were facing bankruptcy. Their exports were in decline because of the squeeze on their own production and because their principal markets (the developed countries) were also being squeezed. The OPEC countries accepted some responsibility for alleviating these burdens. They did so by the established practice of giving aid to the poor. In the late 1970s this aid was flowing at rates between 1 and 3 per cent of their GNP and three OPEC members touched 10 per cent. Such rates compared favourably with an OECD target of 0.7 per cent (achieved in practice by no OECD member except Norway, Sweden, Denmark and the Netherlands). On the other hand the OPEC total was achieved partly by stopping OPEC contributions to the World Bank (on political grounds) and much the greater part of it was given to the more impecunious Arab countries in the Middle East and northern Africa – Egypt excepted after Sadat's agreements with Carter and Begin at Camp David.

From the point of view of the Third World OPEC's munificence had its limitations. In terms of a proportion of GNP, OPEC aid was comparatively generous. But OPEC's GNP was very much smaller than the OECD's (by a factor of about sixteen), so that the volume of OECD aid remained much larger than OPEC's. Furthermore OPEC's aid programme was no more than one more palliative; and alms-giving, on whatever scale, was no substitute for the reform of a fundamentally unbalanced world economic system which was itself contracting. Givers and receivers – OPEC, the EEC or the OECD on the one hand, the UNCTAD majority on the other – assumed a tussle, an adjustment of conflicting interests and divergent fortunes. Theorists had been talking for decades about the interdependence and mutual interests of rich and poor without getting much of a hearing but in 1980 an unofficial group of public personalities under the chairmanship of Willy Brandt produced a report which attempted to restate the problems and propound practical remedies.

The Brandt Report attempted to persuade its readers that the so-called north–south conflict between rich and poor was as dangerous for the whole world as the more obvious east–west conflict between the armed camps of the superpowers. Given the catastrophic properties of nuclear weapons this was an exceedingly difficult point to make. It was common knowledge (of the kind that most people prefer to forget) that millions of poor people were dying because of the world's extreme economic imbalance: some 800 million were living below the harshest poverty line. But it was not at all evident that this constituted a danger as well as a disgrace. The Brandt Report did not set out to repeat familiar humanitarian pleas; it set out to give them an urgent practical turn, to add scare to shame. In the course of doing this it made implicitly a central assertion of great moment by arguing that the common interests of rich and poor were more potent than what divided them. It abandoned UNCTAD's essentially adversary stance. It did not speak for the poor against the rich but for everybody, on the assumption that everybody was sliding down the same slippery slope. It alleged a community of interest running through the entire roll-call of sovereign states.

The report then set about squaring a circle with a tri-cycle. In brief it said: let OPEC's plutocrats lend their surpluses to the poor; let these and other funds be used to end famine in the Third World and develop its agriculture and industry; help the Third World, rescued from stagnation, to buy the manufacture of the industrial world; and help the industrial world, reaffirming its faith in free trade, to expand its business with the Third World (the source of much of its profits in colonial days) and so find the money to pay OPEC for its oil.

At the core of the Brandt proposals was a massive transfer of resources to the poorer countries by the new rich and the old rich of the order of $50,000 million a year by 1985 (at 1980 prices) – an increase in official aid of $8,000 million a year so that this aid would be equivalent to 0.7 per cent of the GNP of the contributing countries by 1985 and 1 per cent by the end of the century. There were two major obstacles to the acceptance of this statistically modest recipe. In so far as it maintained that it would pay the rich to help the poor it seemed to affront common sense. Secondly, it presupposed something like world government, of which there was not the faintest sign, for it was virtually inconceivable that a programme of this nature could be initiated and pursued by a committee or conference of representatives of sovereign states lacking the executive authority to take decisions, to decree action and to tax. The Brandt Report required a politically fragmented world to tackle universal economic problems, for which it did not appear to have the requisite institutions. It was an exercise in persuasion, unbacked by authority or power.

The problem nevertheless persisted much in the terms of the analysis made in the Brandt Report and every improvement in the world's economy enticed the poor into more hopeless positions. Thus a short boom in the rich world in the early 1970s stimulated, as had the Korean War, demand for primary products but when the demand petered out with the boom the producers of these products in the poor world found that they had been seduced into putting more eggs into leaky baskets. The oil price rises of the 1970s, which did not cause but did aggravate the collapse of the boom, forced the poor to borrow, not for investment, but simply to keep afloat. The rich lent money openhandedly as banks, flush with money from oil producers and encouraged (notably in the case of United States loans to Latin America) to lend and so foster the private capitalist sector, searched for borrowers – to whom they might lend at rates far higher than they were paying to their own depositors. So long as this pattern persisted rich banks and other financial institutions got richer still but by the 1980s the debts of the poor to the rich had become so obviously unpayable that one or two statesmen said so openly: President Nyerere in so many words and President Garcia of Peru by unilaterally limiting repayment of Peruvian debt to a fixed percentage of national income. The rescheduling of debts became a pretence, obscuring the fact that these were bad debts of such magnitude that the risk of insolvency was threatening creditors as well as debtors.

Insolvency – particularly the insolvency of states which were halfway between poverty and riches and had borrowed vast sums in order to accelerate the transition and in the expectation of being able to finance these loans out of oil revenues which disappeared with the slump in oil prices – enforced new measures. In 1982 Mexico defaulted on its interest payments. Such a default, and such an example, threatened disaster for many major banks. The IMF lacked the resources to cope with a crisis on this scale but it took the lead in co-ordinating a relief operation with funds contributed by the threatened banks, and it went on to deal with dozens of other indebted states, funnelling fresh funds to them from commercial banks in return for the adoption of rigorous economic reforms. The IMF thereby tided over a crisis. Besides lending sums of its own it extracted five or six times as much money from banks. But the debtors, who discovered that even states forced to beg cannot choose, were obliged to impose domestic economic programmes which, by raising prices and restricting social services, bore heavily on their poorer citizens and which purchased respite rather than lasting relief. During the 1980s African debt (excluding South Africa and the northern fringe) increased by more than $7,000 million a year and passed $200,000 million; in a number of years payments to the IMF exceeded receipts. A plan to convert debt into long-term IOUs found favour with creditors who knew that they would never get their money back but preferred to postpone payment indefinitely rather than write the debts out of their balance sheets. Latin American debt was twice as large as this African debt, with Brazil and Mexico each owing (in round figures) $100,000 million and Argentina $70,000 million. The Third World as a whole, in spite of receiving $50,000 million (net) a year from the World Bank, was piling up $30,000 million of new debt every year: the total exceeded $1,300,000 million. So far from developing, its economies were growing by 1 per cent and its population by 2 per cent.

In 1985 the Secretary of the US Treasury, James Baker, proposed that creditor banks should advance a further $20,000 million and that international institutions should make similar new loans – that is to say, lend money to enable debtors to meet the interest on old loans. This idea was elaborated by his successor, Nicholas Brady, who devised a scheme whereby creditor banks might choose between three ways of relieving their debtors: by exchanging existing bonds for new ones at

65 per cent of their face value, redeemable after thirty years and paying the original rate of interest; exchanging existing bonds for new thirty-year bonds with the face value unchanged but interest reduced and fixed at 6.25 per cent throughout the thirty years; and fresh loans payable in annual instalments over three years. A prototype agreement under the Brady plan was signed with Mexico in 1990.

NOTES

A. VERY SMALL STATES

There were in the world by 1985 over 40 independent states with a population below 2 million and another 30 territories of similar dimensions which were geographically distinct but not fully independent. Virtually all the latter were islands or groups of islands. So were most of the former, which included 9 separate states in the Caribbean and 8 in Oceania. Within this category the states and self-governing territories differed widely among themselves in extent, population and wealth, but with comparatively few exceptions they were too weak to defend themselves against predators attracted by their strategic value or by some exploitable asset – the beauty that appeals to tour operators or the remoteness that appeals to crooks. They can be to all intents and purposes bought by foreign mafias in league with local politicians, since a little money suffices to buy the few votes needed to win elections and power. Against the displeasure or apprehensions of bigger states they have no defence, as the invasion of Grenada by the United States in 1983 showed. In the 1960s proposals were canvassed for the invention of a new status for very small territories about to be released from colonial empires into a nominal independence which they lacked the resources, human and material, to defend; but these discussions came to nothing, largely from fear of giving offence to nascent nations. Association with a great power, or regional associations, offered some protection and hope of betterment. A number of islands, most of them in the Pacific, made association agreements with the United States or France (as Overseas Territories); others became associates of the Commonwealth. The South Pacific Forum, created in 1971, embraced by 1990 11 independent states and non-independent but self-governing territories; established links with other international associations such as ASEAN and the EC and with the Specialized Agencies of the UN; and combined with larger states in the much older (1947) and wider South Pacific Commission which included the United States, Australia, New Zealand, Britain and France. By a treaty of 1985 10 Pacific island states, with Australia, New Zealand and Papua–New Guinea, proclaimed a South Pacific Nuclear-Free Zone; ancillary protocols were signed by the USSR and China, rejected by France and 'not accepted' by the United States and Britain. The treaty imposed a ban on the manufacture, testing and deployment of nuclear weapons in the zone; called on the five principal nuclear powers not to test or locate nuclear weapons in it; but refrained from attempting to prevent warships or aircraft from using international waters or airspace.

The principal island states which achieved independence in the South Pacific were: Western Samoa, independent in 1962 after German and then New Zealand rule; Tonga or the Friendly Islands, a nineteenth-century monarchy which became fully independent in 1970 (and did not sign the 1985 treaty); Fiji, independent in 1970 (see below); the Gilbert and Ellice Islands which were separated in 1975 and became independent in 1979 and 1978 as Kiribati and Tuvalu; Solomon Islands, independent in 1978; and the Anglo–French condominium of New Hebrides, independent in 1980 as Vanuatu after considerable internal dissension and fighting. Vanuatu pushed the concepts of state and nation to the limits: it contained 150,000 people speaking one hundred different languages in what was a geographical, but neither a political nor cultural, entity. In New Caledonia, a French Overseas Territory from 1958, conflict between indigenous kanaks and white immigrants who had become the majority was at least temporarily resolved by the Matignon agreement of 1988 which provided for direct French rule for one year, the division of the territory into three regions (two of them with a kanak majority), considerable economic aid from France and a referendum on independence in 1998. For accepting these arrangements the kanak leader Jean-Marie Tjibaou was assassinated by his more militant compatriots.

Fiji had a similar racial problem. Fiji – an extent of land the size of Wales divided into over six hundred islands embracing 100,000 square miles of water and a population of 750,000 – had made unsuccessful approaches in the nineteenth century to Germany and the United States for protection before finally yielding itself in 1874 to Britain. Under British rule Indians were encouraged to migrate to Fiji where they became the mainstay of the sugar industry and, by the time of independence in 1970, a majority of the population. Tensions between the races were supplemented by tensions between the western and eastern islands and between the generations. In elections in 1987 the ruling Fiji party, which rested upon chiefly paternalism, was defeated by a coalition under Sir Timothy Bavadra who assembled a cabinet with a bare Indian majority. With a considerable part of its small army away in the Middle East in UN service Colonel Siliveni Rabuka was able to stage a coup with a handful of men who invaded the parliament and seized the new prime minister. Bavadra appealed unsuccessfully to the British government and crown (Fiji was a member of the Commonwealth), but Rabuka had covert support from the CIA which looked askance at Bavadra as a neutralist. Rabuka had the support of the Tsakei movement for Fiji-for-the-Fijians, but he progressively lost ground, particularly in the eastern islands, and the beginnings of an Indian emigration exposed the dangers of scaring Indians away from the sugar industry which was Fiji's principal economic resource. Nevertheless elections in 1994 confirmed him in power in spite of his reluctance to hold them. Fiji was a member of the South Pacific Forum. Bavadra died in 1989.

The Bismarck archipelago, annexed by Germany in 1874 and administered by Australia first under mandate from the League of Nations and, after 1947,

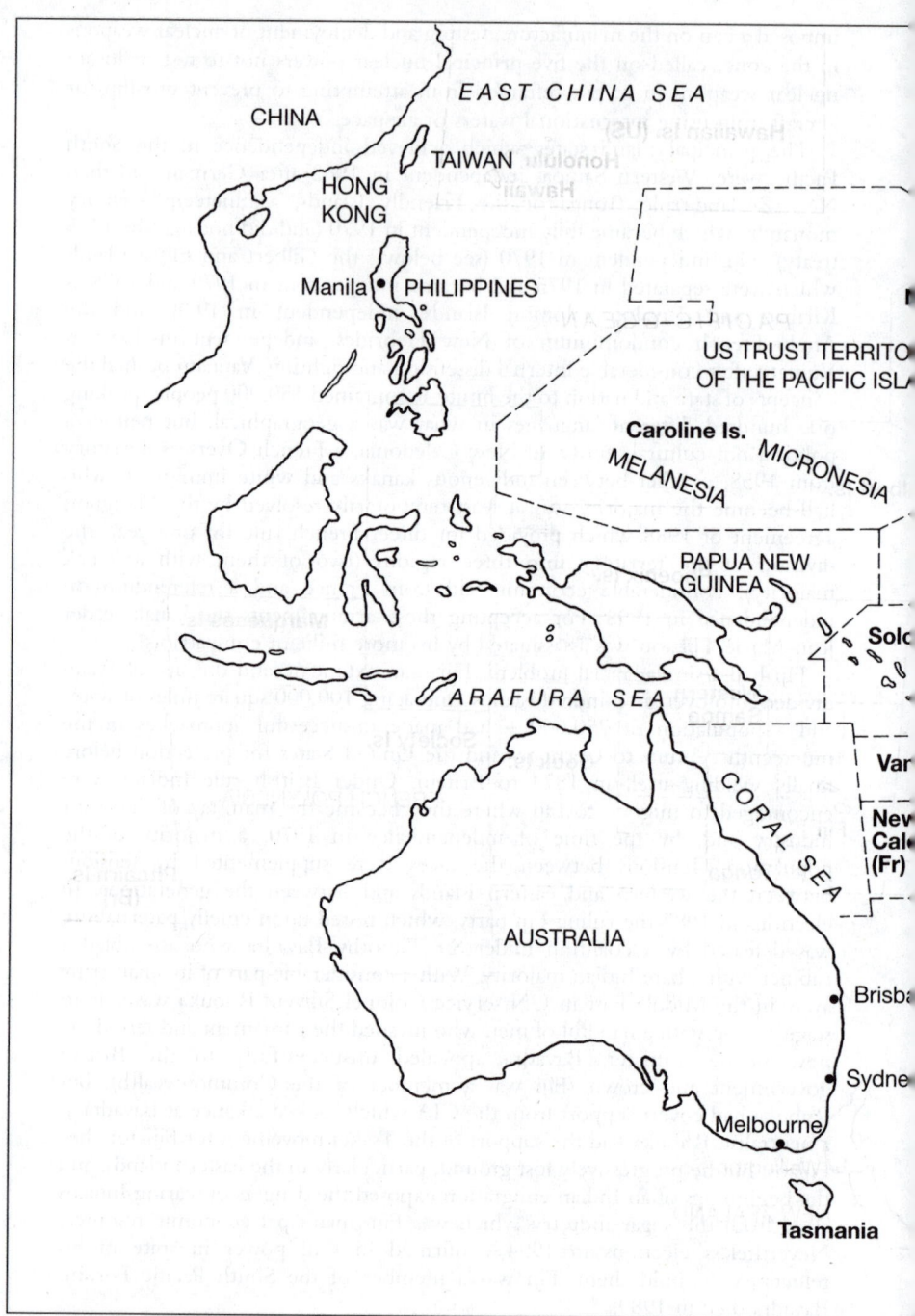

5.2 The South Pacific

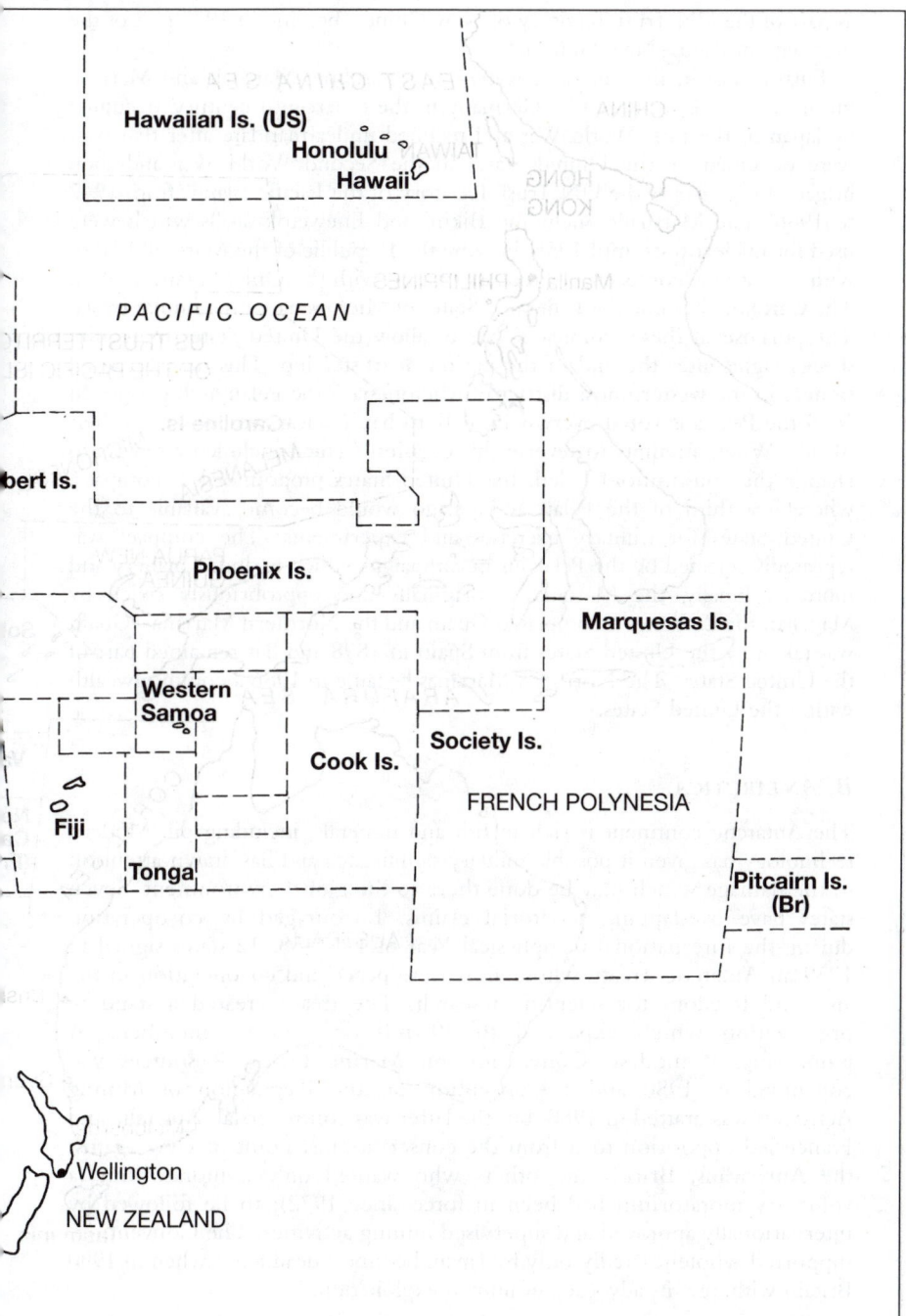

as part of the UN Trust Territory of New Guinea, became in 1975 part of the new state of Papua–New Guinea.

Further north, in equatorial waters, the Caroline, Marshall and Mariana groups of islands, acquired by Germany in the nineteenth century, occupied by Japan in the First World War and retained under mandate after that war, were occupied by the United States in the Second World War and then brigaded together as the UN Trust Territory of the Pacific Islands from 1947 to 1986. The Marshalls, including Bikini and Enewetok atolls which were used for nuclear tests until 1958, became the Republic of the Marshall Islands with a treaty or 'compact' of Free Association with the United States (1982). The Carolines became the Federated States of Micronesia with a similar treaty. The purpose of these 'compacts' was to allow the United States to acquire special rights after the end of the period of trusteeship. This device caused trouble in the westernmost district of Micronesia – the Pelau archipelago. In 1979 the Pelauans voted overwhelmingly to ban nuclear weapons from their islands. When attempts to reverse this decision (which included attempts to change the constitution) failed, the United States propounded a 'compact' whereby a third of the Pelau archipelago would become available to the United States for military exercises and experiments. The compact was repeatedly rejected by the Pelauans in campaigns which included bribery and murders. Finally, the Marianas – originally and opprobriously called by Magellan the Ladrones – comprised Guam and the Northern Marianas. Guam was taken by the United States from Spain in 1898 and has remained part of the United States. The Northern Marianas became in 1986 a commonwealth within the United States.

B. ANTARCTICA

The Antarctic continent is rich in fish and minerals, including oil. Modern technology has given it possible military significance and has drawn attention to the damage which may be done there to the global environment. Seven states have overlapping territorial claims. Encouraged by co-operation during the International Geophysical Year of 1957–58, 12 states signed in 1959 an Antarctic Treaty whose aims were peace and co-operation in the area and freedom for scientific research. The treaty created a standing organization which expanded to 39 full or associate members. A pioneering, if modest, Convention on Marine Living Resources was concluded in 1980 and a Convention on the Regulation of Mining Activities was drafted in 1988, but the latter was controversial. Australia and France led opposition to it from the conservationist point of view against the Americans, British and others who wanted only a moratorium (a voluntary moratorium had been in force since 1972), to be followed by internationally approved and supervised mining activities. The Convention, supported wholeheartedly only by Japan, became a dead letter when in 1990 Britain withdrew its advocacy of limited exploitation.

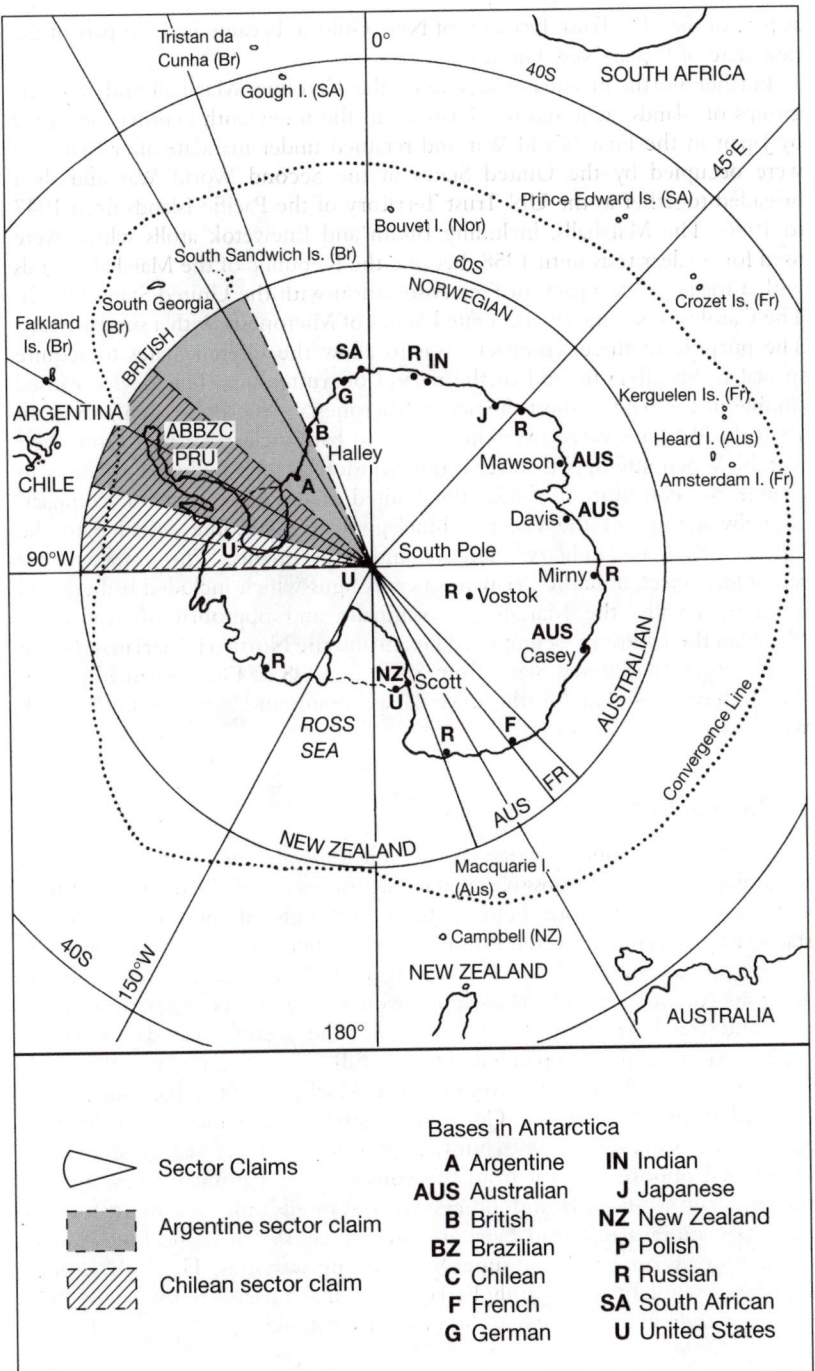

5.3   Antarctica (Source: *An Atlas of World Affairs*, William Boyd)

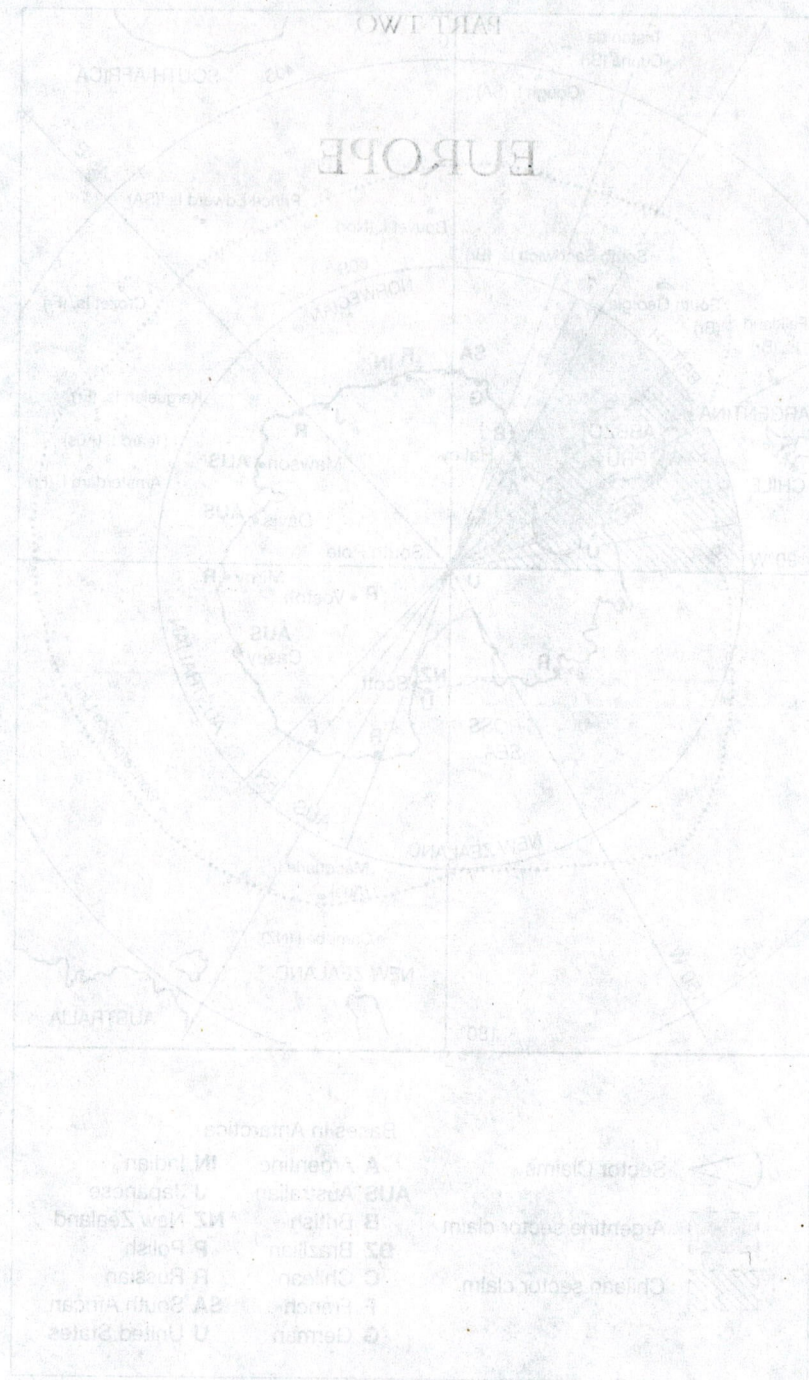

PART TWO

EUROPE

6   WESTERN EUROPE

RECOVERY

Western Europe's recovery after 1945 was unexpectedly robust. Economic recovery, which required the restoration of severely damaged but essentially sound and skilled economies, was powerfully engendered through American financial aid which was itself impelled both by generosity and by fears of the collapse of countries of vital concern to the United States in the Cold War. The Marshall Plan (1947) was, with the North Atlantic Treaty (1949), a crucial factor in the tempo of western Europe's material repair and spiritual reassurance.

When the war ended the countries of western Europe were in a state of physical and economic collapse, to which was added the fear of Russian dominance by frontal attack or subversion. During the war plans had been made for the relief of immediate needs in Europe. The UN Relief and Rehabilitation Agency (UNRRA) was created in 1943 and functioned until 1947: a European Central Inland Transport Organization, a European Coal Organization and an Emergency Committee for Europe were established and merged in 1947 in the UN's Economic Commission for Europe (ECE). These organizations assumed that Europe's ills could be treated on a continental basis, but the Cold War destroyed this assumption and, although the ECE continued to exist and issued valuable *Economic Surveys* from 1948 onwards, Europe became bisected for economic as well as political purposes.

The immediate precursors of American economic aid were the failure of the conference of foreign ministers in Moscow in March and April 1947 and the Truman Doctrine whereby, in March, the United States took over Britain's role of supporting Greece and Turkey and rationalized it in anti-communist terms. In June General Marshall, then secretary of state, propounded at Harvard the plan which bears his name and which offered to all Europe (including the USSR) economic aid up to 1951 on the basis that the European governments would accept responsibility for administering the programme and would themselves contribute to European recovery by some degree of united effort. This American offer required the creation of a European organization; the Russian refusal of the offer, for the USSR and its dependants, turned the organization into a western European one. Sixteen countries established a Committee for European Economic Co-operation which assessed their requirements in goods and foreign exchange for the years 1948–52 and was converted in April 1948 into the more permanent Organization for European Economic Co-operation (OEEC). Western Germany was represented through the three western commanders-in-chief of occupation forces until October 1949 when German representatives were admitted. The United States and Canada became observer members of the organization in 1950 and subsequently co-operation was developed with Yugoslavia and Spain. At the American end the Foreign Assistance Act of 1948 created the Economic Co-operation Administration (ECA) to supervise the European Recovery Programme (ERP). In the following years the OEEC, using American funds, became the principal instrument in western Europe's transition from war to peace. It revived European production and trade by reducing quotas, creating credit and providing a mechanism for the settlement of accounts between countries. While it was a government-to-government and not a supra-national organization, it nevertheless inculcated international attitudes and fostered habits of economic co-operation which survived the ending of the ERP. (It was replaced in 1960 by the Organization for Economic Co-operation and Development – OECD – in which the United States, Canada and Japan were full members and which extended the work of the OEEC into the developing areas of the world.)

The establishment of the OEEC coincided with the signing in March 1948 of the treaty of Brussels by Britain, France, Belgium, the Netherlands and Luxembourg (the last three compendiously referred to as Benelux from the time when they formed a customs union in 1947). This treaty, like the Anglo–French treaty of Dunkirk of 1947, was a military alliance ostensibly directed against a revival of the German threat. It contained in addition provisions for political, economic and cultural co-operation through standing committees and a central organization, and it was also seen by at least some of its promoters as a first step towards a yet broader military alliance with the United States. President Truman, speaking of the need for universal military training and selective military service in the United States, so interpreted it and the leader of the Republicans in the Senate, Arthur H. Vandenberg, proposed and carried a motion in favour of American aid to regional military organizations which served the purposes of American policy: the senator was in essence advocating a military pact between the United States and western Europe, a military counterpart to General

Marshall's economic plan. The North Atlantic Treaty, signed in April 1949, by the United States, Canada and ten European countries, gave the latter for at least twenty years a guarantee of their continuing independence and integrity against Russian attack by formalizing and institutionalizing the American intention to remain in Europe and play the role of a European power. At this date the Russians, like the Chinese fifteen years later, had large and frightening land forces which weighed heavily on all those within their reach but lacked a diversified, modern armament capable of engaging the United States. The North Atlantic Treaty was therefore a way of bringing American air power, including nuclear weapons, to bear in order to inhibit the use of Russian land forces in the area designated by the treaty.

The European members of this new alliance were at first comparatively passive beneficiaries who, in spite of the fact that they provided 80 per cent of its forces in Europe, were dependent on the far more significant American contribution, without which their own contribution was irrelevant to their main needs and fears. Although in terms the treaty was a collective security arrangement, in fact it was, to begin with, more like the protectorate treaties of an earlier age whereby a major power had taken weaker territories under its wing. The treaty created a permanent organization (Nato – the North Atlantic Treaty Organization) for political discussion and military planning and some of its makers and later devotees envisaged the growth of something more than a military alliance – an entente or community or union. But nothing of the kind emerged for a variety of reasons: the enormous disparity between the power of the United States and any other member, the failure of the European members to coalesce into a political unit commensurate with the United States, the breadth of the Atlantic Ocean, the unquestioning addiction of Americans to the sovereignty which seemed to them old-fashioned in others, the revival of European power and confidence, and the waning of the Russian threat halfway through the life of the treaty.

For almost half a century Nato was a principal instrument in the Cold War. Western Europe, with the Atlantic and Mediterranean seas, was its theatre of operations. The European members of the alliance provided the bulk of its forces, the Americans the bulk of the equipment and most of the money. The Europeans were careful to avoid any form of military association among themselves which might seem to weaken the Euro–American link. They were also considerably more reluctant than the Americans to acknowledge that an anti-Soviet alliance entailed an end to hostility with Germans and the integration of West Germany into the alliance. This step – the most portentous in Nato's history and one which was to cause some confusion when the Soviet menace all but disappeared in the 1980s – was precipitated by events thousands of miles away in Asia: the Korean War. Within little more than a year after the signing of the North Atlantic Treaty the outbreak of this war created fresh and substantial calls on the United States' resources and the fear of similar hostilities in Germany. Washington became therefore anxious to convert its allies from passive protégés into junior partners and to build up in Europe itself a counterforce to the Russian armies, distinct from the long-range American air power which, although based in Europe, was under exclusive American command and remained so even when a joint Nato command was created. The allies, however, were still weak. Britain and France, with much of their forces committed outside Europe, could give little help immediately and they were therefore the more easily constrained to accept an American emergency decision to rearm the Germans. The anti-Russian alliance feared by Moscow between two world wars now took shape and at the end of 1950 General Eisenhower returned to Europe as supreme commander of another grand alliance. In the same year Greece and Turkey were invited to co-operate with the allies in the defence of the Mediterranean, although they did not become full allies until early in 1952: their co-operation helped to establish an eastern flank to protect the allied central sector and to threaten the USSR from the south. Early in 1951 a new headquarters came into operation (Supreme Headquarters, Allied Powers in Europe – SHAPE) and a year later, at Lisbon, the Nato council approved a plan to endow this command by 1954 with ninety-six active and reserve divisions, including fifty in the central sector, and 9,000 aircraft. Although these targets were never attained, the Lisbon decisions gave the alliance the shape which it ever afterwards retained, later debate focusing on strategic doctrine, new weapons and their deployment.

With one exception. In 1952 the German problem – that is to say, the status of West Germany as a political entity and its role in Nato's planning and operations – was still unresolved. With the Korean War American pressure to accelerate West Germany's sovereignty and therewith West German rearmament became irresistible, but France in particular was anxious to find a way to prevent the resurgence of autonomous German military power. René Pleven, the French minister of defence, proposed that German units be raised and incorporated in multinational divisions but that western Germany be allowed no separate army, general staff or defence ministry. Adopting the pattern of the European Coal and Steel Community, which had been launched on French initiative and was about to come into existence, Pleven devised a European Defence Community (EDC) with a council of ministers, an assembly and a European minister of defence. The French aim was to minimize the German military unit and at the same time to integrate the German military contribution, both operationally and politically, in an international organization. British participation was all but essential since without it the proposed international organization would consist only of France and Germany with some comparatively trivial makeweights. For France a British commitment was the only way adequately to offset the risks inherent in the rearmament of Germany and the reappearance of a sovereign German state, but no British commitment satisfactory to France was forthcoming during the four years in which the EDC was under debate.

In May 1952 agreements signed in Bonn and Paris created a complex new structure: six continental European states signed a treaty creating the EDC, the three western occupiers of Germany agreed to end the occupation upon ratification of the EDC treaty, and both Britain and the remaining Nato powers entered into separate ancillary treaties promising military aid in the event of an attack upon any of the EDC partners. But the French remained uneasy and undecided. They wanted British membership of the EDC and not a pledge to help it, and they disliked the provision in the EDC treaty permitting the raising of whole German divisions in place of the smaller units proposed by Pleven's plan for incorporation in international divisions. The United States and Britain brought pressure to bear on France, the former by threatening an 'agonizing reappraisal' of American policies if the EDC treaty were not ratified (which was taken to mean the cutting off of American aid to France) and the latter by giving in 1954 a further pledge of military and political co-operation with the EDC (a pledge echoed by President Eisenhower within the limits of his constitutional competence). In August of that year the French parliament finally came to a vote and refused, by 319 votes to 264 on a procedural motion, to debate the ratification of the treaty.

With this vote the EDC and all the Bonn and Paris agreements of 1952 collapsed. There was anger in Bonn, where Adenauer insisted that western Germany must have sovereignty none the less, and in Washington, where Dulles decided ostentatiously to cut Paris out of a tour of European capitals. In London, more constructively if belatedly, Eden set to work to put the pieces together again by diplomatic labours and a more specific pledge than Britain had so far been willing to vouchsafe. By the end of the year the Brussels treaty of 1948 had been expanded to take in the German and Italian ex-enemies and renamed Western European Union (WEU); this WEU took over the non-military functions of the Brussels treaty organization and became militarily an ingredient in Nato; Britain declared that it would maintain on the continent forces equivalent to those already committed to the Supreme Allied Commander in Europe (Saceur), that is, four divisions and a tactical air force; the occupation of western Germany was ended; Adenauer undertook not to produce atomic, bacteriological or chemical weapons, long-range or guided missiles, bomber aircraft or warships, except upon the recommendation of Saceur and with the assent of two-thirds of the council of WEU; West Germany was to become a full member of Nato, and did so formally the following year. One other loose European end was clutched. France and western Germany agreed that the Saar, which France had hoped ever since 1945 to annex in one form or another, should constitute a special autonomous territory embedded in WEU, but the Saarlanders rejected this arrangement by plebiscite in October 1955 and the Saar became a part of western Germany at the beginning of 1957. Thus the first postwar decade closed with Nato in existence to extend American protection over western Europe, Britain as the firmest and most effective of the European members of the alliance, and a nascent German state back in the comity of western Europe.

The weak point was *France*. Whereas Britain had made an energetic recovery from the war and West Germany was on the verge of its economic miracle, France was relapsing into the political instability which had characterized it between the wars, accompanied by economic ineptitude and colonial overstrain. But France too was on the verge of economic reform and advance; was about to shed intolerable imperial burdens; and, with a radical political somersault, to find in Franco–German partnership a new base for French activity in Europe and beyond.

France had been a major European land power and a major imperial power but had failed in the contest with England for sea power. In the nineteenth century the decline of France's position in Europe had been matched by the acquisition of a second overseas empire to replace the territories lost to Britain in the wars of the eighteenth century, but by the beginning of the twentieth century France, slipping back in the demographic and the industrial race and spiritually still divided between the heirs of the Enlightenment and the Revolution and those who accepted neither, was becoming discouraged and unnerved and unresponsive to central government. The awful sacrifices of the First World War and the no less awful humiliation of the Second, separated by incapacity to face up to the problems of the economic crisis or to Hitler's challenge to basic values, brought France low in its own eyes until the exploits of the Resistance and the leadership of de Gaulle revived and personified the French spirit: de Gaulle's identification of himself with France and his constant use of the first person singular were precisely what was needed after the physical and spiritual lesions of a century.

When the war ended the French tried to strike out into a new world with some of the trappings of the old until they found that this would not work. They adopted a constitution and political methods unhappily reminiscent of the defunct Third Republic, made great efforts to retain or recover their empire in Asia and Africa, tried the old game of weakening Germany permanently, and made treaties with their traditional British and Russian allies. But they also revolutionized their foreign policies by joining the anti-Russian western alliance even though it entailed the rearming of Germany, took the lead in devising new political structures suitable to Europe's altered place in the world, accepted the end of empire and – most important – adopted under the lead of Jean Monnet a successful form of central economic planning for the modernization of industry and agriculture. During the war national production had fallen by 65 per cent but within two years of liberation in 1944 and before the inception of the Marshall Plan, 90 per cent of this loss had been recovered. From 1947 the series of Monnet Plans, elaborated and applied by a modest central government department and using Marshall funds mainly for the reconditioning or transforming of industries, allocated resources, determined priorities and master-minded the restoration of the economy sector by sector in association with state-owned enterprises and private businesses. From the date of the second plan (1952–57) socio-economic sectors such as education, research and training were included and led at the next stage to central direction of social policies and a welfare state financed by employers and employed.

The renunciation of empire brought France, over Algeria, to the brink of civil war, from which it was saved in 1958 by the return to power of de Gaulle and the thwarting thereby of a right-wing military plan to seize control of the capital and the state. De Gaulle tamed the generals and colonels, disposed of the politicians and the remaining colonies and, profiting from a rapidly improving economic situation, raised France from a position of pity and scorn to one of independence and attention.

De Gaulle inherited two recent decisions of his predecessors – the decision to become a nuclear power and the decision to join an economic community with avowed political implications (the EEC – see the next section). The first of these decisions was congenial to him. He believed that France could be a major power and he believed that power must be modern. Just as he had been an expert in tank warfare when many of his colleagues were still in favour of the horse, so now a generation later he held that there was a choice between nuclear power and no power, and he held too that beyond a certain point there was little difference between one nuclear power and another: a nuclear power which reached that point became a member of the first league even if it possessed fewer or less sophisticated weapons than other members of the league. The second decision may have been less congenial to him, not so much because he eschewed all unions or harboured antiquated notions about the ability of a country like France to go it alone, but rather because his ideas on the nature of useful unions were different from those of the authors of the treaty of Rome. While aware that the independent states of Europe were no longer what they had been (even if, by becoming nuclear powers, one or two of them could stand up for themselves in the exceptional circumstances when the nuclear argument is brought into play), he did not believe that the minds of Europeans had become supra-national. In his view the vast majority of Europeans still responded to the idea of the nation and he therefore based his European policies on the nation (but not necessarily national sovereignty) with, however, the reservation that the base was a starting point and not a permanently limiting factor. De Gaulle's *patrie* was not identical with the state (*état*) and might be fitted into a partnership or association which recognized and protected national identities. Further, de Gaulle's pragmatic temper led him to insist on the differences between more and less powerful states and held that any association should be governed or guided by a directorate composed of the former – in this case either France, western Germany and Italy, or France and western Germany. Equality between states, with its corollary of one-state-one-vote, he regarded as a vicious pretence, whether portended in the EEC or practised in the UN General Assembly. (The directorate of the five permanent members of the Security Council, however, fitted his theory, subject to putting the right Chinese delegate in the Chinese seat.)

De Gaulle also inherited a position which he found intolerable in a Nato which he found anachronistic. Following his doctrine of directorates, Nato should be directed by the United States, Britain and France, but it was in his view dominated by the United States with a touch of special British influence achieved by British subservience to American policies. Shortly after his return to office in 1958 de Gaulle tried to establish with the United States and Britain a triumvirate within Nato, but his ideas were rejected on the grounds that the formation of an alliance within an alliance would lead to the loss of the other allies. Washington and London, moreover, underrated France at this stage: they derided French nuclear ambitions, did not see how France was to dispose of its Algerian incubus, and failed to see that the status of France was changing and would change more rapidly in the near future.

De Gaulle not only desired the revival of France: he rightly saw that it was happening. He also saw that Europe was changing. The Cold War was over. It might be renewed one day, but for the present the fear of Russian aggression which had prompted the creation of Nato was fading fast. It followed that the Americans could not be expected to stay in Europe indefinitely. Perhaps they would be gone by 1980. Again de Gaulle, taking a long view and a clear view, was probably right, since the development of weapons technology was converting the American presence in Europe from a strategic disposition into a political gesture, and a political gesture is more easily abandoned than a strategic position, especially if, as began to happen after 1960, difficulties with their balance of payments should cause the Americans to ask what their forces were really doing in Europe. Europeans were coming to doubt the automatic immediacy of an American response to a Russian attack since American cities first came under direct threat from Russian intercontinental missiles: either the Russians would not attack or they would do so in a way calculated to avoid an American response.

Immediately after de Gaulle's return Dulles offered France nuclear weapons in return for the right to put launching sites in France. De Gaulle refused and in 1959 withdrew the French contingent from Nato's Mediterranean fleet. He refused to be ruffled by the Berlin or the Cuban crises and was strengthened by both in his view that there was no pressing Russian danger. In 1962 when Kennedy offered France as well as Britain nuclear weapons he again refused and in 1966 he withdrew French forces from all Nato commands. This policy struck some responsive chords in the rest of Europe. The waning of the Russian threat and of the fear of economic collapse, the achievement of the prime purposes of Nato and the Marshall Plan, gave Europeans a new confidence which they translated into a desire to run their own affairs (notwithstanding that for the most part they already did so). Since western Europe had been saved by the Americans from having its affairs run by the Russians, the new mood was anti-American, for it was the Americans and not the Russians whose presence affronted a new nationalism which was further sharpened in the 1960s by resentment and alarm at American economic penetration, the debit side of the American investment which gave Americans control over European enterprises and so over the hiring and firing of labour. De Gaulle's anti-Americanism, rooted in Roosevelt's partiality for Vichy and right-wing French generals and admirals during the war, was not out of tune with Europe's mood – until he gave the impression that he wanted to put an end to the American alliance altogether. For this western Europe was not prepared. The Cold War might be waning but it was not over, it might recur, and so long as there was a doubt there had better be an alliance. De Gaulle himself affirmed the need for an alliance more than once, but his desire to see the Americans at a distance from Europe created the impression that the alliance was itself in the Gaullist view expendable.

France in the 1960s was ready to relinquish, or at any rate relax, an alliance which it had adopted in the 1940s out of economic necessity. The change in France's economic circumstances was at least as important as its change of ruler in 1958 in producing a change in policy. De Gaulle was not untypical of Frenchmen and many other Europeans in wishing to diminish political and strategic dependence on the United States as soon as economic dependence was no longer a fact. At the end of the war de Gaulle and other leading French politicians had wanted France to adopt an intermediate position between the United States and the USSR, but the onset of the Cold War and French military and economic weaknesses forced the French government in 1947 to make a choice and to choose the American side. The need for money and for food determined French policy. The Marshall Plan offered salvation and France took it. But – like the Americans themselves – they saw the programme as a short-term rescue operation and – unlike the Americans – assumed that the consequential alignment would also be reviewed at the end of the short term.

This postwar drift from an intermediate to an aligned policy was facilitated by, and also to some extent a cause of, the dropping of the communists from the government. From their role as national heroes and active partners in the resistance to the Germans the communists had reverted to a suspect, sectarian position in which the interests of Moscow counted for more than national unity and regeneration. Even before the Marshall Plan and its rejection by the USSR their continuance in the government had become next to impossible on account of the Truman Doctrine on aid to Greece and Turkey for the defeat of communism, the decision to abandon discussions with the Viet Minh and fight it, the stern suppression of revolt in Madagascar and of strikes in the nationalized Renault works, and a wage freeze; communists had already been dropped from the Belgian and Italian governments earlier in 1947. At the end of that year the French communists tried to exploit politically serious strikes which had genuine economic sources in the financial policies of the government of Paul Ramadier, but they failed, their representatives were dismissed from the government and the party lapsed into a long period of opposition. Political power shifted to the right and even when it swung back leftward in the mid-1950s it did not re-embrace the communists and France remained for a decade an acquiescent member of the western alliance. By the 1960s France was ready to reconsider its role, and it so happened that, owing to the coup of 1958, its gradual disengagements from the American alliance took place not at the instigation of communists (who remained outside the pale of government) but under the guidance of de Gaulle.

A fourth principal element in de Gaulle's heritage in 1958 – along with France's nuclear programme, the treaty of Rome and membership of Nato – was the *rapprochement* with Germany. The principal architects of this *rapprochement* were Robert Schuman and Jean Monnet on the French side and Adenauer on the German. Adenauer proposed in 1950 a Franco–German union, to which Italy and the Benelux states and possibly Britain too might adhere. De Gaulle, then in retirement, welcomed the idea, and ten years later he turned it to good account by concluding a Franco–German treaty at a time when his relations with the EEC and Nato groups were strained. De Gaulle's return had coincided with a weakening of German–American relations. Adenauer's political attitudes were markedly personalist and the death of Dulles had removed the principal bond between him and Washington. He was moreover suspicious of the Camp David spirit and Eisenhower's attempts to find points of agreement with Khrushchev. Towards Britain Adenauer's feelings had been cold since, after the war, a British officer had found him unfit to be mayor of Cologne in spite of the fact that he had held that office continuously from 1917 to 1933. He did not like Macmillan, was caustic about his attempt to play the role of mediator between Washington and Moscow, and was angered by his visit to Moscow in 1959 to discuss a European settlement – affecting above all Berlin – without prior notice to Britain's German allies who were more closely affected by such a topic than anybody else. Adenauer also resented Britain's aloofness from the EEC and its attempt to block progress by forming EFTA. He was ready to turn to France and, after some initial hesitation, to find a new personal friend in de Gaulle.

In 1959 Adenauer, after ten years as chancellor and now 83 years old, toyed with the idea of accepting the West German presidency. For a man who was Stresemann's senior and had been considered for the chancellorship of the Weimar Republic in 1921 and 1926 the end was approaching and his colleagues considered that the time had come for him to retire to a less active post. But for Adenauer the transfer was only palatable if it were to be accompanied by a transformation of the presidency from an ornamental into an executive office. He was willing to be a president like Eisenhower or de Gaulle but not a president like his own predecessors or his Italian neighbour. When it became apparent that his compatriots did not relish a presidential democracy he decided to remain chancellor. In elections in 1961 his party, the Christian Democratic Union, lost its absolute parliamentary majority and in the ensuing inter-party negotiations for a new government Adenauer was forced to accept a conditional fourth term of office as chancellor, the condition being that he would retire in 1965 at the latest. During meetings with de Gaulle at Rambouillet in July 1960 and in Paris in July 1962 – the latter was followed in September by a triumphant tour of western Germany by de Gaulle – Adenauer opted for a continuing Franco–German understanding in spite of misgivings about de Gaulle's version of European integration and de Gaulle's opposition to a European political union and the inclusion of Britain in the EEC. By 1962 he was further disillusioned with the United States under the new Kennedy administration and with Macmillan's devious approaches to the EEC, and in January 1963 he signed with de Gaulle a treaty which formalized the Franco–German entente and sought to make it the core of European politics, a working alternative to or brake upon Nato, the EEC, the Anglo–American partnership, an American–Russian *rapprochement* or the American–German entente of the 1950s. This treaty was a peak in de Gaulle's diplomacy but he did not remain on it, for the accord was mistimed. Adenauer himself was on his way out and his immediate successors were unenthusiastic about this, among others, of his personal achievements. The Franco–German treaty became almost at once a dead letter, or at best a pointer of uncertain direction and import. De Gaulle failed to achieve the changes which he desired in the way Nato functioned at the top. He was credited with the belief that Britain was Washington's Trojan horse within Europe's walls, but he would have been more correct to see Bonn in this role, for whereas Britain in its chronic ill health was dependent on the United States economically, West Germany remained dependent on the United States for its defence.

By the late 1950s problems over the distribution of power and weapons within the alliance were acknowledged and produced some bizarre schemes. De Gaulle's own proposals in 1958 for a Nato directorate comprising the three powers with extra-continental interests had been made in response to an American request for ideas about this problem, but they found no favour in Washington or London where they were regarded as no more than a French claim to equality with the United States and Britain. A year later, after the installation in Europe of IRBM (Intermediate Range Ballistic Missiles) under a dual control or 'two keys' system, the Supreme Commander, General Lauris Norstad, stressed the need for a multinational nuclear authority, and in 1960 the United States proposed to instal in Europe 300 mobile Polaris missiles on road, rail and river under American control. De Gaulle, asked to accept fifty of these, said he would do so only if France produced its own warheads, thus making French control a condition of acceptance, whereupon the Americans dropped the plan. The Russians objected that it entailed the provision of nuclear weapons for Germany.

These discussions, abortive though they were, showed that Nato could not go on for ever on the basis of an American nuclear monopoly. Either the Europeans would themselves produce a deterrent force roughly equivalent to the American nuclear contribution to the alliance and so turn it into a more equal partnership, or some way must be found of creating an American–European nuclear force. The first solution – a distinct European force – presupposed a European political authority to control it, and although Europeans might have liked to have such a force, they showed no signs of evolving the necessary political authority. The solution must therefore lie on

American–European lines, and there were two schools of thought, the multinationalist and the multilateralist. The multinationalists accepted national sovereign control and aimed at no more than the retractable commitment of national forces to a Nato commander, together with increased participation by all the allies in strategic planning and political consultation. The multilateralists devised a scheme for mixed forces in which nuclear weapons would be operated by units whose personnel would be drawn from different states. The American administration adopted multilateralism in 1962 not long before the British and French governments demonstrated their continuing addiction to, in the British case, multinationalism as the Americans understood it and, in the French case, a multinationalism which excluded the Americans. Since, however, the Americans hoped that multilateralism would provide the answer to the German question – how to give the Germans a satisfactory share in nuclear operations without alarming the Russians – they persisted with it despite the opposition of their other principal allies. They proposed in March 1963 a multilateral force (MLF) of twenty-five mixed-manned surface vessels, each carrying eight Polaris missiles, three-quarters of the cost to be paid by the United States and West Germany.

Western Germans welcomed the scheme as a means to restore the close relations with the United States which had characterized the 1950s; alarmed by the American–Russian *rapprochement* which produced the test ban treaty of 1963, they saw in the MLF a way of securing a special position equivalent to the possession by Britain and France of the independent nuclear deterrent which was denied them. The Russians, for the same reasons, objected stoutly to the MLF and insisted on regarding it as a case of nuclear proliferation. The French ignored it and the British were scornful of its military value but agreed, for political reasons and after strong American pressure, to participate in it. The Italians, Greeks and Turks also agreed to join. At the end of 1964 the new British Labour government produced an alternative scheme without obvious appeal or virtue. Thereafter the MLF wilted because the Americans found that they no longer needed to entice West Germany away from France and because they came to believe that Russian objections were genuine and fatal to the progress of agreements to control further nuclear proliferation. So far as concerned harmony in the alliance they fell back on proposals to give the allies a bigger share in planning committees, and in 1966 such questions were temporarily submerged by a French decision to withdraw from all Nato's military organs and to expel such organs from France. France remained a member of the alliance and continued to insist on the need for it, but would have no part in its operations so long as it remained unreformed.

Restored sufficiently to semi-withdraw from the alliance, France was shaken in May 1968 by an outbreak of revolutionary violence in Paris. The causes were not peculiar to France. All over western Europe there were deep sources of discontent which overlapped and fused: urban squalor, revulsion against the horrors of the war in Vietnam, overcrowded universities and schools, the fight for higher wages in a period of inflating prices. In France de Gaulle's government irritated the young by its tone of paternalism and the liberals by its attempts to direct radio and television and to control the press: the progressive element in Gaullism had grown dimmer during the ten years since de Gaulle had returned to save France from fascism and military rule. The position in French universities and schools was far from being the worst in Europe (in some parts of Italy schoolchildren had to attend on a rota system because there was no room for them all at once), but it was bad enough to inflame a generation which had been attuned to political activism by the Algerian War and was politically better organized than anywhere else in Europe. Since the war the number of university entrants had been quadrupled by the rise in the birthrate and because no government had dared to change the rule that any boy or girl achieving the *baccalauréat* was entitled to go to a university. New universities were being built in Paris and out of it, but they were started too late. The resulting chaos was increased by bureaucratic centralization, the discontent by outdated syllabuses and outdated rules about personal conduct (sometimes enforced by the police). The new university at Nanterre on the edge of Paris became notorious for clashes between students and staff but it was typical rather than unique, and it was trouble at the Sorbonne in the heart of Paris which eventually converted such clashes into something like a revolution. After an occupation of university buildings by students the university authorities called in the police and the police behaved with such brutality that opinion in the capital, not normally on the side of students, swung massively in their favour. The troubles culminated in a night of battle in which the police (this time on government instructions) and the students fought one another for control of the Left Bank while the scenes of violence were relayed to France and beyond by radio reporters roaming the streets. The police won the battle but students continued for a time to occupy parts of the university. At the same time workers in Paris and other cities went on strike, occupied factories and set up action committees which began to look like a new government in embryo. The authority of the legitimate government was badly shaken. The prime minister, Georges Pompidou, was so alarmed that he advised de Gaulle to resign. There was talk of a new French Revolution. But a month later de Gaulle went to the polls and won a sweeping victory.

There were several reasons for this. Although a nucleus among the students had revolutionary political aims, many of them wanted no more than university reform and the strikers were not revolutionary at all. They were not trying to overthrow the government but to get better wages out of it and less unemployment. The leaders of the Communist Party were too much part of the system to want to risk its disruption, were afraid of the more left-wing groups and had no sympathy with students. Above all de Gaulle's nerve held. Although he had to hurry back from a state visit to Romania, he did not let himself be hustled when he got back. Having assured himself by a secret expedition to military headquarters that he had nothing to fear from the army, he correctly weighed up the situation, waited for the university authorities and the trade unions to begin to recover their control in their respective spheres and then, disdaining François Mitterrand's bid to replace him in the presidency, won a sweeping victory by the votes of frightened Frenchmen at the end of June and dismissed Pompidou. He then backed his minister of education, Edgar Faure, who made a radical attack on the problem of higher education in spite of some of his colleagues and of much conservative opinion. Faure introduced joint teacher–student management; abolished the centralized system under which France had in effect a single university and substituted for it sixty-five universities (thirteen in Paris), none of which was to have more than 20,000 students; and further decentralized control within each university by creating joint councils for each nucleus of 2,500 students.

But de Gaulle's days were numbered. One of his aversions was the French senate. One of his preoccupations at this point was the reform of the machinery of government by the creation of regional assemblies. He proposed to link this reform with the abolition of the senate and put the two issues together to the electorate by referendum. But the senate was not unpopular and de Gaulle's use of the referendum, coupled as it was with the implication that the rejection of what was proposed entailed his own resignation, was widely regarded as unfair tactics. A majority voted no. De Gaulle at once resigned. (He died the next year.) The president of the senate, Alain Poher, assumed the functions of the presidency in accordance with the constitution, a presidential election was held, and the Gaullist candidate Georges Pompidou won comfortably in the second round. This change at the top in Paris coincided with a similar change in Bonn.

Konrad Adenauer's long rule in western *Germany* (1952–63) established in Europe a new state which was resuming Germany's economic dominance on the continent. It was built on the economic policies of the western countries (currency reform and Marshall aid) and the division of Germany. Territories lost in the east were less valuable than those retained in the west; refugee labour from the east was opportune for reconstruction and peculiarly mobile. Investment boosted growth without inflation. Growth restored morale. High standards in education, industrial training and discipline, and efficient administration added their quotas. In the four years preceding sovereignty the combined western zones trebled industrial output and raised GNP by two-thirds. The population recovered its prewar numbers and was stabilized in the 1960s at 60 million (eastern Germany's was 17 million). As output rose and unemployment did not, generous provision for public services and a welfare state attracted minimal opposition. Material success was matched by political success on two fronts: internally, a democracy which worked and (unlike Weimar) was respected and popular, and, externally, acceptance in the world's strongest alliance.

In the Adenauer years the west meant, overwhelmingly, the United States. Yet Adenauer himself edged in his later years towards a more precisely European stance, particularly in his relations with de Gaulle, and his successors began to probe Germany's traditional associations in central Europe. Adenauer's task had been to reposition Germany in Europe after the disasters of the Nazi years. His alliance with the United States was a precondition for new relations with France and Russia. The first were firmly established but the latter were interrupted by the building of the Berlin wall in 1961 which not only isolated one part of Germany from the other but also cut western Germany off from the USSR. Adenauer's departure was followed by a short postlude to the Adenauer era with Ludwig Erhard as chancellor until, in 1966, he was forced out of office by his own party. From 1966 to 1969 the Christian Democrats and the Social Democrats governed in coalition with Kurt Kiesinger as chancellor and Willy Brandt as vice-chancellor and foreign minister. These three years constituted a bridge between the Adenauer era and the almost equally long socialist era to come. The grand coalition abandoned Adenauer's attitude of regarding half Europe as virtually non-existent. The sources of this evolution were détente in Europe and Washington's increasing preoccupation with establishing better relations with Moscow without as much regard for German susceptibilities as had been evinced in the past; the abandonment of the pretence that European security and the German problem were inseparable and that no European system could usefully be studied in the absence of German reunification; the growth of eastern European economies leading not only to restiveness against satellite status but also to a desire for the products of the new technology (computers, for example) which western Germany could provide; and to popular appreciation among West Germans of the barrenness of the promise of reunification via a western alliance, the realization that the road to reunification did not run through

Washington. Bonn therefore entered into discussion with eastern European states and established in 1967 diplomatic relations with Romania. This eastern policy quickened after 1969 when Brandt became chancellor. Brandt and his foreign minister, Walter Scheel, opened discussions with East Germany, Poland and the USSR.

Progress was slow because of interlocking complexities. Besides renewing normal relations with wartime enemies (the USSR, Poland, Hungary and Bulgaria) Bonn had to negotiate agreements with Czechoslovakia which was demanding the abrogation of the Munich agreement of 1938 and with East Germany which was demanding recognition as an independent sovereign state. This last issue was complicated by the problems of Berlin, a city divided politically and physically and in which Germany's four principal conquerors still had special rights. Treaties with the USSR and Poland, which included the recognition of the Oder–Neisse line, were concluded in 1970 and ratified in 1972 after other agreements had been concluded. These included: a new four-power agreement on Berlin providing, among other things, for easier rail, road and water communication between West Germany and west Berlin and freer access for west Berliners to East Germany for a wide variety of purposes; a General Relations Treaty between the two German states, a bundle of documents whereby both signatories recognized each other's sovereignty and frontiers and promised to be good neighbours and settle disputes peaceably; and an agreement establishing diplomatic relations between Bonn and Prague and a declaration that the Munich agreement was invalid. Bonn also established diplomatic relations with Hungary and Bulgaria. Both Germanys became members of the UN (1973). A quirky postlude to these transactions was the claim in 1975 by East Germany for the return to Berlin of works of art removed during the Second World War to safer places further west – some 600 paintings by major artists including 21 by Rembrandt, over 200 drawings by Dürer and Rembrandt, Queen Nefertiti and 3,000 other Egyptiaca, and more. But that the Hague Convention of 1954 could be held to apply to these circumstances was a hardly tenable proposition in law.

These various agreements, concluded in 1970–73, resolved much but not all of the business which a peace conference in 1945 might have been expected to settle. Berlin in particular was not purged of its anomalies. The four powers maintained their rights in the city, which remained two cities; the movement of west Berliners to the east became easier but not normal; west Berlin remained constitutionally attached to West Germany but physically contiguous only to East Germany. The reunification of Germany was not ruled out, although its attainment by force was. Brandt's *Ostpolitik* was much criticized by his countrymen and -women and although he had strengthened his parliamentary position at elections in 1971, he lost ground in the ensuing years. In

1974 he was forced to resign the chancellorship by the discovery of a spy at work in his private office. His successor, Helmut Schmidt, inherited a situation in eastern Europe which had been thoroughly transformed in the five years of Brandt's chancellorship. A seal was set on this area of détente when Brezhnev visited Bonn in 1973.

Again by coincidence Brandt's departure was closely followed by change in France. In 1974 Pompidou died. In the first round in the ensuing election the socialist François Mitterrand, backed by the communists, came first but without the necessary margin for election. In the second round between himself and the runner-up, Valéry Giscard d'Estaing, the latter won by the narrow margin of 50.8 per cent to 49.2. The Gaullist candidate came third in the first ballot and so was eliminated together with nine other candidates who received only derisory support. In West Germany Schmidt remained chancellor until 1982 when his government was defeated in the Bundestag and his FDP partners switched their allegiance to give the conservative Helmut Kohl the chancellorship; at elections in 1983 the swing of the electoral pendulum confirmed Kohl in the post which he held into the 1990s, even in 1994 against strong expectations. In 1983 too Mitterrand became president in *France*.

Mitterrand was a politician with intelligence, cunning and patience. Domestically his main task was to reassert the socialists' predominance on the left and as far into the centre as possible. To do so he needed to recover votes which the working classes had given to de Gaulle but were not so ready to give to the general's successors on the right; and he needed also to make inroads into communist areas and make the socialists independent of communist support. After Pompidou's death Giscard d'Estaing challenged and defeated the Gaullist heir apparent, Jacques Chaban-Delmas, thus fragmenting the right. The united left was badly beaten in 1978 and the alliance of socialists and communists dissolved, but in the course of this defeat Mitterrand and the socialists overtook the communists in all but a few communist strongholds. Mitterrand then won the presidency in 1981. He briefly admitted communists to government for the first time since 1947 and initiated expansionist policies which, however, he was forced to abandon by an influx of German goods. He adopted left-of-centre policies not very different from the right-of-centre policies of his predecessors and moved to closer relations with West Germany within a European partnership (the policy spurned by Thatcher). The French economy, although it had by this date a GNP higher than Britain's, had weathered the rough economic climate of the 1970s less well than the West German. Older industries which were in transition had to withstand depression as well as the strains of change; newer industries which France had successfully fostered wavered. After Helmut Kohl's re-election as chancellor in 1987 Mitterrand welcomed a German proposal for closer Franco–German military co-operation and the formation of a joint Franco–German brigade; the two leaders created a Franco–German defence council. Mitterrand was forced by electoral results in 1986 to collaborate with the right, appointed Chaban-Delmas prime minister but outwitted him and was re-elected to the presidency in 1988.

Mitterrand was a pragmatist in the tradition which stretched from Talleyrand to de Gaulle, a skilful national and international politician but uninterested in political ideas and prone in his later years to errors of judgement. In the Gulf War of 1991 he contrived to take an active role against Iraq in spite of some uneasiness in his relations with Maghrib states and the presence of 3 million Muslims in France, but he stumbled over other issues: the coup against Gorbachev in Moscow which he appeared at first to accept, and proposals for admitting ex-Soviet satellites into the EC which he was obliged to ditch for the time being. His touch with his own party became unsure and with his prime ministers ungenerous, even disloyal. Under his guidance the socialist party had abandoned much of its traditional programme without espousing anything else beyond tactical opportunism. Having captured the main place on the left from the communists he failed to unite the left or construct an anti-right alliance with the centre parties or prevent damaging feuding within the socialist party. In 1991 he dismissed the ablest of his prime ministers Michael Rocard and appointed instead Edith Cresson who, besides having unusually little aptitude for the post, was disastrously tactless and was removed in her turn. The French economy was prosperous in parts but unemployment was rising, special groups (farmers, fishermen) were rampantly obstreperous and immigration became a prominent issue with openly racist overtones. Mitterrand himself gave an impression of lassitude, due either to ill health or simply weariness after too many years in office in taxing times. He became aloof, capricious, nepotistic and (like Thatcher whom he otherwise resembled not at all) semi-detached from parliament and parliamentarism.

In 1993 Mitterrand's party was routed at the polls, winning only 19 per cent of the vote, and he faced for the second time a period of government with a right-wing prime minister and cabinet. With the principal leaders of the right – Chirac and Giscard d'Estaing – evading this awkward situation, Edouard Balladur became prime minister. President and prime minister being both aware that the former's term must end in 1995, the latter was content to leave Mitterrand's external policies undisturbed, the more readily since both men were wedded to the general aim of securing for France an elevated role in Europe and the world as an ally and more than an ally of Germany. Unlike Thatcher who distrusted Germany and scorned the European Community, Mitterrand cherished the Franco–German entente at the heart of Europe. He defended it at some cost to French independence and some strain on the French economy since his policies entailed a strong franc and painful anti-inflation measures (to keep monetary union on course), aggravated unemployment, depressed consumer demand and provoked violent popular clamour for protectionist subsidies, tighter immigration controls and narrower rights to citizenship. In his last years the scene had shifted more than Mitterrand perhaps realized, for the reflux of Soviet power in Europe virtually ensured that sooner or later central Europe – *Mitteleuropa* – would constitute a force in European affairs which it had not exercised for half a century and which gave Germany a range of choices alternative to the bilateral Franco–German direction of the first phases of European union.

Balladur showed unexpected skills in his handling of the emotional issues which faced him when he became prime minister. While insisting on strict enforcement of rules on immigration he stood by residence, not blood, as the test for citizenship and defused agitation for breaching the principle of the Loi Falloux of 1850 by which France had for so long regulated strife over religion in state schools and government money for non-state (mainly Roman Catholic) schools. His mild-mannered conservatism and economic dependability reassured the French electorate and he became therefore a candidate for the presidency. But he lacked the gifts needed to succeed against the more extrovert and supple Jacques Chirac and came third in a first round which stirred only modest enthusiasm for any candidate. The socialist Lionel Jospin, thrust into the battle when Jacques Delors ruled himself out, received most votes in a high poll but Chirac won the second. He inherited the inescapable problems of any national leader constrained to play second fiddle to a greater, even if friendly, power together with the specially teasing problems of combining a proud nationalism with a current of internationalism. On the question of whether a free-trade community was incomplete without financial integration Chirac was an integrator but he also shared sentimental nationalism and was perplexed by the immediate economic consequences – with unemployment at 12 per cent of the workforce – of the preconditions for the monetary union to which he remained committed. By temperament he was, like Mitterrand, a man more adept at finding ways round problems than, like Delors, eager to confront them.

Britain's recovery from the war was strong, swift but not sustained. Morale at home and prestige abroad were very high. War losses had been severe but the financial management of the war had been prudent.

Half the costs had been met out of taxation, the other half by selling foreign assets and borrowing. The abrupt end of American Lend-Lease in August 1945, followed by an American loan which was disappointingly low and accompanied by impossible conditions (the convertibility of sterling within one year), presented the new government with a daunting task which was mastered by a combination of wise management and American aid. Marshall funds, although available for four years, were dispensed with after two. They were used primarily to restore war-damaged industries and to fund an ambitious programme of social reform in education, health and social security – but not for the wider and more radical restructure of the manufacturing economy which Britain needed to retrieve the position in the world which it had been losing since the late nineteenth century to Germany, the United States and newer competitors. Taxes remained high but interest rates low; unemployment was low and wages were under control; economic growth averaged 4 per cent a year. Prewar production, the prewar value of foreign assets and the prewar level of personal incomes were all recovered by the end of the decade. Simultaneously education was expanded and a comprehensive social security system was introduced.

But Britain's longer-term industrial decline was not mastered and the country remained therefore the more vulnerable to external vicissitudes and domestic mismanagement. Of the former the most damaging were rises in the cost of imported industrial materials such as were occasioned by the Korean and Vietnam wars and the huge rises in the price of oil stemming from wars in the Middle East in the 1970s. The relaxation of controls and trade liberalization, by removing protection from renascent industries, created from the 1950s onwards serious imbalances on external account and distracted attention – and resources – from a more radical renewal of the British industrial base. This base had always been comparatively narrow – coal and a few prime industries such as textiles and engineering – and responded only sluggishly to the development of new methods, new competitors and altogether new industries. Adaptation there was, but not enough to preserve the economy from the buffets inescapable from a world economy no longer controlled from Britain. In addition Britain was losing its primacy in the provision of financial services and other invisible exports. Only rarely throughout the nineteenth century and the twentieth did Britain achieve a surplus on trade in manufactured goods, but the gap was consistently and comfortably covered by surpluses on invisibles until the Thatcherite 1980s. Conservative policies in the early 1970s created a boom but a speculative boom which diverted profits and savings away from industrial investment into short-term gambles and drove wages up. Attempts to impose wage restraint through income policies were progressively ineffectual and culminated in a miners' strike and the fall of the government. A Labour government addressed itself to labour relations (often another euphemism for keeping wages down) but Harold Wilson recoiled when it became clear that they would split his party. This political pusillanimity strengthened left-wing shopfloor dissidents against the trade unions.

Politically as well as economically Britain was slow to shake off a long past whose characteristic feature was aloofness. Having lost their continental possessions towards the end of the Middle Ages, the English people extolled their island status and fortified it with the Protestant secession and their seafaring gusto. Thereafter the British fought other Europeans rather less than they fought each other, but this amiable trait was more dismissive than pacific and the sense of distance – not always distinguishable from a sense of superiority – was far from extinct in the twentieth century. In 1945 the view from the island focused more easily on the Commonwealth and the United States than on Europe.

The Commonwealth, originally the British Commonwealth but renamed in the light of postwar decolonization, was a descendant of the British empire and a happy historical accident which eventually comprised fifty independent sovereign members. The dissolution of British rule in Asia, Africa and the Caribbean began with the retreat from India and Burma in 1947–48, a step long envisaged but not precisely programmed until the end of the war. This dramatic move had unexpectedly precipitate consequences throughout the colonial empire which (with the notable exception of Rhodesia) was abandoned with good grace and little fuss in the ensuing twenty years. But the emerging Commonwealth was not a power centre, still less an instrument for the exercise of British power. Its headquarters were in London but meetings of its political chiefs sometimes left Britain in a minority – in Thatcher's time in a minority of one – and although the Commonwealth provided Britain with special contacts and opportunities it did not reinforce British power in the world.

Nor did the special relationship with the United States which, like the links with the Commonwealth, was a reality but not one which gave Britain what it thought it was getting. (Washington's only truly special relationship was with Israel.) Before the war Anglo–American relations were poor to bad. During the war they were excellent, but the excellence was due to crisis and to personalities: what was special about the wartime relationship was the unusually close friendship and collaboration between Churchill and Roosevelt. After the war many wartime friendships and channels persisted, facilitated by a common language and by shared experiences and ideals, but old divergences –

commercial, imperial – reappeared and attempts by British prime ministers on visits to the White House or in reaching for the telephone to play the part of Churchill degenerated into backscratching, not without a suggestion on the British side of subservience. The special relationship not only deflected British attention from political developments on the Continent but also gave Continentals – France in particular – grounds for barring Britain from the European Community.

In 1979 Margaret Thatcher became prime minister, the first woman to hold that office. She held it with exceptional dominance for longer than anybody in the century. She was determined, vigorous and hard working – perhaps the most hard working prime minister since William Pitt but not the most intelligent. She was handicapped by her predecessors' failure to grasp serious long-term economic problems, by her poor choice of cabinet colleagues, and by her autocratic temper which, besides mauling local government, depreciated the House of Commons and the democratic order. Thatcher came to power in a recession and left it in another. She enjoyed but squandered two uncommon economic advantages: the yield of North Sea oil which reached its peak in the 1980s, and the proceeds of her programme of denationalization which, whatever might be judged of its merits or its handling, brought the government large sums which were credited to current instead of capital account and enabled it to present illusorily balanced budgets.

Thatcher concentrated on the alarming rate of inflation with a single-mindedness reinforced by a narrow version of monetarist dogmatism which, by a myopic attention to the money supply to the exclusion of other factors, succeeded in reducing inflation but at dire cost. The pound was kept at a high value by high interest rates which stifled industry, extinguished viable as well as deservedly moribund undertakings, reduced manufacturing output by a quarter and therefore diminished the tax base, rendered Britain incapable of financing its public education and social services and raised unemployment to a level unforeseen by the government. During the Thatcher years a nation of savers became a nation of gamblers. Private domestic debt trebled, mortgage debt quadrupled and personal savings as a percentage of national income sank from 16 per cent to 2.5. There was a return to widespread poverty which, although much less personally distressing than the poverty of the great depression between the wars, was painfully felt and evident and was reflected in a variety of telling statistics such as prosecutions for begging or loitering which rose to 4,000 a week. By cutting credit controls and taxes Thatcher fuelled property speculation, encouraged the nation to live beyond its means and created demands on the economy which would be met only by imports which crippled the balance of payments. Vast amounts of money were lent to speculators and lost, money which could have been used in re-equipping or converting flagging industries or starting new ones. Britain became the prime example of sick capitalism as distinct from the healthier capitalism of Germany or Japan.

Thatcher's talents were for combat. She won applause for her response to the Argentine seizure of the Falkland Islands and embarked on a characteristically root-and-branch attack on union power – a power much exaggerated by governments and no less by miners' leaders – but she was frustrated when tempted to handle foreign and Commonwealth affairs in the same mode. Under her rule Britain's relations with the EC were conditioned by an innate dislike of foreigners, imperfect comprehension of foreign affairs and dogmatic hostility to anything beyond minimal co-operation with the Community and minimal abandonment of formal sovereignty. She fell abruptly from power when half the Conservative members of the House of Commons revolted against her style of government, her disdain of her colleagues, her cavalier attitude to the constitution (particularly her manipulating of the cabinet system), the discord within her administration over the EC, gathering economic stringencies and above all the prospect of losing their seats at the coming general election if she remained their leader. Thatcher was succeeded by John Major, whose modesty and lack of evident disqualification was made by Thatcher's performance to seem a recommendation and who scored an unexpected victory in elections in 1991. Major had a tenacious conviction that, the EC having come into existence and the alternative of a free trade area not being on offer, Britain needed to play a leading role at the centre of the EC instead of trying to emasculate it from the sidelines. But the transition from Thatcher to Major did not give the new prime minister a free hand. Thatcher herself, wounded by dismissal from office by her own party, became a rancorous focus for disunity within it. In the new House of Commons, Conservatives were divided into a faction loyal to Thatcherite attitudes, a faction prepared to follow Major in claiming a constructive role for Britain in European affairs and a third faction as much concerned to find fault with the EC as to try to shape its future and understand its workings. Major was hampered by taking office at a moment when the oil revenues which had masked Thatcher's budgetary legerdemain were in decline and were forcing the government to raise instead of cutting taxes if it was to avoid a hazardous level of public borrowing and maintain at least adequate public services. The Conservative victory of 1991 was followed by economic recovery which produced growth of nearly 4 per cent but also laid bare the trap created by the policies of the 1980s. With manufacturing output still inadequate, growth again stimulated imports to satisfy increasing purchasing power but thereby threatened the balance of payments and the value of the currency and obliged the government to raise interest rates (to the dismay of manufacturers) in order to restrain the growth which it was nevertheless earnestly seeking. In foreign affairs Britain was a country grappling negatively with distasteful problems – large ones in Northern Ireland and the European union, lesser ones in Hong Kong.

Thatcher gave xenophobia a political point which it had not enjoyed since a much earlier generation of Tories had loved to hate the first Hanoverians. British opposition to the EC was emotional, irrational, mostly poorly informed and yet popular against a background of resentful disappointment. However convinced Major might be of the inescapability of a British commitment to the EC, he failed to extricate himself or his party from the trough dug by Thatcher who put at risk by her fractiousness London's title to be Europe's financial capital. Thatcher had come to power as the leader of the Conservative Party but she was in spirit the leader of a right-wing section within it. At the beginning of the twentieth century Britain had recoiled from what Lord Salisbury in 1885 called 'the abyss of isolation' but it never (in peacetime) got further than substituting ambiguity for isolation. The Thatcherite minority in the Conservative Party hankered after the unfettered independence which had vanished around 1900 when Conservatives like Balfour and Lansdowne appreciated that neither politically nor economically was it any longer sensible or safe.

EUROPEAN UNION

The European Community was created with two main purposes: to find a solution to Europe's German problem and to make its members richer and more influential in the world as partners than they could be as separate states.

From Bismarck's time to Hitler's Germany was, with only brief intervals, the state which most Europeans most feared. In a European states system Germany was inescapably the most powerful state, for although it was twice defeated in war and might be curbed in the immediate aftermath of defeat, its resources and skills ensured its resurgence. Its neighbours moreover, although they might fear it, depended on a prosperous Germany for their own prosperity. Even before the end of the Second World War a few Europeans, notably Dutch and Belgian, confronting this dilemma of reconciling their economic dependence on German recovery with their fear of German military might, envisaged a new European order in which the sovereign nation state would no longer be paramount. Whereas a states system by its very nature encouraged national aggressiveness, in a partnership each nation – including the German – might develop its resources and preserve its identity and cultivate its pride in a context more conducive to co-operation than aggression. Opinion in the more powerful states – France, Britain – was unsympathetic to anything so radically restrictive of their formal sovereignty and independence, but within a few years of the war's end French leaders became converted to ideas for a western European association which, in economic affairs, would maximize commerce and output and, politically, would circumscribe Germany's predominant power and redirect its ambitions. France abandoned the policy, entertained both in 1945 and 1919, of weakening Germany by partition and disarmament in favour of a policy of supporting and sharing in German recovery. Britain, on the other hand, although – or because – its own economic recovery from the war was at first more emphatic than France's failed to perceive either the economic advantage or the political calculation behind the movement towards a partnership which would make the sovereign state avowedly less than sovereign. British attitudes were partially reversed in the1960s and 1970s when governments of both the right and the left came to fear the consequences of exclusion from a western European economic association, but in the 1980s Thatcher personified and inflated British atavistic aversion to political association with the result that Britain was cast in the role of a disgruntled, even subversive, member of the Community whose development was bedevilled rather than strengthened by Britain's adherence to it.

On the continent the prestige of the nation state suffered during the war. National governments failed to prevent nation states from being battered or their citizens from being killed, tortured and enslaved. In Britain, however, the institutions of the state were not diminished; they remained intact and functioned with remarkable efficiency and fairness. In British eyes the separation from the continent by the English Channel remained axiomatic. The British were uninterested in giving the lead towards that strong and more permanent association which many continentals desired and which Britain, uniquely in 1945, had the strength and prestige to offer. The British still thought of themselves as a maritime power and a world power, only peripherally European, unthinkably less than sovereign. For two and a half centuries Britain had had no land frontiers, since even its troubles in Ireland lay beyond the sea. Its principal preoccupations were the freedom of the seas, the movements of commerce, and peace. The first two of these objects it pursued by maintaining a naval lead over the combined strength of other substantial naval powers and by ensuring so far as possible that the European nations which dominated the world should include a number of land powers of the first rank but only one such naval power. In this

Founder members of the EU in 1957
The Netherlands, Belgium, Luxembourg, W. Germany, France & Italy

Additional members of the EU with dates of membership in brackets
UK, Ireland, Denmark (1973), Greece (1981), Spain, Portugal (1986),
Austria, Sweden & Finland (1995)

1. Further enlargement would lead to the incorporation of central and eastern European
states into the EU: Bulgaria, the Czech Republic, Hungary, Poland, Romania and the
Slovak Republic.
   March 1992 Europe agreements signed with Poland, Hungary and the Czech and Slovak
Republics came into effect February 1994; from 1995 EU entered a free trade agreement
with Estonia, Latvia and Lithuania; Slovenia and Albania may become signatories of EU
agreements.

2. Three southern European states have applied for membership: Turkey in1987 and
Cyprus and Malta in 1990. The association agreement with Turkey was intensified in 1989.

6.1   Growth of the European Community

context the continent of Europe was a place to which negative principles applied: it must not be allowed to distract or threaten Britain, it must not fall under the dominance of one among its principal land powers. British diplomacy was directed to maintaining a balance and preventing a hegemony in Europe; if British diplomacy failed, then British arms had to shoulder the task which, though in a sense negative, was also vital to British interests as they had evolved since the Tudors had laid the foundations for a kind of British power altogether different from the continental imperialism of the Plantagenets. The British therefore developed a state of mind which drew no distinction between the near and the far. Geographers might talk of the 'Far' East and measure the distance to India in thousands of miles, but to many an Englishman Delhi and Singapore and Hong Kong were psychologically no further away than Calais; they were often more familiar, and they were of course more British.

In 1945 Britain's innate inattention to European affairs was enhanced by the fortunes of war and the prospect of peace. During the war every continental European combatant, including the USSR, had been overrun and at some point defeated or almost defeated. Britain had been terribly hard pressed and had been bombarded from the air, but it had not been invaded or occupied or defeated. Its victory vindicated its right to go on as before, since it is the prerogative of a victor to retain its past, whereas its shattered and disillusioned European neighbours were looking for a new start and not for a restoration of the old order of things which had failed. The British and continental attitudes to the past were therefore completely different, and continentals who expected British sympathy for radical political experiment in Europe were overlooking not only Britain's separate historical development but also its postwar psychology, the intent to repair and improve the structure of British life but not fundamentally to alter or find fault with it.

The advent of a Labour government in Britain in 1945 should have revealed the difference, for the Labour Party, although a reforming and not a conservative party, was no less traditionalist than the Conservatives. It consisted of pragmatic radicals and socialists who wanted to make life happier for the lower classes by continuing the gradual and non-revolutionary adaptation of Britain's social structure to modern notions of social justice. It had no intention of overturning the British apple-cart and not much interest in other people's apple-carts. It was a hard-working middle-of-the-road administration which was trying, in exceptionally difficult economic circumstances (aggravated by the end of Lend-Lease and American insistence on the premature convertibility of sterling), to restore the British economy and reform British society and it did not wish to be diverted from these tasks by unprofitable foreign entanglements. The continent was chaotic and impoverished and, as the transfer of Britain's commitments in Greece and Turkey to the United States showed, could better struggle out of its troubles with American rather than British aid. Moreover, the new leaders in Europe were (quite apart from being foreigners) mostly conservatives and Roman Catholics; opponents, it was wrongly thought, of planned economies, uncomfortable partners for British socialists. In so far as they were attracted by federal ideas, these leaders were regarded as unpractical visionaries. For the British the nation state was one of those bits of the past which practically nobody questioned.

Winston Churchill had told the British during the war that they operated in three circles – the Anglo–American, the British imperial and the European – and that this triangularity gave Britain special opportunities and a unique position in the world. Until Harold Macmillan applied in 1961 to join the European Economic Community the European circle was the one which seemed to offer Britain the least. The most important was the Anglo–American. Britain – or at any rate Ernest Bevin, who became foreign secretary in 1945 – saw that the consolidation of Europe under the aegis of a single power could no longer be prevented by British diplomacy or British arms alone, and that if this bugbear of British foreign policy was to be avoided the Americans must be made a European power. Nato was the outward and visible sign of his success, but his endeavours to create an Anglo–American thrust in European affairs, in the place of the expired British power to intervene and rectify, made him suspicious of continental federalists who might hanker after an independent European power to the exclusion of the Americans. Their policies were at best irrelevant, possibly damaging, to his aim of bringing in the New World to create a balance in Europe. Furthermore, those in Britain who were hostile to the United States or wary of its preponderance were not for the most part European federalists. In so far as there was a party in Britain which was thinking in terms of a 'third force' in world affairs, it conceived at this period a third force provided by the Commonwealth rather than by a united Europe. The Commonwealth, together with nuclear weapons and the pound sterling as an international currency, would keep Britain in a separate category.

Britain's change of heart did not begin to occur until some ten to fifteen years after the end of the war and even then it manifested itself much more fitfully and slowly than the comparable revolution in continental thinking which had been imposed by wartime defeats. Britain continued to think of itself as a worldwide, even if no longer an imperial, power – a somewhat uncritical adjectival substitution. One of the most striking consequences of the war was the British departure from India in 1947 (followed more rapidly than was expected by departure from Africa), but this abnegation of empire took place in such an atmosphere of self-congratulation that the attendant loss of power was overlooked. The loss of India was regarded as a victory for British commonsense, which it was, but not as a curtailment of British power, which it was too. For generations Britain had been a world power because it possessed in Asia an area where it could keep, train and acclimatize armies for use in distant parts of the globe, and this reserve of power was at least as important as the command of the seas in making Britain what it was in the world. The departure from India, coupled with the loss of wealth and strength during the war, sapped Britain's staying power in the Middle East and made Australia and New Zealand turn to the United States for their security. (The tripartite Anzus Pact of 1951, to which Britain was not a party, confirmed this lesson of the Second World War. It was one of the world's most equable alliances until the 1980s when the return of anti-nuclear Labour governments in both Australia and New Zealand created difficulties over naval exercises and visits. In 1984 New Zealand banned from its ports all nuclear-powered or nuclear-armed vessels.)

Britain did not, however, draw the conclusion that the end of empire and of defence commitments in Asia, Africa and Australia had converted Britain into a primarily European state. The empire had been replaced by the Commonwealth, a more elevating concept perhaps but one of less substance since it lacked the empire's bonds of allegiance to the British Crown, government by a ruling class which regarded itself as all one kin, mutual comprehension through a prodigal exchange of secret telegrams, and a British commitment to the defence of all its territories. The Commonwealth became an association of monarchies and republics of widely differing traditions and inclinations, requiring above all development capital which Britain could not provide, and pursuing independent and even contradictory foreign policies on the basis that this permissive latitude was a necessary price to pay for a continuing association which was still worthwhile. And so perhaps it was, since the Commonwealth proved to be an international organization which worked up to a point. But it contained within itself racial conflicts which posed tests of statesmanship which the British governments of this period failed to pass. In Rhodesia Britain was credibly accused of dealing softly with rebels because they were white and at home the same government exposed itself to even more serious charges. In 1963 Britain had given Asians in Kenya the right to opt for British citizenship, which many of them took. In 1968 the most important element in this right – the right to enter Britain – was summarily removed from them by a government which, in its ignorance of the true facts and figures about coloured immigration and integration, allowed itself to be panicked into slamming the door against some of its own fellow citizens. This unprecedented act, grounded in colour prejudice in a section of British society and in racial discrimination by the government, made nonsense of the Commonwealth ideal – and was later challenged and condemned in the Council of Europe. Even if Britain had in the past thought of the Commonwealth as a source of political strength, Britain's rulers in the 1960s were finding it more an embarrassment than a support. Community with Europeans seemed all the more real and manageable.

Unofficial pressure groups in favour of a European union had been encouraged, not least by Churchill who spoke more than once during the war of the need for European unity and advocated in a famous speech at Zürich in September 1945 a Council of Europe. These groups organized a convention at The Hague in May 1948 which was held under the sponsorship of many of Europe's leading figures, including Churchill, and which succeeded in persuading the five Brussels powers to set up a Council of Europe consisting in the first place of themselves and Norway, Sweden, Denmark, Ireland and Italy – to which were shortly afterwards added Iceland, Greece, Turkey, West Germany and Austria. The members, besides being European, were required to respect the rule of law and fundamental human rights. The constitution was a hybrid, an assembly without legislative powers yoked to a committee of ministers; the members of the assembly were appointed by national parliaments, in practice in accordance with the party representation in each parliament; the committee of ministers, which was included in the constitution on British and Scandinavian insistence against the more federalist wishes of other members, ensured that any authority which the Council of Europe might exercise should be subject to the control of national ministers responsible to national parliaments. The assembly never acquired any real authority and at the end of 1951 its president, Henri Spaak, resigned in despair.

But another and more substantial initiative was under way. In May 1950 the French foreign minister Robert Schuman proposed a *European Coal and Steel Community* (ECSC). Although multinational in scope, this was an essentially Franco–German venture with political as well as economic aims. It marked the conversion of France to a partnership with Germany which became one of the principal features of postwar Europe. Coal and steel were at the heart of Franco–German economic and military competition and by proposing to place these industries under joint international control France was, openly, burying the hatchet and, privately, acknowledging the foolishness of trying to compete: to this degree it accepted, consciously or not, that in the new Europe it would be a junior partner with Germany and also that being a junior partner was more sensible than trying to be an independent actor. Britain was sceptical on principle and mildly hostile because it hoped that British steel would undersell European steel. There has been argument whether Britain refused to join the ECSC or France made it impossible for Britain to do so; the one view does not exclude the other. In April 1951 six states signed a treaty establishing the ECSC which came into existence in the following year. It consisted of a High Authority of nine individuals acting by majority vote with power to take decisions, make recommendations, make levies on enterprises, impose fines and generally control production and investment in the six countries; a court of justice empowered to pronounce upon the validity of the High Authority's decisions and recommendations; a council of ministers; and an assembly entitled to censure the High Authority and by a two-thirds majority to enforce the resignation of the council of ministers. In the late 1950s, when the demand for coal declined, differences arose between the High Authority and the council of ministers, and the High Authority suffered some attenuation in practice of its supra-national competence.

By the early 1950s the Council of Europe had been joined on the European stage by the Coal and Steel Community and by an incipient Defence Community, and in 1952 Eden proposed the amalgamation of these three bodies and their parallel institutions. The Council of Europe appointed an *ad hoc* assembly to work out a scheme on these lines, to include provision for a directly elected assembly and a European cabinet. This was an attempt to build a political association, tentatively called the European Political Community, on the twin bases of economic and military co-operation and it was to embrace most of the non-communist countries of Europe. The prospective membership was large, even though Sweden's empirical neutralism, Switzerland's doctrinaire neutralism and the unpalatable autocracies of Spain and Portugal might exclude these countries in the shorter or the longer run. But the scheme was stillborn. The demise of the EDC in 1954 killed it and even without this blow it is difficult to believe that the rudimentary institutional economic association so far achieved sufficed to support so ambitious a parliamentary structure. As the nation states of Europe recovered from their postwar blues they became less disposed to abandon their identities, even though they might be prepared for permanent international co-operation in other fields.

The Coal and Steel Community was from the outset a first step rather than an end in itself. For devotees of European union it was one move in a process of creating a series of functional associations which could later be agglomerated. Coal and steel were no longer the prime sources of energy in a world powered by oil and expecting to be powered by nuclear energy. By 1955 the six members of the ECSC had progressed far enough to be ready to convene a formal meeting at Messina where they resolved to form a *European Economic Community* (EEC) and a

European Atomic Community (Euratom). The British attended the conference as observers and then withdrew, partly because they thought the new venture would be a failure, partly because they were opposed to a common external tariff which would be inconsistent with what was left of Commonwealth preferences, and partly because they wanted to limit co-operation to forming a free trade area – which they proceeded to form and which, unlike the EEC, was a failure. (Britain proposed a free trade area in which the EEC would be one member. It would remove internal barriers to trade and industry but not agriculture; have no common external tariff and permit each member to maintain existing preferences; have no obligation to permit free movement of labour or capital, align general economic or social policies, or to create common political institutions. Discussions began at the end of 1957 and collapsed a year later.)

The treaties establishing the EEC and Euratom were signed at Rome in March 1957 and both bodies came into existence at the beginning of 1958 and began to function a year later. In 1967 they were merged with the ECSC. The treaty of Rome created a constitution in four parts: a Council of Ministers, a Commission, a Parliament and a Court. The Court of seven was the guardian and interpreter of Community law. The single-chamber Parliament – its members were elected from 1979 onwards (but later in Britain) – was primarily a debating body. Created at a stroke by treaty between states, it had only embryonic powers of the kind which national parliaments had accumulated through time in their attempts to constrain the executive power. It might reject but not amend the Community's budget and dismiss the Commission in its entirety but not in part; it possessed a limited share in framing legislation but no control over the Council of Ministers. The Commissioners, at first two per member state but after the Community's enlargement only one from each of the lesser members, were fully and exclusively occupied with Community affairs at the Commission's headquarters in Brussels. Most of them had had ministerial experience in their own countries but performed in Brussels functions closer to those of departmental chiefs in a national civil service. (The Brussels bureaucracy was, however, small by comparison with national bureaucracies.) Finally, the Council of Ministers, representing the Community's several member states, was within the terms of the treaty of Rome the Community's dominant organ. It was a replica of the similar Council which had been incorporated at a late stage in the plans for the Coal and Steel Community in order to ensure that that Community should be less a supra-national entity than a national collective, and the EEC's Council was even more clearly superior to the Commission in Brussels than the ECSC's Council was to the High Authority of the Coal and Steel

Community. Consequently the most significant debates and decisions on the EEC's development took place in the Council and revolved not round the distribution of power among the Community's main bodies but round the exercise of power within the Council – which matters might be settled by a majority and which required unanimity. (There were some matters which might be decided by a 'qualified' majority. This was a prescribed majority reached after ascribing to each member's vote a weight commensurate with its economic power.)

The first stage in economic integration, which was to be completed in twelve years (and was), was commercial – a customs union with a common external tariff and, internally, the removal of all tariffs and the elimination of quotas. A second and overlapping stage envisaged a broader economic union with, in particular, a common agricultural policy (CAP); free movement of labour and capital; the homogenizing of social policies, laws (specially company law), and health and safety and other standards; and monetary union with a common currency and a single central bank. Political integration, so far as it might be necessary to secure the Community's economic aims, was an inevitable corollary but the treaty of Rome was silent on this aspect. For some, political integration was an end in itself and a necessary means to strengthen the Community's influence not only in Europe but also in further areas of special concern to Europeans, notably colonial empires on the verge of independence and the Middle East whose oil was western Europe's most crucial import. The parallel aims of Euratom, embodied in the separate treaty creating that community, were to conduct nuclear research, construct nuclear installations, work out a safety code and set up a body to own and secure pre-emptive rights over nuclear raw materials.

The first six members were joined in 1973 by Britain, Ireland and Denmark, in 1981 by Greece and in 1987 by Spain and Portugal. Norway's adherence, negotiated alongside Britain, Ireland and Denmark, was rejected by plebiscite, and Greenland, which joined with its Danish parent, seceded in 1986. Of all these additions Britain's was the most important and the most contentious, delayed and later soured by Britain's own ambivalence but also vetoed for a number of years by France.

Britain's scepticism about the EEC and its unwillingness to entertain so broad an association led it to form in 1959 a European Free Trade Association (EFTA) together with the three Scandinavian states, Switzerland, Austria and Portugal. EFTA was a rival to the EEC. Its aim was to abolish tariffs between its members over ten years but without any attempt to establish a common external tariff or any form of political union. The British believed that de Gaulle – who returned to power in France at this time – would kill off the EEC, but this was a misinterpretation since de Gaulle intended not to undermine the EEC but merely to delay its development while France's economic recovery gathered pace and France's political weight in Europe increased commensurately. De Gaulle's proclaimed objection to British membership on the grounds that it was a stalking horse for American interests had enough plausibility to carry conviction with other members of the EEC in spite of their anxiety to have Britain as a member to offset German and French dominance. By 1960, however, de Gaulle was ready to consider British membership and publicly proposed it. Harold Macmillan, who had repaired Britain's relations with the United States and the Commonwealth after the Suez débâcle of 1956–57, was also anxious to give Britain a stronger European role. His visit to Moscow in 1959 had shown him to be an honest broker in international affairs but not a necessary one; the cancellation of the British Bluestreak missile in 1960 had exposed the difficulties of maintaining an independent nuclear force with a sporadically uneven economy; and in 1961 he responded to de Gaulle's overture and negotiations for Britain's adherence to the treaty of Rome began that year. These negotiations, although laborious, were proceeding to a successful conclusion when they were overborne by a serious Anglo–French misunderstanding. In June 1962 Macmillan visited de Gaulle at the Château de Champs. What passed between them is uncertain and each may well have mistaken the intentions of the other: an interview between a devious man and a silent one can leave much unclear. It would appear, however, that Macmillan, who had made admission to the EEC a centrepiece of his foreign and economic policies, not only played down current difficulties over Commonwealth preferences and agricultural policy but also left de Gaulle with the impression that Britain was prepared to integrate with its continental neighbours in the military sphere. This integration was a matter of the first importance to de Gaulle, who was looking for ways of making Europe independent of the United States but was unlikely to be able to create a credible European defence establishment without British participation.

In France opinion was divided about the admission of Britain to the EEC. Tangled discussions in Brussels over food prices disturbed many Frenchmen and during 1962 the *patronat* became increasingly hostile to British admission. But for de Gaulle these were minor matters if he could get Britain into an association which would be strategic as well as economic. After June 1962 he seems to have felt confident that he could do this. The French elections of November 1962 confirmed his authority by giving him a majority in parliament and a welcome success after a setback in October when his vote in a constitutional referendum had gone down. He was aware of Macmillan's domestic difficulties when the Labour Party came out in October against joining the EEC but he probably saw little reason to suppose that the British government would be defeated. (He may have been wrong. Macmillan might have felt constrained to go to the country before so momentous a step as joining the EEC, and it is not inconceivable that he would have lost an election fought on this issue. The main arguments against the EEC, other than the national distrust of over-association with foreigners, were: that the EEC was a bureaucracy and not a democracy, constitutionally speaking an irresponsible form of government; that the EEC was devoted to free competition as opposed to planning and that entry into it meant abandoning national planning, not for international planning but for *laissez-faire*; that parliament would have to renounce its control over vital aspects of British public business; that the EEC was an inward-looking, Eurocentric organization with un-British traditions such as multiparty government and Roman–Dutch law in which British civil servants would be at a disadvantage; and that the process of government by qualified majority – that is, by giving a power of veto to a combination of one major and one minor partner – was a sure way to create disgruntled factions.)

Towards the end of 1962, when the British and French leaders both seemed intent on getting Britain into the EEC and the American administration was blessing the union, a decision in the Pentagon started a chain of events which led to the rupture of the negotiations. This was the decision to cancel the manufacture of the air-to-ground nuclear missile Skybolt, which the British had contracted to buy from the United States. The decision, taken in November on the grounds of cost, deprived Britain of the instrument with which it had hoped, after the cancellation of Bluestreak, to maintain an independent nuclear force up to 1970. By paying half the development costs Britain might have saved Skybolt and its own nuclear programme, but the price was too big and in December Macmillan went to Nassau in the Bahamas to meet Kennedy and find an alternative. He did not, to de Gaulle's disgust and perhaps surprise, turn instead to France and make the failure of Skybolt the occasion for switching from an Anglo–American to an Anglo–French or Anglo–European nuclear association. This demonstration of where Britain's first allegiance lay led de Gaulle to pronounce, at a gathering of the press on 14 January 1963, the exclusion of Britain from the EEC.

At Nassau Kennedy offered Macmillan the Polaris submarine missile in place of Skybolt. Macmillan accepted. Kennedy made the same offer to de Gaulle. This was a reversal of a decision in the previous year not to offer France nuclear weapons. This decision had been reached after a division between the proponents of such a deal who hoped thereby to improve Franco–American relations (a course initiated by Kennedy at a successful meeting with de Gaulle in June 1961) and its opponents who argued that it would change nothing but give a fillip to nuclear proliferation. De Gaulle refused the offer. Both Macmillan and de Gaulle were insisting on their nuclear sovereignty. De Gaulle refused to accept Polaris; Macmillan, while accepting Polaris and proposing to commit Britain's bomber and tactical air forces and Polaris units to Nato, insisted on ultimate British command and the right of withdrawal. The one attitude was not very different from the other, and both were essentially nationalist. The Americans, however, were looking for a supra-national solution to the problem of nuclear sharing in Nato and they may have believed at Nassau that Macmillan, as opposed to de Gaulle, had agreed to fall in with their plans in return for Polaris.

Macmillan and de Gaulle disappointed each other and the British approach to the Community lapsed. In 1964 the Labour Party returned to power with Harold Wilson as prime minister. Wilson renewed the British application. His approach was markedly different. Macmillan had given the impression that the main purpose of joining was to pull British economic chestnuts out of the fire, and that for Britain the economic benefits of joining would be considerable and would outweigh both the loss of sovereignty and the inconveniences of closer contacts with non-British folk. This prospectus was not an appealing one either in Britain or in the Community, and the procedure adopted to implement it was the bizarre one of sending a senior minister to Brussels to conduct protracted negotiations over details (unkindly dubbed the kangaroo tail syndrome) and without a firm policy decision on the main issue of whether Britain would in the end want to join or not. By contrast Harold Wilson told the British public that the economic disadvantages of joining would be considerable; he argued only that the advantages were in the long term more considerable. He also stated the aim unequivocally: to join the Community, accepting all its rules in advance. He himself, a convert to British membership of the Community, was much influenced by the conflict between the EEC and the United States over the (GATT) Kennedy Round which revealed the vulnerability of an economically isolated Britain in the event of a trade war between Continental western Europe and North America. From the British point of view de Gaulle's stand against the Commission had made joining more attractive, since it had removed to the distant future, perhaps to a never-never land, the federal trappings and aspirations which scared Britain as much as they irritated the French president. The future form of a European entity had become even vaguer than before and the British liked it that way. But the chief obstacle remained. Although five of the six were ready and even eager for Britain to join, de Gaulle could still prevent it. Britain's links with the United States and its economic dependence, its recurring debits on external account, and its continuing commitments in other continents than Europe provided arguments, if arguments were needed, for classing Britain as a thing apart.

The question was whether France would continue to insist on these arguments. They had always, even in de Gaulle's mind, been counterbalanced by others, notably by France's uneasiness about a European Community in which two states – France itself and West Germany – outranked the rest and might one day confront each other on a major issue. Britain therefore had two faces. While on the one hand it was as an American appendage unacceptable, it was simultaneously as a counterweight to Germany very desirable. During 1967–69 the latter aspect began to overhaul the former, partly because of developments within the Community and partly because de Gaulle's resignation in 1969 facilitated a change of emphasis in Paris.

In 1969 de Gaulle once more dropped a hint that the time might be ripe for British membership. He invited the newly appointed British ambassador, Christopher Soames, who happened to be Churchill's son-in-law, for a general discussion of Franco–British relations in which he talked about closer four-power co-operation in Europe (France, Britain, West Germany and Italy) and an adaptation of the EEC to fit Britain in. These were basic Gaullist ideas of long standing but as a result of a chapter of ineptitudes the interview produced a public diplomatic row when the British divulged the tenor of the talks to other governments and allowed the impression to gain ground that France was offering Britain a place in a European directory of major states in return for the suppression of the EEC and perhaps Nato too. Britain's chances of getting into the Community either by the front door or the back seemed therefore to have receded. But not for long. That year de Gaulle resigned the presidency and was succeeded by a man of different temper and outlook, Georges Pompidou. Within a few months a change of government in Britain made Edward Heath prime minister. Heath had been the protagonist of British membership of the EEC at the time of the first application in 1961. In December 1969 the heads of government of the six, at a meeting at The Hague, had committed themselves to British membership and Heath lost no time in confirming that France, also under new leadership, would back a second application. Heath visited Pompidou in Paris in May 1971 and the negotiations for the treaty were completed within a few months. By treaties signed in January 1972 Britain, Denmark and Ireland became members of the EEC on the first day of 1973. The Norwegian government, which negotiated with the EEC in company with these three countries, also agreed to join but its act was disavowed on referendum. In Denmark and Ireland a referendum held in accordance with the constitution endorsed the treaty of accession.

The British case was peculiar. The treaty-making power of a British government requires no popular endorsement. Nevertheless a referendum was held, although not until two and a half years after the date fixed for Britain's accession. This unusual procedure arose out of the equivocations of the Labour Party. The Labour Party was more seriously divided than the Conservatives over the EEC and Harold Wilson, his natural ambivalence sharpened by fears of splitting his party and ruining its electoral prospects, declared that the terms secured by the Conservatives should and could be bettered. He said that when Labour returned to office these terms would be renegotiated. This was in effect a threat to denounce the treaty unless its other signatories agreed to alter its terms. Wilson also promised that revised terms would be submitted to the country as well as the cabinet. Early in 1974 Heath, having narrowly miscalculated his electoral advantage, called an election and lost it, whereupon Wilson's new administration (a minority government until strengthened by a second election in October) opened discussions with the EEC in which the main British effort, successfully accomplished, was to get new terms sufficiently different from those in the treaty to show that the Conservatives should have done better, while the main aim of the EEC negotiators was to concede as much as was necessary to retain Britain as a member but no more. The main issue was the size of the British contribution to the Community's budget, on which the foreign secretary James Callaghan obtained sizeable concessions. The British cabinet approved the new terms with few dissentients. The electorate, certainly confused by elaborate and conflicting economic arguments, probably somewhat bored by this long-drawn-out affair, and not a little impressed by the plea that it would be wrong to undo what a previous government had with all due form and propriety done, said yes to membership in June 1975 by precisely two to one. Thus, twenty-five years after it could have joined a European community on virtually any terms of its own choosing, Britain haggled its way into the Community which had been constructed without it.

At the core of the Community was the Franco–German entente, tentatively but enduringly established by the Coal and Steel Community. Britain's entry into the Community was expected to broaden and strengthen this core but did not do so. If the initiative in creating the entente was French, its affirmation was German. Adenauer's first concern after the formation of the West German state in 1952 was to anchor it to the west by unquestioning association with the United States and membership of Nato, but in his later years – and particularly after de Gaulle's return to power in France in 1958 – the chancellor gave increasing attention to Germany's place in Europe. France was becoming, in Bismarck's phrase, more *Bündnisfähig* – a more worthwhile ally – and at the same time disenchanted with Britain and freed from postwar wrangles with Germany. The plebiscite in the Saar in 1955 removed the last serious Franco–German difference from the political agenda. Britain's unilateral abandonment under American pressure of the ill-designed and ill-executed Anglo–French attack on Egypt at Suez soured Franco–British relations. France was overcoming its reputation for political and economic instability and the firm handling of the Algerian crisis in May 1958 was followed by two meetings between Adenauer and de Gaulle at Colombey-les-Deux-Eglises and Bad Kreuznach which put Franco–German relations on a new footing. Between these meetings a French statement bluntly rejected British proposals (supported by Adenauer's finance minister Ludwig Erhard but not by himself) for a loose free trade area.

There were nevertheless serious divergences between French and German attitudes to the EEC's development. France, with Italian support, proposed in 1959 regular meetings of the six foreign ministers, backed by a secretariat to be established in Paris. The treaty of Rome was silent on political integration and it seemed to the Germans and to the Benelux members of the Community that the French were trying to create in Paris a political organization of an international character distinct from the institutions in Brussels and designed to bypass, throttle or even take over the supra-national economic activities which were proceeding there. These suspicions were sharpened in the next year when the French elaborated their plan and proposed a council of heads of government with a secretariat in Paris. Such a 'union of states', successfully pressed by de Gaulle on Adenauer at Rambouillet in July, was clearly incompatible with federalist ambitions. The whole subject was referred by the six to a special committee (the Fouchet, later Cattani, committee) which discussed two successive plans of French origin and the objections to them. These objections amounted in sum to the contention that the plans left out all the principal features of the EEC's own constitution, since they contained no provision for a parliamentary element or for an independent executive or for eventual decision by majority vote. The subject was allowed to fizzle out. Since it had been raised by France this outcome was a rebuff to France but also evidence that France was willing to accept some rebuffs rather than demolish the Community.

The next dispute arose over the elaboration of a common agricultural policy (CAP). The CAP was one of the Community's most pressing concerns. Its terms were not settled until 1967 (the travails of the CAP are examined later in this chapter). The principal bone of contention in the negotiating stage was the level of uniform prices to be set for cereals. In 1963 Dr Sicco Mansholt, one of the Commission's two vice-presidents and specially charged with agricultural matters, produced a plan for fixing a common price at one fell swoop instead of by instalments. The price was to be lower than the price ruling in Germany (and Italy and Luxembourg), so that French eagerness collided with reservations in Bonn which were all the more stubborn because the direction of affairs passed in 1963 to Erhard who had his eyes on the general election due towards the end of 1965. De Gaulle played on the fact that Erhard, while reluctant to accelerate the CAP, was anxious to establish an agreed EEC position in relation to the coming negotiations under the GATT (General Agreement on Tariffs and Trade). These negotiations, dubbed the Kennedy Round, were part of the series of periodic swapping of tariff cuts between members of the GATT, and the six were proposing to bargain as a single team with their most important trading partners, the United States and EFTA. The American president had been empowered by the Trade Expansion Act of 1962 reciprocally to cut tariffs by as much as half but his powers to do so were to expire on 1 July 1967. The tariffs imposed by the six were mostly bunched in a band ranging from 6 to 20 per cent, whereas American and British tariffs were either much higher or zero. Consequently the EEC Commission and also the French government pressed for the lopping of the higher tariffs (*écrêtement*) rather than general cuts (in which they were eventually forced to yield when the Americans proved adamant). Some Americans and British had expected and hoped that the six would split among themselves over a common approach to the Kennedy Round but at the end of 1963 compromises were accepted in order to permit both the Kennedy Round and the CAP to proceed.

During 1964, however, differences between France and Germany, instead of evaporating, became worse with France still holding up agreement on essential preliminaries to the Kennedy Round to extract Bonn's acceptance of a uniform cereal price but also to prevent it from joining the MLF and so swinging between the European and Atlantic groupings – an attempt to get the best of two worlds which de Gaulle was determined to block. The year ended with another of the compromises for which the Community was becoming noted. The American administration having tacitly abandoned the MLF, Bonn agreed to the uniform cereal price in return for special subsidies (also payable to Italy and Luxembourg) and the postponement of the common agricultural policy from 1966 to July 1967. This was a victory for de Gaulle, but it left Bonn resentful and ready to give a lead against France in the new crisis which de Gaulle's policies evoked in the next year.

The crisis arose out of proposals by Dr Hallstein, the Commission's president, to extend the authority, within the Community, of the Commission and the Assembly at the expense of the Council of Ministers by, first, expediting the switch in the Council from unanimity to majority voting and, secondly, by channelling agricultural levies directly to the Commission instead of to national governments for transmission to the Commission. These proposals were a challenge to de Gaulle's views or at least to his timetable, and some of Hallstein's colleagues warned him that he was going imprudently fast. Moreover, by presenting his proposals first to the Assembly instead of to the Council as the rules provided, Hallstein gave de Gaulle an opportunity to put his foot down with some show of justification. The French ambassador to the Community was withdrawn and France attended no Council meetings for six months. These tactics succeeded. The crisis of the 'Empty Chair' was resolved at the beginning of 1966 in a series of meetings in Luxembourg (not Brussels) which were in effect a governmental negotiation between France and its five partners. By the Luxembourg Compromise the members of the Community accepted informally that any member might insist that a particular proposal which it considered of special interest should remain under discussion until unanimity was reached. France thus secured a veto in circumstances to be judged and defined on each occasion by itself. By accepting this compromise France's partners, although giving little away on paper, acknowledged implicitly that French membership and full co-operation were of the essence of the Community. Thereafter the Luxembourg Compromise receded into the background. It was not invoked by Britain when, in the 1980s, Britain succeeded France in the role of reluctant or recalcitrant partner: by that date a Community which had existed without Britain for decades did not feel its very existence threatened by threats such as those posed by France in the 1960s.

These were constitutional disputes. There were also financial troubles, particularly in connection with the CAP which by the 1980s was absorbing 70 per cent of the Community's budget and dispensing more on buying up surpluses for resale at a loss than on subsidies to farmers. The Community needed an agricultural policy because so many of its inhabitants were engaged in agriculture and because the Community was not self-sufficient in many foodstuffs, but the policy ran away with itself and instead of turning under-production into sufficiency turned it into embarrassingly large surpluses. In addition, the Community was propelled into competition with the United States where also subsidies to farmers were producing surpluses which had to be sold or dumped in the same markets as Community surpluses. Financially, the Community's problem was aggravated by fluctuations in exchange rates among its members. In order to reduce the impact of these ups and downs the Community created a special 'green' currency, which was, however, abused by currency dealers.

The main mechanism of the CAP was an annual fixing of prices by the Commission. These prices were set somewhat above world prices in order to give farmers and farm workers an attractive standard of living. The Commission undertook to buy surpluses at fixed prices and it levied import duties on foreign foodstuffs whenever prices fell below prescribed levels. When this system turned production into surplus the Community was faced with the logical need to reduce its agricultural acreage. In 1974 Dr Mansholt proposed a ten-year programme to this end but it was rejected by the Council of Ministers who, individually, were afraid of losing votes at their next national elections if they endorsed the plan. There was also a second and more respectable objection to the plan. Cutting agricultural production meant putting out of business smaller farmers who not only commanded emotional and sentimental admiration but were often the economic core of whole communities: an end to farming in a given district could impoverish and depopulate the whole district. Whereas an older industrial revolution had involved a switch from one kind of manufacture to another, agricultural revolution threatened to extinguish one economic activity without replacing it by another.

To finance the CAP the Commission had at its disposal the proceeds of agricultural levies and import duties and also 1 per cent of each member's GNP, but these sources of revenue were intended to cover all the Commission's expenditure, not only the CAP. From 1979 the Commission might cover a year's deficit by requisitioning from members a further contribution not exceeding 1 per cent of its receipts from VAT. In the 1980s this further source ceased to bridge the gap, and since the main cause of the imbalance was the CAP members had strong grounds for challenging the scope of the CAP and demanding a radical review of its subsidies to farmers as a condition for increasing the share of VAT to be transferred to the Commission. But in this respectable endeavour they – particularly Thatcher's Britain – so overdid the rhetoric and underplayed the genuine social obstacles to radical change that little was achieved and in 1984 the VAT ceiling was raised without any serious inroads into the CAP's excesses. Renewed attempts to grasp these nettles at the end of the decade were equally fraught. Coincidentally with the latest round of GATT negotiations (the Uruguay Round, 1986–90) the Commission produced proposals for cutting subsidies by 30 per cent over ten years. A number of members, including Germany, objected that this scale was too harmful to their farmers. The United States on the other hand, which had been demanding cuts in EC subsidies of 70 per cent or more, threatened to abandon the Uruguay negotiations and bring the entire GATT process to a halt.

Thatcher retarded the development of economic union, although her objections to it were not the only obstacle. The worsening of the general economic climate in the 1970s, contemporaneous with Britain's entry into the Community, introduced serious currency instability, stunted the Community's economic growth and raised unemployment to 10 per cent or more of the workforce. The initial aims of the customs' union had been achieved ahead of schedule and the complementary integration of standards and other internal economic policies was proceeding at an acceptable pace, but monetary union was set back both by recession and by Thatcher. A series of programmes for assimilating currencies and credit policy – two from Raymond Barre and a third from Gustav Werner – were set aside. The last proposed a ten-year period during which fluctuations in currencies would be progressively restricted, movements of capital would become unrestricted, and national economic and budgetary policies would become subject to communal consultation with, at the end of the period, full monetary union with a single currency and a central Community bank. This plan was revived during Roy Jenkins's presidency of the Commission and particularly through a fresh Franco–German impulse from Valéry Giscard d'Estaing and Helmut Schmidt. The latter formulated a *European Monetary System* (EMS) with strict limits to currency fluctuations, a reserve currency and a central reserve of $50,000 million in gold and foreign exchange.

The EMS, which came into existence in 1979, was created by all twelve Community members but four of them – Britain, Greece, Spain and Portugal – stayed outside the mechanism (the Exchange Rate Mechanism – ERM) devised to operate the system. The function of the system was to keep currencies fixed in terms of one another, subject to variations not exceeding 4.5 per cent (with greater latitude for Italy) and subject also to annual review and, if necessary, readjustment of the fixed values. The system had two main purposes. Managed exchange rates were a compromise between floating and fixed rates, a way of controlling currency fluctuations whose volatility not only discouraged trade by making future prices peculiarly uncertain but also allowed trading rivals to engage one another in competitive inflation for short-term advantage. Secondly, the system was a step towards a monetary union (EMU) with a single currency. This currency was already taking shape as the ecu (European Currency Unit – ECU). Within a year of the EMS agreement loans were being negotiated and bonds issued in ecus; and within ten years ecu loans exceeded in value loans in currencies other than the dollar, yen, Deutschmark or Swiss franc, and the Council of Ministers had instructed the Commission to prepare the next steps towards monetary union. This last decision followed the signing in 1985 of the Single European Act to complete commercial union by the end of 1992.

The 1980s were marked by economic recession, the sharp aversion of the Thatcher régime in Britain to the EC and most of its ways, and the sudden wakening of its strongest member – Germany – in the wake of German unification. In these years EC members were much occupied in seeking a compromise between the Commission and most of the members on the one hand and Britain on the other. Jenkins's successor, Jacques Delors, was an enthusiast for progress towards economic and monetary union and also the development of joint or common external policies. But he was also aware of the value of British membership and the need for British co-operation when progress should entail revision of the treaty of Rome. So long as Thatcher remained in power progress was obstructed, often acriminiously, but not fundamentally endangered.

By the end of the 1980s an extensive relaunch of the EC seemed in sight when a more serious danger arose from an opposite quarter: Germany, and the impact on European union of the dissolution of the Soviet empire in Europe and the precipitate unification of Germany. This impact had two aspects: the weakening of the German mark and a rush of postulants for full or limited membership of the EC. Germany's economic success had been an essential ingredient in the development of the EC's economic and monetary integration although there were also some apprehension and envy over the strength of the mark. With the unification of Germany the mark became not too strong but too weak. It could not for the time being cope with the costs of unification – the conversion of the eastern mark and the investment required to salvage east Germany's derelict industries and agriculture – and at the same time serve as the anchor of the EMS. Partly in response to generous exuberance and partly to win votes Kohl, overriding the German Bundesbank, promised to redeem the east German mark at parity with the western mark. In order to guard against the inflation inherent in this open-handedness and in the funding of the rehabilitation of the east the German government had in theory choices, but a devaluation of the mark was regarded as inconceivable and higher taxation as more burdensome than borrowing. Therefore interest rates had to be raised and, given the standing of the mark throughout the EC, interest rates in other parts of the EC had to be raised too – or at least could not be lowered. Put crudely, the rest of the EC was paying in higher interest charges part of the cost of reviving eastern Germany when what they needed was cheaper money to rekindle their own growth. The Bundesbank was caught between the need on the one hand for higher rates to counter inflation in Germany and, on the other, the need throughout Europe for lower rates to counter recession. The bank gave priority to the former as by its constitution it was bound to do. It was too unenthusiastic about monetary integration which would shift power from itself to a central Eurobank.

The extent of *Community membership* had been an open question from its foundation. The increase in membership from the original six by 50 per cent in 1973 had not seriously impaired its homogeneity or its efficiency. The accession of Greece (1981) and Spain and Portugal

(1986), however, and the implication that the door was open to yet more postulants – Malta, for example, or Cyprus or Turkey – threatened to make it dangerously cumbrous. Turkey's application was fended off on the pretext that the Community was unable to consider it while preoccupied with the completion of the single market by 1993; less publicly the Community was loath to admit a state whose regard for human rights and civil decencies was substandard. In economic terms the Community had been conceived as a self-help organization of advanced industrial states in north-western Europe with, by political accident, the addition of Italy which had been closely associated with western Europe since the fall of Mussolini in 1943. In the 1980s the Community became an association of twelve states, four of which were Mediterranean or Atlantic and more or less economically weaker brethren. This expansion created administrative problems, economic strains (particularly on the funds established to help poorer regions) and political complexity, but pressures for further expansion persisted both from states in the west which were revising their decisions to remain aloof and then from ex-communist states which hoped to find in the EC short cuts to economic repair. To meet the solicitations of the former the Commission had put forward as early as 1984 a plan for a European Economic Area (EEA), a free trade envelope round the EC itself whereby seven states might enjoy limited membership. The states envisaged were Switzerland, Liechtenstein, Austria, Sweden, Norway, Finland and Iceland (many of which had had bilateral free trade agreements with the EC from the 1970s) and at the beginning of 1994 the EEA was formed by all but the first of these with the EC. By that date, however, the EEA had become not an alternative to full membership of the EC but a step towards full membership, for which the governments of all seven had formally applied. Public opinion, however, was less enthusiastic than governments and when the Swiss government put its application to a referendum it was rejected by the Swiss people, mainly on the grounds that the EC's constitution was too undemocratic but also out of fears of damaging Switzerland's permanent neutrality and the equally prized neutrality of the International Red Cross which was a Swiss body based in Switzerland. For eastern states the EC created a European Reconstruction and Development Bank located in London and negotiated a series of association agreements, postponing the question of full membership and the implicit prospect of turning the EC into a regional organization for Europe within the terms of the UN Charter (and to some extent in competition with the ECSC).

These broad issues were perforce debated simultaneously with the most far-reaching review of the Community's internal structure and purposes. The events of 1989–90 in central and eastern Europe coincided inopportunely with this review and raised the question whether the Community should apply the brake or the accelerator to its own development. Germany chose to put its weight behind purposeful integration and, once more in concert with France, to accelerate economic and monetary union, closer co-operation in external affairs and constitutional changes requiring amendments to the treaty of Rome. At the end of 1990 two conferences assembled in that city to consider such a programme and in the same year the five central members of the Community agreed to abolish reciprocal visas and border controls. This so-called Schengen agreement (it was based on a Schengen computer system) was introduced in 1995 by its five originators and Spain and Portugal. They also envisaged a common policy on political asylum and an integrated police force.

These were the preliminaries to the full conference held at *Maastricht* early in 1992. At Maastricht all twelve members of the Community signed a treaty whose main purpose was to give more precision to progress to monetary and economic union and to append to economic policies a series of principles and mechanisms for the co-ordination of policies in other areas. The treaty created a European Union in which the Community and its organs were subsumed.

The European Union was a federation in all but name (a federation being essentially a union established by treaty, not by conquest). The draft of the treaty acknowledged this fact but the word 'federal' was deleted in deference to British susceptibilities. This deletion did not affect the substance since in all political associations, whatever they may be called, the inescapable question is the delineation of powers between authorities at different levels. In this matter the treaty affirmed the principle of 'subsidiarity' which ascribed authority to the member state in all cases of doubt or apparent overlap between the state and the Community or Union. The European Union introduced a European citizenship in addition to citizenship of its several members and gave all citizens the right from 1994 to vote in elections in the state in which they were residing subject to regulations to be unanimously adopted by the Council of Ministers. (These provisions were similar to proposals made by Churchill to Roosevelt for an Anglo–American union: Churchill also proposed a single currency.) A European Council composed of heads of government and the president of the Commission was created, charged to meet once a year and to present to the European parliament an annual report on the state of the Union. The Union was also charged with the development of common policies in areas outside the Community's immediate concern with economic and monetary integration. In these matters the Union was in effect the Community's Council of Ministers acting in spheres from which the

Commission was entirely and the parliament almost entirely excluded. These spheres were foreign and security policies and internal policies relating, for example, to immigration, drugs, terrorism and some other kinds of crime. This part of the treaty amounted to little more than declarations of intention. The treaty did not rescind the Community's competence specifically to negotiate economic agreements on behalf of the Community with states or groups of states outside the Community.

The most concise part of the Treaty of Maastricht dealt with economic and monetary union, of which the first stage was already in place. The second stage, designed to prepare the way for stage three (full implementation), was set to start in 1994. In stage two a European Monetary Institute, an association of members' central banks, was to oversee and try to secure the preconditions for a single currency and a central bank. The criteria for monetary union were precisely defined in terms of convergence in four categories: price inflation, yields on ten-year bonds, budget deficits as a percentage of GDP and gross public debt as a percentage of GDP. Stage three was to be reached by 1999. From the beginning of 1997 the Council of Ministers might decide that a majority of members (seven) had achieved the neccessary convergence and might in that event set a date for the inauguration of a full union by those members. In the absence of such a decision full union would nevertheless be inaugurated from the beginning of 1999, subject to provisions for exemption for members not yet in a condition to join. The first *terminus ad quem* would put pressure on laggards if they wished to avoid the creation of a two-tier Community. This was the only part of the treaty which committed members specifically to anything not contained in the treaty of Rome or the Single European Act. Britain was expressly exempted from its provisions. More generally the treaty also redefined or extended the Community's commitment to harmonizing policies and law in certain areas such as transport, the movements of funds, professional qualifications and the environment. It was no less important for what it excluded from joint action or consideration: for example, the provision of health services and overall Community principles in relation to such services. The main weakness of this part of the treaty was a yawning discrepancy between the programme for economic unity and the timetable incorporated in the programme.

Social services and the grey areas between economic and social policy became a subject of sharp debate, not so much because the members disagreed with one another but because of a rift between Britain and the other eleven. The latter wanted to advance beyond the Social Charter adopted in 1989 but the British Conservatives adamantly opposed this part of the treaty which was therefore recast as a separate protocol not applicable to Britain. At the root of this disagreement was the continued British disposition (much strengthened by Thatcher) to treat labour relations as a form of combat as distinct from the view, which had played a conspicuous role in the postwar German economic miracle and had been assimilated by much of continental Europe, that economic prosperity required a partnership between capital and labour. Thatcher's successor, John Major, refused to have any truck with a Social Chapter or protocol in the hope of asserting his nationalist credentials within the Conservative Party while at the same time reversing Thatcher's disdain of the Community and British alienation from it. The other eleven members were content to put social matters into an optional protocol as the price for keeping Britain in the fold. The protocol itself was a declaration of unexceptionable general principles which might be invoked by the Commission if it were minded to draft relevant directives which might then be approved or not by the Council of Ministers. The protocol covered subjects such as health and safety at work and equal pay for women and men but excluded others such as the right of association among workers, strikes and levels of pay.

The structure of the Community was little affected by the Maastricht treaty. The Council of Ministers, the stronghold of national governments, retained its domination, although within the Council each member's veto was somewhat diluted by the expansion of majority voting in place of unanimity. From 1995 the president of the Commission was to be nominated by member governments after consultations with the parliament and all other commissioners were to be nominated by governments separately after consultations with the president. The parliament's authority received a marginal boost. Besides being entitled to offer opinions and to amend directives it acquired a new right of veto in limited circumstances and it could by a two-thirds majority censure the Commission and so secure its entire dismissal. It was to be elected by the same method in all states in accordance with regulations to be approved by the Council of Ministers. The European Investment Bank also received a modest boost and a new Regional Committee was created with 169 members and vague advisory powers. It seemed unlikely to have more influence than the equally hazy advisory Economic and Social Committee. All EC members except Denmark and Britain ratified the treaty by the end of 1992. It was rejected in Denmark in a referendum by 50.7 per cent but this decision was reversed after Denmark was granted the right (already conceded to Britain) to opt out of important sections of the treaty on the pretence that the treaty itself allowed such manoeuvres. In Britain, where accession was made subject to parliamentary endorsement (although strictly only its fiscal implications required the approval of the House of Commons) Major narrowly won the day for ratification after some equivocation and secret deals with Irish parties in the House.

The conclusion of the treaty of Maastricht was quickly followed by the collapse of the ERM and the emasculation of the EMS. The EC had approved in 1989 a plan (accepted even by Thatcher) for economic and monetary integration and this plan was broadly incorporated in the treaty but without time limits for any stage. The plan endorsed the extension of the EMS with its ERM to all EC members and although Thatcher continued to object to it the logic of events obliged Britain to join the ERM in 1990. But it did so at an unsustainable overvaluation of the pound and at a rate set without consulting its German or other partners who might be called upon to sustain it. Britain simultaneously cut interest rates and so raised doubts about its commitment to reducing inflation and pursuing the convergencies required for implementing monetary union.

The EMS/ERM was at once a mechanism for regulating the exchange values of its various currencies and a step towards the supersession of these currencies by a single currency. As a mechanism it required and provided for periodic readjustments of exchange rates by orderly government agreement rather than by short-term speculators operating in money markets for their own immediate profit. As a stage towards a single currency it encouraged the belief that adjustments should be increasingly infrequent until the need for them faded away: stability became confused with fixity. In its first years the ERM was used, as intended, to make small but frequent adjustments in rates, but from 1987 none; and after the unification of Germany it lost its essential anchor – a central currency not vulnerable to inflation. By 1992 neither the Italian lira nor the pound sterling could be sustained within their prescribed limits. The lira was devalued by 7 per cent and driven out of the system by speculators to whom the interventionist role assigned to the ERM had been surrendered by inaction. The pound was similarly assailed; the German and French governments intervened to bolster it in the markets. But Norman Lamont, Major's chancellor of the exchequer, preferred to raise interest rates by 5 per cent (unavailingly), spurned German overtures for a general readjustment of currencies together with a reduction of interest rates, and seriously underestimated the effectiveness of speculators in the markets. When Lamont's gamble failed the pound fell below its ERM limit, was suspended and then devalued by 17 per cent; the British exchequer lost billions of pounds. There were compensations, particularly for exporters, in this severe and ill-managed adjustment of an overvalued currency but Lamont's attempt to show that he could run the British economy without regard to the Germans was neither sensible nor successful. The ERM itself was discredited but further vagaries among the currencies were stemmed more quickly than expected. At the beginning of 1994 the European Monetary Institute, the Eurobank in waiting, opened its doors in Frankfurt. Integration remained on course but off schedule.

But if the treaty of Maastricht was a next step in the development of the Community which had grown from six to twelve members, it coincided with events which made that Community obsolete – the emancipation of central and eastern Europe from Soviet rule and the wars in Yugoslavia which created yet more independent states and put the EU's collective leadership in a poor light (see Chapter 8). New states seeking admission to the EU outnumbered its existing members and were almost all of them poor agricultural countries whose admission would further bedevil the common agricultural policy and require a radical overhaul of the Union's administration and finances. A major review was planned for 1996. In 1994 Jacques Delors' tenure of the presidency of the Commission came to an end and Jean Santer was chosen to succeed him after Major vetoed Jean-Luc Dehaene as a protest against the Union's accelerating integration. (The two candidates' views were virtually indistinguishable.) In the same year Austria, Sweden and Finland became members; Norway's admission was repudiated by a referendum.

THE SOUTHERN FLANK

Command of the Mediterranean, although disputed during the Second World War, was not in doubt after it and all the European riparian states (and Portugal) joined Nato over the years and then became members – in the case of Turkey an associate member – of the Community. Malta and Cyprus, members of the Commonwealth from 1964 and 1961, also became associates of the Community.

Historical accident – its shift into the Anglo–American alliance against Germany in 1943 – and the uninterrupted predominance of the Christian Democrat Party after the war made *Italy* an original and unquestioned member of Nato and of the Community. Italy's retreat from fascism converted it into a parliamentary democracy but did not make it prosperous. Post-fascist governments inherited a situation in which 2 million out of a working population of 20 million were unemployed and nearly half of Italy's workers were employed (or unemployed) in agriculture. By 1970 this proportion had been reduced to 20 per cent but the fundamental problems of economic overpopulation remained, for which the classic remedy was emigration (not new, since Julius Caesar had founded Narbonne in Gaul for this purpose). The chief haven, the United States, had been progressively closed by quotas and literacy tests. Some 150,000 Italians left home every year for Australia, Canada and elsewhere, but there was still not enough work for those who remained (and who, since the leavers were the younger, became a senescent population). Schemes for helping the impoverished south failed to check the widening gap between the two halves of the country, and the twin problems of surplus manpower and depressed areas became two of the mainsprings of Italy's postwar European policies. Court Sforza persuaded de Gasperi that the only cure was participation in a European confederation and so Italy joined the European Coal and Steel Community as a founder member in 1951 even though it had no coal and, outside Elba, no ironstone. Instead of growing oranges and lemons for the delectation of richer Europeans beyond the Alps Italy would emulate them, even if it had to import iron all the way from Venezuela to feed the modern steelworks being constructed at Taranto; and twenty years later Italy was exporting 20 million tons of steel a year. In the 1950s GNP was doubled by modernizing industry, investing in training and making good use of surplus labour; industrial output rose from a quarter to nearly a half of total output. The discovery of natural gas in the Po valley gave an unexpected boost from 1958 although supplies proved unhappily limited and this uncovenanted benefit lasted only a few years.

Throughout the whole of the post-fascist period up to and including the elections of 1987 the Christian Democrat Party was the largest in parliament. Unrelieved by respites in opposition and unwilling or unable to produce a new generation of political leaders, it evinced a declining sense of purpose, an incapacity to master economic problems and a laxness in regard to standards of public morality which would have ensured its defeat if its main opponent had not been the Communist Party. The socialist left was riddled by fissures but the communists, like the Christian Democrats, maintained their unity – in spite of the shocks of 1956 and 1968 which lost them some middle-class and intellectual support but did not erode their popular following. During the brief pontificate of John XXIII (1958–63) there was a lessening of papal interference in Italian affairs but no abrogation of the Vatican's special right to intervene under the Lateran Treaty of 1929. Pius XII's excommunication of communists was lifted, there was no repetition of the abuse levelled at President Gronchi for visiting Moscow, and although the debate on civil divorce in Italy in 1969–70 led to a recrudescence of clerical intervention, the church-and-state question did not rise above the level of an accustomed, if sometimes exasperating, family dispute. Italian democracy succumbed neither to communism nor clericalism. Nor did it suffer the fate of Greek democracy although a fascist plot contrived by Prince Valerio Borghese was detected in December 1970. Nevertheless by the 1970s Italy was in poor health politically as well as economically. The government inspired little faith, public services were regularly breaking down, corruption was an open and even popular topic of conversation, inflation rose and in 1974, year of the first oil crisis, the deficit on the balance of payments reached an appalling $825 million (of which $500 million was for oil). West Germany and the IMF came to the rescue but Italy looked like being a permanent drain on the EEC and, particularly after the regional elections of June 1975, a political risk as the communist share of the poll edged up towards that of the Christian Democrats and Italy's allies asked themselves what they would do if the communists came into the government (they had last shared in government in 1947) or if communist successes in elections provoked the right into a coup. The problem was evaded, and thereby prolonged, when in the general election of 1976 the Christian Democrats managed to maintain a small lead over the communists but were forced to form a single-party government without a parliamentary majority and dependent on communist abstentions. The fall of this government two years later revived the arguments for and against a coalition of the two main parties, the Christian Democrat and the communist. Opponents included the United States and the Vatican on the one hand and also the extreme left which regarded communist participation in such an alliance as treason. Aldo Moro, chairman of the Christian Democrats, was kidnapped and murdered by extremists intent on making this point.

The Italian Communist Party (PCI) was exceptional among its associates. It was the largest outside the USSR; it played a leading part in Italian national and local politics, whether in power or not; and it had a strong native intellectual tradition which made it much less foreign than other communist parties appeared to be. It was at the centre of the Eurocommunism which became a talking point in the 1970s and attracted in particular the Spanish and French parties (PCE and PCF).

Eurocommunism was an assertion that communist parties need not be blindly subservient to an international movement and particularly not to the Communist Party of the Soviet Union (CPSU); it was an attempt to deny the Stalinist past. It was also an attempt to bolster the central communist tradition against the New Left which, in many instances still further to the left, was representing communist parties as fossilized anachronisms. The Italian party was reverting to the nineteenth-century view that socialism could win substantial victories by democratic means – that fair elections between a plurality of parties could be turned to good account. This view had been submerged in the twentieth century by Lenin's disciplined democratic centralism, by the sight of democrats taking the side of the whites in Russia after 1917, and later by western democrats' evident preference for fascism over communism. But the older view was never quite lost in Italy.

Palmiro Togliatti was an outspoken critic of the USSR even in the 1950s, and at the conference of communist parties in Moscow in 1969 he alone opposed the Brezhnev theses which sought to justify the Russian invasion of Czechoslovakia and which were accepted even by the Spanish and French parties – the latter without reservations. He found an ally for his more nationalist outlook in the Spanish leader Santiago Carrillo who, after opposing Eurocommunism, was converted and wrote its basic text *Eurocommunism and the State*. The PCI supported the so-called historic compromise in Italy, the alliance of all anti-fascist parties. It approved Italy's membership of the EEC by contrast with the ambivalent PCF, which boycotted the European parliament until 1975 and opposed direct elections to it until 1977. (The first such elections were held in 1979.) The PCI also declared in 1974 that it did not require Italy to leave Nato.

The Italian example carried the Spanish and, more hesitantly and temporarily, the French parties into a loose Eurocommunist entente but there was little positive substance to it beyond disengagement from the taint of Russian tutelage. Communists in north-western Europe, including France, had next to no prospect of winning power through the ballot box, mainly because of the existence of sizeable socialist parties. In southern Europe they might hope to do better but if they did they would almost certainly be prevented from taking office by inflexibly anti-communist armies.

The Italian political order collapsed with disconcerting abruptness in the years around 1990. The fundamental principle of Italian politics from 1947 had been the exclusion of communists from government. The Christian Democrat Party had ruled, alone or in coalitions of which it was an essential partner, throughout this period and formed the nucleus of a wider governing class which embraced financiers, industrialists, linkmen with their counterparts in the Vatican, the Mafia and other more or less right-wing interests whose common anti-communism and devotion to their several material interests overrode their differences, subordinated the free play of democratic politics and nourished widespread corruption both by the smaller parties which exacted a price for their part in keeping communists out of office and by the larger parties which needed more and more money to keep their allies content and their leaders in a befitting style of living. With the collapse of communist rule in Europe the system's anti-communist lynchpin disappeared. At the same period economic strains, besides threatening the national economy and the national mood, punctured the political order which was funded by mulcting the public purse. Malpractice, long suspected and including trafficking by political leaders in the allocation of lucrative public contracts, was so widespread – the socialist prime minister Bettino Craxi was as deeply implicated as Christian Democrat leaders – that the system could not be salvaged for want of sufficiently untainted segments. Courageous action by judges and prosecutors against Mafia bosses was followed by criminal charges against eminent politicians and businessmen. Local elections in 1991 and national elections in the following year reduced the reputable remains of the Christian Democrat Party to insignificance. Demands for political reform were only partially met by changes in 1993 with the aim of reducing the number of small parties. With thousands of politicians under police investigation, tens of thousands of public servants in fear of prosecution or dismissal, and all the established parties more or less discredited, new parties appeared with promises of radical cleansing of public life and vaguely miraculous cures for a huge public debt which was growing annually by sums equivalent to 10 per cent of GDP. The general effect was the polarization of politics round an empty centre.

From this complex cataclysm two forces emerged: a new left-wing force whose principal ingredient was the Party of the Democratic Left (formed by the more flexible and moderate segments of the old Communist Party) and a right-wing alliance consisting of an old party, a new party and something which was hardly a party at all. The old party was the Movimento Sociale Italiano (MSI), formed in 1946 as a relic of fascism. The party had operated tactfully within the rules of democratic politics, gradually attracted members and votes from other right-wing groups, became a not unimportant minor party and on occasions provided the parliamentary votes need to keep a Christian Democratic government in office. After nearly half a century its first leader Giorgio Almirante gave way (1987) to Gianfranco Fini who, although ousted temporarily in 1990–93, recovered power in time for the elections in 1994, in which year the party changed its name to Alleanza Nazionale (AN). Fini represented the traditional conservative bulk of the party which also had a radical revolutionary wing. Besides Roman Catholic causes common to many right-wing parties in Europe the AN–MSI preserved such fascist notions as the corporatist state, Italy's destiny as a Mediterranean power and its claims to territories along the eastern shores of the Adriatic. It was stronger in the south and centre than in the north.

By contrast the newly created Northern League had little support in south or centre. The League was an amalgam of lesser Leagues, beginning with the Venetian League founded in 1983. It embodied the conservatism of small businesses against the over-mighty conglomerate and the over-intrusive state. It regarded the centre and Rome as a land of corrupt drones and the south as a land of delinquent layabouts and advocated regional devolution in a loosely federated Italy. Its leaders in

1994 were Umberto Bossi and Gianfranco Miglio, leaders of the Lombard League which had overtaken the Venetian League as the voice of northern Italy. To these two parties were added in 1994 *Forza Italia* and Silvio Berluscone, the first little more than a slogan and the second one of Italy's most successful businessmen, president of Fininvest, the holding company of his own creation whose assets included much of Italy's press and television. Berlusconi's reasons for entering politics on the right (he had earlier supported and befriended politicians of the left) included the demise of the Christian Democrat Party and his wish to step into its shoes, the need to defend his personal business empire (particularly his interests in the media) from any attempt to restrain or regulate it, alarm at the country's crumbling economy, crumbling political order and prospects of communist electoral victories, and the need for some new party to bridge the unbridgeable gap between the two main parties of the right, the AN–MSI and the Northern League. With these and other parties of the right Berlusconi created in January 1994 the League for Freedom (*Polo della Libertà*) which won the elections two months later. His political programme was more rhetorical than precise. It sought to exalt and liberate private enterprise, diminish government, reduce government expenditure, create a million new jobs and introduce a presidential rather than a parliamentary democracy; but its principal appeal was its novelty.

Berlusconi and his allies won 366 of the 630 seats in the national assembly with 43 per cent of the votes cast; left-wing parties won 213 seats with 34 per cent; centre groups a mere 46 seats. The result therefore was a victory for the right but not obviously for democracy or political cleansing. With *Forza Italia* emerging as the largest single party in parliament Berlusconi became prime minister with embarrassingly large promises of lower taxes, more jobs and clean government. To the power which he already had through his newspapers and television he added the endorsement of the ballot box. How far he regarded these two sources of power as compatible was unclear, a question all the more pressing since *Forza Italia* alone had no majority in the assembly where it was dependent on the expressly regional interests of the Northern League and the sometimes disquieting prejudices of the semi-fascist AN–MSI. As a successful tycoon Berlusconi's outlook was European and internationalist but his allies' outlook was either indifferent to these wider issues or the reverse of his. Berlusconi presented himself as a break with an unsavoury past but his own links with that past were not negligible. The most obvious difference between *Forza Italia* and the Christian Democrats was the former's much weaker links with the Vatican and the Italian ecclesiastical hierarchy – links retained, however, by the rump of the Christian Democrats, renamed *Partito Popolare*.

Berlusconi's promises to eliminate corruption, reduce taxes, salvage the economy and reform the constitution all came to nought. He himself came under investigation for corruption through his master corporation Fininvest; he raised taxes and introduced a budget which was condemned by the World Bank and other international bodies for failing to tackle a frightening deficit; and he refused to relinquish control over his newspaper and television empire. (Berlusconi, it could be said, understood the power politics of his own time as had Pompey the Great two thousand years earlier. Pompey risked forfeiting the consulship in Rome rather than lay down the command of his army. Berlusconi risked the premiership rather than divest himself of the power which his media empire gave him.) His term in office lasted seven months, less than the average of the fifty ministries which preceded his from the end of the Second World War. He was forced to resign when the Northern League defected from his coalition. His demands for a dissolution of the assembly and fresh elections (in which he hoped to win more seats) were denied by President Oscar Scalfaro who found in Lamberto Dini, a stop-gap prime minister of unobtrusive competence who was appointed at the end of 1994 and stopped the gap for a whole year. But three years after the collapse of the old political order nothing coherent had taken its place.

At the eastern end of the Mediterranean the western alliance secured a lodgement in Greece and Turkey which was strategically important but also embarrassing because of the weaknesses and mutual hostility of those two countries.

From the end of the Second World War to the end of the Cold War *Turkey* was a key point in the American policy of containing the USSR. It was (with Greece) the specific object of the Truman Doctrine, was included in the Marshall Plan and became a member of the OEEC and Nato. It had a foot in Europe and another in the Middle East without belonging comfortably to either. Turkey was a welcome member of European organizations dominated by the United States but, even apart from the special case of Greek hostility, not so readily acceptable by purely European bodies, the EC in particular. It preferred to hold aloof from Middle Eastern affairs but could not be indifferent to them and was at least tangentially involved by Dulles's schemes for a Middle Eastern extension of the American cordon round the USSR, in the clashes between the old order and the new personified by Nasser and those who emulated him, in the use of Turkish territory by the United States in emergencies such as the landing of American forces in Lebanon in 1958 and in the Gulf War of 1991, in the revolutionary increases in the price of oil in the 1970s, and in the Kurdish problems which Turkey shared with its Middle Eastern neighbours. Turks were also vaguely conscious of being a detached western outpost of a larger Turkish nation most of which had stayed in mid-Asia and been engulfed by Tsarist Russia and the USSR.

Turkey was a new state created out of the ashes of the Ottoman empire by Mustafa Kemal, known as Ataturk, a secular and modernizing parvenu not unlike his contemporary Reza Shah in Iran. Ataturk ruled Turkey from 1923 until his death in 1938. His concept of the state was national, secular, western and parliamentary. He believed that the right place for religious leaders was the mosque and not the government. His friend and comrade-in-arms Ismet Inönu (president 1938–50 and prime minister 1961–65) maintained this tradition but the party political mechanisms of this state functioned in a nation which had not taken them to its heart: the parliamentary stage was significantly smaller than the national with the result that the political parties and their leaders, while treating each other as adversaries, were regarded by many as linked cogs in a system which half the country rejected. The authority of the politicians was correspondingly weakened and their policies tacked uneasily between the basic principle of a secular state and the voters' attachment to an older Islamic tradition – secular parties appealing to Muslim masses. Turkey after Ataturk was a country as deeply riven as Russia in the century after Peter the Great, the modernizing tsar. Politicians and the political system were further weakened when their modernizing ambitions failed to deliver economic rewards for more than a small minority of the people. One consequence was the dominant influence of the army, itself twice modernized (by Ataturk and through the American connection after the Second World War), broadly anxious to maintain the democratic parliamentary system and unwilling to assume the responsibilities of government except sporadically and briefly at critical moments.

With the end of the Second World War a new party, the Democratic Party (DP), a splinter from the Republican People's Party of Ataturk and Inönu, appeared and in 1950 all but extinguished the RPP in parliament. Its leaders, Celal Bayar and Adnan Menderes, became president and prime minister. They enjoyed the first flush of the Marshall Plan, encouraged private industries, subsidized agriculture and sought to curb the political power of the army. The DP's modernizing zeal improved the economic infrastructure and expanded Turkey's industrial sector but at a cost and pace and with a dogmatic hostility to planning which drove the country into bankruptcy, from which it was rescued by the IMF and accommodations with its foreign creditors. Economic mismanagement was compounded by Menderes' increasingly crazy despotism and the latter half of the decade was punctuated by demonstrations and riots, including indiscriminate attacks on Greeks and their businesses. After a carefully prepared coup in 1960 the army arrested all the DP's members of parliament, dissolved the party, put nearly six hundred persons on trial and executed three of them including Menderes. General Cemal Gürsel became head of a ruling National Unity Committee of thirty-eight with Colonel Alparslan Turkes – a Cypriot and something of a cultural, religious and nationalist fanatic – as (temporarily) the power behind the throne. This régime was weakened by plots and coups within the military and produced a new constitution which was greeted at a referendum by a surprisingly large adverse minority. The dissolved DP was revived as the Justice Party – one of a dozen new parties – and came a close second to the RPP at elections in 1961. It regained power in 1965 with its new leader Suleiman Demirel, a self-made engineer, cautious conservative and powerful speaker with a common touch. In or out of office Demirel became the outstanding political figure in Turkey for the next thirty years.

To the perennial problems of economic stability and progress the 1970s added that of internal order, threatened by dissidents of both left and right and by Kurdish separatism (see Note A to Part Three). In coping with these problems Demirel and his principal adversary Bulent Ecevit, who succeeded to the leadership of the RPP in 1972, were hampered by the proliferation of parties at both ends of the political spectrum, notably the Islamic and nationalist Grey Wolves formed by Turkes. Parliamentary majorities became more difficult to secure and military interference more disconcerting and unpredictable. The constitution of 1961 had given the military a special status within the state but military interventions took a perplexing variety of forms from the declaration of emergencies and imposition of martial law to the creation of temporary military authorities alongside and in ill-defined tandem with civilian bodies. Demirel and Ecevit moved in and out of office while the economy plummeted and public order evaporated. After 1977 neither of the two principal parties had a parliamentary majority. Demirel formed a coalition with Turkes but defections from his own party forced him to resign. Ecevit was obliged to introduce martial law in 1978 and to resign in his turn in 1979. Anti-communist measures – the term communist being very loosely interpreted – included the suppression of press and academic freedoms, arrests by the thousand and the recourse to terror and torture as a routine. Disorder was compounded by fatuity when more than a hundred ballots failed to elect a new president in 1980. Disorder also destroyed hopes of economic recovery (which was further retarded by the oil price rises of the 1970s), stimulated the more militant tendencies among the Kurds, turned to political assassination and compelled the army once more to assume direct rule in order to fend off anarchy. All parties were dissolved, local as well as national politicians and thousands besides were arrested; political discussion was banned. General Kenen Evren was proclaimed head of state, order was restored with a ruthlessness rendered acceptable to many by the fears which had preceded the coup; and in 1983 the army installed a new civilian régime.

The new man was Turgut Ozal, another engineer and an economist who had worked in the United States and the World Bank and as an adviser to Demirel who promoted him to the cabinet in 1979 with special powers over the economy. Ozal formed the Motherland Party and won in 1973 a clear victory over all other parties, including one preferred by the army and led by a general. He curbed inflation and rectified the balance of trade and payments by rigorous monetary controls, deflation, high interest rates, low wages and reduced subsidies. The outcome was mixed: on the one hand substantial growth in GNP and exports, useful public work on roads, irrigation and telecommunication and a boost for tourism; on the other hand an explosion in speculation and corruption and a calamitous fall in living conditions for all but a small number of entrepreneurs. The Motherland Party was an uncomfortable amalgam of modernizing westernizers and Islamic purists and nationalists and Ozal was no more successful than his predecesssors in pruning the bureaucracy or imposing taxes and making people pay them. Remittances declined alarmingly as distrustful workers in Germany and elsewhere preferred to keep their money where they were. Financial scandals and allegations of nepotism eroded Ozal's personal standing and public confidence in his doctrinaire market capitalism. He secured in 1987 after laborious negotiations a fresh economic and defence agreement with the United States and readmission to the Council of Europe from which Turkey had been expelled when its constitution was abrogated in 1980. Ozal was gradually replaced by the evergreen Demirel. Ozal moved in 1989 from the premiership to the presidency – the first civilian president for twenty-eight years – and when he suddenly died in 1993 he was followed in each office by Demirel, who appointed Turkey's first woman prime minister, Tansu Ciller.

Demirel faced a worsening Kurdish situation and the Gulf War against Iraq. The latter aggravated the former which was already peculiarly intractable because the Kurds in Turkey refused to become Turks while the Turks refused to contemplate a Turkish–Kurdish state. In Ottoman days Turks and Kurds were alike Muslims (and so allies against the persecuted Armenians) but in Kemalist and post-Kemalist Turkey the religious bond was much less of a counterweight to their ethnic differences and the Kurds achieved neither a distinct share in the new state's governance nor a state of their own: nor were they united in deciding which of these unachievables they preferred. In the aftermath of the Gulf War large numbers of Kurds tried to flee from Iraq into Turkey which repulsed them.

Both Ozal and Demirel were intrigued (as Enver Pasha had been intrigued in the early years of the century) by the opportunities offered by the emergence of new independent Turkish states in central Asia (see Note A to Part Four). Ozal visited them and invited their leaders to Istanbul. For Demirel there were alternative possibilities: an association of Turkey with them and with Iran in an Islamic block, or a Turkish association without and to some extent against Iran which was Shi'ite, theocratic, fundamentalist and not Turkish. Awkwardly Turkey, although it could claim kinship with all these new states except Tajikistan, had no borders with any of them. Turkey's wariness of its near or fairly near neighbours, in particular Russia and Iran, was balanced by their importance as markets for Turkish exports. Hence Turkey's promotion of a Black Sea Economic Co-operation Treaty which was signed by eleven states in 1992, a similar Caspian Sea Organization and an Economic Co-operation Treaty with Iran and Pakistan. The United States remained a firm friend – President Bush visited Turkey in the aftermath of the Gulf War – and would have liked to see it join the EU, if only as a reward for its contributions to the war, but it got no further than associate membership of the WEU in 1992 and a customs agreement with the EU in 1995.

Economic problems did not relent in spite of IMF aid. Foreign debt, which had massively increased in the 1970s and thereafter, was out of control; an alarmingly declining balance of trade was threatening Turkey's ability to borrow; the privatization prescribed by the IMF was in haphazard confusion; budget deficits remained heavy and inflation high; an influx of Kurdish refugees into the main cities increased poverty and disorder. The government intensified hostility against the Kurds, even to the point of invading Iraq on the ground and in the air to destroy Kurdish bases and disrupt supplies. It also resorted to violence against (Shi'ite) Alawis demonstrating against official persecution of their relatively liberal teaching. But a general sense of failure afflicted the ruling True Path Party and, first in municipal elections and later in a general election, the Rafeh or Welfare Party led by Necmettin Erbakan scored startling gains with calls for a return to Muslim traditions, puritan patterns of behaviour and a renewed self-confidence not always distinguishable from abrasiveness (in particular in relation to Cyprus and to Turko–Greek relations generally). At the general election Erbakan won more votes than either of the two main secularist parties: all three leading parties came with 1–2 points of each other and well short of a governing majority.

The fighting against Italians and Germans in *Greece* in the Second World War was sharp but short, but the ensuing occupation was exceptionally severe. The mainly agricultural economy was devastated; bridges and railway stock was almost totally destroyed; three-quarters of

Greek shipping was lost; starvation in Athens was severer than almost anywhere in Europe. The war engendered a civil war which outlasted the wider war by several years and, following atrocities on both sides, seared the Greek consciousness for a couple of generations. A communist attempt to seize power in 1944 was foiled by British troops. The eventual defeat of the communists in 1949 was effected with American aid which included the first use of napalm. However welcome the outcome, Greeks resented British insistence on the restoration of the monarchy and the American support for the political right which characterized the next thirty to forty years. The royal line, inaugurated by a Danish prince nearly a hundred years earlier, was not only tainted with acquiescence in the prewar dictatorship of General Metaxas but, more generally and often unfairly, was suspected of a willingness to subserve the interests of one foreign power or another. King George II returned to Athens after the war in much the same way as Louis XVIII returned to Paris in 1814 – in the baggage train of foreigners – and although his brother Paul who succeeded him in 1947 was widely accepted as a symbol of anti-communist unity so long as a plausible communist threat lasted, the defeat of the communists left him vulnerable to charges of being the tool of the Americans and the Greek right. That defeat was secured by the Yugoslav secession from the Stalinist bloc (Tito needed to mend his fences with the United States) and by the creation of a large, well-equipped and well-trained Greek army with rigorously selected right-wing officers. The continuance of the civil war to the end of the 1940s was an economic and social disaster, partially relieved by American aid, and the major determinant of Greek foreign policy. Greece joined Nato in 1952 and became a cog in the formal alliance of western Europe with the United States against eastern Europe and the USSR, but an uncomfortable partner since the abiding concern of Greek policy was hostility not to the USSR but to Turkey, a valued friend of the United States.

After the civil war and a short period of political instability the conservative governments of Field Marshal Alexander Papagos and Constantine Karamanlis (1952–63) provided tranquillity and the beginnings of economic recovery. Under Karamanlis, who had an unprecedented run of eight years in office, the currency was rescued and tourism, which became a bonanza with the invention of the package holiday, was skilfully developed; but modernization was interpreted as industrialization which, in competition with other industrial countries, had disappointing results and widened the gap between the cities and a neglected countryside. A scarcely disguised Communist Party re-entered the political arena but never got more than 15 per cent of the vote, except in the special circumstances of 1958 when its 25 per cent included a big protest vote against the government's handling of the Cyprus crisis. The centre was fragmented by personal jealousies and out of office until, in 1963 and under Papandreou's leadership, it won more seats in parliament than any other party. During the next few years Papandreou initiated a programme of social improvement, notably in education and health, but he soon got into trouble with the palace, where on Paul's death his young, inexperienced and badly advised son Constantine now reigned, and he roused the hostility of the class of persons who equate all social reform with communism. This hostility was accentuated by the activities of the prime minister's son Andreas, an economist who had been induced by Karamanlis to exchange an American professorship for a non-party professional post in Athens. Andreas Papandreou was a social democrat of the type found in Scandinavia or the left wing of the American Democratic Party (he had been an active campaigner for Hubert Humphrey); his incursion into Greek politics irritated other leaders more particularly because the Centre Union led by his father was not a left-wing party like the British Labour Party so much as a centre or right-centre party like the French Radical Socialists. He became a useful bogey for the right which began to raise the cry that the country was in danger from communism, and in 1965 he was alleged to have become implicated with a left-wing secret society in the army called Aspida – a society which can hardly have been very left-wing if it existed in the Greek army, and a charge which was probably a put-up job (after the coup expected revelations failed to materialize).

So long as king and prime minister worked in harmony there was little chance of a coup, but after an initial amicability these two fell out over the control of the armed forces. The king regarded this control as part of his royal prerogative, to be exercised through a minister of defence acceptable to him. Papandreou, while retaining the palace's man in the post, believed on the contrary that the armed forces should be subordinate to civil control exercised through the prime minister and cabinet. The clash came when Papandreou sought to change the chief of staff and the minister of defence refused to dismiss him or to accept his own dismissal from the government. Papandreou decided to solve the tangle by becoming his own minister of defence but the king refused to appoint him and dismissed him from the premiership. The king had meanwhile been intriguing with some of Papandreou's party colleagues who were induced to abandon him and so bring the government down in parliament. There followed a period of squalid manoeuvres accompanied by minor demonstrations in the streets and a few strikes, until the moderate elements of both the right and the centre came together to put an end to an unedifying spectacle which had been started by the king's unconstitutional behaviour and to hold office pending fresh elections. But the extremists on the right, assuming as did most people, that the Centre Union would win the election, resolved that it should not take place.

This was the immediate cause of the coup of April 1967 by a small group of army officers, a grade below the top ranks and outside the upper-class crust which had normally monopolized these upper grades. They owed their advancement to the expansion of the army after the Second World War and inasmuch as this expansion had been a riposte to a communist threat they were in some sort its product. They were also fanatical, if unintelligent, anti-communist salvationists who, no less than Hitler, saw politics in black-and-white terms, a simple and fierce contest between good and evil. Their coup was therefore not a conservative one but a radical one, although it had at the start the acquiescence and even the support of traditional conservatives (as the Nazis had been supported by aristocrats like von Papen and by conservative industrialists). Whether the coup had also American support is doubtful. American interference in Greek affairs was undoubted and so was American suspiciousness of the centre politicians who were ousted by the coup, but of the variety of possible coups under more or less open discussion at the time the Americans may be presumed to have preferred something more traditional than the one which in fact occurred. Colonels Papadopoulos and Makarezos and Brigadier Pattakos were neither particularly eminent in the service nor particularly appealing to the public. What is certain, however, is that the Americans were widely assumed to have had a hand in the plot and behaved afterwards as though they had. Their support of the régime was, in its turn, conditioned by the balance of American and Russian forces in the eastern Mediterranean and by the continuing war in the Middle East.

After the coup little was heard of the communist threat which was supposed to justify it, and the seizure of the papers of the left-wing party failed to produce any evidence in support. The coup was a simple seizure of power by a handful of military bigots. At the end of the year the king attempted a counter-coup which was as incompetent as the first coup was efficient; after a few hours he fled to Rome. The new régime dismantled the apparatus of the state, purged the armed forces and the Church, annulled Papandreou's social reforms, browbeat the judiciary and established strict control over the press and broadcasting. In addition it resorted to torture and brutality on a big scale and, as an international commission of lawyers subsequently found, as a deliberate instrument of policy. Torture by the police was not unknown in Greece but the extent to which it was practised and approved by the régime appalled the outside world when the facts became inescapable. As a result charges brought against Greece in the Council of Europe by the Dutch and Scandinavian governments for breaches of democratic safeguards were amended to more serious charges under the European Convention of Human Rights and Greece was forced to leave the Council in order to escape suspension from it. The American government, however, viewed these proceedings with no more than embarrassment and after a brief period of indecision gave the régime its accolade in the shape of arms supplies. Greece as a *place d'armes* for Nato seemed more important than Greece as a conforming member of the society of free and democratic nations which Nato was proclaimed to be.

The Greek dictatorship, which began as a triumvirate of colonels (George Papadopoulos, Stylianos Pattakos and Nikolaos Makarezos), was modified in 1971 when the last two were eliminated by the first; in 1973 when, following a mutiny in the navy, the monarchy was abolished; and later in that year when the army transferred effective power from Papadopoulos to Brigadier Dimitrios Ioannides, the chief of the military police. The funeral of George Papandreou earlier in the year had provided the occasion for a portentously large demonstration in Athens against the régime and students in particular continued to harass it with an indignant boldness which cost them dear. The régime collapsed in 1974 under the weight of its bungling in Cyprus where it tried to get rid of Archbishop Makarios and annex the island in the belief that the Americans would prevent Turkey from interfering. The Americans had twice before blocked a Turkish invasion but this time they did not. Makarios escaped, the Turks invaded and the Greek government, having brought upon itself a confrontation with Turkey, ordered a general mobilization for which no preparations had been made. Karamanlis, who had been living in exile in Paris, returned to form a democratic coalition government and consolidated his position by winning for his party a clear majority over all others in elections held at the end of the year. The electorate also voted by two to one to abolish the monarchy.

Karamanlis knew that Greece needed friends: he renewed the campaign to join the EEC and visited Balkan and other communist capitals. He was also anxious to defuse conflicts with Turkey which had become many and acute. He had exceptionally wide popular support and a rare overall majority in parliament. Elections in 1977 reduced but did not destroy this majority and when he ascended to the presidency in 1980 his colleague George Rallis moved smoothly into the premiership.

The most pressing of the Greco–Turkish quarrels lay in Cyprus (see Note C at the end of this Part) but it was not the only one. There was also a complex of disputes over the Aegean. At the time of the Turkish invasion of Cyprus Greece fortified islands in the eastern Aegean which had been demilitarized by the treaties of Lausanne (1923) and Paris (1947). There were also disputes over territorial waters, the continental shelf and air traffic control.

Territorial waters – that is to say, the reach of sovereignty beyond the water's edge – created no troubles so long as both sides observed, as they did, the traditional six-mile limit. On this basis Greece held sovereignty over 35 per cent of the Aegean, Turkey over less than 10 per cent. If, however, the limit were stretched to twelve miles, the area of Greek sovereignty would be almost doubled, the Turkish would be enlarged only slightly; the points of direct contact between international and Turkish waters would be drastically reduced; and Turkish vessels would be obliged to take passage through Greek waters or stay at home. The international community had not endorsed a twelve-mile limit for territorial waters but it did move in that direction when the first international conference on the law of the sea (1958) recommended that for certain defined purposes (customs, immigration and health control) national jurisdiction be extended from six to twelve miles. Neither Greece nor Turkey signed the convention drafted by the conference, but the trend was one calculated to disturb the Turks. (This issue was revived in 1994 when, a new law of the sea coming into force, Greece threatened to exercise its right to a twelve-mile limit which would give it sovereignty over nearly three-quarters of the Aegean.)

The continental shelf, a postwar concept introduced into international law by the United States, caused more friction. The question here was the right to exploit, and the proprietorship of, the seabed beyond the limits of territorial waters. Rights in the shelf carried with them no rights in the waters or the airspace above it. A convention, called the Geneva Convention, elaborated by the 1958 conference, was signed by Greece but not Turkey. It defined the debatable area as the seabed beyond territorial waters if no deeper than 200 metres or capable of being exploited for the extraction of minerals. It provided that islands should confer on their sovereign owners the same rights – and Greece has over 2,000 islands in the Aegean. Exploitation of rights in the shelf was, however, to be subject to freedom of navigation, fishing and scientific research. Where two sovereign states confronted each other in such a way as to create overlapping shelves, the median line should normally constitute their submarine frontier. In the disputes over the Aegean Greece maintained that this convention, being declaratory of international law, was legally unalterable and so not politically negotiable. Greece had some support for this contention in decisions of the International Court of Justice and also for the even more highly contested assertion that islands had continental shelves of their own. All these contentions were opposed by Turkey.

The issue became acute at the end of 1973 when Turkey published a provocative map of the Aegean continental shelf and granted submarine exploration licences to a parastatal corporation. A few months later it despatched the survey ship *Candarli* into the Aegean with a massive escort of warships. This move was quickly, although coincidentally, followed by the invasion of Cyprus and a consequent exacerbation of the conflict over air traffic control.

From 1952 air traffic control over the Aegean had been exercised by Greece; Turkish flights had to be notified to Athens and Turkish pilots in the area had to take instructions from Athens. But during the first stage of the Cyprus crisis Turkey claimed the right to take control over the eastern half of the sea. Greece rejected the claim and declared the Aegean airspace unsafe as a consequence of the conflict. Turkey later extended its claim to cover international as well as Greek and Turkish flights and, in the Greek case, military as well as civilian flights.

The years 1974–75 were a tense period, sporadically lightened by bouts of discussion between Greek and Turkish ministers or lawyers. On both sides the parliamentary opposition was more bellicose than the governments. In February 1976 another Turkish ship, *Sizmik I*, carried out a series of surveys accompanied this time by a single warship only. Nerves began to tingle again, particularly when the Turkish government, under pressure from its own opposition, rejected a Greek proposal formally to renounce force as a way of settling the two countries' disputes. Greek approaches to the Security Council and the International Court provided no alleviation. The Court refused to grant an interim protective order or to accept jurisdiction, while the Security Council passed the buck back to the protagonists with only platitudinous exhortations to settle their problems themselves. This was marginally a victory for the Turks, who preferred bilateral political discussions to international intervention, juridical or political. Karamanlis and Ecevit met in 1978 and instituted regular meetings at lower levels at the rate of three or four a year. They served to prevent the situation from getting worse but there was no substantial lifting of the clouds until the military takeover in Turkey in 1980, at which point the new Turkish government made a concession on another front.

In protest against its allies' inactivity at the time of the Turkish invasion of Cyprus in 1974 Greece had withdrawn from co-operation with Nato. Karamanlis wanted to return to full partnership but Turkey under civilian rule tried to set conditions, in particular by securing in advance a redefinition of Greek and Turkish responsibilities in the Aegean. These preconditions were dropped by the new Turkish government in return for a Greek promise to exclude no Greco–Turkish issue from future discussions. Greece therefore returned to Nato with all these issues open and undecided but avowedly negotiable.

The foremost aim of Karamanlis's foreign policy was to join the EEC. His reasons were political and economic. On the political side nearly all Greeks in 1974 found dependence on the United States obnoxious because they believed that the Americans had not only supported the colonels' dictatorship but had helped it to power in the first place. There was no thought, except among communists who had done badly in the elections, of switching to the Russian camp. But in the EEC Karamanlis hoped to find friends against Turkey (the United States always found it difficult to choose between the two) and a barrier against another lapse into dictatorship in Greece. Economically too Greece needed the EEC. The dictatorship had reversed the favourable economic performance of the 1960s when prices in Greece had been remarkably stable and the booming economy of Germany had provided some relief for Greece's chronic state of economic overpopulation. The dictators, anxious to win support at home, had distributed favours to various classes. In order to do this they had inflated the money supply with the result that the lucky favourites had stimulated an import boom which had seriously upset the balance of payments and produced heavy and largely short-term foreign debts. Inflation reached an acknowledged 35 per cent and was certainly higher.

Furthermore Greece, as a predominantly agricultural country, faced a basic problem which was politically insoluble without the goad of foreign pressures such as an organization like the EEC might provide. Greek agriculture employed 40 per cent of the working population but contributed only 16 per cent of GNP. (By the date of entry to the EEC these figures had changed little: agriculture still provided work for 32 per cent of workers and contributed 14 per cent to GNP.) Greek agriculture, like agriculture in other Mediterranean countries, was as much a way of life as an economic activity. The land was owned in small and often separate parcels (the national average was about twelve acres, most holdings much smaller), Greek convention and testamentary law promoted fragmentation, and the consequent inefficiency drove peasant proprietors into the towns whence they neglected their holdings either through absenteeism or because of the family feuds which were a common consequence of the division of property enforced by law. The produce of the land went primarily to the rich foreign markets of western Europe; the domestic market got what was left over and since demand exceeded supply the Greek consumer had to pay high prices. The trade, foreign and domestic, was in the hands of middlemen who made their profits not only in exports but also at home since they bought the whole of a farmer's crop early in the year before domestic prices had begun to rise and sold it (after satisfying export markets) on a rising market. In order to appease the farmers, who were a substantial proportion of the electorate and also commanded more sympathy than middlemen, governments disbursed subsidies which condoned the structural shortcomings of the industry and paid for them out of the public purse. While economists believed this to be mismanagement, politicians could see no alternative – unless membership of the EEC were to enforce changes in practices which were incompatible with the EEC's own rules. Whether this was what the EEC was for had not been discussed among its members when Greece formally applied for full membership. By a treaty of accession signed in 1979 Greece was to become a full member at the beginning of 1981. Andreas Papandreou's socialist PASOK and the communists boycotted the parliamentary debate on the ratification of the treaty and opposed membership of the EEC, mainly on the grounds that it would inhibit the development of Greek industry and intensify the parasitic relationship of the Greek on the European economy, but when Papandreou came to power later in that year he made no move to annul Greek membership. In the event the economic consequences of joining the EC were disappointing: a big boost for fruit-growers (curtailed after Spain and Portugal joined too) and substantial financial help which failed, however, to prime industrial reorganization, attract foreign investment or alleviate Greece's severe trade and payments deficits.

In 1981 the conservative New Democracy party lost power to PASOK as a consequence of an overwhelming desire for changes and of the personality of Andreas Papandreou who, besides being an eminent economist, had all the trappings (and many of the defects) of a commanding pasha. He understood that Greece was not a modern industrial state nor capable of becoming one but a poor country half-way between the wealthy west and a Third World dependent on richer countries and entitled to special consideration from them. He was, like many Greeks of all political allegiances, aggressively anti-American and promised to close American bases in Greece (but did not). His party had a social programme which amounted to the beginnings of a welfare state which he proposed to finance by making people pay their taxes and by pruning expenditure on public administration. He introduced sensible and moderate social reforms and had some modest and temporary success in reducing inflation and the deficit on external account. But he did not tackle Greece's cumbrously over-centralized and inefficient government machinery, his rule was autocratic and unpredictable, his social policies were fiercely attacked by the Church and the moneyed classes, and this accumulation of obstacles destroyed his government when, in addition, he was assailed by public and private scandals. These scandals, economic failure and its leader's authoritarian handling of a mediocre cabinet caused the collapse of the PASOK government in 1988.

In external affairs Papandreou followed the example of Karamanlis in seeking better relations with Greece's Balkan neighbours but not at first with Turkey. Karamanlis had cultivated Greece's traditional friendship with Yugoslavia without, however, going as far a formal pact and had tried tentatively to promote some form of Balkan entente. He received Zhivkov in Corfu in 1979 and later in that year Greece, Yugoslavia, Bulgaria and Romania all attended a conference on communications in the Turkish capital. Karamanlis became the first Greek prime minister to visit Moscow since 1917 and the first ever to visit Beijing. Papandreou improved relations with Albania but allowed relations with Turkey to rise to nearly boiling point in 1987 over oil exploration in the Aegean. In 1988 the Turkish prime minister made an unprecedented visit to Athens and more temperate relations prevailed.

Papandreou's successor, Constantine Mitsotakis, had little room for manoeuvre. His victory in 1988 was not decisive, he was forced into alliance with incompatible parties united only in their dislike of PASOK, he secured a (narrow) majority only in 1990 and he was plagued by economic crisis – the highest rate of inflation in the EC, a debt approaching 150 per cent of GDP, failure to meet the conditions prescribed for aid from the IMF and serious strikes against his attempts to tackle overmanning in the public sector. The disintegration of Yugoslavia, the revival of the Macedonian question and the opening of the Greek frontier with Albania (which had been sealed with wire fences by the latter's communist rulers) turned the Balkans into quicksands where it was difficult for Greece not to put its feet wrong. In spite of the brutal behaviour of the Bosnian Serbs, Mitsotakis adhered to Greece's old friendship with Serbia. He also cultivated its other old friendship with Romania in spite of the fact that Iliescu was hardly less of a communist than Ceauşescu had been. He staved off international recognition of the Yugoslav Republic of Macedonia as an independent state under the name of Macedonia but at the cost of adding to the nationalist emotions already too plentiful in the Balkans. Refugees from Albania were at first charitably received by Greeks; they were also welcome for their willingness to do dirty and ill-paid jobs. But as their numbers increased, so did thieving and hooliganism by some, with the result that they became widely vilified and many were sent back to Albania unceremoniously and indiscriminately. Frontier questions which had been dormant for more than a generation recovered their relevance together with complaints on the Albanian side of aggressively nationalist behaviour by Greek Orthodox churchmen. In these same years, with the end of the Cold War and the disappointments of EC membership, Greece became of smaller account to its Nato allies and less enamoured of its EC partners who, on their side, were finding Greece embarrassingly expensive. In 1993 Papandreou got his electoral revenge on Mitsotakis whose government was accused of ineffectiveness and nepotism and had no plausible plans for curbing two-figure inflation and unemployment approaching 10 per cent of the workforce. In 1994 Greece asserted its right under the international law of the sea to extend its territorial waters from six to twelve miles but provisionally withdrew after Turkish protests.

The end of the Cold War affected Greece more profoundly than appeared at first sight. The Cold War had cut Greece off from the Balkan region to which geographically it belonged; forced it into uneasy alliance (in Nato) with its age-old Turkish adversary; and annexed it to western European associations (including the EC) to which it was less important than they to it. Like Finland and Portugal it was a small peripheral nation state but, unlike those two, it lacked regional support.

At the other end of the Mediterranean it had become axiomatic to suppose that in *Spain* nothing much would happen until Franco died. Spain emerged from the Second World War in an ambiguous position. General Franco had wanted to join forces with Germany and Italy but had not done so except to the extent of sending a token force to fight with the Germans against the Soviet Union. His price for closer co-operation, which Hitler refused to pay, was a free hand in creating a new Spanish empire in north-west Africa. Spain was denied membership of the UN but within a short time it was wooed by the United States which needed bases in Spain for the Cold War. Franco's conservative and paternalistic Catholic nationalism – he absorbed rather than adhered to the fascist Falange – included an attempt to make Spain economically self-sufficient and brought the country near to collapse in the late 1950s. He was persuaded, against his instincts, to sanction a more international economic policy to Spain's considerable profit – and his own, since he was given the credit for economic recovery and died in office and in his bed in 1975. His death made Spain a kingdom in fact as well as name. Franco had pronounced Spain a kingdom in 1947 but he had no liking for the last king's son, Don Juan, and no intention of relinquishing his own quasi-monarchical control. He therefore ignored Alfonso XIII's act of abdication in favour of Don Juan and gradually prepared the latter's son, Don Juan Carlos, for the succession. Juan Carlos played an equivocal role in an equivocal position, so that on Franco's death his character and political ideas were something of a mystery. His first government was a mixture of Francoists and democrats but within a year he appointed as prime minister the 47-year-old Adolfo Suarez, an able and single-minded politician who fitted no precise faction. This was a skilful move which alarmed conservatives and the army without giving them distinct cause for revolt and pleased moderate groups on the right and the left which believed that Spain needed change. The appointment of Suarez signified the king's intention to accelerate change and also to be seen as a monarch capable of picking his own man. He was fortified by the approval of the Cortes and a referendum in favour of substantial constitutional changes. Within two years of Franco's death the king and Suarez were strong enough to face parliamentary elections with confidence.

Franco's Spain, although politically fossilized, had not been immune to all change. Franco's autarkic economic policies had begun to crack in the 1950s under the weight of their own irrationality and of increasing American intervention in Spanish affairs induced by Washington's policy of diversifying its anti-Russian deployment in Europe. But his modernization was ill regulated and careless of the poorer classes and the new régime inherited a distorted economy. It also inherited the perennial problem of regional discontent, particularly in Catalonia and among the Basques. The Socialist and Communist Parties were legalized, the latter in spite of American and military pressures to the contrary. The socialists, led by Felipe Gonzalez, proved the stronger of the two. The right was torn between those who wanted a new clean Christian Democratic Party and those who wanted a broader party including as many old Francoists as possible. In the first popular test in 1977 the communists and the right fared badly and Suarez's right-centre triumphed. In further elections in 1979 the right was routed, the communists staged a small recovery and the voters opted for two broad centre groups, giving Suarez a slight advantage over Gonzalez. But Suarez's handling of economic affairs was tentative and of the Basque separatists insensitive and indecisive: in the Basque country (Euskadi) violence became serious and Basque extremists carried their violent protest into other parts of Spain (see further p. 349).

In 1981 the constitutional and democratic monarchy sponsored by the king was challenged by an army coup. The army, which had no properly military function, was dominated by officers whose ideas were archaic and anti-democratic. They engaged in sporadic plots which, although ludicrous, were also dangerous until an overt but incompetent attempt to overthrow the régime, which included a spectacular assault on the Cortes, forced the king to disavow its leaders and confront the army as a whole with a choice between himself and their Francoist past and predilections. Suarez, however, did not survive nor did his right-centre bloc. He resigned and was succeeded by Leopoldo Calvo Sotelo, but at the end of 1982 fresh elections brought Gonzalez to power. His supporters included many who wanted Spain to leave Nato (which it had just joined) and he promised to hold a referendum on the question: he subsequently declared himself in favour of continued membership and carried the day in a referendum, although Spain then insisted on the removal of American aircraft by 1992 (Italy agreed to receive them). The Spanish economy flourished vigorously within the EC, enjoying wider markets, foreign investment, a share in world boom and the benefits of relatively cheap labour, but boom brought its familiar accompaniment of speculation and corruption. In the 1990s Gonzalez was weakened by rising unemployment and dissension in his own party, including a split which occasioned the resignation of his deputy, Alfonso Guerra. Having narrowly won a majority in 1986, he lost it after by-elections in 1990 and called a general election in 1993 in an attempt to revive his authority. He survived in office but only with the support of a small right-wing Catalan party. The conflict with Britain over Gibraltar was unfrozen to the extent that the frontier was reopened and the British government, while repeating its commitment to honour the wishes of the Gibraltarians, agreed that sovereignty be an item in future discussions. Talks in 1987 preserved goodwill without altering either party's stance.

Unexpectedly change in Spain was preceded by change in *Portugal* when, in April 1974, a group of middling and junior army officers overthrew the dictatorship and installed a junta of seven under the presidency of General Antonio de Spinola, a returned and critical African governor. A tussle for power followed within the dominant Armed Forces Movement and between the political parties which took shape after the coup. General Spinola resigned in September and fled the country in the following March after becoming involved in an unsuccessful anti-communist coup. A communist bid for sole or predominant power failed. Elections in April 1975 gave the socialists led by Dr Mario Soares 38 per cent of the vote and relegated the communists with 12.5 per cent to third place behind the right-centre Popular Democrats (26 per cent). Portugal seemed many times on the brink of civil war. The Armed Forces Movement and the army itself were split, but a coup in favour of the extreme left misfired and senior officers, alarmed by the prospect of anarchy, combined to support a minority socialist government and the installation of the relatively uncommitted General Antonio Ramalho Eanes as chief of staff.

Elections in 1976 gave no party a clear majority, the most successful being the Socialist Party which won 107 of the 263 seats. In the presidential election which followed General Eanes won 61.5 per cent of the votes distributed among four contenders. Soares formed a minority socialist government which was attacked both from the communist left and by the right representing not only the old régime but also the poor, anti-communist and conservative farmers of the north. Soares resigned at the end of 1977, formed a new coalition but was dismissed in 1978 by the president who appointed a government of technicians and, when this faltered for lack of a parliamentary base, another under Portugal's first female prime minister, Maria de Lurdes Pintassilgo. The president was trying to find a parliamentary coalition or a non-parliamentary alternative which would stem the country's post-revolutionary swing back towards the right, but this swing became more manifest at the end of 1979 when Francisco Sá Carneiro won nearly half the seats in parliament. He became prime minister but found himself at odds with the president who, using his constitutional right to interpret the constitution, blocked a number of measures whereby Sá Carneiro intended to steer the economy back into the paths of free enterprise. President Eanes won a further term of office in 1980. Portuguese politics continued to oscillate between centre-right and centre-left. In 1983 Soares, having failed to win an overall majority, formed a government with the social democrats to his right but this alliance collapsed in 1985 and Soares resigned. He became Portugal's first civilian president for over sixty years with the conservative Anibal Cavaco Silva as prime minister – an economist who had twice served briefly as a minister under Sá Carneiro and won a landslide victory in 1987 with, unprecedently, more than half the votes cast. Minor changes were made to the constitution to expedite the denationalization of industries and the decollectivization of agriculture. In 1991 Soares was re-elected president for five years with convincing popular approval. Poor by western European standards but orderly and enjoying the benefits of EC membership, Portugal shifted leftward in 1995 when the Socialist Party led by Antonia Guterres won nearly half the votes. With Spain, Portugal joined the Western European Union in 1988.

7 CENTRAL AND EASTERN EUROPE

STALIN'S EMPIRE

The partition of Europe after the Second World War was the consequence of a trend and an accident. The trend was the decline, accentuated by war, of the European nation states. Europe was a continent which had functioned in the form of comparatively small and comparatively strong entities, capable of maintaining separate existences because of the industrial sophistication of some and the addiction of all to the principle of self-determination. Thus the stronger European states existed because they were strong, while the weaker ones existed because it seemed right to the stronger that they should. But with the waning of the strength of the strong the basic element in the pattern of Europe disappeared and Europeans ceased to be able, for the time being, to maintain truly independent states. The question was what new forms would be imposed by dependence.

The answer to this question was determined by accident, by the fact that the precipitating cause of the war had lain in the centre of the continent – in Germany – so that the course of war meant a convergence of anti-German forces on the centre from the sides. Despite some plans to the contrary, the Anglo–American and the Russian advances into Germany were in substance separate operations which created separate American and Russian dominances to the west and to the east of Germany. Anglo–American sea power modified this pattern by decreeing that Mediterranean Europe as far as the Aegean should fall into the American and not the Russian sphere. This new distribution of power was recognized by Stalin's abandonment of Greece to Churchill, his refusal to pay attention to the Greek communist revolt or help it, and by the subsequent attachment of Greece and Turkey to Nato. At the other end of Europe Finland fell into the Russian sphere not only because of its strategic importance for the defence of Leningrad but because Anglo–American sea power did not overlap Europe round the north as far as it did in the south. Only Germany and Austria were designated common ground and, as we have already seen, even here the new principle of Russo–American partition prevailed and created in

Countries belonging to Warsaw Pact. (Albania formally withdrew from the Pact in 1968)
1989–90 Collapse of the Eastern Bloc;
Warsaw Pact ceased in 1990;
Comecon ceased in 1991

Czechoslovakia ceases to exist January 1993; new Slovakian and Czech republics come into existence
October 1990 DDR absorbed by the Federal Republic of Germany

Russia cedes independence to Latvia, Lithuania and Estonia in 1991 – final withdrawal of Russian troops completed in August 1994

June 1991 Slovenia and Croatia unilaterally declare independence; Macedonia and Bosnia–Hercegovina follow; Macedonia secedes September 1991
April 1992 Serbia and Montenegro become Federal Rupublic of Yugoslavia

7.1 Eastern Europe

Germany a partition within a partition which assumed crucial and critical significance in world affairs as the focus of the Cold War. In the rest of Europe the Americans and the Russians let each other be, but in Germany they could not – partly because Germany was the cause and prize of war, partly because of its central position and potential power, and partly because the old notion of the nation state remained strong enough to make the division of a state seem much more unnatural than the allocation of whole states to great power spheres of influence.

It has been argued that the division of Europe and the resulting Russian overlordship in eastern Europe were the consequence not of historical accident but of agreement, notably agreement at Yalta by Roosevelt and Churchill to give Stalin a position of power which otherwise he could not have achieved. This argument cannot be sustained. Roosevelt and Churchill conceded at Yalta nothing that it was in their power to withhold. The Russian armies were already in occupation of positions in Europe from which they could not be expelled and Stalin's postwar dominance in eastern Europe derived from his victories and not from any bargain with his allies. The most that Roosevelt and Churchill could do was to try to get Stalin to accept certain rules governing the exercise of the power that was his. This they succeeded in doing by persuading him to endorse a Declaration on Liberated Countries which promised free elections and other democratic practices and liberties. When, later, Stalin ignored the engagements contained in this declaration western governments could do no more than protest. Action in eastern Europe was impossible. Only in western Europe could they do anything and steps such as the formation of a West German state were their riposte and also a further manifestation of a division of the continent which preceded, and was not determined by, Yalta.

Within the Russian sphere Stalin's problem was the nature of Russian control and its mechanisms. He created a satellite empire in which the component states retained their separate juristic identities – separate from each other and from the USSR – but were subjected to Russian purposes by the realities of Russian military power and the modalities of Communist Party and police rule and unequal economic treaties. There was soon little difference between former foes like Hungary, Romania and Bulgaria, and wartime allies like Poland, Czechoslovakia and Yugoslavia. This indifference was manifested at an early stage: in Poland the Lublin Committee, a communist-dominated group of leaders formed in July 1944 from among the Polish resistance and then groomed in Moscow, was established in Warsaw as the government of Poland in order to frustrate the London Poles who had conducted the fight against the Germans from exile; in Romania the king was compelled in March 1945 to appoint a government controlled by communists and to cede to the USSR its Moldavian province (which had been part of Russia from 1812 to 1917). In form the defeated enemies were equated with the allies by the conclusion of peace treaties in Febuary 1947 and of further treaties between each of them and the USSR during 1948 (with the allies the USSR already had treaties dating from the war years). The peace treaties cost Hungary Transylvania which (as in 1920) went to Romania, and a smaller piece of territory awarded to Czechoslovakia; confirmed Romania's loss of Bessarabia and northern Bukovina to the USSR, and southern Bukovina to Bulgaria; and gave the USSR the Petsamo area of Finland and a fifty- year lease of the naval base at Porkkala with an access corridor. The treaties of 1948 provided for mutual assistance against Germany and proscribed any alliance by the one signatory which might be construed as directed against the other.

But Stalin aimed at more than formal arrangements and by the end of the 1940s he had, except in Yugoslavia and Finland, transferred the machinery of government into the hands of obedient communists who were not merely conscious of the realities of Russian power but determined puppet-like to serve it. This transfer involved the suppression or emasculation of non-communist parties and the elimination from the communist ranks of communists who were more national than Muscovite. This process was successfully achieved in the short run, unsuccessful in the longer run in that it failed to secure for Moscow a trouble-free zone of influence round the USSR's European borders. Yugoslavia rejected Russian dominance in 1948, Poland and Hungary kicked against it in 1956 and Romania led a campaign against it in the mid-1960s. By the 1970s it was in evident decay but still sustainable, even though the use of force – which was its ultimate guarantee – was becoming more a memory and a bluff than a plausible practicability. The advent of Gorbachev triggered the collapse in 1989 which had long been impending.

In 1946 Yugoslavia, Czechoslovakia, Bulgaria and Albania had communist prime ministers: Tito, Klement Gottwald, G. M. Dimitrov and Enver Hoxha. In Hungary and Romania the post was occupied by Peasant Party leaders, in Poland and Finland by socialists. All these countries had coalition governments, although only the governments in Prague and Helsinki gave the appearance of a real distribution of power. In Finland the communists were left out of a new government formed after elections in July 1948, in which they fared badly. Elsewhere communist control was intensified during 1947–48, although in Yugoslavia the communist monopoly of power worked against and not for the Russian interest and culminated in June 1948 in the eviction of Yugoslavia from the fraternity of communist states.

Poland's history is a struggle against neighbours, not least against Russia. Polish communism has had its anti-Russian side: Rosa

Luxemburg disputed Lenin's peasant policy, and the Polish Communist Party later deprecated Stalin's campaign against Trotsky. The leadership of the Polish party was wiped out in 1937–38 and the party itself was dissolved by the Comintern when Stalin was laying his plans for his pact with the Nazis. It was resuscitated in 1942 and nourished by the hideous behaviour of the Germans in Poland, which caused some revulsion of feeling towards Russians and communists. The discovery in April 1943 of the Katyn massacre (only implausibly ascribed by the Russians to the Germans) reminded the Poles that, for them, the choice between Russians and Germans was a hopeless one, but the Germans were at that time the present pest from which the Russians were future liberators. At the beginning of 1944 the Russians entered Poland in pursuit of the Germans and in July of the same year they accepted the Curzon Line as Poland's frontier in the east and sanctioned the Lublin Committee, which shortly afterwards became the country's provisional government. In August the people of Warsaw rose against the Germans in the expectation of swift help from the advancing Russians, which however failed to materialize; the victims included many leaders of the resistance, national communists as well as non-communists. The rising, besides being directed against the Germans, was an attempt by the London Poles and their secret army in Poland to establish their authority in Warsaw before the arrival of the Russians.

The looming Polish problem was twofold: what were to be Poland's boundaries and who was to rule? The Russians wished to shift Poland to the west in order to gain territory for themselves in the east and, perhaps, to perpetuate a pro-Russian and anti-German slant by adding German lands in the west to the Polish state; the Russians were also determined to insist on a government which was wholly or preponderantly communist. At the Yalta conference in February 1945 the Americans and the British disputed the Russian plans at length but irresolutely; they recognized the force of Stalin's arguments about the importance of a reliable Poland between the USSR and Germany, they did not regard the Polish question as the most important on the agenda, and they believed that no great harm would come from leaving the composition of the future Polish government vague since they had Stalin's agreement to the broadening of the provisional government and to fair and early elections after the end of the war. Before the Potsdam conference in July the Poles had been put by the Russians in possession of German lands beyond their old western borders and at that conference Churchill's forebodings and remonstrances were uttered in vain. The British and the Americans accepted the accomplished fact provided it were called an interim measure which would be reopened upon the negotiation of a peace treaty with Germany.

Stanislaw Mikolayczyk, chief of the Polish government in exile in London and leader of the Polish Peasant Party, had been added to the provisional government in Warsaw as a deputy prime minister. The other principal figures in the government came from the Lublin group: the communist Boleslaw Bierut as president, the socialist Edward Osobka-Morawski as prime minister and the communist Wladyslaw Gomulka as a deputy prime minister. The promised elections were held in January 1947 to the accompaniment of every conceivable electoral abuse, which failed however to conceal the underlying popular strength of the Peasant Party. This party was said to have won 10 per cent of the vote and 28 of the 444 seats in the parliament but Mikolayczyk felt obliged to seek safety in flight, as did many others. The rump of the Peasant Party was absorbed, into the newly created Democratic Bloc, which took the place of the Communist Party. In the following year the Social Democrat Party was led by Josef Cyrankiewicz, who had taken the place of Osobka-Morawski, into a merger with the Democratic Bloc which thereupon became the United Workers Party. The Americans and the British protested in vain over proceeding which they had no power to rectify. But within the communist leadership the old division between Polish and Muscovite communism, already visible in the days of Rosa Luxemburg, reappeared with Gomulka, now secretary-general of the single party, as leader of a faction which wanted to make communism more Polish, more popular and less subservient to Moscow. The quarrel between Stalin and Tito gave him an occasion to express support for Titoism, as a result of which he was gradually denuded of all his posts and disappeared in the background for the next eight years.

These events coincided with the transformation of the political scene in *Czechoslovakia*. Eduard Beneš, who resumed the presidency after the defeat of Germany, was the principal symbol in central Europe of the wish to conduct a state in accordance with western values and in friendship with the USSR. This formula, if not intrinsically inoperable, was made so by the evident wish of Beneš and his non-communist colleagues to participate in the Marshall Plan which was incompatible with Stalin's requirement of total and exclusive loyalty to Moscow. Beneš survived for a few months only. In February 1948 the minister of the interior, Vaclac Nosek, dismissed eight police inspectors in Prague. The cabinet voted to reverse this step but the prime minister supported Nosek, and the arrival in Prague of a deputy foreign minister of the USSR, Valerian Zorin, suggested outside interest of an unusual kind. Nosek declined to reinstate the policemen and eleven ministers tendered their resignations. They were the non-communist ministers, minus however the socialists who, although they had voted with the majority of the cabinet against Nosek, were reluctant to break their association with the communists in government – despite the fact that three months earlier they had selected a new leader to replace the fellow-travelling Zdenek Fierlinger. Anti-communist demonstrations, chiefly by students, occurred in Prague. Police were brought into the capital from outside. Amid fears of increasing tumult Beneš tried to restore calm by accepting the eleven resignations. Two ministers were killed by falling from windows, one of them – Jan Masaryk – having perhaps been pushed out. In June Beneš resigned. He was succeeded by his prime minister, Klement Gottwald. The mopping up operations consisted of amalgamating all Czech parties or remains of Czech parties with the Czech Communist Party, and likewise in Slovakia, whereupon the single Slovak party was amalgamated with the single Czech party to make the National Front. A similar attempt to extend Russian control in Finland in 1948 was dropped when it ran into difficulties.

Concurrently in *Hungary*, the monarchy having been abolished in January 1946 – nearly thirty years after it had ceased to function – a coalition government was formed with the Smallholders and Socialist parties. The leaders of the former, Zoltan Tildy and Ferenc Nagy, became president and prime minister and it won 56 per cent of the vote in the first postwar election. In the winter of 1946–47 rumours of a plot against the state were put about, trials were staged and the secretary-general of the Smallholders Party, Bela Kovacs, was abducted by the Russians. The Americans and British, who were partners with the Russians in the Control Commission for Hungary, protested but were helpless. In May 1947 the prime minister went to Switzerland to see a doctor and, while he was away, was asked to stay away and resign. A new plot was discovered. Elections in August were patently rigged. Members of non-communist parties fled or were put on trial, their parties were split and partially absorbed, as elsewhere, into a single National Independence Front. Tildy resigned the presidency in July and was succeeded by a complaisant social democrat, Arpad Szakasits.

In *Romania* the Peasant leader Ion Maniu was accused in 1947 of plotting against the state with American and British agents. He and others were tried and condemned, and the Peasant Party was dissolved. In December the king abdicated. In the following February the Social Democrat Party was merged with the Communist Party, and in March this party won 405 of the 414 seats in parliament. But before this election dissension had struck the Communist Party too and one of its veteran leaders, Lucretsiu Patrasceanu, was dismissed from the government, arrested, expelled from the party and, according to rumour, lodged in the Lubianka prison in Moscow. In Bulgaria the leader of the Agrarian Party, Nicola Petkov, was arrested with others in June 1947 and executed. The Fatherland Front, product of the usual socialist–communist merger, appeared in the next year.

During this period of regimentation Moscow's central purpose was to refashion each satellite in accordance with a single general pattern and to attach each of them to the USSR by all means short of incorporation. But there was to be no incorporation. The new fashioned states were to be People's Democracies; they were not to become Soviet Republics. Talk of incorporation was dowsed by Moscow, which made it clear that a People's Democracy was something short of a fully communist state and something different from a Soviet Republic. Communist enthusiasts with visions of admission to an extended Soviet commonwealth were disabused. The reasons for his policy may be found in Stalin's essentially cautious nature; or in his realization that these areas would, if fully absorbed, cease to be a buffer zone; or in the inability of the war-ravaged USSR to make radical changes in its structure in the mid-1940s; or in the knowledge that some of the satellites had enjoyed and expected a higher standard of living and of public administration than the USSR could provide; or in Stalin's desire to avoid unnecessary provocation of the western powers and to hoodwink them by getting the substance of empire without making constitutional changes which they would not stomach.

At the same time Stalin prevented the satellites from making new political associations among themselves. Everybody visited everybody and piled up bilateral treaties of friendship and mutual assistance until almost all the possible permutations had been employed, but schemes of a more radical kind withered quickly. Various such schemes were in the air immediately after the end of the war. Czechoslovak minds reverted to the Little Entente with Yugoslavia and Romania; and Hungary, whose relations with Czechoslovakia were marred by problems of exchanges of population, riposted in some alarm with a plan for a Danubian confederation. (A Danubian conference at Belgrade in July 1948 left the Russians in effective control of half the river. The Americans, British and French, outvoted at every turn, protested that the convention of 1921 remained in force in the absence of a universally accepted substitute. The riparian states were no counterpoise to the USSR so long as they remained separate.) Downstream the notion of a south Slav federation flourished for a while and produced some talk of a Romanian–Hungarian counterweight. The south Slav federation was the least unlikely of these federative ideas, if only because it was sponsored by two leaders of the first eminence, Tito and Dimitrov. But their plans became too grandiose for Moscow's liking and a Yugoslav–Bulgarian union looked uncommonly like a takeover of Bulgaria by Yugoslavia: the chief city of such a union would be Belgrade not Sofia. In June 1947 Tito told the Bulgarians publicly that he looked forward to a monolithic entity of free Balkan peoples. In August Dimitrov, visiting Yugoslavia to sign four pacts, secretly ceded Pirin Macedonia to the Yugoslav Macedonian Republic. In December Traicho Kostov, deputy prime minister of Bulgaria, spoke of a union of all south Slavs in the near future, and a month later Dimitrov spoke in the Romanian capital of a customs union leading to a federation or confederation which would include not only the south Slavs but also the north Slavs (other than those in the USSR) and Hungary, Romania, Albania and Greece. At this point Moscow intervened, summoned Yugoslav and Bulgarian leaders to Moscow and told them that Romania must be left out of their plans, although Albania might later be added to a Yugoslav–Bulgarian state. The Yugoslavs were also told to drop plans for sending troops into Albania. No union of any kind was effected. Dimitrov died in Moscow in July 1949, a year after Yugoslavia was expelled from the communist bloc.

The dispute between the Russian and Yugoslav leaders was conducted by correspondence in the months of March, April and May 1948. The heart of the matter was the refusal of Tito to accept direction from Moscow and his insistence on the right to think out Yugoslavia's own problems in their own context and apply its own solutions in preference to Russian principles and Russian programmes. Tito maintained that *Yugoslavia* was not only separate from the USSR but different, and that communist doctrine and practice were not so rigid as to be unable to take account of the differences. But on his side Stalin had abandoned the notion of diverse roads to socialism and became suspicious of Tito whom he wanted to displace in favour of one of his subordinates (Andriye Hebrang or Sreten Zujovic). He wished to secure in Yugoslavia a régime as obedient as any other in eastern Europe. If Tito were recalcitrant there were other Yugoslavs prepared to toe the Stalinist line. Stalin was intent not on casting off Yugoslavia but on casting out Tito, and the harsh retribution inflicted by Tito on these Stalinists (particularly in Serbia) after the breach suggests that the struggle within the Yugoslav party was acute and Tito's victory a narrow one. The dispute ranged over such topics as the proper organization of a communist state, the role of the Communist Party, agrarian policies, Yugoslavia's laxity in liquidating capitalism and the person of the Yugoslav foreign minister, Vladimir Velebit, whom the Russians accused of being a British agent. Friction was increased by the presence in Yugoslavia of Russian civilian and military advisers who seemed to the Yugoslavs to represent a Russian claim to superiority and to be paid too much; Russian attempts to put pressure on the Yugoslavs included threats to withdraw these experts. Throughout the correspondence the Yugoslavs were evidently anxious to avoid a breach, an attitude which may have strengthened the Russian resolve to exact admissions of error on the points in dispute. But the outcome was a Yugoslav refusal to accept the status of pupil and in June the breach was made public by the eviction of Yugoslavia from the Cominform, the international association of communist parties which had been formed in the previous September to ensure ideological unity and conformity.

Tito was a convinced communist but not an obsequious one. He had envisaged the adherence of a Yugoslav communist republic to the Soviet Union but he had also in his younger days been disturbed by the Russian purges of the 1930s, had been shaken by the Russo–German pact of 1939 and had had experience of Stalin's attitude to lesser communist leaders. In 1948 he abandoned the Stalinist mode of international communism out of national and personal pride; his genuinely national roots distinguished him from communist leaders who had lived longer in the USSR than in their native countries, and his position as a successful communist chief against the Germans had given him self-confidence and a popular following. He was fortunate in the fact that Yugoslavia had no frontier with the USSR and that western aid enabled him to parry communist economic blockade. Yugoslavia became an international anomaly, a communist state dependent on American and other western aid, an ally of Greece and Turkey in the Balkan pact of 1953 and then a protagonist with Nehru and Nasser of neutralism and non-alignment. The Yugoslav secession entailed the end of schemes for a Balkan union round a Yugoslav–Bulgarian core, diplomatic and economic breaches with the remaining satellites, and shifts in Yugoslavia's relations with its non-communist neighbours. It contributed to the defeat of the communist rebellion in Greece and the settlement by partition in 1954 of the problems of Trieste (Italy finally renounced its claim to the whole Free Territory in 1975). It dispelled the myth that a communist government not subservient to Moscow was a contradiction in terms. It promoted a series of witch-hunts in the satellite bloc where the Russians proceeded to eliminate communists who might be sympathetic to Tito or tempted to follow in his footsteps.

The most spectacular consequence of the breach was the trial of Laszlo Rajk in Hungary. The secretary-general of the Hungarian party, Matyas Rakosi, had fallen under suspicion but, having confessed to errors, survived. Then in September 1949 Rajk and other Hungarian communists were brought to trial on a compendium of charges to which they confessed and which included spying for the prewar Horthy régime, for the Nazi Gestapo, for the United States and Britain and, more significantly, for Tito. The trial was an anti-Tito demonstration, all the more forceful in that it ended with the execution of the accused. It was repeated elsewhere. In Bulgaria the veteran communist Traicho Kostov was expelled from the party in March 1949, arrested in June and, along with others, tried and executed in December. The charges ranged from Trotskyism to Titoism; their essence was conspiracy against the state. In Albania Koci Xoxe, who had favoured a union with Yugoslavia, was eliminated by his anti-Yugoslav rival, Enver Hoxha (who survived an inept Anglo–American attempt to unseat him in 1949 and a Russian coup after a quarrel with Khrushchev in 1960, evicted his erstwhile Chinese friends in 1978 and died in 1985). In Poland Gomulka lost his remaining badge of respectability by being expelled, with others, from the central committee of the party after accusations of collaboration with the Pilsudski dictatorship and the Gestapo and of nationalism and deviationism. He was not, however, put on trial. Moscow preferred to strengthen its position in the most important of the satellites by sending the Soviet marshal Konstanty Rokossovsky to Warsaw where he became a Polish citizen and minister of defence. In Czechoslovakia Tito's secession and the Rajk trial were followed by a purge of suspected pro-Titoists and of communists who had spent the war years in London. The victims included the foreign minister, Vladimir Clementis, who resigned in March 1950 but was not made to stand trial. At the end of 1951 communists of the wartime Moscow group, including the party secretary Rudolf Slansky, were put on trial in proceedings which had a distinct anti-Jewish tinge and seemed intended to use Jews as scapegoats for the unpopularities of the régime.

These years were critical for Stalin's hold on the satellites – or, judging by Russian actions, were deemed by him to be critical. The rejection of the Marshall Plan in 1947 had been followed by the Russian blockade of Berlin and by more militant action by communists in western Europe (particularly in France and Italy), but these ventures had failed and their failure coincided with a challenge to Russian rule in eastern Europe which succeeded in Yugoslavia and looked like becoming contagious. Stalin's first response was harsh and practical: he stamped, where he could, on threats to Russian interests. In addition, he added to the apparatus of communist integration in two ways – by creating in Comecon an institution for economic assimilation and by creating the beginnings of military co-ordination as well. These measures, although primarily conceived as retorts to western measures and although not developed during the remainder of Stalin's lifetime, ultimately affected the nature of the relations between the USSR and its satellite neighbours.

Comecon – the Council for Mutual Economic Assistance – was founded in January 1949 as a counter to the Marshall Plan. Its founding members were the USSR, Poland, Czechoslovakia, Hungary, Romania and Bulgaria, which were joined almost at once by Albania and a year later by eastern Germany. It was in form an association of sovereign states: communist propaganda was at this time attacking the Marshall Plan as an American device for overriding European sovereignties. But in Comecon there was for many years no question of any insistence on sovereign rights to dissent, and the dominance of the USSR was underlined by the exclusion of Yugoslavia. For ten years Comecon had no constitution, meagre quarters and a tiny staff in Moscow, and few activities. In so far as it was anything more than an anti-American gesture, it was an adjunct of the Russian policy, chiefly pursued by other means, of using planning to annex the satellite economies to Russian needs and not to develop the area as a whole in the interests of all its parts. The satellite governments were themselves in the process of adopting the rigid communist system of planning by setting national targets and instructing each separate enterprise how much it must contribute to the aggregate (a system which the head of the State Planning Commission in Moscow, Nikolai Voznesensky, was trying to reform until his dismissal by Stalin in March 1949). Until the crises of the mid-1950s Comecon occupied itself modestly with statistical research, technical exchanges and the promotion of bilateral and triangular trade treaties, but from 1956 considerable changes were introduced. Twelve standing commissions were established in different capitals, Yugoslavia and China were admitted as observers, a constitution was worked out and came into effect in 1960, an international executive was inaugurated in 1962, and meetings of these various organs became regular and frequent. Comecon organized aid for Hungary in materials and credits after the revolution of 1956, promoted joint planning and investment, and expanded satellite co-operation on a broad multilateral basis; for example, for the distribution of electric power and in the construction of oil pipelines.

In military matters Moscow exercised control and supervision through officers in the satellite forces who had been trained in the USSR and were regarded as dependable. The despatch of Marshal Rokossovsky to Warsaw was a uniquely overt and elevated gesture but one which had numerous parallels at lower levels. In 1952 this tentacle policy was supplemented by the creation of a military co-ordinating committee with Marshal Bulganin as chairman, and at a conference in Warsaw at the end of that year a combined general staff was instituted with headquarters in Cracow under a Russian general. Military facilities, involving considerable displacements of population, were developed along the Baltic coast, in Poland and eastern Germany and round the Black Sea. The satellites were at this stage contributing something like 1.5 million men to military and security forces, were incurring a commensurate financial burden, and were being obliged to adjust their industries, their industrial revolutions and their economic planning to the military requirements of the bloc as assessed by the USSR. There was, however, no formal multilateral defence treaty until, after Stalin's death, the admission of western Germany to Nato prompted the inauguration in May 1955 of the Warsaw Pact. This treaty, to which the USSR and all its satellites were parties, was expressed as a regional arrangement for self-defence within the meaning of article 52 of the Charter of the United Nations: it was renewed in

1975 and 1985. It created joint organizations with headquarters in Moscow under the command of Marshal Koniev. It incidentally regularized the presence of Russian troops in Romania and Hungary where they would otherwise have had no legal standing after the termination of occupation rights in Austria by the treaty which restored Austrian sovereignty in that year. In general the Warsaw Pact was a formalization of existing dispositions and added little of substance to them. Membership of the alliance became, however, a touchstone of the reliability of Moscow's associates in the Cold War.

AFTER STALIN

Stalin's death in March 1953 stimulated unrest. His death did not initiate unrest: there were strikes in Czechoslovakia, for example, the previous year. But in June 1953 serious disorders occurred in East Germany which the government was unable to bring under control without the help of Russian tanks. In Hungary tens of thousands were confined in camps in a campaign directed in particular against the peasants, the majority of the population. In June 1954 Moscow gave evidence of new thinking about relations with the satellites when it sold to the satellite governments the Russian share in joint companies created after the war for the control of key industrial and commercial enterprises, and at about the same time the liquidation of Beria in the USSR was copied further afield: in Hungary the chief of the security police, Gabor Peter, was sentenced to life imprisonment, and the first secretary of the ruling party, Istvan Kovacs, was forced to confess to unjust arrests and false witness. In 1956 Poland and Hungary gave much sharper testimony to the insecurity of Russian rule.

In June 1956 there were strikes and riots in the *Polish* city of Poznan. They were directed chiefly against low wages for long hours. Industrial unrest coincided with intellectual ferment and with Roman Catholic demonstrations in August at Czestochowa, the home of a specially venerated shrine of the Virgin Mary. A meeting of the central committee of the party in July was attended by Gomulka, and when Bulganin and Zhukov arrived from Moscow in the hope of taking part, they were not allowed in. The first secretary of the party, Edward Ochab, became at some point persuaded that Gomulka must be readmitted to grace and power. During visits to the USSR and China Ochab seems to have convinced the Chinese, but not the Russians, of the need for this reversal. In October the politburo resolved to reconstitute itself, admitting Gomulka and excluding Rokossovsky. On 18 October a powerful Russian delegation arrived consisting of Khrushchev, Molotov, Mikoyan and Kaganovich, and Russian troops in Poland and eastern Germany began to move. Gomulka was included in the Polish team chosen to confront the visitors. These were not allowed to attend a meeting of the Polish central committee and after twenty-four hours they left. Hot words were spoken but rough action was stayed. Gomulka was appointed first secretary on 21 October. On 29 October Rokossovsky left for Moscow. He was followed within a few days by Gomulka. The upshot of these events was that the Russians, who had presumably gone to Warsaw with the intention of checking Gomulka and his party, decided after a quick look to accept them. The alternative, a direct use of Russian forces in Poland, was too risky because Russian forces might well have been resisted by the Polish army and a fight in Poland could have led to serious trouble in other countries. Gomulka was a communist and had no illusions about Poland's need to keep on reasonably good terms with the USSR. He did not propose to take Poland out of the Warsaw Pact or to share power in Poland with non-communists. He could be lived with. If the Russians had initially feared that a new Polish government would go too far in jeopardizing essential Russian interests, they concluded on second thoughts that Gomulka would not exceed the bounds of the tolerable. In December a new treaty gave the Russians the right to retain troops in Poland.

In *Hungary* the revolution which immediately followed the settlement in Poland did exceed those bounds. After the death of Stalin Imre Nagy was wafted into the premiership on a wave of general relaxation and of reaction against old-school hard-liners. In April 1955, however, he was ousted by Rakosi who took over the post himself and held it until July 1956. But Rakosi was unable to stem the rising tide of opposition emanating from the Petöfi circle (the centre of intellectual debate and dissatisfaction) and from the keen dislike of Tito, his neighbour to the south. When even the party newspaper *Szabad Nep* turned against him, the Russians realized that he had become a handicap and put their money on Erno Gerö. But Gerö too was disliked by the malcontents and by Tito, and in October Nagy regained power. He was in fact, if not reinstated, at least confirmed by the Russians, who had been forced to the conclusion that Rakosi, Gerö and communists of that ilk must be expended. Military operations by the Russians against Budapest had been begun on 24 October but the next day Mikoyan and Suslov arrived in Budapest and signified their acceptance of a Nagy régime – but without knowing, perhaps, what kind of government Nagy intended to form. Everything depended on whether Nagy could shape a policy acceptable both to the Russians and to the upsurge of Hungarian nationalism. He announced his government on 27 October and included in it two leaders of the suppressed Smallholders Party, Zoltan Tildy and Bela Kovacs. On the next day a cease-fire was arranged and the day after that the Russian troops began to withdraw from Budapest. Nagy then announced the end of one-party rule and the complete evacuation of Russian troops from Hungary. Nagy had now gone much further than Gomulka and it is probable, though not certain, that he had exceeded the limits of what the Russians deemed tolerable. By 30 October it seemed that the withdrawing Russian troops were preparing to return, or at least were being redeployed with a view to a possible return. The die was cast either on 31 October when Mikoyan and Suslov were told that Hungary intended to leave the Warsaw Pact, or at the latest on the following day when Nagy made a public statement to this effect and declared that Hungary would become a neutral. On that day an alternative government was set up by the Russians under Janos Kádár and when, two days later, General Pal Maleter, Nagy's minister of defence, went to negotiate with the Russians over the withdrawal of their troops, he was kidnapped. Budapest was attacked on 4 November (the day on which Gomulka went to Moscow) and thereafter the revolution was quickly suppressed. Thousands of Hungarians were deported to the USSR or executed. Kádár applied himself to running Hungary within the limits imposed by Moscow: no elections and no quitting the Warsaw Pact. He ruled for thirty-two years. He was on friendly terms with Khrushchev. He proclaimed a general amnesty in 1963, cultivated a comparatively unautocratic style, abated the use of torture and imprisonment, and permitted the discussion and introduction of economic reforms. Under Gerö, a poor economist, Hungary's manufacturing output had at first been increased but the productivity of a greatly enlarged industrial workforce did not improve, there was only pitiful investment in plant or technology, and when the workforce ceased to grow manufacturing growth subsided to zero. Investment in roads, housing, education and research was negligible. By the mid-1970s economists were advocating a drastic abandonment of central economic planning and direction, the unleashing of market forces of supply and demand, and schemes for allocating the profits of particular enterprises between workers, the state and reinvestment. These discussions were open and even broadcast on television. But intellectual barriers were coming down faster than the political. Actual reforms were unheroic and the ruling party unadventurous and, unlike its counterparts in Poland and Czechoslovakia, small.

The suppression of the Hungarian revolution of 1956 was one of those brutal acts of policy which do grave damage to the perpetrator but are undertaken none the less upon the calculation that graver damage would otherwise result. Communist parties lost members on a considerable scale and even communist governments which reckoned that Nagy had gone imprudently far shuddered at the display of Russian might and determination. Before long they found new cause for unease in the Russo–Chinese split. The Chinese were judged to have given Moscow the right advice on Poland and Hungary – that is, to acquiesce in changes and, in Hungary, to use force only after Nagy had given the revolution an anti-communist course – and Zhou Enlai visited Poland

and Hungary early in 1957 to consolidate this advantage and stress the need for good Russo–Chinese relations. Khrushchev's subsequent handling of the quarrel with Beijing disturbed satellite leaders who disliked the way in which Khrushchev insisted on bringing it into the open and making communists take sides. In June 1960 Khrushchev used a congress of the Romanian party to stage a demonstration against the Chinese, and although in November 1961 eighty-one communist parties (Yugoslavia alone abstaining) signed in Moscow a declaration intended to lessen strife and paper over cracks, Khrushchev continued until his fall in 1964 to conduct a public campaign against the Chinese. On this issue *Romania* took the lead first in persisting with attempts to resolve the disputes and then in refusing to take sides. Gheorghe Gheorghiu-Dej, the virtual ruler of the country since the end of the war (and undisputed ruler after the fall in 1952 of Ana Pauker, Vasile Luca and other Muscovite leaders), also challenged the Russians in Comecon where, in 1962, Khrushchev proposed to create a supra-national planning organ with powers to direct investment throughout the bloc and prescribe what should or should not be done in each member state. The Romanians, who wanted a steel mill but were cast for the role of producers of raw materials, invoked the principle of national sovereignty which the Russians themselves had made much of when Comecon was founded. They displayed their dissatisfaction by entering into a separate agreement with Yugoslavia for a hydroelectric scheme at the Iron Gates on the Danube, by proposing that China be a full member of Comecon and by threatening to leave it. Romanian leaders visited Paris, London, Ankara and other non-communist capitals. After Gheorghiu-Dej's death in 1965 his successor as party secretary, Nicolae Ceauşescu, also talked of the dissolution of the Warsaw Pact and of the loss of Bessarabia to the USSR twenty-five years earlier. After a tactful visit by Brezhnev to Bucharest in July 1965 the more implausible Romanian proposals abated as the realities of the situation reasserted themselves. Romania continued to assert its idiosyncrasy by establishing diplomatic relations with western Germany in 1967.

The year 1968 brought a fresh challenge from a different quarter: *Czechoslovakia*. Before the Second World War, the Communist Party of Czechoslovakia was, unlike its neighbours, neither illegal nor underground. It was the second largest political party in the country. It escaped being compromised by the general communist alliance with Hitler in 1939–41 because Czechoslovakia was already occupied by the Germans and its parties banned. It played a patriotic role during the war and emerged after it as the biggest party owing to the proscribing of the collaborationist Agrarian Party. Its leader, Klement Gottwald, was therefore the natural prime minister. He and his colleagues had operated before the war in a democratic system and they now co-operated with other parties in what was at first a genuine coalition of all anti-fascist groups. The first postwar elections endorsed Gottwald's position by giving the communists 39 per cent of the vote in a free election, the highest percentage. Moreover, when in 1948 the non-communist parties tried to undermine communist authority (a legitimate aim) by resigning *en bloc* from the government and forcing President Beneš to instal a government of officials (a dubiously democratic procedure), Beneš backed Gottwald and the scheme collapsed. But so too did genuine coalition government. The communists began to make a mockery of it and to govern increasingly through tyranny and terror. Czechoslovakia became a police state. The Yugoslav secession gave Gottwald a motive and an excuse for tightening his control (there were more executions in Czechoslovakia than anywhere else) and the bleak years of the Cold War with their talk of American action to liberate eastern Europe put a clamp on all meaningful criticism of the government. Although the government was distrusted, it was not seriously opposed; fear of the police stifled talk and prevented organization. Within the governing party itself debate was atrophied. Even the events of 1956 in Poland and Hungary struck no visible spark in Czechoslovakia, which came to be rated as the most docile of the satellites.

Presiding over this inertia was Antonin Novotny, first secretary of the Czechoslovak Communist Party from 1953 and president of the republic from 1957. Novotny was a Czech who despised Slovaks and did not conceal the fact. This became one item in a movement against his leadership which led to his removal from his party post in January 1968, from the presidency in March and from the party in May. Ditched by Moscow he was succeeded in the first office by the first secretary of the Slovak Communist Party, Alexander Dubček, and in the second by General Jan Svoboda. The first and more important of these changes was effected by a special meeting of the central committee of the party called to resolve a deadlock in the presidium. It amounted to a reversal of power within the party, largely instigated by Slovaks.

Dubček, besides being a Slovak nationalist, was an economic reformer, in government relatively humane and in economics relatively liberal: the sort of communist who makes other communists look peculiarly tyrannical and incompetent. He presented Moscow with a conundrum and for several months Moscow could not make up its collective mind what to think or what to do about him. The new government's first measures were vague and without explicit threat to the communists' monopoly of power or to the Warsaw Pact, but its Action Programme published in April and a promise of elections in May put Moscow in a dilemma which it resolved at the end of August by invasion.

Czechoslovakia was an industrial state which was specially hampered by the drawing of the Iron Curtain where it was drawn. Economically its western half belonged with the western world, even if Slovakia did not. The country's economy made a decent recovery from the war but was slack and even stagnant in the 1960s. Its prewar strength in the manufacture and export of consumer goods was undermined by Russian insistence on expanding heavy industries whose products were required by the USSR but had no markets outside the satellite bloc – since in western terms their technology was outdated. Its methods were (like Poland's) dangerously polluting. It produced twice as much steel as it could sell. It became isolated from the rest of the world. The standard pattern of central communist direction and control engendered ossification, and a return to consumer goods was obstructed by new vested interests in heavy industry (which would however be specially vulnerable if disarmament took hold). Czechoslovakia avoided excessive indebtedness to the west on the Polish or Hungarian scale but could not readjust or repair its economy without western capital.

Economists were alive to these dangers. Some mainly ineffective reforms were initiated in 1958 and a more far-reaching programme for decentralizing industrial management, elaborated by Professor Ota Sik, was approved by the Central Committee of the ruling party in 1965. Similar ideas had been discussed in neighbouring countries, including the USSR itself, and more radical measures than Sik's had been adopted in Hungary. There was in 1968 no compelling reason to suppose that Moscow would veto the kind of economic policy which the Dubček government wished to implement.

But pressure for such changes was accompanied by a second kind of ferment. Decentralization of economic management was equated with liberalization of controls and worker participation (or industrial democracy) and these trends overlapped naturally with demands for more freedom generally, notably freedom of expression in the press and on the radio and the democratization of party politics and the parliament – all of which posed more serious problems for the Russian guardians of the established order. Whereas in January 1968 the Russians had apparently decided that Novotny had lost his grip and was expendable, and that Dubček was acceptable in his place (Dubček proclaimed the solidarity of Czechoslovakia with the USSR and visited Moscow immediately after his appointment), a couple of months later Brezhnev and his colleagues were becoming worried by Dubček's programme and perhaps also by the likelihood of Dubček being forced further in a liberal direction by the enthusiasm released by the change of government in Prague. This change had been followed by a considerable relaxation of the censorship, by a number of ministerial changes and by the prospect of political democratization as well as economic liberalization. There was popular pressure on Dubček to make such radical internal reforms that these would be bound to raise questions about Czechoslovakia's external relations. Dubček had taken care to make personal contacts with Russian, Polish and Hungarian leaders before meeting the suspect Romanians and for a couple of months his neighbours evinced no distrust of him or his new course, but from the end of March criticisms began to appear in East Germany and Poland and the Russians had to ask themselves whether the changes in Czechoslovakia were not more portentous than they had at first seemed. The reformers were growing in confidence and were exciting ever greater popular expectations. Their Action Programme, produced at the beginning of April, made radical proposals concerning the reorganization of the respective functions of party and the government, the rehabilitation of victims of the purges of 1949, the position of Slovakia, the revival of the parliament and some freedom for minor parties (within the National Front which the communists would continue to control). Dubček's radicalism, however – not unlike Gorbachev's in years to come – was limited and he relied injudiciously on his ability to remodel the Communist Party and secure its support for his programme. He turned an unwilling ear to warnings of military action by Moscow.

To the Russians the Action Programme was objectionable in itself and doubly objectionable in the wider context of central and eastern Europe. It brought personal and political freedom to the centre of a debate which could hardly be confined to Czechoslovakia. The first major upheaval in the postwar communist bloc had been caused by the example of Yugoslavia in 1948, the second by the examples of Poland and Hungary in 1956. Both had required, and in Moscow's eyes justified, recourse to violent measures including judicial assassination and force of arms. It was vital to prevent Dubček from setting a third bad example. Romania, particularly since Ceauşescu's accession to power in 1965, had been making itself awkward in Comecon and the Warsaw Treaty Organization; it had ceased attending meetings of the latter. Tito was still alive and vibrant. (Both these leaders were to visit Prague in August within a few days of each other and to rapturous acclaim.) Czechoslovakia under Gottwald and Novotny had been a key element in the western sector of the USSR's European dominions, aligned with Poland and eastern Germany. The prospect of Dubček's Czechoslovakia sliding out of this sub-bloc and into alliance with Yugoslavia and Romania was both strategically alarming and an unacceptable display of political independence. Political changes in Prague, themselves a sequel to economic and managerial changes, might in their turn have dangerous strategic consequences. And some in Moscow were becoming alarmed by signs of declining communist authority in the government of Czechoslovakia and even in its police.

If the Russians were becoming uneasy about Czechoslovak actions, so were the Czechoslovaks about possible Russian reactions, particularly in view of the fact that popular pressures had induced Dubček to accelerate to early September the party congress which would presumably acclaim and endorse the reform programme. At the beginning of May Dubček and other leaders of the new course went to Moscow. Two weeks later Kosygin paid a prolonged visit to Prague and so too – at the same time but separately – did Marshal Grechko, accompanied by Marshal Epishev, the chief of the Soviet army's political intelligence. These visits pointed in opposite directions. Kosygin seemed to be seeking an accommodation, the generals to be preparing for action if there were none. In June the forces of the Warsaw Pact held manoeuvres in Czechoslovakia. These had been arranged a long time earlier, but they were considerably enlarged and the Russian tanks which came with them seemed in no hurry to leave. This hint or demonstration by the Russians coincided with the publication of a new liberal manifesto – the Two Thousand Words – which put further reformist pressure on Dubček and sharpened the tension between democratic and counter-reformist elements in Prague. The situation was now so dangerously charged that the French and Italian communist parties tried to mediate and the West Germans, equally alarmed by the turn of events, withdrew their forces from the Czechoslovak border in order to belie rumours that they and their allies were instigating a secession from the Warsaw bloc or were proposing to take advantage of the rifts within it.

Intervention in the affairs of a neighbour was neither new nor ideologically entirely unjustifiable, but the extreme form of military invasion was to be avoided if possible – perhaps to be eschewed altogether on some estimates of the damage that such strong-arm methods might do to international communism and the USSR's place in it. The debate on how and how far seems to have engaged the Russian leaders throughout July and half of August. A first meeting at Warsaw, not attended by the Czechs (or the Romanians), produced a letter warning them that their proposed reforms were tantamount to allowing power to escape from the Communist Party. They were asked to explain their doings. This meeting was followed by a Russo–Czechoslovak meeting held, on the latter's insistence, in Czechoslovakia – at Cierna-nad-Tisou on the Slovak border. It began on 29 July and lasted four days and was immediately followed by yet another meeting, at Bratislava, of all members of the Warsaw Pact except Romania. Before Cierna the Russians issued a threatening statement saying that a cache of American arms had been found on Czech soil and it is difficult to judge whether Cierna was anything more than window-dressing. Bratislava reasserted the military threat. Russian troops moved out of Czechoslovakia. Moscow's propaganda against Prague stopped. But on 20 August the Russians, accompanied by East German, Polish, Hungarian and Bulgarian units, invaded.

It is not possible to say whether the decision to do so had been taken in principle before the July meetings or whether it was taken at short notice after them. If there was a change of plan during August, then the event most likely to have caused it was the publication on 10 August of the new statutes of the Czechoslovak Communist Party which amounted to the ending of democratic centralism and the granting of substantial rights to minor parties. These statutes were to be considered at the party congress a month later and would certainly be adopted unless drastic steps were taken to change the party leadership once more and prevent the holding of the congress. That the Russians invaded at short notice rather than after long and concealed premeditation is suggested by the fact that this political programme was only partially fulfilled. Although the invasion was militarily precise and efficient (the Czechoslovak armed forces offered no resistance and the Dubček government had said they would not do so), its political tactics were palpably confused and its political aims only partially achieved. Dubček was not overthrown. He was seized, flown to Moscow under arrest, possibly drugged and forced to give his assent to a Russian invasion. He did so in the belief that invasion was inevitable and that without his assent it would be uselessly and bloodily resisted by his own people. If, as seems probable, the Russians expected the presidium in Prague to displace Dubček and instal a new government, they were ill informed and their dispositions insufficient. They came as blatant conquerors and although their power was overwhelming they had to negotiate with Dubček and Svoboda.

Twenty years of communist rule had stifled political and cultural life and created economic disaster through, chiefly, sterile bureaucracy on the one hand and unreality on the other – none of the five-year plans had worked as envisaged by the planners. Reformers in the 1960s, besides being anti-Russian, aimed for a better kind of socialism which the Russians would not countenance because reform anywhere in the satellites was necessarily a threat to the entire Stalinist system. The concrete symbol of this imperative was a treaty signed in October which permitted Russian troops to be stationed in Czechoslovakia in undefined numbers. Dubček was gradually demoted, sent to Turkey as ambassador and then recalled to be expelled from the Communist Party. Reform and the reformers were eliminated. From the Russian point of view the invasion was a regrettable necessity and a well-calculated action. Some western tremors notwithstanding, it created no threat to international peace and it did not halt the course of Russo–American détente: there was no more than a brief interruption in the talks that led to the opening of SALT in 1969 and the agreement with Bonn in 1970. As a crisis manager Brezhnev proved to have many of the qualities which had served Kennedy well in the Cuban crisis of 1962. But the invasion forced Moscow to proclaim an extreme doctrine about the limits of sovereign independence within the communist bloc and to make plain that Russo–American détente implied no loosening of the reins of power therein. The invasion and the doctrine unsettled eastern Europe by the violence of the action and the implications of the doctrine, and the use of the ostensibly anti-western Warsaw Pact against one of its own members emphasized the strains prevailing within the bloc twenty years after its consolidation.

These strains had two main sources: nationalism and conflicts of economic interest. Nationalism was endemic throughout the bloc, although weaker in some places than others. Bulgaria, at one end of the scale, endorsed the Brezhnev doctrine in a new constitution in 1971 and so elevated socialist internationalism above traditional nationalism and states' rights. This was an echo of Dimitrov's old-style international communism coupled with Bulgaria's perennial leanings towards Moscow to offset its perennially uneasy relations with its neighbours. (But in the 1970s Bulgaria sought to improve its relations with Romania and Yugoslavia and even with Greece and Turkey too.) Most eastern Europeans, however, were as loath as western Europeans – or, for that matter, Arabs – to subordinate their national identity to supra-national organizations or causes. To some extent the USSR had contributed to this particularism not merely by its heavy-handedness in moments of crisis but also by obstructing regional associations within its sphere (Balkan, Danubian or other). Any association had to be all-embracing and it must of course include the USSR itself.

Of the two principal organs of eastern European integration the Warsaw Pact was little more than an expression and deployment of Russian power. Its forces were commanded by a Russian commander-in-chief and its headquarters were a departmental office of the USSR's high command. It had been created in opposition to Nato and its principal function was to face Nato's forces in Europe. But whatever their avowed purpose, the Pact's forces had other potentialities of which the invasion of Czechoslovakia was an uncomfortable reminder. The defence of eastern Europe included, according to the Brezhnev doctrine, firing shots in anger against enemies within the gates. The doctrine raised questions not about power but about sovereignty. All eastern European countries knew that they had to live with Russian power and observe the limitations which it imposed on their own freedom of action, but they wished at the same time to maintain, even if they might not always exercise, their sovereign rights. This was for the most part a vain aspiration, a kicking against the pricks exemplified by Romania's continued refusal (maintained in spite of a visit to Bucharest by Marshal Grechko in 1973) to take part in Warsaw Pact activities and by Ceauşescu's symbolic visit to Beijing in 1971 and his reception in Bucharest in 1972 of the presidents of the United States and western Germany.

In Comecon, the second quasi-international body, strains were more concrete. Having come to life in the late 1950s after a somnolent start, Comecon was endowed in 1962 with a Basic Plan of International Socialist Division of Labour. This title proclaimed the intention. In 1971 a second basic document was adopted: the Complex Programme for the Development of Socialist Economic Integration, incorporating a long-range programme reaching fifteen to twenty years ahead. (In this year Albania rejoined the organization after a gap of nine years.) Comecon's practical problems were fundamentally no different from those of any international organization trying to reconcile the good of each with the good of all. Its members had divergent views of their interests and the harmonization of these was further complicated by the immense preponderance of power of one member – a complication absent from the similar problems of the EEC. In eastern Europe the division of labour meant two things in particular: that each non-Russian member should concentrate on one or two economic activities prescribed for it by the organization as a whole, and that the resulting exchanges within the group should largely take the form of trading these prescribed manufactures for Russian primary products – notably oil: Russian exports of oil to other members of Comecon rose from 8.3 million tons in 1965 to a projected 50 million tons in 1975. Such a division of effort within a state would seem natural enough but when applied in advance of political unification it entailed a supra-nationalism which members – Romania in particular – distrusted because it removed decisions from national control (the same objections were heard in western Europe) and threatened to reduce particular members to dependence on a single industry or crop, with consequent further loss of political independence. There was also the question of payments, the fear that the terms and benefits of trade with Comecon would be manipulated to the disadvantage of the weaker brethren by arbitrary fixing of unfavourable currency parities. The 1971 Complex Programme acknowledged these fears to the extent of envisaging a common currency or convertible rouble by 1980 after a period in which separate currencies would be equitably and permanently adjusted in terms of one another.

Members of Comecon had a common interest in raising their economic performance and their trade with one another, but particular members also had an interest in trading with, and therefore producing for, countries outside the bloc: some of their needs could be satisfied only by buying from western countries, their trade would expand faster if they dealt with western as well as eastern countries, and there were political advantages in a commercial diversification which would reduce economic dependence on one or two neighbours. The USSR itself set an example which it could hardly denounce in others by concluding with the United States in 1972 an agreement designed to treble Russo–American trade by 1975. (This agreement was terminated by the USSR at the beginning of 1975 after the US Congress had inserted into the Trade Reform Act 1974 an amendment linking the expansion of Russo–American trade with the relaxation of the USSR's emigration policies. The author of the amendment, Senator Henry Jackson, hoped to facilitate Jewish emigration from the USSR but succeeded only in checking trade between the two countries.) From 1973 Comecon engaged in talks with the EEC about agreements between the EEC and particular Comecon members embodying quota reductions and most-favoured-nation clauses. Western countries were particularly attracted by the possibility of increasing their purchases of non-Middle Eastern oil.

But these horizons were not all fair. Increased east–west trade coincided with increased world commodity prices and increased inflation in the west. Eastern Europe, experiencing inflation in its own economies for domestic reasons including higher wages, was faced with the alternatives of importing a further measure of inflation with western goods and raw materials or of cutting back its trade. When Romania, which had redirected half of its foreign trade to countries outside Comecon, was preparing a new five-year plan in 1975, it altered its first intentions by reducing the share of its trade which it proposed to do with the west; but, loath to retreat into the closed circle of Comecon, it began to explore the prospects for more trade with the Third World and for this purpose applied for and was given membership of the conference of non-aligned states convoked in Lima in that year (but held in Colombo the next year). Even Czechoslovakia, consistently the strongest economy and a creditor country within the bloc, was caught in this dilemma, partly because of the vigour of its trade and industry. Its trade with the west, facilitated in some measure by the establishment of diplomatic relations with western Germany in 1973, brought in the world's goods but at the world's inflated prices, and since Czechoslovakia could not abandon incomplete projects which depended on foreign contracts it found itself having to increase the value of its exports to the west by more than 20 per cent if it were to pay for its imports by trade.

In Hungary decentralization in economic affairs troubled the Russians and the more conservative wing of the Hungarian communist establishment, but a visit by Brezhnev to Budapest in 1972 was taken to carry with it continuing Russian support for Kádár and cautious liberalization. Nevertheless Kádár remained harassed. Hungary's import bill rose vastly, with Russian as well as world prices increasing so much that the government, unable to bridge the gap by greater productivity and exports, was obliged to pass price increases on to the consumer. At the eleventh party congress in 1975 Kádár himself survived criticism but his prime minister, Jenö Fock, and other senior personages had to resign and new policies of austerity and recentralization were adopted to cope with an unmanageable deficit in the balance of payments.

Much more serious from the Russian point of view was *Poland*, where economic difficulties led to the overthrow of the government by workers' demonstrations, a phenomenon rare in any quarter of the globe and least expected in an authoritarian communist state. Poland was something of a special case in Stalin's empire, for Poles harboured an atavistic national hostility to Russians whether Tsarist or bolshevik and, as Roman Catholics, were equally hostile to Orthodox Christianity and communism. Increases in food prices provoked at the end of 1970 strikes and riots which the government tried but failed to control by force. Forty-five people were killed and over a thousand injured. Gomułka, who had been brought back to power in the midst of the troubles of 1956, resigned. A new government under Edward Gierek cancelled the price increases, raised wages and social security payments, imported foreign consumer goods (at considerable cost) and purged party and administration by dismissing officials at all levels and placing new men and women in half of the top positions at the centre and beyond. The press was given greater freedom (but was restricted again in 1974) and in 1972 wages and benefits were raised again – the government demanding in return harder and more punctual work – and a new parliament was elected with many new faces but also the standard 99 per cent vote for the communist candidates. By 1973 many economic indicators were so propitious that it was permissible to speak of a boom: industrial and agricultural production up, real wages up, investment up, prices stable. But the cost was a large increase in the import bill and foreign debt, as the government satisfied consumer demand and the needs of industrial modernization by buying abroad and financing its purchases by foreign borrowing and encouraging foreign investment. By 1974 Poland, like Romania, was doing half its trade with the west. Huge price increases, specially for food, led in 1976 to strikes, riots, deaths and heavy prison sentences. A Committee for the Defence of the Workers (KOR) came into existence to help the families of the dead and imprisoned and to protest against the heavy-handed brutality of the police. The price rises were rescinded, wages in industry and prices to farmers were raised, and Gierek persuaded Moscow to send substantial quantities of food and industrial machinery to Poland. This was a policy of repression tempered by concession. It did not work. The wage increases accentuated the demand for imported consumer goods and so increased inflation and the already unmanageable debt to western trading partners. The deaths of 1970 and 1976 were not forgotten; further demonstrations and hunger strikes forced the government to grant in 1977 an amnesty to those prosecuted for the previous year's disturbances; and Gierek, who held his first meeting with Cardinal Wysczynski, was seen to be appealing to the Roman Catholic Church for help. The election of Cardinal Wojtyla to the papacy – the first Polish pope since Poland was Christianized 1,000 years earlier – had an unmeasurable but far from insignificant effect in a country which regarded itself as the jewel of the Counter-Reformation. Pope John Paul II paid an emotional visit to Poland in 1979, which was another year of bad harvests and mounting foreign debt. Gierek's task was becoming beyond his powers and the industrial workers were moving, albeit cautiously, towards another confrontation with the government.

In 1980 they struck again, bringing the shipyards along the Baltic to a halt against a background of unmanageable foreign debts, soaring prices and severe shortages. Earnings on exports were almost wholly absorbed by the charges on a massive foreign debt of over $20,000 million, increasing by $2,000 million a year. Meat was alternatively unobtainable and prohibitively expensive. Wage increases of 10–20 per cent, granted by the government in the summer, threatened to aggravate inflation without making life tolerable for the recipients. There was nobody to blame except the government, but equally there was no constitutional way of blaming the government or seeking redress, even explanation, from it. Consequently the ensuing strikes in August were inevitably and immediately political. Not only could they not be settled by collective bargaining between the workers and their employers; but such bargaining was itself impossible because the workers were permitted no lawful organizations to do the bargaining. The system could solve nothing. If the strikers stayed out the government would have to choose between using force and bypassing the system.

But the government itself was not free to choose. It had to consult the USSR. Gierek flew to see Brezhnev to discover how much rope he was allowed. Neither he nor Moscow wanted a showdown with the strikers (which would rapidly become a showdown with a far wider segment of the population), but the use of force by the USSR could not be ruled out. Gierek had to find out what he might in no circumstances concede, and the agreement ultimately reached with the strikers in the Gdansk yards in August showed what these limits were. In that agreement, signed on behalf of the strikers by Lech Walesa and subsequently copied in similar terms in the Silesian coalfields and other areas, the strikers accepted the leading role of the Communist Party, Poland's socialist system and its membership of the Russian bloc. In return for recognizing these limits Poland's remarkably mature and disciplined opposition leaders won, on paper, astonishing victories: the right to strike, the right to form trade unions independent of the state, wider discussion of the government's economic policies, an abatement of censorship, appointments and promotions on merit and irrespective of party membership, increased wages and pensions, promises on working conditions, housing accommodation and maternity leave, a second day off in the week, and regular broadcasts of Roman Catholic church services. Some of the clauses were vague; others could hardly be implemented without an economic miracle which nobody expected. Nevertheless nobody doubted that something extraordinary had happened. An officially non-existent opposition, using non-violent industrial muscle, had forced a totalitarian government to meet many of its demands, to change the country's constitution and to reintroduce political dialogue where there had been none for a generation. The immediate and most potent cause of the government's defeat by the workers' movement Solidarity was its economic failure, but the workers' demands showed how far discontent spread beyond the purely material sphere to questions of human freedom and dignity such as the right of industrial groups to manage (some of) their affairs for themselves and to express and publish their own views.

Communist control was preserved in diluted form when General Jaruzelski, prime minister from February 1981, was appointed head of the party later in that year. To make assurance doubly sure he then proclaimed martial law. Walesa's control over Solidarity, by contrast, faltered as the government's measures and the arrest of some of its leading members produced divided counsels. When martial law was lifted after a year the main focus of opposition had shifted from Solidarity to the more tractable hands of the Roman Catholic Church, traditionally accustomed and disposed to deal with lay power. An uneasy accommodation was tacitly sealed by a visit by the Pope in 1983, followed by an amnesty for (most) political prisoners which emphasized the army's confidence. But confidence was misplaced, for there were no answers to accumulating economic problems or to an accelerating flight from membership of the Communist Party which was the only recognized instrument of government. The murder in October 1984 of Father Jerzy Popieluszko, one of a number of outspoken priests whose sermons praised Solidarity in patriotic terms, revealed the fires beneath the surface. It revealed also the inadequacy of the government's control over its own agencies: Father Popieluszko was kidnapped and tortured to death by the police.

The extraordinary significance of the Polish protest lay in its source – the industrial workers. In the USSR protest against bad government had been endemic for over a hundred years, back to the days of Alexander Herzen; but this was intellectual protest, as ineffectual as it was honourable and admirable. It reappeared in the postwar USSR and elsewhere, for example Hungary and Czechoslovakia. But only in Poland – and East Germany (1953) – were there, before the late 1980s, any industrial protests on a politically significant scale. Stanislaw Kania, who replaced Edward Gierek as secretary-general of the Polish Workers Party, faced a situation unique in the bloc. For externally little had changed. In spite of the domestic convulsions which brought him to the top Kania had to reckon with the fact that Moscow was still the master. This mastery was moreover intensified by Poland's economic plight. Poland could either pay its debts or feed itself but not both: the meat which had been at the centre of the summer's discontents could either be exported for foreign currency or released into the home market where, as a result of the wage increases granted before and after the strikes, it was more in demand than ever. Coincidentally a bad harvest was adding to import bills for grains and feeding stuffs, and although Poland's principal western trading partners – the United States and West Germany in the lead – were willing to extend fresh loans (within limits), only in Moscow could Poland's new government get credits commensurate with its difficulties. Poland in 1980 was a portent of trouble within the bloc but not, like Yugoslavia when Tito confronted Stalin thirty-two years earlier, a detachable part of the bloc. As the Poles themselves saw, geography ruled out anything like the Yugoslav secession and, by the same token, the massive financial and political aid from the west which sustained Tito from 1948. The degree to which Kania might succeed where Gierek had failed would therefore be settled in Moscow whence alone Kania might get the economic aid or concessions needed to maintain the treaty with the strikers – a compromise which Kania could not honour if Moscow chose not to let him do so. The USSR permitted the August settlement as the least of a number of evils but there was no guarantee of its continuing endorsement, still less of approval for further tinkering with the constitution or with the dominance of the Communist Party.

That the USSR was capable of reoccupying Poland was never in doubt and the examples of Czechoslovakia and Afghanistan showed that the Kremlin did not lack the nerve to use force. There were nevertheless strong reasons for supposing that it was extremely averse to doing so. These reasons derived neither from Poland's fighting spirit, which could be presumed to be as splendid and as ineffectual as it had been in 1939; nor from any American deterrent, for an American military threat was not credible and economic sanctions – as the recent Afghan crisis underlined – were slow to operate and fairly easy to circumvent. What made the USSR recoil from an invasion of Poland was the prospect of finding there nobody to govern the country – no Kádár as in Hungary in 1956, no Hušak as in Czechoslovakia in 1968.

Poland's communists, already identified with economic catastrophe and internal repression, could not affront Polish nationalism by playing a similar role, and so a Russian invasion would entail direct Russian rule – as in Tsarist times. But this was a policy which had been rejected both by Stalin in 1945 when he poured cold water on the idea that the countries of eastern Europe should join the USSR as Soviet Republics (the three Baltic states alone excepted), and by Brezhnev when he rejected in 1970 Gomulka's plea to use Russian forces to save his falling régime. Therefore Moscow adopted a minimal policy. It wrote off the Polish Communist Party and fell back on the Polish army in the hope that Jaruzelski could, by martial law if necessary, keep Poland a one-party state and in the Warsaw Pact. Failing such an outcome Stalin's recipe for the control of central Europe was – in Poland's case at least – in ruins. In 1944–45 Stalin had occupied this area as a prelude to establishing in it governments which would be both subservient to Moscow and wield adequate authority each in its own zone. That done, Stalin could withdraw all or some of his armed forces. But if the authority of these governments were to collapse the system collapsed. The USSR would then have to choose between tolerating a multiparty system and reimposing direct Russian military rule. The appointment of General Jaruzelski as prime minister was a makeshift response to a hopeless dilemma. Polish military rule might avoid the extremes of overt Russian dictation and party political competition, but it could solve neither Poland's main problems nor those of the USSR in Poland. Martial law from 1981 to 1983 delayed an inevitable dialogue with Solidarity while further damaging the economy by angering western countries which, venting on Poland their displeasure with the USSR, refused to renew their loans or to extend their trade.

END OF EMPIRE

In 1985 Gorbachev came to power in Moscow prepared to abandon the Stalinist empire. In 1989 it vanished. Everywhere except in Romania it vanished almost bloodlessly. The sources of these revolutions in central and eastern Europe were similar, their outcomes less so. The East German state disappeared; in central Europe communist rule was decisively rejected and centre-right governments were installed; further east, in Bulgaria and Romania, the old régime with a new face was not so clearly removed.

In *Poland* Jaruzelski's government tried to stave off economic collapse by imposing painful reforms, but since his government lacked popular legitimacy its programme was rejected in a popular referendum in 1987.

Two years later, in April 1989, Jaruzelski, overriding opposition from the military and the communists, legalized Solidarity. He agreed to introduce a multiparty system, remove censorship of the press and broadcasting, and hold elections for the Senate and the Seym (the lower chamber), reserving however for communists and their minor allies two-thirds of the seats in the Seym. Two months later elections were held and the communists routed. After the second round of voting Solidarity had won 99 of the 100 seats in the Senate and the maximum open to it in the Seym (162 out of 460). In the reserved seats many official communist candidates failed to win the 50 per cent of votes required for election on the first ballot. This outcome produced consternation in ruling circles and astonishment practically everywhere. Jaruzelski, having resigned his party post, was elected unanimously to the state presidency by the two chambers of the parliament but the communists failed to form a government acceptable to parliament and their minor allies switched to the side of Solidarity which found itself propelled inexorably, but not wholly willingly, into government. After tortuous negotiations one of its leaders, Tadeus Mazowiecki, was made prime minister, most of the communists in the Seym voting for him. His appointment was welcomed by Gorbachev. But the economy continued to plummet so disastrously that the European Community established a special emergency fund to buy food for Poland and provide three-year subventions from the EC itself, its principal member states, the United States and Japan. During the decade wages lost a fifth of their purchasing power, inflation came near to 500 per cent a year, and Poland's debt to the west of $35,500 million was much the largest of the satellites'. (Hungary's at $13,700 million at the beginning of 1989 was higher per head of population. Others were: East Germany $7,600 million; Bulgaria $5,400 million; Czechoslovakia $4,000 million; Romania $3,900 million but then paid off almost entirely at Ceauşescu's quirky resolve. Yugoslavia owed the west $17,600 million.) In 1990 Jaruzelski extricated himself from a hopeless position by prematurely resigning the presidency which was thereupon contested between Lech Walesa with demands for speedier but unspecified action and, on the other hand, the more prosaic but more reflective Mazowiecki: the man of destiny who had challenged the communist order, against the man who seemed to many better equipped to pick up the pieces. The contest between them was bizarrely bewildered by the incursion of a third candidate, Stanislaw Tyminski, who had spent the previous twenty years in Canada and Peru making himself (so he said) a millionaire and promised to do the same for countless Poles. He beat Mazowiecki into third place on the first round and although overwhelmed on the second by Walesa secured a quarter of the votes cast.

Shock therapy for the economy imposed all but intolerable burdens in order to cope with the huge debts incurred in the 1970s and 1980s to the Paris Club of western lenders and private bankers and to secure a fresh start. The United States, followed by Britain and France, wrote off two-thirds of its debts as a contribution to, and reward for, democracy; the EC concluded a helpful association agreement, and the IMF provided funds in return for drastic reductions in government expenditure, including social schemes and benefits. Hyperinflation was reduced to a less terrifying 60 per cent a year but unemployment doubled and revenue continued to fall. Walesa and the parliament were at odds over remedies and over the distribution of power: Walesa, having played the central role in getting rid of communist rule, found it uncongenial to adapt to parliamentary democracy. Parties proliferated. Of sixty-seven taking part in elections, twenty-nine won seats – pluralism with a vengeance. Mazowiecki's Democratic Union and the (ex-communist) SLD were in first and second place. Walesa failed to get himself made prime minister as well as president. In 1993 the SLD with its allies of the Peasant Party won elections and the latter's leader Waldemar Pawlak became prime minister. But Walesa's unaccommodating personality, insistence on controlling senior appointments and eagerness to be re-elected in spite of declining popularity made for political instability. Pawlak was forced out of office after little more than a year, replaced by the ex-communist Josef Oleksi. Presidential elections in 1995 were narrowed down to a contest between Walesa and the SLD's Alexander Kwasniewski – young, intelligent but a former communist who had held office in Poland's last communist government. During the campaign Walesa recovered much of his lost popularity but in spite of aggressive support from the Roman Catholic hierarchy which focused attention on past ideological battles rather than current economic problems, he was narrowly defeated by opponents who were better organized and more forward-looking.

Poland's external affairs were unusually equable. Walesa recognized Poland's existing frontiers, there were no frontiers with Russia and there were no Germans left in Poland.

In *Hungary*, in the same month as the Polish elections of 1989, the body of the murdered Imre Nagy was brought to Budapest and given a reburial which turned into a vast popular demonstration against the régime. Kádár's thirty-two-year rule had been brought to an end the year before (he died in 1989). He had advocated economic reforms before the advent of Gorbachev in the USSR but after that event his concept of reform was made to seem too lame and was overtaken by a new momentum. His successor as general secretary of the ruling party, Karolyi Grosz, came into collision with his colleagues over the scope and pace of change – notably with Miklos Nemeth, the prime minister, and Imre Poszgay, the régime's economic expert, who wanted not only to accelerate economic reform but also to introduce radical political changes such as the abandonment of the communist monopoly of political power. Elections in a multiparty system were promised for early in 1990 and the party set about salvaging some of its power by changing its name and putting on a new public face: the Hungarian Socialist Workers' Party became the Hungarian Socialist Party. Two main opposition parties emerged: the Hungarian Democratic Forum, a centre-right party akin to the Christian Democrats of western Europe, and the Alliance of Free Democrats, the vehicle primarily of urban intellectuals. Fifty other parties appeared too. The non-communists defeated the communists (who were reduced to 10 per cent of the vote) and formed a coalition which, however, proved uneasy and remained intact for only a few months. It was replaced by a centre-right coalition under the unobtrusive but firm guidance of Josef Antall who, until his death in 1993, was the chief begetter of the least turbulent transition from communism to democracy in the former Soviet satellites. His party won nearly half the seats (with a quarter of the popular vote) but could find no miraculous economic cures and was embarrassed by an extreme right-wing minority – two burdens which contributed to its defeat in 1994 by a resurgent, largely ex-communist, socialist party led by Gyula Horn. Crisis in Yugoslavia sharpened Hungary's worries about Hungarian minorities in Romania and Slovakia as well as Serbia. Horn's government tried to conclude treaties with Romania and Slovakia which would define frontiers in return for guaranteed rights for minorities. Nearly a quarter of Hungarians lived outside Hungary – 2 million in Romania, 1 million in Slovakia, Serbia and Ukraine, smaller numbers in Croatia and Slovenia.

Hungary played an elliptical role in the next act of the year's drama. In May *East Germans* began to desert their state in a mass emigration when Hungary, in a gesture with no special reference to German affairs, removed restrictions on its frontiers with Austria. East Germans on holiday in Hungary found that they had an open road through Austria to West Germany where they had automatic rights of access and citizenship. They took the road at a rate initially of 5,000 a day. This exodus was swollen a few months later when Hungary abrogated an agreement with East Germany and virtually constituted itself a transit area for mass emigration of the discontented and the desperate. Others in East Germany staged huge anti-government demonstrations in Leipzig, Berlin and other cities and under these pressures – and with the example of events elsewhere in central Europe derived from West German television – new escape routes to the west were fashioned through Czechoslovakia, from Poland through East Germany in special trains, and eventually direct from East to West Germany. By the end of the year about half a million had left and they were still leaving at the rate of 2,000 a day.

At the beginning of the year Erich Honecker's inclination had been to hold on to power, if necessary by force, and to treat his people as Deng had treated the Chinese in Tiananmen Square. Honecker's position, however, was in one crucial respect the weakest among communist leaders since the state which he ruled was a Russian creation without a national identity of its own. The USSR was the ultimate source of the ruling party's power (as it was throughout the satellite empire) but also, and uniquely in the German case, the USSR was the source of the state's legitimacy and *raison d'être*. When therefore Gorbachev arrived on a visit to Berlin in October 1989 his appearance in the midst of a crisis not only recalled his appearance in Beijing in similar circumstances six months earlier but also underlined the essential difference that, whereas in Beijing he had been an accidental observer, in Berlin he held a key to future developments. In terms which were at once guarded and unmistakable he warned Honecker against the dangers of not moving with the times and privately made it clear that the East German régime must expect no help from the USSR in suppressing violent demonstrations. He may have gone even further, thwarting Honecker's assumed willingness to use East German forces in the manner of Deng and encouraging Honecker's removal from power. In a flurry of unforeseen events Honecker was replaced as general secretary of the Socialist Unity Party by Egon Krenz, a younger but not much less tarnished member of the inner core of the communist élite; and Krenz, who barely outlasted the year, was himself replaced by Gregor Gysi, under whose guidance the party renounced its monopoly of political power, changed its name and prepared for multiparty elections early in 1990. Hans Modrow, mayor of Dresden and a comparatively respectable communist, became prime minister in an interim administration formed to conduct negotiations with non-communist groups which were pupating into political parties. These parties, campaigning under the management of their West German counterparts, submerged but did not obliterate the communists in elections in which nearly half the votes were cast for the right, fewer than a quarter for socialists and 16 per cent for communists. The Berlin wall was demolished and the city united as men and women from east and west were once more free to pass from one side to the other. Germany itself was reunited in fact while German and foreign politicians argued about the terms and timing of reunification. The West German government, desperate to stem the wholesale migration of East Germans to the west, promised to convert the East German currency into Deutschmarks at a basic rate (there were exceptions) of one-for-one – a measure which entailed an increase in the money supply in West Germany by 14 per cent and corresponding inflation over an indefinable period – but nothing could prevent the collapse of East German industries. Kohl promised to drag East Germany up to the levels of prosperity enjoyed by West Germans. The means to this end were to be investment in the east by the west and by non-Germans. The immediate reality was the halving of industrial production in a single year, a persistent migration of East Germans fleeing not a police state but economic disaster, and unemployment among those who remained at between a third and a quarter of the workforce. A special office, the *Treuhandanstalt*, created to sell into private ownership 14,000 East German enterprises as quickly as possible and at whatever cost in job losses, found many of them so overmanned and environmentally noxious as to be unsellable. Its first chief was assassinated and it was wound up at the end of 1994.

The dissolution of the East German state catapulted the unification of Germany on to the international agenda. Gorbachev briefly tried and predictably failed to secure the neutrality of the newly enlarged state. At a meeting with Kohl in the Caucasus he acquiesced in the new Germany's membership of Nato but won in return the retention of Russian troops in what had been East Germany until the end of 1995, a substantial German contribution to their costs and an undertaking that no German units committed to Nato nor any Nato units stationed in Germany should be deployed in the former German Democratic Republic. In October the unification of the two Germanys was formally completed and the East German state was converted into five Länder of the German Federal Republic. This republic was not prewar Germany. East Prussia remained part of the USSR, the lost lands beyond the Oder–Neisse line remained part of Poland. On the first anniversary of the piercing of the Berlin wall, Germany signed treaties affirming these frontiers.

In central Europe revolution in *Czechoslovakia* came last but in the end fast. Opposition had crystallized round Charter 77 which was a staid statement of basic rights and a specific protest against the persecution of a rock group. It carried 1,250 signatures. It was followed by the more broadly based Civic Forum which sought discussions with the government on the release of political prisoners and the removal from office of certain communists accused of inhuman behaviour in 1968 and later repressions. It was able to promote and control demonstrations involving 1,000–2,000 people. The Communist Party, uneasily aware of its unpopularity and mismanagement, of apathy in its own ranks and of the winds of change blowing from Poland and the USSR, narrowly preferred discussion to force. It discarded Gustav Husak as its general secretary (an office which he had held together with the presidency since 1968) and appointed a group of more conciliatory communists led by Ladislas Adamec to discuss changes with representatives of Civic Forum led by Vaclav Havel. The communists were, however, offering too little and when it became evident that Gorbachev, on a visit to Prague in 1987, preferred Havel to Adamec the latter threw in his hand. Marian Calfa, a member of Adamec's government, assembled a coalition in which communists and their docile allies held only eleven out of eighteen posts. The dour Milos Jakes, who had succeeded Husak in the presidency, resigned in favour of the more emollient Karel Urbanek as a rising tide of popular demonstrations accelerated the dissolution of the old régime, bloodlessly but decisively. At the end of 1989 Havel, who had served more prison terms than any triumphant national hero since Mahatma Gandhi, became president. His position was confirmed by elections in 1990 although it was less solid in Slovakia than in the 'historic lands' of Bohemia and Moravia. Dubček reappeared and was rewarded for his past endeavours and sufferings with the post of chairman of the parliament. Uniquely in the former satellite states the Czech and Slovak communists decided not to change the name of their parties. By agreement with Moscow the 70,000 Soviet troops in Czechoslovakia were to be withdrawn by 1991 (Gorbachev had already agreed to withdraw his 50,000 troops from Hungary by that date).

But Czechoslovakia did not survive its liberation. It split into two states. The Slovak prime minister Vladimir Meciar, communist turned nationalist, asked of his Czech counterpart Vaclav Klaus more than the latter was prepared to concede and found himself, pehaps not unwillingly, obliged to demand independence. Slovakia, a poor country of 5 million inhabitants, slid in economic terms from central to eastern Europe although it formed part of the Visegrad group with Hungary, Poland and the new Czech Republic. (The Visegrad group was more a convenient geographical term than an active political association.) In the Czech Republic, where Havel was persuaded to accept the presidency, reform concentrated on privatizations of smaller state enterprises and the elimination of price controls. Manufacturing and construction contracted sharply and inflation rose to 40 per cent but unemployment was kept within tolerable limits. A Czech–German treaty declared that neither state had any claim on the territory of the other. Like Poland and Hungary the Czech Republic concluded an association agreement with the EC. In Slovakia Meciar, increasingly autocratic and ill-tempered beyond the tolerance of colleagues and political allies, was twice removed from office. But he recovered with a decisive victory in Slovakia's first general election (1994) when he outdistanced twenty other parties with lavish promises on the economy and attacks on Slovakia's Hungarian minority, its Czech neighbours and western capitalism. A more promising sign was the conclusion by

Slovakia and Hungary in 1995 of a Convention on the Protection of National Minorities.

In *Bulgaria* Todor Zhivkov had ruled for even longer than Kádár in Hungary. In 1983 he tried to fortify his position and breast the tide of dissension within and beyond his party by dismissing his more liberal deputy Chudomir Alexandrov, but to no avail. His fall a year later was preceded by an episode peculiar to Bulgaria: the persecution and flight of its Turkish inhabitants. These people, called Pomaks, had increased from about half a million in 1945 to twice that number forty years later. This rate of increase was not in itself startling but it was higher than the increase among ethnic Bulgarians and the Pomaks, being Muslims, were at a disadvantage whenever the government wanted to curry favour with Bulgarian Christians by diverting attention to people who could be stigmatized as undesirable aliens. In 1950 Bulgaria had asked Turkey to grant visas to 250,000 Turks in Bulgaria, and under an agreement of that year 154,000 left Bulgaria for Turkey without their relatives, who were however allowed to emigrate under a second agreement of 1968. At that point the Bulgarian government declared that there were no Turks left in Bulgaria and that all remaining Muslims were the descendants of Bulgarians who had been forcibly converted in the past to Islam. In 1985 these Muslims were required, for the purposes of the census, to adopt Bulgarian names. This decree revealed massive discontent and even panic, and in the face of Bulgarian obduracy Turkey opened its frontier to all would-be migrants without requiring visas. The result was an exodus. Although Bulgaria refused to allow males of military age to leave, at least 300,000 persons did so, whereupon Turkey reversed its earlier decision and reintroduced visas. (Bulgaria evicted some 2,000 Turks into Romania and Yugoslavia, whence most of them proceeded to Turkey. After the revolution of 1989 the remaining Turks in Bulgaria were given the right to choose their own names and many refugees returned in order to claim privatized land but ethnic hostility persisted.) But the final blow to the Zhivkov régime was delivered not by the Pomaks but on the occasion of a conference of environmentalists in Sofia which was turned into the setting for demands for *glasnost*. These were met by police brutality of a kind which had served its purpose in the past but led now to the ousting of Zhivkov by his rattled colleagues. Like Honecker, Zhivkov was arrested and charged with crimes against the state. Under his successor Petar Mladenov the Bulgarian Communist Party changed its name to Socialist, renounced its political monopoly and began negotiations with other parties – the Agrarian People's Party and the Union of Democratic Forces, the latter an assemblage of anti-communist groups – to form a traditional government. These talks having led nowhere, the communists remained in office and contrived to win what appeared to be reasonably conducted elections, but the ensuing popular clamour forced Mladenov to resign the presidency to Zhelu Zhelev, the leader of the UDF (who was re-elected in 1992). Political power wavered between the renamed communists and their opponents. A split in the communist–socialist party and bounding inflation cost the prime minister Andrei Lukanov his majority and then his office. His successor, Dimitru Popov, won a close election in 1991 with the help of a Turkish party and imposed severe economic measures. But in 1994 the left, whose better organization outweighed memories of cast-off ideologies, recovered enough lost ground to win overall control of parliament. The new prime minister, Zhan Videnov, was typical of the new breed of left-wing leaders of a new generation which was in the ascendant throughout much of the former satellite empire: young men with more open minds, better educated and disposed to pursue left-wing aims within a democratic system and a capitalist economy. But they had to find answers to pressing economic problems which deflated right-wing parties in a country which could no longer sell its tobacco to Russia and whose industry was moribund.

In *Romania* the fall of the communist dictatorship was accomplished only with sickening bloodshed but communist power was not extinguished. Nicolae Ceauşescu had used the Romanian Communist Party, never large, as a base for converting Communist Party rule into a personal or family tyranny of intense malignity, supported by extensive and efficient delation, a well-equipped private army, a peculiarly ruthless police and driving megalomania. Like Stalin, whom he admired, Ceauşescu was obsessed with getting things done and damning the human cost. He aimed to modernize and aggrandize Romania by the force, not so much of communist doctrine, as of his own authority and personality, but like many a dictator he failed increasingly to draw the distinction between Romania and himself, with the consequence that he embarked on spectacular enterprises reminiscent of those of the madder Roman emperors. Both at home and abroad he won for a time a measure of approval through anti-Russian policies and gestures, including his refusal to co-operate in the Warsaw Pact or allow foreign troops on Romanian soil. He was rewarded with lavish praise from, among others, George Bush and with a British knighthood and was received effusively in countries as diverse as China, India and Israel. At home, however, he proceeded to denude the economy and even to order the destruction of 7,000 villages whose inhabitants were compelled to migrate in search of a livelihood and were employed in building ostentatious palaces in the capital on starvation wages. Serious riots were first recorded from Brasov in 1987, and at Timisoara in Transylvania at the end of 1989 the army refused to fire on large popular demonstrations which turned out to protest at the persecution of a fearlessly outspoken Protestant cleric, Laszlo Tokes. This spark lit a fire which Ceauşescu, returning from a visit to Teheran, was unable to quench. His security police or private army was opposed by the regular army as well as by large unarmed crowds, and thousands were killed before Ceauşescu and his wife took to flight. They were captured and executed after the merest formality of a trial. Paradoxically Ceauşescu, the least satellite-minded communist leader, paid the grimmest price in the débâcle of the Stalinist empire. Paradoxically again, his removal allowed other communists, who had been plotting a palace revolution, to take power. These had formed a National Salvation Front which declared itself a transitional government, emerged victorious from internationally supervised elections and thrust its chairman Ion Iliescu into the presidency. He was re-elected in 1992. Romanians looking for an alternative to the rump of the *ancien régime* seemed more inclined to turn to what they had known in the past – Antonescu's nationalist dictatorship, for example – rather than to the unfamiliar ideas of liberal and democratic groups which failed to form effective political parties. The disintegration of the USSR impinged on Romania through declarations of independence by the Moldavian SSR with its partly Romanian population and by the Ukraine which had been given a part of Romania in 1939. Attempts to reconstruct the oil, gas and coal industries with a view to making them profitable by 1996 attracted only disappointing amounts of foreign capital.

The year 1989 witnessed revolutions in the grand tradition – the assertion of civic rights and human values – but, like the French Revolution of two hundred years earlier, they were grounded in economic collapse and sustainable only through economic improvement. They were revolutions against tyranny, corruption and incompetence. Since communist parties had been responsible for these evils, the revolutions were also revolutions against these parties and against communism itself. The parties were demoted from their privileged positions and removed from power in most places for (probably) a long time, but they were not eliminated and might be expected to find some place in the spectrum of multiparty politics. The principal features in the overthrow of the Stalinist empire in Europe were economic disaster, Gorbachev, persistent intellectual and popular protest over decades, and the rulers' lack of a plausible claim to legitimacy. The principal consequence was articulation in both senses of the word: the expression of opinion triumphing over suppression, and the re-emergence of the half-buried pattern of sovereign nation states. Gorbachev saw that the Stalinist system did not work. It was an expensive millstone round Moscow's neck and had lost the strategic rationale which had been its prime justification after 1945. Forty years later an American attack on the USSR, which was in any case far less probable than Stalin might himself have imagined, would bypass the satellites by passing miles over their heads instead of moving across their territory. Gorbachev's disenchantment with Stalin's system was not concealed from the regional satraps whose business it was to operate it. They could see that they were no longer wanted. They had become proconsuls in a defunct empire and they could not look to their fellow citizens for support since they had grossly abused their power and, besides destroying basic freedoms, had allowed whole economies to moulder. Although in the 1950s economic growth had reached 10 per cent in the more favoured parts of the zone, by the 1980s it was zero or nearly zero practically everywhere. Hence hardship, indignation, tensions, demonstrations, repression (often extremely brutal) and so to revolution. All the revolutions were swift and thorough: second-rank communists who were thrust into leading positions disappeared almost as soon as they appeared. Most portentous was the standing of Gorbachev, the president of the USSR which had imposed the régimes which were being discarded. Instead of being reviled he was acclaimed by crowds which chanted his name with grateful enthusiasm and seemed to regard the United States and other western democracies as peripheral. Yet Gorbachev, however crucial his role in the transformation of European affairs, was not the only begetter of the revolutions in central and eastern Europe. These were not as sudden as many bewildered observers imagined. They were part of a sequence (to which they owed much) of abortive risings reaching back to the first postwar decade: 1953, 1956, 1968, 1980. All these commotions, including the formation and early achievements of Solidarity in Poland, ante-dated the elevation of Gorbachev and were made by people – some of them out of the ordinary, but many of them people commonly dubbed ordinary – who won attention for their indignation and their values and eventually triumphed on the streets in the tradition of the barricades. Solidarity, for example, ignited a jet of popular pressures which could not be capped. The destruction of the imperial superstructure by the fires of revolution uncovered the perennial alternative: a *Mitteleuropa* articulated into separate states, conceived as nation states but in reality conglomerates of greater or less coherence. In the eyes of the new men and women these states were real and legitimate. Only the old rulers had lacked legitimacy – except in East Germany where the state itself did so too. Revolution had therefore strong nationalist ingredients. Its leaders shared a common plight and common aims and were in that sense linked internationally, but revolution was in no sense supra-national even to the limited extent that supra-nationalism was taking hold in western Europe.

The revolutions did not solve economic problems. By the standards of western Europe the material state of these countries was pitiable. Yet most people in central and eastern Europe were materially better off than most people in Asia or Africa. Their economies suffered from distortion, not from sterility. They possessed agricultural and mineral resources and experienced growth in output, reaching in some places 10 per cent a year – albeit growth from a low postwar base and increasingly in the wrong directions. In Poland's large agricultural sector, four-fifths of it in private hands, output rose throughout the postwar decades but investment was inadequate and the agricultural workforce failed to contribute to GNP commensurately with its numbers or to feed the population. Investment in Polish industry increased output but not productivity; exports declined and foreign debt accumulated. Czechoslovakia, with a lengthy history of good education, good administration and technical skills, with the highest income per head in central Europe, with a flourishing export business and enviable natural resources, was converted under communist rule from the things it did best (notably medium industries such as making shoes) to serving the satellite bloc's appetite for heavy industrial goods and armaments. It lost old skills and concentrated on products which, with the end of the Cold War, were wanted neither at home nor abroad. Hungary, the smallest of this central European trio, possessed rich and well-watered agricultural land and significant mineral resources, although these were not – with the exception of its bauxite – of the first quality. Postwar modernization boosted output and industrial exports, but the economy was neither adequately capitalized nor efficiently managed, returns on investment were low, foreign debt high. In all three states much effort brought little reward. Food and consumer goods became dearer and scarcer. Industrial pollution was the worst in the world. Eastern Europe fared no better. Romania embarked on an ambitious, even ruthless, attempt to modernize its economy on the basis of its coal, oil and other minerals, its chemical, hydroelectric and metallurgical industries, and its underpaid and overworked labour force. Ceauşescu's megalomania turned failure into catastrophe. In Bulgaria a similar, if less ruthless, expansion of industry increased output, much of which was however unsaleable because of its poor quality. In sum, considerable efforts to catch up with western Europe left central and eastern Europe further behind, impoverished, polluted and verging on hopelessness.

8 YUGOSLAVIA AND ALBANIA

FEDERATED YUGOSLAVIA

Tito lived for thirty-two years after his rift with Stalin in 1948. After Stalin's death Khrushchev paid two conciliatory visits to Yugoslavia. The first, in 1955, was tantamount to a confession of error and an apology for the Russian stand in 1948. The second, in June 1956, followed the twentieth congress of the Communist Party of the USSR at which, earlier in the year, Khrushchev had indicated that relations with the satellites needed to be put on a new basis; it followed also the dissolution in April of the Comintern, the body which had pronounced Tito's excommunication. But Russo–Yugoslav relations remained distant and distrustful, if less bitter. Internally Tito, having won the right to tackle his problems in his own way, did so only tentatively, although the introduction of workers' management in industry in 1950 and the abandonment of collectivization in the countryside in 1953 attested a willingness to moderate doctrinal rigidity. The constitution of 1945 had nationalized all industrial, commercial and financial enterprises, limited individual landholdings to sixty acres, and organized the surplus agricultural land into collective farms. The first five-year plan was an immensely detailed, voluminous and bureaucratic blueprint for a Russian-style command economy. It was gradually dismantled in the 1950s, mainly because it was hopelessly cumbersome. In 1950 and again in 1961 industrial control was devolved on to workers' councils with wide powers of management, including the right to allocate investment funds and to decide what to do with profits. In the early 1960s new lines of credit were made available through local banks (as opposed to the central bank) and the system of fixing prices centrally was relaxed. A new constitution in 1963 introduced real but limited decentralization without, however, removing ultimate power from the party or Tito himself. These half-measures failed to give industry the expected stimulus, the economy remained stagnant, inflation rose and so did Yugoslavia's dependence on aid from the IMF and the United States. There was a brief reaction in economic policy but the liberalizing and

8.1   Yugoslavia and its republics until 1991

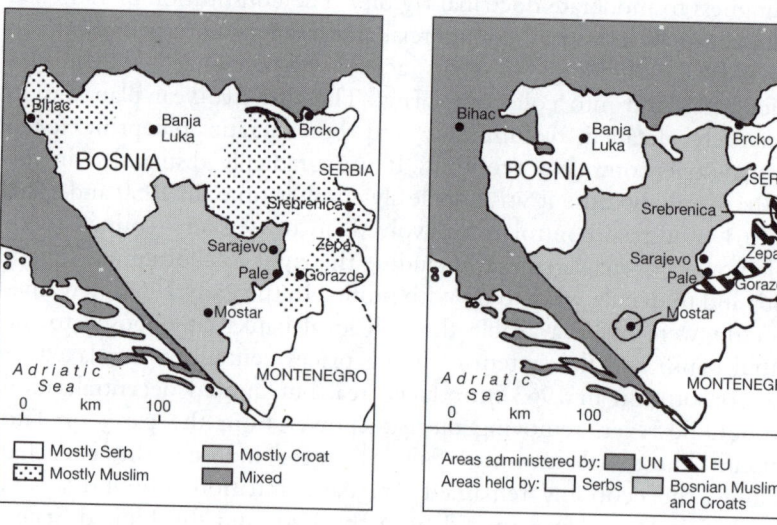

8.2   The Vance–Owen plan
(April 1993)

8.3   The Contact Group proposal
(July 1994)

8.4   The Dayton peace plan (November 1995)

decentralizing strands were – also briefly – resumed. So in fits and starts the economic sector was fragmented into tens of thousands of autonomous units. Whether the psychological stimuli of decontrol and self-management outweighed losses in efficiency and productivity was inconclusively debated both inside Yugoslavia and out. More certainly, these industrial changes created a new base for political power by opening careers for talents in industry besides the traditional ladders of the Communist Party and the armed forces.

Besides the economy Tito's second main concern was the cohesion of the state and the creation of a Yugoslav identity over and above the Serb, Croat and other nationalisms within the Yugoslav federation. Tito, half Croat and half Slovene by birth, was determined to preserve the Yugoslav state which had emerged from the destruction of the Habsburg and Ottoman empires in the First World War. He was aided by the camaraderie of the partisan warfare of the Second World War, during which he had been careful to assemble an ethnically diverse leadership in the Communist Party and army. The integrity of Yugoslavia was not seriously threatened so long as this generation remained in sole charge. By the 1970s, however, a new generation had grown up, which was more nationalist and separatist, partly because the wartime magic was receding and partly because peacetime competition for goods and investment fostered far from forgotten scores. In Croatia agitation for more autonomy went to the length of demands for sovereign independence (but within a Yugoslav confederation) and a separate seat in the UN such as had been granted to Ukraine and Byelorussia. A time of troubles loomed – with Tito close on eighty. In 1971 Tito decided to stamp on his aberrant younger colleagues. He made it plain that he was the judge of when enough was enough, and that in his judgement Croat separatism was unpardonably threatening the integrity and welfare of the Yugoslav state. Once he resolved to intervene he carried the day. Significantly the Croat elder statesman Vladimir Bakarić stood by him. The younger leaders resigned. Tito then repeated this performance in Serbia where the younger politicians were no less nationalist; and in Slovenia and Macedonia (but here dissent was not purely nationalist).

Separatism was an extreme form of the continuing problem of decentralism which called in question the powers of the central government and – an even more sensitive matter – the central organs of the Communist Party. In the 1950s and 1960s both centralism and permanent communist rule were questioned. Centralism was attacked on grounds of efficiency and offended local patriotism. Criticism was open and lively. It raised the question how political authority could be dispersed without denting communist control. The decentralizers argued as technocrats, managers and grassroots politicians. Their arguments either skated round the consequences for communist control or, as in the case of Djilas, accepted the conclusion that the Communist Party had no automatic right to permanent power. They became a double source of offence to the centralizers who, pre-eminently Ranković, not only preferred central to dispersed authority for its own sake but also saw in decentralization a menace to the communist monopoly of power. Tito allowed the debate to flourish until it seemed to become dangerously disruptive. He was not an enemy of limited debate, but he did not intend that either side should win. In the liberalizing atmosphere of the 1950s the first need was to curb the liberalizers. Djilas was persecuted and imprisoned. Ranković remained in office – and was even promoted – until his anti-liberal campaign overstepped the bounds and he was found to be spying on Tito himself, whereupon he was dismissed. At the end of this phase Tito devised a new constitution which created an Executive Bureau or Presidential Commission consisting of two representatives of each federal republic and one from each autonomous district with Tito as permanent federal president. (After his death the presidency was to revolve.) This arrangement was intended to signal a compromise but it left the realities untouched and conflict no more than deadened.

The outcome of the separatist and decentralizing currents of the 1950s and 1960s was to block secession but at the same time to allow a federation with a strong centre to evolve towards something approaching a looser confederation of sovereign entities on the model of Switzerland. And the state remained a communist one. In 1958 the Communist Party had been renamed the League of Communists. This change was taken to indicate that the party might not always be the right organ for running local government bodies or industrial plants. But the party, or League, remained unique, so that the unity of Yugoslavia remained linked with communist rule rather than the development of a Yugoslav nationalism. Older nationalisms survived and flourished and when Yugoslav federalism disintegrated leaders in the several republics – Milosevic in Serbia was only the most conspicuous example – transformed themselves from communists into strident nationalists, reanimating the ethnic and religious hatreds which had served to get rid of alien overlords (Turkish, Austrian) in living memory and were now aligned against one another.

At Tito's death in 1980 Yugoslavia was unique. It was the only communist neutral in the world. But it was also fragmenting and indigent. It was no more cohesive as a communist federation than it had been as a triune kingdom. The task of the tail end of the Karageorgević dynasty had been to compound Serbs, Croats and Slovenes into a Yugoslav state and nation, but this task was no more than half achieved before the Second World War reopened old divisions, particularly between the Serbs and Croats, whose grisly wartime leader Ante Pavelić tried to create a separate Croat state, buying Italian support by ceding marginal Croat lands to fascist Italy (which was looking across the Adriatic for its version of *Lebensraum* – *spazio vitale* – in Albania and Dalmatia). On the other hand, the communist resistance to Italians and Germans produced an aura of Yugoslav confraternity which, after the war, Tito sought to enhance by giving ethnic minorities – Montenegrins, Macedonians, Albanians (in Kosovo), Hungarians – equal or approximately equal status with the three original constituents of the Yugoslav state. But Tito could not give these different groups anything approaching economic equality. Their national and religious differences – within Christianity and, for Muslims, beyond it – were not offset by any economic fruits of federation. Aid from the west which followed the rift with Moscow was used mainly to develop industries. The consequent drift from the countryside to towns was at first a welcome relief for a problem of rural overpopulation insufficiently tempered by emigration, but in the longer run it weakened agriculture without commensurate benefit from industry which faltered when reforms in management and financing were applied too slowly and without turning the right to manage into competent management. Western helpers lost interest when they lost money and when aid to Yugoslavia lost its political edge with the cooling of the Cold War. Foreign debt piled up, so that paying the interest on it became a burden on the Yugoslav economy bigger than the benefits conferred by the loans in the first place. The economic balance spread over a generation showed a net loss. In 1989 the federal prime minister, Ante Marković, a Croat, adopted desperate measures, instituting a convertible currency, commercial banks, a stock exchange and other apparatus of a capitalist market economy. He aimed by strengthening the economy to stave off Serb, Croat and Slovenian nationalism. He won some international support but not enough nor soon enough. Inflation was reduced from 2,000 per cent to zero in a matter of months and exports and the reserve rose, but at the price of catastrophic unemployment and bankruptcies and dire want. Endemic tensions between the several republics were accentuated. In the largest, Serbia, which contained 40 per cent of the population, nationalists led by the flamboyant Slovodan Milosevic – who made up in Serb nationalism what he was in danger of losing as a communist – were intent, first, on annexing the autonomous districts of Kosovo and Vojvodina; second on restructuring the federation in its original tripartite form; and thirdly, in absorbing the newer republics and dominating the federation itself.

The issue in Kosovo went back to the Second World War and to the Middle Ages. During the war promises of autonomy had been used to enlist the co-operation of Albanians against Italians and Germans. But Kosovo, although now inhabited by nine Albanians for every Serb, was the historic heartland of Serb nationalism and the symbol of Serb resistance to the Turks in the fourteenth century – the six hundredth anniversary of the fateful battle of Kosovo which had extinguished Serbian independence fell in 1989. In 1990 the Albanian majority in Kosovo declared the district to be a full republic and the Serbs retaliated by adopting a new constitution which extinguished the autonomy of both Kosovo and Vojvodina. Milosevic's rhetoric and a sweeping victory in multiparty elections in 1990 accentuated Croat and Slovene tendencies to secede from what in their eyes was becoming Greater Serbia. In Croatia, Franjo Tudjman, a communist turned chauvinist, won over 40 per cent of the vote and 70 per cent of the seats in the parliament. In multiparty elections in 1990, Croats and Serbs confronted one another in eastern Slavonia and the Dalmatian district of Knin (two Croat areas with significant Serb populations) and over the fate of Bosnia where Croatia claimed a strip of the republic and the more militant Serbs claimed the whole of it.

DISSOLUTION: SERBS, CROATS AND SLOVENES

From Tito's death or earlier Yugoslavia displayed signals of strains portending either constitutional change towards a looser federation or, in default of adjustment, disruption and war. The driving force was Serbia, the most powerful of the six republics, the largest ethnic group and the one with the largest numbers outside its own republic: 2 million

Serbs in Croatia and Bosnia besides 10 million in Serbia. Serbian leaders also had the support of the Orthodox Church in and beyond Serbia – particularly in Greece and Cyprus – which was to play a significant part in blunting economic sanctions from 1991. Religion was still a large part of nationalism in the Balkans, the wars about to erupt were among other things wars of religion, and the intersection of secular and religious paranoia gave these conflicts a ruthless inhumanity all but banished from most of the rest of Europe. Most important of all to Belgrade was its control of the Yugoslav federal army of 135,000, officered predominantly by Serbs.

The temper of the Serbs was embodied in Slobodan Milosevic who rose through the Serbian League of Communists to become president of the Serbian Republic. He was a Serb nationalist and a skilled politician with minimal loyalty to the idea of Yugoslavia. He extinguished in 1989 the autonomous status of the two regions of Kosovo and Vojvodina – the former nine-tenths Albanian in population (a small group of Albanian communists collaborated with Belgrade in this suppression) and the latter largely Hungarian; he gave military and propagandist help to Serbs in Croatia who were agitating for an autonomous Serbian region; he revived Serb claims to parts of Bosnia; he obstructed the routine advancement of the Croat Stipe Mesic to the federal presidency; and in the Krajina, a district of Croatia, he condoned and encouraged Serb activities reminiscent of Henlein and the Sudeten Germans in Czechoslovakia in 1938. He advocated the maintenance of the Yugoslav federation within its existing borders and thereby won goodwill in the EC and the United States which, while threatening economic sanctions, wanted to enlist his help in getting Mesic smoothly installed as federal president and in pacifying rather than inflaming the position in Kosovo. For states outside Yugoslavia the federation represented stability but for Slovenes and Croats it represented Serb dominance while for Serbs it was an artificial entity in which Serbs lived in non-Serb republics. This last consideration did not apply to Slovenia where there were next to no Serbs. In June 1991 Slovenia and Croatia declared themselves independent, repudiating not only the existing federation but any modification of it.

Slovenia, compact, homogeneous, comparatively remote and with a mere million inhabitants, had made preparations to make good its resolve. Each Yugoslav republic possessed its own defence force distinct from the federal army whose units were isolated from one another and from their supplies; their attempts to take control in Slovenia were repulsed. Milosevic was content to see it go. He was intent on concentrating the federal army against Croatia. This crucial distinction was at first unperceived by the increasingly embroiled EC's negotiators.

Croatia, covering a fifth of Yugoslavia with a population of 4.5 million, contained 600,000 Serbs, mainly at the eastern end (eastern Slavonia) and in the Krajina between the Adriatic coast and Bosnia. It had (as had Serbia) memories of ferocious fighting between Croats and Serbs in the Second World War, abetted by the rival Roman Catholic and Greek Orthodox Churches to which nearly all Croats and Serbs respectively belonged. Although worsted in Slovenia the federal army, simultaneously invading Croatia, won control of a third of it, making and unmaking a series of cease-fire agreements in the course of doing so. The Croats lost eastern Slavonia to the federal forces and the Krajina to the Serbs of that region. In Slavonia the Serbs' devastation of Vukovar won the Croats international sympathy which, Germany apart, they had lacked and helped them to secure international recognition of their independence. The Serb attack on Croatia, although speedily successful, was suspended in the early part of 1992 when both Serbs and Croats turned their attention to Bosnia. At this stage Slovenes, Croats and Serbs had effectively demolished the federation and established separate states. The two latter, however, were not minded to allow Bosnia to become a fourth. (The Krajina was originally a Croat–Serb buffer zone or Hadrian's Wall created in the sixteenth century by the Habsburgs against the Turks, the Serbs being refugees from the Ottoman empire. In modern times the Krajina was extended southward to include Knin and its surrounding area peopled by Serbs more militantly anti-Croat than the Serbs of the Krajina proper had been wont to be.)

International intervention had two motives. The first was to stop the fighting for fear that it would spread to the whole of Yugoslavia and beyond. With this aim in view the UN imposed in 1991, in the first or Serb–Croat phase of the war, an embargo on the delivery of arms to any part of Yugoslavia. In addition the EC, later the EC jointly with the UN and – from 1994 – the EC and UN with a Contact Group of five (the United States, Russia, Britain, France and Germany) tried by diplomacy to negotiate a settlement between the combatants. The fighting, which began in the northern republics, threatened to become widespread because the independence of Slovenia and Croatia was likely to produce similar demands in other republics; because Serbia and Macedonia contained ethnic minorities and Bosnia consisted only of minorities; and because conflict in Yugoslavia attracted nervous attention and inflamed national susceptibilities in Albania, Greece and (to a lesser extent) Bulgaria. The initial aim of international diplomacy was to negotiate a political settlement between Serbs and Croats which would nip further troubles in the bud before they became violent. The settlement might involve recasting the Yugoslav federation – a solution which quickly faded away – or accepting its dissolution and defining the boundaries of a new map. A *troika* of EC foreign ministers was succeeded by a single negotiator, Lord Carrington, who was quickly followed by Lord Owen and ultimately by Karl Bildt. The UN appointed first Cyrus Vance and then Thorvald Stoltenberg to work with Lord Owen until this effort was superseded in 1994 by the Contact Group of five. All these eminent persons strove to secure cease-fire agreements and to devise a generally acceptable territorial apportionment of Bosnia within its existing boundaries but with a new constitution. The cease-fire agreements were adopted and broken with sickening cynicism while the territorial proposals were all dismissed by one faction or another with a flippancy which paid scant tribute to the arduous and mainly sensible labours of their authors.

The second branch of international intervention was the succour of victims of war and the protection of those providing these services. It was pursued by various agencies, including UN agencies, and by a UN force – UNPROFOR – recruited from over twenty countries and despatched to aid and protect the aiders but not to become involved in the hostilities. UNPROFOR was not a peace-keeping force. A UN peace-keeping force was sent to Macedonia and may have helped to keep the peace there, but UNPROFOR was established, first in Croatia and later in Bosnia, to protect aid agencies, to succour victims in various areas where peace had not been secured and to keep out of the fighting. It was despatched for humanitarian purposes but in breach of the UN's normal practice of sending peace-keeping or humanitarian missions only to areas where hostilities had been reliably suspended. UNPROFOR was partly a gamble on getting such a suspension and partly a reaction to the unexpectedly fearful atrocities which accompanied the fighting. It was put into a war situation without the authority or capacity to make war and in the hope that it might perform its tasks in spite of the wars being waged in its zones of operation: this was an operation of a kind never before undertaken by the UN Security Council. In many areas and various ways UNPROFOR was successful but it was always a pawn waiting to be seized by one belligerent or another. Its fortunes were inextricably entangled with the parallel diplomatic attempts to take war out of the situation.

The EC's diplomatic effort to pacify Yugoslavia, grounded in fear of spreading disorder and sharpened by horror at the stupefying brutality used by the combatants, was flawed from the outset. EC members discerned an opportunity to assert – or at least test – their collective power but neither the Community as such nor its several members had given much thought to a crisis which had been in the making for a decade or more. Nor was the EC merely unprepared. It had as a Community no standing outside the territory of its members, no armed forces, no community of purpose and no established machinery for concerting Community foreign policies. It was on the contrary only embarking on the first steps towards creating such machinery and fell into the error of supposing that one way of doing so was to act as though it already had what it hoped to concoct. The EC was not a regional organization for Europe in the terms of article 52 of the UN Charter. Attempts to invest it with authority from the CSCE (an even more embryonic body with likewise no armed forces) were an unconvincing subterfuge, the more so since Yugoslavia was a member of the CSCE with a veto on its more positive actions. The EC pressed for the acceptance of Mesic as Yugoslav president, a three months' suspension of the Croat and Slovene declarations of independence, the withdrawal of all armed forces to barracks and a conference on the future shape of Yugoslavia. It achieved none of these things and although it negotiated a series of Serb–Croat cease-fire agreements, these were routinely disregarded. The fighting in Croatia was halted only because the Serbs had conquered most of what they wanted and Bosnia claimed the attention of both Serbs and Croats.

The EC – and notably Germany – also took the lead in arguing for immediate and unconditional recognition of Croat and Slovene independence. This was a political act. In law a people or nation has no right to secede from a sovereign state by invoking the principle of self-determination although other states may, using established diplomatic criteria, choose to recognize a claim to statehood. Slovenia satisfied the broad criteria of international practice but Croatia did not. An adjudicator consulted by the EC – Robert Badinter, a former French minister – reported that the state of affairs in Croatia did not warrant immediate recognition. Croatia neither offered nor was it required to offer guarantees to its large Serb minorities; Croats were not reminded that if Croatia seceded from Yugoslavia Serbs in Croatia might with equal force claim a right to secede from Croatia; and there was no Croatian government with control over the whole of the territory claimed by the new state. Croatia was granted recognition as an independent state in the slipstream of Slovene independence and in the hope of preventing the spread of war into Bosnia and beyond. The opposite view, expressed by the UN secretary-general and others, was that recognition would do nothing to check Serb–Croat hostilities and would evoke a declaration of independence by Bosnia–Hercegovina and its invasion by Serbs and Croats in temporary alliance. Bosnia–Hercegovina satisfied the criteria for independence even less than Croatia.

Milosevic's position in Serbia came under some threat in 1992. Milan Panic, who had returned from the United States after making a lot of money there and had been appointed prime minister of the Serbian republic, was disenchanted with Milosevic. So too was Dimitri Cosic, an eminent Serb nationalist who had been appointed to the vacant Yugoslav presidency. At a conference in London, Cosic struck a bargain with the Croat president, Franjo Tudjman: Serbia would recognize Croatia's independence and its right to the Krajina in return for internationally guaranteed rights for Serbs in Croatia. This agreement was a setback for Milosevic if only because it had been negotiated by somebody else. He was faced with the choice whether or not to endorse it at the cost of antagonizing the more militant Serb nationalists in Serbia and the Bosnian Serbs led by Radovan Karadzic and his military counterpart General Radko Mladic who had purported to annex the Krajina to Bosnia. Milosevic used a dispute over control of the army to secure the dismissal of Cosic, decisively defeated Panic in a presidential election and affirmed his authority in parliamentary elections – in which, however, Vojislav Seselj, campaigning for the eviction of all non-Slavs from the whole of Bosnia and Kosovo, won 20 per cent of the vote and so presented Milosevic with a challenge as the standard-bearer of the purest nationalists. Seselj and Karadzic, who declared an independent Serb republic in Bosnia in 1992, were for Milosevic rival Serb chieftains who had to be kept in order or outmanoeuvred.

For the time being Milosevic had consolidated his position and became the key figure in the search for peace by negotiation. Outsiders had to choose between seeking a deal acceptable to him or making war on him with economic weapons and perhaps more. Since by this date he had invaded Bosnia with rapid success, the only available deal must include some acceptance of Serb conquests in Bosnia as well as Croatia, both of them now internationally recognized sovereign states. The choice or risk of military war against Serbia raised the prospect of lengthy, costly, bloody, possibly ineffective operations with ill-defined purposes, unforeseeable wider consequences and, most telling, disagreement among strategists on what kind of operations and what weight of force would be appropriate. One of Milosevic's main strengths was his awareness that no such war was at all likely. He was little disturbed by the arms embargo imposed by the UN in 1991 on all parts of Yugoslavia since the ban hurt other parties more than the well-armed Serbs; and economic sanctions, imposed on Serbia in 1992 by the EC and the UN, had yet to make their mark. From mid-1992 into 1995 his central problem was how much of Bosnia to annex in the face, on the one hand, of growing international hostility and sanctions and, on the other, the successes of the Bosnian Serbs whose calculations and ambitions might not accord with his own. Serbs in Serbia were prepared to suffer for Greater Serbia but not necessarily for a Greater Serbia with Radovan Karadzic as its hero. The architects of Greater Serbia were in Belgrade, not Bosnia; they were Milosevic and the Serb Orthodox hierarchy, not Karadzic and Mladic.

PARTITION: BOSNIA–HERCEGOVINA

Bosnia–Hercegovina was distinct from Serbia and Croatia not merely as a separate Yugoslav republic but as a political entity which, within varying boundaries, had been accepted as such formally since the sixteenth century and in practice for many generations earlier. Many of the inhabitants had affinities with Serbs or Croats; others were different because they were Muslims. The Bosnian Serbs accounted for less than a third of the population, the Bosnian Croats for about a sixth, the Muslims for nearly half. All these people were Slavs by race and tongue as a result of invasions in the early Middle Ages by Slavs of various kinds who overlaid Iberian, Celtic and Avar peoples and other trace elements. Among these invaders Serbs and Croats were the most prominent and they penetrated into Bosnia as well as establishing principalities of their own on three sides of it. They arrived in the Balkans as pagans but became Christians, and although they were often allies against the dominant but declining power of the Byzantine empire they were divided by giving their allegiance to rival Byzantine and Roman ecclesiastical authorities. Relatively inaccessible Bosnia became a patchwork of lordships great and small which were relatively independent of outside domination until the conquest of the whole area by the Ottoman Turks after the battle of Adrianople in 1463. The Bosnian church was Catholic, not Orthodox, but was suspected of heresy in Rome where nobody knew much about it. The weakness of the ecclesiastical link, the growth of towns under Turkish rule and the simple good sense of adopting the new rulers' faith contributed to the unusually wholesale conversion of Bosnians to Islam. By about 1600, half of them were Muslims. Peasants trying to get a living with the minimum of interference from superiors, merchants anxious to maintain and extend their commerce in the new environment of Turkish administration, Turkish law and Turkish favours, found little embarassment in putting off one faith and putting on another – or the reverse, when expedient.

From beginning to end of the Turkish domination, Bosnia's fortunes were conditioned by the frequent wars between the Ottoman and Habsburg empires. Its frontiers were fixed by the treaties which punctuated those wars in the eighteenth century. As the next century progressed the Turks lost control and Bosnia was up for grabs. With Serbia and Montenegro becoming independent states and clients of Russia, Austro–Hungary became more worried about the Russians than about the Turks and Bosnia more vulnerable to expansionist neighbours. After the Russian defeat of the Turks in 1877 the European powers gathered at Berlin to redraw the map of the Balkans. They decreed that Bosnia should pass to Austro–Hungarian administration (converted into annexation in 1908 but lost to Vienna at the end of the Great War which started in the Bosnian capital Sarajevo in 1914). Between that war and the next there was wide agreement on the idea of a Yugoslav state but no agreement on the division of powers between its central government and its provinces. Bosnia was, with Croatia, opposed to centralization which meant in effect power to the Serbs. The problem was unsolved when the Second World War broke out and it was no more than suppressed, partially, by Tito who, as a communist, was a centralizer but, as a Slovene–Croat, was not. In Tito's Yugoslavia, Bosnia became one of the federation's more depressed areas. One consequence was increased communal tension; another was the departure of many Bosnian Serbs to Serbia, a flight which made the Muslims much the largest of the three main communities in Bosnia. Tito encouraged the Muslims to think of themselves as a distinct community and they were formally so recognized in 1971. But they were split between communist and anti-communist Muslims: Ilia Izetbegovic, future president of Bosnia–Hercegovina, was a prominent anti-communist leader.

In 1990 the government of Bosnia–Hercegovina (Hercegovina was its small south-western corner) was a coalition formed after elections in that year. It contained Muslims, Serbs and Croats but the Serbs left the coalition – their first step towards a distinct Serb state in as much of Bosnia as they might be able to conquer. When fighting in Croatia began President Izetbegovic was faced with a choice between remaining in a new Yugoslavia shorn of Slovenia and Croatia, or claiming independence as Slovenia and Croatia were doing. He chose the latter. He feared Serb and Croat designs on his republic's territory and wanted to be in a position to appeal for outside help in protecting it. He was accorded international recognition and membership of the UN early in 1992 but not before the UN, reacting to the doubling of Serb regular forces in Serbia (normally around 45,000) and the growth of Serb irregular units, had imposed its embargo on the supply of arms to all parts of Yugoslavia, including Bosnia. Serbia treated these events of the spring of 1992 as a signal for open hostilities by regular and irregular forces. They gained control of half Bosnia in a few days. Their methods included terror and atrocities which changed the nature of the war and of UN intervention which, originally confined to protecting aid agencies and civilian victims in need of food and medicines, became entangled in the fate of a million refugees congregated and besieged in lethal enclaves and clamorous to escape from Bosnia rather than be succoured in it. At this point Serbia and Montenegro together proclaimed a new Yugoslavia consisting of themselves, another new state.

There were now several new states in what had been Yugoslavia and two of them were at war with one another: Bosnia–Hercegovina and the new, reduced Yugoslav federation. What had been a civil war in Yugoslavia had been converted into a war between two sovereign members of the UN and the Security Council could have branded Serbia an aggressor. If the war was in any sense a civil war it was a civil war in Bosnia, not a civil war in Yugoslavia. There would have been no case for applying an embargo to Bosnia had it not already existed. But a one-sided lifting of the embargo implied an engagement in the hostilities which in 1992 no government suggested or seemingly contemplated, partly because to do so must jeopardize UNPROFOR's tasks and put an end to the EC/UN's mediatory diplomacy and partly because the Bush and Clinton administrations set their faces against the kind of UN intervention which must entail active American involvement on the ground. The result was equivocation at the international level and covert arms supplies to all parties including the Bosnian government.

In 1992–94 the EC/UN negotiators and the Contact Group produced a series of plans for partitioning Bosnia into Serb, Muslim and Croat segments within a continuing Bosnian entity. These plans were accepted at one time or another by all the principal parties in Bosnia but never simultaneously and perhaps with more guile than sincerity. They accepted the principle of division but not the divisions proposed. The Muslims regarded them as outrageously unfair invitations to them to commit suicide while the Serbs regarded them as inadequate recognition of their conquests and ethnic entitlements. In Serbia sanctions, tightened during 1993, were causing considerable distress as unemployment engulfed half the workforce and the national economy was starved by the loss of access to international aid and investment. Although the worst effects were for a time evaded with the help of Greece, Cyprus and (for a short time) Russia, Milosevic was forced to consider reining back the Serbs in Bosnia and applying his own sanctions to the Bosnian Serbs who depended on Serbia for fuel for their war machine. In 1994 he closed the border between Bosnia and Serbia after the Contact Group devised yet another set of proposals which allotted 49 per cent of Bosnia–Hercegovina to the Serbs. This was in territorial terms generous but Karadzic and his associates reckoned that they might get even more, claiming in particular a larger slice in the east, access to the Adriatic in the west and half Sarajevo. The Contact Group hoped to secure their acceptance through Milosevic who, besides his economic worries, was distrustful of Karadzic's independently expansionist and defiant attitudes.

The Bosnian Serbs were more emboldened than chastened by international intervention which, on the scale represented by

UNPROFOR, excited their derision. This confrontation with the UN was crystallized during 1993 when the Security Council was pressed by its leading members into designating six safe areas (on the analogy of the safe havens proclaimed in Iraq in 1991 but in very different circumstances and terrain) and authorized UNPROFOR to use force to prevent their bombardment. The Council's members did not, however, provide UNPROFOR, already overburdened, with the means to discharge this new function. These safe areas were peopled by Muslims and threatened by Serbs. They were safe only so long as the Serbs abstained from attacking them. There were six of them: four in eastern Bosnia close to the border with Serbia and strategically important for communications between Serbia and Bosnia; Bihacs in north-western Bosnia athwart a main railway and adjacent to the main pocket of Serbs in Croatia; and Sarajevo. The town of Bihacs and its surrounding area were a Bosnian enclave vulnerable to attack from three quarters: from the west by Krajina Serbs, from the east by Bosnian Serbs and from the north by dissident Muslims at odds with the government in Sarajevo and led by an eccentric Muslim entrepreneur with visions of establishing a city state of his own. In 1995 the Bosnian Serbs took the eastern havens of Srebrenica and Zepa, evicted their non-Serb inhabitants and 400–500 Dutch and Ukrainian troops of UNPROFOR and cleansed by terror and murder fresh areas to be attached to a Greater Serbia. They also showed how little they cared about international intervention. In June 1995 they declared themselves at war with the UN and Nato and, in a flagrant breach of the rules and customs of war, took hostage some 380 members of UNPROFOR units – a step too far which enabled the more circumspect Milosevic to assert his pan-Serb authority and insist on their release. (At the same time he put his extremist adversary in Serbia, Vojislav Seselj, in prison.)

For three years the Bosnian Serbs had known only success. Yet there were adverse factors to which Karadzic paid perhaps too little attention. He depended on Milosevic for supplies of, particularly, fuel but he failed to perceive how far Milosevic was becoming at least as anxious for the removal of economic sanctions on Serbia as for further victories in Bosnia by Karadzic; and the successes and excesses of the Bosnian Serbs (unchecked by the Security Council's decision in 1993 to establish at The Hague a tribunal to consider allegations of war crimes in Yugoslavia) created a revulsion which, notably in the United States, took the form of threatening air warfare against them and turning an increasingly blind eye – and lending a helping hand – to the supply of arms to the Bosnian government by, among other channels, the clandestine and unauthorized use of Nato aircraft.

This last factor was a source of dissension among the external states as well as a threat to the Bosnian Serbs. Peace in the region remained the overriding aim of the EC states but in the United States a conflicting aim – justice for non-Serb Bosnians, particularly Muslims – gained ground. Europeans reckoned that the Serbs had won the war in Bosnia and it was futile to encourage the Muslims or a Muslim–Croat alliance to continue it. France and Britain, which provided the largest contingents for UNPROFOR, determinedly opposed any measures which might provoke attacks on it, necessitate its withdrawal together with the aid agencies and delude the Muslims with vain hopes. Americans on the other hand were impelled towards a policy of helping the Bosnian government to redress the military balance by removing the arms embargo from the Bosnian government and using air power to halt Serb aggression and defend the safe areas. Finding that a resolution to raise the embargo would be defeated in the Security Council, Clinton decided instead to recall American vessels from the patrol in the Adriatic which was enforcing it. The practical effect of this step was small but the wider implications of disavowal of a resolution for which the United States had lobbied and voted in the Security Council were graver. Clinton's decision – which was taken, but not announced, before the Democrats' heavy losses in the mid-term elections in November – meant backing Muslims and Croats against Serbs, risking overt Russian support for the Serbs in return and putting an end to hopes of ending the war by negotiation. As in the civil war in Spain in the 1930s, the Bosnian war was being covertly internationalized. The United States and Russia were providing encouragement and more than encouragement to different belligerents while Islamic states in the Middle East were vying with one another in supplying or promising aid to the Muslims. (Some of these countries had been sending charitable aid as far back as the 1970s. More robust commodities were entering the pipelines.)

American advocacy of air power was no less contentious. Air power meant Nato air power since this was the only force readily available to the Security Council. Its use was contentious on political and tactical grounds. The Russians, on whom the Contact Group relied to persuade Milosevic to compel Karadzic to accept its latest peace plan, found it hard to stomach the introduction of the quintessentially anti-Russian Nato into the Bosnian tangle, regarded it as a ruse to by-pass the Security Council and the Russian veto in it, and were privately indignant about the use of Nato to arm the Bosnian government against their friends the Serbs. Europeans attributed the Americans' fondness for air operations to their reluctance to incur casualties in ground operations. Europeans, and some American strategists and commanders, also contested the efficacy of air strikes on the grounds that targets in Bosnia were difficult to find and still more difficult to hit. Over Bihacs these doubts appeared to be justified. When the Serbs, having countered over-optimistic offensives against them, held the town at their mercy, UN commanders invoked air retaliation but it proved equivocal and was discontinued. But the main impediment to the use of any kind of force was not tactical disagreement or political foreboding but the vulnerability of scattered UNPROFOR units liable to be taken hostage by Bosnian Serbs.

The Croats had bided their time. Whether or not Presidents Tudjman and Milosevic had a tacit agreement to partition Bosnia, the Croat president had a second string to his bow: an alliance with the Bosnian government. (During the Second World War the independent but puppet Croat state established by Italy and Germany had extended over most of Bosnia.) In 1993 the governments of Croatia and Bosnia agreed to form a Muslim–Croat federated state with loose confederal links with Croatia – a significant step towards the bisection of Bosnia between a Serbian state or sub-state and a rump federated with and dependent on Croatia. Tudjman, having unostentatiously reorganized and re-equipped his army, launched offensives in 1995 which recovered the Krajina with unexpected efficiency and speed and voided it of 100,000–200,000 terrified Serbs who, fleeing eastward, constituted the war's biggest forced migration and one of its most atrocious human calamities; and overwhelmed the aid agencies and their UN protectors. The Croats also recovered most of eastern Slavonia which they had lost to the Serbs in 1991. The Bosnian Serbs retaliated against the safe areas, seizing Srebenica and Zepa and slaughtering their inhabitants, and delivering one of the fiercest bombardments of the war on Sarajevo. The principal western powers, having removed their UNPROFOR contingents from their exposed positions, attacked the Bosnian Serbs round the remaining safe areas from the air (using Nato forces) and with the ground artillery of a new Rapid Reaction Force created by France, Britain and the Netherlands. The air operations, which were the largest in Nato's entire history, were undertaken without informing Yeltsin. In substance, the United States overruled its European allies, humiliated Russia, sidelined the UN – and got results. One result was to damage the UN as seriously as the League of Nations had been damaged by Britain and France in the Ethiopian crisis of the 1930s. The United States insisted that military operations be under Nato's control.

But the main result was a prospect of peace. The United States effectively took the field against the Bosnian Serbs without going through the formalities of condemning them at the UN as a threat to international peace and after driving a wedge between them and Milosevic. It convoked the presidents of Serbia, Croatia and Bosnia–Hercegovina to talks at an air base at Dayton, Ohio, and kept them there for three weeks until they accepted terms which were presented to them. The Bosnian government, which had the most reason to reject the terms, accepted them under the threat, implied or explicit, that the American aid which had enabled them to fight on would cease. The Bosnian Serbs, who were not invited to Dayton, accepted the terms with unconcealed repugnance. Milosevic abandoned the Bosnian Serbs' leaders but not the prospect of Greater Serbia. In substance these terms followed earlier proposals for the partition of Bosnia–Hercegovina into two roughly equal halves (a Bosnian Serb republic and a Croat–Muslim federation) within a sovereign Bosnian state with a weak central government and a weakly integrated capital, Sarajevo, under overall Croat–Muslim control. Each of the two halves was to be entitled to raise its own army; it was therefore a state in all but name. Persons convicted of war crimes were to be debarred from holding public office; some fifty persons, including Karadzic and Mladic, had been named by the prosecution at the tribunal established by the UN but only one had been apprehended and arraigned; the rest were beyond the tribunal's reach and had a good chance of remaining beyond it. An international implementation force (I-FOR), substantially American and outside UN control, was raised to police the Dayton agreement.

From the international point of view the wars in Yugoslavia were on balance a fiasco. Much was done to succour victims from the accustomed hazards of war and from the heightened horrors of ethnic vengeance. The first round between Serbs and Croats was brought to a halt through international mediation but also because the combatants had reasons of their own to desist: the spread of the war to Bosnia suspended their conflict for three years. It was also the real test for international intervention.

The chief aim of the main international bodies most of the time was to stop the war. They failed because, initially, they were not prepared to deploy the necessary force and, later, because they could not agree among themselves what kind of force to use. The discredit for these failures fell on the UN as an organization but the blame attached more properly to the principal members of the Security Council: they ducked the issue by trying to use UNPROFOR for purposes for which it was neither intended nor equipped by them and, having recognized Bosnia–Hercegovina as an independent sovereign state, they refused to act as though they had done so. The Security Council was not asked to treat the war in Bosnia as a threat to international peace (which it clearly was) or to authorize consequential measures as prescribed by the UN Charter. The UN was further diminished when the United States, having resolved to take effective action, preferred to do so not by giving a lead in the UN but through Nato whose members could be relied on to mute their objections to American policies.

All external states grievously underestimated the ruthlessness of the Serbs and particularly the Bosnian Serbs. In the Bosnian round of the Yugoslav drama the Serbs made the running until checked by the Croats. Preferring international diplomacy to international action the EC and, for a time, the United States wrongly assessed, first, the readiness of the belligerents to heed reasonable proposals for peace and, later, the capacity or willingness of the supremely cool Milosevic to control the Bosnian Serbs. They recognized Milosevic's skills and influence but failed to perceive that Karadzic was his Achilles' heel. They relied on Milosevic for the one thing which he was least able to deliver. These mistakes prolonged the conflict.

The introduction of UNPROFOR in the absence of a cease-fire was an unprecedented gamble which, despite some promise in Croatia in the first round, went disastrously wrong in the next. The early initiatives of the EC were impelled by good intentions unaccompanied by equivalent good judgement. They were on the contrary ill-conceived and ill-informed.

Finally, the final solution adopted and strenuously pursued by the United States had a strong air of non-finality. It gave Serbia and Croatia much of what they wanted and had acquired illicitly and brutally. Formally it preserved a single Bosnian state; substantially it partitioned Bosnia into two entities waiting to be annexed by Serbia and Croatia. The Muslim community, the largest, retained some leverage in one half of the state, none in the other. About a quarter of a million people were dead as a direct result of wars of aggression and ethnic cleansing. In a world accustomed to talk about 'sending the right signals' this outcome looked terribly perverse.

Much of this mediocre performance could be ascribed to the relative unimportance of south-east Europe compared with, for example, the Middle East whose economic and strategic importance had, almost contemporaneously, predisposed major states to go to war and pay for it. With the partial exception of the war in Korea forty years earlier there was no clear or close precedent for UN intervention in an international war such as, given the recognition of an independent Bosnia–Hercegovina, this war was. More fatefully, American intervention where it mattered – that is to say, on the ground – was never made available.

MACEDONIA AND ALBANIA

Fears of secondary conflicts resulting from the disintegration of Yugoslavia were focused on Albania and Macedonia. For Serbs Kosovo was a magic name which recalled the valour of Serb resistance to the Turks in the fourteenth century but the population of the Kosovo district of Serbia in the twentieth century was overwhelmingly

Albanian. Milosevic's abrogation of its autonomy in 1989 prompted an ineffectual declaration of independence. Parliamentary elections were held in 1992 but the Serbs prevented the parliament from meeting. Albanian leaders in the region disclaimed any intention to use force (they had no arms) or seek union with Albania and the Albanian president Sali Berisha limited himself to expressions of sympathy. Milosevic's power was circumscribed less by Albanians than by the irregular and ruthless Serb nationalist bands within the region and the extremer nationalists in Serbia led by Vojislav Seselj.

In *Macedonia*, whose very name conjured up powerful emotions in Greece and Bulgaria, the internal tensions of these years were internationalized by paranoid reactions over its borders, particularly in Greece where ancient history was invoked to bolster modern nationalism (and win elections). When in 1389 Sultan Murad I had extinguished the Kingdom of Serbia at the battle of Kosovo he anaesthetized for five centuries what was to become the Macedonian Question. The Macedonian name went back to antiquity and to the warrior-kings who conquered Greece as a preliminary to conquering much of the world known to them. The Macedonians of more recent times, however, were Slavs but whether closer kin to Serbs or Bulgars was a much disputed question – the Bulgars pointing to close linguistic affinities and the Serbs to common cultural rituals unknown among the Bulgars. The quickening reflux of Ottoman power in Europe in the nineteenth century resuscitated ancient Serb and Bulgar principalities, revived separate Serb and Bulgar churches anxious to assert themselves against the Hellenic patriarch of Constantinople, and created between Serbia, Bulgaria and Greece tensions which were focused particularly upon Macedonia and the port of Thessalonica. Whereas in political terms the separatism of the emergent Slav states was anti-Turkish, in religious terms it was anti-Greek. After the Russo–Turkish war of 1877–78 the Treaty of San Stefano, imposed by Russia, gave Bulgaria all Macedonia but not Thessalonica. The Treaty of Berlin, by which the other Great Powers insisted on curbing this Russian victory, restored Macedonia to the Turks and divided Bulgaria into two pieces (which succeeded in coalescing in 1885). In the last decade of the century Bulgaria fostered ambitious Macedonian organizations but these quickly split on their tactics and strategy and by the time the progressive decline of the Ottoman empire led to the Balkan wars of 1912–13, Serbia and Greece had discovered a common purpose in preventing Bulgaria from getting Thessalonica. The second stage of the Balkan wars ended with the Treaty of Bucharest by which Macedonia was divided in such a way that Bulgaria received only one-tenth of it, the rest being allotted to Serbia and Greece with the latter securing a larger share than Serbia and also Thessalonica. There was some consequent migration and more after the First World War, all of it more or less voluntary. These movements left only a small Slav-speaking population in Greece but an abiding fear among Greeks of a revival of plans for a distinct Macedonian state which would remove from Greece its most fertile province and 1.5 million of its inhabitants.

After the Second World War plans for a union between Yugoslavia and Bulgaria were scotched by Stalin and Tito's reorganization of prewar tripartite Yugoslavia into a federation in six parts created the Yugoslav Republic of Macedonia. Forty-five years later this Republic, with a population of 2 million, a quarter of whom were Albanians and the rest divided into a dozen categories, had a crumbling economy, virtually no industry or purchasers for its principal crop (tobacco), no currency apart from the worthless Yugoslav dinar and no army. The writ of Kiro Gligorov – an ex-communist who became president in 1992 – ran but feebly beyond the capital Skopje. In 1991 the Macedonians confirmed by plebiscite a declaration of independence but the Albanian part of the electorate boycotted the voting and the Greek government prevented for two years international recognition of the country's independence and maintained for most of 1992–94 an economic blockade of it. The Albanians of Macedonia fared better than their kinsfolk in the Kosovo region of Serbia but were all but excluded from the republic's government and army and hankered after autonomy or independence or union with an enlarged Albania – which Albania was in no mood or condition to promote.

Greek concern about the new state had several sources. The most recent was the Greek civil war when many Slavs had been recruited into the ranks of ELAS. There were also deep-seated suspicions of possible meddling in anything with the name Macedonia attached to it; exaggerated reactions to the appearance in Skopje of ancient but provocative maps showing Macedonia extending over parts of Greece; fears that a Macedonian state on Greece's border might engender claims also for Albanian and Turkish minorities. Finally and most prominently, political leaders in Athens revived romantic illusions about the continuing relevance to modern affairs of the career in the third century BC of Alexander of Macedon whom the Greeks of those days had regarded as more barbarian than Greek. Greece's membership of the EC ensured a degree of attention for chauvinist posturing which might otherwise have gone unremarked. Macedonia was obliged to adopt the unwieldy nomenclature of Former Yugoslav Republic of Macedonia (FYROM) before it was admitted to the UN. It remained uneasily at peace with the help of a small UN peace-keeping force (to which in 1993 the United States contributed units) and in spite of limited outbreaks of violence in 1995. Less volatile than Bosnia it had, however, borders with four unfriendly and more securely established states –

Albania, Bulgaria and Serbia as well as Greece – and was itself landlocked. In 1995 Gligorov narrowly escaped assassination. He had won recognition from Bulgaria, Turkey, Serbia and even Greece but his country was imperilled rather than assured by the triumph in Bosnia of the Serb and Croat states against the idea of a multi-ethnic state such as the Macedonian Republic had to be.

Albania's independence from the Ottoman empire was first recognized in 1912 with a German prince as king (for a few months), much of its territory allotted to Serbia, Montenegro or Greece, and more of it occupied by Italian and French forces during the First World War. On various occasions from 1878 onwards Greece had claimed what it called Northern Epirus, a variable area amounting to about a third of Albania and inhabited by a substantial Greek minority. The size of this minority was unassessable since the only reliable census was by religion, not tongue or nationality. A number of Albanians spoke Greek since the Turks had forbidden the learning or use of Albanian but not of Greek. During the First World War it appeared likely that Albania would be partitioned between Greece, Italy and the new kingdom of Yugoslavia but such plans were vetoed by Woodrow Wilson. A southern boundary, first delineated by an international commission in 1913, was endorsed by the League of Nations in 1921 and accepted by the states concerned, including Greece and Italy, in 1926.

After the First World War Albania was internally chaotic until Ahmet Zogu asserted his rule over most of it and made himself king in 1928 (he died during the Second World War). In 1939 Italy invaded it in order to attack Greece but Albania succeeded with German help in gaining territory from Serbia and Montenegro, if only temporarily. Resistance to the Axis occupiers was led by Enver Hoxha with aid from Britain. Hoxha, who was a middle-class paranoiac partly educated in France and Belgium, established a Stalinist dictatorship. He eliminated Koci Xoxe and other close associates whom he plausibly suspected of being seduced by Tito's schemes to absorb Albania into the Yugoslav federation. Another close associate Mehmet Shehu, a veteran of the Spanish civil war, was prime minister until 1981 when he committed suicide and was attacked *post mortem* as an agent of both the CIA and the KGB. Hoxha became hostile to Khrushchev's USSR and severed relations with it in 1961. He lauded Mao's China until 1978 when he attacked it for revisionism. He isolated Albania which he tried to revolutionize and modernize with ruthless incompetence, wasting its substantial mineral resources and hydro-electric potentialities. He died in 1985. His successor, Ramiz Alia, tried for a while to preserve Hoxha's system in spite of the upheavals in the communist world, but the country became anarchic, large numbers of Albanians fled to Italy,

Greece and Yugoslavia and Alia was obliged to permit a general election in 1991. Communists fared fairly well but the government collapsed and in 1992 Sali Berisha was elected president by the parliament in place of Alia who was arrested on charges of peculation.

Two-thirds of the Albanians in Albania are Muslims of diverse sects. The remainder are mostly Orthodox and Roman Catholic Christians, with the former outnumbering the latter by two to one. Ethnically the principal minorities are the Greeks whose number is put at 40,000 or 100,000 according to political viewpoint; and the Vlachs who look to Romania. There are as many Albanians outside Albania as in it, most of them in Serbia (Kosovo), Macedonia and Montenegro but also in Turkey, Greece and Italy from old migrations and in the United States and Canada from more recent ones.

NOTES

A. NORTHERN IRELAND

Northern Ireland, frequently but wrongly called Ulster, was that part of Ireland which remained part of the United Kingdom when in 1922 the rest of Ireland broke away from the union created in 1800. During the winter of 1921–22 there were two principal issues in debate between the English and the Irish. The partition of Ireland was only the second of these since it was widely regarded as inevitable and the more heated debates, in London and among the Irish leaders themselves, were about the status of the Irish state which, the English insisted and a narrow majority of the provisional Irish parliament agreed, should be a Free State within the British Commonwealth and empire, owing allegiance to the British Crown and presided over by a British governor-general. In Ireland a republican minority pressed its case to the point of civil war but lost, and Ireland, although a separate member of the League of Nations from 1923, did not become a fully independent republic until 1937.

In the north-east a border was eventually agreed in 1925. The province of Northern Ireland inherited from the Act of Union of 1800 the right to elect members to the parliament at Westminster. It had also, under the Government of Ireland Act 1920, a bicameral legislature and an executive of its own with a considerable degree of internal autonomy (extended in 1948). The province was governed from its inception until 1972 by a Protestant oligarchy dominated by landowners but embracing in later years representatives of the more prosperous urban and professional classes, one of whom, Brian Faulkner, became its last prime minister. The power of this oligarchy rested fairly and squarely on the electorate in which Protestants outnumbered Roman Catholics by two to one. The people of Northern Ireland were divided three ways. There was, first, the most striking and emotive religious division, a relic of the antiquated politics of seventeenth-century Europe. There was the division between those who wanted the unification of Ireland, those who opposed it and those who felt no urgent concern about it; and finally there was the division between bosses and workers, rich and poor. These divisions overlapped and the overlapping made the province's politics. The bosses were all Protestants but not all the Protestants were bosses. The partisans of a united Ireland were all Roman Catholics but not all Roman Catholics wanted Ireland united; or not immediately; or not by force. The aim of the dominant class – part of a majority in religious terms but a minority in socio-economic terms – was to remain in control, making such reformist concessions as calculation might enjoin (as to which opinions would differ). The aims of the opposition, and so their tactics, varied and in 1969 the IRA (descendants of the Irish Republican Army formed to evict the English from Ireland) split. The larger group, adopting a Marxist interpretation of the situation, aimed to enlarge its base among the deprived Roman Catholic minority by enlisting the support of the poorer Protestants; this group put socialism first and abandoned violence in favour of propaganda. The other group, called the Provisional IRA, maintained the traditional policy of achieving a united Ireland by the eviction of the English (who had conveniently returned) by the traditional weapon of violence. Their interpretation of the situation was not class conflict but national war. English aims, finally, were negative. The English had lost all will to remain in any part of Ireland and regarded a united Ireland as a natural but probably remote eventuality. They felt an obligation to stand by the established order in Northern Ireland but both Conservative and Labour governments saw the need to push the provincial oligarchy into civil and political reforms, as a matter of expediency and as a matter of justice. Both believed, or acted as though they believed, that no more than this was necessary, a delusion from which they should have been disabused by the events of 1969–72.

These were the last years of the old order and in them the province had three prime ministers. Disorders, particularly on historic or religious festivals, were a normal feature of political life but not normally lethal. In 1966 disorder had been aggravated by a spate of killings by the newly formed (Protestant) Ulster Volunteer Force, but the authorities regained control until in 1969 political murder reappeared. In that and the next year the number of victims was small (thirteen and twenty respectively) but they sufficed to induce the British government to send troops. This seemingly extreme reaction was an aspect of the peculiar dilemma of the province. The local forces of law and order – the Royal Ulster Constabulary supplemented by the B-Special Constables – were composed of Protestants and were regarded by Roman Catholics as instruments of the Protestant sectarian supremacy. Their use to suppress disorders, even their continued existence, were incompatible with a policy of recognizing and rectifying Roman Catholic grievances. Consequently some other force had to be imported. Conversely the dissolution of the B-Specials and the disarming of the RUC, on which the British government insisted in 1970, intensified Protestant apprehensions for their own fate and reinforced their intransigence.

Having intervened thus far the English, anxious to retire again as soon as possible, pressed the provincial executive to introduce reforms to satisfy those among the Roman Catholics who were agitating for civil rights. Successive prime ministers, Terence O'Neill and James Chichester-Clark, were willing in principle but inhibited by their own supporters who regarded reforms as a prelude to further concessions leading to the unification of Ireland with dire consequences for the plight of Protestants and the economy of the northern counties. The hold of such leaders over the Protestant majority was further weakened by the emergence of new leaders whether, like Ian Paisley, eloquent anti-papalists of a kind no longer found outside Northern Ireland or, like William Craig, protagonists of the right of the majority to have its way and fight for it. Nevertheless a number of reforms were enacted. Unhappily they were years too late. They divided Protestants and no longer satisfied Roman Catholics whose attention was being diverted from civil rights to sectarianism by the Paisleyites on the one hand and the Provisionals on the other. Killings multiplied (173 in 1971) and in August a new prime minister, Brian Faulkner, resorted to detention and internment without trial. In one night 342 persons were arrested. A number of them were clearly the wrong persons. More important, not one of them was a Protestant. This was not surprising since the object of the operation was to break the IRA.

But the operation had the opposite effect. It strengthened the Provisionals. It swung Roman Catholic opinion, which had at first welcomed British troops as protection against Protestant militants, over to the IRA and forced the official IRA to join the Provisionals in denouncing the British government, without whose endorsement Faulkner could not have acted. Stories of brutality and torture by the English, subsequently endorsed by the European Commission on Human Rights, added fuel to the flames. The Protestants were both emboldened and alarmed, a fateful combination: emboldened because Faulkner had not ventured to intern even the most militant UVF leaders and alarmed because they felt denuded in the face of increasing IRA violence and the emasculation of their own means of securing law and order. A new Protestant self-help force appeared, the Ulster Defence Association, as a counterpart to the Provisionals. The provincial executive became ineffective and in March 1972 it was suspended. The British government assumed direct responsibility.

Direct rule from Westminster involved Britain in direct confrontation with the opposition and the opposition now was not the Roman Catholic parties or civil rights movement but the Provisional IRA. Nevertheless the Heath government set to work to find a constitutional answer to Northern Ireland's problems. The policy had two pillars and they were incompatible. The one was power-sharing, an insistence on Roman Catholic participation in government at all levels including a restored provincial executive which was to be permanently a coalition and, secondly, democratic endorsement by the province's electors. By 1972 the Protestant majority of the electorate was prepared to endorse no such thing. Protestants as a whole were justly enraged by the murderous activities of the Provisionals, and Protestant militants took to killing twice as many civilians as the Provisionals (who directed their fire more particularly at the British troops). Thus the British government's policy, enshrined in a White Paper of March 1973 and apparently triumphant at a conference at Sunningdale in December, was for a second time foredoomed by unreality, and the addition at Sunningdale of a gesture towards unification by the creation of a Council of Ireland (north and south) only made its rejection doubly certain. The British general election, unexpectedly called by Heath in February 1974, gave the electors in Northern Ireland a chance to show their mind and they decisively rejected the Sunningdale scheme and made it unworkable. The restored executive survived for a while but was brought down in May by Protestant demonstrations and a general strike against power-sharing, leading to the declaration of a state of emergency. London's answers were to fly in more troops and resume in May direct rule: the two things it least wanted to do.

From the winter of 1972–73 both sides within the province had produced groups which were trying to reach out to one another but they were too tentative and submerged to be much heard amid the clash of arms and rhetoric. Truces were arranged but were imperfect and short. Horrors multiplied but produced for the time being more horrors rather than revulsion. The English put a brave face on their helplessness and justified their presence by predicting, with some superficial plausibility, a blood bath if they were to leave. That they would leave one day was obvious to all but themselves: a repetition of the blindness of the French to the realities of their tenure of Algeria. That they should do so sooner rather than later was an argument which the larger part of them was emotionally unprepared to entertain. In 1980 two new prime ministers, Margaret Thatcher and Charles Haughey, were trying to devise a way round the blockage created by the claim of the Protestants of Northern Ireland – and accepted by all British governments – that, being a demographic majority within the province, they had a democratic right to veto constitutional change and, to that extent, to regulate the relationship between Britain and Ireland. Discussions continued after a switch in Dublin from Haughey to Garret Fitzgerald but the two governments were far apart, not because of divergent attitudes to violence but because of their presuppositions. Dublin saw no end to the troubles without significant political change; it did not seek immediate or even early change but the various changes which it advocated at one time or another all pointed in that direction. Thatcher, on the other hand, believed that the first task was the conquest of violence by force and her readiness to consider change, even in the long term, was strictly circumscribed by her view – familiar from the Falklands, although less inflexible in Gibraltar and discarded in Hong Kong – that British policies must be subordinated to the wishes of the majority of the people of the territory concerned. When in 1984 the New Ireland Forum, an association of Irish Republic parties, proffered a series of possible formulas for political manoeuvre pointing towards unification or a condominium in the north, Thatcher dismissed them with marked asperity. Nevertheless in 1985 she and Fitzgerald agreed by the Hillsborough or Anglo–Irish agreement to establish a standing joint body or 'ministerial conference'. The immediate practical purpose of this agreement was to improve police co-operation and border control. Its larger purpose was to marginalize the IRA. The British government, looking to Dublin for help on the first count, was willing to allow the Republic to become associated with the affairs of the province. Irish governments felt constrained to co-operate with the British to some degree since the *status quo* was, from the Republic's point of view, less hazardous than the alternatives:

either, on the one hand, war in the north leading to victory for the Protestants or the extreme right; or, on the other hand, a union which would saddle the Republic with a million Protestant dissidents and bankrupt it. The agreement carried, however, a price for both parties. Revelations of serious miscarriage of justice in British courts, which neither Fitzgerald nor Haughey (who returned to power in 1987) could easily overlook, made the agreement peculiarly and unexpectedly uneasy for the Irish. On the British side the agreement entailed political stagnation in the province and an end to attempts to return it to parliamentary rule so long as the Unionist (Protestant) parties made discussions on devolution conditional on the abrogation of the agreement. The Unionist leaders, James Molyneaux and Ian Paisley, profoundly hostile to Dublin's involvement in the province's affairs in any shape or to any degree, believed that they might nullify the agreement by non-co-operation with the British secretary of state. The British government, although it was as much determined as the Unionists to keep the province in the United Kingdom, never affirmed this resolution convincingly, and although the agreement provided that the 'status of Northern Ireland' would not be altered without the assent of a majority in the province, the Unionists retorted by pointing to the article in the Irish constitution which described the whole island as the territory of a single nation – a provision which had been interpreted by the Supreme Court in Dublin as laying upon the Irish government a duty to reintegrate Ireland. To Unionists the British government had been hypocritical when it signed the Anglo–Irish agreement without securing, or even asking for, the abrogation of this article in the Irish constitution. One effect therefore of the agreement was to marginalize not the IRA but the Unionist parties. These, however, were mistaken in their belief that they could wreck the agreement, since the British refused even temporarily to suspend its more obnoxious manifestations (joint conferences of ministers and officials). The agreement achieved little by way of policing the border which could not have been achieved without the agreement. It failed to stop or reduce the murderous activities of the IRA in Ireland or in England, and it froze – at least for a number of years – attempts to bring direct rule to an end. The IRA's campaign to evict the English by force – a campaign which included an attempt to kill Thatcher and other ministers at Brighton in 1984 and intensified violence with imported weapons and explosives from 1988, and which was met by British attempts to pick off and kill leading IRA activists – kept the conflict alive but with scant prospect of the success sought by the IRA. Despite its name the IRA was not an army and it could not dislodge the British army which needed to send to Ireland no more than 15–20,000 troops (briefly increased to some 30,000 for a special operation in 1972).

From the late 1980s a section of the nationalists which included Gerry Adams, president of Sinn Fein – the dynamo of Irish nationalism – became convinced that their political aims were not being furthered by military action. They also became persuaded that the British government no longer harboured strategic or economic reasons for preserving British power in Northern Ireland, although Britain would not relinquish power against the wishes of a majority of the people of the province. The British, however, were profoundly distrustful of the nationalists and coincidentally John Major's government became dependent on Irish Unionist votes in the House of Commons after the election of 1992 left it with a precarious majority in debates on the Treaty of Maastricht. The Unionists used their parliamentary power to block proposals for joint Anglo–Irish administration in Northern Ireland. In the short term the Unionists were suspicious of London's attempts to stop the fighting through an accord with and concessions to Dublin, attempts which were far from unpopular in Britain where sympathy with the Protestants dwindled as they too took to arms and, in 1992 for the first time, murdered more Catholics than vice versa. In the longer term the Unionists were obliged to contemplate a demographic shift as Catholics looked forward to overhauling the Protestant majority in not much more than a generation. Finally, in the province itself the age-old religious backing for armed conflict was weakening, politics were becoming secularized, leaders could rely less on sectarian demagogy and fears, and northern Protestant distrust of Catholics was beginning to be allayed as political power in the Irish Republic slipped away from an over-mighty and over-secretive church to a secular political class.

In Dublin a revolt against Haughey in 1991, although initially a failure, secured his departure and replacement by Albert Reynolds as leader of Fianna Fail and Taoiseach (prime minister). Elections dented Reynolds's position in the Dail so far that he was obliged to give substantial weight in his cabinet and the ministry of external affairs to the Labour Party led by Dick Spring. This coalition broke up acrimoniously in 1994 and a new government was formed by John Bruton, leader of Fine Gael, in coalition with the Labour Party and the small, new and socialist Democratic Left. Behind these changes Irish leaders persisted in pursuing what they saw as an opportunity for a truce and permanent peace. Hopes were stimulated when a series of meetings between Adams and John Hume, leader of the Social Democratic Labour Party in Northern Ireland and a member of the British parliament, produced an (unpublished) set of proposals for peace. In oblique response the British and Irish prime ministers issued in Dublin a declaration which amounted to an offer of discussions provided the IRA first renounced violence unequivocally and permanently. In what looked like prevarication the IRA asked for clarification, which the two governments refused to give without a prior renunciation of violence, and advanced the proposition that constitutional changes in Ireland should be submitted to the Irish people as a whole and not separately to a plebiscite in the north. This argument was clearly unacceptable to the British and failed, if that was its purpose, to drive a wedge between London and Dublin even though Dublin was convinced, as London was not, that Adams was in earnest and capable of delivering an agreement for peace. Major's cautious procrastination had the general support of the British Labour Party and he succeeded in winning the trust of James Molyneaux, the leader of the province's largest Protestant party, thus diminishing the abrasive and unco-operative Ian Paisley and his unregenerated Protestant faction. When the IRA proclaimed a cease-fire Major's tactics seemed for a while to be justified.

The gradual transformation of the situation rested, however, on a major change which carried with it fresh obstacles to a settlement. As the 1990s progressed it became clear that the driving force was the establishment of continuing and close accord between the British and Irish governments which was regarded by the Protestants in the North as the mechanism for concessions to Irish nationalism and the looming abandonment by Britain of the union with Northern Ireland. For two years governments worked laboriously and secretly on a joint document to be presented to all parties as a basis for discussions about the future of the province. It was published in 1995 and shocked not only Paisley's party but also Molyneaux's Ulster Unionist Party (UUP) because of its proposals for joint Anglo–Irish authorities to co-ordinate such administrative problems as border controls. The British government represented this 'framework' document as no more than a basis for a conference at which all parties might table other proposals and stressed the safeguards for Unionists in the document's insistence on its acceptance by the North's political parties, by a referendum in the North and by the British parliament. But the Unionists, whose bargaining positions rested on their voting power in a finely balanced House of Commons whose term was running out, were genuinely alarmed by pan-Irish elements in the framework document and by the evidently declining sympathy for them in the rest of the United Kingdom where they were portrayed as mock militarists flavoured by Protestant fundamentalism. The enigmatic Molyneaux was replaced by the more outspoken, apparently uncompromising, but intelligent David Trimble as leader of the UUP. Between the main protagonists – the British government and the IRA – there was little trust, so that progress was slow and became slower as an issue crucial to both – the surrender of arms – reached the agenda. John Major, cautious by nature and constrained by conflicting political considerations (the need for radical change versus loyalty to the Union) forfeited the initiative. Like Rhodesia in the 1960s but much nearer home, the fag end of the Irish question posed for the British a problem whose profoundly uneasy emotions cramped the British genius for pragmatism and good sense.

B. THE BASQUES

The Basques are a people of sharp self-consciousness and obscure origins. They have preserved a cultural identity based on an unaffiliated language and romantic historical claims. Straddling the occidental Pyrenees they have been acquiescent in France but effervescent in Spain. Prosperity in Spain's Basque provinces in the 1960s led, as it did in the 1890s, to an influx of non-Basque workers which split local discontent between nationalist separatism and a workers' movement. After violence in the late 1960s, sixteen persons, mostly from the middle class and some of them priests, were put on trial in 1970 before a military court in Burgos charged with murder, bombings, illegal possession of arms and illicit propaganda. The trial lasted several years and gave a boost to the Basque separatist movement, ETA, which in 1973 carried its war into the Spanish capital and there assassinated Franco's prime minister, Admiral Carrero Blanco. After Franco's death the new régime in Madrid offered the Basques as much autonomy as it dared and so divided ETA between acceptors and rejectors. But in spite of some police successes against ETA leaders and police brutality in what amounted to civil war in the 1980s both ETA tendencies survived.

C. CYPRUS

In 1878 Disraeli, prompted by British military opinion, contemplated the seizure of Cyprus but contrived to secure it by diplomacy. On the eve of the Congress of Berlin he entered into an agreement with the Turks to defend their empire against Russia upon being permitted to occupy Cyprus which, however much the Turks might regard it as an anti-Russian base, was for the British of those days the key to western Asia. Britain, contemplating the collapse of the Ottoman empire or alternatively the substitution of German for British influence at the Sublime Porte, was on the look-out for vantage points in the eastern Mediterranean. The occupation of Cyprus in 1878 was followed a few years later by the occupation of Egypt. The British remained in both for the best part of three generations.

Cyprus was annexed by Britain in 1914 on the declaration of war between Britain and Turkey, and British sovereignty was recognized by Turkey by the treaty of Lausanne in 1923. Between the wars British rule was challenged by the partisans of *enosis* or union with Greece but was never seriously threatened. For the British Cyprus was a colony which could gradually be granted the degree of self-government compatible with its usefulness as a staging-post and a base in the British imperial scheme of things. For the Greek section of the population, however, the British themselves and their programme of limited democratic development were simply obstacles to a more natural scheme of things in which all Greeks were being brought together in the single nation state which had been inaugurated by the Greek revolt against the Turks in 1821. The Turkish minority was a spectator of Anglo–Greek conflict, sometimes worried and sometimes not. At the end of the Second World War the British estimate of the value of Cyprus rose as a result of the retreat from Palestine and the weakening of the British position in Egypt. The British assumed therefore that they would and should stay in Cyprus. The Greek Cypriots assumed the opposite. They expected a reward for their loyalty to Britain during the war and they regarded the British withdrawal from Egypt as a step which should logically entail a withdrawal from Cyprus and not a reinforcement of British rule there. Discussions for a constitutional advance towards limited self-government were abortive. To the Greeks they were irrelevant.

In 1946 the exiled bishop of Kyrenia returned to the island and in 1950, at the age of 37, he was elected to the archiepiscopal throne of the autocephalous church of Cyprus, which had been vacant during the years 1937–47. He took the regnal name of Makarios III and he became at the same time the ethnarch or national leader of the Greek community. Makarios was therefore both the head of one of the most venerable of the churches within the Orthodox communion (it had been founded by the Cypriot apostle Barnabas) and also the leader of a modern nationalist movement which had fought the Turks for a century and a half and was now to fight both Turks and British on this last remaining battlefield. He and his fellow enotists considered that the kingdom of Greece had a sacred duty to support them, a belief shared by very many – probably most – Greeks outside Cyprus but embarrassing nevertheless to Greek governments which were reluctant to impair their good relations with Britain and even had reasons for co-operation with the traditional Turkish enemy after the admission of both Greece and Turkey to Nato. In 1950 a plebiscite organized by the church returned the inevitable (but not, as the British deludedly imagined, faked) response in favour of *enosis*; the Greek prime minister, General Plastiras, equally inevitably responded to it in a tone of mixed encouragement and moderation. About the Greekness of Cyprus no Greek bothered to think twice. Four Cypriots in five were Greek by race, tongue and religion, and called themselves Greeks. Whether they supported the left-wing party AKEL or the right-wing KEK they shared a common nationalism, for which they were soon to fight together in the insurrectionary movement of EOKA.

The Turkish section of the population (18 per cent) was no less alive to the Greekness of the island but it naturally drew opposite conclusions from the same facts. Although the government had passed from Turks to British, the Turks knew that they remained the principal and hereditary enemies of the Greeks, and they feared Greek rule. These fears may not have been great in the first postwar years, since the Greek record in relation to Turkish minorities was no bad one, but they were latent fears which could easily be inflamed. The Turkish state, as opposed to the Turks of Cyprus, also feared *enosis* because Cyprus was only 60 km from the Turkish coast, because post war Greece seemed for a short while to be exposed to communism, and because Greece might still cherish the ambition to conquer Constantinople and the coasts of Asia Minor (which it had tried unsuccessfully to do immediately after the end of the First World War). Turkish governments showed, however, little inclination to intervene in Cypriot affairs until encouraged to do so by Britain.

In 1951 the Greek government sought a way of satisfying its Cypriot compatriots and its British allies by offering Britain bases in Cyprus – and also in Greece itself – in return for *enosis*, but Eden was not interested in a solution which would have been more attractive to Britain than the eventual settlement of 1959 and could have saved much bloodshed. In 1953 Eden gravely exacerbated the situation by declaring that there was no question of a British withdrawal. This statement forced Greeks to decide between an indefinite acceptance of the existing position and a resort to violence to change it. At first Makarios tried an intermediate course. In 1954 he went to Athens to try to get the Greek government to raise the Cypriot question at the UN. Eden repeated that no discussion was possible; a junior British minister with a lamentable sense of history and a notable lack of caution declared that Cyprus would 'never' be fully independent; and the colonial secretary affronted the Greek government by advancing as an extra reason for the maintenance of British rule the argument that Greece was too unstable to be allowed safely to extend its sway to Cyprus. The Greek government then raised the question of self-determination for Cyprus at the UN but half-heartedly and without pressing the case, which was shelved. The enotists, disappointed by this effort, reverted to the local scene and organized demonstrations which evoked an excessive British counter-reaction (including measures against schoolchildren). The British in Nicosia and London believed that the *enosis* movement was a bubble blown in their faces by a small and unrepresentative group of irresponsible agitators who had succeeded in cowing the bulk of a law-abiding and pro-British population. This was, at the very least, a serious exaggeration, just as the British appreciation of the value of Cyprus as a base was also seriously mistaken: Cyprus, though valuable as a command headquarters (HQ Middle East Land and Air Forces was transferred there in 1954) and as an air staging point, was a poor country with meagre resources, a hostile population, no adequate naval base (Famagusta being too shallow) and a manifest vulnerability to Russian nuclear attack. The Suez War of 1956 proved that it had some value for the Royal Air Force but no other.

In 1955 Eden and his foreign secretary Harold Macmillan decided to fortify the British position in Cyprus by bringing the Turkish government officially into the matter. The Greek and Turkish governments were invited to a conference in London. The result was the collapse of the Greco–Turkish alliance. In Turkey hideous atrocities were perpetrated against Greek residents, and Greece boycotted Balkan Pact meetings and Nato exercises.

In Cyprus the governor and the archbishop met for the first time, and a British colonial secretary appeared in the island also for the first time. At the end of the year the governor was replaced by Sir John Harding, a field-marshal and former Chief of the Imperial General Staff. The policy entrusted to him was to separate the ethnarch-archbishop from the insurrectionary movement which had come into the open in 1954, to negotiate with the one and to extirpate the other. This policy, which was pursued until the end of 1957, was a total failure because it was based on false premises and poor information and because, at one decisive point, it was abandoned by the British government. The Harding–Makarios negotiations were proceeding early in 1956 towards a promising conclusion when the British government intervened and decreed the deportation of the archbishop to the Seychelles. Eden appeared to have been swayed at this juncture by pressure from the right wing of the Conservative Party, by the dismissal of General Glubb by the king of Jordan (which Eden interpreted as a deliberate slight to his government), and by the failure to manoeuvre Jordan into his prized creation, the Baghdad Pact. These rubs roused in him the assertiveness of a weak man and he resolved to teach his enemies a lesson, foremost among them Makarios and Nasser. But Makarios was released in the following year, much to the annoyance of the Turks and without any compensating advantage since the archbishop refused to return to Cyprus and took up residence in Athens.

The British attempt to quell the insurrection was equally unsuccessful. This revolt was led by Colonal Grivas, an officer of the Greek army and a Cypriot by birth who set himself to evict the British by a combination of military skill, faith and ruthlessness. Grivas had fought on the same side as the British in two wars and was outraged to discover at the end of the second that, despite the UN Charter and frequent British promises in the past, Britain had no intention of allowing the Cypriots to choose how and by whom they should be governed. Like Cavour after the Crimean War Grivas took the view that his compatriots had paid for self-determination with their blood, and when he saw that the British view was different, he set about shedding more blood. He decided, according to his own account, to resort to violence in 1951 but he laid his plans with considerable professional care, carried out an extensive and open reconnaissance of the terrain and waited until 1954 before making his first purchase of arms and setting up a headquarters in a suburb of Nicosia inhabited chiefly by British families. He launched his revolt in April 1955, survived a drive against him at the time of Makarios's deportation in March 1956 and immediately struck back. His main weapons were bushcraft, discipline and terrorism, his victims frequently Greek civilians, his armoury and front-line manpower always small. He provoked the British into retaliatory measures which failed – collective fines on villages, high but ineffective bribes, hangings and torture. He defeated the policy which Harding had been sent to implement.

In the course of the year in which Makarios was in the Seychelles and the Harding–Grivas duel was taking place, the first Greco–Turkish riots also occurred and the British government began, although unintentionally, to transfer the initiative from London to Ankara and Athens. During 1956 Eden produced a plan by which Cyprus would be allowed self-determination after ten years of self-government, but instead of applying his scheme he submitted it for approval to the Turkish and Greek governments and accompanied it with a proposal for a tripartite treaty. The Turkish government rejected the plan and this rejection both killed it and gave Turkey a new commanding position. Later in the same year new constitutional proposals, elaborated by an eminent British judge, Lord Radcliffe, were in the same way submitted to the Turkish and Greek governments. The Radcliffe plan rejected self-determination and mentioned partition. It was accordingly rejected by the Greeks, while Turkey was emboldened to suggest that either half of Cyprus or the whole of it should be annexed to Turkey. The Greeks, thoroughly alarmed, threatened to leave the western camp. In 1957 General Ismay, the secretary-general of Nato, offered to mediate, but although the Turks were willing the Greeks were not. The Turks believed that a majority of the members of Nato were sympathetic to Turkey; the Greeks believed that their cause would prevail in the UN but not in Nato. There was deadlock internationally and continuing disorder and murder locally. The government of Harold Macmillan reviewed Eden's Cyprus policy and, faced with the threat to Nato's eastern flank which had been created by it, decided that Britain no longer needed to be sovereign in the whole of Cyprus. Sovereign bases would do and the Greek and Turkish governments must be brought to accept independence for the rest of the island. For Turkey independence was acceptable, since it automatically excluded *enosis*. Upon Greece independence might be forced, since Greece abominated partition and was afraid that in the absence of a settlement Greeks in Istanbul and other parts of Turkey would be stripped of their property and either killed or expelled.

In December 1957 Harding was replaced by Sir Hugh Foot who produced a new plan: self-government as a colony for a period followed by self-determination, with the proviso that *enosis* would need Turkish approval. The mention of *enosis* was too much for the Turks and demonstrations were organized in Ankara when Selwyn Lloyd (now foreign secretary) and Foot visited that capital. The Foot plan disappeared. It was succeeded by the Macmillan plan which was a further step away from undiluted British rule. Macmillan proposed to introduce representatives of the Greek and Turkish governments alongside the British governor and to create a mixed cabinet and separate Greek Cypriot and Turkish Cypriot local administrations. The last provision was unacceptable both to Makarios and to the Greek prime minister, Constantine Karamanlis. Their rejection of it led to fresh riots in Cyprus in which the Turks seemed to be applying pressure in order to enforce acceptance of the Macmillan plan. The revolution in Baghdad in July 1958 may have inclined the British government, threatened with the loss of its Iraqi ally and the collapse of the Baghdad Pact, to lean further to the Turkish side in Cyprus, and Macmillan set out on a tour in the course of which he slightly modified his plan, failed to reconcile Makarios or Karamanlis to it, but resolved to apply it none the less. Violence increased horribly.

The spreading communal hatred shocked and alarmed the Greek and Turkish governments into an accord. After exploratory contacts at the UN and Nato between their foreign ministers they conferred together in Zürich in February 1959 and announced that they had agreed that Cyprus should be independent. Britain would be accorded sovereign rights in certain areas which would be British military bases. The new state would have a Greek president, a Turkish vice-president with a veto in certain matters, and a cabinet of seven Greeks and three Turks; this 7–3 proportion would be repeated right down the administrative ladder. The Greek and Turkish states would station small armies of 950 and 600 men respectively in Cyprus. This scheme was accepted with the greatest reluctance by Makarios, who declared it unworkable. He was, however, threatened with abandonment by the Greek government and on 1 March 1959 he returned at last to Cyprus. Grivas, infuriated by the politicians' betrayal of the cause of *enosis*, was fêted, promoted and sent back to Athens. Instead of driving the British out and making Cyprus part of Greece, his campaign had ended with the British still in possession of sovereign bases and Cyprus still not part of the Greek kingdom. Cyprus became independent in August 1960, a member of the UN the following month and a somewhat strange member of the Commonwealth in 1961.

The Zürich settlement was an attempt by frightened men to prevent the situation from getting completely out of control. That was its one merit. But Makarios was right in regarding the constitution as unworkable. The Turkish Cypriots were entitled to 30 per cent of the posts in the administration, although they did not have the men to fill so many posts. Each of the five principal towns was to have two separate municipal bodies, though this concession to communal distrust produced such absurdities in practice that it was never implemented. Discussions for an improved system broke down and when the Turkish leaders in the island proposed the extension of the existing arrangements for a year, the Greek majority in parliament rejected the proposal. The constitutional court, upon being appealed to, pronounced both the Greek and Turkish cases to be wrong, and in December 1963 Makarios made proposals which were meant to force the Turks into further discussions but were taken by them to be a breach of the constitution and an attack on their safeguards. Serious fighting developed and attempts by leaders on both sides to arrange and enforce cease-fires broke down. Moreover the progenitors of the Zürich agreement had by now passed from the scene. Both the Karamanlis government in Athens and the Menderes government in Ankara had fallen. The former had been replaced by a government under George

Papandreou which began its existence without a parliamentary majority, while in Turkey a military coup had overthrown the parliamentary system in May 1960, indicted over 400 people and executed fifteen of them including Menderes himself. General Gürsel, who was successively provisional head of state and then in October 1961 president, installed a coalition government, but the new military régime was assailed by an abortive coup from within its own ranks in May 1963 and by the recrudescence of Menderes's Democratic Party under the new name of the Justice Party. In December 1963, when fighting broke out again in Cyprus, the veteran Ismet Inönü had just resigned the premiership and been persuaded to retain it, albeit with a majority in parliament of only four.

Some 200 Turks were killed in this new bout. Turkish jet aircraft flew menacingly over Nicosia, a Turkish naval invasion was thwarted by the American Sixth Fleet prowling in the vicinity, the Greek and Turkish forces in Cyprus took up hostile battle positions, and the British colonial secretary, Duncan Sandys, abandoned his Christmas holiday to fly overnight to Cyprus. After four years of independence Cyprus had brought Greece and Turkey to the verge of war.

Britain tried at first to transfer the Cyprus problem to Nato. The Greek and Turkish governments were willing to accept Nato intervention but Makarios was not; nor were other members of Nato anxious to become embroiled. In February a second Turkish invasion threat was unostentatiously foiled by the American fleet and Britain accepted the need to invoke the UN. The raising of a UN force coincided with a third invasion scare, but by mid-March the Canadian advance party of a UN force reached Cyprus. It was followed by units from Eire, Sweden, Denmark and Finland which, together with British units transferred to UN command, constituted a force of 7,000 under the Indian General Gyani and subsequently under his compatriot General Thimayya and, after his death in 1966, the Finnish General A. E. Martola. This force gradually asserted control, although occasional outbursts continued to occur – and led to occasional reprisals against the Greeks in Istanbul.

In addition to the peace force the UN secretary-general appointed a mediator to seek a political solution, but neither of two incumbents of this office was able to find a solution acceptable to both sides. The United States, alarmed by the consequences of a Greco–Turkish conflict, also took a hand through the former secretary of state Dean Acheson who produced a scheme for *enosis*, excluding an area in north-eastern Cyprus which would go to Turkey. The Turks thereupon asked for a larger area and so converted the plan into partition in a new form. Acheson then revised his plan and proposed that the north-eastern area should merely be leased to Turkey for twenty to twenty-five years. At this point the plan became unacceptable to everybody and the Turks, already disappointed in their hopes of Nato and frustrated by the American fleet, turned tentatively towards the Russians (who may at this time have sensed a possibility of undermining the entire Baghdad Pact since Pakistan was also disappointed by the United States and Iran was internally unstable and internationally plastic). Early in 1965 the UN mediator proposed a demilitarized independent Cyprus, debarred from *enosis*, in which the Turkish minority would be protected by a UN guarantee and a resident UN commissioner. A new government in Ankara rejected the idea.

With the UN force preventing a resumption of civil war, the basic political fact reasserted itself: namely that Turkey, Cyprus's nearest neighbour and a country with three times the population of Greece, was capable of preventing *enosis* but was not capable of achieving the reconquest of the island. Hence the independent status of Cyprus, an independence resulting from a balance of external forces which was countered by an opposite balance of internal forces. Externally power lay with the Turkish state and not with the Greek state; internally power lay with the Greek community and not with the Turkish community. The power of the Greek community was limited by the power of the Turkish state, and the power of the Turkish state was limited by forces which were not native to the area itself. Cyprus therefore was an international problem-child destined, so long as these circumstances prevailed, to independence tempered by ungovernability. The UN stopped the killing but could not resolve the underlying dispute. Two UN mediators tried but failed, whereupon the Turkish and Greek governments insisted on assuming the role but did very little about it.

Year after year, and twice in each year, the mandate of the peace-keeping force was renewed by the UN who feared the consequences of saving money by removing it. Talks between the two communities were started and stopped more than once. Any whiff of *enosis* from the Greek side was met with Turkish talk of double *enosis*, a new name for partition and the attachment of northern Cyprus to the Turkish state. In 1971 Grivas was back in Nicosia. The only new element in the situation was the worsening relations between Makarios and the military junta which had seized control of Greece in 1967. The junta supported Grivas's heirs, now called EOKA B, against Makarios whom they regarded as a troublesome red priest. They tried to force him to change his government (and did succeed in making him remove his foreign minister) and they incited his fellow bishops to bring charges of simony against him for combining the presidential and archiepiscopal offices. They were more and more anxious to score a popular victory by forcing the pace in Cyprus in order to be able to pose as Greek patriots who had united Cyprus with the Greek heartlands. But Makarios was re-elected president in 1973 without opposition, routed the bishops and had them unfrocked for good measure, and struck back at Athens in 1974 by demanding the recall from Cyprus of the officers of the Cypriot National Guard who were doing the junta's bidding and subverting rather than protecting the Cypriot state. At this point, 15 July, the junta acted. Makarios, who had already survived at least one attempt on his life, was attacked in his palace by the National Guard. He escaped in a helicopter with British help and was flown to England. The insurgents proclaimed Nikos Samson president in his place, a choice as unwise as it was unsuitable, as Samson had been a notorious EOKA gunman and had to resign at the end of a week.

The independence and integrity of Cyprus had been guaranteed by Greece, Turkey and Britain. Greece was in the process of destroying both. Turkey saw in Greece's foolhardy action an opportunity to intervene and occupy at least a part of the island. Britain was unwilling to do anything, partly because of the difficulty of finding reinforcements for its units in the British bases but more emphatically because intervening meant intervening against Turkey and so on the side of the Athens junta and the equally unattractive Samson. Consequently Turkey invaded Cyprus five days after the coup against Makarios. A cease-fire was imposed two days later and the next day the junta in Athens collapsed.

The three guarantors met at Geneva. Turkey's attitude was threatening but realistic: either there must be a new constitution acceptable to Turkey or Cyprus would remain *de facto* partitioned. The constitution proposed was a loose confederation not far short of independence for the components, and the proposal was accompanied by an ultimatum. The talks were broken off. The Turks attacked again, occupied 40 per cent of the island in two days and turned 200,000 Greeks into homeless refugees. The American ambassador in Nicosia was murdered by Greeks who took American inaction to betoken sympathy for or even complicity with Turkey. The British too were criticized violently in Cyprus and in Greece for not doing more, as a guarantor, to help Cyprus escape the plight brought upon it by a Greek government. Makarios returned at the end of the year. Cyprus was in effect partitioned but nobody was prepared to say so and its affairs were therefore back to inter-community talks, hampered by the emotions of war, charges and counter-charges of atrocities, the plight of refugees, economic disruption and the unreality of any attempt to restore the integrity and independence of Cyprus with a Turkish army in control of a large part of it. Makarios, who died in 1977, was succeeded by Spyros Kiprianou.

Both the UN, which tried to mediate from 1977, and the Greeks had tacitly to accept federalism as the basis for any possible settlement. In discussions between the leaders of the two communities within Cyprus, and between the Greek and Turkish governments, the Greeks were intent on securing a strong federal centre together with a substantial Turkish withdrawal from the territories which they had conquered. At the end of 1983, however, the Turks forced the pace by declaring the northern part of the island independent – partition in all but name. The UN secretary-general, Javier Perez de Cuellar, who had considerable personal knowledge of Cypriot affairs, contrived to persuade leaders from both sides to discuss a loose federation: the Turks, in possession of 37 per cent of the territory, seemed willing to reduce their share to 29 per cent and to accept a Greek, not a rotating, presidency; the Greeks seemed willing to allot to the Turks 30 per cent of the seats in a lower and 50 cent in a second chamber. Progress was, however, imperceptible. In 1986 the Turkish régime accepted a UN plan for reunification but the Greeks, hoping for better terms, countered with a different plan. Talks continued, punctuated in 1987 by the election of a new president of Cyprus, George Vassiliou, who promised progress towards agreement but could not make any. President Bush tried to mediate in 1991 but the moment was ill chosen. Elections in Turkish Cyprus strengthened the separatists, elections in Turkey itself produced a hung parliament, and the Greeks in Athens and Cyprus were stiffly opposed to a demand by Denktash for an entrenched right to secede from any federation which might be set up.

PART THREE

THE MIDDLE EAST

9 THE ARABS AND ISRAEL TO THE SUEZ WAR

In the seventh century AD the Arabs, united by a single tongue and a single faith, surged out of the Arabian peninsula and created an empire which stretched at its zenith from the Pyrenees, along North Africa, through what was later called the Middle East and deep into central Asia. The successors of Mahomet, or caliphs, failed to preserve the unity of this vast and increasingly polyglot realm but they imposed their religion on non-Arab peoples, so that in the twentieth century there were Muslims under Russian rule after there had ceased to be Arabs under the rule of other Europeans. This Arab empire lost first its unity and then its independence, and for something like 1,000 years Arabs were subject to Kurds, Turks, British and French. But they never lost the enormously powerful links of a common language and a common faith, and when they began to recover their independence these links served to revive and give substance to visions of a renewed unity. The collapse of the Ottoman empire in Asia in 1918 offered to its Arab subjects a prospect altogether different from that offered, by the more gradual withdrawal of the same empire in Europe, to the racially and linguistically divided Christians of the Balkans.

But in 1919 the Arabs were disappointed. The Ottoman empire was virtually partitioned between the British (who already held Egypt and Cyprus) and the French, and one effect of this Balkanization of the Middle East was to foster separate Arab particularisms (Syrian, Iraqi, Jordanian and so on) and dynastic feuds at the expense of Arab unity: the Arab world was more united under the Turks than without them. The new rulers from the west, lured by international politics and by oil back to the scene of their crusading adventures, became obstacles both to Arab unity and to Arab independence, not least because of the deadening effects of their overwhelming power on the Arab will to struggle for these aspirations; a second world war was needed to get the westerners out. French rule was eliminated by the British when the French authorities in Syria and Lebanon declared for Vichy; British rule on the other hand was maintained and even temporarily strengthened in spite of powerful anti-British currents in Egypt and an attempted pro-German coup in Iraq in 1941 by Rashid Ali el-Gailani. The veiled occupation established in neighbouring Iran by Britain and the USSR did not immediately affront the Arabs, and when the war ended the British were therefore the sole surviving target of Arab nationalism, whose temper was sharpened not only by this concentration of one enemy but also by Britain's administration of the mandate over Palestine where, in consequence of Britain's endorsement in 1917 of the Zionist aim of a Jewish National Home (the Balfour Declaration), a new non-Arab and non-Muslim community had gradually taken hold and was claiming the right to be not a home but a state.

Faced with this powerful and stubborn remnant of the imperialist centuries, and with the vigorous new threat of Zionism, the Arabs were exceptionally divided among themselves.

At the level of power monarchs, Saudi and Hashemite were divided by inherited rivalries; more important, there was a rift between an old order, largely monarchical and traditionalist in its views on society and religion, and a new order which, with its beginnings in Arab intellectual movements of the nineteenth century, aspired to modernize religious and political thought and forms and to reduce the huge differences between the style of living of the very rich and the very poor. This revaluation inevitably produced within the Arab world internal conflicts and strains which enfeebled the Arab capacity to remove the British or defeat the Jews. The British were able to leave in their own time, which was slow; and the Jews established their state of Israel.

THE CREATION OF ISRAEL

The British had struggled for a generation to reconcile their pledges to the Jews and to the Arabs (the latter reinforced by Britain's desire to be on good terms with the oil-producing Arab states and to retain bases in the Arab world), but the two were irreconcilable and the attempt to find an accommodation passed imperceptibly into a hand-to-mouth evasion of the most urgent current complications until the whole responsibility of the mandate was abandoned in 1948. At one point Britain tried partition. The Peel Commission proposed (1937) a tripartite division into an Arab and a Jewish state, leaving Britain with a mandate over a reduced area which would include the holy places of Jerusalem and Bethlehem with access to the Mediterranean. But upon closer inspection this scheme proved impracticable and Britain, forced by the approach of war in Europe to choose between the two sides,

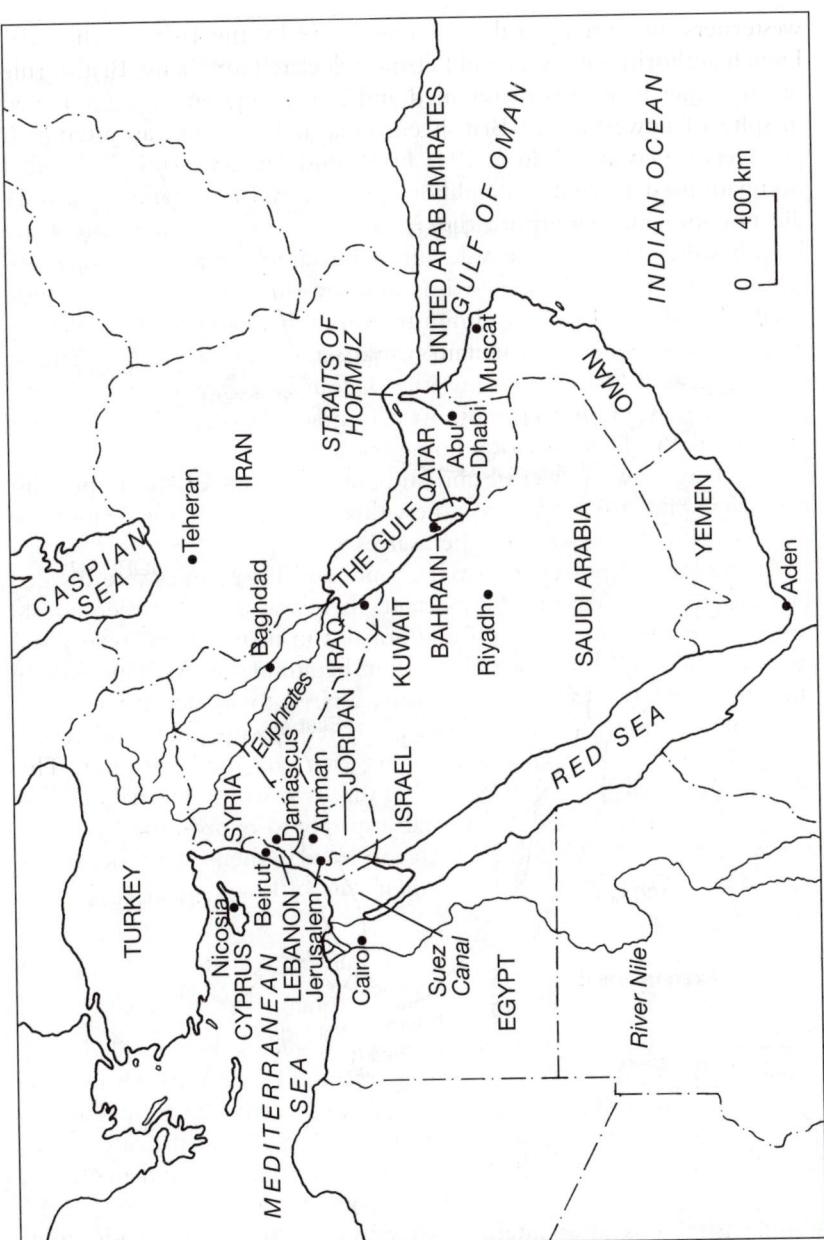

9.1   The Middle East

chose the Arabs and undertook, in the White Paper of 1939, to keep
the Jewish element in the population of Palestine to one-third of the
whole (it had risen since 1919 from 10 to nearly 30 per cent) and so to

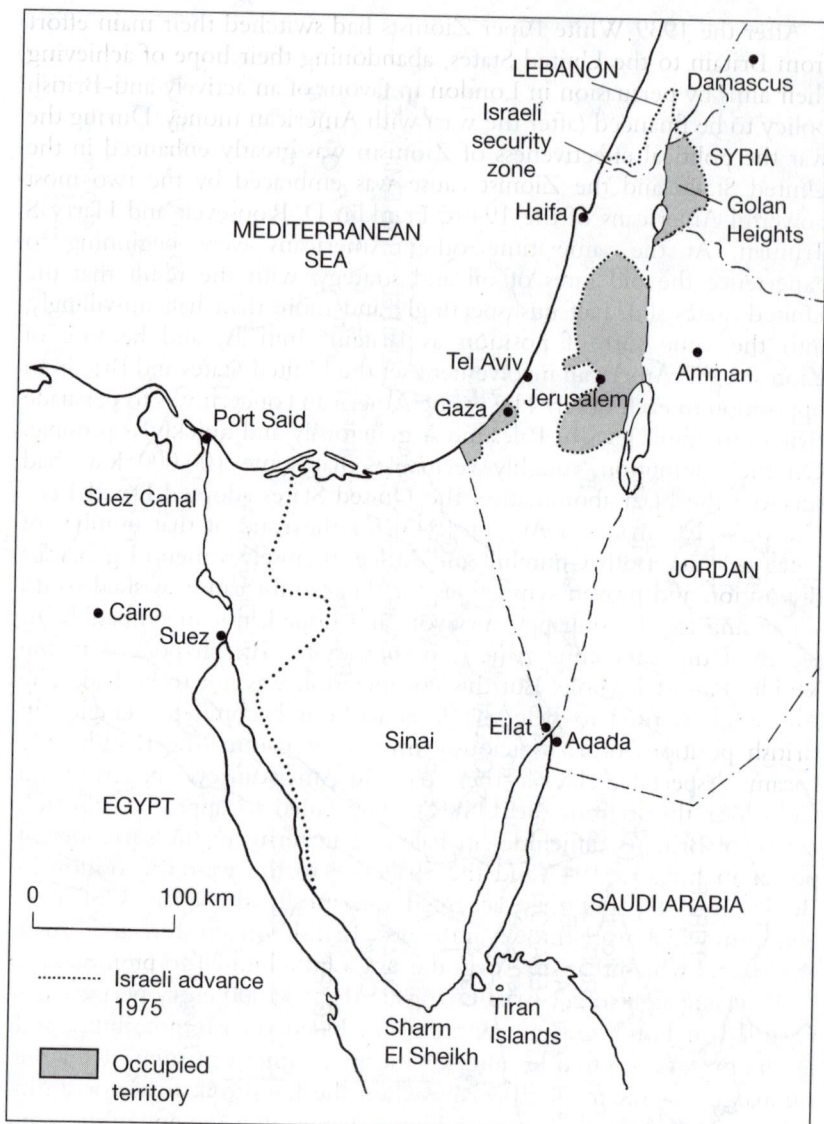

9.2   Israel and its neighbours

stop Jewish immigration after a further 75,000 Jews had been admitted. Thus Hitler cast upon Britain the odium of refusing asylum to Germany's persecuted Jewry because the imminence of a war against Germany made Britain even more sensitive to the need for Arab friendships: grand strategy, as well as oil strategy, dictated the terms of the White Paper.

After the 1939 White Paper Zionists had switched their main effort from Britain to the United States, abandoning their hope of achieving their aims by persuasion in London in favour of an actively anti-British policy to be financed (after the war) with American money. During the war the political effectiveness of Zionism was greatly enhanced in the United States and the Zionist cause was embraced by the two most powerful Americans of the 1940s, Franklin D. Roosevelt and Harry S Truman. At the same time other Americans were beginning to experience the old lures of oil and strategy, with the result that the United States slid, half unsuspectingly and more than half unwillingly, into the same sort of position as Britain. Initially, and because of Zionism, this American involvement set the United States and Britain in opposition to each other. The prime American concern was to persuade Britain to admit Jews to Palestine as generously and quickly as possible. On the assumption, roughly verifiable, that some 100,000 Jews had survived the Nazi abomination the United States adopted David Ben-Gurion's plea, made in August 1945, for the issue of that number of entry permits. Both Churchill and Attlee, themselves men of generous disposition and proven sympathies for the Zionist cause, wished to do something for the unhappy survivors and hoped, not incompatibly, to secure at the same time American support for British policies in the Middle East as a whole. But this co-operation was not to be had. The Americans wanted to help the Jews without becoming entangled in British positions of a suspiciously imperialist nature (the British only became respectable allies in these parts in American eyes in terms of a Cold War threat from the USSR); they failed to appreciate the full extent of Britain's difficulties in Palestine in terms of the surrender of power in India in 1947 and the challenge to the western position in Berlin in 1948; and they accepted uncritically the charges of anti-Semitism which were thrown at the new British foreign secretary, Ernest Bevin, and which that statesman did not a little himself to promote.

The campaign to get Britain to issue the 100,000 entry permits was regarded in London as an extravaganza based on irresponsibility and ignorance and inspired by ulterior motives. American support for the campaign was resented, especially when the Jews took to terrorism in Palestine, as a section of them did immediately after the end of the war. Nevertheless the British government opted for a joint Anglo–American approach to the problem, and in October 1945 a committee of six Britons and six Americans set off to take soundings in Palestine, five Arab states and the European camps where the survivors of Nazism were waiting. Its report, published in April 1946, endorsed the estimate of 100,000 homeless Jews in Europe and the plea for their immediate admission to Palestine; it also rejected partition, recommended the continuance of the British mandate and – besides urging massive Jewish immigration – proposed the abolition of existing limitations on the purchase of land by Jews. The committee had hoped to produce an acceptable package deal, but Truman endorsed only the plea for 100,000 entry permits and the Arabs and the British government rejected the proposals as a whole. The appearance of the committee's report coincided with a Jewish terrorist attack in Tel Aviv which seemed to have no other purpose but murder and which succeeded in killing seven persons. This attack provoked counter-terrorism from the British side which indicated that the British nerve in Palestine, no less than British tempers in Whitehall, was beginning to break. While it was still widely supposed that Jewish atrocities were the independent work of special units (the Stern gang and Irgun Zvai Leumi) undertaken without the approval of the main Jewish defence force (Haganah) or established political bodies (the Jewish Agency and Zionist organizations), there was already evidence that the case was otherwise and that the British authorities were faced with a concerted nationalist attempt to coerce or remove them. The extent of the violence to be used was demonstrated in June 1946 when a part of the King David Hotel in Jerusalem was blown up and ninety-one persons were indiscriminately killed.

After the rejection by Britain of the report of the Anglo–American committee the idea of partition was revived. Ambassador Henry Grady for the United States and the British home secretary, Herbert Morrison, produced in July a plan for two autonomous but not sovereign provinces, and the issue of 100,000 permits a year after the establishment of this hybrid state. Truman rejected the plan and the British government next tried the expedient of a round table conference, but the Jews refused to engage in anything except bilateral discussions with the British. There was this time no American participation, and the repetition by Truman of his support for the 100,000 permits during his campaign for the presidency was not a help. Discussions nevertheless continued sporadically between September 1946 and February 1947. They were abortive. The realities of the situation were reflected by increasing terrorism and the execution by hanging of the young Jew, Dov Grüner, the first victim of British exasperation.

In February 1947 Britain announced that the problem, and possibly the territory too (though this was not clear), was to be transferred to the United Nations. In May a special session of the General Assembly created Unscop (United Nations Special Committee on Palestine) and its eleven members set off for Jerusalem. While they were there, and possibly because they were there, 4,554 refugees aboard the *Exodus 1947* arrived at Haifa after having been collected by the Jewish Agency at Sète in the south of France and furnished with travel documents for Colombia. These pawns in the Zionist game were refused permission to land by the British authorities and were shipped back to Sète whence, again refused permission to land, they were directed with a horrible insensitivity to a German port. In July the toll of innocent suffering was dramatically increased when two British sergeants were hanged by an Irgun band. Unable to solve their problem, or to keep order, or even to defend themselves, the British were now more than ready to go. When Unscop produced, by a majority, a new partition plan of ridiculous complexity (three Arab and three Jewish segments linked in a sort of economic union with Jerusalem under international trusteeship), they said they would do so.

The Unscop plan was adopted with modifications by the UN in November. It was accepted with misgiving by the Jews and rejected by the Arabs. Jewish acceptance was superficial and misleading: the Jews were willing to accept publicly the promise of a Jewish state whatever their reservations about its size and shape, but they did not accept the corollary of a Palestinian Arab state and entered into secret talks for the annexation of the designated Arab sections by the emir Abdullah of Transjordan. The prevention of a Palestinian state was a cornerstone of Israeli policy for a generation and more, beginning with the creation of the Israeli state. The UN plan, whatever its validity (the UN had no authority to make states), was quickly made irrelevant and the Jews won their state by conquest. Fighting began before the end of 1947 with attempts by the Jews to get control of the segments allotted to them. There were also disorders in Arab towns. The British were impotent and lost even the reputation for fairness which they regarded as one of their special contributions to public morality. On 11 December they declared that they would surrender the mandate on 15 May 1948. Fighting increased. The Jews managed to procure arms in substantial quantities, with which to meet the expected invasion by the regular armies of neighbouring Arab states. In April 1948 they contrived at Deir Yasin a massacre which set in train an exodus of refugees who were to constitute, with their as yet unborn progeny, one of the bitterest bones of contention between Arabs and Jews and one of the sorriest human spectacles of the times.

The last British official left Palestine on 14 May 1948. The state of Israel was movingly proclaimed by Ben-Gurion who became its prime minister, embodiment and inspiration; Chaim Weizmann, the veteran leader of Zionism who enjoyed worldwide respect, became the first president despite the fact that, in the previous December, the twenty-second Zionist Congress at Basle had marked the passing of his influence and had even exposed him to some ruderies. Truman immediately recognized the new state and Stalin was not far behind. In its fledgling years Israel was more beholden to the USSR than the United States for practical help, but it quickly observed that the United

States had more to offer in money and arms and from about 1950 alliance with the United States had become the sheet anchor of the new state. In these same years Israel formed a similarly enduring, if less crucial, alliance with Iran which provided asylum for Jews fleeing from Iraq in the late 1940s and recognized Israel in 1950. This alliance was strengthened in the ensuing decades by commercial ties and exchange of military intelligence and, surviving the fall of the shah, contributed to Khomeini's resistance to the Iraqi attack on Iran in 1980. Its cement was hostility to the Arabs which had been a persistent theme in Iran's history and became the determining factor in Israel's fight for survival.

In 1948 five Arab states marched against Israel, but the promptness of their reaction was no index of their keenness for a fight or their effectiveness in it. The Syrians did little and the Lebanese less; the Iraqis retired early and the Egyptians arrived late; the Jewish defence of Jerusalem, which thwarted the Jordanians, combined with the Arabs' incoherence to give Israel the victory. The UN intervened by appointing the Swedish Count Folke Bernadotte to mediate. He effected a truce which lasted from 11 June to 9 July and was followed by more fighting in which the Israelis were decisively victorious. Bernadotte was murdered on 17 September by the Stern gang; he was succeeded by a UN official, the American Ralph Bunche. The Israelis gained further successes in renewed fighting in October but by the end of the year the war was virtually over, Palestine and Jerusalem itself had been partitioned by the verdict of arms, and during the first half of 1949 armistices were effected between Israel and four of its assailants (Iraq being the absent fifth). In the second phase of the fighting the Israelis had secured their sovereign state by defeating the regular forces of Arab states as well as the Palestinian Arabs and they were ready to consolidate their gains.

Arabs, other than Palestinians, never had much appetite for a war with Israel. They were ready to accept an Israeli state but Ben-Gurion and other Israeli leaders, more intransigent, put their faith in military dominance. During the war Ben-Gurion rebuked his generals for not driving more Palestinians out of Palestine and in the two years 1948–50 Palestinian property was systematically wrecked and 700,000 Palestinians were evicted or, in Zionist terminology, transferred. When King Farouk of Egypt tried to open negotiations during the war, his advances were rejected by Ben-Gurion and when, after seizing power in Syria in 1949, Husni Zaim proposed peace and an alliance and homes for 300,000 Palestinians in Syria, Ben-Gurion, who rightly regarded Zaim as unstable, did not respond nor did he pursue these proposals with Zaim's successors. (Later moves from the Arab side, including the PLO, in the 1970s were ignored by Israel and spurned by the United States.) But Israel's militancy failed to secure its medium- or its long-term aims – formal peace with security and recognition. The Arabs, out of their deep humiliation and their just concern for the Arab refugees from Palestine, refused to recognize Israel's existence or make peace with it and continued the war by economic means – notably by closing the Suez Canal to Israeli shipping and to goods going to or from Israel. This action was condemned, fruitlessly, by the Security Council which also, equally fruitlessly, asserted that Israel should either readmit the refugees or compensate them. Since these refugees outnumbered the Jews in Palestine the Israelis argued with much plausibility that they could not be expected to readmit them so long as the Arab nations as a whole continued to proclaim their intention to eliminate the state of Israel. The Arabs themselves did little to integrate the refugees in their places of refuge since, politically, the most important point about a refugee was his refugee status and plight. From 700,000 in 1949 their numbers swelled by natural causes to over 1.25 million.

The creation of the state of Israel was a most extraordinary political phenomenon. Israel came into existence as a result of the tenacious memories of a persecuted people whose misfortunes in various parts of the world had given them an intense addiction to the words of their holy books; as a result of atrocious crimes perpetrated against European Jewry in sight of Europe and the world; and as a result of the exertions of leading Jews who worked with vigour and intelligence to capture a piece of territory not their own, and who would probably have failed to do so if they had not believed that their end was one of those which justifies every means and had not enjoyed the advantages of working under a British umbrella. The state which they founded was as exceptional as its origins. By adopting the principle that the door to it must be open to all Jews everywhere, it became a mixture of tongues and cultures and unequal skills, held together by the unifying power of a Hebrew revival and by a community of race and of hope (though not of religion, since the Jewish religion meant little to many Israeli citizens). The Law of Return of 1950 gave all Jews the right to come to Israel. The Law of Citizenship of 1952, however, conferred on them Jewish, not Israeli, citizenship and Arabs in Israel were designated persons of Arab nationality, not Israeli citizens. These Arabs suffered certain disadvantages – for example, in the acquisition and holding of land. Israel was dependent on external aid, which it received liberally from American and other Jews and, by way of reparation, from the German Federal Republic; and its life was conditioned by external hostility, since the Arabs refused to accept its existence and insisted that, as an imperialist subterfuge for the maintenance of Anglo–American power in the Middle East, it must be dismantled. It adopted, in spite of its economic and military stringencies, a mainly democratic form of government. Externally, however, it developed polices, shaped by its victories on the one hand and its continuing precariousness on the other, which secured no more than the prospect of more victories. To defend its frontiers it retaliated promptly, vigorously and often indiscriminately against every assault great or small through special units under the command (initially) of General Ariel Sharon. This escalation of violence was effective in the short term but at the cost of sharpening the aggressive psychology of the new state, prompting counter-action by Arabs and prolonging a chain of defensive and offensive operations against hardened enemies. That these operations were against Arab states helped to obscure the centrality of the problem of the Palestinians who were without a state.

NASSER AND REVOLUTION

To the Arabs the fact that Jews had suffered at the hands of Christian Europeans through the ages seemed a poor reason for allowing them to expropriate a part of the Arab world and drive a million Muslims out of it. The failure of the armies of five Arab states to prevent the establishment of Israel was a humiliation which could only be met by rounding on the leaders who had so conspicuously failed and by declaring that the verdict of 1948 was only an interim one which would be reversed later. Those – mainly outsiders – who thought that time would heal or at any rate soften the acerbities of this conflict had to retreat to a more pessimistic view as officially sponsored Arab propaganda maintained a vicious anti-Zionism and as the new generation in the refugee camps was consciously nourished on visions of a return to Palestine after a gloriously successful war which would erase Israel from the map. Zionist nationalism and Zionist extremism were matched by Arab nationalism and Arab extremism. Although the situation remained frozen by external factors – in particular by the Tripartite Declaration of 1950 by which the United States, Britain and France undertook to maintain a balance of armaments between Arabs and Jews and to consult together over any infraction of frontiers – basic attitudes within the area of conflict changed little, if at all. What did change as a result of the war of 1948 was power in the Arab states immediately concerned. The defeat of 1948 accelerated the demise of some ancient régimes. In Syria Husni Zaim seized power in March 1949, held it for a few months before being ousted by Sami al-Hinnawi who was in turn ousted after a few months by Adib Shishakli who ruled until February 1954. In July 1951 Abdullah, having transformed his emirate of Transjordan into the kingdom of Jordan, was murdered as he entered the Al Aqsa mosque in Jerusalem; neither then nor later could Palestinians easily forgive him for collaborating with the Jews to enlarge his own principality at the expense of the Palestinians who, in the UN plan accepted by the Jews, were to have a state of their own. More significantly, in the following July the Egyptian monarchy was overthrown by officers who had bitterly resented the unpreparedness and ineptitude of their country's operations in 1948 and now despatched the last representative of Mohamed Ali's line into exile.

The titular leader of the Egyptian revolution of 1952 was General Muhammad Neguib but the real leader was Colonel Gamal Abdel Nasser, who became prime minister in April 1954 and supplanted Neguib as president of the new republic a few months later. Nasser was a young man of 36 who had made a career in the army from origins in what might be called the lower middle class. He had the necessary impetuousness and indignation to make a nationalist revolutionary, but he had also the qualities of coolness, shrewdness and humour which keep the revolutionary sane and afloat after success. He was also sufficiently quick-witted to keep more or less abreast of the problems which assailed him when he found himself required, as head of the most important of the Arab countries, to elaborate and execute policies in world affairs. The first of these problems was the unfinished business of getting rid of the British. In addition he had to find his place in the Arab world, African as well as Asian, and take a stand in the Cold War for or against the west or the Russians or neither.

The British occupation of Egypt began in 1881 with a debt-collecting expedition but its real motive was and remained strategic – to secure control of the eastern Mediterranean (threatened when British influence over the sultan's government in Constantinople was challenged by German influence), of the route to India and the east, and of the Nile valley. Egypt remained formally a province of the Ottoman empire under the hereditary rule of the heirs of Mohamed Ali, but it became in effect semi-independent under the British instead of semi-independent under the Turks. When Britain and the Ottoman empire went to war in 1914 Britain proclaimed a protectorate over Egypt which was converted in 1922, under nationalist pressure, into a treaty relationship. Britain secured in 1936 the right to station 10,000 men in the Suez Canal zone for twenty years, an arrangement which was rendered acceptable to the Egyptians because this base could serve to protect Egypt against the ambitions of Mussolini who was engaged at this very time in conquering Ethiopia. During the Second World War Churchill resolved to maintain and strengthen Britain's Middle Eastern position. The Arabs had shown certain pro-German proclivities which threatened to give Germany dominion in the Middle East after the German conquest of the Balkans and Crete in 1941. The British therefore occupied Syria, Lebanon and Iran and wooed Turkey (unsuccessfully until 1944 when Turkey severed relations with the Axis – it entered the war the following February); they also forced the pro-British Nahas Pasha on King Farouk as prime minister. Nahas hoped that in return the British would give Egypt after the war the unshackled independence which they had promised over sixty times since 1881, but the king, preferring a tougher to a waiting game, dismissed Nahas in October 1944.

At about this time Eden was preparing a postwar British position in the Middle East, based on an association of pro-British Arab states. Eden encouraged the hardened pro-British Nuri es-Said, Iraq's prime minister, to revive the concept of the Fertile Crescent as a political unit with Baghdad as its capital and embracing autonomous Zionist and (Christian) Maronite communities. Eden also promoted a pro-British Arab League, but in September 1944 at Alexandria, Egypt, as well as Iraq, Syria, Lebanon and Transjordan, concluded the Protocol of Alexandria which preceded the treaty of March 1945 whereby these five states plus Saudi Arabia and Yemen constituted the Arab League: this larger issue was not the league designed by Britain. From the beginning Egypt was much the most important member of the League by virtue of its greater population and sophistication and the influence which radiated from Cairo's universities, printing presses and radio. But the British and Egyptian interests in the Arab League were fatally crossed. For Britain the League was to be an ally and a pillar of support; Egypt, however, obsessed with the continuing British occupation not only of the Canal Zone but also of the points in Cairo and Alexandria where the Union Jack had been hoisted during the war, saw the League as an anti-British lever and not as a symbol of partnership. Britain's plight in Palestine helped to draw Arabs together in an anti-British attitude which suited Egypt and belied Eden's hopes. (Palestine, or rather Israel when it came into being, also affected the Arab League by converting it from a regional organization which, in its members' opinion, concerned nobody but themselves into an alliance with wider objectives and therefore international significance.)

After the war Britain appreciated the need for a revision of its treaty relations with Egypt – and also with Iraq and Transjordan – but had no thought of an early departure from the Middle East. British forces were withdrawn from the Nile delta to the Canal Zone, but only tardily; it still seemed axiomatic that Britain, victorious in a war in which British exertions and tenacity in the Middle East had played a major role, was and should be a Middle Eastern power. In addition, the withdrawal from India in 1947 would have made it difficult for the new British Labour government, accused by the opposition of 'a policy of scuttle' in India, to propose further substantial abdications in the Middle East. But the British position had been gravely weakened. Britain owed Egypt £400 million and Iraq £70 million; and these debts were only a part of the economic strains which were intensified by the abrupt ending of American Lend-Lease and were soon to force Britain to ask the United States to take over British responsibilities in Greece and Turkey. In 1948 the strain would be further increased by the Palestine finale and by the need to defend the western position in Berlin by the air lift. Such were the circumstances in which Bevin set about negotiating fresh treaties with Egypt, Iraq and Transjordan. He quickly succeeded in the last case (1946), but King Abdullah was much criticized for allowing British troops another twenty-five years' lease of Jordanian soil. With the Egyptian and Iraqi governments Bevin also succeeded in reaching agreement, but the draft treaties accepted by Sidky Pasha for Egypt and Salih Jabr for Iraq (the Treaty of Portsmouth, 1948) were rejected by their parliaments and people because they did not provide for complete British evacuation.

There followed an uneasy and unproductive period which was broken in 1951 by a clash between two rival concepts. By this time the Korean War had begun and the western allies were anxious to create in the Middle East an anti-Russian alliance similar to Nato. With this end in view the American, British, French and Turkish governments produced plans for a Middle East Defence Organization (Medo) which would include Egypt and the Suez Canal base; the British would evacuate the base but the allies, adapting an idea once produced by Bevin, would have the right to return in certain eventualities. This scheme with its roots in the world situation ran foul of Egypt's conception of a foreign garrison on the Canal as a symbol of indignity, and since Egypt was not moved by the fear of the USSR which animated the west, Egypt rejected Medo. At the same time and for good measure Egypt denounced in October 1951 the Anglo–Egyptian Treaty of 1936, which had five years to run, and began guerrilla attacks on the Canal Zone. In January 1952 anti-British riots in Cairo caused extensive damage.

When therefore Neguib and Nasser seized power a few months later they were confronted not only with the traditional problem of the British, whose occupation of Egyptian soil was a standing national affront, but also with the new problem created by the desire of the west as a whole to enlist Egypt in the Cold War and make the Canal Zone an anti-Russian arsenal and base. Provided they could secure total British evacuation, the new Egyptian leaders were anxious to come to terms with Britain, but they had less positive ideas about an association with the west.

In the two years after the revolution Egypt and Britain settled their outstanding differences. They began not with the Canal Zone but with the Sudan, which had been an Anglo–Egyptian condominium since its recapture from the Mahdists in 1899. Egypt wished to restore the union of Egypt with the Sudan because of the vital importance of the Nile waters and because of ancient pharaonic vistas of the unity of the Nile valley. The British, who had been effective rulers of the Sudan for half a century, insisted that the Sudanese must decide their future for themselves and the Egyptians, wrongly believing that the Sudan would opt for unity with Egypt, so agreed in February 1953. In the event the Sudan chose independence in spite (or because) of Egyptian propaganda and pressure, and became a sovereign state in 1956. Nasser accepted this first reverse in the foreign field with a pragmatic acquiescence which he was to have to display again on future occasions.

Agreement on the Canal Zone was reached in July 1954 (and a treaty signed in October) on the basis that the British would depart in twenty months but would have the right to return if any member of the Arab League or Turkey were attacked by any outside foe except Israel. Iran was not included in the reverter clause in spite of British attempts to put it there. This treaty represented considerable, if sensible, concessions by Britain which were unpalatable to a section of the ruling Conservative Party but were accepted by the government partly on the grounds that the existence of nuclear bombs had turned the base into a death trap and partly – if less so – in response to pressure from the United States where continuing bad relations between Britain and Egypt were regarded as a serious impediment to the Middle Eastern segment of American world policies. But for Egypt the agreement of 1954 with Britain was not a preliminary to alignment with the west. The west on the other hand believed during 1952–54 that better relations with Egypt implied such an alignment. Consequently the better relations soon took a turn for the worse.

In 1955, the year of the Bandung conference of neutralists, Egypt became wedded to non-alignment. The issue did not arise in the first postwar years because alignment meant alignment against the Russians and there was as yet no Russian presence in the Arab world to be aligned against. The traditional Russian spheres of activity in the Middle East were the non-Arab states of Turkey and Iran, and in both the USSR had experienced and apparently accepted rebuffs at the end of the war; claims for the return of Kars and Ardahan (lost to Turkey in 1921) and for the revision of the Montreux Convention governing passage through the Straits of Constantinople were ineffective, while the attempt to subvert the régime in Iran by sponsoring an Azerbaijani republic and a Kurdish bid for independence was defeated by the unexpected astuteness of Qavam es-Sultaneh and the unexpectedly firm reactions of the United States and Britain. The prompt Russian recognition of Israel was not followed up and seemed to have been regarded by Moscow as a blind alley.

When therefore the Medo plan was first presented to Egypt, it seemed foreign and irrelevant to Egyptian interests. It was also vaguely repugnant. The Korean War, which lay behind it, was regarded by most Arabs as no concern of theirs, so much so that all the Arab members of the United Nations except Iraq refrained from voting on it in the General Assembly. Nasser himself came under the influence of Nehru, especially after the latter's visit to Cairo in August 1953 and during K. M. Pannikkar's embassy there in 1953–54. But the principal factor in Nasser's route to neutralism and the 1955 Bandung spirit was provided by the western powers themselves.

IRAQ: THE BAGHDAD PACT

In 1954 the Americans were arming Turkey and Pakistan. They decided to help Iraq in the same way and to brigade these countries and Iran in a new anti-Russian organization which would stretch from the Bosphorus to the Indus. The inclusion of Iraq was enthusiastically welcomed by the British who saw a way of getting a new treaty with Iraq in place of their existing treaty which was due to expire in 1957 and which Bevin's negotiations with Britain's Iraqi friends had failed to prolong. But the inclusion of Iraq was a cardinal mistake. In 1953 Dulles had decided that it was no good trying to get the Arabs to join a Cold War front (the Medo policy) and that the west should therefore construct instead a non-Arab group of friends along the so-called northern tier of countries bordering on the USSR. Iraq was not a necessary member of this group, since it had no frontier with the USSR and was in any case an Arab state. Moreover, the attraction of Iraq into this northern orbit roused the deepest suspicions in Cairo both on account of existing tensions between Egypt and Iraq and as a manoeuvre by outsiders to disrupt the unity of the Arab world. The Americans were proposing to arm Iraq and Pakistan against the USSR, but Egypt interpreted the agreements as a strengthening of the Iraqi side in Arab politics (just as India regarded the arming of Pakistan as a threat to itself in Kashmir).

Egypt and Iraq stood for different brands of Arab unity long before the Egyptian revolution of 1952 made Egypt a socialist republic by contrast with the traditionalist monarchy which lasted in Iraq until 1958. Both before and after 1952 Egypt's leaders wished to be the leaders in the Arab world and planned to use the Arab League for this purpose; they were conscious of Egypt's natural claim to leadership and they sensed the advantages which would accrue to them in their dealings with the outside world if Egypt were the acknowledged head of an Arab commonwealth as well as an independent sovereign state. Their approach to Arab politics was Egyptian first and Arab second, and their plans stopped a long way short of Arab unity in any organic or institutional sense. The Iraqis on the other hand inherited other ideas, notably the vision of a vast Arab kingdom which had been nurtured by the British and French during the First World War and then thwarted by them. Although the sons of the sherif of Mecca – the Hashemites – came to rule in Baghdad and Amman and briefly in Damascus, they had been evicted from Saudi Arabia by the rise of Ibn Saud and had been foiled further north by the mandates system (which did not bother the Egyptians). Nevertheless the vision persisted, embracing in its changing forms – whether Greater Syria or the Fertile Crescent – Iraq, Syria, Lebanon and Palestine. But any union between Iraq and Syria would deprive Egypt of hegemony in the Arab world.

The Hashemite regent of Iraq, Abd ul-Ilah (who never forgot that he ought to have succeeded his father and grandfather as king of the Hejaz), aimed to be king in Syria when he ceased to be regent in Iraq. Had he succeeded, the Hashemites would have formed a powerful bloc, for Abd ul-Ilah's young cousin Feisal became king in Baghdad and his uncle Abdullah was emir, later king, in Amman. After the first Syrian postwar revolution Husni Zaim turned to Iraq, although later in his short period of power he veered towards Egypt and Saudi Arabia; his successor, Sami al-Hinnawi, favoured an Iraqi–Syrian union but by so doing he precipitated his own fall, and his supplanter, Adib Shishakli (1949–54), the strong man of the Arab world before the advent of Nasser, rejected any close association with Iraq because it was monarchical and pro-British. The overthrow of Shishakli in February 1954 by the Syrian army, helped by Iraqi gold and by British well-wishing at the least, revived for Nasser the danger of an Iraqi–Syrian union at the moment when its other principal opponent, Ibn Saud, had just died and the western powers were preparing their new treaty with Iraq. In the event opposition among Syrians to the Iraqi royal family and its British connections, manifested at elections in 1954, prevented the consummation of the union.

The first formal step towards the Baghdad Pact complex was the treaty of 4 April 1954 between Turkey and Pakistan. It was followed by American military aid agreements with Iraq (21 April) and Pakistan (10 May). Turkey and Iraq signed a mutual assistance pact on 24 February 1955. This was the Baghdad Pact itself and it was declared to be open to all members of the Arab League and other states interested in the peace and security of the Middle East. Britain joined on 5 April 1955, and Pakistan and Iran in September and October. But no other Arab state followed Nuri's lead into the western camp. The Arabs rejected Nuri's belief that salvation lay in alliance with old friends in the west – partly because, to a younger generation, Britain did not appear a friend. Syria, by rejecting the pact, swung the bulk of the Arab world to Cairo and away from Baghdad, and in March 1955 a counter-alliance was formed between Egypt, Syria and Saudi Arabia. This was a victory for Nasser over Nuri and the British. It also marked the end of the period of improving relations between Egypt and the west and the beginning of violent polemics within the Arab world. Shortly after signing the treaty of March 1955 Nasser set out for Bandung where, in obedience to the prevailing anti-colonialist and neutralist wind, Iraq was censured for joining the Baghdad Pact and Pakistan for joining its eastern equivalent, Seato (South East Asian Treaty Organization).

Whatever the relative merits of British and American approaches to Middle Eastern problems at this time, the failure of London and Washington to find a common policy was manifest and significant. Britain may have been wise or unwise to use the Baghdad Pact for special purposes of its own in Iraq, and the United States may have been wise or unwise to join only the pact's economic and anti-subversion committees, but the most obvious consequence in Arab eyes was the discrepancy between the two western powers and the attempt of the Americans to look un-British by not signing the pact or joining its central organs. A more forthright American policy might have persuaded the Jordanian and Saudi monarchies to join a pro-western alliance in 1954, but at that date American anti-imperialism (not yet killed in Vietnam) and Anglo–American oil rivalries were still making Washington wary of entanglements with Britain in the Middle East with the result that there was a time lag between the adoption of essentially identical policies first in London and later in Washington.

THE SUEZ WAR

Besides the division within the Arab world Nasser was concerned with the enduring feud between the Arabs and Israel and with the fact that Israeli retaliation against Arab propaganda and provocative raids was being turned against Egypt. Shortly before the signature of the treaty of March 1955 Egypt experienced an exceptionally sharp Israeli attack in the Gaza area. This raid had been ordered by Ben-Gurion who had very recently returned to power from retirement and was to Nasser's knowledge seeking arms from France. This Franco–Israeli association, although never enshrined in a formal alliance, became an additional prime factor in Middle Eastern politics and one of the principal ingredients in the Suez War of 1956. The Tripartite Declaration of 1950, which was supposed to prevent an arms race in the Middle East, was circumvented by Israel which found in France sympathizers ready to help Israel in secret. French motives were various: there was a feeling of obligation to the Jews as a people who had suffered too much; a feeling of admiration for what they had achieved in Israel: a feeling of socialist solidarity between men like Guy Mollet and Ben-Gurion.

These affinities were played upon by skilful Israel lobbyists who found their task much facilitated when the Algerian revolt of 1954 and Nasser's help for the rebels led many Frenchmen to conclude that a blow at Nasser was the right way to solve their troubles in Algeria – or at any rate a necessary pre-condition. In addition some Frenchmen shared Eden's view that Nasser was a menace like Hitler and must be put down before it was too late. French policy, traditionally pro-Arab, was therefore pulled in a new direction. France agreed in 1954 to supply Israel with fighter aircraft.

On his return from Bandung Nasser too began seriously to look for arms. The three signatories of the Tripartite Declaration refused to supply Egypt or Syria with all they requested. Syria turned successfully to the Russians who had become interested in the possibility of playing a more active role in the Middle East since the Gaza raid had exposed Egypt's weakness. But Nasser was reluctant to buy Russian. After trying Beijing (where he got a hint to try Moscow) and then Washington and London once more, he finally took the Russian plunge and announced in September 1955 that he was to receive Czech arms without strings. It was now Israel's turn to be alarmed. By the Czech deal Egypt was to get a wide range of weapons including 80 MiG 15s (the fighters used in Korea), 45 Ilyushin 28 bombers and 115 heavy tanks equal to the best in the Russian army and superior to anything which Israel had. Israel pressed France to revise the agreement of 1954 by supplying Mystère 4 jet fighters instead of Mystère 2s. France complied and Israel received in April 1956 a contingent of the best fighter aircraft in Europe. (Their arrival just after the French foreign minister, Christian Pineau, had paid a successful visit to Cairo evoked an indignant anti-French outburst from Nasser and destroyed what chance there was of the Arab wind prevailing over the Israeli in the French cabinet.)

Like most countries Egypt wanted both guns and butter. Nasser had hoped to get both from the west but had been forced to buy communist arms or go without the guarantee which he was seeking. He then faced the question whether he could get economic aid from the west after accepting military aid from the communist bloc. The answer proved eventually to be no, but it was in doubt for some months. The test was the Aswan High Dam, and it proved that not only France but also Britain and the United States were turning against Nasser.

The high dam was designed to transform Egypt's economy and society by adding 860,000 hectares to the area of cultivable land, making the Nile navigable as far south as the Sudanese frontier and generating electricity to service industrial plants which would provide some of the growing population with a living. It was to cost $1,400 million, including $400 million in hard currency, of which the World Bank would advance $200 million and the United States and Britain $56 million and

$14 million respectively at once and the remaining $130 million between them later. During 1955 negotiations to this end seemed to be proceeding without more than normal hitches. During the first half of 1956, however, they petered out. Britain and the United States decided not to help. Nasser's credit, in both senses of the word, was running down, especially as a result of his purchases of communist arms.

During 1956 he strengthened his enemies by recognizing the communist régime in Beijing – a step which caused special irritation in Congress in Washington, even though the real reason may have been Nasser's fear that the new Russian leaders might, on their visit to London, be persuaded to join the western powers in a new Middle Eastern arms embargo. The Czech arms deal had broken the 1950 embargo, to the general delight of the Arab world which resented the embargo as a clog on their sovereignty, but that embargo could be reimposed if the Russians were looking for inexpensive ways of showing goodwill towards the west, and in that event Beijing would be the only alternative source of supply. In addition the cotton lobby in the American Congress disliked laying out American money to help Egypt to grow more cotton to compete with American cotton. Egypt was blamed for not coming to terms with the other riparian states along the Nile (Sudan, Ethiopia and Uganda) and for pledging for arms purchases money which would be needed to service its foreign loans. In the United States it was argued that it was imprudent to allocate so much American money to a single project since the United States would then have to refuse all other requests for aid to Egypt for many years, leaving the Russians to step in and say yes. But behind all these reasonings lay the plain fact that Washington and London did not like Nasser and thought (like the French for a different and more specific reason) that it would be salutary to administer a snub to him and cut him down to size. This attitude was strongest in Britain where it found vent in the obsessive misinterpretations and miscalculations of the prime minister, Eden, who mistook Nasser for a fascist dictator and thought he could be easily replaced. (Eden's animosity had been sharpened in March by the dismissal of General Sir John Glubb and other British officers from Jordan's Arab Legion, of which Glubb was the commanding officer. Fortuitously this anti-British move by King Hussein coincided with a visit to Cairo by the British foreign secretary, Selwyn Lloyd, and it was therefore erroneously regarded by Eden as a deliberate affront to Britain contrived by Nasser.)

On 19 July 1956 Dulles informed the Egyptian ambassador in Washington that the Anglo–American offer to finance the dam was revoked. The French ambassador in Washington, Maurice Couve de Murville, had predicted that in this event Nasser would retaliate by seizing the revenues from the Suez Canal. On 26 July, in a speech in Alexandria, he did just that.

The Suez Canal was indubitably a part of the Egyptian state but it was also the subject of two, very different, instruments – a concession agreement and an international treaty. The former, granted to Ferdinand de Lesseps by the khedive or viceroy of Egypt, Said Pasha, and confirmed by the Ottoman sultan, conceded the right to operate the canal for ninety-nine years from its opening, which took place in 1869. The concession had passed from de Lesseps to the Universal Maritime Suez Canal Company, which was an Egyptian corporation with headquarters in Cairo and Paris and with a diversity of shareholders including the British government and a host of ordinary French *rentiers*. The concession, which was a valuable one, had twelve years to run in 1956. Thereafter the operating rights would revert to the Egyptian state. Nasser's action amounted to the nationalization of the company's rights, but since he promised compensation it was difficult to maintain that he had done anything illegal or, in the twentieth century, particularly unusual – although the company might well ask where the compensation money was to come from. Nasser would, however, be on the wrong side of the law if he broke the terms of the second relevant instrument. This was the convention made in 1888 between nine powers including the Ottoman empire, which was at that date suzerain over Egypt. The parties engaged themselves to keep the canal open to all ships of commerce or war, in times of war as well as peace, and never to blockade it. Should Nasser fail to keep the canal open he would be in breach of the convention and the signatories would be entitled to take measures to reopen it. There was some ill-judged expectation that if the canal pilots were withdrawn, the canal would cease to function and the right to intervene could be said to have arisen, but in the event the canal continued to function smoothly until bombarded by the British and French, even though the company's pilots were nearly all withdrawn as a result of pressure by outside powers.

The nationalization of the canal company gave Britain and France an excuse for the forcible action which they wished to take against Egypt. The British cabinet allocated £5 million (imperialism on a pittance) and resolved to use force within a week, only to discover that Britain's military preparedness was such that nothing could be ventured before the middle of September or without calling up the reserves. This delay enabled the United States to intervene. Eisenhower and Dulles agreed with the British and French governments in wishing to put the canal under international control but, although they had little love for Nasser, they were opposed to the use of force until all methods had been tried and had been seen to have been tried. Eisenhower, who made his position clear in letters to Eden and in public statements, was temperamentally averse to force and was also convinced that force was inexpedient because it would lead to sabotage of pipelines, would encourage other leaders (for example, Jiang Kaishek and Syngman Rhee) to claim American support for the use of force in their quarrels too, and would turn the uncommitted world against the west. Eisenhower sent a special emissary to London who reported on 31 July 1956 that the British were intent on using force, and thereafter a duel developed between Eden and Dulles, enemies since the crisis of 1954 in Indo-China, with Eden manoeuvring to get American sanction for a forward policy and Dulles side-stepping and playing for time.

First the British and French, with American support, convened in London a conference of the canal's principal users and presented to it a plan for a new operating board to ensure international control of the canal. The conference did not approve this plan unanimously; it was criticized as an unjustifiable infringement of Egyptian sovereignty. Nevertheless the Australian prime minister, Robert Menzies, and four other members, representing the majority view, went to Cairo to present the plan to Nasser, who turned it down and pointed out that the canal was functioning normally. Next Dulles, perhaps merely in order to keep talking and avoid shooting, propounded a Suez Canal Users' Association with the right to organize convoys and take tolls from the vessels in them. This plan appealed to the British who saw a chance of running a convoy through the canal against Egyptian opposition and so putting Egypt in the wrong in American eyes. Dulles, who was consistently pragmatic throughout this phase, then killed the scheme upon becoming suspicious that Britain and France might use it to start shooting; he pointed out that the American government had no power to force the masters of American vessels to pay tolls to the association and not to the Egyptian government. Britain and France then referred the dispute to the Security Council while explicitly stating that they reserved the right to use force. When the Council met on 5 October Egypt proposed negotiations, while Britain and France produced a plan for international control of the canal. Unofficial negotiations outside the Council made substantial progress but Egypt persisted in its refusal to accept international control and the Anglo–French plan was defeated by a Russian veto.

During these months the French became increasingly exasperated with the British. Joint Anglo–French commands had been set up at the beginning of August but the prospects of action by the slowly assembling hosts diminished as the British wavered between their desire to keep in step with the French and their anxiety not to get out of the Americans' good books. By the end of September or early October the French were reverting to their old line of co-operating with Israel, from which they had been distracted by the lure of a joint Anglo–French operation in response to the nationalization of the canal company.

Israel had powerful reasons for wishing to make war on Egypt. Raids into Israel by *fedayeen* based in the Sinai peninsula had become more audacious and more frequent. Land near the frontier was becoming too dangerous to farm and the Israeli government feared outrages even in the centres of its cities. Only a spectacular gesture could end this murderous nuisance. Further, Israel wished to break the Arab blockade of the gulf of Aqaba and so win a sure outlet to the countries of Asia and Africa from the port of Eilat which was languishing at the head of the gulf; even Israel's air links with Africa were insecure. The opening of the straits of Tiran at the entrance of the gulf would compensate Israel for the Egyptian refusal to allow ships bound to or from Israel to use the Suez Canal. But Israel's capacities were not the equal of its intentions. Egypt's new Russian bombers were in a position to bomb Israel's cities and cause panic among the newer immigrants who had not yet become tempered to life in a besieged state. Israel's air force was barely able to defend these cities even with its new French fighters, or to protect Israeli land forces operating in the open desert, and it was quite unable to bomb Egyptian airfields and so prevent the Egyptian air force from taking off. When therefore the French bethought themselves once more of the Israeli attack on Egypt for which they had been supplying arms, they found that the Israelis wanted more than arms. They wanted active French co-belligerence in the form of units of the French air force to be stationed on Israeli airfields for the defence of Israeli cities, and they also wanted British co-belligerence and the bombing of Egyptian airfields by the only locally based bombers capable of the task – which were British bombers using British bases in Cyprus.

The French set about engineering this combined tripartite operation and succeeded. During October French ministers divulged Israel's plans to British ministers; a meeting in Paris on 16 October between Mollet and Pineau, Eden and Lloyd, with nobody else present was particularly important. The British ministers, however, were reluctant to embark on any but the most furtive co-operation with Israel because of the repercussions in the Arab world. But Ben-Gurion insisted on a formal commitment from the British, whom he did not trust, and this he secured at a secret meeting at Sèvres where, on 23 or 24 October, the British foreign secretary joined French and Israeli ministers, subsequently authorizing the signature of a secret tripartite treaty. At this point the Israeli commander-in-chief altered his battle orders, which had envisaged an Israeli raid in force similar to earlier raids but bigger, and proposed instead to commit his forces to open desert upon the assumption that British attacks on Egyptian airfields would give the Israeli troops immunity from air attack.

Israel attacked on 29 October and duly received the anticipated support of Britain and France (the French also prevented the Egyptian fleet from attacking the Israeli coast), but the Egyptian air force was in any event incapacitated since the Russians, who were still in operational control of the llyushin 28s, ordered their pilots out of the battle area. Britain and France also issued an ultimatum to Israel and Egypt requiring both sides to withdraw 16 km from the canal. This was a ruse intended to preserve the fiction that Britain had not colluded with Israel. Neither side paid attention to the ultimatum, Egypt because the canal was 160 km within its own frontiers and Israel because its forces were some way from the canal and were not intended to go there. The Israeli campaign was virtually over on 2 November when its principal objectives – the clearing out of the *fedayeen* bases, the opening of the straits of Tiran and a resounding victory against Egypt – were assured.

A few hours after the initial Israeli attack the Security Council met to consider an American resolution requiring the Israelis to return to their borders. Britain and France vetoed this resolution but the General Assembly, convoked under the Uniting for Peace resolution (a procedure adopted in 1950 on western initiative to enable the Assembly to consider and make recommendations on matters on which the Security Council was stultified by a veto), adopted in the early hours of 2 November an appeal for a cease-fire. At the UN a small group which included the secretary-general, Dag Hammarskjöld, and the Canadian minister for external affairs, Lester Pearson, worked to stop the approaching Anglo–French sea and land attack on Egypt (which, unlike the Israeli, had not yet been launched) and to recover control over an alarming situation by imposing a cease-fire and despatching an international peace force to the area. The Anglo–French attack began with a parachute drop on 5 November. On the next day a seaborne armada from Malta landed troops but on the same day Britain cried halt and France after some hesitation desisted too.

The British decision was the result of an accumulation of pressures, of which one was decisive. Britain, unlike France, was split. The parliamentary opposition, much of the press and a substantial part of the public were opposed to the government's policy. Eden's own party and a majority of the country as a whole supported him consistently except on the use of force, for which there was never a popular majority. The existence of doubts within the government itself was common knowledge. The independent members of the Commonwealth were also split; Australia and, less enthusiastically, New Zealand supported Eden, but Canada and the newer dominions did not (there were as yet no independent African members). This opposition had, however, been foreseen and discounted and for that reason the usual processes of Commonwealth consultation were omitted and Commonwealth governments, like senior advisers in Whitehall and all the pertinent British ambassadors abroad, were kept in the dark. But the decisive reason for calling off the operation was the failure to secure American endorsement and the failure to see what American opposition entailed.

The attack on Egypt caused the biggest financial crisis in Britain since 1945. Britain lost on balance $400 million during the last quarter of 1956; withdrawals were probably half as much again but were partly offset by one or two exceptional influxes which were credited during the quarter. Sterling was healthy and the reserves more than adequate for ordinary purposes, but losses of this order could only be borne for a number of weeks without external aid to preserve the exchange value of the pound. It became clear that Britain would have to borrow to save the pound (which, apart from the war, was not threatened) and that neither the United States nor the IMF would lend the necessary sums until the fighting was called off. Britain could not support an international currency and at the same time conduct independently an aggressive foreign policy. Nevertheless the prophesied run on the pound which panicked Eden and many of his cabinet colleagues into abandoning military operations without warning to their allies was seemingly exaggerated by the chancellor of the exchequer, Harold Macmillan, who was to emerge from the disaster as the next prime minister.

There was in these calculations one other subsidiary element which may have had some effect on some people. This was the entry of the Russians upon the scene. Until 5 November the Russians were too much preoccupied with the suppression of the Hungarian rebellion to take a hand in Middle Eastern affairs, but on that day they proposed to Washington joint action to force Britain and France to desist and threatened vaguely to use rockets against Britain and France. They also indicated that they might allow volunteers to go to the Middle East but statements to this effect were made only after the fighting was over, except in one case when Khrushchev made, at a diplomatic party in Moscow, remarks about volunteers which were not reported in the Russian press. The Russian threat to use rockets – which was taken seriously by British intelligence for a few weeks – was countered by an American threat to retaliate, whereafter no more was heard on this subject. By their intervention the Russians gained a sizeable propaganda victory in the Arab world; it is very improbable that they ever intended anything else.

The Suez War raised Nasser's prestige high. He had secured the leadership in Egypt only in 1954. Suez confirmed him in it and made him a popular leader as well as a military ruler. He had kept his head and his dignity, had emerged intact from an imperialist onslaught and an Israeli invasion, and had demonstrated his power in the Arab world when even Nuri's Iraq had felt obliged to condemn Britain's action and propose Britain's eviction from the Baghdad Pact. Jordan too rejected its traditional British links and the subsidies which went with them, denounced its treaty with Britain and joined instead the Egyptian–Syrian alliance of 1955 (to which Saudi Arabia and Yemen also now belonged). Nasser kept the canal and showed he could work it and got the dam too. The Americans, who might be said to have precipitated the whole affair by abandoning the dam project, had had to come to the support of a régime which they had notoriously ceased to admire and were left in the aftermath without a policy. The Russians were jubilantly claiming all the credit and getting much of it, and they undertook to finance the Aswan Dam in place of the Americans and British.

While the debris was still flying the United States tried to make a fresh start with what came to be called the Eisenhower Doctrine. This venture proceeded from the assumptions that, with the defeat of Britain, it had become necessary for the United States to take some sort of initiative and that the decline of British power had created a vacuum which must be filled by the United States if it were not to be filled by the USSR. Between $400 and $500 million were to be disbursed in two years in the form of economic and military aid to willing recipients who would enter into agreements with the United States authorizing and inviting the use of American arms to protect the integrity and independence of the signatory if threatened with overt aggression from any nation controlled by international communism. Neither Egypt nor Syria was expected to conclude an agreement with the United States on these lines, but a special emissary was sent by President Eisenhower to tour the Middle East and get as many takers as possible. His only success was in Lebanon where, more out of courtesy than enthusiasm, a Christian leader entered into the requisite agreement – and was later to suffer for this decision. King Saud of Saudi Arabia was also polite and paid a visit to Washington but evaded signing any agreement. In Jordan there were anti–American riots. The Eisenhower Doctrine was a new version of the old plan of constructing an anti–Russian front in the Middle East, and its failure was due to the spread of neutralism among Arabs who realized, especially after the lessons of the Suez War, that they were no longer helpless against major outside powers and that the decline of Britain might be followed not by a fresh foreign domination but by none.

REVOLUTION IN IRAQ

During the late 1950s both the Americans and the Russians absorbed this lesson. For the Russians the test came in Syria and then more decisively in Iraq where events seemed to offer opportunities for intervention in a classical mode but brought instead disillusionment. Upheavals in both these countries opened for Moscow the possibility of alliances and even bases in the Middle East from which power could be exercised in imperial fashion. Syria, whence came reports of spreading communism in 1957, entered into economic and military agreements with the USSR, expelled three American diplomats and carried out a purge of the army. Turkey, disturbed by these pointers, concentrated forces on its southern borders. Egypt sent troops which were received in Damascus with acclaim. The tension was temporarily eased but in January 1958 some Syrian officers went to Egypt and asked Nasser to declare a union between the two countries in order to avert a communist takeover in Damascus. The pan-Arab nationalists of the Syrian Ba'ath Party had become alarmed at the growing influence of the Russians and of Syria's principal, and somewhat lonely, communist Khaled Bakdash. Preferring Egyptians to communists they instigated a political move which Nasser, though embarrassed and hesitant, felt unable to reject upon being faced with the argument that an Egyptian refusal would leave no alternative to communism. The creation of a United Arab Republic, consisting of Egypt and Syria, was proclaimed on 1 February 1958. Yemen became loosely attached to it in March.

The Hashemite monarchs in Iraq and Jordan retaliated with an Arab Federation. This union was, however, insubstantial and short-lived, for on 14 July in the same year King Feisal II, the former regent and other members of the royal family, and Nuri es-Said were murdered in a military uprising led by Generals Kassim and Aref. The revolutionaries included communists whose presence at the centres of power alarmed the west and enticed the Russians. The Americans and the British, dismayed by this revolution in the centre of the Baghdad Pact, at once moved forces into Lebanon and Jordan in order to prevent the trouble from spreading, President Camille Chamoun of Lebanon invoking the Eisenhower Doctrine and King Hussein of Jordan the Anglo–Jordanian treaty. (Jordan's request to Britain to intervene was drafted in London.) In Jordan British force saved the monarchy at a time when its fall would have produced turmoil, laying it open to attack from Israel. In Lebanon there was already a civil war in progress which was threatening the religious equilibrium on which the life and prosperity of the country had been based for decades. The Lebanese government had invoked the United Nations in May but by the time of the revolution in Iraq in July the dangers of increasing civil and religious strife had become so great that the marines sent in by President Eisenhower were welcomed by a large majority. In both countries therefore foreign intervention was a stabilizing factor and was not held against the intervenors, especially as they contrived to get themselves out again rapidly with the help of Hammarskjöld. In Lebanon a new president, General Fuad Shehab, a member of an old Maronite family, was able to re-establish the country's traditional equilibrium.

The Baghdad revolution looked at first sight like another link in the lengthening chain of Russian opportunities which included Arab hostility to the Baghdad Pact, the Czech arms deal, the Suez War, the financing of the Aswan Dam, Russian attendance at the Afro–Asian conference in Cairo in December 1957, and a visit by Nasser to the USSR in April–May 1958. Iraq left the Baghdad Pact (which was renamed the Central Treaty Organization – Cento – and moved its headquarters to Ankara). But Kassim was himself no communist, the communists failed to consolidate or improve upon their advantages and Iraq too became more or less a neutralist state where the Russians were as unwelcome as any other major power. The Russians, who had been careful not to offend the nationalists by committing themselves to the communists, swallowed their disappointment and adapted themselves to the temper of the Arab world. They also reverted to more familiar ground. They tried (unsuccessfully) in 1959 to inveigle the shah of Iran away from the western camp, and they began to show an interest in better relations with Turkey after the fall of Menderes in 1960, in which attempt they were later assisted by Turkish disappointment over the American refusal to permit a Turkish invasion of Cyprus. The USSR recognized Kuwait in 1963 and Jordan in 1964, and the presence of Khrushchev at the opening of the Aswan Dam in 1964 reasserted the usefulness of the Russian presence in the Arab world.

The Iraqi revolution also occasioned intervention by China in the affairs of the Middle East or, in Chinese terminology, west Asia. Until this time American and Russian interests in the Middle East had been openly antagonistic. Russian policies in Iran in 1945 on the one hand, and on the other the Truman Doctrine, the Baghdad Pact, American bases at Dhahran in Saudi Arabia and Wheelus Field in Libya and the US Sixth Fleet in the Mediterranean, made the Middle East look like an annexe of the Cold War. In 1956, however, the Americans and the Russians had found themselves in agreement on the need to thwart the British and the French at Suez, and the Russians had even proposed joint Russo–American action; and in the aftermath of the Iraqi revolution of 1958 there appeared some prospect of an open or tacit accord between the two giant powers, an acknowledgement that both had legitimate interests in the Middle East and that some of these interests might be the same. This accord was not to the liking of the Chinese and when an international conference on Middle Eastern affairs was mooted Beijing objected to the holding of such a conference without Chinese participation. The conference never took place. Although Chinese effectiveness in the Middle East was minimal, China began in 1958 to venture there. It gave exclusive support to the Iraqi communists, while the Russians were advocating a front policy; it championed Egypt's gaoled communists, to whom the Russians were turning a deaf ear; it gave some aid to the Yemeni republicans; and, more anti-Israeli than the Russians, it received the Palestinian Arab leader, Ahmed Shuqeiri, in Beijing almost as a head of state and promised him military aid. Zhou Enlai visited Cairo in December 1963 and April 1965. But the Chinese, besides being remote, proved little more popular than other outsiders.

Within the Arab world itself the Iraqi revolution, which was wrongly expected to entail the early disappearance of the monarchy in Jordan, created a new pattern by shifting Iraq from the traditionalist monarchist category into the revolutionary republican one. But the overthrow of the monarchy introduced decades of political instability in which the predominant military power and the civilian political world were split within themselves as well as being opposed to one another. The first military leaders, Abdul Karim Kassim and Abdul Salem Aref, were mutually hostile and divided on many issues, notably relations with Nasser's Egypt and the burgeoning United Arab Republic. Of the main civilian parties the Ba'ath was hostile to Kassim who tried therefore to use the communists to maintain what became an unsuccessful balancing act. The Ba'ath tried to kill Kassim in 1959, secured his downfall in 1963 and elevated Aref to a presidency designed to be more decorative than executive. A year later, however, Aref seized full power with an openly military régime. In 1966 he was killed in an accident and succeeded by his weaker brother, Abdul Rahman Aref. After another two years the Ba'ath returned to power determined to redress the civilian–military balance, albeit under the presidency of the Ba'athist general Hassan al-Bakr. In this endeavour the Ba'ath politicians failed. Under cover of the new president's favour, Saddam Hussein set out to create a powerful modern professional army and turn Iraq into a totalitarian autocracy in the grasp of himself and his extended family. Saddam Hussein's vision of Iraq was not, however, merely clannish for he also aspired to use his personal power base as the motor for a nation state which would become a major actor in the Middle East through military power, economic wealth and nationalist emotions: for some years he courted Shi'ite as well as Sunni Muslims.

THE DECLINE OF NASSER

During the early 1960s Nasser's prestige began to decline from its post-Suez peak. The revolution of 1952 had had a strong socialist strand among officers who were largely of peasant origin and had high, if vague, hopes of the benefits to come from land distribution and agricultural co-operatives, but disappointing results led to a shift to industrial development, nationalization and state controls. In external affairs the fall of the senior (Iraqi) branch of the Hashemite line had brought no gain to Egypt. The union with Syria was not a success: Egypt and Syria had no common frontier; Nasser and the Ba'ath had too little in common beyond a superficial socialism; and the influx of Egyptians into Syria and Nasser's policies of land reform and of forcing all political parties to merge in a single movement or front strained a union which had been a shot-gun marriage from the start; Syrians swung back to the idea that a union with Iraq would suit them better (especially after the Iraqi Ba'ath helped to oust Kassim). In 1961 a short-lived right-wing revolution in Syria brought about the dissolution of the union.

In Yemen, loosely attached to the United Arab Republic from the start, an attempt to overthrow the imamate and instal a republic led to a civil war in which Nasser backed the republican leader Brigadier Sallal without realizing that he was thereby entangling himself in Yemen for several years and to the tune ultimately of 50–60,000 troops. The imam on the other hand was supported by Saudi Arabia, so that the Yemeni civil war developed into a contest between two of the principal Arab states. After two years both sides found the effort unrewarding and in 1965 Nasser travelled to the Saudi capital of Riyadh to meet King Feisal (a relatively progressive member of his house who had displaced his brother Saud the year before), put an end to the civil war and even effect a *rapprochement* between Egypt and Saudi Arabia, the protagonists of the opposing socialist and traditionalist tendencies in the Arab world. Earlier in the same year he had initiated a *rapprochement* with Jordan.

This re-emergence of the theme of Arab unity probably owed at least as much to apprehension about Israel's plans for diverting the Jordan's waters as to war-weariness in Yemen. An American plan in 1955 for an equitable apportionment of these valuable waters had been rejected by the Arabs on political grounds, whereupon Israel had started to construct engineering works which would take water from the Galilee region in the north to the Negev desert in the south. Israel maintained that the quantities to be pumped out of the Jordan's main stream would neither exceed the Israeli quota in the 1955 plan nor leave the lower reaches of the river unduly salty. The Arabs denied both these propositions. They also saw that the Israeli engineering works would come into operation in 1964 and in January of that year they convened a conference in Cairo, attended by traditionalist monarchies and progressive republics alike, in order to concert counter-measures. These measures included the diversion of two of the Jordan's tributaries, the Hasbani in Lebanon and the Banias in Syria – the latter at points within sight and within range from the Israeli frontier; they also included the creation of a unified Arab High Command which, with a useful ambivalence, could be construed either as a means of preventing Israeli attacks on the work on the Banias or as a covert way of keeping the Syrians, the most unpredictable and volatile of the Arab allies, from starting anything on their own. And thirdly, Arab counter-measures included the promotion of the Palestinian Arabs to something approaching sovereign status with a Liberation Organization, an army and a headquarters at Gaza.

This unity was, however, imperfect. To King Hussein the pretensions of the Palestinians amounted to a threat to disrupt the Jordanian state forged by annexing parts of Palestine and to end the monarchy. Syria, Lebanon and Jordan all jibbed at the idea of having Egyptian troops stationed on their soil. Feisal turned out to be a dubious ally. Not only did the agreement of 1965 on Yemen come to nothing, but Feisal began to create a traditionalist or Islamic bloc within the Muslim world. He paid visits to the shah of Iran and to Hussein and ordered arms from the west in alarming quantities. Thus despite pan-Arab conferences at Alexandria in 1964 and Casablanca in 1965 Arab unity and Nasser's role as its leader were in decline in these years.

1967: THE THIRD ROUND

In the decade after the war of 1956 Israel enjoyed its gains without confirming them. The tranquillity on the Egyptian border achieved by the extirpation of the *fedayeen* was maintained and was for a while repeated on Israel's other borders. Eilat flourished, growing from a small town of fewer than 1,000 inhabitants to a flourishing port of over 13,000, trading over a large part of the world and rendering the continuing blockade in the Suez Canal harmless. Israel's warlike spirit was also kept up and, allied with its technical skills, placed Israel in a position to become a nuclear power, should it so choose. Although Arab talk of eliminating Israel went on unabated, it was believed by some to have become a mask, hiding the secret conviction that Israel had come to stay. Israel's reliance on its own might and the divisions of its enemies was reinforced, at second remove, by continuing the alliance with France which had preceded the war of 1956 and by new undertakings from the United States and Britain, as well as France, which followed it. These states declared in March 1957 that they regarded the straits of Tiran as an international waterway and would take action to ensure free passage through them into the Gulf of Aqaba. (The straits were indubitably territorial waters but international law required riparian powers to permit innocent passage for all vessels through such straits if they led to non-territorial waters or to the territory of another state.)

Until 1967 therefore there was some hope that the third round in the Arab–Israeli contest might be postponed *sine die* but by that year growing threats from Syria and from the terrorist organization Al Fatah supported by Syria had changed the climate. An Israeli attack on Syria, leading possibly to a war with both Syria and Egypt, became a matter for practical speculation. The Israeli response to Al Fatah's raids, which began in 1956, was to retaliate in force and by daylight with regular army units against villages in Jordan and Lebanon from which these raids originated. Forays across the Syrian border presented more difficult problems because the lie of the land favoured the Syrians and also because there were no specific limited objectives for the Israeli forces; an expedition into Syria would have been an open-ended *promenade militaire* with no precise stopping place short of Damascus. The brunt of Israel's counter-measures fell therefore on Jordan, notably in a raid on the village of Es Samu in November 1966 in which eighteen Jordanians were killed and much of the village destroyed.

The Egyptian moves which initiated the third round were to a large degree prompted by the state of Syrian politics The Ba'ath Party, founded in Syria in the early 1940s as a pan-Arab party of secular socialism and mild reform – not anti-religious but opposed to the interference of mullahs in politics and to fanaticism of all kinds – had merged in 1950 with the left-wing Arab Socialist Party. It was dissolved (in Syria) in 1958 after the union with Egypt but was in alliance with a group in the army led by General Salah Jadid, an Alawi. The Alawis, a Shi'ite offshoot, one of the principal communities in Syria, strong in the western parts of the country, distrusted by the orthodox Sunnis, by the Druzes and by the Christians, had been favoured by the French during the period of the mandate. The dominant Alawi group in the army established a working partnership with the civilians of the Ba'ath;

this partnership was rendered uneasy by the Marxist and atheistic elements on the left wing of the Ba'ath, but on the other hand the military leaders found the left wing's anti-parliamentary leanings congenial and contrived in any case to secure control of much of the Ba'ath's provincial organization. In 1966 Jadid led a successful military coup but he gradually lost ground to his more subtle colleague and defence minister General Hafiz Assad. In 1970 Syria was placed in a dilemma – the more acute since Syria had no friend in the world except Algeria – by conflict in Jordan between King Hussein and his army on the one hand and, on the other, Palestinians who were encamped in Jordan and suspected of aiming to take over the country. The Palestinians asked for Syrian help which Jadid was willing to give. Assad, however, more circumspect, sent tanks but not aircraft and the Jordanians were able to cripple the unprotected tanks and force them to retreat. As Jadid sought to lay the blame for this humiliation on Assad, the latter carried out a coup, imprisoned Jadid and kept him in prison until he died there more than twenty years later.

Jadid, besides being outwitted by Assad, had been weakened by a scandal caused by the publication of an atheistic article which provoked demonstrations against his government. Anxious to deflect these attacks the government tried to lay the blame on Zionists and Americans, whom it accused of fabricating lies in order to destroy it. The Russians, alarmed that Jadid's fall might entail a conservative reaction, abetted the story of an external threat to Syria. So too for similar reasons did the Palestinians. The cry was taken up throughout the Arab world and even kings Feisal and Hussein were moved to expostulate and promise aid. Israel was said to have moved troops to the Syrian border, a report which was first thought to have reached Syria from Lebanon but may have been given to the Syrians by the Russians. In Israel the Russian ambassador refused an invitation to go and see for himself that there had been no such troop movements.

Nasser had been anything but in the lead during this clamour. He did not like the Syrian government but he could not stand against the tide of Arab feeling and, as in 1958, he was ill informed about the true state of affairs inside Syria and afraid that its government might indeed be about to fall and be replaced by another that he would like even less. His standing in the Arab world had been diminished over the past years, notably by his unsuccessful Yemeni venture; he was being taunted by Jordanians and Saudis for his inactivity; gradually he became convinced that he had to do something. So he took the field. On 16 May he presented a demand for the withdrawal of the UN Emergency Force (UNEF) which had been in Sinai since 1957 and despatched troops to harass UN positions. Two days later, his demand having been rejected by the UN commander who said he had no authority to entertain it,

Nasser repeated it to U Thant. After consulting his advisory committee on UNEF, U Thant complied and on the 23rd UN forces pulled out of Sharm es-Sheikh, leaving Egypt in control of the straits of Tiran. On the previous day Nasser had declared that the straits would be closed to vessels flying the Israeli flag and to contraband of war on whatever vessel (but not to Israeli commerce in non-Israeli ships).

The moves of these days created a clear danger of war between Egypt and Israel. Syria's troubles were overtaken. In 1957 Israel had withdrawn from Sharm es-Sheikh in reliance on western promises to guarantee free passage through the straits, the opening of which had been one of Israel's prime objectives in making war on Egypt in October 1956. Israel had stated that the closing of the straits would constitute a *casus belli* and on the day when the UN troops departed the Israeli prime minister, Levi Eshkol, publicly called on the western powers to implement their guarantee. Washington and London issued statements about international waterways and the British foreign secretary, George Brown, propounded a maritime declaration whose bearing on the crisis appeared, however, remote. De Gaulle proposed four-power talks but the Russians refused. In the early hours of the morning of the 26th the Russian ambassador in Cairo got Nasser out of bed to tell him to go carefully, but the Egyptian moves had caused a wave of enthusiasm and foolish optimism in the Arab world which carried Nasser even further forward. King Hussein arrived in Cairo to make friends with Nasser, a defence pact was signed and a joint command established, Iraq joined the pact a few days later, Egyptian troops moved to Jordan. Israel, which had viewed Egypt's first moves with equanimity, concluded that the danger of war was real, and a few hours after Iraq acceded to the Egypt–Jordanian pact, Israel – after delaying, partly under American pressure, for longer than some members of the cabinet thought prudent – struck the first blow. Victory was swift and total. Although the war has since been designated the Six Day War, Israel defeated both Egypt and Jordan within two days, annexing Jerusalem and occupying all Jordan west of the Jordan River and the whole of the Sinai peninsula. Syria was dealt with later but no less summarily. Nasser, who had moved back to the centre of the stage only to collapse even more humiliatingly than Farouk twenty years earlier, resigned but other scapegoats were found (and executed) and Nasser survived to fall in a later war.

There was much debate over the legal correctness and political consequences of U Thant's decision to withdraw UNEF and to do so speedily. UNEF had been deployed with Egypt's agreement, which was required because the operation was launched under chapter VI and not chapter VII of the UN Charter. But Hammarskjöld had made an agreement with Nasser and the question arose whether, by this agreement, Nasser had abrogated to any degree Egypt's sovereign right to require the removal of the force. On the one hand it has been argued that the effect of the Nasser–Hammarskjöld correspondence was to make the stationing of UNEF on Egyptian soil terminable only by mutual consent. On the other hand it was said that this limitation, if and so far as any had been intended, applied only so long as UNEF was fulfilling its original role of bringing the hostilities of 1956–57 to an end, which role had undoubtedly been completed long before 1967. (It was never part of UNEF's role to keep the straits of Tiran open. This was an obligation of the western powers, if anybody's.) But whatever the true construction of the relevant documents, U Thant had to consider practical matters as well. His Advisory Council was divided, the UN's forces in the field were being forced out of their positions, two of the governments supplying forces indicated that they would withdraw whatever U Thant decided. In these circumstances it is difficult to see how U Thant had any choice. Even if he had prevaricated and delayed, as some of his critics maintained, he would probably have done no more than get some of his own forces killed.

Israel's new conquests made it safer. Although its territory was much larger, its frontiers were shorter and easier to defend. Its troops stood beside the Suez Canal and the straits of Tiran, while in the north the commanding Golan Heights had passed from Syrian to Israeli possession. For many Israelis at this time (but fewer and fewer as time went by) this new territory was not for keeping but for bargaining with. Israel hoped that it now had the power and the counters to force its neighbours to make peace and to recognize a state of Israel with defined frontiers not very different from those existing before the war. Strategic details apart, the one gain to which Israel intended to cling with non-negotiable tenacity was the emotion-sodden old city of Jerusalem. For some months this mood prevailed but it was backed by no effective moves, even though Egypt and Jordan were thought by many observers to have signalled a readiness to negotiate. In August Arab leaders met in Khartoum. Saudi Arabia and Kuwait agreed to make good to Egypt its losses from the closing of the canal and to Jordan its losses from the capture of all its lands west of the Jordan River, which included the revenues provided by tourists and pilgrims visiting Jerusalem. In return Egypt agreed to quit Yemen and to make no fuss about the ending of the ineffective and burdensome boycott which the oil-producing countries had applied against western customers. Unofficial sanction was given for separate and secret discussions by Jordan with Israel. Various peace terms were mooted, including free passage for the Israeli flag through the straits of Tiran and the Suez Canal and eventual Egyptian recognition of Israel. But this piecemeal approach did not satisfy Israel which remained inflexibly intent on a formal peace conference and direct Arab–Israeli negotiations without intermediaries (as opposed to the so-called Rhodes formula by which each side would communicate with the other through a UN or other mediator). Fighting began again. The USSR, which had lost prestige as well as, vicariously, materials of war, decided to rearm Egypt. In October an Israeli destroyer was sunk by Egyptians using Russian weapons. As the exchanges of fire across the canal increased Israel resolved to force Egypt to revert to a cease-fire by massive and deep retaliation but Egypt, instead of complying, called for and got more Russian help. The Russian position in Egypt was fortified; the Americans were correspondingly alarmed.

On the Jordanian front the principal changes effected by the war of 1967 were, in addition to the shifting of the frontier, the influx of a further 250,000 Palestinian refugees into Jordan, a more effective organization of Palestinian guerrilla forces and the conversion therefore of Jordan into a primary target for Israeli attacks. The humiliating collapse of the regular armies of Arab states had intensified the Palestinian belief that it was futile to rely on these states for the recovery of the lands and rights which they had lost in Palestine. They resolved to fend for themselves and, because their cause had a compulsive emotional appeal throughout the Arab world, they had a degree of political influence in Arab capitals out of proportion to their military effectiveness and felt themselves to be in a position to block proposals for an accommodation with Israel which fell short of their irredentist demands. Their principal weapon was the threat to disrupt the kingdom of Jordan where, besides constituting more than half the population, they now also had armed forces. Militarily they were no serious threat to Israel – a further reason why they had to pursue their aims by threatening Jordan – but their guerrilla tactics provoked Israel to retaliations which fell upon the countries which harboured them. In 1968 a section of them took to hijacking aircraft. In July an aircraft of the Israeli line El Al was forced to land at Algiers where its Israeli passengers were held until released through Italian mediation, and in December the Israeli air force destroyed thirteen aircraft on the ground at Beirut in response to an attack by Palestinians on another El Al aircraft at Athens.

In November 1967 Britain had succeeded in getting the support of the United States, the USSR and France for a resolution in the Security Council which condemned the acquisition of territory by force, required Israel to withdraw from its recent conquests and advocated a settlement which would include recognition of Israel and a fair deal for the Palestinian refugees. This resolution (242/67) was endorsed by the Arabs (other than Syria) after some hesitations; it was rejected by Israel. This unaccustomed solidarity among the major powers – long but unsuccessfully sought by France – was facilitated by the fright of Washington and Moscow at finding themselves on opposite sides in a shooting war which, fortunately for them, lasted only six days. Both the superpowers wanted stability in the Middle East. So long as the state of war lasted the Americans were committed, if in uncertain degree, to Israel and therefore debarred from improving their relations with Arab states. By the same token the Russians were able greatly to improve their standing as a Middle East power and their military facilities in the Middle East.

The cessation of American aid to Egypt in 1966 (including free food under Public Law 480) had cast Egypt into the arms of the USSR even before the disasters of the 1967 campaign and so enabled Moscow to bring to a culminating point the capture of vantage points which it had begun after Eden, by adhering to the Baghdad Pact, and Ben-Gurion, by the Gaza raid of 1955, had initiated the Russo–Egyptian alliance. Their gains in this short period gave the Russians an interest in the stability of the Middle East where they had become something of a *status quo* power; and the emergence of an independent Palestinian power, hostile to Arab governments which were friendly to the USSR, and partly nourished by China, threatened to complicate Moscow's diplomatic problems if peace were not made. Moscow also wanted to see the canal reopened for the passage of supplies to North Vietnam and for the flotilla established as a permanent symbol of its expanding reach and pretensions in the Indian Ocean. France, no longer dependent on Middle Eastern oil after the opening of the Algerian and Libyan oilfields, had extricated itself from the Israeli alignment into which the Mollet government had led it and, alarmed by the growth of Russian naval power in the eastern and (prospectively) western Mediterranean, tried to break the Russian monopoly in the Arab world and give Arabs some freedom of diplomatic and commercial manoeuvre by offering to sell arms to Iraq and Libya. As in 1956, so in 1967, France's main concern in the Middle East was the implications of events there upon the balance of power in the western Mediterranean. Britain, still dependent on Middle East oil and still enmeshed in the Persian Gulf, wanted peace on general grounds and particularly for the commercial convenience of reopening the canal. But resolution 242 of November 1967 did not bear fruit.

In 1968–69 a war of attrition was waged on the Suez front to which Israel resolved, early in 1970, to put a stop by making air raids into Egypt to within a few miles of Cairo itself. This tactic failed because it led the USSR to bolster Egypt's defences with Russian missiles and Russian pilots and rocket crews: by the end of the year the USSR had stationed 200 pilots and about 15,000 men in missile crews in Egypt and were manning eighty missile sites in addition to earlier sites manned by Egyptians but also equipped with Russian missiles. Israel was forced to desist. Fighting was once more confined to the canal and its environs. The United States and the USSR pressed their clients to start talking instead of fighting. Israel agreed to talk upon the basis of a withdrawal to pre-1967 limits and Nasser said that Egypt would recognize Israel. Israel agreed that talks might be indirect (through a UN negotiator, the Swedish Ambassador Gunnar Jarring) or direct. In June therefore the American secretary of state William Rogers produced a plan for a cease-fire as a preliminary to indirect negotiations having as their goal an Israeli withdrawal and Egyptian recognition of Israel. Israel disliked this plan, as did the Palestinians; Nasser rejected it. But in July Nasser spent two weeks in Moscow, and Egypt and Israel both accepted the idea of a cease-fire. In August a ninety-day cease-fire and stand-still agreement was concluded: within an area of 50 km on each side of the canal fighting would cease and no fresh units would be introduced, and talks would begin under Jarring's aegis. This agreement was renewed in November and again in February 1971, but it was immediately infringed by Egypt which moved missiles into the stand-still zone. Israel responded by quitting the Jarring talks and demanding massive American aid. It got only part of its demands, waxed indignant with Washington and was forced back to the talks before the year ended.

These talks greatly alarmed the Palestinians who foresaw an Israeli–Egyptian deal, followed perhaps by an Israeli–Jordanian deal, which would leave them out in the cold. They decided to wreck the proceedings. George Habash's Popular Front for the Liberation of Palestine, a small but specially militant group among the dozen or so separate Palestinian organizations, hijacked four American and British airliners, held the passengers and crews hostage and burned the aircraft. In Amman King Hussein, yielding to pressure from his extremist anti-Palestinian advisers, decided on a trial of strength with the Palestinians who were turning his country and his capital into an armed camp and exposing it to enemy attacks. He proved the power of his army which inflicted heavy casualties on the Palestinians but shocked the Arab world by the spectacle of fratricidal war and suffered a political rebuff when peace was restored by the intervention of other Arab states (notably Syria which invaded Jordan) and he had to sign, in the embassy of a foreign state, what was in effect a treaty of peace with the Palestinian leader Yassir Arafat, who was thus accorded the status of a head of state without the inconveniences of having territory to defend and control. The king committed himself to support the Palestinians' aims, though he soon afterwards entered into secret discussions with Israel which were unlikely to produce terms of peace acceptable both to Israel and to the Palestinians.

During the fighting in Jordan Iraqi troops stationed in that country made no move to help the Palestinians. Syria on the other hand sent a small detachment of tanks across the border but withdrew it, probably under Russian pressure induced in its turn by the prospect of Israeli or

American intervention against this intervention. General Jadid's ill-fated gesture recalled his failure in 1967 to defend the Golan Heights against Israel and contributed to the overthrow of the Syrian government later in the year by his rival and defence minister, General Hafiz as-Assad, a leader no less anti-Israel than Jadid but more solicitous about good relations with Egypt and other Arab countries. More important by far, among the consequences of Jordan's internal war, was the sudden death of Nasser from a heart attack caused by his exertions in restoring peace in Jordan. He was succeeded with constitutional smoothness by vice-president Anwar as-Sadat.

DEATH OF NASSER

The death of Nasser removed from the scene the first Egyptian to rule in Egypt since before the days of Alexander the Great. The movement which brought him to power had had complex sources. It sought national emancipation, a spiritual (Islamic) revival, social reform and economic and military modernization. Nasser wanted to rid Egypt of a parasitic monarchy and upper class, to extinguish British domination over Egypt and the Sudan, and to raise the miserable standard of living of the Egyptian people by a more equitable distribution of land, the extension of the cultivable area and the promotion of industries. He acquired the further aims of leading all Arabs against Israel and against Arab régimes deemed reactionary.

The revolutionary movement, in which he was at first only one among a number of leaders, was predominantly but not exclusively a military collectivity in which Nasser soon came to be the leading figure by force of personality and by the elimination of possible rivals. The enemies of the old régime had included elements – the Muslim Brotherhood on the right and the communists on the left – which constituted distinct centres of power. They were suppressed almost as quickly as the pashas of the old order. Political parties were banned. Nasser out-manoeuvred the nominal leader of the coup, Neguib, who was suspected of being insufficiently implacable against some pre-revolutionary élites. The monarchy was abolished; the big landowners were stripped of some of their land; rich business entrepreneurs were hobbled by nationalism; in 1954 Nasser negotiated the removal of British forces from the Canal Zone and British rule from the Sudan. The stage seemed set for the economic reforms which would convert the coup of 1952 into a social and economic revolution.

Egypt's economy was weak at home and abroad. Egypt had too little cultivable land, a one-crop economy (cotton), a stagnant agriculture, no mineral wealth, a very small share of international trade, little industry, little capital with which to develop industries and a population growing at the rate of 3 per cent a year. The relatively small sector of modernized industry was in foreign ownership. There was, however, a native bourgeoisie with some capital resources and Nasser aimed at first to get its co-operation in the development and diversification of the Egyptian economy. But this class had no faith in the new régime and preferred to put its money into unproductive savings at home or abroad rather than venture it in industry. By the end of the 1950s massive unemployment and crushing poverty had been scarcely affected and Nasser turned to other ways. The western refusal to finance the Aswan Dam and the Anglo–French attack on Egypt in 1956 had given him cause to seize the assets of foreign companies, and in the early 1960s he went further and established state control over the greater part of the economy (other than retail trade). He also extended land reform from the modest and largely ineffective first steps of 1952 (when a limit of 200 feddan had been set on individual holdings, a reform which was evaded by various devices such as transferring parts of estates to relations: this reform did not apply to state or religious lands). A five-year plan for the years 1960–65 aimed to increase the gross national product by 7 per cent a year and did in fact increase it by 5.5 per cent, but this improvement was hardly felt by the working population whose numbers increased in the same years by 4 per cent.

Nasser's economic problem was never an easy one and it was made impossible by his foreign policy. He may well have intended in 1954 to devote more of his attention and resources to domestic affairs, but the resolution in that year of his differences with Britain was almost immediately followed by a more aggressive Israeli policy and by Britain's adherence to the Baghdad Pact, which Nasser interpreted as British intervention in Arab politics on the side of adversaries intent on stifling Egypt's revolution. So Nasser found himself increasingly concerned with foreign affairs instead of the reverse. The authors of the revolution had always intended to create a stronger, more efficient and better equipped army than Farouk's. This aim was intensified by the seeming need to defend the country against Israel and the revolution against Nuri's Iraq and its British friends. Nasser's consequent search for arms – and for the means to pay for them – first pushed Egypt's defence spending up beyond all normal percentages of national income and then forced Nasser into borrowing sums which Egypt had little prospect of repaying. Aid and other resources which might have gone into development were appropriated to finance a deficit on external account which increased alarmingly from 1961 onwards. The war in Yemen made matters worse. The continuing war with Israel deprived Egypt of American aid and free food. At the time of Nasser's death Egyptians were materially hardly better off than they had been eighteen years earlier (though there had been some improvement in living conditions in towns); and Egypt itself was in pawn to the USSR.

11 Focus on Lebanon

When Nasser died in 1970 the politics of the Middle East (excluding Iran) were dominated by a bipolar pattern: Israel–United States versus Egypt–USSR. Within a few years they looked very different. The Russian alliance with Egypt was broken. The United States, while not abandoning Israel, was increasingly embarrassed in its role of protector which was barely compatible with the role of peace-maker which it also cherished. The Palestinians acquired a new strength, only to see it gravely jeopardized. The Arab oil exporters gave a startling demonstration of the efficiency of economic sanctions. The Lebanese state was all but destroyed.

Sadat, like Nasser, found himself under pressure from Moscow to reach some accord with Israel on the assumption that Israel was under equivalent pressure from Washington. In fact both the United States and the USSR were oscillating between putting pressure on their clients and acceding at least in part to their demands for aid and arms: by the end of 1971 American aid to Israel, after being reined back for political reasons, was again substantial. After the fighting in Jordan in 1970 and the death of Nasser, Egypt, Syria, Libya and Sudan had agreed to form a new federation of Arab republics, but Sadat's diplomacy was multifaceted. He wished to improve Egypt's relations with the Saudi and Jordanian monarchies and also to effect a reconciliation between Syria and Jordan, which had been within an ace of fighting each other over the Palestinians. He was well equipped for these exercises since he scared the heads of other states less than Nasser had done, and although he made the mistake of assuming and saying that he would bring the Arab–Israeli matter to the point of decision within a year, he contrived not only to establish himself at home but also to consolidate much of the Arab world while at the same time pursuing the policy of edging towards a bilateral agreement with Israel. Having inherited the cease-fire and stand-still of August 1970 and the consequent talks, he offered in February 1971 to open the canal to Israeli cargoes in return for a partial Israeli withdrawal or the convoking of a conference to be attended by the four major outside powers, but neither in this wise nor in the Jarring talks (which Israel again cold-shouldered) was any progress made during the year. In the next year Sadat twice went to Moscow, found he could not get the help he wanted, concluded that Brezhnev had betrayed Egypt by promising Nixon to keep Egypt on short commons, and with wholly unexpected boldness told the Russian specialists and advisers in Egypt to leave. They did so in a matter of weeks, leaving Moscow – which had lost an earlier foothold in Israel – looking for new centres of influence in the Middle East.

THE WAR OF 1973

In 1973 Sadat waxed even bolder. His Arab fence-mending complete and his hopes for a bilateral agreement soured by increased American aid to Israel and increased Israeli settlement in the occupied areas contrary to international law, he decided to take the offensive, and on 6 October Israel and the rest of the world were taken by surprise by an Arab attack on two fronts. The Egyptian army attacked across the canal and pierced Israeli positions, but these successes were nullified when the Egyptians ventured to operate beyond air cover and the Americans rushed aid to Israel by every available military and civil aircraft. The Russians, determined not to be left out, sent aid to Syria, Iraq and Egypt. Nine days after the first shots Israel counter-attacked, found a gap in the middle of the Egyptian front and, crossing the canal westward, surrounded an Egyptian corps. The Egyptian stroke had been stymied and the next day Kosygin arrived in Cairo. In effect the war was over, although there was still some fighting to come. In the north it was even shorter. The Syrians, with the help of Jordanian, Iraqi, Saudi and Moroccan units, attacked the Golan Heights but were held and pushed back after two days.

Stalemate suited the United States and the USSR, the latter as soon as it became clear that the Arabs were not going to win and the former because, once Israel had been saved, the principal American concern was to prevent an Israeli counter-attack which would provoke a more forceful Russian riposte. Kissinger accordingly was invited to Moscow whence he proceeded to Jerusalem, and the United States and the USSR jointly presented to the Security Council a resolution requiring a cease-fire, implementation of UN resolution 242/67 and peace talks under 'appropriate' auspices. Egypt and Israel accepted this resolution and it was adopted by the Council on 22 October. The Russians had privately proposed to the Americans that American and Russian troops be sent to the Middle East but Washington rejected this idea emphatically. It was, however, reintroduced by Egypt either spontaneously or at Russian instigation, and the USSR then publicly supported it. The American reaction was extreme. All American forces throughout the world were brought to the most advanced state of alert, whereupon the USSR retracted. Washington had demonstrated implacable opposition to the arrival of Russian units in the Middle East; but some European allies were disturbed by what they deemed American over-reaction, provoking in return from Kissinger some acid comments on their nerve and reliability.

There were two ways of winding up the war – a conference or diplomatic exchanges. A conference was convened at Geneva at the end of 1973 and was dignified by the presence of Kissinger, Gromyko and the secretary-general of the UN, Kurt Waldheim, but the ensuing year was devoted primarily to personal diplomacy by Kissinger while the conference was held in abeyance like a net under an acrobat. A first Israeli–Egyptian disengagement agreement was reached in January 1974; both sides withdrew and a UN force took station between them. On the northern front negotiations were slower because Israel insisted on proper notification of the numbers, names and fates of Israelis captured by the Syrians, which the Syrians either would not or could not give, and because it was impossible for Israel to withdraw more than a mile or two without endangering its entire strategic position and its settlements within range of the Golan Heights. Nevertheless Kissinger, by persistent shuttling between capitals, succeeded in getting a first agreement here too. (Both agreements were extended in November for six months and then again for another six months.) At the next stage the main crux was the status of the Palestine Liberation Organization (PLO) and its demands. Israel refused to recognize the PLO as anything but a terrorist organization or the Palestinians as anything but refugees. Israel also refused to consider territorial retreat except in the context of a general peace agreement. The PLO's proclaimed position was that it did not accept the existence of the state of Israel, but it was supposed that it might be brought to negotiate on the basis of the borders obtaining before the war of 1967. It insisted on being a principal party in negotiations as opposed to having the Palestinians represented by some Arab government. Israel wanted a series of bilateral arrangements beginning with Egypt (and sidetracking the Palestinians). Sadat was disposed to seek such an agreement provided it was quickly followed by similar agreements between Israel and its other neighbours, so that Egypt might not be accused of breaking Arab ranks or leaving the Palestinians in the lurch. Kissinger wanted an Israeli–Egyptian agreement as quickly as possible; he wanted the United States and not an international conference to be the peace-maker of the Middle East; and he wanted to fortify the American–Egyptian link without too greatly offending Israel or American Jewry: everything else could wait. Kissinger saw an Israeli–Egyptian peace as a stage towards peace in the Middle East and he did not believe that peace in the Middle East could be attained except by stages. Peace moreover was increasingly important for the United States, whose attempts to get away from a policy focused on support for Israel had been encouraged by Sadat's break with the USSR and then made urgent by the Arabs' use of oil sanctions during the war. The weakness of Kissinger's tactics was his emphasis and his dependence on Egypt which he was pressing into a bilateralism offensive to Egypt's Arab allies and into a disregard of the Palestinians whose claims remained none the less more central to the Arab–Israeli conflict than such purely Egyptian–Israeli questions as their common frontiers or the passage of Israeli cargoes and vessels through the Suez Canal. In order to succeed Kissinger had to divorce, and get Sadat to divorce, Egyptian–Israeli issues from wider Arab–Israeli issues and above all from the Palestinian cause. He risked therefore damaging his new Egyptian friend throughout the Arab world.

The Arabs were in no mood to abandon the Palestinians, however much they might wish that the Palestinians had never been heard of. The Palestinians brought trouble and danger, but at an Arab conference in Rabat in October 1974 even their most ruthless enemy, King Hussein, who had been firing on them four years earlier and had forced 15,000 of them to flee from Jordan to Syria, joined with everybody else in acknowledging that the PLO represented Arab Palestine; and from Rabat Yassir Arafat, chairman of the PLO's executive council, proceeded to the UN General Assembly which received him and listened to him as though he were a head of state and adopted a strongly pro-Palestinian resolution affirming the rights of the Palestinians to sovereignty inside Palestine and to be principals at a peace conference. In the winter 1974–75 it seemed likely that the PLO would declare itself a government in exile and be recognized by a majority of the United Nations, despite the lesions which weakened its claim to represent all Palestinians and despite the extremism of splinter groups whose indiscriminate kidnapping and killing shocked most people. (The PLO was formed in 1964, came to prominence in the wake of the war of 1967, and had its headquarters in Jordan until 1970, in Lebanon until 1982 and in Tunisia until 1994. Its strongest component, created in the 1950s, was Al Fatah which came to prominence from 1965 onwards by staging commando operations. The more militant PFLP of George Habash crystallized out of earlier groups in 1967. From it later broke away the PDFLP – People's Democratic Front for the Liberation of Palestine – and the PFLP High Command. Al-Saiqa, the organization of the Palestinians in Syria, was created in 1966 with the support of the Syrian Ba'ath. And there were other groups.)

If the war of 1973 gave the Palestinians a boost and a chance to recover from their drubbing in Jordan in 1970, it also boosted Arab morale and hopes in other ways. Quite apart from showing that the Egyptian army had become a match for the Israelis (at least for a week), it demonstrated Arab power against greater states. The Arab oil producers, banded together in the Organization of Arab Petroleum Exporting Countries (OAPEC, a division of OPEC), shocked the whole of the developed world by cutting off oil and raising its price. These measures were very effective and there was every reason to believe that in another emergency the oil weapon would be used yet more effectively.

For the oil-producing states oil was not merely, or even primarily, a weapon. It was above all a source of revenue and for some states the only or overwhelming source. All of them were concerned to safeguard and increase their oil revenues and in 1971 they had made, at Teheran and Tripoli, agreements to secure these ends for the next five years. They also insisted through OPEC in 1972 that purchasing countries which devalued their currencies should revise the terms of these agreements so that the producers should not lose by the devaluation. They also changed their attitudes towards the way in which they would take their profit. Whereas hitherto they had done so mainly by fiscal measures, by taxes on offtake at the wellhead, they now moved towards participation or part ownership of operating companies and resolved to appropriate 25 per cent of the share capital of major companies, rising to 51 per cent by 1982. They felt moreover that what they set out to get they would get because – even before the war of 1973 – the developed world was worrying about an energy shortage amounting to an energy crisis. This crisis began to push up prices alarmingly in 1970–71 and made the producers more conscious of the advantages of restricting production: restrictions on output would prolong the life of reserves of an irreplaceable asset, put the screws on customers, and yet maintain producers' revenues if prices were raised – as they easily could be in a seller's market – to compensate for a lower volume of sales. Importing countries were vulnerable either way. Restrictions, let alone a stoppage, would jeopardize their industry and their daily life; price increases would disfigure their balance of payments. (The companies, however, were not in the same position as the importing countries. They suffered only one way. Restrictions and stoppages harmed their profits, higher prices did not.)

When the war began the Gulf producers raised their prices by 70 per cent. OAPEC threatened to cut deliveries by 5 per cent a month until Israel undertook to evacuate Arab territory and accede to the legitimate rights of the Palestinians. Saudi Arabia cut deliveries by 10 per cent; to the United States and the Netherlands, Israel's main champions, totally. Iraq nationalized parts of certain foreign companies and Libya talked of total expropriation. As a result of this multifarious, if ill co-ordinated, Arab action oil prices in Iran, Nigeria and elsewhere shot up. Also as a result the EEC expressed its sympathy for the Palestinians, Britain stopped supplies of arms to Israel and Japan reversed its pro–Israeli proclivities. Arab threats did not fully materialize – it was, for example, not possible to boycott the Netherlands when oil despatched to other countries could be transshipped to Rotterdam – and within the Arab world there were some who feared the effects on themselves of harming western economies which were their principal customers. But the point had been made and taken. A long-standing major factor in international affairs had been shaken. Primary producers were no longer the exploited but, where there was co-operation and shortage, could call the tune and use economic leverage for political ends. The change was so startling to public opinion in the industrial world that it was greeted with cries of blackmail. For the Middle East it meant that so long as there was an energy crisis Arabs could not be trifled with.

Thus the Kissinger–Sadat axis was a brittle one, and although Kissinger's stage-by-stage diplomacy was admirably motivated, it was less well grounded in practical terms. The arguments which Kissinger could bring to bear on Sadat without discountenancing him in his own country were limited by the availability of so efficacious a new weapon and by the enhanced prominence of the Palestinians. Nor were the Israelis on their side to be pressured so easily. They had not been defeated. They had on the contrary satisfied themselves that in spite of initial reverses they had been on the way to another victory. But they had been severely shaken by the failure of intelligence services which, in so small a country, were vital to prevent instant obliteration. In elections at the end of 1973 the government parties lost six seats; the gains were on the right but there was increased support too for doves critical of Israel's Arab policies. The chief of staff and other senior officers resigned. Moshe Dayan, ultimately responsible for national security as minister of defence, lost some of his hero's image and was left out of the new government formed in April 1974 when General Yitzhak Rabin succeeded Mrs Golda Meir as prime minister. The spectacle of an American president touring Arab capitals in June was a disturbing one, even though Nixon came to Jerusalem too. The speed with which the USSR more than made up Syria's and Iraq's losses in arms was more disturbing, even though the United States did the same for Israel. At the end of 1974 Syria, by proposing a new joint military command with Egypt, Jordan and the PLO, emphasized the underlying unity among Israel's neighbours and their commitment to the Palestinians. Perhaps Israel had never been so unsettled. This was not a situation in which an American secretary of state could wield too big a stick, certainly not in public view.

Kissinger renewed his shuttle diplomacy early in 1975 with the aim of getting a further Israeli–Egyptian disengagement agreement. At first he was unsuccessful, owing mainly to Israeli obduracy. Sadat, after a meeting in May with President Ford at Salzburg, announced the reopening of the Suez Canal and in September a new agreement was reached. Fresh front lines and force levels were accepted; the keeping of the agreement was to be monitored by a string of early warning stations, some of them manned by American civilians; Israel abandoned the Mitla and Giddi passes and Egypt's oilfields in Sinai, in return for which it was promised massive American aid and the passage of Israeli cargoes (but not vessels) through the canal. This was the crown of Kissinger's exertions and by comparison it seemed to matter little that on the northern front the greater acerbity of the adversaries and Israel's persistence in planting new settlements in occupied territory prevented any equivalent relaxation. Yet it was in this area that the next crisis was about to occur.

CIVIL WAR IN LEBANON

One of the recurrent features of Middle Eastern politics from 1919 onwards was the attempt to forget about the Palestinians. From the time when Emir Faisal and Chaim Weizmann talked at the Paris peace conference without troubling about them the Palestinians had been overlooked or worse, but they refused to disappear or lose their identity. By 1975 there were rather more than 3 million of them, of whom close on half lived in Israel, or under Israeli occupation. Of the remainder 750,000 were in Jordan, 400,000 in Lebanon, 200,000 in Syria. Through half a century they had persistently posed political problems, tugged at consciences (non-Arab as well as Arab) and resorted to violent means – including in the 1970s guerrilla operations, kidnapping, murder and hijacking – in order to draw attention to themselves and their grievances. After their pounding in Jordan in 1970 the centre of these activities was Lebanon.

Lebanon has been called the Switzerland of the Middle East. This meant two things: first, that the Lebanese were highly successful merchants and bankers, good at making money, richer than their neighbours and rich because of the services they rendered rather than from domestic natural resources; and secondly, that the country was a patchwork of communities held together by political skill and tolerance in the service of material self-interest. The patchwork was imposed by geography as much as history, and the cohesion of the patchwork was a condition of the survival of the state. Two parallel ranges of mountains, one close to the sea and the two separated by a narrow valley, cut the country into vertical strips which are again divided by transverse barriers, and within this grid separate communities had preserved their individualities and mutual hostilities. These had been sharpened by religions – Muslim and Christian – and by further divisions within religions: Sunni, Shia and Druze; Maronite, Greek Orthodox and Greek Catholic. They had also been sharpened by varieties of economic experience, for not all Lebanese were rich. The richest group were the Maronite Christians who held in addition the presidency, given to them by the constitution and at a time when Christians as a whole outnumbered Muslims. (A decline in the Christian proportion of the population was one of the ground factors in Lebanese tensions.) The other Christian communities were neither so rich nor so powerful, but so long as Lebanon presented the appearance of a prosperous mercantile community in the guise of a state, by so much was it Christian rather than Muslim.

The largest of the Muslim communities were the Shia, but the Sunni, who monopolized the post of prime minister by constitutional right, were more influential politically. The Shia, many of whom had drifted from being poor countrymen to becoming poor townspeople, felt themselves to be equally neglected by Maronite Christians and Sunni Muslims. They were most exposed to Israeli incursions. The Druzes on the other hand, who still preserved much of the fierce exclusiveness which had marked their origins in the eleventh century, were confident and assertive because, though for the most part poor and frequently indignant, they formed a compact and well-knit society in a hereditary mountain homeland and under a hereditary leader, Kamal Jumblatt, who commanded their loyalties and expressed an appropriate left-wing have-not philosophy.

Into this delicate cat's cradle of criss-crossing confessional and economic tensions had come in 1948 refugees from Palestine who, by 1975, numbered about 400,000, mostly housed (an ironic word) in dreadful camps. They included some five or six thousand active gunmen whose aim was the recovery of lost lands in Palestine by any effective means and the extinction – if that were necessary, as it was assumed to be – of the Israeli state. Since such an aim could never be attained by a few thousand armed men it had to be prosecuted through Arab states which had to be induced to make war on Israel. Since these states were only half willing to do so, and since when they ventured they got beaten, the Palestinians had a survival problem which led them to look even further afield for aid and sympathy, notably to the leading left-wing powers, China and the USSR (although the latter preferred to put its money on governments rather than on movements which were attacking governments), and to universal left-wing opinion which would rally to dispossession and destitution and not be too gravely offended by the use of terror as a weapon justified by desperation. Within Lebanon Palestinians could count on support, not from the political and mercantile élite which regarded them as a nuisance, but from other sections of the population which had to recognize them as fellow Arabs and fellow underdogs. By the 1970s the Palestinians were more than a nuisance. Their militant organizations, encamped on Lebanese soil, regarded themselves as being at war with Israel, which was Lebanon's neighbour, and did not hesitate to strike back when attacked from Lebanon. The Maronites in particular, as conservative Christians, disliked and feared the Palestinian organizations because they were left-wing, militant, Muslim and a threat to the stability of the state which gave the Maronites their wealth and their influence. It was an important coincidence that the Maronite position in Lebanon was already under some threat from the country's changing demographic balance and from the nemesis which (cf. Northern Ireland) haunts a too prosperous exclusivity.

In 1975 a group of Muslim fishermen who felt aggrieved by a concession granted to a group of Christian fishermen staged a demonstration which turned into an affray. Some Palestinians took sides with the Muslims. Christians saw in this episode a writing on the wall: the Palestinians, already a standing incitement to Israeli aggression against Lebanon, were now interfering too in the balance of internal Lebanese politics. Anti-Palestinian incidents followed, caused by the Phalange, a right-wing Christian faction. Violence escalated; the government, always weak in a crisis because of conflicting loyalties implanted in it by the constitution, fell; the army, small and itself divided, failed to keep order. The disorder, which had begun as faction fights, turned into a battle for territory. The Palestinians, who had arms in plenty, became more and more involved (against the wishes of some of their leaders) and their opponents more and more provocative. There were innumerable truces of insignificant duration. Battle having been joined, opportunity beckoned to both sides. A left-wing alliance of Druzes and Palestinians seemed set to take over the country, the Druzes in order to supplant the existing ruling groups and the Palestinians in order to make Lebanon safe for Palestinians and a base against Israel. On the other side the Maronites and their allies sensed the chance to do to the Palestinians in Lebanon what King Hussein had done to them in Jordan: expel as many as possible, kill some leaders and perhaps even, as tempers became uncontrollably bitter, exterminate them with Israel's help.

As the civil war threatened to entail the destruction of the state Syria had to consider what to do. Lebanon had been a part of Syria (as had Palestine) under Ottoman dominion and before, but Syria hesitated to intervene for a number of reasons. If the Syrian army were to go into Lebanon the Israeli army would most probably do so too and a new conflict would be staged, unwelcome to Syria and to Syria's Russian patrons. Nor would Jordan, with which President Assad had been improving relations for some years, take kindly to any move which looked like a step towards re-creating Greater Syria. The same argument applied at least as strongly to Iraq. Syria therefore, though inclined as a Muslim and left-wing state to favour the Druze–Palestinian side, was more concerned to end the fighting and put the Lebanese state back on its feet; and Assad concentrated on finding a way of restoring the constitutional proprieties under a Maronite president (though a new one), even if this entailed attacking the Palestinians. Syrian intervention took therefore an anti-Palestinian and pro-Maronite flavour which outraged other Arab states which, suspicious of Syrian designs, tried to substitute pan-Arab for Syrian management of the crisis by introducing a mixed Arab force and an Arab League mediator. But these moves had little substance. Assad chose his moment to put his own troops in and imposed a provisional settlement whereby the Maronites would retain the presidency *de jure* but the president would no longer choose the prime minister and the Muslims would have equal representation with Christians in the parliament. This was a restoration which amounted to a defeat for the leftish Shi'ites and Druzes (Kamal Jumblatt was killed, probably murdered) who had scented power only to be denied it by foreign intervention; and it was a defeat for the Palestinians, who had made common cause with the Druzes and other anti-Maronite elements and in doing so had backed the losing side and been battered first by the Maronites and then by the Syrians. It was also a defeat for the Russians who, having perforce shifted their pivot in the Middle East from Cairo to Damascus, saw Syria acting with a degree of independence and in a direction which were both unpalatable to Moscow. Yet Moscow gave no help to Jumblatt or Arafat; its influence in the crisis was minimal. Assad, not Arafat, had become the dominant force in the state. Lebanon itself was in ruins and in tutelage. It could eject Syria from its system only at the price of letting loose new conflicts between the private armies of Maronites, Druzes and others, which no purely Lebanese army could control. The Maronites, although rescued by Syria, were no less hostile to it and remained prone to flirt with Israel in spite of the fact that this flirting fatally divided them. The Palestinians, having been attacked by Jordan and by Syria, having lost their stronghold in Lebanon, and conscious that Sadat's sympathy for them was offset by Egypt's contrary interest in peace with Israel, had once more nobody to rely on but themselves. Arafat had looked for support in the wrong quarter. Disillusioned with Arab governments and nervous about leftish breakaways from his own movement, he had formed an alliance with the Arab left and so become a target for Arab governments determined to annihilate the left.

The civil war in Lebanon, which lasted from April 1975 to November 1976, cost 40,000 lives and huge material and commercial damage. The new president, Elias Hrawi, less committed to the Maronite ascendancy in its old form than his predecessor Suleiman

Franjieh, lacked the authority and the armed power to make his
government's writ run. The fighting was brought to a stop by the
Syrians but Assad's position in Lebanon was both taxing and
embarrassing. Syria was reviled by the rest of the Arab world; Iraq cut
off oil (but Saudi Arabia provided credits to fill the gap). Assad had
neither the means nor probably the wish to take the whole
responsibility for Lebanon's affairs. He reverted to negotiation. At
Shtoura in July 1977 Syria, Lebanon and the PLO reaffirmed a
Lebanese–PLO agreement made in Cairo in 1969 whereby the PLO
would withdraw its armed units from Beirut and the frontier zone in
the south. This disarming of the Palestinians was put in train in Beirut
but not in the south where fighting started again between Palestinians
and Lebanese Christians. The Israelis invaded and with their help Major
Saad Haddad established (in a largely Muslim area) a semi-independent
Christian state of Free Lebanon. Neither the Lebanese army, always
helpless in the face of communal strife, nor the UN Interim Force in
Lebanon (UNIFIL) could do anything to counter this creation of an
Israeli puppet state, while in the rest of the country private armies
resumed their battles and assassinations. There were about forty armies
or armed bands in this small state. In 1980 battles between the two
principal Maronite forces resulted in victory for the Phalange and the
Gemayel family over the National Liberal Party of the Chamoun and
other prominent Maronite clans. The victors were the more
determined to remove from Lebanon all Palestinians, civilian refugees as
well as armed units, and they had Israel's armour to help them. A new
no-go area, like Major Haddad's and beyond the reach of the
government, was created in and around Beirut, a Phalangist
autonomous canton.

SADAT IN JERUSALEM: CAMP DAVID

This disintegration of Lebanon was consciously promoted by Israel.
South Lebanon was part of the lands claimed for Israel by Zionist
irridentism, but this claim might have remained muted if the area had
not become during the 1970s a refuge of Palestinian revanchism. Israel
was determined to eradicate the militant arm of the Palestinians by
force, and most Israelis regarded any aggressive action for this purpose
as a legitimate exercise of the right of self-defence. In its earliest days
the threat to Israel's existence came from Arab states. Israel believed
that, by its superior efficiency and American aid, it could always defeat
enemies who would be perpetually at odds with each other. At this
stage the Palestinians were no more than a minor instrument of the
Arab states, useful mainly for propaganda purposes. But they became
the central issue, evolving a fighting force of their own, attracting

international sympathy and commanding the support of Arab states – a reversal of roles which turned them into a more persistent danger to Israel than any Arab government.

Israel's increasingly aggressive response was a recognition of this change. Whereas Arab states could not be destroyed, the political power of the Palestinians might be if enough Palestinians could be killed or dispersed. A change of government in Israel helped to put this reasoning into practice. During 1976 the Labour-dominated coalition struggled in vain with economic problems and disorders on the West Bank. Splits developed within the coalition and within the Labour Party itself. At the UN the United States was increasingly alone in vetoing anti-Israeli resolutions in the Security Council and eventually joined the majority in condemning Israel's ruthless and illegal behaviour in the West Bank. The prime minister Yitzhak Rabin resigned when his wife was accused of a currency offence but his successor Shimon Peres did not survive a general election and in 1977 Menachem Begin, veteran fighter against the British and biblical fundamentalist, formed Israel's first government without the Labour Party. He declared that Israel must keep the West Bank and should accelerate the creation of new settlements there. President Carter, among others, called these settlements illegal. He proposed that the Geneva conference be reconvened (with Russian participation) and that the PLO be invited to attend it on condition that it recognize Israel's right to exist. Begin was willing to see Palestinians at a conference but adamant that he would never at any point negotiate with the PLO, thus allowing Arafat to avoid giving a straight answer to Carter's question. In November Begin, to general amazement, invited Sadat to Jerusalem and Sadat, who was thought to have contrived the invitation, possibly through Ceauşescu, at once accepted. He badly needed peace. He had fought a four-day war with Libya on his other flank; the Egyptian economy was in tatters; and domestic disaffection had turned into riots.

Addressing the Knesset with a directness and modesty which attracted worldwide praise Sadat deplored delays in reconvening the Geneva conference and appealed for peace between Egypt and Israel, Israel's withdrawal to its 1967 frontiers and the right of self-determination and statehood for the Palestinians; he did not mention the PLO by name. Before the year ended talks between Egypt and Israel began in Cairo and Begin visited Ismailia where he spoke of an autonomous Palestinian entity with an Israeli military presence for twenty years. His formal proposals for the West Bank, published at the very end of the year, envisaged continuing Israeli control of security and public order, a choice for the inhabitants between Israeli and Jordanian (but not Palestinian) citizenship, Arab immigration in 'reasonable numbers' to be agreed between Israel, Jordan and the new authority of the autonomous territory, and freedom for Israelis to buy land. Carter approved this response, thus opposing the creation of a Palestinian state.

This was the beginning of the process which ended the war between Egypt and Israel but, contrary to the expectations roused on many sides, did nothing for the Palestinians. Begin, while prepared to abandon valuable forward positions and defences in Sinai (which was demilitarized) had no intention of abandoning control over the West Bank; he proposed on the contrary to accelerate its demographic transformation by encouraging settlements regardless of their strictly military justification; the Jewish population of the occupied West Bank quintupled in three years of his government. Egypt secured a much needed peace and the return of the territory and oilfields in Sinai which it had lost in 1973. But Sadat was violently attacked by almost the whole of the Arab world for failing to make this peace settlement with Israel contingent upon an equitable deal for the Palestinians. He lost his Saudi subsidies and became dependent on American charity.

Carter, having failed to reactivate the Geneva conference, became an eager champion of this strange Israeli–Egyptian concord which he saw as the forerunner in a series of agreements between Israel and all its circumambient enemies. The United States assumed the role of a principal negotiator with Israel and Egypt and at a tripartite conference at Camp David in September 1978 got both Begin and Sadat to subscribe, alongside the bilateral peace terms, what was described as a framework for a Middle East settlement: Israel was gradually to leave the West Bank (and Gaza), creating an autonomous Palestinian entity with a temporary Israeli presence, all other problems to be deferred for five years. Begin and Sadat – like Kissinger and Le Duc Tho some years earlier – were awarded the Nobel Prize for Peace, thus repeating the curious deviation by which the prize was bestowed on warriors who mended their ways rather than on more exemplary men of peace.

The euphoria generated by Sadat's appearance in Jerusalem in 1977 lasted about a year. American pressure on Jordan and Saudi Arabia to applaud what had been done and to further it failed completely. The Arabs were divided into the hostile and the very hostile. Israel continued its economic decline (its rate of inflation became one of the highest in the world) and pursued its tough policies on its northern borders and the West Bank. Begin's cabinet lost its more eminent members and its popular support; General Dayan, who had crossed the floor when Begin won the election of 1977, deserted him over his government's insistence on promoting settlements which had a political but no strategic purpose, and General Ezer Weizmann did so too on similar grounds. But although Begin's abrasive and unyielding manner lost him friends, there were few signs that his basic objectives had become unpopular. The election of the outspokenly pro-Israeli Ronald

Reagan to be President of the United States was a relief for Israelis who had begun to ponder the future policies of an American administration forced to choose between its devotion to Israel and its material and strategic interests in Saudi Arabia and the Gulf. And the war launched by Iraq against Iran in 1980 readvertised the fissures among Israel's enemies as some Arab governments supported Iraq and others did not.

At Camp David Begin had bought off Egypt, a rare diplomatic feat in Israel's history of military triumphs. But he secured his southern flank as part of an expansionist policy in other directions. Israel kept its hold on the occupied West Bank with little difficulty. Although it was obliged to use harsh measures which did it no good in the outside world, internally there was minimal Israeli criticism of either the occupation or its seamier side, and Begin used his freedom of manoeuvre to the south and east to launch a war northward into Lebanon. A narrow electoral victory in 1981 left him dependent in the Knesset on small and myopic religious parties whose impact on policy was both hard and adventurous. Israel's creeping colonization of the West Bank became more brutal and precipitate – the Jewish population was to be increased from 20,000 to 120,000 by 1985 – and the Israeli air force displayed its muscle by destroying a nuclear installation near Baghdad in June and raiding Beirut in July: a left and right against Iraq on the one hand and Syria on the other. The former exploit, designed to pre-empt or delay the emergence of Iraq as an offensive nuclear power, was condemned at the UN even by the United States which delayed, by a few token weeks, the delivery to Israel of new military aircraft. The attack on Beirut was prompted by discord within Lebanon over the presence of Syrian forces – the so-called Arab Defence Force (ADF), created to keep Lebanon in one piece and independent, but a force which was Syrian in all but name. The ADF was challenged by the Christian Phalange which was hostile to Syria and alarmed about the gradual decline of (mainly Maronite) Christian *vis-à-vis* Muslim power in Lebanon. When at the end of 1980 the Syrians evacuated Zahle to the east of Beirut the Phalange forestalled the regular Lebanese army and occupied it. The Syrians supported the army's attempts to oust the Phalange. Fighting spread. Israel saw its opportunity. Besides raids into southern Lebanon it flew demonstrative sorties over Beirut and when Syria then introduced ground-to-air missiles it attacked them. Israel hoped to propel President Bashir Gemayel, who was not only a Maronite Christian but also the son of the founder of the Phalange, to take sides with that militant body and so entrench a Christian and anti-Syrian statelet in parts of Beirut and its surroundings. But President Gemayel was persuaded by Syria and other Arabs to denounce an Israeli–Phalange alliance. Torn between Syria and Israel, the Maronites were on the slippery slope to the loss of their once dominant role in Lebanon.

At the end of 1981 the Saudi prime minister, Prince Fahd, presented a plan for peace which required the withdrawal of Israel to its 1967 borders and implied the recognition of the Israeli state within those limits. This plan was anathema to Israel but was aborted by the Arabs themselves when President Assad refused to go to a conference at Fez to discuss it. King Hussein strongly supported the Saudi initiative, patched up his quarrel with Morocco by his first visit to that country for five years, but displeased President Reagan by his insistence on including the PLO in any peace process notwithstanding that the Palestine National Council refused formally to abate its demand for the elimination of the Israeli state.

Reagan was still clinging to the view that peace could be achieved without the concurrence of the PLO but his commitment to Israel and to his predecessor's Camp David programme received two knocks at the end of 1981. In October President Sadat was assassinated. More of a hero outside his country than in it, Sadat offended many Egyptians by his style as well as his policies. A notable feature of his death was the peaceful transfer of power to Hosni Mubarak, demonstrating the comparative stability of Egypt's political system. The new president, although too cautious to make any abrupt changes of policy, was concerned to garner the full fruits of the Camp David agreement so far as they related to Israel's retreat from Egyptian soil (which was completed in April 1982) but also to mend Egypt's relations with the rest of the Arab world which had been wrecked by Sadat's incautious acceptance of the agreement's vague reference to Palestinian claims. Where Sadat had flaunted western ways Mubarak was decorously Islamic; he relaxed Sadat's (and Nasser's) repression of political opponents, won foreign aid for Egypt's public services and quietly restrained military expenditure. By 1984 he felt strong enough to legalize a revived, if fairly tame, Wafd and allow it to take part in a general election. Reagan, secondly, was obliged to recognize some linkage between different parts of the Middle East and so implicitly to modify his belief that the Palestinian issue could be kept in a tight compartment. Concerned by the extension of the Iraqi–Iranian war to the Gulf he agreed, and narrowly persuaded the Senate to agree, to the sale of AWACS intelligence aircraft to Saudi Arabia (for delivery in 1986) but at the cost of giving Israel still more military aid, an agreement for 'strategic co-operation' and an enhanced capacity for independent adventures. By the same token the American room for manoeuvre in the Middle East was becoming more inhibited even before Israel marginalized all peace plans by launching on 6 June 1982 a full-scale invasion of Lebanon.

ISRAEL'S INVASION OF LEBANON

The ostensible and limited object of this operation was to root out the Palestinians from south Lebanon who were harassing Israel, but there were deeper causes and wider ambitions. For Menachem Begin and Zionists of his kind parts of Lebanon belonged to Israel by divine and biblical right and all Israel's governments since the creation of the state had mooted the destruction of Lebanon by arraying its Christian against its Muslim communities. The spark for the invasion was provided by the attempted assassination of the Israeli ambassador in London. Its scope was immediately made clear as the Israeli forces advanced rapidly from southern Lebanon to the outskirts of Beirut. It was a military success but a political disaster. In the first stage Israel failed to secure the eviction of the PLO from Beirut since the Israeli government shrank from an assault on the capital which would entail heavy Israeli casualties and American displeasure, but in July east Beirut was taken with Christian help and the Palestinian fighters were evacuated by sea. In September President Gemayel was assassinated, probably with Syrian connivance or at Syria's instigation. He was succeeded by his brother Amin. Israel attacked west Beirut. At this point and in revenge for Bashir Gemayel's death, appalling massacres of defenceless Palestinian refugees were perpetrated in two camps at Sabra and Chatila. The perpetrators were Phalangists but Israelis were present at Phalangist headquarters throughout the three days and nights of atrocities which continued ceaselessly with the help of lighting provided by Israelis. Subsequently an official Israeli inquiry, forced upon a reluctant Israeli government by international outcry and internal disquiet, censured a number of highly placed Israelis for their responsibility in these doings.

By invading Lebanon Israel put at stake the virtually unconditional support which successive American governments conceded to it. Whether the invasion alone would have permanently eroded that support is questionable but the massacre made a deep mark on an American view of Israel which, ever since the foundation of the state in the wake of the Nazi holocaust, had permitted a double standard to be applied to the more dubious Israeli actions. In Israel itself the shock was profound and, coupled with continuing complications and deaths in the occupation of southern Lebanon over the next few years, created at least the beginnings of a revulsion against the more unacceptable features of Israeli aggressiveness, a suspicion that Israel's policies were in the long run hopeless, and some disinclination to accept the majority view that only the Palestinians were blameworthy. Above all the invasion's declared aim of peace for Galilee was not attained.

A multinational peace-keeping force of French, Italian and American units, later joined by a minute British contingent, had arrived in

Lebanon before the massacre in order to oversee the evacuation of the PLO and Syrians from west Beirut. Casualties at this stage were around 12,000, including Lebanese civilians (and 350 Israelis). At the same time President Reagan produced another peace plan. Since it included a ban on Israeli annexations, an end to Israeli settlements on the West Bank and an autonomous West Bank state in association with Jordan, it was immediately rejected by Begin. Reagan's simultaneous plea for the mutual withdrawal of Israeli and Syrian forces was ignored by both and was a dead letter from the start. The Arabs produced their own plan which again implicitly recognized an Israeli state (within the 1967 frontiers) but also demanded a Palestinian state with Jerusalem as its capital: this was a refurbishment of the Fahd plan of the previous year (Fahd had become king in June on the death of his brother Khaled). Lebanon's faction fighting continued and the Syrians and Israelis remained in the country. The Palestinian fighters (about 7,500) were dispersed in ten different directions, half of them to Syria, but Syria failed to promote an effective alternative to Arafat as the Palestinians' outstanding leader.

The Syrian position was initially precarious but was gradually consolidated, mainly because neither the Israeli–American front nor Syria's Arab enemies had anything viable to offer. The Israelis were overstretched and unhappy, the Lebanese army impotent, the multinational force decamped when it got caught in the crossfire of Lebanese politics: 239 US marines were killed in a single suicide raid in October 1983 after the US navy bombarded Shi'ite and Druze positions from offshore. Begin resigned in 1983 from a combination of personal domestic grief on his wife's death and political frustration, leaving Israel militarily, politically and morally weakened. President Assad on the other hand, having survived two rounds of risings against him in the first months of 1982 (although only at the cost of massacring his domestic enemies to the number of perhaps 10–20,000) was left locally in control. In Lebanon Amin Gemayel had nobody but Assad to turn to and when in 1983 Gemayel tried to secure an Israeli withdrawal by direct negotiation with Israel his partners in the Lebanese government denounced his efforts and Assad forced him to cancel the deal which he had made. Gemayel's task of piecing together a coalition from his country's warring forces was repeatedly frustrated by one or another of them and Gemayel himself was reduced to insignificance.

The Israeli withdrawal was completed in 1985. The Israelis took with them 1,000 Lebanese captives. To get them back a group called al-Jihad hijacked an American B-727 on its way from Athens to Rome and brought it to Beirut. The American and Israeli governments, emphasizing their refusal to deal with terrorists, refused to negotiate for the release of its passengers and crew but the Shi'ite leader Nabih Berri did so and secured their freedom in return for the release of Israel's 1,000 captives. The Americans contented themselves with threats of retaliation which did not materialize.

Syria failed to find a way to pacify or unite Lebanon. The PLO, buffeted but not eradicated, rebuilt its positions in the camps round Beirut and Sidon and so presented Syria with the possibility of another Israeli invasion to destroy them. This time Syria, abandoning its policy of restoring the Maronite–Sunni ascendancy, promoted an alliance of the Shi'ite Amal Party of Nabih Berri with the Druzes led by Kemal Jumblatt's son Walid, but the Palestinians offered stiff resistance to attempts to dislodge them from the camps and re-established their armed presence in them. The Shi'ite–Druze alliance evicted the Maronites from west Beirut but not the Palestinians. Syria's vexations were compounded by the surreptitious activities of its supposed ally, Iran, which split the Shi'ites between the pro-Syrian Amal and the pro-Iranian Hizbollah whose purpose was to create an Islamic state in Lebanon and use it as a springboard for war with Israel and the recapture of Jerusalem – policies regarded by Syria as impractical and suicidal. Amal and Hizbollah came to blows in west Beirut and the latter attacked the South Lebanese Army which was under Israeli protection. In addition Iran supported fundamentalist Sunnis (Tawhid al-Islam) who raised their own army and used it to fight Syrian units in Lebanon. Among the Maronites Gemayel was from the Syrian point of view a broken reed; by attaching himself to Syria he antagonized half his own community, half the Phalange and half the (Christian) Lebanese army – all bodies where Syria was as much disliked as Israel. The Phalangist Samir Geagea, commander of the Lebanese Forces, challenged Gemayel's authority but the Maronites, fearful of these divisions among themselves, created a Salvation Council as a means of getting rid of Geagea and retaining their hold on east Beirut. Another leading Maronite, Elie Hobeika – a former commander of the Lebanese Forces, who had been displaced by Geagea in 1986 – threw in his lot with Syria, went to Damascus and agreed to a truce with the Shi'ites and Druzes and to Syria's political programme: parity between Christians and Muslims, a reduction in the powers of the Maronite president, and a special relationship between Lebanon and Syria. But these moves failed to restore unity in Lebanon or to stop the fighting. President Gemayel, who had not been consulted, repudiated the agreement with Berri and Jumblatt, and the Maronites disowned Hobeika in favour of Geagea. In 1987 Maronites murdered the Sunni prime minister Rashid Karawi and developed in east Beirut a single-minded and militant hostility to the Syrians, who returned in force.

Lebanon was divided three ways and then four ways. In the south Israel supported its puppet régime under Christian military control. The rest of the country had rival governments headed by Karami's successor Selim al-Hoss (a Sunni) and Michel Aoun (Christian commander of the Lebanese army). Aoun's main aim was the eviction of Syria's 30,000 troops, in which he had help from Syria's perennial foe, Iraq. Fighting increased. In 1989 a rump of the Lebanese parliament elected in 1972 was assembled at Taif in Saudi Arabia. Its aim was to give Lebanon a new start after fifteen years of civil strife, foreign invasion and progressive disintegration. It recognized the main shifts in Lebanese politics by leaving the presidency to the Maronites but giving more powers to the Sunni prime minister and more parliamentary seats to the Shi'ites, the largest single community. President Elias Hrawi dismissed Aoun but his army remained loyal to him. Aoun and Geagea fought one another until Aoun was bombarded into surrender in 1990 by Syrian missiles: the battles between these two Christian forces marked the schismatic distintegration of Maronite power. Geagea withdrew his forces from Beirut leaving Hrawi's government ostensibly, Syria effectively, in control of the capital and most of the country. Israel retained its indirect control in the south through the puppet South Lebanese Army and through its evident readiness to use force – as, for example, in 1993 when heavy bombardment propelled half a million refugees to Beirut in an attempt to compel Lebanon to expel Hizbollah units from its territory. The attempt was unsuccessful and Israel's actions made Hizbollah more popular than it otherwise would have been.

Elections in 1992 demonstrated the strength of the Shi'ites but also their division between the main body of Nabih Barri's Amal and Hizbollah, each with its own private army. Al-Hoss was succeeded as prime minister by Rafiq al-Hariri, a richly successful businessman who mastered inflation, kept taxes and unemployment reasonably low, brought the balance of payments into surplus and secured a measure of foreign aid and investment. He restored some vigour to the urban economy but could neither arrest rural decline nor diminnish the load of 250,000, mostly homeless, Palestinians alongside a Lebanese population of no more than 3.5 million.

KING HUSSEIN'S DIPLOMACY

In the wake of the Israeli invasion of Lebanon King Hussein resumed his efforts to find a settlement between Israel and its neighbours. The rebuff to Israel in Lebanon and the worsting of the PLO by Syria revived the king's hopes of an agreement whereby Israel would cede some territory in exchange for peace. These hopes had long been shared by Shimon Peres, leader of Israel's Labour Party and joint head of the coalition formed in 1984 with Likud, led after Begin's resignation by Yitzhak Shamir; under a power-sharing agreement Peres was prime minister from 1984 to 1986 when he switched posts with Shamir and became foreign minister. These hopes were, however, unreal. Among the terms of such an agreement each side had a non-negotiable item which was unacceptable by the other. Likud, while outwardly ready to discuss some form of autonomy for Palestinians, insisted on keeping control over all territory conquered in and since 1967 and on negotiating only with Jordan. From its earliest days Israel had refused to countenance a distinct Palestinian state and it refused to enter into talks with the PLO or with a mixed Jordanian–PLO team. Hussein on the other hand believed that no peace was possible or sustainable without the participation of the Palestinians and that the PLO was their natural and proper representative. In this view he was supported by western European governments while the United States, obedient to Israeli insistence, interpreted direct talks to mean talks between Israel and Jordan only with no wider international conference. Since one main element in any peace talks must be the cession (or other disposition) of territory occupied by Israel, not to Jordan but to Palestinians, the pattern was lop-sided unless Jordan were to abandon its plea to include the PLO and the PLO were to give Hussein an open mandate to negotiate the fate of the Palestinians independently of themselves. Hussein's task in these circumstances was to reach the conference room by some dexterous fudging and not surprisingly he failed.

He began by reaching agreement with the PLO on the principle of territory-for-peace and on convoking an international conference which would include all five permanent members of the UN Security Council, Israel and a joint Jordan–PLO delegation. This formula buried the PLO in a large assembly but it implicitly rejected Israel's form of direct negotiation and, by making no reference to the Security Council's resolution 242, affronted Israel's demand for prior recognition of Israel's right to exist as a state. Hussein's manoeuvre got no support from other Arab states apart from Egypt (still an outcast because of Sadat's evasion of the Palestinian crux in the Camp David agreement) and it split the PLO. In 1985 Hussein made one of his repeated visits to Washington in order to assure the United States that the PLO would endorse resolution 242. He also asked for American arms. Reagan refused to modify the American–Israeli formula for direct negotiations but alarmed Israel by agreeing to supply Jordan with missiles and aircraft. His rejection of Hussein's plans for a peace conference was the penultimate nail in their coffin and later in the year they foundered completely when even the moderate Arabs such as King Fahd refused to attend an Arab meeting to discuss them.

Hussein blamed the PLO for this setback. Encouraged by their partial recovery in Lebanon the more militant PLO leaders were keen to draw attention to themselves and their cause by startling and violent acts. In one incident in Cyprus they killed three Israelis. Later in 1985 four Palestinians boarded at Genoa the Italian *Achille Lauro* as it set out for a cruise in the eastern Mediterranean. They intended (so they later said) to leave the ship at Ashdod in Israel and there execute some spectacular destruction of Israeli property and lives. They were, however, discovered on board before reaching Ashdod and thereupon – and apparently on the spur of the moment – seized the liner and issued a demand for the release of a number of Palestinians held captive in Israel. One of the hijackers murdered an American passenger. Alarmed by this turn of events Palestinian leaders – notably Abu Abbas, an associate of Arafat, if not always in total accord with him – ordered the hijackers to return the liner to its captain and come ashore in Egypt on being assured that they would thence be conveyed to a safe place (presumably Tunisia). This resolution of the incident was made possible by President Mubarak who provided the means of getting the hijackers off the *Achille Lauro* and out of Egypt. But the Americans, incensed by the murder of the hapless passenger, intercepted the Egyptian aircraft carrying the hijackers and Abu Abbas to Tunisia and, by threatening to shoot it down, forced it to land in Sicily where armed Americans from a Nato unit tried to seize its passengers. This last part of the exploit was foiled by the Italians who arrested and charged the hijackers but allowed Abu Abbas to escape to Yugoslavia. The consequences of these adventures included the fall of the Italian government (subsequently reinstated); the weakening of President Mubarak whose mediation brought him the humiliation of having one of his aircraft brought down by the Americans; the raising of illegal violence to higher levels, not only by the PLO but also by Israel which bombed the PLO's offices on Tunisian soil; the final abrogation of Hussein's peace plans and of his co-operation with the PLO; and a new hiatus in which no party was offering any viable proposals for an approach to peace in the Middle East.

In the next year (1986) the United States tried to reflate the peace process by offering to accept the PLO at a broad international conference if it would accept resolution 242 and forswear violence – a belated conversion to the defunct Hussein plan. But Hussein, increasingly occupied with the war in the Gulf between Iraq and Iran, washed his hands of the PLO; Peres came to the end of his term as prime minister; and the PLO retorted to the American initiative by demanding that the United States publicly endorse the Palestinians' right to self-determination – that is, to a state of their own. In 1987 the Palestinian National Council abrogated the 1985 accord between Arafat and Hussein and affirmed its demand for an independent Palestinian state and separate representation at any peace talks.

As 1987 ended the Palestinians under Israeli occupation surprised everybody, including the Israelis, by staging a rising – *intifada* – on their own account. It began with a traffic accident in Gaza, where the Palestinians' living conditions were peculiarly disgraceful and Israeli police brutality peculiarly barbarous. Stones were thrown, mostly by children, and the Israelis retaliated by opening fire and killing people. The rising spread to the West Bank and Jerusalem and evoked from the startled Israelis a riposte so harsh that Israel's already tarnished reputation was severely damaged throughout the world and even governments protested publicly. Israel failed to suppress the *intifada*. The *intifada* failed to achieve anything more concrete than the violent advertisement of Palestinian grievances. The arrival in the late 1980s of thousands of Jews from the USSR hardened attitudes on both sides: 200,000 immigrants in 1990 and the tide still rising presented Israel with demands for housing and jobs which it could not satisfy. Israel was losing its grip on southern Lebanon and the crisis over Kuwait was to add to its worries by allowing Syria to come in from the cold and win some (limited and grudging) goodwill in Washington by contributing to the American assemblage of armed force in Saudi Arabia against Iraq. And Jordan, Israel's sometime interlocutor, was fading from the picture.

The *intifada* and Israel's handling of it caused King Hussein finally to disengage from the West Bank, leaving Israel with the entire responsibility for its affairs and accentuating the transformation of Jordan itself into an annexe of Palestinian hostility to Israel. Hussein turned to Iraq for support against an increasingly belligerent Israel. The United States, which was by this stage providing Israel with aid up to and beyond $2 billion a year (in despite of laws which forbade aid to governments which flouted human rights), was embarrassed by Israel's behaviour and by the foundering of the peace process and felt driven in 1989 to talk for the first time officially and directly to the PLO. But in Israel Shamir, under pressure from colleagues such as David Levy and Ariel Sharon, reverted to Begin's position at Camp David. The Likud declared that it would trade no land for peace, would continue to expedite Jewish settlement of the West Bank, would never give votes to Arabs in east Jerusalem and would attend no conference so long as the *intifada* lasted. In the next year Sharon increased the pressure for bold Zionist policies by resigning from the cabinet. The Likud–Labour coalition, looking for ways of preventing a peace conference without giving too much offence to the United States, broke up and after lengthy bargaining by both major parties with minor religious parties Shamir was able to form a new government but one uncomfortably dependent on the fanatic fringe. The Labour Party was openly divided between adherents of Peres and of Rabin. Arafat, under similar pressure from colleagues, also took refuge in a regression to extreme positions. The United States tried to use Mubarak as a mediator but with no success.

Iraq's invasion of Kuwait (see the next chapter) deflected attention to more urgent conflicts. It was also one more setback for the always tenuous prospect of Arab unity and peace among Arabs. Arabs were no more united, or more likely to unite, than other kindred. But since the waning of the Ottoman power, and more specifically since the creation of the Arab League in the 1940s, they had made a point of the search for unity and had been correspondingly downcast among themselves and mocked by others when it eluded them. The late 1970s had been a bad time by these standards. Egypt, the most prestigious Arab state, was made an outcast by its Camp David policies and was not even invited to attend the Arab conference at the end of 1980. Syria was cast as another deviant on account of its intervention in Lebanon, supported only by Libya and Algeria in the remote west and by South Yemen in the remote south. In the first Gulf War Saudi Arabia, carrying with it all the Gulf states, supported Iraq against Iran, whereas Syria supported Iran; Jordan took Iraq's side with unexpected directedness. When the war ended King Hussein embarked on an attempt to refashion the politics of the Arab world. His aims were pacification in the Middle East and the preservation of his own kingdom: they were interrelated. To the north and west the king faced enemies in Syria and Israel; to the south was the Saudi monarchy, hereditary enemy of the Hashemites; to the east was Iraq, waxing strong under a leader as ambitious as he was unpredictable but a man for whom the king had feelings of friendship and misplaced trust; and further off was Egypt which was readmitted to the Arab League in 1989. During his arduous but fruitless efforts to inaugurate peace talks between Palestinians and Israel the king had become disenchanted with the United States and, if to a lesser degree, with Britain and the more convinced that the Arabs themselves must take a concerted and purely Arab initiative. At his invitation the presidents of Iraq and Egypt – and of the newly reunited Yemen at Saudi Arabia's back door – met in Amman early in 1990. But a few months later both his main guests torpedoed his plans and endangered his throne, Saddam Hussein by invading Kuwait and Hosni Mubarak by precipitately joining the forces being assembled by the United States in Saudi Arabia. The king denounced the Iraqi invasion and taking of hostages, insisted on an Iraqi retreat and the restoration of Kuwait's independence, and supported UN sanctions; but he continued to import Iraqi oil out of necessity and advocated the continued supply of necessary food and medicines to Iraq. He also continued to champion an exclusively pan-Arab resolution of the crisis. His attitudes were popular in Jordan but hardly anywhere else, and even in Jordan his own position was weakened by the addition of new to old economic strains. In attempting once more to play the role of honest broker he found himself snubbed by President Bush and received by Thatcher with the barest courtesy. The Kuwait crisis cruelly demonstrated that he was the most well intentioned but the least effective statesman in the Middle East.

The Gulf War (see the next chapter), together with the extinction of Russian influence in the Middle East, encouraged the United States to seek a comprehensive peace settlement between Israel and its neighbouring Arab states and between Israel and the Palestinians. With Egypt Israel was already at peace. Between Israel and Jordan there had been no hostilities for more than forty years, although formally the two countries were at war. Neither Egypt nor Jordan had a serious conflict of national interest with Israel, and Jordan had even profited from the creation of the state of Israel by enlarging its territory and was being tempted to enlarge it still further at the expense of the Palestinians. Syria's case was different since Syria was determined to recover territory lost to Israel and, more vaguely, cherished hopes of reconstituting Greater Syria at the expense not only of Palestinians and Lebanon but also of Israel itself. The plight of the Palestinians and their aspirations created bonds between Israel and Jordan but not between Israel and Syria. Israel wanted peace with Jordan and Syria but was adamantly opposed, as it had always been, to the creation of a Palestinian state or recognition of a distinct Palestinian identity. It was prepared to make minimal concessions to the Palestinians in order to meet American pressures and secure peace treaties with neighbouring states but it was not prepared to accept or even discuss the Palestinian state which had been a part of the UN partition plan which Israel had purported to accept in 1948.

The Gulf War had aligned Syria as well as Egypt with the United States and ostracized Jordan and the PLO (which supported UN action to evict Iraq from Kuwait but opposed the direct attack on Iraq and the attempt to depose Saddam Hussein). The United States sought to mollify Arabs by opposing the development of Israeli settlement in the occupied territories and by supporting the principle of land-for-peace. It secured Israel's participation in a peace conference by a mixture of concessions (no independent Palestinian state) and coercion (withholding a $10 billion guarantee for financing the settlement of Soviet Jews). The Likud government was divided on the issue of land-for-peace but united in refusing to admit a separate Palestinian delegation to peace talks and making only vacuous proposals for Palestinian autonomy. After much wrangling a conference opened in 1991 in Madrid whence it was transferred to Washington where disputes over seating arrangements reduced the talking to encounters in corridors. These talks were little more than a charade and the killing of the Hizbollah leader Sheikh Abbas Musavi by Israelis in Lebanon at the beginning of 1992 created fresh, if temporary, obstacles. Later in the year the Israeli Labour Party defeated the Likud in a general election (winning 44 seats in the Knesset to Likud's 32) and formed a government with smaller parties and with Yitzhak Rabin as prime minister.

Israel's main aims remained the same – to come to terms with Jordan and Syria and isolate and divide the Palestinians – but the tactics changed inasmuch as Rabin declared himself willing to discuss with Syria Israel's occupation of the Golan Heights. But this overture, which seemed designed to win a quick deal with Syria, failed when it turned out to mean a thinning out of Israeli forces but not their withdrawal or any cession of occupied territory. On Palestine Rabin proposed a five-year experiment during which a Palestinian local authority would have limited powers over (mostly) social services but little political authority or policing power. Discussions about Jerusalem were ruled out and the position of the PLO in Israeli eyes was left unresolved.

Neither Rabin nor Arafat was in a strong position. Neither had enough support to deliver a meaningful agreement, nor was it clear that either wished to. Arafat and the PLO were gravely devalued by their stance in the Gulf War on top of their persistent failure over years to secure a Palestinian state. Arafat's prestige had been highest in the 1970s when he succeeded in achieving a status approaching that of a head of state and in establishing formal links with the European Community (1974). But he was weakened five years later when the Camp David accords, which were applauded by the EC, gave the Palestinians nothing beyond vague phrases. He recovered some of his political stature when the EC reasserted at Venice in 1980 its support for Security Council resolution 242 but he was a performer on a narrowing stage, diminished not only by more militant groups but also by the *intifada* whose leaders within Israel appeared – to Israelis in particular – to constitute an alternative Palestinian leadership. Arafat therefore was in the early 1990s a leader uncertain of his next move. He was, however, still the Palestinian leader of international consequence; and the Palestinians still outnumbered Israelis in the Middle East and accounted for nearly a fifth of the inhabitants of Israel.

Rabin's electoral victory had left him short of a majority in the Knesset and he would all but certainly lose his parliamentary and electoral edge if he were to make any but the smallest concessions to the PLO. Yet the decline of Arafat and the PLO presented him with an opportunity, if a risky one: to discredit the Palestinians' principal leaders still further without raising up new and more effective ones. Rabin had hoped to find within Israel other Palestinians with whom to negotiate a strictly limited kind of autonomy within a strictly limited area, but these hopes proved vain. He found himself with no alternative to the PLO, under strong American pressure to engage in talks with Palestinians as well as Jordan and Syria, and in a new situation where the alternative to the PLO was not a comparatively accommodating Palestinian leadership but the more militant Hamas – an extremist group unashamed of violence and bent on the extinction of the Israeli state. Hamas, which come into prominence around 1990, gathered strength with each manifestation of compromise or hesitation by Arafat and each display of lawlessness or brutality by Israel (for example, the expulsion in 1992 of more than 400 indiscriminately arrested Palestinians from Gaza into the barren and freezing wastes of southern Lebanon in breach of international law).

Given that Rabin and Arafat were irreconcilably opposed on the main issue of a Palestinian state, no discussions between Israel and the PLO could get under way unless they were diverted into other issues. This diversion was found in secret talks which, engineered by the Norwegian government, produced the Oslo declaration of 1993 which was designed as the thin end of a wedge whose other end was kept diplomatically out of sight. Israel recognized the PLO as representative of Palestinians and the PLO accepted Israel's right to exist and agreed to denounce violence. On this basis the Oslo plan had three stages: first, an Israeli withdrawal from the city of Jericho and the Gaza strip and the transfer of powers to the PLO; second, elections within three months for Palestinian authorities in these areas; and ultimately, discussions over the future of the 140 settlements in the occupied territories and of Jerusalem. In spite of much imprecision and serious obstacles, negotiations for the implementation of the first stage proceeded and Israel withdrew its forces from the two designated areas. But the further stages were bedevilled by distrustful sparring between the two sides, by challenges from Israeli settlers seeking to coerce or unseat Rabin's government, by the hostility of Hamas seeking to discredit Arafat and destroy the PLO, and by Rabin's unwillingness to proceed faster than he felt compelled and Israel's accelerated confiscation of land in the occupied territories for further Israeli settlement. For Israel the cession of Gaza was a good riddance of a destitute and violent area from which it drew no profit and some discredit, an area moreover whose transfer to the PLO would expose Arafat's shortcomings as a possible head of state. The cession of Jericho, on the other hand, was unwelcome since it portended the cession of substantially more of the West Bank to Palestinian government.

A second stage in this peace process was concluded, albeit more than a year behind schedule, towards the end of 1995 when six cities and several hundred small towns and villages on the West Bank were transferred to (limited) Palestinian administration – excluding, however, Hebron, the main city in the south and the scene in 1994 of the killing of twenty-nine Muslims in a mosque under the eyes of Israeli police, where Israeli settlers succeeded in enforcing a mixed régime for it. The new area of Palestinian authority was a patchwork in which the six cities were separate from each other, their supplies of water and other necessities were controlled by Israel, and the Israeli police and army retained rights to operate in the intervening spaces which comprised two-thirds of the West Bank. Palestinian police were denied the right to arrest Israelis. A Palestinian Council and the chairman of a Palestinian executive were to be elected early in 1996. The resulting map was a concatenation of two dispersed and hostile chains of Palestinian towns and Israeli settlements, the whole effectively under Israeli control. For Palestinian refugees, most of them still in camps, there was nothing. On the crucial question whether a Palestinian state lay at the end of the peace process, Arafat and Rabin remained in stark disaccord.

For Rabin the peace process seemed a road to electoral disaster for himself and his party and he probably trod it under compulsion and without the optimism consistently expressed by his foreign minister, Shimon Peres. With more conviction Rabin pursued better relations with Jordan through a declaration of mutual non-belligerence following a peace treaty. The two states had not engaged in hostilities for nearly half a century but Israeli strategists were perennially nervous about the possibility of an attack from Jordan by Jordanians or other Arabs using Jordanian territory. Rabin and Hussein shared a common aversion to Arafat personally, and just as Ben-Gurion had negotiated with Hussein's grandfather for Jordan to appropriate lands allotted by the UN partition plan to a Palestinian state, so Rabin sought Hussein's co-operation against Palestinian claims in Jerusalem by conceding rights of guardianship over Muslim holy places to a religious body in Jordan. Israel also proceeded to expropriate lands around Jerusalem for Israeli settlement, escaping international censure only through the use of the American veto in the Security Council. Rabin stayed these operations only when threatened with defeat in the Knesset by the loss of Arab votes to the Likud.

Rabin's peace with Jordan was a step in his strategy of making peace with Israel's neighbours in order finally to bury the premise, almost fifty years old, that there should be a Palestinian state as well as an Israeli in the old mandated territory of Palestine. But the key to this strategy was not in Jordan but in Syria. Jordan had no valid or plausible claim to territory which had formed part of the mandate but Syria had. Assad's first aim was the recovery of the Golan Heights but Syria's older and more persistent purpose was the reconstitution of Greater Syria. Syria was, like Israel, an irredentist state and its horizons encompassed the whole of Israel and Lebanon and parts of Jordan. And Assad knew, and Rabin feared, that Syria had an alternative to peace with Israel. With help from Iran or Pakistan it might threaten to create a second nuclear power in the Middle East which would eliminate Israel's monopoly and puncture its defence strategy.

In 1995 Israel's survival rested, as it had rested for half a century, on arms and American money. Israel's military expenditure exceeded $7 billion a year and it had, with American encouragement and finance, fortified itself with a nuclear armoury. But American subventions could not safely be expected to continue at current levels of $3 billion a year and Israel's nuclear weaponry was causing Arab and other Muslim states to follow it into a regional arms race and mutual bankruptcy. Yet Rabin's policy seemed no different from that of his predecessor, Shamir, who avowed his aim to spin out peace talks with the Palestinians for ten years. Rabin faced only slowly (if at all) the need to make peace with Syria by surrendering the Golan Heights and so risking the loss of the elections due in 1996 – elections in which by a constitutional amendment of 1992 Israel's prime minister would for the first time be elected by direct popular vote. But his tortured and hesitant search for a way of combining security with peace ended in 1995 when he was shot dead by a young Israeli who had no doubt that his probings for peace were misguided and sinful.

Iran is in traditional parlance a part of the Middle East but it is emphatically not part of the Arab world. It remembers the Arab conquest of the seventh century more resentfully than the Mongol which is more recent by 600 years. It is the unique stronghold of Shia Islam, the tense and high-pitched deviant from Sunni orthodoxy: unique because the Shia, although a majority in Bahrain, Iraq and perhaps Oman too, are everywhere except in Iran subject to a Sunni dynasty or ruling class. Iran is rich in oil but, unlike the Arab world, not only in oil. Its dominant Iranians rule over a heterogeneous variety of races and religions. It has never been subject to western imperialisms although it has been obliged on occasions to humble itself before them, particularly the British and Russian. The world's thirst for oil has given it the means to turn itself into an industrial and military power without equal in its surrounding region.

If history and geography made Iran wary of both the British and the Russians, oil made it particularly suspicious of the former. In the Second World War both powers occupied Iran on the plea of strategic necessity (the Iranians resisting for three days), and enforced the abdication of the founder of Iran's new dynasty, Reza Shah Pahlavi, who was exiled to Mauritius and later died in South Africa in 1944. His son and successor, Muhammad Reza Pahlavi, had had neither the time nor the opportunity to make a mark among his people by the time that the war and the occupation ended. The treaty of 1942 which had sanctioned foreign occupation provided for the withdrawal of the British and the Russians six months after the end of hostilities. The Russians made an attempt to retain influence in the province of Azerbaijan, comparatively rich, traditionally hostile to the central government and situated on the borders of the USSR. They also lent support to Kurdish separatism and may have entertained hopes of establishing a Russian sphere of influence stretching southwards from the Kurdish republic of Mahabad through other Kurdish territories to the Persian Gulf. They were assisted by the Tudeh Party, an amalgam of Marxist communists with an older liberal tradition which had infused the Constitutional Movement earlier in the century.

British troops left Iran punctually in March 1946 but Russian troops had to be manoeuvred out by the Iranian government. The astute elder

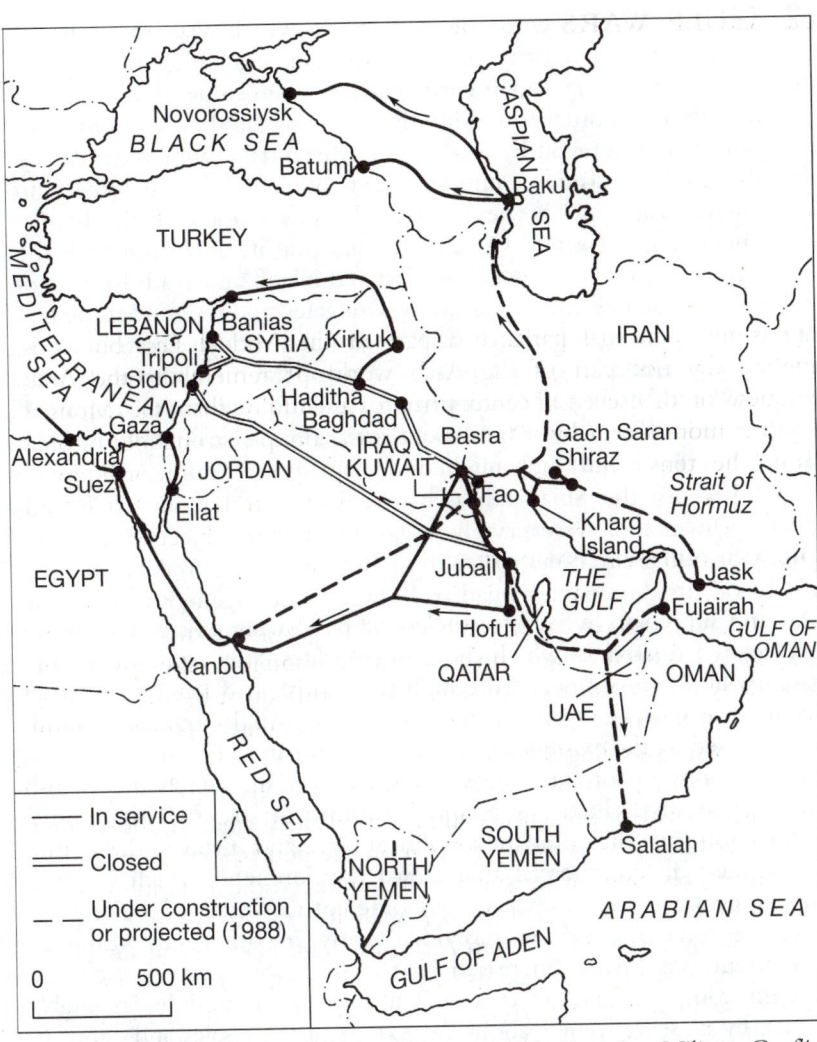

12.1 Major oil pipelines of the Gulf (Source: *The Iran–Iraq Military Conflict*, Dilip Hero)

statesman Qavam es-Sultaneh, who became prime minister in January 1946, visited Moscow in February on the heels of an Iranian complaint to the newly established Security Council and contrived to persuade Stalin that Russian aims in Iran could better be attained through good relations with the Iranian government than by the continued presence of Russian troops in north-western Iran. The Russians, who were at least as keen on an oil agreement with the central government as on fostering separatist movements against it, withdrew their troops only a few weeks late. Qavam entered into discussions about an oil agreement but evaded any conclusive step on the plea that, constitutionally, the ultimate decision lay with the parliament which was about to be elected; he also delayed the elections. Qavam simultaneously played a double game with the Tudeh Party. Having taken some of its members into his cabinet in order to mollify the Russians, he then welcomed (to put it no higher) a revolt by powerful southern tribes which demanded the dismissal of Tudeh ministers and other Tudeh members in prominent positions. The new Majlis (the lower house of the Iranian parliament) duly censured Qavam for engaging in discussion with the Russians for an oil agreement and declared it null and void. In spite of these oblique achievements Qavam was defeated and resigned at the end of the year. Having outwitted the Russians and seen the British depart too, he was embarking on a policy of co-operation with the United States, whence he hoped to get financial aid and diplomatic support in Iran's traditional search for means of keeping both the Russians and the British at arm's length.

OIL AND NATIONALISM IN IRAQ

Qavam's successors were nevertheless to be sharply engaged with the British in a quarrel in which the United States, after some hesitation, gave Britain firm support. Although the British had left Iran in strict accordance with the terms of the 1942 treaty, Anglo–Iranian relations were soured by the existence of the Anglo–Iranian Oil Company which had a monopoly of Iran's proven oilfields and in which the British government itself held a substantial number of shares. This unusual connection gave the company a political tinge and involved the British government in commercial affairs, two developments which were, in Iranian eyes, neither natural nor welcome consequences of the grant of a concession to a private individual early in the century. Iranian discontent was further increased by the suspicion that this rich and foreign company sold oil to the British navy on unduly favourable terms, by its secretiveness about the extent of these sales and about its accounts generally, by its failure to publicize in the rest of Iran the good wages, working conditions and other benefits which it provided for its labour force. On the other hand the company's shortcomings were not all its own fault: its failure to advertise its own virtues was due to the fact that, since it was practically unique in complying with the labour laws, it could not claim credit where credit was due without casting aspersions on other employers.

The concession inherited by the company had been granted in 1901 to a certain W. K. D'Arcy and acquired by the company before the First World War. The Iranian government had bargained to receive a percentage of the company's net profits, but this bargain turned out badly for Iran since the Iranian share was rendered dependent on the

level of taxation in Britain. It was further affected by the slump between
the wars and in 1932 Iran purported to cancel the concession. As a
result of negotiations between Reza Shah and the company a new
agreement, duly ratified by the Iranian parliament, gave the company a
new concession running for sixty years from 1933 (instead of sixty years
from 1901) over a substantially diminished area, and gave Iran royalties
to be calculated on the quantity of oil extracted. By 1950 this output
was 32.5 million metric tons, three times the amount produced in 1938;
Iran was the largest producer in the Middle East and contained at
Abadan the largest refinery in the world.

The revenues from the oil were much needed. After the Second
World War Iran embarked on economic expansion and a plan adopted
by the Majlis in 1947 envisaged the expenditure of $651 million of
which $242 million were to come from oil. But oil revenues in 1947
and 1948 were a disappointment in spite of huge increases in the
company's profits and the company, sensing trouble, entered into
discussions for a revision of the agreement of 1933. In 1949 a
supplemental agreement was concluded: the royalty rate was raised by
50 per cent and the company agreed to pay £5.1 million at once out
of its reserves and thereafter to make further payments out of reserves
annually instead of waiting until 1993, as the 1933 agreement provided.
This agreement was very favourable to Iran but it was also complicated
to the point of incomprehensibility and was unacceptable to a group of
nationalists who wanted to terminate the concession entirely and take
the management and profits of the oil industry into purely Iranian
hands. General Ali Razmara, who became prime minister in 1950,
refrained for a while from pressing the supplemental agreement on the
Majlis. When he recommended its adoption – largely on the grounds
that Iran was short of qualified technicians – the Majlis special oil
committee demanded the nationalization of the oilfields and refinery,
and the prime minister withdrew the agreement. The company,
unexpectedly made aware of terms being offered by the Arab American
Oil Company (Aramco) to the Saudi Arabian government – terms
which were more favourable (in good years) and above all easier to
understand – prepared to reopen discussions and simplify the
supplemental agreement. Razmara produced a series of reports by
experts who opposed nationalization but he was assassinated in March
1951 by a fanatic nationalist belonging to the semi-religious Fidayan-i-
Islam. On the next day the Majlis approved the oil committee's
proposals and a few weeks later it nationalized the oil industry.

General Razmara was succeeded by Husain Ala, a friend and
nominee of the shah who wanted to avoid appointing the chairman of
the oil committee, Dr Muhammad Musaddiq, in which attempt the
shah failed. Dr Musaddiq began at the end of 1951 a spell of power

which lasted tumultuously until August 1953. Musaddiq was a rich and aristocratic hysteric and hypochondriac who appealed to a wide variety of emotions; he was supported by xenophobic chauvinists, by religious fanatics, by communist and non-communist radicals, by the old landowning aristocracy to which he belonged, and by all those who distrusted the shah's attempt to resurrect the authority of the dynasty which had been abased by his father's abdication and the foreign occupation during the war. His political stock-in-trade was a blinkered nationalism which, though it may have been intended as a screen, became the substance of his political actions. If, as seems probable, Musaddiq intended to combine the nationalization of the oil industry with the continued employment of foreign technicians and continuing foreign finance, he was soon defeated by his own extreme professions and supporters. Musaddiq became almost a figure of fun to the world at large, certainly to the western world, but in his own country he roused genuine popularity (quite apart from the bought support of the capital's mobs) and even his political opponents were to acknowledge his success in reducing corruption in public affairs. He failed, however, to retain control over the course of the oil dispute or to retain his own confidence in himself, he became the prisoner of his own rash attitudes, and he underrated the effectiveness of the economic sanctions which the British government was able to deploy against him.

The oil nationalization law expropriated the British company and created a new Iranian company to take its place. Throughout the ensuing controversies Dr Musaddiq insisted on British recognition of the validity of the nationalization decree as a prerequisite for any negotiations on compensation for the British company. The British on the other hand maintained that the decree was illegal and inoperative, that the company's title remained intact, and that the concession agreement of 1933 could not properly be revoked by unilateral act of one party to it even if that party was a sovereign state. Consequently Britain claimed not only compensation for the loss of the benefits secured to the company by the concession agreement but also a further sum representing damages for a wrongful breach of contract.

The controversy involved not only the British company but also the British government which was responsible for the safety of British citizens (who might be endangered by nationalist frenzy) and wished to safeguard its own financial stake in the British company; it was also concerned to stand firm in defence of British rights for fear that weakness in one part of the Middle East might provoke attacks on British interests elsewhere. During 1951 various attempts were made to negotiate with Dr Musaddiq – by the British company, by a special emissary of President Truman and by a British delegation led by a cabinet minister. The American administration chose the British side after some initial hesitation. Its sympathy for the British case was to some extent countered by its desire to attract Iran into the western camp; since the oil dispute threatened on the contrary to separate Iran from the west, there were some Americans who were anxious not merely to mediate but to secure a settlement without necessarily getting the full measure of what London thought fit and proper. But Dr Musaddiq's extravagant histrionics alienated American sympathy and a visit by him to New York (to address the Security Council) and to Washington in October 1951 did him no good and some harm. His flirtations with Moscow and his indebtedness to Iranian communists helped to persuade the Americans to make common cause with the British and to support his internal enemies' plans to overthrow him.

The dispute was not in the last resort ruled by the legal rights and wrongs but by the British government's unwillingness to use force (except to protect British citizens); by the Iranian company's inability to sell Iranian oil in face of obstacles interposed by the British company and the lack of solidarity among producers; and by the economic collapse of Musaddiq's régime which, deprived of its revenue from the British company, was unable to find alternative sources of cash. Having failed to raise loans from the United States or the World Bank, Musaddiq called in the renowned Dr Hjalmar Schacht who judged him, after a few days' acquaintance in September 1952, to be one of the wisest men of the age but was unable to help him. Meanwhile some of Musaddiq's allies were wavering. He was reappointed prime minister in July 1952 after a routine resignation upon the convening of a new Majlis, but the members showed some reluctance to grant him the full powers which he asked for. When the shah refused to give him the war ministry he resigned. But Qavam, who succeeded him, was only able to stay in office for four days and then had to take flight; the mob demonstrated in Musaddiq's favour, the Majlis voted him full powers, and the throne itself seemed to be in danger. Only the army had the capacity to unseat him, and a year later it did.

Musaddiq's triumph had exposed his political dependence on the mob without lessening his financial dependence on the British, which in turn was conditioned by his dependence on the extreme nationalist mullah Kashani who had become president of the Majlis and would countenance no approach to the British. By the end of 1952 Musaddiq and Kashani had fallen out. In the next year Musaddiq triumphed over Kashani and dismissed the Majlis but failed to prevail over the shah. In August the shah dismissed Musaddiq and appointed General Zahedi in his place. Three days later the shah and the general were both in flight, but whereas the shah fled to Rome the general withdrew only a little distance and within a week of his original appointment he returned to put an end to the Musaddiq régime once and for all with not inconsiderable help from the CIA and the British secret service. The shah returned too. The helping hand of the United States was barely concealed and was not forgotten when the shah was once more forced into flight a quarter of a century later.

Peace was soon made in the oil dispute. New agreements were worked out on the basis, ironically, of recognizing the validity of the Iranian nationalization decree. The Iranian oil company remained in being and in possession of the oil. A consortium of eight foreign companies – British, American and French – was created. The British company accepted, not without demur, a 40 per cent share in this consortium and received from its seven associates £214 million for the remaining 60 per cent which they bought between them. The British company was also to receive £25 million, spread over the years 1957–66, from the Iranian government in compensation for its losses since nationalization. The consortium was given effective control over the Abadan refinery and the principal oilfields for twenty-five years, with a series of options of renewal, and undertook to pay the Iranian government 50 per cent of profits. Thus nationalist doctrine was satisfied, the consumers were assured of their requirements and the producers of their revenues. Full diplomatic relations between Iran and Britain were restored early in 1954. Musaddiq spent the next two years in gaol.

THE RULE OF THE SHAH

The fall of Musaddiq was a victory for the shah. Musaddiq's left-wing supporters were persecuted with a thorough ferocity and the shah gradually asserted the paramountcy of the throne, first through military rule which lasted until 1957 and then through a series of prime ministers who were either submissive or turned out. The death of the shah's only brother late in 1954 jeopardized the dynasty and obliged the heirless shah to divorce his second wife and marry a third, who bore him a son a year later. Thus fortified the shah began to implement a policy of land distribution and reform which proved so unpopular with the landowning classes and the Majlis (in which they were well represented) that the shah dispensed with parliament for the two years 1961–63. In 1963 he felt strong enough to hold a plebiscite which confirmed his personal ascendancy and the decline of the power of the provincial notables. The worlds of the urban politician, the tribal chiefs and the educated young remained for diverse reasons disgruntled, but oil revenues increased and Iran's gross national product began to register annual increases of the order of 7 per cent.

In foreign affairs the shah had to decide whether to join the Baghdad Pact and to identify his country with the west. He did so in 1955 after becoming the first recipient of American Point Four aid, and in 1959 he paid a state visit to London and received President Eisenhower in Teheran. After a short period of frigidity, Russo–Iranian relations improved and in 1963 President Brezhnev too was officially received in the Iranian capital, a reminder that the traditional suspiciousness between the two countries had to take account of the fact that Iran had a long undefended frontier with the USSR and looked to northern trade routes for the export of the produce of its northern provinces. The shah undertook not to permit the installation of nuclear missiles in Iran and, without abandoning Cento (as the Baghdad Pact had become) or joining the unaligned group, he moved towards a more independent position in world politics. By 1969–70 he was able to play a decisive role in shaping the political future of the Persian Gulf after the departure of the British and, relishing the role of a crowned entrepreneur, to use mineral wealth and a bounding economy to turn Iran into a military and industrial power of considerable significance. The shah's policy and bent was growth at all costs and the key was oil, although oil was not the country's only resource. (It was rich too in natural gas, other minerals and agriculture and was establishing industry as fast as 50 per cent illiteracy and a wretched educational system would allow.) When in 1973 war in the Middle East gave the oil producers the excuse to push up prices the shah insisted on maximum increases, successfully but against the wishes of more cautious Arabs who hesitated to damage western countries which were their best customers. In two years the Iranian government's revenue per barrel was multiplied by ten and its total annual oil revenues rose from $2.3 billion to $18.2 billion. In the year after the oil price rise of 1973 GNP rose by 42.5 per cent. Spending rose too, particularly defence spending which also multiplied by ten in the half decade and passed the $10 million mark; by 1975 Iran was spending on defence a larger proportion of GNP than any country in the world except Israel. The results of this explosion were not all happy: 1975 saw a deficit in the balance of payments of nearly $1 billion. Waste and corruption flourished commensurately; inflation took hold. Those hurt by inflation and least able to make a profession of corruption had to be compensated and wages were nearly doubled in 1974–75 with the usual cyclic nightmare: demand for goods, inadequate supply, rising prices and increased exports to fill the gaps, further price rises and further wage demands. The shah, who had dealt roughly with the landed aristocracy in the early 1960s, showed signs of imperial displeasure with the new, ostentatious and corrupt rich, and toyed with schemes for handing over half the ownership and profits of industry to the workers. Yet wages remained derisory and Teheran became a shanty town of 5 million for whom housing was shamefully inadequate. His régime's weaknesses were the uncertainty surrounding an autocracy with an infant heir, the opposition of the conservative mullahs, the opposition of radical students and other protesters which even one of the world's most ferocious secret police apparatuses could not mute, and his own refusal to listen to others. He was obsessively concerned with left-wing conspiracies but blind to the threat from clerical radicalism and he became dangerously ignorant of the state of his own country, where the savagery of his police SAVAK (trained by Americans and Israelis) and blatant inequality alongside ill-gotten wealth ensured that when the tocsin sounded thousands of civilians rushed unarmed into the streets, prepared to face his fearsome military machine. Between his return in 1953 and his second flight in 1979 the shah worked a revolution in Iran, using the country's wealth to create prosperity and strength; but the headlong pace and fearful inhumanity of this revolution united conservatives, radicals and liberals against it and so generated a counter-revolution. When in January 1979 the shah asked Dr Shahpur Bakhtiar to assume the premiership, Bakhtiar consented only upon condition that the shah leave the country.

If the shah was the chief engineer of his own destruction, the chief beneficiary was Ayatollah Ruhollah Khomeini, an aged and obdurate cleric incensed by the shah's material values and animated by personal hate born of the shah's refusal to allow him to attend the funeral of one of his sons. Khomeini was the principal spokesman for those who were opposed to secular government and believed that the modernization of Iran in the American mould was sinful. This movement was evident from the early 1960s, broke into serious and harshly repressed riots in 1962–63, and led in 1964 to Khomeini's flight to Turkey and exile – at first at Najaf in Iraq and then in France – whence he continued to animate opposition to the shah. The personalization of this struggle culminated in the flight of the shah and Khomeini's return: the triumph of imam over shah. One month later Bakhtiar too fled the country.

THE TRIUMPHS OF KHOMEINI

Khomeini was all powerful. Other groups opposed to the shah judged it expedient to rally behind Khomeini or keep out of sight. He proclaimed an Islamic republic and instituted a régime even more intolerant than the shah's, although possibly less murderous and less corrupt. He appointed as prime minister Mehdi Bazargan, a liberal Muslim intellectual with a scientific education who had been imprisoned by the shah, but there was in fact no central government. Bazargan was harassed from left to right, by Kurdish and Arab minorities, and by conflicting pronouncements from the ayatollah himself who retired to the holy city of Qum, dominated the scene by sporadic utterance and allowed a kind of religious hooliganism to prevail. Local Islamic committees spent their time rounding up and executing those whom the hazards of denunciation exposed to their indiscriminate wrath. In November radicals in Teheran invaded the American embassy and took fifty-three of its inmates hostage. This coup was partly directed at Bazargan, who finally succumbed, but more overtly against the United States. It incidentally, but portentously, alarmed the Russians who feared a retaliatory American *coup de main* in Iran at a time when their own hold over neighbouring Afghanistan was becoming insecure.

Restored to his throne in 1953 by the Americans, and thereafter showered with American aid of every kind, the shah was widely regarded as an American creature and the American government as accomplices in his misgovernment. Via Egypt and Mexico the shah reached the United States in search of the best medical treatment for the disease which would soon kill him, but many in Iran regarded his arrival in New York as the prelude to another American attempt to put him back on his throne. Moreover the younger militants were at one with the vindictive clerics in wanting to get the shah back to Iran and put him on trial for his alleged crimes against the state. The seizure of the hostages was a move towards this goal, and the anti-Americanism which it exemplified and fuelled was useful to Khomeini in rallying the splintered fragments of Iranian society behind himself. At every turn over the next twelve months the ayatollah supported the hostages' captors, whose urge to lay hands on the shah and to humble the United States overrode the milder counsels of those – Bazargan and, after him, Abolhassan Bani-Sadr, appointed president in January 1980 – who wished to restore normal relations with Washington, if only to unfreeze Iranian assets in American banks and, after Iraq's attack, to get spare parts for its weapons. Bani-Sadr was forced into flight by a fresh wave of terror in 1981 of which his successor was an early victim, one of many.

Imperial powers have been wont to accept with a certain equanimity the disasters which befall their servants in foreign lands. But not so the more responsible American public. There was no evidence that the hostages were ill-treated but the mere fact of their detention was regarded as a slur and disgrace not to be borne. Their fate became an obsession and the situation was further inflamed by the Russian invasion of Afghanistan a few weeks later. On the one hand the United States had an interest in preserving Khomeini's rule because he seemed to be the one man who could prevent Iran from falling to pieces; civil war in this sensitive area was not only to be feared on general grounds but also because it might provide the USSR with a legitimate excuse to intervene under its treaty with Iran of 1921. On the other hand the United States saw politicians like Bani-Sadr as its natural allies and wished to strengthen them against the coalition of extremes represented by the ayatollah and the radicals. Washington calculated that it would reinforce the moderates by imposing severe sanctions, for which however Washington needed European and Japanese collaboration. But this strategy was vitiated by a latent contradiction. Bani-Sadr was trying to negotiate goods-for-oil agreements with Europe and Japan, so that American attempts to get these countries to impose sanctions on Iran hampered Bani-Sadr whom the Americans were hoping to bolster and placate.

For their part Washington's allies had little faith in sanctions but were disposed to fall into line partly in order to show solidarity with Washington against a blatant breach of diplomatic practice and partly also to prevent Washington from resorting to force to free the hostages. Force, they reckoned, would not succeed and it might lead to the closing of the Gulf and the loss therefore of shipments of oil far more important than Iran's. Iran responded to this threat of sanctions by making economic agreements with the USSR, East Germany, Czechoslovakia and Bulgaria. These agreements were economically marginal but, for the west, politically disturbing.

Early in January 1980 President Carter told the US Congress that the use of force to rescue the hostages would almost certainly fail and might lead to their deaths. In April he made the attempt. It was a gamble and it failed. Eight Americans were killed and six helicopters and one troop carrier lost. The hostages were not killed but they were dispersed. American prestige suffered. The allies, in the course of preparing sanctions which were to be an alternative to force, were vexed at not being told that force was after all to be used. Moscow was pleased that the world should have another military incursion to talk about besides Afghanistan. Muslim countries felt obliged to rally round Khomeini, whose personal power was strengthened by this episode, and again at the end of the year when Iran was attacked by Iraq.

SADDAM HUSSEIN

In Iraq the Aref régime had been overthrown in 1968 by a coup which put General Ahmad Hassan al-Bakr in the presidency. This was a victory for the Iraqi branch of the Ba'ath Party and more particularly for Saddam Hussein Takriti, the brutally strong man of the new régime. Saddam Hussein remained half in the background until 1979 when he took open control after a plot of obscure origins (probably by a disgruntled rival clan within the Sunni establishment). He looked south rather than west. His first ambition was to reassert Iraq's position in the Gulf. His chief adversaries therefore were not Syria, Egypt or even Israel but Iran and Saudi Arabia, and since both of these were supported by the United States he turned to the USSR and concluded in 1972 a treaty for, among other things, the supply of arms. But Saddam Hussein did not intend to get hitched to the Russian wagon and three years later he concluded an agreement with France for the supply of a nuclear reactor and set up a nuclear research establishment with a staff of 600 engineers. Iraq, which had signed the Nuclear Non-Proliferation Treaty, maintained that all the safeguards prescribed by the International Atomic Energy Agency were being observed, but Iraq's adversaries – particularly Israel – feared that it was preparing to produce nuclear weapons: the plant where the reactor was being built in France was sabotaged, material destined for Iraq was sabotaged at Toulon, an Egyptian nuclear physicist in Iraqi service was murdered in Paris. Iraq also turned to Italy for help in training naval and air officers and for the delivery of ten vessels of war. Given Iraq's geography these could be used only in the Gulf. Contacts outside the USSR and eastern Europe, particularly with West Germany and Japan, were intensified after displays of Russian power in Aden and Ethiopia in the late 1970s. In 1980 Iraqi nuclear installations were bombed by Israel.

Iraq's development was not purely military. The government embarked also on literacy and other educational programmes, technical training, industrial expansion and an agricultural plan designed to enable Iraq to produce all its food instead of a mere quarter. All these measures were based on oil. Oil provided 98 per cent of Iraq's export revenue and financed 90 per cent of the government's investment. Output reached 2 million barrels a day in 1973 and 2.5 million in 1977, and hit a brief peak of 3.7 million before the end of the decade. Iraq became the world's second exporter. (Exports to the USSR remained comparatively trivial: 220,000 barrels a day in 1973, declining to 70,000 in 1979.)

In 1980 Iraq attacked Iran. The causes of this war included the weakness of Iran after the fall of the shah – a temptation to score off Iran; Saddam Hussein's profound dislike of Ayatollah Khomeini, whom he regarded as a religious lunatic; unease about the consequences of Khomeini's Shia fanaticism and intrigues among the Iraqi Shia, an underprivileged majority who staged serious riots at the end of 1979; possibly a suspicion that Khomeini had been involved in the unsuccessful coup of that summer against his régime; and, finally, the perennial question of the Kurds, a sizeable minority in Iraq (18 per cent) with awkward claims in the oil-bearing region of Kirkuk and an even more awkward propensity for allowing themselves to be used by Iran against Iraq.

Saddam Hussein hoped that he had settled the Kurdish question in 1975 (for its antecedents see the note at the end of this Part). After admitting Kurds to the Iraqi cabinet in 1973 and granting autonomy to the Kirkuk region in 1974 Iraq bought from the shah in 1975 a promise to stop Iranian aid to Kurdish dissidents. But Saddam Hussein could not be sure that the shah's promise would be honoured by Khomeini and in any case he resented the price he had had to pay for it and wanted to go back on his own promise. This was the immediate cause of the war. It concerned the respective rights of Iraq and Iran in the Shatt al-Arab.

The Shatt al-Arab carries the waters of the Euphrates and Tigris to the Persian Gulf. These rivers are in Iraq but the Shatt itself marks the frontier between Iraq and Iran, and half-way along its course of 200 km it is joined by the Iranian River Karun. The Shatt is Iraq's only outlet to the sea and it carries also the traffic of the Iranian ports of Khorramshahr and Abadan in the province of Khuzistan, whose population consists of Iranian citizens of Arab race. The Gulf into which the Shatt debouches leads, not directly to the open sea, but to the Straits of Hormuz which are 800 km from Iraq but easily commanded by Iran. When in 1971 Iran had seized the small islands in the straits (the Tunbs and Abu Musa) which Britain had wished to transfer to two of the Arab emirates, Iraq was unable at that date to do more than break off diplomatic relations with Iran and expel Iranians from Iraq.

The first treaty concerning the frontier between the Iranian and Ottoman empires along the Shatt al-Arab was signed in 1555. It has many successors. No treaty ever satisfied both parties to it and the complexities have been increased by shifts of the terrain, waterways and islands and (eventually) by the discovery of oil. At the beginning of the twentieth century the Ottoman empire had secured control over almost the whole of the Shatt and this position was inherited between the two world wars by the successor state of Iraq. It was, however, challenged by the resurgent power of Iran under the Pahlavi dynasty which claimed that the frontier should run down mid-channel. In 1937 a new treaty considerably improved the Iranian position, notably by making the Shatt freely usable by naval and merchant vessels of both states. After the Second World War, and particularly after the 1958 revolution in Iraq, Iran began to put pressures on Iraq. In 1969 Iran abrogated the 1937 treaty. Iraq riposted by declaring the whole of the Shatt to be Iraqi territorial water but Iran's support for Kurdish revolts against Baghdad forced Iraq in 1975 to accept a deal whereby the shah would abandon the Kurds in return for recognition of the mid-channel frontier. In 1978 the fall of the shah again transformed the situation. Iran lapsed into something like chaos and lost the American support which had been so conspicuous an element in the expansion of Iranian power. Iraq on the other hand had been gathering strength since the coup in 1968. In 1980 Saddam Hussein abrogated the 1975 agreement and invaded Iran.

He miscalculated. The war was not the walkover which he hoped for. Khomeini's Iran did not fall to pieces and Iraq became committed to wearing operations which exposed its weaknesses as well as its ambitions. The Iranians stemmed the Iraqi attack despite the disorganizations of their country and the reluctance of the clergy to give the army – a possible rival for power – a free hand. Saddam Hussein, like Ayub Khan in Kashmir in 1965, failed to get the quick victory which was the only kind of victory worth having. The war entered upon years of ding-dong slaughter. Iraq's gamble on a quick victory, inspired by the post-revolutionary chaos in Iran, a lull in Kurdish turbulence and a treasury filled by the oil price rises of the 1970s, failed and with it Iraq's vision of dominating the Gulf and the Arab world. Khomeini was able to throw thousands of conscripted Iranians into battle with revolutionary and religious ruthlessness, insisting that he would accept no terms for peace short of the overthrow of President Saddam Hussein.

Although it all but won the war in the first few weeks, Iraq then made a fateful pause in the false belief that it had done so. Iran held the main Iraqi offensive and an extension of the front northward to Desful; retaliated by attacking Basra from the sea and targets as far north as Mosul by air; achieved limited successes over the next two years; emboldened Syria to block the flow of Iraqi oil through Syria to the Mediterranean; and proclaimed in 1982 war aims which amounted to a demand for Iraq's surrender – the replacement of the Iraqi régime and substantial reparations for aggression. But Iran's military successes were too modest to match these aims. Year after year small gains were made at hideous cost in death, mutilation and destruction but without denting Iraq's resolve. Iraq, thwarted on land, developed a new twofold strategy: it attacked Iranian oil installations (oil exports were crucial in financing Iran's war effort) and internationalized the war by raising fears of an oil shortage as a consequence of damage to shipping in the Gulf and the closing by Iran of the Straits of Hormuz. Iran's export earnings were significantly curtailed but it reacted cautiously to this escalation of the war and refrained at first from interfering with foreign shipping. In 1985 Iraq launched a second major land offensive – the first since the beginning of the war – but this attack and an Iranian counter-offensive made small impact and confirmed the stalemate on land. Even when Iranian forces crossed the Shatt in the following year and subjected Baghdad to regular bombardment no military decision seemed to be in sight; and a further Iranian offensive against Basra in 1987 was equally inconclusive.

In the Gulf, however, Iraq had some success with its economic warfare and its policy of goading Iran into activities which would invite international intervention. Iran's economy was more susceptible than Iraq's to the loss of oil revenue, since Iran was financing the war largely by exporting oil in excess of the quotas prescribed by OPEC, whereas Iraq – although it had lost one of its main oil outlets at the outset of the war – was sustaining its war effort with subsidies from Kuwait and other Arab states: Kuwait, although traditionally hostile to Iraq which had territorial claims against it, was nervous about possible Iranian subversion among Kuwait's Shi'ites who numbered nearly a third of its population. Iraq's new strategy presented Iran with the problem of scaring Kuwait into desisting from aiding Iraq without at the same time scaring western states into active participation in the Gulf against Iran. This was a particularly delicate operation since the war presented President Reagan with an opportunity to settle old scores with Iran over the taking of American hostages in Teheran and new scores over the humiliating exposure of the arms-for-hostages traffic in which the United States (and Israel) had been engaging. Reagan, hoping to succeed where Carter had failed, hoped to get Iran to secure the release of American captives in Lebanon in return for the supply of American and Israeli weapons. This covert manoeuvre was contrary to the United States' principle of never negotiating with hostage-takers or terrorists. It was later justified on the grounds that its aim was to encourage so-called moderates in Iran, but these moderates were for the most part mythical and in so far as they existed their influence was diminished when the American proposals and furtive visits to Teheran were revealed.

By opening a second front at sea Iraq aimed to inveigle the United States into blockading Iran and even perhaps into hostilities against it. The initiative was taken by Kuwait which, economically dependent on the free flow of oil through the Gulf, asked the United States to protect Kuwaiti tankers by allowing them to fly the American flag (contrary to the Geneva Convention of 1959) and to send a substantial naval force to the Gulf where it was likely to become engaged in hostilities with Iran. This request followed an attack on the US frigate *Stark* which was hit by an Exocet missile (albeit launched by Iraq, not Iran) and it was reinforced by reports of the installation of Chinese ground-to-sea missiles at Hormuz and by a judiciously similar Kuwaiti appeal to the USSR for the use of the Soviet flag which raised American fears of an expanded Russian naval presence in or near the Gulf.

Simultaneously with these moves the Security Council unanimously approved a call for a cease-fire in terms which side-stepped Iran's demand that Iraq be labelled the aggressor in the war. Iran therefore prevaricated, while Iraq stalled in the hope of increased American support. The United States was covertly helping Iraq with intelligence, arms and instruction in the use of them and was believed to be contemplating the invocation of chapter VII of the Charter and military sanctions against Iran. In the Gulf the provision of armed escorts for shipping was only moderately successful but it succeeded in dragging reluctant European forces in too and in exposing Iranian mine-laying. But the United States shrank from outright war on Iran in support of the Iraqi aggressor and the war went on until a cease-fire was accepted by both sides in 1988. The next year Khomeini died. He had established a theocratic state and become the international symbol of active opposition to the seamier side of western civilization, but he had also imposed on his country a tyranny as savage as the shah's and a war which entailed vast slaughter and economic catastrophe (inflation at 30-40 per cent a year, plummeting production, negative investment at home and abroad, loss of oil revenues which constituted 90 per cent of Iran's foreign earnings). An armistice and proposals for an exchange of prisoners indicated the prevailing war-weariness after an exceptionally long war. Neither side had defeated the other. The main issue of rights in the Shatt was resolved only when Saddam Hussein, having precipitated a different crisis, abruptly conceded to the new government of Ali Akbar Hashemi Rafsanjani all that he had fought for so ruthlessly and lethally against Khomeini.

KUWAIT AND THE GULF WAR

On 2 August 1990 Iraq invaded and annexed Kuwait. The sheikhdom of Kuwait was something of an anomaly in the Gulf. Much smaller than Iran, Iraq or Saudi Arabia, it was however more populous and richer than the Gulf's other minor states and was a long way away from them: it was a solitary small state surrounded by larger ones. It had been part of the Ottoman empire under the autonomous rule of a family which established its local rule in the eighteenth century and was still ruling in the twentieth. Kuwait was also the subject of special treaties between the Ottoman and British empires. Unlike the rest of the Gulf Kuwait concerned Britain not because of the piracy which impeded British commerce and prompted British intervention, but because of fears of German or Russian expansion to the Gulf by means of railway concessions and Ottoman favours. To allay these fears treaties were concluded in 1899 and 1913, and after the First World War Kuwait became a British protectorate. In the 1930s the new state of Iraq claimed that Kuwait, as a former part of the Ottoman pashalik of Basra, belonged by right of succession to Iraq. More particularly Iraq laid claim to the islands of Bubijan and Warbah at the head of the Gulf and to the tip of the Rumeila oilfield which, mainly in Iraq, extended beneath the frontier into Kuwait. (These claims were not in themselves substantial. The islands were barren, without oil, partly submerged for part of the year, and no help or hindrance to Iraq's commerce to or from the Gulf. Kuwait's extraction of oil from the Rumeila field amounted to about 1 per cent of its yield. But the claims could be used to stalk bigger prizes in northern Kuwait.) In 1961 Britain left Kuwait, which became fully independent and a member of the United Nations. General Kassim repeated Iraq's traditional claims and the ruler of Kuwait, afraid of an Iraqi attack, appealed to Britain for help. A small British force was expeditiously landed, the Iraqis (who had remained a long way from the frontier) subsided, the British troops were quickly withdrawn and were replaced for a short time by contingents from other Arab states. Two years later Iraq recognized the independence and sovereignty of Kuwait, which became a member of the Arab League.

Upon the Iraqi invasion in 1990 the emir of Kuwait and his family fled, a puppet administration was installed and Kuwait was declared to be a province of Iraq. It was also despoiled. Saddam Hussein's motives were greed and need. He had been rearming on credit after his war with Iran and his suppliers had stopped his credit. He spent or pledged perhaps $100 billion on war in a decade, most of it in purchases in countries bound by UN resolutions (which they had approved) not to supply Iraq or Iran. On the eve of his invasion of Kuwait his debt to non-Arab creditors was about $35 billion, much of it owed to commercial companies but underwritten by governments and so ultimately a charge on the public. Kuwait's wealth was fabulous, Iraq's postwar needs were urgent and Saddam Hussein may have believed that Kuwait was ripe for the taking. In 1986 the emir had disbanded the Kuwaiti parliament and in 1989 he rejected pleas to reinstate it; half the emirate's population were immigrants without citizenship or full civic rights; nomadic bedouin were denied rights because they could not prove fixed residence, as too were Palestinians and others, even if they had been born in Kuwait. But if Saddam Hussein was counting on some sort of a welcome he was greatly mistaken. He was no less mistaken about Arab and international reactions. The invasion was an incontestable act of aggression by one member of the UN against another and by one Arab state against another, and unlike the equally blatant act of aggression by Iraq against Iran a decade earlier the attack on Kuwait was also a threat to the interests of the United States and other countries. The appropriation of Kuwait's oilfields considerably increased Iraq's weight in OPEC and its influence over the pricing of oil worldwide and its flow in and from the Middle East. It might furthermore be a prelude to an attack on Saudi Arabia, which would place virtually all Arab oil under Iraq's control, cause widespread political chaos by unseating the Saudi monarchy and weakening other Arab régimes, and precipitate universal economic instability and recession if, with or without an attack on Saudi Arabia, oil prices doubled or trebled.

Saddam Hussein's action was therefore a miscalculation of massive proportions which created an impressively broad coalition against him and, given the gravity of its possible consequences on the one hand and his own stubbornness on the other, a confrontation not easily to be dispelled without war. Over previous decades Iraq had enjoyed western and Soviet help in creating powerful, partly modernized armed forces. To Iraq's aggression against Iran the United States, Britain and other countries had turned a blind eye; some had given significant direct or indirect aid to the Iraqi war effort, as had a variety of Arab states. Even more culpably leaders in most of these states had disregarded Saddam Hussein's slaughter of thousands of Kurds with chemical weapons; in the United States attempts to chastise Iraq through economic sanctions had been thwarted in Congress and the White House. To Saddam Hussein's delusions therefore about the international effects of his second major act of aggression foreign states had themselves contributed and one of Washington's principal concerns in the months after the invasion of Kuwait was to drive home the message that in American eyes this act was not to be condoned as had Iraq's earlier misdeeds. In addition the context in 1990 was changed as well as the stakes. Since the fall of the shah in Iran the United States had become determined to prevent a similar fate overtaking the Saudi monarchy. The Cold War had come to an end with both sides looking for occasions to display co-operation. Arab governments which had supported Iraq against Iran were angered by Saddam Hussein's attack on one of themselves. They feared his pretensions to leadership or dominion in the Arab world – pretensions partly based on his view of himself as successor to Michel Aflaq, the founder of the Ba'ath who died in 1989 in Paris after long years in exile in Iraq from his native Syria.

The American response was two-pronged: invocation of chapter VII of the UN Charter and a distinct and massive American military expedition into the Middle East. But although the American response was swift and vigorous its motives were confused. The occupation of Kuwait provided justification, even an obligation, to take international action to reverse· the invasion by embargo, extending if necessary to blockade and the implementation of these measures by force, under articles 41 and 42 of the Charter (see Chapter 4). But the occupation of Kuwait was not the prime cause of the distinct American action, and the reversal of the occupation not the United States' sole object. Since similar infractions of international law by Iraq (and others) had evoked no such response the despatch to the Middle East of a quarter of a million troops could not plausibly be attributed to Kuwait's fate alone. The cause of this vast effort was plainly something more than solicitude for international law or for the emir of Kuwait; the cause was fear of further aggression by Iraq with the possible consequences already noted and a general fear of the temper and intentions of the régime in Baghdad. Fears of an imminent Iraqi assault on Saudi Arabia were probably baseless, but in the absence of adequate political and military intelligence and appraisal they could not safely be discounted and President Bush – apparently taken by surprise in spite of some timely warnings – felt unable to pause to weigh the question whether such an attack was probable or merely possible. He committed himself therefore to a display and deployment of force which, while it relieved his immediate fears, hampered him in a variety of ways: it narrowed his room for future manoeuvre, cast doubt on the sincerity of the UN prong of his two-pronged strategy (which was the centrepiece of his appeal for international support), sowed seeds of uneasiness about his aims and methods, and created trouble for himself on the home front both by declaring that this obviously offensive capacity was strictly defensive and by committing the United States to huge expenditure and a humiliating need to tout for foreign contributions. The use of overriding force had weaknesses which, unlike the weaknesses of the UN embargo, were accentuated by the passage of time. The embargo, if it worked at all, would do so *accelerando* after an interval which might be longish: the Achilles' heel of the Iraqi economy was its need for grain and therefore for exports of oil to pay for the grain from the spring or early summer of 1991 to the tune of perhaps $2 billion. The American armed threat on the other hand was a threat with diminishing returns since, first, the supplementary 200,000 men and women despatched by Bush at the end of the year did not remotely double the threat posed by the 250,000 despatched at the beginning of the crisis and, secondly, the implied intention to abandon the strategy of embargo alienated the American public and churches and gave opportunistic allies – Syria, for example – a handle which Americans (and Israelis) much disliked.

Bush's initial and impressive success lay in his rallying of international support both at the UN and for his fighting forces in Saudi Arabia: the active Saudi alliance was a *sine qua non*. His problem lay in keeping either coalition together when his purposes appeared to be different from theirs. The forces assembled in Saudi Arabia comprised, besides the essential co-operation of Saudi Arabia itself, units from Egypt, Syria, Morocco, Britain, France, Pakistan, Bangladesh and more (recalling in some instances Cavour's despatch of Sardinians to the Crimean War as part of his diplomatic effort to stand well where it mattered). But this heartening multiplicity created problems. Iraq's aggression meant different things in different places; the aims of the anti-Iraqi alliance were diverse. The common aim was the restoration of Kuwait's independence. Americans, however – and more exaggeratedly the Thatcher administration in Britain – wanted to take the opportunity to overthrow the régime in Iraq or at least Saddam Hussein himself, to exact reparations for the considerable destruction and suffering perpetrated in Kuwait, to initiate war crimes trials, and to secure the destruction of Iraqi weapons including weapons which, however horrible, were not prescribed under international law and were manufactured and held by a number of states. It was this proliferation of aims beyond the original aim of removing the Iraqis from Kuwait which dictated the doubling of the American forces in Saudi Arabia; it also delayed American preparedness to strike since the forces deemed adequate to defend Saudi Arabia remained for months inadequate to conquer Iraq; and it laid bare Washington's concerns over Saudi Arabia and Iraq rather than the liberation of Kuwait.

To Arabs on the other hand, from King Fahd of Saudi Arabia to President Hosni Mubarak of Egypt, Saddam Hussein's offence was above all his flouting of the principles, proclaimed by himself in 1980, that no Arab state should attack another and that all issues in the Arab world should be settled by the Arabs themselves without seeking or provoking non-Arab intervention. By flouting these principles Saddam Hussein put his fellow Arab leaders in the awkward position of having to choose between two unpalatable courses – acquiescing in an extension of Iraq's unprincipled abuse of power, and allying themselves with the Americans whose behaviour and very presence in force in the Middle East were offensive to many Arabs and Muslims. Even those Arabs who chose the latter course were having misgivings a few months later. In November the king of Morocco revived the notion of an Arab conference; by December Saudi Arabia was secretly discussing through Arab intermediaries a possible revision of the Iraqi–Kuwaiti frontiers; Kuwaitis in exile revived an old proposal for a long lease of Bubijan and Warbah to Iraq; in January Mubarak was to be seen in the unlikely company of Qadafi whom he visited in Tripoli along with Assad; and no Arab was untouched by the prospect, floated by Saddam Hussein and cautiously approved by sundry Europeans and the USSR and China, of an international conference on the Middle East which would embrace Israel's occupation of the West Bank, Gaza and Jerusalem and be held in tandem with a settlement of the Kuwaiti crisis or soon thereafter. (The Arab League had at this date twenty-one members: twenty states and the PLO.)

The initial flurry of diplomatic and military activity was followed by an uneasy pause lasting several months, necessitated partly by the very nature of action under chapter VII of the UN Charter, partly by American determination to launch nothing less than an immediately crushing blow and partly by genuine, if tenuous, hopes of scaring Saddam Hussein out of Kuwait by demonstrations of military might but without bloodshed on an unpredictable scale. Between the Iraqi invasion on 2 August and the recourse on 29 November to article 42 of the Charter (supplementing measures not involving force under article 41) the United States held its disparate coalition together while at the same time accumulating more and more force in the Middle East in the hope of ensuring that, in the event of war, a first blow against Iraq (whether in Kuwait or in Iraq was understandably left uncertain) would achieve its purpose without unbearably long and costly operations. Publicly at least the United States side-stepped the uncomfortable question whether these purposes, whatever they might be, warranted a war whose cost in lives, money, general economic disruption and long-term influence in the Middle East might prove disastrous. It also played down, to the point of trying to expunge, that part of the first Security

Council resolution which – besides requiring complete Iraqi evacuation of Kuwait – required also a negotiated settlement of the crisis: negotiation, for the United States (and Britain) meant negotiation after the end of the crisis and not in order to end it. For his part Saddam Hussein played a militarily weak hand with some dexterity but limited success. He seized foreigners in Kuwait (who numbered several thousands and came from two dozen countries) and transported some of them to Iraq where about 340 were dispersed to likely targets in and outside Baghdad as a shield against armed attack; he prevented foreigners in Iraq as well as those in Kuwait from leaving the country; he threatened to set Saudi oilfields and installations ablaze and to attack Israel if he were attacked by the United States; he tried with minimal success to sow discord among the United States' associates and to inflame anti-American Arab and Muslim emotions; he mended his fences with Iran in the hope of securing a sizeable gap in the UN's economic cordon. His refusal to allow foreigners to leave Iraq or Kuwait and his planting of them in and around sensitive installations (a clear violation of the Geneva Convention on the treatment of civilians) provided ammunition for extended denunciations of his law-breaking and barbarity, but the presence of these possible victims was an embarrassment to their governments which had to present themselves as undeterred as well as outraged and criticized unofficial rescue attempts by assorted emissaries – Kurt Waldheim, Jesse Jackson, Edward Heath, Yasuhiro Nakasone, former US Senator John Connally, British MP Tony Benn – whose motives governments tried obliquely to impugn. Women and children were permitted to leave from September, and in December, by which date the propaganda disadvantages of these violations of international law were outweighing their anticipated strategic advantages, Saddam Hussein declared that all who wished might be home for Christmas.

The period of waiting was, if not ended, given a prospective term when the Security Council approved on 29 November a resolution – the culmination of a sequence of twelve – authorizing the use after 15 January 1991 of any necessary measures to secure the removal of Iraq from Kuwait and the restoration of its former rulers. This explicit legitimization of the recourse to war was meant to frighten Saddam Hussein more than it might disturb the American public, whose vivid remembrance of war in Vietnam and its failures remained a prime ingredient in American policy-making. American diplomacy secured the passing of this resolution by twelve votes to two (Yemen, Cuba) with one abstention (China). The firmness of the Arab with the non-Arab members of the anti-Iraqi alliance was bolstered by American and British somersaults in their attitudes to Syria whose anti-Iraqi credentials were held to outweigh its earlier ostracism as a paymaster of international terrorism; and by subventions to Egypt which, while gravely damaged by truncated tourism and by loss of remittances from Egyptians forced to leave their jobs in Kuwait and Iraq, was compensated by substantial remission of its foreign debts and the provision of large new credits (mainly from the United States and Saudi Arabia). Saddam Hussein, although his hopes of dissolving the array against him were slimmer than he believed, continued to insist that his occupation of Kuwait was irreversible. Bush, refusing to believe this proposition, adopted increasingly exclusive eyeball-to-eyeball tactics, complaining repeatedly that the obstacle to a peaceful implementation of the UN's resolutions was Saddam Hussein's failure to understand the American position. In what he described as a last effort to save the peace Bush proposed talks in Washington and Baghdad on the basis however that such talks must be exploratory only and entail no negotiation. Saddam Hussein prevaricated over the timing and objected that there was no point in talks which amounted to no more than a restatement of known positions. The foreign ministers of the two states met, however, in Geneva. Neither side gave way, the Iraqis showing themselves to be as adamant as the Americans. The secretary-general of the UN, Javier Perez de Cuellar, made a despairing attempt to fend off war, meeting Saddam Hussein in Baghdad two days before the date set by the Security Council for the permissible use of force to secure compliance with its resolutions. On the previous day the US Senate and House of Representatives authorized President Bush, albeit by narrow majorities, to go to war.

These last moves were undertaken under the shadow of the deadline which had been adopted by the Security Council on 29 November at American insistence. The United States had wished to set 1 January as the deadline but had accepted 15 January in response to criticism that sanctions were not being given time to work. Since, however, sanctions were never expected to have a serious effect before the spring or early summer the difference between 1 January and 15 January was inconsequential. The deadline was set not because sanctions were falling short of expectations but because Bush could not afford to keep vast forces inactive in the Middle East: their cost, although more than half underwritten by Saudi Arabia, was alarming and the alliance which sustained them was less than solid and in danger of being exposed as such. The crisis was caused by Saddam Hussein's invasion and purported annexation of Kuwait in violation of the UN Charter. It was made unmanageable by the American deployment of large forces whose primary objects were not the redress of this offence but the defence of Saudi Arabia and American interests there and, secondly, the destruction of the Iraqi régime and its warmaking capacities. These aims, whatever their virtues or legitimacy, were incompatible with a peaceful settlement, within the terms of the UN's resolutions, of the crisis provoked by the Iraqi aggression.

When 15 January was reached the United States wasted no time and opened hostilities during the night of 15–16 January. It did so without informing the secretary-general of the UN in whose name the hostilities were launched. But for six weeks these hostilities were muted while the United States and its allies assembled ever larger forces and attempted to win the war by bombardment at long range and without recourse to the hazards of general land warfare. The Iraqi air forces were reduced to impotence and put to flight, seeking refuge in Iran where they were interned; the navy fared no better; Iraqi ground forces, armour and communications were severely pounded and Baghdad was subjected to destruction greater than anything which it had suffered for 700 years. Iraq countered with largely ineffective missiles aimed at Saudi and Israeli cities and by devastating Kuwait City and barbarously maltreating its inhabitants. As the contest's extreme inequality became obvious Saddam Hussein was driven to attempt negotiations but did so equivocally and obliquely. Bush responded by insisting on unconditional compliance with all pertinent UN resolutions, by inviting Iraqis to revolt against their government and by adding conditions of his own in order to maintain pressures for unconditional surrender. Attempts, notably by the USSR, to broker an Iraqi withdrawal from Kuwait on acceptable conditions were rejected by the United States which evaded their discussion by the Security Council and on 23 February, a date chosen ten days earlier, converted the six-week war into a general onslaught for the defeat of Iraq which would, it was hoped, involve the removal or death of Saddam Hussein but would not lead to Iraq's dismemberment, which Bush explicitly disavowed (but was nevertheless a likely sequel to defeat).

The final stage, lasting less than a week, was for the Americans and their allies little more than a *promenade militaire*, for the Iraqis a massacre of fugitives by unopposed air power, mitigated only by surrender. Victory, mainly by encirclement of the principal Iraqi formations, was swift and cost the allied forces extraordinarily few lives. Kuwait was liberated and an unsavoury régime in Baghdad was humiliated. Some 50,000 Iraqis were killed, perhaps 100,000. The infliction of comprehensive and undeniable defeat on a cruel dictator was very widely welcomed but since this American purpose was neither avowed by the United States nor licensed by the UN, the former stood accused of being two-faced and the latter, made use of rather than used, was weakened.

The financial and political costs of the war were heavy, although the latter were not easily assessable immediately. The US secretary of state and the British foreign secretary toured the Middle East and other rich capitals soliciting contributions and succeeded in collecting sums equal to about four-fifths of their costs. They demonstrated, however, that they had embarked on a war which they could hardly afford and, by side-lining the UN, had been obliged themselves to undertake the business of levying contributions which would otherwise have been performed by the secretary-general. The political costs included evident strain on the Soviet–American détente, but the mutual interests of the superpowers took the strain; also a pronounced accentuation of anti-Americanism from Morocco to Iran, but one possibly more clamorous than enduring. These rising costs were evident enough to the United States to persuade President Bush to launch his attack sooner rather than later in February and to terminate it promptly after Iraq's evident defeat even though at that point the overthrow of Saddam Hussein had not been accomplished. The war did not make the Middle East more stable. American intervention to protect Saudi Arabia exposed the inability of that country and others to defend themselves and introduced into the Middle East the spectacle of rules of behaviour – extending from democracy to irreverence to sexual tolerance – which Arab rulers must find hard to swallow and hard to resist. Opposition to the régimes in Egypt and the Maghrib received a fillip. Although Mubarak profited by the cancellation of a quarter of Egypt's foreign debt of $50 billion and the rescheduling of the rest of it (he won a third presidential term in 1993) the government's brutality, corruption, inefficiency and unpopular pro-western policies and mannerisms encouraged a series of attempts on Mubarak's life. The president himself remained the epitome of personal rectitude but failed to curb or apparently notice the growing corruption in high places and the barely concealed use of terror and torture. Jordan's support for Iraq unified the country but bankrupted it. Saudi–Yemeni hostility was sharpened. Kuwait itself was pushed into an uneasy experiment in tempering established absolutism with a degree of democracy. On the future of Iraq, with or without Saddam Hussein, its neighbours had conflicting ideas and the Americans seemingly none. In the Gulf Iran moved closer to recovering its dominant position. To the north the use of Turkish air bases recalled Turkish claims on northern Iraq. The international arms trade, as lucrative as the oil business, was more stimulated than strangled. Israeli intransigence was, if possible, further entrenched. In spite of missile attacks on its cities Israel, pressed and generously rewarded by the United States, refrained from entering the war and used this forbearance to extract large American subventions: the fact that much of this money was recouped by the United States from Saudi Arabia and Kuwait did not go unremarked in the Arab world. Israel improved its relations with the United States and saw its principal enemy battered and humiliated and the PLO – which had denounced Iraq's aggression against Kuwait but otherwise supported Iraq – discredited and weakened. Finally, most immediately disruptive and most appalling, Bush – having failed to secure the death or removal of Saddam Hussein – publicly opined that the Iraqis themselves would do so and in consequence Kurds in the north and Shi'ite and other groups in the south revolted on the assumption that Saddam Hussein's military power had been destroyed. But it had not and it was used against them ruthlessly. The Kurds in particular suffered in battle and in flight misery and death on a scale unusually bitter and devastating even for them. (Further on the Kurds see the Note at this end of this Part.)

Iraq was attacked and defeated because it had invaded Kuwait and was supposed to be intent on invading Saudi Arabia too, but the latter threat (supposing it to have existed) was not to be removed by substituting one Iraqi régime for another. Iraq's aggressiveness, whatever its scale, raised one of the central issues in international politics in the second half of the twentieth century: how to ensure the supply of essential resources lying outside the territorial confines of powerful and avid consumers. The United States and other powerful states were dependent on Middle Eastern oil but unable to secure it by occupying or dominating the relevant areas in the manner of the Ottoman empire or the Anglo–French mandates system. (The Kuwait war was fought to assert the rule of law forbidding one state to appropriate the territory or resources of another.) The alternative to this outmoded imperialism was to secure national interests through international peace and stability and the operation of market forces. When that order broke down, as it did upon Iraq's annexation of Kuwait, force had to be used and the UN Charter so provided. But the *casus belli* was the infringement of national sovereignty, not the interruption of trade or threats to trade. Consequently legitimate international action in defence of national supplies depended on a simultaneous breach of the law relating to sovereign independence. The United States showed that it could and would fight for its interests. The display of American will, no less than the display of immense technical competence, both logistical and operational, was a major event in international affairs but there was a proviso: that there had to be a legal loophole, which in this case was provided by the violation of Kuwait's sovereignty and independence. How the United States would act in the face of a threat to its interests arising from domestic upheaval and not from international aggression was not only dubious but also peculiarly germane to the Middle East

where conflict (Israel apart) came from internal threats to governments
rather than from hostilities between them. Until Iraq attacked Kuwait
no Arab state had attacked another in seventy years and if the Kuwait
war were to restore that pattern the occasions for American or other
foreign intervention would be limited to aggression by or against Israel.

The greatest winner in the Gulf War was Iran. In the politics of the
Gulf Iran had to look to its Iraqi and Saudi flanks. The war eliminated
Iraq for the time being and forced Saudi Arabia into a controversial
role and inordinate expenditure. The shah's misrule and fall and
Khomeini's revolution had weakened Iran but after the latter's death
Rafsanjani consolidated his relations with the military and won a
comfortable victory in elections for the Majlis in 1992. In that year he
felt strong enough to rekindle conflict over Abu Musa in the Straits of
Hormuz by requiring the nationals of all the Arab Emirates
except Sharjah to present special permits before entering the islands:
Iran and Sharjah were joint sovereigns under an agreement made
before the creation of the Union of Arab Emirates in 1971. He
mollified western governments by helping the release of hostages in
Lebanon and manoeuvred to attract western bankers and industrialists
scrambling for a share in new enterprises and contracts. Yet Iran
remained a regional power unable to exercise its regional power to the
full. Tensions persisted between Rafsanjani and the religious interests
which looked to Ayatollah Ali Khamenei as the keeper of the true
Islamic flame in succession to Khomeini. The economy, so far from
recovering, wilted: foreign debts accumulating and unserviceable, the
currency losing value faster than almost anywhere in the world,
industry more than half closed, foreign investment negligible, and
inflation causing widespread despair to which the government's
counter was shriller anti-western propaganda and blatant support for
extremist régimes (Sudan) or subversive movements (Egypt). Like the
first Safavid shahs 500 years earlier the Pahlavi shahs had tried to re-
create an Iranian empire on the basis of shi'ism and modernization,
only to be unseated by Khomeini who pronounced the two
incompatible. Nevertheless Iran, with or without Khomeini – and
possibly stronger without – remained the spearhead of a cultural
revolution which, with echoes in Arab lands, challenged the western
culture whose leadership had passed to the United States but which had
been attacking the Islamic Middle East since Europeans checked the
Ottoman Turks in the seventeenth century at Vienna.

Iraq's exposure to international intervention did not end with the
war. The Security Council remained seized of two issues: protection of
minorities and destruction of weapons of mass destruction in
accordance with the terms of the cease-fire. The council ordained safe

havens in north and south and no-fly zones where Iraqi aircraft were forbidden to operate. UN teams inspected missile sites, nuclear plants and other places where weapons of mass destruction might be made or stored. They were frequently obstructed, although their missions were reinforced by American air attacks on targets within Iraq. In 1993 the inspectors declared themselves satisfied (prematurely, as it turned out) that proscribed weapons had been destroyed but no continuing monitoring régime had been established to prevent reanimation. In 1994 they reported that Iraq had complied with all the Security Council's requirements but in the council the United States and Britain argued that the economic sanctions imposed to enforce the terms of the cease-fire had also other purposes, such as an explicit renunciation of Iraq's claim to Kuwaiti territory. The sanctions caused considerable privation and suffering in Iraq – infant mortality, for example, trebled – but they did not not unseat Saddam Hussein. The Security Council exempted food and medicines but since it also blocked sales of Iraqi oil (except through the UN) it effectively deprived Iraq of the money needed to buy food or medicines. Some members of the council became uncomfortable about maintaining measures which penalized the innocent. At the end of 1994 Saddam Hussein made an armed demonstration towards his border with Kuwait. It was interpreted in the United States as an imminent fresh invasion but was more probably another misjudged gesture by the Iraqi ruler who was forced to retreat in the face of American military might. But the Russian government publicly dissociated itself from the continuance of sanctions and France and China began to vacillate. In 1995 two of Saddam's sons-in-law fled the country, exposing the dictator's most insidious weakness – the hazards of autocratic family rule when the autocrat's sons, sons-in-law and half-brothers fall out.

The most ambiguous outcome of the Gulf War was its impact on the Saudi state which played host to it. On the one hand the ruling dynasty played a conspicuous and successful role on the international stage, presented in public a united front and demonstrated its immense wealth by undertaking to pay a lion's share of the costs of all the anti-Iraqi allies including the United States. On the other hand two Gulf wars seriously depleted the country's reserves and, concurrently with falling oil prices, halved its revenues and forced the régime to cut its expenditure in 1994 by 20 per cent. In the longer term Saudi Arabia was still the world's largest producer of crude oil and the possessor of a quarter of the world's proved oil reserves. Although not a modern industrial power it was a financial power backed by a large part of the world's most essential industrial commodity. But oil was not enough to secure the stability of a régime whose two main supports, the dynasty itself and the prosperous middle class, were both growing larger but not more coherent: the former counted sixty princes in an upper echelon whose number created opportunities for discord, while the latter was not immune from doubts about the régime. There were pressures for some degree of social reform and financial modernization, particularly from the growing number of Saudis educated abroad; the labouring class, also growing, was increasingly foreign, and the kingdom's close relationship with the United States was potentially unpopular, not least with sections of the Muslim establishment. King Fahd made conciliatory gestures to his Shi'ite minority, proclaimed an amnesty for political offenders, convoked a new (advisory) majlis and instituted regional councils under the authority of princes of the royal house. But he refused to tolerate a Committee for the Defence of Legitimate Rights set up in 1993 and its chairman Muhammad Masari fled to London in the following year. Prodigious overspending, including the costs of two Gulf Wars, presented the government with unfamiliar economic problems. After a heart attack, the king briefly transferred effective powers to his half-brother, Abdullah, at the end of 1995, a shift in power rather than policy and a postponement of whatever crises might be latent within the existing order.

13 THE ARABIAN PENINSULA

SAUDI KINGDOM

The Saudi kingdom is a historical alliance which has lighted on a crock
of gold. In the eighteenth century the Saudi clan and the puritan sect of
the Wahabis formed an alliance which was so far successful that it got
possession, early in the nineteenth century, of the holy places of Mecca
and Medina and pushed its tentacles as far afield as the areas now called
Syria and Iraq. Success was temporary, for with the help of Egyptian
arms the Ottoman sultan threw the Saudis back into the desert whence
they had erupted and for the rest of the century they had to struggle
with neighbouring clans to stay alive. They barely did, but fortune
returned to them with the young Abdul Aziz ibn Saud (c. 1880–1953)
whose sword created a new realm, called from 1932 the kingdom of
Saudi Arabia. The crown of this kingdom belongs to the family of Saud,
all the sons of a deceased monarch having claims over grandsons. Abdul
Aziz was succeeded by four sons: Saud (1953–64), Feisal (1964–75),
Khaled (1975–82) and Fadh; and in 1995 there were in existence some
thirty more princes of this generation and perhaps 6,000 adult princes
related to Abdul Aziz in different degrees. Territorially the kingdom was
nearly coextensive with the Arabian peninsula, bounded by the strategic
waterways of the Persian Gulf, Indian Ocean and Red Sea; but to the
east and south a number of principalities and republics impaired the
tidier pattern which seemed to have been designed by nature but was
not yet accomplished by Saudi man. The departure of the British in
1971 from the Gulf and Aden removed one great power from the area
without substituting another.

Saudi Arabia was a country of great size and, after the Second World
War, vast wealth. But in human terms it was a small state. Its population
in 1990 was no more than 6–7 million, but some estimates make it
much smaller, and about a third of the inhabitants were not Saudis. The
richer it grew the more tempting its riches (which lay in the north) to
a predator. It was therefore driven by prudence to seek a powerful ally
and to build powerful armed forces. The ally – and main supplier of the
hardware for these forces – was the United States.

American penetration of the Middle East began in Saudi Arabia where the American oil company Aramco concluded in 1950 with the Saudi government a fifty–fifty deal which was considerably more generous than the terms traditionally vouchsafed by British and French companies and was designed to enable American companies to compete with and ultimately displace the British. From this commercial vantage point the United States developed a political relationship with the dependably anti-communist Saudi royal house. This alliance lacked the fervour of Washington's commitment to Israel or the thrust of its alliance with the shah of Iran, but it was the one abiding feature of its relations with the Arab world and had special strategic connotations which increased with the growth of Russian naval power. For much of the postwar period Americans contrived to shut their eyes to the fact that the Saudis were deeply opposed to both Israel and Iran, the two principal pillars of the Middle Eastern policies of the United States.

One legacy of the creation of the Saudi kingdom was its opposition to the Hashemite descendants of the sherif of Mecca whom Abdul Aziz had evicted from his temporalities and spiritualities alike. These descendants ruled in Jordan and (until 1958) Iraq. They were also the butt of Nasser's reformist drive and so provided a bond between the conservative Saudi monarchy and revolutionary Egypt. Abdul Aziz's successors were anxious to escape from an isolation which was no longer safe for a country so rich but so under-equipped in human and material terms; and they began in Cairo. The association was, however, half-hearted. Nasser was anti-American but King Saud was not. Egypt and Saudi Arabia took opposite sides in the civil war in Yemen. King Saud's conservative instincts proved stronger than his anti-Hashemite inheritance, so that in Jordan in 1957 he supported King Hussein against a Nasserist opposition. An ambitious ten-year pact between Saudi Arabia, Egypt, Jordan and Syria, concluded in 1957, faltered from the start and quickly went the way of the numerous inter-Arab pacts of these years when Arabs were finding it difficult to decide which of their kin were friends and which were not.

After Nasser's death in 1970 King Feisal, who had been the power behind the throne even before he came to occupy it in 1964, moved towards the leadership which had been slipping from the more exuberant Nasser – particularly after the war of 1967 with Israel and the failures of the Egyptian expeditionary force in Yemen. Feisal had already in the 1960s strengthened Saudi links with the United States and improved relations with the shah; he visited Iran in 1965 and concluded an agreement on the ownership of the possible submarine wealth of the Gulf; and in 1968 the shah went to Mecca. Feisal's leadership in the Arab world was evident at the conference of Arab states at Rabat in 1969, where he succeeded in institutionalizing, by the

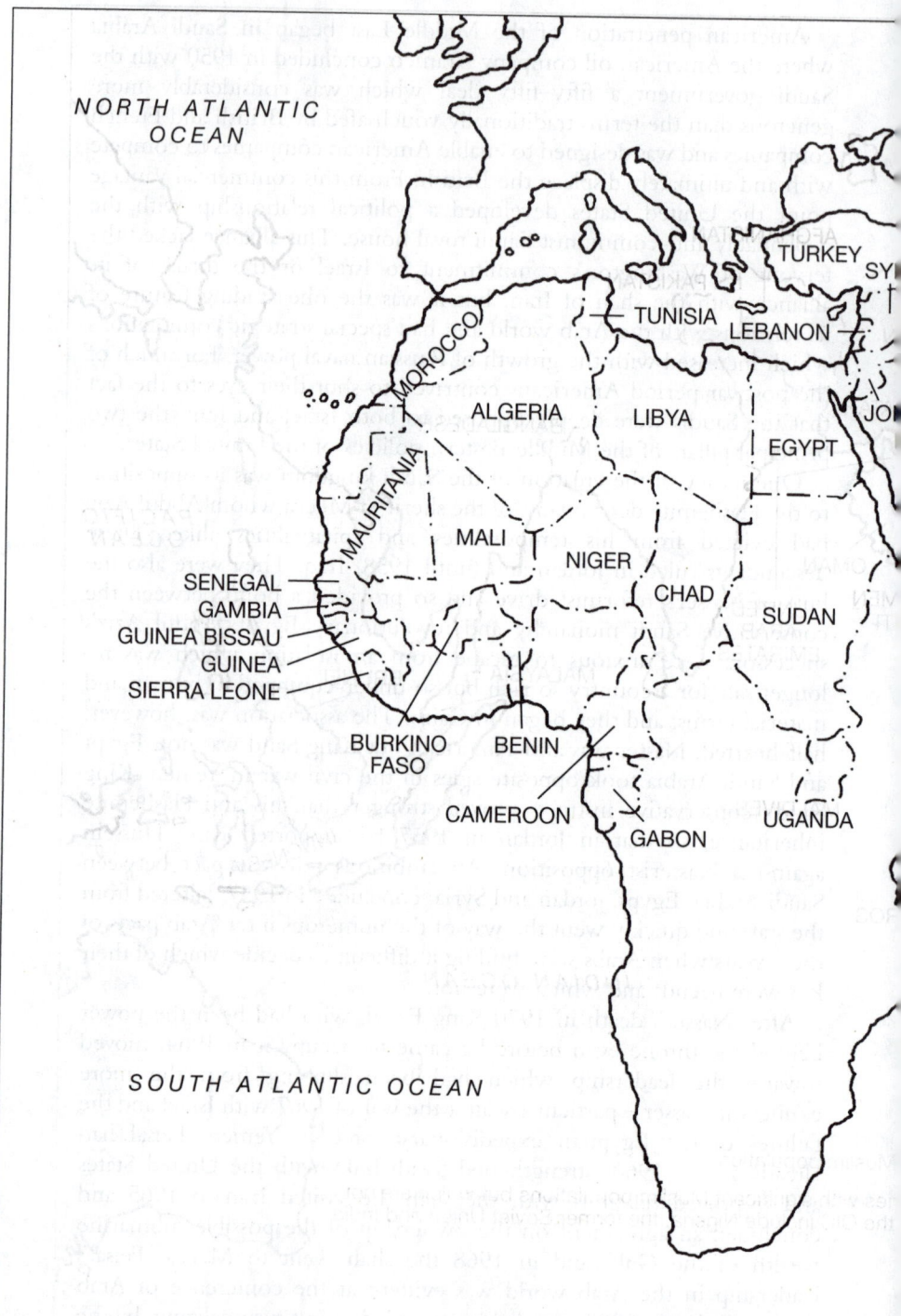

13.1   The Islamic world

AFGHANISTAN

PAKISTAN

N

ATAR

OMAN

MEN
TI

UNITED
ARAB
EMIRATES

MALDIVES

ROS

BANGLADESH

MALAYSIA

BRUNEI

INDONESIA

*PACIFIC
OCEAN*

INDIAN OCEAN

Muslim population

ries with significant Muslim populations but which are not
the OIC include Nigeria, the former Soviet Union and India

creation of a standing Islamic Conference, an idea which he had cherished for some years. Egypt became a natural part of this scheme when Sadat evicted the Russians from Egypt in 1972 and, much more conservative than Nasser, stemmed at the centre of the Arab world the revolutionary current which had characterized it from the early 1950s. (But Egypt, having forfeited its place in the Muslim family as a result of its separate peace with Israel at Camp David, was not invited to the Islamic assembly which, at Taif and Mecca itself early in 1981, brought together forty-two monarchs and other heads of states covering northern Africa and southern Asia from Morocco to Indonesia – an imposing advertisement of slowly maturing Saudi tactics.)

The war of 1973 further enhanced Saudi Arabia's international importance. For the first time an economic weapon – oil – was brandished almost as frighteningly as tanks and aircraft. Saudi Arabia cut its production by 30 per cent, more sharply than any other exporter. By this action it put its anti-Israeli before its pro-American policy. The price of oil shot up from $3 to nearly $12 a barrel in a few months. Saudi Arabia's GNP rose by 250 per cent in a year (and went on rising: oil revenues of $4 billion in that year passed $40 billion four years later). Saudi action had profound political effect as the rest of the world digested the implications of a cutback in oil exports, while at the same time Saudi Arabia's great wealth rose to legendary proportions. By 1980 one-third of the financial reserves of the whole world outside the USSR and its satellites belonged to Saudi Arabia. Nor was this a fleeting boom, for Saudi Arabia possessed also a quarter of the world's known reserves of oil, was producing 9.5 million barrels a day at comparatively low cost and could increase this production by about 70 per cent with little effort.

Saudi Arabia was therefore more of a prize than ever. But the world in which it lived was becoming progressively less tranquil and there were doubts about its own internal stability.

The rule of the royal family, however bizarre or anachronistic in western eyes, was reasonably well assured so long as the princes maintained the historic alliance with Wahabi Islam and established a fresh understanding with the classes, civilian and military, which worked the new institutions engendered by wealth and power. The princely house was on the whole successful in these tasks and in preserving its own discipline. The royal family was extensive, all powerful and very rich. Its stability was maintained by a strict hierarchy. Which prince would be the next king was well known, and which would follow after him. The thousands of lesser princes had their allotted places in what was in effect a *nomenklatura* of aristocrats. Rarely did any of them step out of line. The régime appeared to be impregnable so long as there were sons of Abdul Aziz to occupy the throne, but it was evident that the system would be more difficult to operate, the hierarchy more difficult to define, when the first generation of brothers and half-brothers was followed by the next generation of – far more numerous – cousins. Nevertheless the monarchy had a nasty shock when in 1979 250 Oteiba bedouin seized the Great Mosque at Mecca and held it for two weeks in protest against the worldliness and corruption which were gaining ground in Saudi society – in spite of draconian enforcement of rigorous laws. This incident coincided with riots, stimulated by Khomeini's successes and precepts in Iran, among the Shia minority – a very small minority of 120,000 but a major component of the workforce in the oilfields. Saudi society was also vulnerable by reason of its backwardness and lack of professional and technical skills, a weakness which obliged it to employ large numbers of workers from other Arab countries and from east Asia (Korea, Thailand, the Philippines) and to devote to education as large a share (15 per cent) of government spending as it allotted to defence in the five-year plan for 1975–80 and thereafter.

In external affairs Saudi Arabia was embarrassed by the failure to resolve the Arab–Israeli dispute and the consequent divergence between itself and the United States. Saudi Arabia kept out of the more militant aspects of this dispute but it was in a sense more committed than any state. In politics most things can be compromised, even frontiers, but the Saudi claim against Israel was not political. It was religious: the demand by the custodian of holy places for the return of Jerusalem, another holy place, to Muslim hands (but not to Saudi Arabia itself). No Saudi prince was willing to abandon this claim, which the west consistently failed to take seriously enough, and it united the Saudis with King Hussein of Jordan, whose grandfather had been murdered in the Al Aqsa mosque in Jerusalem. At the conference in Baghdad summoned to protest against Camp David Saudi Arabia tried to moderate the more extreme Arab reactions to Sadat's separate peace with Israel and at the same time retain the leading role which Iraq might snatch out of the heat of general Arab indignation.

In the wider conflict between the superpowers the advent of the Russians to Aden and the Horn of Africa created a new war zone in the Indian Ocean with all eyes trained on the Straits of Hormuz, traversed by 140 ships a day, three-quarters of them oil tankers. Saudi Arabia's traditional border concerns acquired an added international dimension. These concerns were: first, to the south-west, Yemen; secondly, to the south, Oman and the ocean; and thirdly, to the east, the Gulf and all its riparian states.

YEMEN

Yemen, part of the Ottoman empire from 1872 to 1919, was ruled until 1962 by hereditary imams whose rule was one of the least amiable and admirable in the world. Yemen had a coastline stretching northward from the Bab el Mandeb along the Red Sea but no access to southerly waters. By the treaty of Sana in 1934, made between Britain and the Imam Yahya (1918–48), Britain recognized the imam as sovereign and accepted an adjournment until 1974 of territorial disputes arising out of Yemeni claims against the protected sheikhs and the colony of Aden. The Imam Yahya's policy was to temporize but his son Ahmed revived Yemen's claims and argued that constitutional changes in the colony and in the protectorate were a breach of the treaty of Sana inasmuch as they prejudged matters which were to be settled in 1974 and created an anti-Yemeni group intended to disrupt the Yemen which Britain had recognized. Border affrays resulted and the imam concluded in 1956 the treaty of Jidda with Egypt and Saudi Arabia and in 1958 the federal association with the United Arab Republic.

In 1963 the Imam Ahmed died and was succeeded by his son, Muhammad al Badr, who in 1956 had visited Moscow and Beijing. A revolution broke out immediately and a republic was proclaimed by Brigadier Sallal who invoked Egyptian help under the treaty of Jidda. The imam invoked Saudi help under the same treaty, and there ensued a mixed civil and international war, not unlike the Spanish civil war of the 1930s. At Riyadh in August 1965 Nasser and Feisal agreed to discontinue their aid and withdraw their troops; the Yemenis were to instal a coalition and hold a plebiscite at the end of 1966 to decide their country's form of government. But this agreement was abortive, troops were not withdrawn and Nasser later said that Egyptian forces might stay indefinitely. After the war with Israel in 1967 he had to change his mind again and withdraw them. The war ended in 1970 in compromise. The republic prevailed but royalists joined its government.

At precisely this time the adjacent Aden protectorates and colony were being vacated by Britain in the course of its retreat from global power, a retreat which affected the Persian Gulf as well as Aden and the command of the southern entrance to the Red Sea.

The port of Aden was in the possession of the East India Company and then of the government of India from 1839 to 1938, when it became a British colony. It was given a legislative council in 1947 and elected members were introduced in 1955; under a new constitution in 1958 the elected members became a majority of the council. During the period of Indian government (first as part of the Bombay presidency and finally for a few years as a separate province under the direct control of Delhi) Adenis complained that they were a neglected outpost of the

Indian empire. After the transfer of their affairs to the British colonial office nationalist demands for independence appeared and waxed. Important as a port for 2,000 years, Aden also became in the middle of the twentieth century the site for a big new oil refinery and for the headquarters of Britain's Middle East Command. Conservative ministers therefore decided, and stated with incautious boldness, that nationalist aspirations could not be allowed to go to the lengths of independence. The nationalists, led by Abdullah al Asnag, the secretary-general of Aden's trade unions, proceeded to press their views by means of strikes which seriously threatened a base so dependent on native Adeni and immigrant Yemeni labour, and by boycotting the electoral processes with which Britain had hoped to satisfy local political aspirations.

The adjacent Aden protectorate consisted of twenty-three sheikhdoms divided for administrative convenience into a western protectorate embracing eighteen sheikhdoms and an eastern protectorate containing the other five. All these principalities had entered into protection agreements of some kind with Britain between 1839 and 1914, and Britain had performed a useful pacificatory role in this part of the world. After the Second World War Britain negotiated new treaties under which British political officers were appointed to advise the sheikhs who agreed to accept the advice given them except in relation to Islamic law and customs. In the western principalities the appointment of a new sheikh had to be confirmed by the British governor of the colony. The sheikhdoms and the colony constituted a geographically compact, religiously homogeneous area, but the colony differed from its surroundings in being populous, comparatively rich and hostile to the monarchical principle. The nationalists in the colony envisaged a union with the protectorate territories, and ultimately with Yemen also, but not under their existing rulers.

Aden was a small and neglected world of its own until Britain's final departure from the Suez base in 1956. In that year a British minister told the colony's legislative council, in the course of a speech which was particularly patronizing and insensitive in its references to self-government, that the British government foresaw no possibility of changes in Aden's affairs and was confident that this immobilism would be welcomed by a vast majority of its inhabitants. Later in the year the blocking of the Suez Canal by Nasser as a result of the Suez War caused unemployment, strikes and unrest in Aden. At the same time the sheikhs became aware of the threat to their way of life. In 1958 the sultan of Lahej decided to take the road to Cairo, while in 1959 six of the western sheikhs, caught between the devil of Adeni nationalism and the deep seas of Yemeni subversion, formed the federation of Arab emirates which the British had been working for in vain since 1954. This group, renamed the Federation of South Arabia in 1962, was gradually enlarged but never embraced all the sheikhs in spite of British grants fifty times bigger than those provided theretofore.

Britain, intent above all on retaining the Aden base as other bases in Kenya, Egypt and Cyprus vanished or became unreliable, decided in 1963 to attach the colony to this new federation, uniting the sheikhs and the merchant class and circumventing the nationalists. Under the British scheme Britain retained its sovereignty in the colony; it undertook not to extract the whole colony from the federation, although it might withdraw parts; during the seventh year of this symbiosis, but not before or after, the colony might of its own volition secede but if it did it would have to revert to colonial status and could not become an independent state; whereas each normal member of the federation had six seats in the Federal Council, Aden was to have twenty-four. This bizarre concoction intensified nationalist agitation in the colony. The Adeni nationalists refused to co-operate with Britain or recognize the federation, turned to Egypt for help and took to terrorism to accelerate the British departure and ensure the collapse of the rest of the British plans. The federated sheikhs tried to get promises of continuing British military support which the British Labour government was unwilling to give since they would virtually negate the policy of withdrawal and retrenchment and would entangle Britain in Arab feuds: by accepting a military commitment while abandoning political power Britain would get the worst of two worlds. The British government preferred to accelerate departure. Aden and its hinterland became in 1967 the independent People's Democratic Republic of Yemen (PDRY). The departure of the British, who had provided most of the jobs, and the closing of the Suez Canal in the same year plunged the new state into economic distress which accentuated its inherent instability. It was also at odds with its northern neighbour, the Yemeni Arab Republic (YAR), whence it was invaded with Saudi help in 1972. A short war ended with talk – but no more – of amalgamating the two Yemeni states. The island of Socotra, south of Aden, a British colony from 1880 to 1967, became part of the PDRY.

From the Saudi point of view a union, or federation, of the PDRY and YAR would be an advantage only if the resulting entity were right-wing and not left. Saudi Arabia wished to exercise control over the YAR and to seduce the PDRY from Russian influence. In 1976, after the secret meetings in Cairo, Saudi Arabia established relations with the PDRY. This move, a logical step in Feisal's diplomacy but a surprise to those who overrated the role of ideology in international politics, was made possible by the termination in the previous year of the rebellion against the sultan of Oman in his Dhofar province, in which Saudi Arabia and the PDRY had been on opposite sides. Thereafter Saudi Arabia and the USSR vied with each other in offers of aid to the PDRY.

In the YAR the president, Colonel Ibrahim al-Hamdi, was assassinated in 1977 after three years of power and was succeeded by Colonel Ahmad al-Ghashni who was assassinated a year later. Although these murders seemed motivated in the main by tribal feuds there was also some reason to suspect al-Hamdi of trying to play a separate game of his own with the PDRY, and in 1979 the PDRY invaded the YAR in company with discontented refugees from the YAR. In the same year President Abdel Fattah Ismail of the PDRY concluded a twenty-year treaty with the USSR. He was succeeded in 1980 by his prime minister Ali Nasser Mohammed, who wanted to find a way of reconciling the Russian connection with better relations with the YAR and Oman. The two Yemeni states were united in 1990 under the presidency of the president of the YAR. The new state, whose capital was Sanaa, had burgeoning oil revenues (mostly from the north) but the two armies were not merged and in 1994 an attempt to re-establish a separate southern state led to the conquest of the south by the north.

MUSCAT AND OMAN

In the sultanate of Muscat and Oman, where a coup brought a new sultan to power in 1970, the People's Front for the Liberation of Oman and the Arab Gulf (PFLOAG) sustained with Russian and Chinese help a civil war which the sultan, with British and Iranian help, overcame in 1975. Behind Britain and Iran stood the United States.

Between the powerful American fleets in the Mediterranean and the South Pacific lay a gap. For twenty-five years after the Second World War this gap was filled by the British who, up to the mid-1960s, insisted that they would go on filling it. The Defence White Paper of 1957, the first to appear after the fiasco of the Suez War, contained a mixture of old and new ideas. It still envisaged local forces not only in Aden and Cyprus but also in Kenya (which, however, on becoming independent in 1963 granted Britain only limited facilities by an agreement of March 1964); but it also envisaged a carrier group in the Indian Ocean. The Kuwait operation of July 1961 (see below) reinforced arguments in favour of local garrisons for acclimatizing troops, since between a quarter and a half of the men flown to the scene of that action from temperate climes had been quickly prostrated by the heat. On the assumption that such operations remained an inescapable part of Britain's lot in the world, tropical bases seemed essential. The White Paper of 1962 reiterated the need for a British presence to assure stability and with it the need to maintain forces in Aden and Singapore.

But the arguments and assumptions of the framers of Defence White Papers were challenged by those who counted the cost of these establishments (especially after the Labour victory of 1964 in the middle of a financial crisis) and by those who believed that bases in Arab territory created political ill will out of proportion to their military usefulness.

PERSIAN GULF

Finally, the Persian Gulf. The Gulf became in the 1970s what Berlin had been in earlier decades – a point so charged with tension that a dispute involving its navigation or installations might prove to be beyond the control of statecraft.

The Gulf is an arm of the Indian Ocean which here makes a long and narrow penetration into the Middle East, dividing Arabia on the one hand from Iran on the other. This arm of the sea is at its narrowest at the Strait of Hormuz. Southward lies the Gulf of Oman leading to the ocean, northward the Persian Gulf. The whole of the eastern shore of both gulfs is Iranian territory, but the opposite shore counts fourteen sovereigns. At the head of the Persian Gulf is Iraq and next to it Kuwait, separated by 560 km of Saudi coastline from the other principalities – the island of Bahrain, the promontory of Qatar, the Trucial States and lastly the corner state of Muscat and Oman. Kuwait's political and business affinities have been with its Arab neighbours to the north and its sons have gone to Cairo or Beirut for their education, whereas the links of the remaining Gulf states have been rather with Pakistan and India and their sons have gone to school in Karachi.

Britain's initial concern in these waters was to secure a monopoly or dominance over the routes to India and to protect British trade from the lawless depredations of the so-called piratical sheikhs who, at the southern end of the Persian Gulf, lived off piracy and earned for this strip of land the name of the Pirate Coast. From 1820 onwards the British imposed a maritime peace by a series of treaties or truces with these Trucial sheikhs and in 1861 extended the system northwards to Bahrain by an agreement which pledged Britain to protect Bahrain in perpetuity. Later in the century Britain negotiated fresh agreements by which it assumed control over the external policies of the sheikhdoms and also of Kuwait and Qatar. Finally, as pearling and oil made the states in this area increasingly important for their own sake and not merely as pieces of territory adjoining an important trade route, a series of twentieth-century agreements gave Britain exclusive rights in the commercial exploitation of local riches but without any control over or responsibility for internal affairs.

Throughout most of this period of developing British control the landward frontiers of these states, lying vaguely in uninhabited and supposedly worthless country, were of no concern, and the obligations contracted by Britain as a part and consequence of the maritime truce remained for a long time unquestioned since Britain had additional reasons for remaining in the Persian Gulf and was undisturbed by commitments to local rulers in an area which it intended to go on policing by naval and air forces for complementary purposes. These purposes were the continuing need for communications with the east and the new traffic in oil.

In the endless traffic between Europe and Asia the Middle East must either be crossed or circumvented, and in this context the Arab nationalism of the twentieth century affected Europeans in exactly the same way as the Muslim conquests which posed the alternative ventures of the crusades and the voyages round Africa. In the heyday of British power in the Middle East the British travelled freely across it, but after the Second World War the direct central route was lost as one Arab government after another denied to Britain the special rights and facilities which it had previously enjoyed by treaty or by occupation. The withdrawal from Palestine and the Canal Zone, the revolution in Iraq in 1958 and the abrogation of the Anglo–Jordanian treaty at the time of Suez in 1956 eliminated this route and made it necessary to find a southern or a northern detour. The southern route lay through Libya and the Sudan to Aden, but in 1964 the Libyan government sought a revision of its 1952 treaty with Britain and in the same year the Sudan attached virtually prohibitive conditions to overflying rights. Going south meant flying across central Africa or even southern Africa, both routes being politically awkward as well as expensively long. The northern alternative, by Turkey and Iran and the Persian Gulf, remained therefore valuable, even though it could no longer be described as essential since technical developments were opening up a new, if arduous and expensive, route from Britain westward to Singapore.

Communications provided therefore an argument, effective though not conclusive, for staying in the Persian Gulf a little longer provided it was understood that the days of the British presence were already numbered. Oil provided only a weaker argument. The importance, even the growing importance, of Middle Eastern oil to Europe was undeniable, but the policy of ensuring supplies by a physical presence seemed increasingly dubious and anachronistic. Europe, unlike the other major consuming areas, the United States and the USSR, had become dependent on Middle Eastern oil. Britain for example, which before the Second World War imported more than half its oil from the American continent, was by 1950 importing half its needs from Kuwait alone, and it was estimated in 1965 that Europe's annual consumption of 300 million tons would rise in fifteen years to 750 million tons, despite the greater use of natural gas and nuclear power. Important discoveries of oil in the Sahara in 1956 and Libya in 1959 would decrease the proportion of Middle Eastern oil in Europe's consumption but not the total amount of Middle Eastern oil required. Oil was thought likely to remain cheaper than gas or nuclear power or submarine hydrocarbons, and Middle Eastern oil was cheaper than other oil, but this European dependence was not as frightening as it seemed to a number of European statesmen since Middle Eastern producers were on their side heavily dependent on their European customers. By the mid-1960s when a British retreat became a reality the oil revenues of the states concerned exceeded £2,000 million and provided between 70 and 95 per cent of the budgetary income of the several producers. Neither the increasing demand from Japan nor the prospect of China's entry into the market seemed likely to counter the need of the Middle Eastern producers to sell their oil to Europe, and if this were so military bases were as irrelevant to the flow of oil as to the supply of any other commodity.

There was, however, a further argument of a political rather than an economic nature. The British, it was said, had gone to the Persian Gulf to keep the peace and provide the stability without which commerce is endangered and they had done so. A British withdrawal could be followed by disorders which would interrupt the flow of oil. These disorders might be the result of sabotage or of frontier disputes between the states of the region. Sabotage, however, is not a thing which regular military units are particularly suited to prevent, while one of the more usual causes of sabotage is nationalist resentment against the presence of foreign garrisons. Frontier disputes certainly existed: Iran claimed Bahrain; Iraq claimed Iran's province of Khuzistan and also Kuwait; Saudi Arabia had disputed frontiers with some of its smaller neighbours, many of whom were at feud with one another. So long as such disputes persisted Britain might claim to be rendering a service by remaining in the area and performing the policeman's task which it had once exercised over half the world.

In the Trucial States Britain tried but failed to promote a federation as a first step towards release from its perpetual obligations. Britain was not supposed to interfere in the internal affairs of these states but was often obliged to do so because of their mutual disputes, the shortcomings of some of their rulers and the aid which the poorer ones required. The rulers, though no longer in need of protection from the seaward, felt the need for protection against Saudi Arabia and against their own not very numerous subjects. The ruler of Qatar (which had loosely formed part of the trucial system in the early nineteenth century but had left it in the 1860s) asked for British help against internal troubles in 1963 but was refused it. The principal problem, however, was the relationship between the sheikhs and Saudi Arabia. Britain wished neither to let the sheikhs down nor to prop them up indefinitely.

Britain's relations with the Saudi royal house had been traditionally good, notably in the days of King Abdul Aziz ibn Saud, although less so after the accession of his son Saud in 1953 and the substitution of American for British influence in Saudi Arabia. Britain was also embarrassed by the apparently complete imperviousness of the Saudi régime to the slightest touch of modernity, except in the accumulation of oil royalties which were expended by the royal family with lavish uselessness. It did not seem right to abandon states once redeemed from piracy to the mercy of a big neighbour still noted for slavery. Moreover in 1955 a running dispute over the Buraimi oasis led to a minor military confrontation. This oasis, a collection of ten villages partly in the possession of the sheikh of Abu Dhabi and partly in that of the sultan of Muscat, was coveted by Saudi Arabia whose claims had been rejected by Britain for a generation. The dispute, which involved the two powerful forces of honour and oil, was referred to an arbitral tribunal at Geneva but the Saudis, having too little faith in their cause or in the tribunal, bribed witnesses on such a scale that Britain was moved to public protest. The Saudis also sent troops into the villages, whence they were forcibly expelled by the British. Yet Britain was loath to pursue the quarrel, which died down once more.

As in India, so too in the Middle East, the positions of power vacated by Britain were transmitted to no single successor. The post-British régime in the Gulf was the outcome of negotiations between Britain itself, Iran and Saudi Arabia. The shah renounced the Iranian claim to Bahrain which became in 1970 a fully independent member of the United Nations. Populous and wealthy enough to stand on its own feet, it was unwilling to become a member of the proposed new federation of Gulf states, unless it were accorded an equivalently overwhelming representation in the projected Union of Arab Emirates. This federation came into being at the beginning of 1972 but without Qatar and Ras al-Khaimar (one of the seven Trucial States), both of which chose independence. All existing treaties with Britain were abrogated. Iran, pleading strategic exigency, seized the islets of Abu Musa and the Tunbs in the narrows between the Persian Gulf and the Gulf of Oman, causing Iraq to break off diplomatic relations with Iran and expel Iranians from Iraq. The new order rested essentially on agreement between Iran and Saudi Arabia, on the latter's inability to secure at this stage full control over the western shores of the Gulf to match Iran's control on the other side, and on the fortuitous absence of Egyptian and Iraqi voices owing to the former's defeat in 1967 and the latter's continuing war with the Kurds and other internal weaknesses; but none of these factors was necessarily permanent. The assassination of Feisal in 1975 and the peaceful succession of his brother Khaled did not disturb the pattern, but in the same year Iraq began to recover its freedom of action. The Gulf Council, created in 1981 on Saudi initiative, was part of the Saudi effort to counter both Iraqi and Iranian influence: it embraced Saudi Arabia, Kuwait, Bahrain, Qatar, the United Arab Emirates and Oman.

IRAQ

Iraq is the link between the western and eastern sections of the Middle East. More than any other state in the Middle East Iraq looks both ways: westward over west Asia, eastward to Iran and therefore to the Gulf. In 1918 the Ottoman order in the Middle East was dismantled. It was succeeded by an Anglo–French order which was dismantled in its turn after a second world war. The main outcrops of the Anglo–French order were Palestine and Lebanon. They became the battlegrounds of the next phase. Further east, where the Ottoman Turks had confronted Iran, they had no single successor except in so far as Britain ruled the Persian Gulf. With the departure of the British Iraq and Saudi Arabia disputed the primacy in the area, the Iraqis as champions of Arabism and the Saudis as the guardians of Islam's holiest places: a new sultan and a new caliph. And both feared a modern Iran which rejoiced in their quarrels and profited from them.

NOTES

A. THE KURDS

The Kurdish peoples trace their tribal names further back in time than any other peoples in the world and their presence in western Asia for about 4,000 years. In recent centuries they have been divided between Turkey, Iraq and Syria. In the mid-twentieth century they numbered about 30 million, the greater part in Turkey (16 million), Iran (8 million) and Iraq (4 million). There were about 350,000 of them in Syria and perhaps 200,000 in the Armenian SSR and Soviet central Asia. Their fortunes had been in decline since the days of their twelfth-century sultan known to Europeans as Saladin.

Between the wars the Kurds in Turkey were among the opponents of Ataturk's rule. There were Kurdish risings in 1925, 1930 and 1937 but after the Second World War – and particularly during 1950–60 when the Democratic Party was in power – relations eased. There was some recrudescence of racial hostility in the 1960s but the Kurds' worst enemy was the poverty of eastern Anatolia, an area of agricultural decrepitude. In Iran the Kurds were harried by Reza Shah between the wars and then led up the garden path by the Russians who paid diligent attention to Kurdish notables during the war, fostered Kurdish nationalism and supplied a separatist movement with arms. In January 1946 the Kurds proclaimed the independent republic of Mahabad, but for the Russians the Kurdish movement was meant to be no more than an appanage of the separatist Azerbaijani republic which they had contrived in Tabriz at the end of 1945. The Mahabad republic was an expression of genuine non-communist local feelings, whereas the Azerbaijani republic was a communist artefact. Attempts to ally the two were never more than superficially successful although the Kurds, dependent on Russian support, were half tied to the Azerbaijanis. After the withdrawal of Russian troops in May 1946 the Azerbaijanis negotiated, without reference to the Kurds, a favourable agreement with the Iranian government which restored Azerbaijan to the Iranian state as an autonomous province. At this time the Iranian government was a coalition including communists, but later in the year the prime minister, Quavam es-Saltaneh, evicted his communist colleagues in Teheran and sent his army against Tabriz where the communist provincial administration was immediately overthrown with the help of the local populace. The Mahabad republic was now at the mercy of the Iranian army. It was annihilated, its leaders fleeing or being hanged.

The fate of the Kurds in Iraq was largely in British hands when, after taking Baghdad in the First World War, British forces went on to Mosul and installed the Kurdish leader Mahmud Barsandji as governer. The Kurds hoped for an independent Kurdistan as a reward for their anti-Turkish services during the war and in the first postwar years the British were primarily concerned to prevent the return of Mosul to the new Turkish state. The abortive Treaty of Sèvres of 1920 (which provided for an independent Armenia) gave the Kurds no more than a promise of autonomy which the Iraqis, a protected or mandated state on its way to full independence, were determined to limit as far as possible or even to annul. His hopes disappointed, Barsandji declared an independent Kurdistan. The British deported him to India but then (1922) reinstated him as governor in Mosul. A year later he fled to Iran. The Treaty of Lausanne (1923) annulled the Treaty of Sèvres and therewith its provisions for Kurds and Armenians. The new Turkish state claimed Mosul but in 1925 the League of Nations endorsed the British claim that it was part of Iraq. Five years later Barsandji returned to Iraq in an attempt to win some Kurdish advantage out of sporadic clashes between Kurds and Iraqis but he was thwarted by British air power and with the end of the British mandate over Iraq the Kurdish region became a part of the sovereign state of Iraq.

Towards the end of the Second World War a new Kurdish leader, Mustapha Barzani, led a revolt which was crushed: some 10,000 Kurds fled to Iran and Barzani fled to Moscow. The revolution of 1958 in Baghdad destroyed the alliance between the three anti-Kurdish states of Iraq, Turkey and Iran and substituted for the Iraqi monarchy a republic of 'Arabs and Kurds'. Barzani returned from Moscow and was honourably and even munificently received by President Kassim, who looked with favour on the Kurds because of the left-wing tendencies among them. In 1960 a Kurdish Democratic Party was allowed, but the Kurds were divided in their views on Kassim and even their left wing disliked the communist ideas which gained ground in Iraq after the revolution. After some internal dissension the Kurdish Democratic Party became definitely anti-communist, rapidly lost Kassim's favour, was persecuted and officially extinguished, even though the communists also fell from grace at this time.

In 1961 the Kurds rebelled. Kassim claimed that they were receiving military supplies from Britain and the United States. He turned for arms to the USSR, which found it opportune to forget its earlier support for the Kurds in the hope of finding a more useful ally in Kassim. The Iraqi army started a full-scale campaign against the Kurds with napalm bombs and rocket-carrying aircraft. In February 1963 Kassim was overthrown and killed and the pan-Arab nationalists of the Ba'ath Party took control of the government under the presidency of General Aref who engaged in discussions with the Kurds and gave them to understand that they would get something like autonomy. Later in the year, however, the discussions were broken off by the seizure of the Kurdish negotiators and the Iraqi army, supported in the field by the Syrians where the Ba'ath had also come to power, embarked on an even more ferocious campaign against the Kurds than the operations of 1961–62. But the Ba'ath anti-communist nationalism had alienated the communist world which raised the cry of genocide. Early in 1964 a truce was negotiated. It proved fragile, but in 1970 President al-Bakr and his vice-president Saddam Hussein renewed it and Barzani won a fresh agreement on autonomy which, however, again collapsed for a variety of reasons, including the disputed status of Kirkuk. When Iraq and Iran joined forces against the Kurds, Barzani fled to the United States. But fundamentally Iraq–Iranian relations were bad and offered opportunities to the Kurds.

The shah feared the new type of Arab régime, specially Nasser's. The Iraqi revolt of 1958 was therefore unwelcome to the shah and the successes of Aref and Kassim in 1963 even more so. The shah wanted Shi'ite Muslims to have a share in the government of Iraq and if the Shi'ites were to be so treated the Kurds would have to be so too. He refused to help Aref's anti-Kurdish operations even when Kurds took refuge in Iran and he gave Iraqi Kurds surreptitious – and not so surreptitious – help. His *rapprochement* with King Feisal in 1965 encouraged him to think in terms of a Persian Gulf regulated by a Saudi–Iranian compact without the need to include or accommodate Iraq. In 1980 Iraq turned the tables on Iran by supporting Kurdish separatists within Iran and when Iraqi Kurds appealed for support from the exiled Khomeini, Saddam Hussein slaughtered them with chemical weapons. In and after the Gulf War of 1991 the Americans hoped that Kurds and Shi'ites would revolt against Saddam Hussein and the incautious language of President Bush was interpreted by them as incitement to do so. Not for the first time these Kurds misread ambiguous signals emanating from greater powers. Bush's Turkish allies were opposed to Kurdish independence or autonomy anywhere and his Saudi allies were equally opposed to similar recognition for Shi'ites. Saddam Hussein's crushing defeat at the hands of the United States and its allies did not prevent him from savaging Kurds and Shi'ites. The latter's homelands in southern Iraq were drained to make them an easier prey and many were killed by Iraqi forces or driven to flee into Iran.

The Iraqi Kurds were never more than tenuously united among themselves. The Patriotic Union of Kurdistan (PUK) split from the Kurdish Democratic Movement (KDM) in 1975, advocated more active opposition to the central government and refused to support Iraq in its war with Iran, whereas the KDM – and Kurds in Iran – sided with Iraq. After the Gulf War the PUK's leader, Jalal Talabani, responded positively and disastrously to Bush's call to revolt, whereas the KDM and Masud Barzani were more cautious. The two leaders' following among Iraqi Kurds was about equal. Saddam Hussein, dealing with each separately, produced a draft law for autonomy which satisfied neither and in the aftermath of their abortive rising 1.5 million Kurds fled into Iran and half a million to Turkey. In 1992 fresh proposals which included an elected Kurdish assembly and an autonomous executive in an (imprecisely defined) zone were accepted by both the KDM and the PUK but their co-operation was no more than a veneer. International attention and commitment flagged, UN members failed to keep up their contributions to UN aid and supervision, diseases took hold, Iraqi harassment increased and so too did fighting between the Kurdish parties.

In Turkey the Kurdish question, which had been smothered for the best part of two generations, was forced into the open by the Kurdistan Workers Party (PKK) which claimed to represent 20 per cent of Turkey's population, exercised despotic and to some extent corrupt authority in south-east Turkey and provoked an incipient civil war. The Turkish government took the opportunity presented by the Gulf War and its aftermath to pursue PKK forces into Iraq where it won a considerable battle but by 1993 the PKK had recovered sufficiently to renew quasi-military operations which were marked by atrocious behaviour on both sides.

B. THE SHI'ITES

The Shi'ites, who separated from the main Muslim stock over 1,000 years ago, comprise about a tenth of Islam. They revere the Prophet's son-in-law Ali and his two sons Hassan and Hussein as the originators of the true line of divinely inspired imams (leaders). Most of them await the reappearance of the twelfth imam who disappeared in A.D. 880 but some of them await other reappearances. They are widespread in Islam but strongest east of the Euphrates. They dominate Iran where they have had the control of power for 500 years, outnumber other groups in Iraq and Lebanon, and constitute important minorities in Syria, Pakistan and elsewhere. Khomeini's successes boosted their standing wherever they were to be found but also scared non-Shi'ites, particularly Arabs who distrusted Iranians and Sunnis who disliked clerical intervention in government. Their faith is imbued with suffering and persecution, a bottled bitterness directed chiefly against Sunnis whose faithlessness to the true line has been further sullied by western – notably American – modernism and immorality; also against Jews and Baha'is. The legitimacy of violence is more obvious to Shi'ites than to Sunnis or most Europeans.

C. SECTARIAN VIOLENCE

The Muslim Brotherhood, founded in Egypt in 1928 by Hassan al-Banna who died in 1949, asserted what it held to be basic Muslim values and particularly opposed the separation of church and state (as those terms were understood in the west) and foreign influences on Islam. Its prime aim was the reassertion of Islamic law (the shari'a) and the paramount authority of shari'a courts. Hassan al-Banna preached patience but not all his followers achieved it. The Brotherhood murdered the Egyptian prime minister, Nokrashi Pasha, in 1948, was dispersed and dissolved, revived in the anti-British climate of the 1950s, supported the movement which brought Naguib and Nasser to power but was discarded by them, tried to kill Nasser, suffered thousands of deaths in prisons but recovered after Nasser's death. Sadat, who was more partial to religious groups, tried to split the Brotherhood into violent and anti-violent wings but his pro-western policies alienated both and narrowed the gap between them. The Brotherhood was saddled with responsibility for his murder and again suffered arrests on a large scale. The murder in 1990 of Rifaat al-Mahgoub, the chairman of the Egyptian parliament, provoked a similar operation against religious revolutionaries. Similar movements in, for example, Syria, Sudan, Algeria and Pakistan built up parties within the state with the aim of subordinating the constitution to fundamental Muslim tenets, codes, courts and practices. Although banned from party politics in Egypt the Brotherhood was active in middle-class professional bodies and attracted a new class of follower among the social and economic victims of accelerating urbanization. Its links with more militant movements – Gamaa el-Islamiya and Jihad el-Islami – were sufficiently ambiguous to ensure it some toleration from the government.

These conflicts within Islam interlocked with the Arab–Israeli feud and Arab responses to Israel's existence and oppression of Palestinians: a number of anti-Israeli bodies resorted to violence against Israelis. In the 1970s the Patriotic Front for the Liberation of Palestine (PFLP) and its leader, George Habash, took to hijacking in order to draw international attention to the grievances of Palestinians. In 1970 the PFLP hijacked four aircraft, destroyed one of them on the ground at Cairo and three in Jordan, released all those on board and secured in exchange the release of PFLP prisoners from British, German and Swiss gaols. In 1972 Japanese sympathizers opened fire at Lod airport in Israel, killing twenty-five people including three of themselves. In 1973 the Israeli air force forced an airliner to land in Israel in the mistaken belief that Habash was on board. In 1976 the PFLP hijacked a French airliner with a number of Jews on board and forced it to land at Entebbe in Uganda in an attempt to secure the release of Palestinians from Israeli gaols; after most of the passengers had been released by the hijackers the rest were rescued by a daring and efficient Israeli raid on the airport. The 1980s saw no diminution in violent protest but it took a variety of forms, adding the taking of hostages to murder and suicide. In 1983 pro-Iranian Shi'ites attacked public buildings and the American and French embassies in Beirut; among twenty-one persons arrested were three Lebanese. In response hostages – mostly American or French – were seized in Beirut and a Kuwaiti aircraft was seized at Teheran and two American passengers murdered. In 1985 a handful of Palestinians hijacked the cruise ship *Achille Lauro* as a means to get fifty Palestinians out of Israeli gaols; one passenger, an American, was killed. The hijackers surrendered at Port Said where they were put on an Egyptian aircraft for Tunisia which was forced by the American air force to land in Sicily. The Italian government refused to hand them over to the United States and they escaped to Yugoslavia. In the same year a Palestinian group led by Abu Nidal hijacked an Egyptian aircraft which was damaged and forced to land in Malta where an Egyptian attempt to recover it caused fifty-seven deaths; and the same Palestinian group attacked passengers for Israel on Rome and Vienna airports. Shortly afterwards Israel forced a passenger airliner down in Israel again in the mistaken expectation of finding Palestinian leaders and taking them hostage.

In 1986 the United States with British help bombed Libya on the plea that the killer of an American soldier in a Berlin night club had been a Libyan; the United States sought to justify this act as self-defence. In 1988 the Hizbollah in Lebanon captured and hanged an American colonel serving with the UN; a Kuwaiti aircraft was flown by hijackers to Mashad in Iran, thence to Beirut where it was refused permission to land and then to Cyprus where its passengers were released; and an American airliner was destroyed over Lockerbie in Scotland and all its passengers and crew were killed. In 1989 Israel kidnapped from Lebanon Sheikh Abdul Karim Obeid in another tit-for-tat operation. In the 1990s the most noticeable group was Hamas, a small group of Palestinians who, like the even smaller Islamic Jihad, turned to violence to vent their rage against the Israeli state. It attracted widespread, but largely passive, support among Palestinians in Gaza and the West Bank who were losing faith in the PLO and distrusted Arafat's negotiations with Israel. The high-handedness of Rabin's government strengthened Hamas and enabled it to pose as a significant alternative to the PLO in the struggle for an independent Palestinian state. After the massacre in 1994 of twenty-nine Muslim worshippers in a mosque in Hebron under the eyes of the Israeli police, Hamas abandoned its proclaimed policy of killing soldiers but not civilians. Unlike Hizbollah in Lebanon, Hamas was neither Shi'ite nor apparently funded by Iran.

In these years the term Muslim fundamentalism came into general use as one of the more mischievous over-simplifications of the age. A fundamentalist is one who believes the totality of what he is told to believe. Islam, more thoroughly than any other major religion, retained sway over the hearts and minds of its adherents and these were attracted into parties prepared to use violence and justify it, whether the cause were religious, nationalist, social or a compound of these. Muslim fundamentalism was used not only to describe these various and sometimes dangerous groups but to imply that Islam was essentially violent and that parties claiming a special allegiance to Islam were part of a single unified and menacing force.

Islamic fundamentalism, a catch-all phrase coined in the west, stresses the more disruptive aspects of a cultural revolution of long standing. From one standpoint it denotes a conservative reaction against assorted political and social ideas which have been reaching Islam from the west for centuries. These ideas have created intellectual movements which are opposed by traditional establishments and create counter-movements which are characteristically xenophobic and often violent. To outsiders they appear dangerously concerted and vengeful, the more so because of their impact on Middle Eastern politics and their spread to West Africa on the one side and South-east Asia on the other.

PART FOUR

ASIA

The half century preceding the departure of the British from India had seen a series of reforms, evolved by the British and leading logically and explicitly to Indian independence; the growth of divisions within India which led to the partition of 1947; and a growing awareness of world politics, in which nevertheless the new rulers were for the most part but imperfectly versed. In the nineteenth century India, though less closed to the outside world than China or Japan, had a view of the world which was vastly overshadowed by the British presence; and the principal episode of the century was the Sepoy Revolt or Mutiny of a part of Britain's subjects against British rule. This view was changing by the end of the century. The Russian advance towards Afghanistan; Curzon's preoccupation with the north-western frontier and Tibet; the extension of British Indian power into the Middle East; the alliance of Britain with Japan in 1902, followed by Japan's defeat of Russia in 1905; revolutions in Turkey, Iran and China – all these impinged on India and turned the Indian mind or part of it outward. Although in 1947 Jawaharlal Nehru was the only member of his cabinet with any claim to expert knowledge of foreign politics, his colleagues and many others among his compatriots had grown up with the feeling that, if India's special problem was the defeat of the British raj, there were also other problems and other powers to be reckoned with, notably Russia, China and Japan. Even though many Indians misjudged the nature of the problems, deceiving themselves with the belief that their cause was the British presence and their cure would be the British departure, this mistake was a matter of misinterpretation and not of cloistered ignorance. Within India nationalism, which was one of the by-products of the Hindu and Muslim intellectual revivals of the nineteenth century and took visible shape in the founding of the Indian National Congress in 1885, was inevitably anti-British. Like most nationalist movements it came to be divided into a more militant and a less militant faction (led in this case by B. G. Tilak on the one hand and by G. K. Gokhale and then by M. K. Gandhi on the other), but unlike others it was also divided in a more enduring way before the day of victory. As independence approached, the ability of Hindu and Muslim to work together diminished until it proved impossible to maintain a single successor state to the British raj and at independence the two great religious communities feared and hated each other more than they feared or hated the British.

Britain had long envisaged the surrender of empire in India but without formulating a timetable. The Second World War imposed the timetable. At the beginning of that war the viceroy made the ludicrously inept mistake of declaring war on India's behalf, as he was entitled to do, without consulting a single Indian. Congress ministries, in office under the constitution of 1935, thereupon resigned. In 1942, Sir Stafford Cripps was sent by the British cabinet to India to offer it dominion status with the right to secede from the empire, but the exercise of this option and all other internal advances were to be postponed until the end of the war. The Congress, which may have misjudged British intentions and certainly – if recruiting figures meant anything – misjudged Indian sentiment, decided to have no part in the war and to use it to wage a Quit India campaign against the British. In 1944 Britain, now sure of victory in all theatres, appointed a new viceroy, Lord Wavell, and released Indian leaders who had been put in prison. After the general election of 1945 the Labour government sent three cabinet ministers to India to try to get agreement between the Congress and the Muslim League as a preliminary to independence, but relations between these bodies had deteriorated during the war (the reverse of the experience of the First World War) and the British attempt failed. The League and its leader Mahommad Ali Jinnah were convinced that Britain was partial to the Congress; in August 1946 Jinnah inaugurated Direct Action by the League to secure a separate sovereign state for Muslims, and the winter of 1946–47 was marked by violent communal riots. Early in 1947 the viceroy came to the conclusion that no single Indian central authority could be constituted and he accordingly advised the British government either to retain power for at least a decade or to transfer it, fragmented, to the several provinces.

The British government rejected this advice, replaced Wavell by Mountbatten and announced that Britain would abdicate in June 1948. It proposed to resolve the dilemma by neither of the methods recommended by Wavell but to partition India and hand over power to two separate central governments. The 562 princely states, which were not part of British India, were to be cajoled into one or other of the new states. Their relationship with the British Crown was regulated by the doctrine of paramountcy. Britain did not propose to transfer its paramount rights to the new India or to Pakistan but declared that paramountcy would lapse, with the result that each ruler would be free in law to accede to India or Pakistan or neither – and free in practice to accede to India or Pakistan. Junagadh made the impractical choice of acceding to Pakistan with which it had no border; the ruler was forced to see his error, departed for Pakistan and left his state to become part of India. Three states toyed with independence: Travancore, Hyderabad and Kashmir. Travancore's ambitions were quickly seen to be illusory: Hyderabad had first to be blockaded and then (in 1949) invaded. Both became part of India. With Kashmir we shall be further concerned.

Britain's shock tactics were intensified when Mountbatten reported that even June 1948 was too late for the transfer of power. With earlier visits in mind he concluded that violence would become uncontrollable by then, and the British cabinet accepted his view. By advancing the date to August 1947 it left scant time for settling the biggest of all the issues raised, the lines of partition between India and Pakistan. It was apparent that Pakistan must consist of two widely separated areas, each containing large non-Muslim populations. A boundary commission was created, consisting of two Hindu and two Muslim judges with Sir Cyril (later Lord) Radcliffe as chairman. On most contentious points the two Hindu and the two Muslim voices cancelled each other out and Radcliffe was left to take, in two months, a series of detailed decisions on localities which he did not know and had no time to visit. One of the most important of these decisions was an award which gave India access to Kashmir. But the new borders did not serve to contain the peoples within them; probably no borders could have done so. Fear drove millions across them and in the course of this mass exodus millions were slaughtered. Sikhs, quitting the homes in the Punjab in which they no longer felt safe, attacked Muslims moving westward for the same reason; Muslims retaliated; atrocities were multiplied over a wide area reaching to Delhi itself. Probably 2 million men, women and children died, and this horrible feast of violence was capped in January 1948 by the assassination of Gandhi himself, Hindu apostle of non-violence killed by a Hindu fanatic.

In the new state of *India* two men took control – first and foremost Jawaharlal Nehru and with him Vallabhai Patel. Nehru became prime minister and held that office until his death in 1964. Patel, to whom fell the task of consolidating the Indian federation of former British provinces and former princely states, died untimely in 1950. The Congress remained paramount, though divided. In 1950 it chose the right-wing Purshottandas Tandom to be its president but in the following year Nehru enforced Tandom's resignation by resigning from the Working Committee which was the Congress's power house. Nehru remained president of the Congress until 1955 when he handed the office over to a reliable subordinate. In general elections in 1952, 1957 and 1962 the Congress received a slightly rising share of the vote (45–48 per cent) and a consistently massive majority in the federal parliament, despite its internal divisions and despite growing criticisms

14.1 Pakistan, northern India and Bangladesh

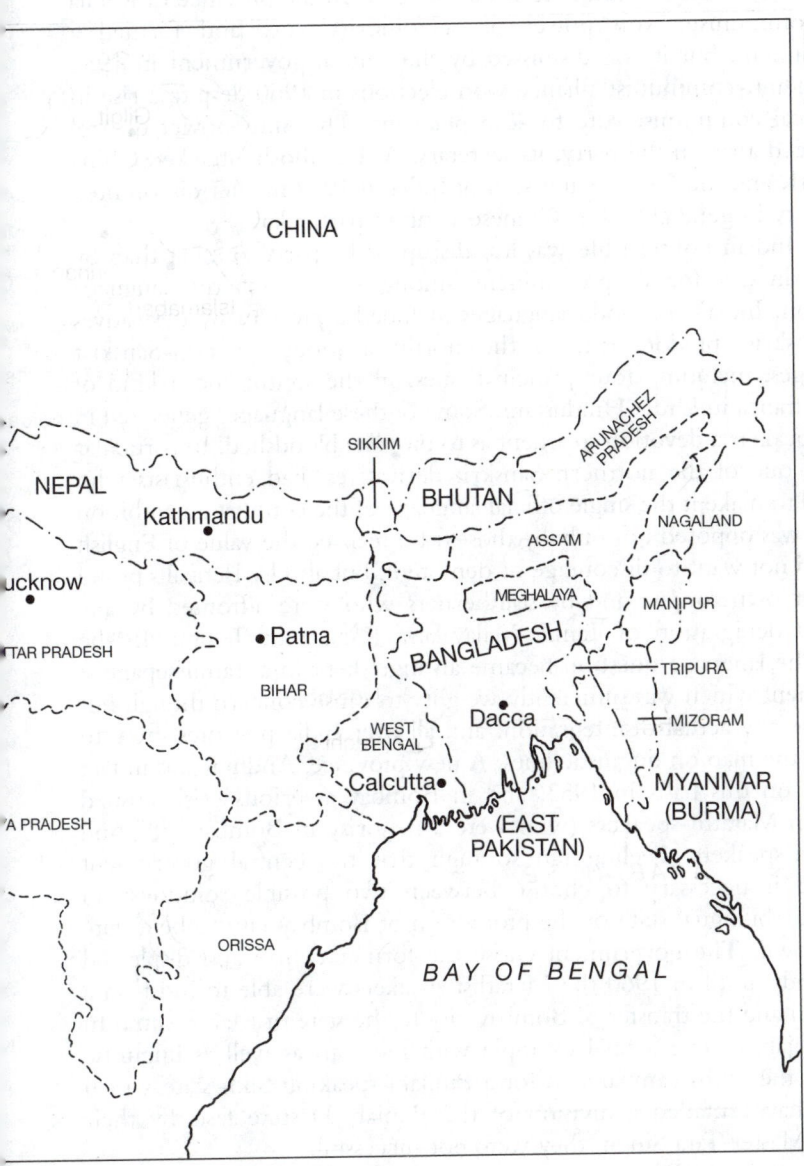

from right and left. Its principal adversary, the Communist Party, won 3.3 per cent of the vote in 1952 and around 10 per cent in the two following elections, but this vote was so concentrated as to be more effective than mere arithmetic would suggest. In the province of Kerala the communists won power on a minority vote and formed a government, but it was dismissed by the central government in 1959 and an anti-communist alliance won elections in 1960 despite a rise in the local communist vote to 42.5 per cent. The Sino–Soviet quarrel produced a rift in the party; its secretary, A. K. Ghosh, attacked China in 1961, and the Chinese invasion of India in 1962 further discomfited the party in general and its Chinese wing in particular.

But Indian political life was less disrupted by party conflict than by other divisive forces, pre-eminent among which was the language question. India's sixty-odd languages included a great many derivatives of Sanskrit, predominant in the north; a group of non-Sanskrit languages, including four principal ones, in the south; and a kind of lingua franca in Urdu/Hindustani. Some of these languages generated in their speakers a devotion so ardent as to provoke bloodshed. In particular Hindi, one of the northern Sanskrit derivatives, had enthusiasts who wished to make it the single official language of the country, an ambition which was opposed not only by those who realized the value of English and did not want to discourage or demote it, but also by Bengalis proud of their own tongue and by southerners who were affronted by any implied denigration of Tamil, Malayalam, Kanada or Telugu. In the south the language question became an ingredient in a Tamil separatist movement which was sufficiently weighty to cause concern though not to cause any actual disintegration, and all over India pressures grew to redraw the map on linguistic lines. A new province, Andhra, was in fact created on this basis in 1953, and in Bombay a serious crisis ensued between Marathi-speakers (who were a majority in Bombay city) and Gujarati-speakers. Feeling ran so high that the central government deemed it necessary to choose between two possible solutions: an avowedly bilingual state or the promotion of Bombay city to be a state on its own. The government chose the former course and displeased everybody, until in 1960 the Marathsi-speakers were able to insist on a partition and the transfer of Bombay city to the state of Maharashtra. In the Punjab, to cite a final example with religious as well as linguistic aspects, the Sikhs campaigned for a Punjabi-speaking Sikh state which would have entailed a division of the Punjab. Despite fasts by their leader, Master Tara Singh, they were not successful.

In external affairs Nehru was determined to remain in the Commonwealth (especially if Pakistan did) while adopting a republican constitution and conducting a foreign policy which might be not merely independent of Britain's but contrary to Britain's. He succeeded in persuading the other prime ministers of the Commonwealth (as the British Commonwealth was significantly renamed at the period) that India might remain a member even though it became a republic. This was a revolutionary step. Without it the expanding postwar Commonwealth, with its strong republican tendencies, could not have taken shape. In 1949 a Commonwealth conference accepted an Indian plan to declare the British sovereign head of the Commonwealth and leave each member free to adopt a monarchical or republican form of constitution. India itself became a republic at the beginning of 1950, and within a few years there were many more republics than monarchies in the Commonwealth. The independence of each member of the British family of nations had been accepted for a generation but none had so far consistently pursued a foreign policy which ran counter to Britain's. This Nehru set out to do without impairing relations with London, and he was substantially successful. Although Britain was a committed protagonist in the Cold War Nehru held that it was none of India's business and that the two camps were behaving with an equally deplorable folly. Having taken a conspicuous part in bringing the Korean War to an end he went on to elaborate a policy and posture of neutralism in the hope of keeping sizeable powers out of the Cold War, of limiting its evil effects and paving the way for its eventual termination. In 1955 he visited Moscow, and India received a return visit from Bulganin and Khrushchev (who made the mistake of making the anti-British speeches which Indians were capable of making themselves but did not countenance in others). In the following year the Anglo–French attack on Egypt and the Russian intervention in Hungary confirmed Indian belief in the wickedness of all major powers, even though Nehru himself was less censorious of the Hungarian episode than of Suez (perhaps because of the latter's implications for Commonwealth solidarity). Nehru's determination that Asia should give the rest of the world an example in sanity caused him to pursue the myth of Sino–Indian friendship with an unrealistic tenacity which, when China attacked India in 1962, gravely weakened his prestige, his policy and his country.

One source of the refusal of many Indians to give the Chinese menace sufficiently serious attention was the all-consuming quarrel with *Pakistan*. Even India's claims against western colonial powers like France and Portugal were emotionally trivial compared with the animosity against Pakistan. (These claims were admittedly small in territorial extent. France ceded Chandernagore – virtually part of Calcutta – in 1951 and its remaining possessions – Pondicherry, Karikal, Mahé and Yanan – in 1954. Portugal adopted in India, as in Africa, the device of converting its colonies into provinces of metropolitan Portugal, but this nominal metamorphosis was of short avail and in 1961

India took the Portuguese territories – Goa, Danan and Dia – by a show of force.) Unlike India, Pakistan lost its father figure early. Jinnah, who had become governor-general on independence, died in September 1948. Moreover Liaqat Ali Khan, Pakistan's first prime minister, was assassinated three years later. For several years Pakistan wasted much of its tenuous substance on barren constitutional disputes while public figures succeeded each other in high offices, corruption became scandalous and the army wondered how long to let it go on. Khwaja Nazimuddin succeeded Jinnah as governor-general and then succeeded Liaqat Ali Khan as prime minister in 1951; in 1953 his own successor as governor-general, Ghulam Muhammad, dismissed him and installed Muhammad Ali Bogra in his place.

These changes, which involved deliberate attempts to preserve a balance between West and East Pakistan, were accompanied by a gradual collapse of authority and by rioting. In 1954 the Muslim League was severely defeated in provincial elections in East Pakistan and the central government, humiliated and jeopardized by this reverse, despatched General Iskander Mirza to East Pakistan as military governor. This appointment was the beginning of the movement towards military rule. In the next year General Mirza became governor-general on the death of Ghulam Muhammad who had been ailing for some time, and appointed Chaudri Muhammad Ali, an able and honourable civil servant, to be prime minister. He, however, resigned in 1956 and was succeeded in 1957 by Firoz Khan Noon, a distinguished veteran. A constitution was finally adopted in 1956, but in 1958 parliamentary democracy came to an end after the deputy speaker of the East Pakistan parliament had been hit on the head with a plank during a debate, so that he died. Martial law was proclaimed with General Ayub Khan as administrator and subsequently as General Mirza's successor as head of the state. Political parties were banned. A new constitution was introduced in 1962 based on the American presidential rather than the British parliamentary system.

Pakistan's instability gave India the excuse to justify its fears of its neighbour on the grounds that there was no knowing what governments in such straits might do next. When unstable government was succeeded by military government, Indian fears were merely transposed into a different key and it was alleged that an efficient junta was necessarily even more of a danger than an inefficient civilian régime. The ill will between the two countries was concentrated on Kashmir, but Kashmir was not its only cause. The massacres of 1947 had given a spectacularly bad start to a relationship which was almost foredoomed in that it arose out of the inability or refusal of India's two communities to live on good terms with each other. The tendency in India to treat partition as an ephemeral aberration was an additional source of irritation in Pakistan. There were also disputes over the distribution of the waters of the Indus and its tributaries, and over the property of those (about 17 million of them) who had fled from one country into the other and found themselves unable to sell the pieces of land which they had left behind. Further, the division into two parts of what had been a single economy produced economic tensions which developed into a trade war and reached a high pitch of acrimony when India devalued its currency in 1949 in step with Britain but Pakistan refused to follow suit until 1955. But worst of all was Kashmir.

The state of *Kashmir* consisted of Kashmir proper; Jammu; an upper tier running from the north-west to south-east and consisting of Gilgit, Baltistan and Ladakh; and a western fringe which included the small territory of Poonch. Kashmir had suffered centuries of oppression by a variety of tyrannical overlords. It was under Afghan rule when it was conquered by the Sikh prince Ranjit Singh in 1819. The same prince shortly afterwards installed Gulab Singh as ruler over Jammu, and Gulab Singh duly added Ladakh and Baltistan to his realm. In the 1840s all these territories, Gilgit in the far north-west still excepted, became part of British India as a result of the two wars between the British and the Sikhs for the mastery of the Punjab. The Sikh princes remained in power as maharajahs of Kashmir and Jammu, and in 1947 the British were turning a blind eye to the notoriously unsatisfactory rule of the rich and incompetent Hari Singh. This maharajah's Muslim subjects, four-fifths of the total, were solidly opposed to his rule and so were many of the Hindu minority. Muslim opposition was divided between the Kashmir Muslim Conference, to which no Hindu could belong, and a larger organization led by the intelligent and open-minded Muslim Sheikh Abdullah which, following the example of the Indian National Congress, included members of both creeds.

As independence approached in 1947 the maharajah delayed and prevaricated, partly because he was toying with the notion of an independent Kashmir and partly because his interests lay in directions other than statecraft. In October the small territory of Poonch purported to secede and was immediately invaded by tribesmen from Pakistan. How far the Pakistani government had foreknowledge of these events has remained obscure. The maharajah appealed to India which refused to come to his help unless he formally acceded to India. This he did just in time to enable Indian troops to be flown into Kashmir to forestall the capture of its capital, Srinagar, by the tribesmen. The maharajah was then removed and, in 1949, deposed, and Nehru made in November the first of a number of promises to hold a plebiscite. Early in 1948 India referred the situation to the Security Council and Pakistan sent regular army units into Kashmir which recovered some of the ground which the tribesmen had lost to the

Indian army. A UN mission proposed a cease-fire, which came into effect on the first day of 1949, and a plebiscite, which was never held. The cease-fire hardened into partition along an adventitious line. Parts of Kashmir became integrated with Pakistan; these were the west including Poonch, and also Baltistan and Gilgit. The rest of the country, including Ladakh, was ruled by Sheikh Abdullah as prime minister under the nominal authority of a member of the princely house as head of state, until in 1953 the Indians, suspicious that Sheikh Abdullah too wanted an independent Kashmir, had him put in prison and replaced by Bakshi Ghulam Muhammad, with whom they proceeded to work out a new constitution for Kashmir in order to make it a fully integrated part of the Indian federation.

During the 1950s a number of abortive attempts were made to supplement the cease-fire by a political settlement. The first UN attempts were immediately unsuccessful. The American Admiral Chester Nimitz, who was appointed to supervise the plebiscite, never even went to Kashmir. A UN conciliation commission (UNCIP) abandoned its task at the end of 1949. An Australian judge, Sir Owen Dixon, was appointed UN mediator but was obliged to announce failure. Talks in 1951 between Nehru and Liaqat Ali Khan also failed. Dr Frank P. Graham took over Sir Owen Dixon's task with the same result. After the dismissal of Sheikh Abdullah in 1953 there was a new round of talks between Nehru and Mahommad Ali but again fruitlessly. Then in 1954, at the time when Kashmir's constitution and status were being altered, a decisive event occurred. The United States and Pakistan entered into an agreement which provided for American military aid to Pakistan. From the American point of view this was an anti-Russian move, part of the American policy of containment or encirclement, but from the Indian point of view it was a powerful reinforcement of India's principal enemy. India deeply resented this American move and Nehru made it the occasion finally and explicitly to go back on his promise to hold a plebiscite in Kashmir. During negotiations in 1954–55, during which the Pakistani governor-general Ghulam Muhammad went to Delhi, Nehru refused to entertain the idea of a plebiscite. At the beginning of 1957 Pakistan asked the Security Council to order the withdrawal of all troops from Kashmir, to send a UN force there, to organize a plebiscite and to require India to abandon the new constitution which was about to integrate Kashmir in India. India opposed UN intervention and the USSR vetoed the resolution endorsing Pakistan's plan. A year later Dr Graham made similar proposals to India, which rejected them.

During the Indo–Pakistani dispute over Kashmir in the 1950s it was often forgotten that Kashmir also had a frontier with *China* and this frontier had, since 1948, been partly in Indian hands and partly in Pakistani. In the late 1950s, however, Indians became suddenly aware of hitherto unsuspected Chinese activities in Ladakh. At the same time the flight of the Dalai Lama to India attracted popular attention, and especially Indian attention, to Chinese activities in Tibet; an incident in the far north-eastern corner of India revealed a Sino–Indian clash at this point; and politics in Nepal split openly into pro-Indian and pro-Chinese strands. China became a factor in the Kashmir dispute between India and Pakistan, and likewise Kashmir became a factor in the evolving Sino–Indian dispute which ran all the way along a frontier of more than a thousand miles and involved, besides Kashmir, the small Himalayan states of Nepal, Bhutan and Sikkim and the much larger and constitutionally anomalous country of Tibet.

Since India and China are the two largest Asian countries their attitudes towards each other are a major element in the politics of Asia. In 1954, in the context of a commercial treaty about Tibet, they proclaimed five principles for the regulation of their mutual and possibly conflicting affairs. These principles, often known as the Panch Shila, were: respect for each other's sovereignty and territorial integrity, non-aggression, non-interference in each other's internal affairs, equality and mutual benefit, and peaceful coexistence. On Nehru's side the Panch Shila reflected certain basic aims of his foreign policy: to eschew war, to create an Asian order resting on the mutual trust and respect of the two leading Asian powers, to obviate anything like a cold war in Asia, and to give the rest of the world an example in international behaviour. That the context should be Tibet was partly accidental but also fitting and prophetic, for it is along common frontiers that the resolutions of states are most frequently tested.

China's subjugation of Tibet, undertaken in 1950, was a logical implementation of the determination to reunite all China's 'five races' under the control of Beijing, and it brought China to the borders of India. Indians had various reasons for viewing this confrontation – a meeting, as it turned out, of ideologies as well as power – with surprising unconcern. There was first of all the determination to get on well with China, a determination which was heightened by the challenge of ideological differences and which therefore obfuscated the difficulties of doing so. Then there was a pervasive if vague belief that bad-neighbourliness was a mark of capitalist, power-minded states rather than a consequence of rubbing shoulders, coupled with a similar belief that many of the troubles of the past in India (including the troubles which the British had had along its frontiers) were part and parcel of an alien imperialism and had automatically been removed with the British. Finally, there was the obsession with Pakistan, so that the division of British India into two states proved not merely a weakness in the sense of a dissipation of material resources but also a weakening of the power to see things in proportion. Whether or not the Chinese were emboldened by this division, it certainly gave them opportunities, not least the opportunity to play for years a hazardous frontier game which many Indians preferred in their myopia not to notice.

The Sino–Indian border runs for about a thousand miles through some of the most daunting territory in the world. To the west Tibet marches with Ladakh, the south-eastern portion of Kashmir which, jutting eastward, thrusts a wedge between India and Tibet. The border then touches Indian territory and makes the turn to the eastward which carried it all the way to Burma by way of Nepal, Sikkim, Bhutan and the North-East Frontier Agency of India (Nefa).

The Chinese claim to regard Tibet as an integral part of China had not seriously been contested by other sovereign states, however uneasy they might have been about this extension of Chinese power. The Tibetans themselves, conscious of their wholly different culture, religion and language, took a different view, partly on the grounds that *de facto* independence since 1911 had ripened into full independence and partly by construing vague and ancient declarations of Chinese respect as formal grants of independence and not the mere courtesies which the Chinese said they were. The Mongol Khan Kublai, a grandson of Jenghiz, who became emperor of China in the thirteenth century and was converted to Buddhism, bestowed favours and rights on a lama who established in Tibet a local dynastic rule of uncertain radius. A hundred years later a schismatic line of so-called Yellow Hat Buddhists appeared and after a further two hundred years supplanted the line installed by Kublai as effective rulers of Tibet. The chief of this line was the Dalai Lama (he claimed spiritual descent from a contemporary of the Buddha in the fifth century BC), and in the seventeenth century he received from the first Manchu emperor in China marks of respect which may or may not have amounted to something approaching sovereignty. In the next century the Chinese entered Tibet to protect the country against Mongols and refused to go away again. They also defended it against a Gurkha invasion later in the same century and consolidated their position during the nineteenth century, aided by the tendency (sometimes described as mysterious) of the boy Dalai Lamas to die just before or soon after reaching the age to assume full powers.

The nineteenth century also witnessed the approach of the British and the Russians, and in 1903 Sir Francis Younghusband rapped on Tibet's southern door, proceeded to Lhasa and so served notice of Britain's unwillingness to leave China a free hand in Tibet. This was the period of Chinese disintegration but any British notion of taking China's place in Tibet was soon abandoned. The thirteenth Dalai Lama fled in 1903 to Mongolia and thence to China, where his reception was disappointing. He returned to Lhasa in 1909 but fled again in 1910, this time in fear of the Chinese and into India. The collapse of the Chinese empire in 1911 seemed to open the way to real independence but at the Simla conference of 1913–14 between Chinese, Tibetans and British the latter proposed a recognition of Chinese suzerainty in return for a Chinese promise of Tibetan autonomy which would include the right to conduct its foreign affairs independently. This proposal, repeated in 1921, was never bindingly adopted. The Simla conference also propounded a frontier between India to the south and Tibet and China to the north and north-east (the so-called McMahon line) in a document which was initialled by the Chinese but never ratified, not because China questioned the line but because its acceptance was linked with a division of Tibet into inner and outer zones and the exclusion of Chinese troops from the inner.

Upon the death of the Dalai Lama in 1933 the Chinese took the opportunity to return to Lhasa. A mission bearing condolences arrived and remained until 1949 when it was ejected as a result of the general collapse of the Kuomintang. Before this exodus China supported a revolt by the regent of Tibet against his ward, the fourteenth Dalai Lama, who was still a child. The Kuomintang patronized and recognized as Panchen Lama a boy who had been discovered in China in 1944 and was still there: the Panchen Lama, another Yellow Hat hierarch with at least temporal superiority over the Dalai Lama in eastern Tibet, was a spiritual and temporal rival of the Dalai Lama. (The previous Panchen Lama and the Chinese had fled to China in 1923 and died there in 1937.) This last throw by the Kuomintang proved useful to the communists, who took over the new Panchen Lama and set him at the head of provisional Tibetan government in exile. He lived in Beijing with a Chinese wife but in 1962 he disappeared after refusing to attack the Dalai Lama. He reappeared in 1979, having apparently been severely ill-treated in the interval, and died in 1989 curiously young. By 1995 the Dalai Lama and the Chinese had discovered the new reincarnation in different boys.

During 1950 there were attempts by the authorities in Lhasa to negotiate with Beijing in Hong Kong, Calcutta, Delhi or wherever contact could be made, but in October the Chinese invaded and soon secured control of the capital and much of the country. Tibet appealed unsuccessfully to the United Nations and the Dalai Lama fled in 1959 to India. The authorities remaining in Lhasa accepted Chinese suzerainty in return for a promise of a measure of autonomy as an Autonomous Region of China. After Mao's death Deng Xiaoping and the Dalai Lama exchanged proposals for discussions about the future of Tibet on the basis of ruling out in advance nothing except complete independence. But in the years which followed, China was much less energetic in pursuing such proposals than in despatching to Tibet enough Chinese to outnumber its 6 million Tibetans.

The Indian government protested to China and deplored the use of force but to the Chinese argument that Tibet was a part of China the Indian government had no reply. Nehru pursued his endeavour to reach amicable and rational solutions to current problems. After discussion in Beijing during 1953 a Sino–Indian agreement on trade and intercourse in Tibet was signed in April 1954. It dealt with the rights of traders and pilgrims, transferred to China postal and other services previously operated by India as Britain's successor, provided for the withdrawal of Indian military units from Yatung and Gyantse, and enunciated the Panch Shila. India also recognized Chinese sovereignty in Tibet. Two months later Nehru and Zhou Enlai met for the first time as the latter returned home to Beijing from the Geneva conference on Indo-China and Korea, and in October Nehru went to Beijing. The Dalai and Panchen Lamas were already in Beijing where they stayed from September to December 1954. Like the British before and after this period, and like the Americans after it, Nehru was more interested in his relations with China than in the status of Tibet or the fate of the Tibetans (who numbered only about 6 million, not all of them in Tibet, and were in the process of being swamped in their own country by politically inspired Chinese immigration). American and British governments periodically expressed sympathy with Tibet so long as they were on poor terms with China but both gave the Dalai Lama the brush-off when they became engaged in improving relations with Beijing – in the American case after Nixon's visit to China, in the British case when Thatcher was trying to get acceptable terms for the cession of Hong Kong.

In the late 1950s there were a number of incidents leading to Chinese complaints of Indian troops crossing into Tibet and Indian complaints of Chinese troops found south of the frontier. These aberrations could be accounted for by the difficulty of knowing exactly where one is in such country and the Indians in particular, anxious to prove that Indian and China could coexist peaceably in Asia, were not on the lookout for more serious or sinister explanations. The possibility that the two sides had radically different ideas of where the frontier ran on the map was evaded. Yet by the mid-1950s the Chinese had entered the Aksai Chin or Soda Plains in Ladakh. This area, between the two mountain ranges of the Kuen Lun and the Karakoram, had long been disputed ground because it had never been agreed which range marked the Sino–Indian border. To the Chinese the Aksai Chin was important because it lay in the path of a road which they wished to build to link the Tibetan capital with their western province of Jinjiang. This road they now proceeded to build. Their operations can hardly have been concealed from the Indians. What is conceivable, and indeed probable, is that knowledge of what was going on did not immediately reach Nehru himself and that both locally and at the centre there was a conspiracy of silence among Indians whose animosities against Pakistan blinded them to the significance and consequences of what China was doing in territory claimed by India.

The Chinese occupation of Tibet had thus become something more than a rounding-off of the traditional domains of the Chinese empire. It was also a step towards Chinese involvement in international affairs. The continued exclusion of China from the United Nations gave China an appearance of detachment which was later reinforced by its diplomatic isolation after the breach with the USSR, but during the 1950s China was prosecuting active interests in central Asia which brought it into contact with the Himalayan states of Nepal, Sikkim and Bhutan and with Kashmir and so with the Indo–Pakistani dispute. The subjugation of Tibet was an extension of Beijing's authority throughout the Chinese empire in more senses than one. Tibet was reduced not only for its own sake but also as a step towards the more effective control of the great province of Jinjiang to the north and north-west of it. This was the purpose of the Tibet–Jinjiang highway and the chief cause of the flouting by China of the Panch Shila.

Jinjiang, conquered by the Manchu dynasty in the middle of the eighteenth century, had borders in the twentieth with Kashmir, Afghanistan, three Soviet Republics (the Kirghiz, the Kazakh and the Turkoman) and Outer Mongolia. It has in the past been one of those provinces where a governor exercises unusual powers by virtue of his very distance from the imperial centre. He was a semi-independent proconsul who had sometimes looked to the Russian rather than the Chinese empire for help in troubles which he could not cope with himself (as, for instance, during Muslim revolts in 1930–34 and 1937). With China in disarray he could expect little from the east and had to turn west; when, however, the Russians became fully occupied by the German invasion in the Second World War, he faced about and became the friend and ally of the Kuomintang, who were in law his suzerains. In 1944 the Russians helped to foment and sustain a revolt in the Ili district of Jinjiang where an Eastern Turkestan Autonomous Republic was proclaimed, but in the Stalin–Jiang treaty of August 1945 Moscow recognized Chinese sovereignty in Jinjiang and promised not to interfere there – a promise which seems to have been inadequately kept. During the last phase of the Kuomintang the Russians tried to extend their pre-war (1939) monopoly of civil aviation in Jinjiang and to re-create a Russo–Chinese partnership in economic opportunities. At the time of the collapse of the Kuomintang the former object had been achieved on paper but not the latter, and after the governor of Jinjiang had gone over to Mao, the Russians opened negotiations with the new régime. In March 1950 agreements were signed for the creation of joint

(50/50) companies to exploit oil and non-ferrous metals for thirty years and to operate civil airways for ten. Mao was evidently in no position to hold out for complete Chinese control in the province, indubitably Chinese though it was in law, but he immediately set about improving its communications with the rest of China by rail and by road. In casting his eyes on the Aksai Chin he was, as in so many other things, following a line of policy which had occurred to his predecessors fifty years earlier. If in 1950 Mao had had to temporize over Jinjiang, in the same year he was successful in Tibet.

The later 1950s were a period of mixed gains and disappointments for the Chinese in external affairs. On the credit side were Zhou Enlai's appearances at the Geneva and Bandung conferences and his visits to Asian capitals; commercial agreements (1957) with Nepal and Ceylon, a Sino–Burmese border settlement and a Sino–Cambodian treaty; the failure of western policies in Laos and South Vietnam, and the progressive abrogation of democracy in Pakistan, Burma, Ceylon and Indonesia. On the debit side were the failure to take Quemoy in 1958, the quarrel with the USSR, the strains of domestic revolution and economic misfortunes. Most surprising perhaps to Beijing was the uninterrupted construction of the Tibet–Jinjiang highway and its implied claim to 31,000 square km of territory claimed by India and without any protest from Delhi. But an incident involving the capture of an Indian patrol by the Chinese in Ladakh brought the issue into the open and in 1958 the Indian government formally expressed surprise and regret that Beijing had not seen fit to consult Delhi about the highway.

In 1959, when Tibetan discontents mounted into a serious anti-Chinese revolt, the Dalai Lama again fled to India from Chinese retaliation and Nehru wrote privately to Zhou to express his concern but received no answer until six months later when a public Chinese reply laid claim for the first time to extensive stretches of Indian territory. In the interval Moscow had agreed to give India financial aid and Khrushchev, before setting out for the United States, had adopted a neutral instead of a pro-Chinese posture. India likewise was maintaining a neutral position as between China and the United States and refusing to take an anti-American line. The Chinese accused India of interfering and stirring up trouble in Tibet, and in the summer the border incidents which had been going on for several years without attracting wide attention produced casualties, publicity and bitterness. In August an Indian policeman was killed at Longju at the eastern end of the Sino–Indian frontier, and in October several Indians were killed in a skirmish in the Changchenmo valley which lies about midway along the north–south border between Tibet and Kashmir and on the Kashmiri side. It was no longer possible to conceal the fact that the dispute was not about who was where on the occasion of a particular clash, but about where the frontier itself was supposed to run. Up to 1960 Nehru refused to discuss the border problem with China. In 1960–61 a conference of officials met at Beijing but failed to produce agreement. Nehru neither pressed the matter nor prepared his armed forces to meet any attack in what had now become a dangerously contested area.

The eastern half of the frontier was of much less concern to the Chinese than the vital Aksai Chin, and Chinese pretensions in the east may well have been regarded by Beijing as a useful lever for the extraction of concessions in the west. To India the eastern areas were more sensitive than the western, for they provided easier access into India itself; included the *Himalayan principalities* which were under Indian protection but would, upon any reversal of alliances, provide a Chinese springboard into India off the southern ledge of the Himalayas; and included also Nefa where the Naga tribesmen were in revolt against Indian rule, were tying up Indian forces and were damaging India's moral standing and prestige as stories of vicious Indian tactics leaked out to the world. The unratified McMahon line, the rebellious Nagas and the weak Himalayan states (where alone frontiers were defined) gave the Chinese a small orchestra of opportunities.

Sikkim, the central and smallest of the three Himalayan states, received from India in 1950 a guarantee of internal autonomy and a subsidy in return for Indian control of its defence and foreign relations. India was allowed to station troops in Sikkim. Bhutan, the easternmost of the three, agreed in 1949 to accept Indian guidance in foreign affairs in return for an Indian promise of non-interference in its internal affairs. In both cases India was continuing the policy of the British. In the background was a Chinese claim to Bhutan rejected by Britain early in the century, and the uncomfortable fact that Sikkim, governed by a minority of Tibetan stock, had been virtually part of Tibet in the eighteenth century.

Of the Himalayan states much the largest, and the only fully independent, was Nepal, the home of the Gurkhas who had provided famous regiments for the armies of Britain and India. It had been a refuge for Hindus fleeing from the Moghul conquest and had become a separate unified state in the mid-eighteenth century. From the middle of the nineteenth century to the middle of the twentieth it was under the dual control of a royal family without power and the less than royal but more powerful family of the Ranas who ruled it much as the mayors of the palace ruled Merovingian France or the shoguns ruled Japan between the fourteenth century and the Meiji restoration of 1867. By the middle of the twentieth century the dominance of the Ranas was threatened by a recrudescence of the royal power and by

Congresses on the model of the Indian National Congresses, of which there were two: the Nepali National Congress founded in 1947 in Calcutta and led by B. P. Koirala and his relatives, and the Nepali Democratic Congress, founded in Calcutta in 1949 by a member of the royal family. In 1950 King Tribhuvana provoked a constitutional change by fleeing first to the Indian embassy in Katmandu and then to India itself. The next year he returned, entered into an unofficial compact with the Ranas, introduced a parliamentary régime and installed a coalition government which included Ranas and Koiralas.

These dissensions were embarrassing to the Indian government whose object was to dominate Nepal politely and keep it out of the news. India recognized Nepal's sovereignty by a treaty of 1950. The Ranas had tended to look to China as a counter against India, and Nehru therefore wanted correct and amicable relations with the king. It was also important for Nehru that the king and the Koiralas should co-operate, since in any clash India's natural sympathies would be with the National Congress rather than the monarch and such a clash might induce it to turn to China.

King Tribhuvana was succeeded in 1955 by King Mahendra who proceeded to visit Moscow and Beijing and to receive in his own capital not only the Indian president and prime minister but also Zhou Enlai. Alive to the possibilities of exploiting his strategic position he sought economic aid from all quarters and concluded with China in 1961 a border agreement which gave Nepal the whole of Mount Everest. He also agreed to the construction of a road by the Chinese from Lhasa to Katmandu. He died in 1972. Over the next twenty years India gradually took over Nepal. Mahendra's son, King Birendra, was regarded by Indians as not very intelligent, his queen and her family as grasping. A Nepali arms deal with China in 1988 alarmed India and when in the following year an Indian–Nepali trade agreement expired, India closed thirteen of the fifteen border points between the two countries, thus imposing an embargo which caused much economic distress in Nepal and fuelled disgruntlement with the king's absolute rule. Riots in Katmandu in 1990 reminded the king of his régime's dependence on India and also impelled him towards a measure of constitutional change which reduced his powers and his divinity. Elections in 1994 were won by a party which professed loyalty to the king and to market economics but called itself Marxist-Leninist. (In these years Gurkha immigration into Bhutan threatened to submerge Bhutanese in their own country and take it either into India or into a greater Gurkhaland centred on Nepal.)

LIMITED WARS

In the same year as its border agreement with Nepal in 1961 China proposed to Pakistan negotiations for the settlement of border disputes in Gilgit and Baltistan. In the next year, 1962, it broached its Indian border questions in a very different way. In October it put troops across the McMahon line, skirting Bhutan to the west and entering a part of India which was peculiarly difficult to reach from the rest of India. (Nefa, bounded to the south by East Pakistan, has a panhandle stretching westward to Darjeeling, whence a thin corridor runs between Nepal and East Pakistan into the Indian province of Bihar.) Nehru, who had consistently admitted that the frontiers were ill defined and needed to be discussed, refused to begin talking unless the Chinese withdrew behind the line. The Indian army in the north-east had been reinforced during 1962 but its intelligence and logistical services were conspicuously bad and when the Chinese attacked in earnest India was humiliatingly defeated and was saved from greater disasters only by the intervention of the Americans or Russians or both – or alternatively because the Chinese had had limited aims and had attained them.

China's vigorous action against India was at variance with its approach to its frontier problems with Pakistan and Nepal, and surprising in the light of its domestic preoccupations. The antecedents of this operation are unclear but there are grounds for supposing that China acted as it did in response to a change of policy by India where counsels were divided. Nehru himself and his army chiefs had been opposed to a forward policy which would precipitate the issue of the disputed frontier by occupying disputed areas, but there was an opposing view which regarded forceful action as opportune and Chinese retaliation as unlikely. If, as may be supposed, Nehru became converted to this view he was quickly and bruisingly undeceived and the riddle of Chinese moderation in victory would be explained on the hypothesis that China wanted to do no more than stop Indian infiltration into the disputed areas (by pushing forward police posts) and put the frontier problem back on ice until India was prepared to negotiate about it. China in fact offered to negotiate but the humiliation of defeat prevented Nehru from accepting: the Indian army had lost 3,000 men killed and 4,000 captured. Following a cease-fire at the end of the year a group of India's fellow neutralists – Burma, Sri Lanka, Indonesia, Cambodia, the United Arab Republic and Ghana – offered to mediate, but they did so in so neutral a spirit that many Indians were indignant, hoping for greater sympathy and support. This attempted mediation produced no resolution and the crisis simply petered out.

The coincidence of this short war with the Cuban crisis prompted speculation about deeper Chinese calculations and more dramatic international pressures behind the scenes. While it is exceedingly unlikely that Moscow took Beijing into its confidence over Cuba, it is more than likely that the Chinese were kept informed by the Cubans.

That being so, the Chinese could have seen in the possibility of war between the USSR and the United States an opportunity to press their claims against India and force out of Delhi a cession of the territory in Ladakh which they had occupied. Even larger motives could be imputed to them: to inflict severer defeats and losses on India, to bring Nehru's government down in chaos, to help Indian communists, to strike at India's economic planning. But for such grand designs there is no firm evidence, nor is there evidence for the surmise that China's advance was stopped by external threats. Once the Cuban crisis had been peacefully resolved the Americans were in a position to help India by bombing Chinese airfields and communications, but it is not known that they threatened to do so. The Russians, with obvious irritation against Chinese action, were willing to supply India with Russian aircraft, but whether they also threatened to cut off China's oil is another unknown.

Nehru's standing in his own country was both weakened and strengthened. He had been forced to seek military aid from the United States and Britain and his neutralism had led India into acute danger – because a neutralist, more perhaps than anybody else, needs to be prepared and strong – but he himself seemed more irreplaceable than ever and he was in no danger of losing office. The British government hoped to use this shock to India to bring about a settlement in Kashmir, but once the Chinese attack had been suspended, the shock to Indian security became less than the shock to Indian pride, so that India was in no mood to compound its differences with Pakistan, whose attitude towards India during the critical weeks had been the reverse of soothing. India moreover believed that Britain was on Pakistan's side. Conversations therefore, though initiated, produced nothing, and at the end of the year relations between the two countries were suddenly inflamed by the theft of a hair of the Prophet from its shrine at Hazratbal, in Kashmir. This incident produced riots in both countries and led Nehru, aware that his life was approaching its end, to make an endeavour to resolve the Kashmir problem. In April 1964 he released Sheikh Abdullah who had talks with both Nehru and Ayub Khan. These were fruitless. In May Nehru died and was succeeded by Lal Bahadur Shastri. By the end of 1964 Pakistan, convinced that nothing short of force would serve, was preparing for war. It feared that the time for success in war was running out. On the other side India was worried by Pakistan's American arms and its overtures to China: Ayub Khan visited Beijing early in 1965. Both governments were suspicious and weaker at home than they had been. Temporarily attention was diverted from Kashmir to a desolate and uninhabited mudflat called the Rann of Kutch. This unattractive area, under water for part of the year and above water for the rest, was regarded by India as part of the state of Kutch which was undeniably part of India, but Pakistan claimed that the border between Pakistan and India ran through the middle of the Rann on the principle that boundaries in water-courses lie in midstream. The dispute, in itself somewhat ridiculous, brought the two countries to the verge of fighting but abated after an offer of British mediation. In 1968 Pakistan got by arbitration one-tenth of its claim. More serious was the arrest of Sheikh Abdullah. Since his release the Kashmiri leader had visited Britain and a number of Muslim countries and he was about to go to Beijing. The Indian government, alarmed by his undiminished independence, took the view that he would be better back in prison.

On 28 August 1965 Pakistan troops crossed the cease-fire line in Kashmir which had been established and kept under UN observation since January 1949. A second attack was launched on 1 September. The Pakistani air force conducted some successful operations but the crucial land attacks were held by the Indian army and on 6 September India retaliated by invading Pakistan itself. Thereafter the fighting rapidly came to a stalemate. China delivered a threatening note to India but took no effective action in support of Pakistan. U Thant went in person to Asia and secured an (imperfectly observed) cease-fire. The USSR offered to mediate if Shastri and Ayub Khan would meet Kosygin in the Uzbek capital of Tashkent. Britain and the United States also urged both sides to stop fighting and there were at least implications that economic aid and military supplies would be stopped if they did not.

Pakistan had presumably hoped to score a quick and decisive military victory as a preliminary to negotiations from a position of strength. Its precise political aims were unknown but probably entailed the cession to Pakistan of considerable areas of Kashmir, including perhaps the central Vale of Kashmir with or without a plebiscite. These hopes were frustrated by the Indian army whose performance surprised all who were still judging it by its failures against the Chinese three years earlier. India's successes in the fighting were matched by stubbornness on the political front. Having agreed to a cease-fire India, which remained in occupation of a slice of Pakistani territory, showed no more inclination than it had after 1949 to proceed to a political settlement. Sentiment apart India had two substantive grounds for its continued refusal to treat with Pakistan. The first was strategic. The only serviceable road for the Indian army to use in order to reach Ladakh ran through the Vale with the result that the abandonment of the Vale would have crippled India in any further encounters with the Chinese in Ladakh. An alternative road skirting the Vale could be constructed but only at considerable cost and over a period of years. Secondly, Indians were genuinely averse to a settlement on a religious basis. A plebiscite in Kashmir meant counting Muslims and Hindus and determining the political future of a territory by reference to the religion of the majority of its inhabitants.

India (unlike Pakistan) was a secular state, firmly committed to a non-confessional view of politics. It could hardly accept any procedure in Kashmir which was also acceptable to Pakistan without betraying this principle and also – a matter of practical urgency – endangering the 50 million Muslims in India whose property and lives could well be jeopardized if they were regarded as adherents of Islam rather than citizens of India.

The brief war in Kashmir enabled India to re-establish its military prestige and to score a modest diplomatic point against the feeble intervention of the Chinese. India was not obliged to cede anything to Pakistan. Pakistan on the other hand failed to secure its objectives and gave its larger neighbour an opportunity to demonstrate the strength of its negative position over Kashmir. China had felt compelled to do something in support of Pakistan and had chosen to do the least. The USSR was embarrassed by a possible renewal of the Sino–Indian conflict and also by the possibility of having to take sides between India and Pakistan. India, intrinsically the more important of the two countries if only because of its size, had been cast by the USSR as a useful adjunct in the Sino–Russian conflict, was in receipt of Russian aid and had been championing the USSR's claim (rejected by China) to be an Asian country and a proper member of Afro–Asian conferences. The USSR therefore had powerful reasons for not offending India, but on the other hand it wished also to have good and improving relations with Pakistan. It disliked Pakistan's recent tendency to look for aid and comfort to Beijing, from whose lukewarm cajolements Pakistan ought in Moscow's view to be weaned. Further, the Kashmir War had disenchanted Pakistan with the United States. Pakistan had accepted Seato and Washington's general alliance system but when the crunch came in Kashmir the Americans had failed to give Pakistan the support which it supposed itself to have bought and paid for. Consequently there was at least a possibility of detaching Pakistan from the American system as well as from the Chinese flirtation. More than that, the similar disenchantment of the Turks, whom the Americans had prevented from invading Cyprus, and the perennial fluidity of Iranian politics, where too the shah might be glad of a friend uncommitted to the Arabs, gave Moscow the exciting prospect of dissolving the Northern Tier. But since Russian diplomacy in Pakistan must not lose sight of more important Russian interests in India, it was essential for Moscow to reduce Indo–Pakistani animosities to the minimum. The Tashkent meeting, which took place at the beginning of 1966, was designed both to illumine the USSR in the role of peace-maker and to clear the complex channels of Russian diplomacy in Asia. The meeting stayed the expiring war and boosted Russian prestige but produced no answer to the basic problem in Kashmir. Shastri died suddenly at the end of the conference.

Britain's position during the Kashmir War was that of a friend who is so impartial that he has become useless to both sides and distrusted by both. India and Pakistan each believed Britain to be committed to the other side under a cloak of pious objectivity. In India Britain's position was made worse when Harold Wilson deplored India's invasion of Pakistan without having previously deplored Pakistan's original act of aggression. (Although Pakistan had attacked in Kashmir and not in other parts of India it transpired that there had been a slight incursion into other Indian territory by Pakistani forces.) The Americans were in a similar position: Pakistanis judged that they had failed to live up to their engagements, Indians that they had done as little as had been expected of them.

In India itself the war in Kashmir, coming the year after Nehru's death, intensified an inevitable debate about his foreign policy. Traditionally the central point of a country's foreign policy is its own security, elaborated in national defence forces and foreign alliances. The weakness of India's foreign policy in the age of Nehru was that it decried these traditional concerns and gave more attention to the exercise of influence upon the conflicts between major powers which affected Nehru's vision but did not directly affect India's independence or integrity. In order to play this role in world affairs India needed an exceptional prestige (to command the attention of the great and to collect a following of the less great, without which India by itself would be of comparatively little account) and exceptional detachment. Nehru personally provided both and so succeeded in winning for himself and his country a position which, if not always popular with the great or the less great, was nevertheless gratefully used by the great on such occasions as the ending of the Korean War and the Indo-China settlement of 1954 when Indians were accepted as impartial chairmen or mediators. But Nehru's detachment, and also his refusal to allow his policy of non-alignment to be prejudiced by arms deals and alliances, was only compatible with India's own prime requirements upon the basis that its relations with its neighbours were good. And this was not the case. Both India's most powerful neighbours were hostile: China and Pakistan claimed territory under Indian control, and the attacks delivered first by the one and then by the other forced India to consider whether a policy of non-alignment between the United States and the USSR was not at least irrelevant, and possibly a hindrance, to the defence of its Himalayan borders and the retention of Kashmir. Could India be non-aligned and safe? Could it, as Nehru had believed, be safer non-aligned than dependent on one great power and forced into hostility to the other? Perhaps non-alignment remained the wisest attitude, but if so, must not India become a nuclear power as well as a neutral one?

Shastri's successor was Nehru's daughter, Mrs Indira Gandhi, and India entered a phase in which internal affairs increasingly overshadowed the world role which had so engrossed and appealed to Nehru. The contradictions within the vast Congress Party led to dissensions and splits which foreshadowed a re-formation of Indian political patterns. After elections of 1967 a number of provinces were governed by unstable coalitions and within a year five of them had been placed under president's rule. The central government faced threats to law and order from strikes, students and the continuing failure either to come to terms with the insurgent Nagas and Mizos in the north-east or to silence them. (The Mizos became India's twenty-third state in 1986.)

BANGLADESH

Pakistan too was suffering internal distractions. Ayub Khan's rule had gone on too long with too few results. In East Pakistan secessionist feeling grew and the leader of the Awami League, Sheikh Mujibur Rahman, was arrested. In West Pakistan as well as East there was resentment against Punjabi domination. Leaders appeared to crystallize political and social discontents which reached sufficient proportions to force Ayub Khan to step down in 1969. He was succeeded by General Yahya Khan who proposed to guide Pakistan along much the same lines but perhaps a little faster. Elections were held in 1970 for a constituent assembly to be charged to produce a constitution within 120 days. Ayub Khan's selective democracy − a form of indirect choice based on local elections and proceeding upwards by stages by the election at each stage of delegates to the next − was set aside and universal suffrage permitted. The result in East Pakistan was an overwhelming victory for Sheikh Mujibur, in West Pakistan a somewhat less decisive victory for Zulfikar Ali Bhutto, who had been Ayub Khan's foreign minister in the years 1963–66 and formed the Pakistan People's Party (PPP) in 1967. Sheikh Mujibur's victory had been predicted, though its proportions had not; they were probably inflated by a hideous cyclone which, among other things, demonstrated the incapacity of the central government to organize relief and so accentuated East Pakistan's conviction that the government in the west did not care about its problems. The success of the Awami League was primarily an expression of Bengali separatism. East Pakistan, the more populous of the two halves of the country, objected to its status (under Ayub and Yahya) as one of five provinces, the other four all being in the west; it wanted a generous measure of autonomy in a loose federation in which the central government's authority would be confined to defence, foreign affairs and some currency matters. The president and Bhutto envisaged a stronger centre. After the elections talks began between Bhutto and Mujibur. The latter was now able to point out that he was the leader of the largest party in parliament. In the background was the fact that the ultimate repository of power was the army and the army was largely western. But recourse to the army to coerce the east could mean the disruption of Pakistan.

The Bhutto–Mujibur talks got nowhere and in the east Mujibur began to act like the head of an independent administration. He was again arrested. The president gambled on stopping secession by jailing the seceders, but he provoked instead full-scale fighting. It lasted two weeks. India, moved as some thought by anti-Pakistan venom but more certainly by fears of an anti-Indian left wing coming to power in West Bengal, and by a flood of Hindu fugitives quoted at 10 million, intervened in arms and the Pakistani forces in the east were forced to surrender: 90,000 men were taken prisoner. American and Russian warships appeared briefly in the Bay of Bengal. India also overran the Rann of Kutch and a slice of Azad Kashmir. Pakistan suffered substantial losses in men and material on land, at sea and in the air. The president resigned and Bhutto took his place. Mujibur Rahman was released to become the prime minister of the new state of Bangladesh.

The outlook for the new state was bleak in the extreme. Mujibur was popular but weak and during 1972 he was away ill in England for two months. Postwar chaos was aggravated by a catastrophic wave of disease and death, and then by general disorder, crime and corruption on such a scale that a distracted government had to proclaim a state of emergency in 1974. A year later Mujibur lost the support of the army and was murdered in a coup which was followed by a struggle for power between sections of the army. The economic activity of the country was in ruins and foreign aid ($1 billion) was quickly swallowed up. China, acting in support of Pakistan, vetoed the admission of Bangladesh to the UN for three years. Relations with Pakistan, the unravelling of the foundered association of what had been West and East Pakistan, and the release of prisoners and return of fugitives were hampered at the outset by Bangladesh talk of war crimes trials (which never took place). After the assassination of a second president General Hussein Mohammad Ershad became president in 1982. He possessed at the outset of his rule the reputation of being neither corrupt nor violently Islamic, but in 1985 he made Islam the state's religion (85 per cent of the population were Muslims) and hopes of honest government evaporated. He introduced a measure of stability but little hope of relief for a country crushed by war, poverty and frequently by nature: appalling floods in 1988, for example, put three-quarters of it under water. Opposition, divided between twenty groups of which the two most prominent were led by the widow and daughter of former presidents (whose mutual animosities were hardly less acute than their hatred of Ershad) was rendered the more ineffectual by its habit of boycotting such elections as the government ordained. After elections in 1986 Ershad lifted martial law but a year later he reimposed it and dissolved the parliament. He won fresh elections in 1988, but mounting popular disorder and waning support in the armed forces (which he failed to appreciate) undermined his position and he was obliged to resign in 1990. He was arrested and charged with corrupt practices. Elections in the following year were a contest between the widow of one president and the daughter of another with victory for Begum Khaleda Zia who became prime minister of a country of outstanding poverty and vulnerability to natural disasters.

For Mrs Gandhi the bisection of Pakistan was no bad thing. India's decisive intervention in Bangladesh was immensely popular in India. She crushed opposition to herself in the Congress Party at elections in 1971, getting 40 per cent of the vote and making her opponents look like clueless has-beens, but she produced no solutions to the prime needs of a huge population growing at the rate of 2.5 per cent a year or to those of India's industry whose rate of growth was in decline. The consequent stresses, together with an ill-conceived scheme for compulsory sterilizations, led to disturbances in different parts of the country and although elections in a number of states went well for her in 1972, food problems were becoming acute in the early 1970s and were alleviated only when the USSR supplied 2 million tons of grain and the United States resumed aid; some basic goods and commodities disappeared altogether. In 1974 India brought Sikkim more closely under its control by turning the Chogyal into a figurehead (in the name of democracy) and making Sikkim an associate state of the Indian republic with representatives in both houses of the Indian parliament. China was upset and so was Nepal. The Chogyal committed suicide two years later.

In 1975 a judge of the High Court started a train of unexpected events by ruling that Mrs Gandhi had offended against the Corrupt Practices Act in the elections of 1971 and by imposing on her the statutory disqualification from political activities for five years. The Supreme Court granted a stay and suspended the disqualification pending appeal but two days later Mrs Gandhi declared a state of emergency, arrested hundreds of her political opponents and introduced a stiff censorship. Mrs Gandhi explained that a conspiracy against progress and democracy had been discovered but no convincing evidence of so serious a threat was provided. She was ousted in 1977 by the Janata, a coalition formed by the austere and elderly Morarji Desai who had been an unsuccessful candidate for the premiership in 1964 and 1966, had been slighted by Mrs Gandhi and had formed a new party in opposition to her authoritarianism and the growing political influence of her favourite son, Sanjay. Desai became prime minister at the age of 82 but the Janata lapsed into squabbling and survived only two years. In 1980 Mrs Gandhi won sweeping victories and resumed her rule. A few months later Sanjay was killed in an air accident. His elder brother, Rajiv, was adopted by their mother as heir apparent.

Mrs Gandhi was assassinated in 1984 by *Sikhs* with whom she had developed something of a personal feud. The Sikhs were a would-be nation in a vast conglomorate federation. They had ruled the Punjab between the decay of the Muslim Moguls after 1700 and the arrival of the British in the mid-nineteenth century. The partition of India in 1947 entailed the partition of the Punjab and the flight of the Sikhs of (Muslim) West Punjab eastward to where the dominant Akali Sikhs in particular hoped to create a Sikh state or quasi-state. They agitated for greater autonomy in Punjab, the incorporation into it of Chandigarh (shared with the neighbouring state of Haryana under central control) and a reallocation of river waters. These were political claims arising out of the Sikhs' religious identity. Mrs Gandhi's hostility was directed to the former which smacked of separatism, not to the latter which gave no offence in a secular and multireligious state. The Sikhs aggravated their offence in Mrs Gandhi's eyes when their party, the Akali Dal, made common cause with the Janata in 1977, and after her return to power she sought to discredit it by allowing extremists among the Sikhs latitude for their wilder ways. Under a militant leader, Sant Bindranwale, they occupied and stocked with arms nearly forty Sikh shrines, including the Golden Temple at Amritsar. Faced with this threat of insurrection, Mrs Gandhi ordered the army to clear the shrines and in the course of the main operation at the Golden Temple 1,000 Sikhs were killed, including Bindranwale who was thus promoted to martyrdom. A group of 300 Sikhs later recovered possession of a part of the Golden Temple in defiance of their own leaders and the army again resorted to force to evict and arrest them. The army's operations, which included the use of tanks, appeared incompetently crass and unnecessarily destructive. In revenge two Sikhs of Mrs Gandhi's bodyguard killed her.

At the invitation of Congress leaders Rajiv Gandhi took over the premiership in a remarkably smooth transition. Known as a retiring man without political ambition or experience he was regarded as a convenient but temporary figurehead whose presence could forestall chaos. Within a year he had become a leader in his own right who impressed Indians and others by a quiet determination and obvious integrity but this unexpectedly auspicious start was not maintained. Although he held his ground at the centre he lost it in many provinces. He had some success in economic affairs; growth rose to 9 per cent a year and consumer goods became more plentiful. He was genuinely opposed to religious factionalism but alienated both Muslims and Hindus. He distanced himself from the crusty barons of the Congress Party but replaced them by personal cronies. He bewailed corruption but failed to deal forthrightly with scandals. His good sense did not suffice to solve the Sikh problem or prevent him from publicly slighting India's Sikh president, Zail Singh. His well-intentioned effort to prevent massacre and civil war in Sri Lanka was not rigorously thought out, but he eased relations with Pakistan and China.

His most urgent problem was Punjab. He sought and gained an accord with the Sikh leader Sant Harchand Singh Longoval, giving Punjab the city of Chandigarh in return for the cession of a number of villages to Haryana and an allocation of river waters vaguely favourable to Haryana. Although Sant Longoval was then assassinated by Sikh militants the agreement stood the test of provincial elections in Punjab which were won by the Akali Dal, campaigning for its endorsement. In Haryana on the other hand the proposed loss of Chandigarh gave great offence to Hindus and led to a crushing defeat for the ruling Congress (I) Party in 1987. Nor did the agreement itself survive. Disputes over its detailed application gave the more militant Sikhs the excuse to disavow their leaders, pull down parts of the Golden Temple on the grounds that it had been polluted by Hindus, and proclaim their separate state of Kalistan (1986). The Golden Temple was recovered from the militants but the underlying conflicts persisted, among Sikhs and between Sikhs and Hindus. Violent disorders occurred from time to time.

The setback to the Congress (I) Party in Haryana was preceded, if less dramatically, in other provinces – West Bengal, Kerala. Gandhi's personal standing was tainted by the resignation of his esteemed defence minister, V. P. Singh, in protest against the government's reluctance to explore financial scandals, particularly those relating to defence contracts with Bofors of Sweden, and continuing dissensions within the ruling party caused further electoral losses at by-elections in 1988. Huge trade deficits, as exports declined and imports rose, added to the feeling that the government was as precariously in control of the country's economy as of its cohesion. On the credit side Gandhi pacified Assam's discomfort over the influx of Muslim refugees from Bangladesh; a fence was built to stem the flood while plans were made to send refugees back. On balance, however, Gandhi's performance in domestic affairs began to look like honourable inadequacy. Abroad he improved relations with China but achieved no concrete settlement of the two countries' frontiers. He reaffirmed India's good relationship with the USSR and made some improvement in relations with the United States; he got credits from the one and technical aid from the other. He was a principal author of the South Asian Association of Regional Co-operation (SAARC), an association of seven states – India, Pakistan, Sri Lanka, Bangladesh, Nepal, Bhutan and the Maldives – which, by deftly avoiding all contentious issues, engineered some useful co-operation in commercial and technical matters and transport: it established a headquarters in Katmandu. In the all-important direction of Pakistan Gandhi met both General Zia and Benazir Bhutto to demonstrate goodwill as a cautious prelude to reducing mutual fears and suspicions. But in 1989, his virtues having been overtaken by his shortcomings, he lost a general election to a motley association of opponents, led by V. P. Singh and stretching from the Bharatiya Janata (BJP) on the Hindu right to assorted communists on the left. The Congress (I) Party fared particularly badly in the northern states but it remained the largest party in the Lok Sabha and Gandhi was quickly re-elected chairman on the motion of the man tipped to supplant him in the event of electoral defeat. V. P. Singh commanded a majority in the Lok Sabha only so long as the extremists of right and left continued to support him. He jeopardized this majority with proposals for positive discrimination in the job market for the lower castes and he lost it in 1990 when he ordered the arrest of Lal Krishna Advani, leader of the BJP, who was conducting a long march across northern India with the aim of pulling down a mosque and building a Hindu temple in its place – a threat to communal peace and even to international peace in the sub-continent. Chandra Shekhar, who was Singh's colleague and adversary in his own party, Janata Dal, took the opportunity to split the party, destroy Singh's majority and succeed him as prime minister by winning from Gandhi a promise of parliamentary support. Gandhi thus steered clear of an immediate return to power and responsibility but his attempt to be recognized as leader in the Lok Sabha of the opposition to a government which he had promised to support failed amid some derision.

This disorderly political scene was the more damaging owing to an accumulation of economic reverses: interest rates at an unprecedented level, repeated devaluations of the rupee, a foreign debt exceeding $70 billion, a population growing so fast that it seemed likely to match that of China in not much more than a hundred years. In 1991 Rajiv Gandhi, India's most solid politician, was assassinated by a Tamil in the course of an election tour in southern India. Although the Congress Party won 225 of 544 seats in the Lok Sabha, the BJP, which had two seats in 1984, won 119 with 20 per cent of the vote. The BJP, whose strength was mainly in the north and west, was an expression of militant Hinduism organized by ideological zealots and led by mendacious demagogues who traded on respectable antecedents as anti-British nationalists to transfer this hostility to Muslims, trade unions and much else. With the Janata Dal reduced to a small group in the Lok Sabha, the Congress Party returned to government with concessions to right and left and with Narasimha Rao as a relatively featureless but unexpectedly enduring prime minister and Manmohan Singh as his finance minister. India's reserve of foreign exchange was entirely exhausted but India was too big to be ignored or bullied. Manmohan Singh, a banker, rescued its finances and won favourable deals from the World Bank and the IMF by steering a course between monetary correctness and China's more hazardous dash for spectacular growth. He devalued the currency, reduced inflation, lowered tariffs, introduced tax reforms, adopted modified market policies, replenished the reserves and made India talked about with more attention than despair. In provincial elections in 1993 the BJP suffered severe losses and split, retaining substantial popularity only among middle classes in Delhi and other big cities. In Punjab Sikh militancy died down. But Rao turned a blind eye to rising charges of corruption in government and the Congress Party and experienced unexpectedly sharp rebuffs at provincial elections in 1994, including losses in his own state of Andhra Pradesh.

THE NEW PAKISTAN AND INDIA

The abiding sore in the sub-continent was the bad relations between its two biggest states. Mrs Gandhi and Zulfikar Ali Bhutto had concluded in 1972 a comprehensive agreement in the course of direct talks in Simla and by the end of that year they had fixed provisional lines of demarcation in Kashmir and begun the business of repatriating prisoners taken in the war over Bangladesh. The Bangladesh crisis had shaken settled attitudes in both countries. It caused Mrs Gandhi to abandon India's traditional non-alignment and conclude a treaty with the USSR to offset the partiality of the United States and China for Pakistan. (Washington had gone so far as to stop aid to India and denounce it for aggression.) Bhutto's main problem was to build a new state. Pakistan had lost not only face and half its territory but also half its domestic market, a major part of its raw materials and manufactures, its overseas markets (which had been supplied by East Pakistan) and its foreign earnings; and the constitutional problem, unsolved since 1947, remained. This last problem was temporarily laid to rest in 1973 by the adoption of a presidential, federal constitution, but in the same year economic problems were exacerbated by disastrous floods.

Bhutto set Pakistan on the nuclear road. He sought French help which was initially promised but then revoked upon American remonstrance. Nevertheless Bhutto succeeded, at first secretly, in starting a programme under the charge of a Pakistani scientist who had acquired the requisite expertise in the Netherlands. Bhutto's expressed attitude was that, since Hindus, Jews and Christians had nuclear weapons, there ought to be an Islamic bomb too. (India achieved a nuclear explosion in 1974: Mrs Gandhi and the pacific Desai both forswore the ambition to develop a nuclear armoury.) Washington tried by pressures and cajolery to stop Pakistan's progress; it cut off economic aid as required by the Foreign Assistance Act 1976, refused to relax debt repayment terms and offered Pakistan modern fighter aircraft if it would apply international safeguards to its nuclear activities. Washington's dilemma was how to make Pakistan a secure ally but without fatal injury to its relations with India.

In 1977 the Pakistani army stepped in once more, installed General Zia ul-Haq as president, hanged Bhutto two years later and inaugurated a cruelly repressive régime. In external affairs Zia continued the programme initiated by Bhutto to give Pakistan the option of becoming a nuclear power; some Pakistanis professed to fear a combined Indian and Israeli operation against Pakistan's nuclear installations. Zia took Pakistan out of the moribund Cento alliance and made it a member of the non-aligned group but the Soviet invasion of Afghanistan in 1979 turned him back to the United States camp and consolidated his personal position. The army, which had put him in power and on which he was wholly dependent, had shown few signs of keeping him as president until the Soviet invasion made an early change inopportune: initially the extension of direct Soviet power to Afghanistan seemed to entail a possible threat to Pakistan also. That fear abated when it became apparent that the USSR was less interested in Pakistan than in Afghanistan itself and Iran and that the puppet régime in Kabul effectively controlled little more than a fifth of the country. Zia thereupon set himself successfully to use the invasion as a lever to recover American military aid and funds. The Soviet action, although a costly blunder, was interpreted in the United States as a dangerous increase in Soviet power which – the shah of Iran having been overthrown – could be checked only by arming Pakistan and using it as a channel for arming opposition to the USSR in Afghanistan. Zia lent himself to this policy with enthusiasm. He received American aid in profuse quantities, notwithstanding much American dislike of Pakistan's nuclear programme and its role in the drug trade and notwithstanding too the Symington Amendment which prohibited aid to states manufacturing nuclear weapons.

By 1983 the remission of fears of Russian aggression against Pakistan revived pressures within the country for an end to military rule, and in an attempt to preserve his position Zia surprisingly announced at the end of 1984 that he would hold a referendum to test popular approval of his policies in general and himself as head of state, with the corollary that, if approved, he would serve for another five years and intensify the

Islamicization of the constitution and the application of the severest Koranic penalties for infringements of religious rules: penalties which delighted many Muslims and shocked others. The votes cast were almost all in his favour. Martial law was removed at the end of 1985 – a show of moderation which, however, was preceded by an act of indemnity for past actions and a substantial increase in the presidential powers, including the extension of Zia's own term of office to 1990.

Zia had done nothing to lessen tensions between Punjabis on the one hand and Sindis and Baluchis on the other and the return to Pakistan in 1988 of Bhutto's daughter and political heir Benazir increased bitterness and uncertainty. The persistent weakness of the Pakistani economy was aggravated by the profligacy of a military régime which allocated 40 per cent of the budget to the armed forces, ran huge deficits on the budget and external account, and allowed foreign debts to escalate ominously: defence costs and the service of the foreign debt together absorbed 85–90 per cent of the budget. Zia was constrained to seek a large loan from the IMF, for which he had to submit his country to severe austerity, higher prices, new taxes and collapsing social services and roads (the health budget was cut to less than one-fiftieth of the defence budget). Zia's problems were multiplied by declining remittances from Pakistanis in the Gulf and other foreign parts, by a sequence of crop failures, by the stream of refugees from Afghanistan, and by the private armies of drug syndicates which operated even in the capital where Pathan immigrants clashed with the non-Pathan inhabitants. Opposition to Zia therefore grew but it was not compact. After her return Benazir Bhutto pulled discordant groups together, but not seamlessly. The first local elections after her return disappointed her followers. She was handicapped as a woman and a Sindi, particularly in the dominant province of Punjab; by a personality in which intelligence was not unmixed with arrogance; and by an inauspicious marriage arranged by her mother. The government had some success in denigrating and marginalizing her PPP until two events in 1988 transformed her position: Zia abruptly dismissed his civilian prime minister Mohammad Khan Junejo who dissented from the army's espousal of total support for the *mujaheddin* in Afghanistan, and Zia was himself killed in an air accident. Two other generals, several senior officers and the American ambassador died with him.

The elections which followed this catastrophe were fought between two consolidated groups: the Islamic Democratic Alliance which consisted of the Muslim League and minor associates, and the PPP and its associates. In a low poll the PPP won much the larger number of seats but not a majority and it failed to win in any province except Sind and the North West. Benazir Bhutto was installed as prime minister and the army made one of its periodic retreats into the background. But not for long. In varying degree the army, the president and a majority of the provincial governors were hostile to Bhutto and awaiting a chance to get rid of her. Her heritage of corruption, drug trafficking, Afghanistan, economic decline, inter-provincial squabbles and growing Islamic fundamentalism was too complex for a government which controlled only two provinces and was ultimately dependent on army officers whom she could neither conciliate nor subordinate. Her government was accused, not without reason, of incompetence, corruption and nepotism and in 1990 she was dismissed by the president, using the pretext of violence in Sind which she had failed to control. (Violence in Sind was endemic between Sindis and Muhajirs, post-1947 immigrants from India.) In fresh elections later in the year she lost to the accompaniment of far from implausible charges of ballot rigging. The victors were the Islamic Democratic Front, an amalgam of nine parties with a strong flavour of religious intolerance and extremism and the backing of the army. The Front and its allies controlled two-thirds of the parliament but by 1993 the new president, Ghulam Muhammad, a Pathan, had quarrelled with his prime minister, Nawaz Sharif, over the right to appoint the commander-in-chief of the army and the prime minister was dismissed by the president but reinstated by the Supreme Court. In the ensuing elections Bhutto, who had improved her relations with army leaders, got her revenge on Nawaz Sharif and became once more prime minister. The Islamist Jamaat-i-Islam fared surprisingly badly.

The United States collaborated with Pakistan's support for anti-communist *mujaheddin* but was alarmed by its apparent determination to become a nuclear power, an ambition forcefully proclaimed by Ali Bhutto in 1966. India had tested a nuclear device in 1974 and was believed to be developing its research on plutonium warheads with Chinese help, but the United States accepted that India was constructing no nuclear weapons. By the Pressler Amendment of 1986 both countries were debarred from American economic and military aid unless the president certified to Congress that they had no nuclear weapons. In 1990 President Bush refused to give this certificate in the case of Pakistan which, a few years later, was believed to possess half a dozen uranium 335 warheads, to be building a plutonium reprocessing plant and to be importing plutonium.

During the 1980s Pakistan's conflict with India had been over-shadowed by the war in Afghanistan but it was not obliterated and it was in two respects sharpened: by India's Sikh problems and by Kashmir once more. Indians accused Pakistan of inciting the more militant Sikhs in India's province of Punjab and moved troops to stop Sikhs from crossing into India from Pakistan. Pakistan concentrated its own forces on the same border. In 1985 fighting broke out between India and Pakistan over possession of a glacier north of Kashmir. India had the advantage; mediation was proposed in 1987; but more fighting followed and in 1989 the central Kashmiri issue was once more inflamed. Sheikh Abdullah having died in 1983, his mantle had fallen on his son Faruq Abdullah (a lightweight character too close to New Delhi for the taste of most Kashmiri Muslims) and his son-in-law Ghulam Mohammad Shah, and the family disagreements of these two compounded political feuds and led to disorders in l989–90 in which a number of people were killed. India blamed Pakistan for the disorders, dropped Faruq and resorted to direct rule from Delhi. In Kashmir the anti-Indian and increasingly violent Jammu and Kashmir Liberation Front gained popular support while in reply the Indian authorities fell back on harsh reprisals, including the random torture of innocent victims. These disorders inevitably involved Pakistan and even Iran and Saudi Arabia which vied with one another in their support for Pakistan as a way to demonstrate the purity of their Islamic credentials. In 1993 the temple of Hazrat Bal in Srinigar, which possessed a hair from the Prophet's beard, was surrounded by Indian troops upon rumours of a plan by Muslims to remove the hair. Muslims were outraged by this show of force which non-Muslims deplored as unwisely provocative. Kashmiri dissidents were divided between those wanting independence and those preferring incorporation in Pakistan, a division which Indians were glad to observe and possibly foment. In 1994 Rao retaliated against Pakistani meddling in Kashmir by claiming the return to India of Kashmiri territory annexed to Pakistan.

In the half century after the end of the Second World War the world at large had wondered at economic miracles in Germany and Japan but had found little else deserving vigorous applause. The world had perhaps overlooked a political miracle in India where daunting economic, technical, administrative and sectarian problems had been handled, most of the time, in a domestic framework to the considerable credit of India's governing classes – and their predecessors. But this achievement was still more remarkable than secure.

15 THE INDO-CHINESE PENINSULA

The most disputed postwar arena in South-east Asia was Indo-China. This area, put together by the French, contained the protectorates of Annam and Tongking and the colony of Cochin-China (Annamite by race, Chinese by culture, and together called the three Kys) and the protected kingdoms of Luang Prabang or Laos, and Cambodia (Thai by race, Hindu by culture). The Khmer forebears of the modern Cambodians ruled an empire which, at its peak in the twelfth century, reached from sea to sea and included the southern parts of Burma, Siam, Laos and Annam. In Laos invading Thais had established an ascendancy in the thirteenth century. By the nineteenth Cambodia and Laos were threatened by Annam but were saved by the French.

The French established themselves in Asia later than the other European powers. Their defeats in the Franco–Prussian War of 1870–71 on their own soil had deprived them of territory, valuable resources, population, prestige and self-respect, and although recovery was surprisingly rapid, the acquisition of a new colonial empire in the 1880s was in some sense a compensation. It took the form of entering the competition for whatever might be made out of the disintegration of the Ottoman and Chinese empires. Tunisia was virtually annexed as an extension of French power in Algeria, Madagascar was occupied, and in Asia Annam, Tongking and Cochin-China were made the nucleus of an Indo-Chinese dominion which might provide a way into southern China and was extended by taking the kingdoms of Cambodia and Luang Prabang under the French wing. The acquisition of Annam and Tongking led to an unpopular war with China, while the western and Hindu-ward move brought France into competition with the British in Burma, where they were checked by the viceroy, Lord Dufferin (who feared – or perhaps invented – a French threat to India), and also into a more enduring hostility with the independent kingdom of Thailand. But French power remained substantially unshaken until 1940.

During the Japanese occupation in the Second World War the three Kys became the autonomous state of Vietnam and upon the Japanese withdrawal Ho Chi Minh, the leader of a communist-dominated nationalist coalition, proclaimed the independent republic of *Vietminh*.

15.1   Vietnam, Cambodia and Laos

Again, as in Indonesia and Korea, practical reasons of convenience dictated the course of events with more far-reaching effects than were contemplated at the time. The British took control south of the 16th parallel and the Chinese north of it; both withdrew during 1946 but the former had first cleared the way for the return of the French who arrived to find Ho Chi Minh in control of a sizeable area in the north and presiding over a government. The Annamite emperor Bao Dai, having abdicated in 1945, had accepted the post of chief counsellor to Ho.

The French began by recognizing Ho's government as autonomous within the French Union but refused to accept his demand for the union of the three Kys. Towards the end of 1946 French policy stiffened as a result of right-wing pressures. In November Haiphong was bombarded and in December the French in Hanoi were attacked by the Vietminh, convinced that France now intended to overthrow Ho: some forty Frenchmen were killed and 200 abducted. This was the beginning of a war which lasted seven and a half years. After some hesitation the French decided to give up further negotiation with Ho and to switch instead to Bao Dai, to whom they offered the union of the three Kys which they had refused to Ho. The so-called Bao Dai experiment was an attempt to separate communists from other nationalists and to preserve with Bao Dai's complaisance a general French overlordship throughout Indo-China. An agreement signed in Along Bay in June 1948 embodied this policy (which was to be reflected in the south in the next decade by the Americans with President Diem as the anti-communist nationalist) and a new state of Vietnam was formally constituted in June 1949. Bao Dai, who had been in France during the negotiations, returned to his own country as head of state. Vietnam was proclaimed an Associated State in the French Union in December, and Laos and Cambodia were given the same status. Among those who refused to co-operate with Bao Dai was the Roman Catholic leader and future president of South Vietnam, Ngo Dinh Diem.

But the Vietminh did not give up and the victory of the Chinese communists across the northern border transformed the situation. The attempt of the French to retain power by the device of the semi-independent Associated State led to protracted bickering between themselves and Bao Dai and failed in the end to achieve its aim. France had not been able to nerve itself to take the extreme but simple step taken by the Attlee government in India, and the history of the next few years suggested that nothing less was of any avail. In 1950 five months' conferring at Pau about constitutional advances exacerbated all parties, while in Indo-China the Vietminh took the field under its able general Vo Nguyen Giap. A Vietminh offensive was strikingly successful. Marshal Juin was sent to the scene. Bao Dai also returned. Pierre Mendès-France and others began to say that it was time for

France to clear out of Asia. The French government, which had been resolutely opposed to admitting that the situation was in any sense international, relented to the extent of accepting American economic and military aid; to fight for French supremacy brought neither sympathy nor success, and the only way to continue the fight was to call it a fight against communism and invoke the aid of anti-communist friends. On this new basis General de Lattre de Tassigny was appointed high commissioner and commander-in-chief in December 1950. With a new general to raise morale (in which he was very successful), with American aid and with the negotiation of a political settlement of the status of the Indo-Chinese states within the French Union, the French made their final effort.

But the Vietminh struck first and showed that it was capable of waging open war as well as guerrilla operations. Its attack was renewed in October 1951 and a few weeks later the French lost de Lattre who was invalided home to France and died there in January. Bao Dai's Vietnamese army was coming into existence with extraordinary slowness, and the private armies which complicated the picture were not being reduced to his control. Chinese aid to the Vietminh was increasing, although the invasion which could have brought the Americans in never materialized. During 1952 French losses in men, material, morale and prestige were considerable, and in 1953 the Vietminh carried the war into *Laos*, threatened the royal capital of Luang Prabang and forced the French to divert forces from Vietnam to the Plain of Jars in Laos. In *Cambodia* King Norodom Sihanouk raised the political stakes by asking for a status in no way inferior to that of India and Pakistan in relation to Britain, and then put the French in great embarrassment by temporarily decamping to Thailand. Amid this scene of disintegration the French could do no more than offer to review the constitution of the French Union and begin yet another round of negotiations with Bao Dai as well as the Laotian and Cambodian monarchs. (Sihanouk had succeeded to the throne in 1940 at the age of 18. In 1955 he abdicated in favour of his father who reigned until 1960 with Sihanouk as prime minister. In 1956 Zhou and Sihanouk signed a non-intervention agreement. Sihanouk visited Beijing, Moscow, Prague and Belgrade; also Madrid and Lisbon. He adopted a policy – the reverse of Burma's – of the widest diplomatic contacts and aid from as many places as possible. He seemed, however, to have a preference for China, which he again visited in 1958 and 1960. In 1960 he resumed the top place in the state with the title of Head of State.)

There was by this time nothing left for the French in Indo-China except a need to salvage pride, but since the French have quite enough to be proud of without retaining distant satrapies, they became increasingly fed up. Moreover they were now preoccupied with the revival of German strength, beside which the strength of the Vietminh seemed both inconsiderable and irrelevant to France's position in the world. France had failed to brace itself to give its Indo-Chinese territories independence after 1945 and had entrusted their affairs to generals who, often unrealistic and at odds with one another and their civilian colleagues, had fought and lost an old-style colonial war. If the Vietminh mattered to anybody, it mattered to the Americans who were already paying for the war indirectly and saw it as a new-style anti-communist war. It required only a climactic event to make France acknowledge that what it really wanted in Indo-China was to get out. This event occurred at Dien Bien Phu.

Dien Bien Phu was a small garrison or camp in a bowl in the north-west. Its possession was important in relation to the Vietminh's threats to neighbouring Laos, which were themselves important inasmuch as they demonstrated French inability to protect a protectorate and at the same time diverted French forces from the defence of the strategically and politically central Red River delta and Hanoi. Dien Bien Phu had changed hands more than once during the war. It was taken by the French in November 1953 and after some hesitation they decided to stay there.

General Navarre, now in command of French and Vietnamese forces which considerably outnumbered the Vietminh, believed that if he could force the enemy to battle he could inflict upon them a major defeat and permanently reduce their operations to minor guerrilla scale; he believed that the Achilles' heel of the Vietminh was in numbers and that the Chinese or Russians, though willing to supply arms and equipment, would not send fighting men across the borders for fear of American retaliation. Dien Bien Phu was therefore to provide the setting for the battle which would cripple the Vietminh.

Early in 1954 agreement had been reached among all the principal powers concerned to hold an international conference on Indo-China and Korea, and as the preparations went ahead it became increasingly clear that the outcome at Dien Bien Phu would have a powerful influence on the course of negotiations. It also became clear that the French, so far from delivering a knock-out blow, were being surrounded and pounded by an unexpectedly large force and were in danger of having to capitulate. What was not so clear in these circumstances was whether France's allies would be well advised to make a special effort and intervene. The Americans, who had been opposed to any such involvement, began to have second thoughts and to propound the view that the loss of Indo-China would be a fatal blow to the fortunes of all South-east Asia and even further afield. In a famous speech in January Dulles threatened 'massive retaliation' as a means of stopping communist expansion and aggression, and at the end of March he appealed for united international action to prevent the imposition of communism on South-east Asia. Congressional leaders and allies were sounded on joint intervention in Indo-China and retaliation against China itself in the event of a Chinese counterstrike. The response was unfavourable.

The Korean War had left the United States with little appetite for Asian adventures and the allies had not recovered from their distrust of MacArthurism which they detected reviving in the views of Admiral Radford, the naval chief of staff, who advocated American air strikes. Eisenhower was prepared to give his consent to Radford's policy if Congress were to agree and the United States were not the only intervenor, but General Ridgway, chairman of the chiefs of staff, opposed intervention on the grounds that it would force the Chinese to enter the war as they had done in Korea. Eisenhower, who had campaigned for the presidency on a promise to stop the war in Korea, may have known that his conditions for intervening were most unlikely to be met, but he allowed Dulles to pursue the question of allied co-operation. Dulles discussed intervention in London and Paris on 11–14 April and returned to Washington under the impression that he had secured agreement for a general conference to devise a plan but Eden, with whom his personal relations were bad, immediately denied this interpretation of their talks and refused to send a representative to a preliminary discussion. Dulles returned to Paris later in the month with a proposal for unilateral American air intervention and Eden, who was also in Paris on his way to Geneva, flew back to London where a Sunday meeting of the cabinet refused to give its endorsement. Eden communicated this decision to Georges Bidault, the French foreign minister, at Orly airport on his way to Geneva and so acquired the reputation of being the man who saved the world from being plunged into a new world war by American temerity at Dien Bien Phu. It seems, however, truer to judge that American intervention had already been dismissed because of the opposition of the American chiefs of staff (Admiral Radford alone excepted) and American congressional and public opinion. Within a very short time the view that Indo-China was essential to the free world had been temporarily dropped and the fate of Vietnam was once more being treated as a local affair.

The Geneva conference, convened to discuss Korea and secondarily Indo-China, opened on 26 April. Dien Bien Phu fell on 7 May. The French government also fell and Mendès-France took the place of Laniel (who resigned on 13 June) with a promise to reach a settlement in Indo-China by 20 July or resign. On 23 June Mendès-France and Zhou Enlai had a long private discussion in Berne before the latter set off for Beijing via India and Burma during a break in the conference.

At the same time Churchill and Eden visited Washington to discuss a variety of topics and repair the damage to Anglo–American relations caused by the Dulles–Eden misunderstandings of April. Soon after the conference resumed three armistice agreements were signed. Vietnam was partitioned roughly along the 17th parallel (a compromise choice), the Vietminh agreed to withdraw from Laos and Cambodia, and three armistice commissions were constituted with Indian, Polish and Canadian members to supervise the implementation of the agreed terms. The conference marked above all the defeat of France and its withdrawal from all the states of Indo-China. It purported to proclaim the creation of three new independent states: Laos and Cambodia, which were to be safeguarded in this condition against their hereditary enemies in Vietnam and Thailand by guarantees by China, France, Britain and the USSR; and Vietnam which had won independence but not perhaps integration.

In *Laos*, although the Geneva settlement provided for the withdrawal of Vietminh forces, there were other forces to keep revolt going. The Pathet Lao, created in 1949 by Ho as an adjunct of the Vietminh with a member of the Laotian royal family, Prince Souphanouvong, among its senior leaders, had established itself sufficiently to be able to retain control of the two northern provinces of the country. In 1956 this prince visited Beijing and Hanoi and in the next year he negotiated a coalition with his half-brother Souvanna Phouma (who was prime minister) on the basis that Laos would be neutralized. But this coalition lasted only to 1959 when Souphanouvong and other Pathet Lao leaders were arrested. (Later coalitions in 1962 and 1973 were equally evanescent.) The United States, having adopted the domino theory that communism would engulf the whole of South-east Asia if it won a victory in any part of the region, was not prepared to tolerate neutralism and decided to exclude not only the communist prince but also the neutralist. It began to pour money and other aid into Laos with an abandon which produced much corruption and waste and two years of civil war. A new government in Laos, making the most of North Vietnamese incursions, asked for a UN mission and a UN force. Hammarskjöld visited Laos in person and sent a special representative to observe and report and so gain time for temperatures to fall, but in December 1959 General Phoumi Nosavan staged a successful coup which had the opposite effect. It also evoked, a few months later, another coup led by Captain Kong Le, a rather naïve symbol of the irritation of ordinary men who were fed up with feuding and corruption. Souvanna Phouma declared his support for Kong Le, became prime minister once more and contrived briefly to reconcile the general and the captain, again on a neutralist ticket. But his solution did not last, largely because the Americans were able to rebuild a right-wing front under General Nosavan and Prince Boun Oum, a distant relative of the reigning family and mediatized ruler of Champassak (in the south). Souvanna Phouma fled to Cambodia, whence he was transported in a Russian aircraft to confer with his half-brother and Kong Le. He was also courted by Boun Oum and Nosavan but preferred to set off on a world tour. Meanwhile Kong Le had inflicted a defeat on Boun Oum and Nosavan. In 1961 the three princes met in Switzerland and agreed that Laos should become a neutralized state without military alliances. In Washington the new President Kennedy was more inclined than Eisenhower had been to accept a neutral Laos, partly through disenchantment with the Laotian right and partly because of the failure of direct military intervention at the Bay of Pigs in Cuba. War was brought to an end when the United States and China reached agreement behind the scenes on a neutralized Laos and the evacuation of foreign troops — that is, American troops which China feared and the United States felt to be engaged in a worthless cause. But Laos continued to be used by Ho for supplies into South Vietnam, the authority of the Pathet Lao spread and the American forces were replaced by an active communist army — part Laotian and part Vietnamese — which probably numbered 60,000 or more. In 1963 the Laotian coalition government dissolved and the country became virtually partitioned with the Pathet Lao ruling in the north-east and Souvanna Phouma the rest. From 1964 the United States reversed its policy of withdrawal, gradually built up very large ground forces and used its air forces in operations by the Laotian government against the Pathet Lao. Laos became an important American theatre of war, ancillary to the waging of the war in South Vietnam and to the protection of American forces of about 50,000 in Thailand. The Americans dropped a greater weight of bombs on Laos than had been dropped on Germay from beginning to end of the Second World War.

In Vietnam the 1954 Geneva settlement gave Ho Chi Minh half the country and the expectation of the rest in less than two years if the terms of the settlement were fully accepted and implemented. The formal agreement concluded at Geneva was an armistice agreement signed by generals on behalf of France and the Vietminh. It drew a line (roughly the 17th parallel), imposed a cease-fire, and made arrangements for the regroupment of military forces and the resettlement of civilians on either side of the line. The conference also produced a number of declarations, including a final declaration propounded but not signed. The United States and South Vietnam dissociated themselves in particular from this declaration's provision for elections throughout Vietnam by the middle of 1956; they believed that such elections were bound to transfer the whole of the country to communist rule since the part north of the armistice line contained a majority of the population — which was in addition likely to vote with that 90 per cent solidarity characteristic of authoritarian régimes. The government in South Vietnam, which had been established in Saigon by the authority of Bao Dai before the Geneva conference, considered itself bound by nothing that was settled there.

For the French the armistice agreement of 1954 was a means of escape. They were able to conclude a war which they were losing. Thereafter they watched with an excusable but irritating smugness as the Americans repeated in the next phase many of the mistakes which they had made between 1945 and 1954. For the Russians and the Chinese the settlement was a political arrangement which they pressed upon Ho, the Russians because they wanted France to stifle the nascent European Defence Community and the Chinese because they wanted to remove western forces and influence from a country on their borders; both may have supposed that Ho would soon rule as effectively in Saigon as in Hanoi and were faced with fresh problems when this did not happen.

For the Americans the Geneva settlement marked the end of a French presence in Asia which, however obnoxious at one time to Americans on general anti-colonialist principles, could have been rendered useful in anti-communist terms. Having decided in 1954 not to buttress French rule any longer, the United States sought an alternative anti-communist and anti-Chinese force. They disapproved of the Geneva settlement because it not only failed to constitute such a force but also threatened to accelerate Chinese communist expansion by giving Ho the whole of Vietnam in two bites – the north by the armistice agreement and the south through elections: they regarded Ho as a satellite and discounted his chances of becoming the Tito of Asia. They resolved therefore to maintain the independence of the anti-communist régime established by Bao Dai in the south, and also to create a new anti-communist alliance to check China in Asia as Nato had checked the USSR in Europe and to facilitate in future the united action which Dulles had tried and failed to organize for the relief of Dien Bien Phu. Accordingly the South-East Asia Collective Defence Treaty (the Manila Pact, establishing a South-East Asia Treaty Association – Seato) was signed in September by three Asian and five non-Asian states: the Philippines, Thailand, Pakistan, the United States, Australia, New Zealand, Britain and France. These signatories bound themselves to take joint action in the event of aggression against any one of them in a designated area, and to consult together in the event of threats from action other than armed action (namely, subversive activities). The designated area was the general area of South-east Asia including the territory of the signatories and the general area of the South-west Pacific to 20 degrees 30 minutes north; the area included therefore Vietnam, Cambodia and Laos but not Taiwan or Hong Kong.

Seato never became an impressive organization. No purely Asian state of substance joined it except Pakistan which, distant from China and not really concerned about Chinese expansion in South-east Asia or anywhere else, joined for ulterior reasons – to please the United States and get American support against India. France became an increasingly cynical member; Britain became an increasingly embarrassed one, balancing the obligations of a loyal ally (with a special interest in the area so long as it retained obligations to Malaysia) against the wish to keep out of Vietnam and the temptation to criticize American mistakes there. Seato was no more than the United States writ differently. Its purpose was to ensure the independence of South Vietnam but it could not save that ill-governed country from the dilemma of either collapsing or surviving as an American protectorate. It was dissolved in 1975.

Ho had accepted the Geneva armistice reluctantly and under Russian and Chinese pressure. It may be surmised that only the prospect of elections in 1956 persuaded him to come to terms with France when he was entitled to expect not only the surrender of Dien Bien Phu but the collapse of the entire French position in any event and in the same year. He soon saw that his hopes were to be falsified and that the 17th parallel was another armistice line destined to harden into a political frontier. It was moreover a line of graver consequence than the division between the two halves of Korea, for in Vietnam the south fed the north and the perpetuation of the division entailed economic problems as well as disappointment over the denial of reunification. In the period of uncertainty between the signing of the armistice and the date fixed for the elections which were never held Ho initiated a programme of agrarian reform and sought also to expand industry and exploit North Vietnam's mineral wealth, but the agrarian reform, modelled on Chinese collectivization, provoked a peasant rising which provoked a reign of terror which got out of hand and caused at least 50,000 deaths. Industrialization required help from advanced countries such as the USSR and Czechoslovakia rather than China, and after the Sino–Russian split in the 1950s Ho had to balance between Moscow and Beijing. At first he allowed relations with the latter to deteriorate and in 1957 he received Voroshilov in Hanoi, but this partiality displeased some of his colleagues and may even have caused a threat to his own position. Ho himself had associations of a lifetime with the USSR and Muscovite communism, and General Giap, backed by the North Vietnamese army, expressed the traditional Vietnamese distrust of China, but other leaders were reputedly pro-Chinese, including Truong Chinh whose strength lay in the Lao Dong Party (founded in 1951 and essentially a communist party although it embraced a few non-communist notables). Ho spent two months in Moscow in 1959, returning to Hanoi via Beijing. By this time fighting had been resumed in the south and the problem of getting rid of the French was succeeded by the far bloodier problem of getting rid of the Americans. He lived just long enough to see it happen, dying in 1969.

American policy during the Eisenhower and Kennedy administrations rested on two main propositions: first, that South-east Asia mattered to the United States and second, that a communist victory anywhere in this area would be followed by communist victory everywhere – the domino theory. This theory, which turned out not to be valid, seemed the more plausible in the 1950s and 1960s owing to the strength of the communists in the huge country of Indonesia and the introduction there of Sukarno's Guided Democracy (see Chapter 16). The American resolve to stem the anticipated communist flood assumed, third, that it could be attained without committing American forces to battle on the Asian mainland. The crisis in Vietnam during the Johnson administration was precipitated when this assumption was seen to be untrue because the Diem government in South Vietnam was unequal to the task of defeating the north. The domino theory invested Vietnam with a signal significance beyond its intrinsic importance. The Americans, besides portraying Diem's régime with marvellous inaccuracy as a democracy, misjudged their man.

South Vietnam started its independent career comparatively peacefully and prosperously in spite of the arrival from the north of nearly a million refugees (two-thirds of whom were Roman Catholic Christians). Its prime minister, Ngo Dinh Diem, a northerner and a Roman Catholic, eliminated Bao Dai, who was deposed in 1955, and a year later inaugurated a republic with himself as president. Although the last French troops did not leave until April 1956, American aid was transferred from the French to the Diem government from the first day of 1955. Diem was an anti-French intellectual, not particularly pro-American and in no way a democrat. He treated South Vietnam as a personal or family demesne which he administered through a network of secret societies, nepotistically, intolerantly and unintelligently. He antagonized the Buddhists, the dominant religious group, and the hill peoples who, although only a small minority of the population, inhabited more than half of the countryside in which subversive movements might be sustained. In the cities too Diem and his relatives, who included five brothers, became increasingly unpopular. The cities became generators of inflation and vice while the countryside became an open field for the settlement of private grudges and for extortionate demands which drove the peasants into the arms of the communist opposition.

A first coup against the régime in November 1960 failed but three months later the opposition was strong enough to attack Diem's palace from the air. At about the same time Ho decided to give material aid to the forces gathering against Diem in the countryside. Despite

American pressure Diem had refused to supply food to North Vietnam in the mistaken belief that the northern régime was about to collapse under the weight of peasant revolts. Thus Ho was given an economic as well as a political motive for reopening the war. At the time of the Geneva armistice the communists in the south, many of whom retreated to the north, had concealed their arms against a possible resumption of fighting. An active communist opposition to Diem had come into existence under the name of the Vietcong – originally a term of opprobrium, like Whig or Tory. In 1960 a National Liberation Front was formed and in 1962 the International Control Commission, an observer group established at Geneva eight years earlier, reported that North Vietnam was intervening in civil war in the south in support of the National Liberation Front.

In 1963 Buddhist hostility to the Diem régime reached a climax. Buddhist monks publicly burned themselves to death in gruesome protest against persecution. The authorities retaliated by looting and sacking pagodas and torturing monks. The Americans cut off aid to Diem and instigated a coup against him. In November Diem and his most markedly unpopular brother, Ngo Dinh Nhu, were killed, the régime collapsed and General Duong Van Minh ('Big Minh') became the head of the first of a series of transient military governments. In the next eighteen months half a dozen coups took place. General Minh was forced into the background within two months of Diem's end. General Nguyen Khanh, more militant in relation to North Vietnam but no more secure in Saigon, was opposed by Buddhists and students demanding an end to military rule. He survived a coup against him in September 1964 but was displaced early in 1965 by the yet more militant general Tran Van Minh ('Little Minh'). From the latter part of 1965, however, the rising men were General Nguyen Cao Ky, the most militant of the generals to gain power and the most determined to carry the war beyond the 17th parallel, and General Nguyen Van Thieu.

The United States had two incompatible aims. It was determined to prevent communist government in Vietnam and it was determined not to become directly involved on the Asian mainland. The failure of France's Bao Dai experiment, which the United States supported, and of Bao Dai's non-communist successors created a dilemma which the United States resolved by rating its first aim above its second, with disastrous results. It still believed that its choice lay between non-intervention and modest military intervention and, choosing the latter, became progressively involved in one of the major conflicts of the second half of the century. During the early 1960s the Americans became both alive to the weaknesses of their policy of supporting weak South Vietnamese régimes and more committed to it. One of Kennedy's early decisions on taking office in 1961 had been to increase

American aid in men and material but without committing combat troops. He accepted the advice of General Maxwell Taylor, a former chairman of the US chiefs of staff, to build up American forces and by 1962 American aircraft were flying combat missions and the CIA was conducting more or less underground operations. The pretence that the Americans were in Vietnam as no more than advisers had worn desperately thin. During 1963 and 1964 the Vietcong, aided by the misrule of Diem, the confusion among his successors and by North Vietnam and China, extended its control, until the South Vietnamese and American forces were in danger of suffering the same fate as the French in Algeria and the British in Cyprus and of being driven into a few fortified footholds. The policy, copied from Malaya, of isolating villagers from guerrillas was unsuccessful because the differences between Vietnamese villagers and Vietnamese guerrillas were insignificant compared with the differences between Malay villagers and Chinese guerrillas in Malaya. The Americans, seeing that the Vietcong was winning, decided further to increase their military effort and their control over the direction of the war, and as a result the position of the Vietcong rapidly worsened. Thereupon the government of North Vietnam, seeing that the Vietcong was losing, began to send regular divisions to its rescue. The war became, with scant disguise, a war between the United States and North Vietnam. In the south the Americans aimed to subjugate the whole country. Ultimately they failed and, with hindsight, it could be said that they would have been wiser to have adopted General Salan's strategy of holding the centres of population from which no enemy could have dislodged them, and of securing and sealing off the rich, easily defensible and ethnically distinct Cochin-China. Instead American forces were multiplied year by year: from 23,000 at the end of 1964 to 390,000 two years later and 550,000 by the beginning of 1968. When American withdrawals began in July 1969 these forces had lost 36,000 dead. At their peak the combined American, South Vietnamese and allied forces numbered 1.25 million, backed by a mighty air force. Against them General Giap proved himself a brilliant guerrilla commander who knew how far to follow and how far to deviate from the strategic teachings of Mao Zedong.

The war was extended to the north in 1964. In July of that year the US destroyer, the *Maddox,* operating with South Vietnamese sea and land forces against North Vietnam, was twice attacked in the Bay of Tonkin. President Johnson used this episode, tendentiously explained, to obtain from Congress authority to use US forces in open naval combat. At the end of the same year Johnson became the elected instead of merely the accidental president of the United States. He was seduced by visions of a quick victory by the air bombardment of a fourth-rate foe. In February 1965 a successful attack on an American garrison at Pleiku near the armistice line was followed by sharp American bombing reprisals. American ground forces were also now committed to the battle against the north. But the quick victory was prevented by Russian aid to Ho, notably for the anti-aircraft defence of Hanoi, and as the American bombardment became fiercer, and napalm, poison gases and defoliants were introduced, so did opposition to the war within the United States develop vociferously and violently. The home front, mobilized by television, took the part not just of an enemy vilely massacred but of its own kith and kin involved in these horrors.

As the American war effort was intensified President Johnson also began to bid for peace. In April he offered to treat unconditionally with North Vietnam for a cessation of hostilities on the basis that South Vietnam would be an independent and neutral state; and he offered $16,000 million in aid to South-east Asia, including North as well as South Vietnam. But he was not willing to accept the Vietcong as a party to the negotiations. In reply Ho enunciated a set of conditions which were not irreconcilable with the American proposals and in July Johnson implicitly but not explicitly opened the door of the negotiating chamber to the Vietcong as well. He was, however, unwilling to stop all bombing immediately (a bombing pause in May lasted only a few days) or to agree to withdraw all American troops before negotiations began. At the beginning of 1966 Johnson conferred with Ky and Thieu in Honolulu. Bombing was interrupted for several weeks and negotiations seemed not improbable, but they remained in the event elusive and the war went on, increasing in weight and horror. Vietcong adherents and others were slaughtered by the most up-to-date weapons. They were hunted by and thrown out of helicopters, tortured, raped and murdered in cold blood. In a celebrated instance at My Lai in 1967, which later led to a criminal trial and conviction in the United States, 300 civilians were killed by a US army unit.

By the end of 1967 the United States had forces of 470,000 in Vietnam and was bombarding the north heavily from the air. But American casualties were severe and at a meeting with Kosygin at Glasboro in New Jersey, Johnson tried unsuccessfully to enlist Russian help to end the war. In South Vietnam Ky fell foul of the Buddhists and the Ky–Thieu government split and was re-formed as a Thieu–Ky government (which lasted until 1971 when Thieu prevailed). A truce was declared for the festival of Tet in January 1968 and in Hanoi the foreign minister, Nguyen Duy Trinh, made pacific statements but his remarks, while welcome to some in Washington, were also regarded there as a feint to cover an impending offensive. Both interpretations were valid, for counsels were divided in Hanoi too. Optimistic hawks believed that they could seize Saigon and other southern cities, win the active support of the urban populace and cause Thieu's armies to collapse. The attack was launched, caught the Americans unprepared but failed in all its main aims except in the capture of Hue by the NLF which held it for two months. The attack was, however, successful in an unexpected way. Although it did not destroy the Thieu régime it caused a loss of nerve in Washington. Two months later Johnson announced that American bombing was to be substantially curtailed and that he himself would not stand for re-election as president. In May peace talks between the United States and North Vietnam opened in Paris. Later in the year South Vietnam and the Vietcong joined the discussions. But they led nowhere. North Vietnam had discovered that it could survive the American onslaught and maintain the supply of men and material to the south by using routes and methods which, being comparatively primitive, were never wholly disrupted by aerial bombardment. North Vietnam had also discovered that the Americans really wanted to get out and either turn the war over once more to the South Vietnamese (whose army was brought up to a strength of over 1 million and given the most modern equipment) or end it. Johnson could neither win a victory, since he was not prepared to use nuclear weapons, nor negotiate a peace, since North Vietnam preferred waiting to negotiating. He was pursuing incompatible aims – to get out and to secure the existence of a separate, non-communist South Vietnam. North Vietnam was not convinced that it need concede the latter in order to get the former, and when the Democrats went down to defeat in November the new president, Richard Nixon, found himself in the same dilemma. Although he had pledged himself to end the war, he immediately sanctioned the extension of American bombing to Cambodia in another vain hope to find the quick victory.

So *Cambodia* was brought into the field of war. In March 1970 its ruler, the neutralist Prince Sihanouk, was unseated by a coup by his own prime minister, General Lon Nol. Sihanouk had ensured comparative peace and quiet from 1955 to 1965 but forfeited American friendship by his neutralism and his friendship with Zhou Enlai. Cambodia, like Laos, had been used by North Vietnam without much regard for its neutrality and Sihanouk had been reproached by some of his compatriots for putting up with too much infringement of the country's rights. The new Cambodian government asked for American arms to defend itself against intruders. It got instead American armies, accompanied by South Vietnamese forces (hereditary enemies) and by all the devastation of which the American military machine was capable. The American command in Vietnam was justifiably anxious to counter the use of Cambodia by the North Vietnamese and unjustifiably convinced that it would capture vast communist stores and major headquarters. The operation was militarily successful, though doubtfully necessary; politically it was at best irrelevant. It was stopped by the US Congress which refused to vote funds for troops or advisers in Cambodia and, in 1973, by the same means stopped Nixon's resumed bombing of the country. Lon Nol protested that he had never been told that American and Vietnamese forces were to enter Cambodia.

Although American bombing of North Vietnam was resumed in 1972 these operations were the last lashes of the frustrated American giant. The withdrawal proceeded steadily and only 25,000 Americans remained at the end of the year. Since 1970 the governments of the United States and North Vietnam had been secretly in contact through Henry Kissinger and Le Duc Tho and in January 1973 representatives of the United States, both Vietnams and the Vietcong met in Paris and agreed on a cease-fire, to be internationally supervised, and on the creation of a Council of National Conciliation in Vietnam to prepare elections. These agreements endorsed the American retreat but did not bring peace. There was heavy fighting in the latter part of 1973 and throughout 1974, with Thieu expressing confidence that his new army of a million men could win the war. He expected American aid but did not get it and proclaimed himself let down by his ally. His army and his power distintegrated in the face of 200,000 North Vietnamese and 100,000 Vietcong adversaries. In 1975 he resigned and Big Minh returned from the shadows to surrender what authority was left in Saigon. The war was over. About 2 million people had been killed. In Laos, which had been invaded in 1971 by South Vietnam with American air support, North Vietnam won decisive victories and fighting came to a stop in 1973 on the basis that yet another coalition would be constructed and all foreign troops withdrawn. The new government came laboriously into existence in 1975 and perished the same year. The monarchy perished too. Prince Souphanouvong, now president, and the Pathet Lao had won the internal battle. Laos became a dependency of Vietnam under the name of the Lao People's Republic. In spite of its name its first elections were not held for fourteen years.

For Cambodia the end of one war meant the beginning of another. Sihanouk, upon being displaced by the Americans and Lon Nol, had shifted to the left and allied himself with the Khmer Rouge, thereby transforming a comparatively insignificant faction of about 3,000 active members into a militant body ten times that size. Its leader, Saloth Sar, better known as Pol Pot, was a zealot of about fifty years who had spent time in left-wing circles in Paris and later in China during the Cultural Revolution. His revolutionary programme included the abolition of religion and money, the creation of a rural populist communism and the extinction of cities; his revolutionary performance killed a million people in a sickening reign of terror after the United States abandoned Lon Nol and the Khmer Rouge took Phnom Penh a few days before the North Vietnamese entered Saigon. The Americans bombed

Cambodia (now Kampuchea) once more in 1975 when the Khmer Rouge captured the US intelligence vessel *Mayaguez* but the war was now between Kampuchea and Vietnam. It was provoked by the Khmer Rouge which attacked Vietnam in debatable border lands and in indisputably Vietnamese territory. Vietnam retaliated with a successful invasion of Kampuchea at the end of 1978 and installed a subservient government with Heng Samrin as president and Hun Sen as foreign minister, later also prime minister; both of them were Khmer Rouge renegades with a heavy responsibility for past atrocities. Hun Sen introduced some beneficial reforms, including land distribution and encouragement for small businesses, and permitted wider freedom of speech and religion, but his rule was brutal and corrupt. The Khmer Rouge offered little resistance, retreating to the north-west where it was pursued by the Vietnamese invaders. Kampuchean troops and refugees crossed indiscriminately into Thailand, whence the troops were able to filter back and fight again.

The war between Vietnam and Kampuchea was of prime concern to China for reasons both ancient and modern. China's traditional enmity towards an independent Vietnam had been overlaid during the 1960s and 1970s when the Sino–Soviet split created competition between China and the USSR. China helped the Vietcong in order to pre-empt Russian aid and influence, but this amity did not last. The Cultural Revolution, derided by the Vietcong, severely curtailed Chinese external activity, and Nixon's visit to China and the subsequent Sino–American *rapprochement* were bitterly resented in Vietnam. North Vietnam's victory over the United States, the unification of north and south in 1975, this new state's large army, its alliance in 1978 with the USSR (when it became also a full member of Comecon), and its conquest of Kampuchea – all these factors ignited China's hostility. In particular, Vietnam's invasion of Kampuchea seemed to pose a threat to the hegemony in Indo-China which China had pursued for a thousand years, while Vietnam's links with the USSR betokened a dependence which, occasioned by the exhaustions of war, economic mismanagement and a sharply rising population, heralded a Russian dominance in Indo-China even more distasteful to China than Vietnamese or American dominance. Russian influence in Vietnam was part of a threat manifest also in northern Asia and Afghanistan. There were also two other irritants: territorial conflict in the South China Sea and the treatment of Chinese in Vietnam.

There were about a million Chinese in Vietnam, most of them in the south. These Hoa, as they were called, were economically successful and commensurately unpopular. Their nationality was open to argument and their loyalties to doubt. In 1977 many Hoa, including some from the remote south, trekked into China in what the Vietnamese described as migration and the Chinese as expulsion.

Whether or not incited in the first place by China, this trek acquired a momentum of its own and continued for two or three years, after which it subsided. Conflict in the South China Sea, however, was more enduring. Old disputes were fanned by the belief that there might be oil under the waters of the Gulf of Tonkin. This belief enhanced the value of contested groups of islands to which both China and Vietnam laid claim. In 1974 a Chinese evicted a Vietnamese force from the Paracel islands whereupon South Vietnam occupied the Spratly islands – small, numerous, uninhabited and further south. Both these groups were within the 200-mile offshore limit recognized by China and Vietnam. In 1988 the growing Chinese navy took possession of some of the Spratly islands and, in 1994, the Mischief Reef in the same archipelago and in spite of Philippine claims. Malaysia and Brunei also had claims in the area.

China's manifold preoccupation with Vietnam gave Kampuchea a powerful friend. China strongly berated Vietnam's aggression and Deng Xiaoping, on a visit to the United States, threatened to teach Vietnam a lesson. The USSR came to the help of Vietnam by vetoing its condemnation by the Security Council and, when Chinese troops invaded northern Vietnam, sent a flotilla which included a missile carrier into Vietnamese waters. But China's demonstrative riposte across its frontiers with Vietnam – which did much damage but also cost the invaders unexpected casualties – did not stop Vietnamese forces from sweeping across Kampuchea as far as the Thai border. More persuasive in Hanoi were the burden of keeping 250,000 troops in Kampuchea, the evaporation of Soviet aid, and the difficulties of restoring relations with the United States so long as 230 American soldiers missing during the war remained unaccounted for and Vietnam continued to occupy Kampuchea. By 1987 Vietnam had resolved to get out of a troublesome and very costly situation. It announced that its forces would begin to withdraw at once and would all be evacuated by 1990. They left in 1989.

Simultaneously discussions began for a replacement. Sihanouk and Hun Sen had a number of meetings in France and Indonesia, encouraged by China (to a limited degree, for China was not wholly averse to a war which was a drain on Vietnam's resources and on the popularity of its government at home). But neither Sihanouk nor Hun Sen nor both of them together could construct an effective alternative régime without the Khmer Rouge which had substantial armed forces as well as Chinese – and some American – support: the United States, like China, was prepared to give at least one cheer to any enemy of Vietnam and had been urging Sihanouk since the early 1980s to make common cause with the Khmer Rouge and with Son Sann, a former prime minister under Sihanouk and the leader of the third effective anti-Vietnamese force. (Sihanouk and Son Sann each commanded perhaps 10,000 armed men, the Khmer Rouge 30,000 at the very least. Hun Sen's army amounted to about 40,000, including a number of Vietnamese who remained in or returned to Kampuchea because they saw small prospect of a decent living in Vietnam.)

Partly at Australian prompting the idea of overthrowing Hun Sen was replaced by attempts to unite all the main factions in a Supreme National Council in which Hun Sen would have six seats and the three opposition groups two each. At a late stage Sihanouk, the chairman-designate, insisted on being the thirteenth member of this Council, thus raising his contingent to three. This obstacle was surmounted by proposing that Hun Sen be deputy chairman to Sihanouk with a seat of his own additional to his six nominees. The plan envisaged a gradual return to normal political activity after eighteen months of supervision and administration by the UN, but Hun Sen's distrust of Sihanouk and his Chinese backers, and his fear of Khmer Rouge preponderance, made him prevaricate in the hope of getting some international recognition for his own régime without being forced into a coalition. A UN force was despatched to lay the base for elections by preparing electoral registers, assuring voters that it was safe to vote, organizing the return of some 250,000 refugees from Thailand and keeping the various parties to their engagements. UNTAC (UN Transitional Authority in Cambodia) came close to being an interim government for most of Cambodia and so a new kind of UN intervention. Although small – about 20,000 persons – it was markedly successful in its main tasks. It could not prevent Sihanouk from moodily retreating once more to Beijing or the Khmer Rouge, in occupation of a tenth of the country where it supported itself by sales of timber and precious stones through corrupt Thai entrepreneurs and army officers, from threatening to disrupt and then boycotting the elections. But these elections were duly held in 1993 in a generally orderly way and with 90 per cent of voters taking part. Contrary to expectations Hun Sen's Cambodian People's Party with 38 per cent of the vote was defeated by its rival Funcipet led by Sihanouk's son, Prince Rannarith, which won 45 per cent. Sihanouk, mindful of the continuing strength of the Khmer Rouge, exerted himself to bring Funcipet and the CPP together even though he had to concede to the latter an equal share in government and a blocking power in the parliament. Having appointed his son prime minister with Hun Sen as deputy he reassumed the regal style which he had laid aside in 1955. But the Khmer Rouge could not be written out of the plot and a year later it re-created enough chaos to bring about the disintegration of the new government and the destruction of UNTAC's work, or most of it. As in Angola the UN relied perforce on a minimum of goodwill and good faith and was powerless when these failed.

One persistent aspect of the travails of Vietnam after its victory over the United States was a stream of refugees trying to escape from the miseries of life in Vietnam under an incompetent and harsh régime compounded by an American economic embargo. They went by sea, many of them being attacked, despoiled, raped and killed on the way. Most of them came from the north of the country but the tally of refugees from the south increased during the 1980s. Around 57,000 of them reached Hong Kong where they were most unwelcome and were detained and confined in large pens in insanitary and degrading conditions. All but a fifth of them were adjudged by the colonial government of the already overcrowded island to be economic refugees (a new term) and not genuine political refugees whose return to Vietnam was precluded by international conventions. In 1989 the government of Hong Kong announced a scheme for the voluntary return of these refugees with one aircraft a month, but few were willing to go. At the end of the year fifty-one men, women and children were sent back against their will in a night-time operation which occasioned such an international outcry that the scheme had to be abandoned. In the next year the British and Vietnamese governments and the UN High Commissioner for Refugees concluded an agreement for the repatriation of refugees 'not opposed' to it. Britain, while it did not abjure its claim to be entitled to repatriate economic refugees by force, enlisted the help of the UNHCR in persuading them to go and hoped that many would do so after discovering that their treatment in Hong Kong was as bad or worse than anything which they might have to suffer in Vietnam. The raising of the American embargo in 1993 gave a fillip to the Vietnamese economy which, albeit belatedly and from a wretchedly low level, set out to emulate other east Asian economic 'tigers'. Diplomatic relations between the United States and Vietnam were fully restored only in 1995.

16 SOUTH-EAST ASIA AND ASEAN

The term 'South-east Asia' is used to describe the countries which lie between India, China, Australasia and the open expanses of the Pacific Ocean. Diverse by race, religion and wealth they had before the Second World War one nearly common feature: with the solitary exception of Thailand all were ruled by foreigners. The British, French, Dutch, Americans and Portuguese had spread over the area and appropriated varying amounts of it. This state of affairs was viewed with dissatisfaction within the region and also by the Japanese, whose New Order had been expanded into the Great East Asia Co-Prosperity Scheme under the direction of a special ministry in Tokyo. When war brought the Japanese to South-east Asia they came as anti-imperialist and pro-nationalist liberators, promising to remove European overlords, an operation which proved in the circumstances astonishingly easy. Three days after the attack on Pearl Harbor the Japanese sank the British warships *Prince of Wales* and *Repulse* (10 December 1941); Singapore fell in February 1942 and Corregidor in May; western dominance finished.

It was succeeded by a very short Japanese phase in the course of which the new overlords, like Napoleon in Germany, found that nationalism is not a commodity that can be turned on or off at will. Some Japanese genuinely believed in the co-prosperity theme and wished to help the peoples of South-east Asia, but more were simply new imperialists who quickly alienated the local nationalists. As the fortunes of war turned against the Japanese, so did the nationalists, preparing to achieve their ends partly by services rendered to the former colonial powers and partly by a new strength that would not be overborne by war-weary Europeans. In Burma and the Philippines the end was easily achieved, in Indonesia less easily. In Malaya independence was delayed by an insurrection that was more communist than nationalist. Indo-China (see the previous chapter) was fated to suffer the long war which assumed international proportions and whose settlement posed further international problems in the successor states of Vietnam, Laos and Cambodia.

Throughout this region, the Philippines apart, the predominant power as the war ended was the British, represented until 1946 by the

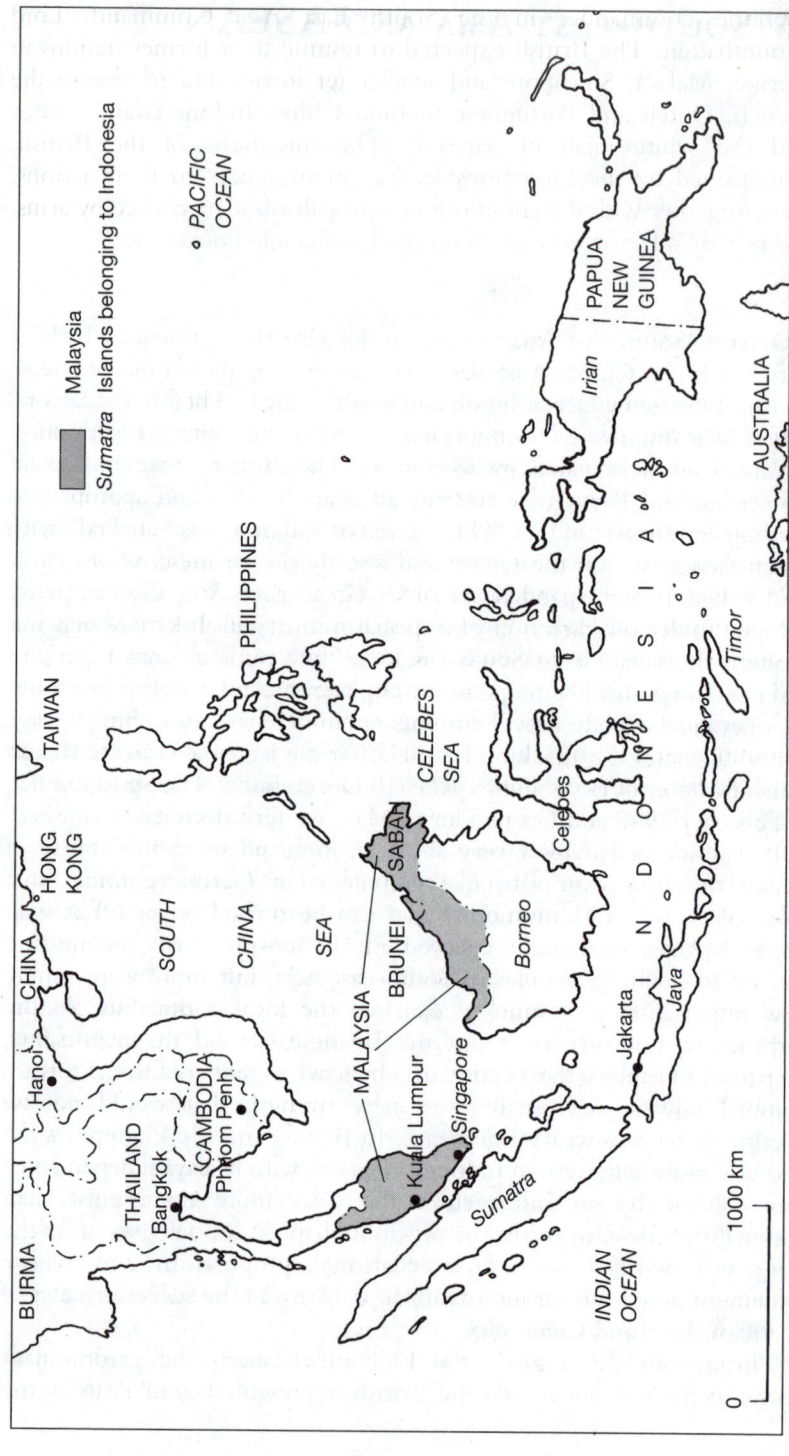

16.1 South-east Asia

supreme commander in the South East Asia Command, Lord Mountbatten. The British expected to resume their former stations in Burma, Malaya, Singapore and smaller territories and to restore the French, Dutch and Portuguese in Indo-China, Indonesia and Timor and the white rajah of Sarawak. The intentions of the British, conditioned by the limitations set by circumstances to their actions, were, together with the ambitions of nationalist leaders backed by arms, the principal factors in a situation of considerable uncertainty.

MALAYSIA

In *Malaya*, a conglomeration of principalities plus small British colonial territories, even less of a unitary state than Burma, the British had no easy way out. The naval base on the adjacent island of Singapore (itself a British colony) created arguments for staying which did not exist in Burma, and the strength of the Chinese (in Malaya a substantial minority and in Singapore a majority) made many Malays less resentful of British rule than apprehensive of the local Chinese, whose leaders – replacing prosperous prewar leaders compromised by their collaboration with the Japanese invaders – had taken to the jungle and to communism. Among the active opponents of the Japanese the largest group had been Chinese and most of these were communist rather than adherents of the Kuomintang, but by the very fact of their racial and doctrinal distinctiveness they could not claim to be a nationalist movement like the AFPFL in Burma. Their bid for power was of a different nature.

An initial and premature attempt was checked by the British after some hesitation. The British then tackled the racial problem and proposed a Malayan Union in which citizenship would be obtainable by any person who had lived ten years in the country. The Malays opposed this plan which would have made citizenship, and so political power, available to a large section of the Chinese population. The British thereupon proposed in its place, in February 1948, a Malayan Federation in which the powers of the Malay sultans were greater than in the proposed Union and the opportunities for the Chinese to become citizens were restricted.

In that year the Chinese communist insurrection began and in June an emergency was declared which was to last for twelve years during which the insurgents baffled 50,000 troops, 60,000 police and a home guard of 200,000. In 1950 General Sir Harold Briggs, the Director of Operations, realized that the key to the situation lay in the silent support given to the insurgents, often out of terror, by the great mass of the population, and he therefore made plans to assemble and protect the people in resettlement areas where they would be immune from blackmail and would cease to supply the insurgents with food. General Sir Gerald Templer, who arrived as governor in 1952 in succession to the murdered Sir Henry Gurney, continued this policy and developed at the same time political measures designed to bring the Malay, Chinese and Indian communities together as a prelude to independence. Malaya became independent on 31 August 1957 under a constitution which established a revolving presidency tenable by the Malay sultans in turn. Malaya entered into a defence agreement with Britain but did not become a member of Seato and refused in 1962 to allow Malayan territory to be used by British units which might be called upon to help Thailand in the event of a threat to that country from Laos.

Singapore, with a population three-quarters Chinese, was set on the road to independence by the usual British process. A new legislative council with an unofficial and elected majority was installed alongside a nominated executive council. This machinery of government then evolved into a legislative assembly and a council of ministers under a chief minister, most of whose colleagues were chosen by the assembly, the governor retaining certain reserved powers. Full independence came in 1959 subject to the retention of British rights in the naval base. The prime minister, Lee Kuan Yew, was anxious for closer links with Malaya but since a union would place the Chinese in a majority the Malays were reluctant unless the union were at the same time extended to other and less Chinese territories. Such territories existed in Sarawak, North Borneo and Brunei.

In 1946 the rajah of *Sarawak*, Sir Charles Brooke, ceded the principality which his family had held since 1841 to the British government and in the same year Britain assumed in North Borneo the rights and responsibilities which had belonged before the war to the British North Borneo Company. Both Sarawak and *North Borneo* became crown colonies. *Brunei*, a third territory along the northern side of the island of Borneo (most of which formed part of Indonesia), resumed its prewar status as a British protectorate.

In 1963 Sarawak and North Borneo (now Sabah) joined Malaya and Singapore to form a Greater Malaysian Union, despite protests from the Philippine Republic which had claims in North Borneo, and from the Indonesian Republic which regarded the scheme as a plan to create a western–orientated state capable of checking Indonesian growth and ambitions. Brunei refused to join at the last moment and a revolt against the sultan, who was believed to favour accession, gave the Indonesian President Sukarno the idea that the federation was unpopular and might be destroyed by inexpensive guerrilla operations, persistence and propaganda; and he might be able to add the rest of Borneo to Indonesia. The ensuing Indonesian confrontation forced Malaysia to appeal for British and Australian military help and frustrated the British intention of getting out of the region after creating Malaysia. The confrontation abated when Sukarno was demoted by his army in 1965–66, but Malaysia was a contrived constellation whose components had been forced into federation for external and adventitious reasons and upon the assumption that the territories tacked on to Malaya could not exist on their own. After its creation Malaysia's largest ethnic group was Chinese (43 per cent) with Malays the next largest (40 per cent). In the Malayan peninsula Malays were 50 per cent and Chinese 37 per cent of the population.

Brunei had rejected the assumption that the smaller territories in the region could not exist on their own and Singapore rebutted it. Brunei, vestigial survival of an empire which had once reached to the Philippines, fabulously rich in relation to its size and population and virtually the private estate of its autocratic sultan, attained internal self-government in 1971 and complete independence from Britain in 1984 when it joined the UN, the Commonwealth and ASEAN (see below). Singapore, having joined Malaysia at its inception, broke away. Its socialist People's Action Party (PAP) was opposed by the conservative Malayan Chinese Association on the issue, among others, whether Malaysian politics should be conducted essentially on class lines or racial. Both parties were, like the island itself, largely Chinese but the PAP wanted politics throughout the federation to be organized round the needs and interests of economic classes. Across the water in Malaya Tunku Abdul Rahman, who was a conservative Malay nobleman as well as party leader and prime minister, was alarmed by the attitudes of the PAP and its leader Lee Kuan Yew and was determined to ensure that no Chinese Singaporean should become prime minister of the federation. He decided to evict Singapore from it. Lee Kuan Yew, doubting Singapore's independent viability, had no wish to secede but tensions between the two territories and their two leaders convinced him that an early and reasonably amicable divorce was imperative. His doubts about the island's economic prospects quickly dissolved. Sinapore developed an outstandingly successful mixed economy. Foreign capital flowed in; income per head trebled in the first decade after the breach in 1965; and the PAP reaped the benefits by remaining in power for a generation. Singapore had for several years uneasy relations with Indonesia, inflamed by incidents such as the execution in Singapore of two Indonesian soldiers and the refusal to recognize Indonesia's conquest of East Timor in 1975 (see the note at the end of this Part), but disagreements were contained and eased within the framework of ASEAN.

In *Malaysia* – thus constituted with the Malay states, Sarawak and Sabah but without Brunei or Singapore – the principal political problem was relations in the Malay peninsula between the rural Malays, numerically superior, and the urban Chinese, economically predominant. Elections in 1969 were tumultuous and the army had to be used to restore order. The prime minister, Tunku Abdul Rahman, resigned in the wake of this setback to hopes of peaceful communal development. He was succeeded by Tun Abdul Razak who enjoyed the fruits of an impetus to the economy which brought benefits to Malays through economic growth rather than economic redistribution at the expense of the Chinese. Although the second Development Plan (1971–75) made a faltering start and was temporarily upset by the world oil crisis in 1973–74 Malaysia's diverse economy successfully withstood the buffets of the mid-1970s. Politically, the risk of racially polarized conflict was avoided. Tun Abdul Razak began by reviving the United Malayan National Organization (UMNO) which had been created in 1946 as a focus for Malay nationalism against the British and the assertion against the Chinese of the Malay nature of Malaya. Next he reconciled factions within UMNO and coaxed it into alliance with the Malayan Chinese Association (MCA – founded in 1949) and the Malayan Indian Congress (founded in 1954). By the early 1970s all parties of consequence in Malaysia were associated in a multicommunal Alliance, in which UMNO was the largest component.

In external affairs Tun Razak modified Malaysia's pro-western stance in the direction of non-alignment. He visited the USSR and China and after 1975 he favoured the inclusion of Vietnam in ASEAN, but he died in 1976. His successor, Dato Hussein Onn, who was prime minister until 1981 when his health gave way, faced economic deceleration, renewed threats to communalism from Malay nationalism (and assertive Muslims), allegations of corruption and fears – unrealized – of secession by Sabah, but economic recovery helped his National Front of ten parties to convincing victory in elections in 1978. His deputy prime minister and successor, Muhammad Matathir, was a more controversial politician who had first made his mark by criticism of Tunku Abdul Rahman and later adopted a robust stand against what he regarded as undue and unnecessary deference to British interests. After becoming prime minister Matathir's political fortunes were erratic. His position was tainted by rumours of financial scandals, in spite of which he won unexpectedly clear victories in federal and all state elections in 1986. In the next year he almost lost the presidency of UMNO and in 1988 the Supreme Court declared UMNO an unlawful organization on grounds of procedural delinquencies, but Matathir circumvented this damaging judgment by going to war with the judiciary. The Chief Justice was suspended and then dismissed; five other senior judges were suspended and two of them dismissed; the Court nevertheless confirmed its original ruling. On the political stage Matathir was opposed by two of his most eminent former colleagues, Tungku Razaleigh Hamzah and Datuk Musa Hitam. The latter accepted an ambassadorial appointment in 1989 and in a general election in 1990 Matathir defeated Razaleigh's challenge. The following years of strong and steady economic growth (8 per cent a year) entrenched Matathir as leader of UMNO and the ruling coalition of fourteen parties dominated by UMNO. In elections in 1995 some local opposition persisted in Sabah and the Muslim PAS retained its leading position in Kelantan in the north. But the principal opposition party, the Democratic Action Party, was severely defeated even in its own stronghold, Penang.

Beyond the British possessions in Burma, Malaya and Singapore lay in 1945 the former Dutch empire in Indonesia to the south and the French empire in Indo-China to the east. When the British returned to South-east Asia after the defeat of the Japanese, their actions were consciously or unconsciously dominated by their decision to leave India. South-east Asia (with the exception perhaps of Singapore) was for Britain an adjunct of India and it was hardly conceivable that the British could for long pursue in South-east Asia policies that were plainly at variance with their chosen course in India. No such considerations affected the Dutch and the French whose positions in Indonesia and Indo-China were not regulated by any other positions in Asia. For them the question was the basic question which the British put to themselves in India – whether to go or to stay – and not the secondary question which confronted the British in Burma – whether to stay in spite of the departure from India; but since the factors affecting the British in India were quite different from those affecting the Dutch and the French further east, the first answer of the latter was also different. The Dutch and French proposed to revert more or less to their prewar positions and they looked to their British allies to help them.

The return of the Dutch to *Indonesia* was generally assumed – or feared – in 1945 but for purely practical reasons the British supreme commander was unable to undertake extensive operations and dealt directly with the nationalist leader Sukarno in the key island of Java. Dutch forces arrived to take the place of the token British occupation but found themselves opposed by a comparatively well-organized and well-equipped movement under leaders of ability and sophistication whose foreign contacts and travels had given them an insight into the wider forces affecting the continuance of European rule in Asia. In this situation the British tried to mediate and by the Linggadjati agreement of November 1946 the Dutch recognized the *de facto* authority of the self-proclaimed Indonesian republic in Java, Sumatra and Madura and agreed to withdraw their forces from these areas; a union between the Netherlands and Indonesia was to be created by the beginning of 1949.

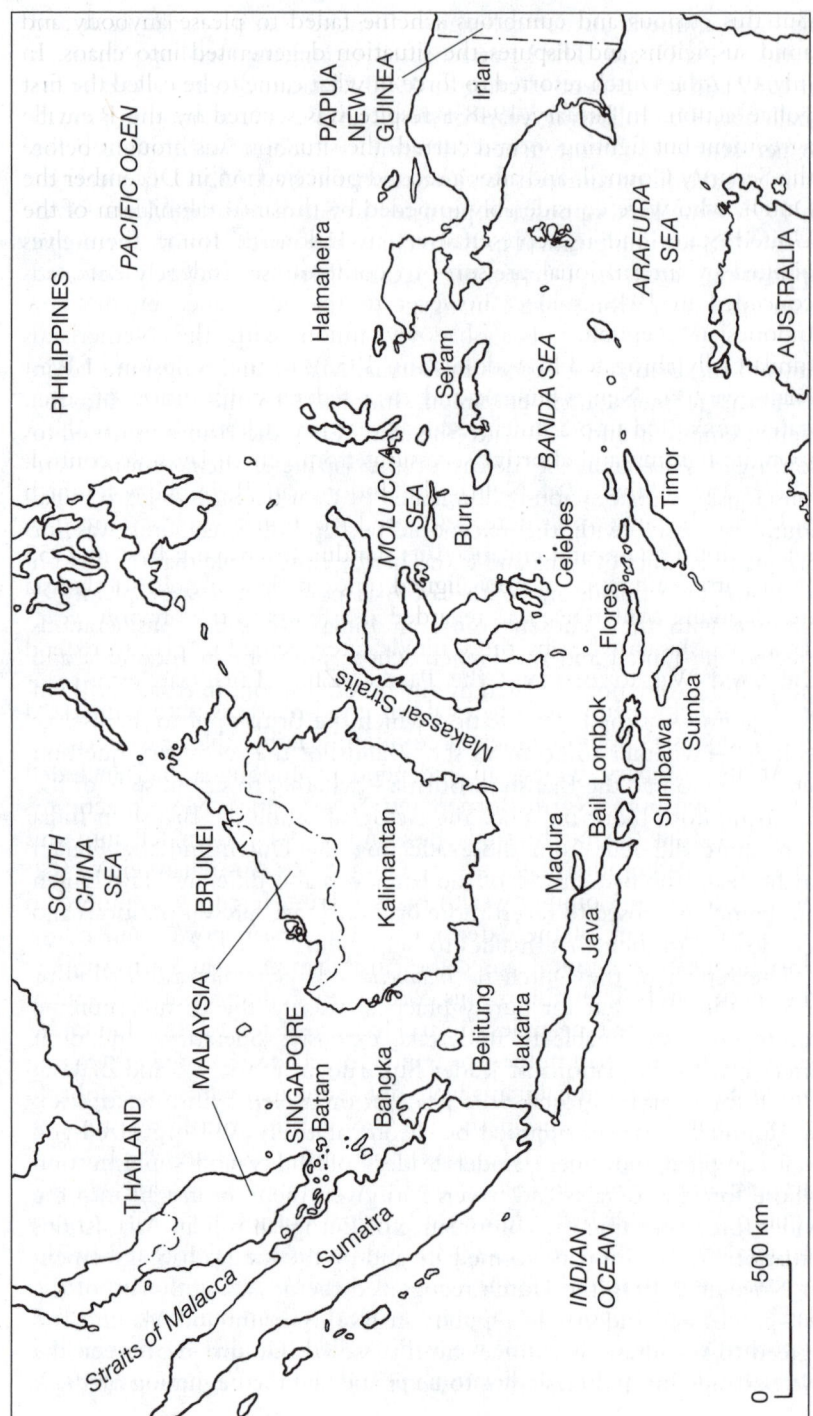

16.2 Indonesia

But this curious and cumbrous scheme failed to please anybody and amid suspicions and disputes the situation degenerated into chaos. In July 1947 the Dutch resorted to force in what came to be called the first police action. In January 1948 a respite was secured by the Renville agreement but fighting soon recurred, the situation was brought before the Security Council, and after a second police action in December the Dutch, who were considerably impeded by the anti-colonialism of the United States and its representatives in Indonesia, found themselves obliged by international pressures to compromise. Independence was conceded in 1949, subject however to the acceptance by the new Indonesian Republic of a shadowy union with the Netherlands (unilaterally abrogated by Indonesia in 1956); to the exclusion of West Irian (western New Guinea); and to a federal constitution of seven states, converted into a unitary state a year later, although not without sporadic fighting and abortive secessions from central Javanese control. Under the influence of Nehru the Indonesian Republic, of which Sukarno became president, adopted a non-aligned stance in international affairs and enhanced its standing by playing host in 1955 to the first conference of non-aligned states at Bandung. In south Asia the Bandung conference was regarded as a retort to the previous year's Treaty of Manila whereby the United States seemed to plan to extend the Cold War to Asia and the Pacific. Zhou Enlai's appearance at Bandung helped Indonesia to establish good relations with China and gave Indonesia's non-alignment a communist rather than a western tilt.

At its visionary widest an independent Indonesia had included Malaya, Singapore, North Borneo and New Guinea. More practically, it included all the Netherlands East Indies. West Irian fell into this category but the other half of New Guinea was Australian, either part of the Australian Commonwealth or as a Trust Territory administered by Australia. Fears of the extension of Indonesian power and claims throughout New Guinea caused the Australian government to support the Netherlands' retention of this corner of its empire, but the logic of events – measured in miles from Holland – forced the Dutch to abandon in 1962 a last outpost to which they had succeeded in clinging on the plea that the inhabitants were not Indonesians and were unlikely to get a fair deal from Indonesia. The prolongation of this question for thirteen years after independence kept nationalist spirits on the boil and strengthened the position of Sukarno, the country's most eloquent and popular nationalist, in spite of the gradual collapse of parliamentary democracy under his rule and economic recession (aggravated by the eviction of the Dutch and the nationalization of their enterprises). In 1959 Sukarno, supported by the army and the Communist Party (PKI), transformed the state into what he called Guided Democracy, a euphemism for personal dictatorship and a bravura performance in which however he overplayed his hand. His determination to secure West Irian drove him to closer relations with the USSR and China and earned him therefore the distrust of the United States which regarded him as a conduit through whom the communist powers might get control of South-east Asia's largest state. The PKI encouraged Sukarno in his second major external venture, opposition to the creation of a Malaysian federation which was strongly anti-communist and suspected by Sukarno of being a device for preserving British colonial influence. Sukarno backed a revolt in Brunei in 1962 against its merger with Malaysia – it was suppressed by the British in a few days – and kept up confrontation with Malaysia by operations across its borders and in protests at the UN and other international meetings from its creation in 1963 until the end of his rule.

Indonesia was, by virtue of its size (over 13,600 islands) and potential wealth, a power of a completely different order from any other state in the region of South-east Asia. It could be a counter to Chinese influence and expansion or an adjunct to them. The PKI under its leader K. N. Aidit was one of the most effective in the world and pro-Chinese. There was therefore constant speculation about the balance of power between the communists (who had failed in 1948 to win a monopoly of power by a coup at Madiun in Java) and the army and about the likely outcome of a struggle between the two when Sukarno, frequently reported to be a sick man, should die. In September 1965 a false report of his death became the signal for a communist putsch. Six senior generals were murdered after the cruellest mutilation but others escaped and the coup failed. In a counter-coup, widely believed to have had American aid, something like half a million communists (including Aidit) or supposed communists were killed and the army established its control over Sukarno and the country. Bung Karno, or brother Karno, the creator of Indonesia, was gradually stripped of his powers: he resigned in 1967 and died in 1970. His successor T. N. J. Suharto, a peasant's son in general's clothing, was a deft political operator who turned Indonesia into a centralized economy run by himself and his very extended and extensively enriched family. Suharto was elected president for six terms stretching to 1998. He kept vast numbers of opponents and presumed opponents in detention (they were later transported to a convenient offshore island) but the national economy prospered with American and Japanese help. Although earnings from the export of rubber fell, Indonesia's oil and its enviable variety of other minerals more than redressed the balance. By the 1970s and 1980s the economy as a whole was growing at the rate of 7–10 per cent a year. This growth was checked in the late 1980s by declining revenues from oil and by the rising cost of imports from Japan and debt service to it, and the currency was devalued. In 1989 Indonesia resumed formal relations with China after a breach lasting twenty-four years and Suharto visited Moscow after an equally long period of tepid and distant contacts. The same year saw the beginning of secessionist violence by Muslim fundamentalists in northern Sumatra.

Thailand had the unique experience of escaping the European conquest which embraced the rest of South-east Asia. It was left as a buffer between the British who advanced from India into Burma and the French who advanced from Annam and Tonkin into Laos and Cambodia. Unlike Burma, Laos and Vietnam it had no frontier with China and it had succeeded in assimilating a substantial part of its Chinese population of 4 million (12 per cent of the total), but it shared the general apprehensiveness of all South-east Asia concerning the revival of Chinese power and welcomed an American alliance as an insurance. It became the one genuinely Asian member of Seato (the Philippines and, even more so, Pakistan being but peripheral members of an alliance termed South-east Asian). The army, which had imposed limits to royal absolutism in 1932, ruled directly from 1938 to 1973 through a succession of strong men – Marshals Pibul Songgram, Thanarat Sarit and Thanom Kittikachron – with an interlude of civilian rule (1944–46) under Nai Pridi Panomiong who, on being worsted by Pibul, retired to China and there revived his wartime Free Thai movement which did not this time amount to much. Rich in rubber, tin, teak and rice Thailand stood in less need of foreign aid than most of its neighbours but it became expensively embroiled in the wars to the east. Threatened in 1962 by communist units on its borders with Laos it was offered and accepted the help of American troops. Two years later it permitted the establishment of American air bases for use in the war in Vietnam, and this counter-aid to the United States was gradually expanded to six major bases accommodating forces of 50,000. By the 1970s the strains of war, and the awareness that the war was being lost by the Americans, fostered discontent which was accentuated by the general corruption, inefficiency and uncertain policies of military rule. The American alliance and the consequent involvement in Vietnam were increasingly criticized and after riots in 1973 civilian rule was restored. This change marked a partial reversal of external policies as well as a desire to clean up public life which had been corrupted by oligarchic government and the inflation and vices attendant on a foreign military presence. The American forces began to leave in the same year but another product of the war remained in the shape of the Patriotic Front which, with some Chinese support, had for ten years maintained guerrilla groups a few thousand strong among the Meo peoples who (numbering in all about 300,000) lived astride the Thai–Laotian border. In 1976 an increasingly bewildered civilian régime was again replaced by the military behind a military–civilian façade, whose main worry was the advent of Vietnamese power to the borders of Thailand.

Vietnam's conquest of Kampuchea (Cambodia), for long the buffer state between Thailand and Vietnam, threw thousands of refugees into eastern Thailand and forced Thailand to look for help to counter Vietnamese expansion and restore Kampuchea's independence. Thailand's friends and allies in South-east Asia, less directly involved and linked in an unmilitary alliance (see below), did not provide the kind of support required by Thailand, which turned to China and the United States. Both supplied arms, the latter (by an agreement of 1987) in large quantities and on condition that these arms might be available for use by the United States as well as Thailand. With economic growth around 10–12 per cent a year, industrial investment and output growing, tourism also growing and external payments in balance, Thailand prospered although it could not rival the economic explosion of Hong Kong or Taiwan. Within its monarchical framework it was governed by interlocking and mutually beneficial associations of the army and the business community, headed by generals but marginally constrained by a parliamentary constitution and the residual powers and considerable prestige of the monarchy. In 1988 General Prem Tinsulanonda gave way at the age of 82 to General Chatichai Choonhaven but four years later the choice of General Suchinda Kraprayoon as the next prime minister without parliamentary approval caused riots, led by ex-General Chamlon Srimuang, on a scale which prompted King Bhumipol to intervene, snub the military and instal a civilian prime minister. Besides quarrels between generals the régime was marked by outstanding peculation and, mainly in the west and south, disturbances of various kinds caused by Muslim or communist insurgents or organized opium dealers.

The Philippines (about 700 islands), overrun by the Japanese during the war, were recovered by the United States in 1945 and given their independence in the next year. The terms of the transfer of sovereignty included a lease of bases to the United States for ninety-nine years, reduced in 1966 to twenty-five years from that date. The central problem of the new government was authority. The country had become a gangland in which different groups, including the police, had become laws unto themselves. Violence was endemic and rose in a crescendo at or near election times. Prominent among the forces in the state were the Hukbalahaps, originally anti-Japanese guerrillas supported during the war by the Americans but, the war having ended, now expendable or worse: their leader Luis Taruc proclaimed himself a communist. He surrendered in 1974 but his movement persisted in the north where its New People's Army (NPA) governed substantial areas with some support from the local population. The NPA's sphere was near enough to the American air and naval base at Clark Airfield and Subic Bay to cause the United States some uneasiness and ensure support for a series of presidents judged capable of containing the insurgents. There was also dissidence in the south where an established Muslim majority found itself becoming a minority and formed the Moro National Liberation Front. In the capital, Manila, ostentation flourished alongside the debris of war.

In 1962 President Diosdado Macapagal exhumed the Philippines' claim to Sabah (North Borneo) which forms one side of the Sulu Sea, otherwise enclosed by Philippine islands. He broke off relations with Malaya and refused to recognize the new state of Malaysia of which Sabah had agreed to become part. But this issue was allowed to drop by his successor Ferdinand Marcos (president, 1966–86) who, although elected as a reformer, became preoccupied with attempts to retain the presidency beyond the constitutional term and a fabulous enrichment of himself and his family. A referendum in 1973, following the imposition of martial law in 1972, allowed him to prolong his hold on office. Martial law was annulled in 1981 and Marcos unsurprisingly won a presidential election in the same year. His rule was tyrannical and his wife, who became a political personage in her own right, was no better. In 1983 Benigno Aquino, the most eminent opponent of the régime (who had been imprisoned in 1973 to keep him from winning an election and was released in 1979 to seek medical treatment in the United States), was assassinated as he set foot on his native soil from an aircraft, killed by a hired gunman who was himself immediately shot dead. A subsequent inquiry pointed the finger at a group which included the chief of army staff, General Fabian Ver, and two other generals, but they were officially exonerated. With fresh elections due in 1986 and increasing unrest verging on civil war, the Americans were forced to ask themselves whether Marcos was not a rapidly wasting asset, but the more vulnerable he became the more the Americans found themselves trapped between him and what was to them an even more unpalatable, increasingly left-wing alternative. Having lost Vietnam and its base at Camranh Bay to allies of Moscow they were fearful of seeing the Philippines with their even more important bases go the same way. They sustained Marcos until he became unsustainable and then switched their support to Aquino's widow, Corazon, who campaigned against Marcos for the presidency and forced him to decamp two weeks after a further re-election – and the Americans to change sides – under the pressures of huge peaceful demonstrations and a few highly placed desertions to her side. Marcos died in the United States in 1989.

Mrs Aquino, although president by popular acclaim, was none the less dependent on the army, personified by Juan Ponce Enrile and Fidel Ramos. Her vice-president Salvador Laurel had been her rival for the presidency. She inherited a twenty-year-old insurrection in the south, an external debt of $25 billion, a debt service which absorbed a third of foreign earnings, declining domestic production and an agreement with the IMF which was on the verge of being abrogated for non-compliance with its terms. She negotiated a new agreement but recovery was sluggish. Discord in the insurgent provinces persisted and discord within the government soon made its appearance, particularly from army officers critical of her inability to repress the NPA and the Moro Front. Forced to dismiss Enrile she found herself the more dependent on Ramos; Vice-president Laurel deserted her for Enrile at local elections in 1987. A series of revolts advertised her uncertain control, prevented her from embarking on much needed land reforms and increased her dependence on American goodwill. In spite of earlier support for those demanding the closure of American bases, she stood by the agreement permitting their presence until 1991 while obtaining Washington's assent to quinquennial reviews, which were a way to raise the rent. She concluded with Reagan an agreement whereby the two presidents undertook to use their best endeavours to ensure payment of $962 million for two years, but the US Congress approved only $365 million for one year. Any extension of the bases beyond 1991 would require endorsement by two-thirds of the Philippines' senate and possibly a referendum. The bargaining power of the Philippines was far from inconsiderable since the bases were not wholly replaceable except at staggering cost and their evacuation would impose on the United States a thorough and painful strategic review. A new draft treaty was rejected by the Philippine senate in 1991 in spite of the considerable sacrifice in jobs and revenues at a time of severe and worsening poverty. President Aquino, after initially contesting the Senate's resolution, required the withdrawal of the Americans within three years. In 1992 Aquino was succeeded by Ramos who conducted a more coherent administration and directed a modestly more prosperous economy.

The countries of South-east Asia other than those in Indo-China formed in 1967 the Association of South East Asian Nations (*ASEAN*). The first members of this association were Indonesia, Malaysia, Singapore, Thailand and the Philippines. They were joined in 1984 by Brunei. Their aims were to attack poverty, disease and other social ills; improve their commercial and economic weight in the world; secure their independence by reducing the sources of domestic upheaval and the temptations to outsiders to meddle in their affairs; and keep foreigners away. There were to be no joint military planning or exercises nor any other kind of military collaboration. This was a mutual aid association, economic and social but not without some silent hopes that it might curb Indonesia's potential capacity to overawe the region.

ASEAN was not the region's first postwar essay in co-operation. In 1961 Thailand, Malaya (as it then was) and the Philippines formed the Association of South-east Asia which, however, disintegrated as a result of the rival claims of the last two to Sabah. In 1963 these two and Indonesia projected a different tripartite association which never came to birth. There were at this period two major obstacles: tensions arising out of the creation of Malaysia (achieved in 1963) and the pro-communist attitudes of Sukarno in Indonesia which was much the biggest, richest and most populous state in the region. The removal of Sukarno by the Indonesian army was a necessary pre-condition for the creation of a broad South-east Asian association, as too was Indonesia's willingness to collaborate. Sukarno's successors were willing. Confrontation between Indonesia and Malaysia (see above) faded away and the quarrel between Malaya and the Philippines, although it produced a diplomatic rupture, also died down after Sabah voted in 1967 to join Malaysia: President Marcos visited Kuala Lumpur a year after his inauguration. Relations were again soured that same year after a mysterious massacre of Muslim soldiers on Corregidor (possibly by their own officers) and by a brief revival of the Philippines' claim to Sabah, but the claim was put on ice by Marcos who, although unwilling to take the unpopular step of finally abjuring it, rated it below the new regional solidarity. President Aquino later made a formal reununciation.

The British departure from South-east Asia and the Americans' war in Vietnam tested this solidarity. The announcement in 1968 of imminent British withdrawal from Malaysia and Singapore prompted these two states to conclude with Australia, New Zealand and Britain a new agreement (to which Brunei later adhered) replacing the existing Anglo–Malayan defence agreement and then in the same year to join their partners in ASEAN in proposing that South-east Asia be declared a zone of peace, freedom and neutrality. If these two steps were barely consistent with each other, the inconsistency was the price paid for transition from one colonial world to a new system of regional co-operation for the better securing of national independence. The American war in Vietnam was an affront to that independence, but after the Tet offensive at the beginning of 1968 the United States began planning the retreat which was eventually assured by the treaties signed in Paris in 1973. For the ASEAN states the most disturbing sequel to the United States' failure in Vietnam was Nixon's *rapprochement* with China and the opening given to China to exercise in South-east Asia the power role abandoned by the United States, with the further problem of deciding whether the prime enemy in the region was now communist China or communist Vietnam. The threat from Vietnam was made the more acute by Vietnam's invasion and domination of Kampuchea.

Members of ASEAN were divided. At the UN in 1970 Malaysia and Singapore had voted to give the Chinese seat in the Security Council to the communist régime in Beijing, the Philippines had voted against and Indonesia and Thailand had abstained. Most members of ASEAN had hoped to add Vietnam to their association in spite of its communist régime, but by invading Kampuchea Vietnam had offended against one of ASEAN's basic tenets – respect for national sovereignty and independence. Vietnam's treaty with the USSR was another count against it since another of ASEAN's tenets was its resolve to keep major powers away. The dilemma was most acute for Thailand. Thailand wished to act in concert with its ASEAN partners but it was more directly threatened than they were by Vietnamese expansion into Kampuchea and was well aware that ASEAN had no military might to oppose to Vietnamese aggression, whereas China had. In fighting terms the anti-Vietnamese effectives were the Khmer Rouge and China. For other ASEAN members on the other hand, in particular Indonesia, China was the main threat to the region in the long term and China's incursion into northern Vietnam in 1979, a sinister omen. In these circumstances ASEAN's survival was a tribute to the statesmanship of its leaders and their perception of its value to them, but the strains within it were not relieved until the Vietnamese withdrawal from Kampuchea and the Russian withdrawal from Vietnam at the end of the decade.

The allies were disturbed by reports during 1989 that Singapore was preparing to grant to the United States the right to establish military bases in breach of ASEAN's commitment to keep the region a zone of peace. They were assured that any such base would be used for repairs and maintenance only.

After a quarter of a century of an unostentatious co-operation ASEAN was enticed into expanding its scope and size. By 1993 regular talks on security were expanded into an ASEAN Regional Forum embracing fifteen states and the European Union. In economic affairs the emergence, at Australian initiative, of APEC (Asian Pacific Economic Co-operation) portended a wider zonal organization which might overshadow ASEAN or absorb it in a new and vast free trade area.

17 AFGHANISTAN

The identity of the modern state of Afghanistan began to be formed by Ahmed Shah in the eighteenth century after the assassination of Nadir Shah of Iran whose empire had extended over Pathans, Turkomans, Uzbeks and Hazars (descendants of the Mongols) who lived between the deserts of eastern Iran and what was to become the north-west frontier of the British in India. In the nineteenth century the British, after defeating the Sikhs, extended their rule westward and came into collision with Amir Abdur Rahman of Afghanistan who, having consolidated his position after a time of troubles, was moving in the opposite direction. In 1893 the Durand line, named after the foreign minister of India and running through Pathan country, was drawn but the nature of this line was not precisely defined and successive Afghan governments denied that it was ever meant to be an international frontier.

Between eight and nine million Pushtu-speaking Pathans live on either side of the Afghan–Pakistani frontier. On the eve of Pakistan's independence in 1947 and after it Afghanistan tried to persuade first Britain and then Pakistan to agree to the creation of an independent Pathan or Pushtu state which would not, however, have included the Pathans living in Afghanistan (who were alleged to desire no change) but would stretch from Chitral in north-western Kashmir down to Sind and might embrace parts of Baluchistan and Sind and even Karachi. Pakistan rejected the notion. For some years there was border fighting – associated in particular with the faqir of Ipi, a persistent thorn in Pakistan's flesh – and a series of domestic protests and flurries. An offer by the shah of Iran to mediate in 1950 was accepted by Pakistan but never eventuated. Shortly afterwards the dispute died down, but it continued to affect relations between Afghanistan and Pakistan and, together with the former's tradition of friendly relations with the USSR, was a factor in keeping Afghanistan out of the negotiations for the Baghdad Pact, sponsored by the west. From 1953, when it became the first non-communist recipient of Russian aid, Afghanistan moved into the Russian sphere, but its dependence on the USSR remained discreet for twenty-five years. Afghanistan stayed off the international map.

Modern Afghanistan has been governed by the Pathans of the south-east with the reluctant acquiescence of other races, among whom the more important were the northerners who, although quick to stigmatize the Pathans as idlers, found in the comparative richness of their own lands and herds compensation for their disproportionately low political influence. Afghanistan was a country with few natural resources and a medieval fiscal system. Since the parliament refused to impose any but the lightest taxes on land, public revenue was derived from customs duties which were necessarily small. Smuggling was a major economic activity. Muhammad Zahir Shah, who came to the throne in 1935, was well disposed to modest progress but obstructed by a parliament which he, unlike his neighbour the shah of Iran, was not strong enough to dismiss or manipulate. He was deposed in 1973 while on a visit to Europe, the monarchy was abolished and a republic inaugurated under the presidency of one of his relations, Muhammad Daud Khan. During the 1960s foreign aid (Russian, but also Chinese and French) was used mainly for road building. Some of this activity, notably the road and tunnels built by Russian engineers from Mazar-i-Sharif near the frontier with the USSR and over the Hindu Kush to Kabul, had obvious strategic significance, particularly alarming to Pakistan.

Muhammad Daud had been helped to power in 1973 by a section of the (more or less) communist People's Democratic Party (PDP) which came into existence in 1965 and almost immediately split into two parts. The one – the Khalq – led by Nur Muhammad Taraki and Hafizullah Amin was broadly rural and Pathan. The other – Parcham – led by Babrak Karmal was predominantly urban and stronger among Tadjik and other non-Pushtu-speaking peoples. Parcham supported Muhammad Daud and was rewarded with places in his cabinet. But Muhammad Daud was trying to play off east against west and vice versa. He sought aid from the shah of Iran (and possibly the United States), persecuted both Parcham and the Khalq and in 1977 put their leaders in jail. Both factions, however, had made headway in the army and in 1978 the tables were turned and Muhammad Daud was ousted. This was a victory chiefly for the Khalq which, after a brief spell of co-operation with Parcham, despatched Karmal and other Parcham leaders into dignified exile as ambassadors. When later summoned to return home they preferred not to.

Taraki and Amin quickly came to grief. Their impulsive reforms enraged landowners and the clergy and precipitated rioting in which many were killed (including incidentally the American ambassador whose role behind the scenes, if any, has remained a mystery). In what may have been an abortive coup against the government some fifty Russian advisers in Herat, close to the Iranian frontier, were gruesomely murdered. Taraki's star waned and Amin tried to recover control

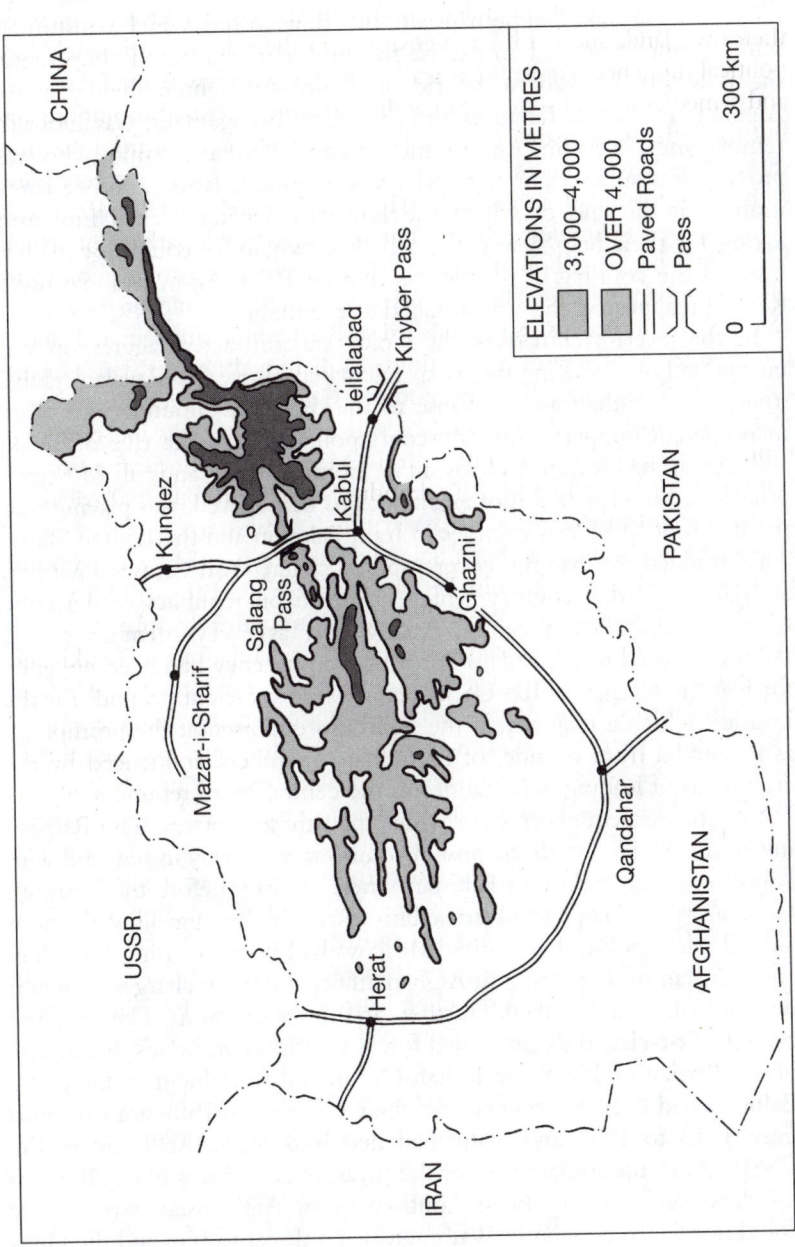

17.1 Afghanistan

by ordering ferocious razzias in the countryside and killing his enemies by the thousand. Alarmed by these disorders, by seething Islamic opposition provoked by Amin's persecution and the example of neighbouring Iran, and perhaps also by events in Iran which constituted both a temptation and an excuse for an American armed coup there, the Russians resolved to get rid of Amin and tighten their grip on Afghanistan. But as Taraki, with Russian encouragement, was about to remove and kill Amin, Amin removed and killed him. Amin, who was proving less subservient than Moscow required, lasted another three months at the end of which the Russians, having tricked him into asking for their help against the rebellion which he could not master, invaded the country in the closing days of 1979. Amin was executed, Karmal reappeared and was installed as president.

In the feverish climate of the Cold War, this act of aggression was interpreted in Washington as a calculated move in Russian global strategy rather than as a response to the dangerous flounderings of an incompetent puppet. From Moscow's point of view the rule of Taraki and Amin had created chaos in a frontier zone and, if Moscow's proclaimed interpretation of events was to be believed, was playing into the hands of Moscow's enemies in Iran, Pakistan and the United States. This troubled area had moreover a special characteristic. It was a Muslim and largely Turkic country which bordered on republics of the same texture within Soviet central Asia where, as every member of the Politburo could recall, *bashmak* or robber insurgency had been endemic for half the lifespan of the USSR. Here were sufficient grounds for the invasion which a majority of the Politburo endorsed at the prompting, as it seemed from outside, of its military members, influenced by the arguments of leading personalities in the central Asian republics.

But the consequences were wider than these motives. The Russian invasion coincided with the post-revolutionary disarray in Iran and with continuing uncertainty in Pakistan where, shortly before the invasion, General Zia ul-Haq had advertised his insecurity by cancelling elections and all party political activities. Moscow had been accusing both Iran and Pakistan of interfering in Afghan affairs and these charges, whether or not true, might herald further acts of aggression. The invasion brought first-class Russian armed forces to within an hour's flying time of the Persian Gulf and the Indian Ocean and to Baluchi country: the Baluchis had been fighting against their Iranian and Pakistani overlords from 1973 to 1976 and many had fled into Afghanistan and to the USSR. After the initial phase of the invasion two-thirds of the Russian invaders were seen to be in south-western Afghanistan where they could not fail to be seen in Washington as a threat to Iran and the Gulf. Movements of American vessels of war to the Gulf had in Moscow the same message in reverse.

The United States chose to react by playing up the invasion's frightening implications. This was not difficult since the invasion was a blatant act of aggression. It was also necessary from the American point of view because American prestige and influence had been severely dented by the continuing captivity of the hostages and also by the fall of the shah; the shah in person had been portrayed as a major bastion of American strength and Washington was accused – by, in the first place, dismayed Americans themselves – of having failed to stand by and salvage a powerful friend. The American position in and around the Gulf was weakened by this charge, even though Arabs secretly rejoiced over the discomfiture of a non-Arab monarch whom none of them loved. President Carter had also a legitimate interest in switching the international limelight from his own dilemma over the hostages to the enormity of Russian behaviour in Afghanistan and he was strongly supported by the Muslim world. The Islamic Conference, the institution created in 1969 by the Saudi monarchy, held an unprecedented emergency meeting at which Khomeini inveighed against the USSR as almost as satanic as the United States and all Arab members joined in condemning the invasion. The Gulf states were more emphatic than the anti-Israeli Confrontation Front (Syria, Algeria, Libya, South Yemen and the PLO), but even the latter could not forbear to curse.

In terms of action, however, President Carter's range was limited. Earlier in the year he had announced, with special reference to the Gulf, the formation of a Rapid Deployment Force of 100,000, but no such force yet existed. Nor was it clear what it would do if it did. The president was therefore thrown back on to long-term measures such as stopping the export of grain to the USSR (unpopular with American farmers who sold the grain to Argentinians whence it found its way to the USSR) or trying to orchestrate international displeasure by a boycott of the Olympic Games in Moscow, which annoyed the Russians but did them little harm and evoked a mixed response from the rest of the world. In geopolitical terms the principal immediate consequence of the invasion was that the USSR had to commit some 90,000 troops to pacify a country already in its sphere of influence and see this large force fail to regain convincing control over it.

From Moscow's point of view the invasion made the immediate situation both worse and better. The substitution of Karmal for Amin softened and improved the régime's domestic standing but the invasion boosted opposition by endowing it with American funds, which rose to something like $1 billion a year, and with modern American missiles and other weapons. Karmal's government reintroduced religious instruction in schools, allowed religious programmes on television and even built new mosques, thus winning the allegiance of younger mullahs. It also wooed tribal leaders. Its opponents, collectively known as the *mujaheddin*, controlled about two-thirds of the country but were seriously divided between seven groups based in Peshawar and a further eight based in Iran. The lion's share of American aid, channelled through Pakistan and allocated with Pakistan's advice, went to Muslim Sunni fundamentalists who were distrusted by Shi'ite groups operating in the western parts of the country with Iranian support, by more intellectual or secular leaders, and by partisans of the ex-king (who lived in Rome and showed no wish to leave it). A small number of individuals enriched themselves by trafficking in arms and drugs, while much larger numbers of brave men fought and died in frightful warfare which was kept going by the superpowers.

From 1982 the UN issued annual pleas for a cease-fire, for a timetable for the expeditious withdrawal of Russian forces, and for agreements between the Pakistani and Afghan governments to stop aid to the *mujaheddin* through Pakistan and to return the thousands of Afghan refugees who had fled to Pakistan. Pakistan and the United States were unco-operative so long as the *mujaheddin* appeared to have a good chance of overthrowing the communist régime in Kabul, and the Reagan administration was also reluctant to forgo the advantages of exploiting the discomfiture which Moscow had brought upon itself by invading.

The advent of Gorbachev to power in Moscow transformed the situation, if only slowly. Gorbachev was determined to get out of a venture which was disastrously costly in money and lives. He wished to preserve the communist régime in Kabul, but if he had to choose between saving it and evacuation he would choose the latter. In 1986 he offered an immediate token withdrawal and total withdrawal as soon as hostilities against Kabul were abandoned. A year later he dropped the condition and announced total withdrawal within ten months. In the interval he tried to strengthen the Afghan government by jettisoning Karmal in favour of Muhammad Najibullah, a tough Pushtu police chief, who became the general secretary of the ruling People's Democratic Party and later head of state. Najibullah had the dual task of controlling the factions which made up the PDP and of enticing the opposition (or some of it) into a broad coalition with a newly elected parliament and under a new constitution. But his conditions, which included the reservation of a leading role to the PDP and a special relationship with the USSR, were comprehensively unacceptable to the *mujaheddin* who were at this point given doubled American aid and more and bigger weapons. The Russian invaders departed on schedule but the anticipated collapse of the government which it left behind did not occur and the *mujaheddin* failed to take Jellalabad and other key points in the east. Najibullah contrived both to stay in power in the capital and in much of the country and gradually to come to terms with a sufficient number of the guerrilla groups to prepare the ground for peace under a broad-bottomed government. But in 1992 he lost his nerve and took refuge in a UN encampment, leaving his various adversaries to fight among themselves. Their divisions were as much ethnic as ideological: Pathans, Uzbeks, Tajiks competing for power and for aid from Pakistan or Iran or Saudi Arabia or the United States. The new president Burhanuddin Rabbani, himself a Tajik, courted his fellow Tajik leader Ahmed Shah Masud and the Uzbek Abdul Rashid Dostum (who exercised firm control in the north around Mazar-i-Sharif). But the next year Rabbani appointed his principal adversary, Gulbuddin Hekmatyar, prime minister whereupon Hekmatyar, who had useful help from Pakistan and Saudi Arabia, and – in spite of being a Sunni Muslim – Iran, forged a new alliance with Masud and Dostum and began shelling Kabul. The UN laboriously persuaded the principal factions into a broad coalition involving the resignation of Rabbani but the emergence of a new faction, the Taliban, destroyed the plan. The Taliban, mostly youthful Pathans from the south, united by their impatience with the feuds of everybody else, seized Kandahar and advanced on Kabul, routing Hekmatyar's forces on the way and giving new heart to Rabbani who withdrew his resignation. Thwarted before Kabul the Taliban proceeded to Herat which fell to them without a struggle and returned thence for a renewed onslaught on the capital.

18 KOREA

The Korean War left North and South deeply hostile to each other and condemned to decades of misrule. The military inferiority of the South was mitigated by American reinforcement – a UN force commanded by an American general remained in the country – and a vastly and ultimately spectacularly better economic performance. The UN voted annually in favour of the reunification of the country by way of free elections but these resolutions were without effect. So were talks between North and South in 1971–73 and in the 1980s. In North Korea Marshal Kim Il Sung held uninterrupted sway. Opposition dared not show its head. As he grew older (he was born in 1912) he became more autocratic and intent upon securing the succession for his son Kim Jong Il. In South Korea Synghman Rhee's ruthless rule, buttressed by a defence treaty of 1953 with the United States, ended in 1960 when he was forced to resign and flee to Hawaii where he died in 1965 at the age of 90. A military coup in 1961 carried General Park Chung Hee to power, and such was the inflexibility of South Korean political life that Park continued in power for nearly twenty unlovely years. They were marked by considerable economic success (checked in and immediately after 1973) and continuous, if ineffectual, protest against harshness and corruption. In 1963 martial law was lifted and Park and his principal colleagues transformed themselves into civilians. Park was elected president in 1963 and again in 1967 and 1971 and sent a contingent to fight with the Americans in Vietnam. In 1971 one of two United States divisions was removed. Martial law was reintroduced in 1972 with such extreme disregard for human rights that President Carter announced that the remaining American division would also be withdrawn – a decision reversed by Reagan when it appeared that North Korea's large armed forces were being made even larger. In 1979 Park was assassinated by the chief of his intelligence services. The new president, Choi Kyu Hwa, bade fair to inject a measure of democracy but was quickly rendered powerless by a group of officers led by General Chon Doo Hwan. Demonstrations which turned into a revolt in Kwangju in May 1980 were isolated and brutally repressed (perhaps 2,000 were killed) and after nine months as a civilian façade for military rule Choi resigned and was succeeded by Chon who was invited by Reagan to visit Washington. In 1983 a bomb, all but certainly placed by North Koreans, killed twenty-one people in Rangoon, including four South Korean cabinet ministers. President Chon, the main target, escaped. This outrage disrupted talks which had begun for the unification of the two Koreas. They were, however, resumed in 1984–85. The results were meagre: a few dozen visits in either direction to see relatives but no reduction in frontier fortifications or military exercises. The choice of Seoul for the 1988 Olympic Games induced North Korea to ask for some participation. The International Olympic Committee offered to hold five events in the North but on terms which North Korea did not accept.

During the 1980s South Korea emerged from a period of economic development behind strong protectionist cover to become a flourishing industrial power in a worldwide and mainly liberal economic order. Reunification of the peninsula and better relations with ideological enemies offered further economic gains without, however, being essential to them. When Chon's presidency came to an end in 1988 the succession might be by fiat, coup or election. Chon chose the last and his choice demonstrated the vast change which had come over South Korea from a country at war with North Korea to one in competition with Japan and other economic giants. Chon's friend and chosen successor, General Roh Tae Woo, won the election with 37 per cent of a large turnout and against a divided opposition. The imminence of the Olympic Games may have helped him inasmuch as South Koreans were anxious to avoid disturbances or bloodshed during the Games and were therefore drawn to the established régime rather than to the uncertainties of radical change. Six months of extreme violence had preceded the election, alarming the United States as well as Roh who adopted a conciliatory tone directed particularly to the middle classes and middle-aged who were showing signs of sympathy with the seething indignation of radical youth. After his victory Roh established diplomatic relations with China and Russia (on a visit to Seoul in 1992 Yeltsin apologized for the shooting down of Korean Airlines flight 007 in 1983) but he had strained trade relations with the United States, resisted an American request to quadruple his contributions to the costs of American forces in South Korea and refused to send South Korean forces to the Gulf War in 1991. He engaged in talks, eventually fruitless, with North Korea, approved the admission of both Korean states to the UN and set out to conciliate some of his domestic opponents. These were three: Kim Yong Pil, an ex-officer who had played a leading role in the 1961 coup but broke with Park in 1973 and re-emerged in 1980 as leader of the New Democratic Republican Party; Kim Young Sam, leader of the Democratic Reunification Party; and Kim Dae Jung, leader of the largest opposition group, the Party for Peace and

Democracy (later the Democratic Party). The first two of these accepted amalgamation with Roh's Democratic Justice Party, jockeyed for the role of Roh's favoured successor and formed with him the Democratic Liberal Party. Kim Young Sam succeeded Roh in 1992 – the first civilian president for thirty years, a cautiously conservative reformer pledged to expose corruption in government and business and an assiduous visitor to Asian neighbours. After a pause in the early 1990s the South Korean economy resumed its vertiginous growth, albeit with corresponding growth in wages and deficits on foreign account. But the sequel to the shift from military autocracy included much publicized revelations of disturbingly vast financial malpractice which had accompanied the country's transformation into one of the world's most productive economies.

In North Korea the government declared in 1985 that it would adhere to the Nuclear Non-Proliferation Treaty but seemed intent nevertheless on countering with nuclear weapons South Korea's superior non-nuclear armoury. Deprived of Soviet support by the turn of events in Moscow, North Korea was the more susceptible to American and Japanese pressures and agreed in 1991 to permit inspection by the International Atomic Energy Agency (IAEA) but then refused to sign the agreement unless the United States withdrew its forces from South Korea. Kim Il-Sung hoped for American and Japanese recognition as part of the price for abandoning his nuclear possibilities. He played a teasing game with the IAEA, agreeing to some of its requests, going back on some or his promises, finally driving the IAEA to refer the *impasse* to the Security Council and President Clinton to threaten economic sanctions or even armed intervention.

Clinton's paramount concern was over North Korea's nuclear capacity and intentions but his forcefulness worried South Korea – and Japan and China – which feared that it might provoke the wayward Kim into starting a war, nuclear or not. Kim Il-Sung tried to divert attention from the nuclear issue by making overtures to South Korea for an economic union or even reunification, neither of which seemed likely in the near future. The nearest he got to internationalism was the Tuman River Free Economic Zone in which he was a partner with China, Russia, Mongolia and South Korea. In 1994, on the eve of a first meeting with his southern counterpart, he suddenly died leaving his son as a semi-designated successor who was assumed to be the new head of state but was not formally so appointed, perhaps on account of divisions in the army. After the Great Leader's death, Clinton secured an agreement by a judicious combination of threats of military action, carrots and concessions. North Korea agreed to stop the construction of new reactors capable of producing weapons-grade plutonium, not to reprocess its spent fuel rods and to allow regular inspection by the IAEA in return for the provision by the United States by the year 2003 of two reactors of insignificant military capacity. The United States agreed also to open diplomatic relations, supply oil and remove obstacles to trade and investment. It quietly dropped demands for inspection of North Korea's alleged stock of nuclear weapons.

NOTES

A. CENTRAL ASIA

From the collapse of the Soviet Union and the Russian empire in Asia emerged five new sovereign states with a population, unevenly distributed, of 50 million, the great majority of them Turkish by race and language and Muslim by religion, the largest minority being some 8 million Russians mostly in the northern half of Kazakhstan. Previously called with political ambiguity Russian Turkestan, this block of states, all of which declared independence in 1990 with the borders which they had had as Soviet Republics, had frontiers with Russia, China, Pakistan, Afghanistan and Iran but none with Turkey. None had access to the sea. In the northern part of the area was the huge state of Kazakhstan, the size of India and twice the size of the other four republics combined, but with a population of only 17 million of whom 40 per cent were Kazakhs and 40 per cent were Russians. To the south were the two lowland states of Turkmenistan and Uzbekistan and the much smaller Tajikistan and Kirghizia – renamed Kyrgystan – in the high mountains to the east of Uzbekistan. Most populous of the five and the most homogeneous ethnically was Uzbekistan. There were sizeable Uzbek minorities in the states to the east of it and in the area as a whole Uzbeks were much the most numerous (the Russians came second). The entire area was predominantly Turkish except Tajikistan which was 60 per cent Iranian; and predominantly Sunni with some Shi'ites in the southern belt. Kazakhstan, the richest in natural resources, in first place oil, alone marched with Russia. Its oil reserves were believed to be so copious as to challenge the dominance of the Middle East. Turkmenistan, specially rich in gas, oil and other minerals, marched with Iran and Afghanistan; Uzbekistan, also endowed with oil and gas, with Afghanistan; Tajikistan with Afghanistan, Pakistan and China; Kyrgystan with China. All these new states had considerable but underdeveloped resources alongside weak economic infrastructure, weak administrative machinery, rapidly increasing populations, disputes between and within themselves and artificial boundaries laid down by Moscow in the 1920s and 1930s when they had become Soviet Republics. After independence their governments ceased to call themselves communist but remained autocratic: in Turkmenistan, for example, elections in 1992 gave almost 100 per cent of the vote to the former communist Saparmurad Nigazov. They also ceased to be off the map. Iran and Turkey eyed new areas of possible influence. Visits were exchanged between leaders of the four Turkish states and President Ozal of Turkey. There were similarly calculated courtesies between them and Iran which took the lead in establishing in Teheran a Caspian Council (Iran, Russia, Azerbaijan, Turkmenistan) and an Economic Co-operation Organization (Iran, Turkey, Pakistan, Afghanistan and the five new states). These bodies were suspected of cloaking semi-imperial Iranian ambitions. A Black Sea Economic Co-operation Treaty, promoted by Turkey, was signed in 1992 by eleven, mostly Asian states. Non-Muslim China and India also displayed eagerness to establish commercial links and regular communications from telephone services to air schedules. Russians and others feared the spread of specifically Islamic parties, which they dubbed fundamentalist. In Tajikistan, where 20 per cent of the population was Uzbek, ethnic and religious conflicts amounted in 1992 to civil war. President Rakhmon Nabijev was forced to flee but was restored by Russian and Uzbek intervention after a brief interval in which the Islamic Renaissance Party, which had support from the communist Nazibullah régime in Afghanistan, formed a coalition government pledged to more religion in a nevertheless secular state. On Nabijev's return 100,000 or more refugees fled into Afghanistan (which contained a Tajik population of 3 million).

B. Sri Lanka

Ceylon gained its independence from Britain in 1948 as a consequence of the British departure from India. In 1972 it became Sri Lanka and a republic.

It was governed alternatively by two main families and the parties which they formed: the United National Party led in turn by D. S. Senanayake, his son Dudley Senanayake and his nephew Sir John Kotelawala, and the Sri Lanka Freedom Party, founded by Solomon Bandaranaike, a defector from the UNP, and after his murder in 1959 by his widow Sirimavo. The UNP governed from 1948 to 1956, from 1965 to 1970 and from 1977 onwards; the SLFP the rest of the time. The most important election campaign was that of 1956 in which Bandaranaike defeated the UNP by running on a combined racial and religious ticket, pro-Sinhala and pro-Buddhist. His success was followed by persecution of the Tamil minority (a quarter of the population) and a permanent increase in the political influence of the Buddhist establishment which had subsequently to be courted by the UNP as well as the SLFP. Alongside the contest for power between these two parties opposition was provided by a variety of left-wing parties of which the most notable was the Lanka Sama Samaj Party, founded in 1935 as an expression of anti-colonialism and a member of the Trotskyist international until 1963 when it joined Mrs Bandaranaike's government (this alliance was dissolved in 1975). Mrs Bandaranaike also brought the pro-Russian Communist Party into government and since other left-wing groups had lost ground her coalition was expected to win the elections of 1965. That it did not was mainly due to an economic situation which neither main party had contrived to control but which had deteriorated alarmingly during the SLFP's years of office in the 1960s.

After independence, as before, the economy of Sri Lanka was heavily dependent on the export earnings of tea, coconut and rubber estates owned by British companies. The revenue from these products fell steadily while at the same time the import bill rose, the largest single item being for food. This was a typical Third World situation in which the economy was dominated by increasingly less remunerative dealings with the former metropole and an increasing inability to feed a growing population without ruinous expenditure of foreign currency or, when that failed, foreign borrowing. The population was rising fast. It roughly doubled between independence and 1975. In the 1960s alone the external debt quadrupled. Unemployment rose from around 40,000 at independence to something near to 700,000 when, at the next turn of the electoral wheel, Mrs Bandaranaike and her associates handsomely won the elections of 1970 and re-formed the coalition that had been defeated five years earlier. In April 1971 it was severely shaken by a carefully prepared, well-armed peasant rising.

This rising, which came as a surprise, was organized and directed by the Janatha Vinukhti Peramuna (JVP) or Popular Liberation Front, formed in 1965 as a splinter from the ailing Maoist Communist Party. It regarded the SFLP and UNP as two virtually indistinguishable aspects of a post-imperialist and neo-colonialist bourgeoisie which was prevented by self-interest from tackling the country's basic economic problems and social ills. The emphasis of its doctrine and its practice was on the peasants and it attempted to use the peasants to strike a direct blow at the government and the system. There was later the usual insoluble squabble about who struck the first blow, but there was no doubt about the outcome. After seven weeks the government had prevailed but the challenge to it was such that the number of people killed in the process of asserting its authority ran into tens of thousands. A most unusual constellation of states supported Mrs Bandaranaike materially or orally. They included the USA, USSR and China; India and Pakistan; Britain, Australia and Egypt. Mrs Bandaranaike maintained until 1977 the state of emergency proclaimed in 1971 but was then defeated by the UNP under Junius R. Jayawardene who became prime minister in that year and president the next. This swing to the right was marked by a dash for growth to emulate the fortunes of Singapore, Hong Kong and Taiwan, but economic ambitions were wrecked by acute racial and religious conflict in which the government became involved with the Tamil minority, once the country's rich ruling class, 18 per cent of its population and mostly Hindu.

The Tamils constituted a majority in the north and parts of the east. Distinct in race and religion they sought autonomy or independent statehood from the Sinhalese Buddhist majority which had established a powerful central government but had failed to use its power to allay Tamil fears and separatism. By the 1980s time for an accommodation was running out. The more militant Tamils formed the Liberation Tigers of Tamil Eelam and civil war began. Both sides committed atrocities. India became involved because Rajiv Gandhi could not ignore the outrage felt by the Tamils of Tamil Nadu in southern India. In 1986 the governments of India and Sri Lanka agreed that the Tamils of Sri Lanka should be granted autonomy but they could not agree on the area of the autonomous province. Talks therefore collapsed and the next year Jayawardene decided to extirpate the Tigers. He failed, and Gandhi intervened to stop massacres of Tamils by the Sinhalese army. Jayawardene was obliged to let the Indian army into Sri Lanka to protect Tamils. But the Tigers, who were set on independence and therefore opposed and ignored the Indo–Sri Lankan agreement, refused to surrender their arms as envisaged by that agreement with the result that the Indian force, whose original complement of 15,000 was

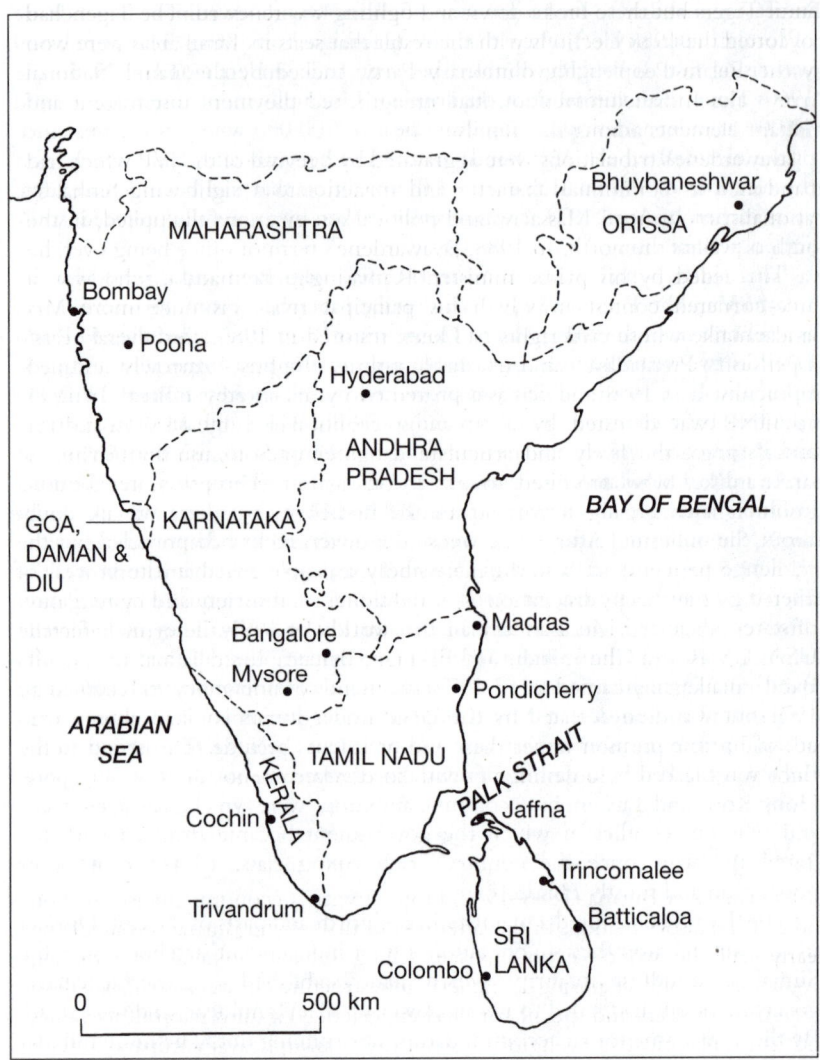

18.1 Southern India and Sri Lanka

doubled within a few months, found itself trying to enforce the surrender of Tamil arms instead of protecting Tamils. To do so it had to abandon its peace-keeping role and use force whereupon India was accused of planning to detach northern Sri Lanka and annex it to India. In 1988 Gandhi declared that he would withdraw his force the next year. For his part Jayawardene was accused of opening the way to Indian imperialism and also of having made undue concessions to Tamil autonomy. As the Indian forces prepared to depart – the last of them left early in 1990 – the government engaged in talks with the

Tamil Tigers but these broke down and fighting was renewed. The Tigers had boycotted the 1988 elections with the result that seats in Tamil areas were won by the Eelam People's Revolutionary Party, backed by the Tamil National Army. The Indian intervention had strengthened the more intransigent and militant elements among the Tamils.

Jayawardene's tribulations were aggravated by a revival of the JVP which had abandoned its communal character and turned into a right-wing Sinhalese nationalist movement. Massacres and political assassinations multiplied, in the south as well as the north. In 1988, Jayawardene's term of office being over, he was succeeded by his prime minister, Ramasinghe Premadasa, who won a three-cornered contest in which his principal rival was once more Mrs Bandaranaike whose civil rights had been restored in 1986 after several years' suspension. Premadasa, an autocratic political fighter, narrowly escaped impeachment in 1991 and was assassinated two years later by a Tamil. In 1994 the UNP was defeated by a left-wing coalition led by Mrs Chandriga Kumaratunga, the lively and articulate daughter of Solomon and Sirimavo Bandaranaike, who promised tough action against corruption, retribution against right-wing gangs terrorizing and killing helpless peasants, and talks with Tamils. She inherited a surge in economic growth and low democratic growth (below 1.5 per cent) and also the melancholy reality of enviable natural wealth afflicted by the abrupt dislocation of a traditional rural society and by religious ethnic conflicts as vicious as any in the world. In 1995 the army inflicted serious reverses on the militant Tamil Tigers and captured Jaffna, the Tamil capital, incidentally turned hundreds of thousands of non-militant Tamils into refugees and indicated that Mrs Kumaratunga's government had chosen the path of harsh repression rather than − or at least as prelude to − negotiations which must lead to a loosening of centralized government.

C. Myanmar

British rule over Burma (1886–1942) brought with it economic transformation and cultural disorientation. In the latter part of that period Burma was supplying nearly half the world's exported rice, selling valuable minerals and timber, producing oil and acquiring a modern infrastructure. It was also invaded by foreign marketeers and speculators and experienced a subversion of its values and its monarchical and religious institutions − an upheaval comparable with Ataturk's revolution in Turkey with the added infelicity that the engineers of change were alien. Long governed by an absolutism tempered mainly by anarchy, Burma received a veneer of constitutional and economic modernization both striking and superficial which fostered, besides anti-colonial resentments, a conflict between more modernization in an opening world and a retreat into conservative traditionalism.

In wartime Burma a group of nationalists without much knowledge of the outside world, pro-Japanese at the beginning of the war but later anti-Japanese and supplied with British arms, were associated together in the Anti-Fascist People's Freedom League (AFPFL). As the war came to an end there was doubt among the British whether to recognize and deal with the AFPFL. The government in London was opposed to recognition but the supreme commander, who was faced with the problems of installing a new administration without adequate resources of his own, favoured it. The Labour victory in the British general election of July 1945 and the precipitate surrender of the Japanese in August operated to produce a decision to recognize. The new British governor of Burma, who wanted to arrest the nationalist leader Aung San and nearly did, was replaced, the AFPFL was treated as an embryonic government of Burma, and although Aung San and other leaders were assassinated in July 1947 their surviving colleagues achieved their goal of independence from Britain on 4 January 1948. There was no fighting. The British, strongly influenced by their own pledge to leave India and also by the belief that it was not possible to use the Indian troops of South East Asia Command against the Burmese, retired, leaving the AFPFL to struggle with huge war damage, its own internal divisions and with the hill peoples round the central Burmese plain whose traditional distrust of Rangoon created a string of separatist problems. The AFPFL included communists who not only split off from the main association but also split among themselves and waged separate campaigns against government; assassination had robbed Burma of a number of coming leaders during 1947 and 1948; revolt in Arakan along the west coast, added to troubles in the east (from the Karens spread across central Burma, the Shans – a Thai people – wedged between Thailand and China, and the Kachins in the far north) threatened independent Burma with disruption. The Shans and Kachins were mostly Christian. There were also Muslims in most parts of Burma, the majority of them in Arakan where, known as Rahingya, they had been since the seventh century AD.

In 1950 a new danger was introduced with 4,000 Kuomintang troops who crossed into Burma from China under General Li Mi and gave rise to fears that the new Chinese régime would follow its retreating enemies. But Beijing made a point of not blaming Rangoon and after a while General Li Mi and his followers were removed by air to Taiwan by the Americans. Nevertheless Burma felt it wise to edge politely and circumspectly away from its British and American connections. Although the Chinese in Burma numbered only about 300,000 Burma had a frontier with China and this frontier was straddled by the Kachins (300,000 in China, 200,000 in Burma and a small number in India). In 1953 the Burmese government intimated that it did not wish to renew the expiring Anglo–Burmese defence treaty or retain the British military mission which had been in the country since independence; it also informed Washington that it wanted no more American aid. Zhou Enlai visited Rangoon on his way back to Beijing from Geneva in 1954 and the Burmese prime minister, U Nu, returned the visit later in the year. Further meetings led to the initiation of frontier discussions which were completed when General Ne Win displaced U Nu (who returned to power in 1960 but was again displaced by Ne Win in 1962 and this time imprisoned for six years).

Ne Win was a general with a taste for philosophical and socialist discourse which he combined with the suppression of parliament and the imprisonment of his political opponents. His seizure of power was, in the first place, a response to the threat of disintegration posed by the country's dissident movements. The demands of the Shans in particular (in later years the most

Glossary

Current Name	Former Name
Myanmar	Burma
Yangon	Rangoon
Tanintharyi	Tenasserim
Bago	Pegu
Magway	Magwe
Ayeyarwady	Irrawaddy
Dawei	Tavoy
Pathein	Bassein
Bamar	Burman
Kayin	Karen
Mawlamyine	Moulmein
Hpa-an	Pa-an

Name change: On 18 June 1989 the name of the country was changed to the Union of Myanmar. The English spelling of many towns, divisions, states, rivers and nationalities was changed.

0   500 km   ▨ Mountains

18.2   Burma (Myanmar)

persistent dissidents and the most successful drug dealers) for a new federal and so looser constitution created a fear that the refugee Kuomintang forces in the north-east might, with American help, launch an invasion of China which would turn Burma into a battlefield (like Korea), while the existence in Burma of at least three different communist rebel movements made a Chinese invasion of Burma in support of communists against the government in Rangoon not improbable. The government's failure to remove these dangers of disintegration and foreign intervention were the chief causes of its removal by Ne Win and the army. The new régime, being overtly military, looked less feeble but it too failed to eradicate the various rebellions in spite of a policy of tough measures combined with declarations of willingness to talk with rebel leaders. It watered down U Nu's socialism but also watered away democracy: political parties were suppressed and universities closed in 1964. Burma's traditional fear of China (and of its friends in Burma, many of them of Chinese descent as well as communist ideas) persuaded Ne Win to visit the United States but the American war in Vietnam was all but universally regarded in Burma as unjustifiable and ominous interference in Asian affairs.

In economic matters Ne Win was even less successful than his predecessors in restoring Burma's economy. Under the Japanese occupation the acreage under rice had been halved and although this area was increased after the end of the war production was ruined by civil wars and general anarchy. Exports fell to one-tenth of their prewar figure and by 1966 domestic consumption was rationed. Desperate measures, such as the remission of all agriculture rents, did little good. The overall decline of the country's economy left farmers with nothing to buy even when they were paid for their produce, so that they had no incentive to produce more than they could themselves eat. Rice, once the source of Burma's wealth, was no longer worth growing, let alone exporting. Inflation accompanied by static wages impoverished the bulk of the population. Grandiose enterprises turned to dust, democratic forms were supplemented by intimidation and violence, poverty and disease killed hopes as well as people.

In 1977 an attempt on Ne Win's life was followed by discrimination against Muslims, torture and killing. A massacre in Arakan caused flight into Bangladesh. Ne Win assumed nevertheless a new term of office in 1978. He had become autocratic, secretive and out of touch and in 1981 he made way for General Sam Yu who ruled in his shadow rather than in his place. The various peripheral wars continued in the north-east and south-east, with the Burmese communists joining in where they were welcome. The economy deteriorated to such an extent that Burma graduated downward to the status of a Least Developed Country (LDC – eligible for specially soft international loans) and in 1987 the state cancelled all notes below a certain face value, thus extinguishing the savings of countless people. Discontent spread from the poor to the middle classes and the army. Demonstrations were brutally repressed by the police, and General Sein Lwin, noted for his ferocity, was thrust into the presidency which Ne Win had resumed but now once more abandoned. Continuing riots forced Sein Lwin to resign after a few weeks and were not abated by the installation of the civilian Maung Maung who lasted only two months. The army resumed power and dissolved parliament. When fresh elections were promised, some two hundred parties appeared but only those acceptable to the government were licensed. Aung San's daughter, Aung San Suu Kyi, returned from exile in England to lead a National League for Democracy but was arrested. The League claimed a massive victory in 1990 but the official results of the elections mirrored the government's wishes rather than the voters' choices and the government remained unmoved. Aung San

Suu Kyi was put under house arrest until 1995 when she was unexpectedly and unconditionally released. In 1989 the country's name was changed from Burma to Myanmar.

D. TIMOR

Timor, the last major island in the chain stretching between continental Asia and Australia, was divided by the Dutch and Portuguese in almost equal sections. The eastern and Portuguese half was detached from Macao at the end of the nineteenth century to become a separate colony, which it remained after the western half had become part of the Indonesian republic. The overthrow of the military régime in Lisbon in 1974 was expected to lead to the engulfing of eastern Timor in Indonesia, but opposition to Indonesia was manifested by two groups, the one pro-Portuguese and the other demanding independence. The latter was left-wing. Civil war broke out in 1975, the small Portuguese force left and at the end of the year Indonesia annexed and invaded the territory, devastating it with modern (American and British) arms and killing about half the inhabitants and incarcerating many of the survivors in camps. Portugal and Australia implicitly acquiesced. The rest of the world hardly noticed. The president and vice-president of the independence movement FRETELIN were assassinated in 1979. Continuing resistance throughout the 1980s was met by ruthless ferocity powered by weapons supplied to the Indonesians by western arms-makers under licence from western governments.

PART FIVE

AFRICA

19 NORTH AFRICA

THE MAGHRIB

In the 1950s and 1960s Africa produced a phenomenon of unparalleled extent – the emancipation from foreign rule of enormous areas in the form of independent sovereign states. This process had worldwide repercussions and overshadowed all other African affairs. It was on the whole unexpected. When the Second World War ended there were only three fully independent states in Africa: Ethiopia, Liberia and South Africa. The next ten years were a decade of preparation for the liquidation of the French, British and Belgian empires, and a further ten years later most of Africa was free. Since France, Britain and Belgium came to accept with versatile swiftness the need to go, these decades witnessed struggles over timetables rather than principles. Compelled by calculation rather than by force the imperial powers abandoned with unexpected ease vast areas whose governance they had acquired in the previous century with equal facility. The process of decolonization was, however, halted in the southern tip of the continent by the stubbornness of the Portuguese in Angola and Mozambique and by the ruthless determination of the semi-independent white settlers of Southern Rhodesia. The self-preservative resistance of the latter, and the refusal of the former to calculate in the same way as the French, the British and the Belgians, were decisively influenced by the existence still further south of the South African stronghold of white supremacy, where the white minority was comparatively much larger (one in four) than elsewhere in Africa and was at the same time fortified by riches, by modern technical power, by having nowhere else to go, and by a doctrinaire racialism which permitted extremities of repressive injustice and cruelty. There were also other pockets of Portuguese or Spanish rule.

The artificiality of dealing with political and economic affairs in terms of geographical categories is nowhere better illustrated than in Africa. The northern fringe of the African continent has been made by history a part of the Arab–Islamic civilization and has been more conscious of affinities with the Middle East than of its ancient economic or current political links with the rest of Africa. Moreover the European overlordship exercised from Casablanca to Suez through protectorates, unequal treaties, military agreements and direct annexation was different in kind from the colonial empires established by Europeans south of the Sahara. But North Africa is at the same time part of Africa; the ethnic line between the Arab–Berber and Bantu races runs through the Sudan and gives trouble there which other states on either side of the line cannot altogether ignore; Egypt and Morocco have played prominent parts in African affairs and conferences and associations; the more reticent Tunisia had a leading voice in the Congo's early troubles; Libyan ambitions and arms have troubled not only Central but also West Africa. The desert is no longer the barrier that it used to be since the aeroplane and above all the radio have enabled men to transcend it; language is no more a barrier between Arab Africa and Bantu Africa than it is within these two areas; and religion provides points of contact between Muslims and Christians on both sides of the divide. The North therefore, though still distinct from the rest of the continent in special and enduring ways, will be classed here as more African than Asian with the sole exception of Egypt whose postwar African role has been consistently subordinated to its Asian and not the other way round.

The whole of the North African coast came under European domination – French, Italian, British – during the nineteenth century or soon afterwards. The British sphere in Egypt has been considered elsewhere in this book but the rest of the area, though preponderantly Arab and Muslim, will be considered here in its geographical African context. So too will the Sudan, the country where the dividing line between Arab and African, Muslim and non-Muslim, cultures is most evident.

The French invaded Algeria in 1830 and declared it a part of metropolitan France in 1848. As a consequence of their occupation of Algeria they became involved in the two neighbouring monarchies of Tunisia to the east and Morocco to the west, then under the nominal suzerainty of the Ottoman sultan. By the Bardo Treaty of 1881 they established a protectorate over Tunisia, a poor and relatively small country with a population, in 1945, of only 3 million. In Morocco the evanescence of the Ottoman sultan's power coincided with the decay of the Moroccan sultan's own authority, thus creating an invitation and an excuse for foreign intervention, but French power was not so easily substituted owing to the ambitions of other European states. In the first years of the twentieth century France obtained a free hand in Morocco by conceding the same to Italy and Britain in Libya and Egypt, by defeating a claim by Germany (which accepted compensation in central Africa) and by allowing Spain to appropriate the northern strip whence the Arabs had invaded Spain 1,200 years earlier. The Treaty of Fez with the sultan in 1912 crowned these diplomatic successes and established a protectorate of the same kind as the French protectorate over *Tunisia*, although the whole country was not brought under effective French control until the 1930s. Tangier became an international zone and a fiddlers' paradise. The Italians reaped the reward of their complaisance when they were allowed in 1911 to take Tripolitania from the Turks on the eve of the Balkan wars.

During the Second World War the whole of northern Africa became a battlefield or, in the case of Morocco, a military rear area. Equally important were the political concomitants of the war, especially the Atlantic Charter, the eviction of France from the Arab countries of Syria and Lebanon, and the appearance on the scene of the Americans, including President Franklin D. Roosevelt himself, who had a much publicized interview with the Sultan of Morocco, Muhammad V ben Yusuf. The Tunisian nationalist leader Habib Bourguiba was released from a French prison in 1943, went to Cairo in 1945 and thence to the United States, and settled in his native country once more in 1949. The Moroccan leader Allal al-Fasi, who had been in prison from 1937 to 1946, also went in 1947 to Cairo and then settled temporarily in Tangier. In 1947 the newly created Arab League established a Maghrib Bureau (Maghrib means 'West' in Arabic; the Maghrib comprises at least Morocco, Algeria and Tunisia, sometimes Tripolitania as well), thus institutionalizing the Arab interest in the affairs of North Africa which was to become deeper and more effective in the coming years.

Both France and Italy had been defeated during the war. Both, however, ended up on the winning side. In North African terms Italy paid the price of defeat while France won the rewards of victory; Italy lost its African colonies while France was reinstated in the Maghrib. In the Maghrib the French, aware of the need for changes, sought modifications within the framework of the Bardo and Fez treaties, but the nationalists aimed to terminate the protectorate status altogether. In all three territories the French had encouraged immigration, so that there was a considerable French population settled on the land or in business alongside the French administrators who ran the government of Algeria and had also come to do much the same in Morocco and Tunisia in spite of the sovereignty of sultan and bey. French education had nurtured an élite which appreciated French culture, as the French intended, but also became attracted to the idea of independence, which the French had not foreseen. These modernizing nationalists found themselves allied, if on this point only, with traditionalist malcontents who resented the French presence from a conservative and Muslim standpoint. Caught between these currents the bey of Tunis wavered ineffectually until he was more or less captured by the French, to his eventual discomfiture, while the much younger sultan of Morocco wavered more purposefully, attached himself to the nationalist movement and was exiled by the French, to his eventual benefit.

The governments of the Fourth French Republic were all coalitions which contained ministers who wanted to meet nationalist movements more than half-way and other ministers who did not want to make life unpleasant for the settlers. This was an impossible combination which rendered France ineffective, allowed Tunisia and Morocco to achieve their aims and drove the largest group of the settlers, the Algerian, to revolt against the government of France. The first French plan was also the first French failure. The French Union, conceived in 1946, created the new category of Associated States with Morocco and Tunis, and also Vietnam, Laos and Cambodia, in mind, but whereas the Asian trio accepted this new status the African pair refused it. There were, however, reasonable hopes of a settlement with Tunisia after the return of Bourguiba in 1949 and until his arrest early in 1952. Robert Schuman, prime minister of France, spoke in 1950 of the ultimate independence of the protected states, and internal reforms of a democratic nature were seriously discussed; Bourguiba's Neo Destour Party was represented in the bey's government. But many nationalists regarded the reforms as inadequate except possibly as a first step, whereas the French regarded them as a long step with no immediate next step in sight. By the end of 1951 the dialogue had turned into rivalry for the support of the bey who was himself so uncertain of his better course that he appeared sometimes to be a nationalist and sometimes a puppet of the French who were pressing him to accept the projected reforms.

At the beginning of 1962 a tougher line gained ground in Paris. Bourguiba was arrested in January and the prime minister, Muhammad Chenik, was dismissed and arrested in March. The bey accepted the French programme and a number of nationalists fled to Cairo. In Tunisia constitutional changes which had been meant to please the nationalists were imposed by authority against their will. The attempt to reach agreement bilaterally had failed, for the acquiescence of the bey was unimportant by contrast with the disagreement of the nationalists who now carried the debate into the international sphere. The next two years merely confirmed the failure and exposed its consequences. Then in 1954 Pierre Mendès-France, after his blitz peace in Indo-China, insisted that the Tunisian problem must be solved no less radically. He flew to Tunisia accompanied by the right-wing Marshal Juin to propose full internal self-government. He fell from power in February 1955 before his initiative had borne fruit, but his successor Edgar Faure pursued the negotiations and in June Bourguiba accepted the French proposals as a step towards independence and returned to Tunisia. Nine months later, on 20 March 1956, Tunisia became fully independent and concluded a treaty with France which included provision for the stationing of French troops in the country.

The *Moroccan* case was not very different. A period of genuine negotiation revealed to both sides the gap between the French programme of democratic gradualism and the nationalists' determination to get independence very soon. The principal difference between the Moroccan and Tunisian cases lay in the temper of the ruler. Muhammad V ben Yusuf had shown signs of allying himself with the Istiqlal (Independence) Party at the end of the war, and by so doing he had destroyed the basis of government in Morocco which, during the seven-year term of office of General Noguès as resident-general, had rested upon the good personal relations between the two men – the *dialogue sultan–résident*. A second major factor was the isolation of the French settlers from the outside world during the war years 1940–42, an isolation which had larger consequences in Morocco than Tunisia because the French community in Morocco was also isolated from the surrounding Muslims by Lyautey's prewar policy of siting new French towns away from the traditional centres of Moroccan life. From 1947 to 1951 Marshal Juin was resident-general, but in spite of his somewhat forbidding presence a new *dialogue* Paris–Fez was initiated and the sultan visited the French capital in 1950. The French, however, believed that they had an alternative to treating with the nationalists, whom they and some Moroccans were tempted to write off as uncharacteristic and irresponsible townees, of less significance than traditional personages like the pro-French, and anti-sultan, pasha of Marrakesh, el Glaoui. The sultan was persuaded – he subsequently said coerced – into signing in 1951 decrees initiating the reforms which the French were prepared to introduce, but the consequent agitation in Morocco and elsewhere in the Arab world caused him to swing away from the French and for a year there was increasing uncertainty and disorder, culminating in December 1952 in violent and barbaric anti-white outbursts in Casablanca. In the following February the sultan was sent into exile. His absence, however, did not serve to restore order or strengthen French rule, while in Spanish Morocco an assembly of notables refused to recognize the exiled sultan's uncle, Muhammad ben Arafa, whom the French had placed on the throne. In 1955, following the settlement with Tunisia, the sultan was brought back and before the end of the year France had agreed to concede full independence. It took effect on 2 March 1956. Both Morocco and Tunisia became full members of the Arab League in October 1958.

By this time the revolt in *Algeria* against France, a far more serious problem, had begun. Hostilities which opened in the Aurès mountains in 1954 were at first regarded as a fresh instalment of familiar colonial troubles but they developed into a war which involved the flower of the French army, the full panoply of military rule and censorship and terrorism and torture, three separate white French challenges to the authority of Paris, the fall of the Fourth Republic and the achievement by Algeria – uniquely in Africa until Zimbabwe – of independence by force of arms. The situation was without any parallel in the rest of Africa. The European population had been a part of the country for much longer than any other settler community, it was nearer to the mother country, and in the main cities it was as numerous or almost as numerous as the Muslims; its services to Algeria were conspicuous. Juridically too the situation was peculiar since Algeria was constitutionally a part of metropolitan France, so that Frenchmen who failed even to envisage a severance were maintaining an unreality which was nevertheless no fiction. The legal position contributed to a stubborn psychological attitude which caused outsiders to wonder why it was that Frenchmen alone in the world were unable to see that their days as rulers in Algeria were numbered. The French, moreover, like the British in the Middle East after the retreat from India, hated the thought of further surrenders after the collapse of their empire in Indo-China; just as Africans were encouraged by the ending of French and British rule in Asia, so in reverse were the French and British influenced in their African policies by their postwar experiences in Asia, making in Algeria and Egypt in particular mistakes of timing and understanding which they might have avoided if they had not felt that their descent from the first rank was proving too hasty for dignity or safety.

In Algeria, additionally, senior French officers developed in the shadow of the defeat in Indo-China a sense of mission so strong that it distorted their sense of proportion and led them in the end to jettison their oaths of allegiance. These officers convinced themselves that they belonged to the gallant and prescient category of the saviour with the sword, that they alone appreciated the full import of the communist threat to civilization, and that theirs was the honourable destiny of leading the resistance to the hosts of darkness and opening the eyes of woolly-minded sluggards to the dangers and responsibilities of the twentieth century. This apocalyptic determinism was accompanied by an almost equally passionate emotion which was local and practical instead of cosmic and visionary. The day-to-day business of administering large tracts of Algeria had become the responsibility of the army, and in the course of governing their localities officers had acquired a proficiency, knowledge and sympathy for the people in their care which, they rightly judged, would not easily be supplied by anybody else.

There was a foretaste of revolt at Sétif in 1945. Like a similar rising in Madagascar in the same year, this revolt was suppressed with brutality. France in 1945 was in no mood to do things by halves; for the killing of one hundred Europeans about six thousand Algerians were killed in retaliation. The nationalist leader Ferhat Abbas was arrested and the

French community was given the incentive and the excuse to arrogate to itself an authority which belonged rightly to the government in Paris. The weaknesses and divisions of the governments of the Fourth Republic allowed this authority to be exercised and enhanced until the return to power of General de Gaulle in 1958 put it to a test which it failed. There was, however, no effective nationalist threat to French rule during the decade between the Sétif rising and the revolt in the Aurès mountains at the end of 1954 which forced France to deploy half a million troops. During these years French ministers and governors-general tried to temper the repression of nationalism with economic advancement and democratic reform, but they failed to mollify the nationalists whose aim was not reform but independence and they antagonized the European community whose preoccupation with repression left little room for anything else.

This dilemma was illustrated vividly during the first stage of the revolt. The defeat of the Faure government in November 1955 had been followed by a general election and the installation of a minority government led by Guy Mollet. The new prime minister attempted to find a way to bring the fighting to an end. Yet when he went to Algiers he was pelted with garbage by Europeans, while approaches to leaders of the insurrectionary Front de Libération Nationale (FLN) were entirely unproductive. Mollet appointed General Catroux, a widely respected and liberally minded proconsul, to the governor-generalship, but Catroux resigned his office within a week and without leaving France. In May 1956 Mendès-France resigned from the Mollet government on the grounds that it was not doing enough, while the prime minister probably felt that if he ventured more boldly he would provoke a trial of strength between Paris and the Europeans in Algeria, which Paris would not win. In October 1956 the weakness of Paris was dramatically illustrated, and its position with regard to the FLN seriously damaged, when five FLN leaders, including Ahmed Ben Bella, were kidnapped when returning by air from a meeting with the sultan of Morocco. The aircraft in which they were travelling, which was registered in France and piloted by a Frenchman although operated by a non-French company, was diverted from its destination to Algiers without the knowledge of the government in Paris. During the next eighteen months political attitudes remained irreconcilable, the French army and the FLN secured positions in which neither could defeat the other, terrorism increased on both sides and spread to Paris and other cities in France, torture became a regular instrument of government and was seen to be, and any lingering intention of applying the new compromise constitution called the Algerian Statute of 1947 was finally abandoned. The impasse seemed to be complete, politically and militarily.

In this same period Morocco and Tunisia were drawn more closely into the conflict. The sultan of Morocco had been deeply offended by the kidnapping of Ben Bella who had been his guest but an hour or two earlier, and Bourguiba was angered when the French, irritated by the FLN's freedom to use Tunisian and Moroccan soil as an asylum where they could rest and refit without fear of attack, delivered an air attack on Sakiet in Tunisia in February 1958 and killed seventy-five people; Bourguiba threatened to stop supplies to French units in Tunisia. The Moroccan monarch and the Tunisian president had met at Rabat at the end of 1957 and had offered to mediate but the French, deceived into optimism by some recent successes in the field, had declined. Bourguiba persisted in his attempts to find a peaceful solution, not without some regard to the increasing links between the FLN and Egypt with which he was on bad terms as a result of an attempt on his life for which he blamed Nasser. There was talk of a Maghribian federation to include an independent Algeria, Morocco and Tunisia.

On 28 May 1958 the last civilian prime minister of the Fourth French Republic, Pierre Pflimlin, resigned, a victim of the Algerian war for which he and five predecessors could find no solution. On 13 May Algiers had rebelled against Paris. On 30 May, so it was planned, this government would seize power in Paris by a coup. Almost the whole of Corsica, the stepping stone, had accepted the rebel régime and half the commanders of the military regions in France were believed to be disloyal. Only one obstacle to success in the capital remained – a Frenchman of enormous prestige and outstanding political skill. On 1 June General de Gaulle was invested with full powers. On 4 June he flew to Algiers.

It is only possible to guess what policy de Gaulle had in June 1958. He may have had no fixed policy but he probably saw the inevitable end. What happened was that, by a mixture of authority and ambiguity, he imposed himself upon the situation and gradually acquired the power to impose a solution upon it. This took him nearly four years. By doing enough to retain the initiative, but not too much to reveal himself, he prevented potentially hostile groups from acting against him until it was too late. He began by patching up relations with Tunisia and Morocco, agreeing to withdraw French forces from both countries (except from the Tunisian naval base at Bizerta). He next moved from Algeria many senior officers who, even were they minded to object to their postings, could not gainsay the legitimacy of an order from *the* general. General Salan, a prime rallying point for disaffection and leader of the May *putsch*, was permitted to remain temporarily in his command, but he was relieved of his civilian functions which were now once more divorced from the supreme military command. After these preliminary moves, and with cautious deliberation, de Gaulle prepared his first major statement on the future status of Algeria and made his first bid for peace with the FLN. In September 1959 he formulated a choice (not unlike his offer to France's colonies in western and central Africa in 1958) between independence, integration with France and association with France, the choice to be made within four years from the end of hostilities, defined as any year in which fewer than 200 people had been killed in fighting or by terrorism. This pronouncement precipitated a second white revolt. On 24 January 1960 the European community showed that it would oppose even de Gaulle rather than accept the independence of Algeria. The revolt was a failure. The French government, acting energetically in Algeria and at home, showed in return how the authority and power of Paris had grown during the past eighteen months. But to Algerians de Gaulle's offer in 1959 was no more than a half-way house, a solution short of true independence and by now unacceptable.

Support for de Gaulle within France, more widespread and more positive in 1960 than in 1958, was partly due to a feeling that the war had gone on too long and partly to restiveness over the methods which were being used to wage it. Henri Alleg's book *La Question* focused attention on the use of torture by units of the French army. The trial of Alleg in 1960, followed by the disappearance and (as it was correctly surmised) murder of the French communist university lecturer Maurice Audin, the trial in 1961 of the Algerian girl Djamila Boupacha, protests by the Roman Catholic cardinals occupying French sees, and a manifesto signed by 121 leading intellectuals, all contributed to turn French opinion against the French community and the French army in Algeria. Towards the end of 1960 the leaders of the January revolt were themselves put on trial. But there was still one white rebellion to come. It came in April 1961. It was led by four generals and it lasted four days. Two of the four generals, Salan and Jouhaud, were subsequently sentenced to death *in absentia* and the other two, Challe and Zeller, who surrendered, to fifteen years' imprisonment – all sentences being eventually reduced. Out of the failure of this rebellion rose the Organisation de l'Armée Secrète (OAS) which resorted to terrorism and, by creating among the European population fears of reprisals by an independent Algerian government, provoked – as independence became inevitable – an exodus which deprived the country of much needed skills in administration, education and other public services.

De Gaulle's victory on the white front was not at first accompanied by any improvement in the nationalist quarter. In September 1959 the FLN had proclaimed a provisional Algerian government with Ferhat Abbas as prime minister and the imprisoned Ben Bella as his deputy, and Ferhat Abbas had left Tunisia for Cairo which was to be the seat of government for the time being. De Gaulle, true to his general disposition, temporized after the defeat of the white revolt of January 1960; and this inactivity caused the Algerians to turn for help to Moscow and Beijing. During 1960 moreover it became apparent that the movement of opinion among non-combatant Algerians was towards the FLN and its unequivocal demand for independence and not towards any middle position between the FLN and the Europeans (as de Gaulle may have hoped at the time of his pronouncement of September 1959). De Gaulle began therefore to move more purposefully towards negotiation with the FLN. A first secret encounter at Melun in June was a failure but after discussions between de Gaulle and Bourguiba, between FLN leaders and Georges Pompidou (still at this time a private banker) and between the FLN and Moroccans, Tunisians and Egyptians, a conference opened at Evian in May 1961. But suspicions and difficulties proved at this stage too great; the latter included the FLN's claim to be recognized as a government, the right of the imprisoned Ahmed Ben Bella to appear at the conference, guarantees for the French who might wish to remain in Algeria, continuing French rights in the naval base at Mers-el-Kebir, Saharan oil, and the conditions under which the proposed referendum on the status of Algeria would be held.

This conference failed to reach agreement but in July de Gaulle, in a televised speech, unequivocally accepted Algerian independence. In the same month, however, Franco–Tunisian relations suffered a sharp relapse when Bourguiba, concerned about Tunisian rights in the Sahara, demanded a complete French evacuation of Bizerta (effected in October 1963 after Tunisia had laid its complaints before the United Nations, thus accentuating de Gaulle's dislike of that organization); and the FLN adopted a more assertive line when Yusuf Ben Khedda succeeded the more moderate Ferhat Abbas at the head of the provisional Algerian government, supported Bourguiba over Bizerta, and delivered some forceful speeches at the conference of non-aligned states in Belgrade in September. Still in the same month the OAS made an unsuccessful attempt on de Gaulle's life, OAS activities increased throughout France as well as Algeria, and there were rumours of the proclamation of a dissident French republic under General Salan in northern Algeria. In October Ben Khedda proposed a new round of negotiations.

The second Evian conference took place in March 1962. On 18 March a cease-fire agreement was signed. The conference also agreed on the terms for holding a referendum and, on the assumption that the result would be a vote in favour of independence, further agreed (among other things) that French troops would be withdrawn progressively over three years except from Mers-el-Kebir which France was to be permitted to occupy for at least fifteen years; that France might continue its nuclear tests in the Sahara and retain its airfields there for five years; that France would continue its economic activities in the Saharan oilfields; and that French technical and financial aid to Algeria would continue undiminished for at least three years. On 3 July 1962 Algeria became an independent sovereign state for the first time in history. But its leaders did not hold together. In the ensuing quarrels Ben Bella, returning to the scene after six years' absence in prison, won power but alienated colleagues and followers by moving too fast, by trying to reorganize the FLN on communist lines and by trying to play a leading and radical part in African and Afro–Asian affairs to the neglect of urgent domestic problems. In June 1965 he was overthrown when he was on the point of ousting his minister of defence, Colonel Houari Boumédienne, who succeeded him at the head of the government. Ben Bella was imprisoned until 1978 and a fugitive until 1990.

Boumédienne's policies included, in domestic affairs: a new structure of government, state capitalism, the nationalization of natural resources, vigorous exploitation of oil and gas deposits, and industrialization; in external affairs: cautiously good relations with the USSR, continued collaboration with France, a Maghrib entente, and active association with the Arab states against Israel. He ruled through a Council of the Revolution which was created at this point; he himself was its chairman but its further membership was undisclosed. He created in 1968 regional authorities for economic and social affairs and made them elective a year later. Dissatisfaction – sharply evinced by an abortive army revolt led by Colonel Tahar Zbiri – focused on the failure to partition all big estates at once or give workers in industry as much control in management as many of them wanted, but by and large the new government became quickly and firmly established. In the foreign field Boumédienne concluded a series of agreements with France for the development and nationalization of mining and other industries, securing both French aid and Algerian control. He also secured the return to Algeria of 300 works of art removed by the French. In 1967 France evacuated its remaining land bases in Algeria and next year the naval base at Mers-el-Kebir. Boumédienne visited Moscow soon after his accession to power, broke off diplomatic relations with Britain over Rhodesia, declared war on Israel and sent troops to fight it, and broke off relations with the United States. When a first pan-African cultural festival was held in Algiers in 1969 it was attended by the president of the USSR and the French foreign minister as well as a concourse of African and other notables.

Algeria went briefly to war against Morocco in 1963 over their common frontier and as a result of the removal of French control over both of them. The conservative monarchy of Morocco and the socialist Algerian republic were not made to be friends, even in the absence of an ill-defined frontier, but talks between King Hassan and Boumédienne in 1968–70 led to a settlement which fixed the border and also provided for joint exploitation of iron ores at Tindouf on the basis that Tindouf itself was in Algeria. This agreement was distinctly favourable to Algeria. Morocco signed it in the hope of getting Algerian support against Mauritania in the disposal of Spanish Mauritania (Rio de Oro).

Boumédienne ran a one-party state for thirteen years during which the FLN dissolved into cliques, became corrupt and lost much of its authority to the army. He was removed in 1978 and replaced after an interval by Colonel Chadli Benjedid who failed to stem corruption, remedy poverty or dispel dissatisfaction. Riots in 1988 and open army intervention in government ensued as an ageing élite struggled to govern a country where three-quarters of the population were under 25 and nearly half under 14. In 1989 a new constitution described Algeria as a multi-party democracy but in reality it was becoming a splintered country in which parties old and new and the army were decomposing and losing their authority and legitimacy. The Front Islamique du Salut (FIS), formed in 1989, scored successes in local elections in 1990 and in effect won a general election the next year when the FLN won only sixteen seats in the first round and cancelled the second. Chadli was forced to resign and an improvised Haut Conseil d'Etat persuaded Muhammad Boudiaf, a hero of the war of independence, to return as president from Morocco where he had taken refuge after being condemned to death by Ben Bella in 1964. Boudiaf was a devout Muslim and a champion of non-religious, non-military government but he was assassinated within a few months of his inauguration and the army installed General Lamine Zeroual as president in his place. Boudiaf and, after him, Zeroual were in a quandary between trying to crush or split the FIS which, while Islamic in its insistence on more rigorous Islamic rules and behaviour, did not aim to establish an Islamic theocratic state on Sudanese or Iranian lines. The FIS was legally dissolved and thousands of its adherents were arrested but Zeroual was inclined to side with the conciliators rather than the eradicators so far as he felt able to do so in the face of hawkish army colleagues and rivals. In 1994 he held talks with a number of party leaders, excluding the FIS, but these parties were unwilling to agree measures for a return to democratic order without the co-operation of the FIS and later in the year a conference convened in Rome was attended by the FIS and consequently boycotted by the government. It produced a National Contract which the FIS accepted and the government rejected. In the next year and with the FIS banned, Zeroual, civilianized, was re-elected. His victory lay less in his share of the vote than in holding an election at all. The country was in a state of war or anarchy, hundreds of people were being killed every day and the business community, the IMF and Algeria's foreign creditors were engulfed in gloom.

Between Morocco and *Mauritania* there was a common bond so long as Spain retained its African possessions but an underlying conflict over the eventual fate of one of these: Rio de Oro. Spain had ceded Spanish Morocco to Morocco in 1956, retaining however the towns of Melilla and Ceuta and three other small enclaves whose population was mainly Spanish. In 1957 Spain sent troops to Ifni in south-west Morocco but ceded it to Morocco in 1958 as far south as latitude 27° 40'. There remained the Canary Islands, which stayed part of Spain, and Rio de Oro, claimed by both Morocco and Mauritania. These rival claimants were united in opposition to the Spanish presence in a territory where the world's richest phosphate deposits had been found in 1945. Morocco, heavily dependent on the export and therefore the world price of phosphates, would have liked to acquire the whole of Rio de Oro but was prepared to concede a substantial part of it to Mauritania rather than see the creation of a new state, possibly with a Hispanophile monarch and extensive Spanish aid and tutelage. Mauritania wished to thwart the creation of a Greater Morocco. Spain, having announced its departure in 1974, wished for a referendum which would produce a majority for a new independent state. Morocco fought off this prospect by getting the UN to ask the International Court of Justice who had owned Rio de Oro before the Spanish got there. The Court answered in 1975 that the status of Rio de Oro must be determined on the basis of self-determination and not by reference to past history.

While the Court was still deliberating Morocco reached a preliminary agreement with Mauritania over the exploitation of the phosphates and in 1975 King Hassan personally led 350,000 Moroccans a few miles into Rio de Oro in a demonstration designed to force Spain to negotiate with Morocco and Mauritania and not hand over to any third authority. Spain agreed at the end of the year to transfer Rio de Oro to the two African claimants jointly and they, five months later, partitioned it: two-thirds to Morocco and one-third to Mauritania.

Algeria looked askance at a settlement which had been made without its intervention and barred its access to the Atlantic. It resolved to support the Polisario (Popular Front for the Liberation of Saguiet el-Hamra and Rio de Oro) which had been in existence since 1973 and claimed that Rio de Oro should belong neither to Morocco nor Mauritania. In February 1976 a *Sahrawi Arab Democratic Republic* was proclaimed (in Libya). Morocco and Mauritania joined forces against the Polisario but the latter found the effort too great for its penurious economy and in 1978 the army removed President Ould Daddah and gave up the fight. The war went on because the Polisario's tactics and resources, effective against Mauritania, were not effective against Morocco. Whereas the Polisario was able to inflict on Mauritania military defeats whose economic consequences Mauritania could not take, Morocco was not so easily disposed of. Equally, however, Morocco was not able to drive the Polisario from the field.

The conflict had wider dimensions. The Polisario had Algerian support and Russian arms. Morocco received aid from France and the United States: King Hassan was in American eyes a bastion against communism in north-west Africa. In the period after the fall of the shah of Iran, when Washington's loyalty to its friends was being questioned, the United States judged it expedient to promise the king increased military and other aid, if only to allay this anti-American current. (The promises, once given, were only tardily executed.) The king further courted the west by his readiness to despatch modest forces to troubled areas such as the Shaba province of Zaïre. Algeria on the other hand, although suspiciously left-wing, had the makings of a more solid ally. It was to Morocco what India was to Pakistan in western geopolitical calculation: strategically and economically a worthwhile friend but politically dubious. The death of Boumédienne in 1979 marginally shifted the equation, for Boumédienne's successor was the more cautious Benjedid Chadli who was expected to play Sadat to Boumédienne's Nasser.

Chadli consolidated, in a lower key, his personal position as Boumédienne's successor and sketched a vision of a federated Greater Maghrib. The first instalment was a treaty with Tunisia in 1983 to which Mauritania later adhered. With Morocco, however, Algeria's relations remained difficult on account of the two countries' difference over the western Sahara and were further strained when Colonel Qadafi surprisingly visited Rabat in 1983 and bartered away his support for the Polisario in exchange for a Moroccan promise to keep out of the imbroglio in Chad. But Morocco's attempt to detach Libya from the Polisario by placating Qadafi displeased the United States and in 1986 King Hassan denounced his treaty with Libya in order to repair his relations with Washington.

In the same year Hassan celebrated twenty-five years on the throne of Morocco. He combined ostentatious wealth and unconcealed autocracy with political astuteness. He commanded the trust of his army and exploited the nationalism which animated Moroccans of all parties and classes. All patronage, political and religious, was in his hands. Political parties, which had been allowed from 1977 to take part in elections, accepted nominal democracy at the price of good behaviour. Muslim fundamentalism was curtailed in spite of not inconsiderable popular support. Left-wing parties, when not destroyed, were penetrated by the secret police. His policy in the Sahara was not questioned, but its success encouraged opponents of the royal autocracy whose voices had been muted out of patriotism. In 1993 he conspicuously displayed his confidence by opening a huge new mosque and permitting the first elections for nine years (for two-thirds of the seats in parliament). Opposition parties made some gains but not enough to disturb the king.

In the Sahara Morocco was chiefly concerned to keep the Polisario away from its Atlantic provinces. Although Hassan rejected a plea by the OAU to talk with the Polisario he appeared willing to back a referendum which, it was assumed, would give him those disrupted areas which mattered to him at the cost of letting others become an independent Sahrawi republic. By the mid-1970s the war was a stalemate in Morocco's favour, for whereas the Polisario made no significant gains Morocco not only contained it but was slowly pushing it back by the inglorious but effective strategy of building walls of sand stretching along 2,000 km of desert. This moving line, punctuated by forts 5 km apart and manned by 100,000 troops, created enclaves which were then peopled by Moroccan immigrants. But the war was ruinously expensive and in 1987 Hassan agreed to a meeting with Chadli in the hope of stemming Algerian aid to the Polisario. The Polisario was allowed to retain its quasi-government in Algiers but lost Algerian funds and by the end of the decade its leaders were split on whether to continue the struggle and, if so, how. By 1992 it seemed a spent force. A UN team of 2,000 observers and peace-keepers (MINURSO) was assembled to supervise a referendum offering a choice between independence and incorporation in Morocco but powerful, if unobtrusive, American support and Algeria's internal turmoil enabled Hassan to harden his face against any form of Sahrawi autonomy and stifle the UN presence and plans for a referendum.

Morocco was simultaneously freed of its anxieties on the Mauritanian front when Mauritania, although it adhered to the Algerian–Tunisian treaty of 1983 and recognized the Sahrawi African Republic, was beset with troubles at home and to the south. Economic stringencies, including an encroaching desert, exacerbated class and ethnic tensions. Mauritania comprised paler and darker (Beydane and Haratine) Moors and black Africans. The Beydane constituted a ruling caste but the Haratine, descended from slaves, were twice as numerous while the black Africans – about a third of the total – were akin to the Fulani, Wolof and other peoples across the border in Senegal. Under the brutal rule of Colonel Masouiya Ould Taya, who came to power by force in 1984, thousands of Mauritanians were killed in a series of massacres in 1989–92 and tens of thousands fled into Senegal. The arrival of these (non-Muslim) refugees provoked a counter-expulsion of some 70,000 and created for a while a threat of war. (There were about ten Mauritanians in Senegal for every Senegalois in Mauritania.) Ould Taya espoused a multiparty system and used it in a string of elections in 1992 to reinforce his autocratic rule as the various parties quarrelled with each other.

In 1988 a fresh attempt was made by the UN secretary general, Javier Perez de Cuellar, to resolve the Saharan problem. Morocco and the Polisario accepted a UN plan for a cease-fire, to be followed by a referendum to choose between independence and incorporation with Morocco. This agreement was completely without substance. Within a few weeks the fighting had started again and within a year Morocco concluded with Algeria, Tunisia, Libya and Mauritania a treaty for an Arab Maghrib Union which left the Polisario out in the cold. The new union was essentially geographical and not the ideological bloc which Qadafi had from time to time tried to form. Its brittle cohesion was quickly tested when Morocco (like Egypt) responded to the American call for the despatch of troops to Saudi Arabia to confront Iraq's seizure of Kuwait, while Algeria and Tunisia were more critical of the United States and their peoples more openly sympathetic to Saddam Hussein's denunciations of the United States and Israel.

Popular emotions were becoming more open and influential in both countries. In Tunisia, where Zayn al-Abdin Ben Ali succeeded the venerable but senile Bourguiba in 1987, political parties were allowed to emerge, although no specifically religious party might be formed: the government feared Islamic fundamentalists. In 1994 it won all the seats. In Algeria riots in 1988 preceded the formulation of constitutional changes which were put to the people in a referendum and adopted. Political opposition was legalized within limits and civil rights, including the right to strike, were recognized. As in Tunisia and for the same reasons no party devoted to a particular ideology or limited to a certain race or religion might be licensed, although this provision was in practice overlooked with the licensing of the Front Islamique and the Berber Rassemblement. Under the new constitution a party winning half the votes in a (multimember) constituency took all the seats, a party winning less than 10 per cent of the votes got none.

LIBYA AND CHAD

The postwar fate of Italy's North African colonies was to be resolved by the four principal victorious powers by September 1948, failing which the problem would be transferred to the United Nations. The British and Italian governments devised a plan, called after their foreign ministers the Bevin–Sforza plan, whereby an independent Libyan state would come into existence at the end of ten years and during the interval its existing three parts would be under British or Italian tutelage: the Fezzan (which was coveted by the French as an addition to Tunisia) under Italian trusteeship, Tripolitania in the middle under British administration for two years and Italian trusteeship for eight, and Cyrenaica further east under British trusteeship for the whole ten years. This plan was objectionable to the Arabs and to the Russians, who proposed a five-year UN trusteeship (they also proposed UN trusteeships of five and ten years for Eritrea and Italian Somaliland). The political committee of the Assembly accepted the Bevin–Sforza plan but the Assembly itself just failed to endorse it and there followed a period of negotiation and investigation by a UN commissioner (Dr Adrian Pelt), as a result of which the whole of Italy's North African empire was converted into a tripartite constitutional monarchy under the emir of Cyrenaica, Muhammad Idris as-Sanusi, as King Idris of Libya. The new state came into being on the first day of 1952 and shortly afterwards entered into an agreement with Britain which secured essential economic aid and arms to Libya and military rights to Britain in the small post of El Adem. The Americans subsequently established a larger base at Wheelus Field. In 1969 the king was ousted by army officers, who were disgruntled by corruption in high places. Their leader was Colonel Qadafi who disdained the conventions by interfering wherever possible to help (notably the Palestinians) or harry (oil companies, communists, Israel and established governments, especially monarchies). Within the Arab world Qadafi projected a series of unions or federations which were insubstantial and increasingly derided: with Egypt and the Sudan in 1969, extended to Syria in 1970, but never effective; with Egypt and Syria again in 1971, equally without consequence; with Egypt in 1972 upon the eviction of the Russians from Egypt but followed in 1973 by complete diplomatic rupture; with Tunisia in 1974, the most short-lived of these attempts at fusion since it was denounced by Bourguiba two days after it was proclaimed; with Syria in 1980; and with Chad at the beginning of 1981. In the last case union appeared to be a euphemism for at least partial annexation. Qadafi's contempt for conventions led him to organize the killing of political opponents who had fled to foreign countries, even using his diplomatic missions for this purpose.

Within Africa Qadafi not only played a leading part in the troubled affairs of *Chad* but was suspected of much wider ambitions. Huge, arid and underpopulated (about 4 million inhabitants), athwart the dividing line between Arab and Bantu-speaking Africa, with six international frontiers and presumed wealth in uranium, gold, oil and other precious commodities, Chad was endemically unstable and a standing temptation. Its first president, François Tombalbaye, a representative of the educated southern élite, was unsympathetic towards the Muslims of the north whose opposition to his government he insisted on regarding as banditry. It was also a threat to the tourist traffic which he hoped to encourage. By 1966 he was challenged by FROLINAT, a northern liberation movement with headquarters in Libya. He turned for help to France with which Chad, like other former French colonies, had a defence agreement. (France had some 6,000 troops in West and Central Africa in these years, distributed over thirteen states.) This help began with supplies but expanded in 1968–70 to direct intervention with paratroops and aircraft which secured Tombalbaye's position until 1975 when he was assassinated. His successor Felix Malloum gradually lost the war against FROLINAT. Meanwhile the leaders of FROLINAT's several armies fell out. The French withdrew, Qadafi backed Goukouni Oueddei who was willing to cede – or to promise to cede – the northern Ouzou strip and its supposed uranium deposits to Libya. His rival Hissen Habre formed a brief alliance with Malloum before turning on him and seizing power in the capital Ndjamena in 1979. Libya, the Sudan and Niger engineered a superficial accord between Oueddei and Habre but in 1980 Libya, driving its Russian tanks across the desert with surprising speed and efficiency, enabled Oueddei to put Habre to flight. In 1981 Qadafi and Oueddei announced a union of their two countries. Qadafi then agreed to pull his troops out in return for a French undertaking to do likewise, but when the French withdrew the Libyans nevertheless remained in occupation of the Ouzou strip. Elsewhere Habre regained and consolidated his position but his impoverished government was weakened by falls in the price of cotton which fuelled popular discontent. A peace conference at Brazzaville in 1984 was unsuccessful but thereafter Habre's fortunes revived, particularly from 1986 when Libya, having failed in a fresh offensive, abandoned Goukouni and Goukouni himself split with one of Habre's other opponents, Acheikh Ibn Ommar. In 1987 Habre's forces won a convincing victory in the north although without recovering the Ouzou strip – thus vindicating President Mitterrand's policy of giving aid to Habre but deprecating large-scale hostilities against Qadafi. After this victory many of Habre's enemies either came to terms with him or fled to the Sudan, but in 1989 a coup against him (which failed) demonstrated the turmoil beneath the surface. The plotters, led by an old and close associate Idriss Deby, resented favours dispensed to erstwhile enemies, personal or tribal, in order to seal this reconciliation and soften some of the asperities of Habre's harsh rule. They had help from Libya and in 1990 they tried again and were successful. France, which had intervened on several occasions to protect Habre, refused to do so again although it had a force of about 1,300 in the country. American patronage and his Israeli guards were not enough to save Habre, who fled to Cameroun. Deby, no democrat, disappointed the French who began after a few years to have second thoughts. Chad's continuing troubles were used by the United States, Eygpt and Iraq –

strange conjunction – to make trouble for Qadafi in Libya.

When Reagan took office in 1981 the new administration in Washington, understandably irritated but extravagantly obsessed by Libya's often outrageous behaviour, sent warships into the Gulf of Sirte and shot down two Libyan aircraft. The United States also imposed economic sanctions and in 1983 and 1984 sent AWACS intelligence aircraft to Egypt to deter Qadafi from adventures in that direction: Qadafi had threatened to march into Egypt and he had a hand in an abortive plot to overthrow President Nimeiry by dropping bombs on Khartoum as a prelude to a coup by Sudanese officers. In 1986 the Americans delivered a massive air and naval attack on Tripoli and Benghazi (see also Chapter 4) which failed to kill Qadafi and probably strengthened his position in his own country. In a smaller attack in 1989 two Libyan aircraft were destroyed after the United States had accused Libya of building a plant for the manufacture of poison gases.

Africans were alarmed by Qadafi's activities and by his visions of a great Islamic Saharan empire which would point a menacing finger at the wealth of Zaïre, overawe the Sudan and Uganda, and cut a swathe westward from Chad through Mauritania, Mali and Niger to the Atlantic in Senegal, the Guinea and Ivory Coasts and the Bight of Benin. The murder of President Luiz Cabral of Guinea-Bissau, the overthrow of President Sangoule Lamizana of Upper Volta, a coup which failed against President Seyni Kountche of Niger (who endured almost annual coups after seizing power in 1974) – all these events in 1980 were ascribed to Libyan machination. So too was an abortive coup in the Gambia which was suppressed by prompt Senegalese intervention. The remoter Central African, Cameroon and Gabon Republics also felt a Libyan current blowing their way and privately wished that France might send troops back to Chad. The Gambia, Senegal, Ghana and Gabon severed relations with Libya and Nigeria threatened to follow suit.

Yet Libya was in no position to create an empire. Its population numbered only 3 to 4 million, its army about 40,000. During the crisis in Chad in 1980 Egypt mustered more than twice that force on its border with Libya alone and breached the frontier with impunity. What Qadafi had was an alarming political style and oil. He used the oil partly to twist the arms of customers to whom he chose to sell at half price or less, and partly to buy Russian arms. But in the 1980s Libya's oil production fell to a quarter of its peak in the previous decade, and although Qadafi turned Libya into the largest *place d'armes* of Russian origin outside the USSR and central and eastern Europe much of this weaponry was either unsuited to his foreign ventures or left inadequately tended in places where it quickly deteriorated. His interventions in Chad were inconsequent. After declaring himself the champion of the

Muslims he helped non-Muslims to power. The uranium which he was supposed to covet had no proven existence. (The Ouzou strip had been assigned to Chad by a Franco–British agreement of 1899. An unratified Franco–Italian agreement of 1935 assigned it to Libya. Tombalbaye allegedly sold it to Libya in 1973 by an agreement of which no record has been produced. In 1989 the OAU persuaded both countries to refer their dispute to the International Court of Justice.)

Although hostile to Islamic fundamentalism Qadafi supported Iran against Iraq. During the Gulf War of 1991 he denounced both Saddam Hussein's invasion of Kuwait and the intervention of the United States and its western allies against him. After the war he began to repair his relations with Egypt where President Mubarak was receptive in the hope of breaking earlier links between Libya and Sudan (which sheltered Egyptian fundamentalists) and of getting an interest in Libyan oil.

In 1988 an American airliner was destroyed in flight over Scotland. The dead included many Americans. There were grounds for suspecting that this crime was the work of Iranians retaliating against the destruction of an Iranian airliner over the Persian Gulf by American fire. But the American, British and French governments suspected two Libyans, whom they identified, and demanded their extradition for trial in the United States or Scotland. The relevant international convention – the Montreal Convention of 1971 – provided that in the absence of applicable extradition treaties there was no obligation on the suspects' country to hand them over but the three governments used their weight in the UN to procure in 1991 a resolution of the Security Council demanding that Libya should do so. Two years later the same three states secured the imposition of economic sanctions against Libya when it did not comply. The consequent economic distress in Libya strengthened Islamic opposition to Qadafi's secular régime.

Europeans were interested in Africa long before they occupied it. In the century after the death of the Prophet Mahomet North Africa mounted the greatest threat to European Christendom which it has ever faced, and although that threat was parried by the Frank Charles Martel and by the Byzantine emperor Leo the Isaurian, Spain remained for centuries partly under an alien rule buttressed on occasions by fresh support from Berber Africa. When the Christians finally drove the Muslims out of the Iberian peninsula, their momentum carried them on into Africa, and Spanish and Portuguese adventurers, still finding it too difficult to turn east, explored and circumnavigated Africa and made Cape Horn a station on a new route to the east. In more modern times Africa became a place where Europeans got things: slaves for plantations in the west, food for industrialized countries whose peoples were leaving the land for the factory, precious minerals like gold and copper and diamonds and uranium. At first only the coastal areas were exploited, but later the rumoured wealth of the interior tempted organized expeditions to follow in the steps of adventurers and missionaries. Although checked at first by the unexpected strength of African kingdoms, white power eventually prevailed – especially when the motives of curiosity and enrichment were reinforced by inter-white competition. So, in a final phase of European penetration, Africa was partitioned by official emissaries, part soldiers, part administrators, making territorial claims and fighting for them because traders demanded protection and each European state was afraid that others would take what it did not annex for itself. Areas, large but vaguely defined, were assigned to adventurers backed by, or acting in the name of, European states which had been taught by centuries of history to think first and foremost in territorial terms.

In the last quarter of the nineteenth century it became apparent that Europe had acquired a great part of Africa in a very short space of time, and that there was some danger of fights between the European states concerned as a result of the untidy and uneven distribution of the spoils. The Europeans proceeded to settle these matters reasonably amicably among themselves. There were many wars in Africa during the colonial

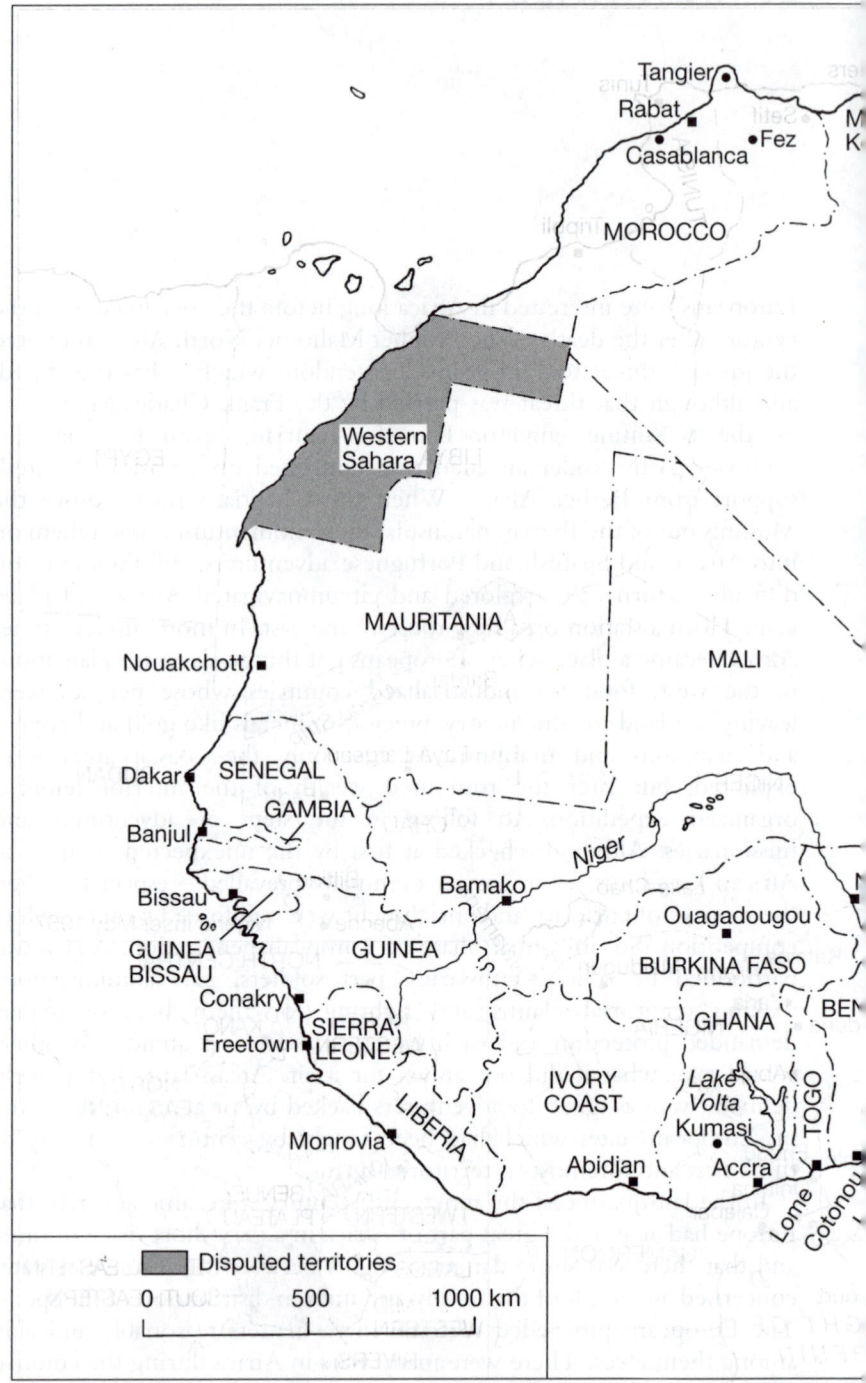

20.1 North–west Africa (with inset of Nigeria at the time of the Biafra W

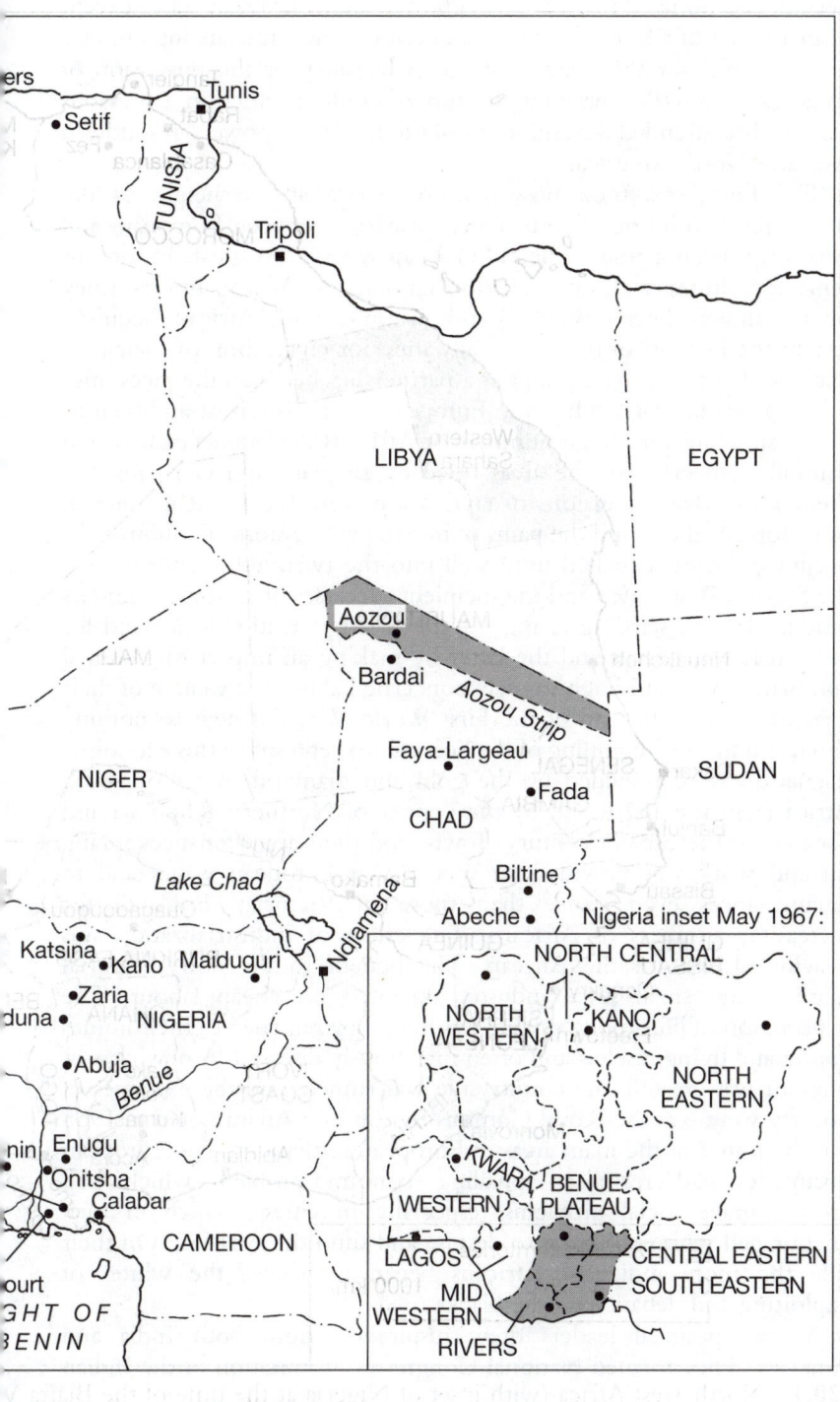

Setif

Tunis

TUNISIA

Tripoli

LIBYA

EGYPT

Aozou

Bardai

Aozou Strip

Faya-Largeau

Fada

NIGER

SUDAN

CHAD

Biltine

Lake Chad

Abeche

Nigeria inset May 1967:

Katsina

Kano   Maiduguri

Zaria

Ndjamena

duna

NIGERIA

NORTH CENTRAL

NORTH
WESTERN

KANO

NORTH
EASTERN

Abuja

Benue

KWARA

BENUE
PLATEAU

nin   Enugu

Onitsha

Calabar

WESTERN

CENTRAL EASTERN

CAMEROON

LAGOS

SOUTH EASTERN

ourt

MID
WESTERN

GHT OF
BENIN

RIVERS

period, but none of them was fought between European states except as an adjunct of a European war, and even the most menacing disputes (for example, for the control of the Nile valley or the possession of Angola) were settled without the sort of conflict which in a previous century had attended the ambitions of the European powers in southern Asia and North America.

The Europeans took possession of Africa at the height of the Industrial Revolution. The technical disparity between Europeans and Africans was enormous. The cultural gap was no less great. Europeans remained almost without exception ignorant of African history (they often assumed there was none) and customs, while Africans acquired few of the benefits of the technically superior civilization of their new masters. There was no attempt at a partnership between the races until nearly a century later when the Europeans were in retreat and seen to be. Meanwhile very large numbers of Africans died unnecessarily and painfully, especially in the areas ruled by Belgians and Germans; the French and British versions of civilization were less lethal in spite of wars, forced labour and the pains of mineral concessions. Economically, occupied Africa stagnated until well into the twentieth century when the Second World War and the incipient successes of nationalist leaders produced some startling changes – the first by creating a demand for African raw materials and the latter by making an impact on colonial authorities who, although vaguely concerned about the welfare of their territories from the end of the First World War, did next to nothing about it until the beginning of the Second. Exceptions to this economic stagnation were provided by the gold and diamond mines of South Africa from the 1870s and by the copper of Northern Rhodesia and Katanga in the present century. Towns and their appurtenances sprang up and work was provided (at special non-European wages and in circumstances even worse than those of Victorian England) for increasing numbers of Africans. The effect of industrialization was heavily adverse for the African. The benefit to the new industrial workers was small since industry operated on cheap labour. The expectation of life among miners was low; they returned to their homes young and dying, useless and often infectiously diseased. A new class of migrant was created, the countryside was ruined, and the extremes of poverty were – as the Royal Commission on East Africa noted in 1955 – to be found in the main areas of European settlement. The European occupation had created an appalling economic problem, which itself created grave social problems (especially in cities), which in turn encouraged white observers to despise and shun the Africans. On their side the more indignant Africans began to accuse the whites of exploiting and debauching the blacks.

African political leaders drew inspiration from both India and America. They formed National Congresses in imitation of the Indian

National Congress; many of them were attracted by Gandhian ideas of passive resistance; and the independence of India in 1947 had an effect in Africa which had not been foreseen by the colonial powers. From the American continent, and notably the Caribbean, Africans gained confidence and dignity and a habit of meeting together. A first pan-African conference was held in 1900, followed by a second in Paris during the peace conference of 1915. These first meetings were dominated by West Indian blacks but the sixth, held at Manchester at the end of the Second World War, was attended by the principal African leaders – Kenyatta, Nkrumah, Akintola, Nyerere, Banda. It voiced demands for independence which would have seemed totally unreal five years earlier. A mere ten years later West Africa was leading the way to independence from European rule.

In Accra, the capital of the British colony of the Gold Coast, there were riots in 1948 started by ex-servicemen in protest against high prices. The colonial authorities were taken by surprise, and a commission of inquiry produced a radical report which said in effect that the colony's newly proposed constitution was out of date before being introduced: to give Africans a majority of the seats in the legislative council was no longer enough. A new commission of Africans, with an African judge as chairman, was appointed to devise a new constitution. These developments coincided with the appearance of a new nationalist leader, Kwame Nkrumah who, aged probably 38, had returned to his country from the United States in 1947 and was determined to press for independence more energetically than older leaders such as J. B. Danquah. Nkrumah demanded immediate self-government. He seceded from Danquah's party, formed the Convention People's Party, earned the distinction of a prison sentence and won electoral victories in 1951, 1954 and 1956. After the first of these the British governor, Sir Charles Arden-Clarke, summoned him from prison and made him chief minister, thereby adopting the view that the colonial problem was not how to prolong colonialism, but how to make the best use of what time was left, to swim (not drift) with the tide and not uselessly to fight against it.

The British method, in the Gold Coast and elsewhere, was gradually to increase the elective and African elements in legislative and executive councils. Legislative councils in British territories progressed from assemblies dominated by appointed officials to assemblies containing a majority of elected members, and at the same time the governor's executive council was similarly transformed by the introduction into it of the leaders of the main party in the legislature. The governor retained at first extensive reserve powers but at a later stage this association between the colonial authority and nationalist movements was carried a stage further by converting the nationalist leader from the governor's chief minister into a prime minister of a self-governing territory. At this point the territory was on the verge of independence, and when independence was granted the governor disappeared. If the territory decided to remain in the Commonwealth, it might accept the British queen as its titular head with a governor-general as her representative on the spot, or it might become an independent republic within the Commonwealth but without any direct link with the British Crown. The new state would in any event establish diplomatic relations with the British government (as distinct from the British Crown) through representatives called high commissioners if the Commonwealth link were maintained, or ambassadors if it were not.

Given this method, the main problem was to regulate the pace. This was inevitably an uneven exercise in compromise between an irresistible flood and a removable power. The Gold Coast became self-governing in 1955, independent under the ancient name of Ghana in March 1957 and a republic in 1960. It led the way for Africans into the Commonwealth. Nkrumah, influenced by India's example, calculated also that the Commonwealth association would be a help to new states coming naked into international society.

The second West African state to become independent was French *Guinea* (1958). The French had been slower than the Spaniards or the Portuguese to enter Africa and less successful at first than the British or the Dutch. When in the seventeenth century they followed the new fashion their motive was emulation rather than any large expectation of gain. In the eighteenth century, however, they were participating in the slave trade on a big scale, extracting at one time as many as 100,000 slaves in a year, and after the abolition of the trade in 1815 and of slavery itself in 1848 (fourteen years after a similar prohibition in the British territories) French merchants turned from human goods to ivory and rubber and French explorers and missionaries began to penetrate inland and so stimulated ambitious dreams of a vast and compact empire stretching from the west coast to the Nile and from Morocco to the equator. In the event France acquired 3 million sq. km by the end of the century, something between a fifth and a quarter of the whole continent. Almost the whole of the great African bulge came under French rule; it was formed into the governor-generalship of French West Africa in 1895 and comprised eventually eight separate colonies and, after 1945, the two trust territories of Togo and Cameroon. Four equatorial territories were similarly federated in 1910. Only in the valleys of the Niger and the Nile were the French worsted by the British. In 1900 French advances from Morocco in the north, from Senegal in the west and from the Congo in the south, where the great explorer de Brazza left his name, converged at Lake Chad. France also acquired in the First World War the enormous island of Madagascar;

most of the German Cameroons, leaving a small slice which passed to the British who administered it with the eastern province of Nigeria; and half of Togoland, which was also originally German and was divided in 1919 between France and Britain as the owners of the adjacent territories of Dahomey and the Gold Coast.

After the fall of France in 1940 French West Africa opted for Vichy until the invasion of north-west Africa by the Americans and the British in 1942. An attempt by the Free French and the British to seize Dakar, the capital of Senegal, in 1940 failed. In Equatorial Africa, however, the governor-general Félix Eboué, a native of the Caribbean, took the Gaullist side in August 1940 and introduced a number of imaginative social and political reforms. At a conference at Brazzaville in 1944 the Africans were promised more participation in mixed Franco–African councils, more decentralization and a wider franchise.

The first French constitution of 1946 was liberal from the colonial point of view but it was rejected by the French people and the second constitution of that year was less far-reaching. It created the French Union, comprising the French Republic, Associated States and Associated Territories (there were also Overseas Departments and Overseas Territories, which were part of the French Republic). All the West and Equatorial African territories became Associated Territories with representatives in the French National Assembly and the Council of the Republic. Representative assemblies were also established in each territory with a grand council at federal level. From 1947 to 1954 France had a succession of predominantly conservative governments, but in 1956 the Mollet government, which included Gaston Defferre as minister for overseas territories, introduced a *loi cadre*, which was intended to lead to a substantial degree of internal autonomy by way of universal franchise, elected councils and the Africanization of public services. The *loi cadre* marked the abandonment of the policy of integration or assimilation in favour of a freer federation in which the African territories, while still associated with France, would increasingly order their own affairs and develop their own services and personalities. The *loi cadre* was elaborated by a series of decrees in 1957 which endowed the twelve West and Equatorial African territories with assemblies elected on a common roll and with councils of government elected by the assemblies. Considerable powers were reserved to the governors, the West or Equatorial high commissioners, or the metropolitan government, but nationalist demands had secured the elimination of special votes for whites and every constituency in every territory had a majority of black voters. The Rassemblement Démocratique Africain, the principal nationalist party which operated throughout French West Africa, won the ensuing elections in Guinea, Soudan, Ivory Coast and Upper Volta.

These political changes were, however, offset on the economic side where French policy remained integrationist and, in return for French aid and guaranteed markets in France, proposed to make French Africa a highly protected area serving French metropolitan economic interests. In the following years Africans became increasingly restless about economic policy in the franc zone. They claimed that a system which allowed free trade in the zone but erected barriers round it profited the stronger members rather than the weaker, impeded economic growth in the African territories, and prevented them from diversifying their economies. In addition the *loi cadre* accentuated differences among African leaders. Some, of whom Felix Houphouet-Boigny of the Ivory Coast was representative, seemed well satisfied with the French proposals, while others, among whom Sekou Touré of Guinea was the most eminent, suspected the *loi cadre* of being not a stage to sovereign independence but a device to postpone it indefinitely.

When de Gaulle returned to power in 1958 he faced the West Africans and Equatorial Africans with a choice: either autonomy within a *communauté* in which France would clearly retain control of the economic levers, or independence – which was a polite term for expulsion into a francless world. All but Guinea made the first choice. Guinea became an independent sovereign state in October 1958, humiliatingly discarded by its French mentors and forced to look to communist powers for the wherewithal to sustain its independence. The association was not a happy one and Touré was forced to change course (see Chapter 24) before he died in the United States in 1984 after a heart attack, his position already undermined by discontent which came to a head in a coup which put Colonel Lansana Conte in his place to cope with a legacy of repression, corruption and bankruptcy. Conte's position was strengthened by the failure of a counter-coup but he faced persistent popular grievances grounded in tribal rivalries and the standard IMF medicine administered to impecunious countries and swallowed with revulsion by its chief victims, the poor. Ethnic tensions, serious trade deficits augmented by falls in the price of bauxite, loss of IMF support and the decline and collapse of basic public services so weakened Conte that he was obliged to concede elections in 1993 which he won by the narrowest of margins.

France's other West African colonies, which had chosen autonomy within the projected *communauté*, did not long remain content with their status. In 1959 Senegal, Soudan, Upper Volta and Dahomey decided to federate under the name of Mali and ask for independence. The two latter changed their minds under French pressure and although Senegal and Soudan persisted the resulting federation lasted only a couple of months, after which Senegal too withdrew leaving Soudan with the new name of Mali; it moved into the Guinea–Ghana orbit.

The idea of a Mali federation had been unpopular not only in Paris but also among French African states further south, especially in the Ivory Coast where Felix Houphouet-Boigny riposted by forming with Niger and the detached Upper Volta and Dahomey an association called the Council of the Entente. The Entente then asked for independence by the beginning of 1960, which was conceded. Houphouet-Boigny wished to fortify the Entente by introducing dual nationality in it, but it was weakened by coups in Upper Volta and Dahomey in 1961 and by persistent suspiciousness in these and other states (including Togo which joined in 1966) of Houphouet-Boigny who was regarded as unduly hostile to Nkrumah and too partial to Tshombe in the Belgian Congo (see the next chapter). The net result of these centripetal and centrifugal forces was to solidify the several French colonies as independent sovereign states.

The independence of the Entente countries in 1960 deprived the *communauté* of any meaning although in theory it continued to exist as an association comprising France, Madagascar and the Equatorial states. These latter – Gabon, Chad, Ubangi-Shari and French Congo (the last two renamed the Central African Republic and Congo-Brazzaville) – produced a plan for a federation which was however abortive; they too became independent as separate sovereign states in 1960. In 1963 the Senegal River states (Senegal, Mali, Guinea and Mauritania) planned an association which was however frustrated by political differences: the notion was revived in the late 1960s. The enduring association was the much bigger and looser OCAM (see Chapter 5).

The huge British territory of Nigeria also became independent in 1960, followed in 1961 by Sierra Leone, the British colony wedged between Guinea and Liberia, and in 1965 by the Gambia, the northernmost and last remaining British territory in West Africa, a narrow enclave within the embrace of Senegal. Of the trust territories in West Africa the French Cameroons and French Togoland became independent republics, while the British Cameroons and British Togoland were attached to Nigeria and Ghana respectively.

Spain and Portugal were slower to divest themselves. The Spanish island of Fernando Po in the Bight of Biafra – renamed Macias Nguema island – together with Rio Muni, an enclave in Gabon, became the independent state of *Equatorial Guinea* in 1968 and shortly thereafter a family autocracy characterized by economic calamity and mass murder. President Obiang Nguema, who replaced Francisco Macias Nguema in 1979 and relied for his protection on his Moroccan guards, was not much of an improvement on his predecessor but was propelled in the 1990s towards a multiparty order. The islands of Príncipe and São Tomé, 200 miles west of Gabon in the Gulf of Guinea, which had first seen the Portuguese in the fifteenth century, became independent in 1975 and moved in the same democratic direction when President Aristide Pereira was trounced in elections in 1991 by Antonio Mascarenhas Monteiro but were engulfed by accelerating unemployment, inflation and foreign debt. A coup in 1995 by junior officers against President Miguel Trovoada failed in spite of suspicions of compliance from prime minister Carlos da Graça.

In Portuguese Guinea, or *Guinea-Bissau,* a liberation movement (PAIGC: African Party for the Independence of Guinea and the Cape Verde Islands) founded in 1956 by Amilcar Cabral, started a rising in 1959 and full-scale war in 1963, in the course of which the Guineans gradually established their control over the greater part of the country. In 1970 Portugal was implicated in an invasion of the neighbouring state of Guinea where PAIGC had its headquarters. An inquiry by the UN found that 350–400 invaders had been landed by ships manned by white men; the Security Council censured Portugal. In 1973 Cabral was murdered in the Guinean capital, Conakry. Guinea-Bissau and the Cape Verde islands became separate states in 1975 with plans for a union which was not consummated. The islands, ten of them 320 km from the mainland, provided sustenance for no more than a third of their inhabitants. They were isolated and kept politically stable by the tight control of a ruling group until the early 1990s when Cape Verde adopted a democratic constitution. In Guinea Bissau Joao Vieira turned multiparty democracy to his advantage by winning an election in which opposition parties were so numerous that none had a chance of winning.

Nearly every West African state experienced after independence one or more military coups. The tale of these coups is depressing. Twenty years after the main emancipation year (1960) the only original rulers still in power were Leopold Senghor in *Senegal,* Sekou Touré in Guinea and Houphouet-Boigny in Ivory Coast. Senghor – poet and scholar (the first African *agrégé*) before he was a statesman, a Christian leader in an overwhelmingly Muslim country, a leader in whom charisma was an addition to and not a substitute for intelligence, a president who abdicated voluntarily and peacefully after twenty-one years in power – made way in 1981 for Abdou Diouf who was confirmed in his position by a convincing electoral victory two years later but failed to stem sectarian rivalries, escape the control of elderly and conservative local bosses, or retrieve an economy overburdened by unwanted groundnuts and afflicted by drought. Unemployment rose, living standards fell, regional feuds and Islamic fundamentalism reared their heads. External debt charges rose to approximately half the national income, recourse to the IMF entailed painful austerity, and ventures such as phosphate plants, the Dakar shipyards and the Senegal River Plan languished as did agriculture. Foreign aid cushioned these failures. Senegal enjoyed strong French and American support and Diouf was able to win further elections in 1988 by hook or by crook but the disorder which followed obliged him to impose martial law and he was further weakened by economic decline, new austerity measures, the devaluation of the CFA franc (see p. 619) and signs of French reluctance to bail him out unless he gave Senegal's democracy more reality. At the next elections (1993), Senegal's first multiparty contest, Diouf won only a disappointing 58 per cent, further reduced the next year. Senegal's foreign relations with its neighbours Mauritania and the Gambia were uneasy, leading in the first case to undeclared war (see the preceding chapter).

The *Gambia* was a stranger neighbour since it was entirely embraced within Sengal. This smallest of the British colonies in West Africa led at first an equable existence after independence. With an economy which enabled it to dispense with British aid after 1967, the Gambia experienced nothing worse than rubs in its attempts to work an 'association' with Senegal until in 1981 President Dawda Jawara was forced into the humiliating expedient of calling in Senegalese troops in order to escape eviction. He accepted a confederation with Senegal – by which the Gambia hoped to have the benefits of independence and of partnership with a larger economy. But relations were soured by the prevalence of Gambian smuggling and Senegalese suspicions of Gambian intrigues among malcontents in Senegal's southern province of Casamance where the mainly Wolof speakers maintained a secessionist movement (which appeared to have support also from Mauritania and the adjacent Guinea-Bissau). In 1989 Senegal renounced plans for the integration of the Gambia with Senegal. In 1994 Jawara's long and honest rule as prime minister and then president ended when an army coup led by Lieutenant Yahya Jammeh forced him to flee. His departure ended the longest stint of democratic government in independent Africa and perhaps also the dominance of Gambia's largest ethnic group, the Mandinka.

South of Senegal and south again of the Guinea Coast (Guinea and Guinea-Bissau) *Sierra Leone* was ruled by the brothers Margai until 1966 – Sir Milton until his death in 1964 and the more radical Sir Albert until his plans for constitutional changes cost him the elections of 1966. The army intervened briefly, chiefly in order to keep Siaka Stevens from power, but in 1968 Stevens became prime minister and in 1971 president of the newly proclaimed republic of Sierra Leone. Stevens turned the country into a one-party preserve, an economic catastrophe and a paradise for corrupt tycoons who sucked it dry while waiting for the aged Stevens to depart. He resigned in 1985 after seventeen years' rule and was succeeded by General Joseph Momoh with promises to end the corruption and misrule which had characterized the government of which he had himself been a member for over a decade. Momoh had difficulty in dissociating himself from the unsavoury pillars of the old régime, while the business world became a battlefield between old-established Lebanese operators and immigrant Israelis (backed in some cases by South Africa). In 1991 Momoh tried to strengthen his position by an overt anti-Liberian stance when southern parts of the country were invaded by adherents of Charles Taylor (see below) but he was unable to resist the current trend towards a multiparty system. Six months before his term of office was due to end he was forced by junior officers to flee. The youthful leader of this coup, Captian Valentine Strasser, jailed some of Momoh's ministers but kept others, combined the appeal of another new broom with doses of brutality but failed to set a date for the elections which were supposed to bring his transitional authority to a democratic conclusion. He secured the cancellation of about a fifth of the country's foreign debt to a dozen creditors in return for promises of a return to civilian rule by 1996. But he failed to establish his authority in eastern and southern parts of the country where dissidents, with help from neighbouring Guinea and Liberia, seized the principal mines and took native and foreign hostages. Contrary to repeated predictions, Strasser survived with help from Ghana and Nigeria but the country was atomized by the indiscipline of an army suddenly enlarged tenfold, the forays of irregular forces intent on opportune loot and the no less opportunist manoeuvres of (partly) foreign organizations dabbling in a vastly expanded arms trade.

The eccentric history of *Liberia*, virtually an American preserve, was disrupted in 1980 by the violent removal of the dominant Tolbert family and the ruling True Whig Party by Sergeant (later General) Samuel Doe. Doe's support was predominantly tribal and although he survived for ten years, protected by an Israeli-trained guard and the Israeli secret service, he never established his authority throughout the country. Economic disaster supplemented tribal animosities and in 1990 invasion from the north-east started a chain of massacres and civil war in which tens of thousands were killed and about a million made homeless. The United States, which deplored Doe's demolition of the Tolbert régime but came to terms with him and used Liberia as an intelligence centre, cold-shouldered him when his rule became blatantly tyrannical and his prospects dim. Five West African states assembled a military force (ECOMOG), mainly Nigerian but commanded at first by a Ghanaian officer, to stop the fighting between the rebels and Doe (beleaguered in the capital) and between rival rebel sections fighting one another. But ECOMOG failed to protect Doe who was seized, tortured and killed after taking refuge with it. Doe's executioner, Prince Johnson, an alcoholic psychopath, seized control temporarily in the capital but most of the country was dominated by Charles Taylor, who had served in Doe's government before being charged with embezzlement and driven into temporary exile in the

United States. Taylor recruited exiled and discontented Nigerians and Ghanaians, thwarted ECOMOG's efforts to make peace, invaded Sierra Leone and fostered suspicions that ECOMOG was being used by Nigeria less to make peace than to pursue quasi-imperial ambitions in West Africa. (Nigeria met four-fifths of ECOMOG's costs and supplied two-thirds of its 15,000 troops.) ECOMOG slowly prevailed against Taylor with the apparent aim of forcing him to come to terms with the relatively feeble government in Monrovia protected by ECOMOG under the titular and increasingly phantom presidency of Amos Sawyer and (later) David Kpormakpor. The task of reconciliation was aggravated as factions split and multiplied and ECOMOG's Nigerian and Ghanaian principals grew evidently worried about the cost of their intervention. A comprehensive but fragile cease-fire was negotiated at the end of 1994, mainly by Rawlings, and converted the next year into a power-sharing agreement. Taylor had the most reason to be satisfied with it and therefore to honour it. Liberia was, with Ethiopia, an outstanding warning of the faults and fissures in the fabric of many African states where mutually hostile ethnic groups contended against one another with an eye to sympathetic kin beyond the state's borders.

Ivory Coast, Liberia's neighbour, was by contrast the most successful of all West African states in the first decades after independence. Superposing good management on good fortune (a benign climate and no real desert) its government designed, first and foremost, to intensify and diversify the rural sector; secondly, to give measured encouragement to industrialization, including opportunities for private enterprise within a state-controlled system and assurances to foreign capitalists wishing to invest and earn a fair profit; and thirdly, to avoid heavy spending on defence and other forms of ostentation. As a result the national product grew steadily at the rate of 3 per cent a year and the average annual income of the country's 8 million people rose from less than $100 before independence to about $800 by 1980. But this average increase owed more to the rapid enrichment of the few than to the spread of prosperity to the many, and after 1980 growth faltered. Oil began to flow in 1984 but with limited effect on the economy, offsetting recession rather than enhancing growth. Collapsing cocoa prices, due to over-production worldwide, undermined the commodity which still in the 1980s provided (with coffee) half of Ivorian foreign earnings, and this shortfall combined with high borrowing at high cost to create an economic crisis. In 1987 Ivory Coast suspended payment on its foreign debts (which amounted to 150 per cent of GDP) and exports of cocoa; two years later it halved the price paid to cocoa growers. Private businesses were owed more than $1 billion by the state and private banks were collapsing. Loans from the World Bank expiring in 1989 were not renewable and the country was not poor enough to qualify for special debt remissions. Cuts in public services and wages – the IMF demanded cuts up to 75 per cent in salaries in the public sector – provoked strikes and demonstrations but failed to rectify the country's deficits. As Houphouet-Boigny, confessedly 80 but nearer to 90, approached the end of his natural span he took few steps to ensure an orderly succession. A constitutional amendment introduced in 1980 had designated the vice-president as heir but nobody was appointed to that office and five years later Houphouet-Boigny announced that he would remain in office until he died. In 1990 he so far bowed to domestic anxieties and to the new wind of change in Africa as to sanction the formation of more than one political party, but his main concern in his declining years was the construction at unconscionable cost of a vast church in his native village and getting the Pope to sanctify it, which the Pope did. Houphouet-Boigny was re-elected president. Some two dozen parties, newly established, contrived to field only one candidate who nevertheless received 18 per cent of the votes cast in elections held under the control of, and not without manipulation by, the ruling party. Rises in cocoa and coffee prices in the mid-1980s were not sustained and austerity and bureaucratic pruning caused some distress. Houphouet-Boigny died in 1993 after thirty-three years in power and leaving no designated successor. The president of the national assembly, Henri Konan Bédié, moved deftly to win the succession but not without splitting his party. He inherited a position of strength in one of Africa's more stable societies but chose to adopt an assertive style, even disfranchising non-Ivorian residents who constituted a third of the electorate. In 1995 he was comfortably re-elected, most of his adversaries ducking the contest at the last moment.

In *Ghana*, the pioneer of West African liberation, the personal leadership of Nkrumah turned into a febrile autocracy which passed step by step beyond a struggle to maintain the unity of the new state and a struggle to modernize it, and became perverted into a struggle to assert and preserve the authority of Nkrumah himself. Upon the attainment of independence Nkrumah had three principal aims. He wished to propagate his vision of African unity by accelerating the independence of other territories and consolidating the energies of Africans and Africanism in the service of African dignity and African effectiveness in the world; but his methods and his personality did not always commend him to other African leaders, whose own pan-Africanism was filtered through their several national experiences and ambitions. Like Nasser's pan-Arabism, Nkrumah's pan-Africanism was suspect. The exuberance and vigour which had contributed so much to make him the first ruler of a sub-Saharan African republic were drawbacks in a new situation which demanded the arts of diplomacy rather than the *élan* of leadership.

Nkrumah's second ambition was to make Ghana a modern industrial state. This ambition was impeded by a complex of regionalism, conservatism and political jealousies and also by a serious drop in the world price of cocoa, Ghana's principal source of revenue. The Convention People's Party, whose main strength was in the coastal areas, was opposed by chiefs whom it regarded as a divisive and reactionary force and by the United Party led by K. A. Busia and J. E. Appiah. Above all his showpiece failed him. He revived an abandoned British plan to build a dam across the river Volta and he concluded an agreement with an American corporation to build a smelter which would buy and refine Ghana's large deposits of bauxite. The World Bank would lend the money to build the dam on the strength of Ghana's prospective profits from the sale of bauxite but the contract with the American Kaiser corporation failed to provide that the bauxite to be used in the smelter should be Ghanaian bauxite and the corporation imported bauxite from elsewhere. The dam was built but Ghana's share of the profits from the smelter was minimal and a country which had started its independent existence with enviable reserves began to slide into debt. Other foreign companies enticed Ghana into expensive projects of which the most notorious was the enlargement of the Accra–Tema road into a four-lane highway with all the appurtenances of a British motorway. To these extravagances Nkrumah contributed by grandiose building. Bloated economic activity carried with it a grievous load of corruption. Nkrumah's failure to meet his promises of a better life for all was compounded by an increasingly dictatorial and suspicious manner. The illiberality evidenced, for example, by the removal of the chief justice after the delivery of an unpalatable verdict in a treason trial, coupled with increasing corruption and spendthriftness in the administration, destroyed Nkrumah's attempts to get foreign friends and finance and created an opposition to him in the army which, by a bloodless coup in 1966, unseated him while he was away on a visit to China.

Third, in the world at large Nkrumah wished to follow a policy of non-alignment. His circumstances were exceptionally favourable. Upon independence Ghana's resources and reserves – at £200 million the latter were larger than India's – gave it a material base for a measure of independence which might the more easily be maintained since the Cold War had not at that date reached Africa. But in practice Ghana under Nkrumah never achieved non-aligned status in the eyes of outsiders. For half the period 1957–66 it appeared pro-western, for the other half pro-eastern. Ghana remained in the Commonwealth, was visited in 1961 by Queen Elizabeth II and its leader welcomed in Washington by President Eisenhower, but the crisis in the Congo and particularly the murder of Lumumba, for which he blamed the west, turned Nkrumah against the western world, while the elaboration in theory and practice of his 'scientific socialism' with the inauguration in 1962 of a one-party system turned the western world against him. In contradistinction to Nasser's evident pragmatism, Nkrumah became increasingly ideological and illogical, evolving a form of economic planning akin to communist models but nevertheless dependent on western capital for its success. The coup which removed him was managed by the CIA.

Nkrumah wasted Ghana's enviable heritage and belied his own vision and ideals. He reduced the country to bankruptcy and initiated the corruption which no subsequent régime proved able to master. With one two-year interlude his fall was followed by four military governments of varying incapacity. The first of these, led by Colonel Ankrah, found that Nkrumah's extravagances had run the reserves down to £4 million and the external debt up to £279 million. Ghana also had a deficit on external account of £53 million. Colonel Ankrah and his colleagues made some improvements to this dismal tale before relinquishing power in 1969, but economic growth remained below 1 per cent a year, the population was increasing at an annual rate between 3.5 and 4 per cent, the standard of living had been cut by a severe devaluation and the first attempts of the new prime minister, Dr Kofi Busia, to secure an alleviation of the external debt were only moderately successful. By the beginning of 1972 Busia had been defeated by his predecessors' legacy, by an imprudently liberal policy which permitted an inflow of foreign goods which crippled the balance of payments, and by a drastic devaluation which doubled the external debt at a stroke. He was ousted while away in London and succeeded by any army committee calling itself the National Redemption Council under the chairmanship of Colonel I. K. Acheampong. This well-intentioned group struggled, not always harmoniously, with economic problems which were being aggravated by steep rises in the price of oil and other essential imports and by consequent rises in the cost of living. Ghana's creditors gave it some relaxation on the repayment of debts but not as much as it hoped for. By 1978 Acheampong's rule had become corrupt and ineffective. He was ousted by General F. W. K. Akuffo, who lasted less than a year against a rising tide of indignation. He too was removed and, together with Acheampong and six others, executed after a coup led by Flight-Lieutenant Jerry Rawlings, who promised to end corruption, reverse economic decline and restore civilian rule. In his three months in power he succeeded only in the last, completing an electoral process which had already been set in motion and which installed Dr Hilla Liman as president of a country with a collapsing economy and aggravated social tensions. At the end of 1981 Liman was ousted and Rawlings returned, struggled against the economic slide and sporadic plots and riots, pulled away from his more radical associates and introduced substantial deflation and a number of devaluations in order to get help from the IMF. Having attacked Liman for being willing to deal with the IMF and having failed to get economic aid elsewhere Rawlings switched to a policy of IMF-led growth and had some qualified success: inflation fell from 100 to 20 per cent, the budget was balanced and growth reached 6 per cent. But revenues from cocoa and timber declined (gold sales, however, went up); foreign debt remained a throttling burden; the cost of imported oil, which had quintupled in the decade preceding Rawlings's return, absorbed nearly half the country's foreign earnings; the devaluation of the cedi in 1983 from 2.75 to 1.62 to the dollar caused great distress; promises of an early return to civilian rule and of popular consultation were not kept; the ruling Provisional National Defence Council was a clique with good intentions but no sound constitutional standing; opposition in politics or in the administration became a road to jail but this treatment did not deter Rawlings's enemies at home and abroad (mostly in Ivory Coast and Togo) from keeping up their intrigues against him. The justification for these tribulations and stringencies would be the attraction of foreign capital but the issue – and the outcome therefore of Rawlings's economic strategy – remained uncertain. He had forfeited the support of radicals, introduced measures which hurt the many more than they hurt the few at the top, and risked the disgruntlement of junior officers which had kindled more than one coup in Ghana's history. Such a coup in 1989, although unsuccessful, was the more ominous for coming from his own Ewe people.

By 1990 Rawlings was being nudged towards constitutional reform by international forces, foreign finance and the regional fashion for more democracy, while remaining himself innately averse to a multiparty system or at any rate to more parties than two. Unexpectedly in 1991 he announced a programme for a return to civilian rule and at the end of the following year he was re-elected president by a majority of two to one. Parliamentary elections were boycotted by a cynical and divided opposition whose refusal to enter the multiparty fray gave Rawlings 189 seats in an assembly of 200 and in effect turned Ghana into a one-party state with a multiparty constitution. It was also a state with rosier economic expectations than it had been when Rawlings seized power for the second time eleven years earlier. With the help of about $2 billion of foreign aid, government spending had been axed, inflation had been reduced to 15 per cent, growth in the agricultural sector had averaged 2.5 per cent since the early 1980s, new sources of revenue had been developed in, for example, fishing and there was even talk of Ghana emulating the Tigers of East Asia in Taiwan, South Korea and Hong Kong. On the other hand continued economic progress (at least to the next elections due in 1996) depended on rising foreign aid and rising rates of growth, whereas the economy remained heavily dependent on cocoa and gold, corruption in the private sector persisted, there was no domestic and disappointingly little foreign money for investment, strikes were on the increase, and forecast budget and trading surpluses were unfulfilled. But the balance of these political and economic gains and losses, together with discord in Nigeria (see below), encouraged Rawlings to play a stronger role in African affairs as chairman of ECOWAS (whose francophone members trusted Ghana more readily than Nigeria), as elder statesman in the eyes of new men in Sierra Leone and Gambia and as respected visitor to a string of southern African states.

In *Togo*, where the first president Sylvanus Olympio was assassinated in 1963 and the second Nicholas Grunitzky displaced a few years later, Gnassingbe Eyadema maintained himself in power against substantial – largely regional and ethnic – unpopularity and frequent plots. The president's political strength lay in the north whence came also a large proportion of the army's officers. Matters turned to open crisis when, after riots in 1991 and a conference lasting two months to ventilate accusations of fraud and murder in high places and proposals for constitutional reform, the president rounded on his prime minister, Joseph Kokou Koffigoh, dismissed the army's commander-in-chief (both Ewes) and cancelled projected elections. The army, whose attitude had been uncertain, rallied to the president in alarm at spreading disorder – some of which was allegedly incited by the president himself to keep the army on his side. Further serious riots in 1993 prompted half a million Togolese to flee to Ghana and Benin, whereafter Eyadema allowed elections to proceed and was declared the winner.

In adjacent Dahomey, renamed *Benin* in 1972, a country living under the shadow of Nigeria and without jobs or resources, the military were in and out of government, displacing one civilian president and installing another until in 1970 they forced three past presidents to work together on a rota system which proved to be no more than a two-year interlude before military rule. In 1972 Colonel Matthieu Kerekou seized power, survived a series of attempts against his life – including one in 1977 allegedly launched from Togo with French, Moroccan and Gabonese help – was re-elected in 1984 and (narrowly) in 1989, made a tour of communist capitals including Beijing (1986) and then turned cautiously back to the French connection. The opposition to his rule was divided, consisting of young people without jobs, junior officers without promotion, unions, and ethnic rivals in the south. He saw his country become a small exporter of oil and also an entrepôt for drugs from Asia to the west. But by 1990 his days were numbered and Benin led the way in West Africa to multiparty systems. Kerekou, like Kaunda in Zambia a year later, accepted his dismissal peacefully, and was succeeded by Nicéphore Soglo who released political prisoners, abstained from strong-arm rule and introduced economic changes calculated to win foreign (mainly French) aid. Conversely, however, his economic reforms put large numbers of people out of work and nearly a hundred parties appeared. Soglo, whose ruling coalition was mainly Fon, southern and Christian, had only a patchy electoral victory in 1995 and he lost to a revived Kerekou in 1996.

Behind the arc of states stretching from the mouth of the Senegal river almost to the mouth of the Niger lay three large, arid and landlocked countries: Mali, Burkina Faso (formerly Upper Volta, the smallest of the three) and Niger. In *Mali* Modibo Keita, one of the architects of the abortive Malian federation, was ousted in 1968 by General Moussa Traore who ruled for twenty-three years. In his search for foreign aid he vacillated between the communist states of Europe and Asia (which he visited in 1986) and the readier, if unpalatable, succour of the IMF. Mali had trouble from the Tuareg, Berber desert-dwellers who lived also in Burkina Faso, Niger, Algeria, Libya and Chad. As decolonization loomed in the 1950s the Tuareg hoped to secure with French help a state of their own. After independence – and largely on account of drought – they dispersed into neighbouring countries, notably Libya where they enlisted in Qadafi's wars in Chad whence, however, they began to return to Mali around 1989–91, causing fighting in Mali and also Niger. Traore was trying to bring this fighting to an end when he was ousted by his army. (He and his chief of staff and two senior ministers were condemned to death.) A transitional government concluded with Algerian help an agreement with the Tuareg which conceded them regional autonomy but did not satisfy a more militant group; fighting went on with increased viciousness. An outburst of democracy in Mali produced twenty political parties to contest local elections in 1992. A series of civilian administrations (Younnoussi Traore, Abdoulaye Sekou Sow) grappled with the consequences of bad government, droughts, corruption and crippling debt.

Part of Mali's pre-independence heritage was a border dispute with *Burkina Faso* which was aggravated when Captain Thomas Sankara took power there in 1983 (ousting a three-year-old civilian régime which had succeeded the military régime of General Saye Zerbo). Sankara was attractively efficient, young and honest but his authority rested on an amalgam of discordant elements. He developed good relations with Benin and projected a union with Ghana but this trend alarmed Mali which launched a small, short and ill-conducted war in 1985. Two years later Sankara was murdered and succeeded by a close friend and colleague Blaise Compaore, who, less radical than Sankara and more autocratic, presided over a multiparty régime in which his party held the biggest stick and did not shrink from brandishing it against unions, students and others with views of their own.

Niger too suffered plots and disorders, generally instigated by military dissatisfaction with the efficiency or honesty of civilian rule. These coups were effected with the minimum of violence and amounted to an accepted, if not very satisfactory, way of changing a government which had failed but which had no ready-made successor in a one–party state without adequate machinery for changing the government in any other way. Niger's first ruler, Hamami Diori, had won his position with strong French help against a rival, Djibo Bakary, who leaned rather to the left and also to the Islamic north of the country. During the civil war in Nigeria Diori supported the central government in defiance of France's support for Biafra and he paid the price in 1974 when the army turned him out: a drought in the previous year, extreme even by Niger's standards, had weakened his position. His successor Seyni Kountche remained president until his death in 1987. He was followed by Colonel Ali Seybou who would, it was hoped, display a less offensive lifestyle than Kountche but did not. A new constitution in 1992 was a recipe for stalemate between president and prime minister should they fail to agree. They did, and President Ousmane Mahamane, a Hausa, was forced into competition with prime minister Amadou Hama for support from the second largest ethnic group, the Songhai. Niger's Tuareg, comprising a tenth of the population which felt excluded from jobs and education, were disgruntled and dissident but failed to muster an active rebel force of more than two thousand. Niger's uranium made it a cynosure of French attention so long as uranium was believed to be a scarce mineral.

Skipping for the moment West Africa's most important country, Nigeria, the southward route leads to three more ex-French colonies and another landlocked empire. In *Cameroun* – a relatively successful adjuster of linguistic and ethnic tensions – France intervened to sustain the friendly government of Ahmadu Ahidjo against a revolt whose causes were mainly economic. Ahidjo resigned suddenly in 1982, to be succeeded by Paul Biya who proved to be disappointingly ineffective. Sharp falls in the prices of Cameroun's principal exports – coffee, cocoa, cotton – produced crippling debt, left public servants unpaid, encouraged corruption and smuggling, dictated recourse to the IMF and austerity, and stimulated political in-fighting which Biya failed to master. He refused to appoint a prime minister until forced to do so. He won a narrow victory in less than perfectly conducted parliamentary elections in 1992 and was re-elected president after deferring elections while opponents squabbled over the selection of a single candidate. In 1995 Cameroun joined the Commonwealth.

In *Gabon* hopes turned to disillusion. Potentially one of the richest states in Africa, Gabon was ruled by Presidents Leon M'Ba (deposed in 1963 but restored by French troops) and Albert Bongo who changed his first name to Omar upon conversion to Islam in 1973 and completed the country's transition to a one-party state. In spite of natural wealth which included a quarter of the world's known store of manganese Gabon was struck by economic crisis in the 1970s and 1980s, so that the government had to levy a forced loan on salaries and reschedule its foeign debts as the only way to avoid default. Serious riots in 1990 forced Bongo to introduce a multiparty system faster than he would have liked. After elections in which his party, the Parti Democrate Gabonais, comfortably outpaced its challengers he formed a new government of six parties, the PDG retaining the principal ministries but taking only a minority of places in the cabinet, and he himself was re-elected president in 1993 – by the proverbial whisker and against a challenger with support from Cameroun.

In *Congo-Brazzaville* the first president, Fulbert Yalou, was overthrown by revolution, France refusing to intervene to save him in a situation which it recognized as one of genuine and well-founded popular indignation. Alphonse Massemba-Debat and then Marien Ngouabi took the country to the left but the latter was assassinated in 1977 and the former shortly afterwards executed. President Denis Sassou-Nguesso of the Parti Congolais du Travail kept both pro-Chinese and pro-Russian parties down but not wholly out. His ill-planned use of public funds and extravagant spending on projects of little merit, coinciding with falls in oil production and the oil price, debilitated the economy, encouraged a self-defeating drift to the towns, exacerbated ethnic rifts and increased imports and foreign debt. When the army refused to quell riots in 1985 the president saved himself only by the use of the presidential guard and six years later he was cut down to size, albeit escaping retribution and keeping his position and some of his powers. Elections in 1992, local and then presidential, led to the installation of Pascal Loussaba as president in an evolving but turbulent democracy for which he sought American support against (mainly northern) opposition parties favoured by France.

In the landlocked *Central African Republic* a coup by Colonel Jean-Bedel Bokassa, similar in its origins to military coups in Burkina Faso, Benin and Niger, led however to enormities when the colonel declared himself emperor and went to insane lengths of ostentation and cruelty. France, after become embarrassingly associated with one of the period's most indecent tyrannies, assisted with its overthrow in 1979. An attempt by Bokassa in 1983 to return failed. After a short transitional period André Kolongba, a dependant and relative of President Mobutu of Zaïre, ruled for twelve years. He resisted the trend towards a multiparty state for as long as he could, received aid from Iraq until the Gulf War of 1991, only partially replaced it with aid from Taiwan, and most importantly lost the favour of France which still had influence and 3,500 troops in a country which it regarded as strategically valuable. Unable to stave off elections any longer, Kolongba was eliminated in the first round, tried to cancel the second but was replaced by André-Félix Patasse in 1992.

Nigeria, huge and diverse, was a twentieth-century creation. The colonies of Lagos and South Nigeria were united by the British in 1906. The former was the home of the Yoruba whose society was based partly on tribal chiefs and partly on cultivated cities. The dominant people in the latter were the Ibo who had neither chiefs nor cities but were therefore the more easily united as a nation and displayed a thrusting vigour which carried many of them long distances from their homes in eastern Nigeria and into the western and northern parts of the federation created by the British when, in 1914, they joined northern Nigeria to the two southern colonies.

The north was in a state of flux when the British reached it at the end of the nineteenth century. Early in that century the Islamicized Fulani had established their authority over a medley of pre-existing Hausa states of very varying size and had so created the empire of Usman dan Fodio and his descendants. Conflict between Fulani and Hausa and others persisted and the British assumed the task of mastering it as a prelude to uniting all their territories in this part of the world, but the federation which they created was an uneasy amalgam of states and cultures without a national consciousness or unity. After the Second World War the southern half of the country pressed for independence, while the north hesitated for fear of being reduced to dependence on the south with its coastal windows on the outside world's commerce, skills and manners. The north even toyed with the idea of a separate state with access to the sea either down a river corridor along the Niger or through Dahomey or the Cameroons on the one side or the other. Before independence, riots with racial and religious overtones gave manifest warning of the dangers which the new state would run, but the advantages of unity seemed to make the experiment worthwhile and in 1960 Nigeria became independent with a constitution based on three regions. The federal premiership went to a northerner, Sir Abubakr Tafewa Balewa. The eastern leader, Namdi Azikiwe, became president and his party co-operated with the principal northern party in the government at the centre. It was Azikiwe's belief that a country as big and heterogeneous as Nigeria could only be governed by a coalition of all its main groups within a federal structure, but the western leader, Chief Obafemi Awolowo, preferred to form an opposition in the federal parliament.

The accord between east and north did not last long, nor did the cohesion of the west. The Ibos annoyed their northern partners by trying to win parliamentary seats in the northern region. In the west Chief Akintola challenged Awolowo and succeeded in getting him and other leaders tried and imprisoned. Akintola was not satisfied with Awolowo's policy of forming an opposition at the federal level. He preferred a policy of alliance with the north, displacing the Ibos and getting for his own people a share in the power, perquisites and finances of central government. In 1963 a fourth, mid-western region was formed. In 1964 a census showed a total population for the whole country of 55.6 million, of which 29.8 million were in the north – more than half. This demographic preponderance of the north increased the attractions of an alliance with the leaders of that region and, together with charges that the census had been faked, also increased political tensions. Elections in the western region in 1965 produced allegations of flagrant malpractice and fighting. In the next year more serious fighting began in the north. The country's uneasy tripartite constitution broke down with, first, a mainly Ibo insurrection; then a coup which provoked an attempt to create an independent Ibo state; and eventually successive redivisions of the country into more and more constituent states.

In January 1966 a group of young Ibo officers rebelled, partly in protest against bad government and partly as a demonstration against the north and its allies in the west. The federal and northern prime ministers, Sir Abubakr Balewa and the Sardauna of Sokoto, were murdered along with Chief Akintola and many more, but senior officers stepped in quickly to arrest further developments and the junior officers. A comparatively senior officer, General Johnson Ironsi, was proclaimed head of state with the mission of keeping the country from breaking up and providing competent and honest administration. General Ironsi was an Ibo. He was well intentioned but not otherwise well equipped for the delicate task of holding the country together and when in July he incautiously suggested that a unitary constitution might be better for Nigeria than a federal one, northern fears of attempted Ibo dominance flared up into another coup and he was murdered in his turn. He was replaced by Colonel Yakubu Gowon, a northern Christian, who released a number of imprisoned civilians, including Awolowo, and expressed the usual hopes for an early return to civilian rule. In September a conference at Lagos agreed on a loose fourfold federal structure, but it coincided with a counter-massacre of Ibos in the north (there had been a similar, though less severe, massacre in May) and the flight of the survivors back to the eastern region.

After the first coup in 1966 Lieutenant-Colonel Odumegwu Ojukwu had been appointed military governor of the eastern region.

He now concluded that the salvation of the Ibos could only be secured by secession from Nigeria. Dissuaded from taking this step at a conference in January 1967 at Aburi in Ghana, he nevertheless proclaimed in May the independent state of Biafra and civil war began. Attempts at mediation by the OAU (at Kinshasa in 1967, Algiers in 1968 and Addis Ababa in 1969) all broke down. The federal army, expanded from 10,000 to over 200,000 men, gradually prevailed in spite of the toughness of Biafran resistance. While Britain and the USSR supplied arms to the federal government, France supplied Biafra (thereby perhaps somewhat prolonging the war) and two of France's closest African associates, Ivory Coast and Gabon, recognized the secessionist state – as too did Zambia and Tanzania – in attempts to stop the killing. The suffering in Biafra, overcrowded and virtually besieged, was appalling and widely reported throughout the world. By January 1970 it had become decisive. In that month Biafra was forced to capitulate and, as such, ceased to exist. Gowon proved a statesmanlike victor. He preached reconciliation and practised it. Nigeria's economic growth was resumed at a great pace. It became one of the world's largest producers and exporters of oil, besides being rich in coal, tin and other minerals, and in agriculture. But the government was none the less overspending its revenues and internal problems were not washed away by this wealth. The rich got richer and the poor mostly stayed poor or, with inflation, got poorer. The huge army and a large police force of 30,000 became notorious for corruption, as did the more affluent sectors of civilian life. There were strikes and unemployment. The constitutional problem was unsolved. After the civil war the country was divided into twelve states (later increased to nineteen) but Gowon incautiously indicated that there might be more, on an ethnic basis which, if seriously applied, could produce up to 300–400 states. The census of 1973 returned a total population of 79.76 million, a growth of 43.5 per cent since 1963 and an average annual increase double that of the fastest growing populations in the world. Nobody believed the figures although some people in some areas liked them. The fact that increases varied widely and suspiciously from state to state forced Gowon to say that they would not be used as a basis for any political decisions. By 1974 it was announced to the relief of many Nigerians that the return to civilian rule promised for 1976 would be postponed. Gowon's too evident inability to grasp the nettle of corruption led to his displacement in 1975 while he was out of the country. He was allowed to go with his family to England. His successors Generals Murtala Muhammad (assassinated after a few months) and Olesugun Obasanjo were more vigorous in their attacks on corruption and other problems. The latter, who refrained from styling himself president, promised to restore civilian rule in 1979 and after prolonged constitutional debate which produced the bulkiest constitution in the world Alhaji Shehu Shagari became president of a new federation of nineteen states.

Nigeria was a giant among African states in more than size. It had wealth. It was among the first half-dozen oil producers in the world and had a GNP at the end of the 1970s of about $35 billion ($460 per head) and a growth rate of 6–7 per cent a year. At its peak in 1974–76 oil production exceeded 2 million barrels a day. But this bonanza engendered an orgy of activity which got out of control. Nigerian oil, exceptionally expensive to produce, happened to come into world markets in large quantities in the same period as Alaskan and North Sea oil, with the result that demand slackened, the price had to be cut and production held back to 1.5 million barrels: the balance of payments, which had shown a surplus of over 4 million naira in 1974, went suddenly into the red in 1977, reserves fell by two-thirds and inflation rose to 40 per cent a year. But Nigeria's credit remained high and it was able to borrow large sums to put it back on to sounder courses. Its basic problem was not how to balance its books but how to apply its enviable revenues. It was at this stage an immensely wealthy country whose inhabitants nevertheless saw nothing of this wealth but continued to live in a colonial, mainly agricultural, society which was using no more than half its cultivable land and, as a result of neglect, inefficiency and incompetent investment, was importing food instead of exporting it. The country was also frighteningly lawless and blatantly corrupt; its ethnic divisions and mistrusts had not evaporated; its army (although reduced from 230,000 to about half that number in 1980) was an expensive and unsettling legacy of civil war. But it was a measure of the new civilian government's resources and optimism that it felt able to produce a five-year plan for 1981–85 which aimed to make Nigeria self-sufficient in food and manufactured goods by the end of that period.

Both the outgoing military régime and its civilian successor intended Nigeria to play a positive part in African affairs, perhaps in world affairs too. It forced its way on to the Security Council in 1976 against the candidature of a fellow African state (and neighbour) nominated by the OAU. It put a lot of energy and money into the 1976 Festival of African Arts in Lagos, a prestige jamboree which partially rebounded because of organizational chaos and scandals about overspending. It so dominated the Economic Organization of West African States (ECOWAS) that this organization would have been boneless without Nigeria. Lagos became a necessary port of call for British, American and other politicians trying to resolve the Rhodesian crisis. As a member of the OAU, the Commonwealth and OPEC, as well as the UN, Nigeria was active in a unique sample of international organizations.

On the last day of 1983 the army returned to power. Major–General Muhammad Buhari became head of state in place of President Shagari. There were many reasons for an upheaval which displeased few Nigerians: blatantly rigged elections in the previous August; even more blatant corruption; soaring food prices; an economy no longer buoyed up by oil whose price had been cut by 10 per cent and production by nearly 25 per cent; interminable constitutional wrangles; religious disorders in the north, notably in Kano where thousands were killed in suppressing riots by the Muslim Maitatsine at the end of 1980 and again in Borno and Gongola states in 1982 and 1984 respectively. The new government was akin to its predecessor, conservative and pledged to produce prosperity without corruption. For some, including junior officers waiting their turn and Muslims of a puritan turn of mind, it was too like the Shagari régime with too many of the same businessmen and chiefs in the seats of federal and provincial power. It was threatened within a few months by rumours of fresh coups which were suspected by some of being fabricated in order to enable the government to forestall action from within the army's middle ranks, dissatisfied by dilatory measures against the profiteers of the previous régime. When the government did take action it bungled. The most notorious of the alleged profiteers, Umaru Dicko, had taken refuge in London where he was kidnapped by three Israelis and a Nigerian with ill-concealed Nigerian government involvement and more than a little suspicion of official Israeli participation: this coup was foiled when British police extracted Dicko from a packing case in which he was about to be flown to Nigeria.

The Buhari government was an all-round failure. It denounced corruption without abating it, savaged criminals without reducing crime, intensified economic austerity without any countervailing benefits and forwent needed aid from the IMF (to bolster its foreign creditworthiness) by refusing to devalue the naira. To its neighbours it was no improvement on its predecessor. Both expelled large numbers of immigrants who had flocked to Nigeria to find work and both reduced ECOWAS to near inanition: ECOWAS was either a way of sharing wealth and enterprises in the region or it was nothing, and a few improvements in transport systems and freer trade did not suffice to alter the complexion of the area as a cluster of separate and disparate economies whose inequalities and inefficiencies could not be improved by a dominant Nigeria beset by political and economic instability. Further afield the Buhari government irritated its partners in OPEC when, like Britain and Norway, it cut its oil prices but without giving OPEC notice in advance: oil revenues had sunk to half their 1980 value and the Buhari government was nervously aware of the part played by oil in the decline and destruction of the Shagari government. Within two years Buhari was evicted by one of the principal members of his governing council, General Ibrahim Babangida, a recurrent figure in Nigeria's political crises but one who had hitherto avoided the first place and its responsibilities.

Babangida's main tasks were to rescue the economy, restore civilian rule within a reasonable time and prevent Nigeria's regional conflicts from turning into a religious war between north and south, Muslims and Christians. Nigeria's economic problems arose not from lack of resources but from its over-optimistic use of a single source of wealth (oil) whose value had suddenly plummeted for reasons outside Nigeria's control. The collapse of the oil boom stunned Nigeria the more forcibly because it had been riding so high on the crest of the wave that was now spent. With the end of the boom the naira was floated and lost 70 per cent of its value. Inflation, wage freezes and penal interest rates hit both the poor and the flocks of *nouveaux riches* born of the boom. Average income per head dropped in five years from over $1,000 to $250. Babangida needed to restore confidence within Nigeria and to recapture international confidence and funds, but his attempts to do so evoked strong criticism from those – including Obasanjo – who inveighed against an economic policy which made the poor poorer and gamblers richer without reversing industrial recession. As a prelude to the restoration of civilian rule a constituent assembly was elected in 1987 with, however, strictly limited competence: it was precluded from considering the federal principle, the two-party system, the ban of former politicians, or religion. Babangida's confidence in himself was shown by two striking acts in 1989. When an electoral commission approved six political parties from which the government would select two, Babangida rejected all six and licensed two others of his own devising. Secondly, he intervened decisively in the succession in the sultanate of Sokoto. The death in 1988 of the aged sultan provoked a politico-religious tangle which developed into an issue in relations between the sultanate and the federal government. A son of the deceased sultan was proclaimed sultan in his father's place but the government interposed a veto and appointed instead an eminent Muslim, Ibrahim Dasuki, a direct descendant by a junior line of Usman dan Fodio. No less disturbing than these economic, constitutional and religious problems was one other: demography. A population of 118 million in 1990 was projected to reach more than twice that total within thirty years in the absence of birth controls, of which there was only the most modest prospect. This was a problem widespread in Africa but peculiarly alarming in Nigeria which would on existing trends become one of the most crowded countries in the world and drastically short of food, schools, hospitals and other crucial services.

A return to civilian rule was repeatedly promised and postponed. In 1991 Babangida cancelled a presidential election for the third time but later in the year elections were unexpectedly held for governorships and assemblies in the several states which now numbered twenty-nine: the newly licensed Social Democratic Party won majorities in sixteen of the assemblies and the National Republican Convention the largest number of governorship. Demands for Babangida's resignation multiplied as a comprehensive break-down of law and order loomed with riots in Lagos and other cities (chiefly against price rises) and with looting and religious clashes in the north and north-east. Babangida's term of office was extended to August 1993 but powerful personalities inside Nigeria and its principal foreign creditors – the United States, Japan, Germany – were no longer concealing their opinion that he ought to go. The foreign debt had reached $30 billion, corruption was blatant, Nigeria was becoming a major international drugs entrepôt and its very existence as a single state was once more being questioned. A presidential election in 1993 was cancelled when it appeared that the winner would be Chief Moshood Abiola, millionaire Yoruba leader of the SDP. Abiola fled to England. Babangida's plans for a fresh election were rejected by the SDP. The SDP and NRC offered to join an interim government with the military but their offer was rejected by Babangida. Apprehensions among senior officers caused splits as a result of which Babangida was discarded and replaced by his deputy General Sani Abacha, with Ernest Shonekan as a figurehead who proved too independent and resigned after a few months. In 1994 Abacha, perhaps out of weakness, sanctioned a Nigerian occupation of the Bakassi peninsula in Cameroun which commanded the approaches to the Calabar Channel and Cross river, areas rich in oil. Notions of Nigeria as a stabilizing force in West Africa and an economic dynamo vanished as it accelerated towards political disintegration, religious discord and general lawlessness. The execution in 1995 of the writer and political activist Ken Saro-Wiwa with eight other protesters from the oil-bearing Igoni lands after an unsatisfactory trial and on the eve of a Commonwealth conference evoked worldwide condemnation but no effective international action. Although Nigeria was abnormally dependent on revenues from oil, the United States and other purchasers were unwilling to deprive themselves of it by an international ban.

A quarter of a century of political instability, economic disaster and misrule left West Africa with the independence which it had won from the colonial powers but few of the anticipated fruits of independence. Spirits were low, expectations reduced, disappointments bitter, tempers frayed. A number of the states of West Africa hoped to alleviate their economic plight through federal associations among themselves. Sixteen of them created in 1975 an Economic Community of West African States (ECOWAS), a half-way house between national sovereignties and the continental OAU. The first aims were to increase trade between the members and to fund co-operative developments in agriculture, communications and education, with further economic co-operation to follow. Nigeria's affluence gave the members hope but Nigeria's power gave them pause, more especially the francophone members, six of which created a community within a community: CEAO – Communauté de l'Afrique de l'Ouest. (CEAO had counterparts in Central Africa in CEEAC – Communauté Economique des Etats de l'Afrique Centrale – and UDEAC – Union Douanière et Economique de l'Afrique Centrale – whose names sufficiently indicated their ambitions. All six members of the latter belonged to the former, whose ten members included English and Portuguese states as well as French speakers.) In 1948 France had instituted a franc zone by undertaking to buy local currencies at the fixed rate of 50 CFA francs for one French franc, an increasingly expensive way of securing political influence and an increasingly disruptive arrangement as the several currencies diverged from one another. The French franc became extravagantly undervalued, labour became more and more expensive, French investment was diverted to anglophone countries where real rates of exchange prevailed, and a busy trade grew up in smuggling goods to francophone countries in return for local currencies convertible into French francs at the fixed rate which remained unchanged for nearly half a century. In 1994 the new French government of Edouard Balladur devalued the CFA franc by half in terms of the French franc (and the associated Comoro franc by one-third). The aims and hopes of this devaluation were increased investment and liquidity and – a necessary precondition – a change of mood among creditors who might be induced to write off part of their debts. Some debt was written off but creditors already committed to cancelling the debts of the former Soviet satellites in Europe were reluctant to tackle African problems with equal vigour. Inflation running at 30–40 per cent in most of the zone was halted and then reduced to around 10 per cent. Overall growth in West African states was strong at 4–5 per cent but negligible in central Africa where growth depended too greatly on mere optimism. Domestic liquidity rose, mainly through repatriation of capital which had fled abroad, but the money remained in banks instead of being invested, partly because would-be borrowers recoiled from the rates of interest demanded by the banks and partly because the banks distrusted the applicants.

The members of both ECOWAS and CEAO concluded also defence agreements: ECOWAS in 1981, Mali, Guinea-Bissau and Cape Verde abstaining; and CEAO by stages from 1977, Togo joining the original six. The signatories of CEAO's defence pact were divided between some who regarded it as a first step to getting rid of French troops from West Africa and others who were far from ready to approve such a step. These defence arrangements were of much less consequence than economic agreements. The economic agreements were on the other hand hampered by the inescapably worldwide nature of the problems which they addressed.

West Africa's political problems were largely home grown and blamed on the new rulers. The colonial past bore some responsibility, but a receding one. Colonialism had by its very nature destroyed the ruling class and its institutions. Even where the colonial rulers practised indirect rule and so made use of the indigenous system, this system and its protagonists were downgraded. Moreover the new nationalists who led the successful campaigns for independence – unlike the leaders of the National Congress in India which had been founded in the nineteenth century – had minimal experience of government and precarious authority. Their authority was personal. It derived from charisma or rank, so that the choice for the people, so far as they had a choice, lay between the demagogue and the general. Some were good and some were bad, but all lacked systematic political backing and they were obliged therefore to rely on their wits or their swords. The natural outcome was either an entrenched tyranny or constant shifts. In West Africa, which was in this respect distinct from most of the rest of the continent, there was more instability than tyranny. Freedoms were curtailed or abused on the plea that the autocrat or the one-party state would be more efficient, but such régimes failed to deliver the goods.

West Africa – and in this respect its lot was repeated through most of sub-Saharan Africa – saw a modest economic advance in the 1960s, checked in the 1970s and reversed in the 1980s. In the course of the twentieth century Africa's population was multiplied by nearly ten (from 100 million to an expected 1,000 million or not much less) and it moved into cities. Migrants became refugees. The demand for food created by the first of these trends was magnified by the second, as the migrants became non-producing consumers of food. The countryside became denuded, the cities insanitary and dangerous. The food deficit was further aggravated by the post-independence view that the road to prosperity was industrialization. The application of this nostrum created more debt than success, increased Africa's manufacturing output only marginally (it remained about 1 per cent of world output), exposed the lack of adequate managers and fanned the corruption inherent in a chase after contracts. Aid was misused and often the misuse was encouraged by ignorant or greedy donors. Industrial and commercial failures were visited on other sectors: roads, schools and universities decayed and health deteriorated. Only half the children went to school; half the teachers were unqualified. Corruption in business was barely severable from corruption in government which extended to bullying the judiciary and the press. Wars, drought, famine and adverse terms of trade in a world governed by the wealthier countries made matters worse. Some states were woefully small (nine had fewer than 1 million inhabitants at independence) and some were unmanageably big. Nearly all were multilingual, multiethnic, multireligious, multicultural – not the unities which they seemed to be on the map or which outsiders took them to be. Their boundaries, at first decreed sacrosanct, gradually ceased to be so. Foreigners turned their backs on investment in Africa and ignored its tourist attractions. Yet much of the continent possessed excellent agricultural land, valuable minerals and plentiful sources of energy. The economic balance sheet was not wholly negative but the remedies for centuries in an economic backwater, followed by decades of bad policies and bad government, required harsh sacrifices which must fall mainly on the numerous poor, making them not only poorer but also ill-disposed to trust their governments. The IMF's recipe for salvation was two-edged. On the one hand these institutions provided funds but on the other they required devaluation of currencies, liberalization of trade and increases in productivity which more often than not could be achieved only by sacking people, raising prices and cutting services. This was medicine which cured the community by killing its members. Plans to offset the painful side-effects were, again more often than not, either feeble or side-stepped. The World Bank, in a review in 1989 of the well-known ills of poor agricultural performances, industrial decline and excessive debt, noted their political as well as their economic sources – the precariousness of the rule of law, the absence of a free press. The Bank and the IMF had previously tied aid to economic plans and economic performance, setting terms which spelt severe hardship or worse for small business and individuals. These institutions, by identifying good government as a central economic requirement and equating good government with democracy, boosted a shift throughout Africa away from military or one-party régimes to multiparty states (Sudan was the outstanding exception when even Malawi succumbed) but how far this shift might affect either economic fortunes or political behaviour remained to be seen.

The instability of the African state, and therewith its disappointing performance, were due to more than the shortcomings of its leading personalities. The post-independence population explosion negated economic growth and powered poverty, corruption and crime in swollen cities. It shifted the balance between town and countryside at a rate which cities could not handle. It contributed to failures amounting often to collapse, in education at all levels, health services and standards, transport and other public services. Ill conceived and ill managed economic policies produced crippling external debts out of all accepted proportion to GDP or export revenues. But beside these handicaps the state itself lacked the essential attributes of a state: definition and authority. In 1964 the OAU resolved to sanctify the borders delineated by and inherited from the colonial powers in order to avoid disputes likely to lead to war, but this sanctification was also petrification and most African states harboured in consequence ethnic conflicts within themselves or ethnic minorities with kin across the border: they were arenas, not agoras. Political groups or parties tended to be factional rather than national, armies alone being more nearly national. The multiplication of political parties in the name of democracy – encouraged, sometimes imposed, from outside – tended to institutionalize ethnic conflicts to the advantage of an oppressively dominant group determined to monopolize power. A multiplicity of parties was a necessary but not a sufficient ingredient of democracy.

West Africa became independent without anything that could be called an international crisis. East Africa was to follow suit in the next few years. But the independence of the Belgian Congo produced not only internal chaos and civil war but also one of the principal international crises of the postwar period, in which major powers were brought to the brink of a confrontation. African states were divided among themselves, and the United Nations was called upon to play expected and unexpected roles in the course of which it was attacked by some of its principal members, its secretary-general was killed and its very existence was called in question. The sources of this catastrophe were, first, the hurriedness with which the Belgians took their departure from a colony which they had hardly at all prepared for independence; secondly, the very great size of the Congo and its ethnic and tribal diversity; thirdly, the revolt immediately after independence of the Force Publique or army, whose mutinous conduct left the new central government powerless; next, the attempt to detach the rich southern province of Katanga and make it a separate state; and finally, the fact that the United Nations was required to perform a multiplicity of barely consistent tasks and was hampered in them by the inadequacies of its own machinery and by the hostility and independent actions of certain governments, notably the Russian and the British.

The Congo is a large country of over a million square miles inhabited by many different tribes, some of which constitute tribal federations. The most prominent of these are the Bakongo in the west, including the capital of Leopoldville, and also in the neighbouring territories of Angola and Congo-Brazzaville; the Baluba in southern Kasai and northern Katanga; and the Balunda in southern Katanga. In spite of its great size the Congo touches the ocean only at the end of a corridor between Congo-Brazzaville and Angola, and its modern history begins with the ventures of explorers along the Congo River. It became famous as a result of H. M. Stanley's expedition in 1874 and was at once an object of international competition among the principal European powers. Britain, already well endowed and also busy elsewhere, did not enter the lists on its own account but pushed the claims of its old friend,

Portugal. France and the three-year-old German empire were not so altruistic and in 1876 an international conference reached a temporary compromise by creating the International African Association which was to act as a sort of composite cultural mission to enlighten the darkest heart of Africa and find out what was there. This conference met, not by accident, in Brussels and was convened by the Belgian king Leopold II.

The International African Association did not eliminate the ambitions and manoeuvres of the European powers, and in the early 1880s there was widespread fear of war as a result of their jealousies and claims in West Africa from the Niger down to the Congo. In 1884 a conference in Berlin, engineered and presided over by Bismarck, sorted things out on the basis that there was enough for everybody and no need to fight: Europeans would do best to recognize one another's possessions and tacitly sanction in advance further acquisitions of anything not yet under a European flag. The Congo was handed over to the Congo International Association which was the International African Association under a new name, and in effect Leopold II in person. The king thus became the largest private estate owner in the world, although he did not yet know how rich his property was. His obligations were to extirpate the slave trade, permit free trade and secure free passage for all on the Congo River. He defaulted in his first obligation until an outcry in Britain and elsewhere compelled him to abolish the trade – and to resort to forced labour instead. His administration of his estate became one of the most notorious scandals since Cicero's denunciation of the proconsulship of Verres, and in 1908 the Congo was transferred from Leopold to the Belgian state which despatched a governor-general to rule with the help of a Belgian civil service. The interests of Africans were to have priority over the exploitation of the domain but this principle was not in practice found to involve any advancement for Africans or any but the most rudimentary mingling of the races.

Shortly before the First World War the mining of copper in Katanga began and brought with it two potent changes. Katanga gradually became twice as rich as the whole of the rest of the Congo; should the Congo ever become independent the Katangans would be in a position to demand first place in the state or to leave it and set up on their own. The second and more immediate change was the transformation of the territory into a booming partnership between the administration, Belgian finance houses and the Roman Catholic Church. And so it remained for nearly half a century. Some attention was paid to the economic wellbeing of the Africans but no political activity was tolerated and education above the elementary level was reserved for very few. In so far as Belgians thought about the future they imagined a slow advance by Africans to a level at which a new form of association might have to be devised, but a transfer of power to Africans did not enter into their calculations and so no steps were taken to train even an élite. So far as any practical man needed to see the Congo would remain as shut off from the rest of the world as Japan before the Meiji restoration.

This view began to be questioned in the 1950s. The missionaries became alive to the pressures of nationalism and uneasy therefore about the assumptions on which the territory was being ruled by the church and its associates. In Brussels, where a left-wing coalition came to power in 1954, the liberation of French and British territories could not be ignored; de Gaulle's offer of autonomy in 1958 was made at Brazzaville just across the river from Leopoldville. African leaders, some of whom met one another for the first time at the Brussels Fair of 1958, began to espouse independence in place of a somewhat less gradual evolution, and in December 1958 many of them attended the Pan-African Conference in Accra where they received encouragement from fellow Africans and were reinforced in their determination not to get left behind. A few days after the end of the Accra conference there were riots in Leopoldville.

The Belgians had recently appointed as governor-general Maurice van Hemelrijk who believed that an acceleration was the only possible policy. He advocated a parliament for the Congo by the end of 1960 and independence in 1963 (reduced, by the end of 1959, to independence in 1960), but van Hemelrijk was forced by conservative protests in Belgium to resign in September 1959. His successor, Auguste de Schryver, found he had to go even faster. Tribal fighting began before the end of the year and in January 1960 the Belgians assembled in Brussels a conference at which, to their surprise, the Africans unitedly asked for immediate independence. The Belgians agreed to leave the Congo at the end of June. Never had so much been conceded so quickly. The frontiers of the new state were to be the same as those of the colony, and when Sir Roy Welensky suggested that Katanga should leave the Congo and join the Rhodesian federation, the Belgian government was much displeased. The internal structure of the new state, whether it should be military or federal, was however left undecided, while the distribution of top posts was also unsettled pending elections which were held in May. The Belgians made some attempt to remedy their sins of omission by running a crash training programme in Brussels for Africans and by enlisting Ghanaian help in the training of more Africans in the Congo.

At this time the most prominent of the Congolese politicians, and the favourite of the Belgians, was Patrice Lumumba, one of the founders in 1958 of the Mouvement National Congolais which was the main non-tribal party in the Congo. Lumumba and his principal associates, who included Cyrille Adoula and Joseph Ileo, demanded the

Africanization of public services and professions with a view to ultimate independence. At the Accra conference Lumumba struck up a friendship with Nkrumah and, although he became more extreme in his demands during 1959 and was eventually imprisoned, he retained his good standing with the Belgians until the eve of independence. He wanted a strong centre rather than a diffuse federation, in contradistinction to his opponents who were essentially the modern equivalents of tribal and local chieftains: Joseph Kasavubu, the more sedate and aristocratic leader of Abako which was founded in 1950 among the Bakongo with the aim of restoring the old Kongo empire but which became converted to the idea of a Congo state provided it was a federal one; Moise Tshombe, the rich middle-class *évolué* and leader of Conakat through which the Balunda aimed to exercise power in Katanga either within a federal Congo or independently; and Jason Sendwe, whose Balubakat eyed the Balunda to the south with considerable suspicion and prevented Conakat from speaking for the whole of Katanga. Lumumba's tendency to quarrel with other leaders in his own party, and a change of attitude in Brussels, led to attempts to create an anti-Lumumba front, but the attempts failed, Lumumba's party emerged from the elections as the biggest and Lumumba was installed as prime minister of a broad coalition government with Kasavubu as president.

Independence was proclaimed on 30 June 1960 and the official celebrations lasted for four days. Forty-eight hours later there occurred the first mutinies in the Force Publique which sparked off a train of terrible disasters. The soldiers seem to have expected that independence would instantly improve their pay and open their way to the officer grades, which were entirely filled by white men. When nothing happened, they decided to replace their officers. This indiscipline was accompanied by a certain amount of violence and some rapes, much multiplied in the telling. An African sergeant, Victor Lundula, was appointed commander-in-chief and the situation seemed to have been brought under control in a couple of days. But fresh mutinies and more serious violence quickly followed; there was, however, no killing until after Tshombe's proclamation of the independence of Katanga on 11 July.

The mutiny of the Congolese army was the source of all the Congo's woes. It deprived the Congolese government of power and authority. It caused a panic among Europeans, and the Belgians announced that, with or without Congolese consent, they would return to protect their nationals. Tshombe unilaterally asked for Belgian help and declared Katanga independent. The long-term consequences of this action will be examined below. The immediate consequence was to humiliate and infuriate Lumumba who became suspicious of a Katangan–Belgian plot to subvert the independence of the new state and disrupt it by detaching its richest province. Subsequent events successively strengthened his suspicions. Although he remained for a time willing to discuss with the Belgians the maintenance of law and order, he was not willing to do what Tshombe had done and ask for their help in suppressing the mutinies. Since there was no other force immediately available to him, this rift between Lumumba and the Belgians enabled the mutinies to spread and take hold. At the same time the rift between Lumumba and Tshombe made Lumumba more anxious to maintain than to disarm the Congolese army with the result that when UN forces arrived to restore order Lumumba dissuaded them from taking the essential step of disarming the mutinous units.

These were not the only consequences. The Belgians, who occupied Leopoldville with parachute troops on 11 July, in effect abrogated their recently signed treaty of friendship with the Congo and switched to Tshombe, whom they provided with a Belgian armed force and commander-in-chief. They thereby annulled not only the Belgian–Congolese association but also the unity of the Congo. This unity was already precarious and the attempts to re-establish it form one of the two principal strands in the complex story of the next three years. The other strand is the conflict over the powers and duties of the UN forces which began to arrive in Leopoldville on 14 July to restore order and displace the Belgians – but without agreement on whether the restoration of law and order included the subjection of Katanga to the authority of the central government.

On the domestic political front Lumumba and Kasavubu began working in harmony, but in September they broke with each other and Kasavubu dismissed Lumumba and appointed a new government. The parliament supported Lumumba who maintained, with some justice, that the president's action was illegal. In the ensuing crisis Hammarskjöld's representative in Leopoldville, Andrew Cordier, closed airports and radio, thereby giving an advantage to Kasavubu by denying to the more popular Lumumba the opportunity to state his case in different parts of the country or make his voice heard on the air. This action was bitterly resented within the Congo and beyond and led to fierce Congolese attacks on the UN with Russian support. Lumumba remained in Leopoldville in his official residence. Cordier's successor, Rayeshwar Dayal, refused to assist him, and he became in effect not the prisoner but the protégé of the UN (which was now attacked for being pro-Lumumba) until he fled from Leopoldville at the end of November, hoping to reach Stanleyville by car. He was overtaken a few days later with the help of aircraft and was lodged in jail in Thysville, whence he fled to the Katangan capital of Elisabethville on 17 January, only to be captured again and murdered.

Shortly after the Lumumba–Kasavubu rift in September power in Leopoldville was seized by Colonel Joseph Mobutu, the chief of staff of the army (the commander-in-chief, Victor Lundula, was not in the capital). Mobutu ejected both the parliament and the Russian and Czech embassies and, when attempts to reconcile Lumumba and Kasavubu failed, declared for Kasavubu who acquiesced in the colonel's virtual usurpation. Mobutu was the principal ruler in the capital until the end of the year. He introduced a new constitution and some degree of order and efficiency, but although supported by the west he failed to establish the sort of military régime which other soldiers were successfully establishing in Asian and Arab states at this time. The provinces did not respond and the resources available to him – communications, trained personnel – were totally inadequate. Moreover the army itself was divided; General Lundula and the forces in Orientale province remained pro-Lumumba, as did many politicians and, so far as was ascertainable, the popular favour.

In February 1961 it was apparent that the Mobutu–Kasavubu team had failed and a new government was appointed under Ileo. It lasted six months until August when Ileo was succeeded by Adoula, who remained prime minister until 1964. During this period the Lumumbists, led by Antoine Gizenga in Stanleyville, and the Katangans, led by Tshombe and Godefroid Munongo in Elisabethville, were separate and often separatist factors in the situation. Various attempts were made to bring all three sections together but hopeful moves in the direction of Stanleyville usually caused Elisabethville to shy away, and vice versa.

The Stanleyville secession was never formalized in the same way as the Kantangan. It developed after the break between Lumumba and Kasavubu and gathered strength after the murder of Lumumba when the Russians seemed to be on the verge of recognizing the Stanleyville party as the government of the Congo and of supplying it with arms. But if the Russians were expected to secure a foothold in Africa by exploiting the emotions roused by Lumumba's death, they miscalculated; their faces were as white as any Belgian's, their intervention was in principle as unwelcome, and instead of pleasing they shocked Africans by the ruthlessness with which, upon their arrival in Orientale, they proposed to set about the anti-Lumumbists. They were seen off by a small party of UN troops, having badly damaged their reputation in Africa. Gizenga meanwhile equivocated over his relations with Ileo, watching the state of Ileo's relations with Tshombe, but when he was appointed deputy prime minister to Adoula he accepted the post on the understanding that Adoula would use force to put an end to the Katangan secession. He accompanied Adoula to the conference of non-aligned nations in Belgrade in September 1961. The failure of the operations against Katanga, about to be related, revived his suspicions and he returned to his own base in Stanleyville, thereby re-creating the tripartite pattern. Attempts to induce him to return to the capital having failed, he was brought there under arrest in January 1962, lodged in jail and expelled from the government.

The Katangan secession was a formal, if illegal, state of affairs with a precise beginning and a precise end. It was proclaimed by Tshombe on 11 July 1960 and renounced by him on 21 December 1961. Its inception thus coincided with the mutiny of the Force Publique and Belgian intervention, and it was accompanied by an appeal to Belgium and a refusal to allow Kasavubu and Lumumba, the federal president and prime minister, to go to Elisabethville. As a result of these moves Katanga was for the time being protected from the chaos developing in other areas, and while the UN was trying to re-establish order in Leopoldville province the Belgians did so in Katanga. They also provided Tshombe, at his request, with administrative services, and – the most important aspect of the partnership – they ran the mines and paid royalties to Tshombe instead of to the central government. These payments were a direct breach of the pre-independence agreement (the *loi fondamentale*) which had been signed by the Belgian government and accepted by, among others, Tshombe; but they enabled Tshombe to enlist and pay an army of foreigners with which to oppose his Congolese adversaries and, if need be, the UN. One of the most pregnant eventualities of the first weeks of independence was this alignment of Katanga – wealthy, relatively isolated, strongly led and able to command Belgian aid – against the rest of the Congo and its UN supporters.

The UN came to the Congo in response to three appeals by Kasavubu and Lumumba on 10, 12 and 13 July 1960. They appealed first for technical aid, including aid in the organization and equipment of security forces. In their second and third messages they appealed for help against Belgian aggression. Hammarskjöld, taking the initiative under article 99 of the Charter, on 13 July asked the Security Council to consider technical aid for the Congo and the problem of law and order. The Council authorized the despatch of military aid to the Congolese government with the proviso that force should not be used by UN troops except in self-defence. The council was divided on whether to require the Belgians to withdraw, Britain, France and China abstaining from the vote. Hammarskjöld immediately asked African states north of the Congo for military and other help, an appeal later extended to certain European and Asian members. The first troops reached the Congo on 14 July and four days later a force of 4,000 had been assembled. The air lift was provided by the United States, Britain and the Soviet Union. Differences of opinion about the functions of the UN troops arose at once. General Alexander, in command of the Ghanaian contingent, immediately began disarming the mutineers of the Force Publique. Had he been allowed to continue to do so, all might have been very different, but Lumumba insisted on stopping the process, partly because he was suspicious of outside interference and partly because he wanted to use the Force Publique against Katanga.

The Security Council met again on 20 July and resolved unanimously that the Belgians should withdraw and that other states should refrain from aggravating the situation; it confirmed the authority already given to Hammarskjöld and commended what he had done. The Belgians began to withdraw on the 20th and by the 23rd they had left the Congo again – except Katanga. Ghana and Guinea then threatened to withdraw their troops from the UN force and place them at the disposal of the Leopoldville government in order to help evict the Belgians from Katanga and force the province to accept the authority of the Kasavubu–Lumumba régime. Consequently the UN had to consider most urgently whether its own force was entitled to enter Katanga to achieve these objects and, if not, whether it should now be empowered by new resolutions to do so. Hoping to by-pass this difficulty Hammarskjöld flew to Brussels and the Congo to secure the entry of UN units by negotiation and announced on 2 August that a first contingent would do so in three days' time. But the Katanga authorities said they would resist and Ralph Bunche, despatched to Elisabethville to discover if they really would, advised postponing the move. Rather than use force, and uncertain whether he was empowered to in these circumstances, Hammarskjöld returned to New York to seek further instructions from the Security Council. This check was accounted a victory by the Katangans. It further infuriated Lumumba who decided that the UN had let him down and that he should seek African support for a campaign against Katanga.

The Security Council's third meeting took place on 8 and 9 August and (France and Italy abstaining) repeated the injunction to the Belgians to leave immediately, authorized the entry of UN forces into Katanga and said once more that these forces were not to be used to influence the internal conflict. The last part of the resolution was but doubtfully consistent with the second, since the essence of the conflict was the authority of Leopoldville over Elisabethville, and the entry of UN troops into Katanga, even if designed only to compel the departure of the Belgians, could not fail to have an effect on the balance of domestic Congolese forces which were in conflict. Hammarskjöld, fortified by this resolution, returned to the Congo and to his policy of getting UN forces into Katanga bloodlessly. He entered Katanga with a token force but refused to take a representative of Lumumba with him, thus further antagonizing Lumumba who became convinced that the secretary-general was embroiled in a plot against him. For his part Hammarskjöld became convinced that Lumumba's main aim was to get rid of the UN presence. Tshombe meanwhile used the breathing space afforded by these quarrels to consolidate his position. A fourth meeting of the Security Council on 21 August produced an ominous sign when a resolution of strong support for Hammarskjöld was opposed by the Soviet Union and Poland.

Throughout August the situation in the Congo deteriorated with every prospect of a clash between the Congolese and Katangan armies. A conference of thirteen African states in Leopoldville failed to give Lumumba the support he wanted and advised against an attack on Katanga, but Congolese troops were taking matters into their own hands and contrived a massacre of Baluba who were trying to establish a state of their own between the two principal river forces. (They had already suffered a massacre by the Balunda since Tshombe could not establish his predominance in Katanga as a whole without this radical alteration of the numerical balance.) At this point Kasavubu's resolve to get rid of Lumumba, whose unpredictability had grown alarmingly and whose reputed pro-communism was, though unfounded, embarrassing, gave Tshombe a second breathing space; the projected attack on him was called off, and his Belgian troops took the opportunity to move north and establish a second secessionary state in Kasai under the short-lived presidency of Albert Kalonji. A separate state backed by the Belgians had come into existence in the south, while in the north-east the Russians were beginning to play with the idea of another separate state backed by themselves.

The Congo seemed to be about to break up into three large and warring units, two of which would be foreign bases, and a number of smaller ones. The situation was aggravated upon the opening of the UN General Assembly in September, when Khrushchev arrived in person to attack the secretary-general and two rival Congolese delegations competed with each other for seats in the Assembly and the ears of its members.

This session of the Assembly was notable for the admission of seventeen new African members, and for the part they played. They refused to support the Russian attack on Hammarskjöld and joined with the western bloc to isolate the communist states. But they did not agree with prevailing western views on the Congo, nor did they remain solidly united among themselves. In Britain, France and the United States the Katangan case, propagated by an active and lavishly supplied promotion lobby which tended to put the interests of its clients before the truth, made many converts in political and business circles on the plea that Katanga was an oasis of peace and sanity in an otherwise barbarous and increasingly communist Congo. This thesis found no acceptors among Africans who unanimously condemned Tshombe and his ways, while being divided on what to do. One group turned against the UN and reverted to Lumumba's plan for a joint African invasion of Katanga. Another group remained attached to the idea of UN action, though dissatisfied with the action apparently contemplated; this group became a pressure group at the UN with the object of persuading the secretary-general and other members that a policy of reducing Katanga by negotiation was hopeless and should be replaced by direct action. A third group of Africans, consisting of recently emancipated French colonies, placed its faith for a time in the Mobutu–Kasavubu alliance and the gradual radiation of law and order from Leopoldville out into all the provinces. The decline of Mobutu, the murder of Lumumba and the installation of Ileo provided a depressing background to events in the last months of 1960 and the beginning of 1961, during which successful attempts were made by Hammarskjöld to get the Russians and the Belgians to stop their independent bolstering of Stanleyville and Elisabethville.

On 21 February 1961 the Security Council explicitly authorized the UN forces to use force in the last resort to prevent a civil war. It did not, however, authorize the use of force against Katanga, or for the removal of the Belgians, or to secure a political solution. This development marked a return to better relations between Hammarskjöld, the independent African states and the west – or at any rate the United States where Kennedy had just assumed the presidency – but it alienated not only Tshombe but also the Kasavubu–Ileo government which suspected the UN of being in western pockets, disliked any increase in its pretensions and now began to draw closer to Tshombe. At a conference at Tananarive in March Tshombe persuaded Ileo to accept a plan for a loose confederation of sovereign Congolese states. No Lumumbist had attended the conference, whose proceedings they regarded as a western ruse to give legitimacy to Tshombe by manipulating the ex-French colonies in his favour. Ileo, however, quickly repented of having moved so far in the direction of Katanga, set about repairing relations with Gizenga and the UN and, at a further conference at Coquilhatville, ordered and effected the arrest of Tshombe.

In July the Congolese parliament assembled at Lovanium, the university town near Leopoldville, in order to patch up a grand coalition. Tshombe was released, Ileo gave way to the honest and respected manager of men Adoula (a sort of Congolese Eisenhower), and Gizenga joined the government. On the UN side the replacement of Dayal, who was weighed down by the unhappy past, by the Tunisian Khiari and the Ghanaian Gardiner helped to improve relations between the government and the UN. But the grand coalition was not achieved. Tshombe was now odd man out.

There followed the military operations against Katanga. By now Katanga was in African terms a formidable power, equipped since the beginning of the year with men, supplies and even aircraft from Belgium, France, South Africa and Southern Rhodesia. A few specified Belgians had been removed as a result of laborious negotiations, but the course of these discussions and of events convinced UN representatives on the spot that Tshombe was only playing for time and had no real intention of dismissing the Belgian and other mercenaries or of coming to terms with Leopoldville. These suspicions received blatant confirmation at the end of August. UN forces seized about 100 foreign officers who had been declared to be undesirable aliens by the Adoula government. Tshombe acquiesced but the local UN representative, Conor Cruise O'Brien, agreed that in order to minimize personal affronts the details should be handled by the Belgian consul in Elisabethville, who guaranteed the voluntary departure of the officers but then failed to honour his word. Moreover other foreign officers not in Elisabethville remained untouched. This attempt to evict some officers was doubly unfortunate for the UN since its failure encouraged Katangan obduracy while its legality was strongly attacked by British representatives in the Congo and New York and Britain became for a while as useful an ally for Katanga as Belgium or France.

In the face of this rebuff the UN had to decide whether to take further steps or to acquiesce. Khiari and O'Brien, believing that Munongo was the real kernel of resistance, hoped to separate Tshombe from Munongo and to get Tshombe to Leopoldville where Hammarskjöld would come to talk him into a more amenable and reasonable frame of mind. But Tshombe refused to be enticed and O'Brien, believing that he had Khiari's authority to use force to round up the foreign officers in Katanga and put an end to the Katangan secession, planned a second military operation, which included the virtual kidnapping of Tshombe. It failed. The Katangans were better prepared and the British government, by refusing to allow Ethiopian jet aircraft to fly over Rhodesia to the Congo, presented the Katangans with a decisive advantage. UN forces captured the post office and radio station at Elisabethville, but Tshombe took refuge with the British vice-consul and then in Rhodesia. The UN operation had produced bloodshed which, much exaggerated in the reporting, shocked and antagonized those who thought that a peace force must achieve peace without force, and Tshombe had been let slip. Hammarskjöld arrived in Leopoldville to find an utterly unexpected situation – confusion and stalemate in Katanga, stubborn and effective hostility from Britain and, if in smaller degree, from the Belgian and American governments also. He determined to seek out Tshombe and talk to him. He left for Ndola in Rhodesia by air on 17 September and was killed when his aircraft crashed *en route*.

This appalling calamity – for Hammarskjöld was one of the half-dozen outstanding personalities in postwar international affairs – was followed by a cease-fire and the return of Tshombe to Katanga. After the appointment of U Thant to serve the rest of Hammarskjöld's term of office, the Security Council returned to its familiar dilemma:

whether to try to reconcile Adoula and Tshombe or to bring the latter to heel. On 24 November the Council, Britain and France abstaining, authorized the new secretary-general to use force to expel foreign mercenaries and political advisers from Katanga, thus implicitly endorsing O'Brien's policy although he himself was removed from the scene. UN forces were still in Elisabethville but they were awkwardly placed since Tshombe's mercenaries were manifestly keen to provoke a next round of fighting, in which they would surround and destroy the UN forces. The UN representatives in Katanga decided to take action in order to prevent their scattered units from being picked off piecemeal before the arrival of reinforcements, but they were again handicapped by Britain which, having promised to deliver a supply of 100-lb bombs, succumbed to right-wing pressures and cancelled its undertaking. Fighting was inconclusive and when Tshombe agreed to meet Adoula, U Thant ordered a new cease-fire. On 21 December 1961, at Kitona, Tshombe renounced secession, and the Kitona agreement was endorsed by the Katangan assembly two months later.

But the troubles of the Congo were not so easily assuaged. During most of 1962 Adoula and Tshombe were engaged in a series of fruitless discussions concerning the implementation of the Kitona agreement, which was opposed in Katanga by Munongo and European secessionists. Tshombe, who interspersed his meetings with Adoula with secret visits to Welensky in Rhodesia, seemed unable to make up his mind. He eventually departed for a prolonged stay in Europe. Although Adoula took three Katangans into his cabinet in April 1963 no genuine reconciliation was brought about. The central government continued to be oppressed by problems of economics and law and order, and early in 1964 a more serious revolt broke out in Kwilu under the leadership of Pierre Mulele, who had recently made a trip to China. This outbreak was the more ominous because the UN troops were departing. The last planeload left at the end of June, the fourth anniversary of independence.

The balance sheet of these four years of international endeavour, sometimes co-operative and sometimes competitive, is not easily struck. At the very beginning the UN, through sticking to the letter of the law, desisted from disarming the mutinous Force Publique, a major source of continuing disorder. There were also mistakes and misunderstandings in the handling of relations with the legally appointed, if personally difficult, prime minister Patrice Lumumba, and the resulting suspiciousness affected not only the UN's own standing and operations but also relations among African states and relations between them and other states. Given the circumstances, however, the misunderstandings could easily have been worse. In the Katanga operations the UN earned the hostility of certain powers and suffered from the criticism that UN forces had come to secure order but had shed blood in the name of peace. Apart from the fact that all police operations must envisage the use of force as an ultimate sanction, the blame for the ambiguities over Katanga belonged primarily to the members of the Security Council whose instructions were not at first sufficiently precise; the reverses suffered by the UN in terms of public sentiment demonstrated above all the need to provide the secretary-general with more effective consultative and executive machinery. In the last resort, however, a joint operation such as the UN operation in the Congo was bound to run into difficulties as soon as there appeared among the members any serious discrepancy of aim.

The successes of the UN were considerable. It achieved almost immediately its first aim of displacing (except from Katanga) the Belgians who had returned when the Force Publique mutinied. Its intervention also prevented intervention by individual states on their own account and enforced in one case, the Russian, a retreat; fears that Africa would become a new theatre for the Cold War were allayed. The UN also succeeded to a remarkable degree in keeping the Congolese economy going, providing elementary administrative services, and preventing famine and epidemics. It could take credit for staving off civil wars in the Congo which would almost certainly have been worse but for the UN presence and became worse when that presence was removed. Finally, when the UN forces departed in June 1964 Katanga had not seceded. But it was still, as it was bound to be, richer than the whole of the rest of the country, economically so dominant that its leaders were in a position either to play a separatist hand or to claim a dominant position in the central government.

The central government's lack of authority became patent immediately after the departure of the UN forces. Civil war started again. The Adoula government was first reconstituted and then replaced by a new administration under Tshombe, who returned from Europe as the UN departed. Tshombe tried to form a broad coalition and to enlist the support of the OAU, but Gizenga (whom, among others, he released from prison) formed a new opposition party and the OAU – in spite of creating a reconciliation commission under Kenyatta's chairmanship – failed to find a remedy for the Congo's ills. Tshombe was too greatly disliked to be able to control the situation except by arms and he began, soon after his return, to enlist a new force of (mainly) Belgian and South African mercenaries. This force was quickly successful and the rebels, who had taken Stanleyville in August, found themselves threatened with the loss of all their principal strongholds. They had meanwhile taken hostages, and in an attempt to rescue these hostages Belgian parachute troops were flown in October in American aircraft from the British island of Ascension to Stanleyville, which was recaptured from the rebels. About 200 hostages were killed in spite of this operation (or, as some maintained, because of it), in addition to some 20,000 Congolese who lost their lives in the furies of this rebellion. Africans outside the Congo were divided between those who took the easy course of denouncing the Belgians, Americans and British as imperialist recidivists and those who, stifling their dislike of Tshombe, defended the right of the Congolese government to ask for outside help if it wanted to. Tshombe, who had already in July been refused an invitation to the OAU's conference of heads of state in Cairo, was kept out of the conference of non-aligned states in the same city in October. Upon his arrival in Cairo he was escorted to a hotel and kept there until he decided to return to Leopoldville.

Both in Leopoldville and in Stanleyville, the Congolese leaders were divided among themselves. Rebel unity did not survive reverses and President Kasavubu's appointment of Tshombe as prime minister had not betokened any real reconciliation. Although Tshombe was successful in elections in April 1965, Kasavubu shortly afterwards dismissed him. The president failed, however, to construct a new government without Tshombe and in November the army, in the person of General Mobutu, stepped in, dismissed the president and established military rule. With this revolution the army, which alone had had real power since independence, assumed responsibility as well. Mobutu, who had once been a Lumumbist, defeated Mulele's revolt in 1966 and next year survived an attempt, assisted by mercenaries, to restore Tshombe. He broke up the political map by reducing the number of provinces from twenty-one to twelve and then to eight, and he reduced the likelihood of secession by nationalizing the assets of the Union Minière. He repaired relations with Belgium, which he visited in 1969, concluded financial and technical agreements and played host to King Baudouin in Kinshasa. Although he had to imprison large numbers in the process, he brought a semblance of peace and some prospect of economic improvement to an exhausted country. In 1970 he became president for seven years. He felt strong enough at home to give his attention to wider African problems, meeting President Ngouabi of Congo-Brazzaville (relations between the two countries began a much needed improvement) and also Presidents Kaunda and Gowon. He introduced civilians and younger men into senior government posts in place of Adoula who fell ill and Bomboko and Nendaka, two other senior figures, who were sacked. He consolidated his personal power, renamed the country Zaïre and set about turning it into the Iran of Africa with himself as its shah.

In 1977 the Lunda peoples of Zaïre and neighbouring Angola, organized as the Front de Libération Nationale Congolaise (FNLC), staged a revolt in Shaba (ex-Katanga). The west assumed that Cubans, who were helping to train refugees from Zaïre in Angolan camps, had promoted the revolt but Castro swore that he had on the contrary tried to restrain it and there was no evidence to gainsay him. The wealth of the province – cobalt, uranium, diamonds and copper – made it a matter of concern to the outside world and in spite of the embarrassment of going to the aid of a régime as corrupt as Mobutu's France transported, largely at Saudi Arabian expense, 1,500 Moroccan troops to stiffen a Zaïrean army which was widely regarded as useless. A year later a second revolt took place. One hundred and thirty Europeans were killed and in order to save further lives 700 French and 1,700 Belgian paratroopers were flown to Zaïre in American aircraft. Having evacuated 2,500 Europeans they were replaced by a mixed African force recruited from Morocco, Senegal, Ivory Coast, Togo and Gabon. This African intervention was to be coupled with a wider international effort to rescue Zaïre from the bankruptcy into which it had been plunged by the collapse of the world price of copper, the closing of its principal rail outlet (to Benguela in Angola) and Mobutu's malpractices. But aid from the IMF and the EEC was dependent on internal reforms and resident IMF supervision, and since Mobutu postponed the one and cold-shouldered the other the plans came to nothing. Some sections of the economy prospered but the state piled up huge debts and was shored up by western countries which valued its minerals and superficial stability. Over a quarter of a century in which Mobutu ruled most Zaïreans became progressively worse off and the life of the average Zaïrean ended at 40, while Mobutu himself and his family became prodigiously and ostentatiously rich. For most of this period Mobutu seemed irremovable but suddenly appearances changed. By 1990 bazaars and universities were seething, the economic infrastructure collapsing, prices soaring and civil servants striking for a fivefold increase in their salaries. Serious disorders culminated in a massacre at Lubumbashi. Mobutu promised constitutional changes but granted none. He espoused a multiparty system. Hundreds of parties appeared and he proposed to legitimize three of them; but these were mere manoeuvres. The reality was two governments in the capital, provincial governors who were virtually independent pro-consuls, and everywhere lawlessness. By promoting one prime minister after another, Mobutu weakened all of them and when the Supreme Council appointed a prime minister (Etienne Tshiseki) Mobutu installed another. He kept the subservient national assembly in being beyond its proper term. The economy continued to dissolve, prices soared, salaries were unpaid, education, transport and other services collapsed, swindlers flourished openly, great numbers of people lived in fear, parts of the army were out of control. In serious rioting, possibly incited by the government itself, shops and factories and private property were ransacked and Mobutu was forced into seclusion but not out of office. He kept much of the army paid, using the assets and printing presses of the central bank to ensure its loyalty as well as maintain his imperial style. He recovered some of his frayed international standing by mediating in Angola, by being helpful when the Rwandan crisis spilled over into Zaïre and by trading on international fears of accelerating chaos in so important a country as Zaïre. In the 1990s the World Bank was forecasting a different disaster: that within forty years Zaïre's population would treble.

The Sudan

The history of the Sudan in modern times is an alternation of subjection and self-rule. From 1820 to 1881 it was a part of the Ottoman empire in name and of its autonomous Egyptian pashalik in fact. From 1881 to 1898 it was ruled by the Mahdi and his successors. (The Mahdi – Muhammad Ahmad Abdullah, who proclaimed himself Mahdi in 1881 – died in 1885.) From 1899 to 1955 it was subject in theory to an Anglo–Egyptian condominium, in practice to Britain. In 1956 it became an independent state, and then a member of the Arab League and the United Nations.

The Anglo–Egyptian condominium reflected no Anglo–Egyptian agreement about the Sudan. The administering powers agreed in little and did not bother to co-operate. After the Anglo–Egyptian treaty of 1936 the British allowed a greater degree of Egyptian penetration, but although the Sudan was coveted by Egyptian governments it attracted few Egyptian administrators. Sudanese nationalism gathered strength in the 1930s under the leadership of Ismail al-Azhari who became secretary of the Graduates General Congress, formed in 1938. Owing to the active role of Britain and the comparatively negative role of Egypt Sudanese nationalism was at first more anti-British than anti-Egyptian and indeed looked to Egypt to help displace the British. This predominantly anti-British streak was intensified during the Second World War when nationalists raised the issue of the Sudan's postwar status and received unsympathetic answers from the British who were fully occupied with present and pressing realities in the campaigns against the Italians in Ethiopia and against the Italians and the Germans in North Africa. The Egyptians sensed the opportunities in this situation and welcomed the formation by Azhari of the Ashigga party which aimed at the union of the Sudan with Egypt. He had the support of Ali el-Mirghani, one of the Sudan's two principal religious leaders. As a direct response a second party, the Umma Party, was created to work for the independence of the Sudan under the leadership of the Sudan's other principal religious leader, Abd al-Rahman al-Mahdi, the

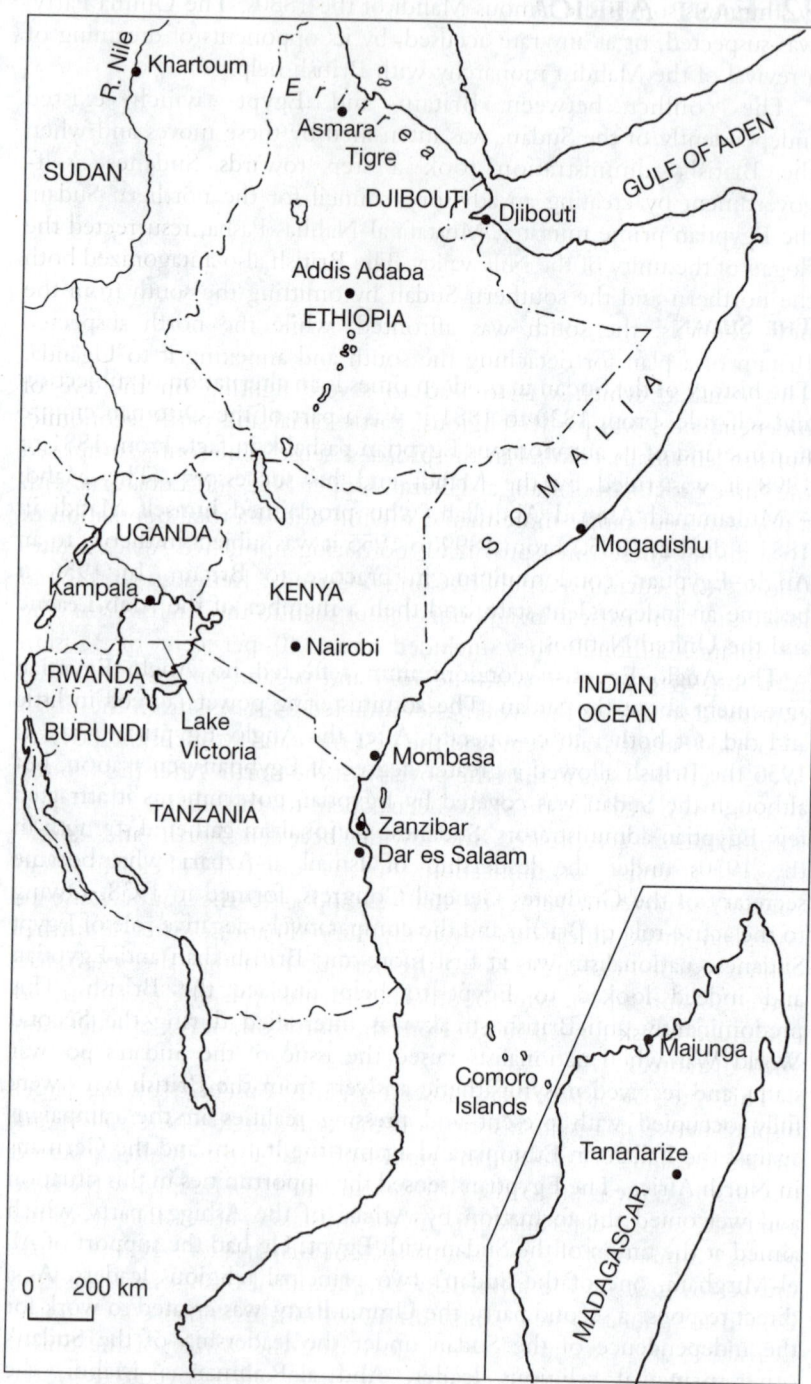

22.1 The Horn and East Africa

posthumous son of the famous Mahdi of the 1880s. The Umma Party was suspected, or at any rate accused, by its opponents of dreaming of a revival of the Mahdist monarchy with British help.

The conflict between Britain and Egypt, which existed independently of the Sudan, was intensified by these moves and when the British administration took a step towards Sudanese self-government by creating an advisory council for the northern Sudan, the Egyptian prime minister, Mustafa al-Nahhas Pasha, resurrected the slogan of the unity of the Nile valley. The British also antagonized both the northern and the southern Sudan by omitting the south from the new councils; the south was affronted, while the north suspected Britain of a plan for detaching the south and annexing it to Uganda. This conflict, which was to lead to severe fighting on the eve of independence, was partly religious, partly racial and partly economic, but fundamentally cultural in the broadest sense. The south comprised the three provinces of Bahr al-Ghazal, Upper Nile and Equatoria with a population of rather more than 3 million out of a total population of 10 million. These three provinces, bordering upon five African states (the Central African Republic, Congo, Uganda, Kenya and Ethiopia) looked to their African neighbours rather than to the Arab north. The inhabitants, although they included some 40 per cent of Muslim Africans, were predominantly negroid and pagan; some of the tribes continued to display the weaknesses inflicted by slavers from the north who had systematically removed the best men and women; the poverty of the south was worse than the poverty of the north. After the war the British, who had in the past been open to the charge that their administration accentuated the divide between north and south, adopted a policy of unification, but a conference held in 1946 contained no southern representatives and gave the south cause to complain that unification was really subordination. After a further conference at Juba in 1947 northerners claimed that the south had agreed to unification, while southerners denied this interpretation and continued to ask for separation from the north or a form of federation with guarantees.

This internal dispute was, however, partly obscured by the larger dispute between Egypt, insisting on the unity of the Nile valley, and Britain, insisting on the Sudan's right to decide for itself after a period of self-government under the British aegis. In 1947 Egypt accepted the principle of self-determination in the mistaken belief that it would produce a union of the two countries, but Britain and Egypt were still unable to agree on the immediate constitutional changes and in 1948 Britain introduced a new constitution without Egyptian concurrence. Legislative and executive councils covering the whole of the Sudan were established but the elections for them were boycotted by the pro-Egyptian parties. After anti-British demonstrations Azhari was arrested. Egypt was at this time embroiled in the Arab–Israeli war and its immediate aftermath.

In 1951 the British governor appointed a commission to consider further constitutional advances. The Sudan was plainly moving fast along the road through self-government to self-determination, and Egypt thereupon carried out a threat, uttered in 1950, to abrogate the condominium agreement of 1899 and the Anglo–Egyptian treaty of 1936. King Farouk was declared king of Egypt and the Sudan. But in 1952, as a result of the revolution in Egypt, he became king of neither.

A few months earlier Britain had produced a self-government statute for the Sudan which provided for the creation of a council of ministers, a house of representatives and a senate, and reserved to the governor only external affairs and ultimate emergency authority in domestic affairs. The new Egyptian government, whose chief, General Neguib, was half Sudanese, accepted the self-government statute, improved it (from the Sudanese point of view) and persuaded all the principal Sudanese parties to issue a pro-Egyptian declaration. The modified statute curbed the governor's authority by introducing a five-man international commission and by reducing special powers originally reserved to him in the south. It was embodied in the Anglo–Egyptian agreement of February 1953 and was intended to lead to independence not later than January 1961. Elections, preceded by vigorous Egyptian campaigning, were held at the end of 1953 and gave an overwhelming victory to Azhari's refurbished party, now called the National Unionist Party. Azhari became prime minister at the beginning of 1954. At the end of the same year Neguib was out.

The disappearance of Neguib was not the only cause of the ebbing of the pro-Egyptian tide in the Sudan. The tide had in fact not been flowing as strongly as it seemed, since the voting in 1953 had been more anti-British than pro-Egyptian. Two years after the elections the Sudan shook off both the British and the Egyptians and declared its independence, but before doing so it experienced mutiny and rebellion in the south. Disappointed by the new government's allocation of posts and by the persistent rejection of its demand for secession or a federal association, the south exploded in August 1955 as the result of an incident in which an army section turned upon its Arab officer. Disaffection spread rapidly and was only mastered when the prime minister invoked the aid of the British governor and moved substantial forces from the north into the south. The south, submitting, claimed that it had done so in reliance upon British mediation and was subsequently let down. Repression was certainly thorough and probably brutal. Southerners were executed in their hundreds and deported to the north in their thousands and fled to neighbouring African territories in their tens of thousands. The south became a closed area whence reliable information was difficult to obtain and was replaced by gruesome reports.

The revolt accelerated independence which was claimed by Azhari before the end of the year and inaugurated on the first day of 1956. It was followed by the break-up of Azhari's party, a coalition under him and then a coalition without him. New parties formed. Abdullah Khalil, who succeeded Azhari in 1956, grappled for two years with economic problems (lightened by an aid agreement with the United States in 1958) and with reviving fears of Egypt. In 1958 Egypt sent troops into two disputed areas on the Egyptian–Sudanese border, but this tactless demonstration recoiled on Cairo at elections in the same year which gave the Umma Party the largest number of seats. Nevertheless Khalil was afraid of an alliance between Azhari and Nasser and he was suspected of being privy to an army coup by which he and his government and the constitution were all swept away. General Ibrahim Abboud took power with a supreme council of officers.

Military rule lasted for six years, during which the situation in the southern provinces deteriorated drastically. The Sudanese government pursued a policy of Arabization and looked with extreme suspicion on the foreign Christian missionaries whom it suspected of being not merely anti-Muslim but also active agents of a western policy of separatism. The Missionary Societies Act of 1962 restricted their activities and induced a number to leave; the surviving 300 were expelled in 1964 after the resumption of full-scale insurrection by the separatist Anya Nya movement and brought with them to the outside world tales of extensive massacres by government troops. In October 1964 General Abboud's military régime was overthrown and after a few weeks during which General Abboud ruled in conjunction with a civilian administration he himself resigned. Ser el-Khatim Khalifa formed a transitional coalition of intellectuals and re-emergent politicians which was replaced early in 1965 by a more normal coalition of politicians in which the Umma Party had the largest voice – a position which it confirmed and enhanced at elections in April. A southerner, Clement Ngoro, held the ministry of the interior and north–south negotiations took place both in the Sudan and in Nairobi with a number of leading southern exiles. A further conference planned for February at Juba was postponed owing to a refusal by the Anya Nya to surrender in advance, but discussions were resumed in Khartoum in March. With the south demanding a federal constitution and northern politicians inveighing against separatist plots little was achieved. Fighting continued and southern opinion moved increasingly away from federation and towards independence with both sides seeking support from neighbouring states. The Sudanese government exerted itself to improve its relations with Ethiopia, Uganda and the Central African Republic, and the Sudanese African National Union appealed to African opinion through the Organization for African Unity and its Liberation Committee in Dar-es-Salaam.

In 1969 a group of military officers led by General Gaafar Muhammad Nimeiry seized power and established a more left-wing and pro-Egyptian government. The coincidence of similar coups in Somalia and Libya gave rise to projects for a union of all these countries with Egypt and the creation therefore of an Arab bloc of north-east African states. In 1970 this plan was further expanded to attract South Yemen and Syria, thus linking it with the defunct union of 1958 with Egypt as the central point and common factor, but these foreign schemes came to nothing. In the Sudan they were attacked by critics who argued that they were diverting attention from more urgent domestic problems, especially the dissatisfied south, where civil war dragged on until 1972. Nimeiry survived a series of attempts to unseat him which, in 1975 and 1976, were at least assisted, if not instigated, by Libya and Ethiopia. In 1977 the Sudan moved close to war with Ethiopia as a result of the help which it was giving to the Eritrean revolt against Addis Ababa. At home a reconciliation between Nimeiry and his principal opponent Sadiq al-Mahdi, marked by the latter's election to a vice-presidency of the Sudanese Socialist Union, brought Muslim fundamentalists into the government and thereby split the Muslim Brotherhood between those prepared to share power in this way and others who insisted on holding out for the imposition of comprehensive theocratic rule. The coalition crumbled in 1979. Nimeiry refused to sever relations with Egypt after Camp David.

Throughout the early 1980s Nimeiry's fall was confidently predicted. His troubles in the south revived, his economy went from bad to worse, students and other malcontents demonstrated against him, his suppliers – mainly Saudi Arabia and the United States – began to ask themselves whether they would not be better without him; and his borders were crossed by hordes of destitute and starving refugees from Ethiopia and Chad. He reacted by imposing and savagely implementing Islamic laws and penalties, calling himself Imam (although a third of his countrymen and -women were not Muslims) and by granting valuable concessions to Saudi and other tycoons. Sadiq al-Mahdi opposed the severity of the application of the Shar'ia and was put in prison in 1983 but released the next year. By 1985 about a fifth of the population was starving, maladministration had brought public services and basic works (irrigation, for example) to a standstill, the special security police were justly loathed, the foreign debt service exceeded the value of exports, foreign loans were obtainable only at ruinous cost, the south was again in arms against Khartoum and to all intents and purposes autonomous.

In April Nimeiry, away on a visit to the United States, was removed by
the army after what amounted to a strike by the professional classes – a
pre-emptive coup by the military to forestall more radical action. The
succeeding military government proved only a stop-gap while
traditional political parties argued their way to a civilian coalition with
Sadiq al-Mahdi as prime minister. But al-Mahdi, weak and inconsistent,
gave hostages to fortune by espousing (contrary to his campaign
rhetoric) Islamic fundamentalist policies and fell out with his allies of
the Democratic Unionist Party by failing to stop – or not trying hard
enough to stop – the war in the south. He was accused of sabotaging
peace moves. The DUP left his government in 1988 and he became
dependent on the National Islamic Front. The army feared defeat in the
south, was impatient with his dithering and fearful of economic
collapse. The war, resumed in 1983, had cost at least half a million dead
and 1.5 million refugees. The external debt had risen to $7 billion;
discussions with the IMF broke down; food riots became frequent. In
1989 the army resumed overt power under General Omar el-Bashir, a
more circumspect but hardly less convinced Islamicist than his
predecessor, but the army had by this time become permeated by the
NIF and el-Bashir was a figurehead masking the real power of Hassan
al-Turabi, the NIF's leader, who succeeded in claiming for himself and
his party a perverse monopoly of Islamic orthodoxy. Besides its appeal
to army officers, the NIF had made converts among businessmen and
in the public services in reaction against the striking incompetence and
outrageous corruption of earlier governments. But the new régime
failed to check inflation, attract foreign investment or stop the war in
the south. Its sympathy for Iraq soured its relations with Saudi Arabia
although its ideological credentials won it a visit from the Iranian
President Rafsanjani. In the south the fall of Mengistu in Ethiopia
threatened the Sudan People's Liberation Army supplies and
communications with the outside world, exposed half a million refugees
to starvation and strafing while they struggled to find escape routes to
and through Kenya, and caused defections from the SPLA's leadership
over the question whether to accept anything less than full sovereignty
for the south. The government in Khartoum, widely vilified abroad for
its atrocious conduct of the civil war, anxious to arrest its economic
slide and alarmed by proposals for international intervention to create
safe havens for southern peoples threatened with genocide, agreed to
meet the SPLA's leader John Garang at Abuja in Nigeria with or
without representatives of the SPLA splinter group formed by Riak
Machar (largely among the Nuer as distinct from Garang's main power
base among the Dinka). It discarded in 1993 its military trappings,
converting el-Bashir and others into civilians but leaving al-Turabi's role
unaltered (but somewhat impaired by a serious accident). As civil war

and economic decline continued, Sudan also became increasingly isolated with no friends among its neighbours and worsening relations with Egypt, particularly after Sudanese complicity in an attempt on Mubarak's life during a visit to Addis Ababa in 1995.

THE HORN

The ancient *Ethiopian* kingdom, an amalgam of races, languages and religions, retained its independence even in the nineteenth century and therewith a certain immunity from the modernizing trends which normally accompany both colonialism and anti-colonial movements. Subjected for only five years in the twentieth century (1936–41), it resumed its hallowed way of life under its astute monarch Haile Selassie, who had first mounted the throne in 1930 at the age of 38. The emperor was a cautious innovator whose reforming proclivities were in advance of the landed feudatories and wealthy churchmen who were the most eminent and ultimately the most powerful people in his realm. An attempted coup at the end of 1960, followed by army mutinies in March 1961 and a reported conspiracy in August of that year, suggested the existence of some discontent among the educated young and junior officers who looked to Prince Asfa Wossen for more spirited progress, but the emperor re-established his control with awe-inspiring ease and continued his policy of slow internal advance. He also cultivated good relations with the emerging states of Africa (perhaps as an insurance against traditional Muslim hostility towards his predominantly Christian empire); in 1958 his capital became the headquarters for the UN Economic Commission for Africa and in 1963 the Organization for African Unity came to birth there. For both these organizations the emperor caused spacious quarters to be built.

Ethiopia's special external concern was with the neighbouring and non-Bantu *Somalis* who, unlike the Ethiopians, were not only conquered by Europeans in the nineteenth century but also partitioned among them. The Somalis, Muslims but not Arabs, appeared in the Horn of Africa towards the end of the European Middle Ages and subsequently joined in Muslim attacks on the Christian kingdom of Ethiopia. They were subjected in the latter part of the nineteenth century, first by the French in the 1860s and then by the British and the Italians (who also took Eritrea – so named by the Italians in that century – flanking the Red Sea to the north of Ethiopia) in the 1880s. The British colony of Kenya extended northwards over a predominantly Somali area, and Ethiopia appropriated in its Ogaden province territory to which the Somalis laid claim. Relations between Ethiopia and the Somalis were therefore inherently bad and British relations with both were uneasy, since the Ethiopians suspected Britain of partiality to Somali claims against Ethiopia, while Somalis found Britain unsympathetic to their claims against Kenya.

In 1935 the Italians, dissatisfied with the barrenness of their part of Somaliland and their imperial pretensions in general, exploited an incident at Wal Wal in the disputed Ogaden in order to conquer Ethiopia. They were suspected of toying with the idea of a Greater Somalia which would annex British Somaliland, but their defeat by the British in 1941 revived the independent Ethiopean empire (to which was added Eritrea in 1952 after a period of British trusteeship) and left the Somalis still subject and divided. At the end of the war Ernest Bevin proposed at one moment a Greater Somalia consisting of British and Italian Somaliland and parts of Ethiopia, but this notion antagonized Ethiopia without profiting the Somalis. Discussions on the Ethiopian–Somali frontier proceeded sluggishly until 1959 when a conference in Oslo with Trygve Lie as arbitrator produced a compromise agreeable to neither side. In 1960 Italian Somaliland, to which the Italians had returned in 1950 to administer a ten-year trusteeship, became independent, and as this date approached the British, who had become nervous of Egyptian interference in British Somaliland, hurried their own colony forward so that it could be joined with Italian Somaliland to make the independent republic of Somalia – large but poor, racially mixed, ill-prepared, and distrusted and menaced by its Ethiopian and Kenyan neighbours.

At the Lancaster House conference on Kenya in 1962 the Somalis asked unsuccessfully for a plebiscite in the Northern Frontier District of Kenya (an area of over 260,000 sq. km) and its union with Somalia. Later in the same year Kenyan politicians discussed with Somalis an East African federation which would embrace not only Somalia and the British East African territories but also Ethiopia; in the event of such a development the Kenyans made it plain that they intended to keep the whole of the Northern Frontier District for themselves. In December a boundaries commission recommended that the district be divided into two regions, both to be included in the new Kenyan state. This recommendation, which was accepted by the British government, produced riots and a rupture of diplomatic relations by Somalia. Kenya was able to get other African states on its side and a Kenyan delegation walked out of an Afro–Asian conference at Moshi in Tanganyika in February 1963 when the Somalis raised the border issue. At a further conference in Addis Ababa in May a number of Africans chided the Somalis for again raising the question.

In the same year open hostilities broke out between Somalia and Ethiopia. The Somalis did not accept the treaty of 1897 by which Britain had ceded part of British Somaliland to Ethiopia; between

Italian Somaliland and Ethiopia there was no established border. On independence the Somalis had refrained from challenging a neighbour which possessed American equipment. Ethiopia too, conscious of the racial divisions within its own borders, avoided a clash. But the Somali claim against Kenya alarmed Ethiopia owing to its affinity with Somali claims on Ethiopia's Ogaden province. Fighting developed unofficially along the border during 1962. In the next year the Somali president, Abdurashid Shermake, visited the USSR, Egypt, India, Pakistan and Italy. He got little help or encouragement. Kenya's independence at the end of that year saw also the formal conclusion of a pre-arranged Kenyan–Ethiopian defence treaty, and a few months later open fighting began between Ethiopia and Somalia. The Russians offered their mediation and a deputy foreign minister of the USSR went to Mogadishu, thereby evincing Russian concern, if not for the Somalis, at any rate about possible American or Chinese influence in the Horn. More effective mediation was proffered by the president of the Sudan and the king of Morocco, and after talks in Khartoum hostilities were suspended. But the underlying problem was not resolved and further discussions between Kenya and Somalia in 1965 were abortive. Sporadic fighting continued until 1967 when a new Somali government asked President Kaunda to mediate. Diplomatic relations were restored the next year.

In French Somaliland a movement in favour of amalgamation with the Italian and British territories was circumvented by a vote in the assembly at Djibuti to continue as an Overseas Territory of the French Union, but in 1975 France decided to leave (retaining however an armed force of 4,000) and in 1977 the independent Republic of Djibuti came into existence. It included the port of Djibuti and the west shore of the Bab el-Mandeb, was coveted by Somalia and vital to the external commerce of Ethiopia, and was divided between Afars who had kin in Ethiopia and Eritrea and Issas who were Somalis. After the fall of Mengistu in Ethiopia Djibuti adopted (1992) a multiparty constitution which did not, however, alter the essentially ethnic nature of its politics. President Hassan Gouled Aptidon, an Issa, was accused of manifold atrocities against Afars which lost him French support and isolated him from Djibuti's neighbours in east Africa and the Arabian peninsula.

The Horn of Africa, besides its regional feuds, acquired international significance in the context of the Cold War. It provided at the southern end of the Red Sea a bridge from Africa into Asia hardly less important than the complementary passage through Sinai, and eastward it faced the wider waters of the Indian Ocean. The United States established influence and bases in Ethiopia. The USSR responded in Somalia where it supplied arms for the Somalis to use against their neighbours in return for storage and communications facilities, intelligence posts and a large air base. In 1974 these manoeuvres were inflamed and transformed by revolution in Ethiopia.

The Ethiopian emperor Haile Selassie was pressed by friends and relatives to abdicate in 1973 when he reached the age of 80. He was old and much burdened – his heir had a stroke in 1973; his alliance with Israel crumbled in that year of war in the Middle East; American cordiality was waning; students were rioting and workers striking – but he refused to resign. He made a move to loosen the lid which he had clamped on his country but the past blew up in his face. A group of army officers mutinied. This mutiny led to a further revolt by privates, NCOs and junior officers, to the dethronement of the emperor (who later died in prison, possibly murdered or maltreated) and to numerous political assassinations. The deadening influence of the aged, if once noble, emperor had produced intolerable strains (between regions and classes and even within the army), conspiracies, corruption, misgovernment and stagnation, all of which were worsened by the drought and famine which afflicted much of Africa in the early 1970s and by the contrasting affluence of the court and nobles. The first mutiny was followed by a series of short-lived governments and by the formation of a loose amalgam of groups called the Dergue (or Committee) whose membership seemed to be fluid and was largely unknown but which became the real power in the capital.

The dominant figure in the new régime was Colonel Haile Mariam Mengistu, a clever manoeuvrer who defeated his main rivals in the Ethiopian People's Republican Party with the help of the socialist Meison which he then routed. Mengistu and his Alyotawi Seded secured control in Addis Ababa but not over the whole country and, most particularly, not in Eritrea in the north nor in the disputed lands in the south. For the next fifteen years and more Mengistu was at war on several fronts, and without foreign help he would have lost one or all of these wars. He got help from the USSR and its Cuban surrogates when, in 1977, Moscow decided to back Mengistu, jeopardizing and forfeiting its alliance with Somalia in order to displace the United States in Ethiopia. Cuban arms saved Harar and Diredawa for Ethiopia and repulsed the Somalis who had been on the verge of conquering the Ogaden. This relief enabled Mengistu to hold his own against the Eritreans until, once more with Russian and Cuban help, he could mount a counter-offensive. For the time being Mengistu was saved – and the USSR committed to supporting the Christian Amharic empire of Haile Selassie in its new colours under the Dergue. Besides his conflicts with Eritreans and Somalis Mengistu faced revolts in Tigre in the north-west and by the Oromo or Galla peoples of the south (the latter constituting about half the country's population).

Somalia was switched by these events from Russian to American protectorship. But it too was a gravely divided country. In the last third of the nineteenth century French, British and Italian colonizers had asserted patchy rule over a network of pastoral clans and sub-clans which reasserted themselves after independence in 1960. In 1969 President Shermake was assassinated. He was succeeded after an interregnum and a military coup by Colonel Siad Barre. Barre belonged to the Merehan clan which was a section of the Darode peoples who were sandwiched between the Issas to the north and the Hawieh to the south. He strengthened Somalia's links with the USSR and the Arab world, promulgated an eccentric socialist programme, benefited after 1977 from American favours and invaded Ethiopia with groundless hopes of American military support. His forces were quickly evicted by the Ethiopians with Cuban and Soviet help and his own position was seriously undermined by his misconceived and humiliating adventure. He became a harsh and nepotistic autocrat, lost the support of allied sub-clans and in spite of a rally to his leadership against an Ethiopian invasion in 1982 became enmeshed in a civil war which accelerated the disintegration of the state and caused famine, rapine and the flight of 2 million terrified and starving people. (At one point the Red Cross was devoting one-third of its worldwide resources to Somalia and twenty other humanitarian agencies were involved there.) One of Africa's few homogeneous countries was transformed under Barre's rule into a maze of hostilities sharpened by a flood of undiscriminating American aid which, although often allocated to inappropriate and uncompleted projects, created nevertheless pots of gold.

The northern, formerly British, part of the country purported to secede as the Republic of Somaliland but its president Muhammad Egal won no international recognition and little foreign aid. His position was threatened by tribal rivals but, so long as they remained divided among themselves, not by Somali leaders to the south. There, clan leaders who had joined forces against Barre resumed mutual hostilities while much of their forces descended into brigandage. The most prominent of these leaders were Muhammad Farah 'Aideed' who (like Barre) was an Italian-trained soldier and policeman, had been kept in prison by Barre for years but had been released when his talents were needed by Barre against Ethiopia. He narrowly missed becoming Barre's successor when through a tactical error he allowed Mogadishu to fall into the hands for his rival, Ali Mahdi Muhammad, who assumed the title of president. To compensate for this failure Aideed sought allies among other groups, including Islamic zealots who rallied to him when the United States singled him out as a special enemy and tried to kill him.

There were in Somalia two separate but interlocking calamities. The one was the chaos and mayhem which followed the collapse of Barre's régime; the second was drought, famine and disease due in the first place to natural causes. The first gravely aggravated the second. The first was felt mainly in the capital and along the coast, the second most severely a hundred miles inland. The first required international mediation or more forceful intervention to pacify and disarm the warring factions; the second required food, medicine and protection for those bringing these things. The worst of the famine was over by the middle of 1992 but relief agencies continued to be robbed of 10–20 per cent of their supplies. Following a series of Security Council resolutions in 1992 a UN peace-keeping force (UNOSOM) was despatched but, contrary to the wishes of the secretary-general, was not given the authority to disarm factions. A UN arms embargo had little effect since arms were already plentiful. UNOSOM, consisting initially of five hundred Pakistanis and designed to reach a total of 3,500, was ineffectual and towards the end of the year President Bush offered to supplement it with 3,000 Americans. At the same time – and perhaps with an eye on his presidential campaign – Bush prepared an independent force of 25,000 to succour the starving millions by easing the flow of aid but set a limit of a few months for this operation. It provided some relief in a limited area for the aid purveyors but upon being brought to an end in 1993 it dumped the unresolved political and security problems on the UN. Essential American support was ensured for a while by giving the United States the two posts of special representative of the secretary-general and second-in-command of a reconstituted UNOSOM with 28,000 troops (including 8,000 Americans) and the authority to enforce as well as keep peace. In this expanded UN force the Pakistani element was overtaken by the American (with 2,600 Italians and 1,100 French in third and fourth places) but the United States also retained the relatively small but uniquely powerful US Rapid Deployment Unit with 1,000 troops and helicopter gunships under independent American command. The first American arrivals had been welcomed by Aideed whom the United States appeared at this point to favour against Ali Mahdi, but during 1993 the United States abandoned the aim of bringing war lords together in favour of eliminating Aideed politically and, if necessary, personally. The secretary-general's first special representative, Muhammad Sahoun, an Algerian, resigned in protest against American militancy. Confusions multiplied. Twenty-five Pakistanis were killed in an affray for which the Americans blamed Aideed; he, however, described the episode as an attack on a broadcasting station in his area of Mogadishu and implied that the Americans had been trying to kill him. The new and baffled President Clinton sent a special unit which arrested UN leaders instead of Aideed. Clinton then sent a larger force while at the same time promising to withdraw it after six months.

When he duly did so, no progress had been made towards reuniting Somalia either by negotiation, by force or by any combination of the two and the various parties – excluding, however, Aideed – began to toy with the idea of a new and loosely federated Somalia. The departure of the Americans was followed sixteen months later by the departure of UNOSOM, leaving Somalia much where it had been three years earlier – divided between two principal tribal coalitions manoeuvring against one another and pretending to ignore the precariously detached fiefdom of President Egal in the north. Fighting went on.

In Ethiopia Mengistu's government was able to cow the Western Somali Liberation Front and to check, but not defeat, the Eritrean People's Liberation Front (EPLF). Sudan and Saudi Arabia tried to unify minor Eritrean groups – three of which signed an agreement in Jeddah at the beginning of 1983 – as a prelude to a broader accord with the EPLF whose leftward inclination they distrusted, but these moves were ineffectual and the EPLF demonstrated its vitality by inflicting serious defeats on Ethiopian government forces in 1983 and again in 1988. The United States reclassified the EPLF as a democratic liberation movement instead of a Marxist insurrection. The revolt in Tigre waxed and waned with the fortunes of the Eritreans and the government suffered serious defeats in Tigre in 1989. Russian aid was tapering off and Mengistu felt obliged to offer Eritrea autonomy in a federation. A coup against him misfired but the army, on which he depended, had had enough of war and pressed for any agreement with Eritrea short of independence. Talks began in 1989 with the former President Carter as intermediary. In 1991 Mengistu fled and the capital and central government were taken over by 'Meles' Zenawi, now an ex-communist, as chief of the Ethiopian People's Revolutionary Front (EPRDF: its main components were the Ethiopian People's Democratic Movement, the Tigray People's Liberation Front and the Oromo People's Democratic Organization). Chaotic elections followed a year later, less chaotic ones in 1994. The EPRDF retained dominance although the Oromos and others, wary of Amharas, wanted secession or at least a loose Ethiopian federation. Afars were similarly divided. The EPRDF swept the board in central and regional elections in 1995 which were boycotted by the principal opposition parties. In Eritrea the EPLF acted as though Eritrea were already independent, which it became in 1993 with Issayas Aferworki as president of a state of 3 million people about equally divided between Islam and Christianity, speaking a dozen languages, short of food for a decade or more to come, struggling half-heartedly to devise a new-fangled market economy, and a prey to the rival intrigues of Sudan and Saudi Arabia. In the rest of Ethiopia the new government adopted a policy of devolution over five years to a number (originally fourteen but reduced to ten) of regions whose separate ethnic identity threatened, however, to turn devolution into fragmentation – and therefore to bring devolution to an early halt. For all the peoples of this region the horrors of seemingly endless wars were compounded by famines so lethal that the outside world put them at the top of the news and was briefly stirred to raise its support for impossibly overworked and underfunded relief agencies. As the Cold War receded the superpowers found themselves at one in wishing to end the wars and reduce the other horrors of the region by promoting a loose federation of Eritrea with Ethiopia which might be expanded to include Djibuti and Somalia, possibly also the Sudan, Uganda and Kenya – all of them members of the International Authority for Drought and Development.

UGANDA, TANZANIA, KENYA

The preponderant power in east and south central Africa had been the British, but the first Europeans to arrive were the Portuguese. Diaz touched land in East Africa on his journey from the Cape to India and back again. On this coast the Portuguese came in touch with the Arab world, establishing ports of call and repair where Arabs were already trading: Kilwa, Zanzibar, Mombasa, Malindi. The nature and extent of Portuguese rights were vague and fluctuating, and they were gradually reduced by the Arabs and then by the British and the Dutch until only the territory of Mozambique remained to them.

The British interest in East Africa has been twofold. To the dominant power in the Indian Ocean the coastal territories were a natural bait, while Britain also became a continental African power in the course of establishing control over inland regions which were deemed necessary to safeguard British strategic interests at the Cape and in the valley of the Nile. A northern thrust from the Cape by-passing the Boer republics and continuing into Rhodesia combined with a southern thrust from Egypt and the Sudan to create the strategic importance of Uganda and the route of access to it through Kenya. In a particularly colourful chapter of the history of exploration and great power competition Britain acquired Uganda and Kenya at the end of the nineteenth century, adding Zanzibar in 1890 (in exchange for giving Heligoland to Germany) and Tanganyika in 1919 as a mandated territory after Germany's defeat in the First World War.

British power was not at first territorially deployed. Preferring, as usual, the indirect approach Britain chose to exert influence on the coast through the Sultan of Zanzibar, an Arab potentate who was also Sultan of Muscat in Arabia but had moved his capital to Zanzibar in 1840. (Zanzibar and Muscat were separated again in 1861, two sons of a deceased sultan setting up separate states and dynasties as the result of mediation by the Viceroy of India, Lord Canning.) Inland, British governments left the business of expansion to commercial concerns until the last quarter of the nineteenth century when the failure of the companies chartered to exploit parts of Africa (found to be only doubtfully rewarding), coupled with the competitive expansion of European powers in Africa, induced Britain to shift to policies of territorial annexation. The failure of the British East Africa Company, for example, led Lord Rosebery's cabinet to endorse a British protectorate over Uganda, the prime minister's doubting colleagues being overpersuaded by strategic arguments about the intentions of other European powers, particularly Italy and Belgium, in the regions of the headwaters of the Nile. After the capitulation of the Italians at Adowa in 1896 to the Emperor Menelek II, Britain feared that Ethiopia would make an alliance with France or with the Mahdists in the Sudan, and Lord Salisbury pressed the construction of the Uganda railway as an adjunct of his policy of reconquering the Sudan. No less important was the contest with Germany, whose determination to become an African power had been made manifest at the Berlin Conference of 1884–85.

The Germans had staked out claims in South-West Africa and then in East Africa to the embarrassment of Britain, which disliked German expansion in Africa but was in need of Germany's friendship in Europe and elsewhere during the period of bad Anglo–French relations from 1880 to 1904. Britain consequently acquiesced in the German occupation of South-West Africa, while taking the precaution of securing Bechuanaland against possible German (or Boer) aspirations which might interfere with the railway from the Cape northwards. In East Africa Britain likewise acquiesced in a German presence and gave up using the Sultan of Zanzibar to make things difficult for the Germans in Tanganyika, while at the same time leasing from the Sultan in 1887 a coastal strip 16 km long (including Mombasa) and developing British power in Uganda and Nyasaland so that the emerging German empire might be contained within limits compatible with essential British interests. The First World War eliminated the German factor, but the early part of the twentieth century also saw, in Britain's more northerly colony of Kenya, an enterprise which was later to produce its own problems. This was the development of Kenya by white settlers. While the removal of great power complications enabled the Foreign Office to dismiss East Africa from its mind, the emergence of new complications of a different nature does not seem to have troubled the mind of the Colonial Office until too late, for whereas the Devonshire Declaration of 1923 had affirmed the primacy of African interests, the settler community assumed, and was allowed to go on assuming until the eve of independence, that its own powers and privileges were not threatened within the foreseeable future. Even in the 1950s both races believed that a white government's devotion to the principles of self-determination and majority rule would stop short at putting a substantial white community under black rule.

East African independence followed West African but – partly because it came later – was achieved by a telescoped sequence of the established pattern of evolution from nominated to mixed councils and so to the fully elected parliaments which accompanied self-government and presaged independence. Tanganyika became independent in December 1961, Uganda in October 1962 and Kenya in December 1963, but in spite of their geographical closeness the circumstances of the three territories were very different.

The special features of *Uganda's* progress to independence were provided by the composite political structure of the country. The Uganda protectorate included a number of monarchical entities, of which the most important was Buganda under the rule of its kabaka, Frederick Mutesa II. Others were Bunyoro, Toro, Ankole and Busogo. Moreover, between Buganda and Bunyoro there was a territorial feud of long standing. One consequence of the existence of these principalities was that Uganda had a relatively strong nationalism of a traditionalist kind which was antagonistic both to British colonial rule and to the more democratic and anti-monarchical forms of nationalism. The comparative weakness of these latter strands tended to cast the colonial administration for a progressive role in opposition to the conservatism of the kabaka, who was concerned to preserve his traditional powers and the separate identity of his principality. In 1953 a British minister let fall in London an ill-timed remark about an East African federation which was taken by the kabaka – and many other Africans – to betoken a British scheme to create a large new political unit for the better preservation of white rule. A central African federation was being formed at this time and East Africans feared that the white settlers in Kenya were to be given throughout East Africa the powers which Southern Rhodesia's white minority was in the process of confirming in its own country and extending to Northern Rhodesia and Nyasaland. A quarrel ensued between the kabaka and the governor, Sir Andrew Cohen, and the kabaka was despatched into exile for failing to observe the Uganda agreement of 1900, by which he was obliged to accept British advice on certain matters.

The kabaka's exile lasted until 1955. A commission under Sir Keith Hancock worked out a new constitution which the kabaka accepted by the Namirembe agreement and by which he agreed to turn Buganda into a constitutional monarchy and recognize it as being an integral part of Uganda; the Buganda parliament or lukiko was to send representatives to sit in the Uganda parliament (though it began by refusing to do so). The kabaka was thus restored, but the principality to which he returned was a budding democracy within a larger democracy and the British aim to create an independent Uganda which was a unitary parliamentary democracy and not a federation had been significantly advanced.

Britain failed, however, to resolve Buganda's feud with Bunyoro. This quarrel went back to 1893 when the British under Lugard and the then kabaka had made war on Bunyoro. Buganda had taken from Bunyoro five counties and two parts of counties and this transfer had been sanctified by Britain in the Uganda agreement of 1900, since when Britain had turned a deaf ear to all Bunyoro complaints. In 1961, however, Britain appointed a commission of privy councillors (the Molson Commission) to try to resolve the quarrel before independence. The commission recommended a compromise which was only partially implemented by the British authorities. Buganda was confirmed in its possession of four counties which were allotted to it by the commission, but a decision on the rest of the disputed territory was postponed. In 1964, after independence, the government of Uganda conducted the plebiscite which had been recommended by the Molson Commission and which conclusively restored these areas to Bunyoro.

Constitutional change in Uganda had begun in 1950 when the legislative council was given an equal number of official members (that is, colonial servants) and unofficial members. Of the latter eight were African, four Asian and four European in recognition of an ethnic problem which did not, however, impede the advance to independence owing to the fact that the European settlers were too few to think of retaining political power and the Asians judged it expedient at an early date to conciliate rather than antagonize the African majority. A similar balance of official and unofficial members was established in the executive council in 1952. In 1954 the size of the legislative council was doubled. In 1961 Uganda was given self-government and Benedikto Kiwanuka became its first prime minister, but Bugandan separatism continued to delay full independence. The lukiko, which had petitioned the British Crown in 1957 for greater autonomy and refused in the following year to play its allotted role in a general election, even declared Buganda independent. The Uganda People's Congress, which had been formed by Milton Obote to fight for independence, entered into an alliance with the Bugandan home rule party, Kabaka Yekka, in order to win a parliamentary majority. Obote took Kiwanuka's place. Independence followed in October 1962. In the following year Uganda became a republic in the Commonwealth and the kabaka accepted the ornamental office of president. But the alliance between Obote and the kabaka did not last. Early in 1966 rumours of scandals in high places threatened to weaken Obote's position. He set up an inquiry and then assumed emergency powers, dismissed and arrested a number of his ministers and dismissed the president. Two months later he introduced a new constitution which precipitated a fresh clash between himself and the lukiko and made himself president. Recurrent rumours of Bugandan plans to resort to force induced, or enabled, him to act first. The kabaka's palace was sacked and the kabaka, barely escaping, was driven once more into exile. No kabaka resided in Buganda until 1993.

Tanganyika followed in form a similar course through official–unofficial partnership in government to self-government and independence, but the British authorities tried to give a special multiracial twist to events by espousing the equality of races as opposed to the equality of individuals. This concept was enshrined in a constitution of 1955 which provided that the electors in each constituency should all elect one member of each of the three races, but a party formed to apply it, the United Tanganyika Party, never secured much support and was eclipsed by the Tanganyika African National Union (Tanu) formed by Julius Nyerere in 1954. In 1957 Nyerere, who had been asking for independence in twenty-five years' time, was appointed a member of the legislative council together with Rashid Kawawa, but he soon resigned in order to press the pace and urge independence by 1969. In elections spread over a period in 1958–59 Tanu won all the seats which it contested, and in 1960 it extended its victories throughout the country. The multiracial experiment was given up after a change of governor and of colonial secretary in 1959, and Tanganyika attained independence in 1961. It chose to be a republic and a member of the Commonwealth.

One of the most important elections in the history of British decolonization was held not in any colony but in Britain itself. In 1959 Harold Macmillan was returned to power by the British electorate and one of his first undertakings after this refreshing experience was a journey through Africa which made manifest a new attitude to colonial affairs. Macmillan had felt the 'wind of change' and had determined to let it blow him along. His new colonial secretary, Iain Macleod, was immediately faced with the most difficult of all the East African situations, Kenya.

In *Kenya* unofficial members of the legislative council began to be given ministerial appointments immediately after the end of the Second World War. They were, however, not black but white, representatives of the British settlers who had been coming to Kenya since the beginning of the century and had been acquiring and developing, in good faith and with intelligent toil, the excellent agricultural land in what came to be called the White Highlands. This community became also politically powerful. It hoped either to rule Kenya in lieu of the colonial authorities or to share in governing a multiracial state on a scale appropriate to its wealth and sophistication rather than to its numbers. It was in other words an aristocracy with scant prospects in a democracy and it was suddenly faced with the problems, more familiar to historians than to farmers, of the aristocracy which is required by events to come to terms with a non-aristocratic future. In 1953 this community and the colonial régime were faced with a savage outbreak among the Kikuyus who lived in and around Nairobi and had long nourished grievances against the white settlers as well hostility to black neighbours.

The Kikuyu and Luo were Kenya's two main ethnic groups, led respectively by Jomo Kenyatta, one-time student of anthropolgy at London University and president of the Kenya African Union (who had returned to Africa in 1947) and Oginga Odinga. The British refused to allow them to form political parties in central Kenya and promoted as a counter-weight an association of smaller tribes (Kadu). Shortly after Kenyatta's return the Kikuyu formed a secret society called Mau Mau, whose activities – known to the authorities but not made much of – were the militant expression of a deep-seated nationalist or xenophobic movement. Mau Mau administered oaths and performed secret rites and cherished apocalyptic fancies, all of which were anti-European and anti-Christian. With time the society became extreme in its ambitions and barbarous in its practices. It took to murder – mostly of other Kikuyu – and finally developed a campaign of violence and guerrilla warfare. The government declared an emergency, called for military reinforcements from neighbouring territories and from Britain, arrested thousands of Kikuyu including Kenyatta (who was sentenced in 1954 to seven years' imprisonment for organizing Mau Mau), and gradually suppressed the rising. It also initiated a programme for the psychological reorientation of the detainees, although in other departments it succumbed to the infectious passions roused by the Mau Mau and became responsible for ugly beatings of detainees, notably at the Hola camp where gross inhumanities and murders were disclosed at a coroner's inquest in 1959. The African victims of Mau Mau numbered some 8,000; the number of Europeans killed was sixty-eight. Suppressing the insurrection took the British five years and 50,000 troops.

The British government realized that Mau Mau could not be made an excuse for abandoning constitutional advance and in 1956, the year of the termination of the emergency, it introduced changes in the legislative council. The guiding principle was that of multiracialism or partnership between the races, a theory of government which found almost no support among the Africans who demanded a majority in the legislative and executive councils and refused to accept parity with the European elected members in the former or a minority of seats in the latter. To Africans multiracialism was a device for giving the whites unfair shares.

The Africans also insisted on being given a date for independence, whereas the British government, unwilling to accept the evidence of an accelerating tempo throughout Africa or to affront the Kenyan settlers by naming a date acceptable to Africans, tried to proceed towards independence at an unrevealed pace without losing control of the situation. In 1959 this policy was abandoned (along with the current colonial secretary) and Kenyatta was conditionally released; in 1961 he was given full freedom of movement and then allowed to stand for and be elected to the legislative assembly.

A constitutional conference was held at Lancaster House in London in 1960 and shortly afterwards the leader of government business in the assembly, Ronald Ngala, was upgraded to chief minister. But the African political leaders, who were now clearly destined to take power in the near future, were divided among themselves. Many of them represented tribes rather than a nation and they failed therefore to create the single unified independence movement which was characteristic of other emergent African countries. The weaker tribes combined to oppose the stronger Kikuyu and Luo and to press for a federal constitution in which important powers would reside in regions rather than with the central government. Constitutional conferences became contests between the Kenya African Democratic Union (Kadu), the proponents of regionalism led by Ngala, and the Kenya African National Union (Kanu), which objected that too much regionalism would make the constitution unworkable and that Kenya had neither the money nor the trained administrators to be able to afford the complications and duplications of a federal system. The advantage lay with Kanu, under the leadership of Kenyatta, who became prime minister in June 1963, and the British government was even forced to amend in September constitutional proposals of a federal nature which had earlier been accepted by all parties. Kanu's success in elections in May enabled it to face the British government with the choice between revising the constitution on the eve of independence or seeing it changed immediately afterwards. The British preferred to give way in the hope of sparing Kenya a constitutional crisis on the morrow of independence, even though the cost of concession was a not implausible charge of bad faith from Kadu. Kenya became independent on 12 December 1963. It became a republic in 1964, whereupon Kenyatta swiftly removed the constitutional safeguards for minorities and regional rights and, by absorbing Kadu into Kanu, created a centralized one-party state. In 1969 it lost the ablest of its younger leaders when Tom Mboya was assassinated. Before this Kenyatta had removed his principal left-wing rival Oginga Odinga from the vice-presidency and then put him in detention.

Kenya's delayed independence had been a source of anxiety in Tanganyika and Uganda. Kawawa and Obote went to London in 1963 to try to persuade Britain to hurry Kenya into independence. In June the leaders in the three countries declared their intention to federate. Uganda, however, had reservations about a federation since Obote feared that Buganda would insist on being a full member of such a union instead of a component part of Uganda. Kenyatta subsequently disclaimed any serious intention to federate, declaring that the plan had been no more than a means of pressing Britain to hasten Kenya's independence. Perhaps Nyerere alone was wholehearted. The proceedings were certainly dilatory. Obote failed to attend a meeting held in Nairobi in September to discuss the scheme, and the Kenyans established an awkward system of being represented by Kanu at one meeting and Kadu at the next. There were also genuine difficulties such as the location of the federal capital, the choice of a federal president (the promotion of Kenyatta, the obvious candidate, would unleash an inopportune contest for the Kenyan presidency), the division of powers and other constitutional matters, opposition from Ghana to any regional associations likely to hamper Nkrumah's pan-African schemes, and Tanganyika's trend to the left in its domestic and its external policies. In the end Tanganyika, regretfully but firmly, insisted on a decision and the scheme thereupon collapsed.

The three East African territories made some attempts to integrate their economies. Under British rule they had had a common currency and certain common services (for example, posts, railways, medical services) and had constituted a common market. Tanganyika and Uganda complained, at intervals and with some justice, that Kenya took the lion's share of the resulting benefits and in 1960 the Raisman Commission was appointed to report on these dissensions. It recommended that the links should be retained subject to some reorganization in favour of Tanganyika and Uganda. In 1964 Kenya offered further concessions in order to prevent the disruption of this partial union and in 1968, after a fresh inquiry and report by a Danish expert, the three states signed a Treaty of East African Co-operation. It became a dead letter as the three partners pursued divergent social and economic policies and their relations became strained, and it was finally dissolved in 1977.

Kenya under Kenyatta became a relatively stable country run by a progressively more corrupt black élite. Nairobi became a magnet for the rural unemployed who congregated in shanty towns which were sporadically destroyed by the authorities. Kenya's relations with its neighbours deteriorated. Its rumbustious capitalism grated with Tanzania's possibly more salubrious but less successful socialism and these ideological differences turned in the 1970s into bickerings, mutual expulsions and the closing of the frontier. With Uganda a border dispute initiated by Idi Amin, followed by charges of Kenyan complicity in the Entebbe rescue operation (see below: the Israeli rescuers used Nairobi on their way back) almost led to war. Kenya's traditional friendship with Ethiopia against the Somalis temporarily lost its charms when the Dergue deposed the emperor Haile Selassie. Kenyatta's senescence was another source of worries. His family and familiars were expected to make a bid to retain power and affluence when he died and prepared to do so, but in the event (1978) vice-president Daniel Moi (a Kalenjin) stepped smoothly into his place with promises of continuing the country's good fortunes and erasing its less estimable features.

Moi displayed unexpected authority and gave Kenya a second instalment of comparative political stability. He withstood a semi-covert challenge from Charles Njonjo (one of Kenyatta's more ebullient colleagues and a standard-bearer of the Kikuyus) and he survived an attempted coup in 1982, but his position was sapped by unappeased Kikuyu resentment at his elevation and by economic recession. He transformed Kenya into a one-party state, harassed independent bodies (lawyers, for example) and resorted to strong-arm rule, imprisonment and torture, while he himself became a flagrantly rich man surrounded by mediocre cronies. Kenya's ambition to be the Ivory Coast of the east – a thrustingly successful capitalist state – lost its momentum, unemployment rose, wages fell, landless cultivators flocked to the capital which became one of the more dangerous as well as one of the more corrupt cities in Africa, and by the mid-1980s this corruption had led to financial scandals, collapsing banks and secret police intrigue. The importance of Kenya's ocean coast (a tropical paradise) attracted the United States which built bases there, dispensed lucrative contracts and provided funds for strategic and less serious ancillary services. Kenya's relations with its neighbours were poor, particularly with Uganda: each country harboured refugees from the other. The murder in 1990 of the (Luo) Robert Ouko, foreign minister and possible successor to Moi, and the arrest in 1991 after accumulating accusations of gross peculation of the former economics minister Nicholas Biwott, increased clamour against the government but not coherence in the opposition. Serious disorders approaching civil war ravaged the Rift Valley as Kikuyu and Luo were killed or driven out of their homes by Kalenjin, Moi's people. Ethnic divisions were compounded by class divisions as Moi manoeuvred between new rich landowners and tycoons on the one hand and less wealthy traders and speculators in rapidly growing towns. By 1992 elections could no longer be deferred, Kanu suffered severe losses, fifteen ministers lost their seats; but the opposition lost their opportunity and much of their credit. Foreign governments suspended aid as reports of corruption and breaches of human rights multiplied. Moi defiantly rejected the IMF's terms for an aid programme and engaged in teasing negotiations for a revised agreement. Kenya provided a striking example of new kinds of foreign intervention and their limitations. Outside powers, operating through the IMF or World Bank, brought pressures to bear and imposed conditions for their aid, forcing Moi to hold elections but doing little or nothing to reduce corruption or inject anything more than the forms and parlance of democracy.

Tanganyika became *Tanzania* by an act of union with Zanzibar concluded in 1977 but foreshadowed by earlier events. In Zanzibar, still a British dependency, the Afro–Shirazi Union staged a successful coup at the beginning of 1964. This party, which represented the bulk of the African population of about 200,000 (the Arabs numbering some 44,000) had suffered a setback in elections in 1961 which had given control to a coalition of the Zanzibar National Party and the Zanzibar People's Party; and in 1963 the British, ignoring warnings from Tanganyika, had transferred power to the sultan and the Arab minority with the result that Zanzibar had carried into independence an inbuilt racial conflict in which the scales had been artificially tipped in favour of a minority. The Afro–Shirazi Union wanted neither the sultan nor any kind of Arab dominance. Its leaders – Abeid Karume, Abdullah Hanga, Othman Shariff – led what was essentially an African, anti-Arab revolt and proclaimed a republic with the first two as president and vice-president. Their allies in this reversal of power included a minor Arab party, the Umma Party led by Abdul Rahman Muhammad Babu, the local correspondent of the *New China News Agency*, and a curious soldier of fortune styled Field-Marshal John Okello, who was said to have fought in Cuba but was soon sent into exile when the value of his services proved less than the embarrassment of his presence. The Afro–Shirazis and their friends were quickly branded by a startled world as tools of China.

Alarm was felt in Dar-es-Salaam too and on 17 January Nyerere, remembering perhaps Nasser's anti-communist pre-emptive action in Syria in 1958, sent his home minister and some armed police to Zanzibar, thus acquiring control of the situation. On 20 January troops in Dar-es-Salaam mutinied; a second mutiny occurred at Tabora in central Tanganyika on the following day. Similar acts of insubordination took place in Uganda and Kenya on the 23rd and 24th. In all these places order was restored with the help of British troops who were still stationed in Nairobi (they were not due to leave until the end of the year). The mutineers asked for better pay and the replacement of British officers by Africans. They were using violence to protest to their leaders, not attempting to overthrow them. But in the light of events in Zanzibar rumours circulated of a widespread communist plot, reinforced by Okello's presence in Dar-es-Salaam on the eve of the first disturbance. The British left a week after their arrival. In April Tanganyika and Zanzibar were joined together to make the new state of Tanzania with Nyerere as president and Karume as senior vice-president. The association proved most unhappy for Nyerere. He was denied all influence in Zanzibar and even access to it. Karume established an autocratic régime which, whether or not it was under Chinese control, certainly produced some extreme scandals such as the much publicized forcing of teenage girls into marriage or concubinage with the island's new élite. In 1969 Karume was able to exact from Nyerere the return to Zanzibar of political enemies who were thereby consigned to their deaths. This humiliating concession to brutality by a statesman of an uncommonly humane nature may have been due to Nyerere's fear that Zanzibar might add to his other current troubles, for he was facing discontent in the south, among former members of trade unions (which he had suppressed) and from women's organizations, junior officers and disappointed civil servants. Karume was assassinated in 1972. The formal unification of Tanganyika with Zanzibar and Pemba was effected in 1977. Tanzania's stability (despite poverty) and its influence in African affairs (for example, in the protracted Rhodesian crisis) were due to the character and intelligence of President Nyerere who remained in office until 1985 but bequeathed to his successor, Ali Hassan Mwinyi, an economy which, partly by bad management and partly from bad luck and its own inherent weaknesses, was faced with famine, rationing, a declining standard of living and unpayable foreign debts.

Although Nyerere resigned the presidency he did not retire from politics and he remained chairman of Tanzania's ruling party, Chama cha Mapindusi (CCM); he was re-elected to this post in 1987 for a further five years. He criticized his successor's economic policies. Mwinyi turned for help to the IMF and secured large credits at the price of devaluing the currency by 50 per cent and accepting the standard IMF package of funds in return for austerity – the policy which Nyerere had always opposed. Mwinyi also had problems in Zanzibar. He had been president of Zanzibar from 1984 when he succeeded Aboud Jumba until he became president of Tanzania a year later. Zanzibar's new president, Abdul Wakif, was a figurehead and the Zanzibaris feared outright annexation; they took to the streets and people were killed. After elections in 1990 Mwinyi appointed a new prime minister, John Malecela, who became thereby the outstanding contender for the presidency in succession to Mwinyi until he was forced by dissatisfied aid agencies to resign. The emergence of many parties helped to prolong the rule of the CCM but Mwinyi was scathingly attacked by Nyerere. The CCM picked Ben Mkapa to succeed Mwinyi but chaotic elections in 1995 reflected little credit on multiparty democracy and left the new president uncertain how to construct a majority in the new parliament.

Kenyatta and Nyerere brought stability, if of different kinds, to their countries – virtually unchallenged in office, the former becoming more remote and unpredictable with age and giving rise to some apprehension about what would happen when he died; the latter continuing to guide Tanzania along the road of self-reliance proclaimed by the Arusha Declaration which in 1967 pledged the country to a village socialism and participatory democracy within the framework of a single-party state. In *Uganda* things were different. Obote was overthrown by an army coup while he was attending a Commonwealth conference in Singapore in January 1971. In coming to power he had antagonized many traditionalists, particularly in Buganda; he had then scared the property-owning classes by mildly left-wing pronouncements, and intellectuals by authoritarian scorn. After surviving at attempt on his life in 1969 he tried to curb the power of the army and its commander, Idi Amin, but failed to do so. Amin took his place. Hailed at first as a good sound army type (and boxing champion) with a respectably British background, Amin instituted a reign of terror, especially after an unsuccessful attempt by Obote in September 1972 to invade Uganda from Tanzania with a force of about 1,000 men. (Obote had gone first to the Sudan but had had to leave as a result of the pacification of the south by Nimeiry.) Amin was not reluctant to play a part in international affairs. He proclaimed himself a friend of Israel but then changed sides and became a strong partisan of the Palestinians. In 1976 Uganda became a centre of world attention when an Air France airbus, hijacked at Athens by the Palestinian PFLP in an attempt to secure the release of Palestinians from Israeli and other jails, landed at Entebbe. After its 150 non-Israeli passengers had been released, the remaining 100 were rescued in a spectacularly daring and efficient Israeli commando operation.

Amin made a particular stir by giving notice to Uganda's 70,000 Asians to leave the country within three months, though he later exempted those of them who were Ugandan citizens. Since most of them were British citizens Amin's move greatly embarrassed the British government which, having foolishly failed to take up an earlier offer by Obote to discuss this problem, now found itself caught between its evident obligation to allow British citizens into Britain and the clamour against allowing them in if they were numerous and black. Amin also evicted the British military mission and High Commissioner whom he accused of complicity in Obote's attempted counter-coup. In the following year he began the takeover of hundreds of foreign, mostly British, businesses while throughout the country the roll of Ugandans, distinguished and undistinguished, who disappeared increased gruesomely, to 500,000 by the time he was overthrown in 1979. Amin was what ordinary people call mad. (Coincidentally 1979 saw the disappearance of two other mad monsters, the emperor Jean-Bedel Bokassa of Central Africa and President Macias Nguema of Equatorial Guinea. All three had received more support than was decent from the three antecedent colonial powers.)

Amin's fall was effected by a Tanzanian invasion, the only practical step by any African state to rid the continent of a barbarous tyranny. A Tanzanian force, accompanied by a Ugandan National Liberation Army, seized the capital without difficulty and installed a provisional government under a former vice-chancellor of Makerere university, Dr Yusufu Lule. After a mere two months Lule was deposed by his National Consultative Council and was replaced by Godfrey Binaisa, a lawyer who had held ministerial office under Obote. Binaisa was himself deposed by the army and placed in detention while a pro-Obote Council took over the government of the country. These changes, together with Nyerere's known partisanship of Obote who soon reappeared in Uganda, presaged the latter's return to power, and at the end of 1980 his popularity in the northern half of the country and the active support of the civil and military authorities carried him back to the presidency by a narrow and disputed margin. His rule, which lasted until 1985, was disastrous and hardly, if at all, less bloody than Amin's. With ethnic feuds unabated and the army out of hand, mass slaughter was reduced only by mass flight – mainly to Sudan and Zaïre. Those who ousted Obote were soon fighting one another. The ultimate victor was the National Resistance Army of Yoweri Museveni which, having defeated Obote's forces and the succeeding military government of Tito Okello, put an end to organized opposition by the defeat of General Basilio Okello in the north (but mopping up continued for three years). Museveni became the new president at the age of 40 with a reputation for decency and political acumen, an appalling legacy of savage destruction and no obvious way to convert foreign sympathy into economic aid. He was also committed to the one-party state which, although increasingly unfashionable, commanded much assent on Uganda's hazardous road to recovery. His personal position was somewhat compromised by his supposed involvement in 1990 in the invasion of Rwanda by Tutsi exiles in Uganda: he was by inclination and kinship partial to the Tutsi. The invasion was a fiasco. Nevertheless elections in 1994 for a constituent assembly to draft a new constitution gave him clear, if by no means overwhelming, popular approval. His chances of re-election in 1996 were helped by remarkable (around 10 per cent) economic growth and Uganda's lawfulness in East and Central Africa's many scenes of frightful lawlessness. Against him were campaigns for a multiparty unitary state or, contradictorily, a decentralized federal Uganda.

23 AFRICA'S DEEP SOUTH

THE LEGACY OF CECIL RHODES

Towards the end of the nineteenth century the British presence in Cape Colony, hemmed in and menaced by the German occupation of South-West Africa on the one hand and the Boer republics across the Orange and Vaal rivers on the other, caused the British to venture northwards in order to rule out a junction between these two potential enemies and to secure a passage for a railway to the north through British territory. In 1844 the British government declared a protectorate over Bechuanaland, a huge and largely desert area to the north of Cape Colony, but for the rest of the century it was a British citizen rather than any British government who directed the British advance in this sector. That citizen was Cecil Rhodes and one of the principal reasons why he was able to direct policy was the fact that he was able to finance it.

Rhodes struck north from the Cape Colony into Bechuanaland with his eyes on the Zambezi River – and possibly even the Nile. His company, the British South Africa Company, was chartered by the British government in 1889 to administer Bechuanaland and he lost no time in pushing on. In 1896–97 he fought and conquered the Matabele and the Mashona and so became the ruler of what was later to be called Southern Rhodesia, but also in 1896 the failure of the Jameson raid into the Transvaal wrecked his ambition to rule in Johannesburg too and resulted in a gradual reassertion of British governmental control over policy towards the Boers: in 1899 it was the British government and not Rhodes who made the Boer War. From Southern Rhodesia he continued across the Zambezi and, by questionable methods and with questionable legality, won concessions from the sovereign or litunga of Barotseland and other African chiefs. (Lewanika, the litunga of Barotseland, had given a concession to a fortune-hunter in Johannesburg who sold it to the company. He then gave further concessions to the company in the hope of enlisting British support against the Matabele monarch, Lobengula. It seems very unlikely that the parties to the concession agreements were talking about the same

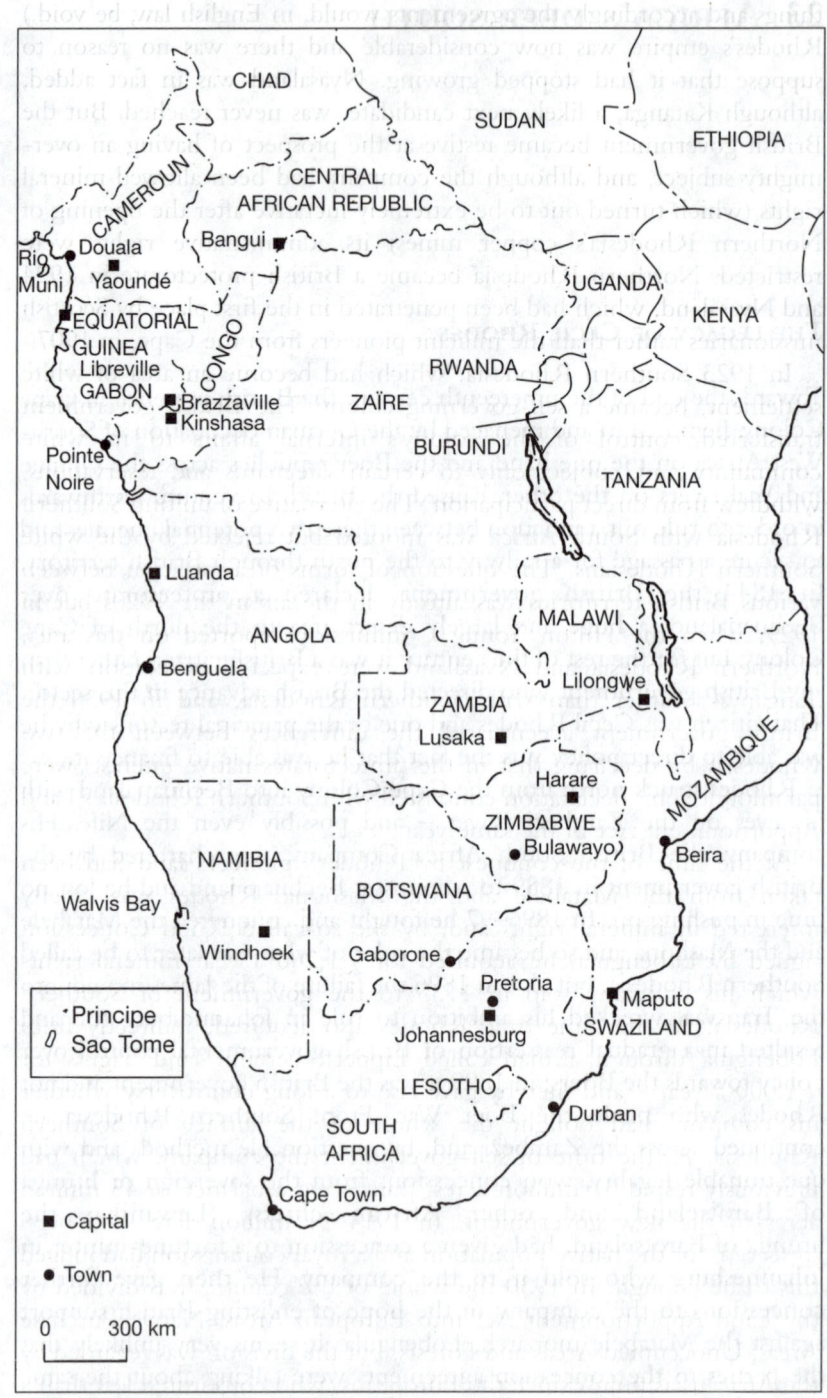

23.1   Equatorial and southern Africa

things and accordingly the agreements would, in English law, be void.) Rhodes's empire was now considerable and there was no reason to suppose that it had stopped growing. Nyasaland was in fact added, although Katanga, a likely next candidate, was never reached. But the British government became restive at the prospect of having an over-mighty subject, and although the company had been allowed mineral rights (which turned out to be extremely lucrative after the opening of Northern Rhodesia's copper mines) its administrative rights were restricted. Northern Rhodesia became a British protectorate in 1924 and Nyasaland, which had been penetrated in the first place by Scottish missionaries rather than the militant pioneers from the Cape, in 1907.

In 1923 Southern Rhodesia, which had become an area of white settlement, became a self-governing colony. The British government transferred control of the colony's internal affairs to its white community and, subject only to certain safeguards and reservations, withdrew from direct participation. The alternative of uniting Southern Rhodesia with South Africa was mooted but rejected by the white Southern Rhodesians. The question of forms of association between various British territories was already in the air in the 1920s but in 1929, when the Hilton Young Commission reported on this area, Northern Rhodesia and Nyasaland were expected to consort with Tanganyika rather than with Southern Rhodesia, and in 1930 the British government accentuated the differences between the two Rhodesias by declaring that in the protectorates native interests were paramount. This declaration contrasted with Southern Rhodesia's Land Apportionment Act of the same year.

At the time of the conquest by Rhodes's pioneers land had been taken from the Matabele and the Mashona. Rhodes was chiefly interested in mineral rights and, by the so-called Rudd Concession signed by Lobengula, he acquired for £1,200 a year mineral rights which his company sold in 1933 to the government of Southern Rhodesia. At the same time Rhodes also acquired, indirectly from Lobengula through a man called Lippert, certain 'land' rights for £1,000 a year – and thereby gave rise to a long controversy whether his company had bought the whole of the surface of Southern Rhodesia. At the time of self-government the company, which had previously resold 31 million acres, transferred a balance of 45 million acres to the new government. In 1914 21 million acres had been 'reserved' for the native population and a royal commission had judged this to be enough. In 1930 the whole of the country was divided by the Land Apportionment Act into European Areas, Native Purchase Areas, Unoccupied Areas and Forests, but the division was regarded by the Africans as unjust since the Europeans, who constituted less than a fifth of the population, were allotted a slightly larger share of the whole, and all the towns; in this area no African might own land. The 1930 division therefore sharpened instead of allaying resentments about how the land had been acquired in the first place, and showed that in Southern Rhodesia native interests would be anything but paramount. The principle of racial division of the land was also a foretaste of further racial irritants on a similarly unjust and discriminatory basis.

After a preliminary conference of officials from the two Rhodesias at Victoria Falls in 1936 the Bledisloe Commission was appointed in 1938 to consider closer union between them and Nyasaland. The commission rejected the idea on the grounds that divergences in native policies and the hostility of the Africans made it impracticable. It recommended no more than a Central African council consisting of the governors of the two protectorates and the prime minister of Southern Rhodesia to co-ordinate matters of common concern. This pale and tepid council was created in 1944. But the protagonists of closer union were neither discouraged nor defeated and in nine years they secured the creation of the Central African Federation. Throughout these years Southern Rhodesia was pursuing a policy of association with Northern Rhodesia after having rejected the alternative of association with South Africa. It was not contemplating a separate existence on its own, although there was among the whites a group which wanted dominion status for Southern Rhodesia within the Commonwealth.

The white leaders in Northern and Southern Rhodesia realized that their interests were not the same and were wary of each other. Roy Welensky in the north, where the riches lay, suspected Godfrey Huggins in the south of wishing to take over Northern Rhodesia for the benefit of the more numerous white population of Southern Rhodesia. On the other hand Southern Rhodesia, although poorer, was freer in terms of British control over its internal affairs; it had a quasi-independence which the whites in Northern Rhodesia (excluding the colonial administrators) coveted for themselves and wished to secure through the Southern Rhodesian back door. These attitudes were clearly to the fore at a conference at Victoria Falls in 1949 which, with no African present, became a tussle between Huggins and Welensky, with the latter falling back on the position that there should be no federation without a referendum. The northerners were at this stage suspicious of the federal idea, but the conference put it on the map and made it the central talking-point of the next few years. This conference was followed in 1951 by a conference in London of officials from the three territories and the two British departments of state concerned (the Colonial Office and the Commonwealth Relations Office). Their report, while again recognizing African opposition to federation, hoped that it would evaporate under the impact of the economic advantages which they listed. At a further conference in the same year at Victoria Falls the politicians (who included the two British secretaries of state as well as the two governors), senior officials and white Rhodesian leaders accepted the greater part of the officials' arguments. This conference was also attended by Africans from Northern Rhodesia and Nyasaland, but they were less impressed by the economic advantages of a federation than by the fear of coming under the rule of the white minority in Salisbury.

A change of government in London from Labour to Conservative shifted the balance of forces in favour of federation. Whereas Labour ministers had come to look favourably on federation but were unwilling to go forward with it without first discovering more about the wishes of the African population, Conservative ministers were more emphatic in their approval, considered that it was impossible to discover what the Africans really thought, ascribed African opposition to irascible and ignorant factiousness, and believed that it was in any case the duty of government to do what was best even if some people did not yet see how good it was. Yet another conference assembled, this time at Lancaster House in London (the maternity ward for emergent constitutions). It was boycotted by the Africans of the two northern territories, though the Southern Rhodesian delegation included Joseph Savanhu and Joshua Nkomo who acted as uneasy camp-followers until the end of the conference. The outcome was a federal constitution with temporarily significant reservations. The federation was to have three separate territorial administrations and assemblies as well as a federal government and parliament, the retention of the British protectorate over the northern territories, the proviso that the federation should not be granted dominion status without the consent of a majority of the population, and the creation of an African Affairs Board with blocking powers designed (unsuccessfully as it proved) to prevent racially discriminatory legislation. The federation came into being on 1 August 1953. Huggins became the first federal prime minister with Welensky as his deputy and a cabinet of six; Huggins also became the leader of a newly formed Federal Party with branches on both sides of the Zambezi. In Southern Rhodesia Huggins was succeeded by Garfield Todd as prime minister and leader of the United Party.

The *Central African Federation* lasted ten years. It was accepted with misgiving by some whites in Southern Rhodesia who feared cheap black labour and rightly opined that the connection with the British protecting power in the north would act as a drag on their plans for dealing with the racial situation. But on the whole federation was welcomed by the whites who believed that within it they would conserve their privileged standards of living, excitingly expand the material things which they were already doing well, and somehow find a way of fitting a small African middle class into the existing order. The idea of partnership, which was written into the constitution and salved a lot of uneasy consciences in London, meant at best a half share in power for Africans in a barely discernible future. Like the Belgians in the Congo they regarded the African as an economic man who could be satisfied with a material competence (however assessed) and, apart from a few over-educated eccentrics and professional trouble-makers, had no real interest in politics. Nationalism they grievously underestimated, and also the force of human indignation over inequality and injustice. Ideas were not their stock-in-trade and so they failed to realize that ideas were at the root of the refusal of the Africans to accept their rule. From the prime minister downwards they insulted and humiliated their black fellow citizens with remarks about living in trees and with the practical application of segregation in public places like post offices and restaurants, and so quickly confirmed the Africans' conviction that partnership was to be not an endeavour but a pretence.

The life of the federation was violent as well as short. The Africans began the violence. Most of them were never interested in the economic aspects of federation or never understood them; the more sophisticated among them understood and saw that they were largely bogus. Despite a distinct current of non-violence, in the African national movements the partisans of violence, aided by circumstances and by exaggerated white reactions, became increasingly prominent. An incident provoked by thieves in the night at Cholo in Nyasaland in August 1953 caused the death of eleven Africans and injuries to many more and led the authorities to magnify the prevalence of hooliganism and subversive activities. It was also a curtain-raiser for more significant events in Nyasaland.

Nyasaland was included in the federation because the British insisted, unwilling to keep it on their hands as a separate dependency with little hope of becoming anything but a drain on the British exchequer. It was an African country with a white population of only 1 in 500, no settlers or industry, its people dependent on Southern Rhodesia and South Africa for the jobs which did not exist and had not been created in their own country. It had been nurtured by the Church of Scotland in much the same way as Paraguay by the Jesuits. It had also a charismatic figure beyond the horizon in Dr Hastings Banda who had spent most of his life learning and practising the profession of medicine in the United States, Scotland, Liverpool, Tyneside and Ghana.

Dr Banda arrived in Nyasaland in 1958 to lead an anti-federation crusade in conjunction with nationalists of a younger generation who were glad to enlist and serve under this more famous elder figure. As in many similar cases certain differences of aim and outlook were easily submerged by the single paramount objective of seceding from the federation and establishing an independent sovereign state. In January 1959 there occurred a secret and apparently confused palaver between nationalist leaders which was afterwards represented as a murderous conspiracy to kill a number of Europeans and seize power. Banda was not present at this meeting and either knew nothing of it, or did not mind what happened at it, or knew all about it and chose to turn a blind eye. Banda himself was a man of considerable, even violent, oratorical powers who preached non-violence, but the situation was one of growing violence and jumpiness and in February the governor of Nyasaland asked for Rhodesian troops to help keep order. The government of Southern Rhodesia sent 3,000 troops and took the opportunity to declare a state of emergency in its own territory. The governor of Nyasaland followed suit a week later. Between 2,000 and 3,000 Africans were rounded up in the federation, the African National Congresses in all three territories were dissolved and Banda and his chief associates were among those lodged in jail. The British colonial secretary, Alan Lennox-Boyd, said that he had evidence of an impending massacre.

These dramatic events evoked some scepticism as well as alarm, and the British government appointed a commission under a High Court judge, Sir Patrick Devlin, to verify the allegations with which these emergency measures had been justified. The commission was unable to find evidence of any murder plot and specifically exonerated Banda. It found that the governor of Nyasaland had got into a position where he had either to abdicate or reach for emergency powers and troops, and that in consequence Nyasaland had become temporarily a police state in which it was unsafe to express approval of the policies of the Congress to which the great majority of Africans adhered. This report, by exploding a myth propagated by authorities in London as well as Salisbury, gave a fillip to the anti-federation campaign and discredited the bona fides of their opponents. Shortly after its appearance the Colonial Office was placed in the charge of a more liberal member of the Conservative government, Iain Macleod, and the prime minister, Harold Macmillan, began a series of devious moves designed to give British colonial policies a leftward shift. These moves were to include his tour of Africa, his 'wind of change' speech at Cape Town during that tour, the appointment of the Monckton Commission, and Macleod's first visit to Kenya — all in 1960.

The federal constitution of 1953 had left the territorial constitutions untouched. It had also made clear that the federation could not hope for independence or dominion status until they were touched. In 1958 there were elections for the federal and the Southern Rhodesian parliaments. In the former Welensky, who had succeeded Huggins (now Lord Malvern) in 1956, scored an overwhelming victory; few of the qualified Africans bothered to register or vote, partly because they disliked the existing two-roll franchise and partly because they were afraid of the police. In Southern Rhodesia Todd was forced to resign by his cabinet colleagues, who refused to accept a wage increase for Africans which had been recommended by the Labour Board. More important, they regarded the comparatively liberal Todd as an electoral Jonah and although he was allowed a seat in the new cabinet formed by Sir Edgar Whitehead he was soon dropped and in the elections his new United Rhodesia Party won no single seat. Whitehead scraped in, although the Dominion Party led by Winston Field, campaigning in favour of independence by 1960, got more votes. This eclipse of Todd was in part a white reaction to the reanimation of the African National Congress by George Nyandoro and Robert Chikerema in 1957 and was also the first of a series of moves which placed the premiership in the hands of ever more extreme politicians who were gradually forced to be more explicit about the basic white demand for independence from Britain as the only way of ensuring that power would remain in white hands for as long as anybody cared to look into the future. In Northern Rhodesia a new constitution preceded elections in 1959 which gave victory to the United Federal Party (the new name for the local version of the Huggins–Welensky party); the party was, however, almost universally rejected by Africans in spite of the support it received from the whole press. These elections showed that the enchantments of federation were not working even before the alleged murder plot and the judgments of the Devlin Commission.

From about 1960 there was in effect a race between the forces which wanted to extract independence for the federation out of the British government and the forces which wanted to break up the federation and establish black rule in its several parts. The British government, caught in this crossfire, resorted to the expedient of sending a commission to look into the situation. Welensky opposed this manoeuvre in private but acquiesced in public. The Africans boycotted it, as did the British Labour opposition. The commission, presided over by Lord Monckton, an eminent Queen's Counsel and former Conservative minister, produced an ambivalent report in which a majority extolled the principle of federation but judged it unworkable. The nature of the commission's work inevitably raised the question of the right to secede from the federation, although the white leaders in

Rhodesia and their friends in Britain hotly contended that the commission had no power to consider the matter and that the British prime minister had promised that it would not be raised; a majority concluded that there was no legal right to secede but that, as a matter of practical politics, the issue should be placed on the agenda of a federal review conference and that Britain should declare its willingness to permit secession after the passage of a defined trial period of federation. This was the commission's most important recommendation since it placed federation on probation. The commission's report marked in fact the turning of the tide in favour of the break-up of the federation, even though the commission gave more attention and space to reforms designed to make it work (such as parity between Europeans and Africans in the federal parliament, a broader franchise, and immediate advances towards self-government in Northern Rhodesia with an African majority in the legislative council and an unofficial majority in the executive council).

A federal review conference assembled in London at the end of 1960. The federal constitution did not require such a conference at this date but in the judgement of the British government it had become necessary. Yet it achieved nothing and was adjourned *sine die*, to be followed by a Northern Rhodesia constitutional conference which produced bitter backstage fighting in which Welensky (who, like Cecil Rhodes, relied partly on a group of members of the Westminister parliament) was worsted by Macmillan and Macleod. Each side suspected the other, perhaps correctly, of being about to use force and Welensky called up troops. The constitutional proposals which escaped from this mêlée would have made it impossible for Welensky to secure control of the Northern Rhodesian parliament. But the proposals, when embodied in a White Paper, were not allowed to pass unscathed and the British government made some slight concessions to Welensky. In so doing it angered Kenneth Kaunda and other leaders of the United National Independence Party (Unip) who accused the British of tinkering behind the scenes with the agreed outcome of the conference. There were violent demonstrations in Northern Rhodesia and the proposals were reversed. These swingings and swayings reflected the divided mind of the British Conservative Party and cabinet; it was by this time reasonably clear that the federation was doomed, although nobody was prepared to say so; and although Welensky and his white followers had fervent friends in London, there was a growing body of opinion in the party which recognized the greater expendiency of being friends with Kaunda and the many African states which stood behind him. A third set of proposals was eventually propounded which ensured that neither Unip nor the Federal Party could get a parliamentary majority and was in any case too complicated to be understood by anybody but a constitutional maniac. It was rejected by Kaunda.

During this same period another conference went through similar travail to produce a new constitution for Southern Rhodesia. Its proposals, which guaranteed a larger minority parliamentary representation to Africans but also removed nearly all the residual powers of the British government, were accepted by Nkomo at the conference but repudiated by him immediately afterwards because he came to believe, again probably correctly, that if he accepted the increased number of seats, the British government would immediately grant independence to Southern Rhodesia under its white rulers who would then either arrest or even reverse the advance towards African majority rule. The British government had hoped to find a formula containing a large enough element of agreement to enable it to give independence with a good conscience and get out of a hopeless situation, but African insistence on the magic formula of majority rights – echoed after all by many in Britain itself – baulked its efforts and kept it on the hook.

The following year, 1962, saw an abortive British Plan to carve up Northern Rhodesia by elevating Barotseland into a separate state under its traditional and conservative ruler, the litunga, and by giving a further slice of it to Southern Rhodesia. The only effect of this bizarre and anachronistic notion was to create bad blood between the nationalists and traditionalists. R. A. Butler was then appointed to a special office of Central African Affairs. His business was to assuage the internal feuding in the Conservative Party and to wind up the federation. By the end of the year Nyasaland had been promised self-government and the right to secede from the federation. In Northern Rhodesia Unip was asked to join the government in 1962 and the right to secede from the federation was acknowledged in 1963. With the secession of Northern Rhodesia there was no federation left. It expired on the last day of 1963.

Nyasaland and Northern Rhodesia, independent from July and October 1964, became the republics of Malawi and Zambia (both of them within the Commonwealth) and with Banda and Kaunda as their presidents. Banda developed into a conservative dictator, intolerant of opposition but successful both in surviving threats of civil war and in developing Malawi's agricultural economy to the point where he was able to raise in 1978 a Euroloan of £14 million. Under a plausible plea of economic necessity he made a commercial agreement (1967) and established conventional relations with South Africa and he supported Renamo in Mozambique until about 1986. Autocratic, sharply anti-communist, morally intolerant, Banda had some of the traits of the South African Nationalists. As age dimmed his wits – he was a little older than the century – his authority passed to Cecilia Kadzamisa and her uncle John Tembo who became Banda's heir apparent in a situation of deepening uncertainty. In 1993 a referendum on constitutional change, followed by brain surgery on Banda in South Africa, strengthened opposition to a régime approaching its conclusion. When elections came, Banda's party held some seats in the central segment of the country but north and south were swept by two different parties and the winner in the south – the United Democratic Front – emerged with a clear parliamentary majority. Its leader, Bakili Muluzu, became president with, however, a need to conciliate the north. The aged Banda and his closest associates were charged with murder and other crimes.

Zambia's fortunes were diametrically different. Zambia became independent with healthy resources and exports and a favourable balance of payments. First among its assets was copper, but it was rich also in other minerals, water and agriculture. A boom in the later 1960s fortified hopes for prosperity together with progress in education and other social services; but the 1970s were a bitter disappointment particularly after the closing of the frontier with Rhodesia in 1973. The Tanzam railway (see Chapter 24) failed to come up to expectations. Copper prices slumped in 1975; some mines became unprofitable; whites began to leave the Copperbelt; blacks lost their jobs and their troubles were compounded by a shortage of maize. Kaunda's popularity waned and there were plots or at least rumours of plots against him. With the ending of the war in Rhodesia it became apparent that Zambia's misfortunes could not wholly be blamed on that war. One of Africa's potentially wealthier countries and the recipient of a fair share of western financial and technical aid, Zambia had lapsed into perilous economic straits because development had been grievously mismanaged. The poor got poorer and lacked basic goods; agriculture was neglected while city building and an élite of city dwellers flourished. Kaunda, his personal rectitude unsullied, was none the less criticized for picking mediocre advisers, for refusing to face economic facts and for presiding over a saddeningly inept decline in his people's fortunes. To secure re-election in 1978 he resorted to disqualification of rivals in advance and in 1980 he took emergency measures against an alleged conspiracy to overthrow him. To secure essential foreign credits through the IMF he cut subsidies on foodstuffs with the inevitable result that prices soared and there were serious riots in the Copperbelt in 1986. Whether or not these riots were inflamed by South African agents South Africa applied economic and military sanctions against Zambia which had become a principal headquarters and base for the ANC. Economically Zambia's dependence on South African routes and ports had been intensified by disorders in Mozambique and Angola which atrophied the alternatives. In 1987 the South African air force attacked Lusaka and killed a number of Zambians and in 1989 Kaunda restricted the ANC's presence in Zambia, disowned its militant tactics and embarked on a dialogue with South Africa's in-coming president F. W. de Klerk. Mounting popular discontent, allied with growing uneasiness in the army and the churches, forced Kaunda to abandon in 1990 his opposition to the multiparty system which he had abrogated early in his rule. Parties had in fact sprung up and the one-party system was dissolving of its own accord. Elections in 1992 produced an overwhelming victory for the Movement for Multiparty Democracy (MMD). Unip won all the seats in the eastern provinces but few anywhere else. Kaunda immediately resigned and Frederick Chiluba became Zambia's second president. Kaunda returned to politics in 1995 by resuming the leadership of Unip as Chiluba's coalition disintegrated and his popular support crumbled.

In *Southern Rhodesia* the constitution of 1961 had left only one political issue between the Southern Rhodesia and British governments: independence. Winston Field had envisaged an independent Rhodesian dominion consisting of Southern Rhodesia and most of Northern Rhodesia, with Barotseland and Nyasaland as separate states under high commissioners appointed by the United Kingdom, but this unlikely scheme had been overtaken by events. In 1963 he began negotiations with Britain on the bare issue of independence for Southern Rhodesia and was confronted with five conditions: unimpeded progress towards majority rule in Southern Rhodesia; guarantees against regressive amendments of the 1961 constitution; an immediate improvement of African rights; an end to racial discrimination; and a basis for independence acceptable to the people of the country as a whole. Ian Smith, who ousted Winston Field in 1964, tried to meet the last and most intractable of these conditions by assembling an *indaba* of chiefs who, having been fêted by the government and even given nice trips abroad, said exactly what was expected of them – but without convincing anybody in Britain that this was a demonstration of the popular will. The advent of a Labour government in London at the end of 1964 caused despondency among whites in Salisbury and increased the demand for a unilateral declaration of independence, but the new British government, beset by a major economic crisis and possessing only the thinnest parliamentary majority, decided to keep talks with Salisbury going for the sake of talking and gaining time. There was, however, no basis for agreement since the British government was pledged to conditions which ran counter to the fundamental demand of the great majority of the white community in Rhodesia: the indefinite preservation of their minority rule. As a last resort the British government proposed the appointment of a royal commission under the Rhodesian chief justice, Sir Hugh Beadle, to examine ways of testing the popular will but this proposal was rejected by Smith and on 11 November 1965 the Smith government declared the country independent. The government was thereupon dismissed by the governor, Sir Humphrey Gibbs. It remained, however, the effective government of Rhodesia within its borders.

The British government was determined not to use force except in the event of a major breakdown of public order. Superficially the situation was one in which previous British governments had not hesitated to use force, but the Rhodesian case was in substance different in two decisive ways: first, because Rhodesia, although technically a colony, had been administered and in effect governed by its own white population for more than forty years and not by Whitehall, and secondly because the rebellious government was white and not black, so that a recourse to force in what the Rhodesian whites had made a racial issue would bitterly have divided opinion in Britain and could even have faced British army officers with a test of obedience which they might not have passed. The British government was left therefore with two courses, negotiation and economic coercion.

For a year it pursued both courses with the emphasis on the first. It immediately took fiscal measures against Rhodesia, imposed oil sanctions and secured the passage of a Security Council resolution against the supply of arms to Rhodesia and requesting an international economic boycott. Over forty countries besides Britain complied with this request. In April 1966 the Security Council, at British request, authorized the use of force to implement oil sanctions and in December, again at British request, it imposed mandatory sanctions over a wide assortment of commodities. Eighteen months later this ban was made total. But despite hardships, which took time to take effect or become apparent, the Smith régime was able to maintain itself thanks to the South African government (which supplied credit and goods and facilitated the export of Rhodesian products), thanks to Rhodesia's capacity to retaliate economically against Zambia (which was dependent on Rhodesia for coal for its copper mines and in other ways), and thanks also to the British government's reluctance to intensify its economic measures so long as it could go on hoping for a negotiated settlement. Rhodesia's exports were substantially blocked, its reserves were depleted, its government had to resort to internal loans, it became virtually a dependency of South Africa, but the régime was not compelled to capitulate and the economy, though damaged, adjusted itself.

In applying sanctions Britain was intent not on destroying the Rhodesian economy or even the Smith régime but on forcing Smith to come to terms. These tactics and aims did not commend themselves to most of the Commonwealth which suspected the British government of being bent on a deal with Smith even to the extent of betraying its predecessor's and its own pledges. At a Commonwealth conference in Lagos in June Britain succeeded in buying time but not in retrieving the trust of its African partners in the Commonwealth, and at a further conference in London in September Britain was obliged, by a nearly unanimous show of resolution by members from all continents, to promise to ask the Security Council to apply selected mandatory sanctions if negotiations between London and Salisbury did not produce a return to legality by the end of the year. By the first anniversary of the unilateral declaration of independence the futility of these negotiations had become manifest, and Britain faced a lengthy economic battle which, if pursued, was bound to develop into a contest between the overwhelming majority of the United Nations on the one hand and South Africa and (less resolutely) Portugal on the other hand, whose policy was to keep the battle going until Rhodesia's opponents got bored and gradually ceased to enforce sanctions. Two attempts to exorcize the problem by personal negotiations between Harold Wilson and Ian Smith – the one on board HMS *Tiger* in 1966 and the second on board HMS *Fearless* in 1968 – failed, principally owing to the obduracy of Smith (or his more extreme lieutenants), but the return of the Conservatives to power in Britain in 1970 revived the prospect of a settlement – and rekindled the mistrust of the rest of Africa.

An initial and unofficial exploration of the chances for a settlement by Lord Alport produced a pessimistic answer but Sir Alex Douglas-Home, back at the Foreign Office in London, opened a dialogue with Smith through Lord Goodman and produced a plan for a return to constitutionality subject to his being satisfied that it was acceptable to the people of Rhodesia as a whole. A commission under Lord Pearce went to Rhodesia in 1972 to find the answer to this question and reported that the majority of Africans rejected the plan. The British government retired once more into the background. Inside Rhodesia guerrilla operations, which had been prematurely and unsuccessfully started a few years earlier, increased but were contained by Smith's forces with South African help. Attempts by the OAU to heal splits among black Rhodesian leaders were unsuccessful but the collapse of Portuguese rule in Mozambique transformed the situation by dealing strategic and psychological blows to the white Rhodesians (the frontiers to be defended became several hundreds of miles longer) and by posing new policy problems for South Africa. The leaders of neighbouring countries – Zambia, Botswana, Mozambique, Zaïre, Tanzania – wanted the overthrow of Smith without a war. They were willing to talk, and to try to persuade black Rhodesians to talk, about a short transitional period before majority rule and about guarantees for the lives and property of whites. They were encouraged by the shift in South African policies, notably by a speech at the UN in which the South African representative spoke plainly about the existence of *apartheid* in South Africa, deplored it and looked forward to a time (unspecified) when it would be removed. This surprising statement was subsequently qualified, presumably for internal consumption, by Vorster, who said that it did not mean one-man-one-vote or a black parliament. Vorster's balancing act was plain for outsiders to see and Kaunda was not deterred from sending an envoy to talk with Vorster who seemed as keen as Kaunda and his colleagues to compass a change of régime in Rhodesia without bloodshed. In December 1974 Smith released the two principal detained nationalist leaders, Joshua Nkomo and Ndabaningi Sithole, as a preliminary to a cease-fire and a possible constitutional conference, and in the following September Smith, at Vorster's prompting, attended a conference with nationalists on a bridge over the Victoria Falls, Vorster and Kaunda themselves also attending. The conference was a failure, technically because of a dispute about the terms of reference already agreed between Vorster and Smith, substantially because Smith was not prepared to discuss any constitutional proposals except his own, which did not then include a transfer of power to the majority. The notion of a constitutional conference was kept alive by Smith and Nkomo but the prospects were dimmed both by white intransigence and by renewed splits among black leaders who had been temporarily persuaded into a show of unity. The collapse of the Vorster–Kaunda initiative left little in prospect except war. In Rhodesia the stalemate continued but around it two big changes had occurred. For Vorster the Rhodesian situation represented a wolf at the door, but the wolf had changed its colour. The failure of the Victoria Falls conference completed the alienation of Vorster from Smith. The main danger for South Africa was no longer the disappearance of white rule but its persistence in hopeless circumstances and the manner of its eventual undoing. Smith was no longer a shield but an Achilles' heel. Secondly, Russo–Cuban intervention in Angola must bring the United States too into a field of forces from which nevertheless it would have preferred to keep out. The chief American concern, besides the removal of Cuban armies from Africa and checking the spread of Russian influence, was stability which in Rhodesia was unachievable without the eclipse of white power.

Smith did not see the writing on the wall. In the first place economic sanctions were a farce. The Beira naval patrol, which was maintained by the British for ten years (until Mozambique became independent in 1975) at a cost of at least £100 million, prevented the delivery of crude oil to Rhodesia through Beira, but nothing was done to stop the supply of refined products through Lourenço Marques or by other routes.

Portugal and South Africa made little secret of their assistance to Smith to circumvent sanctions. American compliance was imperfect: the import of Rhodesian chrome and other minerals was legalized in 1971. And it subsequently became clear that Smith was plentifully supplied with oil, mainly by British and associated companies which were assiduous in devising means to help Smith – in which they were able to count on the benevolence of British officials and ministers who turned a blind eye to traffic which was too bulky to be invisible. (Even when this discreditable story became public knowledge no prosecutions were launched against companies or individuals who, in breach of the law, had conducted a foreign policy at variance with their government's commitments and ostensible objectives.)

Smith was also encouraged by the dissensions among his black adversaries. His tactics were to exploit these differences with the aim of saving white control through a deal with one or more of the black leaders, preferably Nkomo. It was a classic policy of divide and rule. But Smith aimed too high, clinging too long to white minority rule and then, when forced to come to terms with Bishop Muzorewa and others, drove a bargain too good to stick.

The failure of the Victoria Falls conference was followed by negotiations between Smith and Nkomo. These broke down by the early part of 1976, mainly as a consequence of Smith's obdurate rejection of majority rule. During these months guerrilla operations, although still relatively ineffective and confined to border areas, increased and Robert Mugabe emerged as the one leader whom most guerrillas trusted: he alone was uninterested in wheeling and dealing with the whites. Mugabe eschewed the politicking which was second nature to Smith and he also divined, correctly as it turned out, that the conflict was one which would be settled by other means. Smith, unable to distinguish one kind of left-winger from another, was over-confident of his ability to represent Mugabe as a Muscovite stooge and over-contemptuous of the guerrillas.

The collapse of Portuguese rule and the Russo–Cuban intervention in Angola obliged the United States to review its policies. These had been predicated on a stability which no longer existed. Kissinger's first look at southern Africa in the early 1970s had led him to the conclusion that Washington should and could improve its relations with both black and white régimes in order to prevent the USSR from doing likewise; and he assumed that the régimes which he saw there would remain there. The collapse of Portuguese rule undermined this policy. By cutting all but the South African outlets for Rhodesia's commerce it rendered Rhodesia both more dependent on South Africa and less attractive to South Africa as an ally; it dramatically reduced Smith's chances of survival. To the east it not only closed Rhodesia's routes to the sea but also opened a new guerrilla front. To the west it created in Angola an international crisis by which the USSR secured a toehold in southern Africa. The immediate American ripostes, all of them misconceived, were to switch aid from the Portuguese to the two Angolan liberation movements which failed and to instigate a South African invasion which also failed. When Kissinger returned to Africa in 1976 he announced a surprisingly sharp pro-black turn by declaring himself in favour of majority rule not only in Rhodesia but in Namibia and South Africa too. This pronouncement was, however, only the rhetorical prelude to an uneasy alliance with South Africa to put pressure on Smith. Vorster stopped military aid to Smith and allowed his surviving economic channels to the outside world to become clogged. Kissinger went to South Africa for a tense meeting with Smith who was forced to yield to superior power. He accepted majority rule within two years together with a transitional plan which included a black majority in the cabinet and a black prime minister.

Within Rhodesia the guerrilla war became a major factor during 1976. It spread beyond the borders into the heart of the country; its strains on the economy and manpower became visible and painful; it gave a boost to white emigration. But in his own mind Smith was far from surrender. He insisted on interpreting majority rule to mean not black but responsible majority rule; that is to say, he rejected one-man-one-vote. While announcing the end of white rule he contrived to give the impression that white control would continue. He succeeded in tearing up the Kissinger plan. Kissinger had implied to both sides more than had been agreed by either. He allowed Smith to think that the details which he sketched had been affirmed by the African Front Line presidents and by the guerrillas' leaders, when this was not the case. The Front Line presidents complained in their turn that they had been misled by Kissinger; they and the Patriotic Front regarded as negotiable terms which Smith held to be part and parcel of the bargain which had been forced upon him by Kissinger. It was therefore easy for Smith to declare himself released from that bargain unless no letter of it were altered. The Kissinger plan was exposed as a set of imprecise undertakings which fell far short of agreement. A conference of all interested parties in Geneva failed to rescue the plan. So did a subsequent British diplomatic trot round the continent.

At this point the deadlock seemed worse than ever but the guerrilla leaders had meanwhile agreed to co-operate, militarily and politically, as the Patriotic Front. Brittle though it was this alliance was to win the war. Nor were external powers prepared to remain inactive. Britain, the United States and South Africa were committed to replacing the Smith régime by something else. So too were the Front Line presidents, whose countries – particularly Zambia and Mozambique – were suffering severely from the economic consequences of the war and from Rhodesian retaliatory strikes against guerrilla camps in their territory.

On assuming office in 1977 President Carter appointed Andrew Young to handle African affairs. London too had in David Owen a new foreign secretary, and after a joint tour of African capitals these two produced a fresh plan for ending UDI by the transfer of power to the black majority after a very brief resumption of British authority and elections to be held under international supervision. The plan failed chiefly because Smith was outraged by its provision for one-man-one-vote and the prospect of losing white control of the armed forces and security. Smith flew to a secret meeting with Kaunda, once more in search of an alliance with Nkomo – which again eluded him. He turned therefore to a second best version of the same tactics: a deal with other African leaders, an alternative way of pre-empting a victory for Mugabe. After acrimonious bargaining Smith, Muzorewa, Sithole and Chief Jeremiah Chirau were able to announce agreement on a scheme for sharing power between whites and blacks. Smith now accepted one-man-one-vote but secured the white position by entrenched rights in the constitution. It was a political move to defeat the militant Patriotic Front. Britain and the United States endorsed it with a reservation which made nonsense of it: they approved it provided the Patriotic Front were brought in. But it was a plan to keep the Patriotic Front out.

After much heart-searching the Rhodesian whites approved the new constitution in a referendum in 1979. Elections with a commendable turnout gave Muzorewa the victory and he became prime minister of the aptly named Zimbabwe-Rhodesia. But this settlement was unconvincing. The whites remained obviously in control for an indefinite period and the new government failed to introduce even eye-catching reforms to alleviate, much less remove, racial discrimination.

Fortuitously the Rhodesian elections coincided with the elections in Britain which returned the Conservatives to power under a prime minister suspected of having rather more sympathy for Smith than most British politicians. Mrs Thatcher appeared from her public statements to favour recognition of the Muzorewa government and the immediate removal of sanctions. But British pragmatism prevailed. The Patriotic Front was by now too effective to be affronted lightly; it could even be said to be the winning side. At a Commonwealth conference in Lusaka Britain continued to withhold recognition from the Muzorewa–Smith settlement until one more attempt could be made to reconcile the warring parties. Britain prepared a new constitution, purged of the more blatant pro-white provisions of the 1978 document, and presented it to both sides at a conference in London at the end of 1979 on a take-it-or-leave-it basis. The new foreign secretary, Lord Carrington, was able to threaten to recognize Muzorewa if the Patriotic Front did not accept his constitution, while on the other side Mugabe and Nkomo, the rival leaders of Zanu and Zapu, were under strong pressure from the Front Line presidents to reach agreement. This combination prevailed. The whites and Muzorewa, although bitterly disappointed and angry, had no alternative to acceptance because what they had constructed was unsound. The conference, contrary to initial expectations, reached agreement. A British governor went to Salisbury for four months. Elections, fair and free in the view of some, the reverse in the view of others, gave Mugabe and Zanu a decisive and overall majority and he became the prime minister of the independent state of *Zimbabwe*.

Independence disposed of the conflict between black and white but sharpened that between the Shona and the Ndebele, personified by Mugabe and Nkomo. The next few years deepened the personal and tribal rifts. Nkomo was expelled from the government in 1982 and the Ndebele, far from being reconciled to a junior partnership in the new state, were progressively alienated and alarmed: disorders were met by methods which were certainly ruthless and seemed racial. After elections in 1985 which confirmed Zapu's hold over Matabeleland Mugabe, intent on firm one-party rule – constitutionally Zimbabwe was not a one-party state but in practice it was – initiated a reconciliation with Nkomo which was, however, barely more than superficial. Hopes of moulding Zimbabwe into a truly united state crumbled as the Ndebele minority became permanently alienated and mistrustful, numerous enough to disrupt the country but too few to create on their own a constructive society in a hostile environment. The hopes invested in Mugabe himself were clouded as internal violence led to illegal detentions without trial, rampages by army units and the diversion of food from, and consequent starvation in, politically selected areas. Not all Mugabe's senior colleagues withstood the temptations of power and in 1988–89 scandals forced five ministers to resign and one to kill himself. Mugabe's unconcealed preference for state socialism in a one-party state was unpopular not only with western opinion and governments but also within his own party where each section feared that in a one-party state another section of the party would become the single ruling group. An impressive economic plan which initially attracted international (specifically western) support was rendered precarious by drought, political instability, poor transport and distribution, a flight from the land to overcrowded towns with inadequate housing and no jobs, an intolerable foreign debt service, declining foreign revenues as world prices fell, and inflation which went up and down but a national product which went only down: all of which weakened the personal position of the man whose pre-eminence had made him peculiarly vulnerable to the twists and turns of fortune.

Two special factors aggravated this situation. The first was war in Mozambique. Zimbabwe formed in 1982 a Special Task Force to defend the oil pipeline to Beira on the Mozambican coast against the Renamo guerrillas, and this comparatively modest venture led Zimbabwe deeper into Mozambique's civil war. In 1986 Renamo declared itself at war with Zimbabwe, thus bringing hostilities to Zimbabwe's eastern province of Manicaland and leading to the creation in 1989 of the Zimbabwean Unity Movement by disgruntled members of Zanu – chief among them Edgar Tekere whom Mugabe had dismissed from his government. Secondly, discontent was nourished by the sluggishness of the land settlement programme, one of the major promises of independence. Families evicted from their lands after 1890 – the more massive evictions had taken place in the period 1930–60 – were promised recovery, but the new state also undertook to pay compensation in foreign currency to evicted white owners at a rate applicable to sales between a willing seller and a willing buyer. In spite of British contributions to this programme the cost was burdensome and progress slow. A first plan to settle 162,000 families in five years reached only a third of its target, partly because of practical difficulties such as the need to make new roads, partly because too little land was made available for the landless, and partly because the government insisted that newcomers must first renounce their title to land (if any) which they were leaving – and they were reluctant to do so. After ten years no more than a third of the land programme had been carried into effect; many of the new settlers quickly sold their allotted land and found themselves as poor as before or poorer. The government announced more vigorous action which, however, it could take only by watering down its obligations to white owners and introducing some form of compulsory purchase.

In spite of these travails and in spite of unemployment, poor public services, lack of foreign exchange and austerity measures, Mugabe's leadership was not seriously challenged. Nkomo became a back number and Mugabe's Zanu-PF won 147 of 150 seats in elections in 1990, 148 five years later. The World Bank and the IMF endorsed a five-year plan for investment in both the public and private sectors with a social fund for the relief of unemployment in return for promises of drastic pruning in the administration and ruthless refusal to bail out failed private businesses.

ANGOLA AND MOZAMBIQUE

Until 1974 Africa's southern cone was flanked by the Portuguese territories of Angola and Mozambique. The Portuguese record in

Africa was first in, last out: the first colonists and the last imperialists. This small European country first ventured into strange continents 500 years ago and it acquired in Africa an estate twenty-three times its own size. Bartolomeo Diaz rounded the Cape of Good Hope in 1487, followed by Vasco da Gama in 1498. The Guinea trade, of which Portuguese Guinea was a relic, flourished during the sixteenth century but the Portuguese were gradually pushed out of West Africa by the Dutch, the British and the French, and they were to find their biggest prizes further south. Luanda, the capital of Angola, was founded in 1576 by a grandson of Diaz; it was held by the Dutch for a short time in the seventeenth century but was recaptured by the rich Brazilian adventurer Salvador de Sá who became its governor and restored Portuguese rule. In the seventeenth and eighteenth centuries Angola provided slaves for Brazil. The trade was abolished in 1836 and the institution of slavery itself supposedly from 1858. During the intensest phase of the European scramble for Africa Portugal was supported sporadically but ineffectually by Britain, but by nobody else. Germany coveted Portuguese territories and hoped to succeed to them by lending Portugal more money than it could repay and then foreclosing. A Portuguese dream of linking Angola with Mozambique by a land strip never materialized. Although Mozambique was settled in the twenty-five years before the First World War, the Portuguese evinced only tepid interest in Africa, while their laxity in the suppression of slavery earned them a scandalous notoriety, especially after the publication in 1906 of H. W. Nevinson's *A Modern Slavery*.

Major risings in Angola in 1922 and 1935 were ruthlessly suppressed but also nurtured nationalist movements. Lisbon began to see that its rule was threatened less by white home-rulers than by black nationalism, even though the nationalists were in the short term hopelessly hampered by illiteracy, tribal divisions and the overwhelming might of the Portuguese police and army. In 1952 Portugal's colonies were renamed overseas provinces and the 1950s saw a burst of material progress in the shape of public works and development plans. The change of name enabled Portugal to claim that it was under no obligation to submit reports to the UN under article 73(e) of the Charter, Angola and Mozambique having become not non-self-governing territories but provinces of Portugal. In this contention Portugal had the support of its American and British allies and its Brazilian kinsmen, but in 1963 a resolution of the General Assembly urging Portugal to accelerate self-determination was supported by the United States and Britain; only Spain and South Africa voted against it; France abstained.

The Portuguese practised a policy of assimilation as a means of gradually expanding the franchise, but increasing immigration from

Portugal in the mid-1950s altered social and economic patterns, reinforced the colour bar and reduced many Angolans to unemployment or semi-employment. Tens of thousands of Angolans were in exile, mostly in the Congo, by 1960, and the nationalist movements established their headquarters in Leopoldville. Confidence was shaken by a dispute over the election of a northern chief, in which the Portuguese authorities acted in accordance with the letter of the law but with dubious wisdom and probity. The emancipation of the British and French West African colonies and of the Belgian Congo brought the modern African problem to Portugal's Angolan doorstep and in February 1961 there were riots in Luanda, partly stimulated by the adventures of a Captain Galvao, who seized the liner *Sta Maria* and was expected in Luanda to appear off the coast in the role of liberator. These riots were suppressed but in March there were further disorders in the north and an invasion from across the Congolese frontier. The Portuguese were taken by surprise, suffered casualties of about 1,400 in killed and wounded and disastrously responded to barbarities with further barbarities. African casualties were of the order of 20,000 and a stream of hideously wounded refugees – 20,000 in 1961 and a further 150,000 in the next three years – into the Congo blackened the Portuguese name, while the cost of Portuguese authority became a serious strain on Portugal's economy.

In Mozambique a guerrilla liberation movement, Frelimo, began an armed rising in 1964 and secured control over parts of the country, but it suffered in 1969 the assassination of its leader, Dr Eduardo Mondlane. The economy was hit by the blockading of Beira to implement sanctions against Rhodesia, but the Portuguese showed no signs of giving up. They embarked instead on a vast scheme to develop Mozambique's economy in association with South African, European and American concerns (which would thereby acquire a vested interest in Portuguese rule). This scheme, which centred round the building of a hydro-electric barrage at Cabora Bassa, would also entail the addition of a million persons to the white population. The political nature of the plan was barely concealed and, considering that Mozambique already had six hydro-electric schemes and needed no more, hardly could be. The political implications and lobbying by African states caused some of the participants to think again and – in the cases of Sweden and Italy – to withdraw.

By 1970 Portugal's colonial wars had forced it to allot half of its budget and over 6 per cent of its gross national product to military spending, to raise the largest armies ever raised in Portugal's history and to conscript its young men for four years. The absence of a politically active public opinion shielded the government from what had been one of the strongest, perhaps even the strongest, of the decolonizing forces in London, Paris and Brussels, but ten years of costly and unsuccessful fighting in three widely separated territories sapped the will of the Portuguese armies. The liberation movement had survived and grown, winning popular support, extending their geographical control and getting arms from abroad. In response the Portuguese tried bombing and barbed wire, but bombing was inefficient against guerrillas and barbed wire failed to separate them from the populace because the latter (unlike their counterparts earlier on in Malaya) were on the side of the former. In Guinea-Bissau the governor, General Spinola, becoming convinced of the hopelessness of the struggle, resigned in 1973 and went back to Lisbon as an outspoken opponent of continuing colonial rule. His actions and writings contributed to the overthrow by the army in 1974 of the military dictatorship founded in 1926 and the Armed Forces Movement which won control in Lisbon declared itself in favour of independence for Portugal's African colonies. This was immediately achieved in Guinea-Bissau and Mozambique, and Vorster, bowing to the accomplished facts, publicly accepted Frelimo's victory and wished the new government well. Both he and Samora Machel, Mondlane's successor and first president of the new state, knew that a quarter of the black labour in South Africa's mines consisted of Mozambican migrants, a constraint upon both governments. In Angola there was no single equivalent to Frelimo and when on the eve of independence Holden Roberto's FNLA attacked Agostinho Neto's MPLA South Africa clandestinely invaded Angola in support of a third movement, Jonas Savimibi's Unita. But Neto appealed for help to Fidel Castro and the South Africans were forced for the time being to retreat although they continued openly to arm Savimbi who succeeded in winning control over a substantial part of the country.

Washington's overriding aim was the eviction of the Cubans. Although Washington and Pretoria were at one in supporting Unita, that support undermined the Angolan government's ability to dispense with the Cubans and Pretoria was simultaneously negotiating with President dos Santos while helping Savimbi to throttle him. Militarily South Africa's incursions were dubiously valuable to it. Tracts of Angola were occupied, devastated and then abandoned. But in 1985 Angola, fortified by fresh Russian military equipment, took the offensive against Unita, and South Africa accepted the challenge to rescue it at the cost of escalating the conflict. Both sides failed. South Africa suffered another rebuff. A joint Unita–South African air onslaught on the strategic base at Cuito Cuanavale in 1987–88 was thwarted and the South African air force humiliated. Unita, although battered in the south, was re-established in Zaïre by the United States, but this was only a tactical move as talks were begun in London between South Africa,

Angola and Cuba with American blessing (but with Unita left out in the cold) and were continued in Brazzaville where agreement was reached on the withdrawal of South African and Cuban troops from Angola. South African troops left during 1989, as did half the Cubans that year and the remainder by mid-1991.

The end of seventeen years of warfare seemed in sight. The arrival of the Cubans in 1964 to protect the government in Luanda and the American and South African aid to Unita to prevent it taking and keeping power had produced no win for either side. In the 1980s Reagan, in defiance of the Clark Amendment of 1975 cutting off aid for Unita, had given huge quantities of aid to Savimbi whom Reagan dubbed a democrat. But by 1991 Savimbi still controlled no more than a fifth of the country and was ready to try and improve his position by the ballot box. The civil war was brought temporarily to a halt by the Bicesse Accords of that year mediated by Portugal, the United States and the USSR; and the UN, which had supervised the Cuban withdrawal in 1989–91 through UNAVEM I, was asked to oversee the implementation of these accords (UNAVEM II) and, subsequently, general and presidential elections. By the accords Savimbi recognized dos Santos as president and the two leaders agreed to withdraw their troops to fifty specified points and to fuse them into a joint army of 50,000 – a quarter of their combined strength. Political parties were formed by the dozen and elections were held under UN supervision in 1992. Thanks largely to UNAVEM they were peaceable, fair and strikingly popular – 90 per cent of those entitled to vote did so. But contrary to his hopes and expectations, Savimbi found that he was not the winner. Unita won 34 per cent of the votes against the MPLA's 54 per cent and although dos Santos fell narrowly short of the 50 per cent needed for confirmation as president, Savimbi won only 40 per cent. Savimbi restarted the war which, with 1,000 or more deaths a day, surpassed the horrors of all earlier campaigns. UNAVEM, with a mere 400 persons (increased to 1,000 for the elections) and starved of funds and other resources, was powerless to prevent the destruction of its hard-won achievements. Attempts in 1993 to stop the fighting again produced fresh agreements but Unita refused to yield territory won in the latest operations unless a refurbished UNAVEM were first introduced into the country. This the Security Council refused to promise unless Unita first accepted the new agreements.

Angola was devastated for so many years because it was rich. The war in which each side spent $1–2 billion a year on arms was financed by oil and diamonds – oil which enabled the government to borrow abroad (thus mortgaging its future) and diamonds which Unita sold to the international cartel managed by De Beers. A sub-plot in Angola's drama was provided by the comparatively small province of Cabinda which had been of little account before oil was discovered there in 1966. Its population of 100,000 included a secessionist movement. It was coveted by Zaïre but Mobutu was bought off by Agostinho Neto in 1978. Savimbi's occasional attempts to enlist Mobutu's help against the Angolan government in Luanda failed because Unita had no substantial or continuous presence in the province but separatist Cabindan groups maintained hostilities against the government in Luanda.

In *Mozambique*, as in Angola, South Africa played a double game for much of the 1980s and beyond. It concluded in 1984 at Nkomati an agreement whereby it undertook to cease supplying the insurgent Renamo (or MNR) and end its economic warfare in return for Mozambique's abandonment of the ANC, but in spite of a confirmation of these promises later in the year Renamo continued to receive aid from South Africa. President Botha (as he had become under a new constitution) had an interest in stifling the ANC and in pacifying southern Mozambique but no interest in inhibiting Renamo's activities further north so long as these did not force President Machel back into the arms of the USSR. Pretoria's approaches to Mozambique before the Nkomati agreement were prompted both by the presence of the ANC there and by generally worsening relations which had provoked talk of war and visits of Russian warships. For South Africa Renamo was a useful thorn in Machel's side. But in spite of aid from South Africa and from Portuguese sympathizers it failed to create anything like a rival government. Equally, however, Machel's government failed to suppress it and this failure was due not only to outside aid to Renamo but also to two internal factors: the fact that Mozambique was a large country divided into numerous small areas suspicious of one another and of central authority, and the unpopularity of Machel's government which rode roughshod over traditional customs and culture in its headlong attempts to modernize the country by fiat and dogma. Machel was killed in an air crash in South Africa, probably accidental, in 1986. His successor Joaquim Chissano was, at 49, a representative of a younger generation which criticized many of Frelimo's old guard for military incompetence and personal corruption. Chissano preferred diplomacy to war, if only because his government might contain Renamo but could not extinguish it so long as South Africa chose to sustain it. On the government's side morale and discipline were low as the cost of living rose and the pacification of the country appeared to be out of reach. Under the impact of the interlocking problems of civil war and economic collapse Chissano steered a reforming course, away from communist centralism and towards a multiparty system, with elections promised for 1991.

Communist Mozambique abandoned the one-party state sooner than its non-communist neighbours in Malawi, Zambia and Tanzania.

By 1990 Mozambique was ready for talks with South Africa, if only to recover from recent setbacks, and the Roma Agreement, negotiated with the help of President Kaunda, established a cease-fire and a plan for the reduction of forces and their concentration in specified areas. Renamo later denounced the agreement on the grounds that Zimbabwean forces had not been withdrawn from Mozambican territory; and whereas the government was weakened by a (failed) coup against Chissano in 1991, Renamo was strengthened when its leader, Alfonso Dhlakama, was received by the Pope and Portuguese and other European leaders. Fresh agreements were signed and the UN agreed to field 7,000 observers (ONUMOZ) but Renamo remained obstructive as its position steadily improved in northern and central areas which were alienated from the government not only by their miserable poverty but also by distrust of its predominantly southern (Shangaan) complexion.

SOUTH AFRICA

Bantu peoples reached the southern tip of the African continent in the eighteenth century and there met the Dutch, clashed with them and were gradually subdued and turned into hewers of wood and drawers of water. The Dutch, who had followed the Portuguese to the Cape and were at first confronted only by the Hottentots, enjoyed a brief dominance until, their home country having become a part of the French revolutionary empire, they became fair game for the British in the Napoleonic wars and lost their base at Cape Town which they had founded in 1652. In the 1830s they packed up and trekked and set up new independent states – successfully beyond the Orange River and the Vaal but unsuccessfully in Natal where, although they defeated the Zulus, they were again pushed on by the expanding British who annexed Natal in 1843. The British colony of the Cape grew politically, economically and demographically; it was granted constitutions in 1853 and 1872, receiving on the latter occasion the right to a prime minister of its own.

In the latter part of the nineteenth century South Africa achieved an entirely unexpected prominence in British thinking for two quite separate reasons. The first of these was the discovery of its wealth – diamonds in the 1860s and gold in the 1880s. The second was the conviction that South Africa was a vital link in Britain's imperial communications and that therefore Britain must keep firm hold not only of Cape Town and Simonstown but also of the hinterland. This second preoccupation turned the Dutch, or Boers, into potential enemies, since they might invite other European powers to occupy the sensitive hinterland or cause trouble for the British in Cape Colony by creating an alliance with their kinsmen in the colony. Moreover attempts by the Boers to reach the coasts in Portuguese territory and to trade with the outside world through Portuguese ports were a further source of friction since these activities threatened the monopolistic profits of British traders at the Cape. The British therefore began to expand more purposefully than before. They took Basutoland from the Boers and made it a colony in 1868. They annexed Kimberley, the diamond city, in 1871 and the Transvaal in 1877. Brought up against the Bantu they fought a series of wars (notably the Zulu War of 1879) and in 1880–81 they went to war with the Boers, whom they incidentally accused of maltreating the Bantu. A British Liberal government, perplexed at this same time by the graver practical and moral dilemmas of the Irish question, was divided about the right attitude to adopt towards the Boers, and after the battle of Majuba in 1881 adopted a policy of conciliation, restored the Boer republics of the Transvaal and Orange Free State and so implicitly recognized the existence of a separate and valid Boer nationalism.

Cecil Rhodes became prime minister of Cape Colony in 1890 and for a few years his voice rather than London's was the decisive one. Within the colony his policy was one of co-operation with the Boers for the paternal rule of the Africans, but any extension of this policy to form an Anglo–Boer entente in a wider field was prevented by British fears about the political and commercial policies of the Boer republics. Rhodes developed and paid for the Cape–Bechuanaland–Zambezi axis which put the Boer republics in jeopardy, countered the German threat from South-West Africa and forestalled a possible anti-British alliance between the Germans and the Transvaal. In 1895 he over-reached himself. With the connivance of a part of the British government he stimulated a rising against President Kruger of the Transvaal and sent Dr Jameson raiding into Boer territory to sustain it. The result was a rebuff for the raiders and also for the British government when the German emperor sent Kruger his moral support by telegram. But although Kruger triumphed and Rhodes fell, the British government, which now resumed the dominant role, adopted Rhodes's basic thesis that the first British aim must be to prevent the emergence of a Boer South African republic comprising the Boer republics and the British colonies. By fighting and winning the Boer War in 1899–1901 Britain postponed this eventuality but did not succeed in scotching it, for it came to pass in 1961. Britain also earned the hatred of the wounded Boers but failed to appreciate its depth.

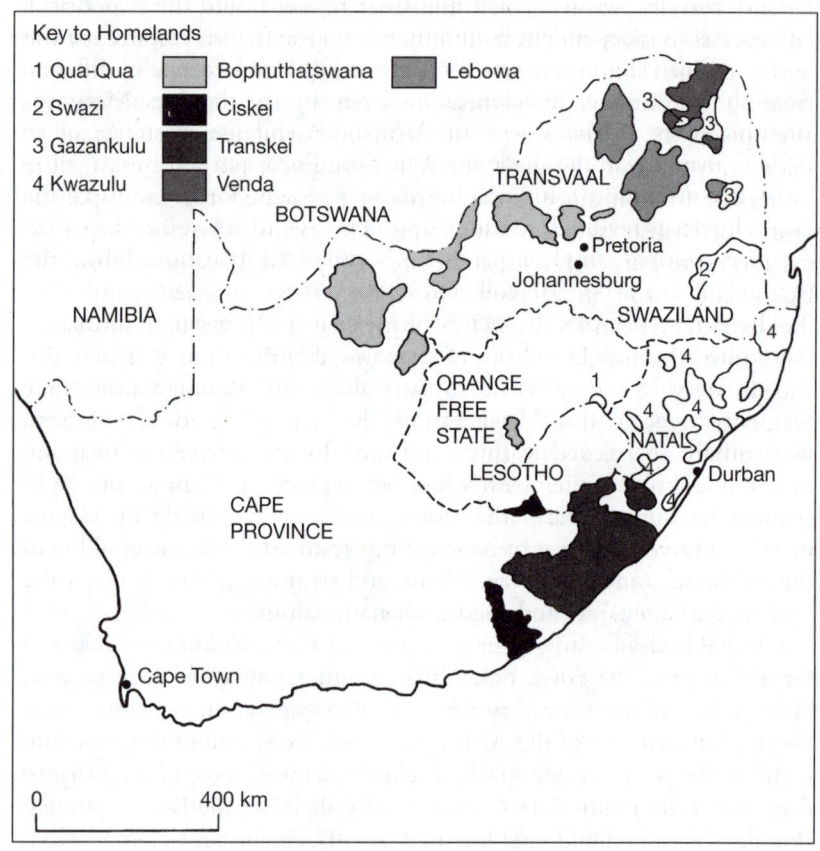

23.2 South Africa and the former Homelands

After the Peace of Vereeniging in 1902 the British pursued two barely compatible lines of policy. On the one hand they busied themselves being fair and generous to the defeated Boers, often forgetting the Africans in the process. The Transvaal and Orange Free State were granted self-government in 1906–7, and the constitution of 1908 provided for an all-white parliament which, while meant as a symbol of Anglo–Boer reconciliation, looked in retrospect more like a pledge of African exclusion. On the other hand the British had entrenched the policy of British supremacy by insisting on the extinction of the sovereignty of the republics and the dominance of the English language, so that the Boers planned for the day when the balance of power in South Africa would shift in their favour and they would be able then to turn the tables on the British. By creating the Union of South Africa in 1910 the British government tried to confederate the two self-governing Boer republics and the two British colonies as an independent dominion within the British empire, but this South African dominion was fundamentally if unperceivedly different from the three other old dominions by reason of the persistent non-British culture of the Boers (or Afrikaners) and the existence of an African majority in the shadows. When the European war broke out in 1914 the dominion joined in hostilities and sent forces overseas; the Boers however revolted, refusing – as they were to refuse again in 1939 – to be counted simply a part of a South African nation. They saw themselves as a people or Volk and not as a state.

The persistence of a distinct Afrikaner nationalism which ultimately triumphed through the electoral victories of the Nationalist Party was accompanied by the growth of an African nationalist movement which was frustrated by the Afrikaners. The Afrikaners had to resort to increasingly stringent measures and extreme ideologies in order to preserve and justify their newly won supremacy, and by the middle of the century South Africa had become an oligarchy ruled by a particular racial group without the consent of the majority of the governed who – to make matters worse – were a different colour.

The Afrikaners' attitude to colour was an extreme instance of a common prejudice. It had played a part in their decision to trek out of Cape Colony, where the British were putting more liberal ideas into practice and where the coloureds (the offspring of white alliances with black servants) were given the vote in 1853 subject to certain educational tests – a right which they retained until 1956. But the African was an essential factor in the state since his labour was needed in industry and could be had cheaply owing to the scarcity of work in the native reserves (which, comprising one-tenth of the surface of the state, were supposed in theory to accommodate all the Africans but were incapable in fact of supporting more than half of them). Colour became therefore equated with economic degradation; a black man was a labourer who must and could be had for a low wage. And economic degradation led to social degradation as the Africans were crowded into slummy locations which bred self-disgust and crime. Whereas in Europe the victims of the Industrial Revolution were the more unfortunate members of society, in South Africa the revolution which exalted mining and heavy industry above agriculture found its victims in a separate society, alienated and increasingly hostile. Natal possessed an additional element in the Asians who first migrated there in search of work in the middle of the nineteenth century.

The Nationalist Party was built up by General James Hertzog who was supported by the Labour Party (the representatives of white labour) and opposed by the British section of the electorate and by the moderate Afrikaners led by Louis Botha until his death in 1919 and then by Jan

Smuts. Hertzog was prime minister from 1924 to 1939. He and Smuts joined forces in a coalition government in 1933 but upon the outbreak of war in Europe in 1939 they again separated and Hertzog drifted back towards D. F. Malan who had founded a new Nationalist Party rather than follow Hertzog into alliance with Smuts. The Nationalists refused, however, to ally themselves with the fascist Ossewa Brandwag led by Oswald Pirow. In the first postwar elections, held in 1948, the United Party was ousted by the Nationalist Party, which embarked on an extended period of power under a series of prime ministers – Malan, J. G. Strijdom, H. F. Verwoerd, B. J. Vorster, P. W. Botha – who turned South Africa into an independent republic outside the Commonwealth and a police state based on racial discrimination.

The policies of the Nationalist Party were based partly on the fears and hatreds of a minority faced by a majority of a different race and colour, and partly on a theory which was elaborated in this context. To the Afrikaner the African, especially the urban African, was an uncivilized barbarian unfit for public responsibility or private intercourse; or alternatively the African, especially the rural African, was a tame, happy and rather lazy servant whose chief desire was to remain in his undemanding, if servile, state. The mingling of the races moreover was biologically evil. Logic therefore, as well as prejudice, dictated the attempts to separate the races which collectively amounted to the policy of *apartheid*. Although the races must remain in some contact with one another for a transitional period, the aim was the segregation of the black Africans in separate territories – dubbed Bantustans – which would be racial enclaves, economically self-sufficient, within a single South African polity. This theory and its elaboration were equally flawed. Scientists, including South African scientists, rejected the biological assertions at the basis of the policy of *apartheid*; economists demonstrated that the native areas could never become viable except at a cost unacceptable to the white population, and that the white community could not do without black labour: the African was to be returned to a semi-industrialized countryside while the white urban centres nevertheless craved for his services. The theory also overlooked the strength of human feelings of righteous indignation, and the political awareness of Africans who knew what was happening elsewhere in Africa. Finally, many of the protagonists of *apartheid* probably underestimated the extent to which their policies forced them to rule by violence and fraud and would attract worldwide attention and condemnation. Prisons filled, floggings became a habitual part of the process of government, executions took place at the rate of one a day, the mail ceased to be private, spies and informers multiplied, an inevitably brutalized police became a law of assault unto itself, and the régime felt itself obliged to build up a white citizen army with a potential strength of 250,000 in addition to a part-time commando force of 50,000 (based on four years' compulsory service) and a police force of 20,000 and police reserve of 6,000.

On the other side the African nationalist movement also took to violence. A Native Nationalist Congress, later renamed the African National Congress, had been formed in 1912. The object of its first leaders – chiefs and then lawyers – was to secure a place for the black African in South African society without upsetting any applecarts, but both the leaders and the rank and file began to have doubts about their prospects during the Hertzog period. The Nationalist victory in 1948 sharpened African apprehensions. The Congress leaders pursued a policy of strikes and civil disobedience to draw attention to demands which they formulated in 1955 at a conference at Kliptown. The government's response to the Kliptown Charter was to charge the leaders with treason. One hundred and fifty-six persons were indicted in a trial which lasted four years and ended with the acquittal of all the accused. In 1957 an increase in bus fares produced a wholesale bus boycott in which Africans, by trudging long miles to work every day, demonstrated their poverty and their discipline, but violence also began to grow and in 1958 outbreaks in the Transvaal were fiercely repressed. In 1959 a section of the African National Congress broke away to form the Pan-Africanist Congress led by Robert Sobukwe.

On 21 March 1960 a crowd of Africans, probably 3,000–4,000 strong but subsequently inflated by reports to much larger dimensions, converged on the police station at Sharpeville. It was one of a number of similar crowds which were doing the same thing all over the country in protest against the pass laws which required adult Africans always to have their identity documents upon them. In organizing this demonstration Robert Sobukwe had told Africans to leave their passes behind and present themselves at police stations in an orderly manner and unarmed. Newspaper reporters said afterwards that the crowds at Sharpeville and elsewhere were in cheerful holiday mood. At Sharpeville, where a small police force of seventy-five was nervously on duty, somebody fired a gun and within minutes dozens of Africans had been killed and many more wounded; the wounds of the victims were, according to hospital testimony, almost all in the back. At Langa in Cape Province there was a similar scene and altogether that day eighty-three Africans were killed and about 350 wounded. Sharpeville became a household word throughout the world. Statesmen of all colours denounced the bloodshed, flags were flown at half mast and formal tributes of silence were paid to the dead. In South Africa a state of emergency was declared as the protests against the pass laws continued, the militia was called up and thousands of Africans were apprehended and put in prison.

Nevertheless the principal African leaders – Chief Albert Luthuli (shortly afterwards awarded the Nobel Peace Prize), Nelson Mandela, Walter Sisulu – continued the struggle by non-violent means. A successful three-day strike in 1961 was followed by preparations for wholesale non-co-operation throughout the country and Mandela went underground in order to organize the campaign. The government was gradually placing most of the responsible leaders behind bars. As their leaders, their activities and their organizations were proscribed both Congresses produced more militant offshoots in the Sword of the Spirit and Poqo (founded in 1961 and 1962 respectively, the latter using murder as a weapon). In 1963 the government arraigned Mandela, Sisulu and others in a further mass trial at Rivonia and sentenced them to long terms of imprisonment.

The government did not hesitate to use force but it preferred, like any government, to provide its actions with legislative cover. Its legislative programme was extensive and cannot be considered in detail here. Its principal features, although framed within a single grand design, may be reviewed under separate headings.

In the first place there were a number of statutes affecting the individual in his private capacity. Sexual relations between Africans and Europeans had been a criminal offence since 1927. In 1950 this prohibition was extended to relations between Europeans and coloureds, and in 1949 the Prohibition of Mixed Marriages Act (unlike many other Acts) did precisely what its title indicated – not without reference to the marriage of Seretse Khama of the Bamangwato to an English girl. Also in 1950 the government enacted the Popular Registration Act, described by Malan as the basis of *apartheid*, which prescribed registration and classification by race. The position of the individual before the law was also affected by the Native Laws Amendment Act 1957, which gave the authorities power to exclude Africans from such places as churches on the grounds that they might cause a legal nuisance (the Dutch Reformed Church protested and then complied); by the Separate Amenities Act 1953, which declared that separate amenities for blacks and whites need not be equal, as the courts had tended to insist; by the General Law Amendment Acts of 1961, 1962, and 1963 which stringently reduced the rights of the individual against the police, introduced house arrest for periods up to five years and sanctioned the indefinite detention of suspects; and by the Bantu Laws Amendment Act 1963, which legalized the eviction of any African from any urban area no matter how long he had lived there and turned him into a squatter on sufferance who was allowed to work but not to settle.

A second category of statute fortified and formalized Nationalist control over the machinery of the state. In 1949 separate Indian representation in parliament was abolished and the government then launched its attack on the coloured franchise in the Cape province. Its Separate Representation of Voters Bill 1951, removing the Coloured voters from the common roll to a separate one, was declared unconstitutional by the courts because it took away rights which were entrenched by the South Africa Act 1909, and could not therefore be validly removed without a two-thirds majority of the two houses of parliament sitting together. The government thereupon brought in the High Court of Parliament Bill 1952 to give a committee of members of parliament power to override a decision of the Court of Appeal, but it too was voided by the courts. Malan then dissolved parliament, increased his majority but failed to win the two-thirds majority necessary for his purposes. He won over some United Party members in the lower house but not enough and then summoned a joint session of the two houses of parliament to consider a bill to amend the South Africa Act. This move also failed in spite of a measure of United Party support and Malan was then replaced by J. G. Strijdom who abandoned Malan's legalistic tactics in favour of packing the judiciary and altering the constitution of the senate. By these means he secured, first, the passage by the requisite two-thirds majority of the South Africa Act Amendment Act 1956 (which validated the ill-fated Separate Representation of Voters Act 1951) and, secondly, judicial endorsement of the Senate Act 1956.

This long tussle stimulated extra-parliamentary but still mainly non-violent opposition to the government among liberal whites as well as among Africans. The Torch Commando, the Black Sash movement and the bus boycott were its principal manifestations; but the principal consequences were the extinction of the lesser opposition parties (the Liberal and Progressive Parties), the increasing resort to house arrest and detention without trial, and a vigorous use of the communist bogey to brand all opponents of *apartheid* as subversive agents of an international conspiracy. The South African Communist Party had, after some internal disagreement, taken up the cause of the black worker which the Labour Party refused to espouse, but the communists had been as much hampered in South Africa as elsewhere by the twists and turns of Moscow and they had not greatly prospered. In 1950, however, the Malan government secured the passage of the Suppression of Communism Act which was important, not because it proscribed communist activities and the relatively feeble Communist Party, but for its definition of communists and for the powers given to the executive to deal with such people. A communist was defined as any individual whom the minister of justice chose to call a communist, and in so determining the minister was to take the widest possible view since any person seeking to promote political, industrial, social or economic change by unlawful acts or omissions, or the threat of them, was to be considered a communist, as were various other broad categories of people. The Act was in fact an instrument for eliminating every kind of opposition in parliament, trade unions, schools, universities and elsewhere, and at the same time a piece of propaganda meant to exaggerate the communist menace and so win support for the Nationalist régime.

In the field of labour Africans had been forbidden to form unions since 1937. A ban on strikes imposed during the war remained in force after it, and the Native Labour (Settlement of Disputes) Act 1953 repeated these bans on unions and strikes and kept labour relations under white control.

A further category of statute was concerned with the special subject of education. The Bantu Education Act 1953 showed the government's awareness of the crucial fact that children do not naturally adopt racial antagonisms (juvenile loves and hates being otherwise motivated), and also of the crucial need to give Africans a qualitatively different education if they were to form a permanently separate and inferior community in a society under white minority control. The Act created a new division of the Department of Native Affairs which was to direct and finance Bantu education and which would displace the religious missions which were running most of the schools for Africans. Some churches handed over their schools willingly, others unwillingly; the Anglicans preferred to close their schools in the Transvaal and the Roman Catholics decided to carry on and bear the cost themselves. In the new nationalized schools education was geared to the place of the African in South Africa; that is, an unskilled worker in industry or a docile rustic in the special territories to be allotted to him. A further Bill to permit the exclusion of Africans from universities had to be withdrawn in 1957, but two years later the university college of Fort Hare was taken over by the state and separate (but well-equipped) university colleges for Africans were created. Africans remained entitled to apply for admission to white universities if they could show that they were qualified to attend courses not available elsewhere, but in practice such pleas were always found by the authorities to lack substance and the liberalism of the universities themselves was thwarted by executive action.

The kernel of the Nationalist programme was the territorial separation of the races. The South Africa Land Act 1913 had allocated 87 per cent of land to whites and left open the question what to do with the remaining 13 per cent which was regarded as a labour reservoir. The Group Areas Act 1950 prescribed racial segregation in terms of residence, business and ownership of land. One of its immediate effects (completed by the Natives Resettlement Act 1954) was the wholesale removal of Africans from areas of Johannesburg which were needed by the whites. Some of these areas were slums, others were not; to the Africans they were areas where they had been able to own property. This Act was followed in 1951 by the Bantu Authorities Act which sought to re-create chieftainship as a basis for the management of the African population. Chiefs, appointed and paid by the government, were to exercise authority in tribal areas, which would be grouped in regions, which would be grouped in territories, over which the Minister for Native Affairs would have ultimate control. This scheme was carried a stage further by the Promotion of Bantu Self-government Act 1959, which (besides depriving the Africans of the right to elect whites to represent them in parliament) consolidated the African reserves into eight Bantustans. Six territorial authorities were established by the end of 1962, including one for the Transkei (an economically non-viable area of 40,000 sq. km with a population of 2 million in the north-east of Cape province) which was equipped with a cabinet of six under Chief Kaiser Matanzima and a legislature of four paramount chiefs, sixty chiefs and forty-five elected members. Meanwhile the Tomlinson Commission, appointed by the government, had reported in 1955 that the cost of implementing the Bantustan policy would be £104 million spread over ten years and that the African population in 1960 was likely to be 21.5 million. The government promised £3.5 million for the development of the Bantustans and established in 1959 a Bantu Investment Corporation with a capital of £500,000 to develop Bantu industries.

South Africa's domestic racial policies became the chief determinants of its external relations and led eventually to something approaching isolation. Even with Britain, a close associate and recent ally in the Second World War, relations became awkward as, for example, over the conclusion of a new alliance in which one of the sticking points was special guarantees over the treatment of black employees in the British naval base at Simonstown. The Simonstown agreements, concluded in 1955, were the fag end of several years of discussion in which neither Britain nor South Africa got what it wanted from the other. Britain wanted South African co-operation in the defence of British interests in the Middle East. South Africa wanted a broad alliance of (white) states similar to Nato for the defence of South Africa against communist attack or subversion and, second, the transfer of Simonstown to South Africa sovereignty. Both Britain and South Africa expected too much. The agreements of 1955, one of which – on staff talks – was secret, transferred Simonstown to South Africa in return for its indefinite use by Britain in peace and war and its use by Britain's allies in war, and in return also for the expansion and modernization of the South African navy with materials to be supplied by British shipyards. Beyond these provisions everything was vague: agreement in principle on the desirability of an African Defence Organization (which never came into being) and South African agreement, also in principle only, to contribute to the defence of Africa and the Middle East gateways to it (a meaningless phrase). If Britain had felt uneasy but no more in its dealings with the governments of Malan and Strijdom, with Verwoerd (who succeeded Strijdom in 1958) and his foreign minister, Eric Louw, the gap was wider. It was widened by events such as the Sharpeville massacre of 1960 and the exclusion of South Africa from the Commonwealth.

It so happened that the Sharpeville calamity fell shortly before a meeting of Commonwealth prime ministers at which South Africa intended to seek assurances that it would be accepted as a continuing member of the Commonwealth if it became a republic. This change of status had been a prominent item in the Nationalist programme and it had not been expected to give rise to any difficulties, since the Commonwealth had already adapted itself to the presence of republics. Formally, however, the change required endorsement and there were in the Commonwealth a number of members to whom South African racial policies were so repugnant that they might grasp at an excuse to withhold their assent. Verwoerd was unable to attend the conference owing to an attempt on his life by a white farmer, and South Africa was therefore represented by Louw. The question of *apartheid* was with difficulty kept off the formal agenda on condition that individual prime ministers might express themselves to Louw in private. Louw, so far from making the impression which he had hoped to make on his colleagues, was forced in these conversations to appreciate the strength of their feelings and to accept a final communiqué which hinted that an application to remain in the Commonwealth, unaccompanied by reform of South African racial policies, would be rejected. Ghana, which had already voted by referendum to become a republic, was accepted as a continuing member. The Commonwealth therefore side-stepped the unpleasant and unprecedented step of expulsion and left the initiative with South Africa. A referendum was held later in the year with the expected republican result and South Africa thereupon left the Commonwealth.

Sharpeville also had repercussions at the United Nations. The General Assembly had been concerning itself with *apartheid* for some years. During its 1952–53 session it had appointed a commission to consider the racial situation in South Africa and this commission had reported that the situation was harmful to peaceful international relations. In succeeding sessions the Assembly regularly passed resolutions asking South Africa to review its policies in the light of the UN Charter and deploring its failure to do so. After Sharpeville the

Security Council resolved that the situation in South Africa might endanger international peace and security, called on South Africa to mend its ways and instructed the secretary-general, in consultation with the South African government, to take measures to uphold the Charter. Hammarskjöld went to South Africa in January 1961 but his visit was as fruitless as the Assembly's resolutions. The Assembly repeated its censures in 1961 and then in 1962 asked members to break off diplomatic relations with South Africa, close their ports to South African shipping and their airports to South African aircraft, prevent their own ships from calling at South African ports, boycott all South African products and suspend exports to South Africa. The Assembly also appointed a special committee to keep the situation under review. This committee produced a series of reports and in August 1963 the Security Council recommended the suspension of all sales of military equipment. Throughout these proceedings the African members were pressing for UN action partly out of genuine attachment to the idea of collective international action as embodied in the Charter and partly because they had no prospect of achieving anything effective on their own against the vastly superior military and economic strength of South Africa. They were the more embittered by the refusal of Britain and other powers which were capable of exerting economic pressures but reluctant to do so.

The campaign for economic sanctions against South Africa developed in the 1960s as the wave of independence came up against its most formidable obstacle in Africa's deep south. Here was a white minority fortified by an authoritarian and non-egalitarian theology totally different from the secular libertarianism and egalitarianism which, even if subconsciously, formed the policies of the principal European colonial powers. This minority moreover was strong and rich, and it was not a colonial offshoot but a people at home with no thought of going anywhere else. The new African states found themselves powerless to do anything about a problem to which they had given an absolute priority. They were compelled therefore to turn to those who could do something about it – in the first place Britain and secondly the United States. They hoped that these countries would be not only outraged by the enormities perpetrated by the South African régime but also convinced that *apartheid* was a danger to international peace; they wanted international economic sanctions which would either force the Nationalists to parley with the Africans and give them a share of political power, or would cause a disintegration of the régime and the emergence of something new which could in no event be worse.

The British and Americans were not prepared to agree that the practices of *apartheid* were a threat to international peace within the meaning of article 39 of the Charter. Their reasons were many. Both countries were preoccupied with problems which seemed to them more urgent. The United States gave unhesitating priority to Vietnam and to Latin American affairs at least. Britain was more intimately involved because of the old Commonwealth connection and Britain's much longer acquaintance with Africa, but in Britain's financial situation even the relatively small profit on trade with South Africa could not lightly be forgone; British exports to South Africa earned about £150 million a year, 4 per cent of total British exports, while imports from South Africa cost £100 million, 2 per cent of the import bill, leaving a favourable balance of about £50 million. In addition about two-thirds of South Africa's gold went into western reserves and was largely handled, at a profit, by the London market. In both Britain and the United States there were vested financial interests and pressure groups which, though they might not be so influential with government as was sometimes supposed, were nevertheless marginally effective owing to the articulate assiduity with which they made the case against intervention. British investments in South Africa were valued at about £1,000 million. Britain was also held back by fears of reprisals against its dependencies of Bechuanaland, Basutoland and Swaziland, by the residual importance of the right to use the naval base at Simonstown, and by a postwar mood of getting out of things rather than into them. Public opinion was largely unaroused; the committed minorities were believed to be too small to have electoral significance.

The most weighty of these various considerations was the existence of other absorbing international commitments which made both London and Washington hope that an upheaval in South Africa could be postponed, but almost equally important were doubts about the effectiveness of sanctions, about their legality in the circumstances which had arisen, and about the immediate consequences of applying them. The South African economy had not only been expanded rapidly but had also become more self-sufficient; at the same time oil and other materials had been accumulated. There was no agreement among experts on the period during which South Africa would stand a virtual siege, nor was there agreement on the degree of effectiveness of the naval blockade which would be necessary for a short, and possibly for a long time. On the legal side it was argued that the preconditions for such action laid down in article 39 of the Charter (namely, a threat to international peace) existed and had indeed been acknowledged by the members of the Security Council who formally resolved in 1963 that the policies of the South Africa government 'disturbed' international peace. But the substitution of the word 'disturb' for the word 'threaten' was a precise gauge of the political temperature, for it enabled members of the Council to say one thing and yet avoid the consequences of their words. Britain and the United States could no longer bring themselves to deny the substance of the allegations made by the African states but they were determined all the same not to be embroiled with South Africa and so they insisted on using a word which was not part of the language of article 39. Nor was this simply hypocrisy, although it may have been unwise inasmuch as it postponed an issue which was getting worse and not better. It was not simply hypocrisy because those who had the power were unwilling to use it so long as they could see so little of its probable effects. Sanctions would require a blockade and a blockade would lead to a war. Within South Africa the collapse of the régime could produce an anarchic situation, bloodshed and misery. Ought they to take this responsibility? Since they did not want to anyhow, the question was not too difficult to answer.

The refusal of the British and Americans to be moved by the arguments in favour of sanctions on the grounds that *apartheid* constituted a threat to peace was, however, not the end of the matter. South Africa was also vulnerable on a completely different count. It was the mandatory power in *South-West Africa* and was accused of violating the mandate agreement by, among other things, introducing *apartheid*. If this charge could be established, South Africa could be required to mend its ways in South-West Africa or expose itself to legitimate intervention upon its failure to comply; and *apartheid* would be condemned in at any rate one context. The Africans went to law.

South-West Africa, huge, rich and under-populated, the most barren expanse in Africa except for the Sahara, came under German rule at the end of the nineteenth century because nobody else wanted it. In the 1870s the Cape Colony had expressly considered and rejected annexation and even the great Berlin share-out of 1884 left South-West Africa as *res nullius*. In the concluding years of the century, however, the Germans moved in, first as allies of the Bantu Herrero against the Nama (an aboriginal people second in antiquity only to the Bushmen) and then as settlers and farmers. Rebellions by the Herrero and later by the Nama led to savage repressions amounting to genocide, and by 1907 the territory had been devastated, subdued and handed over to German farmers together with the surviving Africans who became little better than slaves. In the First World War it was conquered by General Botha and a South African army in a matter of weeks, but after the war it was – together with all other conquered German possessions outside Europe – placed under international mandate instead of being treated as the spoils of the conqueror. The mandate was entrusted by the Principal Allied and Associated Powers to King George V to administer through his government in South Africa but without any transfer of sovereignty.

The mandate agreement provided, among other things, for the suppression of trade in slaves, arms and liquor; prohibited the establishment of military bases and the military training of Africans; guaranteed freedom of religion; and required the mandatary to submit annual reports to the Permanent Mandates Commission of the League of Nations and to forward to the League complaints and petitions from the inhabitants. In its execution of the mandate the South African government treated the new order as tantamount to annexation. Attempts were made to convert South-West Africa into a fifth province of South Africa and although these attempts were successfully resisted by the League the character of the territory was profoundly altered by an influx of white South Africans who acquired land cheaply and became a ruling class complete with an elected parliament. In the north areas were reserved for Africans. White immigration was prohibited in these areas which were ruled by a governor-general appointed by South Africa acting through African chiefs who received salaries if they were complaisant but were deposed if they were not.

After the Second World War Smuts repeated pleas for the integration of the territory with South Africa on the grounds of propinquity and economic advantage; he denied that, as the mandatory, South Africa was under an obligation to negotiate a trusteeship agreement with the United Nations; and he argued that the inhabitants of the territory were happy and prosperous and wanted to be part of South Africa. Smuts was obliged to drop his attempt to secure integration and he agreed to submit to the UN reports on conditions in the territory, but after 1948 the Nationalist government stopped the reports and also allotted six seats in the South African lower house and four in the senate to representatives of South-West Africa's 53,000 European inhabitants (the non-Europeans numbered 400,000).

The International Court of Justice ruled in 1950, in an advisory opinion (which was elaborated on subordinate points in further opinions of 1955 and 1956), that the mandate was still in force, that South Africa was not obliged to place the territory under trusteeship, but that it was obliged to submit reports and transmit petitions. From this time the Herrero, Nama and Damara tribes made their views known to the world through their spokesman, the Rev. Michael Scott, who presented on their behalf a picture of their conditions and wishes very different from the official South African line. South Africa offered to negotiate with the United States, Britain and France, as the survivors of the Principal Allied and Associated Powers of the First World War, an agreement to place South-West Africa under the administration of the International Court, but this novel idea was rejected by a special committee of the United Nations and was withdrawn in 1955. In 1957 the UN appointed a good offices committee of three (Britain, the United States and Brazil) which visited Pretoria the next year and recommended that either the Trusteeship Council or a new body consisting of former League members should receive reports from the mandatory, but South Africa maintained that the mandate had lapsed and that it was gratuitously continuing to administer the territory in the spirit of the mandate. South Africa next suggested a partition whereby the southern part of the territory would be annexed to South Africa and the northern part (where there were no whites) would be under South African trusteeship and administratively part of South Africa. The General Assembly rejected this scheme. The territory was in practice progressively absorbed into South Africa's administrative system and became subject also to the politics and practices of *apartheid*.

In 1959 a report by the UN Committee on South-West Africa comprehensively condemned South Africa's execution of the mandate and in the following year the Organization for African Unity decided to initiate a substantive case before the International Court, claiming that the mandate agreement had been violated. Ethiopia and Liberia, former members of the League and the senior independent states of the continent, thereupon petitioned the court accordingly and engaged an eminent American lawyer, Ernest Gross, to argue their case at The Hague. In 1960 the court accepted jurisdiction by the narrow margin of eight judges to seven. After protracted and expensive proceedings the court in July 1966 non-suited the plaintiffs and, in spite of its preliminary decision in 1960, ruled by the president's casting vote (making eight against seven) that they had no standing in the matter before the court. This refusal to entertain the substance of the case caused universal astonishment and was considered all the more unsatisfactory in that the outcome was plainly due to accidents of death and ill-health, without which the judgment on this point would have gone the other way. The immediate effect, besides jubilation in South Africa and some discrediting of the role of law in international affairs, was to return the South-West African issue to the arena of politics (where it was bound to end up in any case).

In October 1966 the General Assembly, brushing aside the notion of further recourse to the court, resolved by 114 votes to 2 with two abstentions that the mandate was terminated and the UN must assume the administration of the territory. A special committee was appointed to consider how the second part of this resolution could be made effective. Essentially the majority in the Assembly was searching for a course of action in which Britain (which had abstained on the resolution) and the United States (which had voted for it) would participate. The committee and a subsequent special meeting of the General Assembly in 1967 were divided between those who asserted, without contradiction, the enormity of South Africa's proceedings and those who asked, without reply, what was to be done about it. In 1968 the UN created a Council for Namibia (as the territory was henceforward called). This Council was not in a position to do anything. In 1970 the Security Council declared South Africa's occupation of Namibia illegal on the ground that it contravened the terms of the mandate. In effect the UN annulled the mandate. The Council's western members abstained but cast no veto.

While the case before the International Court was still in progress the South African government had appointed the Odendaal Commission to make recommendations about South-West Africa. This move was prompted by more than international interference, for in 1960 a South-West Africa People's Organization (Swapo) had been formed and had declared itself at war with South Africa, initiating disorders in the territory of unpredictable consequence. The Odendaal Commission reported in 1964 in favour of dividing the territory into one white and ten black areas and spending £80 million over five years on communications, water and other schemes. The black areas would have certain local government powers but matters such as defence, internal security, frontiers, water and power would be reserved to the central authorities. The habitable land and the known minerals fell within the white area. The consideration of this report was postponed pending the decision of the International Court.

The collapse of the judicial offensive against South Africa and the increasingly evident refusal of major powers to enter on an economic war with it coincided with the aggravation of the Rhodesian independence crisis and also with the granting of independence to black régimes in Britain's three protected territories of Bechuanaland, Basutoland and Swaziland which lay on or within South Africa's borders. South Africa was simultaneously strengthened and faced with problems of external policy. The South African government aimed to preserve white supremacy by economic and military power and did not doubt its ability to do so. It had to consider whether white supremacy could be preserved beyond its own borders and, secondly, how it should conduct relations with established black states. White supremacy was strengthened by the waning of the old feud between the Afrikaner and English-speaking minorities. The Afrikaner Nationalist victory at the elections of 1948 seemed to have become irreversible. Moreover the rapid growth of the South African economy blurred the old dividing line between an Afrikaner country party and an English urban one. Afrikaners, while continuing to constitute a country party, were also penetrating in larger numbers into the upper reaches of the industrial plutocracy and sharing the bounding standard of living enjoyed by that class. This change in the fortunes of a part of the Afrikaner community created a possible dilemma for the future, when the affluent Afrikaners might have to choose between the rigour of their racial doctrines and the maintenance and the expansion of this standard of living by consenting to the lifting of job reservations. This prospect gave some comfort to the more optimistic liberals who hoped that economic pressures would enforce a gradual abandonment of the practices of *apartheid*. Externally the South African government appeared to conclude that white rule in Rhodesia and the Portuguese territories could not be indefinitely maintained and that the weaker black states could be induced by necessity to respond to a show of amicability and so breach the unanimity of black Africa's hostility. South Africa supported the illegal Smith régime in Rhodesia not so much because it believed in the permanence of white rule in that territory but because of the force of South African white popular sympathy for it and in order to show that the economic measures threatened against South Africa could not work even against the very much weaker Rhodesian economy. At the same time Verwoerd and Vorster pressed Smith to come to terms with Britain. Vorster at least hoped that the creation of a moderate black régime would forestall the victory of black nationalists friendly to the ANC, but he misjudged the obduracy of the Rhodesian whites. In the longer view South Africa, like the USSR after 1945, strong but beleaguered, would look out towards a hostile world over an expanse of satellites which would one day include also a black Rhodesia, Angola and Mozambique. The model was Malawi.

In the late 1960s the Nationalist Party was strong enough to afford internal dispute without risking the loss of power. A hard-line or *verkrampte* section took issue with the enlightened or *verligte* wing (a comparative term) but was severely defeated at the polls. The *verligte* majority proceeded to develop its economic relations with African states, concluded an economic agreement with the Malagasy Republic in 1970 and was encouraged by the response of such diverse African leaders as Houphouet-Boigny and Busia, Kenyatta and Bongo. South Africa also saw in 1970 an opportunity to gain credit by obtaining from the new British government a badge of respectability in the form of a reversal of the Labour government's refusal to sell it arms. Vorster perceived and played on the Conservatives' continuing vision of world affairs in primary Cold War colours and their reluctance to consummate Britain's withdrawal from military vantage points in Asia and the Indian Ocean.

This British withdrawal from the world beyond Europe had continued during the 1960s. Partly because of the natural reluctance to withdraw and partly too because the British were filling a gap which the Americans were in no hurry to man, British military power had remained in evidence in the arc stretching from the Red Sea to Singapore. But after the departure from Aden in 1968 the retention of forces in the Persian Gulf seemed more than ever anomalous and transient, and in 1968 economic stringencies led the British Labour government to announce that British forces in the Gulf, and also east of it, would be withdrawn as early as 1971. This acceleration of earlier intentions alarmed Malaysia and Singapore. Singapore's prime minister, Lee Kuan Yew, went to London to contest it: he got a little money and a nine months' extension. Discussions, hitherto largely academic, for a defence agreement between Malaysia, Singapore, Australia and New Zealand were begun.

The British decision to withdraw in 1971 instead of a few years later seemed at the time precipitate. Yet in another sense it was a belated recognition of the obvious and its significance rested more in the delay than in the haste, for the delay ensured that it should coincide with the growth of Russian naval power. Thus a British withdrawal which, earlier, would have forced the Americans, Asians and Australasians to fill the gap now offered the alternative prospect of Russian naval power taking the place of British naval power in the east. Although impeded by the closing of the Suez Canal by the 1967 war, the Russians were able to extend their gains in the Middle East to the Indian Ocean where a small Russian naval force became a familiar feature, supported by bases in Socotra, Mauritius and as far east as the Andaman Islands in the Bay of Bengal. Whether or not this extension was prompted by conflict with China or by a natural tendency of major powers to be active everywhere, it redoubled American opposition to the British withdrawal and caused Britain itself, which saw withdrawal as humiliating rather than sensible, to look around for a different policy. It cast east-of-Suez strategies into a Cold War, anti-communist mould.

During the British election campaign of 1970 Sir Alec Douglas-Home, the putative foreign secretary, said that a Conservative government would reverse the Labour government's policy of not selling arms to South Africa. Conservatives were not partisans of *apartheid* but wished to be friends with South Africa in spite of *apartheid* and because they could see no state nearer to the Indian Ocean that was both strong and anti-communist. After the Conservative victory one of Sir Alec's first acts was to repeat and set about implementing this declaration. He was supported by the new prime minister, Edward Heath, who resisted all arguments against this course, largely, it seemed, on emotional grounds: that election promises must be kept and that British policy decisions must rest in British hands (a proposition that nobody had denied). While Heath treated arguments by African leaders as attempts to push Britain around, Douglas-Home treated African politics as a department of the anti-communist Cold War. On the other side it was argued that a Russian naval threat, should it develop, could not in any event be met by a relatively small British–South African naval force; that the focus of such a threat, essentially a threat to the oil traffic, would not be in the southern oceans but round and about the southern shores of the Middle East; and that the British alignment with South Africa – a political demonstration – would create for the Russians political opportunities in Africa far greater than any military benefits likely to accrue to the anti-Russian side. Just as conflict in the Middle East and the ensuing arms race in that area and the American attachment to Israel had created a Russo–Arab front, so would a British alliance with South Africa throw African countries, however reluctantly, into the arms of the USSR; the USSR could look forward in the 1970s not merely to a growing role on the high seas but in addition – and once more because of western mistakes – to a political windfall on the African continent. Within South Africa these external arguments kept alive the illusion that South Africa would be welcomed into a southern counterpart of Nato in a global anti-communist strategy. Yet South Africa, mainly for technological reasons, was becoming unnecessary for any such strategy and, mainly for political reasons, was an unacceptable ally. The Anglo–South African agreement for the use of the Simonstown naval base (1958) was allowed by the British to expire in 1975.

For South Africa the 1970s were a decade of alarms surmounted. At the end of them South Africa appeared stronger and more independent in spite of the collapse of Portuguese rule in Mozambique and Angola, the transfer of Rhodesia to black rule and the abrupt cessation of Iranian oil supplies. At home a boom in the price of gold gave South Africa a windfall. But Angola's independence had accentuated the problem of Namibia while race relations, including the special aspect of controlling black labour, remained unresolved and looked insoluble.

Angola's independence in 1974 gave Namibia a frontier with a sovereign black state, gave Swapo bases within easy reach of that frontier and reanimated international concern over the territory. Since growing disorder inside Namibia would oblige South Africa to make its occupation even harsher and so exacerbate foreign criticism and the threat of international sanctions, South Africa abandoned its policy of annexation for the alternative of transferring authority to an anti-Swapo alliance likely to be responsive to South Africa's needs and susceptibilities. Discussions in the Turnhalle in Windhoek during 1975–77 led to the formation of the Democratic Turnhalle Alliance (DTA), an interracial association of racially based parties. It was opposed by the multiracial Namibian National Party (a coalition of five parties) and by white segregationists. The Turnhalle conference produced a plan for transition to independence at the end of 1978 together with a draft constitution so complex, and a fragmentation of the territory so excessive, that it sharpened suspicions that its main object was to leave all effective power in Pretoria. The plan, whose essential aim was to put an acceptable Namibian government in place before elections and before independence, was stillborn because it was unacceptable to the UN which was concurrently elaborating other plans. Vorster, having abandoned annexation, abandoned the Turnhalle alternative. He substituted no new policy but formally annexed the port of Walvis Bay, which had never been part of the mandated territory, and twelve offshore Penguin islands.

The initiative passed to the UN which, by Security Council resolution 435 of 1978, adopted its own programme and also established a 'contact group' consisting of the United States, Canada, Britain, West Germany and France to negotiate with South Africa. The UN programme required a cease-fire, release of prisoners, return of fugitives to Namibia but retention of Swapo guerrillas in their bases in Angola, withdrawal of all but 1,500 of the 20,000 South African troops, the despatch of a UN force, elections for a constituent assembly, supervision of the whole process by a Special Representative of the UN and by a UN Transitional Assistance Group (UNTAG), and so to independence on the last day of 1978. South Africa accepted this plan provided that Walvis Bay were expressly excepted, and an effective cease-fire supervened before the withdrawal of South African forces began. Swapo was worried by the deployment of the remaining South African forces which it wished to see concentrated in the south, and about the ill-defined co-operation between the South African and UN representatives. While discussion on these points was still in train South Africa delivered heavy attacks on Swapo's bases in Angola and Swapo broke off the talks. The murder of Chief Clemens Kapuuo, an enemy of Swapo and pillar of the DTA, and the South African Administrator's harsh repression which followed, all but wrecked the negotiations. They were however resumed after strong Zambian and Tanzanian pressure on Swapo to keep talking and the UN was able to convoke a conference of all interested parties in Geneva at the beginning of 1981. Besides the protagonists – South Africa and Swapo – the five western and six black African states attended as observers. The conference was brought to naught by South African design. Namibia's independence was further delayed by two changes: the succession of P. W. Botha to the leadership of South Africa and the arrival of Ronald Reagan in the White House in Washington. Botha conducted for some years a more vigorous regional policy than his predecessor; Reagan tied progress in Namibia to the removal of the Cubans from Angola.

The result was stalemate. The DTA was flagging, its leader Dirk Mudge resigned from the council of ministers, and a new association of anti-Swapo groups – the Multi-Party Conference (MPC) – came into being. In 1984 South Africa declared itself ready to give independence to a government of national unity but a conference at Lusaka attended by Swapo and the MPC produced no unity, Swapo refusing to co-operate with the MPC except on the basis of a precise timetable for independence. In 1985 South Africa devolved some powers to a transitional government but the key to the situation was outside the territory in Angola. As part of the Brazzaville Agreement of 1988 for the withdrawal of troops from Angola the South African and Angolan governments agreed on a joint military patrol of the Angolan border with Namibia and the implementation of Security Council resolution 435. The execution of this agreement and the preparation and supervision of elections were undertaken by the UN (UNTAG). In elections in 1989 Swapo got 57 per cent of the vote and 41 seats out of 72 in the constituent assembly, fewer than the two-thirds which would have given it virtually unbridled power in framing the constitution. The DTA got 29 per cent of the vote and 21 seats. The remaining seats were divided among small and resentful offshoots of Swapo which had defected or been driven out during the years in exile (when dissidents had been treated with deplorable severity and cruelty). A constitution was quickly adopted in 1990 and Swapo's leader Sam Nujoma was elected Namibia's first president. Swapo proved itself genuinely conciliatory, made no attempt to monopolize power or posts, succeeded in combining a multiparty system with stability, and respected the independence of the judiciary and the press.

In 1978, partly as a result of half-veiled scandals and partly because he was ill, Vorster moved to the presidency of the South African Republic and was succeeded as prime minister by Botha. Vorster had ruled for twelve years. He was heir to an insoluble problem and handed it on unresolved but not unchanged. The task of South Africa's leaders was to defend not a state but a state-within-a-state and a racial system hated by many at home, reviled throughout the continent and abhorred in the world at large. Vorster tried to shift South Africa's regional and global positions but did not significantly improve them. In terms of power these two positions were diametrically different. Regionally South Africa was a dominant power but globally it was of secondary importance at best. Even at the height of the Cold War South Africa failed to persuade more than a handful of conservatives (mostly in Britain and the United States) that South Africa mattered more to them than they did to it. South Africa was rich in resources but dependent on the outside world for capital to exploit these resources, vulnerable therefore to financial pressures or sanctions, and geographically and strategically more marginal than it liked to think. So long as defending South Africa meant defending *apartheid* South Africa found few friends and those whom it did find were not so much genuine as (Israeli and Taiwanese arms dealers, for instance) merely mercenary. Vorster perceived the hopelessness of trying to ally white South Africa with the principal western powers and addressed himself to the region, particularly after the abdication of Portuguese power in Angola and Mozambique and the collapse of white rule in Rhodesia exposed his landward flanks. In an attempt to come to terms with the transition in Rhodesia he sought the co-operation of President Kaunda to pre-empt a victory for the armed guerrillas in Rhodesia and he offered independent Mozambique friendship and economic aid. But in Angola he made war. He did so ineptly. His invasion of Angola in 1975 had no clear political or military aims and if, as seems probable, he counted on western support because of the Cuban presence he was wrong or at least easily deceived about the extent and decisiveness of (mainly American) help. South African casualties, although not severe, shocked South Africans who had not been told that they were engaging in war. This setback was compounded in 1977 when other clandestine operations of a different kind came to light. The government's Department of Information had been authorized to engage in deception and bribery in order to win friends and disarm criticism and in the course of its campaigns much of the money provided for these purposes had been misused or appropriated by officials of no mean standing.

Finally in the Vorster years, a special event both symbolized South Africa's bedevilled situation and further soured its relations with the west. It became a nuclear power, but its added military might was irrelevant to its particular problems and alienated its potential friends. In 1977 preparations for a nuclear test in the Kalahari desert, about which the United States was informed by the USSR, led to strong protests from the principal powers. South Africa denied that the Kalahari site was designed for a nuclear explosion but the denial was not widely believed and South Africa continued to refuse to adhere to the Nuclear Non-Proliferation Treaty. The United States, which had been supplying South Africa with enriched uranium, adopted in 1978 legislation prohibiting such sales to any state which did not permit inspection by the International Atomic Energy Authority. (Reagan later found ways of circumventing this ban.) In 1979 a phenomenon inexplicable except as a nuclear explosion was observed by an American intelligence satellite over the ocean between South Africa and the Antarctic. In 1987 South Africa abruptly changed its attitude to the Non-Proliferation Treaty and agreed, under threat of expulsion from the IAEA, to adhere to the treaty. By this time it was clear that South Africa had at least the capacity to make and use nuclear weapons, even though it had no rational use for them. Against its neighbour they were unnecessary, against its own townships and internal enemies unusable.

P. W. Botha inherited therefore a situation which was not only intrinsically daunting but also tarnished by recent failures within and beyond South Africa's frontiers. He ruled (1978–89) for nearly as long as Vorster. He won the leadership by a narrow margin and would not have got it at all if his principal rival and Vorster's obvious successor, Connie Mulder, had not been eliminated by the scandals and divisions which precipitated Vorster's retreat. Botha appeared relatively a soft-liner but more in the sense of having doubts and hesitations than in the sense of having firm policies of his own. He was a politician from the Cape and not from the boilerhouse of Afrikanerdom in the Transvaal. He had held the defence portfolio for eleven years and had the respect and confidence of the military; although victorious within the Nationalist Party he was not strong enough to offend the hard-liners. Stubborn and cautious but increasingly domineering, he was determined that South Africa should know the firm smack of government and should stand on its own feet rather than being beholden to allies who laced their friendship with offensive homilies about its internal affairs. Serious riots in Soweto in 1976 (met with brutal police reaction) and the emergence of the Black Consciousness movement led by Steve Biko until he was beaten to death by police in 1977 showed that the will to live dangerously and fight against fearful odds had not been extinguished. These trends illuminated the refusal of a new generation to come to terms with the whites; here were two communities with no way of talking to each other and not much wish to do so. Nevertheless Botha accelerated the abatement of petty *apartheid*: football teams and audiences, for example. He toured the Homelands or Bantustans (see the note at the end of this Part). He hoped to conciliate and detach from the black majority a small black bourgeoisie with a marginal share in the country's prosperity – an exercise in the use of economic resources for political ends – and tried also to prevent Indians and Coloureds from making common cause with the black peoples by offering the former limited internal autonomy. Separate assemblies of whites, Indians and Coloureds would be created; a mixed electoral college with an overall white majority would choose a president who would appoint three prime ministers, who would pick their own cabinets; and there would be also a multiracial council of ministers with a white majority. There was no place for black Africans in this scheme since their numerical preponderance would have destroyed the commanding position which, on arithmetical principles, the whites were able to use against the Indians and Coloureds. But these were gestures which changed nothing of substance. The Homelands had become evidently nonsensical and the constitutional tinkering was first spurned by Indians and Coloureds and then only half-heartedly accepted.

Other varieties of federal and confederal solution were canvassed. They were more academic than practical. Talk of partition into a black state and a white was revived, but no line of demarcation conceivably acceptable to both sides could be devised; fairness, in terms of an equitable division of wealth, could not be achieved in any way likely to secure peace between them. Chief Gatsha Buthelezi, who had revived the Inkatha movement (founded in 1928) and aimed to turn it into a vehicle for negotiating a non-violent settlement with the whites, proposed a National Convention to discuss non-violent progress towards majority rule in a unitary state and under a constitution accepted by all four racial groups. This proposal had the merit of all plans which prefer talking to fighting, but it seemed to offer little prospect of agreement between blacks and whites in their prevailing moods. Nor was it clear how far Buthelezi, a Zulu chief, could speak for non-Zulus or for the younger generation.

In elections in 1981 Botha lost votes but not seats; his plans were checked but not abandoned. In the following year the right wing of the Nationalist Party, led by Dr Andreas Treurnicht, having failed to gain control of the party left it, leaving Botha freer to proceed with his constitutional changes. These were approved by the white electorate by a 2–1 majority in a referendum at the end of 1983 and accepted, after some hesitation, by the Coloured and Indian communities who were thereby embraced in the new tricameral legislature with 85 and 45 representatives to 178 whites. The blacks were entirely excluded but were promised some relaxation of the minor indignities of *apartheid* and some participation in the local administration of their own communities. The black response was overwhelmingly cynical, as was that of the outside world. The banned ANC intensified its activities both within South Africa and on its borders.

In regional affairs Botha was as ambivalent as Vorster. He liked to show the flag and did not shrink from firing the gun. On the other hand he resuscitated the concept of an association of all states in southern Africa. Verwoerd had toyed with the prospect of an African commonwealth stretching up to (and even possibly including) Zaïre and under South African economic control. Botha's version was called a 'constellation' but it was too nebulous to attract serious attention and was regarded by black states as no more than a design to increase their economic dependence on South Africa. South Africa's economic strength served to scare rather than unite the area. Distrust and indeed aversion to South Africa were profound and black states were more intent on lessening their dependence on an overmighty neighbour than on going into partnership with it. They created therefore a countervailing Southern African Development Co-ordinating Council (SADCC) of nine black states which, at conferences at Arusha in 1979 and Lusaka and Maputo in 1980, evolved ambitious plans for regional economic co-operation and secured promises of $670 million from outside sources – priority being given to husbandry and communications. SADCC was an essay in regionalism by sovereign states with a population of 60 million, considerable natural resources but inadequate capital, technology or managerial skills. Its members suffered from sporadic drought, direct attacks from South Africa, increasing defence expenditure and an overall decline in output of 15–20 per cent in the first half of the 1980s. First steps were modest but the organization gained international regard and at its fifth congress in Harare in 1986 thirty-seven states were represented. So long, however, as South Africa was out of bounds, SADCC could not itself generate sufficient resources to attract the foreign capital which it needed. In 1994 South Africa joined and SADCC changed its name to Southern Africa Development Community – SADC – twelve states from Tanzania to South Africa plus Mauritius with an aggregate GNP of $150 billion, most of it South African.

For most of the 1980s South Africa lacked settled policies. In spite of its military and economic dominance in the region it displayed little more than a patchwork of inconsistent expediencies. The Nkomati agreement with Mozambique was initially ambivalent and then dead; the forward policy in Angola was abandoned, resumed and abandoned again; none of Pretoria's shifts in Namibia advanced its interests. Displays of strength had incommensurate results. They coincided with disappointments and half-measures at home. The Homelands, most of them still on paper, had became a monument to dogmatic inflexibility pointing down a dead end; the brutality of the police was failing to quell protests or riots and was fuelling outrage abroad; constitutional changes in 1984 were criticized by whites on both of Botha's flanks as well as being rejected by non-whites; and the president himself wobbled between going forward and going back to pure repression. He announced in 1985 a relaxation of the rules of *apartheid* governing sex and marriage but not the restrictions imposed by such centrepieces of *apartheid* as the Group Areas Act or the Bantu Education Acts. He envisaged some devolution of authority to blacks locally but was adamantly opposed to any black share in power at the top, even in the restricted form conceded to the Coloureds and Indians. As violence led to scenes reminiscent of Sharpeville in 1960 and Soweto in 1976, South Africa's principal trading partners – including this time the United States – took fright and began once more to talk about economic sanctions. Western capitalists began to run down their investments, the rand collapsed temporarily and the white business community in South Africa expressed forthright dismay when Botha's riposte was more rod than olive branch.

In 1985 South Africa seemed to be on the edge of an explosion. It had so often been there before that there was a stubborn insistence, among whites and their foreign friends, to regard the crisis as exaggerated: white power could be preserved by superior force, superior intelligence and changes which would include no concessions on the Group Areas Act, the Reservation of Separate Amenities Act or the constitutional dominance of the white vote. Yet there was by this date a new situation which was being fashioned by more than the steady accumulation of pressures and discomforts, and the greatest change was the public acceptance – albeit by a small section of the whites – that there must be negotiation with the ANC. The biggest obstacles to this change of mind had been, first, the almost universal (white) belief that there was no need to talk to the ANC or any other black body and, second, the equally widespread belief that the ANC was merely a stalking horse for the communists and black power a prelude to rule from Moscow. The second belief, buttressed by the presence among the ANC's leaders of a sizeable minority of communists, remained strong but the first had become a subject for discussion. During 1985 white leaders of industry and banking held a formal meeting with ANC leaders outside South Africa and at the end of the year the creation of a new Congress of South African Trade Unions emphasized the government's problems in retaining control over organized labour. The government imposed censorship over the reporting of black riots and police counter-brutalities but although these measures impeded foreign journalists they did not stop the disorder or render the principal black townships less ungovernable. The banned ANC and the precariously legal United Democratic Front were the principal beneficiaries of these cracks in white solidarity and government confidence, although they themselves, being amalgams of groups of different origin and temper, were not unmarked by suspicions and jealousies which might be expected to increase as the black cause gained in strength and optimism. Nevertheless the overriding change of these years was the arrival at the top of the political agenda of the questions how, when and within what limits to treat with the blacks.

This shift owed something to the internationalization of the affairs of Africa's deep south – the internationalization which was anathema to white South Africans – and reluctance on the part of governments of major white states which would have been happy to leave South Africa to order its own affairs if they had thought it capable of doing so. But as the white state-within-a-state lost moral authority and relied more and more on force, as the whites themselves became more nervously divided, and as the blacks remained capable only of harassing the white régime and incapable of supplanting it, so chaos seemed a not impossible prospect. For countries which looked to South Africa for valuable minerals or corporate profits chaos had to be avoided. Mouthing abhorrence of *apartheid* was not enough. Nor were partial or token economic sanctions. The UN had imposed in 1977 a mandatory ban on the supply of arms to South Africa (which fostered South Africa's manufacture of relatively small weapons but left it dependent on clandestine suppliers of heavier equipment and advanced technology). Otherwise Mrs Thatcher and to a lesser degree President Reagan espoused a policy of business as usual on the assumption that prosperity would soften racial asperities and enable the whites to keep control of a situation which, if it was deteriorating under white rule, would deteriorate faster without it. From this standpoint Reagan was able to see Unita and Renamo as romantic freedom fighters, but in 1985 the US Congress approved additional sanctions and it seemed probable that the next president, whoever he might be, would not veto severe measures adopted by the Congress. Thatcher, who saw the ANC as nothing more than a terrorist organization, tried to thwart but succeeded only in snubbing her Commonwealth partners when they voted in favour of tougher economic measures. (She was again isolated at a later Commonwealth conference in 1989.) By way of compromise the Commonwealth despatched an Eminent Persons Group to South Africa. Simultaneously, but perhaps not coincidentally, South African attacks on Botswana, Zimbabwe and Zambia scotched any usefulness which it might have had, and a mission by the British foreign secretary, Sir Geoffrey Howe, in the following year met blunt and resentful rebuff: both missions hoped to set a diplomatic ball rolling by securing the legalization of the ANC and PAC and the release of Nelson Mandela, imprisoned since 1963. Nevertheless, South Africa's efforts to extinguish the ANC having failed, a new international agenda which included discussion and accommodation with the ANC was necessarily taking shape.

So too, painfully, slowly and arguably too slowly, were minds changing within the white South African community. Broadly speaking the 1970s had witnessed a surprising level of confidence; the 1980s questioned this confidence. In the 1970s whites, who had become used to facing external criticism, fancied that they saw the threats to their position diminishing. The ease with which economic sanctions against Rhodesia were defeated by western corporations with governmental connivance reduced to insignificance the threat (never strong) of similar action against South Africa. The late 1970s saw too an economic recovery unforeseen a few years earlier. Growth exceeded forecasts, partly export-led but also nourished by domestic consumer demand. The boom in the price of gold filled the country's coffers and enabled it to pay off foreign debts; taxes were cut; the balance of payments improved. South Africa was not merely surviving but prospering in spite of internal tensions and external vilification. It was becoming less of a 'windfall' economy (although the gold boom belonged to that category), more self-sufficing and independent. Its principal weaknesses were its dependence on imported capital and fuel, shortage of skills, and the uncertainties surrounding the black labour force. Capital continued to flow in spite of the political risks, and the development of oil from domestic coal advanced so rapidly that the cessation of critical supplies of oil after the Iranian revolution of 1979 imposed no more than minor restrictions on private consumption. But the labour market, skilled and unskilled, posed more serious problems. The shortage of skills compelled the abandonment of many promising projects and increased unemployment. The unionization of black labour was potentially a source of effective militancy, by strikes or violence; it was unclear how the whites could keep control of black labour. Whether unemployed or organized in labour unions the black urban proletariat could become allied with traditional black nationalism.

The 1980s darkened this side of the picture. GDP declined; foreign investment virtually ceased; inflation rose above 20 per cent a year; the economy was ceasing to generate profit to finance its future; the reserves were running down. Discontent was manifested in five by-elections in 1985 when the ruling party lost votes to the hard right in very different constituencies. The ANC had been badly harmed without being smashed: the raids on neighbouring countries which were designed to scare them into disowning and expelling the ANC did not do so. Attempts to control the black townships by tens of thousands of arrests and the declaration of a state of emergency shook confidence, particularly the confidence of the business community. Even South Africa's undoubted military strength was less reassuring than it had seemed in earlier years. South Africa, although neither a world power nor a superpower, was a fully modernized power capable of spending about 2,000 million rand (about £1 billion) a year on defence and vulnerable only to a direct attack assisted by a superpower, or to a prolonged guerrilla war assisted by a major internal rebellion. Yet it was dependent on outside sources for the maintenance and further development of its more sophisticated equipment; its expanding forces were short of officers, NCOs and instructors; and its perimeter was uncomfortably long.

To retain the goodwill, or negate the ill will, of the west South Africa boasted invaluable minerals which, it believed, the west could not do without. These minerals were not only remarkable in themselves but, in a number of cases, irreplaceable except by recourse to the USSR as the only other known major repository. In manganese, essential and irreplaceable in steel making, South Africa dominated world production and reserves; in vanadium it was a major producer and possessed enormous reserves; in uranium it had considerable reserves and although other countries also had reserves most of these had no exportable surpluses, so that western Europe and Japan could be seriously upset by any cut in supplies from South Africa (and Namibia); in platinum, whose diverse uses embraced the chemical, electrical, glass and petroleum industries, the United States was dependent on South Africa. To some extent this dependence might be lessened by stockpiling and the search for substitutes but the haphazardness of nature in endowing these two areas – southern Africa and the USSR – with such large shares of the world's known deposits of a variety of strategic minerals was a significant factor in international affairs and in the way in which white South Africans assessed their prospects in the world.

In these confused and dangerous cross-currents two facts stood out. The Afrikaner political and intellectual élites were losing their faith in the *apartheid* which had sustained them since the Nationalist Party came to power in 1948. They were less dogmatic, more flexible. But although they shed their ideology they were not ready to shed power – or, in so far as some of them foresaw the need to loosen their monopoly on power, were restrained by an electorate which envisaged no such loss. The blacks on the other hand were unimpressed by a shift in attitude which seemed to them to betoken no change in the realities of everyday life or the realities of power. Events spoke louder than the conferences where a few leaders on both sides were beginning to know one another. The death of Steve Biko was only the most flagrant and famous of a stream of incidents which were creating martyrs for the black cause and making a peaceful settlement less and less likely. Each martyr was a figure to whom his surviving compatriots owed a debt which they felt could not be compromised. The generation of the under-thirties equated South African capitalism with exploitation, *apartheid* and police rule; and it was uninterested in any dialogue with the whites, even liberal whites.

This situation was changed by one man: F. W. de Klerk. Botha's political failings, his unattractive personality and ill health combined to embolden his senior colleagues to get rid of him. Of the candidates to succeed him de Klerk was the most conservative but he transformed the political atmosphere. He stopped trying to appease the hard white right; he set out to curb the security forces, military and police, which were acting as though they were a law unto themselves; he stressed, in contra-distinction to Botha's style, the role of cabinet and parliament; he abandoned the policy of destabilizing governments in neighbouring states; he permitted protests and orderly demonstrations; he initiated secret talks with Nelson Mandela; he released from jail Walter Sisulu and other incarcerated black leaders; and he released Mandela himself. These measures indicated a determination, probably irreversible, to negotiate in effect a treaty with black South Africans, represented first and foremost by the ANC, and to give South Africa a new constitution.

Nelson Mandela was an even more striking figure. With Oliver Tambo and Walter Sisulu he had been one of a trio of black youth leaders. In 1963 he had been sentenced at the Rivonia trial to imprisonment which seemed likely to last as long as his life. He was vice-president of the ANC, the president being Tambo who had escaped imprisonment but was at this date seriously ill in Sweden. The ANC was the creation of a liberal black élite. Although its popular base expanded in the 1950s and onwards its leaders remained select, urban and middle class. It looked to the USSR for financial help and became closely associated with the South African Communist Party in spite of some distrust of that white organization. It preferred non-violent tactics but endorsed violence when non-violence got it nowhere. Unlike, for example, Black Consciousness (a product of the hopeless 1970s), the ANC did not disdain white support and was not dogmatically anti-white. On emerging from prison after twenty-seven years Mandela was revealed as a man of extraordinary composure and dignity. If de Klerk had created a new situation, Mandela was the man without whom their new starting point would most likely be of no avail.

Both men had the task of finding common ground for working out a new constitution. Both realized that beginning to talk was an initial portent and the merest prelude to an eventual agreement which might lie some way ahead. Both had the no less daunting problem of securing adequate consent, each from his own constituency, by marginalizing extremists. The ANC was insistent upon one-man-one-vote. The government had to find a way of reconciling this recognition of democratic majority rights with safeguards for the apprehensive white minority, part of which was viciously hostile to any relaxation of white rule. The ANC accepted a multiparty system and a mixed economy (so too did the SACP) but was wedded to much more government control over the economy than white politicians or businessmen approved. Public order was an immediate issue. De Klerk tried to get Mandela and the ANC to forswear violence, which they were unwilling to do so long as the only actual changes in the country were the release of

Mandela and others and the president's evident sincerity and willingness to talk. The ANC agreed, however, in 1990 to suspend violence in return for the gradual release of all political prisoners and the termination of the state of emergency which gave the authorities extensive powers of arbitrary arrest and indefinite detention. The government limited its removal of the state of emergency by excepting Natal which was the scene of violence of a special kind – violence between blacks. While the ANC was countrywide the undoubted representative of the black community as a whole, in Natal this position was contested by the Inkatha, the movement led by the Zulu chief Gatsha Buthelezi. Inkatha had solid support in rural Natal but was losing ground in the towns to the ANC (or, before the ANC was legalized, to the United Democratic Front which was the ANC in disguise). The resulting contest took the form of hideous violence between Zulu and Zulu in which thousands were killed. The violence spread beyond Natal, in particular to the Transvaal where it pitted Zulus against non-Zulus. The ANC complained with more than a little plausibility that the police were helping Inkatha at least negatively by standing by instead of intervening to keep or restore order, even perhaps positively by encouraging violence in order to impede the government's talks with the ANC. Police revelations (suspected as a fabrication) of a revolutionary conspiracy between blacks and communists were assumed to have the same disruptive purpose, but de Klerk and Mandela kept to their common purpose. De Klerk declared that membership of the Nationalist Party should become open to blacks, thus creating the prospect of two mass parties – the NP and the ANC (already open to whites) – as the principal political pillars of a non-racial democratic South Africa in which the white minority would not be sequestered in a permanently minor party.

The internal evolution of South Africa provided in these years the main focus for the debate on the efficacy of international economic sanctions which, before the Second World War, had been focused on the external threat to Ethiopia from Italy. The more this question was generalized the more intractable did it become since the question itself was intrinsically complex and its several practical applications peculiarly divergent. In general the efficacy of sanctions rested on two entwined sets of calculations: their economic consequences, and their impact on the political will of various groups within the country assailed by them. In the particular case of South Africa sanctions aggravated the existing economic problems of an expanding and sophisticated economy, largely dependent on the international economy, in a period of worldwide economic stringency. Commercially, sanctions obliged South African enterprises to redirect their business, notably away from the United States and towards West Germany and Japan, but this forced search for new markets was in itself no bad thing, and with the aid of subtle and not so subtle evasions of the obstacles imposed by sanctions, South Africa's trade suffered from sanctions no serious net loss. Particular losses were mitigated, if not fully compensated, and South Africa learned to live with the marginal loss at least in the short term. Financially, the withdrawal of foreign companies gave South African companies the opportunity to buy, often at bargain prices, the plants and operations left behind by the departing foreigners; some individuals became rich out of these transactions. More serious, if also more ambivalent, was the reduction in the flow of foreign funds. In common with other countries South Africa experienced in the mid-1980s a payments crisis which it surmounted by rescheduling its debts but which was nevertheless a disturbing reminder of its dependence on foreign capital. In purely economic terms this dependence on foreign creditors was offset by the creditors' knowledge that, if they refused to extend their loans, they risked losing their money and their chief bargaining counters. Psychologically, however, the shock was sharp and one element in it was the feeling that sanctions had aggravated the situation and might do greater damage in the future. This psychological reaction had political repercussions, first among financiers and industrialists and so among politicians who took note of the fact that the white business community needed to participate in the world economy and feared isolation. While the poorer whites might be driven by sanctions to support extreme and racialist policies and parties, the business classes on the other hand were driven by these sanctions in the opposite direction and this trend gave force to arguments outside South Africa for continuing or even intensifying sanctions. Within the space of a few months in the second half of 1989 the OAS, the Commonwealth and the UN not only asserted the efficacy of existing sanctions but also treated them – and the promise of their remission in return for fundamental changes in South Africa – as a major instrument for the social and constitutional reforms which de Klerk seemed irreversibly to have espoused but was introducing with tantalizing slowness.

The main business of the years 1991–94 was the elaboration and inauguration of a new constitution, non-racial and democratic. A Convention for a Democratic South Africa (CODESA) was convoked. It included more than twenty parties. The context was past passions, persistent violence and fears of civil war to come, counterbalanced however by an overriding commitment by the leadership of the Nationalist Party and the ANC to continue their struggle without violence and without fateful delays. The essence of this struggle was the balance between majority and minority rights, the safeguards therefore whereby the (white) minority might enjoy a blocking power short of a veto. The principle of majority rule, meaning black rule, was not negotiable but the limiting conditions of its exercise were negotiable. The negotiations were clouded by the existence of numerous groups besides the two protagonists: white extremists of varying degree; the PAC and other black groups (some of them within the ANC) distrustful of Mandela's drive; and the Inkatha Freedom Party (IFP) whose power was overrated both by its leader, Mangosuthu Buthelezi, who envisaged an autonomous Zulu domain, and by de Klerk who hoped for a time to construct with the IFP a right-wing alliance in the new non-racial political system.

On the black side Mandela's position was never seriously threatened. The ANC's first national conference for thirty-three years – in Durban in 1991 and attended by 2,234 delegates – tested and confirmed his authority at a point where blacks feared that their bargaining power was in decline as the outside world planned to remove the economic sanctions which had played a large part in converting de Klerk to conciliation and democracy. During CODESA's proceedings Mandela's personal dominance crucially preserved unity in the ANC in the face of differences over the handling of the discussions, the tone of propaganda and the use of violence which many in the ANC believed to be justified by disclosures of aid given to Inkatha in money, arms, training and even incitement to fighting by blacks against blacks. On the opposite side, de Klerk's position was strengthened by a referendum in which whites supported with unexpected emphasis – 69 per cent of a turnout of 89 per cent – accelerated power-sharing under a new constitution; and by his opponents' recourse to violence which, by its failures, discredited them, dissolved their alliance and killed off any prospect of a right for whites to secede from the South African republic. De Klerk, however, was weakened against the ANC by a refusal by the United States to come out in favour of white demands for a blocking power equivalent to a veto. He was further weakened by evidence that he was much less than fully in control of his security and armed forces or alternatively that he had connived at breaches of the law and of police instructions.

The ANC responded to pressures on de Klerk by producing precise timetables of its own which envisaged an interim government by mid-1992, general elections in the first quarter of 1993 for a constituent assembly, a mixed cabinet by the middle of that year and a new constitution in operation one year later. By insisting that timetables were as important as substance the ANC forced de Klerk to reconsider and substantially abandon his hopes of keeping many of the more intransigent whites in the Nationalist Party or of forming an alliance with the IFP to serve as a counterweight or alternative to his current alliance with the ANC. The ANC did not get all that it was asking for but it gave up fewer of its aspirations than did the government. The ANC successfully jostled the government.

It did so partly because it held a stronger hand from the moment that de Klerk had felt obliged to release Mandela from prison and partly because the opposition to CODESA's initiatives was more damaging to the government than to the ANC. This opposition came from whites and blacks, united by their common aim to secure the greatest possible degree of devolution or even the right to secede but divided over the means to do so – particularly over a resort to force. They combined uneasily in a Freedom Alliance which embraced the Afrikaner Volksfront led by General Viljoen, the Conservative Party led by Ferdie Hertzenberg, the AWB of Eugene Terre-Blanche, Buthelezi's IFP, the presidents of two Homelands – Bophuthatswana and Ciskei – and another twenty smaller groups. Having exerted little or no influence on the constitutional debate, these groups confronted the question whether to participate in elections or try to wreck them. The two more important leaders – Viljoen and, at the eleventh hour, Buthelezi – decided to participate. Viljoen parted company with the AWB and other extremists after an affray in Bophuthatswana when Chief Lucas Mangope invited the Afrikaner Volksfront to help him stay in power and the AWB took the opportunity to join in riotously. Mangope was overthrown and Viljoen turned his back on violence and formed a new party with defectors from the Conservative Party. In Ciskei, Brigadier Oupo Gqoso resigned after conflict over police pay and pensions and both these Homelands were in effect taken over by the South African army. Buthelezi, who possessed the greatest capacity for disrupting the elections, was in a cleft stick, incapable of winning more than a fractional vote even in Zulu areas and equally incapable of extracting from Mandela and de Klerk more than a fraction of the demands which he had been making over the terms of the constitution or the timing of the elections. A week before the elections he reversed his refusal to take part in exchange for face-saving declarations concerning the status of his nephew, the Zulu king (who had shown signs of being irked by his uncle) and promises by Mandela and de Klerk of international mediation over autonomy for KwaZulu-Natal. These promises remained unfulfilled. Whether Buthelezi's aim was autonomy or independence remained uncertain.

Elections were held in April 1994. At the end of 1992 nineteen of CODESA's participants had agreed on a programme which provided for a Transitional Executive Council operating alongside the existing Nationalist cabinet; elections by adult suffrage in April 1994; a federal state with nine regions, a president and up to two vice-presidents (nominated by parties winning more than 20 per cent of the vote); an executive of not more than 27 to which parties might nominate members in the proportion of one for every 5 per cent of votes won; a legislature consisting of an assembly of 400 (half elected from national lists and half from regional lists) and a senate of 90, ten from each province; regional legislatures with, depending on population, 30 to 100 members; a Bill of Rights and a Constitutional Court. Senior military leaders had declared their loyalty to these arrangements. The government and the ANC had agreed that there would be a single army, no dismissals from the civil service and no confiscation of land. The conclusion of these arrangements was due to the persistence of Mandela and de Klerk and to the fact that each needed the other because they feared anarchy. They were, however, tactically divided because de Klerk wanted time in which he might stultify his extremists, while Mandela could not afford time for fear of losing ground to his own extremists.

The elections were largely undisturbed by violence but seriously disrupted and eventually prolonged by administrative shortcomings. They were declared to be substantially free and fair and this verdict was accepted by the principal parties in spite of widespread suspicions of grave misdemeanours by Inkatha supporters in KwaZulu-Natal. The ANC scored a sweeping victory but fell narrowly short of a two-thirds majority − 62.6 per cent of the vote, 252 seats in the assembly. The National Party won 20.4 per cent and 81 seats, Inkatha 10.5 per cent and 43 seats. Three other parties won seats in the assembly but failed to qualify for places in the cabinet of which the ANC occupied 18, the National Party six and Inkatha three. In seven of the provincial assemblies the ANC held majorities but the National Party won the Western Cape (with the help of the Coloured vote) and Inkatha won KwaZulu-Natal.

24 RUSSIANS, CUBANS, CHINESE

What the Russians knew about Africa when the Second World War ended was little. A few nineteenth-century travellers in the Nile valley, a comic opera invasion of Somaliland and Ethiopia in the 1880s and some marginal activity by the Orthodox Church were all that tsarist Russia had to put against the stores of knowledge acquired by western traders, explorers, missionaries and colonial governors; and since almost none of Africa was independent Russia had no diplomatic missions to observe and report about it. The Bolshevik revolution gave a spurt to African studies but these were quickly submerged under the pressure of more urgent cares. As the Second World War ended Stalin put in a bid for a share in Italy's conquered colonies, but it failed. In the 1950s African studies were expanded, an Africa Institute was founded in Moscow (1959) under the direction of an eminent historian, I. I. Potekhin, and the USSR began to send representatives to Afro–Asian conferences. In 1960 the People's Friendship University, later renamed Lumumba University, was founded in Moscow.

African leaders were by definition anti-colonialist and so anti-western. Most of them also professed to be socialists. This was for Moscow a promising background, but it proved less fruitful than expected for a number of reasons: the anti-colonial struggle was unexpectedly short and peaceable; African leaders, beginning with Nkrumah, espoused non-alignment; the familiar paths to London and Paris remained well trodden; Soviet aid never matched western aid. Even African socialism created no strong association with the socialist Mecca. It was a nebulous concept derived at least as much from western socialism as Russian communism. The principal vehicle had been the French and not the Soviet Communist Party and although the Rassemblement Démocrate Africain (RDA) – established in October 1946 at Bamako – had in Gabriel d'Arboussier a secretary close to communist thinking, its president Felix Houphouet-Boigny was hostile to it. Another founder of the RDA, Sekou Touré, was typically unhelpful to the USSR. Evicted from the French trade union federation of the CGT he had joined its African counterpart the CGTA (later UGTAN), in which communists were in a minority, and he became the first of a number of African presidents to evict Russians from his country.

Moscow's first venture into African politics south of the Sahara was a reaction to an opportunity. Guinea, pitched into independence in 1959 upon refusing to join de Gaulle's '*communauté*', was born in desperate straits. A poor and thinly populated state it had nevertheless a strategic position on the south-west corner of upper Africa, a port nearer than any other to mid-Atlantic, about a third of the world's bauxite and massive grievances against the west. Moscow was quick with diplomatic recognition, credits (140 million roubles in 1959), offers of trade and Czech arms. Sekou Touré was invited to Moscow. Khrushchev promised to visit Conakry and Brezhnev did. Daniel S. Solod, who had served with credit as ambassador to Syria and Egypt, was transferred to Guinea with the purpose of turning Conakry into a centre of Soviet influence and activity in *West Africa*.

Mali and Ghana fell more or less fortuitously into this picture. Mali (or Soudan, as it still was) was the poor relation of the short-lived union with Senegal which collapsed in August 1960 under the weight of its incompatibilities. Huge, arid and landlocked it had nothing to offer an ally but Modibo Keita, a socialist of the Sekou Touré brand, was brought by Guinea into the Soviet sphere by a similar friendlessness. He was offered credits of 40 million roubles and Bamako airport was made available to Aeroflot. Ghana, already independent three years earlier, was offered 160 million roubles at this time, although Moscow had at first shown only perfunctory interest in Ghana and sent no ambassador there until 1959. Nkrumah was a strong Commonwealth man, candid about his need for western aid and – in Soviet eyes – a typical product of the wrong section of the African bourgeoisie which was ready to do deals with the British. But Nkrumah had also been the first to hold out a helping hand to Sekou Touré, he gave aid and a home to the truly revolutionary Union Populaire du Cameroun, and he became bitterly anti-American after the murder of his friend Lumumba. He was a pendulum swinging in the right direction· for Moscow. Like Sekou Touré and Modibo Keita he was to receive the Lenin Prize.

Yet within a few years there was nothing left of this pro-Russian constellation in West Africa. Upon the fall of Nkrumah in 1966 the régime of General Ankrah expelled all Soviet experts (about 1,000 of them), the bulk of the Soviet, Cuban and Chinese diplomatic missions and the entire East German trade mission; relations with the USSR degenerated into a slanging match. Modibo Keita survived until 1968 (when he was removed by the army and kept in prison until he died eight years later) but the Russian connection languished almost from the start, Russian aid came to a full stop and although a Russian group was reported in the 1970s to be prospecting for minerals in northern and eastern Mali the common belief was that the USSR was more interested in possible landing grounds for aircraft or paratroopers. The initial, and most promising, association with Guinea turned sour even more swiftly when, in December 1961, Solod was told to pack his bags and leave Guinea, accused of complicity in a communist-inspired conspiracy against Sekou Touré. Moscow sent no less a personage than Anastas Mikoyan to try to patch things up but he failed and in 1962 Sekou Touré took a first step towards a more neutral position by accepting $70 million of American aid. During the Cuban missiles crisis he refused to allow the Russians to use Conakry airport, which they had built. Ironically the first jet aircraft to land there was French.

The main consequence of these Russian failures in West Africa was the fillip which they gave to new thinking in Moscow about Africa. The Russian ventures in Guinea, Ghana and Mali were in origin partly fortuitous and partly ideological. Their principal instrument, however, was economic: money plus expert advice. Russian credits (which were made available on exceptionally easy terms) were to be used to finance mutually approved capital projects, meet the balance on trading accounts so far as Russian imports were not paid for by Guinean bauxite or Ghanaian cocoa, and contribute to the training of West Africans in the USSR. Moscow sought in this way to detach these states from the western economic embrace, to secure control of their exports and to plant Russian experts in them. But the experts turned out to be more numerous than welcome – 3,000 in Guinea alone at their peak. They tended to recommend showy rather than useful works – such as a big stadium and theatre in Conakry – and they promoted collectivization of agriculture which was so unpopular that it led in Guinea to unrest which the government had to suppress. Excessive quantities of Russian goods were imported; bills and resentment piled up.

Worst of all from Moscow's point of view was the growing feeling that the Russians were interested only in driving crassly selfish bargains from a position of strength. This was not entirely fair. Over the 1960s as a whole the USSR paid rather more than the world price for Ghanaian cocoa. The USSR was for Ghana an assured and steady purchaser and when world prices fell Russian purchases stayed the fall by reducing the supply to other markets. The Russian contract kept prices in Ghana's traditional markets more buoyant than they would otherwise have been. In Guinea on the other hand the Russian appetite for bauxite ran far ahead of anything resembling generosity or tact. After deposits at Kindia had been prospected by Russian experts a so-called joint company was formed; it was owned and run entirely by Russians. This company's profits were allocated in the first place to a clearing account to offset Guinean purchases of Russian arms and other goods but were juggled in the worst neo-colonial manner. In addition the USSR contrived to pay a beggarly $6 a tonne for bauxite all the way to 1976 when the price was raised to $16 – still a mere two-thirds of the world price. The ill will so generated was compounded by the Russian refusal to help Guinea build a fishing fleet or to part with any of its own catch in Guinean waters, a foolish niggardliness in sharp contrast with the East Germans and Cubans who were willing to hand over some – in the Cuban case the whole – of theirs. In sum Guinea, Ghana and Mali served no practical Russian purpose and did less than nothing for the USSR's image in Africa.

But they were a testing ground, for in these years Moscow was developing a new, more pragmatic style in the aftermath of Stalin's death in 1953. Stalin's death sanctioned an intellectual ferment which Khrushchev's temperament encouraged. The rigid dogmatic categories into which Stalinist communist thinking had been driven were loosened and Khrushchev, an unintellectual old man in a hurry, was more than willing to stir the pot without caring too much where the spirit of inquiry might lead. Africa, always towards the bottom of any Russian agenda, was not a precipitating cause of this new mood, whose main foreign source was probably the new rulers' determination to find ground for ending the quarrel with Tito. But there were implications for African policy too. The search for friends in Africa need not be confined to leaders whose credentials could pass the tests of Stalinist orthodoxy, which in fact no leading African did. Nationalists, provided they were moving in a socialist direction, became acceptable friends and allies. The path of 'non-capitalism' – a usefully vague phrase which was a favourite with Khrushchev – became respectable. It was in this context that the links with Sekou Touré, Nkrumah and Keita were made without doing violence to basic Russian postulates, and although these particular links led only to disappointment the policies underlying them survived. In consequence Moscow had a vastly broader field for diplomatic manoeuvre.

Moscow was also becoming much better informed. The African Institute was expanded particularly in its economic and social sections, and on the death of Potekhin in 1965 the directorship went to V. G. Solodovnikov, an economist (and, from 1976, ambassador in Zambia). Solodovnikov was also put at the head of a Co-ordinating Council for African Studies, created by the Academy of Sciences in 1966. By 1970 the USSR had 350 scholars working on African ethnography, economics, law, philology, history and geography; twenty-two of them reached the rank of DSc. When Solodovnikov left the Africa Institute its standing was such that he was succeeded by the son of the foreign minister, A. A. Gromyko. There was similar development in Leningrad and in universities in eastern European countries.

Whether or not for these reasons, there was a marked difference in Moscow's handling of the crises and opportunities created by the two big civil wars of the 1960s, in the Congo and Nigeria. In the former the USSR began by endorsing UN intervention, partly because

Lumumba – Belgium's own choice to lead the new state – fitted Moscow's identikit of an African socialist of the right sort; but when Lumumba was ousted and later killed Moscow attacked Hammarskjöld's handling of the situation and threatened unilateral military intervention in support of Lumumba's heir, Antoine Gizenga, who, however, lacked adequate support in the country or on its borders (Sudanese hostility sealed the fate of his revolt in Orientale province); his rebellion created a temporary alliance between Kasavubu and Tshombe, which was detrimental to Russian aims; and the final outcome – the establishment of Mobutu, the USA's great black hope – was accompanied by the forcible closing of the Russian embassy in Kinshasa. The USSR retired hurt by its own miscalculations. It had also offended the great majority of the members of the OAU which was brought into existence in 1963 largely as a consequence of this war.

By contrast Moscow considerably strengthened its position in Nigeria during the civil war in that country. This was a victory for pragmatism over ideology. The USSR had been slow to pay attention to Nigeria. Although a Nigerian ambassador went to Moscow a few months after independence in 1960, Moscow did not return the compliment until 1964. Its instinctive sympathies did not lie with the dominant North, or with a general such as Johnson Ironsi who was regarded as a member of a feudal, pro-British class. But disillusion with civilian leaders in West Africa led Moscow to see virtue in military men who might be expected to provide more stable government and to prove themselves more predictable and enduring friends. Theory too accommodated itself to the practical necessities by depicting the military as a popular and progressive force, sweeping away bourgeois capitalist hangovers. Gowon, Ironsi's successor, was not a northerner and showed some signs of making an alliance with Awolowo, whom he released from prison. (Awolowo had found favour in Moscow's eyes immediately after independence when he became the first leader of the opposition in the federal parliament. Moscow now hoped, wrongly, that Gowon might make him prime minister.) Moreover, as Ojukwu moved towards the Biafran secession it became clear that an overwhelming majority of the OAU opposed him. Moscow did not propose to offend them. It decided to back – and arm – Gowon.

The negotiations were carried out quickly, professionally and the more easily because Nigeria was already negotiating with the USSR for economic aid. Nigeria's development plans were so extensive that it could not get all the aid it required from the west and so, overcoming its profound anti-communism, it turned also to the USSR and eastern Europe. In 1967 a small five-man team of Russian experts spent four weeks in Nigeria travelling over many parts of the country and studying possibilities for a steel industry (for which, with other projects, the

USSR eventually advanced $140 million at the end of 1968). But with the outbreak of civil war Gowon's problems had been reduced to one: getting arms. Britain and the USA refused to give him what he wanted and he turned to the Russian ambassador, Alexander Romanov, an adept and sensitive negotiator who did not try to drive an over-hard bargain (Lagos was in any case ready to pay cash) and understood that although Nigeria could not be made communist it could be made friendly. Within weeks of the beginning of hostilities Russian and Czech arms began to arrive and Russian support for Gowon persisted firmly and openly. This was a policy which achieved its limited aims. Successive Nigerian governments, all of them strongly anti-communist, preserved the equable relationship with the USSR which was established during the civil war.

For ten years after the end of the Second World War the USSR failed to find a point of entry into *northern Africa*. Opportunity came in a curiously roundabout way. Nasser wanted arms which he could not get from the west. He did not know Stalin's successors but on his way to Bandung he unexpectedly met Zhou Enlai at Rangoon airport. He explained his predicament and Zhou suggested he try Moscow. Zhou promised to put in a good word for him. After discussions in Cairo and Prague the Russians sent Shepilov to Cairo to give Nasser to understand that he could do what he liked with the Egyptian communists: it would be no part of an arms deal that these should benefit from the pact. When a year later the Americans declined to help with the financing of the Aswan Dam Nasser declared that he would go ahead all the same. He counted on Russian help. This was a chance for the USSR and after some hesitation Moscow plunged to the extent of offering finance for the first stage. When France and Britain encouraged Israel to invade Egypt and joined in the attempt to overthrow Nasser Moscow's position was consolidated.

This marriage of convenience ran smoothly enough so long as it was personified by Nasser and Khrushchev. The USSR provided Nasser with transport for his expedition to Yemen in 1962 and two years later Khrushchev attended the celebrations at Aswan. This was a few months before his fall. Already his colleagues were charging him with undue adulation of Nasser. Doubts were strengthened when Nasser refused to back the USSR against China in the row over whether the former was an Asian power and so eligible to attend a second Bandung Conference. During the Israeli attack on Egypt in 1967 Cairo accused the Kremlin of giving only feeble military and diplomatic support. When Nasser died in 1970 the Russian presence in Egypt had become as uneasy as it was large. In 1972 Sadat demanded the immediate withdrawal of 20,000 Russian advisers and experts. They left in seven days. Sadat appropriated all their installations and equipment.

Nasser and Khrushchev were both exceptionally plain-spoken heads of state. They were able to cut through a lot of the suspicions inevitable in a dialogue between two states so unfamiliar to each other as Egypt and the USSR. Their successors lacked this knack. Sadat suspected the Kremlin of intriguing with Ali Sabry and others of his domestic adversaries. He suspected Brezhnev of wanting to do a global deal with Nixon which would incidentally stop aid to the Arabs from either superpower for the struggle against Israel. He resented having to make two begging trips to Moscow in one year and getting little in return. For his part Brezhnev suspected Sadat, whom he rightly regarded as much less radical than Nasser, of unsavoury transactions with rich Saudis and Americans. In the 1973 war the USSR, swallowing its pride in an attempt to recover lost ground, rescued and rearmed Egypt, but this was only an interlude in Sadat's progression away from Nasser's Russian alliance and to the Camp David powwow with Carter and Begin. In 1976 Sadat abrogated the treaty he had made with the USSR in 1971. The Kremlin had lost its greatest prize in Africa.

The USSR stumbled in the Sudan too. Moscow was prompt to recognize and welcome Sudanese independence; it classified Abboud, when he took power in 1960, as a progressive soldier and was further encouraged by the Nimeiry coup of 1969. Nimeiry gave the communist Muhammad Ahmed Mahgoub a cabinet post, and the coincidence of similar coups in Somalia and Libya created the pleasing illusion of a left-wing bloc of these three countries with Egypt. But the next year Nimeiry expelled Mahgoub from the Sudan and, when he came back and stirred up a revolt, had him executed. This was particularly embarrassing for Moscow, which acclaimed the coup in the belief that it had already succeeded.

More important than Sudan and hardly less alluring than Egypt was Algeria. Mers el-Kebir was almost as attractive a naval base as Alexandria and, by its position westward in the Mediterranean, offered opportunities for outflanking Nato – a counter to Nato's gains when Greece and Turkey adhered to the western alliance. But Russo–Algerian relations were never easy. Moscow was half-hearted about the FLN, whose leaders seemed to stand at the wrong end of the socialist spectrum. It hesitated to offend the French Communist Party and its Algerian offshoot, both of which were hostile to the FLN. It hesitated too to offend de Gaulle so long as he was a thorn in the side of Washington. Russian recognition of the FLN as a government did not come until 1960 (China recognized it in 1958), and Russian backing remained restrained until the failure of the French generals' coup in 1961, followed by the second Evian conference in 1962, set the seal on the FLN's success. Moscow then gave Ben Bella enthusiastic support until his fall in 1965 when, with almost indecent haste, it sought Boumédienne's friendship with equal zeal.

Algeria's problem was to get the best of two worlds, the USSR and France, and to sup with a long spoon with both. It expanded its trade links with the USSR by a series of commercial agreements, sent hundreds of young men to study in the USSR and accepted some 2,000 Russian technicians and as many military advisers. But it employed as many French and other technicians; made a point of honouring heterodox leaders like Tito and Ceauşescu; and took the line that both the Russian and American fleets should get out of the Mediterranean. Algeria became a prime example of the USSR's main difficulty in Africa. Ideological sympathies led to military aid but not much else. The Russian and Algerian economies were not complementary; their several needs and resources did not mesh. Morocco could supply the USSR more copiously than Algeria with the phosphates which it needed for its chronically sick agriculture; but Morocco was ideologically opposed to the USSR and also at odds with Algeria.

In *East Africa* the USSR had an unhappy time. In Tanzania Nyerere made no secret of the fact that he distrusted Moscow as much as Washington. In Kenya the USSR made a bad tactical mistake when it backed Oginga Odinga who, at independence in 1963, seemed to be Kenyatta's favoured number two but overplayed his hand and was disgraced in 1966. Although Kenya had no arms agreement with the USSR, Russian and Czech arms arrived in the country and were found on Odinga's land. Kenyatta accused Moscow of organizing and financing a conspiracy against him and Odinga of being a communist agent. Russian and other communist diplomats were expelled. The best that Moscow could do in these parts was to keep on better terms with Uganda, which it did even with the atrocious Amin (to whose court a deputy foreign minister, Alexei Zakharov, was sent as ambassador in 1972). But Uganda, coastless, was the least promising of East African states in geopolitical terms.

Compensation was found further north. In 1963 the USSR made a first agreement with Somalia, offering credits and grants up to $35 million. The connection was strengthened after the coup of 1969 which brought Colonel Siad Barre to power. The payoff was the development of a naval base at Berbera, an air base at Hargeisa, storage for missiles and other weapons, and facilities for intelligence and telecommunications. In 1972, when the USSR's positions in Egypt and the Sudan were becoming dubious, Marshal A. Grechko, the Russian minister of defence, paid a visit to Somalia. He was followed two years later by President Podgorny, who concluded a ten-year treaty of friendship and collaboration which, among other things, expunged the Somali debt to the USSR – an unusual piece of generosity. By 1977 Russian military aid had reached $250 million, there were 2,000

Russian military technicians in the country and the value of the bases and equipment available for Russian use was about $1 billion. The Russian stake was considerable.

All this had more to do with the Indian Ocean than the continent of Africa. Russian preoccupation with these waters is ancient. Before 1917 Russian access to what is a segment of the world's main maritime highway was blocked by the Ottoman empire and Persia. Thereafter a variety of Arab states, and their western sponsors, stood in the way, until in 1955 the arms deal with Egypt opened a door into the Middle East and so, via the Red Sea, to the Indian Ocean. And in 1971 Aden, at the further end of the Red Sea, became a part of the independent and left-wing state of South Yemen – a strategic outpost of enormous significance if, under Russian tutelage, it could be fitted into a more comprehensive group of client states. As its positions in Egypt and the Sudan degenerated the USSR became commensurately concerned to counter the American dominance in Ethiopia which, since its virtual annexation of Eritrea in 1952, commanded 1,500 km of the coast of the Red Sea (the opposite coast being under the control of Washington's even stauncher friend, Saudia Arabia).

During the 1960s moreover the USA evolved the A3 Polaris missile with a range of 4,000 km. If launched from vessels in the Indian Ocean this new weapon would be able to hit Soviet cities. Moscow's first reaction was a proposal to declare the Indian Ocean a nuclear-free zone. When this idea fell flat Admiral S. G. Gorshkov, the father of the modern Russian navy, sent a token force from Vladivostok to the Indian Ocean and established during the 1970s a permanent patrol there. Thus the development of Russian naval power, one of the outstanding features of the postwar contest between the superpowers, took effect in waters where much of the United States' imported oil and two-thirds of western Europe's were at risk.

In the mid-1970s events offered Moscow a chance greatly to enlarge its sphere of influence in this part of the world. In 1974 Haile Selassie fell. In the ensuing confusion outsiders moved gingerly. The United States, already shifting from its Ethiopian base at Kagnew to a new strongpoint on Diego Garcia in the Indian Ocean, continued for a while to provide Ethiopia with financial aid, arms and training. The USSR was pulled two ways: its Somali ally was bitterly hostile to Ethiopia, but the chance for Moscow to supplant the Americans there was very appealing. Moscow had long been mindful of Ethiopia's significance. In 1959, when the country was firmly in the American camp, Moscow had sent the head of the African department of the foreign ministry, A. V. Budakov, to be ambassador in Addis Ababa. It had extolled Haile Selassie's part in the creation of the OAU, depicted his antiquated and repressive régime as progressive, and given him a comparatively generous measure of aid. After his fall Moscow found itself possessed of a weapon which Washington did not have. It might arm the new régime: Washington would not.

Arming Ethiopia must put the Somali alliance at risk, but Moscow hoped to have it both ways. In 1977 it resolved to take the plunge. Mengistu had emerged as the man to back but he would have to be backed against foreign as well as internal enemies – in particular against Somalia. Moscow turned to Castro to persuade Siad Barre to drop his quarrel with Ethiopia and join a left-wing constellation to be formed round Ethiopia, Somalia and South Yemen.

Castro's sympathies lay with the Somalis (and the Eritreans) rather than the Ethiopians: there had been a Cuban training mission in Somalia since 1974. Yet he accepted Moscow's charge. But predictably he failed to persuade Barre to forgo the golden opportunity to annex the long-disputed Ogaden from Ethiopia. Castro thereupon obliged Moscow a second time. He placed his armies at the disposal of Moscow and Addis Ababa to defeat his erstwhile Somali friends. First contingents were flown from Cuba to the USSR and thence to Africa. Later units followed by air and by sea from Angola. They saved Harar and Diredawa from capture by the Somalis, halted the Somali advance and then utterly repulsed it. The Russian contribution to these operations was a gigantic air lift via Aden which, beginning in October 1977, comprised 550 T54 and T55 tanks, at least 60 MiG 17s and 21s and 20 MiG 23s, SAM 2 and 3 ground-to-air missiles, BM 21 rocket launchers, 152 and 180 mm artillery, self-propelling anti-aircraft systems, armoured personnel carriers and much else. For the USSR this was not only a strategic move but also a display of strength, efficiency and confidence. It was all the more remarkable since never before, even to protect its positions in Egypt, had the USSR resorted to the direct use of force in Africa. But having saved Ethiopia from Somali attack the USSR did not permit an Ethiopian counter-attack upon Somalia itself – a measured restraint probably dictated by a determination not to provoke the United States too far and to avoid open superpower confrontation. But even with Russian aid Mengistu could do no more than stave off his enemies and ask for more aid. By the time Gorbachev came to power Mengistu was clearly more expensive than he was worth. He was told that Russian aid and Cuban troops would fade away.

The defeats inflicted on the Somalis enabled Mengistu to hold his own against the Eritreans until, once more with Russian and Cuban help, he could mount a counter-offensive. For the time being Mengistu was saved – and the USSR committed or, as some thought, entangled. Mengistu's Ethiopia, no less than Haile Selassie's, was a patchwork empire dominated precariously by its Christian Amharic minority. Besides its conflicts with Eritreans and Somalis the Dergue was opposed by the Oromo peoples of the south (themselves about half the country's population) and the Tigrayans to the north-west. At the Islamic conference at Taif early in 1981 eleven members of the OAU joined in unanimous condemnation of the repression of Muslims in the Horn. Neither as an ally nor a dependency was Ethiopia likely to prove a stable asset of Russian world power.

Upon the fall of Salazar in Lisbon in 1974 the USSR sent V. G. Solodovnikov to Zambia as ambassador. Solodovnikov had held no diplomatic post. He was Moscow's leading African academic. He was reputedly given high rank in the KGB. Approachable, genial but a notable absentee from the diplomatic cocktail party round, he had the task of making Lusaka a centre of Russian influence and intelligence such as Cairo and Conakry had failed to become. Solodovnikov was as much concerned with Zapu in Rhodesia and Swapo in Namibia as with Zambian affairs. At the same date the USSR established a surprisingly large embassy in Botswana.

During the rest of the decade the USSR pursued a low-risk policy in *southern Africa*, in contrast with its forward policy in the Horn. It gave minimal support to insurgent movements in Rhodesia and Namibia, enough to keep them going and keep them sweet but no more. In Angola it contrived to interfere only by proxy. But it became increasingly alert. In 1976 General (later Marshal) S. L. Sokolov travelled as far south as Mozambique in the course of an African tour. Next year a group of no fewer than eleven Russian generals was spotted in southern Angola and another year on General V. I. Petrov visited Angola and Mozambique. The calibre of these explorers was notable. Sokolov was a member of the highest group in the military hierarchy, a group of four. Petrov was the number two in the ground forces, had seen service in the Far East and had been the controlling genius of the operations in Ethiopia and the Horn. The group of eleven included specialists in planning, training, supply, air transport and radio and electronic intelligence; both the ground forces and the air forces were represented. Their surveys led to no operations. They were, however, symptomatic of the Kremlin's awareness that such operations might one day become desirable.

A little earlier, and a little further off the map, a prestigious delegation from the Africa Institute visited (in 1976) Madagascar and Mauritius to testify to Russian interest in these parts. In 1975 Madagascar took a sharp turn to the left with the advent to power of Didier Ratsirake, while in the Comoros Ali Soilih (murdered in 1978) also took his country to the left. So did France-Albert René in the Seychelles in 1977.

But the weightiest developments in southern Africa in the 1970s were the sequel to the collapse of Portuguese rule. In Mozambique Samora Machel established, against virtually no opposition, a one-party state which was distinctly left-wing in its domestic affairs but non-aligned in relations to external power blocs. In **Angola** the fight for the succession was more closely contested and produced one of the strangest intrusions in the whole history of the continent: the arrival in force of the Cubans. This astonishing move was prompted by Fidel Castro's personal temperament but would have been impossible without Russian aid and sanction. It gave Moscow a vicarious instrument in a zone where it was reluctant to operate directly and (like Washington) was probably still too ignorant to operate to good effect. It led also to the use of Cuban forces in the Horn.

As a result of American economic sanctions against Cuba Castro had become by this date critically dependent on the USSR: a series of economic disasters on the home front put him at Moscow's beck and call. But Moscow would hardly have called him into Africa if the idea of doing things there had not been his in the first place. Africa always beckoned to Castro. Cuba was too small for him; his attempts to spread his revolution in Latin America were failures; and when he recalled that 'African blood flows in our veins' and spoke of Africans as his brothers and sisters he was expressing something that was real to him, as well as romantic. He wanted to help. Within two years of coming to power he sent instructors to help train guerrillas in camps in Ghana. Two years later, in 1963, a second training mission went to Algeria. But here training spilled over into something more. When Algeria became involved in border fighting with Morocco Castro sent Ben Bella three shiploads and an aircraft load of equipment, accompanied by Cubans who knew how to operate it. These Cubans would almost certainly have become involved in the fighting if it had not stopped just before they reached the firing line. The combat troops were then withdrawn but the training mission remained until the overthrow of Ben Bella in 1965.

Castro was also involved in the Congo. Advisers and a small fighting force of about 200 men with Che Guevara in person went to the aid of Lumumba's heirs, but Mobutu's coup at the end of 1965 put an end to their adventure. A few of the advisers crossed the Congo River to Brazzaville but when, next year, the Cuban mission in Ghana was withdrawn upon the fall of Nkrumah, Castro's African expeditions seemed to have proved as futile as those he sent to Latin America.

But he did not lose interest. The toehold at Brazzaville was matched by another in Conakry – the nearest African capital to the Caribbean and a centre for training guerrillas to fight the Portuguese in Guinea-Bissau. Castro also organized an internal security force and bodyguard for Sekou Touré and others for Presidents Alphonse Massemba-Debat in Brazzaville and Siaka Stevens in Sierra Leone. (In the 1970s he branched out into South Yemen, Oman's Dhofar province and the Golan Heights.)

Castro made contacts with liberation movements in all Portugal's African colonies during the 1960s. The Angolan scene was peculiarly perplexing. Dozens of movements came into being but these sorted themselves out into three major forces: Agostinho Neto's MPLA, Holden Roberto's FNLA and Jonas Savimbi's Unita. The FNLA looked consistently the most effective but Roberto was anti-communist and closely linked with Mobutu. He got aid from various quarters: Zaïre, China, Romania, Libya and the USA (which stuck to Portugal to the last minute but switched the bulk of its aid to the FNLA early in 1975). So the USSR backed Neto, but only sporadically; in the year before the fall of Salazar Russian aid had dwindled so far that Neto turned in desperation to Scandinavia and Cuba. But six months after the Lisbon coup Moscow reinstated its aid and, when the FNLA attacked the MPLA in March 1975, in effect saved Neto by rushing supplies and weapons to him by air via Conakry and Brazzaville and also by sea. By October China had retired from the game, unable to match Russian aid to the MPLA which – according to the CIA – had by then received from the USSR and eastern Europe arms worth $80 million. Romania also retreated, but the USA, after some hesitation, persisted and wasted a lot of money.

These successes of the MPLA settled nothing because in October, four weeks before the date fixed for independence, South Africa invaded the country. In November the MPLA declared itself a provisional government; other Africans rushed to recognize it and to denounce the South African invasion and its American sponsors; and Nigeria gave Neto $20 million. Nevertheless Neto's position was precarious. This time he was saved by Castro.

Castro and Neto had become personal friends. Cuban aid had helped to fill the gap when Moscow turned cold on Neto in 1973–74. Several hundred Cuban advisers arrived in MPLA camps in April and September 1975 and when Neto appealed to Castro early in November for combat troops to fight the South Africans he did not appeal in vain. The first unit arrived by air two days later. It consisted of eighty-two men in civilian disguise, the forerunners of what was to become an army of 20–30,000. The first seaborne reinforcements arrived at the end of November and this first ocean crossing of nearly 10,000 km was followed by forty-one more during the ensuing six months of active hostilities. At one point no fewer than fifteen ships were simultaneously at sea on an eastward course, the biggest procession of men and material across the Atlantic since the Americans had sailed to North Africa and Europe to make war on Hitler. The seaborne effort was supplemented by an air lift in which Cuba's antiquated Britannias, after landing at Barbados, flew to Guinea-Bissau and Brazzaville. When the Barbados staging point was cut out, under US pressure, by a refusal to refuel the aircraft, an alternative was improvised in the Cape Verde islands.

The Cuban expedition to Angola was a gamble that, on balance, came off. Its failure would have put paid to the Cuban presence in Africa and perhaps to Castro's rule in Cuba itself. But success was not complete. The MPLA achieved no final victory over its domestic rivals, particularly Unita which remained in being in southern Angola. Nor did the expedition develop as Castro intended. He hoped to have his troops back home in six months and was still expecting to accomplish this as late as March 1976. In the event his overseas force, which absorbed a considerable proportion of Cuba's entire army of 160,000, was obliged to remain in Angola indefinitely in order to keep Neto and (after his death in 1979) his successor, José Eduardo dos Santos, in power. Even though the Russians paid the bills, Castro's quixotry imposed strenuous demands and sacrifices on thousands of Cubans – fighting, dying or suffering bereavement back home. Unlike the Piedmontese sent by Cavour to flounder in the mud of the Tchernaya, they could not comfort themselves with the thought that they were paying for the independence of their own country.

Nobody has been generous to Africa, and the Russians have been among the least generous. If economic aid is a major diplomatic instrument, then the Russian use of it has been surprisingly feeble: they have not even kept up with the Joneses. At the end of the 1970s the total of Russian overseas development aid was equivalent to 0.02 per cent of estimated GNP. Even Italy did better than that. (The more generous givers were Sweden, the Netherlands, Norway and France, with respectively 0.99, 0.85, 0.82 and 0.6 per cent. Britain managed 0.38, the USA 0.22, Switzerland 0.19 and Italy 0.09 per cent. Although there are conflicting ways of taking these measurements they do not materially affect these relative positions.) Of the net flow of aid to Africa from all sources, national and international, the USSR provided less than 3 per cent. Economic aid therefore was treated by the USSR, not as a major tool, but as a minimum political investment.

On the other hand the effects of Russian aid may have been greater than such figures suggest because it was concentrated in relatively few countries. Certain recipients stood out: Egypt, Algeria, Morocco, Guinea, Somalia. They were picked for political reasons and liable to be dropped for the same reasons. Within them the USSR – like other donors – showed a preference for big industrial projects rather than, for example, rural development: in Egypt, for example, the Aswan Dam, the Helwan steel works, the Nag Hammadi aluminium works. Russian technicians were similarly concentrated; of some 34,000 in Africa in 1977 only 5,000 were south of the Sahara.

The proportion of loans to grants in Russian aid was about 2:1. Rates of interest were low at 2.5 per cent or even nil; but repayment dates were comparatively short, normally ten to twelve years where a western loan would be repayable over thirty or forty years. The USSR proved even more reluctant than other creditors to cancel or reschedule debts, although Somalia was so favoured in 1974. Repayment was normally in goods at fixed prices, a practice which led to much grumbling by borrowers who were thus prevented from reaping the benefits of rising world prices.

In terms of trade the USSR was doing a mere 2 per cent of its trade with African countries, but this insignificant proportion was not entirely inconsequent since the USSR was running a surplus on its trade with Africa which was more than sufficient to offset the deficit on its trade with the west. For Africa the USSR had little commercial importance: only Egypt and Guinea ever did more than 10 per cent of their trade with the USSR.

Military aid fell into two broad categories. There was the provision of material and training given as part of the business of making friends in likely quarters; and there were special efforts made for a particularly strategic gain. The statistics of military aid showed therefore jumps at crucial points. During the 1970s, up to and including 1976, Russian and eastern European military aid to Africa averaged $300–350 million a year. It was fairly widely spread: more than half of Africa's air forces acquired Russian aircraft. In 1977, principally on account of Angola, military aid reached $1,500 million and in 1978 a further $1,000 million was expended on Ethiopia alone. These were, in Russian terms, big figures, a reflection of Moscow's preparedness to take a plunge in Africa and pay for it.

Invitations and grants to students are another kind of aid. Its pitfalls are well known. Students are frequently disappointed in their hosts and the Russians were no more successful than westerners in avoiding the slights inflicted by colour prejudice. The USSR opened its arms to foreign students as far back as 1922 when the first Mongolians were welcomed to Moscow, but substantial programmes for students were developed only after the Second World War. In the first postwar decade nearly all of them came from the European satellites but with the decolonization of Asia and Africa, and the Cold War's competition for the goodwill of the Third World, steps were taken to attract students from these countries and Latin America too.

Few knew any Russian. They spent therefore a preliminary year learning the language (for which 4,500 teachers were recruited) and doing a course in 'scientific socialism'. The second half of this programme was sometimes resented but perhaps less so than westerners liked to imagine. When Kenyan students went on strike in Baku in 1965 and asked to be sent home their grievances included indoctrination; but communism was part of what the USSR had to offer and many students approached this course with at least a modicum of initial curiosity. Of the substantive courses the most popular were in medicine and engineering.

Foreign students in the USSR were offered 100 roubles a month – twice the grant given to a Soviet citizen – plus cheap lodging at about 2 roubles a month, free medical treatment and an annual vacation grant of 100 roubles for spending in the USSR. About half of all Third World students were at the Patrice Lumumba University and more than half of them came from Africa. According to Russian sources there were in these years some 50,000 foreigners studying in universities and technical colleges, not counting others doing vocational courses at lower levels or postgraduate work at higher levels. This total was supplemented by universities and specialist schools in eastern Europe – for example, schools in Budapest and Berlin set up to train Africans in trade unionism and journalism respectively. In purely numerical terms this was a useful and welcome contribution to one of Africa's most urgent needs but (again merely numerically) it was only a minor addition to more familiar educational routes. In the 1960s and 1970s Britain alone had twice as many African students as the USSR and all eastern Europe combined, and – to take a particular example – at the end of ten years of independence Nigeria had about 15,000 students studying abroad and nearly all of them were in Britain.

The foreign student in eastern Europe, like his domestic fellow, was granted much less freedom than the student in the more permissive west; his daily round was bleaker, although East Germany provided some of the bright lights and carefree style sought by young people who have embarked on a foreign adventure; he was more supervised, even segregated, and experienced more than the usual unpleasantness when claiming the attention of local girls. In 1963 Ghanaian students demonstrated in Moscow's Red Square after the death of one of them in a punch-up, and in a particularly bothersome incident 250 Egyptian students were flown from the USSR to the USA after complaints about their living conditions and indoctrination courses. There was in other words a debit side to the account.

For the immensely larger number of Africans who stayed perforce at home the USSR gradually built up its foreign broadcasts. At the end of the 1970s it was broadcasting to the world at large for 2,000 hours a week in eighty languages (China was using only half that number). But Africa, excluding North Africa which was served by the distinct Middle East service, was not a prime target. In 1979 it was allotted 147 hours per week, the same as in 1966. The leading languages in use by the African service were French, Swahili and Hausa which occupied 66.5 hours, leaving 80.5 hours for eleven other languages. Apart from its African service the USSR was broadcasting in English round the clock and many of these programmes could be heard in Africa. Among other eastern Europeans broadcasting to Africa the East Germans were the most prominent. The substance of the programmes was the familiar mixture of news laced with music, but the music was not the latest thing and probably turned more sets off than listeners on. Africans seemed to like rather more western decadence than Russian (or Chinese) broadcasters could bring themselves to provide. The quality of news and feature programmes improved considerably after the 1950s when unfamiliarity with the target areas and their culture gave rise to much factual error and poor judgement, but – to the western ear at least – these broadcasts remained crassly slanted and the Russian and Chinese broadcasters devoted a tedious amount of time to abusing one another.

All these general adjuncts of foreign policy reflected the marginality of Africa in the Russian scheme of things. Africa was worth some study and some effort and some expenditure but most of the time it was not a prime or even secondary field of action. The attention devoted to it indicated that Moscow had in mind that one day it might be. There were exceptions – areas and occasions where the effort assumed much more significant proportions (Egypt, Ethiopia) – but the motive in these cases could be found in the Middle East rather than the African continent itself. Meanwhile Moscow accumulated knowledge through its diplomatic missions, its secret services, the journalists of Tass and Novosti, and academic bodies in Moscow and elsewhere. This work coincided with the striking increase in Russian capabilities which characterized the Brezhnev years. Under neither Stalin nor Khrushchev was the USSR a truly world power. Stalin turned it into a nuclear power sufficiently menacing to be a counterpoise to the USA in a bipolar world. Khrushchev inherited this position and, by the failure in the Cuban missiles crisis, demonstrated that Moscow's reach was still limited. But in the 1970s the continuing growth of Russian power and technology put the USSR on the map everywhere. Thenceforward the question, in relation to each part of the map, was what it might decide expediently to do there.

Gorbachev saw that what was being done in Africa was both expensive and fruitless. The commitment in Ethiopia, like the commitments in Vietnam and Afghanistan, was a ruinous folly. Military activities cost too much and bought no political gain. Even the relatively small subsidies given to the ANC were money wasted so long as the ANC relied on violence to destroy *apartheid*: the USSR must continue to support the ANC and the South African Communist Party but not necessarily their methods. Gorbachev preferred to put Soviet weight behind the Front Line states. He improved relations with Zimbabwe where his predecessors had made the mistake of backing Nkomo against Mugabe. He reduced the Russian presence to the political, rhetorical and covert, and waited for others to make mistakes instead of making them himself.

Throughout these decades the Chinese played a sporadic *disobligato* to Russian and Cuban performances in Africa. China's achievements in Africa have been prodigiously exaggerated. Africa is even more remote from China than from the USSR. By the end of the 1970s China still had no direct link with Africa by air or sea (Aeroflot was flying to twenty-nine African airports). Trade, totalling about $400 million a year, was mutually negligible.

Yet China arrived in postwar Africa not far behind the USSR and at the same point: Egypt, where a Chinese embassy was opened after the Bandung conference in 1955. The main motives – international recognition, admission to the UN and to friends in the Third World – were extraneous to Africa. After the break with the USSR came the desire to make difficulties for Moscow, notably by outbidding its revolutionary fervour at a time when the USSR was shifting towards more pragmatic diplomacy. But the rapid spread of independence undermined Beijing's emphasis on subversion; and although Zhou Enlai visited ten African countries in 1963–64 (the revolts in Kenya, Tanganyika and Uganda curtailed his tour), the Cultural Revolution caused an almost complete withdrawal for a time: in 1966 China's eighteen ambassadors were all recalled. To Africans the Chinese appeared to be nice people from a long way away who behaved more agreeably than Russians or Americans but had much less to offer.

China recognized Sudan and Ghana in 1956 and 1957 but had to wait for three and two years respectively for counter-recognition. It was the first to recognize the FLN as the provisional government of Algeria in 1958. It was prompt to offer aid to Sekou Touré: Guinea is one of the few states in Africa with a sizeable Chinese population (4,000 – the others are Mauritius, Madagascar and South Africa). But Chinese credits for Guinea ($26 million), Ghana and Mali ($19.5 million each) were a long way short of Russian offers. Terms, however, were extraordinarily generous: no interest, repayment over long periods, even (in the case of Kenya in 1961) no repayment at all. But when the Cultural Revolution reduced aid to a trickle only about 15 per cent of Chinese credits had been taken up. They were even more marginal than the Russian.

To the smiling face of communism was added revolutionary zeal. From Conakry China supported a group of Ivorian exiles and, until his assassination in 1960, Felix Moumié's exceptionally savage Union des Populations du Cameroun. China gave Lumumba £1 million, supported Gizenga until he joined the Adoula government and then helped to train guerrillas for Mulele and Soumaliot. In 1963 it intervened in Burundi's troubled waters where Tutsi refugees from

Rwanda were plotting a return to their country. The Chinese backed the feudal Tutsi but the expedition was a disaster and the Burundi, suspicious of Chinese meddling in their own hardly less turbulent politics, evicted the newly arrived Chinese diplomatic mission. A bigger mission in Congo-Brazzaville, established in 1964, also came to grief after a military coup there in 1966.

China's steadiest friend in Africa was Tanzania. The Chinese arrived there by the back door as a result of the revolution in Zanzibar in 1964. One of its leaders, Muhammad Babu, had good enough connections with China to raise a loan of $14 million and when Zanzibar was united with Tanganyika (paradoxically Babu opposed the union) the Chinese won the goodwill of Nyerere who, ever cynical about Moscow and Washington, happened at this time also to be on cool terms with Britain. Zhou Enlai and Nyerere exchanged visits in 1965 and the head of the African department of the Chinese foreign ministry was sent to Dar-es-Salaam as ambassador. Then came the Tanzam railway, China's showpiece in Africa.

There were at independence no rail links between Tanganyika and Zambia. Kaunda was particularly anxious to create a passage for Zambian minerals avoiding Portuguese territory, and both he and Nyerere saw the railway as a means to develop neglected areas in their two countries. China, which in 1964 offered aid for building railways inside Tanzania, next year extended the offer to an international line. Both African presidents were seeking western aid and turned to China only when they failed to get it. A first agreement was signed in 1967 after a visit by Kaunda to Beijing. Final agreement followed in 1970 and work began that year.

The Tanzam railway is a single-line track of nearly 2,000 km with ninety-one double-line stations. It was finished two years ahead of schedule and on completion became the property, in equal shares, of the Tanzanian and Zambian governments. Fifteen thousand Chinese were employed in its construction. They set a good example at work and behaved themselves when off. There was some propaganda byplay but after complaints by the African governments this was restricted to handing out the works of Mao – a latterday version of the distribution of Bibles by Christian missionaries. The building of the railway did the Chinese a lot of good for a time, but maintenance was poor, services degenerated and the Dar-es-Salaam terminal became clogged. For these shortcomings the Chinese were not responsible but their repute suffered none the less when the railway ceased to be one of the wonders of the modern world.

Chinese penetration of Africa was facilitated by French recognition of the communist régime in Beijing at the beginning of 1964. Most of francophone Africa followed suit. On their return to Africa after the Cultural Revolution the Chinese quickly took over embassies previously occupied by the régime on Taiwan, often in countries where entry could have been difficult if the Taiwanese had not got there first. Chinese broadcasts to Africa, which began in Arabic at the time of the Suez crisis in 1956, were gradually expanded but did not match the Russian either in hours per week or in their range of languages. In the 1960s and 1970s the Chinese instigated no revolutionary change, largely because they picked the wrong people to support or supported them too feebly; a guerrilla leader in search of arms was better advised to ask Moscow for them. But China showed interest in Africa and established a presence unlikely to be expunged. In 1982, nearly twenty years after Zhou's continental tour, a new prime minister, Zhao Ziyang, visited ten African states but Africa remained in Beijing a marginal and specialist concern.

NOTES

A. RWANDA AND BURUNDI

Rwanda and Burundi, formerly German colonies, were placed under Belgian mandate after the First World War. The population consisted of the Hutu; the Tutsi, supposedly descended from warrior invaders who had arrived in the sixteenth century and become a semi-divine ruling minority; and the pygmy Twi, a distinct but very small minority despised by both Hutu and Tutsi. Hutu and Tutsi spoke the same language, intermarried and looked much alike, but were differentiated by occupation: the Hutu were mainly agriculturalists who grew bananas and other crops, the Tutsi raised cattle. This distinction was enlarged during the eighty years of colonial rule when the Tutsi were treated as and turned into a superior administrative caste. In 1959 the Hutu of Rwanda revolted. The Tutsi monarch, or Mwami, who had succeeded on the mysterious death of his half-brother, was deposed in 1961 and fled to Uganda. A republic was proclaimed under Grégoire Kayibanda, a Hutu. In 1962 the Belgian mandate was terminated, Rwanda and Burundi ceased to be jointly administered, Rwanda became an independent republic and Burundi an independent (Tutsi) monarchy.

In *Burundi* the ruling Tutsi were viciously assailed by Hutu in 1965 and the monarchy was overthrown a year later but the Tutsi retained power until 1972 when the Hutu rose again, many Tutsi were slaughtered and some 200,000 fled to Tanzania. In 1987 a new and conciliatory president, Pierre Buyoya, allowed the Hutu a bare majority in his cabinet and the post of prime minister. He may have hoped by these concessions to improve his case for foreign aid but he risked hostility from both Tutsi (a mere 16 per cent of the population) who feared that concessions were the thin end of the wedge, and from Hutu who branded the Hutu prime minister a collaborator. Buyoya, having survived an attempt on his life and an invasion from Tanzania, was succeeded after elections in 1993 by the Hutu Melchior Ndadye, who was killed by Tutsi officers. He was succeeded by Cyprien Ntaryamisa, who was killed in the attack which killed President Habyarimana of Rwanda as he returned from peace talks in Tanzania. After the explosion of violence in Rwanda in 1994 Burundi's army (Tutsi) overshadowed a feeble coalition government and at least condoned scattered massacres which amounted to a suppressed and one-sided civil war.

In 1964, shortly after independence, the Hutu of *Rwanda* perpetrated a genocidal massacre of Tutsi and many Tutsi fled to Uganda. The Hutu of Rwanda were divided among themselves as well as from the Tutsi: the division was partly regional but also over relations with the Tutsi. President Kayibanda, a southern Hutu, was overthrown in 1973. He was replaced by General Juvenal Habyarimana from the north who repelled a Tutsi invasion from Uganda in 1990 with French help but at the price of agreeing to share power with the Tutsi. To support himself against rival Hutu who distrusted his attitudes he created a Hutu youth movement which became in a few years a sinister force supplied with arms by the government and incited by the radio to massacre Tutsi and the more moderate Hutu. The former were regarded by many Hutu as a fifth column ready to join with Tutsi exiles in Uganda to exterminate the Hutu but in 1993 Habyarimana, largely in response to foreign pressures, agreed at a conference at Arusha in Tanzania (implementing an earlier agreement at Kinshasa in Zaïre) to create a two-nation government. This agreement alienated and alarmed those Hutu who wanted a purely Hutu state and felt safe in no other. Habyarimana consequently dithered until threatened with economic sanctions by the World Bank and the IMF. In 1994, returning from a further conference at Dar es Salaam in Tanzania, he was killed, together with his fellow president of Burundi, when his aircraft was destroyed as it prepared to land at his capital.

This was a signal for instant conflagration and a Tutsi invasion. Half a million Hutu were gruesomely massacred and two or three times that number fled into neighbouring countries, mainly Zaïre. International attempts to protect refugees were frustrated by the refusal of governments (except the French) to provide either forces or funds in aid of African states which were willing to intervene. A French attempt to succour refugees in the south-west corner of Rwanda was discontinued when the relatively small force involved came under military threat. Civilian aid agencies were unable to persuade the refugees in camps on Rwanda's borders with Zaïre, Burundi and Tanzania to return to their homes and untended crops since these camps contained not only a million or more destitute refugees but also armed and vengeful Hutu whose single prospect was to fight their way back to power by marshalling the hapless refugees at gunpoint against the Tutsi. The UN and other agencies were left with a choice between abandoning the refugees who were dying in thousands or continuing supplies of food in the knowledge that much of it would be appropriated by the gangs which dominated the camps. An attempt by the new Tutsi government to persuade refugees back to their homes by appointing as prime minister the Hutu Pasteur Bizimanga fell short of the needful reassurance. In Rwanda the Tutsi, having suffered perhaps a million deaths by violence, became a yet smaller dominant minority, more embattled, more vulnerable and more intransigent. After a year the government of Zaïre resolved to force the unwelcome Hutu encamped on its territory back to Rwanda but many of them preferred, however hopelessly, to take flight into the hills and wastes of Zaïre. The wars and massacres of these years, although not entirely ethnic, were the clearest instance of genocide since the adoption of the Genocide Convention in 1948.

B. THE MALAGASY REPUBLIC AND THE INDIAN OCEAN

The island of *Madagascar*, lying off the Mozambique coast, came under French rule at the end of the nineteenth century. In 1947 a serious revolt against the postwar restoration of that rule was rudely repressed, but independence was nevertheless not long delayed. Within the island there was a conflict of interest between the Hova dominant minority, which wanted independence at the earliest possible moment, and a larger nationalist group under the leadership of Philibert Tsirinana, which preferred a short delay in order to be sure of gaining power for itself and not for the Hovas. This group succeeded in its aims. Madagascar became the Malagasy Republic in 1958, remaining in the French Union. Tsirinana became president and full independence followed in 1960. Tsirinana steered a pro-French course and became a prominent supporter of black dialogue with South Africa. Although re-elected in 1972 he was forced that year to hand over real power to General Gabriel Ramanantsava, the nominee of the army, to whom he resigned the presidency after a referendum which approved the army's action. Ramanantsava was displaced three years later by Colonel Richard Ratsimandrava who was assassinated within a week and succeeded by Captain Didier Ratsiraka. French troops left the island in 1973. Ratsiraka nationalized French concerns without compensation and disputed with France the ownership of the Iles Glorieuses, Juan de Nova, Bassas da India and Europa in the Mozambique Channel. The riots which removed Tsirinana in 1972 were repeated and savagely defeated in 1985. Asians were indiscriminately assaulted and many fled. Outside the capital mob rule prevailed. Elections due in 1988 were postponed to the end of 1989 when Ratsiraka retained the presidency but grave riots broke out again and after ruling seventeen years Ratsiraka was replaced by Albert Zafy. A new constitution in 1992 reduced the presidential powers enjoyed by Ratsiraka but produced an assembly in which twenty-five parties were represented. The party of the prime minister, Francisque Ravony, had only two seats. A plan by the World Bank and the IMF to rescue the economy, encumbered with $5 billion of foreign debt, was so stringent that it was rejected by the assembly and the president. Madagascar was a country in which the population but little else was rising.

At the northern end of the Mozambique Channel lie the *Comoros*. Mayotte, the southernmost of them, remained by its own choice French territory when the rest of the group became independent in 1976. Their first president Ali Soilih, a man of the left, was murdered two years later and the islands became a base for right-wing mercenaries and their Belgian commander Bob Denard, dubiously styled colonel, who gained control of this small fiefdom after various adventures in Africa and Asia and maintained his rule by useful contacts with French and (increasingly) South African covert agencies, by the ruthless savagery of his small private army and by adopting the Muslim faith. The Comoros became a staging post for the supply of South African and Saudi military aid to the insurgent Renamo guerrillas in Mozambique less than 300 km away across the water. Denard removed Soilih's successor President Ahmed Abdallah in 1975, restored him in 1978 and had him killed in 1989. Denard returned to the attack in 1995 when he once more invaded the islands, seized President Said Djohar and transferred one of his friends from prison to the presidency. French troops reversed these events a few days later.

To the north-east of the Comoros is the British island of Aldabra, famous for its giant tortoises, and beyond that the *Seychelles*. This group, about 1,500 km from the African coast and twice as far from Bombay, was a British colony from which Britain detached the Aldabra and Chagas groups (mostly uninhabited) to form in 1965, for strategic purposes, a new colony called the British Indian Ocean Territory (BIOT). The rest of the *Seychelles* were given a constitution in 1967, revised in 1970, and became in 1976 a member of the Commonwealth. In return for the islands detached to the BIOT, Britain built on Mahé, the capital island, an airport to help the Seychelles to become a tourist area. The first president of the new state, James Mancham, was soon ousted by his prime minister, Albert René, more to the left and in external affairs non-aligned. René survived an invasion by white mercenaries connived at by South Africa in 1980. A second such attempt in 1983 was, however, stopped by South Africa. In 1986 René pre-empted a coup fostered by the CIA with British and South African assistance. He subsequently improved his relations with these enemies, chiefly in order to expand the islands' tourist industry.

Yet further to the north-east and nearer to the Asian shores of the ocean the *Maldives* – 1,200 small islands rising only a few feet above water and lying some 1,300 miles south of Sri Lanka – became independent of Britain in 1965, granting to Britain a lease for twenty-six years on Gan where the British were building an air base. Britain also bought for £3 million the island of Diego Garcia from its colony of Mauritius (east of Madagascar) before giving Mauritius independence in 1966. In 1970 Britain and the United States resolved to build a joint naval base on Diego Garcia and in 1971 the island was secretly leased to the United States. Its inhabitants were removed, mostly to the slums of Mauritius, and were given £650,000 (later raised to £4 million) against promises never to go back. A few years later Mauritius laid claim to Diego Garcia. The claim was subsequently toned down but not abjured. Mauritius was a French-speaking British colony with a racially mixed population in which Indians were in a majority. Originally acquired in order to pre-empt France, it was turned into a sugar plantation when its strategic importance evaporated and was worked by imported Indian labour. In time

Indians came to own more than half the land and were opposed politically by French-speaking Creoles who had the support of a small Chinese minority (about 3 per cent of the population). Mauritius was an economic success and it hoped to increase its prosperity by economic links with the Seychelles and the Malagasy Republic. It was wary of its neighbour Réunion, still a French dependency, and of French plans for a francophile association of Réunion, Mayotte and the Comoros. An attempt in 1988 against the life of the prime minister Aneerood Jugnauth, a low-caste Hindu, failed.

An Indian Ocean Commission for economic co-operation on the lines of ASEAN brought old adversaries closer: Madagascar, Mauritius, the Seychelles. Increased air traffic, investment and trade by South Africa suggested that South Africa was displacing the EC as the region's main external prop. Proposals for a demilitarized zone were advanced with an eye to reducing American military activity on Diego Garcia which had been used by the United States in the Gulf War of 1991.

C. BOTSWANA, LESOTHO, NGWANE

The three territories of Bechuanaland, Basutoland and Swaziland became protectorates of the British Crown in 1884, 1868 and 1890 respectively. Their transfer to South Africa, which had been envisaged in the South Africa Act 1909, was mooted by Malan during the Commonwealth conference of 1949, during a visit to South Africa by the Commonwealth relations secretary, Patrick Gordon-Walker, in 1950, again in 1951, 1954 and 1956. But the developing policy of racial *apartheid* and the intensification of police rule in South Africa eliminated such possibility as there might have been for British compliance. The Tomlinson Report included all three territories among South Africa's Bantustans but Tomlinson himself acknowledged a few years later that there was no likelihood of their transfer. In 1960 a transfer under the terms of the South Africa Act became an impossibility, since that Act required an order by the king in council upon the receipt of addresses from both houses of the South African parliament – an impossible procedure after South Africa became a republic. More important, the British parliament had been given assurances, frequently repeated, that there would be no transfer without a debate at Westminster and consultation of the wishes of the inhabitants.

But Britain's neglect of the territories had left them economically dependent on South Africa; Britain's traditional predilection for conservative chiefs retarded the advance of self-government without reconciling the traditionalists who, fearful of the newer type of nationalist, found some attractions in South African schemes for separate African territories based on the authority of chiefs. British weakness in regard to South Africa was demonstrated when Seretse Khama, the hereditary chief of the Bamangwato in Bechuanaland, married an English girl in 1949. Under pressure from the outraged South Africans the British banished both Seretse and his uncle, the highly talented and efficient Tshekedi Khama, on the plea that the tribe was split on the question whether to accept Seretse with a white wife as chief. Seretse remained in exile until he renounced in 1956 his claims and those of his descendants. He subsequently formed the Bechuanaland Democratic Party and after a general election in 1965 (held under a new constitution introduced in 1961) became prime minister. Bechuanaland – huge, underpopulated and largely desert – became an independent state under the name of Botswana in 1966 but its nominal independence was even more constricted than that of most poor countries, for its poverty made it dependent either on outside aid or on its big neighbour, and in the absence of substantial aid it seemed doomed by past neglect and present stringency to become a satellite of South Africa with all the consequent threats to the moderate and democratic rule which Seretse and his party hoped to give it. Under Seretse, president until his death in 1980, Botswana allowed the Rhodesian guerrillas to set up training camps and joined the group of Front Line states whose aim was to secure majority rule in Rhodesia with the least possible extension of violence. Seretse was succeeded by Quetumile Masire who continued to provide tolerant and moderately democratic government, albeit essentially one-party rule in an ostensibly multi-party state. It was one of the few places in Africa where people in general became better off.

In Basutoland, later Lesotho, a landlocked territory of 2 million people, the paramount chief Constantine Bereng, later King Moshoeshoe II, who succeeded to his office in 1960 after an education in England, led a moderate nationalist party (the Maramotlou Party) which was sandwiched between the traditionalists in the Basuto Nationalist Party led by Chief Leabua Jonathan and the more forthright nationalists of the Basuto Congress Party led by Ntsu Mokhele. In 1965 the Nationalists defeated the Congress but then lost ground. The opposition to them, suspicious of Nationalist links with South Africa where Chief Jonathan was regarded as the lesser evil, pressed Britain to strengthen the powers which the paramount chief would have as monarch after independence and to conduct fresh elections before independence, but the British government was unmoved by these pleas and in 1966 Basutoland became independent Lesotho with a narrow balance of domestic political forces and a tenseness in the political atmosphere which boded ill for Chief Jonathan's government and suggested that if it ran into difficulties in maintaining its authority it would have no choice but to ask South Africa for help. Riots which occurred at the end of the year gave Jonathan the opportunity to exact from King Moshoeshoe II a pledge to keep out of politics and to confine him to his palace. In 1968 some chiefs who had sided with Jonathan appeared to be switching their allegiance to the king and when Jonathan seemed to be losing the elections held in that year he cancelled the proceedings, suspended the constitution and so kept for the time being his office of prime minister. The king retreated into exile but returned in 1970. Jonathan survived until 1986 by which date South Africa, concluding that Jonathan was no longer the best bet from its point of view, applied economic pressures and encouraged the army to take over. General Justine Lekhanya, who had been dismissed by Jonathan in 1964, put an end to the twenty-year rule of the Nationalist Party. The new government was more cautious in relation to South Africa, less accommodating to the ANC. The king was dispossessed by the military in 1990, went into exile again and was replaced by his son but retained partisans in and outside the army. In 1994 the comparatively left-wing Congress won a comfortable victory in a general election but was attacked by a section of the army demanding the ex-king's recall. Diplomatic activity by (the new) South Africa, Botswana and Zimbabwe resolved this conflict. King Moshoeshoe was reinstated as a constitutional monarch with Ntsu Mokhehle as prime minister, but was killed in a car accident a year later.

Swaziland, later Ngwame and the smallest of the three territories and the last to attain independence, had an elderly paramount chief, Sobhuza II, who was advised by a singularly conservative council. Between 1960 and 1973 he experimented with a modest degree of democracy. At elections in 1964 his Mbokodo Party formed an alliance with the United Swazi Party, the organ of the white farming and business community of 10,000, and routed an assorted opposition of divided and ill-prepared nationalists. Under the constitution prepared for independence in 1968 the monarch was given a dominant position: he could nominate enough members of the parliament to block measures which he did not like, he was not bound to act on ministerial advice in all matters in which a constitutional monarch is normally so bound, he was given control over Swaziland's principal asset (its minerals) subject only to advice from his traditional council and not from his minister, and his party was entrenched in power by an electoral system in which the towns, and therefore the nationalists, were submerged in large rural constituencies. Control over the minerals was, however, taken away from him on the eve of independence. Opposition was weakened because a number of nationalists joined the Mbokodo Party for opportunist reasons. The death of Sobhuza II in 1982 at the age of about 83 was followed by multiple intrigues in and around the royal family and by a regency under a queen mother who was quickly deposed in favour of another queen mother. Under South African pressure the ANC was proscribed. A new king, Mswati II, aged 18, was installed in 1986. He opposed partisans of democracy and the ANC but when he tried to repress his adversaries by widespread arrests and treason trials he was rebuffed by the acquittal of those indicted.

D. THE HOMELANDS OR BANTUSTANS

By 1990 four Homelands had been established and six more were projected:

Transkei: a geographically almost coherent agglomeration of 2.5 million Xhosa-speaking peoples, self-governing from 1963 and 'independent' in 1976; 45,000 sq. km of (largely neglected) agricultural and timber-bearing land; 1.7 million Xhosa-speakers outside the Homeland lost their South African citizenship. Transkei broke off relations with South Africa – it had none with any other state – over the latter's refusal to incorporate East Griqualand in the Homeland and over the dumping of unwanted Xhosa-speakers in it. In 1979 South Africa stopped the main source of this forcible migration: the bulldozing of Crossroads in Cape Province. Transkei's first president, Kaiser Matanzima (subsequently imprisoned), gradually lost ground to the left personified by General Bantu Holornisa and Chris Hani, chief of staff of Umkhonto na Swize and general secretary of the South African Communist Party (assassinated in 1993).

Bophuthatswana: a Homeland for about half of the Tswana peoples not in Botswana which, since the migrations to the goldfields in the nineteenth century, has contained only a minority of them; 'independent' in 1977; over 40,000 sq. km in seven different segments without a capital but with the world's largest platinum mines.

Venda: an almost coherent Homeland of 6,500 sq. km incapable of supporting its 1.3 million Vhavenda (kin of the Shona); an area of poor agriculture but, being only 10 km from Zimbabwe, some strategic importance; 'independent' in 1979.

Ciskei: a poor and unruly area of 5,500 sq. km and 500,000 Xhosa-speakers who crossed the Kei in the nineteenth century; claimed by Transkei; self-governing from 1972. The Ciskei and the 'white corridor' leading to East London (where the jobs were) were mutually dependent – and hostile.

The designated Homelands comprised *Kwazulu*: a Homeland intended for 4 million of the 5.5 million Zulus and comprising 31,000 sq. km in ten (originally forty-eight) segments; includes one-third of Natal, of which the Zulus claim the whole. Chief Buthelezi, the outstanding Zulu personality (although formally inferior in rank to King Goodwill Zwelithini), refused to accept independence. *Lebowa* and *Qwagwa*: for the North and South Sotho; small and arid; the former self-governing from 1972. *Kangwane*: north of Ngwane; 3,700 sq. km in two segments for 250,000 Swazis (one-third of the total). *Gazankulu*: 6,300 sq. km in four segments for 350,000 Shangaan; arid but possibly covering minerals. *South Ndebele*: extracted from Bophuthatswana; the North Ndebele, destined for Lebowa, might join it.

By 1990 all four established Homelands were demonstrating their rejection of government policy and support for the ANC. In Bophuthatswana there were demonstrations against President Lucas Mangope. In Venda President Frank Ravele was forced out of office. In Ciskei Lennox Sebe, president for life, was ousted by the local army which renounced independence and aligned itself with the ANC. In the oldest of the Homelands, Transkei, a similar coup had removed two chief ministers in 1987. The embryonic Homelands also renounced their projected independence and, including even Kwazulu, declared their support for the ANC. All the Homelands ceased to exist as such upon the adoption of South Africa's new constitution in 1994.

PART SIX

AMERICA

25 CANADA

Canada, the world's third largest country, has played a singularly benign part in international affairs. It fought in two World Wars not for itself but for others. It was unusually secluded in geographical and political terms with to all intents and purposes only one neighbour and no serious external disputes. It had considerable resources and a relatively small population. It had been, after Britain, the senior member of the old British Commonwealth and became in the postwar multiracial Commonwealth the most attentive of the older (white) members to the needs and susceptibilities of the newer. To Canada's historical, political and sentimental attachment to Britain was juxtaposed its economic, strategic and geographical attachment to the United States. The shift in importance from the former to the latter was accelerated by the Second World War and the Cold War and recognized by the Ogdensburg Declaration of joint Canadian–United States responsibility for the security of North America, by the wider commitments entailed by membership of Nato, by participation in the Early Warning System for nullifying surprise Soviet missile attacks on North America and by the creation in 1958 of North American Defence Command (NORAD). By sending troops to the Korean War, Canada associated itself with the view that the paramount issue in world affairs was the fight against the Soviet Union and international communism. Yet it did so with reservations. Many Canadians resisted the simplistic demonology of the Cold War. Some Canadian leaders toyed with non-alignment and queried the usefulness of Canadian troops in Europe: at one point the government unilaterally halved their number (after observing the total withdrawal of French forces from Nato commands). The conduct of the war in Vietnam by the United States was widely condemned and Canadians were quick to feel hurt when the United States overlooked – as, for example, in the Cuban missile crisis – the need to tell its continental allies what was going on.

Canada was a prominent supporter of the abortive provisions in the UN Charter for an international force and a military staffs committee and Canada's was the main voice in insisting that Nato be committed by article 2 of the North Atlantic Treaty to economic as well as military purposes. More effectively, Canada worked to mollify the asperities created by the British assault on Egypt in 1956 and the tensions caused by *apartheid* in South Africa (when these proved irresoluble Canada aligned itself with those members who judged that the Commonwealth would be better off without South Africa).

If, as was supposed, President William H. Taft had coined in 1911 the expression 'special relationship' to describe US–Canadian relations he was not far wrong, particularly in economic terms. Economic links grew throughout the century as the Canadian economy expanded in old ways and new: in grain, lumbering and established mineral (chiefly nickel) undertakings, and additionally in electronic and chemical industry, petroleum research and extraction, and the exploitation of radium deposits which enhanced a nuclear programme initiated during the war. These activities were on such a scale that Canadian capital did not suffice for them and was supplemented by money from the United States to the point where control of half Canada's manufacturing, mining and carbon fuel industries passed into US hands. For the United States Canada was a neighbour of vast space and unmeasured, untapped resources, which might replace United States resources in food and raw materials that were threatening to become inadequate through natural exhaustion and population increase. The United States might the more easily concentrate its economic efforts in the manufacturing sector if it were assured of supplies of Canadian primary products. For Canadians on the other hand the flow of US capital was unwelcome as well as welcome. Economically it was not only welcome but essential. Yet it was also a source of resentment and apprehension for those Canadians who complained of a loss of control over their own assets and even scented a loss of national identity. This imbalance began to be reduced in the 1970s and 1980s when Canada became the fourth largest investor in the United States – behind Japan, Britain and the Netherlands. Each of the two countries was the other's biggest trading partner and the two together constituted the world's biggest bilateral trading partnership. A free-trade agreement which came into force at the beginning of 1989 provided for the elimination of all tariffs and other obstacles to trade over the next ten years. Nevertheless this commingling did not remove all the acerbities. Nixon's surprise decoupling of the US dollar from its fixed gold price in 1971, accompanied by a 10 per cent surcharge on US import duties, hurt Canadian businesses and angered Canadian politicians. This shock was quickly followed by the sharp rises in the price of imported oil caused by a war in the Middle East in which Canada had no part but the United States, by promising to rearm Israel after its initial setbacks, had. Canada's eastern provinces, being importers of oil, were badly hit and although the oil-bearing western provinces benefited from higher prices they were embroiled in disputes over the

25.1 North America

transport of Alaskan oil to the United States by sea or through pipelines through Canada. In this segment of economic affairs nationalist considerations tended on both sides to prevail over the prospective value and virtue of internationalism.

Canadian politics of the first postwar decades were dominated by two Liberal leaders: Lester Pearson as minister of external affairs and then prime minister for twenty years (1948–68) and Pierre Trudeau as prime minister with one short break for sixteen (1968–84). Both were pronounced champions of an international as distinct from an American regional role for Canada. Both were notable and more effective for what they said than for what they achieved: Canada's resources were in a worldwide context modest but these leaders made Canada's voice significant. Pearson, who spent a lifetime in the public service, conformed the more easily with the Cold War consensus and with Washington's priorities in the era of Truman, Eisenhower, Kennedy and Johnson. Trudeau, almost a newcomer to high office when he first became prime minister, was faced with lesser presidents and was by nature more cynical or disdainful. He had travelled widely in four continents. In 1970, soon after succeeding Pearson, he recognized the communist government of China and led the way for other states to do so too. Trudeau's hold on power was less secure than Pearson's. He nearly lost the election of 1972, recovered his majority in 1974, lost office in 1979, and regained it in 1980 but as the choice of the eastern provinces only – the Liberals won few seats in the west. He was also uncomfortably placed in Canada's great domestic drama, the status of Quebec, for he was a determined federalist as well as himself a Québecois.

The rapid increase in Canadian prosperity posed a special problem in Quebec, the seat of Canada's French heritage and culture. Four-fifths of the inhabitants of the province were French-speakers, most of them rural conservatives. In spite of their numbers they owned, however, only one-fifth of its productive economy and their average income was substantially below that of English-speakers. From 1944 to 1959 the province was dominated by Maurice Duplessis (prime minister also in 1936–39) who presented himself as a traditional Québecois while associating also, and not always openly, with English-speaking Canadian and US capitalists who were contributing to Quebec's economic transformation. He was succeeded in office by the Liberal Jean Lesage who espoused the modernization of Quebec with a more determined policy of securing the benefits for the French-speaking majority, but in 1966 the Liberals were unexpectedly defeated, losing working-class voters who had come to the conclusion that modernization was benefiting the middle classes only. A visit by de Gaulle in 1967 enthused separatists. It is unlikely that de Gaulle wanted to disrupt Canada. What he did want was to recruit Quebec to the international, mostly African, francophone community and by ending a public speech in Montreal with the old and emotive slogan 'Vive le Québec libre' he both dramatized Canada's rumbling domestic discords and forced his affronted hosts in Ottawa to issue a deprecating rejoinder. He went home without visiting Ottawa. Much ill will was generated over the next few years over whether the province of Quebec or the government of Canada should be invited to francophone meetings in Africa.

In 1968 – the year when Trudeau became prime minister – a new party, the Parti Québecois (PQ), was formed to campaign for independence for Quebec. More extreme than the PQ, the Front de Libération de Québec (FLQ) advocated and practised violence in order to win the independence which the PQ aimed to win peacefully. The FLQ consisted of a tiny handful of men and women who, invoking Che Guevara and Mao Zedong, committed a number of bank robberies and other acts of violence without attracting widespread support or, until 1970, much attention. In elections in that year the PQ won a quarter of the votes cast but only seven seats out of 108: the winners were the Liberals under Robert Bourassa. The FLQ resorted to more serious violence, kidnapping the British Trade Commissioner James Cross and the province's Minister of Labour Pierre Laporte, whom they strangled. The central government applied the War Measures Act and outlawed the FLQ but gave the kidnappers a safe exit to Cuba in return for the release of Cross after two months' confinement. The security forces, whose ineffective intelligence had allowed the FLQ a surprisingly long lease of life, succeeded in extinguishing it by 1972. The separatists' cause became entirely limited to the non-violent route which however got them nowhere.

The problem reverted to constitutional debate. A conference in 1971 between the federal government and Canada's ten provinces produced the Victoria Charter, a compromise which was rejected by Quebec, and in 1976 the PQ won elections in the province. Its leader, Jean Levesque, became provincial prime minister, visited France and introduced a series of measures giving the French language a special status (the federation as a whole had been declared officially bilingual by statute in 1969). But the separatist cause was set back when a referendum in Quebec proposing negotiations for 'sovereign-association' was defeated in 1980.

Quebec's nationalism impinged on federal positions when Trudeau's Conservative successor, Brian Mulroney, tried to woo Liberal voters in Quebec in 1984 by supporting constitutional changes (the Meech Lake amendment) designed to increase the influence of provincial governments in senior judicial and senatorial appointments, immigration and financial affairs. In provinces other than Quebec the amendment was not popular and it failed in 1990 to win the necessary endorsement of a majority of the provinces. This rebuff to the Québecois produced for the first time a poll showing a majority in the province in favour of sovereign independence. Mulroney's next attempt to solve the problem by giving Quebec a special status in the federation, although backed by all ten provincial prime ministers (the Charlottestown Accord), was rejected in plebiscites in five provinces as well as Quebec. These constitutional setbacks, combined with economic recession, higher taxes and unemployment and an air of inconsequence at the centre, destroyed Mulroney's standing to the point where he was compelled to resign.

Mulroney was succeeded in 1993 by Canada's first female prime minister, Kim Campbell, who led her party to a defeat verging on extermination. Her party won two seats and the Liberals returned to power with Jean Chrétien as prime minister and the familiar problems of relations with the United States (in the guise of NAFTA – see Chapter 27); balancing the public accounts, particularly financing generous social services from a relatively narrow tax base; and, still, Quebec where, however, the PQ won in 1994 provincial elections with so small a margin of the popular vote that its promise to proceed to a referendum on independence appeared rash – the more so when the Inuit and other indigenous peoples began to suggest that the right of the province to secede from the federation implied a similar right for them to secede from the province. A referendum in 1995 gave the federalists victory by 1 per cent, an inconclusive result.

Canada's last constitutional link with Britain, derived from the British North America Act of 1867, was demolished in 1982 when at Canada's request the British parliament enacted an entrenched charter of rights and simultaneously extinguished its own remaining powers to legislate for Canada. This measure had been anachronistically delayed by disputes within Canada over the allocation between the centre and the province of the rights to be relinquished by Britain; by differing judicial decisions on whether the central government might petition the British parliament without the consent of all the provinces (which Trudeau failed to get at two constitutional conferences); and by Trudeau's determination to secure the enactment in Britain of what amounted to a Canadian Bill of Rights. In Quebec, Lévesque objected to the enactment of such rights by either the British or the Canadian parliament and they remained ineffective in Quebec except in so far as they were already or in future might be enacted by the provincial legislature.

In 1949 Newfoundland became part of Canada after a referendum in which a narrow majority chose confederation with Canada in preference to penurious self-government.

26 SOUTH AMERICA

South America in the nineteenth century was isolated from world politics not – as were Africa and much of Asia – by the muffle of European imperialism but by the heritage of post-colonialism. At the same time the South American republics were largely isolated from one another as well as from the rest of the world. The twentieth century witnessed an accelerating reversal of this pattern, accompanied and complicated by spasmodic attempts to assimilate the democratic and industrial revolutions which were the hallmark of the experiences, and to some extent the successes, of western Europe and the United States of America: to implant, that is, democratic political forms and democratic social values in resistant, narrow oligarchies, and to develop manufacturing industries where trade in primary products had hitherto sufficed for the needs of the ruling classes.

In a century and more after independence South America had become a byword for political instability and could, but for widespread ignorance of its affairs, have become no less a byword for social immobility. It was notorious for civil wars, revolutions, coups, political assassinations and short-lived constitutions, while at the same time it entrenched extreme social and economic injustice. Its basic needs were the reverse of its experience; namely, political stability and social and economic change. It was in these respects not unique but its ills had been aggravated by time until they posed a daunting dilemma: was it possible to get social and economic change without revolution? Was it possible to get political stability without perpetuating social and economic stagnation? The underlying conflict between the few and the many was not mediated by a middle class such as had tided western Europe over the bar of oligarchy without intolerable violence.

The government of South America after the end of Spanish rule devolved upon a social élite consisting of big landowners supported by the Roman Catholic Church and by a military caste aspiring to the same social status. By 1945 the traditional power of the upper classes had been destroyed nowhere south of the Panama isthmus and ten years later only in Bolivia (1952). Nevertheless the oligarchy's props were weakening. Within the religious and military establishments there were growing

26.1   South America

doubts about the immutability and propriety of its monopoly of power and profit and some concern, expressed with varying degrees of success and sincerity, over the plight of the rural poor, the growing urban proletariat and the suppressed Indians. An awareness of gross inequalities was stirring consciences and fears and so enlisting in the service of the underdog those two powerful political forces, indignation and the recalculation of expediencies. Churches began to shift their attention and their political weight somewhat to the left, and there appeared in the armed forces officers with some of the instincts of populists and a taste for demagogy.

Through South America a great part of the population was extremely poor, illiterate, unproductive and virtually outside the state. Many states were not merely run by an oligarchy but also owned by it in the sense that the land and what grew on it and what lay beneath it were the private property of a small number of individuals: in various countries 60 to 90 per cent of agricultural land was owned by a tenth of the population. Quite apart from abstract notions of fair or unfair shares, this distribution was the cause of great inefficiency. Many landowners, possessing more land than they needed to cultivate, left much of it untilled and untended, but were firmly opposed to any redistribution to other proprietors who might be more disposed to cultivate it. (Forcible redistribution produced the opposite evil of a multitude of economically intractable plots: in Colombia, for example, more than two-thirds of the land was uncultivated while much of the cultivable area had been broken down into holdings of a few acres.) Bountifully supplied with cheap labour the big landowners had no need to invest their profits in their land, modernize their methods or increase production.

The rural poor remained poor to the verge of destitution and beyond, and either endured short lives of useless and hopeless misery or drifted to towns where they were not much better off, since they were not fitted to take jobs. Public education in South America as a whole was so meagre that less than a tenth of the population completed a primary course and illiteracy rates of 50 per cent were not uncommon, 90 per cent not unknown. If they were Indians, the peasants who went to the towns invited incomprehension and ridicule by their strange speech and attitudes. Moreover, if the peasants had little to offer, the towns too had little to offer on their side. Industry cannot flourish in places where half the population is too poor to buy its products. South American industries were handicapped by the lack of a domestic market with the result that the South American countries continued to import goods which they could have been making for themselves. Hence the towns to which the peasants migrated, so far from needing their labour, already contained unemployment; in some of them a third of the inhabitants might be unhoused. And this unemployment was growing not only for economic reasons but also because the population explosion was greater than anywhere else in the world: the population was in the process of doubling itself in periods of a quarter century.

There existed therefore in most parts of South America a revolutionary situation. This situation was accentuated by awareness of it, since an assumption that a pattern cannot last much longer is itself a potent factor for change. The forces making for change were the relatively passive rural and urban proletariats and the active leaders who might emerge from the established social groups or the nascent middle class and who, out of dissatisfaction with the existing state of affairs, might make common cause with the masses some or all of the way towards revolution.

The growth of a South American middle class had been stunted by the slow pace of industrial development just as industrial development had been stunted by the self-sufficiency of a ruling class capable of maintaining its standard of living by exporting primary products and using the proceeds to import all the necessities and luxuries which it wanted from the outside world. This economic pattern had social and political consequences, since a small middle class is less likely than a large one to adopt distinctive social habits and political aims of its own, and in South America the small middle class, imprisoned within the oligarchic system, was relatively effortlessly seduced into gravitating into and so preserving the upper reaches of a system which it had no power to subvert. Here again South America was not unique, except perhaps in degree. The standing of the middle class was, however, altered by the Second World War which deprived South America of its habitual imports and so promoted industrial development. After a pause in the immediate postwar years the demands created by the Korean War gave a fresh fillip to industrial expansion. These wars, events external to the sub-continent's own affairs, thus altered its economic course, although they did not do so to the extent of seriously altering its social hierarchy. The industrial middle class prospered in some centres, notably in Brazil, but for the most part it came nowhere near to supplanting the traditional ruling class in the exercise of political power.

The rise of a middle class is commonly associated with the growth of democracy as the middle class, annexing a part of the power wielded by a landed aristocracy, points the way to a yet further expansion of the political system. This process was reinforced in South America by a strand in its history of European origin. While South America owed its independence to the circumstances of the Napoleonic Wars which sundered it from Spain and Portugal as well as France, and whereas the waging of wars of liberation favoured the establishment of autocratic and military rule, the revulsion from European dominion also carried with it ideas of the Enlightenment so that liberated South America was imbued with a rejection of autocracy as well as rejection of distant, foreign rule. But this was not a dominant theme, and it succeeded only sporadically in making serious modifications to a quasi-feudal pattern which was supported by the most powerful element in the state, the armed forces.

After the wars of liberation early in the nineteenth century the new South American states engaged in few wars among themselves and faced no enemy threat from beyond the sub-continent. Their armies assumed therefore a domestic political role rather than the function of national defence. Officers, with the conservatism natural to their caste and the social ambitions which became a substitute for more serious military employment, saw themselves as the guardians or godfathers of the state – and the state as the same as the status quo. Unduly inefficient or corrupt politicians would be replaced by others or by a period of direct military rule, and in the exercise of this regulative function the officer class pictured itself as acting for the public weal, a habit of mind which could lead to surprising results if the army's interpretation of the public weal should change. And around the middle of the twentieth century there were indications of such a change. As armies expanded, they acquired officers from somewhat different backgrounds, so that the awareness of the need for social change and mild support for it found their way into the military establishment. The consequences of this shift were, however, ambivalent. On the one hand the insemination of a radical element into the military mind was an encouragement for the proponents of change, but on the other hand an establishment ready for no more than a small dose of change could easily be frightened back into a more rigid conservatism, if the pace of change began to outstrip the progressive officers' own cautious assessment of the need. The addition of a military element to the progressive forces could affect the pace of change in one of two ways: either by increasing it and so alienating progressive officers and bringing on a clash with the forces of conservatism, or by reducing it to a level acceptable to these officers and so alienating the more intrepid elements and producing a clash at a different point along the political spectrum.

Alongside this regrouping of domestic forces old and new, whose outcome was likely to differ from one state to the next, there was a search for new models, for political forms to replace the existing forms which were being judged and found wanting. This search led the more inquiring minds to consider two foreign models, each of which seemed to have something to contribute: western democracy with its emphasis on freedom and human rights, and communism with its reputation for economic growth. The South American intellectual who could discover how to get the best from both these two worlds would perhaps have found the synthetic short cut to prosperity and justice. But he too was faced with a multiple dilemma. He knew that, for economic reasons, South America needed foreign aid; he knew that, for historical reasons, South America wanted to steer clear of foreign aid; and he had to face the fact that if, in his search for the synthetic short cut, he looked for aid and inspiration to both the western and the communist strongholds, he would be met in each with the argument that the two models were antithetical and that he must make up his mind between them before expecting help from the guardian of either. Like Asians and Africans, although for different reasons, he would need to be non-aligned while soliciting the favours of those whose line he refused exclusively to take.

South America's approach to the outside world reflected the needs of countries whose economies were in a process of transition without adequate domestic resources or machinery, against a background of ill-regulated foreign aid in a previous generation. Towards the end of the nineteenth century and at the beginning of the twentieth the principal capitalist nations had lent money to South America to excess and at very high rates of interest to the mutual dissatisfaction of both sides – of the South Americans who felt that they had been exploited, and of the lenders who resented many a subsequent default and the expropriation of a number of enterprises which they had built up. But South America could not do without foreign capital if it was to implement development programmes, close trade gaps and meet its large (and largely short-term) foreign debts. The capital required was not being produced at home. The poor were too poor to save and were getting poorer, the rich frequently invested or banked their wealth abroad, and domestic banks and other financial institutions had not developed the habits or the mechanisms for providing credit for industrial expansion. The reformers, who appreciated the need for foreign aid but were required by their environment to be anti-foreigner as well as anti-establishment, hoped to be extricated from this dilemma by getting aid from international agencies instead of foreign governments, but these agencies proved disappointing since they applied strict financial rules to their lending, insisted on currency stabilization and a reasonable prospect of profit as preconditions, and hesitated to come to the aid of petitioners with limited financial credentials and limited economic expertise: the postwar years in South America were peppered with too many economic blunders as wartime profits were dissipated on luxuries and industrialization was pursued fitfully and without discrimination.

In so far as common features could be found in this vast sub-continental area's external affairs they were two: attitudes towards the northern part of the continent and foreign debt.

South America differed from North America not only by its Latinity but also by its fragmentation. While South America became divided into many states, North America did not. From the days of the Liberators, when the sub-continent was Balkanized by post-imperial wars and politics, there were hankerings for the sub-continental solidarity of the great Spanish viceroyalties. A surviving cultural community and a certain geographical compactness and isolation promoted in time an inter-American system, which was enlarged in turn into a pan-American (as opposed to a Latin American) organization. But the contrast between south and north could not be eradicated. In the north the United States had shown an amazing capacity to accommodate a hotch-potch of races and preserved its unity in spite of the disruptive social and economic forces which produced the Civil War, while Canada succeeded in keeping its British and French populations under a single political roof. These two states therefore were much bigger and more powerful than any South American state, and the South Americans became fearful of US preponderance and imperialism. Canada, which might have served as a makeweight, was reluctant to join any organization which might involve it in the conflicts between the United States and the Latin states.

The United States fed Latin American fears and suspicions. During its years of expansion the United States was uncertain of its attitudes towards its neighbours in Central America and the Caribbean, and during its years of emergence as a great power it frequently acted as though these states were something less than sovereign. Just as Britain in the twentieth century found it hard to think of Middle Eastern countries as independent or deserving independence, so the United States in an overlapping period felt much the same way about a group of states which were supposed to have a special impact on North American vital interests. When President Monroe forbade the expansion of European territorial dominions in the New World, Spain and Portugal had already lost theirs; the British, French and Dutch, who had arrived too late to get more than the pickings left over from the Hispano–Portuguese partition, had little interest in challenging Monroe's unilateral declaration and the one serious attempt to do so – France's attempt to turn Mexico into a new Habsburg empire during the American Civil War – was a catastrophic failure. By this time the United States had itself annexed one-third of Mexico, had evinced interest in an isthmian canal and was about to toy with the idea of acquiring the large islands of Cuba and Hispaniola (the latter containing the Dominican and Haitian republics).

The Monroe Doctrine, enunciated in 1823, was the basis of a policy of turning America into an island by purchases (Louisiana, Florida,

Cuba, Alaska) and by barring all European powers from recovering their possessions or extending their influence in the continent. It was inspired equally by fears of Russia in the north-west and other Europeans to the south-east. For more than a century the doctrine required little exertion on the part of the United States, mainly on account of the state of Anglo–American relations, and it was not seriously called in question until, in the Second World War, the remaining French and Dutch colonies came within the legal grasp of Nazi Germany and, in the Cold War, the Russians dared to establish a base in Cuba. Britain never made any attempt to enlarge its West Indian empire and during the decades after the breach between Britain and the new United States British naval power served to buttress rather than challenge the Doctrine. Geography has placed no islands between the British Isles and the seaboard of the United States, so that the initial post-independence period was one of estrangement but not of conflict; it is hard to believe that conflict could have been avoided, especially during the Civil War, if the North Atlantic had been dotted with islands which the two powers would have competed to occupy. In the event the possession of a common language and common traditions was not countered by territorial disputes (except over Canada which, illuminatingly, failed to get very robust support from London), and when by the end of the century the power of the United States had grown to significant proportions, the goodwill between the two countries prevailed over occasional conflicts and led in the next century to an alliance which pulled Britain out of a European and into an American orbit. This alliance might not have been achieved but for Britain's abnegation of the role which it had once been in a position to play in Central America.

In the middle of the nineteenth century, when the United States first began to think of a canal across Nicaragua to link the Atlantic and the Pacific, British assent and co-operation seemed essential. Britain had territories and claims along the nearby coast (in British Honduras, the Mosquito Coast and the Bay Islands); the United States negotiated favourable treaties with Nicaragua and Honduras. From these positions the two countries negotiated the Clayton–Bulwer Treaty of 1850 whereby they agreed that neither should acquire exclusive control over any canal or special privileges in it; and further that neither should occupy or fortify or colonize or assume or exercise any dominion over any part of Central America. This treaty, concluded at a time when the United States was the weaker of the two parties, became an obstacle to later plans to go ahead with the construction of a canal without British collaboration and to acquire dominion over its course and banks, but by the end of the century British interest in this part of the world was small compared with British interest in the Middle East and southern Asia and in 1901 the Clayton–Bulwer Treaty was replaced by the Hay–Pauncefote Treaty which reaffirmed the principles of neutrality and free and undiscriminating use, but otherwise removed the limitations set by the earlier treaty. The United States then entered into discussions with Colombia for a grant of territory at Panama. A treaty was negotiated but rejected by the Colombian senate, whereupon in 1903 the United States promoted a revolt at Panama and the secession of the area from Colombia. A new Panamanian republic was created and, for $10 million and an annual rent of $250,000, granted to the United States perpetual sovereignty over the area to be called the Canal Zone and also the right to intervene in Panama's internal affairs. The United States made use of this right by despatching troops on a number of occasions.

More important to the United States than Panama was the large island of Cuba, closer to the United States than any other Latin American country with the sole exception of Mexico. For strategic reasons the United States tried on various occasions in the nineteenth century to buy Cuba from Spain but without success. In 1868 the Cubans rose against Spain and waged war for ten years. They were defeated but rose again in 1895. Feelings in the United States were enraged by the cruelty which the Spanish authorities used to defeat the revolt and by concern for American investments, but the government took no action until, in February 1898 and in circumstances which have remained unexplained ever since, the battleship USS *Maine* was sunk in harbour at Havana. Washington delivered an ultimatum and, although the terms were largely accepted and the war virtually over, declared war on Spain. Fighting, which extended to the Pacific, lasted three months and ended with the complete defeat of Spain and the cession to the United States by the Treaty of Paris of the Philippine Islands, Guam and Puerto Rico in return for $20 million. Cuba passed in effect under the tutelage of the United States and so remained until 1933. In 1901 the United States asserted by the Platt Amendment (an amendment to the Army Appropriation Bill) that it would not withdraw its military forces unless its right to intervene for the preservation of good government were embodied in the Cuban constitution. US forces were stationed in the island in 1906–9 and 1912–13 in support of corrupt military régimes, and a naval base was built at Guantanamo.

Within a few years of the Cuban War the United States intervened in 1905 in the Dominican Republic. Fearful of European intervention as a consequence of the default of the Dominican government on its debts, President Theodore Roosevelt formulated the Roosevelt Corollary to the Monroe Doctrine, by which the United States arrogated to itself the right to intervene in Latin American countries in order to keep governments in order. US forces reappeared in 1916 and for the next eight years the country was under the direct military rule of the United States with a US officer as president. A similar occupation of the neighbouring Haitian republic, also intended to forestall European creditors, lasted from 1915 to 1934. On the mainland the United States intervened openly in the Mexican revolution and civil war by a naval bombardment in 1914 and an unsuccessful army expedition in 1916–17; and US forces, sent to Nicaragua in 1911 to support a favoured president, kept the liberal opposition quiet until 1933 when, upon their departure, the dictatorship of Anastasio Somoza was inaugurated.

This US policy of direction and control, sustained by sporadic military descents, was abandoned by President Franklin D. Roosevelt and his Secretary of State, Cordell Hull. In his first inaugural address Roosevelt promulgated a good neighbour policy based on non-intervention, and his undertaking was repeated at the Pan-American Conference of 1936 at Buenos Aires. The Platt Amendment was repealed. The right to intervene in Panama's internal affairs was abrogated by treaty. The withdrawal of US marines from Haiti was accelerated. The president accepted the right of the Mexican government to nationalize oil properties within its territory. Latin American hopes were also raised by the passing of the Reciprocal Trade Agreements Act of 1934 which gave the president power to reduce tariffs by as much as 50 per cent, and by the establishment in the same year of the Import-Export Bank for the lending of US public funds to foreign governments.

But this amelioration of the intra-American atmosphere did not produce all that Washington desired. Before the Second World War, as after it, the United States wanted to enlist Latin American sympathies and support for wider matters, but largely failed. Before the outbreak of war Hull, mindful of Germany's attempt in the First World War to get Japan to attack the United States through Mexico, tried but failed to persuade Latin America that Nazism and fascism were present dangers against which the whole American continent should take joint precautions. At Havana in July 1940 the American states agreed that no non-American state should be allowed to take over any piece of American territory, but this new and anti-German enunciation of the Monroe Doctrine was to be upheld by joint American action; the United States was given no invitation to intervene on its own against any external threat, but rather to supply its neighbours with the arms and equipment to do so. Consequently, one result of Washington's fears for the defence of the hemisphere was the strengthening of the military class throughout the area. Small Brazilian and Mexican forces were sent overseas during the war, but American military supplies affected the structure of politics within Latin America far more than they affected the course of the war.

After Pearl Harbor, Mexico, Colombia and Venezuela broke off relations with Japan, Germany and Italy, and all the Central American and Caribbean republics declared war. When the end of the war was in sight the American states, meeting at Chapultepec in Mexico in February 1945, declared themselves in favour of collective defence against internal as well as external threats (this arrangement was placed on a more permanent footing by the Inter-American Treaty of Reciprocal Assistance, concluded at Rio de Janeiro in 1947), but Washington's concern to create a continental anti-communist alliance or defence system found little response among states which were still used to thinking of the United States and not the USSR as the prime intervenor in their internal affairs and were much more worried by postwar economic problems than by communism. The military classes, which were the most immediately affected by joint defence schemes, were less interested in co-operation than in strengthening their own forces. They looked to the United States for more, and more modern, equipment. The United States on the other hand became increasingly concerned with communism and increasingly trapped between policies of pre-empting communism by economic aid to its neighbours and interfering in their affairs to suppress communists or anybody who looked like a communist from Washington.

In 1951 the Mutual Security Act was extended to Latin America and from 1952 the United States concluded a series of bilateral defence agreements. If the United States (and Latin American civilians) had qualms about this further reinforcement of armies which operated most often as domestic political forces, Washington felt itself constrained by the veiled threat that armies unable to satisfy their needs in the United States might go shopping elsewhere. An inter-American police academy was established at Fort Davis in the Canal Zone in 1962 for the study and practice of techniques of counter-insurgency, but otherwise inter-American military co-operation was not noticeably fruitful or popular.

The conferences at Chapultepec and Rio were the seventh and eighth in a series of inter-American conferences which had been inaugurated in Washington in 1889. The ninth of these meetings was held at Bogatá in 1948 and created new institutions and continuing machinery for pan-American consultation and action (Canada still, by its own wish, excluded until 1989). Obstensibly pan-American, but in some minds a Latin counter-weight to northern power, the Organization of American States had for its objects the maintenance of peace within the area of its members, the peaceful settlement of their disputes, joint action against aggression and co-operative development of economic, social and cultural interests. For Latin America this association with the United States was welcome chiefly on account of the prospect of an economic outpouring like the Marshall Plan for Europe, but the prospect proved a mirage since Washington saw Europe and Latin America in a completely different light: Europe had been ravaged by war, was in danger of immediate economic collapse and was believed to be a defenceless bait for further Russian advances. These arguments to the hearts and heads of US policy-makers applied little or not at all to Latin America, even had these policy-makers been as much concerned with good neighbourliness as Franklin D. Roosevelt in the 1930s. Latin America had not a high priority in postwar Washington. Disenchantment grew on both sides. When war broke out in Korea, in Washington's eyes an anti-communist war in which all should be prepared to stand up and be counted, Colombia alone sent troops. The Korean War sharpened Washington's global anti-communist susceptibilities and clarified its priorities in Latin America where it abetted and organized the overthrow of the left-wing President Jacobo Arbenz Gúzman (see Chapter 27) and so showed how much more importance it attached to the suppression of communism than to the principle of non-intervention or to good relations with its Latin neighbours. Washington also displayed its belief that its actions in the southern half of the continent need not be trammelled by great power considerations: the reaction of the USSR could be regarded, at this date, as minimal or impotent. The final lesson of the Guatemala affair was, or should have been, the impossibility of drawing a line between radical programmes which were social and therefore good and radical programmes which were communist and therefore malignant. The United States evicted in Guatemala a reforming administration whose actions were not in themselves noxious but whose motives and further intentions were suspect. Latin America did not accept the anti-communist justification put forward in Washington.

In 1958 Vice-President Richard Nixon, on a tour of Latin American countries, was received with insults and even violence which enlightened official and popular opinion north of the Rio Grande about some Latin American attitudes towards the United States. In two of the countries in his itinerary, Peru and Venezuela, dictators had recently been displaced, but not before they had received from President Eisenhower the Legion of Merit: Nixon found himself the target of popular indignation against the US approval of dictators. In 1960 Eisenhower himself undertook a Latin American tour as part of a calculated attempt to improve relations. In the interval between the two visits the United States agreed to the creation of an inter-American Bank and an inter-American Fund for Social Development (the latter to finance unprofitable schemes which the World Bank was debarred from helping); Washington had previously frowned on these institutions. Although the eleventh inter-American conference, due to be held at Quito in 1959, was postponed, the foreign ministers of the American states met once in that year at Santiago and twice in the following year at San José. The first of these meetings was mainly devoted to denouncing Trujillo and Batista, matters on which it was not difficult to get a wide consensus. In 1960 a Venezuelan allegation that Trujillo had instigated an attempt to murder President Betancourt produced an inquiry, a condemnation of Trujillo and the eviction of the Dominican Republic from the OAS, but when the United States proposed economic sanctions and internationally supervised elections in the Dominican Republic its associates drew back for fear of setting a precedent for intervention in their own affairs (Trujillo was himself murdered in 1961). There was also a clash between the United States and the rest on whether Cuba had become decently or dangerously left-wing. The conference was on the whole anti-Cuban but was not prepared to express itself at all directly. A proposal for inter-American mediation between the United States and Cuba gained no ground. Nevertheless the Organization of American States was functioning and seen to be functioning and the discords within it were not acute, although on the perennial problem of economic co-operation an attempt by President Kubitschek of Brazil to obtain United States aid for an 'Operation Pan-America' had produced little response in Washington.

The next year, however, President Kennedy sounded a new note. Addressing himself to Latin America as no president of the United States had done since Franklin D. Roosevelt, Kennedy proposed an Alliance for Progress, massive and long-term co-operation between the United States and its Latin neighbours for the improvement, at the expense of the United States, of their economies on condition that they would also introduce certain fundamental reforms. The United States would provide $20 billion over ten years to pay, in effect, for economic and social reform. Although Latin American governments preferred bilateral to multilateral aid, and although they presumably preferred aid without strings to aid tied to reform, the Alliance for Progress was the sort of intervention which they had been seeking ever since the inauguration of the Marshall Plan for Europe, as opposed to the sort of intervention which they habitually feared or said they feared. Later in the same year, at Punta del Este in Uruguay, all the members of the OAS except Cuba subscribed a charter giving effect to Kennedy's proposals.

The Alliance for Progress was a bold psychological stroke, important for its impact upon the citizens of the United States as well as on the peoples further south, but it was imperfectly thought out in advance and the practical results had proved disappointingly meagre by the time of its author's death. The reasons for the disappointment were many. First, Kennedy's death came all too soon after his inauguration; there had been little time to get results. But, secondly, there were more material reasons. The sum pledged was large but it was questionable whether even this sum could do more than hold a deteriorating situation. It was expected moreover that government funds would attract private spending, but the volume of the private spending did not come up to expectations, and some of the government funds had to be diverted to unforeseen short-term purposes owing to a fall in Latin American revenues from primary products. Thirdly, there was some confusion about the purposes and priorities of the Alliance from Washington's point of view. Was it designed primarily to alter social structures, or to alleviate immediate poverty, or to promote a state's economic progress, or to facilitate interstate economic integration, or to combat communism? Unless priorities were established it was difficult to know where to begin or what to approve. Finally, there were misunderstandings and pitfalls. The misunderstandings transpired when the question was put whether funds were to become available immediately or only conditionally upon the introduction of social changes. The pitfalls were disclosed when the question arose of giving aid to a reactionary or unconstitutional régime. Cutting off aid meant damaging the hopes of the blameless poor. Continuing aid might mean financing backward régimes instead of encouraging progressive ones.

A second sub-continental common factor was *debt*. Early in the century many South American countries had found it easy to borrow abroad. Superficially this relationship had for a time seemed mutually satisfactory but the foreign borrowing masked the need for domestic economic restructuring and much of it was ill spent – for example on armies (as harmful to the economy and political stability as the Roman legions had been, and with no one to fight). The Second World War eliminated export markets. Postwar industries tended to produce unexportable surpluses. Domestic markets were small because impoverished and intra-America trade meagre. Hyperinflation, unmanageable debts and – in the 1990s – calls for funds for eastern Europe alienated lenders whose reluctance to help on economic grounds was reinforced by political disincentives: insurrections, a state of affairs in which half the population was officially described as poverty-stricken, and the drug trade. In some areas the drug trade, its controllers and covert and corrupt supporters, its private armies and labouring dependants, were the only flourishing part of the local economy. Its activities involved not only the zones of production in Colombia, Bolivia and Peru but also its trade routes and sources of finance in Argentina, Brazil, Venezuela, Panama, countless islands and beyond.

Economically most of South America remained in the shadow of foreign domination. Industrialization was weak or delayed for many reasons: ruling élites opposed import tariffs and quotas which infant industries needed but rich customers did not like; the several domestic markets were too small and attempts to combine them rudimentary; industries tended to excess capacity, their products to high costs; technology was either local and behind the times or foreign and therefore strange and expensive. Transport and other public services were poor or non-existent, except near big cities. Half the population was drawn into cities of 20,000 or more, overcrowded, dirty, violent and a prey to property speculators. The economic gap between town and country was widening. The land provided status, wealth and a way of life for the few; the towns failed for the most part to become engines of new economic activity. The need for land reform was manifest and a number of countries passed relevant laws, but little practical action followed except in the wake of revolutions. Workers on the land were ill paid, under-employed and growing in numbers in spite of the drift to the towns. Without credit or adequate communications or, in many areas, water, small proprietors would get neither seeds nor fertilizers nor machinery for what was economically productive land.

The 1990s saw a revival of schemes mooted in the 1960s for alleviating South America's economic weaknesses and therewith its social ills through regional association. The Andean Pact, formed in 1966 by Colombia, Bolivia, Peru and Venezuela, was reanimated in 1989 with the object of establishing a common market and customs union by 1995 and thereafter an economic union. Further south Brazil, Argentina, Uruguay and Paraguay created in 1991 Mercosur with similar aims and dates. (Similarly in Central America, Caricom, created by thirteen states in 1964, was revived with a projected economic union by 1993.) Although these plans might be dismissed as paper fantasies they were also a sign of the times in which regional co-operation, with or without political or constitutional reforms, was regarded as the way of salvation.

External concerns complicated but did not create South America's domestic problems over *forms of government* which wavered between oligarchy, tyranny and democracy. During the 1950s there was a clatter of falling dictators beginning with the fall of Juan Domingo Perón in Argentina in 1955. The following three years saw the withdrawal of Manuel Odria in Peru and the displacement of Presidents Gustavo Rojas Pinilla and Marcos Perez Jimenez of Colombia and Venezuela. The trend in these years was for the military to retire from sight. They could not abdicate real power since power inevitably was theirs. The exercise of that power in politics had become a habit partly because, with no military enemy in view, they had nothing else to do and partly because of the defects of the rich ruling classes. The officer who is not required to prepare for a war can either opt for society or for politics. Many opted for the best society and nothing more; others for society and a share in the political dominance of the upper classes; others again, having opted for politics, became dissatisfied – and, although they might despise democracy, wished to make the conduct of public affairs more efficient and less corrupt and to make a modest start with social reform. This last group came to be dubbed *nasseristas* and it was their influence which was felt in the late 1950s and 1960s and spread the notion that both national politics and national armies should become both more professional and more socially conscious. But if the first impact was the destruction of old-style dictators and a diminution of the prestige of old-style officers, it did not follow that government by civilians was to be the new rule, for the civilians might fall short, in efficiency or integrity, of what was desired of them, or exceed expectations and tolerance by going ahead too fast with social reforms. In either event the military would return.

On the one hand the personal autocracy of the *caudillo*, interested only in his own power and unencumbered with ideology of any kind, came to be represented south of the Panama isthmus only by Alfredo Stroessner of Paraguay who survived until 1989 in a backwater, a secluded fiefdom of Brazil (which got electric power from Paraguay); on the other hand one-party rule corresponded with a search for stability, a way of fending off a too liberal parliamentary system which threatened anarchy and a recourse to violence which spelled disaster. Two things had failed. Violent revolution, rural or urban, achieved no success outside Cuba. Equally thwarted were those who had hoped for the advance of the bourgeois liberalism or social democracy familiar to Europeans of the nineteenth and twentieth centuries. Developments of this kind, although espoused by a small cosmopolitan élite, had too few roots in the political soil of South America where the enduring traditions were those of pre-Napoleonic Spain. South America did not revolt against Spain. Napoleon's subjection of Spain had simply caused Spanish power in the New World to evaporate. The Liberators were therefore successors rather than liberators and they assumed, for themselves and for generations to come, the centralist powers and attitudes which the Spanish crown had exported to Spanish presidencies and vice-royalties in South and Central America. This was an autocratic tradition, often liberally interpreted, but not a democratic one. The new men of the second half of the twentieth century, most but not all of them military, were with rare exceptions not concerned to alter the system. They were concerned to take for themselves the positions of power and direction which had hitherto been monopolized by a different or more restricted ruling class.

From the early 1970s the *guerrilla activity* characteristic of much of South America in the previous decade declined. The death of Che Guevara in Bolivia in 1967 had been a blow to morale. The development of anti-guerrilla techniques under the surveillance of US officers enabled governments to defeat movements which had never been numerically strong and were only fleetingly united. Protest found different and less romantic forms. The guerrilla *foco* was replaced by urban kidnappers – the Uruguayan *tupamaros* and their like who in Argentina, Brazil and elsewhere specialized in the capture of diplomats as hostages. One of the more notable exploits of the *tupamaros* was the capture of the British ambassador in Montevideo, but although they held him for eight months they gained nothing by doing so. Urban tactics were no improvement on the *focos* and in elections in Uruguay in 1971 the Frente Amplio or Broad Left ran third to the two traditional parties of Blancos and Colorados. Juan Bordaberry, installed as president in 1972 after intricate and disputed calculations, was – like his predecessors – more concerned with the task of finding for Uruguay, once rich on its exports of wool, an economic future in a world which had invented rayon. He was unseated by the military but in 1980 the Uruguayans, with unexpected temerity, rejected a new constitution proffered to them by this régime in anticipation of a more obedient vote and from 1985 Uruguay returned to the peaceful civilian ways for which it had long been noted.

Optimists hoped that such trends would become increasingly characteristic of the closing years of the century but no trend could be counted on to persist, nor could generalizations applicable to the whole sub-continent easily be found. Two crucial elements were economic growth and institutional reform. Without them prospects for democratic rule and social justice must be hazy. In the early 1990s economic growth was evident in many countries but precarious and uneven. Most – Brazil was the principal exception – sought to handle their immediate financial troubles by privatizing state enterprises and properties in order to raise money to service foreign debts, give budgets a healthier complexion and stave off tax increases: Argentina's plan to denationalize its state oil corporation was the most extensive measure of this kind, in Chile and elsewhere similar transactions were pursued, but in Peru and Venezuela they were suspended. The obstacles were not only the wayward will of ruling classes but also the weakness of the financial and administrative institutions needed to formulate and carry through complicated economic manoeuvres.

Brazil, a country perpetually on the edge of becoming more influential than it is, the fourth or fifth largest country in the world and second in the Americas only to Canada, became an independent state as an empire which lasted from 1822 to 1889 when the second emperor was displaced by a bloodless revolution led by a professor of mathematics. It possesses enormous natural resources and a vast internal market, occupies one-third of Latin America and accounts, with Mexico and Argentina, for nearly two-thirds of its products. According to the census of 1970 its population had grown by a third in ten years and had reached more than 93 million. Thanks to its vast and varied resources it was able to make economic progress both during and after the First World War. In 1930 Getulio Dornelles Vargas, a mild liberal, was put in power by the army which was dissatisfied with the traditional conservative oligarchy. He began the modernization of Brazil. He strengthened the central government, nurtured industry and introduced state economic planning. He became both more autocratic and more popular. The army grew suspicious of him on both counts. It feared his new power base in the poorer classes and in 1945 he was forced out of office. But he was returned to power in 1950 by popular vote and with the support of a new type of younger army officer. The army, hoping to control him and reluctant blatantly to flout a fair election, allowed him to take office but by 1954 it had had enough of him and was about to evict him – on the grounds of incompetence – when he committed suicide.

The next decade saw the running down of the Vargas era under three presidents. The election of 1955 was won by Juscelino Kubitschek with Julio Goulart as his vice-president. The armed forces, divided in their attitudes towards this result, interposed no objection and indeed acted to allow the successful pair to take over. They thereby demonstrated the new president's dependence upon them, and he in return tacitly acknowledged the relationship by raising pay and providing funds for more and better military equipment. Some officers, like their counterparts in other countries, wanted more than this and hoped for a mildly reforming administration which would attack corruption, incompetence and the grosser manifestations of social injustice, but they were not willing to endorse any very radical measures nor – in the Brazilian case – did they have the satisfaction of observing a good administration at work, for President Kubitschek embarked on extravagant enterprises (such as the building of the new capital, Brasilia) and in the course of an energetically misguided administration opened up huge new opportunities for private speculation and peculation.

At the next election, in 1961, Jânio Quadros, with Goulart still as vice-president, succeeded, but he resigned after seven months leaving Goulart to take his place under the eyes of a divided and increasingly dubious army and a solidly conservative navy and air force. In these circumstances President Goulart's powers were strictly limited and he quickly reached the limits. Where Quadros had preferred to quit, Goulart chose to forge ahead; he rode for a fall. He proposed a wide extension of the franchise, land reform and a neutralist foreign policy. He appealed to the people against the armed forces and congress, accepted communist support at home, established relations with the USSR and opposed the eviction of Castro's Cuba from the OAS (but joined the United States naval blockade of Cuba during the missile crisis). These measures won him some popularity but also evoked fear and hatred, enabled communists to infiltrate into central and state administrations and trade unions and, as different provinces lined up on different sides, produced a threat of civil war, at which point the army intervened.

Having evicted Goulart the army leaders forced the congress, after a pause in which constitutional propriety was observed by recognizing the president of the chamber of deputies as president, to instal General Humberto Castelo Branco as president for the remainder of Goulart's term. The general was succeeded by Marshal Costa e Silva and he by Generals Garrastazu Medici, Ernesto Geisel and João Batista de Oliveira Figuieredo. Political parties were banned with the exception of two new ones of no independent vitality. The régime turned a hard face towards trade unions, peasants and students. It made little progress with land reform and halted literacy programmes which seemed to it dangerous: Brazil remained a country of unused land and unused talent. Faced with growing opposition it resorted to strong-arm methods, including wide-spread and appalling cruelty to the political prisoners who were thrown in increasing numbers into the jails.

On the other hand the military régime claimed credit for its handling of the economy. Having inherited galloping inflation it not only reversed the trends of the lean 1960s but achieved for a while (1968–74) a growth rate of 10 per cent a year, much in excess of anything seen in the previous twenty years. The main instrument was exports, which were quadrupled, assisted by heavy borrowing from foreigners impressed by Brazil's natural resources and disciplinary government. But 1974 was a year of reckoning in Brazil as elsewhere in the world. The healthy world markets which had taken Brazil's exports became unhealthy. Internal developments financed from abroad slowed down and Brazil turned to import controls and import substitution. The beneficiaries of the boom began to grumble, its victims to assert themselves. The latter were numerous. In spite of the boom nearly half the population was worse off than before – 40 per cent were sharing less than 10 per cent of the national income. Regional as well as class tensions were aggravated, as the under-privileged north-east waxed indignant about the fat cats in the south. With the advent in 1979 of President Figuieredo the government set about the dual task of restructuring the economy and at the same time humanizing the régime. Censorship was greatly relaxed, torture probably diminished and a wide, though not total, amnesty decreed for political prisoners. In the economic sphere special effort was put into the search for alternatives to foreign oil: prospecting for offshore oil (which proved disappointing), the search for substitutes and a nuclear programme. The last raised the usual questions about whether Brazil had it in mind to develop nuclear military power as well as a nuclear-powered economy.

Brazil in the 1970s was not the leader in nuclear development in South America. Its old adversary Argentina was ahead. Brazil had not signed the Nuclear Non-Proliferation Treaty but had signed, with reservations, its regional equivalent, the Treaty of Tlatelolco (1967). Argentina signed neither. (Mexico, also in the nuclear race, signed both.) In 1975 Brazil, after a rebuff from the United States, turned to West Germany for help in accelerating its nuclear programme. In return for part-payment in Brazilian uranium, Bonn agreed to supply Brazil with eight reactors, a fuel-processing plant and uranium enrichment technology. Brazil's aim was to have no fewer than seventy-three reactors by the end of the century. Although the main purposes of this programme were industrial and civilian, the processing plant and the technology to be supplied by West Germany pointed to a military capability. The safeguards embodied in the agreement were stringent but, in the nature of the case, less than total. Brazilian and Argentinian programmes created the possibility of a nuclear arms race in South America similar to those between India and Pakistan or Israel and Iraq.

But Brazil's attitude to its traditional opponent was not bellicose. Brazil's great size, undeveloped resources and comparatively stable government allowed Brazilians to look forward to the day when one of South America's classic feuds would be ended for ever by Brazil's irreducible superiority. Questions remained about the buffer lands of Paraguay and Uruguay but so far as Brazil was concerned these were matters which could be confronted with assurance and left to time.

Brazil's size made it a giant among Latin American states. Its potential wealth made it a potentially mighty giant. Its Portuguese background differentiated it from its Spanish-speaking neighbours and gave it a more cosmopolitan flavour which was accentuated by its awareness of its African origins. Brazil was both a part of South America and turned its back on South America. It was no less detached from North America, determined to assert its independence from the United States and not to play second fiddle to Washington, even when the two countries shared conservative and capitalist outlooks. It sought closer links with the rising stars of Africa and Asia – Nigeria and Japan – and also with Iraq, another country with well-grounded ambitions and the supplier for the time being of the greater part of Brazil's imported fuels. Its weak points lay in the excessive growth of its population and its foreign debts, conducing the one to social, the other to economic nightmares, to grave unemployment and unrest and to inflation. As the dreams of the 1970s turned thin, congressional and provincial elections in 1982 gave the government unaccustomed reverses; the IMF insisted in 1983 on painfully classic measures as conditions for its help; and in 1985 the military deemed it prudent to take a back seat after twenty-three years of political power. Tancredo Neves was elected president but immediately collapsed and died, leaving a little-known deputy and successor, José Sarney, to cope with daunting economic and social problems. Sarney abandoned the monetarist recipe for economic rehabilitation, partly because its purely economic dictates were bound to foment misery and likely to create a revolution, partly because Brazil's inherent strengths enabled it – or seemed to enable it – to negotiate with its creditors for relief and with the IMF for funds to support a moderately expansionist economic plan. In spite of its huge debts Brazil was able to pay the interest on them punctually. It was also able to imply that without outside aid the Sarney régime would founder, leaving the creditors without hope of interest or the repayment of their capital and confronting the world with the abhorrent possibility of either a reversion to military rule or a régime much to the left of Sarney's. But Sarney lacked political anchorage. His government was a coalition in which the largest partner, the Brazilian Democratic Movement, gave the president full support only when he least needed it. The Movement scored an overwhelming victory in federal and provincial elections in 1986 and then, with the introduction of painful economic measures, bickered with Sarney. These measures excited opposition without stemming inflation which was running at 200 per cent a year and growing to astronomic four-figure proportions. Elections in 1989 gave a clear majority to Ferdinando Collor de Mello who, at the age of 40, won nearly twice as many votes as his main rival. Collor, governor of the north-eastern province of the country, made the Sarney régime look boring and incompetent, was fortunate in having to contest the second round of the election with the extremer candidate of the left and had spent on his campaign a great deal of money. He introduced severe measures which curbed inflation at the price of curbing everything else too. Bank accounts were frozen for eighteen months, thus removing about $80 billion from the economy; a new currency was introduced; public services were drastically reduced; tax changes shifted money from the private to the public sector. Unemployment soared, businesses collapsed and the weaker classes were driven to penury and despair. And Collor turned out to be outrageously corrupt. After he and his cronies had stolen about $1 billion he was suspended and impeached and replaced by vice-president Itamar Franco pending elections in 1994. In the interim the parties of the right recovered from these discreditable upheavals sufficiently for Fernando Cardoso (of the Social Democratic Party, a minor component of the broad right) to win largely on the basis of his successful anti-inflation measures as minister of finance. In the same year the Supreme Court acquitted Collor of the charges against him by the narrowest possible margin.

Argentina, the second in extent of South America's major states, achieved its independence from Spain at the cost of losing territories which formed the new states of Uruguay, Paraguay and Bolivia. It was only gradually and painfully unified but from the latter part of the nineteenth century it prospered with the development of its lands, largely by immigrants from Europe. The government was in the hands of upper-class conservative and moderately radical politicians who competed for power and sporadically permitted an admixture of democracy, especially between 1916 and 1930. The country's population, its agriculture and its railways expanded steadily and with the introduction of refrigeration it became one of the world's major exporting countries before the First World War; but it lacked minerals and the capacity to develop industry alongside agriculture and commerce. It was severely hit by the slump of 1929 and suffered, partly in consequence, a revolt in 1930 in which army officers (including a 35-year-old Juan Domingo Perón) had the support of the possessing and the dispossessed classes. This event was followed by a period of autocratic conservative rule, buttressed by the army and by renewed prosperity. At the same time urbanization and immigration were producing a more politically conscious and socialist working class.

During the war the presidency was held by the ultra-conservative Ramón Castillo, first as acting president from 1940 to 1942 and then as president for a further year. Castillo refused to break off relations with the Axis powers, partly because he had fascist leanings himself and partly in despite of the United States, towards which Argentina's governing classes were traditionally hostile. In 1943 the army again intervened in politics, this time in order to remove the reactionary régime of landowners whose tenure of power since 1930 had proved that they had learnt nothing. The revolution was more than a transfer of power. Its leaders had a programme, or at least a series of general aims, which epitomized powerful desires in one section or another of Argentine society and mentality. One of these aims was Argentina for the Argentinians: a rejection of the role of foreigners in the Argentine economy who promoted useful enterprises but took all the profits away, and a rejection too of the Argentine ruling class itself which aped foreign, mainly French, fashions and spent half its time in Paris. Another aim, going beyond the assertion of Argentine pre-eminence in the sub-continent, was to make Argentina not only an independent power in the world but a great one whose voice would be heard and heeded in international affairs generally. And thirdly, the new men – or many of them, including Perón – wanted social justice at home, by which they meant a less blatant maldistribution of wealth and a less extreme concentration of power in the capital at the expense of the provinces.

A series of generals occupied the presidency in the next two years. In 1944 Argentina broke off relations with Germany and Japan, less out of a change of heart than from a well-calculated need to stand well with Washington as the war in Europe approached its end. In the next year Argentina declared war. (All the independent states of South and Central America and the Caribbean declared war on Germany and its associates between 1941 and 1945. Argentina was the last to do so.) During this period Perón held a number of official positions, developed good relations with union leaders and great popularity with the *descaminados* (the shirtless or destitute), but earned the jealousy of many of his brother officers. Dismissed in 1945 he created a new political party which won elections in 1946. Perón became president partly through this expression of the popular will but no less through the wave of extra-parliamentary enthusiasm displayed by the crowds of *descaminados* who invaded the political arena. Throughout this contest Perón was openly and vituperatively opposed by the US ambassador.

Perón ruled for nine and a half years which were crammed with legislative and other measures designed, mainly by authoritarian methods, to turn Argentina into a modern and just country. Banks and other enterprises were nationalized, foreign concerns were bought out with wartime profits, public services and popular education were expanded, industrialization was accelerated, the centralized buying and selling of agricultural products was introduced in order to cushion the effects of price fluctuations. But the pace was too hot for the economy and for the propertied classes, and Perón also gave grave offence to the Church. Although the programme was trimmed in the later years Perón made too many mistakes and too many enemies, and in June 1955 the navy and the air force, with the Church in the background, tried to bomb him out of the presidency. The army, however, not only remained loyal to him during the abortive coup but did not feel obliged to replace him by any other leader after it. Perón dismissed some of his ministers who were most obnoxious to the conservatives and set about safeguarding his position by organizing and arming the *descaminados*. This latter move alarmed the army which turned against him and, in September 1955, removed him by a coup which placed another general in his place.

The oligarchy had closed ranks in order to put an end to economic policies which were harming the country and social reforms which were harming themselves. They had the support of intellectuals who, although they might approve Perón's social aims, disapproved of his authoritarian methods, particularly censorship; for Perón's advocacy of social reform did not include freedom of expression. The fall of Perón halted the reform movement and, as in Brazil, raised the hopeless question of what to do when the need for reform is urgent but the readiness of the powers that be for change is restricted. The supporters of Perón – and of the vivacious Eva Duarte, whom Perón had married just before his triumph in 1942 but who had died of cancer in 1952 – did not disappear. They represented a force which, if it could not be wooed away to some other movement, must either be disfranchised or be allowed to resume the *peronista* course.

Meanwhile the army ruled. It first installed General Eduardo Lonardi in the presidency, in which he was later succeeded by General Pedro Aramburu pending elections in 1958. In that year Arturo Frondizi, a representative middle-class politician and leader of the Radical Civic Union, became president with army consent. Perón's followers remained numerous and the Radicals split over political alliance with them, Frondizi opting for an alliance through which he hoped to seduce *peronistas* to his party. This gamble failed conspicuously. The persistence of *peronista* feeling was demonstrated at elections in 1960 when, with the *peronistas* barred from putting up candidates of their own, a million abstentions were recorded. In 1962 the *peronistas* were allowed to vote once more. Officers of the three services, angered by Frondizi's miscalculation, discarded him and took counsel among themselves on the basis that the first article of government was the continued exclusion of Perón (in exile in Spain) and his like. The navy, Perón's stoutest opponents, insisted on the simple course of not having elections, and they were supported by a section of army officers known as the *gorillas*. Other army officers, however, backed by the air force, preferred to revert to the system of holding elections as required by the constitution but with the *peronistas* disfranchised. Fighting ensued and the latter group won. New elections were held but could not disguise *peronista* strength which could always be manifested either by pro-*peronista* votes or by abstentions. In the next year the moderate army group made Arturo Illía president. The armed forces had refused to accept the electoral verdict of 1962 and had failed to find a constitutional way of excluding Perón. They had therefore resorted to unconstitutional manoeuvres and had in the process further exposed the divisions in their own ranks. In 1966 they removed Illía and substituted Juan Onganía. In 1970 they removed Onganía and substituted Roberto Levingston who was removed in his turn in 1971 to make way for General Alejandro Lanusse. None of these governments could master inflation or keep order. Disorder increasingly took the form of kidnapping for publicity and ransom – for example, the seizure and murder by a Trotskyist group of Fiat's director-general in Argentina.

In 1972 Perón, now 77 years old, returned amid speculation and apprehension. His arrival was undramatic and his stay short. He endorsed the candidacy of Hector Cámpora in the approaching presidential election and went back to Spain after a few weeks. Cámpora won the election in 1973 with 49.6 per cent of the vote and his party – the Frente Justicialista, alias *peronista* – won twenty out of twenty-two governorships and control of both parliamentary chambers. But Cámpora was unable to control the violence and kidnapping by groups which claimed a share of the Peronist mantle but were disavowed by the Frente, and after a brief interval Perón reappeared, Cámpora resigned, the Frente nominated Perón and his second wife for the presidency and vice-presidency and they received 61.8 per cent of the vote. Perón introduced heavy taxes, bearing specially on the rich but also – in the case of VAT at 16 per cent – on other classes too. He died in July 1974, leaving his widow to face the problems of inflation and public order, in dealing with which she was no more successful than her predecessor. She and her government were removed in 1976 by the army which put General Jorge Rafael Videla in her place but could think of nothing more useful to do than imprison and kill its critics.

The military régime, which continued under the presidencies of Generals Roberto Eduardo Viola and Leopoldo Galtieri, lasted until 1982 when it was undone by the humiliating outcome of the *Falklands War*. Under its aegis and by its agents some 30,000 Argentinians were tortured and killed, many of them being dumped alive into the sea out of aircraft. Inflation exceeded 250 per cent a year, the foreign debt was sextupled, production and industry collapsed. Galtieri hoped to redeem this record by capturing the Falkland Islands on the 150th anniversary of their occupation by Europeans. After talks at the UN early in 1982 which perpetuated Anglo–Argentinian stalemate, Galtieri sent a civilian unit, under naval protection, to South Georgia in the Antarctic to dismantle a disused British whaling station and then proceeded to invade the Falklands. The Argentinian expedition, which landed on 1 April, was immediately successful and Galtieri himself visited the capital, Port Stanley, three weeks later. But the British despatched a force to regain possession of South Georgia and then attacked the Argentinians on the Falklands in a war for which Argentina was militarily and psychologically unprepared.

Over the ownership of the Falklands there has been a long dispute which neither bilateral diplomacy nor international machinery had been able to dispel. The islands were discovered by John Davis at the end of the sixteenth century and named over a century later after Lucius Cary, Viscount Falkland, one of the most illustrious victims of the English Civil War. The French colonized East Falkland in 1764, the English West Falkland in 1765. The Spanish bought the former and seized the latter a few years later but then yielded West Falkland to the British who promptly abandoned it. The Spanish subsequently abandoned East Falkland. In 1829 a new colony was established in the name of the United Provinces of Rio de la Plata, successors in title to the Spanish empire in South America and predecessors of the Argentina republic. Britain registered a verbal protest. It had established in 1820 a symbolic presence which was destroyed in 1833 by the United States, whereupon the British came back, evicted the few remaining Argentinians and established a colony embracing both East and West Falkland. Argentina persistently challenged the British title which Britain offered on several occasions to submit to the International Court of Justice, an offer always rejected. Argentina also claimed sovereignty over South Georgia, the South Orkneys and other territory in the South Pacific and Antarctica. This claim clashed with Chilean as well as British claims, of which the latter dated from early in the twentieth century.

In 1979 the British Foreign Office, uneasily but not urgently alive to the dangers, elaborated a plan for relinquishing sovereignty over the Falklands and leasing them back. This plan was presented to a committee of the cabinet which gave it scant attention although the approach of the 150th anniversary of the British reoccupation in 1833 was a likely source of trouble: a full-scale invasion was not envisaged even after General Galtieri's promotion to the presidency at the end of 1981 was followed by rumours and suspicious acts. After bilateral ministerial talks in New York at a relatively low level in the following February the Argentine government issued a brusque declaration which the talks' pallid conclusions did not seem to warrant, and in the next month a party of Argentinian scrap dealers was sent to South Georgia under naval protection to dismantle an abandoned British station. This probe was not unconnected with the well-advertised withdrawal in the previous year, on grounds of expense and against the advice of the Foreign Office, of the British survey vessel *Endurance* – a symbol of the British presence whose removal led Argentinians to believe that Britain's position throughout the South Pacific was no longer greatly prized in London. Two weeks after the arrival of the scrap dealers in South Georgia Argentina seized the Falklands, undeterred by last-minute attempts by the United States to prevent this blatantly illegal act of aggression. President Galtieri was able to make a triumphant visit to Port Stanley, the islands' capital.

The British government's failure to assess Argentinian intentions until too late left it with few choices. These were negotiation from a position of weakness or war. Ostensibly it professed to be using force to secure negotiation but once a naval armada had been assembled and despatched there was little prospect of anything except war:

peace-makers were up against a time limit and entrenched national obduracies. The British government may have toyed with the notion of a deal whereby – provided the invading Argentinian force were voluntarily removed – Britain would reintroduce its rule for a short time only and then transfer sovereignty to Argentina; but the dominant British resolve was not to treat with the enemy but to beat him, and in this mood negotiation – and all third-party negotiators – tended to appear as impediments. The outcome upon which the British government was resolved was a clear-cut victory, and negotiation is apt to produce something much less satisfying. The British expedition was sent into action swiftly but only precariously protected – its air cover was inferior to that provided for any British squadron since the *Prince of Wales* and *Repulse* had been sunk by the Japanese in 1941. It won a resounding victory, but by a whisker and not without grievous loss. It recovered the Falklands, caused the fall of a hateful dictatorship and saddled Britain with the defence of the islands for an indefinite future and at greatly increased cost.

In the aftermath of this victory one episode stood out. This was the sinking of the Argentinian cruiser *General Belgrano* on 2 May. The reasons were various: the loss of life (368 dead); the circumstances in which the action was authorized; the consequent suspicion, in Britain, that the specially constituted war cabinet had not exercised proper control over naval operations; and the allegation, again in Britain, that the torpedoing of the *Belgrano* was an item in the determination to torpedo current peace negotiations. The last two charges appear to be mutually incompatible. The ensuing controversy was of little interest outside Britain.

The attempts to keep the peace while the British force was sailing south consisted, first, of some hectic diplomacy by the United States secretary of state, Alexander Haig, whose main aim was to persuade President Galtieri (and his associates in the military junta) that the British force was no mere bluff and would recover the Falklands by force if no agreement was reached between the adversaries before the expedition reached the islands. This attempt had failed by 29 April and the chief consequences of the failure were two: the United States, which had failed to condemn Argentinian aggression as the UN Charter required, abandoned its neutrality and came out in Britain's favour; and secondly, the war cabinet in London abated its opposition to the Royal Navy's pleas for more hostile action. For professional but also political reasons (the economic axe was threatening the navy more severely than the other services) naval chiefs wished to demonstrate their unique value to the nation. Proposals for an attack on South Georgia having been rejected, the navy wanted to bag a scalp in the shape of a major Argentinian warship, preferably the aircraft carrier

25 de Mayo or alternatively the *Belgrano*. The collapse of the first stage of Haig's diplomacy cleared the way and on 2 May the war cabinet, at a strangely slapdash meeting at Chequers, changed the rules of engagement laid down earlier and so authorized an attack on either vessel. By the prevailing rules Britain had publicly given notice that it would engage and sink any Argentinian vessel within a twelve-mile 'exclusion' zone round the Falklands. On 2 May this rule was changed to permit action anywhere outside Argentinian territorial waters although, unlike the defining of the exclusion zone, the new rule was not made public until 7 May.

The sinking of an enemy ship in war is not normally an event to cause surprise. The sinking of the *Belgrano* became controversial on two grounds: first, because it was allegedly unnecessary, and, second, because it destroyed the second of two attempts to preserve the peace by Peruvian diplomatic intervention.

The case for the sinking of the *Belgrano* was that it posed a threat so long as it was at sea and regardless of its position or course. The case against the sinking was that the *Belgrano* was known to have changed course for port, was not – as the defence secretary and the prime minister told the House of Commons and the prime minister repeated a year later – closing rapidly on the British force, and had ceased to pose any threat. (In a strange *volte face* seven months later the chief of defence staff said that the only threat came from the *25 de Mayo*.) By a stream of contradictory and sometimes false statements British ministers dug pitfalls for themselves, revealing a disconcerting ignorance of important details and some tardiness in correcting mistakes after the true facts had become available.

The second matter of controversy was the connection, if any, between the sinking of the *Belgrano* and the proposals for peace which were being canvassed by the UN's secretary-general and the government of Peru. Alexander Haig subsequently wrote that both he and the Peruvian president, Fernando Belaunde Terry, had received from both belligerents general approval of these proposals before the sinking of the *Belgrano* and were working on the details, all of which were known to British officials. The British government denied this account. The British foreign secretary, Francis Pym, flew to Washington on 1 May and reported to London on the state of negotiations late on 2 May, after the sinking. That negotiations were in progress was known to all. That the sinking of the *Belgrano* put an end to them is sufficiently evident. That the British war cabinet instigated the sinking as a means to abort them is an imputation supported by no evidence. Thatcher, determined on the arbitrament not of politics but battle, was hostile to the UN/Peruvian initiative but chose to leave Galtieri to block it, which he did. Galtieri misjudged the delicate balance in Washington between pro-British and pro-Argentinian sympathizers. He also underestimated the crucial importance of the naval intelligence which Regan was able to pass to Thatcher.

Three days after final British victory in June Galtieri was dismissed. Civilian rule was restored after a short interval and the ensuing elections were won by the Radical Party, whose leader Raoul Alfonsin became president with the tasks of repairing the economy, investigating the disappearance of thousands of victims of the military régime, keeping the military in order, and repairing relations with Britain without abandoning Argentina's claim to the Falklands.

Alfonsin had decency but little else in his favour in politics. He was distrusted by the military whose overweening style survived the Falklands fiasco; by the Church, by the unions and eventually by the populace which had acclaimed him but turned away from him when prices began to double every month. He survived scattered military mutinies at the cost of giving in to demands to abandon trials of officers accused of enormities against civilians during the military régime. At mid-term elections a revived Peronist party, named the Justicalista party, won sixteen out of twenty state governorships and in the presidential election of 1989 the Peronist candidate Carlos Menem, a flamboyant campaigner of Syrian extraction, easily won the popular vote. During his campaign Menem praised Castro and Stroesser indiscriminately; deployed the familiar Peronist rhetorical style; made extravagant promises to the poor but seemed to have no idea how to redeem these promises; and was fierce in his resolve to recover the Falklands. The economic state of the country on the morrow of the election was so alarming that Alfonsin was persuaded to resign in order that Menem might be inducted in June instead of November. He quickly authorized exploratory talks for the resumption of diplomatic and commercial relations with Great Britain (which proceeded slowly in the face of British lack of enthusiasm) and embarked on an economic programme similar to that of President Collor in Brazil. He ditched controls over prices and wages; the former rose steeply and the latter were halved. Businesses and employment collapsed. So too did the currency. Retailers took to pricing their goods in US dollars whose value quintupled in two months. Hyperinflation was reduced to inflation at about 60 per cent a year but then rebounded into the thousands per cent. Menem tried to sell off state enterprises but had difficulty in finding buyers. His government could raise no money by issuing bonds since nobody would or could buy the bonds. He reduced the armed forces by a third and gave senior officers pardons for their crimes against human rights. Mid-term elections in 1993 for half the seats in the lower house left the *peronistas* without a majority and Menem with the task of amending the constitution to enable him to run for a second successive term in 1995 as the man who had killed inflation, made the rich richer without making the poor either poorer or richer, and shown an unnerving ability to change his policies to fit his circumstances at the shortest notice. Relations with Britain took a turn for the better with an agreement over oil which was believed to exist superabundantly round the Falklands. The two governments agreed to divide the proceeds of exploitation in one area 50/50 and in another 75/25 in Britain's favour. On implications for sovereignty both governments expressed themselves decisively, openly and in diametrically opposite senses.

Chile's history in the nineteenth century had been one of progress only briefly interrupted by civil war in 1891 and based on rich mineral resources and successful wars against Peru and Bolivia. In the twentieth century Chile had its share of political instability and inflation, its difficulties being accentuated by competition from artificial nitrates. The landed oligarchy lost its hold on power after the revolt of 1891 and thereafter Chile moved painfully towards a more democratic order. In 1938 elections gave power to a popular front which included communists and was not debarred from taking office by the armed forces. A period of orderly civilian government, conservative rather than radical, followed. In 1964 two progressive forces competed for the succession to President Jorge Alessandri – a popular front led by the socialist Salvador Allende and a Christian Democrat Party, an offshoot of the conservatives founded in 1941 and led by Eduardo Frei Montalva. The latter won and established a progressive, anti-communist government, overtly akin to the Christian Democrat parties of Europe but ostensibly more radical.

By the end of his term Frei had to his credit a considerable advance in education, a noticeable drop in illiteracy, some industrial development and new housing, and the introduction of graduated taxes. But inflation was not mastered, wages remained low, the number of landless peasants high – so high that they began to occupy lands, from which many were evicted by force and some killed. Conscious of the size of his economic problem Frei tried to avoid a direct clash with the oligarchy or the United States, but his gradualism offended on the one hand the more militant elements and on the other the more chauvinist. In 1970, in a three-cornered presidential election, Frei's designated successor, Radomiro Tomic, came bottom of the poll and amid general surprise Allende narrowly defeated the conservative Alessandri.

Allende's coalition had in broad terms a clear programme but it had no clear popular or parliamentary majority, its six constituent parties quickly became an uneasy team and it had the support of only some of the country's senior military figures. The coalition's broad aim was to create a socialist state on an established democratic base; the means were to be state control over foreign and domestic capitalist power centres, extensive nationalization, land reform leading to higher agricultural productivity, redistribution of wealth through tax reforms, full employment and an acceleration of Frei's educational and social programmes. These reforms, particularly the redistribution of wealth to create new purchasing power and the stemming of the outflow of profits to the United States and other foreign parts, would generate the necessary finance – an imprecise and optimistic calculation. Of the principal government parties the oldest was the Radical Party, a typical nineteenth-century, progressive, anti-clerical party which had emerged from the possessing classes (landed and mine-owning) and was in the throes of debating whether it was Marxist or not. The Chilean Communist Party was the largest and best organized communist party in South America and was predisposed to co-operation with other parties in a Popular Front.

Allende's Socialist Party had started as a diverse group of intellectuals and had succeeded in attracting votes from small businesses and the teaching and other professions but was divided on the legitimacy of the use of force. Allende had some of the talents of the conjuror but after a promising start they failed him. In his first year the price of copper went up, temporarily; wages rose while prices remained stable, also temporarily; inflation was held back, again temporarily. Expropriation of US copper companies and domestic banks was popular, as was land reform until it impinged on medium as well as big proprietors. By 1972, however, evident economic strains caused rifts in the government between those who wanted to move faster and those who urged a touch on the brakes. Allende tried but failed to bargain for support from the Christian Democrats. He persuaded senior officers to join his cabinet but without conciliating enough of them. By 1973, when he was more than half-way through his term of office, his economic difficulties became critical with a lurch into serious inflation and food shortages. He also alienated many by a style of living out of keeping not only with socialism but also with a usage observed by Frei, Alessandri and other presidents who had followed a Chilean tradition of living as simply in office as out of it. The parliament refused to vote tax increases or funds for subsidies or welfare, and the government's alternative – printing the money – made things worse. Restlessness in the navy and army culminated in a coup in which General Augusto Pinochet Ugarte seized power with US help. Allende committed suicide. Pinochet consolidated his success in a referendum in 1980 which gave him the presidency for a further eight years from 1981. His régime restored economic order by putting Allende's measures into reverse. It also earned for Chile international opprobrium for its fearful barbarity.

Churchmen became uneasy about their association with an appalling régime; military and police chiefs became uneasy about keeping order against rising indignation and outrage; businessmen wondered whether they might not do better without Pinochet than with him. The threat to Pinochet came from these groups which wanted Pinochet's Chile without Pinochet.

In 1988 Pinochet failed to obtain a renewal of his rule beyond its term in the following year. Some twenty parties entered the lists in the ensuing elections but the effective choice lay between going back to Frei or back to Alessandri. The centre and left united round Patricio Aylwin Azocar, himself a man of the centre-right, and defeated a splintered right. Pinochet's long reign, the longest in Chile's history, came to an end. Aylwin, who assumed office early in 1990, inherited a foreign debt of $16 billion, a decline in the price of copper which accounted for half of Chile's foreign earnings, and discontent among a population which had seen the gap between rich and poor widen over Pinochet's sixteen years in power and wages halved. But Pinochet, an ardent modernizer not unlike the shah in Iran, also bequeathed to his successor an economic legacy not wholly bad and Aylwin was able to hold unemployment at politically tolerable levels while achieving an annual growth rate of 5 per cent or more. The attitude of the army remained, however, ambiguous and the drift into cities foreshadowed continued instability. At elections in 1993, Frei's son defeated Alessandri for the succession to Aylwin.

One item which Aylwin did not inherit was Chile's ancient quarrel with Argentina at the southern tip of the American continent where the Beagle Channel leads from the Atlantic Ocean to the Pacific. The question between them was where the one ocean begins and the other ends. Their dispute concerned islands lying at the eastern, or Atlantic, end of the channel. In 1977 an arbitral award gave these islands to Chile but Argentina refused to accept the award and staged naval and air demonstrations, to which Chile responded in kind. In 1979 both sides agreed not to prosecute their claims by force and to accept the mediation of the Vatican. In 1980 Chile again sent armed forces to the disputed area. The quarrel was settled in 1984, substantially in Chile's favour.

The three major states of South America, while occupying almost the whole of the south and east of the sub-continent, left room for seven other successors of the Spanish empire (joined in 1966 by the new state of Guyana – see the note at the end of this Part). In the north-west, where Bolivar had hoped to create a single grand Colombia, three states resulted: Colombia and its two offshoots, Venezuela and Ecuador. Colombia and Venezuela possessed exceptional wealth in two very different forms, cocaine and oil. To the south Peru and Bolivia (formerly Upper Peru) also possessed considerable mineral wealth: Bolivia lost its coastal regions to Chile in the nineteenth century.

In *Venezuela* the ruling establishment was displaced in 1945 by a coup from within itself: an alliance of army officers frustrated by the professional stagnation of their class with the mildly radical Accion Democratica led by Romulo Betancourt. Its slogans were modernization and change. It lasted three years. It achieved an initial measure of land reform but its education policies gave offence to the Church which regarded education as an ecclesiastical preserve and it was divided over economic policy: its more cautious members preferred to modernize agriculture rather than encourage industries which would create a politically significant urban proletariat. It was overthrown in 1948 by a counter-coup led by General Carlos Delgado Chalbaud which inaugurated a military dictatorship which lasted for ten years, proscribed political parties and trade unions, suppressed the free press and sponsored grandiose building programmes while leaving the gap between rich and poor to widen. The most effective of the dictators of this period was Marcos Peres Jimenez (1952–58) who benefited from increasing oil revenues but failed to cope with recession in the late 1950s. A military revolt which failed was followed by strikes, fighting, the flight of the president, a transitional régime under Admiral Wolfgang Llarazabal and elections which brought the Accion Democratica and Betancourt back to power. There followed a period of civilian rule under Betancourt and five successors from, in turn, Accion Democratica and its chief rival COPEI, the Venezuelan equivalent of Christian Democracy. All these presidents were peacefully elected and served five-year terms. This equable régime was interrupted when Peres returned to power but reverted to autocratic rule under the cloak of democracy, became embroiled in corruption, was impeached and forced to resign in disgrace. After two interim presidencies, ex-president Rafael Caldera, who had fallen out with his party COPEI and formed a new one, broke the mould by defeating in 1993 candidates from both COPEI and Accion Democratica.

Venezuela's mainly civilian governments secured political stability. They enjoyed considerable revenues from oil, iron and other natural resources; kept the loyalty of the army and – if sometimes grudgingly – the business community; reached a compromise with the Church over education; was required to cater for a comparatively homogeneous and small (12 million) population; but produced disappointingly meagre social changes, did little for labour on the land, created new jobs in manufacturing and service industries which went largely to skilled foreigners, and fell victim to the familiar boom-to-bust sequence characteristic of states whose fortunes were linked to the international price of oil. In a single decade in the 1970s this price rose tenfold. Oil was nationalized (a few years ahead of the end of existing concessions to foreign companies) but economists were divided over whether to use oil wealth to diversify the economy or to invest ever more heavily in oil and petrochemicals. In spite of economic growth at rates of 10–15 per cent, huge foreign debts were incurred and large enterprises were begun which depended for their completion on something over which Venezuela had no control. The oil price rebounded and domestic prices soared amid complaints of inefficiency in the nationalized businesses, wastefulness and corruption in the highest places, neglect of education and health and persistent poverty. Popular discontent reached the level of riots in 1989. There was an increase in guerrilla dissidence but attempts by an amalgam of communists, non-communists and disgruntled officers to make common cause under the banner of a Liberation Army came to grief: the communists in particular being sceptical of the value of guerrilla activities.

In external affairs successive presidents tried unsuccessfully to resolve border disputes with Colombia and Guyana and then played them down. Betancourt and his first successors made a point of alliance with democratic régimes against dictatorships but in the expansionist and ambitious 1970s Peres, who restored diplomatic relations with Cuba in 1974, cast Venezuela for a leading role in the South American–Caribbean region and the Third World. It had been a founder member of OPEC in 1960 and joined LAFTA (Latin American Free Trade Area) six years later. Less happily, it was becoming by the 1990s involved in the world of the international drug trade whose centre was in neighbouring Colombia.

Before the Second World War *Colombia* enjoyed a reputation for political stability and measured economic progress. The latter was based on coffee, cocoa, a variety of manufacturing enterprises and a manageable foreign debt. The ruling class was divided into two political parties, conservatives and liberals, and shared power on a roster system until menaced by left-wing insurgents and the dictatorship of Rojas Pinilla (1953–58). The 1940s and 1950s were a time of ferocious violence, *la violencia*; in 1949 the province of Marquetalia, not many kilometres south-west of Bogota, declared itself independent and was recovered in the 1960s only with the help of the United States. The Cuban revolution led to hopes of aid for the flagging revolutionaries, but these hopes were falsified and the Colombian revolutionaries fared only marginally better than their Venezuelan comrades. At elections in 1974 the liberal candidate Julio Cesar Turbay Ayala beat the conservative and an assortment of candidates of the left. He was succeeded by Alfonso Lopez, the first president of the Frente Nacional, an alliance of liberals and conservatives formed to defeat both the *violencia* and military dictatorship. In 1978 the conservative Belisario Betancur beat a divided opposition but was reduced by further elections in 1983 to governing with a hostile parliament. Facing the Caribbean and Central America as well as the Pacific and South America, Colombia was drawn into the politics of the area to its north and drifted into measured criticism of Reagan's policies in and around Nicaragua. By formal agreements with rebel forces Betancur put an end to civil wars which had been going on for thirty-six years but his pacific policies did not command universal approval and a spectacular attack on the Supreme Court, in which many were killed before the attackers were forced to surrender or kill themselves, was variously interpreted as a last act of desperation by the insurrectionary M-19 movement or the recrudescence of civil war on a fiercer scale.

A more serious threat appeared in a different quarter. The trade in cocaine became Colombia's most successful enterprise and its chiefs established themselves not only as men of enormous wealth ostentatiously displayed but also as a state within the state in the north-western parts of the country, including the cities of Medellín and Cali where they ruled with the help of Israeli, British and other mercenaries and provided more jobs and better pay than did the government in its own sphere of competence. The economic empire of these drug barons depended on their markets in the United States where, for a time, US agencies were less alive to the dangers of flooding the market with poisonous drugs than to the opportunities of taking part of the profits to finance US policies against Nicaragua: many blind eyes were turned until the murder in 1989 of Luis Carlos Galan, likely to be the next president of Colombia, concentrated attention on the slide, not into anarchy, but to the capture of the state by a sinister cartel whose single aim was money. President Virgilio Barco promised tough action and President Bush promised lavish help, but the former was reluctant to accept the latter's offer of military support which was distasteful to many Colombians. Some 12,000 persons were arrested but hardly anyone of much consequence. The United States pressed for the extradition of senior malefactors but the barons replied with a threat to kill judges, magistrates and children for every operator extradited. Since more than fifty judges had been murdered the threat was plausible. In 1987 the Supreme Court of Colombia invalidated an extradition treaty of 1979 with the United States. In elections in 1990 the main themes were appeals for better prices for cocoa and other exports and promises to do no deals with drug barons. The winner was the young Liberal candidate Cesar Gaviria whose protestations on the last issue seemed unlikely to survive the harsh test of practical choices.

Southward from Colombia the Andean republics of Ecuador, Peru and Bolivia displayed the troubled swings between civilian and military rule and the scarcely less troubled problems of joint military–civilian alliances. *Peru* had an extra-parliamentary governing class and a parliamentary world of political parties; it also had a modern or modernizing economic sector and an impoverished economy distinct from this sector. These political and economic worlds were themselves divided. The governing class, held together by a common conservative outlook and often by marriage, comprised army officers and large landowners and the more eminent bankers and businessmen. The army, however, contained a radical wing which did not subscribe to all the values of the conservative establishment. The outstanding political party was APRA, founded in the 1920s with a semi-socialist programme, regarded by conservatives as deeply dangerous and tainted with a willingness to resort to violence. By 1945 it was forswearing violence and moving towards the centre, partly because its stance in the 1920s and 1930s condemned it to nearly permanent opposition and partly in order to compete for votes with the less radical Accion Popular. These parties manoeuvred for power against one another and also with, but at arm's length from, the holders of military and economic power who were frequently in two minds about whether to use their power indirectly through a political party or directly by occupying the presidency and other public offices. The mainstays of the economy were agricultural (cotton and sugar) but oil was becoming increasingly important. Their development required foreign, in effect United States, capital which was as unpopular as it was necessary. Nearly half the population were Indians, deprived and despised, and another third mestizos.

In 1948 a military coup led by General Manuel Odria ousted the APRA government of José Luis Bustamente (president from 1946) and forced the Aprista leader Victor Raul Haya della Torre to flee for safety to the Colombian embassy where he remained for several years before departing into exile in Italy. After holding power for eight repressive years, Odria allowed elections to precede the transfer of power to Manuel Prado, a member of one of Lima's most exalted banking families, but a surprising number of votes were cast for Fernando Belaúnde Terry, an architect by profession and leader of Accion Popular, who was presented as a new force between the old APRA left and the military–oligarchic complex. When presidential and congressional elections came round again in 1962 Belaúnde stood against Odria and Haya della Torre. None of the candidates won outright; the army stepped in to block both Odria and Haya della Torre and declared the elections fraudulent for no discernibly adequate reason. After an interval of rule by a military junta the elections were rerun and Belaúnde was declared the winner in a close contest between Accion Popular and

APRA. Belaúnde was no conservative but from the army's point of view he was a safe man and political opinion within the army had been shifting for an accumulation of reasons: contempt for corruption in the civilian business élite, a desire for modernization in the armed forces and also outside them, alarm at the persistence of guerrilla risings, the drift of the destitute to Lima and riots in the capital, seizures of land in the interior. The elections of 1963 registered, if still tentatively, a new political configuration with a section of the army willing to back significant change and Belaúnde willing to enlist military support for a progressive programme.

Risings in the 1950s had been scattered and easily suppressed. Their leaders were divided and, as in Venezuela, the communists among them were averse to guerrilla operations – in this case partly because the most notable guerrilla chief, Hugo Blanco, was a Trotskyist and partly because the communists were split between Russian and Chinese factions. But in the 1960s insurrection revived, so much so that in 1965 the United States intervened with a special military corps, much resented by the Peruvian military. Belaúnde's government failed to defeat or appease the rebels. His economic programme, another failure, forced him to devalue the currency in 1967 by 40 per cent. Most conspicuously and ultimately fatal to him was his inept handling of the oil problem.

The International Petroleum Company (IPC) was the most prominent example of the importance of foreign capital for Peru and correspondingly offensive to Peruvian pride. Earlier governments had tried to milk the company without going to the lengths of expropriating it (and having to run it). Its presence was the one issue on which the far right, the far left and government could speak in unison, laying blame for poor economic performance on foreigners. The company had been established in Peru soon after the First World War under an agreement which was probably made in breach of Peruvian law but was nevertheless plain as between the company and the Peruvian state. The resulting legal imbroglio gave rise to a series of disputes with the United States as well as the company. Belaúnde struck a deal with the company on outstanding disputes but so secretively and open to misinterpretation that its formal adoption by the Act of Talara was greeted with scandalized outrage and was discussed by a conclave of generals who decided that Belaúnde must go. By the time Belaúnde had been forced to seek a coalition with APRA and in doing so had split his own party as well as forfeiting his military support. He fled the country. The succeeding régimes wrecked the Peruvian economy.

This time the army did not look for a civilian partner but decided to shoulder the entire responsibility of government. General Juan Alvarado Velasco, with the enthusiastic support of a group of radical colonels and the more measured support of disillusioned conservatives, took power with promises of economic and social reform, greater efficiency and greater integrity, but his government came under immediate strain from the inherited IPC issue and, opting for complete expropriation, lost its less radical component. It further antagonized the United States by declaring a 200-mile fishing limit and firing on a United States vessel in the course of enforcing it. Social reform, which was presumed to include an extension of political power and economic wellbeing to the poorer classes, was another source of uneasiness between the radicals and others; it was also dependent on economic growth from, in particular, exports of copper and other minerals which failed to come up to expectations.

The Peruvian course from 1968 was an experiment, under strongly centralized and military direction, to introduce economic planning and a measure of social reform within a capitalist framework (from which however foreign capital had been extruded) with the emphasis on technological and managerial innovation in the service of industry and exports. Velasco himself was a nationalist hostile to US penetration of Peru, a man of the mild right but born outside the dominant oligarchy. A number of his army and civilian colleagues and supporters were radicals as hostile to the oligarchy as to the United States. The régime had therefore an innate instability which afflicted it when its first achievements gave way to ideological division and were clouded by economic strains. In spite of initially broad agreement on a programme of nationalization (which included the International Petroleum Company) and land reform, Velasco quickly lost the support of the principal political groups which had helped him to power and felt compelled to seek instead a popular base outside the political establishment. His government moved therefore to the left. In doing so Velasco lost more than he gained, particularly because land reform, although far from insignificant, fell short of expectations and so prevented him from winning peasant and student suffrages. In 1973 both these groups were joining in riots against the government which stiffened their right-wing enemies within it. Conservatives, encouraged by the Pinochet coup in Chile in September, were further alarmed by the economic consequences of world recession which, in 1974, hit Peru at a time when the president fell seriously ill. The piling up of foreign debt in earlier years and a sharp deterioration in the balance of payments were countered by disagreeable measures which were blamed on left-wing incompetence. A bloodless conservative coup removed Velasco in 1975 and promoted his prime minister General Francisco Morales Bermudez to the presidency. But the military were tiring of grappling with the tasks of government and taking the blame for its failures. A return to civilian rule seemed expedient and after an election in 1980 Belaúnde returned to the presidential palace from which he had fled twelve years earlier. Not much had changed in the interval.

Belaúnde failed to curb inflation or the insurgent Sendero Luminoso (a small army of about 5,000 recruited chiefly among the rural poor and led by a rigid Marxist and visionary mestizo Abimael Guzman Reynoso, called President Gonzalo) and in 1985 APRA captured the presidency and Peru announced that it would limit debt-service charges to a fixed percentage of the national product. The new president, Alan Garcia, set a centre-left course flavoured with anti-Washington gestures at home and in Central America. But the filthy war went on with multiple murders, torture and unaccountable 'disappearances' of people whom the army or police did not like. Economic distress went on too: inflation at incalculable levels, a large budget deficit, and the suspension by the World Bank and the Inter-American Bank of their lines of credit. Garcia was soon at odds with the army. He proposed to give it one instead of three places in his cabinet, to reform military administration and cut the army's share (40 per cent) of government spending, and to investigate the massacre of 300 persons in a Lima prison allegedly by the army. The armed forces and the political right, with covert United States support, undermined Garcia's government and then overthrew it. Few regretted it, for Garcia had alienated the poor as well as infuriating and alarming the rich.

Garcia left Peru in even worse shape than it had been when he was elected president. With his own party APRA discredited and his most eminent opponent Belaúnde offering nothing much, the country hankered for a new man. In the elections it got a choice between two: Mario Vargas Llosa, well-known man of letters, and Alberto Fujimori, politically unknown businessman of Japanese descent. Vargas won a slight but insufficient lead in the first round but after a peculiarly vicious interval Fujimori soared to a convincing victory. Vargas, who had moved from communism to the far right, failed to attract the Peruvian *homme moyen* and scared many potential supporters by visions of austerity to come. Fujimori promised solutions to Peru's main problems without being too specific about the means. Once elected he reversed Garcia's refusal to honour foreign debts, reversed Peru's decline in output, reduced inflation from the astronomic to the merely large, and put Gonazalo in prison. He also suspended the constitution, dispensed with parliament, interfered with the judiciary, arrested political opponents and suppressed the press. In 1995 he was re-elected, obliterating the former UN secretary-general, Javier Perez de Cuellar. Fujimori's unexpected electoral successes pointed to changes in Peruvian society which had gone unnoticed and were submerging the simplicities of the old party alignments.

Bolivia's social configuration was rather different from its neighbours' because Victor Paz Estenssoro, after an unsuccessful attempt in 1943, had led in 1952 a successful revolt in which non-communists,

communists, Trotskyists and junior officers, appalled by corruption and inefficiency, injustice and poverty, had participated. His régime nationalized Bolivia's extensive tin mines, broke up large estates, introduced universal suffrage and elementary social services, and attempted to diversify the economy in order to reduce the country's dependence on mining. The régime quickly ran into economic trouble by trying to do more than its resources permitted and was forced to adopt severe stabilization measures which cut down its programmes and so produced political trouble in the form of disputes between the different components within the new order. But it ran none of the risks which undid President Arbenz of Guatemala because land-locked Bolivia was not so close to the United States as Guatemala and its possessing classes were not American corporations.

President Paz, who was succeeded by Hernán Siles Zuazo in 1956, returned to office in 1960 but was exiled in 1964 after a coup by General Alfredo Ovando Candia. In 1967 the existence of guerrilla activities was admitted and made world famous by the arrest and trial of the French Marxist writer, Regis Debray (committed to prison for thirty years) and the death in Bolivia of Che Guevara. The government was weakened by splits and by conflict between Ovando and the titular president René Barrientos. In 1969, a few months after the death of Barrientos, Ovando took power, installed a purely military régime and embarked on policies partly borrowed from General Velasco's populist military régime in Peru. A year later an abortive right-wing coup assisted by the United States was followed by a coup from the left led by General Juan Torres, but in 1971 the pendulum swung again as Colonel Hugo Banzer displaced Torres. Banzer's rule was a stereotype of military repression, censorship and brutality but it observed constitutional forms to the extent of permitting an election in 1978 for the succession to Banzer. The army's candidate, Juan Pereda Asbun, was unexpectedly defeated by the veteran Siles, who then went on hunger strike when he was not allowed to take over. The Bolivian Congress annulled the election but a military rising brought about the premature retirement of Banzer and the installation of Pereda in his place. Pereda, however, lasted only five months and in a rerun election in 1979 Siles defeated Pereda, Banzer and a number of other candidates. Siles offered the army a series of concessions and guarantees but he was not allowed to take office. After a two-week interlude when Colonel Alberto Natusch Busch seemed to have gained control, Lidia Gueiler Tejida became the country's first female president. She was expelled in her turn in 1980 by a coalition of right-wing officers and prosperous drug traffickers who, with Generals Luis Garcia Meza Tejada and Celso Torrelio Villa as presidents, inaugurated a fresh régime of terror and torture. Fifteen years of military rule ended in 1982 with the return to power of Siles, half-heartedly applauded by the United States. The growth of the drug business caused the United States to send troops and detectives to help suppress it – an intervention much resented, particularly by coca farmers. Siles was followed once more by Paz who was expected to be followed by Banzer who, however, was defeated in 1993 by General Gonzalo Sanchez de Lozada campaigning with an Indian vice-presidential partner. Hyperinflation was tackled by methods which produced hyperunemployment.

In *Ecuador's* milder political climate the military seized power in 1976 but held it only briefly, making way in 1979 for a left coalition led by Jaime Roldos Aquilera who was killed in an air accident two years later. Political power oscillated between left-centre and right-centre, but no party commanded a parliamentary majority and the austere rectitude of market economics provoked popular discontents and riots. Ecuador disputed with Peru an area believed to contain valuable minerals and oil. This dispute was settled in 1942 by a treaty which was denounced by Ecuador in 1960. The borderlands remained unruly. In 1992 President Fujimori of Peru on a visit to Quito accepted a proposal for mediation by the Vatican but in 1995 Peruvian forces provoked serious armed clashes. In the same year financial scandals caused the flight of the vice-president and apprehensions about the future of civilian rule as fresh elections due in 1996 approached.

Mexico, like Canada, is a country which may be destined for ultimate embrace in a North American economic zone dominated by the United States. Between Mexico and the United States, however, there is a historic feud which is more easily forgotten in the latter than in the former. Mexico lost Texas to the United States in 1836. Ten years later US troops occupied Mexico City. Half of what was then Mexico became the south-western United States. In 1862 a French debt-collecting expedition (with British and Spanish encouragement) turned into an imperial adventure and created a short-lived Mexican empire with an Austrian archduke as emperor. This régime was followed after a short interval by the dictatorship of Porfirio Diaz which lasted from 1876 to 1910 with only one four-year interlude. His remorseless, modernizing rule ended in revolution and seven years of civil war in which the United States intervened on the side of counter-revolution, and a war between Mexico and the United States was only narrowly avoided. The long dictatorship of Diaz was succeeded by the even longer rule of a single party called, eventually, the Partido Revolucionario Institucional (PRI) and committed to social mobility and economic reform. But the years eroded the party's revolutionary zeal, it never grasped the nettles of taxing the rich and reforming a notoriously inefficient agriculture, and it became the political vehicle of the well-to-do. In 1938 President Lazaro Cardenas nationalized oil reserves and the oil industry. Presidents succeeded one another decorously and constitutionally and by the end of the Second World War Mexico was a stable country under civilian rule. Relations with the United States had proceeded to a dignified wariness.

Mexico declared war on Germany, Italy and Japan in 1942 and sent an air contingent to the Philippines. After the war it played a leading part in establishing an inter-American security system: by the Treaty of Chapultepec (which is in Mexico) the American states agreed to mutual consultation if the borders of any one of them were infringed and envisaged joint action up to and including the use of force; the Treaty of Rio in 1947 made more precise provision for the peaceful settlement of disputes and collective defence; and these agreements culminated in the creation in 1948 of the Organization of American States (OAS). Foreign leaders paid attention to Mexico. All United States presidents from F. D. Roosevelt onwards visited it and so did all French presidents from de Gaulle onwards. In regional affairs Mexico ostracized Cuba in 1962 but reversed this attitude in 1975; it joined Venezuela and other American states in attempts to restore Cuba to inter-American respectability and to settle the war in Nicaragua; it condemned Pinochet's Chile and sympathized – sometimes sided – with left-wing movements in Central America; its professed aims included the reduction of United States and Russian influence south of the Rio Grande; it refused to accept a US military mission (1951) and openly criticized US actions and judgments in Guatemala, Cuba and Nicaragua; it was a sponsor of the Treaty of Tlatelolco in 1967 which essayed to ban nuclear weapons in the sub-continent; and, principally as a consequence of its oil wealth, it established links with other oil producers, notably Nigeria, while refusing characteristically to become a member of OPEC.

From around mid-century the Mexican economy was transformed by a policy of industrialism, urbanization and an extended internationalization designed to reduce the country's dependence on foreign economies while at the same time securing the external finance required to promote and sustain industrial growth. Employment in manufacturing industry passed the level of employment in agriculture. Foreign earnings grew with oil exports but these accounted for no more than a third of the total in which cotton, coffee and sugar were substantial earners. Growth, already substantial during the war years, continued and enabled Mexico for a time to take an unprecedented demographic explosion in its stride. It was accompanied by a shift from military to civilian rule but not any relaxation of the entrenched single-party régime of the PRI. During the 1970s, however, progress was assailed from many quarters: world recession and domestic inflation, inadequate creation or investment of domestic capital, inadequate training or recruitment of skilled labour, a serious decline of food production, heavy external indebtedness and debits on external trade, increasingly obvious maldistribution of wealth, particularly in the south where most of the Indian population lived without a sight of the benefits of economic growth: something like nine-tenths of the country's wealth was owned by fewer than half a million persons out of a total of 85 million.

These failings were partially masked by oil. By 1980 Mexico was producing 2.5 million barrels a day, exporting more than half of it (and half of these exports to the United States), had proved reserves of 60 billion barrels and actual reserves three or four times larger, and ranked sixth among the world's oil giants. But the lure of oil enticed

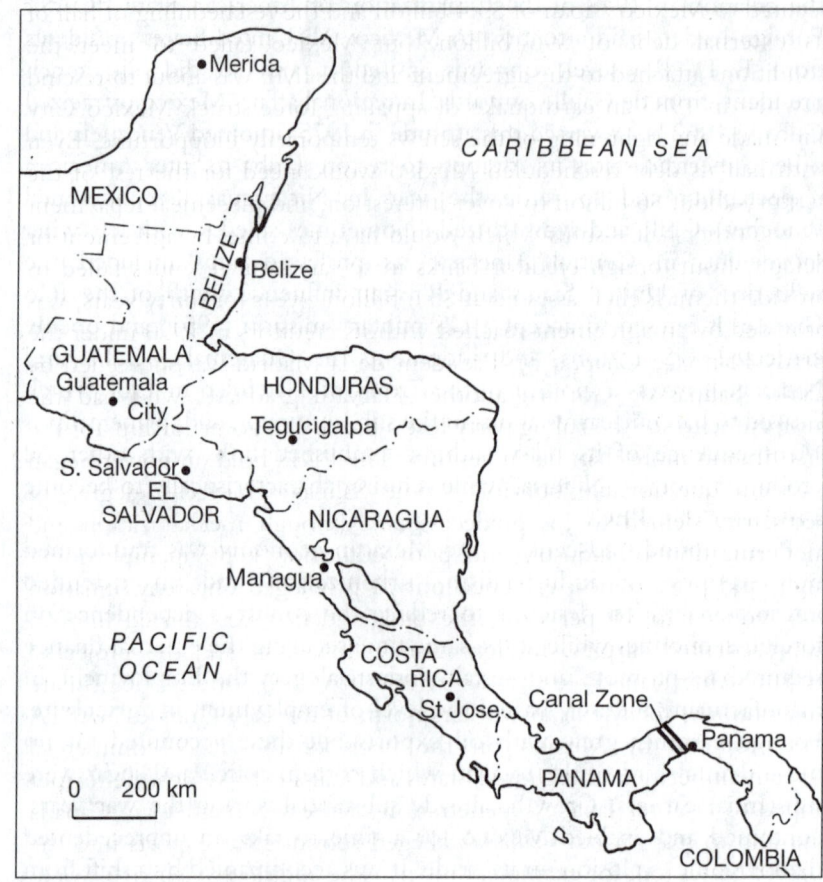

27.1   Central America

governments into an extravagant pursuit of growth and, when the oil boom stopped booming, into debt. Distress and discontent, which exploded shockingly with a massacre of young demonstrators at Tlatelolco in 1968 on the eve of the Olympic Games in Mexico City, combined with economic disappointments and general unease to compel a thorough review. This produced in 1982 a devaluation of the currency by 70 per cent, the nationalization of the banks, a reduction of real wages by nearly half over the next five years, cuts in education, health and other public services and a culture increasingly attuned to short-term speculation rather than long-term growth. Miguel de la Madrid Hurtado, a gifted politician and economist and Harvard graduate, elected president in 1982 at the age of 47, faced a collapsing peso, rising unemployment, clamps on public capital spending and oil-fed corruption. He was able to strike a bargain with the IMF which secured to Mexico a loan of $3.4 billion and the rescheduling of half of its external debt of $96 billion, but Mexico failed to meet the conditions attached to the agreement and the IMF was about to rescind it when in 1985 an earthquake of appalling force struck Mexico City and made any tightening of the screws temporarily inopportune. Even with half its debt rescheduled Mexico would need for the rest of the century about $6 billion to cover interest on, and piecemeal repayment of, the other half – sums which would have to come, by agreement or default, from foreign creditor banks in so far as oil revenues failed to provide them. Relief, assessed at $3.6 billion a year for thirty years, was provided by an agreement reached with its creditors in 1990 under the Brady Plan (see Chapter 5). President de la Madrid was succeeded by Carlos Salinas de Gortari, another Harvard graduate, who had the misfortune to see the ruling party lose a provincial governorship for the first time in more than half a century. The PRI's hold on power began to seem questionable but for a time Salinas achieved a degree of popularity denied to his predecessors. Although mechanization and modernization of industry and poor education kept unemployment high, and privatization enriched only a privileged oligarchy, inflation was reduced to 10 per cent, the budget deficit was turned into an unprecedented surplus, foreign money flowed in and the projected North American Free Trade Area (NAFTA) offered legitimate opportunities for Mexican workers north of the Rio Grande. NAFTA was proposed by Mexico on the heels of the 1989 commercial agreement between the United States and Canada by which the former hoped to counter Japanese penetration of the US market via Canada and also to stop or cut Canada's alleged subsidies for exports into the United States (on, for example, timber). The addition of Mexico would create an economic zone with a population of 370 million. But the US Congress and public were uneasy about the extension to Mexico – and perhaps thereafter to Chile and other Latin American countries – sensing that it might encourage US industries to move south into Mexico and Mexican job-seekers to flood contrariwise into the United States. President Bush delayed action on the proposal during his election campaign in 1992 to the chagrin of the Mexican government whose economic programme, including growth at 6 per cent a year, depended heavily on US investment in Mexico but in the following year Clinton succeeded in securing Congressional approval for it. Initially Mexican exports to the United States grew faster than trade in the opposite direction but its benefits were unevenly distributed geographically and barely apparent to smaller businesses.

The Mexican régime was badly bruised by two events in 1994: open revolt in the southern province of Chiapas and the assassination of the PRI's next candidate for the presidency, Luis Donaldo Colosio. Chiapas had been in ferment for many years. At the core of its troubles were the destitution and despair of recently emancipated serfs and landless peasants, mostly Mayas. Their sense of grievance and isolation was accentuated by their speaking little or no Spanish and their poverty by the collapse in 1989 of the price of coffee. Their plight won the sympathy and active support of the Roman Catholic clergy, led after 1960 by the Bishop of San Cristobal, Samuel Ruiz; it generated self-styled Maoist groups which put their faith in guerrilla war; and, most importantly, it created the evocatively named Zapatista National Liberation Army led by the white or half-white Subcomandante Marcos. The growth of opposition from the late 1980s caused dissension and splits, particularly on the usefulness of violence about which Marcos was ambiguous or at least discreet. A sharp but short bout of violence early in 1994 was followed by a truce and by government proposals which were almost universally rejected in Chiapas as the inadequate response of an untrustworthy régime. Almost simultaneously Colosio was assassinated. The Mexican government nervously closed financial markets and the United States government quickly and publicly extended a credit of $6 billion.

A few months later the PRI retained the presidency and a majority in the parliament although with a bare majority of the popular vote and amid open cynicism about the workings of Mexican democracy. The new president, Ernesto Zedillo Ponce de Leon, although belittled as less exciting than his predecessors, seemed a safe choice in troubled times. The economy, however, was more troubled than safe and confidence in the government's handling of it low. Zedillo's recourse to the IMF entailed promises to limit wage increases and redress the budget deficit by privatizing public enterprises − promises easier to make than fulfil when the trade unions were clamouring for better pay and the purchasers of state industries were few and mean. A spate of imports was playing havoc with the balance of payments, an unexpectedly sharp devaluation of the peso at the end of 1994 was interpreted as panic, and interest rates rose to 40 per cent. Devaluation was accompanied by the issue of dollar-linked securities (in exchange for short government bonds) but the total of these *tesobonos* quickly exceeded the country's foreign reserves. To prevent the collapse of the peso, of United States banks involved in Mexico and of United States exports to Mexico, Clinton contrived with the IMF and the Bank for International Settlements a supporting guarantee of $50 billion to which the United States contributed $20 billion, secured on Mexican oil reserves. But the Mexican crisis called in question the ability of international finance to meet such a crisis and therewith the willingness of financiers worldwide to support Latin American governments pursuing economic policies dependent on foreign loans and investment. Although primarily a domestic and a Mexican–United States event, the collapse of the Mexican peso dismayed all Latin American countries where economic growth was desperately needed for its own economic sake and as a precondition for stable democratic government. In Mexico itself, economic disaster threatened political stability as the gap between poor and rich widened and that between poverty and starvation closed. This dual instability threatened the future of a country whose problems, although massive, were also simple because they proceeded not from inherent poverty but from the mismanagement of potential wealth and the sclerosis of almost a century of one-party rule.

When the Second World War ended, Central America south of Mexico was in the course of transformation – becoming part of a wider world, looking to world markets and scrutinized by international corporations. This revolution had consequences for land ownership, the structure of politics and the organization of labour which the ruling élites failed to master or even perceive. They found themselves opposed by liberals and by more strenuous radicals who were, however, divided against one another partly on class lines and partly by opposing views on the morality and expediency of violence. In decades of strife the firepower and traditional authority of the élites prevailed to the extent of thwarting insurrections, so that the 1990s witnessed a return to a more pacific political order based largely on weariness.

As Mexico narrows southward almost all of its land border is shared with *Guatemala*. (British Honduras, independent as Belize from 1981, has a short border with Mexico as well as a longer, disputed border with Guatemala.) Of three republics of comparable size in Central America Guatemala lies to the north, Nicaragua to the south and Honduras marches with both. For over a hundred years (1838–1944) Guatemala was ruled by four military dictators with short intervals of civil governments in between. In 1944 the army was divided and a group of junior officers supported a comparatively left-wing candidate, Juan José Arévalo, who was president until 1950. During this period the division in the army continued and was represented by the rivalry between two majors, Francisco Xavier Arana and Jacobo Arbenz Gúzman. The former was assassinated in 1949 and the latter won the election of 1950. This victory caused alarm in the United Sates where the new régime was regarded (disputably) as pro-communist or a forerunner of communism and therefore as presenting a threat to the Panama Canal; and also (correctly) as an enemy of foreign capitalists and especially of the United Fruit Company which, as the owner of a tenth of the country's land, exercised even more economic power in Guatemala than the Anglo–Iranian Oil Company in Iran, and stood in the way of essential land reform.

The Guatemalans, 10 million people of whom over half were of Indian stock, suffered from extremes of poverty, disease and social inattention, which were rendered the more intolerable to the thinking section of the people by the prosperity of a small minority and by the economic omnipresence of foreign enterprises which owned not only an abundance of land but also the railways, docks and public utilities. President Arbenz accelerated his predecessor's reform programme; he nationalized uncultivated land and supported strikes against foreign concerns. These moves were interpreted in Washington as the beginnings of a fully fledged communist policy, notwithstanding that communists were greatly outnumbered by anti-communists in the Guatemalan government, parliament and civil service and that power lay with an anti-communist army. So long as Truman was president the United States waged veiled economic war but refused to go further. Eisenhower and Dulles, however, resorted to (vicarious) armed force. At the tenth inter-American conference at Caracas in March 1954 Dulles tried to get a condemnation of the Arbenz régime but discovered that no other state accepted Washington's interpretation of events in Guatemala. The conference passed a general resolution condemning communist domination of any American state but refused to single out Guatemala for special mention, and Dulles thereupon left Caracas abruptly and turned to conspiring with disaffected Guatemalans who were preparing to invade their country from Honduras and Nicaragua. These two countries, having complained of communist incursions from Guatemala, were themselves provided with arms from the United States, while Washington tried to prevent Guatemala from getting arms by appealing to its allies not to supply them and by intercepting communist shipments from Europe. The invasion took place in June and was successful. The Guatemalan government appealed to the Security Council but was frustrated by a proposal to refer the matter to the OAS. This attempt to deflect the matter was vetoed by the USSR but a week later President Arbenz was forced to resign. The leader of the invasion, Colonel Carlos Castillo Armas, succeeded him and retained office until he was assassinated in 1957. The annual rate of growth of the Guatemalan economy fell from 8.5 per cent (over 1944–54) to 3 per cent and under Castillo's successors Guatemala lived in a state of suppressed civil war.

Castillo was succeeded by General Miguel Ydígoras Fuentes. In an election in 1957 Ydígoras came second, but after some disorder and a second election in 1958 he was duly installed. His power was confirmed by a fraudulent election in 1961 but riots, repressed by the army, demonstrated a dependence which he was not wise enough to acknowledge. Military opinion turned against him on the grounds that he spent too much money on himself and too little on the armed forces.

In 1962 the army defended him against an attempted coup by the air force, but he then fatally antagonized the army by allowing ex-president Juan Arévalo to return to Guatemala and campaign for the presidency. Arévalo, who seemed certain to win a fair election, was regarded as a reincarnation of Arbenz, and in 1963 the army under Colonel Enrique Peralta ejected Ydígoras and suspended the constitution. In 1965 the army tolerated the election of a comparatively liberal president, Julio Cesar Mendez Montenegro, but he was no more able than his more oppressive predecessors to pacify the country. A motley and frequently divided opposition, consisting of communists, Trotskyists and guerrillas who were neither, staged unsuccessful risings and resorted to murder and kidnapping when their risings failed. US forces helped the government. A US ambassador was among those assassinated.

In the 1970s nominal control was shared between the army and civilian parties with left-wing parties proscribed. A number of officers went into business and became rich, thus earning the jealousy of other officers and attracting charges of corruption. Elections in 1974 were contested by three candidates, of whom the youngest – Colonel Efrain Rios Montt, a passionate Protestant – won but was shouldered out of the way by his elders. The military were divided between hardliners who saw only guerrillas and on the other hand a more enlightened minority who saw also widespread poverty and injustice and, at the centre, corruption. Faction was added to violence, corruption and economic collapse. Abuses of power became so gross that in 1977 Carter stopped US aid (it was resumed in 1981 by Reagan who had other priorities and values). Elections in 1978 gave the presidency to Romeo Lucas Garcia, a right-wing general of peculiar brutality who eradicated villages but not guerrillas. He was ousted in 1982 by Rios Montt who promised to end private murder, civil war and financial scandals but was himself ousted in the next year with these promises unfulfilled. He was followed by General Oscar Mejis Victores who devoted himself to the suppression of the left in his own country but showed less interest in conflicts beyond his borders, for which indeed Guatemala had limited resources to spare. A nominal reversion to civilian rule in 1985 left the army unfettered, President Vinicio Cerezo being little more than a figurehead who might be saddled with the blame for economic chaos. He edged away from Washington's stark anti-Sandinista policy in Nicaragua and collaborated with other Central American presidents in seeking peace with the Nicaraguan régime, reconciling President Duarte of Salvador with President Ortega of Nicaragua and ending the Contras' use of Costa Rica as part of a pincer attack on Nicaragua. In elections in 1990 a dozen candidates took the field, ranging from extreme right to centre-right. Rios Montt, declared constitutionally ineligible by the Supreme Court, backed Jorge Elias

Sarrano, a Protestant and the candidate of middle-class business, who won on the second round with fewer than half of the electorate voting. In 1993 he clashed with his vice-president and with the military and in the ensuing confusion a new president, Ramiro de Leon Carpio, a civil rights leader, emerged. He failed to stem the excesses of ordinary criminals and psychopathic officials and officers. Civil war subsided but did not end. Its roots had been planted in the mid-1950s by young people who, under mainly Christian leadership, set out to establish rural communities on reclaimed land but were converted by official persecution into a partly missionary, partly militant movement. It was brutally attacked by the army which deployed 40,000 troops, destroyed 4–5,000 villages, killed tens of thousands of people and rendered at least half a million homeless. After the most punishing attacks in 1980–84 the militants were reduced to about 2,000, the visionaries lost their illusions of bringing justice to Guatemala and the country was split between bitter indignation and brutal repression with, however, some hope of milder times ahead with the defeat at the turn of 1995–96 of Rios Montt's latest nominee for the presidency.

On the Pacific coast of Central America south of Guatemala the small and once wealthy state of *El Salvador* was wrecked by civil strife. A standard military dictatorship under General José Maria Lemus was modified in 1961 by the installation of a middle-of-the-road government under General Julio Rivera which adopted a programme of social reform and economic investment which, if inadequate in the eyes of the left, nevertheless made a substantial step away from the narrow conservatism of past rulers. But the modest hopes raised by this change proved illusory and during the 1970s extreme violence erupted between the government and left-wing guerrillas of the Farabondo Marti Liberation Front (FMLF), the militant arm of a broad political left. The inauguration in 1977 of President Carlos Humberto Rovero was boycotted by the clergy in protest against the violence and torture practised by the régime itself and in 1979 a coup by junior officers installed a new government pledged to moderate agrarian reform and the nationalization of selected financial and commercial concerns. But this mildly right-wing government failed to control extreme right-wing terrorism. The murder in 1980 of the archbishop of San Salvador typified growing anarchy and a polarization to extremes which left the government stranded in the middle. The United States, fearing that the extreme left might prevail, supported the government as a lesser evil and tried to persuade Guatemala and Honduras to agree to intervene should the government collapse and the extreme left take power but neither country was keen to denude its domestic position by sending troops abroad.

With the election of Reagan in 1980 El Salvador became a piece in a positive, even obsessive, US policy of overthrowing the government of Nicaragua in particular and helping any Central American government of the right. The FMLF was stigmatized as a terrorist organization and Reagan certified, in the face of ample evidence to the contrary, that the Salvadorean government's record in human rights was satisfactory and that its army was under government control – an equally implausible statement. But in spite of increased American aid President Napoleon Duarte was unable to suppress the Front. He was also unable to break free from the extreme right which in elections in 1982 won more seats than the president's party of the less ferocious right. Robert d'Aubuisson was elected president but was persuaded under pressure from Washington to give way to the more respectable and neutral Alvaro Mangana. Another election two years later was similarly indecisive as between Duarte and d'Aubuisson. Mounting guerrilla successes during 1984 were countered by more American aid in money, helicopters and miscellaneous weaponry. If, as seemed to be the case, Duarte genuinely wished to treat with the insurgents and introduce some reforms he was thwarted by continuing civil war of a peculiarly nasty kind in which the greater proportion of the killings were the work of right-wing squads belonging to the army or more loosely connected with it. In 1985 Duarte prevailed over the extreme right and so won room to manoeuvre for peace with the left, provided the terms were acceptable to the holders of ultimate power, the army. But he remained uncomfortably wedged between two extremes, hardly an independent force, his authority more nominal than real, a spent force (in health as well as politically) and the ineffectual arbiter of a war which cost his small country at least 50,000 lives in twelve years. The guerrillas, in spite of maintaining a threat, failed to incite a broad popular revolt and were reduced to a few thousand against an army of 50,000 with US aid amounting to $1 million a day. The military and political wings of the left were divided on whether to negotiate with Duarte; on the right the military and their paramilitary accessories or death squads were out of control and perpetrated indiscriminate massacres. Attempts from 1986 to arrange a cease-fire came to nothing. Elections in 1989 were won by d'Aubuisson's Arena Party which installed Alfredo Cristiani as president but lost its parliamentary majority in 1991. The death squads were no more responsive to Cristiani than they had been to Duarte. US aid was increased but failed to ensure either of its professed aims – stability and democracy. In 1991–92 UN intervention secured a cease-fire, disengagement of armed forces and an agreement for the creation of a new police force and the absorption of the guerrilla units into the regular army. The

FMLF gradually demobilized under UN supervision. Elections in 1994 gave Arena a comfortable win, supervised and endorsed by the UN.

Internationally much the gravest commotion in Central America occurred in *Nicaragua* following the collapse in 1979 of the dictatorship of the Somoza dynasty which had lasted for nearly fifty years. The Somoza dynasty stemmed from General Anastasio Somoza Garcia who seized power with the help of the United States in 1930. He and his son Luis, who succeeded him in 1956, ruled in alliance with the prosperous landowning and business classes but after the latter's death in 1967 his brother, Anastasio Somoza Debayle, alienated and even scandalized these allies by the brutality and corruption of his rule, notably his appropriation of large sums out of foreign aid sent to Nicaragua after a serious earthquake in 1972. Armed resistance had begun in 1961 by guerrillas calling themselves the Sandinista Front after Colonel Augusto Cesar Sandino, assassinated in 1934 after leading a rising which was suppressed with the help of troops from the United States. The Sandinistas remained for many years few in number since they failed to win widespread support among the peasantry or the urban working class, but Anastasio II's misrule earned them more sympathy among the professional classes and the Roman Catholic clergy and the murder in 1978 of the editor of Managua's principal newspaper *La Prensa* was the signal for serious attempts to put an end to one of Latin America's most outrageous dictatorships.

President Peres of Venezuela took the lead and tried to persuade President Carter to intervene. Carter was torn between, on the one hand, his revulsion against the tortures and other abuses prevailing in Nicaragua and his fears that Somoza's excesses would lead to the opposite extreme and a second Cuba in Central America; and, on the other hand, his reluctance to intervene in the internal affairs of a neighbouring sovereign state, particularly to intervene with force. While these issues were being debated and various forms of economic pressure discussed, the chance to replace Somoza by a moderate and democratic régime slipped by. Washington's indecision hardened the determination of the Sandinistas and their allies, domestic and external (Cuba supplied arms through Costa Rica), and in 1979 Somoza was forced to flee.

The victorious alliance began immediately to fall apart. Promises to hold quick elections and instal a multiparty system were shelved. The Sandinistas' clerical and professional allies dribbled away. President Reagan began a covert war which eventually put about 12,000 men – the Contras – into the field against the new government. These Contras, dubbed by Reagan freedom fighters, consisted of old and newer opponents of the Sandinistas reinforced by mercenary bandits of no fixed political views except that they were anything but democrats. Their main bases were in Honduras where the United States built a large air base and increased its aid to the Honduran government tenfold. These exertions failed, however, to overthrow the Nicaraguan government; provided it with excuses to maintain and tighten its authoritarian régime and further postpone the promised elections; alienated anti-Sandinista elements which reprobated Reagan's support for the Somozistas in Honduras and Florida; and gradually converted Nicaragua's neighbours from auxiliaries in Reagan's war to active proponents of a peace which would include the recognition of the Sandinista government.

The attack on Nicaragua, although mainly directed from Honduras in the north, had also a pincer arm in Costa Rica in the south. The former's enthusiasm for aiding the US cause waned towards the end of the 1980s, more particularly after the election in 1989 of Raffaello Leonardo Callejas who wanted to get out of the war. On the Costa Rican front the principal figure was Eden Pastora, a social democrat who had joined forces with the Sandinista Front in the early 1970s, quickly changed his mind and broke with it, was then persuaded to rejoin it in 1976, became its commander-in-chief under the sobriquet of Comandante Zero two years later, again defected soon after the victory of 1979 to become leader of the anti-Sandinista forces in Costa Rica, and finally in 1988 abandoned his opposition to the Sandinista government out of disgust with Reagan's tactics. His shifts and changes symbolized the confusion and perplexities among anti-Somozistas who were half disenchanted with the Sandinistas but no less disheartened by United States policies. The governments of other Central American countries oscillated in the same way. Anything but left-wing, they nevertheless came round to preferring the Sandinistas to Reagan's war.

Washington's rooted dislike of any left-wing government in Central America was sharpened in the case of Nicaragua by the fact that, however exaggerated Washington's propaganda about their links with Cuba and the USSR, the Sandinistas both needed and got arms from Cuba. This Cold War aspect of the upheavals following the fall of Somoza became a ruling, and to some considerable extent a blinding, element in the United States after the advent of Reagan. The Reagan administration took office resolved to give high priority to Central America as a critical zone in the global conflict of the superpowers. Its endemic troubles, economic and political, had been exacerbated during the preceding decade. Its economic growth, which had been nurtured on an expanding middle class, was brought to a halt by the general world recession with the result that this class became divided in its political allegiances and the poorer classes became poorer: the middle class was drawn on the one hand to the old oligarchies which represented, even when they could no longer command, the settled order dear to middle-class enterprise and money-making but, on the other hand, this class was alienated from the oligarchs by their selfish brutality and by the suspicion that their days were numbered. The poor had been emboldened by glimpses of a juster order and hardened by the renewed hardships of the late 1970s. The consequent upheavals were, however, regarded in Washington as something more sinister and any inclination by liberal America to welcome Central America's belated revolutions was overlaid by the conviction that they were more than half provoked and extensively supplied by communist Cuba and the USSR. Nicaragua was seen as a Soviet satellite in the making, all the more dangerous than Cuba for being on the mainland. Reagan was therefore easily persuaded that he must ensure the defeat of the guerrillas in El Salvador, stifle any other leftish movement which might arise in the area and above all overthrow Nicaragua's Sandinista régime. The means were military, covert and open, and they failed. Besides organizing, training and paying for anti-Sandinista forces in Honduras and Costa Rica, the United States navy staged demonstrations off the Nicaraguan coasts in 1983 and in the next year mined Nicaraguan ports on the Atlantic and Pacific oceans; this mining in contravention of international law was censured by the Security Council (but the United States vetoed the resolution) and condemned by the International Court of Justice. Further hostile action was impeded by the US Congress which refused funds requested by the president for the Contras – a refusal circumvented by a number of subterfuges such as selling arms to Iran and diverting the proceeds or part of them to the Contras with, it can hardly be doubted, the connivance of the president.

In 1983 four states – Mexico, Venezuela, Colombia, Panama – formed the Contadora Group, an outer ring of states concerned about the manifold disorders in Central America. Their purpose was to bring about a general pacification in Central America on the basis of recognition of all governments actually in place, self-determination and the reduction of armaments, but the group was distrusted in Washington on account of its even-handedness which would recognize and probably validate the position of the Sandinistas in Nicaragua. Under pressure from Washington, Honduras, Guatemala and El Salvador refused to attend a conference which the group proposed to convene in 1985. But the group persevered and produced in 1986 a detailed programme (the Carabellada Declaration) for talks between Central American governments. This programme was irremediably opposed to Reagan's insistence that peace must be achieved through talks between the Sandinistas and the Contras and that the former must first introduce political reforms within Nicaragua. To all appearances Reagan put the removal of the Sandinistas above peace, while the

Contadora Group envisaged a pacification which might leave the Sandinistas in power. Washington described the Contadora proposals as a threat to the region and to the United States.

But Reagan recoiled from the open military intervention which alone could achieve what he wanted and his support for the Contras was hampered by the revelation, on the eve of mid-term elections in 1986, of the Iran arms deals. The United States was able to keep the Contras in the field in Honduras but in spite of a strength of 15–20,000 and of lavish modern equipment they took no sizeable town in Nicaragua. Honduras, which harboured them and was well paid for doing so, received no support from any other Central American government, while on the southern front Costa Rica abandoned the war. The misgivings of the outer ring had reached the inner ring.

Costa Rica had been a commitedly pacific state which, after a century and more of peculiar turbulence, had gone to the lengths of abolishing its army in order to have more to spend on public services and utilities. But it remained poor and dependent on the United States and after the Sandinista victory in 1979 it responded to Washington's appeal to play a part in containing and eradicating communists in Central America. This attitude was, however, reversed when in 1986 Alberto Monge was succeeded by Oscar Arias who produced a plan (for which he was later awarded the Nobel Prize for Peace) which was subscribed by five Central American presidents including Nicaragua's. It proposed that presidential, national and local elections should be held in Nicaragua in February 1990, preceded by the removal of press censorship and of obstacles to multiparty political activity. This scheme was viewed askance in Washington and with premature optimism elsewhere. Nicaragua took quick action by releases from jail, by allowing *La Prensa* to reappear and by annulling its state of emergency and its ban on political parties – but it was criticized for not doing more. Other signatories squabbled over whether they too had done what was required of them in their countries. Nevertheless the Arias plan was a portent, if only because it pointed in a direction which doomed Reagan's policies and forced his successor George Bush to review and ultimately slide away from them. In 1989 the five presidents pressed their initiative by demanding the disbandment of the Contras and their reintegration into Nicaraguan society.

In 1990 Nicaraguans were allowed their first elections since 1984. Then they had given the victory to the Sandinistas, but this time they did not. Although the Sandinistas remained the largest single party they were beaten by a coalition led by Mrs Violeta Chamorro, widow of the murdered editor of *La Prensa*. Although the US attack by means of the Contras had been an expensive failure, US sanctions had wrecked the Nicaraguan economy and Nicaraguans wanted a new government which would end the sanctions and ensure them a decent standard of living. Moreover, the Sandinistas' good record in such matters as health and education was offset by economic incompetence, corruption and in a broader sense by the mournful consequences of taking on the United States in a war which had cost tens of thousands of lives. Although over half the land and over a third of the principal industries remained in private hands government had nationalized the banks and taken control of wages, prices and imports with disastrous results. Foreign aid, most of it from the USSR and East Germany, was considerable but wasted. Real wages fell during the 1980s by nine-tenths of their value, industrial output declined by a fifth each year and domestic product per head sank to $300 a year, the lowest in the region. Mrs Chamorro formed a government in which place was found for some Sandinistas but none for any of the Contra leaders. Scrapping between Sandinista and Contra bands became endemic, fuelled by steep price rises and unemployment. Chamorro's election, coinciding with Reagan's departure from the White House, allowed the war to come to an end. Throughout Central America the Gorbachev revolution in the USSR – Gorbachev's policies of retrenchment and the ending of the overriding confrontation of the superpowers – was inducing the United States to take a more relaxed view of the left in Central America and to look less indulgently on the excesses of its right-wing associates. And communism began to seem a lesser menace than cocaine. By the 1990s Central American governments were turning their attention to economic co-operation, although commercial agreements – with, for example, Mexico – were of limited significance since the volume of trade within the region was not great. A veneer of peace was spread over the area.

In *Panama* President Arnulfo Arias Madrid set a record of three depositions from office (1941, 1951, 1968). His successor on the third occasion was Colonel Omar Torrijos Herrera, a popular mestizo who had army support and engineered a boom by encouraging foreign banks to do business in Panama and foreign commerce to make the fullest use of its free port. He negotiated the treaties of 1977 by which the United States agreed to hand over the Panama Canal in 1999 on condition that it be permanently neutral and that the United Sates have permanently the right to defend it. US troops would be withdrawn in the year 2000. The canal was becoming less strategically important to the United States but remained economically vital to Panama whose economy rested on the rent paid by the United States, the commerce of the free-trade zone, the traffic in flags of convenience, and off-shore banking in the Canal Zone. In 1981, after he had relinquished office but was still commander-in-chief, Torrijos was killed in an air crash. From 1982, following a military coup, a sequence of lustreless generals ruled and power accumulated in the hands of General Manuel Noriega Morena, second in command of the army. Noriega was a likely president, but in 1984 Arias was re-elected against the wishes of the army. Funded by the CIA in spite of his known role in the drug trade, hailed in the United States as a democrat, Noriega remained in the background organizing the torture and murder of his opponents, enjoying an income said to be $100 million a year from making Panama the chief channel for the traffic in drugs between Colombia and the United States, helping the United States to arm the Nicaraguan Contras and rising to be chief of the army. He was dropped by the United States when scandals became too open and he was publicly accused of drug trafficking, electoral frauds and murder. In 1988 Washington urged President Eric Delvalle to dismiss Noriega from his army post, whereupon Noriega dismissed Delvalle from the presidency. In the next year he cancelled the election which had all but certainly elected Guillermo Endara Galimany to that office. He was indicted on drug charges in the United States and Washington tried also to bribe him to leave Panama and live in Spain, but he refused to go. All else having failed Bush resolved to invade Panama and seize Noriega, which he succeeded in doing in an operation which – depending on the point of view – was either hilarious or humiliating. Noriega was taken to jail in the United States. Endara was installed in his place but lost the presidency in 1994 to Noriega's political heir.

28 CUBA AND THE CARIBBEAN

The island of *Cuba*, the largest of the Antilles, thrusts its western end
into the jaws of the Gulf of Mexico almost midway between the
peninsulas of Florida and Yucatan, from which it is separated by
channels about 160 km wide. Cuba is therefore of all the West Indian
islands also the nearest to the mainland of north and central America.
Its affairs were a special concern of the United States from the middle
of the nineteenth century onwards. Its liberation from Spain at the end
of the century proved (like the liberation of Arab lands from Turkish
rule after the First World War) to be no more than a change of masters,
and it entered upon a period of colonial rule without the benefits of a
colonial administration. Far ahead of its Caribbean neighbours in
educational standards and facilities, and as well-endowed with a middle
class as the most advanced Latin American country, it endured
nevertheless a record of bad government uninterrupted from its
liberation up to and including the Castro régime. The abrogation in
1933 of the Platt Amendment (see p. 772) coincided with the end of
the odious rule of Gerardo Machado, which had rested to some extent
on US support. Machado had just transferred the presidency to Manuel
de Cespedes but a revolt by non-commissioned officers (including
Sergeant Fulgencio Batista) and students overturned the régime and
inaugurated a period of twenty years during which a number of
presidents held office. Batista, who ruled from 1940 to 1944, refrained
at first from infringing a constitution which prescribed four-year terms
with a ban on immediate re-election, but in 1952 he made himself
permanent dictator and introduced a reign of terror. On 26 July of the
next year, a date which gave its name to a movement, Fidel Castro led
an attack on the Moncada barracks in an unsuccessful attempt to
supplant Batista. After eighteen months in prison Castro emerged to
prepare in Mexico a second attempt, and in December 1956 he led an
invasion band of eighty-four which was swiftly defeated. The survivors,
who numbered only twelve, escaped to the Sierra Maestra, where they
turned from the tactics of a *coup de main* to guerrilla warfare which
they waged for two years. In March 1958 Batista attacked the growing
forces of rebellion, but his campaign was a failure and served only to

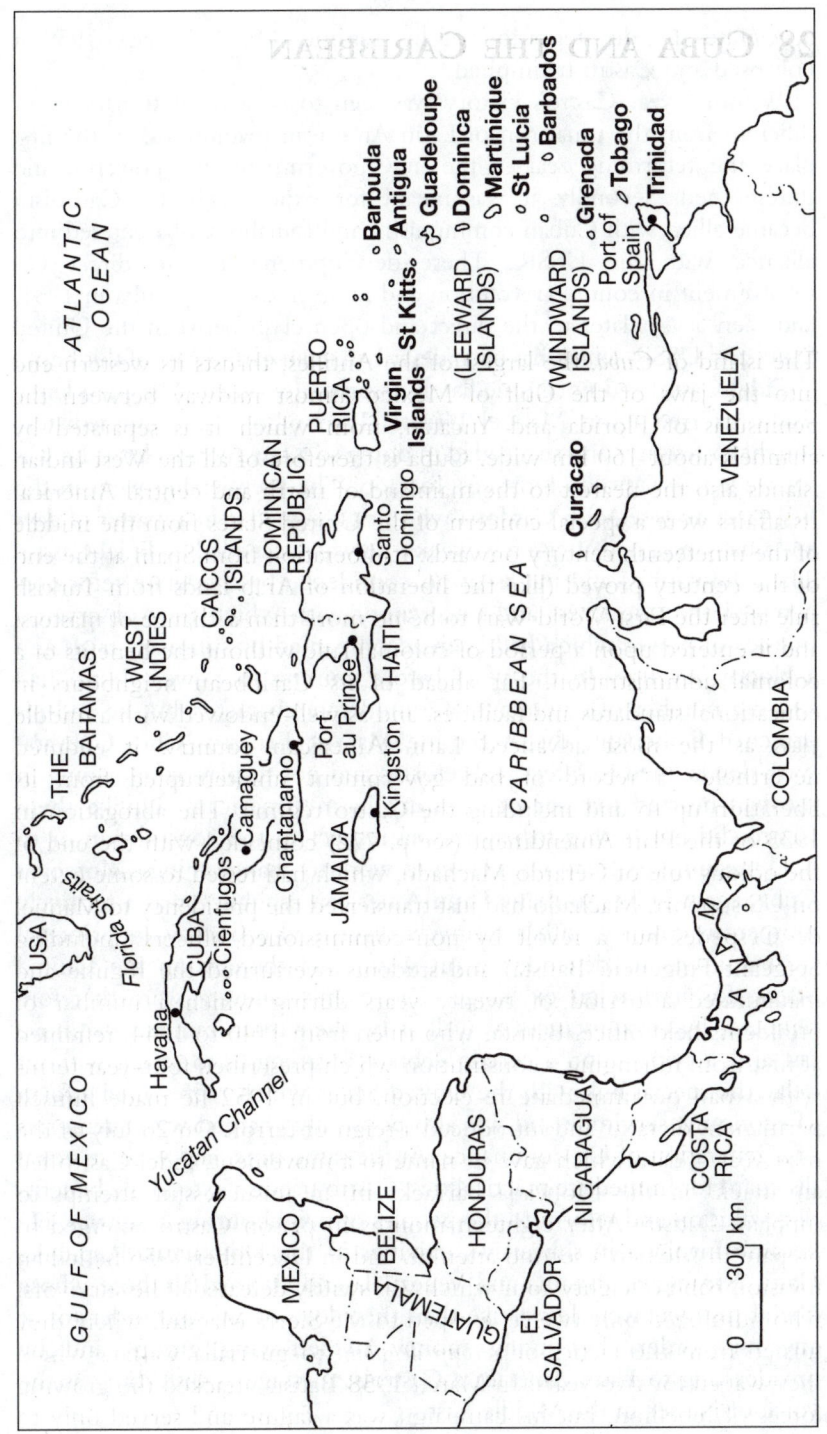

28.1　The Caribbean

accelerate the disintegration of his régime. On 1 January 1959 it collapsed and Castro triumphed.

Within a year Castro's victory was seen to be a revolutionary event different from the usual run of Latin American revolutions. In the first place the reforming zeal of the new government was powerful and unrestrained. Secondly, it was meant for export. Thirdly, Castroism became allied with Cuban communism, and fourthly Cuba entered into alliance with the USSR. These developments led to direct US involvement in counter-revolution and in an invasion of Cuba in 1961, and then a year later to the direct and open clash between the United States and USSR which has been described in an earlier part of this book.

Fidel Castro did not immediately assume any office. The presidency was conferred upon Manuel Urrutia Lleo who almost immediately sought to resign it and succeeded in doing so a few months later; he was succeeded by Osvaldo Dorticós Torrado. The premiership went in the first place to José Miró Cardona who first resigned it in a matter of days and finally resigned after a few weeks; he was succeeded by Fidel Castro. These hesitations and marks of no confidence by moderate reformers betrayed the uneasiness and uncertainty with which they observed a transfer of power which had been accompanied by summary trials and a bloody revenge. In the years before coming to power Castro had issued a number of statements of a moderate character, but he and his principal lieutenant, the Argentine Marxist Ernesto 'Che' Guevara, were determined to effect real reforms and not to play at reforming in the half-hearted manner of so many Latin Americans. Unlike them Castro did not bother to pay formal respect to the constitution or to hold meaningless elections. He set to work to change things. Moreover Cuba, again unlike so many Latin American republics, was a relatively prosperous country with a relatively diversified economy. Its principal weakness was its dependence on sugar, and thereby on the United States, for its foreign exchange. Castro and Guevara were therefore tempted, by circumstances as well as by their temperaments, to move fast and, relying upon a certain economic strength, to attack without delay the special link with the United States which represented a form of economic servitude and was also politically charged with memories of a generation of US dominance. At the same time, and with an equal disregard for immediate practicalities, Castro tried to extend the benefits of revolution and reform to the peoples of neighbouring countries. He became involved in subversive activities in the Dominican Republic, Haiti, Nicaragua, Panama and Venezuela and alarmed all those whose liberal instincts were less active than their love of law and order (or at any rate order) by giving money to left-wing groups and by broadcasting to Latin America a Castroist message as disturbing as the voice of Nasser in the Middle East.

Castro's economic policy called for foreign aid and credits and for foreign customers to replace the United States. He made the obvious move. In February 1960 Mikoyan visited Havana and concluded a trade agreement with Cuba which, among other things, enabled Cuba to buy Russian oil. Later in the same year Guevara made a tour of eastern European countries. In May Cuba established diplomatic relations with Moscow and in June it began buying arms from the USSR and other communist governments. At the same time Castro attacked the US rights in the Guantanamo naval base and its economic influence through the sugar trade. In June US and British refineries refused to refine Russian oil, thereby giving Castro an excuse to take them over. The Eisenhower administration in Washington, which had been watching events with restraint, hoped to bring joint American pressure to bear but at San Jose in August an inter-American conference of foreign ministers, while condemning Russian and Chinese intervention in Latin America, refused to refer to Cuba by name. Washington decided to act on its own. It stopped further purchases of Cuban sugar and, after Castro had retaliated by nationalizing US property, imposed a complete commercial boycott and in January 1961 severed diplomatic relations.

From early in 1960 the United States had been helping and encouraging Cuban refugees in two places, both under the aegis of the Central Intelligence Agency. In Florida exiled politicians were formed into a political committee which hoped to become a government of Cuba; it included men of very different views who were only held together by their common opposition to Castro and by their US managers. In Guatemala a force was being trained against the day when it would return to Cuba in small bands and start guerrilla warfare; at first there was no thought of US military participation in the adventure, but as time went on the original tactics were transformed from piecemeal infiltration to a single invasion thrust with US air cover, and the exiles were further allowed to assume that the United States would back their ground operations with force rather than let them fail. When Kennedy was apprised of these activities immediately after his successful campaign for the presidency, he was troubled by them, but between that date in November 1960 and the launching of the attack on the night of 14/15 April 1961, he never translated his doubts into prohibition. The operation was already in train; the chiefs of staff, his most awe-inspiring advisers, were in favour of it (but his secretary of state Dean Rusk was not); he did not want to let down several hundred Cubans, with whom he sympathized, nor did he know what to do with them if they were to be disbanded; the expectation that Castro would soon have Russian jet aircraft would make it very difficult, if not impossible, to sponsor such an invasion at a later date. Kennedy reaffirmed the ban on the involvement of United States forces but overrode the opposition of

Senator William Fulbright and a few other civilian advisers who advocated the containment of Cuba in preference to direct action which would be contrary to the charter of the OAS and all too consonant with Washington's reputation for imperialism and hypocrisy in its dealings with its southern neighbours.

The invasion was a complete failure. A force of 1,400 men – nine-tenths of them semi-trained civilians – supported by B-26 bombers operating from Nicaragua with Cuban pilots landed in the *Bay of Pigs* only to discover that, contrary to the assurances of the CIA, the US administration was not prepared to back them up and that expected risings in Cuba itself were not materializing. The Cuban government's riposte was more effective than had been anticipated and after forty-eight hours all was over. For good measure Castro imprisoned several thousands of his fellow citizens, thus seizing the opportunity to silence, demoralize and in some cases extinguish his opponents. His prestige beyond Cuba was greatly increased.

Nevertheless his position was in other respects unhappy. The economic measures taken by his government had, by its own admission, been ill conceived. A modish passion for industrialization led to the construction of factories for the manufacture in Cuba of articles which could be imported from abroad at less cost than the cost of the raw materials for their manufacture. In the countryside peasants displayed the worldwide dislike of their kind for co-operatives, the nationalization of land, and the enforced cultivation of crops destined for sale at fixed low prices. The middle class, which had been a more active ingredient in the revolution than the peasantry or the urban working class, became antagonized when nationalization was extended from foreign to domestic enterprises, and when the new régime took to the bad old ways of its predecessors in putting political opponents away in noisome jails. By 1962 there was an economic crisis, a general refusal to work by the peasants, food rationing and widespread disillusion, discontent and poverty.

At the end of 1961 Castro declared himself a Marxist. The Cuban Communist Party was a distinct entity which had at the outset little or nothing to do with Castro's 'July 26' movement. It was not a party with mass support and in the Batista period it had preferred back-stage political manoeuvres which secured it a humble position on the periphery of the ruling constellation. Before the Castroist victory in 1959 there were in both the Castroist and the communist camps some who encouraged a *rapprochement* or even a fusion and others who opposed any such move. During 1961 the former groups prevailed and the parties effected a considerable degree of integration. Cuba became in consequence a part of the communist world in terms of international politics, but Castro did not become either an orthodox communist (he seemed to care little about communist teaching) or a captive of a communist machine. As independent in his way – and within the limits imposed by Cuba's need for foreign friends – as Mao or Tito, he also remained domestically the leader of a government and a movement which were only partly communist. Leaders of the Cuban Communist Party were promoted to office and influence but in March 1962 the most eminent of them, Anibal Escalante, was dismissed and fled to Czechoslovakia shortly before Castro set out for an extended visit to the USSR. The two parties had fallen out, the Castroists prevailed and Moscow backed the winners.

Moscow's support for Castro had become an economic burden with exciting political and strategic possibilities. Keeping Castro afloat economically was probably costing the USSR more than it had bargained for in financial terms; keeping Castro afloat politically in the face of the US determination to destroy him could be an even more costly and risky policy, but it was also an exceptionally tempting one since an effective alliance between the USSR and Cuba would give the Russians a foothold in Latin America with all its unpredictable revolutionary possibilities, and a base within 160 km of the United States to offset the bases with which the USSR was ringed by its antagonists. According to Cuban – and Chinese – sources the idea of sending to Cuba the missiles and jet aircraft which were first observed by United States reconnaissance in October 1962 was Khrushchev's. However that may be Cuba, which Castro had made an economic and political ward of the USSR, was converted by Khrushchev into an armed pawn. Whereas Castro had wanted to equip Cuba to defend itself against a second attack like the one in the Bay of Pigs, Khrushchev despatched weapons of an altogether different significance. The upshot has already been discussed (see pp. 33–5). In the narrower Latin American context the results were that the OAS (which had expelled Cuba from its ranks at the beginning of the year) approved the deployment of force by the United States against the approaching Russian ships; that a number of Latin American states contributed to the blockade of Cuba instituted by the United States; that Cuba was left out of the reckoning as Kennedy and Khrushchev moved to resolve the crisis; but that Castro was given an outstandingly splendid reception when he revisited Moscow in April 1963 and won assurances of continuing Russian favour and support.

The *missile crisis* of 1962 involved for Castro a threat of extinction inasmuch as the rebuff to the USSR might carry with it an implicit freedom for the United States to work its will in Cuba and remove a government which had connived at, or even possibly instigated, an attempt to alter the balance of power in the American hemisphere more radically than at any time since the eighteenth century. But time seemed to show that Kennedy's triumph over Khrushchev was not to be so interpreted. Castro himself felt secure enough against renewed intervention by the United States to engage, towards the end of 1963, in a plot against the government of Venezuela (for which the OAS in July 1964 declared him an aggressor and recommended the severance of diplomatic and commercial connections, a step which only Mexico then refused to take); and Castroism was also one of the ingredients in a rising in Andean Peru in 1965 in which desperate underdogs tried, with the help of Castroist and communist supplies and moral support over the radio, to force their plight upon the attention of their government. But the ostracism of Castro in 1964 was a reminder that he was still there to be ostracized, and if he was still there part of the reason was that his state was under the protection of a major power.

The Russians had been physically beaten out of the Caribbean but they retained a protective or intrusive capacity of another order. The assertion of the Monroe Doctrine had revealed its limitations. There were to be no foreign bases in the American continent but the American hemisphere's politics could no longer be sealed off from wider international politics. The United States had been harping for some time on the dangers of international communism in Latin America, but this ideological approach had missed the main point which was the opening of Latin America to the processes of non-ideological international politics, in much the same way as the Middle East and the rest of Asia and then Africa had become international magnetic fields as soon as both the major world powers decided to exercise there the powers which nobody could prevent them from exercising. In the first postwar decades it had been assumed that two areas in the world were immune from this interplay: Latin America in the penumbra of the United States, and Moscow's satellite empire in Europe in the irresistible tow of the USSR. The Cuban crisis showed that this assumption was at least an exaggeration. In the case of Latin America Khrushchev, with Castro as his eye-opener, scented that the time had come to question the unquestionable. This capacity was one of his strengths as a politician. Characteristically, however, he followed his instinct with more enthusiasm than caution and so exposed himself to a stinging tactical defeat. Characteristically too Washington, with the restraint which democracy imposes on action and even sometimes on thinking, had not come within sight of any comparable policy in Europe, thus avoiding any comparable defeat. But the principal implication of the Cuban crisis was the demolition of the theory that major powers have back gardens round which it suffices to put up notices against trespassers. Although interference in such areas remains preternaturally hazardous, the areas themselves are not enclosures where special conditions are bred, but areas where the various winds of international politics blow in different force. This revelation marked a stage in Latin America's international history as well as in the appreciation of great power politics.

For Cuba itself the experience was bitter. During the period of United States dominance up to 1933 Cuba had been a quasi-colonial territory which was permitted to contract no foreign alliances and harbour no foreign bases. Castro reversed this situation to the extreme extent of making an alliance with the USSR and turning Cuba into a Russian base. He did so on a calculation of Cuban interests which turned out to be a miscalculation, since the USSR showed by retreating in the face of Washington's challenge that Cuba's interests played very little part in Moscow's calculation of its own interests and possibilities in the western hemisphere. Although the USSR had defied the Monroe Doctrine it had not done so in support of Cuban nationalism or Cuban ambitions. Castro, who had hoped to become the Latin American equivalent of Tito, Sukarno and Nasser, at once the personification of a new national dignity and the regional leader of revolution, found himself instead the ruler of an island which had become an international curiosity rather than an international fulcrum and which was beginning to look rather bedraggled as a consequence of economic and administrative muddle. For Guevara, the Argentinian pan-American for whom Cuba's revolution (not Mexico's or Bolivia's) was to have been the real beginning of Latin America's revolution, the reassessment necessitated by the Russian retreat was even bitterer than it was for Castro, and in the course of it the two men fell out and Guevara disappeared, taking with him much of the Cuban revolution's international favour which he tried to inject into South America until he met his death in Bolivia in 1967.

Castro's adventures in Africa have been described in Chapter 24 of this book. They increased and publicized his dependence on the USSR and they coincided with the global inflation and recession of the late 1970s. In the first part of that decade the Cuban economy expanded at the rate of 10 per cent a year, but this rate fell to 4 per cent for the rest of the 1970s. Food, housing and transport became scarce and dear; the price of fuel became prohibitive for large numbers of people; many workers were having to travel four hours a day to and from their work. Stricter regulations against small businesses created a large and economically stranded sub-proletariat. Nepotism, inequalities and stifling controls unaccompanied by commensurate benefits aggravated its plight. In 1979 a cyclone added its toll, the sugar crop failed and the tobacco crop was badly blighted. The weight of Washington's economic blockade was increasingly felt. In April 1980 10,000 Cubans invaded the Peruvian embassy in Havana demanding to go to Peru and hoping to go to the United States. Some of them were common criminals or would-be political refugees, but the bulk were Cubans who had had enough of being in Cuba. Castro made the best of a bad job by allowing them to go, but they were a startlingly bad advertisement for his conduct of his country's affairs. His plight worsened a few years later when Gorbachev, who visited Cuba in 1989, cut Russian aid and trade to and from eastern Europe ceased altogether. Public services collapsed, industries closed for want of fuel as well as orders, wages and salaries went unpaid, emigration accelerated (over the whole Castro period, Cuba lost about 15 per cent of its population). The ultimate symbol of Cuba's sorry state was Gorbachev's removal of 11,000 troops while US forces remained in Guantanamo.

But there was no revolt. Castro even contrived to turn the miseries of his people to his own advantage by permitting and even facilitating their exodus to the United States. The Clinton administration inherited a policy of admitting Cuban refugees with open arms but faced with a new wave of 50,000 it resolved to intercept them (or such of them as did not drown on the way) and transport them to the Guantanamo base. This, however, was already encumbered with 14,000 Haitian refugees and the cost of their indefinite upkeep. In this quandary Clinton was forced into talks with a Cuban régime which showed no signs of collapse. Castro's single purpose in such talks was the removal of the stifling economic sanctions which were Clinton's main weapon against him, short of armed force.

In one sense Castro was a revolutionary leader who failed because his revolution was throttled by the United States but in a deeper sense the revolution failed because it turned optimism into pessimism or at best resignation. Before Castro, Cuba was a relatively prosperous but ill-governed Latin American state whose business was selling sugar to the United States. It was also a relatively new state liberated from Spain towards the end of the nineteenth century but ruled by a narrow, inefficient and increasingly corrupt élite which purloined the fruits of development and failed to maximize them. Castro's revolution appeared to complement the revolution against Spain. It excited hopes. But its increasingly dogmatic elements fired few emotions and cut the ground from under the country's economic base; and when the revolution failed to fulfil hopes the revolution itself had failed. What the USSR supported until 1989 was no longer a revolution but a sadly bankrupted and disconsolate state. Yet in 1993 Castro was re-elected president for another five years.

Apart from Cuba and the neighbouring large island of Hispaniola (divided between Haiti and the Dominican Republic) the Caribbean was colonial, mainly British, ground until after the Second World War. Decolonization was the main topic of the first postwar decades. A federation of British colonies in the Caribbean had been under discussion since the middle of the nineteenth century. Such a federation was created in 1958 but came to grief in 1961. There have been divergent views about the causes of this failure. On the one hand it has been alleged that the delay in bringing it into being after the end of the Second World War allowed animosities to develop between the islands and that the blame for this delay rested upon the British government which, regarding West Indian leaders as much more left-wing than in fact they were, slowed down the pace towards independence and therefore federation. On this view there was a genuine desire for federation, at the popular as well as the political level, despite the acknowledged difficulties – great disparities in the size and wealth of the different islands and the great distances between them. The alternative view is that, except perhaps in the smaller islands, there was no popular interest in federation or support for it and that the federation of 1958 rested therefore on a narrow base of political and trade-union leaders who were liable to be undermined by a popular revulsion; and that such a revulsion was in the event stirred up by other leaders who saw in it a way to take power. This view stresses the ways of thinking of the Caribbean peoples, islanders with a keen sense of belonging to their particular islands but little sense of community with others beyond the horizon (a habit accentuated by the fact that under colonial rule each colony had been treated by the British Colonial Office as a distinct object in a series of bilateral relationships with Britain).

Shortly after federation, and before its economic benefits had had a chance to make their impact, the prime minister of Jamaica, Norman Manley, was challenged by his rival, Alexander Bustamente, to hold a referendum to see whether Jamaicans wished the federation to continue or not. The result was a narrow defeat for Manley and the federation, which Jamaica left in January 1962. Trinidad-Tobago followed suit. Both became independent members of the Commonwealth and the OAS, as did Barbados in 1966. The smaller islands became associate members of the Commonwealth, self-governing except in relation to foreign and defence policies in which Britain retained a share of responsibility. There were six of these associates: Antigua with Barbuda which progressed to independence and UN membership in 1981; Grenada which became an international celebrity in 1983 (see p. 832); Dominica; St Lucia; St Vincent; and St Kitts–Nevis–Anguilla. The last proved unhappily assorted and Anguilla rebelled against the government on St Kitts. It achieved a *de facto* separation which Britain, despite some sympathy with Anguillan complaints, refused at first to endorse but accepted in 1980 when Anguilla became formally a separate state, Britain assuming responsibility for its defence, foreign affairs and internal security with a say in its senior appointments. Grenada, Dominica, St Lucia and St Vincent formed a regional assembly to consider a federation of the Windward Islands. Four British colonies remained colonies: Montserrat, the British Virgin Islands, the Turks and Caicos Islands, and the Cayman Islands – some of them becoming favoured haunts of tax-evading bankers and their Third World associates in the drug business. On the mainland British Honduras, which changed its name to Belize in 1973, moved only slowly to independence under the shadow of Guatemala's claim to the whole territory. It became in 1981 an independent member of the Commonwealth protected by a small British force until in 1991 Guatemala formally renounced its claim and Britain terminated its commitment to its defence.

In 1968 a modest economic association, Carifta (Caribbean Free Trade Association), was formed by all Commonwealth states and territories except British Honduras and the Bahamas; the former joined three years later. The weaker members claimed that Carifta benefited the stronger members more than themselves. All were worried by British accession to the EEC since they had, under the Commonwealth Sugar Agreement and similar schemes, an assured market and stable prices for their sugar, fruit, rum and other produce. With the exception of the solitary oil producer, Trinidad and Tobago, they were harder hit by rising oil prices in the 1970s which aggravated endemic unemployment and anti-government sentiment. In 1973 four states – Trinidad and Tobago, Guyana, *Jamaica* and Barbados – created a Caribbean Community (Caricom), membership of which was open to all Commonwealth countries which chose to join within a year. The participation of Jamaica, which had torpedoed the federation of 1958, was vital. This new association, which replaced looser meetings of heads of government, had a small population but was nevertheless economically overpopulated: its two largest members, Jamaica and Trinidad (whose perennial prime minister, Eric Williams, died in 1981), counted only half a million people between them but a quarter of the population of working age was unemployed and the average income was less than $500 a year. Caricom languished until resuscitated in the late 1980s with proposals for an economic union by 1993.

Much the largest of the British possessions was *Jamaica*. In 1961 Jamaicans, asked to choose in a referendum between independence and a West Indian federation, chose independence. This was the policy of the Jamaica Labour Party (JLP) led by Alexander Bustamente – and subsequently by Donald Sangster, Hugh Shearer and Edward Seaga. The pro-federation People's National Party (PNP) of Norman Manley (in office 1957–62) resigned and the JLP governed for the next ten years. This was a period of largely unhappy post-colonial reconstruction. The economic base was broadened from its traditional reliance on sugar to embrace also tourism, bauxite and manufacture, but the shift from agriculture to manufacture increased unemployment to 20–30 per cent of the working population at a time when the habitual escape – emigration to Britain – became all but barred. Hence pressures and conflicts, notably in overcrowded Kingston, and discontent directed against the economic establishment: a small oligarchy, foreign capital and – in the commercial field – Lebanese, Syrian and Chinese minorities. Discontent mounted to violence not far short of civil war (major riots in 1968) and a helpless JLP government lost the 1972 election. The PNP, now led by Norman Manley, Jr, returned and remained in office until the end of 1980. Like Castro, Manley wanted to break away from Washington's economic stranglehold and, in rebellion also against the terms on which the IMF did business with impecunious developing countries, to get aid from the USSR and eastern Europe. As a result he got himself a bad name in the west without compensating benefits from the east. The island was also hit by declining revenues from bauxite and other exports. In 1980, in an election which claimed at least 500 lives, the comparatively right-wing JLP triumphed with an economic recovery programme based on conventional capitalist remedies with US approval. The new prime minister, Edward Seaga, made a quick tour of Washington and other western capitals and closed the Cuban embassy but he was not able to stem the slump in Jamaica's main exports – sugar, bananas, bauxite – or reduce a foreign debt which was eating up half of Jamaica's export earnings. With the economic infrastructure collapsing and the poor getting poorer Manley won the elections of 1989. He secured fresh loans from the IMF but had to agree to cut subsidies, devalue the currency and keep inflation below 9 per cent a year. The soaring prices occasioned by the first part of this programme led to high wage-claims which endangered the undertaking on inflation and therefore the whole package. Economic decline, rising inflation and higher taxes weakened Manley who gave way to P. J. Patterson who all but eliminated the opposition in elections in 1993.

For a generation after independence *Trinidad and Tobago*, besides being at the other end of the Caribbean, presented a contrast to Jamaica in its style and fortunes. Its two principal communities, of African and East Indian origin, coexisted without serious conflict. It pursued unselfish policies of lending to its neighbours and subscribing for World Bank loans. But after 1980 declining oil revenues and high interest rates dented its prosperity and stability. Meanwhile it had lost its status as a developing country, while its attempts to diversify its economy and reduce its dependence on oil and sugar ran into protectionist obstacles from the United States and the European Community. As world prices for its exports collapsed, unemployment rose, a series of devaluations forced up prices, wages were frozen, and crime and disorder increased.

In these circumstances the rule of the People's National Movement came to an end in 1986 when the National Alliance for Reconstruction, led by Robbie Robinson (of African origin), won a convincing victory. The new government accepted the standard IMF remedies.

With the exception of Jamaica and perhaps Trinidad the several new states of the Caribbean had electorates of eighteenth-century proportions to go with their constitutions of twentieth-century design. Consequently elections were all too easily conducted with threats and promises – a promise to abolish taxes, for example. This circumstance did not make for good or stable government. It also robbed electoral verdicts of much significance, so that generalizations about a leftward or rightward trend were uncommonly deceptive and misleading. The poverty of so many of the islands, coupled with Castro's appeal and meddling, favoured parties with radical programmes. Nevertheless, where a valid distinction could be drawn between right and left, there was in the 1970s a tendency to choose the more right-wing parties. In *Antigua*, for example, Vere Bird's Labour Party recovered in 1976 the power it had lost in 1971 to a more radical rival; but its return was facilitated by its promise to abolish income tax and by its predecessor's too extreme corruption. In the same year J. M. G. Adams's *Barbadian* Labour Party defeated the more left-wing party of Errol Barrow, and in Trinidad and Tobago the comparatively conservative Eric Williams won another term of the office which he had held since 1956. In 1979 *St Vincent* and *St Kitts–Nevis* also shifted to the right and so perhaps did *Dominica*, where the government of Patrick John was forced to resign after revelations concerning peculiar dealings with South Africa. Many of these small states were obliged to search for unorthodox sources of revenue but Patrick John's scheme for leasing a substantial part of his island to the South African government was too eccentric and portentous to pass muster.

The contrary left-wing trend, supported by Cuban aid and inspiration, won some success in *Grenada* where Eric Gairy's United Labour Party, re-elected in 1976, failed to cope with the island's economic distress or to maintain a semblance of honest government. It was removed by force by Maurice Bishop, an admirer of Castro. Other Caribbean parties sympathetic to Castroism remained in the wings, their fortunes still latent but their presence significant enough to induce Washington to quadruple its financial aid to the region between 1975 and 1980. A scare in 1980 over the supposed arrival in Cuba of a Russian brigade reinforced this beneficence and also the feeling that the Caribbean was being drawn once more into world politics as it had been at the time of the Cuban missiles crisis in 1962. The election of Reagan added a touch of gunboat diplomacy to the prevailing dollar diplomacy, its most notable exploit being the invasion of Grenada in 1983.

Bishop had begun his rule by making overtures to the United States but he was snubbed in Washington and found his development plans blocked by US votes in international financial institutions. He obtained nevertheless favourable mention from the World Bank in 1982 for his rural schemes and extensions of public services and he released in the same year some of the political prisoners whom he had put in jail upon coming to power. His party, the New Jewel Movement, was an amalgam of two roughly equal sections, the one led by himself and the other by Bernard and Janet Coard who were regarded as closer in temperament and ideology to Castro (or even, in the eyes of their enemies, to Pol Pot). By 1983 these two groups had become openly hostile and were murdering each other. Bishop was assailed from two sides – his more left-wing colleagues on the one hand and, on the other, the United States which had decided to unseat him and began using blatantly exaggerated propaganda against him. The Coard faction murdered him. A US force landed and destroyed the Coards and their adherents, using a variety of excuses to cover a rehearsed operation: that the airport being built by British contractors was not, as alleged, intended to boost tourism but to serve the purposes of Castro and his allies; that the lives of a group of US nationals engaged in research in Grenada were at risk; and that a substantial Cuban military force was already on the island. In spite of being rehearsed the United States adventure was bungled; intelligence was poor; the invaders took ten days to conquer a small defenceless island and suffered over a hundred casualties, inflicted mostly on themselves. There were no Cuban forces on Grenada and official statements were embarrassingly and blatantly untrue. But the operation was far from unpopular. Washington secured endorsement of its action from neighbouring Caribbean governments, although most Commonwealth governments, including the British, were piqued by what they regarded as an illegal and politically unnecessary act. Washington got away with it because the operation was brief, welcome to the inhabitants of Grenada and in tune with prevailing anti-left-wing sentiments. A new political party was put together under Herbert Blaize and captured in 1984 all seats in the parliament except one. Blaize died in 1989 and in elections a few months later the ruling National Democratic Congress failed by one seat to secure a majority of the parliament's fifteen members. Relics of Bishop's party won no seat and a mere 2 per cent of the vote. Nicholas Braithwaite became prime minister. He participated in desultory talks about a federation of the Windward Islands.

Haiti's independence was won by slave revolts against Revolutionary and Napoleonic France after being a French colony from 1697. French rulers set an example of atrocious behaviour which was not lost on their wards and successors. Until 1843 the island of Hispaniola comprised a single state but in that year it was divided into Haiti and the Dominican Republic. Haiti was occupied by the United States from 1915 to 1934. Society and politics revolved round a black majority and a mulatto élite which, after briefly combining to get rid of French rule, treated one another with savagery. Dictators were common but not always or entirely vicious. The main characteristics of the régime of Colonel Paul Magloire (1950–56) were extravagance, jollity and corruption. His place was taken by Francois Duvalier (Papa Doc) who had been in hiding in Magloire's time but emerged to win an election in 1957 and survived, first, by being a better plotter than his rivals and, secondly, by recruiting, clothing and equipping, mainly from the slums, a counter-army, the fearsome Tonton Macoute. He won the support of the army and the Church, extracting from the Vatican a concordat (1966) placing the appointment of bishops in his hands. He also had support from the United States until Washington connived at a ludicrously incompetent invasion by his enemies in Florida. After this fiasco Washington switched back to Duvalier. His rule became rough, ruthless and crazed. He died in 1971 and was succeeded by his son Jean-Claude (Baby Doc). Whereas Papa Doc had been the leader of a black middle-class coterie which hijacked the state and treated it as private property while at the same time extolling negritude, the national voodoo religion and the cruder ideas and practices of European fascism, Baby Doc found his friends among the rich mulattos. He was less intelligent than his father and perhaps less ruthless but he was evicted in 1986 after over-straining the indulgence of the Roman Catholic Church, the professional classes and the young (here, as in many countries, habitually referred to as students). The Duvalier régime was followed by instability, two elections in 1987 and 1990 marred by violence, and the eviction in 1990 of another strongman, Prosper Avril. A third post-Duvalier election gave Father Jean-Bertrand Aristide a convincing victory after a campaign in which he promised to make life better for oppressed and undernourished underdogs and the electorate allowed itself to hope that he might be allowed to do so. Aristide was a fearless left-wing priest who denounced his own ecclesiastical hierarchy as well as the Duvaliers. From the United States he got good marks and some aid as a reward for democracy but bad marks for being on the left. Within a year he was ousted, nearly killed and fled. Army and police chiefs took control – Generals Raoul Cédras and Philip Biamfy and Colonel Joseph François – and the Tonton Macoute reappeared under the name of *attachés*. By the Washington Accord of 1992 the United States undertook to reinstate Aristide if he would commend a military blockade, to which he reluctantly agreed. By the next year's Government Island Accord, the new Clinton administration broadly endorsed its predecessor's policies while also trying to side-step military action by persuading the Haitian triumvirate to allow Aristide back without seriously curbing military rule. This manoeuvre failed and a swelling tide of refugees from Haiti to the United States stiffened Clinton's resolve. He secured the imposition of UN sanction and UN approval for the use of all measures necessary to restore democracy, and he assembled a naval force to invade the island. In an attempt to evade the use of force, Clinton sent ex-president Jimmy Carter to negotiate an unopposed landing which Carter, flanked by Senator Sam Nunn and General Colin Powell, achieved but at the cost of dropping demands for the triumvirate to leave Haiti, promising them an amnesty and allowing them to remain in power for a month. But after a brief period of suspense they left and Aristide returned, restored by the United States where it was hoped that he would soon be replaced by a less radical figure more to Washington's liking. But when Aristide's term expired in 1995 he was succeeded – he himself being debarred from two consecutive terms – by René Préval who was even more suspect in the United States.

Clinton's explicit aim in Haiti was to restore democracy. While laudable in itself and likely to stem inhuman atrocities, this attempt by one state to change by force the government of another was in breach of the UN Charter and was not rendered less so by Clinton's success in getting twenty other UN members to provide small token adjuncts for a US invasion. Even if inhumanity constituted grounds for international intervention (see Chapter 4) that right was most questionable where the defence of human rights could be effected only by the overthrow of the government of a sovereign state. Clinton's Wilsonian purpose, placing democracy at the centre of his foreign policies, was a moral purpose not clearly supported by law.

In the other half of Hispaniola the *Dominican Republic*, ruled by Rafael Trujillo for more than thirty years until his assassination in 1962, was dominated for the next thirty by his political heir, Joaquin Balaguer, a shrewd and comparatively unobtrusive politician who succeeded in the 1990s in mastering inflation and increasing overall production but not in reducing unemployment (which was running at about a quarter of the workforce) or in the easing the lives of the poorer half of the population or redressing the country's balance of trade. Balaguer was persistently opposed by Juan Bosch who won the first post-Trujillo election but was removed by the army after a few months. A bout of civil war was ended by US intervention and the beginning of the first of Balaguer's six presidencies. When these rivals were both in their eighties, a third man entered the political lists – José Francisco Gomez – who, in spite of the disadvantages of being black and born in Haiti, pushed Bosch into third place in elections in 1994 and evidently defeated Balaguer too although those counting the votes ruled otherwise.

At the northern edge of the Caribbean, north-east of Cuba and south-east of Florida, the Commonwealth of the *Bahamas*, an independent member of the Commonwealth from 1973, flourished as a principal ancillary of the seamier side of international finance and of the drugs trade. Sir Lynden Pindling, chief minister 1967–73 and prime minister 1973–92, rose to power as the champion of the black majority against the white merchant élite (the Bay Street Boys). His governments were periodically assailed by scandals, notably in 1984 which he survived and in 1987 when he felt obliged to call an election which his Progressive Liberal Party won after a confused and ill-tempered contest which, however, was adjudged by US observers to be fair enough. The opposition Free National Movement, accused of being equally corrupt and élitist to boot, won a few seats from the PLP but Pindling won 31 out of 49. He enjoyed the support, with diminishing enthusiasm, of the United States. The Bahamas was probably the only country in the world where over half the members of parliament were millionaires.

NOTE

GUYANA AND SURINAM

Guyana, once British Guiana, is an almost equally divided bi-racial country in which the descendants of Indians imported by the British (after the abolition of slavery in 1838) slightly outnumber the Africans while a small Roman Catholic minority with European origins can in these circumstances be electorally important. The Indians are predominantly rural, the Africans urban. The postwar leaders of these two groups were at first divided more by temperament than policy. The Indian leader, Cheddi Jagan, a Rooseveltian liberal with a socialist American wife, joined forces with Forbes Burnham, a shrewd politician of African descent. Both men moved to the left but the former moved faster than the latter. While Jagan allowed himself to be pushed by conflict into more extreme attitudes, Burnham exercised more self-control and successfully subordinated the expression of opinions to the capture of power. They became therefore the leaders of opposing parties, separated at first by tactical considerations but increasingly racial in their confrontation. In 1961 Jagan won an election, but his radicalism, and especially his plans to unionize labour, alarmed the European planters, while local American interests were reporting him to be a dangerous communist. This view of Jagan's impact on the colony was sharpened by riots necessitating the despatch of British troops and by Jagan's leanings towards Castro's Cuba, an inclination partly temperamental but partly also economic: Cuba was one of Guyana's principal markets for its surplus rice and this market had been cut off after Castro's revolution.

With independence in the offing the United States and to some extent Britain were anxious to ensure the victory of Burnham over Jagan. Burnham cultivated the small Roman Catholic party and won the election of 1964 after the British had conveniently altered the constitution by introducing proportional representation. On a straight vote Jagan could and almost certainly would win; under proportional representation he could not. This election was followed in 1966 by independence, whereupon British Guiana became Guyana and joined the UN but not, owing to a border dispute with Venezuela, the OAS. In the next few years the victorious Burnham began to move to the left, impelled chiefly by the needs of the urban unemployed Africans and by general anti-American and anti-planter ideological imperatives. Burnham's perennial rival, Cheddi Jagan, and his People's Progressive Party suffered a series of defeats too overwhelming to be credible and the alternative opposition, the Working People's Alliance, retired from the elections of 1980 after its leader was assassinated. A new constitution in 1978 established a presidential form of government. Burnham died suddenly in 1985. His successor Desmond Hoyte maintained the dominant position of Burnham's People's National Congress, subject however to increasingly trenchant accusations of sharp electoral malpractice and a steep decline in the country's economic fortunes. Revenues from sugar and bauxite flagged, the currency was devalued, food and electric power became scarce, wages were cut. Hoyte was obliged to postpone the elections more than once because it was impossible to guarantee proper procedures.

East of Guyana Surinam became independent of the Netherlands in 1975. From 1980 to 1988 it experienced military rule which was reasserted in 1990 by a coup which Surinam's neighbours in South America and the Caribbean deplored and denounced. In the following years internationally supervised elections placed Colonel Desi Boutresse and his civilian allies in control and in a presidential election their candidate, Ronald Venetiaan, won with promises to reduce the army by two-thirds. He granted an amnesty to various opposition groups in return for the surrender of their arms. Boutresse resigned as commander-in-chief in 1992. French Guyana remained a Department of France. A thousand miles westward of Surinam the Netherlands Antilles and Aruba were, by an agreement of 1986, promised independence in 1996.

APPENDIX: KEY TO UNITED NATIONS MISSIONS (MAP 4.1)

MINURSO: UN Mission for the Referendum in Western Sahara
ONOMOZ: UN Operation in Mozambique
ONUC: UN Force in the Congo
ONUCA: UN Observer Group in Central America
ONUSAL: UN Observer Mission in El Salvador
UNAMIC: UN Advance Mission in Cambodia
UNAMIR: UN Assistance Mission to Rwanda
UNAVEM: UN Angola Verification Mission
UNBOG: UN Border Observation Group – Thailand/Kampuchea
UNCI: UN Commission for Indonesia
UNCIP: UN Commission for India and Pakistan
UNCOK: UN Commission on Korea
UNDOF: UN Disengagement Observer Force – Golan Heights
UNEF: UN Emergency Force – Middle East
UNFICYP: UN Peace-keeping Force in Cyprus
UNGOMAP: UN Good Offices Mission in Afghanistan and Pakistan
UNIFIL: UN Interim Force in Lebanon
UNIIMOG: UN Iran/Iraq Military Observer Group
UNIKOM: UN Iraq/Kuwait Observer Mission
UNIPOM: UN India/Pakistan Observation Mission – Kashmir
UNITAF: UN Unified Task Force – Somalia
UNMIH: UN Mission in Haiti
UNMOGIP: UN Military Observation Group in India and Pakistan
   – Kashmir
UNOGIL: UN Observer Group in Lebanon
UNOMIL: UN Monitoring Force in Liberia
UNOMUR: UN Observer Mission in Uganda/Rwanda
UNOSOM: UN Operation in Somalia
UNPREDEP: UN Prevention and Development Force – Macedonia
UNPROFOR: UN Protection Force – Croatia/Bosnia
UNSCOB: UN Special Committee on the Balkans
UNSCOP: UN Special Committee on Palestine
UNTAC: UN Transitional Authority in Cambodia
UNTAG: UN Transitional Assistance Group – Namibia
UNTEA: UN Temporary Executive Authority – West Irian
UNTSO: UN Truce Supervision Organization – Palestine
UNYOM: UN Observation Mission in Yemen

INDEX